W9-ARX-026

Mechanical and Electrical Equipment for Buildings

Science Building, The Madeira School, Greenway, Va.

Photograph by Norman McGrath.

Architects:
Arthur Cotton Moore, Associates, Washington, D.C.

Mechanical and Electrical Consultants
Flack & Kurtz, Consulting Engineers, New York, N.Y.

JOHN WILEY & SONS
New York • Chichester • Brisbane • Toronto • Singapore

Mechanical and Electrical Equipment for Buildings

SIXTH EDITION

WILLIAM J. MCGUINNESS
Professor of Architecture Emeritus,
Pratt Institute

BENJAMIN STEIN
Visiting Associate Professor
The Technion, (Extension Division), Israel

JOHN S. REYNOLDS
Professor of Architecture
University of Oregon

This book was printed and bound by Halliday Lithograph Corporation. It was set in 10/12 Laurel by York Graphic Services. The designer was Maddy Lesure. Linda Sadovnick supervised production. The manuscript was edited by Joan Knizeski and Ellen MacElree.

Library of Congress Cataloging in Publication Data:

McGuinness, William J.
Mechanical and electrical equipment for buildings.

Includes bibliographies and index.
1. Buildings—Mechanical equipment.
2. Sanitation, Household. I. Stein, Benjamin, joint author.
II. Reynolds, John, 1938—joint author. III. Title.

TH6010.M25 1979 696 79-12979
ISBN 0-471-58432-0

Printed in the United States of America

10 9 8 7 6

preface

Change and Progress

Since the publication of the first edition of *Mechanical and Electrical Equipment for Buildings* in 1937, under the distinguished authorship of Professors Charles Merrick Gay and Charles De van Fawcett of the University of Pennsylvania, each succeeding eight-year interval between editions has resulted in what has essentially been—except for unchanging basics—a new book. This edition is no exception. It has been written during a period that marked the beginning of changes that may exceed any that have occurred in the past four decades. It describes the problems that we all face and reports the very considerable early progress in solving them. We look forward to new criteria, an awareness of the need for greater efficiency, a lesser use of energy, and the development of renewable energy sources. In the years ahead, we can expect a more controlled and thoughtful involvement in the design and use of building equipment.

Objectives of the Book

Since equipment is a major element in the design of buildings, this textbook is intended for architects and engineers. Written for architectural and engineering students, it nevertheless offers information and reference material for practicing professionals. A strong trend in modern architectural treatment is the casual acceptance of equipment as a design element, together with esthetics, function, and structure. For example, the inclination to hide equipment behind furring is diminishing. Use of this textbook is facilitated by a knowledge of college-level mathematics and physics, but expertise in engineering is not presumed.

The Sixth Edition

Three new introductory chapters stress the interrelationship of energy, climate, the site, and architectural design. Reduced use of energy in any form and the increased utilization of renewable energy sources are intrinsic themes throughout this book. Passive and active solar designs are featured. The recovery and use of energy from refuse and trash take on new importance. Lower lighting levels, without diminution of visual performance, are promoted, as is the use of local task lighting in conjunction with reduced general overhead lighting. A discussion of natural daylighting introduces the latest recommendations of the Illuminating Engineering Society. Energy-saving standards of the American Society of Heating, Refrigerating and Air-Conditioning Engineers are also introduced. The elevator chapters are much expanded. They include elevator control during fires in high-rise structures and an introduction to materials-handling in commercial buildings. The choice between sprinklers and fire-resistant compartmentation through the use of special walls and floors introduces a subject for consideration. Advantages of the characteristic of slow thermal transmission through heavy masonry construction are also discussed. The control of peak electrical power demand and the minimization of power loss in wiring are considered as well as the recent preference for the use of the variable air volume method over the reheat and dual-duct methods of air conditioning. Advantages of the recycling of water, ventilating air and, indeed, of buildings are also examined. Annual budgets for the minimal overall use of energy in buildings take on a new dimension. An expanded acoustics section that includes higher values of sound transmission loss through a multitude of listed floor and wall constructions may help us to reduce noise levels in our busy world. We have also included a few guidelines for the transition to the Système Internationale (SI), the Metric System.

William J. McGuinness
Benjamin Stein
January 1980 John S. Reynolds

acknowledgments

The authors are grateful for the overall reviews and support they have received from Dean Olindo Grossi, Norman Kurtz, John I. Yellott, Donald S. Duncan, G. Z. Brown, Robert S. Harris, Phyllis R. Naylor, and Jim McGuinness. For reviews in specialized disciplines we thank: William J. Cavanaugh—acoustics, John Van Dusen—vertical transportation, Howard Haynes—lighting, Bill Donaldson—sewage treatment, Dick Koral—HVAC and plumbing, Joe McCann and Dick Hall—fire protection. We appreciate the technical editing done by Jim Richards. For correspondence and preparing the manuscript we thank Lila Stein. Valuable assistance in proofreading and other production items was given by Emily Stein and J. Y. Stein. Editing of the manuscript was aided considerably by Carol Venolia. For typing and other assistance our appreciation goes to Elsie Cohen, Jean Dobkin, Sharron Peel, Tisha Egashira, Mike Clark, Michael Powell, and the staff of the School of Architecture and Allied Arts at the University of Oregon. To the numerous architects, professional societies, trade associations, and manufacturers who have aided us we express our thanks. We are grateful to our wives, Lillian McGuinness, Lila Stein, and Nancy Reynolds, for their encouragement and assistance.

contents

Part One ENERGY AND ENVIRONMENT

Part Two INDOOR CLIMATE CONTROL

Part Three WATER AND DRAINAGE

Part Four FIRE SAFETY

Part Five ELECTRICITY

Part Six ILLUMINATION

Part Seven SIGNAL EQUIPMENT

Part Eight TRANSPORTATION

Part Ten APPENDICES

Mechanical and Electrical Equipment for Buildings

PART ONE

energy and environment

1

Energy Sources for Buildings

Today buildings in the United States are frequently labeled as energy gluttons. Our society uses more energy per capita than most other countries, even some with "high standards of living" that have colder winters than ours. Although buildings are not our only users of energy—automobiles, industry, and agriculture are some others—buildings and their energy-related equipment are the subjects of this book. One way to begin is to examine the path that architecture has followed, from simple shelters that modified the surrounding climate, to sophisticated and tightly controlled internal climates that strive to ignore the conditions outside.

1.1 Energy and Architectural History

Few books on architectural history deal with the influence of changing fuel sources on the development of building. Style—including the proportion of spaces and the elements of the facade—has often been more influential, particularly on exterior appearance. Structural innovation is more visible and more permanent than fuel sources in architecture, and therefore easier to trace back to its earliest forms. Most buildings before the latter part of this century relied heavily on the sun for light and sometimes for heat, and on breezes or massive construction to temper the hot portions of the day. These energy-related distinctions were relatively subtle, particularly among the buildings within a given climatic region. The size and placement of windows were energy-related; yet, the size, placement, and, particularly, the shape of the windows were more clearly attributable to prevailing customs of proportion and materials than to the buildings' energy supply. The amount of sun or breeze admitted through the window, or the amount of building heat lost through it, has less often been cited as an influence on window design.

The impact of energy on buildings in the past has been both considerable and visible, and it is becoming so again. Space heating in buildings beyond that provided by the sun began with the burning of available fuel (Figs. 1.1, 1.2, and 1.3). It then proceeded by the gradual use of more imported fuels and less visible equipment, until the obvious architectural impacts nearly disappeared (Figs. 1.4 and 1.5). Today, solar collection devices, such as large windows and flat-plate collectors (Fig. 1.6), have reintroduced the significant architectural impact of heating; their large glass areas and the slopes of collectors combine to produce one of the greatest visible influences of any heating system on building form.

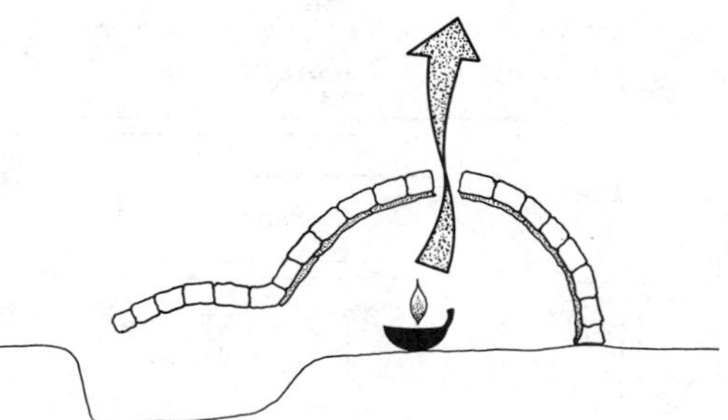

Fig. 1.1 The portable lamp (as used in the igloo) has a slight architectural impact: a storage place is needed for fuel; the design must admit the necessary combustion air (to feed oxygen to the flame) and must allow ventilation for the gases of combustion; there is also a residue of soot on interior surfaces.

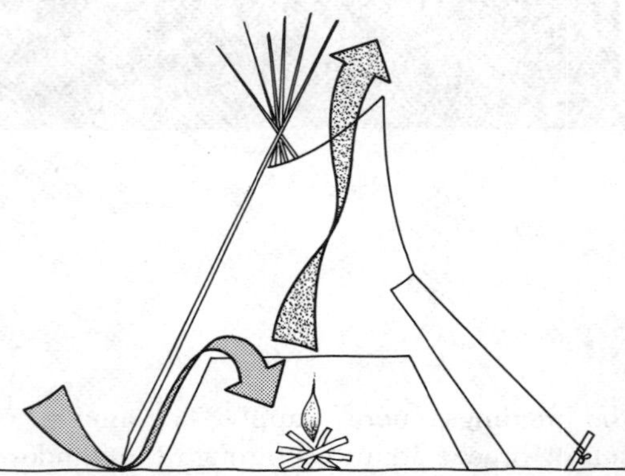

Fig. 1.2 The indoor fire (as used in the tipi) introduces two evident architectural impacts: the adjustable smoke flap *to encourage venting of smoke by prevailing winds and to minimize the entrance to rain, and the* interior liner *that forces the cold combustion air to rise along the sides of the tipi, gaining some warmth, before it moves across the occupants on its way to the fire. Again, some fuel storage is needed, and there is a residue of soot indoors.*

Fig. 1.3 The fireplace and the more efficient wood stove increase the impact of heating on architecture. Their permanent location within a space both allows a massive and visible chimney and determines the placement of other furnishings within the space. As the enclosed fire burns hotter and heats more space, the needs for fuel storage and for providing combustion air intensify. Photo by William Johnston.

Fig. 1.4 The furnace (or boiler) begins the trend toward less visible, more automatic heating. (a) *A central large "machine" can be located in a basement along with its fuel storage area. A single chimney exhausts gasses, and combustion air can infiltrate the basement without causing undue discomfort upstairs. An extensive duct-work system is threaded through floors and walls to simple grille openings in each room.* (b) *The use of piped-in fuel, such as natural gas, eliminates on-site fuel storage, and furnace sizes diminish with technical developments. With the electric furnace, even the chimney disappears along with the gasses of combustion. These plans show a basement remodeling that resulted from changing fuel sources.*

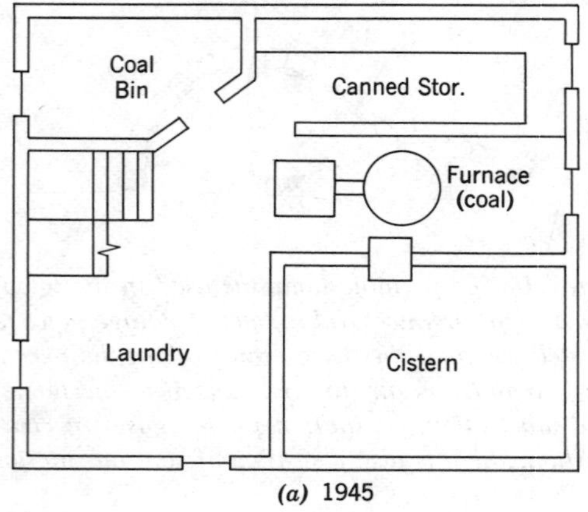

(a) 1945

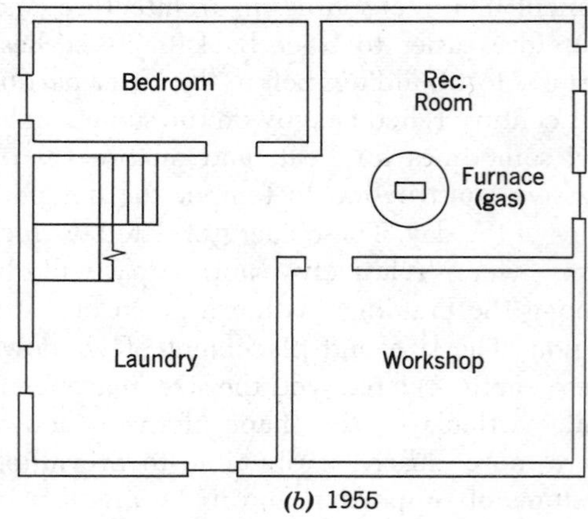

(b) 1955

Fig. 1.5 Radiant heat in ceiling or floor, provided through pipes or cables, reduces the visible architectural impact of heating to a thermostat on the wall—and occasional cracks in the heating surfaces!

This is not the only sudden and recent shift in long-term trends concerning energy in buildings. A new awareness of the sources, characteristics, and limitations of energy supplies is resulting in new directions in building design, away from many practices of the recent past.

Fig. 1.6 Solar energy for heating involves the visible impact of large south-facing glass areas—sometimes windows *directly admitting sun to spaces, sometimes* collectors *from which the sun's heat is transfered to a large storage place. Bishoprick House, Salem, Oregon. Photo courtesy of Portland General Electric Company.*

1.2 Trends, Recent Shifts, and Some Challenges

(a) ***A Trend from Local, Renewable Energy Sources to Imported, Nonrenewable Ones.*** Renewable fuel sources are those that are available indefinitely but that arrive at a relatively fixed rate; the influx of solar energy varies from day to day, but on the average it continues at a steady rate. A woodlot will produce a limited amount of wood per year, but it will do so for centuries if properly managed. A popular analogy for renewable fuel sources is a fixed-but-steady monetary income, such as a salary with no raises.

Using the same analogy, nonrenewable fuel sources are like savings accounts that draw no interest—once spent, they are gone. The nonrenewable resources we utilize can be bought and used in large quantities all at once, which makes possible many processes that are more difficult to attain with low-concentration, steady renewable resources. Thus, the use of nonrenewable fossil and nuclear fuels in high-temperature and high-concentration processes—for example, in power-generating plants—is widespread and is likely to continue until resource depletion is more closely approached.

Plentiful, locally available fuels such as wood and solar energy are cheap and convenient up to a point. As population density increases, however, such supplies dwindle per capita, and locally available quantities can become inadequate. As in the case of firewood in urban areas, prices can also rise. At that point, the conveniences of an established distribution network and a precisely measured heat content, as found with electricity, natural gas, and fuel oil, become attractive. The pollution that was formerly evident with some local fuels, such as woodsmoke, can be relocated to a distant generating plant. Eventually, of course, the problems of pollution must be faced, wherever they occur.

Even where local fuels are plentiful and "free," imported fuels have often supplanted them. Figure 1.7 shows one of thousands of residential solar water heaters formerly in use in Florida. When automatic washing machines were introduced, their greater appetite for hot water taxed the capacity of these heaters, which had been sized for laundry practices that reused hot soapy water rather than discharging it after one washing cycle. Natural gas utilities promoted the fast temperature recovery of their water heaters, and solar water heating was replaced by natural gas. In another climate, Eskimo families have turned, with government subsidies, from burning seal oil—a locally available, renewable resource—to imported and nonrenewable kerosene for home heating.

Shifts back to renewable and local fuel sources, such as the sun in areas like Florida and California, are occurring rapidly. For example, California has banned new hookups for natural gas swimming-pool heaters, creating an early and heavy demand for solar pool heaters. Although the shift in the United States to solar heating is so far slight, it is nonetheless highly visible. Figure 1.8 shows the increase in the number of known solar-heated buildings in the United States between 1940 and 1976 (the last year in which Dr. William A. Shurcliff compiled his now-famous directory of solar buildings). A similar growth is currently felt in the wood stove industry. Government subsidies are expected to add solar heating to thousands of buildings in the near future.

Buildings designed today may outlast the supplies of the fuels that support them. This fact provides two major challenges for designers:

1. To design our buildings so that they can eventually be weaned away from nonrenewable fuel dependence.
2. To use only a "fair share" of locally available renewable fuels, recognizing that such sources are limited even though they are continually available.

For example, with supplies stretched tight by increasing density, it is tempting to erect a larger solar collector to intercept sunlight that would otherwise be utilized by the neighboring building. The temptation grows stronger as the building is designed to rely more heavily on the sun.

Fig. 1.7 Solar water heaters are widely used and are available in many versions. This example in Miami, Florida, includes a storage tank above the collector, which is enclosed in a chimneylike form. When the lower collector is warmer than the upper storage tank, water circulates without the need for a pump. Photo by M. Steven Baker.

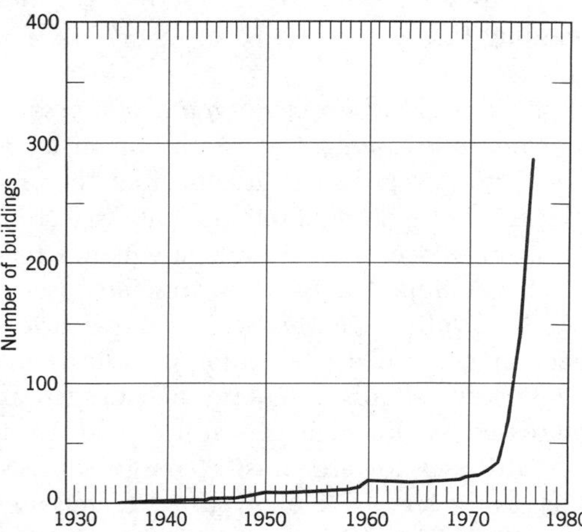

Fig. 1.8 This graph shows the rapid increase in the number of solar-heated buildings in the United States during the 1970s. In 1976, the 146 new solar buildings exceeded the total number of such buildings in all preceeding years. William A. Shurcliff, Solar Heated Buildings: A Brief Survey, *13th edition, 1977.*

(b) The Trend from Labor to Energy-Intensive Practices Within Buildings. All of us know the convenience of energy-driven motors that do what we otherwise would do manually. As appliances that replace human labor become widespread, we design them into our buildings, often without provision for the now-obsolete human-powered practices they replace.

Figure 1.9 illustrates the effective but time-consuming and labor-intensive practice of solar clothes drying. The use of the energy-intensive alternative, the mechanical clothes dryer, was increasing U.S. energy consumption for this purpose by about 10% per year in the 1970s. A designer who considers weather unpredictability, cultural expectations, and energy scarcity can provide an outdoor clothesline that is both visible and easily accessible from the mechanical dryer (see Chapter 3 for further discussion).

Fig. 1.9 *Compare the simple solar clothes dryer with the collectors above. This time- and labor-intensive solution is very different from the energy-intensive mechanical clothes dryers; energy consumed in such appliances was increasing 10% per year in the early 1970s. Photo by Douglas Boleyn.*

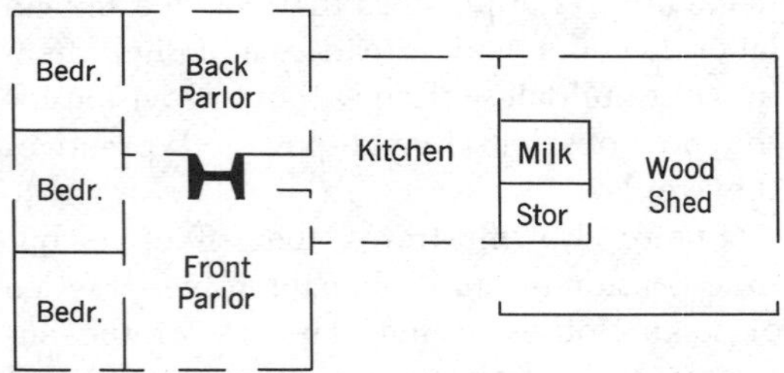

Fig. 1.10 *Labor-intensive heat regulation. The house dependent on fireplaces or wood stoves also depends on someone to tend the fire. The warm area near the fire in this early Oregon farmhouse was used for social purposes; the colder spaces at the extremities served for sleeping areas and for storage of food and fuel. Based on a plan drawn by Philip Dole.*

Another trend to energy-intensive practices is evident in the thermal control of buildings. Figure 1.10 shows the plan of a pioneer house that is dependent on fireplaces and the kitchen's wood stove for heat. Spaces containing fireplaces could be kept warm by adding wood to the fire or by stirring it, requiring periodic labor on the part of the inhabitants. Spaces without fireplaces are less controlled, with a temperature somewhere between that of the fireplace rooms and that of the outdoors. Today, the common approach is to provide thermal control via a thermostat, only occasionally adjusted by users. One such device often governs several rooms, keeping all of them near one temperature, whether or not they are occupied. Another labor-intensive versus energy-intensive design solution is shown in Figure 1.11. In this example, the labor-intensive approach to snow removal has significant consequences for architectural form.

Recent shifts to labor-intensive practices are scarce; yet, in facing a future with a continued increase in world population and nonrenewable resource depletion, it seems obvious that labor-intensive alternatives must be made available. The designer's challenge is to make these alternatives attractive and rewarding to use.

(c) ***The Trend From Energy Awareness to Unconscious Energy Use.*** With human labor being greatly reduced and the fuel that replaces it being brought into our buildings from far away, it seems a natural consequence that we have slipped into an unconscious use of energy. Until recently, with energy prices declining, the cheap, seemingly endless, and unobtrusive supply of energy encouraged an increased use by each person it served. Even our attitudes toward on-site energy sources, such as the sun, have evolved to exclude user awareness (Fig. 1.12).

With the apparent depletion of nonrenewable energy resources, a "scarcity ethic" is developing. This encourages increased self-reliance in the face of eventually decreasing imports, and brings "user control" of

Fig. 1.11 *Two approaches to dealing with snow at entrances. The labor-intensive solution* (a) *utilizes the stair (or ramp); as snow accumulates, "ground level" simply moves up this inclined surface. The person climbing toward the entry, or using a shovel to clear the upper treads, has a labor intensive task. In* (b), *heated water is piped below the sidewalk to turn snow into water. As temperatures drop below freezing, this solution becomes a particularly energy-intensive procedure.*

on-site energy sources back to architectural design considerations. Thus another challenge to the designer: to engage the users in indoor environmental control without making them slaves to the building. For example, a certain minimum level of thermal performance could be possible independent of daily human intervention.

1.3 Buildings and the Demand for Energy

The construction, operation, and demolition of buildings in the United States accounts for about one-third of this country's energy use. Figure 1.13 graphs the approximate percentage of gross-energy use for various purposes in our society. Transportation, 25% of gross usage, is linked with energy used in buildings; this is particularly evident in the layout of residential areas, which essentially require frequent automobile use. The automobile accounts for about one-half of transportation's share of energy usage. The goal of renewable energy sources for buildings may conflict with the goal of reduced automobile usage, since solar energy is a somewhat thinly spread resource. This tends to limit the density of buildings; decreasing density usually results in the increasing use of transportation, as people travel further between places of home and work. This conflict has several possible resolutions that may include a merging of work and home sites, a substitution of communication for transportation, or a provision for transportation to be powered by renewable resources.

Fig. 1.12 *Varying provisions for user awareness in schools. The famous early solar-heated school at Wallasey, England* (a, *and* b) *depends on custodians and users to make the adjustments that alternately admit or exclude direct sun, control the extent of diffused light and outdoor air admitted, and insulate against loss of stored heat at night. (Photos by Reg McDonald.) In a similar climate, a windowless school in the Pacific Northwest* (c) *excludes the unpredictable outdoors. Here the users' consciousness is directed away from energy. (See Chapter 8 for another school's approach to users, the sun, and energy.)*

Fig. 1.13 (opposite) *Percent and quality of energy use in the United States. The categories of water and space heating, air conditioning, refrigeration, and lighting are those most influenced by architectural design. "Quality" is related to terminal process temperatures; "low-grade" uses, at lower temperatures, can be fueled by a variety of energy sources. As temperatures rise, the variety of possible fuels diminishes. At the top of the "high-grade" uses, electricity is produced by the expenditure of much lower-grade fuel. (See also Fig. 1.17.) From E. Lazslo and J. Bierman (Eds.),* Goals in a Global Community, *published by Pergamon Press, Inc., Fairview Park, Elmsford, N.Y. Copyright © 1977 by State University of New York.*

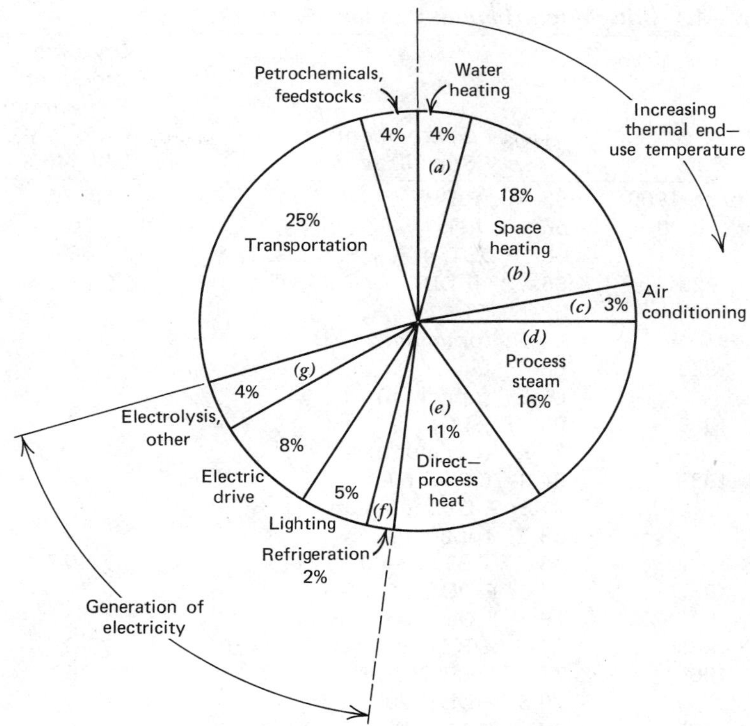

Functional Energy Uses	Percent of 1968 U.S. Energy Use	Terminal Process Temperature Required, in Centigrade
(*a*) Water heating	4.0	50–85
(*b*) Space heating	17.9	40–75
(*c*) Air conditioning	2.5	77–96[a]
(*d*) Industrial process steam	16.7 (87%)	100–200
	(13%)	200–250
(*e*) Industrial direct heat	11.5 (80%)	<350
	(20%)	>350
(*f*) Refrigeration	2.2	150–175
(*g*) Selected materials for refining (conventionally by electrolysis)	1.2	>350
Other:		
Clothes drying	0.3	20–65
Cooking	1.3	100–250

[a]These temperatures are the range for solar air conditioning.

1.4 Past Increases in Energy Usage

The last few decades in building design have seen a dramatic increase in recommended levels of heating and lighting for buildings. The evolution of higher recommended indoor temperatures is shown in Table 1.1. Lighting levels have risen even more rapidly, as a look at past editions of this textbook (Table 1.2) reveals. (Footcandles are defined in Chapter 18.) A reassessment of these lighting levels is underway, as shown by later scaled-down recommendations of the Federal Energy Administration. These recommended increases and their impact on energy consumption

Table 1.1 ***Criteria for Thermal Comfort, Since 1900***

Date	Environmental Specifications	Dry Bulb Temperature (in Fahrenheit), for RH = 40%
Prior to 1900	65–70° F DBT	65–70
Early 1900	56° F WBT	70
1914	68° F DBT, 40% RH	68
1923	66–72° F DBT, 19–61% RH	66–72
1923	62–69 ET 64 ET (optimum)	68
1925	63–71 ET 66 ET (optimum)	71
1929	66–75 ET 71 ET (optimum)	77
1939	64.8–76.0° F ET 71.8° F ET (optimum)	78
1941	68 ET (optimum)	74
1938–1956	73–77 DBT, 25–60% RH	73–77
1960	77.6° F DBT, 30% RH 76.5° F DBT, 85% RH	77
1965[a]	73–77 DBT, less than 60% RH	73–77
1965[b]	77° F DBT, 70% RH 79.5° F DBT, 20% RH	78
1975[c]	72° F DBT, 30% RH, Winter[d] 78° F DBT, Summer	

[a]ASHRAE Standard.

[b]For lightly clothed subjects engaged in sedentary activity. Reprinted by permission from *Criteria for Thermal Comfort* by R. G. Nevins, Institute for Environmental Research, Kansas State University.

[c]ASHRAE Standard 90-75.

[d]Suggested maximum RH if humidification is provided.

reached a peak in the years around 1970, which we could call the "Pre-Energy Awareness" period in recent U.S. history. One example of the extent to which increased lighting has been carried is seen in an electric utility headquarters building, renovated and enlarged in the mid-1960s (see Fig. 1.14). Among its features are:

- 300 to 350 footcandles (abbreviated "fc") of light in office areas.
- Up to 550 fc in display areas.
- Up to 600 fc in conference and demonstration rooms.
- 500-watt (abbreviated "w") luminaires in each of 288 window sills, for nighttime facade lighting.

During the design phase of this building, the architect calculated that construction costs would be reduced by about $1,000,000 if lighting levels were cut to 150 fc. This savings would be evident in the lower number of luminaires and in the reduced size of cooling equipment necessary to remove the surplus heat. The utility, however, expressed its interest in leading the trend to higher lighting levels, and retained the 300-plus footcandle level. Some of the heat emitted by this lighting is captured and used when needed to heat the office building ("heat-of-light" systems are discussed in Chapter 21, Section 21.18). Yet this high a lighting level produces a need for cooling, not for additional heating, in a typical office building on most winter days. The lighting thus provides surplus heat for almost all of

Table 1.2 ***Trends in Recommended Minimum Lighting Levels (in Footcandles)***

Category	*From* Mechanical and Electrical Equipment for Buildings		*Federal Energy Administration (1976)*
	2nd Edition (1945)	*5th Edition (1971)*	
Offices			
Accounting, bookkeeping	30	150	
Regular office work	20	100	50
Conference rooms	10	(Not listed)	30
Corridors, stairs	5	30 (But not less than one-fifth the level of the adjacent area)	10
Schools			
Auditoriums	10	30	
Classrooms, regular deskwork	20	70	
Drafting, drawing	30–50	100	
Sewing	50–100	150	
Libraries			
Reading room	20	70	

Fig. 1.14 *An office building for an electric utility, designed in the mid-1960s when energy seemed in plentiful supply. Extraordinarily high interior lighting levels as well as (now unused) 500-w exterior night-lights in* each *window sill are remnants of a less energy-conscious era. Photo by Stan A. Adams.*

the year and requires that still more energy be spent to remove surplus heat. This utility has since discontinued the lighting of its facade at night but continues to provide very high lighting levels for its interiors.

1.5 The Role of Architecture in Future Energy Consumption

There are many differing projections of future U.S. energy consumption, most of which assume continued, though less rapid, growth in demand. These views of energy supply-and-demand are constantly shifting, and this year's estimate does not assure next year's performance, nor certainly the next decade's. Consider the impact of the 1973 Oil Embargo both on the reality of our subsequent consumption and on the revised forecasts for the future (Fig. 1.15). In Figure 1.16, the predicted impact of an "active conservation program" is shown, including improved energy efficiency for automobiles, homes, and office buildings. Clearly, energy efficiency in buildings can have an impact on our future consumption patterns.

A more detailed look at building energy efficiency is found in many parts of this book, especially Part Two (Indoor Climate Control) and Part Six (Illumination). The relationship between indoor climate and illumination is likely to be particularly influential on building energy consumption, as the office building cited above demonstrates. Electricity is one of

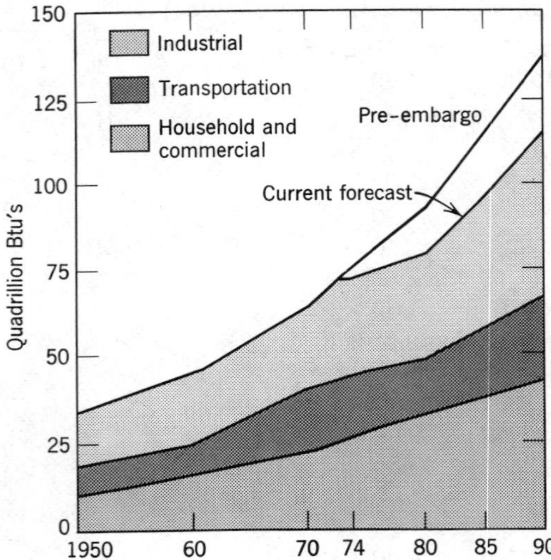

Fig. 1.15 U.S. energy consumption, past and future. Energy for buildings is expected to continue to form a large portion of our national energy usage. Based on Federal Energy Administration (FEA), 1976 National Energy Outlook.

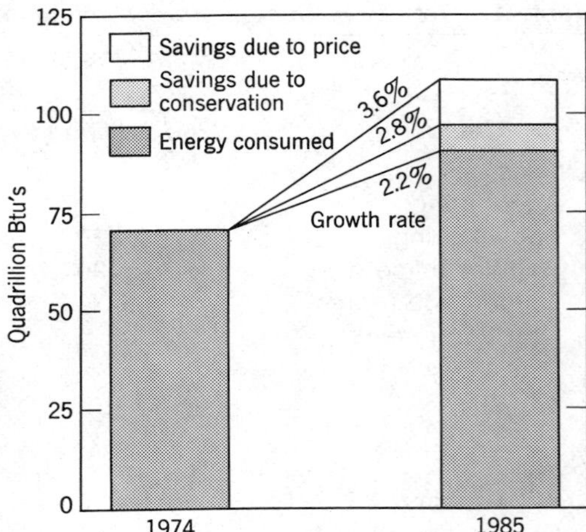

Fig. 1.16 Potential for energy conservation. Savings due to higher prices are reflected in the forecast of Figure 1.15. In addition, an active conservation program is predicted to be capable of further *reducing our annual energy-demand growth rate to 2.2%. FEA,* 1976 National Energy Outlook.

the keys to this relationship, for the following reasons.

1. Consumption of electricity is expected to rise about twice as fast (5.4% annually) as the overall energy demand, and more and more we are expected to use electricity in place of other energy forms. Part of the reason for this is seen in Section 1.9; for some energy sources, conversion to electricity before distribution to buildings is the most convenient option.
2. Electricity is almost the only source of illumination in buildings, other than daylight; the past annual growth rate (about 7%) of electricity consumption was boosted by the steeply increased illumination recommendations and the cooling that accompanied them.
3. Electricity is a convenient and versatile energy form; it not only serves the high-temperature and highly concentrated (or "high-grade") energy tasks, for example, lighting and motive power, but it can also serve the low-temperature, simpler (or "low-grade") tasks, such as cooking, heating water, and space heating for buildings (see also Fig. 1.13). "All electric" buildings have become commonplace, despite the fact that they are subject to paralysis in blackouts, as is any building dependent on a single energy source.
4. Electricity delivers only about one-third of the total energy that goes into its production; the other two-thirds are usually lost as waste heat at the generating plant.

Thus, the relationships among greater energy demands, electricity, lighting, and interior climate are complex. As we increase our use of electricity, we doubly increase our use of the nonrenewable fuels that serve almost all of today's electrical generating plants.

The building designer interested in slowing our energy-demand growth rate can begin with a careful utilization of electricity; low-grade energy tasks might be better served by nonelectric fuel sources. As an example, consider the furnace in a residence (Fig. 1.17).

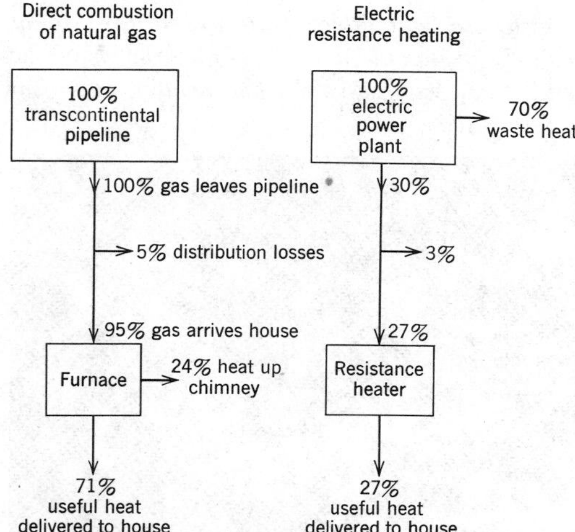

Fig. 1.17 *Efficiency comparisons. For lower-grade heating tasks, more of the energy available in fossil fuels is utilized by the furnace, which burns the fuel directly. The electric furnace using resistance heaters is fully efficient in the home, but the higher-grade energy it receives has produced waste heat at the generating plant. John C. Fisher,* Energy Crisis in Perspective, *John Wiley & Sons, 1974.*

The low-grade alternatives to electric-resistance furnaces include not only fossil fuel-burning furnaces but also renewable heat sources, such as wood and solar energy.

Although the low efficiency of electrical generation is obvious, the central generating plant has some advantages over many fuel-burning furnaces. Air-pollution control is more efficient at a central plant, and the massive concentration of its waste heat could be coupled with certain industrial processes or piped to nearby buildings for space heating. (However, since buildings need this waste heat only in winter, a separate, additional cooling solution for the summer must also be provided: two cooling systems are obviously more costly to build.)

The higher-grade tasks for which electricity is suitable also need careful consideration. The higher the level of electric lighting, the more cooling is needed; depending on the climate, every unit of energy expended in lighting will require another half-unit of energy for removing the surplus heat thus generated. The annual energy growth rate for the cooling of buildings is about 10%—more than three times as fast as our overall energy growth shown previously in Figure 1.15.

One way to reduce the electricity consumed by lighting is to substitute daylight for electric lighting. This is most applicable at the perimeters of buildings; yet, even the interiors of low-rise buildings can be served with daylight for the general or overall illumination, using small individually controlled electric lights only where and when needed (Fig. 1.18).

As with many such substitutions, the designer must consider the trade-off: will more glass area to admit daylight produce greater heat loss on winter nights and undesired heat gain on summer days? (Calculations of heat gain and heat loss are presented in Part Two.) In Chapter 3 techniques are discussed for protecting glass against heat loss with insulating shutters and for designing windows to minimize summer solar gain. See in particular Figure 3.12, page 69.

1.6 Energy in Building Construction

In the United States, the total energy expended in *constructing* new buildings, as a percentage of the total used for all purposes, appears relatively small (Fig. 1.19). The decisions that designers make regarding the building envelope and the structural and mechanical systems generally have greater energy consequences in building operation over many years than in the shorter construction period. Most of the energy embodied in construction is invested in the manufacture of materials and components (Fig. 1.20).

Some building types are much more "energy intensive" in the construction phase than others, as shown in Figure 1.21. Single-family residences are relatively low in "energy intensiveness," compared to the total energy invested in them (from Fig. 1.19); such residences constitute almost one-third of the total square footage of all new buildings constructed in the years for which these statistics are compiled. The single-family residence and the closely related garden apartments (two- to

Fig. 1.18 Combining general daylight with specific-task electric light. Alvar Aalto's library at Mount Angel Abbey (Oregon) utilizes a central north-facing skylight to provide daylight on two levels of the interior. The individual electric lights (b), *each with a pull chain, can be used when reading fine print makes more light desirable.*

four-family residential and low-rise residential) are low in energy intensiveness, primarily because they utilize so much wood in their construction, and wood is a low-energy material. A comparison of the approximate total energy embodied in selected materials, delivered to the job site, is shown in Table 1.3.

Until recently, building material choices were rarely influenced by the amount of energy invested in those materials. Now that such information is becoming more widely available, energy content can be included with other factors—esthetics, durability, fire resistance, labor intensiveness, and installed cost—that a designer considers in choosing materials for a building.

A related consideration is the monetary cost of the energy-consuming equipment in a building. Table 1.4 indicates the portion of recent construction costs that are assigned to the building elements discussed in this text: plumbing, HVAC (heating, ventilating, air

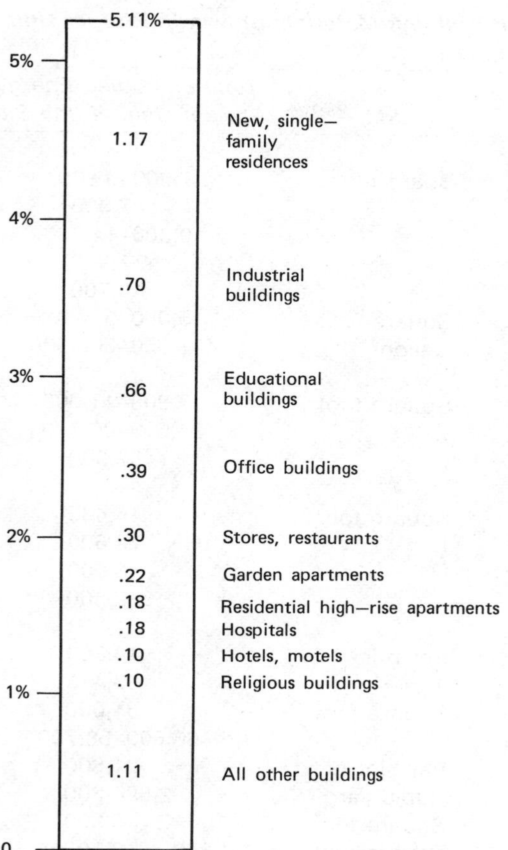

Fig. 1.19 Energy in new building construction, as a percent of U.S. total energy consumption. This includes energy to manufacture, transport, and erect new buildings. From B. M. Hannon, R. A. Stein, B. Z. Segal, D. Serber, Energy Use for Building Construction, *for U.S. Energy Research and Development Administration (ERDA), 1977.*

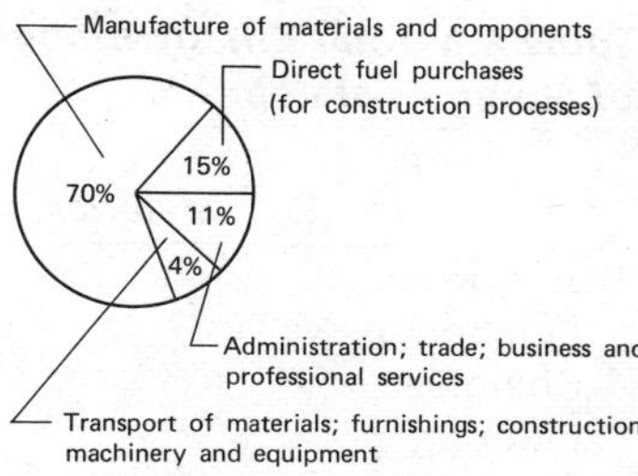

Fig. 1.20 Energy embodied in building construction goes largely into the manufacture of building materials and other components. Based on B. M. Hannon, R. A. Stein, B. Z. Segal, D. Serber, Energy Use for Building Construction, *for ERDA, 1977.*

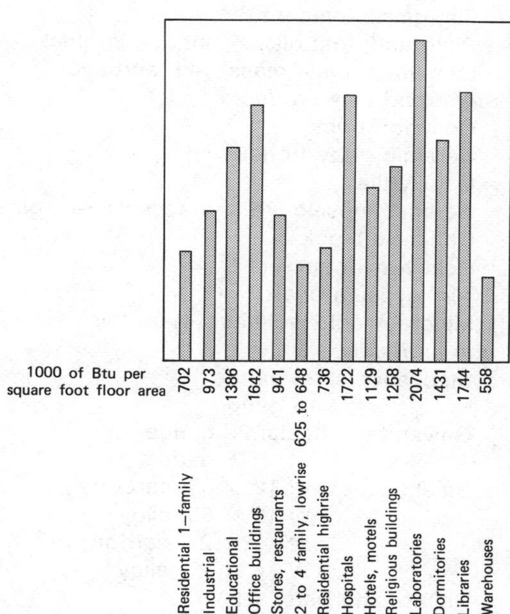

Fig. 1.21 Energy embodied per square foot of floor area, for various building types. These comparisons of 1000 Btu per unit area are for the building types shown previously in Figure 1.19. From B. M. Hannon, R. A. Stein, B. Z. Segal, D. Serber, Energy Use for Building Construction, *for ERDA, 1977.*

conditioning), and electrical equipment. The thermal and other environmental control systems for buildings are already a major influence on construction costs, as well as on energy consumption in operating the building. As the techniques and equipment for reducing energy consumption are developed through such steps as recovery of waste heat and the substitution of solar energy for fossil fuels, the percentage of mechanical and electrical system cost will probably become an even greater influence on the cost of constructing new buildings.

1.7 Energy in Building Operation

The typical building will consume more energy in its lifetime of operation than will be used in its construction. The estimated energy

Table 1.3 ***Total Energy Embodiment in Selected Building Materials Shown in Btu per Unit of Material, at Job Site***[a]

Material	*Unit*	*Total Embodied Energy Btu per Unit, at Job Site*
Wood products:		
Lumber	Board foot	7,600–9,800
Shingles		7,300
Flooring		10,300–14,300
Mouldings		17,900
Glu-lam		16,700
$\frac{3}{8}$-in. plywood (softwood)	Square foot	5,000–5,800
Paints	Gallon	437,000–508,500
Asphalt roofing:		
Rolls	Square foot	7,800–11,000
Shingles		25,600–29,700
Mineral-surfaced insulating board siding		67,500
Glass:		
Flat glass: double strength	Square foot	15,430
Flat glass: tempered		72,600
Plate and float glass, $\frac{1}{8}$-in. to $\frac{1}{4}$-in. thick		48,000
Laminated plate glass, $\frac{1}{4}$-in. and over		212,500
Stone and clay products:		
Common brick	Per brick	14,300
Ceramic glazed brick		33,413
Quarry tile	Square foot	51,000
Ceramic mosaic tile and accessories, glazed		63,600–68,700
Concrete block	Per block	31,800
Ready-mix concrete	Cubic yard	2,594,300
$\frac{3}{8}$-in. gypsum board	Square foot	5,300
Mineral wool insulation, $4\frac{1}{2}$-in. thick	Square foot	8,300
Iron and steel:		
Steel sheets: 22 gauge	Square foot	29,400
16 gauge		58,800
Galvanized sheets: 22 gauge		49,800
16 gauge		98,500
Steel shapes: W 12 × 65, carbon	Lineal foot	1,217,800
W 12 × 65, alloy		1,749,200
WT 6 × 27, carbon		543,344
WT 6 × 27, alloy		780,400
Reinforcing bars: #2		2,600
#8		41,800
Welded wire mesh: 2 × 4, 14/14	Pound	3,900
2 × 12, 8/8		25,400
Pipe, carbon steel: $\frac{1}{2}$-in. diameter	Lineal foot	21,900
6-in. diameter		489,700
Stainless steel sheets: cold rolled	Pound	138,300
hot rolled		80,800
Aluminum:		
$\frac{1}{4}$-in. plate	Square foot	420,700
1-in. plate		1,680,300
$\frac{1}{8}$-in. sheet		174,800
Standard shapes: 8 I 8.81	Lineal foot	811,800
6 I 5.10		469,900

[a](Data from Hannon, Stein, Segal, Serber, *Energy Use for Building Construction,* for ERDA, 1977.)

Table 1.4 *Construction Costs for Mechanical and Electrical Equipment*

Building Type	*Portion of Total Construction-Dollar Cost in Percent*			
	Plumbing	*Mechanical*	*Electrical*	*Elevators*[a]
Apartments	10	7	7	4
Auditoriums	4	12	10	
Banks	5	10	10	
Churches	5	11	8	
College:				
Classrooms and administrative offices	6	16	11	
Laboratories	8	16	8	
Department stores	8	11	9	3
Hospitals	9	16	12	3
Libraries	4	13	11	2
Offices	5	12	11	4
Restaurants	8	13	10	
Retail stores	5	10	10	
Schools:				
High schools	7	11	10	1
Elementary schools	6	11	11	1
Supermarkets	6	9	14	

Source. (From 1976 Means, *Building Construction Cost Data,* Copyright © 1976, Robert Snow Means Company, Inc.)

[a]Elevator costs obviously vary considerably with building height.

needs for the construction, operation, and demolition of an office building in Albany, New York, are shown in Figure 1.22.

There are several proposals to establish an upper limit for building energy consumption in operation. In their simplest form, such regulations would specify the maximum allowable energy input per building-area unit per year: British thermal units per square foot per year, or megajoules per square meter per year. A more complicated set of regulations to define—but relatively easy to apply to the design of walls, roofs, and so on—would establish maximum overall thermal transfer values. This means that the designer would be required to choose a combination of wall materials, for instance, those that did not allow more than a specified rate of heat transfer. These more detailed regulations would also include HVAC-system performance (indoor temperatures, controls, ventilation rates, humidity control, zoning, pipe and duct insulation, equipment-performance ratings, etc.), hot water service, electrical distribution systems, and lighting. Details of one such set of proposals, ASHRAE Standard 90-75 (American Society of Heating, Refrigerating and Air Conditioning Engineers), are found throughout Part Two.

The impact of more energy-efficient buildings can be estimated in several ways. The American Institute of Architects (AIA) estimated in 1976 that a high-priority national energy conservation program could save at least 30% of the energy now used in existing buildings, and 60% of the energy that would have been used by new buildings had they not been designed with energy conservation in mind. The AIA estimated that by 1990 the implementation of such a program would save as much energy as our nation could expect from the daily output of any one of our 1990 domestic energy sources. A more specific example is shown in Figure 1.23.

As energy used in operating buildings becomes more expensive and more difficult to obtain, its influence on the design process will become evident over a wider range of the designers' choices. We can expect buildings to be placed not only to take advantage of natu-

Randy Cook

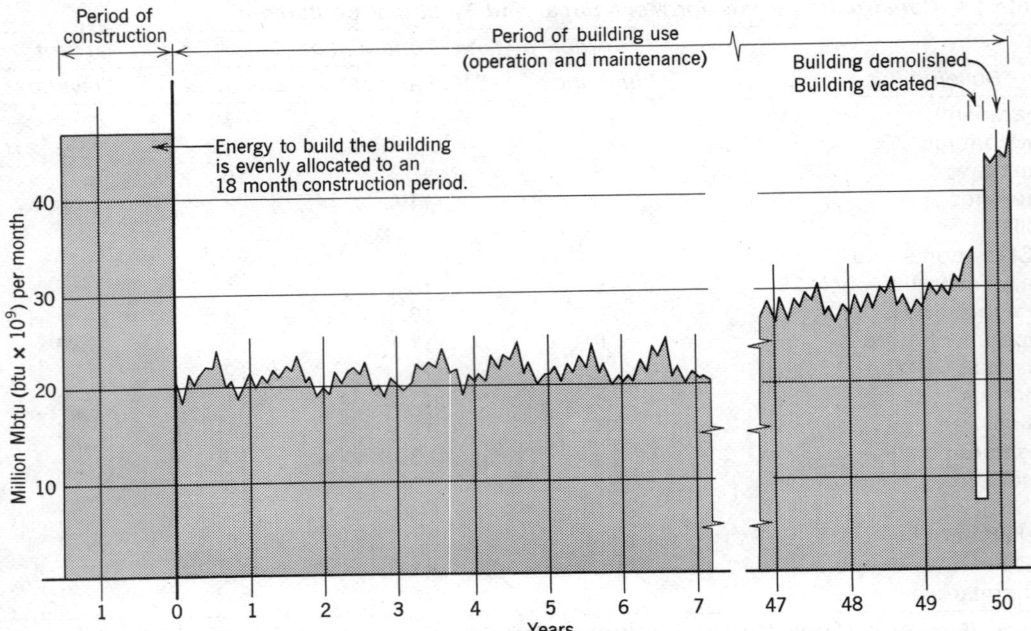

Fig. 1.22 *Energy use in constructing, operating, and demolishing an office building. This shows the approximate energy used over an estimated 50-year life span for a 650,000 sq ft office building in Albany, New York. Energy in construction includes total energy embodied in the materials and is shown as constant through the construction period. The peaks and valleys of energy use in each year of operation correspond to the expected weather; peaks during the extremes of winter cold and summer heat, valleys in the milder spring and fall. As the building ages and energy leaks become more numerous, the consumption is assumed to rise gradually. From* Architecture and Energy, *by Richard G. Stein. Copyright © 1977 by Richard G. Stein. Used by permission of Doubleday & Company, Inc.*

ral energy sources, but to be so conscious of the energy they consume that significant efforts will be made to utilize formerly wasted heat. It is possible that even a single-family residence will be so thoroughly insulated and so equipped to recover otherwise wasted heat that a furnace (or other space-heating equipment) will become unnecessary in the future. For many brightly lighted office buildings, this has already occurred.

1.8 Energy in Building Demolition

Compared to construction and operation, relatively little energy is invested in a building's demolition (see again Fig. 1.22). Yet, another look at Figure 1.20 is a reminder that about 70% of the total energy invested in construction is embodied in a building's materials and components. The more of a building that can be recycled, the more energy can be recovered. At present, the recovery of usable materials from demolition is limited. The cost of labor is high, and the cost of energy is relatively low; it is currently easier, quicker, and cheaper to reduce a building to rubble and haul it to a landfill.

Designing for the recycling of buildings is a two-part balancing act. First, the designer should provide enough flexibility to prolong the useful life of the building, by enabling it to easily adapt to changed usage. Flexibility can be expensive to implement physically and can result in a bland "sameness" throughout the structure. The latter characteristic is easier for the designer to change than the former. Second, the design can allow for demounting

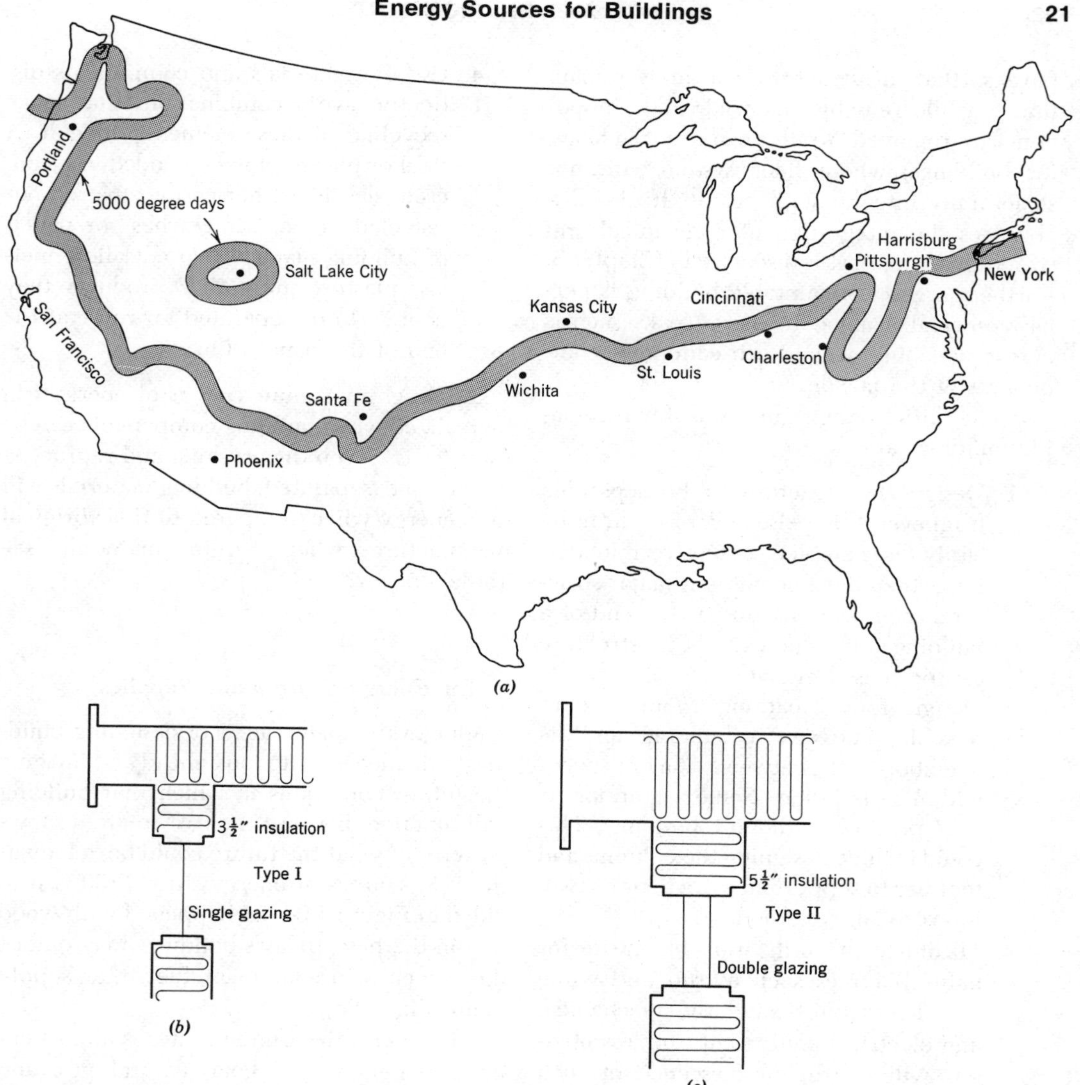

Fig. 1.23 *Estimated energy for constructing and operating a residence. These figures are based on a one-story, flat-roofed house with a 1500 sq ft floor area. It is located in an area with about a 5000 degree-day heating season, near the band shown on the map* (a). *(Degree-days are discussed in Chapter 5.) In areas above the line, yearly energy consumed—and saved, by the more efficient house—would be even greater.*

	Energy Embodied	*Annual Energy Demand*	*Demand over 20 Years*
Figure (*b*), Type I:	169 million Btu	109 million Btu	2180 million Btu
Figure (*c*), Type II:	179 million Btu	77 million Btu	1540 million Btu
Difference between types:	10 million Btu more to build Type II	32 million Btu less to operate Type II yearly	640 million Btu less to operate Type II for 20 years

The annual energy demand includes heat lost to cold air infiltrating the house. The extra energy used in building the $5\frac{1}{2}$ in. insulation and the double glazing is quickly repaid—in about one-third of one heating season! Based on B. M. Hannon, R. A. Stein, B. Z. Segal, D. Serber, Energy Use for Building Construction, *for ERDA, 1977.*

parts so that the structure can safely remain intact, while reusable materials and components are removed. Yet this can result in heavier buildings, where floor systems are not structurally integral with beams. It also discourages the integration of mechanical and structural systems, as is discussed in Chapter 3. Furthermore, a demountable building is perhaps especially subject to energy leaks, such as cracks widening around self-contained components of the facade.

Some initial design guidelines for recyclable buildings are:

1. Design the structure to be separable from everything else and, of itself, to be easily disassembled. Extensive remodeling is then possible without major structural modifications, and at the end of a building's life, elements of its structure can be reused elsewhere.
2. Design for "breathing room" where possible, between a building and its neighbors, or between major spaces within a building. Some expansion is thus possible without rebuilding. This could include designing the columns and footings to support an extra floor or two, for vertical expansion.
3. Maximize the utilization of on-site (or natural) forces such as sun and wind. The less sophisticated the mechanical and electrical equipment, the less obvious will be the obsolescence of such equipment with the passing of time.
4. Use the materials and components distinctly; avoid combinations that make recycling of these elements difficult. A steel or plastic pipe embedded in a concrete slab is neither easily repaired nor recycled; some "sandwiches" or panels of building materials do not allow metals, plastics, and other products they contain to be separated for reuse at the end of the panel's life.

Although maximum savings of energy can be realized when building components can be used "as is," even the crushing and reprocessing of some (separated) building materials will save energy when compared to their original manufacture when virgin material; see Table 1.5.

1.9 Energy Sources and Supplies

One of the many facets of designing buildings is the necessity to look ahead and imagine the future conditions in which that building will function. Because the past so often shows patterns of what the future could be, a look at the U.S. sources of energy since 1850 is provided in Figure 1.24. As happened with wood fuel in the past, today's buildings may outlast the currently predominant fuels that supply them (Fig. 1.25).

At present, the United States is almost entirely dependent on coal, natural gas, and oil—much of the oil is imported. There are

Table 1.5 ***Recycled Material Compared to Virgin Material, Percent Savings***[a]

Material	*Point of Impact*	*Percent Saved*	*Ease of Recycling*
Aluminum	Molten aluminum	96	Present technology limited by separation and collection.
Plastics	Molten polymer	96	Extremely difficult; no satisfactory technology known.
Steel	Molten steel	52	Present technology limited by impurities, separation, and collection.
Paperboard	Pulp	50	Present technology for separated waste; no satisfactory technology for mixed waste.

[a](Data from R. Berry and H. Makino, "Energy Thrift in Packaging and Marketing," *Technology Review*, February 1974. Copyright © 1974 by the Alumni Association of the Massachusetts Institute of Technology.)

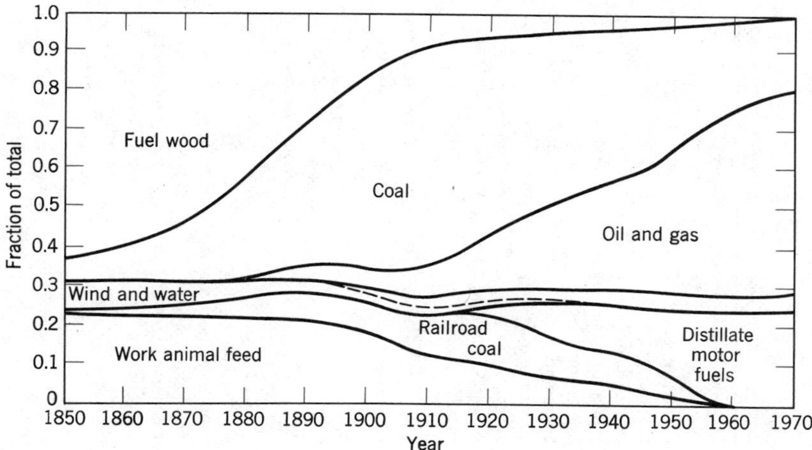

Fig. 1.24 U.S. fuel input, 1850 to 1970. An illustration of the progression from dependence on the renewable fuels of wood and work animals to the fossil fuels of coal, then to oil and gas. Wind and water power changed their form from "mills" to hydroelectricity between 1890 and 1940., Similarly, though not shown here, much fossil fuel is now converted to electricity before use. From J. Fisher, Energy Crisis in Perspective, *John Wiley & Sons, 1974.*

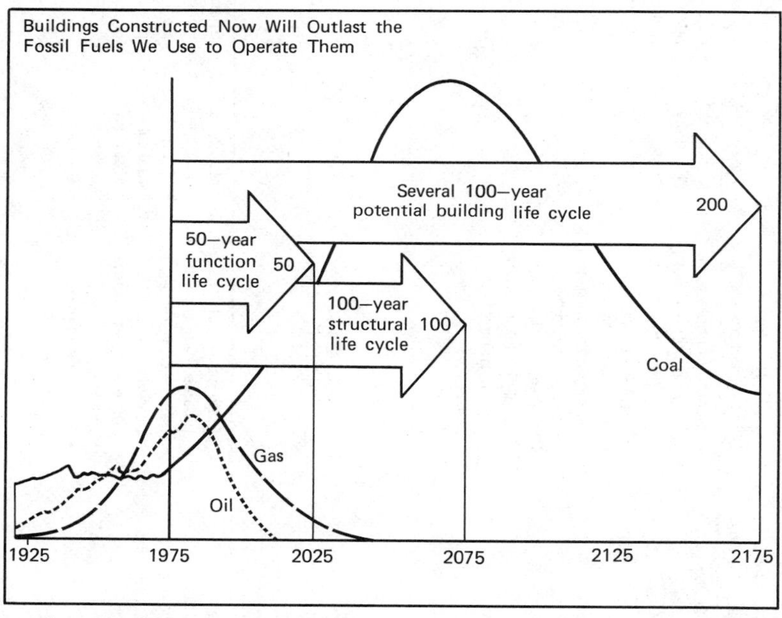

Fig. 1.25 Life cycles of buildings and of fossil fuel reserves. This graph shows our relative dependence on three major energy sources. In 1925, we used mostly coal, some oil, and very little natural gas. In 1975, natural gas ranked first, followed by oil, then coal. By 2025, we are predicted to be almost entirely dependent on coal. From the Office of the State Architect of California, Building Value, *1976.*

many other possible energy sources, even though they are at present minor contributors to our energy supply. A summary of energy sources, their characteristics, and their impacts on architecture and the environment is presented in Table 1.6. Several widely discussed, but still uncertain, energy sources are omitted from this table. These include nuclear

Table 1.6 ***U.S. Energy Sources***

I. Nonrenewable Sources

Nonrenewable Source	*Estimated U.S. Reserves and Resources in Quads (10^{15} Btu)*	*Notable Characteristics*	*Form in Which Supplied to Consumer*	*Typical Architectural Utilization as Fuel (not including electricity)*	*Architectural Impacts*	*Environmental Impacts*
Coal	**9000** | 38000[a] 28000 U.S. 1974 use: 15 Est. U.S. 1985 use: 23[b]	• Least flammable fossil fuel to transport, but bulky (less fuel value for its weight) • Coal dust presents mining, transport, and storage problems • Coal gasification can supply cleaner fuel to electric generating plants	• Mostly as electricity • Some as coal	• Some central steam plants (for campuses and other building complexes)	• Combustion air, stack, air-pollution control equipment, and cooling tower or pond • District heating possibilities for waste heat from steam plants • Large and "dusty" fuel storage and delivery areas • Ash collection and disposal	Recovery • Strip mining's long-term ecological disruption, including acid drainage to streams • Underground mine safety problems, and subsidence of surface over mines Transport • For slurry pipelines, water pollution Use • Air pollution (particularly with high sulfur coal) • Thermal pollution Waste • Landfills for ash disposal
Crude oil	**300** | 2700[a] 1700 | 12000 U.S. 1974 use: 21 Est. U.S. 1985 use: 30[b]	• Flammable but less explosive than natural gas, readily transported • Stored on site where used • Fuel-price manipulation by cartels	• Mostly as gasoline • Much as fuel oil • Much as electricity • Some for nonenergy usages	• Space heating • Some water heating • Some central steam plants • Some large motors	• Combustion air, flue (and waste-heat recovery opportunities) • Oil storage tank and delivery area	Recovery • Off-shore oil spills • Saltwater intrusion and surface subsidence near oil fields Transport • Oil spills, especially on water • Pipeline causes ecological disruption, especially during construction Refining • Air and water pollution

						Use • Air pollution • Thermal pollution
Natural gas	**300** | 2200[a] 180 | 4000 U.S. 1974 use: 21 Est. U.S. 1985 use: 22[b]	• Explosive, harder to transport and to store • High fuel content, burns relatively cleanly • Distributed by piped networks (except as propane)	• Mostly as natural gas • Much as electricity • Some for nonenergy usages	• Space heating • Water heating • Cooking • (Some lighting, cooling) • Some central steam plants	• Combustion air, flue (and waste-heat recovery opportunities) • Meter accessible to utility • Care in locating gas lines, especially underground on site • (For propane, outdoor storage tank and delivery area)	Transport • Pipeline causes ecological disruption • For liquified natural gas, potential for catastrophic explosion (spreading icy vapor cloud, which is highly explosive) • For bottled gas, hazard of explosive contents under pressure on site with buildings Use • Air pollution (though less than with oil, coal) • Thermal pollution
Uranium	**220** | 600[a] 5700 | 4,400,000 U.S. 1974 use: 2 Est. U.S. 1985 use: 7[b]	• Light-water reactors now operating use very little of the energy contained in the enriched fuel • Fuel enrichment itself consumes large amounts of energy • An especially controversial source, involving both environmental and arms-control issues • Fuel-price manipulation by cartels	• Almost all as electricity	• (Radioactive materials utilized in some medical and scientific equipment)	• (Radioactive materials require heavy shielding for delivery, storage, use, and waste pick-up areas) • "Fallout shelter" designation for qualifying spaces	Recovery • Mining problems similar to coal; less bulky, but with additional problem of radioactive tailings Transport • Radiation shielding and security needs for nuclear wastes Refining • Large amount of energy consumed in fuel enrichment • Radiation shielding and security needs for waste reprocessing Use • Potential for catastrophic radiation release • Radiation shielding and security needs • Thermal pollution

Table 1.6 *U.S. Energy Sources (Continued)*

Nonrenewable Source	*Estimated U.S. Reserves and Resources in Quads (10^{15} Btu)*	*Notable Characteristics*	*Form in Which Supplied to Consumer*	*Typical Architectural Utilization as Fuel (not including electricity)*	*Architectural Impacts*	*Environmental Impacts*
						Waste • Long-term radiation shielding and security needs
Geothermal energy	**2500** >10,000,000[a]	• Renewable, from earth's core, but millions of years after heat is removed from outermost layers of rock	• Electricity • Hot water	• Direct heating of buildings	• "Hot springs" water often highly corrosive • Source usually must be very near to point of use	• Unknown impacts of large-scale removal of heat from outer crust • Subsidence of surface is possible where water is removed without recharging to the earth

II. Renewable Sources

Renewable Source	*Estimated Maximum U.S. Arrival Rate in Quads (10^{15} Btu) per Year*	*Notable Characteristics*	*Form in Which Supplied to Consumer*	*Typical Architectural Utilization as Fuel (not including electricity)*	*Architectural Impacts*	*Environmental Impacts*
Solar radiation	47,000[c]	• Thinly spread, with large daily and seasonal variations in most of the United States		• Daylighting • Space and water heating • Some cooling	• Window and skylight areas for daylighting • Internal arrangement of thermal-storage materials in passive solar buildings • Large collectors, storage volumes • Less densely built complexes for solar buildings	• Ecological disruption where large land areas are converted to solar collection • Soil depletion where organic waste is not returned as fertilizer • River ecosystem disruption by hydroelectric dams
	Est. U.S. yearly capture[c] by year 2000					
Fuel wood	3[c]		• Wood fuel	• Some cooking (wood and gas)		
Farm waste	6[c]	• Methane gas highly explosive	• Methane gas • Gas or oil			
Photosynthesis fuel	15[c]					

Solar heating and cooling of buildings	9[d]		• Heated air or water	
Photovoltaics	7[d]		• Electricity	
Hydropower	9[c] U.S. 1974 use: 2 Est. U.S. 1985 use: 3[b]	• Varies with seasonal river flows	• Electricity	
Wind power	6[d]	• Varies considerably between sites	• Electricity	
Waste paper and plastic incineration	2[e]		• Electricity	• Air pollution • Thermal pollution
Tidal power	3[c]		• Electricity	• Ecological disruption of estuaries

[a]From J. Fisher, *Energy Crisis in Perspective,* John Wiley & Sons, 1974; based on U.S.G.S. Circular 650, *Energy Resources of the United States,* 1972.

Known Recoverable Reserves	Undiscovered Recoverable Resources
Known Submarginal Resources	Undiscovered Submarginal Resources

[b]Federal Energy Administration, *1976 Energy Outlook.*

[c]From C. Starr, "Energy and Power," *Scientific American,* September 1971.

[d]William J. Jewell, *BioMass Fuels—Past, Present and Future,* for International Solar Energy Society, American Section Conference, Denver, Colo., 1978.

[e]R. Berry and H. Makino, "Energy Thrift in Packaging and Marketing," *Technology Review,* February 1974. Copyright © 1974, by the Alumni Association of the Massachusetts Institute of Technology.

fusion, nuclear fission with breeder technology, oil shale, and solar energy from sea-thermal gradients. The uncertainties inherent in these sources include a combination of technical, political, and environmental problems (especially with the nuclear breeder and with oil shale) and involve doubt about whether more energy would eventually be produced by some of these sources than is required to develop them.

Similarly, the known total U.S. quantities of natural gas, crude oil, coal, and uranium in the ground are greater than the portion called *recoverable* reserves; the numbers of Quads (10^{15} Btu) of this most easily tapped portion for each of the nonrenewable energy sources are shown in boldface type in Table 1.6. The other numbers for each source are estimates of the Quads of energy that are either difficult to utilize ("submarginal") or are as yet undiscovered (see note below Table 1.6). For the known "submarginal" resources shown, the effort of retrieving them currently surpasses their value. For example, why use the energy of a barrel of oil in order to remove a barrel of oil from a well?

Despite the significant total Quads of solar radiation that the United States receives each year, we utilize relatively little for energy in buildings. The high cost of using such a plentiful but thinly spread resource has discouraged its use, but as the submarginal, nonrenewable resources must be worked to replace the dwindling recoverable portions, the difference in cost between solar and nonrenewable energy sources will shrink.

Probably even before widespread utilization of solar energy is achieved, the United States is expected to shift away from reliance on gas and oil to coal as its major energy source. This is evident from a comparison of Figures 1.24, 1.25, and Table 1.6. At present, electricity is the dominant form by which the energy of coal is delivered to buildings. Thus part of the prediction for electricity's rapid growth (Section 1.5) can be traced to the United States' relatively plentiful coal reserves. The waste heat and other environmental impacts associated with the production of electricity are significant and reinforce the necessity of careful and appropriate use of this versatile energy form by the designers of buildings.

The environmental impacts listed in Table 1.6 show only the most obvious or large-scale impacts for each energy source. Although each source has environmentally damaging consequences when used, the relatively mild impacts of some sources make them more environmentally attractive than others. The details of such consequences of energy utilization are not treated further in this text, except for a summary of air pollution in the United States and its impacts on buildings and their occupants, in Chapter 2, Section 2.4.

References and Bibliography

1. *The Architecture of the Well-Tempered Environment,* 1969, R. Banham, The University of Chicago Press, and The Architectural Press, London.

2. *The Poverty of Power: Energy and the Economic Crisis,* 1976, B. Commoner, Knopf, distributed by Random House.

3. *Energy Crisis in Perspective,* 1974, J. Fisher, John Wiley & Sons.

4. *Energy Use for Building Construction,* 1976, B. Hannon, R. G. Stein, B. Z. Segal, and D. Serber, Center for Advanced Computation, University of Illinois at Urbana-Champaign.

5. *Building Value, Energy Guidelines for State Buildings,* 1976, Office of the State Architect, Sacramento, Calif.

6. *Architecture and Energy,* 1977, R. G. Stein, Anchor Press/Doubleday.

2

Climate and the Site

A designer's early decisions in site planning will influence the later choices of both the building's mechanical and electrical equipment, and its overall consumption of energy. If the site is seen as a collection of resources (sun, wind, water, plants) and also as a part of the environment we all share, then the buildings we design can approach self-sufficiency of energy supply, without limiting the availability of local energy resources for neighboring buildings. The use of on-site resources not only can reduce the amount of energy needed to maintain the interior climate, but it also can produce outdoor spaces that become especially pleasant to use. Such spaces can direct winter sun to a glass wall while blocking the wind, or funnel the summer breeze through shade to an open window. Site planning is greatly influenced by economic considerations, zoning regulations, and adjacent developments, all of which can interfere with the design of a site to utilize the sun and the wind. Integration of all these concerns at the site-planning stage is the first step in adapting a building to its climate. This chapter looks at climate as an influence in the planning of the environment near buildings.

2.1 Different Climates, Different Buildings

Within the wide range of North American climates, four general regions can be identified (Fig. 2.1). Table 2.1 summarizes the seasonal characteristics of these four climate types and some of their implications for site and building design.

One example of how the information in Table 2.1 can be applied (and manipulated) in the planning of a building and its site is a house designed by Frank Lloyd Wright more than 30 years ago (Fig. 2.2). The direct gain of solar heat through its windows in winter makes this an early example of "passive solar heating" (by contrast, "active solar heating" includes collectors and a separate storage volume for solar heat). This house, known as the Solar Hemicycle, was built in 1948 near Madison, Wisconsin, between the cool and temperate climate zones. Winter heating is the dominant thermal influence in this area. The house stands on a hilltop site, particularly vulnerable to winter winds. In construction, earth was scooped from in front of the south face of the house and bermed against the entire curved north wall, almost to roof level.

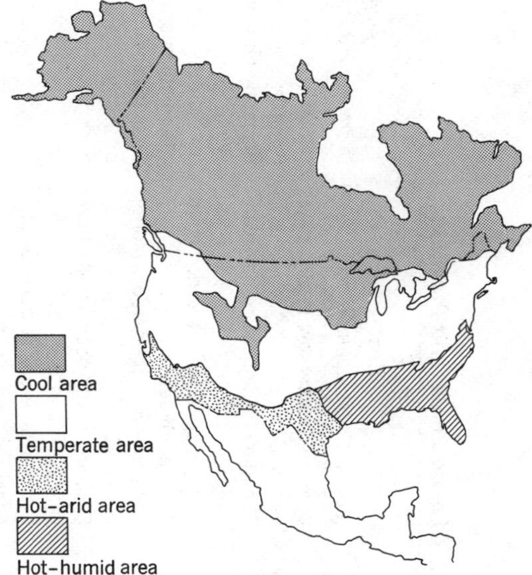

Fig. 2.1 Regional climate zones of the North American continent. From Victor Olgyay, Design with Climate: Bioclimatic Approach to Architectural Regionalism, *Copyright © 1963 by Princeton University Press. Reprinted by permission.*

Table 2.1 *U.S. Climate Zones and Planning*

	Cool (*about 45° N. Latitude and above*)	*Temperate* (*from about 35 to 45° N. Latitude*)	*Hot Arid* (*below about 35° N. Latitude, primarily Southwest*)	*Hot Humid* (*below about 35° N. Latitude, primarily Southeast*)
Climate characteristics	Winter sun very low in sky, days very short. Summer sun high in sky, days very long. Long, cold winters are dominant thermal influence. Fall somewhat shorter, often drier than spring. Summer cool (occasional short, very hot periods). Marked differences between the seasons.	Winter sun low in sky. Summer sun high enough to significantly impact roof surfaces. Winter and summer each critical in thermal influence, especially where summers are humid. Seasons about equal in length and distinctly different.	Summer sun very high in sky. Greatest impact on roofs, also on east and west walls. Winter sun relatively high in sky, days longer than at northern latitudes. Long hot summer is dominant thermal influence. Short and sunny winter. High daily temperature range from hot afternoon to cold at dawn.	Summer sun very high in sky. Greatest impact on roofs, also on east and west walls. Winter sun relatively high in sky, days longer than at northern latitudes. Summer very long, with slight seasonal changes and relatively constant temperature. High humidity with heat is dominant thermal influence.
Early building responses	Compact buildings huddled together, or sharing common walls. Minimal window areas; rooms for social usages kept warmer than bedrooms. Attention to protection from winter wind.	Buildings with flexible perimeter spaces, such as porches screened in summer, glazed in winter. Wide diversity of buildings and local climate conditions. Some underground space for winter food storage, summer escape from heat.	Buildings sharing east-west walls, with heavy walls to absorb heat by day, release heat in evening. Wind ventilation not favored, to avoid overdrying. South-facing terraces to provide a warmer, early morning place in winter. Courtyards with deciduous plants for summer outdoor shade.	Emphasis on shade, with openings to breeze both day and night. Elevated floors, ceiling-to-floor windows and open plans facilitate air passage; lightweight roofs (sometimes ventilated below) shed sun without storing heat. Little more than a roof, a privacy screen, and a platform for much of the year.
Desirable site characteristics	Protected from winter winds, but with access to winter sun. Access to morning sun for most of the year. Shielding to west and northwest in summer. Avoid the bottom of slopes where cold air collects, or the top where winds carry off heat. Dry enough for underground spaces.	Protected from winter storm winds, but with access to both summer breeze and winter sun. Shading in summer both to east and west, with deciduous trees over roofs. Dry enough for underground spaces but able to support summer plants.	Protected from summer sun, especially to west, but with some access to winter sun. Protected from winds. Not too dry for deciduous plants near buildings, not too damp for underground spaces.	Plentiful shade and access to breezes. Quiet. Limited access to winter sun. Trees over roof for shading.

Building location on site	Outdoor space to south (and east) for control of sun access and for summer outdoor activity. Share heated walls with neighbors, especially where they are upwind in winter. Use of underground space. Locate to provide shade from west in summer.	Outdoor space to south and north for control of winter access to sun and to facilitate summer passage of breezes. A variety of outdoor space orientations for seasonal outdoor activities. Use of underground space. Locate to provide summer shade east, west, and above.	Outdoor space to north, south, and perhaps east, for seasonal outdoor activities. Use of underground space. Summer shade is especially important to west and overhead.	Outdoor space on all sides, for breeze. Shading especially important to east, west, and above. Wall openings away from noise sources. Floors somewhat elevated above ground.
Plan and volume characteristics[a]	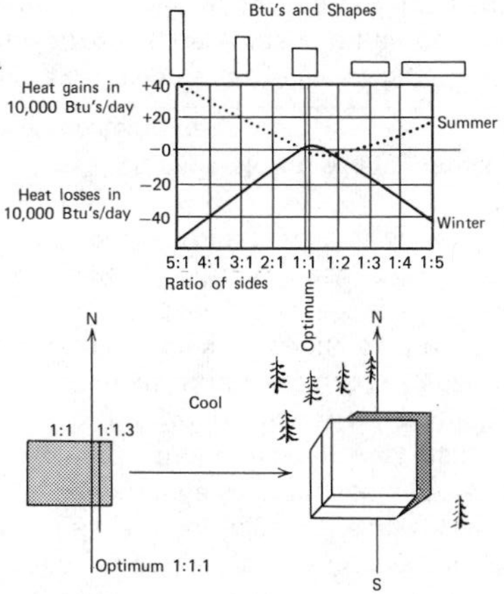 Btu's and Shapes Heat gains in 10,000 Btu's/day: +40, +20, −0 Heat losses in 10,000 Btu's/day: −20, −40 Summer, Winter 5:1 4:1 3:1 2:1 1:1 1:2 1:3 1:4 1:5 Ratio of sides Optimum N, S Cool 1:1 1:1.3 Optimum 1:1.1	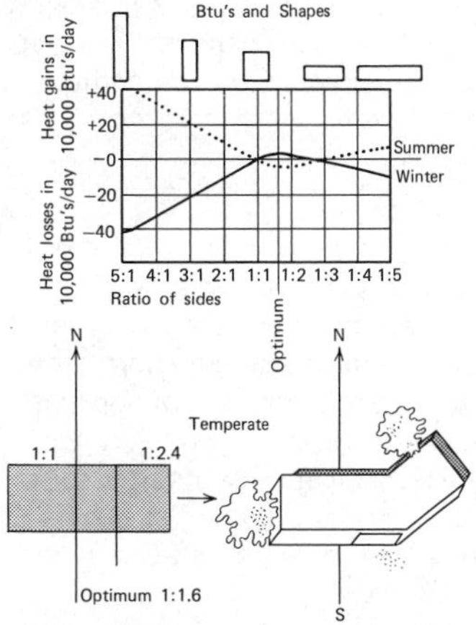 Btu's and Shapes Heat gains in 10,000 Btu's/day: +40, +20, −0 Heat losses in 10,000 Btu's/day: −20, −40 Summer, Winter 5:1 4:1 3:1 2:1 1:1 1:2 1:3 1:4 1:5 Ratio of sides Optimum N, S Temperate 1:1 1:2.4 Optimum 1:1.6	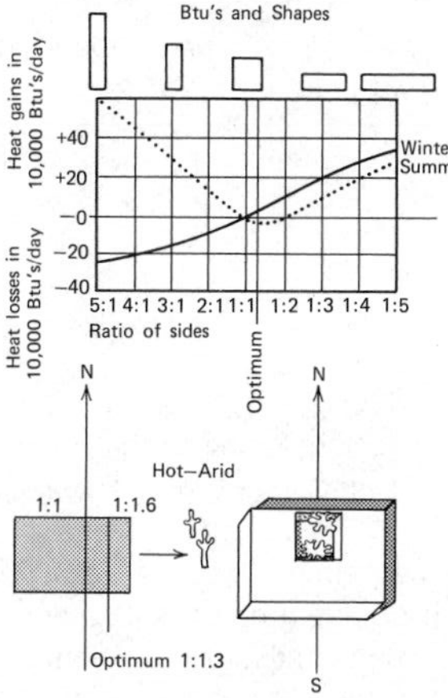 Btu's and Shapes Heat gains in 10,000 Btu's/day: +40, +20, −0 Heat losses in 10,000 Btu's/day: −20, −40 Winter, Summer 5:1 4:1 3:1 2:1 1:1 1:2 1:3 1:4 1:5 Ratio of sides Optimum N, S Hot–Arid 1:1 1:1.6 Optimum 1:1.3	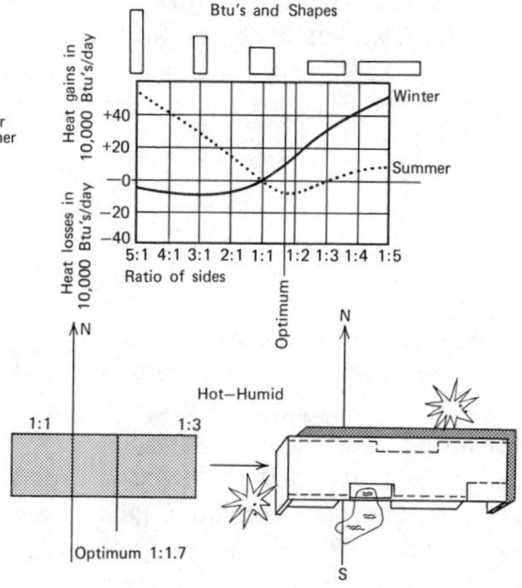 Btu's and Shapes Heat gains in 10,000 Btu's/day: +40, +20, −0 Heat losses in 10,000 Btu's/day: −20, −40 Winter, Summer 5:1 4:1 3:1 2:1 1:1 1:2 1:3 1:4 1:5 Ratio of sides Optimum N, S Hot–Humid 1:1 1:3 Optimum 1:1.7
Roofs[b]	Summer solar gain about twice that of east or west wall. Winter solar gain less than half that of south wall. Dominant influence: winter snow, wind. Attics helpful as insulation in winter, and readily lose heat on cool, clear summer nights if ventilated. Steep pitch to south for optimum winter solar gain.	Summer solar gain more than twice that of east or west wall. Winter solar gain about half that of south wall. Attics helpful in winter, but must be well ventilated in summer to lose heat in high humidity areas. Moderately steep pitch to south for optimum winter solar gain.	Summer solar gain more than twice that of east or west walls. Winter solar gain more than half that of south wall (at lower latitudes, almost equal). Dry, cool, clear nights facilitate loss of heat from flat roofs. Roof terraces serve evening, early morning activities for most of the year.	Summer solar gain more than twice that of east or west walls. Winter solar gain more than half that of south wall (at lower latitudes, almost equal). Roof must protect spaces below from heat and is helped by ventilation both over and under its surface. Lightweight construction to avoid heat storage.

Table 2.1 *U.S. Climate Zones and Planning* (Continued)

	Cool (*about 45° N. Latitude and above*)	*Temperate* (*from about 35 to 45° N. Latitude*)	*Hot Arid* (*below about 35° N. Latitude, primarily Southwest*)	*Hot Humid* (*below about 35° N. Latitude, primarily Southeast*)
Skylights	Particularly useful for this zone's minimum perimeter buildings, but skylights must cope with snow, high winds, and low temperatures. Double glazing and insulating shutters inside are both recommended.	Should be insulated at night in winter and shaded against direct summer sun. Avoid snow accumulaton. Skylights that can open can aid summer ventilation.	Direct sun can be selectively admitted in winter. Insulating shutters inside, seasonal sun shades outside.	Avoid direct sun for almost all of the year. Skylights that can open can aid breeze passage all year round.
Wall orientations	*Summer:* Shading of west wall most critical (east morning gains follow cool nights); south wall solar gain also high, about three-fourths that of east or west walls. *Winter:* South wall by far best for solar gain, but subject to freeze-thaw cycles on sunny cold days. Low sun can create glare problems. (East and west wall gains less than one-third that of south wall.) Wall facing storm winds needs protection.	*Summer:* Shading of east and west walls critical. (South walls gain about two-thirds that of east and west walls.) *Winter:* South wall clearly best for solar gain, with east or west wall gain about one-third that of south wall. Potential freeze-thaw and glare problems on south walls, though less severe than in cool regions.	*Summer:* Shading of west wall critical (east morning gains follow cool nights). South and north wall gains are less than one-half those of east or west walls. *Winter:* South wall best for solar gain, about twice that of east or west walls.	*Summer:* Shading for east and west walls critical; north and south wall gains are less than one-half those of east or west. At lower latitudes, north wall surpasses south wall gains. *Winter:* South wall best for solar gain, about twice that of east or west walls.

[a]From Victor Olgyay, *Design With Climate: Bioclimatic Approach to Architectural Regionalism,* Copyright © 1963 by Princeton University Press. Reprinted by permission. These results were based on an only moderately insulated house with 40% of its south wall (and 20% of its other three walls) as single-glazed windows. With today's better insulated ceilings and walls, and with summer shading over south glass, further elongation on the east-west axis becomes optimum for all zones.

[b]Roof solar gain refers to horizontal surfaces.

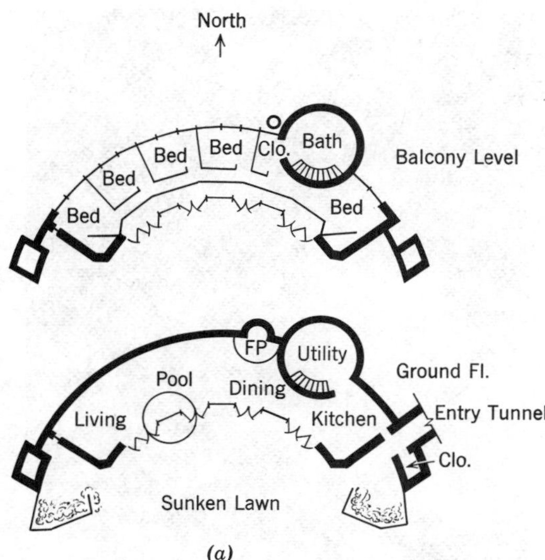

This gives winter wind protection from northeast to northwest and provides further insulation behind the almost windowless north wall. The north wall is made of stone, which absorbs and stores the winter solar heat that comes in through the floor-to-ceiling, southeast-to-southwest-facing glass. The concrete floor also stores solar heat in winter. The impression that this house and site are sun col-

Fig. 2.2 *An early passive solar-heated home, Frank Lloyd Wright's Solar Hemicycle near Madison, Wisconsin (between the cool and temperate zones), designed in the early 1940s, built in 1948.* (a) *A representative floor plan.* (b) *An early view of the northeast berm wall.* (c) *The passive solar-heated interior. Photos: Ezra Stoller, Copyright © ESTO.*

lectors is heightened by an entrance tunnel at the northeast end of the house, which leads from the parking area through the berm wall and onto the sunny, protected south terrace and front door.

This house is longer in the east-west direction (about 1 : 3) than the recommended "optimum configuration" (about 1 : 2, between the cool and temperate region; see Table 2.1). This elongation is typical of passive solar designs, which are able to store and use the winter solar heat gain, and thus profit from having long south-facing glass walls to act as collectors. Had the large windows been well insulated at night, this house would be a very up-to-date example of passive solar heating.

In the Solar Hemicycle house, protection from summer overheating is provided by an overhang along the south glass walls, the cool thermal mass of the north stone/berm wall combination (which receives no direct sun in summer), and the high windows in both north and south walls, which allow warmed air to rise and escape. Over the years since these early photographs were taken, the growth of some plants around the south rim of the scooped pocket has diminished the summer "heat trap" effect of this pleasant outdoor space.

Fig. 2.3 *Example of a courtyard in an arid zone. Attention to rainwater collection, deciduous plants, and massive walls creates a microclimate that moderates the extremes of the arid zone's range between daytime heat and nighttime cold.*

2.2 Climates Within Climates

The Solar Hemicycle encourages a small pocket of warm air to form by day within its site—a microclimate that differs slightly from the overall climatic region. Another way to achieve desirable microclimates is via the courtyard, a feature common to buildings in hot-arid regions (Fig. 2.3). In a courtyard, scarce rainwater is channeled into a small area where it can nourish deciduous plants; these then provide shade to the courtyard floor and walls in summer. The wind that threatens to dry out both plants and people is diminished by the enclosing walls. These massive, shaded walls absorb heat slowly by day, so the courtyard temperature rises more slowly than the air temperatures elsewhere. With sundown, the outside air temperature drops swiftly, but it drops less quickly in the courtyard where the walls radiate their stored heat to the space.

When many buildings are clustered, the number of local climate effects created is multiplied. In the case of large cities, these changes have combined to make both the winters and the summers warmer and have also created other climatic differences that are particular to urban areas (Table 2.2). The most obvious reason for the city's year-round warmth is its concentration of *heat sources:* the air conditioners, furnaces, and internal combustion engines in cars and buildings. The *rain* that falls on the city and countryside can be an effective cooling mechanism, especially as water evaporates from wet surfaces; but the materials of streets and buildings are usually designed to shed water quickly and thoroughly, so evaporative cooling for these surfaces is minimized.

The city also blocks the *wind* or diminishes

Table 2.2 *Average Changes in Climatic Elements Caused by Urbanization*

Element	*Comparison with Rural Environment*
Contaminants	
Condensation nuclei and particulates	10 times more
Gaseous admixtures	5 to 25 times more
Cloudiness	
Cover	5 to 10% more
Fog, winter	100% more
Fog, summer	30% more
Precipitation	
Totals	5 to 10% more
Days with less than 5 mm	10% more
Snowfall	5% less
Relative humidity	
Winter	2% less
Summer	8% less
Radiation	
Global	15 to 20% less
Ultraviolet, winter	30% less
Ultraviolet, summer	5% less
Sunshine duration	5 to 15% less
Temperature	
Annual mean	0.5 to 1.0°C more
Winter minima (average)	1 to 2°C more
Heating degree days	10% less
Wind speed	
Annual mean	20 to 30% less
Extreme gusts	10 to 20% less
Calms	5 to 20% more

Source. (H. E. Landsberg, "Climates and Urban Planning," *Urban Climates,* World Meteorological Organization, 1970. WMO Technical Note No. 108, p. 372.)

its overall cooling action by channeling it into narrow streets. The geometry of high vertical walls and narrow streets also increases the summer heat in cities as the high sun is reflected downward and is absorbed by the often rocklike street and building surfaces. In winter, however, this geometry puts most urban surfaces at a solar disadvantage, since the low sun strikes only the upper portion of south-facing walls.

A more subtle climate influence is *contaminated air;* small particles in the city's air can keep some sunlight from reaching the city; yet, it can help to keep the city's heat from radiating outward. These particles also form the nucleus of fog droplets; Table 2.2 shows that up to twice as much fog occurs in the city in winter as in the surrounding countryside.

The city thus changes its climate from that of its surroundings—to the city's decided disadvantage in summer. In winter, the city's warmth and its reduction of wind velocity reduces, but does not eliminate, the need for heating buildings. The typical means of providing additional winter heating (by fossil fuel-burning furnaces and power plants) contribute to airborn particles and urban fog. A switch to solar heating would diminish this pollution—and with less air pollution, more sun would reach a solar collector.

Site and urban-planning responses to these urban climate characteristics can sometimes lead in contradictory directions. For example, the provision of greenways within cities would bring softer ground surfaces cooled in summer by shading, breezes, and evaporation.

Its winter impact might be an increase in fog via evaporation of retained water; or perhaps fog would be discouraged by the increased wind speeds facilitated by such greenways. The summer impact of the greenway is also not entirely positive; increased evaporation could mean higher relative humidity. On balance, the greenway within a city seems a positive and esthetic step in ameliorating the urban climate, but the complexity of such a question deserves attention.

While considering how the sun, wind, and other natural forces can be utilized on a site for the benefit of a building, it is important to remember the need to protect the access of others to these same resources. This is effectively illustrated in Garrett Hardin's "Tragedy of the Commons" (*Science,* December 13, 1968, and often reprinted since). In his example, the commons are meadows publicly owned and shared by many herders. Each herder realizes that his personal wealth will increase as animals are added to his herd, so all herders increase their livestock. But the meadow capacity is not increased; overgrazing occurs, and the commons become unable to support any animals as a result. The following discussions of the resources of sun, air, and water found on a site are influenced both by the "private" needs of a building and the "public" patterns of these resources, which should remain accessible to all.

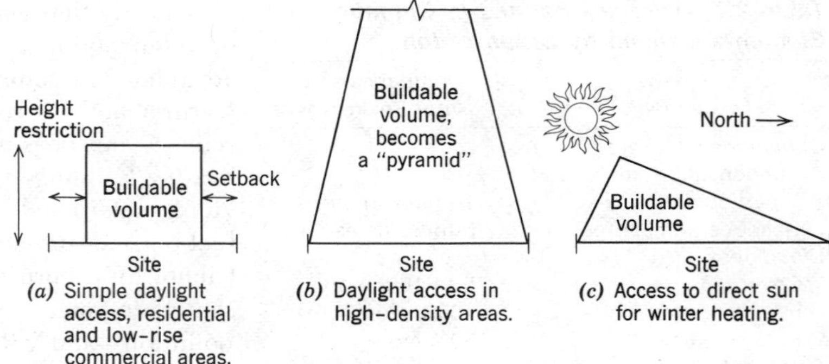

(*a*) Simple daylight access, residential and low-rise commercial areas.

(*b*) Daylight access in high-density areas.

(*c*) Access to direct sun for winter heating.

Fig. 2.4 *Access to light. Some methods of compromising private optimums (such as maximum rentable floor space) with public optimums (such as daylight at street level).*

Fig. 2.5 *Some results of zoning for "pyramids" in densely built areas.*

2.3 Direct Sun and Daylight

(*a*) ***Access to Light and Sun.*** The value of daylight (and air) to buildings has long been recognized in zoning laws, which require that minimum distances (setbacks) be maintained between a building and the property line in lower density areas. Height restrictions often accompany these setbacks, defining a maximum "buildable volume" in which a building can grow (Fig. 2.4). As buildings become taller and density increases, daylight reflecting down between them is diminished; in response, the maximum buildable volume approximates a pyramid (Figs. 2.4 and 2.5).

When *direct sun* in winter is desirable at the ground floor of each site, this pyramid changes shape and its volume decreases (Fig. 2.4*c*); this is due to the low angle of the winter sun, which is readily blocked by taller objects south of a given window. This most-restricted pyramid is at present rare, but various proposals to guarantee access to direct winter sun for solar collection are under development on federal, state, and local levels. The most restrictive feature of this solar-access pyramid is the low slope of its northern face, usually corresponding to the altitude of the sun above the horizontal at about two hours from noon on December 21. This allows four hours of access to direct sun on even the shortest day.

(*b*) ***Charting the Sun.*** There are many approaches to describing the position of the sun relative to a point on the earth. Presented here are charts of the sun's position for two latitudes (Fig. 2.6). These charts are actually graphs superimposed on the eye of an observer. At the vertical center line of each chart, the observer is looking due south. The horizon (a horizontal plane at observer eye level) is the line at the bottom of the chart.

Fig. 2.6 *Sun charts. Illustrations* (a) *through* (g) *describe the development of the charts shown in* (h) *and* (i)*, for two latitudes that cross the United States* (j). *From Edward Mazria and David Winitsky,* Solar Guide and Calculator, *Center for Environmental Research, University of Oregon, 1976, and Edward Mazria,* The Passive Solar Energy Book, *Rodale Press, Emmaus, Pa., 1978.*

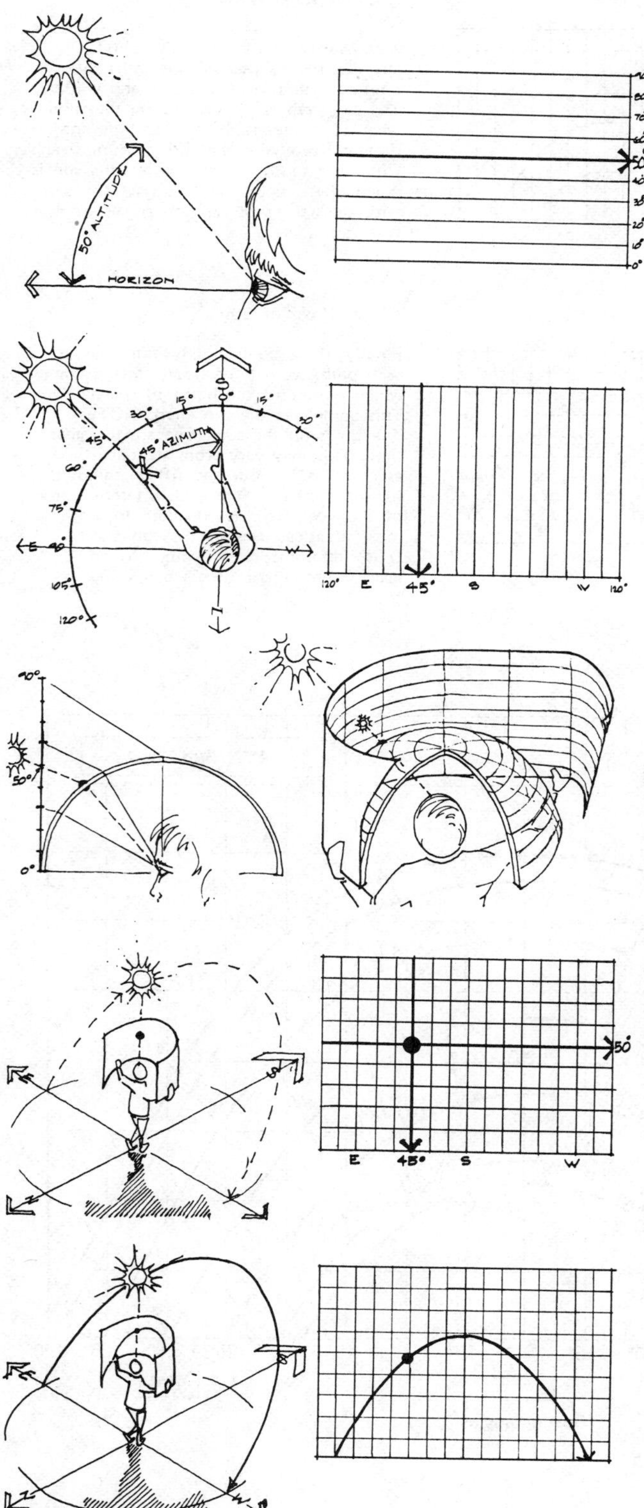

(a) Altitude

Two coordinates are used to locate the position of the sun in the sky. They are called the altitude and the azimuth (also called the bearing angle).

Solar altitude is the angle, above the horizon, of the position of the sun measured from the horizon. The horizontal lines on the chart represent altitude angles in 10° increments above the horizon.

(b) Azimuth

Solar azimuth is the angle, along the horizon, of the position of the sun measured to the east or west from true south.

(c) Skydome

The grid on the chart thus represents vertical and horizontal angles of the whole skydome. It is as if there was a clear dome around the observer, and then the chart was peeled off of this dome and laid flat.

(d) Sun Position

With the altitude and azimuth angles, the sun can be located at any point in the sky.

(e) Sun Path

By connecting the points of the location of the sun throughout various times of the day, the sun's path for that day can be drawn.

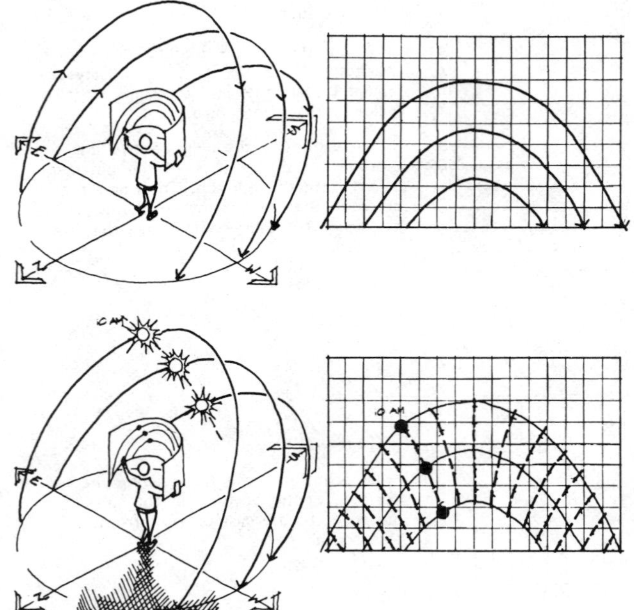

(*f*) Seasonal Paths

Thus, we can plot the sun's path for any day of the year. Those paths shown represent summer, fall/spring, and winter. The sun path is greatest during the summer when it reaches its highest altitude and rises and sets with the widest azimuth angle from true south. During winter, the sun is much lower in the sky and rises and sets with the narrowest azimuth angles from true south.

(*g*) Times of Day

Finally, if we connect the hour lines on each path, we get the heavy dotted line which represents the hours of the day. This completes the sun chart. NOTE: The times on the sun chart are for solar time. This may vary from standard time as much as an hour and fifteen minutes for different locations and different times of the year. This is adequate for most practical uses of the sun chart. It is important to remember to at least use standard time when using the charts.

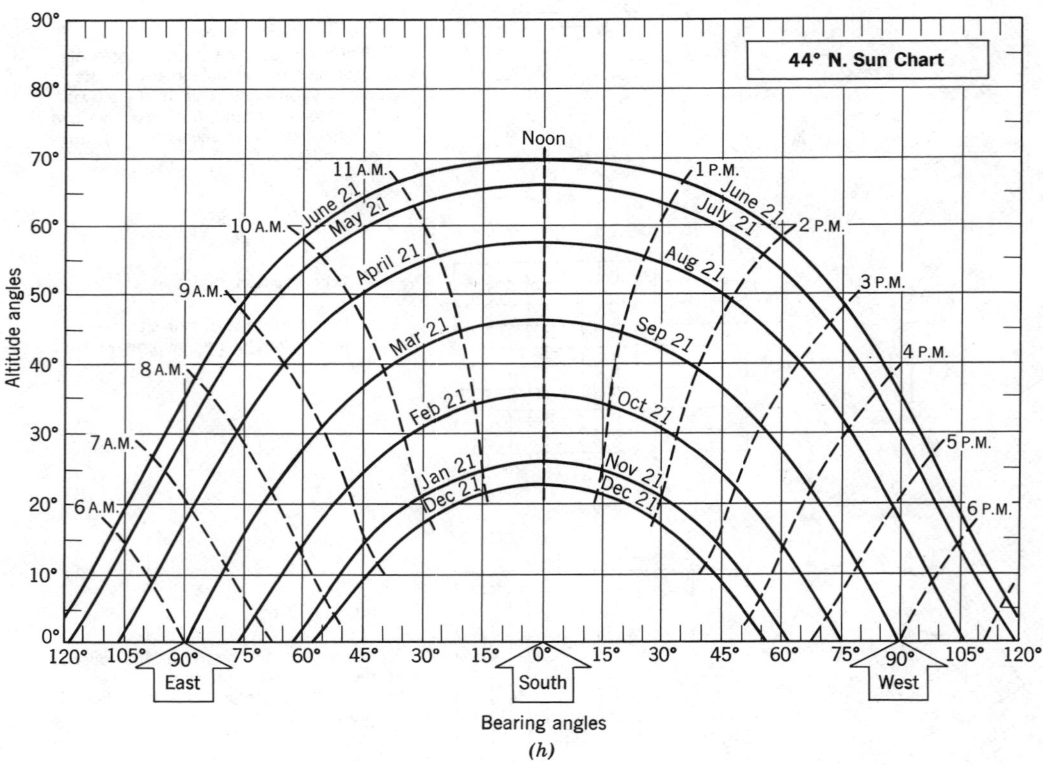

(*h*)

Fig. **2.6** (*Continued*)

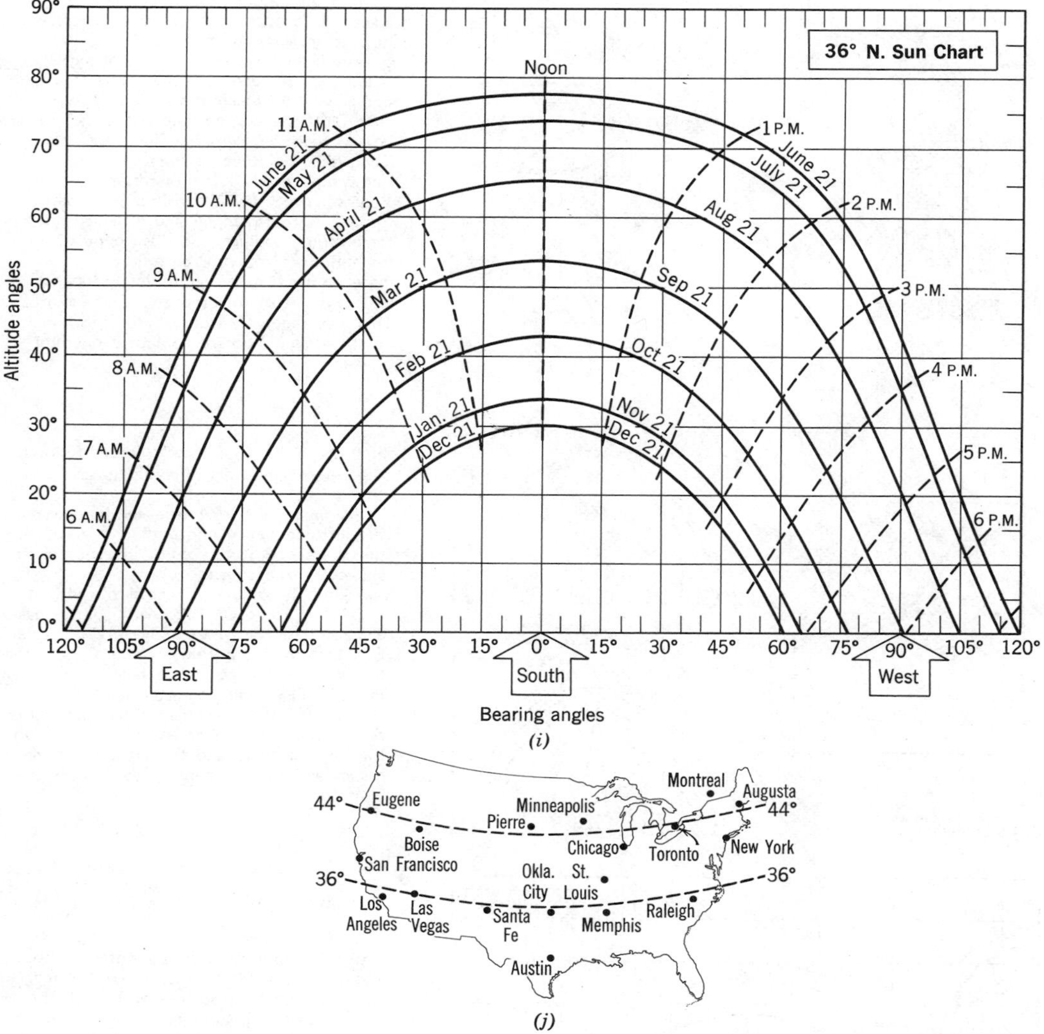

(i)

(j)

(*c*) ***Skylines and Winter Sun.*** The actual skyline as seen from a given place can be charted to determine access to direct sun at any time of year (Fig. 2.7). Such an analysis should precede the planning of the location of a solar collector. Obstructions within the four "best" hours (10 A.M. to 2 P.M.) are particularly serious and should be minimized in collector siting.

At the same time, consideration of the neighbors' access to direct sun is appropriate. This can be checked by charting another skyline, somewhere along the northern boundary of your site. The building you are designing should be included on this skyline and modified, if necessary, to preserve solar access for your neighbor.

(*d*) ***Controlling Solar Reflections.*** The use of highly reflective (or "mirror") glass to reduce heat gain in office buildings, and the rapid spread of solar collectors on walls and sloped roofs, has increased the frequency of annoying solar reflections from buildings (Fig. 2.8). The further the sun's rays are from being perpendicular to any surface, the more of the sun's light is reflected, rather than absorbed, by that surface. Thus, the intensity of the reflection will be greatest when the sun's rays

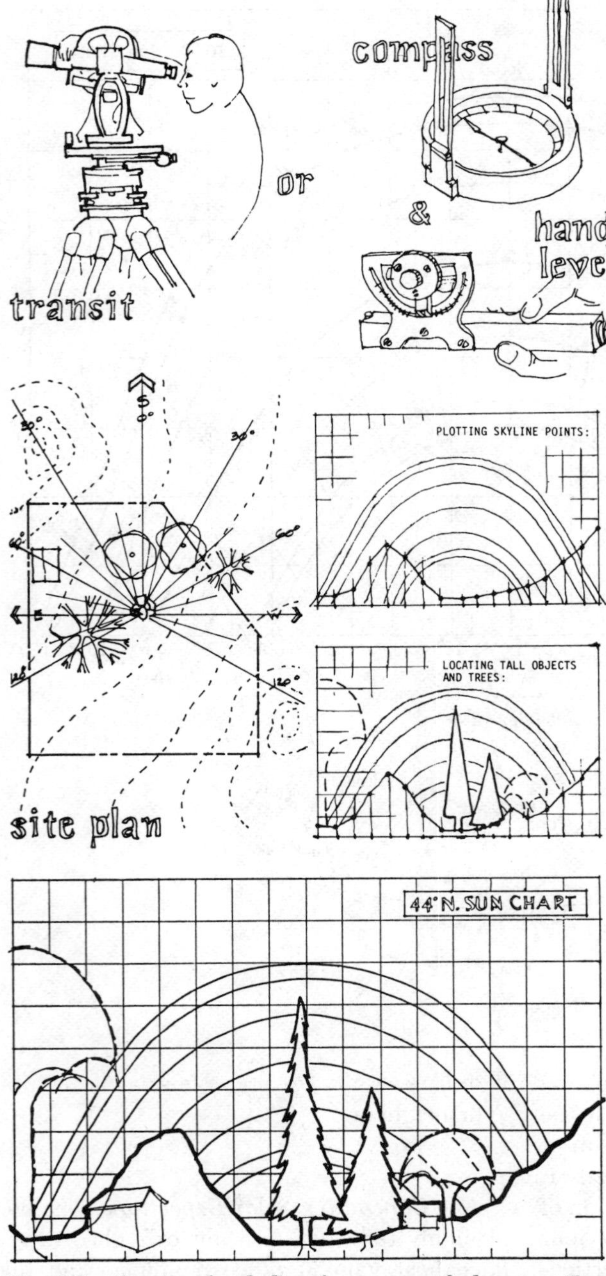

To find the amount of direct sun that a specific location receives, it is necessary to plot obstructions that block the sun as seen from that point. This plotting is called the "skyline" and it is done in a similar way in which the sun's path was plotted. If the skyline is low, that is with no tall trees, adjacent buildings, abruptly rising hills, or other obstructions, the following procedure is unnecessary.
You will need: (1) a transit or compass for the bearing readings and a hand level for altitude readings, and (2) a copy of the sun chart for your latitude.
Place yourself at the approximate location where you wish to know the solar radiation This may require a ladder if the location is on the second floor or roof of a building–to–be.

Find the altitude of the skyline as follows:

- Determine which direction (bearing) is true south.
- Aiming your level true south, determine the altitude (angle above the horizon) of the skyline. Plot this point on your chart above 0° true south.
- Similarly, find and record the altitudes of skyline for each 15° bearing, both east and west of south, to at least 120°. This is a total of 17 altitude readings. Plot these readings for the respective bearing angles on the sun chart and connect them.
- For isolated tall objects that block the sun such as tall evergreen trees, find both the bearing angle and the altitude for each object and record them at the appropriate point on the chart.

- Finally, plot the deciduous trees in the skyline with a dotted line. These are of a special nature because they will block the sun during spring through fall and let most of the sun pass through when their leaves are gone, late fall through early spring.

This completes the skyline. The open areas on your completed chart are those times when the sun will reach this specific location.

Fig. 2.7 *Charting the skyline from a specific location. From Edward Mazria and David Winitsky,* Solar Guide and Calculator, *Center for Environmental Research, University of Oregon, 1976, and Edward Mazria,* The Passive Solar Energy Book, *Rodale Press, Emmaus, Pa., 1978.*

are nearly parallel to a surface. Fortunately, this is the poorest time for solar collection through that surface; curtailed reflections need not mean curtailed collections.

The path or direction of the reflection is a "mirror image" of the sun ray's path. Figure 2.9 shows the intensity (percent of sunlight striking a surface that is reflected) and the

Fig. 2.8 Reflections. Mirror-glass windows in this newer office building in Milwaukee, Wisconsin, cast strong reflections on the north-facing wall of the older hotel next door. Although the winter heat thus reflected could be welcome, the glare is intense. In summer, the older building is particularly disadvantaged.

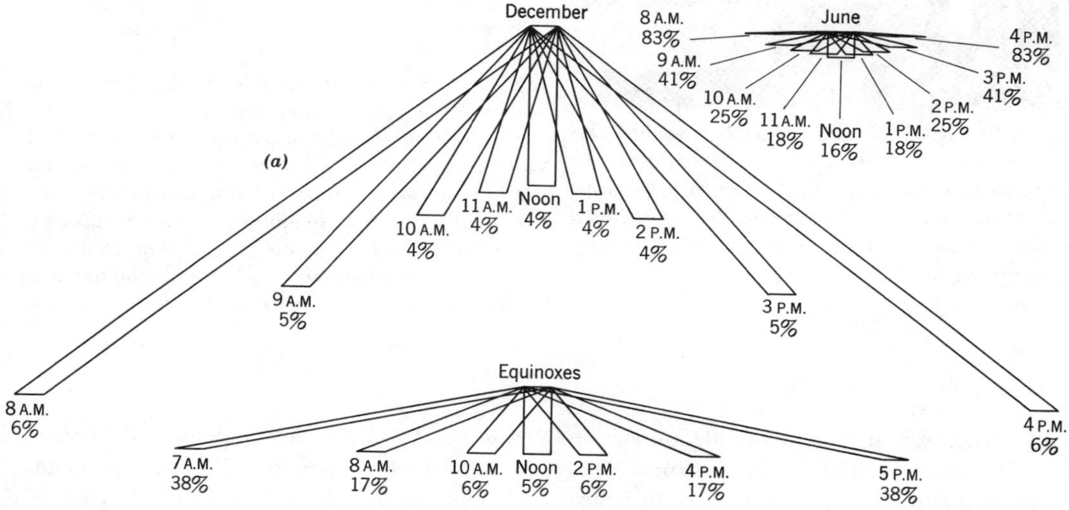

Fig. 2.9 Reflections in plan view (a). *These plans show the sizes of the reflections from a vertical south-facing wall (at 44° N. Latitude) whose east-west length is four times its height. Although the December reflections reach furthest from this vertical glass wall, they are the weakest in intensity; only 8% of the sunlight striking the glass is reflected at 8* AM *and 4* PM. *Conversely, the June reflections plunge quickly into the ground, but at great intensity (82% at 8*AM *and 4*PM). (b) *Reflections at noon, in section, for solstices and equinox. Copyright © Michael D. Lee, Anchorage, Alaska. Reprinted by permission.*

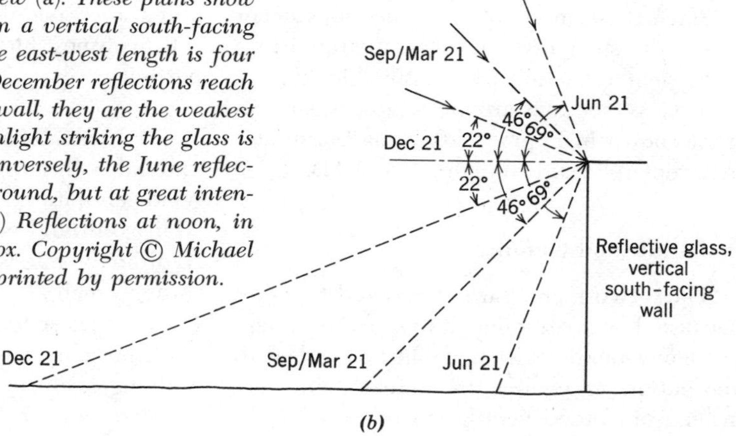

Fig. 2.10 *External architectural shading. This combination of overhangs and vertical fins prevents the more intense reflections of early morning and late afternoon by intercepting the sun before it strikes the glass surfaces. For a similar example, see Figure 6.2, page 252.*

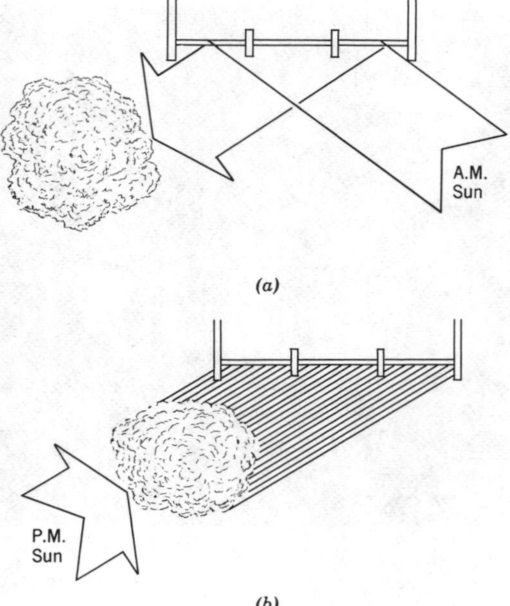

Fig. 2.11 *Selective protection from reflections. The tree standing west of this south window wall does not interfere with solar access during the best hours for solar gain (around noon), nor does it prevent early morning sun from entering the windows* (a). *The reflections of early morning are intercepted by the tree, before they can escape to annoy elsewhere. In* (b), *the late afternoon sun is blocked by the tree, before either heat gain or reflections can occur.*

path (on horizontal ground) of reflections from a vertical south wall at 44° N. Latitude. Such an analysis can be used to minimize the most intense solar reflections.

Since these most intense reflections occur where the sun's rays are nearly parallel to a wall, such reflections are easily blocked by external projections around windows, such as those shown in Figure 2.10. Foliage can also intercept reflected sunlight (Fig. 2.11).

2.4 Air and Sound

These two factors are considered together because they are so difficult to separate. There are many buildings that could be opened to ventilation or cooling by breezes, but rely instead on forced ventilation because of the noise that would accompany the breeze through an open window. Polluted air is another potential deterrent to "natural" ventilation. Almost any device that reduces sound will also reduce the velocity of the breeze, as is true of most filtering devices to remove dirt particles.

(*a*) ***Noise.*** The building in Figure 2.12 is unusual both in providing an opportunity for wind ventilation—air moves freely below it as well as around and over it—and in the extraordinary intensity of noise that is carried through such air. Especially for this example, it seems easier to understand the impact of a city's noise on a building than to imagine the building itself as a contributor to the noise problem.

Fig. 2.12 A series of apartment buildings straddle the approach ramps to New York's George Washington Bridge. These buildings were the scene of a study linking noise level with reading disabilities. S. Cohen, D. C. Glass, & J. E. Singer, "Apartment noise, auditory discrimination, and reading ability in children." Journal of Experimental Social Psychology, *1973, 9, pp. 407–422.*

Many of the noise complaints against buildings involve air-conditioning equipment (the compressive refrigeration cycle and its year-round utilization as a heat pump are described in Chapter 5). Noise is generated both by the compressor and by the great quantity of exterior air that must be rapidly pushed through outdoor coils (further heating outdoor air so that the indoors may be cooled). When densely packed buildings are forced to rely on mechanical cooling, this closeness makes the noise of these systems even more annoying. The machine's appetite for outdoor air is so enormous that attempts to surround it with noise shields can greatly hinder its efficient operation and shorten its life.

In residential neighborhoods, the greater distance between buildings might be expected to lessen these difficulties. Yet the much lower ambient (or background) noise level of residential areas is one of their more desirable characteristics, and an intruding compressor can trigger lawsuits from annoyed neighbors.

Two characteristics of cities (see Section 2.2) also contribute to increased noise at street level: *hard surfaces* reflect rather than absorb sound, and *parallel walls* intensify noise between them rather than dissipating it (see Chapters 26 and 27). Although building surfaces are generally hard for durability in weathering, softer and multiplaned materials (such as plants) are desirable from a public noise-reduction viewpoint. Their impact on measured noise may be slight, but the perception of softer surfaces and the sounds made by the plants themselves can contribute to an environment that seems less noisy.

(b) Air Pollution. While one of the deterrents to operable windows is the dirt that they could admit to the interior, air pollution is a threat to people and their buildings in other ways. Figure 2.13 summarizes the impacts of air pollution in the United States and indicates hope for improvement from the enforcement of the 1970 Clean Air Amendments. Buildings (and the electric generating plants that supply them) are major contributors to the pollution of the "fresh air" needed for ventilation. It is encouraging that pollution levels of particulates and carbon monoxide have declined since 1970. Sulfur oxides and hydrocarbons have declined slightly in this five-year period, and nitrogen oxides have actually increased. The latter effect is due mainly to increased emissions from electric generating plants, as well as to more automobiles.

Carbon dioxide is not a pollutant, but a necessary and widely debated constituent of air. Will the amount of carbon dioxide continue to increase in our atmosphere primarily as a result of the burning of fossil fuels? (Coal is projected as a prime future U.S. energy source, see Table 1.6, pages 24–27.) If carbon dioxide does increase, will it warm the earth by discouraging our outward radiation of heat, as does the glass cover on a solar collector? Or will particulate matter (that accompanies carbon dioxide from the furnaces of buildings and

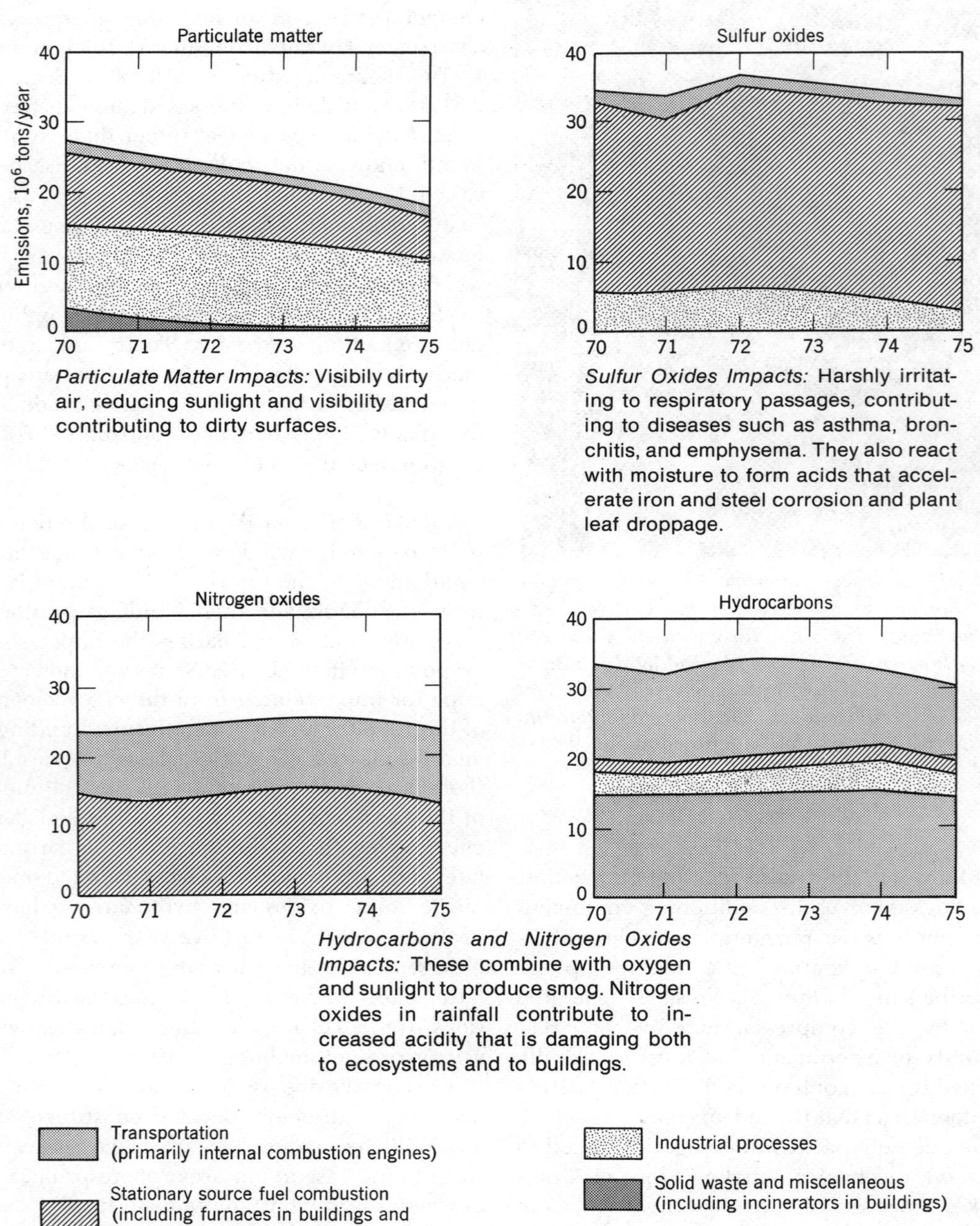

Fig. 2.13 Air pollution in the United States, 1970 to 1975. Calculated total emissions of pollutants by source category. U.S. Environmental Protection Agency, Trends in the Quality of the Nation's Air, *1977.*

Fig. 2.13 (*Continued*)

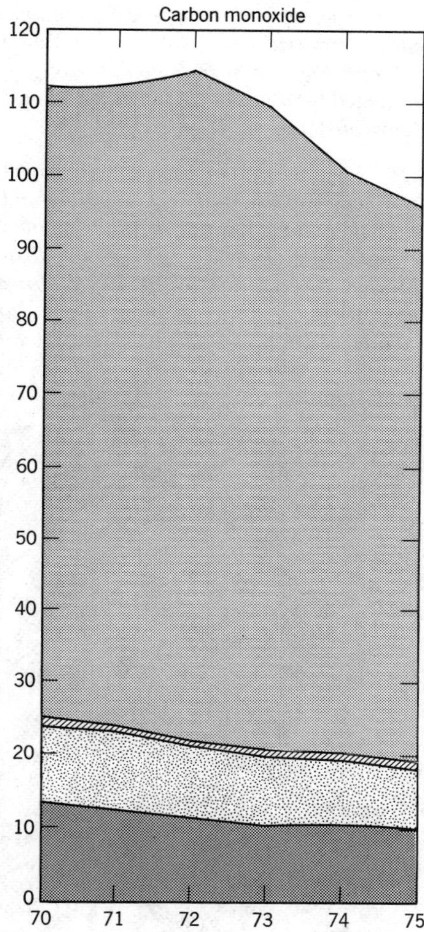

Carbon Monoxide Impacts: Tends to cause suffocation by interfering with the blood's ability to supply oxygen to the body; as a result, heart and respiratory systems are forced to work harder. Symptoms include headache, loss of vision, decreased muscular coordination, nausea and abdominal pain.

generating plants) reflect incoming sunlight and nullify the warming effect of carbon dioxide?

(*c*) ***Ventilation and Cooling.*** A distinction must be made between two common architectural introductions of outdoor air: ventilation and cooling. *Ventilation* involves the provision of "fresh air" to interiors to replenish the oxygen used by people and to help carry away their by-products of carbon dioxide and bodily odors. Ventilation is desirable all year round; recommended minimum rates for providing fresh air are found in Table 4.3, page 101. *Cooling* (with outdoor air) replaces heated indoor air with cooler outdoor air. Thus cooling by breezes is a seasonal opportunity, limited to times when the outdoor air temperature is lower than the indoor desired temperature. Cooling can require far greater quantities of air than ventilation, and its influence on building siting and window size and placement is considerable. Figure 2.14 illustrates the differing approaches to these two usages.

Two examples of contemporary buildings that use the wind appear in Figures 2.15 and 2.16. The first served as a summer exhibition space in Montreal, Canada, and relied on pre-

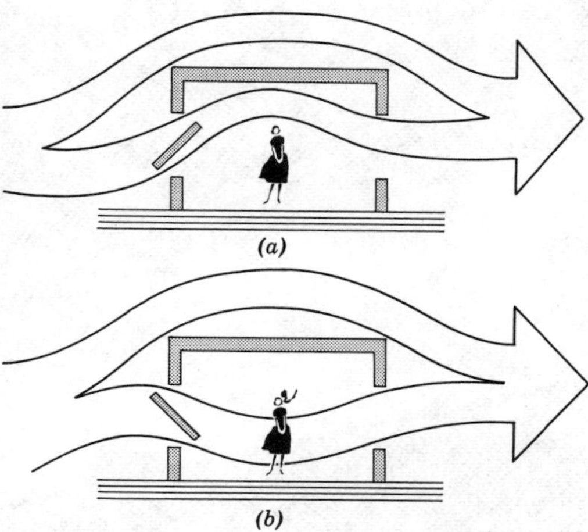

Fig. 2.14 *Ventilation, with and without cooling. The window's placement and size can influence the stream of air within a space, providing either ventilation alone or adding a cooling function to moving air.* (a) *Ventilation: the pivoting window sends the breeze toward the ceiling, where it replaces inside air. It has minimum contact with people below.* (b) *Ventilating and cooling: this position of the window directs the breeze toward the floor, where it encounters people and provides a direct cooling effect from air motion. Adapted from Victor Olgyay,* Design with Climate: Bioclimatic Approach to Architectural Regionalism, *Copyright © 1963, Princeton University Press. Reprinted by permission.*

Fig. 2.15 Cooling with the wind. The air inside this Canadian exhibition hall is heated by lights and crowds of people. It then rises and is sucked out of the large clerestory opening by the prevailing winds. Note that these winds do not blow into this high opening, but encourage outward flow of the overheated interior's air. African Place, at Expo '67, Montreal; John Andrews, Architect.

Fig. 2.16 Ventilation and cooling options. These faculty offices and seminar rooms in England (a) have the choice of admitting a small amount of winter ventilating air or a large amount of air to directly cool the interior (b). The prefabricated window unit (c) is hoisted into position during construction (d). Leicester University's Attenborough Building; Arup Associates, Architects, Engineers and Quantity Surveyors, London.

(a)

Fig. 2.16 (Continued)

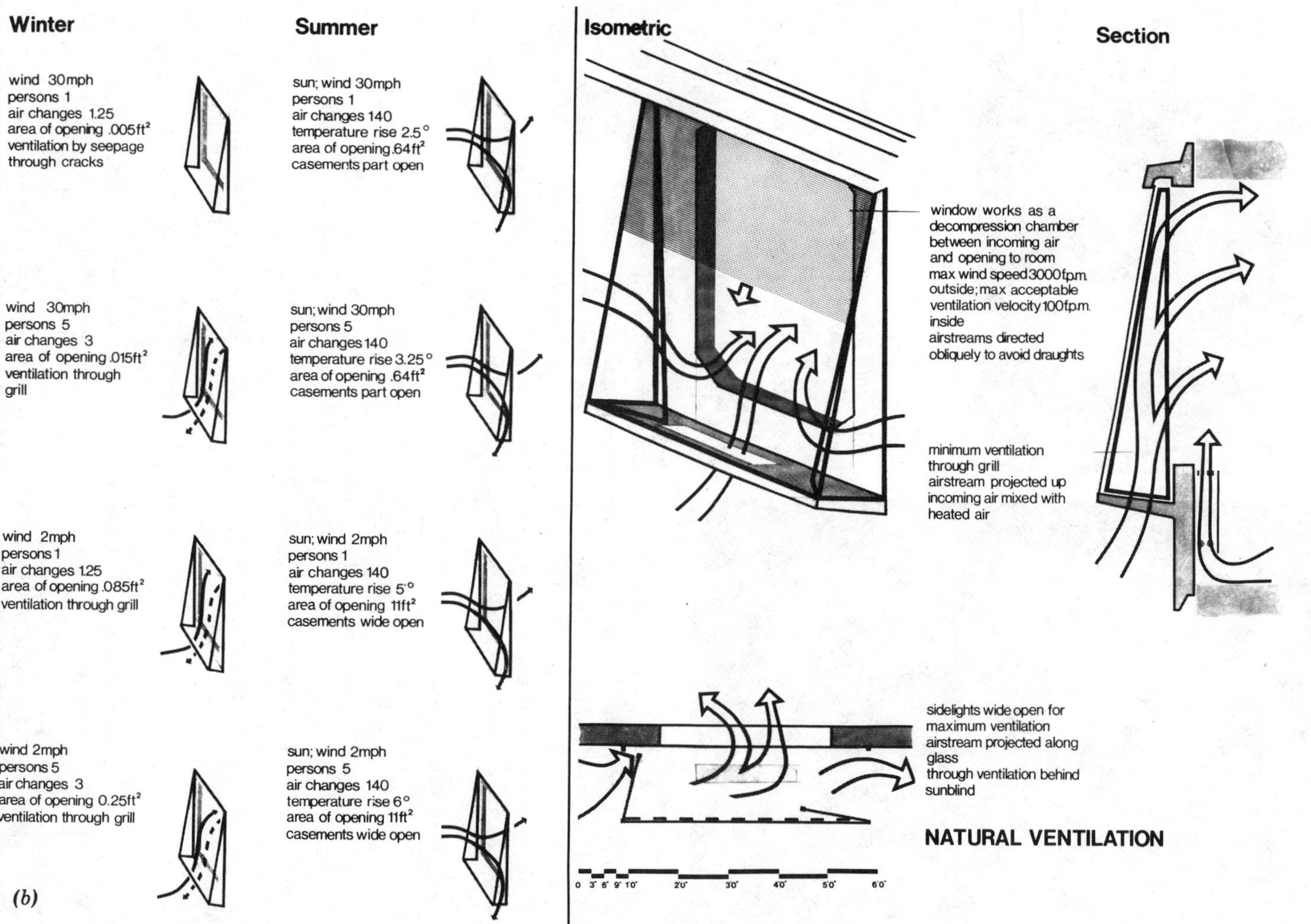

(b)

Fig. 2.16 (*Continued*)

vailing winds to remove the air heated by display lighting and crowds of people. The second is a faculty office building in England, whose windows can be manipulated to provide either a small amount of ventilation air or a more thorough scouring for cooling purposes.

The building that relies on prevailing wind for cooling must be sited with attention to wind direction. Again, one building should not be erected to obstruct the next building's access to breeze. Obstacles upstream from intake openings, or downstream near the outlets, can substantially reduce the velocity—and thereby the cooling effect—of the wind. But although wind cools people primarily by speeding the evaporation of sweat from skin, it can itself become an irritant at high velocity (Table 2.3), and manual controls of openings are a necessary part of natural ventilation equipment.

Table 2.3 *Wind Velocity and Comfort*

Wind Velocity	*Possible "Lower Temperature" Comfort Sensation (between 80 and 90° F; larger numbers correspond to high humidity areas)*	*Probable Impact*
Up to 50 fpm	No change in comfort sensation	Unnoticed
50 to 100 fpm	2 to 3° F lower	Pleasant
100 to 200 fpm	4 to 5° F lower	Generally pleasant, but causing a constant awareness of air movement
200 to 300 fpm	5 to 7° F lower	From slightly drafty to annoyingly drafty
Above 300 fpm	More than 5 to 7° F lower	Requires corrective measures if work is to be efficient and health secured

Source. (Adapted from Victor Olgyay, *Design with Climate: Bioclimatic Approach to Architectural Regionalism,* Copyright © 1963, Princeton University Press. Reprinted by permission.)

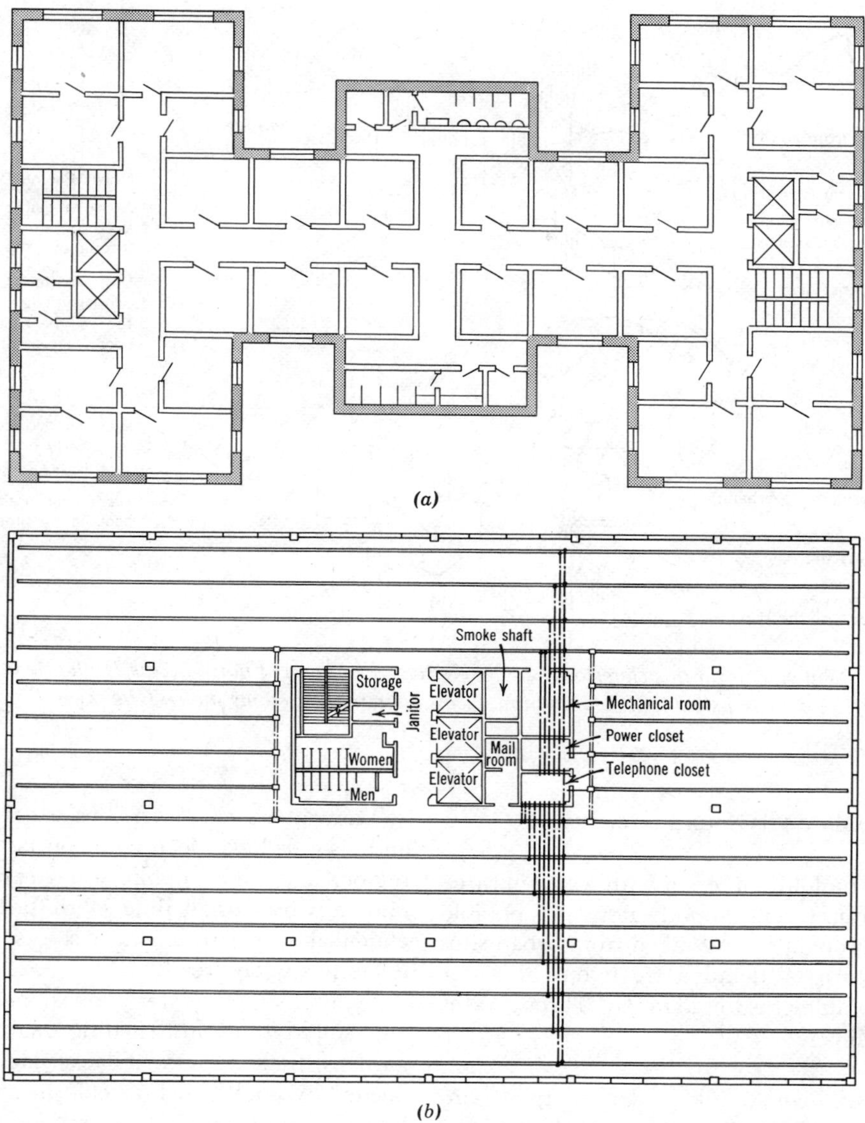

Fig. 2.17 *In contrast to building plan* (a) *that uses daylight and wind ventilation in each office, office building* (b) *receives cooler, filtered air, is less subject to the noise of the city, typically provides both constant light and temperature throughout, and provides for more rentable floor space on the site. It also allows less daylight to reach street level, requires much more electricity (though probably less heating fuel), and thus contributes more heat (and possibly noise from mechanical equipment) to the city all year round.*

(d) Wind, Daylight and Sun. When daylight and wind ventilation are desired, they combine to limit the width of buildings. This is particularly evident in multistory office buildings, where increasing urban density and reliance upon electric lighting and cooling have changed the form of buildings considerably in the past few decades (Fig. 2.17). Another impact of combined wind and sun is shown in Figure 2.18, where a designer weighs a decision about orienting a clerestory window.

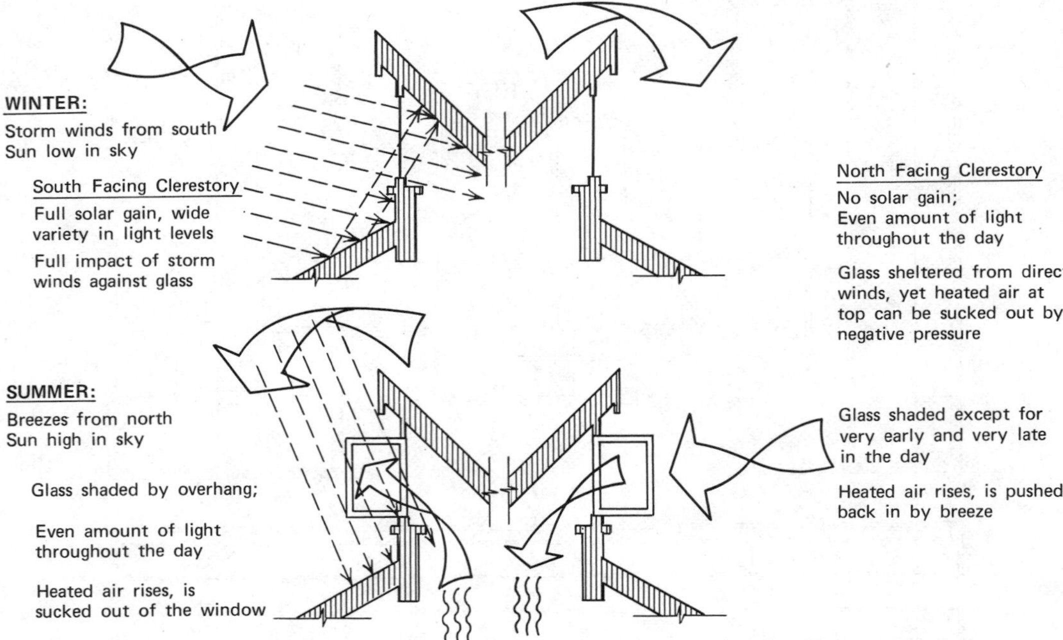

Fig. 2.18 Which way for this clerestory? Some relative advantages of north versus south orientation for a clerestory window and shed roof combination. (Seasonal wind directions shown here prevail in the Pacific Northwest.)

2.5 Rain and Groundwater

Most buildings interact with water in three forms: rainwater, groundwater, and potable water (brought to and taken from urban sites by utilities). A detailed treatment of water within buildings is found in Part Three, Water and Drainage.

(*a*) ***Rainwater.*** Like solar energy, rainwater is a diffuse, intermittent, and often seasonal resource. As a nonpotable source of water, it is most often collected and used where other water sources are scarce, or of poor quality. Rain also has an influence on building design; heavy rains and pitched roofs have long been found in the same locales. Overhangs may extend further beyond walls exposed to storm winds; even gutter and downspout details can become a design feature, as shown in the examples in Chapter 12. A building that reflects the combined influences of daylight, wind, and rain is shown in Figure 2.19.

Rainwater's characteristics can be useful, and cisterns have been common both in arid regions and where public water supplies are unusually hard or alkaline. With the advent of chemical water softeners, cisterns have become less widespread.

Unfiltered rainwater seems particularly well-suited to use in irrigating small lawns or gardens, both because it lacks additives unneeded by plants, such as chlorine or sodium fluoride, and because it can reduce the user's demand on the public water supply on the hottest summer days. Cisterns located above the irrigated area have the advantage of replacing a pump with simple gravity flow. The rainbarrel at the bottom of a downspout was an example of this approach, though of limited capacity.

Some information about rainfall is shown in Figure 2.20, and about cisterns in Figure 2.21. To calculate the average annual volume of water capturable by a cistern, multiply the annual rainfall by the roof area contributing to the cistern (keep units consistent). The aver-

Fig. 2.19 *Rain, wind, sun, daylight, and design. This covered outdoor tennis facility at the University of Oregon, Eugene, was designed to ward off the rainbearing south winds in winter. Direct sun is also unwelcome; instead, north skylight is admitted along with reflected light from roof surfaces. Cool outdoor temperatures are maintained, appropriate to strenuous activity. Gutters at the lower edge of each roof plane carry away rain; the courts stay dry for all but a few days of the year. Unthank, Seder, Poticha Architects.*

age number of gallons or liters of irrigating water needed per unit area vary by crop and location; local agricultural extension offices are a source for this information.

In areas of moderate building density, rainwater not stored is commonly discharged onto the ground's surface or into drainage lines under surrounding lawns. In these cases, care is necessary to prevent one user's solution to rainwater disposal from becoming another's problem. Buildings that divert the normal course of surface runoff from rainfall or that add greatly to runoff at a concentrated spot can cause erosion or flooding of a neighboring lower area. As density increases, rainwater is more likely to be promptly discharged to storm sewers.

(*b*) ***Groundwater.*** Generally avoided by designers where possible, groundwater is a threat to foundations and underground spaces. This avoidance carries over into site planning, and marshy places are usually unwelcome near buildings. The previously mentioned urban dryness is intensified as ponds are drained and streams are piped away.

On the other hand, architects have used groundwater as a heat sink, discharging building heat to groundwater in summer and re-

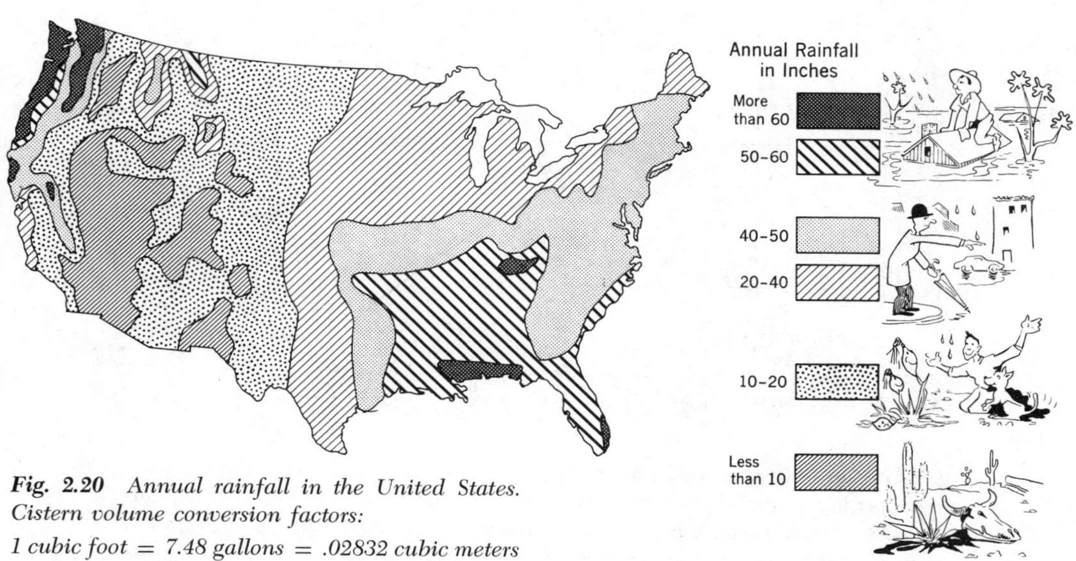

Fig. 2.20 *Annual rainfall in the United States. Cistern volume conversion factors:*

1 cubic foot = 7.48 gallons = .02832 cubic meters

1 cubic meter = 1000 liters = 264.2 gallons

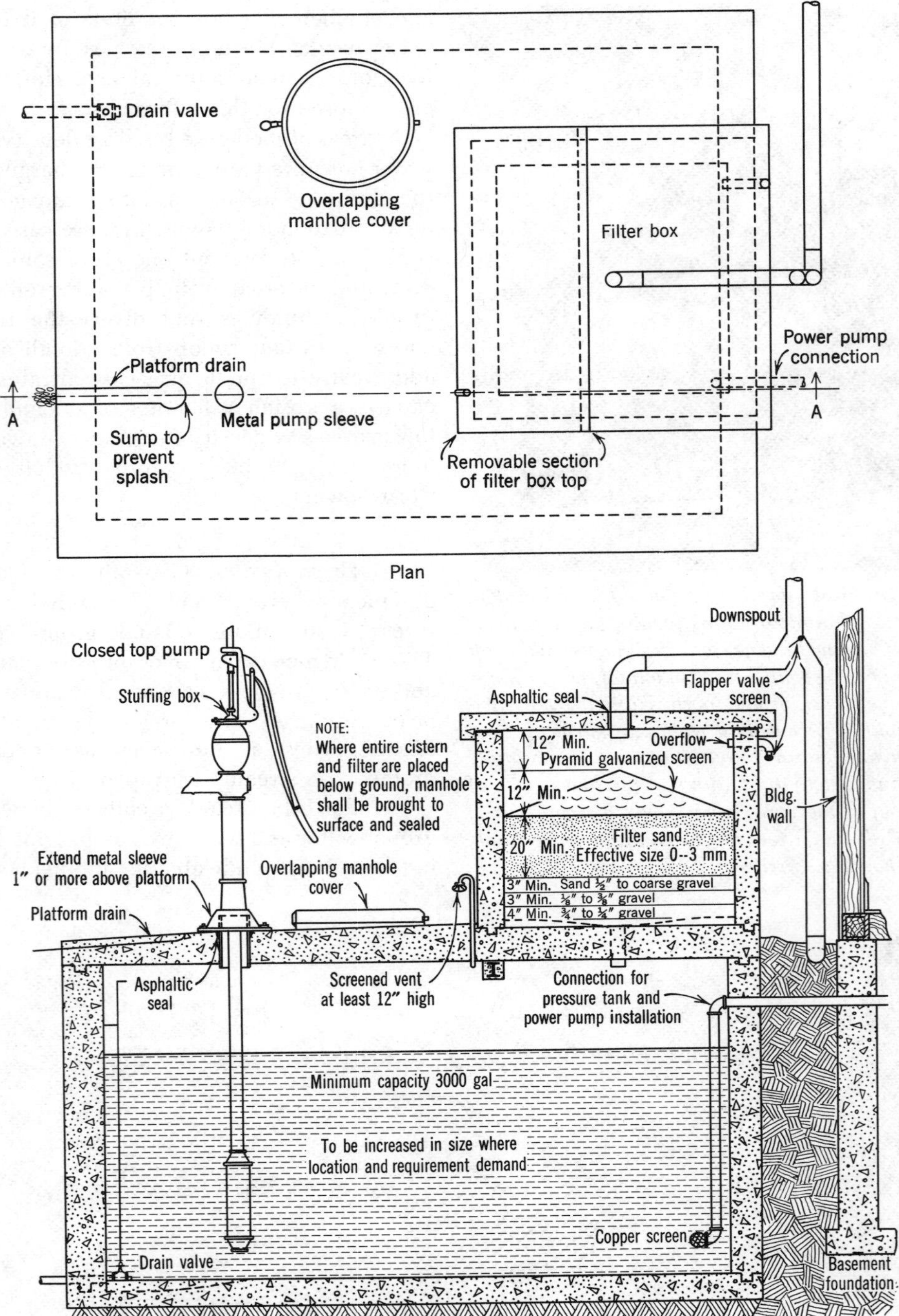

Fig. 2.21 Rainwater cistern with a sand filter. This cistern is also equipped with an access opening, a drain, and a "flapper valve." The flapper valve is a manually operated switch that allows the first and dirtiest water of each rainfall to bypass the filter and cistern entirely. Adapted from U.S. Public Health Service, Individual Water Supply Systems, *1950.*

moving the groundwater heat for the building's winter benefit. The quantity of groundwater available to act thus as a heat sink varies with subsurface conditions, and little information is yet available to indicate its limits. More buildings discharging more heat will eventually raise groundwater temperatures; but how many degrees, and for how long, remains unknown in most cities.

The ability of water to conduct and store heat, which encourages its use as a heat sink, also makes it a thermal enemy of heat storage tanks or bins in solar buildings. There is little point in collecting solar energy to store in an underground tank for later use if groundwater is allowed to rob the tank of its heat.

Fig. 2.22 Fixed overhang sun control. This south-facing corridor in Oregon is in the open air all year round. The low winter sun fills the space, and some of its heat is stored by the tile floor. However, in cold weather the offices that it connects are disadvantaged by exposure to cold air. In summer, little of the corridor is exposed to sun. Sun control in the spring and fall is identical.

2.6 Plants

Plants have several roles: they are part of the water cycle; they turn carbon dioxide into oxygen by day; they provide organic matter suitable for composting (eating, in some cases); and they tell time both by growth and by change with the seasons. Our associations with plants are mostly pleasant ones, and they contribute to our enjoyment of the places where they grow.

Plants are also of immediate practical value to buildings, as they enhance privacy, slow the winter wind, reduce the glare of strong daylight, or prevent summer sun from entering and overheating buildings. In this latter role, plants are particularly noteworthy, because they enhance a feeling of coolness when breezes rustle or sway their leaves, and especially because they respond more to *cycles of outdoor temperature* than to those of *sun position*. Unlike fixed sunscreens on buildings, plants thus provide deepest shade in the hottest weather.

To illustrate this contrast, a fixed sunscreen (here an overhang for south-facing windows) is shown in Figure 2.22. Such sunscreens block the sun for some portion of the year, centered on June 21; that is, maximum shade is provided at the summer solstice. A typical approach to such sunscreens in the U.S. temperate zone is to shade at least half of a south window from March 21 to September 21, or equinox to equinox. Yet, March is on the average a colder month than September; March 21 is the last day of winter, while September 21 is the last day of summer. Full solar radiation is more welcome in early spring than in early fall, yet sun position is identical at these times. In contrast to fixed sunscreens, deciduous plants do most of their shading from the middle of June to early October, giving windows access to solar radiation through much of the spring (Figs. 2.23 and 2.24).

Deciduous trees have a potential solar disadvantage in that certain species (such as some oaks) hold onto their leaves well into the heating season, a tendency increased by fertilizing or irrigating near the tree. Others have a dense

Fig. 2.23 *Deciduous vines, temperature, and sun position. The sun's path through the sky is identical in late May and late July (see Fig. 2.6) when photos* (a) *and* (b) *were taken. Identical lower sun paths occur in November and January, photos* (c) *and* (d). *This deciduous vine responds more to the temperature of its Oregon climate than to the sun's position at 44° N. Latitude, which makes it particularly useful as a potential sun-control device. From J. Reynolds,* Solar Energy for Pacific Northwest Buildings, *Center for Environmental Research, University of Oregon, 1976.*

Fig. 2.24 *A deciduous tree and the equinoxes. The sun position (4:45* PM *sun time) is identical on September 21* (a) *and March 21* (b)*; yet, this hour's average temperature (Eugene, Oregon) is 75° F in September, 53° F in March. The second-floor overhang provides identical shading, while the tree shades only in September. From J. Reynolds,* Solar Energy for Pacific Northwest Buildings, *Center for Environmental Research, University of Oregon, 1976.*

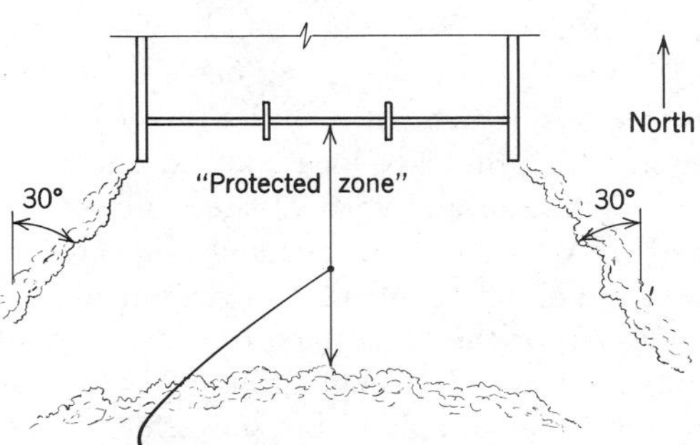

Fig. 2.25 *Protecting access to winter sun. Shown here are the size and shape of a minimum lawn or terrace to the south of solar collecting surfaces. Deciduous plants within this "protected zone" should be avoided, unless they are very low growing or are a reliably early-defoliating species. Summer sun protection for these south windows is best provided by flexible architectural controls, such as awnings or hanging screens.*

branch structure, blocking a surprisingly high percentage of solar radiation even when bare. Agricultural extension services in most areas can provide information on late-defoliating trees. For solar collection, avoiding trees or large shrubs within the area shown in Figure 2.25 is recommended.

References and Bibliography

1. "The Climate of Cities," *Scientific American,* August 1967, W. P. Lowry.

2. *Solar Control and Shading Devices,* 1957, A. Olgyay and V. Olgyay, Princeton University Press.

3. *Design with Climate: Bioclimatic Approach to Architectural Regionalism,* V. Olgyay, Princeton University Press.

4. *Plants, People and Environmental Quality,* 1972, G. O. Robinette, U.S. Department of the Interior, National Park Service, Washington, D.C.

3

Energy and Architectural Design

The previous chapters have discussed the total energy consumption of a building's mechanical and electrical equipment and have stressed the use of energy sources from the site itself. Before moving to the parts of this book that treat this equipment in detail, a closer look is provided here at the impact of energy conservation on architectural design. This chapter deals first with the design of the exterior envelope, then moves inside to consider specifics such as thermal and solid waste, and then to more general questions of how people relate to energy inside buildings.

Heat, light, sound, and water are important elements in the design of spaces. Their impact on the user's comfort is obvious, but their esthetic role is often subtle. Consider, however, a large fountain deprived of its water, along with the coolness, the sparkle, and the splash. What is a fireplace without firelight, heat, crackling sounds, and a smoky aroma?

Whether provided through mechanical and electrical equipment or by more direct use of the surrounding climate, *heat, light, sound,* and *water* deserve a central role in designing—along with color, texture, materials, and forms. *All* these elements can be blended to give each place its distinctive character. The users of spaces have five measurable senses of perception; utilizing all of them when designing for people can result in particularly successful architecture. (Some detailed information on human response to heat, moisture, and air motion are found in Chapter 4; to light, in Chapter 18; to sound, in Chapter 26.)

3.1 Schematic Design

One of the designer's first concerns is to recognize the resources that exist on and around a site (or in a place) and to decide how best to integrate these resources into the final design, while making the design a successful addition to the larger patterns of its surroundings. Schematic site plans are typically used as a kind of inventory; overlaid "bubble diagrams" can test possible design arrangements that can relate rooms and functions to their surroundings in the plan. Sun and wind conditions (in both summer and winter), noise sources, and water run-off patterns can be included in this schematic plan. Typically, however, the site plan-bubble diagram approach has been limited to the *surface* conditions of a site. Conditions of privacy and accessibility, of view, heat, light, air motion, sound, and water, all change with vertical distance from the surface (Fig. 3.1).

To minimize energy consumption in constructing and using buildings, and to integrate them with their surroundings, the conditions best suited to various functions should approach the characteristics of the "layer" of the site in which they are located. Consequently, both horizontal and vertical site analyses are needed. A lecture hall, needing both an isolated and closely controlled environment, is a candidate for the subsurface layer. Electrical equipment, which benefits from cool environments, also is suitable to the subsurface layer.

One building that derives its form from these layers is Boston's City Hall (Fig. 3.2). Activities with the most frequent public interaction are located near the skylit, high-ceilinged lobby on surface levels, whereas special ceremonial functions are elevated to distinctive forms in the near surface layer. The city offices with less frequent public contact occupy several floors in the sky layer, and storage and mechanical functions are in the subsurface layer.

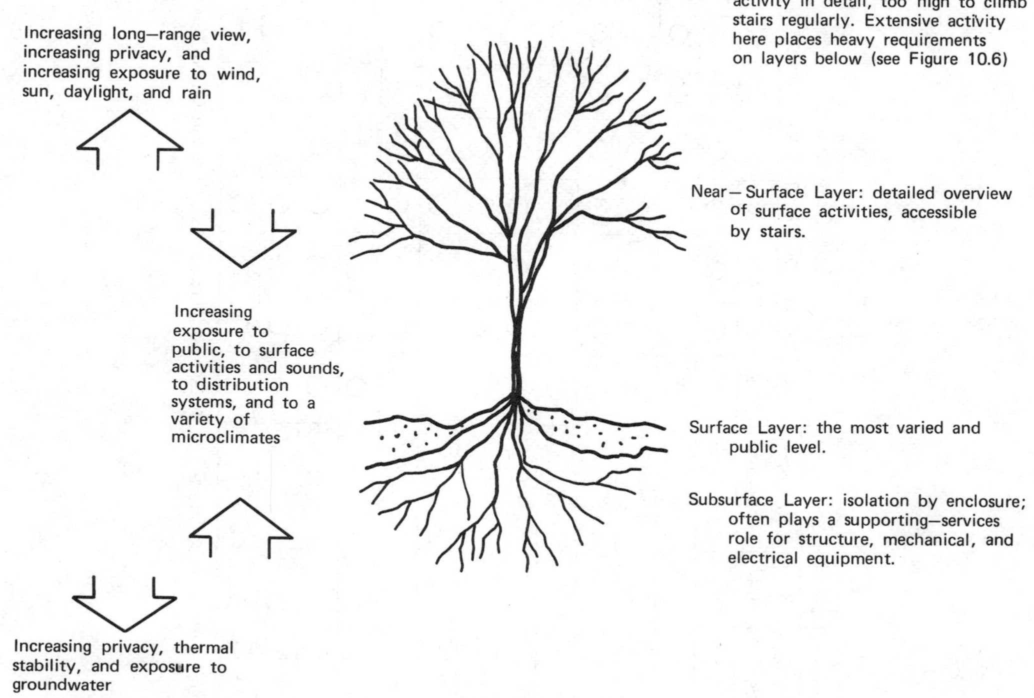

Fig. 3.1 *Vertical characteristics and layers for site analysis.*

Sky Layer:
least frequent public contact: public works, housing, health administration, parks and recreation, building, redevelopment.

Near–Surface Layer:
mayor and council offices, council chambers, reference library, news conferences, exhibits.

Surface Layer:
most frequent public contact: entry and lobby, complaints, elections, licensing, assessing, health registration.

Subsurface Layer:
parking, mechanical equipment, data processing, inactive files, and storage.

Fig. 3.2 *Layers and form: Boston City Hall, 1969. Kallmann, McKinnell and Knowles, Architects; photo by M. D. Ross.*

Table 3.1 ***Residential Functions and Categories***

Thermal- and Moisture-Control Range	Function	Acoustic Category			Sun Orientation[a] (for Temperate and Cool Zones)							
		Sound Source	*Neutral*	*Sound Sensitive*	*NW*	*W*	*SW*	*S*	*SE*	*E*	*NE*	*N*
Major sources of heat and moisture (see also Section 3.7)	Sun porch		O			O	O	O	O			
	Laundry-utility	O			O					O	O	O
	Kitchen	O					O	O	O	O		
	Bathrooms	O			O	O	O	O	O	O	O	O
Inactive, lightly clothed users	Library-study			O	O						O	O
	Living	O		O		O	O	O	O			
	Dining	O		O		O	O	O	O	O		
Insulated users	Bedrooms			O			O	O	O	O	O	O
Active users	Play areas	O				O	O	O	O			
	Workshops	O			O						O	O
Minimal difference from outdoor conditions	Garage	O			O	O	O	O	O	O	O	O
	Terraces		O			O	O	O	O	O		
	Drying yards		O			O	O	O	O			

← Increasingly Critical Thermal Control →

[a]Copyright © 1974 by Jeffrey Ellis Aronin, AIA, FRIBA, 389 Woodmere Blvd., Woodmere, L.I., New York, 11598. As printed in *Climate and Architecture*, New York, Reinhold Publishing Corp., 1953, 304 pp., illus.

Another design procedure for sorting through the influences on a building's schematic layout is shown in Table 3.1, where residential functions are related by categories of thermal range, moisture, acoustics, and recommended orientation to the sun. There are ambiguities in this procedure; the library-study is shown as a particularly critical area for thermal control, yet it is not recommended for south exposure. This recommendation results from the even quality of north light as opposed to glare from the direct sun; yet, a sunny library can often be successful. Southern exposures are recommended for the spaces with prolonged occupancy; thus, a laundry room is more suited to north exposure, even though clothes-drying yards perform best on south exposures. A particularly prized view, a noise source at any orientation, or simple personal preference can obviously produce a layout quite different from the general recommendations of Table 3.1.

3.2 The Building Envelope

The envelope of a building is not merely a set of two-dimensional exterior surfaces; it is a transition space—a theater where the interaction between outdoor forces and indoor conditions can be watched. Some of these interactions include the ways in which sun and daylight are admitted or redirected to the interior, the channeling of breezes and sounds, and the deflection of rain. This transition space, which forms the envelope, is a place where people indoors experience something of what the outdoors is like at the moment, and where people outside get a glimpse of the functions within. At entries, where there is a space created in the transition from one environment to another, a person will be most aware of the difference between outdoors and indoors; an example of entry as space, not just surface, is shown in Figure 3.3.

The envelope also has a fourth dimension; it changes with time. The seasons have a marked effect on the entry in Figure 3.3, and a more subtle effect on the east-facing balconies of the apartments in Figure 3.4. The year-round usable volume of these apartments is increased by making the balcony into a sun porch, as has been done for many apartments in Figure 3.4. Not all buildings encourage such flexibility; an unchanging envelope can be symbolic of stability and is considered appropriate for some governmental and religious monuments. Generally, the more that users are involved in decisions about how much outside to bring inside, the more changeable the building envelope will be. Further examples appear in Figures 3.15 and 3.16.

3.3 Components of the Envelope

Called by their familiar names, the basic components of the envelope are windows, doors, walls, and roofs. On closer inspection we find that windows can include skylights, clerestories, screens, shutters, drapes, blinds, diffusing glass, and reflecting glass—an array of components that determines more exactly how the envelope does its job of making the transition between inside and outside. In his book *Intentions in Architecture,* Christian Norberg-Schulz suggests that a component also be thought of by its function in the exchange of energies: as a *filter, connector, barrier,* or *switch.*

> In general, we define a *connector* as a means to establish a direct connection, a *filter* as a means to make the connection indirect (controlled), a *switch* as a regulating connector, and a *barrier* as a separating element. . . . An opaque wall thus serves as a filter to heat and cold, and as a barrier to light. Doors and windows have the character of switches, because they can stop or connect at will.[1]

This approach to the choice of components can be illustrated by two different concepts of envelope design: the *open frame* and the *closed shell.* In harsh climates (or where unwanted external influences such as noise or intruding activities abound), the designer frequently conceives of a building's envelope as a closed shell and proceeds to selectively punch holes in it to make limited and special contacts with the outdoors. In the hot-humid regions (or where external conditions are very

[1] Christian Norberg-Schulz, *Intentions in Architecture,* Copyright © 1965, M.I.T. Press, Cambridge, Mass.

Fig. 3.3 *The envelope is more than a surface. This south-facing entry to an architect's office in Oregon becomes a microclimate that buffers the transition between the indoors and outdoors. In* (a), *three kinds of entry conditions are visible: the awning (over the windows of a restaurant), the gable roof with bare rafters (over the planting in the architect's entry), and the arcade, a second story carried out over a covered walkway, that links shops. The architect's entry is shown in detail on March 21* (b). *The change of seasons* (c) *brings deep shade to this entry in summer and early fall. Wilmsen & Endicott, Architects.*

Fig. 3.4 Envelopes change with time. This east-facing building envelope shows changes between seasons and between years. (a) *December 21, 8:30* A.M. *the sun is low and shining from the southeast; shading occurs not from balcony overhangs but from vertical balcony-divider walls.* (b) *June 21, 8:30* A.M.; *the sun is high and nearly due east, so shading occurs only from balcony overhangs and from blinds at the railings.* (c) *An August morning, several years later, showing the conversion of balconies by many of the occupants. Cascade Manor, Eugene, Oregon; John Graham & Associates, Architects.*

close to the desired internal ones), the envelope begins as an open structural frame, with pieces of building skin selectively added to modify only a few outdoor forces. The open-frame or the closed-shell approach to envelope design, when combined with material availability and the influence of local culture, can produce a distinct regional architecture (Fig. 3.5).

With a wide range of energy sources, building materials, and mechanical equipment available, it is possible to design buildings anywhere that are connector-dominated, despite the climate. The consequences from the resulting energy consumption can be severe. By contrast, if defending against outdoor conditions becomes an overriding consideration, then barrier-dominated envelopes can occur in any climate. The resulting fitness for human usage—and the potential of using solar heat directly—can be reduced. The designer's combinations of connectors, filters, and barri-

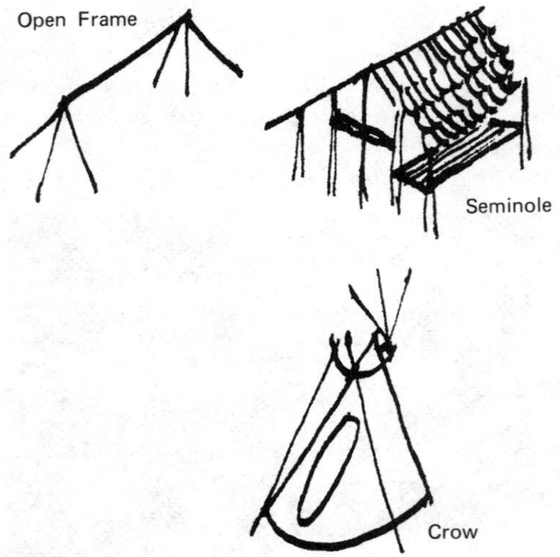

Hot Humid Climate: To the open frame, a barrier roof of local plant materials is added to reject rain and sun. A raised floor avoids damp earth and its creatures and allows breezes to pass over and under its users.

Temperate Climate: This open frame is wrapped in light–filtering animal skin, doubled near the ground. Wind and rain are rejected; protection against cold is provided by users' clothing (blankets) more than by the envelope. The switch at the crown controls smoke (see also Fig. 3.11). Portability of shelter is a cultural factor here.

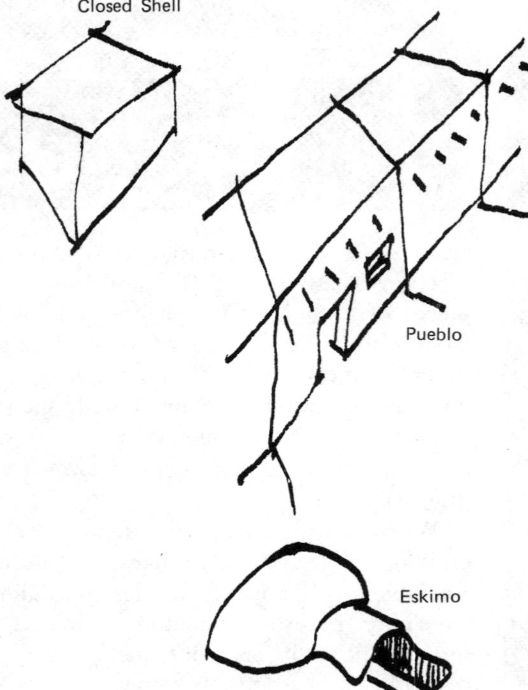

Arid Climate: The closed shell of mud block is a barrier to wind and sunlight; it filters heat by both delaying and reducing its impact on the interior. Some light and heat are admitted directly by small connectors: the door and window, typically south–facing.
By early morning, the cold interiors are abandoned in favor of rapidly–warming south terraces.

Cold Climate: The igloo's closed shell of ice is a filter to light and heat, a barrier to wind. Holes for entry and for smoke are allowed, but sparingly. Fur–bearing hides hung inside can increase thermal comfort for users.

Fig. 3.5 *Open frame and closed shell: climate, materials, and culture. The influence of climate is clear, but material availability and cultural patterns are also strong determinants in the choice of envelope design in these examples.*

ers (and the switches that allow these elements to respond to changing conditions) are basic to the design of building exteriors and can give them the liveliness that makes a building an attractive addition to its neighborhood.

1. *Connectors* are strong indicators that something outside is welcome inside. They are characteristic of regional architecture in milder climates, but sun connectors are dominant in solar-heated buildings anywhere. Connectors, being open to outside influences, are often one position of a switch that in other positions becomes a filter or a barrier. Or, as in Figure 3.6, a connector is sometimes followed by a combination of filters and barriers somewhere further inside the space that forms a building's envelope.

2. *Filters* represent decisions about how much or what kind of outdoor condition is to be admitted. They are found in some form in all building envelopes and in all climates, and they include a wide variety of types. Because they admit desired amounts or qualities of light, air, and sound, they offer an opportunity for an enhanced awareness of selected outside conditions from inside the building. For example, the stained glass of a church selects the blue of a north sky, or the warm reds and oranges of a sunset; the *texture* of the sky's cloud patterns is not admitted to the interior—only its color comes through. Often the filter is one of the positions of a switch, as in the case of the windows in this famous building by Corbu (Fig. 3.7).

3. *Barriers* are more drastic in their complete severance of the outdoor-indoor relationship. They are characteristic of regional architecture in harsh climates, but are also common to spaces needing a tightly controlled environment (such as an auditorium). Barriers to rain are an almost universal building feature; barriers to wind are at least seasonally common in all climates, except hot-humid ones. Barriers to sun are more likely to be one position of a switch, unless a building is suffused with electric light or other plentiful sources of internal heat that make solar heat

Fig. 3.6 *Connectors for a corridor. An old cannery building in the mild climate of San Francisco is converted into a series of shops; a corridor is given a near-outdoors environment by the connectors at the right. Across the corridor, filters give greater internal climate control to the shops beyond. The Cannery; Joseph Esherick & Associates, Architects; photo by Richard A. Cooke III.*

Fig. 3.7 *Filters in various positions of a switch. Le Corbusier's Pavillion Suisse, 1930–1932, at the University of Paris; photo by Nicolai Shur.*

Fig. 3.8 Barriers, but not to heat. These barrier-dominated exteriors, typical of Haida villages in coastal British Columbia and southern Alaska, shed rain, wind, and light. The walls are filters, however, to heat that passes with relative ease through the thick planks. (In winter, the narrow spaces between the houses can be enclosed, producing a more heat-conserving "row-house" configuration.) The central fires of each house are fed by plentiful wood fuel, and firelight is preferred over daylight for illuminating the elaborate masks used in Haida ceremonies. Photo courtesy of the Milwaukee Public Museum.

permanently unwelcome. In practice, cultural influences often override those of climate; barriers to sun are erected even in cold, damp environments, such as those of the Canadian Pacific (Fig. 3.8).

3.4 Envelopes and Energy Leaks

With our increased awareness of the dwindling supplies of nonrenewable fuels, greater attention is being given to the flow of heat in buildings. One option for reducing the amount of energy consumed in warming a space is to increase the thermal resistance of its envelope. This barrier-wall approach requires more material (such as insulation) in construction, and the extra labor of carefully installed vapor barriers; but it achieves conservation without making demands on the users to change their living patterns. The detailed procedure for evaluating a wall as a heat barrier is presented in the next chapter, but preliminary observations can be made for two situations.

(*a*) ***Existing Buildings.*** These buildings can first be weatherstripped—a process of plugging obvious cracks with filler material. They can also be retrofitted with insulation, often by filling the cavities between wall studs or ceiling/floor joists.

Attention is needed to a vapor barrier at the interior surface of the insulation; water vapor will readily enter walls, ceilings, and floors from the interior spaces. As the vapor cools upon approaching the colder exterior, it eventually condenses inside the wall, causing water damage to materials through staining, disintegration, or degradation of their insulating value. As a building envelope is better insulated, its interior surface temperature *rises* but its exterior temperature *drops;* the lower the temperature somewhere within a wall, the more likely it is that condensation will occur. Thus it is necessary to stop moisture from entering the wall from the interior space, particularly in well-insulated construction.

Another typical retrofit technique is to add a second layer of glass to the window openings. This provides an air space between the panes of glass, increasing the typically low thermal resistance of window areas.

There are major changes that can be made in existing buildings, such as replacing windows with insulated walls or increasing the thickness and insulation of walls or roofs. This involves either a costly new exterior surface, with changes in the frames of all openings, or a new interior surface with changes at electrical outlets, counters or fixtures adjacent to exterior walls, and again at all window and door frames.

(b) New Buildings. Better opportunities for dramatic increases in the thermal resistance of walls and roofs are provided in new buildings. The long-term energy savings achieved by using a $5\frac{1}{2}$-in. rather than a $3\frac{1}{2}$-in. framing with wall-cavity insulation has been shown in Figure 1.23.

One example of conservation by such attention to the envelope began in Arkansas, where a construction system known as the "Arkansas House" was developed in the 1960s. The central heating systems of these houses, illustrated in Figure 3.9, use substantially less energy than conventionally constructed homes. This trend, if continued, could lead to the disappearance of central heating in "superinsulated" residences, with such heat supplied by cooking, appliances, lighting, and body heat.

Regulations to ensure greater energy efficiency in buildings typically deal with the building envelope, as will be shown in the ASHRAE Standard 90-75 recommendations encountered later in this text. Such a preoccupation with barriers could eventually lead to an elimination of windows along with the central furnace. Another approach to energy conservation involves life-style changes: using less energy for appliances, lighting, and heating, and relying more on the sun for space heating. The variability of solar heat and the emphasis on the user's life-style combine to make the switch an important element in the envelopes of such user-dependent environmental control systems.

3.5 Switches and Users' Choice

When window area oriented toward the winter sun is increased, a different kind of residential envelope results. The house in Figure 3.10 was designed around principles similar to those used in Frank Lloyd Wright's Solar Hemicycle (see Fig. 2.2, page 33), with the important addition of switches that can make the window more of a barrier when appropriate. The house is provided with a large amount of thermal mass, so that some of the heat admitted on a sunny day can be stored for nighttime warmth. The users of this house must often base their actions of the moment on what will be needed later, by manipulating thermal switches. This is called "thermal sailing," and is similar to actions of outdoor workers in the far north, who learn to unbutton their coats in the cold early hours of the workday *before* they begin to sweat and rebutton their coats in the relative warmth of the late afternoon before the rapidly falling temperatures near dusk can chill the skin. Misjudgments in passive solar-heated homes can result in extraordinarily high temperatures on a sunny day or nights without stored solar heat. The relationship between user and envelope is a continuing and educational one; it can also be largely automated when desired. A much earlier example of thermal sailing is shown in Figure 3.11.

The contrast between the limited window areas of the Arkansas Houses and the glass expanses of passive solar-heated buildings raises the question of the relative thermal benefit of south-facing glass. The answer is complicated; it varies from one area to another, depending heavily upon the *average insolation* (or solar radiation) available on south-facing surfaces, as well as on *average outdoor temperatures*. Given these figures, a

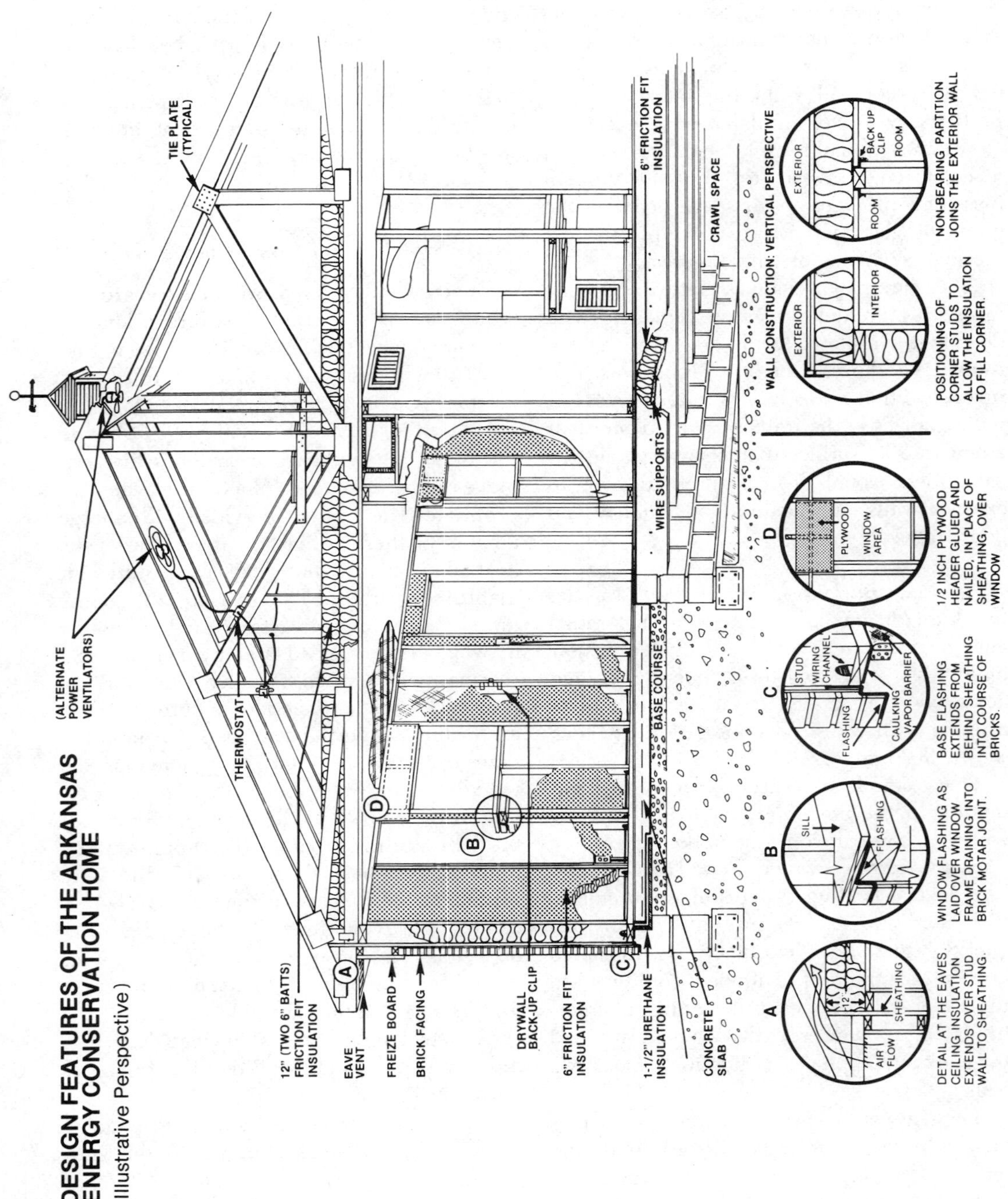

Fig. 3.9 *The Arkansas House. Using thicker walls and a deeper ceiling cavity to accommodate much more insulation, these houses also limit window area to 8% of the floor area; they are so tightly sealed that humidity control is essential. This scheme was a cooperative effort between an electric utility, an air-conditioning engineer, and a government (HUD) construction analyst. From Owens-Corning Fiberglas,* Energy Conservation Ideas to Build On, Report #1: The Arkansas Story. *Courtesy Owens-Corning Fiberglas, Toledo, Ohio. Further details of this construction method are shown in Figures 4.21 through 4.24, pages 156–158; see Section 4.23.*

Fig. 3.10 *Thermal sailing in New Mexico.* (a) *The north wall is mostly a barrier;* (b) *the entire south wall can be a connector to sunlight, or* (c) *it can be isolated to varying degrees from the outside by operating switches. David Wright, designer; photos by Edward Mazria.*

comparison can be made, as graphed in Figure 3.12 for Oregon's Willamette Valley. The heat-gain advantages of south glass are much more evident on *clear* days (data for south walls at 40° N. Latitude are found in Table 4.27, page 170) than on *average* days. Oregon's relative winter cloudiness might lead to the conclusion that south glass anywhere else must produce a higher net gain, but Oregon's winters are relatively mild. What is clear from the graph in Figure 3.12 is that insulated switches used with south glass have an advantage over the simple connectors to light.

The use of daylight to supplement or replace electric light involves a more familiar switch: the one that controls electric lights. It seems obvious that electric lights near windows should be controlled locally so that,

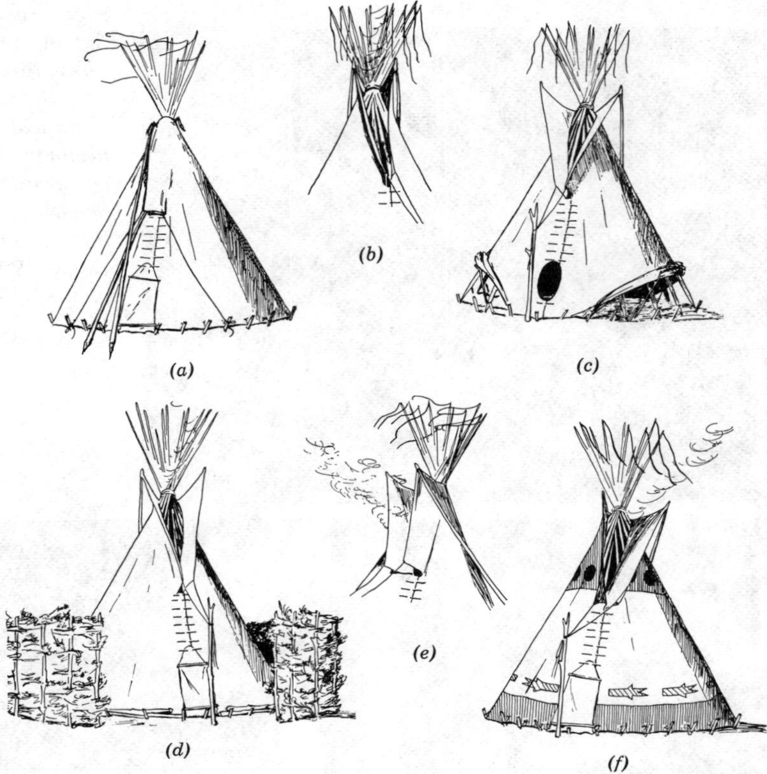

Fig. 3.11 Thermal sailing on the Great Plains. Six adjustments to the lightweight, translucent, and portable tipi are shown; the tipis in these diagrams are east-facing, with their backs to the prevailing westerly winds. (a) *In severe rainstorms, the smoke flaps can be closed.* (b) *For ordinary conditions of west wind, the smoke flaps block the wind, thus creating a suction at the opening to draw out the smoke.* (c) *For hot weather, breezes are admitted under the cover at the ground.* (d) *For extremely cold weather, a temporary windbreak can be added.* (e) *For the unusual northeast wind, or* (f) *southwest wind, the smokeflaps are manipulated to block this wind, thus encouraging smoke draw-out, as in* (b). *From* The Indian Tipi: Its History, Construction, and Use, *by Reginald and Gladys Laubin. Copyright 1957 by the University of Oklahoma Press.*

when daylight is sufficient, they can be turned off. Many buildings are electrically lighted with all controls at a central point to reduce the installation cost. The removal of these switches from the lighted areas usually results in wall-to-wall electric lighting whenever anyone uses the space, even if one of those walls provides a plentiful supply of daylight. Designers of electric lighting systems must recognize two basic conditions of use: without *and with* daylight; the luminaires and their controls should be chosen and located accordingly. A careful look through the windows in Figure 3.13 shows a separate recessed lighting layout at the perimeter of most floors of this

Fig. 3.13 (opposite) *Perimeter electric lighting. The recessed incandescent lights just inside these windows contrast with the fluorescent lighting further inside. The warm color of incandescent lights mixes well with the cooler color of daylight; their radiant heat can be welcome on cold, dark winter days; and they can be swtiched off in the warmer, brighter days of summer and fall. With today's energy considerations, they are seldom used. Bank of America, San Francisco; Wurster, Bernardi &Emmons and Skidmore, Owings & Merrill, Joint Venture Architects; photo by Richard A. Cooke III.*

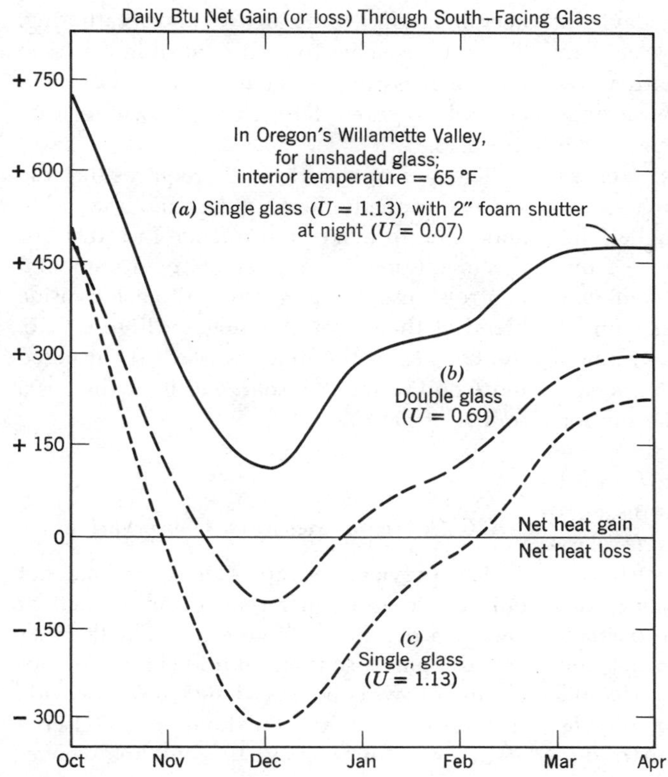

***Fig. 3.12** South-facing glass in Oregon: heat gain or heat loss? The average daily heat gains through unshaded south-facing glass surfaces are compared to their average daily heat losses; net gain or loss for an interior kept at 65°F is evident.* (a) *Single glass, but with a 2-in. foam shutter closed for nondaylight hours;* (b) *double glass;* (c) *single glass. For clear days, even single glass has a big net gain. (See solar heat gain factors in Chapter 4, Table 4.27, page 170; U-values are defined in Section 4.15.) From John Reynolds,* Solar Energy for Pacific Northwest Buildings, *Center for Environmental Research, University of Oregon.*

office building; perimeter areas with daylight are given a different electric light source than the core areas of the interior. (Another designer's mixing of daylight and electric light was seen in Fig. 1.18, page 16.) One of the subjects of conservation regulations, such as ASHRAE Standard 90-75, is the provision of local control for electric lighting installations.

Windows are also subject to glare from direct or reflected sunlight—a problem that can conflict with the desire for heat gain or daylighting. This can be especially troublesome when the sun is low in the sky, appearing near eye level. Rooms open to the sun for heat-gain purposes are not alone in this problem, as is shown in Figure 3.14. Sun control devices that encourage heat gain but discourage glare are typically located *inside* glass surfaces, where sunlight can be converted to heat and kept within the space. Where the sun's heat is unwelcome, filters or barriers should be located *outside* the glass so that the heat of intercepted sunlight can be carried off by outdoor air. This exterior location enlivens a facade, but it can also cause problems of dirt accumulation, wind damage, and weathering. Three approaches to solar control in the same city, contrasting a reliance on switches to control by fixed filters, are shown in Figure 3.15.

The user-operated switch requires that the occupants of a building understand how, why, and when to make adjustments. In return for this investment in rudimentary environmental controls, users are given an outside-inside blend of their own choosing. In Figure 3.16, the choices of the office workers seem based more on the sun as a source of light than as a source of heat.

3.6 Conscious Resource Conservation

The previous examples illustrate that not only can the designer reduce the amount of energy a building will consume, but the user can also play a substantial role via the manipulation of switches. The designer can provide further opportunities for the user to conserve at many points beyond the building's enve-

Fig. 3.14 Sun versus view. A restaurant on the California coast has connectors that give both a view of the Pacific and blinding dinnertime sun. As diners arrive, switches are employed to block the sun but to leave open some ocean vistas in other directions.

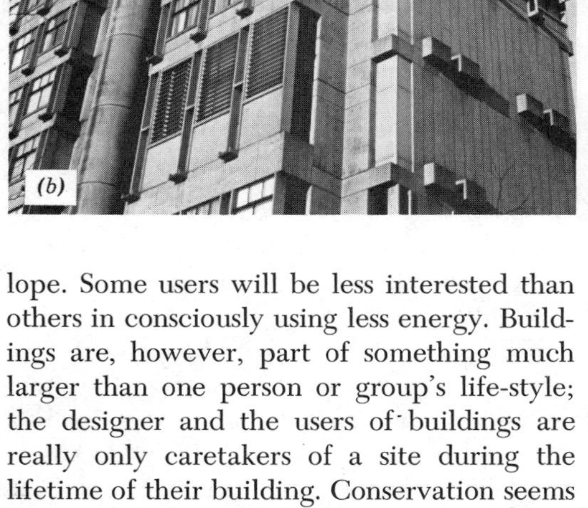

Fig. 3.15 *Sun control for Boston offices.* (a) *Switches both inside and outside the glass are evident, and the windows themselves are operable. Awnings are more likely to be fully extended as the windows sit higher up in the canyon of the street. (New England Merchants' Bank; Shepley Rutan & Coolidge, Architects; demolished 1966).* (b) *Three-dimensional filters in the form of overhangs dominate the south facade (right) of this building at Boston University; the west windows (where overhangs are less effective) have internal switches, shown here in a variety of positions. (Law and Education Building; Sert, Jackson and Gourley, Architects.)* (c) *The two-dimensional filter of reflective glass sheathes all faces of this office tower, sending a beam of reflected sunlight to the neighborhood below. The switches in this case are thermostats; users are not involved in the outside-inside decisions. John Hancock Headquarters; I. M. Pei & Partners, Architects; photo by Stephen Tang.*

lope. Some users will be less interested than others in consciously using less energy. Buildings are, however, part of something much larger than one person or group's life-style; the designer and the users of buildings are really only caretakers of a site during the lifetime of their building. Conservation seems to be in everyone's long-term best interests; the designer should provide the opportunity to conserve so that users can play their caretaker role more responsibly.

One example of the number of conservation options a designer can provide is found in the familiar practice of bathing. Consider the person in a shower or tub; Table 3.2 describes a number of design opportunities and tradeoffs in attempting to conserve heat and water in this situation. Similar analyses could be

Fig. 3.16 Which is spring, which is summer? The awnings for this office building (a) *in Eugene, Oregon, are joined by overhangs and side walls (or fins) to provide sun control for these south-facing windows. The summer sun, at a high angle* (b), *is readily blocked by the overhang; the awnings can be rolled up with less risk of direct glare into the users' eyes. Moreland-Unruh-Smith, Architects.*

done for other widespread practices that involve resource consumption, including food preparation, waste recycling, elimination of body wastes, personal transportation, laundry, and space heating. (If some of the user's opportunities in Table 3.2 seem unlikely and the designer's opportunities farfetched, consider the reality of affluent Californians in the late 1970s who stood in plastic garbage cans while showering, then siphoned this collected water outdoors through garden hoses in an attempt to keep their plants alive through a two-year drought.)

There are at least two challenges to the designer who wishes to encourage the user's involvement in conservation: to keep the user aware of consumption and to provide attractive alternatives to the high-energy conveniences that are, at first, hard to give up. To keep the user aware of consumption, the de-

Table 3.2 *Conservation in Bathing: Some Objectives, Opportunities, and Trade-Offs* (*Arranged for each objective in approximate order, with the least-conscious life-style change listed first.*)

User's Opportunities	*Designer's Opportunities*	*Trade-Offs*
Objective: Use Less (Hot) Water		
1. Reduce flow through shower head.	Choose and locate controls to encourage frequent and exacting regulation; install a restriction on maximum flow; install a simple flow meter on supply to shower head.	Exacting controls probably will increase first cost, as will a flow meter. (Maximum-flow restrictors are free from some utilities.)
2. Use less water in bathtub.	Design a scale on side of tub to indicate volume below water line; select a tub with lower capacity (i.e., overflow located lower in the tub).	A water-volume scale can also act as a signal to the overweight: the more you displace, the greater the problem!
3. Use a fixed-capacity solar water heater, seasonally.	Design an unshaded, wind-sheltered place for the collector/storage device (typically a pillow, transparent on top and black underneath, that is refilled after each use with cold water). If refill is not automatic, place fill control valve in obvious location.	The solar device and its plumbing will increase first cost. Where water can be solar heated above 110°F, automatic cold water mixing valve should be installed as well.
4. Spend less time in the shower.	Install a timer device on *one* of the bathroom lights or on the shower stream itself. A view of a clock from the shower enclosure could be provided, or install an x-minute "hourglass."	Timer devices increase first cost. The signal could inform rather than annoy or endanger the user (by sudden deprivation of water or light).
5. Separate cleansing from relaxing functions of bathing.	Provide a "sponge-bath" cleansing facility as well as a well-insulated soaking tub (in which water is not often changed).	Conservation of water is more likely to occur than conservation of heat, unless tub is covered between soakings.
6. Use a cooler water temperature.	Insulate bathroom surfaces to raise winter radiant temperatures; install a dial thermometer on the shower stream; choose controls that do not completely cut off cold water to shower or tub; select and locate water heater with an easily accessible and adjustable thermostat.	The thermometer will increase first cost. Lower setting on water heater means less heat lost through pipe and tank walls, and (since less cold water is mixed with hot, stored water at the shower head) quicker depletion of hot water, shortening the shower.
7. Don't bathe as often.	Select and design for a lavatory, counter, and floor area that are ample for "sponge baths," encouraging frequent partial body cleansing.	Increased size of lavatory and counter area will increase first cost. Impacts on personal hygiene vary.

(Continued)

Table 3.2 ***Conservation in Bathing: Some Objectives, Opportunities, and Trade-Offs (Arranged for each objective in approximate order, with the least-conscious life-style change listed first.) (Continued)***

User's Opportunities	*Designer's Opportunities*	*Trade-Offs*
Objective: Recover Otherwise Wasted Heat and Water		
1. Capture waste hot water (from any source) in central location where heat can be utilized.	Select and locate a holding tank to achieve heat exchange, probably with the fresh cold water enroute to the water heater. Tank must allow for occasional removal of sludge from bottom; controls on tank are necessary to exclude cold waste water.	The tank and its temperature-selective control will increase first cost. Possible contamination of incoming fresh cold water from hot water tank.
2. Capture less-polluted (gray) water at any temperature, for irrigation.	Select and locate a large holding tank to allow gravity feed of irrigated area and to allow occasional removal of sludge from bottom. Select controls at fixtures to encourage user to store only the relatively clean waste water; design outflow to discourage drinking of this water by those unfamiliar with its source.	The tank, its plumbing, and controls will increase first cost. Possible health problems from utilizing gray water unless it is treated.
3. Keep shower or bath water in tub until heat exchange with room air is achieved.	Choose and locate the tub or shower to facilitate cleaning; provide for user to stand out of collecting water while showering; encourage passage of room air over standing water; locate bathroom so that warmed, moist air can rise to other, cooler rooms.	Standing water leaves scum on surfaces; more moisture in air will result, with varying impacts on comfort.

signer can locate meters for water, electricity, and fuel consumption where they will be frequently seen, and where resulting conservation practices will occur. Perhaps an ideal residential meter location would be in a place visible from the dishwasher, clothes dryer, and bill-paying desk, as well as from the outside (for utility personnel).

Conservation proposals such as ASHRAE Standard 90-75 recommend individual metering for tenants of apartment buildings to avoid feelings like, "I've paid for it already, so I'll be sure to use my share." Further refinements in metering and supply are likely; one indicating meter is shown in Figure 3.17. Some utilities install signal lights that warn customers of peak demands on the utility network; during those times the cost of energy is made more expensive. A growing array of automatic electric controls exists to disconnect nonessential loads during peak periods (and thereby lower the customer's electric bill); this practice is further discussed in Chapter 14.

A more substantial challenge is to design lower-consumption alternatives to be both available and attractive to users who want to keep the conventional air conditioner, clothes dryer, dishwasher, garbage disposal, and trash compacter for occasional use. These alternatives typically require more space, time, and labor to use, and may have special orientation needs. Some basic objectives for conservation alternatives are:

a. The alternative should be visible and directly accessible from the convenience appliance that it could replace.
b. The alternative should be visually attractive.
c. Other pleasant activities should be integrated with the alternative, as appropriate; this can increase the rewards for choosing the alternative.
d. The alternative should advertise its resource savings, such as a comparative energy-consumption chart located in view of the convenience appliance.

The opportunities for personal attention to conservation are numerous in residences and small shops, but what of larger office buildings? There are few such opportunities for an office worker who is placed behind sealed reflective-glass windows, with ceiling lighting controlled from a locked panel at some distant

Fig. 3.17 Indoor electric "meter" for conscious conservation. This panel is installed flush with an indoor wall surface, usually near the thermostat, where it can be readily seen by the consumer. (It does not *replace the meter, which is still located outside for utility personnel.) It shows the* cost *of electricity being used in a residence at any given moment, and it is, of course, adjustable to reflect increasing electric rates. (It also tells time of day, when not indicating cost.) Courtesy of Alex Funk, for Fitch Creations; Alex Funk, inventor; R. B. Fitch, developer.*

(a)

point. A provision of more switches could produce personal incentives: a combination of adjustable external sun shades, internal glare controls, and locally controlled overhead and task (desk) electric lights would provide substantial opportunities to reduce personal consumption. Figure 3.18 illustrates a step in this direction for an office building in California. Specifics of the energy savings for a St. Louis office building using similar modulating shading devices are detailed in Table 19.8 (Case 3, page 807). A particularly intriguing example of a switch for office buildings is found in Chapter 8, Section 10.

Reinforcing signals, as well as opportunities, can be provided to the office worker. For example, with "variable volume" air supply systems (Chapter 8), the greater an office's load on the cooling system, the more cooled air is pumped into the space. A series of signal lights can be installed at each variable volume orifice (the more lights on, the greater the load you are placing on the system), or recording meters can be utilized to compare the consumption of various offices. Alarm devices, signaling high consumption, are shown in Chapter 14, Figure 14.12, page 526.

Fig. 3.18 *An office building for Sacramento's climate. A variety of approaches to envelope and equipment distinguishes this design, developed by the staff of the California State Architect's office. From the exterior* (a), *the east facade (and its counterpart on the west) is brightened by roll-down fabric shades. These disappear daily, when direct sun no longer threatens to overheat the perimeter offices. (Sun control on the south facade is a simple trellis.) The interior court* (b) *is roofed to admit south sun in winter for passive solar heating; movable louvers can close out direct sun whenever overheating is a problem. The range of heating and cooling devices is shown in* (c), *including solar collectors for domestic water heating, rockbed and building mass to store "coolth" (obtained from venting with Sacramento's cool night air), and vertical canvas tubes with fans to reduce the courtyard's air stratification. Courtesy of the Office of the State Architect, Sacramento, California.*

3.7 Harnessing Thermal Waste

Some "spillage" of energy is inevitable; electric lights produce more heat than light, cooking-range tops send much of their heat

(b)

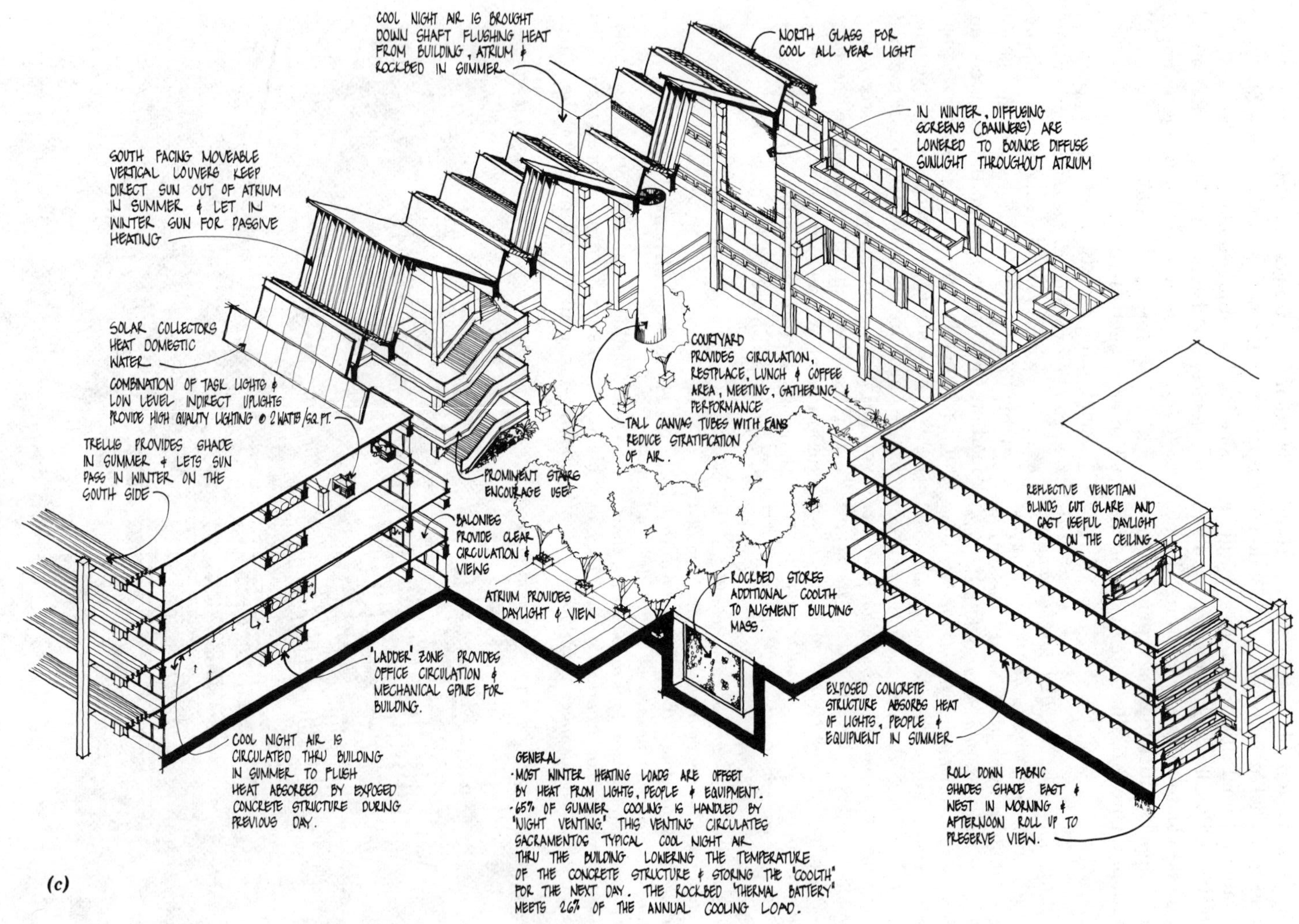

(c)

Fig. 3.18 (*Continued*)

directly to the room, and heated water loses some of its energy through pipe and tank walls. There are opportunities to make some use of otherwise wasted heat; these opportunities can have design consequences as well as an impact on the cost of constructing—and operating—a building.

As buildings become larger (or the temperature and/or quantity of the waste products increases), the benefits from utilizing waste become so substantial that products are devised to recover these resources. Some of the mechanical devices for heat exchange between exhaust and incoming fresh air, for example, are shown in Chapter 5. Analysis of expected energy savings can show approximately how much time will elapse before these devices pay for themselves in reduced fuel bills.

The goal of capturing waste can result in a rearrangement of appliances within spaces, altered layouts of the spaces themselves, or even wholly new spaces as shown in Figure 3.19. Some of the more common opportunities for heat exchange in dwellings are shown in Figure 3.20; estimates for the energy consumption of some common residential appliances are shown in Table 3.3. The functions of the kitchen, laundry, and bathroom are of special interest in residential waste harnessing. In Table 3.1, the orientation and thermal-comfort grouping of spaces was discussed; there, the sources of heat and moisture present in the kitchen, laundry, and bathroom set them apart from other spaces. This is a result of both their need for moisture-resistant materials and the moisture, itself, which is sometimes unwelcome in other spaces. These functions also have an uncommonly large appetite

Fig. 3.19 *This greenhouse utilizes clothes dryer exhaust; it also substantially reduces heat loss from the living room window, which serves as the greenhouse's rear wall. Paul Barnhart, Designer.*

Table 3.3 *Residential Energy Usage*

Reason for Energy Expenditure or Specific Appliance Expenditure	*Approximate Percent of Annual U.S. Residential Energy Consumption*[a]	*Approximate Average Monthly Kilowatt-Hours for Specific Appliances*[b]
Space heating	57	(Varies widely)
Water heating	15	400
Refrigeration: 14 cu ft refrigerator/freezer	6	96
with frost-free feature		154
Other appliances	6	
(not including use of hot water): dishwasher		36
clothes washer		8
Cooking: range with oven	5	70
Air conditioning	4	(Varies widely)
Television: black/white, solid state	3	10
color, solid state		36
Food freezing: 15 cu ft freezer	2	100
with frost-free feature		147
Clothes drying: 5 loads per week	2	83

[a] From Stanford Research Institute, *Patterns of Energy Consumption in the U.S.* (data for 1968).
[b] From the Washington State Energy Office, 1976.

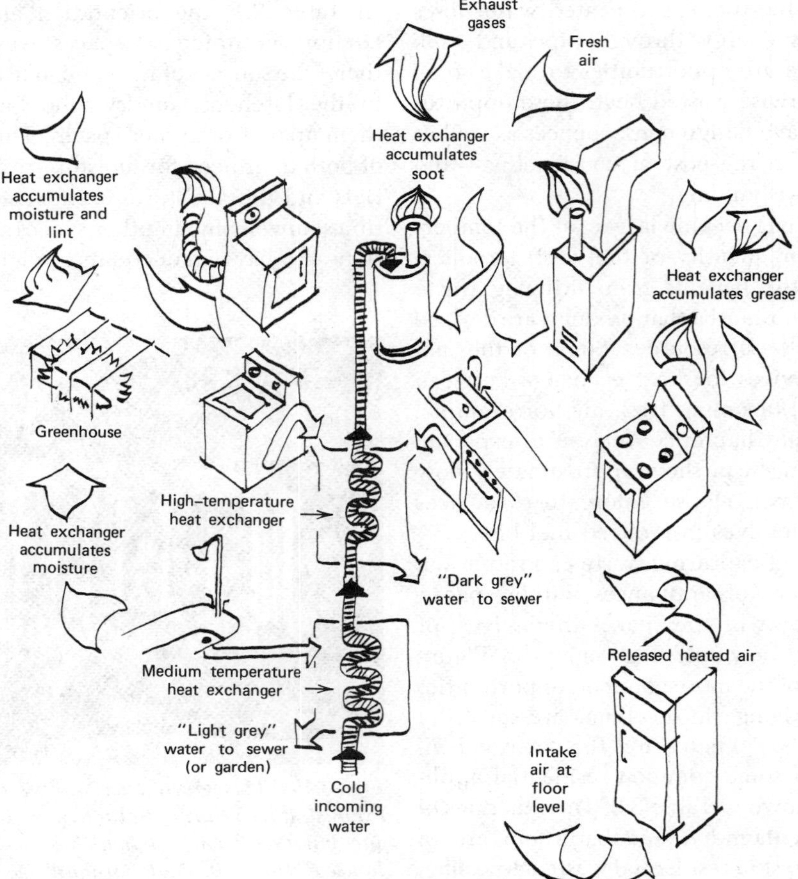

Fig. 3.20 Some opportunities for heat exchange in dwellings.

for outdoor air (for the purposes shown in Fig. 3.20) and can be sources of intense heat for short periods of time.

As both air exchangers and heat sources, these functions present not only opportunities for harnessing waste heat (primarily by exchanging outgoing heat with incoming colder air or water), but for *storing* heat for use by other spaces later on. In this role of accepting heat for shorter periods and releasing it later, they resemble passive solar-heated spaces, and thermal storage materials (such as masonry) can become appropriate for their surfaces. This is particularly true for the kitchen, already recommended for south exposures and, therefore, capable of receiving solar heat directly as well as substantial heat from cooking.

3.8 Recycling Solid Waste

This is another conservation influence to which the kitchen is especially subject. The separation of dry organic garbage (paper), wet organic garbage, and inorganic wastes such as glass, plastics, aluminum, and other metals is encouraged by community recycling efforts in many locations. In the home, wet organic garbage can be composted for garden use, and dry organic garbage might fuel the fireplace. Thus, the kitchen needs direct access to a compost pile, as well as a place (which is cool and dry, to discourage odors) to store various waste metals, plastics, and glass until recycling time.

The Clivus Multrum (see Chapter 9) is one built-in way to deal with wet organic waste from both kitchen and bathroom, but typically where a Clivus Multrum has not been installed, a place for food wastes awaiting composting is also required. Since these wet wastes are particularly odorous, they also need cool temperatures and isolation from the kitchen. In addition to being roomy, easily accessible, closed off, and cool, these spaces must be able to be cleaned. The energy consuming and antirecycling garbage compacter is clearly an easier approach for the designer as well as the user, but it fails the challenge of resource conservation.

The outdoor compost pile has its own characteristics to challenge the designer. At its best, it is a frequently turned, quite warm, damp, well-aired source of rich humus (and red worms) for gardens; odors are noticeable only while the pile is turned. At its worst, it is a source of unpleasant odors and a breeding place for vermin. Where odors are not objectionable, the heat generated in a frequently fed and tended compost pile could be welcome against the exterior walls of residences. Clearly, these walls must have nonorganic exterior materials!

In larger buildings, solid-waste collecting and processing systems are often installed to deal with volumes of "specialized" wastes; paper from an office building is one example. There are two basic approaches to these systems:

1. Utilize the waste to recycle both organic and inorganic materials and to recover heat from nonrecyclable materials.
2. Ship the waste elsewhere as quickly as possible.

The first approach includes composting systems, which are available (if not yet common) for multistory buildings; an obvious problem in urban areas is the distribution of compost from so many families to so few gardens. Another example of the first approach is the incinerator—a potential source of both heat and air pollution; Table 5.1, page 178 lists the heat benefits obtainable from incinerating various wastes.

A major problem is the need for separate waste storage, collection, and processing for various categories of garbage. At best, without separation, otherwise recyclable materials are utilized only for heat, as when paper is incinerated. At worst, they become part of our society's growing solid-waste disposal problem.

The second approach, of quickly shipping these wastes elsewhere, is particularly likely to generate large quantities of unrecoverable solid waste. Garbage disposal units in apartment kitchens feed the sewer, which lessens the building's solid-waste collection problems, but not those of the municipal sewage treatment plant. For the wastes that cannot be flushed, compacters are often used to reduce the volume (therefore the cost) of shipped wastes; once compacted, later separation to recover the material is more difficult. Yet the amount of space taken up by the collection and storage of unseparated wastes is typically less than that required for separated wastes, and space becomes more precious as density increases. Storage and loading areas for a downtown apartment tower are likely to be far smaller than for a suburban apartment complex.

An example of the second approach to waste handling is shown in Figure 3.21, where *pulping* of waste is used to greatly reduce its volume. These systems grind waste into pulp in the presence of water, making a readily transportable slurry. At the loading docks, this slurry enters a water press where about 90% of the water is squeezed out, reducing the volume to about one-fifth of that of the original wastes. This water is reused and replenished as required. Such pulping systems are used not only for general refuse, as illustrated, but also for the destruction of documents and for food service wastes. Pulping systems have limitations; they should not be used to handle metal or plastics, so that the recovery of at least these materials remains a possibility. Pulping systems are replacing incinerators in urban areas; their advantages in reduced air pollution must be weighed against the possible heat contribution from incineration.

Once shipped from buildings, unsorted garbage must undergo energy-intensive processing if its resources are to be recovered. This

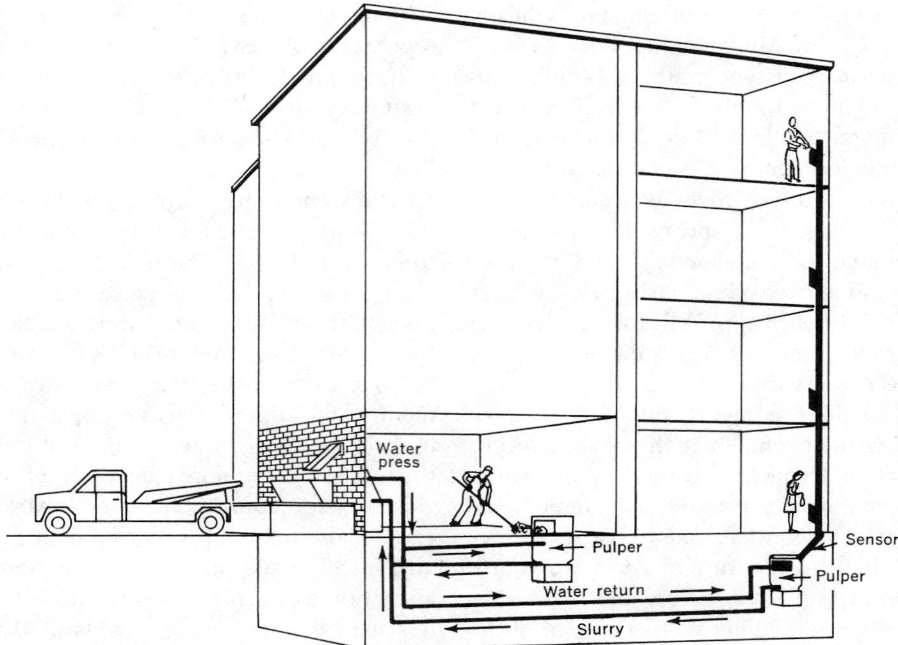

Fig. 3.21 *A pulping installation for apartment refuse. The pulper is located to deal directly with solid waste; the slurry then transports the waste to the press at the loading dock. The pulper can operate manually or automatically by means of a sensor as shown. One water press can serve two or more pulping units.* Wascon Systems, Inc.

procedure typically involves shredding the mixed garbage, then blowing it through massive "air classifiers" that separate the organic (burnable) wastes from metals and glass. Burnable wastes can then be used, under controlled combustion, to generate electricity, and the glass and metals (which have been further magnetically separated into ferrous and nonferrous metals) can be recycled. Figure 3.22 illustrates a proposal for one such facility for the recovery of solid wastes.

3.9 The Building Interior

A look at the interior of a building and its dependance on mechanical and electrical equipment can serve to illustrate some basic design choices. This summary introduces the remainder of this text, which will provide a closer look at heat, water, electricity, light, and sound.

Away from the perimeter and its visible interaction with climate and other outside forces, the interior is potentially an isolated environment. As the designer manipulates heat, air, light, sound, and water to best match the environment of these interior spaces with their function, the following choices must be made.

(*a*) ***Mechanical-Structural Integration or Integrity.*** The similarity of these two technical support systems has intrigued designers ever since mechanical systems began to require substantial volume for distribution, as in air-duct systems. As the complexity and size of the mechanical distribution systems was *increasing* with technological development (typically, more air is required to cool a space than would be required for simply heating it), increased strength of materials was *reducing* the size of the structural system. The "uncluttered" floor areas between the more widely

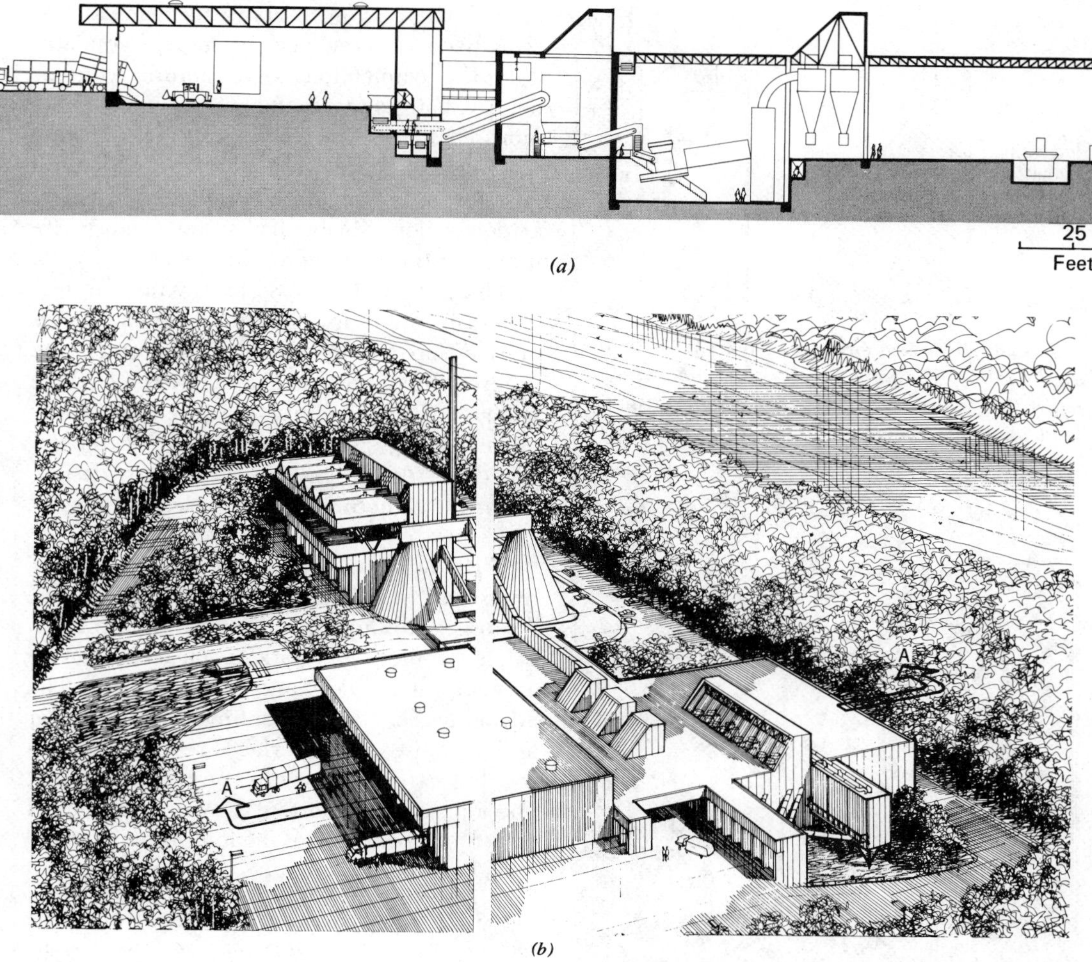

Fig. 3.22 Solid-waste resource recovery project. The flow of initially unseparated garbage through a proposed facility is shown in section (a); *the rendering* (b) *shows the recovery plant in the foreground, the electric generating station beyond. Cone-shaped storage bins hold shredded refuse, awaiting conversion to steam for the turbines. Clean Communities Corporation, Haverhill, Mass.; Camp Dresser & McKee, Inc., Engineers, reprinted from* Architectural Record, *June 1975, Copyright 1975 by McGraw-Hill, Inc., with all rights reserved.*

spaced columns became desirable for flexibility in spatial layout. Keeping the mechanical systems at or within these columns allowed these floor areas to remain clear, so mechanical-structural integration was given further impetus. With the new expectations for cooling, the refrigeration cycle's cooling tower often moved to the roof, taking the air-handling machinery with it. This further encouraged the merging of systems, for one system was growing wider as the other diminished (see Fig. 3.23). Thus, a fixed column cross section became possible, consisting mostly of the structural column at its base and the air duct at its top. (One of the responses to this opportunity, expressed as a dual-duct supply and return system, is shown in Chapter 7, Fig. 7.4, page 280, and Fig. 7.5, page 281.)

Yet the functions of these systems are very different; compared to the dynamic on-off air, water, and electrical distribution systems, the structural system is static—gravity never

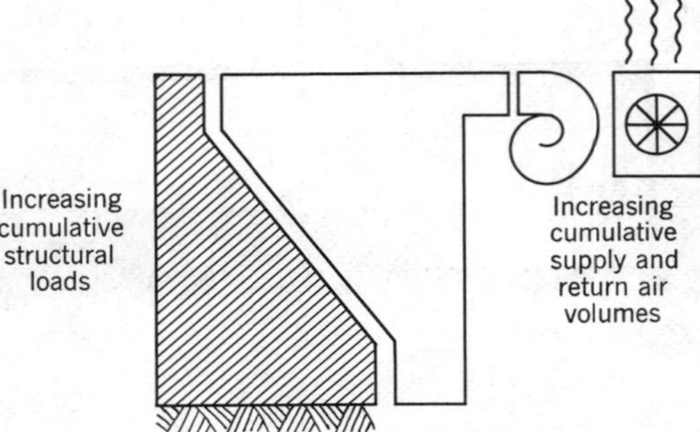

Fig. 3.23 Technical support system. With rooftop air handling, the total air duct size decreases toward the ground. Conversely, the total structural load increases toward the ground.

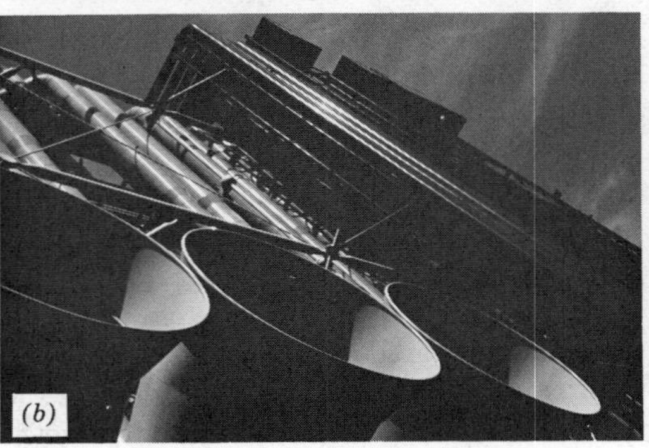

ceases. The moving parts in mechanical systems need maintenance far more frequently than the connections of structural components. Changes in occupancy can mean enormous changes in mechanical systems, requiring entirely different equipment; structural changes of such magnitude usually occur only at demolition. Mechanical systems can invite user adjustment; structural systems rarely do.

Thus, while it is possible to wrap the mechanical systems in a structural envelope, it is of questionable long-term value, given the differing life spans and characteristics of these systems. The probability of future change suggests that the *mechanical* system be the exposed one, despite the appeal to many designers of the structural system's cleaner lines.

(*b*) ***Concealment and Exposure.*** The pipes, ducts, and conduits that take the necessary resources to and from the interior are often carried within a network of spaces unseen by anyone but builders and repair people. The advantages of concealment include: less water and air noise, fewer surfaces requiring cleaning, less care necessary in construction (leaks, not looks, are important), and more control over the appearance of the interior ceiling and

Fig. 3.24 *Centre Georges Pompidou, Paris. Some views of the mechanical support systems. The design competition for this complex was announced in 1971, but subsequent lawsuits and budget cuts delayed its opening until January 1977.* (a) *The view from a noisy and congested street; an open square on the other side is thus protected by the building.* (b) *Some air exhaust ducts, seen from the street.* (c) *The facade from the open square. Piano + Rogers, Architects; photos by John Tingley.*

Fig. 3.25 Diversity within a uniform module. This wooden ceiling module serves both the office and the corridor space; the fluorescent office luminaires spread light evenly at desk height (a), *whereas the incandescent spotlights send sharp patterns across the otherwise plain walls of the corridor* (b). *The lighting is deliberately and starkly different between these interior places.*

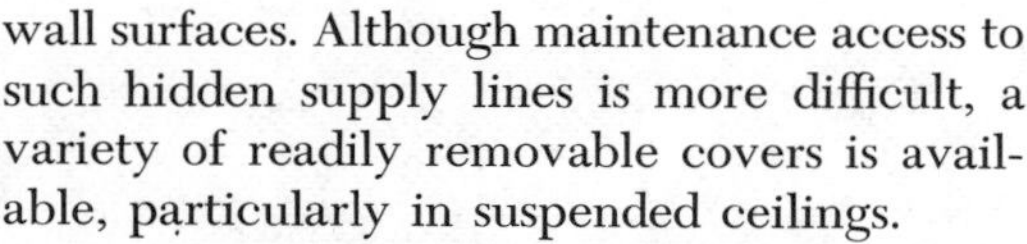

wall surfaces. Although maintenance access to such hidden supply lines is more difficult, a variety of readily removable covers is available, particularly in suspended ceilings.

On the other hand, the exposure of these supply networks provides an honest and direct source of visual (and occasionally acoustical) interest. Exposure in corridors and service areas, and concealment in offices is an approach used in many office buildings (see Fig. 3.25). Flexibility is usually encouraged by exposure; changes can be easily made when not accompanied by a need for neatly cut holes in concealing surfaces. However, flexibility from movable partitions requires constant ceiling heights, which is a feature of the suspended-ceiling approach.

One of the more spectacular examples of exposed mechanical (and structural) systems is shown in Figure 3.24—the result of a design competition for a museum of modern art, reference library, center for industrial design, center for music and acoustic research, and supporting services, in downtown Paris.

When users are invited to play an active role in adjusting conditions inside, exposure of the switches they manipulate is helpful. Not only are users reminded of their opportunities by seeing these mechanisms, but user interaction is encouraged; adjustments are sometimes discovered that the designer had not anticipated.

(c) ***Uniformity and Diversity.*** The advantages of uniformity are most evident in rapidity of design and construction that, through mass production and speed, often brings lower first costs. As mentioned earlier (also in Chapter 1), the flexibility in office arrangements that are accompanied by uniform ceiling heights, light placement, grille locations, and so on, can extend a building's usable life span. However, uniformity is not always attractive to users, and diversity is often encouraged at a more personal level, with office furnishings, for example. A more thorough approach to diversity can provide stimulus to the user who spends many hours away from the variability of the exterior climate.

If offices must be uniform in ceiling lighting, air handling, and size, the corridors that connect them and the lounges, or other supporting service spaces, can be deliberately different. Figure 3.25 illustrates diversity within a simple uniform ceiling module as one example of this distinction. Diversity requires that the designer be complete and detailed about creating places; it gives the builder a more complex and interesting task; and it can provide orientation and interest to the users. The attractiveness of diversity is evident in most collections of retail shops, where light and sound, and sometimes heat and aroma, are used to distinguish one shop from the next.

The following chapters provide a detailed look at the design elements of heat, water, light, and sound, and the energy forms, principles, and equipment that are associated with them. Although sometimes implicit, the underlying principles of these first three chapters continue to apply to the remainder of the text:

1. Design to utilize on-site sources first in order to provide a degree of building self-sufficiency. Then add any needed equipment.
2. Design so that a building's responsibility to the public—its environment, resources, and cultural patterns—is as clear as its responsibility to its users of the moment.

PART TWO

indoor climate control

4

Comfort, Heat Loss, Heat Gain

4.1 Metabolism

Food taken into the body may be thought of as a fuel that is subject to a low-grade burning process sufficient to maintain a body temperature of 98.6° F (37° C). There is a wide variation in metabolic rates, dependent on physical activity (see Table 4.1). The values in that table are expressed in *met* units. For an average size man, the met unit corresponds approximately to 360 British thermal units per hour (Btuh); women tend to have maximum levels about 30% lower. The met represents the average heat produced by a sedentary man. For a person sleeping, the heat given off is 0.7 × 360 = 252 Btuh. A basketball player in action generates and loses 7.6 × 360 = 2136 Btuh. For any specific value of heat generation and loss such as 400 Btuh (sedentary office work), which is the basis for Figure 4.1, heat is lost from the body by several different methods. They vary in rate, dependent largely on dry-bulb (air) temperature, the tempera-

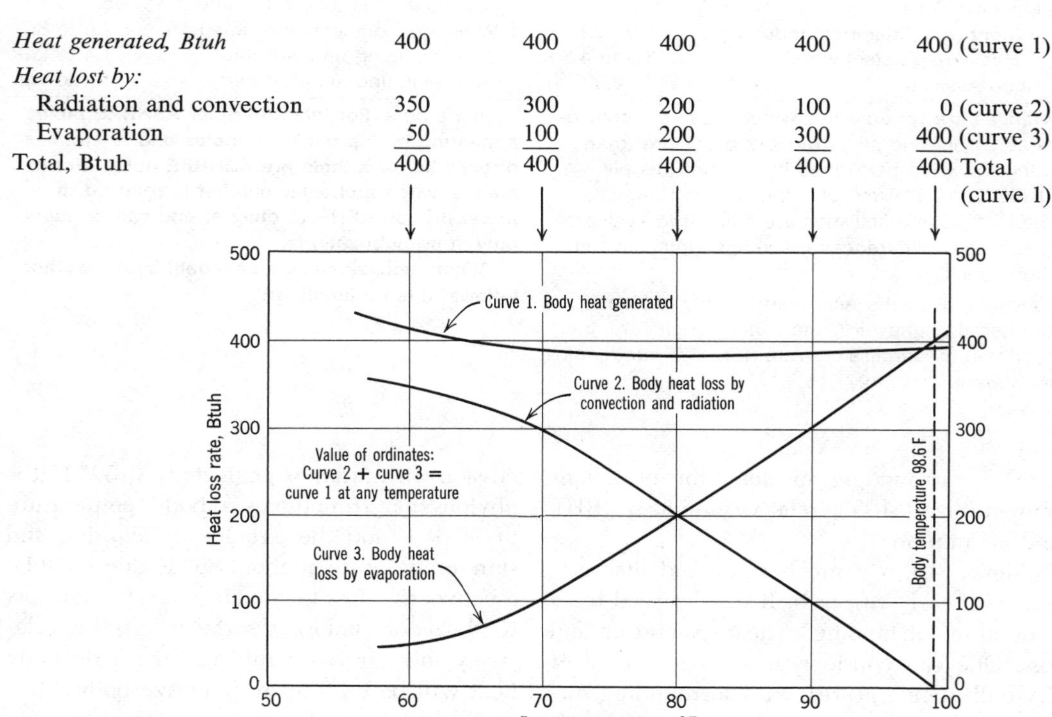

Heat generated, Btuh	400	400	400	400	400 (curve 1)
Heat lost by:					
Radiation and convection	350	300	200	100	0 (curve 2)
Evaporation	50	100	200	300	400 (curve 3)
Total, Btuh	400	400	400	400	400 Total (curve 1)

Fig. 4.1 Heat generated and lost (approximate) by a person at rest (RH fixed at 45%).

Table 4.1 ***Metabolic Rate at Different Typical Activities***

Activity	Metabolic Rate in Met units*	Activity	Metabolic Rate in Met units*
Resting		Teacher	1.6
Sleeping	0.7	Watch Repairer, Seated	1.1
Reclining	0.8	Vehicle Driving	
Seated, quiet	1.0	Car	1.5
Standing, relaxed	1.2	Motorcycle	2.0
Walking		Heavy vehicle	3.2
On the level mph		Aircraft flying routine	1.4
2	2.0	Instrument landing	1.8
3	2.6	Combat flying	2.4
4	3.8	***Domestic Work, Women***	
Miscellaneous Occupations		House cleaning	2.0 to 3.4
Bakery (e.g., cleaning tins, packing boxes)	1.4 to 2.0	Cooking	1.6 to 2.0
Brewery (e.g., filling bottles, loading beer boxes onto belt)	1.2 to 2.4	Washing by hand and ironing	2.0 to 3.6
Carpentry		Shopping	1.4 to 1.8
Machine sawing, table	1.8 to 2.2	***Office Work***	
Sawing by hand	4.0 to 4.8	Typing	1.2 to 1.4
Planning by hand	5.6 to 6.4	Miscellaneous office work	1.1 to 1.3
Foundry Work		Drafting	1.1 to 1.3
Using a pneumatic hammer	3.0 to 3.4	***Leisure Activities***	
Tending furnaces	5.0 to 7.0	Stream fishing	1.2 to 2.0
Garage Work (e.g., replacing tires, raising cars by jack)	2.2 to 3.0	Calisthenics exercise	3.0 to 4.0
General Laboratory Work	1.4 to 1.8	Dancing, social	2.4 to 4.4
Machine Work		Tennis, singles	3.6 to 4.6
Light (e.g., electrical industry)	2.0 to 2.4	Squash, singles	5.0 to 7.2
Heavy (e.g., steel work)	3.5 to 4.5	Basketball, half court, intramural	5.0 to 7.6
Shop Assistant	2.0	Wrestling-competitive or intensive	7.0 to 8.7
		Golf, swinging and walking	1.4 to 2.6
		Golf, swinging and golf cart	1.4 to 1.8

*Ranges are for activities which may vary considerably from one place of work or leisure to another or when performed by different people. 1 met = 58.2 W/m^2; 50 kcal/hr · m^2; 18.4 Btuh/ft^2. Some activities are difficult to evaluate because of differences in exercise intensity and body position.

Author's note. **For this and other ASHRAE tables appearing in this book, footnotes and references directly below a table are ASHRAE notes. For instance, when a chapter number is referred to there, it is an ASHRAE chapter and can be found only in its publications.**

When, occasionally, a comment by this author follows, it is so identified.

ture of surrounding surfaces, mean radiant temperature (MRT), relative humidity (RH), and air motion.

Figure 4.1 is a much simplified diagram, based on early research. It can be used for a general understanding of heat generation and loss. One can consider an extreme range of dry-bulb temperatures, with surrounding surfaces at about the same temperature as the air in each case, and a RH of 45%. For this discussion, air motion is neglected. At 60° F it is obvious that, from the usual body temperature of 98.6° F and the average of clothing and skin temperature of about 85° F, there will be a convective loss to the air and a radiant loss to the cool enclosing surfaces. At this relatively low air temperature, very little body heat will be used up to vaporize body moisture in the form of perspiration from the skin or respiration from the lungs. Figure 4.1 indicates that, at 60° F, the body loses about seven times as much heat by convection and radiation as by evaporation.

Consider now the other extreme condition in which the air temperature is 98.6° F and surrounding surfaces are about the same. *No* heat can be lost from the body either by convection or radiation because heat will flow only in the direction of an environment of *lower* temperature. In this situation, all of the body heat is lost by evaporation, as is shown in Figure 4.1. It is, of course, possible for humans to survive in temperatures over 100° F, in which condition the body heat balance can only be preserved by evaporation.

Returning now to a reasonable room temperature of 75° F with enclosing surfaces nearly the same, it is seen in Figure 4.1 that evaporation plays a minor role and that body heat loss by convection and radiation is about twice that by evaporation. The most recent research, culminating in ASHRAE Standard 55-74 (discussed later), has established that at normal temperature, RH has an even smaller effect than was previously thought.

4.2 Thermal Equilibrium and Comfort

It becomes apparent that the thermal environment may be *regulated* to permit an easy and comfortable rate of loss for the heat that is generated by humans for any given activity to avoid body chill in one case and heavy respiration in the other.

The term *environmental* comfort has taken on a broad meaning. Some discussions have enumerated more than 15 factors that affect humans pleasantly or adversely. They include items such as esthetics and acoustics. Factors that can be controlled by *air-conditioning systems* include:

(a) Temperature of the surrounding air.
(b) Mean radiant temperature of the surrounding surfaces.
(c) The relative humidity of the air.
(d) Motion of the air.
(e) Odors.
(f) Dust.

Of these, the first four are related to the *thermal* environment. They can and should be adjusted so that occupants of conditioned space can experience a thermal equilibrium by which bodily heat loss is adjusted to the heat generated by the level of activity in the space.

4.3 Regulation of the Thermal Environment

Figure 4.2 illustrates heat loss by convection. At any selected air temperature in a space, with fixed air motion, air particles in contact with the body, or with clothing heated by the body, are warmed. They become less dense and rise, to be replaced by cooler particles that follow in turn. If the air temperature rises as a result of this, the optimum, then the selected temperature must be restored within limits of 1 or 2 degrees. In winter, this may be done by controls that reduce the output of the heating system. In mild weather, cool outdoor air may be introduced. In summer, mechanical cooling may be necessary. The most recent standards call for a temperature related to the activity but not to the season. For this reason, year-round air conditioning is essential under conditions of dense occupancy and significant activity.

Heat loss by radiation is the transmission of energy through the air from the warmer human body or its clothing to the cooler surfaces that are "seen" by the warmer sources. This expression is chosen to emphasize that the energy cannot go *around corners* or be

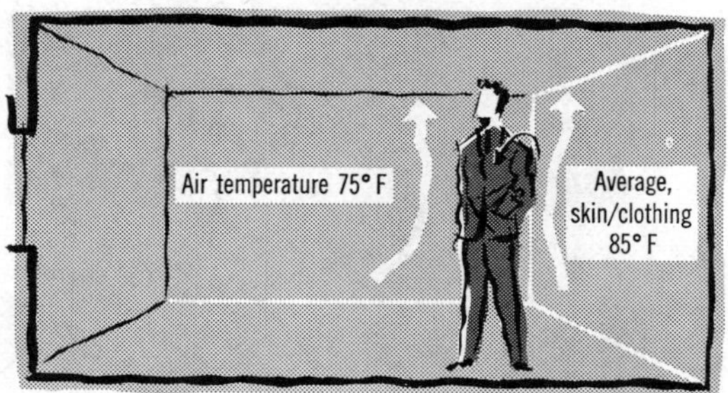

Fig. 4.2 *Heat loss by convection. Differential between the average skin and clothing temperature and the room air temperature is usually about 10° F.*

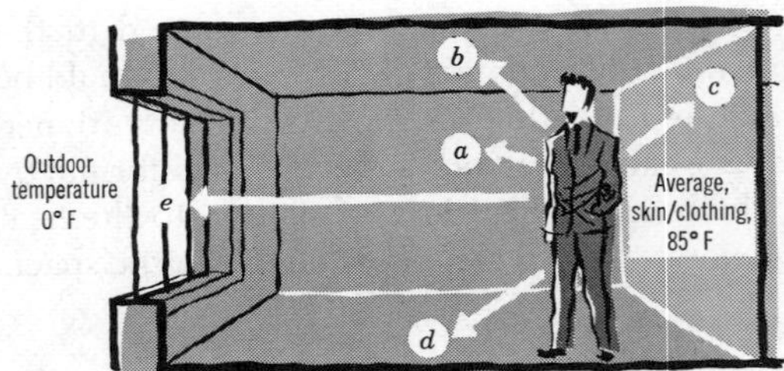

Fig. 4.3 Heat loss by radiant transmission to surrounding surfaces. Mean radiant temperature, MRT, reflecting the thermal nature of people's indoor surrounding environment is usually between 70 and 80° F. Since either is less than the usual average of skin and clothing temperature of about 85° F, heat radiates from *the body to its surroundings. In the winter condition illustrated, surfaces* a, c, *and* d *are about at room air temperature (72° F). The mean, however, is affected by a higher temperature* b, *due to lights, and a lower one,* e, *reflecting outdoor conditions.*

affected by air motion. The surface temperature of the human body and its clothing less the MRT of the surrounding surfaces is the determining differential in this radiant transmission (see Fig. 4.3). Although the MRT usually lies between 70 and 80° F and tends to stabilize at the room air temperature, it is sometimes affected by large glass areas, degree of insulation, lights, and so on. Recent trends have tended to stabilize the MRT at a temperature close to that of the air in the spaces. These trends include more insulation, lower lighting levels, and smaller glass areas.

The use of circulated air brings several advantages to occupied rooms. The motion tends to unify the thermal conditions necessary for comfort. Also, as air gently passes the occupants, it carries away accumulated warm air and the humidity given off by the body. This results in a feeling of freshness (see Fig. 4.4). The rate of circulation relates principally to the heat to be carried to or from the room and it generally results in air speeds that lie between 70 and 10 feet per minute (fpm). In setting dry-bulb standards it is usual to take into account the slightly cooling effect of moving air. Methods of introducing and removing air from rooms are under continual study.

As early as the 1920s RH was studied to determine its effect on comfort. It was thought that humidity had a strong influence on the regulation of body heat loss and comfort. Subsequent to 1938, these findings were, at times, challenged and, although the early recommendations were carried in official reference literature until 1961, recent experiments at the Institute for Environmental Research, at Kansas State University, have established new standards. Figure 4.5 illustrates the range of RH found to be satisfactory, provided dry-bulb temperature and MRT are in proper adjustment.

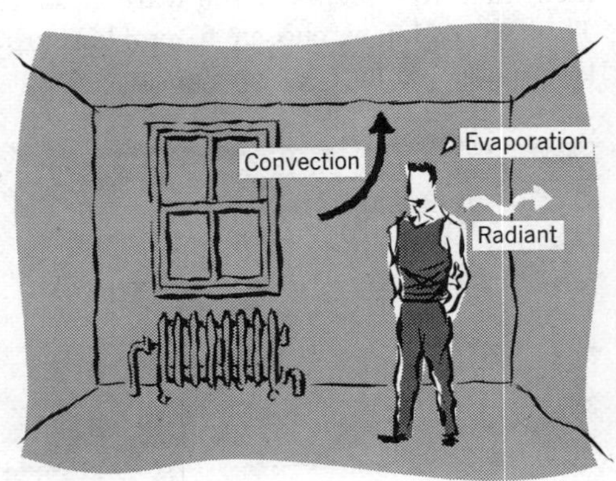

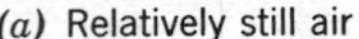

(a) Relatively still air

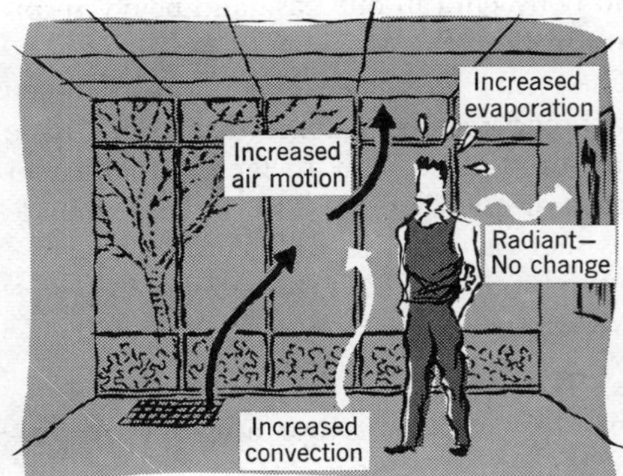

(b) Air in motion

Fig. 4.4 Increased air motion slightly accelerates heat loss by convection and evaporation. Air motion is desirable within the limits of ASHRAE Standard 90-75. It improves the general feeling of comfort by assuring uniform conditions in the space and by clearing out stuffy spots.

Fig. 4.5 (a) *and* (b) *Equal comfort. Within the range of 20 to 60%, relative humidity (RH) has little effect on comfort.*

4.4 Criteria for Thermal Comfort

It is interesting to learn from Table 1.1, page 12, that comfort studies have been in progress since before 1900. It will be noticed that there had been a gradual increase in the recommended temperatures. Reasons for this are thought to include changes in clothing habits, working conditions, and improvement in heating and cooling systems. Near the end of the list in Table 1.1, we find reference to the ASHRAE Standard adopted by the American Society of Heating, Refrigerating and Air Conditioning Engineers in 1975, and replacing their earlier standards. This new Standard is ASHRAE Standard 90-75. Its principal theme is energy conservation. Therefore, we are not surprised that it calls for a reduced indoor temperature in winter and other energy-saving changes.

	Winter	*Summer*
1965	77° F DBT, 70% RH	79.5° F DBT, 20% RH
1975	72° F DBT, 30% RH	78° F DBT

In regard to winter operation, the 5 degree reduction in dry-bulb (air) temperature is accompanied by a recommended upper limit of 30% RH. This low RH limit is another energy-saving item, since it would require additional energy to achieve the 70% RH suggested in the 1965 Standard. The summer cooling-plus-dehumidification process is similarly energy-saving. Sections 5.3.2.2 and 5.4.2 of Standard 90-75, quoted here by ASHRAE permission, indicate the intention of these design standards.

5.3.2.2 INDOOR DESIGN CONDITIONS

Winter

The *recommended* design condition is 72F (22° C) dry bulb. If humidification is provided, it *shall* be designed to a *maximum* relative humidity of 30 percent.

Summer

Where comfort air conditioning is required or

used, the *recommended* indoor design condition is 78F (25.5° C) dry bulb. The actual design relative humidity within the comfort envelope as defined in ASHRAE Standard 55-74 "Thermal Environmental Conditions for Human Occupancy" *shall* be selected for minimum total HVAC system energy use.
Although the *recommended* design points are established above, the HVAC system design *should* permit operating at minimum energy levels within the boundaries of the comfort envelope.
Due to internal heat gains or losses, it *may* be more energy efficient to operate at other than the design points for major portions of the year.

5.4.2 Humidity Control. If an HVAC system is equipped with a means for adding moisture to maintain specific selected relative humidities in spaces or zones, a humidistat *shall* be provided. This device *shall* be capable of being set to prevent new energy from being used to produce space relative humidity above 30 percent RH. Where a humidistat is used in an HVAC system for controlling moisture removal to maintain specific selected relative humidities in spaces or zones, it *shall* be capable of being set to prevent new energy from being used to produce a space relative humidity below 60 percent RH.

The decade of the 1970s was a time for diligent and intensive action in revising, among other things, standards for reduced energy use in places of human occupancy. The work of specialists, technical committees, and the staff of the American Society of Heating, Refrigerating and Air Conditioning Engineers culminated in much new data including four significant documents. They comprise:

Copyright year	*Publication*
1973	ASHRAE Standard 62-73, *Standards for Natural and Mechanical Ventilation.*
1974	ASHRAE Standard 55-74, *Thermal Environmental Conditions for Human Occupancy.*
1975	ASHRAE Standard 90-75, *Energy Conservation in New Building Design.*
1977	*ASHRAE Handbook and Product Directory, 1977 Fundamentals.*

This great fund of information and guidance is available for our use. It is recommended to readers as in-depth reference material. Only a few of the highlights can be reported in the scope of this book. In this section on the subject of environmental comfort, it can be said that comfort, always an important consideration, would appear to be a bit less important than in the past. However, comfort, together with energy conservation, will affect building design and possibly our choice of indoor clothing. The *ASHRAE Handbook of Fundamentals, 1977* presents a table that assigns ratings in "clo" units for "Individual Insulation Values of Men's and Women's Garments." In *comfort line* charts these personal insulation values are considered along with items that include ambient air temperature, clothing insulation (Icl), RH, and air movement. As part of the research findings of Fanger at Kansas State University and later in Denmark, a number of such comfort charts are presented in Chapter 8 of the *ASHRAE Handbook of Fundamentals, 1977.*

The comfort chart shown in Figure 4.6 includes an indication of the findings and conclusions of ASHRAE's Technical Committee 55, regarding comfort. Conditions at the edges or within the hatched area of Figure 4.6 should afford the greatest comfort. (We saw how the two conditions in Fig. 4.5 fall into very acceptable places in Fig. 4.6.) Temperature is optional and RH falls well within the 20 to 60% bracket. The energy recommendation of Standard 90-75, however, may make these conditions of RH a little more difficult to achieve.

4.5 Indoor Humidity in Winter

For many reasons other than thermal comfort, the RH should not fall below 20%. One of these reasons is that humidity has a strong effect on the wood of furniture, paneling, and other interior equipment and finishes. The

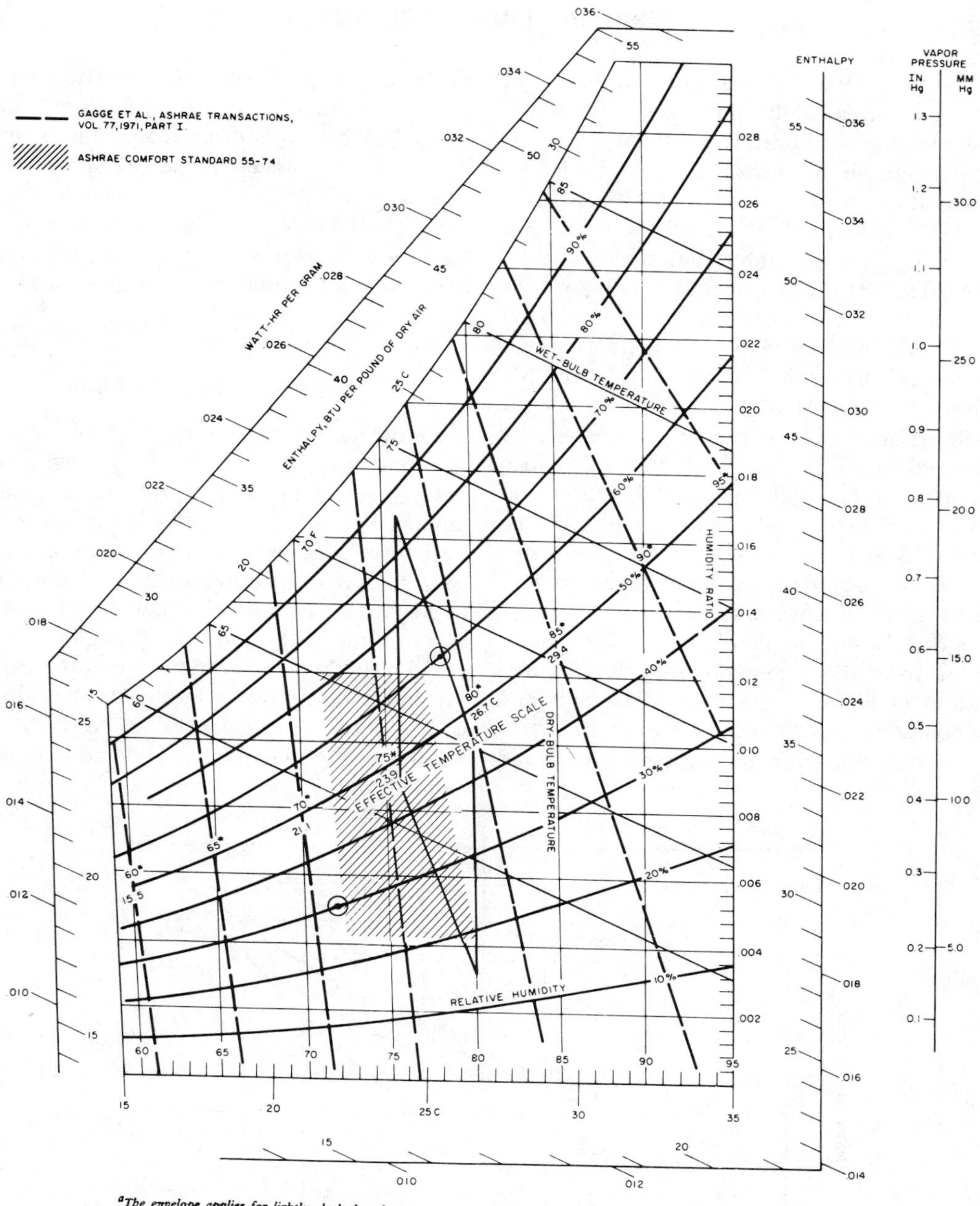

[a]The envelope applies for lightly clothed, sedentary individuals in spaces with low air movement, where the MRT equals air temperature

Fig. 4.6 *Comfort zone recommended in ASHRAE Comfort Standard 55-74. It generally applies to altitudes from sea level to 2134 m (7000 ft) and to the most common special case for indoor thermal environments in which the mean radiant temperature is nearly equal to the dry-bulb (air) temperature and the air velocity is less than 0.2286 m/s (45 fpm). For this case, the thermal environment is well specified by the two variables shown: dry-bulb air temperature and RH. Copyright © by the American Society of Heating, Refrigerating and Air Conditioning Engineers, Inc. Reprinted by permission from* ASHRAE Handbook of Fundamentals, 1977. Note. *From Table 1.1, page 12, it is seen that ASHRAE Standard 90-75, Energy Conservation in New Buildings, suggests winter dry-bulb temperature of 72° F and 30% RH. For summer, a temperature of 78° F is named, with a suggested minimum RH of 60%. Both of these conditions, circled on the chart by this author, will be found to be slightly outside the hatched comfort zone above. We will see the use of heavier indoor clothing in winter and lighter clothing in summer.*

shrinkage of wood lateral to the direction of the grain often results in unsightly cracks and the loosening of furniture joints. Another reason is that skin becomes rough and dry in low humidities.

Without humidification, the RH in houses is generally low in winter, but seldom below 20%. The RH is often found to be 30 or 35%, which falls within the range of the new ASHRAE Standard 90-75. Architects sometimes feel that humidification to raise the RH above these levels in winter is undesirable for two reasons. Condensation on glass, especially on single glass, can occur (see Fig. 4.7). Introducing heat below the glass reduces this possibility but there are practical limits to the humidity that can be tolerated by cold surfaces. The second reason is that vapor barriers, despite efforts toward the best workmanship, are not always completely impervious. Vapor pressure at the higher humidity levels often causes moisture to get through barriers and condense within exterior walls in winter.

Modern houses are increasingly airtight and the moisture produced by domestic operations, as shown in Table 4.2, is usually retained in the building to afford a reasonable humidity. This is emphasized by the fact that exhaust fans are often used in bathrooms while showering, their purpose being to *reduce* concentrations of moisture that cause copious condensation on mirrors and other glass surfaces.

4.6 Coping with Special Conditions

Of the numerous building materials, glass is the most sensitive to weather changes and poses some of the major problems in maintaining thermal standards. It has taken some time for architects and air-conditioning engineers to develop solutions to the problems that large glass areas can create in winter (Fig. 4.8) and in summer (Fig. 4.9).

Admittedly the condition in Figure 4.8*a* is extreme and aggravated by the fact that there is no controlled passage of air through the room. Even with carefully designed inlet and

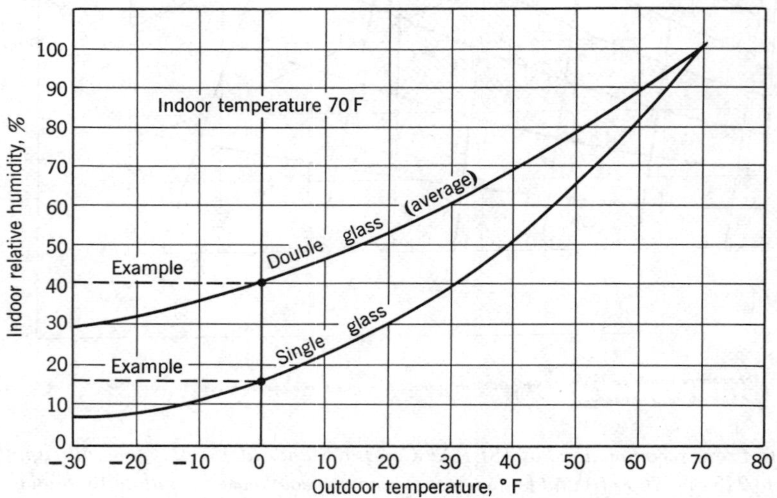

Fig. 4.7 Condensation on glass (indoor temperature 70° F). Example: At 0° F outdoors, relative humidities above 15 will cause condensation on single glass. Double glass will permit about 40% RH. Convectors or warm air registers placed below the glass to warm it will raise considerably the practical limits of relative humidity that can be maintained without condensation on the glass. The recommendation of ASHRAE Standard 90-75 that indoor relative humidities be no greater than 30% in winter reduces the probable occurrence of condensation that was formerly common on single glass subjected to high indoor humidities. The additional precaution of warming the glass by forced warm air can also be accomplished by directing the warm air from above *instead of from below (see Figs. 6.24 and 6.25a, page 272).*

Table 4.2 ***Moisture Production for Various Domestic Operations***

Operation				Pounds of Moisture
Floor mopping (8 ft × 10 ft kitchen, 0.03 psf)				2.40
Clothes drying indoors				26.40
Clothes washing				4.33
Cooking	From food	From gas		
Breakfast	0.34 plus	0.56 equals		0.90
Lunch	0.51 plus	0.66 equals		1.17
Dinner	1.17 plus	1.52 equals		2.69
Dishwashing				
Breakfast				0.20
Lunch				0.15
Dinner				0.65
Bathing				
Shower				0.50
Tub				0.12
Human contribution, family of four			per hr	0.46
Gas refrigeration			per hr	0.12
House plants, each			per hr	0.04
Humidifier, when used			per hr	2.00

Source. From *Research in Home Humidity Control,* by S. C. Hite and J. L. Bray, Purdue University Engineering Experiment Station Bulletin Research Series No. 106, p. 24.

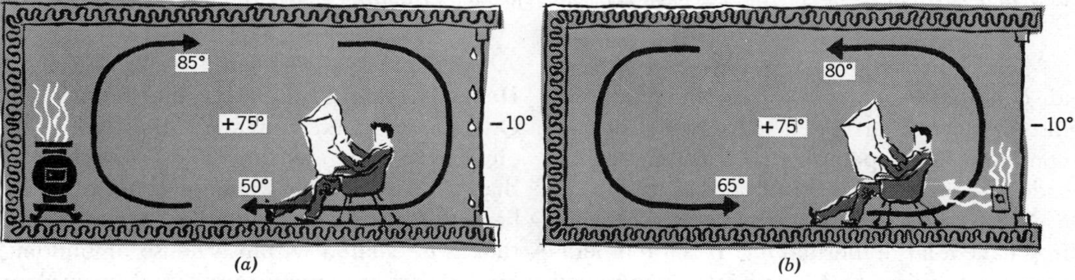

Fig. 4.8 *Dealing with outdoor conditions. Convection is inevitable. It can work for or against you. The temperatures are approximate.* (a) *The stove is not merely in the wrong place. It* accelerates *the "down-slip" of cold air from the glass.* (b) *The convector strip moves air up to warm the glass and provides local radiant warmth. The location of the newly popular woodburners should be carefully chosen.*

outlet registers and grills, chilly conditions can be experienced in the vicinity of glass. Most designers now consider that every linear foot of glass should be blanketed by heat from below.

It is not strictly true that for comfort an air-cooling system can be contrived to deal with *any* condition of glass-plus-sun in critical summer weather. It is not just a question of reducing the cooling load, which is important in itself, but of avoiding uncomfortably over-hot solar-radiant areas next to glass. Many architects now feel that glass should be avoided or minimized in the path of the sun's rays or that the impact of solar energy in *summer* should be absorbed on the *outside* of the conditioned space. (See Fig. 4.9.)

A sudden increase in the number of people occupying a space already adjusted to proper thermal conditions can put a tax on sluggish

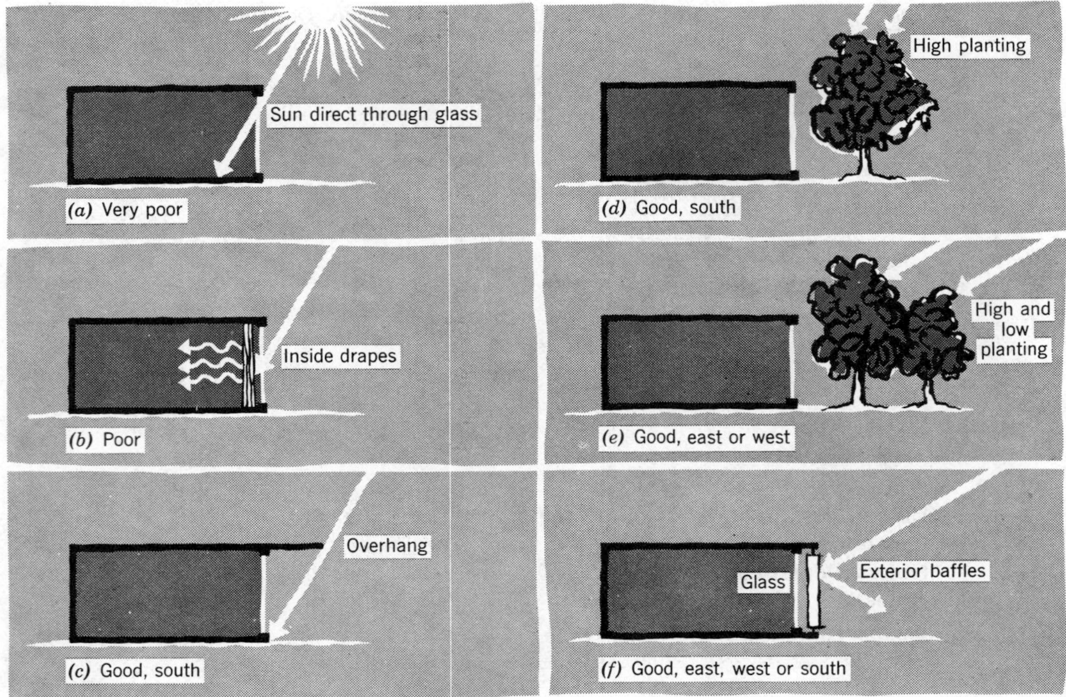

Fig. 4.9 (a) *to* (f) *Indoor comfort in summer. Sun on shaded and unshaded glass. For applications of* (c) *above, see Figures 6.1 and 6.13, pages 251 and 262. For* (d) *and* (e), *Figures 6.4, 6.17, and 6.19, pages 254, 266, and 267. For* (f) *see Figure 6.3, page 253. For reflective insulating glass, see Figure 4.27, page 165.*

air-conditioning systems. It is essential that installations be arranged to prevent the uncomfortable temperature increase caused by sudden increase in population density.

Finally it is important to know that buildings have a large thermal *lag*. If the temperature in a building that is normally maintained at 70° F when occupied is turned down to 50° F for a week or more, the MRT gradually approaches this lower value. Despite a rapid return of air temperature to 70° F, it may be days before the MRT returns to its former (higher) value. During that period the occupants will experience a radiant chill that may not be compensated for by an even further raise in air temperature.

4.7 The Recycling of Air

Conventional titles for this subject could include Air Cleanliness, Ventilation for Health, Comfort and Safety, and Removal of Odors and Contaminants. Yet the term recycling of air suggests the rather new concept that air is a reusable commodity. It has long been the custom to add outdoor air to the air that is circulated within a house or building for heating or cooling purposes. But outdoor air is not always the "fresh" air that it used to be.

For the rare once-through systems (as for hospital operating rooms), outdoor air must conform to requirements of ASHRAE Standard 62-73, which incorporates, for the first time, a quantitative definition of *acceptable outdoor air* and specifies conditions under which the amount of outdoor air may be reduced, thereby taking advantage of advancements in air-cleaning technology. The requirements apply to levels of:

Particulates.
Sulfur.

Fig. 4.10 *Air filtration devices.*

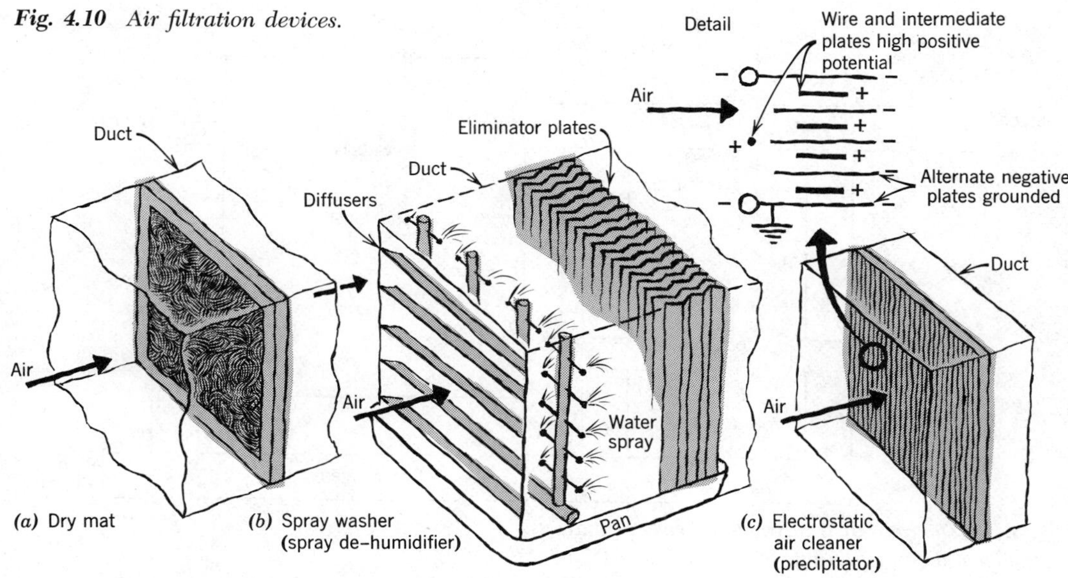

Carbon monoxide.

Photochemical oxidant.

Hydrocarbons, not including methane.

Nitrous oxides.

Odor.

When within the prescribed levels, the air is termed acceptable outdoor air. In many major cities these levels are met when this outdoor air is subjected to the minimal treatment, including such processes as are seen in Figure 4.10.

In *recirculated* air systems, the requirements for quality apply to the air entering the *occupied space*. When air in *that* part of the system qualifies and has had additional *special* treatment, the actual rate of outdoor air can be reduced to as little as 15 or 33%. Section 5.0 of the Standard governs these reductions (see Fig. 4.11). The following section of ASHRAE Standard 62-73 is quoted by permission of ASHRAE:

Section 5.0 Recirculation

The requirements for ventilation quantities given in Section 6.0[1] are for 100% outdoor air when the outdoor air meets the specifications for air quality given in Section 3.0. Except for areas where recirculation is prohibited by other codes or standards having precedence, the outdoor air requirements may be reduced to 33% of the specified required ventilation air quantity if adequate temperature control is provided, in addition to filtering equipment, so that the maximum allowable concentration of particulates entering the space is less than that specified in Table I. If, in addition, high efficient adsorption or other odor and gas removal equipment is employed, so that the air entering the space has been purified to meet the requirements of Sections 3.1 and 3.3, the outdoor air requirement may be reduced to 15% of the specified required ventilation air quantity. *In no case shall the outdoor air quantity be less than 5 cfm per person.*

Large savings in energy accrue from such reduction in the use of outdoor air. In winter, to quote approximate values, outdoor ventilation air at a low (0° F) temperature would need a 120° F rise to join the 120° F air delivered to heat the space. In contrast, recirculated air leaving the space at 75° F would require a rise of only (120–75° F), a 45° F differential.

In summer, if indoor cooling were provided, outdoor ventilation air at 95° F would

[1]The sections and tables referred to in this quotation are those of ASHRAE Standard 62-73.

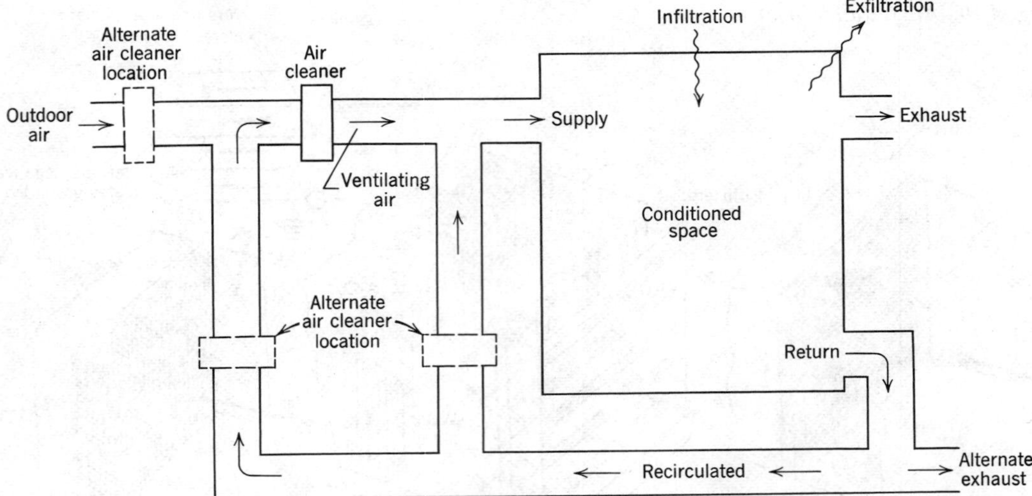

Fig. 4.11 Diagram of a recirculated system indicating location and alternate locations of air-cleaning equipment. Copyright © by the American Society of Heating, Refrigerating and Air Conditioning Engineers. Reprinted by permission from ASHRAE Standard 62-73.

require cooling to 60° F for delivery to the conditioned space, a differential of 35° F. Recirculated air would only require cooling from 75 to 60° F, a 15° F differential. These values can be clarified by reference to Figure 5.54, page 433.

In mild seasons, late fall and early spring, densely occupied buildings that require cooling can be cooled by *high* flow rates of the relatively cool outdoor air. Quality of the air entering the occupied space is *still* a matter of concern. Air-cleaning equipment, the location of which is shown in Figure 4.11, must "quality condition" the air to the requirements of Standard 62-73. The values in Table 4.3 are typical of requirements for outdoor air. In ASHRAE Standard 62-73, this kind of schedule has been expanded to 12 pages and, in the *Handbook of Fundamentals, 1977*, four pages are included. Values are subject to great reduction under the conditions of Standard 62-73, Section 5.0, Recirculation, quoted earlier.

The entire subject is under expanded and continuing study. Additional changes may be expected.

4.8 Heat Loss, Thermal Value of Walls and Roofs

For energy conservation, walls, and roofs, and sometimes floors—if there is outdoor space below—must be resistant to the rapid transmission of heat. Slow passage of heat also results in warmer, more comfortable inside surface temperatures. Insulation is highly essential. Vapor barriers are needed to prevent room moisture from penetrating outward to colder parts of roofs and walls where it condenses or freezes. Tight construction retains warm air and resists the entry of cold air during windy times. Finally, the proportion of glass to insulated opaque walls and roofs should be studied if fuel economy and human comfort are to be achieved.

4.9 Importance of Heat Conservation

One of the requirements for the comfort of people occupying indoor spaces during cold weather is a constant temperature of room air reasonably higher than that of the outdoor air. Heat supplied to the room for this purpose is constantly dissipated by transmission losses through the surfaces of the enclosure. It is lost also by the escape of warm air through minute openings, such as the cracks between window sash and frames. The air is forced out by cold

Table 4.3 *Outdoor Air Requirements*[a]

Application	*Smoking*	*Cfm per Person*[b] *Recom-mended*	*Cfm per Person*[b] *Mini-mum*[c]	*Cfm per Sq Ft of Floor*[b] *Mini-mum*[c]
Apartment				
Average	Some	20	10	......
DeLuxe	Some	20	10	......
Banking space	Occasional	10	7½	......
Barber shops	Considerable	15	10	......
Beauty parlors	Occasional	10	7½	......
Brokers' board rooms	Very heavy	50	20	......
Cocktail bars	....	40	25	......
Corridors (supply or exhaust)	....	......	.....	0.25
Department stores	None	7½	5	0.05
Directors' rooms	Extreme	50	30	......
Drug stores[e]	Considerable	10	7½	......
Factories[d,f]	None	10	7½	0.10
Five and Ten Cent stores	None	7½	5	......
Funeral parlors	None	10	7½	......
Garages[d]	....	......	.....	1.0
Hospitals				
Operating rooms[f,g]	None	......	.....	2.0
Private rooms	None	30	25	0.33
Wards	None	20	10	......
Hotel rooms	Heavy	30	25	0.33
Kitchens				
Restaurant	....	......	.....	4.0
Residence	....	......	.....	2.0
Laboratories[e]	Some	20	15	......
Meeting rooms	Very heavy	50	30	1.25
Offices				
General	Some	15	10	0.25
Private	None	25	15	0.25
Private	Considerable	30	25	0.25
Restaurants				
Cafeteria[e]	Considerable	12	10	......
Dining room[e]	Considerable	15	12	......
Schoolrooms[d]	None	......	.....	......
Shop, retail	None	10	7½	......
Theater[d]	None	7½	5	......
Theater	Some	15	10	......
Toilets[d] (exhaust)	....	......	.....	2.0

[a] Taken from present-day practice.

[b] This is contaminant-free air.

[c] *When minimum is used, take the larger of the two.*

[d] See local codes which may govern.

[e] May be governed by exhaust.

[f] May be governed by special sources of contamination or local codes.

[g] All outside air recommended to overcome explosion hazard of anesthetics.

outdoor air infiltrating through similar openings on the opposite (windward) side of the room. Loss of warm air may occur also when controlled ventilation operates to change the air in the room at established rates. Since valuable energy must be used to offset these losses, a careful study of infiltration, ventilation, and transmission rates is part of every architectural design.

Among the criteria for the selection of exterior construction, thermal transmission is one of considerable importance because every square foot of material carries a permanent upkeep cost for fuel over the years. The index for comparison is the *U*-coefficient of transmission. The method of establishing this value for selected walls is developed in the following sections. For the purpose of interpreting the relative merits of the walls shown in Figure 4.12, the *U*-coefficient is defined here as the number of British thermal units per hour that pass through 1 sq ft of wall, floor, or roof under actual conditions at the building when the difference between the inside and outside air temperature is 1° F under a steady rate of heat flow.

A few examples of walls are illustrated in Figure 4.12. Undue consideration for fuel savings *only* might suggest the elimination of all glass and the use instead of insulated walls such as *A* or *D*, which have minimum heat losses. Although this might not be architecturally desired, it is nevertheless essential that, in the case of opaque walls, adequate insulation be used. Its cost is quickly repaid by fuel savings. The fuel-saving advantage of double and triple glazing should be considered. The proportion of glass to opaque materials must obviously be judged by architectural qualities in addition to those relating to heat. It should be pointed out, however, that while glass is quick to lose heat under critical conditions,

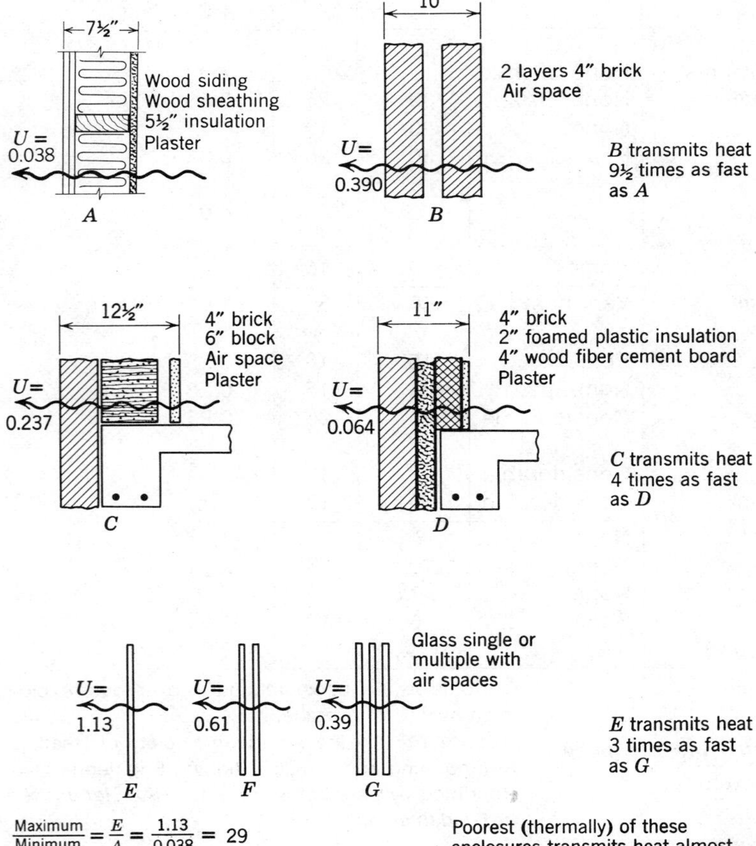

***Fig. 4.12** Approximate comparison by* U-*coefficients of the heat-transmitting rates of some opaque and transparent enclosing walls.*

including the absence of sunshine, it is most receptive, when correctly oriented, to the passage of solar energy into the building during sunny hours (see Chapter 6). When glass is used in this manner, the nightly use of thermal shutters should be considered. See Figure 3.10*c*, page 67.

4.10 Nature of Heat Flow

Beginning with the combustion of fuel in boilers or furnaces, heat flows by various methods to warm the occupied spaces and thence minimally to outdoors by transmission through exterior room surfaces or by the loss or expulsion of warmed air through openings in the building. The analysis and evaluation of the transmission through a combination of building materials leads to the finding that the rate of heat flow is related to the passage of heat through these assembled materials by conduction, convection, radiation, or combinations of them (see Fig. 4.13).

It is evident that a reduction in the rate of heat loss can be achieved by the use of insulating materials having slow conduction rates and, conversely, high thermal resistance.

4.11 Heat Flow through Homogeneous Solids

Conductivity (unit conductance), is designated as k and defined as the number of British thermal units per hour that flow through 1 sq ft of material, 1 in. thick, when the temperature drop through the material under conditions of steady heat flow is 1° F. Conductivity is established by tests and is the basic rating for a material. When conductance C is referred to, in a homogeneous material, it is for a thickness other than 1 in. The other conditions remain the same.

Figure 4.14 compares the conductivities of a dense and a light material. It also shows the method of computing the thermal resistance, R, of 1 in. of material. R is the reciprocal of the conductivity. This is $1/k$ and is stated as the number of hours needed for 1 Btu to flow through the material. For thicknesses other than 1 in., the conductance C decreases (k/x) as the thickness increases. The resistance R increases directly (x/k) with the increase in thickness. In each case x stands for the thickness of the material in inches.

4.12 Air Spaces

Air spaces may be introduced into the structure to reduce the U-coefficient and to aid in retarding heat flow from the building. Unlike conventional batt or "fill" insulation, the resistance of an air space is not related primarily to thickness, but is determined by many factors including position of the space, direction of heat flow as shown in Figure 4.13, and the nature of the surfaces lining the space. Radiant transmission across the space is much reduced by the use of shiny material such as aluminum foil. It is sufficient to use such material on only one of the sides. Using it on both makes for very little improvement because the reflective material will accept *or emit* only a small amount of heat relative to that transmitted by the common rough building materials. Thicknesses of the air space are generally evaluated for 0.5 and 0.75 in. (see Table 4.4). The increase of thickness does little to improve the resistance except in spaces in floors with heat transmission downward, in which case there is little convection. Table 4.4 shows the many variables involved, not the least of which is emissivity.[2] For our purposes, we may consider that reflective lining has an emissivity of 0.05 (effective thermal emission) and common building materials have an emissivity of 0.82. An example of the value of *reflective* lining is seen in the following instance. For a $\frac{3}{4}$-in. vertical space through which the heat flow direction is horizontal, and assuming an air space temperature of 50° F, with a temperature drop across the space of 30° F, one finds that for an emissivity of 0.05 (reflective), the thermal resistance is 2.77. For an emissivity of 0.82 (nonreflective) the thermal resistance is 0.94 under the same conditions. The resistance to heat flow of the reflective-lined air space

[2] Also known as emittance.

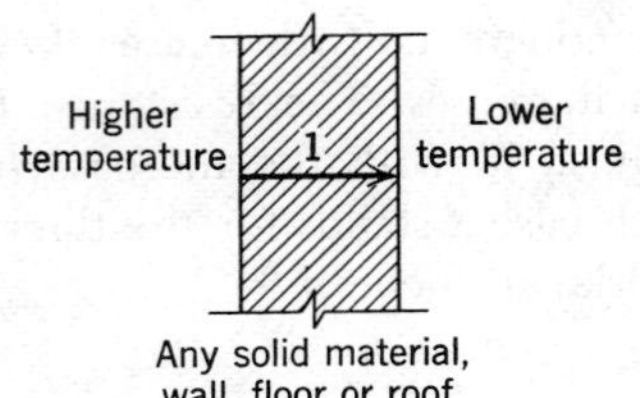

A single solid material illustrates the transfer of heat from the warmer to the cooler particles by conduction (1).

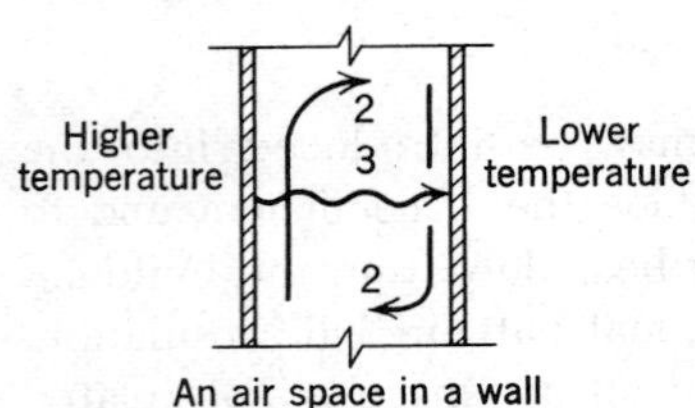

As air is warmed by the warmer side of the air space it rises. As it falls down along the cooler side it transfers heat to this surface (2). Radiant energy (3) is transferred from the warmer to the cooler surface. The rate depends upon the relative temperature of the surfaces and upon their emissive and absorptive qualities. Direction is always from the warmer to the cooler surface.

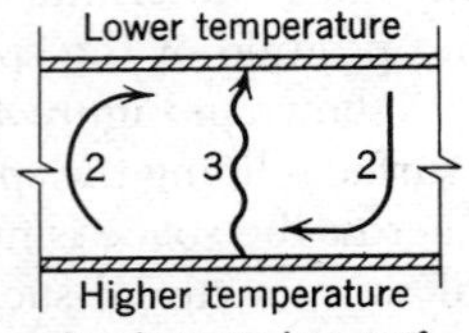

The convective action (2) in the air space of a roof is similar to that in a wall although the height through which the air rises and falls is usually less. The radiant transfer is up in this case because its direction is always to the cooler surface.

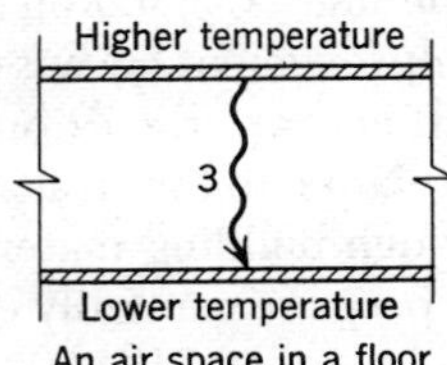

When the higher temperature is at the top of a horizontal air space the warm air is trapped at the top and, being less dense than the cooler air at the bottom, will not flow down to transfer its heat to the cooler surface. This results in little flow by convection. The radiant transfer in this case is down because that is the direction from the warmer surface to the cooler.

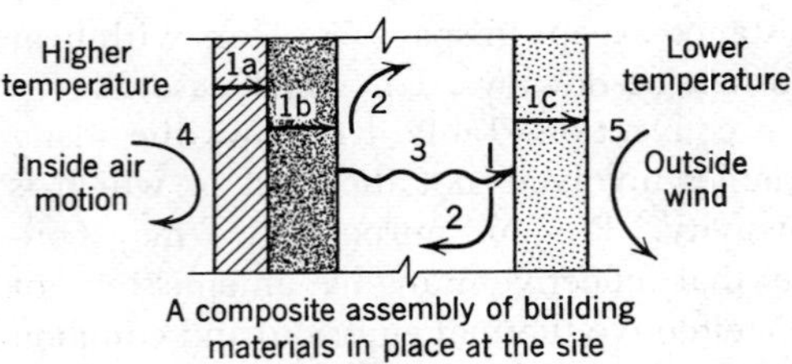

This example of a wall in place illustrates the several methods by which heat is lost through a composite assembly of materials. Conduction at varying rates in different materials is accounted for in 1a, 1b, 1c. Convection currents (2) and radiation (3) carry the heat across the air space.

Heat is conducted from the room air by warm air currents that strike the inside wall. Heat is conducted away from the exterior surface of the wall by the action of the wind.

Fig. 4.13 *Nature of heat flow through materials, air spaces, and assembled structures. Thermal action is identified by the following numbers.*

1. *Conduction*
2. *Convection*
3. *Radiation*
4. *Inside surface conductance*
5. *Outside surface conductance.*

with respect to the nonreflective-lined air space is 2.77/0.94 = 2.95, which is almost three times as effective in reducing heat flow. It is important to know that a sheet of reflective foil in direct contact on both of its sides with other materials has no significant value in heat flow retardation. It *must* face an air space. In *ASHRAE 1977 Fundamentals,* for the first time, air spaces are evaluated for resistances *only*. Earlier issues (1972) quoted *both* conductances *and* resistances. In an energy-conscious era, this is significant. We now "think *resistance*" instead of conductance. Although it is always important to evaluate and include air space resistance where it occurs, another change appears. We no longer *plan*

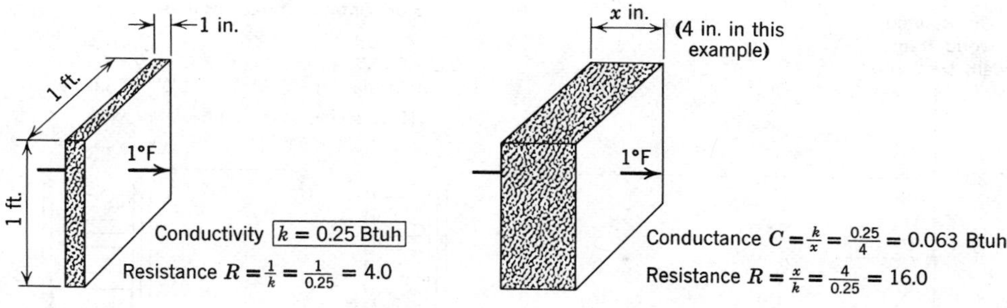

Glass Fiber Insulation Board

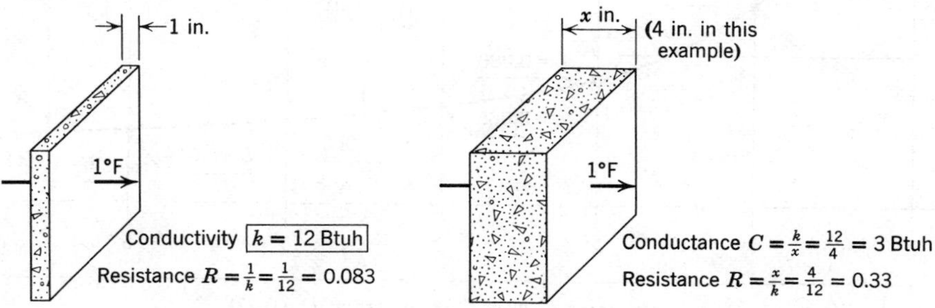

Sand and Gravel Concrete

Fig. 4.14 Example of conductivities (k) *for 1-in. thickness, conductances* (C) *for any thickness (4 in. in this example), and resistances* (R) *for glass fiber, a material of low conductivity, and concrete, a material of high conductivity. See Table 4.6, page 110.* Note. *Standard unit of area 1 sq ft. Standard unit temperature differential 1° F.*

air spaces where, instead, insulation of the same thickness can be employed. The increased resistance value is indicated in the caption of Figure 4.15. Turn to page 132 for Section 4.13.

For easy reference. Tables 4.4 to 4.19 are listed below and grouped on pages 107-131.

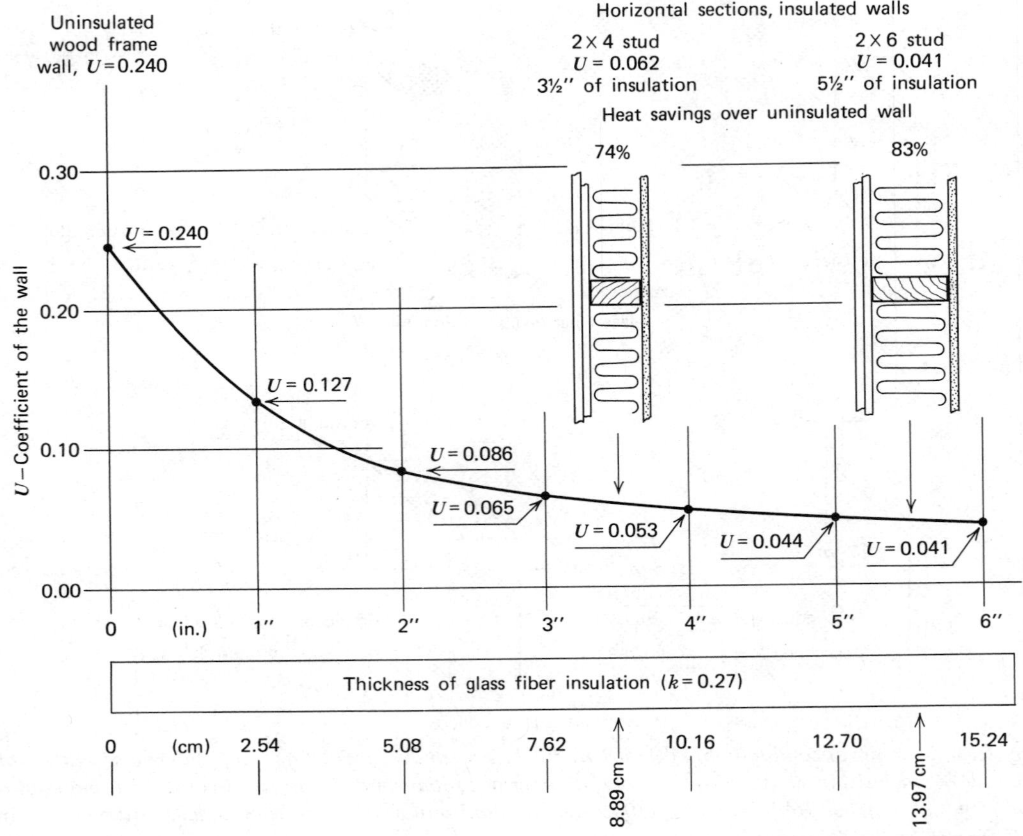

Fig. 4.15 Approximate values of heat savings when exterior frame walls are insulated. A trend is to entirely fill the void. One inch of insulation has an R*-value of about two to four times that of a 1-in. air space.*

In the illustration, parallel SI units of thickness are shown. Let us compare conventional and SI units:

Conventional (nominal) stud size	*2 × 4 in.*	*2 × 6 in.*
Actual *lumber size*	*1½ in. × 3½ in.*	*1½ in. × 5½ in.*
Exact *SI conversion*	*3.81 cm × 8.89 cm*	*3.81 cm × 13.97 cm*
Possible *future SI stud*	*4 × 9 cm*	*4 × 14 cm*
Possible future insulation thickness	*9 cm*	*14 cm*

Note. *In the past, nominal,* whole-number, *conventional dimensions have usually been chosen for the convenience of the skilled trades. In the future, simple* whole-number *SI dimensions may be used for the same reason. Thus, we may expect to see a 4 × 9 or a 4 × 14 stud. Centimeters, that is!*

Table 4.4 *Thermal Resistances of Plane[a] Air Spaces[d,e]*

All resistance values expressed in (hour) (square foot) (degree Fahrenheit temperature difference) per Btu Values apply only to air spaces of uniform thickness bounded by plane, smooth, parallel surfaces with no leakage of air to or from the space. Thermal resistance values for multiple air spaces must be based on careful estimates of mean temperature differences for each air space. See the Caution section, under Overall Coefficients and Their Practical Use.

Position of Air Space	Direction of Heat Flow	Air Space Mean Temp,[b] (F)	Air Space Temp Diff,[b] (deg F)	0.5-in. Air Space[d] Value of E[b,c] 0.03	0.05	0.2	0.5	0.82	0.75-in. Air Space[d] Value of E[b,c] 0.03	0.05	0.2	0.5	0.82
		90	10	2.13	2.03	1.51	0.99	0.73	2.34	2.22	1.61	1.04	0.75
		50	30	1.62	1.57	1.29	0.96	0.75	1.71	1.66	1.35	0.99	0.77
		50	10	2.13	2.05	1.60	1.11	0.84	2.30	2.21	1.70	1.16	0.87
Horiz.	Up	0	20	1.73	1.70	1.45	1.12	0.91	1.83	1.79	1.52	1.16	0.93
		0	10	2.10	2.04	1.70	1.27	1.00	2.23	2.16	1.78	1.31	1.02
		−50	20	1.69	1.66	1.49	1.23	1.04	1.77	1.74	1.55	1.27	1.07
		−50	10	2.04	2.00	1.75	1.40	1.16	2.16	2.11	1.84	1.46	1.20
		90	10	2.44	2.31	1.65	1.06	0.76	2.96	2.78	1.88	1.15	0.81
		50	30	2.06	1.98	1.56	1.10	0.83	1.99	1.92	1.52	1.08	0.82
		50	10	2.55	2.44	1.83	1.22	0.90	2.90	2.75	2.00	1.29	0.94
45° Slope	Up	0	20	2.20	2.14	1.76	1.30	1.02	2.13	2.07	1.72	1.28	1.00
		0	10	2.63	2.54	2.03	1.44	1.10	2.72	2.62	2.08	1.47	1.12
		−50	20	2.08	2.04	1.78	1.42	1.17	2.05	2.01	1.76	1.41	1.16
		−50	10	2.62	2.56	2.17	1.66	1.33	2.53	2.47	2.10	1.62	1.30
		90	10	2.47	2.34	1.67	1.06	0.77	3.50	3.24	2.08	1.22	0.84
		50	30	2.57	2.46	1.84	1.23	0.90	2.91	2.77	2.01	1.30	0.94
		50	10	2.66	2.54	1.88	1.24	0.91	3.70	3.46	2.35	1.43	1.01
Vertical	Horiz.	0	20	2.82	2.72	2.14	1.50	1.13	3.14	3.02	2.32	1.58	1.18
		0	10	2.93	2.82	2.20	1.53	1.15	3.77	3.59	2.64	1.73	1.26
		−50	20	2.90	2.82	2.35	1.76	1.39	2.90	2.83	2.36	1.77	1.39
		−50	10	3.20	3.10	2.54	1.87	1.46	3.72	3.60	2.87	2.04	1.56
		90	10	2.48	2.34	1.67	1.06	0.77	3.53	3.27	2.10	1.22	0.84
		50	30	2.64	2.52	1.87	1.24	0.91	3.43	3.23	2.24	1.39	0.99
		50	10	2.67	2.55	1.89	1.25	0.92	3.81	3.57	2.40	1.45	1.02
45° Slope	Down	0	20	2.91	2.80	2.19	1.52	1.15	3.75	3.57	2.63	1.72	1.26
		0	10	2.94	2.83	2.21	1.53	1.15	4.12	3.91	2.81	1.80	1.30
		−50	20	3.16	3.07	2.52	1.86	1.45	3.78	3.65	2.90	2.05	1.57
		−50	10	3.26	3.16	2.58	1.89	1.47	4.35	4.18	3.22	2.21	1.66

(*Continued*)

Table 4.4 *Thermal Resistances of Plane[a] Air Spaces[d,e] (Continued)*

Position of Air Space	Direction of Heat Flow	Air Space: Mean Temp,[b] (F)	Air Space: Temp Diff,[b] (deg F)	0.5-in. Air Space[d], Value of $E^{b,c}$: 0.03	0.05	0.2	0.5	0.82	0.75-in. Air Space[d], Value of $E^{b,c}$: 0.03	0.05	0.2	0.5	0.82
		90	10	2.48	2.34	1.67	1.06	0.77	3.55	3.29	2.10	1.22	0.85
		50	30	2.66	2.54	1.88	1.24	0.91	3.77	3.52	2.38	1.44	1.02
		50	10	2.67	2.55	1.89	1.25	0.92	3.84	3.59	2.41	1.45	1.02
Horiz.	Down	0	20	2.94	2.83	2.20	1.53	1.15	4.18	3.96	2.83	1.81	1.30
		0	10	2.96	2.85	2.22	1.53	1.16	4.25	4.02	2.87	1.82	1.31
		−50	20	3.25	3.15	2.58	1.89	1.47	4.60	4.41	3.36	2.28	1.69
		−50	10	3.28	3.18	2.60	1.90	1.47	4.71	4.51	3.42	2.30	1.71

Position of Air Space	Direction of Heat Flow	Air Space: Mean Temp,[b] (F)	Air Space: Temp Diff,[b] (deg F)	1.5-in. Air Space[d], Value of $E^{b,c}$: 0.03	0.05	0.2	0.5	0.82	3.5-in. Air Space[d], Value of $E^{b,c}$: 0.03	0.05	0.2	0.5	0.82
		90	10	2.55	2.41	1.71	1.08	0.77	2.84	2.66	1.83	1.13	0.80
		50	30	1.87	1.81	1.45	1.04	0.80	2.09	2.01	1.58	1.10	0.84
		50	10	2.50	2.40	1.81	1.21	0.89	2.80	2.66	1.95	1.28	0.93
Horiz	Up	0	20	2.01	1.95	1.63	1.23	0.97	2.25	2.18	1.79	1.32	1.03
		0	10	2.43	2.35	1.90	1.38	1.06	2.71	2.62	2.07	1.47	1.12
		−50	20	1.94	1.91	1.68	1.36	1.13	2.19	2.14	1.86	1.47	1.20
		−50	10	2.37	2.31	1.99	1.55	1.26	2.65	2.58	2.18	1.67	1.33
		90	10	2.92	2.73	1.86	1.14	0.80	3.18	2.96	1.97	1.18	0.82
		50	30	2.14	2.06	1.61	1.12	0.84	2.26	2.17	1.67	1.15	0.86
45°		50	10	2.88	2.74	1.99	1.29	0.94	3.12	2.95	2.10	1.34	0.96
Slope	Up	0	20	2.30	2.23	1.82	1.34	1.04	2.42	2.35	1.90	1.38	1.06
		0	10	2.79	2.69	2.12	1.49	1.13	2.98	2.87	2.23	1.54	1.16
		−50	20	2.22	2.17	1.88	1.49	1.21	2.34	2.29	1.97	1.54	1.25
		−50	10	2.71	2.64	2.23	1.69	1.35	2.87	2.79	2.33	1.75	1.39
		90	10	3.99	3.66	2.25	1.27	0.87	3.69	3.40	2.15	1.24	0.85
		50	30	2.58	2.46	1.84	1.23	0.90	2.67	2.55	1.89	1.25	0.91
		50	10	3.79	3.55	2.39	1.45	1.02	3.63	3.40	2.32	1.42	1.01
Vertical	Horiz.	0	20	2.76	2.66	2.10	1.48	1.12	2.88	2.78	2.17	1.51	1.14
		0	10	3.51	3.35	2.51	1.67	1.23	3.49	3.33	2.50	1.67	1.23
		−50	20	2.64	2.58	2.18	1.66	1.33	2.82	2.75	2.30	1.73	1.37
		−50	10	3.31	3.21	2.62	1.91	1.48	3.40	3.30	2.67	1.94	1.50

		90	10	5.07	4.55	2.56	1.36	0.91	4.81	4.33	2.49	1.34	0.90
		50	30	3.58	3.36	2.31	1.42	1.00	3.51	3.30	2.28	1.40	1.00
45°		50	10	5.10	4.66	2.85	1.60	1.09	4.74	4.36	2.73	1.57	1.08
Slope	Down	0	20	3.85	3.66	2.68	1.74	1.27	3.81	3.63	2.66	1.74	1.27
		0	10	4.92	4.62	3.16	1.94	1.37	4.59	4.32	3.02	1.88	1.34
		−50	20	3.62	3.50	2.80	2.01	1.54	3.77	3.64	2.90	2.05	1.57
		−50	10	4.67	4.47	3.40	2.29	1.70	4.50	4.32	3.31	2.25	1.68
		90	10	6.09	5.35	2.79	1.43	0.94	10.07	8.19	3.41	1.57	1.00
		50	30	6 27	5.63	3.18	1.70	1.14	9.60	8.17	3.86	1.88	1.22
		50	10	6.61	5.90	3.27	1.73	1.15	11.15	9.27	4.09	1.93	1.24
Horiz.	Down	0	20	7.03	6.43	3.91	2.19	1.49	10.90	9.52	4.87	2.47	1.62
		0	10	7.31	6.66	4.00	2.22	1.51	11.97	10.32	5.08	2.52	1.64
		−50	20	7.73	7.20	4.77	2.85	1.99	11.64	10.49	6.02	3.25	2.18
		−50	10	8.09	7.52	4.91	2.89	2.01	12.98	11.56	6.36	3.34	2.22

[a]See Chapter 20, section on Factors Affecting Heat Transfer across Air Spaces.

[b]Interpolation is permissible for other values of mean temperature, temperature differences, and effective emittance E. Interpolation and moderate extrapolation for air spaces greater than 3.5 in. are also permissible.

[c]Effective emittance of the space E is given by $1/E = 1/e_1 + 1/e_2 - 1$, *where* e_1 and e_2 are the emittances of the surfaces of the air space (See section B of Table 1.)

[d]Credit for an air space resistance value cannot be taken more than once and only for the boundary conditions established.

[e]Resistances of horizontal spaces with heat flow downward are substantially independent of temperature difference.

[f]Thermal resistance values were determined from the relation $R = 1/C$, *where* $C = h_c + Eh_r$, h_c is the conduction-convection coefficient, Eh_r is the radiation coefficient $\cong 0.00686\ E\ [(460 + t_m)/100]^3$, and t_m is the mean temperature of the air space. For interpretation from Table 2 to air space thicknesses less than 0.5 in. (as in insulating window glass), assume $h_c = 0.795\ (1 + 0.0016)$ and compute R-values from the above relations for an air space thickness of 0.2 in.

Based on National Bureau of Standards data presented in Housing Research Paper No. 32, Housing and Home Finance Agency 1954, U. S. Government Printing Office, Washington 20402.

NOTE: The values indicated are circled by the author and are discussed in Section 4.12, *Air Spaces.*

Table 4.5 ***Surface Conductances and Resistances for Air***

All conductance values expressed in Btu/(hr · ft² · F). A surface cannot take credit for both an air space resistance value and a surface resistance value. No credit for an air space value can be taken for any surface facing an air space of less than 0.5 in.

SECTION A. Surface Conductances and Resistances[a,b,d]

Position of Surface	Direction of Heat Flow	*Surface* Emittance: Non-reflective $\varepsilon = 0.90$ h_i	R	Reflective $\varepsilon = 0.20$ h_i	R	Reflective $\varepsilon = 0.05$ h_i	R
STILL AIR							
Horizontal	Upward	1.63	*0.61*	0.91	*1.10*	0.76	*1.32*
Sloping—45 deg.	Upward	1.60	*0.62*	0.88	*1.14*	0.73	*1.37*
Vertical	Horizontal	1.46	*0.68*	0.74	*1.35*	0.59	*1.70*
Sloping—45 deg.	Downward	1.32	*0.76*	0.60	*1.67*	0.45	*2.22*
Horizontal	Downward	1.08	*0.92*	0.37	*2.70*	0.22	*4.55*
MOVING AIR		h_o	R	h_o	R	h_o	R
(Any Position) 15-mph Wind (for winter)	Any	6.00	0.17				
7.5-mph Wind (for summer)	Any	4.00	0.25				

SECTION B. Reflectivity and Emittance Values of Various Surfaces[c] **and Effective Emittances of Air Spaces**

Surface	Reflectivity in Percent	Average Emittance ε	*Effective* Emittance E of Air Space: One surface emittance ε; the other 0.90	Both surfaces emittances ε
Aluminum foil, bright	92 to 97	0.05	0.05	0.03
Aluminum sheet	80 to 95	0.12	0.12	0.06
Aluminum coated paper, polished	75 to 84	0.20	0.20	0.11
Steel, galvanized, bright	70 to 80	0.25	0.24	0.15
Aluminum paint	30 to 70	0.50	0.47	0.35
Building materials: wood, paper, masonry, nonmetallic paints	5 to 15	0.90	0.82	0.82
Regular glass	5 to 15	0.84	0.77	0.72

[a]For ventilated attics or spaces above ceilings under summer conditions (heat flow down) see Table 6.

[b]Conductances are for surfaces of the stated emittance facing virtual black body surroundings at the same temperature as the ambient air. Values are based on a surface-air temperature difference of 10 F and for surface temperature of 70 F.

[c]See also Chapter 2, Table 4.

[d]See Fig. 1 for additional data.

Author's note. The values indicated are circled by the author. They are discussed in Section 4.13.

Table 4.6 ***Thermal Properties of Typical Building and Insulating Materials—(Design Values)***[a]
(For Industrial Insulation Design Values, see Table 3B). These constants are expressed in Btu per (hour) (square foot) (degree Fahrenheit temperature difference). Conductivities (*k*) are per inch thickness, and conductances (*C*) are for thickness or construction stated, not per inch thickness. *All values are for a mean temperature of 75 F, except as noted by an asterisk (*) which have been reported at 45 F.* The SI units for Resistance (last two columns) were calculated by taking the values from the two Resistance columns under Customary Unit, and multiplying by the factor 1/*k* (*r*/in.) and 1/C (*R*) for the appropriate conversion factor in Table 18.

Description	Customary Unit						SI Unit	
				Resistance[b] (*R*)			Resistance[b] (*R*)	
	Density (lb/ft^3)	Conductivity (*k*)	Conductance (*C*)	Per inch thickness (1/*k*)	For thickness listed (1/*C*)	Specific Heat, Btu/(lb) (deg F)	$\frac{(m \cdot K)}{W}$	$\frac{(m^2 \cdot K)}{W}$
BUILDING BOARD:								
Boards, Panels, Subflooring, Sheathing								
Woodboard Panel Products								
Asbestos-cement board	120	4.0	—	*0.25*	—	0.24	1.73	
Asbestos-cement board 0.125 in.	120	—	33.00	—	*0.03*			*0.005*
Asbestos-cement board 0.25 in.	120	—	16.50	—	*0.06*			*0.01*
Gypsum or plaster board 0.375 in.	50	—	3.10	—	*0.32*	0.26		*0.06*
Gypsum or plaster board 0.5 in.	50	—	2.22	—	*0.45*			*0.08*
Gypsum or plaster board 0.625 in.	50	—	1.78	—	*0.56*			*0.10*
Plywood (Douglas Fir)	34	0.80	—	*1.25*	—	0.29	8.66	
Plywood (Douglas Fir) 0.25 in.	34	—	3.20	—	*0.31*			*0.05*
Plywood (Douglas Fir) 0.375 in.	34	—	2.13	—	*0.47*			*0.08*
Plywood (Douglas Fir) 0.5 in.	34	—	1.60	—	*0.62*			*0.11*
Plywood (Douglas Fir) 0.625 in.	34	—	1.29	—	*0.77*			*0.19*
Plywood or wood panels 0.75 in.	34	—	1.07	—	*0.93*	0.29		*0.16*
Vegetable Fiber Board								
Sheathing, regular density 0.5 in.	18	—	0.76	—	*1.32*	0.31		*0.23*
0.78125 in.	18	—	0.49	—	*2.06*			*0.36*
Sheathing intermediate density 0.5 in.	22	—	0.82	—	*1.22*	0.31		*0.21*
Nail-base sheathing 0.5 in.	25	—	0.88	—	*1.14*	0.31		*0.20*
Shingle backer 0.375 in.	18	—	1.06	—	*0.94*	0.31		*0.17*
Shingle backer 0.3125 in.	18	—	1.28	—	*0.78*			*0.14*
Sound deadening board 0.5 in.	15	—	0.74	—	*1.35*	0.30		*0.24*
Tile and lay-in panels, plain or acoustic	18	0.40	—	*2.50*	—	0.14	17.33	
0.5 in.	18	—	0.80	—	*1.25*			*0.22*
0.75 in.	18	—	0.53	—	*1.89*			*0.33*

(Continued)

Table 4.6 *Thermal Properties of Typical Building and Insulating Materials—(Design Values)*[a] (Continued)

Description	*Customary Unit*						*SI Unit*	
				Resistance[b] *(R)*		*Specific Heat,*	*Resistance*[b] *(R)*	
	Density (lb/ft³)	*Conductivity (k)*	*Conductance (C)*	*Per inch thickness (1/k)*	*For thickness listed (1/C)*	*Btu/(lb) (deg F)*	*(m·K)/W*	*(m²·K)/W*
Laminated paperboard	30	0.50	—	*2.00*	—	0.33	13.86	
Homogeneous board from repulped paper	30	0.50	—	*2.00*	—	0.28	13.86	
Hardboard								
Medium density	50	0.73	—	*1.37*	—	0.31	9.49	
High density, service temp. service underlay	55	0.82	—	*1.22*	—	0.32	8.46	
High density, std. tempered	63	1.00	—	*1.00*	—	0.32	6.93	
Particleboard								
Low density	37	0.54	—	*1.85*	—	0.31	12.82	
Medium density	50	0.94	—	*1.06*	—	0.31	7.35	
High density	62.5	1.18	—	*0.85*	—	0.31	5.89	
Underlayment 0.625 in.	40	—	1.22	—	*0.82*	0.29		*0.14*
Wood subfloor 0.75 in.		—	1.06	—	*0.94*	0.33		*0.17*
BUILDING MEMBRANE								
Vapor—permeable felt	—	—	16.70	—	*0.06*			*0.01*
Vapor—seal, 2 layers of mopped 15-lb felt	—	—	8.35	—	*0.12*			*0.02*
Vapor—seal, plastic film	—	—	—	—	Negl.			
FINISH FLOORING MATERIALS								
Carpet and fibrous pad	—	—	0.48	—	*2.08*	0.34		*0.37*
Carpet and rubber pad	—	—	0.81	—	*1.23*	0.33		*0.22*
Cork tile 0.125 in.	—	—	3.60	—	*0.28*	0.48		*0.05*
Terrazzo 1 in.	—	—	12.50	—	*0.08*	0.19		*0.01*
Tile—asphalt, linoleum, vinyl, rubber	—	—	20.00	—	*0.05*	0.30		*0.01*
vinyl asbestos						0.24		
ceramic						0.19		
Wood, hardwood finish 0.75 in.			1.47		*0.68*			0.12
INSULATING MATERIALS:								
Blanket and Batt								

Mineral Fiber, fibrous form processed								
from rock, slag, or glass								
approx.[e] 2–2.75 in.	0.3–2.0	—	0.143	—	*7*[d]	0.17–0.23		*1.23*
approx.[e] 3–3.5 in.	0.3–2.0	—	0.091	—	*11*[d]			*1.94*
approx.[e] 3.50–6.5	0.3–2.0	—	0.053	—	*19*[d]			*3.35*
approx.[e] 6–7 in.	0.3–2.0		0.045		*22*[d]			*3.87*
approx.[d] 8.5 in.	0.3–2.0		0.033		*30*[d]			*5.28*
Board and Slabs								
Cellular glass	8.5	0.38	—	*2.63*	—	0.24	*18.23*	
Glass fiber, organic bonded	4–9	0.25	—	*4.00*	—	0.23	*27.72*	
Expanded rubber (rigid)	4.5	0.22	—	*4.55*	—	0.40	*31.53*	
Expanded polystyrene extruded								
Cut cell surface	1.8	0.25	—	*4.00*	—	0.29	*27.72*	
Expanded polystyrene extruded								
Smooth skin surface	2.2	0.20	—	*5.00*	—	0.29	*34.65*	
Expanded polystyrene extruded								
Smooth skin surface	3.5	0.19	—	*5.26*	—		*36.45*	
Expanded polystyrene, molded beads	1.0	0.28	—	*3.57*	—	0.29	*24.74*	
Expanded polyurethane[f] (R-11 exp.)	1.5	0.16	—	*6.25*	—	0.38	*43.82*	
(Thickness, 1 in. or greater)	2.5							
Mineral fiber with resin binder	15	0.29	—	*3.45*	—	0.17	23.91	
Mineral fiberboard, wet felted								
Core or roof insulation	16–17	0.34	—	*2.94*	—		20.38	
Acoustical tile	18	0.35	—	*2.86*	—	0.19	19.82	
Acoustical tile	21	0.37	—	*2.70*	—		18.71	
Mineral fiberboard, wet molded								
Acoustical tile[g]	23	0.42	—	*2.38*	—	0.14	16.49	
Wood or cane fiberboard								
Acoustical tile[g] 0.5 in.	—	—	0.80	—	*1.25*	0.31		0.22
Acoustical tile[g] 0.75 in.	—	—	0.53	—	*1.89*			0.33
Interior finish (plank, tile)	15	0.35	—	*2.86*	—	0.32	19.82	
Wood shredded (cemented in								
preformed slabs)	22	0.60	—	*1.67*	—	0.31	11.57	
LOOSE FILL								
Cellulosic insulation (milled paper or								
wood pulp)	2.3–3.2	0.27–0.32	—	*3.13–3.70*	—	0.33	*21.69–25.64*	
Sawdust or shavings	8.0–15.0	0.45	—	2.22	—	0.33	*15.39*	
Wood fiber, softwoods	2.0–3.5	0.30	—	*3.33*	—	0.33	*23.08*	
Perlite, expanded	5.0–8.0	0.37	—	*2.70*	—	0.26	*18.71*	

(Continued)

Table 4.6 ***Thermal Properties of Typical Building and Insulating Materials—(Design Values)***[a] **(Continued)**

Description	*Density (lb/ft³)*	*Conductivity (k)*	*Conductance (C)*	*Customary Unit: Resistance*[b] *(R): Per inch thickness (1/k)*	*Customary Unit: Resistance*[b] *(R): For thickness listed (1/C)*	*Specific Heat, Btu/(lb) (deg F)*	*SI Unit: Resistance*[b] *(R): (m·K)/W*	*SI Unit: Resistance*[b] *(R): (m²·K)/W*
Mineral fiber (rock, slag or glass)								
approx.[e] 3.75–5 in.	0.6–2.0	—	—		11	0.17		*1.94*
approx.[e] 6.5–8.75 in.	0.6–2.0	—	—		19			*3.35*
approx.[e] 7.5–10 in.	0.6–2.0	—	—		22			*3.87*
approx.[e] 10.25–13.75 in.	0.6–2.0	—	—		30			*5.28*
Vermiculite, exfoliated	7.0–8.2	0.47	—	*2.13*	—	3.20	*14.76*	
	4.0–6.0	0.44	—	*2.27*	—		*15.73*	
Roof Insulation[h]								
Preformed, for use above deck								
Different roof insulations are available in different thicknesses to provide the design *C* values listed.[h] Consult individual manufacturers for actual *thickness of their material*			0.72 to 0.12		*1.39 to 8.33*		— —	*0.24 to 1.47*
MASONRY MATERIALS:								
Concretes								
Cement mortar	116	5.0	—	*0.20*	—		*1.39*	
Gypsum-fiber concrete 87.5% gypsum, 12.5% wood chips	51	1.66	—	*0.60*	—	0.21	*4.16*	
Lightweight aggregates including expanded shale, clay, or slate; expanded slags; cinders; pumice; vermiculite; also cellular concretes	120	5.2	—	*0.19*	—		*1.32*	
	100	3.6	—	*0.28*	—		*1.94*	
	80	2.5	—	*0.40*	—		*2.77*	
	60	1.7	—	*0.59*	—		*4.09*	
	40	1.15	—	*0.86*	—		*5.96*	
	30	0.90	—	*1.11*	—		*7.69*	
	20	0.70		*1.43*			*9.91*	
Perlite, expanded	40	0.93		*1.08*			*7.48*	
	30	0.71		*1.41*			*9.77*	
	20	0.50		*2.00*		0.32	*13.86*	
Sand and gravel or stone aggregate (oven dried)	140	9.0	—	*0.11*		0.22	*0.76*	

Sand and gravel or stone aggregate (not dried)	140	12.0	—	*0.08*			*0.55*	
Stucco	116	5.0	—	*0.20*			*1.39*	
MASONRY UNITS								
Brick, common[i]	120	5.0	—	*0.20*	—	0.19	*1.39*	
Brick, face[i]	130	(9.0)	—	*0.11*	—		*0.76*	
Clay tile, hollow:								
1 cell deep ... 3 in.	—	—	1.25	—	*0.80*	0.21		*0.14*
1 cell deep ... 4 in.	—	—	0.90	—	*1.11*			*0.20*
2 cells deep ... 6 in.	—	—	0.66	—	*1.52*			*0.27*
2 cells deep ... 8 in.	—	—	0.54	—	*1.85*			*0.33*
2 cells deep ... 10 in.	—	—	0.45	—	*2.22*			*0.39*
3 cells deep ... 12 in.	—	—	0.40	—	*2.50*			*0.44*
Concrete blocks, three oval core:								
Sand and gravel aggregate ... 4 in.	—	—	1.40	—	*0.71*	0.22		*0.13*
... 8 in.	—	—	0.90	—	*1.11*			*0.20*
... 12 in.	—	—	0.78	—	*1.28*			*0.23*
Cinder aggregate ... 3 in.	—	—	1.16	—	*0.86*	0.21		*0.15*
... 4 in.	—	—	0.90	—	*1.11*			*0.20*
... 8 in.	—	—	0.58	—	*1.72*			*0.30*
... 12 in.	—	—	0.53	—	*1.89*			*0.33*
Lightweight aggregate ... 3 in.	—	—	0.79	—	*1.27*	0.21		*0.22*
(expanded shale, clay, slate ... 4 in.	—	—	0.67	—	*1.50*			*0.26*
or slag; pumice) ... 8 in.	—	—	0.50	—	*2.00*			*0.35*
... 12 in.	—	—	0.44	—	*2.27*			*0.40*
Concrete blocks, rectangular core*[j]								
Sand and gravel aggregate								
2 core, 8 in., 36 lb[k]*	—	—	0.96	—	*1.04*	0.22		*0.18*
Same with filled cores[j]*	—	—	0.52	—	*1.93*	0.22		*0.34*
Lightweight aggregate (expanded shale, clay, slate or slag, pumice):								
3 core, 6 in., 19 lb[k]*	—	—	0.61	—	*1.65*	0.21		*0.29*
Same with filled cores[l]*	—	—	0.33	—	*2.99*			*0.53*
2 core, 8 in., 24 lb[k]*	—	—	0.46	—	*2.18*			*0.38*
Same with filled cores[l]*	—	—	0.20	—	*5.03*			*0.89*
3 core, 12 in., 38 lb[k]*	—	—	0.40	—	*2.48*			*0.44*
Same with filled cores[l]*	—	—	0.17	—	*5.82*			*1.02*
Stone, lime, or sand	—	12.50	—	*0.08*	—	0.19	*0.55*	

(Continued)

Table 4.6 *Thermal Properties of Typical Building and Insulating Materials—(Design Values)*[a] **(Continued)**

Description				Customary Unit			SI Unit	
				Resistance[b] (R)			Resistance[b] (R)	
	Density (lb/ft^3)	Conductivity (k)	Conductance (C)	Per inch thickness (1/k)	For thickness listed (1/C)	Specific Heat, Btu/(lb)(deg F)	(m·K)/W	(m^2·K)/W
Gypsum partition tile:								
3 × 12 × 30 in. solid	—	—	0.79	—	*1.26*	*0.19*		*0.22*
3 × 12 × 30-in. 4-cell	—	—	0.74	—	*1.35*			*0.24*
4 × 12 × 30 in. 3-cell	—	—	0.60	—	*1.67*			*0.29*
METALS								
(See Chapter 37, Table 3)								
PLASTERING MATERIALS								
Cement plaster, sand aggregate	116	5.0	—	*0.20*	—	0.20	*1.39*	
Sand aggregate 0.375 in.	—	—	13.3	—	*0.08*	0.20		*0.01*
Sand aggregate 0.75 in.	—	—	6.66	—	*0.15*	0.20		*0.03*
Gypsum plaster:								
Lightweight aggregate 0.5 in.	45	—	3.12	—	*0.32*			*0.06*
Lightweight aggregate 0.625 in.	45	—	2.67	—	*0.39*			*0.07*
Lightweight agg. on metal lath 0.75 in.	—	—	2.13	—	*0.47*			*0.08*
Perlite aggregate	45	1.5	—	*0.67*	—	0.32	*4.64*	
Sand aggregate	105	5.6	—	*0.18*	—	0.20	*1.25*	
Sand aggregate 0.5 in.	105	—	11.10	—	*0.09*			*0.02*
Sand aggregate 0.625 in.	105	—	9.10	—	*0.11*			*0.02*
Sand aggregate on metal lath 0.75 in.	—	—	7.70	—	*0.13*			*0.02*
Vermiculite aggregate	45	1.7	—	*0.59*	—		*4.09*	
ROOFING								
Asbestos-cement shingles	120	—	4.76	—	*0.21*	0.24		*0.04*
Asphalt roll roofing	70	—	6.50	—	*0.15*	0.36		*0.03*
Asphalt shingles	70	—	2.27	—	*0.44*	0.30		*0.08*
Built-up roofing 0.375 in.	70	—	3.00	—	*0.33*	0.35		*0.06*
Slate 0.5 in.	—	—	20.00	—	*0.05*	0.30		*0.01*
Wood shingles, plain and plastic film faced	—	—	1.06	—	*0.94*	0.31		*0.17*
SIDING MATERIALS (On Flat Surface)								
Shingles								
Asbestos-cement	120	—	4.75	—	*0.21*			*0.04*

Wood, 16 in., 7.5 exposure	—	—	1.15	—	*0.87*	0.31		*0.15*
Wood, double, 16-in., 12-in. exposure	—	—	0.84	—	*1.19*	0.28		*0.21*
Wood, plus insul. backer board, 0.3125 in.	—	—	0.71	—	*1.40*	0.31		*0.25*
Siding								
Asbestos-cement, 0.25 in., lapped	—	—	4.76	—	*0.21*	0.24		*0.04*
Asphalt roll siding	—	—	6.50	—	*0.15*	0.35		*0.03*
Asphalt insulating siding (0.5 in. bed)	—	—	0.69	—	*1.46*	0.35		*0.26*
Wood, drop, 1 × 8 in.	—	—	1.27	—	*0.79*	0.28		*0.14*
Wood, bevel, 0.5 × 8 in., lapped	—	—	1.23	—	*0.81*	0.28		*0.14*
Wood, bevel, 0.75 × 10 in., lapped	—	—	0.95	—	*1.05*	0.28		*0.18*
Wood, plywood, 0.375 in., lapped	—	—	1.59	—	*0.59*	0.29		*0.10*
Wood, medium density siding, 0.4375 in.	40	1.49	—	*0.67*		0.28	*4.65*	
Aluminum or Steel[m], over sheathing								
Hollow-backed	—	—	1.61	—	*0.61*	0.29		*0.11*
Insulating-board backed nominal 0.375 in.	—	—	0.55	—	*1.82*	0.32		*0.32*
Insulating-board backed nominal 0.375 in., foil backed			0.34		*2.96*			*0.52*
Architectural glass	—	—	10.00	—	*0.10*	0.20		*0.02*
WOODS								
Maple, oak, and similar hardwoods	45	1.10	—	*0.91*	—	0.30	*6.31*	
Fir, pine, and similar softwoods	32	0.80	—	*1.25*	—	0.33	*8.66*	
Fir, pine, and similar softwoods 0.75 in.	32	—	1.06	—	*0.94*	0.33		*0.17*
1.5 in.		—	0.53	—	*1.89*			*0.33*
2.5 in.		—	0.32	—	*3.12*			*0.60*
3.5 in.		—	0.23	—	*4.35*			*0.75*

NOTES FOR TABLE 4.6

[a] Representative values for dry materials were selected by ASHRAE TC4.4, Insulation and Moisture Barriers. They are intended as design (not specification) values for materials in normal use. For properties of a particular product, use the value supplied by the manufacturer or by unbiased tests.

[b] Resistance values are the reciprocals of C before rounding off C to two decimal places.

[c] Also see Insulating Materials, Board.

[d] Does not include paper backing and facing, if any. Where insulation forms a boundary (reflective or otherwise) of an air space, see Tables 1 and 2 for the insulating value of air space for the appropriate effective emittance and temperature conditions of the space.

[e] Conductivity varies with fiber diameter. (See Chapter 20, Thermal Conductivity section, and Fig. 1) Insulation is produced by different densities; therefore, there is a wide variation in thickness for the same R-value among manufacturers. No effort should be made to relate any specific R-value to any specific thickness. Commercial thicknesses generally available range from 2 to 8.5.

[f] Values are for aged board stock. For change in conductivity with age of expanded urethane, see Chapter 19, Factors Affecting Thermal Conductivity.

(Continued)

NOTES FOR TABLE 4.6 (Continued)

[g] Insulating values of acoustical tile vary, depending on density of the board and on type, size, and depth of perforations.

[h] The U.S. Department of Commerce, *Simplified Practice Recommendation for Thermal Conductance Factors for Preformed Above-Deck Roof Insulation,* No. R 257-55, recognizes the specification of roof insulation on the basis of the *C*-values shown. Roof insulation is made in thicknesses to meet these values.

[i] Face brick and common brick do not always have these specific densities. When density is different from that shown, there will be a change in thermal conductivity.

[j] Data on rectangular core concrete blocks differ from the above data on oval core blocks, due to core configuration, different mean temperatures, and possibly differences in unit weights. Weight data on the oval core blocks tested are not available.

[k] Weights of units approximately 7.625 in. high and 15.75 in. long. These weights are given as a means of describing the blocks tested, but conductance values are all for 1 ft^2 of area.

[l] Vermiculite, perlite, or mineral wool insulation. Where insulation is used, vapor barriers or other precautions must be considered to keep insulation dry.

[m] Values for metal siding applied over flat surfaces vary widely, depending on amount of ventilation of air space beneath the siding; whether air space is reflective or nonreflective; and on thickness, type, and application of insulating backing-board used. Values given are averages for use as design guides, and were obtained from several guarded hotbox tests (ASTM C236) or calibrated hotbox (BSS 77) on hollow-backed types and types made using backing-boards of wood fiber, foamed plastic, and glass fiber. Departures of $\pm 50\%$ or more from the values given may occur.

Author's Note. The *k* value of face brick is circled by the author. It is discussed in Section 4.14.

Table 4.7 ***Coefficients of Transmission (U) of Frame Walls***[a]

These coefficients are expressed in Btu per (hour) (square foot) (degree Fahrenheit difference in temperature between the air on the two sides), and are based on an outside wind velocity of 15 mph.

Replace Air Space with 3.5-in. R-11 Blanket Insulation (New Item 4)

Construction	1 Resistance (R) Between Framing	1 At Framing	2 Between Framing	2 At Framing
1. Outside surface (15 mph wind)	0.17	0.17	0.17	0.17
2. Siding, wood, 0.5 in. × 8 in. lapped (average)	0.81	0.81	0.81	0.81
3. Sheathing, 0.5-in. asphalt impregnated	1.32	1.32	1.32	1.32
4. Nonreflective air space, 3.5 in. (50° F mean; 10° F temperature difference)	1.01	—	11.00	—
5. Nominal 2-in. × 4-in. wood stud	—	4.38	—	4.38
6. Gypsum wallboard, 0.5 in.	0.45	0.45	0.45	0.45
7. Inside surface (still air)	0.68	0.68	0.68	0.68
Total Thermal Resistance (R)	$R_i = 4.44$	$R_s = 7.81$	$R_i = 14.43$	$R_s = 7.81$

Construction No. 1: $U_i = 1/4.44 = 0.225$; $U_s = 1/7.81 = 0.128$. With 20% framing (typical of 2-in. × 4-in. studs @ 16-in. o.c.), $U_{av} = 0.8(0.225) + 0.2(0.128) = 0.206$ (See Eq 9).

Construction No. 2: $U_i = 1/14.43 = 0.069$; $U_s = 0.128$. With framing unchanged, $U_{av} = 0.8(0.069) + 0.2(0.128) = 0.081$.

[a]See text section Calculating Overall Coefficients for basis of calculations.

Table 4.8 *Coefficients of Transmission (U) of Solid Masonry Walls*[a]

Coefficients are expressed in Btu per (hour) (square foot) (degree Fahrenheit difference in temperature between the air on the two sides), and are based on an outside wind velocity of 15 mph.

Replace Furring Strips and Air Space with 1-in. Extruded Polystyrene (New Item 4)

Construction	1 Resistance (R) Between Furring	1 Resistance (R) At Furring	2
1. Outside surface (15 mph wind)	0.17	0.17	0.17
2. Common brick, 8 in.	1.60	1.60	1.60
3. Nominal 1-in. × 3-in. vertical furring	—	0.94	—
4. Nonreflective air space, 0.75 in. (50° F mean; 10° F temperature difference)	1.01	—	5.00
5. Gypsum wallboard, 0.5 in.	0.45	0.45	0.45
6. Inside surface (still air)	0.68	0.68	0.68
Total Thermal Resistance (R)	$R_i = 3.91$	$R_s = 3.84$	$R_i = 7.90 = R_s$

Construction No. 1: $U_i = 1/3.91 = 0.256$; $U_s = 1/3.84 = 0.260$. With 20% framing (typical of 1-in. × 3-in. vertical furring on masonry @ 16-in. o.c.) $U_{av} = 0.8(0.256) + 0.2(0.260) = 0.257$

Construction No. 2: $U_i = U_s = U_{av} = 1/7.90 = 0.127$

[a]See text section Calculating Overall Coefficients for basis of calculations.

Table 4.9 ***Coefficients of Transmission (U) of Masonry Walls***[a]

Coefficients are expressed in Btu per (hour) (square foot) (degree Fahrenheit difference in temperature between the air on the two sides), and are based on an outside wind velocity of 15 mph.

Replace Cinder Aggregate Block with 6-in. Light-weight Aggregate Block with Cores Filled (New Item 4)

Construction	1		2	
	Resistance (R)			
	Between Furring	*At Furring*	*Between Furring*	*At Furring*
1. Outside surface (15 mph wind)	0.17	0.17	0.17	0.17
2. Face brick, 4 in.	0.44	0.44	0.44	0.44
3. Cement mortar, 0.5 in.	0.10	0.10	0.10	0.10
4. Concrete block, cinder aggregate, 8 in.	1.72	1.72	2.99	2.99
5. Reflective air space, 0.75 in. (50° F mean; 30° F temperature difference)	2.77	—	2.77	—
6. Nominal 1-in. × 3-in. vertical furring	—	0.94	—	0.94
7. Gypsum wallboard, 0.5 in., foil backed	0.45	0.45	0.45	0.45
8. Inside surface (still air)	0.68	0.68	0.68	0.68
Total Thermal Resistance (R)	$R_i = 6.33$	$R_s = 4.50$	$R_i = 7.60$	$R_s = 5.77$

Construction No. 1: $U_i = 1/6.33 = 0.158$; $U_s = 1/4.50 = 0.222$. With 20% framing (typical of 1-in. × 3-in. vertical furring on masonry @ 16-in. o.c.), $U_{av} = 0.8(0.158) + 0.2(0.222) = 0.171$.

Construction No. 2: $U_i = 1/7.60 = 0.132_4$; $U_s = 1/5.77 = 0.173$. With framing unchanged, $U_{av} = 0.8(0.132) + 0.2(0.173) = 1.40$.

[a]See text section Calculating Overall Coefficients for basis of calculations.

Table 4.10 ***Coefficients of Transmission (U) of Flat Masonry Roofs with Built-up Roofing, with and without Suspended Ceilings***[a,b] ***(Winter Conditions, Upward Flow)***

These Coefficients are expressed in Btu per (hour) (square foot) (degree Fahrenheit difference in temperature between the air on the two sides), and are based upon an outside wind velocity of 15 mph.

Add Rigid Roof Deck Insulation, C = 0.24 (R = 1/C) (New Item 7) *Construction* *(Heat Flow Up)*	*1*	2
1. Inside surface (still air)	0.61	0.61
2. Metal lath and lightweight aggregate plaster, 0.75 in.	0.47	0.47
3. Nonreflective air space, greater than 3.5 in. (50° F mean; 10° F temperature difference)	0.93*	0.93*
4. Metal ceiling suspension system with metal hanger rods	0**	0**
5. Corrugated metal deck	0	0
6. Concrete slab, lightweight aggregate, 2 in.	2.22	2.22
7. Rigid roof deck insulation (none)	—	4.17
8. Built-up roofing, 0.375 in.	0.33	0.33
9. Outside surface (15 mph wind)	0.17	0.17
Total Thermal Resistance (R)	4.73	8.90

Construction No. 1: $U_{av} = 1/4.73 = 0.211$
Construction No. 2: $U_{av} = 1/8.90 = 0.112$

[a]See text section Calculating Overall Coefficients for basis of calculations.

[b]To adjust U values for the effect of added insulation between framing members, see Table 5 or 6.

*Use largest air space (3.5 in.) value shown in Table 2.

**Area of hanger rods is negligible in relation to ceiling area.

Table 4.11 ***Coefficients of Transmission (U) of Wood Construction Flat Roofs and Ceilings***[a] ***(Winter Conditions, Upward Flow)***

Coefficients are expressed in Btu per (hour) (square foot) (degree Fahrenheit difference in temperature between the air on the two sides), and are based upon an outside wind velocity of 15 mph.

Replace Roof Deck Insulation and 7.25-in. Air Space with 6-in. R-19 Blanket Insulation and 1.25-in. Air Space (New Items 5 and 7)

Construction (Heat Flow Up)	1 Resistance (R)		2	
	Between Joists	At Joists	Between Joists	At Joists
1. Inside surface (still air)	0.61	0.61	0.61	0.61
2. Acoustical tile, fiberboard, glued, 0.5 in.	1.25	1.25	1.25	1.25
3. Gypsum wallboard, 0.5 in.	0.45	0.45	0.45	0.45
4. Nominal 2-in. × 8-in. ceiling joists	—	9.06	—	9.06
5. Nonreflective air space, 7.25 in. (50° F mean; 10° F temperature difference)	0.93*	—	1.05**	—
6. Plywood deck, 0.625 in.	0.78	0.78	0.78	0.78
7. Rigid roof deck insulation, c = 0.72, (R = 1/C)	1.39	1.39	19.00	—
8. Built-up roof	0.33	0.33	0.33	0.33
9. Outside surface (15 mph wind)	0.17	0.17	0.17	0.17
Total Thermal Resistance (R)	$R_i = 5.91$	$R_s = 14.04$	$R_i = 23.64$	$R_s = 12.65$

Construction No. 1: $U_i = 1/5.91 = 0.169$; $U_s = 1/14.04 = 0.071$. With 10% framing (typical of 2-in. joists @ 16-in. o.c.), $U_{av} = 0.9\,(0.169) + 0.1\,(0.071) = 0.159$

Construction No. 2: $U_i = 1/23.64 = 0.042$; $U_s = 1/12.65 = 0.079$. With framing unchanged, $U_{av} = 0.9\,(0.042) + 0.1\,(0.079) = 0.046$

[a] See text section Calculating Overall Coefficients for basis of calculations.

* Use largest air space (3.5 in.) value shown in Table 2.

** Interpolated value (0° F mean; 10° F temperature difference).

Table 4.12 ***Coefficients of Transmission (U) of Metal Construction Flat Roofs and Ceilings*[a] *(Winter Conditions, Upward Flow)***

Coefficients are expressed in Btu per (hour) (square foot) (degree Fahrenheit difference in temperature between the air on the two sides), and are based upon an outside wind velocity of 15 mph.

Replace Rigid Roof Deck Insulation ($C = 0.24$) and Sand Aggregate Plaster with Rigid Roof Deck Insulation, $C = 0.36$ and Lightweight Aggregate Plaster (New Items 2 and 6)

Construction (Heat Flow Up)	*1*	2
1. Inside surface (still air)	0.61	0.61
2. Metal lath and sand aggregate plaster, 0.75 in.	0.13	0.47
3. Structural beam	0.00*	0.00*
4. Nonreflective air space (50° F mean; 10° F temperature difference	0.93**	0.93**
5. Metal deck	0.00*	0.00*
6. Rigid roof deck insulation, $C = 0.24$, ($R = 1/c$)	4.17	2.78
7. Built-up roofing, 0.375 in.	0.33	0.33
8. Outside surface (15 mph wind)	0.17	0.17
Total Thermal Resistance (R)	6.34	5.29

Construction No. 1: $U = 1/6.34 = 0.158$
Construction No. 2: $U = 1/5.29 = 0.189$

[a]See text section Calculating Overall Coefficients for basis of calculations.

*If structural beams and metal deck are to be considered, the technique shown in *Examples 1 and 2,* and Fig. 3 may be used to estimate total *R*. Full scale testing of a suitable portion of the construction is, however, preferable.

**Use largest air space (3.5 in.) value shown in Table 2.

Table 4.13 *Determination of U-Value Resulting from Addition of Insulation to the Total Area[e] of any Given Building Section*

Given Building Section Property[a,b]		Added R[c,d,e]						
		$R = 4$	$R = 6$	$R = 8$	$R = 12$	$R = 16$	$R = 20$	$R = 24$
U	R	U	U	U	U	U	U	U
1.00	1.00	0.20	0.14	0.11	0.08	0.06	0.05	0.04
0.90	1.11	0.20	0.14	0.11	0.08	0.06	0.05	0.04
0.80	1.25	0.19	0.14	0.11	0.08	0.06	0.05	0.04
0.70	1.43	0.18	0.13	0.11	0.07	0.06	0.05	0.04
0.60	1.67	0.18	0.13	0.10	0.07	0.06	0.05	0.04
0.50	2.00	0.17	0.13	0.10	0.07	0.06	0.05	0.04
0.40	2.50	0.15	0.12	0.10	0.07	0.05	0.04	0.04
0.30	3.33	0.14	0.11	0.09	0.07	0.05	0.04	0.04
0.20	5.00	0.11	0.09	0.08	0.06	0.05	0.04	0.03
0.10	10.00	0.07	0.06	0.06	0.05	0.04	0.03	0.03
0.08	12.50	0.06	0.05	0.05	0.04	0.04	0.03	0.03

[a]For U- or R-values not shown in the table, interpolate as necessary.

[b]Enter column 1 with U or R of the design building section.

[c]Under appropriate column heading for added R, find U-value of resulting design section.

[d]If the insulation occupies previously considered air space, an adjustment must be made in the given building section R-value.

[e]If insulation is applied between framing members, use Eq 9 to determine average U-value.

Table 4.14 ***Determination of U-Value Resulting from Addition of Insulation to Uninsulated Roof Deck***

U-Value of Roof without Roof-Deck Insulation[a]	*Conductance C of Roof-Deck Insulation*					
	0.12	*0.15*	*0.19*	*0.24*	*0.36*	*0.72*
	U	*U*	*U*	*U*	*U*	*U*
0.10	0.05	0.06	0.07	0.07	0.08	0.09
0.15	0.07	0.08	0.08	0.09	0.11	0.12
0.20	0.08	0.09	0.10	0.11	0.13	0.16
0.25	0.08	0.09	0.11	0.12	0.15	0.19
0.30	0.09	0.10	0.12	0.13	0.16	0.21
0.35	0.09	0.11	0.12	0.14	0.18	0.24
0.40	0.09	0.11	0.13	0.15	0.19	0.26
0.50	0.10	0.12	0.14	0.16	0.21	0.30
0.60	0.10	0.12	0.14	0.17	0.23	0.33
0.70	0.10	0.12	0.15	0.18	0.24	0.35

[a]Interpolation or mild extrapolation may be used.

Table 4.15 *Heat Losses, Edge of Concrete Slabs*

(a) Heat Loss of Concrete Floors at or Near Grade Level per Foot of Exposed Edge

	Heat Loss per Foot of Exposed Edge, Btuh	
Outdoor Design Temperature, F	*Recommended 2-in. Edge Insulation*	*1-in. Edge Insulation*
−20 to −30	50	55
−10 to −20	45	50
0 to −10	40	45
Outdoor Design Temperature, F	*1-in. Edge Insulation*	*No Edge Insulation*[a]
−20 to −30	60	75
−10 to −20	55	65
0 to −10	50	60

[a]This construction not recommended; shown for comparison only.

(b) Floor Heat Loss to be Used When Warm Air Perimeter Heating Ducts Are Embedded in Slabs[a] Btuh per (linear foot of heated edge)

	Edge Insulation		
Outdoor Design Temperature, F	*1-in. Vertical Extending Down 18 in. Below Floor Surface*	*1-in. L-Type Extending at Least 12 in. Deep and 12 in. Under*	*2-in. L-Type Extending at Least 12 in. Down and 12 in. Under*
−20	105	100	85
−10	95	90	75
0	85	80	65
10	75	70	55
20	62	57	45

[a]Factors include loss downward through inner area of slab.

Table 4.16 ***Coefficients of Transmission (U) of Windows, Skylights, and Light Transmitting Partitions***

These values are for heat transfer from air to air, Btu/(hr·ft²·F). To calculate total heat gain including solar transmission, see Chapter 28.

Part A—Vertical Panels (Exterior Windows, Sliding Patio Doors, and Partitions)—Flat Glass, Glass Block, and Plastic Sheet

Description	*Exterior*[a]		
	Winter	*Summer*	*Interior*
Flat Glass[b]			
Single glass	1.10	1.04	0.73
Insulating glass—double[c]			
0.1875-in. air space[d]	0.62	0.65	0.51
0.25-in. air space[d]	0.58	0.61	0.49
0.5-in. air space[e]	0.49	0.56	0.46
0.5-in. air space, low emittance coating[f]			
$e = 0.20$	0.32	0.38	0.32
$e = 0.40$	0.38	0.45	0.38
$e = 0.60$	0.43	0.51	0.42
Insulating glass—triple[c]			
0.25-in. air spaces[d]	0.39	0.44	0.38
0.5-in. air spaces[g]	0.31	0.39	0.30
Storm windows			
1-in. to 4-in. air space[d]	0.50	0.50	0.44
Plastic Sheet			
Single glazed			
0.125-in. thick	1.06	0.98	—
0.25-in. thick	0.96	0.89	—
0.5-in. thick	0.81	0.76	—
Insulating unit—double[c]			
0.25-in. air space[d]	0.55	0.56	—
0.5-in. air space[c]	0.43	0.45	—
Glass Block[h]			
6 × 6 × 4 in. thick	0.60	0.57	0.46
8 × 8 × 4 in. thick	0.56	0.54	0.44
—with cavity divider	0.48	0.46	0.38
12 × 12 × 4 in. thick	0.52	0.50	0.41
—with cavity divider	0.44	0.42	0.36
12 × 12 × 2 in. thick	0.60	0.57	0.46

Part B—Horizontal Panels (Skylights)—Flat Glass, Glass Block, and Plastic Domes

Description	*Exterior*[a]		
	Winter[i]	*Summer*[j]	*Interior*[f]
Flat Glass[e]	1.23	0.83	0.96
Single glass			
Insulating glass—double[c]	0.70	0.57	0.62
0.1875-in. air space[d]	0.65	0.54	0.59
0.25-in. air space[d]	0.59	0.49	0.56
0.5-in. air space[c]			
0.5-in. air space, low emittance coating[f]			

Part B—Horizontal Panels (Skylights)—Flat Glass, Glass Block, and Plastic Domes (*Continued*)

Description	Exterior[a]		
	Winter[i]	*Summer*[j]	*Interior*[f]
$e = 0.20$	0.48	0.36	0.39
$e = 0.40$	0.52	0.42	0.45
$e = 0.60$	0.56	0.46	0.50
Glass Block[h]			
11 × 11 × 3 in. thick with cavity divider	0.53	0.35	0.44
12 × 12 × 4 in. thick with cavity divider	0.51	0.34	0.42
Plastic Domes[k]			
Single-walled	1.15	0.80	—
Double-walled	0.70	0.46	—

Part C—Adjustment Factors for Various Window and Sliding Patio Door Types (Multiply *U* Values in Parts A and B by These Factors)

Description	*Single Glass*	*Double or Triple Glass*	*Storm Windows*
Windows			
All Glass[l]	1.00	1.00	1.00
Wood sash—80% glass	0.90	0.95	0.90
Wood sash—60% glass	0.80	0.85	0.80
Metal sash—80% glass	1.00	1.20[m]	1.20[m]
Sliding Patio Doors			
Wood frame	0.95	1.00	—
Metal frame	1.00	1.10[m]	—

NOTES FOR TABLE 4.16

[a] See Part C for adjustment for various window and sliding patio door types.

[b] Emittance of uncooled glass surface = 0.84.

[c] Double and triple refer to the number of lights of glass.

[d] 0.125-in. glass.

[e] 0.25-in. glass.

[f] Coating on either glass surface facing air space; all other glass surfaces uncoated.

[g] Window design: 0.25-in. glass—0.125-in. glass—0.25-in. glass

[h] Dimensions are nominal.

[i] For heat flow up.

[j] For heat flow down.

[k] Based on area of opening, not total surface area.

[l] Refers to windows with negligible opaque area.

[m] Values will be less than these when metal sash and frame incorporate thermal breaks. In some thermal break designs, *U*-values will be equal to or less than those for the glass. Window manufacturers should be consulted for specific data.

Table 4.17 ***Coefficients of Transmission (U) for Slab Doors Btu per*** $(hr \cdot ft^2 \cdot F)$

	Winter			Summer
	Solid Wood,	Storm Door[b]		
Thickness[a]	*No Storm Door*	*Wood*	*Metal*	*No Storm Door*
1-in.	0.64	0.30	0.39	0.61
1.25-in.	0.55	0.28	0.34	0.53
1.5-in.	0.49	0.27	0.33	0.47
2-in.	0.43	0.24	0.29	0.42
	Steel Door[14]			
1.75-in.				
A[c]	0.59	—	—	0.58
B[d]	0.19	—	—	0.18
C[e]	0.47	—	—	0.46

[a]Nominal thickness.

[b]Values for wood storm doors are for approximately 50% glass; for metal storm door values apply for any percent of glass.

[c]A = Mineral fiber core (2 lb/ft^3).

[d]B = Solid urethane foam core with thermal break.

[e]C = Solid polystyrene core with thermal break.

Table 4.18 ***Infiltration and Air Changes***

(a) Infiltration Through Double-Hung Wood Windows

Expressed in cubic feet per (hour) (foot of crack)[d]

	Pressure Difference (Inches of Water)				
Type of Window	*0.10*	*0.20*	*0.30*	*0.40*	*0.50*
A. Wood Double-Hung Window (Locked)	25	50	25	100	125
(Leakage expressed as cubic feet per hour per foot of sash crack; only leakage around sash and through frame given)					
1. Nonweatherstripped, loose fit[a]	77[d]	122[d]	150[d]	194[d]	225[d]
2. Nonweatherstripped, average fit[b]	27[d]	43[d]	57[d]	69[d]	80[d]
3. Weatherstripped, loose fit	28[d]	44[d]	58[d]	70[d]	81[d]
4. Weatherstripped, average fit	14[d]	23[d]	30[d]	36[d]	42[d]
B. Frame-Wall Leakage[c]					
(Leakage is that passing between the frame of a wood double-hung window and the wall)					
1. Around frame in masonry wall, not caulked	17[d]	26[d]	34[d]	41[d]	48[d]

Table 4.18 ***Infiltration and Air Changes (Continued)***

2. Around frame in masonry wall, caulked	3[d]	5[d]	6[d]	7[d]	8[d]
3. Around frame in wood frame wall	13[d]	21[d]	29[d]	35[d]	42[d]

[a]A 0.094-in. crack and clearance represent a poorly fitted window, much poorer than average.

[b]The fit of the average double-hung wood window was determined as 0.0625-in. crack and 0.047-in. clearance by measurements on approximately 600 windows under heating season conditions.

[c]The values given for frame leakage are per foot of sash perimeter, as determined for double-hung wood windows. Some of the frame leakage in masonry walls originates in the brick wall itself, and cannot be prevented by caulking. For the additional reason that caulking is not done perfectly and deteriorates with time, it is considered advisable to choose the masonry frame leakage values for caulked frames as the average determined by the caulked and non-caulked tests.

[d]Multiply by 0.0258 for 1/s per metre of crack.

(b) Air Changes Occurring Under Average Conditions in Residences, Exclusive of Air Provided for Ventilation[a]

Kind of Room	*Number of Air Changes per hour*
Rooms with no windows or exterior doors	0.5
Rooms with windows or exterior doors on one side	1
Rooms with windows or exterior doors on two sides	1.5
Rooms with windows or exterior doors on three sides	2
Entrance halls	2

[a]For rooms with weatherstripped windows or with storm sash, use two-thirds these values.

Table 4.19 ***Resistance (R) Values of Some Available Fiberglas Insulation Products for Residential Construction***

R-Values	*Nominal Thickness in Inches*
9	$2\frac{3}{4}$
11	$3\frac{1}{4}$
13	$3\frac{5}{8}$
19	6[a]
30	9
38	12[a]

Source. Courtesy of Owens-Corning Fiberglas Corporation.

[a]Used in the Arkansas House, Figure 3.9, page 000.

NOTE: Table 4.13 gives another schedule of *R*-Values from *R*-4 to *R*-24. When using Table 4.19 and also Table 4.13, other approximate thicknesses may be determined by interpolation and extrapolation.

4.13 Effects of Air Motion

When a wall or roof is in place to enclose a room under conditions resulting in heat loss from the space, the gentle motion of the nominally *still* air within the room and the more active motion of the wind outside of the room both act to increase the rate of heat loss. The room air is of course, higher in temperature than the inside surface of the room. The convection currents within the room cause the warmer air particles to collide wth the cooler surface. The resulting surface conductance, called h_i (i for interior), is least on the floor and increases slightly for the walls and ceiling. The outdoor air temperature is less than that of the outside surface of the structure. When the wind blows these cooler air particles against this warmer exterior surface, the heat loss rate is increased. This conductance factor is called h_o (o for outside). The factors are for the number of British thermal units per hour passing through 1 sq ft of surface for 1° F difference in temperature. Thickness is not involved.

In using Table 4.5*A*, one will most often use the value for nonreflective surfaces (common building materials), since few of the outside or inside surfaces in buildings are highly reflective. Emissivity (or emittance) (E) is therefore listed as 0.90. It is seen that, for still air (interiors), h_i is 1.46 for walls. The value of h_o is found to be 6.00.

4.14 Transmission Through Building Units

It is only coincidental when building materials or products are produced in exactly 1-in. thickness. Determining the conductance factor C of *homogeneous* materials not 1 in. thick was examined in Section 4.11 and Figure 4.14. For instance, the conductance C value for a 4-in.-thick wall of face brick[3] is $k/x = 9/4 = 2.24$ Btuh for 1 sq ft when 1° F temperature difference exists.

Table 4.6 lists not only conductivities k but also conductances C for homogeneous materials of thicknesses other than 1 in. Their resistances, $1/k$ and $1/C$, are also tabulated. In addition, the table shows C values for products such as hollow concrete block. The C values of composite products of two or more materials may be obtained from literature of manufacturers who usually publish these values. They may also be calculated as shown in Figure 4.16. They do not include the effect of surface conductances because the future placement of these products in a specific structure is not known. For completeness of presentation, the example in Figure 4.16 begins with k values, but whenever convenient the resistance values may be read from Table 4.6 and then added to find the total resistance.

4.15 Overall Coefficients of Heat Transmission

The rate of heat flow through a roof, wall, or floor is known as the overall transmission coefficient U. This and some other coefficients that affect it are defined as follows, for conditions of steady-state flow of heat:

U = *Overall Coefficient of Heat Transmission.* The British thermal units per hour flow-

[3]The k value of face brick is 9 (see Table 4.6).

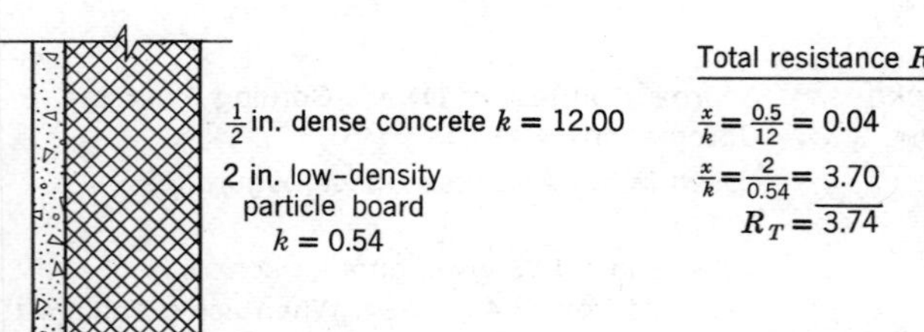

Fig. 4.16 Method of computing resistance and conductance of a manufactured composite building product. Conductance C is a reciprocal of the total resistance R_T. k-values from Table 4.6, page 111.

ing from air to air through 1 sq ft of roof, wall, floor, or other building component in place in the structure under actual conditions for a difference of 1° F in temperature between the air on the inside and the air on the outside. It can apply to a combination of materials or to a single material such as glass.

k = *Conductivity.* The Btuh rate of heat flow through 1 sq ft of a homogeneous material 1 in. thick for a 1° F difference in temperature between its two surfaces.

C = *Conductance.* The Btuh rate of heat flow through 1 sq ft of a homogeneous material or a combination of materials for 1° F difference in temperature between the exterior surfaces; for the thickness of construction stated, not necessarily per inch of thickness.

a = *Air-space conductance.* The Btuh rate of heat flow through 1 sq ft of area for 1° F difference in temperature between the bounding surfaces. It is affected by position and by the emissivity E of the surfaces.

h = *Film or surface conductance coefficient.* The rate of heat flow in Btuh through 1 sq ft of surface due to the motion of air against the surface, for 1° F difference in temperature. h_i is the symbol for the inside film coefficient and h_o the symbol for the outside film coefficient. These coefficients reflect the speed at which the air strikes the surface.

E = *Emissivity (sometimes ϵ for emittance).* The effective thermal emission (or absorption) of the surfaces bounding an air space.

R = *Thermal resistance.* Its value is obtained from the reciprocal of heat transfer as expressed by coefficients such as U, C, h, or a. It may be expressed as hours per British thermal units for the standard square foot of surface and 1° F temperature difference. For example, a wall with a U-coefficient of 0.25 would have a thermal resistance R of $1/0.25 = 4.0$.

The U-values in Tables 4.7 to 4.12 and modified for insulation as indicated in Table 4.13 may be used directly in heat loss calculations. These values are developed from basic data that appear in Tables 4.4, 4.5, and 4.6. The U-values of other constructions may also be calculated from these tables. Space limitations permit only a few of the most common U-values to be reprinted in this book. The designer who is continuously engaged in heat loss problems and who deals with a variety of structures is referred to the *Handbook of Fundamentals, 1977* of the American Society of Heating, Refrigerating and Air Conditioning Engineers, from which much of this information has been adapted.

In computing U-values it is evident that component heat transmissions are not additive. The value of the overall coefficient is always less than that of any of its parts. Greater thickness, more parts, insulation, and air spaces all serve to lower the overall coefficient of transmission. To arrive at this coefficient, it is necessary to add the resistances of the various elements, including film coefficients. This results in a total resistance, R_T, which may be expressed in total hours for the passage of 1 Btu through the construction. The reciprocal of this total resistance, stated as British thermal units per hour, is the overall U-coefficient of transmission.

Tables in *ASHRAE Handbook of Fundamentals, 1977* emphasize the importance of filling voids with blanket insulation. In frame construction of flat roofs, it was often previously considered sufficient to provide minimal rigid "deck" insulation above the plywood deck. Another item of interest appears in the ASHRAE tables. Overall transmission (U-value) at wood joists (or studs) can be different *at* the wood structural members than between them. Energy conservation now prompts a comparison of these two paths of heat loss. From the *approximate* values shown in Figures 4.12*a* and 4.15, we see that the value at the wood structural element was not shown. Table 4.11 for a flat roof in frame construction takes into account both improved insulation and values at joists. This suggests an analysis of Table 4.11 as a basis for the following example:

EXAMPLE 4.1. From Table 4.11, tabulate the U-coefficients for Construction 1 and Construction 2 (which adds R-19 insulation to replace two other items of lesser insulative value). Tabulate also the values at joists and *between* joists. Include the weighted *average* value in each case for between-joist and at-joist U-coefficients. Calculate the savings effected by the change to Construction 2.

SOLUTION

	Construction 1	Construction 2
Between joists	$U = 0.169$	$U = 0.042$
At joists	$U = 0.071$	$U = 0.079$
Weighted average	$U = 0.159$	$U = 0.046$

Energy savings, from Construction 2 over Construction 1, are

$$\frac{0.159 - 0.046}{0.159} = 0.711 \text{ (about 71\%)}$$

COMMENTS. When space permits, insulation can be added to reduce U-coefficients to 0.050 or less and save 70 to 80% of the heat loss through uninsulated or *poorly* insulated construction. The Arkansas House (Fig. 3.9, page 66) employs R-38 insulation above the living story, twice as much as is indicated in this example (see Table 4.19).

For the purpose of planning and decision-making in the early stages of architectural design, Table 4.13 can be useful. A quick approximation is obtainable. This is illustrated by the following example:

EXAMPLE 4.2. A wood-frame, uninsulated-roof construction has a U-coefficient of 0.25. What will be its approximate U-coefficient if R-20 (about $6\frac{1}{2}$-in. thick) insulation is added.

SOLUTION. From the value in the left-hand column of Table 4.13, go right to the vertical column at R-20. It is found that the revised coefficient is 0.04. The heat saving in Btuh per square foot will be

$$\frac{0.25 - 0.04}{0.25} = 84\% \text{ energy saving}$$

Although satisfactory for the use stated, note that, for accuracy, the average of U-values at and between joists should be considered as well as the adding or eliminating of air spaces in the construction. However, with the insulation thicknesses of 6 in. or more, the effect of these two items is relatively small. The present availability of water-resistant rigid fiber insulations should encourage the insulation of *masonry* walls. This could become as common as the present universal use of mineral wool batts in stud and joist spaces in frame construction. Figure 4.12*d* illustrates this.

4.16 Glass, Windows and Doors

As seen in Figure 4.12, the U-coefficients of single, double, and triple layers of glass are about 1.13, 0.61, and 0.39, respectively. Multiple layers with air spaces between are sometimes referred to as *insulating* glass. Transmission through *windows* is affected by *net* glass area and type of frame (see Table 4.16).

Reflective glass, so important in resisting heat *gain* (see Section 4.27), has little special value in resisting heat *loss*, the critical value of which is commonly calculated for the hours of darkness when sun is not a factor.

The U-coefficients of exterior doors and storm doors are listed in Table 4.17.

4.17 Heat Loss at Edge of Slabs

The edges of concrete slabs laid on the ground or on a gravel base-layer should be insulated. Suggested details are shown in Figure 4.17. The heat loss is not expressed as a U-coefficient but instead as the loss in Btuh per *linear* foot (lin ft) of perimeter *edge*. Also it is not evaluated for a temperature difference between indoor and outdoor air. It is, however, related to *outdoor* air temperature under conditions of normal winter occupancy. Thus, for an outdoor design temperature of 0° F, Table 4.15 indicates a heat loss of 40 Btuh/lin ft of exposed edge, when 2-in.-thick insulation is used. Rigid urethane insulation is usually chosen. Its k-value is 0.13.

4.18 Infiltraton and Ventilation

Winter winds blow cold outdoor air into indoor spaces through cracks around windows and doors on the windward side of the house. After this air has been warmed by heating

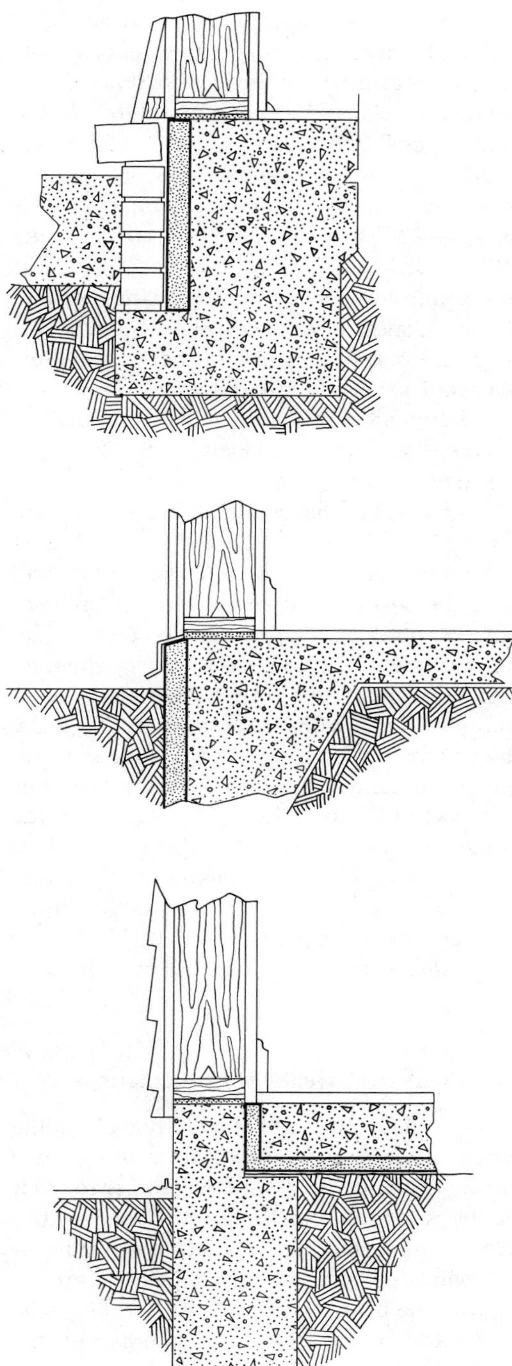

Fig. 4.17 Three appropriate details for rigid edge-insulation of slabs on grade. Courtesy of Owens-Corning Fiberglas Corporation. Reprinted by permission from Guide to Constructing an Energy-Efficient Home.

units near these locations, the warmed air leaves through similar cracks on the leeward side, where there is usually a reduced air pressure on the outside, creating a suction. This process is costly in energy because fuel is required for the warming operation. Losses due to infiltration are part of the heat loss calculations. Weather stripping reduces the rate of air infiltration.

The basis for calculating hourly heat loss due to infiltration is the number of cubic feet per hour (cfh) of outdoor air that must be heated from outdoor design temperature to room temperature. There are two methods of calculating this rate in cubic feet per hour, as shown in Table 4.18. The crack method, Table 4.18(*a*), is the more accurate of the two but requires tests or good judgment concerning the air pressure difference between outdoor and indoor conditions. This is measured in "inches of water" in a manometer device—a rather special engineering measurement. From pressure differences 0.10 to 0.50 in., the cubic feet of air entering per foot of crack can vary by as much as three to five times.

For a long time the air change method [Table 4.18(*b*)] was considered quite satisfactory for *residences*. It is still valid, though a bit on the generous side in Btuh estimating.

EXAMPLE 4.3. Using the air change method, find the rate of infiltration in cubic feet per hour of outdoor air entering a room that is 10 ft × 15 ft × 8 ft high, with openings on one side.

SOLUTION. Table 4.18(*b*) shows that, for rooms with windows on one side only, the rate of probable infiltration should be one air change per hour. Therefore, the rate would be $10 \times 15 \times 8 \times 1 = 1200$ cfh. After the rate of flow of outdoor air into the space is established, the calculation for hourly heat loss as a result of infiltration is as follows: The heat required to raise 1 lb of air 1° F is 0.24 Btu. This is the value of specific heat for air. The density of air may be assumed to be represented by an average value of 0.075 lb/cu ft. If, for example, the outdoor and indoor temperatures are 0° and 70° F, respectively, the hourly heat loss in warming the air in the room described in this example would be

$$1200 \times 0.075 \times 0.24 \times 70 = 1512 \text{ Btuh}$$

Since the product of the two constants, density and specific heat, is $0.075 \times 0.24 = 0.018$, the expression usually used for heat lost by infiltration is

$$\text{cfh} \times 0.018 \times \Delta t = \text{Btuh} \qquad (1)^4$$

where Δt is the temperature difference between indoor and outdoor air.

The use of the *crack* method in the "Arkansas Story" is a very accurate way of showing the great reduction in infiltration due to tight weather stripping, caulking, sealing of frame-cracks, and the use of a storm sash. Table 4.22(*c*), page 149, shows an infiltration heat transfer factor of 22.1 for the Energy Conservation Construction House as compared to a factor of 34 for the Minimum Property Standards House, a saving of 35% in energy use.

For a 3 × 4 ft double-hung window, the length of crack between sash and frame and between upper and lower sash would be

$$\begin{aligned} 3 + 3 + 4 + 4 = {} & 14 \text{ between sash and frame} \\ & \underline{\ 3} \text{ where sash meet} \\ & 17 \text{ lin ft} \end{aligned}$$

4.19 Indoor and Outdoor Temperatures

In the design for a controlled environment, indoor temperature is only one of many items to be considered. Sections 4.1 to 4.7 deal with the various conditions involved in the total design. Since hourly heat loss rates from buildings with known *U*-values depend on the temperature differential between inside and outside air, some thought must be given to the selection of appropriate temperatures. The range of outdoor design temperatures in the United States extends from plus 40°F in southern Florida to minus 50°F in Montana, as shown in Figure 4.18. This map has been included to facilitate a general grasp of winter conditions throughout the United States. More detailed data may be found in Table C2, Appendix page 1299.

If one considers the example of a heated residence, where the indoor temperature is usually 72° F, the design differential in Florida will be 32° F and in Montana 122° F. It is reasonable to expect that the heating system in the Montana residence should have four times the output capacity of that in the Florida residence. The comparative yearly fuel use is additionally related to the number of degree-days in the two locations. See Example 5.3, page 183.

It should be understood that the temperatures given in Figure 4.18 are not the lowest ever recorded in each locality. These will often be several degrees lower. The recommended outdoor temperature for heating design represents the concensus of many heating designers about the correct temperature assumption for a satisfactory system. To make the system much larger, to cope with an infrequent critical day, is usually unnecessary. For instance, in Boston, Massachusetts, where the design temperature is 0° F, the temperature has been found to drop to minus 8° F once in 40 years. If, on these few days, the house temperature cannot be kept at 72° F, the slight discomfort will have to be tolerated.

4.20 Typical Heat Loss Calculations

The hourly rate of heat loss from a building when the indoor temperature is maintained during critically cold outdoor temperatures is the basis for the selection of a proper central heating unit. The net or minimum rating of the boiler or furnace must match or exceed slightly the heat loss rate of the building. The convectors or air registers, depending on the kind of systems used, must be of proper size in each space to make up for the heat loss in that space. Spaces are sometimes grouped to give the heat loss in a selected "zone" of control, with its own separate circulator pump, mo-

[4]The term Δt is commonly used as a symbol for temperature difference.

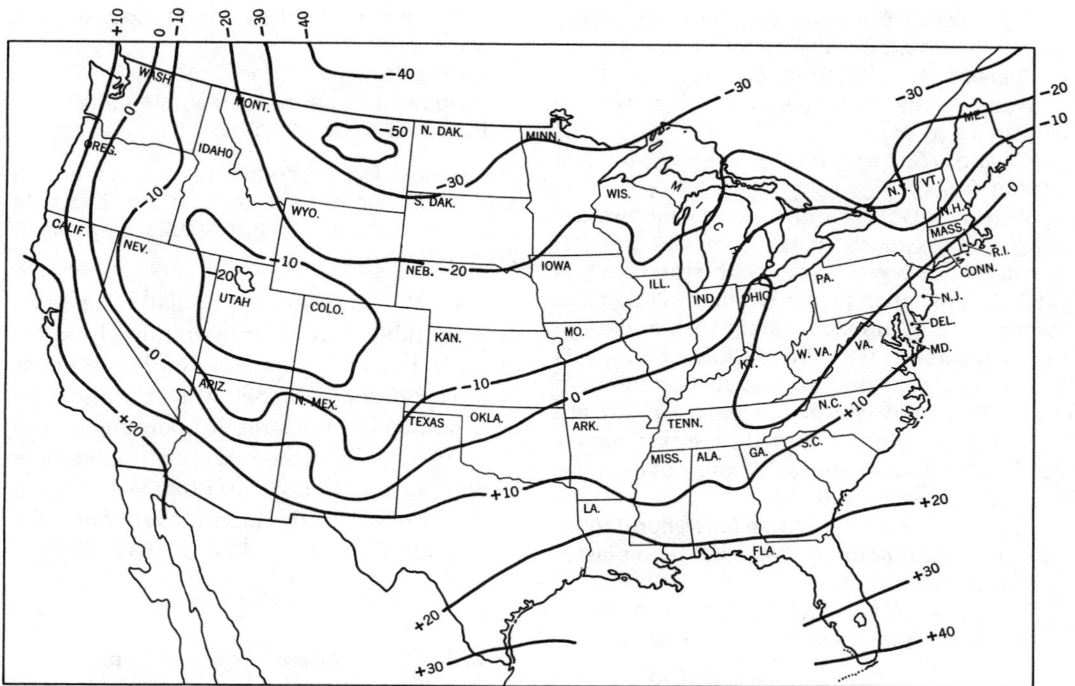

Fig. 4.18 Suggested outdoor temperatures for heating design. See also Table, Appendix C2, page 1299.

torized valve, fan, or even a separate heating unit.

Prior to the 1930s, little use was made of insulation as we know it today. Houses were frequently uninsulated. Storm sash, however, had an earlier history. These sash provided a second layer of glass that reduced heat loss by transmission. They also reduced cold air infiltration, which was excessive in the days of unweatherstripped windows.

In Example 4.4, excerpted from *ASHRAE Handbook of Fundamentals, 1977,* typical heat loss calculations are presented. In addition, the value of insulation is sharply emphasized. This is done by two sets of calculations. The first, Example 4.4(a) is for an old-style house with storm sash and unweatherstripped windows and no insulation. The second, Example 4.4(b), is for the same house with its storm sash and unweatherstripped windows, but with insulation in walls, attic walls, and roof; also in floors subjected to downward heat loss to colder space below. See Figure 4.19 and Tables 4.20 and 4.21.

EXAMPLE 4.4(a).[5] Calculate the heat loss rate at design conditions of the residence (Fig. 4.19) located in the vicinity of Syracuse, New York. From Table 1, Chapter 23, design outdoor conditions are −10° F and 11 mph wind speed. Indoor temperature is assumed to be 75° F. The attic is unheated. Assume ground temperature to be 50° F under basement and garage floors, and 32° F adjoining basement walls. Estimate infiltration losses by the air change method. No wall, ceiling, or roof insulation is to be considered in this problem, but all first- and second-floor windows, except in the garage, are to have storm sash. The building is constructed as follows [overall heat transfer coefficients in Btuh/(ft^2)(°F) are in parentheses]:

Walls: Brick veneer, building paper, wood sheathing, studding, metal lath, and plaster (0.29). Walls of dormer over garage are the same, except wood siding is used in place of brick veneer (0.26).

[5] Except for references to Figure 4.19 and Tables 4.20 and 4.21, all other references are to chapters and tables in *ASHRAE Handbook of Fundamentals, 1977.*

Attic Walls: Brick veneer, building paper, wood sheathing on studding (0.42).

Basement Walls: 10-in. concrete (0.10).

Roof: Asphalt shingles on wood sheathing on rafters (0.44).

Ceiling (Second Floor): Metal lath and plaster (0.74).

Windows: Double-hung wood windows averaging 70% glass (0.45; from Chapter 22, *U*-value for wood windows with storm sash is 0.53 × application factor; by interpolation this factor is 0.85). Steel casement sash in garage and basement (1.13; from Chapter 22, *U* is 1.13 for all glass and the application factor is 1.00). French doors in the dining room are 50% glass, no storm doors (0.85; from Chapter 22, *U* is 1.13 for all glass; by interpolation the application factor is 0.75).

Floor (Bedroom D): Maple finish flooring on yellow pine subflooring; metal lath and plaster ceiling below (0.26).

Floor (Basement and Garage): 4-in. stone concrete on 3-in. cinder concrete (0.10).

SOLUTION. Calculations for this problem are in Table 4.20, and a summary of the results is in Table 4.21. The values in Column F of Table 4.20 were obtained by multiplying together the figures in Columns C, D, and E. The heat losses are calculated to the nearest 10 Btuh. See reference notes for Table 4.20 for further explanation of the data.

Attention is called to the summary of heat losses (Table 4.21) for the uninsulated residence. As storm windows are used in this instance, the *glass* and *door* transmission heat losses of 19.8% are relatively small. The infiltration losses of 14.0% are also comparatively small, because the storm windows are equivalent to weatherstripping. In this problem, the wall, ceiling, and floor transmission losses comprise 66.2% of the total.

EXAMPLE 4.4(b). Calculate the heat loss of the residence in Fig. 4.19, based on the same conditions as in Example 4.4(a), but having construction improved or insulated to obtain overall heat transfer coefficients in Btuh/(ft²)(F) as follows:

Walls: 0.13
Walls of Dormer over Garage: 0.12
Attic Walls: 0.28
Walls Adjoining Garage: 0.18
Basement Walls (Recreation Room): 0.10
Roof: 0.53
Ceiling (Second Floor): 0.15
Windows [Same as in Example 4.4(a)]
Floor (Bedroom D): 0.18

SOLUTION. The procedure for calculating the heat losses is similar to that for Example 4.4(a). A summary of the results is given in Table 4.21(*b*).

Energy savings due to insulation are evident in Table 4.21. The operating[6] total for the heat loss of the uninsulated house under design conditions is 99,890 Btuh. For the insulated house, the comparative operating total is 61,170 Btuh. The saving due to insulation is 99,890 − 61,170 = 38,720 Btuh. With respect to the heat loss of the uninsulated house, the savings are 38,720 ÷ 99,890 = 0.39; 39%.

4.21 Attics, Basements, Crawl Spaces and Slabs

Attics are much less popular than formerly, when they served as storage areas. Thus, heavy insulation directly above the top floor is now most common. When basements are used, their exterior walls should be well insulated. Loss above outdoor grade in basement walls is calculated in the usual way. Below grade, the loss is very small. Old standards commonly suggested values of 4 Btuh/sq ft for walls and 2 Btuh/sq ft for floors in below-grade space regardless of the outdoor *design* air temperature. Crawl spaces should be ventilated and provided with maximum insulation in the floor joists above. For concrete slabs on grade, edge insulation, already discussed, is standard. Common practice in our new energy-oriented

[6]The operating total heat loss for an entire house is the design total less one-half the value of the total infiltration in all rooms. The reason for reducing the infiltration for the entire house is the fact that the wind affects, at one time, only the rooms on the windward side (about one-half) of the house. The boiler or furnace is selected on this basis. The heating unit in any room, however, must be adequate enough to offset the heat loss by infiltration when, in changing wind patterns, *that* room is on the windward side of the house.

design is to *enclose* all living volume with insulated constructon above, below, and on all sides. Refer to Figures 3.9, page 66, and 4.20.

4.22 Heat Transfer Factors

These are values that are part of simplified heat loss (and heat gain) computations. The intricacy of the ASHRAE calculation methods, even when applied to a simple residence, was evident in Example 4.4. The calculations on large buildings become very complex. Fortunately, there are simpler methods for those of us who must deal with the problem of evaluating the heat loss of *residences*. These methods are sufficiently accurate and most useful to architects and contractors when the services of a consulting engineer are not required. They are used in the "Arkansas Story,"

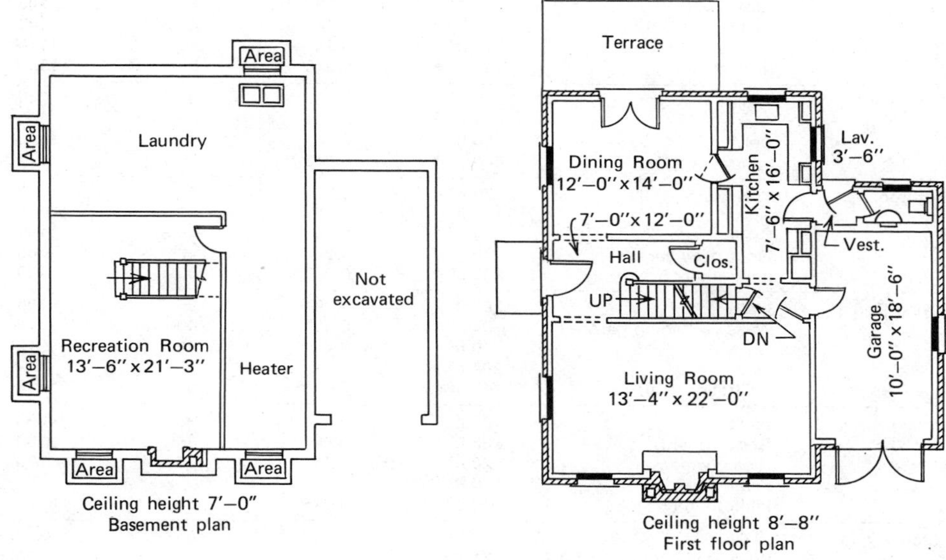

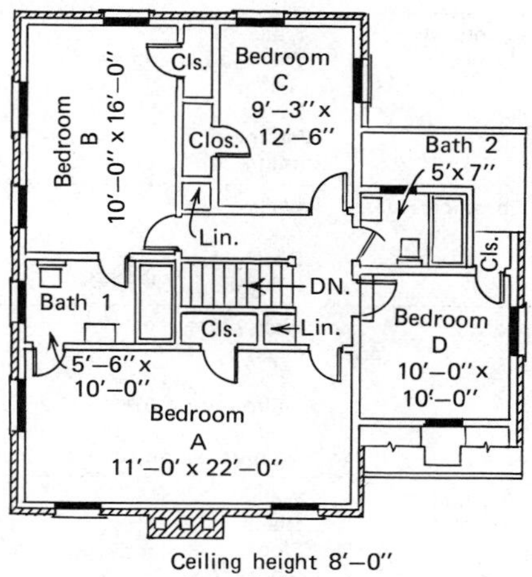

***Fig. 4.19** (Example 4.4). Floor plans of residence. Copyright © by the American Society of Heating, Refrigerating and Air Conditioning Engineers, Inc. Excerpted by permission from* ASHRAE Handbook of Fundamentals, 1977.

For details of heat loss calculations, see Table 4.20 and for summaries of room-by-room losses for uninsulated and insulated construction, see Table 4.21.

Table 4.20 [Example 4.4(a)] *Heat Loss Calculation Sheet for Uninsulated Residence (Fig. 4.19)*

A *Room or Space*	B *Part of Structure or Infiltration Air Changes*	C *Net Exterior Area and Air Volume*	D *U-value Coefficient, (Btuh)/(ft²)(F)*	E *Temp. Diff.,*[a] *F*	F *Heat Loss, (Btuh)*	G *Totals, (Btuh)*
Bedroom A and Closet	Walls	238 ft²	0.29	85	5870	
	Glass	40 ft²	0.45	85	1530	
	Ceiling	252 ft²	0.74	44.8[d]	8350	
	Infiltration (1)[g]	2016 cfh[b]	0.018[c]	85	3080	18,830
Bedroom B and Closet	Walls	156 ft²	0.29	85	3840	
	Glass	40 ft²	0.45	85	1530	
	Ceiling	170 ft²	0.74	44.8[d]	5630	
	Infiltration (1)[g]	1360 cfh[b]	0.018[c]	85	2080	13,080
Bedroom C and Closet	Walls	114 ft²	0.29	85	2810	
	Glass	27 ft²	0.45	85	1030	
	Ceiling	129 ft²	0.74	44.8[d]	4280	
	Infiltration (1)[g]	1032 cfh[b]	0.018[c]	85	1580	9,700
Bedroom D and Closet	Walls	118 ft²	0.29	85	2910	
	Glass	20 ft²	0.45	85	770	
	Ceiling	110 ft²	0.74	44.8[d]	3640	
	Floor over garage	110 ft²	0.26	40[e]	1140[p]	
	Infiltration (1)[g]	880 cfh[b]	0.018[c]	85	1350	9,810
Bathroom 1	Walls	30 ft²	0.29	85	740	
	Glass	14 ft²	0.45	85	540	
	Ceiling	55 ft²	0.74	44.8[d]	1820	
	Infiltration (1)[g]	440 cfh[b]	0.018[c]	85	670	3,770
Bathroom 2	Walls	79 ft²	0.26	85	1750	
	Glass	9 ft²	0.45	85	340	
	Ceiling	35 ft²	0.74	44.8[d]	1190	
	Floor over garage	35 ft²	0.26	40[e]	360	
	Infiltration (1)[g]	280 cfh[b]	0.018[c]	85	430	4,070
Living Room	Walls	267 ft²	0.29	85	6580	
	Walls (adjoining garage)	94 ft²	0.39[f]	40[e]	1470	
	Glass	50 ft²	0.45	85	1910	
	Floor	294 ft²				
	Infiltration (1.5)[h]	3745 cfh[b]	0.018[c]	85	5730	15,690
Dining Room	Walls	166 ft²	0.29	85	4090	
	Glass (doors)	35 ft²	0.85	85	2530	
	Glass (windows)	20 ft²	0.45	85	770	
	Floor	168 ft²				
	Infiltration (1.5)[i]	2140 cfh[b]	0.018[c]	85	3270	10,660
Kitchen and Entrance to Garage	Walls	96 ft²	0.29	85	2370	
	Walls (adjoining garage)	51 ft²	0.39[f]	40[e]	800[p]	
	Glass	18 ft²	0.45	85	690	
	Door	17 ft²	0.51	40	350	
	Floor	125 ft²				
	Infiltration (1.5)[j]	1595 cfh[b]	0.018[c]	85	2440	6,650
Lavette and Vestibule	Walls	82 ft²	0.29	85	2020	
	Walls (adjoining garage)	85 ft²	0.39[f]	40[e]	1330[p]	
	Glass	9 ft²	0.45	85	340	
	Door	19 ft²	0.51	85	820	
	Floor	30 ft²				
	Infiltration (1.5)[k]	383 cfh[b]	0.018[c]	85	590	5,100
Entrance Hall	Walls	39 ft²	0.29	85	960	
	Door	21 ft²	0.38	85	680	
	Ceiling[r]	87 ft²	0.74	44.8[d]	2990	
	Infiltration (2)[l]	1110 cfh[b]	0.018[c]	85	1700	6,330
Garage	Walls	167 ft²	0.29	50[e]	2420	
	Glass	53 ft²	1.13	50	3000	
	Doors	44 ft²	0.51	50	1120	
	Infiltration (1.5)[m]	2360 cfh[b]	0.018[c]	50	2120	
	Floor	29 ft[s]	0.81	50	1180	
	Gain adjoining rooms				−5100[p]	4,740
Recreation Room[q]	Walls	220 ft²	0.10	43	950	
	Glass	8 ft²	1.13	85	770	
	Floor	287 ft²	0.10	25	720	
	Infiltration (1)[n]	2010 cfh[b]	0.018[c]	85	3080	5,520
					TOTAL	113,950

Table 4.20 [Example 4.4(a)] *Reference Notes for Calculations*

[a] The indoor-outdoor temperature difference is [75 −(−10)] or 85 F, except where otherwise noted.

[b] Volume of infiltration, cfh = (number of air changes) × (floor or ceiling area) × (ceiling height).

[c] From Eq. 6.

[d] The ceiling heat losses are calculated by estimating the attic temperature and then calculating the loss through the ceiling using the proper temperature difference. This unheated attic is not ventilated during winter months. The attic temperature is estimated from Eq. 2 to be 30.2 F when the outdoor temperature is −10 F and room temperature is 75 F. The temperature difference is then (75 − 30.2) or 44.8 F. For the insulated residence, the attic temperature becomes 4.6 F and the temperature difference, 75 − 4.6 = 70.4 F.

[e] Temperature in garage selected to be 35 F.

[f] Coefficient for wall adjoining garage calculated on the basis of metal lath and plaster on both sides of studs [U = 0.39 Btuh/(ft^2)(F)].

[g] Two-thirds of value from Table 1, Chapter 21, for storm windows or weather-stripping.

[h] Exposed on two sides; weatherstripped windows offset by fireplace. Use 1.5.

[i] Window on one side weatherstripped but double-doors are hard to close tightly. Hence conservative value of 1.5.

[j] Assuming kitchen vent door to vestibule usually open, allow full table value of 1.5.

[k] Value in Table 1, Chapter 21, increased to 1.5 by nearby outside door in the vestibule.

[l] Value in Table 1, Chapter 21.

[m] Two sides exposed; large doors but large volume. Use value 1.5, as given in Table 1, Chapter 21.

[n] Two small unweatherstripped windows in protected location, but fireplace, indicate one change.

[p] Heat losses from these rooms into garage are heat gain for garage.

[q] Neglect heat loss to basement, as losses from boiler, piping, etc., will probably keep basement near, if not above, 75 F.

[r] Upstairs hall ceiling figures with downstairs. Heat should be provided downstairs for both.

[s] Linear feet of exposed edge.

Author's note. In the above notes, all references to chapters and tables are to those in *ASHRAE Handbook of Fundamentals, 1977.*

Table 4.21 [Example 4.4(a) and (b)] ***Comparison of Heat Losses for Uninsulated and Insulated Residence***

(a) Summary of Heat Losses of Uninsulated Residence (Btuh)

Room or Space	Walls	Ceiling and Roof	Floor	Glass and Door	Infiltration	Totals
Bedroom A	5870	8350	—	1530	3080	18,830
Bedroom B	3840	5630	—	1530	2080	13,080
Bedroom C	2810	4280	—	1030	1580	9,700
Bedroom D	2910	3640	1140	770	1350	9,810
Bathroom 1	740	1820	—	540	670	3,770
Bathroom 2	1750	1190	360	340	430	4,070
Living Room	8050	—	—	1910	5730	15,690
Dining Room	4090	—	—	3300	3270	10,660
Kitchen	3170	—	—	1040	2440	6,650
Lavette	3350	—	—	1160	590	5,100
Entrance Hall	960	2990	—	680	1700	6,330
Garage	−1180[a]	−1500[b]	1180	4120	2120	4,740
Recreation	950	—	720	770	3080	5,520
Design Totals	37,310	26,400	3,400	18,720	28,120	113,950
Operating Totals[c]	37,310	26,400	3,400	18,720	14,060	99,890
Percentages[d]	37	27	3	19	14	100

[a]Wall heat loss of 2420 Btuh minus wall heat gains of 1470, 800, and 1330 Btuh.

[b]Heat gains of 1140 and 360 Btuh.

[c]Based on 0.5 computed infiltration.

[d]Based on operating totals.

(b) Summary of Heat Losses of Insulated Residence (Btuh)

Room or Space	Walls	Ceiling and Roof	Floor	Glass and Door	Infiltration	Totals
Bedroom A	2620	2770	—	1530	3080	10,000
Bedroom B	1720	1870	—	1530	2080	7,200
Bedroom C	1260	1470	—	1030	1580	5,340
Bedroom D	1300	1220	790	770	1350	5,430
Bathroom 1	330	610	—	540	670	2,150
Bathroom 2	870	280	250	340	430	2,170
Living Room	3630	—	—	1910	5730	11,270
Dining Room	1830	—	—	3300	3270	8,400
Kitchen	1430	—	—	1040	2440	4,910
Lavette	1520	—	—	1160	590	3,270
Entrance Hall	430	960	—	680	1700	3,770
Garage	−580[a]	−1040[b]	1180	4120	2120	5,800
Recreation	950	—	720	770	3080	5,520
Design Totals	17,310	8,140	2,940	18,720	28,120	75,230
Operating Totals[c]	17,310	8,140	2,940	18,720	14,060	61,170
Percentages[d]	28	13	5	31	23	100

[a]Wall heat loss of 1080 Btuh minus wall heat gains of 680,370 and 610 Btuh.

[b]Heat gains 790 and 250 Btuh.

[c]Based on 0.5 computed infiltration.

[d]Based on operating totals.

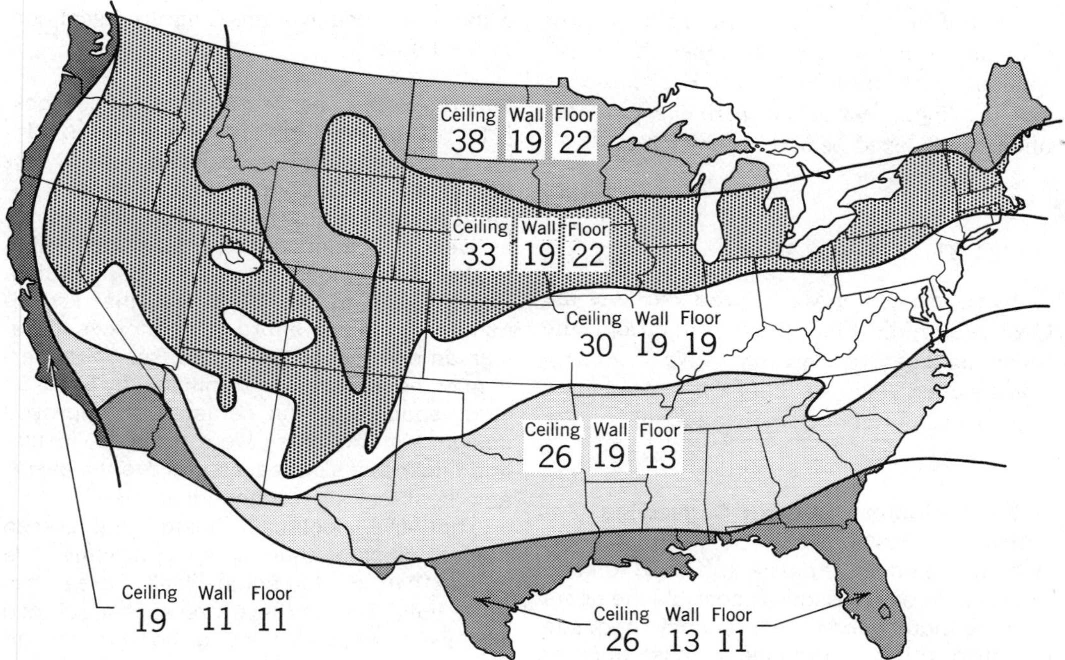

Fig. 4.20 *Recommended insulation* R-*values for ceilings, walls, and floors for indicated zones of the United States. Courtesy of Owens-Corning Fiberglas Corporation. Reprinted by permission from* Guide to Constructing an Energy-Efficient Home. k-*value of glass fiber insulation, 0.27.*

which follows in Section 4.23 and compares Energy Conservation Construction (Table 4.22) with the less efficient Minimum Property Standards (Conventional) Construction, (Table 4.23).

A heat transfer factor for transmission through building components such as glass, doors, walls, or ceilings is the product of the *U*-coefficient and the temperature difference between indoors and outdoors. Therefore, for an item such as glass, if the area of glass in the house is known, then the product of the glass area in square feet times the selected heat transfer factor is the heat loss through the glass in British thermal units per hour. It is not necessary to select a *U*-coefficient, but only to select a heat transfer factor by use of an established specification of the type of construction. Values for these heat transfer factors are found in Tables 4.22*c* and 4.23*c*. When chosen, it is then only necessary to multiply the heat transfer factor by the area of the building surface (e.g., wall and doors) to establish the heat loss. See Tables 4.22*b* and 4.23*b* for these heat loss values.

Therefore, in our *simplified* method, we do *not* use *U*-coefficients but only the corresponding heat transfer factors based upon the type of construction. The analysis below, however, indicates how they are established:

Item	*For Energy Conservation Construction See Table 4.22c*	*For Minimum Property Standards (Conventional) Construction See Table 4.23c*
	$U \times \Delta t$ = HTF	$U \times \Delta t$ = HTF
Glass	0.63 × 70° F = 44.3	1.13 × 70° F = 79.1
Doors	0.073 × 70° F = 5.1	1.13 × 70° F = 79.1
Wall	0.051 × 70° F = 3.6	0.083 × 70° F = 5.8
Ceiling	0.024 × 70° F = 1.7	0.051 × 70° F = 3.6

NOTE. In the tables mentioned in the above headings, you will find the heat-transfer-factor values marked for identification.

Heat losses by infiltration and by slab-edge loss are not analyzed in terms of square feet, as in the categories above. In our simplified method of heat loss, we use the linear feet of door and window crackage times the corresponding heat transfer factor. For slab-edge loss, the linear feet of slab *perimeter* is multiplied by *its* listed heat transfer factor.

4.23 The Arkansas Story

By permission of Mr. Stewart Byrne of the Owens-Corning Fiberglas Corporation, the following digest of their *Report No. 1: Energy Conservation Ideas to Build On* is presented. It is an account of outstanding energy savings.

A Demonstration in Energy Conserving Home Construction

With the competitive desire and need to keep the cost of homes as low as possible the home building industry has, until recently, had little incentive to inform prospective customers of the fact that heating and cooling costs for the life of the mortgage exceed the cost of preventing much of the energy loss. While low construction cost is still important, reduced energy consumption and costs have become essential to the nation and a marketable home benefit for the builder.

A forecast by Townsend-Greenspan made in the Fall of 1974, anticipated that between 1972 and 1985 the cost of No. 2 (residential) fuel oil would increase 613% and electricity 157%. No 1985 forecast on natural gas is available but natural gas for utility use was expected to rise 409% by 1981. Because of the unpredictable pressures on fuel prices these forecasts may change, but the trend is clear, energy for heating and cooling is going to increase by a much greater order of magnitude than we have experienced to date.

Homeowners and renters are already protesting loudly at today's increases in energy costs which at this point are only averaging 33% above 1973. If homes continue to be built to conventional pre-energy-crisis low insulation standards and energy costs continue rising beyond 1985, few will be able to afford their home heating and cooling costs by the turn of the century. The Arkansas study has indicated that by building homes designed for energy conservation we can avoid the need for such a life style of austerity and discomfort without necessarily increasing the initial cost of the home.

Why the Home was Developed. In Little Rock, Arkansas an electric utility, interested in adding customers to utilize existing capacity together with an air conditioning engineer dedicated to making heat pumps provide economy as well as comfort and a HUD construction analyst with innovative ideas in construction and a desire to provide homes that low income families can afford, joined forces to design an energy conserving house. Thirty percent of the framing was eliminated by using 6″ studs spaced 24″ on centers and improved design, window areas were reduced, insulation thicknesses increased and every measurable heat leak traced and eliminated.

Thirty-five electrically heated and cooled homes, most of them three bedrooms, two baths, 1040 to 1200 square feet in area have been built. Ten of these are each equipped with two meters so heating, cooling and air conditioning use can be measured separately from lighting and equipment. Homeowners vouch for their greater comfort and the quietness of these energy conserving homes. See Figures 4.21 and 4.22.

Owens-Corning Fiberglas has been asked for the facts by many of its customers. This is what we found out, this is The Arkansas Story.

NOTE. Wherever the expression "FHA, MPS, or Conventional Homes" is used in this report, it refers to homes built with 2″ × 4″ studs 16″ o.c. insulated with 6″ Blown Insulation (R-15) in the ceiling, $3\frac{1}{2}$″ R-11 batts in the walls and 1″ Styrofoam perimeter insulation around the slab. These homes had a window area of 11–12% of the total living area. Exterior doors were $1\frac{3}{4}$″ hollow core with storm doors. The houses had gable louvers without soffit vents.

The Objectives of the Energy Conservation Design

ASHRAE heat loss calculations used as the guide for thermo-dynamic heat loss calculations in utilizing cheap and plentiful energy are now under review by that association. We would not presume to anticipate the outcome of their review but bring up this point because one of the departures in the Arkansas project was the development of a new heat loss calculation chart based not just on thermodynamics but also on physiological factors.

Table 4.22a *Planning and Specification Form, Energy Conservation Construction*

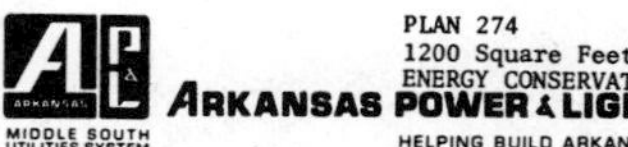

PLAN 274
1200 Square Feet
ENERGY CONSERVATION CONSTRUCTION

Residential Electric Comfort Conditioning

For ______________________ of ______________________
No. Street City

Built by ______________________ of ______________________
No. Street City

Comfort Conditioning By ______________________ of ______________________
No. Street City

Design Conditions

Winter		Summer
5 °F	Outdoor Temperature	100 °F
75 °F	Indoor Temperature	75 °F
70 °F	Temperature Differential	25 °F
____ %	Indoor Relative Humidity	____ %

Structural Characteristics

Gross Exposed Wall Area	1331 sq. ft.	Glass Area	5.6 % Floor Area
Ceiling Area Outside Dim.	1200 sq. ft.	Total Heat Loss	16159 BTUH
Floor Area Inside Dim.	____ sq. ft.	Unit Heat Loss	13.5 BTUH/Sq. Ft.
Total Glass Area	____ sq. ft.	Total Heat Gain	10,490 BTUH

Structure Faces ____

Equipment to be Installed

Heat Pump & Supplementary: Compressor 19,200 BTUH, 600 CFM, And Supplementary ____ BTUH
Manufacturer & Model No. ______________________

Resistance Heating: Type ______________; Capacity ____ BTUH And ____ CFM
Manufacturer & Model No. ______________________

Cooling Equipment: Capacity ____ BTUH And ____ CFM
Manufacturer & Model No. ______________________

Filter Equipment: Type Electrostatic
Manufacturer & Model No. ______________________

Humidifier: Type ______________ Rating ____ Gal./Day
Manufacturer & Model No. ______________________

Dehumidifier: Type Mechanical Rating 4 Gal./Day
Manufacturer & Model No. ______________________

Attic Ventilation: Rating 790 CFM; Control Thermostatic
Manufacturer & Model No. ______________________

Estimated Cost of Operation

	DESIGN NO. 1	DESIGN NO. 2	DESIGN NO. 3
Heating Season of ___ Degree Hrs. ___ KWH @ ___¢ =	$____.___Yr.	$____.___Yr.	$____.___Yr.
Cooling Season of ___ Degree Hrs. ___ KWH @ ___¢ =	$____.___Yr.	$____.___Yr.	$____.___Yr.
Heating & Cooling	$____.___Yr.	$____.___Yr.	$____.___Yr.
Estimated Cost of Total Electrical Usage of This Home Appliances, Lighting, Heating & Cooling With Levelized Billing	$____.___Mo.	$____.___Mo.	$____.___Mo.

Calculation of BTUH heat gain and loss are based on thermal control as follows: Insulation of *R* ____ ceiling, *R* ____ walls, *R* ____ floor; vapor barrier type ____ ceiling, ____ walls, ____ floor; ground cover type ____; kitchen exhaust ____ manual, ____ auto.; other ______________________

Any change in thermal protection will affect the equipment size and estimated operating cost.

Date ______________ Signature ______________

Table 4.22b ***Quantities, Heat Loss, Heat Gain, Energy Conservation Construction (Continued)***

Schedule 1. Load Calculations

1	Name of Room or Space		MBR	BATHS	DINING	LR	BR-2	BR-3		
2	Ceiling Height		9.5	9.5	9.5	9.5	9.5	9.5		
3	Linear-Ft. of Exposed Wall	(a)	28.5	28.5	37	30	11	25.5		
		(b)								
4	Gross Exposed Wall Area	(a)								
		(b)	271	76	352	285	105	242		
5	Window and Door Area	(a) Windows	10	0	21	15	9	9		
		(b) Doors			21	21				
6	Net Exposed Wall Area	(a)								
		(b)	261	76	310	249	96	233		
7	Window & Door Crackage—Linear Feet		16	0	51	39	15	15		
8	Ceiling Area (Outside Dimensions)		192	114	360	214	175	145		
9	Floor (Outside Dimensions)	(a) Sq-Ft Area								
		(b) Ln-Ft Perimeter (Slab)	28.5	8	37	30	11	25.5		
Heat Loss										
10	Of Windows And Doors	Sq-Ft (Line 5a) × 44.3 HTF	443	0	930	665	399	399		
		Sq-Ft (Line 5b) × 5.1 HTF			107	107				
11	Of Net Exposed Wall	Sq-Ft (Line 6a) × HTF								
		Sq-Ft (Line 6b) × 3.6 HTF	940	274	1116	896	346	839		
12	By Infiltration	Ln-Ft (Line 7) × 22.1 HTF	354	0	1127	862	332	332		
13	At Ceiling	Sq-Ft (Line 8) × 1.7 HTF	326	194	612	364	298	247		
14	At Floor	Sq-Ft (Line 9a) × HTF								
		Ln-Ft (Line 9b) × 22.7 HTF	647	182	840	681	250	579		
15	Of Structure	Lines 10 + 11 + 12 + 13 + 14	2710	650	4732	3575	1625	2396		
16	Equipment Requirements	WATTS								
		BTUH								
		CFM (Actual)	78	19	137	103	47	69		
Heat Gain										
17	Of Windows And Doors	Sq-Ft (Line 5a) × 15.8 HTF	158	0	332	237	142	142		
		Sq-Ft (Line 5b) × 1.9 HTF			40	40				
18	Of Net Exposed Wall	Sq-Ft (Line 6a) × 1.3 HTF	339	99	403	324	125	303		
		Sq-Ft (Line 6b) × HTF								
19	By Infiltration	Ln-Ft (Line 7) × 7.9 HTF	126	0	403	308	119	119		
20	At Ceiling	Sq-Ft (Line 8) × .7 HTF	134	80	252	150	123	102		
21	At Floor	Sq-Ft (Line 9a) × HTF								
		Ln-Ft (Line 9b) × 0 HTF								
22	Sensible	Lines 17 + 18 + 19 + 20 + 21	757	179	1430	1059	509	666		
23	Equipment Requirements	BTUH								
		CFM (Adjusted)	90	21	169	150	75	95		

				Design No. 1	Design No. 2	Design No. 3
				TOTAL		
				9.5		
				140		
				1331		
				64		
				42		
				1225		
				136		
				1200		
				140		

				2836		
				214		
				4411		
				3007		
				2041		
				3179		
				15628		
				453		

				1011		
				80		
				1593		
				1075		
				841		
				4600		
				600		

Schedule 2. Window & Door Schedule

Symbol	*Dimensions*	*No.*	*Type*	*Sq. Ft. Area Ea./Total*	*Ln. Ft. Crackage Ea./Total*
A	3-0 × 5-0	2	S.H.	15/30	19/3
B	3-4 × 3-0	1	S.H.	10/10	16/1
C	3-0 × 3-0	2	S.H.	9/18	15/3
D	2-0 × 3-0	1	S.H.	6/ 6	12/1
				/64	/9
				/	/
1	3-0 × 6-8	1	Therma-Tru	21/21	20/2
2	2-8 × 6-8	1	"	21/21	20/2
				/42	/4
				/	/
				/	/
				/	/
				/106	/13

Schedule 3. Solar Areas of West and South Elevations

Symbol	*West*		*South*	
	No.	*Total Area*	*No.*	*Total Area*
A	1	15		0
C	2	18		0
Totals		33		0

Use The Larger Of These Two Total Areas To Calculate Solar Gain

Table 4.22*b* continued on the following page.

Table 4.22*b* Continued
Summaries, Heat Gain and Heat Loss

Gains—Size Equipment On Cooling Load

	Design No. 1
Sensible Gain (Line 22, Schedule 1)	4600
Latent Gain, 30% Sensible Gain	1380
Solar Gain (Schedule 3), 29.2 × 33 sq. ft.........	965
(Assuming 35% shading coefficient.)	
People __6__ (Min-5) × .300 × 1.80..............	3240
Subtotal..	10185
Duct Gain __3__ % of Subtotal....................	305
Total BTUH Heat Gain	10490

Losses—Design Room CFM On Heating Load

Structural Loss (Line 15, Schedule 1).............	15688
Duct Loss __3__ % of Structural Loss	471
Total BTUH Heat Loss	16159

Energy Conservation Construction

Ceiling *R*-38

Walls *R*-19

Floor *R*-10.7 (Slab)

Storm Windows

Insulated Doors

Indicate North

OUTLINE PLAN OF STRUCTURE

Table 4.22c *Heat Transfer Factors (HTF), Energy Conservation Construction*

	Heat Transfer Factors (HTF)						
	Heat Loss				*Heat Gain*		
Design Temperature Differential (TD)	*75°*	*70°*	*65°*	*60°*	*30°*	*25°*	*20°*
Windows (Weather stripped)							
Single Pane	84.8	79.1	73.5	67.8	31.8	26.5	21.2
Single Pane and Storm Windows	47.5	44.3	41.2	38.0	19.0	15.8	12.7
Insulating Glass ($U = 0.61$)	45.8	42.7	39.7	36.6	18.3	15.3	12.2
Triple Glazed	31.5	29.4	27.3	25.2	12.6	10.5	8.4
Doors (Weather stripped)							
Hollow Core	84.8	79.1	73.5	67.8	31.8	26.5	21.2
Hollow Core and Storm Door	47.5	44.3	41.2	38.0	19.0	15.8	12.7
Solid Core ($1\frac{3}{4}''$)	36.0	33.6	31.2	28.8	14.4	12.0	9.6
Solid Core ($1\frac{3}{4}''$) and Storm Door	23.3	21.7	20.2	18.6	9.3	7.8	6.2
Therma-Tru *R*-13.8		5.1				1.9	
Infiltration (Doors and Windows)							
Without Storm Sash or Door	36.5	34.0	31.6	29.2	14.6	12.2	9.7
With Storm Sash or Door	23.7	22.1	20.5	19.0	9.5	7.9	6.3
Walls							
Frame Standard Sheathed No Insulation	18.2	16.9	15.7	14.5	7.3	6.1	4.9
Brick Standard Sheathed No Insulation	20.0	18.6	17.3	16.0	8.0	6.7	5.3
Brick or Frame Standard Sheathed: 2″ Insulation (R-7.0)	8.6	8.1	7.5	6.9	3.5	2.9	2.4
$3\frac{1}{2}''$ Insulation (R-11)	6.1	5.8	5.3	4.9	2.5	2.1	1.7
6″ Insulation (R-19)	3.9	3.6	3.4	3.1	1.6	1.3	1.1
2″ Insulation (R-7.0) Cavity, Impregnated Vermiculite Poured	7.5	7.0	6.5	6.0	3.0	2.5	2.0
$3\frac{1}{2}''$ Insulation (R-11) Cavity, Impregnated Vermiculite Poured	5.5	5.2	4.8	4.4	2.2	1.9	1.5
6″ Insulation (R-19) Cavity, Impregnated Vermiculite Poured	3.7	3.4	3.2	2.9	1.5	1.2	1.0
8″ Solid Wall Brick or Stone: No Insulation	50.3	46.9	43.6	40.3	20.1	16.8	13.4
Gypsum Wallboard, Furred With 2″ Insl. (R-7.0)	8.5	7.9	7.4	6.8	3.4	2.8	2.3
4″ Brick & 8″ Concrete Block: No Insulation	32.3	30.1	28.0	25.8	13.0	10.8	8.9
Blocks, Impregnated Vermiculite Poured	23.3	21.7	20.2	18.6	9.3	7.8	6.2
Standard Aggregate Concrete Block: 6″ Block No Insulation	42.6	39.8	36.9	34.1	17.1	14.2	11.4
8″ Block No Insulation	38.3	35.7	33.2	30.6	15.3	12.8	10.2
12″ Block No Insulation	35.2	32.9	30.5	28.2	14.1	11.7	9.4
6″ Block Impregnated Vermiculite Poured	27.0	25.2	23.4	21.6	10.8	9.0	7.2
8″ Block Impregnated Vermiculite Poured	25.5	23.8	22.1	20.4	10.2	8.5	6.8
12″ Block Impregnated Vermiculite Poured	24.8	23.1	21.1	19.8	9.9	8.3	6.6
Lightweight Aggregate Concrete Block: 6″ Block No Insulation	28.3	26.4	24.5	22.6	11.3	9.4	7.5
8″ Block No Insulation	25.8	24.1	22.4	20.7	10.3	8.6	6.9
12″ Block No Insulation	22.9	21.4	19.9	18.3	9.2	7.7	6.1
6″ Block Impregnated Vermiculite Poured	17.6	16.4	15.2	14.0	7.0	5.9	4.7
8″ Block Impregnated Vermiculite Poured	16.6	15.5	14.4	13.6	6.6	5.6	4.5
12″ Block Impregnated Vermiculite Poured	16.1	15.0	13.7	12.9	6.4	5.4	4.3
Ceiling							
Pitched Roof: 6″ Insulation (R-19.0) Standard Ventilation	3.9	3.6	3.4	3.1	2.1	1.8	1.6
8″ Insulation (R-25.3) Standard Ventilation	3.0	2.8	2.6	2.4	1.6	1.4	1.2
10″ Insulation (R-31.7) Standard Ventilation	2.4	2.3	2.1	1.9	1.3	1.1	1.0
12″ Insulation (R-38.0) Standard Ventilation	2.1	1.9	1.8	1.7	1.1	1.0	0.8
Hipped or Flat Roof: 6″ Insulation (R-19.0) Standard Ventilation	3.4	3.2	2.9	2.7	3.1	2.6	2.1
8″ Insulation (R-25.3) Standard Ventilation	2.7	2.5	2.3	2.1	2.4	2.0	1.6
10″ Insulation (R-31.7) Standard Ventilation	2.3	2.1	1.9	1.7	1.9	1.6	1.3
12″ Insulation (R-38.0) Standard Ventilation	1.8	1.7	1.6	1.5	1.7	1.4	1.1
Pitched, Hipped or Flat Roof: 6″ Insulation (R-19.0) Controlled Ventilation	3.4	3.2	2.9	2.7	1.5	1.3	1.0
8″ Insulation (R-25.3) Controlled Ventilation	2.7	2.5	2.3	2.1	1.4	1.2	0.9
10″ Insulation (R-31.7) Controlled Ventilation	2.3	2.1	1.9	1.7	1.0	0.8	0.6
12″ Insulation (R-38.0) Controlled Ventilation	1.8	1.7	1.6	1.5	0.8	0.7	0.5
Floor							
Double Wood and Crawl Space: 2″ Insulation (R-7)	4.8	4.5	4.2	3.9	1.5	1.2	1.0
$3\frac{1}{2}''$ Insulation (R-11)	3.3	3.1	2.9	2.6	1.3	1.1	0.9
6″ Insulation (R-19)	2.0	1.8	1.6	1.5	0.8	0.6	0.5
$1\frac{1}{2}''$ Urethane R-10.7		22.7				0	
Slab on Ground: 1″ Insulation (R-3.5) Standard Duct System	66.8	62.3	57.9	53.4	.0	.0	.0
2″ Insulation (R-7) Standard Duct System	35.8	33.4	32.6	31.8	.0	.0	.0
1″ Insulation (R-3.5) Perimeter Duct	199.3	186.9	173.7	160.2	.0	.0	.0
2″ Insulation (R-7) Perimeter Duct	107.4	100.2	97.8	95.4	.0	.0	.0

NOTE: (A) For additional factors, use the American Society of Heating, Refrigerating and Air-Conditioning Engineers, Inc. Guide & Data Book.

(B) All above factors include correctly applied vapor barriers.

(C) Above factors are based on blanket insulation; adjust for blown material.

(D) Boxes indicate *selected* HTFs.

Simply explained, even though a home may be warm, if its occupants' feet feel cold they will tend to raise the thermostats which wastes energy. A recent research project by the Electricity Council Research Center, England has substantiated this phenomenon. Studies on the life support systems for the astronauts have provided additional information on the air movement and humidity requirements for human comfort.

Although the developers of the Arkansas project were unaware of these independent studies, their research led to the same conclusions which explains why the owners of these new homes are so complimentary regarding the comfort they are experiencing in addition to the energy savings and costs.

The objective in Little Rock was to gain maximum control over the interior environment of the home. In other words, to isolate the interior environment from the variable exterior environment. By erecting a thicker and more continuous than usual insulation barrier to do this they were able to provide the desired comfort and the fuel saving. What was so ingenious about the way this was accomplished was that the new design is reported to cost no more to construct than the same size energy leaking conventional homes.

That the basic heating for these homes is electrical does not diminish the value of the specifications for other areas of the country. Due to the shortage of natural gas, it appears many new homes are going to have to be electrically heated. Gas lines are not being extended to the new development areas in many parts of the country. Due to the relatively air tight construction and polyethylene vapor barrier envelope, air consuming flame type heating and cooling was not considered in Little Rock.

Evolution of the Energy Conserving Home

The test homes ranged from 1040 to 1200 square feet of living area with three bedrooms and two full baths. Accurate measurement of the heating and cooling was recorded separately. The second meter recorded the total electric consumption of each home. Heating and cooling amounted to 12% to 32% of total energy consumption. Compared to the same size conventional homes, the *total* averaged monthly electric costs of the test homes were about 37% less.

Origin of the Concept. Built to new minimum construction standards for energy conservation homes issued by the Little Rock area office of HUD, and designed by Mr. Frank Holtzclaw, HUD Construction Analyst, in cooperation with the Arkansas Power and Light Company, these new homes were the culmination of 12 years of testing and experimentation.

During the late 1950's a series of research studies on owner satisfaction of conventional homes around Little Rock equipped with heat pumps indicated some were unhappy with the high energy costs and dissatisfied with the comfort provided. House by house inspection of the homes of the dissatisfied owners revealed that the problems were not due to the equipment but to insufficient or improper insulating, uninsulated ducts, construction changes after the equipment had been installed and lack of humidity control.

Mr. Harry Tschumi, President of Harry Tschumi Company, air conditioning wholesale distributor consulted with the late Mr. Les Blades, then heating and air conditioning coordinator for Arkansas Power and Light Company. Together they started testing the effects of changing the insulation thickness, insulating doors, using double glazed windows and measured the results. The conclusions were that a super insulated home with as small as practical heat pump would prove the most economical application and provide the most comfort. The problem was that they couldn't convince home builders or buyers to invest in the extra materials. Then came the fuel crisis with rocketing energy costs and a demand for lower utility costs.

The late Les Blades working with Frank Holtzclaw and using the earlier findings came up with an ingenious solution, a house design that is reported to have equal or greater strength yet uses less wood, more insulation and includes a number of cost saving innovations. It is claimed this construction costs no more than the conventional home construction it is designed to replace.

Details of the Super-Insulated Energy-Conserving Home

Wall Construction. The Little Rock HUD designed energy conservation home is built with

Table 4.23a ***Planning and Specification Form, Minimum Property Standards***

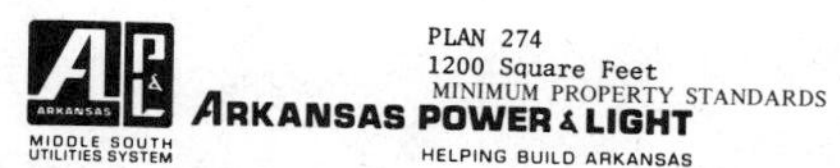

PLAN 274
1200 Square Feet
MINIMUM PROPERTY STANDARDS

Residential Electric Comfort Conditioning

For ______________________ of ______________________
No. Street City

Built by ______________________ of ______________________
No. Street City

Comfort Conditioning
By ______________________ of ______________________
No. Street City

Design Conditions

Winter		Summer
5 °F	Outdoor Temperature	100 °F
75 °F	Indoor Temperature	75 °F
70 °F	Temperature Differential	25 °F
____ %	Indoor Relative Humidity	____ %

Structural Characteristics

Gross Exposed Wall Area	____ sq. ft.	Glass Area	10.7 % Floor Area
Ceiling Area Outside Dim.	1200 sq. ft.	Total Heat Loss	46,542 BTUH
Floor Area Inside Dim.	1154 sq. ft.	Unit Heat Loss	38.8 BTUH/Sq. Ft.
Total Glass Area	124 sq. ft.	Total Heat Gain	27,576 BTUH

Structure Faces ____

Equipment to be Installed

Heat Pump & Supplementary: Compressor ____ BTUH, ____ CFM, And Supplementary ____ BTUH
Manufacturer & Model No. ______________________

Resistance Heating: Type ______________; Capacity ____ BTUH And ____ CFM
Manufacturer & Model No. ______________________

Cooling Equipment: Capacity ____ BTUH And ____ CFM
Manufacturer & Model No. ______________________

Filter Equipment: Type ______________________
Manufacturer & Model No. ______________________

Humidifier: Type ______________________ Rating ____ Gal./Day
Manufacturer & Model No. ______________________

Dehumidifier: Type ______________________ Rating ____ Gal./Day
Manufacturer & Model No. ______________________

Attic Ventilation: Rating ____ CFM; Control ____________
Manufacturer & Model No. ______________________

Estimated Cost of Operation

	DESIGN NO. 1	DESIGN NO. 2	DESIGN NO. 3
Heating Season of ___ Degree Hrs. ___ KWH @ ___¢ =	$____.___Yr.	$____.___Yr.	$____.___Yr.
Cooling Season of ___ Degree Hrs. ___ KWH @ ___¢ =	$____.___Yr.	$____.___Yr.	$____.___Yr.
Heating & Cooling	$____.___Yr.	$____.___Yr.	$____.___Yr.
Estimated Cost of Total Electrical Usage of This Home Appliances, Lighting, Heating & Cooling With Levelized Billing	$____.___Mo.	$____.___Mo.	$____.___Mo.

Calculation of BTUH heat gain and loss are based on thermal control as follows: Insulation of *R* ____ ceiling, *R* ____ walls, *R* ____ floor; vapor barrier type ____ ceiling, ____ walls, ____ floor; ground cover type ____; kitchen exhaust ____ manual, ____ auto.; other ______________________

Any change in thermal protection will affect the equipment size and estimated operating cost.

Date ______________ Signature ______________________

Table 4.23b *Quantities, Heat Loss, Heat Gain, Minimum Property Standards* (*Continued*)

Schedule 1. Load Calculations

			MBR	BATHS	DINING	LR	BR-2	BR-3		
1	Name of Room or Space		MBR	BATHS	DINING	LR	BR-2	BR-3		
2	Ceiling Height		9	9	9	9	9	9		
3	Linear-Ft. of Exposed Wall	(a)								
		(b)	28.5	8	37	30	11	25.5		
4	Gross Exposed Wall Area	(a)								
		(b)	271	76	352	285	105	242		
5	Window and Door Area	(a)								
		(b)	20	0	59	51	18	18		
6	Net Exposed Wall Area	(a)								
		(b)	251	76	293	234	87	224		
7	Window & Door Crackage—Linear Feet		32	0	72	58	30	30		
8	Ceiling Area (Outside Dimensions)		192	114	360	214	175	145		
9	Floor (Outside Dimensions)	(a) Sq-Ft Area								
		(b) Ln-Ft Perimeter (Slab)	28.5	8	37	30	11	25.5		
	Heat Loss									
10	Of Windows And Doors	Sq-Ft (Line 5a) × HTF								
		Sq-Ft (Line 5b) × 79.1 HTF	1582	0	4667	4034	1424	1424		
11	Of Net Exposed Wall	Sq-Ft (Line 6a) × HTF								
		Sq-Ft (Line 6b) × 5.8 HTF	1456	440	1700	1356	505	1300		
12	By Infiltration	Ln-Ft (Line 7) × 34 HTF	1088	0	2448	1972	1020	1020		
13	At Ceiling	Sq-Ft (Line 8) × 3.6 HTF	692	410	1296	770	630	522		
14	At Floor	Sq-Ft (Line 9a) × HTF								
		Ln-Ft (Line 9b) × 62.3 HTF	1776	498	2305	1869	685	1589		
15	Of Structure	Lines 10 + 11 + 12 + 13 + 14	6594	7348	12416	10001	4264	5855		
16	Equipment Requirements	WATTS								
		BTUH								
		CFM (Actual)	130	27	245	198	84	116		
	Heat Gain									
17	Of Windows And Doors	Sq-Ft (Line 5a) × HTF								
		Sq-Ft (Line 5b) × 26.5 HTF	530	0	1563	1352	477	477		
18	Of Net Exposed Wall	Sq-Ft (Line 6a) × HTF								
		Sq-Ft (Line 6b) × 2.1 HTF	527	160	615	492	183	470		
19	By Infiltration	Ln-Ft (Line 7) × 12.2 HTF	390	0	878	708	366	366		
20	At Ceiling	Sq-Ft (Line 8) × 1.8 HTF	346	205	648	385	315	261		
21	At Floor	Sq-Ft (Line 9a) × HTF								
		Ln-Ft (Line 9b) × 0 HTF								
22	Sensible	Lines 17 + 18 + 19 + 20 + 21	1793	365	3704	2937	1341	1574		
23	Equipment Requirements	BTUH								
		CFM (Adjusted)	172	42	297	236	102	152		

Table 4.23b *Quantities, Heat Loss, Heat Gain, Minimum Property Standards* (*Continued*)

				Design No. 1	Design No. 2	Design No. 3
				TOTAL		
				9		
				140		
				1331		
				166		
				1165		
				222		
				1200		
				140		

				13131		
				6757		
				7548		
				4320		
				8722		
				40478		
				800		

				4399		
				2447		
				2708		
				2160		
				11714		

Schedule 2. Window & Door Schedule

Symbol	*Dimensions*	*No.*	*Type*	*Sq. Ft. Area Ea./Total*	*Ln. Ft. Crackage Ea./Total*
A	3-0 × 5-0	4	S.H	15/60	19/76
B	3-4 × 3-0	2	S.H	10/20	16/32
C	3-0 × 3-0	4	S.H	9/36	15/60
D	2-8 × 3-0	1	S.H	8/ 8	14/14
				/124	/182
				/	/
1	3-0 × 6-8	1	H.C	21/21	20/20
2	2-8 × 6-8	1	H.C	21/21	20/20
				/42	/40
				/	/
				/	/
				/	/
				/166	/222

Schedule 3. Solar Areas of West and South Elevations

Symbol	*West*		*South*	
	No.	*Total Area*	*No.*	*Total Area*
A	2	30		
C	4	36		
Totals		66		

Use The Larger Of These Two Total Areas To Calculate Solar Gain

Table 4.23*b* *Continued*
Summaries, Heat Gain and Heat Loss

Gains—Size Equipment On Cooling Load

	Design No. 1
Sensible Gain (Line 22, Schedule 1)	11714
Latent Gain, 30% Sensible Gain	3514
Solar Gain (Schedule 3), 83.5 × ____ sq. ft.	5511
People __6__ (Min-5) × .300 × 1.80	3240
Subtotal	23979
Duct Gain __15__ % of Subtotal	3597
Total BTUH Heat Gain	27576

Losses—Design Room CFM On Heating

Structural Loss (Line 15, Schedule 1)	40471
Duct Loss __15__ % of Structural Loss	6071
Total BTUH Heat Loss	46542

Minimum Property Standards

Ceiling *R*-19

Walls *R*-11

Floor *R*-3.5

Single Pane Windows

Hollowcore Doors

Indicate North

OUTLINE PLAN OF STRUCTURE

Table 4.23c *Heat Transfer Factors (HTF), Minimum Property Standards*

	Heat Transfer Factors (HTF) Heat Loss				Heat Gain		
Design Temperature Differential (TD)	*75°*	*70°*	*65°*	*60°*	*30°*	*25°*	*20°*
Windows (Weather stripped)							
Single Pane	84.8	79.1	73.5	67.8	31.8	26.5	21.2
Single Pane and Storm Windows	42.0	39.2	36.4	33.6	16.8	14.0	11.2
Insulating Glass	45.8	42.7	39.7	36.6	18.3	15.3	12.2
Triple Glazed	31.5	29.4	27.3	25.2	12.6	10.5	8.4
Doors (Weather stripped)							
Hollow Core	84.8	79.1	73.5	67.8	31.8	26.5	21.2
Hollow Core and Storm Door	47.5	44.3	41.2	38.0	19.0	15.8	12.7
Solid Core (1¾″)	36.0	33.6	31.2	28.8	14.4	12.0	9.6
Solid Core (1¾″) and Storm Door	23.3	21.7	20.2	18.6	9.3	7.8	6.2
Infiltration (Doors and Windows)							
Without Storm Sash or Door	36.5	34.0	31.6	29.2	14.6	12.2	9.7
With Storm Sash or Door	23.7	22.1	20.5	19.0	9.5	7.9	6.3
Walls							
Frame Standard Sheathed No Insulation	18.2	16.9	15.7	14.5	7.3	6.1	4.9
Brick Standard Sheathed No Insulation	20.0	18.6	17.3	16.0	8.0	6.7	5.3
Brick or Frame Standard Sheathed: 2″ Insulation (*R*-7.0)	8.6	8.1	7.5	6.9	3.5	2.9	2.4
Brick or Frame Standard Sheathed: 3½″ Insulation (*R*-11)	6.1	5.8	5.3	4.9	2.5	2.1	1.7
Brick or Frame Standard Sheathed: 6″ Insulation (*R*-19)	3.9	3.6	3.4	3.1	1.6	1.3	1.1
Brick or Frame Standard Sheathed: 2″ Insulation (*R*-7.0) Cavity, Impregnated Vermiculite Poured	7.5	7.0	6.5	6.0	3.0	2.5	2.0
Brick or Frame Standard Sheathed: 3½″ Insulation (*R*-11) Cavity, Impregnated Vermiculite Poured	5.5	5.2	4.8	4.4	2.2	1.9	1.5
Brick or Frame Standard Sheathed: 6″ Insulation (*R*-19) Cavity, Impregnated Vermiculite Poured	3.7	3.4	3.2	2.9	1.5	1.2	1.0
8″ Solid Wall Brick or Stone: No Insulation	50.3	46.9	43.6	40.3	20.1	16.8	13.4
8″ Solid Wall Brick or Stone: Gypsum Wallboard, Furred With 2″ Insl. (R-7.0)	8.5	7.9	7.4	6.3	3.4	2.8	2.3
4″ Brick & 8″ Concrete Block: No Insulation	32.3	30.1	28.0	25.8	13.0	10.8	8.9
4″ Brick & 8″ Concrete Block: Blocks, Impregnated Vermiculite Poured	23.3	21.7	20.2	18.6	9.3	7.8	6.2
Standard Aggregate Concrete Block: 6″ Block No Insulation	42.6	39.8	36.9	34.1	17.1	14.2	11.4
Standard Aggregate Concrete Block: 8″ Block No Insulation	38.3	35.7	33.2	30.6	15.3	12.8	10.2
Standard Aggregate Concrete Block: 12″ Block No Insulation	35.2	32.9	30.5	28.2	14.1	11.7	9.4
Standard Aggregate Concrete Block: 6″ Block Impregnated Vermiculite Poured	27.0	25.2	23.4	21.6	10.8	9.0	7.2
Standard Aggregate Concrete Block: 8″ Block Impregnated Vermiculite Poured	25.5	23.8	22.1	20.4	10.2	8.5	6.8
Standard Aggregate Concrete Block: 12″ Block Impregnated Vermiculite Poured	24.8	23.1	21.1	19.8	9.9	8.3	6.6
Lightweight Aggregate Concrete Block: 6″ Block No Insulation	28.3	26.4	24.5	22.6	11.3	9.4	7.5
Lightweight Aggregate Concrete Block: 8″ Block No Insulation	25.8	24.1	22.4	20.7	10.3	8.6	6.9
Lightweight Aggregate Concrete Block: 12″ Block No Insulation	22.9	21.4	19.9	18.3	9.2	7.7	6.1
Lightweight Aggregate Concrete Block: 6″ Block Impregnated Vermiculite Poured	17.6	16.4	15.2	14.0	7.0	5.9	4.7
Lightweight Aggregate Concrete Block: 8″ Block Impregnated Vermiculite Poured	16.6	15.5	14.4	13.6	6.6	5.6	4.5
Lightweight Aggregate Concrete Block: 12″ Block Impregnated Vermiculite Poured	16.1	15.0	13.7	12.9	6.4	5.4	4.3
Ceiling							
Pitched Roof: 6″ Insulation (R-19.0) Standard Ventilation	3.9	3.6	3.4	3.1	2.1	1.8	1.6
Pitched Roof: 8″ Insulation (R-25.3) Standard Ventilation	3.0	2.8	2.6	2.4	1.6	1.4	1.2
Pitched Roof: 10″ Insulation (R-31.7) Standard Ventilation	2.4	2.3	2.1	1.9	1.3	1.1	1.0
Pitched Roof: 12″ Insulation (R-38.0) Standard Ventilation	2.1	1.9	1.8	1.7	1.1	1.0	0.8
Hipped or Flat Roof: 6″ Insulation (R-19.0) Standard Ventilation	3.4	3.2	2.9	2.7	3.1	2.6	2.1
Hipped or Flat Roof: 8″ Insulation (R-25.3) Standard Ventilation	2.7	2.5	2.3	2.1	2.4	2.0	1.6
Hipped or Flat Roof: 10″ Insulation (R-31.7) Standard Ventilation	2.3	2.1	1.9	1.7	1.9	1.6	1.3
Hipped or Flat Roof: 12″ Insulation (R-38.0) Standard Ventilation	1.8	1.7	1.6	1.5	1.7	1.4	1.1
Pitched, Hipped or Flat Roof: 6″ Insulation (R-19.0) Controlled Ventilation	3.4	3.2	2.9	2.7	1.5	1.3	1.0
Pitched, Hipped or Flat Roof: 8″ Insulation (R-25.3) Controlled Ventilation	2.7	2.5	2.3	2.1	1.4	1.2	0.9
Pitched, Hipped or Flat Roof: 10″ Insulation (R-31.7) Controlled Ventilation	2.3	2.1	1.9	1.7	1.0	0.8	0.6
Pitched, Hipped or Flat Roof: 12″ Insulation (R-38.0) Controlled Ventilation	1.8	1.7	1.6	1.5	0.8	0.7	0.5
Floor							
Double Wood and Crawl Space: 2″ Insulation (R-7)	4.8	4.5	4.2	3.9	1.5	1.2	1.0
Double Wood and Crawl Space: 3½″ Insulation (R-11)	3.3	3.1	2.9	2.6	1.3	1.1	0.9
Double Wood and Crawl Space: 6″ Insulation (R-19)	2.0	1.8	1.6	1.5	0.8	0.6	0.5
Slab on Ground: 1″ Insulation (R-3.5) Standard Duct System	66.8	62.3	57.9	53.4	.0	.0	.0
Slab on Ground: 2″ Insulation (R-7) Standard Duct System	35.8	33.4	32.6	31.8	.0	.0	.0
Slab on Ground: 1″ Insulation (R-3.5) Perimeter Duct	199.3	186.9	173.7	160.2	.0	.0	.0
Slab on Ground: 2″ Insulation (R-7) Perimeter Duct	107.4	100.2	97.8	95.4	.0	.0	.0

NOTE: (A) For additional factors, use the American Society of Heating, Refrigerating and Air-Conditioning Engineers, Inc. Guide & Data Book.

(B) All above factors include correctly applied vapor barriers.

(C) Above factors are based on blanket insulation; adjust for blown material.

(D) Boxes indicate *selected* HTFs.

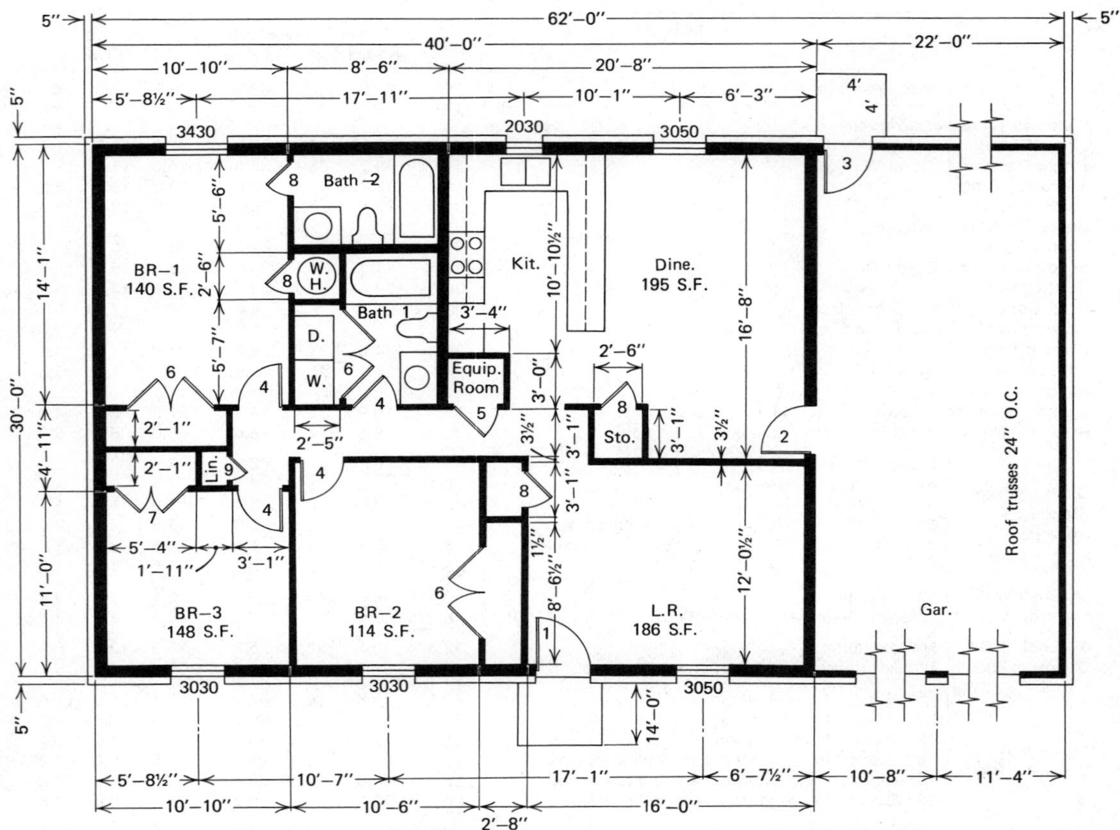

Fig. 4.21 Plan, The Arkansas House, Type 274. Courtesy of Owens-Corning Fiberglas Corporation.

6″ stud walls, 24″ on center. This allows for compression of 6″ of fiber glass insulation in the $5\frac{1}{2}$″ walls. But there are other design differences, too. Since the ceiling is insulated with 12″ of insulation the sheathing is attached to the vertical truss member flush with the wall all the way to the top of the insulation.

The studs in the house corners are set so one is flush to the outside sheathing and the other starting the connecting wall is at right angles 6″ from and parallel to the sheathing. It will be seen that instead of leaving an uninsulated box in each corner of the house this permits the corner to be insulated. Similarly, where a partition meets an outside wall a single stud is used. At the center bottom of each stud through which wiring is going to be passed a hole is augered or vee cut. This avoids having wiring interfere with insulation installation. One other major difference in wall construction is that the total window area of the house must be restricted to 8% of the living area. In addition, these must be double windows or windows suitably insulated with a storm window. Boxing around windows is simplified to permit insulation to be placed between the front and back 2″ headers. The sole plate is set on a full bed of caulking. The single top plate serves to tie the wall laterally. The second story studs and/or roof truss, also 24″ on center are required to rest on top of studs. Use is made of metal perforated tie plates at wood joints and back-up clips are used to hold the drywall. A polyethylene vapor barrier completely covers both studs and insulation.

Many people reviewing these specifications ask if they couldn't retain their standard 2″ × 4″, 16″ o.c. standard wall and just add foam insulation to the exterior. A quick calculation will show that the old 16″ o.c. wall contains a greater area of semiconductive studs and unless the inside vapor barrier is better than the exterior insulating foam water will condense in the wall nullifying the thermal resistance of the insulation. Just based on thermal resistance the 6″ insulated stud de-

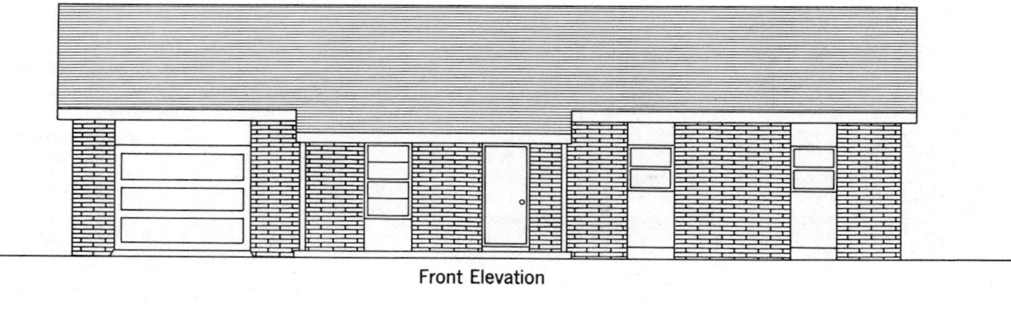

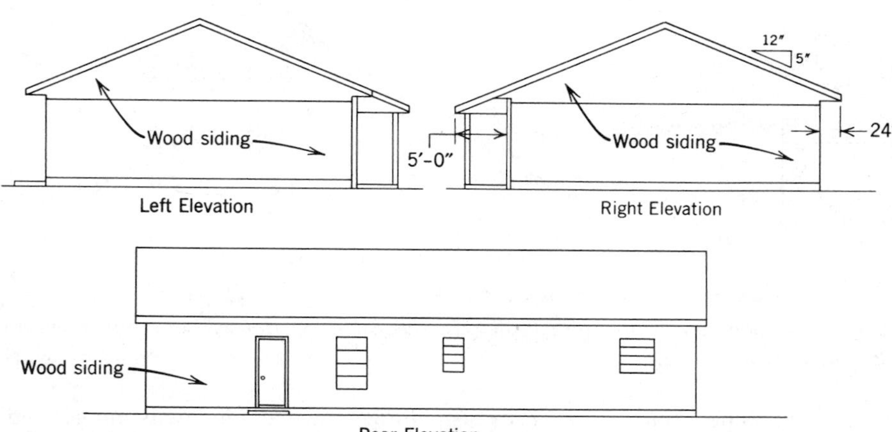

Fig. 4.22 *Elevations, The Arkansas House, Type 274. Courtesy of Owens-Corning Fiberglas Corporation. Type 274 establishes a fixed living area. The plan may be reversed as indicated by comparing Figures 4.21 and 4.22. Size of garage does not affect thermal calculations.*

sign 24″ on centers is 30% more efficient and much less expensive to build.

Crawl Space and Slab Insulation (see Fig. 4.23). While it is true, that if sufficient temperature difference exists, hot air rises it is also true that heat flows to cold. Where walls and ceiling are well insulated the heat will be attracted to and dissipate through the coldest surface even if this is the floor.

Since people are quickly chilled if their feet feel cold this tends to make them desire a higher thermostat setting. For both energy management and comfort the floor areas must be maintained as close as possible to room temperature. This, it has been found, requires 6″ of insulation over crawl spaces and a minimum of $1\frac{1}{2}$″ polyurethane rigid foam insulation around the perimeter of the slab. Because dampness can chill and require greater use of energy to operate a dehumidifier, or can raise the humidity level higher than desirable for comfort by overloading the dehumidifier the complete area under the slab must be provided with a vapor barrier. Likewise, the finish grade in a crawl space must be fully covered with a vapor barrier and since another vapor barrier is installed above the floor insulation the crawl space must be ventilated.

Ceilings. Savings estimates justify the installation of 12″ (R-38) Friction Fit Batts in these Little Rock homes. As with the walls and floor, a complete vapor barrier under the ceiling insulation is specified. To permit easier and safer installation of the insulation the designers have specified an inspection catwalk.

The Savings Provided by the Arkansas Energy Conservation Design

The basic heating and cooling of these homes has been with electrical heat pumps or, on the smaller homes, requiring one ton or less,

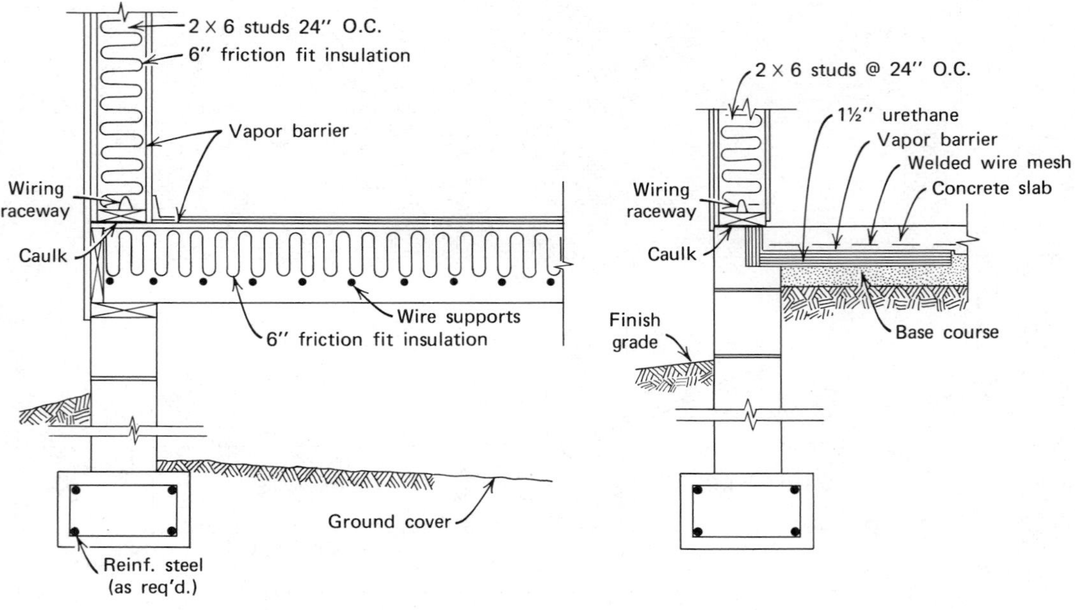

(a) Crawl space construction (insulation) *(b)* Section thru slab

Fig. 4.23 *Crawl space and slab-on-grade details, the Arkansas House. Courtesy of Owens-Corning Fiberglas Corporation.*

where small enough heat pumps are not available, an electric furnace and compressor. On larger homes, other energy sources can be utilized and savings achieved providing capacity does not exceed loads. Resistance type heating in the ceiling or baseboard was not used, despite its low initial cost because it raises the effective surface temperature and does not provide the continuous slow air flow, humidity control and air cleaning necessary for ideal comfort.

When these houses were built, Tschumi was unable to find commercially available fossil fuel heating units of small enough capacity to provide the desired continuous low heat operation necessary to comfort and economy . . . in other words, for these 1200 sq. ft. homes in Little Rock, units with a bonnet capacity of 15,000 to 18,000 Btuh. See Figure 4.24.

Whatever the energy source, heating equipment should be sized to run with minimum cycling and temperature variation and with continuous fan operation.

If flame-type heating and/or water heating is used in a house built to these standards, the firing unit must either be installed outside the conditioned space or be provided with adequate outside air through 6 inch or larger ducts to prevent fumes being sucked into the living area by the turning on of vent fans such as those installed in the bathrooms, in the clothes dryer, or over the stove.

It is the responsibility of the builder to obtain expert advice and specifications before installing any fossil fuel equipment in or attached to the house. This also applies to fireplaces. While these are not recommended in

Heat Loss Comparison Chart[1]
Plan #274 1/1/75

	Energy Conservation Construction		*FHA Minimum Property Standards*
Floor	3,179		8,722
Walls	4,411		6,757
Ceiling	2,041		4,320
Windows & Doors	3,050		13,131
Infiltration	3,007		7,548
Sub Total	15,688		40,478
Duct Loss 3%	471	15%	6,072
Total Btuh Heat Loss	16,159		46,550

Total Heat Loss Reduction = 65%

[1]From Arkansas Power and Light Load Calculation Forms at 70°TD Heating and 25°TD Cooling.

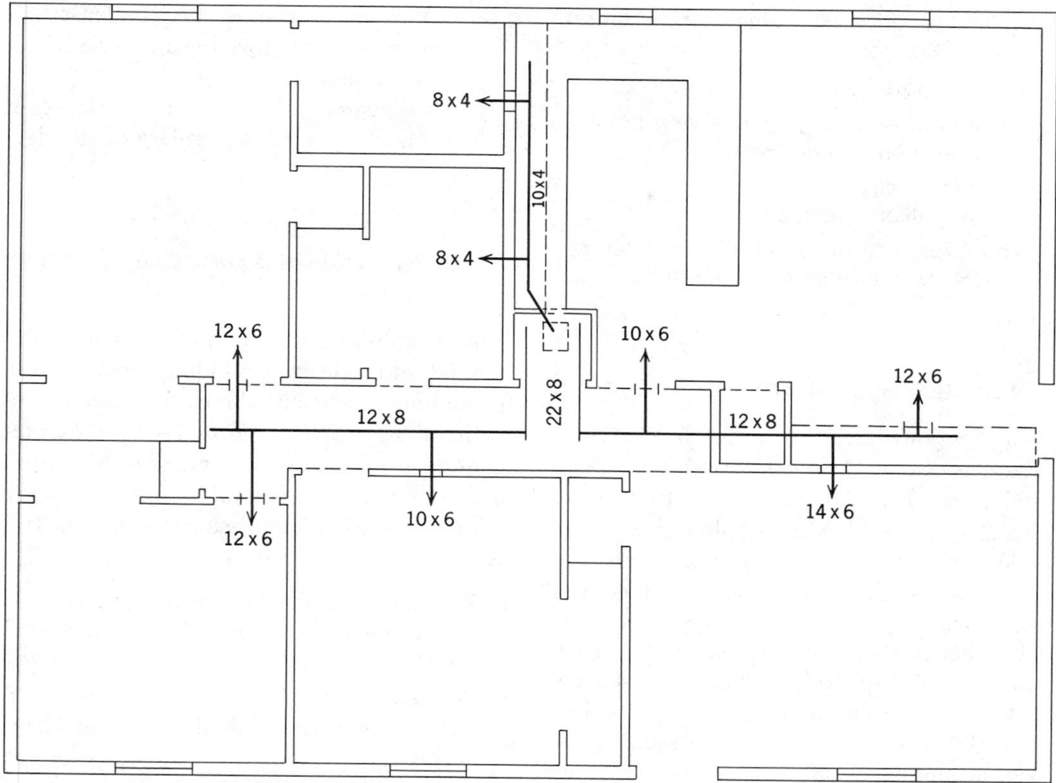

Fig. 4.24 *Air-conditioning duct layout, the Arkansas House. Sizes shown are for ½-in.-lined metal ducts (Fiberglas duct boards can be used). Courtesy of Owens-Corning Fiberglas Corporation.*

an energy conserving home, it is recognized some owners would prefer to have them and pay the cost and energy penalty.

When reports came out of Little Rock that homes were being heated and cooled for about $130 annually, many of Owens-Corning's customers were incredulous and started asking our representatives if it was true and if so, how could such savings be accomplished.

As of March 1, 1976, the following one year's metered *heating and cooling costs* on four homes have been reported from Arkansas:

House Size Sq. Ft.	*Size Heat Pump*	*6 months Htg. Cost*	*6 months Clng. Cost*	*Total 12 Month Cost*
1200	1½ ton	73.26	70.15	143.41
1200	2 ton	34.98	76.25	111.23
1200	2 ton	54.68	91.73	146.41
1344	1½ ton	42.38	96.32	138.70

Credits

Arkansas Power and Light Company
Fred Clark
Bob G. Haynie

Harry Tschumi Company
Harry Tschumi, President

HUD, Little Rock Regional Office
Andy Watts, Acting Deputy Area Director
Frank Holtzclaw, Senior Construction Analyst

Whited Builders
Bill Whited, President

Rex Rogers
Home Construction

NASA
Jim Waligorra, Thermal Physiologist
Dr. Bill Schumate, Physiologist
Dr. Federson, Physiologist

Institute for Environmental Research

Dr. Frederick H. Rohles, Jr., Professor and Director

Hendershott Company

Owners of an energy conserving home
Mr. and Mrs. John Heiser

Modern Realty
Jim Walker, President

The Electricity Research Council Centre
D. A. McIntyre and I. D. Griffiths

4.24 Residential Heat Gain

In Section 4.22, the use of heat transfer factors is discussed, together with their use in establishing heat *losses*. Careful study of Tables 4.22 and 4.23 will complete the general understanding of the simplified method of calculating both heat loss and heat *gain*. It begins with areas, linear footage (edges), and cubic feet per minute (for infiltration). These values are multiplied by the heat transfer factors for the corresponding items. For heat gain, the process starts out the same. However, a number of differences must be noted:

(a) The heat gain is based upon a temperature difference of 25° F, since summer indoor air is designed for 75° F and outdoor air is 100° F. The *loss* is related to a winter indoor temperature of 75° F and outdoor temperature of plus 5° F (70° difference). This is the reason for different heat transfer factors as seen in Tables 4.22*c* and 4.23*c*.

(b) The cooling load is burdened in summer by a minimum of five occupants each of whom gives off about 650 Btuh.

(c) Solar effect in summer is two-fold. In addition to the 25° F *transmission* gain, the direct rays of the sun on glass, sometimes reduced by a shading coefficient of 35% (see Table 4.22*b*), constitute an extra gain.

(d) Because humidity builds up from cooking, bathing, clothes washing, and perspiration from people, 30% is added to the sensible gain. This is an approximate value. *Sensible* gain relates to air temperature. *Latent* gain relates to the moisture that must be condensed to prevent uncomfortable increase of indoor humidity.

(e) It is assumed that there is no edge-gain of heat through the insulated slab-edge.

4.25 Nonresidential Heat Gain Calculations

In establishing the cooling loads for commercial and industrial buildings and, indeed, for unusual residences, the classic methods set forth in the *ASHRAE Handbook of Fundamentals* are essential. To illustrate the nature of these methods, a *few* tables from the many in the *Handbook* have been included in this text, beginning on page 167:

The above tables are taken from *ASHRAE Handbook of Fundamentals, 1967*. The reason is that the 1977 Edition has been much expanded and is assembled as a cross-referenced *package*. Its "Procedure for Calculating Space Design Cooling Load—Summary of Load Sources and Equations" governs the whole process and is quite beyond the scope of this book. The tables that we have selected from the 1967 Edition, though briefer, are for fundamental values that are little changed. They are especially suitable for a simple way of describing *methods* of calculation.

The previous sections of this chapter have applied to a rather special problem—the cooling load for a residence. The cooling load for a residence is by no means typical of the diverse problems in heat gain that would be encountered by the average architect or consulting engineer. It has been stated that the major components of the cooling load include:

1. Transmission through walls and roof.
2. Transmission through glass.
3. Occupants.
4. Infiltration or ventilation.
5. Appliances.
6. Lights.

Transmission through structure and shaded or unshaded glass is determined by a number of factors (Fig. 4.25). Primary is the effect of the sun at the specific latitude, the month and day and the time of day. See Table 4.27 for

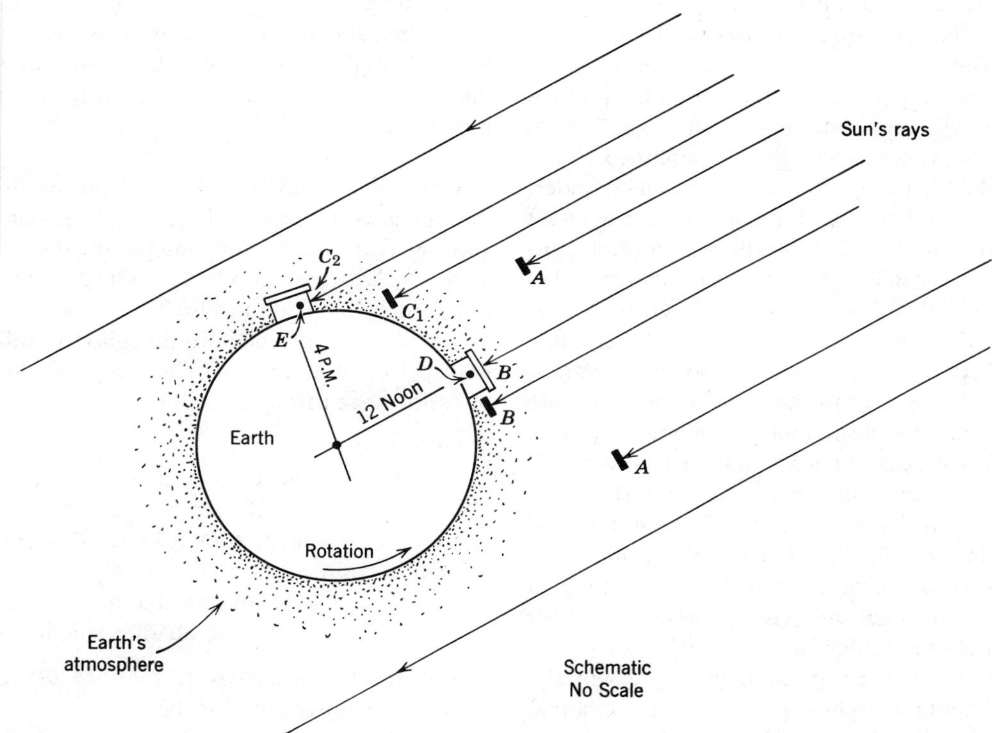

Fig. 4.25 Solar radiation and transmission. See Example 4.5 and 4.6.

Approximate value of solar radiation impingement on 1 sq ft of surface normal to the sun's rays			Approximate Btuh/sq ft
A	*Outer space*		*450*
B	*Earth's surface (or roofs), 12 noon*	*40° N. Latitude Midsummer*	*250*
C_1	*Earth's surface, 4 P.M.*	*40° N. Latitude Midsummer*	*240*
C_2	*West wall (not quite normal), 4 P.M.*	*40° N. Latitude Midsummer*	*210*

Value of heat penetrating to inside of structure at:

D	*Through 4-in. concrete roof at noon*	*(Example 4.5)*	*5.1*
E	*Through the same at 4 P.M.*	*(Example 4.5)*	*8.2*
E	*Through window with blind at 4 P.M.*	*(Example 4.6)*	*135.0*
E	*Through window, full outside shading*	*(Example 4.6)*	*40.0*

40° N. Latitude, which is part of a series of *ASHRAE* tables for various latitudes. The radiant solar effect in outer space is about 450 Btuh/sq ft of area. Under the conditions shown in Figure 4.25 this is moderated by the earth's atmosphere to 250 Btuh/sq ft at the earth's surface. The greater travel of the sun's rays through the atmosphere at 4 P.M. reduces this to 240 Btuh/sq ft and the rate of impingement on a west wall, which is, of course, not quite normal at 4 P.M. to the sun's rays, is 210 Btuh/sq ft. It is seen that 40 to 135 Btuh/sq ft in this situation represents approximately the value that gets through glass with some shading (Example 4.6). Solid construction interposes a more protective barrier with a thermal time lag, and only passes through 5 to 8 Btuh/sq ft under conditions explained in Example 4.5. From these figures it can be understood that, because of the heat-reducing effect of structural roofs or walls, temperature differentials can be established for a range of 0 to 50° N. Latitude and, in general, for the hottest weather, as explained in the footnotes of Table 4.26 (Item 1, *Applications*). So, for transmission through roofs and walls, temperature differential and summer *U*-coefficients can be used; whereas, for glass that relates more directly to the sun, one starts with the Solar Heat Gain Factor (SHGF) of Table 4.27 and applies the Shading Factor (SF) from Table 4.28. To this product must be added the heat passing through the glass by reason of the air temperature difference on both sides of the glass. Their sum represents what is essentially an instantaneous heat gain from the combination of solar energy and normal transmission.

EXAMPLE 4.5. Making use of Tables 4.25 and 4.26, find the heat transmitted through a roof of 4-in. concrete (lightweight aggregate), on gypsum board with 1-in. insulation, $\frac{1}{2}$-in. plaster and $\frac{3}{4}$-in. acoustical tile, but no suspended ceiling, at 12 noon and at 4 P.M. sun-time in summer at 40° N. Latitude. The roof is exposed to the sun.

SOLUTION. Referring to Table 4.26, it is noted that, while insulation changes the temperature differential for light wood roofs, it has no significant effect on this heavy construction roof. The temperature differential at 12 noon is found to be 46° F and at 4 P.M., 74° F. The summer *U*-coefficient in Table 4.25 is 0.11. (It would be 0.17 if the roof were *without* insulation.)

At 12 noon $0.11 \times 46 = 5.1$ Btuh/sq ft
At 4 P.M. $0.11 \times 74 = 8.2$ Btuh/sq ft

The fact that transmission through the roof at 4 P.M. is greater than at 12 noon is evidence of the *time lag* of heavy construction. Warmed all day by the sun, it gives off its greatest heat to the interior, not when the sun is most intense normal to its surface, but later, when the heat has had time to pass through (see Fig. 4.25).

EXAMPLE 4.6. Find the rate of heat passing through a west window of regular sheet single glass, shaded by light Venetian blinds, at 4 P.M. at 40° N. Latitude on July 21st. Outdoor temperature is 90° F. Find also the value under the same circumstances if the glass is fully shaded from the outside. The basic relationships for heat through glass are:

Total heat gain
= Solar heat gain
+ Heat gain due to outdoor—indoor temperature difference

Solar heat gain
= Solar heat gain factor
× Shading coefficient

Substituting from Tables 4.27 and 4.28, the *solar* heat gain is found to be

$$216 \times 0.55 = 119 \text{ Btuh/sq ft}$$

The heat gain due to outdoor-indoor temperature difference is the product of the summer *U*-coefficient and the temperature difference. Summer coefficients are slightly smaller because of the lower outdoor wind velocity. A value of 1.06 is used here instead of the usual winter value of 1.13. Temperature difference is 15° F (90–75). Substituting in the first formula above,

$$1.06 \times 15 = 16 \text{ Btuh/sq ft}$$

Total heat gain = 119 + 16 = 135 Btuh/sq ft

When full outdoor shading is provided, the *solar* gain is reduced by about 80% (see Fig. 4.25):

$119 \times 0.20 =$ 24 reduced solar gain
16 gain due to temperature difference
40 Btuh/sq ft

Of the six major components of heat gain, transmission through structure and glass are illustrated by Examples 4.5 and 4.6. Gain from appliances and occupants is found in Tables 4.29 and 4.30. Table 4.3 establishes ventilation rates that affect heat gain. Tables 4.1 and 4.3 included some reference to the rate of metabolism of sedentary and active occupants as well as the requirements of outdoor air for odor reduction. As summarized in this chapter, they are the basis for evaluating these heat-producing items. Table 4.29 is an excerpt from a much larger compilation of heat-producing cooking devices listed in *ASHRAE Handbook of Fundamentals, 1967*, but it indicates a few of the many cooking units that must be considered. Finally, general-area lighting if its heat is not exhausted directly, often adds a cooling load component at the rate of about 3 watts per sq ft of floor area in the space. In British thermal units per hour, this amounts to 3×3.41[7] (Btuh/w) = 10.23 Btuh/sq ft.

Now consider a large, multiple-use building, where, during normal daytime hours of occupancy, there could be many heat-producing activities in diverse and widely separated parts of the building. They might include:

(*a*) A number of small classroom or lecture spaces, all seats occupied, glass to the east and the need for ventilation and possibly lights; occupancy beginning at 9 A.M.
(*b*) Banquet facilities with kitchens and pantries actively preparing food for long periods before the dinner hour.
(*c*) A large, top-story cocktail lounge commanding a view to the *west* (maximum solar impingement) with glass, floor to ceiling below a roof that transmits the heat of the day accumulated during many daytime hours, occupants at maximum density, and the need for ventilation (at least 25 cfm per person).
(*d*) An interior gallery that opens at 12 noon. Gain components, though not including the sun, derive from dense occupancy, ventilation, general lighting, and much special accent lighting.

[7]One watt = 3.41 Btuh.

The need for many zones to cope speedily and in full capacity with the separate cooling problems is evident. The zones could, at certain hours, present additive loads. In establishing the full cooling capacity for the building and the capacity of the several zones, heat gain calculations may have to be made for several critical periods; perhaps for every hour.

It should not be overlooked that occupants, ventilation, and cooking all introduce humidity, known as latent gain. Energy is required to condense this moisture in order to prevent the RH from rising above a reasonable limit often set at 60% RH (see Fig. 4.6, page 95).

Industrial, commercial, and public buildings frequently have problems that relate more to occupancy and activity than to the effects of sun and the outdoor air temperature and RH. Evaluation of these items is seen in the following example.

4.26 Nonresidential Heat Gain Components

EXAMPLE 4.7. An analysis of heat gain from lights, ventilation, and people is desired for a small one-story office building. Based upon Figure 4.26 and Table 4.24, calculate the percentages of the grand total heat gain load that are represented by:

1. Heat gain from lights.
2. Sensible and latent heat gain from ventilation.
3. Sensible and latent heat gain from people.

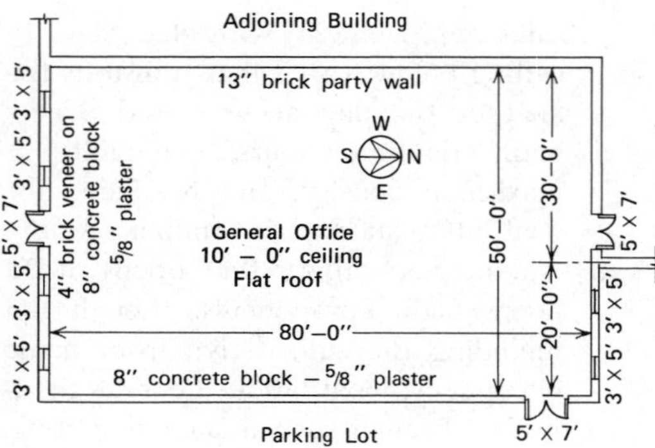

Fig. 4.26 (Example 4.7) *Plan of one-story office building. Copyright © by the American Society of Heating, Refrigerating and Air Conditioning Engineers, Inc. Reprinted by permission from* ASHRAE Handbook of Fundamentals, 1977. *For heat gain values, see Table 4.24.*

Table 4.24 (Example 4.7) *Summary of Calculations for a One-Story Office Building. See Figure 4.26*

Outdoor Conditions	*94 F db*	*77 F wb*	*0.0161 Humidity Ratio*
Indoor Conditions	*78 F db*	*65 F wb*	*0.0102 Humidity Ratio*
Difference	*16 F*		*0.0041*

	Sensible Cooling Load at 3 pm, Btuh	
Roof and Exposed Walls		
Roof	26,190	
South wall	2,641	
East wall	11,310	
North wall	1,005	
Doors in south wall	672	
Doors in east wall	636	
Fenestration Areas		
South windows	2,719	
North windows	695	
Party Walls		
West and north wall	4,118	
Doors in north wall	291	
Internal Sources		
Tungsten lights	12,000	
Fluorescent lights	50,410	
People	16,480	
Ventilation and Infiltration		
Ventilation	22,440	
Infiltration	1,180	
Total		152,787
	Latent Cooling Load at 3 pm, Btuh	
People	21,675	
Ventilation	36,410	
Infiltration	1,915	
Total		60,000
Grand total load		212,787

SOLUTION

Item	Btuh		Percentage
1. Lights	Tungsten	12,000	
	Fluorescent	50,410	
		62,410	29
2. Ventilation	Sensible	22,440	
	Latent	36,410	
		58,850	27
3. People	Sensible	16,480	
	Latent	21,675	
		38,155	18
			74%

Each item is based upon the grand total of 212,787 Btuh.

NOTE. It is evident that lighting and ventilation represent more than one-half of the cooling load. Part Six in this book deals with energy savings resulting from natural daylighting and control of artificial lighting.

Section 4.7 points out the permissible reductions in ventilation rates when special air cleaning is provided for outdoor air and recirculated air.

4.27 Reflective Insulating Glass

With its research and development beginning well before the energy-conscious era of the early 1970s, reflective, insulating glass has attained a high degree of performance and acceptance.

Included among the many products of this type that are available for use is a recently announced item of PPG Industries known as Solarban 490, Twindow unit. It is a double-glazed unit said to have the highest energy-saving potential of their extensive inventory. It consists of two panels of clear glass separated by a dry air space. A gold-toned, metallic-reflective coating is applied to the inner surface of the outdoor glass panel. Transmitted light is a soft bluish-gray. On a typical summer day, the assembly is said to reduce the peak solar heat gain by 86% compared with the conventional $\frac{1}{8}$-in. clear glass (see Figs. 4.27 and 4.28).

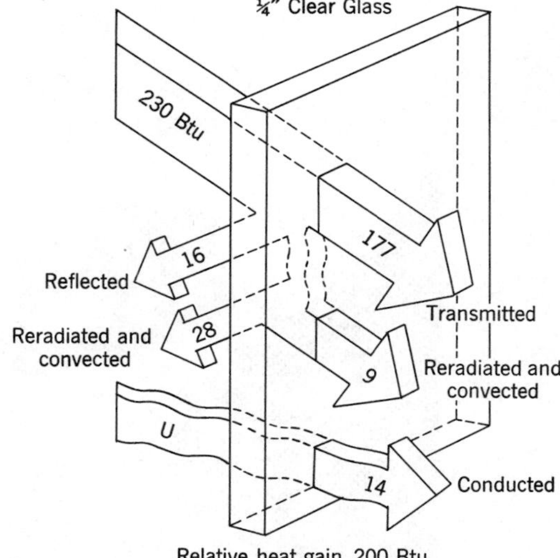

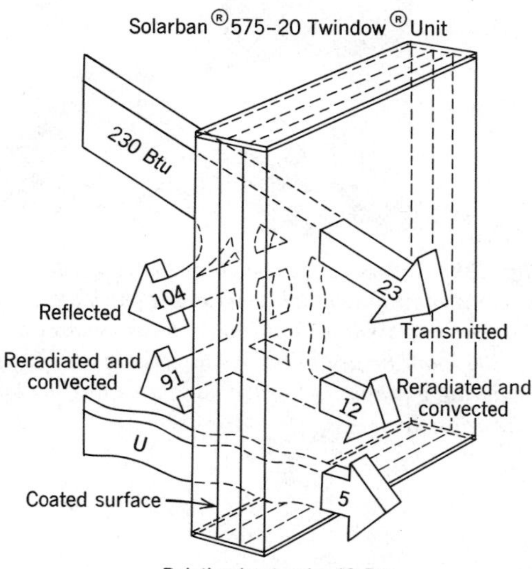

Fig. 4.27 Reflective glasses. Solarban Twindow Units are high-performance, reflective-coated, insulating (double) glass units. They offer significant environmental-control benefits due to their ability to reduce solar heat gain and heat transfer by conduction. The values indicate that a major share of the solar heat *is turned away, by reflection, reradiation, and convection. Courtesy of PPG Industries, Inc.*

AUTHOR'S NOTE. For dealing with external reflections of the sun, the reader is referred to Figure 2.8, page 41; and Figure 2.9, page 41.

Fig. 4.28 *Reflective glass wall of the Hyatt Regency Dearborn Hotel in Dearborn, Michigan, mirrors its spacious environment while reducing entry of the sun's heat, which reduces the energy needed for cooling. More than 150,000 sq ft of bronze-toned Solarban Twindow double-glazed units by PPG Industries sheath the 800-room luxury hotel. Architects Charles Luckman Associates of Los Angeles selected the reflective insulating glass units to save energy as well as to enhance the building's appearance. Courtesy of PPG Industries, Inc.*

Table 4.25 ***Summer Coefficients of Heat Transmission U of Flat Roofs Covered with Built-Up Roofing***

Btu per (hour) (square foot) (F deg difference between the air on the two sides)

Type of Roof Deck (Ceiling not shown)	*Thickness of Roof Deck, Inches*	*Type of Suspended Ceiling*	*Insulation on Top of Deck (Covered With Built-up Roofing), Inches*						
			None	½	1	1½	2	2½	3
Flat metal roof deck	—	None	0.67	0.35	0.23	0.18	0.15	0.12	0.10
		Gypsum Bd. and Plaster (½ in.)	0.32	0.22	0.17	0.14	0.12	0.10	0.09
		Acoustical Tile (¾ in.)	0.23	0.18	0.14	0.12	0.11	0.09	0.08
Concrete (Lightweight aggregate on gypsum board)	2	None	0.27	0.20	0.15	0.13	0.11	0.10	0.08
		Gypsum Bd. and Plaster (½ in.)	0.18	0.14	0.12	0.10	0.09	0.09	0.08
		Acoustical Tile (¾ in.)	0.15	0.12	0.11	0.09	0.08	0.08	0.07
	3	None	0.21	0.16	0.13	0.11	0.10	0.09	0.08
		Gypsum Bd. and Plaster (½ in.)	0.15	0.12	0.11	0.09	0.08	0.08	0.07
		Acoustical Tile (¾ in.)	0.13	0.11	0.10	0.08	0.08	0.07	0.06
	4	None	0.17	0.14	0.11	0.10	0.09	0.08	0.07
		Gypsum Bd. and Plaster (½ in.)	0.13	0.11	0.10	0.08	0.08	0.07	0.06
		Acoustical Tile (¾ in.)	0.12	0.10	0.09	0.07	0.07	0.06	0.05
Gypsum slab on ½″ gypsum board	2	None	0.32	0.22	0.17	0.14	0.12	0.10	0.09
		Gypsum Bd. and Plaster (½ in.)	0.21	0.17	0.13	0.11	0.10	0.09	0.08
		Acoustical Tile (¾ in.)	0.17	0.13	0.12	0.10	0.09	0.08	0.07
	3	None	0.27	0.19	0.15	0.13	0.11	0.10	0.08
		Gypsum Bd. and Plaster (½ in.)	0.19	0.15	0.13	0.11	0.10	0.09	0.08
		Acoustical Tile (¾ in.)	0.15	0.12	0.11	0.09	0.08	0.08	0.07
	4	None	0.23	0.17	0.14	0.12	0.10	0.09	0.08
		Gypsum Bd. and Plaster (½ in.)	0.17	0.13	0.12	0.10	0.09	0.08	0.07
		Acoustical Tile (¾ in.)	0.14	0.12	0.11	0.09	0.08	0.08	0.07
Wood[a]	1	None	0.40	0.26	0.19	0.15	0.13	0.11	0.09
		Gypsum Bd. and Plaster (½ in.)	0.24	0.18	0.15	0.12	0.11	0.09	0.08
		Acoustical Tile (¾ in.)	0.19	0.15	0.13	0.11	0.10	0.08	0.07
	2	None	0.28	0.20	0.16	0.13	0.11	0.10	0.08
		Gypsum Bd. and Plaster (½ in.)	0.19	0.15	0.13	0.11	0.10	0.09	0.07
		Acoustical Tile (¾ in.)	0.16	0.13	0.11	0.10	0.09	0.08	0.07
	3	None	0.21	0.16	0.13	0.11	0.10	0.09	0.08
		Gypsum Bd. and Plaster (½ in.)	0.16	0.13	0.11	0.10	0.09	0.08	0.07
		Acoustical Tile (¾ in.)	0.13	0.11	0.10	0.09	0.09	0.08	0.07

[a]Wood deck 1, 2, and 3 in. is assumed to be 25/32, 1⅝ and 2 25/32 in. thick, respectively. The thermal conductivity k is assumed to be 0.80.

Table 4.26 ***Total Equivalent Temperature Differentials for Calculating Heat Gain Through Roofs***

Description of Roof Construction[a]	Outside Air Dew Point (F)	Water Layer Thickness (in.)	Sun Time A.M. 2	4	6	8	10	12	P.M. 2	4	6	8	10	12	λ	
Light Construction Roofs—Exposed to Sun																
1″ Wood[b]			−5	−6	−6	16	52	81	94	84	56	20	5	−4	0.95	2
2″ Insulating board + 1″ wood[b]			2	−3	−6	−1	20	50	75	85	77	53	26	7	0.82	4
2″ Concrete or 2″ gypsum Plank			−2	−5	−6	11	41	69	85	83	63	32	13	1	0.83	3
2″ Wood[b]			1	−3	−6	1	24	54	77	85	75	48	23	6	0.82	3
Medium Construction Roofs—Exposed to Sun																
2″ Insulating board + 2″ concrete or 2″ gypsum plank			10	3	−1	−1	13	36	60	75	75	60	38	19	0.69	5
2″ Gypsum or 2″ concrete + 4″ rock wool + ½″ plaster			13	5	0	−1	9	30	54	71	75	64	43	22	0.66	5
2″ Wood[b] + 4″ rock wool + ½″ plaster			29	19	12	6	7	16	32	49	61	63	54	39	0.48	7
Heavy Construction Roofs—Exposed to Sun																
4″ Concrete			9	3	−1	5	23	46	65	74	68	49	31	15	0.64	4
6″ Concrete			20	12	7	6	16	32	49	61	63	54	40	26	0.48	5
2″ Insulating board + 4″ concrete			27	19	14	10	14	25	39	51	57	54	45	32	0.38	6
2″ Insulating board + 6″ concrete			32	27	22	18	17	23	31	41	47	48	44	35	0.26	7
Roofs Covered with Water—Exposed to Sun																
Light Construction	60	6	3	−1	−4	−6	−6	−1	6	13	17	17	13	7		
		1	−6	−9	−11	−12	−6	4	15	21	22	17	8	0		
		0	−8	−10	−12	−12	−4	7	17	23	22	16	5	−3		
	70	6	7	4	1	−1	0	4	11	18	21	21	17	12		
		1	0	−3	−5	−5	0	10	19	25	26	21	12	5		
		0	−2	−4	−6	−5	2	12	21	26	26	19	9	2		
Heavy Construction	60	6	6	3	−1	−3	−4	−1	4	9	13	15	13	10		
		1	1	−3	−6	−8	−6	1	8	15	18	17	11	6		
		0	−1	−5	−7	−9	−5	2	10	16	19	16	10	4		
	70	6	11	8	5	2	2	4	9	14	18	20	18	15		
		1	6	3	0	−2	0	6	14	20	23	21	16	11		
		0	5	1	−1	−2	1	8	16	21	23	21	15	9		

[a]Includes $\frac{3}{8}$-in. felt roofing with or without slag. May also be used for shingle roof. [b]Nominal thickness of the wood.

Explanation: $\left\{\begin{array}{l}\text{Total heat transmission from solar radiation}\\ \text{and temperature difference between outdoor}\\ \text{and room air. Btu per (hr) (sq ft) of roof area}\end{array}\right\} = \left\{\begin{array}{l}\text{Equivalent temperature differential}\\ \text{from above table}\end{array}\right\} \times \left\{\begin{array}{l}\text{Heat transmission coefficient}\\ \text{for summer Btu per (hr)}\\ \text{(sq ft) (F deg)}\end{array}\right\}$

1. *Application.* These values may be used for all normal air conditioning estimates; usually without correction, in latitude 0 deg to 50 deg north or south when the load is calculated for the hottest weather. Note 3 explains how to adjust the temperature differential for other room and outdoor temperatures.
2. *Attics.* If the ceiling is insulated and if a fan is used in the attic for positive ventilation, the total temperature differential for a roof exposed to the sun may be decreased 25 percent.
3. *Corrections. For temperature difference when outdoor maximum design minus room temperature is different from 20 deg.* If the outdoor design temperature minus room temperature is different from the base of 20 deg, correct as follows: When the difference is greater (or less) than 20 deg add the excess to (or subtract the deficiency from) the above differentials.

For Outdoor Daily Range of Temperature other than 20 deg, correct Equivalent Temperature Difference as follows:

Outdoor Daily Range	Medium Construction	Heavy Construction
For each 1 deg difference *less* than 20 deg	*Add* $\frac{1}{4}$ deg	*Add* $\frac{1}{2}$ deg
For each 1 deg difference *greater* than 20 deg	*Subtract* $\frac{1}{4}$ deg	*Subtract* $\frac{1}{2}$ deg
Maximum Correction	3 deg	5 deg

For Light Construction, apply no correction.

Light Colors. Credit should not be taken for light colored roofs except where the permanence of the light color is established by experience, as in rural areas or where there is little smoke.

For solar transmission in latitudes other than 40 deg north, and in other months. The table values of temperature differentials will be approximately correct for a roof in the following months:

North Latitude		South Latitude	
Latitude (deg)	Months	Latitude (deg)	Months
0	All months	0	All months
10	All months	10	All months
20	All months except Nov, Dec, Jan	20	All months except May, June, July
30	Mar, Apr, May, June, July, Aug, Sept	30	Sept, Oct, Nov, Dec, Jan, Feb, Mar
40	April, May, June, July, Aug	40	Oct, Nov, Dec, Jan, Feb
50	May, June, July	50	Nov, Dec, Jan

For other months, light colored roofs, or both, use the total equivalent temperature differential from Equation 30. For room air temperature less than 75 F, add the difference between 75 and room air temperature; if greater than 75, subtract the difference.

Table 4.27 *Solar Position and Intensity; Solar Heat Gain Factors for 40 Deg North Latitude*

Date	Time A.M.	Solar Position Alt.	Solar Position Azimuth	Direct Normal Irradiation, Btu	N	NE	E	SE	S	SW	W	NW	Hor.	Time P.M.
					Solar Heat Gain Factors, Btu									
Jan 21	8	8.1	55.3	141	5	17	111	133	75	5	5	5	13	4
	9	16.8	44.0	238	11	12	154	224	160	13	11	11	54	3
	10	23.8	30.9	274	16	16	123	241	213	51	16	16	96	2
	11	28.4	16.0	289	18	18	61	222	244	118	18	18	123	1
	12	30.0	0.0	293	19	19	20	179	254	179	20	19	133	12
			Half Day Totals		59	68	449	903	815	271	59	59	353	
Feb 21	7	4.3	72.1	55	1	22	50	47	13	1	1	1	3	5
	8	14.8	61.6	219	10	50	183	199	94	10	10	10	43	4
	9	24.3	49.7	271	16	22	186	245	157	17	16	16	98	3
	10	32.1	35.4	293	20	21	142	247	203	38	20	20	143	2
	11	37.3	18.6	303	23	23	71	219	231	103	23	23	171	1
	12	39.2	0.0	306	24	24	25	170	241	170	25	24	180	12
			Half Day Totals		81	144	634	1035	813	250	81	81	546	
Mar 21	7	11.4	80.2	171	8	93	163	135	21	8	8	8	26	5
	8	22.5	69.6	250	15	91	218	211	73	15	15	15	85	4
	9	32.8	57.3	281	21	46	203	236	128	21	21	21	143	3
	10	41.6	41.9	297	25	26	153	229	171	28	25	25	186	2
	11	47.7	22.6	304	28	28	78	198	197	77	28	28	213	1
	12	50.0	0.0	306	28	28	30	145	206	145	30	28	223	12
			Half Day Totals		112	310	849	1100	692	218	112	112	764	
Apr 21	6	7.4	98.9	89	11	72	88	52	5	4	4	4	11	6
	7	18.9	89.5	207	16	141	201	143	16	14	14	14	61	5
	8	30.3	79.3	253	22	128	225	189	41	21	21	21	124	4
	9	41.3	67.2	275	26	80	203	204	83	26	26	26	177	3
	10	51.2	51.4	286	30	37	153	194	121	32	30	30	218	2
	11	58.7	29.2	292	33	34	81	161	146	52	33	33	244	1
	12	61.6	0.0	294	33	33	36	108	155	108	36	33	253	12
			Half Day Totals		153	509	969	1003	489	196	146	145	962	
May 21	5	1.9	114.7	1	0	0	0	0	0	0	0	0	0	7
	6	12.7	105.6	143	35	128	141	71	10	10	10	10	30	6
	7	24.0	96.6	216	28	165	209	131	20	18	18	18	87	5
	8	35.4	87.2	249	27	149	220	164	29	25	25	25	146	4
	9	46.8	76.0	267	31	105	197	175	53	30	30	30	196	3
	10	57.5	60.9	277	34	54	148	163	83	35	34	34	234	2
	11	66.2	37.1	282	36	38	81	130	105	42	36	36	258	1
	12	70.0	0.0	284	37	37	40	82	112	82	40	37	265	12
			Half Day Totals		203	643	1002	874	356	194	171	170	1083	
June 21	5	4.2	117.3	21	10	21	20	6	1	1	1	1	2	7
	6	14.8	108.4	154	47	142	151	70	12	12	12	12	39	6
	7	26.0	99.7	215	37	172	207	122	21	20	20	20	97	5
	8	37.4	90.7	246	29	156	215	152	29	26	26	26	153	4
	9	48.8	80.2	262	33	113	192	161	45	31	31	31	201	3
	10	59.8	65.8	272	35	62	145	148	69	36	35	35	237	2
	11	69.2	41.9	276	37	40	80	116	88	41	37	37	260	1
	12	73.5	0.0	278	38	38	41	71	95	71	41	38	267	12
			Half Day Totals		242	714	1019	810	311	197	181	180	1121	

July 21	5	2.3	115.2	2	0	2	1	0	0	0	0	0	0	7
	6	13.1	106.1	137	37	125	137	68	10	10	10	10	31	6
	7	24.3	97.2	208	30	163	204	127	20	19	19	19	88	5
	8	35.8	87.8	241	28	148	216	160	29	26	26	26	145	4
	9	47.2	76.7	259	32	106	194	170	52	31	31	31	194	3
	10	57.9	61.7	269	35	56	146	159	80	36	35	35	231	2
	11	66.7	37.9	274	37	39	81	127	102	42	37	37	255	1
	12	70.6	0.0	276	38	38	41	80	109	80	41	38	262	12
			Half Day Totals		211	645	986	850	347	197	177	176	1074	
Aug 21	6	7.9	99.5	80	12	67	82	48	5	5	5	5	11	6
	7	19.3	90.0	191	17	135	191	135	17	15	15	15	62	5
	8	30.7	79.9	236	23	126	216	180	40	22	22	22	122	4
	9	41.8	67.9	259	28	82	197	196	79	28	28	28	174	3
	10	51.7	52.1	271	32	40	149	187	116	34	32	32	213	2
	11	59.3	29.7	277	34	35	81	156	140	52	34	34	238	1
	12	62.3	0.0	279	35	35	38	105	149	105	38	35	247	12
			Half Day Totals		161	503	936	961	471	202	154	153	945	
Sep 21	7	11.4	80.2	149	8	84	146	121	21	8	8	8	25	5
	8	22.5	69.6	230	16	87	205	199	71	16	16	16	82	4
	9	32.8	57.3	263	22	47	195	226	124	23	22	22	138	3
	10	41.6	41.9	279	26	28	148	221	165	30	26	26	180	2
	11	47.7	22.6	287	29	29	77	192	191	77	29	29	206	1
	12	50.0	0.0	290	30	30	32	141	200	141	32	30	215	12
			Half Day Totals		116	300	803	1045	672	221	117	116	738	
Oct 21	7	4.5	72.3	48	1	20	45	41	12	1	1	1	3	5
	8	15.0	61.9	203	10	49	173	187	88	10	10	10	43	4
	9	24.5	49.8	257	17	23	180	235	151	18	17	17	96	3
	10	32.4	35.6	280	21	22	139	238	196	38	21	21	100	2
	11	37.6	18.7	290	23	23	70	212	224	100	23	23	167	1
	12	39.5	0.0	293	24	24	26	165	234	165	26	24	177	12
			Half Day Totals		83	143	610	989	783	245	84	83	535	
Nov 21	8	8.2	55.4	136	5	17	107	128	72	5	5	5	14	4
	9	17.0	44.1	232	12	13	151	219	156	13	12	12	54	3
	10	24.0	31.0	267	16	16	122	237	209	50	16	16	96	2
	11	28.6	16.1	283	19	19	61	218	240	116	19	19	123	1
	12	30.2	0.0	287	19	19	21	176	250	176	21	19	132	12
			Half Day Totals		61	71	442	884	798	267	62	61	353	
Dec 21	8	5.5	53.0	88	2	7	67	83	49	3	2	2	6	4
	9	14.0	41.9	217	9	10	135	205	151	12	9	9	39	3
	10	20.7	29.4	261	14	14	113	232	210	55	14	14	77	2
	11	25.0	15.2	279	16	16	56	217	242	120	16	16	103	1
	12	26.6	0.0	284	17	17	18	177	253	177	18	17	113	12
			Half Day Totals		49	54	380	831	781	273	50	49	282	↑
					N	NW	W	SW	S	SE	E	NE	Hor.	← P.M.

Table 4.28a ***Shading Coefficients—Single Glass with Indoor Shading by Venetian Blinds and Roller Shades***

Type of Glass	Nominal Thickness[a]	Solar Trans.[a]	Type of Shading				
			Venetian Blinds		Roller Shade		
					Opaque		Translucent
			Medium	Light	Dark	White	Light
Regular Sheet	$\frac{3}{32}$ to $\frac{1}{4}$	0.87–0.80					
Regular Plate	$\frac{1}{4}$ to $\frac{1}{2}$	0.80–0.71					
Regular Pattern	$\frac{1}{8}$ to $\frac{9}{32}$	0.87–0.79	0.64	0.55	0.59	0.25	0.39
Heat-Absorbing Pattern	$\frac{1}{8}$	—					
Grey Sheet	$\frac{3}{16}$, $\frac{7}{32}$	0.74, 0.71					
Heat-Absorbing Sheet	$\frac{7}{32}$	0.51					
Heat-Absorbing Plate	$\frac{1}{4}$	0.46					
Heat-Absorbing Pattern	$\frac{3}{16}$, $\frac{1}{4}$	—	0.57	0.53	0.45	0.30	0.36
Grey Sheet	$\frac{1}{8}$, $\frac{7}{32}$	0.59, 0.45					
Grey Plate	$\frac{13}{64}$, $\frac{1}{4}$	0.52, 0.45					
Heat-Absorbing Sheet, Plate or Pattern	—	0.44–0.30					
Heat-Absorbing Plate	$\frac{3}{8}$	0.34	0.54	0.52	0.40	0.28	0.32
Grey Plate	$\frac{3}{8}$	0.33					
Heat-Absorbing Sheet, Plate or Pattern	—	0.29–0.15	0.51	0.50	0.36	0.28	0.31
Grey Plate	$\frac{1}{2}$	0.24					

[a]Refer to manufacturer's literature for values.

[b]For vertical blinds with opaque white and beige louvers in the tightly closed position, SC is 0.25 and 0.29 when used with glass of 0.71 to 0.80 transmittance.

Table 4.28b ***Shading Coefficients for Insulating Glass[c] with Indoor Shading by Venetian Blinds and Roller Shades***

Type of Glass	Nominal Thickness, each light	Solar Trans.[d]		Type of Shading				
				Venetian Blinds		Roller Shade		
						Opaque		Translucent
		Outer Pane	Inner Pane	Medium	Light	Dark	White	Light
Regular Sheet Out Regular Sheet In	$\frac{3}{32}$, $\frac{1}{8}$	0.87	0.87	0.57	0.51	0.60	0.25	0.37
Regular Plate Out Regular Plate In	$\frac{1}{4}$	0.80	0.80					
Heat-Absorbing Plate Out Regular Plate In	$\frac{1}{4}$	0.46	0.80	0.39	0.36	0.40	0.22	0.30
Grey Plate Out Regular Plate In	$\frac{1}{4}$	0.46	0.80					

[c]Refers to factory-fabricated units with $\frac{3}{16}$, $\frac{1}{4}$, or $\frac{1}{2}$ in. air space, or to prime windows plus storm windows.

[d]Refer to manufacturer's literature for exact values.

Table 4.29 *Recommended Rate of Heat Gain from Commercial Cooking Appliances Located in the Air-Conditioned Area*[a]

Appliance	Capacity	Overall Dim., Inches Width × Depth × Height	Miscellaneous Data (Dimensions in Inches)	Manufacturer's Input Rating: Boiler hp or Watts	Manufacturer's Input Rating: Btu/Hr	Probable Max. Hourly Input Btu/Hr	Recommended Rate of Heat Gain, Btu/Hr: Without Hood: Sensible	Without Hood: Latent	Without Hood: Total	With Hood[b]: All Sensible
			Electric, Counter Type							
Coffee brewer	240 cups per hr									
Per burner				625	2,130	1,000	770	230	1,000	340
Per warmer				160	545	300	230	70	300	90
Automatic		27 × 21 × 22	4-burner + water htr.	5,000	17,000	8,500	6,500	2,000	8,500	1,700
Coffee urn	3 gal.			2,000	6,800	3,400	2,550	850	3,400	1,000
	5 gal.			3,000	10,200	5,100	3,850	1,250	5,100	1,600
	8 gal. twin			4,000	13,600	6,800	5,200	1,600	6,800	2,100
Deep fat fryer	14 lb fat	13 × 22 × 10		5,500	18,750	9,400	2,800	6,600	9,400	3,000
	21 lb fat	16 × 22 × 10		8,000	27,300	13,700	4,100	9,600	13,700	4,300
Dry food warmer, per sq ft of top				240	820	400	320	80	400	130
Egg boiler	2 cups	10 × 13 × 25		1,100	3,750	1,900	1,140	760	1,900	600
Griddle, frying, per sq ft of top				2,700	9,200	4,600	3,000	1,600	4,600	1,500
Griddle-Grill		18 × 20 × 13	Grid, 200 sq in.	6,000	20,400	10,200	6,600	3,600	10,200	3,200
Hotplate		18 × 20 × 13	2 heating units	5,200	17,700	8,900	5,300	3,600	8,900	2,800
Roaster		18 × 20 × 13		1,650	5,620	2,800	1,700	1,100	2,800	900
Roll warmer		18 × 20 × 13		1,650	5,620	2,800	2,600	200	2,800	900
Toaster, continuous	360 slices/hr	15 × 15 × 28	2 slices wide	2,200	7,500	3,700	1,960	1,740	3,700	1,200
	720 slices/hr	20 × 15 × 28	4 slices wide	3,000	10,200	5,100	2,700	2,400	5,100	1,600
Toaster, pop-up	4 slice	12 × 11 × 9		2,540	8,350	4,200	2,230	1,970	4,200	1,300
Waffle iron		18 × 20 × 13	2 grids	1,650	5,620	2,800	1,680	1,120	2,800	900
			Steam Heated							
Coffee urn	3 gal.			0.2	6,600	3,300	2,180	1,120	3,300	1,000
	5 gal.			0.3	10,000	5,000	3,300	1,700	5,000	1,600
	8 gal. twin			0.4	13,200	6,600	4,350	2,250	6,600	2,100

(Continued)

Table 4.29 ***Recommended Rate of Heat Gain from Commercial Cooking Appliances Located in the Air-Conditioned Area[a] (Continued)***

Appliance	*Capacity*	*Overall Dim., Inches Width × Depth × Height*	*Miscellaneous Data (Dimensions in Inches)*	*Manufacturer's Input Rating*		*Probable Max. Hourly Input Btu/Hr*	*Recommended Rate of Heat Gain, Btu/Hr*			
				Boiler hp or Watts	*Btu/Hr*		*Without Hood*			*With Hood[b]*
							Sensible	*Latent*	*Total*	*All Sensible*
Steam Heated										
Steam table per sq ft of top			With insets	0.05	1,650	825	500	325	825	260
Bain marie per sq ft of top			Open tank	0.10	3,300	1,650	825	825	1,650	520
Oyster steamer				0.5	16,500	8,250	5,000	3,250	8,250	2,600
Steam kettles per gal. capacity			Jacketed type	0.06	2,000	1,000	600	400	1,000	320
Compartment steamer per compartment		24 × 25 × 12 compartment	Floor mounted	1.2	40,000	20,000	12,000	8,000	20,000	6,400
Compartment steamer	3 pans 12 × 20 × $2\frac{1}{2}$		Single counter unit	0.5	16,500	8,250	5,000	3,250	8,250	2,600
Plate warmer per cu ft				0.05	1,650	825	550	275	825	260

[a]Heat gain from cooking appliances located in the conditioned area (but not included in the table) should be estimated as follows:

1. Obtain *probable maximum hourly input* by multiplying the manufacturer's hourly input rating by the usage factor of 0.50.
2. If appliances are installed without an exhaust hood, the estimated latent heat gain is 34 percent of the *probable maximum hourly input* and the sensible heat gain is 66 percent.
3. If appliances are installed under an effective exhaust hood, the estimated heat gain is all sensible heat and can be calculated from equations (18) and (20) in the text.

[b]For poorly designed or undersized exhaust systems the heat gains in this column should be doubled and half of the increase assumed as latent heat.

Table 4.30 ***Rates of Heat Gain from Occupants of Conditioned Spaces***[a]

Degree of Activity	*Typical Application*	*Total Heat Adults, Male, Btu/Hr*	*Total Heat Adjusted,*[b] *Btu/Hr*	*Sensible Heat, Btu/Hr*	*Latent Heat, Btu/Hr*
Seated at rest	Theater—matinee	390	330	225	105
	Theater—evening	390	350	245	105
Seated, very light work	Offices, hotels, apartments	450	400	245	155
Moderately active office work	Offices, hotels, apartments	475	450	250	200
Standing, light work; or walking slowly.	Department store, retail store, dime store.	550	450	250	200
Walking; seated Standing; walking slowly	Drug store, Bank	550	500	250	250
Sedentary work	Restaurant[c]	490	550	275	275
Light bench work	Factory	800	750	275	475
Moderate dancing	Dance hall	900	850	305	545
Walking 3 mph; moderately heavy work	Factory	1000	1000	375	625
Bowling[d] Heavy work	Bowling alley Factory	1500	1450	580	870

[a]*Note.* Tabulated values are based on 75°F room dry-bulb temperature. For 80°F room dry-bulb, the total heat remains the same, but the sensible heat should be decreased by approximately 20 percent, and the latent heat values increased accordingly.

[b]*Adjusted total heat gain* is based on normal percentage of men, women, and children for the application listed, with the postulate that the gain from an adult female is 85 percent of that for an adult male, and that the gain from a child is 75 percent of that for an adult male.

[c]Adjusted total heat value for *sedentary work, restaurant,* includes 60 Btu per hour for food per individual (30 Btu sensible and 30 Btu latent).

[d]For *bowling* figure one person per alley actually bowling, and all others are sitting (400 Btu per hour) or standing (550 Btu per hour).

References and Bibliography

1. *Load Calculation for Residential Winter and Summer Air Conditioning,* 4th ed., Manual J. NESCA, National Environmental Systems Contractors Association, Arlington, Va., Copyright © 1975.

The following publications of the American Society of Heating, Refrigerating and Air Conditioning Engineers:

2. ASHRAE Standard 62-73, *Standards for Natural and Mechanical Ventilation, 1973.*

3. ASHRAE Standard 55-74, *Thermal Environmental Conditions for Human Occupancy, 1974.*

4. ASHRAE Standard 90-75, *Energy Conservation in New Building Design, 1975.*

5. *ASHRAE Handbook and Product Directory, 1977, Fundamentals.*

5

Heating, Cooling, Ventilation

5.1 Energy Sources for Heating

Chapter 1 introduced this subject. There, the groundwork was laid for a continual awareness of the limits of our nonrenewable resources, the adoption of renewable sources, and the ever-present challenge of greater efficiency in the utilization of any and all of our very precious and scarce natural assets.

The energy situation is "wide open." It is with the greatest difficulty that firm opinions can be formed about future trends. The current chapter deals largely with the methods of using *conventional* fuels and systems for heating, cooling, and ventilation. This approach is justified by the fact that, at least for the present, backup systems are essential even with the use of renewable sources such as solar energy.

As we watch the evolving energy scene, it would be well to start with an inventory, which can be expanded as advances are made. Energy sources can include the following.

(*a*) ***Natural Materials.*** Wood can be a renewable energy source as a product or by-product of the proper management of our forests. One sees a large increase in the sale and use of freestanding woodburners. Their use is common in the timber regions of the Pacific Northwest. We may return to the early English custom of lighting a fire only in the room where it is needed and closing the doors of other rooms. The large central power plant of the University of Oregon at Eugene burns *hogged fuel,* wood chips from the lumber industry. Peat from the bogs of Ireland, though not renewable, is still a prime source for heating in that country. Natural oils such as whale oil and seal oil served earlier generations in Alaska and in the Arctic generally (see Fig. 1.1, page 3).

(*b*) ***Fossil Fuels: Gas, Oil, and Coal.*** These fuels are, of course, the sources where the greatest change in availability and use is most evident. For several recent years on Long Island, in the New York area, gas has not been available for *new* installations. Depending on constantly varying prices, oil and electricity are the alternatives. At the same time, great piles of coal were seen in reserve at one power plant of the Consolidated Edison Company awaiting approval for use. Con Ed serves New York City and Westchester County. Consulting engineers are being contacted by their clients concerning plans for the design of new coal-fired boiler plants and for the conversion of oil and gas plants to the use of coal. For further information, refer to item No. 1 in the references at the end of this chapter.

(*c*) ***Hydroelectric Power.*** Within limits this could be considered a renewable source. However, there may be a limit to the burden we can place on our natural waterways. As a percentage of the total energy use in the nation, the amount contributed by hydroelectric energy is relatively small; yet the Columbia River, for example, is already dotted with a multitude of such plants.

(*d*) ***Nuclear Plants.*** At the time of this writing, environmentalists, and the general public, largely oppose the rapid increase in nuclear installations. Fusion looms as a future possibility.

(*e*) ***The Heat Pump.*** This popular energy *saver* is not new. Developed many years ago, it was not frequently adopted during the years that provided us with cheap energy sources. Most suitable in moderate and warmer climates, it is, at 40° N. Latitude, about twice as efficient as straight electric-resistance heating. It refrigerates cold outdoor air in winter to an even lower temperature. It steps up to a higher temperature, by means of a refrigeration cycle, the heat that is drawn from the outdoor air. Its low-cost operation was evident in the Arkansas Story, page 144, and it is described later in this chapter, page 234.

(*f*) ***Solar Energy.*** Developed several decades ago by Dr. Maria Telkes of the Massachusetts Institute of Technology, the use of solar energy is our great hope for the future. For houses, where the energy demand for heating is relatively light, it can replace 40 to 60% of the energy required for a season's heating. We had better keep our atmosphere clear or solar effect could be dimmed. In densely built cities where collector plates could be shaded by later high-rise construction, legal rights-to-the-sun may have to be established. In New Mexico a controlling law has already been passed. For large power demand the collector plate area could be bigger than we presently care to contemplate. But we have made a start.

(*g*) ***Geothermal Energy.*** The earth's crust is much thinner than most of us realize. Whether or not the enormous heat of the earth's center can be trapped and utilized on a large scale is not yet established. In the Cerra Prieto region of Mexico's Baja California, however, the most recent and largest geothermal plant in the Americas is in operation. Steam at 670° F, 7500 ft below the surface, is piped to supply the energy needs of nearly a half million people. Japan, Italy, and New Zealand have major geothermal projects. For many years, the town of Rekjavik in cold Iceland has been liberally supplied with geothermal heat.

(*h*) ***Trash.*** Table 5.1 of the Incinerator Institute of America rates type 0 trash as having a heating value of 8500 Btu/lb. This value is 58% of the comparative value of anthracite coal, as listed in Table 5.3. Values of trash and rubbish range between 8500 and 6500 Btu/lb. Recovery of a large portion of this energy is now a possibility. Recovery plants are now in operation. One, completed recently in Westbury, Long Island, N.Y., recovers energy and also eases the problem of finding diminishing locations where garbage can be used as land fill. Before this concept was introduced, efficient incineration (Fig. 5.1, Type III in Table 5.2) had already effected a combustion that assured minimal air pollution. Now further adaptive design can retrieve much of the heat of the trash and rubbish by recovery in a heat exchanger that produces steam. The basic arrangement is seen in Figure 5.2 and engineers' conclusions as to feasibility are summarized in the legend of that illustration.

5.2 Fuel Consumption Rates, Hourly and Yearly

(*a*) ***Hourly Rates.*** When outdoor temperatures in winter drop below 65° F (20.6° C), heating systems usually begin to operate. The hourly rate of burning fuel depends on the hourly heat loss from the interior space. If the fuel burner and the boiler or furnace are selected to run continuously at the outdoor, critical winter design-temperature, then they will cycle (run intermittently) at higher outdoor temperatures. The equipment, however, *is* selected on the basis of the maximum winter demand rate and therefore relates to the calculated heat loss at the design temperature.

EXAMPLE **5.1.** Calculate the rates of burning the several fuels (or the rate of using electricity) to make up the hourly heat loss, under design conditions, of a residence. Its maximum hourly heat loss is 159,840 Btuh. For fuel values refer to the data in Table 5.3.

SOLUTION. If, for instance, coal were used, the statement would be:

lb/hr × Btu/lb heat value × efficiency
= Btuh heat loss

Table 5.1 ***Types, Composition, Heating Values of Wastes***

Type	Typical Mixture	Incombustible Solids, %	Moisture, %	Heating Value, Btu/lb
0	Trash, a mixture of highly combustible waste such as paper, cardboard, cartons, wood boxes, and combustible floor sweepings, from commercial and industrial activities. The mixtures contain up to 10% by weight of plastic bags, coated paper, laminated paper, treated corrugated cardboard, oily rags, and plastic or rubber scraps.	5	10	8500
1	Rubbish, a mixture of combustible waste such as paper, cardboard cartons, wood scrap, foliage, and combustible floor sweepings, from domestic, commercial, and industrial activities. The mixture contains up to 20% by weight of restaurant or cafeteria waste, but contains little or no treated papers, plastic, or rubber wastes.	10	25	6500
2	Refuse, consisting of an approximately even mixture of rubbish and garbage by weight. (Common to apartment and residential occupancy)	7	50	4300
3	Garbage, consisting of animal and vegetable wastes from restaurants, cafeterias, hotels, hospitals, markets, and like installations.	5	70	2500
4	Human and animal remains, consisting of carcasses, organs, and solid organic wastes from hospitals, laboratories, abattoirs, animal pounds, and similar sources.	5	85	1000
5	By-product waste—gaseous, liquid, or semi-liquid—such as tar, paints, solvents, sludge, fumes, etc., from industrial operations.	—	—	[a]
6	Solid by-product waste, such as rubber, plastics, wood waste, etc., from industrial operations.	—	—	[a]

Source. Incinerator Institute of America.

[a]Btu values must be determined by the individual materials to be destroyed.

Transposing

$$\text{lb/hr} = \frac{\text{Btuh heat loss}}{\text{Btu/lb} \times \text{efficiency}} \qquad (1)$$

(Other efficiency statements are similar.) Applying values to this and to statements for the other fuels, the rates are

$$\text{Coal} \quad \frac{159{,}840}{14{,}600 \times 0.70} = 15.6 \text{ lb/hr}$$

$$\text{Oil} \quad \frac{159{,}840}{141{,}000 \times 0.75} = 1.51 \text{ gal/hr}$$

$$\text{Gas} \quad \frac{159{,}840}{1052 \times 0.75} = 203 \text{ cu ft/hr}$$

$$\text{Electricity} \quad \frac{159{,}840}{3.41 \times 1.00} = 46{,}600 \text{ watts (46.6 kilowatts)}$$

The foregoing results are based on the assumption that the boiler and its piping are enclosed within the useful volume of the house as is most usual. If they were in cold basements, or if the ducts or pipes ran through unheated space, more fuel would be used and

Table 5.2 ***Classification of Incinerators***

Type	*Description*	*Feed*	*Miscellaneous Criteria*	*Burning Rate pounds per hour*	*Type of Waste Consumed*
I	Portable, packaged, completely assembled	Direct	Not over 5 cu ft storage	Not over 25	2
IA	Portable, packaged, or job assembled	Direct	Primary chamber volume, 5–15 cu ft	25–99	0, 1, 2
				25–74	3
II	Single chamber, served by one vertical flue functioning both as a chute for charging waste and to carry the products of combustion to atmosphere. This type installed in apartment houses or multiple dwellings up to five stories high	Flue	More than 2 sq ft burning area	—	2
IIA	Multiple chambers, served by a vertical chute for charging wastes from two or more floors above the incinerator and a separate flue for products of combustion	Chute	More than 2 sq ft burning area	—	1, 2 (not industrial)
III	—	Direct	—	100 and over	0, 1, 2
IV	—	Direct	—	75 and over	3
V	Municipal type, rated in tons per hour or tons per 24 hours	—	—	—	0, 1, 2, 3 or combination of all four
VI	Crematory and pathological incinerators	—	—	—	4
VII	Special designs for specific by-product wastes	—	—	—	5, 6

Source. Incinerator Institute of America.

Table 5.3 ***Approximate Heat Values of the Three Fossil Fuels and the Thermal Equivalent of Electricity. Approximate Efficiencies of Burner-Boiler and of Electrical Heating***

Fuel	*Heat Value*	*Efficiency, %*
Anthracite coal	14,600 Btu/lb	65–75
No. 2 oil	141,000 Btu/gal	70–80
Natural gas	1052 Btu/cu ft	70–80
Electricity	1 watt = 3.41 Btuh	95–100

the efficiency would be less. The rates established set the values by which the fuel-burning apparatus is selected. For instance, if oil were used, a nozzle that would inject oil at the rate of about $1\frac{1}{2}$ gal/hr should be tried.

These rates are for design (extreme) conditions and are not typical of the lower average rate of operation throughout the winter.

Note: The hourly heat loss of 159,840 Btuh in Example 5.1 indicates a large house. Thermal budgets in future planning must be judged not only by Average Thermal Transmittance (U_0) (see Section 8.7, page 333), but also by the total rate of energy use. We may tend to live in smaller houses or in the more energy-saving multiple dwelling units.

(*b*) *Yearly Fuel Consumption.* An important determinant of annual fuel use is the number of degree-days at the geographic location of the building. For any one 24-hour day, the number of degree-days is the difference between 65° F and the mean outdoor temperature for the day. Their total for the year, as shown in Table C.1, Appendix page 1295 and graphically in Figure 5.3, is a measure of the effect of the local climate on fuel demands. The choice of using only temperatures

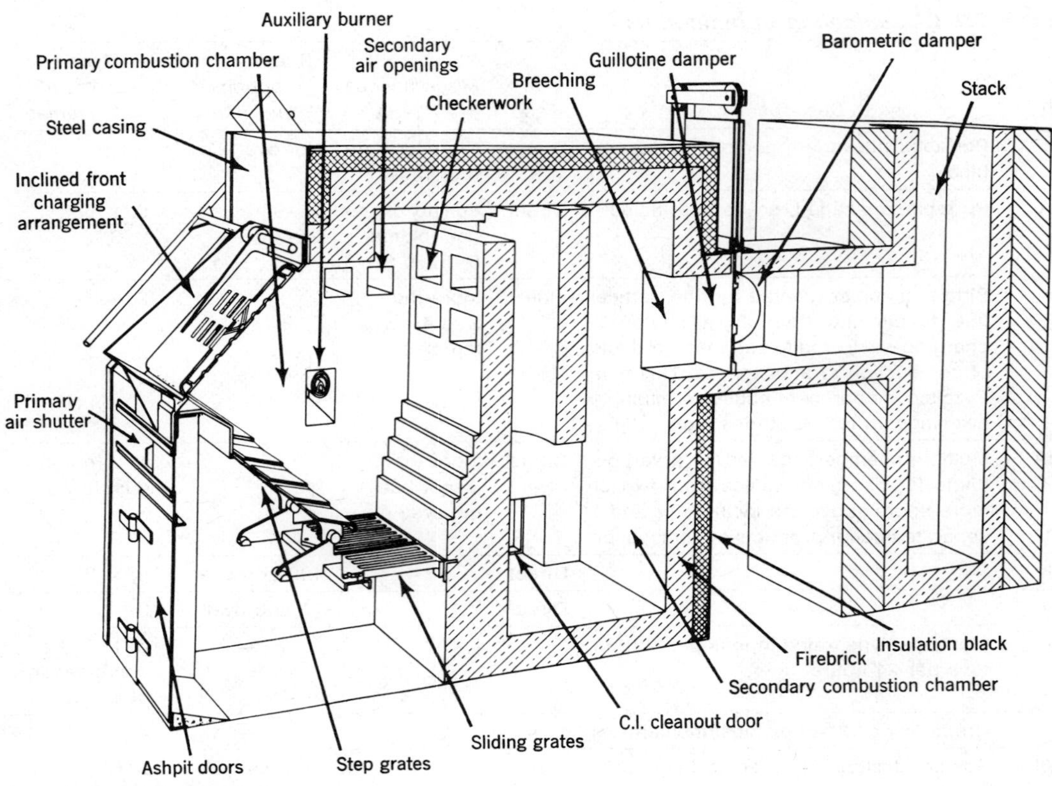

Joseph Godor Incinerators

Fig. 5.1 *Modern direct-fed, multiple chamber incinerator, Type III* (*Table* 5.2). *Suitable for the incineration of waste types 0, 1, and 2* (*Table 5.1*).

under 65° F evolves from findings by heating engineers that buildings seldom require heating until outdoor temperatures drop below 65° F. Next in importance is the size of the building as measured thermally by the hourly heat loss under critical conditions as described in Chapter 4. Finally the kind of fuel, its heat value per unit (see Table 5.3), and the overall efficiency of the boiler and the heating systems have their effect. Fuel consumption calculations thus include:

Calculated hourly heat loss of the building.

Degree-days annually.

Average winter temperature (see Table C.1, Appendix page 1295).

Heating value per unit of fuel.

Efficiency of the heating system.

Fig. 5.2 (opposite) *Incineration with heat recovery. Reprinted from "Heat Recovery from Solid-Waste Incineration,"* Technical Letter, *Vol. 24, No. 6, May 1974, by Syska & Hennessy, Engineers, 110 West 50th Street, New York, New York, 10020. For full information the reader is referred to the above* Technical Letter. *Its conclusions, excerpted from the* Letter, *follow:*

Conclusions

Incineration with heat recovery is now a real possibility for universities, hospitals, large office buildings, and residential complexes. Where it is feasible, it will save fuel and pay for itself in a short time.

The feasibility in each individual case must be demonstrated by a detailed study which includes among other parameters: a qualitative and quantitative evaluation of the waste, collection methods and costs, heat load profiles of the complex including recoverable heat integration into them, equipment location, as well as all pertinent labor and environmental factors, legal codes, and ordinances.

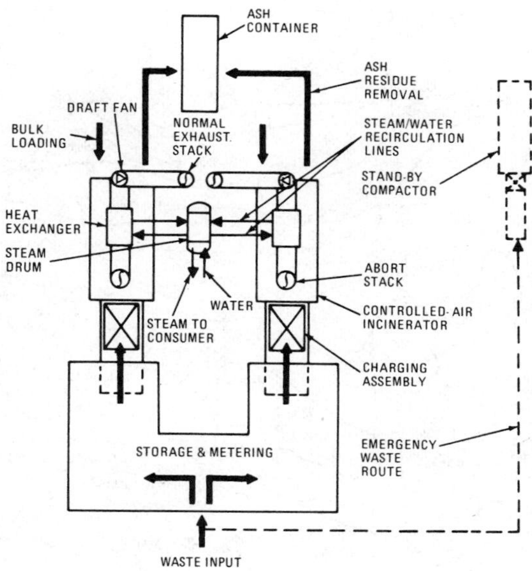
ASH CONTAINER
ASH RESIDUE REMOVAL
DRAFT FAN
BULK LOADING
NORMAL EXHAUST. STACK
STEAM/WATER RECIRCULATION LINES
STAND-BY COMPACTOR
HEAT EXCHANGER
STEAM DRUM
ABORT STACK
WATER
STEAM TO CONSUMER
CONTROLLED-AIR INCINERATOR
CHARGING ASSEMBLY
STORAGE & METERING
EMERGENCY WASTE ROUTE
WASTE INPUT

PLAN VIEW

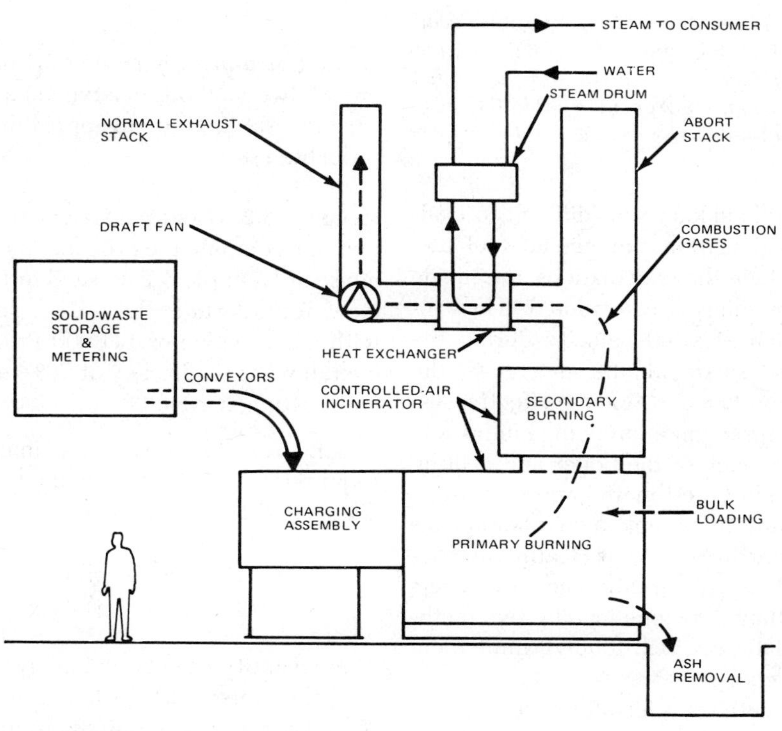
STEAM TO CONSUMER
WATER
STEAM DRUM
NORMAL EXHAUST STACK
ABORT STACK
DRAFT FAN
COMBUSTION GASES
SOLID-WASTE STORAGE & METERING
HEAT EXCHANGER
CONVEYORS
CONTROLLED-AIR INCINERATOR
SECONDARY BURNING
CHARGING ASSEMBLY
BULK LOADING
PRIMARY BURNING
ASH REMOVAL

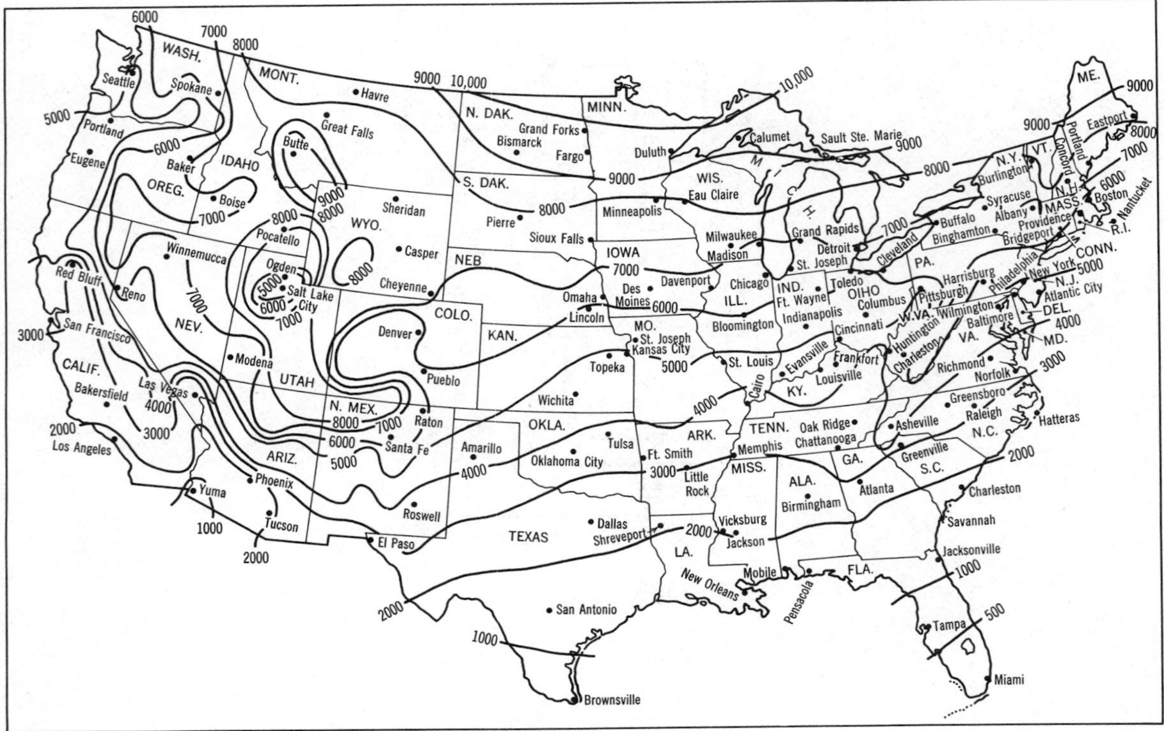

Fig. 5.3 Normal number of degree-days per year. Adapted from Building Systems Design. *For a list of average yearly degree-days in cities in the United States and Canada, see Appendix Table C1, page 1295. The list also includes average winter temperatures.*

Many factors unknown or difficult to evaluate affect calculations for annual fuel use, especially when the calculations are made before the building is constructed. They include variation of wind velocity during the winter, actual solar gain through glass, the tightness of construction, and the effectiveness of weatherstripping in minimizing infiltration. Even the efficiency of the boiler and heating system can only be estimated approximately. So variant are these unknown components that the American Society of Heating, Refrigerating and Air Conditioning Engineers (ASHRAE), though recommending two methods of calculation (which follow), limit them only to *residences*.

The two methods referred to are:

The Calculated Heat Loss Method.

The Degree-Day Method.

(*c*) ***Calculations, Annual Fuel Use.*** Applying the two methods in a typical example, one can arrive at limits or an approximate range of probable use.

EXAMPLE **5.2.** Using the Calculated Heat Loss method, compute the probable fuel use for the house in Example 5.1, located in Long Island, New York. Assume the use of No. 2 fuel oil with a heat value of 141,000 Btu/gal and an overall winter efficiency of 70% for the boiler and heating system.

SOLUTION. The quantity of fuel or energy required for heating may be found by the Calculated Heat Loss method from

$$F = \frac{XN}{EC} \tag{1}$$

where

F = Quantity of fuel or energy required (in the units in which C is expressed).

X = Average heat requirement for the period under consideration, British thermal units per hour.

N = Number of heating hours in estimate period (for an Oct. 1–May 1 heating season, 212 days × 24 hr = 5088).

E = Efficiency of utilization of the fuel over the period, expressed as a decimal; not the efficiency at peak or rated load condition.

C = Heating value of one unit of fuel or energy.

The value of X is usually found from the equation

$$X = \frac{H(t - t_a)}{t_d - t_o} \tag{2}$$

where

H = Calculated heat loss including infiltration loss, British thermal units per hour, based on t_o and t_d.

t = Average indoor temperature maintained during heating period, Fahrenheit.

t_a = Average outdoor temperature through estimate period, Fahrenheit (for U.S. cities with an Oct. 1–May 1 heating season—see Table C.1, Appendix page 1295).

t_d = Indoor design temperature, Fahrenheit (usually 70° F).

t_o = Outdoor design temperature, Fahrenheit.

so that the value of F becomes

$$F = \frac{H(t - t_a)N}{E(t_d - t_o)C} \tag{3}$$

Substituting directly in Equation 3, the probable annual fuel use is found to be

$$F = \frac{159{,}840 \times (70 - 43) \times 5088}{0.70(70 - 0) \times 141{,}000} = 3170 \text{ gal/yr}$$

EXAMPLE 5.3. Using the Degree-Day method and the same fuel and efficiency suggested in Example 5.2, find the probable annual fuel use for the same house. Use values listed in Tables 5.4 and 5.5.

SOLUTION. The general equation for calculating the probable fuel consumption by the Degree-Day method is:

$$F = U \times N_b \times D \times C_j \tag{1}$$

Table 5.4 ***Unit Fuel Consumption Constants***

Based on 0° F outdoor temperature, 70° F indoor temperature

	Utilization Efficiency (%)		
	60	*70*	*80*
Fuels and units	*Unit fuel consumption per degree-day per 1000-Btuh design heat loss*		
Gas in therms[a]	0.00572	0.00490	0.00429
Oil in gallons[b]	0.00405	0.00347	0.00304
Coal in pounds[c]	0.0476	0.0408	0.0357

[a] One therm is equal to 100,000 Btu.

[b] Based on a heating value of 141,000 Btu per gallon.

[c] Based on a heating value of 12,000 Btu per pound.

Table 5.5 ***Correction Factors for Outdoor Design Temperatures***[a]

Outdoor Design Temp F	*−20*	*−10*	*0*	*+10*	*+20*
Correction factor	0.778	0.875	1.000	1.167	1.400

[a] The multipliers in Table 5.5, which are high for mild climates and low for cold regions, are not in error as might appear. The unit figures in Table 5.4 are per square foot of radiator or thousand Btu heat loss per degree day. For equivalent buildings and heating seasons, those in warm climates have lower design heat losses and smaller radiator quantities than those in cold cities. Consequently, the *unit* figure in quantity of fuel per (square foot of radiator) (degree-day), is larger for warm localities than for colder regions. Since the northern cities have more radiator surface per given building and a higher seasonal degree-day total than cities in the south, the total fuel per season will be larger for the northern city.

where

F = Fuel consumption for the estimate period.

U = Unit fuel consumption, or quantity of fuel used per (degree-day) (*building load unit*). See Table 5.4.

N_b = Number of *building load units* (when

available, use calculated hourly heat loss instead of actual amount of radiation installed).

D = Number of degree-days for the estimate period, Table C.1, Appendix page 1295.

C_j = Temperature—correction factor from Table 5.5.

Substituting directly in Equation 1, the probable annual fuel use is found to be

$$F = 0.00347 \times 159.8 \times 4811 \times 1.00 = 2670 \text{ gal/yr}$$

A general value of 3000 gal ± would represent the expected yearly consumption of oil for this house based on an average of the results of Examples 5.2 and 5.3.

Similar calculations can be made for other fuels. When quantities have been established for each, local unit prices per pound, gallon, etc. can be applied to establish the probable yearly *cost* of fuel for heating.

(d) Energy Requirements for Heating Commercial and Institutional Buildings. The idea of calculating the annual requirements of a specific fuel for the isolated item of *heating* modern buildings of this type is largely an outmoded concept. The parameters are many and diverse. A few items that make the problem different from that of houses can be noted. Dense human occupancy often provides body heat that is much more than that required to heat a space, even in critically cold weather. By means of heat pumps, thermal gain is frequently interchanged with the lesser energy content of cooler areas. It is nonetheless a credit item in the annual *heating* fuel bill. Computers and other business machines are similar credits. Until recently, high footcandle levels of general office lighting had effects that displaced some of the fuel that would otherwise operate heating units in such buildings. This is evident in Example 4.7, page 163. Smaller windows and the effective heat barrier of reflective insulating glass are again changing the scene. Fossil fuel heating now becomes only one item in the general energy picture. The yearly energy study in commercial and institutional buildings has become a trade-off budget in which each item must justify its value and cost in energy.

5.3. Combustion, Chimneys, and Fuel Storage

As fuels burn to produce heat they require oxygen to support the combustion. Since oxygen is only about one-fifth of the volume of air, reasonably large rates of air flow are required. The air should be drawn in from outdoors at a position close to the fuel burner or led to this location by a duct. For residences and other small buildings a louver about twice the cross-sectional area of the flue should prove satisfactory. It should be arranged to remain open at all times. If an attempt is made to draw this air from the general space of the house, modern tight construction may retard the effort. If the air rate *is* sufficient by the use of this scheme, it is likely to result in undesirable acceleration of cold infiltration flowing in to replace the air that is used.

The most important combustible element in the chemical makeup of fuels is carbon. It may be burned well or poorly. When burned poorly it can cause great energy losses and sooty operation. For success, much depends upon the proper selection of well-designed boilers, furnaces, and burners. This, however, is only part of the story. Adjustments of primary and secondary air rates of flow and of draft (flow of air and gases through the boiler) are important responsibilities of the engineer and the heating contractor. Carbon may burn to carbon monoxide (CO) or more completely to carbon dioxide (CO_2) with greater heat production. Flue gases should be analyzed and the percentage of carbon dioxide measured. The best economy and the cleanest and most efficient combustion occur when the CO_2 content of the flue gases most nearly approaches the values in Table 5.6 for Maximum Theoretical Stoichiometric Percent CO_2 (see also I=B=R standards, Table 5.7). The architect should require that these tests be made with the adjustments necessary to effect the most efficient performance. Flue gas temperatures should be taken. Temperatures lower

Table 5.6 ***Approximate Maximum Theoretical (Stoichiometric) CO_2 Values, and CO_2 Values for Various Fuels with Different Percentages of Excess Air***

Type of Fuel	Stoichiometric Percent CO_2	Percent CO_2 at Given Excess Air Values		
		20%	40%	60%
Coke	21.0	17.5	15.0	13.0
Anthracite	20.2	16.8	14.4	12.6
Bituminous coal	18.2	15.1	12.9	11.3
No. 1 and 2 fuel oil	15.0	12.3	10.5	9.1
No. 6 fuel oil	16.5	13.6	11.6	10.1
Natural gas	12.1	9.9	8.4	7.3
Carbureted water gas	17.2	14.2	12.1	10.6
Coke oven gas	11.2	9.2	7.8	6.8
Mixed gas (natural and carbureted water gas)	15.3	12.5	10.5	9.1
Propane gas (commercial)	13.9	11.4	9.6	8.4
Butane gas (commercial)	14.1	11.6	9.8	8.5

Table 5.7 ***Conditions Set by the Institute of Boiler and Radiator Manufacturers for the Rating of Boilers***

Some of the requirements that must be met by the testing procedure before I-B-R Ratings on cast-iron boilers are approved are:

	Type Oil	Minimum Overall Efficiency	CO_2	Maximum Flue Gas Temperature	Maximum Smoke Reading
Natural draft or induced draft—bare boilers or boiler-burner units	Light	70%	10%	600° F	#2
	Heavy	75%	$12\frac{1}{4}$%	600° F	#4
Natural draft or induced draft—boiler-burner units only	Light	70%	$12\frac{1}{4}$%	600° F	#2
Forced draft—boiler-burner units only	Light	75%	10%	600° F	#2
	Light	75%	$12\frac{1}{4}$%	600° F	#2
	Heavy	75%	$12\frac{1}{4}$%	600° F	#4

Efficiency: The overall efficiency of a boiler must be not less than that specified. This requirement has served to improve the performance of boilers since the establishment of the I-B-R Testing and Rating Code.

Carbon Dioxide (CO_2) *in the Flue Gas:* During oil-fired tests, the oil burner must be set to produce the specific percentage of carbon dioxide in the flue gas. This requirement was established so that all boilers are rated on a comparable basis and to prevent anyone from obtaining ratings that are unrealistic by testing the boiler with an exceptional oil boiler under ideal conditions at a combustion efficiency that could not be reproduced in the field.

Flue Gas Temperature: The flue gas temperature limitation insures the user of safe and economical operation. The temperature is measured in a thoroughly insulated stack.

Draft Loss Through Boiler (*oil-fired*): The draft loss through the boiler, or the difference between the draft in the stack and the draft in the combustion chamber, must not exceed a specified value. This requirement is included because excessive draft losses invariably lead to trouble when boilers are connected to poor chimneys or in poor draft areas.

than 600° F indicate that heat is properly retained in the boiler instead of going up the chimney (see the note on *efficiency* that follows the numerical tabulation in Table 5.7).

It is important that chimneys with their high-temperature flue gases be safely isolated from combustible construction to prevent the possibility of fire. Conventional standards for houses call for a terra cotta flue lining surrounded by 8 in. of brick, with an additional 2 in. of space between the brick and any wood. The space is usually filled with incombustible mineral wool. The size of flue will be dictated by the specification for the boiler or furnace selected for use. Its height (see Fig. 5.4*a*) had been traditionally 35 to 40 ft. The function of providing a draft, for which chimney height was an important consideration, is no longer as necessary as it previously was—because fans are used now. For example, oil is injected under pressure, accompanied by air, and forced in by a fan. Often a draft adjuster in the breeching (smoke pipe) that carries the flue gases to the chimney is arranged to open slightly to *reduce* the normal stack draft. If increased draft should ever be required, an induced draft fan that puts a suction on the flue side of the fire is usually chosen instead of greater stack height. Draft hoods above gas burners prevent downdraft from blowing out the flames.

Prefabricated chimneys (see Fig. 5.4*b*) are replacing with increasing frequency the bulkier and heavier field-built masonry. They offer a number of advantages and may be easily supported on a normal structure.

The storage space to be allowed for coal or oil depends upon the proximity of the supplier and the space available at the building. For oil, when more than 275 gal are stored, it is common practice to use an outside tank buried in the ground (see Fig. 5.5). It is often set on a concrete slab and strapped down to the slab. This prevents the tank from sinking when full or rising in flotation bouyancy that might be caused by adjacent ground water when the tank is empty. The tank, usually of steel, receives two coats of asphalt emulsion to inhibit rust. Tubing for the gauge and for the supply and circulating lines are of copper, and the fill and vent lines are of wrought iron with swing joints to accommodate possible slight settlement of the tank. Oil deliveries are often made on the basis of the degree-days elapsed since the last fill-up of the tank. Thus the customer is relieved of the chore of checking the gauge and ordering periodically.

5.4. Warm Air Heating

The use of air for heating has certain advantages. The motion of air in the space helps to assure uniform conditions and reasonably

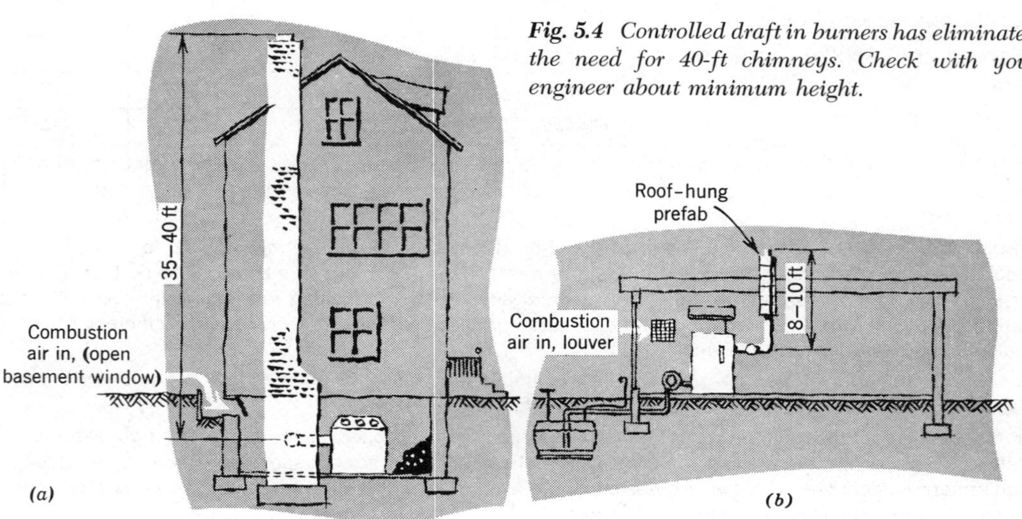

Fig. 5.4 *Controlled draft in burners has eliminated the need for 40-ft chimneys. Check with your engineer about minimum height.*

equal temperatures in all parts of the house. It is possible to clean both the recirculated air and the outdoor air by means of filters and other special air-cleaning equipment. Air may be circulated in nonheating seasons. Fresh air may be introduced to reduce odors and to make up the air exhausted by fans in kitchens, laundries, and bathrooms. Central cooling can be incorporated or introduced if ducts are designed originally for this. Cooling sometimes calls for greater rates of air circulation. Humidification may be achieved by a humidifier in the air stream and, if cooling is included in the design, dehumidification is accomplished in summer. For both heating and cooling, a good arrangement is to place the supply registers in the floor, below areas of glass. This is important for winter operation. Return grills should be on interior walls at high locations. This is especially advantageous if cooling is part of the scheme. High grills pick up the warmer air for recooling at the equipment. Many systems are planned to circulate air at all times with an arrangement to warm or cool it as required.

In planning warm air systems, good balance is achieved if the heating furnace is located reasonably close to the center of the house. After the system is designed, a furnace must be selected. It should be capable of burning fuel at a rate suitable to make up the hourly heat loss in the house. The rate of air delivery must be correct to transmit this heat to the house at the air temperature *rise* that is planned. Finally, the motor and blower must be powerful enough to overcome the friction of air against metal in both the supply and return duct system as well as the friction of air flowing through the furnace, filters, registers, and grills (see Fig. 5.10, page 191). Minor adjustments can be made at the furnace to adapt to the demands of the system and the house.

Some of the system components are discussed below.

(*a*) ***Furnace.*** Figure 5.6 represents a typical furnace, which embraces within its housing the fan (blower), motor, filters, oil burner, and heat-transfer surfaces. A humidifier can be added to this assembly. Arrows indicate the direction of the air. In passing through the blower the air enters at the end of the cylinder opposite the pulley and is forced into the warming chamber by a cylindrical impeller unit. See also Figure 5.7.

(*b*) ***Ducts.*** These are illustrated in Figure 5.8. They are constructed of sheet metal or glass fiber—either round or rectangular. Ductwork will conduct noise unless the following suggestions are followed:

Do not place the blower too close to a return grill.

Fig. 5.5 *Details, fuel oil storage tank.*

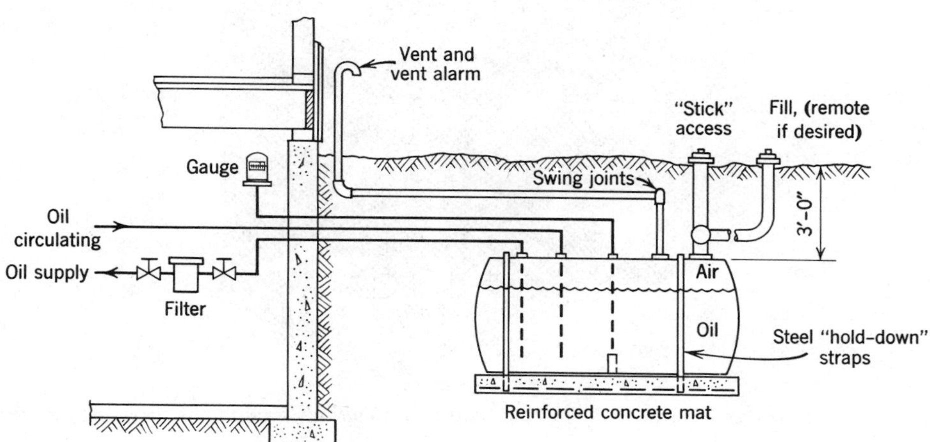

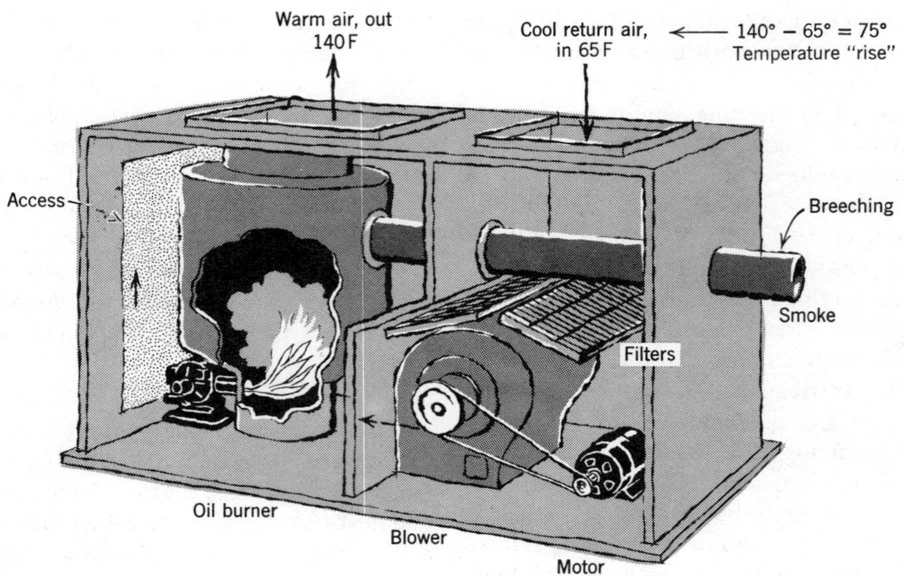

Fig. 5.6 Principle of the warm-air heating furnace. Other styles have the elements "in-line" and replace the U-path flow of air with straight-line flow, vertically upward (Fig. 5.38), horizontal, (Fig. 5.39), or vertically downward (Fig. 5.13a) (counterflow).

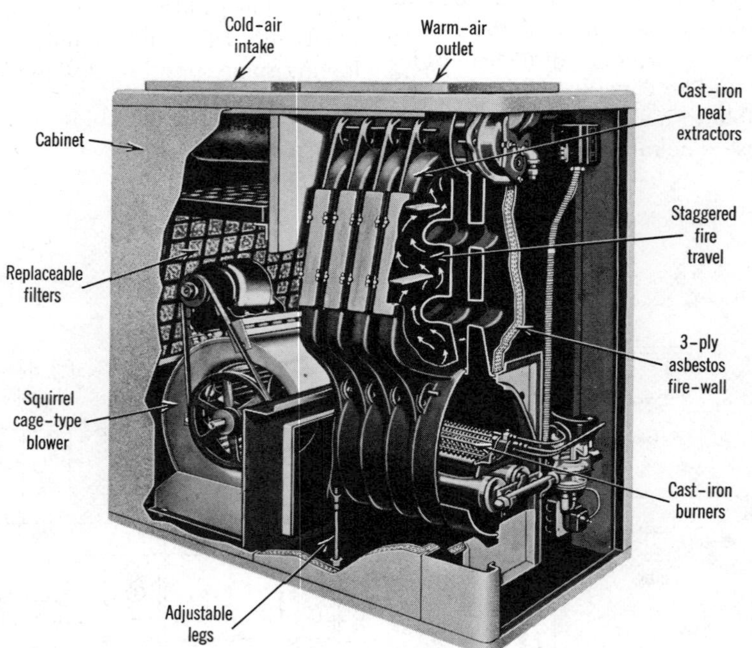

Fig. 5.7 Gas-fired furnace for warm-air heating. Fan blows cold air around cast-iron enclosure within which gas-fired element is located. Thatcher.

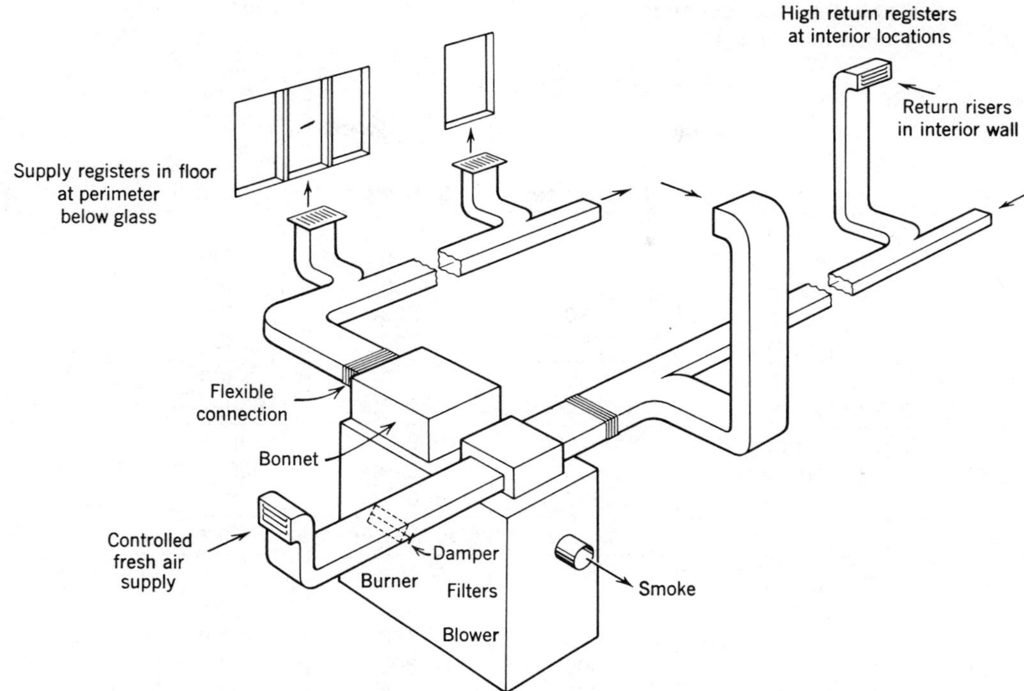

Fig. 5.8 Conventional warm air furnace and ducts. This system with supply registers in the floor under glass and interior, high return registers is suitable for heating or *cooling. Furnace and ducts in basement of this basement-and-one-story house. In two-story houses, supply and return registers should be in the same relative positions in each story.*

Select quiet motors and cushioned mountings.

Do not permit connection or contact of conduits or water piping with the blower housing.

Use canvas-asbestos flexible connection between bonnet and ductwork.

Duct sizing. Sizes may be selected on the basis of permissible air velocity in the duct.

EXAMPLE 5.4. The main duct in the low-velocity, warm air system of a residence delivers 1600 cubic feet per minute (cfm). Select a size for this duct.

SOLUTION. Table 5.8 indicates that 800 ft per minute (fpm) would be an acceptable velocity. The area of the duct in square inches would be

$$\frac{1600 \text{ cfm} \times 144 \text{ (sq in./sq ft)}}{800 \text{ (fpm)}} = 288 \text{ sq in.}$$

A 20 × 14 in. (280 sq in.) duct is acceptable and will operate without undue noise or friction.

(*c*) ***Dampers.*** These will be necessary to balance the system and adjust it to the desires of the occupants (see Fig. 5.9). Splitter dampers are used where branch ducts leave the larger trunk ducts. Each riser can have its flow controlled by an adjustable damper in the basement at the foot of the riser. Labels should indicate the rooms served. Some codes require dampers of fire-resistant material actuated by fusible links in order to prevent the possible spread of fire through a duct system (see Chapter 13).

(*d*) ***Registers.*** Supply registers (Fig. 5.12, page 193) should be equipped with dampers and should have their vanes arranged to disperse the air and to reduce its velocity as soon as possible after entering the room. A common

Table 5.8 ***Recommended and Maximum Duct Velocities for Conventional Systems***

Designation	Residences	Schools, Theaters, Public Buildings	Industrial Buildings
	Recommended Velocities, Feet per Minute		
Outdoor air intakes[a]	500	500	500
Filters[a]	250	300	350
Heating coils[a,b]	450	500	600
Cooling coils[a]	450	500	600
Air washers[a]	500	500	500
Fan outlets	1000–1600	1300–2000	1600–2400
Main ducts[b]	700–900	1000–1300	1200–1800
Branch ducts[b]	600	600–900	800–1000
Branch risers[b]	500	600–700	800
	Maximum Velocities, Feet per Minute		
Outdoor air intakes[a]	800	900	1200
Filters[a]	300	350	350
Heating coils[a,b]	500	600	700
Cooling coils[a]	450	500	600
Air washers[a]	500	500	500
Fan outlets	1700	1500–2200	1700–2800
Main ducts[b]	800–1200	1100–1600	1300–2200
Branch ducts[b]	700–1000	800–1300	1000–1800
Branch risers[b]	650–800	800–1200	1000–1600

[a]These velocities are for total face area, not the net free area; other velocities in table are for net free area.

[b]For low-velocity systems only.

method is to provide vanes that divert the air half to the right and half to the left. When a supply register is in the corner of a room, it is best practice for the vanes to deflect all the air in one direction, away from the corner. Return grills are of the slotted type in walls and of the grid type in floors. All registers and grills should be made tight at the duct connection. See Table 5.9*a* for selection of registers based upon output and recommended face velocity.

(*e*) ***Controls.*** The burner is started and stopped by a thermostat, which is placed in or near the living room at a thermally stable location that is protected from cold drafts, direct sunlight, and the warming effects of nearby warm air registers.[1] A cut-in temperature of between 80 and 95° F is selected for the fan switch in the furnace bonnet. After the

[1]Similar considerations govern thermostat locations in hydronic (water) systems or in those using electricity.

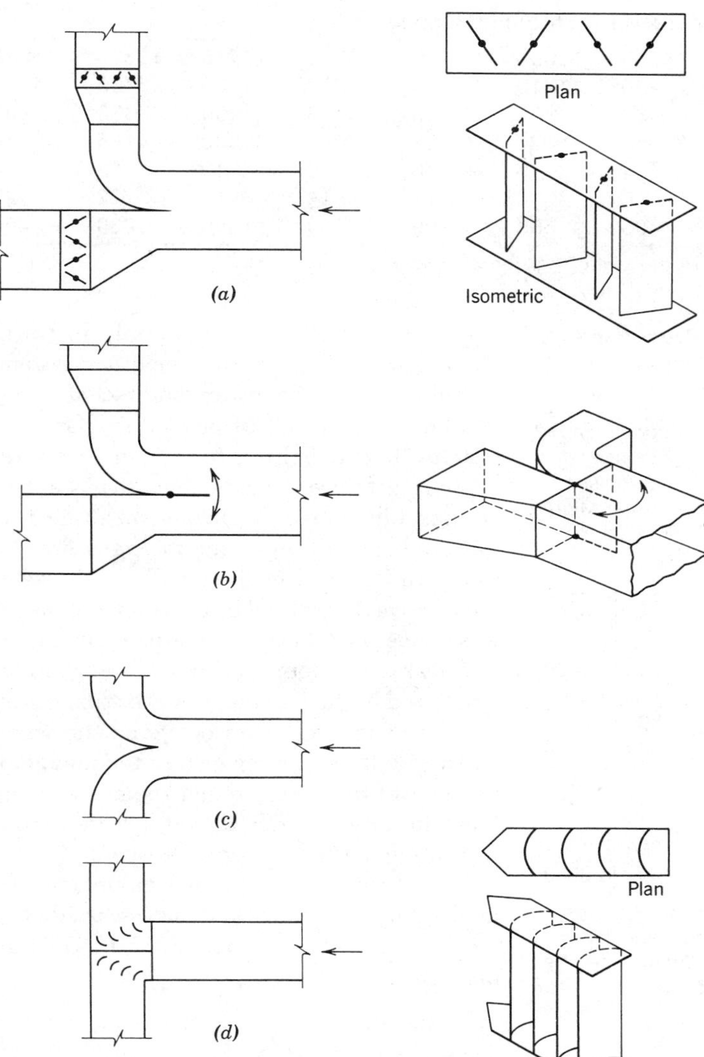

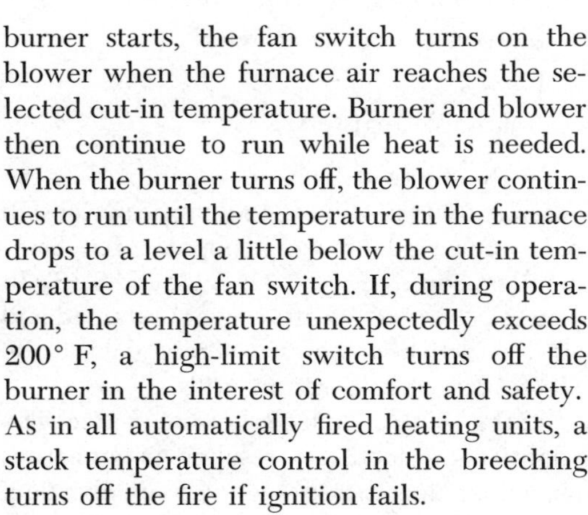

Fig. 5.9 *Air controls in ducts.* (a) *Air adjustment by opposed-blade dampers.* (b) *Air adjustment by splitter damper.* (c) *Conventional turns in ducts.* (d) *Right-angle turns with turning vanes. Method* (d) *is more compact.*

burner starts, the fan switch turns on the blower when the furnace air reaches the selected cut-in temperature. Burner and blower then continue to run while heat is needed. When the burner turns off, the blower continues to run until the temperature in the furnace drops to a level a little below the cut-in temperature of the fan switch. If, during operation, the temperature unexpectedly exceeds 200° F, a high-limit switch turns off the burner in the interest of comfort and safety. As in all automatically fired heating units, a stack temperature control in the breeching turns off the fire if ignition fails.

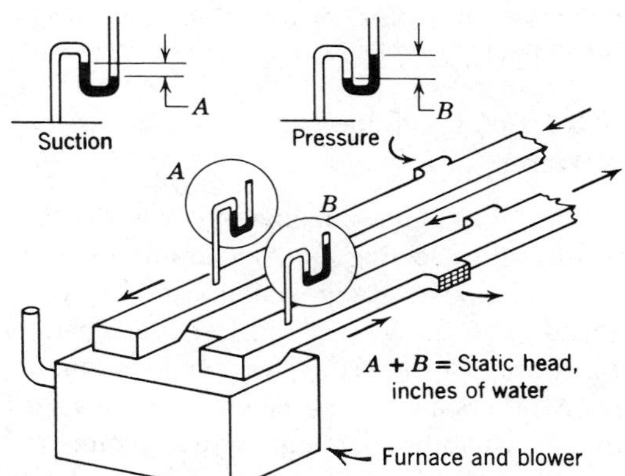

Fig. 5.10 *The static head is the pressure, measured in inches of water, available to overcome friction in the entire system.*

Table 5.9*a* *Characteristics of 2¼ × 12-Inch Floor Register*

Heating Btuh	3045	4565	6090	7610	9515	11415	13320	15220
Cooling Btuh	855	1280	1710	2135	2670	3200	3735	4270
CFM	40	60	80	100	125	150	175	200
T.P. loss	0.009	0.015	0.027	0.037	0.050	0.080	0.105	0.134
Vertical throw, ft	3	4	5	6	8	10	12	14
Vertical spread, ft	6	8	10	11	14	17	22	25
Face velocity, fpm	280	420	565	705	880	1050	1230	1400

Source. Excerpt from the *Catalog* of the Lima Register Company.

Table 5.9*b* *Recommended Delivery Face Velocities for Various Applications (Registers)*

Application	*Recommended Velometer Velocities, Feet per Minute*
Broadcasting studios	500
Residences	500 to 750
Apartments	500 to 750
Churches	500 to 750
Hotel bedrooms	500 to 750
Legitimate theaters	500 to 1000
Private offices, acoustically treated	500 to 1000
Motion picture theaters	1000 to 1250
Private offices, not treated	1000 to 1250
General offices	1250 to 1500
Stores, upper floors	1500
Stores, main floors	1500
Industrial buildings	1500 to 2000

NOTE. The sound caused by an air outlet in operation varies in direct proportion to the velocity of the air passing through it. The air velocity can be controlled by selecting outlets of proper sizes. The outlet velocities recommended above are within safe sound limits for most applications.

NOTE. For residences (500–750 fpm), any of four low-velocity registers of Table 5.9*a* qualify (280, 420, 565, 705 fpm). For stores (1500 fpm), a higher-velocity register of Table 5.9*a* (1400 fpm) would be permissible.

5.5 Resourcefulness in the Design of Warm Air Systems

Flexibility and imagination are the key qualities in selecting and developing *any* mechanical or electrical system. In the years ahead as in the years past, frequent changes are always expected. They include among many others, new architectural trends, a shift in the availability of energy sources, improved construction methods, and newly developed equipment. These influences apply to warm air heating as they do to any other system.

We can trace the history and recent acceleration in the speed of change in warm air systems. It may help us to remain alertly receptive to future developments that we cannot possibly foresee. In 1900 warm air heating systems began to supersede the open fire. An iron furnace stood in the middle of the basement. It was hand-fired by coal. Surrounding it was a sheet metal enclosure. An opening in its side near the bottom admitted the cool air that gravitated to the basement. A short duct from the top of the enclosure delivered the warm air by gravity to a large grill in the middle of the floor of the parlor. Other rooms including those in upper stories shared a little of this warmth when doors were left open.

Very gradual changes had culminated by midcentury to produce systems essentially like the one described in Section 5.4. The improvements included:

Automatic firing of oil or gas.

Operational and safety controls.

Ducted air to and from each room.

Blowers to replace gravity.

Filters.

Adjustable registers.

In the 1960s it became apparent that the basement was beginning to disappear. Subslab perimeter systems were popular for basementless houses (Figs. 5.11, 5.12, and 5.13). In general, the above features were retained and air was delivered upward across glass, to be taken back at high-return grills.

In the 1970s oil and gas became scarce. The adoption of electricity resulted in numerous changes. Necessities such as combustion,

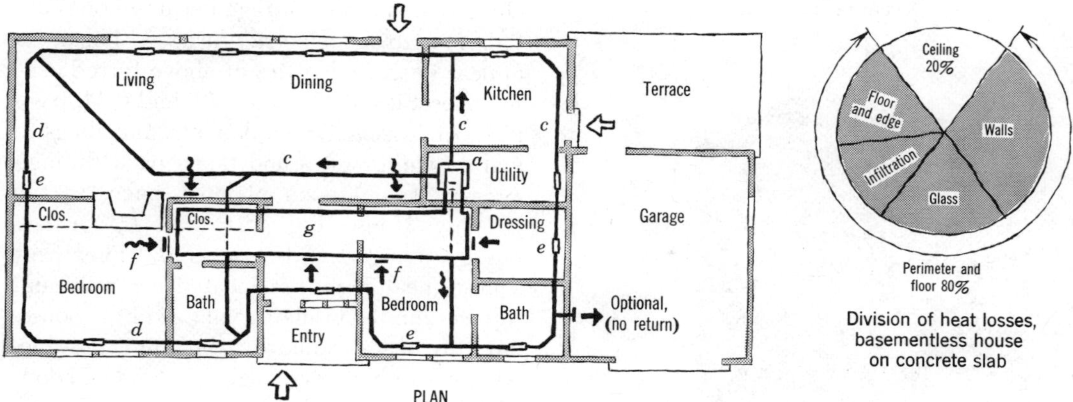

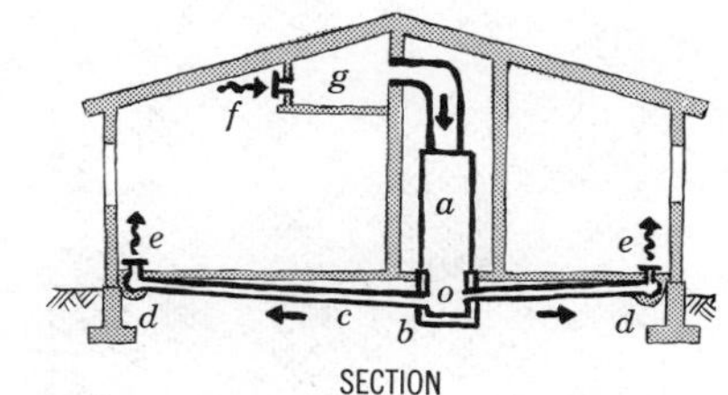

Fig. 5.11 Forced warm-air, perimeter loop system. Adaptable for cooling. No returns from kitchen, baths, or garage.

Legend (a) *Downflow air furnace (Fig. 5.13).* (b) *Supply plenum.* (c) *Eight-inch (plus) subslab supply ducts (encased in concrete).* (d) *Eight-inch perimeter duct (encased in concrete).* (e) *Floor register, adjustable for direction and flow rate (Fig. 5.12).* (f) *Return grill.* (g) *Return plenum.*

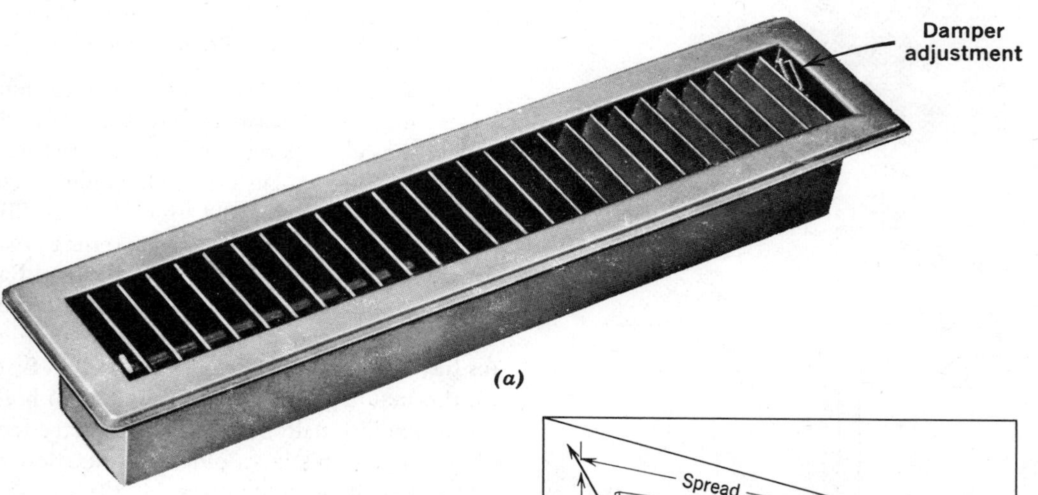

Fig. 5.12 Floor registers and their action. (a) *A $2\frac{1}{4} \times 12$ in. floor register (diffuser). One of many sizes and shapes. It has diverting vanes for "spread" and an adjustable damper. See Table 5.9 for characteristics.* (b) *Concept of spread and throw. By aspiration (suction), cooler room air is induced to join the stream of warm air, resulting in a bland and pleasant air stream that crosses the room.*

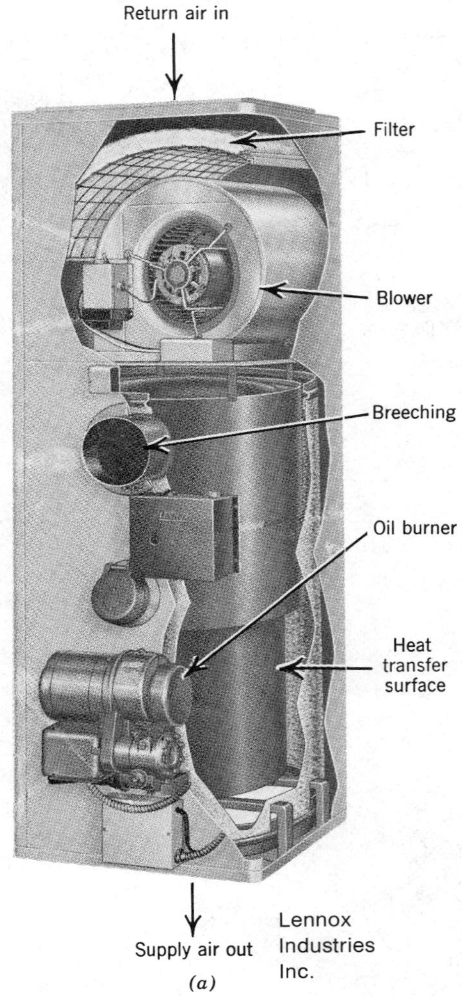

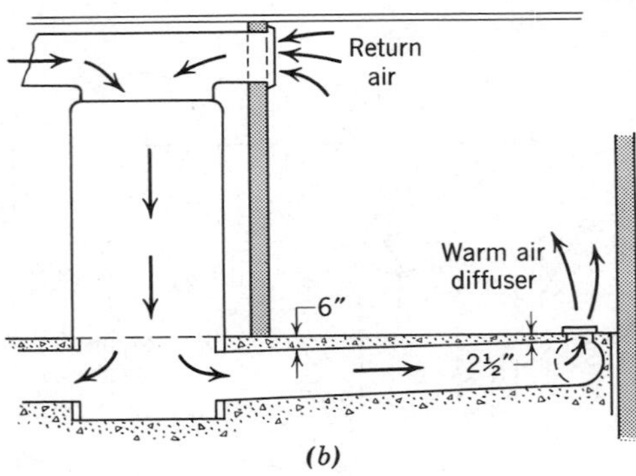

Fig. 5.13 (a) *Downflow (counterflow) furnace and its relation to the perimeter system.* (b) *Airflow pattern. From* Warm Air Perimeter Heating, Manual 4, *the National Warm Air Heating and Air Conditioning Association.*

chimneys, and fuel storage became nonessential. Horizontal electric furnaces began to appear in shallow attics or above furred ceilings (see Figs. 6.22, page 270, and 6.24, page 272). Air was delivered down from ceilings to warm exterior glass and taken back through door-grills and open plenum space. In the Arkansas House (Fig. 4.21, page 159) a new energy-saving trend appeared. The heat source (heat pump) is located centrally and fully *within* the insulated volume of the house. No stray heat escape from the unit is possible. Also, because of very small windows and double glazing, short ducts can deliver warm air from the inner side of each room, since the warming of the double glass is less essential. Air returns to the unit through open grills in doors and at the heat pump enclosure.

There are no fixed rules about designs. The principles of efficiency, comfort, cost, and conformity to the architectural design are paramount.

5.6 Hot Water and Steam Boilers

(*a*) ***Ratings.*** The Hydronics Institute publishes ratings for hot water and steam boilers. The ratings also distinguish between cast iron and steel boilers. The cast iron boilers have I=B=R ratings and the steel boilers SBI ratings. The initials, respectively, refer to two organizations, The Institute of Boiler and Radiator Manufacturers and The Steel Boiler Institute, both now combined in the Hydronics Institute. The ratings are expressed in British thermal units per hour (Btuh). Steam boilers are, additionally, evaluated in square feet of radiation. This is an older classification in which 1 sq ft of radiation is 240 Btuh.

When interpreting the ratings, it is important to select a boiler whose net (useful) rating matches the calculated critical heat loss of the house or building.

(b) Equipment Suppliers' Responsibility. ASHRAE Standard 90-75, *Energy Conservation in New Building Construction,* in *its* Section 6.2, which follows, requires suppliers of HVAC equipment (which includes boilers) to furnish certain information when requested. You will note in Section 6.2 that input and output ratings are included.

6.0 HEATING, VENTILATING AND AIR-CONDITIONING (HVAC) EQUIPMENT

6.1 Scope

This section deals with equipment and component performance and manufacturers' documentation, in accordance with the criteria for effective utilization of energy established in this Standard, for HVAC-systems equipment and components used in new buildings.

6.2 Responsibility of Equipment Suppliers

Suppliers of HVAC equipment *shall* furnish, *upon request* by prospective purchasers, designers, or contractors, the input(s) and output(s) of all HVAC equipment. These *shall* be based on equipment in new condition, and *shall* cover full load, partial load, and standby conditions, as required, to enable determination of their compliance with this Standard. This includes performance data under modes of operation and at ambient conditions necessary to make the analysis outlined in Section 10 of this Standard.[2]

(*c*) ***Boiler Types, Steel and Cast Iron.*** Boilers and their accessories comprise a wide inventory. A few selected types are shown here.

1. *Oil-Fired steel boiler, Figure 5.14.* A refractory chamber receives the hot flame of the oil fire. Combustion continues within the chamber and the fire tubes. Smoke leaves through the breeching at the rear. Water, *outside* the chamber,

[2]The above two subsections of the nine in Section 6.0 are reprinted here by permission of the American Society of Heating, Refrigerating and Air Conditioning Engineers, Inc.

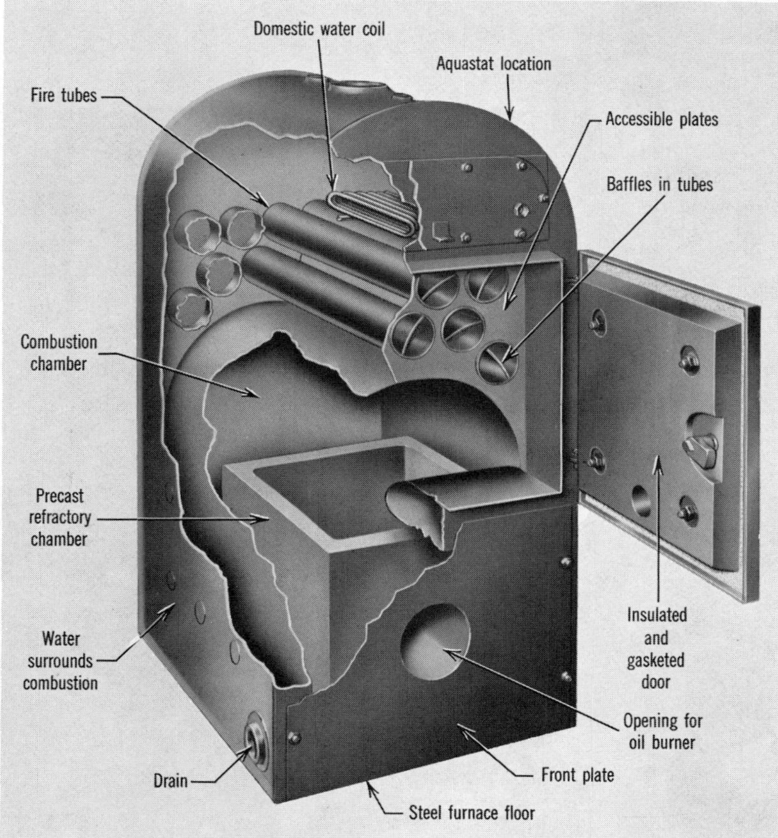

Fig. 5.14 *Steel boiler with horizontal fire tubes. Adapted for oil-firing and suitable for steam or hot-water heating. Federal.*

receives the heat generated in the combustion chamber. If the domestic hot water coil is connected for use, a larger capacity boiler is selected. An aquastat (water thermostat) turns on the burner whenever the boiler water cools off, thereby maintaining a reservoir of hot water ready for heating the house.

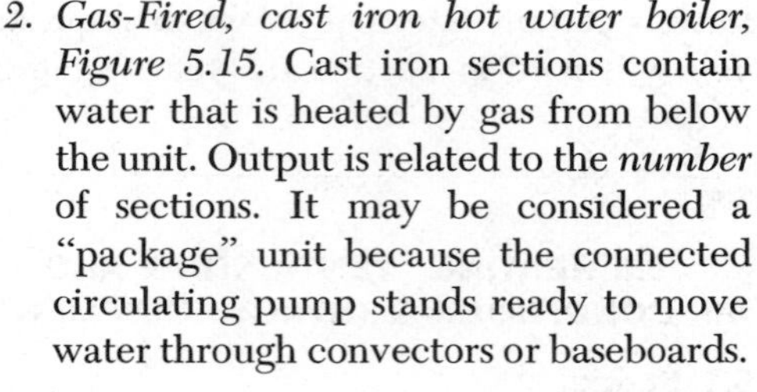

2. *Gas-Fired, cast iron hot water boiler, Figure 5.15.* Cast iron sections contain water that is heated by gas from below the unit. Output is related to the *number* of sections. It may be considered a "package" unit because the connected circulating pump stands ready to move water through convectors or baseboards.

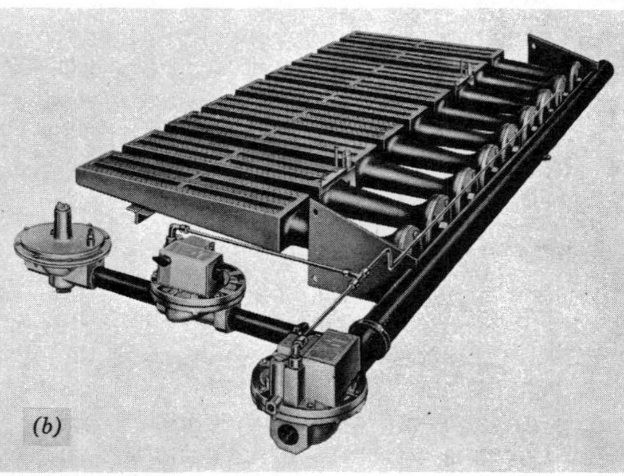

Burnham Corporation

Fig. 5.15 (a) *Gas-fired cast-iron sectional boiler for hot water heating. Circulating pump is seen at left, below gas piping.* (b) *Gas-burning unit.*

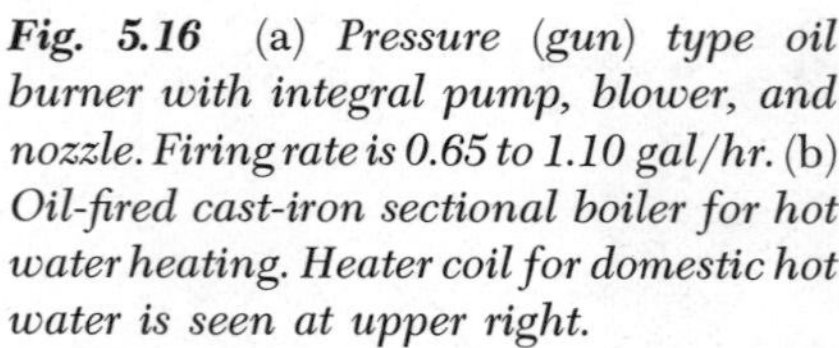

Fig. 5.16 (a) *Pressure* (*gun*) *type oil burner with integral pump, blower, and nozzle. Firing rate is 0.65 to 1.10 gal/hr.* (b) *Oil-fired cast-iron sectional boiler for hot water heating. Heater coil for domestic hot water is seen at upper right.*

Burnham Corporation

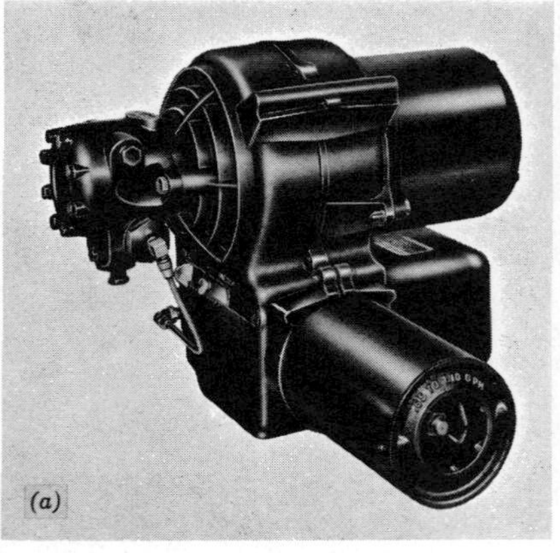

The American Gas Association (AGA) sets standards for gas-fired equipment.

3. *Oil-Fired, cast iron hot water boiler, Figure 5.16.* Primary and secondary air for combustion may be regulated at the burner unit. Flame enters the refractory chamber and continues around the outside of the water-filled cast iron sections.
4. *High-Output, package-type steel boiler, Figure 5.17.* For large buildings using steam as a primary heating medium, one or several such boilers may be used. Direct use of steam may be seen in Figure 5.47, where it supplies preheat and reheat coils and also a humidifying unit. The relative lightness of this boiler type compared to the older styles with ponderous masonry bases (boiler settings) makes this package type suitable for use on upper floors of tall buildings. Figure 7.17, page 293, shows two such boilers on the thirteenth floor of the Fox Plaza Building.
5. *Convertor, steam to hot water, Figure 5.18.* When, in a building using primary steam boilers, secondary circuits using hot water for heating are required, a convertor is used. It is considered a heat exchanger. Figure 7.17, page 293, shows downfeed steam supply for the two boilers on the thirteenth floor to two such converters, one for hot water heating in the apartments and one below the garage ceiling for hot water heating in the commercial area. The secondary (hot water) circuits are not detailed in that illustration. A convertor may also be used to transfer heat from steam to *domestic* (service) water.
6. *Electric boilers, Figures 5.19a and b.* The two boilers illustrated are for hot water heating and are of high-capacity output.

Fig. 5.17 *Package-type, fire-tube steam boiler. Capacity is 600,000 to 3,000,000 Btuh; adaptable for oil, gas, or both. Complete with controls and all fittings.*

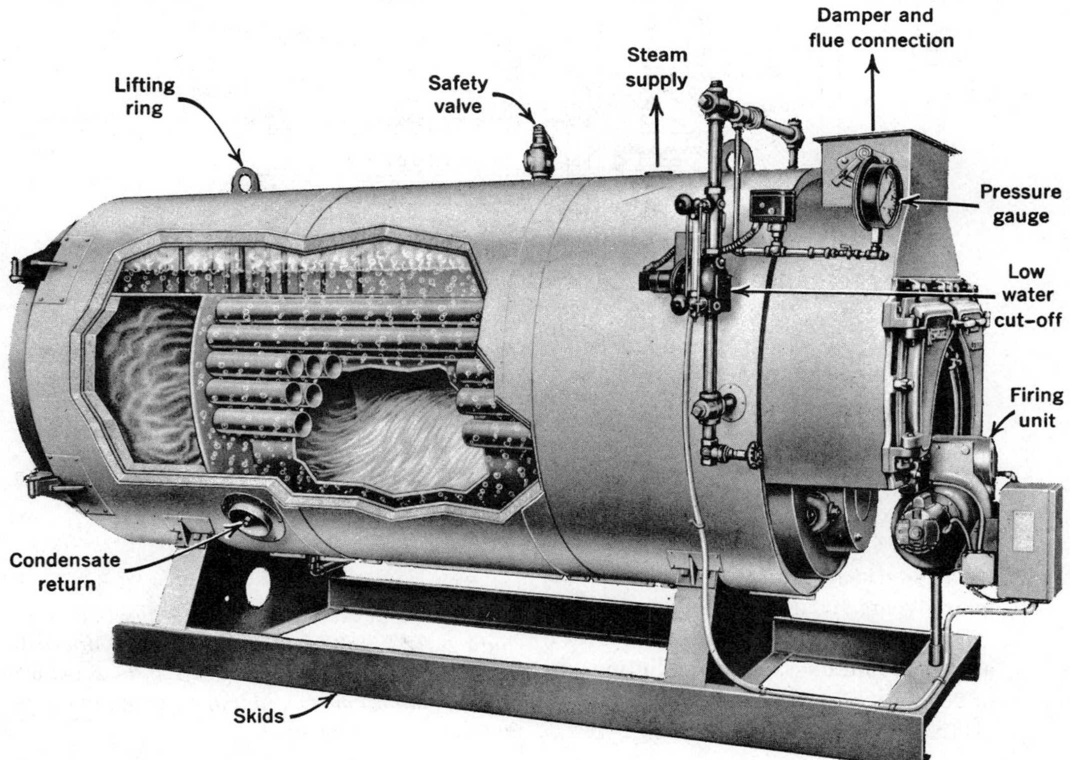

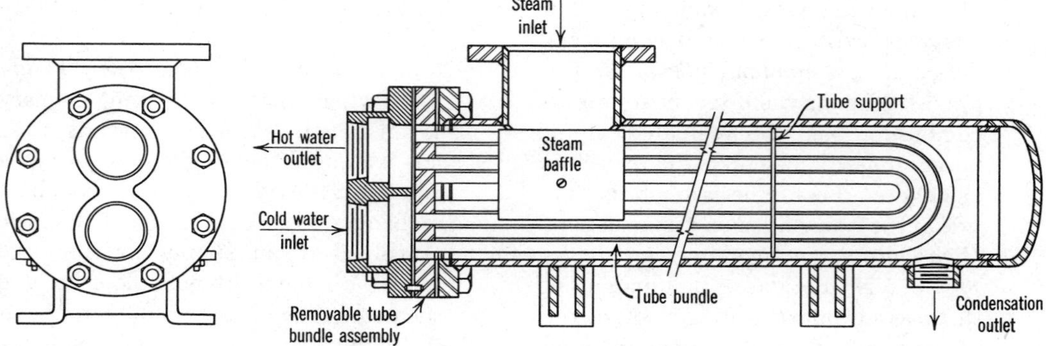

(*a*) Section illustrating the principle of heat transfer from steam to water.

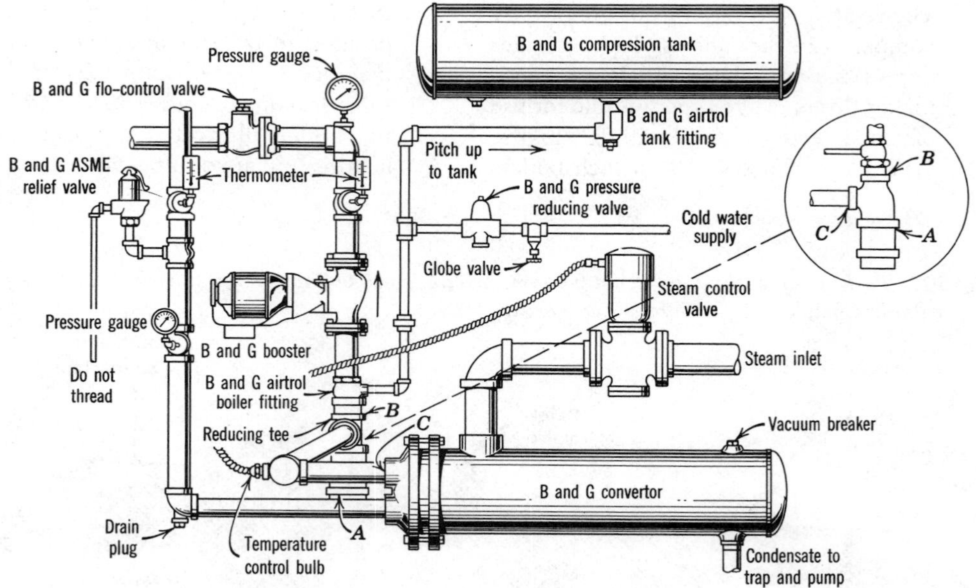

(*b*) A convertor connected to steam supply and equipped with all devices necessary for a complete hot water heating system.

Fig. 5.18 Conversion. Transferring heat from steam to hot water. Courtesy of ITT Bell and Gossett.

Electric *steam* boilers are also commonly used in large buildings. Section 7.27, page 297, refers to their use in the all-electric, Georgia-Pacific Building in Portland, Oregon.

7. *Coal-Fired steam boiler, Figure 5.20.* For many years, coal-fired boilers of this type were used in steam-heated houses and small buildings. As less-polluting use of coal is achieved, we may see a return of a variation of this equipment.

Fig. 5.19 (opposite) *Electric hot water boilers.* (a) *Ratings, 12 to 180 kw (41,000 to 614,160 Btuh).* (b) *Ratings, 180 to 1800 kw (614,160 to 6,141,600 Btuh). Similar boilers with appropriate fittings are available for steam systems.*

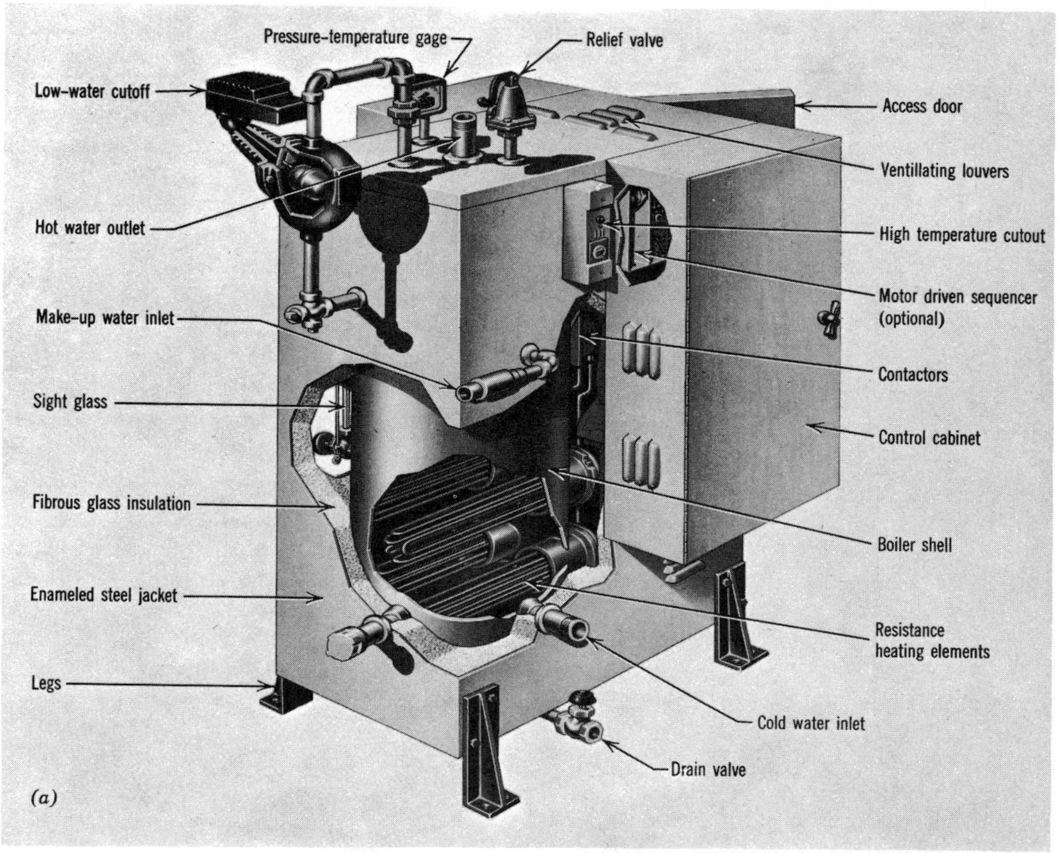

Pressure–temperature gage
Relief valve
Low–water cutoff
Access door
Ventillating louvers
Hot water outlet
High temperature cutout
Motor driven sequencer (optional)
Make–up water inlet
Contactors
Sight glass
Control cabinet
Fibrous glass insulation
Boiler shell
Enameled steel jacket
Resistance heating elements
Legs
Cold water inlet
Drain valve

(a)

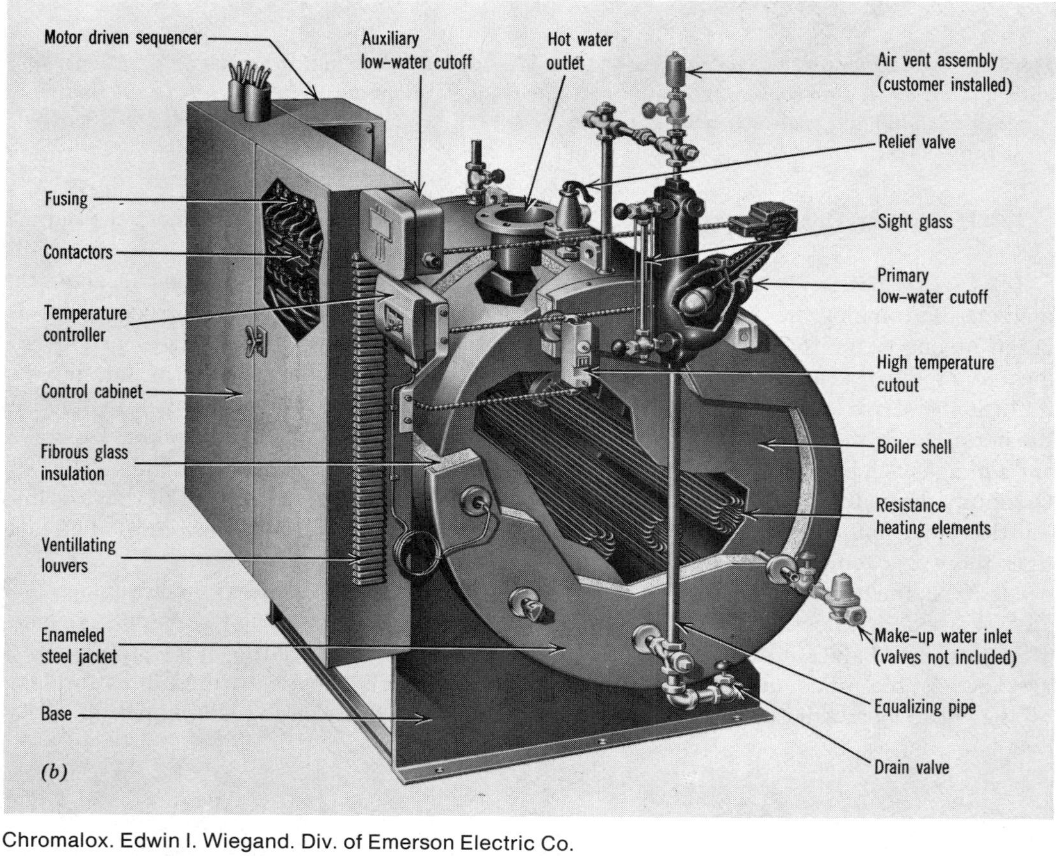

Motor driven sequencer
Auxiliary low–water cutoff
Hot water outlet
Air vent assembly (customer installed)
Relief valve
Fusing
Sight glass
Contactors
Primary low–water cutoff
Temperature controller
High temperature cutout
Control cabinet
Boiler shell
Fibrous glass insulation
Resistance heating elements
Ventillating louvers
Make–up water inlet (valves not included)
Enameled steel jacket
Equalizing pipe
Base
Drain valve

(b)

Chromalox. Edwin I. Wiegand. Div. of Emerson Electric Co.

Fig. 5.20 *Steel boiler for steam heating (note gauge glass on left side to indicate water level). Hopper-type automatic stoker feed for coal-firing. The current adequacy of coal resources in the United States may spur a return to individual coal-fired boilers. Pacific.*

5.7 Hot Water Heating Systems

(*a*) ***Circuit Types.*** There are four principal methods of arranging the piping for the circulation of hot water to the heating elements located in the spaces to be heated. Figure 5.21*a* is the series loop system usually run at the perimeter of the house. The water flows to and *through* each baseboard or fintube in turn. Obviously the water at the end of the circuit is a little cooler, but since, in all hot water systems the water temperature drop seldom exceeds 20° F, the *average* temperature can usually be used in selecting the baseboard or other elements. Valves at each heating element are not possible, since any valve would shut off the entire loop. Adjustment is by a damper at each baseboard, which reduces the natural convection of air over the fins. This is usually satisfactory. There is no general rule about the maximum allowable length of water circuit, but for long runs, the pipe size can be increased or *several* loops used in parallel.

The one-pipe system shown in Figures 5.21*b* and 5.22 is a very popular choice. Special fittings, described in subsection (b) that follows, act to divert part of the flow into each baseboard. A valve may be used at each one to allow for reduced heat or a complete shut-off to conserve energy, an advantage that the loop system does not provide. The one-pipe system uses a little more piping and is thus not as economical to install as the loop system in which piping is minimal.

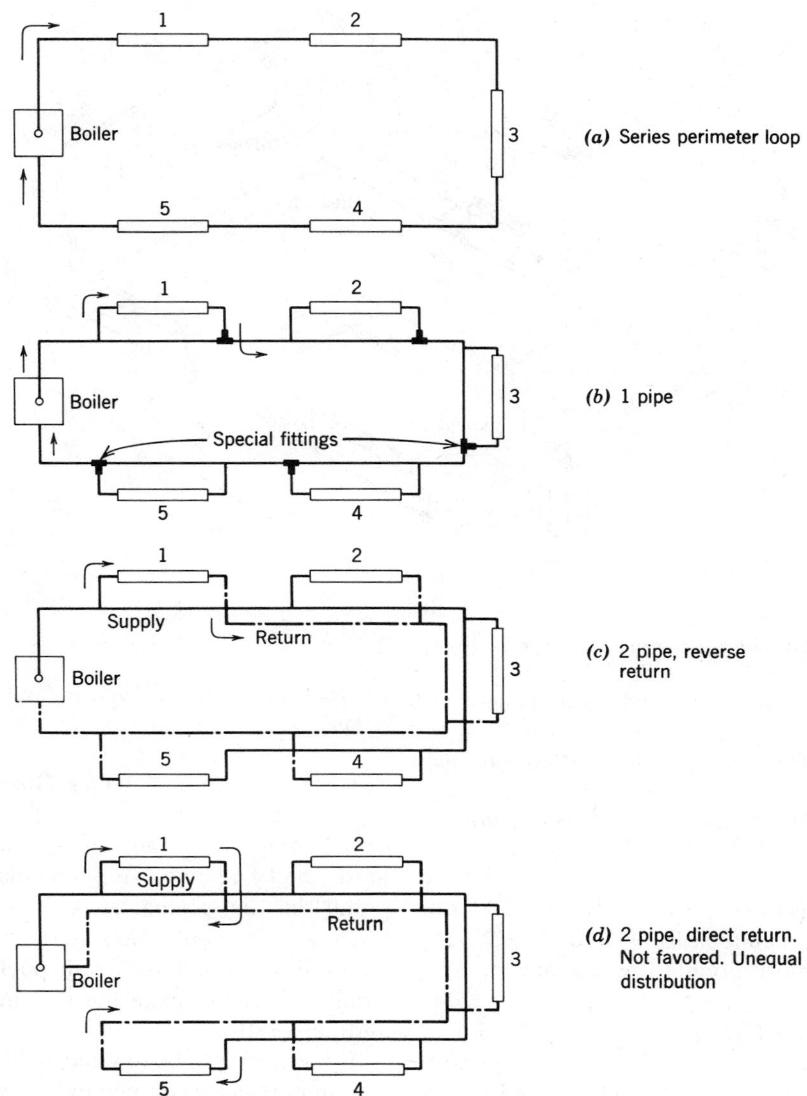

Fig. 5.21 *Plan views showing principles of piping for water distribution to heating elements (baseboard convectors shown here) in hot-water-heating systems. For simplicity, controls are not shown.*

The two-pipe reverse-return of Figure 5.21*c* may be considered the classic method, though not too often employed and indeed not always necessary. Water nearly at boiler temperature is supplied to each baseboard without being cooled by passing through a previous baseboard as in Figure 5.21*a* or accepting the cooler return water as in Figure 5.21*b*. Equal fricton, resulting in equal flow, is achieved through all baseboards, Nos. 1 to 5, by *reversing* the return instead of running it directly back to the boiler. This equality is effected by equal lengths of water flow through any baseboard together with its lengths of supply and return main. More pipe is required for this than for the systems shown in Figure 5.21*a* or *b*.

Figure 5.21*d* shows a method that is not favored because the path of water through baseboard No. 1 is much shorter than that

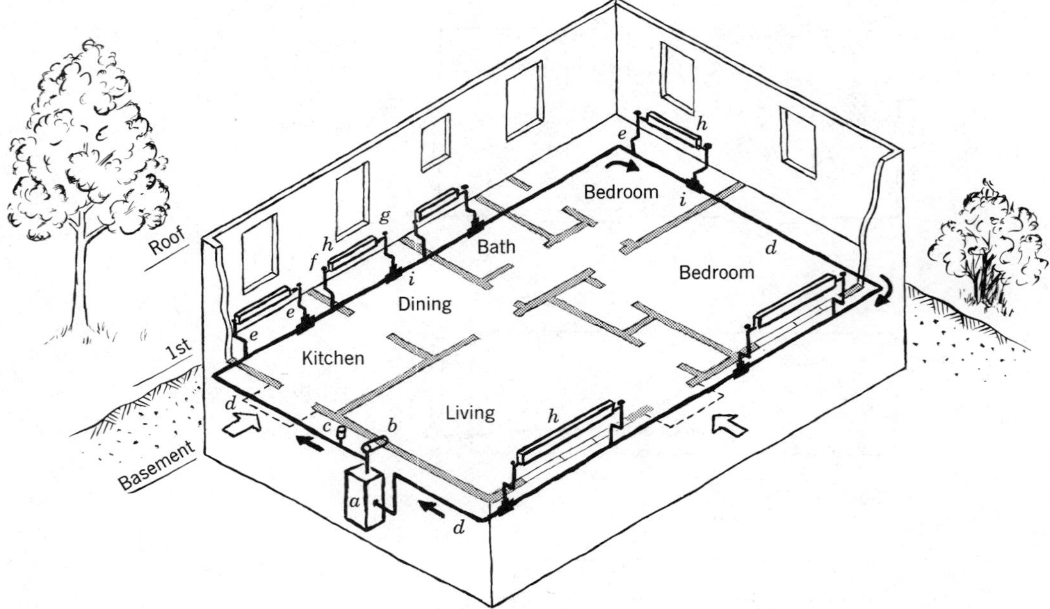

Fig. 5.22 *Hot water, one-pipe system, type* (b) *of Figure 5.21.*

Legend
(a) *Boiler*
(b) *Compression tank*
(c) *Circulator (pump)*
(d) *Hot water main*
(e) *Runout (branch)*
(f) *Control valve*
(g) *Air vent*
(h) *Baseboard heating unit*
(i) *Special return fitting*

through the others, especially No. 5. Baseboard No. 5 could easily be undesirably cool, since it is short-circuited by the others.

(b) Special Fittings. In Figures 5.21*b* and 5.22, which show the usual single circuit, one-pipe system, hot water is carried in the main and diverted to the several convectors. A number of special tee fittings have been devised to accomplish this. One kind (Figs. 5.23 and 5.24) constricts the main, forcing water into the convector supply branch. A venturi-type jet reduces the pressure in the return fitting, inducing flow of water out of the convector branch return. The first convectors receive slightly warmer water than the later ones in the circuit, but this difference is negligible.

When greater pressure differential is required for downfeed (against gravity) or high-output upfeed, arrangements as shown in Figure 5.24*b* and *c* are used instead of the conventional one of Figure 5.23 or Figure 5.24*a*.

(c) Air Vents and Water Drains. Except for the necessary air cushion in the upper part of the compression tank above the boiler, air must not be allowed to accumulate at high points in the piping or at the convector branches. Air vents such as those of Figure 5.25 relieve these possible air pockets, which would otherwise make the system air-bound and inoperative.

If a system is to be drained and left idle in a cold house, water trapped in low points could freeze and burst the tubing or fittings. Operable drain valves must be provided at such locations and, of course, at the bottom of the boiler as shown in Figure 5.26.

(d) Hydronic and Electrical Controls. Hot water heating systems are fully automatic and operate as controlled by methods described in Figure 5.26. Makeup water is added as required, the air level in the tank is regulated by the air control fittings, and the circulator and burner operate as controlled by the aquastat and thermostat. If air vents in the *piping* are not automatic they will require periodic manual "bleeding" of unwanted air.

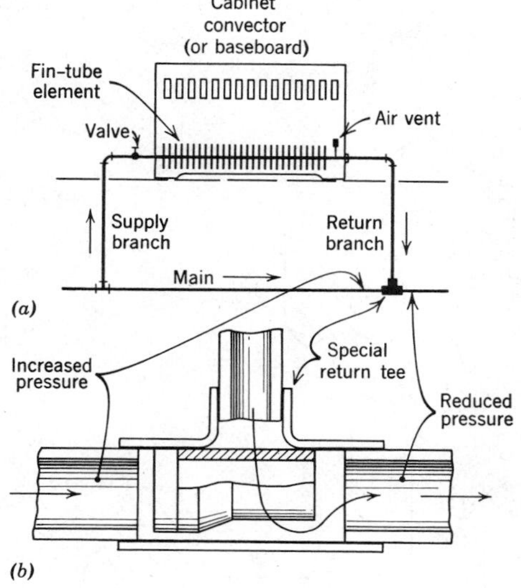

Fig. 5.23 (a) *Special fitting for one-pipe systems. Venturi-jet-type tee used here on the return branch connection to the main. It induces flow through the convector by retarding the flow to force water into the supply branch and producing a jet to reduce the pressure in the main following the return branch.* (b) *Illustrative sectional view to describe the function. No scale. Check with manufacturer as to dimensional details of the tee fitting.* (c) *Manufacturer's photo. Courtesy of ITT Bell & Gossett.*

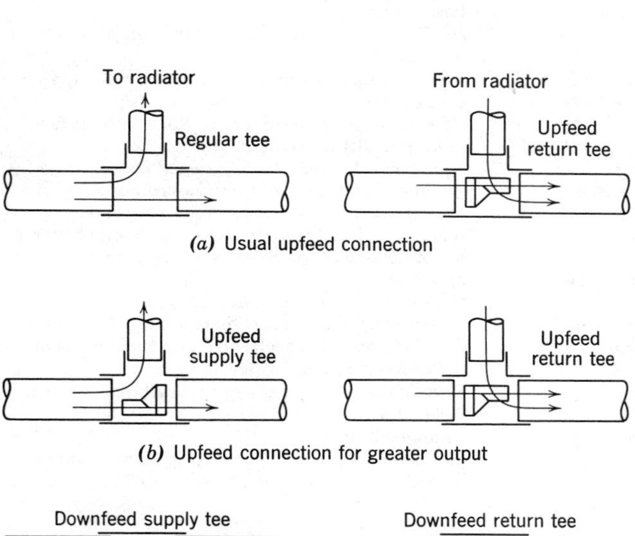

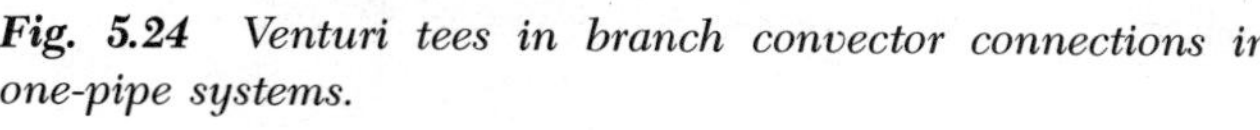

(c) Required downfeed connection

Fig. 5.24 *Venturi tees in branch convector connections in one-pipe systems.*

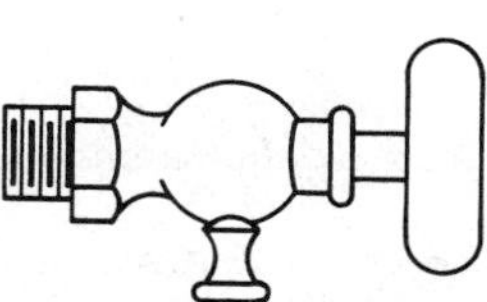

(a) Manual, for convectors

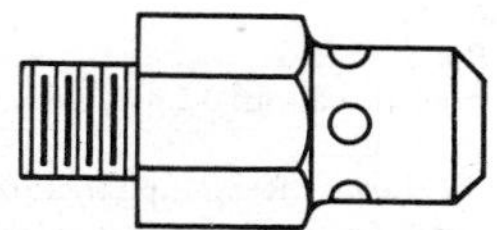

(b) Automatic, for convectors

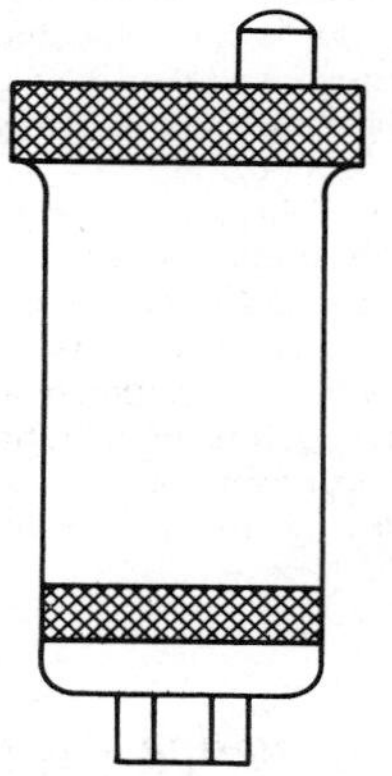

(c) Automatic, for piping

Fig. 5.25 *Air vents. Types* (a) *or* (b) *would be located at positions* (g) *of Figure 5.22.*

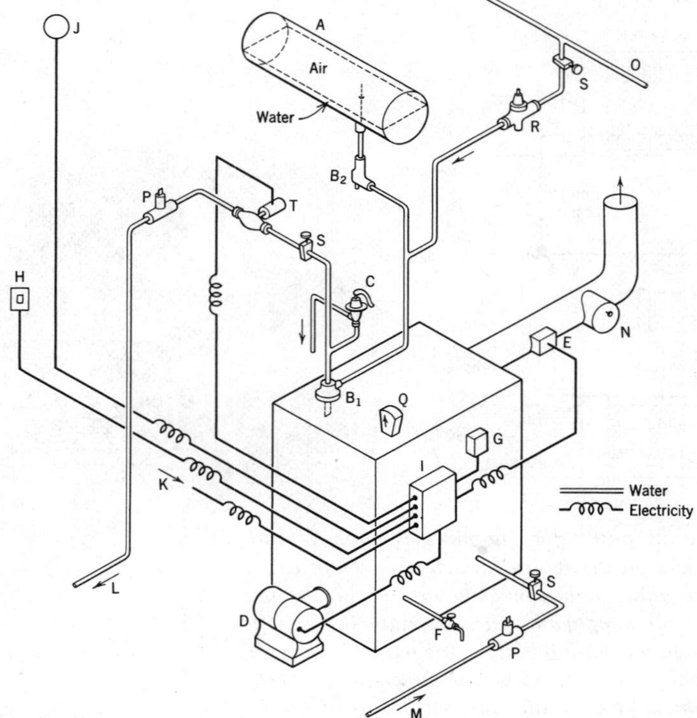

Fig. 5.26 Hydronic and electrical controls; an oil-fired boiler for heating by hot water.

A *Compression Tank.* Accommodates the expansion of the water in the system.
B *Air Control Fittings.* Vent out unwanted air in the boiler and maintain the level in the compression tank.
C *Pressure Relief Valve.* Usually set for 30 psi. Initial cold pressure about 12 psi. Relieves excessive system pressure.
D *Oil Burner.* Responds to aquastat or thermostat.
E *Stack Temperature Control.* Senses stack temperature and stops oil injection if ignition has not occurred.
F *Drain Valve.* At low point in the water system.
G *Aquastat.* Maintains temperature of boiler water by starting the oil burner when temperature of water drops below the aquastat's setting. Sometimes set at about 180° F.
H *Remote Switch.* At a safe distance from the boiler so that the plant can be turned off in case of trouble during which the boiler cannot be approached.
I *Junction Box and Relays.* General control center.
J *Thermostat.* When the room temperature drops below its setting, it turns on both the oil fire and the circulating pump.
K *Electrical Power Source.* Operates from a separate individual circuit at the power panel.
L *Hot Water Supply.* Copper tubing to convectors or baseboards.
M *Hot Water Return.* Copper tubing from convectors or baseboards.
N *Draft Adjuster.* Regulates the draft (combustion air) over the fire.
O *House Cold Water Main.* From which water is fed automatically into boiler.
P *Flow Control Valves.* Prevent casual flow of water by gravity when the circulator is not running.
Q *Temperature/Pressure Gauge.* Indicates water temperature and pressure. Sometimes supplemented by immersion thermometers in supply and return mains.
R *Pressure Reducing Valve.* Admits water into the system when the pressure there drops below about 12 psi. Has a built-in check valve to prevent backflow of boiler water into the water main.
S *Shutoff Valves.* Normally open. Can be closed to isolate the system and permit servicing of components.
T *Circulator.* Centrifugal circulating pump that moves the water through the tubing and heating elements.

(*e*) ***Circulating Pump.*** A typical centrifugal pump is shown in Figure 5.27. It is selected to overcome the friction-of-flow in the piping and fittings and to deliver water at a rate sufficient to offset the hourly heat loss of the house or building. Calculations that lead to its selection on the basis of its rated performance follow in Section 5.8.

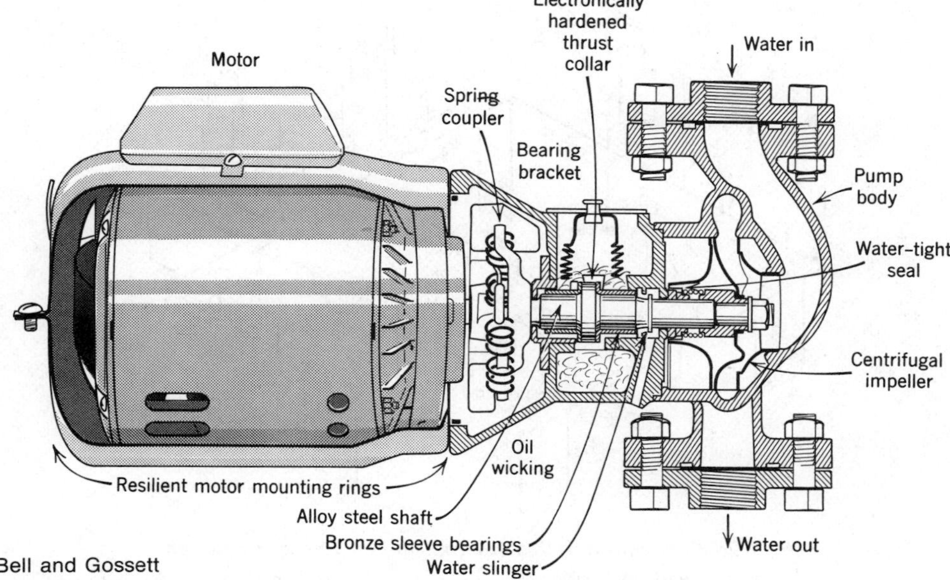

Fig. 5.27 *Centrifugal circulator for hydronic heating systems and for the circulation of* domestic *hot water. Impeller and pump body shown in section.*

5.8 Hydronic Heating Design and Zoning

For hot water (hydronic) heating systems the following design concepts and design procedure are recommended.

(*a*) *Design Concepts*

Total equivalent length. The length of the longest circuit through which the water is circulated plus a length equivalent to the resistance offered by the fittings, boiler, etc., is the total equivalent length (TEL).

Pressure drop in the pipe. This drop due to friction, expressed in milinches of water per foot of pipe, is the difference in pressure caused by friction in 1 ft of pipe and represents the static height of water in thousandths of an inch capable of being sustained by this difference in pressure.

Total friction head. Expressed in feet, this head is the column of static water that could be sustained by the difference in pressure in the entire system owing to friction. Thus, if a system were 300 ft long and had a unit frictional resistance of 300 milinches/ft, the total friction head would be $(300 \times 300)/(1000 \times 12) = 7.5$ ft. Check this by Table 5.11.

Required flow. The required flow is the water flow in gallons per minute to be circulated to make up the hourly heat loss in the building. It is determined by the hourly heat loss and the selected drop in the water temperature.

Pump rating. The pump size is selected on the basis of the required flow and the total friction head (Fig. 5.29).

Required volume of expansion tank. This is related to the volume of water in the system and the overall rise in temperature from cold water supply temperature to boiler water operating temperature.

(*b*) *Design Procedure*

The procedure may be outlined as follows:

1. Add to the length of the longest circuit the length equivalent to fittings, as taken from Table 5.10, to determine the *total* equivalent length of pipe in the longest circuit. A 90 deg elbow produces approximately the same friction as a

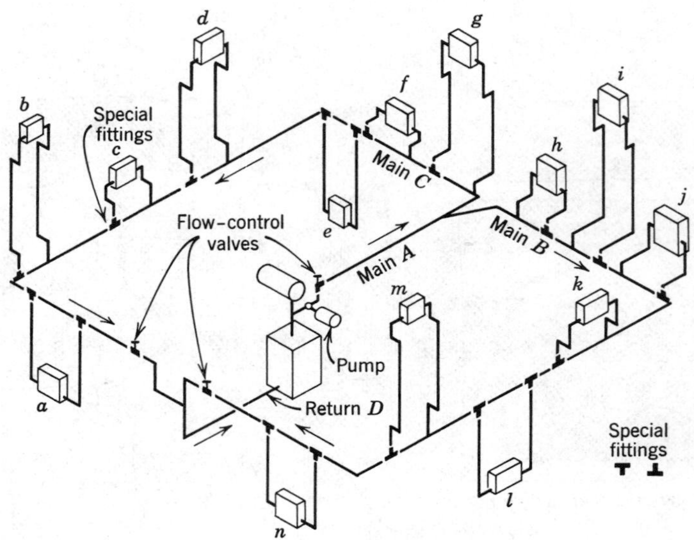

Fig. 5.28 (*Example 5.5*) *Two-circuit, one-pipe forced circulation hot-water heating system.*

Convector requirements in 1000 Btuh (Mbh)

					Through Main *A* and Return *D* 79 Mbh
Main *B*	*h* 4	Main *C*	*a* 3		
	i 3		*b* 4	Through Main *B* 40 Mbh	
	j 9		*c* 8	Through Main *C* 39 Mbh	
	k 5		*d* 6		
	l 4		*e* 5		
	m 6		*f* 4		
	n 9		*g* 9		
Total	40 Mbh	Total	39 Mbh		

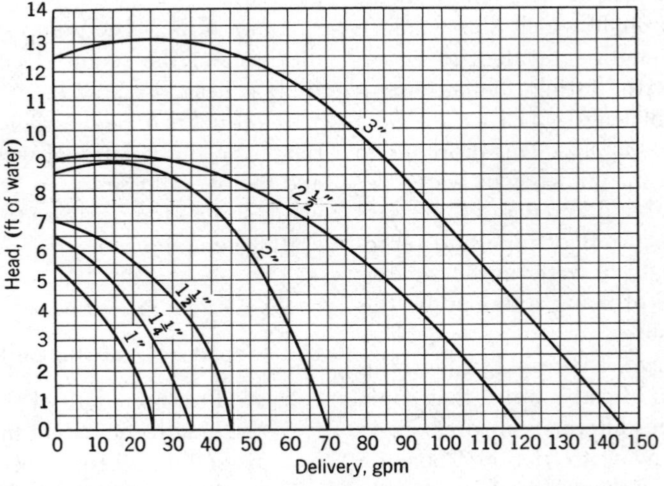

Fig. 5.29 Performance chart for circulating pumps (boosters) commonly obtainable for hot water heating. Check manufacturer's catalog for specific selections.

Table 5.10 ***Copper Elbow Equivalents***[a]

Fitting	Copper Tubing
Elbow, 90°	1.0
Elbow, 45°	0.7
Elbow, 90° long turn	0.5
Elbow, welded, 90°	0.5
Reduced coupling	0.4
Open-return bend	1.0
Open-gate valve	0.7
Open-globe valve	17.0
Angle radiator valve	3.0
Radiator or convector	4.0
Boiler or heater	4.0
Tee, percent flowing through branch:	
100	1.2
50	4.0
25	20.0

[a] The friction in one 90 deg standard elbow is approximately equal to the friction of a length of straight tube of the same nominal size and 25 diam long. Hence, one elbow equivalent in feet of tube equals 25 diam (in inches) divided by 12.

straight pipe of the same diameter, 25 diam long. It is, therefore, necessary to assume an average size of pipe for the system, which is later checked.

2. The rate at which the water is to be circulated is then chosen. High velocities reduce the size of pipe but increase the cost of the pump and its operation. A cooling of the water or drop of 20° F has proved satisfactory and economical in average systems, on the basis of a reasonable velocity from the pump and a logical relationship between the flow and the size of pipe. A mean temperature of 215° F has been found practicable for a system using forced circulation and a closed pressure tank.
3. At the temperatures used in heating, water has a specific heat of 1.0 and a weight of about 8 lb/gal. This is an approximate *round number* to work with in preference to the *actual* 8.33. The flow in gallons per minute to produce the total heat required in the convectors is then:

$$\frac{\text{Total heat}}{\text{Drop} \times 60 \times 8} = \text{gpm}$$

or, for a temperature drop of 20° F,

$$\frac{\text{Total heat}}{9600} = \text{gpm}$$

4. To select the pump—the rate of flow and the pressure head must be known. From Table 5.11 a friction head and pressure head are chosen that will permit the transmission of the required heat and flow of water through the total length of piping without unreasonably high or low friction and pressure. The friction head in milinches per foot of pipe is found by dividing the total pressure head in milinches by the total equivalent length of pipe; 300, 250, and 200 milinches/ft are frequently satisfactory in systems of medium size. Higher-friction heads, while permitting smaller pipe, cause high water velocity accompanied by noise. Lower-friction heads require larger pipes and cause sluggish water flow.
5. With the friction head in milinches per foot and the heat loads, the proper size of pipe for each section of the system may be determined from Table 5.11 or 5.12.
6. With the rate of flow in gallons per minute and the pressure head in feet the circulating pump may be chosen from Figure 5.29.
7. When the entire water volume of the system is known: boiler, tubing, heating units, etc., its expansion for a temperature rise of 165° F (215° F − 50° F) may be calculated. The compression tank must accommodate this.

(*c*) ***Design Sequence.*** Calculations for the design are illustrated by the following example:

EXAMPLE **5.5.** Design piping for a two-circuit, one-pipe forced circulation system (Fig. 5.28). In passing through the longest circuit to one of the upstairs convectors the water travels

Table 5.11 ***Pipe Sizing Table for Mains, Forced Circulation Hot Water Systems***

Section A

Booster Head Pressures, Feet	*Total Equivalent Length of Pipe in Feet*								
2	40	48	60	68	80	96	120	160	240
2½	50	60	75	86	100	120	150	200	300
3	60	72	90	103	120	144	180	240	360
3½	70	84	105	120	140	168	210	280	420
4	80	96	120	137	160	192	240	320	480
4½	90	108	135	154	180	216	270	360	540
5	100	120	150	171	200	240	300	400	600
5½	110	132	165	188	220	264	330	440	660
6	120	144	180	206	240	288	360	480	720
6½	130	156	195	223	260	312	390	520	780
7	140	168	210	240	280	336	420	560	840
7½	150	180	225	257	300	360	450	600	900
8	160	192	240	274	320	384	480	640	960
8½	170	204	255	291	340	408	510	680	1020
9	180	216	270	308	360	432	540	710	1080
9½	190	228	285	325	380	456	570	760	1140
10	200	240	300	342	400	480	600	800	1200
10½	210	252	315	360	420	504	630	840	1260
11	220	264	330	377	440	528	660	880	1320
11½	230	276	345	394	460	552	690	920	1380
12	240	288	360	411	480	576	720	960	1440

Section B ***(Based on 20° F Temperature Drop)***

Pipe Size, Inches	*Main Capacities (in Thousands of Btu)*								
	Pressure Drop in Pipe in Milinches per Foot								
	600	*500*	*400*	*350*	*300*	*250*	*200*	*150*	*100*
½	19	18	16	15	13	12	10	9	7
¾	41	37	33	30	28	26	23	20	15
1	80	71	64	59	53	48	42	37	31
1¼	170	160	140	130	118	102	90	78	63
1½	260	240	210	185	175	156	140	121	94
2	500	450	410	360	322	294	261	227	182
2½	810	750	670	610	551	523	460	385	310
3	1600	1400	1300	1150	1000	900	800	680	550
3½[a]	2300	2100	1850	1650	1500	1350	1190	1020	825
4[a]	3200	2900	2600	2300	2100	1950	1700	1350	1140

Source. ITT Bell & Gossett.

[a] Trunk main capacities only. Fittings are not made larger than 3 in.

NOTE. The figures shown in these tables apply to both steel pipe and Type L copper tubing, as capacity differences are not sufficient to cause design errors.

Table 5.12 ***Pipe Sizing Table for Risers, One-Pipe, Forced Circulation Hot Water Systems with Special Fittings***

(Based on 20° Temperature Drop

Capacity of Risers with Two Fittings (In Thousands of Btu)

	Pipe Size Inches	Milinches 600	500	400	350	300	250	200	150	100
		Upfeed Risers—First Floor (See Note 1)								
A	½	23	22	19	18	17	16	14	12	10
	¾	43	41	37	33	30	28	26	22	20
	1	80	73	64	60	55	50	45	39	32
	1¼	180	140	120	110	100	93	80	74	62
		Upfeed Risers—Second Floor (See Note 2)								
B	½	16	15	14	13	11	10	10	8	7
	¾	31	28	25	24	22	21	18	15	13
	1	58	52	45	43	37	33	32	28	25
	1¼	122	108	92	90	79	72	68	59	50
		Upfeed Risers—Third Floor (See Note 2)								
C	½	14	12	11	10	9	8	8	7	6
	¾	26	24	23	21	19	18	16	14	12
	1	47	43	38	36	34	31	29	28	25
	1¼	99	91	81	77	70	66	59	56	46
		Downfeed Risers (See Note 3)								
D	½	16	15	14	12	11	9	8	For less than 200-milinch resistance, base calculations on pump with higher head pressure	
	¾	33	30	26	24	20	18	14		
	1	58	52	43	41	34	29	25		
	1¼	117	106	86	83	69	59	49		

NOTE. The figures shown in these tables apply to both steel pipe and Type L copper tubing, as capacity differences are not sufficient to cause design errors.

Capacity of Risers with One Fitting (In Thousands of Btu)

	Pipe Size Inches	Milinches 600	500	400	350	300	250	200	150	100
		Upfeed Risers—First Floor								
E	½	16.5	15	13	12	11	10.6	10	9.2	8
	¾	29	27	25	24	21	19	18	17	15
	1	50	48	44	41	37	35	33	31	28
	1¼	95	88	78	76	69	62	55.6	48	40
		Upfeed Risers—Second Floor								
F	½	11	10	9	8	7	7	6	6	4
	¾	20	19	17	16	14	13	12	11	11
	1	34	32	29	28	25	24	22	21	18
	1¼	70	68	59	57	51	49	45	43	36
		Upfeed Risers—Third Floor								
G	½	9	8	7	7	6	6	6	5	4
	¾	18	16	14	14	12	12	11	10	9
	1	31	29	28	27	24	22	21	20	18
	1¼	63	60	56	52	48	45	43	41	36

Source. ITT Bell & Gossett.

(Continued)

Read these notes carefully before sizing risers:

NOTE 1: First-floor upfeed risers—Capacities shown in the table are based upon horizontal branches not more than 3 ft long, with stubs 18 in. long, or a total of 9 ft of pipe. Six elbows, one valve and one union ell, and 1 C.I. radiator are added for the equivalent length. For each additional 10 equivalent ft of pipe, move 2-milinch columns to the right.

NOTE 2: Second- and third-floor upfeed risers—Capacities shown are based upon horizontal branches not more than 3 ft long, with risers 10 ft high and 20 ft high, respectively. Eight elbows, one valve and one union ell, and C.I. radiator are added for the equivalent length. For each additional 10 equivalent ft of pipe, move 2-milinch columns to the right.

NOTE 3: Downfeed risers—Capacities shown are based on a drop of 7 ft to the *center of the radiator,* with not over 3 ft total in horizontal branches, six elbows, one valve and one union ell, and 1 C.I. radiator. For every additional 2 ft of vertical drop, move one column to the right in milinch table.

On downfeed jobs the main *must* be pitched up and a vent installed on end of main.

through boiler, 2 flow-control valves, convector valve, convector, 134 ft of pipe, 3 tees, and 18 ells. This summary would be from *working* sketches. Copper tubing is to be used. The position of special fittings is indicated in Figure 5.28 by heavy lines.

SOLUTION

1. From Table 5.10, friction heads in terms of one elbow are:

1 boiler	4
1 convector	4
1 convector valve	3
3 tees	12
18 ells	18
2 flow-control valves	40 (not in Table 5.10)
	81 elbows

   Tees along the run of the main other than those serving one convector are not considered as adding resistance to the flow. Assuming average pipe diameter to be 1 in., (81 × 25 in.)/12 = 169 ft. Then 134 + 169 = 303 ft total equivalent length.
2. Select 215° F as the average temperature and a 20° drop.
3. Rate of flow through the system: 79,000 Btuh/9600 = 8.2 gpm through main *A* and return *D,* of which (by a similar calculation) 4.15 gpm passes through main *B* and 4.05 gpm passes through main *C.*
4. Select 250 milinches/ft as the pressure drop. For 303 ft equivalent length the pressure head will be 6½ ft (Table 5.11).
5. From Table 5.11 it is found that main *A* and return *D* must be 1¼ in. Branch mains *B* and *C* will both be 1 in.

   The sizes of risers (from Table 5.12) are as follows:

Second Floor (Table 5.12, F)

b	4 Mbh*	½ in.
d	6	½
g	9	¾
i	3	½
m	6	½

*Mbh = 1000 Btuh.

First Floor (Table 5.12, E)

c	8 Mbh	½ in.
f	4	½
h	4	½
j	9	½
k	5	½

Basement (Table 5.12, D)

a	3 Mbh	½ in.
e	5	½
l	4	½
n	9	½

6. Reference to the pump performance curves in Figure 5.29 indicates that a 1½-in. pump will deliver 8.2 gpm against a friction head of 6½ ft.
7. Compression tank may be selected when volume of water in system is known.

NOTE. Tables 5.11 and 5.12 may be used for either steel pipe of Type L copper tubing, as capacity differences are not sufficient to cause design errors.

Fig. 5.30 Details of heating baseboard and finned-tube elements (see Table 5.13). Courtesy of Burnham Corporation.

(*d*) ***Selection of Baseboard Convectors.*** The following example is typical of the method of selecting convectors.

EXAMPLE **5.6.** Select a Heatrim R-750 baseboard convector (Table 5.13) for a 14 ft long living room wall. The required capacity, 9000 Btuh, styles to be as shown in Figure 5.30, the

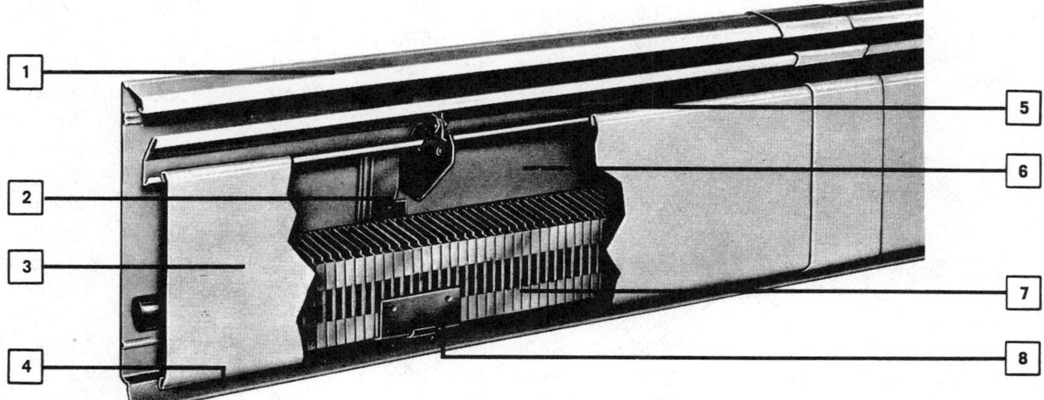

1—enclosure top and back panel—is of one-piece steel construction for strength and durability. Smooth surface won't collect dust and lint. Lower portion is extended ¾" out from wall to cover any openings between wall and finished flooring and to let panel stand unaided while being fastened to wall.

2—snap-in brackets—are die-formed of one-piece heavy gauge steel to support front panel and element where required; can be located at any point along back panel.

3—enclosure front panel—is of flat, dent-resisting steel; firmly snaps onto brackets either way (there is no top or bottom edge) and is easily removed for cleaning when necessary.

4—air intake space—allows unrestricted circulation of air for maximum heating efficiency and permits easy cleaning underneath element.

5—pivot-mounted damper—allows positive fingertip control for easy opening and closing from any point along its length; can be set in any position to regulate heat flow. Damper is standard equipment . . . no extra charge.

6—enclosure interior—gray prime coat provides background so that brackets and element blend in.

7—heating elements—two high-output elements are available, each with heat-reflecting interlocked aluminum fins (vented for maximum performance) bonded to seamless copper tubing for quick response, high I=B=R approved ratings.

8—element slide cradle—of lead-coated metal (terneplate) for lubricant slickness; snaps onto element and rides in bracket for noiseless linear movement of element.

two high-capacity elements

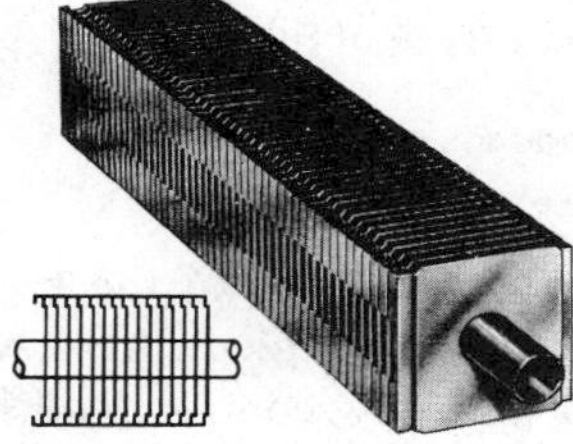

E-500 (½-in. nominal copper tubing)

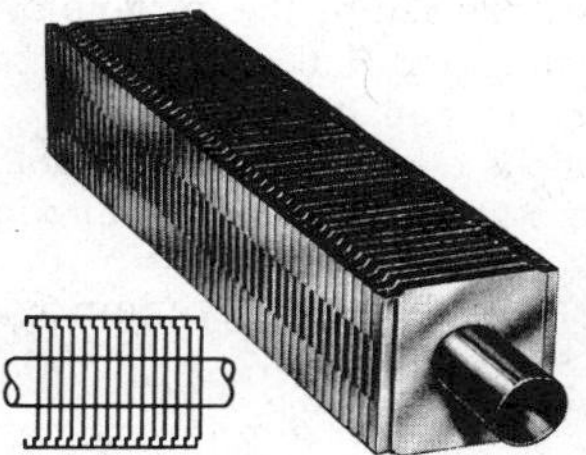

E-750 (¾-in. nominal copper tubing)

Interlocked box-fins . . . 2⅛ inches square . . . tubes have one end expanded for easy coupling with unexpanded end of next element.

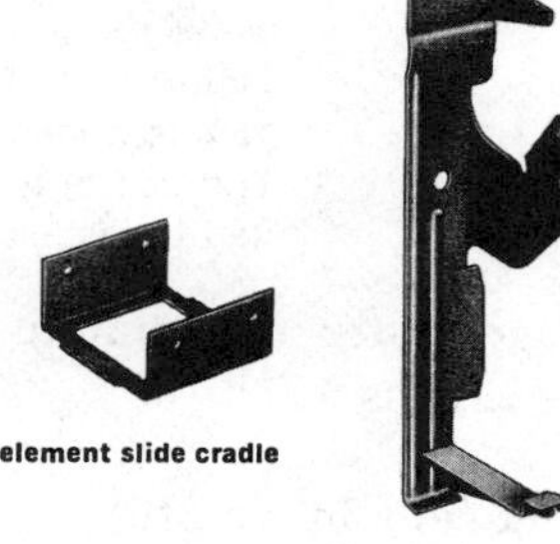

element slide cradle

support bracket

Table 5.13 ***Manufacturers Data for Hot Water Baseboard Convectors***

approved I=B=R water ratings[a]
model R-750 Heatrim—capacities Btu/hr per lineal foot

number of lineal feet[b]	*water flow rate 500 lbs/hr average water temperature °F*					
	170°	*180°*	*190°*	*200°*	*210°*	*220°*
1	500	560	630	690	760	820
2	1000	1120	1260	1380	1520	1640
3	1500	1680	1890	2070	2280	2460
4	2000	2240	2520	2760	3040	3280
5	2500	2800	3150	3450	3800	4100
6	3000	3360	3780	4140	4560	4920
7	3500	3920	4410	4830	5320	5740
8	4000	4480	5040	5520	6080	6560
9	4500	5040	5670	6210	6840	7380
10	5000	5600	6300	6900	7600	8200
11	5500	6160	6930	7590	8360	9020
12	6000	6720	7560	8280	9120	9840
13	6500	7280	8190	8970	9880	10660
14	7000	7840	8820	9660	10640	11480
15	7500	8400	9450	10350	11400	12300
16	8000	8960	10080	11040	12160	13120
17	8500	9520	10710	11730	12920	13940
18	9000	10080	11340	12420	13680	14760
19	9500	10640	11970	13110	14440	15580
20	10000	11200	12600	13800	15200	16400
21	10500	11760	13230	14490	15960	17220
22	11000	12320	13860	15180	16720	18040
23	11500	12880	14490	15870	17480	18860
24	12000	13440	15120	16560	18240	19680
25	12500	14000	15750	17250	19000	20500
26	13000	14560	16380	17940	19760	21320
27	13500	15120	17010	18630	20520	22140
28	14000	15680	17640	19320	21280	22960
29	14500	16240	18270	20010	22040	22780
30	15000	16800	18900	20700	22800	24600

[a]Approved I=B=R water ratings shown above for American-Standard Heatrim Panels (with Model E-750 element) are based on a water flow of 500 pounds per hour with a pressure drop of 0.047 inches of water per lineal foot and a water flow of 2000 pounds per hour with a pressure drop of 0.530 inches of water per lineal foot. As allowed by the Institute of Boiler and Radiator Manufacturers (I=B=R) Testing and Rating Code for Baseboard Type of Radiation, 15% is added to water heat capacity.
The use of I=B=R ratings at water flow rates of 2000 pounds per hour is limited to installation where the water flow rate through the baseboard unit is

average water temperature is 210° F, and water flow is 500 lb/hr.

SOLUTION. In the 210° F column of the left section of Table 5.13 a 12-ft-long baseboard convector is chosen for its capacity of 9120 Btuh. It would fit in the 14 ft length of wall.

(*e*) **Zoning** (Fig. 5.31). Based on the one-pipe principle in multiple circuits, each having its own pump and flow-control (check) valves, hot water systems are very well suited to zoning. This installation comprises three separately heated areas, first, second, and third floors. Each can be heated to different tem-

(*Courtesy of Burnham Corporation*)

water flow rate 2000 lbs/hr average water temperature °F					
170°	180°	190°	200°	210°	220°
530	590	670	730	800	870
1060	1180	1340	1460	1600	1740
1590	1770	2010	2190	2400	2610
2120	2360	2680	2920	3200	3480
2650	2950	3350	3650	4000	4350
3180	3540	4020	4380	4800	5220
3710	4130	4690	5110	5600	6090
4240	4720	5360	5840	6400	6960
4770	5310	6030	6570	7200	7830
5300	5900	6700	7300	8000	8700
5830	6490	7370	8030	8800	9570
6360	7080	8040	8760	9600	10440
6890	7670	8710	9490	10400	11310
7420	8260	9380	10220	11200	12180
7950	8850	10050	10950	12000	13050
8480	9440	10720	11680	12800	13920
9010	10030	11390	12410	13600	14790
9540	10620	12060	13140	14400	15660
10070	11210	12730	13870	15200	16530
10600	11800	13400	14600	16000	17400
11130	12390	14070	15330	16800	18270
11660	12980	14740	16060	17600	19140
12190	13570	15410	16790	18400	20010
12720	14160	16080	17520	19200	20880
13250	14750	16750	18250	20000	21750
13780	15340	17420	18980	20800	22620
14310	15930	18090	19710	21600	23490
14840	16520	18760	20440	22400	24360
15370	17110	19430	21170	23200	25230
15900	17700	20100	21900	24000	26100

equal to or greater than 2000 pounds per hour. Where the water flow rate through the baseboard is not known, the I=B=R rating at the standard water flow rate of 500 pounds per hour must be used.

[b]These ratings are based on active (finned) Heatrim lengths. Difference between active length and total length of the standard Heatrim heating elements is $2\frac{15}{32}$ inches. Elements are unpainted.

Non-ferrous fins on Model E-750 elements measure $2\frac{1}{8}$ x $2\frac{1}{8}$ x 0.008 inches, spaced 52 fins per foot.

peratures as called for by thermostats in each separate apartment. For example, if the thermostat serving the second floor (zone *B*) calls for heat, it turns on pump *B*. Flow-control valves *B* open, admitting hot water from the boiler header to main *B*. Flow-control valves *A* and *C* remain closed, preventing flow in mains *A* and *C*. Any or all of the zones may operate at one time. The boiler keeps a supply of hot water continually ready to supply any zone upon demand. This is achieved by an aquastat immersed in the boiler water. When the boiler water drops below the prescribed temperature it turns on the firing device, such

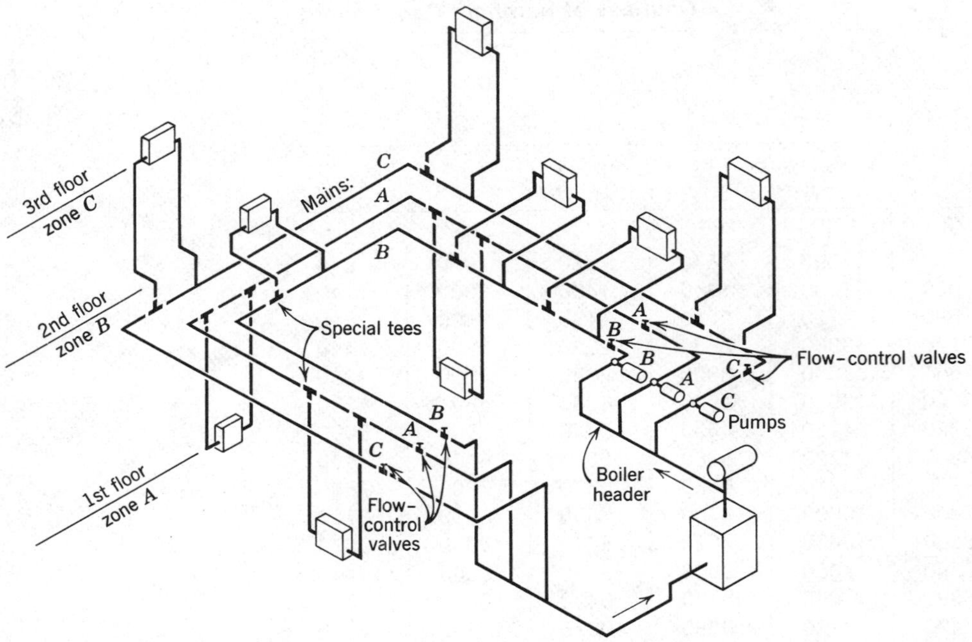

Fig. 5.31 *Three-zone, multicircuit, one-pipe system. Each convector has connections to* one pipe.

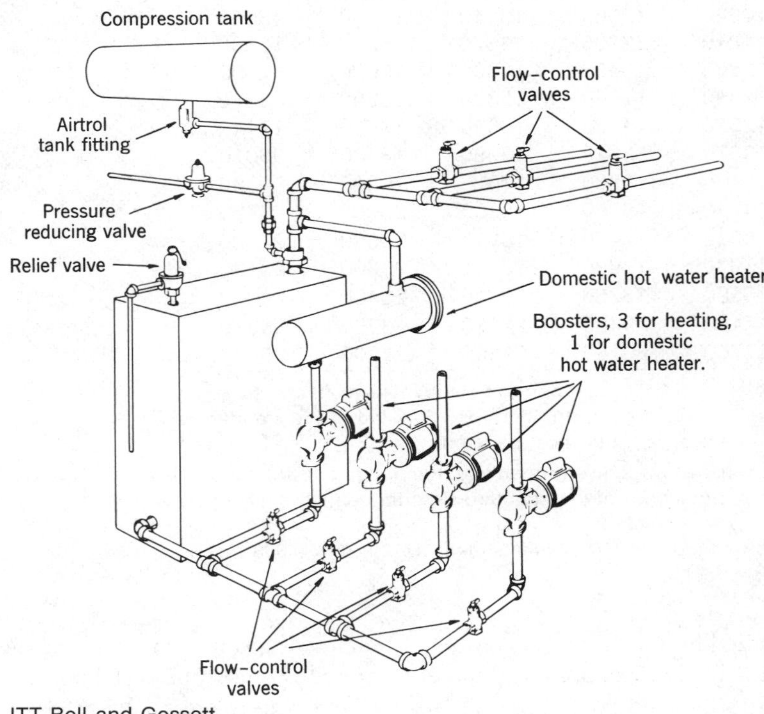

ITT Bell and Gossett

Fig. 5.32 *Boiler, piping, and water controls suitable for the three-zone, one-pipe system in Figure 5.31. Each one-pipe circuit should be provided with two flow-control valves and a circulator on the supply or return pipe. The terms booster, pump, and circulator are interchangeable for the same piece of equipment.*

as an oil burner or gas burner, which brings the water up to temperature. If an overhead main supplies downfeed, as in the first floor of this installation, special downfeed supply and return fittings are necessary. For the second- and third-floor zones, one special return tee is sufficient. If the designer elects to use also a special upfeed *supply* tee of the venturi type, higher outputs of the convectors will result. For similar zoning arrangements at a boiler, see Figure 5.32.

5.9 Electric-Resistance Heating

Each of the three heating systems shown in Figures 5.33, 5.34, and 5.35 can be operated by a thermostat in each room. This feature permits a method of operation that can contribute to the saving of energy. Temperatures in rooms that are not in use can be lowered or the room units turned off by means of a switch. Thus the rooms actually occupied would be the only ones to draw full energy. Baseboard and drop-in units have a fast response. Unoccupied rooms such as bedrooms during daytime hours can be brought to temperature promptly when occupied. Although mean radiant temperatures may drop a bit during hours of nonoperation, reasonable comfort can be expected with the fast pickup of resistance units.

The electric-radiant-ceiling method of Figure 5.35 has largely supplanted the earlier radiant heating that employed hot water coils in ceilings or floors. It has found use in many high-quality apartments.

Fig. 5.33 *Electric residential baseboard in place. Courtesy of Hunter Division, Robbins and Myers, Inc.*

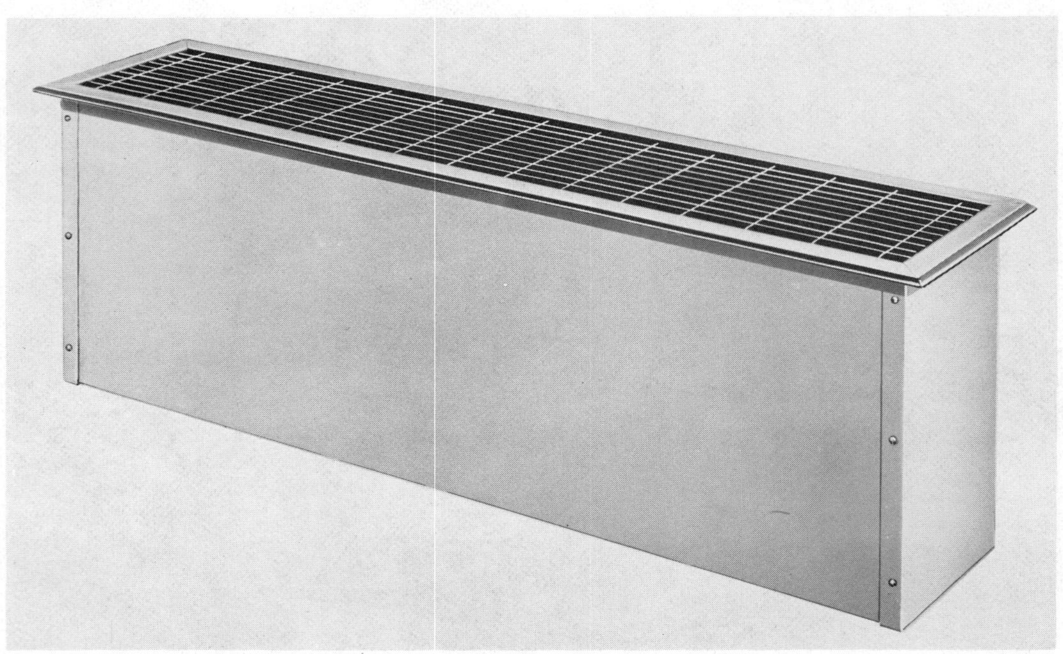

Fig. 5.34 (opposite) *Floor-type, drop-in convective electrical-resistance units. Ratings of this type:*

Watts	Btuh	Width	Length
350	1194	$7\frac{3}{4}$ in.	14 in.
750	2559	$7\frac{3}{4}$ in.	30 in.
2000	6824	$7\frac{3}{4}$ in.	62 in.

Courtesy of Chromalox, Edwin L. Wiegand, Div. of Emerson Electric Co.

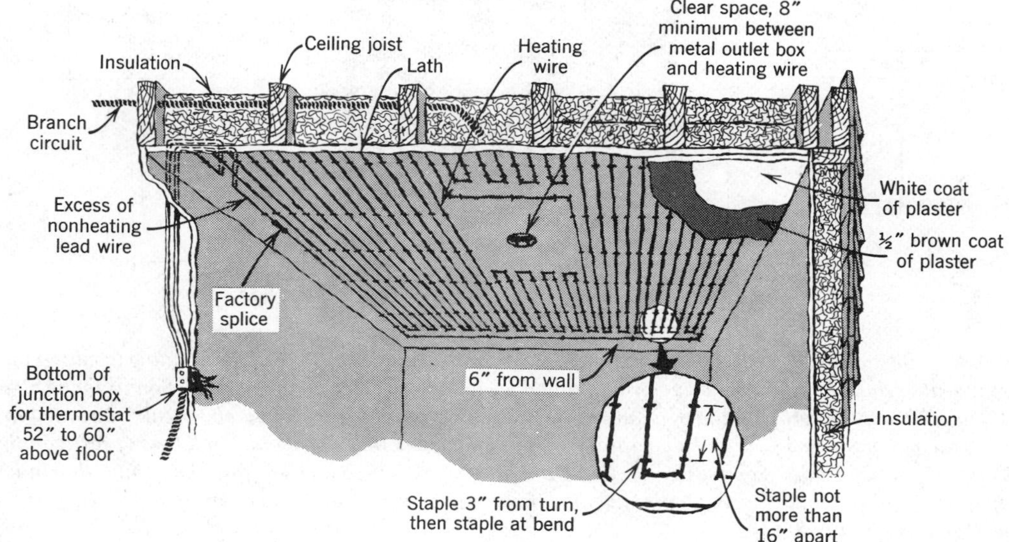

Fig. 5.35 *Typical installation of radiant-heating cable ready for completion of plaster ceiling. Courtesy of General Electric Company.*

5.10 Refrigerated Cooling for Houses

(*a*) ***Air-to-Air Cooling.*** Unlike the technique in large buildings and in district cooling, where distance makes the use of chilled water most convenient for thermal transmission, houses are cooled by a rather simple arrangement of the refrigeration cycle. Reference is made to Section 5.11 and to Figure 5.42, which illustrates the circuit of a refrigerant in compression, condensing, and evaporation, in which the condenser heat is carried away by water and the evaporation process draws heat out of water in another circuit to produce *chilled* water. Thus the heat is *moved* to a heat rejection location outdoors. Figure 5.36 is a schematic diagram of an air-to-air (in distinction to a water-to-water) refrigeration device. Air instead of water can be used to cool the condenser, and indoor air can be cooled directly by passing it over the evaporator coil in which the refrigerant is expanding from a liquid to a gas. Thus heat is moved from the indoor air to the outdoor air by the step-up action or heat pumping nature of the refrigeration cycle. When indoor air is cooled directly in this manner by the expanding refrigerant, the process is usually known as *direct expansion.* One recognizes the assembly just described as the conventional *window unit.* This has progressed to a through-the-wall unit.

To allow more flexibility in the placement of the heat rejection side of the cycle, air-to-air equipment is now obtainable in which the condenser and its somewhat noisy com-

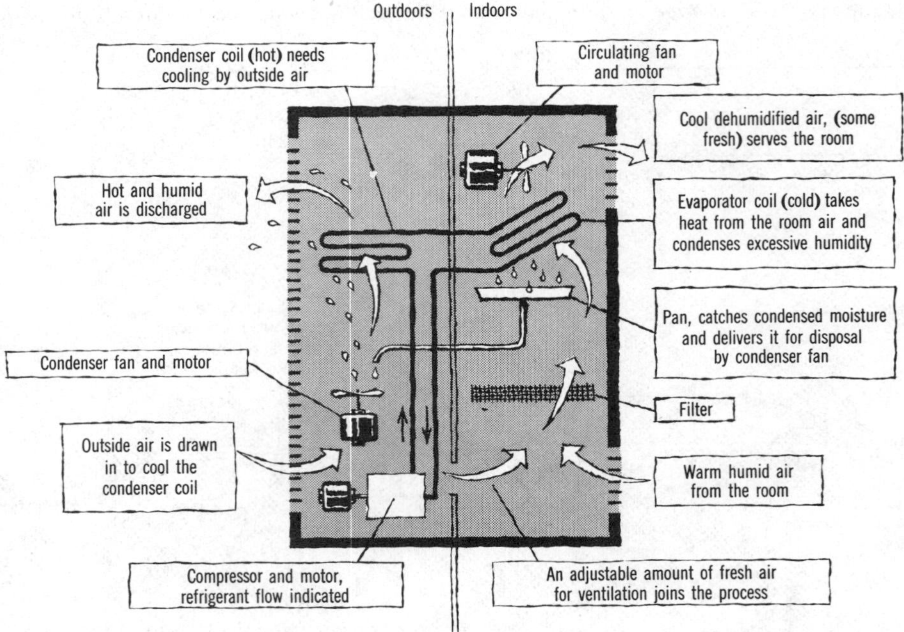

Fig. 5.36 *Schematic diagram of the operation of a through-the-wall, air-to-air, conditioning (cooling) unit. Direct heat-exchange occurs between air and the processes of evaporation or condensation of the refrigerant. The unit is quite self-contained, requiring only access to outdoor air and an electrical connection that powers the motors of two fans and a compressor. The usual capacities are about 1 or 2 tons of refrigeration (12,000 or 24,000 Btuh). Separation of the outdoor compressor-condenser function from the indoor evaporator (cooling) unit (Fig. 5.37) permits the latter to be placed in an indoor central air-cooling system (See Figs. 5.39 and 5.40).*

panion, the compressor, can be placed in a remote location outside of the house and beyond the exterior wall (see Fig. 5.37). The compressor-condenser unit is placed on a concrete slab. This arrangement allows the placement of the evaporator (direct expansion) coil in the air stream of the warm air heating system. Figure 5.38 shows such a direct expansion unit in an upflow cooling-heating system, and Figure 5.39 shows a similar evaporator in the duct circuit of a warm air system employing horizontal, in-line *stowaway* heating elements that can be used compactly in basement or attic.

(b) Cooling/Heating by a Combined Hydronic and Air System. The building firm of Levitt & Sons has used various systems in its projects over the years. The first Levittown, New York house had radiant hot water heating with pipes imbedded in floor slabs. More recently, in an attempt to incorporate air conditioning in low-cost homes, they have found what they feel is the best system from the standpoints of performance and standardization.

The new system combines a perimeter hot water heating pipe with an overhead air-handling system. A boiler having a tankless coil supplies domestic hot water. The heat output supplies both the perimeter loop and a coil in the air-handling unit of the duct system. The total heating load is met by the combination of radiant heat generated by the perimeter loop and heated air from the overhead air-handling system.

The perimeter loop consists only of $\frac{1}{2}$ or $\frac{3}{4}$-in. copper tubing imbedded 4 in. below the top of the floor slab to kill the cold slab effect. It has the capacity to maintain a 35° F differ-

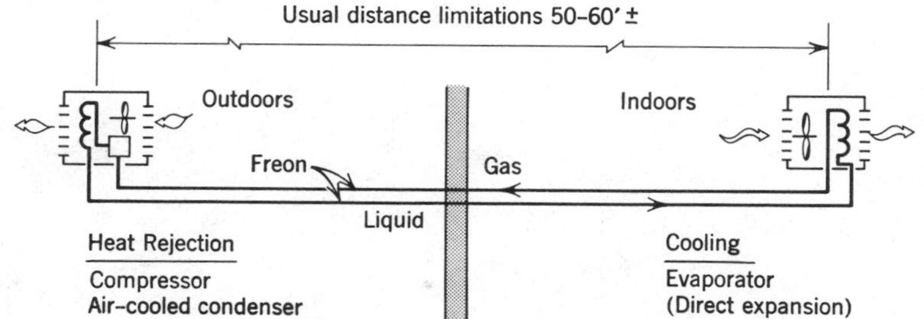

Fig. 5.37 *Schematic diagram of an air-cooled conditioner comprising remote outdoor heat rejection and indoor cooling coil suitable for placement in central air stream (See Fig. 5.39). Progressive Architecture.*

Fig. 5.38 *Cooling/heating air-handling unit. Removal of front panel reveals up-flow circuit. At lower left is air intake and filter. Adjacent are fan and motor. Above these are seen gas-burning elements. At the top is a direct expansion cooling coil. American Furnace Division, The Singer Company.*

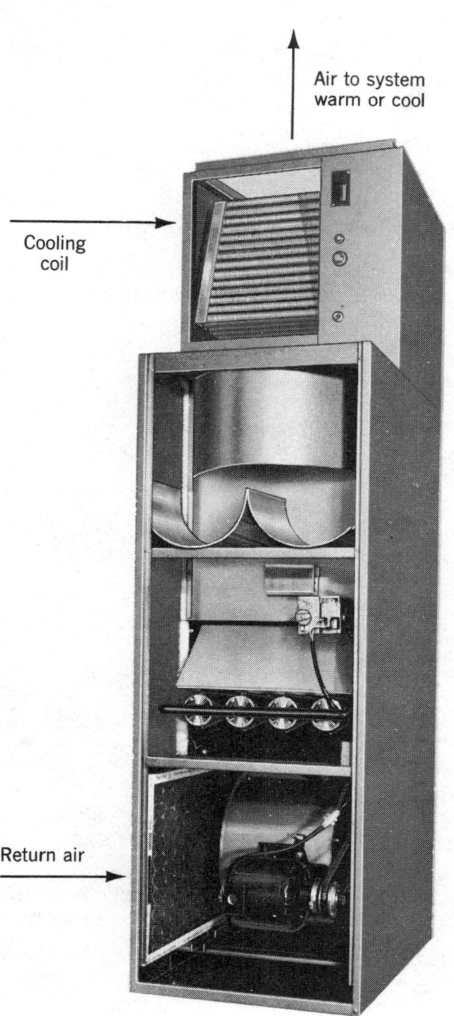

ential between inside and outside temperature at the perimeter.

The air-handling unit and overhead duct system, with supply outlets in each room and central return, is used throughout the year. Its cooling coil is connected to an adjacent outdoor condensing unit (See Figs. 5.40 and 5.41).

Standardization was one of the goals the Levitt organization aimed at with this approach to heating and cooling. Mechanical engineer John Liebl, designer of the system, says that its major advantage is the fact that it provides year-round comfort that works with all types of slab, on any terrain, using any fuel. This standardization simplifies design and construction costs.

An important advantage of the system from a performance viewpoint is that problems of short cycling are minimized, as part of the heating load is carried by the loop, and air discharge can be kept at about 120° F, or 20° F lower than with a conventionally ducted system.

With respect to comfort, it is claimed that not only does the radiant effect eliminate any cold slab feeling but also the temperature variation from floor to ceiling in test houses has not exceeded 4° F.

5.11 Central-Station Air Conditioning

(*a*) ***Scope.*** The first three chapters have set new standards for indoor comfort under varying outdoor climatic conditions. These criteria guide us toward designs that use minimal energy and, accordingly, reduced equipment. Yet in very large, densely occupied buildings, critical conditions may obtain that would re-

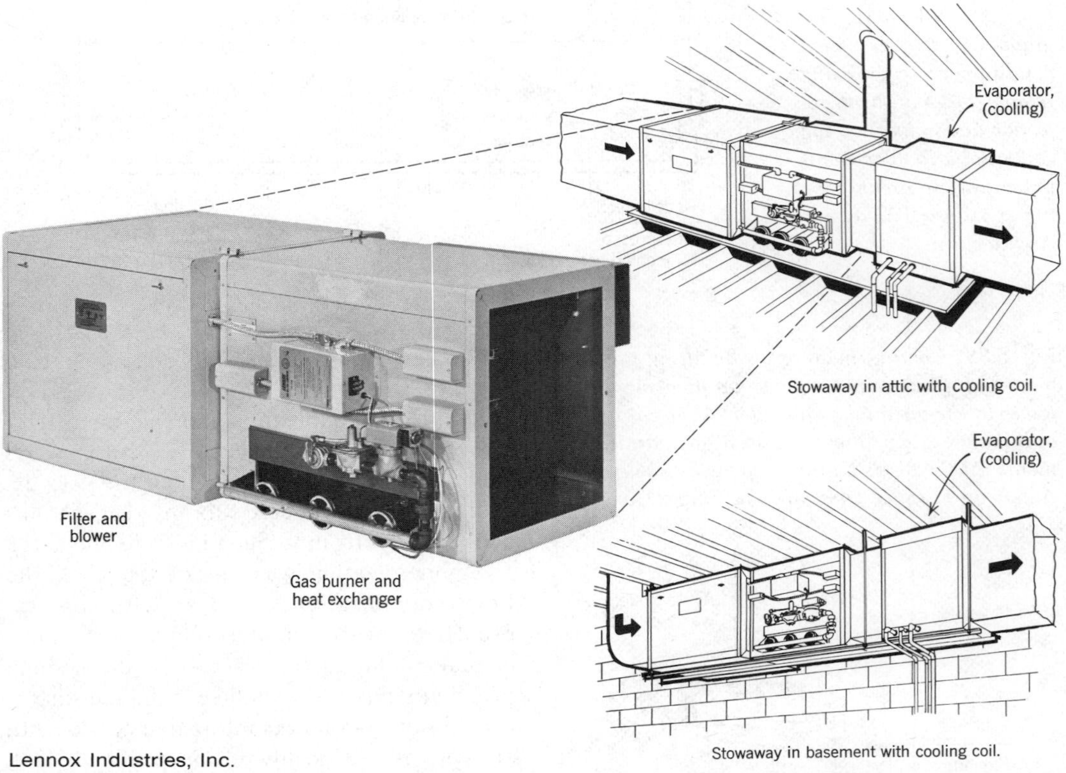

Evaporator,
(cooling)
Stowaway in attic with cooling coil.
Evaporator,
(cooling)
Filter and
blower
Gas burner and
heat exchanger
Stowaway in basement with cooling coil.

Lennox Industries, Inc.

Fig. 5.39 Compact, horizontal-flow combinations for heating and cooling by air. The small pipe between the two refrigerant *pipes of the cooling unit is a water drain that carries away the condensed moisture from the recirculated and outdoor air. Heating unit will require gas and flue connections. Refrigerant tubing connects to outdoor compressor-condenser unit.*

Typical supply grille
Heating coil
Return air duct
Oil fired boiler
Air handler
Condenser
¾″ perimeter copper
loop—underslab

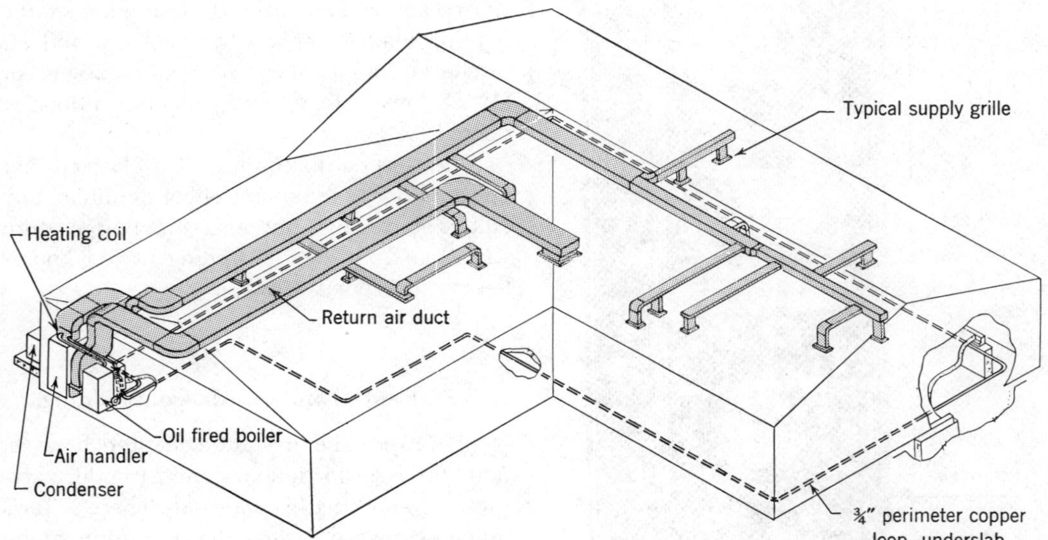

Fig. 5.40 Perimeter loop below slab, combined with overhead air system, provides heating and cooling. Courtesy of Levitt & Sons. Design by mechanical engineer John Liebl. Reprinted from Air Conditioning, Heating and Ventilating.

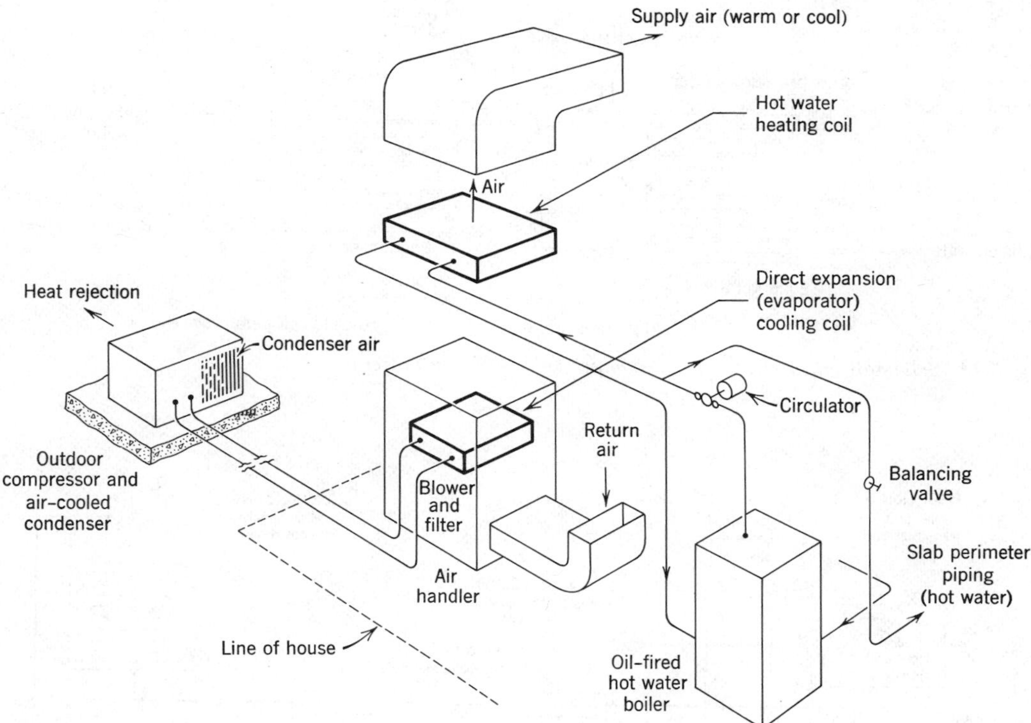

Fig. 5.41 Expanded and partial view of the heating and cooling system shown in Figure 5.40, illustrating schematically the circulation of water, air, and refrigerant. Heating or cooling coils operate, as called for, to warm or cool the circulated air. This is a combination of air and water as thermal media, together with hydronic warming of the slab perimeter.

quire relief. An example would be a very large theater for the performing arts. In solving such problems, a knowledge of the classic methods of air conditioning is essential, to be used with energy-conscious discretion. Complete air conditioning requires a source of heating such as a boiler or furnace, a method of cooling such as the refrigeration cycle, equipment for introducing moisture and for removing it, filters or air-washing devices to clean the air and, for proper air distribution, blowers, ducts, and registers. Fresh air is introduced on the intake side of the conditioning equipment so that it may be treated before entering the space. An equal or slightly smaller amount of indoor air is exhausted from points of odor concentration such as kitchens, toilets, and smoking rooms.

(b) Cooling by Compressive Refrigeration. Since principles of heating have already been discussed in Sections 5.1 to 5.9, it is now necessary to consider the means for producing cool air, or the chilled water by which air may be cooled. Occasionally ground water is obtainable at temperatures low enough for direct use, but generally the use of a refrigeration machine or other special cooling device is necessary. As shown in Figure 5.42 the compressive refrigeration cycle is a scheme for transferring heat from one circulated water system (chilled water) to another (condenser water). The means for doing this is the liquification and evaporation of a refrigerant, usually Freon, during which processes it respectively gives off and takes on heat. The heat that it gives off must be disposed of (except in the heat pump) but the heat that it acquires is drawn out of the circulated water known as the chilled water, which is the medium for subsequent cooling processes.

Freon, a gas at normal temperatures and

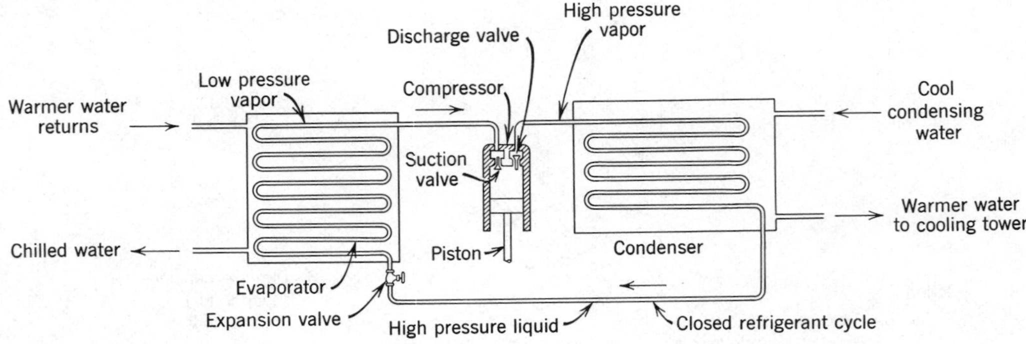

Fig. 5.42 *Schematic arrangement of a compressive refrigeration cycle.*

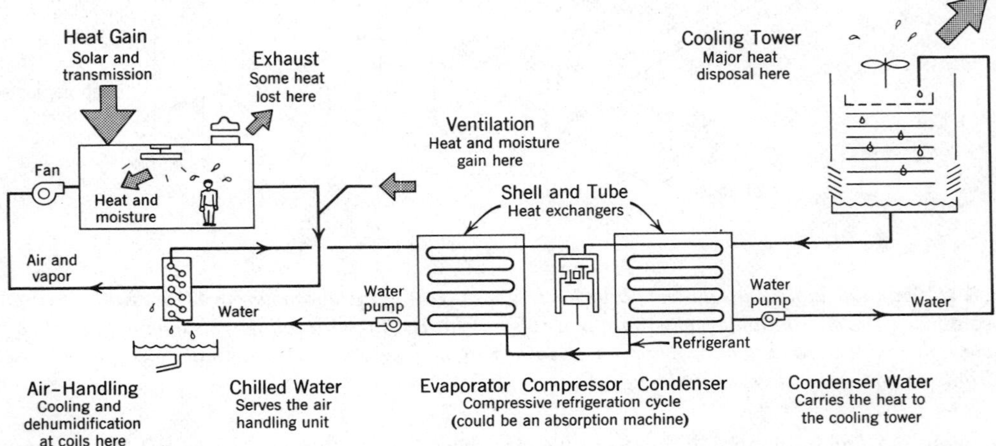

Fig. 5.43 *In most large central systems, water is the vehicle that extends the effects of the refrigeration cycle. Circulated water carries heat from the room via the air-handling unit to the refrigeration unit, and a similar water circuit carries it to the cooling tower for disposal. This diagram includes only the cooling processes of an air-conditioning system. For equipment, see Figures 5.44 [compressive-refrigeration machine (water chiller)], 5.45 [absorption machine (water chiller)] and 5.48 (cooling tower).*

pressures, must be compressed and liquified to be of service later as a heat absorber. In order to liquify it (see Fig. 5.42), it is first compressed to a high-pressure vapor; then, by means of cool water, latent heat is extracted from the Freon, which condenses it to a liquid. This product, high-pressure liquid Freon, is a potential heat absorber since, when it is released through an expansion valve, it springs mechanically to gaseous form. In this change of state it must take on latent heat, by drawing heat out of the circulated water of the chilled water system. It may be said that the refrigeration cycle pumps the heat out of the chilled water system into the condenser water system. Indeed, by special (reverse cycle) arrangements of the water systems, a *heat pump* is the result. So the refrigeration cycle is the basis for the cooling in air-conditioning systems that have fossil-fuel sources of heat, and the basis for both heating and cooling in the heat pump (see Section 5.14).

The piston in the schematic drawing of Figure 5.42 suggests a reciprocating compressor, a type often used in smaller compressor sizes. Larger chilled water "packages" usually

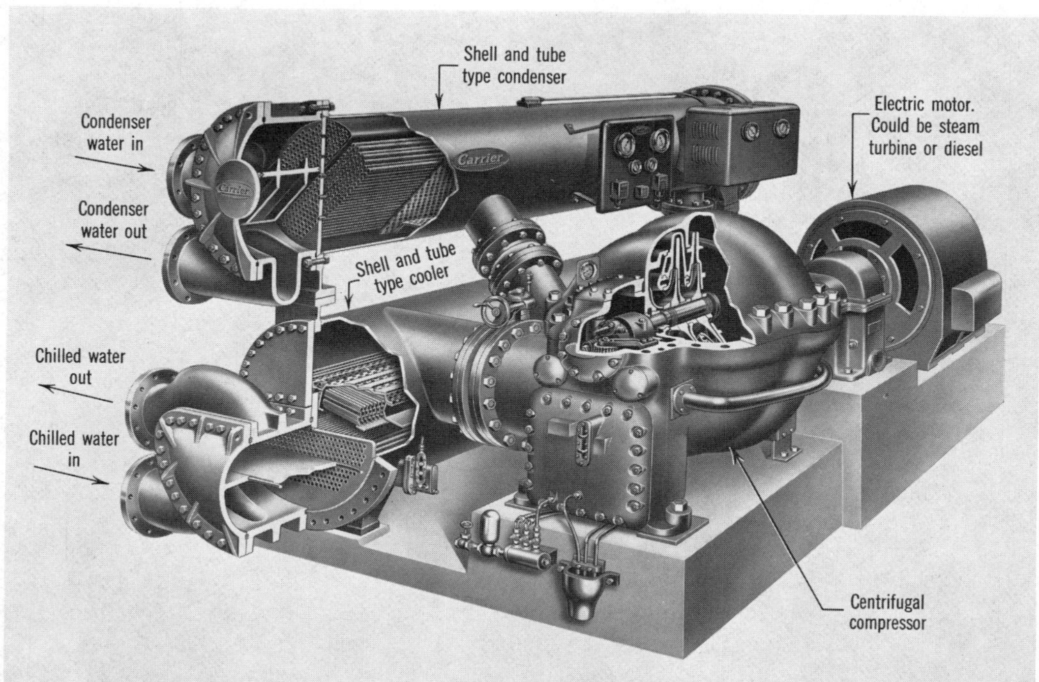

Fig. 5.44 *Centrifugal compressive-refrigeration machine (water chiller). Courtesy of the Carrier Corporation.*

employ centrifugal compressors such as the type of Figure 5.44. Application of the principle of Figure 5.42 is seen in its final form as a piece of tangible equipment, in Figure 5.44. Compressor, condenser, and cooler (evaporator) are usually assembled as a unit package. When the compressor is steam-turbine driven, the exhaust steam is often used as an energy source for an auxiliary absorption-refrigeration machine, the principle of which is described in subsection (d). These two devices make an efficient combination, and the use of steam for cooling can employ, for summer operation, the steam plant that is used for heating in winter. Economic studies to compare the relative costs of steam, electricity, or other power sources should be made independently for each project.

(*c*) ***Unit of Refrigeration.*** A ton of refrigeration is the cooling effect obtained when 1 ton of 32° F ice melts to water at 32° F in 24 hr. Since the latent heat-of-fusion of ice is 144 Btu/lb, the cooling effect or rate of 1 ton of refrigeration (2000 lb) is taken as $144 \times 2000 = 288{,}000$ Btu/day of 24 hr or 12,000 Btuh. The requisite capacity of a refrigerating machine in tons may therefore be found by dividing the total heat gain in a building in Btuh by 12,000.

(*d*) ***Cooling by Absorption.*** Another form of water chiller is the refrigeration absorption machine (see Fig. 5.45). Suitable for installations up to 1000 tons capacity, the external connections of this device are similar to those of the refrigeration machine shown in Figure 5.44. It produces chilled water and has a cycle of hot condensing water that must be cooled. Its motive power is steam,[3] but instead of driving a turbine to run a compressor, the steam is used in a regenerative process to strengthen a salt solution. Although similar in external connections, its interior functioning is very different from the compressor-type refrigeration machine and the steps of the proc-

[3]See the legend of Figure 5.46 where hot water from solar collectors is suggested as an alternative for steam, as an input to the regenerative process.

Fig. 5.45 An absorption machine for producing chilled water. Carrier Corporation. Courtesy of Ingersoll-Rand.

ess are explained in Figure 5.46. The absorption machine has become very popular. It is often economically competitive with the compressive machine, has fewer moving parts, is quieter, and demands somewhat less attention. The cooling principle, shown in step 1 of Figure 5.46, consists of drawing vapor away from a body of water. The change of state from water to vapor requires the acquisition of latent heat. This heat is drawn from the sensible heat of the remaining water, cooling it. The uses of chilled water in cooling coils to produce cool air are described in subsection (e) that follows. Water circulation pumps suitable to serve either a compressive or an absorption machine are used.

(e) ***Principles of Central Cooling.*** In larger buildings and those with varied and diverse occupancy, it is usually preferred to centralize the refrigeration plant. The condenser is cooled by water circulated to an outdoor cooling tower and the evaporator produces chilled water. The latter is then pumped to wherever it is needed in the building. Figure 5.43 shows this arrangement, illustrated here apart from the heating elements that are actually part of the general climate system. In this illustration, chilled water is pumped to one air-handling unit where the circulated air for one room is cooled and its moisture partially condensed. In practice, the water is usually pumped to many air-handling stations, each serving many rooms. The most important single factor affecting such designs is the fact that air ducts are very large and water piping for the same thermal flow is very small. For this reason, the air-handling centers are located as close to the serviced areas as possible. A balance must be achieved, however, and ducts

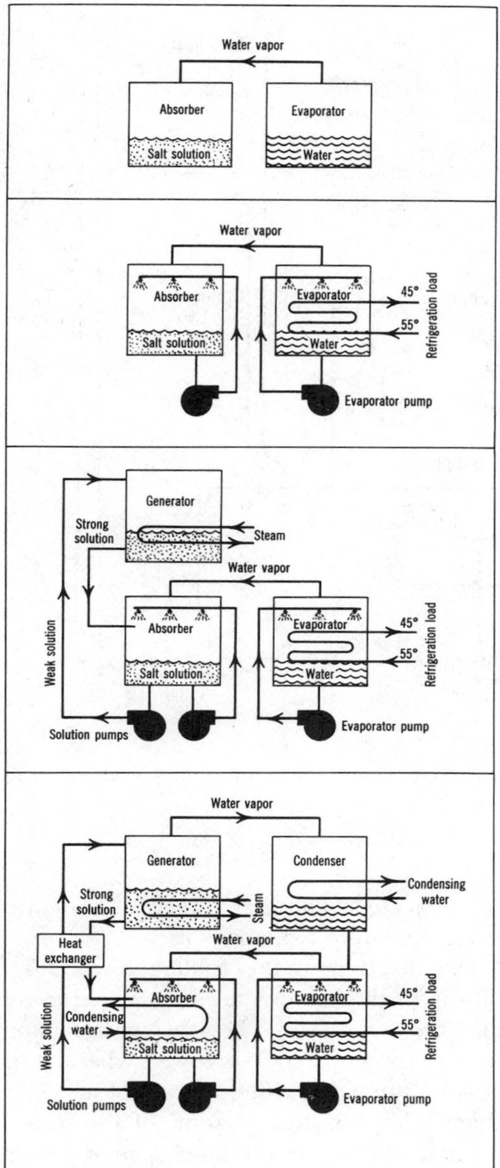

1 Evaporator and absorber

Consider two connected, closed tanks with a salt solution (lithium bromide) in one and water in the other. Just as common table salt absorbs water on a damp day, the salt solution in the absorber soaks up some of the water in the evaporator. The water remaining is thereby cooled by evaporation.

2 Evaporator coil and pump added

This refrigeration effect is utilized by putting a coil in the evaporator tank. Water from this tank is pumped to a spray header which wets the coil. The spray's evaporation chills water in the coil as it circulates to the refrigeration load. Solution pumped to spray in absorber raises efficiency.

3 Solution pumps and generator added

In an actual operating cycle, the salt solution is continuously absorbing water vapor. To keep the salt solution at proper concentration, part of it is pumped directly to a generator where excess water vapor is boiled off. The reconcentrated salt solution is returned to the absorber tank where it mixes with the solution sprayed to absorber in step 2.

4 Condenser and heat exchanger added

Water vapor boiled off from the weak solution is condensed and returned to the evaporator. A heat exchanger uses the hot, concentrated salt solution leaving the generator to preheat the cooler, weak solution coming from the absorber. Finally, condensing water circulating through the absorber and condenser coils removes the waste heat.

Fig. 5.46 *Absorption-refrigeration cycles. For external view of machine, see Figure 5.45. Connected systems are steam, condensing water, and chilled water (refrigeration load). Courtesy of the Carrier Corporation.*

Author's comment. *You will note in the legend of Figure 7.22, page 298, that high-temperature water instead of steam can be used in the generator of an absorption machine. Future planning holds the possibility of using hot water from solar collectors as input to the generator.*

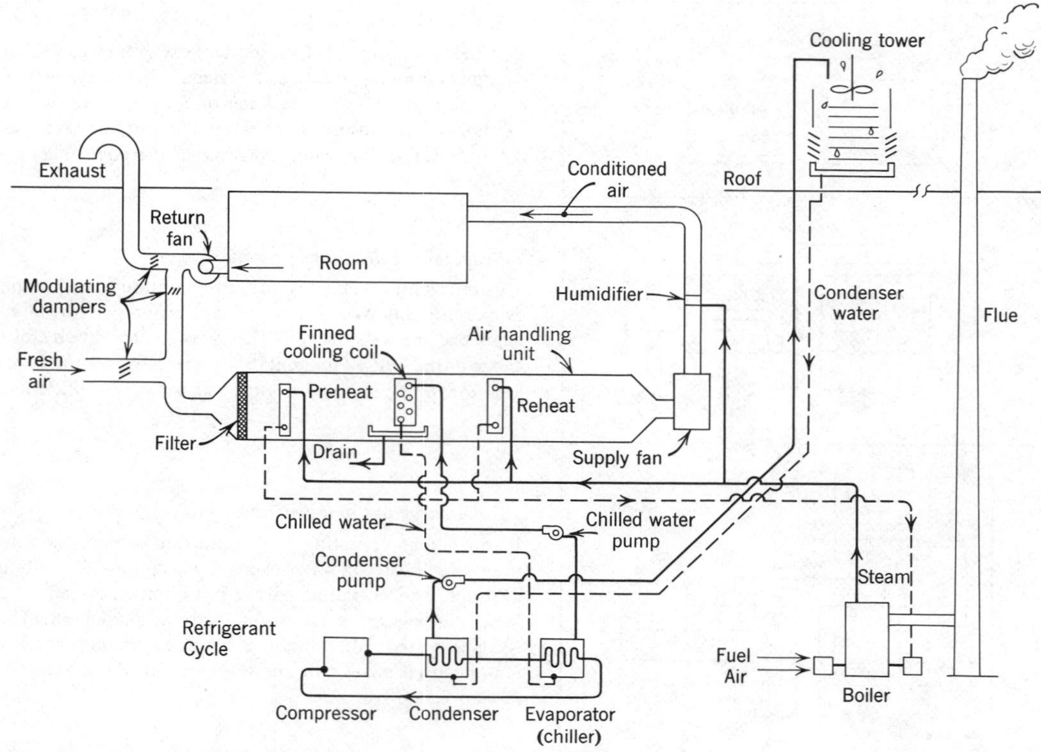

Fig. 5.47 *Schematic diagram of some of the principal components of a central-station, air-conditioning plant for year-round operation.*

from a single air unit may serve several stories. Advantages of central cooling include the significant one that fan and compressor noise does not enter the space. If, additionally, air velocities at registers are kept low, quiet operation will result.

(*f*) ***Central Station Conditioning.*** When heat is needed in winter the air circulated through the room or other space must be heated. Figure 5.47 shows a somewhat more developed sketch of the air-handling unit, indicating the position of the cooling coil in the air stream and the addition of preheating and reheating coils, supplied with steam from a steam boiler. Automatic controls start the steam flow or, by dampers, place the heating coil in the air stream. The reheat coil is available to raise the air temperature if required without increasing the absolute humidity. An example of reheat is seen in cycle No. 3 of Figure 5.53, page 232, in which saturated cool air is reheated to a suitable relative humidity and temperature for indoor use.

The function of the cooling tower (Fig. 5.48) is to dispose of the heat carried away from the condenser. The condenser-cooling water is pumped to the tower where it is dropped through a rising current of air. This effects a vaporization of some of the water. Latent heat necessary to this change of state is drawn from the remaining water, cooling it for reuse in the condenser. Since some of the water is lost in vapor, makeup water must be added at the tower. Water in this circuit, because of the vapor process, has a very corrosive action and needs to be specially treated.

5.12 Psychrometry

(*a*) ***Air, Moisture, and Heat.*** Air and water vapor are the media by which air-conditioning systems operate. The qualities and character-

Fig. 5.48 A cooling tower serving a large building. Condenser water is delivered to this unit by pumps. Courtesy of Ingersoll-Rand Corporation.

istics of these two gases must be understood if we are to gain a working knowledge of air-conditioning processes. Water vapor in varying amounts always exists in air, and the regulation of this moisture content and of the temperature of the vapor-air mixture is the problem posed to the designer of air conditioning. The principles of physics and their development in *ASHRAE* publications are indispensable to a deeper understanding of these matters. A few brief definitions and statements of physical relationships given here will serve as an introduction to the subject.

Dry-Bulb Temperature (*DB*). The temperature of the air-water vapor mixture measured in the normal way with a Fahrenheit thermometer.

Wet-Bulb Temperature (*WB*). The temperature shown by a thermometer with a wetted bulb rotated rapidly in the air to cause evaporation of its moisture. In dry air the moisture evaporates and, in acquiring latent heat, draws heat out of the thermometer to produce a large *wet-bulb depression* (difference between dry- and wet-bulb temperatures). This is an index of low relative humidity. Slow evaporation when the air is already moisture-laden results in a small wet-bulb depression and indicates a condition of high relative humidity. See Figure 5.49.

Relative Humidity (*RH*). The ratio of the partial pressure of the actual water vapor in a mixture to the pressure of a saturated mixture *at the same temperature.* The quantity of moisture showing 40% RH at 40° F would only produce about 13% at 75° F.

Dew Point (*DP*). The temperature at which an air-water vapor mixture will become satu-

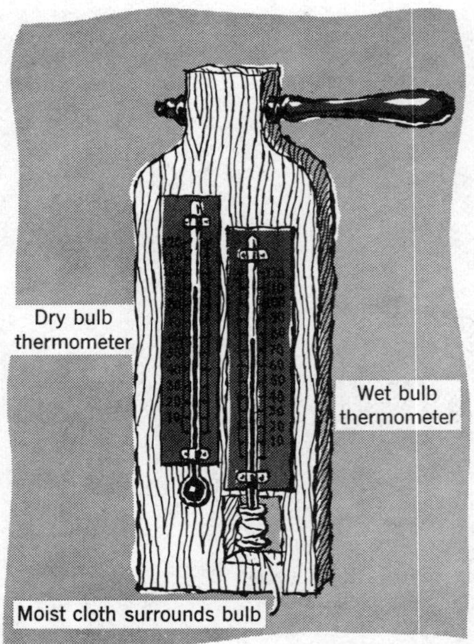

Fig. 5.49 Principle of the sling psychrometer.

rated and begin to yield drops of condensed water. The moisture collecting on the exterior of an uncovered cold-water pipe indicates that the pipe surface temperature is below the dew point of the surrounding mixture. This phenomenon, usually called *sweating*, would more correctly be called *condensing*.

Humidity Ratio (*HR*). The weight of the actual water vapor in a mixture per pound of dry air.

Enthalpy. The total heat in the mixture measured above 0° F and including the latent heat of the water vapor.

Specific Heat. The number of British thermal units required to raise 1 lb of a substance 1° F. For air 0.241 Btu may be used and for water vapor 0.444.

Latent Heat. A term used to express the energy involved in a change of state.

Density. For approximate calculations 0.075 lb per cu ft may be used as the density of air.

(b) The Psychrometric Chart. The qualities of mixtures of air and water vapor are summarized graphically in the psychrometric chart, Figure 5.50. Its use is explained in Figure 5.51. If, for a selected location (New York City, for example), outdoor design conditions are selected (see Table C.2, Appendix, page 1299), other data may be found from Figure 5.50 as explained in Figure 5.51. Dry- and wet-bulb temperatures of 90° F and 76° F, respectively, will fix values as shown in Figure 5.51 for dew point, humidity ratio, relative humidity, and enthalpy. For comparison, indoor conditions are also shown.

The psychrometric chart is the working diagram of the air-conditioning engineer. Assume the simple case of a once-through system, using 100% outdoor air, all of which is discarded. This method is not as economical as when most of the air is recirculated, but it is used in special cases such as for hospital surgical suites and laboratories with many fume hoods (Fig. 7.6, page 282). For every pound of "dry" air (it is not really dry, but the values are based upon the weight of the air alone) that is cooled and dehumidified, $39.6 - 28.3 = 11.3$ Btu must be extracted. Similarly, for every pound of air so treated, $0.0162 - 0.0093 = 0.0069$ lb of condensed moisture must be disposed of. The conditioned air must be introduced to the space at conditions of temperature and relative humidity less in value than those of the indoor conditions so that the entering air may "soak up" heat and moisture and leave through the return grills no worse than the design characteristics of 75° F and 50% relative humidity. So another set of conditions lower in dry bulb and humidity ratio will be established, and the values will depend on the rates at which the air is introduced and the amounts of heat and moisture to be absorbed by the air passing through. This heat and moisture is, of course, the sensible, and latent heat gain of the space to be conditioned.

Each pound of moist air contains heat that is, by custom, measured above the value of 0° F. It consists of the sensible heat of the air and the water vapor, and the latent heat of the water vapor. Using 1061 Btu as an average

Fig. 5.50 The Psychrometric Chart. Courtesy of the Carrier Corporation.

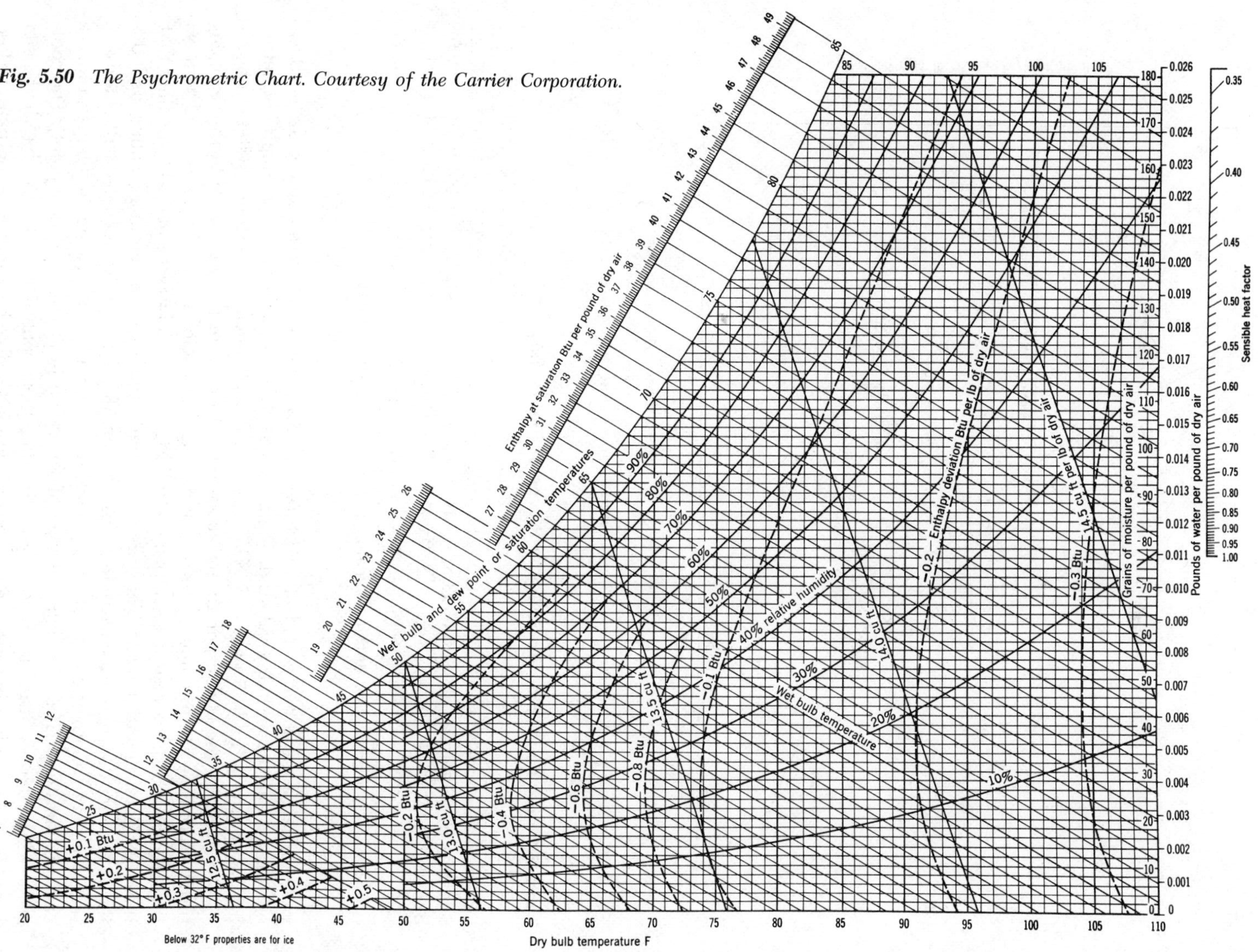

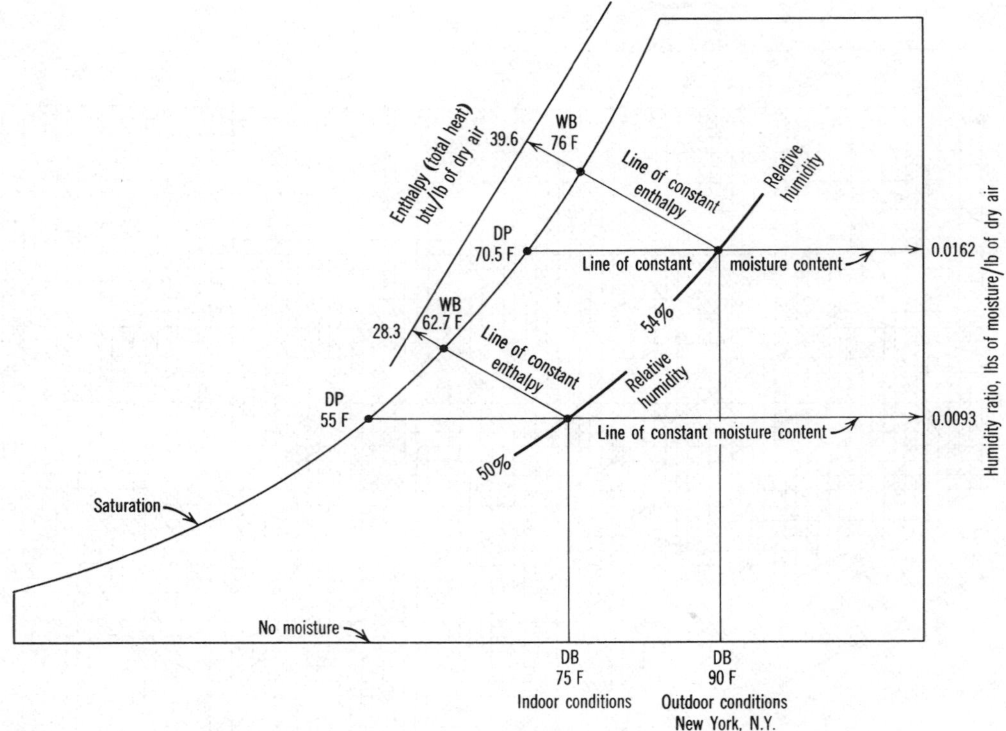

Fig. 5.51 *Use of the Psychrometric Chart. Typical indoor and outdoor conditions in summer.*

Table 5.14 ***Verification of the Enthalpy Values of Figure 5.51***

Item	*Indoor Conditions*		*Outdoor Conditions*	
Latent heat, vapor	$0.0093 \times 1061^a =$	9.90	$0.0162 \times 1061^a =$	17.25
Sensible heat, vapor	$0.0093 \times 0.444 \times 75 =$	0.31	$0.0162 \times 0.444 \times 90 =$	0.65
Sensible heat, air (1 lb)	$1 \times 0.241 \times 75 =$	18.10	$1 \times 0.241 \times 90 =$	21.70
	Enthalpy (Total heat)	28.30		39.60

[a]Latent heat of vaporization, British thermal units per pound (approximate).

value of latent heat per pound of moisture, one may check the enthalpy values of Figure 5.51 as shown in Table 5.14.

The cycles of the mixture of air and moisture between the conditioning equipment and the conditioned space bear a resemblance to those of the refrigeration cycle. They are similar in the fact that they each carry heat from one location to another, but different because only the moisture passes through a change of state while the air itself is subject only to *sensible* cooling or heating as evidenced by its change in temperature. In most cases, for economy, the room air is recirculated and only a small amount of fresh outdoor air is added and stale air exhausted. For simplicity, let us consider again a once-through system using all outdoor air for the purpose of further explaining the cycles in both summer and winter. Figure 5.52 shows schematically the lines that trace the possible summer cooling and dehumidifying steps. Air is cooled without loss of moisture until it reaches a temperature at which it is saturated (its dew point). The cooling then continues. The air that had been losing sensible heat in step 1 continues to lose it in step 2, but additionally, moisture is condensed, which also requires the extraction of

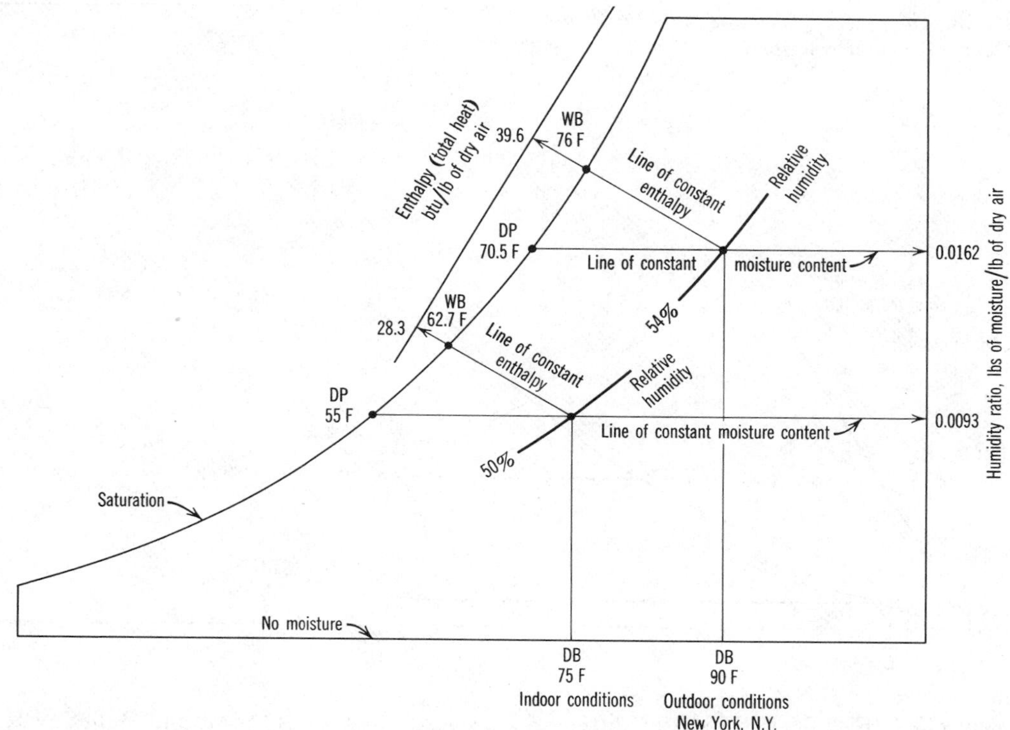

Fig. 5.52 *The process of cooling and dehumidifying outdoor air, summer conditions.*

heat. When step 3 begins (it is known as a reheating process), the remaining smaller amount of moisture and the air are both heated to a condition at which the air temperature and moisture content are slightly below the conditions to be maintained in the room. The changes in the heat content of the air and moisture at the various stages are measured along the enthalpy line. Now refer to Figure 5.53.

Reference to the psychrometric chart shows that air at low temperatures in winter always has a humidity ratio so low that this moisture content would be unsatisfactory when the mixture was warmed to acceptable room temperature. Moisture must be added. Often this is in the form of a warm spray. For our purpose, however, consider an adiabatic spray (no gain or loss of total heat) accomplished by first warming the outdoor air and vapor to a predetermined temperature, in this case 67° F (step 1). Then it is sprayed with water at room temperature to saturation. During this process (step 2) the added water that is evaporated mechanically draws sensible heat from the air, cooling it. The water acquires an equal amount of latent heat for its change of state to vapor. Thus the value at the enthalpy scale does not change. The saturated mixture is then heated (step 3) with no addition of moisture until, at 70° F, it has a reasonably suitable relative humidity. Again the heat changes are read on the enthalpy scale. The action in Figure 5.52 is known as cooling and dehumidifying with a reheat process. Figure 5.53 illustrates a preheat, spray, and reheat arrangement.

5.13 Cooling System Design

The use of the psychrometric chart in computing the relationships of heat and air flow when planning the cooling processes of a central station air conditioner is illustrated in the following example.

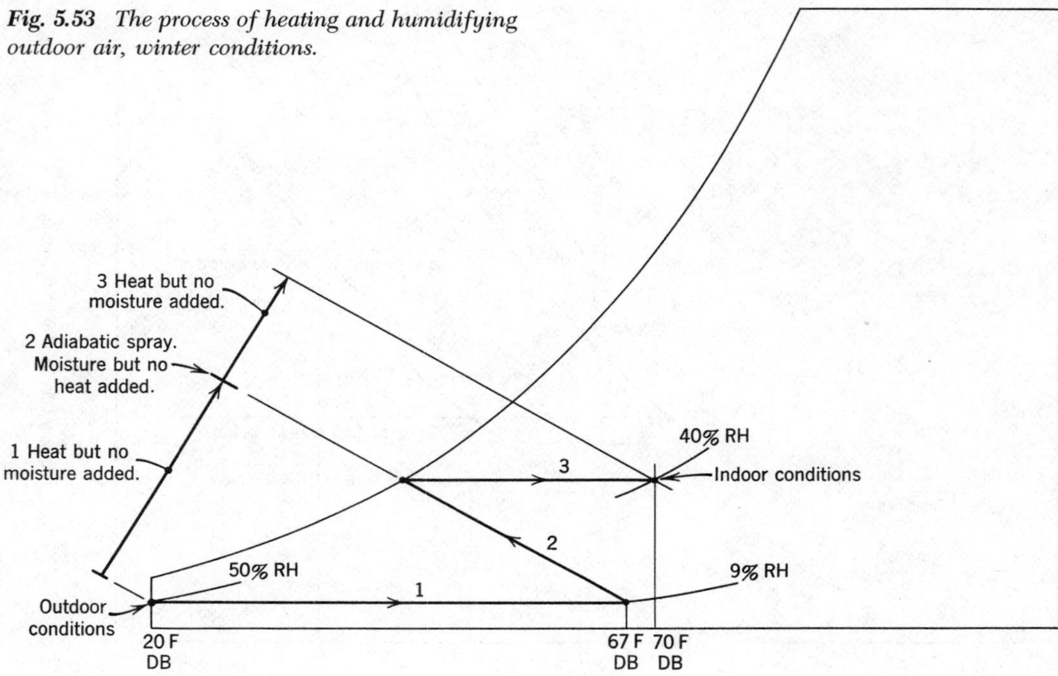

Fig. 5.53 The process of heating and humidifying outdoor air, winter conditions.

EXAMPLE 5.7 (Fig. 5.54). Find the refrigeration required in a room used for public gatherings such as cocktail parties and dances. Establish flow rates for air and chilled water. The design requirements are as follows:

Room conditions (summer)	75° F DB, 50% RH	
Number of occupants	80 people	
Activity	Dancing	
Ventilation required	25 cfm per person[4]	
Conditions, outdoor air	95° F DB, 75° F WB (New York City, for buildings with occupancy other than residential).	
Heat Gains in the Room	Sensible Heat (SH)	Latent Heat (LH)
80 people dancing (see Table 4.30, page 000)		
80 × 305	24,400	
80 × 545		43,600
Total transmission and solar gain	76,000	
	100,000	43,600
	Room Sensible Heat (RSH)	Room Latent Heat (RLH)
		100,000
Total heat gains in room		143,600 Btuh (RSH + RLH)

[4]Subject to reduction if special air cleaning is provided for outdoor air and for recirculated indoor air (see Fig. 4.11, page 100).

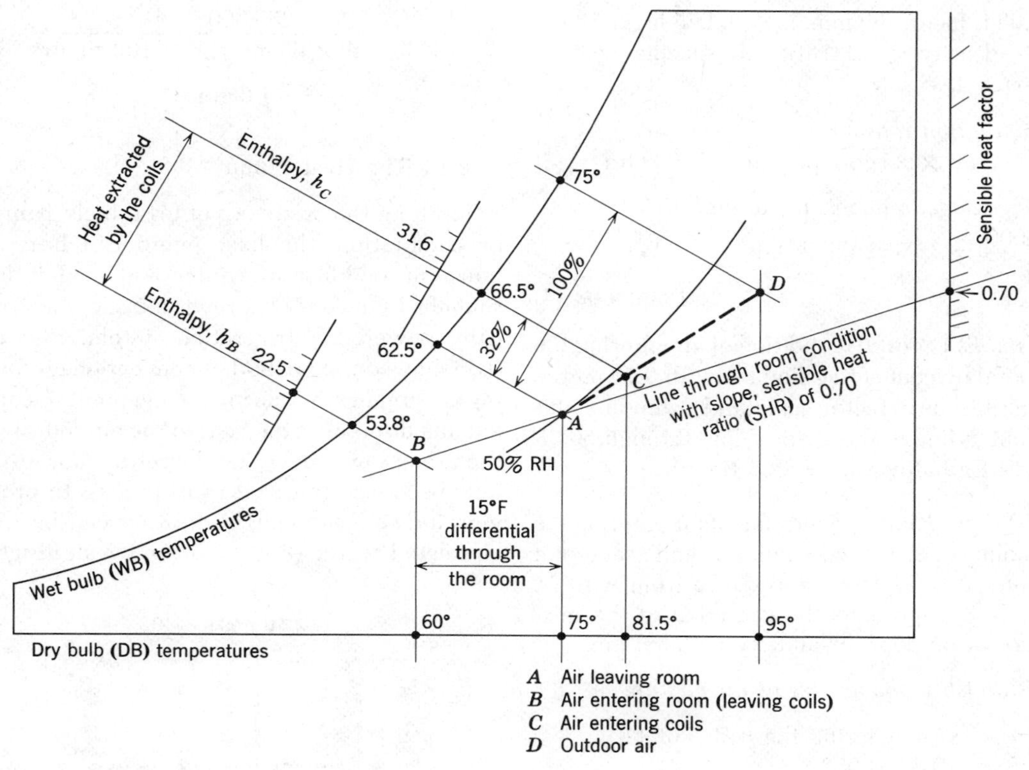

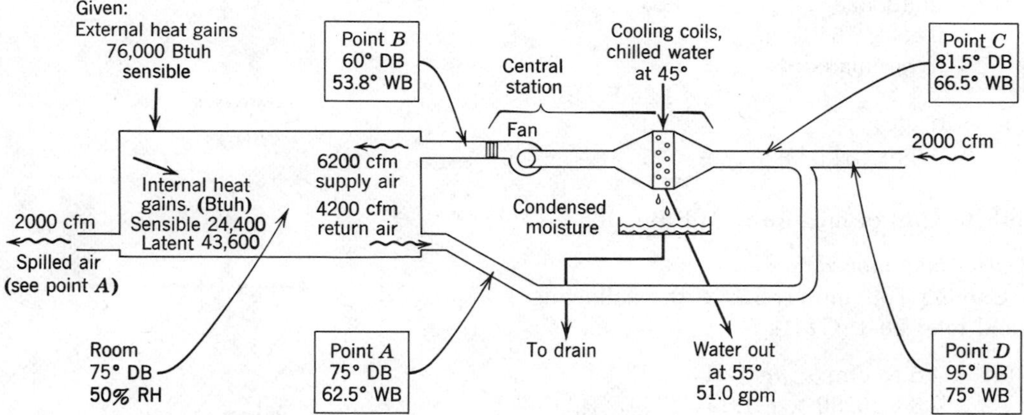

Fig. 5.54 *(Example 5.7) Air-handling and cooling calculations for a room for public gatherings. A central-station installation is used.*

SOLUTION. Sensible heat ratio

$$(\text{SHR}) = \frac{\text{RSH}}{\text{RSH} + \text{RLH}} = \frac{100{,}000}{143{,}600} = 0.70 \qquad (1)$$

Point A. Worst condition of air in the room when returned for reprocessing 75° F DB, 50% RH, 62.5° F WB.

Point B. 60° F DB. Air leaving the cooling condition (assuming a 15° F differential through the room).

Air quantity required

$$\text{cfm} = \frac{\text{RSH}}{1.08 \times \Delta t} = \frac{100{,}000}{1.08 \times (75 - 60)} = 6200 \text{ cfm} \qquad (2)$$

1.08 is factor (60 min/hr × 0.075 lb/cu ft avg air density × 0.241 Btu/lb specific heat of air) = 1.08

Ventilation rate

$$80 \times 25 \text{ cfm/person} = 2000 \text{ cfm} \quad (3)$$

Percentage, outdoor air to total air

$$\frac{2000}{6200} = 32\% \text{ at } 95°\text{ F DB, } 75°\text{ F WB, (Point D)} \quad (4)$$

Establish Point B (condition of air entering the room) by connecting Point A with 0.70 on the sensible heat factor scale and extending this sloping line to the vertical line through 60° F DB. Enthalpy (h_B) = 22.5 Btu/lb.

Establish Point C (condition of air entering the cooling coil). Connect Points *A* and *D* and plot Point *C* = 32% of the distance from *A* to *D*. Point *C* represents the condition of the mixture of outdoor air and recirculated air.

Establish temperatures of air mixture

Air	Air entering the coil (Point *C*)	
DB	0.32 × 95° F (outdoor)	= 30.5
	0.68 × 75° F (recirculated)	= 51.0
		81.5° F DB
WB	0.32 × 75° F	= 24.0
	0.68 × 62.5° F	= 42.5
		66.5° F WB

Enthalpy (h_c) of mixture = 31.6 Btu/lb

Heat to be removed

Cooling coil must remove the following grand total heat (GTH):

$$\begin{aligned} \text{GTH} &= 4.5 \times \text{cfm} \times (h_c - h_B) \\ &= 4.5 \times 6200 \times (31.6 - 22.5) \quad (5) \\ &= 255{,}000 \text{ Btuh} \end{aligned}$$

4.5 is a factor (60 min/hr × 0.075 lb/cf average air density) = 4.5.

Refrigeration required

$$\frac{255{,}000}{12{,}000 \text{ Btuh/ton}} = 21.2 \text{ tons} \quad (6)$$

Chilled water flow rate

Water enters coil at 45° F and leaves at 55° F (10° F differential):

$$\frac{255{,}000}{8.33 \text{ lb/gal} \times 60 \text{ min/hr} \times 10 \text{ Btu/lb}^5} = 51.0 \text{ gpm} \quad (7)$$

5.14 The Heat Pump

Without the onsite use of fossil fuels, flame or combustion, the heat pump can heat a house or building in winter and cool it in summer by means of a reverse-cycle, electrically powered, refrigeration system. Figure 5.36 showed the method of conventional *cooling* in summer by an indoor evaporator coil and the disposal of the heat to the outdoors by the process of condensing the refrigerant. It is only necessary to reverse this process to provide indoor *heating*. The heating can be by circulated water (Fig. 5.55) or by air (Figs.

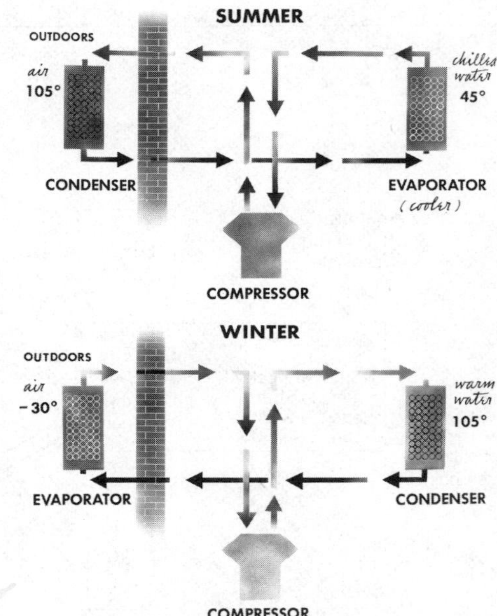

***Fig. 5.55** An air-source heat pump that produces chilled water or hot water for indoor cooling or heating. By an arrangement of valves not shown in this simple diagram, the routing of the refrigerant can be reversed to switch the evaporative and condensing processes between coils that remain fixed. Courtesy of York Division, Borg-Warner Corporation, York, Pa. For air-to-air residential units, see Figures 5.56 and 5.59.*

[5]Specific heat of water = 1 Btu/lb for each 1° F.

Fig. 5.56 (a) *Air-to-air heat pump, its controls, an indoor view, and its operating characteristics. See also Figure 5.56*(b).

Capacity, Btuh	*RB581B/c*	*RB781B/c*
Heating	9,500	14,000
Cooling	10,000	15,000

Courtesy of General Electric Company.

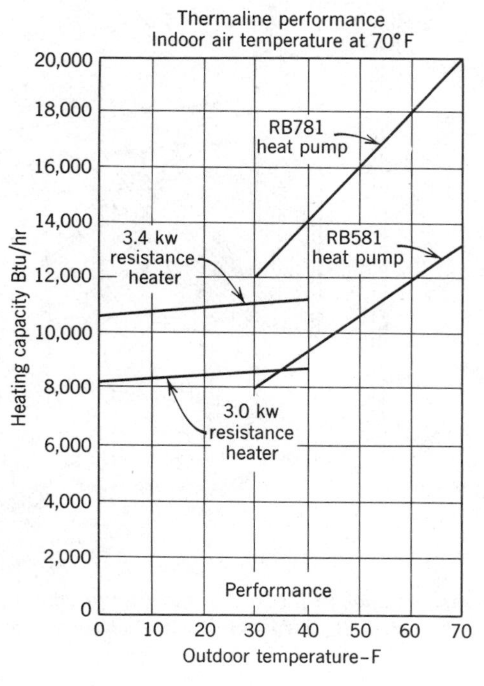

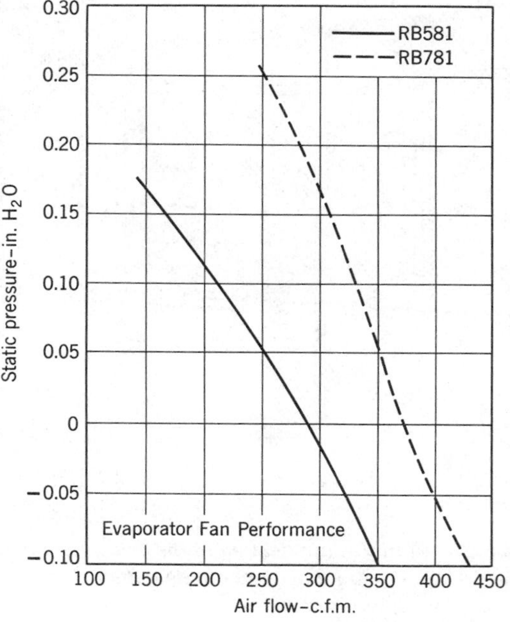

Fig. 5.56 (b) *This heat pump in a living room is quite self-contained. It has an electrical connection and access in a separate chamber to outdoor air. It heats the room in winter and cools it in summer. Indoor air is handled in and out through the edges. Also it is the same unit as pictured in Figure 5.56*(a). *Courtesy of General Electric Company.*

Fig. 5.57 (a) *In this application of the heat pump in a motel, a panel set forward of the exterior wall line allows the pump to inhale and exhale outdoor air around the panel edges. In summer it discharges warm air and, in winter, cool air.*

5.56, 5.57, 5.58, and 5.59). The latter method is called an air-to-air system. At outdoor temperatures below 32° F, the outdoor coil tends to ice up. Then the supplementary (more expensive), electric resistance method takes over. In New York City (vicinity of LaGuardia Airport), for instance, the *average* winter temperature is about 43° F. The resistance coils do not operate very often. Outdoor air is the heat *source.*

The smaller of the two units of Figure 5.56*a* is about right for the heating, *and cooling,* of a single room. The CHP9 unit of Figure 5.58 is suitable for a house. Although the *type* of equipment varies, acceptance of the principle is now widespread. Its economy is seen in its adoption for the Arkansas House. There, the central, indoor air-circulating unit is served by a *remote* outdoor compressor plus a coil that alternates for condensing (summer) and evaporation (winter). An architect on Long Island, New York, recently specified two attic-located heat pumps for a client and subsequently called for three in a three-zone system in his own newly planned home. He reports the installation cost to be less than that of a central fossil-fuel system. We can increase the inherent economy of the method in terms of obvious energy-saving assets, by using only one of several zones at a time when daily occupancy trends permit. Pickup response is fast. In Arlington, Virginia, a large house now in construction will use six units.

5.15 The Induction Method

The principle of this high-velocity system of air conditioning and ventilation is illustrated in Figure 5.60 and its component parts are shown and identified in Figure 5.61. A small amount of outdoor air, representing the required *ventilation-air,* is tempered (cooled or heated, humidified or dehumidified), depending on the season, and delivered at high

Fig. 5.57 (**b**) *Interior cabinet detail and depth serve the heat pump. Room air is taken back as shown and discharged (after cooling or heating) upward across the glass surface. Courtesy of General Electric Company.*

Manufacturer's description of equipment and its performance

High Performance Heat Pump Units Feature Low Operating Cost and Energy Conservation

Lennox single package CHP9 series heat pump units are designed primarily for outdoor residential installations. Six models are offered with a wide and varied heating-cooling capacity range. Units can be installed with ducts extended through a wall in a crawl space, basement, utility room or attic. Installation on a slab at grade level or on a roof will save valuable interior floor space. The compact single package unit contains all refrigeration components (indoor and outdoor unit), air movers, air filters and optional additive electric heat in one complete package. Optional electric heaters are available in several sizes to supplement the heating capacity, space is provided in the unit for simple field installation. Indoor supply and return air openings are both at the same end of the cabinet. Outdoor air outlet is located at the opposite end. Outdoor coil air enters the unit through louvered top panel and both side panels. Powerful direct drive fan(s) discharges air through the outdoor coil quietly and efficiently. Multispeed indoor blower provides a choice of supply air flow. Large indoor and outdoor coils ensure maximum air contact and heat transfer. All components are located within the unit for ease of service. Cabinet is constructed of heavy gauge galvanized steel with a five-step paint finishing process for long lasting protection. Units are shipped completely assembled, piped, pre-wired and precharged ready to install. In addition, units are test operated at the factory insuring continuity of controls and wiring. Installer has only to locate unit, connnect duct work, mount thermostat and make power supply connections to complete a low cost installation.

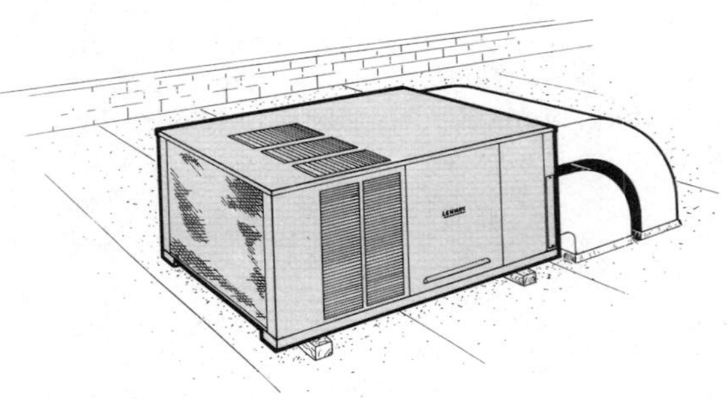

Rooftop installation

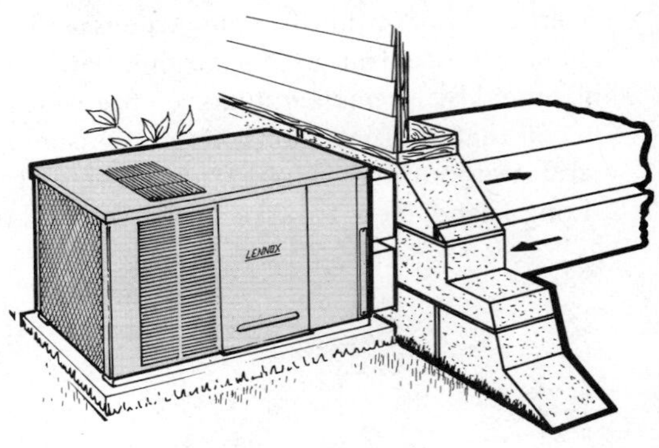

Unit on slab at grade level

NOTE—Specifications, Ratings and Dimensions subject to change without notice.

Fig. 5.58 Series Single Package Heat Pump—Horizontal, CHP9.

23,000 to 56,000 Btuh cooling capacity.

25,000 to 63,000 Btuh heating capacity.

Optional electric heat.

Courtesy of Lennox Industries, Inc.

velocity through a small circular duct. At the cabinet it is attenuated (slowed down and silenced) and blown through jets to induce room air into the flow pattern. The fresh air plus the recirculated air then passes over finned tubes within which is circulated cool or hot water as required. The water heater can be a conversion unit that warms the water by heat transfer from the building steam. The finished appearance is seen in Figure 5.62. The system is frequently used at the perimeter of buildings to provide clean-air ventilation and to offset the adverse seasonal effects of glass. Reference is made to this method in Section 7.9, page 297, concerning its use in the Georgia-Pacific Building, and also in the legend of Figure 7.12, page 289, referring to its use in the World Center Towers, as a low strip-element under the unique fenestration.

5.16 Incremental Heating-Cooling Units

(*a*) ***Description.*** Perhaps the simplest of all air-conditioning devices is the through-the-wall conditioner that was shown in Figure

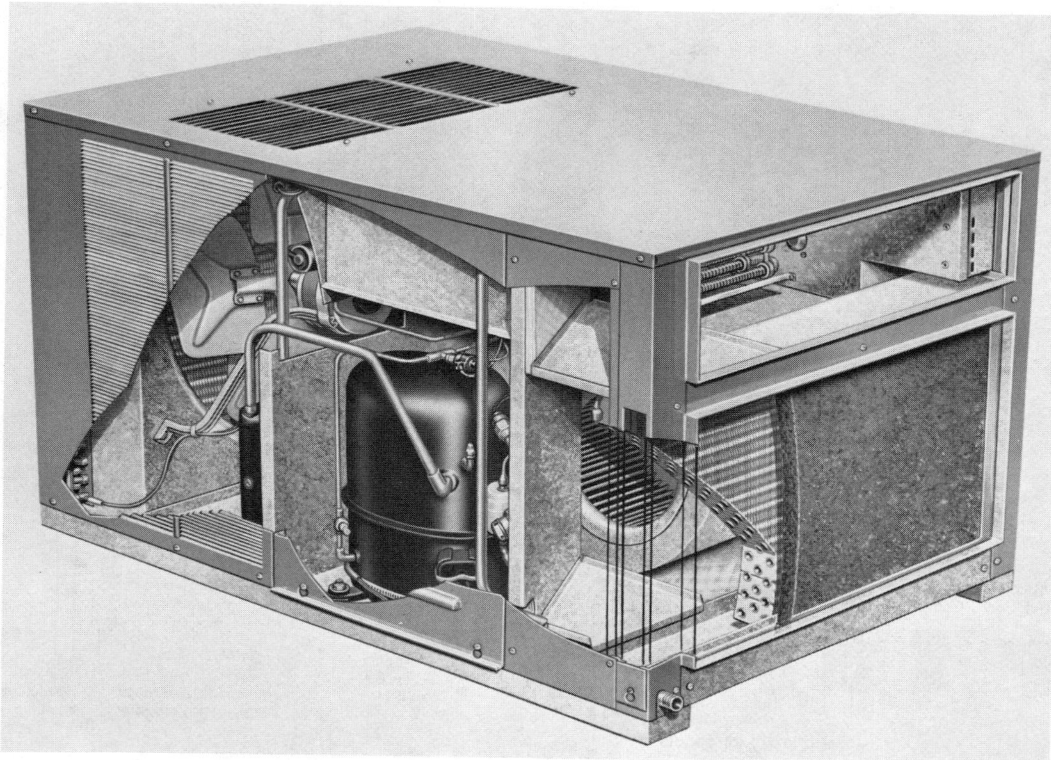

Fig. 5.59 *Series Single Package Heat Pump—Horizontal, CHP9.* Indoor air circuit (right end): *Return air from the house is drawn into the unit through the (lower) duct opening and is passed over the (cutaway) filter and coil. After being heated (or cooled) by the coil, the conditioned air is delivered by the cylindrical blower to the house duct-system through the upper opening.* Outdoor air circuit (left end): *Outdoor air is drawn into the left compartment through the grills on the sides and top of the cabinet. The large fan then blows this outdoor air across the coil in the left end of the cabinet. Refrigerant compressor is seen at the center of the unit. Courtesy of Lennox Industries, Inc. See also Figure 5.58.*

5.36. It uses nonducted air taken from the room and blown over the evaporator coils and returned directly to the room, passing up across the glass or generally upward toward the ceiling to avoid causing draughts on occupants. Air for condensing the refrigerant is taken from outdoors and blown across the coils where it picks up the heat rejected by the unit and is then returned to outdoors. A baffle separates the two circuits except for a small opening to admit fresh outdoor air for ventilation. Through-the-wall units have largely replaced window units, which operate on exactly the same principle. With the addition of a backup heating coil, they are now adapted for use in large buildings and are known as incremental units.

A good example of one of these individual self-contained conditioners is given in Figure 5.63. It is typical in principle of most conditioners of this kind. Note that the compressor is placed on the outdoor side of the baffle, which helps to muffle the indoor sound that it makes. These units have been quite popular for use in apartment houses and motels and are being adapted for use in some office buildings. They are advantageous for the following reasons: Control is in the hands of the occupant or tenant, relieving the management of complaints and making every room an indi-

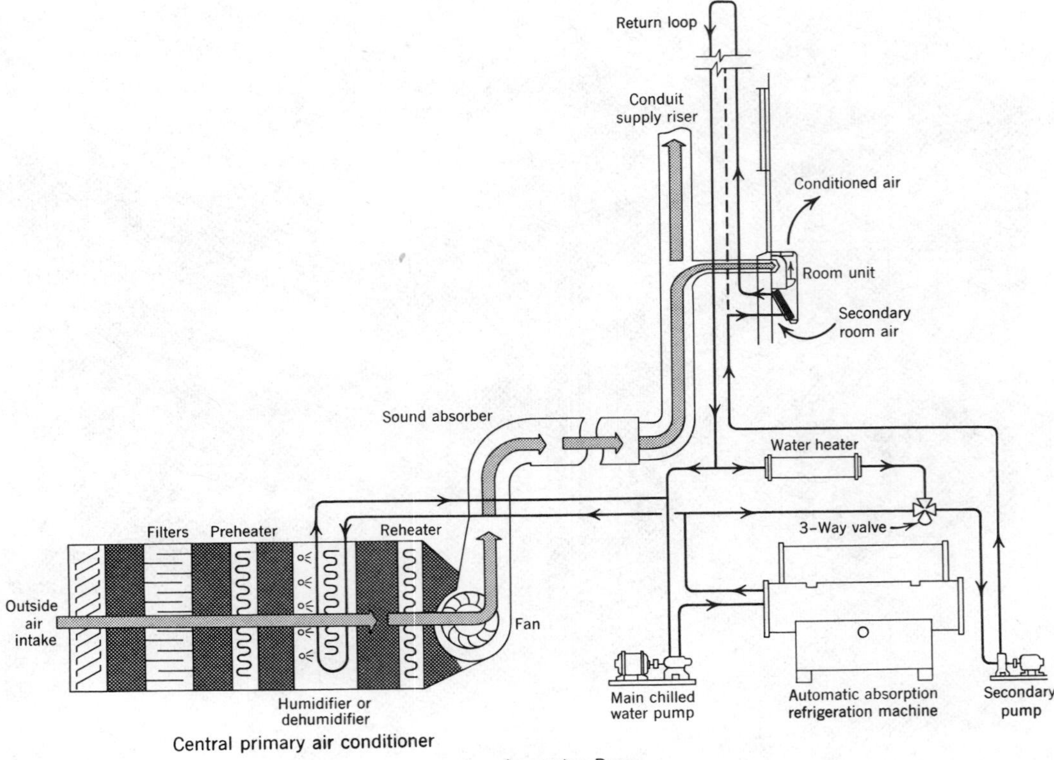

Fig. 5.60 General distribution systems of the high-velocity induction method of air conditioning. Conditioned fresh air (once-through, not *recirculated) provides ventilation, controls air quality and humidity, and induces, through jets, the secondary circulation of room air. This air flow at the cabinet is warmed or cooled to selected temperature by hot or chilled water at controlled rates. Courtesy of the Carrier Corporation.*

vidual zone. If a unit needs servicing it is easy to remove the defective element and insert another. Cooling towers, central chillers, pumps, and piping for chilled and condenser water are all avoided, saving space and making unnecessary the services of a resident operating engineer. Among the disadvantages, the following may be noted: A compressor so close to the occupied space will always create some sounds while a remote central chiller would be inaudible in the space to be conditioned. Large outdoor grills sometimes present problems in architectural design.

(*b*) ***Design Coordination.*** In Figure 5.64, a smooth adaptation of a through-the-wall, all-electric incremental unit to the architectural design of the facade is clearly seen. The unit, powered entirely by electricity, is even more self-sufficient than that shown in Figure 5.63 because it is independent of a central steam system in the building.

5.17 Dual-Duct, High-Velocity Systems

If air is delivered at a velocity of 3000 or more fpm (feet per minute) instead of the more usual 1000 to 1500 fpm, ducts can be much smaller. Obviously fans must be more powerful and problems are faced concerning noises that could be caused by this fast-moving

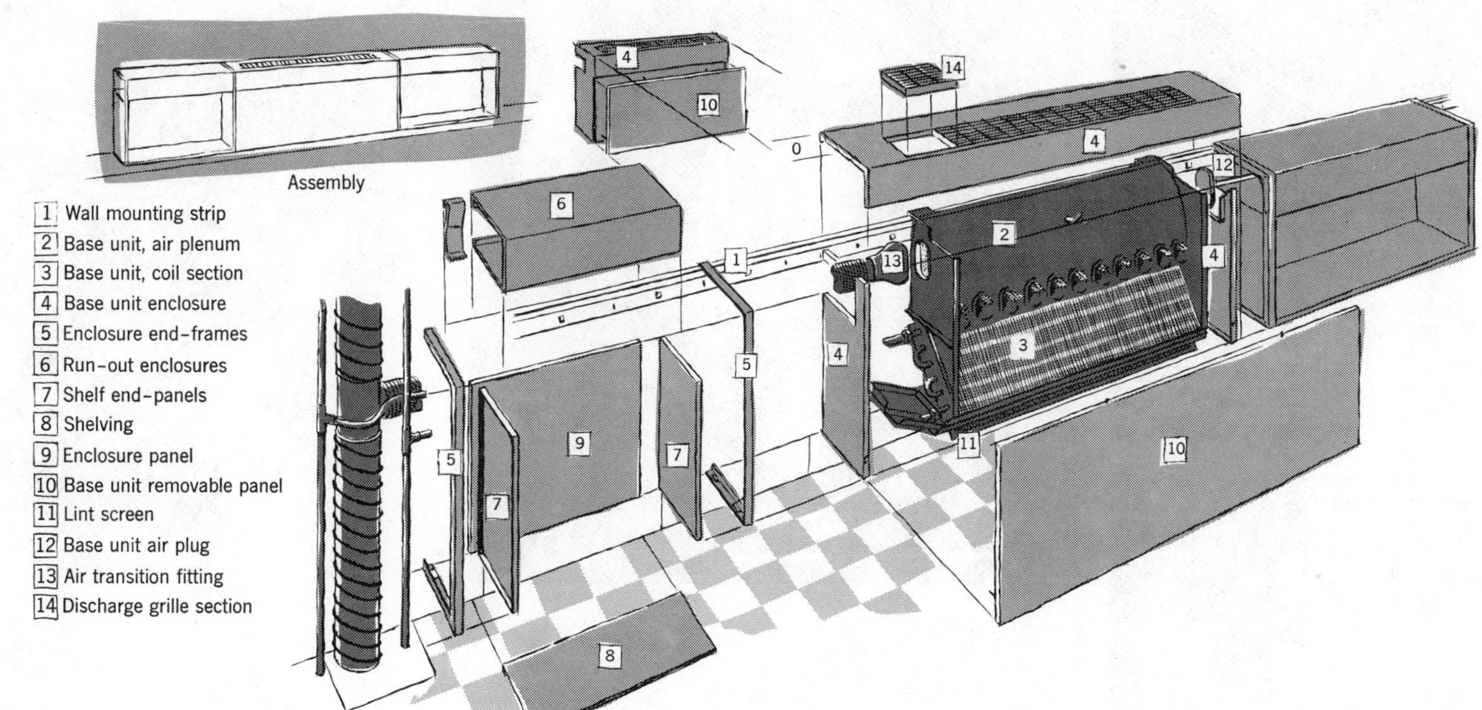

Fig. 5.61 *High-velocity induction system. Conditioned outdoor air for ventilation and to induce circulation of room air is brought in through a single high-velocity duct. It is attenuated and silenced in the chamber* (2) *and then, through jets in the front of this plenum, it induces flow of room air, which is heated or cooled at finned coil* (3) *Courtesy of Carrier Corporation.*

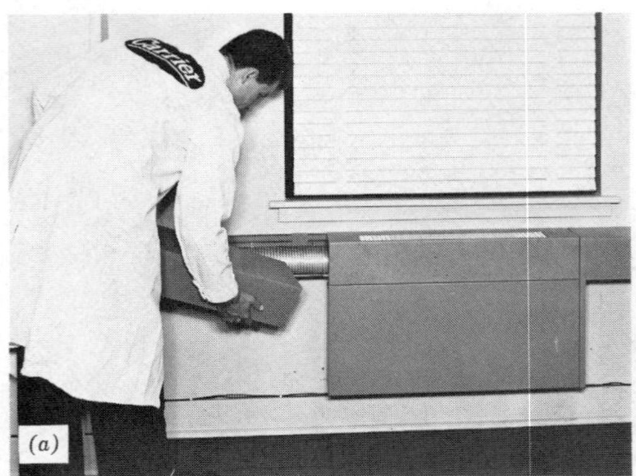

Fig. 5.62 *Finishing stages of the installation of a high-velocity induction cabinet.* (a) *Placing the lateral branch of the high-velocity duct that conveys ventilation air to the central induction unit, through which room air is circulated and conditioned.* (b) *Induction unit assembled as part of bookshelf arrangement.*

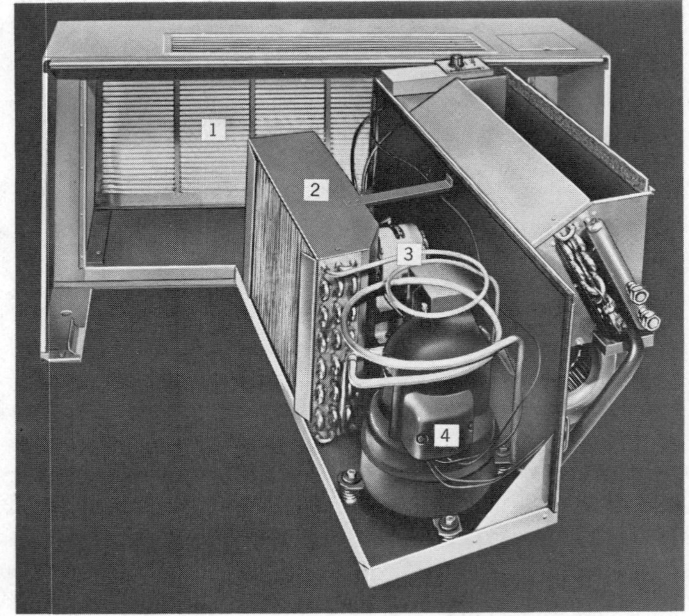

(a) *Equipment assembly swung out from the wall sleeve.*

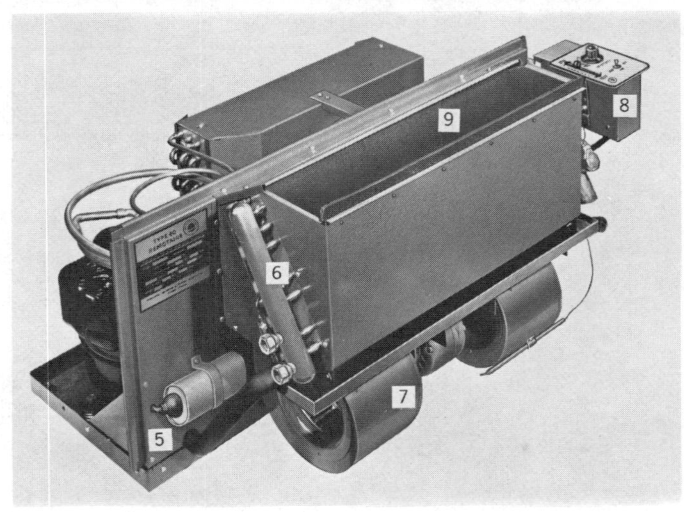

(b) *View of the mechanism of the unit seen from the room side.*

air. These are solved by routing the air through a box performing a function similar to an automobile muffler. These units lined with acoustically absorbent material reduce the sounds to acceptable levels before the air is discharged into the room.

The dual-duct, high-velocity system (see Figs. 5.65, 5.66, 5.67, and 5.68) solves a lot of general problems that had existed in tall buildings as well as in buildings of moderate size. In addition to the much smaller size of ducts, it has the unique characteristic of making both heating and cooling available in different (often closely adjacent) parts of a building. The need for this can occur when, on a reasonably mild day of about 40° F, sun through east or west glass can make cooling necessary while the north (sunless) glass can lose heat so rapidly that heating will be required. Another instance is at midwinter on a very cold cloudy day. All perimeter elements are providing heat. Concurrently, in a crowded interior conference room or auditorium, the heat and moisture gain from people

Fig. 5.63 (opposite and below) *An air-to-air unit applying the principle diagrammed schematically in Figure 5.36. The addition of steam coil (6) just above the cooling coil makes this unit operable for both summer and winter comfort. Control of both is at panel (8). Steam and condensate pipes of the building serve coil (6). Numbered parts are further identified below. Courtesy of the Singer Company.*

Manufacturer's Statement of Design Features

1. Wall sleeve with integral flashing pan and anodized aluminum louver permits standardized installation regardless of application. This assemply becomes "part of the building." The flashing pan provides a solid base for the chassis and a moisture-proof drain surface for masonry protection.
2. Air-cooled condenser has an extra-large surface area for maximum cooling capacity with minimum power consumption. Design complies with ARI requirements governing cooling performance under high outside temperatures and undervoltage conditions.
3. Condenser fan is equipped with a slinger ring (SR), for atomizing condensate over the warm condenser. With the SR unit there is no condensate drip, under the severe conditions set up by the ARI code on condensate disposal.
4. Hermetically sealed motor-compressor assembly is spring mounted and located in the wall for quiet operation.
5. Removable Chassis is sturdily constructed, with gussetted corners for added strength, and designed for balanced component arrangement. Chassis center for gravity is within the wall sleeve. Insulated bulkhead is a sound and thermal barrier and isolates the compressor and condenser from the room.
6. Universal header-type steam-heating coil, directly above the evaporator coil, has provision for either right or left and upfeed or downfeed connections. Concealed flared copper tube connections are made inside the cabinet. A separate coil of serpentine construction is furnished for hot water applications, in place of the steam coil. The heating coil is *not* immediately under the discharge grill, contributing to a lower temperature of the exposed metal cabinet and grill.
7. Quiet operating evaporator fan assembly consists of 2 double-inlet, statically and dynamically balanced, large, slow-speed centrifugal fans mounted directly on the motor shaft. Slow fan speed provides superior heating performance.
8. Built-in control panel is always located on the right hand side of chassis. Access is through a flush-mounted door, from the top or front, dependent on cabinet style.
9. Conditioned air outlet has adjustable louvers in the cabinet, to permit adjustable air deflection with top discharge models. Front outlet models with fixed cabinet grill are also available. On the top outlet models, the adjustable grills are removable by taking out four screws, a handy feature allowing easy cleaning out of any foreign matter that might fall through the grill. Fresh and return air are filtered and mixed before passing over *both* the heating and cooling coils with both models.

Refer to Fig. 5.63.

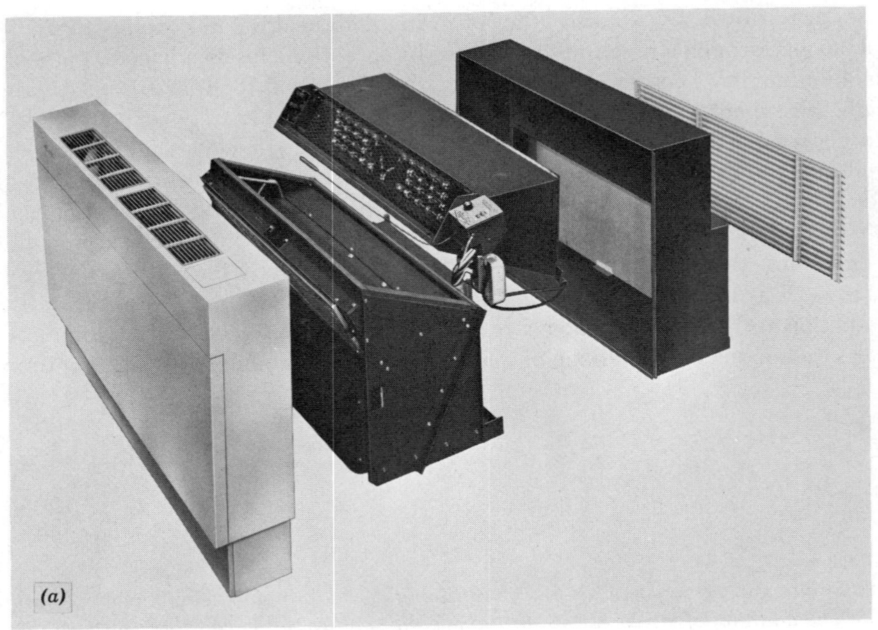

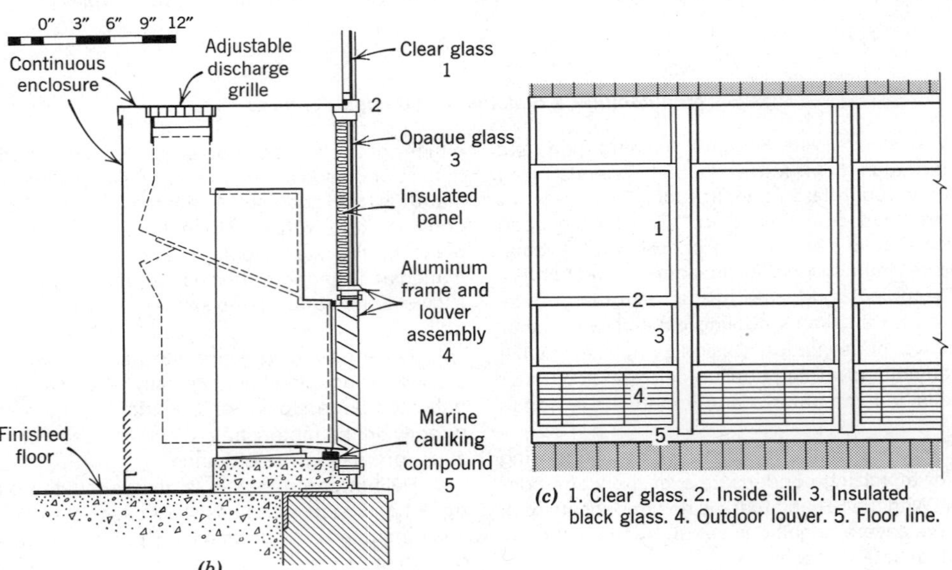

Fig. 5.64 *Incremental (terminal) air-conditioning unit.* (a) *Exploded view of the unit: left to right—room cabinet, electric-cooling chassis, room air circulation and electric-resistance heating element, wall box, and outside louver.* (b) *Section.* (c) *Facade treatment, Travelers Insurance Building, Boston, Mass., Kahn and Jacobs, Architects. Courtesy of Climate Control Division, The Singer Company.*

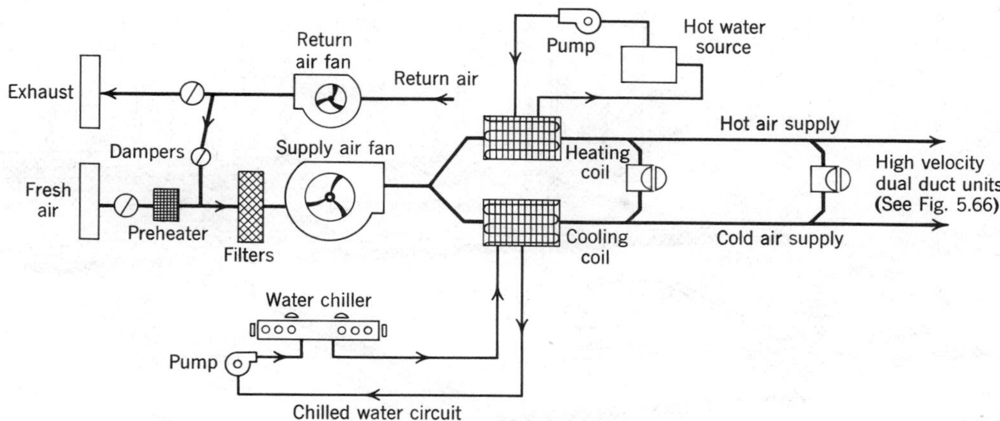

Fig. 5.65 *Principle of the dual-duct, high-velocity method. Hot and cold air is supplied at high velocity through small ducts. Return air is drawn back by its own fan at low-velocity (larger) ducts. Fresh air, exhaust, and recirculation are all provided for. Courtesy of Anemostat.*

Fig. 5.66 *High-velocity, dual-duct unit. Terminal mixing and attenuation (pressure and sound reduction) unit for dual-duct, high-velocity systems. Pneumatically controlled from a thermostat, it blends and delivers air at a selected temperature. These constant volume units provide accurate constant volumetric delivery at each outlet of the system, even though the pressure in the hot and cold ducts may vary widely. The units have simple automatic controls, consisting of a thermostat and a pressure sensitive membrane, both acting through a system of simple mechanical linkages. Courtesy of Anemostat.*

can require cooling to maintain comfortable conditions. These selections can be made when cool air is delivered at all seasons together with warm air at room temperature in summer and warmer than this in winter. This is done by a thermostat and implemented by the high-velocity attenuation and blending units (Fig. 5.66), each of which delivers air at a temperature called for by the local thermostat that controls it. Thus small zones are created that deal with localized heat gains or losses.

The small high-velocity ducts are a great advantage in tall buildings, permitting vertical risers of many stories without excessive bulk. Return air at normal velocities and pressures is often exhausted through ceiling grills into a plenum between ceiling and the floor structure above it, or by systems of return ducts of

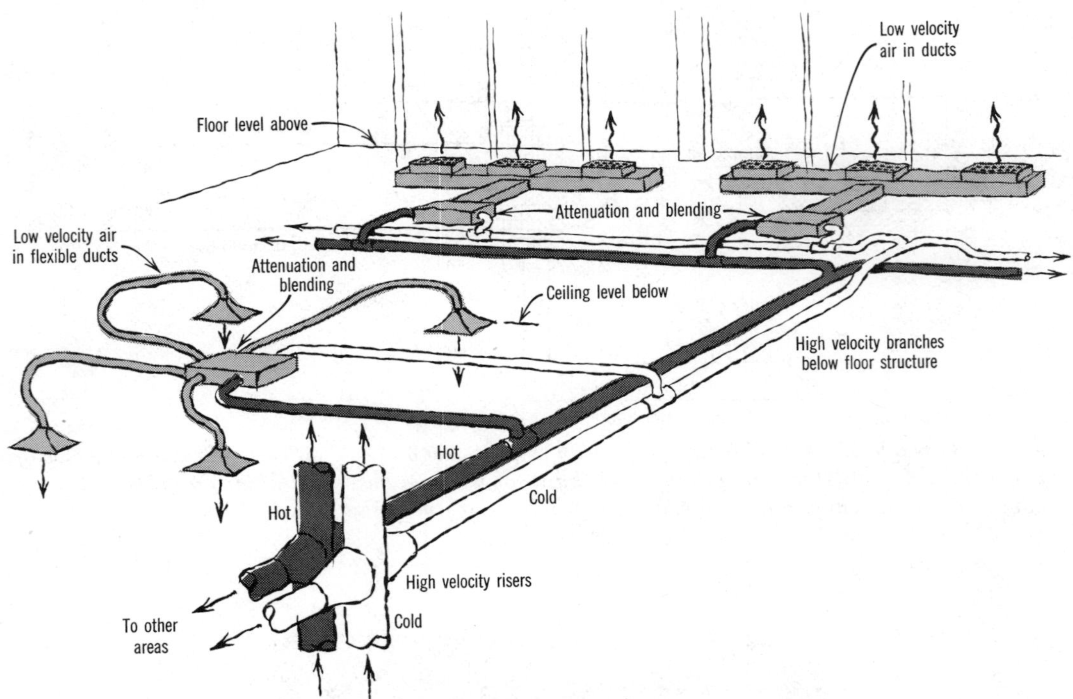

Fig. 5.67 *Distribution of high-velocity, dual-ducted air for use at perimeter (above) and interior spaces (at left) through attenuation boxes (see Fig. 5.66). Air streams are mixed and diffused after reduction of sound and velocity. Each blending unit serves a sseparately controllable zone. Three zones are shown here. Courtesy of Anemostat.*

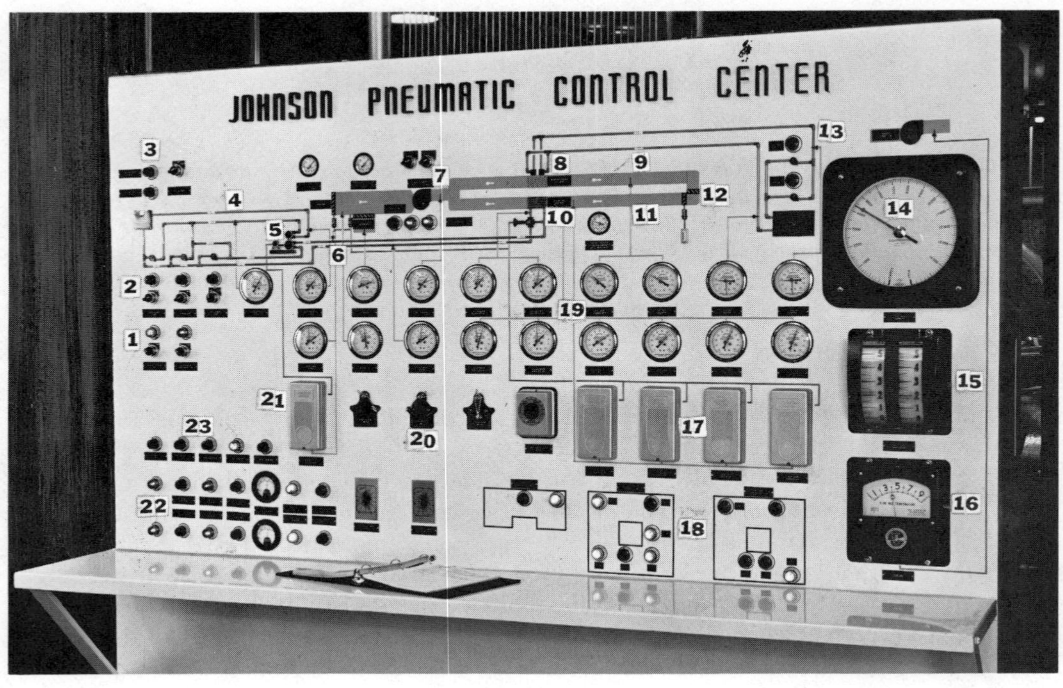

conventional size. Vertical return ducts, also of conventional size, carry the air, usually through the building core, to the return air fan. From there it is recirculated or exhausted. A characteristic of modern air-conditioning systems is the flexibility possible in the flow rates of fresh outdoor air, recirculated air, and exhausted air.

Round ducts offer small frictional resistance and are commonly used in this system. The flexibility of those serving the interior areas allows for office changes with easy relocation of ceiling diffusers. Two common methods of air delivery are the ceiling diffuser, which creates a plane of conditioned and induced air at the ceiling without draughts on occupants, and the floor registers near exterior glass for delivery of air to cope directly with solar heat gains or losses.

The central control panel (see Fig. 5.68), similar in principle to such panels for any modern mechanical system, illustrates the interrelation of the various elements and the methods of checking and controlling them. These parts include boiler, refrigeration, cooling tower, and air handling, as well as the controlling elements in and near the conditioned spaces.

Examples of the successful integration of the dual-duct, high-velocity system into architectural designs are seen in the following illustrations: The International Building, Figure 7.2, page 278; The Blue Cross-Blue Shield Building, Figure 7.4, page 280; and The Medical Towers Building, Figure 7.9, page 284.

The energy considerations of ASHRAE 90-75 tend to favor the variable air volume method (VAV), now gaining in popularity and described in Section 8.10, page 340.

5.18 Ventilation

(*a*) ***Elimination of Negative Pressures.*** Buildings that house industrial processes often have exceedingly high exhaust rates that make the buildings difficult to heat because of high negative pressures and the resulting high rates of cold air infiltration. Figure 5.69 illustrates a makeup unit that supplies warm air in sufficient quantities to cancel this effect. The unit shown is obtainable in ratings of 6000 to 50,000 cfm and input gas ratings of 360,000 to 7,800,000 Btuh.

(*b*) ***Energy Transfer.*** Figures 5.70, 5.71, and 5.72 show a mechanism whose function is to recover the sensible and latent energy of air exhausted from heated buildings and to cool and dehumidify the fresh air entering a cooled building. Seals and laminar flow of air through the wheels prevent mixing of exhausted air and incoming air. A further precaution in the process purges each sector of the wheel briefly, by using fresh air to blow away any unpleasant residual effects of the exhausted air on the wheel surfaces. Fresh air is used for this purpose. Carryover of exhaust air qualities, except those of heat and moisture, is negligible. This system offers great economies in the conditioning of makeup air.

(*c*) ***Infiltration and Natural Ventilation.*** Not many high-rise buildings recently constructed are equipped with operable sash windows that can be opened in summer for natural ventilation. Also, concerning floor levels in these buildings, each level constitutes a sealed fire-barrier to prevent upward flow, during a fire, of smoke, flame, or superheated

Fig. 5.68 (opposite) *Dual-duct, high-velocity control center. The operating engineer no longer has to make trips to the remote equipment, dials, and gauges. All are assembled in a single panel. Pneumatically operated center for supervision and control. Dorr-Oliver Building, Stamford, Connecticut. (1) Exhaust-fan controls. (2) Pump controls. (3) Cooling-tower controls. (4) Refrigerant circuit. (5) Compressors. (6) Chilled-water circuit. (7) Fans (Nos. 1 and 2). (8) Heating coils. (9) Hot air (hot "deck"). (10) Cooling coils. (11) Cold air (cold "deck"). (12) Air blending. (13) Fuel-oil system. (14) Fuel-oil gauge. (15) Draft over the fire. (16) Flue temperature. (17) Automatic-damper controls. (18) Lights come on to indicate operation of major air zones. (19) Twenty dials (temperatures): outdoor air, recirculated air, hot air, cold air, chilled water, condensed water, etc. (20) Switches: winter-summer, day-night, clock operation. (21) Chilled-water temperature regulator. (22) Blower control. (23) Air compressor pilot lights. (Compressed air operates this* pneumatic *control system.)*

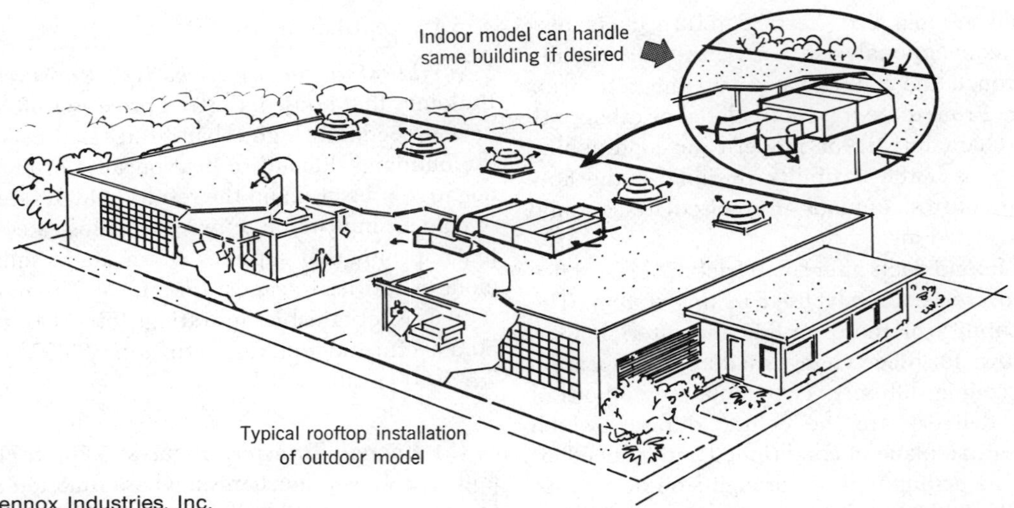

Lennox Industries, Inc.

Fig. 5.69 Elimination of negative pressures. Indoor heaters or weatherproof outdoor heaters can be installed for warm-air delivery to balance the exhaust rates and thus cancel the effect of infiltration, permitting the building's main heating system to maintain comfortable air temperatures throughout the building.

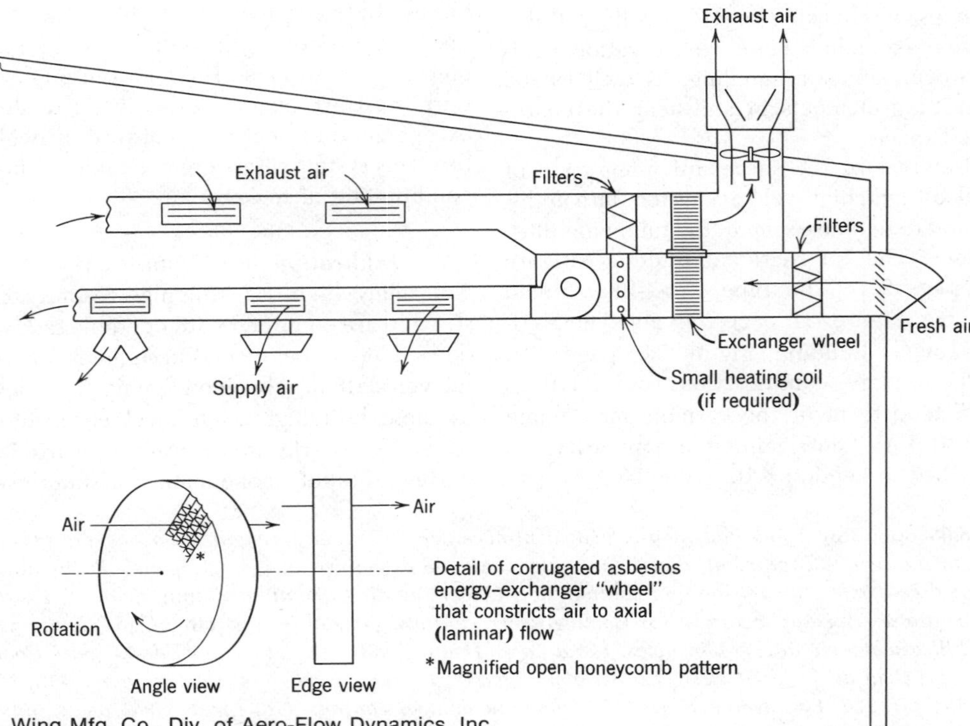

L. J. Wing Mfg. Co., Div. of Aero-Flow Dynamics, Inc.

Fig. 5.70 Energy conservation. By rotating a heat (and moisture) absorbing wheel (cylinder) between the hot and moist exhaust air and the cool and dry fresh air, 70 to 90% of the energy that would be lost by the exhaust is returned to the heating system through transfer to the incoming fresh air.

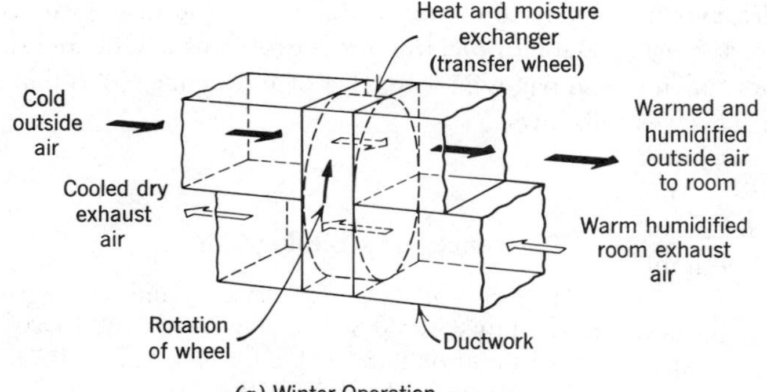

(*a*) Winter Operation

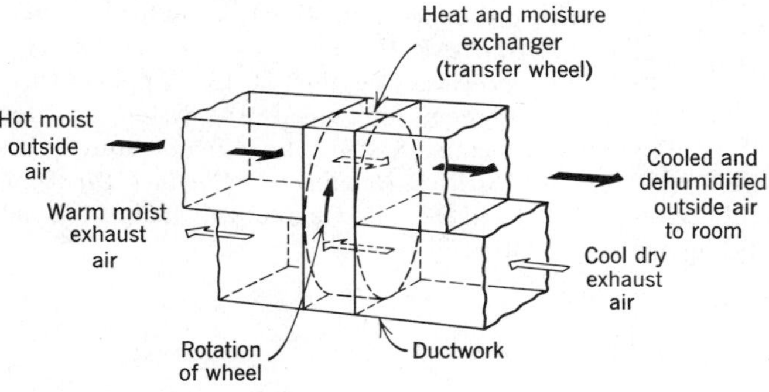

(*b*) Summer Operation

L. J. Wing Mfg. Co., Div. of Aero-Flow Dynamics, Inc.

Fig. 5.71 *Energy transfer wheel acts not only to warm the incoming fresh air in winter but also to cool the incoming air in summer, if the exhausted air is from a cooled building. The wheel surface is impregnated with lithium chloride, which absorbs moisture and transfers it to the other air stream. The wheel delivers moist air in winter and dry air in summer.*

Fig. 5.72 (below) *Multiple unit installation. In this illustration, exhausted room air passes through the upper chambers and the incoming fresh air through the lower chambers. Wheels rotate at 8 to 10 rpm.*

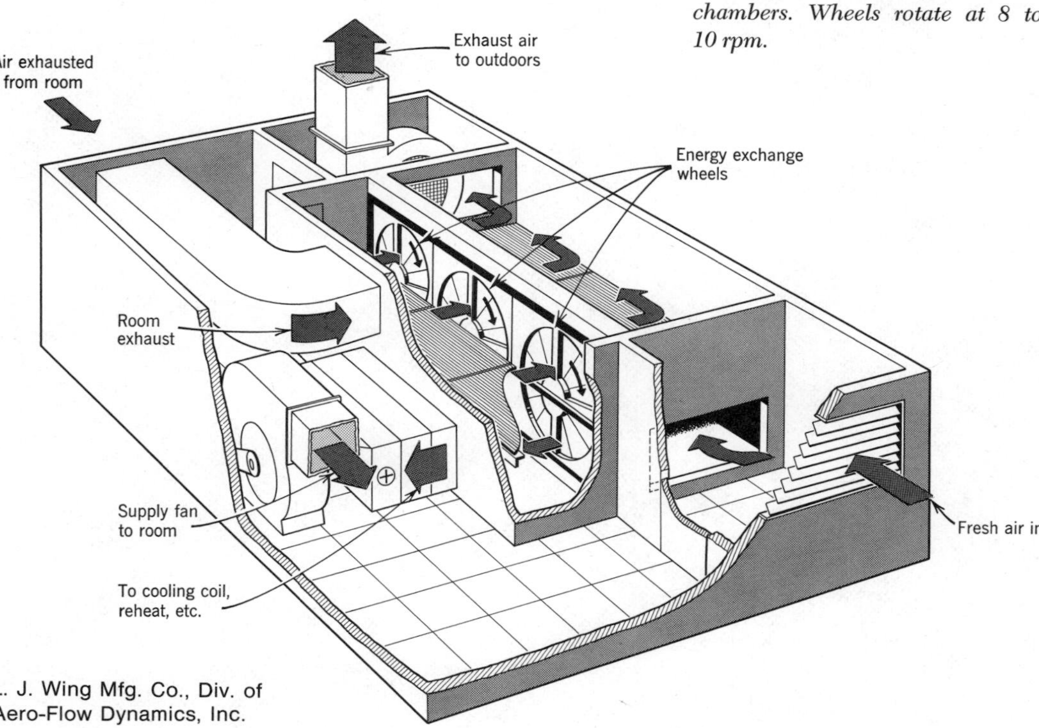

L. J. Wing Mfg. Co., Div. of Aero-Flow Dynamics, Inc.

air (see Chapter 13). A few decades ago, it was customary to compensate for a *stack effect*—upward flow in winter of normal indoor heated air through the building. This was done by adjusting for more heat in the lower stories and less in the upper stories. Since fire barriers at floor levels will, of course, be increasingly essential, stack effect will be minimized.

We are, however, heading toward natural summer ventilation as cooling and mechanical ventilation become greater drains on our energy budgets. Operable sash permit more air-leakage than sealed glazing. Thus in winter, we may again, as in the 1950s, need to deal with greater cold infiltration through cracks at top stories on the windward faces of tall buildings. This subject as well as designs for natural summer ventilation by openings on windward and leeward faces of a building are likely to create challenges in the near future. Data for meeting these problems will be found on pages 21-1 to 21-4 of Reference No. 3 that follows.

References and Bibliography

1. *Technical Letters,* Syska & Hennessy, Engineers, 110 West 50 Street, New York, N.Y. 10020. "Coal Fired Boilers," Vol. *24,* No. 4, March 1974. "Heat Recovery from Solid Waste Incineration," Vol. *24,* No. 6, May 1974. "Coal Conversion of Oil and Gas Burners," Vol. *27,* No. 3, September 1977.

2. *ASHRAE Handbook & Product Directory, 1976, Systems:* Chapters 11–13, "Applied Heat Pump Systems." "Forced Air Systems." "Hot Water Heating Systems, Low Temperature."

3. *ASHRAE Handbook & Product Directory, 1977, Fundamentals,* Chapter 21, "Infiltration and Ventilation."

6 Environmental Planning

In its broadest sense, the general term, environmental planning, is the whole essence of architecture. We plan people's indoor environment. A person's relationship to the site is also necessarily the subject of planning. Neighbors are important. Much consideration is due to other people and to the buildings that adjoin any project. Regional planning is now so important that architects frequently adopt the title "Architect and Planner." In the later years of the renowned Walter Gropius, as the world became crowded, his answer to a student's question was significant. *Question:* "Dr. Gropius, how do you feel about modern architecture?" *Answer:* "I think it's great but it is getting harder and harder to find a place to put it."

Finally, Nature has long since given us an environment. It is a good environment and one that our limited talent cannot duplicate. As we "put" our architecture and its buildings in various new locations, we find, inevitably, that Nature is our neighbor. She must be respected. This chapter is devoted largely to the interrelationship of Nature and the indoor environment.

6.1 Solar Shading in Summer

(*a*) ***Shading by Structural Elements.*** A new influence is affecting the facades of buildings. Frequently they are being designed to intercept *exteriorly* the rays of the sun in summer. Of the many heat-contributing sources, direct solar heat gain is one of the greatest causes of discomfort to occupants. An example of its interception is seen in the south elevation of the Amityville Public Library, Figure 6.1. The unperforated east wall (right) intercepts early morning sun by masonry, whose thermal mass delays by many hours low-rate transmission to indoors. Midmorning (low elevation) sun is intercepted by wide vertical pilasters as are the midafternoon rays at a comparable solar angle. The overhang, as shown, takes care of the nearnoon hours during the hottest days of summer.

Figure 6.2*a* and *b* illustrate the use of dramatically wide external vertical and horizontal concrete baffles, where the shading effect is clearly seen in the almost total exclusion of the sun on the south elevation.

Fig. 6.1 *Solar shading. Amityville Public Library, Amityville, New York, Bentel & Bentel, Architects, FAIA. This elevation faces due south. Shadow of flagpole establishes time as 11 A.M. This September photograph shows glass mostly in shade. Design is such that glass is in full shade at noon on June 21st. Two-foot-deep pilasters and wing walls of the same depth obstruct the sun's lateral rays at low altitudes morning and afternoon in midsummer. Such environmental planning that places a barrier between the fierce rays of the sun and indoor space can be most effective. It can reduce the required refrigerant cooling load, cut it in half (Fig. 6.3b), or make cooling unnecessary (Section 6.3).*

Fig. 6.2 Shading by horizontal and vertical baffles. The Environmental Building at the University of California at Berkeley. In the mild climatic condition of the Bay Area, temperatures are moderate. Yet, direct rays of the sun can be the cause of indoor discomfort and undue heat gain. (a) *South elevation, morning. Small sloping streaks of direct sunlight fall only on concrete walls. Glass is entirely in shade.* (b) *Close-up of the top corner between south elevation (right) and west elevation (left). Details of the relationship of structural components are clarified.*

(*b*) *Powered Louvers to Diminish Heat Gain.* If a building is arranged to intercept the intense rays of the sun before they pass through its glass walls instead of afterward, the air-conditioning heat-gain load can often be cut in half. In approximate terms, the external shading rejects about 80% of the fierce attack of solar energy while the internal shading accepts and reradiates 80% of it. The outside louvers have a chance to cool off in an occasional breeze, but the inside drapes are part of a heat trap, and they constitute a system of hot-weather radiant heating that discomforts those who must work near perimeter surfaces.

Just how fierce is the sun? Compare it to the traditional heating effect of cast-iron steam radiation. Each square foot of such radiation produces 240 Btuh. The amount of energy passing through 1 sq ft of unshaded glass on an east wall in the morning is often evaluated at over 200 Btuh. (The longer wall of the pictured building faces east.) It can be rather frightening to think of an entire wall, blanketed *in summer* by a heat-producing source, which is potentially almost as powerful as cast-iron radiation. The rejection of a maximum amount of this solar energy begins to appear almost mandatory.

Figure 6.3*a* shows the building of the Fidelity Federal Savings and Loan Association of Glendale, California at the time of its completion. In spite of the advantages of full air-conditioning, heat-resistant glass, and the use of fully closed inside drapes, employees moved their desks as far as possible away from the hot exterior walls. Two 50-ton compressors running continuously failed to keep the building comfortable even on days when the outside temperature was only 73° F. The heated drapes were found to be an unwanted radiant-heating system that effectively cancelled the cooling efforts of the air-conditioning plant. The temperature between the glass and drapes was 120° F with thermometers facing *in*. (The working temperature of radiant surfaces for heating in winter is often set at 115° F.)

After spending about $2000 on an experiment with metallic-lined drapes, which aided little, and another $2000 to replace one of the compressors that broke down from overwork, the management knew that a major change was necessary.

A contract was let to install exterior, automatically controlled, power-operated sun louvers on the east and south walls (see Fig. 6.3*b*). (The north and west walls are solid brick.) As soon as the louvers were installed, relief was apparent. One of the two cooling units kept indoor climate cool when outside temperatures were in the 90s. Employees near windows were perfectly comfortable. Much light but little heat was reflected into the building. As the sketch shows, the louvers are not fully closed, even when the sun's rays are in a plane perpendicular to the glass (see Fig. 6.3*c*). Without attendance, the louvers turn to exclude the sun as its relative position changes through the day. They open fully when the sun no longer shines on the controlled facade, or when cloudy conditions prevail. Louvers were manufactured by the Lemlar Manufacturing Company of Gardena, California.

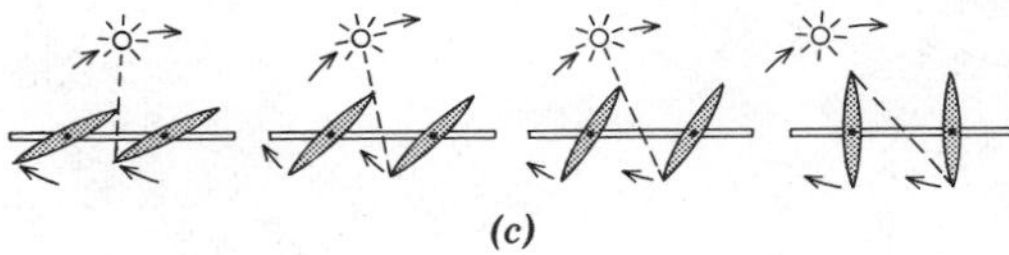

(c)

Fig. 6.3 *Reduction of solar heat gain to lessen the air-conditioning load.* (a) *Bank building prior to installation of exterior, automatically controlled louvers. Long side, east, short side south. Interior drapes and refrigerant cooling did little to relieve discomfort of occupants due to excessive heat from the sun.* (b) *Automatic louvers with solar time control and photoelectric override solve the problem. Rate of energy use for cooling is reduced to less than one-half that of former operation.* (c) *The movement of the louvers is controlled by the position of the sun.*

There are numerous other methods of external shading. On the south, slatted shades or grills do a good job, but this is an easy orientation to control. On east and west glass, it is difficult to beat an operable louver if one is to have a view through these surfaces at *any* time. A city location such as this precludes the alternative possibility of high trees for full shade.

Finally this example gives cause to consider whether, in the best planning, there is ever any justification for accepting the sun's rays through glass in summer and intercepting them only by *interior* shading devices.

6.2 White Roofs and Desert Cooling (Fig. 6.4)

(*a*) ***Evaporative Cooling.*** In hot arid regions, a method of cooling simpler than that of the compressive-evaporative refrigeration cycle can be very effective. Employing one electric motor instead of the three seen in Figure 5.36, page 218, it saves a great deal of energy, though it does use a little of our increasingly precious and scarce resource—water.

Long ago, at about the time that refrigeration units began to be adopted for use in houses, the air cooler, then known as the desert cooler, was popular. It is still very much in use in these earlier installations. Often chosen for new energy-saving jobs, it is now also finding use in more humid regions. In the latter instance, the units are sometimes used for ventilation only.

The Krysan Residence (Fig. 6.4) uses three air coolers. When asked about their use and quality of performance, the response of the Krysans was: "We use our desert coolers frequently, one on the roof and separate units for the front and back of the house. As an in-

Fig. 6.4 *Desert Cooling. The Krysan Residence, Palm Springs, California. View looking directly west. Shadow of the front light-pole indicates the time to be 11* A.M. *Entire east (front) facade is fully shaded from the morning sun by natural planting. Roof overhang on the south puts the few (high) windows largely in shade. Masonry walls delay thermal transmission by many hours. The far (west) end of the house has awnings, and is benefitted by additional shade later as the sun sets behind the mountains. The white roof, typical of hot climates, is heat-resistant, and the living room and bedrooms are on the cool north side. The "desert cooler" on the roof (left) serves the central area of the house. Two through-wall units, one on the east wall and one on the west wall, complete this three-unit evaporative cooling system.*

stance, we are at the moment, (November 23rd), using the one at the back of the house since it is 80° F outside. They do a great job when the humidity is low. In the matter of comfort, we prefer them to refrigeration. We are lucky to have both. Some newer houses have refrigeration only." We will also add a comment: This off-season use of only one of three units is much more conserving of energy than the full use of the central refrigeration system when only partial cooling is desired.

How It Works. Evaporative cooling is an ancient method of lowering air temperature. As water is evaporated to vapor, heat is drawn from the air, reducing its temperature; Figure 6.5 illustrates the equipment. A blower draws outdoor air in through grills, passing it through pads kept moist by recirculated water.

The cooled air is then delivered directly to the indoor space. The effect of the gently moving cool air is to cool the body and, additionally, produce further cooling by evaporation of body moisture. The thermal-evaporative cycle of the cooler is shown in Figure 6.6. Outdoor air at 105° F and 10% relative humidity can be considered as too *unbalanced* a condition to provide comfort for humans. When outdoor air is as dry as this, the adiabatic process of the air cooler results in air for indoors that is quite the same as that provided by refrigerated cooling. The indoor condition of 78° F and 50% relative humidity thus produced are the same as those usually chosen as design standards for refrigerated cooling.

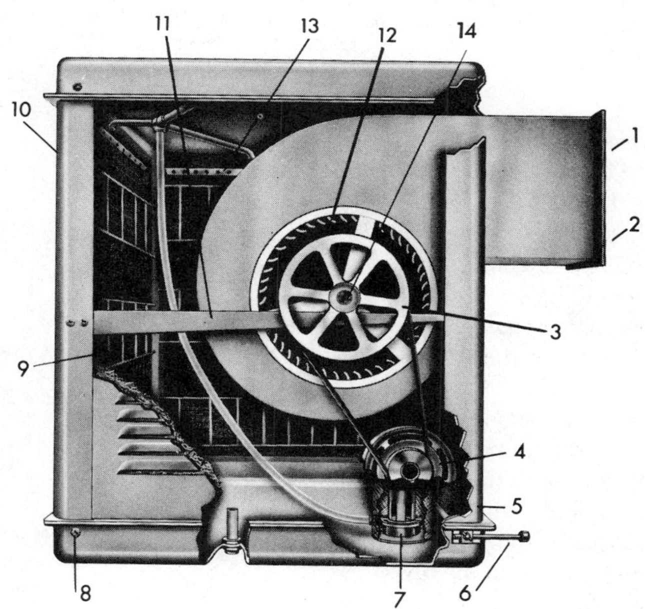

1	Control for air volume	8	Bolts
2	Switches	9	Snap–shut type pad frames
3	Blower pulley (quiet action)	10	Blower mount support
4	Motor	11	Blower wheel
5	Weather resistant exterior finish	12	Even–drip water trough
6	Supports	13	Bronze bearings
7	Water pump, plastic impeller	14	Blower shaft

Fig. 6.5 *Air cooler. Side removed to show operation and to identify parts. Pump, lower right, circulates water from the bottom pan to three horizontal perforated tubes at the top that drip to moisten the pads. Water makeup is automatic. Humidification can, by choice, be discontinued and the blower used for air delivery only. Courtesy of Champion Cooler Corporation.*

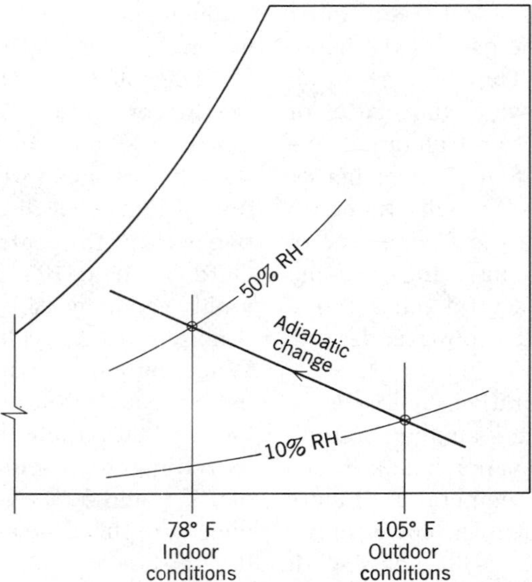

Fig. 6.6 Once-through cycle of outdoor air through an air cooler. Values transcribed from the psychrometric chart (Fig. 5.50, page 229). Hot dry air can be humidified adiabatically to indoor conditions that fall within, or close to, the optimum comfort envelope shown in Figure 4.6; and without high-energy-consuming refrigeration.

Outdoor conditions throughout about one-half of the United States (see Fig. 6.7) lend themselves well to the use of the cooler. In regions approaching the east and west oceans, humid conditions increase. There, more frequent air changes should be as recommended in Figure 6.7. In certain instances, the water pump can be turned off and the system used for ventilation only.

Air introduced into the indoor space must be exhausted to assure proper operation of the system. By selecting the room through which the air is exhausted, cool air can be routed as desired in any chosen path from unit to relief opening. Thus people become involved, a trend set forth in Chapter 1. An indoor view of a typical cooler is seen in Figure 6.8.

EXAMPLE **6.1.** A slab-on-grade, one-story ranch type house is to be provided with air coolers. The open planning of the house is such that three coolers in through-wall locations can serve all areas. Based upon data in Figure 6.7 and for *normal* conditions, calculate the cubic feet per minute required for each unit and select a suitable FD unit from Figure 6.8.

Floor area of the house	2000 sq ft	
Ceiling height	8 ft	
Regional location	Zone 2	
Desired number of units	3	

SOLUTION

Volume of the house	2000 × 8 = 16,000 cu ft	(1)
Total cfm rate, entire house	16,000 ÷ 2 = 8,000 cfm	(2)
Cfm per unit	8,000 ÷ 3 = 2,700 cfm	(3)
Unit selected (Fig. 6.8)	FD 30H-2	(4)

(b) White Roofs. In reflecting heat away instead of absorbing it, which increases the temperature of rooms below, white roofs are effective. Their use in hot climates is illustrated by the photograph in Figure 6.4 of a house in Palm Springs, California.

To evaluate the advantage of white over black roofs, a test was made recently in an-

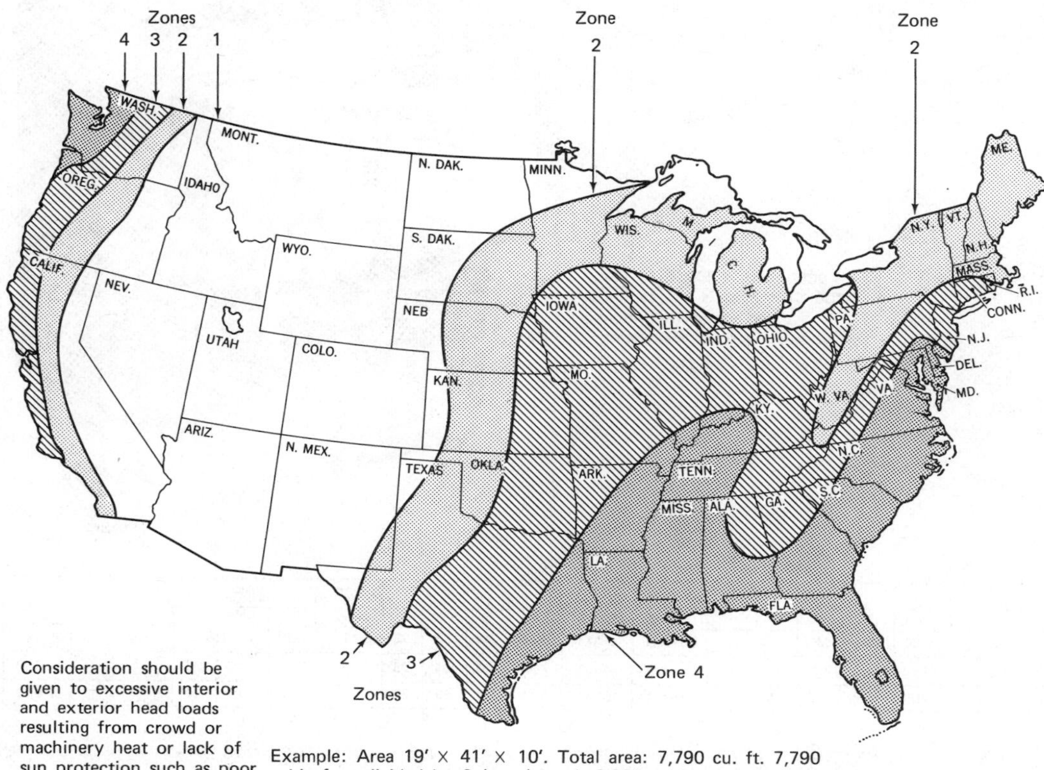

Consideration should be given to excessive interior and exterior head loads resulting from crowd or machinery heat or lack of sun protection such as poor insulation

Example: Area 19′ × 41′ × 10′. Total area: 7,790 cu. ft. 7,790 cubic feet divided by 2 (zone) totals 3,895 cu ft. Size Evaporative Cooler needed: 4,000 CFM.

1. Determine the cubical content of the area to be cooled by multiplying the width times length times height.

2. Check map above and find the zone in which you are located. Divide the cubical contents of the area to be cooled by the number of air changes indicated in the air change table.

Table of Air Changes

Zone		Normal	Excessive
1	change air each	3 minutes	2 minutes
2	change air each	2 minutes	$1\frac{1}{2}$ minutes
3	change air each	$1\frac{1}{2}$ minutes	1 minute
4	change air each	1 minute	$\frac{2}{3}$ minute

Fig. 6.7 *Manufacturer's recommendations for the selection of units with respect to regional zone. Air changes per minute and cfm ratings are established. Courtesy of Champion Cooler Corp.*

3000 CFM 4,200 CFM and 4700 CFM Models for cooling and ventilating three to five average-size rooms.

MODEL	CFM	H. P.	SPEED	BLOWER WHL.		PULLEY		Wt.	CABINET DIMENSIONS			Lower Opening to Cooler Bottom	THROAT DIMENSIONS			ASPEN PADS	BELT SIZE
				Diam.	Width	Drive	Blower		Ht.	Width	Depth		Ht.	Width	Depth		
FD30H-2	3000	⅓	2-VC	12	12	2½AX½	7AX1	148	34	28	28	19	10¾	24¼'	11	3 size 22x25	4L-440
FD42H-2	4200	⅓	2-VC	16	16	2¾AX½	10AX1	173	40	34	28	25	10¾	24¼	11	2 size 22x32 1 size 28x32	4L-500
FD47H-2	4700	½	2-VC	16	16	3AX½	10AX1	178	40	34	28	25	10¾	24¼	11	2 size 22x32 1 size 28x32	4L-500

Fig. 6.8 *One model of the many* air coolers *produced by the Champion Corporation for use at windows or through a wall. Air is drawn in through three outdoor slotted sides. It is passed over moistened pads to lower the air temperature by humidification, and then delivered into the space. Double deflecting grills, operating controls, and a plug-in electric connection are seen on the room side.*

other hot region—Israel. The results as summarized in Figure 6.9*a* indicate a room temperature 4° C (8° F) cooler under the white roof. An interesting feature of the test is that the roof surface temperature of the *discolored* white section is 10° C (17° F) higher than the freshly painted white area. It indicates clearly that frequent repainting would be essential in crowded cities and industrial areas with a heavy outfall of soot.

6.3 Passive Solar Planning

In utilizing the sun, the first principles are to exclude it from interior space in summer and to accept it interiorly (with adjustable drapes), for warmth in winter (see Fig. 6.10). In that late afternoon photograph, we find that the west masonry wall (right) has, all afternoon, intercepted the sun. Heavy masonry reduces and delays heat transmission, which can be tolerated better during cooler evening hours.

Plans of the Grossi residence are shown in Figure 6.11, and overhangs that shade the south glass (*fully* in midsummer) are seen in Figure 6.12. In that south view, the concentration of glass for winter heat pickup is evident. Figure 6.13, a cross-sectional study, indicates the planning needed to achieve suitable dimensions which, together with seasonal solar altitudes, make the best exclusion (or use) of the sun's energy. Small, operable windows on the north provide cross ventilation.

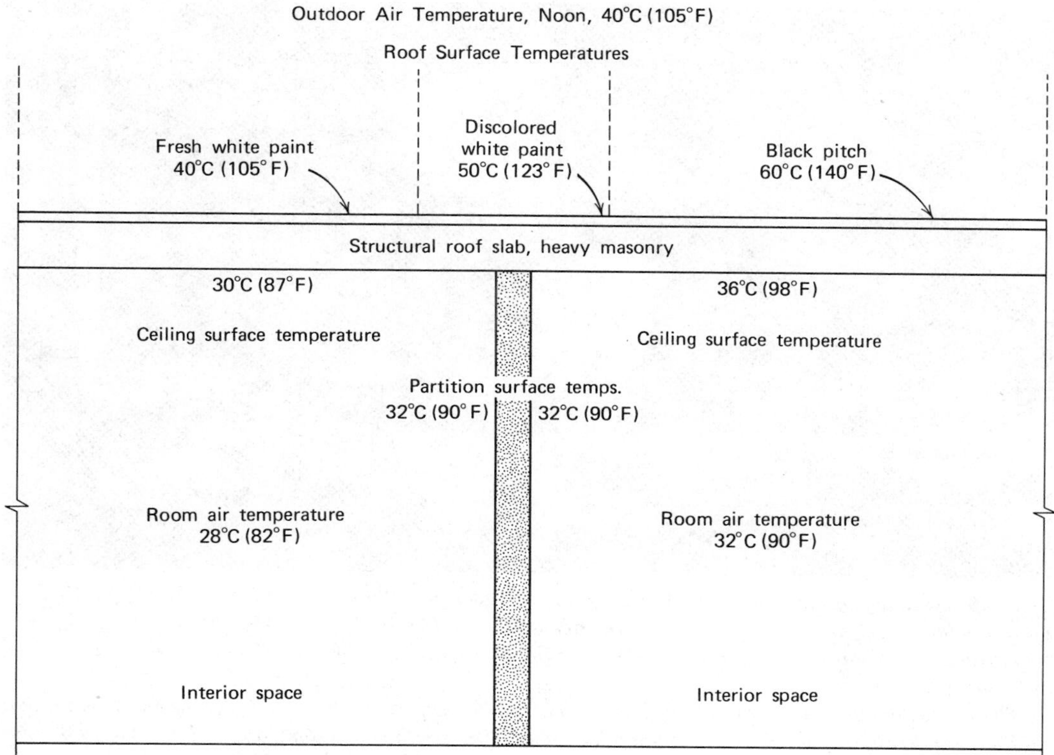

(a) Section Through Two Rooms, Residential Occupancy

Fig. 6.9 (a) *Record of an informal study of the effectiveness of white versus black roof surfaces in resisting the sun's heat.* Location: *Israel at noon on the summer's hottest day, 40°C (105°F).* Results: *White surface 10°C cooler than the discolored white surface and 20°C cooler than the black surface.* Room air temperatures: *Room below white roof 4°C cooler than that below black. It is significant that the Fahrenheit temperature of the cooler room (82°F) was only 4°F warmer than the 78°F American minimum newly set by ASHRAE's Standard 90-75. Notable, also, is the Fahrenheit 23°F (105 − 82) difference between outdoor air and comfortable cooler indoor air. For graphical conversion from Celsius to Fahrenheit, see Appendix Figure D2, page 1304.* (b) *Preparation for measuring the effect of the sun on freshly painted white roof.*

Fig. 6.10 *North elevation, The Grossi Residence, Manhasset, New York, Olindo Grossi, FAIA, Architect. A passive solar house. An analysis of its performance follows in Figures 6.11 to 6.16. Although located at 41° N. Latitude, for convenience, thermal values for the house are taken from Table 4.27 (40° N.L.). Value differences are minimal.*

The house does not have, or need, refrigerant cooling. Note open windows that provide cross ventilation for breezes passing through. The photograph, taken at 5 P.M. (see Fig. 6.14) on a day in June, shows the sun in its orbit just beginning to strike the north wall where the roof overhang puts the second-story glass in full shade. During the hot afternoon, the windowless west wall (right) has provided a barrier to direct high-value solar gain. Earlier, during the middle hours of the day, overhangs (Fig. 6.13) shading both stories, eliminated direct solar gain through glass. Early morning summer sun on the east wall (Fig. 6.11) penetrated only three small windows located there to provide cross ventilation.

The cool earth of the half-story berm against the north wall (Fig. 6.13) together with only minimal and glancing late-afternoon solar impingment on wall and lower-story glass assure a cool area in the living room.

Winter performance is described in illustrations that follow.

In Figure 6.14, a comparison is made regarding sun travel, summer and winter. Azimuths (angles from the south) are more than twice as great in summer as in winter. East and west low-elevation sun is *not* wanted in summer. In winter, building surfaces facing east or west have little value for solar gain. Thus, in *both* seasons blank walls are best.

Winter conditions are summarized in Figures 6.15 and 6.16. The daylong contact of sun's rays with the first-story floor, rear wall, and end walls as the sun travels its orbit are shown. The concrete floor slab, masonry walls, *and the earth berm* of the lower story have a delayed irradiation effect that is most valuable. In the evening, as required, a conventional heating system, little used during the day, takes over. Attention is called to the evaluation of the *rate* of solar heat gain in the caption of Figure 6.16.

See Figure 2.2, page 33, for Frank Lloyd Wright's passive "Solar Hemicycle" House.

6.4 Utilization of Natural Growth

We have seen in Figure 2.23, page 54, how deciduous ivy can shade a building facade

Fig. 6.11 Plans. Except as needed for cross ventilation (operable sash), glass is concentrated on the south to receive the sun's rays in winter and by shading to exclude them in summer (see Fig. 6.13). Air movement through the lower-story glass wall on the south is provided for by open doors. The house was designed and built in 1947.

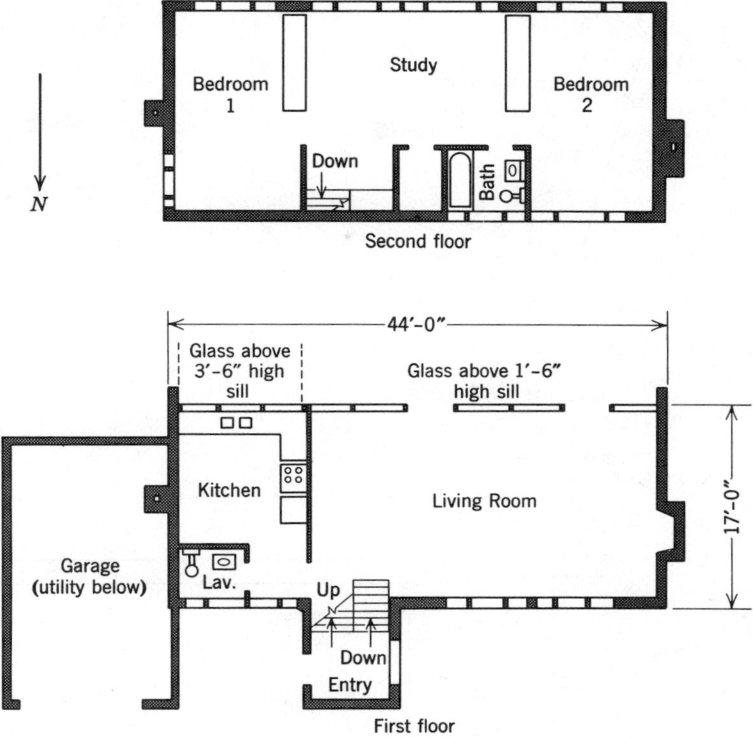

Fig. 6.12 South elevation. This picture was taken in September. In point of time, this is half way between the full shading of June (see Fig. 6.13) and the full solar acceptance of December sun. The vertical shadow of the outer edge of the upper-story open sash is slightly west of south. This "sundial" indication fixes the time at 11:30 A.M.

In this cooler month (and still with open-sash ventilation), a little sun is pleasant. On the second floor it passes through the lower half of the glass. On the first floor, the lower one-third of the six-ft-high glass panels accepts the sun. The kitchen glass (far end) is still largely in shade, which is well-arranged to concur with the cooking of a noon meal.

As the seasons change, there is a variation in the use of filters—operable sash and light interior drapes for shading when desired. Before the winter season begins, the deciduous tree in the foreground (right) will have lost its leaves, affording direct, low-elevation passage of the sun's rays through the south-elevation glass.

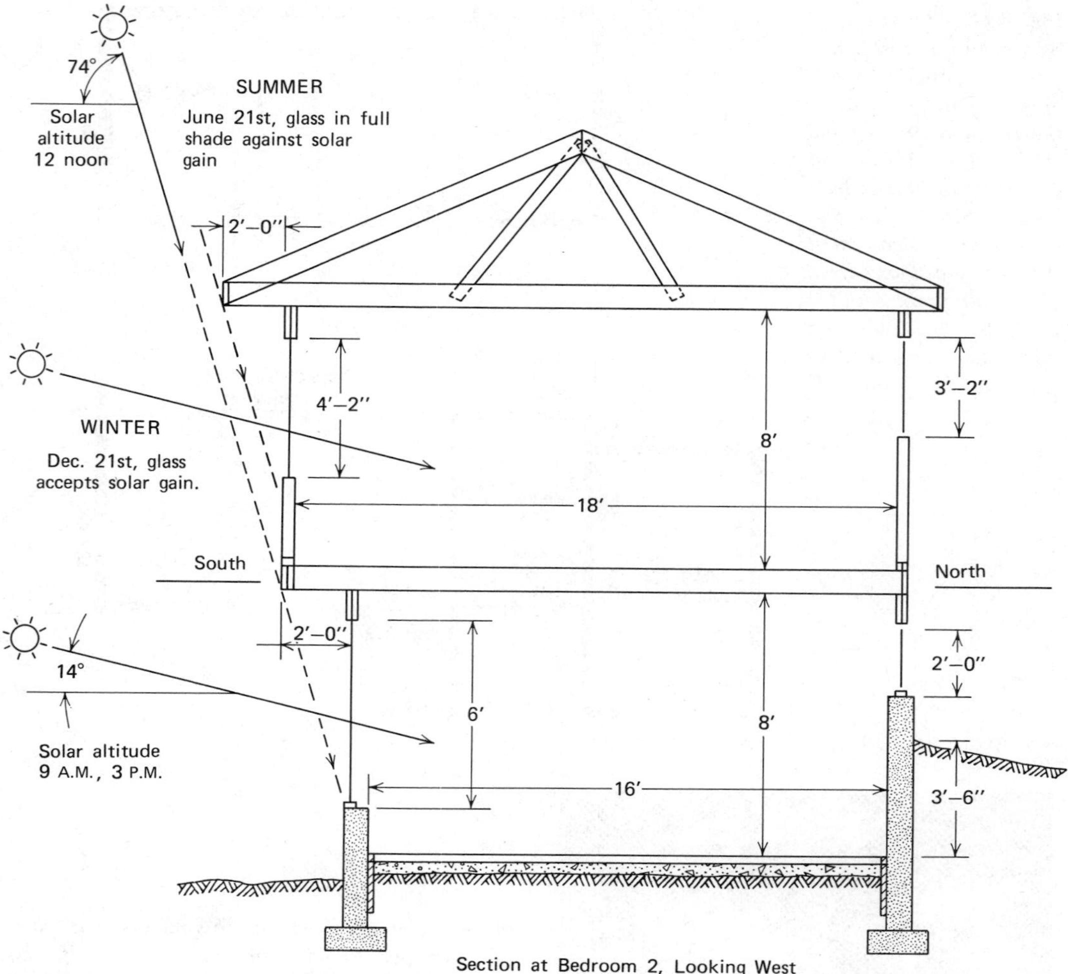

Fig. 6.13 *Cross section of the Grossi Residence. For clarity in showing solar altitudes, glass and overhangs for shading,* architectural *details have been omitted and only the structural frame and its dimensions shown. Highest solar altitude, 74°, June 21st, noon, and the lowest (effective) altitude 14°, December 21st at 9 A.M. and 3 P.M. are shown.*

in summer and allow the sun to shine through to warm it in winter. Natural growth can contribute significantly to cooling in summer and heating in winter. A case can be made for the effectiveness of deciduous *trees.*

Occasionally a large expanse of glass is desirable for an attractive view. Such was the case in planning the living room of a house on Manhasset Bay, Long Island, New York (see Fig. 6.17). Unfortunately the view was largely toward the west. If the glass were unshaded exteriorly in summer, the heat in the room could have been intolerable, despite the available output of an adequate refrigerant zone of cooling that served the room. The architect and engineer were authorized to call for the planting of a row of very tall deciduous trees. When in leaf their branches provided complete shade for the glass and, in turn, for the entire interior of the room. The lowest of the branches did not obscure an eye-level view of the distant shore nor of the summer yacht regattas on the bay.

With the falling leaves of autumn, the sun began to penetrate the trees to warm the room. In midwinter at 3:00 P.M., the sun's rays shone beyond the 33 ft length of the room to irradiate the lower half of the far wall in addition to the floor (Figs. 6.18 and 6.19). The conventional zone for cooling and, in winter, heating, operated infrequently during daylight hours. The room was graced with a very pleasant and natural ambiance.

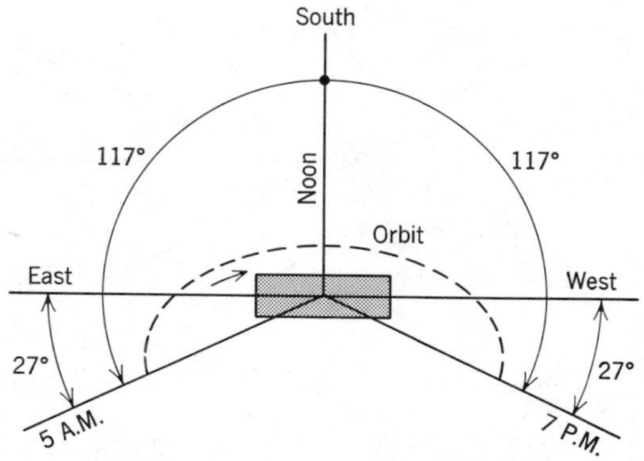

(a) Summer, June 21st, 40° North Latitude

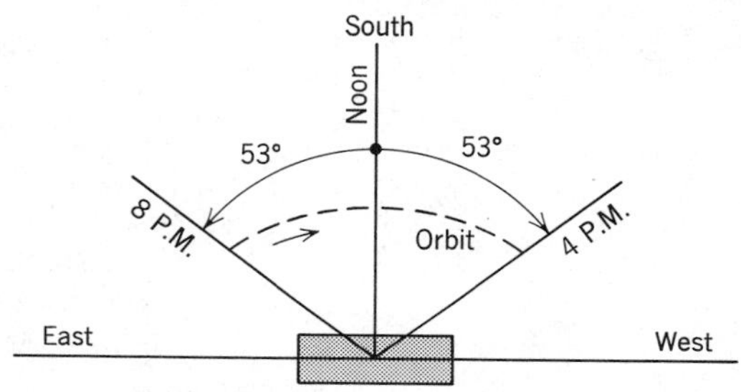

(b) Winter, December 21st, 40° North Latitude

Fig. 6.14 *Plan view of passive solar house with respect to the orbit of the sun during daylight hours, summer and winter. It is seen in* (a) *that, in summer, the sun actually shines on the north elevation early in the morning and late afternoon but only with glancing rays. (see Fig. 6.10). East and west walls are vulnerable to low-altitude summer sun effects and should have minimal glass or none. Midday positions of the sun are at high altitude when horizontal overhangs are most effective. The term* azimuth *is used for angles from the* south. *For winter see Figures 6.15 and 6.16.*

Fig. 6.15 (overleaf, top) *Section and first-floor plan, noon on December 21st. On sunny cloudless days, the sun irradiates most of the first-floor area and much of that on the second floor.*

Definition of irradiation, Webster's New Twentieth Century Dictionary, Unabridged, Second Edition, *1976—"To heat with radiant energy." On fine days in winter this house uses very little fuel for heating during the day and, because of reradiation, less at night than other houses of the same size.*

Fig. 6.16 (overleaf, bottom) *Morning and afternoon sun. Although a little less intense, the altitude of the sun is lower—low enough to warm the floor and the interior of the berm wall and west wall in the morning, with similar action toward the east in the afternoon.*

A rough calculation can be made in order to compare the hourly rate of daytime thermal inflow of energy on sunny days in winter with the conventional calculated heat loss *at sunless nighttime hours, with a temperature difference of 70°F, (70° indoors − 0° outdoors).*

Table 4.27 shows a solar-heat-gain factor on December 21st to be 151 Btuh/Sq ft at 9 A.M. and 3 P.M. With a noon value of 253, we can assume an average of about 200 on the south glass. Of this value, we can estimate that about 80% of this energy enters the house through south glass. The area of glass on the south in lower and upper stories is 230 + 120 = 350 sq ft. Setting up an equation, we have:

$$\text{Solar heat gain factor} \times 0.80 \times \text{glass area} = \text{energy inflow rate} \tag{1}$$

Substituting values,

$$200 \times 0.80 \times 350 = 56{,}000 \text{ Btuh.} \tag{2}$$

The nighttime heat loss *calculation referred to is 55,558 Btuh. The close comparison is entirely coincidental. Nevertheless, the inflow is a significant item. It evaluates free solar energy.*

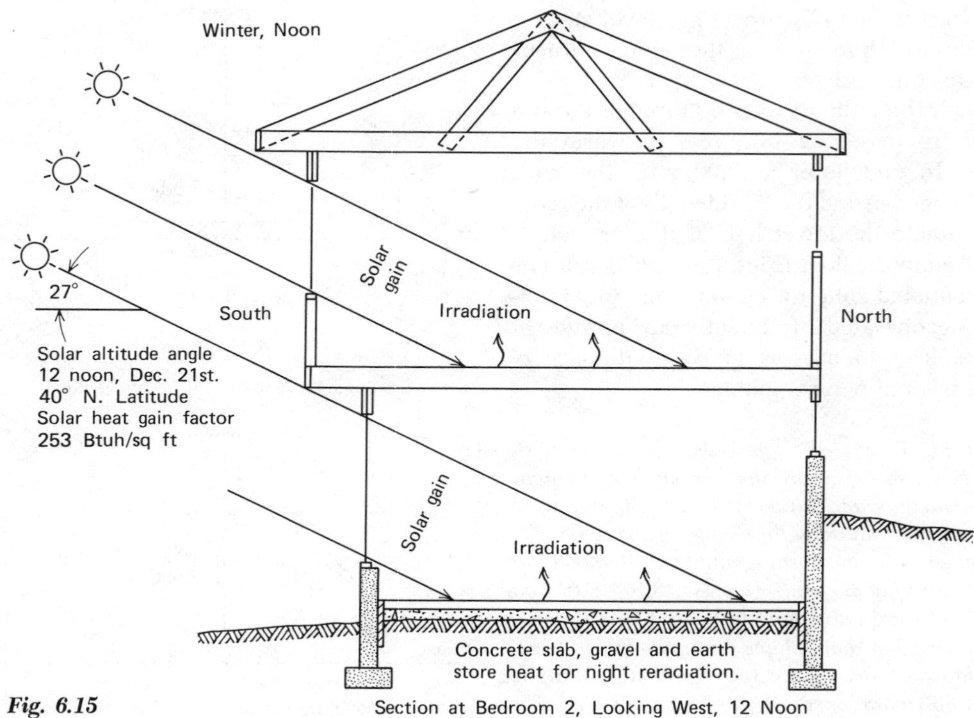

Fig. 6.15 Section at Bedroom 2, Looking West, 12 Noon

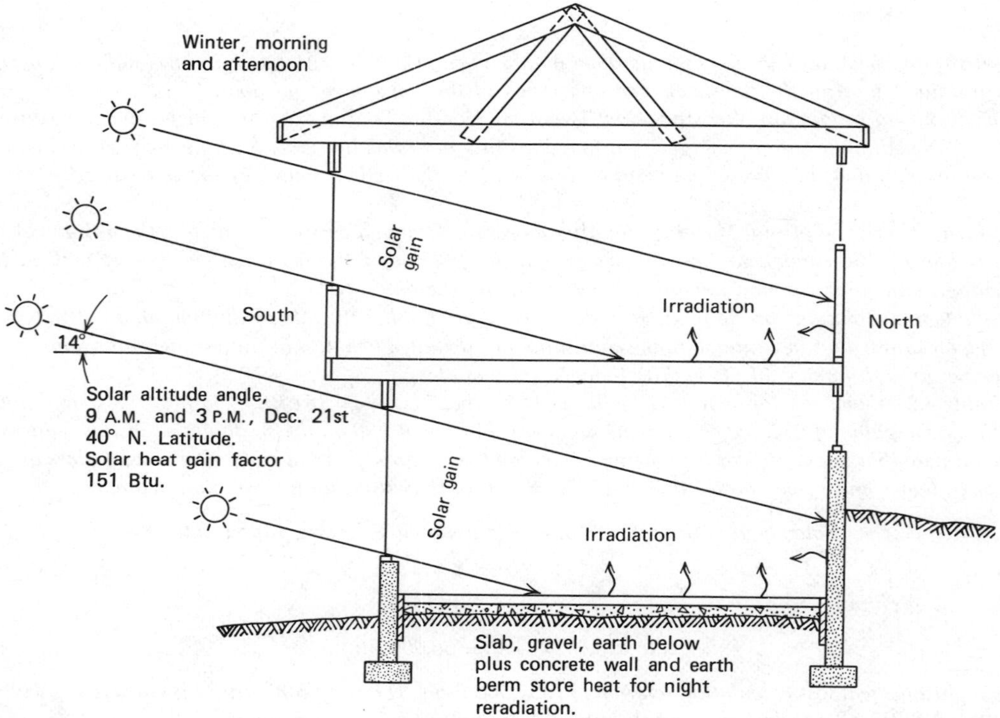

Fig. 6.16 Section at bedroom looking west, 9 A.M., 3 P.M.

South
Solar azimuth 0.0°
12 Noon Dec. 21 40° N. Latitude
Line of glass
42′
High window sills
in kitchen
16′
Irradiated floor area at noon
neglecting (minimal) partitions
and the shadows of 9 mullions
of glass wall
Inside surface, north wall
First-Floor Plan

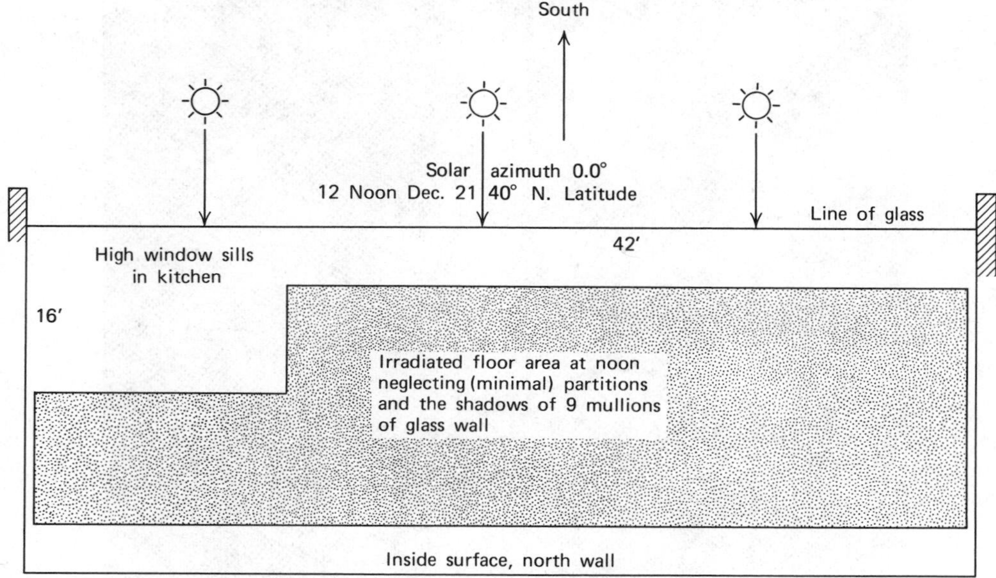

Fig. 6.15

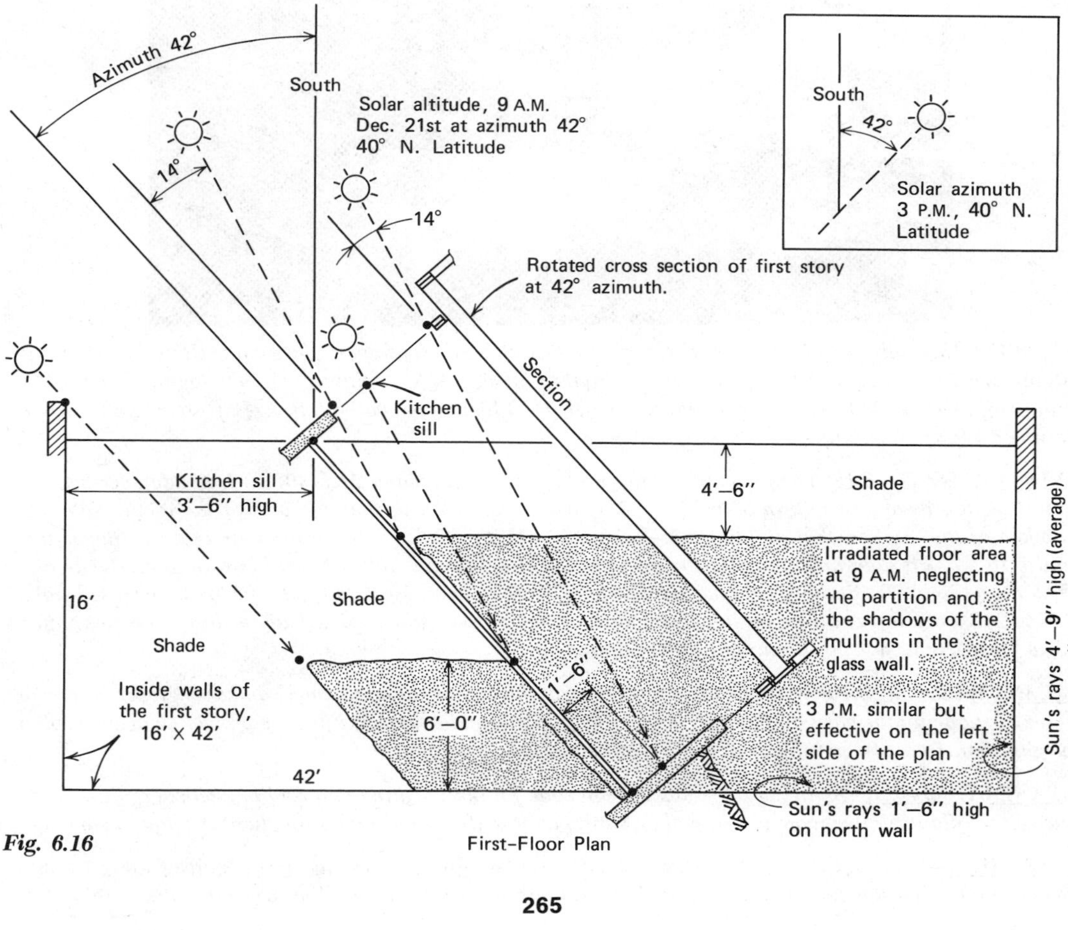

Azimuth 42°
South
Solar altitude, 9 A.M.
Dec. 21st at azimuth 42°
40° N. Latitude
14°
14°
South
42°
Solar azimuth
3 P.M., 40° N.
Latitude
Rotated cross section of first story
at 42° azimuth.
Section
Kitchen
sill
Kitchen sill
3′–6″ high
4′–6″
Shade
Irradiated floor area
at 9 A.M. neglecting
the partition and
the shadows of the
mullions in the
glass wall.
16′
Shade
Shade
Inside walls of
the first story,
16′ X 42′
1′–6″
6′–0″
3 P.M. similar but
effective on the left
side of the plan
Sun's rays 4′–9″ high (average)
on west wall
42′
Sun's rays 1′–6″ high
on north wall
First-Floor Plan

Fig. 6.16

Fig. 6.17 *Thermal control winter and summer with a difficult southwest glass wall. House overlooking Manhasset Bay on Long Island, New York, Olindo Grossi, FAIA, Architect. The photograph shows the corner of an 18 × 33 ft living room with a 3-panel, 18-ft-high glass wall (18 ft × 18 ft) facing 3° west of southwest (48° Azimuth).*

<u>WINTER</u> *(See Fig. 6.18) The photograph, taken at 3 P.M. on December 21st (40° N. Latitude), verifies the sun direction as almost within 6° (48° − 42°) Azimuth parallel to the long axis of the room. Note the shadow of the mullion that dividies the glass panels. With a 3 P.M. solar altitude of 14°, sun floods the entire 18 × 33 ft floor area and also the lower half of the far end wall. At that hour, the solar-heat-gain factor is a (high) 205 Btuh/sq ft value. See Table 4.27. From noon until 3 P.M., the factor is increased from 177 to 205 with rays covering an increasing area of the room's surfaces. The leafless deciduous trees offer little obstruction, nor do the drapes unless drawn.*

<u>SUMMER</u> *(See Fig. 6.19) With the owner's approval, the architect and the mechanical engineer specified summer shade for the glass by planting several 35-ft-high decidious trees. Effective protection against solar penetration, from its first threat at noon until sundown, was provided.*

<u>STANDBY POWER</u> *The living room has its own zone for both heating and cooling, yet energy use from the central plant for this room is minimal on sunny winter afternoons and throughout the day in summer.*

<u>THE VIEW</u> *Manhasset Bay is one of the principal centers for yacht racing in this part of Long Island. When the orientation for such a view is desired, there are ways of handling westerly sun.*

WINTER
Solar altitude
14°
33′
3 P.M. Dec. 21st 40°N. Latitude
Leafless trees do not obstruct solar gain
Zone of solar gain
18′
View of the bay
Irradiation
Vertical Section of Living Room

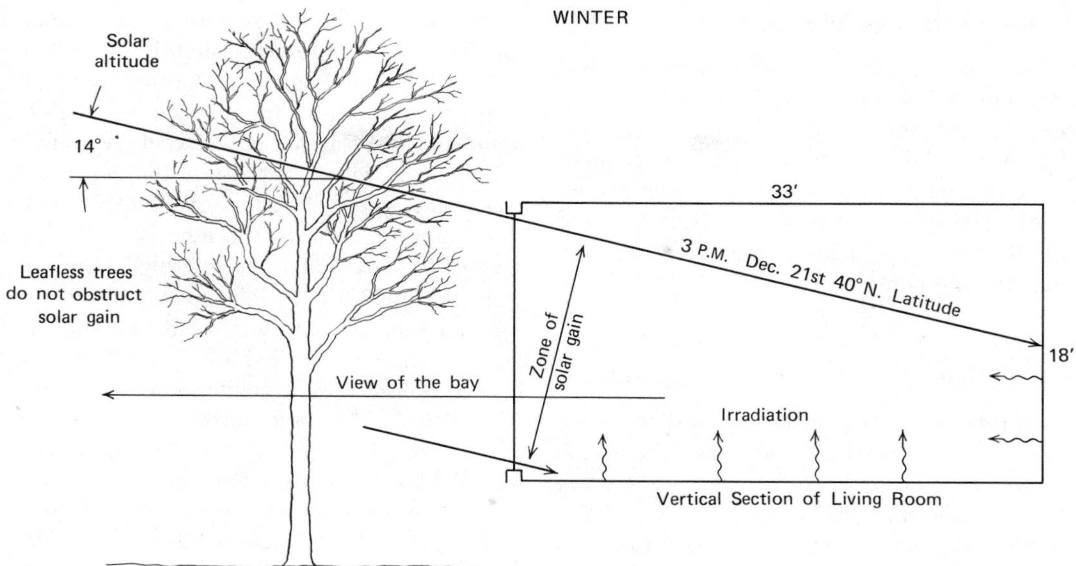

Fig. 6.18 House on Manhasset Bay. Solar gain in living room. Winter.

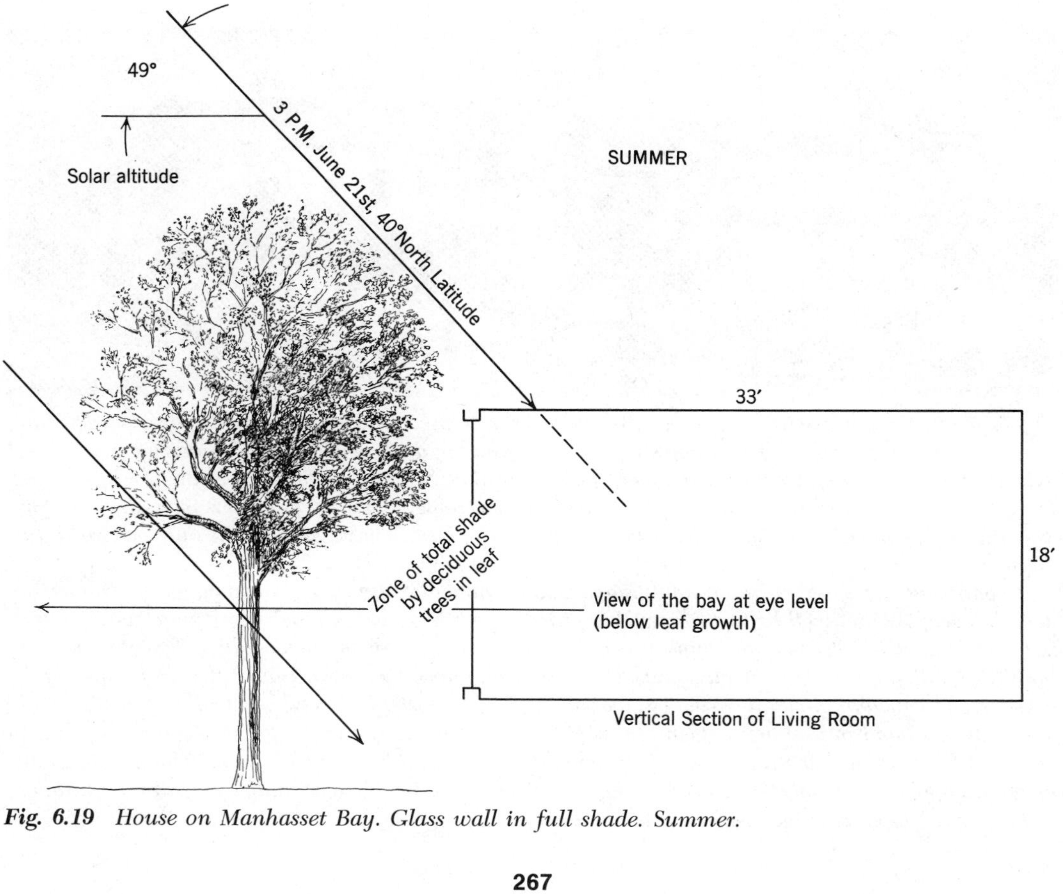

49°
3 P.M. June 21st, 40°North Latitude
SUMMER
Solar altitude
33′
Zone of total shade by deciduous trees in leaf
18′
View of the bay at eye level (below leaf growth)
Vertical Section of Living Room

Fig. 6.19 House on Manhasset Bay. Glass wall in full shade. Summer.

6.5 Windowless Buildings

It is quite evident that glass is frequently a problem. Sometimes it can be omitted. Buildings or large sections thereof can be enclosed by opaque walls. An example of a department store designed in this way is shown in Figure 6.20. The problem is a complex one and should have careful study in each instance by engineering consultants.

6.6 Natural Heating and Daylighting

If the sun's energy is to be used directly for heating in winter, it can also contribute to the illumination of the heated space. The Vincent Smith School in Port Washington, New York, typifies the trend to buildings that put the sun to work. Figures 6.21 and 6.22 illustrate the fact that first-story sliding glass and south glass of the clerestory fenestration contribute materially to daytime (class hours) winter heating. The benefit applies to slightly more than half of the building area. Natural daylight does much to relieve the electric lighting demand not only in the areas receiving direct solar gain in winter but also in the two north classrooms where clerestory glass supplements the sliding glass doors in the north wall. Figure 6.23*a* shows this top-lighting. The east wall (Fig. 6.23*b*) is largely blank against low summer sun, as is the west wall (not shown). Summer shading for the four south classrooms is accomplished by a conventional overhang (Fig. 6.23*c*). The two north rooms are not affected by direct sun rays of the sun, though well lighted by natural sunlight.

The four heating-ventilating units in Figure 6.22 start up as solar heat needs to be supplemented. They typify the new location of many heating plants. Fully automatic, they are located in furred-in areas above the room ceil-

Fig. 6.20 *No windows. Alexander's Department Store at Roosevelt Field Shopping Center, Garden City, New York. Although this chapter emphasizes the importance of dealing with glass for maximum solar heat gain in winter and a minimum in summer, the case of the windowless building cannot be overlooked. Department stores are often best designed this way. Schools sometimes are. Parameters including function, esthetics, and thermal interchange must be examined. The last of these (thermal) is our concern in this discussion.*

Department stores and classrooms fall into a similar thermal category. During daylight hours they are densely occupied and well lighted (though, admittedly, lighting levels are becoming more conservative). See Table 1.2, page 13. The space gains of people-load and lighting-load are usually sufficient to heat the building by day in winter. Often they exceed this state and must be cooled. Hopefully this is done by the low-energy method of circulating cool outdoor air. Glass, no matter how well handled in summer, would add additional and instantaneous heat that is not needed. Quite differently, the transmission of heat in through heavy masonry walls is minimal by comparison and usually delayed by 8 or 10 hours. Merchants do not need to have shoppers distracted by views of the exterior. Children at study can wait for relaxation, to be enjoyed later in windowed recreational area.

Fig. 6.21 The Vincent Smith School, Port Washington, New York, Budd Mogensen, AIA, Architect and Planner. South elevation, construction phase. Passive solar heat in winter serves the four classrooms and (through clerestory glass) also warms the student commons area (see Fig. 6.22). Natural daylighting is characteristic also for these spaces and for the two north classrooms (Figs. 6.22 and 6.23 (a) *and* (b).

ing level. Unit C in Figure 6.22 heats and ventilates classrooms 1 and 2 as shown in Figure 6.24. Multiple-zone systems like this facilitate energy saving by the operation of one or several zones only as needed. Room registers from and to the unit appear in Figure 6.25.

6.7 Equipment on the Roof

Services that connect to an active large building are numerous. *Entering* services can comprise: electricity, oil, gas, and water. These, together with those *leaving* the building (sewage and storm water), obviously relate to lower levels of the structure. Fresh air for ventilation and *used* air, rejected by exhaust fans, frequently pass through the building envelope near the point of use. This can be at open mechanical stories adjacent to air-handling equipment. On the roof in tall buildings as seen in Figure 10.3*b*, page 383, one finds a multitude of items including roof-access penthouse, elevator machinery, cooling towers, water storage tanks, and chimneys. On the roofs of *smaller*, low-rise buildings (Fig. 6.26), rooftop facilities can include chimneys, fan exhausts, plumbing vents, security lights, and roof-access stairs.

The use of outdoor equipment is increasing. Such equipment is necessarily weather-resistant. Locations can be selected. They can include space adjacent to the building or on a roof. Often such locations are preferred to the use of indoor space, which is costly to build and can be used for more suitable or productive purposes.

In Figure 6.27 zonal installations for heating-cooling-ventilation are shown in rooftop locations. Each of these units includes compressor, condenser, and evaporator coil. The latter is in a duct, which also houses a filter, blower, electric heating coil, and provides an adjustable fresh air intake on the suction side of the blower.

6.8 Cause and Cure of the London Fog

The architectural evolvement from 17th Century rural England (Fig. 6.28) to 19th Century crowded London (Fig. 6.29) had environmental overtones. At the start of the 20th Century air pollution in London was at its peak. By the year 1932, the condition described in Figure 6.29 was beginning to abate. In a book published in that year, *The Modern English House*, a prominent English architect,

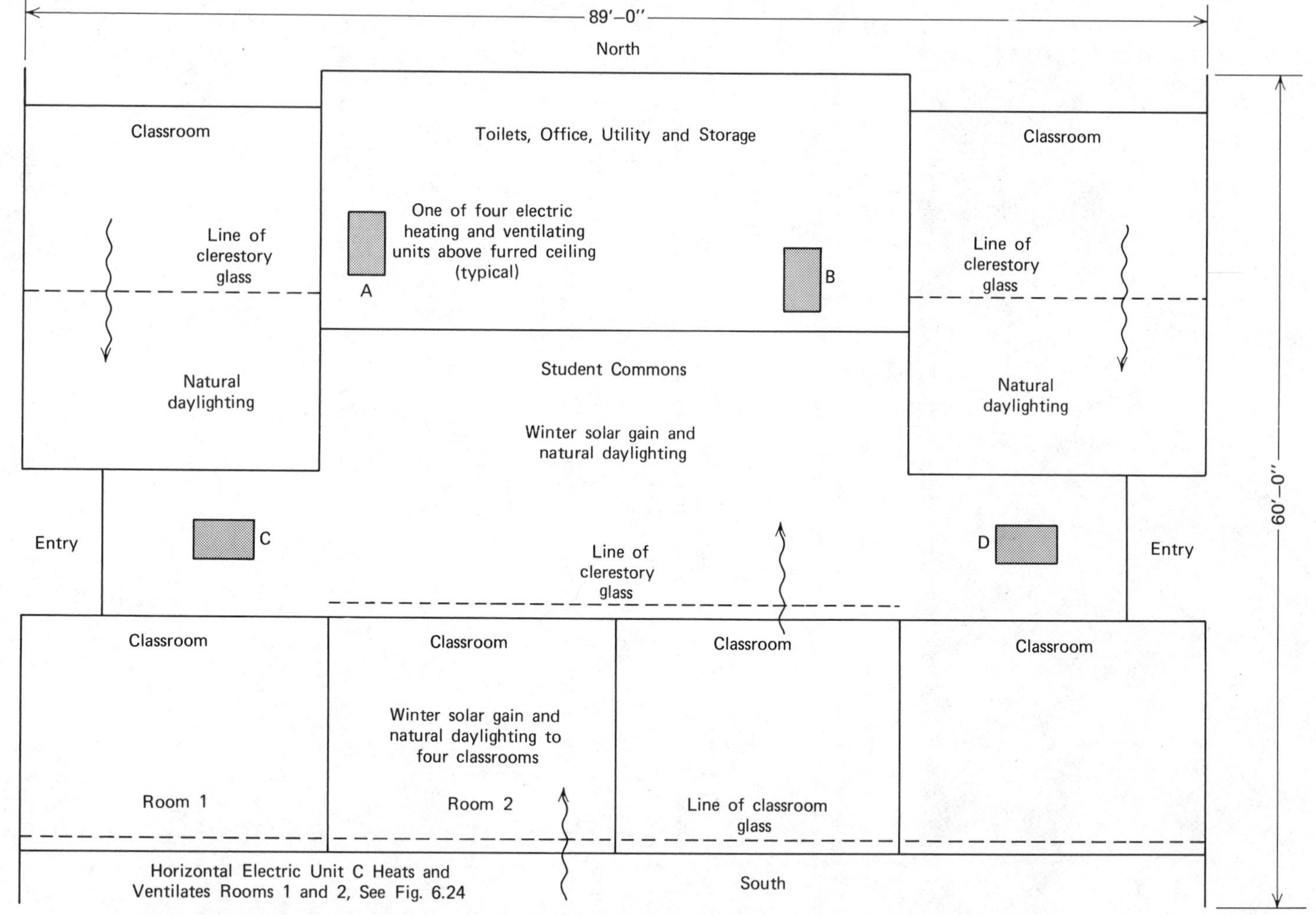

Fig. 6.22 *Line drawing depicting space allocations. The four south classrooms and the student commons receive solar warmth and natural daylighting. The two north classrooms are lighted from above by clerestory fenestration in addition to their first-story glass as shown in Figure 6.23* (a).

Shown also are the four climate control units, each of which adds some fresh air to the circulated warm air.

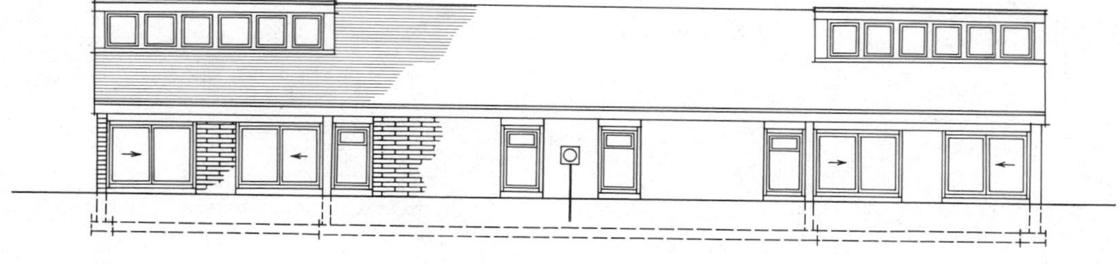

(a) North Elevation

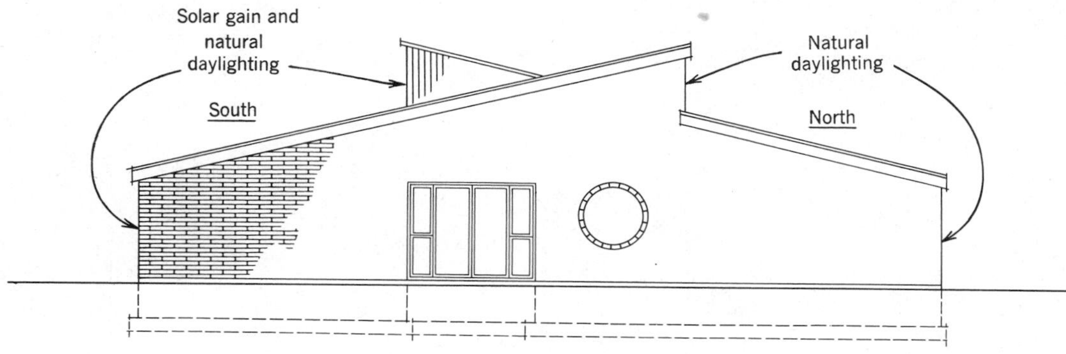

(b) East Elevation

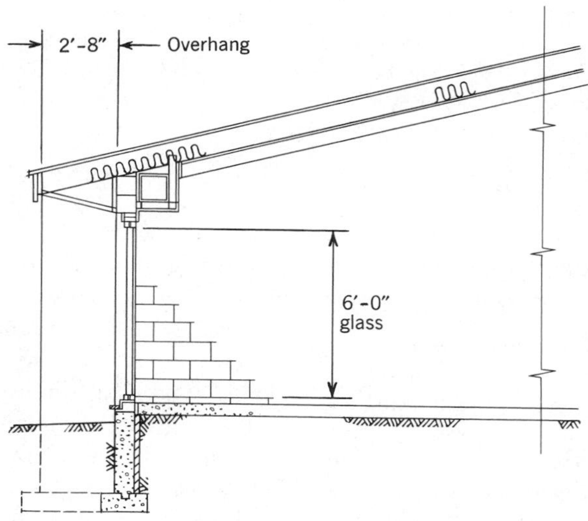

(c) Section Through a South Classroom

Fig. 6.23 (a) *North elevation. The two north classrooms benefit from clerestory natural lighting, which supplements that from the first-story glass panels.* (b) *East elevation. Notice solar gain through south glass, first story and clerestory.* (c) *Conventional overhang above south glass obstructs hot rays of the sun in summer.*

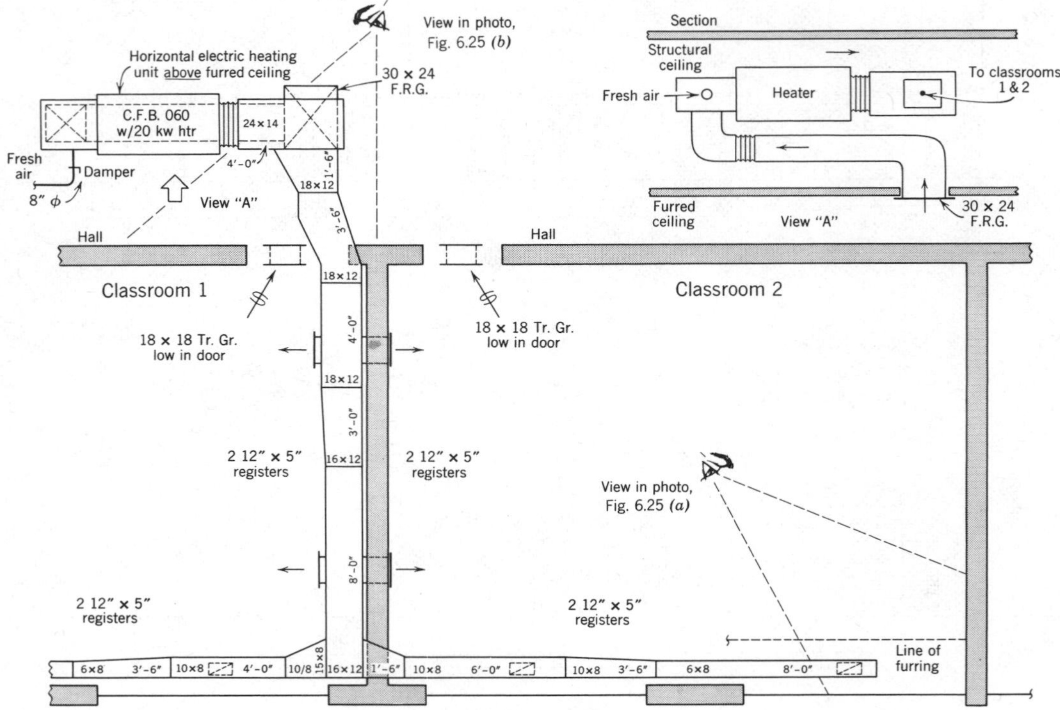

Fig. 6.24 Zone C duct layout, an excerpt from the contractors working drawings. Heating-ventilating unit C (see Fig. 6.22 for location) serves classrooms 1 and 2. In each room, warm air is delivered down *across cool glass through two 12 × 5 in. registers. (see Fig. 6.25*a*). Warm air is also delivered horizontally into the space through two 12 × 5 in. registers in the face of the common duct. Return air leaves through 18 × 18 in. grills in the classroom door and is drawn into the unit through the 30 × 24 in. grill flush with the furred ceiling (see Fig. 6.25*b*). Courtesy of R. Ferina, Inc., Contractor.*

A. Randal Phillips, made the following statement.[1]

It is noteworthy also how coal fireplaces have been reduced in number. Now, in the house of average size, the common arrangement is to have only one coal fire on the ground floor—in the sitting-room; the dining-room and study (if the latter is provided) being heated by gas or electric fires. Similarly, on the first floor, gas or electric fires have taken the place of coal fires, and it is now possible to build gas fires into the wall with special flue blocks for taking away the products of combustion.

[1]Reprinted by permission of Charles Scribner's Sons, Inc.

As of 1979, owners of older London houses are walling up many of their fireplaces and increasing their acceptance and use of gas and electric "fires." In new installations, efficient central heating is often chosen.

In the greater London area, the Solid Fuel Advisory Service is promoting the use of "smokeless coal"—anthracite with minimum smoke and maximum heating value. By law, wood fires are not permitted in Greater London, though in Paris the fragrance of wood smoke is often perceptible on chilly days. The English Parliament's "Clean Air Act" of 1956 has been effective. The Greater London Council has systematically banished the use of soft coal and all other smoky fuels.

Fig. 6.25 *Construction photographs.* (a) *One of the two 12 × 5 in. down-flow registers in classroom 2.* (b) *The 30 × 24 in. opening that will receive the grill through which return air is drawn into unit C. Classroom 1 is seen through the door opening.*

References and Bibliography

1. *Other Homes and Garbage, Designs for self-sufficient living.* Leckie, Masters, Whitehouse and Young, 1975 Sierra Club Books, The Sierra Club, San Francisco.

Fig. 6.26 *Rooftop jungle in New York City. Rearview of one-hundred-year old* brownstones *recently adapted to town-house conversions show the dingy evidence of decades of coal-burning in fireplaces and furnaces. The contrast with the clean new building in the distant background is impressive. We* can *plan better. Improved combustion and better rooftops are possible. Helicopters and the view from adjacent tall buildings make the roof a new subject for good design.* *(Continued)*

Fig. 6.26 (*Continued*)

The photograph presents a potpourri of elements. They comprise: sloped-roof access over stair shafts, gravity-vent cupolas over these stair shaft roofs, gravity vents over obsolete vertical air shafts, groups of four vents from newly installed exhaust fans in kitchens and bathrooms, two chimneys in the foreground and an additional sheet metal chimney (lower right), gooseneck plumbing vent stacks (upper left), and security lights, the need for which is made clear by the iron-barred windows (lower left).

If buildings can be planned for reuse, as is encouraged in Section 1.8 (page 20), and subsequently recycled with good coordination between architect and engineer, we may do a better job of meeting our new design challenge—the roof.

Fig. 6.27 (a) *Conditioning equipment on the roof. For the heating, cooling, and ventilation of public rooms in a motel, five or more self-contained weather resistant units can be used. Each unit delivers warm or cool air to the space below through short duct-runs in furred ceilings. The air is delivered through ceiling registers. Each unit thus constitutes a decentralized zone. Unlike some* central-*station installations, the zones can operate separately and turn off when not in demand. Hence energy is saved. Units are identified as white in color and space-exhaust fans as silver-domed elements on black bases. Appearance of the equiment is not unpleasant and no surrounding parapet is needed to mask them.*

(b) *Operation. Each heating-cooling-ventilation unit recirculates air to the space below. The suction side of the blower draws air from the space and admits fresh air (see short down-turned gooseneck inlet at the left unit). By the action of the blower, the air to be conditioned in the unit passes through a filter, an evaporative cooling coil or, in winter, an electric-resistance heating coil. It is then delivered to the ducts below. The dark areas indicate the grills of the compressor-condenser assembly. The hot air from this cooling process is discharged through these side grills. This constitutes the so-called heat rejection process common to all refrigerant cycle cooling systems.*

(c) *Exhaust fans. At areas of "people concentration" such as dining rooms, conference rooms, lounges and bars, used air is drawn away to prevent odors and stagnaton from building up. The choice of locations for such exhausting is an important consideration.*

Critique: *Referring to* (a) *in this illustration. In the distant background one sees deciduous trees in leaf, which identify the photograph as taken in a warm-weather season. Consideration of energy conservation might suggest that a better design would have masked the glass area against large heat gain from the sun.*

(a)

(b)

(c)

Fig. 6.28 *Energy shortage and air pollution of other days. Buildings of the seventeenth century in rural England. In later years, because of deforestation, wood sometimes became scarce both as fuel and as a structure material. Soft bituminous coal for use in fireplaces and stone for building purposes became the trend. The tall chimneys, each stack serving a fireplace, discharged sulphurous smoke. As congested cities developed, continuation of the custom led to conditions that, prior to corrective measures, became critical* (see Fig. 6.29). *For increase in condensation nuclei, particulates and fog caused by urbanization, see Table 2.2, p. 35.*

Fig. 6.29 *London houses built in the nineteenth century. With acute vision, one can discern 102 chimney pots within two city blocks, opposite Clapham Common. The first (end wall) group of stacks comprise the fireplace smoke exhaust of six rooms. Marching away to the south, the party walls each enclose 12 stacks of two adjacent six-fireplace houses. Soft coal from the mines of England, Scotland, and Wales, transported by the (then) newly developed railroads, made the problem of air pollution acute. In times of high humidity the famous "London Fogs," (normal mist and abnormal pollution), sometimes limited visibility to 6 ft. In the early 1900s steps were begun to correct the problem. By 1930 gas and electric "fires" had largely replaced the use of coal. In more recent times, regulations prescribe the use of "smokeless coal" if fireplaces are used. Now, the contribution of* residential *combustion to the air pollution problem in London is largely eliminated. As we start to use coal in central plants in the United States, the need for care is evident.*

7

Climate Control Systems in Buildings

7.1 Exterior Ducts The International Building

As the popularity of the glass facade of the modern office building developed, pipes and ducts were gradually cleared away from the building exterior. Earlier years had seen the inevitable disappearance of steam and condensate risers at the exterior columns. They were replaced by risers in the core, near the elevators, often hot *and* chilled water risers, branching laterally in the space below each floor and turned up to fan-coil units below the glass. Systems of high-pressure, high-velocity, conditioned air increased in popularity, also taking the form of risers in the central core with supply and return runouts in the space above the ceiling at each story. The concomitant architectural form was a horizontal glass strip with uniform, pipeless columns set back from the building line. Between these facade strips of glass were bands of opaque glass, metal, or masonry; behind these were the floor structure, with horizontal pipes and ducts below it, and sometimes a sill-high space above the floor. There was a strong horizontal accent. Because heat gain and heat loss are maximum at the building perimeter as a result of the effects of the sun in summer and low temperature in winter, the pipe and duct runouts continued at large dimension from the core to the outside curtain of the building. As the intricacy of the mechanical services increased, the depth of the structure-plus-equipment layer attained about one-third of the total, overall floor-to-floor height. High-velocity air systems with central units were beginning to replace the formerly popular systems that used air-handling rooms at each story. This development stressed the importance of the vertical riser. It began to be difficult to get branches out of the central core past elevators, stairs, and other risers.

Much needed had been a relief for the central core, the relocation of risers at the perimeter (where they are most needed) and a reduction in the length—and consequently in the size—of pipe and duct branches. Not least in importance is the avoidance of bulky ceiling crossovers. This is effected by moving large assemblies of building "arteries and veins" to the surfaces that they serve. It was logical to place at the perimeter the parts of the system that have to deal with the effects of sun, shade, and temperature change in the several perimeter zones, leaving at the core a separate network to handle the more stable interior areas.

In their design for the International Building in San Francisco, Architects Anshen & Allen maintained the horizontal-strip effect, yet have succeeded in supplying and returning most of the air for perimeter areas through vertical ducts at the perimeter. This solution has been accomplished by shafts at the corners of the building. A general description of this unusual building may aid in visualizing the scheme. Almost square and supported by 16 columns, the building has four 16-ft cantilevered floor-slab extensions at each level, terminated at their sides by an extension of the

Fig. 7.1 *The International Building, San Francisco, California. Anshen and Allen, Architects; Eagelson, Engineers (Charles Krieger, E.E.). Mechanical Designers. Courtesy of* Progressive Architecture.

Fig. 7.2 *The International Building. In this ingenious scheme, the major supply arteries for conditioned air have been located in alternate corners. In diagonally opposite locations, supply risers are placed in large square enclosures. Although nonstructural, each enclosure is emphasized as a distinct vertical design element. Each encloses both hot- and cold-air ducts, which supply two separately controlled orientations on all of the twenty-one stories. Conditioned air originates at an intermediate floor, the third. In the opposite two corners, similar ducts return much of the air to the equipment story. The balance is returned through duct risers in the core.*

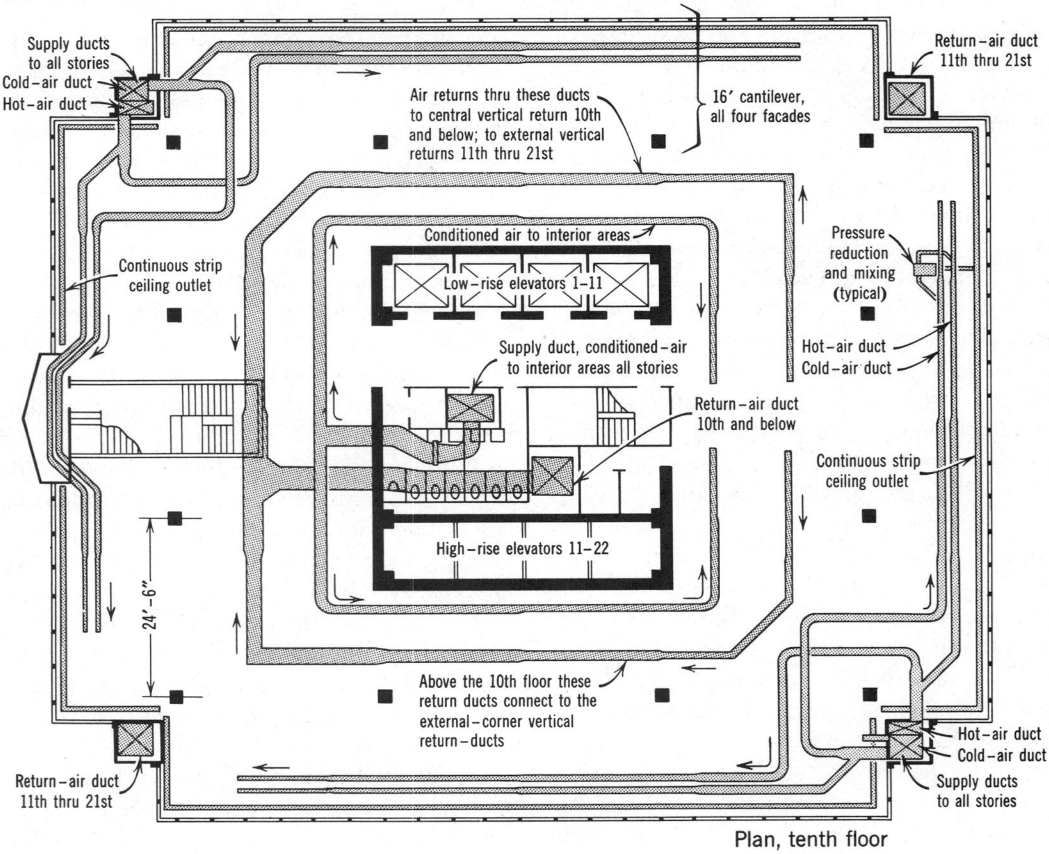

Supply ducts to all stories
Cold-air duct
Hot-air duct
Air returns thru these ducts to central vertical return 10th and below; to external vertical returns 11th thru 21st
16′ cantilever, all four facades
Return-air duct 11th thru 21st
Conditioned air to interior areas
Low-rise elevators 1-11
Continuous strip ceiling outlet
Pressure reduction and mixing (typical)
Supply duct, conditioned-air to interior areas all stories
Hot-air duct
Cold-air duct
Return-air duct 10th and below
Continuous strip ceiling outlet
High-rise elevators 11-22
24′-6″
Above the 10th floor these return ducts connect to the external-corner vertical return-ducts
Hot-air duct
Cold-air duct
Return-air duct 11th thru 21st
Supply ducts to all stories

Plan, tenth floor

Fig. 7.3 The International Building. Evident are the alternate horizontal glass and opaque strips, the columnless perimeter, the 16-ft cantilevered projections, and one of the four corner duct-enclosures. The apparent transparency of the duct-enclosure is caused by the reflection of the sides of the cantilever overhangs in its shiny surface.

exterior column lines. At the internal corners thus created, shafts carry supply and return ducts to serve the four exterior air-conditioning zones. Air-handling equipment and a 750-ton refrigeration plant are located on the second and third floors (these floors are just below the terrace level and are least desirable for renting). Steam is purchased from a utility company and is taken from street mains.

Hot air and cold air are supplied to all stories through vertical ducts in the shafts at diagonally opposite corners of the building. Each corner duct branches to serve two zones, separately controlled. Pressure reduction and blending are done by equipment in the hung ceiling, and from these points air flows to strip-grill diffusers directly above the glass on the four sides of the building. Local controls assure comfort to personnel in each area. Interior zones on each floor are supplied by a riser duct in the building core, which branches at each floor to a loop just outside the line of elevators. The loop serves ceiling diffusers.

Between the perimeter loop and the interior loop just described, a return loop collects air for return to the second-, third-floor central station. These return loops on the eleventh to the twenty-first floors are picked up by external return risers on alternate exterior corners. From the tenth floor and down, the loops are picked up (*as shown*) by an interior return riser that extends down through the core in front of the blank faces of the high-rise elevators. In order to provide a clear space between the elevator banks on the main floor (fourth or terrace), the two core ducts-risers are offset at the ceiling of that story.

In summary, perimeter air for all stories is supplied through corner ducts. Central air for all stories is supplied through a core duct. All return air above the tenth floor is carried down through the return ducts at the *other* two corners. Return air from the tenth floor and below is carried down by a return duct in the core. For simplicity in this discussion, some minor variations have been omitted. The cooling tower is on the roof and the core is the location chosen for risers of chilled water, condenser water, steam, and condensate (see Figs. 7.1, 7.2, and 7.3).

7.2 The Blue Cross-Blue Shield Building

In distinction to the form of the International Building (Fig. 7.1), yet similarly employing exterior surfaces to enclose ducts, the Blue Cross-Blue Shield Building is unique (Figs. 7.4 and 7.5).

The closely spaced and strongly emphasized vertical elements are both structural and utilitarian. Two out of every three of the verticals are structural and *all* the verticals enclose ducts. Each pair (one hot, one cold) of high-velocity, round air ducts at the corresponding structural columns constitutes a vertical air supply system. At each floor they feed an attenuation and mixing chamber. At these locations, temperature is controlled as desired. The return ducts between pairs of hot-cold

Fig. 7.4 Blue Cross-Blue Shield Building, Boston, Massachusetts. Anderson, Beckwith and Haible and Paul Rudolph, Associated Architects; Stressenger, Adams, Maguire and Reidy, Mechanical and Electrical Engineers. The two-story, Y-shaped forms are structural columns that divide at mezzanine level and continue to rise in pairs to form the exterior skeleton frame. Hollow channels on the exterior of each pair enclose, individually, a hot-air supply duct and a cold-air supply duct. These round, high-velocity ducts join for mixing and velocity reduction in attenuation boxes, located between columns at each floor. Conditioned air is discharged upward from a window sill grill above the box. A mullion between each pair of structural columns originates at the second-floor level and extends to the mechanical story at the roof. This mullion encloses a return-air duct, which draws air through grills in the sills of the two adjacent windows on each story. Thus the air is delivered at the exterior, accomplishes its mission at that surface, and returns in the same vertical plane to the suction side of fans on the roof.

supplies complete the parallel, equally spaced pattern. This *vertical* distribution to each floor is at right angles to the position of the *horizontal* dual branch ducts of the International Building.

7.3 Mechanical Services in Special Buildings

In a chemistry building at Berkeley, California, the possible complication of ducts and pipes at ceilings was avoided by a unique solution.

Maintaining a 12-ft floor-to-floor height, and at the same time accommodating the required quantity of horizontal piping and ducting would substantially reduce the usable building volume. Therefore, 16 hollow, exterior, reinforced-concrete, structural columns were used to enclose pipes and ducts connected to short lateral branches, thereby minimizing crossovers and avoiding spaghetti patterns on the ceilings.

These are truly multipurpose columns. Together with the slimmer interior columns, they support the building. In the late afternoon when the balconies (between adjacent columns) have ceased to exclude all direct sunlight, the deep exterior columns take over as totally effective sunshades. Finally, they enclose exhaust ducts from all exterior (and some interior) fume hoods (*FH on the illustration*). These ducts are exhausted by fans behind the masonry grill at the roof level. There are 185 such units, of which 175 are fume-hood exhaust fans. Four of the exterior columns are *pipe-station columns*. Two of these are shown in the illustration (Fig. 7.6), and one is detailed. The piping services include domestic hot and cold water, industrial hot and cold water, distilled water, demineralized water, natural gas, vacuum, compressed air, hydrogen sulphide, nitrogen, oxygen, steam, acid waste, vent pipes, and roof leader pipes. There is access to all exterior columns at all floors.

This is a once-through system, with no air recirculated. The pressure in the halls is

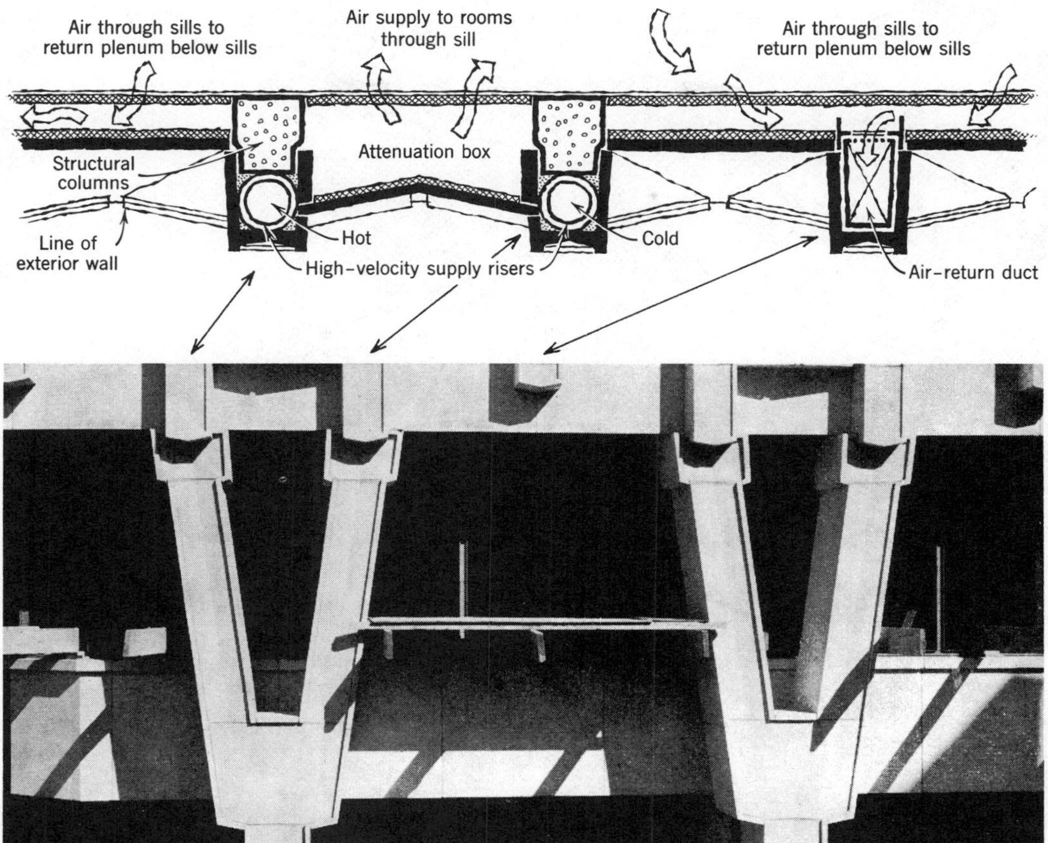

Fig. 7.5 *Dual-duct, high-velocity distribution ducts and return duct integral with the facade of the Blue Cross-Blue Shield Building. Photo shows building in construction.*

greater than that in the rooms, preventing the flow of air or gases back out into the corridors.

Air intakes at ground level supply all fresh air. It is partially conditioned and distributed through a central vertical duct with branches at each floor in all corridors. From these, the rooms are supplied. This air is regionally tempered in response to local controls before it enters the rooms. Air is exhausted through fume hoods by ducts in the exterior columns. Air in some interior rooms is exhausted through fume hoods to ducts that are assembled at one end of the stair. Roof exhaust fans also connect to these ducts.

Campus steam is used as a general power source. It is converted to 275° F primary hot water. Pumps circulate secondary hot water through the various hot water systems. Hot water is used for both heating and cooling; the latter is done by absorption-type water chillers.

Summary: All fresh air is brought in at ground level. It is partially conditioned and delivered to rooms by means of a central riser and ceiling ducts in corridors. After reheat and delivery to rooms, it is exhausted through fume hoods to ducts in hollow exterior columns, and in the case of interior rooms to a central group of exhaust ducts. All exhaust fans are on the roof. Four exterior columns are also *pipe station* columns. See Figures 7.6, 7.7, and 7.8.

7.4 Downfeed Air Distribution

Although heavy equipment such as boilers and water chillers frequently retain their tra-

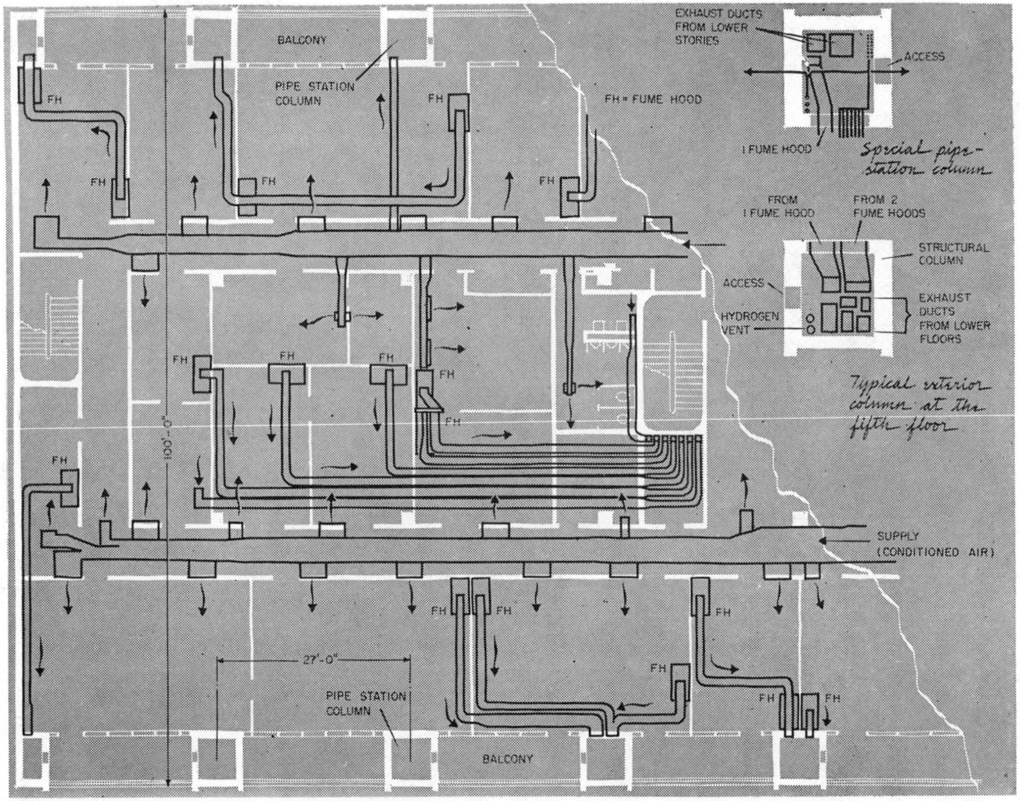

Fig. 7.6 Chemistry Building, Unit 1, University of California at Berkeley, Anshen & Allen, Architects. Mechanical design by Bayha, Weir & Finato. Preliminary rendering and layout of supply and exhaust air shown in four of seven bays at typical floor—the fifth.

The elevation indicates that the balconies exclude direct sunlight through glass. The burden of exhaust from fume hoods is relieved by routing those located in outside rooms to exterior hollow structural columns. Hoods in interior rooms are ducted to exhaust stacks adjacent to the stairway. Hollow columns provide distribution space for piping. Lateral crossover of mechanical services at ceilings is minimized.

Fig. 7.7 Chemistry Building, Unit 1, at Berkeley. The hollow, reinforced concrete columns have a dual function. They are structural supports for the building and also provide channels for the vertical distribution of mechanical services. In the late afternoon when the sun is low they acquire a third role. They become sunshades to put the balcony areas in full shade.

Fig. 7.8 Behind the decorative masonry screen at the roof edge are exhaust fans that pick up the vertical ducts in the hollow columns to relieve the building of gases produced by a multitude of chemical research projects.

At this hour the sun is intercepted by masonry walls that form an effective barrier against heat transmission to the indoors. This relieves a cooling system that must cope with the burden of people, lights, and heat-producing research.

ditional location in basements, freedom of choice can be exercised in placing the lighter air-handling units. Downfeed air systems are not unusual.

For the Medical Towers Building (see Fig. 7.9), an all-air system was chosen to supply heat and also handle the larger problem of air conditioning. It is a dual-duct (one hot and one cold), high-velocity system. Steam boilers and water chillers in the basement serve the two air-handling, central-station units on the top floor. Steam is converted to hot water for use there. Air distribution is downward at the end walls by pairs of ducts, each set supplying one-quarter of the building through dual branches on each story directly under the floor above. At terminal points in the ceiling near offices, attenuation boxes slow down the air and quiet it before delivery to the rooms through ceiling air diffusers. At other points in the ceiling, return air passes into a plenum between the ceiling and the structural floor above, and thence to vertical chases in the core back to the units on the top floor.

This kind of system may really be said to create a new form. All distribution lies neatly in planes: vertical ones at the ends of the structure and horizontal ones below each floor.

Each air-handling unit on the top floor has a capacity of 87,000 cfm of conditioned air, totaling 174,000 cfm. In each unit 13,000 cfm of fresh air is brought into the system, filtered, preheated and precooled to be mixed at 50° F saturated, for summer and winter operations. The fresh air represents about 15% of the air handled. The supply air is then mixed with 74,000 cfm of return air, the total mixture passing over mat filters. About 40% of it then passes through activated-carbon filters for the removal of odors. This is most important in a medical building, where odors from laboratories and other medical processes are likely to accumulate. Total air of 87,000 cfm is then taken into a fan and divided between a hot deck and a cold deck of coils, depending on the requirements in the spaces below. After distribution through the vertical downfeed ducts and the perimeter ducts, flexible ducts carry the air to high-velocity units over the

Fig. 7.9 *Medical Towers Building, Houston, Texas, Goleman and Rolfe, Houston, with Skidmore, Owings and Merrill, Consulting Architects; Bernard Johnson and Associates, Consulting Engineers. Air conditioning for the tower section of this building originates in two air-handling units in the tall top story. Downfeed vertical supply ducts fit into vertical planes at the ends of the building. Below each floor in a horizontal plane are the lateral distribution branch ducts. A dual-duct, high-velocity system, it comprises pairs (one hot, one cold) of downfeed supplies and dual horizontal branches for each quarter of the building. At offices, all air is distributed down from the ceiling after attenuation and blending. Return air at low velocity flows through ceiling plenums to vertical ducts at the core for return to the penthouse units, which are supplied with steam and chilled water from the basement.*

ceiling of the respective doctors' suites. The general design is based on maintaining 75° F and 50% RH. In the basement there are two low-pressure fire-tube boilers, which supply steam for domestic hot water and for conversion to hot water for use in air-handling units. Centrifugal compressors aggregate 600 tons of refrigeration. The lower portion of the building is supplied by a separate, single-duct, low-velocity all-air system.

7.5 The World Trade Center Towers

In any design, the architect must know how much space in the building is to be allotted to mechanical equipment and to the pipes and ducts that serve the heating, ventilating, and air-conditioning systems.

Not only is the *amount* of space significant, but the judgment of the mechanical engineer as to the number, position, and size of the several HVAC service centers must be considered (see Fig. 7.10). Concurrently, the development of the structural system must merge with architectural and mechanical planning. Uniquely, in these towers the tenant space in typical floors is very wide (see Fig. 7.11), in distinction to the usual 20- to 30-ft column spacing that is often seen in typical plans (see Figs. 7.2, page 278; 7.6, page 282; 10.6*b*, page 389; and 13.7, page 500). In achieving such long spans between core and perimeter, the need for greater structural strength at core and perimeter is essential. The multicolumn exterior wall panels are especially important in resisting vertical, lateral, and other forces in this tall and unusual building (see Fig. 7.12). For this reason, since the vertical air distribution at exterior walls (used in the International Building, the Blue Cross-Blue Shield Building, and the Chemistry Building at Berkeley) would not be appropriate, vertical air ducts are located in the core.

By a very special design, the horizontal ducts and pipes run from core to perimeter and from core to tenant areas through the truss-joist construction shown in Figure 7.13. Thus, an efficient and minimum 12-ft story height, floor-to-floor, is maintained with an adequate floor-to-ceiling height in occupied space.

Two-story mechanical floors, one of which is seen in Figure 7.14, have wider exterior columns due to a larger unbraced height than occurs in typical stories.

7.6 Multipurpose Buildings

An excellent example of how mechanical systems are adapted to the architectural program, arranged for the shortest and most efficient distribution circuits, and to some extent affect the general architectural design is seen in Figures 7.15, 7.16, and 7.17, illustrating the planning and design of the Fox Plaza Building in San Francisco.

The project is unusual in that it includes four major building types in one structure:

(*a*) Underground garage for the storage of cars.
(*b*) Commercial center, at ground level, including a bank, women's specialty store, and other uses.
(*c*) Ten floors of offices.
(*d*) Sixteen floors of apartments.

The mechanical level is located between the office portion of the building and the apartments above. In this way, the main runs of all the equipment—heating, air conditioning, electrical, etc.—are directed both upward and downward, in the shortest possible distance. The spatial requirements of offices and the spatial requirements of apartments are quite different; thus the total design layout of each of the two uses differs—floor-to-floor heights are not the same, window treatment is not the same, nor are the heating, cooling, electrical, elevators, and other services alike. The placement of the mechanical level between the offices and the apartments not only facilitates an efficient operation but also provides for a definite visual separation between the two functions.

Quite unusual is the placement of the steam boilers on the thirteenth floor instead of in the conventional location of the basement. Adjunct to the general thirteenth floor center,

Fig. 7.10 *The illustration shown here in Figure 7.10*(a) *is furnished through the courtesy of the Port Authority of New York and New Jersey.* (a) *The World Trade Center Towers in New York City looking south.* (b) *A model of the World Trade Center complex looking west. The twin towers rise to a height of 1350 ft.*

The Port of New York Authority, more recently The Port Authority of New York and New Jersey.
Minoru Yamasaki & Associates, Architects.
Emery Roth & Sons, Architects.
Skilling, Helle, Christiansen & Robertson, Structural Engineers.
Jaros, Baum & Bolles, Mechanical Engineers.
Joseph R. Loring & Associates, Electrical Engineers.

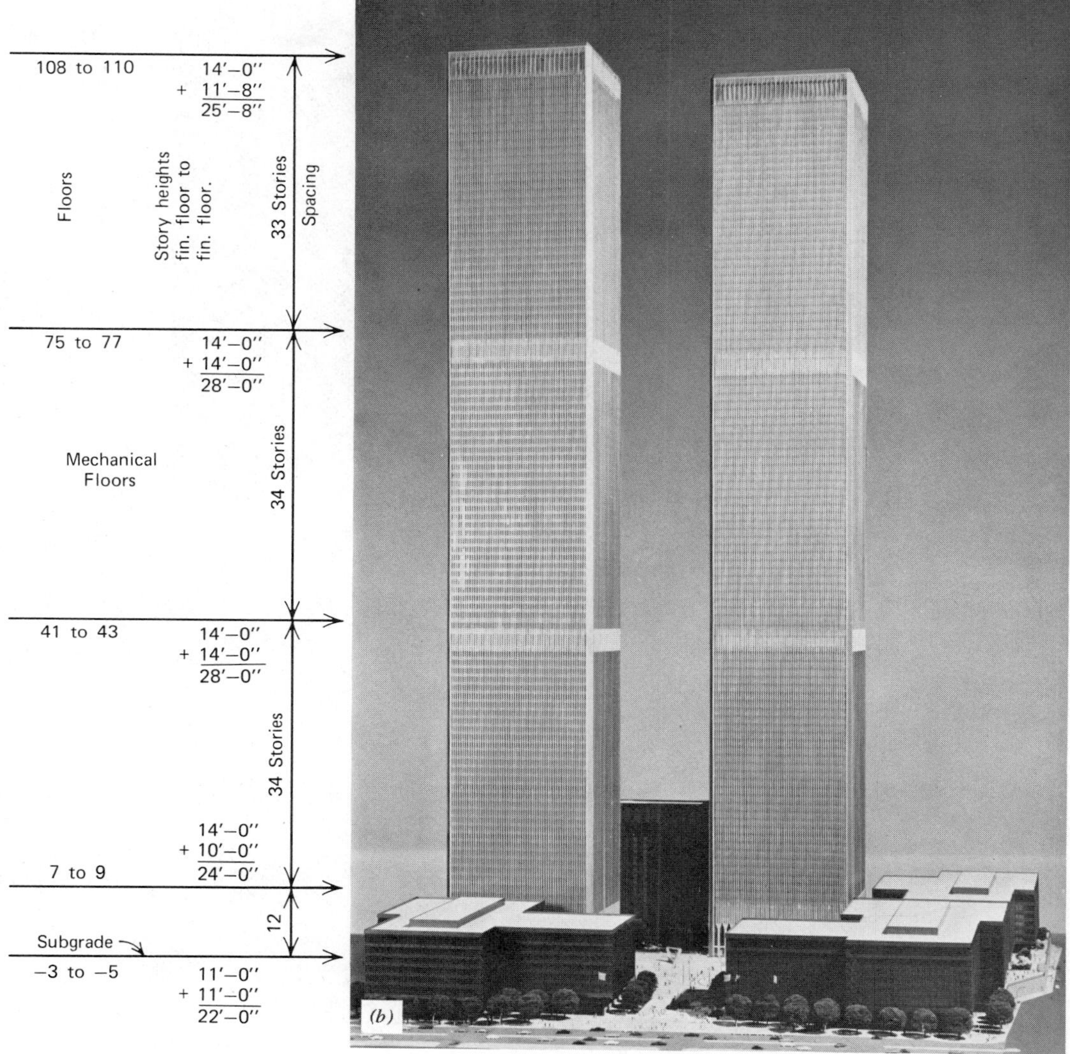

Fig. 7.10 *(Continued)*

Design coordination between the jurisdictional authority, the architects, and the several engineering consultants is most complex in buildings of this magnitude. In each of the twin towers the mechanical floors, of which there are four above grade and one below, are each two stories high (24 to 28 ft, finish floor to finish floor). At the intermediate level of each of these two-story stations, there is not, outside the core, a structural floor. The omission of this affords two-story height for equipment and for horizontal runs of ducts and pipe.

The planned volume of the space required for equipment is a vital consideration in the early design stages. The mechanical stories above grade total 105.7 ft in height. Of the 1350-ft full height of each tower, this comprises 105.7/1350.0 = 0.078 or 7.8%. The subgrade mechanical story, vertical distribution in the core, and other miscellaneous mechanical space increase this percentage considerably.

From intermediate mechanical floors such as the 41st, the distribution of air-conditioning services is both up and down; from the 41st down to the 25th and up to the 57th—16 floors in each direction.

(a)

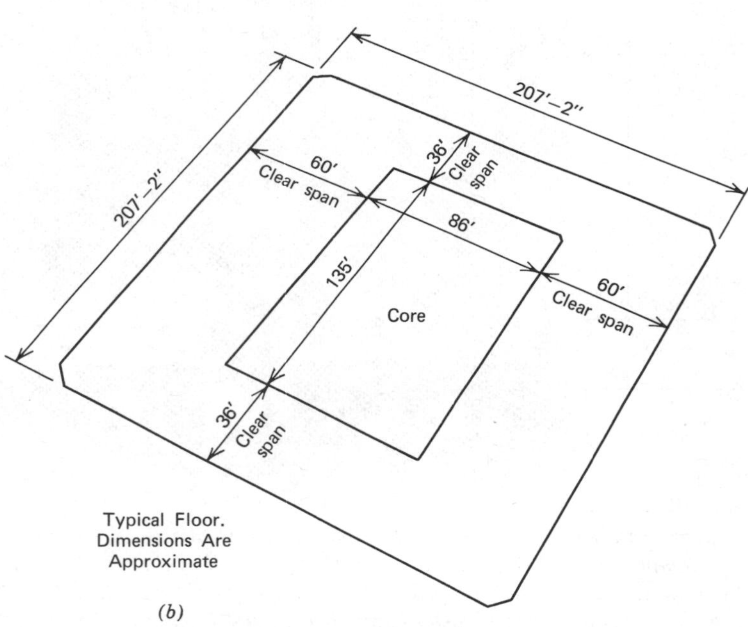

(b)

Fig. 7.11 (a) *Typical floor of the south tower. Photograph taken from an upper level of the north tower. At the corners of the central core are seen the four construction cranes that hoist and place the structural components. In this design, the omission of columns between the core and the perimeter structural wall is unique.* (b) *Plan of typical floor. Deck spans as great as 36 and 60 ft are quite unusual. Great flexibility of later planning in tenant spaces is achieved. The structural floor comprises large, preassembled panels. They consist of custom designed truss-joists carrying steel decking. The joists are in* pairs. *The pairs are spaced 6 ft, 8 in. apart on centers. This spacing affords adequate width for the air ducts that run between them and connect the vertical ducts in the core with the system of air delivery units in the ceiling of the tenant space. Photos, Figs. 7.11 to 7.14, by William McGuinness.*

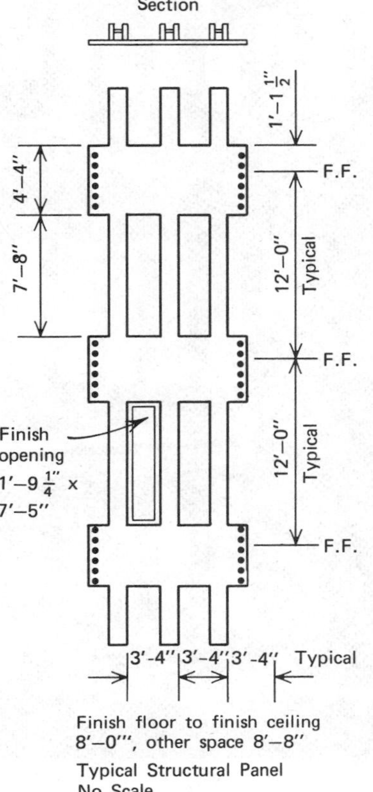

Fig. 7.12 *Mill-fabricated steel panels characterize the perimeter. The staggered position of the multistory units, the verticals on 3 ft, 4 in. centers, and the broad continuous bands attest to the adequate design of the enclosing structure to accommodate the stresses in this unusual building. There are no inert "curtain walls" or spaces for ducts within this all-structure exterior wall. Angles, seen at the left wall, support and tie in the composite floor deck. Just above the partially completed floor one sees at the left a low horizontal sill-high strip. Against this strip will be placed a 1 ft, 3 in. high × 1 ft, 6 in. wide air-conditioning enclosure. This will house induction units that will supply cool or warm air around the perimeter as required to offset outdoor conditions. Four peripheral-air supply ducts, one for each facade (11- or 12-in. round), pass between truss-joists from core risers to perimeter distribution system.*

only a small amount of auxiliary equipment is located on the roof and some in a small portion of the garage. Residential areas have hot water heating, offices have dual-duct, high-velocity heating/cooling, and commercial tenants are supplied with hot and chilled water for their individual climate design requirements.

7.7 Draft Barriers at Perimeter

It has long been generally acknowledged that large exterior glass areas cool the interior air in winter. The vertical layer of cool, denser air thus created then drops to the floor and blankets it like a chilly carpet. Unless one has witnessed tests using smoke and recording thermometers, the speed and the attending discomfort of this phenomenon is seldom fully comprehended.

Fig. 7.13 (opposite) *The open nature of the truss-joist construction, somewhat unusual in tall buildings, permits horizontal ducts emanating from the vertical risers in the core to reach the perimeter. The space between joists and the triangular openings within the joist make this possible. A similar arrangement also within the joist depth serves a distribution duct system in the* tenant *areas. There, rectangular and flexible round ducts form a grid that delivers air down through lighting fixtures in the ceiling. Return air is taken back through plenums and the core return ducts to mechanical floors.*

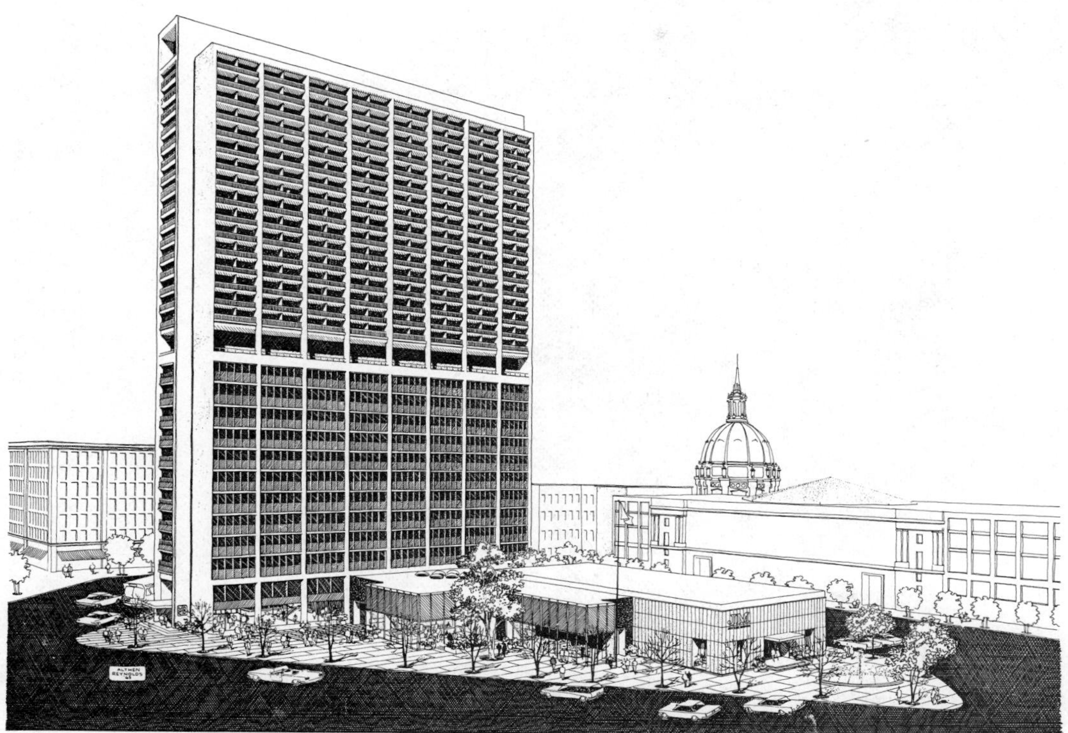

Fig. 7.15 *The Fox Plaza Building, San Francisco, California, Victor Gruen Associates, Inc. Architects and Engineers. Top mechanical story encloses cooling towers and shelters domestic hot water generator-storage units for apartments. Upper stories consist of 16 floors of apartments. Intermediate (thirteenth floor), mechanical story is the location of water chillers, pumps for circulation of chilled water, and of cooling tower water. Hot and cold high-velocity air ducts for downfeed to lower-story conditioning originate there. It is also the location of steam boilers and convertors (steam to hot water), for fan-coil units in residential stories above and similar convertors for hot water for coils in lower-story office air units. Adjacent, also, on the thirteenth floor, is the domestic hot water generator-storage receiver for the offices below. Below the thirteenth floor climate-making center are 10 stories of offices and 2 stories of commercial area. Courtesy of* Progressive Architecture.

Fig. 7.14 (opposite) *Mechanical floors are usually seen as accent bands in modern buildings. The World Trade Center Towers are no exception. Just above the two large working platforms at the extreme left, notice that, at this mechanical floor, the columns become wider because of the greater story height (28 ft). The height of the unglazed openings between these columns is 17 ft in contrast to the 7 ft, 5 in. of typical stories (see Fig. 7.12). Behind this exterior wall of the mechanical story is a 4-ft, open-air perimeter walk. On its inner side is a wall of air louvers, which constitutes a "breathing wall." It permits the intake and exhaust of air as required by the operation of air-handling units.*

The nine bands of flexible fabric wrapped around the left tower enclose the operation of fireproofing the interior steel structure. Columns and spandrels will be covered by aluminum-clad insulation units.

Fig. 7.16 *The Fox Plaza Building in construction. Clearly seen are the air-handling units on the thirteenth floor. Also visible are downfeed ducts that supply high-velocity hot and cold air from these units to the office stories below. Other ducts return the air at normal velocities to the air-handling units for reconditioning. Upper residential stories in this equable climate of San Francisco are supplied with heating but not cooling. Office cooling deals with people-concentration and lights. Photo by Morley Baer.*

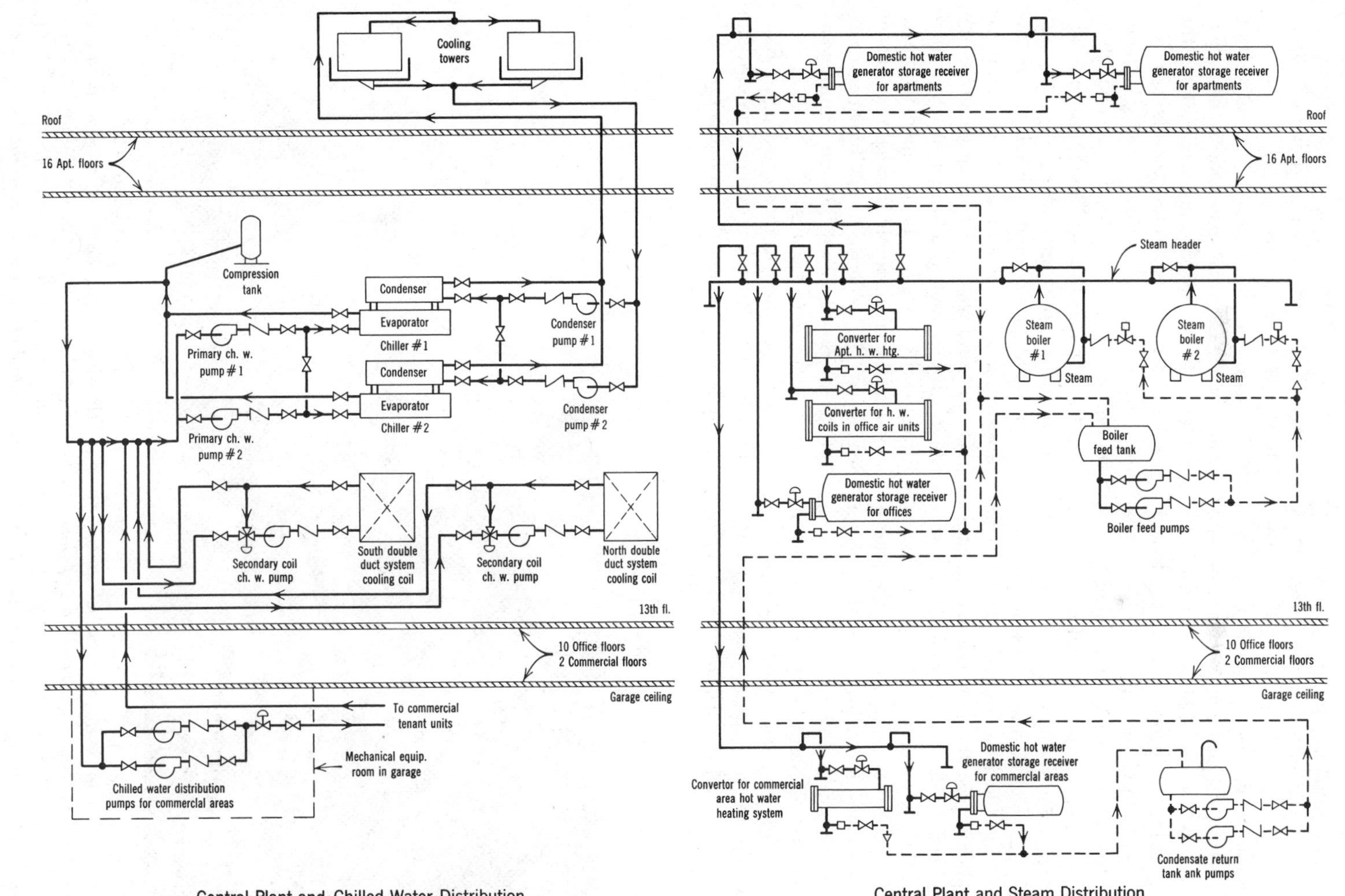

Fig. 7.17 *Mechanical systems for heating, cooling, and domestic hot water for the Fox Plaza Building. Design and these schematic diagrams are by mechanical engineer Howard Wolfberg.*

Under the worst conditions, it is necessary for this cold layer to cross the space and reach a thermostat on an interior wall before a below-the-glass heating element takes over. When it does, its problem is a dual one—to reverse the down-slip of air *and* to warm the space. It is seldom properly adapted to correctly meet both challenges.

Admittedly, the use of a modulated heating medium—water or air—provides a more continuous operation to maintain a warm upflow of air at glass, yet during even brief intervals between these heating periods, the cold air slides quickly past the nonoperative heaters.

A successful solution, and one of the first to provide glass with its own insurance against drafts, has been designed by I. A. Naman & Associates, Consulting Engineers of Houston, Texas. See Figure 7.18.

In this first of Naman's several similar applications, the electric draft barriers have been used in the new home office building of the Great Southern Life Insurance Company in Houston. Architects were Skidmore, Owings & Merrill, associated with Wilson, Morris, Crain & Anderson.

This supplementary system goes into action instantly when cool air descends to meet thermostats placed at 25-ft intervals along the electric heating strips below the glass. It permits no cool air to pass this barrier. The heaters provide 50- to 75-w/lin ft, just enough to stop the downflow of air. Temperature in the space is separately regulated.

The building is dramatically adapted to this new feature. North and south walls are of glass; east and west walls are of concrete (see photo). In concept, the former unsteady performance of glass has been tamed to correspond to the more stable thermal qualities of masonry. The electric draft barriers draw 100 kw of power. Two 500-ton compressors serve the air conditioning.

Should it be thought that a heating problem is unusual in a "warm" Southern state, it can be said that the winter design temperature for

(*a*)

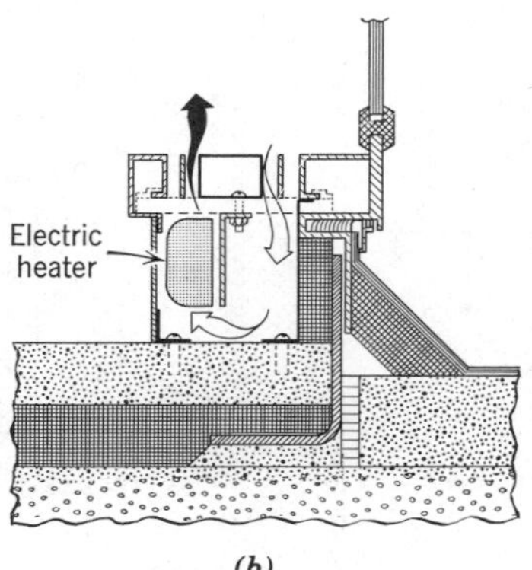

(*b*)

Fig. 7.18 (a) *Great Southern Life Insurance Company, Houston, Texas. East and west walls are of concrete. North and south (end) walls are of glass and are equipped at the sill with electric draft barriers to reverse the down-slip of cold air.* (b) *Electric draft barrier section. Power, 50 to 70 w/lin ft. Winter design temperature of 20° F indicates a need for this heat supplied below glass. Note glass in shadow on south end, minimizing* hot *interior areas in summer adjacent to fenestration. Architects, Skidmore, Owings & Merrill, associated with Wilson, Morris, Crain & Anderson. Engineer, I. A. Naman & Associates.*

Houston is commonly taken as plus 20° F. The region has 1276 degree-days.

In this custom design, the architects wish to acknowledge the collaboration of the manufacturer, the Edwin J. Wiegand Company of Pittsburgh and of the Houston Lighting & Power Company.

7.8 District Services. Airport Planning

(*a*) ***High-Temperature Water and Chilled Water for Airports.*** Although long-distance steam distribution has been used for almost a century, the development of the *high-temperature water* (HTW) principle is measured in several decades. Offering many advantages (though steam is still frequently chosen for city distribution), circulated high-temperature, high-pressure hot water in closed systems has found great favor in Air Force bases and airports, and for groups of buildings such as hospital complexes and those of college campuses. Water will not flash into steam if kept at sufficiently high pressure. It may then be circulated by pumps through supply and return mains and through branches to heat exchangers, which operate conventional low-pressure hot water systems, generate steam, and perform numerous other thermal tasks. Pressures are of the order of 400 psig (pounds per square inch, gauge) and temperatures are about 300° F. During its circuit the water will sometimes lose about 150° F and 60 psig in pressure. A section shown in Figure 7.19 illustrates a common arrangement.

High-temperature water has a number of advantages over steam, which favor it for certain installations. It is a two-pipe system and the temperature drop in the *supply* main is often as little as 10° F. By the use of reasonably high water velocities, mains can be reduced to almost one-half the size of those required for steam distribution. Simplicity results from the lack of need for steam traps and pressure-reducing valves. The pipes need not pitch to low points as in the case of steam but may follow the contours of the ground. Although installation costs are greater, operational costs are less than for steam. Feed water treatment is negligible and corrosion is at a minimum. Underground problems of expansion and insulation are the same as in other subterranean systems. Large sweep-type loops accommodate expansion between fixed points, and underground piping is imbedded in special thermally efficient insulative fill.

(*b*) ***John F. Kennedy International Airport.*** In planning the heating and cooling for complete air conditioning in the principal buildings of the Kennedy International Airport, the original concept was to have the facilities located in utility space in each of the buildings. However, the advantages of centralizing the basic equipment for both heating and cooling soon became apparent. Under the guidance of Mr. Charles Broder, Mechanical Engineer of the Port of New York Authority,[1] planning moved in this direction. A central

[1]The name has recently been changed to The Port Authority of New York and New Jersey.

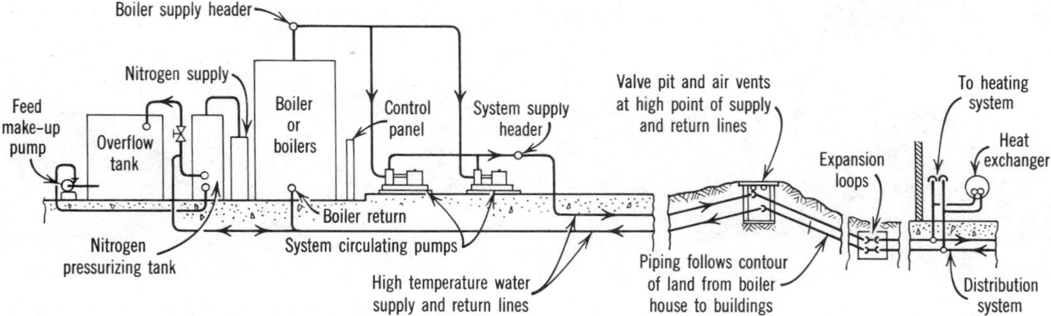

Fig. 7.19 Typical arrangement of a high-temperature-water system. Reprinted from High-Temperature Water Systems, *Industrial Press. By courtesy of author Owen S. Lieberg, Consulting Engineer,.*

heating-cooling plant was designed by Architects Skidmore, Owings, and Merrill and engineered by Seelye, Stevenson, Value, and Knecht. In the selected scheme (see Figs. 7.20 and 7.21), all facilities except air handling and ducts were assembled in the central plant. The elimination of a multitude of stacks, boilers, fuel storage, water chillers, and cooling towers at each building was considered to be a great step forward in the release of valuable space and for cleanliness, control, and architectural freedom. Moreover, it proved to be economically advantageous.

Four La Mont-type International Boiler Works Company high-temperature water boilers supply the heating needs and also serve the Carrier absorption-refrigeration machines. These were specially adapted to use this high-temperature water instead of steam. Thus the boilers are active throughout the year, burning gas or oil. The latter fuel is stored in the 210,000-gal spheroid tank adjacent to the plant. Boilers, chillers, and cooling towers are all in close proximity.

Statistics for the initial stage of the plant are as follows:

(*a*) *Hot water.* 160,000,000 Btuh, 1,140,000 lb/hr pump capacity, 160-ft head, 380° F flow, 240° F return.
(*b*) *Chilled water.* 6210 tons, 16,800 gpm pump capacity, 150-ft head, 55 to 45° F cooling range.
(*c*) *Condenser water.* 24,600 gpm pump capacity, cooling tower capacity for rated flow from 102.4 to 85° F with wet bulb at 78° F. See Figures 7.22, 7.23, 7.24, and 7.25.

At the ends of the four zones for water distribution, namely the various buildings of the airport group, chilled water and hot water are used in the coils of air-handling units from which conditioned air is circulated for complete climate control. Automatic electronic control and monitoring equipment was used (see Fig. 7.26 for the panel of the Arrivals Building). Checked periodically, this equipment is all but *completely* automatic in its regulation of the temperature and humidity throughout the building. The initial stage as described here has been subject to expansion as air travel has increased.

Fig. 7.20 *John F. Kennedy International Airport. The central heating and cooling plant is to the left of the central pool. It is identified as an L-shaped building with a spherical oil-storage tank in its foreground. Courtesy of Port Authority of New York and New Jersey.*

Fig. 7.21 *Heating and cooling plant at John F. Kennedy International Airport, New York, New York, initial stage. At left, spherical fuel oil storage tank. Four stacks serve the high-temperature water boilers. Five of the nine absorption refrigeration machines appear through glass at the right. Six cells identify the two cooling tower structures in the rear.* Architects: *Skidmore, Owings, and Merrill.* Engineers: *Seelye, Stevenson, Value, and Knecht. Courtesy of Port Authority of New York and New Jersey.*

7.9 All-Electric Systems

Many buildings are powered exclusively by electricity for their systems of heating, ventilating, and air conditioning. Notable among those buildings, as listed by the Electric Heating Association, are:

John Hancock Building, Chicago, Illinois. Skidmore, Owings and Merrill, Architects/Engineers.

John F. Kennedy Center for the Performing Arts, Washington, D.C. Edward Durell Stone, Architect, Syska & Hennessy, Inc., Mechanical & Electrical Engineers.

The First National Bank of Chicago Building, Chicago, Illinois. C. F. Murphy Associates and the Perkins and Will Partnership, Architects and Engineers.

Georgia-Pacific Building, Portland, Oregon. Skidmore, Owings & Merrill, Architects/Engineers. (See Fig. 7.27.)

To space-condition the Georgia-Pacific Building's 430,000 sq ft of rentable area, the designers selected an all-electric heat recovery

Fig. 7.22 Night view of the refrigeration-absorption machines in the heating-cooling plant at the John F. Kennedy International Airport. Distinctive colors identify the piping connections to the machines. Chilled water, condenser water and, in this case, high-temperature water instead of steam, comprise the three circuits connected to each of the nine machines. Courtesy of Port Authority of New York and New Jersey.

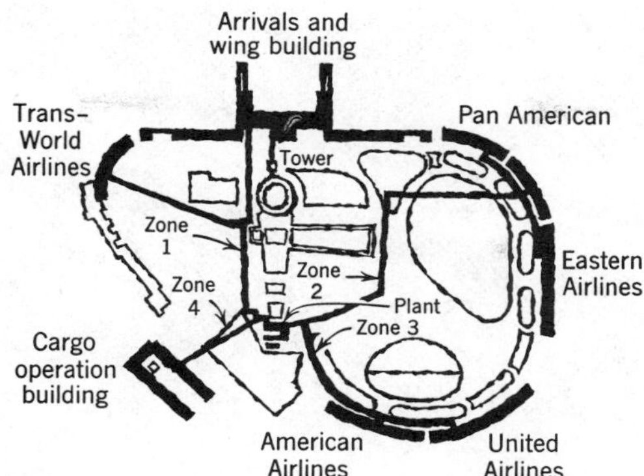

Fig. 7.24 *Control terminal area of the John F. Kennedy International Airport showing location of heating and cooling plant and the four zones of distribution for high-temperature water and chilled water. Reprinted by permission from "Heating and Air Conditioning a Civilian Airport," by Charles Broder, M.E., in* High Temperature Water, *a Symposium Bulletin published by the American Society of Heating, Refrigerating and Air-Conditioning Engineers, Inc.*

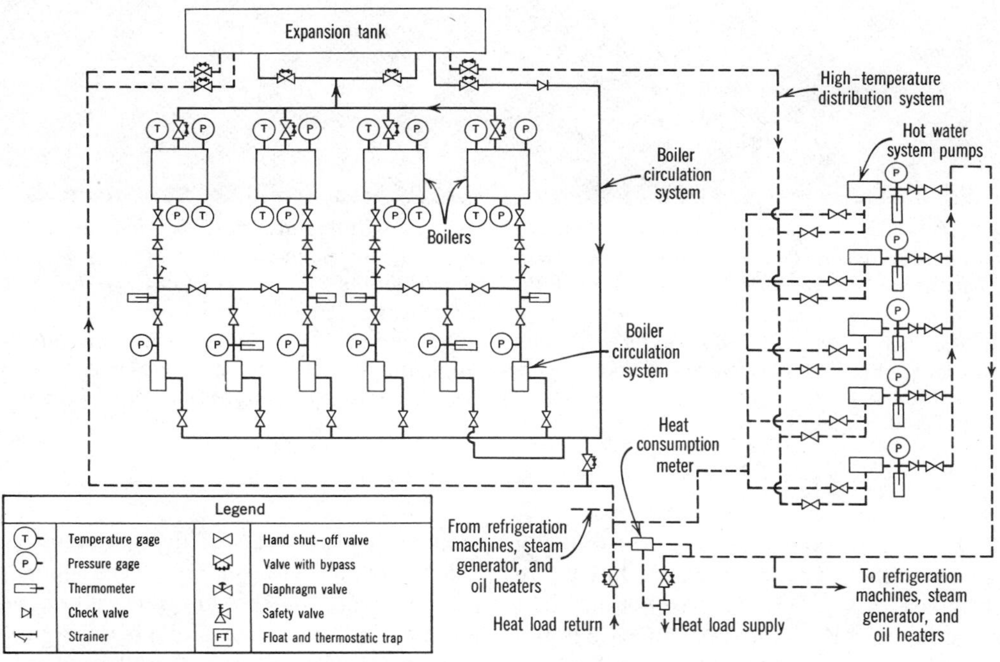

Fig. 7.25 *Piping diagram for high-temperature water system showing circulation through boilers and expansion tank and connection to heating load and refrigeration load. John F. Kennedy International Airport, New York. From "Space Heating at Idlewild" by Charles Broder, M.E., in* Industry Power.

Fig. 7.23 (opposite) *Ground-level view of the central heating-cooling plant. Boiler stacks to the left, cooling towers to the right; water vapor above the cooling towers indicates evident operation of the absorption chillers. The heat rejected above the towers is a measure of the heat removed from the buildings served. Photo by William McGuinness.*

Fig. 7.26 Automatic panels control operation of the Arrivals Building at New York's John F. Kennedy International Airport. Panels at right: central circular dials indicate temperatures of various rooms and of air leaving the local conditioners. Reset of the latter is possible at the dial. Square recorders at top register permanently these temperatures and those of domestic hot water. Panels at left: Motors are started by switches. Their operation is indicated by lights. Panels at left also control all lighting inside and outside of the building. A trouble alarm calls attention to possible difficulties. Controls by Johnson Service Company and IBM's T.E. Division later acquired by Simplex Time Recorder Company. Courtesy of Progressive Architecture..

system, centrally controlled from the building engineer's office. Two centrifugal chillers are in the base of the building—an 806-ton unit for cooling and heat recovery and a 930-ton unit for cooling only. In winter the heat recovery machine transfers enough interior warmth from lights, people, and business machines to heat perimeter areas down to 40° F outside and to temper ventilation air. Below 40° F outside, the building's heat recovery capacity is supplemented by electric hot water boilers.

High-velocity, air-handling equipment is located in a twenty-fifth-floor mechanical room. Induction units at the perimeter provide individual control for offices along the window wall. In the interior, reheat units supply conditioned air through the light fixtures to six separately controlled zones on each floor. All circulated air passes through high-efficiency filter banks and is humidified by steam from electric boilers.

7.10 Total Energy

(*a*) ***Electrical Power Generation at the Site.*** It has been found that, *where conditions*

Fig. 7.27 Georgia-Pacific Building, Portland, Oregon. Skidmore, Owings & Merrill, Architects/Engineers.

are favorable, electricity for power and light can be generated economically by a system that also supplies the building with heating in winter and cooling in summer. It utilizes a fuel such as gas or oil, and is independent of the local electric utility company. Experience has indicated that installation costs are greater than for the more conventional systems that use separate services of electricity and fuel, but the annual operating cost is low enough to pay for the excess installation cost in only a few years. Operational savings continue thereafter.

The success of this method, which has been developed largely in the decade of the 1960s, is evident by its use in hundreds of commercial and industrial buildings and in many schools. It is interesting that total energy should be making significant progress concurrently with all-electric space conditioning (see Section 7.9) and district heating and cooling (see Section 7.8). This strongly suggests the need for expert engineering analysis of conditions before selecting the most appropriate method for the thermal and power services in a building. The conditions include type and size of building, its geographic location, occupancy characteristics, relative costs of oil, gas, and electricity, and whether or not cooling as well as heating is essential.

(*b*) ***On-Site Power Generation Is Not New.*** Prior to 1900 and in the early years of this century, nearly all large buildings and groups of buildings were supplied with direct current generated on or near the premises. The motive power was usually in the form of steam-driven reciprocating engines with belt connections to direct current generators. There were very natural reasons for this local operation. Direct current cannot be transformed to different voltages and must be generated and distributed at the voltage used in the building. At these relatively low voltages, power loss in the distribution system is great and distance adds immeasurably to the loss. With the development and use of alternating current machines, utility companies were able to establish central power stations from which electricity could be transmitted great distances to the user at high voltage. There it was transformed down to domestic voltages for use. Since, at high voltage, power losses are very low, this system became universal. During the 1920s and 1930s, owners were removing their private power generators and accepting utility service with savings in operating expense.

(*c*) ***How Total Energy Developed.*** There are valid reasons for the return of power generation to the site and they relate to the technical advances of recent times. In earlier installations when steam, produced from coal-burning boilers, was used, there was little or no energy salvage from the steam, which was mostly wasted. The newer fuels as used directly in reciprocating engines or turbines have residual heat value that can be recovered for purposes of heating or cooling. The techniques for this recovery have been perfected by experience with nuclear power, rocketry, aviation, and other space-age developments. For total energy to be successful, there has to be a reasonably steady demand in the building for the power generated and also for the heat recovered. It is easily understood that many buildings have this need. Lighting and the demand for power by computers, electrical business machines, electrical cooking, and the great multitude of electrically powered devices often make the call for power nearly constant throughout the year. Similarly the exhaust heat recovery from the engines or turbines that power the generators is in demand for either heating or cooling at any time of the year. Cooling, formerly something of a luxury, is now frequently essential. It also fills the gap of energy demand that formerly existed in summer when heating was not required.

(*d*) ***How It Works.*** An extremely clear diagrammatic exposition of the two principal systems for total energy is found in Figure 7.28, reprinted here through the courtesy of Educational Facilities Laboratories whose study into the feasibility and use of total energy *for schools* is described in Section 7.10(e).

In both systems, one using a turbine and the other a reciprocating engine to operate the

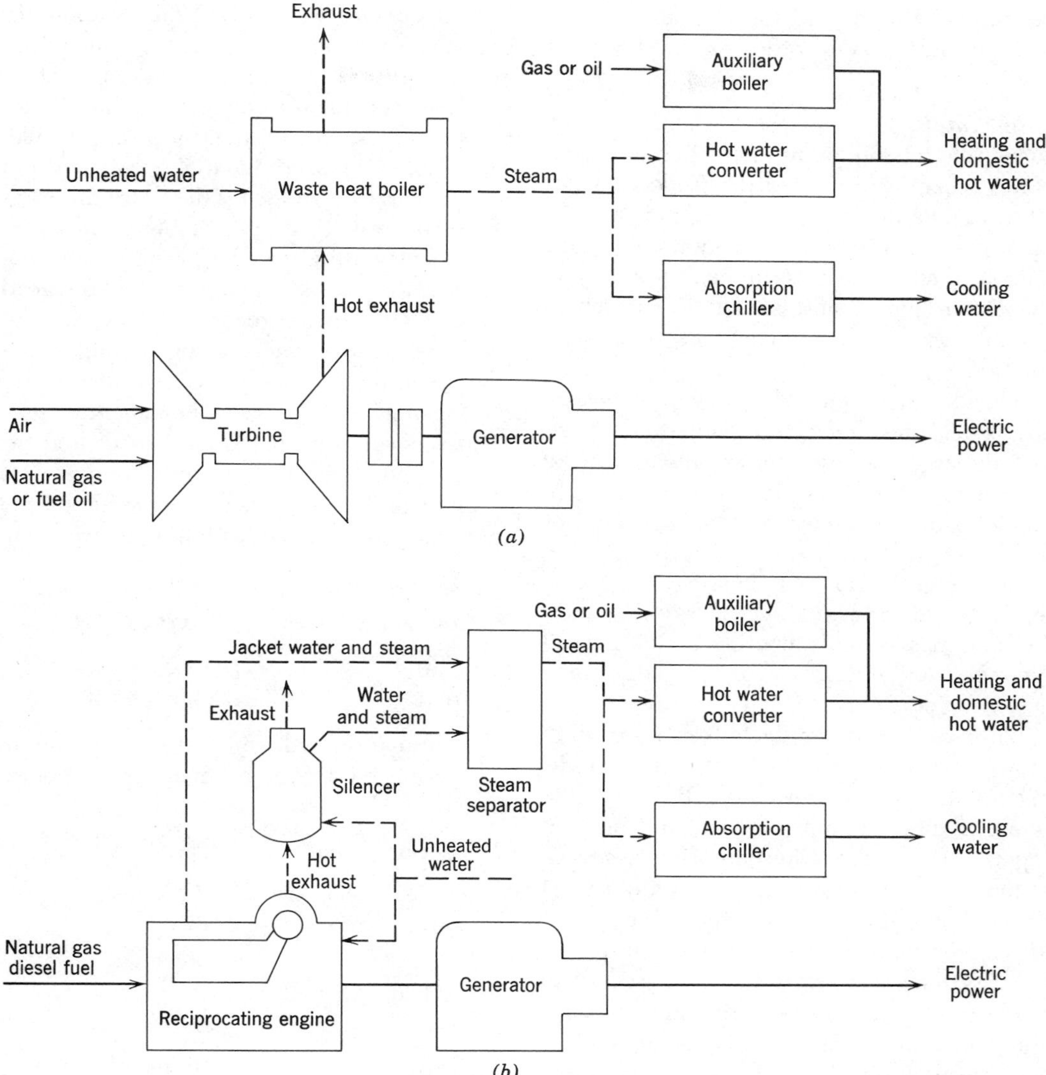

Fig. 7.28 Total energy systems. (a) *Using turbine.* (b) *Using reciprocating engine. Reprinted by permission from* Total Energy, *Educational Facilities Laboratories.*

generator that supplies electric power, heat is reclaimed to produce steam or hot water. The steam or hot water is then used for heating or, by use of an absorption chiller, for cooling. When a turbine is used, the fuels are natural gas or fuel oil. The heat is recovered by passing the hot turbine exhaust through a waste heat boiler to produce steam. Fuel for a reciprocating engine is natural gas or diesel fuel. Both the jacket cooling water and the hot engine exhaust are passed through heat exchangers that utilize the heat to produce steam or hot water for heating or cooling as already described. In both systems, an auxiliary boiler, fired directly by gas or oil, stands ready for operation to help maintain a balance in the system.

The unique feature of total energy is the use of the recovered heat. This is essentially free energy since it would otherwise have to be purchased and paid for separately in the form of electricity or other fuels.

(e) ***Total Energy for Schools.*** Many changes in school buildings and in school programs recently prompted the Education Facilities Laboratories (EFL) to support a study in depth of the feasibility of total energy for schools, methods of determining feasibility in specific instances, and guidelines for school administrators and engineers in developing solutions.

The changing demands, many of which are now established facts, are numerous. They include large energy requirements that make the school of two decades ago seem quite obsolete. In the modern school, lighting has been upgraded and cooling has largely become as high a priority as heating. Electronic teaching aids add to these energy demands. Evening classes, multifunction buildings, and summer programs put many schools into a new category.

The study was conducted as a research project by Fred S. Dubin Associates, Consulting Engineers. Norman Kurtz, P.E., was the research director and prepared the original report. EFL's interest in sponsoring this work stemmed not only from the foregoing school demands but also from the experience of rapidly increasing costs of mechanical and electrical systems in schools and, with regard to total energy, the less objective and inescapably conflicting commercial interests of electric utility companies and the manufacturers of electric generators. The report included a consideration of college buildings as well as schools.

The foregoing conditions and criteria that evaluate schools and college buildings, as possibly suitable for the use of total energy, can also be applied to other building types. These can include commercial and industrial buildings as well as offices and motels.

(f) ***An Example of Suitability in a Motel.*** It has been seen that total energy is appropriate under special conditions that relate strongly to the type of building, its occupancy, and its use. One suitable and successful application was found at the Midtown Holiday Inn at Montgomery, Alabama. Its urban location makes it a center for public gatherings, functions, entertainment, and meetings. Unlike some motels where residential occupancy and the services to care for residents are the principal concern, this inn combines these requirements with a very large and diverse occupancy by visitors. The result is that the need for electric power generation and its by-products, heating and cooling, is quite constant. The motel is served for its electric requirements by two gas engine-generator sets. Heat recovered from the engines is used for heating, air conditioning, and domestic hot water. The several areas at the motel and the space they occupy are as follows:

174 guest rooms	65,000	(see Fig. 7.29)
Commercial area	13,000	
Rooftop supper club	6,000	
Total	84,000	sq ft

The population with which the air-conditioning system must cope on many special occasions is:

Meeting room	425
Dining room	94
Coffee shop	82
Bar	60
Small meeting rooms, 4 @ 20	80
	761 people
Food service	2000 meals per day

The impressive problem of keeping these visitors cool or warm, and supplied with food, hot water, and proper illumination is handled by one fuel, gas. It is used in two 450-hp, engine-generator sets with 300-kw generators operating at 480 v (see Fig. 7.30). Recovered heat from the engines is used to produce steam. Two steam heat exchangers are used to generate hot water for heating and for domestic needs. An additional heat exchanger heats the swimming pool when necessary. A gas-fired boiler stands by as an auxiliary to supply additional steam for peak loads. Cooling is

Fig. 7.29 Midtown Holiday Inn, Montgomery, Alabama. Equipment room for total energy installation is at the left of the swimming pool.

provided by an absorption machine with a gas-engine-driven compressor as a standby. The principal equipment consists of:

Two Caterpillar G-379 engine-generator sets, 450 hp, 480 v (see Figs. 7.30, 7.31, and 7.32).

Auxiliary gas-fired steam boiler, 5,500,000 Btuh.

Three heat exchangers for heating, hot water, and swimming-pool heating.

Absorption machine for cooling, York EK-13, 125 tons.

Standby for cooling, gas-engine-driven compressor, 100 tons.

Cooling tower on roof, 350 tons capacity.

Air-handling units, large meeting room, 3 at 4000 cfm each (15% fresh air), dining room, 4000 cfm plus exhaust fan. Kitchen, 3600 cfm exhaust fan.

Although not related to the use of total energy, the air-handling figures are an indication of the problem of dealing with large and instant crowd gatherings. Moving cooled and dehumidified air into, and the vitiated air out of, these spaces is a task not only involving the need to provide comfort conditions but also requiring control of air speed and sound.

Resident guests control their own environment by means of three-pipe water systems serving fan-coil units in each room. Both hot and chilled water are available when needed and the third pipe carries away the return water. Fan-coil unit blowers in guest rooms operate at 200 or 400 cfm depending on room size. In both the guest rooms and the public

Fig. 7.30 Interior view of equipment room of the natural gas total energy system at Midtown Holiday Inn.

spaces, heating and cooling are available simultaneously when outside temperatures are between 45 and 65° F, at which time both chilled and hot water are circulated. When outside temperature is over 65° F, only chilled water is circulated. When outside temperature is below 45° F, only hot water is circulated.

The mechanical room which houses the principal components of the system is a separate and isolated building seen at the left in Figure 7.29, and in full view in Figure 7.34. The advice of acoustical consultants has reduced the escaping sound to a whisper.

Either of the generators at the Midtown Holiday Inn can carry the entire load of the building, permitting alternate operation of the generators and the assurance of one as a standby in case of mechanical difficulty. Concerning the dependability of gas service, the inn experienced only one 20-minute outage in the first year of its operation.

The net additional installation cost for the total energy system over the originally designed conventional system was $102,400. The net annual savings in operation by the use of total energy instead of a conventional system was $13,162 for a test period of one year.

(g) ***System Components.*** The differences in the equipment needed for a total energy system and a conventional one are generally limited to the equipment room. No change is needed in the piping for chilled and hot water, the fan-coil units, the absorption machine, or the air-handling units. The manufacturer's engine-generator set that was used at the Midtown Holiday Inn is pictured in Figure 7.31 together with its heat recovery equipment and control panel. Figure 7.33 shows in schematic

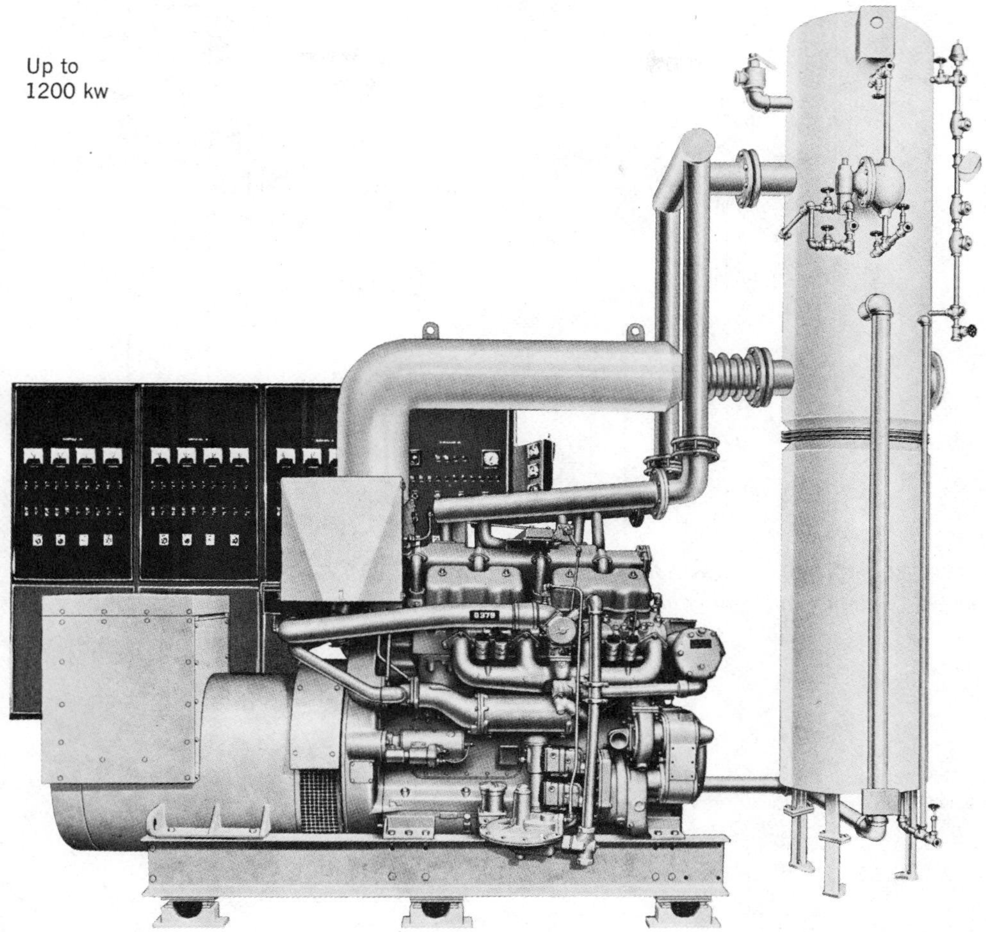

Fig. 7.31 *Manufacturer's illustration of the G 379 engine-generator sets, one of which is seen in the foreground (right) of Figure 7.30. In the above illustration are seen: control panel in background, generator at left, engine in center, and heat recovery unit at right. Note vibration-isolator mountings below dunnage beams.*

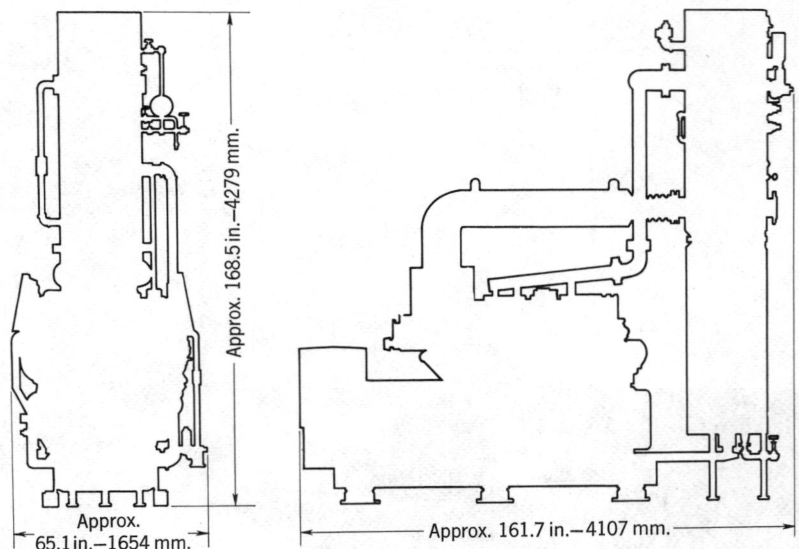

Fig. 7.32 Manufacturer's dimensional silhouette of the G 379 engine-generator-heat-recovery assembly for use in planning layout of equipment room.

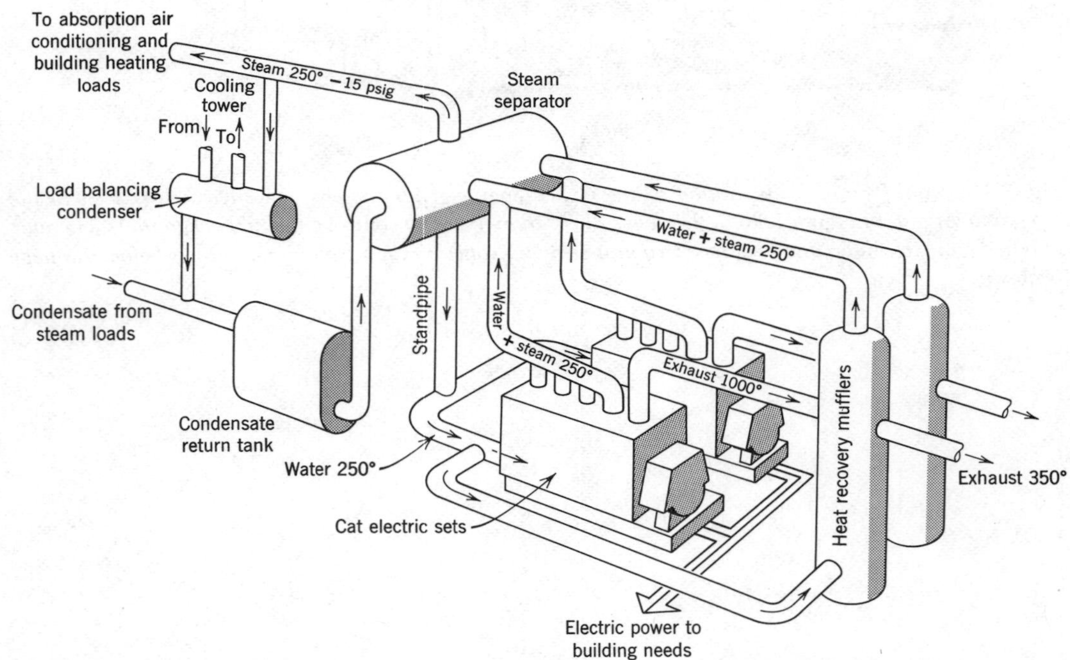

Fig. 7.33 Flow diagram for a typical system recovering both jacket water heat and exhaust heat.

Fig. 7.34 Exterior view of equipment room, housing entire total energy plant. Visible elements are louver for ventilation and combustion air, exhausts, and cooling tower on top.

form the piping and a few additional devices commonly used. They comprise a steam separator, load balancing condenser, and condensate return tank. After the steam has passed through the heat exchangers to produce hot water or through the absorption chiller for chilled water, the rest of the distribution system is as it would normally be in any design suitable for the specific building.

Although the terms just mentioned are typical, it must be understood that there are many reasonable or necessary variations within the design of an efficient total energy system.

References and Bibliography

1. "Engineering for Architecture 1977"
"Engineering for Architecture 1978"
These titles refer, respectively, to the fourth and fifth annual mid-August issues on "Engineering for Architecture" published in *Architectural Record,* a McGraw-Hill publication. These are excellent current studies of the interrelationship of architectural design, structure, and buildings services.

2. *Progressive Architecture.* Reinhold Publishing Company, Inc., a subsidiary of Penton/IPC. Periodic articles on subjects relating to the coordination of architectural and engineering design in the planning and construction of buildings.

8

Solar Energy and Energy Conservation

8.1 Energy and the Future

By this time, there is hardly a person who has not been made acutely aware of the energy situation on our planet. A few still expect it to go away but for the most part, clear-thinking people are prepared to deal with the situation and expect changes which, as time advances, may prove to be more extensive than we now anticipate.

Those of us who will be involved with mechanical and electrical equipment in buildings face a special challenge. It is now our mandate to use *less* equipment. For the equipment that is still basically essential, there must be greater efficiency.

Buildings, of course, have not been the only cause of wasted energy. Transportation, though already more efficient, bears part of the burden. Even further than the consideration of buildings and vehicles, essentially we face the prospect of a changed life-style. Personal effort, manual involvement, and the sacrifice of some of our accustomed creature comforts may lie ahead.

Some have wondered whether in the next few centuries (if not decades) we can maintain the technologically extravagant industrial complex that has served us so well in the times of plentiful energy. *Some* kind of decentralization might have to be considered. For instance, to manufacture a specific product mostly in one city, and then use tremendous energy for the transportation and distribution of the product nationally (and internationally), in a pattern of overlapping radial spider webs, could be a subject for study.

We can't all go back to the farm or to a local shoemaker or blacksmith but we can be ready for *change* when we find out what it has to be. Statistics and estimates are a guide (see Fig. 1.24, page 23, and Fig. 1.25, page 23). Gas and oil, discovered in quantity only 100 years ago, are now insufficient for our needs, to the extent that we must dot our continental shelves at sea with oil rigs.

Nature and the sun had laid down the makings (natural growth) of our fossil fuels millions of years ago. The sun still supports life on the earth's surface (see Fig. 8.1). The African violet in that illustration, typical of all vegetation, positions its own collector-plates at the correct angle and maintains a proper ratio between the area of the *solar panels,* its leaves, and the *power demand,* the blossoms. Nature may guide us in this kind of solar collection and in some other problems.

Fig. 8.1 *Solar pickup. Thirty-one collector plates sustain five blossoms.*

8.2 "Sundown House," The Sea Ranch, California

A strong trend in current thinking stresses the utilization of climatic and site conditions as a foremost consideration in designing for year-round comfort.[1]

In designing his Sundown House, David Wright AIA, who practices environmental architecture, has achieved an almost totally passive solar house. Partially underground, it has no active solar panels for space heating and only a small (two panel) unit for the heating of domestic hot water. His house does, indeed, approach 100% passive solar design as will be seen in Figures 8.2, 8.3, 8.4, 8.5, 8.6, 8.7, and 8.8. His own description of the Sundown House design is our best guide to understanding its planning and performance:

A visually sensitive and florally fragile northern California coastal meadow site provided the opportunity to design a new kind of environmentally responsive structure. The solution is a passive solar heated and cooled home that intrudes minimally on the landscape.

Careful analysis of site and climatic characteristics resulted in a sunken, earth-bermed, sod-roofed structure, planted with native grasses and ice plant, that blends with the meadow environment and is barely visible from neighboring houses. The building materials of native wood, glass, and concrete, used in hard edged planes, compliment the moulded earth form of the structure. The low profile and south-oriented plan shape is tailored to screen prevailing winds and weather while taking advantage of solar exposure to provide most of the space heating/cooling and the domestic hot water.

Several unique features were developed with this design. The sloping solar glass collects the sun's heat which stores passively in the dense structural material—thick concrete walls and brick on sand floors—by the greenhouse effect, radiating back into the living area during sunless periods. Heavy insulation on all exterior surfaces, including under the floor and against the surrounding earth berm, isolate the heat storage mass from exterior loss and gain factors.

[1]In Appendix E, page 1315, forms will be found for the evaluation of passively solar-heated buildings.

In summer, as the sun moves to a higher angle, a system of vents and insulating shades exhausts the solar heated air, inducing outside air movement and effectively cooling the interior. Openable skylights provide additional natural illumination to the north area and assist in ventilation and cooling.

The very economical insulative shades, which are used to trap heat in winter as well as for cooling in summer, were designed by the architect. Three layers of aluminized polyester were sewn together forming multiple radiant barriers, and were quilted in 8 inch squares providing further insulative air pockets. They are covered on each side by colored cotton material to make them visually attractive.

The natural convection solar collector panels which provide domestic hot water are integrated into the weather skin, and require no pumps or heat exchanger. The two story tower facing to the west provides views of waves and whales, and frames the sunsets.

The house, with a generous and flexible living area, was purposefully kept small. The plan makes use of the energy conservation aspect of decreasing surface-to-volume ratio in order to minimize heat loss surfaces to the exterior, resulting in more efficient compact living spaces while at the same time creating a design of visually exciting spatial usage.

The house calculates at 94% solar heated, based on 30-year data:

1977–1978	90%
1976–1977	98%

Provides all cooling and ventilation.

8.3 Heavy Masonry Saves Energy

(*a*) ***Mass Masonry is a Passive Solar Material.*** Brick, stone, concrete, masonry blocks, baked clay, adobe (sun-dried brick), and even rammed earth have special thermal qualities.

Dependent upon their density and mass, as well as thickness of building elements (walls, roofs, etc.), they perform a rather special job of storing heat energy for later use and slowing down, often by many hours, the passage of heat from the warm side to the cool side. Density, of course, is measured by the pounds per cubic foot of the material. The rating of the several materials in place, as in a wall, is

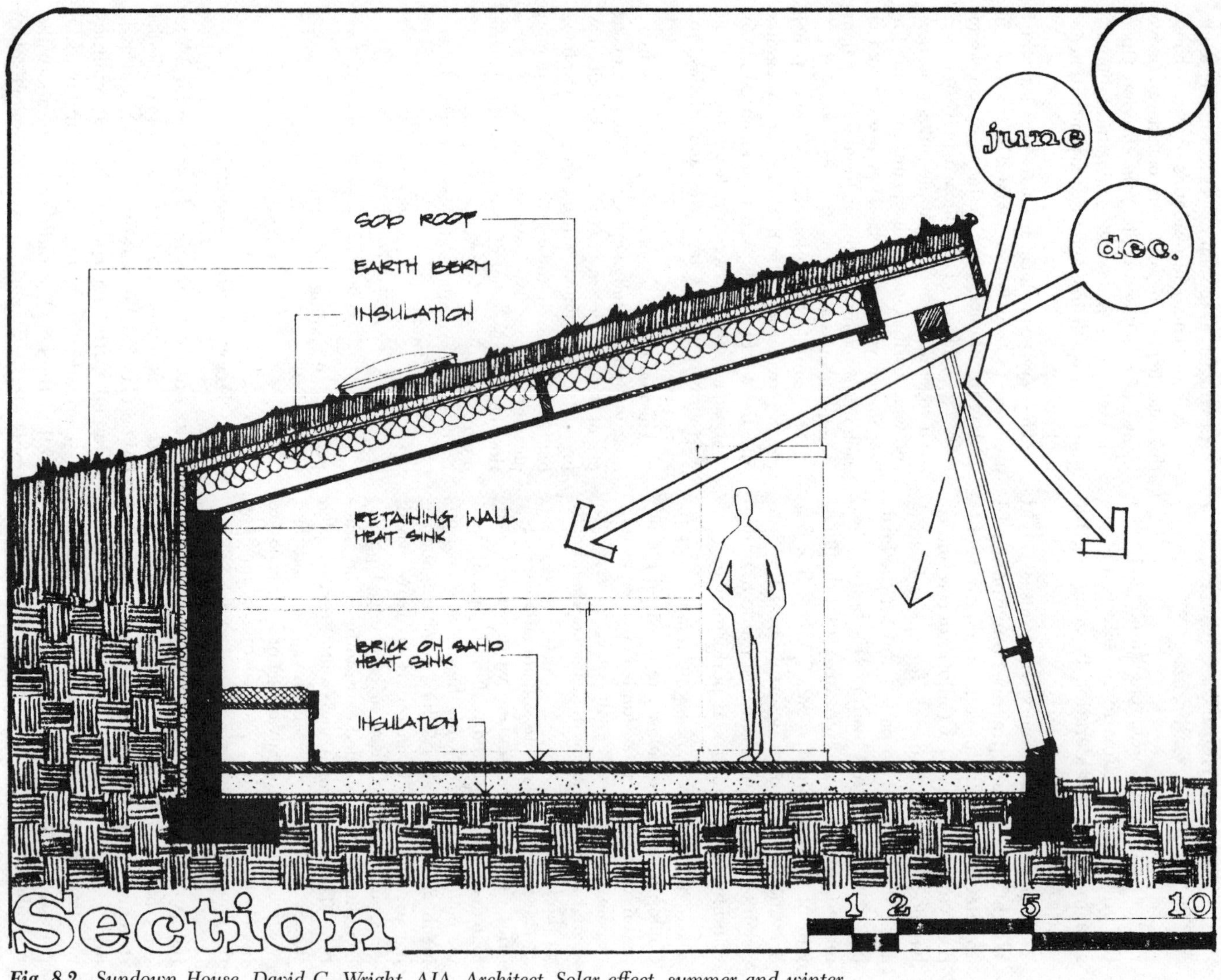

Fig. 8.2 Sundown House, David G. Wright, AIA, Architect. Solar effect, summer and winter.

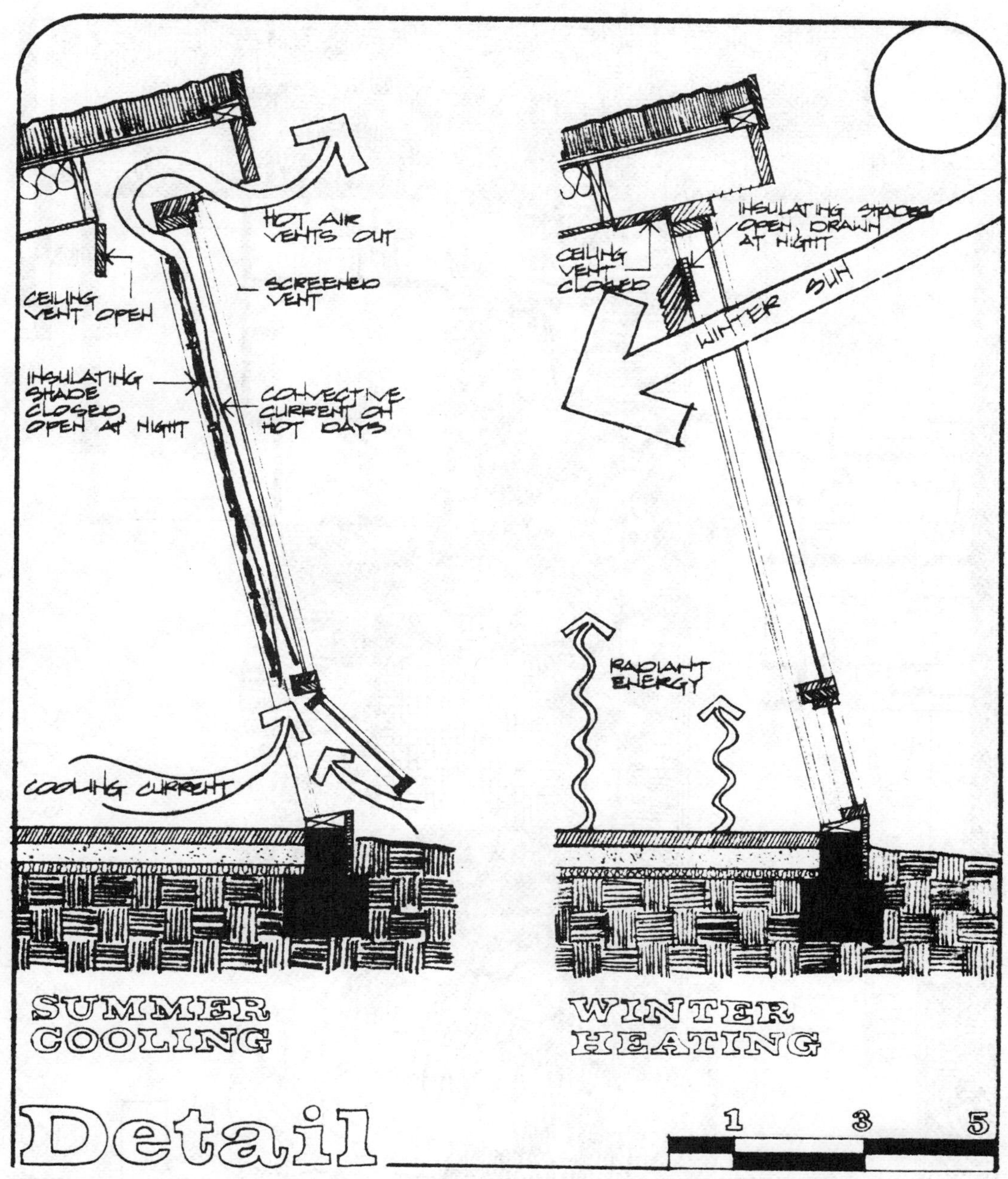

Fig. 8.3 Insulating shades, together with openable vent and low openable sash, control the indoor climate.

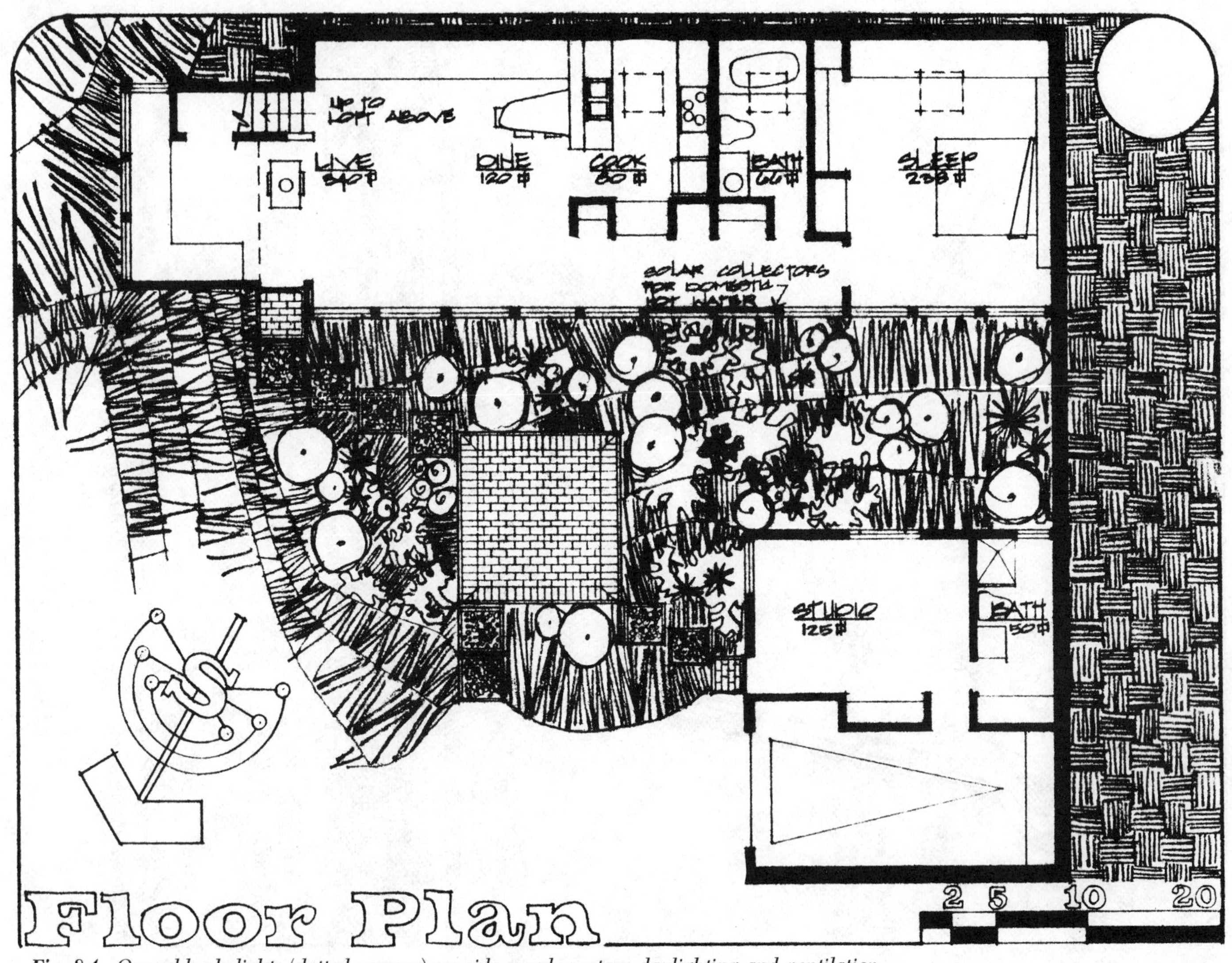

Fig. 8.4 Openable skylights (dotted squares) provide supplementary daylighting and ventilation.

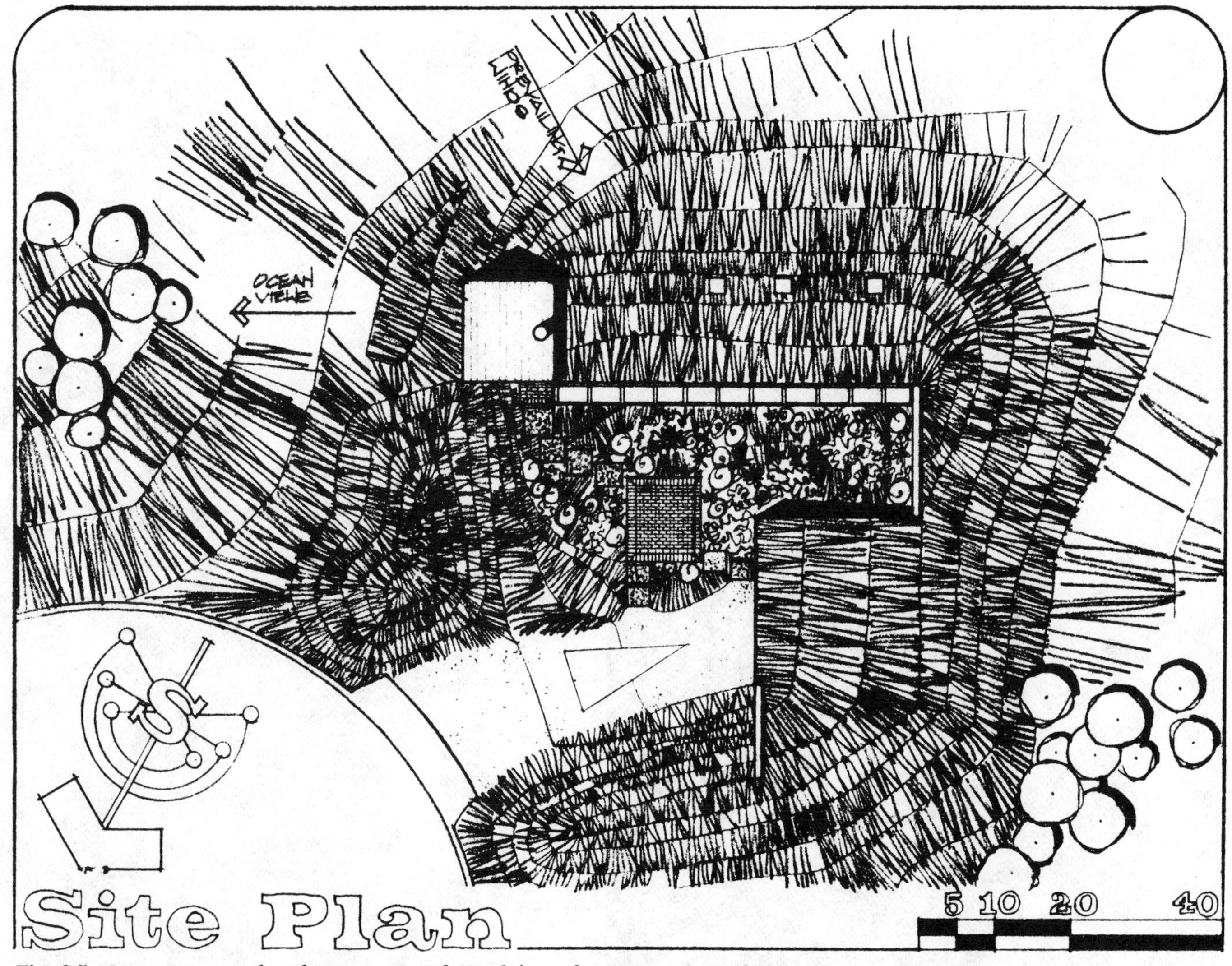

Fig. 8.5 *In environmental architecture, David Wright's* south *arrow with its daily and seasonable paths-of-the-sun replaces the traditional architectural* north *arrow.*

Fig. 8.6 Sundown House, David G. Wright, AIA, Architect. View looking north.

Fig. 8.7 The tower that commands a view of the Pacific.

Fig. 8.8 Openable skylights. They light and ventilate spaces (left to right) for sleeping, bathing, and cooking.

quoted as pounds per square foot (psf). The approximate weight of brick is 120 lb/cu ft and, of concrete, about 150 lb/cu ft. Thus a 1-ft-thick wall of brick would weigh about 120 psf and, of concrete, about 150 psf. As we look with favor on increased thickness of fibrous insulation, we start to consider greater thickness and greater weight (pounds per square foot) of masonry as important in the tasks for which masonry is advantageous.

We have all known for many years that great thicknesses of adobe, often two or more feet thick, have been used in locations such as Mexico, Arizona, California, and New Mexico. The heavy white masonry buildings of the desert areas of the Middle East are also often characteristic of the regions, subject to hot days and cool nights. When the morning sun starts to warm the outside of these massive walls it is not at all unusual to find that the *inside* of the wall does not begin to appreciably transmit the stored heat of the wall until early evening. This delay of 12 hours is not at all uncommon. At that time the outdoor air becomes cooler as the wall continues to gently warm the interior during the cool hours of the night.

Thus during the day, the masonry has eliminated the need for mechanical cooling by performing as such walls have done for centuries, long before refrigerated cooling was developed. When morning comes, the cool mean radiant temperature (MRT) of the cooled-off indoor surface of the wall begins the daytime, indoor cooling process as, at night, it provided a gentle warmth. The time has now come for us to evaluate and quantify these natural and useful phenomena.

Quite ahead of its time as a passive solar installation, the Wallasey St. Georges School in Wallasey, England, is unique. The Brick Institute of America brings us the story of the Wallasey School, including the effectiveness of its masonry, in the illustration and legend of

Fig. 8.9 St. George's School, Wallasey, England. St. George's, built in 1961, provides an example of solar heating on an institutional scale; in this case, for 300 students. It is a long two-story building, its south wall almost entirely glass. Even though the area is frequently cloudy, the masonry envelope of the building can store heat for use for days at a time. The ground floor is a 6-in. concrete slab set over 4 in. of crushed rock, while the intermediate floor is a 9-in. concrete slab. The roof has 7 in. of concrete insulated on top by 5 in. of polystyrene. Partitions between classrooms are made of 9-in. brick, as are the outer walls. Over an entire year, solar energy provided 50% of all heating, and the heat generated by lights and students provided the rest. Reprinted by permission from The Use of Solar Energy Heating Systems in Brick Buildings. *The Brick Institute of America.*

Figure 8.9. The control of sun and air in this school was described in Figure 1.12*a* and *b*, page 10.

(*b*) ***Steady-State Versus Delayed Transmittance.*** Traditionally, the calculations for heat *loss* have been made based on the assumption that steady-state heat flows from a fixed warm interior temperature to a fixed cooler exterior temperature. The *U*-factors and temperature difference (ΔT) are the determinants. The *U*-factors that we are furnished with are evaluated for steady-state flow. Yet in foregoing subsection 8.3(a) we saw that, in the case of heavy masonry, the heat transmittance is often delayed by many hours. This discrepancy has usually resulted in the selection of somewhat oversized heating plants and some inefficiency in operation.

In the matter of heat *gain*, ASHRAE standards emphasize the time delay of heat flowing into a building through various materials including masonry. This is done by adjusting the ΔT by setting *Total Equivalent Temperature Differentials* (TETD). They reflect the nature of heat storage in masonry (see Table 4.26, page 168). Excerpts from that table illustrate a comparison between the TETDs for light and heavy construction for roofs exposed to the sun:

Total Equivalent Temperatures for Heat Gain

	8 A.M.	*4* P.M.	*10* P.M.	
Light construction:				
2-in. wood	1° F	85° F	23° F	TETDs
Heavy construction:				
2-in. insulating board and 6-in. concrete	18° F	41° F	44° F	TETDs

The difference in the thermal transmittance as affected by the TETDs is clear. Aside from the fact that the *U*-coefficients of the light and heavy constructions can differ, the thermal *pressure* as one parameter of actual heat gain is double for the light construction (as compared to heavy masonry construction) at 4 P.M. (85/41) and half (23/44) at 10 P.M. See Example 4.5, page 162.

(*c*) ***The M-Factor of the Masonry Institute Committee.*** The increasing awareness of the influence of masonry on *U*-value correction, especially for heating design, prompted the committee in 1974 to undertake serious research and study. Engineers of the several organizations of the committee took the matter under advisement, and an independent firm of consulting engineers, Hankins and Anderson, was retained to make an objective investigation and report. The results are published by the committee in a nine-page report called *Mass Masonry Energy*. Its subtitle is *The Use of Mass to Save Energy in the Heating and Cooling of Buildings*.

Weather conditions in 10 cities were considered, together with their effect on 10 different walls and the various possible orientations that might occur.

The *M*-factor for the correction of *U*-values is shown by graphs in the report. For details the reader is directed to the publication *Mass Masonry Energy*, in which energy savings are indicated. The report is considered conservative and it is expected that studies will continue. Member organizations of the Committee are:

Brick Institute of America.

International Masonry Institute.

International Union of Bricklayers.

Laborers International Union.

Mason Contractors Association.

National Concrete Masonry Association.

Portland Cement Association.

(*d*) ***Masonry in High-Rise Buildings.*** Concurrent with the new consciousness of the thermal qualities of masonry, its use has become evident in a number of new high-rise buildings. In planning the American Telephone and Telegraph Building in New York City, Architect Philip Johnson will be featuring granite. Architects Collaborative will use precast concrete for almost 50% of the new Fiduciary Trust Building in Boston. We look back at the choice of masonry for large buildings by earlier master architects. Frank Lloyd Wright's Guggenheim Museum and Le Corbusier's Unité d'Habitation in Marseilles are examples.

8.4 The Madeira Science Building, Greenway, Virginia

The features of this design by Arthur Cotton Moore Associates are best described by the following article:[2] See Figures 8.10, 8.11, 8.12, 8.13, and 8.14.

SUNNY SIDE UP

As Wolf Von Eckardt, the architecture critic of *The Washington Post,* put it, "After years of talk about applying solar heating to a large building, Madeira and Arthur Moore just went and did it." The client "did it," as well, by the interest in preservation of natural resources

[2]Reprinted from the February 1976 issue of *Progressive Architecture*. Copyright © 1976, The Reinhold Publishing Company. The article was authored by Jim Murphy.

Photograph by Norman McGrath

Fig. 8.10 *The Madeira School, Greenway, Virginia, Arthur Cotton Moore, Associates, Architects; Flack + Kurtz, Consulting Engineers. Collector roof faces southeast. The pipe manifolds referred to in Figure 8.11 are seen at the ridge.*

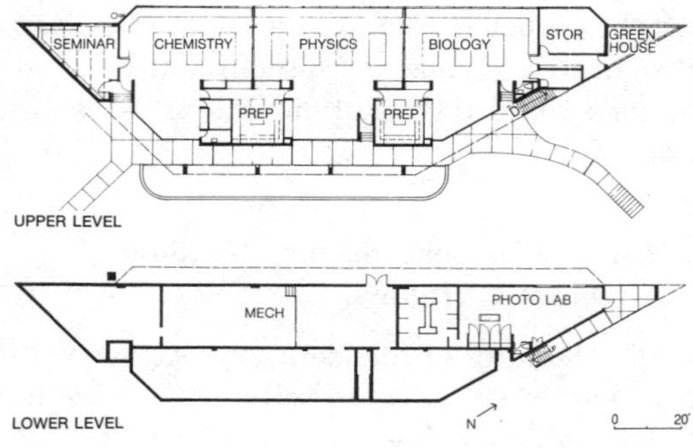

Pipe manifolds supply and withdraw the special oil used to transfer heat from the collector to water storage tank. Appearing along the ridge and under the roof soffits, they form conscious but tasteful design elements.
Arthur Cotton Moore Associates
Washington, D.C.

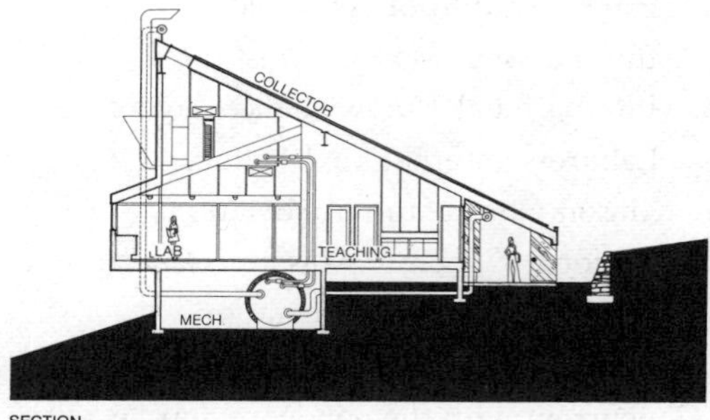

Fig. 8.11 The Madeira School, plans and section.

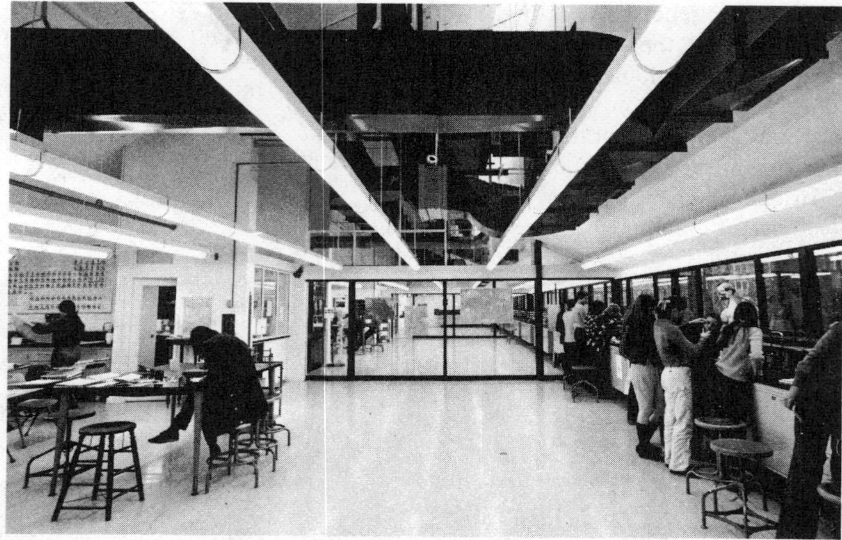

Photo: Arthur Cotton Moore Associates

Fig. 8.12 Upper level, laboratories and prep rooms.

expressed in early meetings. What they did, collectively, was create a striking example of how solar energy can define form.

The major design element of the new science building at the Madeira School—located in the Virginia suburbs of Washington, D.C.—is obvious. A solar collector roof, coupled with long overhangs and careful orientation, led the design process. Program input also stressed flexibility, that venerable programmatic virtue, visual control, and maximum amounts of laboratory and support space. Another parameter was the lighted campus path system, high in priority since an incident in Madeira's recent past. Other constraints followed the normal pattern of most building programs; budgets and practical enclosed volumes shaped some of the decisions.

A parallelogram in plan, the building is set into an unwooded area between older and newer campus buildings. It is sited with the collector roof facing southeast at as steep an angle as the interior volume could practically justify. The upper ridge edge of the single-pitch roof is a skylight, and the lower roof edge overhang forms a shelter over the path. Glazed openings between the laboratories and the path open the facility to views from pathway to lab, and vice versa. At the east end point is a greenhouse.

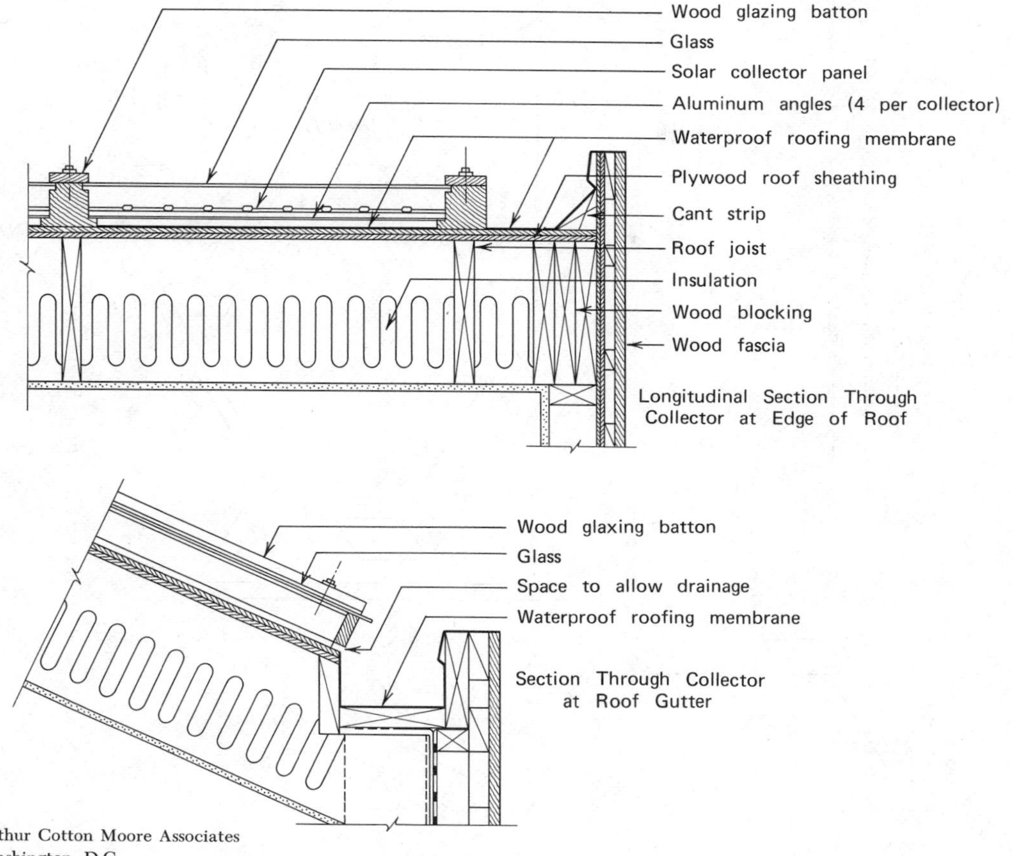

Fig. 8.13 *Architect's details, collector panels and roof construction.*

Inside, one large space, subdivided by sliding glass doors and flanked by prep rooms, houses the laboratories. Through the upper space pass light from the skylight. HVAC ducts, and lighting fixtures. The ducts have been painted in what, by now, has become an accepted fashion—but here, as a celebration of the solar heat source, the colors seem more a part of one overall statement.

More than just a formal design element, the parallelogram shape was chosen to minimize the impact of the solar roof's size (4600 sq ft). By bringing its edge down lower than the adjacent grade, the scale is reduced even farther. The lighted path nestles under this leading edge, and joins the ridge skylight to create dramatic strips of light at night.

The heart of the design, the collector, uses a special oil to capture the heat and transfer it to a 10,000 gallon water tank. This eliminates a major problem of corrosion prevalent in water-and-anti-freeze collectors. The system is to provide 60 percent of the building's heat, backed up by an oversize boiler in the adjacent gymnasium. Also linked to the system is the swimming pool, which can be heated during the spring and fall with excess heat generated. Domestic hot water can be provided for the neighboring gym during those seasons, as well, alleviating some of the necessity to fire the boiler. The building's temperature can be sustained for two or three sunless days without help from the boiler.

Located as it is near the seat of federal government, the Madeira project, albeit for a private girls' school, would have seemed of natural interest to *somebody* in Washington. Not so, Moore reports. Attempts to get any kind of input from those government agencies charged with helping research and develop

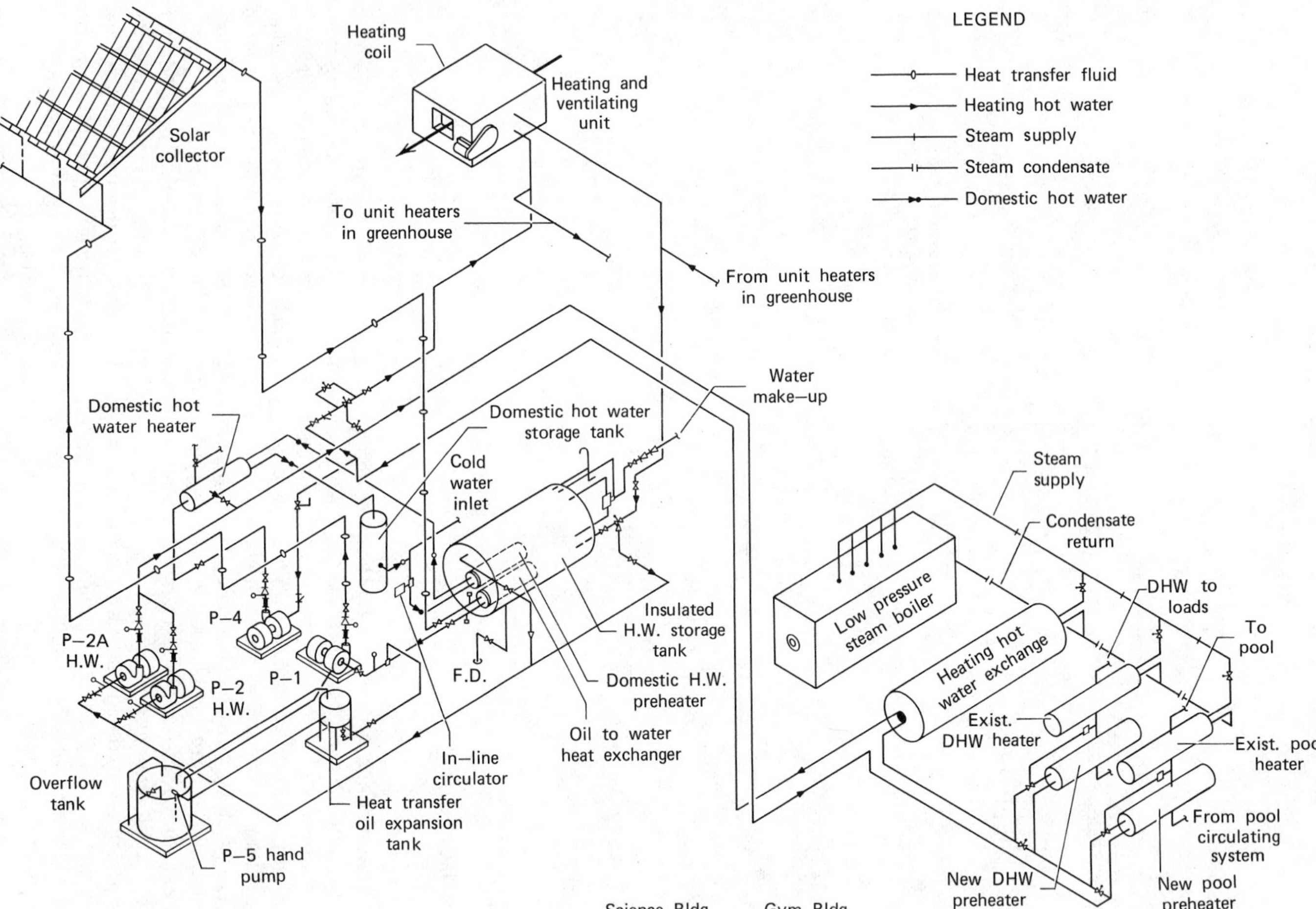

Fig. 8.14 *Solar-heating system, schematic. Madeira School, Flack + Kurtz, Consulting Engineers.*

alternate energy sources were met with a deafening silence. No phone calls returned, and certainly no help offered.

Madeira's science building got built despite the agencies, despite those who may be foot-dragging in the interest of other power lobbies. As Von Eckardt said, "The Madeira science building is no earth-shaking technological breakthrough." It is, however (as he noted), a delightfully unpretentious, positive, and timely contribution. Its design skillfully avoids making more of the technology than it deserves, a very real temptation for most designers. While it is obviously a solar heated building, it is a quiet one, getting quieter as the cedar siding turns gray with age.

Even if large flat-plate collectors are not to be *the ultimate* energy answers, even if they're not new or earth-shaking, it is, as P/A has said many times before, encouraging to see energy-conscious programs turned into realities. When they also turn out to be good architecture, esthetically and functionally, it's doubly rewarding. Science at Madeira has a new teacher. It doesn't talk much, but it says a lot.

Data

Project: science building, The Madeira School, Greenway, Va.

Architects: Arthur Cotton Moore Associates, Washington, D.C.

Program: science facilities for private girls school of 300 in grades 9–12. Client stressed flexibility and preservation of natural resources.

Site: open area on wooded 400-acre site 12 miles from downtown Washington, D.C.

Structural system: steel columns, wood framing and roof, concrete block foundation walls.

Mechanical systems: solar heating, backed up by boiler in adjacent building. Solar collector uses special heat-transfer oil to collect heat, which is stored in a 10,000 gallon hot water tank. Hot water and forced air heating.

Major materials: exterior, glass over brass collector plates over silicone membrane roof, red cedar siding; glass greenhouse; interior, gypsum board walls and ceiling, vinyl asbestos tile and ceramic tile floors, sliding and fixed glass partitions.

Consultants: structural, James Madison Cutts; mechanical and electrical, Flack + Kurtz Consulting Engineers.

Client: The Madeira School.

General contractor: Commercial Industrial Construction, Inc.

Cost: $591,000, $60/sq ft (includes equipment, landscaping, greenhouse, and energy system).

Photography: Norman McGrath, except as noted.

(*a*) ***Engineering.*** A discussion of the Madeira School solar system, designed by Flack + Kurtz, Consulting Engineers, appeared in the July 1976 *NEWSLETTER* of the Virginia Solar Energy Association, John G. Lewis Jr., AIA, Chairman and Editor. It is reprinted here by permission of Mr. Lewis, who wrote the article based upon data supplied by Mr. George Rainer of the office of Flack + Kurtz.

Introduction

The Madeira School, located in McLean, Virginia, includes a science facility of approximately 10,000 sq. ft. of gross area planned by the architectural firm of Arthur Cotton Moore of Washington, D.C. and consulting engineers, Flack + Kurtz of New York City. The science area is mostly on one level, with mechanical facilities containing the solar system underneath, taking advantage of the sloping site. Conceptualized by the architect in 1973, whereby the roof itself was to be pitched and act as the solar collector, the project was to be designed without federal assistance in the most cost effective manner with present state-of-the art solar collector systems for heating only. It was also intended to serve as a teaching tool.

In the final design, the roof angle *ultimately* assumed an inclination of 27 degrees with the horizontal, facing in a southerly direction, shielded in summer by trees and designed for the collection of solar energy during the winter months. As ultimately designed, the solar collector provides approximately 50% of the seasonal heating requirements including comfort heating, service hot water, and pool heating.

Engineering Philosophy and Approach

The design of the solar collector for the Madeira School was approached as a challenging, somewhat unusual, but by no means staggering heating project for which available components had to be properly integrated to provide a heat transfer, storage, and retrieval system.

Flack + Kurtz had designed many radiant ceiling systems; solar collectors are the same type of heat transfer mechanism used in reverse. Given the simple premise, the major tasks were the following:

1. Find components that could operate at the temperatures to be encountered.
2. Select system components that were compatible with each other; and most importantly,
3. Come up with a solution that was inexpensive enough to get built.

Close cooperation with the architect throughout the planning process was continually maintained. Collector orientation, slightly to the east of South was dictated by the site which was available for the new science facility. The collector inclination was dictated by architectural/esthetics and economics; integrating the collector with the roof structure was determined to look better and cost less early on. Raising the roof angle to an optimum of 55° would have caused excessive wasted space at the ceiling as well as extra cost. 27° above the horizontal was considered as a workable compromise.

General System Description

The solar collector is built-up out of prefabricated aluminum panel with integral fluid passages made by Olin Brass and called "Roll Bond". This is installed above a layer of silicone membrane waterproofing over an insulated (6 in. fiberglass batt) wooden roof. Each collector panel section is 27 in. × 104 in. in length and is connected (conventional socket fittings electric resistance welded) to the next collector in series and to the distribution steel header by means of a flexible neoprene connection. The total area of the collector consists of some 102 panels comprising a net metal area of 3550 sq. ft. and a gross collector area of 4600 sq. ft. The collector panels are supported on aluminum angle spacers (4 per collector panel) which create an air space above the membrane. Though the silicone roofing membrane is capable of withstanding temperatures up to 400°, it was thought that providing some air circulation behind the plates was the better part of wisdom. Approximately 2 in. above the panels is one double strength (1/8 in. thickness) piece of glass mounted on wooden supports and held down by wooden battens (as indicated in Fig. 8.13). The 1/8 in. double strength glass is used everywhere except for the lower five rows of glass which are 1/4 in. thick plate to support any snow load which might accumulate by sliding down the collector surface and piling up at the bottom.

The collector glass is lapped as in greenhouse construction. Condensation under the glass is eliminated by permitting a light amount of circulation under the glass from air entering through weep holes at the bottom and a shielded opening at the top. The total circulating rate of liquid through the collector is 9 lb./sq. ft. of collector surface per hour. The collection medium is a special heat transfer oil which is non-toxic and has a flash point in excess of 375° F. The heat transfer is achieved through a tube bundle installed inside a 10,000 gal. steel water storage tank which serves as reservoir for two to three days storage capacity for the heating system. The tube bundles were manufactured by the Alstrom Co. and are of conventional design, but are specially sized for the tank.

The design heating load is 500,000 BTU/hr. peak design capacity. The building has a fairly large air load because in order to use the relatively low temperature of the water from the solar collectors, a lot of air was needed to do the heating. The B.O.C.A. code requires a percentage of outside air of the total air distributed so a little more air was required than that which would have been used for an energy conserving design.

The space heating system itself consists of an all-air circulating system with 2-row hot water preheat and 4-row reheating coils in the air handling system. This fan air handling system and the unit heaters were manufactured by Trane. A small green-house, which is integral with the facility, is heated by means of a hot water unit heater. An auxiliary boiler located in the adjacent gym building fired by No. 2 oil is utilized during extended periods of sunless days. Water temperatures developed in the solar collectors under normal operating conditions are 100 to 150° F. A heat exchanger to provide pool heating is included in the installation. See Figure 8.14.

Description of System Components

Collectors. During the early months of 1974, the products available for collector panels were researched. Materials included carbon steel, copper, and aluminum. Configurations included tubes clamped, soldered, or spot-welded to a flat plate, as well as double sheets with integral passages formed by extrusion, inflation, or spot-welding. The most important element in collection efficiency was determined to be in the close contact between collector plate and fluid passages. A surface with integral passages was decided on; aluminum "Roll-Bond" as manufactured by Olin Brass was selected as the least costly one.

At one point during the investigation, the installation of pre-assembled collectors on top of a finished roof was considered. It was thought at first that a self-contained product, consisting of a sandwich panel of insulation, heat absorbing surface, and glass in a watertight box would be more economical; however, it became a problem to make pipe connections to and between collector boxes, and to cover the spaces between collectors in a tight and esthetic way, and the built-up configuration was reverted to.

Black Surface. Selective coatings were evaluated as to effectiveness, cost, and longevity. No definitive information on coating life was available; also the cost of available coatings was much too high in proportion to the added collection efficiency. It was therefore decided to use a flat black paint, and an alkid flat black enamel manufactured by PPG was used.

Storage. A 10,000 gallon steel storage tank as commercially available for fuel oil or water was specified. The tank was specified to be constructed of 5/16 in. thick steel, coated with red lead paint inside and out, with all necessary pipe connections and heat exchanger installation flanges welded at the factory (Buffalo Tank). A foamed-in-place urethane insulation $1\frac{1}{2}$ in. thick 150 foam SS24-44 (K=1.5), as manufactured by Whitco Chemical was applied and covered with Pyrocrete fireproofing. The tank was mounted on cradles placed on the mechanical room floor slab.

Piping. Black steel piping was specified; connections to aluminum sub-headers was by means of neoprene flexible connections fastened by means of double clamps. Connections between aluminum sub-headers and collectors, and between collectors were electrically welded. The flexible connections provided expansion capability and served as dielectric separators between dissimilar metals.

Bell & Gosset close-coupled pumps were provided.

Heat transfer medium. Present collector systems either utilize water which contains a certain proportion of glycol to prevent freeze-ups during winter operation, or provisions are made for drain-down of the system in case of extreme weather conditions. Free oxygen will react with aluminum, hence special water treatment is required and the pH of the water must be maintained neutral to prevent electrolytic action. It was decided therefore to utilize a non-toxic heat transfer oil which has been widely used in snow melting systems, and has a high flash point (375° F.). It is expected that the heat transfer oil which is formulated for a long life will not interact with the aluminum. Pump gland leakage is collected in a reservoir for reintroduction into the circulating systems of the collector. The heat transfer oil is Caloria HT43 as manufactured by Exxon.

System Operation

Collector temperature as sensed by a thermostat installed within the collector is compared to storage tank water temperature. Whenever collector temperature is higher than storage temperature, the heat transfer oil pump is energized and solar energy is collected through the heat exchanger tube bundle inserted in the storage tank. Tank water is used directly for heating by means of the heating coils in the air handling system and in the green-house unit heaters.

When the science building requirements are satisfied, surplus heat is directed to the gym building where this energy is used for pool heating and service water preheating for the gym building.

When the solar collector is deficient in heating capacity, the steam boiler in the gym building heats the heating water through a heat exchanger as required before returning it to the science building.

A separate tube bundle in the storage tank preheats the service hot water for the science building.

During the summer it may be necessary to operate the oil pump at night for heat dissipation.

Controls are of Barber-Colman manufacture.

Costs

The construction cost for the solar collector and accessories were as follows:

		$ per sq. ft.
Solar Panels, Supports, and Labor	14,500	3.15
Tanks and Heat Exchanger	14,000	3.05
Insulation	8,000	1.75
Piping, Pumps, and Controls	18,500	4.00
Heat Transfer Oil	1,000	–.20
Glass and Framing	13,000	2.85
	69,000	15.00

This comes to a system cost of $15. per sq. ft. of collector area. The cost of roofing materials not required should be taken as a credit against this unit cost.

8.5 The Solar System at Walt Disney World

By courtesy of Don Seely, Chief Mechanical Engineer of the Reedy Creek Utilities Co., a subsidiary of Walt Disney Productions at Walt Disney World in Orlando, Florida, material has been furnished for the following description of this unique installation (see Figs. 8.15, 8.16, 8.17, and 8.18). It is best explained by an engineering paper written and delivered by Harvey C. Jones, Director of the Reedy Creek Utilities Co.

Introduction

Walt Disney conceived of his Florida Project as a showcase of experimental prototypes for tomorrow's living, to show and operate in a real world environment, systems with promise for the future. When the Arab oil embargo brought the pending energy shortage in focus, Reedy Creek Utilities Co. and WALT DISNEY WORLD started actively looking at alternate energy sources. Discussing solar application with ERDA early in 1975, ERDA suggested we explore more advanced systems that were being conceived and tested. The AAI Corporation, in Baltimore, had done considerable work with stationary parabolics, focusing into movable collectors. Reedy Creek Utilities Co. and AAI Corporation started mutual exploration of ways to build a prototype solar system at WALT DISNEY WORLD. We required office space and AAI revealed a concept of horizontal parabolic mirrors used for roof and thereby cutting the cost of the building. It was this concept that was added to the office building requirement, the genesis of the existing project.

The building would be energy conserving and the systems would be optimized for air conditioning. The building is used five days a week and it was decided the storage systems would permit a weekly cycle, i.e., collection seven days a week in order to get high solar output. The system uses an absorption chiller. Components were to be long-life in keeping with the design requirements of the rest of WALT DISNEY WORLD. The solar system would be completely showcased, to include storage tanks, normally placed below ground. The only redundant system would be chilled water from the WALT DISNEY WORLD distribution system for air conditioning.

Construction was essentially complete in mid-November 1977 and solar collection started at that time. The system heat rise to date is within design, but as of this writing has not received sufficient sun to reach the required chiller operating temperature for complete check.

General

The project involves a two-story concrete and block office addition to Reedy Creek Utilities

Fig. 8.15 *The innovative solar heated and cooled office building of the Reedy Creek Utilities Co., Inc. at WALT DISNEY WORLD as guests gather for dedication on January 11, 1978.*

Co., Inc., Central Energy Plant. This insulated building contains a solar equipment room which is open and color coded for high visibility. Sixteen modular solar collectors are joined together along the ridge lines to form the insulated roof. The longitudinal axes of the reflective surface runs in an east-west direction. The solar roof replaces the usual trusses, corrugated decking, and insulating concrete roof. This reduces the costs associated with conventional roofing. Each 8 ft. by 32 ft. (2.44 m by 9.75 m) modular collector is a solar concentrator that consists of a concave cylindrical reflective surface and an absorber. The absorber uses water as a heat transfer medium and is kept in focus with the reflective surface by a linear actuator, controlled by a sun sensor/focusing device.

Design Philosophy

The array was sized to be 100% of the roof of the two-story building. Above ground storage was used to showcase components needed for solar energy utilization. Balancing valves were used in lieu of reverse return in order to save a long piping run (120′) and its associated heat loss. Flow rate through the collector system is variable between 60 GPM and 120 GPM or 3.75 to 7.5 GPM per absorber. The efficiency curve from the prototype collector is relatively flat. Operating experience will compare pump energy requirements with collector efficiency at various flows, thereby establishing an optimum. The control system was designed as a 24-hour a day, seven days a week system, utilizing temperature differentials between storage and collector to control shutdown. The collector bars are positioned by photo electric cell inputs.

The mirrored panels are designed as a $7\frac{1}{2} \times 32$ foot self supporting roof structure supporting a 40 psf live load. The panels are a three inch sandwich of aluminum, isocyanurate foam, and aluminum, formed on a vacuum die, bonded, and then glazed with 3 inch strips of mirror on the concave side. They are factory constructed with appropriate supporting saddles and collector mechanism, and field assembled as major components.

The absorber is an aluminum extrusion with a grooved lower face to which a U-shaped copper tube is swaged. The receiver housing temperature foam for insulation, and a glass cover. The receivers are each supported by two arms, and are aligned parallel to the mirrored panels. The sixteen receivers are mechanically linked by tie bars for

Fig. 8.16 Florida's Governor Reuben Askew and astronaut Gordon Cooper, now of Walt Disney Productions, inspect the solar system roof as Reedy Creek Utiilties Director, Tom Jones, explains during the dedication ceremony at WALT DISNEY WORLD on January 11, 1978.

movement by one drive unit operating through a torque tube to move each side accurately.

Problems & Opportunities

The COP of .7 of the absorption chiller is low. A low temperature Rankine cycle powered heat pump under DOE development may surpass this sufficiently to make it worthwhile to retrofit our project.

Photovoltaic cells integrated into the movable bar to collect the concentrated sunlight would appear beneficial. This is another retrofit potential.

Electrical demand for this system is 60 KWHs to operate pumps which on a good solar day will produce 1.3 MBTU thermal.

The system as built is aircraft type construction. The interface between the building and the roof is machine shop produced. In the next generation, more cost effective construction methods, perhaps an upside down

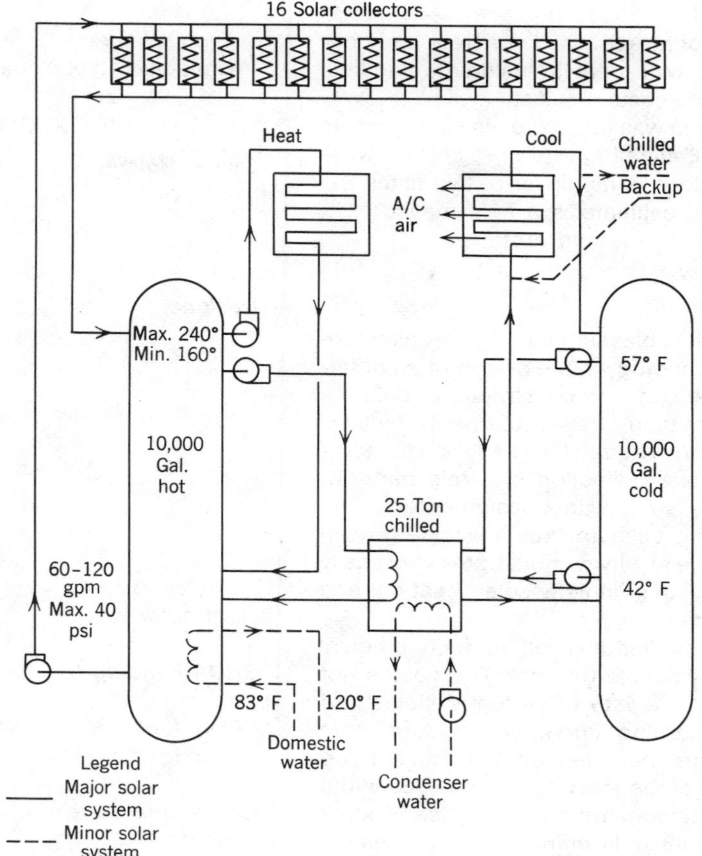

Key Word Abstract

Application: Cooling, heating, hot water hydronic system.
Collector Type: Fixed horizontal parabolic trough mirrored roof with moving absorber.
Collector Manufacturer: AAI Corporation.
Collector Area: 3840 sq ft ($356.75\ m^2$).
Storage Capacity: 10,000-gal, hot water steel tank (37,854 liters) plus 10,000-gal, chilled water steel tank (37,854 liters).
Total Btus: 400×10^6 Btu/yr.

Fig. 8.17 *Schematic diagram of the system. For clarity, the piping at the 16 collector tubes has been shown in simplified form in this illustration. Actually the water passes through each collector tube twice before returning to the hot water storage tank. For additional data see tabulation at the end of this section, page 330.*

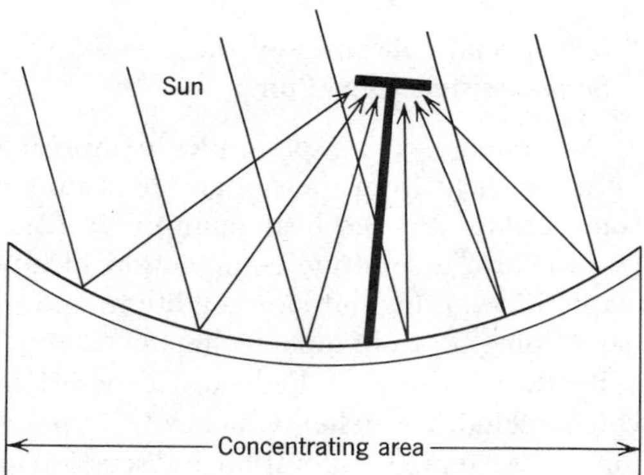

Fig. 8.18 *The stationary mirrored panels concentrate the sun's heat on the absorber, which is an aluminum extrusion. Above this is swaged a U-shaped copper tube, (see Fig. 8.16).*

barrel arch of pre-stressed concrete and concrete tiltup wall panels including saddles, would reduce costs significantly.

This system was the first of its kind and was hand-built. Estimating and cost control were not satisfactory. A multiplier of two times the conventional estimate is a safe measure, in building a first of a kind prototype.

Summary

The facility is a pleasing blend of architecture and new technology in the design of an otherwise simple commercial structure, pointing the way to cutting costs in the future by integrated design. The facility is a long life structure. The solar collection system's performance to date is well within design criteria. The concentrating collector has a sizable growth potential. The facility demonstrates very easily and visually a complete solar heating and cooling system.

The system lends itself to technological advances by simple retrofits. The cost is not now competitive with fossil fuel systems and assuming learning curve construction cost improvements, becomes cost effective if you assume generous fossil fuel price escalation. Any solar demonstration must clearly state that solar energy is diffuse and, at present, expensive to collect.

This system, providing $\frac{1}{1200}$ of the energy at WALT DISNEY WORLD, is not an answer in itself but represents a major start into alternative energy sources.

Climatic Data

Degree Days	Heating 733	Cooling 3226
Avg. Temp. (°F)	Winter 61.1 (16.17° C)	Summer 75.3 (24.05° C)
Avg. Insulation	Winter 1162.35 Btu/Ft/Day (315 Langleys/Day)	Summer 1752.75 Btu/Ft/Day (475 Langleys/Day)

Solar Energy System

Manufacturer: AAI Corporation, Baltimore, Maryland

Application Heating 100% Cooling 80% Hot Water 100%

Collector

Type: Modular cylindrical concentrator
Area: 3,840 Square Feet (356.75 m²)

Storage

Type: Water
Capacity: 10,000 Gallons Hot (37,854 liters)
10,000 Gallons Chilled

Owner Builder:	Reedy Creek Utilities Co., Inc. Subsidiary of Walt Disney Productions
Designer:	Architectural—WED Enterprises, Glendale, California Mechanical—Sudtell Engineering, Glendale, California Solar System—AAI Corporation, Baltimore, Maryland
Contractor:	Buena Vista Construction Co.
ERDA Funding	75% of solar component
ERDA Technical Management:	NASA Marshall Space Flight Center, Alabama
Operational Date:	November 1977
Building:	Type—General Office Area—5,625 square feet (522.58 m²) conditioned
Location:	WALT DISNEY WORLD, P.O. Box 40, Lake Buena Vista, Florida 32830
Latitude:	28.4° N

8.6 Special Efficiency of the Solar-Assisted Heat Pump

Two energy-saving principles comprising the use of heat from the sun and heat drawn from outdoor air (the heat pump) now combine to form a sensitive combination ideally adapted to varying outdoor conditions and to maintaining a stable indoor climate.

By the courtesy of Fedders Corporation, which includes Mueller Climatrol Corporation and Airtemp Corporation, a discussion of the functional characteristics and operation of this system follows (see Figs. 8.19, 8.20, 8.21, and 8.22):

How the System Works

The four heat stages, A to D, relate directly to Figure 8.19.

(*A*) *First Heat Stage.* The solar system functions as the initial heating source for the home between outdoor temperatures of 65° F and 47° F. Only the blower of the indoor air handler and two small water circulators are in operation. The solar heated water is pumped from the collectors (1) or the storage tank (2) into a hydronic coil (3) over which indoor air is circulated.

(*B*) *Second Heat Stage.* At 47° F outdoor temperature, the system automatically switches to heat pump (4) operation which continues to be the sole source of heating as long as the outdoor temperature stays above the first balance point . . . the outdoor temperature at which the increasing heat loss of the house is exactly equal to the declining output of the heat pump, usually between 35° F and 28° F.

(*C*) *Third Heat Stage.* Below the first balance point, stored solar heated water is used in conjunction with the heat pump. Solar heated hot water is pumped from the collector panels (1) or the storage tank (2) into a hydronic coil (3) over which indoor air, preheated by the heat pump (4) is circulated by the blower throughout your home.

(*D*) *Fourth Heat Stage.* When storage temperature is below the requirement of the hydronic coil, electric resistance heaters (5) in the indoor air handler take over the heating function in conjunction with the heat pump. This may occur after cloudy overcast days or on extremely cold days. These occasions represent a small percent of the total heating dollars.

Special Overriding Conditions

To help assure owner comfort and the most economical use of solar energy, the four stage heating sequence described above is subject to override by two special conditions.

(1) When storage water temperature is less than 100° F, the system will automatically advance to the next heating stage.
(2) When storage tank temperatures are 180° F or higher, an automatic control sequence will cause the system to revert to economical direct solar heating no matter what outdoor temperatures register. This can happen on bright sunny winter afternoons when the heat pump would normally be providing the heating.

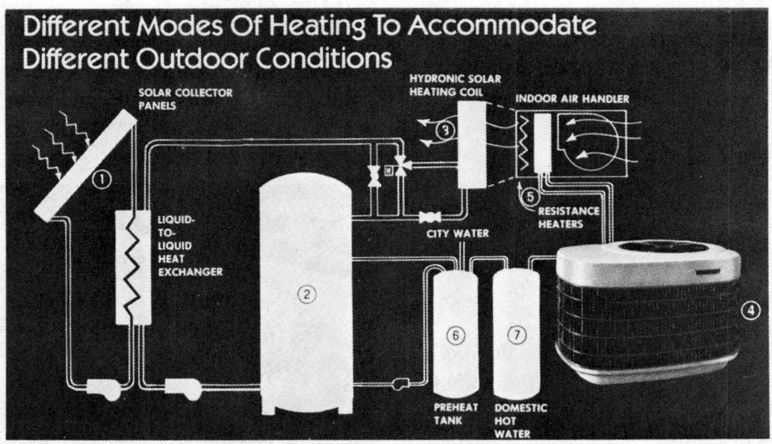

Fig. 8.19 *Components and assembly of the Climatrol Solartrol System. The numbers 1 to 7 identify equipment, the operation of which is described in Section 8.6. Courtesy of Fedders Corporation and Mueller Climatrol Corporation.*

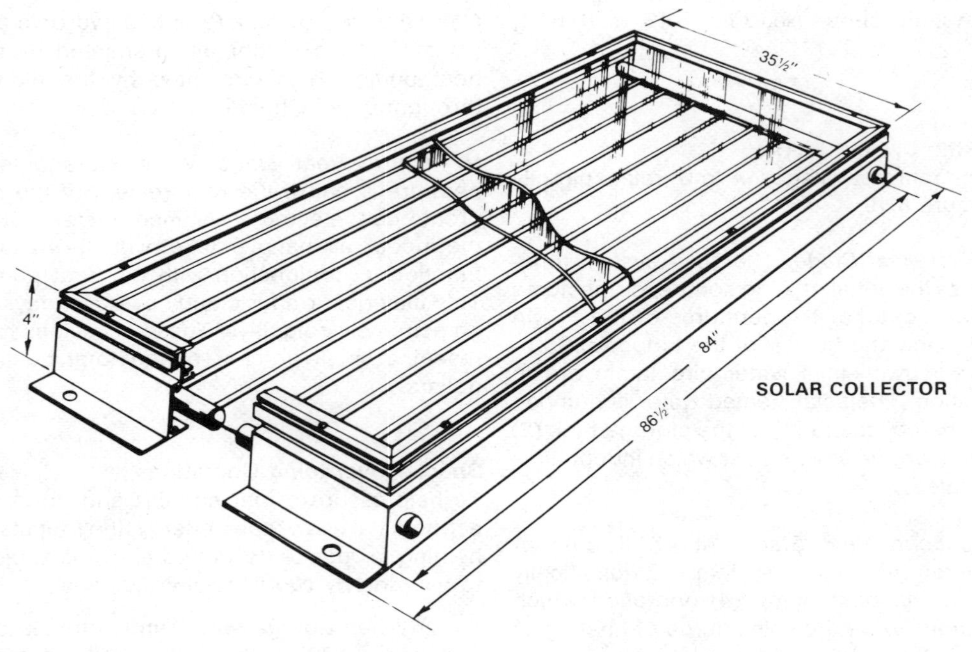

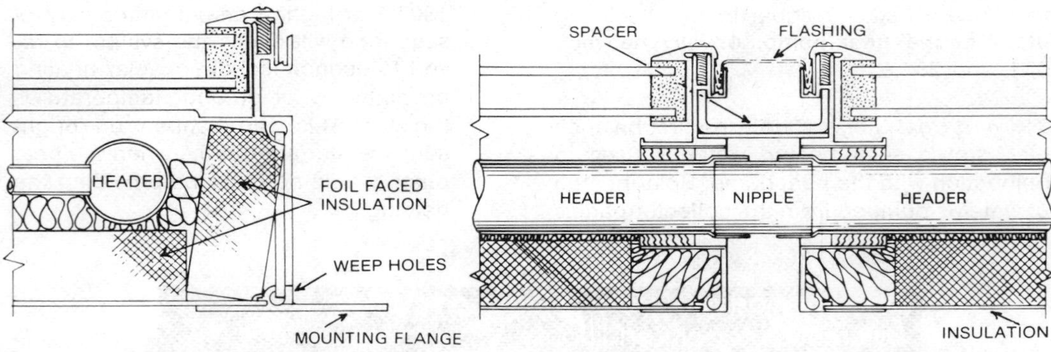

Fig. 8.20 Cutaway section of solar collector (top), cross section of header (bottom left), and method of connecting manifolds of adjacent solar collectors (bottom right).

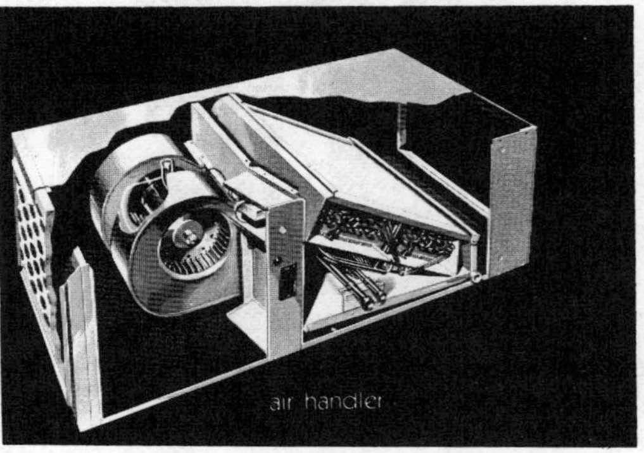

Fig. 8.21 Manufacturer's description of the indoor air handler of the heat pump. Its coil, coupled with the balance of the refrigeration cycle (part 4 of Fig. 8.19, which is located outdoors), *is a condenser coil in winter, providing heat to the circulated air stream. In summer it is an evaporative coil, cooling the air stream.*

SIMPLIFIED PLAN OF CLIMATROL SOLAR-ASSISTED HEAT PUMP INSTALLATION

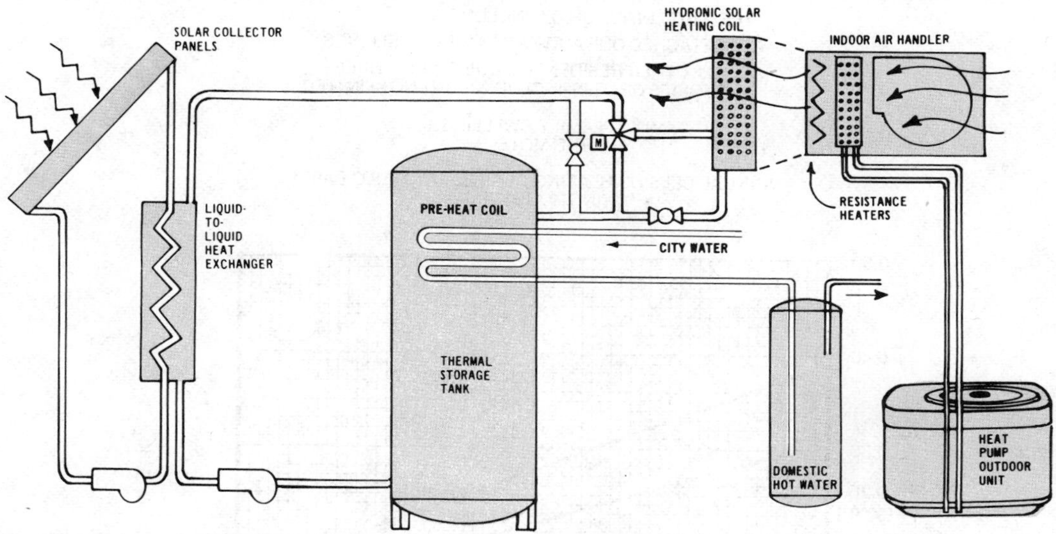

Fig. 8.22 Developed diagram showing details of three coils, the heat pump indoor heating/cooling coil in the air handler, the hydronic solar-heating coil in the continuing air stream of the air-duct system of the house, and the preheat coil in the thermal storage tank. The preheat coil heats the city water before its routing to the domestic hot water tank. In warm weather, when the solar system is inoperative for house heating, the heat pump unit (right) operates to provide cooling through the air-duct system of the house. Reprinted by permission from the "Airtemp Application Guide for Contractors, Builders."

The Bonus of a Domestic Hot Water Supply
In addition to heating your home economically, this system can also satisfy most of your domestic hot water requirements for things like washing, laundry, etc. Solar heated water is circulated through a small heat exchanger in a preheat tank (6) where it heats up supply water from the well or water main. The heated water is then drawn into the domestic water tank (7) where its temperature is raised, as required, to 135°–140° F.

8.7 Average Thermal Transmittance (U_o), a New Concept

ASHRAE Standard 90-75 has established new criteria by which a *weighted average* of the thermal transmittance of the enclosing walls of houses may be evaluated. Walls generally consist of opaque construction plus windows and doors. In preference to considering the *U*-coefficients of these separate elements, it is now usual to design so that the *average* thermal transmittance through the entire (gross) wall, including windows and doors, does not exceed a prescribed upper limit. Thus, the *areas* of the several enclosing structural assemblies have an influence and various *combinations* can be studied. Since roofs and floors above crawl spaces are now often very heavily insulated, the thermal value of the side enclosures is a significant item for special consideration. Let us consider the Arkansas House described in Chapter 4. Figure 5.3, page 182, tells us that Little Rock, Arkansas, the location of the house, experiences about 3000 degree-days annually. Now refer to Figure 8.23. For the house, which is a type A1 dwelling, the upper limit of average thermal transmittance, U_o, for walls should not exceed 0.26.

An example can serve to establish whether the Arkansas House based upon energy con-

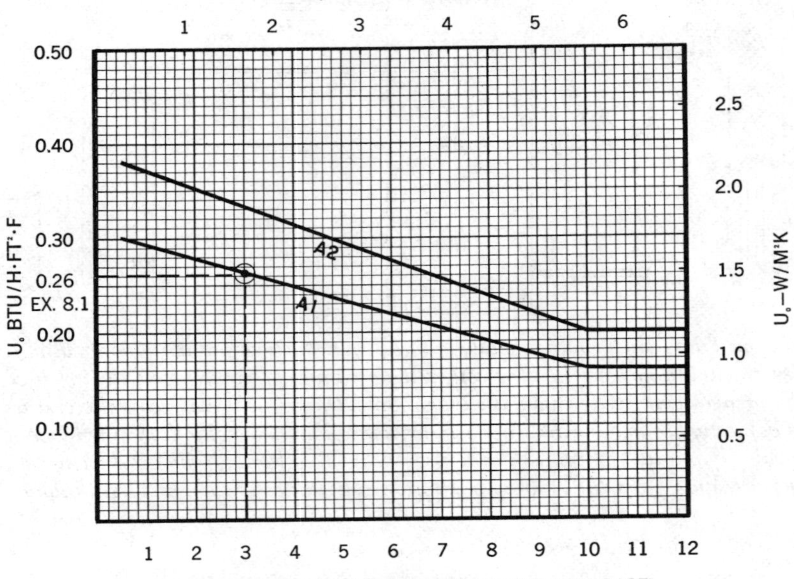

Fig. 8.23 U_o *Walls—Type "A" Buildings. Copyright © by the American Society of Heating, Refrigerating and Air Conditioning Engineers, Inc. Reprinted by permission from ASHRAE Standard 90-75.*

servation construction or the other Arkansas House based upon earlier minimum property standards comply with the 0.26 upper limit for U_o.

EXAMPLE 8.1. Using the following equation 1 and Figure 8.23, find the value of U_o for the two houses mentioned above. The equation and Figure 8.23 are copyrighted by the American Society of Heating, Refrigerating and Air Conditioning Engineers, Inc. and excerpted by permission from ASHRAE Standard 90-75.

Equation 1 Formula for Determining Combinations

$$U_o = \frac{U_{wall}A_{wall} + U_{window}A_{window} + U_{door}A_{door}}{A_o}$$

(See note below.)

where

U_o = the average thermal transmittance of the gross wall area, Btu/h · ft² · F (W/m²K)

A_o = the gross area of exterior walls, ft² (m²)

U_{wall} = the thermal transmittance of all elements of the opaque wall area, Btu/h · ft² · F (W/m²K)

A_{wall} = opaque wall area, ft² (m²)

U_{window} = the thermal transmittance of the window area, Btu/h · ft² · F (W/m²K)

A_{window} = window area (including sash) ft² (m²)

U_{door} = the thermal transmittance of the door area, Btu/h · ft² · F (W/m²K)

A_{door} = door area, ft² (m²)

NOTE. Where more than one type of wall, window and/or door is used, the U × A term for that exposure shall be expanded into its sub-elements, as:

$$U_{wall_1} A_{wall_1} + U_{wall_2} A_{wall_2}, \text{ etc.}$$

SOLUTION. Before applying equation 1, we need to tabulate the areas and U-coefficients for both houses based upon data from Table 4.22, page 145, Table 4.23, page 151, and Section 4.22, page 139.

Areas, *Square Feet*

	Energy Conservation Construction	*Minimum Property Standards Construction*	
Walls (net)	1225	1165	
Windows	64	166 (Windows	
Doors	42	+ doors)	
Walls (gross)	1331	1331	
U-coefficients			
Walls	0.05		0.08
Windows	0.63		1.13
Doors	0.07		1.13

Substituting in equation 1 for the Energy Conservation Construction,

$$U_o = \frac{0.05 \times 1225 + 0.63 \times 64 + 0.07 \times 42}{1331} = 0.079 \tag{2}$$

Substituting in equation 1 for Minimum Property Standards Construction,

$$U_o = \frac{0.08 \times 1165 + 1.13 \times 166}{1331} = 0.211 \tag{3}$$

Both houses qualify with U_o values that do not exceed the 0.26 of Figure 8.23. However, the Minimum Property Standards Construction rates as 0.211/0.079 = 2.67, which is almost three times as thermally transmissive as the more efficient Energy Conservation Construction for the same gross wall area. The combinations of U-values and areas were better for that house.

A much more complex routine, but with similar objectives, applies to other building categories including very large and nonresidential structures, and those that are mechanically cooled. The values established are referred to as Overall Thermal Transfer Values (OTTV).

8.8 Modular Boiler Systems Are More Efficient

In a publication of the General Services Administration, Public Buildings Service, *Energy Conservation Guidelines for Existing Office Buildings, Second Edition,* February, 1977, the use of modular boilers is recommended. The qualities that make this new kind of heating plant an energy saver make it suitable, not only for retrofit of existing buildings, office-types or other kinds, but also for large *new* commercial buildings.

Attention is especially directed to the first of the following four items excerpted from Section 6, "Heating," of the 1977 GSA publication mentioned above:

---

(6) HEATING

Use modular boilers for heating and select units so that each module operates at optimum efficiency.

Replace inefficient boilers, e.g., single pass, with new, unless heat can be reclaimed from the stack.

Select boilers that operate at the lowest practicable supply temperature while avoiding condensation within the furnaces.

Add circulation pumps to hot water boilers to permit greater design temperature differentials. (Boiler inlet temperature should not be below the gas condensation point; i.e., 135 degrees F. for 1000 Btu/ft^3 natural gas and 15% excess air).

Fortunately, the modular system has been under development for 10 years, with a multitude of successful installations, and is ready to meet the energy challenge.

By the courtesy of Hydrotherm, Inc., a manufacturer of modular equipment, their statements and illustrations (Figs. 8.24, 8.25, 8.26, and 8.27) of their *Multi-Temp* system are presented here:

THE MULTI-TEMP PLANT

What It is

Multi-Temp is a cast iron modular heating

Fig. 8.24 Modular boilers, the Multi-Temp system. Human scale emphasizes the compactness of this heating plant with an input of a million and a half Btuh. The units come on in sequence, (each operating at full efficiency), as the buildings demand increases. A bank of twelve or more such units would not be unusual in a large building. Hydrotherm, Inc.

Fig. 8.25 The development of efficient horizontal cast-iron boiler sections, with maximum heat transfer surface, is a new departure from earlier vertical sections.

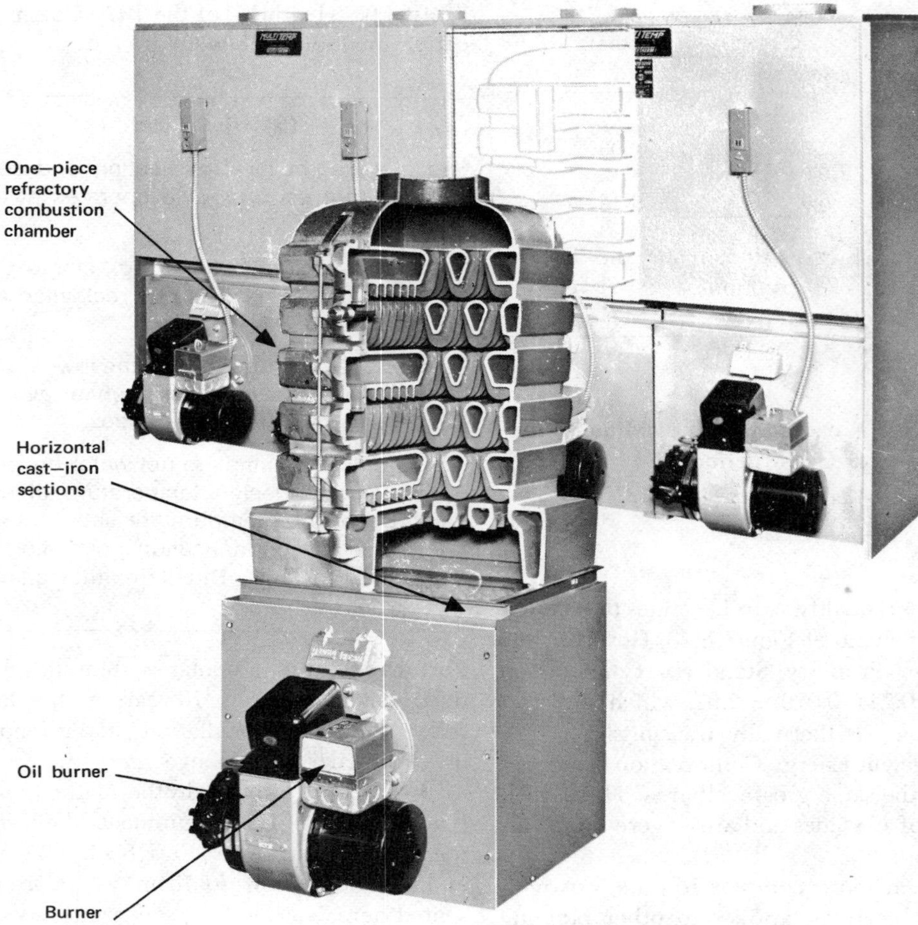

One—piece refractory combustion chamber

Horizontal cast—iron sections

Oil burner

Burner control

CUT-AWAY OF HEATING MODULE

385,000 Btuh Input

Factory assembled with burners and all oil controls

20″ x 32″ x 48½″ High.

Shipping Weight:	644 lbs.	Floor Loading:	179 Lbs./sq. ft.
Water Content:	7.25 gals.	Pressure Drop ($\Delta T=20°$)	1.7′ WC
Fire Side Surface:	49 sq. ft.	Water Side Surface:	42 sq. ft.
Horsepower:	9.1 HP	Pressure Rating:	100 psi ASME

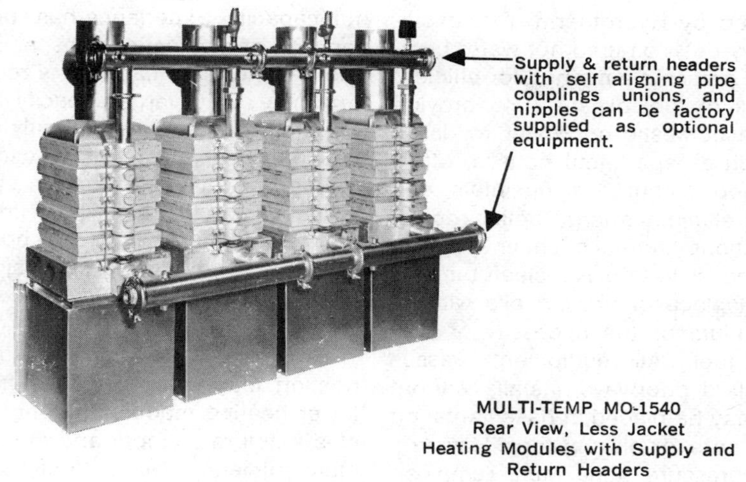

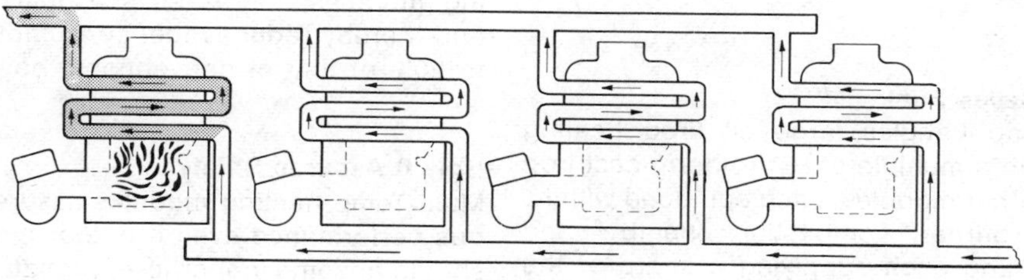

Fig. 8.26 Assembled bank of units. Direction of water flow is shown in the lower diagram. Shaded tone indicates heated *water from the left unit that has started up. Parallel piping of headers creates a team effort that increases as the building load increases.*

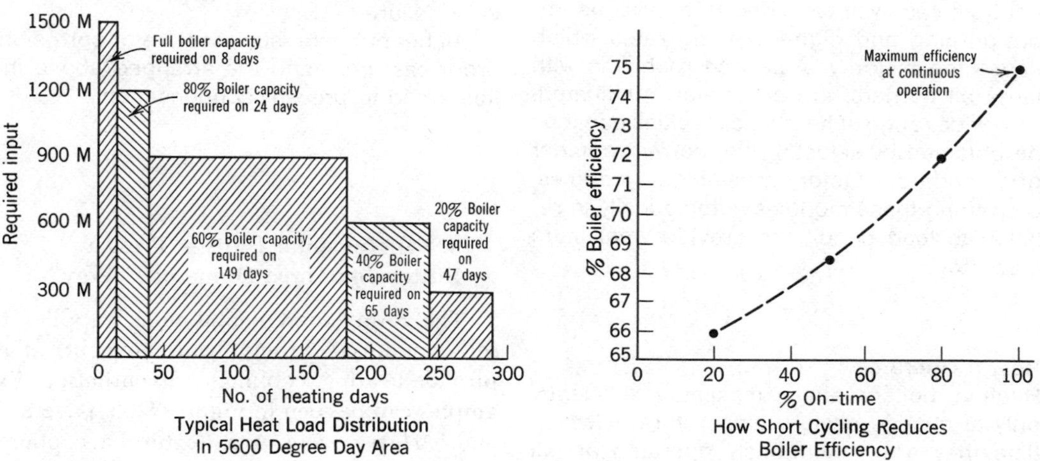

Fig. 8.27 One large boiler versus many small ones. Whether in mild weather with few modules working, cooler weather with several working, or extreme winter weather with all working, the modules actually operating are contributing with minimal short cycling (right-hand diagram). Short cycling of a single large boiler could drop the general efficiency to the lower levels of the 66 to 75% efficiency range.

plant developed by Hydrotherm, Inc. exclusively for heavy-duty, oil-fired hot water heating systems. It presents a method for efficient utilization of light fuel oils (#2) to provide both heat and domestic hot water for large structures such as apartment houses, office buildings, schools, churches, hospitals, etc.

Multi-Temp eliminates large boiler rooms, heavy foundations, reduces chimney costs and makes it possible to utilize clean burning #2 fuel oil in place of heavier oils without paying a premium or the necessity of preheating the fuel. All equipment passes through standard doorways, installs without rigging, and may be placed in basements, on intermediate floors, or on roof tops. Having a 100 psi ASME pressure rating, Multi-Temp can serve high-rise buildings without the use of steam convertors.

How It Saves Fuel

Multi-Temp handles large oil fired heating loads with a multiple of Hydrotherm cast iron boiler burner modules, each equipped with its own oil burner, combustion chamber and controls and each supplying a share of the heating load. By limiting the maximum firing rate of each heating module to 2.75 gal/hr., uncomplicated standard oil burners can be employed. Safe ignition and flame supervision is accomplished with standard primary controls including the modern cadmium sulfide cell. Electronic safety and delayed ignition controls, as well as special provisions for pre-purging and high-low firing rates which cause the majority of service problems with large oil burners, are completely eliminated.

A wide range of heating capacities can thus be obtained by selecting the correct number of standard, factory-assembled modules. Step-firing these modules automatically in relation to load conditions provides maximum efficiency.

How It Operates

Heating boilers attain maximum efficiency only at outside design temperature at which time they are continuously operated at full rated input. Because outside design temperatures are rarely reached during the heating season, a heating system rarely operates at full capacity. Experience has shown that during 90% of the heating season, 60% or less of the heating plant capacity is required.

Single unit, large-capacity boilers must therefore operate intermittently in short combustion cycles, the length of cycle depending on load conditions. Although cycling can be reduced with "high and low" firing rates, neither condition permits the boiler to reach rated efficiency during the major part of the heating season resulting in low "seasonal efficiency".

With the Multi-Temp system, the actual heating load is matched by firing more or fewer heating modules in long operating cycles at full rated input and therefore at maximum efficiency. This in combination with low standby and pick-up losses resulting from the low mass and small water content of the heating modules, increases seasonal efficiency and thereby reduces fuel consumption by 15 to 20% of that of conventional boiler plants.

How It Assures Safety

Multi-Temp heating modules assure continuous performance even if one of the modules should become inoperative. Standby capacity is thus built into every Multi-Temp heating plant and need not be provided by additional, redundant equipment. The small amount of water in each heating module is contained in comparatively small tubular passages. This prevents build up of large, potentially destructive hydraulic forces, even in the event of relief valve failure.

All heat transfer surfaces are of corrosion-proof cast iron, and are arranged above the flame bed to prevent condensation.

8.9 Incremental Units—"Press a Button for Three Hours of Energy"

The term "incremental units" applies to self-contained through-wall elements that provide heating, cooling, and ventilation. Examples can be seen in Figures 5.63, page 243, and 5.64, page 244. Architectural acceptance and design integration of the essential exterior grill (see Fig. 5.64, page 244), impressive upgrading of the engineering design and the quality of materials in excess of industry minimum standards and, most importantly, the development of controls for comfort of occupants and for energy conservation—all these have combined to give this type of air conditioning a new place.

An eight-story office building at 499 Warren Street in Syracuse, New York, demonstrates an especially suitable application. Architect Gordon P. Schopfer and his engineer, Edward A. Fassler, chose unit conditioners for this building in preference to a central-station ducted system for the all-year comfort of office occupants. With the collaboration of Remington Air Conditioning, Division of Remington Corporation, Auburn, New York (more recently a division of The Singer Company), this design team has produced a unique installation that is essentially a custom job.

Tenants in these offices, about 85% of which are peripheral, make use of the self-contained units for heating, cooling, and ventilation. Electrically powered, each contains a hermetically sealed cycle of compressor, condenser, and evaporator. Room air is recirculated, cooled, and dehumidified with the addition of controlled fresh air admitted through a motor-operated and gasketted damper. Heat is available through a separate coil supplied with hot water from central boilers. Integral thermostat and unit controls permit any desired selection of comfort level. Occupants in different offices have the choice of heating *or* cooling at the same time. Two limiting devices operate automatically, one to supply heat in any case when room temperature drops below 55° F and the other to make the cooling cycle inoperable when the outdoor temperature is below freezing.

To this point, the foregoing description is reasonably typical of the service in many high-cost apartment houses and possibly a few office buildings. A number of additional, distinctly new features were responsible for making this building quite different and perhaps the forerunner of a strong new trend. That is why this installation shows a marked improvement in control and quality, and why the architect and engineer felt justified in selecting it in preference to the central ducted system so common in office buildings.

The predominantly important feature is Remington's triple over-riding dual-control system (TODC). Following a schedule established by the building owner (hours can be varied), a clock device turns all units off at 5:30 P.M. and on at 7:00 A.M. This shutdown represents a saving of energy. Units start and each operates at its setting (for cooling or heating) of the previous afternoon to prepare the offices for 9:00 A.M. occupancy. At 5:30 P.M., however, when all units are turned off, they are all immediately reset by an electric impulse over the regular power wiring. No additional electric or pneumatic controls are required for this—an economy in installation cost. When units are thus reset through the special TODC panel within each unit, a single tenant may turn on his or her conditioner for full operation by pressing a button. This will be the *only* conditioner operating in the building, unless others are similarly activated. At 9:00 P.M. and 12:00 midnight, inpulses again turn off and reset all units previously operating. The individual occupants may again press their buttons to continue the service, but if everyone has gone the entire system shuts down. The occupant is only slightly inconvenienced by the need to push a button at three-hour intervals during an evening or on a weekend or holiday.

By building extra quality into through-wall units, reasonably long life may be expected. Those at Warren Street have many special features. One solves the common problem of sleeves that rust out. By use of heavier gage bonderized and zinc-coated steel, with special plastic finish where it is exposed to outdoor air, such corrosion is virtually eliminated.

Qualified and approved service agencies that operate under the supervision of the manufacturer can be responsible for repairs, or a trained mechanic at the building can make them. It is considered mandatory that a number of operable elements be kept available at the site for instant replacement. By means of accessible latches and two or three easy plug-type disconnects, the entire cooling element can be taken out of the cabinet in 30 seconds.

Only 5 minutes are needed for a complete replacement.

The elimination of the usual central system ducts from core to office at each story cut the height of this building by about 6 ft.

8.10 Variable Air Volume (VAV) and Other Energy-Saving Features

This method of air conditioning and air distribution received considerable impetus during the 1960s. Under the proper conditions, control of air flow-rates to suit the requirements of space demands can effect energy savings. By supplying only as much conditioned air as is required, the system shows savings over the reheat method and the dual-duct method, both described in Chapter 5. Although both of these systems are often still appropriate and much used for certain applications, the advantages of VAV are apparent. For summer cooling operation by the reheat method, air is cooled to a lower temperature than required. At specific locations electric elements or heating coils in local branch ducts reheat the cooled air to suitable temperatures. Duplicate use of energy is evident. A similar conflict is seen in the dual-duct system, where cool air and warm air are delivered (both processes using energy), and then the two thermal media are blended with partial cancellation, wasting part of the overall energy input.

At a national meeting of engineers in Denver in July 1969, four papers on the newly developing VAV method were presented. The chairman who introduced the speakers, concluded his introduction with the following statement: "So significant do I consider the varible volume system that within the next ten years, I honestly feel that 75% of the systems designed for office buildings, laboratories and school-type applications will be a form of the variable air volume system. Many of us have been afraid to initiate the design of a variable volume system simply because of lack of knowledge and experience by other people in this area."

Developments and acceptance of the system in the succeeding years has been rapid.

Photo by Jack Horner

Fig. 8.28 The Denver Office Tower (Anaconda Tower), Denver, Colorado. Skidmore Owings & Merrill, Architects, Flack + Kurtz, Mechanical-Electrical Engineers, Oxford-Anschutz Development Company. Photograph of an early model. The Tower Building (right) reflects part of the Hotel (left). Commercial area is seen as a low-rise element in the center foreground.

(*a*) *The Denver Square Office Tower* (*Anaconda Tower*). An excellent example of this system is presented in the following summary of energy-conserving features, including the uses of VAV. It is presented by Flack + Kurtz, Mechanical and Electrical Engineers, to explain their mechanical engineering concepts and design of the mechanical and other features for the Denver Square Office Tower (Anaconda Tower) shown in Figure 8.28. Typical air distribution is seen in Figure 8.29*a*, *b*, and *c*.

ENERGY CONSERVATION FEATURES—DENVER SQUARE OFFICE TOWER (ANACONDA TOWER)

The Denver Square Phase I Office Tower demonstrates that a modern, attractive, high-performance and functional office building can be constructed with the incorporation of energy conservation principles. The building will consume less energy than would have resulted from conformance to the currently prevalent energy conservation guidelines, ASHRAE Standard 90-75, and now being used as the energy code in many parts of the country. The anticipated energy consumption of 69,000 BTU/sq. ft./year is approximately 30–60% less than the energy consumed by comparable modern buildings in the Denver area whose gross energy consumption varies between 100,000 and 225,000 BTU/sq. ft./year.

These low levels of energy consumption are to be achieved by an integrated design approach which is reflected in the architectural design of the building as well as the mechanical and electrical systems. This was all achieved without resorting to negativism and self-denial type design criteria. The building will be bright and airy with good views and standard comfort levels; however, the design will maximize the use of natural elements where possible and protect the enclosed envelope of the building at times when the environment is hostile to it.

The exterior of the building incorporates double glazing with a reflective coating. This is in contrast with most office buildings in Denver which have single glazing and no reflective coating. The double glazing reduces the heat transmission through the glass by approximately 45% and the reflective coating reduces the impact of the sun on the cooling requirements by approximately 40%. Through the year, this generates significant savings in the heating and cooling energy requirements. The double reflective glazing allows the building to have a reasonable amount of glass on the exterior for views and natural brightness while still minimizing the heat loss and heat gain at different times of the year through the exterior skin of the building.

The lighting design in this building will be more efficient than many buildings in the Denver area and will correspondingly have fewer KW consumed in lighting energy. Most buildings consume 3-1/2 to 4 watts per sq. ft. of electricity for lighting whereas this building will provide 60 to 70 footcandles with approximately 2 watts per sq. ft. This not only reduces significantly the electrical energy consumption, but also in the cooling months, provides a substantial reduction in the cooling requirements. This is achieved by using standard 2′ × 4′ commercial lens fluorescent luminaires. Each fixture has within it two lamps rather than the customary four. When operating in this manner, the fixture is 15% more efficient than when operating in the four-lamp mode. Smaller rooms require more lumens per sq. ft. to provide equivalent lighting levels to large spaces. Therefore, each fixture is equipped to accept the future installation of an additional ballast and third lamp. In the event that larger spaces get subdivided into smaller rooms, the 60 to 70 footcandle levels can be maintained with the addition of lamps. Experience has indicated, however, that the lighting level provided by the 2′ × 4′ with two lamps is quite sufficient for the major part of the building.

The heating, ventilating and air conditioning system incorporates a variety of features which distinguish it from most of the present office buildings of the generic high-rise type found in Denver and similar climatological areas throughout the country. The typical office floor will be air conditioned by a variable volume air distribution system with both the perimeter areas and the interior of the floor on the same system. This system has the capability of utilizing 100% fresh air when the outdoor conditions are compatible with comfort requirements; e.g., outdoor temperatures between 50 and 60 degrees will provide free cooling to the entire building. This is particularly important in Denver where the sunny side of the building needs cooling at fairly low outside temperatures. The more commonly found office building air conditioning systems have separate interior and perimeter systems in which the 100% fresh air feature is limited to the interior system. The more commonly found perimeter systems, such as induction or fan coil, are minimum fresh air systems, which in Denver would require mechanical cooling as the temperature rises above 45 degrees F., particularly on the sunny side of the building.

The variable volume feature in the air handling system allows us to reduce the amount of air being distributed through the building during periods of low cooling requirements. This is reflected in the fan horsepower expended in pushing the air through the building. Since these fans operate whenever the building is occupied (60 to 70 hours per week), and there are many periods of reduced loads, this can add up to significant energy savings. Most of the commonly found office

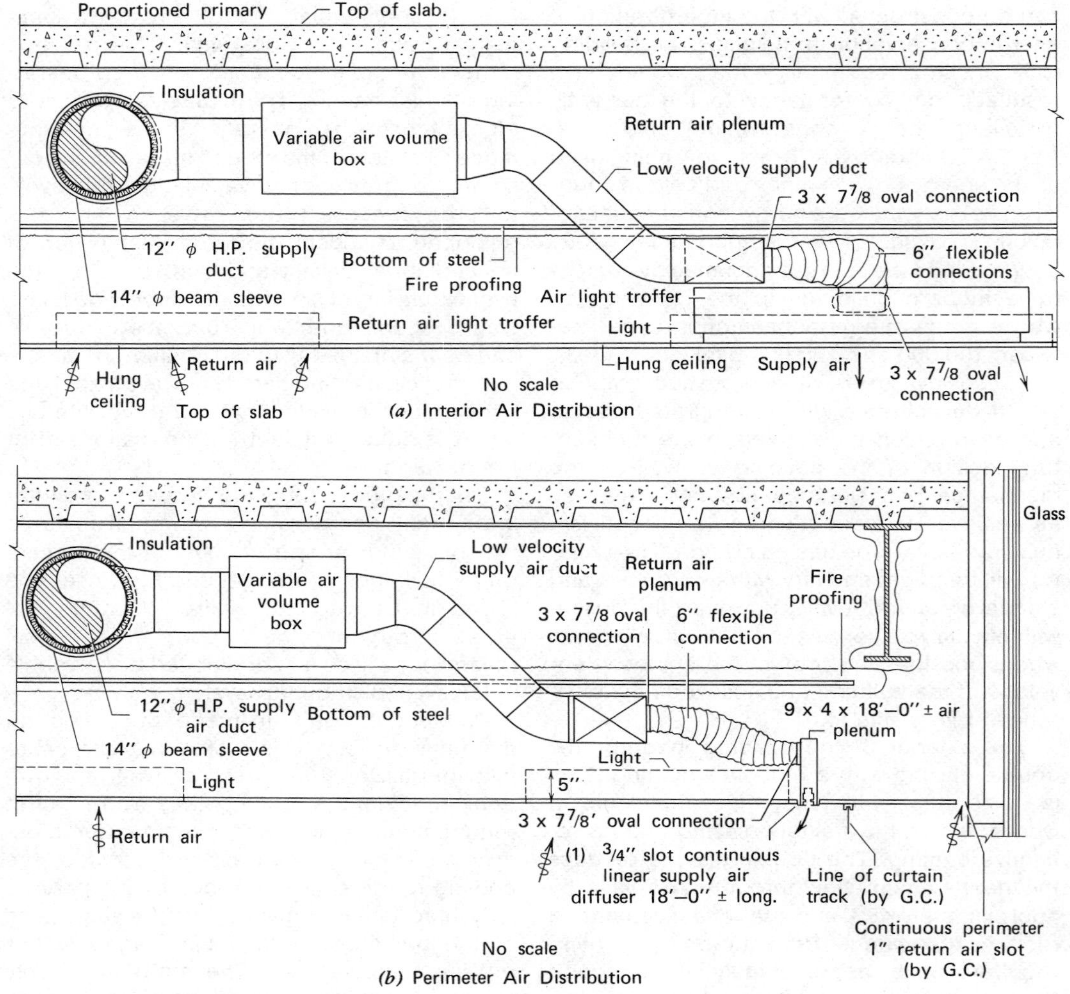

Fig. 8.29 (a) *Air distribution, primary and induced, to interior spaces on typical floors.* (b) *Distribution, primary air only, through $\frac{3}{4}$ in. slot-type continuous linear supply air diffuser in 18-ft lengths, that skirts the building perimeter inside of the curtain track.*

A note on the construction drawings states, "All supply air boxes, shall be provided with individual wall-mounted thermostat unless otherwise noted. Coordinate locations with tenant layout." High-pressure primary air is delivered to all boxes. Low-pressure return air is taken from ceiling plenums and returned to central equipment for reconditioning and the addition of fresh air. (c) *Variable air-volume, induction box detail. Not to scale. After conditioning, primary air (return air plus some fresh air) enters at the left. Its volume is controlled by the pneumatic valve. Local air in the ceiling plenum is induced to join the air stream by the action of the jet element. Air at required temperatures is discharged at the right to the air light-troffer. This unit serves* interior *spaces.* Perimeter *VAV units receive the same primary air, which is similarly volume-controlled but the perimeter unit does* not *add induced air. Perimeter slots discharge volume-controlled primary air only. In cold weather the perimeter is served by hot water convector strips at the sill below glass.*

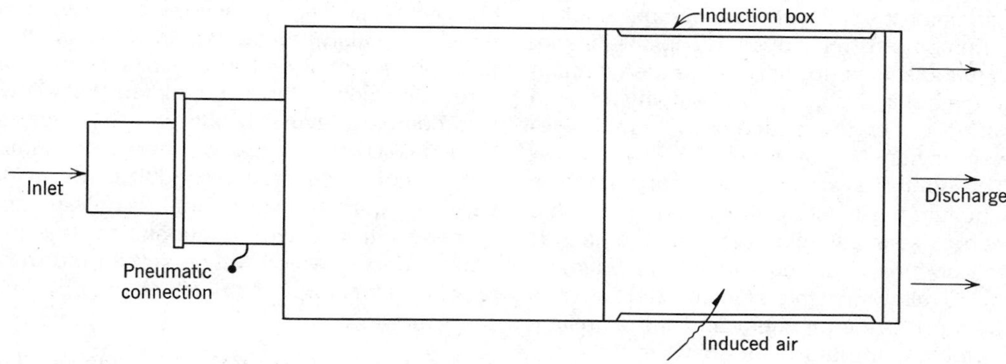

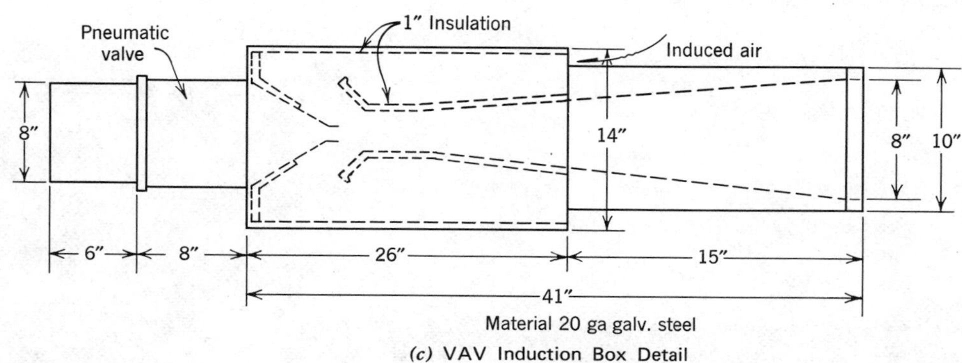

(c) VAV Induction Box Detail

building systems installed until now are fixed air volume systems which do not have this capability. Also recognizing the extensive periods of partial cooling load requirements, the refrigeration machines have been selected to closely match the anticipated demands on the system during partial loads.

The peak cooling demands have been divided 1/3, 2/3 in the selecting of the refrigeration machines and their auxiliaries. Since mechanical equipment operates most efficiently when it is working near its full capacity, selection of refrigeration machine allows for a good load matching performance. It also allows for the use of small auxiliaries (pumps and cooling tower fans) which can consume as much as 30 to 60% of the energy consumed by the refrigeration plant itself. This is often overlooked in the energy consumption of refrigeration plants.

The system design allows the building to respond very rapidly to changes in the exterior climatic conditions. The ability of the building to be "flushed" with fresh air at times when the outdoor conditions are attractive, avoids the need for turning on mechanical refrigeration when exterior sun loads on the building could otherwise cause overheating. The dry climate in Denver also allows for the fan systems to incorporate the feature of enthalpy control which will allow the building to bring in fresh air at temperatures in excess of 60 degrees when its characteristics are superior to recirculated air from the building in terms of offsetting cooling requirements.

The light air troffer device used for air distribution in the building is of the heat reclamation type which will allow the heat from the light fixture to be picked up in the return air to be utilized for heating in the winter and in the summer can be expelled outside the building without being imposed upon the room cooling load. This arrangement allows for efficient reuse of the energy in the winter; however, in the summer it effectively reduces the cooling requirements of the supply air to the space. This reflects itself not only in reduced cooling tonnage but in the quantity of air which has to be supplied to the space with high horsepower air handling systems.

The control system to be incorporated in the building will allow the operators to be quite precise in adjusting the system demands to the comfort requirements of the building. This will allow for the efficient starting and stopping of the equipment and the adjusting of various temperature set points in the heating and cooling media operating throughout the building. This will avoid the tendency of many mechanical systems installed in the 60's and early 70's of overcompensating for excessive heating and cooling; e.g., reheat systems, to satisfy space comfort needs. It is through the many energy conservation features incorporated in this building that allows it to compare favorably with the GSA goal of 55,000 BTU/sq. ft./year of energy consumption which is generally recognized as being very stringent for other than single-tenant, owner-controlled type occupancies. It is this type of design which will be setting the trend for energy responsive office building design in the years to come.

(b) A Schematic VAV Installation. The suitability of the variable air volume principle is not confined to its use in high-rise buildings. As seen in Figure 8.30, one-story buildings can

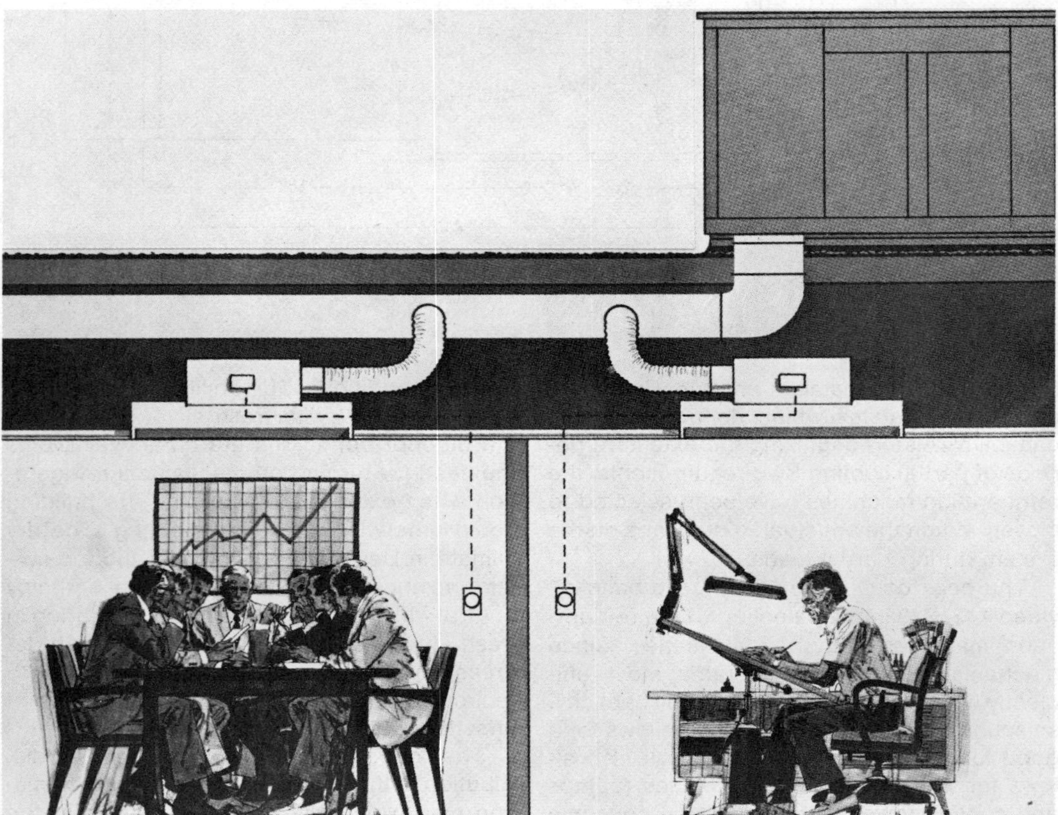

Fig. 8.30 A variable air volume system with equipment located on the roof. Wall thermostats control the volume of air delivered to each space through ceiling luminaires. The conference group of seven, producing about 3500 Btuh, will draw a greater rate of cooling air than the individual designer whose body heat loss could be about 500 Btuh. The duct extends further to serve other spaces but only in proportion to their demands. Economy of energy prevails. The space between ceiling and roof can be used as a return plenum. Courtesy of the Trane Company, La Crosse, Wisconsin.

be well served by rooftop equipment. Extending that general scheme, downfeed air delivery is feasible in a building of several stories.

References and Bibliography

1. *Solar Heating and Cooling,* Kreider and Kreith, McGraw-Hill Book Co., New York.

2. *Energy Conservation Design Guidelines for Office Buildings,* 1975. Prepared by Dubin-Mindell-Bloome, Associates, P .C., in Cooperation with AIA Research Corporation, Heery and Heery, Architects. General Services Administration, Washington, D.C.

3. *Solar Energy and Building,* 1975, S. V. Szokolay, The Architectural Press, London Halstead Press Division, John Wiley & Sons, New York.

4. *Mass Masonry Energy,* 1977, Masonry Industry Committee, The International Masonry Institute, Washington, D.C.

5. "Insulation and the *M* Factor," Mario J. Catani, P.E., *ASHRAE Journal,* June 1978.

PART THREE

water and drainage

9

The Recycling of Water

9.1 Today's Water Supply

Among officials of municipal public works departments, a slogan of long standing had been, "Today's water supply is tomorrow's sewage problem." It had the essence of truth. With great effort and expense, millions of gallons of potable water were delivered to a city, only to be continuously polluted by human wastes. Transformed into sewage, it was then run out to foul the nearest river or, in more recent times, it was purified—also at great effort and expense.

Adequate sources of water are now in short supply. Fresh water lakes, deep wells, and some less-polluted rivers now often yield insufficient water for this extravagant and wasteful cycle.

We face the scheme of making the purified sewage effluent quite a lot purer and drinking it. Thus, the slogan is reversed and becomes "today's sewage flow can be tomorrow's water supply." Terms such as "sewage *disposal*" are used less often and are replaced by "water reclamation" and "recycling of water." Human sensibility is such that we are reluctant to think that we are drinking purified sewage effluent. As in all major social changes, there is a need for capable leaders who can not only implement a process but also assure us of its safety and dependability. A. J. Clayton and P. J. Pybus are such leaders. Their excellent article in the September 1972 issue of *Civil Engineering-ASCE* told of their work in this field in Africa, and offered encouragement to engineers in the United States who are already involved in similar projects. The introduction to that article is reprinted here.

WINDHOEK RECLAIMING SEWAGE FOR DRINKING WATER[1]

A. J. Clayton, City Engineer, Windhoek, South West Africa

P. J. Pybus, Senior Engineer, Stewart, Sviridov & Oliver, Johannesburg, South Africa

From time to time people in the U.S. talk about reclaiming drinking water from sewage. Yet the prevailing view here is that before this can be done, scientists and engineers will have to do much more research on the long-term effects of imbibing such water.

But the people of this capital city of South West Africa have been drinking water reclaimed from sewage for almost four years now—without any apparent problems. Harmful viruses are removed by the process. Plagued by a chronic water shortage, Windhoek has been running a 1.2 mgd (0.0526 m^3/s) reclamation plant since 1969. Reclaimed sewage now furnishes more than 13% of Windhoek's potable water supply. The people have accepted the new water source without complaint.

The article that followed the above introduction was an early commentary on a change that will affect the practice of architects and planners. As it now appears, the movement to reclaim sewage for drinking water is well under way and could change the nature of buildings, their mechanical systems, and their plumbing fixtures. Regional planning will also

[1] Reprinted with permission from the September 1972 issue of *Civil Engineering—ASCE,* official monthly publication of the American Society of Civil Engineers.

be affected, since the recycling may take place in small, newly planned communities and even on college campuses, as will be seen in succeeding sections of this chapter. Space for local treatment plants may be required in the master plans now on the drawing boards or those soon to be contemplated. In this fast-moving development, the possible, complete digestion of sewage at an individual house or even directly at an individual toilet fixture has been seriously considered. At our present time of writing, the speed of change is so great that any appraisal of the current situation must be considered to be subject to additional changes. If we remain alert to the changes, we can be sure of progress and success in our designs. In the following sections of this chapter, a review is made of conditions (and recent developments) that have led to what might be considered the present "state of the art."

9.2 How It All Began

In the 1700s Manhattan Island (Fig. 9.1) was farm country. Typical of all areas that later developed into large cities, its needs were then minimal. Potable water was available in shallow wells and from some springs and streams. These sources were largely unaffected by the minor ground pollution of widely separated dry-pit privies (outhouses) that received human wastes. The 1800s saw the development of some paved city streets at which time the natural streams were enclosed in pipes, called *storm sewers;* these pipes led the rainfall to the many waterways surrounding the island. Thus far, nature had not been seriously hurt. Then in the later 1800s, flush toilets appeared. How natural it seemed to connect them to the already established "storm sewers." It was only necessary to rename them "combined" sewers. These now carried both storm water and so-called "sanitary" (really polluted) drainage to the rivers. Fast-flowing rivers are natural sewage treatment plants, and surprisingly they operated as such for many decades, doing a fair job of keeping pollution reasonably in check. With the prospect of future sewage treatment plants, separate "sanitary" sewers were built. And, indeed, there were also some remaining (and some newly built) storm sewers that were not combined and that did not carry the wastes from toilets. In cities where this confused pattern of sewer systems still exists (this

Fig. 9.1 Pollution problems in big cities. The Ward's Island Plant in New York City marked a beginning in coping with the problem of polluted waterways. Major parts of two of the five boroughs were switched from the "discharge-to-rivers" system to the efficiency of central treatment.

In the center of the aerial photograph is Manhattan Island, identified by Central Park. The island is flanked in the foreground by the Hudson River and, on its other side, by the East River, identified by Roosevelt Island. To the north (upper left), Manhattan is separated from the Borough of the Bronx by the Harlem River. Numerous other plants in New York City have joined this fight for cleaner waterways.

includes most of the larger and older cities), it is extremely difficult and expensive to sort out and reroute sewers so that *only* sanitary drainage goes to treatment plants and *all* storm drainage to waterways or into the ground.

The cost of separating the combined public sewers was reported in "Problems of Combined Sewer Facilities and Overflows, 1967" by the Department of the Interior, Federal Water Pollution Control Administration WP-20-11. In only six of fifty states was there no evidence that combined sewers were used. These states were: Arizona, Hawaii, Louisiana, New Mexico, Oklahoma, and Utah. In the forty-four other states, in which combined sewers *were* used, the "bottom line" totals of comprehensive state-by-state estimates were:

Complete separation of combined sewers (public)	\$30,391,645,000
Plumbing changes to affected buildings	18,378,259,000
Total separation cost	\$48,769,904,000

An example of large-volume public sewage treatment is the Ward's Island Sewage Treatment Works in New York City (see Fig. 9.2). Handling over two hundred million gallons per day (mgd), it serves large parts of two city boroughs, Manhattan and the Bronx. Along the shorelines, "intercepting" sewers were built to head off the multitude of street sewers previously emptying into the waters surrounding the city. They lead the sewage through deep under-river tunnels to the island plant.

Only two-tenths of 1% of sewage is solid material. Sewage, therefore, represents the rate of daily water supply with only a small amount of impure solid material added that pollutes the whole. This is a clear reason for the separate handling of storm water for dispersal to the ground or to waterways rather than permitting it to swell the volume of sewage that requires treatment.

Ward's Island uses the activated sludge process—one of the more efficient methods of treatment. Grit settles out at chambers in Manhattan and the Bronx. Together with other intercepted solids such as pieces of wood it is trucked to outlying districts where it is dumped as "garbage fill" along with garbage from other sources. In the preliminary settling tanks, other heavy solids drop during an hour's pause of the sewage flow. The remaining highly polluted fluid flows through the aeration tanks for a 3-hour trip in the presence of a biologically active culture. As a result of bacterial action, accelerated by air pumped in along the way, digestion takes place and the fluid can be made 95% pure before it is discharged at two points to the river. The resulting floc, digested and much purified, is collected in the final settling tanks. Earlier practice of dumping floc at sea will stop in 1981.

9.3 Three Kinds of Water

Before classifying the degrees of water pollution arising from *human occupancy*, it is necessary to say that the subject of *industrial wastes* is rather special. Although highly important to ecologists, engineers, *and* architects, it is so extensive and diverse that it is not feasible to include it in this discussion. It is sufficient to say that industrial water pollution should receive expert attention with a view to complete treatment, prior to the return of the purified wastes to nature.

Figure 9.3 illustrates the type of installation, overwhelmingly common in older cities, that discharges the kind of mixture that is fed through combined sewers to municipal sewage treatment plants. Descriptive terms are being applied to the three grades of waste water. They are: storm water, grey water (wastes from laundries, wash basins, sinks, showers, bath tubs), and black water (the water-plus-human waste that is flushed out of toilets and urinals).

Although cities are making some progress in improving their existing facilities, a distinct objective statement is needed about the planning that should apply to new buildings in new communities. This planning should include:

1. Storm water to be recharged to the earth.
2. Grey water to be collected separately

Fig. 9.2 *Sewage treatment works, Ward's Island, New York City. Activated sludge process.* (A) *Manhattan grit chamber.* (B) *Manhattan sewage tunnel.* (C) *Bronx sewage tunnel.* (D) *Laboratory and Administration.* (E) *Power plant.* (F) *Pump and blower building.* (G) *Preliminary settling chambers.* (H) *Aeration chambers.* (I) *Final settling chambers.* (J) *Sludge storage building.* (K) *and* (L) *Pure water discharge.* (M) *Dock for sludge boats.*

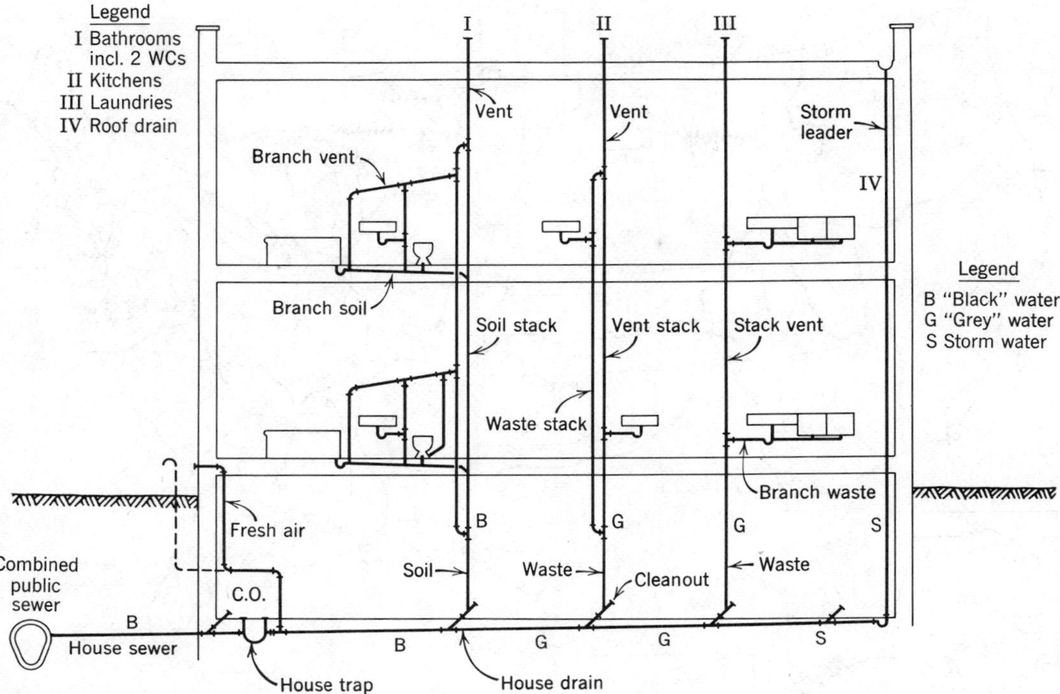

Fig. 9.3 Plumbing section of a "water pollution factory" in a big city. Before the construction of municipal sewage treatment plants, all of the polluted wastes from city residences such as this one were led to the nearest stream or river, polluting in turn the major waterways and the ocean. The plants have relieved this problem to some *extent. However, "combined" sewer facilities, in time of heavy rain, deliver more polluted (type B) water to a plant than it can handle. Much of it has to be bypassed and left untreated; it then finds its way to major river basins. In older major cities this problem of combined sewers may never be fully corrected. In planning new communities our challenge is clear. Storm water must be returned to the earth. Black and grey water can then be treated with greater ease and the purified effluent returned to the earth. Eventually the separation of black and grey water may be anticipated; this could further ease the problem. In the future, piping and fixtures within buildings could be different.*

and given the lighter treatment sufficient for its purification.

3. Black water to be collected separately and then treated, as required, in city plants, locally, or at each building site.
4. Alternatively, eliminating the incidence of black water by the use of composting processes, such as Mullbänk and Clivus Multrum, both described in Section 9.8 of this chapter.

These are goals. We await to see their attainment by these or other methods. Some have already been achieved.

9.4 River Water for Drinking

Those of us who have camped and hiked in wilderness areas cherish the recollection of a draft of fresh clear water from a spring or stream. But let's get back to "civilization." Rivers, including many of the largest ones, can be sources for drinking water. A river can also function for sewage treatment. It operates to the same purpose as process "*H*" in Figure 9.2, where sewage flows for great distances through aeration chambers in which river turbulence is simulated by compressed air pumped into the chambers.

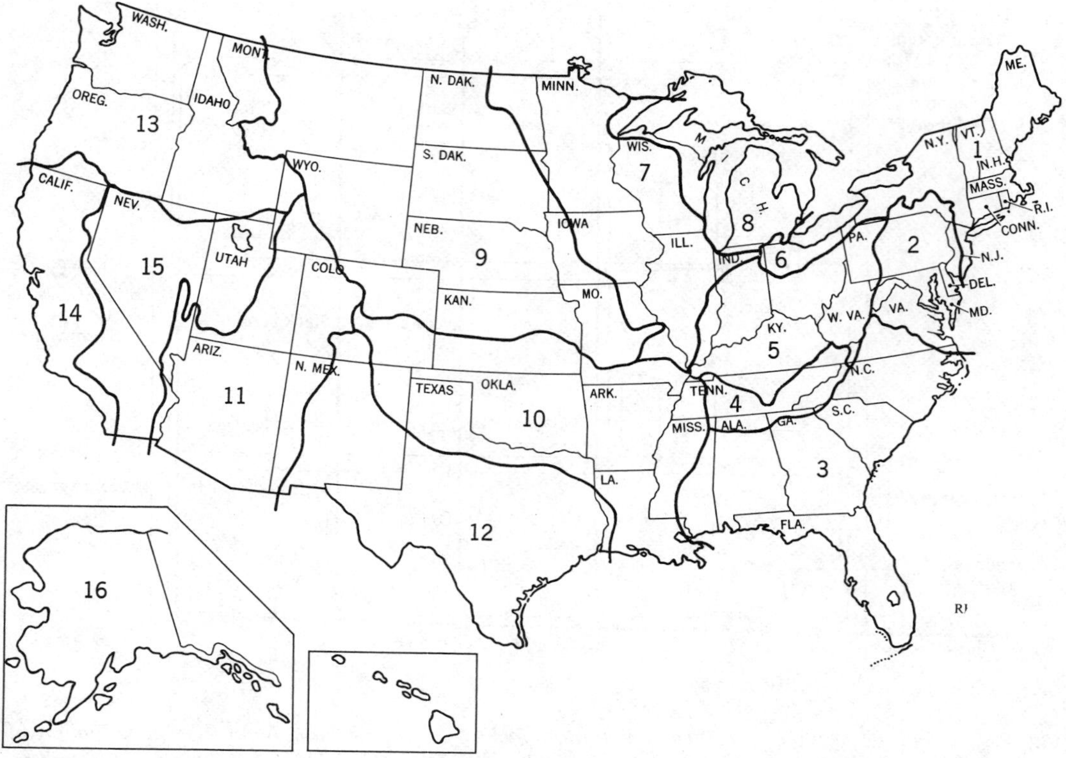

Fig. 9.4 *Major river basins of the United States. Reprinted by permission from "Problems of Combined Sewer Facilities and Overflows, 1967" of the Department of the Interior Federal Water Pollution Control Administration WP-20-11.*

Legend

1 Northeast	*6 Lake Erie*	*11 Colorado River*
2 North Atlantic	*7 Upper Mississippi*	*12 Western Gulf*
3 Southeast	*8 Western Great Lakes*	*13 Pacific Northwest*
4 Tennessee River	*9 Missouri River*	*14 California*
5 Ohio River	*10 Southwest-Lower Mississippi*	*15 Great Basin*
		16 Alaska

Author's note: *These are the river basins that we have been polluting. In nature's pattern, they should be fed by the underground flow of rainwater to the nearby streams. The rainwater is a reasonably pure product and would be futher purified by the natural flow through the ground on which it falls. Our deep wells have sometimes lowered groundwater levels, often reversing the gradient of natural underground flow to streams. Thus, untreated wastes that are dumped into streams and rivers can act to pollute not only the waterways but also the surrounding earth. We owe nature the service of cleaning up the water we have used and putting it back into the earth where it belongs, and whence it can be retrieved and reused.*

Drinking water supply or sewage digestion? Unfortunately, we have been asking our rivers to provide both (see Fig. 9.4). This is a very confusing challenge for a river. In Figure 9.5, we see a nine-process treatment used to make polluted river water suitable for the drinking water supply of a large city in the midwest. Sewage treatment plants on that river are now replacing the direct dumping of raw sewage. This is pleasant news for local

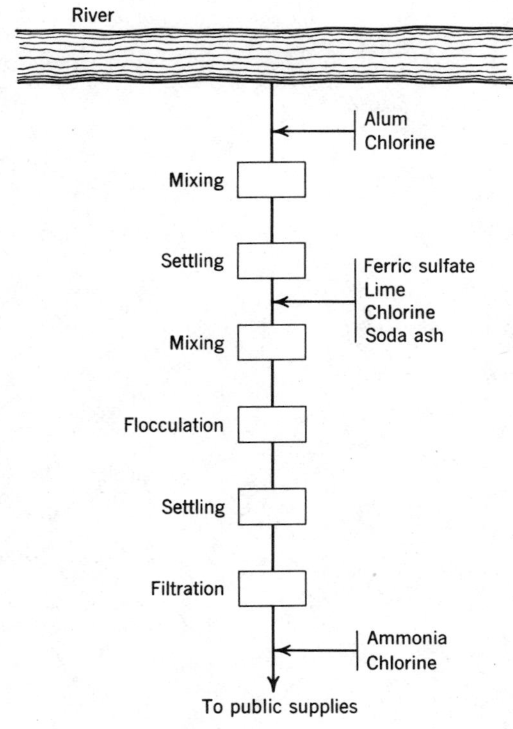

Fig. 9.5 An example of city water treatment. Treated water for a major city using river water as a source. From Water Quality and Treatment, *2nd ed. American Water Works Association, Inc.*

residents who felt it necessary to receive inoculations before water skiing on the river, lest they fall in and drink some of the river water before it went through the nine-process cleanup.

Doesn't it seem logical to tie directly into the ten or more processes of sewage treatment shown in Figure 9.2 and, with a bit more "polishing up," drink the reclaimed water? It does to this writer who, at the time of the completion of the Ward's Island Plant, was offered a clear glass of the purified water by a very proud plant operating engineer. Although declined at the time, the gesture inspired a bit of confidence in the expertise of our scientists.

Conditions *do* improve! In the New York area numerous shad were recently seen in the Hudson River.

9.5 Sewage Treatment in the Front Yard

In suburban areas, where sewers and treatment plants have not yet been built, private on-site treatment of sewage from residences is still (under control) a valid procedure. Methods of treating the sewage and returning the partially purified effluent to the ground are shown in Figure 9.6. A broader coverage of this subject, including design standards, will be found in Chapter 11. Locating the system in the front yard is advised, because a connection to a future sewer is thereby facilitated.

The increasingly serious attitude of health officials cannot be better expressed than by using the following unabridged quotation from *Standards for Individual Sewage Disposal Systems,* which states the requirements of the Nassau County Department of Health in Mineola, New York. Nassau County is a fast-developing suburban area, directly adjacent to the boundary of greater New York City.

Introduction

This Manual has been prepared by the Division of Environmental Health of the Nassau County Department of Health for the guidance of consulting engineers, architects and land surveyors interested in the design and installation of individual sewage disposal facilities within Nassau County. It outlines the type and extent of reports and plans to be submitted in application for approval of facilities, supplies design criteria, gives recommendations for types and arrangements of facilities and describes the policies and procedures of the Department pertaining to realty subdivisions and individual sewage disposal facilities.

General

The individual sewage disposal facility is a system designed to treat sewage from one building plot and dispose of it on the site itself, usually by some type of soil absorption system consisting of a septic tank followed by either leaching pools or tile fields. The individual disposal system cannot be considered a satisfactory permanent method of sewage disposal as compared to a connection to an approved

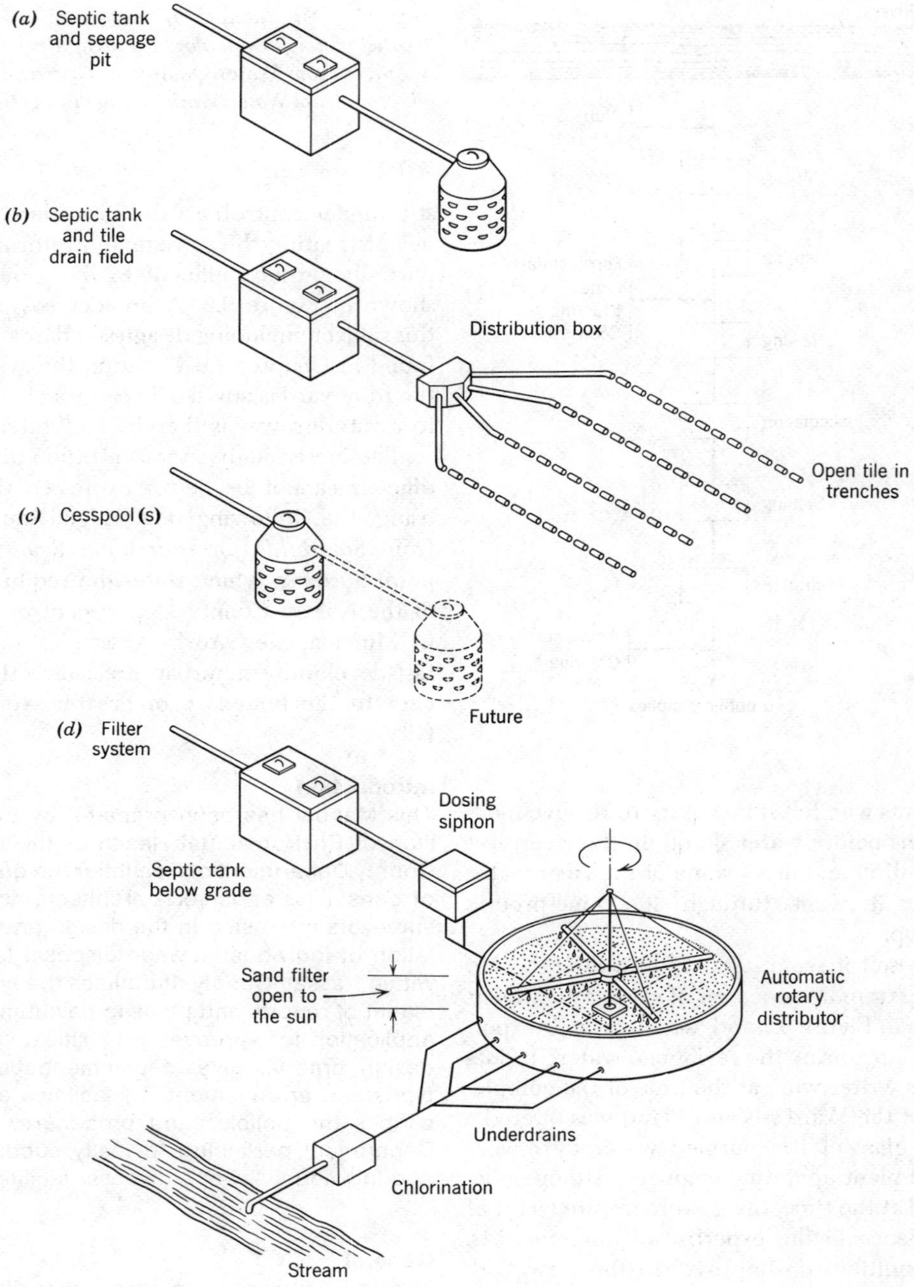

public sewerage system. It is this Department's policy to encourage connection to a sewerage system rather than approve the construction of individual disposal systems.

Individual wells used for drinking purposes will not be approved for any realty subdivision or individual disposal facility where a public water supply is available.

The success of an individual disposal system depends on many factors including the presence of adequate absorptive soil on the site, amenable ground water conditions, ade-

Fig. 9.6 (opposite) *Private on-site sewage treatment. This is* not *the best way to clean and recycle water. These methods are quickly going out of favor, except where there is no better alternative and much open space between dwellings. Type* (b) *which employs a tile drain field, is usually the better method, though it is more expensive to install than* (a). *Types* (a) *and* (b) *may be approved by health authorities, but* (c) *is fast becoming illegal because it puts raw sewage in direct contact with the earth. If there is a sanitary sewer in the vicinity, officials will invariably decline any permission for private treatment. Type* (d), *the filter method, sometimes used in low and marshy ground adjacent to a stream, requires careful design and frequent inspection and servicing. If and when grey water is separated and treated apart from the black water from flush toilets, it is reasonable to assume that it will be permissible to treat these less polluted grey wastes as shown in* (a) *or* (b). *In this event, the black water would have to be handled for full purification by some digester process not yet developed for small private installations.*

quate design and construction and proper operation and maintenance of the system by the occupants of the building. The life span of an individual disposal system is limited even when designed properly, constructed in accordance with the approved plans and carefully maintained. Ultimately the soil becomes clogged and the system will fail. Careless use and insufficient maintenance will definitely shorten the life span of the system even further.

The present day trend towards population movement to suburban areas, the development of small lots in order to reduce land costs, the decrease in available sites with desirable soil and water conditions for individual systems, and the increasing use of water, make necessary a more detailed review of the proposed individual disposal systems. Individual disposal systems were originally designed for rural areas where considerable land was available and failures of such systems did not present a public health threat of the same magnitude as would exist in a built-up residential development. Since rural conditions no longer exist within the County, consideration should be given for complete replacement of the original disposal system should it become necessary before sewers are available.

Small lots *do* pose problems (refer to Fig. 9.7). Consider now the possible need for a 36 ft × 50 ft drain field plus septic tank on a smaller (40 ft × 100 ft) suburban lot with minimum clearances (house to disposal field and property line to disposal field). Planning becomes difficult. Another look at Manhattan Island (Fig. 9.1) indicates the kind of congestion that lies ahead for suburbs that later become built-up cities.

9.6 Recycling at New York Institute of Technology

A new method of sewage treatment, known as the Pasveer Oxidation Stream, employing a principle not unlike that shown in Figure 9.8, has been adopted at New York Institute of Technology. Serving the new 450-acre campus at Old Westbury, Long Island, New York, it provides an on-campus sewage treatment, which returns the purified effluent to the ground through 48 leaching wells located under the athletic field. The groundwater, thus restored, aids in providing a contributing source of water for 400-ft-deep wells, distantly located, that furnish part of the water supply for the campus buildings.

During the development of the campus in the 1960s and 1970s, studies were made of the sewage problem that would develop with the growth of the college. At the beginning, there were a few small existing buildings later converted for administration offices and classrooms. These were, and still are, served by septic tanks and leaching fields. Presently, they constitute less than 20% of the sewage flow of the expanded building complex and may in future years be connected to the campus plant.

There was no public sewer near the campus, and the health authorities ruled out the acceptability of using septic tanks for the numerous additional buildings that were contemplated. Accordingly, the college retained Curtis & Davis, Architects, and S. A. Bogen, Engineers (now Bogen Jenal, Engineers, P. C.), and consultants to design a treatment plant.

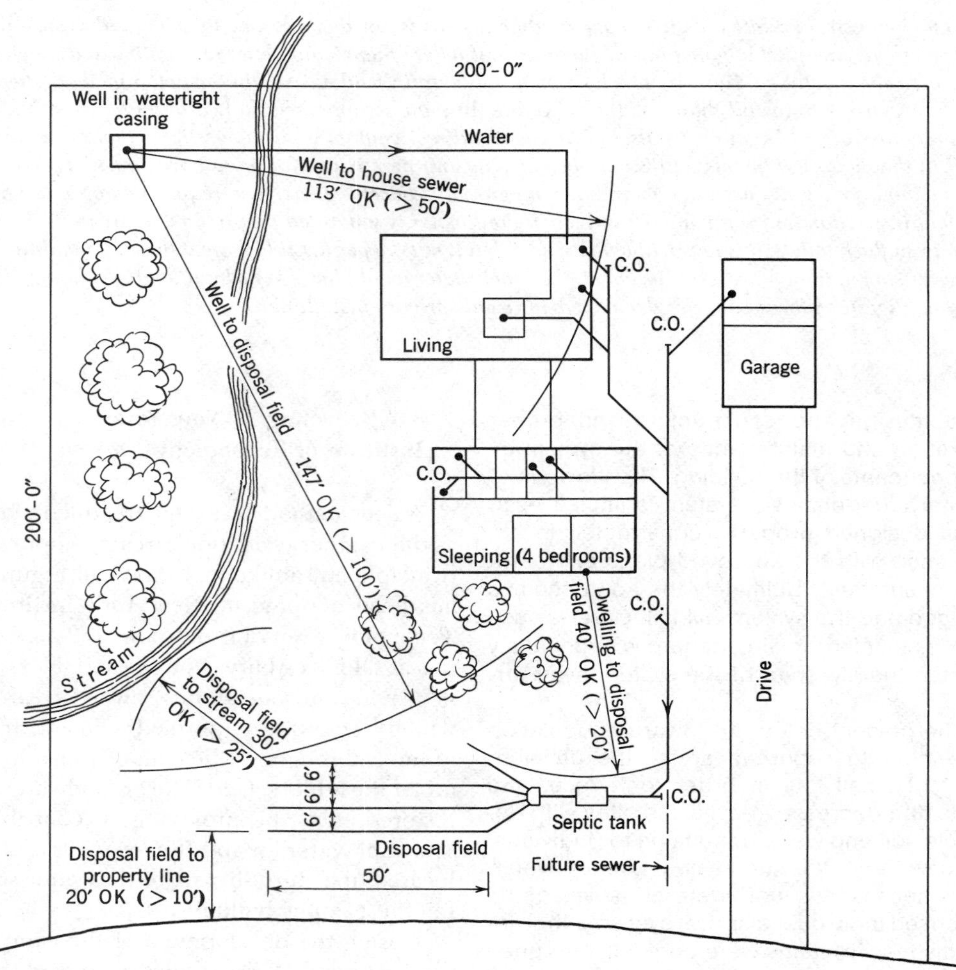

Fig. 9.7 Protection of wells and streams. Health authorities are becoming increasingly alert to the pollution of wells and streams that is caused by the effluent from relatively inefficient private sewage treatment systems. Admittedly, it is somewhat unusual to find, as shown here, a well, a stream, and a tile drain field on a private lot. However, as indicated, the minimum standards of distance-separation (now being increased in many codes) would require a 200-ft. square lot.

Figures 9.9 to 9.13 illustrate the plant, the location of which is shown in Figure 9.12. In a proposal to Dr. John J. Theobald, Executive Vice President of the Institute, Samuel Adams Bogen, P. E., included several paragraphs that are significant to our understanding of the process. They are:

Process

The oxidation stream process is a modified form of the activated sludge process and may be classified in the extended aeration group. It is not a mechanically aerated lagoon. The biological removal and conversion of organic solids occur simultaneously in the aeration basin under aerobic conditions. Aerobic conditions are always maintained by induction of oxygen into the liquid with a highly efficient mechanical aerator. The surface aerator also has excellent propulsion capabilities which keep all solids in suspension.

Operation

The oxidation stream is operated as a closed

Fig. 9.8 *A sewage treatment plant in the Austrian Alps. It is, of course, nothing of the sort. However, this natural, remote, turbulent sparkling mountain stream has the elements for effective sewage digestion—sunshine and a turbulent flow that aerates the water and provides an adequate oxygen content.*

Fig. 9.9 *A quiet lagoon on a country campus. Unlike the wild mountain stream shown in Figure 9.8, this is, in fact, a sewage treatment plant. It does not appear "turbulent" in this photograph, but in operation it becomes so under the action of a mechanical rotor; see Figure 9.10a and b. This action classifies it as an "oxidation stream." It provides the beginning of a water recycling process on the Old Westbury Campus of the New York Institute of Technology.*

Fig. 9.10 (a) *Rotor with its plastic "greenhouse" cover.* (b) *The rotor induces oxidation in this oval, circulating stream.*

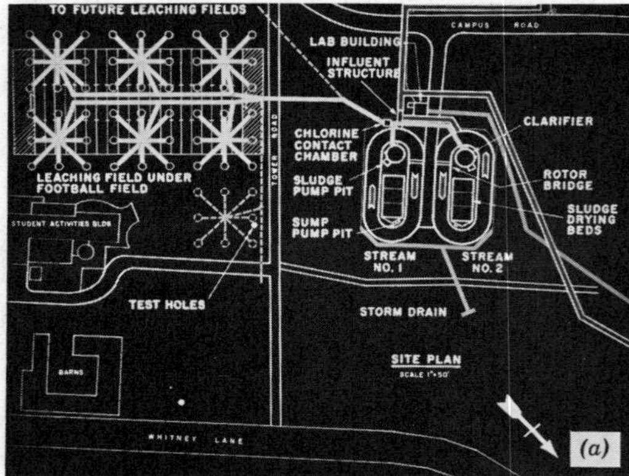

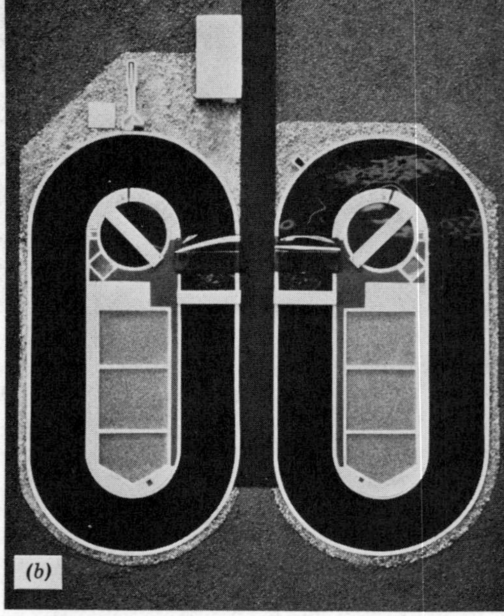

Fig. 9.11 Sewage treatment plant, New York Institute of Technology. (a) *Layout of the design by S. A. Bogen, Engineers, now Bogen Jenal, Engineers, P.C.* (b) *Plan of the two streams. Each of these units includes the stream, rotor, clarifier with adjacent sludge pump pit, and three drying beds. Because of its odorless operation, the proximity of the plant to the Student Activities Building poses no problem.*

system and the net growth of volatile suspended solids will increase so that it will periodically be necessary to remove some sludge from the process. Wasting of excess sludge lowers the concentration in the stream and keeps the metabolism more active. The excess sludge formed in the process is sufficiently broken down so that the water binding properties, present in sludges produced from other processes, is not present. The surplus sludge may be discharged without offensive odors for direct drying on open sludge drying beds.

Origin of the Method

The oxidation stream [the Pasveer method] was developed in Holland by the Netherlands Research Institute for Public Health Engineering (T.N.). The oxidation stream was found to be very successful and has been placed in operation in many localities in the United States.

Recently, after the plant had been operating very successfully for over eight years, interviews with Mr. Bogen, its designer, and with Mr. William J. Donaldson, Director, Building and Grounds Department, under whose jurisdiction the plant is operated, yielded some interesting comments about its design and operation.

Mr. Bogen:

The design had provided for a 4,330 equivalent population and a 340,000 gallons per day flow. It is now operating at less than 1/4 of its design capacity. Storm water is, in all cases, run off separately to sub-surface dry wells and does not enter the plant. Neither does ground-up garbage enter the plant. Three authorities have jurisdiction over the plant. They are:

Nassau County Department of Health
New York State Health Department
Village of Old Westbury

The installation is the first of its type in New York State. There are hundreds of such installations in Europe; twelve or more in the United States and a number in Canada. The process is considered simpler than other methods in that no compressed air is used. [As it is in the Ward's Island Plant; see Fig. 9.2.]

[Mr. Bogen feels that] with a more sophisticated adjustment of the final chlorination, the effluent, instead of being recharged to the ground, could be mixed directly with the local potable water supply. Presently though safe and potable, it is too heavily chlorinated to be pleasant and *palatable.* This system [the

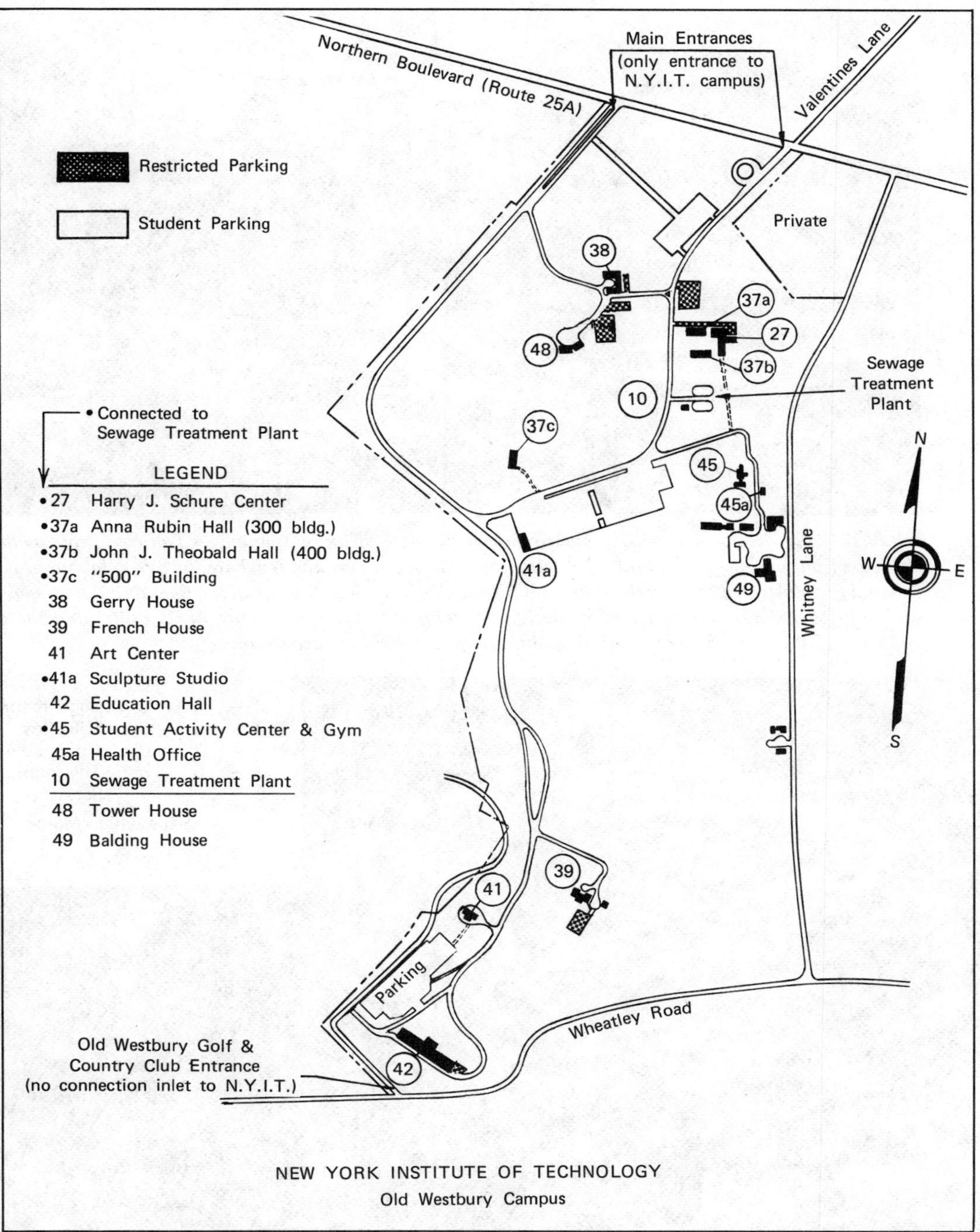

Fig. 9.12 Campus planning. The decision to locate the treatment plant at a position central to all the buildings it serves was a very appropriate choice for efficiency. Surrounded by heavily wooded areas (see Figs. 9.9 and 9.13), it is not a prominent landmark, though it offers a pleasant appearance to the occasional viewer. Although only half of the campus buildings are presently connected to the treatment plant, they are the largest ones and are responsible for 80% of the total campus sewage flow. A frequently used walkway (dotted lines on the map) connects the north academic center with the student activity and athletic center. This path passes just east of the woods that surround the plant. At certain points the plant can be seen through the trees, but no odor betrays its presence.

Fig. 9.13 The "Tech" sewage treatment plant at the time of its completion. A feature of this method is the low visual profile of the plant. The only projections above ground level are the office/laboratory at the left and the plastic covers over the rotors. Of the two lagoons, only the one in the foreground has been used, and only to a small percentage of its design capacity. The unused balance of its rating and the entire capacity of the second lagoon are available for future campus expansion.

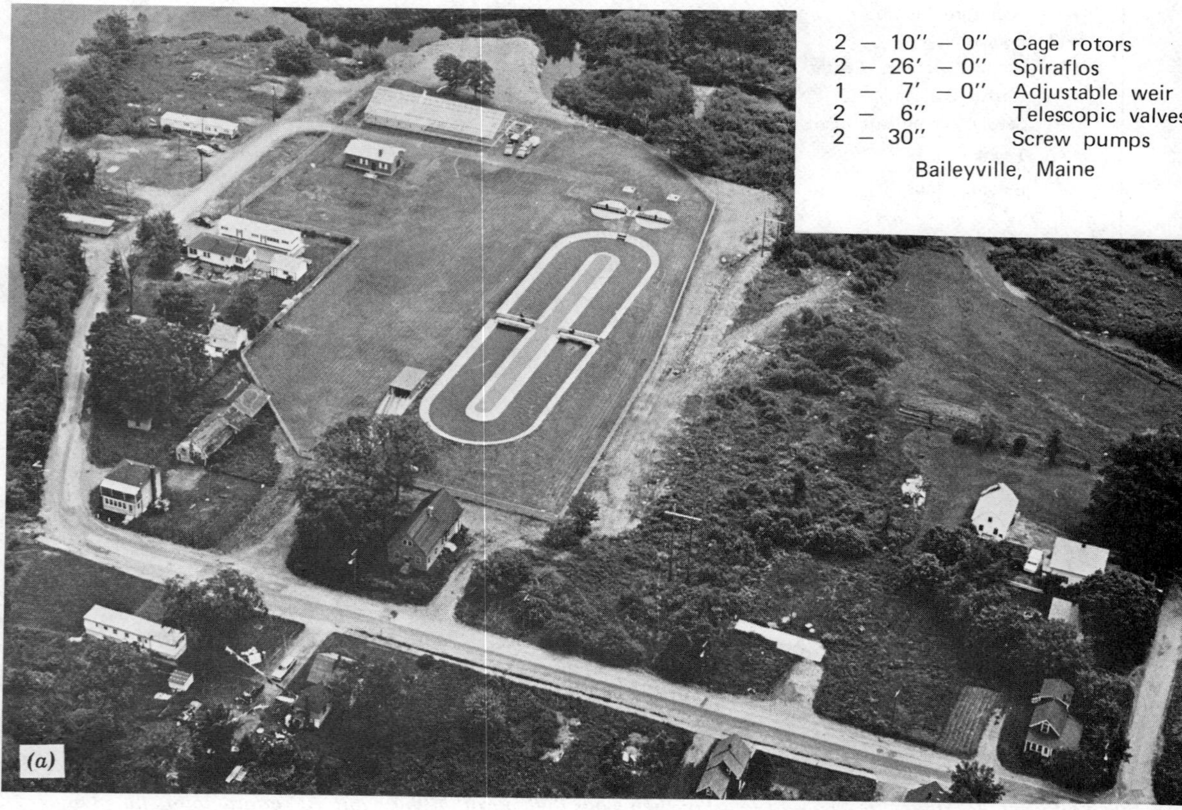

2 – 10″ – 0″ Cage rotors
2 – 26′ – 0″ Spiraflos
1 – 7′ – 0″ Adjustable weir
2 – 6″ Telescopic valves
2 – 30″ Screw pumps

Baileyville, Maine

(a)

Pasveer Method] appears to be most suitable for communities up to 15,000 to 20,000 people.

Mr. Donaldson:
The plant is operated by a full-time accredited operator. There are two assistants and one relief operator. Periodic inspections are made by the Nassau County Department of Health. Buildings served (see Fig. 9.12) represent 80 percent or more of the total campus load. Future leaching pits can be installed if a third lagoon is ever added. Though most of the storm drainage from roofs and paved parking areas goes separately to dry wells, *some* storm run-off goes, *also separately,* to wooded areas where it is quickly absorbed by tree roots, leaving no ponding. Sludge, dramatically small in quantity, has been removed from the drying beds only twice during five years of operation, and buried to form a fertilizing mulch.

The two lagoons of the New York Tech plant are shown in Figure 9.13. The design was honored for "Engineering Excellence" by an award to S. A. Bogen, Engineers, presented by the Consulting Engineers Council.

Other installations, also using equipment of the Lakeside Equipment Corporation, are shown in Figure 9.14*a* and *b*.

9.7 The Santee Water Reclamation Plant

The expression "total use of water" and similar terminology such as waste-water rec-

Fig. 9.14 (a) (opposite) *and* (b) *Other installations using the Pasveer Oxidation Ditch. A spokesperson for the Lakeside Equipment Corporation of Bartlett, Illinois, makes the following comments about the treatment method: "Distinguishing characteristics of interest to architects and planners, aside from low capital and operating costs, are the plant's low profile, the odor free and nuisance free operations which enable dwelling units to be placed nearby* [*see Fig.* 9.14a], *and the potential variability in ditch geometry* [*see Fig.* 9.14b]."

lamation, closed-water management, and water renovation relate to a dramatic new development that may revolutionize the conservation of our water supplies.

The impetus that led to a most significant achievement in water reuse was best summarized by Mr. Ray L. Stoyer.[2]

The changing economics of water supply and pollution control are moving us inescapably toward the establishment of closed water management systems—systems in which we can use our water over and over until it is lost to evaporation, transpiration or other unavoidable "leaks." Realization of this truth came to a small community in Southern California when it was faced with a situation in which the cost of both water supply and waste disposal would become prohibitive in the near future unless a better approach to the management of these vital functions could be devised and implemented.

We are indebted to Mr. Martin O. Poe, Plant Superintendent, for the updated information included in Figures 9.15 to 9.18 and for the following discussion.

The Santee district is a region where rainfall is less than 15 in. per year. It has no local water supplies available and water has to be imported 300 miles from the Colorado River. In a few years, the district is expected to face the necessity of extending its already lengthy California aqueduct an additional 500 miles to the Columbia River. At the same time it was contemplated that the waste water resulting from this flow might have to be thrown away by discharging it into the Pacific Ocean. Adequate sewage treatment could process this fluid waste for secondary uses such as irrigation and recharging of ground water. Considering further that many municipal supplies for potable water were taken from rivers often heavily polluted with sewage (see Fig. 9.5) and, after treatment, pumped into the domestic water mains. Santee officials decided to attempt to perfect a closed system that would make use of purified waste water for many secondary uses and hopefully for ultimate reuse as potable water.

Contrary to common opinion that such a concept might not be acceptable to the public, an effective public relations program concurrent with the technological advances has won over the citizens of the district.

The project involved building a sewage treatment plant and utilizing seven pits left over from a prior unrelated operation consisting of surface mining of sand and gravel. After partial purification of the sewage at the plant by the first two stages of the conventional activated sludge process, the effluent is discharged to form lakes adjacent to the plant. This provides tertiary (oxidation) treatment after which the purer effluent is pumped to a filter area at the north end of the complex where it is further purified by flow through sand and gravel. Chlorination is administered at one point. The water then flows through a series of seven lakes Nos. 7–6–5–4–3–2–1, in that order.

The appearance of new lakes in this semiarid region created an initial interest that was rapidly augmented by a very clever program. At first the lakes were fenced in. Then the fences were removed. The seven lakes were next made available for boating. They were stocked with fish and careful studies were made which indicated that the fish were healthy and flourishing. Fishing was permitted, but for a while all fish had to be returned to the water. Anglers were later permitted to keep and consume the fish. Finally swimming was permitted, a use which connotes pure water that is in no way harmful to health. The overflow from the last lake, No. 1, discharges into Sycamore Canyon Creek and is used for the irrigation of a golf course and the recharging of ground water. It is thought that in the near future these ground waters may be pumped in controlled fashion into the municipal water supply mains.

Technological progress and social acceptance came about during the decades of the 1960's and 1970's. The project was monitored by a committee whose members represented:

United States Public Health Service
California State Water Pollution Control Board
California State Department of Public Health
California Department of Water Resources
California Department of Fish and Game
San Diego Department of Public Health
Santee County Water District.

These agencies agree that the project presents an ideal opportunity for studying the

[2]During the early development of the reclamation project Mr. Stoyer was General Manager of the Santee County Water District.

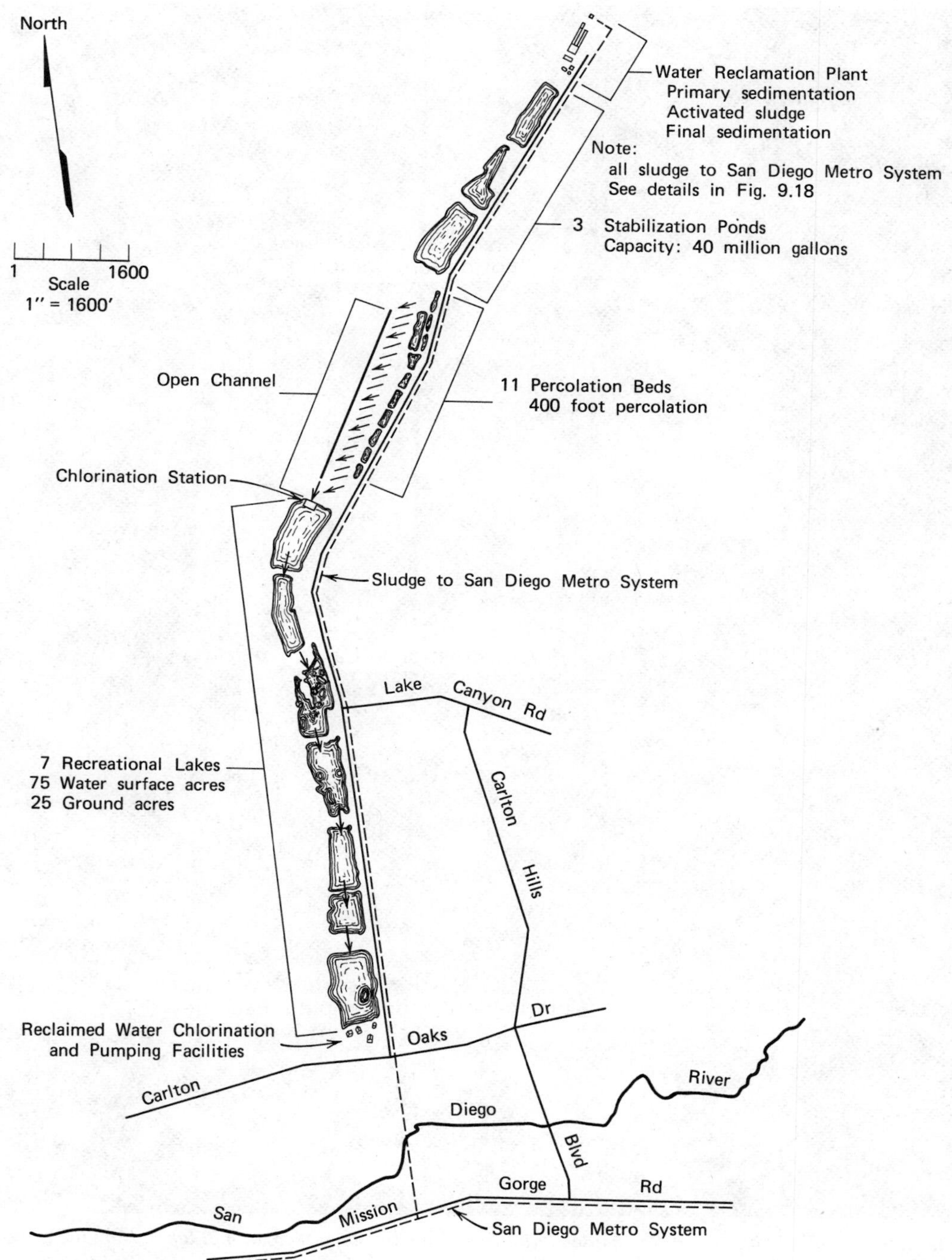

Fig. 9.15 *Santee Water Reclamation Plant and Santee Park and Recreational Facilities, Padre Dam Municipal Water District. Raw sewage from the community of Santee enters the treatment plant, which is located at the top of this diagram. The process then proceeds southward to the point where reclaimed water is pumped to customers. Sludge does* not *enter the San Diego River but is pumped to the San Diego Metro System. There is also a provision by which the sludge can be treated aerobically at the plant. For details of the plant see Figure 9.17.*

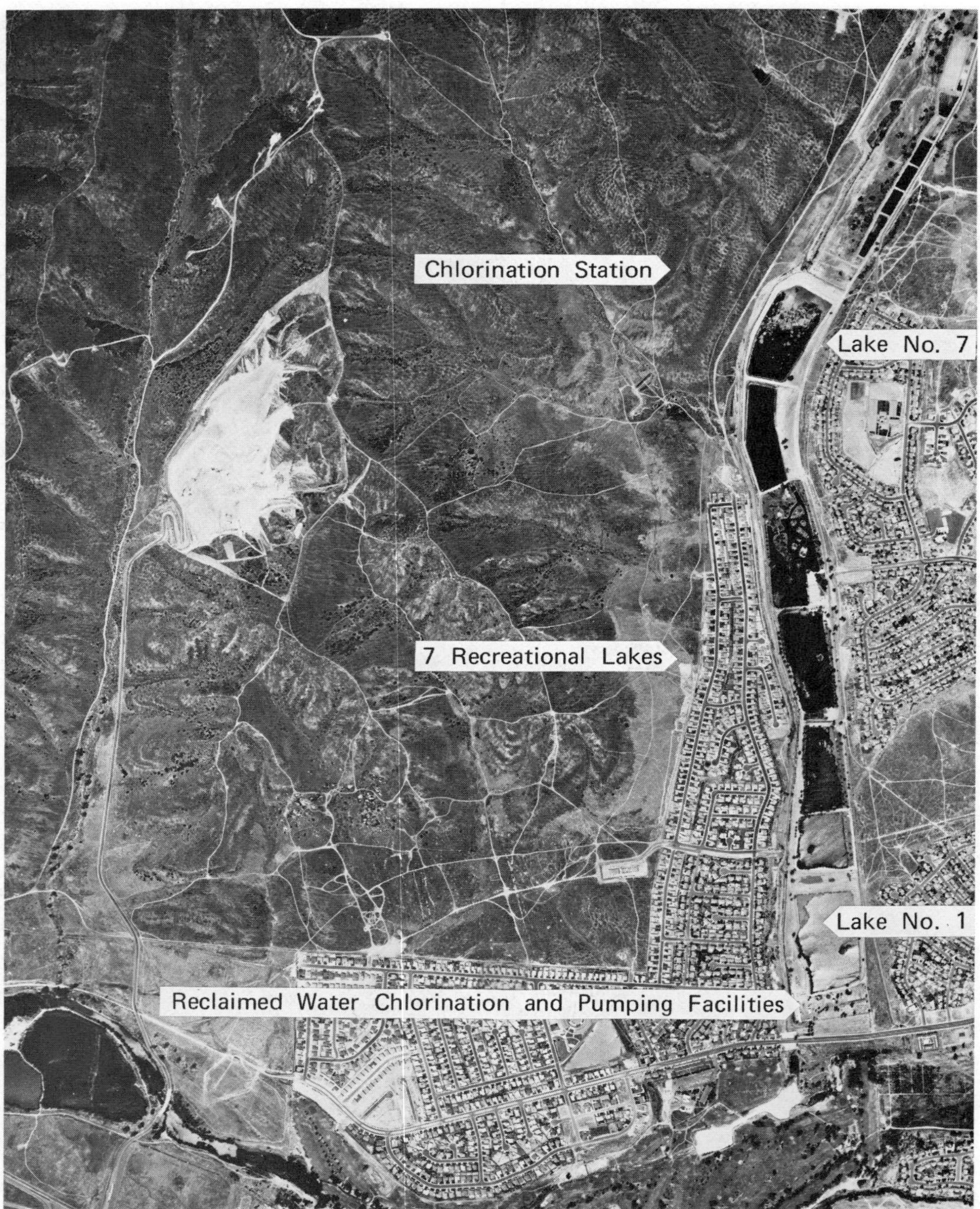

Fig. 9.16 Aerial photograph of the Santee Plant including the seven recreational lakes. Note the location of the San Diego River at the bottom of the photograph. A comparison can be made with the plant diagram shown in Figure 9.15. The water reclamation plant and the stabilization ponds do not appear in this aerial picture but can be seen in the diagram in Figure 9.15.

reuse of waste water for recreational and recharge purposes. It is expected that surveillance of the operation of the system will continue.

9.8 Can We Eliminate "Black Water"?

As an exercise in classifying drains carrying black, grey, or storm water, the diagram below the photograph in Figure 9.19 is helpful. Storm water *only* is carried by the drains at the extreme left and right. At the central storm drain, the addition of lavatory waste puts it in the grey category. The two remaining (larger size) drains carry away both grey and black water, which puts both of these drains in the black category. It is not for us to be critical of this installation that undoubtedly has been functioning well for many years. Indeed, it is much closer to the kind of separation that we look for than is the plumbing "section" of Figure 9.3. In the arrangement in Figure 9.3, *all* of the water is black because of only *two* flush toilets, and this black drainage is much increased whenever it rains.

(*a*) ***Grey Water.*** Dr. John H. Timothy Winneberger has made an in-depth study of grey water treatment. We are pleased to quote from his book *Manual of Grey Water Treatment Practice:*[3]

When considering reduction of wastewater flows, the biggest water-wasting device that first comes to mind is the water flush toilet. Almost everyone has participated in the use of about five gallons of drinking-quality water to flush out of sight, objects of no more consequence than a cigarette butt. Altogether, waters wasted in flushing toilets approach almost half of all water used in a household.

With the recent advent of esthetically acceptable toilets which do not use water at all, there was a need for a term to refer to wastewaters emanating from other household devices. These wastewaters have been collectively termed *grey water* and our body wastes are termed *black*. The day is coming when western man will recognize "black wastes" to be what they really are: legitimate parts of natural cycles. They could be used to man's advantage, and perhaps soon they will be regarded as a valuable resource. Meanwhile, they remain "black" and, in efforts to rid ourselves of those unwanted materials without fouling immediate environs, it seems easier to handle black wastes separately of grey.

(*b*) ***The Mullbänk Toilet.*** Dr. Winneberger speaks of "esthetically acceptable toilets that do not use water at all." The Mullbänk (formerly Ecolet) toilet is an answer to this requirement. Its function, appearance, and adaptability are illustrated in Figures 9.20 to 9.22. This toilet is further explained by its distributing company in the following list.

FUNCTION OF THE MULLBÄNK BIOLOGICAL COMPOST TOILET

1. The Mullbänk is a toilet with a controlled humidification process.
2. It has been developed in Sweden and is marketed by Recreation Ecology Conservation of the United States, Inc., 9800 West Bluemound Road, Milwaukee, Wisconsin, 53226.
3. Odor-free, it can be used continuously without discomfort.
4. It has a continuous, uniform, biological decomposition when in operation and is mainly activated by mesophilic microorganisms.
5. Organic kitchen refuse can be received, as well as newspaper and waste paper, provided that the paper is first torn up (limited input of kitchen waste, etc.).
6. The Mullbänk enables the hygienic handling of waste at the source. No discharge of waste or chemicals to the earth.
7. It produces a fertilizer and a soil amendment.
8. Domestic water consumption is reduced, which implies reduced treatment of sewage water.
9. It will operate in a wide variety of temperature ranges.

[3]From John H. Timothy Winneberger, *Manual of Grey Water Treatment Practice*, reprinted by permission of Ann Arbor Science, publisher and copyright holder, Monogram Industries, Inc.

SANTEE
COUNTY WATER DISTRICT

1 — Available but not in use
2 — Not available
3 — In use

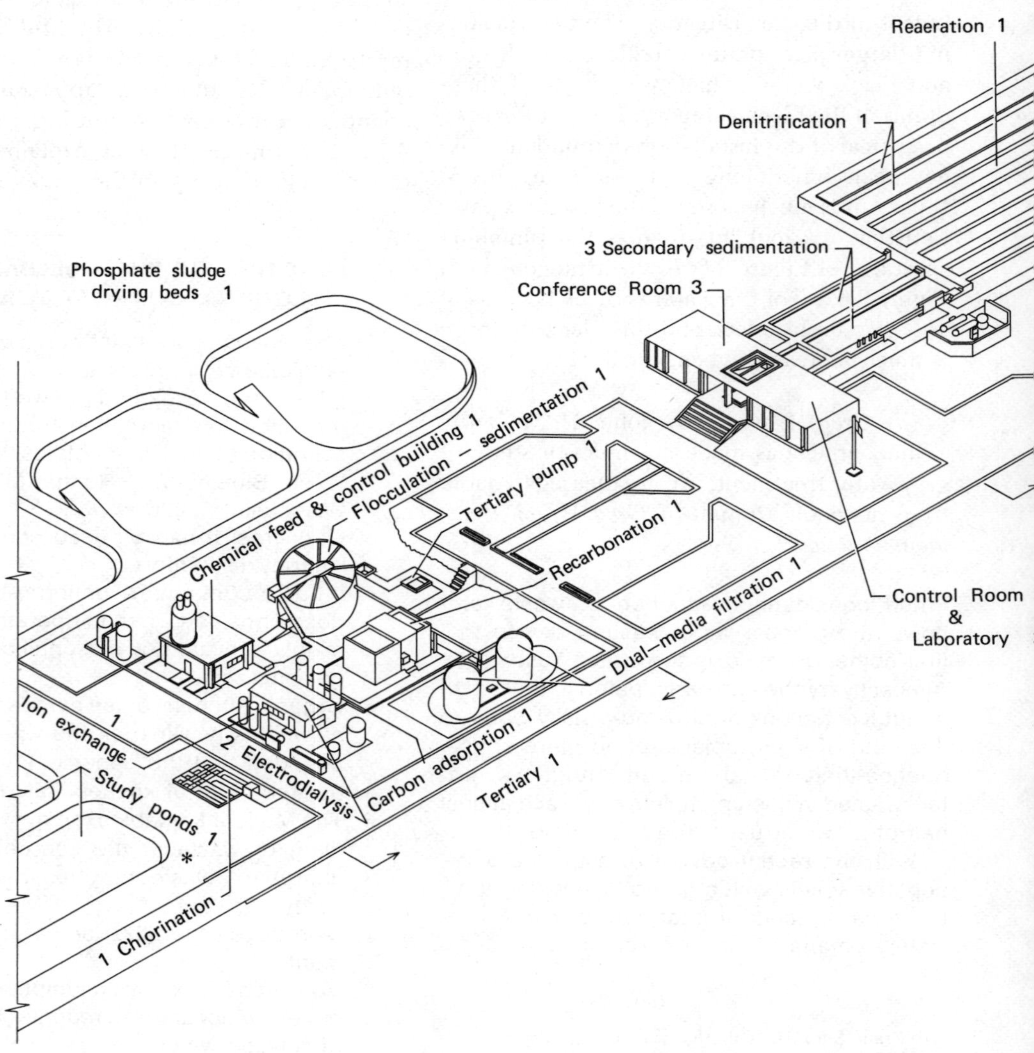

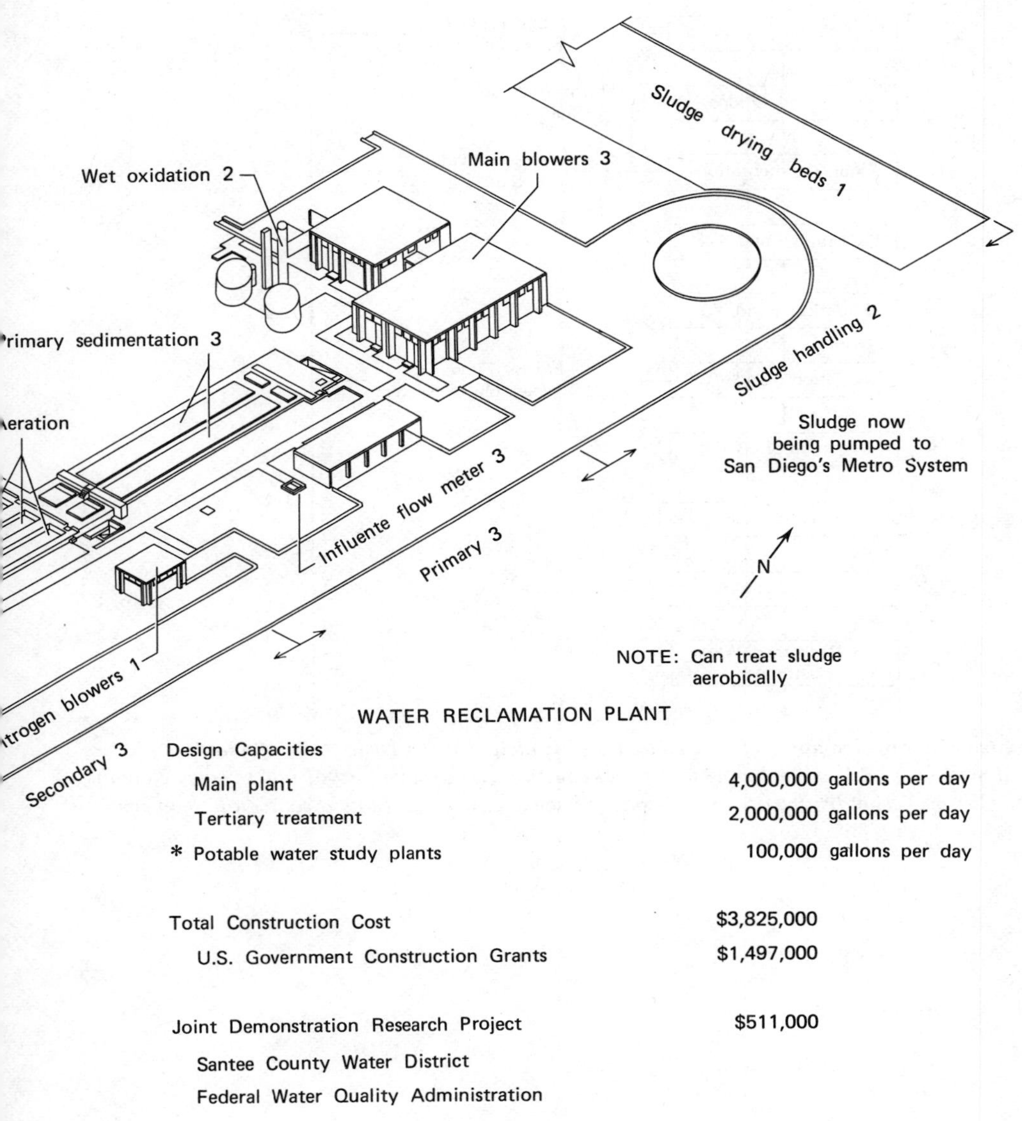

Fig. 9.17 *Details of the reclamation plant. The item "Potable Water Study Plants," marked with an asterisk*, indicates the continuing study and preparation for future distribution of drinking water that is reclaimed from sewage. The plant is not burdened by storm water, which is recharged to the ground where it occurs locally in the community.*

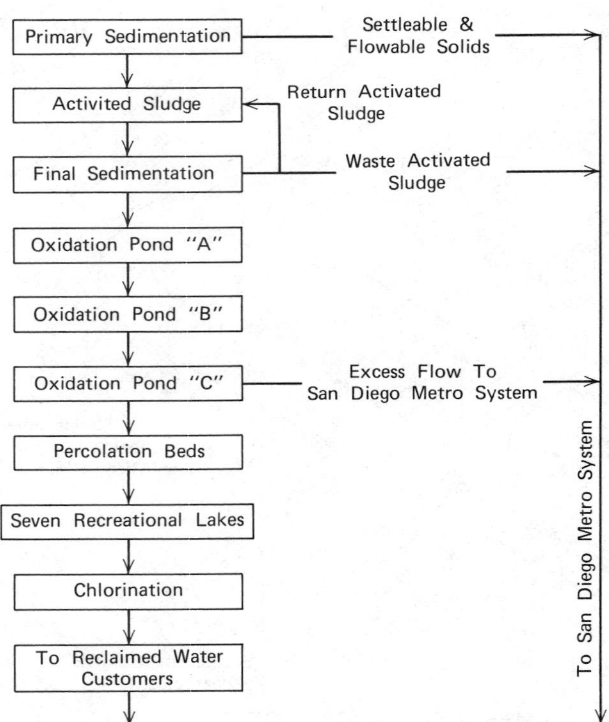

Fig. 9.18 *Santee Water Reclamation System, Padre Dam Municipal Water District. This "laundry" process cleans up and makes reusable the valuable water that constitutes the major part of raw sewage. Note the title: "Water District." The district does not* dispose of sewage, *it operates as a source of clean water.*

Fig. 9.19 (opposite) *Residence of the 27th Vicar of the Parish Church at Bibury, (established* A.D. *1086) in the English Cotswolds.* (a) *It is evident that the plumbing was added after the construction of the residence, perhaps several hundred years later. In England the drainage system of an older building often appears on the* outside *of the building. It seemed inappropriate to ask the Vicar's permission to inspect his indoor facilities, so we have taken the liberty of assigning probable uses and examining this conveniently visual "printed circuit" of drains. We can learn several things about planning that would apply to any system.*

1. *Offsets of the vertical piping are made with "easy bends" and are pitched down in the direction of the flow.*
2. *Vents are located at high outdoor points.*
3. *Stacks carrying so-called "black" water are of larger pipe size than those carrying "grey" water.*

(b) *We can also apply a very new analysis to the system. It is found that there are two each of black, grey, and storm drains. In this very sparsely settled region of the Cotswolds, the dispersal of storm water to the ground and the private septic tank treatment of black plus grey water can be satisfactory. In planning* new *systems the possible separate treatment of B, G, and S waters is becoming an important consideration.*

(a)

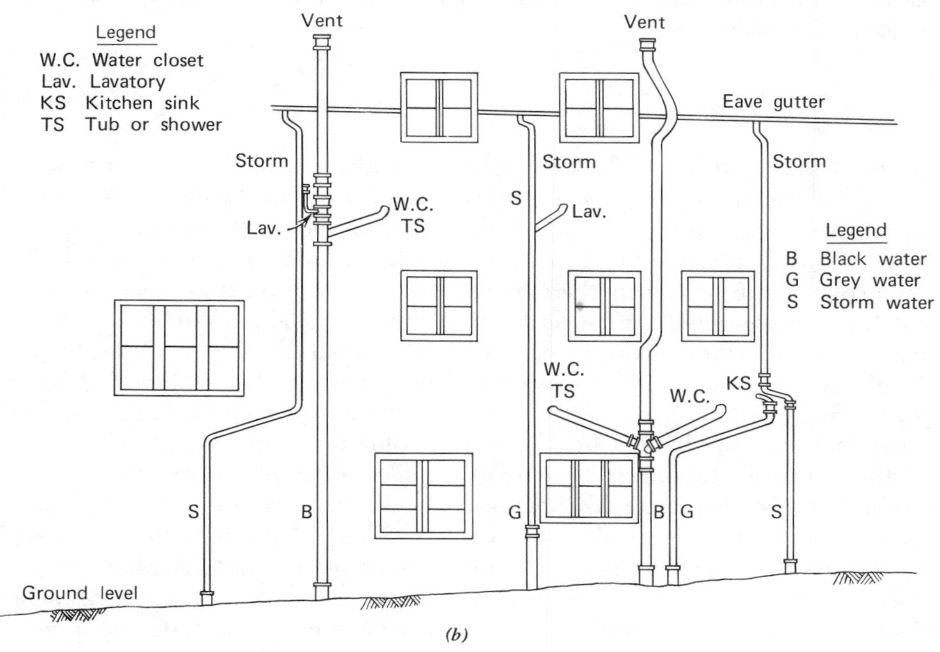
Legend
W.C. Water closet
Lav. Lavatory
KS Kitchen sink
TS Tub or shower
Vent
Vent
Eave gutter
Storm
Storm
Storm
S
W.C.
TS
Lav.
Lav.
Legend
B Black water
G Grey water
S Storm water
W.C.
TS
W.C.
KS
S
B
G
B
G
S
Ground level

(b)

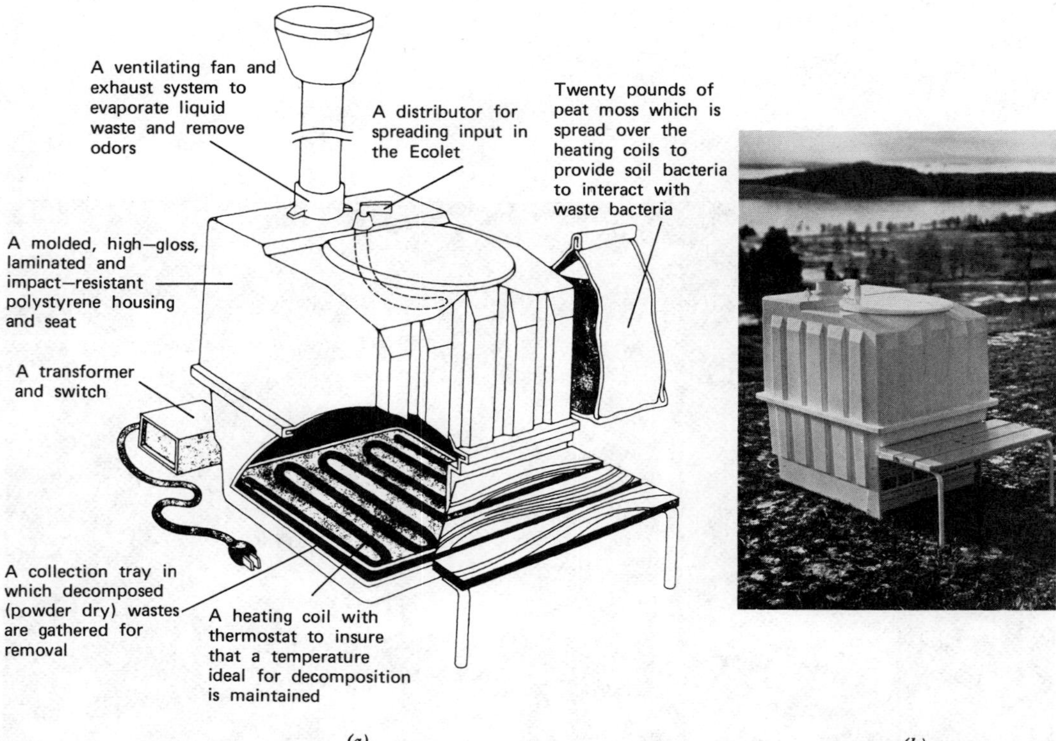

Fig. 9.20 (a) *"Just what the world needs. A toilet you can't flush." A statement from the literature of the manufacturer and distributor of the Mullbänk (formerly Ecolet) toilet. One of the few successful methods of locally digesting human wastes instead of routing them to the general drainage system. It eliminates this major source of pollution and permits separate treatment of "grey water."* (b) *A Mullbänk toilet model exhibited near waterways that it* cannot *pollute. By courtesy of Recreation Ecology Conservation of the United States, Inc., Milwaukee, Wisconsin.*

(c) ***The Clivus Multrum System.*** This method for the treatment of organic wastes is self-contained. It uses no water. Products of the toilet and kitchen are retained in an impervious container from which there is no effluent. It preserves nutrients by conversion to a rich humus suitable for gardens. No external supply of energy or chemicals is needed to effect the decomposition.

In the Multrum, the bottom of which is sloped, the organic wastes slide down at a rate slow enough to be thoroughly decomposed before reaching the storage chamber. A draft, maintained by natural convection, ensures that there is aerobic action in the process and that kitchen and bathroom are odor free.

Bath, dish, and laundry washwaters are not handled by the Multrum. Health authorities differ in their requirements for the treatment of this effluent. At least, it must be disposed of in such a way as not to endanger human health or water quality. With toilet wastes eliminated from the effluent leaving the house, this danger is greatly reduced. The company has prepared a paper describing several promising washwater treatment methods. See *References and Bibliography* at the end of this chapter. One method is seen in Figure 9.23.

Clivus Multrum is the exclusive U.S. licensee of the Swedish Multrum system, invented over 30 years ago and commercially available for the past decade in Scandinavia

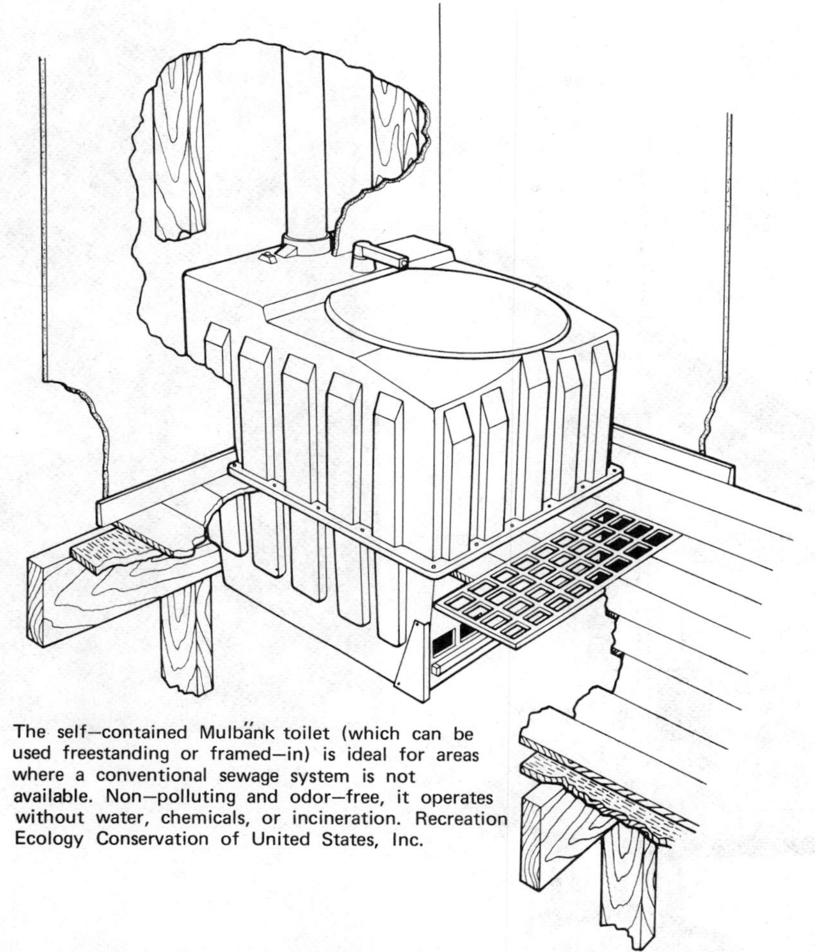

Fig. 9.21 *Details of a "framed-in" application.*

Fig. 9.22 *Adapting to design conditions. A "framed-in" Mullbänk toilet in a finished setting. The vent pipe is concealed behind the paneling. The architect must allow for access and intake ventilation below the finish floor.*

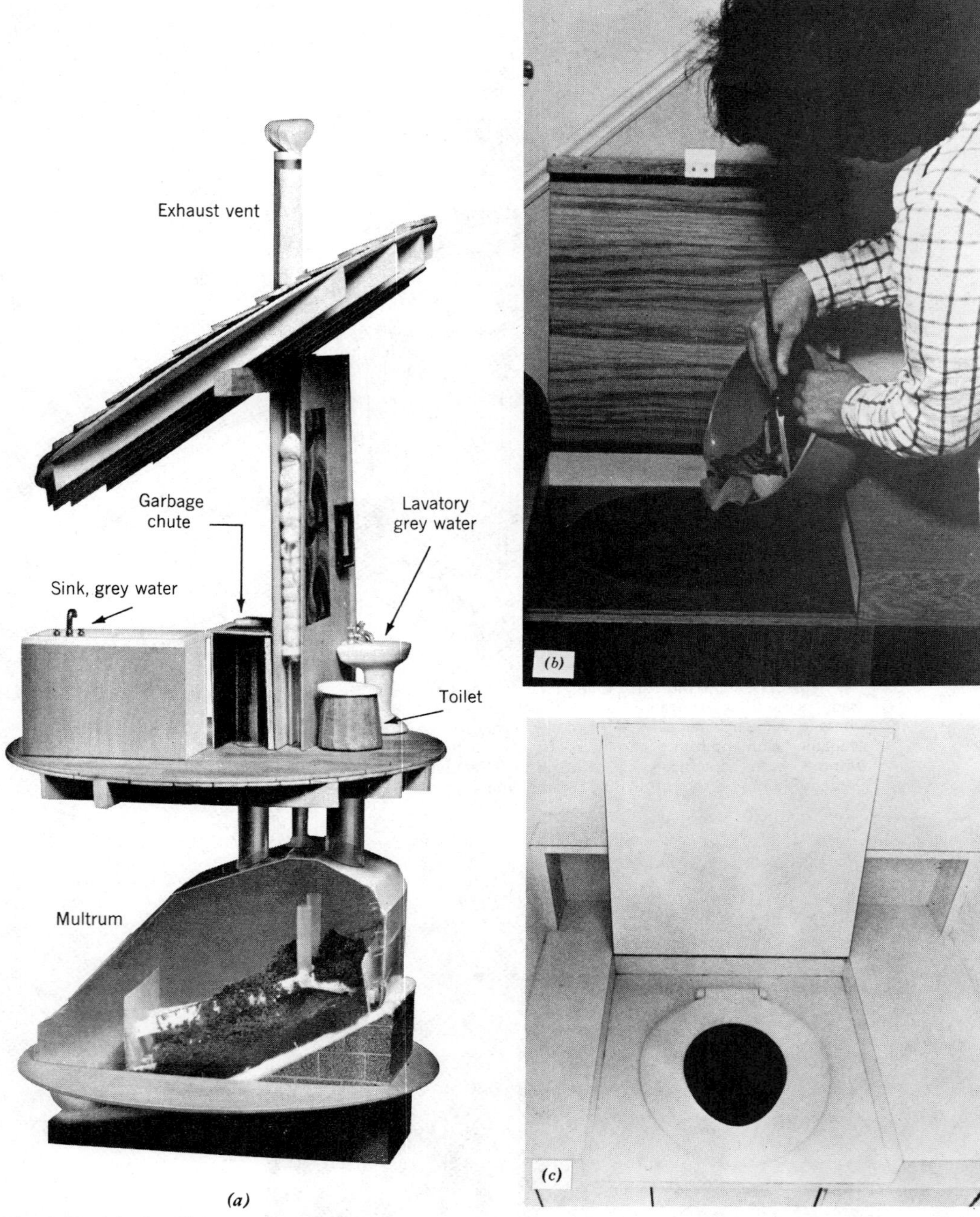

Fig. 9.23 (a) *The Clivus Multrum (inclining compost room) method of organic waste treatment. Using no water, the system provides a compost chamber (4 ft × 8 ft floor space and 7 ft high) that receives garbage and human wastes under controlled conditions. The end product is a humus. Household washwater (grey water) from kitchen sink, lavatory, laundry, and bath/shower is handled separately by a washwater filter, see* (g). (b) *Kitchen garbage being dumped down a chute. For the composting process, garbage is necessary. Alternate use of leaves, sawdust, or shredded hay is possible. Cans, glass, plastic, and similar materials must* not *be introduced.* (c) *Toilet seat and cover. This toilet is* not *flushed with water.* (d) *The vent stack exhausts from the top of the Multrum chamber. This causes a negative pressure (suction) in both chutes,*

where over 2000 units are in use. It was introduced into North America in 1974, and units have now been installed in most U.S. States and Canadian provinces. It is, in principle, adaptable to all types of buildings, urban and rural. At present, however, it is most readily incorporated into domestic and recreational structures. These include new and existing houses, campground and forest comfort stations, vacation cottages, and small commercial facilities.

The system was introduced in 1976 to The Committee on Science and Technology, Subcommittee on the Environment and the Atmosphere, House of Representatives, by Mr. Carl Lindstrom, Director of Research and

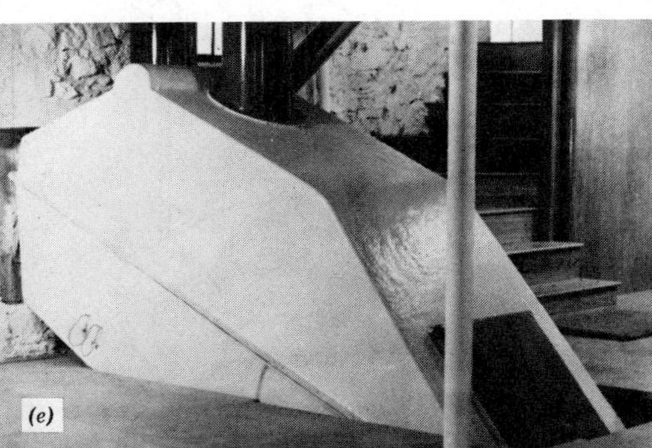

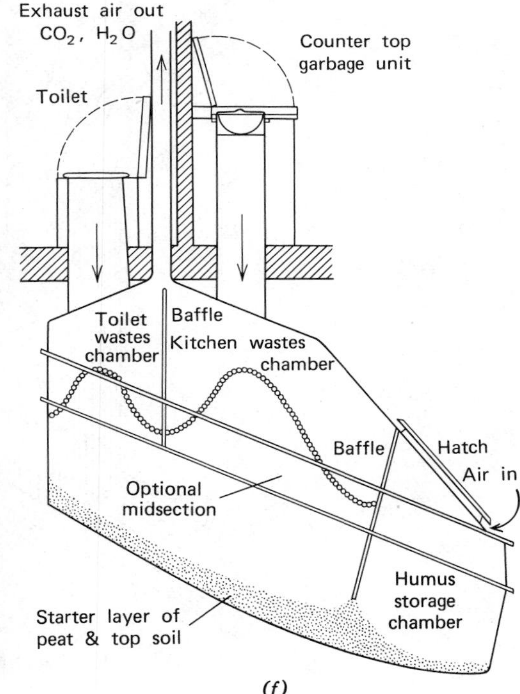

effectively preventing odors from entering kitchen or bathroom. The vent carries away innocuous carbon dioxide and water vapor. It creates less odor than a conventional plumbing vent. (e) *This is a Multrum in the basement of a house, using gravity and vertical chutes to introduce the wastes. Left chute toilet; right chute garbage. The vent, set diagonally at the upper right, continues up through the roof, see* (d). The hatch cover is at the lower right. (f) *Section through the Multrum illustrating the process. The humus that finally reaches the storage chamber represents only 5 to 10% of the volume originally put in. This end product is safe for gardens because of the long retention in the container* (2 to 4 *years*). *This continuous*

(*Continued*)

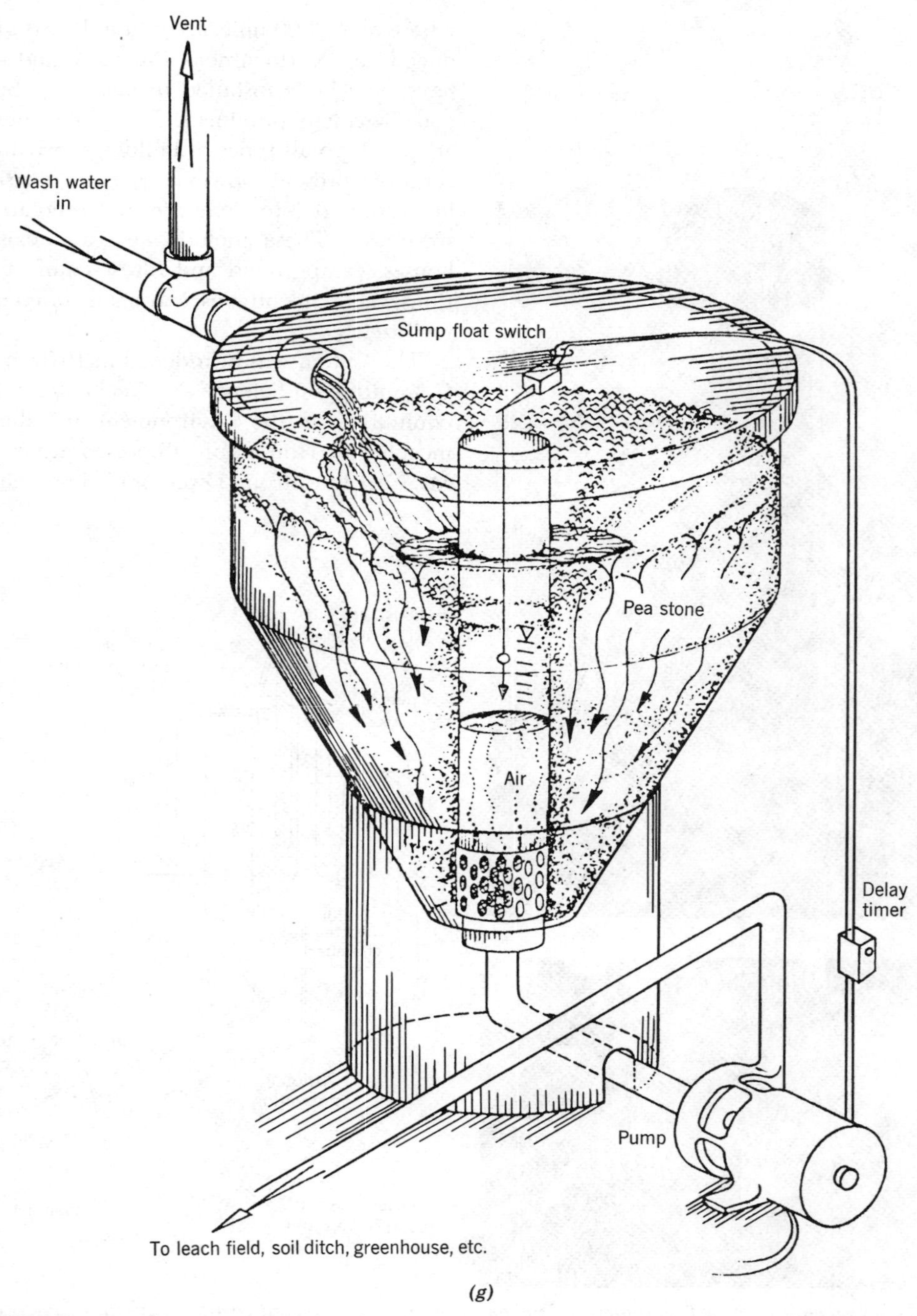

(g)

Fig. 9.23 *(Continued)*
process will generate 3 to 10 gal of humuslike soil per person per year. (g) *The Clivus Multrum Washwater Treatment System. The grey waters (from sink, washbasin, laundry, bathtub, and shower) that are so carefully diverted away from the Multrum composting chamber can be separately treated by the filter. Because of the aerobic conditions within the filter, the effluent is virtually odor free. This is adantageous since the effluent is dispersed to subsoil leach fields. There, evapotranspiration and infiltration occur.*

Development, Clivus Multrum USA, Inc. Mr. Lindstrom had previously served for two years as Environmental Counselor to the Swedish Embassy in Washington, D.C. On May 9, 1977 the Congressional Record-Senate included an article by Abby Rockefeller. Its title is "Private Profit and Public Waste: The Connection." Published earlier in the *Compost Science Journal of Waste Recycling*, it summarized what Ms. Rockefeller had learned about inept waste management in the course of trying to obtain official and public recognition of Clivus Multrum.

(*d*) ***The Microphor Flush Toilet.*** The recycling of water from sewage is not the only way to ease the water supply problem. Obviously, the use of less water at fixtures also helps. This reduces the volume of sewage produced. A notable item for this purpose is the Microphor flush toilet for use in *conventional* plumbing systems. Its 2-qt flush is an answer to the 5-gal flush deplored by Dr. Winneberger. The nature of this toilet is seen in Figures 9.24 and 9.25. The Microphor Company of Willits, California, issues the following comments about its product.

1. Water Savings—Case History

An Interstate Highway comfort station installed 14 low flush toilets into an existing, heavily used comfort station. The toilets were monitored one year for a water use study. Using the original fixtures, the comfort station would have used over $4\frac{1}{2}$ *million* gallons of water. After changing to Microphor Low Flush Toilets, water usage dropped to 280 *thousand gallons.*

This reduction in water usage represents a water savings of over 4 million gallons.

Similar results have been obtained in other locations.

2. How It Works

The Microphor Low Flush Toilet is activated by pressing the flush lever. Water and waste are immediately deposited into the lower secondary chamber which is built into the structure of the toilet.

The toilet has a complete water wash of the bowl. The bowl refills with water to form a seal.

At this point, the secondary chamber is pressurized with compressed air and the waste and water is deposited into the normal sewer line.

Microphor toilets can be used with any sewage system.

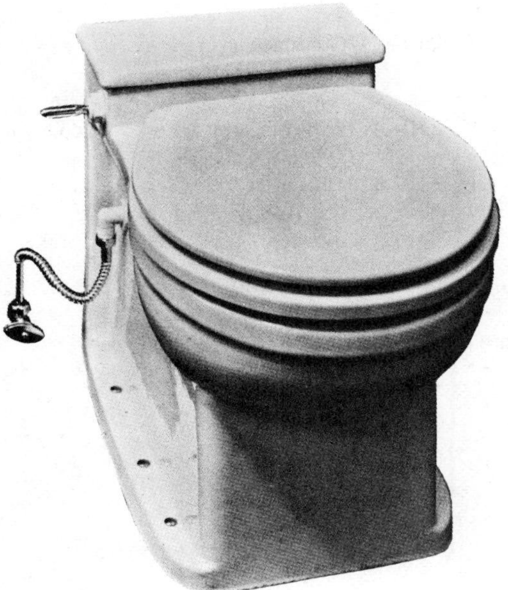

Fig. 9.24 *Reduction of both water use and sewage volume. The Microphor toilet uses 2 quarts of water for each flush instead of 5 gallons. Water use is thus one-tenth the amount used for a conventional toilet, with comparative reduction of resulting sewage volume (for details, see Fig. 9.25 and Section 9.7).*

Partial List of Microphor Users

California Department of Parks and Recreation
California Department of Transportation
U.S. Environmental Protection Agency
U.S. Department of Interior, National Park Service
College of the Virgin Islands
Dodge Ridge Ski Resort
United States Bureau of Land Management
Multnomah County, Oregon
Golden Gate Bridge, Highway and Transportation District
Pre-School Facility
U.S. Corps of Army Engineers
Idaho Department of Transportation
New Mexico Highway Department
Hotels
Homes
Motels
Office Buildings

Cutaway View of Low Flush Toilets

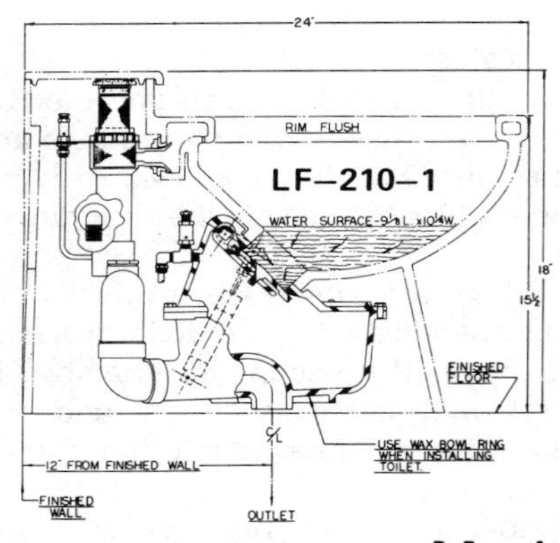

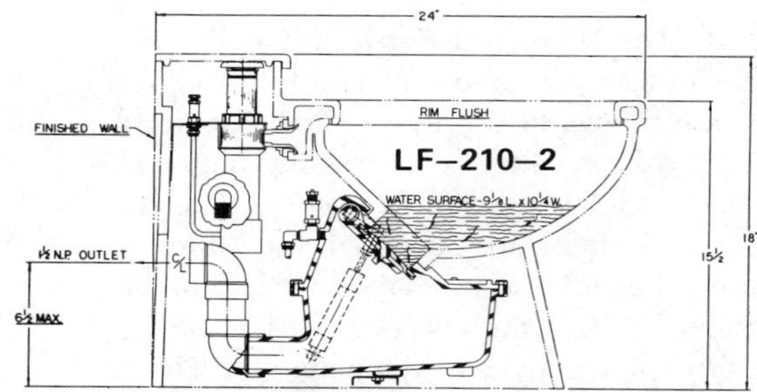

Discharge pipe must rise a minimum of 3" before discharging in a horizontal or downward direction. Discharge line must be connected at the hopper with a no hub coupling.

Model Information

Floor Mounted – Downward Discharge

LF–210–1 Our standard downward discharge toilet with integral valve.

LF–210–3 With Microphor remote valve.

LF–210–5 With Microphor remote valve and Adapter (For existing through wall flush activators).

Floor Mounted – Rear Discharge

LF–210–2 Our standard rear discharge toilet with integral valve.

LF–210–4 With Microphor remote valve.

LF–210–6 With Microphor remote valve and Adapter (For existing through wall flush activators).

Above units available with vandal-proof lid, Part No. 90–44954

SPECIFICATIONS SUBJECT TO CHANGE WITHOUT NOTICE

Fig. 9.25 *Details of the Microphor toilet, with the manufacturer's partial list of users.*

The Microphor Company presents answers to a number of questions commonly asked about the fixture and its performance.

Question Where does the required air supply come from?

Answer From an air compressor, although in remote areas, a compressed air bottle could be utilized.

Question Would I need a large compressor for a home installation?

Answer No, a small $1/4$ or $1/2$ H.P. compressor with a small air tank would be sufficient to operate up to three toilets. In practice, the installations have also demonstrated a lower noise level than a conventional toilet.

Question Does the low flush toilet require different plumbing?

Answer No. The water supply and the sewage outlet remains the same as any standard toilet.

Question Is two quarts of water sufficient to carry waste down the sewer line?

Answer Yes. Extensive testing and monitoring of existing installations have been done and all results show no problems in the waste carrying ability of two quarts of water. Also, most installations will have additional water flow; i.e., sinks, showers, washing machines, dishwashers.

Question Where is the water tank on the Microphor toilet?

Answer None is required because of the small water requirement. The water line out of the wall hooks directly to the toilet.

Question What is the effect of low water usage toilets on a septic tank?

Answer Beneficial! Concentrated and longer retention time for solids breakdown, along with effluent with less particles and lower volume of liquid, should greatly extend leach field life.

Question How does the Microphor toilet compare in price with standard toilets?

Answer There is an extremely wide variance of prices for standard toilets. Microphor low flush toilets are in the top half of the price range.

Many items interact in our concern about the environment; one important consideration is the conservation of energy. Both the Mullbänk and Microphor methods use *some* energy—Mullbänk for warmth and exhaust ventilation, Microphor for air compression. In the overall picture, these items must be evaluated and considered.

References and Bibliography

1. *Problems of Combined Sewer Facilities and Overflows*, 1967, Federal Water Pollution Control Administration, U.S. Department of the Interior, by the American Public Works Association.

2. *Water and Waste Water Technology*, Mark J. Hammer, John Wiley & Sons.

3. *Our Polluted World—Can Man Survive?* John Perry, Franklin Watts, Inc.

4. "Storm Water Control Looks Like Costliest Fight Yet," *Engineering News Record*, March 31, 1966.

5. *Manual of Grey Water Treatment Practice*, John H. Timothy Winneberger, Ann Arbor Science.

6. *The Treatment of Household Washwater in Homes Equipped with the Clivus Multrum Organic Waste Treatment System* by Margaret Fogel, Ph.D., and Carl Lindstrom, P. E. Clivus Multrum USA, Inc., 14a Eliot Street Cambridge, Mass. 02138.

7. *Standards for Individual Sewage Disposal Systems*, Nassau County Environmental Management Council, Mineola, N.Y. 11501.

10

Water Systems

10.1 Distribution and Pressure

In cities having municipal supply, water is distributed through street mains at pressures varying from about 50 pounds per square inch (psi) at the main to about 70 psi. For low-rise buildings of two or three stories in height these pressures are adequate to act against the static pressure of water standing in the vertical piping, to overcome the frictional resistance of water flow in the pipes and still to deliver water at the pressure required to operate plumbing fixtures. The flow-pressure required at the fixtures varies from 5 to 20 psi, depending on the type of fixture—for example, the basin faucet, shower head and faucet, water closet fixture, or others.

(*a*) ***Static Pressure.*** The pressure exerted at the bottom of a stationary "head" of water is related directly to its height. One cubic foot of water weighs 62.4 lb. Consider a "cube" of water 1 sq ft and 1 ft high. Its weight (62.4 lb) rests on a bottom area 1 sq ft (144 sq in.). The *static* pressure at the bottom is $62.4/144 = 0.433$ psi. Reciprocally, 1 psi of pressure will *sustain* a static (stationary) column of water $1/0.433 = 2.3$ ft in height. When fixture pressure and pressure lost in friction-of-flow in pipes are considered, the problem becomes more complex. Example 10.4 in Section 10.7 illustrates this problem when calculating pipe size. For our purpose in discussing upfeed and downfeed distribution, the relationship of heights and static pressure is to a large extent a controlling factor.

(*b*) ***Upfeed Distribution.*** It is safe to assume that in two-story residences and small low buildings of moderate water demand there is seldom any difficulty in achieving the proper flow-pressure at fixtures by the use of an upfeed system. Pressure at the fixtures is usually more than required. When this causes an inconvenient splash, as at lavatory basins, the pressure-at-faucet can be reduced by partially closing the valves on the water supply branches.

Let us examine the system shown in Figure 10.1. In cold climates, water in the service entry pipe (or tube) must not freeze. The pipe must therefore be below the *frost line* of frozen ground. This could vary from 2 to 7 ft, depending on geographical location. The onset of winter in such climates requires the closing and draining of pipes supplying the hose bibbs (and other external piping) by means of a *stop and waste* valve. Houses left *unheated* in cold winter weather must be entirely free of water that could freeze and burst the pipes. Note the drain valve at every low point in the system. House shut-off controls are usually located at the main, the curb, and within the house.

Meters have recently taken on a new role. Beyond their use in measuring the water quantity for which the occupant is to be charged, they now have a restrictive function. In California during water shortages, they were used to indicate water use in excess of 35 gallons per day (gpd) per person, beyond which limit fines were imposed and in some cases water supply reduced by valves under the control of the water company.

Treatment is most often performed to reduce hardness that could clog piping and equipment, or to neutralize acidity, a source of corrosion. During short periods when the treatment tanks are valved off for backwashing or other servicing, the *bypass shut-off valve* is opened.

From this point on, the water continues under pressure to

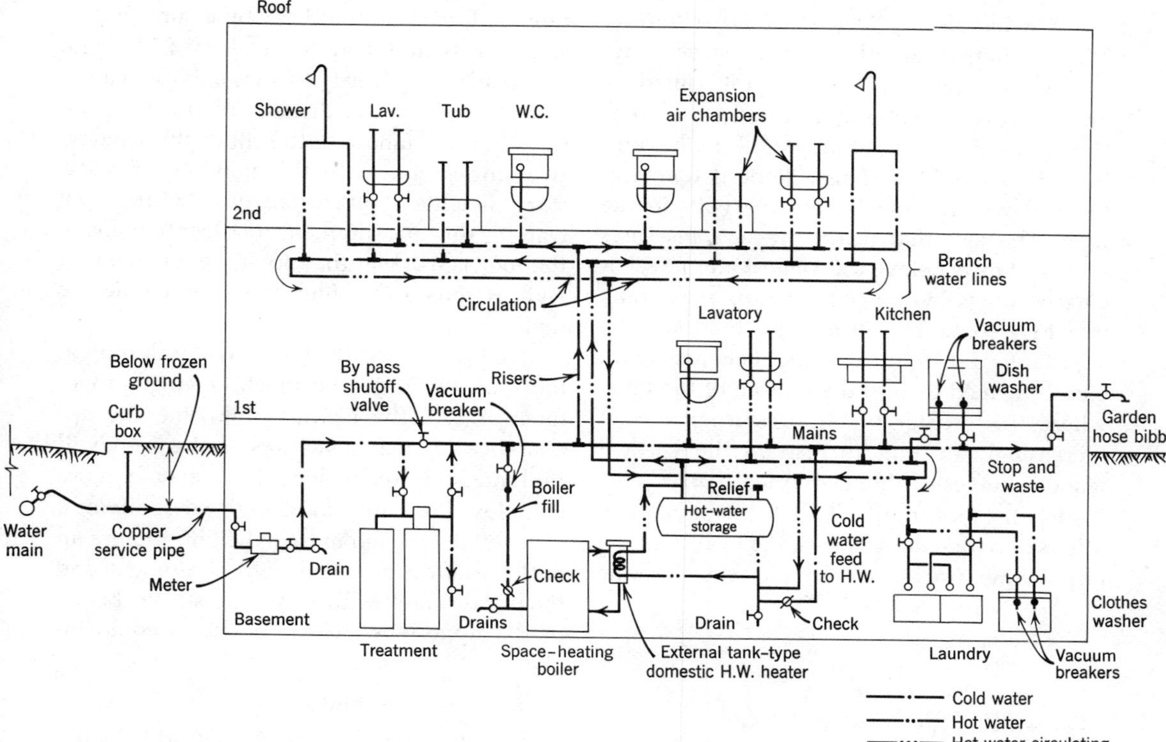

Fig. 10.1 *Upfeed water distribution by pressure in street mains. A schematic section of the water services in a residence. See ASHRAE Standard 90-75, Section 7.0 (page 416):*

Section 7.3.2 for combined space heating and domestic (service) water heating.
Section 7.3.3.1 for insulation of unfired hot water storage tanks.
Section 7.3.3.2 for insulation of hot water recirculation piping.

Note. The above are ASHRAE numbers.

1. Supply make-up water to the house-heating boiler as required.
2. Supply water to, and pressurize, the cold water mains and branches, including the garden hose bibbs.
3. Supply water to, and pressurize, the domestic hot water system through the hot water heater, hot water storage tank and the mains, branches and circulating lines.

The air-filled expansion chambers on cold water *runouts* absorb and reduce the shock of so-called "water hammer" when faucets are shut off. On hot water runouts they perform the same function, *plus* they allow for the expansion of the hot water as it increases in volume with increasing temperature. Vacuum breakers prevent back-flow of polluted water into pipes carrying potable (hot *or* cold) water. Water from polluted or unpleasant locations such as dishwashers, clothes washers, and boilers is thus isolated.

In the legend for Figure 10.1, the word *schematic* is significant. The entire diagram shows all parts lying in a two-dimensional *plane*. It is quite evident that in a real house a three-dimensional system would exist. For instance, economy of piping would suggest that, if possible, kitchen and lavatory be placed back-to-back, as well as the two upstairs bathrooms.

(*c*) ***Principles of Downfeed Distribution.*** Water pumped directly from the street main (or from a basement "suction tank" filled by gravity from the main) is lifted to a roof-storage tank. In cold climates, the tank water in such exposed locations is kept at temperatures above freezing by heating coils in the tank. The fact that water pressure increases with distance below the tank water level is clearly shown by the construction of the tank (see Fig. 10.2). The iron rings, tensioned by adjustable threaded clamps, become more closely spaced toward the bottom of the tank, where the greater water pressure makes it increasingly difficult to restrain the vertical wooden staves of this cylindrical *barrel.*

It must be admitted that this tank and others like it do not improve the beauty of the city skyline. If it were desirable to enclose the tank and its structural base by a surrounding screen, it would have to be about 24 ft high. The reader may verify this visually by scaling from the photo the distance vertically from the sill of a window to the sill of the window two stories above. In this building of 12-ft-story heights, that distance is 24 ft. Now transfer this measurement to a location above the roof, starting at the top of the low parapet wall. A screen that high would just hide the tank.

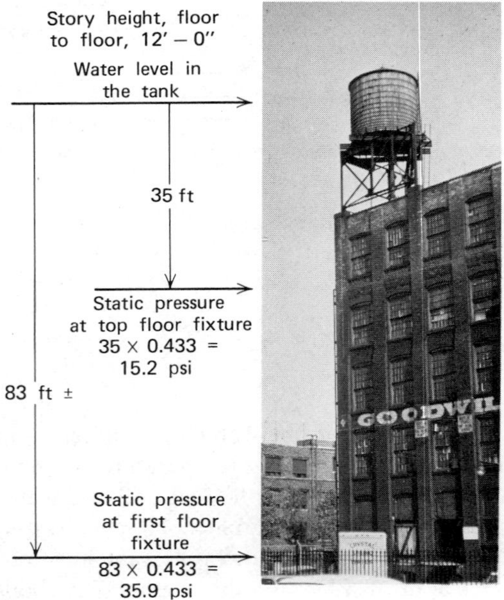

Fig. 10.2 *High-rise building of the 1870s. Downfeed distribution. Although upfeed distribution from a street main has been attempted in buildings higher than two or three stories, it was evident then (and is now) that street main pressures that may drop below the usual 50 psi, together with heavy use, would result in very low pressure at fixtures in upper stories. Therefore, water is pumped from the main to elevated wooden roof tanks high enough to assure reasonable pressure at the top story and ample pressure at the bottom of the downfeed run.*

Architecturally this rectangular lump on the roof would not be much of an improvement. Yet in the century following the appearance of such structures as the Goodwill Building, our technology has become more complex. Presently, most high-rise buildings, often 60 stories and more in height, require an entire rooftop crowded with equipment and technical facilities to serve the stories below or the uppermost *zone.* The items could include:

Water storage tanks.

Two-story penthouses over elevator banks.

Chimneys.

Numerous plumbing vents.

Exhaust blowers.

Air-conditioning cooling towers.

Cantilevered rolling rig to support scaffold for exterior window-washing.

Perimeter track for window-washing rig.

Many of these units can be seen in photographs of the roof of the Chase Manhattan Building (see Fig. 10.3*a*, *b*, and *c*).

Thus in the 1960s and 1970s it was common to find that tall buildings often appeared with a *band* or *screen* two stories (or more) high above the structural roof. It might be said that it all began with the need for elevated water-storage tanks. Other two-story elements increased the need for such bands or enclosures. They are: enclosed hoisting machinery over high-speed elevators and the cooling towers of air-conditioning systems. One sees fewer tall buildings with tapered spires such as those seen on New York's Chrysler Building and Empire State Building, which were both built in earlier decades.

Fig. 10.3 (a) *High-rise building of the 1970s. Downfeed distribution. Uniquely, the elevated wooden outdoor water storage tank has not changed very much in a hundred years (see Fig. 10.2). Its function is the same, with the tank's water level several stories above the top floor plumbing fixtures. As part of what might be called "Rooftop City," the tank has joined the contemporary equipment complex, surrounded by the visual barrier of a louvered "skirt."* (b) *The Chase Manhattan Building in New York City. Skidmore, Owings and Merrill, Architects. A helicopter view. Photo by Sky Service.* (c) *Architects inspect the equipment on the roof of the Chase Manhattan Building. Ahead of and past the ladder is seen one of the nine cooling towers, the top of which is about level with the top of the water storage tank as seen in* (a) *which indicates its elevation.* Note. *Other equipment seen in* (a) *includes perimeter track for a window washing rig and penthouses that enclose elevator hoisting machinery.*

10.2 High-Rise Downfeed Systems

(a) Pipe Routing and Equipment. As described in the previous section, when pressures from city mains are insufficient to maintain adequate fixture pressure at the top story under conditions of probable flow, water is pumped to wooden (or steel) elevated tanks for gravity downfeed. The lower part of the tank is often a reserve space to hold a supply of water for a fire hose system. In this case, only the water in the upper part is available for use as domestic (service) water. The amount stored must be enough to supplement what the pump will deliver during the several daily hours of high demand that occur in most buildings. The pump then continues, often for several hours, to replenish the house supply that had become partially depleted during the busy period. The suction tank is a buffer between the system and the street mains. It usually holds enough reserve to allow the pumps to make up the periodic depletion in the house tank. It refills automatically by gravity flow from the street main that, consequently, will not suffer as much drop in pressure as if it were connected directly to the suction side of the house pumps. Neighboring water users are protected from the adverse effects of sudden demands within adjacent large buildings. See Figure 10.4*a* and *b*, Figure 10.5*a* to *d*, and Figure 10.6*a* and *b*.

House tanks and suction tanks are sometimes of steel plate and are divided vertically in half, each half having identical piping and controls. This permits cleaning out one-half at a time during hours of low demand without shutting down the entire system. One full-capacity pump operates with an equal standby pump for alternate use. Since there is no suction lift below the pump or any fixture pressure at the top of the delivery leg (house tank supply), the head against which the pump works is the sum of the distance from the suction tank water level to the top of the house tank and the feet of head equivalent to the friction loss in the tank supply pipe. For this kind of service the vertical piping is of the order of 3 or 4 in. in diameter for large buildings. Sizes are established by a formal engineering design.

The house supply is fed by a short pipe from the house header to the cold water header that circles the top story and connects to many downfeed cold water risers. For simplicity, Figure 10.4*a* shows only two of them and also omits many valves and controls. Figure 10.4*b* is even more simplified. The hot water circuiting originates as cold water at the house tank header and takes quite a long route. Descending to the bottom of the hot water heater, it rises to seek its own level at the hot water header, becoming available there for hot water downfeed on demand. All of this occurs as flow below the general pressurizing effect of the house tank. In effect, when there is a cold- and hot-water demand on a story near the top of the building, the cold water makes a short trip down to the faucet, while the hot water involves flow in three vertical pipes instead of one. The general scheme just described, with tank above and heater below, is used in multiple form for very tall buildings. The zones are quite independent. Their only common service is the general suction tank. By this zoning method, problems of pipe expansion, excessive pipe sizes, and high pressures in lower stories are minimized. It is usual to have $2\frac{1}{2}$ stories or about 35 ft as the minimum pressure head above the top fixture served by any zone tank. The static pressure created at the fixture is thus $35 \times 0.433 = 15$ psi. If, during flow, not too much pressure is lost in friction, flushometers (flush valves), may be possible at this level, though flush tanks, because of their lower fixture-pressure demand, must often be accepted. Quite the opposite problem occurs at the bottom of the zones where excessive pressures must be valved down at the fixtures. In zone 1 of Figure 10.4*b*, first floor fixtures are below a head of $24\frac{1}{2}$ stories or about 149 psi of static pressure. It is obvious that pressure-reducing valves must be used, and that fixture control valves must be throttled.

(b) Pipe and Tube Expansion. The range of temperature from about 70° F (normal indoor air temperature) to that of service hot water (which often exceeds 160° F) is an index of the expansion of pipes and water as their temperatures rise from shut-down status (70° F) to operating status (160° F). See Table 10.1. The longitudinal elongation of pipe, though negligibly small in houses, can be appreciable in a tall building. Two methods of

Fig. 10.4a (opposite) *Downfeed water distribution, a schematic section. Part of the water services for a 10-story building. Hot water circulation is through the hot upfeed in two directions at the hot water header and down to the tank through the two downfeed hot water risers. For details of one type of steel house tank, a typical centrifugal house pump, expansion joints, and expansion in hot water riser, see Figure 10.5*a *to* d.

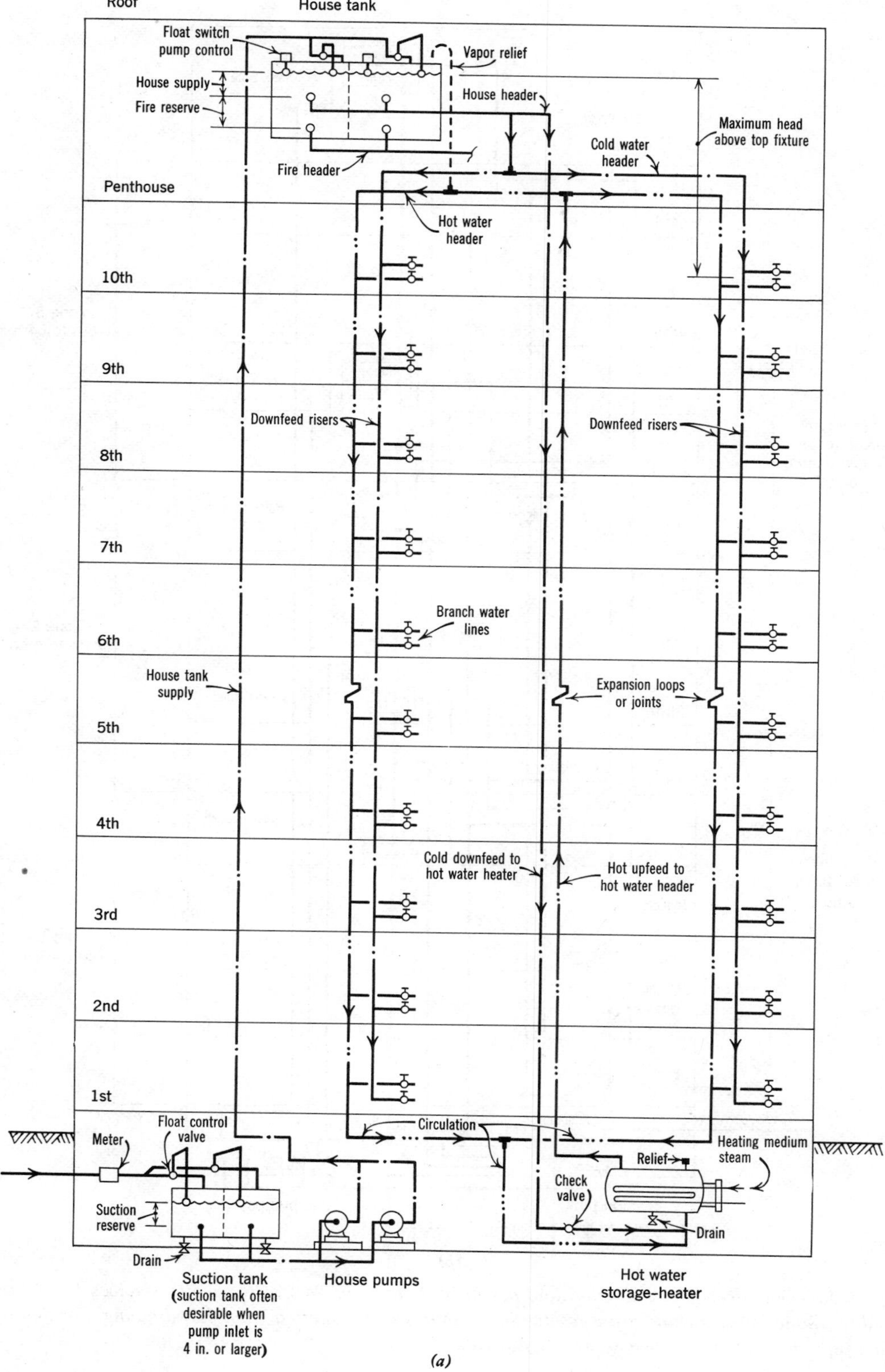
Roof
House tank
Float switch
pump control
Vapor relief
House supply
Fire reserve
House header
Maximum head
above top fixture
Cold water
header
Fire header
Penthouse
Hot water
header
10th
9th
Downfeed risers
Downfeed risers
8th
7th
Branch water
lines
6th
House tank
supply
Expansion loops
or joints
5th
4th
Cold downfeed to
hot water heater
Hot upfeed to
hot water header
3rd
2nd
1st
Float control
valve
Circulation
Meter
Heating medium
steam
Relief
Check
valve
Suction
reserve
Drain
Drain
Suction tank
(suction tank often
desirable when
pump inlet is
4 in. or larger)
House pumps
Hot water
storage-heater

(a)

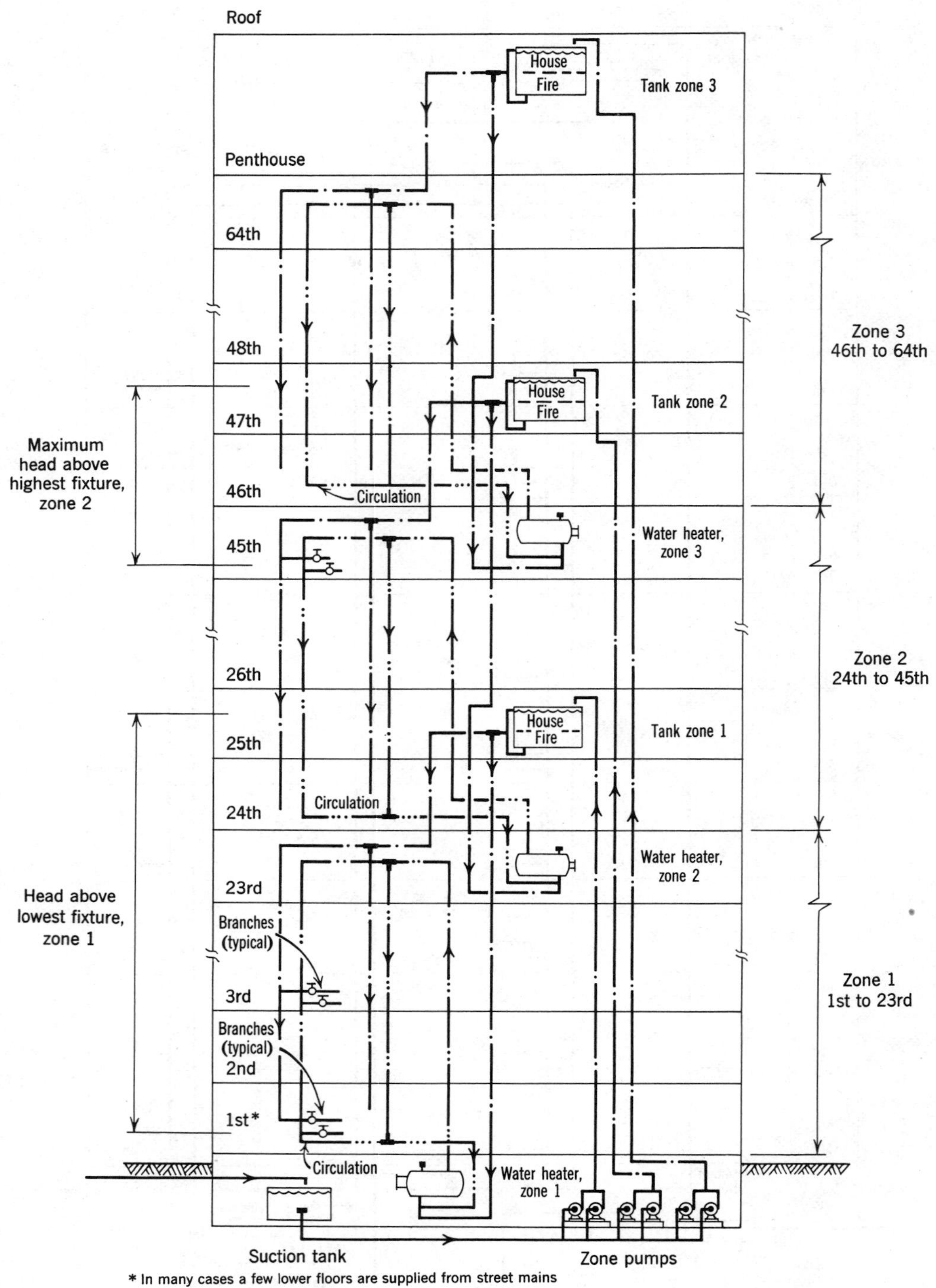

(b)

Fig. 10.4b Downfeed water distribution, a schematic section. Part of the water services for a zoned building. Zone tanks include a fire reserve but standpipes are omitted from this drawing. For detail of steam-type domestic water heater, see Figure 10.24.

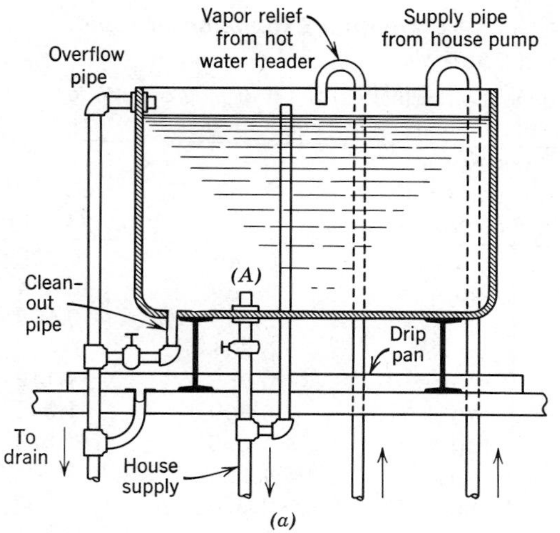

Fig. 10.5a House tank in elevated position for downfeed by gravity. Sediment in tank is drawn off through clean-out pipe and is prevented from entering house supply by pipe projection at (A). Water for fire reserve could be provided by additional piping or a separate tank.

Fig. 10.5b Centrifugal house pump. Commonly used to fill elevated house tank in top story or at intermediate mechanical floor. Electric motor responds to float switch pump control at the tank.

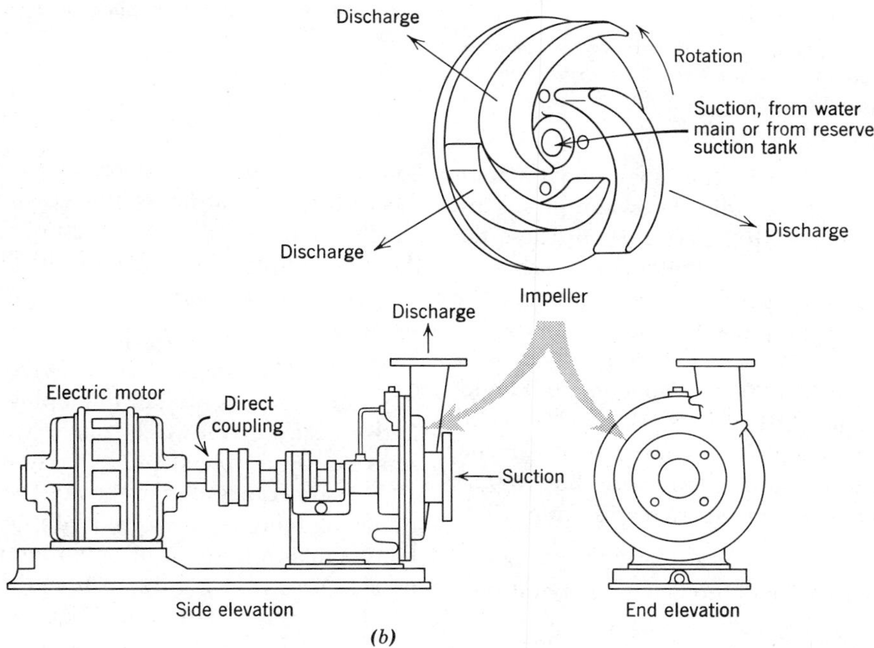

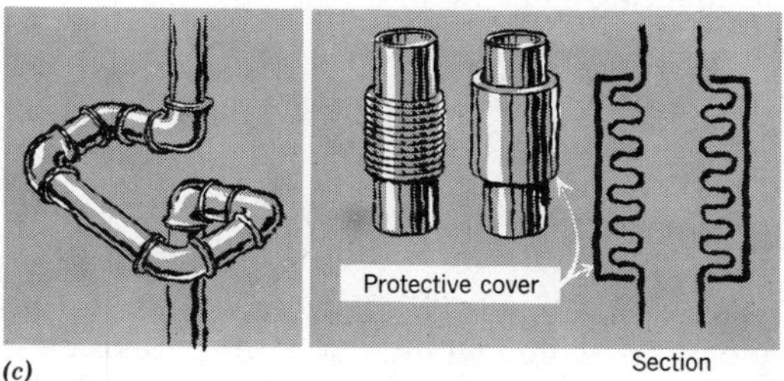

Fig. 10.5c Accommodation for the expansion of hot water piping or tubing. Left: Expansion joint of pipe and fittings. Right: A manufactured product.

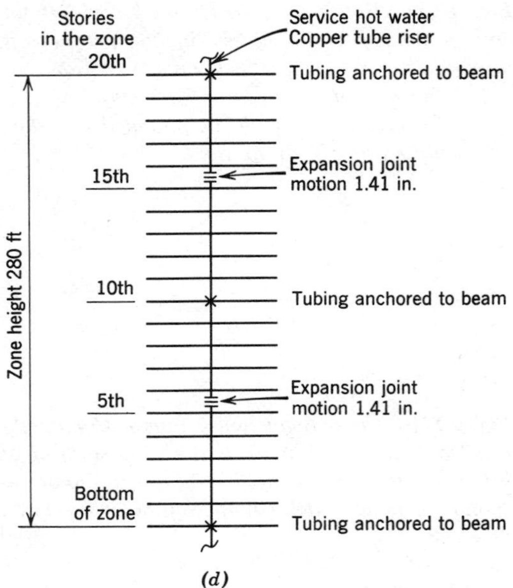

Fig. 10.5d (Example 10.1) Suggested scheme for locating points of anchorage and expansion for service hot water tubing in a 20-story zone.

Table 10.1 *Thermal Expansion of Pipe and Tubing*

Elongation in inches/100 ft of pipe or tube for various increases in temperature.

Increase in Temperature, in Fahrenheit	*Steel Pipe*	*Copper Tubing*
20	0.149	0.222
40	0.299	0.444
60	0.449	0.668
80	0.601	0.893
100	0.755	1.119
120	0.909	1.346
140	1.066	1.575
160	1.224	1.805
180	1.384	2.035
200	1.545	2.268

NOTE. Special care must be taken to adapt for the expansion of *plastic* pipe, the elongation of which is about five times that of copper tubing under the same conditions.

allowing freedom for this longitudinal motion in long runs of expanding hot water piping are shown in Figure 10.5*c*. The use of these devices precludes the buildup of excessive stresses in the metal of the pipes and the tendency of the pipes to buckle laterally.

EXAMPLE **10.1.** A 20-story zone in a tall building has a height of 280 ft. What will be the increase in length of a copper tube carrying "service hot water" (domestic hot water), when its temperature increases from 70 to 160° F?

SOLUTION. The difference in temperature is 90° F.

Elongation per 100 ft for 90° F increase
= 1.01 in.

(Refer to Table 10.1.)
$2.8 \times 1.01 = 2.82$ in.

There are a number of ways of providing for this expansion; one is shown in Figure 10.5*d*. It consists of accepting the motion at two locations, which would make the expansion in each case 1.41 in. Equidistant anchorage to fix the tubing is provided at the bottom, the tenth floor and the twentieth floor. The support of the vertical riser at floors other than those at which it is anchored could consist of clamps of the type illustrated in Figure 10.17*a* or *b*, supported on springs.

(*c*) ***An Alternative: Direct Upfeed Pumping.*** For buildings too tall to employ the limited pressure of the street main, a new constant pressure upfeed pumping system (Fig. 10.7) is sometimes used to replace the conventional method of downfeed supply from a house tank on the roof or on an intermediate floor as shown in Figure 10.4*a* and *b*.

This sophisticated equipment can deliver water at rates varying from what is needed for two or three faucets to full building demand, while maintaining at each outlet a pressure within 2 psi of the design pressure for that outlet.

The installation indicated in Figure 10.7*a* and *b* is a triplex pump group. According to demand, one, two, or three pumps will operate. Since each is of the variable speed type, virtually an infinite number of delivery rates can be achieved within the zero to maximum

Fig. 10.6a *Pan Am Building, New York City. Emery Roth and Sons, Architects: Walter Gropius, Pietro Belluschi, Associated Architects. The distribution principles of the building in Figure 10.4*b *can result in the "mechanical stories" as used in the Pan Am Building. Stories of this type may also contain air-conditioning compressors, water chillers, and blowers.*

Fig. 10.6b *Cross-sectional "mock-up," first floor. Typical of planning in many high-rise buildings, this is one of the early models used in developing space concepts for the Pan Am Building. Equipment risers have a way of being* vertical with no *(or few) offsets. The first floor and typical upper floor set the pattern. In addition to a space for water risers, allocations must be made for sanitary and storm drainage, elevators, ducts, electrical and telephone risers, and fire stairs. Equipment planners must aid in establishing rooftop areas, mechanical stories, and the general building horizontal cross section. Courtesy of Emery Roth & Sons, Architects.*

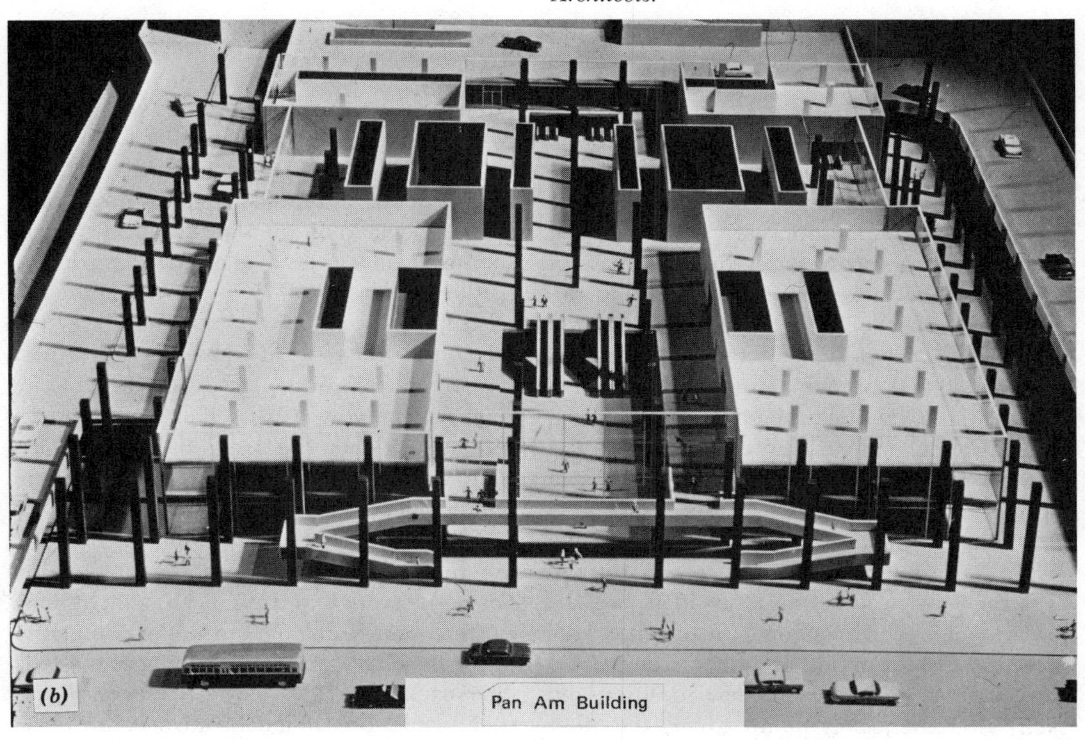

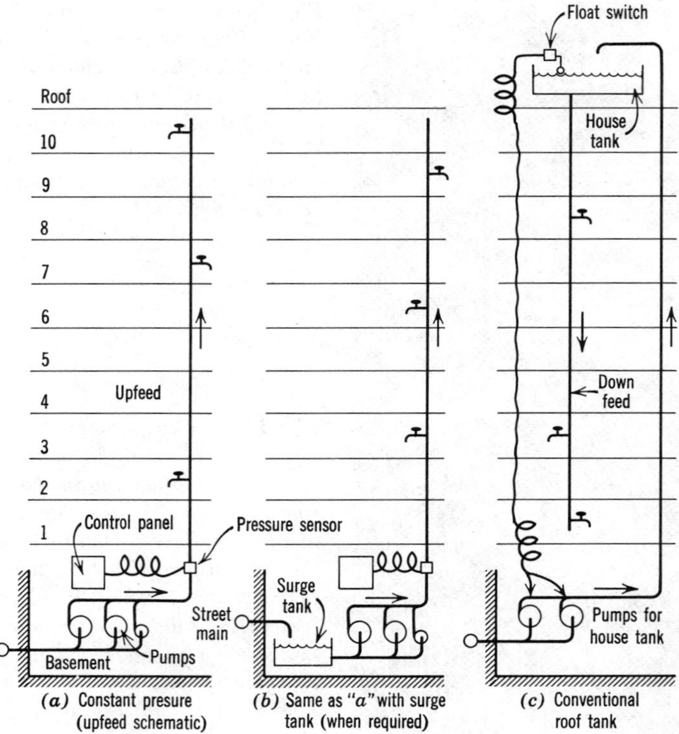

Fig. 10.7 Constant pressure upfeed pumping (a) *and* (b), *compared to gravity downfeed from house tank on roof* (c). *By permission of* Progressive Architecture.

design rate. The pumps operate in sequence. When a very small rate of demand occurs, the smallest or "jockey" pump starts in response to a low voltage impressed on its motor. This and all other operations are triggered and adjusted by the pressure sensor at the base of the riser. The jockey pump continues to run until it has reached its maximum delivery rate at which time the first of the larger pumps cuts in, joined by the third when required. Sequential operation of the three pumps, each increasing in delivery as called for by the sensor, meets the requirement for increasing supply at nearly constant pressure.

It would appear that the second of the two larger pumps would run less frequently and get the least wear. This is avoided and the wear equalized by assigning "lead" and "lag" positions. For a period of 24 hours one of the large pumps will hold the lead position and start after the jockey pump, giving the other large pump a smaller burden. Next day the rested pump takes over for the more active assignment. All of this is automatic.

At full operation this triplex unit can put a suction demand on the street main that could seriously reduce the available water pressure in the neighborhood (Fig. 10.7*a*). Therefore, city officials sometimes require that the system feed on a surge tank, filled by casual flow from the street main, independent of the building requirements (Fig. 10.7*b*). This requirement is often made when the design indicates a maximum building demand in excess of 400 gpm (gallons per minute).

Obvious advantages of upfeed pumping are: the elimination of the house tank and the heavy structure that transmits its weight down to the footings and, of course, the elimination of the necessary periodic cleaning of the tank. A shortcoming is its lack of reserve storage that would be an inconvenience during an electrical power failure. However, minimum flow during this kind of emergency can be arranged by having a diesel or other independent standby motor drive one of the pumps.

10.3 Community and Private Distribution

In big cities, departments of water supply or public works generally assume responsibility for water sources and, to some extent, for the *quality* of water delivered to the customer. Biological purity is, of course, the prime concern. In small communities, remote settlements, or private estates, local authorities or private owners and their technical advisers take over this responsibility. Planning for the quality and quantity of potable water and other water supplies can be aided by the kinds of considerations suggested in Tables 10.2, 10.3, and 10.4.

Table 10.2 ***Water in Various Physical States—Consumed, Utilized, or Controlled In or Around Buildings***

Use	*Desired Quality*
A. Consumed	
1. Drinking and cooking	Potable
2. Bathing	Potable
3. Laundering	Soft
4. Irrigation and watering of livestock	Unpolluted
5. Industrial processes	As required
6. Vapor to increase the relative humidity of air	
B. Circulated	
1. Hot water for heating	Note: Make-up water should be soft or neutral and, for swimming, potable
2. Chilled water for cooling	
3. Condenser cooling water	
4. Swimming pool water	
5. Steam for heating, later condensed	
C. Generally static	
1. Water stored for fire protection	No special requirement
2. Water in fire standpipes	
3. Water in sprinkler piping	
D. Controlled	
1. Vapor condensed to reduce relative humidity of air	

NOTE. For water uses in Section A, above, flow is often continuous. Section B comprises uses for which flow other than circulation is intermittent or at a relatively low rate, the water added to the systems being known as "make-up water." Items C2 and C3 call for piping to provide adequate though infrequent flow in emergencies. Obviously, Item D1 relates only to moisture condensed out of the air and involves no supply.

(a) *Community Systems.* The sources of water for most large cities are rivers and lakes and, increasingly, recycled water. A small percentage of the total public water systems in the United States use groundwater from deep wells. Figure 10.8 shows this kind of installation for a small community on Long Island, New York. The water is taken from the ground by multistage turbine pumps at depths of several hundred feet. The water is delivered to submerged hydropneumatic tanks at a pressure of about 80 psi. As water is demanded in the houses, the air under pressure in the upper part of the tanks forces it through the mains. No treatment of this water is necessary

Table 10.3 ***Chemical Analysis of Water from a Private Well in Virginia***

Quality		*Parts per Million (ppm)*
Total hardness	as $CaCO_3$	30
Calcium hardness	as $CaCO_3$	20
Alkalinity (Methyl Orange)	as $CaCO_3$	27
Alkalinity (Phenolpntalein)	as $CaCO_3$	0
Free Carbon Dioxide	as CO_2	13.5
Chlorides	as Cl	6
Sulfates	as SO_4	4
Silica	as SiO_2	19
Phosphates—normal	as PO_4	0
Phosphates—total	as PO_4	0.5
Iron—total	as Fe	1.6
Total dissolved solids		66
Turbidity or sediment		present
pH[a]		6.6

Source. From a report by Olin Water Service.

[a] NOTE. A low pH is an index of acidity. A value of 7.0 is *neutral.*

Table 10.4 ***Water Quality Problems and Their Correction in Private Systems***

Item	*Cause*	*Bad Effect*	*Correction*
Hardness	Calcium and magnesium salts from underground flow	Clogging of pipes by scale, burning out of boilers, and impaired laundering and food preparation	Ion-exchanger (Zeolite process)
Corrosion	Acidity, entrained oxygen and carbon dioxide (low pH)	Closing of iron pipe by rust, destruction of brass pipe	Raising the alkaline content (Neutralizer)
Pollution	Contamination by organic matter or sewage	Disease	Chlorination by sodium hypochlorite or chlorine gas
Color	Iron and manganese	Discoloration of fixtures and laundry	Precipitation by filtration through manganese zeolite (Oxidizing filter)
Taste and Odor[a]	Organic matter	Unpleasantness	Filtration through activated carbon (Purifier)
Turbidity[a]	Silt or suspended matter picked up in surface or near-surface flow	Unpleasantness	Filtration

[a]NOTE. These problems are not common in private systems using deep wells.

and none is provided. A neutral pH value and low mineral content make this a very satisfactory water. Quite a few systems of this type do not have any treatment; but some, where there is evidence of bacterial impurity, provide chlorination to correct this condition.

Impurities injurious to health are not usually tested as part of the general chemical analysis. County or other official health offices have strict jurisdiction over the use of waters that may be polluted. They will often provide analyses to determine the *biological* qualities of such waters and will allow their use or require treatment for correction of impurities. Troubles generally arise when water sources are polluted by defective cesspools, septic tanks, and other sewage disposal systems, or by the proximity of organic matter or grazing animals. Detergents are now a source of contamination where sewage disposal and wells are close together. As areas become congested, the separation of sewage disposal centers and approved water sources becomes an important matter of public health.

(*b*) ***Private Systems.*** Farms and remote estates usually have their own private water systems. In rural and suburban areas where the progress of building is faster than the development of municipal water supplies or water companies, private sources must be sought. The elaborate treatment necessary for the use of surface waters or those in dug wells of the old type makes the use of driven or drilled wells preferred. Water from these sources usually has at least the advantages of purity, coolness, and freedom from turbidity, odor, and unpleasant taste. There is the possibility that any of these *may* be encountered in addition to either acidity or hardness. Table 10.4 lists the possible problems with their causes, effects, and correction.

When clients plan to build in a remote location, the architect and engineer will want to advise them about water problems. Quality-corrective measures can always be taken and pumping equipment purchased, but the amount of water that can be had from the ground and the depth and cost of wells are

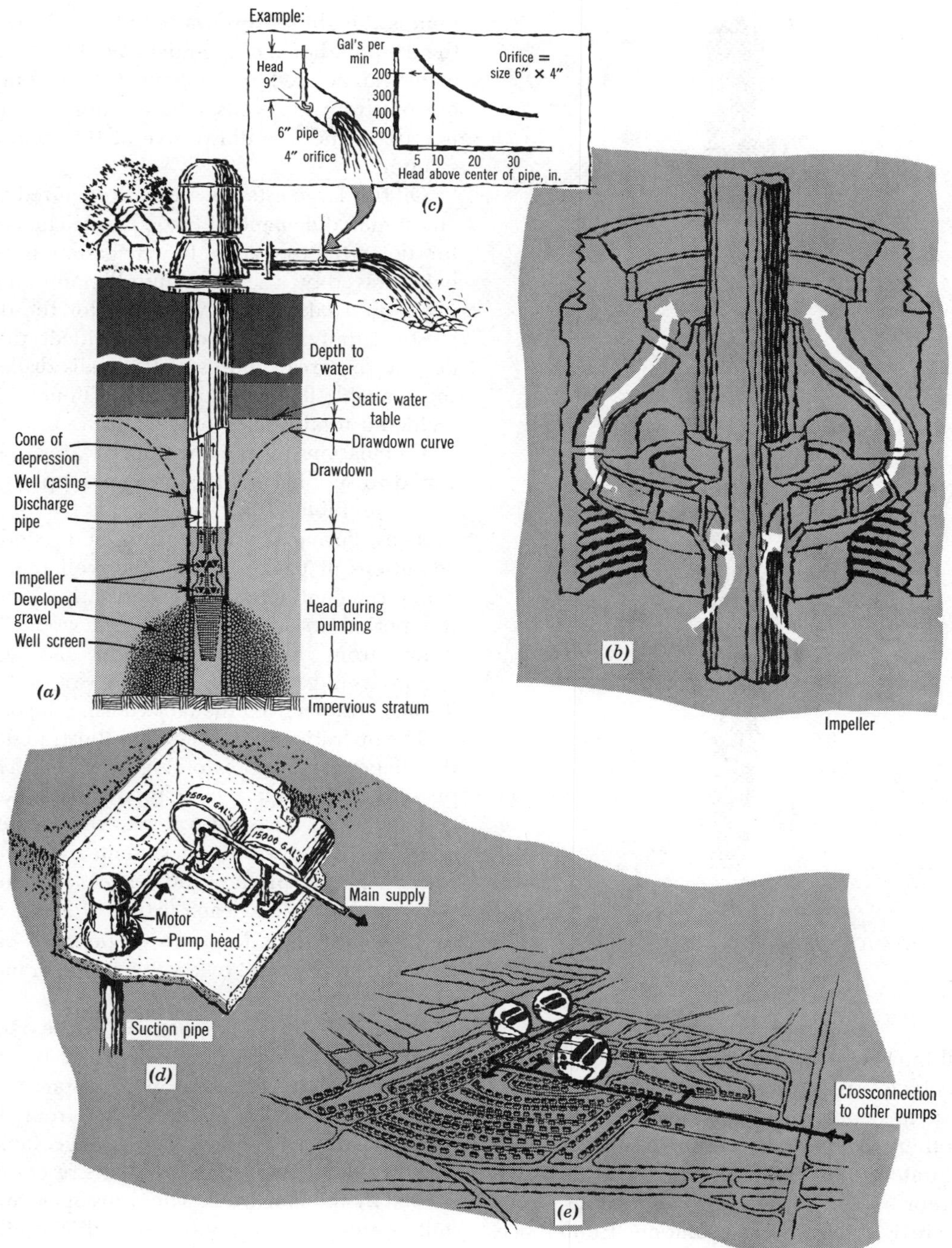

Fig. 10.8 (a) *A Turbine well-pump.* (b) *Its operation.* (c) *Measurement of its capacity.* (d) *and* (e) *Its use in supplying a small community with groundwater.* (f) *A Jacuzzi multistage lineshaft turbine well-pump. Jacuzzi Bros., Inc. Capacities of turbine pumps range from 50 to 16,000 gpm (gallons per minute). By permission of* Progressive Architecture.

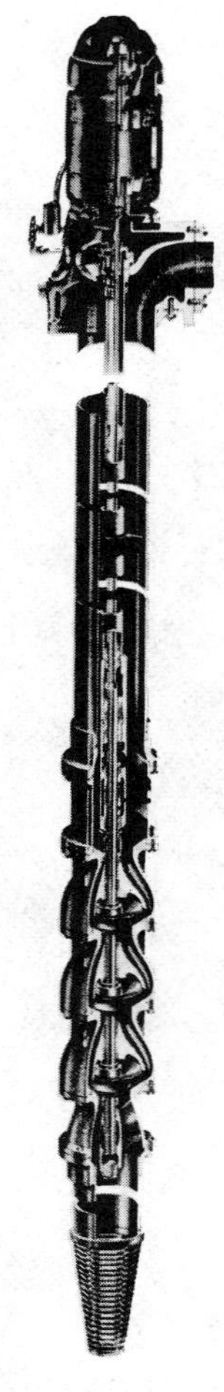
(f)

Fig. 10.8 (*Continued*)

all-important considerations. There are some problem areas where wells several hundred feet in depth will yield as little as 5 gpm or nothing. The cost of drilling a number of exploratory wells may sometimes be excessive. Unfortunately, when such difficulties occur, there is often no easy solution. Conferences with neighboring owners, state and federal geologists, and with local well-drillers are all helpful. Many regions, of course, yield plentiful water, but the cost of the probable depth of the well should be considered.

A solution to the low yield of some wells is seen in the plan to run the well pump or pumps at night to store water in tanks. During the daylight hours of high demand, the stored water can be used to supplement the low rate of flow from the wells. The example in Figure 10.9*a* and *b* is illustrative of this kind of solution.

On this large estate water was required for an estimated demand of 30,000 gpd. This was for domestic use only, the irrigation needs being solved by a separate installation pumping from a lake. Wells were dug for the domestic supply. It was quickly evident that, despite the great depths of the wells drilled, the available flow would be small. Four wells yielded a total rate of only 25 gpm.

Calculations for the amount of water to be stored to supplement this meager supply are shown in Table 10.5.

Thus, during the 14 daytime hours under conditions of peak demand, the well pumps would run continuously. Concurrently, during this period an additional 9000 gal would be drawn from the tanks. At night, the well pumps would run for 6 hours to restore the 9000 gal drawn from the tanks during the day.

The operation of the system is illustrated in the diagram and notes of Figure 10.9*a*. The pressure of 75 psi in the hydropneumatic tank is sufficient to raise the water to the greatest height in the distribution system, overcome friction in the piping, and leave a residual pressure available at each fixture of about 10 to 15 psi as needed. Excessive pressure can always be moderated by a valve in the branch supply of the fixture.

Pressure in this tank is assured by the operation of the air compressor actuated by the pressure switch. The float switch starts the centrifugal pump to deliver water from the storage tanks as needed. The storage tanks, piped together and acting as a single reservoir, are fed by the four wells. The wells operate in unison, singly, or in groups, depending on the

Fig. 10.9a (opposite) *Water control and distribution center for the Kinloch Estate. Bentel & Bentel, Architects, FAIA. The plant is located in a hillside and forms a story below the bathing and dressing pavilion, shown in Figure 10.9*b.

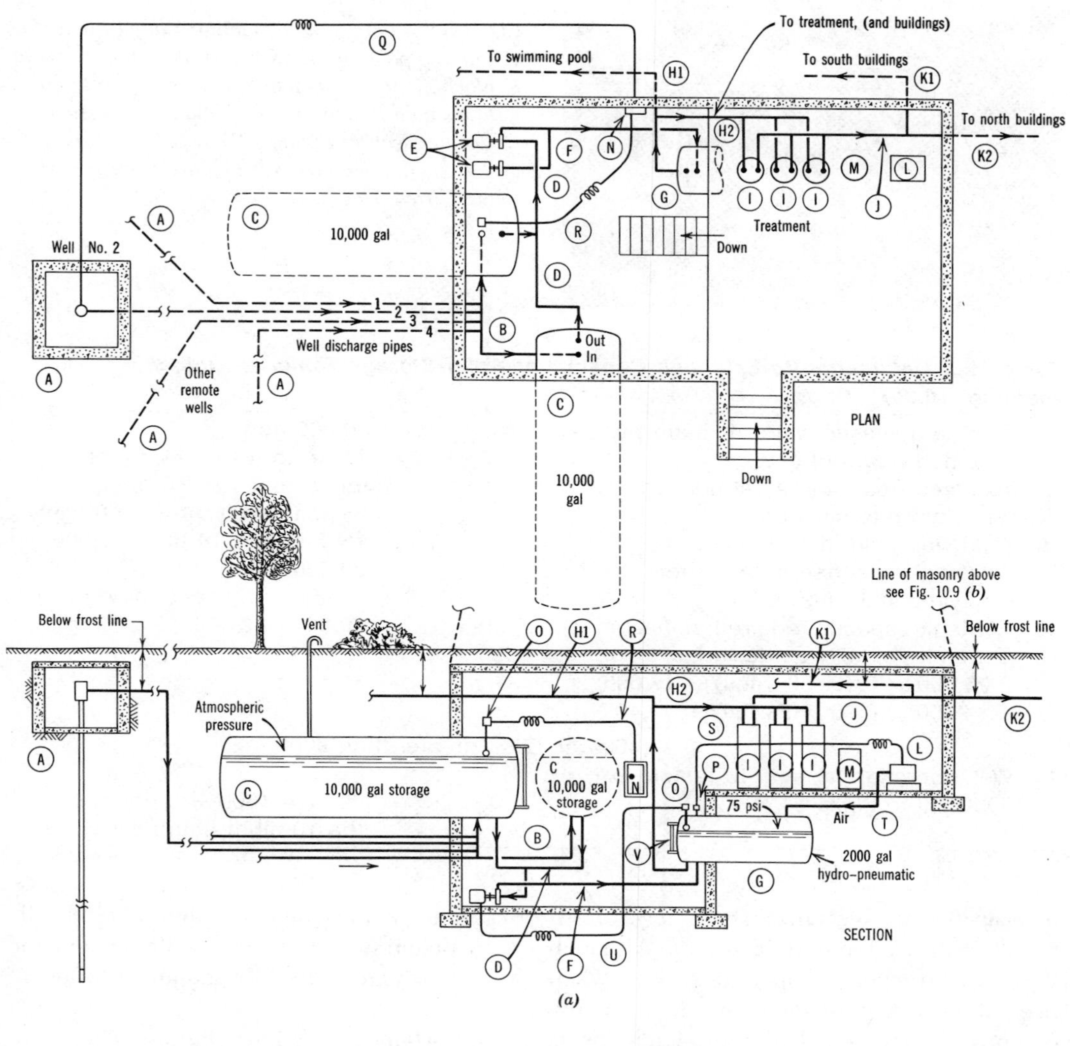

(a)

Legend

(A)	Wells
(B)	Well water header for 4 wells
(C)	Well water storage, vented tanks, (atmospheric pressure)
(D)	Pump supply pipe
(E)	Electric centrifugal pumps (one is a standby)
(F)	Pump discharge to hydro-pneumatic tank
(G)	Hydro-pneumatic tank, 75 lbs. per sq. in.
(H1) (H2)	Discharge pipes, pressurized well water
(I)	Water softeners, (ion exchangers)
(J)	Treated well water under pressure
(K1) (K2)	Branch mains to buildings (treated well water)
(L)	Air compressor maintains 75 psi in tank
(M)	Sodium chloride for regeneration
(N)	Electrical control panel
(O)	Float switches
(P)	Pressure switch (actuates air compressor)
(Q)	Control wiring to well pumps
(R)	Control wiring, float switch to panel
(S)	Control wiring, pressure switch to compressor
(T)	Compressed air piping
(U)	Control wiring, float switch to centrifugal pump
(V)	Gauge glass indicates water level

This schematic diagram shows system components and connections. For clarity, details (valves, drains, check-valves etc.) are not shown. All control wiring, here simply indicated, operates through control panel (N). See text for operation.

Fig. 10.9b One of two pavilions adjacent to the swimming pool and serving residents and guests at Kinloch. Below this pavilion is a separate story with its own downhill access that houses the control center seen in Figure 10.9(a) Exterior tubing to and from the center is all below grade. Bentel & Bentel, Architects, FAIA.

Table 10.5 *Calculation to Establish Capacity of Water-Storage Tanks at Kinloch (see Fig. 10.9a)*

1. Potable domestic water demand per day	30,000 gpd
2. Assumed hours of use	7 A.M. to 9 P.M. 14 hours
3. Assumed hours of virtual nonuse	9 P.M. to 7 A.M. 10 hours
4. Well-yield rate per hour	25 gpm × 60 min = 1500 gph
5. Total well yield in 14 hours	1500 gph × 14 hr = 21,000 gal
6. Total water needed in 14 hours	30,000 gal
7. Well yield in 14 hours	21,000 gal (see 5 above)
8. Net-tank capacity required, minimum	30,000 − 21,000 = 9000 gal
9. Net-tank capacity as designed 20,000 × 0.80 (80% full) = 16,000 2000 × 0.70 (70% full) = 1,400 17,400 gal, O.K., greater than 9000	
10. Well operation at night to restore 9000 gal to tanks	9000 ÷ 1500 = 6 hours of the 10 night hours available.

level in the storage tanks. If the level drops rapidly, all well pumps can run. Although they are arranged to operate all day when needed, in periods of minimum demand only one or two may be called upon. Each has its own power supply, but controls emanate from panel *N*.

Swimming pools should be supplied with pure, potable water. Biological tests showed that this well water was safe. Thus, the fill-line to the swimming pool is connected from *this* control center rather than from the supply of lake water used for irrigation. Since swimming pool water is separately recirculated and provided with its own purification treatment at a location adjacent to the pool, this supply line for make-up water to the pool would have infrequent and off-hour use. Water for domestic use in the buildings is passed through one of three ion-exchangers to provide softening and to make the water more suitable for washing, bathing, and cooking. Periodically the calcium precipitate can be flushed out and the tanks regenerated by sodium chloride.

(*c*) ***Above-Grade Distribution.*** The previous examples of community and private distribution call for pipes located below the frost line. In warmer climates this precaution may not always be necessary (see Fig. 10.10).

10.4 Well Pumps

A number of types of pump are available. They include, among others, the turbine pump, the submersible pump, and the jet pump. In all selections, the advice of the well-driller and pump manufacturer should be sought. In general, however, the turbine is commonly chosen for high-capacity systems with very deep wells. One of this type is illus-

Fig. 10.10 Water distribution in the Florida Keys. With ground level at about "plus 3 ft" above sea level, water can be delivered in pipes above the ground. A glimpse of the Gulf of Mexico is seen at the left. The Straits of Florida are just behind the bushes at the right. Three utility services (water, electric power, and telephone) appear to converge in this one-point perspective as they advance to serve the next Key. The choice between underground or overground services can be influenced by the nature of the terrain and the climate.

trated in Figure 10.8*f*. The submersible and the jet pump are quite suitable for residences and small buildings of moderate water-use demand. Of these two, the deep-well jet type is described here.

Unlike simple suction pumps that will lift water from water tables less than 25 ft deep, the deep-well jet can operate to pump water from well-depths of 25 to over 100 ft. By use of the *venturi* principle, by which a jet submerged below the groundwater level produces a reduced pressure, water is drawn into a circulation system. The addition of this water puts the system under pressure and delivers the excess water to a hydropneumatic storage tank (see Fig. 10.11). The tank contents are then pressurized by the air chamber, against which the water is forced. When a faucet or other control in the house is opened, water flows from it as a result of the pressure in the tank. Check valves in the piping prevent a reverse flow of the water. The foot valve at the base of the well acts as a check to prevent flow back into the ground.

See Figure 10.12*a* to *d* for illustrations of actual equipment available for this method of pumping. The multistage pump is equipped with two centrifugal impellers that circulate and force water through the jet element where the additional water is drawn in. The sets of air/water "Hydrocel" chambers, a recent development (see Fig. 10.13), which often supplant the conventional hydropneumatic tank, provide cushioning and pressuring for the system.

Depending on the motor horsepower, the discharge pressure, and the suction lift or depth to water, this equipment can deliver between 195 and 1620 gallons per hour (gph), also expressed as between 3.5 and 27 gallons per minute (gpm). When it is considered that a lavatory basin faucet flows at 3.0 gpm and a shower at 5.0 gpm[1] (see Table 10.9), a relationship can be sensed between demand and supply. Of course, the well pump cannot draw and deliver water at a rate greater than the groundwater will permit without excessive "*drawdown*" of the ground water-table. This rate can only be established by trial pumping and rate-of-flow evaluation over a significant time period.

10.5 Piping, Tubing, Fittings and Controls

(*a*) ***Piping, Tubing, and Fittings.*** The conveying of water through buildings to locations of use implies the design of a system of piping or tubing, efficient for its purpose, easily maintained, and interfering as little as possible with the interior architectural form. It may be assumed that, except in basements, utility rooms, and at points of access to controls, the system will usually be concealed. Stud and joist construction provides space for concealment but, in fireproof buildings, vertical and horizontal furred spaces must often be provided.

The corrosive effects of water and the resistance of metals to corrosion are usually matters for the attention of chemists and met-

[1] Now reduced by ASHRAE Standard 90-75; see page 416.

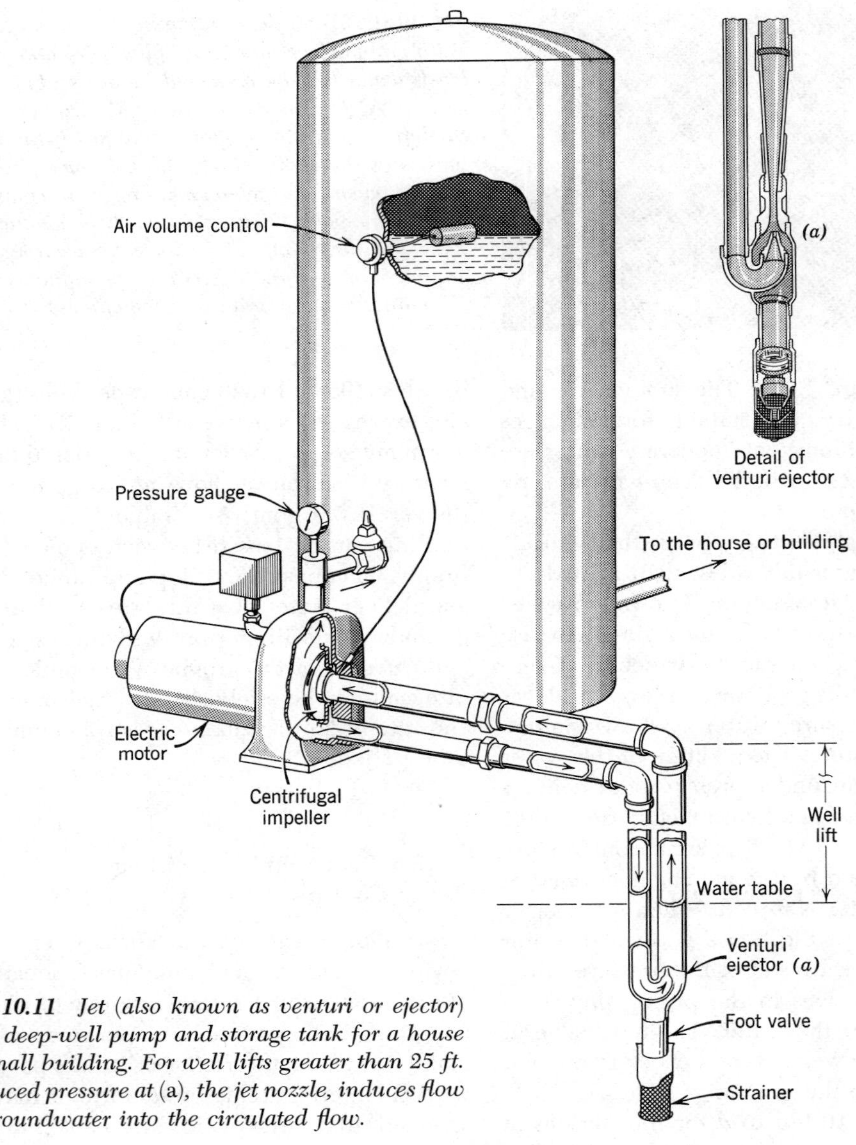

Fig. 10.11 *Jet (also known as venturi or ejector) type deep-well pump and storage tank for a house or small building. For well lifts greater than 25 ft. Reduced pressure at* (a), *the jet nozzle, induces flow of groundwater into the circulated flow.*

Fig. 10.12 (opposite) *Details of the deep-well jet pump.* (a) *Photograph of a multistage jet pump housing and equipment. At the top is the on/off electrical switch actuated by pressure settings. It controls the direct-connected electric motor at the left. Impellers are enclosed in the pump housing at the right. Circulating connections to and from the jet are seen at the right and the pump discharge at the top. Note the pressure gauge readings up to 100 psi within which range the pump can be set to operate.* (b) *Well casing and circulating lines. Jet element is seen at the bottom of the left (larger) of the two pipes.* (c) *Cutaway section of the pump.* (d) *Pumping capacity in gallons per hour (gph) under various conditions. Note the two pressure ranges, 20 to 50 psi and 30 to 60 psi. Jacuzzi Bros., Inc.*

(a)

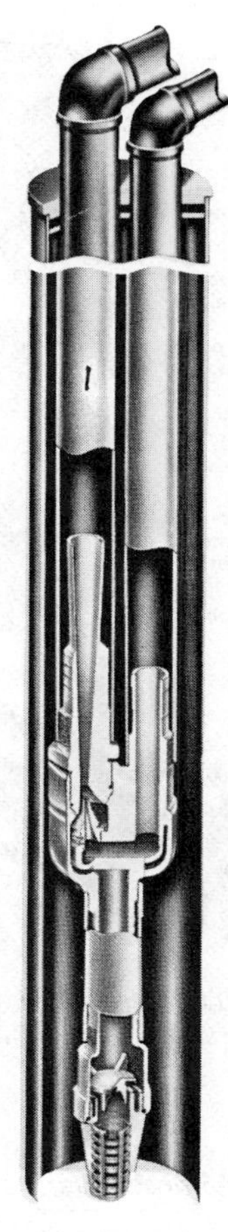

(b)

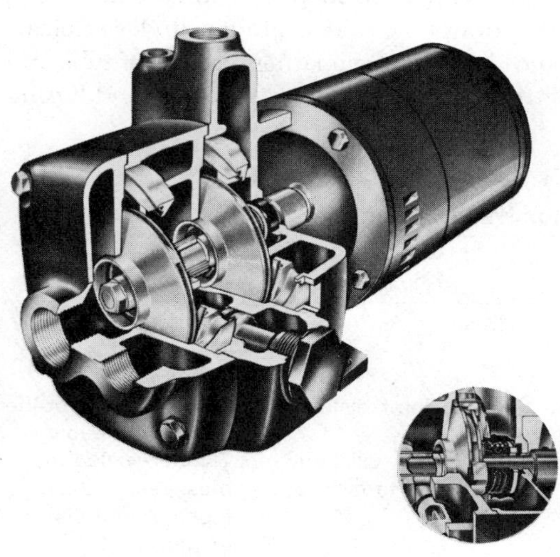

(c)

CHOOSE THE CORRECT JH FROM THESE CHARTS
DEEP WELL (Down to 120 feet)

The JH in the ½, ¾, 1 and 1½ horsepower rating, matched with the appropriate injector and pipe sizes, will pump this amount of water at the indicated lift or depth to the water in the well:

If suction lift or depth to water is	JH ½ HORSEPOWER	JH ¾ HORSEPOWER	JH 1 HORSEPOWER	JH 1½ HORSEPOWER
	Produces these gallons per hour between 20-50 lbs. discharge pressure		Produces these gallons per hour between 30-60 lbs. discharge pressure	
30 feet	795 G.P.H.	990 G.P.H.	1140 G.P.H.	1620 G.P.H.
40	680	875	1000	1470
50	575	735	875	1300
60	445	630	745	1200
70	360	495	620	920
80	310	385	530	820
90	255	315	435	700
100	220	275	340	590
110	195	240	295	540
120		205	250	480

(d)

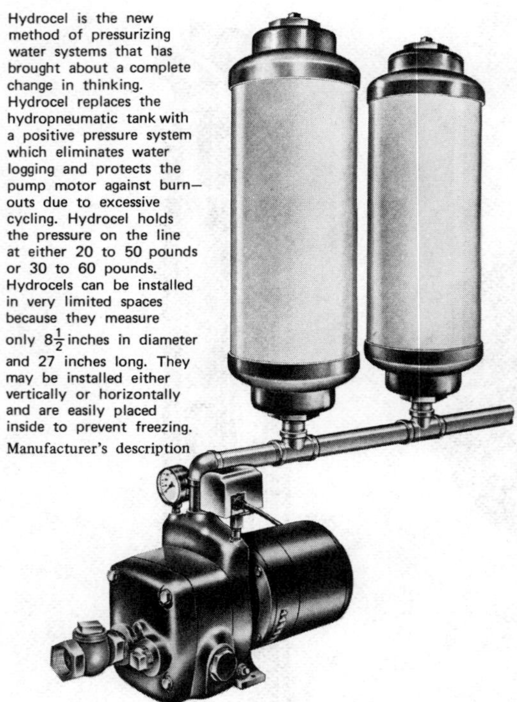

Fig. 10.13 Pump with "Hydrocel" tanks. See manufacturer's description. Jacuzzi Bros., Inc.

allurgists. In general, however, public or private treatment should be provided to correct corrosive qualities. Theoretically, when this is done, it is sometimes suitable to use a cheaper piping material—steel; yet, prudence suggests that a better material be selected. In the nonferrous group, red brass and copper tubing are effective in corrosion resistance. Copper tubing is a very popular choice. It is less expensive than brass, assembles more easily, and is not subject to dezincification, which is the attack by acids on the zinc in brass. For use in handling aggressive waters, plastic is often a good choice. Like copper it is light in weight and assembles with great ease.

For ferrous pipes and "iron pipe size" brass, threaded connections are used. The external, tapered thread on the pipe is covered with pipe compound and makes up tight against the internal tapered thread of the coupling or other fitting. The solder-joint connection in copper depends on capillary attraction that draws the solder into a cylinder of clearance between the mating surfaces of tube and fitting. This occurs after polishing and fluxing

Table 10.6 *Characteristics of Pipe and Tubing for Water Services*

Kind of Pipe	*Material or Manufacture*	*Connections*	*Qualities*	*Notes*
Steel	Butt welded to 2 in. diameter, seamless, large sizes	Threaded	Basic	Should be used only when water is not corrosive
Brass, red	85% copper 15% zinc	Threaded, "IPS," (iron pipe size)	Corrosion-resistant	Bulky because of the need for threading
Copper tube, type "K"	Seamless, hard or soft temper	Soldered fittings	Corrosion-resistant and easy to fabricate	Thinner-walled than brass; easy to put together and dismantle
Copper tube, type "L"	Seamless, thinner walls than type "K," hard or soft temper	Soldered fittings	Corrosion-resistant and easy to fabricate	Thinner-walled than brass; easy to put together and dismantle
Plastic[a]	See Table 10.7	Solvent cement weld[b]	Very easy to fabricate	Not subject to electrolytic corrosion
Nickel silver and chrome	Copper, nickel, and zinc, steel, and chromium	Threaded	Corrosion-resistant	Special applications
Galvanized steel	Zinc-coated steel	Threaded	Moderately corrosion-resistant	Suitable for mildly acid waters

[a]Upper limit of temperature, hot water, 180°F.

[b]For ABS and PVC (Table 10.7).

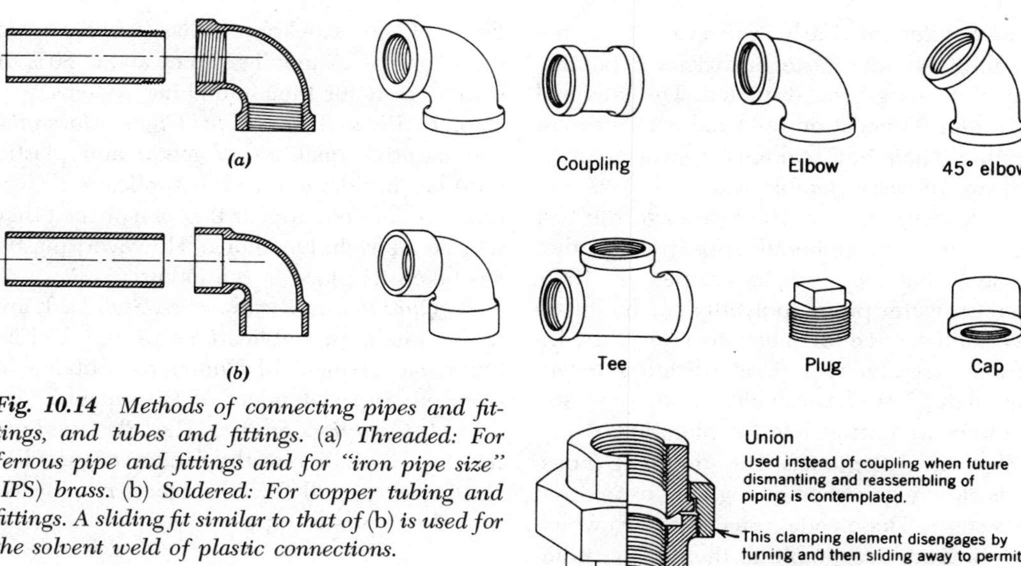

Fig. 10.14 Methods of connecting pipes and fittings, and tubes and fittings. (a) *Threaded: For ferrous pipe and fittings and for "iron pipe size" (IPS) brass.* (b) *Soldered: For copper tubing and fittings. A sliding fit similar to that of* (b) *is used for the solvent weld of plastic connections.*

Fig. 10.15 Examples of threaded *pipe fittings for ferrous or brass pipe. A few of the many fittings used in water piping. These and all common fittings are also available for solder-joint (copper) or solvent-weld (plastic) connections and usually for transition from one system or material to another.*

the surfaces and placing the parts together in final position. They are then heated, and molten solder is applied to the circular opening where the fitting-edge surrounds the tube, with a small clearance. It is then drawn into the cylindrical connection by capillary action. Solders are tin-lead or tin-antimony alloys. This kind of joint permits the advantageous setting up of an entire tubing assembly without turning the parts (as in threaded installations), and before the soldering commences. For the same strength, copper tubing may have thinner walls because no threads need to be cut into it. Its smooth interior surface offers less friction to flowing water. While threaded- and solder-joint connections are the most common in small work, there are many other types. Ferrous pipes in the larger sizes are often welded or connected by bolted flanges (see Figs. 10.14, 10.15, and Table 10.6).

(*b*) ***Plastic Pipe.*** Most of the plastic pipes and fittings now produced are synthetic resins. These do not appear in nature but are derived from such materials as coal and petroleum.

Rapid increase in the development, acceptance, and use of plastics for water piping, fittings and, indeed, for drainage systems (see Chapter 11) suggests a separate discussion of this *family* of materials.

Table 10.7 *Suitable Choices of Material for Plastic Piping in Water Services*

Symbol	*Material*	*Cold Water*	*Hot Water*
PE	Polyethylene	√	
ABS	Acrilylonitrile-Butadiene Styrene	√	
PVC	Polyvinyl Chloride	√	
PVDC	Polyvinyl Dichloride	√	√[a]

[a] Developed recently for this special use. Other plastic materials not currently approved for hot water piping.

Source. Courtesy of the Plastics Pipe Institute.

1. Selection of Material. The chemistry of plastics is quite intricate and the material can appear in a great variety of forms, a few of which, especially suitable for water piping, are listed in Table 10.7. The specifier is well advised to seek information from official and established sources.

2. *Sanitation.* Early concern about the possibility of odor, taste, or toxicity in potable water has long been dispelled. The National Sanitation Foundation tests and certifies plastic pipe. Their NSF seal must appear on pipes that are to carry potable water.

3. *History.* Almost 100 years ago the first plastic material, celluloid, appeared. Another product, Bakelite, was developed in 1905. Used for water piping polyethylene has given successful service for more than 30 years. Intensive research and standardization beginning about 1945 has established many plastic materials as appropriate for plumbing.

4. *Code Acceptance.* The plumbing industry is closely governed by regulations referred to as *codes.* These codes usually govern which materials are acceptable to the local jurisdiction. In 1972 it was estimated that approximately 4,500 state, county, and city codes allowed the use of plastics piping for all or a portion of the plumbing systems. However, prior to specifying or installing plastics piping, check with local plumbing officials.

Codes which are written by associations of plumbing officials or plumbing contractors are called *model codes.* The following model codes approve the use of plastics piping for all or a portion of the plumbing system:

BOCA Basic Plumbing Code
Building Officials and Code Administrators International

National Standard Plumbing Code
National Association of Plumbing, Heating, Cooling Contractors

Southern Standard Plumbing Code
Southern Building Code Congress

Uniform Plumbing Code
International Association of Plumbing and Mechanical Officials

5. *Effects of Temperature and Vibration.* Most of the materials used for piping are thermoplastics and have the quality of repeated softening under the application of heat. PVDC material can carry water at 180° F but plastic pipe should not be subjected to temperatures higher than this. Expansion (see note to Table 10.1) is great and affects the piping design. Quite shockproof, plastic has proved itself for use in mobile homes about 80% of which use it for their plumbing systems.

6. *Position Relative to Other Materials.* Comparative qualities of metal and plastic must be considered for each application. Currently it does not appear that one or the other will be an exclusive choice. However, plastic has taken its place in the industry.

7. *Quality and Performance.* Standards are to be found in publications of the United States Department of Commerce obtainable from the Superintendent of Documents.

8. *Information Source.* The Plastics Pipe Institute, a division of the Society of the Plastics Industry, will furnish objective information about plastic pipe.

(*c*) ***Valves and Controls.*** A good system utilizes many valves. It is usually desirable to valve every riser, the branches that serve bathrooms or kitchens, and the runouts to individual fixtures. This facilitates repairs at any location with a minimum of shutdown within the system. Treatment equipment will have a valved bypass (see Fig. 10.1). Pumps and other devices that may need repair should be disconnectable by unions (Fig. 10.15) after valves are closed.

The gate valve (Fig. 10.16*a*), with its retractable leaf which is machined to seal tightly against two sloping metal surfaces when closed, offers the least resistance to water flow when open. It is usually chosen for locations where it is left completely open most of the time. The compression-type globe valve (Fig. 10.16*b*) has for its usual purpose the closing or throttling of flow near a point of occasional use. Faucets are usually of the compression type, as are drain valves or hose connections. They are similar to the angle valve (Fig. 10.16*d*). When it is necessary to prevent flow in a direction opposite to that which is planned, the introduction of a check valve (Fig. 10.16*c*) will accomplish this. The hinged leaf swings to permit flow in the direction of the arrow but closes against attempted flow in the other direction.

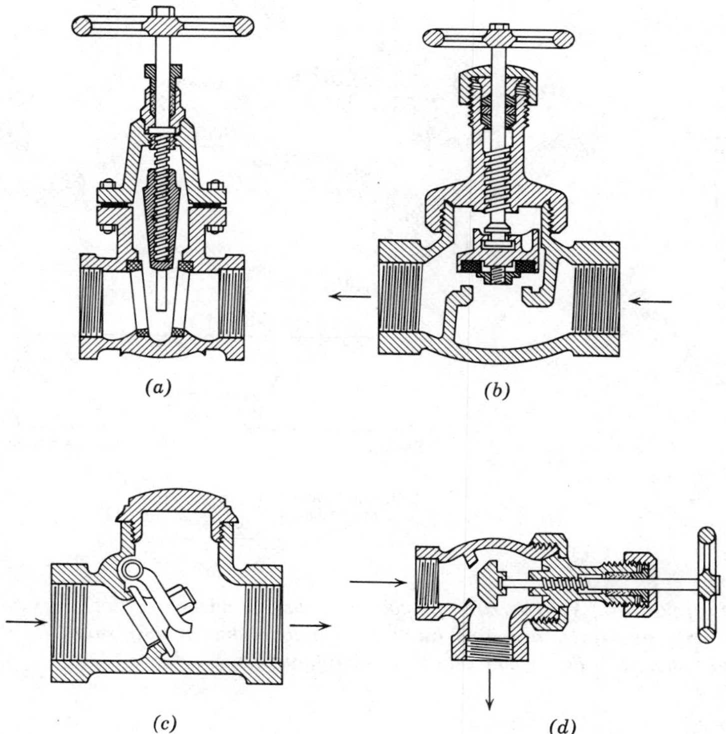

Fig. 10.16 *Typical valves for water systems.* (a) *Gate valve.* (b) *Globe valve.* (c) *Check valve.* (d) *Angle valve.*

(*d*) ***Pipe Supports.*** If a piping system of conventional dimensions were to stand alone without a building to rest on, it would quickly collapse. Quite heavy because of its metallic nature and water content, it needs closely spaced supports (Fig. 10.17). Vertical runs of 1-in. piping should be supported at every story, but larger sizes may extend for two stories. Horizontal pipes should be supported at intervals of 10 ft. Closer spacing, 6 to 8 ft, is preferred for sizes $\frac{1}{2}$ in. and smaller. Horizontal copper tubing should always be supported at closer spacing than steel. Adequate positioning of horizontal runs is important to assure correct pitch and drainage. Hangers are adjustable for this purpose.

(*e*) ***Shock and Hot Water Expansion.*** Water systems can be noisy. When faucets are shut off abruptly, or automatically when released, as in public buildings, the force exerted by the decelerated flowing water shakes and rattles the pipes. Lengths of vertical pipe about 2 ft long at the fixture branches (Fig. 10.18*a*) will usually solve this problem. They trap a certain amount of air that absorbs the impact of the water with some resilience. A somewhat better and a more controllable device is a "rechargeable air chamber" (Fig. 10.18*b*). By closing the valve and draining the water through the hose bibb while the petcock above is open to admit air, the chamber may be refilled with air. Closing the petcock and hose bibb and opening the valve completes the service operation and reconnects the device with the water system. Rechargeable chambers are used on branch lines adjacent to groups of fixtures. Access for servicing them must be provided. When this method is chosen, the smaller pipe extensions (Fig. 10.18*a*) are usually omitted. Perhaps the best method is the use of a special shock absorber (Fig. 10.18*c*).

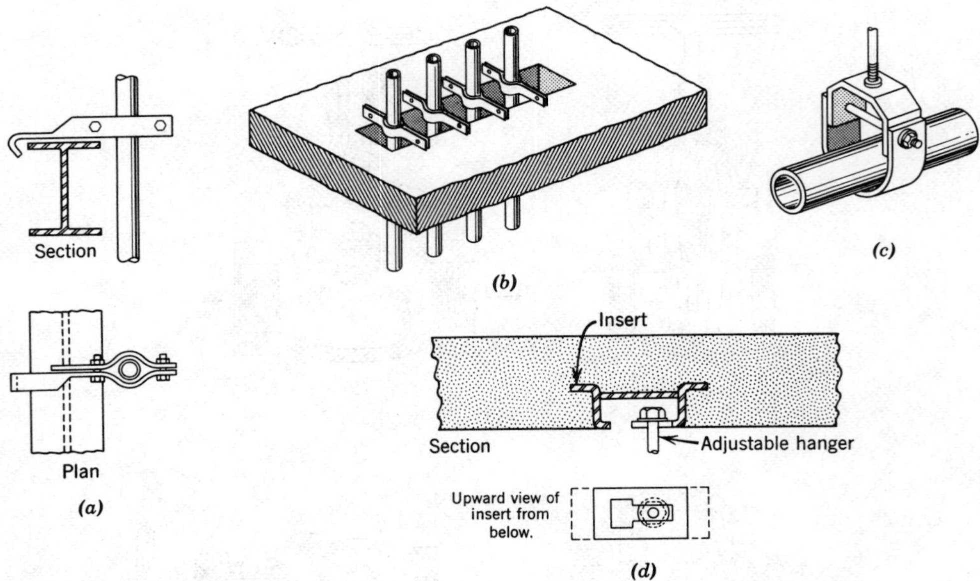

Fig. 10.17 *Pipe supports.* (a) *Vertical riser supported at steel beam.* (b) *Vertical riser group supported at slot in concrete slab.* (c) *Horizontal pipe hung from slab above by adjustable-length clevis hanger.* (d) *Typical metal insert in soffit of concrete to receive hanger rod.*

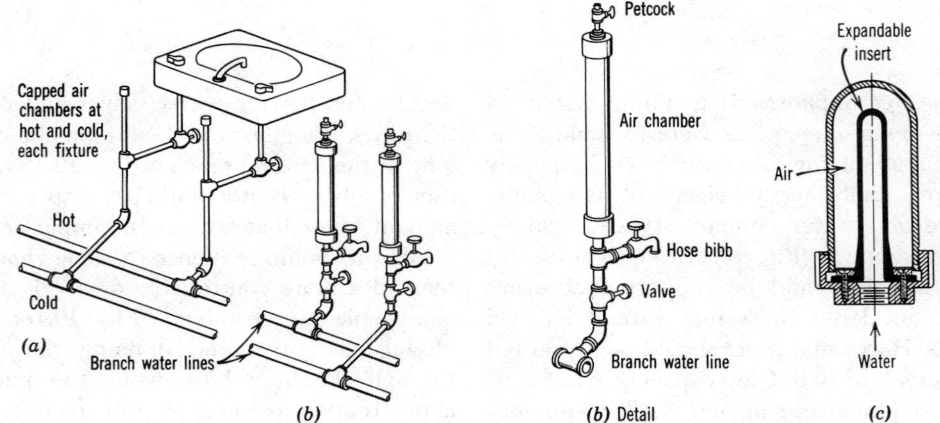

Fig. 10.18 *Shock relief and expansion chambers. Air chambers to cushion the shock of water hammer when fixture faucets are shut off abruptly. They also permit hot water to expand instead of periodically forcing open the hot-water-emergency pressure relief valve at the heater or tank.* (a) *Capped air chambers at each supply pipe of each fixture.* (b) *Rechargeable air chambers on hot and cold branch water lines. When used, the individual fixture chambers are omitted.* (c) *Special shock absorber.*

Air cushions also protect the relief valve against frequent operation with the resulting leakage of hot water, as the hot water periodically expands and contracts in closed systems.

(*f*) ***Condensation or "Sweating."*** The moisture that is always present in air often condenses on the exterior surface of cold pipes. Dropping off the pipes, it creates an unpleasantly wet condition and disfigures finished surfaces. Groundwater in some parts of the United States is 50° F and colder. A pipe carrying such water might have a surface temperature of about 60° F. The psychrometric chart (page 229) indicates that at a summer air temperature of 85° F condensation will occur on this pipe when the relative humidity exceeds 40%. All cold water piping and fittings should be covered. Glass fiber ½ or 1 in. thick is commonly chosen for this purpose. A tight vapor barrier on the exterior surface of the covering prevents the moisture-laden air from penetrating the insulation to reach the colder surface. The insulation provides another advantage of equal importance; it retards heat flow from the warmer air to the water, thus preventing the water from becoming disagreeably warm.

(g) ***Heat Conservation.*** Pipes carrying domestic hot water should be insulated to conserve the fuel used to heat the water and to assure a correct water temperature at the point of use. An efficient covering is a ½ in. thick glass fiber. Parallel hot and cold water piping, even though insulated, should be separated by 6 in. or more to prevent heat interchange.

Small storage tanks and heaters are often manufactured with integral insulation. Uninsulated products should receive insulative covering after installation. For energy-saving standards for hot water piping and hot water storage tanks see Section 10.8.

10.6 Hot Water Heating and Distribution

Water expands and becomes lighter when heated, as may be seen from Figure 10.19*a*. If heat is applied to the lower loop of a glass tube, both ends of which have been inserted in an inverted bottle containing water, the water will move from *A* to *B* and will rise through the tube *BC* into the bottle. It here becomes cooled and drops through the tube *DA* to *A*, is again heated, and rises in the tube *BC*, thus completing the circulation. Since the movement depends on the difference in weight between the two columns of water, the velocity and consequent efficiency of the circulating system increase with the temperature of the water and the height of the circuit. Hot water supply systems therefore usually consist of a heater with a storage tank, piping to carry the heated water to the farthest fixture, and a continuation of this piping to return the unused cooled water back to the heater. A constant circulation is thereby maintained, and

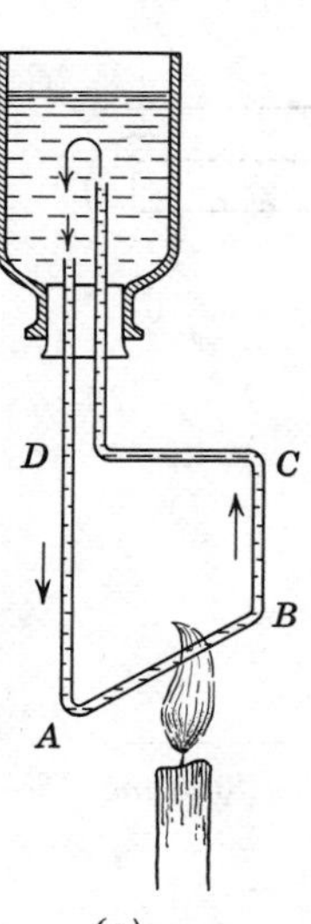

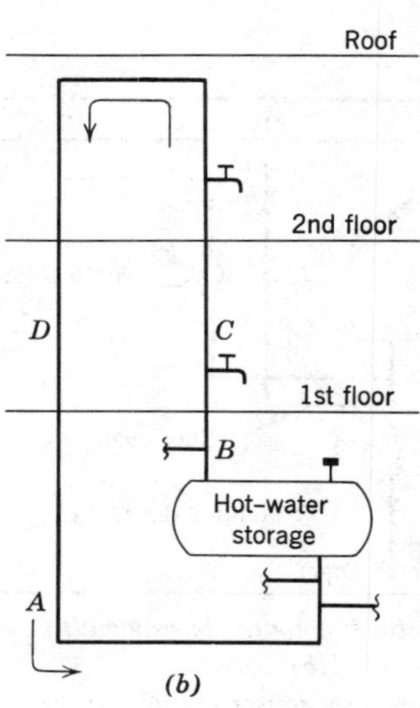

Fig. 10.19 (a) *Principle of hot water circulation by gravity, and* (b) *its application to hot water service. During periods of no-demand there is sufficient incidental cooling between* C *and* D *so that dense (less warm) water at* A *forces the lighter (hot) water at* B *to rise for speedy availability at each faucet. For other applications of the principle of vertical circulation by gravity see Figures 10.1 and 10.4*a *and* b. *The candle flame is, of course, symbolic of other sources of energy, including solar (see Chapter 8), oil, gas, steam (see Fig. 10.4*a*), and electricity. Conformance to ASHRAE Standard 90-75 concerning improved pipe insulation may reduce the effectiveness of gravity circulation.*

hot water may be drawn at once from a fixture without first draining off through the faucet the cooled water that would be standing in the supply pipe if there were no return conduit for its escape. Because heat increases the corrosive action of water on metals, copper tubing is usually the best choice for use in hot water and hot water circulating systems.

This principle of constant circulation of hot water to make it promptly available in branch water lines may be examined in Figure 10.1. Heated water from the tank rises to the hot water main and the branch hot water lines. At the ends of these horizontal pipes the slightly cooler water falls through the circulation piping to the bottom of the tank. Two other circulation processes can be found in Figure 10.1, and all three operate regardless of demand by the opening of faucets. In addition to the circulation that serves the fixtures, there is a second route of water motion from the top of the heating coil to the top of the tank and from the bottom of the tank to the bottom of the coil. The third and separate circulating process takes place when the boiler water circulates around the outside of the coil and then drops to be reheated by the boiler. Upon demand by the opening of any hot water faucet, the pressure in the cold water main opens the check valve below the tank and the cold water forces the hot water in the piping and the upper part of the tank toward the open faucet. Figure 10.4*a* and *b* illustrate a similar system of circulation, but the source of heat is a steam-actuated storage heater instead of the external tank-type heater adjunct to a residential hot water heating boiler, as used in Figure 10.1. For several methods of heating domestic water see Figures 10.22 to 10.25.

There are many buildings that do not have the advantage of height as do those in Figures 10.1 and 10.4*a* and *b*. Low, long rambling buildings such as some large one-story residences, schools, and factories lack sufficient height to set up good hot water circulation by gravity. They also diminish the flow by friction in long pipe runs. For such buildings the forced circulation scheme of Figure 10.20 is good. Three independent aquastats, devices that create an electrical signal impulse when water temperature drops, control this very positive and efficient system. Aquastats *A*, *B*, and *C*, respectively, sense the temperatures of the water in the heater, the tank, and the end

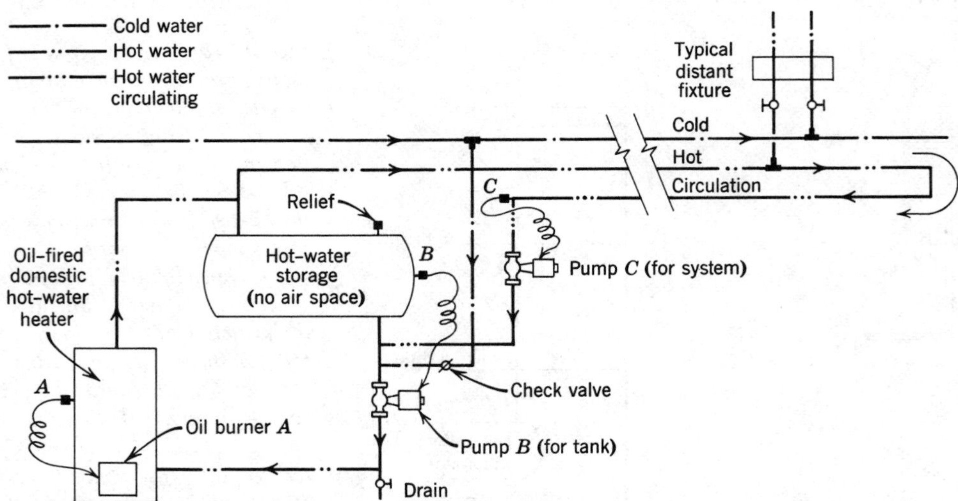

Fig. 10.20 *Forced circulation of domestic hot water is especially needed in long, low buildings. See ASHRAE Standard 90-75, Section 7.0 (page 418):*
Section 7.3.3.1 for insulation of unfired hot water storage tanks.
Section 7.3.3.2 for insulation of hot water recirculation piping.

of the circulation-return main. As needed, they turn on the oil or gas burner, the tank-circulating pump, and the system-circulating pump. Fixtures remote from the tank are as close to hot water as the length of their hot water runout pipes. Water is usually available at full temperature in 5 to 10 sec. Trial aquastat settings in degrees Fahrenheit could be *A* 180, *B* 160, *C* 120.

For relatively small installations and compact fixture groupings, the water may be heated directly and the storage tank omitted. This is done by an internal or external tankless coil (see Fig. 10.22*a* and *b*), its length being great enough to effect a fast and thorough heat exchange between the boiler water and the domestic water. Unlike the short tank-type coil (Fig. 10.1) that operates at a slow recovery rate by the gravity circulation action of the water in it and in the tank, the tankless type operates under full flow whenever hot water faucets are opened. Temperature blending valves are commonly employed to moderate the delivery temperature of the hot water (see alternate piping in Fig. 10.21). Temperature-pressure relief valves are essential in these as in all hot water systems.

The tankless heater, because it meets the full demand without utilizing some stored water, requires a large heat input rate.

EXAMPLE 10.2. A tankless heater delivers 6 gpm at 100° F rise in temperature. How many Btuh are required?

SOLUTION. Average weight of water: 8.33 lb/gal.

Specific heat of water:

1 Btu/(lb)(1° F)

Heat required:

6 gpm × 60 min/hr × 8.33 lb/gal

× 1 Btu/deg × 100° F = 300,000 Btuh.

A large house could have an hourly heat loss requiring a boiler that would deliver about 150,000 Btuh for heating only. During periods of operation a 6 gpm tankless heater delivering hot water while the house was being heated would approximately triple the demand on the boiler. Some heating systems turn off automatically when the domestic hot water demand reduces the boiler-water temperature. This favors the domestic water over the general heating. If, for instance, 200 gal of stored hot water would serve this house during

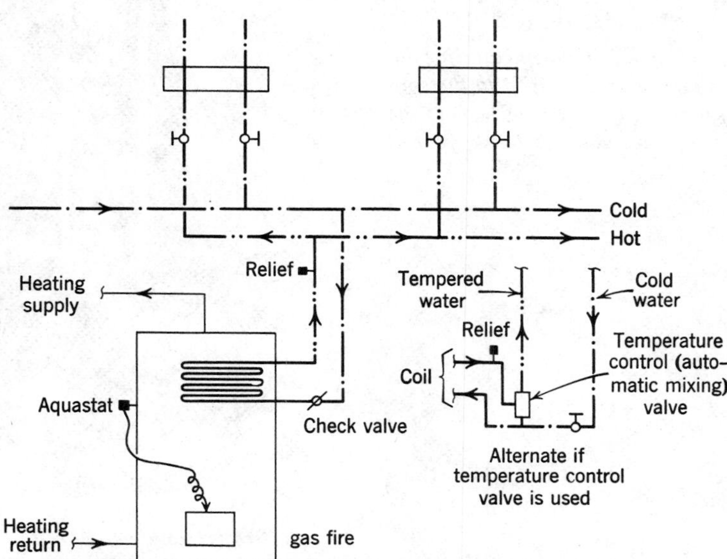

Fig. 10.21 Domestic hot water from a built-in tankless coil in a hot water heating boiler. This method, suitable for small, compact installations, provides no circulation.

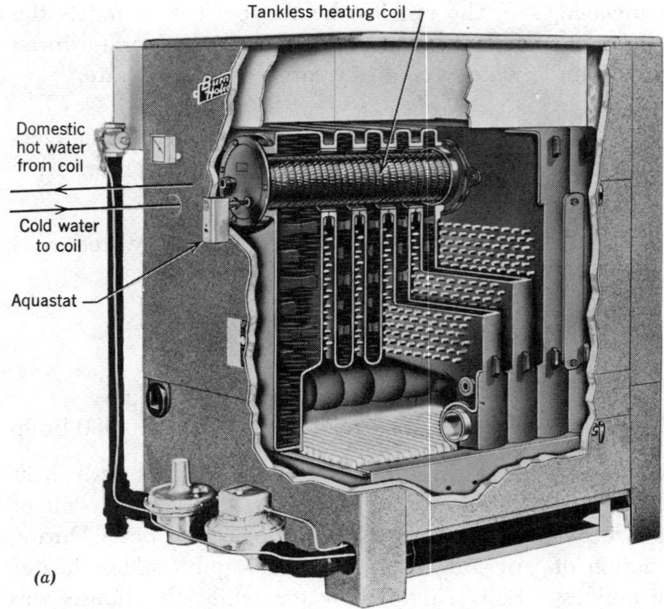

Fig. 10.22a Internal tankless heating coil for domestic hot water immersed in the jacket water of a gas-fired hot water heating boiler. Approximate capacity range 3 to 15 gpm at 100° F rise. See ASHRAE Standard 90-75, Section 7.3.2 for combined space heating and domestic (service) water heating (page 419).

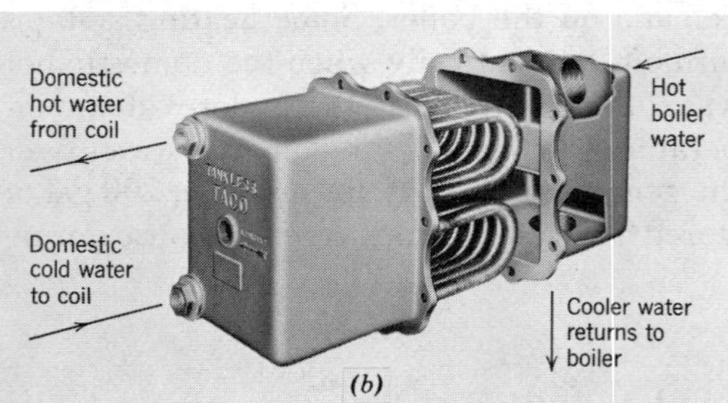

Fig. 10.22b External-type, tankless heater for domestic hot water. Boiler water is piped to the unit and circulates by gravity, transferring heat to the coil. Approximate capacity range 3 to 15 gpm at 100° F rise.

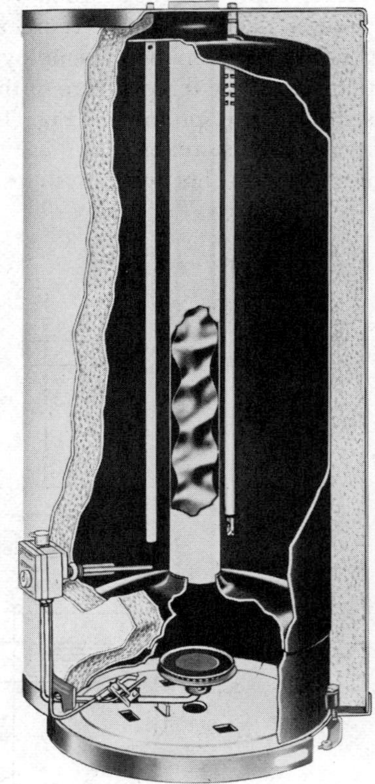

Fig. 10.23a (Example 10.3) *A. O. Smith Conservationist Gas Residential Water Heater, cutaway view. For capacities and other engineering data see Figure 10.23b.*

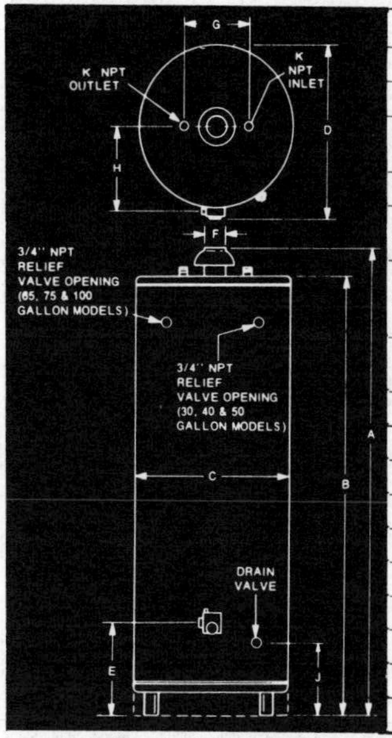

Models	PGC-30	PGCL-30	PGC-40	PGCL-40	PGC-50		PGC-65	PGC-75		PGC-100	
Gases	NAT. & L.P.	NAT. & L.P.	NAT. & L.P.	NAT. & L.P.	NAT.	L.P.	NAT. & L.P.	NAT.	L.P.	NAT.	L.P.
Cap. in gal.	30	30	40	40	50	50	65	75	75	100	100
Input BTU/Hr.	30,000	30,000	32,500	32,500	44,000	40,000	50,000	80,000	75,500	90,000	80,000
Actual Efficiency	79%	76%	76%	76%	77%	77%	75%	70%	70%	70%	70%
Actual Recovery GPH @ 100° Rise	28.4	27.3	29.6	29.6	40.6	36.9	45.0	67.2	63.4	75.6	67.2
A.G.A. Recovery GPH @ 100° Rise 70% Efficiency	25.2	25.2	27.3	27.3	36.9	33.6	42.0	67.2	63.4	75.6	67.2
Approx. shippng wt. lbs.	122	119	136	141	164		268	289		355	
DIMENSIONS IN INCHES A	59-5/8	50-1/8	57-3/4	54-7/8	63-1/2		56-5/8	62-5/8		70-5/8	
B	55-3/8	43-1/8	53-3/8	47-1/2	56-1/4		53-1/4	59-1/4		67	
C	18-1/4	20-1/4	20-1/4	20-1/4	20-1/4		25-3/8	25-3/8		26-3/8	
D	21-3/4	23-3/4	23-3/4	23-3/4	23-3/4		28-3/4	28-3/4		29-3/4	
E	13-3/8	13-3/8	13-3/8	13-3/8	13-3/8		17-3/4	17-3/4		17-3/4	
F	3	3	3	3	3		4	4		4	
G	8	8	8	8	8		16	16		16	
H	10-5/8	11-5/8	11 5/8	11-5/8	11-5/8		14	14		14-3/8	
J	9-1/4	9-1/4	9-1/4	9-1/4	9-1/4		13-1/4	13-1/4		13-1/4	
K	3/4	3/4	3/4	3/4	3/4		3/4	1		1-1/4	

A. O. Smith Corporation reserves the right to make product changes or improvements at any time without notice.

NOTE: To compensate for the effects of high altitude areas above 2,000 feet, recovery capacity should be reduced approximately 4% for each 1,000 feet above sea level.

AOSmith

A. O. Smith Corporation
Consumer Products Division
Kankakee, Illinois • Stratford, Ontario, Canada

WH-1035C Printed in U.S.A. RM25M37

(b)

Fig. 10.23b *Capacities and performance characteristics of Conservationist gas water heaters shown in Figure* 10.23a.

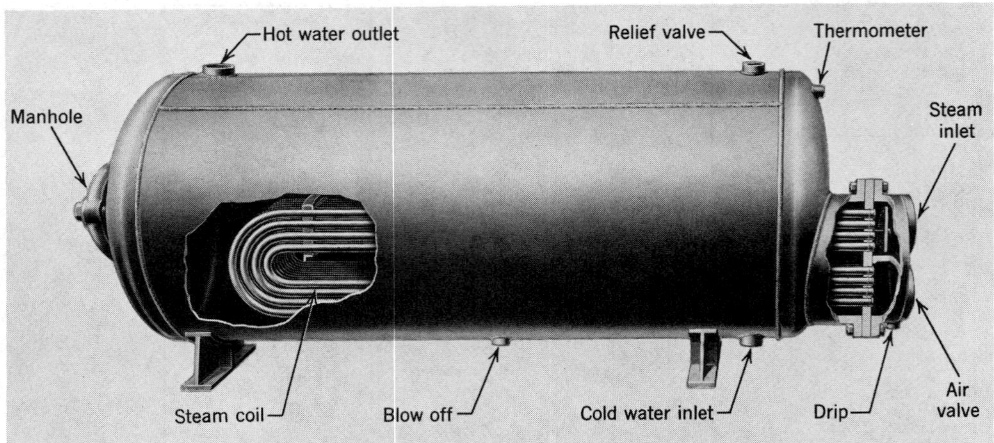

*Fig. 10.24 Storage tank heater and storage for domestic hot water for large-demand applications. Steam coil submerged in tank (see Figs. 10.4*a *and* b*). Capacities 100 to 10,000 gph (gallons per hour), varying by length of coil, for 140° F (40 to 180° F) temperature rise.*

its peak demand, a tank-type smaller coil could be used and it would fill the tank in 3 hours. This would be at the rate of about 55,500 Btuh, which would call for an increase in boiler size of only about one-third. When this joint demand is troublesome, independent domestic hot water heaters are used. They can be electric, oil-fired, gas-fired, or solar. The largest demands are often met by the storage-tank heater (Fig. 10.24), the use of which is shown in Figure 10.4*a* and *b*.

Under certain provisions of ASHRAE Standard 90-75, service (domestic) water heating must not, in some instances, be dependent on year-round operation of space-heating boilers (see Section 10.8). When this regulation applies, independent water heaters are used. One type, a gas-fired residential unit, is shown in Figure 10.23.

An example of the process of selecting the proper size of such a unit is presented with reference to Table 10.8 of the Department of Housing and Urban Development and Federal Housing Administration (HUD-FHA).

EXAMPLE 10.3. Select a natural gas water heater for a four-bedroom house with 2½ baths (two full baths and one powder room). The *minimum* requirements of HUD-FHA are shown in *Column A* below. From Figure 10.23*b* we select two trial units (40 gal and 50 gal) that *might* comply; see *Columns B* and *C*. The analysis is as follows:

	Column A HUD-FHA Minimum (Table 10.8)	*Column B*	*Column C*
		Conservationist Water Heater, A.O. Smith (Figure 10.23b)	
		PGCL-40, NAT.	*PGC-50, NAT.*
Storage (gal)	40	40	50
1000 Btu input	38	32.5	44
1-hr draw (gal)	72	67.3	86.9[a]
Recovery (gph)	32	27.3	36.9

[a]1-hr draw = tank capacity + 1-hr recovery 50 + 36.9 = 86.9 gal

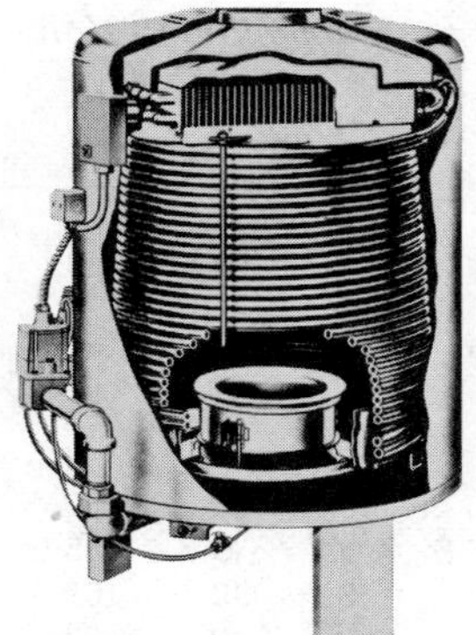

Designed to handle big jobs — large restaurants, laundries, apartment buildings, plant process operations, etc. Normally installed in conjunction with a storage tank and circulating pump to provide an ample supply of 1-temperature or 2-temperature hot water. (Not recommended for instantaneous use.)

Fig. 10.25 *A. O. Smith Burkay Water Heater Models, BC-300, BC-399, BC-420. Cutaway section with manufacturer's statement of inputs and suitable uses. Manufacturer's energy statement—"All Burkay models meet the requirements of ASHRAE 90-75 Standard for efficiencies and bear this label:*

A REPRESENTATIVE PRODUCT SAMPLE OF THIS MODEL HAS BEEN TESTED IN ACCORDANCE WITH ASHRAE STANDARD 90-75.

SOLUTION. PGC-50, NAT. is selected. Although PGCL-40, NAT. complies in storage capacity, it is inadequate in the three other categories. The selected unit is $63\frac{1}{2}$ in. high and $20\frac{1}{4}$ in. in diameter.

For commercial and industrial installations, the water heaters shown in Figures 10.24 and 10.25 are suitable. For information about demand and equipment selection in these more intricate systems, the reader is referred to Chapter 37, "Service Water Heating," in *ASHRAE Handbook, 1976 Systems.*

10.7 Sizing of Water Pipes

There must be sufficient pressure at fixtures to assure the user of a prompt and generous flow of water. Municipal ordinances often state that the flow must be enough to keep the fixtures clean and sanitary. The convenience of the user and the objectives of sanitation are consistent with each other and have resulted in prescribed pressures that must be maintained at the various fixtures to assure the proper flow rates. These pressures and flows are listed in Table 10.9.

Fixture pressures vary from 5 to 20 psi for fixtures other than hose bibbs. Since the pressure in street mains is usually about 50 psi, it is possible to assure the correct fixture pressure provided the water does not have to be lifted to too great a height and that much pressure is not lost by friction in distribution piping that is too long in *developed length* (actual distance of water flow), or interposes too many fittings such as elbows and tees, or is too small in diameter.

The pressure components and their total in an upfeed system actuated by street main pressure are as follows:

Proper fixture flow pressure	A
Pressure lost because of height	B
Pressure lost by friction in piping	C
Pressure lost by flow through meter	D
Total street main pressure	E

In a design, three of the above items, A, B, and E, are known and are reasonably constant. The value of A is found in Table 10.9. Street main pressure, E, is a characteristic of the local water supply. Item B, the pressure lost due to height, is found by multiplying the height in feet by 0.433—a factor that is the pressure of *one* foot of *head* above the main. Item D, the pressure lost in flow through the water meter, depends on flow (gallons per minute) and pipe size (see Fig. 10.28), neither of which is yet known. Therefore, the value of Item D is *estimated.* Later it must be checked and a recalculation made if necessary.

The selection of a pipe size is facilitated

Table 10.8 ***HUD-FHA Minimum Water Heater Capacities for One-and Two-Family Living Units***

Number of Baths	*1 to 1½*			*2 to 2½*				*3 to 3½*			
Number of Bedrooms	*1*	*2*	*3*	*2*	*3*	*4*	*5*	*3*	*4*	*5*	*6*
Gas[d]											
Storage—gal	20	30	30	30	40	40	50	40	50	50	50
1000 Btu input	27	36	36	36	36	38	47	38	38	47	50
1-hr draw—gal	43	60	60	60	70	72	90	72	82	90	92
Recovery—gal/hr	23	30	30	30	30	32	40	32	32	40	42
Electric[d]											
Storage—gal	20	30	40	40	50	50	66	50	66	66	80
Kilowatts—input	2.5	3.5	4.5	4.5	5.5	5.5	5.5	5.5	5.5	5.5	5.5
1-hr draw—gal	30	44	58	58	72	72	88	72	88	88	102
Recovery—gal/hr	10	14	18	18	22	22	22	22	22	22	22
Oil[d]											
Storage—gal	30	30	30	30	30	30	30	30	30	30	40
1000 Btu input	70	70	70	70	70	70	70	70	70	70	70
1-hr draw—gal	89	89	89	89	89	89	89	89	89	89	89
Recovery—gal/hr	59	59	59	59	59	59	59	59	59	59	59
Tank Type Indirect[a,c]											
I-W-H-rated gal in 3 hr, 100° F rise		40	40		66	66[c]	66	66	66	66	66
Manufacturer-rated gal in 3 hr, 100° F rise		49	49		75	75[c]	75	75	75	75	75
Tank capacity in gal		66	66		66	66[c]	82	66	82	82	82
Tankless Type Indirect[b,e]											
I-W-H-rated gpm, 100° F rise		2.75	2.75		3.25	3.25[c]	3.75	3.25	3.75	3.75	3.75
Manufacturer-rated draw in 5 min, 100° F rise		15	15		25	25[c]	35	25	35	35	35

[a]Boiler-connected water heater capacities (180° F boiler water, internal or external connection).

[b]Boiler-connected heater capacities (200° F boiler water, internal or external connection).

[c]Also for 1 to 1½ baths and 4 B.R. for indirect water heaters.

[d]Storage capacity, input, and recovery requirements indicated in the table are typical and may vary with each individual manufacturer. Any combination of these requirements to produce the stated 1-hr draw will be satisfactory.

[e]Heater capacities and inputs are minimum allowable. Variations in tank size are permitted when recovery is based on 4 gph/kw at 100°F rise for electrical. A.G.A. recovery ratings for gas heaters, and IBR ratings for steam and hot water heaters.

by Figure 10.26. Pipe diameter is determined by the point of intersection of a horizontal line representing flow in gallons per minute and a vertical line expressing friction loss in psi/100 ft of pipe length. To select a pipe size, one needs to know the probable flow and the *unit*-friction loss in the pipe and fittings.

Flow is found by assigning the fixture units listed in Table 10.10. The sum of these is an index of the demand flow that can be found in Figure 10.27. These curves, based on experience, indicate that flow does *not* increase in direct proportion to an increase in fixture units. In larger installations, there is less likelihood that many fixtures will be operating concurrently.

Establishing the desired friction loss is done as follows: value C (pressure lost by friction in piping) is divided by the *total equivalent length* (TEL) of the piping. This length is the

Table 10.9 ***Proper Flow and Pressure Required During Flow for Different Fixtures***

Fixture	*Flow Pressure*[a]	*Flow gpm*
Ordinary basin faucet	8	3.0
Self-closing basin faucet	12	2.5
Sink faucet—$\frac{3}{8}$ in.	10	4.5
Sink faucet—$\frac{1}{2}$ in.	5	4.5
Bathtub faucet	5	6.0
Laundry tub cock—$\frac{1}{4}$ in.	5	5.0
Shower	12	5.0
Ball cock for closet	15	3.0
Flush valve for closet	10–20	15–40[b]
Flush valve for urinal	15	15.0
Garden hose, 50 ft, and sill cock	30	5.0

[a]Flow pressure is the pressure psig in the pipe at the entrance to the particular fixture considered.

[b]Wide range due to variation in design and type of flush-valve closets.

sum of the *developed length* (DL) (total linear distance of water travel) plus the length equivalent to the fittings. For instance, Table 10.11 shows that a 90-deg ell causes a friction loss equivalent to that of 3 ft of pipe in a 1-in. diameter pipe run. Obviously the number and style of fittings must be estimated, and the *size* of fittings assumed. This is a puzzling but common engineering procedure that sometimes requires several recalculations.

EXAMPLE 10.4. Using the following data, some of which have been arrived at by the assumptions referred to above, find the proper size for a metered water supply main.

Street main pressure (minimum):	50 psi
Height, topmost fixture above main:	30 ft
Topmost fixture type:	water closet with flush valve
Fixture units in the system:	85
Developed length (DL) of the piping to the highest and most remote fixture:	100 ft
Pipe length equivalent to fittings (estimate)[2]:	50 ft
System uses predominantly:	flush valves

SOLUTION. From the minimum street main pressure subtract the sum of the fixture pressure, the static head, and the pressure lost in the meter.

This sum is

		psi
A—fixture pressure (Table 10.9)		20.0
B—static head 30×0.433		12.9
D—pressure loss in meter (estimated, Fig. 10.28)		8.0
		40.9
Subtract A + B + D	E	50.0
from E	A + B + D	−40.9
	C =	9.1

[2]Often *estimated* as 50% of the DL.

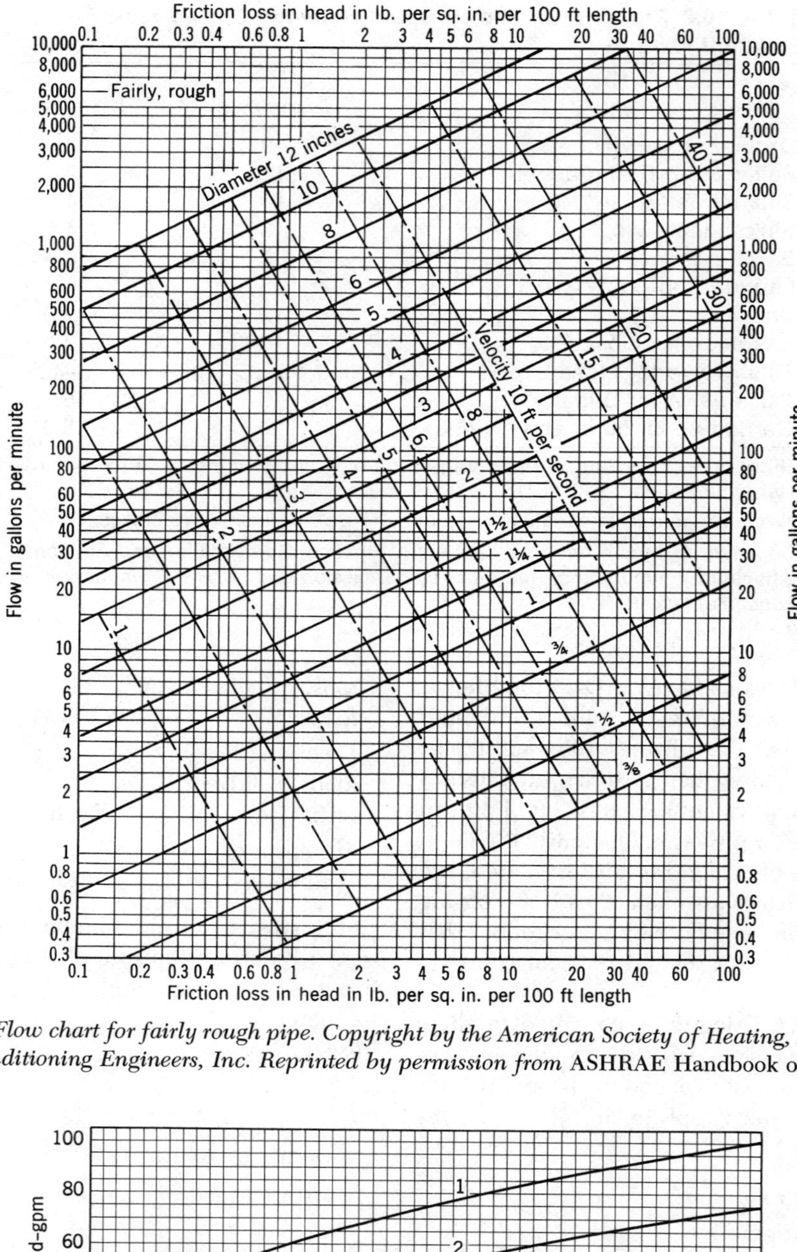

Fig. 10.26 *Flow chart for fairly rough pipe. Copyright by the American Society of Heating, Refrigerating and Air-Conditioning Engineers, Inc. Reprinted by permission from* ASHRAE Handbook of Fundamentals, 1972.

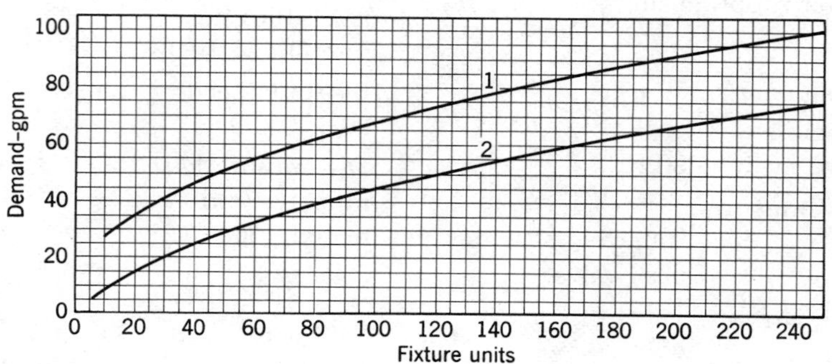

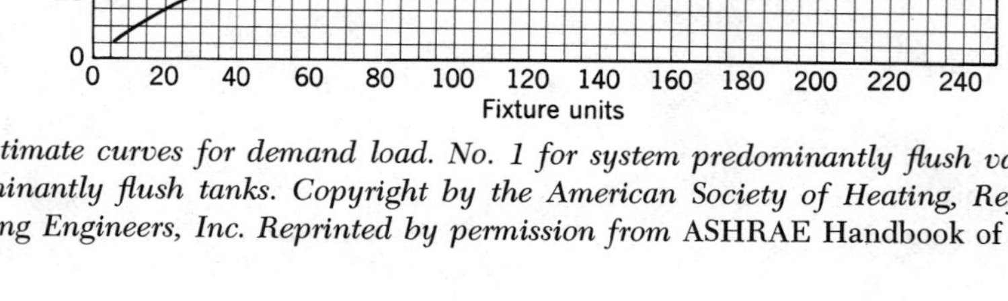

Fig. 10.27 *Estimate curves for demand load. No. 1 for system predominantly flush valves. No. 2 for system predominantly flush tanks. Copyright by the American Society of Heating, Refrigerating and Air-Conditioning Engineers, Inc. Reprinted by permission from* ASHRAE Handbook of Fundamentals, 1972.

Table 10.10 ***Demand Weights of Fixtures in Fixture Units***[a]

Fixture or Group[b]	*Occupancy*	*Type of Supply Control*	*Weight in Fixture Units*[c]
Water closet	Public	Flush valve	10
Water closet	Public	Flush tank	5
Pedestal urinal	Public	Flush valve	10
Stall or wall urinal	Public	Flush valve	5
Stall or wall urinal	Public	Flush tank	3
Lavatory	Public	Faucet	2
Bathtub	Public	Faucet	4
Shower head	Public	Mixing valve	4
Service sink	Office, etc.	Faucet	3
Kitchen sink	Hotel or restaurant	Faucet	4
Water closet	Private	Flush valve	6
Water closet	Private	Flush tank	3
Lavatory	Private	Faucet	1
Bathtub	Private	Faucet	2
Shower head	Private	Mixing valve	2
Bathroom group	Private	Flush valve for closet	8
Bathroom group	Private	Flush tank for closet	6
Separate shower	Private	Mixing valve	2
Kitchen sink	Private	Faucet	2
Laundry trays (1–3)	Private	Faucet	3
Combination fixture	Private	Faucet	3

From NBS Report BMS79 *Water-Distributing Systems for Buildings.*

[a]For supply outlets likely to impose continuous demands, estimate continuous supply separately, and add to total demand for fixtures.

[b]For fixtures not listed, weights may be assumed by comparing the fixture to a listed one using water in similar quantities and at similar rates.

[c]The given weights are for total demand. For fixtures with both hot and cold water supplies, the weights for maximum separate demands may be taken as $\frac{3}{4}$ the listed demand for the supply.

The pressure lost in 100 ft (DL) of piping plus the 50 ft of piping equivalent to the pressure lost by friction in the fittings is 11.2 psi. Total equivalent length (TEL) is 150 ft. The above numerical accounting assures 20 psi at the critical fixture. The unit-friction loss, psi/100 ft of pipe, will be $9.1 \times 100/150 = 6.1$. Figure 10.27, Curve 1, shows that for a flush-valve system with 85 fixture units, the demand (probable flow) will be 64 gpm.

In Figure 10.26, the horizontal and vertical lines, representing 64 gpm and 6.1 psi/100 ft, cross at a point that indicates (sloping line) that a 2-in. diameter main is a correct choice. Figure 10.26 is, of course, developed from basic engineering relationships of flow, friction, pipe diameter, and pipe length.

Now find the actual pressure loss in the 2-in. meter for a flow of 64 gpm. Figure 10.28 shows that this is 4.0 psi. Since this is *less* than the 8.0 psi estimated, the pressure at the *fixture* will be slightly higher than planned. When a final system layout is made, the fittings are tabulated and the length in piping equivalent to fittings is found. If this differs

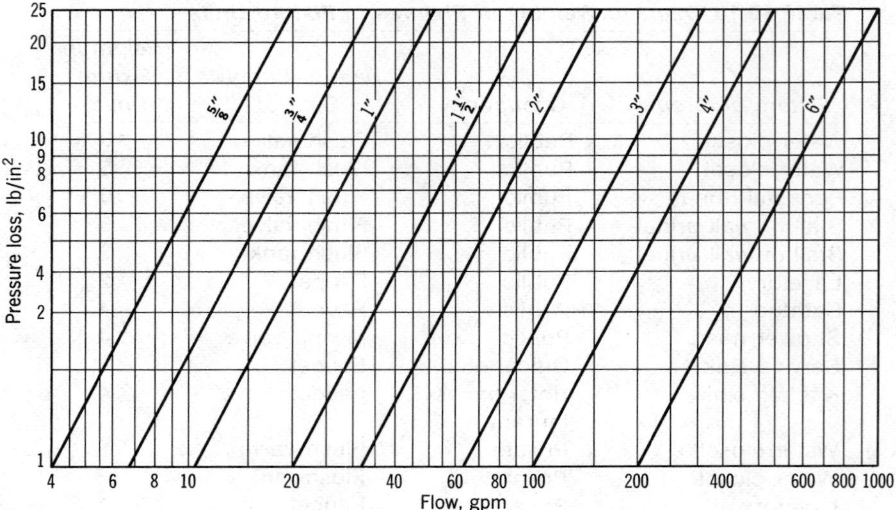

Fig. 10.28 *Pressure losses in water meters. Reprinted by permission of ASHRAE from the* Heating, Ventilating, Air Conditioning Guide, 31st ed.

greatly from the 50 ft estimated in Example 10.4, a recalculation is made.

10.8 Conservation of Energy and Water

ASHRAE Standard 90-75 affects many phases of building systems and equipment in which reductions of energy must be achieved. One of these is service water heating. By permission of the American Society of Heating, Refrigerating and Air Conditioning Engineers, we are privileged to present, in full, Section 7.0 of Standard 90-75, which covers this subject. Following the appearance of the Standard in 1975, there has been an encouraging move toward compliance. Author's comments about early action in this regard follow ASHRAE's Section 7.0.

7.0 SERVICE WATER HEATING

7.1 General

Hot water for domestic, sanitary and swimming pool purposes *shall* be generated and delivered in a manner conducive to saving heat energy.

7.2 Scope

7.2.1 The purpose of this section is to provide criteria for design and equipment selection that will produce energy savings when applied to service water heating.

7.2.2 It is not the intention of this section to develop either a procedure or method for designing a hot water distributing system, other than to offer criteria which, when applied, will reduce unnecessary uses of energy. For a more complete systems design guide, the reader should consult the 1973 ASHRAE HANDBOOK & Product Directory, Systems Volume, Chapter 37, Service Water Heating.

7.3 Water Heaters, Storage Tanks, Boilers, and Piping

7.3.1 Performance Efficiency

7.3.1.1 All automatic, electric, storage water heater(s) *shall* have a stand-by loss not exceeding 6 W/ft² (64 W/m²) of tank surface area. Effective January 1, 1977 all automatic, electric, storage water heaters *shall* have a stand-by loss not exceeding 4 W/ft² (43 W/m²) of tank surface area. The method of test of stand-by loss *shall* be as described in Section 4.3.1 of ANSI C72.1-72 Household Automatic Electric Storage-Type Water Heaters.

Table 10.11 ***Allowance in Equivalent Length of Pipe for Friction Loss in Valves and Threaded Fittings***[a]

Diameter of Fitting, inches	90-Deg Standard Ell, feet	45-Deg Standard Ell, feet	90-Deg Side Tee, feet	Coupling or Straight Run of Tee, feet	Gate Valve, feet	Globe Valve, feet	Angle Valve, feet
	Equivalent Length of Pipe for Various Fittings						
$\frac{3}{8}$	1	0.6	1.5	0.3	0.2	8	4
$\frac{1}{2}$	2	1.2	3	0.6	0.4	15	8
$\frac{3}{4}$	2.5	1.5	4	0.8	0.5	20	12
1	3	1.8	5	0.9	0.6	25	15
$1\frac{1}{4}$	4	2.4	6	1.2	0.8	35	18
$1\frac{1}{2}$	5	3	7	1.5	1.0	45	22
2	7	4	10	2	1.3	55	28
$2\frac{1}{2}$	8	5	12	2.5	1.6	65	34
3	10	6	15	3	2	80	40
$3\frac{1}{2}$	12	7	18	3.6	2.4	100	50
4	14	8	21	4.0	2.7	125	55
5	17	10	25	5	3.3	140	70
6	20	12	30	6	4	165	80

[a]From NBS Report BMS66 *Plumbing Manual.*

7.3.1.2 All gas- and oil-fired, automatic storage heaters *shall* have a recovery efficiency (E_r) not less than 70 percent and a stand-by loss percentage (S) not exceeding

$$S = 4.3 + 67/V$$

where

V = rated volume in gallons

(S = 4.3 + 0.25/V, *where* V = volume in m^3)

Effective January 1, 1977 all gas- and oil-fired automatic, storage heaters *shall* have a recovery efficiency (E_r) not less than 75 percent and a stand-by loss percentage (S) not exceeding

$$S = 2.3 + 67/V$$

where

V = rated volume in gallons

(S = 2.3 + 0.25/V, *where* V = volume in m^3)

The method of test of E_r and S *shall* be as described in Section 2.7 of ANSI Z21.10.3-74—Gas Water Heaters, vol. III, Circulating Tank, Instantaneous and Large Automatic Storage Type Water Heaters.

EXCEPTION. For oil-fired units CF = 1.0; Q = total gallons of oil consumed; and H = total heating value of oil in Btu/gallon.

7.3.2 Combination Service Water Heating/Space Heating Boilers. Service water heating equipment *shall not* be dependent on year-round operation of space heating boilers; that is, boilers that have as another function winter space heating. Excepted from this provision until January 1, 1977 are systems with service/space heating boilers having a stand-by loss in Btu/h (W) less than

$$\frac{25 \text{ pmd} + 250}{n} \qquad \frac{(7 \times 10^6 \text{ pmd} + 73)}{n}$$

where pmd is the probable maximum demand in gallons per hour (m^3/s) (determined as per Chapter 37 of the 1973 ASHRAE HANDBOOK & Product Directory, Systems Volume) and n is the fraction of the year when the outdoor daily mean temperature is more than 64.9F (18°C).

EXCEPTION. After January 1, 1977 excepted from this provision are systems with service/space heating boilers having a stand-by loss Btu/h (W) less than

$$\frac{13.3 \text{ pmd} + 400}{n}$$

$$\frac{(3.7 \times 10^6 \text{ pmd} + 117)}{n}$$

The stand-by loss is to be determined for a test period of 24 hr duration while maintaining a boiler water temperature of 90F (32°C) above ambient.

7.3.3 Insulation

7.3.3.1 Heat loss from unfired hot water storage tanks *shall* be limited to a maximum of 15 Btu/h · ft^2 (47 W/m^2) of external tank surface area. The design ambient temperature *shall* be no higher than 65F (18°C).

7.3.3.2 For recirculation systems, piping heat loss *shall* be limited to a maximum of 25 Btu/h · ft^2 (79 W/m^2) of external pipe surface for above-ground piping and a maximum of 35 Btu/h · ft^2 (110 W/m^2) of external pipe surface for underground piping. Maximum heat loss *shall* be determined at a ΔT equal to the maximum water temperature minus a design ambient temperature no higher than 65°F (18°C).

7.4 Temperature Controls

Service water heating systems *shall* be equipped with automatic temperature controls capable of adjustment from the lowest to the highest acceptable temperature settings for the intended use. (See Table 1 from ASHRAE 1973 HANDBOOK & Product Directory, Systems Volume, Chapter 37.)

7.5 Shut Down

A separate switch *shall* be provided to permit turning off the energy supplied to electric service water heating systems. A separate valve *shall* be provided to permit turning off the energy supplied to the main burner(s) of all other types of service water heating systems.

7.6 Pump Operation

Circulating hot water systems *shall* be arranged so that the circulating pump(s) can be conveniently turned off (automatically or manually) when the hot water system is not in operation.

7.7 Conservation of Hot Water

7.7.1 Showers used for other than safety reasons *shall* be equipped with flow control devices to limit total flow to a maximum of 3 gpm (1.9 × 10^{-4} m^3/s) per shower head.

7.7.2 Lavatories in restrooms of public facilities *shall:*

a. Be equipped with outlet devices which limit the flow of hot water to a maximum of 0.5 gpm (3.2 × 10^{-5} m^3/s)
b. Be equipped with devices which limit the outlet temperature to a maximum of 110F (43°C)
c. Be equipped with self-closing valves that limit delivery to a maximum of 0.25 gal of hot water (9.5 × 10^{-4} m^3)

7.8 Swimming Pools

7.8.1 Heated swimming pools *shall* be equipped with controls to limit heating water temperatures to no more than 80F (26.5°C).

EXCEPTION. Pools used for therapeutic purposes are exempt from this requirement.

7.8.2 Uncovered (unenclosed) heated pools *shall* be controlled so that electric or fossil-fueled pool water heating systems are inoperative whenever the outdoor air temperature is below 60F (16°C).

References Used in Section 7

ASHRAE HANDBOOK & Product Directory, Systems Volume, Chapter 37, ASHRAE, New York, NY, 1973.

ANSI C72.1-72 "Household Automatic Electric Storage-Type Water Heaters," ANSI, New York, NY, 1972.

ANSI Z21.10.3-74 "Gas Water Heaters, vol. III, Circulating Tank, Instantaneous and Large Automatic Storage Type Water Heaters."

Author's Comments on Selected Sections of ASHRAE 90-75 Indicative of Early Compliance

7.3.1.1 Automatic Electric Storage Water Heaters

Most manufacturers are beginning to design for the 1977 requirement of 4 watts (w)/sq ft standby loss, because few of their earlier products met even the pre-1977 standard of 6 w/sq ft. At this time manufacturers already list some products that conform to the later and more stringent requirement.

7.3.1.2 Recovery Efficiency, Gas- and Oil-Fired Automatic Storage Heaters

The same concept (as above) applies to gas- and oil-fired heaters. Although in this case fewer firms have been able to accomplish the requirements of the Standard on a broad product line base, most major manufacturers do offer a limited number of products that meet the 1977 requirements, and some firms offer a relatively complete line of domestic residential water heaters.

Fig. 10.29 Insulating existing hot water storage tanks. ASHRAE Standard 90-75, Section 7.2.3.1 specifies maximum standby heat loss allowed for unfired hot water storage tanks. It is likely that those who are serious about energy conservation will insulate to achieve losses well within the permissible maximums set by the standard.

7.3.2 Combination Service Water Heating/ Space Heating Boilers

This section does not rule out boilers with service water heating coils. It does, however, require that boilers with service water heating coils installed in accordance with ASHRAE 90-75 have a substantially improved performance. Systems having very large heating loads and very small domestic water heating loads may have difficulty in meeting the requirements set forth in 7.3.2 and, therefore, the choice may be for a separate summer water heater to solve the problem.

7.3.3.1 Tank Insulation

The limitation of heat loss from unfired tanks will involve thickness and type of insulation. Assuming a nominal k value of 0.24 for insulation, a temperature difference of 115° F (180° F − 65° F), and neglecting the resistance of the metal shell, the loss could be (for 1 in. of insulation):

$$115° \text{ F} \times 0.24 = 27.6 \text{ Btuh/sq ft}$$

Two inches of insulation are suggested to bring this loss below the 15 Btuh/sq ft level.

Ambient temperature of 65° F leads us to visualize much cooler boiler rooms than we have known in the past (see Fig. 10.29).

7.3.3.2 Piping Heat Loss

Table 10.12 is an approximate guide to the

Table 10.12 ***Heat Loss Btu/Hr/Ft of Pipe at 70° F Air and 140° F Water***

Pipe Size	*Bare Copper Tubing*	*½-in. Glass Fiber Insulated Copper Tubing*
0.75	30	17.7
1	38	20.3
1.25	45	23.4
1.5	53	25.4
2	66	29.6
2.5	80	33.8
3	94	39.5
4	120	48.4

value of a 1/2-in. glass fiber insulation on copper tubing. On the average, the heat loss compared to that of bare tubing is cut by about one-half for the conditions given. In the evaluation of *actual* conditions of loss in a given installation, the determinants are tube size, insulation conductivity (k), the temperature of the hot water in the tube, and the mandatory ambient temperature of 65° F.

7.7 Conservation of Hot Water

Items such as controlled flow rate in showers, a rate for lavatories, a maximum hot water temperature for lavatories plus a self-closing valve that limits each delivery to a controlled amount of hot water are not new. However, the change under Standard 90-75 is that these items are now mandatory in all installations and that their rates must be evaluated—this was not universal in the past.

Characteristic of the concern evidenced by most trade associations, the Plumbing Manufacturers Institute, based in Chicago, is now compiling a list of manufacturers that produce equipment that complies with Section 7.7.

Note: New ideas and further developments concerning energy conservation can be expected in the near future.

References and Bibliography

1. *ASHRAE Handbook, 1976 Systems,* American Society of Heating, Refrigerating and Air Conditioning Engineers, Chapter 37, "Service Water Heating."

2. *Water Quality and Treatment,* 3rd ed., American Water Works Association, New York, N.Y.

3. *Desalination,* Ecodyne, Unitech Division, Union, N.J.

4. *Architectural Graphic Standards,* 1970, 6th ed., Ramsey and Sleeper, John Wiley & Sons, N.Y., pages 588–601.

5. *Time Saver Standards,* 1966, 4th ed., John Hancock Callender, McGraw-Hill, N.Y., pages 812–820.

6. *Jacuzzi Series JH Water Systems,* Bulletin JJH—3 Jacuzzi Bros., Inc., Little Rock, Ark.

7. *Conservationist and Burkay Heaters,* Bulletins GW 69 B, GW 23 B and CH 8207 P, A. O. Smith Corporation, Kankakee, Ill.

11

Plumbing Systems

11.1 Codes and Systems

(*a*) ***The National Standard Plumbing Code.*** In all technical enterprises that affect the public, local authorities, villages, towns, and cities adhere to their prerogative in controlling any facility that preserves the health and safety of human beings. Plumbing is such a facility. Yet, more and more, two national associations with broad countrywide experience in design and construction are setting national standards. These standards are available for adoption by municipalities and other "administrative authorities." The National Standard Plumbing Code—cosponsored by the National Association of Plumbing-Heating-Cooling-Contractors and the American Society of Plumbing Engineers—presents a consensus of the best practices. It is adopted and used by a multitude of administrative authorities. The code is a great unifying influence on diverse local ordinances.

By permission of the two cosponsoring organizations, we include in this section, and in later sections of this chapter, excerpts from the National Standard Plumbing Code. For an overview of the history of the code, we present here, its Foreword.

FOREWORD

Since its founding as the National Association of Master Plumbers in 1883, The National Association of Plumbing-Heating-Cooling-Contractors has maintained a serious interest in plumbing standards, codes and good plumbing design practices.

The Association published the NAMP "Standard Plumbing Code" in 1933 and furnished revised editions until 1942. NAPHCC participated in the development of special standards for wartime plumbing and later was represented on the National Plumbing Code Coordinating Committee, whose work ultimately resulted in the adoption of A40.8 as a standard or model plumbing code in 1955.

NAPHCC served as a sponsor in the early 1960's of the project which attempted to update the 1955 document. This project was operated through the procedures of the American National Standards Institute. However, the A40.8 revision project was not completed because consensus could not be achieved.

In order to provide local and state governments, code administrative bodies and industry with a modern, updated code, NAPHCC published the "National Standard Plumbing Code," in 1971, following the format and sequence of the A40.8 to provide for maximum convenience of users.

With this revision (June 1973) the American Society of Plumbing Engineers has joined this effort by endorsing the National Standard Plumbing Code. ASPE representatives now serve as members of the Code Committee and participate in the continued updating program.

In this book, we call this chapter "Plumbing Systems." In the National Code the general subject of plumbing includes water supply and storm drainage. Since we have already discussed water in Chapters 9 and 10 and have assigned storm drainage to Chapter 12, this chapter is devoted largely to sanitary drainage and private sewage treatment.

(*b*) ***Systems.*** While maintaining all of the health and safety requirements prescribed by codes, it is still possible to expect a great variety of system-design solutions, depending on the designer. The location of the facilities,

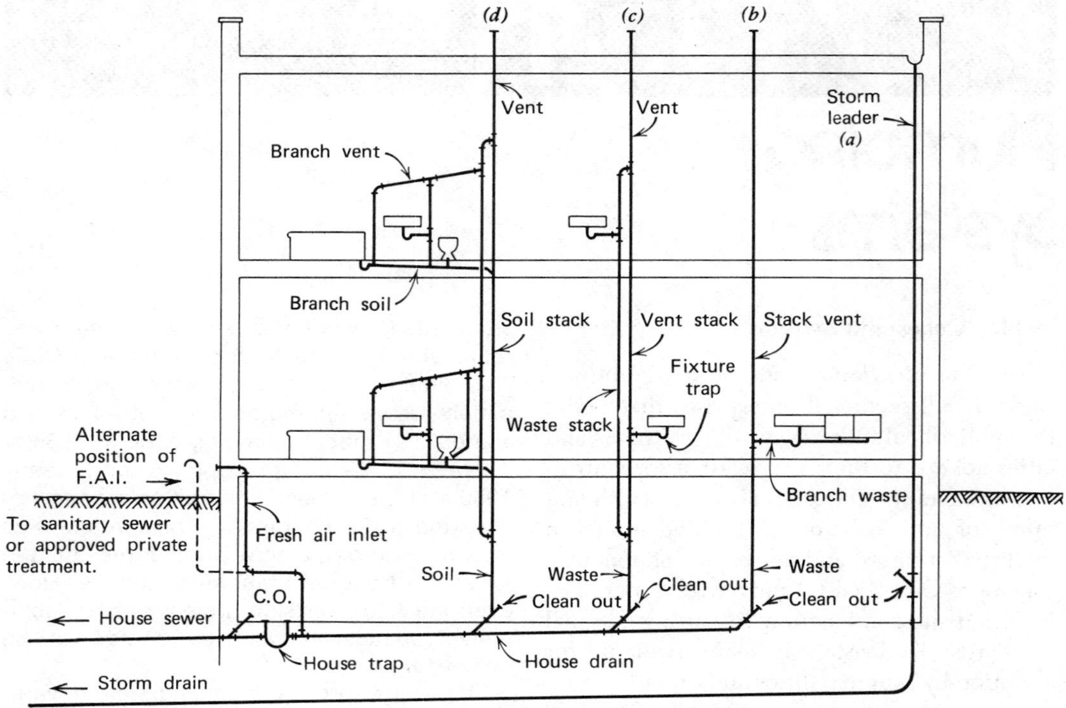

Fig. 11.1 Sanitary drainage system with separate disposal of storm water. Further separation and treatment of black water *(human wastes) and of* grey water *(balance of household wastes) await later study and development. The* house trap, *long a standard feature, is not now mandatory under the National Standard Plumbing Code but may be required by a regional, "administrative authority."*

the routing of the general piping, the location of the cleanouts, and other related topics are subject to considerable freedom of choice. Furthermore, conditions appearing at the site and reasonable construction changes proposed by the plumbing contractor must be considered. If approved by the architect and consulting engineer, these proposals may result in greater efficiency and greater economy of installation. The plumber is a licensed practitioner and, therefore, an accredited and informed consultant to the architect-engineer design group.

Certain essential elements are part of *all* plumbing systems, whether for a simple house or a complicated institutional or commercial building. Figure 11.1 shows the plumbing section of a small house. The same house in a city location was seen in Figure 9.3 (page 353). There it had the great disadvantage of *combined* sanitary and storm drainage, resulting in pollution and in treatment problems. Figure 11.1 corrects this situation by separating the two drainage systems.

Elements of the *sanitary* system, which is our interest in this chapter, include:

Soil stack	House drain[1]
Waste stack	House sewer
Vent stack	Cleanout (C.O.)
Branch vent	House trap
Branch soil	Fresh-air inlet (F.A.I.)
Branch waste	Fixture trap

These and numerous other terms contribute to a plumbing *vocabulary* with which the architect must be familiar. In the following section

[1]The terms "house" and "building" are interchangeable, depending on the size and use of the structure.

and in later sections of this chapter, the *functions* of major plumbing components are described. In later *illustrations* the reader will find applications that are somewhat more complex than those seen in Figure 11.1.

11.2 Traps, Vents, and Air Gaps

(*a*) ***Traps.*** The only separation between the unpleasant and dangerously unhealthy gases in a sanitary drainage system and the air breathed by room occupants is the water caught in the fixture trap after each discharge from a fixture. Sufficient water must flow, especially in water closets, so that this residual water is clean. Traps are of steel, cast iron, copper, plastic, or brass, except those in water closets and urinals which are often of vitreous china cast integrally with the fixture. The deeper the seal, the more resistance there is to siphonage, but the greater the fouling area; therefore, a minimum depth of 2 in. and a maximum of 4 in. with not more than 1 in. loss of seal are common standards. All traps should be self-cleaning, that is, capable of being completely flushed each time the trap operates so that no sediment will remain inside to decompose.

There are a few exceptions to the rule that each fixture should have its own trap. Common exceptions include two laundry trays and a kitchen sink connected to a single trap, not more than three laundry trays using one trap, and three lavatories on a single trap. In the case of the laundry trays and sink, the sink is equipped with the trap and is set nearest to the stack. (See stack (*b*) in Fig. 11.1.)

Traps are usually placed within 2 ft of the fixture and should be accessible for cleaning through a bottom opening, which is otherwise closed by a plug. Overflow pipes from fixtures are connected into the inlet side of the trap. In long runs of horizontal pipe, so-called "running traps" are used only near the drains of floors, areas, or yards and should be provided with hand-hole cleanouts.

When fixtures are very infrequently used, the water in traps can evaporate into the air, breaking the seal of the trap. In contemplating the possible frequency of use, this fact should be kept in mind by the designer. Evaporation to a dangerous degree rarely occurs, except in the case of floor drains where it becomes a real hazard. Trapped drains of this type, employed to carry away the water used in washing floors or drained from heating equipment, may often lose the water seal between infrequent operations. If these drains are connected to soil piping, the results can be dangerous. Many authorities are reluctant to approve floor drains in soil systems, requiring instead that they be separately connected to a dry well. In either case the use of a special hose bibb, affording a source of water directly above the drain, is a wise precaution. It can be used easily to refill the trap of the drain (see Fig. 11.2).

(*b*) ***Vents.*** For the admission of air and the discharging of gases, soil and waste stacks are extended through roofs, and a system of air vents, largely paralleling the drainage system, is also provided. As in the case of drainage stacks, the ventilating stacks extend through the roof or vent through the drainage stack. The functions of venting are often misunderstood. It is true, of course, that one important purpose is to ventilate the system by allowing air from the fresh-air inlet (or from the sewer if there is no house trap or fresh-air inlet) to rise through the system and carry away offensive gases. This provides some purification for the piping. However, several other purposes are served by the vent piping. The introduction of air near the fixture (and in the case of loop venting, at the branch soil line) breaks the possible siphonage of water out of the trap. Under other circumstances, namely when drainage fluids descend to a fixture group through the soil stack, the foul gases would bubble through the trap-seals of that group. The vent system provides a local escape for these gases. Comprehensive experiments have shown that loop venting—which permits air and gases to pass in and out of the soil or waste branch instead of at each fixture, as in the case of continuous venting (individual fixture venting)—is fully effective in prevent-

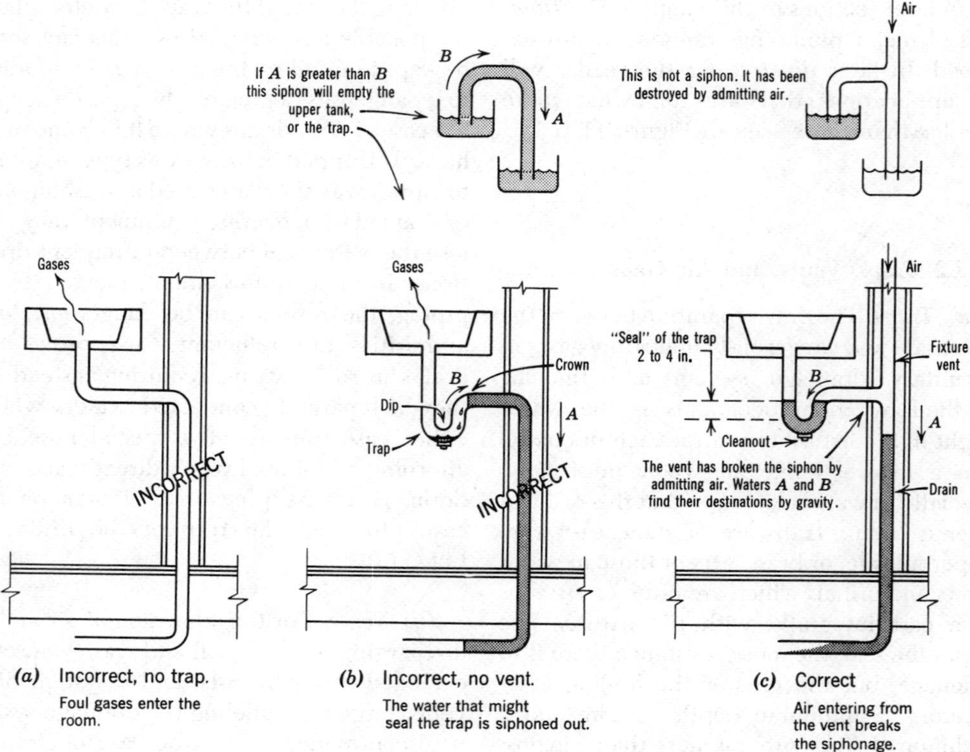

Fig. 11.2 *The function of a trap and one of the several functions of a vent (preventing siphonage).*

ing the siphonage of trap-seals or their penetration by gases (see Fig. 11.3).

(c) ***Air Gaps and Vacuum Breakers.*** Every plumbing fixture is supplied with pure water at one point and discharges contaminated fluids at another. The proximity of sewage to potable water at fixtures is inescapable. It is possible that sewage could accidentally be siphoned into a pipe carrying potable water. Consider a faucet improperly placed with its outlet below the rim of a fixture. If the fixture overflow is plugged and the fixture bowl full, the faucet can easily project into the foul drainage water. If, in this circumstance, the water piping is drained while the faucet is open, contaminated water could be drawn by suction into the water piping.

In water closets served by flushometers (flush valves) the water supply unavoidably enters the bowl below the rim. A vacuum breaker is placed in the flushometer. It closes with water pressure but opens to admit air if there is suction in the water pipe. This prevents siphonage in much the same way that a vent prevents trap siphonage (see Fig. 11.4). The use of vacuum breakers at dishwashers and clothes washers can be seen in Figure 10.1 (see page 381).

Fig. 11.3 (opposite, top) *Gas relief through vents. Gases pressurized by hydraulic action or by expansion due to putrefaction have a path of escape through the vent system and will not enter the rooms.*

Fig. 11.4 (opposite, bottom) *Backflow preventers. Unsanitary fluid wastes cannot be siphoned into the potable water piping.*

11.3 Piping, Fittings, and Accessories

(*a*) ***Piping and Fittings.*** The principal materials used for soil and waste piping and for

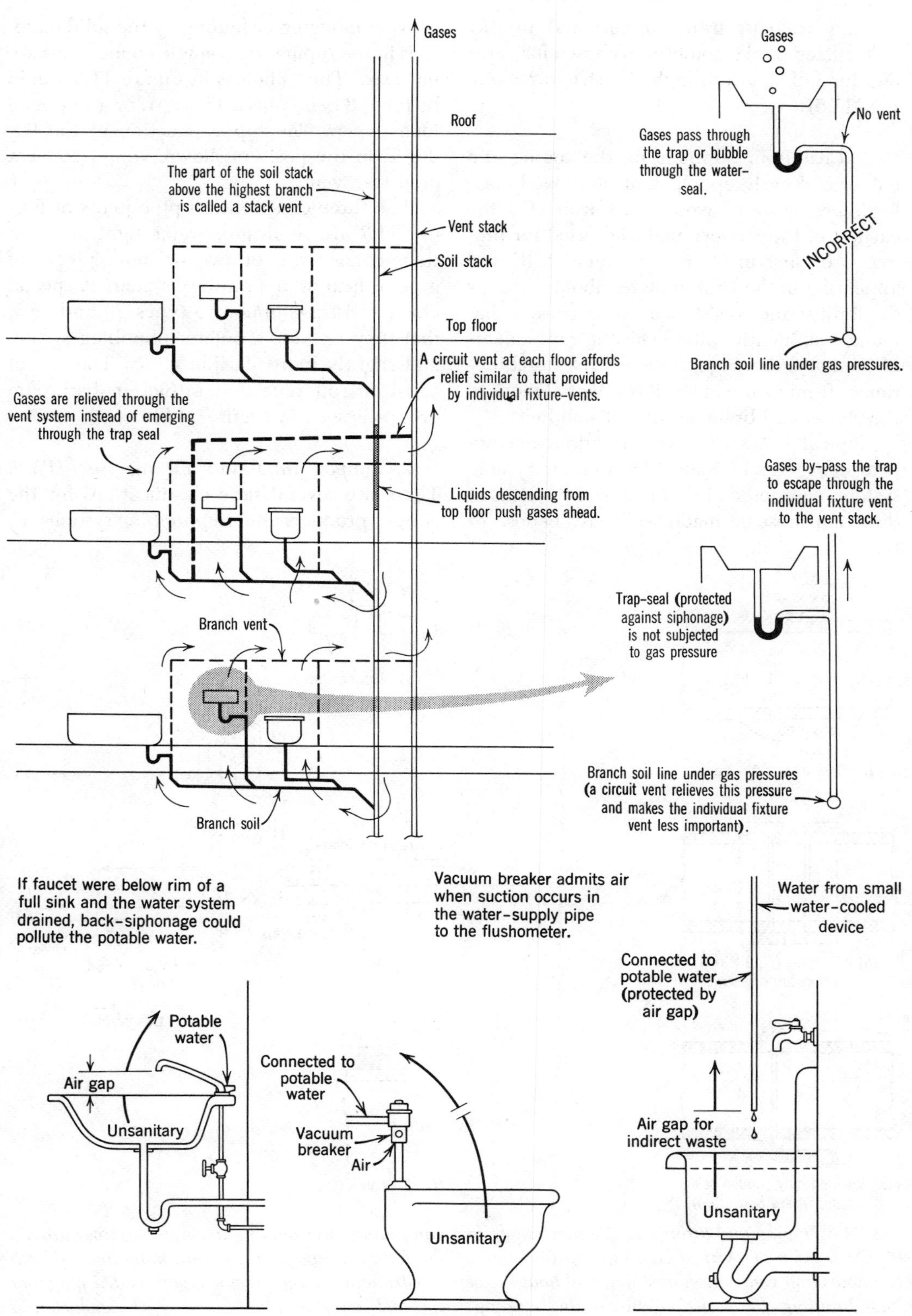

(a) Air gap (b) Vacuum breaker (c) Indirect waste

venting are cast iron, copper, and plastic. Galvanized steel is sometimes chosen for vents and for tall stacks in high-rise structures (see Fig. 11.5).

1. *Cast Iron.* Supplanting the tubing and culverts of early epochs that employed clay, lead, bronze, and wood, cast iron was the earliest of the modern materials used for piping. Used first in Germany around 1562 and appearing in the United States about 1813, its durability and resistance to corrosion has made it eminently suitable for the components of sanitary drainage systems. Its suitability ranges from its use in small residential work to the stacks and branches of tall buildings.

Typical fittings for sanitary drainage appear in Figures 11.5 and 11.6. In sanitary flow systems composed of *any* material, changes in direction must be made with easy bends. To prevent clogging or fouling by the solid materials in the piping, right-angle connections are not used. Thus, choices in Figure 11.5 would be for 1/8 bend plus a 45-deg Y, or a 1/4 bend long sweep. The top connection of the 90-deg T, in the position shown, would connect *only* to a vent.

The three cast-iron soil pipe joints of Figure 11.7 are semirigid, water tight, and gas tight connections of two or more pieces of pipe or fittings in a sanitary system. A special characteristic common to types (*b*) and (*c*) is that they provide a quieter plumbing system and slightly more flexible joints. The use of cast-iron soil pipe and fittings in bathroom groups is seen in Figure 11.8.

2. *Copper Tube and Fittings for DWV.* There are several tube classifications for the copper products used in *plumbing* systems; K,

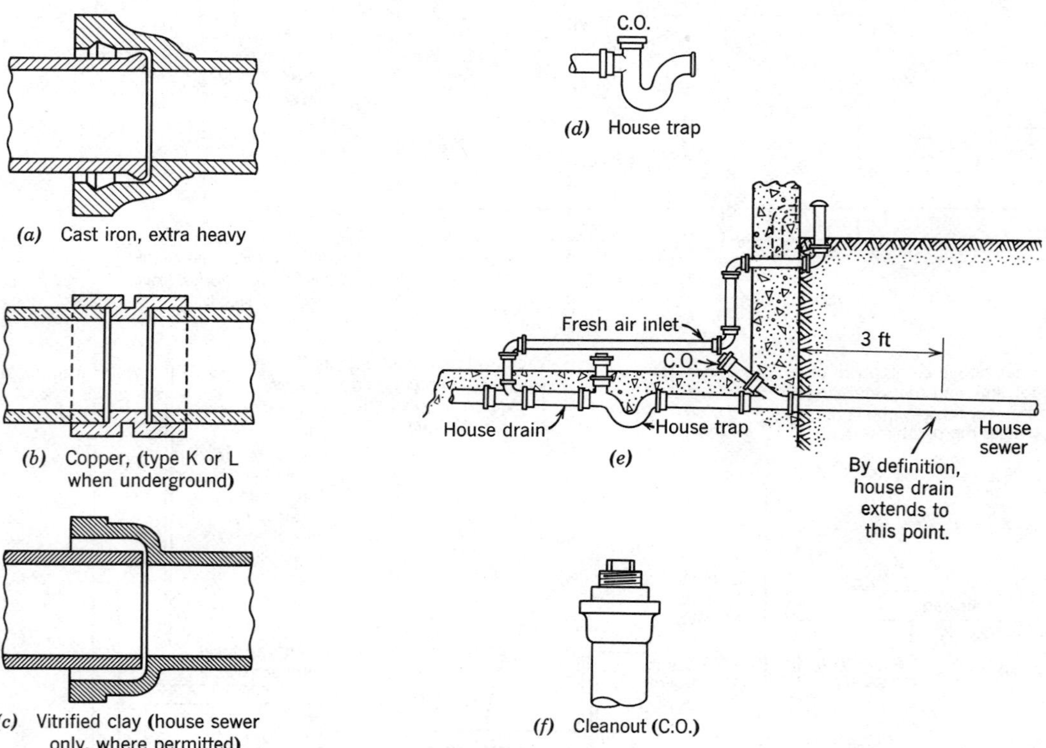

Fig. 11.5 Piping and fittings. (a) *Connection of cast iron piping.* (b) *Coupling to connect copper tubing.* (c) *Connection of vitrified clay piping.* (d) *Detail of house trap fitting.* (e) *House drain, house trap with cleanouts and vent (fresh air inlet) and house sewer.* (f) *Cleanout showing removable threaded plug. For large buildings the terms* building *drain,* building *sewer, and so on, supplant "house." The inclusion or omission of the house trap depends on the local code. Note: Plastic connection is similar to* (b).

Fig. 11.6 Cast-iron fittings. Principal types and method of flashing at roofs.

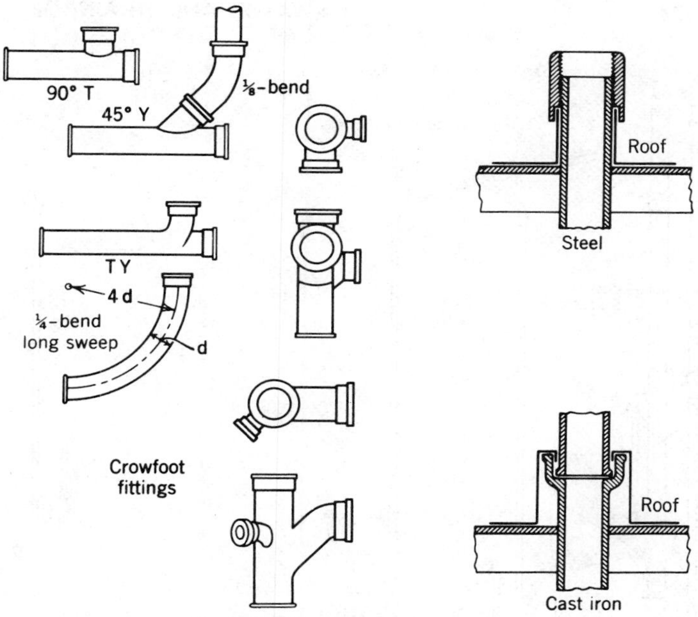

Fig. 11.7 The various joints presently being used to connect cast-iron soil pipe and fittings. Courtesy of Cast Iron Soil Pipe Institute.

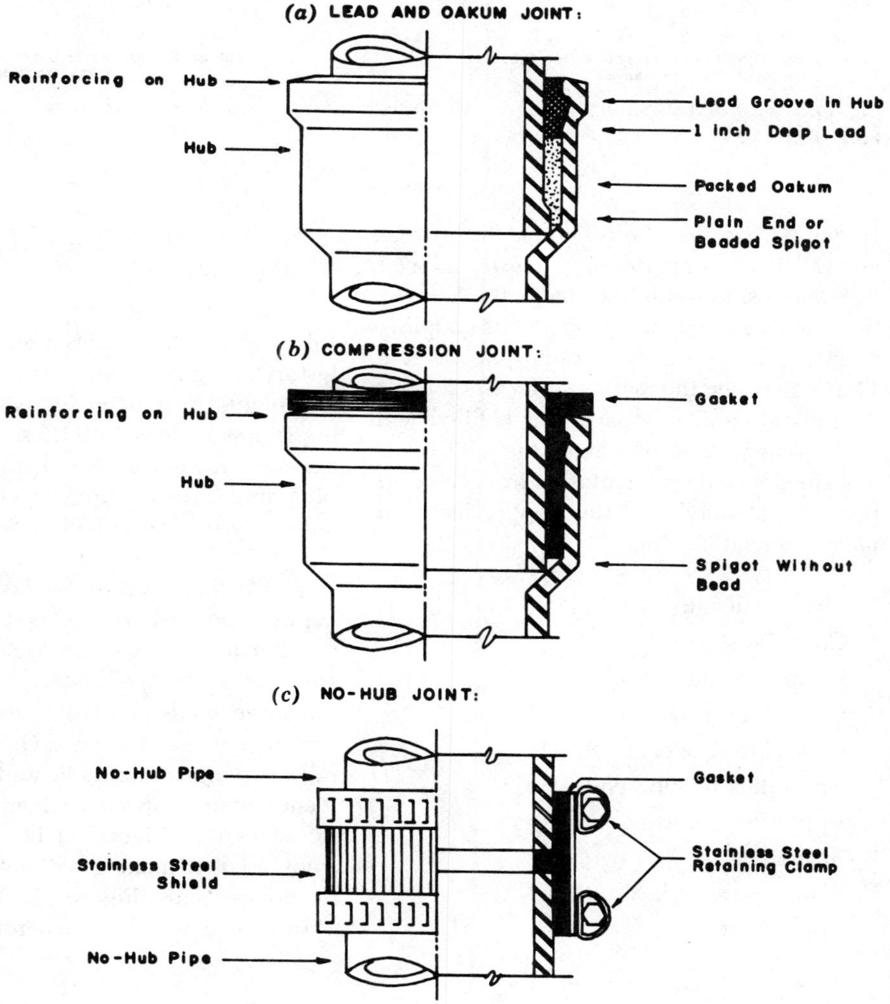

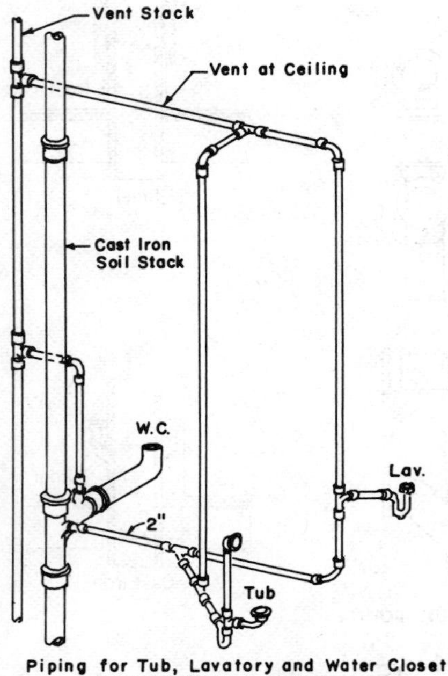

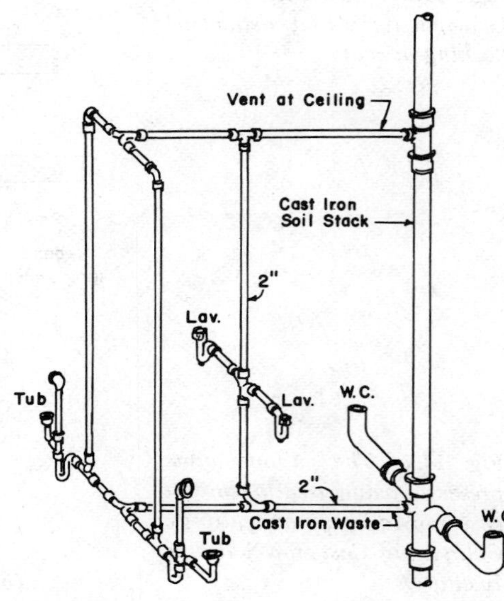

Fig. 11.8 Two typical piping arrangements for water closet, lavatory, and tub. Courtesy of Cast Iron Soil Pipe Institute.

L, and M are the choices for *water* systems and DWV for use in *drainage, waste,* and *vent* installations (as the initials indicate). Connections between copper tubing and its couplings or fittings are made by a sliding fit (see Fig. 11.5*b*). Between the mating surfaces, there is a cylindrical capillary space that is filled with solder. The process of making the joint is best described by a direct quotation from the Copper Tube Handbook of the Copper Development Association, Inc.

Measure length of tube
Cut tube square
Ream cut end
Clean tube end
Clean fitting socket
Apply flux to tube end
Apply flux to fitting socket
Assemble
Remove excess flux
Apply heat
Apply solder
Allow joint to cool.

Like cast iron, copper is a material that has a history of use in ancient installations. Updated and highly developed in recent decades, its use is now widespread. Illustrations of its use are seen in Figure 11.9, in the prefabricated plumbing "tree" of Figure 11.10, and in the Sovent system described in Section 11.7.

3. *Plastic Materials for DWV.* Along with copper and cast iron, plastics are also very suitable for sanitary drainage systems. They comprise a *family* of materials. Table 11.1 lists the three kinds of plastics most suitable for drainage, waste, and vent. One of these materials, acrylonitrile-butadiene-styrene (ABS), is identified and further evaluated by the labeling shown in Figure 11.11. One of several steps used in making a "solvent-weld" connection is seen in Figure 11.12, as is a method of support in wood frame construction. Figure

Fig. 11.9 *Vent and soil stacks (center) with copper waste and soil branches below the structural slab in a fireproof building. Horizontal branches join the stack at the two outstanding hubs of the double T-Y drainage fitting of the cast-iron stack. Sheet metal sleeves in the slab create openings for the verticals. Duct opening is for the ventilation of the toilet room on this story. A hung ceiling will be necessary to conceal the branches and the trap. Courtesy of Copper Development Association.*

11.13 shows assembled bathroom piping in place, and Figure 11.14 illustrates the lightness of the material and its adaptability to prefabrication.

(*b*) ***Accessories.*** Among the many special devices that can form part of a plumbing system, a few are described here; they include floor drains, backwater valves, ejectors, and interceptors.

1. Floor Drains. When floors in buildings must be washed down after such operations as food preparation and cooking, floor drains are usually necessary. Since they are sometimes connected to sanitary drainage systems and, in long periods of disuse, might lose their trap-seals by evaporation, special precautions are necessary to preserve the trap-seal and avoid odors and unsanitary conditions in the room (see Fig. 11.15). The National Standard Plumbing Code describes the trap-seal in the following way.

7.16 Floor Drains

7.16.1 Floor Drain Trap and Strainer

Floor drains shall have a minimum water seal of 3 inches and shall be provided with removable strainers. The open area of strainer shall be at least two-thirds of the cross-section area of the drain line to which it connects. Floor drain trap seals subject to evaporation shall have a water seal of 5 inches or shall be fed from an approved plumbing fixture or by means of an approved automatic priming device or by any other arrangement that is approved by the Administrative Authority.

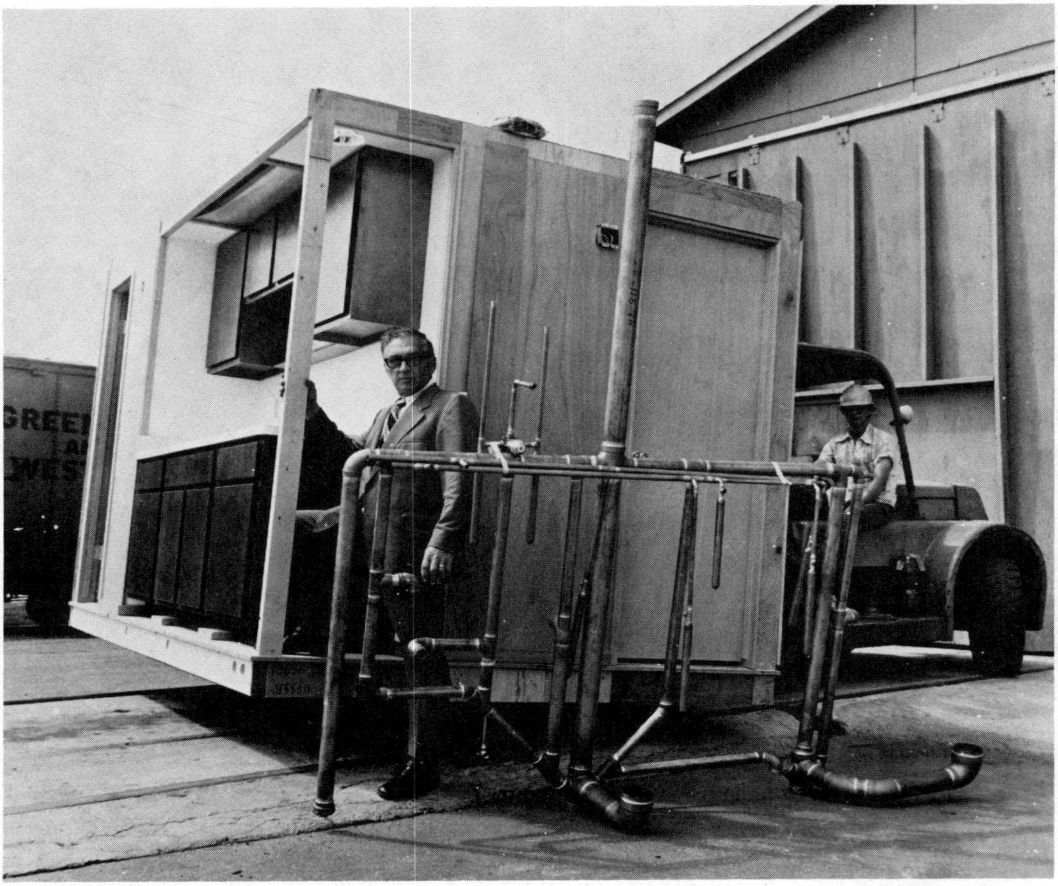

Fig. 11.10 Preassembly. Raleigh Fisher, Chief Mechanical Designer for Wausau Homes, Inc., is shown with a Wausau copper plumbing "tree," which includes use of Copper Drainage Tube DWV. Wausau is one of the nation's largest builders of manufactured housing. Photograph courtesy of Copper Development Association.

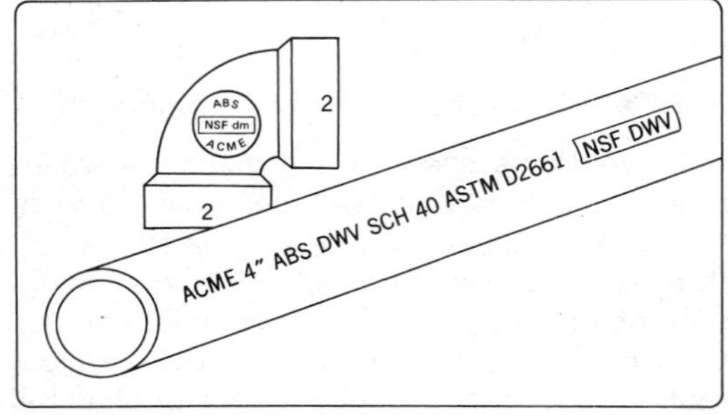

***Fig. 11.11** Typical identification symbols on plastic pipe. Courtesy of Plastics Pipe Institute.*

ACME	The name of the manufacturer.
4 in.	Diameter of the pipe.
ABS	Acrylonitrile-Butadiene-Styrene, the material.
DWV	Suitable for drainage waste and vent.
SCH 40	Schedule 40. This identifies the wall thickness of the pipe.
ASTM D2661	"Standards Number" assigned by the American Society for Testing Materials.
NSF DWV	Tested by the National Sanitation Foundation Testing Laboratory. The pipe meets or exceeds the current standards for sanitary service.

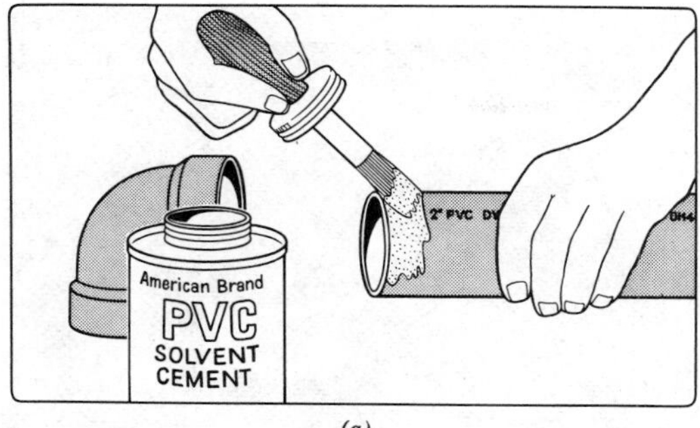

(a)

Fig. 11.12 Details in the use of plastic pipe. (a) One of the steps in making a "solvent weld" of a plastic pipe to a plastic fitting. (b) In wood frame construction, plastic pipe assemblies can be supported by metal straps nailed to the wood joists. Flexibility of the plastic material suggests that the supports be more closely spaced than in the case of metal piping. Courtesy of the Plastics Pipe Institute.

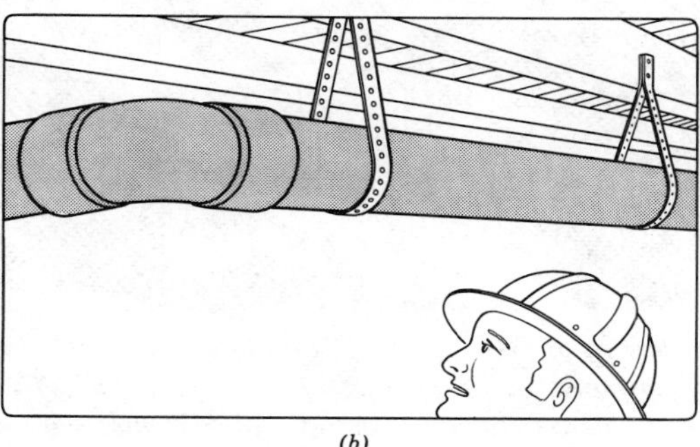

(b)

Table 11.1 *Suitable Choices of Material for Plastic Piping in DWV (Drainage, Waste and Vent) and Sewer Systems*

Symbol	Material	DWV	Sewer
ABS	Acrilylonitrile-Butadiene Styrene	√	√
PVC	Polyvinyl Chloride	√	√
SRP	Styrene Rubber Plastic		√

Source. Reprinted by permission of *Progressive Architecture.*

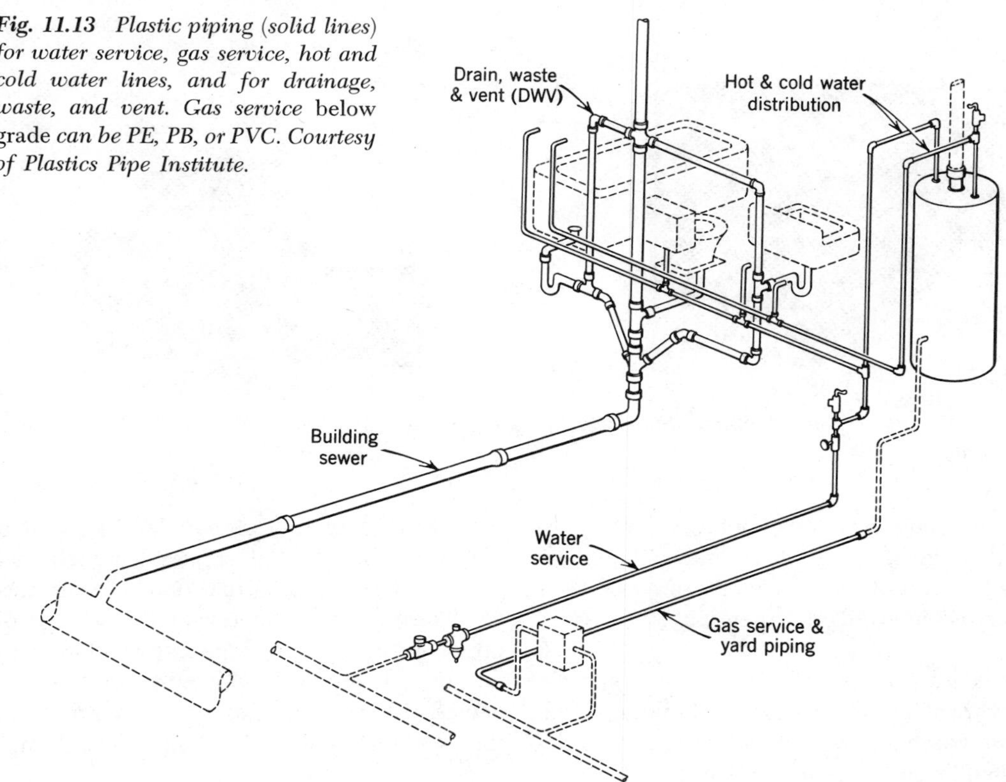

Fig. 11.13 Plastic piping (solid lines) for water service, gas service, hot and cold water lines, and for drainage, waste, and vent. Gas service below grade *can be PE, PB, or PVC. Courtesy of Plastics Pipe Institute.*

Fig. 11.14 Plastics lend themselves to preassembly of sections of DWV piping. Materials used for DWV are: polyvinyl chloride (PVC), either tan or white, and acrylonitrile-butadiene-styrene (ABS), black. The architect or engineer should check structure that may be cut to accommodate piping. Notice in this picture that the notched joists are deeper and more closely spaced than other floor beams. Courtesy of Plastics Pipe Institute. Photo by Richards Studio.

Fig. 11.15 Floor drain. Josam Manufacturing Company.

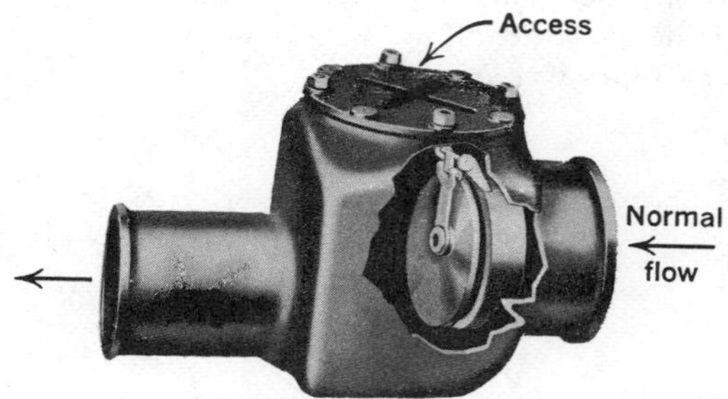

Fig. 11.16 Backwater valve.

After reading the above quotation, one is justified in feeling that floor drains should be avoided, except when mandatory by code or otherwise unavoidably essential.

2. *Backwater Valves.* When plumbing fixtures are at a low elevation in a building, often in the basement, or otherwise located as described in the following quotation, a backwater valve is sometimes used (Fig. 11.16). It is placed in the low soil or waste branch that connects to the piping that drains *upper* stories. The reason is to prevent the sewage of upper stories from backing into the low sanitary branch and flowing out through the connected fixtures. The following excerpt from the National Standard Plumbing Code applies:

6.6 Drainage Below Curb Level

6.6.1 Fixtures Subject to Backflow

Drainage piping serving fixtures that are located below the elevation of the curb at the point where the building sewer crosses under the curb or property line, and above the crown level of the main sewer, may drain by gravity into the main sewer, and shall be protected from backflow of sewage by installing an approved type of backwater valve.

6.6.2 Fixture Branches Subject to Backflow

Backwater valve shall be installed only in that branch or section of the drainage system which receives the discharge from fixtures located as stated in 6.6.1.

6.6.3 Materials for Backwater Valves

Backwater valves shall have all bearing parts of corrosion-resistant material.

6.6.4 Construction of Backwater Valves

Backwater valves shall be constructed so a mechanical seal against backflow will be provided.

6.6.5 Diameter of Backwater Valves

Backwater valves, when fully opened, shall have a capacity not less than that of the pipes in which they are installed.

6.6.6 Location of Backwater Valves

Backwater valves shall be installed so their working parts will be accessible for service and repairs.

In section 6.6.1 of the code quoted above, note the word *may*. A preferred solution often uses a sump pit and an ejector (see Fig. 11.17 and also Fig. 11.29, which show this solution in place in a building).

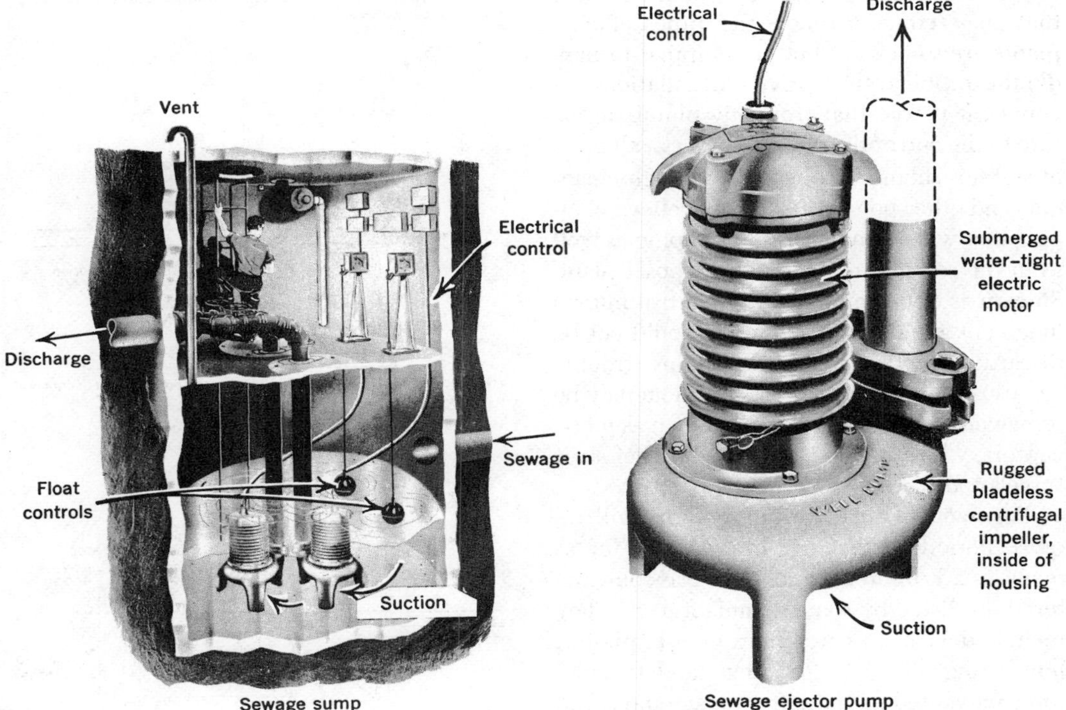

Fig. 11.17 Sump and ejector. Submersible-type centrifugal pump for raising sewage to a higher level. This principle, shown here for an outdoor subgrade sewer installation, may be used in basement applications within buildings. Venting must be carried to roof. Weil Pump Company.

3. Sewage Sumps and Ejectors. Whenever subsoil drainage, fixtures, or other equipment are situated below the level of public sewers, a sump pit or receptacle must be installed into which the drainage from the low fixtures may flow by gravity and from which the contents are then lifted up into the building sewer. Sewage ejectors may be motor-driven centrifugal pumps (see Fig. 11.17) or they may be operated by compressed air. The latter have no revolving parts within the receptacle. An air compressor is started when the float within the sump reaches a certain level, and air at a pressure greater than 0.433 psi for each foot of lift is delivered into the space above the liquid. The air pressure closes the inlet and opens the outlet check valves, expelling the contents of the sump and elevating it to the sewer.

4. Interceptors. Sanitary drainage installations ultimately discharge their waste matter into private or public sewage treatment plants that attempt to digest or cope with anything that may come through the pipes. Public plants are somewhat better equipped to handle this problem than private installations. Of course, it is true that from any plumbing fixture to the end of the disposal process, all parts of systems should be openable through cleanouts and other points of access to relieve clogging that will often occur in the piping as well as in the septic tank or public disposal plant. Since it is quite impossible to control human judgment about what should or should not be discarded into the plumbing drains, trouble can usually be expected. This problem may be somewhat reduced by devices known as interceptors, which catch foreign matter before it travels too far into the system.

It is evident that the interceptors will require periodic servicing. Interceptors for as many as 25 different kinds of extraneous material are listed by some manufacturers. They include devices to catch hair, grease, plaster, lubricating oil, glass grindings, and troublesome unwanted material from many industrial processes. One of the few interceptors that is sometimes needed in homes and more often in institutional kitchens is the grease interceptor.

By passing the waste from a kitchen sink through the circuitous path of the grease interceptor, the grease floats to the top where it is trapped between baffles while the more fluid wastes pass through at a lower level. There are special reasons for removing grease; one reason is that it congeals within piping and physically retards the sewage digestion process (see Fig. 11.18*a* and *b*).

Fig. 11.18 (a) *One type of institutional interceptor, a grease trap. Choice of three locations—adjacent to sink, on floor below, or in a pit.* (b) *Cutaway view with identification of component parts. Josam Manufacturing Company.*

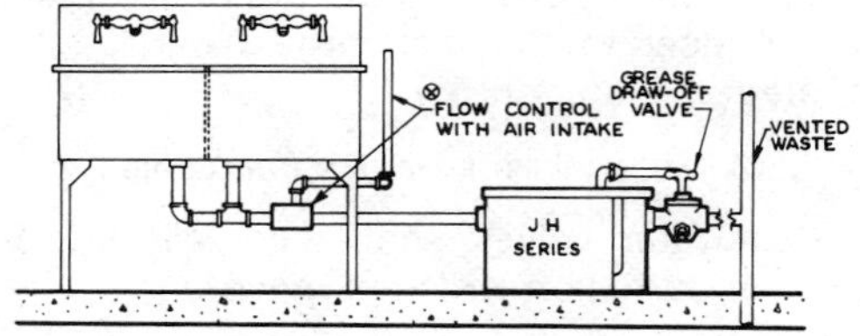

JH Series Semi-Automatic Interceptor installed on floor, servicing a double compartment sink.

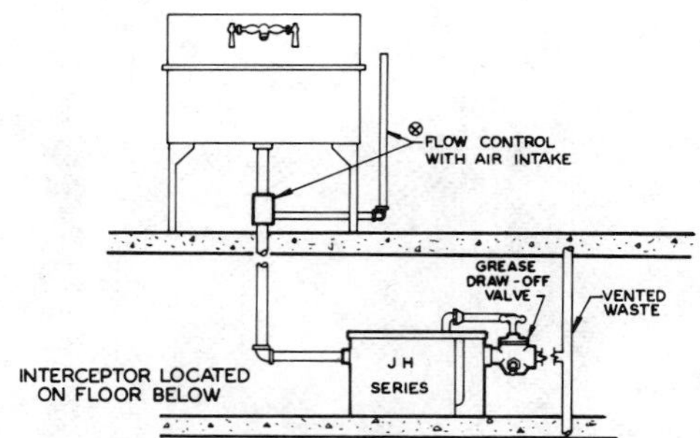

JH Series Semi-Automatic Interceptor installed on floor, servicing a sink on floor above.

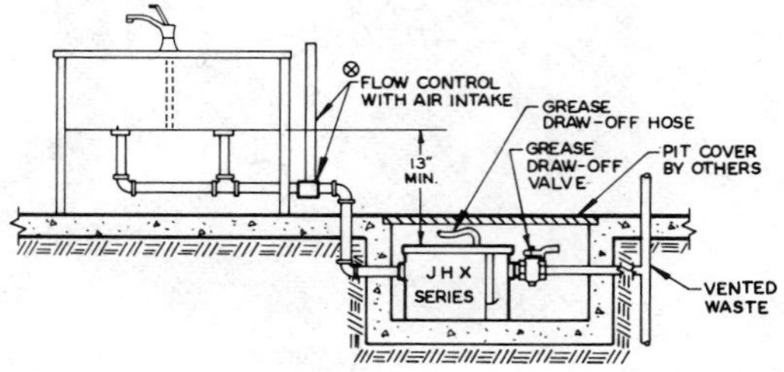

JHX Series Semi-Automatic Interceptor installed recessed below floor, in pit, servicing a double compartment sink.

(a)

11.4 Examples of Installations

In residential work, the piping assemblies may often be considered as a "flag" form, as seen in Figure 11.19. The mast is the soil stack, the horizontal top of the flag is the branch vent, the bottom is the soil or waste branch, and the outer edge is the vertical pipe of the last fixture. In frame construction, the flag usually fits into a 6-in. partition. Fixture branches project from the surface of the flag. There is considerable advantage in "back-to-back" planning of baths and kitchens; this allows the piping assembly to pick up the drainage of fixtures on both sides of it. When all of the fixtures are nearly on the same level as in Figure 11.19, it is unnecessary to have a separate vent stack standing beside the soil stack, as is often the case in multistory construction. In this illustration, and generally in one-story construction, the upper part of the soil stack forms a vent called a stack vent to which the branch vents connect. A separate major vertical vent would be called a vent stack.

Multistory construction, especially in office buildings, needs to be flexible and free of random partitions that would interfere in the periodic replanning of interior spaces and the relocating of dividing partitions. The use of

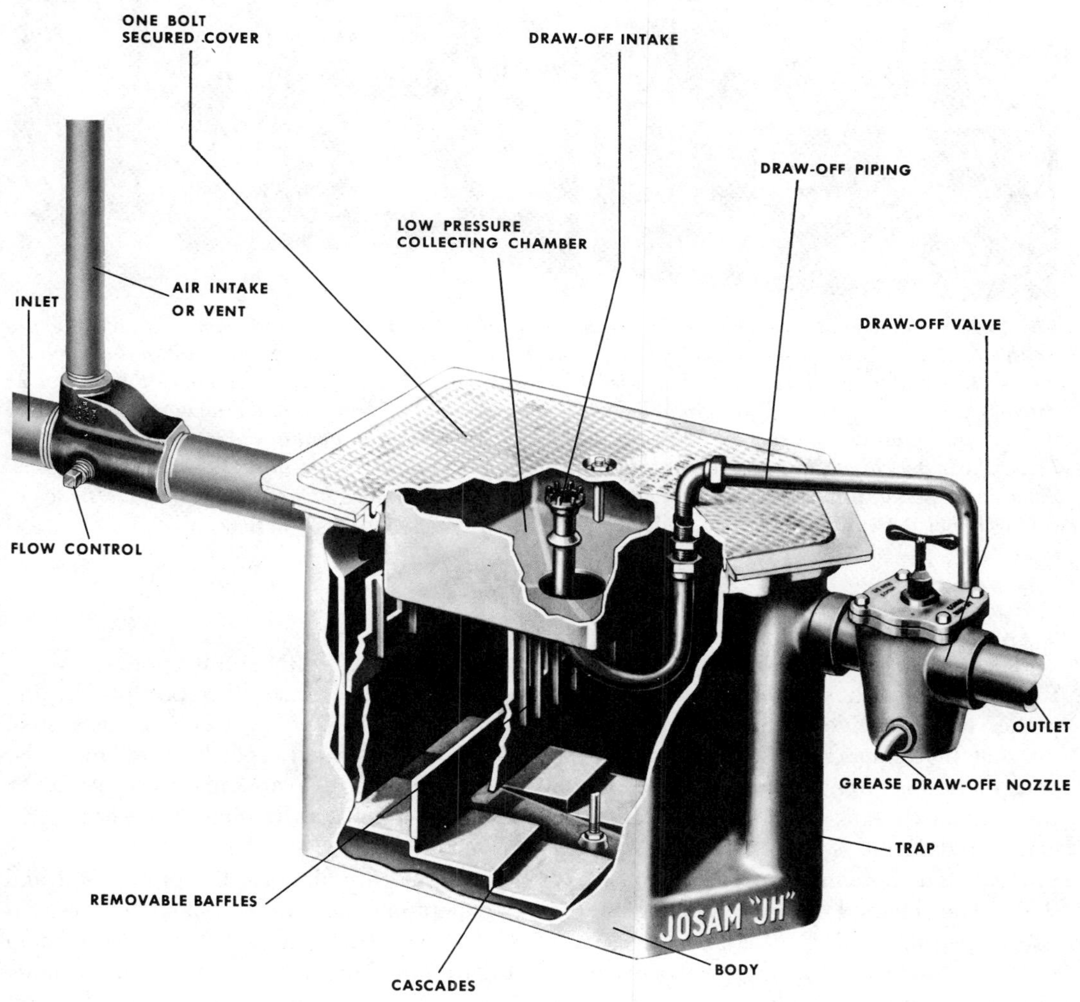

(b)

Fig. 11.19 Drainage and vent piping in a frame residence. This installation in copper "drainage waste and vent tubing" (DWV) serves two bathrooms at the upper level behind the 6-in. stud partition and a kitchen sink and laundry tray at the lower level, which are on this side of the partition. In the bathrooms the roughing serves, from left to right, a lavatory, water closet, and bathtub and a lavatory, shower, and water closet. Bathtub and shower traps can usually be accommodated within the joist depth. The bend below the water closets, however, often leads to a horizontal branch exposed or furred-in below the joists. Some codes permit this branch from a water closet to be 6 to 10 ft long before joining a vent. The water piping is not yet entirely in place. Courtesy of Copper Development Association.

"cores" has become a solution to this problem. Risers of the various systems are grouped in planes that coincide with permanent partitions of block masonry. This island of fixed construction is often at the central section of the building, freeing the surrounding areas for access to daylight. A hole in the floor for each pipe is often chosen in preference to a slot or shaft. This method interferes less with the floor construction (Fig. 11.20).

Offices often need a single lavatory or an executive's complete toilet room at locations away from the central core of the building. "Wet" columns with a full complement of plumbing pipes make this possible. If the pipes are to center on a column in a steel building, structural coordination must be sought early in the planning if the pipes are to clear the structural framing of the floor (Fig. 11.21).

In some installations, the branch soil and waste piping perforates a floor and crosses below the slab to join the stack as seen in Figure 11.9. Tubing has been developed, however, that sits above the structural slab, obviating the need for hung ceilings below (see

Fig. 11.20 Risers in a fireproof, multistory building. Pipes, tubes, conduits, and ducts virtually enclose toilet rooms and utility spaces. Ventilation ducts and a master, 5-in. copper hot water riser are just left of center. Soil and vent stacks with hot and cold water supplies, all of copper, are seen to the right of this group. At the left in lighter tone is the galvanized steel feeder conduit and the distribution circuit conduits, of the same metal, for a local electrical control panel box. Note that some pipes and tubes are supported at this floor by bolted clamps. After testing and before pipes are enclosed, covering will be completed. Copper Development Association.

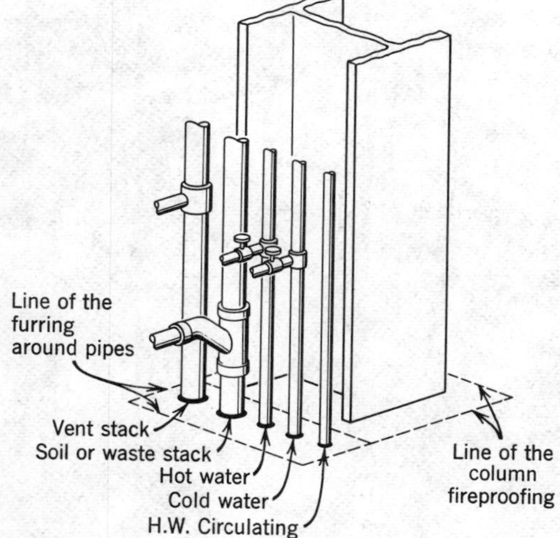

Fig. 11.21 Piping at a "wet" column. In large office buildings there are usually several of these remote from the core and out in the general office area.

Fig. 11.22). A light-weight concrete fill is cast to cover the tubing, raising the floor by 5 or 6 in. This can create a raised floor in the toilet room, which is not the best planning, so the higher floor level is usually carried throughout the floor of the entire story, forming a convenient space into which to place the electrical conduit at a time later than would have been required if it were placed in the structural slab. This affords some freedom in construction because conduits in the structural slab conflict with reinforcement and, furthermore, they must be planned and placed earlier.

11.5 Design, Residential

The sanitary drainage requirements for a residence, the plans of which are shown in Figure 11.23, are the basis for an illustrative example.

EXAMPLE 11.1. Design, lay out, and size the piping for the sanitary drainage system for the house shown in Figure 11.23.

SOLUTION. The first step is to identify the locations where hot and cold water is needed at fixtures and where soil or waste drains must be provided. Figure 11.24 illustrates how this is done. A plan layout for the drains in both levels follows and is seen in Figure 11.25. A photograph showing the paths of water flowing *in* and sewage flowing *out* is the subject of Figure 11.26. Obvious—but interesting in comparing sizes of water pipe and drainage piping—is the $1\frac{1}{4}$-in. water service entering and, after the addition of solids and wastes from the kitchen and baths, the need for a 4-in. soil stack to carry away the sewage.

Fig. 11.22 *An example of plumbing roughing for two lavatory rooms in a fireproof office building. A lavatory and water closet in each room are served by soil and waste branches below and vent branches above. Hot and cold water tubing with air chambers can be seen. The extensions of the water tubing above the two flushometer connections appear to connect into the horizontal vent branches but they do not. They are capped and merely touch the bottoms of the vent branches. Note that soil branches are above the structural slab. A fill of 5 or 6 in. will be necessary to cover the tubing. All vertical tubing will be within the masonry block used to enclose the cubicles.* Copper Development Association.

Drainage	Water
A *Branch soil*	F *Hot*
B *Lavatory waste*	G *Cold*
C *Branch vent*	H *Flushometer supply* (1¼ *in.*)
D *Water closet vent*	I Capped *air chamber*
E *Lavatory vent*	J Capped *air chamber* (*flushometer*) (*I and J absorb expansion and shock*)

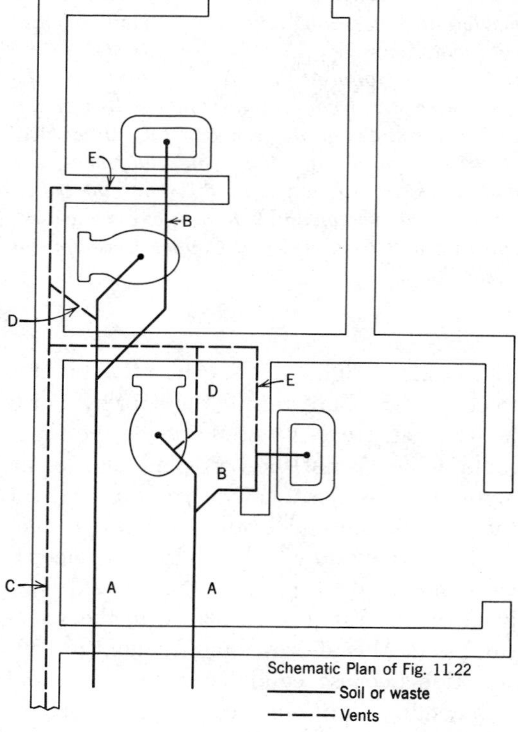

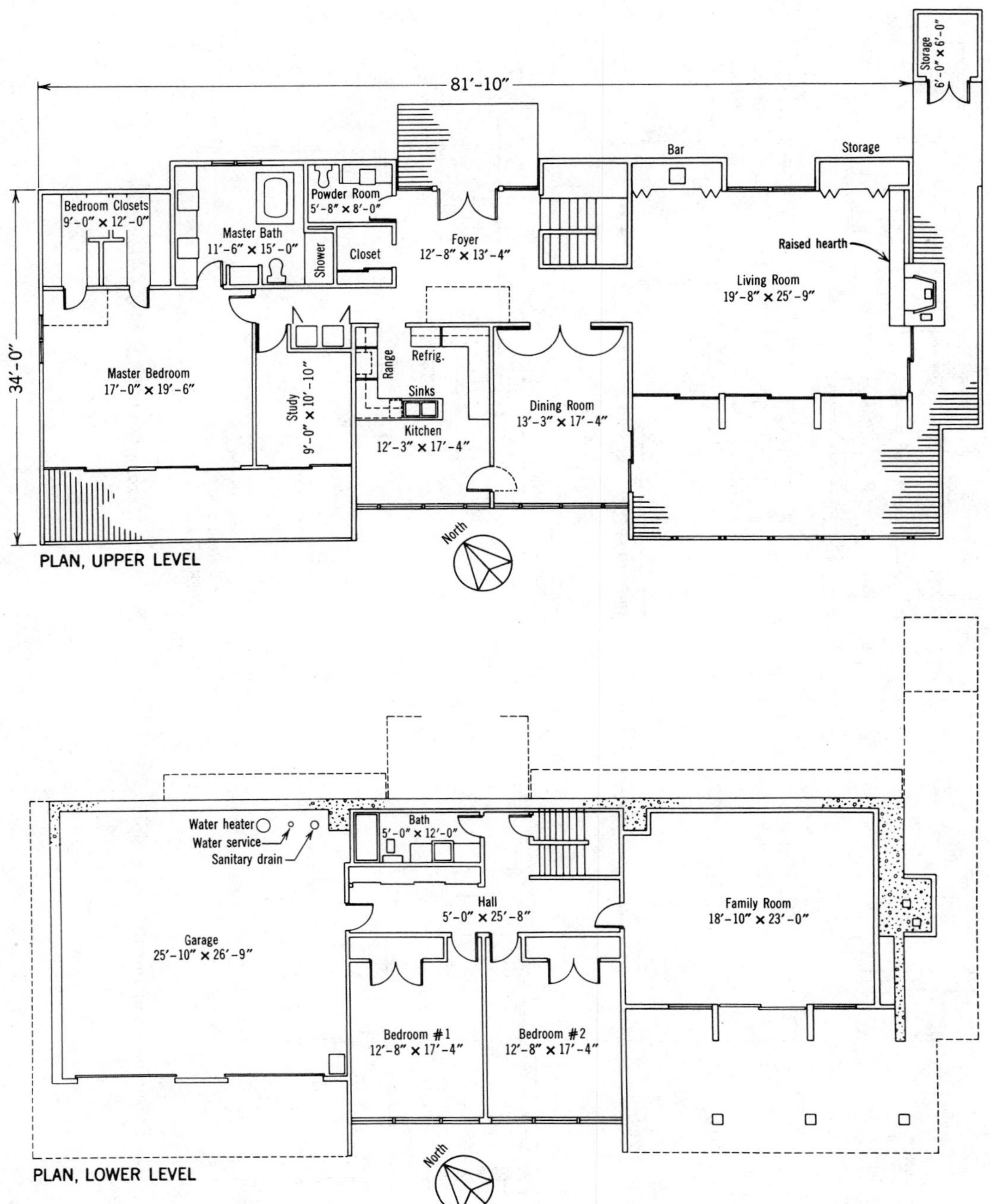

Fig. 11.23 *(Example 11.1) House on Long Island, Budd Mogensen, Architect and Planner. Floor plans to be used in the solution of the example.*

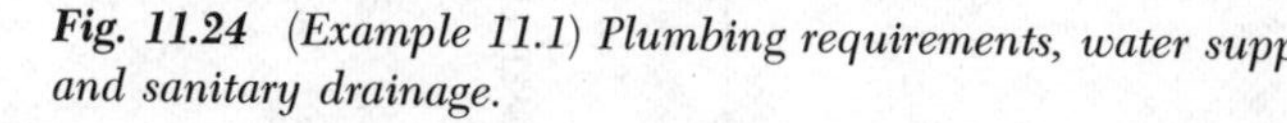

Fig. 11.24 (Example 11.1) Plumbing requirements, water supply, and sanitary drainage.

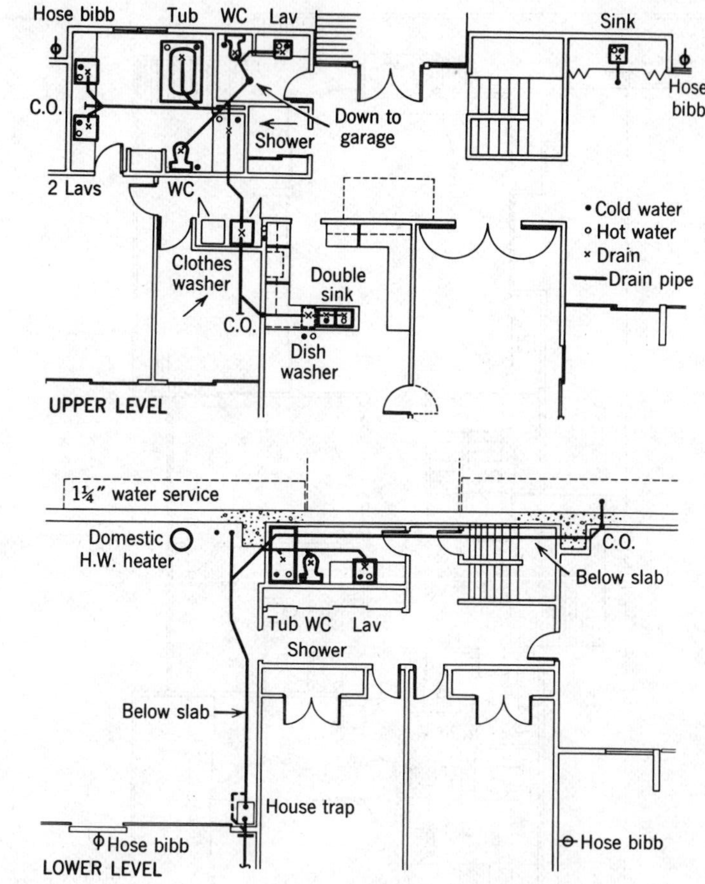

Fig. 11.25 (Example 11.1) Sanitary drainage plan.

Fig. 11.26 *(Example 11.1) At garage floor—water service, $1\frac{1}{4}$ in. copper, soil stack from upper level, 4 in. plastic with cleanout. The copper tubing overhead for water distribution is yet to be connected to heater and service. Heater will stand where pail is seen. Heater is electric in this all-electric house.*

Significant also is the fact that *water* can change direction through 90-deg fittings (Fig. 11.27), while *sewage* flow requires 45-deg changes of direction or other "easy bends." The drainage plan, Figure 11.25, and the plumbing section, Figure 11.28, make this clear.

From these foregoing illustrations a "plumbing section" is developed (Fig. 11.28). The local administrative authority usually requires this to be submitted for its approval. Sizes of all piping are determined from Tables 11.2, 11.3, 11.4, 11.5, and 11.6. Drainage fixture units (d.f.u.) for the system that we are designing are summarized in Table 11.7 from data in Table 11.3. Runouts from *individual* fixtures should be the same size as the fixture trap. Branch runouts from *groups* of fixtures are sized by the drainage-fixture units of the group. Although Table 11.5 permits a 3-in. branch for not more than two water closets, it is common practice to use a 4-in. runout for *every* water closet and not less than a 4-in. branch for every group of water closets. Since Table 11.5 allows 160 fixture units for a 4-in. branch, this size is acceptable for any branch to which a water closet is connected and, as can be seen, for any *stack*. The house drain cannot be less than 4 in. and, because (at a 1/4-in. fall per ft) a 4-in. house drain (Fig. 11.4) will carry 216 d.f.u., that size is chosen and is more than adequate for our 28 d.f.u. system (see Table 11.7). A water closet must have a 2-in. vent. Table 11.6 shows that a 2-in. vent, for vent-lengths not exceeding 150 ft, will serve 20 fixture units. This size is acceptable for branch vents and stack vents. For individual fixture vents, the vent size is usually the

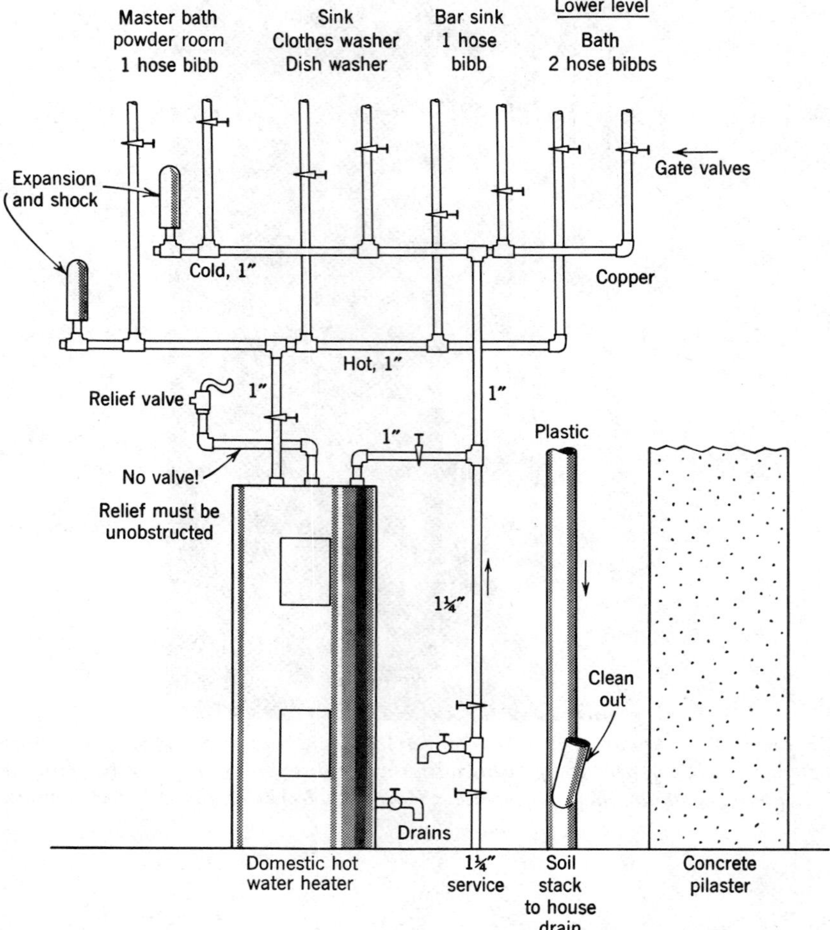

Fig. 11.27 (*Example 11.1*) *Functional diagram of possible water distribution. Soil stack serves all fixtures on upper level except bar sink. Drainage from bar sink and lower level bathroom is in 4 in. branch soil line below garage slab and there joins the stack shown above. See Figure 11.25 and plumbing section, Figure 11.28.*

same as the size of the fixture runout. By most codes vertical vents that penetrate the roof increase to a 4-in. size to prevent blocking by icing in freezing weather.

It will be evident to the reader that in residential applications and in other relatively small buildings certain fairly standard minimum sizes such as a 4-in. soil stack and 2-in. vent are usually adequate. Tables 11.2 to 11.6 are good references but, for large buildings, these standards are highly essential.

Fig. 11.28 (opposite) (*Example 11.1*) *Plumbing section.* Tables: *Trap sizes, Table 11.2; Building Drain, Table 11.4; Horizontal Fixture Branches, Table 11.5; Vents, Table 11.6. Also, see discussion in this chapter.* Note: *When every fixture is vented individually, as in this example, the method is known as* continuous venting. *In larger systems, batteries of fixtures may be vented by a* loop *or* circuit *vent. This reduces the piping of the vent system (see Fig. 11.29).*

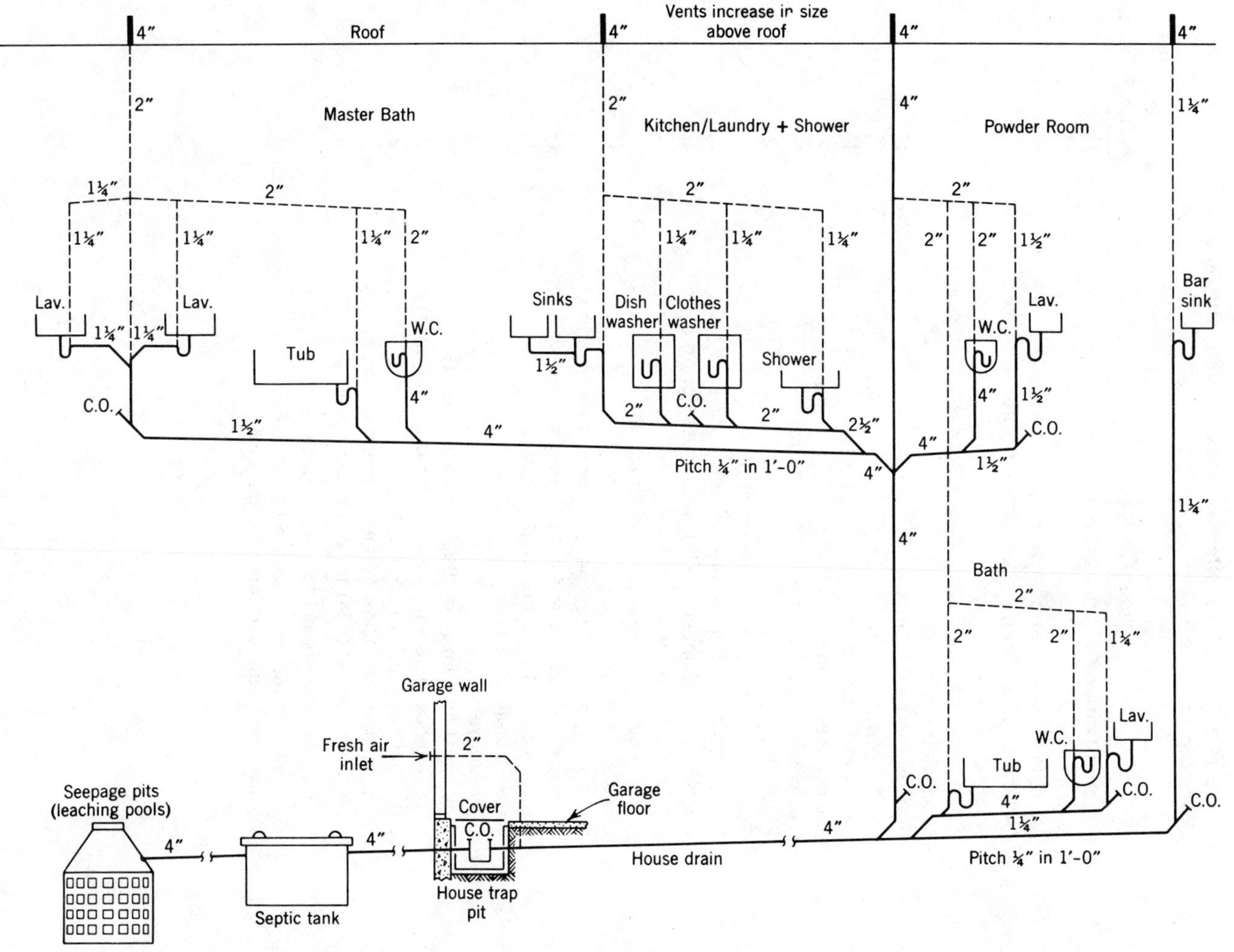
Vents increase in size above roof
Roof
4"
2"
1¼"
1½"
2½"
Master Bath
Kitchen/Laundry + Shower
Powder Room
Lav.
Tub
W.C.
C.O.
Sinks
Dish washer
Clothes washer
Shower
Bar sink
Pitch ¼" in 1'-0"
Bath
Garage wall
Fresh air inlet
Seepage pits (leaching pools)
Septic tank
Cover
House trap pit
Garage floor
House drain

Table 11.2 ***Size of Nonintegral Traps for Different Type Plumbing Fixtures***

Plumbing Fixture	*Trap Size in Inches*
Bathtub (with or without overhead shower)	$1\frac{1}{2}$
Bidet	$1\frac{1}{4}$
Combination sink and wash (laundry) tray	$1\frac{1}{2}$
Combination sink and wash (laundry) tray with food waste grinder unit	$1\frac{1}{2}$[a]
Combination kitchen sink, domestic, dishwasher, and food waste grinder	2
Dental unit or cuspidor	$1\frac{1}{4}$
Dental lavatory	$1\frac{1}{4}$
Drinking fountain	$1\frac{1}{4}$
Dishwasher, commercial	2
Dishwasher, domestic (nonintegral trap)	$1\frac{1}{2}$
Floor drain	2
Food waste grinder—commercial use	2
Food waste grinder—domestic use	$1\frac{1}{2}$
Kitchen sink, domestic, with food waste grinder unit	$1\frac{1}{2}$
Kitchen sink, domestic	$1\frac{1}{2}$
Kitchen sink, domestic, with dishwasher	$1\frac{1}{2}$
Lavatory, common	$1\frac{1}{4}$
Lavatory (barber shop, beauty parlor or surgeon's)	$1\frac{1}{2}$
Lavatory, multiple type (wash fountain or wash sink)	$1\frac{1}{2}$
Laundry tray (1 or 2 compartments)	$1\frac{1}{2}$
Shower stall or drain	2
Sink (surgeon's)	$1\frac{1}{2}$
Sink (flushing rim type, flush valve supplied)	3
Sink (service type with floor outlet trap standard)	3
Sink (service trap with P trap)	2
Sink, commercial (pot, scullery, or similar type)	2
Sink, commercial (with food grinder unit)	2

Source. National Standard Plumbing Code.

[a]Separate trap required for wash tray and separate trap required for sink compartment with food waste grinder unit.

Table 11.3 ***Drainage Fixture Unit Values for Various Plumbing Fixtures***

Type of Fixture or Group of Fixtures	*Drainage Fixture Unit Value (d.f.u.)*
Automatic clothes washer (2-in. standpipe)	3
Bathroom group consisting of a water closet, lavatory, and bathtub or shower stall:	
Flushometer valve closet	8
Tank type closet	6
Bathtub[a] (with or without overhead shower)	2
Bidet	1
Clinic Sink	6
Combination sink-and-tray with food waste grinder	4
Combination sink-and-tray with one $1\frac{1}{2}$-in. trap	2
Combination sink-and-tray with separate $1\frac{1}{2}$-in. traps	3
Dental unit or cuspidor	1
Dental lavatory	1
Drinking fountain	$\frac{1}{2}$
Dishwasher, domestic	2
Floor drains with 2-in. waste	3
Kitchen sink, domestic, with one $1\frac{1}{2}$-in. trap	2
Kitchen sink, domestic, with food waste grinder	2
Kitchen sink, domestic, with food waste grinder and dishwasher 2-in. trap	3
Kitchen sink, domestic, with dishwasher $1\frac{1}{2}$-in. trap	3
Lavatory with $1\frac{1}{4}$-in. waste	1
Laundry tray (1 or 2 compartments)	2
Shower stall, domestic	2
Showers (group) per head	2
Sinks:	
Surgeon's	3
Flushing rim (with valve)	6
Service (trap standard)	3
Service (P trap)	2
Pot, scullery, etc.	4
Urinal, pedestal, syphon jet blowout	6
Urinal, wall lip	4
Urinal, stall, washout	4
Urinal trough (each 6-ft section)	2
Wash sink (circular or multiple) each set of faucets	2
Water closet, tank-operated	4
Water closet, valve-operated	6
Fixtures not listed above:	
Trap size $1\frac{1}{4}$ in. or less	1
Trap size $1\frac{1}{2}$ in.	2
Trap size 2 in.	3
Trap size $2\frac{1}{2}$ in.	4
Trap size 3 in.	5
Trap size 4 in.	6

Source. National Standard Plumbing Code.

[a] A shower head over a bathtub does not increase the fixture unit value.

Table 11.4 ***Building Drains and Sewers***[a]

Diameter of Pipe, Inches	Maximum Number of Fixture Units That May Be Connected to Any Portion of the Building Drain or the Building Sewer Including Branches of the Building Drain — Fall per Foot			
	$\frac{1}{16}$ in.	$\frac{1}{8}$ in.	$\frac{1}{4}$ in.	$\frac{1}{2}$ in.
2			21	26
$2\frac{1}{2}$			24	31
3		36[b]	42[b]	50[b]
4		180	216	250
5		390	480	575
6		700	840	1,000
8	1,400	1,600	1,920	2,300
10	2,500	2,900	3,500	4,200
12	2,900	4,600	5,600	6,700
15	7,000	8,300	10,000	12,000

Source. National Standard Plumbing Code.

[a]On-site sewers that serve more than one building may be sized according to the current standards and specifications of the Administrative Authority for public sewers.

[b]Not over two water closets or two bathroom groups.

Table 11.5 ***Horizontal Fixture Branches and Stacks***

Diameter of Pipe, Inches	Maximum Number of Fixture Units that May Be Connected to: Any Horizontal Fixture Branch[a]	Stack Sizing for 3 Stories in Height or 3 Intervals	Stack Sizing for More than 3 Stories in Height: Total for Stack	Total at 1 Story or 1 Branch Interval
$1\frac{1}{2}$	3	4	8	2
2	6	10	24	6
$2\frac{1}{2}$	12	20	42	9
3	20[b]	48[b]	72[b]	20[b]
4	160	240	500	90
5	360	540	1,100	200
6	620	960	1,900	350
8	1,400	2,200	3,600	600
10	2,500	3,800	5,600	1,000
12	3,900	6,000	8,400	1,500
15	7,000			

Source. National Standard Plumbing Code.

[a]Does not include branches of the building drain.

[b]Not more than two water closets or bathroom groups within each branch interval nor more than six water closets or bathroom groups on the stack.

Stacks shall be sized according to the total accumulated connected load at each story or branch interval and may be reduced in size as this load decreases to a minimum diameter of $\frac{1}{2}$ of the largest size required.

Table 11.6 ***Size and Length of Vents***

Size of Soil or Waste Stack	Fixture Units Connected	Diameter of Vent Required (Inches)								
		$1\frac{1}{4}$	$1\frac{1}{2}$	2	$2\frac{1}{2}$	3	4	5	6	8
		Maximum Length of Vent (Feet)								
Inches										
$1\frac{1}{2}$	8	50	150							
$1\frac{1}{2}$	10	30	100							
2	12	30	75	200						
2	20	26	50	150						
$2\frac{1}{2}$	42		30	100	300					
3	10		30	100	100	600				
3	30			60	200	500				
3	60			50	80	400				
4	100			35	100	260	1000			
4	200			30	90	250	900			
4	500			20	70	180	700			
5	200				35	80	350	1000		
5	500				30	70	300	900		
5	1100				20	50	200	700		
6	350				25	50	200	400	1300	
6	620				15	30	125	300	1100	
6	960					24	100	250	1000	
6	1900					20	70	200	700	
8	600						50	150	500	1300
8	1400						40	100	400	1200
8	2200						30	80	350	1100
8	3600						25	60	250	800
10	1000							75	125	1000
10	2500							50	100	500
10	3800							30	80	350
10	5600							25	60	250

Source. National Standard Plumbing Code.

Table 11.7 ***(Example 11.1) Drainage Fixture Units***

	Units
Bar sink	2
Kitchen sink and dishwasher	3
Lavatory	1
Water closet	4
Clothes washer	3
Master bath, lavatory, water closet, tub	6
Extra lavatory	1
Shower	2
Lower bath, lavatory, water closet, tub and shower	6
	28

Values are from Table 11.3
Hose bibb drainage to ground
Roof drainage to dry wells

11.6 Typical Design, Office Building

EXAMPLE 11.2. Select sizes for drainage and vent piping for the plumbing in an office building for which the fixtures are shown in Figure 11.29.

SOLUTION. Individual fixture branches shall not be less than the size indicated in Table 11.2 for the minimum size of trap for each fixture. Selected fixture units from Table 11.3 are applied to each section of the piping and totaled for each branch and stack, and for the building drain and the building sewer. An example of a fixture-unit summary and sample sizes of individual branches that connect into a typical branch of the men's toilet group on any floor are shown in the following table.

Fixtures	*Units per Fixture*	*Total Fixture Units*	*Diameter, Fixture Branch, Inches*
1 service sink	3	3	3
3 lavatories	1	3	$1\frac{1}{2}$
3 urinals, washout	4	12	2
3 water closets, valve operated	6	18	4
Total fixture units, Men's Toilet Branch		36	

Reference to Table 11.5 indicates that a 3-in. horizontal fixture branch is inadequate for the above group because it will handle only 20 fixture units and not more than two water closets. A 4-in. pipe is selected. Its capacity of 160 fixture units will be more than enough for the 36 needed here. The same table shows that the soil stack can be 4 in. in diameter (it is run thus for its entire height). Its capacity of 90 fixture units per story is sufficient for the 64 that connect in at each T-Y connection.

According to Table 11.4, the building drain and the building sewer at their pitch of $\frac{1}{4}$ in. in 1 ft should be 5 in. in diameter. Their capacity for 480 fixture units exceeds the $350\frac{1}{2}$ placed upon them. The vent stack at a 70 ft length and 338 fixture units could be $2\frac{1}{2}$ in., but 3 in. is a better choice. This is increased to 4 in. as it passes through the roof.

11.7 The Sovent System

Although opinion may vary about the relative merits of continuous or loop venting, either system, properly designed, will be effective in preventing the siphoning out of traps or in relieving air pressures that could cause foul gases to bubble through the traps into the occupied space. It is interesting to discover that a newly developed system, especially suitable for high-rise buildings, eliminates the vent stack completely with equal effectiveness. This is done by changing the nature of the effluent (discharge of wastes and sewage from the fixtures) instead of coping with the pressures and suctions that normal effluent would cause (see Figs. 11.30 and 11.31).

The effects of *normal* liquid effluent descending in a soil or waste stack were illustrated in Figures 11.2 and 11.3. Although the effluent seldom fills the stack completely, it definitely has the action of a *plunger*, pushing gases down ahead of it and creating a suction (negative pressure) above it. In a vented system, gases below the plunger find their escape through branch vents and the vent stack in the lower stories. Negative pressures above the plunger are relieved in a vented system by air that is drawn from the vent pipes above. Thus siphonage of fixture traps above the plunger is prevented. It is evident that, if the effective-

Fig. 11.29 (opposite) (*Example 11.2*) *Plumbing section, office building, in general confirmity with National Standard Plumbing Code. Circuit vents serve branch soil lines. House trap and fresh-air inlet are omitted from building drain. Some codes require continuous venting (see Fig. 11.28).*

Notes:

a. Relief vent not required on top floor.

b. Men's and women's toilets on 3rd floor are typical and would be repeated on 1st, 2nd, 4th, and 5th.

c. Drinking fountains on 5th and 4th floor would be repeated on 1st, 2nd, and 3rd.

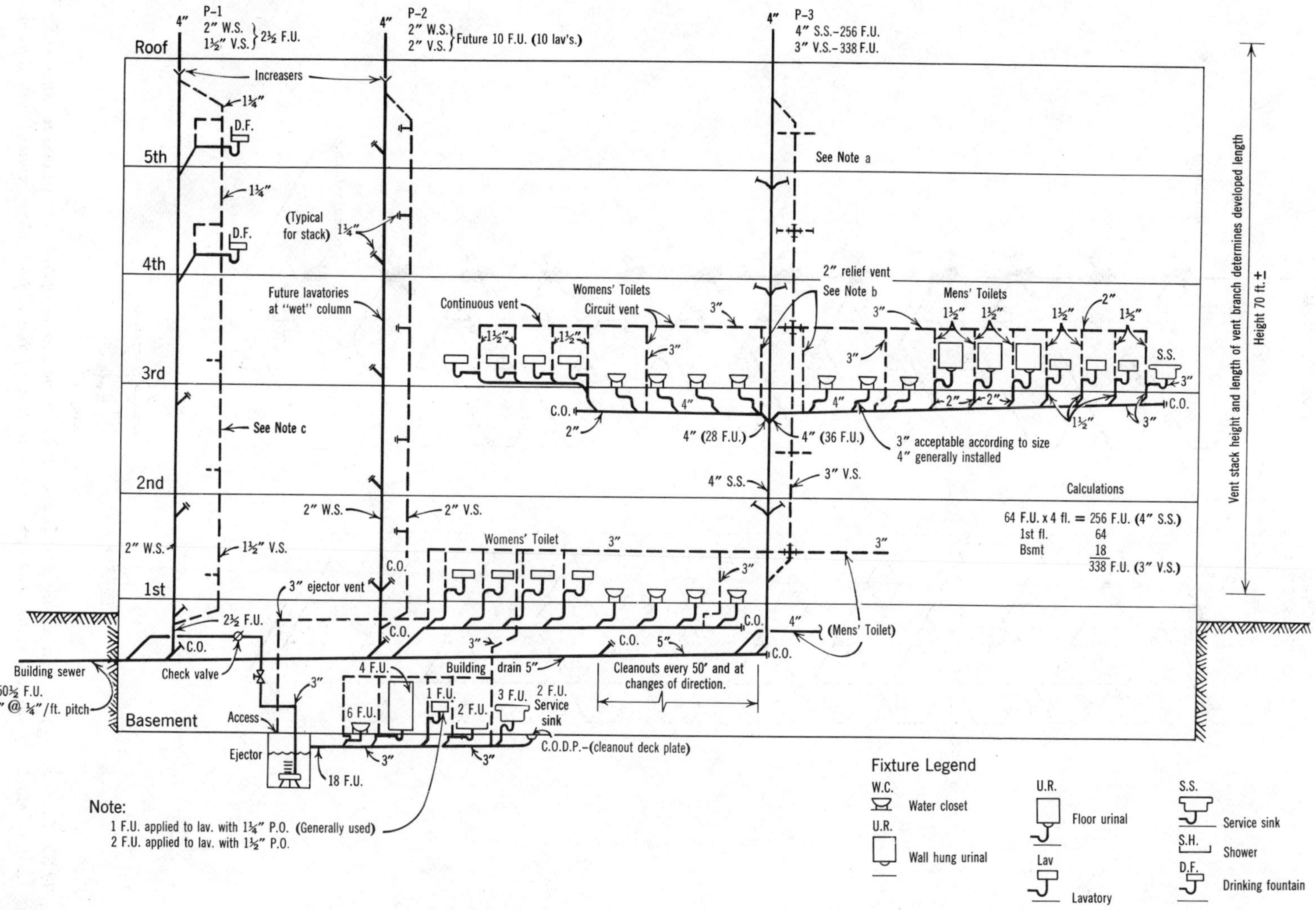

4″ P-1 2″ W.S. 1½″ V.S. 2½ F.U.
4″ P-2 2″ W.S. 2″ V.S. Future 10 F.U. (10 lav's.)
4″ P-3 4″ S.S.-256 F.U. 3″ V.S.-338 F.U.
Roof
5th
4th
3rd
2nd
1st
Basement
Increasers
1¼″
D.F.
(Typical for stack) 1¼″
Future lavatories at "wet" column
See Note a
See Note c
2″ relief vent
See Note b
Womens' Toilets
Circuit vent
Continuous vent
Mens' Toilets
1½″
2″
3″
S.S.
C.O.
4″ (28 F.U.)
4″ (36 F.U.)
3″ acceptable according to size
4″ generally installed
4″ S.S.
3″ V.S.
2″ W.S.
2″ V.S.
1½″ V.S.
Calculations
64 F.U. x 4 fl. = 256 F.U. (4″ S.S.)
1st fl. 64
Bsmt 18
338 F.U. (3″ V.S.)
Womens' Toilet
3″ ejector vent
(Mens' Toilet)
2½ F.U.
Check valve
5″
Building drain 5″
Cleanouts every 50′ and at changes of direction.
Building sewer
350½ F.U.
5″ @ ¼″/ft. pitch
4 F.U.
6 F.U.
1 F.U.
2 F.U.
3 F.U.
2 F.U. Service sink
Access
Ejector
18 F.U.
C.O.D.P.-(cleanout deck plate)
Vent stack height and length of vent branch determines developed length
Height 70 ft. ±
Note:
1 F.U. applied to lav. with 1¼″ P.O. (Generally used)
2 F.U. applied to lav. with 1½″ P.O.
Fixture Legend
W.C. Water closet
U.R. Wall hung urinal
U.R. Floor urinal
Lav Lavatory
S.S. Service sink
S.H. Shower
D.F. Drinking fountain

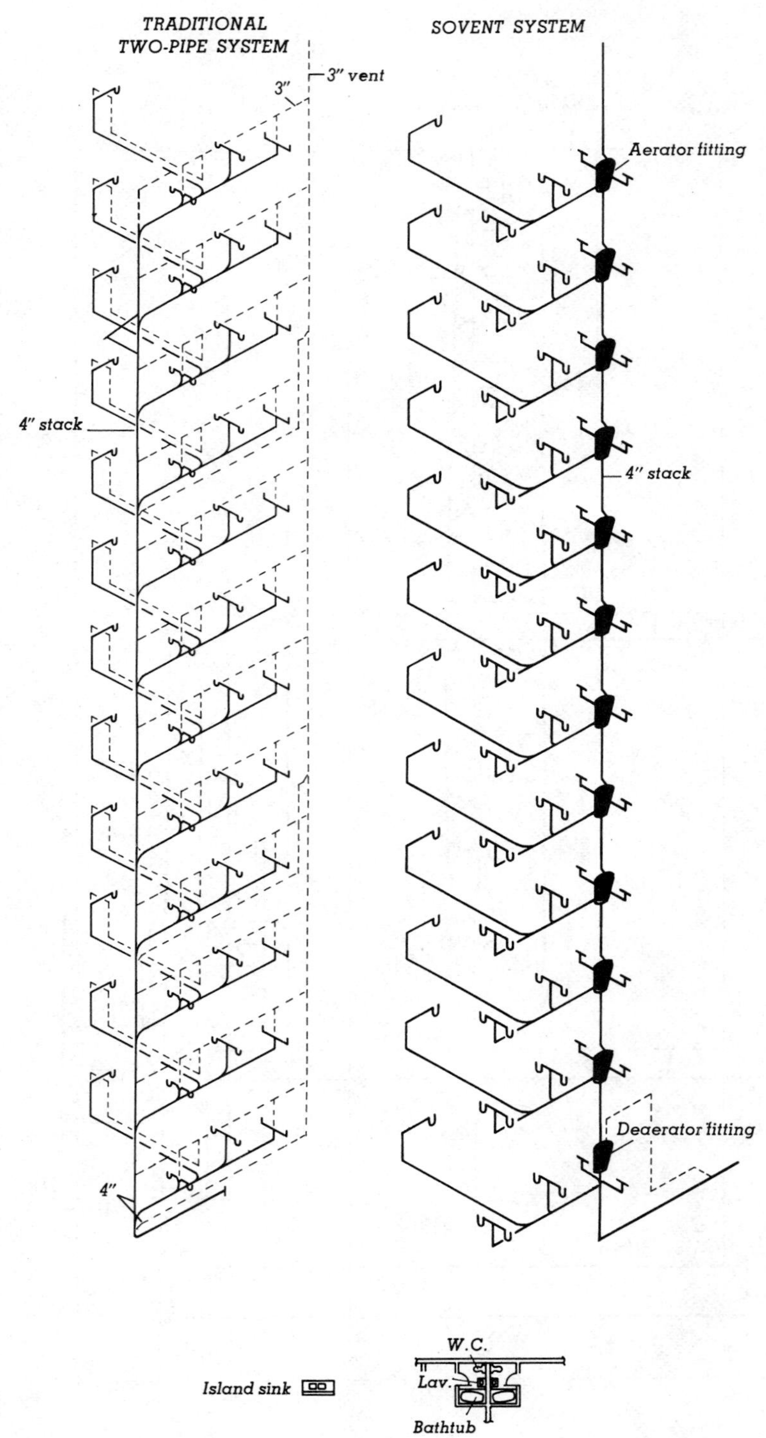

Fig. 11.30 *Cost-saving potential can be seen by the choice of the Sovent system over a two-pipe system for this 12-story stack serving an apartment grouping. Courtesy of the Copper Development Association.*

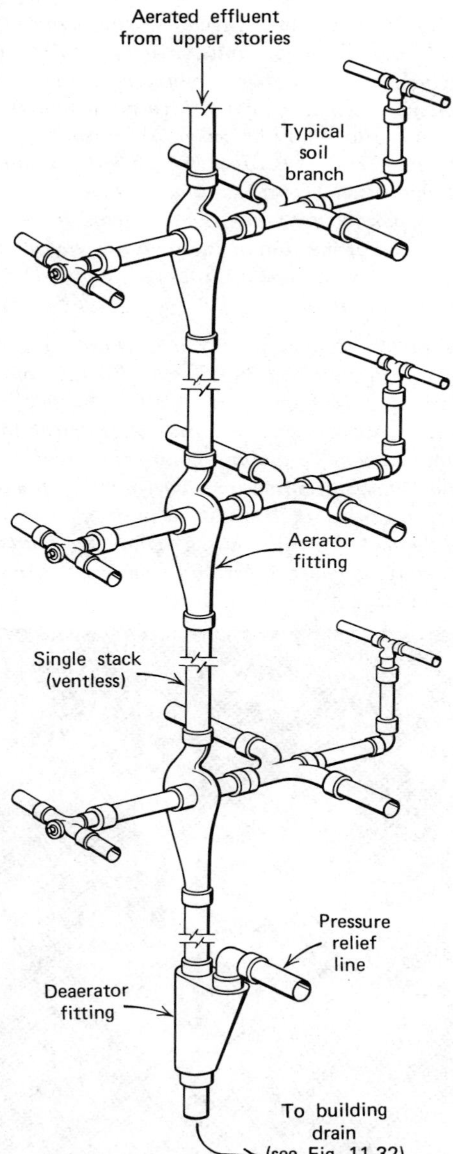

Fig. 11.31 *The Sovent drainage stack consists of aerator fittings, joining the horizontal branches to the stack at each floor level and a deaerator fitting at the bottom of the stack. The stack is open to the atmosphere above the roof at the top. Courtesy of Copper Development Association.*

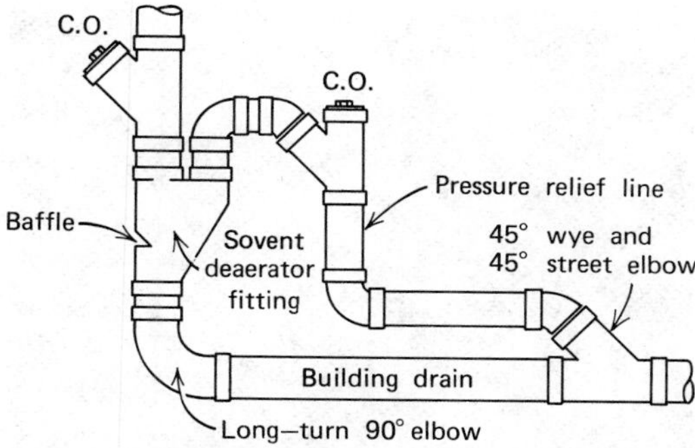

Fig. 11.32 *The deaerator consists of an air separation chamber having an internal nose piece, a stack inlet, a pressure relief outlet at the top, and a stack outlet at the bottom. The deaerator fitting at the bottom of the stack functions in combination with the aerator fittings above to make the single stack self-venting. The deaerator is designed to overcome the tendency that would otherwise occur for the falling waste to build up excessive back pressure at the bottom of the stack when the flow is decelerated by the bend into the horizontal drain. Courtesy of Copper Development Association.*

ness of the "plunger" can be reduced, the negative and positive pressures created by it will be also reduced. If their values can be brought down below the holding power of the several inches of water in the trap, no vents will be necessary. In the single stack *Sovent* system illustrated in Figure 11.31, this is done by dealing with the normal liquid effluent at each floor. By aerating it there, a *foam* is produced that lacks the stack-filling tendency of the liquid effluent. Thus, by creating a *soft* plunger, pressure variations in the single stack are minimized.

Tests have shown that the positive and negative pressures produced by *normal* liquid effluent during its descent and relieved by the vent piping are often about 5 to 12 in. water gauge. Obviously, if the vents were not provided, the 2 to 4 in. of water seal in the traps would be vulnerable to penetration by gases from pipes under positive pressure or siphonage of water seals into pipes that may be under negative pressure.

Figures 11.30, 11.31, 11.32, and 11.33 illustrate the components and the action of the

Fig. 11.33 *This free-standing exhibit clearly illustrates how aeration is accomplished by the offset path of flow provided by the aerator fitting. Soil and waste connections at the story shown are typical in that they all drain to the vertical axis of the stack. The branches, in some parts slightly larger than in a traditional system, provide greater air circulation at design capacity flow than would the branches in a traditional system. Courtesy of Copper Development Association.*

Sovent system. Effluent, already aerated and descending from upper stories, is diverted in the stack at each lower story. The aerator fitting there affords a passage for this diverted flow and also an air space into which the effluent from the local branch soil or waste can drop. Here it spatters, mixing with the air to form a rarified mixture of air and liquid. Tests show that this mixture does not produce pressures, positive or negative, of more than 1 in. water gauge. Thus, a trap-seal of 2 in. or more is safe against siphonage or penetration.

At the foot of the single stack the aerated effluent is compacted—a process which is

Fig. 11.34 (opposite) *Already serving more than 35,000 U.S. dwelling units, chiefly in high-rise structures, all-copper Sovent plumbing also is making important gains in the three-story apartment building sector—and has even been installed in a one-family house. Because the Sovent plumbing drainage system is self-aerating, it eliminates the usual separate vent pipe—thereby achieving economy, speed of installation, more usable space, and greater design freedom. Courtesy of Copper Development Association.*

Fig. 11.35 *Copper Sovent was chosen for the 21-story Hyatt Hotel in Phoenix because of its proven record of quick, trouble-free installation, reports C. A. Pedersen of Precision Mechanical, Inc., Tempe, Arizona, plumbing contractors, seen here against the superstructure of yet another famous Hyatt atrium lobby. Hand-held in the photograph is a typical deaerator fitting. Photo by Markow Photography, Phoenix, Arizona. Courtesy of Copper Development Association.*

aided by a baffle in the path of the flow in the deaerator fitting (see Fig. 11.32). Air piling up at this point could cause pressures in the stack at the first floor if not relieved. An air-discharge pipe provides this relief of air from the deaerator fitting to the upper part of the building drain, above the liquid flow.

The Sovent system was invented by Fritz Sommer of Switzerland who tested it in a 10-story drainage test tower. Since its introduction in 1962, it has been installed and used in more than 100 buildings in Europe and Africa. Canada used the Sovent method in the Habitat apartments at the 1967 Montreal World's Fair. It was first used in the United States in 1968 at the Uniment Apartments in Richmond, California; there it was tested and approved before going into service. Thus Sovent was first granted U.S. code acceptance in 1968 in Richmond, California. Following this success, its code acceptance grew rapidly during the early 1970s. Today, Sovent is accepted by the major model codes used nationwide, as well as by 300 major municipalities and other large jurisdictions (see Figs. 11.34 and 11.35).

Fig. 11.36 Luxor toilet. Fixture dimensions and roughing dimensions. Note the 30 psi water pressure required for this rather special fixture. It exceeds the usual maximum of 20 psi. Courtesy of American Standard.

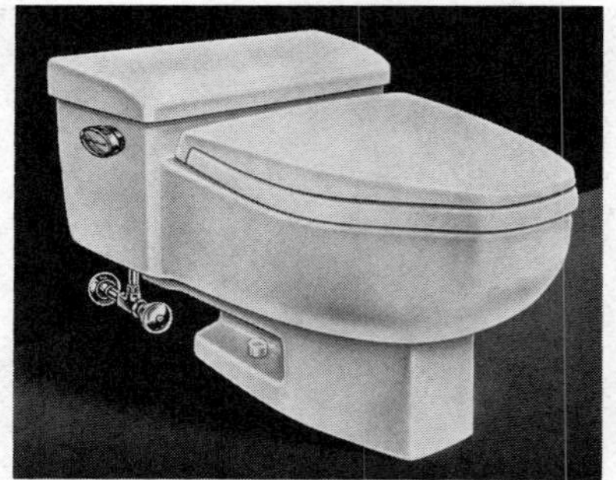

LUXOR TOILET

2003.010
2003.036

VITREOUS CHINA — ONE PIECE

WITH ½" S.P.S. SUPPLY FURNISHED

PLUMBER NOTE: THIS COMBINATION IS DESIGNED TO ROUGH-IN AT MINIMUM DIMENSION OF 12" FROM FINISHED WALL TO C/L OF OUTLET.

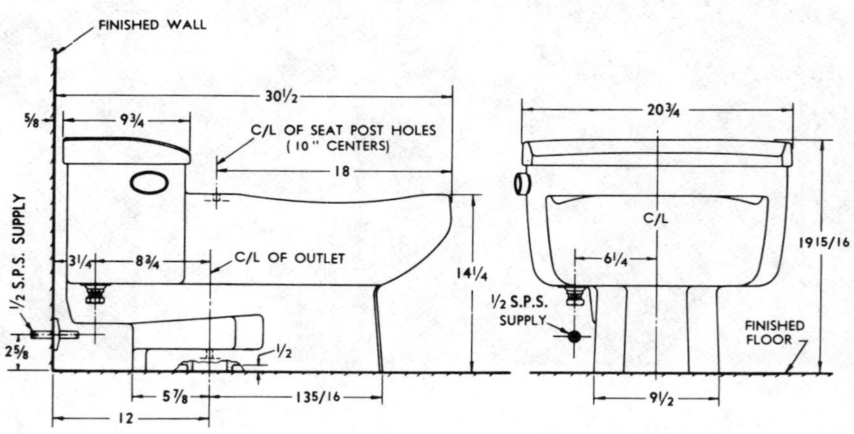

NOTE: 2003.010 - FURNISHED WITH VENT-AWAY TOILET VENTILATOR.
2003.036 - LESS VENT-AWAY.

NOTE - **30 P.S.I. MIN. WORKING PRESSURE REQUIRED AT WATER CONTROL**

NOTE: ½" supply pipe included with toilet.

IMPORTANT: Dimensions of fixtures are nominal and may vary within the range of tolerances established by Commercial Standards CS20.
These measurements are subject to change or cancellation. No responsibility is assumed for use of superseded or voided leaflets.

11.8 Plumbing Fixtures and Roughing

The selection of plumbing fixtures is usually a joint decision of the architect and the client. An example of appropriate selections is seen in Figures 11.36, 11.37, and 11.38. The fixtures—a toilet, a double kitchen sink, and a lavatory basin are choices that would be suitable for the house on Long Island in Figure 11.23.

The illustrations are typical of catalog offerings of the manufacturer and distributor. They are used by the architect in the layout of bathrooms and kitchens for reference to size, spacing, and height. Upon delivery, the fixtures arrive very carefully packed in crates. Architects will often request that the distributor supply three copies of the catalog information: one copy is for the client; one is for their own use; and one is for the plumbing contractor.

EXPLORER SINKS

STAINLESS STEEL — SELF-RIMMING
STRAINERS

3258.233
3258.241

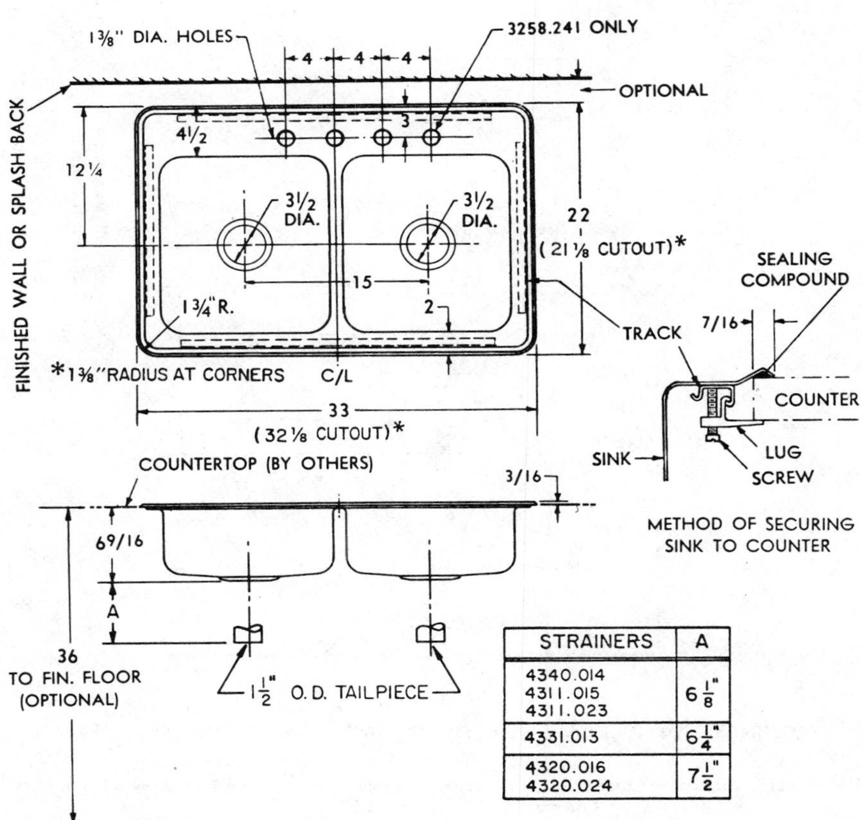

STRAINERS	A
4340.014 4311.015 4311.023	6 1/8"
4331.013	6 1/4"
4320.016 4320.024	7 1/2"

Fig. 11.37 Explorer sinks. Fixture dimensions and roughing dimensions. Courtesy of American Standard.

Fig. 11.38 Merrilyn lavatory. Fixture dimensions and roughing dimensions. Courtesy of American Standard.

MERRILYN LAVATORY 0140.178

VITREOUS CHINA — FOR COUNTERTOP INSTALLATION (POP-UP DRAIN)

2248.045 FITTING — 2303.154 SUPPLIES

4401.014 "P" TRAP

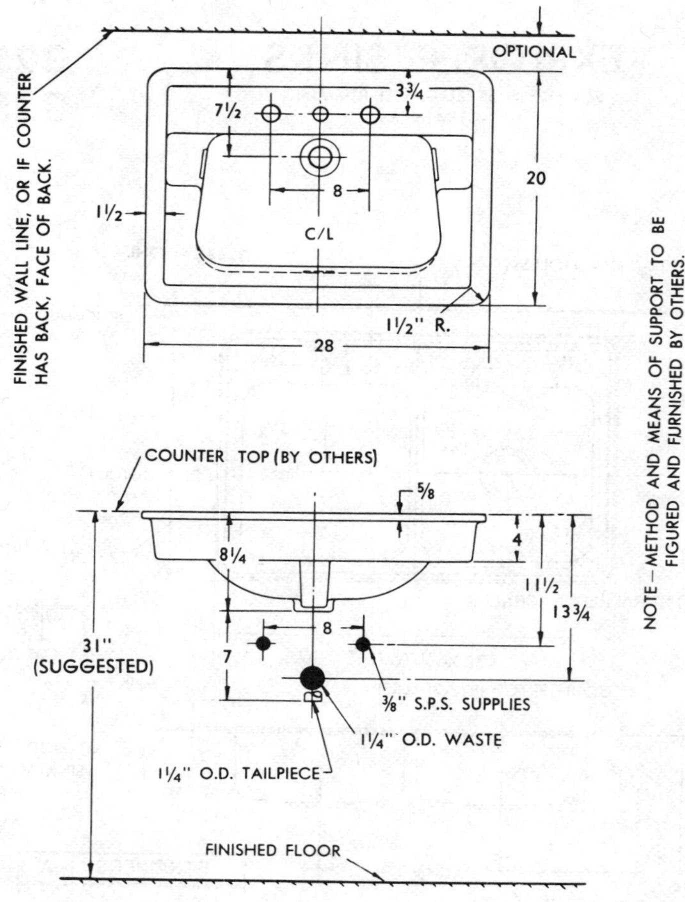

NOTE: FITTINGS NOT INCLUDED WITH FIXTURE AND MUST BE ORDERED SEPARATELY.

IMPORTANT: Dimensions of fixtures are nominal and may vary within the range of tolerances established by Commercial Standards CS20.

These measurements are subject to change or cancellation. No responsibility is assumed for use of superseded or voided leaflets.

Fig. 11.39 Roughing dimensions set the position of pipes. Photograph of the double lavatory installation of piping for the master bath of the house in Figure 11.23. In this early stage, the plastic waste branches are seen in the 2 × 6 stud wall. Copper tubing will serve the fixtures. All pipes will be capped and the system tested. Dimensions of Figure 11.38 were followed and they will be exactly correct to receive two of the Merrilyn fixtures of Figure 11.38. Expansion and shock are not provided for by extensions of the water supply lines, since they can be absorbed by centrally located expansion chambers such as shown in Figure 11.27.

The plumber's interest and responsibility relate to the so-called "roughing dimensions." Since the fixtures are not uncrated and installed until all plumbing piping is complete, tested, and wall surface construction in place, the roughing dimensions are most important. For the location of the water and waste branches that will protrude through the finished wall, the plumber conforms to the manufacturer's "spec sheet" (Fig. 11.38). See also Figure 11.39. A similar procedure applies to other fixtures (Figs. 11.36 and 11.37). Roughing for the washroom of a school is shown in Figure 11.40.

11.9 Minimum Facilities

In order to avoid substandard planning of plumbing facilities, the National Standard Plumbing Code sets minimal limits to the ratio of plumbing fixtures to occupants (see Table 11.8). The health and convenience of people living and working in various types of buildings are thus protected. It is important to keep in mind that the standards are *minimal.* The need for more generous planning of such facilities is often an important consideration.

11.10 Private Sewage Treatment

Nobody likes to think about sewage. Accordingly, the expertise of dealing with it effectively has lagged. However, for the past several decades the problem has become inescapable, and some trends and improvements have been apparent. In Chapter 9, the problems of treatment in large, long-established cities were discussed. There, combined sanitary and storm flow create difficulties that in many cases cannot be fully solved. Where new

Fig. 11.40 Roughing in place for four lavatories in the washroom of a school. At this stage, the waste branches have been capped and the system tested against possible leakage. Using materials different from those of Figure 11.39, we observe that the waste branches are of cast iron, the vents of galvanized steel, and the water lines of copper with soldered fittings. Roughing dimensions have been followed. Vertical capped expansion and shock tubes are seen as extensions of the entering water pipes that serve the hot and cold water branches. The layout course of masonry behind the piping assembly is the position of the partition marking the adjacent room. At the right will be seen a projecting masonry block. The near end of this block is the location of another block wall that will enclose and "fur in" the roughing assembly.

communities are planned with the separation of sanitary and storm drainage, more effective central treatment can be achieved. As new developments occur prior to the installation of such community sewers and treatment plants, traditional methods of on-site treatment are possible. To be most effective, they require fairly large lots to accommodate the underground components. Most desirable also is the foresight to plan for connection to a future sanitary sewer. With the exception of the outmoded cesspool, two systems—the septic tank plus one or more seepage pits and the septic tank plus drainage trenches—do a fair job. Administrative authorities will, in most cases, approve of these systems under strict regulations. Connections to future sanitary sewers are always anticipated.

(*a*) *What* Not *to* *Do.* The only pleasant thing that can be said about Figure 11.41 is that the photograph was taken long ago. It represents a technique that, fortunately, is fast disappearing.

(*b*) *Methods.* As shown in Figure 11.42, methods (*a*) and (*b*) are acceptable. The type chosen and the sizes of its components will depend on the number of people served, the permeability of the soil, the elevation of the groundwater level (water table), the size of the lot, and the proximity of wells that might be vulnerable to pollution by the products of the treatment system. The installation should be designed, of course, to treat only sanitary drainage. Roof and area storm drainage should be recharged to the ground directly through dry wells.

Table 11.8 ***Minimum Number of Plumbing Fixtures***

Type of Building Occupancy	Type of Fixture: Water Closets		Urinals		Lavatories		Bathtubs or Showers	Drinking Fountains	Other Fixtures
Assembly—places of worship	*No. of Persons* 150 Women 300 Men	*No. of Fixtures* 1 1	*No. of Persons* 300 Men	*No. of Fixtures* 1	1			1	
Assembly—other than places of worship, (auditoriums, convention halls, theaters)	*No. of Persons* 1–100 101–200 201–400	*No. of Fixtures* 1 2 3	*No. of Persons* 1–200 201–400 401–600	*No. of Fixtures* 1 2 3	*No. of Persons* 1–200 201–400 401–750	*No, of Fixtures* 1 2 3		1 for each 1000 persons	1 slop sink
	Over 400, add 1 fixture for each additional 500 men; add 1 for each 300 women		Over 600, add 1 fixture for each 300 men		Over 750, add 1 fixture for each 500 persons				
Dormitories—school or labor, also institutional	Men: 1 for each 10 persons. Women: 1 for each 8 persons		1 for each 25 men. Over 150, add 1 fixture for each 50 men		1 for each 12 persons. (Separate dental lavatories should be provided in community toilet room. A ratio of 1 dental lavatory to each 50 persons is recommended.)		1 for each 8 persons; for Women's dormitories, additional bathtubs should be installed at the ratio of 1 for each 30 women. Over 150 persons add 1 fixture for each 20 persons.	1 for each 75 persons	Laundry trays, 1 for each 50 persons. Slop sinks, 1 for each 100 persons.
Single dwellings	1 for each dwelling unit				1 for each dwelling unit		1 for each dwelling unit		
Dwellings—multiple or apartment	1 for each dwelling unit or apartment				1 for each dwelling unit or apartment		1 for each dwelling unit or apartment		
All other structures for human habitation shall be equipped with									

(Continued)

Table 11.8 ***Minimum Number of Plumbing Fixtures (Continued)***

Type of Building Occupancy	Type of Fixture					
	Water Closets	*Urinals*	*Lavatories*	*Bathtubs or Showers*	*Drinking Fountains*	*Other Fixtures*
sufficient sanitary facilities. Plumbing fixtures shall be made of durable, smooth, nonabsorbent and corrosion-resistant material and shall be free from concealed fouling surfaces.						
Industrial—factories, warehouses, foundries and similar establishments	*No. of each sex* — *No. of Fixtures* 1–10 — 1 11–25 — 2 26–50 — 3 51–75 — 4 76–100 — 5 1 fixture for each additional 30 employees	Where more than 10 men are employed: *No. of Men* — *No. of Urinals* 11–30 — 1 31–80 — 2 81–160 — 3 161–240 — 4	*No. of Persons* — *No. of Fixtures* 1–100 — 1 to 10 persons Over 100 — 1 to 15 persons	1 shower for each 15 persons exposed to excessive heat or to occupational hazard from poisonous, infections or irritating material.	1 for each 75 persons	
Institutional—other than hospitals or penal institutions (on each occupied story)	1 for each 25 men 1 for each 20 women	1 for each 50 men	1 for each 10 persons	1 for each 10 persons	1 for each 50 persons	
Hospitals—						Minimum one slop sink per floor for first 50 beds; one additional for each additional 50 or major fraction thereof
individual room	1		1	1		
wards	1 for each 8 patients		1 for each 10 patients	1 for each 20 patients	1 for each 100 patients	

Hospitals— waiting rooms employees	1 Same as public buildings	Same as public buildings	1 Same as public buildings		Same as public buildings	
Penal institu-tions—prisoners	1 in each cell 1 in each exercise room	1 in each exercise room	1 in each cell 1 in each exercise area		1 on each cell block floor, 1 in each exer-cise area	One slop sink per floor
Penal institu-tions—employees	Same as public buildings	Same as public buildings	Same as public buildings		Same as public buildings	
Public buildings offices, business mercantile, stor-age and institu-tional employees	*No. of each sex* — *No. of Fixtures* 1–15 — 1 16–35 — 2 36–55 — 3 56–80 — 4 81–110 — 5 111–150 — 6 1 fixture for each addi-tional 40 employees	Urinals may be provided in men's toilet rooms in lieu of water closets but for not more than $\frac{1}{3}$ of the required number of water closets	*No. of Employees* — *No. of Fixtures* 1–15 — 1 16–35 — 2 36–60 — 3 61–90 — 4 91–125 — 5 1 fixture for each addi-tional 45 persons		1 for each 75 persons	1 slop sink per floor
Schools— elementary secondary	*Boys* — *Girls* 1/30 — 1/25 1/40 — 1/30	 1/25 boys 1/15 boys	*Boys* — *Girls* 1/35 — 1/35 1/40 — 1/40	In gym or pool shower rooms, 1/5 pupils *of a* class	 1/40 pupils 1/50 pupils	
Working men in temporary facili-ties	1/30 working men	1/30 working men	1/30 working men		1 fixture or equiv-alent for each 100 working men	

Source. The National Standard Plumbing Code.

Author's note. Some codes prohibit the installation of drinking fountains in rest rooms.

Fig. 11.41 This is a cesspool. *A "cesspool" is described in two nontechnical dictionaries as* (1) *"A cavity in the earth to receive and retain the drainage and sewage from sinks, toilets, etc.," and* (2) *"A filthy smelly place." The characteristics of the installation shown here that make it unacceptable are:*

1. *The effluent of raw sewage (as in all cesspools) seeps into the soil.*
2. *On this small lot the pit is unavoidably too close to the house.*
3. *The unsealed joints of the clay tile pipe will leak close to the foundation. Polluted effluent can seep into the basement through a poorly constructed foundation wall.*
4. *Fragile clay tile breaks easily and could make leakage worse.*
5. *Loosely laid and poor quality perforated concrete blocks can collapse and make the pit ineffective.*

Fig. 11.42 *Private sewage treatment.* (a) *This method is most suitable in porous soil and where the groundwater level is low.* (b) *This method finds its best use in less pervious soils and where the groundwater level is high.* (c) *This cesspool disposal is now discouraged and in many locations is illegal.*

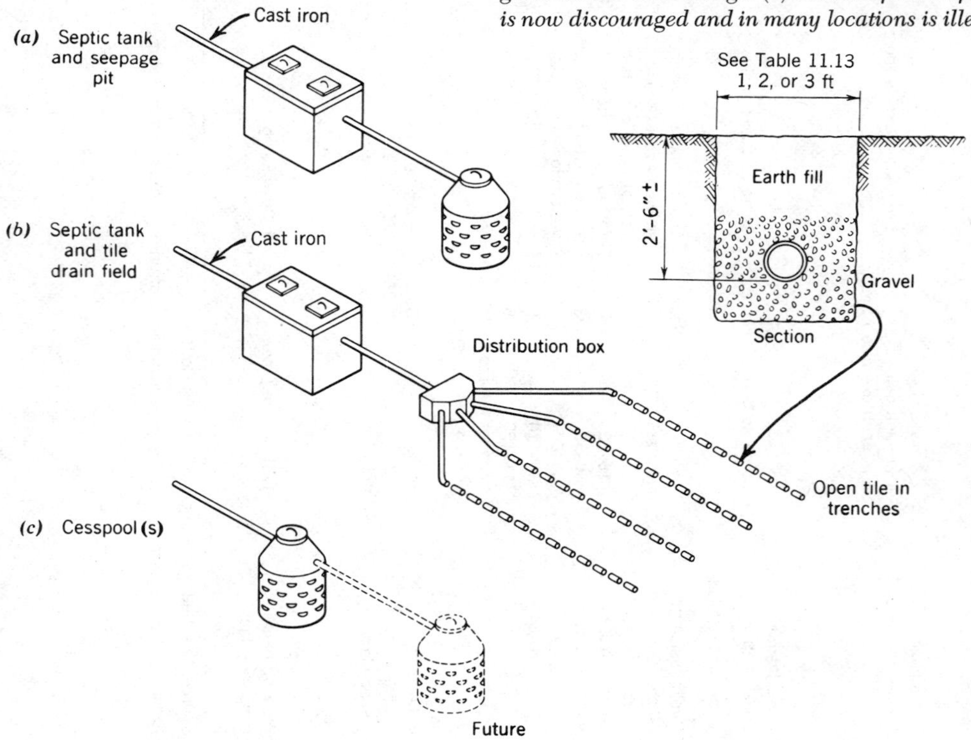

(a) Septic tank and seepage pit

Cast iron

(b) Septic tank and tile drain field

Cast iron

Distribution box

See Table 11.13
1, 2, or 3 ft

2'-6"±

Earth fill

Gravel

Section

Open tile in trenches

(c) Cesspool(s)

Future

It will be observed in Figure 11.42 that the seepage pit dispersal of the partially purified septic tank effluent usually requires considerable depth for the pit. It often extends 10 or 15 ft below the surface. The tile drain field on the other hand is shallow, often not exceeding a depth of about 3 ft. When soil is porous, seepage pits (*a*) are most appropriate. For less pervious soil, the drain field (*b*) extends the discharge of effluent to a much larger area. Another criterion in the choice between pit and drain field is the level of the groundwater. Since a seepage pit must *not* extend down into the groundwater level, the shallow drain field is usually necessary when the water table is high.

(*c*) ***Available Products.*** Crude methods of construction, such as those of Figure 11.41, have long been supplanted for use in acceptable systems by very efficient prefabricated components (see Figs. 11.43 and 11.44). Connections between the house or building and the septic tank should be of extra heavy cast-iron pipe with fluid-tight, approved joints. The same quality of connection is used from septic tank to seepage pit or to the distribution box of a pit-group or to a group of tile drains. Tightly sealed septic tanks and one-piece seepage rings are available as prefabricated products.

(*d*) ***The Septic Tank and Seepage Pit.*** The septic tank has the capacity to retain several days of sanitary flow. The action of anaerobic (active without air) bacteria digests the solids to a compact sludge. So small is the volume of this sludge that it needs to be pumped out only once in several years. The liquid effluent, about 70% purified, then passes to the seepage pit where other bacteria below the vegetation provide further purification. The digestion is entirely biological.

Fig. 11.43 Prefabricated elements for private sewage treatment. As a dry well, (c) *is also used for dispersal of storm drainage to the earth. In specifying actual products, always refer to the manufacturer's dimensions.*

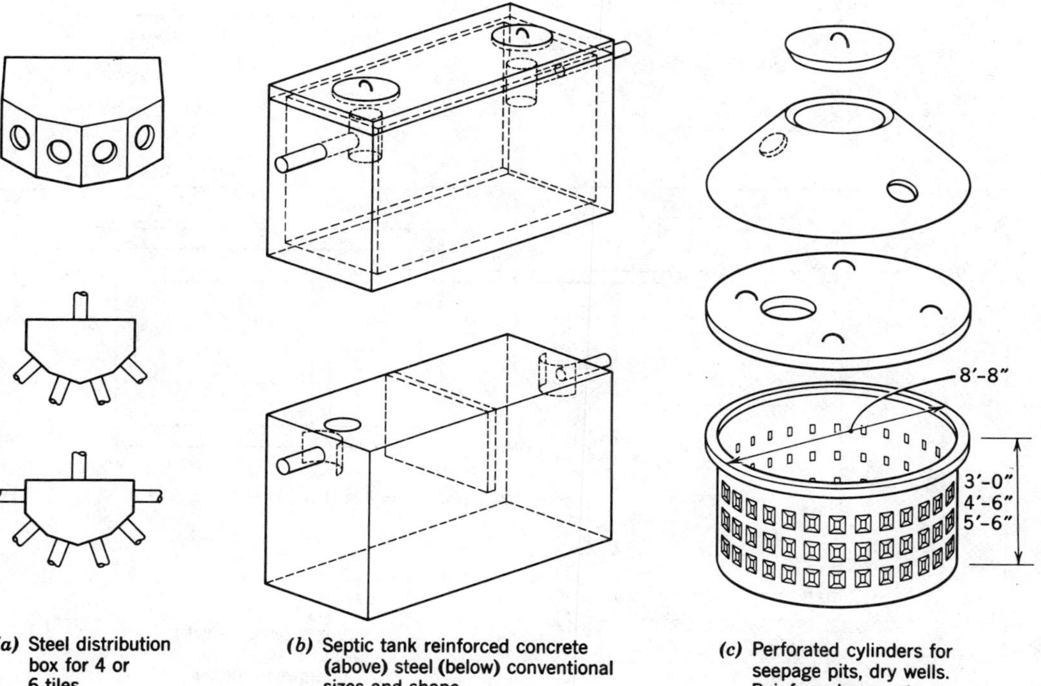

(*a*) Steel distribution box for 4 or 6 tiles

(*b*) Septic tank reinforced concrete (above) steel (below) conventional sizes and shape

(*c*) Perforated cylinders for seepage pits, dry wells. Reinforced concrete

Fig. 11.44 Precast elements used for recharging to the soil the partially purified effluent of septic tanks *or separately collected storm water. In the former use, they are known as seepage pits and, in the latter, dry wells. They are often made of prestressed concrete of high strength (4000 psi ultimate). Here shown in the manufacturer's yard, they are, right to left, two perforated rings with conical top, cylindrical extender and concrete cover for access, alternate perforated cone, and typical extenders for greater depth. Rings are usually available in 8 and 10 ft diameters and heights of 3, 4, and 5 ft.*

Fig. 11.45 (Example 11.3) Plan and section of septic tank and seepage pit for four-bedroom house. Pit is suitable when the earth is absorbent and the water table low (below pit-bottom). Drawing is not to scale.

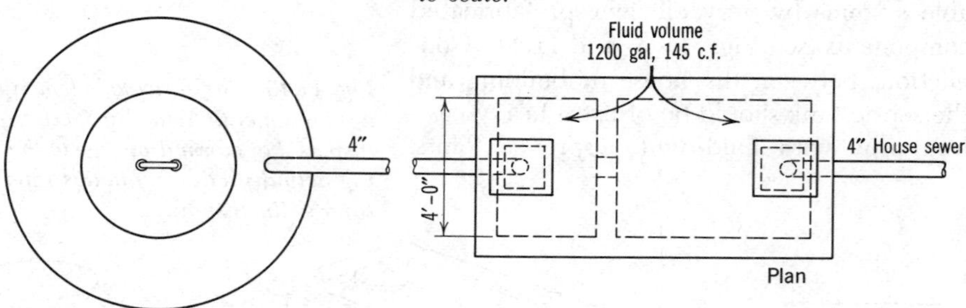

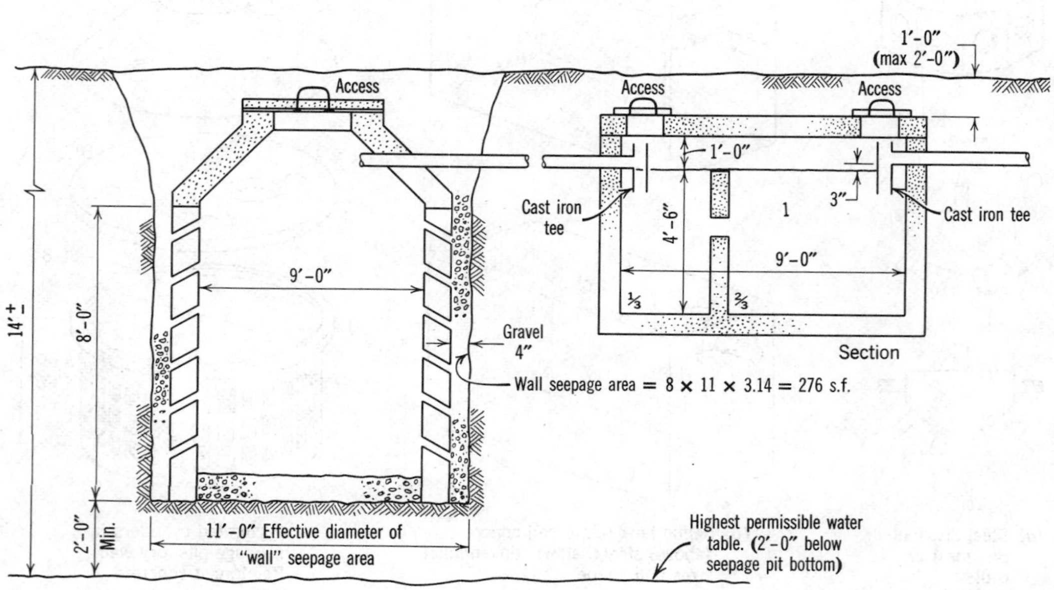

EXAMPLE 11.3 (Fig. 11.45). Design a septic tank seepage pit system for a suburban residence under the following conditions.

Bedrooms	4
Occupants	8
Soil (test)	3 min for a 1-in. drop
Depth to water table	14 ft

SOLUTION. Daily flow

(Table 11.9) is 75 gpd × 8 people	600 gpd
Septic tank capacity (Table 11.10)	1200 gal
Fluid volume of the tank 1200 ÷ 7.48 gal/cu ft	160 cu ft
Pit surface required (Table 11.11) 45 × 6	270 sq ft

Establish dimensions for the septic tank.
Dimensions from the chosen tank (Fig. 11.45) are:
4.5 × 4.0 × 9.0 = 162 cu ft volume
OK, 162 > 160 (that is required).
Check the cylindrical effective seepage area of the selected pit (Fig. 11.45):
8 × 11 × 3.14 = 276 sq ft
OK, 276 > 270 (that is required).

It is interesting to know that the requirement of 1200 gal capacity of tank was increased from 750 required by earlier standards. Private sewage treatment standards are improving.

(*e*) ***Tile Drain Field.*** In some regions, especially near the coastline and major bodies of water, the water table is quite high. Example 11.3, Figure 11.45, indicates that when the water table is closer than 14 ft to the surface, it could become difficult to fit a seepage pit into the relatively thin layer of available dry earth above the water table. In this case several shallower seepage pits may be substituted. Often, however, a tile drain field is preferred. Before designing it, the usual tests are made to determine how long it takes for water in a test pit to drop 1 in. at the proposed level of the disposal field. Table 11.12 lists the tile length required for various degrees of earth permeability as measured by the test. For certain soils of poor absorption, even if seepage pit depth were available, it is necessary to make the choice of a tile-drain-disposal field because the number and size of seepage pits would be too great.

Some sanitarians are of the opinion that for fall-rates of more than 6 min/in. of fall in test pits, tile drain fields are more appropriate and economical than seepage pits. Drains of square-edge agricultural tile, 4 in. or more in diameter which are laid in shallow trenches, are covered with gravel. The ends of the tiles do not meet, being separated by a $\frac{1}{4}$-in. opening. The effluent runs out of these spaces and stands in the interstices of the gravel until it seeps into the earth. In effect, the gravel provides spaces that act as a dry well to receive the fluids and accommodate them until they slowly sink into the ground.

EXAMPLE 11.4 (Fig. 11.46). Design a tile drain field for the conditions stated in the previous Example 11.3.

SOLUTION. Table 11.12 indicates that for a test result of 3 min for a 1-in. drop, a 2-ft-wide trench should have 18 ft of length for each 100 gpd of sewage. Since there will be 600 gpd, the required length of 2-ft-wide trench will be 6 × 18 = 108 ft. Depth and spacing for a 2-ft-wide trench are found in Table 11.13 and are reflected in the design layout of Figure 11.46. The tile length of 120 ft meets the 108 ft requirement. A three-prong layout was selected; yet, subject to the designer's judgment, the nature of the property, and the approval of the local administrative authority, a number of varying schemes can be considered. The length of any single tile-run may not exceed 100 ft.

Table 11.9 *Sewage Flows According to Type of Establishment*

Type of Establishment	*Gallons per Day per Person*[a]
Schools (toilet and lavatories only)	15
Schools (with above plus cafeteria)	25
Schools (with above plus cafeteria and showers)	35
Day workers at schools and offices	15
Day camps	25
Trailer parks or tourist camps (with built-in bath)	50
Trailer parks or tourist camps (with central bathhouse)	35
Work or construction camps	50
Public picnic parks (toilet wastes only)	5
Public picnic parks (bathhouse, showers, and flush toilets)	10
Swimming pools and beaches	10
Country clubs	25 gal per locker
Luxury residences and estates	150
Rooming houses	40
Boarding houses	50
Hotels (with connecting baths)	50
Hotels (with private baths—two persons per room)	100
Boarding schools	100
Factories (gallons per person per shift—exclusive of industrial wastes)	25
Nursing homes	75
General hospitals	150
Public institutions (other than hospitals)	100
Restaurants (toilet and kitchen wastes per unit of serving capacity)	25
Kitchen wastes from hotels, camps, boarding houses, etc. Serving three meals per day	10
Motels	50 gal per bed space
Motels with bath, toilet, and kitchen wastes	60 gal per bed space
Drive-in theaters	5 gal per car space
Stores	400 gal per toilet room
Service stations	10 gal per vehicle served
Airports	3–5 gal per passenger
Assembly halls	2 gal per seat
Bowling alleys	75 gal per lane
Churches (small)	3–5 gal per sanctuary seat
Churches (large with kitchens)	5–7 gal per sanctuary seat
Dance halls	2
Laundries (coin operated)	400 gal per machine
Service stations	1000 gal (first bay) 500 gal (each add. bay)
Subdivisions or individual homes	75
Marinas—flush toilets	36 gal per fixture per hour
urinals	10 gal per fixture per hour
wash basins	15 gal per fixture per hour
showers	150 gal per fixture per hour

Source. National Standard Plumbing Code.

[a]Unless otherwise indicated.

Table 11.10 ***Capacity of Septic Tanks***[a]

Single Family Dwellings Number of Bedrooms	*Multiple Dwelling Units or Apartments—One Bedroom Each*	*Other uses; Maximum Fixture Units Served*	*Minimum Septic Tank, Capacity in Gallons*
1 to 3		20	1000
4	2 units	25	1200
5 or 6	3	33	1500
7 or 8	4	45	2000
	5	55	2250
	6	60	2500
	7	70	2750
	8	80	3000
	9	90	3250
	10	100	3500

Extra bedroom, 150 gal each
Extra dwelling units over 10, 250 gal each
Extra fixture units over 100, 25 gal per fixture unit

Source. National Standard Plumbing Code.

[a] NOTE. Septic tank sizes in this table include sludge storage capacity and the connection of domestic food waste disposal units without further volume increase.

Table 11.11 ***Effective Absorption Area in Seepage Pits for Each 100 Gal of Sewage per Day***[a]

Time in Minutes for 1-in. Drop	*Effective Absorption Area (Square Feet)*
1	32
2	40
3	45
5	56
10	75
15	96
20	108
25	139
30	167

Source. National Standard Plumbing Code.

[a] Table 16.5.7 in the National Standard Plumbing Code.

Table 11.12 ***The Tile Length for Each 100 Gal of Sewage per Day***[a]

Time in Minutes for 1-in. Drop	*Tile Length for Trench Widths of:*		
	1 ft	*2 ft*	*3 ft*
1	25	13	9
2	30	15	10
3	35	18	12
5	42	21	14
10	59	30	20
15	74	37	25
20	91	46	31
25	105	53	35
30	125	63	42

Source. National Standard Plumbing Code.

[a]Table 16.5.4 in the National Standard Plumbing Code.

Table 11.13 ***Size and Spacing for Disposal Fields***

Width of Trench at Bottom (Inches)	*Recommended Depth of Trench (Inches)*	*Spacing Tile Lines*[a] *(Feet)*	*Effective Absorption Area per Lineal Foot of Trench (Square Feet)*
18	18 to 30	6.0	1.5
24	18 to 30	6.0	2.0
30	18 to 36	7.6	2.5
36	24 to 36	9.0	3.0

Source. National Standard Plumbing Code.

[a]A greater spacing is desirable where available area permits.

NOTE. Maximum length 100 ft.

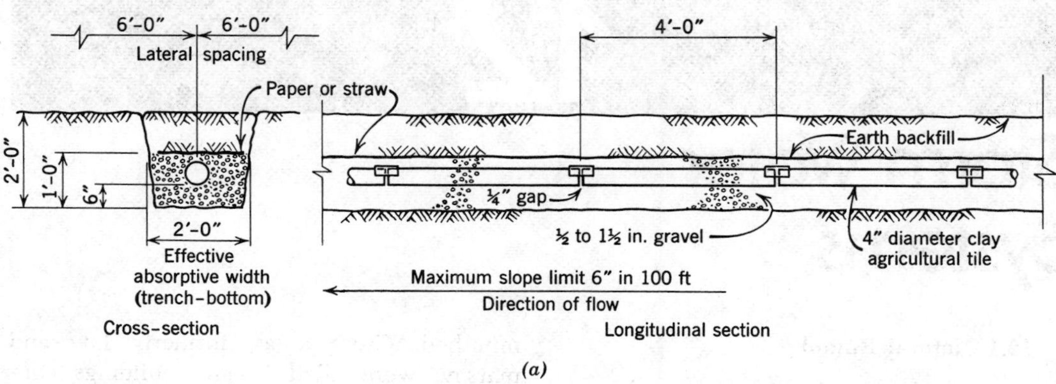

(a)

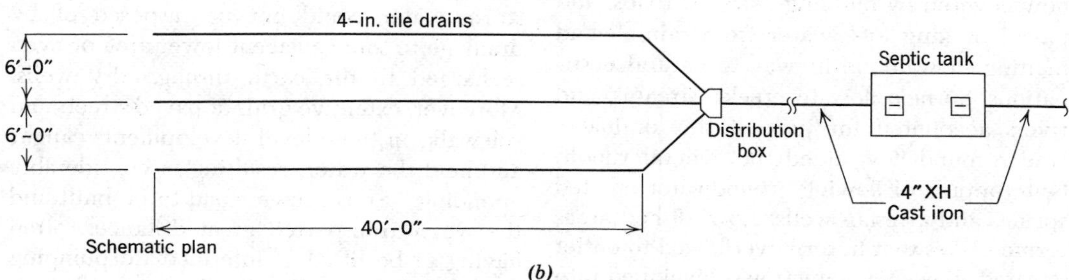

(b)

Fig. 11.46 *(Example 11.4) Tile drain field for a four-bedroom, eight-person house. Although the drawings are not to scale, the dimensions would indicate a required area of about 20 × 70 ft on the lot. When it is considered that it is best not to have the elements run below walks, drives, or other paved areas, sewage treatment on a small lot demands considerable space. For tests that show a time interval of more than 3 min for a 1-inch drop, the space requirements are even greater. When the soil is dense impervious clay, the solution is most difficult.* (a) *Transverse and longitudinal sections.* (b) *Schematic plan.*

References and Bibliography

1. *National Standard Plumbing Code, 1975,* published by the National Association of Plumbing-Heating-Cooling Contractors, 1016 20th Street N.W., Washington, D.C. 20036.

2. *Cast Iron Soil Pipe and Fittings Handbook,* The Cast Iron Soil Pipe Institute, 1824–26 Jefferson Place N.W., Washington, D.C. 20036.

3. *Copper Sovent Single Stack Plumbing System Design Handbook,* Copper Development Association, 420 Lexington Avenue, New York, N.Y. 10017.

4. *Modern Piping with Plastics,* Plastics Pipe Institute, 250 Park Avenue, New York, N.Y. 10017.

12

Storm Water Systems

12.1 Natural Runoff

Prior to the development of rural areas, now covered by buildings, streets, roads, and paved parking lots, water from rainfall and melting snow found its way to natural destinations. Surface flow to creeks, streams, and rivers accounted for part of this drainage. Underground flow aided the general runoff. Outcropping of flowing groundwater created springs and artesian wells. Low, dished areas formed lakes that in turn overflowed to outlet streams. Flat areas sometimes developed into swamps or marshes.

At a time when there was a choice of locations for towns and villages, sites next to rivers were usually chosen. The waterways provided transportation and water was supplied from the river or from adjacent wells. As streets and roads were built, slopes could be arranged whereby the rain falling on these areas and flowing onto them from roofs of buildings could run to the river. At interior parts of the country, high ground was favored for building sites and growing communities. Obviously, swampy or marshy ground would not be chosen, but it did provide terminal locations for the storm water that ran off the high ground. In the course of this natural flow, much of the water was drawn by evaporation to the clouds. The rest, conforming to topographic river basins continued along to seek its way to the sea.

12.2 Growth of Cities

Unfortunately, as building increased, desirable locations were soon in short supply. The possibility of selecting high, dry ground diminished. Great areas, formerly low and marshy, were filled in and buildings constructed, often on piles. From such locations storm water could not be disposed of by drainage to some adjacent lower area or even recharged to the earth through dry wells. Moreover, extensive grids of paved streets and sidewalks in these level developments caught and held the water, resulting in considerable "ponding." Storm sewers had to be built and the water transported great distances, often having to be lifted at intermediate pumping stations before reaching its destination, which might be a remote river.

12.3 Reasons for Drainage

Whether the collection and disposal of storm water is easy or difficult, it is nonetheless important and cannot be ignored. Rainwater collecting on roofs, if not diverted, will run down walls and can cause wall and window leaks, spill on people who are approaching entry doors, cause settlement by washing the soil away from under foundations, subject basement walls to unnecessary groundwater pressure and possible leakage, and erode surrounding ground, often disfiguring landscaped areas.

For buildings and houses with flat roofs, leaders can be interior and concealed by the structure or partitions. Sloping roofs usually require gutters and leaders. Functionally, they can be omitted sometimes in low, basementless, one-story structures with wide overhanging roofs. A gravel-filled trench skirting the perimeter and directly below the edge of the eaves catches the water flowing off the roof. Some designers do not like the appearance of

Fig. 12.1 Gutters and leaders are not always essential, provided doorways, walls, foundations, and landscaped lawns are not subjected to rain concentrations.

conventional gutters and leaders. There are undoubtedly many ingenious ways of avoiding or modifying their use, provided proper drainage is effected (see Fig. 12.1).

Reasons for keeping paved areas clear of water are obvious. Puddles are avoided and contiguous areas of earth and grass are not subjected to erosion and a soft saturated condition. Areas to be drained—including roofs, balconies, terraces, and pavements—are usually connected into a storm drainage system and the water discharged to a stream, dry well, nearby gravel bed, recharge basin, or storm sewer.

12.4 Storm Water Disposal

The easiest way to dispose of storm water is to put it into the ground. This is possible under certain favorable conditions: with absorptive soil such as sand or sand and gravel, with a sparsely built community (it doesn't work well in crowded cities), and with sufficient dry ground to accept the water. The last item connotes a reasonably low groundwater table and the absence of flooded or swampy conditions.

A splash pan at the foot of each leader (Fig. 12.2*a*) is the simplest method. It will lead the water a few feet from the house but will accommodate only a relatively low rate of flow. A gravel-filled pipe is somewhat more effective (Fig. 12.2*b*). When the soil is not very permeable (for instance, clay), it is best to use a dry well with extended area and many perforations through which the water can be discharged to the ground.

Footing drains are often used to collect and lead away groundwater that accumulates around the foundations. This reduces the likelihood of basement wall leakage. These drains are most necessary when higher ground near the buildings increases the flow of groundwater against underground walls. Figure 12.3 illustrates this and also shows how storm water from drains and roofs may be led to a surface absorption area of rock and gravel beyond a head wall where the general storm drain outcrops. Where there is sufficient property area and slope to allow this method, it is sometimes chosen. It has the advantages of easy maintenance and service. Also, one can observe whether or not it is functioning correctly.

In new suburban developments, where storm sewers do not exist, recharge basins are

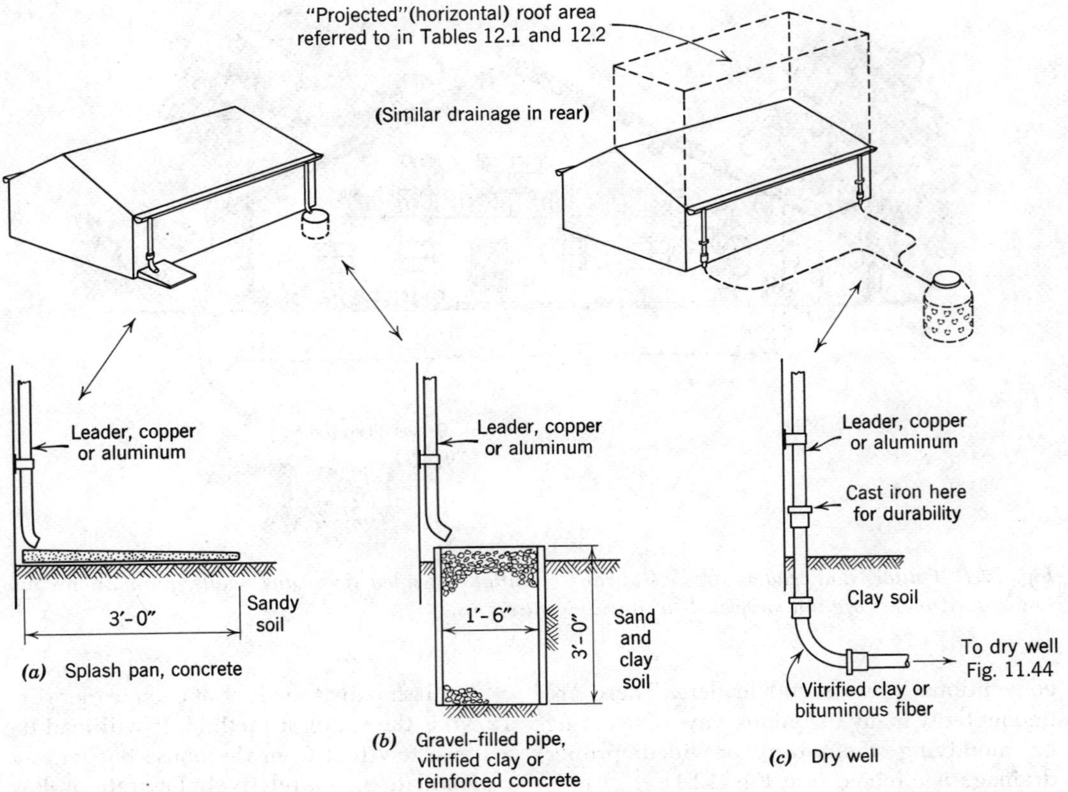

Fig. 12.2 Roof drainage for houses. Gutters and leaders are sized with the aid of Tables 12.1 and 12.2. Method (a) *is suitable for low rates of flow introduced into very pervious soil. When denser soil is encountered* (b) *is used to get the water into the ground and thus avoid surface erosion. For heavy flow or to lead the water further from the structure* (c) *may be used with one or several dry wells.*

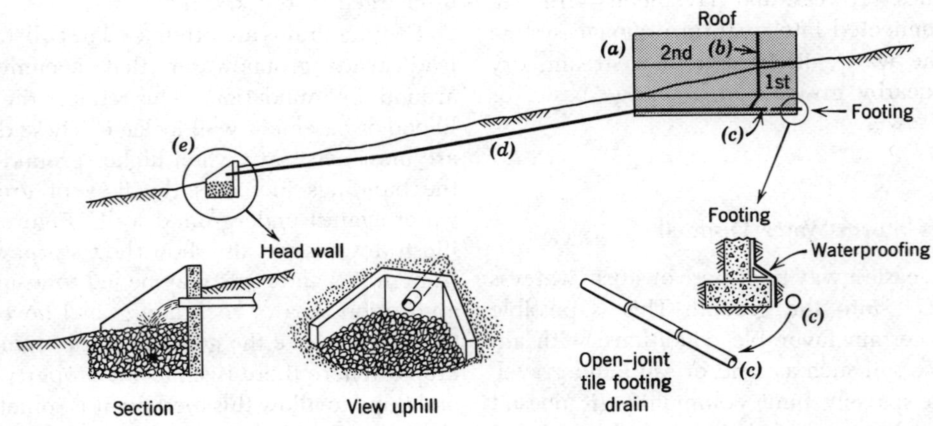

Fig. 12.3 Disposal of storm water on the site but remote from the house or building. When a wall is against a hill it is usually subjected to pressure of groundwater during storms. Open-joint clay tile accepts this water and carries it away. (a) *Position of building.* (b) *Roof drainage.* (c) *Footing drains.* (d) *Tight-joint clay tile or bituminous fiber pipe.* (e) *Flow through stone and gravel returns the water to the earth. Head wall is appropriate in lieu of a dry well if the site permits.*

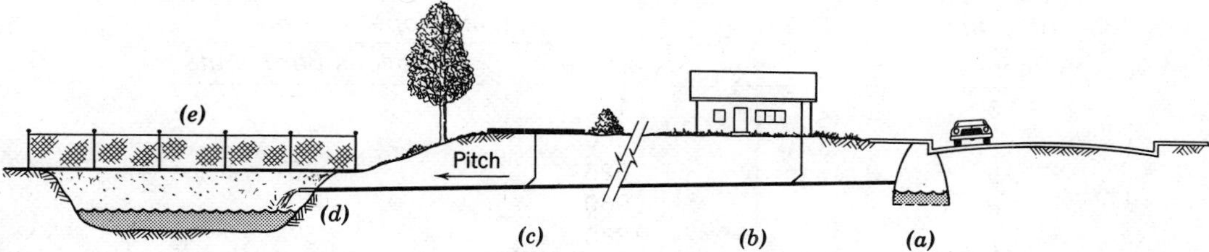

Fig. 12.4 Recharge basins in suburban communities. When topography, goundwater level, and porosity of the soil permit, developers are sometimes required to install systems that collect storm water and carry it from (a) *catch basins at street curbs,* (b) *the roofs of all houses, and* (c) *paved areas to* (d), *a recharge basin that receives the water and returns it to the ground. For the safety of children, a fence* (e) *must prevent unauthorized access to the basin.*

sometimes required to deliver storm water to the ground. Water from numerous roofs, paved areas, and curb catch basins is collected and piped to an open, unpaved pit where it sinks into the earth. This method is not recommended in areas of dense impervious clay soil (see Fig. 12.4).

The softness and other qualities of rainwater are often desirable in arid regions and where public supplies are hard and alkaline. Cisterns, as shown in Figure 2.21, can be used to filter and retain roof drainage for later use.

12.5 Gutters and Leaders for Houses

In selecting sizes for these drainage items, it is necessary to know the area of roof to be drained and the recorded maximum rate of rainfall in the geographic region. For houses with sloped roofs, the area as defined in Tables 12.1 and 12.2 is the *projected* area as shown in Figure 12.2. The maximum regional rate of rainfall for Tables 12.1 and 12.2 is for one selected value of 4 in./hr for a 5-min duration and a 10-year return period. This value is the approximate mean value of rainfall in the United States, calculated using the extreme range of 1.7 to 8.4 in./hr (see *Author's note,* Table 12.1). For the values of specific regions, one may consult Appendix A of the National Standard Plumbing Code, which lists the rates in about 200 cities. Adjustment can be made as stated in Tables 12.1 and 12.2 by using the ratio of the 4-in. rate specified in the table and the regional rate. For instance, a gutter for a fixed cross-sectional area in a region with an 8-in. rate will drain a roof area of 4/8 (or 1/2 the size of the table value).

EXAMPLE 12.1. Select a gutter and two leaders for the front half of a house, as shown in Figure 12.2. Rainfall rate is 4 in./hr for a 5-min duration and a 10-year return period. Projected roof area is 700 sq ft, and the slope of the gutter is $\frac{1}{16}$ in. in 1 ft of length.

SOLUTION. From Table 12.1, we find that a semicircular gutter with a 6-in. diameter will be adequate. It is chosen.

Since two leaders will be used, each will drain 350 sq ft. Table 12.2 shows that a 2-in. leader can be used. For this gutter-leader combination, we specify the detail of Figure 12.6*a*. Verifying the equivalent rectangular cross-sectional area, we have:

Gutter

Required:

$$\pi r^2 \times \frac{1}{2} \quad (1)$$

$$3.14 \times 3^2 \times \frac{1}{2} = 14.1 \text{ sq in.} \quad (2)$$

Rectangular section:

$$3.5 \times 4 = 14.0 \text{ sq in., OK} \quad (3)$$

Leader

Required:

$$\pi r^2 \quad (1)$$

$$3.14 \times 1^2 = 3.1 \text{ sq in.} \quad (2)$$

Rectangular section:

$$2 \times 3 = 6.0 \text{ sq in., OK} \quad (3)$$

(minimum obtainable size)

Table 12.1 ***Size of Roof Gutters***[a]

Diameter of Gutter[b]	*Maximum Projected Roof Area for Gutters* $\frac{1}{16}$*-in. Slope*[c]	
Inches	*Square Feet*	*Gallons per Minute*
3	170	7
4	360	15
5	625	26
6	960	40
7	1380	57
8	1990	83
10	3600	150

Source. National Standard Plumbing Code.

[a]This table is based on a maximum rate of rainfall of 4 in./hr for a 5-min duration and 10-yr return period. When maximum rates are more or less than 4 in./hr, the figures for drainage area shall be adjusted by multiplying by 4 and dividing by the local rate in inches per hour. See Appendix A.[d]

[b]Gutters other than semicircular may be used provided they have an equivalent cross-sectional area.

[c]Capacities given for slope of $\frac{1}{16}$ in./ft shall be used when designing for greater slopes.

[d]*Author's note.* The wide variation in inches per hour is indicated by the three examples listed below and selected from Appendix A of the Code (not reprinted in full here).

	Inches per Hour
Juneau, Alaska	1.7
New York, N.Y.	6.6
Mobile, Ala.	8.4

Rates given are intensities for a 5-min duration and 10-yr period, from Technical Paper No. 25, Rainfall Intensity-Duration-Frequency Curves, U.S. Department of Commerce, Weather Bureau.

Table 12.2 ***Size of Vertical Conductors and Leaders***[a]

Size of Leader or Conductor[b]	*Maximum Projected Roof Area*	
Inches	*Square Feet*	*Gallons per Minute*
2	544	23
$2\frac{1}{2}$	987	41
3	1,610	67
4	3,460	144
5	6,280	261
6	10,200	424
8	22,000	913

Source. National Standard Plumbing Code.

[a]This table is based on a maximum rate of rainfall of 4 in./hr and on the hydraulic capacities of vertical circular pipes flowing between one-third and one-half full at terminal velocity, computed by the method of NBS Mono. 31. Where maximum rates are more or less than 4 in./hr, the figures for drainage area shall be adjusted by multiplying by 4 and dividing by the local rate in inches per hour. See Appendix A.[c]

[b]The area of rectangular leaders shall be equivalent to that of the circular leader or conductor required. The ratio of width to depth of rectangular leaders shall not exceed 3 to 1.

[c]See *Author's note,* Table 12.1.

12.6 Storm Drain Piping in a Building

EXAMPLE 12.2. Select sizes for vertical conductors and horizontal storm drains for the building in Figure 12.5. Roof, balcony, and courtyard areas are as shown in Figure 12.5; rainfall rate is 4 in./hr (table value) and the pitch of horizontal drains is $\frac{1}{4}$-in. slope in 1 ft of run.

SOLUTION. Sizes selected and shown in Figure 12.5 may be verified in Tables 12.2 and 12.3.

12.7 Planning

Arrangements for draining roofs, balconies, and exterior-paved areas in a multistory, flat-roofed building are shown in Figure 12.5. Buildings, other than houses, that have *sloping* roofs often require special study. When the roof design results in dished areas (Fig. 12.7), storm water *must* be drained away at the low points of the roof surface. If this is not done, there might be one of two unfortunate results. Accumulated water could, in a high wind, be blown off in great sheets to the discomfort and injury of nearby pedestrians. Much more serious than this, overloading of the roof structure could result. Water 2 ft deep could impose a load of 124.8 lb/sq ft of roof surface. In buildings that could, in certain areas, be designed for a roof live-load of 40 lb/sq ft, this would constitute a load of three times the design value. Structural deformation or failure might result.

In single-slope roofs, the low edge, of course, creates the line at which a gutter can collect storm water and route it away through vertical drains. Solutions are shown in Figures 12.8, 12.9, 12.10, and 12.11.

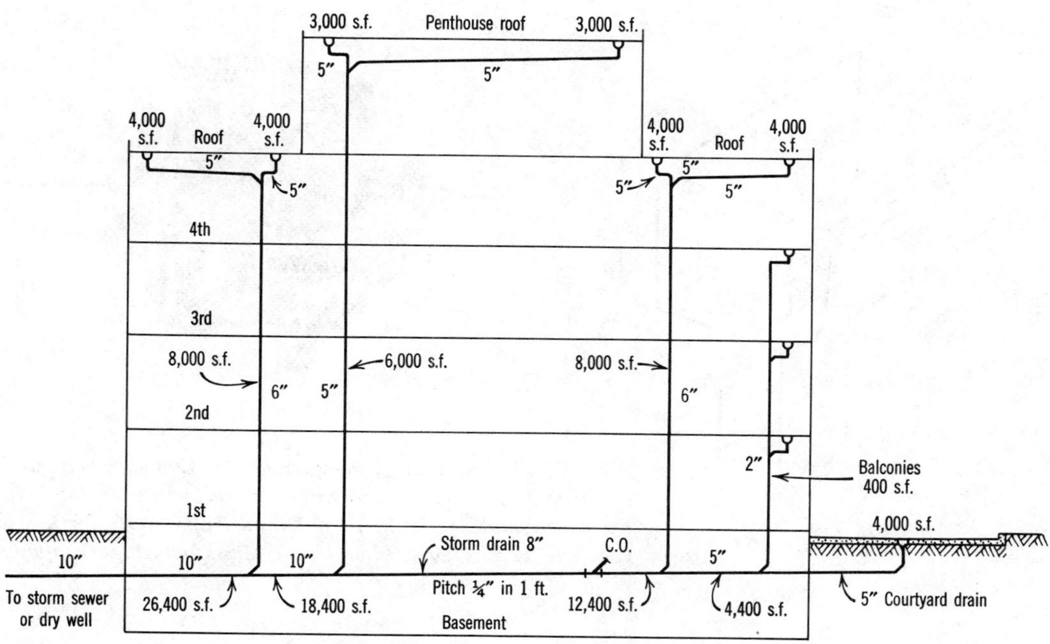

Fig. 12.5 *(Example 12.2) Separate storm drainage. Areas drained and corresponding sizes of vertical leaders and horizontal drains from Tables 12.2 and 12.3. Storm drain piping within a building needs insulative covering with a vapor barrier on the outside. This prevents condensation (sweating) on the pipes when, in winter, warm moisture-laden air in the building could otherwise reach the pipe surface, cold from carrying icy water within, condense there, and cause wet, dripping conditions on the pipes.*

Table 12.3 *Size of Horizontal Storm Drains*[a]

Diameter of Drain, Inches	Maximum Projected Area for Drains of Various Slopes					
	$\frac{1}{8}$-in. Slope		$\frac{1}{4}$-in. Slope		$\frac{1}{2}$-in. Slope	
	Square Feet	Gallons per Minute	Square Feet	Gallons per Minute	Square Feet	Gallons per Minute
3	822	34	1,160	48	1,644	68
4	1,880	78	2,650	110	3,760	156
5	3,340	139	4,720	196	6,680	278
6	5,350	222	7,550	314	10,700	445
8	11,500	478	16,300	677	23,000	956
10	20,700	860	29,200	1,214	41,400	1,721
12	33,300	1,384	47,000	1,953	66,600	2,768
15	59,500	2,473	84,000	3,491	119,000	4,946

Source. National Standard Plumbing Code.

[a]This table is based upon a maximum rate of rainfall of 4 in./hr. Where maximum rates are more or less than 4 in./hr, the figures for drainage area shall be adjusted by multiplying by 4 and dividing by the local rate in inches per hour. See Appendix A.[b]

[b]See *Author's note,* Table 12.1.

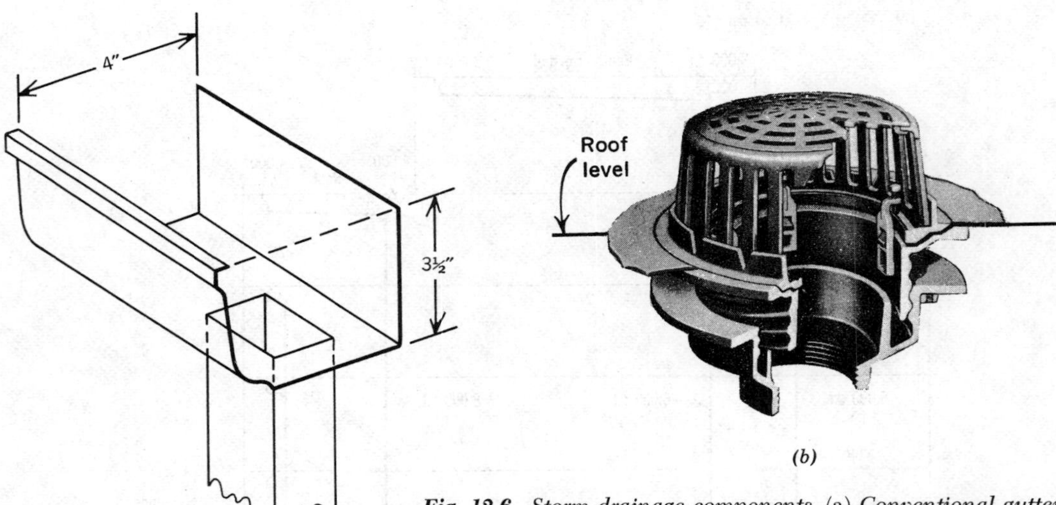

Fig. 12.6 *Storm drainage components.* (a) *Conventional gutter and leader for houses. Sizes vary by manufacturer. Suitable for use in Example 12.1.* (b) *Roof Drain. Appropriate for use in Example 12.2, Figure 12.5. Josam Manufacturing Company.*

Fig. 12.8 (opposite) *Hayward Field West Grandstand at the Eugene Campus of the University of Oregon. Built in 1976. The Amundson Associates, Architects, Springfield, Oregon. It is the track and field facility for the University. In addition to the architectural and esthetic reasons for sloping the roof to the rear of this attractive and functional structure, the design allows for the collection of storm water. The water collects at a horizontal gutter at the rear roof edge and drains through six leaders.*

Fig. 12.7 *Airport terminal, roof drainage. Adjacent to the pilasters that support radial roof girders in this circular building, one sees roof drains that carry storm water to a storm sewer. The water must, of course, be collected at low points. The roof is an* almost *flat "hat form." The cantilevered brim (right) and the crown (left) have slight downward slopes that meet at the low points above the pilasters.*

Fig. 12.9 Portion of the roof gutter and four of the six leaders at the Hayward Field West Grandstand. Storm drainage from all downspouts is collected below ground and is led to the street where it empties into an 18-in. storm sewer.

Fig. 12.10 Collection of storm water at the low *cantilevered edge of a sloping roof with its integral glazed roof-light panel. This shed affords weather protection over the parking area of a motel. Water that collects at the overhanging edge-gutter of the shed is led back through gooseneck branch storm drains. There it flows down through dual-purpose hollow steel pipes. They double as structural columns and storm leaders. Each of these downspouts serves about four panels of roof area.*

Fig. 12.11 Close-up of a "gooseneck" horizontal storm branch. It joins the hollow steel structural column that supports the steel girder. The gutter that is drained by the branch is integral with the edge of the overhanging roof. Below grade and above the footing, the water is diverted laterally through a subgrade branch to a remote dry well.

References and Bibliography

1. *Architectural Graphic Standards,* 6th ed., Ramsey and Sleeper, John Wiley & Sons, New York, pages 102–105.

PART FOUR

fire safety

13

Fire Protection

13.1 Planning

Among the many responsibilities placed upon the architect by society, and more particularly by his license to practice, is the protection against loss of life by fire. The preservation of the structure and its contents against fire damage or destruction, though secondary in importance is, nonetheless, of serious concern to owners and others having a financial interest in property. The experience of municipal fire departments, volunteer fire companies, and of the insurance companies who pay for loss of life and property is an invaluable aid to designers of new buildings. The National Fire Protection Association, of which the American Institute of Architects is one of the organization members, maintains a comprehensive and constantly improved set of standards for use in planning to reduce fire hazards and in designing equipment for fire fighting. Acceptance of their recommendations by the writers of building codes and by practicing architects has done much to reduce fire losses.

Although the inclusion of fire-fighting equipment in buildings is always desirable and increasingly mandatory, good practice begins with the design of the structure. This is necessarily affected by the permitted building density in the locality and by the flammability of the building and its expected contents. Some of the materials and equipment that must be considered are:

a. Fire resistance of the selected structural type and material and the contents of the building.
b. Limitations of volumes within fire-resistant barriers in otherwise vulnerable buildings.
c. Precautions against perforations of approved and required barriers.
d. Exits and fire tower stairs.
e. Protection against fires caused by defective electrical systems.
f. Protection against fires caused by lightning.
g. Detection and alarm systems.
h. Standpipe and hose systems within and near buildings.
i. Automatic sprinkler systems.
j. Automatic smoke and heat venting.
k. Smoke and heat shafts.
l. Control of air-conditioning ducts.
m. Communication in high-rise buildings.
n. Elevator control.
o. Fire command station in high-rise buildings.

Of the above 15 items, those that relate to electricity, signal systems, and elevators will be found in Parts 5, 7, and 8, respectively (they include items e, f, g, m, n, and o).

13.2 High-Rise Fire Safety

In terms of the rate of actual *fatalities* due to fire, high-rise structures have not been the nation's greatest killer. Yet, as the life of any one human being is of infinite value, it has been our challenge in the past decade to assess in retrospect our earlier performance in the design of these structures. Beginning about 1970, the many unexpected high-rise fires have made this challenge inescapable. Large cities responded promptly to the need for research. In one city, the New York City Rand Institute[1] published in September 1975, *Fire Protection and Local Government: an Evaluation of Policy—Related Research.* By permission of the parent organization and publisher of the report, The Rand Corporation of Santa Monica, California, we present an excerpt from the 268-page document. It describes the fire incidents that are the cause of our concern and some of the inadequacies that caused or aggravated the fires.

HIGH-RISE FIRE SAFETY

High-rise buildings present a variety of unique problems, and it should not be surprising that fire safety is one of them. A number of recent fires in tall buildings, some quite spectacular, have caught the attention of the public and of public officials. With that attention came concern: A large percentage of urban residents spend much of their time in tall apartment houses, offices, hotels, hospitals, and schools. More recently, the suburbs have increasingly experienced the high-rise building, and their fire departments are neither trained nor equipped for operations in tall buildings.

The real problem of high-rise fire safety must be put into perspective. On the one hand, high-rise fatalities account for a miniscule fraction of the 12,000 fire deaths per year in the United States. On the other hand, the potential for major disaster is staggering when one considers that the estimated 130,000 occupants and visitors in one day at New York City's World Trade Center exceeds the total population of Albany, New York. A number of psychological factors are at work also. The thought that these imposing creations of modern technology are not completely immune to fatal fires is difficult for people to accept.

To some extent, a useful way to consider the high-rise problem is to *define* a high-rise building in terms of fire protection. Such a definition evolved at the May 1971 International Conference on Fire Safety in High-Rise Buildings (U.S. General Services Administration, 1971a), which characterized a high-rise building as one that was:

1. *Too tall to be completely accessible to fire-fighting equipment from the ground.* This problem occurs anywhere from 6 to 12 stories, but usually 8 to 10 stories, depending upon the reach of aerial ladder equipment available. It means that the fire must be fought internally above these levels.
2. *Too tall to make complete evacuation of occupants feasible.* This occurs at about 25 stories, where complete evacuation would take five minutes theoretically and about 50 to 150 percent longer than this in reality. It means that provision for safety and life support for the occupants must be made within the building.
3. *Tall enough to make possible chimney or stack effects for air and smoke flow.* This is to some extent true in all buildings, but it becomes especially pronounced in buildings about 8 stories high. This means that provision must be made for the control of air flow and products of combustion.

Modern design and construction methods have been accused of increasing the fire hazards in high-rise buildings since World War II, largely because of the feeling that there were fewer fires in the first high-rise structures built between the two world wars. Whether they are responsible for increased hazards, modern design and construction have certainly given rise to changes in the types of fire problems encountered:

- Light construction methods do not offer the same inherent protection that older plaster or concrete provided.
- Nonintegrally constructed floors and walls may separate under heat distortion, providing paths for fire and smoke.
- False ceilings containing electrical and various other building services provide

[1]The New York City Rand Institute is a nonprofit research institution founded, as its Articles of Incorporation state, ". . . primarily to conduct programs of scientific research and study, and provide reports and recommendations, relevant to the operations, planning or administration of the City of New York." The Institute was established in 1969 by the City of New York and The Rand Corporation as a center for the continuing application of scientific and analytic techniques to problems of urban life and local government. It is governed by a Board of Trustees appointed jointly by New York City and The Rand Corporation.

hidden locations where fires can start and propagate.

- Punch-through installations of telephone and other services provide holes that violate designed structural integrity.
- Central air conditioning systems can act as a passage and distribution system for smoke.
- Use of plastics for trim and covering of interior structure creates the potential for smoke and toxic fumes.
- Furniture styling and materials change the nature of the fire load in terms of ignitability, fuel, and smoke potential.

The Cause of the Concern: The Recent Fires

The flurry of activity regarding the high-rise fire problem was, to some extent, instigated by two fatal fires in New York City office buildings; both were new buildings, not yet fully occupied.

One New York Plaza (*New York*). In August 1970, a fire originating in a concealed ceiling on the 33rd floor of this 50-story office building raged for five hours before being brought under control; by then, it had killed two persons, injured 30 more, and caused about $10 million damage. The contents of the 33rd floor were completely demolished by the 1500° F heat; the fire spread to the 34th floor by means of various service openings through the floor, and then to the 35th floor through air-conditioning duct openings. Structural beams were bent on two of the floors; the sprayed asbestos had fallen off in some places and been removed in others to accommodate partitions and services. Smoke spread to most floors through the air return fans—the supply fans were shut down by smoke detectors—with concentration sufficient to require evacuation. Two men died and a third was severely injured when the doors of the elevator they had taken to the fire floor would not close again; a group of firefighters were trapped until they chopped their way out of the elevator they were taking to the fire floor but that stalled at the 4th floor.

919 Third Avenue (*New York*). In December 1970, a fire that originated from cutting torch sparks on the 5th floor of this 47-story office and display building—the Carpet Center Building—burned for 35 minutes before being brought under control; by then, it had killed three persons, injured 20 more, and caused about $2-1/2 million damage. The contents of the 7,500 square-foot compartment in which the fire originated were destroyed, and beams were distorted because of the failure of sprayed asbestos protection. Employees using an elevator to descend were rescued by firemen when their elevator stopped at the fire floor; three workmen died when their elevator also stopped at the fire floor somewhat later. Smoke spread quickly to the 4th floor, to the 11th floor air-conditioning outlets, up to the 8th floor in the stair tower, into the central core above the 22nd floor, and again up to the 33rd floor in the stairwell.

These fires resulted in investigations, tests, and code revisions for tall office buildings in New York. In Chicago, however the effort concentrated on, but was not limited to, residential high-rise fires: Two of its most serious high-rise fires occurred in the Hawthorne House and in the Conrad Hilton Hotel.

Hawthorne House (*Chicago*). In January 1969, a fire originating in the living room of a 36th floor apartment of this 39-story apartment building took four lives, destroyed the kitchen and living room of that apartment, and damaged several other apartments on the floor. Firefighters were first hindered in reaching the fire by evacuees using the elevators to escape, later by a "blowtorch" effect (wind entering the broken windows in the fire apartment caused the corridor to fill with flames, intense heat, and smoke).

Conrad Hilton Hotel (*Chicago*). In January 1970, a fire that started in a pile of chairs in the 9th floor elevator lobby of this 25-story hotel—its 2,600 rooms make it one of the world's largest—took two lives and injured 36 persons; the dead and most of the injured were deaf and unable to hear alarms and instructions. About $100,000 damage occurred, mostly in the corridors.

More spectacular and frightening in their impact were recent fires in tall buildings in other parts of the world, the most notable being two fires in buildings in São Paulo, Brazil.

Andraus Building (*São Paulo*). In February 1972, a fire originating in a storage well outside the 4th floor of this 31-story department store and office building spread through four floors internally and through another 24 floors externally; most of the fuel load was consumed in 90 minutes, but the fire's extinguishment continued throughout the night. Sixteen persons died, over 375 others were injured, and over $2 million damage was incurred, including damage to nonadjacent buildings by radiation and wind-driven flames. Helicopters rescued 350 persons from a rooftop heliport, while 200 persons were trapped but safe in a stairwell. LP gas throughout the building was involved in the fire.

Joelma (Crefisul) Building (São Paulo). In February 1974, a fire started in a window air conditioner on the 12th floor of this 25-story office building and raged for 1 hour and 40 minutes before burning itself out; of the 601 employees in the tower at the time the fire started, 179 died, 300 more were injured, and damage was over $3 million, including all contents of the offices from the 12th floor up. Of the dead, 90 died on the roof waiting to be rescued by helicopters that could not land, and 40 jumped to their death, 30 of these after the fire was out. In this incident the only stairway was unenclosed, there were large amounts of combustible materials, and fire protection facilities were lacking. Early reports indicated the presence of large amounts of LP gas distributed throughout the building, but later reports made no mention of this.

In both of these fires, the structural features, lack of fire protection, and presence of LP gas and other combustibles are unlike situations in the United States. This is also true of the fire in the 13-story Taiyo Department Store in Kumamoto, Japan in November 1973, in which 103 persons died and 119 were injured; and the fire in the 21-story Tae Yon Kak Hotel, in Seoul, Korea in December 1971, in which 163 died and 60 were injured.

13.3 High-Rise Tests and the Law

In the City of New York, the Mayor's Committee sponsored actual fire-safety tests. These tests were a guide to a new city fire law, Local Law No. 5. By permission of Syska & Hennessy, Engineers, their *Technical Letter* is reprinted here in full. It is one of a series of publications for the guidance of architects, planners, and building owners issued periodically by their firm of consulting engineers. This letter reviews the high-rise building tests and the law.

A REVIEW OF THE NEW YORK CITY FIRE SAFETY TESTS & LAW

Fire in high rise buildings has been a widely discussed topic among design professionals, owners, and in the press ever since a series of fires occurred in buildings previously considered "fire safe." To circulate some basic information on the subject, Syska & Hennessy published an issue of the Technical Letter in September 1970 entitled "Mechanical and Electrical Aspects of Fire Protection in Large Buildings." This Letter discussed the stages in the development of a fire, the building systems which are vital during a fire, the problems of partial occupancy, fire prevention, building code requirements, possible additional safety measures, and possible future developments.

The urgency of the problem generated a number of investigative groups, and a second Technical Letter, in January 1972, discussed the appointment of the Mayor's Committee on Fire Safety in High Rise Office Buildings, in New York City in February 1971; the Federal Government General Services Administration Conference in April 1971; and the City of Chicago Conference in May 1971. After a brief review of the two conferences, the Letter detailed the case history of the Mayor's Committee and summarized the recommendations included in its report of November 15, 1971.

This Technical Letter reviews the results of tests under actual fire conditions sponsored by the New York City Mayor's Committee, and summarizes the provisions of the "Fire Safety Requirements and Controls" law which resulted from the Committee's investigations.

Source. Syska & Hennessy, *Technical Letter,* Vol. 23, No. 1, May 1973.

New York Fire Tests

Since a number of questions concerning the feasibility of pressurizing stairwells was unresolved, the Committee arranged for a testing program under actual fire conditions, utilizing the Port of New York Authority's 22 story office building at 30 Church Street which was scheduled for demolition. The Polytechnic Institute of Brooklyn conducted the tests under contract with the City. The Insurance Industry provided additional financial support and advice; the National Bureau of Standards furnished instrumentation and manpower; and industry furnished equipment. This remarkable cooperative effort was organized, financed, and completed in the short period of four months.

The tests were directed primarily to the feasibility of pressurizing stairs in high rise office buildings as a means of maintaining smoke free conditions to permit safe evacuation of occupants and as an aid to fire fighters. A few measurements were also made with regard to the exhausting of smoke and other products of combustion from the fire area, using a stairwell to simulate a smoke shaft; and observations were made in connection with the spread of smoke and heat through ducts and ceiling plenums.

Preliminary tests were conducted with supply and exhaust units operating. These were followed by more tests using cold smoke. Finally four tests were run at three different locations on the 7th and 10th floors using smoke and heat generated by large scale fires. Temperature, pressure, oxygen, carbon monoxide, and smoke measurements were made.

The report of these tests[a] concludes that:

- Pressurization of stairwells is feasible.
- Smoke exhaust systems can remove smoke and heat from fire areas without excess temperatures in exhaust shaft.
- Maximum temperatures and pressures can be reached within four minutes of ignition.
- Maintenance of air supply and return can increase temperature, pressure, rate and extent of fire growth, as well as spread of smoke, gases, and heat.
- Fire-fighter controlled means of ventilation is essential.
- Rapid spreading of toxic concentrations of gases can take place through ducts and ceiling plenums.
- Attention must be given to fire stopping of and elimination of combustibles from plenums.
- Smoke and/or other products of combustion will activate elevator electronic touch buttons (see Syska & Hennessy publication "Elevator Operation Under Fire Conditions").

As a result of these tests, continued discussions within the Committee, and subsequent City Council Committee hearings, the original Committee recommendations were modified and Local Law No. 5, "Fire Safety Requirements and Controls"[b] was passed by the City Council on December 22, 1972, and approved by the Mayor on January 18, 1973. A summary of the law's provisions follows:

[a]Copies of the report, entitled "Fire Tests, Analyses and Evaluation of Stair Pressurization and Exhaust in High-Rise Office Buildings," are available from Professor Paul R. DeCicco, Polytechnic Institute of Brooklyn, 333 Jay Street, Brooklyn, New York 11201.

[b]Available from The City Record, 31 Chambers Street, New York, New York 10007.

Summary of New York Fire Safety Law

For All Office Buildings Occupied by 100 Persons Above or Below Street Level or More Than a Total of 500 Persons in the Entire Building:

Fire Safety Plan

1. Plan for fire drill and evacuation procedures, including appointment of Fire Safety Director, fire brigade and floor wardens.
2. Signs required at elevators and stairs.

For Buildings 100 Feet or More in Height:

Fire Stair Doors

Doors to interior stairs shall not be locked except:

1. On street floor to prevent access to stairs.
2. On stair side if every four stories or less, door is openable.
3. Where failsafe electric door lock, activated by any detector, is provided.

Class E Fire Alarm System

1. Combination fire alarm signal system

consisting of interior fire alarm and two-way voice communications system (direct wire, radio, or carrier current), with at least one station per floor. Activation of alarm station will identify its floor location at the Fire Command Station, mechanical control center, and the regularly assigned location of the fire safety director, sound alarm, and automatically notify Fire Department.

2. Fire Command Station, located in lobby of building containing the following:
   a. Communication Capability (two-way voice) between Fire Command Station and:
      - Floor Warden Stations
      - Mechanical Control Center
      - Elevators and Elevator Machine Rooms
      - Air Handling Control Rooms
   b. Audible Alarm Signal and means to silence.
   c. Visual display system to indicate activation location by floor.
   d. Means to control sounding devices on any floor or throughout building.
   e. Means to transmit Fire Alarm to Fire Department.
   f. Display lamps indicating on/off of air handling systems, unless provided in Mechanical Control Center.
   g. Means to test display lamps, alarms and connection to Fire Department.
   h. Individual manual operation of each air supply and each return or exhaust system unless provided in the Mechanical Control Center and in the room containing the affected fans.

Detectors

1. On each Elevator Landing. Activation shall stop all fans, activate smoke exhaust or stair pressurization fans, return affected elevators to terminal landing, and automatically sound alarm and notify Fire Department.
2. At the return air shafts at each floor. Activation same as above except for elevators.

Compartmentation

(Applies to floors located more than 40 feet above grade served by multi-floor air conditioning system.)

1. Areas to be compartmented by one hour separations, into spaces not to exceed 7500 square feet.
2. Sub-division of area may be increased to 15,000 square feet, provided such area is completely protected with space detectors.
3. Where total area exceeds 10,000 square feet, at least one fire separation shall be two hour rated and create areas of refuge.
4. In existing office buildings, existing one hour separations may be accepted in lieu of the two hour separation.

Smoke and Heat Control

(Applies to buildings served by multi-floor air conditioning system.)

1. Smoke shaft to exhaust heat and smoke to outdoors.
2. Existing buildings may provide pressurization of all interior enclosed stairs (except fire tower) in lieu of smoke shaft.
3. Systems to be activated automatically upon activation of any detectors.

Elevators

1. Elevators serving fire floor to be recalled and retained at terminal floor by activation of landing detectors or Command Station Control.
2. Three elevators servicing each floor shall be equipped with firemen's service.
3. Interlock wiring for elevator hoistway doors will be required to be resistant to high temperatures.

Sprinklers

1. Show rooms exceeding 7500 square feet, located more than 40 feet above grade, shall be sprinklered. Use of domestic water permitted.
2. Sprinklers may be designed in accordance with new NFPA standards (Chapter 8 of NFPA No. 13) which permits Hydraulic pipe sizing. Use of Fire Stand Pipe is permitted.
3. Special provisions permit the following exemptions, when floor is completely sprinklered.
   - Compartmentation.
   - Smoke shaft and stair pressurization.

• Detectors on each elevator landing.

Group E Buildings, Less Than 100 Feet in Height:

1. Two-way voice Communication System, from Fire Command Station to:
   • Warden Station each floor.
   • Mechanical Control Center.
2. Means to transmit fire alarm to Fire Department.

Extensions of the Law

Although the law was limited to office buildings, many of its findings are applicable to other types of high rise buildings. The Mayor has ordered that applicable measures be extended to hotels, apartment buildings, schools and universities, and hospitals, as examples. The Committee is now studying appropriate legislation. Owners of buildings not covered by the law should seriously consider implementing relevant recommendations at this time in order to provide appropriate fire protection for their occupants regardless of present legal requirements.

Implementation of the law should result in a significant reduction in the severity of fires and consequently in fewer injuries and deaths and less property damage. How rapidly this is achieved will depend upon the cooperation of the owners of existing buildings. The law provides for periods of three to fifteen years for compliance, depending upon the specific requirements.

Although prompt action is important, owners of existing buildings should proceed carefully since necessary alteration requirements depend to a great degree on the systems now in each building. The law provides that alternate systems be accepted if judged equivalent to those prescribed by law. Thus, each existing building is an individual case requiring an individual engineering study to ensure that the owner gets the maximum degree of protection for minimum cost and minimum disturbance to tenants.

Extra copies of this and other Technical Letters are available. Written permission must be obtained to reproduce any part of this Technical Letter. Please direct all correspondence to Cathy Ramsey, Syska & Hennessy, Inc., 110 West 50 Street, New York, New York 10020.

13.4 Owner/Builder Reaction

New York City's Local Law No. 5, "Fire Safety Requirements and Controls" was signed by the mayor in January of 1973. Applying essentially to high-rise buildings, it was a response to the problems posed by the frightening, fatal, and destructive fires of the early 1970s (see the report of the New York City Rand Institute, and Syska & Hennessy's, *Technical Letter,* both cited above). In spite of the wish of owners and builders to conform in the interest of life safety, the tremendous cost of changes in existing buildings promises to make full adherence to the law a lengthy process.

Concerning *new* buildings, *The New York Times,* February 25, 1973, quotes the opinion of an architect:

> "I'd say that probably every high-rise office building now being designed in the city will have the sprinklers," commented Richard Roth Jr., an architect active in the office field. "It's just not practical to plan a building with the floor-space requirement in the new code."

The contention, of course, is the option stated in Item 3 of the "Sprinklers" section of the law (page 488) permitting the exemption of compartmentation, smoke shaft, and stair pressurization and detectors on each elevator landing when a floor is completely sprinklered. For a long time now, we have seen the increase of open "landscape"-type office floors with few or no partitions. It would seem likely that the choice might often be for sprinklers that would obviate "compartments" enclosed by fire walls and self-closing fire doors.

The adaptation of *existing* buildings to the law would undoubtedly require that the cost of changes be passed on in amortized fashion to tenants. In the period of transition, the time allowed for adherence, which varies from 3 to 15 years depending on the item, is a reasonable demand. There are over 800 such existing buildings in New York City that do not comply. For more than half of them, working drawings and applications have already been filed to begin some of the changes.

Some sources have estimated the cost of sprinklering existing buildings to be about $1 per square foot of floor area. As an example, the cost of installing sprinklers in the Pan Am Building (Fig. 10.6, page 389) could be $2.4 million. Estimates of the cost of bringing *all* of the high-rise buildings in New York City into conformity with the law range upwards to $500 million.

The law, opposed by the commercial real estate industry as unconstitutional and unnecessary, was reaffirmed in December 1978 by the Court of Appeals, New York State's highest court.

13.5 Water and Standby Power

It is no secret that water resources in the United States are in short supply and may remain so for some time. It is one thing to postpone the washing of a car until tomorrow and another when a building fire needs thousands of gallons of water in fifteen or twenty minutes. Fire insurance underwriters constantly remind us of the importance of the need for sufficient pressure, adequate supply, and dependability. So the terms "secondary supply," "reserve supply," and "dual sources" become familiar terms. A very large industrial plant may settle in a suburb or in a small community where the local water company cannot entirely cope. In such cases, a reserve underground storage of several hundred thousands of gallons or an adjacent artificial *lake* would not be unusual.

When such reserves must be delivered by pumping during a fire, an electrical power outage could be a tragedy; then standby power must cut in. Diesel-powered pumps take over. These units and their fuel storage should be separately housed in fire-resistant enclosures, remote from the possible locations of fires in the main buildings.

(*a*) ***Standpipes and Hoses.*** Fire companies with their apparatus find difficulty in fighting fires from the street in tall structures. Standpipes and hoses with a separate water reserve or upfeed pumping are extremely valuable in *any* building but become highly essential in tall buildings. Figure 13.1 shows such a system, which is intended for use by building personnel until the fire engines arrive and thereafter by the trained staff of the fire department. It is not practical to store enough water on the roof for a protracted fire-fighting period and it is usually assumed that a half hour's supply will be more than enough to provide for the short period it takes the fire engines to arrive. When the system is used by the fire department, its pumps are attached to the street siamese to deliver water from street hydrants or the building's "secondary source" (see Fig. 13.2). The check valve closest to the siamese in use opens and the check valves at the tank close to prevent the water from rising in the tank to no avail. After the engines are disconnected from the siamese, the water between the siamese and the adjacent check valve drains out through the ball drip so that it does not freeze.

The overhead tank is considered a most dependable source, but it sometimes requires a height that is architecturally undesirable. In this case upfeed fire pumps operating automatically to deliver water to higher stories from lower suction reserve tanks may be used. Another alternate in this case is a pneumatic tank used to deliver water by the power of the air that is compressed in the upper portion of the tank.

The water zones as shown in Figure 10.4*a* and *b*, page 385, are also generally followed in planning for fire protection. A *fire* standpipe zone will usually coincide with the service water zone. Fire standpipes and their hoses are usually located at or near fire stairs from which personnel or fire fighters can approach a fire (see Fig. 13.7).

(*b*) ***Sprinklers.*** Automatic sprinkler systems consist of a horizontal pattern of pipes placed near the ceilings of industrial buildings, warehouses, stores, theaters, and other structures where the fire hazard requires their use. These pipes are provided with outlets and sprinkler heads so constructed that temperatures of 135 to 160° F will cause them to open automatically and emit a series of fine water sprays (see Fig. 13.4).

Sprinkler systems include (1) wet pipe, ordinarily with water constantly filling both the

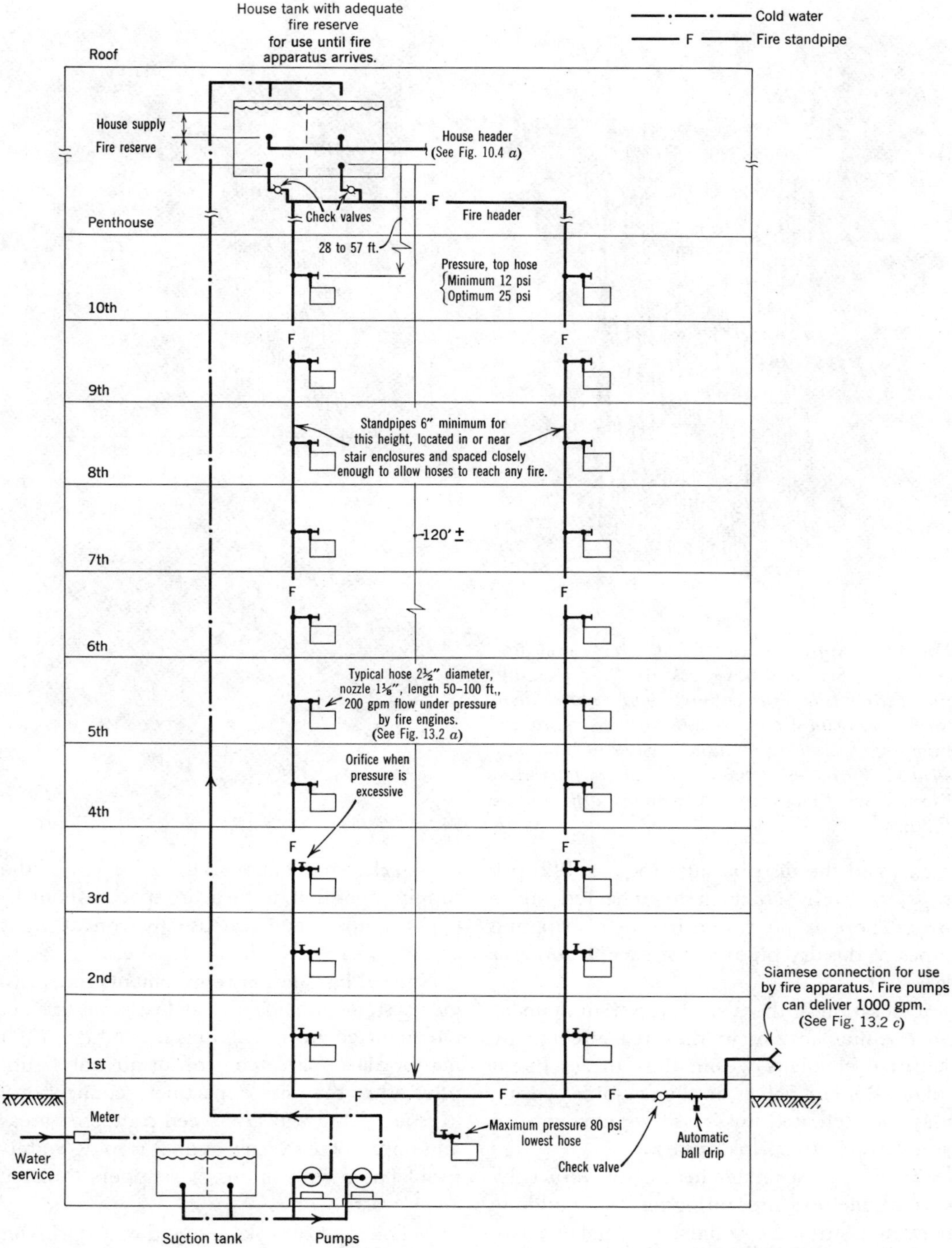

Fig. 13.1 Fire protection. Schematic section; part of a standpipe and hose system. Gravity tank downfeeds to hoses for use by building personnel. Fire department uses siamese connection upon arrival to pressurize the standpipe system. Fire fighters can then approach the floor level of the fire through fire stairs. There they can use the standpipe hoses. In some codes and under certain conditions, sprinklers may make use of the fire standpipe system. See New York City Local Law No. 5, "Sprinkler" section, Item 2, page 488. Also note the reference in the Transamerica story to the "combined wet standpipe—sprinkler risers," page 498.

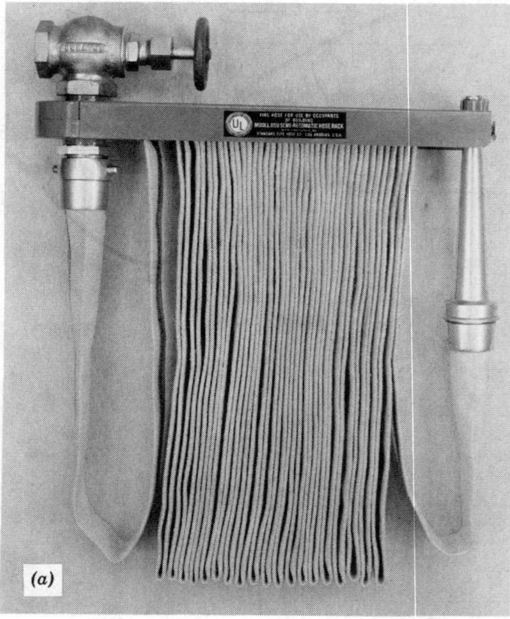

(a)

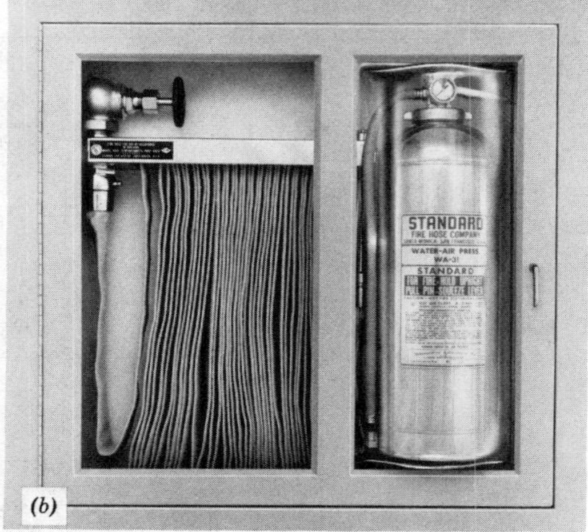

(b)

(c)

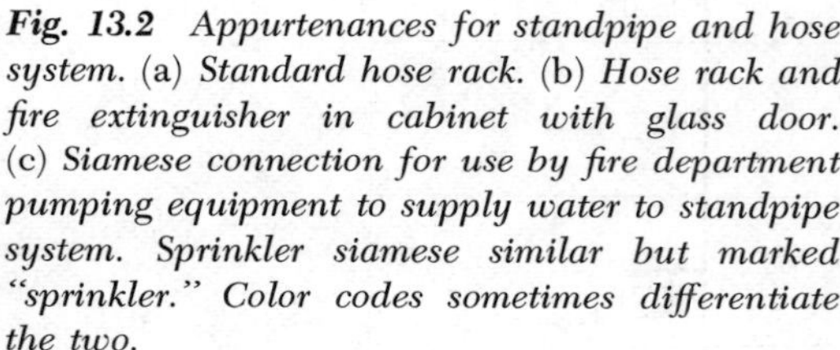

***Fig. 13.2** Appurtenances for standpipe and hose system.* (a) *Standard hose rack.* (b) *Hose rack and fire extinguisher in cabinet with glass door.* (c) *Siamese connection for use by fire department pumping equipment to supply water to standpipe system. Sprinkler siamese similar but marked "sprinkler." Color codes sometimes differentiate the two.*

mains and the distributing pipes, and (2) dry pipe, generally confined to unheated buildings. There is no water in the distributing pipes of the dry-pipe system except during a fire.

Operation of the wet-pipe system depends on opening nozzles in the area affected by sensitive elements within the nozzles themselves. Remote valves, in the dry-pipe system, may be actuated by sensitive elements to admit water to sprinkler heads.

Spacing of sprinkler heads is governed by several factors: fire rating of the building, construction of the ceiling, spacing of joists, type of occupancy, and total area (see Table 13.1 for the relative fire hazard for various occupancies as related to sprinkler installations). The coverage of one sprinkler head varies from about 200 sq ft per sprinkler for light hazard occupancy to about 90 sq ft for extra hazard conditions.

Nozzles are set about 8 to 12 ft apart on the supply pipes that, in turn, are spaced about 10 to 14 ft apart and are usually run at right angles to exposed beams or panels.

Special installation requirements for sprinkler systems include: (1) at least one fire department connection on each frontage; (2) a master alarm valve control for all water supplies other than fire department connections; (3) special fire walls between protected areas and unprotected areas; and (4) sloping waterproof floors with drains or scuppers to carry away waste water.

When gravity tanks are used with sprinkler systems they should provide an adequate reserve for this purpose and, in any case, enough to operate 25% of the sprinkler heads for 20 min. As in the case of standpipe and hose systems, this gives the fire company a chance to arrive and take over.

A typical automatic sprinkler system is

Table 13.1 ***Relative Fire Hazard for Various Occupancies, as Related to Sprinkler Installations***

1-7 Classification of Occupancies

1-7.1 Occupancy classifications for this standard relate to sprinkler installations and their water supplies only. They are not intended to be a general classification of occupancy hazards.

1-7.2 Light Hazard Occupancies

1-7.2.1 Light Hazard—Occupancies or portions of other occupancies where the quantity and/or combustibility of contents is low and fires with relatively low rates of heat release are expected.

Light Hazard Occupancies include occupancies such as:

Churches
Clubs
Educational
Hospitals
Institutional
Libraries, except large stack rooms
Museums
Nursing or Convalescent Homes
Office, including Data Processing
Residential
Restaurant seating areas
Theaters and Auditoriums excluding stages and prosceniums

1-7.3 Ordinary Hazard Occupancies

1-7.3.1 Ordinary Hazard (Group 1)—Occupancies or portions of other occupancies where combustibility is low, quantity of combustibles is moderate, stock piles of combustibles do not exceed eight feet and fires with moderate rates of heat release are expected.

Ordinary Hazard Occupancies (Group 1) include occupancies such as:

Automobile Parking Garages
Bakeries
Beverage Manufacturing
Canneries
Dairy Products Manufacturing and Processing
Electronic Plants
Glass and Glass Products Manufacturing
Laundries

1-7.3.2 Ordinary Hazard (Group 2)—Occupancies or portions of other occupancies where quantity and combustibility of contents is moderate, stock piles do not exceed 12 feet and fires with moderate rate of heat release are expected.

Ordinary Hazard Occupancies (Group 2) include occupancies such as:

Cereal Mills
Chemical Plants—Ordinary
Cold Storage Warehouses
Confectionery Products
Distilleries
Leather Goods Manufacturing
Libraries-Large Stack Room Areas
Mercantiles
Machine Shops
Metal Working
Printing and Publishing
Textile Manufacturing
Tobacco Products Manufacturing
Wood Product Assembly

1-7.3.3 Ordinary Hazard (Group 3)—Occupancies or portions of other occupancies where quantity and/or combustibility of contents is high, and fires of high rate of heat release are expected.

Table 13.1 ***Relative Fire Hazard for Various Occupancies, as Related to Sprinkler Installations (Continued)***

Ordinary Hazard Occupancies (Group 3) include occupancies such as:

Exhibition Halls
Feed Mills
Paper and Pulp Mills
Paper Process Plants
Piers and Wharves
Repair Garages
Tire Manufacturing
Warehouses (having moderate to higher combustibility of content, such as paper, household furniture, paint, general storage, whiskey, etc.)[a]
Wood Machining

1-7.4 Extra Hazard Occupancies

1-7.4.1 Extra hazard occupancies or portions of other occupancies where quantity and combustibility of contents is very high, flammable liquids, dust, lint or other materials are present introducing the probability of rapidly developing fires with high rates of heat release.

Extra Hazard Occupancies include occupancies such as:

Aircraft Hangars[b]
Chemical Works (Extra Hazard)
Cotton Pickers and Opening Operations
Explosives and Pyrotechnics
Woodworking with Flammable Finishing

1-7.5 Under favorable conditions and subject to the approval of the authority having jurisdiction, a reduction of requirements to the next less restrictive occupancy classification may be applied to the following occupancies:

Cold Storage Warehouses
Cotton Picker and Opening Operations
Feed Mills
Leather Goods Manufacturing
Machine Shops
Mercantiles
Metal Working
Paper and Pulp Mills

[a] For high-piled storage as defined in 4-1.3.8, see Appendix for separately published NFPA standards relating to water supply requirements, particularly Indoor General Storage, No. 231, 1974, and Rack Storage of Materials, No. 231C, 1975.

[b] For Aircraft Hangars refer also to NFPA 409, Standard on Aircraft Hangars, 1975.

Source. Reprinted with permission from NFPA, No. 13, "Installation of Sprinkler Systems," Copyright © 1976, National Fire Protection Association, 470 Atlantic Avenue, Boston, Mass. 02210.

shown in Figure 13.3. The building, a printing and publishing plant, is in the category of "Ordinary Hazard, Group 2" (see Table 13.1). The sprinkler design results in a nozzle spacing such that one nozzle (sprinkler head) takes care of 125 sq ft of floor area. This places it between the extremes of 90 and 200 sq ft per sprinkler previously mentioned in this section. Standards of the National Fire Protection Association (NFPA) are constantly being re-

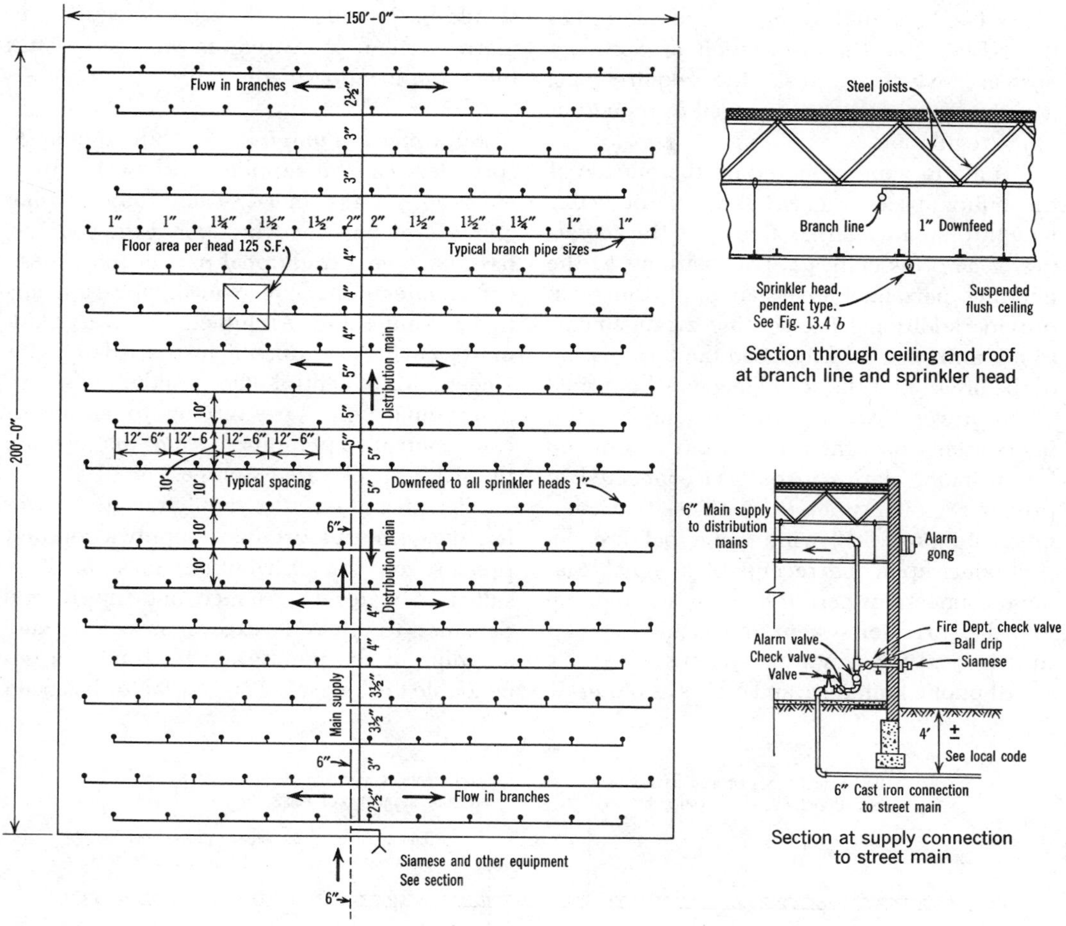

Fig. 13.3 *Plan of sprinklered industrial building. Sprinklers (and standpipes) may use water from street mains when pressure is adequate. Either system may use pneumatic or gravity tanks. When the latter is used to supply both systems, an independent sprinkler reserve occupies the bottom, and the fire standpipe supply occupies the top. Auxiliary fire-engine feed by siamese should be provided in all cases. Auxiliary sources and standby pressurization may be required if street main adequacy is questioned.*

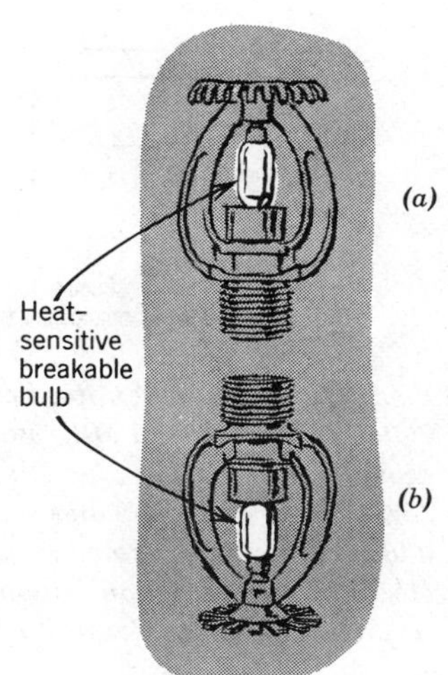

Fig. 13.4 *Sprinkler heads. These are of the quartzoid bulb type. The bulb is transparent and contains a colored liquid. At 136° F the bulb breaks and releases a water stream.* (a) *Upright. This type of sprinkler head is used above piping when piping is exposed. It is safer against damage by workers than the pendent type. The heads should be located at the highest level, just below the structural roof—a location where the hottest gases will accumulate.* (b) *Pendent. This type of sprinkler head projects through a finished ceiling when piping is concealed. Heads can be plated, polished, or colored. The use of upright sprinkler heads is seen in Figure 13.15. Pendent heads are shown in Figures 13.3, and 13.8.*

viewed and updated. Designers are referred to the NFPA No. 13, *Installation of Sprinkler Systems,* which controls the requirements (briefly referred to here), as well as numerous other regulations.

An alarm gong mounted on the outside of the building (see Fig. 13.3) warns of water flow through the alarm valve upon the actuating of a sprinkler head. This warning to the building personnel gives an opportunity to provide additional fire-fighting arrangements to minimize loss and to speed the termination of the fire so that the sprinklers may be turned off to prevent excess water damage to building contents after the fire is out. It is common for sprinkler alarms to be also connected to private regional supervisory offices that communicate promptly with municipal fire departments upon the receipt of a signal. Siamese connections permit fire engines to pump into the sprinkler system in a manner similar to that used for standpipe systems.

All public buildings, and others as required, should be provided with a fire detection and alarm system, indicating in the custodian's office the location of the fire.

(c) ***Upfeed Pumping.*** In tall buildings, sprinklers can be supplied with water from elevated storage tanks. These tanks, rising above large buildings throughout the country, have become a traditional part of the American architectural scene. Now their use appears to be diminishing. Architects—such as Skidmore, Owings & Merrill, in their design for General Mills' central office building in suburban Minneapolis—are turning to automatically controlled pumping systems that obviate the need for elevated storage.

The purpose of the familiar water tower has always been fourfold: to supply a constant pressure on the distribution lines; to store sufficient water to balance out supply and demand; to prevent excessive starting and stopping of the pump; and to provide a dependable fire reserve. This last factor has been

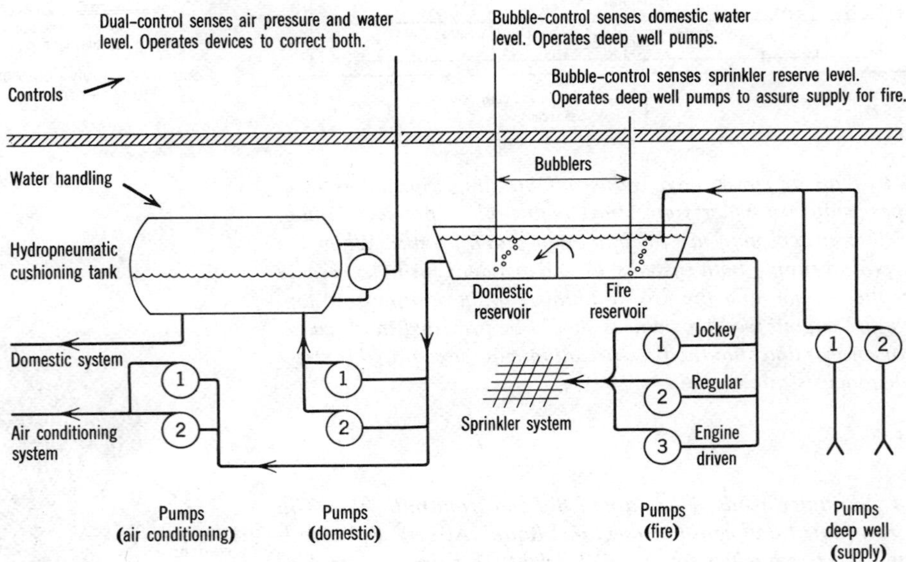

Fig. 13.5 Diagram of upfeed pumping system for supply of sprinklers and building demands for domestic water. In the General Mills Building, referred to in Section 13.5, a large subsurface concrete water tank provides the secondary water source that fire insurance carriers look for. Thus, the wells and their pumps constitute a primary water source. If the well yield fails, the concrete tank supply can be tapped. An additional resource provides engine-driven pumping that is operative in the event of a power outage at an electric utility company. Automatic controls and piping for the storage tank reserve system are not shown in this illustration. Courtesty of Progressive Architecture.

of critical significance in the calculation of fire-insurance rates.

Principal objections to the use of tanks have been their unsightly appearance—lowering property values in their vicinity, the ever-increasing cost of steel and steel-construction work, the problem of freezing, and—in the case of large buildings—their tremendous weight. It is possible that for the General Mills building the fire underwriters would have required at least 50,000 gal of residual water in the tank for emergency purposes, which would mean about 100,000 gal of elevated storage. The alternative chosen was a reinforced-concrete structure placed underground to one side of the building and covered with from 3 to 5 ft of earth. Small vents rising from this reservoir blend in with the lawn and landscaped shrubbery above.

A comparison of costs would be difficult because of the number of factors involved. Generally, however, the saving in steel tended to make the overall cost comparable to that of an elevated tank of the same capacity. The saving is made possible chiefly by refined automatic-pump control. The underground reservoir eliminates the problem of appearance and weight, but the other usual advantages of an elevated tank—reliability in case of fire, a minimum starting and stopping of motors, and the maintenance of pressure while balancing supply and demand—must be equalled in the automatic circuitry. This is not as simple as it might seem at first. There are factors such as fluctuations of demand, friction within the pipes, elevations, starting surges from the pumps, and pressure-flow characteristics of the pumps themselves that must be met. The combinations of these problems, varying as they do from one installation to another, undoubtedly account for the continued use of elevated tanks to a great extent. Yet, these are problems that can be handled efficiently on the drafting board of a controls engineer, rather than in steel mills and contractors' offices where the expense is much greater. Hence, the trend toward the use of more sophisticated pump control.

It will be seen from Figure 13.5 that a continuous flow from the deep-well pumps, through both domestic and fire reservoirs, prevents the water from becoming stale and rancid. The fire reservoir is given the necessary priority over the domestic reservoir by means of a simple weir. Even if the domestic reservoir were completely empty, the fire reservoir would remain full. Pressure for the sprinkler is supplied by a small 20 gpm jockey pump. Signals from the sprinkler system bring in a 750 gpm main pump. If this should fail, a diesel-engine-driven pump of equal capacity automatically takes over.

The circuitry of two 1000-gpm deep-well pumps and two 200-gpm domestic pumps was designed by engineers in the office of Skidmore, Owings & Merrill. Design of the controls was made by the Automatic Control Company of St. Paul, Minnesota. Three sensing units govern the operation of the pumps, bubble-control units in each of the two reservoirs, and a dual-control unit that regulates supply for the pressure tank. All three are connected to a large central cabinet located in an underground pump room next to the reservoir. Pumps are controlled through the cabinet.

The bubble control uses a small air compressor within the central cabinet to send a flow of air through a 1/4-in. tube to the reservoir. Back pressure on this flow, which varies with the level in the reservoir, operates pressure switches within the cabinet.

The hydropneumatic tank is used, not for water storage as is sometimes mistakenly supposed, but to store air under pressure that will balance out surge from the two domestic pumps and reduce the frequency of starting and stopping the pumps. It is a hybrid of the closed system where several pumps are sequenced automatically to supply even pressure. Its advantage is that only two pumps are used.

One of the disadvantages in the past has been the difficulty in maintaining the correct ratio of 60% air to 40% water. Tanks supplied with water from deep wells become air-bound as water stored in them gives up its absorbed air. The dual-control installation in the General Mills system eliminates the need for manual adjustment of this ratio by employing two sensing devices within a single control. A drop in air pressure in the tank sends signals to start the pump. A rise in water level sends signals to stop it.

The signals from the dual-control and the bubble-control units are processed in the central cabinet for correct time delay through motor-driven relays. The central control system also alternates the pumps to give them even wear, or runs them together if the demand requires. In the event of low suction, it shuts the pumps off to prevent motor damage. At the same time, it sends an alarm to the office of the maintenance engineer, indicating the location of the trouble.

Normally, the system runs without human attendance to satisfy the heavy demands of air conditioning, fire control, and domestic water supply in this modern, rurally isolated office building.

(*d*) ***Sprinklers in High-Rise Buildings.*** In the planning of new, tall buildings, there has indeed been a strong move toward total sprinklering throughout. By permission of the Copper Development Association, we include their story of one such installation, in which full sprinklering is part of an overall safety program.

THE TRANSAMERICA BUILDING

A steel framed, Type I, fire resistive, 48-story office tower, the Transamerica Building, San Francisco, California, is, in external form, a pyramid (see Fig. 13.6). Square in plan, succeeding floors decrease in area due to the sloping exterior walls. Rentable floor areas range from 22,000 to 3,000 square feet. A central service core contains elevators, mechanical equipment, wash rooms and stair towers. A modular grid locates partitions, lighting fixtures, air diffusers, and utility jacks.

Two independent diesel driven fire pumps, rated to deliver 750 gallons per minute at 275 pounds per square inch discharge pressure, take suction from city mains at 50 pounds per square inch flowing pressure.

Should the city mains fail—due to an earthquake or similar natural or man-made disaster—a 5,000 gallon closed tank in the building's basement provides a secondary supply source. The tank piping is equipped with a check valve on the system side and a back pressure valve on the city supply.

Maximum sprinkler demand per floor is 186 gallons per minute based on a design density of 0.10 gallon per minute applied over a design area of 1,860 square feet. Fire pump rated capacities anticipate supplying two $2\frac{1}{2}$-inch fire department hose lines from the combined wet standpipe—sprinkler risers in addition to the sprinkler demand.

The fire pumps supply two 6-inch, welded steel, express risers; one in each stair tower (see Fig. 13.7); running the full building height. Local 6-inch risers supplied off the express risers serve each of three 16-story zones. Pressure reducing valves on the local risers limit system pressures to from 50 to 150 pounds per square inch.

At each floor level (see Fig. 13.7) a 2-inch Type M copper looped main connects both local risers. The 2-inch main serves $1\frac{1}{4}$-inch branch lines run to the exterior walls. All lines are above the suspended ceilings. Branch line spacing and sprinkler location are coordinated with the modular grid.

Anticipating interior partition relocation, $\frac{3}{4}$-inch riser nipples terminate in tees with one outlet stubbed and capped for future use. The $\frac{3}{4}$-inch tee accommodates an arm-over and drop which locates the sprinkler in plan and elevation. Fully recessed, $\frac{1}{2}$-inch orifice, 165F, pendent sprinklers serve typical office spaces. Exposed pendent sprinklers protect equipment rooms, wash rooms, and service areas (see Fig. 13.8).

13.6 Fire Load and Compartmentation

(*a*) ***Fire Load.*** When a fire occurs after a building is complete and occupied, fire fighters must cope not only with the building structure as they find it but also with the "occupancy fire load." The nature of the building is established by its design and construction, but the actual occupancy fire load varies greatly. It is difficult to control the decisions of tenants and other occupants in regard to their selection of furniture and finishes. Yet many business firms in choosing their office equipment are now avoiding materials, including some plastics and other products

Fig. 13.6 The Transamerica Building, San Francisco, California, William Pereira & Associates, Los Angeles, California, Architects. This corporate headquarters building features an overall safety program built around an all-copper sprinkler system. Copper Development Association.

that are either flammable or that produce noxious smoke or even explosive gases when under exposure to intense heat. When designing equipment such as sprinkler systems it is often a wise decision to assume a somewhat greater hazard than might be customarily expected.

(*b*) ***Compartmentation.*** In fire protection circles the term "fireproof" has become a dirty word. Older buildings of monolithic poured-in-place concrete slabs employing stone or gravel aggregate provided excellent fire resistance as did partitions of brick, concrete, block, or clay tile. Now structures are lighter. We must deal with the fact that compartment walls and structural slabs above and below required "compartmented" areas need to be

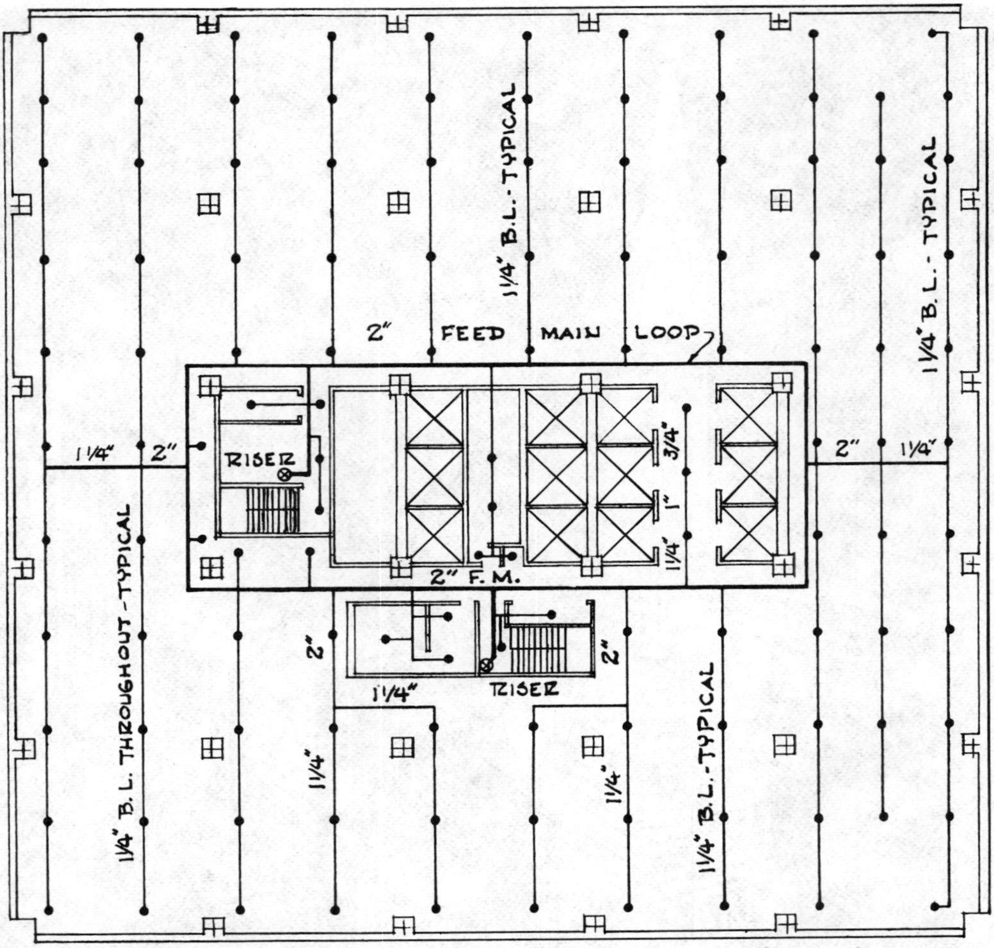

● ½″ flush mounted pendent sprinklers
Scale: 1″ = 20′-0″

Fig. 13.7 *Reflected sprinkler plan, sixth floor, Transamerica Building, San Francisco, California.*

Fig. 13.8 *When changes in the location of partitions require the relocation of sprinkler heads. $\frac{3}{4}$-in. runouts can be connected to capped trees provided for this purpose on the branch lines. A $\frac{1}{2}$-in. line drops through the hung ceiling to a pendent-type sprinkler head.*

properly "fire resistant." Standards of the NFPA call for certain of these enclosing materials or assemblies to have one- or two-hour *fire ratings.* When in place, these fire-rated enclosing partitions, floors, or walls, unfortunately, are subject to numerous perforations.

Fig. 13.9 *Access doors.* (a) *Typical door for access to controls behind a wall.* (b) *Milcor fire-rated access door. This type is chosen when an opening perforates a "fire-rated" wall such as the 1- or 2-hr rated separations that would enclose a compartmented area as discussed in the "Compartmentation" section of New York City Local Law No. 5 (see page 488). Compartmented areas must have no vulnerable openings that could spread fire to other areas. The only opening would be to a smoke exhaust system that removes smoke and heat from the fire area. Inryco/Milcor.*

They include holes for sheet metal ducts, for pipes and conduits, and for access doors. Fire dampers in ducts (see Section 13.7), plugged openings around pipes passing through pipe sleeves, and the use of fire-rated access doors solve these problems (see Fig. 13.9). Tight architectural specifications and diligent inspection at the site during construction are imperative.

13.7 Control of Air Ducts

Air-conditioning systems with supply and return ducts serving a number of stories in a multistory building can spread fire rapidly from the fire floor to several other floors. Control devices before and after the circulating fan or blower can prevent this. Heat- and smoke-sensitive devices can act promptly to stop the blower as soon as the return duct carries evidence of the fire to the sensing elements; this is shown in Figure 13.10.

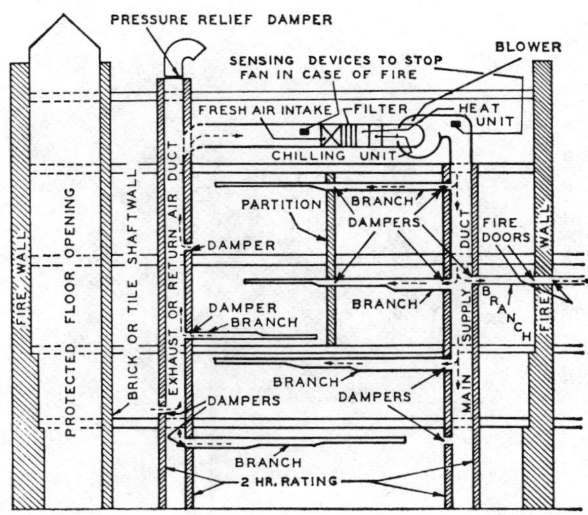

Fig. 13.10 *Typical installation of an air-conditioning system in a building, illustrating protections of penetrations of fire walls, partitions, and shaft enclosures. Reproduced with permission from NFPA 90A,* Installation of Air Conditioning and Ventilating Systems, *Copyright © 1976. National Fire Protection Association, 470 Atlantic Avenue, Boston, Mass. 02210.*

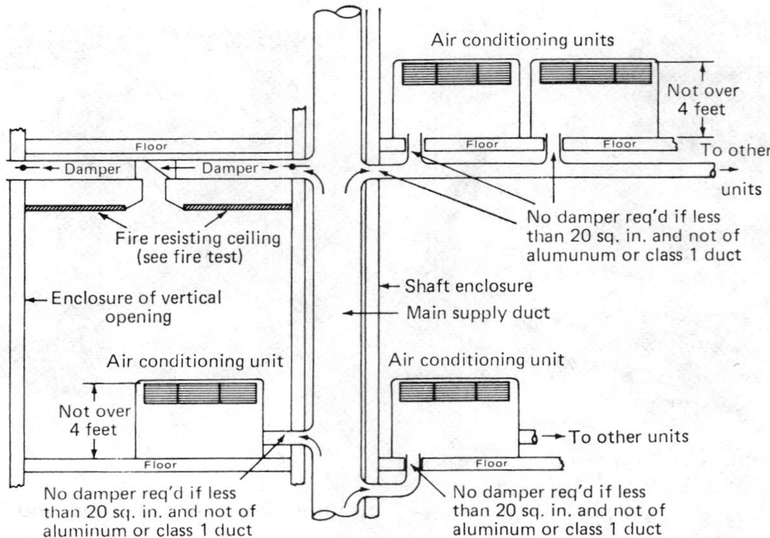

Fig. 13.11 Typical fire damper requirements and examples of locations where dampers are not required. Reproduced with permission from NFPA No. 90A, Installation of Air Conditioning and Ventilating Systems, *Copyright © 1976. National Fire Protection Association, 470 Atlantic Avenue, Boston, Mass. 02210.*

Another precaution is necessary. A fire confined to a single floor can quickly burn through the thin wall of the metal duct, enter a supply or return duct, and spread to other floors. To prevent this, fire-rated dampers in the duct, as shown in Figure 13.10, that are activated by a fusible link can close against such a fire spread. Figure 13.11 shows several cases where, for small duct branches and for certain duct metals, such dampers may *not* be required.

13.8 Smoke and Heat Venting

In attacking a fire, one of the first procedures of fire fighters is to create a vent at a high point in the structure to relieve smoke and hot gases. This permits access to the fire at the level of its occurrence. When *automatic* venting devices open to achieve this relief as in Figure 13.12, delay is avoided since firemen do not have to take the time to chop an opening. This kind of vent often takes the form of "Big Smoky," illustrated in Figure 13.12 and 13.13.

Vent *shafts* in multistory buildings have not

Fig. 13.12 *In commerical and industrial one-story buildings that are extensive in area and are equipped with many vent units, two functions are served.*

1. *The vent closest to the fire opens to relieve the immediate area below, permitting fire fighters to approach* that *point with safety.*
2. *Fire fighters on the* roof *are promptly alerted to the* location *of the fire and can attack it from above.*

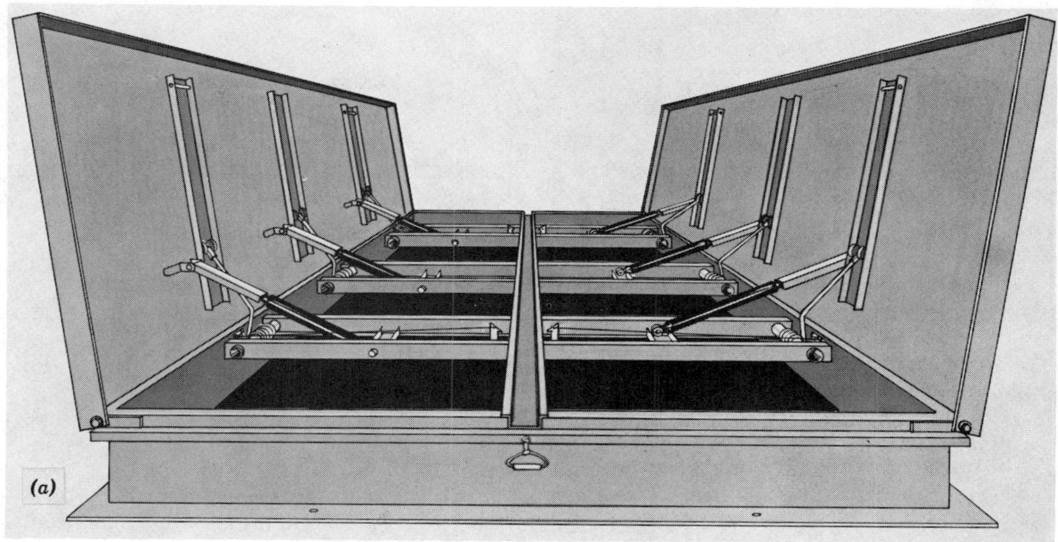

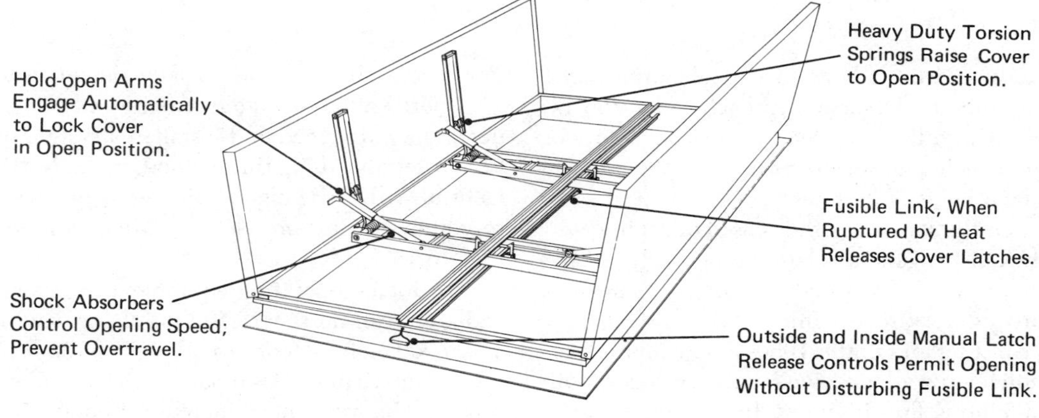

Fig. 13.13 *Together with sprinklers, venting is a highly effective function in automatic fire fighting. Among the many suitable uses of this type of venting, an important one is the prompt exhaust of heat and smoke through the roofs of large one-story industrial and commerical buildings (see Figs. 13.12 and 13.14).* (a) *"Big Smoky" automatic heat and smoke vent, Inryco, Inc., Milwaukee, Wis.* (b) *Operating mechanisms of a heat and smoke vent.*

been too common, and especially not in high-rise structures. Their use in the future is expected to increase so that smoke and heat may be quickly removed and exhausted through the roof.

13.9 Fire Protection for Single-Story Buildings and Low Structures

Although high-rise structures need special study, they are far outnumbered by one-story buildings and other relatively low buildings throughout the world. Frame houses, for instance, should have smoke and heat detectors and alarms. Fire extinguishers and good planning for easy egress are also important, as is an escape plan for the occupants. The subject of fire signals and detectors for residences is discussed in Chapter 22. Protection of low industrial buildings is seen in Figure 13.14.

Buildings for special uses require expert

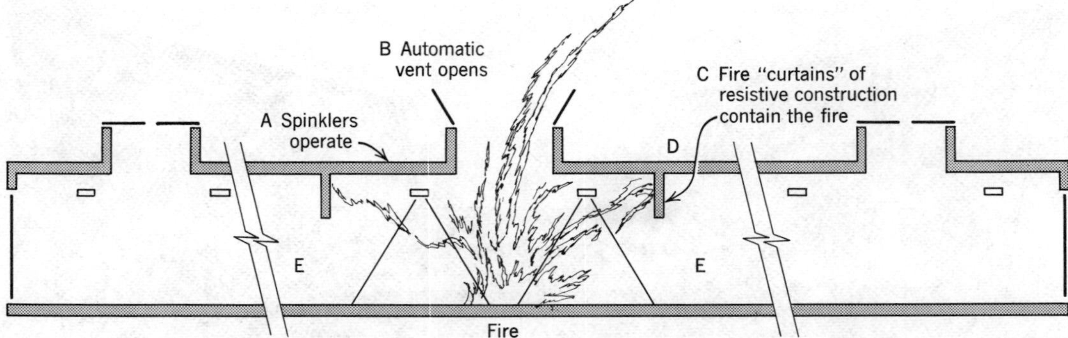

Fig. 13.14 Multiple fire protection of large-area industrial buildings. Schematic cross section, no scale. Three elements combine to fight the fire.

(a) *Sprinklers designed for the occupancy hazard*
(b) *Automatic venting relieves the specific area*
(c) *Fire curtains minimize spread to other areas*

Locations (d) *and* (e) *provide safe positions for fire fighters. Owners are beginning to consider multisystem installations employing water, venting, and partial compartmentation.*

study and the advice of capable professional consultants. Diverse approaches, far too numerous to list, may lead to unusual solutions. One such solution, for the protection of airplane hangars, follows.

Of the several methods used in dealing with fires in hangars, a relatively new approach—the use of a special detergent foam—has proved very successful.

In England in the 1950s a new high expansion foam was conceived and used successfully in fighting fires in mines. Its introduction into the United States by private developers in collaboration with the U.S. Bureau of Mines proved its effectiveness against fires of a similar nature in this country. Years of research and improvement established the ideal foaming agent and led to a wide variety of uses in many other fields of fire fighting. The patents were acquired in 1963 by Walter Kidde & Company. Among the many unique features of the system, which has been brought to a high degree of perfection, is the expansion rate of the foaming process. One thousand gallons of bubbles are produced from 1 gal of water. This 1000-to-1 rate is most dramatic when compared with the 10-to-1 rate of the familiar protein foam that fire fighters have long used effectively, especially on oil fires.

Albert Kahn Associated Architects & Engineers, Inc., designed and installed a system of fire protection, using this method, in the North Central Airlines Hangar Building at the Metropolitan Airport in Detroit, Michigan (see Fig. 13.15).

The foam-generating equipment in this installation can fill the 38,400 sq ft hangar with 1,400,000 cu ft of foam to a height of 36 ft in less than 12 min. Automatic devices, upon sensing abnormal heat increase, operate the foam generators, open roof vents, start the smoke control exhaust fans in the vents and transmit a fire alarm signal to the Airport Fire Department. Foam discharge is delayed 30 sec while evacuation sirens sound to permit occupants to leave the fire area. Manual "override" controls can allow personnel to start the system in the event of failure of the automatic controls, or stop it if the fire is small and controllable by other methods.

The high expansion foam is created by wetting a nylon net with a mixture of water and a special detergent soap concentrate. A large blower directs an air current through the net, producing an avalanche of foam.

Suds, which blanket the fire, deal with it in

Fig. 13.15 *Detergent foam discharged from four units, two of which are seen in the illustration, after 3 min of operation. This installation at the North Central Airlines hanger building at Detroit Metropolitan Airport smothers fire, prevents its spread, and will not harm airplanes or machinery. Fire fighters can breathe inside the foam. The system can handle combustible liquid fires. Note also the grid of upright sprinklers directly below the roof surface.*

several ways. The water in the suds converts to steam, absorbing the heat of the fire. The expansion of the foam into steam reduces the oxygen content to about 7%, which is insufficient to support active combustion. A cooling effect is achieved by the wetting action of the breaking bubbles. The movement of air currents toward the fire to replace the rising hot gases draws the foam to the center of the fire. There it blocks the air flow and cuts off the supply of oxygen.

The fire, thus contained and diminished, can be approached by fire fighters for further control. Personnel advancing through the cooler sections of the foam are safe because the foam in these areas is 99% air and can support human life.

Aircraft are not harmed by the foam. Delicate machinery that might be injured by high velocity streams of water is undamaged and left quite clean when the foam is rinsed away. After all, it *is* a soap.

The structure, in this case an open steel frame with a metal roof deck, is protected from excessive temperatures that might weaken it and cause it to collapse.

References and Bibliography

1. The following are publications of the National Fire Protection Association, 470 Atlantic Avenue, Boston, Mass. 02210.

NFPA No.	*Date*	
13	1976	*Installation of Sprinkler Systems*

14	1976	*Standpipe and Hose Systems*
80	1977	*Fire Doors and Windows*
90A	1976	*Installation of Air Conditioning and Ventilating Systems*
90B	1976	*Warm Air Heating & Air Conditioning Systems*
101	1976	*Life Safety Code*
204	1968	*Guide for Smoke and Heat Venting*
		National Electric Code
		Fire Protection Handbook, 14th ed.

NOTE: The above publications are frequently updated.

2. *Fire Protection and Local Government: An Evaluation of Policy-Related Research*, prepared under a grant from the National Science Foundation, Swersey, Ignall, Corman, Armstrong, and Weindling. The New York City Rand Institute, published by The Rand Corporation, Santa Monica, Calif.

3. "Automatic Heat and Smoke Venting," Robert J. Lyons, Director of Product Design, The Bilco Company, West Haven, Conn., *Progressive Architecture*, April 1972.

PART FIVE

electricity

14

Principles of Electricity

14.1 Electric Energy

Historically, usable energy was most often produced by burning a natural fuel such as coal or oil. The resultant heat energy was used directly as heat and light or converted by machines into other desired forms of energy such as motion. Only within the last century, however, has this heat been in turn used to create another form of usable energy—electricity. Even the recent partial substitution of nuclear for fossil fuels has only affected the heat production portion of this process. Beyond that point, the heat is utilized in the same manner to drive generators that produce electricity. It is well to remember, therefore, that in terms of natural resources electricity is an expensive form of energy, since the efficiency of heat-to-electricity conversion, on a commercial scale, rarely exceeds 40%.

Electricity constitutes a form of energy itself, which occurs naturally only in unusable forms such as lightning and other static discharges or in the natural galvanic cells, which cause corrosion. The primary problem in the utilization of electric energy is that, unlike fuels or even heat, it cannot be stored and therefore must be generated and utilized in the same instant. This requires an entirely different concept of utilization than, for example, a heating system with its burner, piping, and associated equipment.

The bulk of electric energy utilized today is in the form of alternating current (a-c), produced by a-c generators, commonly called alternators. Direct-current (d-c) generators are utilized for special applications requiring large quantities of d-c. In the building field such a requirement is found in elevator work. Smaller quantities of d-c, furnished either by batteries or by rectifiers, are utilized for telephone and signal equipment, controls, and other specialized uses.

14.2 Unit of Electric Current: The Ampere

When electricity flows in a conductor, a certain number of electrons pass a given point in the conductor in 1 sec. Numerically, an ampere of current flows in a conductor when 6.251×10^{18} electrons pass a given cross section in 1 sec. Current, or amperage, is abbreviated *amp*, *amps*, or simply *a*. When current is used in an equation, it is usually represented by the letter I or i. (On 120 v service, the ordinary 100 w lamp filament carries about 0.833 amp; the motor for a desk calculator, about 1.00 amp.)

It is convenient to establish an analogy between electric systems and mechanical systems as an aid to comprehension. Current, or amperage, is a measure of flow and, as such, would correspond to water flow in a hydraulic system (see Fig. 14.1). The correspondence is not complete, however, since in the hydraulic system the velocity of water flow varies, whereas in the electric system the velocity of propagation is constant and may be considered instantaneous; hence, the need to utilize the electric energy the instant it is produced.

14.3 Unit of Electric Potential: The Volt

The electron movement described above and its concomitant energy, which constitutes

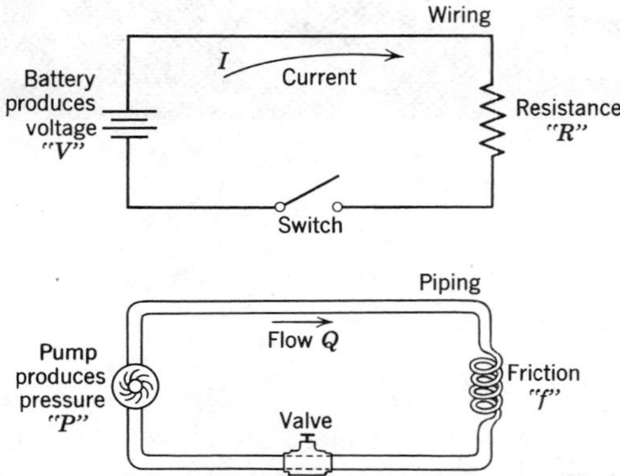

Fig. 14.1 Electric-hydraulic analogy. The circuits show that voltage is analogous to pressure, current to flow, friction to resistance, wire to piping, and switches to valves.

electricity, is caused by creating a higher positive electric charge at one point on a conductor than exists at another point on that same conductor. This difference in charge can be created in a number of ways. The oldest and simplest method is by electrochemical action, as in the battery. In the ordinary dry cell, or in a storage battery, chemical action causes positive charges (+) to collect on the positive terminal and electrons or negative charges (−) to collect on the negative terminal. Assume for the moment that nothing is connected to the battery terminals. There is a definite force attraction, or tendency to flow, between the electrified particles concentrated at the positive and negative terminals. *Potential difference or voltage* is the name given to this electromotive force (emf). This force is analagous to pressure in a hydraulic or pneumatic system. Just as the pressure produced by a pump or blower causes water or air to flow in a connecting pipe, so too the potential (emf, voltage) produced by a battery (or generator)

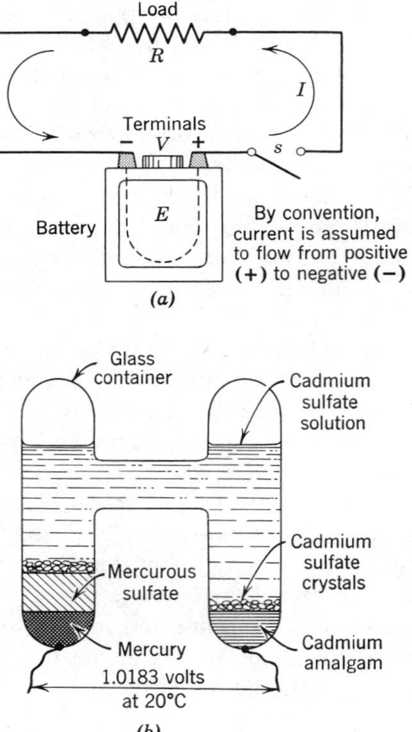

Fig. 14.2 (a) *Current flows in the electric circuit as a result of the voltage (potential difference)* V *that exists between the terminals of the battery. By convention, current is shown as flowing from positive* (+) *to negative* (−) *in the circuit (and from* − *to* + *inside the battery).* (b) *Weston cell used as the international voltage standard.* (c) *Modern lead-calcium storage cell. Such cells are compact, long-lived, and reliable. Applied individually and in banks, they supply normal and emergency power for control and communication applications. Courtesy of Gould, Inc., Industrial Battery Division*

causes current to flow when the terminals between which a voltage exists are connected by a conductor. The higher the voltage (pressure), the higher the current (flow) for a given resistance (friction) (see Fig. 14.2). Other means of producing voltage, both direct (d-c) and alternating (a-c), are discussed in Section 14.8. The unit of voltage is the volt, abbreviated "v."

14.4 Unit of Electric Resistance: The Ohm

The flow of fluid in a hydraulic system is impeded by friction; the flow of current in an electric circuit is impeded (resisted) by resistance, which is the electrical term for friction. In a direct-current circuit (d-c) this unit is called resistance and is abbreviated R; in an alternating-current circuit (a-c) it is called impedance and is abbreviated Z. The unit of measurement is called the *ohm.* (It is interesting to note that the felicitous scientific convention of naming units after persons whose work is closely related to the field is here too followed. Thus the ampere, volt, and ohm are derived from André Ampere, Alessandro Volta, and Georg Ohm).

Materials display different resistance to the flow of electric current. Metals generally have the least resistance and are therefore called conductors. The best conductors are the precious metals—silver, gold, and platinum—with copper and aluminum only slightly inferior. Conversely, materials that resist the flow of current are called insulators. Glass, mica, rubber, oil, distilled water, porcelain, and certain synthetics such as phenolic compounds exhibit this insulating property and are therefore used to insulate electric conductors. Common examples are rubber and plastic wire coverings, porcelain lamp sockets, and oil-immersed switches.

14.5 Ohm's Law

The current I that will flow in a d-c circuit is directly proportional to the voltage V and inversely proportional to the resistance R of the circuit. Expressed as an equation, we have the basic form of Ohm's law that

$$I = \frac{V}{R} \tag{14.1}$$

In a-c circuits, the same relation holds true except that instead of d-c resistance we use the a-c impedance. Ohm's law is frequently written in another form, that is

$$V = IR \tag{14.2}$$

which states the mathematical relationship that volts = current × resistance. This form has no logical basis; therefore, we recommend that the reader remember the form of equation 14.1, which clearly states the physical situation; that is, as a result of voltage V, a current I is produced that is proportional to the electric pressure V, and inversely proportional to the electric friction R.

A few examples will illustrate the application of Ohm's law. The examples chosen are applicable to both a-c and d-c, since the devices are all purely resistive. This will be more fully explained in the subsequent discussion on alternating current.

EXAMPLE 14.1. An incandescent lamp having a hot resistance of 66 ohms is put into a socket that is connected to a 115 v supply. What current flows through the lamp?

SOLUTION

$$I = \frac{V}{R} = \frac{115}{66} = 1.74 \text{ amperes}$$

(These figures correspond to a normal 200 w general service incandescent lamp.)

EXAMPLE 14.2. A bathroom heater draws 11 amperes at 120 v. What is its hot resistance?

SOLUTION

$$R = \frac{V}{I} = \frac{120}{11} = 10.91 \text{ ohms}$$

(These figures correspond to a common 1320 w portable heater.)

EXAMPLE 14.3. A household electric water heater is rated 220 v and 20 amperes. What is the unit's resistance when drawing this amount of current?

SOLUTION

$$R = \frac{V}{I} = \frac{220}{20} = 11 \text{ ohms}$$

We mentioned hot resistance in the above examples. Some resistances are higher when they are hot, at their operating temperatures than when cold, at room temperature. A typical example of this is the tungsten filament lamp that when first turned on takes, for a fraction of a second, 10 to 15 times the current that flows when the filament is hot.

14.6 Circuit Arrangements

To extend somewhat the reader's facility with electric circuit concepts, we will discuss here the two basic electric circuit arrangements, that is, series and parallel. These concepts are the same for both d-c and a-c. As above, we will use purely resistive circuits so that circuit calculations will be applicable to both d-c and a-c. In other than purely resistive circuits, a-c circuit calculations are different, and much more complicated, than their d-c counterparts.

*(a) **Series Circuits.*** In a series arrangement the elements are connected one after the other—in series. Thus, the resistances and voltages add. This is indicated graphically in Figure 14.3*a*. An electric circuit may be defined as a complete conducting path that carries current from a source of electricity to and through some electrical device (or load) and back to the source. A current can never flow unless there is a complete (closed) circuit. A practical series circuit can be represented by the diagram of Figure 14.3*b*. It should be obvious that due to the arrangement of the components, *the current is the same in all parts of the circuit.* (The student can check the values shown by applying Ohm's law.) A somewhat more complicated circuit is shown in Figure 14.3*c*. As stated above, in any series circuit the total resistance R is the sum of the resistances around the circuit. Thus

$$\text{Total resistance, } R = r_{ab} + r_{bc} + r_{cd} + r_{da}$$

Hence the total circuit resistance is

$$0.01 + 0.02 + 1.15 + 0.02 = 1.20 \text{ ohms}$$

It is customary to refer to connection points on such wiring diagrams by letters as *a, b, c, d,* and so on. The battery voltage may then be called $V_{ab} = 120$ v; the voltage across the load resistance, $V_{cd} = 115$ v; the resistance of the two wires $r_{bc} + r_{da} = 0.04$ ohm. The positive and negative terminals of the battery are shown.

EXAMPLE 14.4. The battery in Figure 14.3*c*. is rated at 120 v, the line resistance (both wires) is 0.04 ohm, the battery internal resistance is 0.01 ohm, and the load resistance is 1.15 ohms. Determine (a) current flowing in the circuit, (b) the voltage across the load (V_{cd}).

SOLUTION

(a) The current flowing is

$$I = \frac{V}{R} = \frac{V_{ab}}{r_{ab} + r_{bc} + r_{cd} + r_{da}}$$
$$= \frac{120}{0.01 + 0.02 + 1.15 + 0.02}$$
$$= 100 \text{ amp}$$

(b) The voltage drop across the load is $V_{cd} = I \times R_{cd} = 100 \times 1.15 = 115$ v

A practical application of series circuits is found in incandescent lamp street lighting circuits. In order to maintain equal brightness of lamps in long runs, regardless of distance from the source and attendant voltage drop, a series arrangement of lamps (with variable voltage source) is used. Refer to Figure 19.21 to see the effect of reduced voltage on the light output of incandescent lamps. To overcome the serious problem mentioned in the caption to Figure 14.3*d*, special cutout lamp bases are used that bypass the lamp when it burns out.

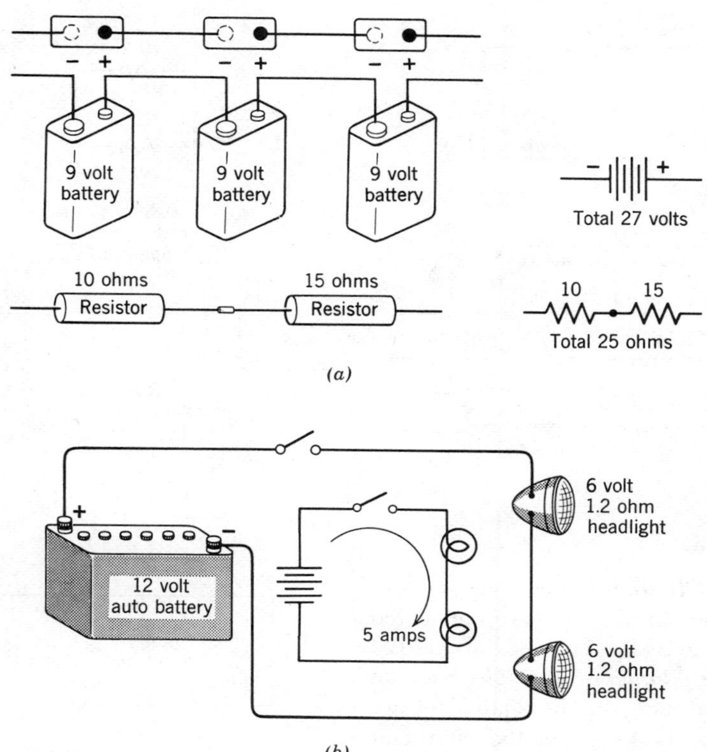

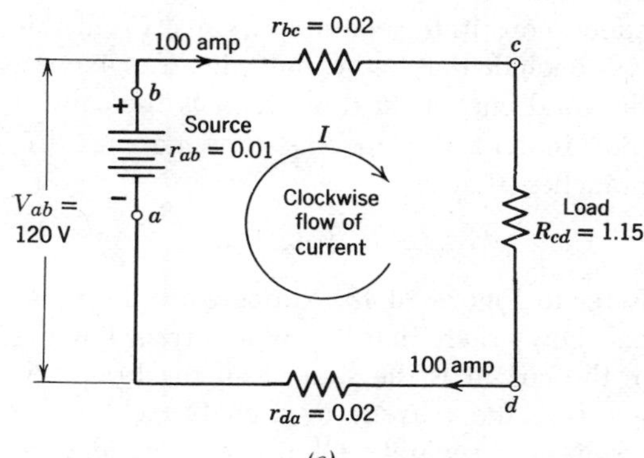

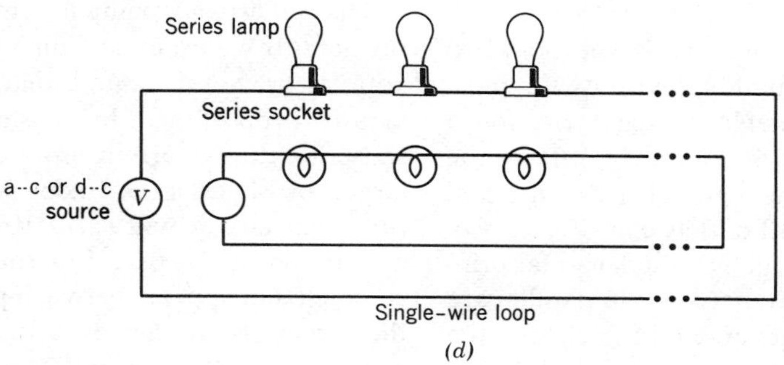

Fig. 14.3 (a) *Physical and graphic representation of series connection of batteries and resistors.* (b) *Physical and graphic representation of a possible d-c circuit.* (c) *A simple series circuit.* (d) *Physical and graphic representation of a series lamp circuit. Loss of one lamp can disable the entire circuit. Furthermore, the point of fault is not obvious, necessitating individual testing of lamps.*

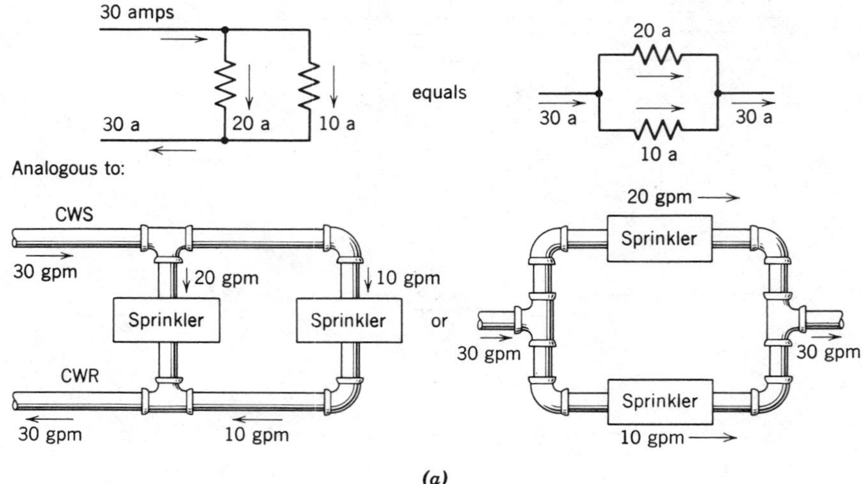

(*b*) ***Parallel Circuits.*** When two or more branches or loads in a circuit are connected between the same two points, they are said to be connected in *parallel* or *multiple.* Such an arrangement and its hydraulic equivalent are shown in Figure 14.4*a*. From the circuit of Figure 14.4*b* it should be apparent that multiple loads are across the same voltage and, in effect, constitute separate circuits. From this we conclude that in the multiple arrangement, the total current in the circuit is the sum of the individual currents flowing in the branches, that is

$$I_T = I_1 + I_2 + I_3$$

Refer to Figure 14.4*b*. Notice from the number shown there that the total current flowing in the circuit is the sum of all the branches, but that the current in each branch is the result of a separate Ohm's law calculation. Thus, in the 10-ohm load a 12-amp current flows, and so forth.

The parallel connection is the standard arrangement in all building wiring. A typical lighting and receptacle arrangement for a large room is shown in Figure 14.5. Here the lights constitute one parallel grouping, and the convenience wall outlets constitute a second parallel grouping. The fundamental principle to remember is that loads in parallel are additive for current, and that each has the same voltage imposed.

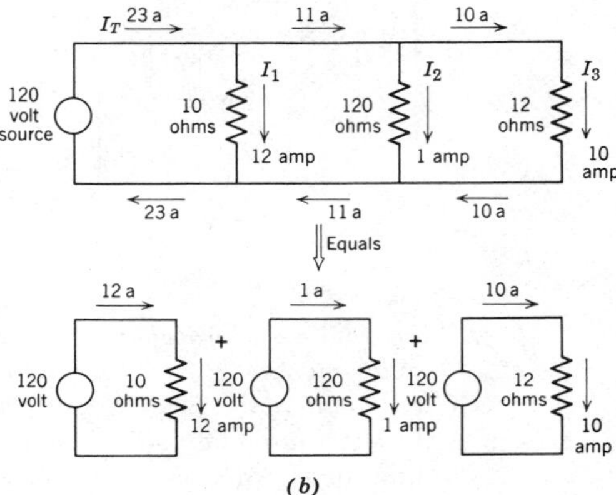

Fig. 14.4 (a) *In a parallel connection the flow divides between the branches, but the pressure is the same in each branch.* (b) *Note that loads connected in parallel are equivalent to separate circuits superimposed into a single connection. Each load acts as an independent circuit unrelated to, and unaffected by, the other circuits.*

One additional point is important to appreciate. If we examine Ohm's law again, we note, as previously stated, that current is inversely proportional to resistance. Thus as resistance drops, current rises. Now look at the circuit of Figure 14.5. Under ordinary conditions that circuit will carry 10 amp and will operate normally. But, if by some mischance, a connection appears between points *a* and *b*, the circuit is *shortened* so that there is no resistance in the circuit. The current rises

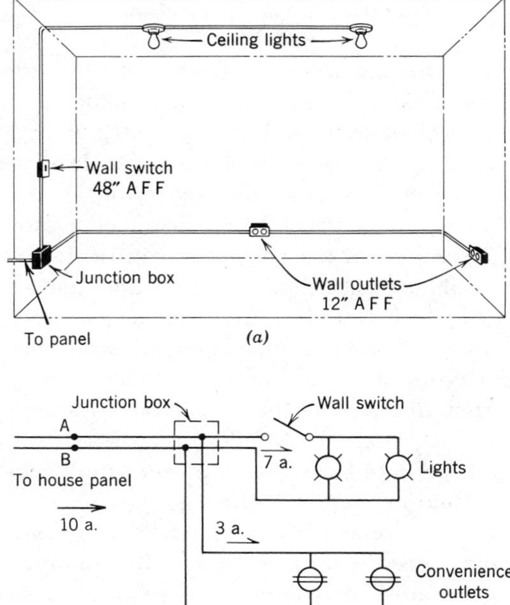

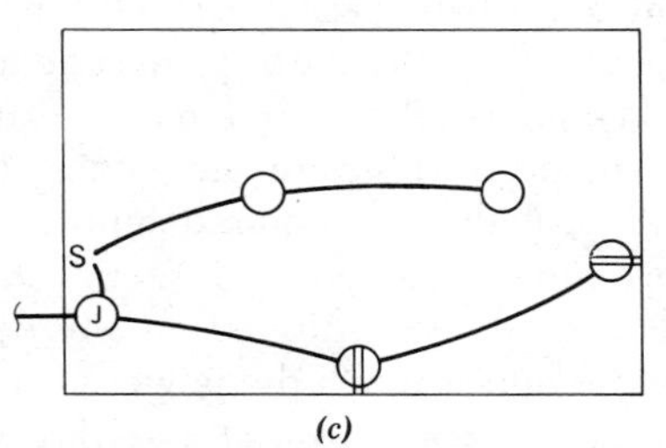

Fig. 14.5 *Parallel groupings of lights and wall outlets are in turn connected in parallel to each other. A circuit is shown* (a) *pictorially,* (b) *schematically, and* (c) *as on an architectural plan.*

instantly to a very high level, and the condition constitutes a short circuit. If the circuit is properly protected, the fuse or circuit breaker will open, and the circuit will be disabled. If not, excessive current will probably start a fire.

14.7 Direct Current and Alternating Current (d-c and a-c)

Whenever the flow of electric current takes place at a constant time rate, practically unvarying and in the same direction around the circuit, it is called direct current. The term *"d-c"* originated as an abbreviation for "direct-current," but because of the oddity of an expression such as "direct-current voltage" or the repititiousness of "direct-current current," the terms universally accepted are "d-c voltage" and "d-c current." The d-c voltages of 1.5 v positive polarity, and 1.0 v negative polarity are shown in Figure 14.6*a*.

Whenever the flow of current is periodically varying in time and in direction, as indicated by the symmetrical positive and negative loops or sine waves in Figure 14.6*c*, it is called an alternating current. The distance along the time axis spanned by a positive and a negative a-c loop is called one cycle of time. Modern a-c systems in the United States operate at 60 cycles per second, or 60 *hertz* (after H. Hertz). This means that current at 60 hertz (abbreviated Hz) is delivered to the consumer. When alternating current is rectified (changed to direct current), it appears as in Figure 14.6*d*. This current is called fully rectified, in contradistinction to that of Figure 14.6*e*, which is half-rectified. Filtering is utilized to create a low ripple wave form, approximately as seen in Figure 14.6*f*.

The a-c circuits differ from d-c circuits in a number of important respects and, since normal a-c current supply is 60 Hz, it is important to understand a-c terminology and usage. Instead of resistance the corresponding parameter in an a-c circuit is impedance, which is also measured in ohms. Depending on the device, it can be markedly different from the d-c resistance. Thus for an a-c circuit, the equivalent to equation 14.1 in Section 14.5 is

$$I = \frac{V}{Z} \tag{14.3}$$

where Z is the symbol normally used for impedance. We will not go into a-c circuit calculations partially because they utilize complex numbers and vector notation but primarily because such calculations are not especially useful to the reader. What *is* useful and important is an understanding of power and energy in both d-c and a-c circuits. This is discussed in Section 14.9.

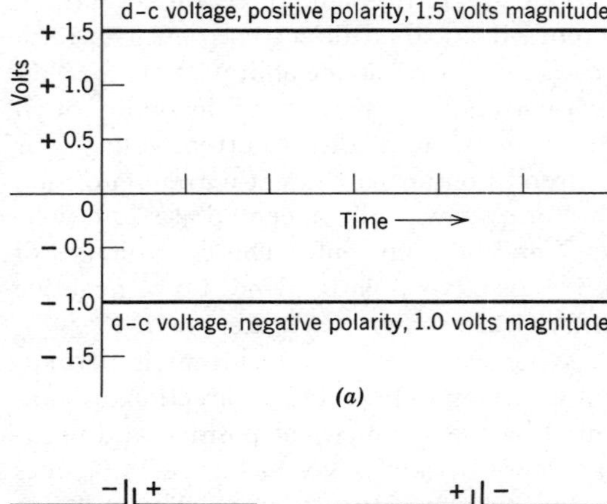

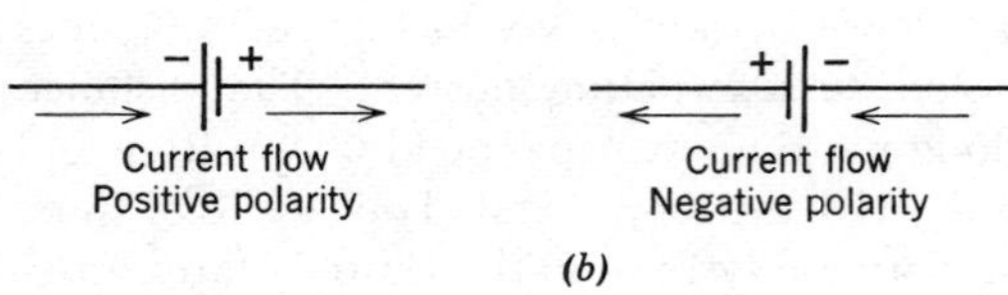

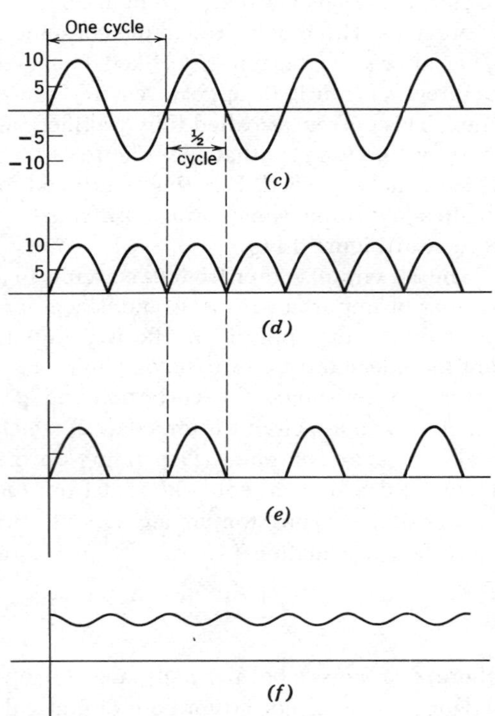

Fig. 14.6 (a) *Graphic representation of d-c voltages with positive and negative polarity.* (b) *Circuit symbol representation of a battery source. The longer bar is positive.* (c) *Alternating current (a-c).* (d) *Fully rectified a-c.* (e) *Half-rectified a-c.* (f) *Fully rectified and filtered a-c approaches the wave form of true d-c as shown in* (a) *above.*

14.8 Electric Power Generation

(*a*) ***Direct Current.*** With respect to generation of large amounts of power, photoelectric, piezoelectric, and thermoelectric effects including solar cells can be ignored, leaving the battery and the d-c generator as the sources of d-c current. Since the d-c generator is in reality an a-c generator with a device (commutator) attached, which rectifies the a-c to d-c, the battery is still the only major direct source of direct current. (There are some special types of generators that produce direct current *directly,* but their use to date has been extremely limited.) A fuller discussion of the application of batteries for power supply can be found in Section 16.29.

The d-c generators are used where accurate motor speed control is desired, for example, for elevators, or where d-c is required on a larger and more sustained basis than would be economical for batteries. Of course, a rotating d-c generator, unlike a battery, must be driven to produce power. The prime mover can be a motor, engine, turbine, or any other device that will provide the required input power. Another source of d-c power is rectification of a-c that can be accomplished on any desired scale to provide as much d-c power as there is available a-c power. Various types of rectifiers are in commercial use, including selenium, silicon, and copper-oxide types.

(*b*) ***Alternating Current.*** Alternating current is produced commercially by an a-c generator, generally called an alternator. The prime mover, as in the case of the d-c units mentioned above, may be any type of engine or turbine. The process by which electricity is produced is illustrated in Figure 14.7. It is based on the fundamental discovery in 1831 by Michael Faraday of the principle of electromagnetic induction. Put briefly, this principle states that when an electrical conductor is moved in a magnetic field, a voltage is induced in it (see Fig. 14.7*a*). The direction of movement determines the polarity of the induced voltage, as shown. If the conductor is formed into a coil and rotated in the magnetic field, a voltage of alternating polarity is pro-

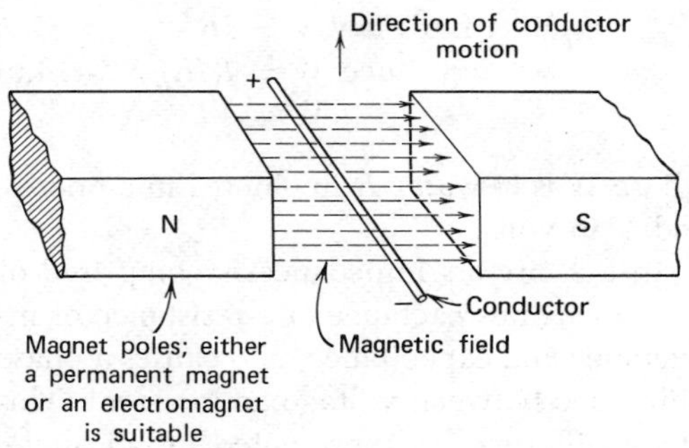

Fig. 14.7a The action fundamental to all generators is illustrated here. When a conductor of electricity moves through a magnetic field, a voltage is produced in the conductor, with polarity as shown.

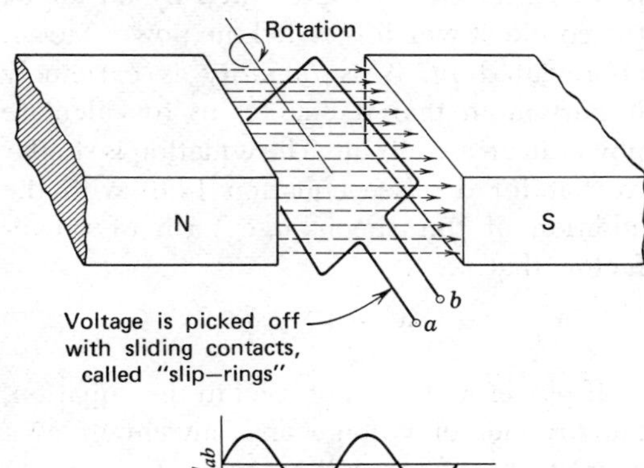

Fig. 14.7b Rotating a coil in a magnetic field produces an alternating sinusoidal voltage at terminals a b due to the alternating polarity (see Fig. 14.7a).

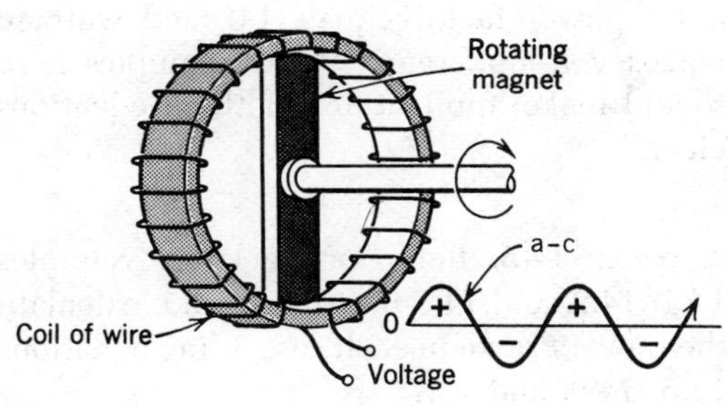

Fig. 14.7c It does not matter whether the conductor moves and the magnetic field is stationary, or vice versa, as long as there is relative motion between the two. The illustration shows the field rotating.

duced, that is, alternating current. It does not matter whether the conductor or the magnet moves; the motion of the conductor and the field with respect to each other produces the voltage (see Fig. 14.7*b* and *c*). It is only one step (of development) from this rudimentary a-c generator to the large, powerful alternator that produces a-c in a modern power plant. The frequency of the voltage generated is a function of the machine design (number of poles) and the speed at which it is driven. Normal generator frequency in the United States is 60 Hz.

14.9 Power and Energy

It is important, indeed imperative, that the distinction between power and energy be clearly understood, since all too frequently the terms are incorrectly used interchangeably. *Energy* is the technical term for the more common expression—work. In terms of power, it is the product of power and time, that is

$$\text{energy or work} = \text{power} \times \text{time}. \quad (14.4)$$

In practical terms, energy is synonymous with

fuel and therefore also cost. Thus energy can be expressed as barrels (tons) of oil, cubic feet (cubic meters) of gas, tons of coal, kilowatt-hours of electricity usage, and dollars of operating cost. The concept of energy efficiency of structures can be stated in terms of annual usage of oil, gas, and electricity or alternatively in terms of dollars of total fuel cost. In technical terms, energy is expressed in units of Btu (calories), foot-pounds (joules), and kilowatt-hours.

Power is the rate at which energy is used or, alternatively, the rate at which work is done, since energy and work are synonymous. The term *power* implies continuity, that is, the use of energy at a particular rate, over a given, generally considerable, span of time. The concept of power necessarily involves the factor of time since it is, as stated above, the *rate* at which work is done. Thus multiplying power by time yields energy. Typical units of power in the English system are horsepower, Btu per hour, watt, and kilowatt. In the metric or SI system the corresponding units are joules per second, calories per second, watts, and kilowatts. In physical terms, power is also the rate at which fuel (energy) is used. Thus power can also be expressed as gallons (liters) of oil per hour, cubic feet (cubic meters) of gas per minute, and tons of coal per day.

14.10 Power in Electric Circuits

The unit of electric power is the watt (w). A larger unit of 1000 watts is the kilowatt (kw). The power input in watts to any electrical device having a resistance R and in which the current is I is given by the equation

$$W = I^2R = I(IR) \qquad (14.5)$$

where W is wattage. This is true for both d-c and a-c circuits. However, since the resistance of an item is generally not known, but the circuit voltage and current *are* known, it would be preferable to be able to calculate power using these two quantities. This can be done, but is different for d-c and a-c.

In d-c circuits

By Ohm's law $V = IR$

$$\text{and, since } W = I(IR), \quad W = VI \qquad (14.6)$$

where W is in watts, R in ohms, I in amperes, and V in volts.

In a-c circuits impedance is comprised of resistance and reactance (a-c resistance of inductance and capacitance) and causes a phase difference between voltage and current. This phase difference is represented by an angle, the cosine of which is called the power factor, abbreviated *pf.* This quantity is extremely important in that it enables us to calculate power in an a-c circuit. The equation is similar to that for d-c (see equation 14.6) with the addition of this special a-c term of power factor, that is

$$W = VI \times \text{pf} \qquad (14.7)$$

If power factor is not used in the equation, the product of voltage and current gives a quantity known as volt-amperes. In a purely resistive circuit, such as one with only electric heating elements, impedance equals resistance, power factor equals 1.0, and wattage equals volt-amperage. A few examples here should make applications of these equations clear.

EXAMPLE 14.5. Referring back to Examples 14.1, 14.2, and 14.3 in Section 14.5, calculate the power drawn in each case, using equations 14.5, 14.6, and 14.7.

SOLUTION. Since in all cases the devices are purely resistive and therefore have a unity (1.0) power factor, it does not matter whether the circuit is a-c or d-c.

From Example 14.1, $R = 66$ ohms, $I = 1.74$ amperes, and $V = 115$ volts

1. In a d-c circuit, we would use equation 14.6

$$W = VI = 115\ (1.74) = 200 \text{ w}$$

2. In an a-c circuit, we would use equation 14.7

$$W = VI \times \text{pf} = 115 \times 1.74 \times 1.0 = 200 \text{ w}$$

3. In either a d-c or an a-c circuit, we can use equation 14.5

$$W = I^2R = (1.74)^2 \times 66 = 200 \text{ w}$$

Similar solutions are applicable to the data of Examples 14.2 and 14.3; these are left to the reader to work out, following the above procedure. (The heater of Example 14.2 is 1320 w; that of Example 14.3 is 4400 w or 4.4 kw.)

EXAMPLE 14.6. Using the data given in Example 14.4 and Figure 14.3*c*, determine (a) the power lost in the wiring and (b) the power input to the load.

SOLUTION

(a) The total line loss is

$$W = I^2R = I^2(r_{bc} + r_{da})$$
$$= (100)^2 \times 0.04 = 400 \text{ w}$$

(b) The power input to the load is

$$W = I^2R = I^2R_{cd} = (100)^2 \times 1.15$$
$$= 11{,}500 \text{ w or } 11.5 \text{ kw}$$

Alternatively, we can find this power by multiplying voltage and current. The voltage on the load is

$$IR = 100 \times 1.15 = 115 \text{ v, and}$$
$$W = VI = 115 \times 100 = 11{,}500 \text{ w}$$
$$= 11.5 \text{ kw}$$

EXAMPLE 14.7. Refer to Figure 14.5. Assume that the two ceiling lights are 150 w each, incandescent. Also assume the load connected to one convenience outlet to be a 10-amp hair dryer and blower, with a power factor of 0.80. Calculate the current and power in the two branches of the circuit, and the total circuit current, assuming a 120 v a-c source.

SOLUTION

(a) In the circuit branch feeding the lights we have

$$\text{Power} = VI$$
$$300 \text{ w} = 120 \text{ v} \times I$$
$$I = \frac{300}{120} = 2.5 \text{ amperes}$$

If we wished to calculate circuit resistance (which is equal to the impedance, since the load is purely resistive), we have:

$$Z = R = \frac{V}{I} = \frac{120}{2.5} = 48 \text{ ohms}$$

This figure is of little practical use to us, and is calculated simply to show technique.

(b) In the second branch we have a 10-amp, 0.8 pf load.

Power in watts = volts × amperes × power factor

$$W = 120 \times 10 \times 0.8 = 960 \text{ w}$$

But the circuit volt-amperes are

$$\text{va} = 120 \times 10 = 1200 \text{ va}$$

This latter figure is significant in sizing equipment.

(c) To calculate the total current flowing from the panel to both branches of the circuit, we must combine a purely resistive current (lamp circuit) with a reactive one (dryer circuit). The exact value of current is the square root of the sum of the squares of the two branch currents. However, as an approximation, the currents may simply be added arithmetically. This yields a result that is higher than actual and is, therefore, on the safe side when sizing equipment. Hence,

approximate total current = 2.5 + 10 = 12.5 amperes

Actual current is 12.1 amperes; our error in approximating is 3.2%, which is acceptable for most uses. One further example at this point will demonstrate the importance of power factor in normal situations.

EXAMPLE 14.8. The nameplate of a motor shows the following data: 3 hp, 240 v, a-c, 17 amperes. Assume an efficiency of 90%. Calculate the motor (and, therefore, circuit) power factor.

SOLUTION

$$1 \text{ hp} = 746 \text{ w}$$

Therefore

$$3 \text{ hp} = 3 \times 746 = 2238 \text{ w output}$$

$$\text{efficiency} = \frac{\text{output}}{\text{input}}$$

so

$$\text{power input} = \frac{2238}{0.9} = 2487 \text{ w}$$

But for a-c,

$$\text{power} = \text{volts} \times \text{amperes} \times \text{power factor}$$

so

$$2487 = 240 \times 17 \times \text{power factor}$$

and

$$\text{power factor} = \frac{2487}{240 \times 17} = 0.61$$

Note the large difference between volt-amperes and watts.

$$VI = 240 \times 17 = 4080 \text{ va}$$
$$P = \text{as above} = 2487 \text{ w}$$

Where P designates power.

14.11 Energy in Electric Circuits

Since power is the rate of energy use, it follows, as stated above, that energy = power × time. Therefore, the amount of energy used is directly proportional to the power of the system and to the length of time it is in operation. Since power is expressed in watts or kilowatts, and time in hours (seconds and minutes are too small for our use), we have for units of energy: watt-hours (wh) or kilowatt-hours (kwh). Obviously, one watt-hour equals one watt of power in use for one hour, and one kilowatt-hour equals one kilowatt in use for one hour.

EXAMPLE 14.9

(a) Find the daily energy consumption of the appliances listed if they are used daily for the amount of time shown.

Toaster (1340 w)	15 min
Percolator (500 w)	2 hr
Fryer (1560 w)	$\frac{1}{2}$ hr
Iron (1400 w)	$\frac{1}{2}$ hr

(b) If the average cost of energy is \$0.06 per kilowatt-hour, find the daily operating cost.

SOLUTION

(a)

Toaster	$1340 \text{ w} = 1.34 \text{ kw} \times \frac{1}{4} \text{ hr} = 0.335 \text{ kwh}$
Percolator	$500 \text{ w} = 0.5 \text{ kw} \times 2 \text{ hr} = 1.00 \text{ kwh}$
Fryer	$1560 \text{ w} = 1.56 \text{ kw} \times \frac{1}{2} \text{ hr} = 0.78 \text{ kwh}$
Iron	$1400 \text{ w} = 1.4 \text{ kw} \times \frac{1}{2} \text{ hr} = 0.70 \text{ kwh}$
Total	2.815 kwh

(b) The cost is

$$2.815 \text{ kwh} \times \$0.06/\text{kwh} = \$0.1689$$

or approximately 17 cents

Clearly the power being used at any specific time during the day by a residential household varies considerably. A meter that measures instantaneous kwh usage and reads in cost per hour is shown in Figure 3.17, page 75. If we were to graph the power in use for a typical American household during a normal weekday, the plot might look something like that in Figure 14.8. The average power demand of the household is obviously much lower than the maximum. The ratio between the two is called the overall load factor and runs between 20% and 30% for a typical household. The energy used by this household for the 24-hr period shown is represented by the *area* under the curve of Figure 14.8 that can be determined by integration only, since it varies continuously. (Note that this house is partially air-conditioned, with window units, which accounts for a considerable portion of the base load.) That this is exactly what a kilowatt-hour meter does will be explained in Section 14.14, which deals with electrical measurements.

EXAMPLE 14.10. It has been estimated that the average power demand of an American household is 1.2 kw. Calculate the monthly electric bill of such a household, assuming a flat rate of \$0.06 per kilowatt-hour.

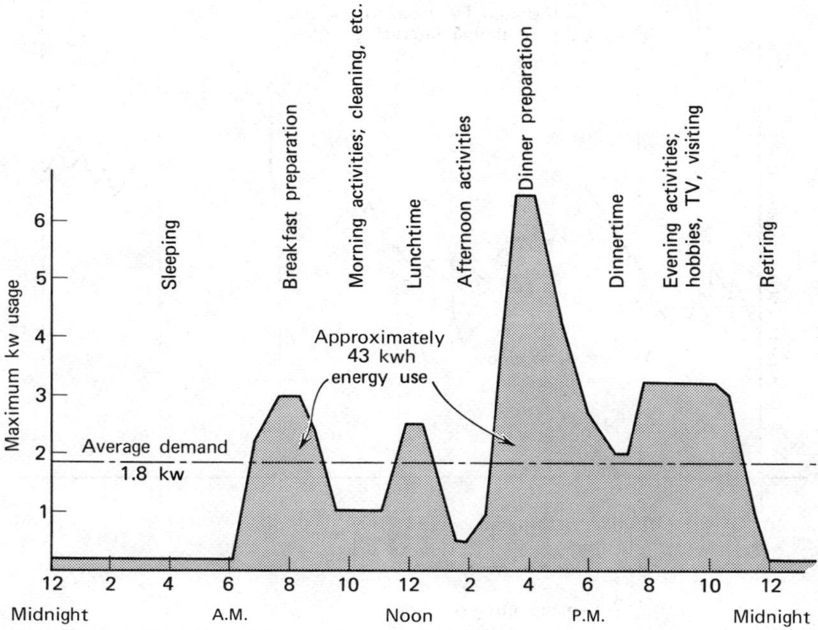

Fig. 14.8 Hypothetical graph of power usage for a typical American household. Electric cooking is assumed. Area under the curve represents energy usage. Maximum kw demand (vertical axis) is based on a 15-min integrated demand, thus eliminating spikes in demand, such as those caused by starting a refrigeration (air-conditioning) compressor. This curve has a 24-hr use of approximately 43 kwh, giving an average demand of 1.8 kw and a load factor of 27.5%.

SOLUTION

Monthly energy consumption

$$= 1.2 \text{ kw} \times \frac{24 \text{ hr}}{\text{day}} = \frac{30 \text{ days}}{\text{month}} = 864 \text{ kwh}$$

Electric power bill

$$= 864 \text{ kwh} \times \$0.06/\text{kwh} = \$51.84$$

We stated in Example 14.10 above that the bill was based on a "flat" rate of 6 cents per kilowatt-hour. In actual fact the rate structure of most utilities is not so constructed. Generally, the tariff *decreases* with increasing use, thus encouraging larger use of electric power. This type of tariff was designed in the halcyon days of low cost fuel, when the utility's costs were in large measure those of transmission, distribution, and administration. As recently as 1973, one large Eastern utility estimated that even if fuel were *free*, their tariff could not be lowered more than 10%. That this situation has changed is well known. Residential electric rates have been increased but, with few exceptions, have not been restructured to encourage either efficient use of electric energy and the facilities for supplying this energy, or reduction in energy usage. One technique, long standard in industrial and commercial user tariffs but almost never applied to residential users, is the levying of a charge for power (kw) in addition to the normal energy (kwh) charge. This *demand charge* is primarily useful in encouraging users to reduce their peak loads. In so doing, energy use is also reduced somewhat.

14.12 Electric Demand Charges

As stated, electric utility companies normally levy a kw demand charge on all but individual residential customers. Varying with the individual utility company involved, this monthly charge runs between \$2 and \$10 per

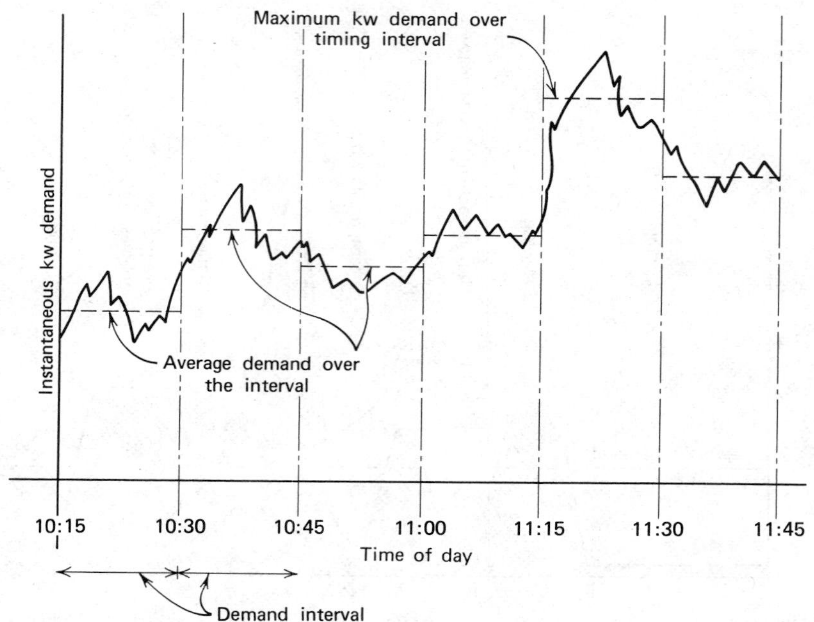

Fig. 14.9 Typical instantaneous load curve for a facility. The utility demand meter records the average demand in each period (here 15 min). The maximum interval demand—in this case between 11:15 and 11:30—is used as the basis for monthly demand billing.

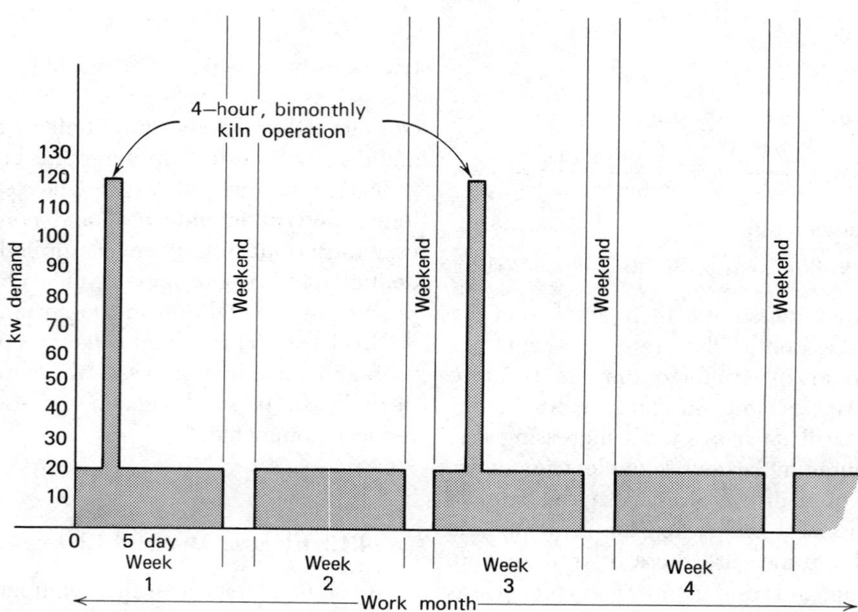

Fig. 14.10 Illustrated here is a case of extremely poor load factor (energy use efficiency) caused by large, short duration loads. Minor variations in load are too small to show on this time scale. Monthly demand billing would be based on the 120 kw maximum demand. Since energy is represented by the area under the curve, it is apparent that the large power peaks do not utilize a proportionately large amount of energy.

kilowatt of maximum average demand over the demand interval for that month. Demand intervals vary, being either 15, 30, or 60 min (see Fig. 14.9), with most companies averaging the two highest figures for the month. Some companies utilize the highest kw demand for any month to establish the demand charge for the *entire year,* thus penalizing those users whose *yearly* load factor is low. Normally, however, only the monthly load factor is considered. The load factor is a measure of the efficiency with which energy is utilized; a low load factor indicates low efficiency, for which such a user is penalized by a high demand charge. The reasoning behind the imposition of a demand charge and the significance of load factors can best be demonstrated by an example.

Assume that a pottery manufacturer, whose average 8-hr daily load is 20 kw for lighting, pottery wheels, and the like, operates two 50 kw electric kilns twice a month for a 4-hr period each time (see Fig. 14.10). Further assume an energy rate of \$.03 per kwh. The total monthly *energy* bill for the operation of the two kilns would be

$$\text{Cost} = 2 \times 50 \text{ kw} \times 8 \text{ hr} \times \frac{\$.03}{\text{kwh}} = \$24$$

Thus, were it not for the demand charge, the utility company (which is required by law to supply maximum customer demand) would have to provide and maintain 100 kw of generation, transmission, and distribution facilities in return for a payment that is the equivalent of 1.11 kw of average continuous load, that is

$$\text{equivalent continuous load} = \frac{\$24}{720 \text{ hr/month} \times .03} = 1.11 \text{ kw}$$

This user's load factor can be calculated readily. By definition,

$$\text{load factor} = \frac{\text{average power demand}}{\text{maximum power demand}} \qquad (14.8)$$

But, since for a given time interval the average power demand is equal to the energy used divided by the time involved, that is

$$\text{average power demand} = \frac{\text{kwh energy use}}{\text{hours of use}} \qquad (14.9)$$

we have the general expression

$$\text{load factor (LF)} = \frac{\text{kwh}}{\text{maximum demand} \times \text{hours use}} \qquad (14.10)$$

For the case under consideration, the *monthly* load factor is

$$\text{LF} = \frac{23 \text{ days} \times 8 \text{ hr} \times 20 \text{ kw} + 2 \text{ days} \times 4 \text{ hr} \times 100 \text{ kw}}{120 \text{ kw} \times 720 \text{ hr/month}}$$
$$= \frac{3680 + 800 \text{ kwh/month}}{86400 \text{ kwh/month}}$$
$$= .052 \text{ or } 5.2\%$$

This is obviously a very poor load factor, which results in a high demand charge. Assuming a \$3.50 per kw demand tariff, this pottery manufacturer would be billed, monthly, an additional:

$$\text{demand charge} = 120 \text{ kw} \times \$3.50 = \$420.00$$

Since the energy bill is only

$$\text{energy cost} = 4480 \text{ kwh} \times 0.03 = \$134.40$$

this manufacturer is paying heavily for an inefficient use of energy. (They would be well advised to use a fossil fuel-fired kiln, rather than one using high-grade energy, like electricity. Generally, to use heat to create electricity in order to re-create heat is wasteful of resources.)

Although the illustration selected is somewhat extreme in its inefficient use of energy, it is not uncommon to find demand charges of the same order of magnitude as energy charges. Obviously, it is impossible to eliminate demand charges entirely without ceasing operation, but it is certainly possible and frequently very simple to reduce them. Such a step is in the interest of the user, the utility, and the public at large: the user—for simple economic reasons; the utility—to permit more efficient use of their facilities; and the general public—by avoidance of unneccesary power plant construction and concomitant inefficient use of fuel, and overall reduction in fuel use. This last item is a secondary benefit of demand control, which will be explained in the following section.

14.13 Electric Load Control

Methods of load control vary greatly in complexity and in degree of automation, but basically perform the same task—efficient utilization of available energy to produce a high load factor. This results in a lowering of demand charges and, secondarily, in a reduction of energy costs. (An ancillary, but important, benefit is the maximum utilization of electrical power equipment, which normally runs underloaded. This results in smaller equipment, lower first cost, and less space utilization.) The control devices and systems are variously referred to as load shedding control, peak demand control, peak load regulation, power use control, and permissive load control. Nomenclature such as "energy-demand control" and "energy management" are somewhat misleading, since these devices are primarily concerned with control of *power*, with energy savings being an important, but secondary, benefit. We will use the expression "electric load control" for simplicity and avoid knowingly using a term that refers to a particular manufacturer's equipment.

(a) Level 1—Load Scheduling and Duty-Cycle Control. This level is the simplest and most obvious, and it is applicable to all types of facilities. The installation's electric loads are analyzed and then scheduled to restrict demand. Thus large loads can be shifted to off-peak hours and controlled to avoid coincident operation. The user can also take advantage of special-night and weekend-utility rates for loads that do not require immediate operation, such as battery charging and transfer pumping. Control can be entirely manual or automated by use of a duty-cycle controller. This device is essentially a program clock (see Section 22.28) with switching for a number of circuits, or loads. Typical applications of this device are control of HVAC loads, lighting loads, and process loads in small commercial, institutional, and industrial buildings.

A typical unit is illustrated in Figure 14.11. The usefulness of these units lies in

1. Eliminating energy waste by shutting down units when not required.
2. Automatic control such as preheat and precool, which results in lower power and energy levels.
3. Establishment of efficient equipment scheduling without continual manual supervision.

These devices are not, strictly speaking, load controllers in that no cognizance is taken of the actual continuous electric loads. Instead, the devices operate on a preset duty-cycle relying entirely on a prior analysis of the loads. Although such an analysis is a necessary first step in all levels of electric load control, its efficacy is limited since many of the loads are automatically controlled. Thus compressors controlled by pressure switches, fans controlled by thermostats, and pumps controlled by float switches cannot be scheduled with this type of duty-cycle controller and coincident operation cannot be prevented. Demand charge savings of 10 to 15% or more are common with these devices; energy charge reduction of this magnitude is also possible.

(b) Level 2—Demand Metering Alarm. If in conjunction with a duty-cycle controller some type of continuous demand metering is installed which will go into alarm when a predetermined demand level is exceeded, a basic load control system will have been established. The load analysis discussed above would have to be extended to determine load priorities so that when the preset maximum demand load is exceeded and the alarm sounds, loads can be shed (disconnected) manually in a predetermined order of priority and, subsequently, reconnected also in order of priority. This type of control is practical only for a limited size installation inasmuch as most of the load switching activity is manual. Also, to avoid excessive alarming, the facility operator may be inclined to set the alarm point

Fig. 14.11a (above) *The heart of a duty-cycle controller is the program drum and associated controlled contacts. The terminal strips are wired to the drum contacts and then to the controlled loads. Courtesy of Pacific Technology, Inc.*

Time of day	Position	Home	Fan #1	Fan #2	Pump #1	Duct heater	Fan #3	Pump #2
		G	F	E	D	C	B	A
	1	■						
	2		■					
	3			■				
	4				■			
	5					■		
	6		■				■	
	7			■				■
	8				■			
	9					■		
	10		■				■	
	11			■				■
	12				■			
	58		■					
	59			■				
	60				■			

Fig. 14.11b (left) *A table of the type shown is indispensable when using this type of controller, because it permits the planning of the duty cycles and the related placing of pins; it also provides a record of the drum contact arrangement. Courtesy of Pacific Technology, Inc.*

higher than it should be to effect all possible savings. A unit typical of this type is illustrated in Figure 14.12. This unit operates on utility supplied energy pulses.

(c) ***Level 3—Automatic Instantaneous Demand Control.*** This type of control (also called "rate control") is, in effect, an auto-

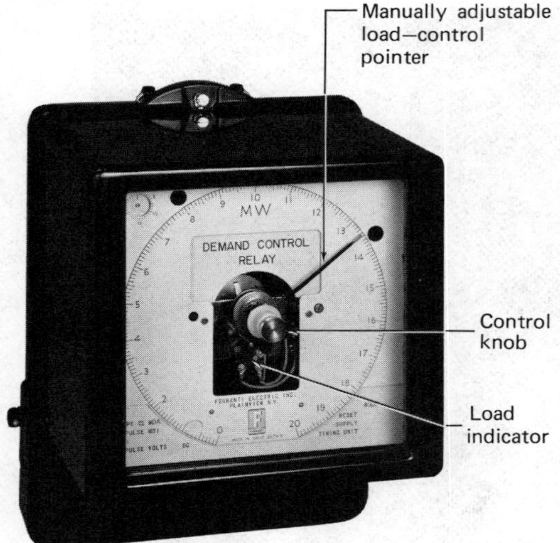

Fig. 14.12 This alarm unit has two basic components: the load indicator that acts in response to utility load information and the control pointer that is manually set at the maximum permissible demand load. The load indicator resets automatically at the beginning of each new billing interval. Courtesy of Ferronti Electric, Inc.

mated version of the level-2 system described above. The unit accepts instantaneous kw load information from the utility system either in the form of continuous current readings or in the more sophisticated form of pulsed energy data. It then compares this information to the preset kw limit (rate control) and acts automatically to disconnect and reconnect loads as required. These units *do not* recognize the utility's metering interval of 15 or 30 min, but act continuously on the basis of load comparison data. For this reason, these units are also referred to as load comparator controllers. Examination of Figure 14.13 will make the unit's operation clear.

The first step in setting up this system is to separate the contollable ("sheddable") loads from those that must remain uninterrupted. Depending on the type of facility, the two lists that follow are typical.

"*Sheddable*"

Nonessential lighting	Domestic hot water heating
Ventilation fans	Sewage ejectors with appropriate level controls
Space heating	Transfer pumps
Comfort cooling	Electric snow melting
Noncritical batch process equipment	Any device with flywheel effect
Electric boilers	

"*Nonsheddable*"

Essential lighting	Process equipment
Elevators	Material handling equipment
Refrigeration compressors	Office machinery

The nonsheddable loads are fed directly from the power line. The sheddable loads are fed via a panel of control relays that respond to on/off instructions from the demand controller. The facility's operator will then analyze the sheddable loads and determine a minimum coincident kw load for this group. This figure, when added to the total coincident of the nonsheddable loads, becomes the kw rate control or the *set point*. To avoid excessive cycling of loads a variable width kw band around the set point is used (see Fig. 14.13*b*). Note that the controller acts to reduce maximum loads (peaks) and fill in low points (valleys). Although theoretically the energy use with or without the controller is the same, in actual practice energy savings of 15% and more are common. Most commercial units permit considerable flexibility in field adjustments in order to arrive at optimum operation. Some units, like the one illustrated in Figure 14.14, have special provisions to overcome some of the limitations inherent in rate control systems. These are:

1. Excessive cycling due to too narrow an operating band or insufficient time delay between control commands.
2. Inability to change or "weight" load priority structure, resulting in excessive cycling of certain loads and insufficient cycling of others.
3. Excessive off-time due to the absence of limit timers.
4. Inability to readily adapt to varying load

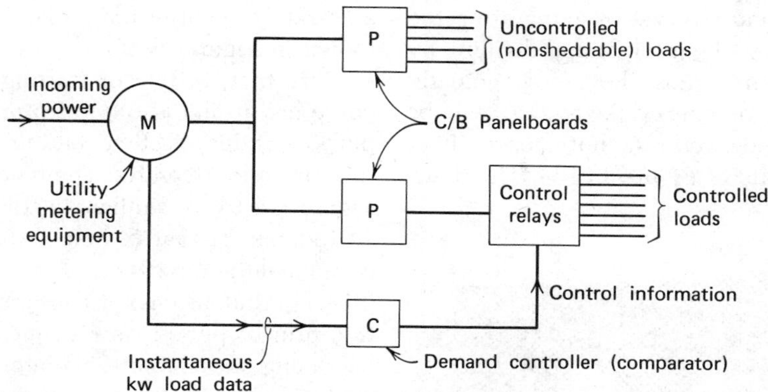

Fig. 14.13a Block diagram of a system of automatic electric power control. The demand controller receives instantaneous load data from the metering equipment, compares it to preset limits, and disconnects and reconnects controllable loads automatically to keep load within these limits.

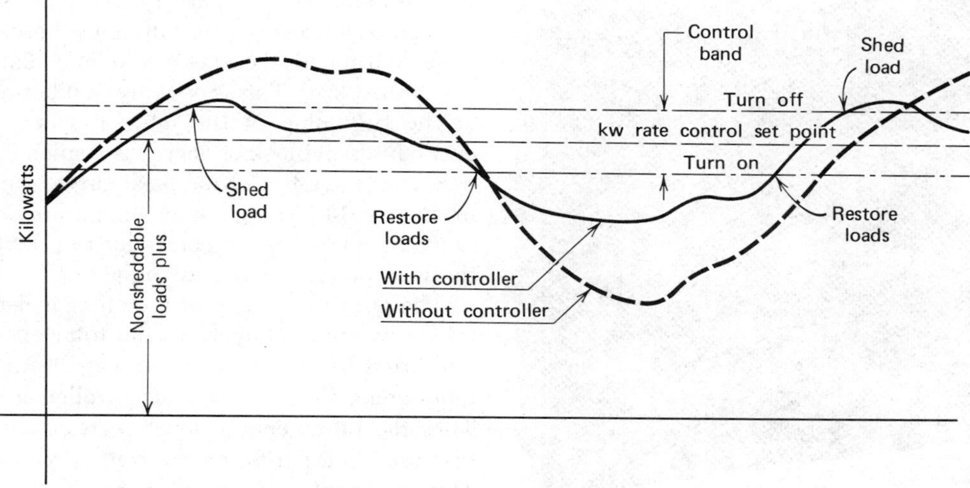

Fig. 14.13b Action of the demand controller is graphically illustrated. The preset maximum demand is shown at the center of the band as rate control. See text, Section 14.13(c).

patterns, resulting from variable production schedules, time schedules, changes in weather, and so on.

5. Nonrecognition of the block of energy available in a given timing interval.

As a result of these limitations, this system is most useful in applications where operating modes do not change frequently and the facility is not very large. Thus stores, supermarkets, warehouses, small industrial facilities, and commercial installations are well served with this level system if they have at least 20% sheddable loads and their connected electric load is at least 150 kilovolt-amperes (kva).

(d) Level 4—Ideal Curve Control. This controller operates by comparing the actual rate of *energy* usage to the ideal rate, and controls kw demand by controlling the total *energy* used within a metering interval (refer to Fig. 14.15). A constant rate of energy use over a demand interval would show as the set of repeated straight lines in Figure 14.15*a*. The utility company determines the demand over the demand interval by integrating the kwh energy over the interval and dividing by the interval time. Thus, the user is actually given a block of energy (kwh) that can be utilized at any desired rate, not necessarily at the constant rate of Figure 14.15*a*. The desirable rate of energy use is the "Ideal Curve" shown in Figure 14.15*b*.

Fig. 14.14 This demand control unit operates on the rate control principle, keeping the demand within a band around the set point. Courtesy of Pacific Technology, Inc.

Note that, in lieu of starting at zero, the curve has an initial offset. To understand the purpose of this offset we must understand how the controller operates. The block diagram of Figure 14.16 is similar to that of Figure 14.13*a*, except that in lieu of the utility supplying continuous data on kw demand, it supplies continuous *energy* consumption data in the form of pulses, plus a time signal at the beginning of each metering interval. The controller compares the actual energy usage to the preset "ideal" curve and acts to shed load when the difference reaches a preset minimum. Thus, if the ideal curve were to start at zero as indicated in Figure 14.15*a* (no offset), the controller seeing zero difference between the actual and ideal curves would immediately act to shed load. This is obviously undesirable at the beginning of the interval when the entire interval block of energy is available for use. Thus the offset of the ideal curve, shown in Figure 14.15*b*, prevents premature load shedding, allowing the controller to establish the load pattern before taking action.

The numerical value of the offset is determined by subtracting from the total energy permitted in the demand interval the maximum energy use of the noncontrolled loads. Thus the offset energy level corresponds to that used in a period by the *controlled* loads. This is shown in Figure 14.15*b*. A typical

Fig. 14.15a Graph of cumulative energy use over a demand interval, corresponding to a constant kw demand. (Energy use is the time integral of power; i.e., kwh = kw × time).

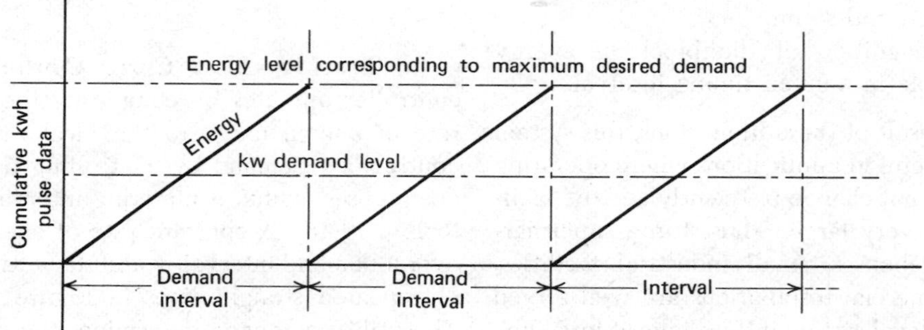

usage curve is shown in Figure 14.15*c*. Shed points can be preprogrammed for each load independently according to a predetermined priority, and priorities can be readily adjusted and rescheduled. Loads are normally shed only toward the end of the interval when the permissible energy total is approached and all loads are restored at the beginning of each interval. Thus during each interval loads are off for only a few minutes at most. Controllers operating on the "ideal curve" principle are considerably more flexible than the kw rate controller (described in the level-3 system above) and are applicable to facilities of widely divergent load size, but with at least a 300 kw connected load. These include office buildings, hospitals, schools, and industrial facilities. As with other controllers, the principal savings will be in demand charges, but almost always with considerable economies in energy billings. The modest cost of these units usually gives a very rapid pay-back period and accounts for their wide use.

An interesting application of these units is to utilize them to operate a standby power plant to supply peak power demands in lieu of

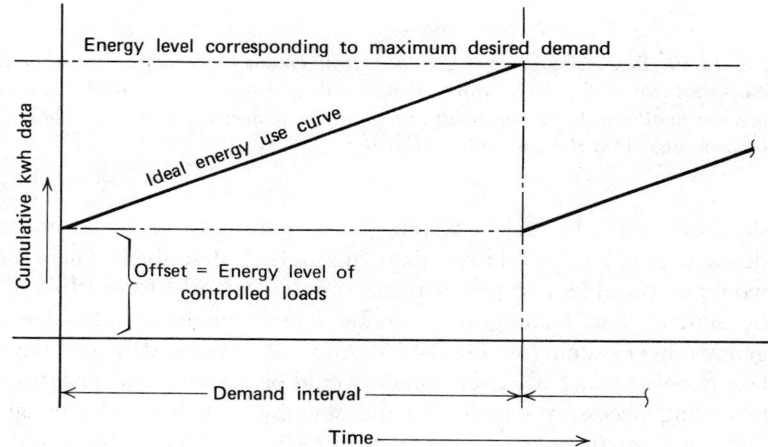

Fig. 14.15b *By modifying the constant kw energy curve of* (a) *with an offset, an "ideal" curve is established to which actual energy use will be compared by the controller, in deciding whether to shed or add loads. The kwh pulses representing this curve are generated within the controller.*

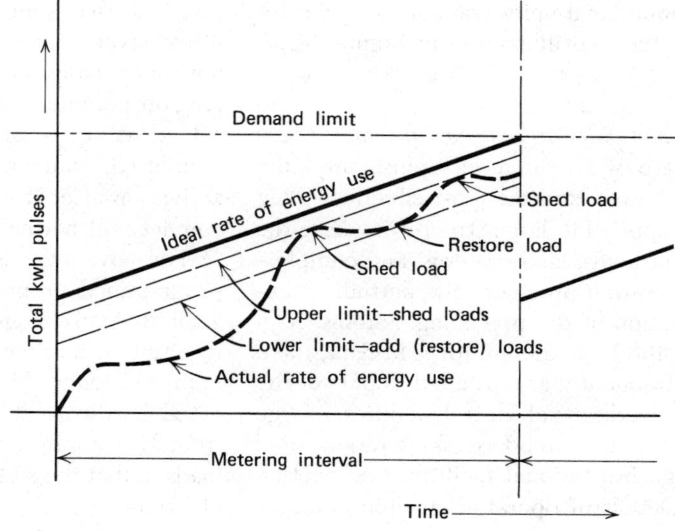

Fig. 14.15c *Graph of action of an ideal rate controller. Note that toward the latter part of the period the pattern has been established, and the actual rate of energy use corresponds closely to the ideal.*

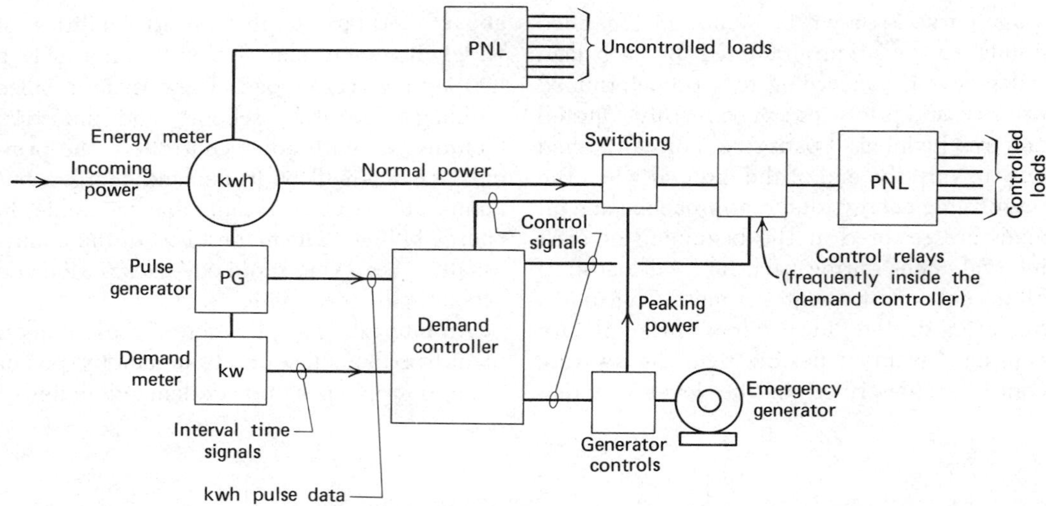

Fig. 14.16 Block diagram of a possible arrangement for demand control with an "ideal curve" controller. Note that cumulative kwh information in pulse form must be available from the utility. If interval signals are not available from the utility, they can be generated in the controller. The control relays are often incorporated into the controller enclosure.

shedding loads. In such a system, maximum demand could be set lower than otherwise would be possible, and power peaks supplied by starting and loading the standby emergency power system (see Fig. 14.16). An auxiliary benefit of such an arrangement would be providing necessary exercise to the standby unit. Such an arrangement would require the approval of the utility, since many utilities prohibit use of standby generation except under emergency conditions. Typical of the many commercial units available in this category are the two illustrated in Figure 14.17.

(*e*) ***Level 5—Forecasting Systems.*** These systems are by far the most sophisticated, the most expensive, and the most effective. They are best applied to large structures where the number of loads, load patterns, and complexity of operation precludes the partially manual operation of the preceding systems. As a result of the large amount of load data, these systems frequently are installed as part of the computerized central control facilities in large industrial plants, modern sky-scrapers, and very large institutional facilities (see Section 22.39). Details of operation are too complex for our needs, but the basic operation can be described. These units operate by continuously forecasting the amount of energy remaining in the demand interval, based on kwh pulse data received. They then examine the status and priority of each of the connected loads and decide on a course of action. Loads, which in other systems are classified noncontrollable, are in this system controlled, because of the accuracy and rapidity of the control function. A pneumatic compressor, for instance, that supplies process air might in other level control systems be classified as noncontrollable, despite the fact that it has long off periods. With computer control, such a load would be classified as "delayable" or "inhibited," since a 30-sec or 1-min delay in activation after the pressure switch closes its contact will normally be acceptable.

The advantage of these systems is that if programmed properly they can make small, accurate load changes throughout the interval, resulting in minimum load cycling and maximum efficiency. Use of this type of sophisticated hardware does, however, require extremely accurate analysis of the nature of all loads so that the system's capabilities are fully utilized.

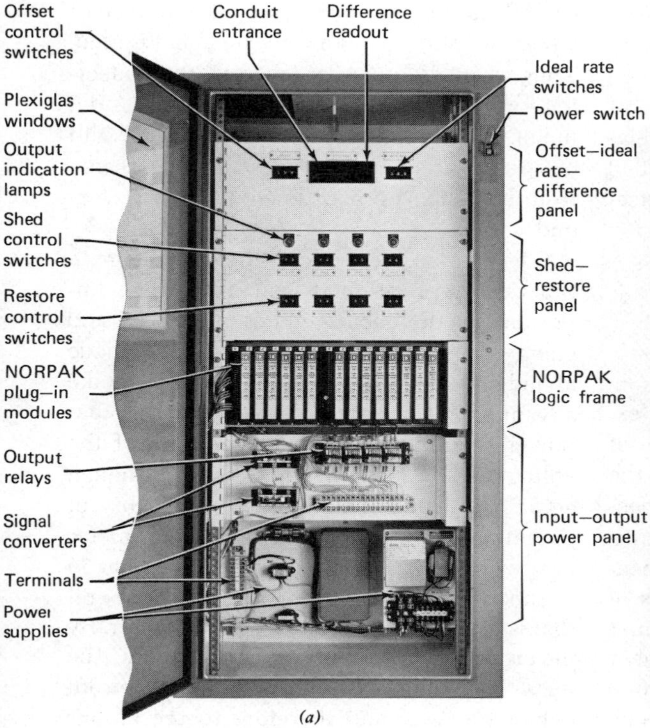

Offset
control
switches
Conduit
entrance
Difference
readout
Ideal rate
switches
Power switch
Plexiglas
windows
Output
indication
lamps
Offset—ideal
rate—
difference
panel
Shed
control
switches
Shed—
restore
panel
Restore
control
switches
NORPAK
plug—in
modules
NORPAK
logic frame
Output
relays
Signal
converters
Input—output
power panel
Terminals
Power
supplies
(a)

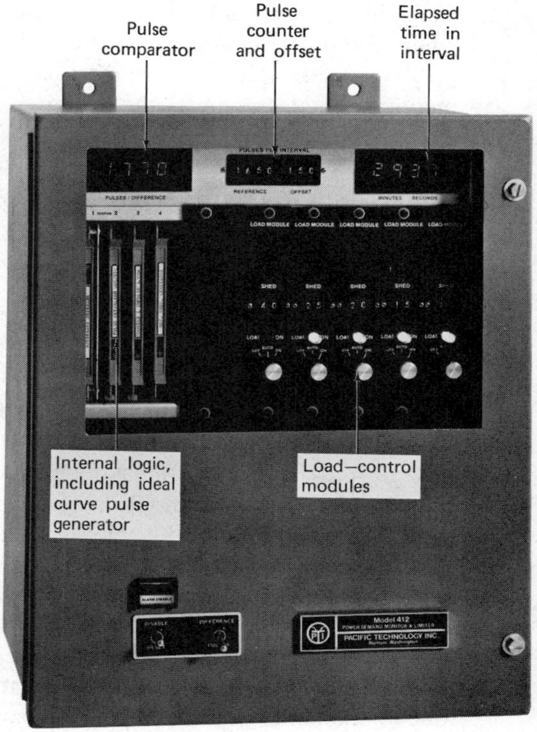

Pulse
comparator
Pulse
counter
and offset
Elapsed
time in
interval
Internal logic,
including ideal
curve pulse
generator
Load—control
modules
(b)

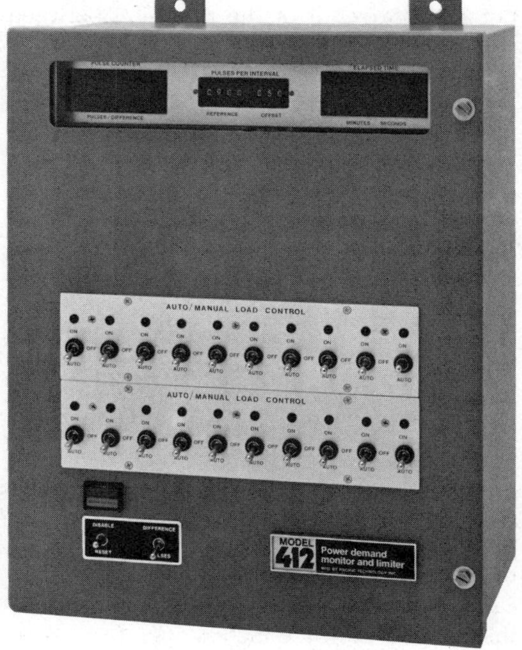

Fig. 14.17 (a) *A typical industrial demand controller, of the ideal-curve type.*
Courtesy of Square D Company, Tradename NORPAK[R] (b *and* c) *These particular demand controllers are also of the ideal-curve type and are applicable to loads in excess of 1000 kw. Unit* (b) *contains load-control modules that are utilized when load priorities are frequently changed. Figure* (c) *shows the same unit with standard auto/manual load switches, used where the above flexibility is not required. Courtesy of Pacific Technology, Inc.*

AUTO/MANUAL LOAD CONTROL
AUTO/MANUAL LOAD CONTROL
MODEL
412
Power demand
monitor and limiter
(c)

(f) Utility Rate Structures. Although actual rates vary from one utility to another, the two most common systems are

1. Energy charge based on usage, and demand charge based on maximum interval demand.
2. Energy charge based on a factor comprising energy and demand, and demand charge as in (1) above.

Generally, greater savings will be realized

from the installation of a demand controller where the utility charges are structured as in (2) above. Although each case must be studied individually, it has been the experience of most users that these controllers have a relatively short pay-back period, plus the added advantages of fuel conservation and ecological desirability.

14.14 Electrical Measurements

In the preceding sections we have explained the fundamental electric quantities of voltage and current, and have defined the units involved as volts and amperes, respectively. As is true of all other physical quantities that are to be used in practical application, the need existed for a simple means of measuring these quantities. This need was met by the development of the meter movement illustrated in Figure 14.18. Everyone at one time or another has felt the repulsion between like poles of two magnets held close together and, conversely, the attraction between opposite poles. This principle is used in the basic meter movement. It causes a deflection of the pointer as a result of the repulsion between the field of a permanent magnet and an electromagnet. The electromagnet is formed when current flows in the coil, and its strength is proportional to the amount of current flowing. Thus, a strong current causes a larger deflection of the needle and, therefore, a higher reading on the dial. A spring (see Fig. 14.18) provides restraining torque on the pointer. To make this very sensitive basic unit usable for large currents (it is intrinsically a microammeter, sensitive to millionths of an ampere), we simply divert, or shunt away, most of the current, allowing only a few microamperes to actually flow in the meter coil.

To use the same unit as a voltmeter, we put a large resistance (multiplier) in series with the meter, again limiting the current flowing to a few microamperes. The scale is then calibrated in volts. All d-c meters are made in this fashion. Most a-c meters operate on basically the same principle, except that instead of a permanent magnet an electromagnet is used. Thus when the polarity reverses, the deflecting force remains in the same direction. A d-c meter connected to an a-c circuit simply will not read, since inertia prevents the needle from bouncing up and down 60 times a second.

The measurement of current and voltage in practical application is generally not as important as the measurement of power and energy, as the preceding section has made abundantly clear. To measure power, we take advantage of the fact learned earlier that power is proportional to the product of the voltage and current in the circuit. Although actual construction is complex, the theory of operation of a wattmeter is simple. The meter has two coils; a current coil that is similar in connection to an ammeter, and a voltage coil that is similar in connection to a voltmeter. By means of the physical coil arrangement, the meter deflection is proportional to the product of the two, and therefore to the circuit power. The meter can be calibrated as desired, depending on the size of the shunts and multipliers. The schematic arrangement is shown in Figure 14.19. To measure energy, the factor of time must be introduced, since as we know

$$\text{energy} = \text{power} \times \text{time}$$

D-c energy meters are available but are not of general interest because of the rarity of d-c power. A-c watt-hour meters are basically small motors, whose speed is proportional to the power being used. The number of rotations is counted on the dials, which are calibrated directly in kilowatt-hours. A diagram of the basic construction of an a-c kilowatt-hour meter is shown in Figure 14.20*a*. As can be seen from Figure 14.20*b*, the kilowatt-hour energy consumption and the maximum interval kw demand can be read directly from the dials. (The illustrated unit is of the standard, not energy pulse, type.) If the numbers involved are too large or a meter is used with current transformers between it and the line, or for calibration reasons, a multiplying factor

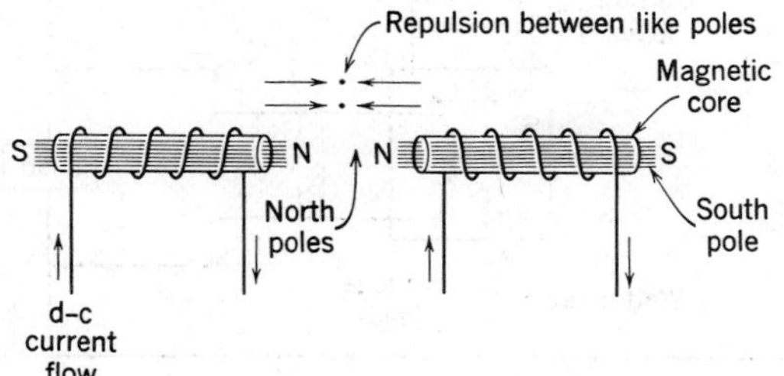

Fig. 14.18a *Diagram showing basic electromagnetic principal and interaction between electromagnets. Any iron core becomes an electromagnet when current flows in a coil around it, as shown.*

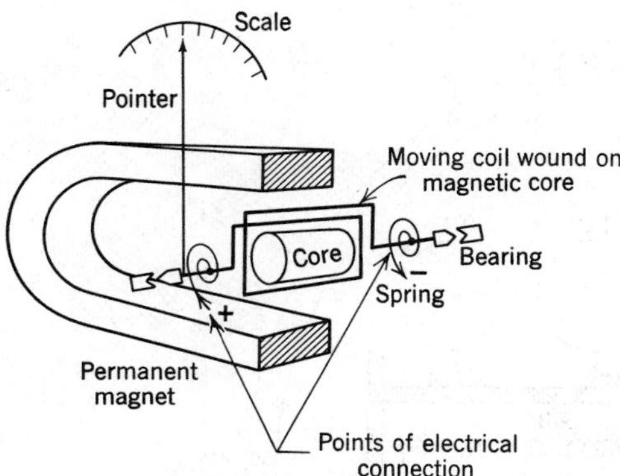

Fig. 14.18b *The principal of the electromagnet is used in this basic meter movement. Current flowing in the movable coil forms an electromagnet whose field interacts (see above) with the permanent magnet's field, causing a deflection proportional to the current flow. Courtesy of Westinghouse, Relay-Instrument Division*

Fig. 14.18c *Illustrated are two typical modern, $4\frac{1}{2}$ in. (11.2 cm) sq switchboard instruments. Their mechanisms and special coil suspension systems give them high accuracy (1%), sensitivity, and shock resistance. Courtesy of Westinghouse, Relay-Instrument Division*

Fig. 14.19 Schematic arrangement of wattmeter connections. Note that the current coil is in series with the circuit load, whereas the voltage leads are in parallel (see also Fig. 14.20).

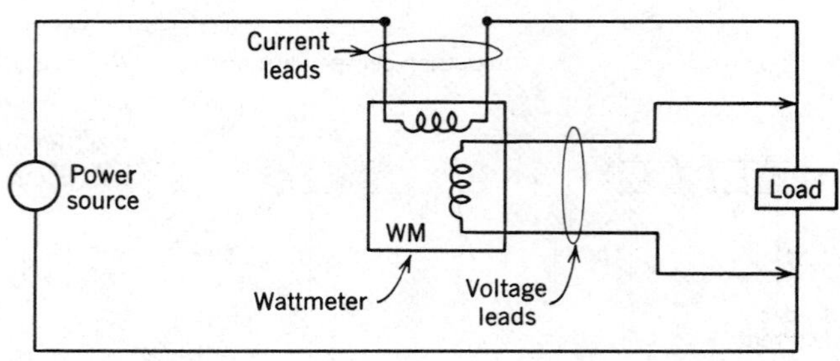

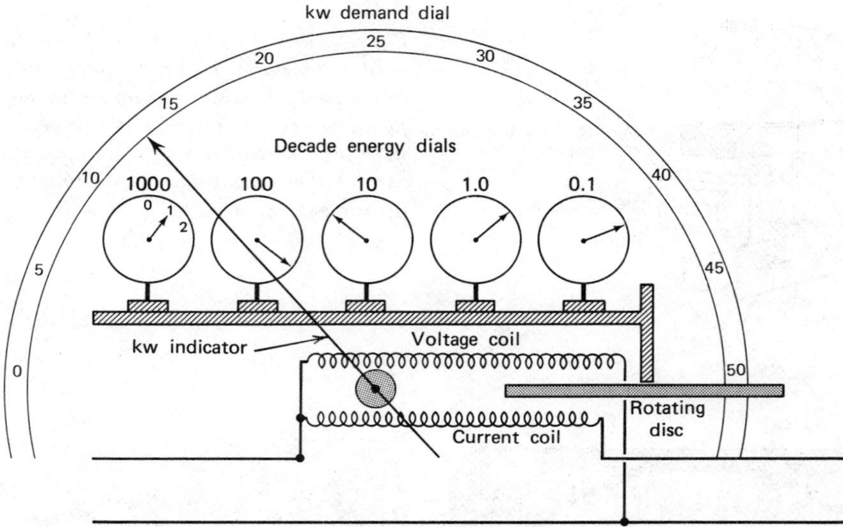

Fig. 14.20a Typical induction-type kwh meter with kw demand dial. Decade dials register total disc revolutions that are proportional to energy. Disc speed is proportional to power. See equivalent meter in (b).

Fig. 14.20b Typical socket-type kwh meter with demand register. Courtesy of Westinghouse, Meter Division

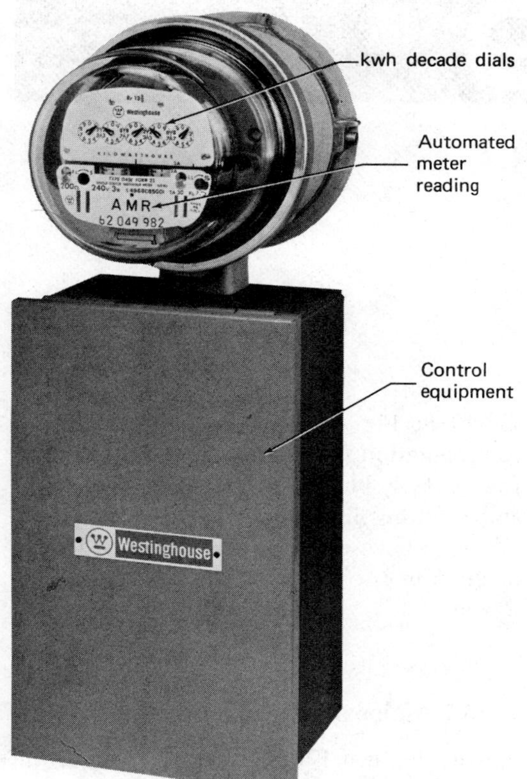

is required to arrive at the proper kilowatt-hour consumption. This number is written directly on the meter nameplate, and we multiply the meter reading by it to get the actual kilowatt-hours. In the absence of such a number, it can be assumed that the meter reads directly in kilowatt-hours.

A special type of kilowatt-hour meter is illustrated in Figure 14.21. This instrument is equipped with an electro-optical automatic meter reading system that is activated from a remote location. The meter data is transmitted electrically to a data processing center where it may be used by the utility to prepare subscribers' bills, to prepare customer load profiles and to study, in combination with other such data, area load patterns, equipment loading, and so on. The customer can use instantaneous data to control loads as explained at length in the preceding section. Of course, the most obvious characteristic of this equipment (whether it is an advantage depends on one's viewpoint) is the elimination of the traditional (and very expensive) meter reader.

Fig. 14.21 *The illustrated encoding register (meter) for automated meter reading is mounted on a control cabinet that contains electronic equipment capable of performing a number of control functions. These include supply of pulse data for demand metering (see Section 14.13), automated metering functions including meter checking and data transmission to a central point, control of selected load, and related functions. Courtesy of Westinghouse, Meter Division*

15

Electrical Systems and Materials: Wiring

15.1 System Components

The major components of a building's electrical power system are illustrated in Figure 15.1*a*. They can be arranged in three major categories, that is: wiring, power-handling equipment, and control and utilization equipment. In the first category, we include conductors and raceways of all types; in the second, transformers, switchboards, panelboards, large switches, and circuit breakers; and, in the last, actual utilization equipment such as lighting, motors, controls, and wiring devices. After some discussion, which is applicable to all electrical materials, this chapter will discuss in detail the items in the first of these three categories, that is, the wiring system. Chapter 16 will cover most of the remaining items in the other two categories with the exception of lighting equipment, which will be discussed in the lighting section. Signal equipment including telephones, intercom, and control will be covered separately.

Referring to Figure 15.1*a* note that the power system proceeds from the service point to the utilization point in a series of descending steps—at each distribution point the branch capacity decreases. This is not so obvious on Figure 15.1*a* which, being a single-line diagram, does not differentiate by line weight between heavy and light (large and small) conductors. If this were done, and assuming, as is usually the case, that the heavier the conductor the greater the amount of power being carried, the single-line diagram of Figure 15.1*a* would be shown as in Figure 15.1*b*. This "size" differentiation, that is indicative of the amount of power handled, is more clearly shown in Figure 15.2 which is a pictorial representation of a system similar to that of Figure 15.1, but in somewhat greater detail and omitting items beyond the panelboard. We shall have occasion to refer back to this diagram in our discussions of the components shown.

15.2 National Electric Code

The National Electric Code (NEC) of the National Fire Protection Association (NFPA) defines the fundamental safety measures that must be followed in the selection, construction, and installation of all electrical equipment. This code is used by all inspectors, electrical designers, engineers, contractors, and the operating personnel charged with the responsibility for safe operation. Having been incorporated into OSHA (Occupational Safety and Health Act) it has, in effect, the force of law. The reader of this book should obtain for personal use the latest edition of the NEC from the NFPA at 80 Batterymarch Street, Boston, Massachusetts 02110. Frequent references will be made to this code.

In addition to the National Code, many large cities such as New York, Boston, and Washington, D.C., have their own electrical codes that, though similar to the NEC, contain numerous special requirements.

In order to assure a minimum standard of intrinsic electrical safety for electrical equipment, a single agency was needed to establish standards for, and to actually test and inspect, electrical equipment. Such an organization is

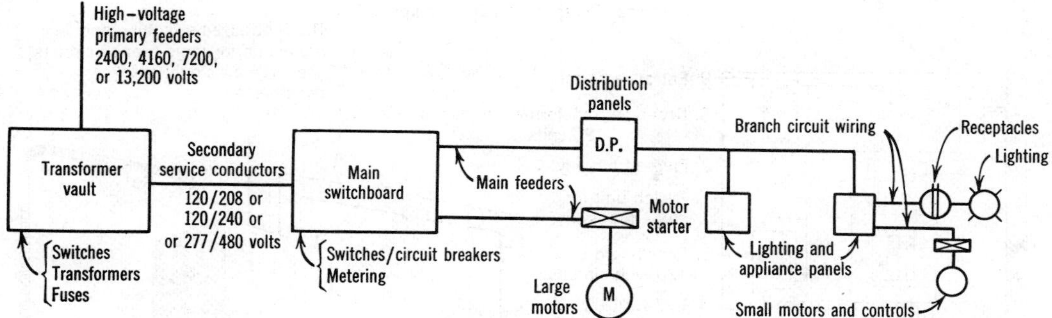

Fig. 15.1a Single-line diagram of a typical building electrical system, from the incoming service to the utilization items at the end of the system. This type of diagram is also referred to as a "block" diagram, since the major components are shown as rectangles, or blocks. When this same type of information is presented showing the spatial relations between components, it is called a "riser diagram"; when electrical symbols are used in lieu of the blocks, it is called a "one-line" or a "single-line" diagram.

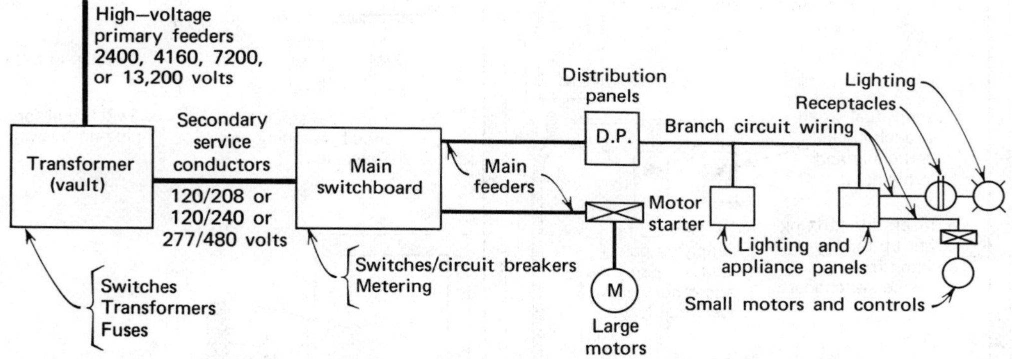

Fig. 15.1b Were the connecting conductors between the major system components drawn to reflect size and thereby power handling capacity, the system of Figure 15.1a would appear as shown here.

the Underwriters Laboratories, Incorporated, which publishes lists of inspected and approved electrical equipment. These listings are universally accepted, and many local codes state that only electrical materials bearing the Underwriters Laboratories (UL) label (of approval) will be acceptable.

15.3 Economics of Material Selection

The selection of electrical materials involves not only choosing a material or assembly that is functionally adequate and, where necessary, visually satisfactory, but also the consideration of economic factors. This is necessary since, in most instances, there is available a multiplicity of equipment that will fulfill the construction need. In such cases, economic factors often decide the issue. The decision is relatively simple when the various materials differ only slightly from each other and a straightforward first-cost comparison is all that is required. Often, however, the choice is not so simple, since the materials may vary considerably in characteristics other than functional suitability, and a more detailed cost study is required. As the reader is no doubt keenly aware, such economic analyses are frequently of great importance in comparisons

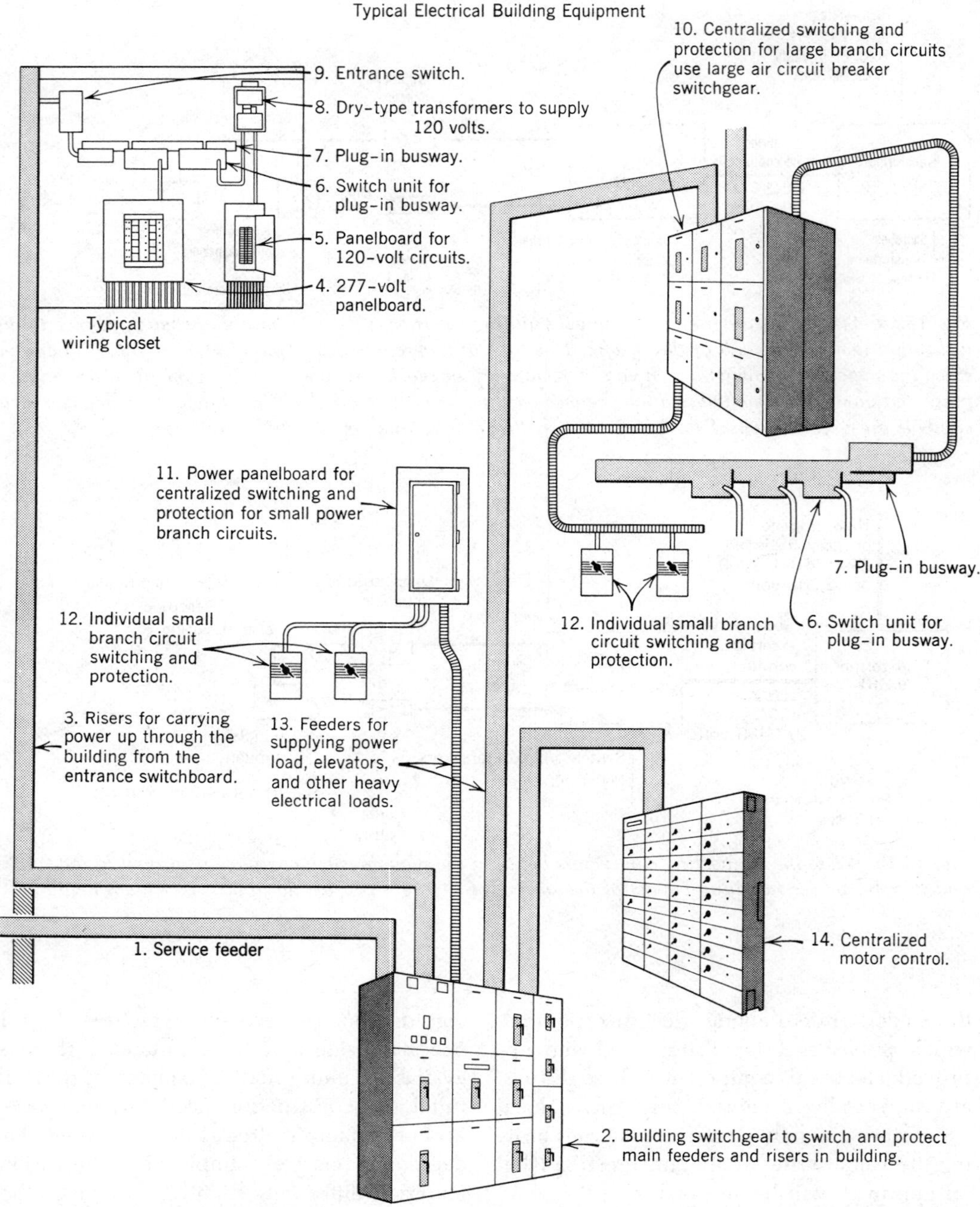

Fig. 15.2 *A building's electrical system shown pictorially, with the power capacity being indicated, more or less, by the size of the conductors. Note that this diagram does not extend beyond the local panelboard. Courtesy of General Electric*

between various HVAC systems. This is less so when dealing with electrical systems, since from an energy point of view electrical systems are more passive than HVAC systems, and it is energy costs that are often the decisive factor in economic analyses. We are speaking, of course, of life-cycle equipment costs (over the life of the *structure*) or annual owning and operating costs, including equipment amortization costs. The type of analysis used depends on the situation. However, such comparisons are only useful when both the initial cost and the operating costs will be borne by the same individual, that is, an owner-operator. Obviously, in the case of a speculative building venture only first cost is considered.

It is often quite difficult to perform such cost analyses since accurate data on life and maintenance costs for electrical equipment may not be readily available. Still, even if a formal analysis cannot be done, the principle involved must be considered in the selection of all electrical material, bearing in mind its particular importance in the case of energy-consuming equipment such as lighting and motors. Detailed economic discussions, on these items will be found with the related technical discussion.

15.4 Energy Consideration

As mentioned above, energy costs are a major factor in economic analysis. However, energy considerations are at least as important, in and of themselves. This aspect will be examined in detail in the discussions both of building energy budgets and of their components, including lighting, elevators, and electric motors in all other systems. At this point the reader should be aware that in addition to purely electrical characteristics, electrical equipment possesses also economic and energy "ratings."

15.5 Electrical Equipment Ratings

All electrical equipment is rated for the normal service it is intended to perform. These ratings may be in voltage, current, duty, horsepower, kw, kva, temperature, enclosure, and so on. Ratings related to the specific equipment will be discussed in detail in the sections below. The ratings that are specifically and characteristically electrical are those of voltage and current.

(*a*) ***Voltage.*** The voltage rating of an item of electrical equipment is the maximum voltage that can safely be applied to the unit continuously. It frequently, but not always, corresponds to the voltage applied in normal use. Thus, an ordinary wall electrical receptacle is rated at 250 v maximum, though in normal use only 120 v is applied to it. The rating is determined by the type and quantity of insulation used and the physical spacing between electrically energized parts.

(*b*) ***Current.*** The current rating of an item of electrical equipment is determined by the maximum operating temperature at which its components can operate properly continuously. That in turn depends on the type of insulation used. As a case in point, consider an electric motor. The current flowing in the motor windings causes a power loss (I^2R), which generates heat. If the windings are insulated with varnished cotton braid, with a maximum safe operating temperature of 65° C, the maximum permissible current (to which the hp rating of the motor is directly related) is that current which will produce this operating temperature. If these same windings are insulated with a silicone or glass compound with a maximum operating temperature of 150° C, obviously more current can be safely carried and the hp rating is consequently larger. Thus we see that, although a motor is rated in horsepower (or kw where SI units are used), a transformer is rated in kva and a cable, as will be discussed below, is rated in amperes; the actual criterion on which all these ratings is based is maximum permissible operating temperature.

15.6 Interior Wiring Systems

Returning now to the principal subject of

this chapter, we believe it will be helpful to survey the different types of interior wiring systems before commencing a detailed discussion of components. The function of any wiring system is to conduct electricity from one point to another. When the primary purpose of the system is to distribute electrical energy, it is referred to as an electrical power system; when the purpose is to transmit information, it is referred to as an electrical signal system. In this chapter we will deal with electrical power systems.

Due to the nature of electricity, its distribution within a structure poses basically a single problem: how to construct a distribution system that will *safely* provide the energy required at the location required. The safety consideration is all-important, since even the smallest interior system is connected to the utility's powerful network and the potential for damage, injury, and fire is always present. The solution to this problem is to isolate the electrical conductors from the structure except at those specific points, such as wall receptacles, where contact is desired. This isolation is generally accomplished by insulating the conductors and placing them in protective raceways. The principal types of interior wiring systems in use today are: exposed insulated cables, insulated cables in open raceways, and insulated conductors in closed raceways.

(a) Exposed Insulated Cables. In this category would be included (using the NEC nomenclature) NM ("Romex"), AC ("BX"), MI (mineral insulated), ALS, and CS. The latter two types are representative of the many metal-jacketed types available. This type of wiring system relies upon the construction of the cable itself for protection both of and from the "hot" conductors, since raceways are not required in the installation (see Sections 15.11 and 15.12).

(b) Insulated Cables in Open Raceways (Trays). This system is specifically intended for industrial application, and it relies upon both the cable and the tray for safety.

(c) Insulated Conductors in Closed Raceways. This system is the most general type and is applicable to all types of wiring in all types of facilities. It can be further subdivided into two major subcategories.

1. *Field Assembled Systems.* We include here wiring in conduit, surface raceways, and underfloor ducts. In general, the raceway is installed first and the wiring pulled in or laid in later. The raceways themselves may be

a. Buried in the structure; for example, conduit in the floor slab or underfloor duct (see Sections 15.22, 15.23, 15.28, and 15.29).
b. Attached to the structure; for example, all types of surface raceways, including conduit and wireways suspended above hung ceilings (see Sections 15.33 and 15.34).
c. Part of the structure; for example, cellular concrete and cellular metal floors (see Sections 15.31 and 15.32).

2. *Preassembled Systems.* In this category are included factory-assembled cables and prewired raceways whose use is generally limited to branch circuits (see Section 15.34).

(d) Combined Conductor and Enclosure. This category is intended to cover all types of factory-constructed busway and busduct, plus a few special types. Application of this construction is generally to feeder circuits, though light-duty busduct is available for branch circuit use as well (see Sections 15.18, 15.19, and 15.20).

The discussion which follows will be devoted to details of the components of the above systems and their application to the wiring systems described above.

15.7 Conductors

Electrical conductors (wiring) are the means by which the current is conducted through the electrical system, corresponding to the piping in the hydraulic analogy. By convention, a single insulated conductor No. 6

Table 15.1 *Physical Properties of Bare Conductors*

Size (AWG or MCM)	Area (Circular Mils)	Diameter (Inches)		Diameter (Millimeters)		d-c Resistance Ohms/1000 ft at 77° F, 25° C (Bare Copper)
		Solid	Stranded	Solid	Stranded	
16	2580	0.0508	—	1.29	—	4.10
14	4109	0.0641	—	1.63	—	2.57
12	6530	0.0808	—	2.05	—	1.62
10	10,380	0.1019	—	2.59	—	1.02
8	16,510	0.1285	—	3.26	—	0.64
6	26,240	0.162	0.184	4.11	4.67	0.41
4	41,740	0.204	0.232	5.18	5.89	0.26
2	66,360	0.258	0.292	6.55	7.42	0.16
1	83,690	0.289	0.332	7.34	8.43	0.13
0 (1/0)	105,600	0.325	0.373	8.26	9.47	0.10
00 (2/0)	133,100	0.365	0.418	9.27	10.62	0.081
000 (3/0)	167,800	0.410	0.470	10.41	11.94	0.064
0000 (4/0)	211,600	0.460	0.528	11.68	13.41	0.051
250 MCM	250,000	0.500	0.575	12.70	14.61	0.043
300 MCM	300,000	0.548	0.630	13.92	16.00	0.036
400 MCM	400,000	0.632	0.728	16.05	18.49	0.027
500 MCM	500,000	0.707	0.813	19.56	20.65	0.022

Source. Extracted from the National Electrical Code, except for millimeter dimensions.

AWG (American Wire Gauge) or larger, or several conductors of any size assembled into a single unit, are referred to as a cable. Single conductors No. 8 AWG and smaller are called wire.

The standard of the American wire and cable industry for round cross-section conductors is the American Wire Gauge. All wire sizes up to No. 0000 (also written No. 4/0) are expressed in AWG. The AWG numbers run in *reverse* order to the size of the wire, that is, the smaller the AWG number, the larger the size. Thus No. 10 is a heavier wire than No. 12 and lighter (thinner) than No. 8. The No. 4/0 size is the largest AWG designation, beyond which a different designation called MCM (thousand circular mil) is used. In this designation, wire diameter *increases* with number; thus, 500 MCM is a heavier wire (double the area) than 250 MCM.

A circular mil is an artificial area measurement, representing the square of the cable diameter when the diameter is expressed in mils (thousandths of an inch). Thus a solid conductor $\frac{1}{2}$ in. in diameter is 500 mils in diameter, or 250,000 circular mils in area, or 250 MCM:

$$\text{MCM} = \frac{\text{CM}}{1000} = \frac{\text{diam}^2}{1000} = \frac{(500)^2}{1000} = \frac{250{,}000}{1000} = 250$$

Outside of the United States where the metric system is in general use, conductor sizes are given simply as the diameter in millimeters (mm). Table 15.1 gives dimensional and stranding data for the commonest wire sizes, and includes the millimeter equivalent of each size—this will prove useful in interfacing American gauges and metric sizes. The d-c electrical resistance data is also shown.

15.8 Conductor Ampacity

Conductor current carrying capacity or *ampacity,* is determined as explained above, by the maximum operating temperature that its insulation can stand continuously. Heat is

Table 15.2*a* *Allowable Ampacities of Insulated Copper Conductors (Not More Than Three Conductors in Raceway)*

Size AWG MCM	*Temperature Rating of Conductor*[a] 60° C (140° F) Types T, TW	75° C (167° F) Types RHW, THW, THWN, XHHW	90° C (194° F) Types SA, RHH, THHN, XHHW	110° C (230° F) Type AVA
14[b]	15	15	25[b]	30
12[b]	20	20	30[b]	35
10[b]	30	30	40[b]	45
8	40	45	50	60
6	55	65	70	80
4	70	85	90	105
3	80	100	105	120
2	95	115	120	135
1	110	130	140	160
0	125	150	155	190
00	145	175	185	215
000	165	200	210	245
0000	195	230	235	275
250	215	255	270	315
300	240	285	300	345
350	260	310	325	390
400	280	335	360	420
500	320	380	405	470
600	355	420	455	525
700	385	460	490	560
750	400	475	500	580
800	410	490	515	600
900	435	520	555	—

Source. Extracted from the National Electrical Code.

[a] These ampacities relate only to conductors described in Table 15.4.

[b] The ampacities for Types RHH, THHN, and XHHW conductors for sizes AWG Nos. 14, 12, and 10 are the same as designated for 75° C conductors in this table.

generated as a result of the current flowing and the conductor resistance (I^2R). This heat is dissipated into the environment. The operating temperature therefore depends on the amount of current, the wire resistance, and the environment. Thus, for a given environment (open-air or enclosed), ampacity increases with increasing conductor size (since resistance decreases; see Table 15.1). Also, for a given environment and conductor size, ampacity varies directly with maximum permissible insulation temperature. These facts are clearly shown in Table 15.2*a*. If more than three conductors are placed in a conduit, the increase in temperature requires that the conductors be derated by the amount shown in Table 15.2*b*. When conductors are placed in an enclosed raceway, the heat generated is not as easily dissipated as it would be if the conductor were in free air,

Table 15.2*b* *Current-Carrying Capacity Derating Factors*

Number of Conductors in Raceway	*Derating Factor*
4 to 6	0.80
7 to 24	0.70
25 to 42	0.60
43 and above	0.50

Source. Extracted from the National Electrical Code.

Table 15.2c ***Correction Factors for Ambient Temperatures above 30° C (86° F)***

Ambient Temp. °C	°F	60° C (140° F)	75° C (167° F)	85° C (185° F)	90° C (194° F)	110° C (230° F)	125° C (257° F)	200° C (392° F)
40	104	0.82	0.88	0.90	0.90	0.94	0.95	—
45	113	0.71	0.82	0.85	0.85	0.90	0.92	—
50	122	0.58	0.75	0.80	0.80	0.87	0.89	—
55	131	0.41	0.67	0.74	0.74	0.83	0.86	—
60	140	—	0.58	0.67	0.67	0.79	0.83	0.91
70	158	—	0.35	0.52	0.52	0.71	0.76	0.87
75	167	—	—	0.43	0.43	0.66	0.72	0.86
80	176	—	—	0.30	0.30	0.61	0.69	0.84
90	194	—	—	—	—	0.50	0.61	0.80
100	212	—	—	—	—	—	0.51	0.77
120	248	—	—	—	—	—	—	0.69
140	284	—	—	—	—	—	—	0.59

Source. Extracted from the National Electrical Code.

Table 15.3 ***Typical Ambient Temperatures***

Location	*Temperature*	*Minimum Rating of Required Conductor Insulation*
Well ventilated, normally heated buildings	30° C (86° F)	(See note below)
Buildings with such major heat sources as power stations or industrial processes	40° C (104° F)	75° C (167° F)
Poorly ventilated spaces such as attics	45° C (113° F)	
Furnaces and boiler rooms (min.)	40° C (104° F)	75° C (167° F)
(max.)	60° C (140° F)	90° C (194° F)
Outdoors in shade in air	40° C (104° F)	75° C (167° F)
In thermal insulation	45° C (113° F)	75° C (167° F)
Direct solar exposure	45° C (113° F)	75° C (167° F)
Places above 60° C (140° F)		110° C (230° F)

NOTE. 60° C for up to and including No. 8 AWG copper and 75° C for over No. 8 AWG copper.

and the temperature rise is greater. Thus, the current rating in free air is much higher than for the same wire in conduit. The free-air table is not reproduced here since it is not generally applicable to interior wiring systems. Conversely, if the ambient temperature is above 30° C, the current capacity must be reduced by the factors given in Table 15.2*c*. These factors apply to Table 15.2*a*. See Table 15.3 for typical ambient temperatures.

15.9 Conductor Insulation and Jackets

Most conductors are covered with some type of insulation that prevents them from contacting other wiring or the grounded raceway, and also prevents accidental contact with the wiring. Insulation also serves as a physical shield to the conductor against heat, water, and so on, although outer jackets usually accomplish this function. Using our hydraulic analogy, conductors are analogous to pipes; when pressure is high, pipes are heavier; when voltage is high, insulation must be thicker. Insulation is rated by voltage, for example: 300, 600, 1000, 3000, 5000, and 15,000 v. If insulation is used above its rating, it may break down causing short circuits and arcing, with the possibility of fire. Ordinary building wiring is usually rated for 300 v or 600 v. The common types of building wire insulation are listed in Table 15.4 with the associated trade names, code letters, maximum temperatures, and special provisions. Table 15.5 gives dimensional data on commonly used building wire.

Various materials are utilized for outer coverings on wire and cable. Lead provides moisture protection. Neoprene gives moisture, corrosion, and abrasion protection. Metal wire or tape coverings of bronze or steel protect against rodent attack or physical damage. These materials and many more are used singly and in combination to provide the specific

Table 15.4 ***Characteristics of Selected Insulated Conductors for General Wiring***

Trade Name	*Type Letter*	*Maximum Operating Temperature*	*Application Provisions*
Moisture- and heat-resistance rubber	RHW	75° C 167° F	Dry and wet locations
Thermoplastic	T	60° C 140° F	Dry locations
Moisture-resistant thermoplastic	TW	60° C 140° F	Dry and wet locations
Heat-resistant thermoplastic	THHN	90° C 194° F	Dry locations
Moisture- and heat-resistant thermoplastic	THW	75° C 167° F	Dry and wet locations
Moisture- and heat-resistant thermoplastic	THWN	75° C 167° F	Dry and wet locations
Moisture- and heat-resistant cross-linked thermosetting polyethylene	XHHW	90° C 194° F	Dry locations
		75° C 167° F	Wet locations
Silicone-asbestos	SA	90° C 194° F	Dry locations
Asbestos and Varnished cambric	AVA	110° C 230° F	Dry locations only

Source. Extracted from the National Electrical Code.

Table 15.5 ***Dimensions of Rubber-, Asbestos-, and Thermoplastic-Covered Conductors***

Size	*Type RHW*[b]		*Types T, THW*[a]*, TW*		*Types THHN, THWN*		*Type AVA*		*Type XHHW*	
AWG[c] *MCM*	*Approx. Diameter (Inches)*	*Approx. Area (Square Inches)*	*Approx. Diameter (Inches)*	*Approx. Area (Square Inches)*	*Approx. Diameter (Inches)*	*Approx. Area (Square Inches)*	*Approx. Diameter (Inches)*	*Approx. Area (Square Inches)*	*Approx. Diameter (Inches)*	*Approx. Area (Square Inches)*
14	0.204	0.0327	—	—	—	—	—	—	—	—
14	—	—	0.162[a]	0.0206[a]	—	—	0.245	0.047	0.129	0.0131
12	0.221	0.0384	—	—	—	—	—	—	—	—
12	—	—	0.179[a]	0.0251[a]	—	—	0.265	0.055	0.146	0.0167
10	0.242	0.0460	0.168	0.0224	0.153	0.0184	—	—	—	—
10	—	—	0.199[a]	0.0311[a]	—	—	0.285	0.064	0.166	0.0216
8	0.311	0.0760	0.228	0.0408	0.201	0.0317	—	—	—	—
8	—	—	0.259[a]	0.0526[a]	—	—	0.310	0.075	0.224	0.0394
6	0.397	0.1238	0.323	0.0819	0.257	0.0519	0.395	0.122	0.282	0.0625
4	0.452	0.1605	0.372	0.1087	0.328	0.0845	0.445	0.155	0.328	0.0845
2	0.513	0.2067	0.433	0.1473	0.388	0.1182	0.505	0.200	0.388	0.1182
1	0.588	0.2715	0.508	0.2027	0.450	0.1590	0.585	0.268	0.450	0.1590
1/0	0.629	0.3107	0.549	0.2367	0.491	0.1893	0.625	0.307	0.491	0.1893
2/0	0.675	0.3578	0.595	0.2781	0.537	0.2265	0.670	0.353	0.537	0.2265
3/0	0.727	0.4151	0.647	0.3288	0.588	0.2715	0.720	0.406	0.588	0.2715
4/0	0.785	0.4840	0.705	0.3904	0.646	0.3278	0.780	0.478	0.646	0.3278
250	0.868	0.5917	0.788	0.4877	0.716	0.4026	0.885	0.616	0.716	0.4026
300	0.933	0.6837	0.843	0.5581	0.771	0.4669	0.940	0.692	0.771	0.4669
350	0.985	0.7620	0.895	0.6291	0.822	0.5307	0.995	0.778	0.822	0.5307
400	1.032	0.8365	0.942	0.6969	0.869	0.5931	1.040	0.850	0.869	0.5931
500	1.119	0.9834	1.029	0.8316	0.955	0.7163	1.125	0.995	0.955	0.7163

Source. Extracted from the National Electrical Code.

[a]Dimensions of THW in sizes Nos. 14 to 8. No. 6 THW and larger is the same dimension as T.

[b]Dimension of RHW without outer covering is the same as THW.

[c]No. 14 to No. 8, solid; No. 6 and larger, stranded. Refer to Table 15.1 for equivalent sizes in millimeters.

construction best applicable to the specific installation.

15.10 Copper and Aluminum Conductors

The use of aluminum wiring has increased because of its inherent weight advantage over copper, with concomitant lower installation costs. Economy usually lies with copper in small- and medium-size cable, since weight is not a problem and the smaller conduit required for the smaller copper conductors generally makes them cheaper. In the larger cable sizes the aluminum weight advantage offsets the economy of smaller copper size and smaller conduit, and generally proves less expensive, particularly in areas of high labor cost such as urban areas.

Aluminum and copper both exhibit the low electrical resistivity necessary for a good electrical conductor. The introduction of aluminum on a mass scale was delayed until approximately a decade ago as a result of difficulties inherent in splicing and terminating. These difficulties—which can be overcome with the use of proper equipment, techniques, and workmanship—stem from aluminum's cold-flow characteristic when under pressure (causing joints to loosen) and aluminum's oxide. This oxide, which forms within minutes on any exposed aluminum surface, is an adhesive, poorly conductive film that must be removed and prevented from reforming if a successful, long-life joint or termination is to be effected. If this is not done, the oxide causes a high-resistance joint with consequent excessive heat generation and possible incendiary effects. Instances of improper terminations increased with the rise in aluminum use, particularly in branch circuit work. Furthermore, when used in branch circuits, even if properly installed initially, aluminum can create problems when wiring devices are replaced by unskilled homeowners.

As a result of a number of unfortunate incidents, some localities in the United States have banned the use of aluminum wire in branch circuitry. Heavy feeders are normally installed by experienced and skilled contractors, and the risk of a poor joint is minimized. The writer, who has had many years of experience in this electrical field, recommends that the use of aluminum conductors be restricted to sizes not smaller than #4 AWG, and that installation be permitted only by contractors who certify expertise in the specialized techniques involved. Also, local codes and the electrical inspectors should be consulted. All references in this text, including all tables and illustrations, are to copper conductors. The following sections are devoted to a brief description of the principal building wire types.

15.11 Flexible Metal Clad Cable (BX)

Among the most common types of cable in the first classification of Section 15.6(a) above—that is, cables run without raceways—is the NEC type AC metal clad cable, commonly known by the trade name "BX." It is an assembly of wires, normally plastic insulated, bound together with a tape or braid, and then wrapped with a spiral-wound interlocking strip of steel tape (see Fig. 15.3). Such cable must have an internal metallic bonding strip in contact with the armor for its entire length. The assembly is installed as a unit, usually by simple U-clamps or staples holding it against beams, walls, ceilings, and columns. This type of installation is frequently used in residences and in the rewiring of existing

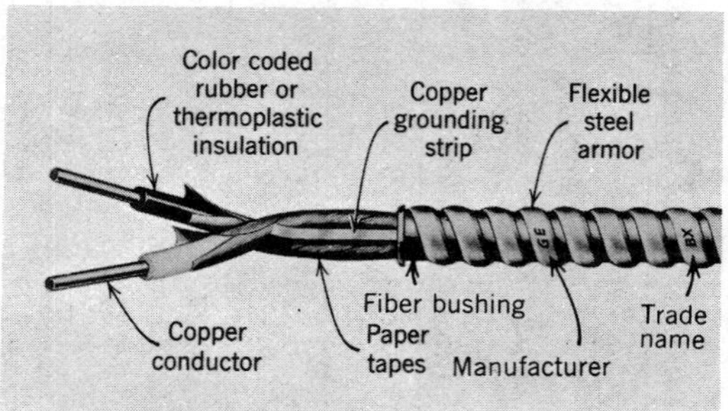

Fig. 15.3 Type AC flexible armored cable (BX). The bushing is installed to protect the wires from the sharp metal edges of the cut armor. Courtesy of General Electric, Wire and Cable Department

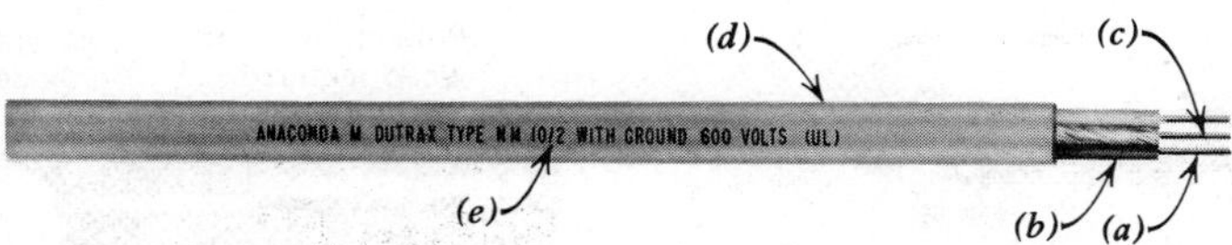

(a) Uncoated copper conductor
(b) Color–coded plastic insulation
(c) Fiber–covered ground wire
(d) Mositure–resistant, flame–retardant plastic jacket
(e) Cable indentification; 2 conductor, No. 10 AWG

Fig. 15.4 *Construction of typical NEC type NM cable. The cable is a 2 conductor, No. 10 AWG with ground, insulated for 600 v. Also shown are the manufacturer, cable trade name, and the letters* (UL), *which indicate listing of this product by the Underwriters' Laboratories, Inc. The ground wire* (c) *may be bare or covered, and the entire cable may be obtained flat (illustrated), oval, or round. Courtesy of Anaconda*

buildings. Such armored cable can be pulled into place through existing spaces in back of plastered surfaces, under floor joists, or between studding in the walls. Special couplings, box connectors, and other fittings are made for BX and to interconnect BX with rigid conduit systems. Use of this type AC cable is generally restricted to dry locations. For specific application details see NEC Article 334, "Metal Clad Cable."

15.12 Nonmetallic Sheathed Cable (Romex)

In application, the NEC types NM and NMC, also known by the trade name "Romex," are similar to type AC (BX). However, not having the physical protection of metallic armor, use is restricted to small buildings, that is, residential and other structures not exceeding three floors above grade. This cable type comprises an assembly of two or more plastic-insulated conductors and a ground wire, all covered with a flame-retardant, moisture-resistant plastic jacket. In the case of type NMC, the jacket is also corrosion resistant (see Fig. 15.4). The plastic jacket, in lieu of the armor on type AC, makes type NM easier to handle but more vulnerable to physical damage. For application details and restrictions see NEC Article 336, "Non-metallic Sheathed Cable." Typical installation technique is shown in Figure 15.5.

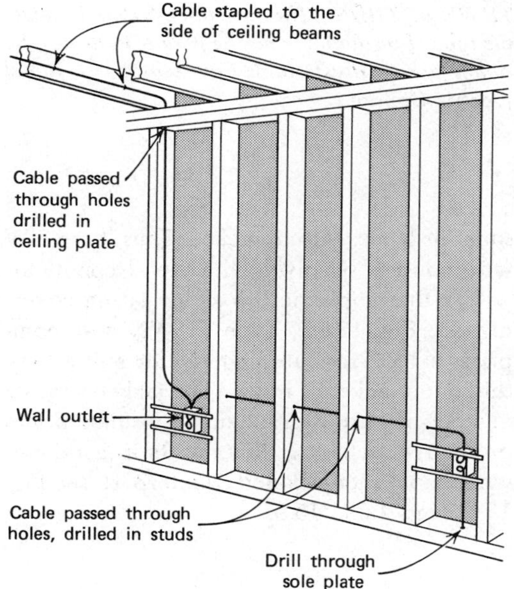

Fig. 15.5 *Typical wiring technique using types NM (Romex) or AC (BX), in wood stud-type construction. With metal stud construction the cables are passed through precut openings in lieu of field-drilled holes.*

15.13 Conductors for General Wiring

Under this heading (Article 310) the NEC lists the wire types that are generally used and installed in raceways, and referred to by the term "building wire." The most common types are listed with relevant characteristics in Table 15.4. These wires consist of a copper conductor covered with insulation, and in

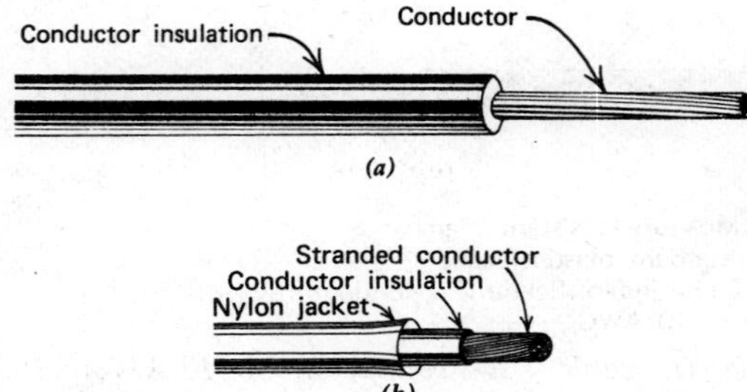

Fig. 15.6 (a) *Typical construction of unjacketed building wire such as types T and TW; see Table 15.4. Conductors normally are solid through No. 8 AWG, and stranded in sizes No. 6 AWG and larger; see Table 15.1.* (b) *The illustrated construction is typical for any nylon jacketed cable such as THWN or THHN. (The first three letters indicate the type of insulation, and the final N indicates the nylon jacket). Illustrations Courtesy of ITT, Royal Electric Division*

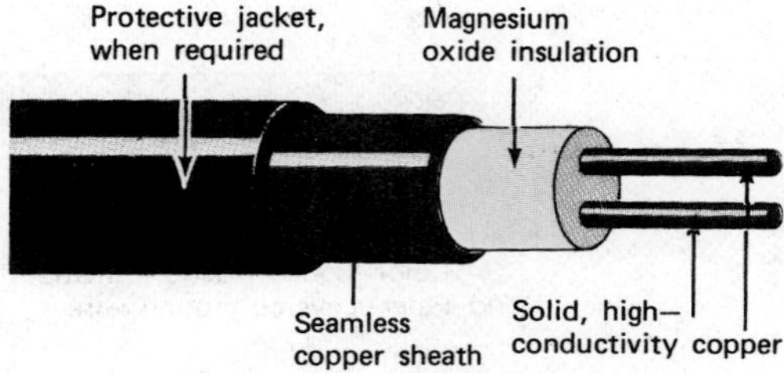

Fig. 15.7a Mineral insulated cable. This 600 v, self-contained wiring system finds its best application in adverse conditions where conventional wire and conduit systems are not adequate. These include extremes of heat and cold, and moisture-laden or explosive-gas environments.

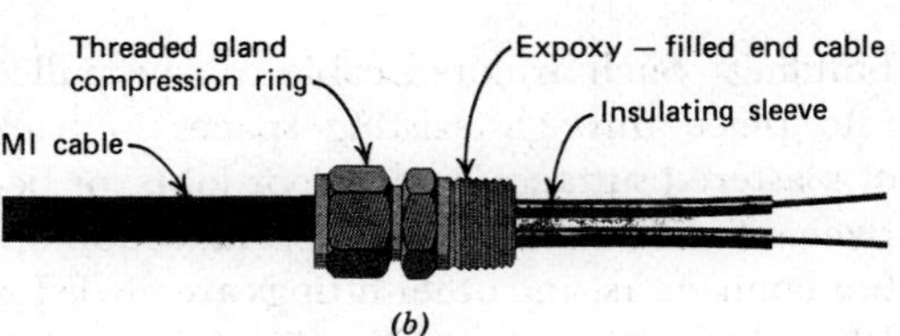

Fig. 15.7b MI cable termination. To maintain the integrity of the MI cable, terminations are made with compression rings, glands, and sealing compound. Sleeves are installed on the exposed bare conductors to provide insulation between the wires, and between the wires and the metal box. Courtesy of General Cable Corporation

some instances with a jacket. Thus, type TW wire consists simply of the metal conductor with a thermoplastic (PVC) insulating covering (see Fig. 15.6*a*). Type THWN wire comprises a PVC-insulated conductor and a very thin outer jacket of nylon. The jacket serves a number of purposes, including insulation and protective jacketing; this results in a *thinner* wire than its nonjacketed counterpart (see Fig. 15.6*b* and Table 15.5).

15.14 Special Cable Types

Although most building wiring is accomplished with plastic-insulated 300 v and 600 v conductors of the types described in the preceding sections, applications often require the use of special cables. In the following paragraphs and illustrations some of these are discussed.

15.15 Mineral Insulated (MI) Cable

This construction, which is covered in NEC Article 330, comprises an integral assembly of copper conductors, mineral insulation, and outer copper jacket that serves as a water and gas seal and a continuous ground. Because of its unique construction (see Fig. 15.7), it requires special fittings for terminations. Despite its relatively high cost, it is often the best solution to a difficult problem. The mineral insulation is flame proof and cold resistant, and the entire construction is explosion proof, lightweight, nonaging, and self-contained—making raceways unnecessary. There are no application limitations except under highly corrosive conditions and that limitation can be avoided by use of an appropriate plastic jacket over the copper outer covering.

15.16 Jacketed Cables

These cables comprise an assembly of two or more individually insulated conductors,

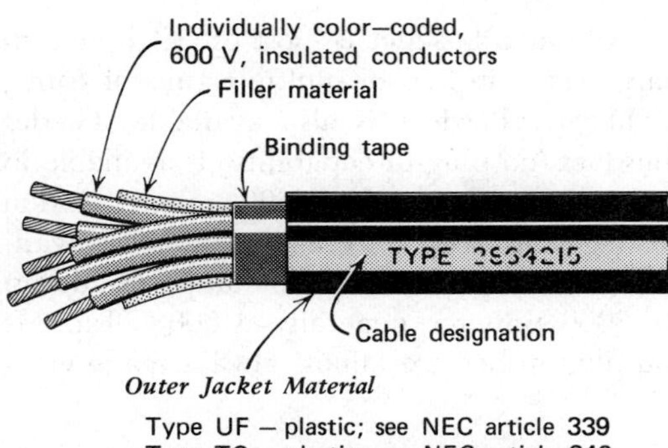

Outer Jacket Material

Type UF – plastic; see NEC article 339
Type TC – plastic; see NEC article 340
Type ALS – aluminum; see NEC article 331
Type CS – copper; see NEC article 332

Fig. 15.8 *All 600 v, jacketed cables have this basic construction. The outer jacket material determines the type of support necessary: metal-jacketed cables may be clamped at appropriate intervals; plastic-jacketed cable must be continuously supported by tray, a messenger cable, or the earth.*

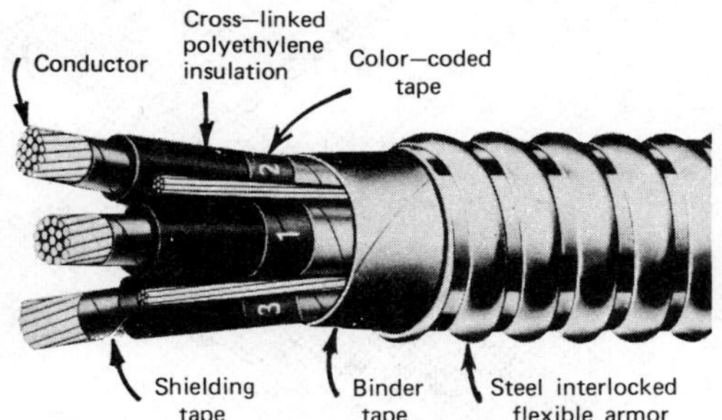

Fig. 15.9 *High voltage, 3 conductor interlocked steel-armored cable. This type of construction carries its own raceway and finds application indoors, in industrial and large commercial facilities. The thickness of insulation is dictated by the voltage rating of the cable. Courtesy of General Electric, Wire and Cable Department*

cabled together and covered with a plastic or metal jacket. Metal clad cables, due to inherent rigidity, may be installed with cable clamps. Plastic-jacketed cables require continuous support. The nature of the jacket material is determined by the cable's application. Cables intended for underground use must be designated type UF. All special-purpose cables may be used for ordinary wiring within the application limitations of their insulation (see Figs. 15.8 and 15.9 for typical jacketed cable construction).

15.17 Service Entrance Cables

These special cables and their installation requirements are discussed in Sections 16.1 to 16.4, which cover electric service.

15.18 Busway

When it is necessary to carry large amounts of current (power) the usual alternatives are to use several conductors in parallel or a single large conductor. The former solution becomes expensive with the increasing size and number of paralleled cables, particularly where tap-offs are necessary. However, the single large cable becomes increasingly inefficient in wire sizes above 750 MCM, because large cables require more cross section per ampere than small ones. This is not the case with flat conductors (called busbars), leading to their use for high-current-carrying application. Busbars are described by dimensions; thus, a bar may be $\frac{1}{4}$ in. × 2 in. (.64 × 5 cm), $\frac{1}{4}$ in. × 4 in., $\frac{5}{16}$ in. × 6 in. (.8 × 15 cm), and so on. As a rule of thumb, the current-carrying capacity of *copper* bus is 1000 amperes/sq in. of cross section (150 amperes/sq cm). Thus, a $\frac{1}{4}$ in. × 2 in. copper bar is $\frac{1}{2}$ sq in. in area and will safely carry 500 amperes. Bus is normally constructed of solid copper and, when assembled with other bars in a metal housing, it is referred to as busduct or busway.

The bars in a busduct, whether bare or insulated, are rigidly assembled by bolting them to insulating supports that are then connected to a stiff metal housing (see Fig. 15.10). A wide variety of fittings and joints are available to enable buswork to be installed with angles, bends, tap-offs, and curves (see Fig. 15.13).

Busduct is specified by material, number of buses (normally three or four, plus ground bus if required), current capacity, type, and voltage. In addition, maximum voltage drop is often specified. Thus, a typical brief description of a busduct would be: Copper busduct, 4-wire, 1000 amp, low-impedance type, 600 v; or aluminum busduct, 3-wire, 2000 amp, plug-in type, 600 v—both with a maximum full-load voltage drop of 2.5% per 100 ft at 90% power factor.

"Plug-in" refers to a design that allows devices such as switches, circuit breakers, and so on, to be directly plugged into the busduct, similar to a common plug insertion into a receptacle (see Fig. 15.11). Low impedance refers to a design that is specifically intended to give minimum voltage drop.

Although busduct is most often applied in large sizes, in lieu of multiple runs of cable, light-duty busduct is also available. Feeder busduct (no plug-in capability) is available in current ratings of 400 to 4000 amperes, and in voltages up to 15 kv. Plug-in busduct is available in ratings as low as 100 amperes and up to 3000 amperes, generally at 600 v. Tap-offs on plug-in bus are almost always made via a

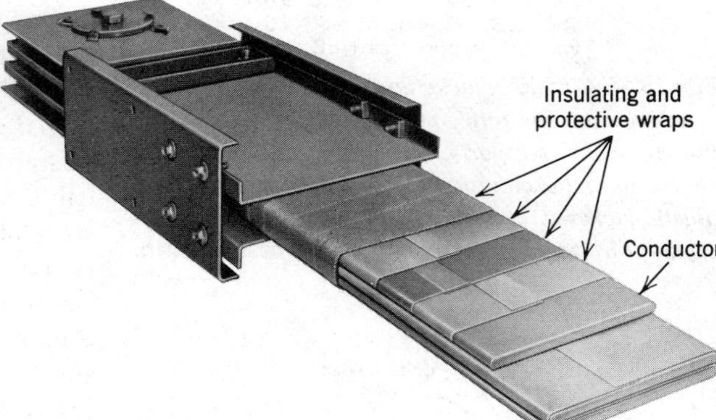

Fig. 15.10 "I-line" busduct is an assembly of molded and insulated conductors rigidly fastened in an aluminum housing. Housing is sectional, and of aluminum to reduce hysteresis and eddy current losses. Design is inherently weatherproof. Heat loss is by contact radiation and conduction. Reproduced by permission of The Square D Co.

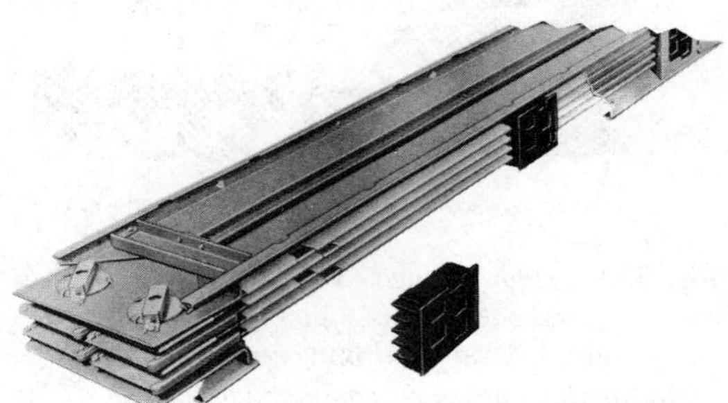

Fig. 15.11 Construction of plug-in busduct. Plug-ins are spaced every 12 in. on alternate sides to facilitate connection of plug-in breakers, switches, transformers, or cable taps. Note that bars are insulated over their entire length and are clamped rigidly at plug-ins with spacer blocks of insulating material. Housing is of sheet steel with openings for ventilation. Cover plate is not shown. Reproduced by permission of The Square D Co.

Fig. 15.12 A sectional view of this type of busduct shows the tight assembly of conductors within the metal housing. This design, unlike the ventilated type, can be mounted in any position, since heat dissipation is by radiation. The 8 sets of cable shown have the same current-carrying capacity as the busduct shown. Reproduced by permission of The Square D Co.

Fig. 15.13 A typical installation of compact design busduct. Note that the individual busducts are supported by channels hung from the ceiling and that the same hangers support more than one level of bus. Turns are easily made in the same plane (horizontal angle) and in two planes (vertical to horizontal). Reproduced by permission of The Square D Co.

circuit-protective device (see Fig. 15.14); that is, power is tapped off via a fuse or circuit breaker plug-in that in turn feeds an electric circuit. In very light-duty busduct the device itself (usually lighting) is fed from the duct directly. This type of bus is often referred to as lighting track, and it is discussed in Section 15.20. Busduct with ventilated housings are intended for indoor use. Designs are available in weatherproof housing for outdoor application. A common design of busduct utilizes insulated bars tightly bolted within a metal frame with no interior ventilation, relying instead on the housing to act as a heat sink (see Figs. 15.10 and 15.12). This design may be used outdoors with only a weatherproofing treatment. Busway is covered in NEC Article 364.

15.19 Cablebus

This construction is similar to ventilated busduct, except that it uses insulated cables instead of busbars. These cables are rigidly mounted in an open space-frame. The advantage of this construction is that it carries the ampacity rating of its cables *in free air,* which is much higher than the conduit rating, thus giving a high amperes-per-dollar first cost figure. Its principal disadvantage is bulkiness. Figure 15.15 shows a construction with 6 cables. Units are available with 3 to 18 cables, sizes 250 through 1500 MCM. Corresponding electrical ratings are from approximately 400 to 6000 amp in voltage with ratings of 600, 5000, and 15,000 v. Cablebus is covered in NEC Article 365.

As an example of the type of study that must be made when considering an item as relatively simple as an electrical feeder, refer to Table 15.6 and the accompanying illustration. Note that when considering first cost alone, the advantage lies in cablebus; adding the energy-loss consideration shifts the advantage to cable tray and interlocked armor cables (see Fig. 15.19). No general conclusion should be drawn from Table 15.6 regarding costs. A change in feeder length, number of taps, hours of operation, energy rates, or any

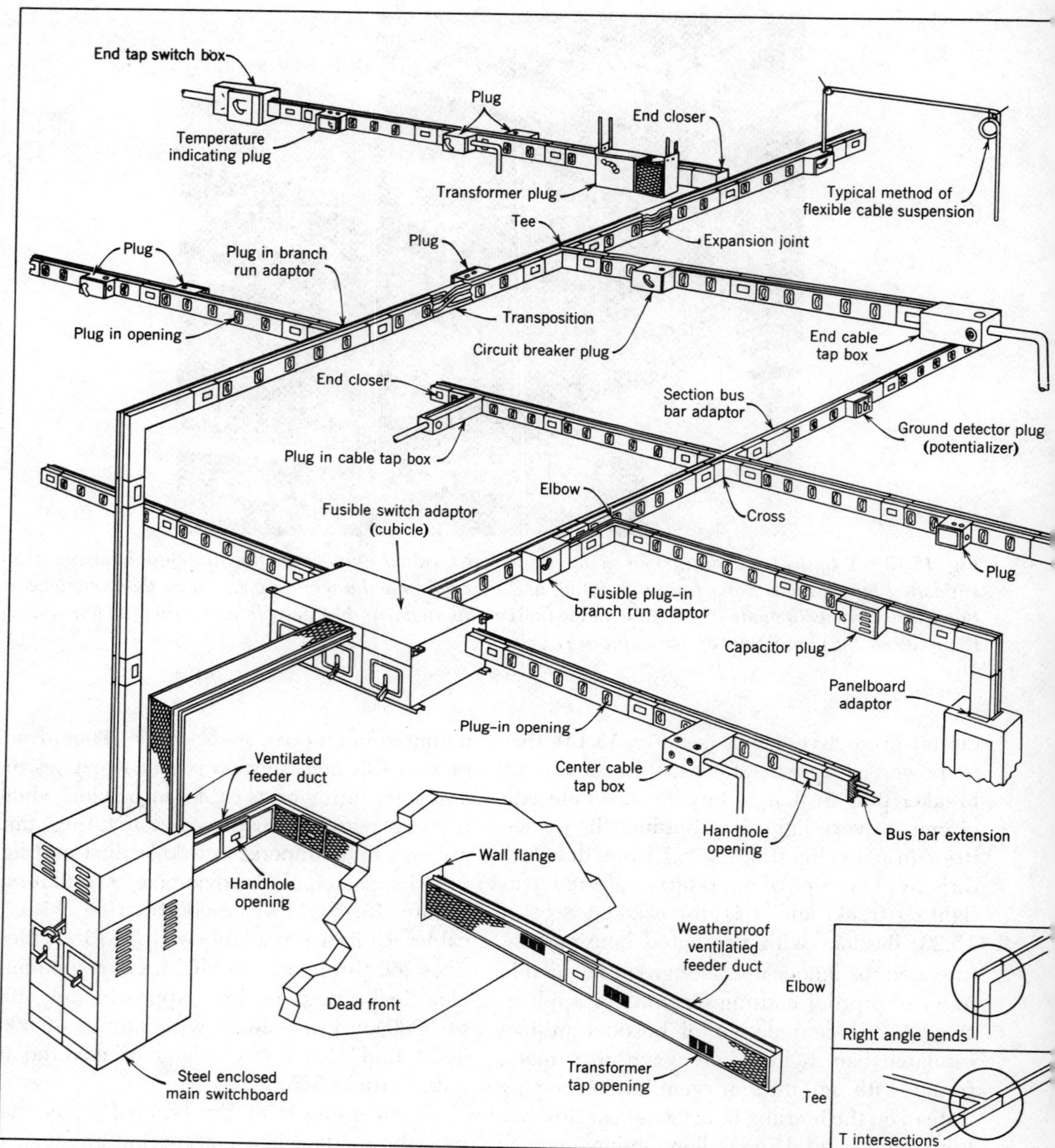

Fig. 15.14 *Typical busduct system. Note the use of feeder and plug-in duct, and the wide variety of devices that can be used with a plug-in busduct system. Courtesy of General Electric*

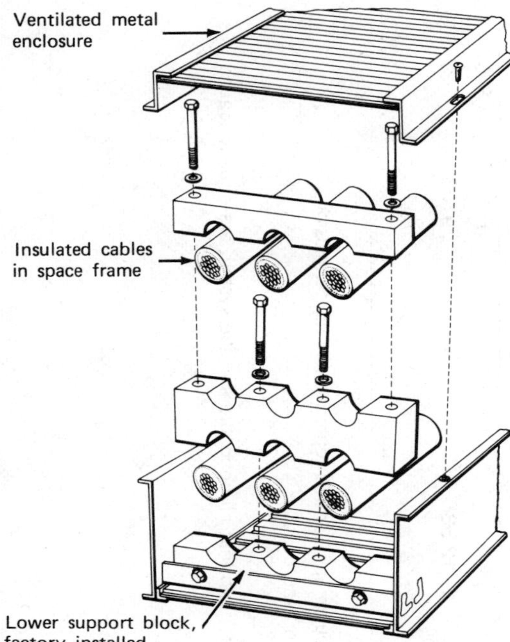

Fig. 15.15 *Typical cablebus construction. The enclosed insulated cables have a free-air ampacity rating, yielding a high ampere-per-dollar factor for this assembly. Courtesy of Husky products, subsidiary of Burndy Corp.*

of the other factors can shift the advantage to a different system. The point of our study is to demonstrate clearly that life-cycle costs and first costs often yield entirely different results and, therefore, that this type of study is required before an intelligent decision can be made. (Life-cycle cost is taken to mean the present value of all costs over the installation's life cycle—in this case, 20 years for the system.)

Two additional items are worthy of note.

1. The very factors that yield lower first cost operate to yield higher operating cost. The smaller copper sizes in busduct and cablebus, permitted by high-temperature insulation and good ventilation, cause increased power loss because of their higher resistivity.
2. If the heat loss from the busduct or cablebus can be used to advantage, the related energy cost can be credited, instead of being considered a total loss, and life-cycle costs can be changed considerably.

15.20 Flat Cable Assemblies and Lighting Track

Two special construction assemblies that act as light-duty (branch circuit) plug-in busways are widely used. The first—referred to in NEC Article 363 as "Flat Cable Assemblies"—is field constructed; the second, known in the industry as lighting track is factory prepared and field mounted. Unlike the heavier plug-in busways, both are specifically intended to directly feed utilization equipment which, as one of the names shows, is generally lighting equipment.

(*a*) ***Flat Cable Assemblies.*** A specially designed cable consisting of two, three, or four conductors, No. 10 AWG, is field installed in a rigidly mounted standard $1\frac{5}{8}$ in. square structural channel. Power tap devices, installed where required, puncture the insulation of one of the phase conductors and the neutral. Electrical connection is then made to the pigtail wires that extend from the tap devices. This connection can extend directly to the device or to an outlet box with a receptacle, which then acts as a disconnecting means for the electric device being served. In this fashion lights, small motors, unit heaters, and other single-phase, light-duty devices can be served without the necessity of "hard" (conduit and cable) wiring. Figures 15.16 and 15.17 illustrate the application and installation of this equipment.

(*b*) ***Lighting Track.*** This is a factory-assembled channel with conductors for one to four circuits *permanently* installed in the track. Power is taken from the track by special tap-off devices that contact the track's electrified conductors and carry the power to the attached lighting fixture. The tracks are generally rated at 20 amperes, and, unlike FC cable assemblies, they are restricted to 120 v. A typical design is shown in Figure 15.18. Application of track lighting is covered in Part Six of this book.

Table 15.6 ***Comparison of Feeder Installations 2000 Amperes, 208 Volts***[a]

Feeder System Description[a,b] 1	*Material Cost* 2	*Labor Cost*[c] 3	*Total First Cost* 4	*Power*[d,e] *Loss per 100 ft (Kilowatt)* 5	*Annual Energy Loss*[f] *(Kilowatt-hour)* 6	*Annual Energy Cost*[g] 7	*Life-Cycle Energy Cost*[h] 8	*Total Life-Cycle Cost*[i] 9
Cablebus: using 4 sets of 350 MCM, XHHW copper	$19,028	$6615	$25,643 100%	5.91	$25,547	$1566	$19,766	$45,409 121%
Wire and conduit: 4 sets of 750 MCM, XHHW copper, $3\frac{1}{2}$ in. rigid conduit	16,752	12,270	29,022 113%	2.90	12,536	769	9703	38,725 103%
Busduct: 2000 amp, 600 v, 3ϕ, 3 w, copper	31,587	5118	36,705 143%	5.60	24,207	1484	18,729	55,434 148%
Cable tray: aluminum ladder type, with 4 sets of 750 MCM armored cable	19,224	8642	27,866 109%	2.90	12,536	769	9703	37,569 100%

[a]Equipment rating is 600 volts.

[b]Data in columns 1 to 4 and the illustrations are taken from a study published by Husky Products, Inc., and is reproduced with permission.

[c]Overall labor rate used is $15 per hour; this includes both journeymen and foremen, and overhead costs.

[d]Data in columns 5 to 9 developed by the author.

[e]Based on published resistivity data for cable and bus—assuming 80% demand (1600 amperes) and that all conductors in the system are equally loaded.

[f]Based on 80% demand, 12 hours per day, 360 days per year.

[g]Using $.035 per kilowatt-hour as the combined net rate, including demand charges.

[h]Using 20-year life cycle, 8% fixed capital cost, and 3% annual escalation in energy cost.

[i]Sum of columns 4 and 8.

Table 15.6 (*continued*)

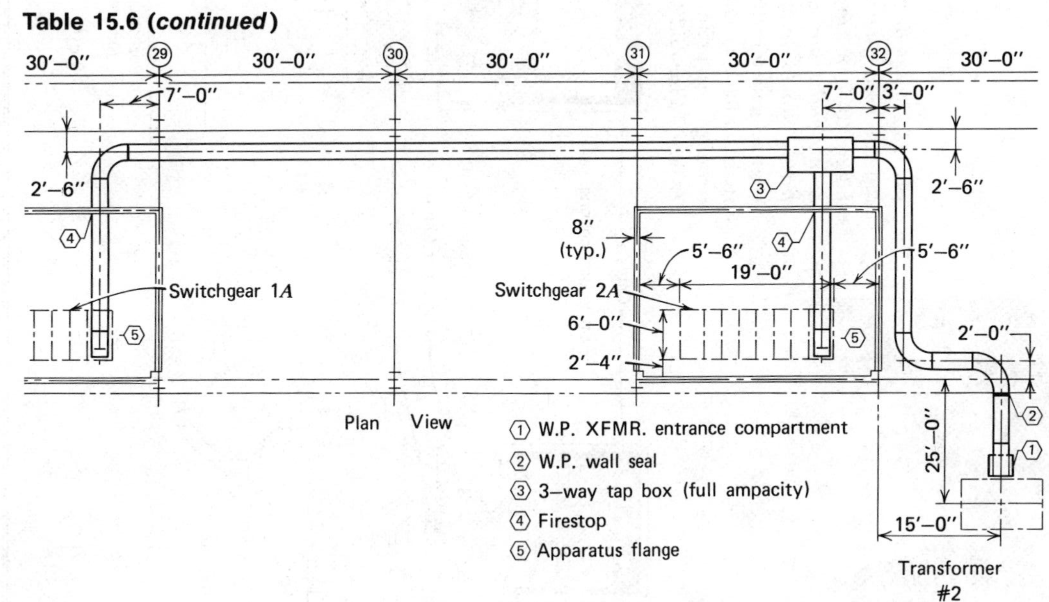

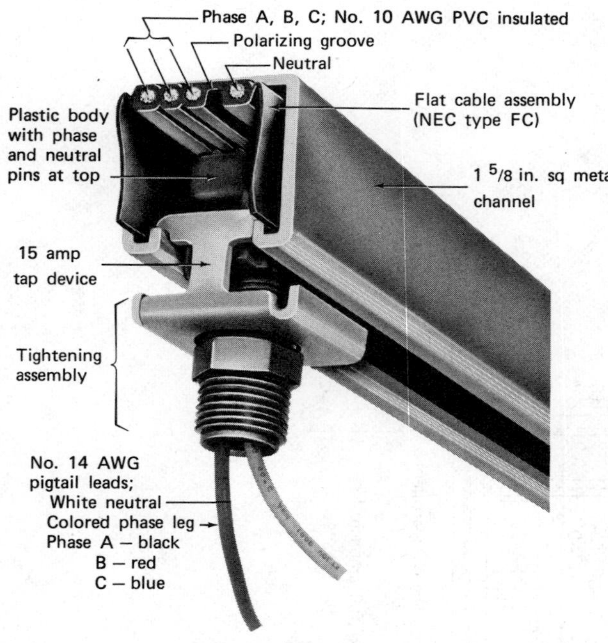

Fig. 15.16 *Flat cable assembly installation. The unit illustrated is a 4 conductor, FC cable. Only a single phase may be tapped, thus limiting the application to single-phase 120 v or 277 v devices. If the tap is removed the pin holes in the PVC insulation "heal," maintaining the integrity of the insulation. Courtesy of Chan-L-Wire/Wiremold Co.*

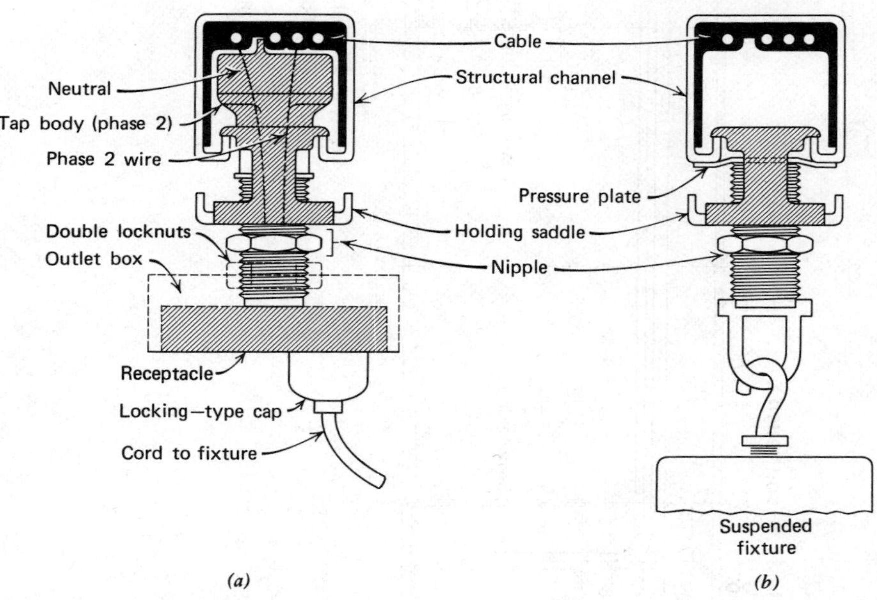

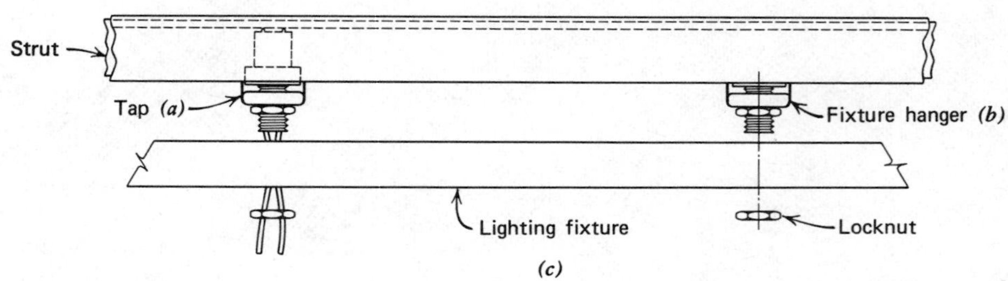

Fig. 15.17 *Application of FC cable. The tap can feed directly to a device as in Figure 15.16 or can energize a receptacle in an outlet box as in* (a). *A lighting fixture can be hung from the FC cable channel with a fixture hanger and hook as in* (b), *or with a combination of a tap and a hanger as in* (c). *Courtesy of Chan-L-Wire/Wiremold Co.*

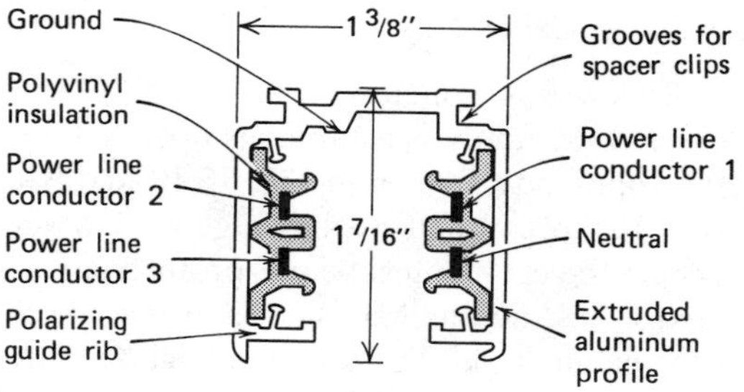

Ground
1 3/8″
Grooves for spacer clips
Polyvinyl insulation
Power line conductor 1
Power line conductor 2
Power line conductor 3
1 7/16″
Neutral
Polarizing guide rib
Extruded aluminum profile

(c)

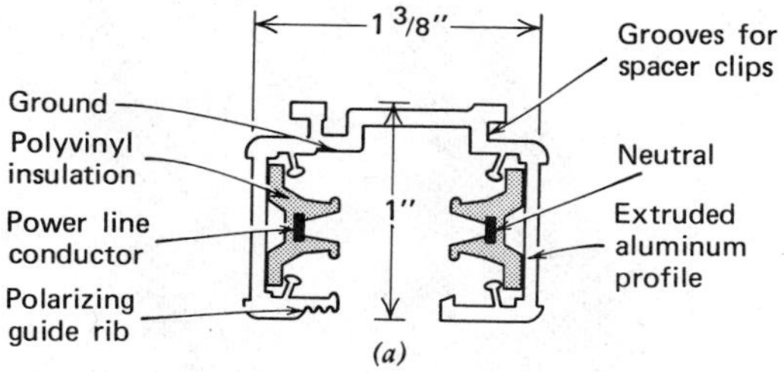

1 3/8″
Grooves for spacer clips
Ground
Polyvinyl insulation
Neutral
1″
Power line conductor
Extruded aluminum profile
Polarizing guide rib

(a)

(b)

(d)

Spring-loaded Ground
Circuit 2 on reverse side
Circuit 1
Circuit 3 on reverse side
Neutral
Polarizing Guide Rib
Locking Ribs (two more on other side)
Circuit Dial
NOTE: Contactors are extended ONLY after switch is turned On.
Track Light Mount
(e)

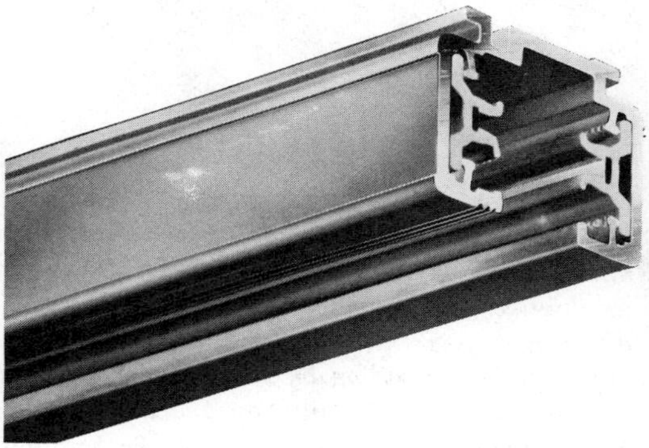

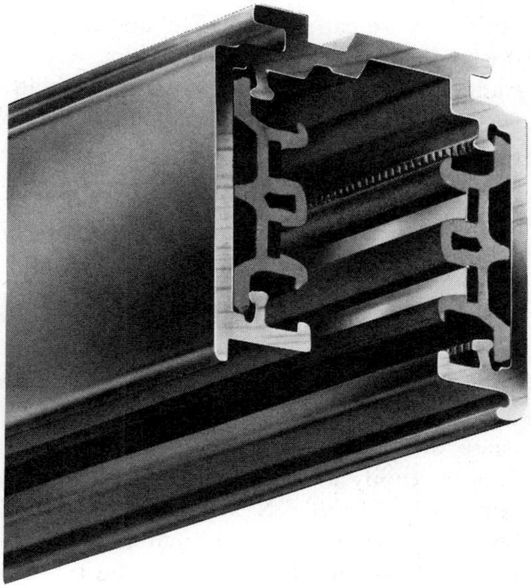

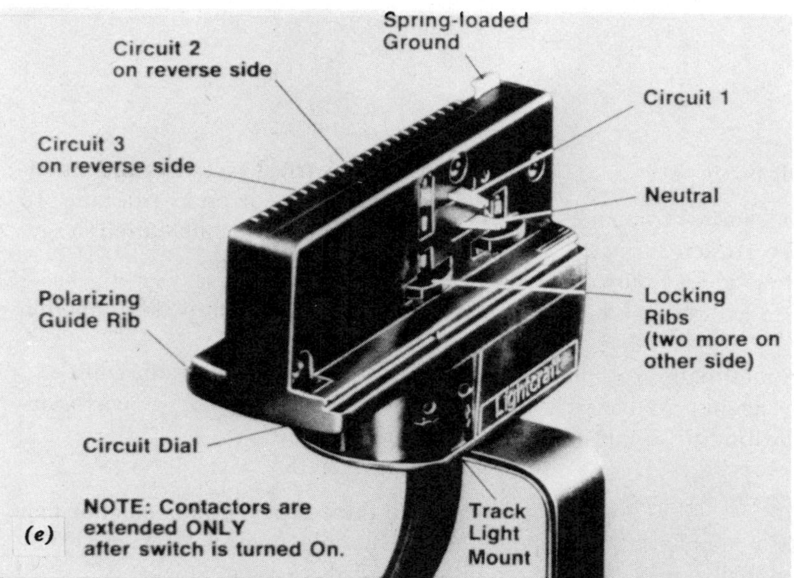

Fig. 15.18 *Lighting track. The electrified conductors are permanently installed in the aluminum track, which is grounded for safety. Tracks and insert plugs are available in single-circuit design* (a *and* b) *and multiple-circuit design* (c, d, *and* e). *Accessory fittings permit taps, joints, feeds, and connections at any point on the track. Courtesy of Lightcraft Track-Line, Nutone Division, Scovill*

15.21 Cable Tray

This system, which is covered in NEC Article 318, is simply a continuous open support for approved cables [see Fig. 15.19 and Section 15.6(b)]. When used as a general wiring system the cables must be self-protected, jacketed types such as MI, ALS, and the special tray cable, type TC. In industrial facilities where only competent maintenance personnel have access to the cable, large-size normal building wire can be used. The advantages of this system are free-air rated cables, easy installation and maintenance, and relatively low cost. The disadvantages are bulkiness and the required accessibility.

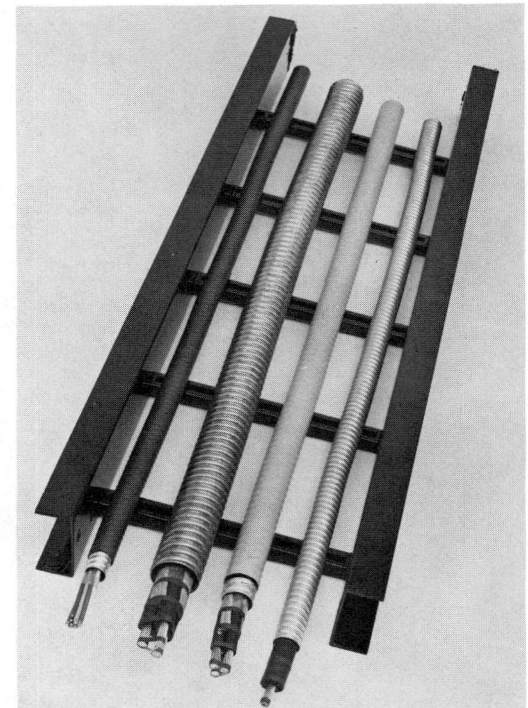

Fig. 15.19 *Note that cables in open trays must have integral mechanical protection. Cables shown are interlocked armor, including two with moisture- and corrosion-resistant jacket. Cable tray is supported from the ceiling by hangers and trapezes, similar to supports for busway and ductwork, as seen in Figure 15.13. Courtesy of Husky products, subsidiary of Burndy Corp.*

15.22 Closed Raceways

The following sections deal with closed wiring raceways, which will complete our discussion of raceways. We will not go into the details of construction and application, because of space limitation, and because that data is readily available from manufacturers and the applicable NEC articles. We will, however, provide enough material for the reader to become familiar with the types of raceways, their common applications and limitations, and, where applicable, comparative characteristics.

15.23 Steel Conduit

The purpose of conduit is to:

a. Protect the enclosed wiring from mechanical injury and corrosion.
b. Provide a grounded metal enclosure for the wiring in order to avoid shock hazard.
c. Provide a system ground path.
d. Protect surroundings against fire hazard as a result of overheating or arcing of the enclosed conductors.
e. Support the conductors.

For these reasons the NEC generally requires that all wiring be enclosed in a rigid metallic conduit. Metal electrical raceways and associated fittings must be corrosion resistant. To this end, steel conduit is manufactured in several ways, among which are:

a. Hot-dip galvanized (dipped into molten zinc).
b. Sherardized (coated with zinc dust).
c. Enameled (coated with a corrosion-resistant enamel).
d. Plastic covered.

There are three types of steel conduit that differ basically only in wall thickness. They are, in order of decreasing weight:

a. Heavy-wall steel conduit, also referred to simply as "rigid steel conduit"; it is covered in NEC Article 346.
b. Intermediate Metal Conduit, usually referred to as IMC; this is covered in NEC Article 345.
c. Electric Metallic Tubing, normally known as EMT or thin-wall conduit; this is covered in NEC Article 348.

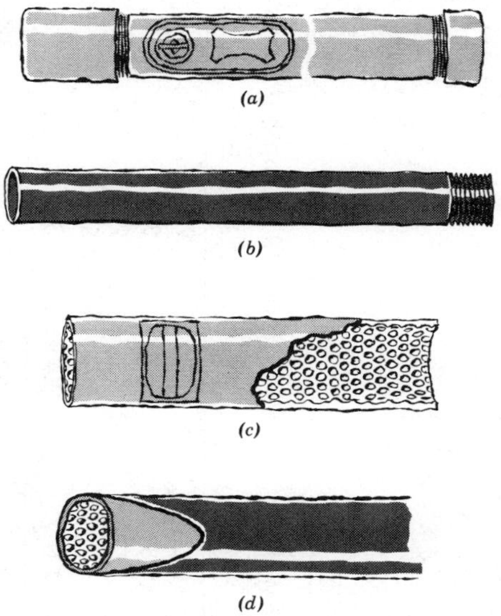

Fig. 15.20 *Steel conduits:* (a) *Galvanized, heavy wall, rigid;* (b) *black enameled;* (c) *EMT thin wall;* (d) *plastic-coated conduit, for use in highly corrosive atmospheres.*

The differences are clearly shown in Tables 15.7 and 15.8. Several types of heavy-wall conduit plus EMT are shown in Figure 15.20. For persons working on projects using SI dimensions, the equivalent millimeter (mm) sizes of conduits are given in Table 15.9.

Rigid conduit and IMC use the same fittings and are threaded alike. They are also identical in application, except that IMC is not permitted in hazardous areas (explosive atmospheres). As a result of its thin wall, EMT does not lend itself to threading; instead, it uses set-screw and pressure fittings. EMT may not be used in hazardous areas and must be approved for embedding in concrete. EMT above 2 in. and IMC and rigid conduits in all sizes have substantially the same OD (outside diameter). The thinner walls of EMT and IMC yield a larger ID (inside diameter) and, therefore, easier wire pulling. This combination of lower weight and easier wire pulling gives EMT and IMC a distinct labor cost advantage over rigid conduit, which is enhanced further in jobs with a great deal of field bending and handling of conduit. Generally, no conduit smaller than $\frac{1}{2}$ in., nominal trade diameter, is used. The nominal trade sizes are $\frac{1}{2}$, $\frac{3}{4}$, 1, $1\frac{1}{4}$, $1\frac{1}{2}$, 2, $2\frac{1}{2}$, 3, $3\frac{1}{2}$, 4, 5, and 6 in., although the last two sizes are made only in rigid conduit and are not often encountered. Conduit is made in standard lengths of 10 ft and includes one coupling per length. Ordinary steel pipe may not be used for electric purposes, and all electric steel conduit is distinctively marked as such.

When steel conduit is installed in direct contact with the earth, it is advisable to use hot-dip galvanized type and to coat the joints with asphaltum. If the earth is very wet, the complete conduit system should be coated

Table 15.7 ***Comparison of Steel Conduit Diameters***

3/4" Trade Sizes

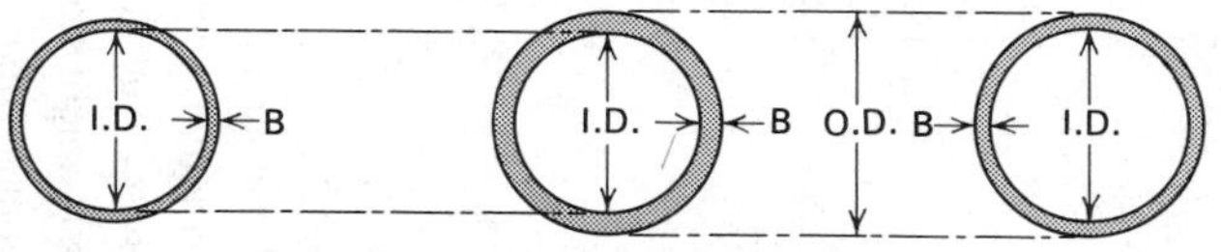

EMT		Rigid Steel		IMC	
O.D.	0.92	O.D.	1.050"	O.D.	1.029"
I.D.	0.82	I.D.	0.824"	I.D.	0.887"
B	0.05	B	0.113"	B	0.071"

Table 15.8 ***Comparative Dimensions and Weights of Metallic Conduit***

Nominal or Trade Size	*Outside Diameter (Inches)*				*Inside Diameter (Inches)*				*Weight per 10 ft Length (Pounds)*[a]			
	RS[b]	*IMC*[c]	*EMT*[d]	*AL*[e]	*RS*	*IMC*	*EMT*	*AL*	*RS*	*IMC*	*EMT*	*AL*
$\frac{1}{2}$	0.84	0.82	0.71	0.84	0.62	0.69	0.62	0.62	7.9	5.7	2.9	2.7
$\frac{3}{4}$	1.05	1.03	0.92	1.05	0.82	0.89	0.82	0.82	10.5	7.8	4.4	3.6
1	1.32	1.29	1.16	1.32	1.05	1.13	1.05	1.05	15.3	11.2	6.4	5.3
$1\frac{1}{4}$	1.66	1.64	1.51	1.66	1.38	1.47	1.38	1.38	20.1	14.4	9.5	7.0
$1\frac{1}{2}$	1.90	1.88	1.74	1.90	1.61	1.70	1.61	1.61	24.9	17.6	11.0	8.6
2	2.38	2.36	2.20	2.38	2.07	2.17	2.07	2.07	33.2	23.5	14.0	11.6
$2\frac{1}{2}$	2.88	2.86	2.88	2.88	2.47	2.61	2.73	2.47	52.7	39.3	20.5	18.3
3	3.50	3.48	3.50	3.50	3.07	3.23	3.36	3.07	68.3	48.3	25.0	23.9
$3\frac{1}{2}$	4.00	3.97	4.00	4.00	3.55	3.72	3.83	3.55	83.1	56.1	32.5	28.8
4	4.50	4.47	4.50	—	4.03	4.22	4.33	—	97.2	62.5	37.0	—
5	5.56	—	—	—	5.05	—	—	—	127.6	—	—	—
6	6.63	—	—	—	6.07	—	—	—	172.6	—	—	—

Source: All data courtesy of Allied Tube & Conduit Corp.

[a]Standard length including one coupling.

[b]Standard heavy-wall rigid steel conduit.

[c]Intermediate-weight steel conduit.

[d]Electric metallic tubing.

[e]Aluminum.

Table 15.9 ***Metric Equivalents of American Conduit Sizes***

Conduit Size (Inches)	*Dimensions (Millimeters)*	
	OD	*ID*
3/8	16.5	9.5
1/2	20.5	12.7
5/8	23.0	15.9
3/4	26.5	19.0
1	33.0	25.4
1 1/4	42.0	31.75
1 1/2	48.0	38.1
1 3/4	52.0	44.5
2	59.7	50.3
2 1/2	76.0	63.5
3	89.0	76.2
3 1/2	101.5	88.9
4	114.0	101.6

with an asphalt compound. Figures 15.21 and 15.22 illustrate some typical methods of installing conduits and raceways.

The selection of conduit size depends on the number and diameter of the wires that may be drawn into the conduit without injuring the wire. The number and radius of bends in the conduit, as well as its total length, affect the degree of abrasion to the wiring insulation. Long, straight pulls may be made through as much as 200 ft of continuous conduit without bends. The NEC states that no wire shall be spliced, connected, or tapped and then drawn into the conduit so that the connection is within the conduit itself. All such connections shall be made within connection boxes. Also, no wires should be installed until the conduit system has been inspected and approved.

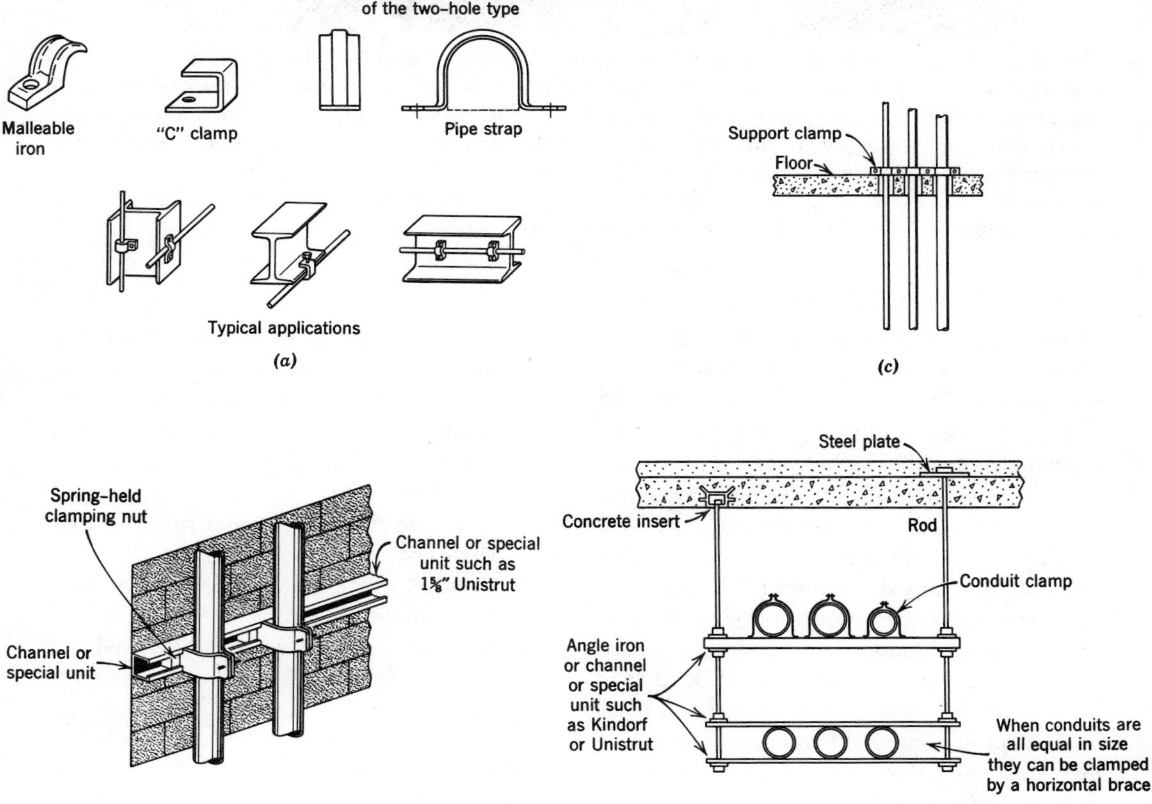

Fig. 15.21 *Conduit supports and methods.* (a) *Pipe straps and their application.* (b) *Vertical and horizontal supports using special support channels and pipe clamps.* (c) *Vertical support at floor openings provided by special conduit clamps that take the vertical load.*

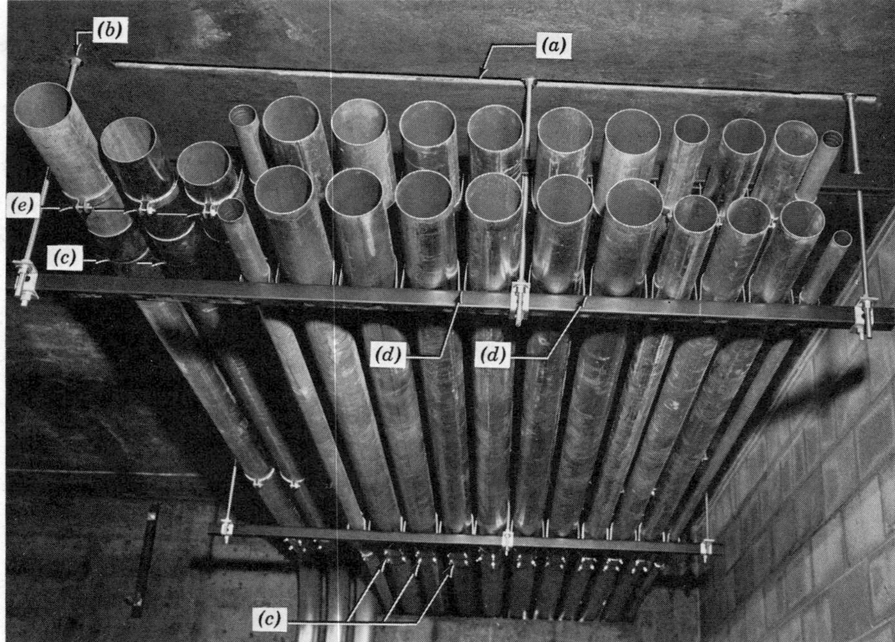

Fig. 15.22 *Typical overhead conduit bank installation. Note that due to field conditions the insert* (a) *for hanger rods was inadequate, and an additional insert* (b) *was added. This conduit bank uses EMT which has a pipe wall thickness of approximately one-third that of heavy-wall rigid conduit. The resulting weight difference in a large bank such as this is very pronounced. EMT is suitable here since the overhead location protects the pipe from severe physical abuse. EMT joints are made with set-screw fittings* (c). *Note that individual conduits are supported by clamps above* (d) *or below* (e) *the trapeze channel. Compare this photo to Figure 15.21*b *above. Courtesy of Republic Steel Corp.*

For structural reasons, conduits in concrete slabs are run close to the bottom surface (in the portion of the slab in tension) or near the central portion. If a great number of conduits must be embedded, it may be necessary to increase the slab thickness. In many instances, the structural slab is covered with a concrete topping, in which conduit may be installed without affecting slab integrity. In all cases, local building codes should be consulted for limitations on embedded conduits. In any event the top of any conduit shall be at least $\frac{3}{4}$ in. below the finished floor surface in order to prevent cracking. When heavy trucking is expected, this allowance should be increased to $1\frac{1}{2}$ in. minimum.

In general, the following rules should be observed and included in all specifications for conduit work in concrete slabs:

1. Conduits shall have an OD no greater than $\frac{1}{3}$ of the slab thickness as measured at its thinnest point.
2. Conduits running parallel to each other shall be spaced not less than three times the OD of the largest conduit center-to-center.
3. Conduits running parallel to beam axis shall not run above beams.
4. Conduit crossings shall be as near to a right angle as possible.
5. Minimum cover over conduits shall be $\frac{3}{4}$ in.

15.24 Aluminum Conduit

The use of aluminum conduit has increased greatly in recent years because of the weight advantage that aluminum has over steel, being even lighter than EMT. The savings in labor cost more than offsets the additional cost of the material itself. In addition, aluminum has better corrosion resistance in most atmospheres; it is nonmagnetic, giving lower voltage drop; it is nonsparking; and, generally, it does not require painting.

Its major drawback is its deleterious effect on many types of concrete, causing spalling and cracking when embedded. Although manufacturers can demonstrate many cases of embedding in concrete without harmful effect, it is a procedure that should be avoided unless the concrete additives are rigidly controlled. It is also inadvisable to bury aluminum in earth, with or without asphalt or other coating because of the rapid corrosion often encountered. Other difficulties frequently encountered are freezing of threaded joints, due to thread deformation, and difficulty in obtaining electrical contact with grounding straps. With the exceptions noted above, aluminum conduit may be used in all locations where steel conduit is used.

15.25 Flexible Metal Conduit

This type of conduit construction—which consists of an empty spirally wound interlocked armor raceway—is known to the trade as "Greenfield" and is covered in NEC Article 350. It is used principally for motor connections or other locations where vibration is present, where movement is encountered, or where physical obstructions make its use necessary. The acoustic and vibration isolation provided by flexible conduit is one of its most important applications. It should always be used in connections to motors, transformers, ballasts, and the like. Flexible conduit is available in galvanized steel, brass, bronze, and aluminum, in all standard trade diameters, and with requisite fittings. Typical applications are shown in Figure 15.23.

15.26 Liquid-Tight Flexible Metal Conduit

This item is covered in NEC Article 351 and is of the same construction as flexible conduit, except that it is covered with a liquid-tight jacket. It is not intended as a general purpose conduit, its use being restricted to connections where flexibility and imperviousness to liquid is required. Although made by a number of manufacturers, the trade name "Sealtite" is most often applied to this product. Figure 15.24 shows a typical application.

15.27 Nonmetallic Conduit

A separate classification of rigid conduit (NEC Article 347) covers raceways that are formed from such materials as fiber, asbestos-cement, soapstone, rigid polyvinyl chloride (PVC), and high density polyethylene. These materials when utilized as an electric raceway must be so labeled and must be resistant to moisture and chemical corrosion.

For use above ground, this conduit must be flame retardant, tough, and resistant to heat distortion, sunlight, and low temperature effects. For use underground the last two requirements are waived. Generally, nonmetallic conduit may be used without restriction in nonhazardous areas, within the physical limitations of the material involved. Thus plastic conduit has a temperature limitation, asbestos-cement has considerable physical strength limitations, and so on. As a result of these limitations, PVC conduit is the material of choice for indoor exposed use and asbestos-cement, fiber, and PVC plastic for outdoor and underground use. A separate ground wire *must* be provided, since the ground provided by a metallic conduit is absent.

15.28 Metal Surface Raceways

These raceways are covered in NEC Article 352. Surface metal raceways and multi-outlet assemblies may be utilized only in dry, nonhazardous, noncorrosive locations and may generally contain only wiring operating below 300 v. Such raceways are normally installed in exposed condition and in places not subject to physical injury.

The principal applications of surface metal raceways are:

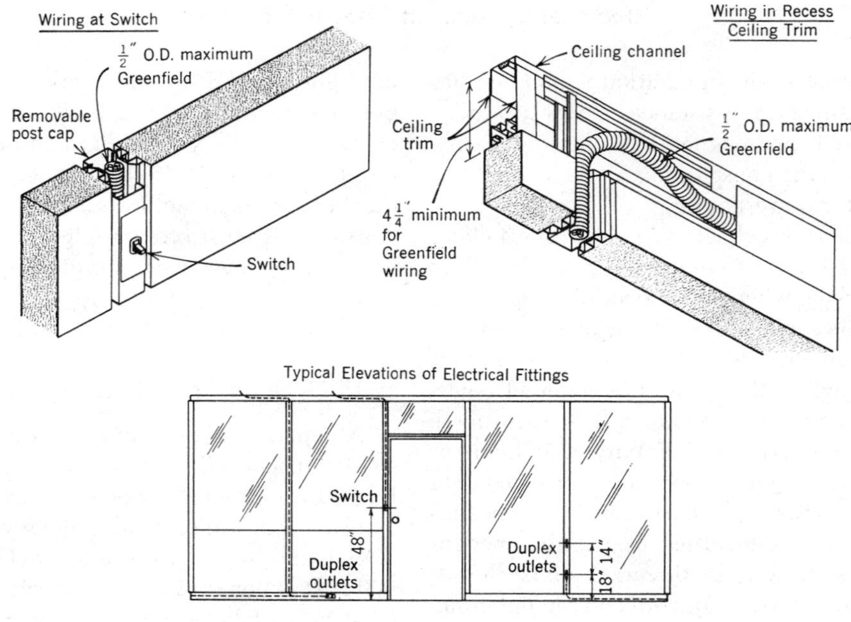

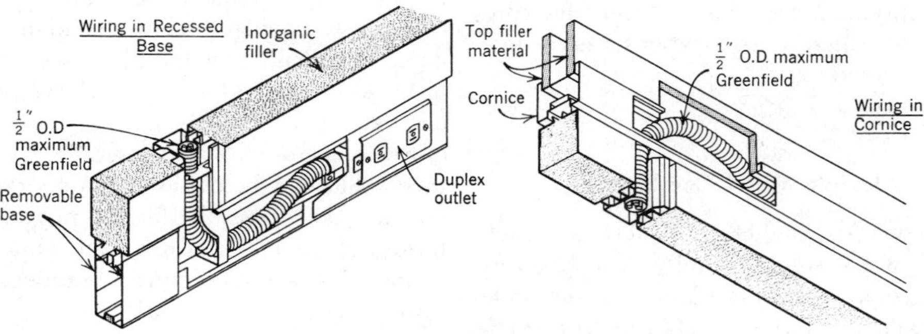

Fig. 15.23 (above) *Flexible conduit is particularly useful in wiring metal partitions.*

Fig. 15.24 (left) *This is a particularly good application of liquid-tight flexible conduit since it provides weatherproofing and acoustical isolation of the noise-producing transformer. Courtesy of Electri-Flex Company*

a. Where the architecture does not permit recessing (see Fig. 15.25).
b. Where economy in construction weighs very heavily in favor of surface raceways (see Fig. 15.26).
c. Where outlets are required at frequent intervals, and where rewiring is required or anticipated (see Fig. 15.27).
d. Where access to equipment in the raceways is required (see Fig. 15.28).

Fig. 15.25 *The exposed wood members require the use of an unobtrusive surface raceway. A small flat raceway feeds receptacle outlets into which the elaborate hanging fixtures are plugged. Courtesy of Wiremold Co.*

Fig. 15.27 (below) *Surface outlet finds ready application in areas where requirements change frequently and are usually heavy, as on test benches. Courtesy of Wiremold Co.*

Fig. 15.26 *A split raceway under a single cover handles both power and telephone conductors. By placing it above the heating unit as shown, or below a fan coil unit, no obstacle is created and no room space is lost. Multisection units are available for running power and signal conductors; where used, the wiring must be segregated and the compartments color coded and maintained in the same relative position throughout the installation. Courtesy of Wiremold Co.*

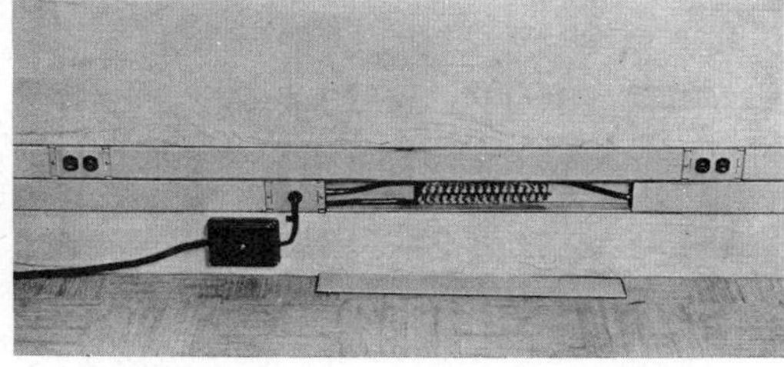

Fig. 15.28 *Parallel power and telephone raceways with separate covers are shown. Note easy access to telephone terminal board mounted* inside *the telephone raceway. In a conduit job, such a terminal board would require the installation of a separate cabinet. Where desired, a single cover can be installed as in Figure 15.26. Courtesy of Wiremold Co.*

e. Where rewiring existing installations to avoid the extensive and expensive cutting and patching required to "bury" a raceway (see Fig. 15.29).

Fig. 15.29 *Although surface raceways may be unsightly in some installations, they are entirely unobjectionable in areas with exposed piping, as in this school rewiring job. Courtesy of Wiremold Co.*

To satisfy the requirement for convenience outlets at frequent intervals, multioutlet assemblies consisting of prewired outlets in a specially designed surface duct are used. Outlet spacing is normally available in 12, 18, and 24 in. sizes. See Figure 15.30, which also gives some dimensional data for two of the many surface raceway shapes available. Multioutlet assemblies are covered in NEC Article 353.

Wireways and troughs are sheet metal enclosures with hinged or removable covers, which may be utilized to carry up to 30 conductors, rated 600 v maximum in dry, nonhazardous, noncorrosive locations and where not subject to physical abuse. If properly gasketed such troughs or wireways may also be used outdoors. Splices and taps are permitted in such raceways, making them particularly useful for feeding groups of starters, switches, and other equipment that may be nippled into the trough and its wiring spliced at that point. Another advantage of this type of raceway is that the hinged or removable cover exposes all the conductors to full view and access for replacement, inspection, addition of other wiring, and so on. Although available in such standard sizes as $2\frac{1}{2} \times 2\frac{1}{2}$, 4×4, 4×6, and 6×8 in. and 1, 2, 3, 4, 5, and 10 ft lengths, special sizes are available.

Fig. 15.30 *Two popular surface raceways with dimensional data are shown in* (a) *and* (b). *A special use of Wiremold No. 2100 is shown in* (c) *where it is used with prewired snap-in receptacles to form a multi-outlet assembly. Courtesy of Wiremold Co.*

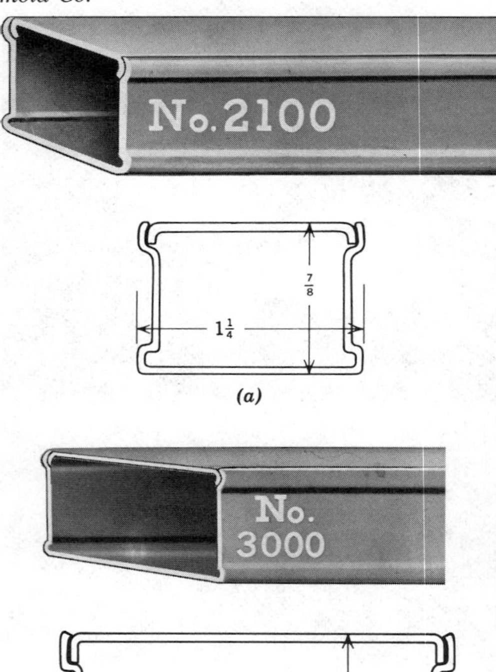

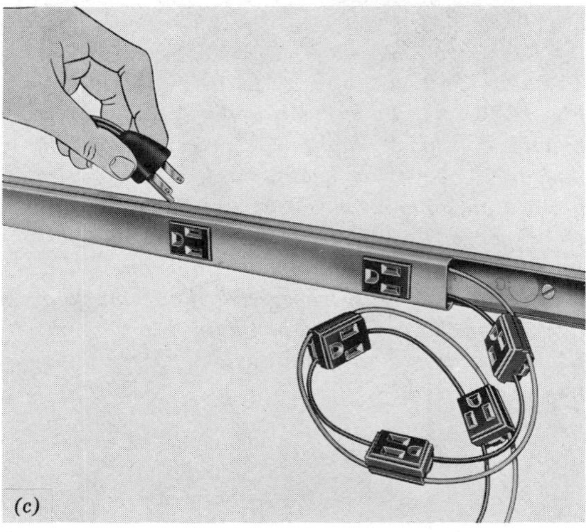

15.29 Floor Raceways

The NEC recognizes three types of floor raceways:

Underfloor raceways—Article 354

Cellular metal floor raceways—Article 356

Cellular concrete floor raceways—Article 358

All three types are applicable to all types of structures and none may be used in corrosive or hazardous areas. The fundamental difference between them is that underfloor raceways are added on to the structure, whereas cellular floor raceways are part of the structure itself—and therefore have a pronounced effect on the building's architecture.

15.30 Underfloor Duct

These raceways, which may be installed beneath or flush with the floor, are covered in NEC Article 354. They find their widest application in office spaces, since their use permits placement of power and signal outlets immediately under desks and other furniture, regardless of furniture layout. Where such underfloor raceways are not employed, and it is desired to place an outlet *on the floor,* one of the following methods is necessary:

a. Channel the floor and install a conduit in the chase, connecting it to the nearest wall outlet. Patch the chased portion.
b. Drill through the floor and run a conduit on the ceiling below to an outlet below.
c. Drill through the floor twice and connect the new outlet to an existing outlet via a conduit on the ceiling below. This, as (b), is expensive and disturbs the occupant below.
d. Install a surface floor raceway.

Since all of these have obvious major disadvantages, underfloor (UF) duct systems were widely employed until the introduction of what may be called over-the-ceiling ducts (in apposition to under-the-floor ducts); these systems are discussed in Sections 15.33 and 15.34. The reason that alternate systems were developed is economic—underfloor duct systems are expensive and frequently underutilized, giving an unsatisfactory return on investment. However, before discussing the relative merits of systems, an understanding of what UF duct systems are, and how they are assembled and utilized, is in order.

Underfloor duct systems are available in two basic designs—single level and two level (refer to Fig. 15.31). In a single-level system all the system components are on the same level—the feeder ducts from the panels, the distribution ducts with inserts for floor outlets and, most important, the junction boxes. As the number of parallel distribution ducts grows with the size of the open floor area and the density of the furniture layout, the need for more feeder ducts and consequently larger and more complex junction boxes also grows. A modern office floor layout almost invariably requires a triple duct system (power, telephone, and signal). Since power and telephone wiring must be separated by metal barriers, the junction boxes become complex, and consequently large and deep. The depth of these boxes obviously controls the amount of concrete fill required. A simple single-level system requires a minimum concrete fill of $2\frac{1}{2}$ in., while a complex one can easily require an inch more. With such a concrete fill requirement, a two-level system with its great flexibility becomes desirable.

Refer now to Figures 15.32 and 15.33, which illustrate a two-level system. Note that the fundamental difference is that the feeder and distribution ducts are on different levels thus eliminating the necessity for complex junction boxes; this gives the system unlimited feeder capacity and thereby obviates the necessity for supplemental conduit feeds, as in the single-level system. The drawback of this system is the conduit fill required. A *minimum* of $3\frac{5}{8}$ in. is required, though additional slab thickness throughout can be avoided by depressing part of the slab to accommodate the feeder ducts, as shown in Figure 15.34.

In both systems, the distribution duct can be provided with preset inserts or can be furnished flush (without inserts) for use with afterset inserts. Although preset inserts can be

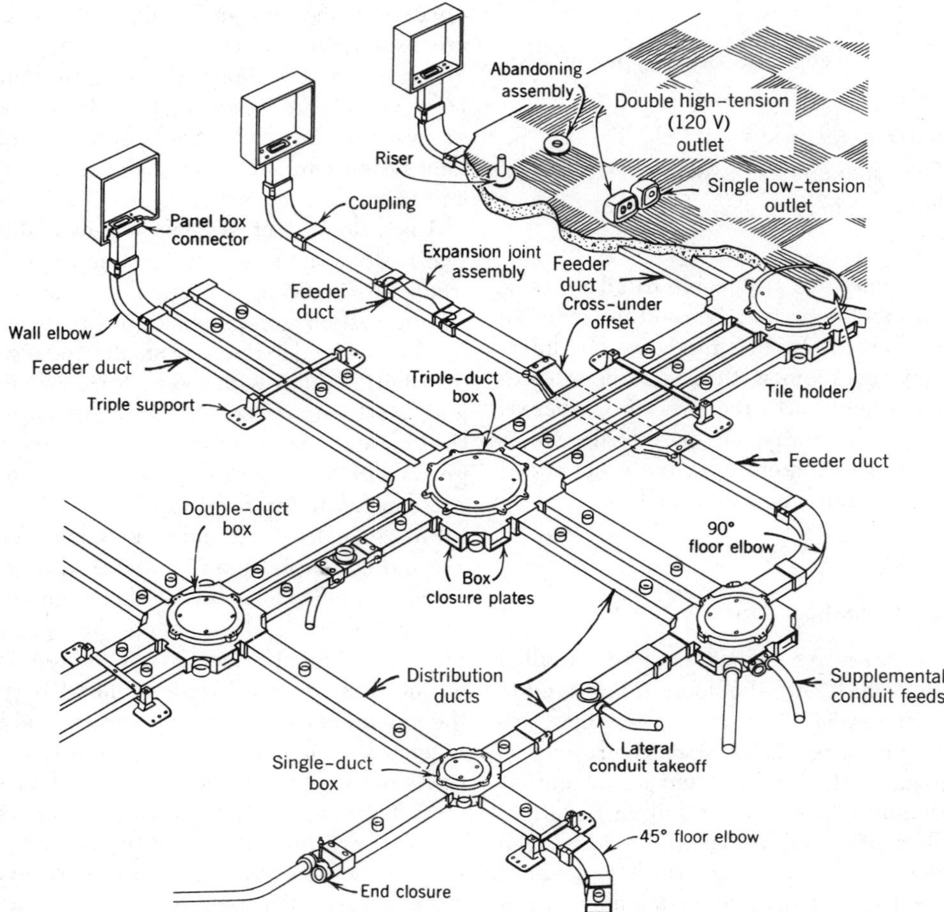

Fig. 15.31 *Single-level underfloor duct system. Since all ducts are mounted at the same elevation, special junction boxes are required to accommodate the varying number, size, and type of duct. Junction boxes may be obtained to accommodate either one, two, or three ducts on each side, with closure plates sealing unused openings. Duct supports are available for one or more cells, while a wide variety of accessory components promotes flexibility in layout and variety in service facilities. When feeder duct capacity is inadequate to supply the needs of all the distribution ducts (which serve the outlets), supplementary conduit feeds are run to the junction boxes.*

furnished with any desired spacing, 2 ft is most common. Afterset inserts are placed at the desired location by core drilling through the concrete and duct and placing the insert. The choice of whether to use preset or afterset depends on the number of inserts to be activated. It has been estimated that the cost of placing one afterset insert, including drilling, is approximately equal to the cost of four preset inserts, not including outlet activation costs.

In a single-level system the ducts are arranged in a grid pattern, with spacing selected to provide desk coverage. Thus, if desks are 48 in. wide with a 4 ft aisle between, an 8 ft lateral duct spacing would provide adequate coverage. Since feeder ducts—particularly for telephone service—rapidly become filled, cross-connecting distribution and feeder ducts must be supplied. Also, the location of telephone closets has a pronounced effect on the pattern and spacing selected. With a two-level

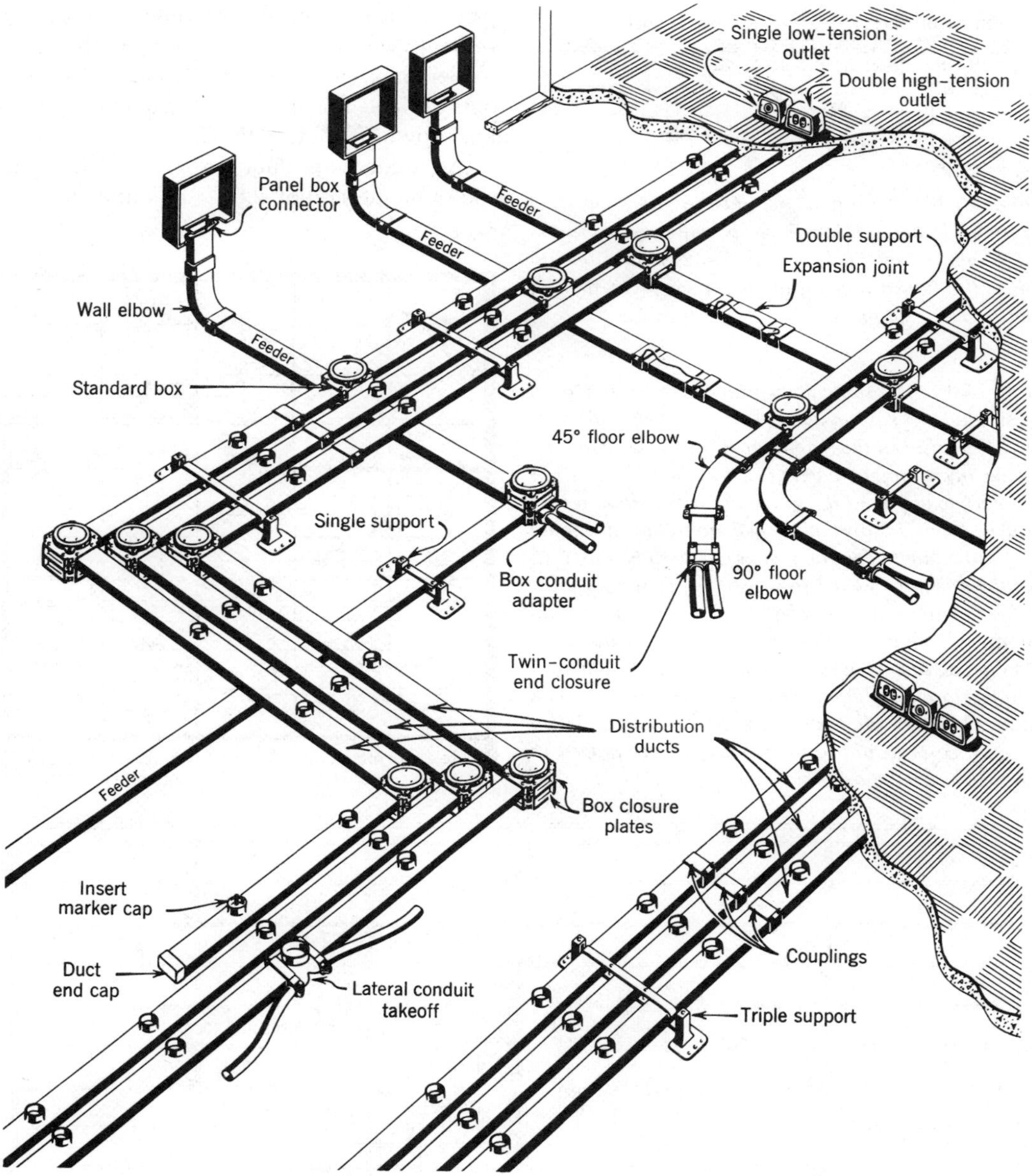

Fig. 15.32 Two-level underfloor duct system, using standard size duct only. Feeder ducts are run below the distribution (outlet) ducts resulting in simple standard junction boxes, (see Fig. 15.33), unlimited feeder capacity, and complete separation of power and signal wiring.

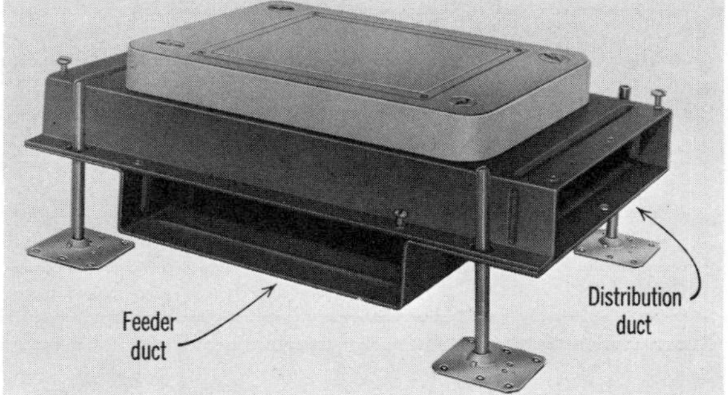

Fig. 15.33 A typical two-level junction box demonstrates the simplicity of the two-level system. Courtesy of The Square D Co.

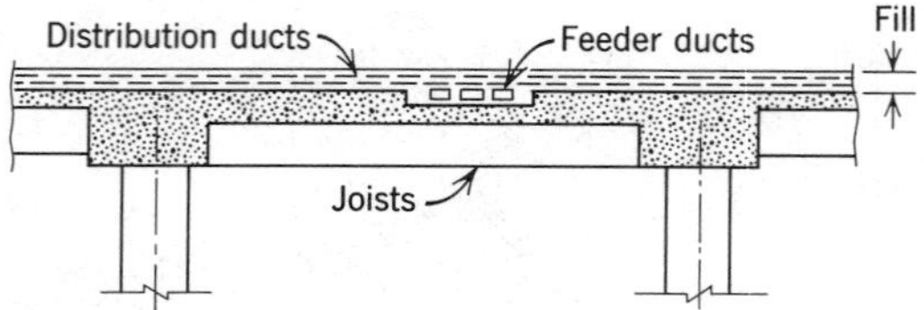

Fig. 15.34 Setting a two-level underfloor duct system. To avoid thickening fill, a depression in the slab can accept feeder ducts. Ducts would be run near the bay center to avoid the negative steel of joists, near columns.

system, distribution ducts would be run longitudinally at 6 to 8 ft spacing, with lateral feeder ducts as necessary. Comparative layouts of ducts for single- and two-level systems are shown in Figure 15.35.

All underfloor duct systems use basically the same method of setting an outlet once an

Fig. 15.35 The same floor plan can be supplied by a single-level system (a) *or a two-level system* (b) *or* (c). *Note that in the single-level system* (a) *conduits have been run to each junction box (only stubs shown) to augment the generally inadequate feeder duct system. In this layout, even with the additional conduits, the telephone feeder capacity is only 60% of the recommended 1 sq in./100 sq ft of floor area. The location of the telephone closet has a pronounced effect on the entire layout. Thus, if a two-level system is restricted by core design to a single closet, the layout shown in* (b) *results. A much better layout with less duct, shorter runs, and lower overall costs would result if closets were placed in the core corners as shown in* (c).

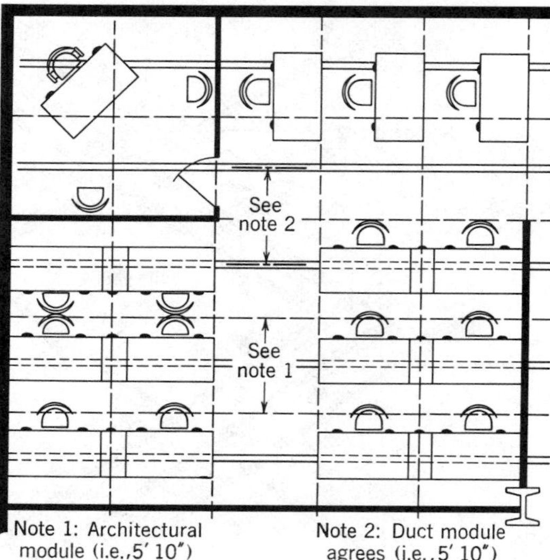

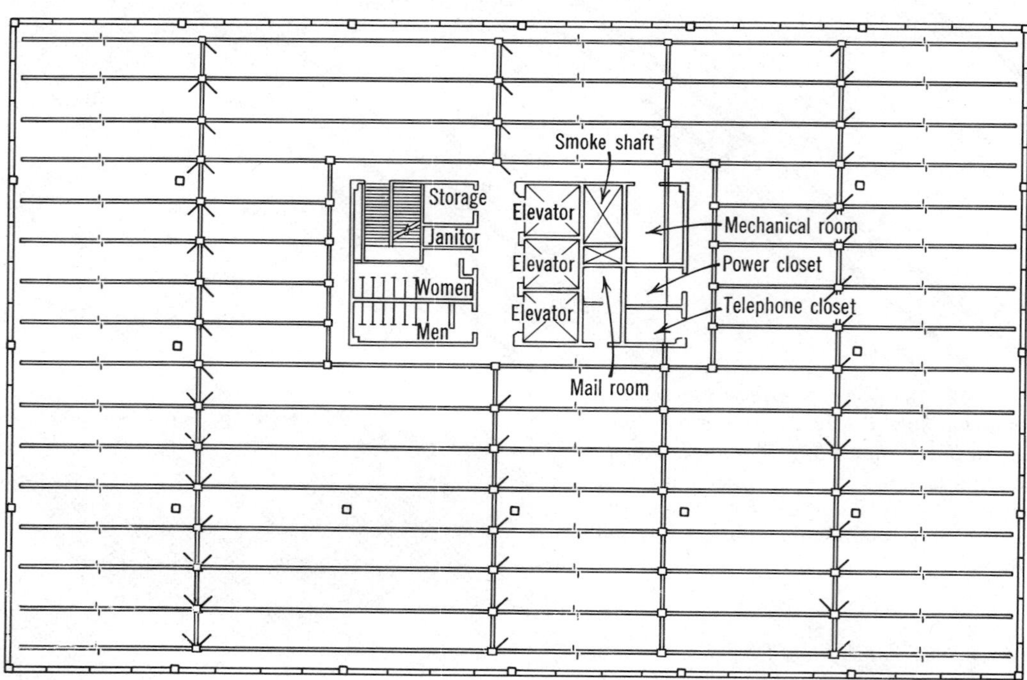

(a)

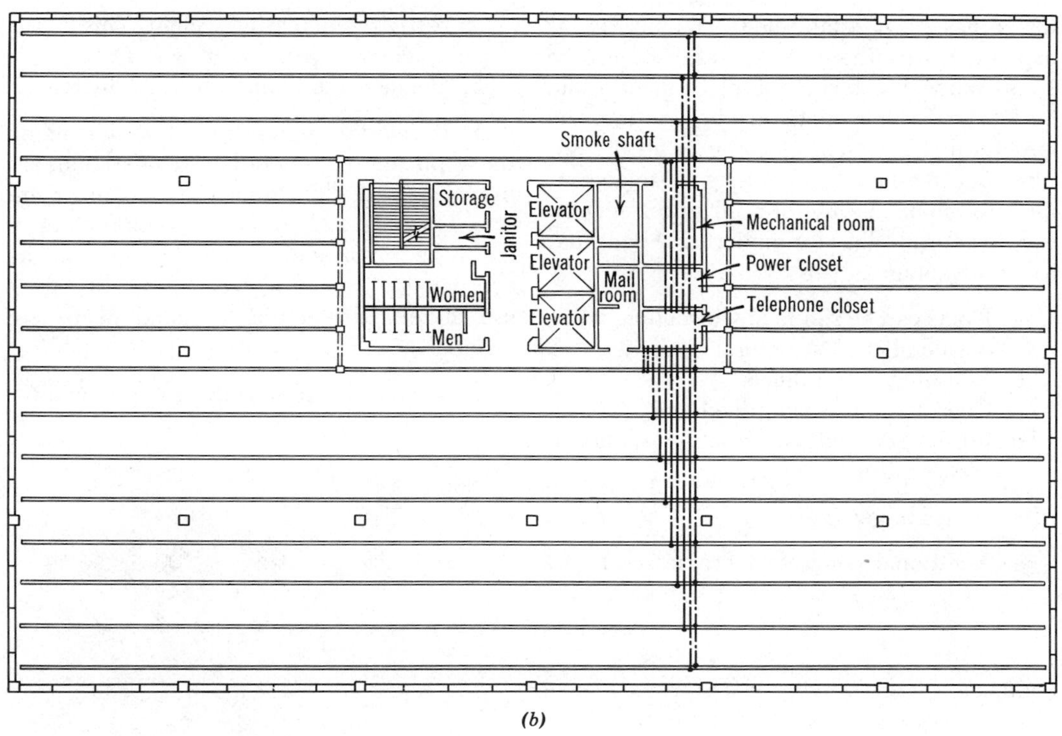
Smoke shaft
Storage
Janitor
Elevator
Elevator
Elevator
Mail room
Women
Men
Mechanical room
Power closet
Telephone closet

(b)

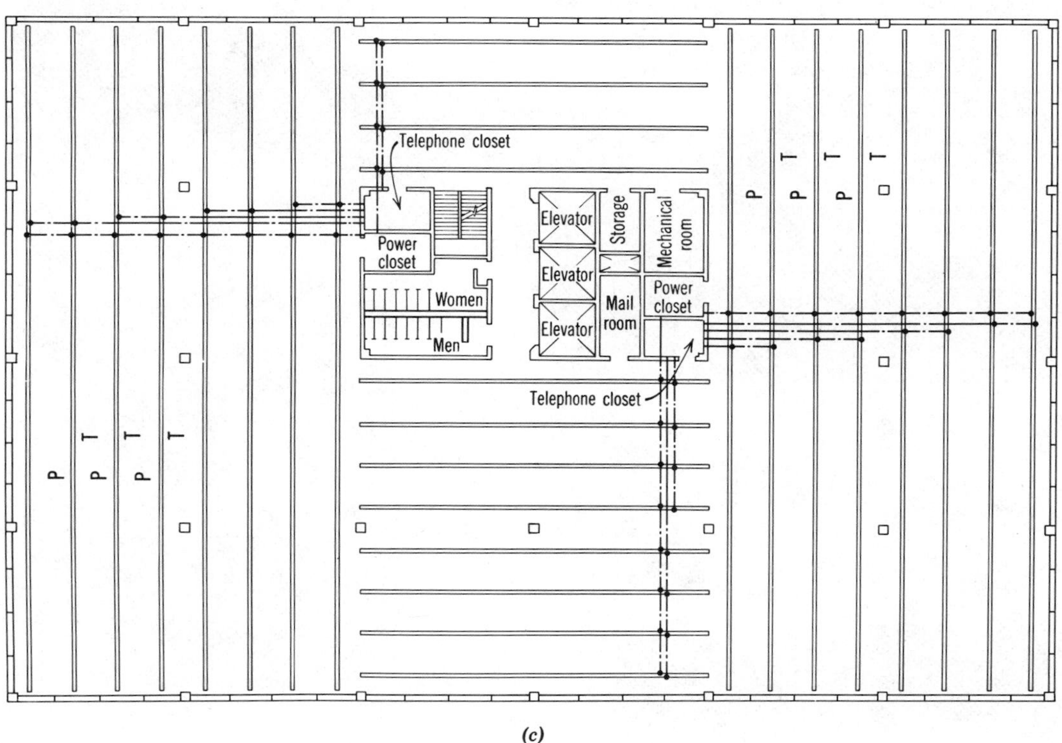
Telephone closet
Power closet
Women
Men
Elevator
Elevator
Elevator
Storage
Mechanical room
Mail room
Power closet
Telephone closet
P
P
P
T
T
T

(c)

insert has been established. The inserts, as explained, are either preset or afterset. Figure 15.36 shows the technique of connecting an outlet. Underfloor ducts may be cast into the structural slab in lieu of being in fill or topping, but the slab must be designed to accommodate them. The use of a fill or topping on the structural slab for underfloor duct has these advantages:

a. Ducts can be run in any direction, without conflict to structural elements.
b. Finishing is simplified.
c. Coordination is simplified.
d. Formwork and construction sequence are simplified.

The disadvantages are:

a. Additional concrete increases costs directly by increasing weight. This is particularly expensive in seismic designs.
b. Height of building may be increased.

In conclusion, some general comments on the application of underfloor duct systems are in order. Underfloor duct systems are *expensive.* They can add 50% to the building's electric system cost, without consideration of the construction costs involved. To justify their use, therefore, the building should meet these criteria:

a. Open floor areas, with a requirement for

Fig. 15.36 *Placing an outlet in a preset duct insert is a multistep operation. The concrete is chipped away* (a), *the insert cover removed* (b), *an outlet collar placed* (c), *and the outlet affixed* (d) *after pulling the wire. Telephone outlets are similarly set.*

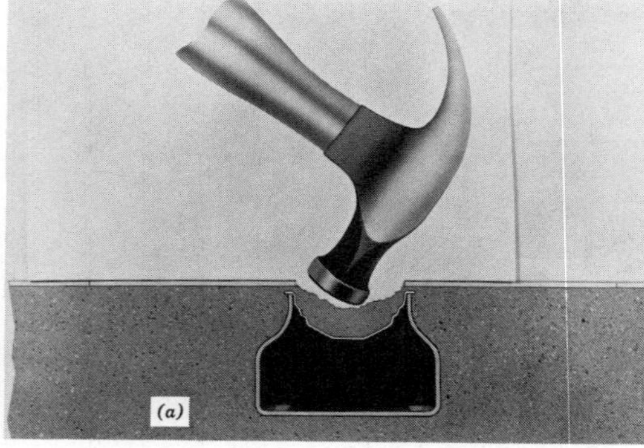

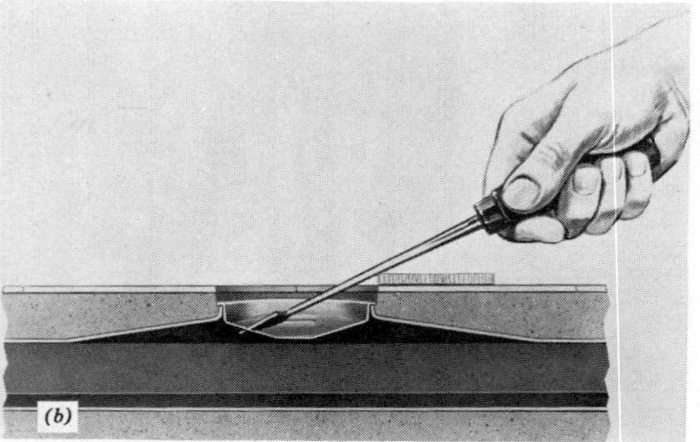

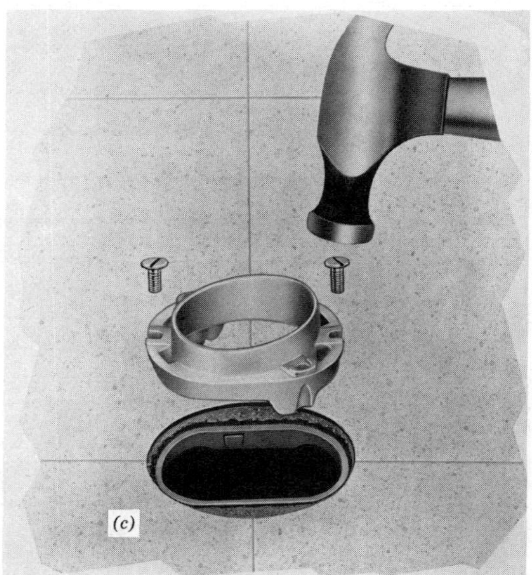

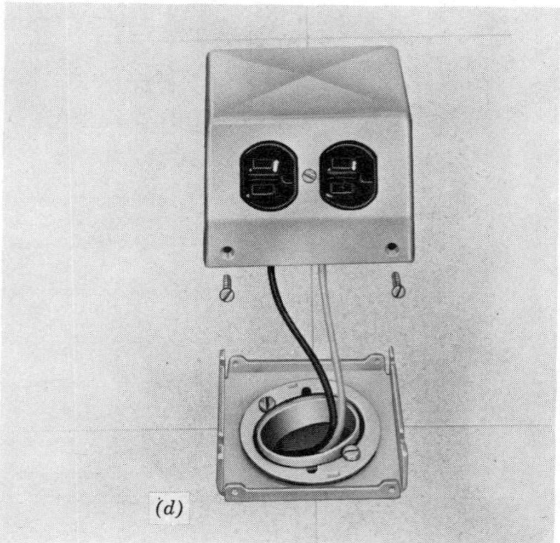

outlets at locations removed from walls and partitions.

b. Outlets from ceiling systems unacceptable.
c. Frequent rearrangement of furniture and other items requiring electrical and signal service.

The facilities that may meet these criteria are prestige office buildings, museums and galleries, high-cost merchandising areas, and selected areas in industrial facilities. Bear in mind that even in high-cost office construction, underfloor duct systems are difficult to justify economically unless furniture layout will change. For fixed layouts, conduit fed floor boxes are the economical choice. In doubtful cases, alternate arrangements can be planned and an intelligent choice made after costs and impact on the building structure are studied.

15.31 Cellular Metal Floor Raceway

The underfloor duct system described above is best applied to known furniture layouts and to rectilinear arrangements. Random arrangements, such as those found in office landscaping, require a fully accessible floor—if indeed the floor is to be used for electrification. This is best provided by a cellular (metal) floor that is an integrated structural/electrical system. The floor can be fully or partially electrified. A floor designed with two or three electrified cells adjacent to several cells of structural floor, as shown in Figure 15.37, will give sufficient coverage for all purposes. One of the many structural element designs available is shown in that figure. Furthermore, the electrified cells can be arranged to feed lighting outlets in the floor below. This practice, albeit economical, may introduce electrical control problems inasmuch as the electric panelboard controlling the lighting circuits thus fed would be located on the floor above.

Figure 15.37 shows a floor design with electrified cells fed from a trench header, as well as a cell detail. Figure 15.38 shows another floor detail plus actual installation photos of the inserts during construction and after completion. Note that the inserts shown in Figure 15.38 are of the type that rest *between* cells and are fed from the sides of the cells. The advantage of this arrangement is that the insert sits very low and therefore the amount of concrete used is the minimum required structurally; compare Figure 15.37*b*. The technique for placing an afterset insert is clearly illustrated in Figure 15.39. As with underfloor ducts, the choice between preset and afterset inserts depends on the number of outlets that will be activated and the frequency with which changes will be made. Also notice that the cellular floor is part of the structural system and is designed accordingly. The electrical wiring is fed into the cells from header ducts and/or trenches that run perpendicular to the floor cells and constitute a system of underfloor duct in themselves. The header ducts in turn are fed from lighting panels, and signal and telephone cabinets in much the same manner as normal underfloor ducts.

All floor cell sections and the crosswise header ducts (for feeder and branch circuits) are tack-welded to the main floor beams and to each other. These welds assure electrical grounding of the raceways. Three types of wiring systems generally run in separate floor cells and header ducts: general lighting and appliances, telephones, and signal systems—although the latter two may be combined in a single cell if the signal system voltage is low. A complete range of outlets and fittings is available, together with installation tools and accessories.

The use of cellular floor construction for wiring is restricted to noncorrosive, nonhazardous areas. No electric conductors may be installed in any cell or header that contains any service other than electric. The usefulness and capacity of a cellular floor system is controlled by the capacity of the feeder raceways that may be header or trench ducts. Use of a trench-header feed eliminates the need to core a hole through the intersection of header and cell, avoiding damage to floor and finish. This convenience is gained at the price of a more difficult installation and usually at higher cost.

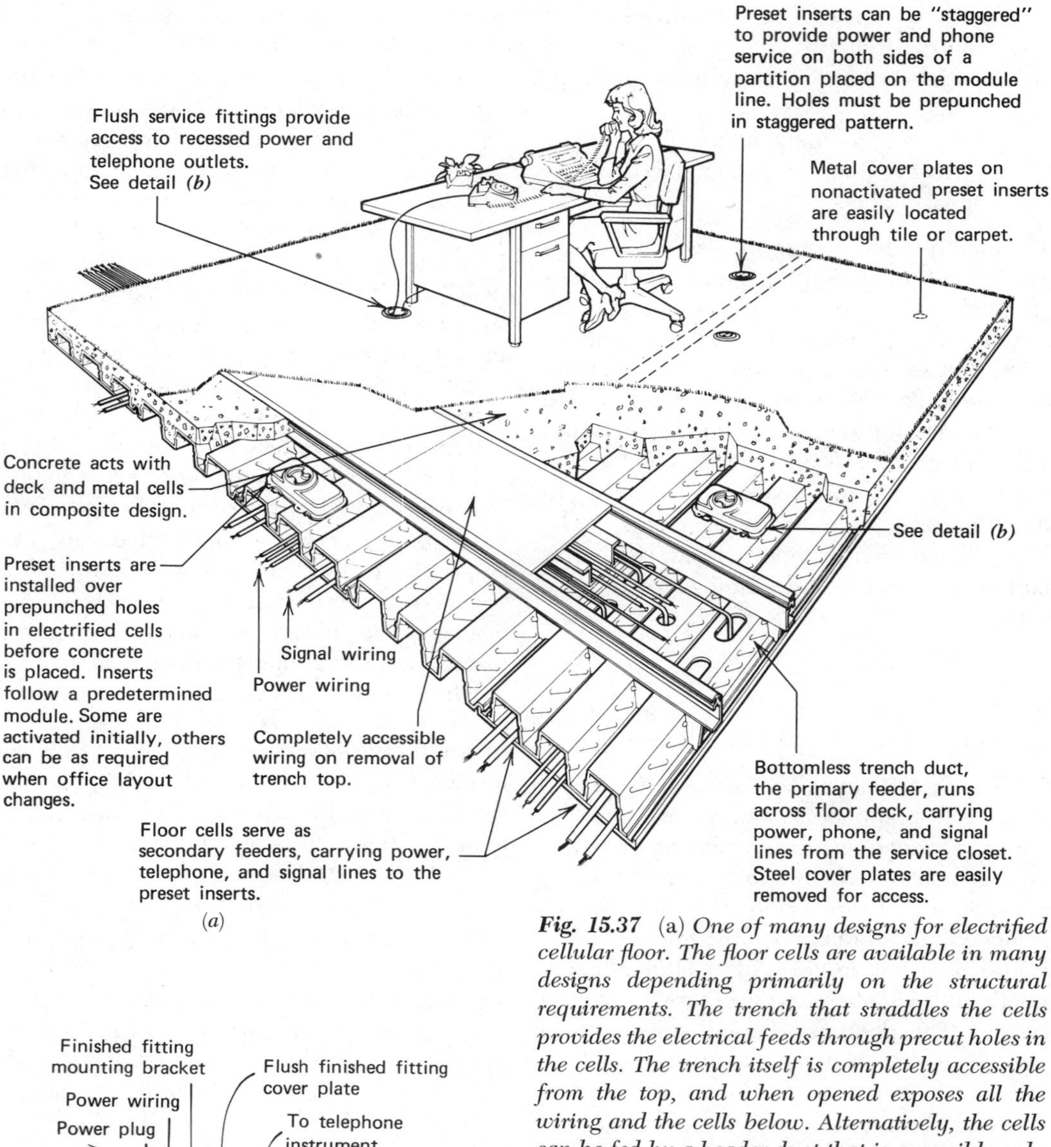

Preset inserts can be "staggered" to provide power and phone service on both sides of a partition placed on the module line. Holes must be prepunched in staggered pattern.
Flush service fittings provide access to recessed power and telephone outlets. See detail (b)
Metal cover plates on nonactivated preset inserts are easily located through tile or carpet.
Concrete acts with deck and metal cells in composite design.
See detail (b)
Preset inserts are installed over prepunched holes in electrified cells before concrete is placed. Inserts follow a predetermined module. Some are activated initially, others can be as required when office layout changes.
Signal wiring
Power wiring
Completely accessible wiring on removal of trench top.
Bottomless trench duct, the primary feeder, runs across floor deck, carrying power, phone, and signal lines from the service closet. Steel cover plates are easily removed for access.
Floor cells serve as secondary feeders, carrying power, telephone, and signal lines to the preset inserts.

(a)

Finished fitting mounting bracket
Flush finished fitting cover plate
Power wiring
Power plug
To telephone instrument
Duplex Receptacle
Telephone connector
Duplex receptacle mounting plate
Base plate
120 volt power cell
Telephone/signal cell

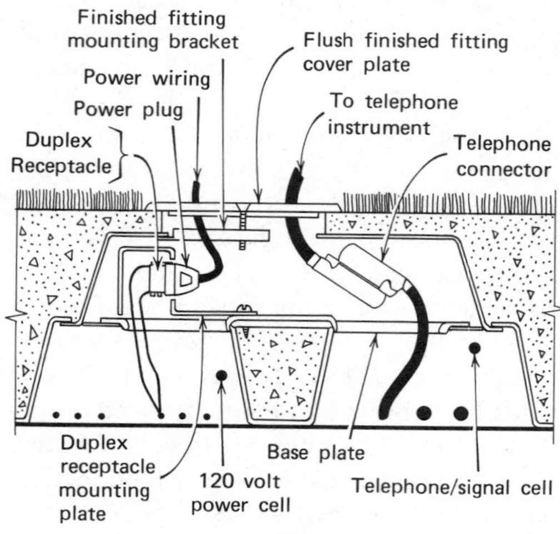

(b)

Fig. 15.37 (a) *One of many designs for electrified cellular floor. The floor cells are available in many designs depending primarily on the structural requirements. The trench that straddles the cells provides the electrical feeds through precut holes in the cells. The trench itself is completely accessible from the top, and when opened exposes all the wiring and the cells below. Alternatively, the cells can be fed by a header duct that is accessible only via handholes.* (b) *Activated preset insert. Note that the insert straddles two cells and takes wiring through prepunched holes in the tops of the cells. Power and signal wiring are completely separated at all times by metal barriers. If desired, a standard surface fitting can be mounted on the floor in lieu of the flush plate shown. When an insert is to be deactivated, the flush cover plate is simply replaced with a blank plate. Courtesy of INRYCO, Inc. and The Square D Co.*

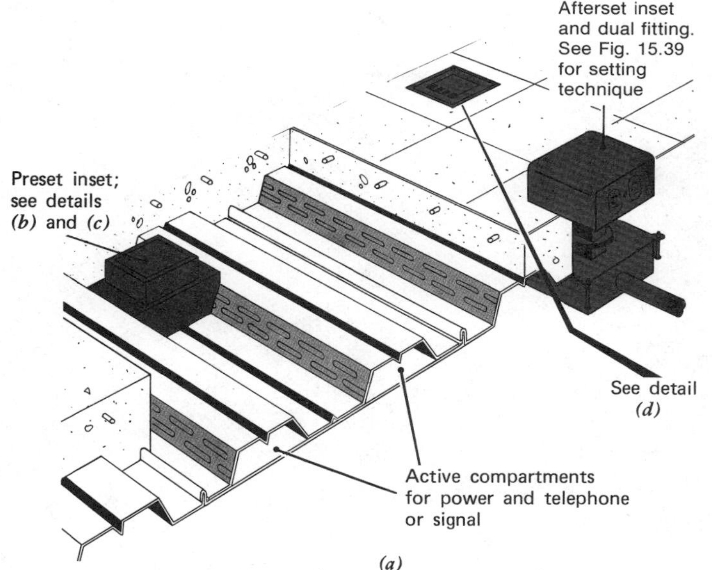

(a)

Fig. 15.38 *Floor section showing schematically both preset and afterset inserts* (a). *An actual field installation is shown in* (b) *and* (c), *while* (d) *shows the completed floor outlet. Courtesy of Rollform Products, Inc.*

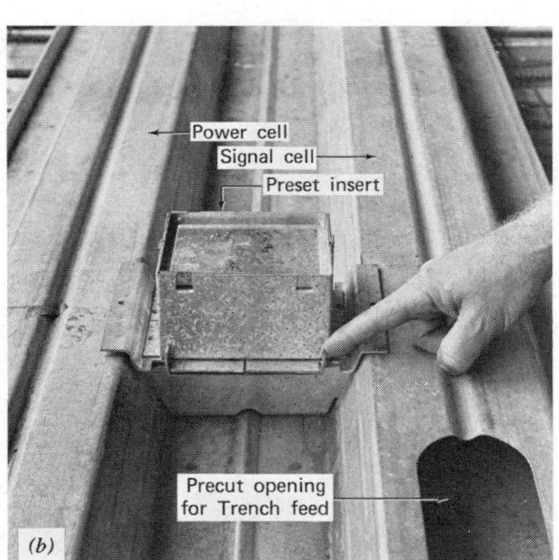

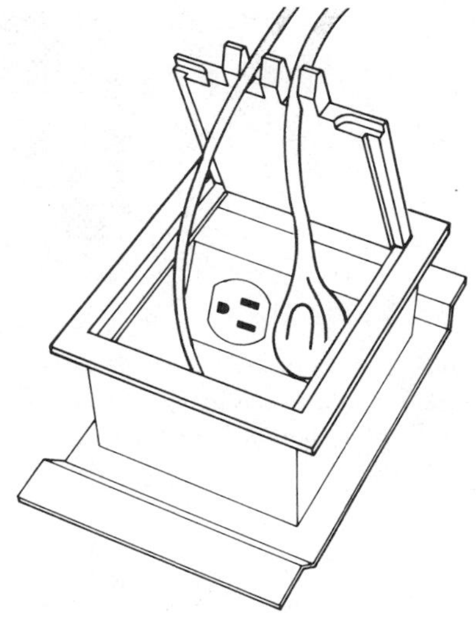

(d)

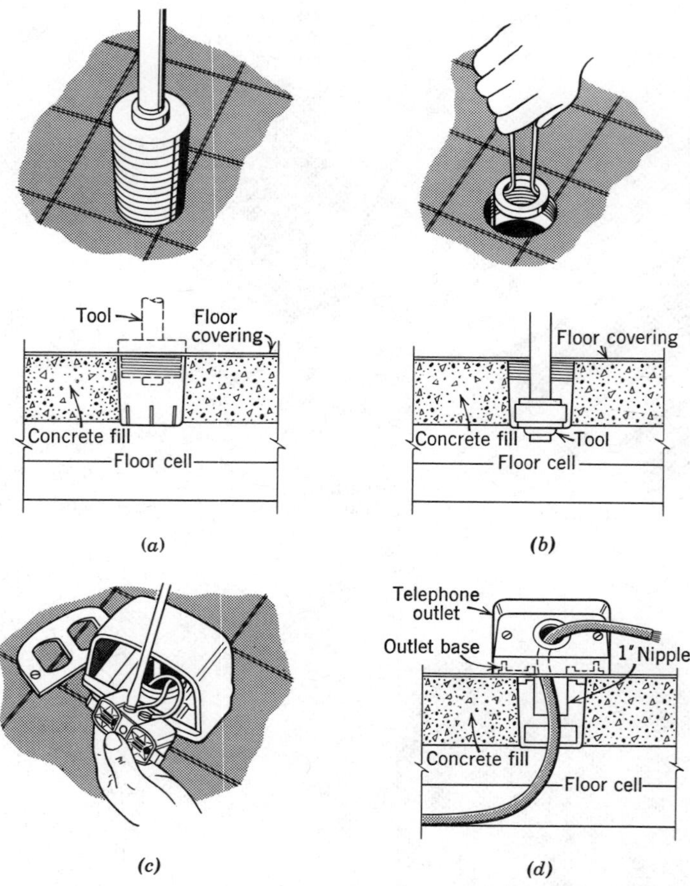

Fig. 15.39 *Afterset inserts and floor outlets in cellular metal floors are simply installed.* (a) *Hole is drilled through floor and into cell using core drill.* (b) *Fittings are set in cell for attachment of floor device.* (c) *The wire is pulled through and attached to the device;* (d) *the floor box is set, leaving the device ready for use. Telephone outlets are similarly installed.*

15.32 Precast Cellular Concrete Floor Raceways

This structural concrete system is similar to a cellular metal floor in application (see Fig. 15.40). A cell is defined in NEC Article 358 as a "single, enclosed, tubular space in a floor made of precast cellular concrete slabs, the direction of the cell being parallel to the direction of the floor member." Feed for these cells is provided, as with metal cellular floor construction, by header ducts. Although header ducts are normally installed in concrete fill above the hollow core structural slab, a header arrangement with feed from the ceiling below is also entirely practical. Like the metallic cellular floor, the cells can be used for air distribution and even for piping, although these items are generally installed in a hung ceiling. Integrated ceiling designs for lighting, HVAC, and acoustical treatment below the slab are discussed in Section 21.17.

15.33 Ceiling Raceway Systems

The need for electrical flexibility in facilities with limited budgets coupled with the

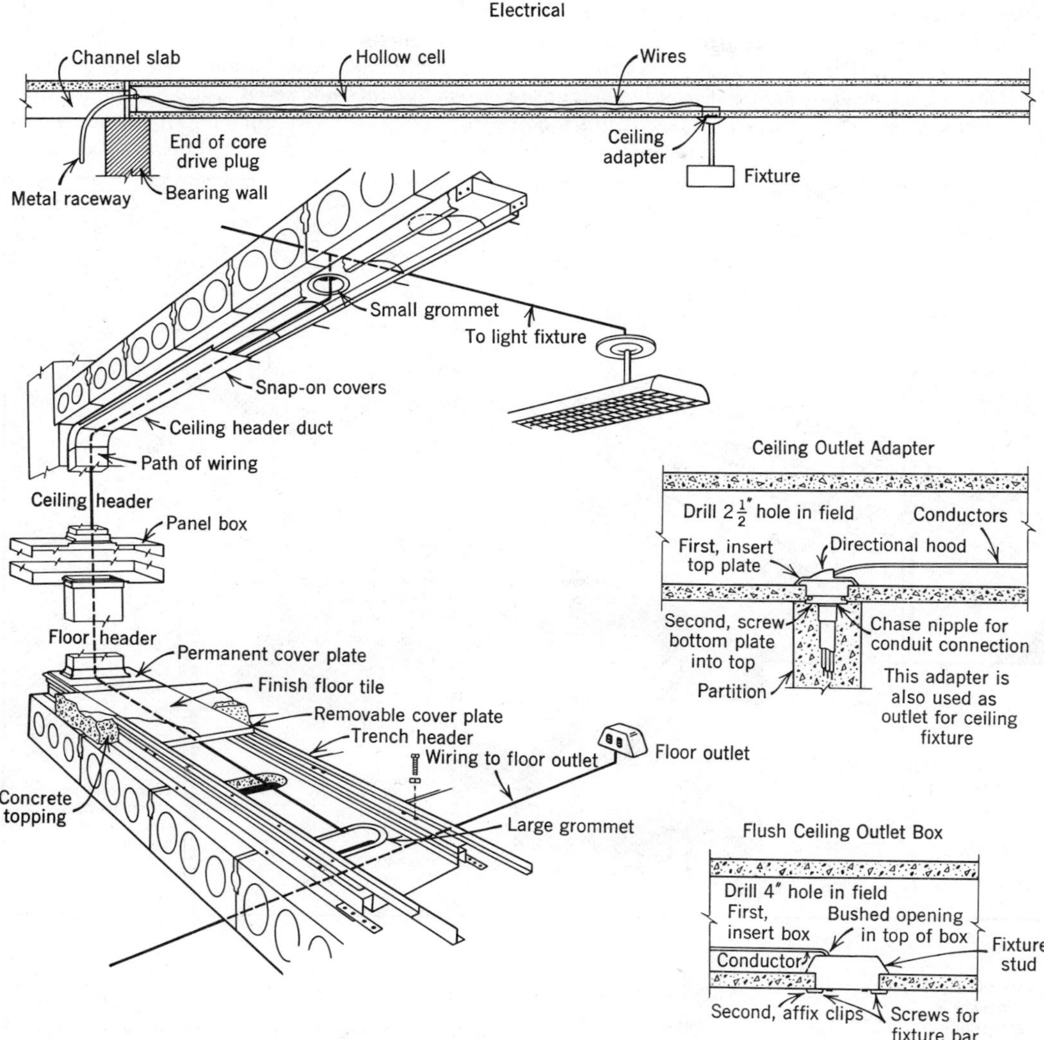

Fig. 15.40 *Precast cellular concrete slabs. This construction is the concrete equivalent of cellular metal construction. Note that headers are run on the floor* and *on the ceiling, to handle floor electrification and lighting, respectively.*

high cost of underfloor electrical raceway systems encouraged the development of equivalent over-the-ceiling systems. These systems are actually more flexible than their underfloor counterparts, since they energize lighting as well as provide power and telephone facilities; furthermore, they permit very rapid changes in layouts at low cost. This last characteristic is particularly desirable in stores where frequent display changes necessitate corresponding electrical facility changes. Development of these systems was made possible by changes in the NEC; it now allows: wireways to be run inside hung ceilings, fixtures to be fed by lengths of unsupported flexible cable, vertical runs of unsupported raceways (floor-to-ceiling system), splices inside wireways, and plug-in type connections in accessi-

Fig. 15.41 Overhead raceway system fed from a central service core. The system is a conventional "tree" system; that is, the main feed at the panel subfeeds to headers that in turn subfeed to distribution ducts, which in turn subfeed to the utilization devices, which here are lighting fixture and ceiling-to-floor poles (Tele-Power poles). The detail shows the method of tapping the laterals (distribution duct) to the feeders (header duct). Depending on the size and layout of the system, laterals may serve as subdistribution only (foreground) or subdistribution plus feed to utilization devices (see insert). All numbers and nomenclature in this and subsequent related figures refer to Wiremold equipment. Courtesy of Wiremold Co.

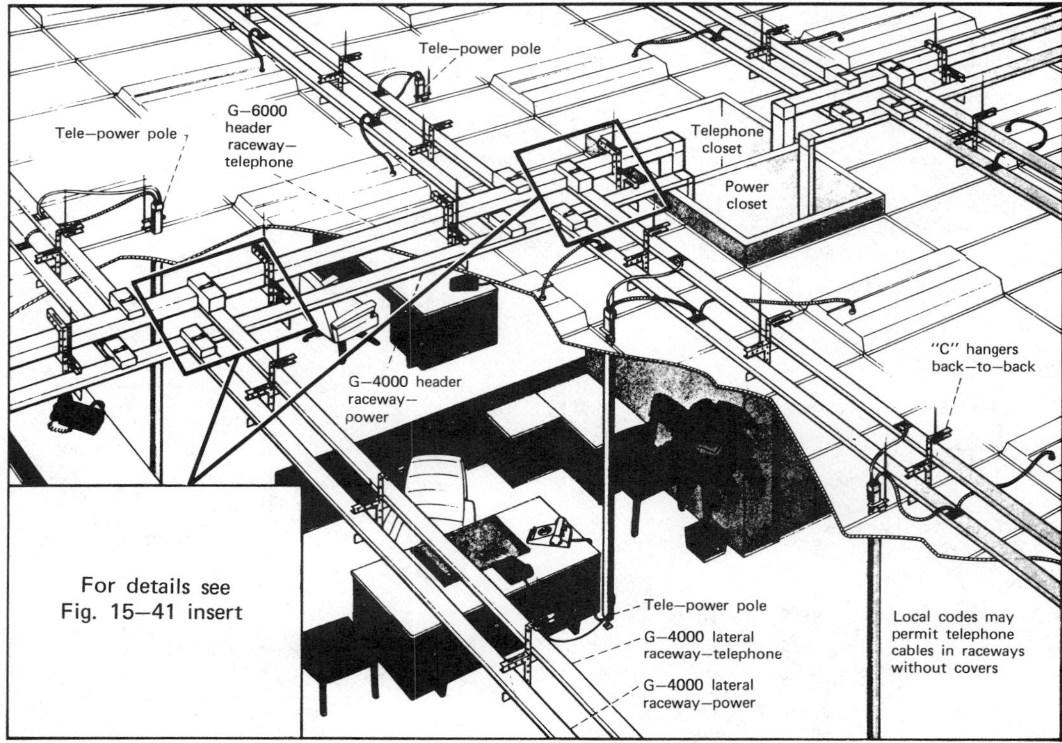

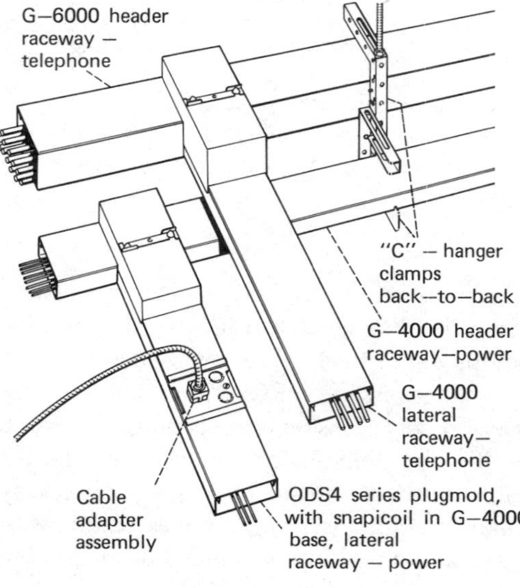

ble ceiling spaces. (Hung ceilings with lift-out tiles are classified in this respect as accessible spaces.) In addition to the extreme flexibility of this electrical layout made possible by the ceiling raceway system, it has the additional advantage that it itself, not being cast in concrete like its underfloor counterpart, can be altered at will. Thus, not only layout changes (as mentioned above) but also changes in the utilization of existing spaces can readily be accommodated. This is a particularly important characteristic in merchandising and educational facilities where spaces originally planned for one use have their application completely changed during the course of the building's life.

Fig. 15.42 A single distribution duct with receptacle outlets, placed at 24-in. centers, feeds a large number of recessed troffers in this brightly lighted merchandising area ceiling. Hard wiring of this density of lighting fixtures would be considerably more expensive initially, and much more so in alterations. Courtesy of Wiremold Co.

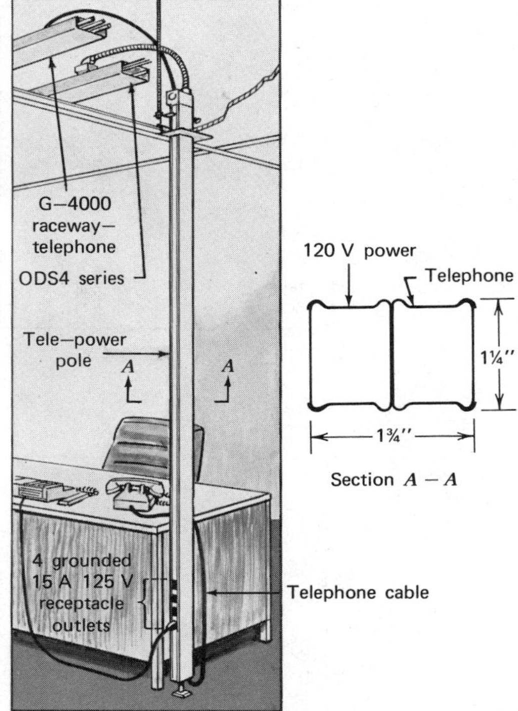

Fig. 15.43 Detail of one design (of many) of Wiremold's Tele-power pole. Other designs have different dimensions, outlets, bases, and colors. The power compartment is prewired with four single or two duplex 15a, 125 v outlets, with top connection as desired. In this case the power compartment is wired to a flexible armored cable terminating in a special polarized plug. The communications compartment is generally unwired. Pole lengths are available to match ceiling height requirement. Courtesy of Wiremold Co.

The systems vary among manufacturers but are essentially the same, and similar to underfloor systems. The entire system is shown clearly in Figure 15.41. Header ducts (wireways) connect to electrical panels and telephone cabinets in the power and telephone closets, respectively. Telephone headers are normally of larger size than the power header and can carry other low-voltage signal equipment as well. Distribution ducts (laterals) tap onto the headers. These laterals may act as subdistribution wireways (foreground) or may feed fixtures and poles directly. Power feeds from these laterals are made by means of plugs and receptacles, thus eliminating the cost involved in the "hard" wiring of fixtures; this allows the desired flexibility. Note in Figure 15.42 how easily fixtures are connected. By using polarized, color-coded receptacles and plugs, power laterals can feed fixtures at 277 v and poles at 120 v without confusion. Poles carrying telephone and 120 v power down to the desk level can be fed from laterals above, below, in, or on the ceiling as desired (see Figs. 15.43 and 15.44). The end result in a hung ceiling office area (Fig. 15.45) or an open ceiling area (Fig. 15.46) is certainly less elegant than an underfloor system, but for most users it is satisfactory. The technique shown in Figure 15.46 of creating a black cavity at the ceiling is undesirable from the lighting viewpoint, as is discussed in Chapters 19 and 20. It creates high brightness ratios and reduces working plane illumination. Figure 15.47 shows the system's versatility in open floor plans.

15.34 Prewired Ceiling Distribution Systems

In an effort to reduce the on-site labor cost of systems such as those described in the preceding section, a number of manufacturers have produced prewired plug-in type equipment that reduces the field assembly time

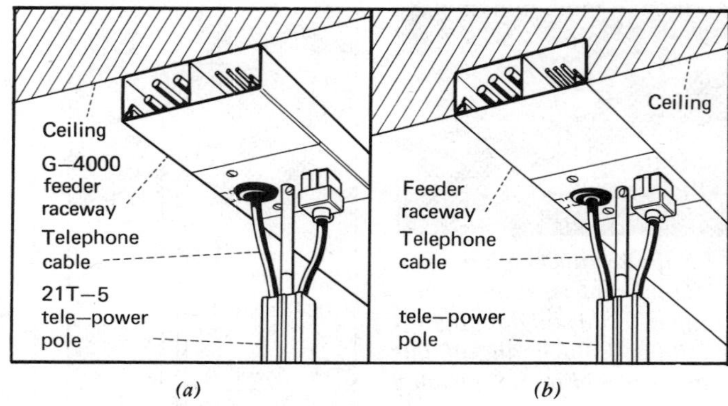

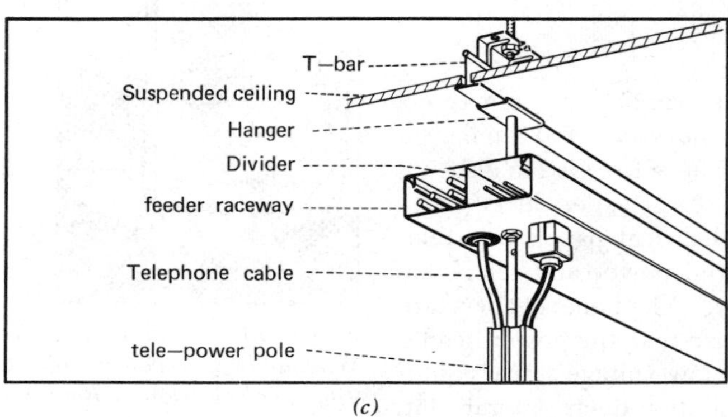

Fig. 15.44 *Ceiling distribution raceways from which service poles are fed may be mounted on* (a), *in* (b), *or below* (c) *the ceiling, in addition to the more common above-the-ceiling installation of Figure 15.43. Courtesy of Wiremold Co.*

Fig. 15.45 *A service pole in this consulting engineer's office feeds a portable light-table and other electrical accessories. The communication compartment has been left unwired. Courtesy of Wiremold Co.*

drastically. The reasoning behind this is that factory labor cost is considerably lower than field labor cost, and therefore a considerable savings can be effected. An ancillary benefit is that a plug-in type system is simply and rapidly altered if required. Figure 15.48 shows schematically how such a system could be employed to feed a group of recessed fluorescent troffers in a typical hung ceiling installation. Figure 15.49 demonstrates the extreme simplicity of field connection and Figure 15.50

Fig. 15.46 In this economical installation the absence of a hung ceiling is minimized by painting the entire ceiling.

Fig. 15.47 Two illustrations of the ceiling raceways system's versatility, in (a) *a learning center and* (b) *in an open plan office.*

illustrates the factory-made system components. Similar type plug-in equipment is available for receptacle systems, low voltage switch, clock and program equipment, and sound and signal equipment. To take full advantage of the potential labor-cost savings in this system, field wiring must be minimized. This is best accomplished by having utilization equipment, including lighting fixtures, factory equipped with appropriate plug-in connectors.

15.35 Boxes and Cabinets

In this category are included pull boxes, splice boxes, and outlet boxes. Splice boxes, as the name suggests, are placed in raceway runs at points where splices or taps must be made;

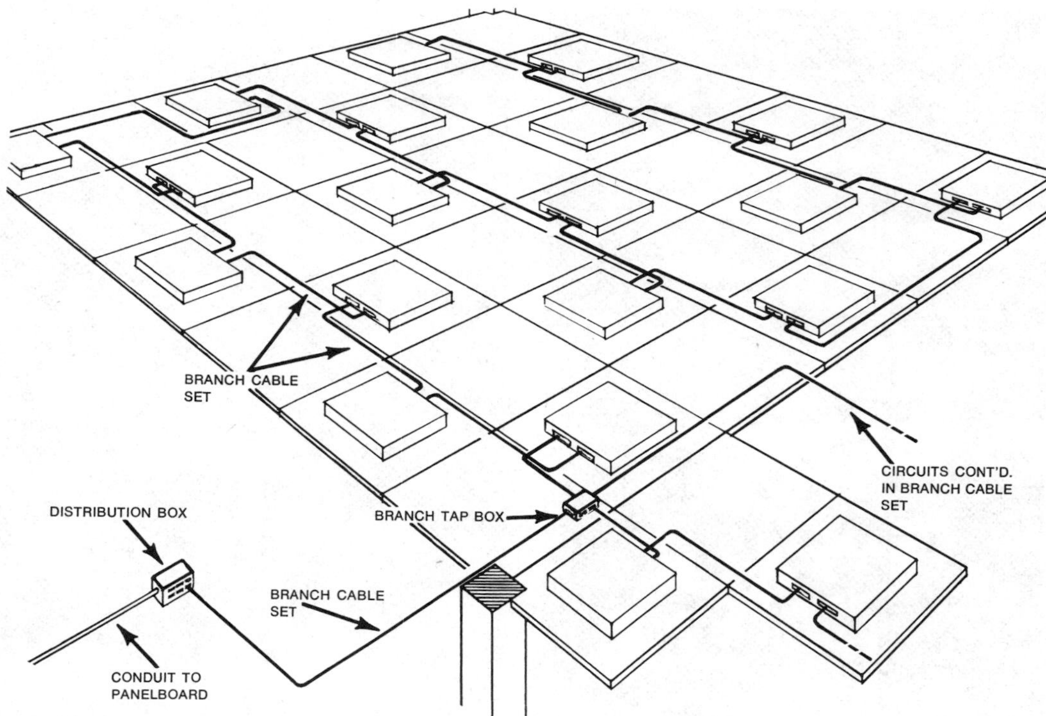

Fig. 15.48 Detail of one design (of many) of Wiremolds' Tele-power pole. Other designs have different dimensions, outlets, bases, and colors. The power compartment is prewired with 4 single, or 2 duplex, 15A, 125 v outlets, with top connection as desired. In this case, the power compartment is wired to a flexible armored cable, which terminates in a special polarized plug. The communications compartment is generally unwired. Pole lengths are available to match ceiling height requirement. Courtesy of Wiremold Co.

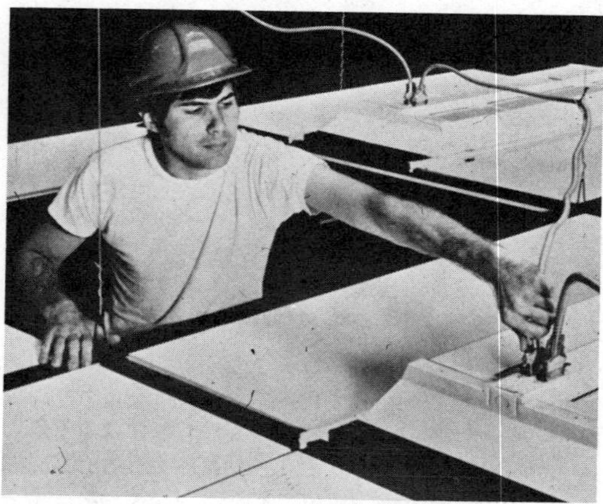

Fig. 15.49 Electrical worker using plug-in feature of prewired ceiling distribution system. Courtesy of Wiremold Co.

the NEC prohibits having splices inside conduits. (Splices are permitted in wireways and troughs with removable covers.) Pull boxes are placed in conduit runs where it is necessary to interrupt the raceway for a wire pulling point. As explained in Section 15.23, this depends on the pulling friction in the system. The size of pull boxes depends on the number and size of incoming conduits, the direction in which conduits leave, and whether or not splices will be made in the box. Minimum sizes based on the above data are specified in the NEC. When a box is equipped with a hinged door(s) and contains some equipment other than wiring, such as a terminal board, it is referred to as a cabinet. All boxes must be equipped with tightly fitting, removable covers. A typical pull box is shown in Figure 15.51.

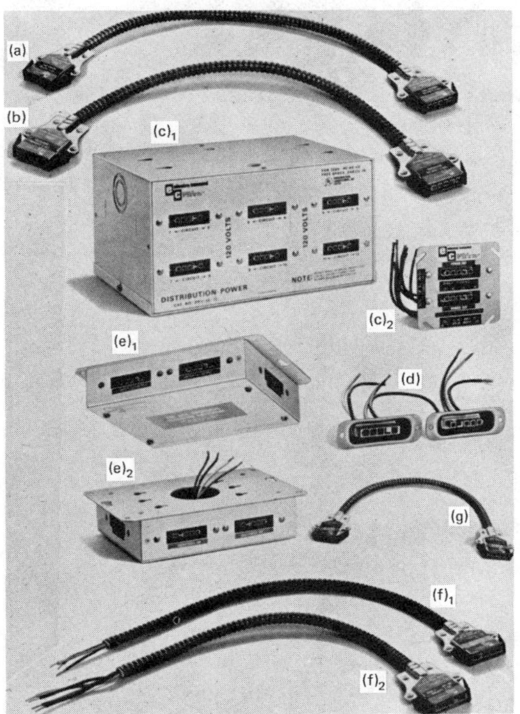

Fig. 15.50 *Some of the equipment in the prewired distribution system is illustrated here; refer to Figure 15.48 for application.* (a) *and* (b) *Branch circuit cables;* (c) *distribution boxes;* (d) *power in/out adapters;* (e) *tap boxes;* (f *and* g) *tap cables. Courtesy of Day-Brite Lighting, Division of Emerson Electric Co.*

Fig. 15.51 *A typical pull box, with EMT conduit entrances. The conduit is fastened to the box by means of a two-part connector; one section clamps on to the EMT, and the second section is installed from inside the box, screwing into the EMT connector on the conduit, thus making a rigid joint. (With threaded rigid conduit, conduits are fastened with locknuts on both sides of the box wall.) Courtesy of Republic Steel Corp.*

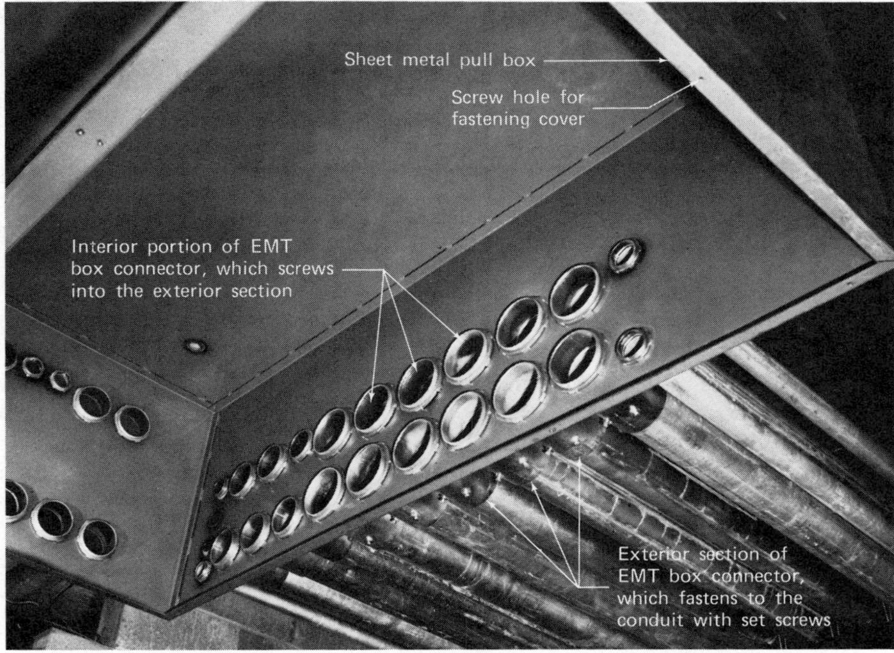

16

Electrical Systems and Materials: Service and Utilization

16.1 Electric Service

Having discussed in the preceding chapter conductors and raceways, the first of three major categories of electrical equipment, we turn now to the two remaining categories, that is, power-handling and utilization equipment. The first step is to examine the means by which electric service is brought into a structure. Public utility franchises require only that service be available at the private property line. Thus service is normally tapped onto the utility lines at a mutually agreeable point at or beyond the property line. The service tap may be a connection on a pole with an *overhead service drop* to the building, an *underground service lateral* to the building, or a connection to an underground utility line with a service lateral to the building. All electrical construction work on private property is normally at the owner's expense.

Under certain conditions the owner can influence the type of construction utilized by the electric utility company in conveying the electric service to the site. This is often the case in large tract developments and in places where owners are willing to share some of the cost of better grade construction. Also, in many areas, the utilities themselves have instituted "beautification" programs in an effort to decrease the objectionable appearance of much of their equipment. Service from the utility line to the building may be run overhead (OH) or underground (UG) depending on the following conditions:

a. Length of service run.
b. Type of terrain.
c. Budget limitations.
d. Utility company voltage.
e. Site and nature of electric load.
f. Importance of appearance.
g. Local practices and ordinances.
h. Maintenance and service continuity.
i. Weather conditions.

16.2 Overhead Service

The principal advantage of overhead electric lines is low cost. Depending on terrain and other factors the cost of overhead as compared to underground installation is in the range of 10 to 50% (the latter when compared to direct burial cable installation). This accounts for the overwhelming majority of installations being overhead. In recent years special techniques in underground installation have lowered that cost, making it a feasible alternative. Where the service run is several hundred feet or more, voltages higher than utilization level may be involved. This weighs heavily in favor of overhead lines, particularly with voltages exceeding 5000 v. Similarly, when terrain is rocky and the electrical load is heavy, the cost of an underground installation rises sharply. Since overhead lines are easily maintained and repaired, and faults easily located, service continuity with overhead lines is usually quite good. In areas where there are extreme

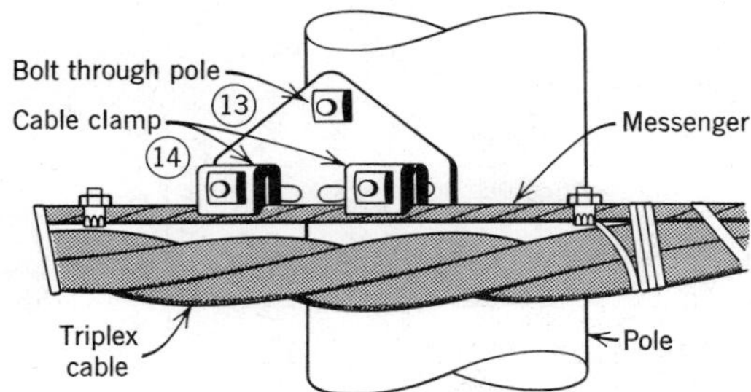

Fig. 16.1 Preassembled aerial messenger cables are carried by steel messenger cables clamped to the poles.

weather conditions, called heavy loading areas, where combinations of snow, wind, and ice increase the possibility of outages, underground lines are preferred if service interruptions involve hardship or financial loss.

Overhead cables are of several types: bare, weatherproof, or preassembled aerial cable. Bare copper cable supported on porcelain or glass insulators on crossarms is normally used for high voltage (2.4 kv and higher) lines. Secondary circuits at 600 v and below are generally run on porcelain spool secondary racks with 1/c weatherproof cable as the conductor. Preassembled aerial cable consists of three or four insulated cables wrapped together with a metallic tape and suspended by hooks from the poles. This type of construction may be used up to 15 kv. It often proves to be more economical than crossarm or rack installation (see Fig. 16.1).

A typical detail of an overhead electric service entrance to a multiple residence is given in Figure 16.2.

16.3 Underground Service

The advantages of underground electric service are attractiveness (lack of overhead visual clutter), service reliability, and long life. The principal disadvantage is high cost. To overcome this, utilities frequently use direct burial techniques which, by eliminating the raceway, reduce costs considerably. Since direct buried cable cannot be pulled out if it faults, as would be the case with a raceway installed cable, it is recommended that the decision of which technique is to be used be based on the consideration of these data:

a. Cost premium for underground raceway installation, including handholes if required (see Section 16.4).
b. Record of outages for direct burial installation.
c. Cost and availability of repair service (utilities frequently will repair customer-owned underground service laterals, *for a fee.*
d. Impact of electric service outage in terms of time delays, inconvenience, necessity to dig up lawns and paved areas, and cost impact in the case of a commercial facility.

16.4 Underground Wiring

Exterior installations are generally in connection with service—either directly from the utility or a subfeed between buildings. The methods available for underground wiring are:

a. Direct burial; see Figure 16.3.
b. Installation in Type II, direct burial duct; see Figure 16.4.
c. Installation in Type I, concrete encased duct, see Figure 16.4.

The first alternative offers low cost and ease of installation, with the disadvantage stated above. The second offers median cost but little strength; therefore, only installations on undisturbed earth and/or under light paving are recommended. The last (c) offers high strength and permanence, but at the highest price of the three.

Nonmetallic duct (conduit) intended for underground electrical use is commercially available in two wall thicknesses. NEMA (National Electrical Manufacturers Association) Type II with a heavy wall provides the physical protection required and is suitable for direct burial installation with no concrete encasement. Type I is manufactured with a thinner wall and is intended for encasement in a

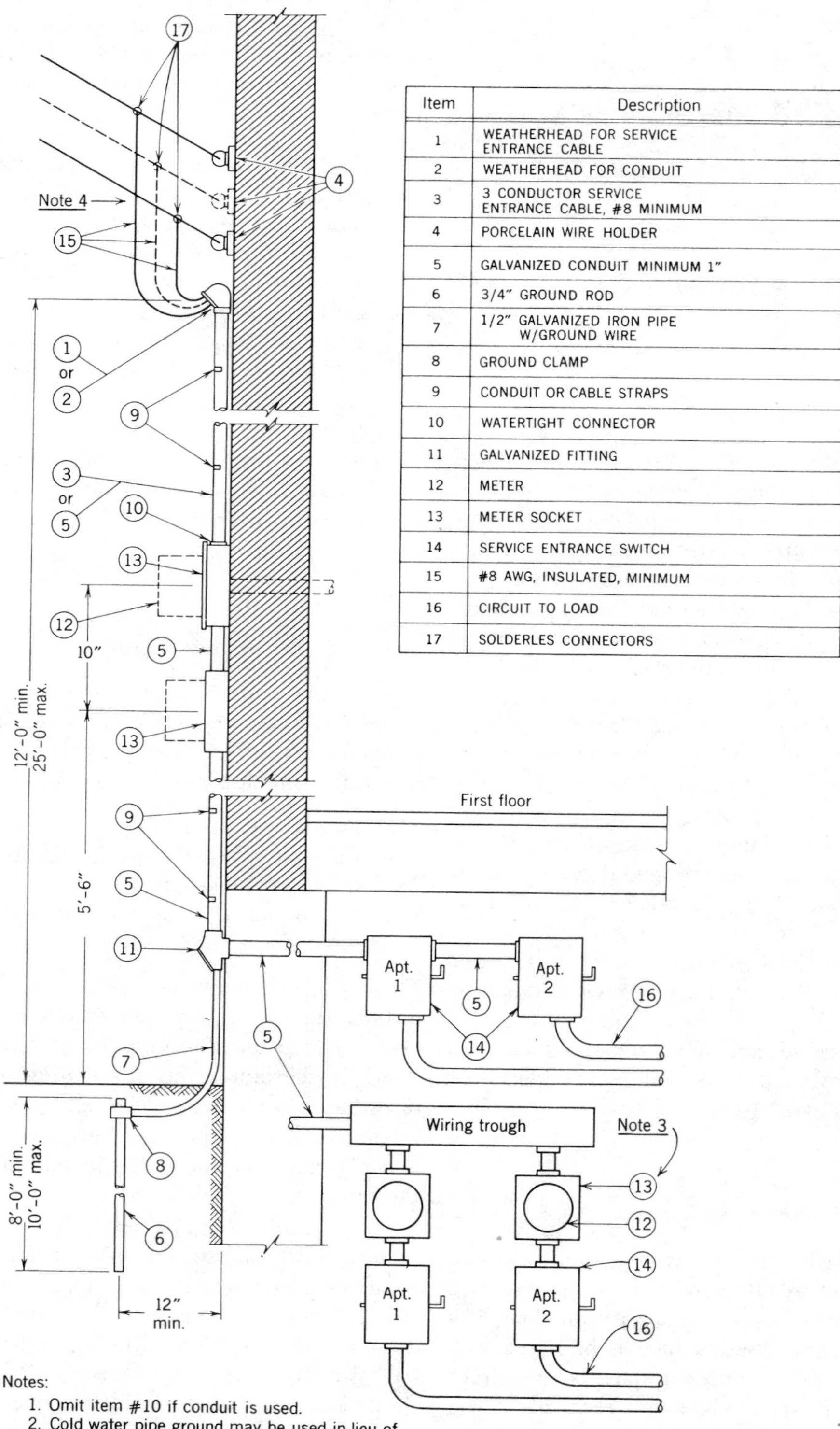

Item	Description
1	WEATHERHEAD FOR SERVICE ENTRANCE CABLE
2	WEATHERHEAD FOR CONDUIT
3	3 CONDUCTOR SERVICE ENTRANCE CABLE, #8 MINIMUM
4	PORCELAIN WIRE HOLDER
5	GALVANIZED CONDUIT MINIMUM 1"
6	3/4" GROUND ROD
7	1/2" GALVANIZED IRON PIPE W/GROUND WIRE
8	GROUND CLAMP
9	CONDUIT OR CABLE STRAPS
10	WATERTIGHT CONNECTOR
11	GALVANIZED FITTING
12	METER
13	METER SOCKET
14	SERVICE ENTRANCE SWITCH
15	#8 AWG, INSULATED, MINIMUM
16	CIRCUIT TO LOAD
17	SOLDERLES CONNECTORS

Notes:

1. Omit item #10 if conduit is used.
2. Cold water pipe ground may be used in lieu of ground rod.
3. Meters may alternatively be placed inside the building.

Fig. 16.2 (opposite) *Detail of typical overhead electric service to a multiple residence. Note that meters are usually placed on the exterior of the buildings. If that is objectionable, they can alternatively be installed inside, provided access is available to the utility's meter readers.*

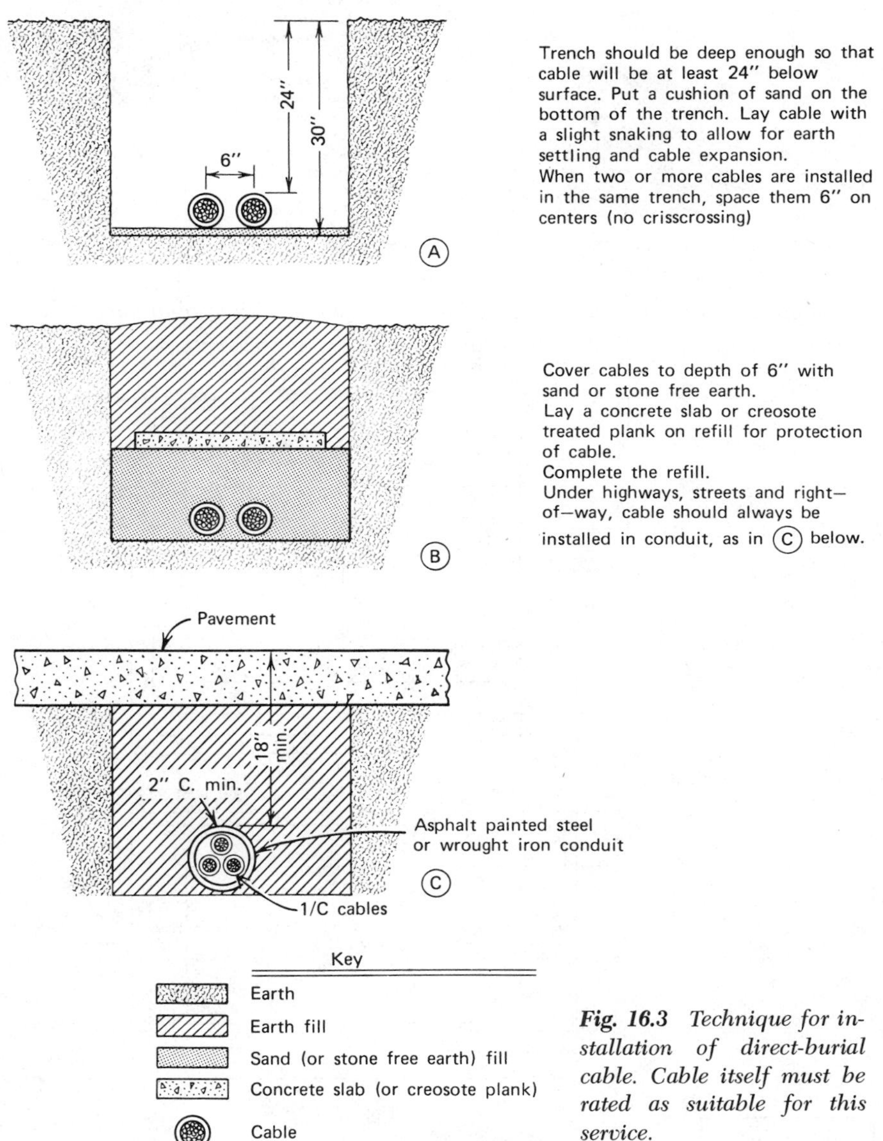

Fig. 16.3 *Technique for installation of direct-burial cable. Cable itself must be rated as suitable for this service.*

minimum of 2 in. of concrete. Common trade names for asbestos-cement and fiber ducts are "Transite" and "Orangeburg." Plastic conduit is referred to as PVC, or simply as plastic. Nonmetallic conduit is most frequently used without concrete encasement for low voltage and signal wiring and with encasement for high voltage wiring. It offers several advantages over steel conduit for underground work, such as lower cost and freedom from corrosion.

When underground electric wiring is duct installed and the run extends over several hundred feet (the exact distance depending on

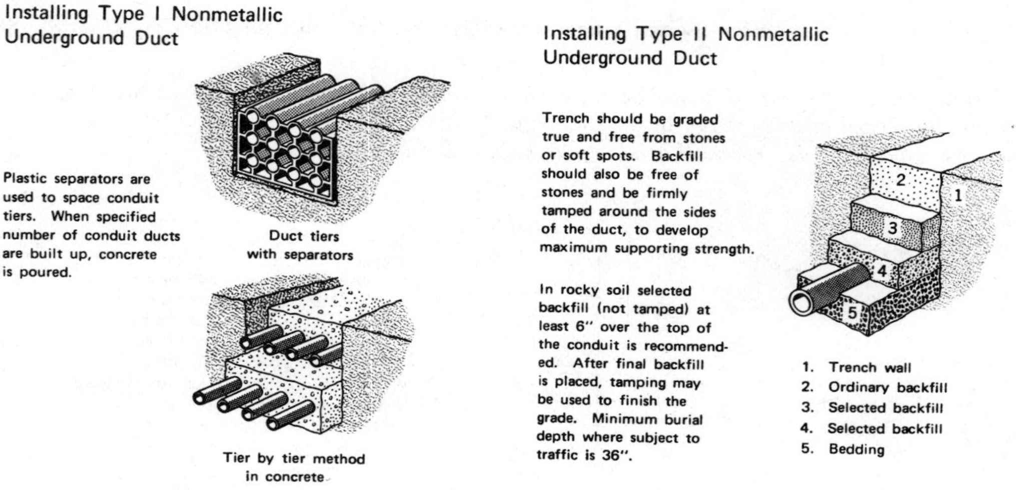

Installing Type I Nonmetallic Underground Duct
Plastic separators are used to space conduit tiers. When specified number of conduit ducts are built up, concrete is poured.
Duct tiers with separators
Tier by tier method in concrete
Installing Type II Nonmetallic Underground Duct
Trench should be graded true and free from stones or soft spots. Backfill should also be free of stones and be firmly tamped around the sides of the duct, to develop maximum supporting strength.
In rocky soil selected backfill (not tamped) at least 6″ over the top of the conduit is recommended. After final backfill is placed, tamping may be used to finish the grade. Minimum burial depth where subject to traffic is 36″.
1. Trench wall
2. Ordinary backfill
3. Selected backfill
4. Selected backfill
5. Bedding

Fig. 16.4 Installation methods for Type I and Type II nonmetallic underground duct.

Fin. grade
Set manhole frame and cover flush with final grade
See specifications for cover details
Varies
Cast iron frame & cover
Masonry & grout
Continuous hot dip galvanized insert flush to wall on each side
6″
1′-6″
3′-0″
For size & location of duct windows see contact drawings
Pulling hooks hot dip galv.
6″
6″
1½″ galv. iron sleeve for ground rod filled with watertight compound
6″
Y
6″
Section A–A
Cont. insert
6″
A
4″
4″
4″
A
5″
X
Hot dip galvanized pulling hooks
#4 rods at 6″ o.c.
2″ cl. typ.
6″
Plan
3″ facility symbol location
2″
½″ raised letters
3″
Non penetrating pick holes
Cast iron frame & cover

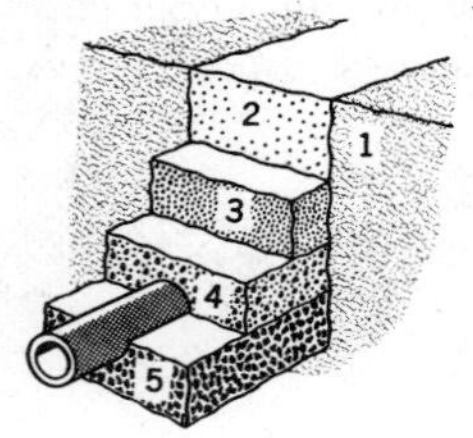

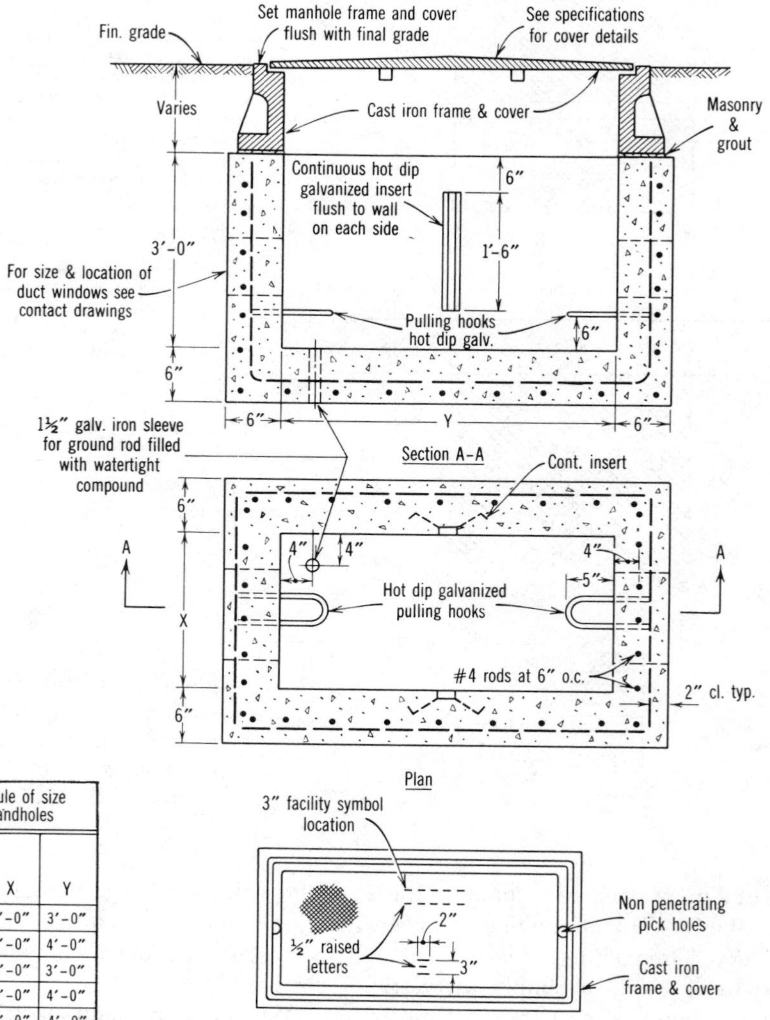

Schedule of size of handholes

Size	X	Y
A	2′-0″	3′-0″
B	2′-0″	4′-0″
C	3′-0″	3′-0″
D	3′-0″	4′-0″
E	4′-0″	4′-0″

Fig. 16.5 Typical handhole detail. Note wall insert for cable support, hooks for cable pulling, and ground rod inside the handhole.

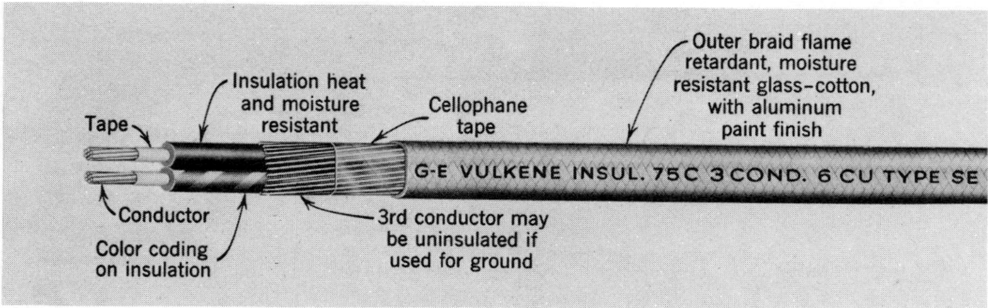

Fig. 16.6 *Detail of construction of a three-conductor service entrance cable, Type SE. Courtesy of General Electric, Wire and Cable Department.*

pulling tension), a pulling handhole or manhole is necessary. Handholes are used for low voltage power and signal cables, and runs with a small number of cables. Manholes are used for high voltage cables and where large duct banks must be accommodated. A typical detail of a field-poured concrete handhole is given in Figure 16.5. Precast handholes and manholes are readily available in most standard sizes and are usually cheaper than field-formed and poured units.

Cable used in underground wiring must be specifically manufactured and approved for that purpose. Type SE is the basic service entrance cable, constructed with moisture and flame resistant covering (see Fig. 16.6). When provided with moisture proofing for underground use, the designation is SE type U, or simply USE. Type USE cable may not be buried directly in the earth unless equipped with an outer moisture-resistant jacket such as neoprene, designated for direct burial use. Underground cable for other than service runs is classified type UF (underground feeder) and comprises one or more rubber or plastic insulated conductors with an outer jacket that is designed to protect the cables from moisture and corrosion when buried directly in the earth. These cables are available up to No. 4/0 AWG.

16.5 Service Equipment

Referring back to Figure 15.1, page 537, the reader will note that interposed between the high voltage incoming utility lines and the secondary service conductors is a block labeled "transformer." This item is required whenever the building voltage is different from the utility voltage. It may be pole or pad mounted outside the building, or installed in a room or vault inside the building as will be discussed below.

16.6 Transformers

A transformer is a device that changes or *transforms* alternating current of one voltage to alternating current of another voltage. A transformer would typically be used to step down an incoming 4160 v service to 480 v for distribution within a building; another transformer would be used in a local electric closet to step down the 480 v to 120 v for use on receptacle circuits. Thus one of the ratings of a transformer is its voltages. A transformer rated 4160 v to 120 v transforms (steps down) 4160 v to 120 v; a transformer rated 120/480 v transforms (steps up) 120 v a-c to 480 v a-c. Transformers cannot be used on d-c. A transformer usually has two sets of terminals; an input set called the *primary* terminals and an output set called the *secondary* terminals. Unfortunately, there is some ambiguity in the use of the terms "primary" and "secondary." High voltage is usually called primary and low voltage, secondary. In a step-down transformer—for example, 4160/120—this is fine, but in a step-up transformer—for example, 480/4160—this is confusing, since the primary of the transformer is connected to secondary (low) voltage and the secondary of the trans-

Table 16.1 *Typical Transformer Data*

Transformer Type	*Cooling and Insulating Medium*	*Voltage*[a] *Primary*	*Voltage*[a] *Secondary*	*Max. Capacity*[b] *of 3-Phase Bank*	*Usage*
General purpose, dry-type	Air	120, 208, 240 480, 600	120, 208, 240 480, 600	750 kva	General purpose, light and power circuits, indoor.
Load center	Air Silicone Oil	2400 4160 7200 11000 13200 13800	120 208 240 480 600	2000 kva	Building service, unit substations, load center indoor and outdoor.
Distribution	Silicone Oil	2400 4160 7200 13200 13800	120 208 240 480 600	750 kva	Pole, pole platform, or concrete pad outdoors.
Substation	Oil	2400 4160 7200 12470 13200 22000 34000	480 600 2400 4160	Above 500 kva	Outdoor substations for groups of buildings or large single buildings.

[a]These are nominal voltages. Taps are often available to change the voltage ±10%.

[b]Larger ratings are available on special order.

[c]Older models utilize askarel liquid.

former yields primary (high) voltage. It is well to remember that ordinarily 120, 208, 240, 277, and 480 v are called secondary voltages, and 2400, 4160, 7200, 12,470, and 13,200 are primary voltages.

Transformers are available in single-phase or 3-phase construction (see Section 17.3 for an explanation of these terms). Transformer power capacity is rated in kva (kilovolt-amperes). For a single-phase unit, this figure is the product of the full load current and the voltage. Since the voltages are different on primary and secondary, so are the currents—because the kva remains constant. Thus a 100 kva 2400/120 v transformer will carry at full load:

$$\text{primary current} = \frac{100{,}000 \text{ va}}{2{,}400 \text{ v}} = 41.6 \text{ amperes}$$

$$\text{secondary current} = \frac{100{,}000 \text{ va}}{120 \text{ v}} = 832 \text{ amperes}$$

Two other characteristics by which transformers are specified are cooling medium and noise level. The transformer coils are either oil-immersed, askarel-filled, or dry. Askarel is a synthetic nonflammable liquid used in lieu of oil. When overheated askarel releases dangerous gases; as a result legislation has been passed to severely restrict its use. To replace askarel, manufacturers now produce silicone liquid-filled units. These are used where the excellent electrical characteristics of liquid-filled units are desired, without the problem of a flammable oil-filled unit. Obviously, there is a price premium to be paid (see Table 16.5). Some pertinent transformer data are given in Tables 16.1 and 16.2. The dimensional data in Table 16.2 vary considerably from one manufacturer to another, and therefore the table is useful only for a concept of bulk volume and weight.

The physical size of a transformer of given kva rating and voltage depends on the type of insulation used. In order of decreasing physical size and increasing operating temperature we have, for dry transformers, class A, B, F, and H insulations that represent organic, inorganic, and silicone insulating materials, respectively (see Table 16.3). Although class H insulation transformers can withstand 150° C rise, many users specify class H insulation with 115° C rise or even 80° C rise; that is, a better grade of insulation is used with an underrated transformer. The reason for this is threefold:

Table 16.2 ***Typical Dry-Type Transformer, Dimensions and Weights***

Kva Output Continuous 80° C Rise	Approx. Dimensions (Inches)			Approx. Weight (Pounds)
	Height	Width	Depth	
Primary 2400/4160 Y v with (4) 2½% Taps Below Normal—Secondary 120/240 v				
SINGLE-PHASE				
3	15¼	8⅝	7¾	97
5	15⅝	11¼	9⅞	135
10	19⅛	11¼	9⅞	235
15	30⅛	22⅜	18⅛	325
25	34⅛	25⅞	20⅛	375
37.5	36⅛	30⅛	22⅛	500
50	36⅛	30⅛	22⅛	600
75	40⅛	36⅛	25⅜	900
100	40⅛	36⅛	25⅜	1100
167	46⅞	38¾	25½	1400
250	51⅞	44¾	35	2050
Primary 480 v with (4) 2½% Taps Below Normal—Secondary 208Y/120 v				
THREE-PHASE				
45	34⅛	33¼	20⅜	580
75	36⅛	41⅛	22⅜	820
112.5	40⅛	45⅞	25¾	1175
150	40⅛	50⅞	25¾	1500
225	58¾	62¼	35	2630
300	58¾	62¼	35	3100
500	70½	69	39¼	4100

a. Longer life.
b. Higher overload capacity.
c. Lower operating cost.

A class H, 150° C transformer operated at full load *continuously* (an unusual situation) has a shortened life—estimated by some experts on the basis of accelerated aging tests to be between 3 and 10 years. The same class H insulation transformer rated at 80° C has a life expectancy of over 100 years. With respect to

Table 16.3 ***Electrical Insulation Temperature Ratings (Based on 40° C Ambient)***

Insulation Class (System)[a]	*Insulation Type*	*Average Conductor Temperature Rise*	*Ambient Temperature*	*Hot-Spot Temperature Gradient*	*Total Maximum Temperature*
A (55°)	Organic	55° C	40° C	10° C	105° C
B (80°)	Mica, glass	80° C	40° C	30° C	150° C
F (115°)	Asbestos	115° C	40° C	30° C	185° C
H (150°)	Silicones	150° C	40° C	30° C	220° C

[a]The modern terminology for insulation class uses "system" in lieu of "class." Thus class A is a 55° system, and so on.

operating cost, the same situations obtain here as with the high-temperature busduct insulation discussed in Section 15.19 and studied in Table 15.6. The truism that one gets nothing for nothing can be expressed here as: in return for the smaller, lighter, cheaper, and hotter transformer (class H, 150° C), we have higher losses and correspondingly higher operating cost, and shorter life.

Refer to Table 16.4 which compares two dry-type transformers, class H; one is insulated for 80° C and the other for 150° C. The operating cycle is chosen to be representative of commercial use. With such a cycle, even the 150° C unit will probably last 30 years—the specified life cycle. The current difference in first cost between the two units is about $1000. Table 16.4 tells us that the lower energy waste of the 80° C unit will repay the first-cost differential in 4.1 years. Beyond that the advantage is entirely on the side of the 80° C unit. Also, with a total permissible hot-spot temperature of 220° C (428° F), the location of the 150° C rise unit must be very carefully chosen, since it can create a serious heat generation and radiation problem.

In summary, then, a transformer is specified by type, phase, voltages, kva rating, sound level, and insulation class. Thus, 112.5 kva, 3-phase, 480/120–208 v, air-cooled, indoor, dry-type transformer with class H insulation and 115° C rise, 45 db (decibel) maximum sound level is an adequate transformer description. Sound ratings of transformers, as well as installation techniques and acoustical treatment, are fully discussed in Part Nine, "Acoustics."

16.7 Transformers Outdoors

A service transformer bank is necessary, as explained above, when the facility utilization voltage is different than the utility voltage (see Fig. 16.7*a*). Also, the designer occasionally opts for a step-up, step-down arrangement when the service run is so long that the conductor cost when run at a low voltage would be excessive. Raceway cost increases also due to large-size cables. In such instances, the cost of the double transformer installation must be more than offset by the savings in conductor cost—only then is the installation (shown in Fig. 16.7*b*) economically justifiable.

The advantages of an outdoor transformer installation are:

a. No building space required.
b. Reduced noise problem within building.
c. Lower cost.
d. Ease of maintenance and replacement.
e. No interior heat problem.
f. High impulse level—a characteristic of liquid-filled units, which is generally important only in large, high-voltage units.

On the other hand, it is frequently easier to find space indoors (preferably in a basement) than to find a suitable exterior spot; noise may be more disturbing from the available exterior spaces, such as courtyards, than from a basement. Costs can run high if long secondary runs are required; heat can often be handled by louvers or areaways adjoining a basement or even made use of profitably if the transformer load is fairly constant. Also, since exposure to direct sunlight decreases the unit

Table 16.4 ***Transformer Operating-Cost Comparison, 750 kva, Dry Types***

750 kva Dry-Type Transformer Class H Insulated	Operating Cycle[a] Load (Percent)	Operating Cycle[a] Time (hr/day)	Efficiency at Stated Load[b] (Percent)	Loss[c] (kw)	Annual Energy Use (kwh)	Annual Energy Cost[d]	Life-Cycle Energy Cost[e]
80° C rise	100	2	98.8	9.0	6,570		
	75	8	99.0	5.18	15,126		
	50	4	98.9	3.71	5,417		
	25	2	98.3	2.86	2,088		
	0	8	—	2.2	6,424		
Totals		24			35,625	$1,247	$19,491
150° C rise	100	2	98.4	12.0	8,760		
	75	8	98.7	6.45	18,834		
	50	4	98.5	4.46	6,512		
	25	2	97.8	3.09	2,256		
	0	8	—	2.2	6,424		
Totals		24			42,786	$1,498	$23,406

[a] Operating cycle is representative of a typical commercial building.

[b] Efficiency figures are averaged from manufacturers' published data.

[c] Loss at 0 load is core loss of 2200 w.

[d] Based on electric energy cost of $0.35 per kwh, including demand.

[e] Using 30-year life,. 8% fixed capital cost, 3% annual escalation.

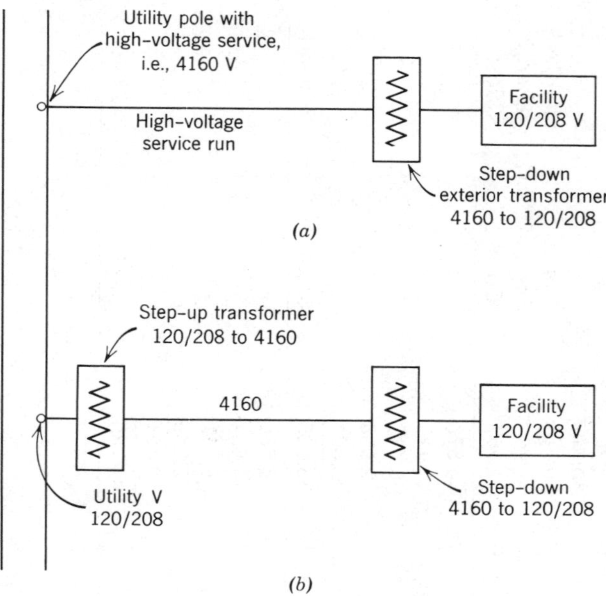

Fig. 16.7 Service transformer arrangements. (a) *High-voltage service with step-down service transformer at the facility, and* (b) *low-voltage service, with transformation at both ends of the service run.*

rating by increasing its temperature, a shaded spot may be difficult to find. Furthermore, an exterior transformer, except where pole mounted, is a questionable choice in an area with a high incidence of vandalism, regardless of sturdiness of construction. Finally, appearance of an exterior unit may be objectionable. This latter point has received much attention from manufacturers, and numerous designs have been developed that minimize the appearance problem (see Fig. 16.8). The most popular type of exterior transformer installation is the pad mount. It has all the above listed advantages in addition to extreme simplicity of installation—it is simply set on a concrete pad. Figure 16.9 gives some typical data for pad-mount units.

16.8 Transformers Indoors—Heat Loss

When the characteristics discussed above indicate an indoor transformer installation, special consideration must be given to the transformer type, as well as to heat- and noise-generating properties. The noise problem and its treatment are discussed in detail in Part Nine, Acoustics. Dissipation of transformer heat can be a problem if not carefully considered. Between 1 and $1\frac{1}{2}$% of the transformer's rating, depending on type, is converted to heat at full load. Table 16.4 states that for a 750 kva, dry-type, 150° C rise unit, 12 kw or 41,000 Btu/hr of heat loss is generated at full load! Losses are lower for 80° C rise units, as indicated. Liquid-filled units have approximately the same losses as 80° C rise dry units. Unless the heat can be used, sufficient ventilation—either natural or forced—must be provided to keep the ambient temperature from exceeding 40° C.

Ventilation by natural convection is most desirable, necessitating the location of the transformer room on an exterior wall (with an areaway if the room is located below grade). The size of the *free* area required for a ventilation opening is 3 sq in./kva of capacity, plus an additional 1 sq in./kva for switchgear losses, if any. If a louver is used in the open-

Fig. 16.8 Pad-mounted exterior transformers are neat, compact, and, if sited properly, unobtrusive. (a) (left) *A well-packaged unit that hinges open at the center to completely expose interior cables, for ease of installation and maintenance.*

Fig. 16.8 (b) (below) *Large units can be partially screened by shrubbery.*

Fig. 16.8 (c) (below) *When the size is such that visual screening becomes a problem, consideration should be given to a structural screen such as a decorative brick wall, which also provides some much needed sound screening. Courtesy of General Electric.*

ings, the size of the opening must usually be doubled (total: 8 sq in./kva), since most louvers have a 50% free area. For good convection, it is desirable to divide the louvered areas in half, placing one-half near the ceiling, and the remaining half near the floor. To provide for equipment removal, if the areaway is large enough, it may be useful to add louver area between the upper and lower openings and make the full louver removable. A bird screen is also desirable.

Since outside air temperature will vary, it is advisable to use a temperature-controlled, adjustable louver. In severe cold climates, heat loss from the electrical equipment may not be

Transformers
Pad Mounted, Three Phase

Liquid Immersed, 75-500 Kva, 18 Kv and Below, 65°C Rise, Type CTP

Fig. 16.9 Typical pad-mount transformer data. Courtesy of Westinghouse Electric Corp.

Dimensions and Weights

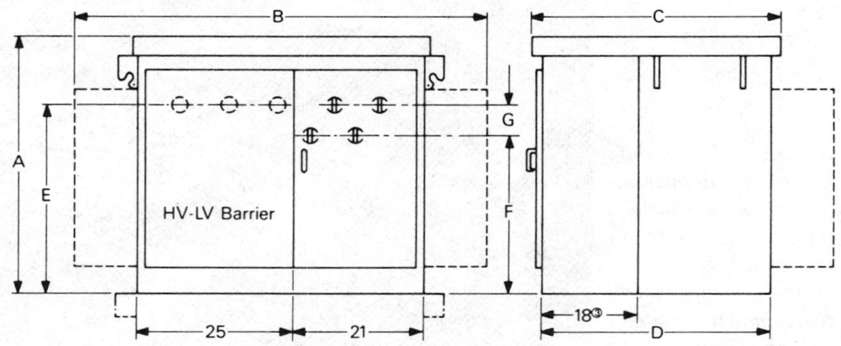

Transformer Dimensions

Kva Cont. 65°C	Dimensions in Inches A	B②	C	D③	E	F	G	Net Weight, Lbs.
75-112½	40	53	42	38	24	22	6	2300-2600
150	40	53	52①	38	24	22	6	2900
225-300	44	53	58①	44	30	27	8	3700-4200
500	52	67①	58①	44	30	27	8	5100

① Dimensions include coolers.
② For loop feed switched units, refer to Westinghouse.
③ Certain combinations require an additional 6 inches, refer to Westinghouse.

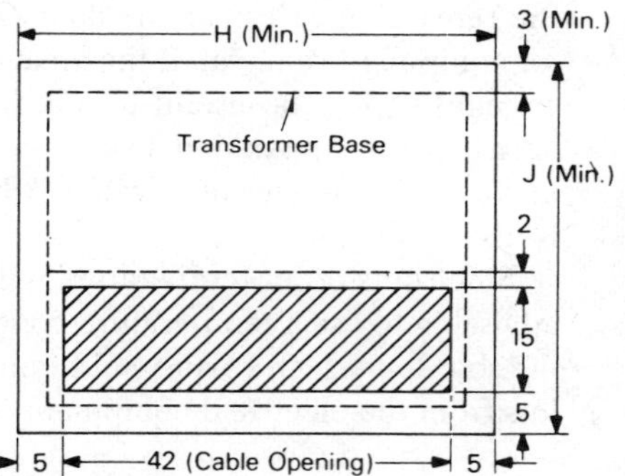

Recommended Pad

Recommended Pad Dimensions

Dimensions in Inches

Kva	H①	J①②
75-112½	52	44
150	52	44
225-300	52	50
500	52	50

① To fully extend beneath coolers, add:
11 inches to "J" for 150 through 500 Kva
22 inches to "G" for 500 Kva

② Dimensions based on 18 inch deep cable compartment. For 24 inch compartment, add six inches to "J" and cable opening depth.

sufficient in the winter to warm the room. In such instances, a unit heater should be installed in the room with a thermostat set at 55° F.

16.9 Transformers Indoors—Selection

When transformers are installed indoors they are subject to stringent NEC regulations that are designed to make the installation intrinsically safe. These regulations are detailed in NEC Article 450 and no purpose is served in duplicating them here. Instead, we wish to present the reader with the essential considerations involved.

(*a*) Oil-filled transformers present a fire hazard indoors, because flammable oil can spread from a tank leak or rupture. To prevent this, oil-filled transformers must be installed in a fire-resistant vault, the construction of which involves a heavy cost. Advantages offsetting this cost are the oil-filled transformer's small size, low weight, low first cost, low losses, long life, excellent electrical characteristics, low noise level, and high overload capacity. Despite these, the vault requirement has had the effect of restricting oil-filled units to industrial facilities and other structures where electrical considerations require its use.

(*b*) Noninflammable liquid-filled units have most of the advantages of oil-filled units and do *not* require a vault. They do, however, require a sump or catch basin of sufficient capacity for all the contained liquid. This and a relatively high first cost are the negative aspects.

(*c*) Dry-type units are the transformers of choice in the majority of indoor installations, despite shorter life, higher losses, high noise level, greater weight, and larger size than the liquid-filled units. The principal advantage is ease of installation and almost unrestricted choice of location. As explained above, by using an underrated transformer (Type H, 80° C rise), losses can be reduced and life extended. Also, for a price premium, noise level can be reduced. When a dry-type transformer is equipped and modified so that its characteristics are equal to that of liquid-filled units, its installed and owning costs are in the same range as the liquid units. Still, most users find the lack of restrictions on placement a sufficient reason for its choice. Table 16.5 gives a comparison of installed costs for these three classes of transformers.

16.10 Transformer Vaults

A transformer vault is basically a fire-rated enclosure, provided because of the possibility of transformer case rupture and an oil fire. However, this must not be construed as an inference that transformers are hazardous or

Table 16.5 ***Relative Installation Costs for a 300 to 1000 kva Transformer***

Transformer Type	*Temperature Rise*[a] *(° C)*	*Basic Impulse Level (BIL)*	*Transformer Cost*	*Construction Cost*	*Total Relative First Cost*
Oil filled	80	95	1.00	50–100[b]	1.50–2.00
Silicone	80	95	1.40	20–40[c]	1.60–1.80
Dry, ventilated	80	95	1.65	—	1.65
	80	50	1.50	—	1.50
	150	95	1.35	—	1.35
	150	50	1.20	—	1.20

[a] Transformers of equal temperature rise have approximately equal life and equal losses.

[b] Cost of vault depends on local labor costs and size of transformer. The relative cost decreases with increasing transformer size.

[c] Cost of catch basin. As in the case of a vault, relative cost depends on labor rates and transformer size.

delicate devices prone to faults. On the contrary, transformers are extremely tough, sturdy, long-lived, capable of sustaining large and prolonged overloads, and are indeed among the most reliable elements of an electrical system. However, faults do occur, and an oil-filled transformer is a potential fire hazard.

Transformer vaults should be located, if practicable, where they can be ventilated to the outside air without flues or ducts. The combined net area of all ventilating openings (gross area less screens, louvers, etc.) should be as explained in the preceding section, not less than 3 sq in./kva of transformer, but in no case less than 1 sq ft. Further ventilation recommendations plus details of enclosure construction materials, fire rating, door and sill details, and other relevant and important information are provided by NEC Article 450.

16.11 Service Equipment Arrangements and Metering

To summarize the preceding discussion and to proceed with our study of electric power-handling equipment, refer to Figure 16.10. Electric service connection will be overhead or underground. Where building voltage is different than utility voltage a transformer is required. This transformer may be pole or pad mounted outside the building, or installed in a room or vault inside or outside the building.

Metering must be provided at either the utility or the facility voltage, and at either the service point or inside the building. The choice is generally left to the owner with the understanding that inside meters must be readily accessible to utility personnel. If high voltage service is purchased, then the transformers and all equipment beyond the service connection must be furnished by the owner. Conversely, if low voltage is purchased, all equipment necessary to provide low voltage is furnished by the utility. Obviously, the electric service rates for low voltage service are *higher* than for high voltage service in order to compensate the utility for the cost of providing and maintaining the step-down transformer and associated equipment. Therefore, it is often advisable for the owner of a facility (other than a small, residential facility) to investigate the economics of purchasing power at high voltage. Since many owners are not equipped to maintain high voltage equipment, arrangements can sometimes be made to pay the utility to provide and maintain transformers, while taking the cost advantage of high voltage service.

Generally, for a single-use building or a building where electric energy is included in the rental charge only, a single meter is necessary. Provision for such metering in the form of current transformers and wiring is made in the main switchboard as in Figures 16.21 and 16.22, or the meter may be independently mounted as in Figure 16.11. In both cases, the meter is furnished and installed by the utility company. Where submetering is required, such as in apartment houses, banks of meter sockets are installed to accommodate the multiple meters. Recently promulgated federal regulations forbid master metering in new multiple-dwelling constructions, because it encourages energy waste. A typical assembly of such socket meters is shown in Figure 16.12 (see also Fig. 16.2, above). Automatic energy

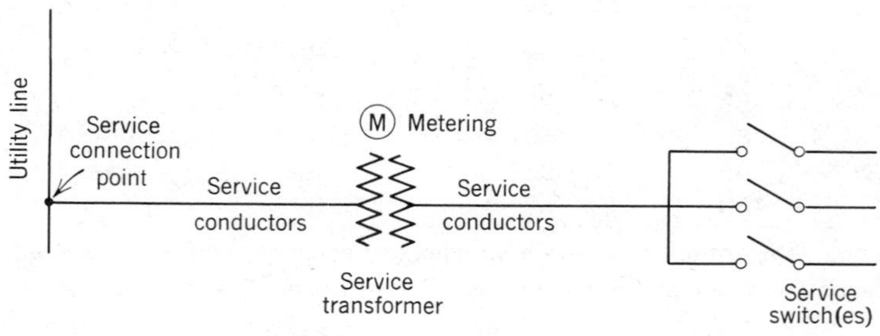

Fig. 16.10 Typical electric service arrangement.

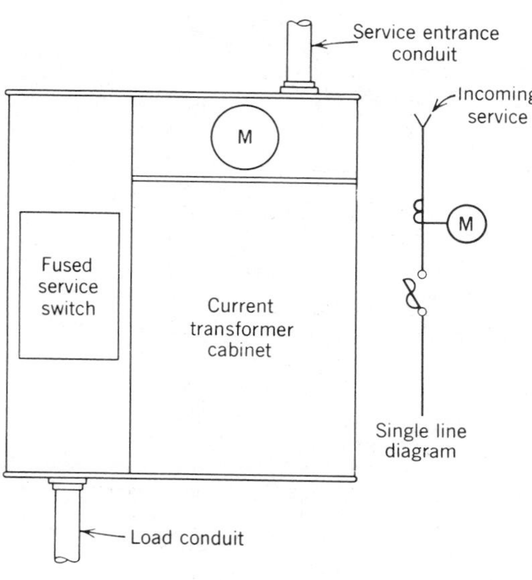

Fig. 16.11 Typical separate (interior or exterior) meter and service disconnect arrangement for low-voltage service of 400 amps and larger. The assembly includes current transformers, meter, and single-service switch.

Fig. 16.12 Typical modular metering equipment. The cabinet at the left contains the main service disconnect—in this case a circuit breaker. Adjacent, and fed from this, are two sections of meter pan: one 8 and one 4. Below each meter socket is the circuit breaker that is the main protection for the apartment involved. More sections can be added, as needed. (Note the incoming service conduits and the outgoing load conduits.) Complete electrical and physical data are available in manufacturers' catalogs. Courtesy of The Square D Co.

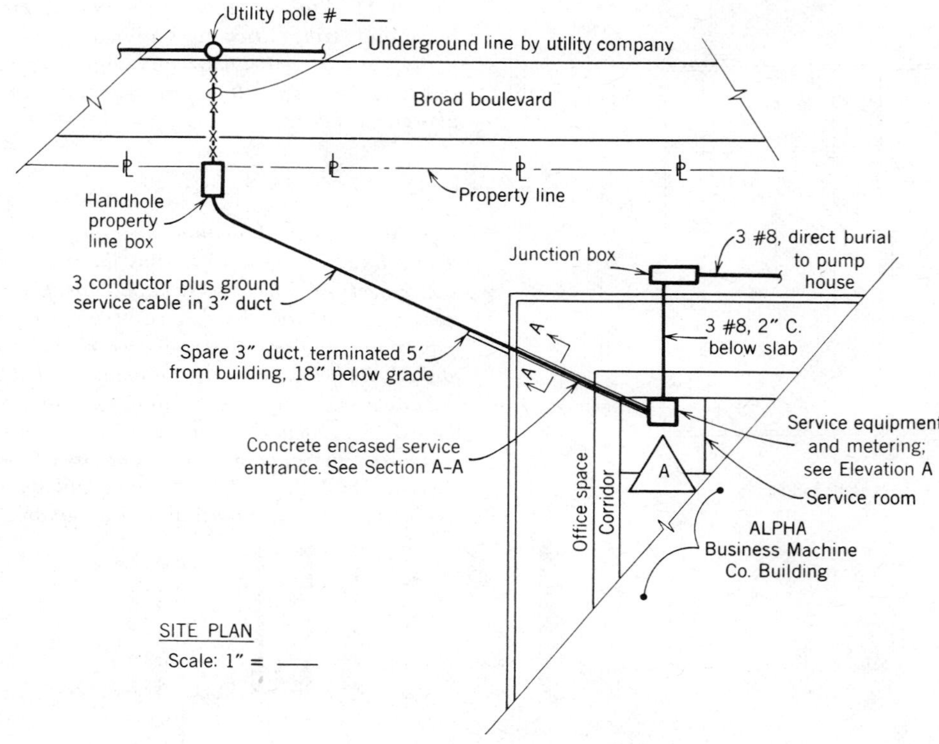

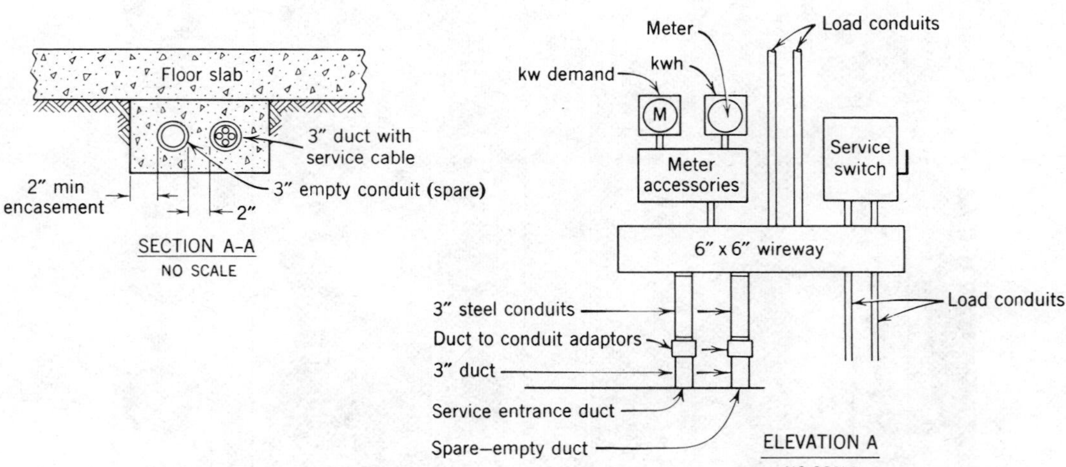

Fig. 16.13 Typical electrical plan of underground service to an industrial building, with details.

metering is discussed in Section 14.14. A low voltage underground service detail as it would appear on a set of contract drawings, including relevant details, is given in Figure 16.13. Note that here the service switch and meters are separately mounted. Meters are always installed *electrically* ahead of the service switch so that they cannot be disconnected.

16.12 Service Switch

The purpose of the electric service switch is to disconnect all the electric service in the building except emergency equipment. Thus, in the event of fire, no electrical hazard will face fire fighters. It is therefore obvious that this disconnecting apparatus must be located at a readily accessible spot near the point the service conductors enter the building. If such a location is not feasible, service conductors may be run in concrete encasement under the building and will be considered "outside the building" up to the point at which they emerge from the floor in the building (see Fig. 16.13). At that point, the service switch must be installed. The service switch, or more accurately, the "service disconnecting means," may comprise one to six properly rated switches. These are frequently assembled into a switchboard. Before discussing switchboards, a description of switches, circuit breakers, and fuses, the components of which switchboards are constructed, is in order.

16.13 Switches

An electrical switch is a device intended for on/off control of an electrical circuit and is rated by current and voltage, duty, poles and throw, fusibility and enclosure. The current rating is the amount of current that the switch can carry continuously and interrupt safely. Certain light-duty switches are not intended to interrupt rated current and are clearly so labeled. Switches intended for motor control are also rated in horsepower. The voltage rating of a switch is, as for other electrical equipment, by voltage class. Thus a switch is rated 250 v, 600 v, or 5 kv as required. Switches intended for normal use in light and power circuits are called general-use safety switches and are rated ND for *normal duty*. Switches intended for frequent interrupting are rated HD for *heavy duty* (see Fig. 16.14). Conversely, switches intended to be opened under load only occasionally, such as service switches, are rated LD for *light duty*.

An examination of Figure 16.15 should clarify what is meant by the number of poles and throws of a switch. Unless otherwise noted, a switch is assumed to be single throw. Since the NEC states generally that the grounded neutral conductor of a circuit should not be broken, most switches carry the neutral through unbroken, by means of a solid link within the switch. This gives rise to the term *solid neutral* (SN) *switch*. Switches are available in 1, 2, 3, 4, and 5-pole construction. Poles are indicated by a "P"; thus 3-pole is written "3P," and so on.

A switch may be constructed with or without provision for fusing. If provided, the switch is fusible; if not, the switch is nonfus-

Fig. 16.14 *Typical heavy-duty (H.D.), 3-pole fusible switch in industrial-type NEMA 12 enclosure. Note the handle and door interlock that prevents opening the cover while the switch is in the "closed" position; note also the gasket that provides the dust and drip protection. Courtesy of The Square D Co.*

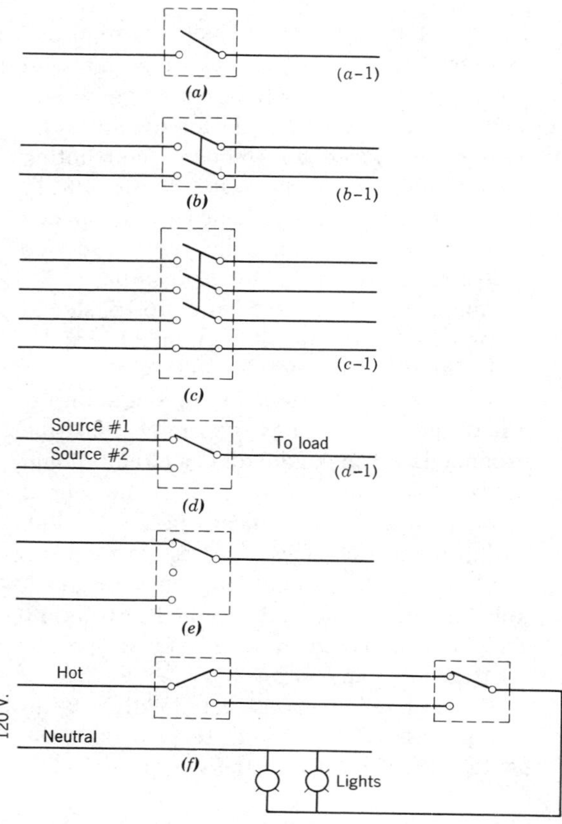

(*a*) Single-pole single-throw switch.

(*a-1*) Single NO contact.

(*b*) Two-pole single-throw switch.

(*b-1*) Two NO contacts.

(*c*) Three-pole and solid-neutral (3P and SN) switch.

(*c-1*) Three NO contacts.

(*d*) Single-pole double-throw switch (also called, in small sizes, a 3-way switch).

(*d-1*) One NO and one NC contact.

(*e*) Single-pole double-throw switch with center "off" position (in control work called a hand-off automatic switch).

(*f*) Use of two single-pole double-throw (3-way) switches for switching of a lighting circuit from two locations.

Fig. 16.15 *Typical switch and contact configurations. In* (a) *to* (d) *the equivalent contact configurations are shown to illustrate how a contactor performs the switching function. Note that switches are always shown* open. *Contacts are shown in their de-energized state, either NO or NC. The operating coil of the contacts is not shown, since it is not relevant to the circuit.*

ible. All separately enclosed switches must be in an appropriate cabinet. The National Electrical Manufacturers Association has standardized the nomenclature and application of enclosures for all electrical control equipment of which switches are only one item. These are detailed in Section 16.15. Summarizing the above, then, we could adequately describe a switch thus: switch, HD, 3P & SN, 200A/150AF (fuse), 600 v, in NEMA 12 enclosure.

16.14 Contactors

A contactor is a switch. Instead of a handle-operated, movable blade and a fixed grip, a contactor uses two *contact* blocks of silver-coated copper, which are forced together to make (close) or are separated to break (open) the circuit. The common wall light switch is a small mechanically operated contactor. A relay is a small electrically operated contactor (see Fig. 16.44*a*, page 633). Most contactors are operated by means of an electromagnet that causes the contacts to close (or open). The reverse action is by spring and/or gravity. Contactor terminology is somewhat different from that of a switch. Its condition when deenergized is its *normal* state. Thus, a contactor whose contacts are open when the coil is not energized is *normally open* (NO). One with normally *closed* contacts is NC. Small contactors are called relays. Units intended for motor control are called motor starters and are discussed in Section 16.22. Current, voltage, and number of poles have the same significance for contactors and relays as for switches.

The great advantage of contactors over switches is their facility for remote control. Switches must be manually thrown—or at best with a motor. However, the magnetic contactor is inherently a remotely controlled device, making it ideal for a myriad of control functions. They are controlled by push-button or automatic devices such as float switches, thermostats, pressure switches, and so on. Since control can be both remote and automatic, the application of relays and contactors is universal in remote control of lighting, heating, air conditioning, all motor control, sequence switching, automatic load transfer, and the like.

16.15 Special Switches

Many special types of switches are available. Most of these types are beyond the scope of this book, except for the following which we will discuss briefly.

(*a*) ***Time Controlled Switches.*** These devices are all basically similar, comprising a precision low-speed miniature drive motor (timer) to which some type of electric contact-making device is connected. The units differ essentially in the complexity of the electric contact arrangements. Figure 16.16 shows two of these units and their applicability to control problems. All of the units depend on the accuracy of the power line frequency for their own accuracy. Some units provide spring-wound, reserve power motors to carry over power interruptions. This feature is particularly useful in areas subject to power outages, however short, since each outage will show up directly as a switch delay. This type of timed switch mechanism is the heart of all time-dependent program systems. One such unit, used in demand/load control, was discussed in Section 14.13(a) and shown in Figure 14.11. Other similar units are discussed in Section 22.28, which is devoted to programmed signal systems.

(*b*) ***Remote-Control (RC) Switches.*** A contactor that latches mechanically after being operated is known as a remote-control switch. These devices are extremely useful in lighting control and energy-conservation techniques. A discussion of low voltage control for individual lighting circuits is found in Section 16.26.

(*c*) ***Automatic Transfer Switch.*** This device, which is an essential part of all standby power arrangements, is basically a double throw switch—generally 3-pole—so arranged that on failure of normal service it *automatically transfers* to the emergency service. When normal service is restored, it automatically retransfers to it. The control devices are voltage sensors that sense the condition of the service and operate the switch accordingly. Auxiliary devices can be built on to the basic switch, the most common of which are emergency generator starting equipment. Figure 16.17 illustrates a typical unit. Refer to Section 17.22 for a discussion of emergency power arrangements.

16.16 Equipment Enclosures

As mentioned above, design requirements for control equipment enclosures is standardized by NEMA in its standard ICS 1-110. Control equipment includes switches, circuit breakers, contactors, relays, and similar devices. The proper nomenclature, enclosure description, application, and figure references are found in Table 16.6. Enclosures are illustrated in Figure 16.18. It is important to note that there is no enclosure described as WP or weatherproof. Equipment intended for outdoor use should be specified:

a. In a Type 3R enclosure to protect against rain.
b. In a Type 3S enclosure to protect against wind-driven rain and sleet.
c. In a Type 4 enclosure to protect against the above plus splashing and condensation.

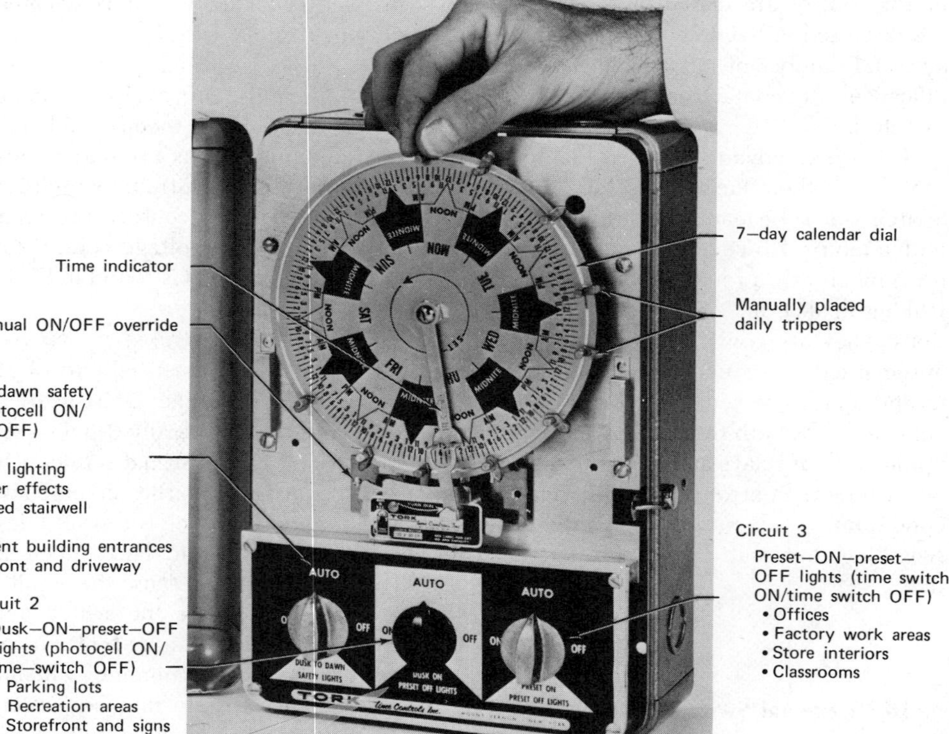

Fig. 16.16 *Time switches vary in complexity. The basic switch comprises a drive motor, with attached 24-hr clock face on which one or more on/off trippers are placed and adjusted manually. The unit can usually also be operated manually. Switch* (a) *is arranged for three different types of control functions that are applicable as indicated to different types of load. This unit's dial is a 7-day calendar type that permits setting a different schedule for every day of the week. The dial is* not *astronomical* (solar), *since all the turn-on functions are either photocell or time-of-the-day controlled. Solar dials are useful when photocell control is not available, since sunrise and sunset times vary with the season.* (b) *Illustrates a repetitive timer, also called a program time switch. The illustrated unit has a 24-hr dial and space for 96 trips to permit up to 48 on/off operations. By ganging control wheels on a common shaft, a program drum, similar to the program drum in Figure **14.11a**, can be constructed. Courtesy of Tork, Inc.*

Note also that the Type 12 industrial enclosure is similar to Type 1, except that it is gasketed for dust and drip resistance and therefore is well applied in *all* "dirty" indoor environments, including commercial and institutional spaces. Its use is *not* restricted, as the name implies, to industrial areas.

16.17 Circuit-Protective Devices

In order to protect insulation, wiring, switches, and other apparatus from overload and short circuit currents, it is necessary to provide automatic means for opening the circuit. The two most common devices employed to fulfill this function are the fuse and the circuit breaker (c/b).

(*a*) ***Fuses.*** The fuse is a simple device consisting of a *fusible* link or wire of low melting temperature that when enclosed in an insulating fiber tube is called a cartridge fuse, and when in a porcelain cup is known as a plug fuse. Figure 16.19 shows common types of fuses. Plug fuses, such as those normally used in a residence, are rated 5 to 30 amp. Cartridge fuses are made in sizes 5 to 600 amp. Table 16.7 lists the standard fuse sizes. When a fuse is subjected to excess current, the energy

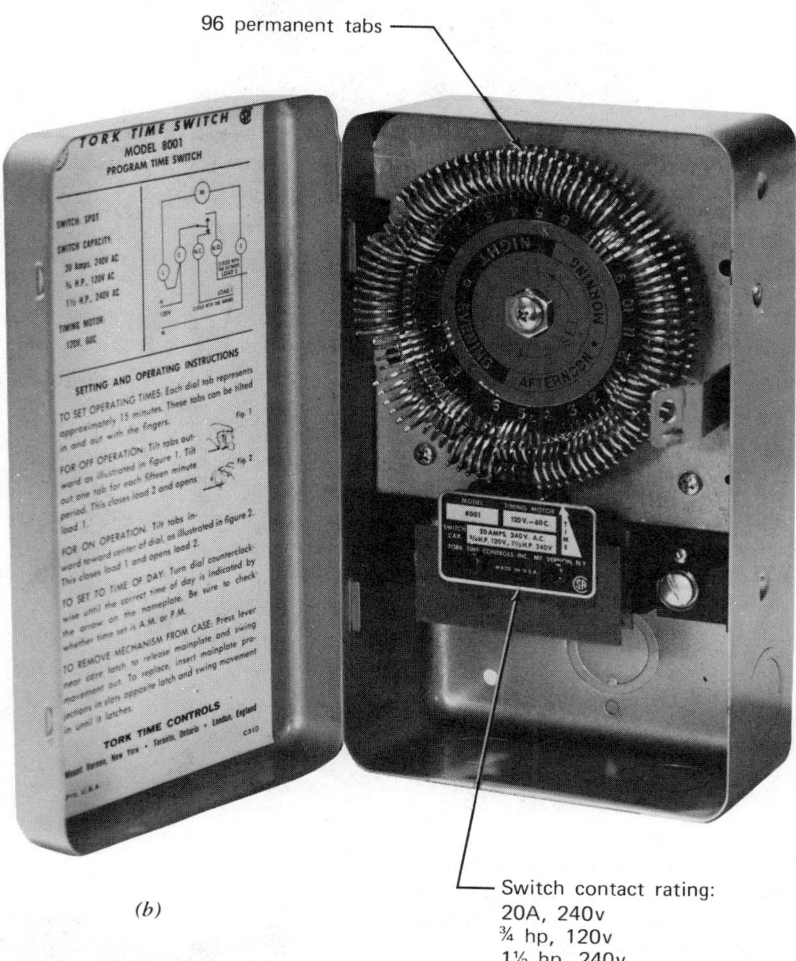

(*b*)

loss in the link (I^2R loss) generates heat and melts it.

Fuses are applied not only by current rating but also by interrupting capacity, which is the capacity to safely interrupt a short circuit current of given magnitude. Fault (short circuit) current if not quickly interrupted can cause explosive damage, fire, and loss of life and property. This is particularly true in areas with large power systems and is the reason for the almost universal use of the current-limiting fuse.

(*b*) *Circuit Breakers.* A circuit breaker is an electromechanical device that performs the same protective function as a fuse and, in addition, acts as a switch. Thus it can be used in lieu of a switch-and-fuse combination to both protect and disconnect a circuit. Most circuit breakers are equipped with both thermal and magnetic trips. The thermal trip, which acts on overload, is similar in action to a thermostat. Heat generated by excessive current causes an element to move and trip the latching mechanism of the breaker, thus opening the breaker contacts and thereby the circuit. The magnetic trip comprises a coil with a movable core. On short circuits, magnetic forces actuate the core which trips the circuit breaker latch. Both the thermal and the magnetic action have inverse time characteristics, that is, the heavier the overload the faster the trip action. A special type of circuit breaker designed to protect against ground faults in addition to overloads and short circuits (line faults) is illustrated in Figure 16.20.

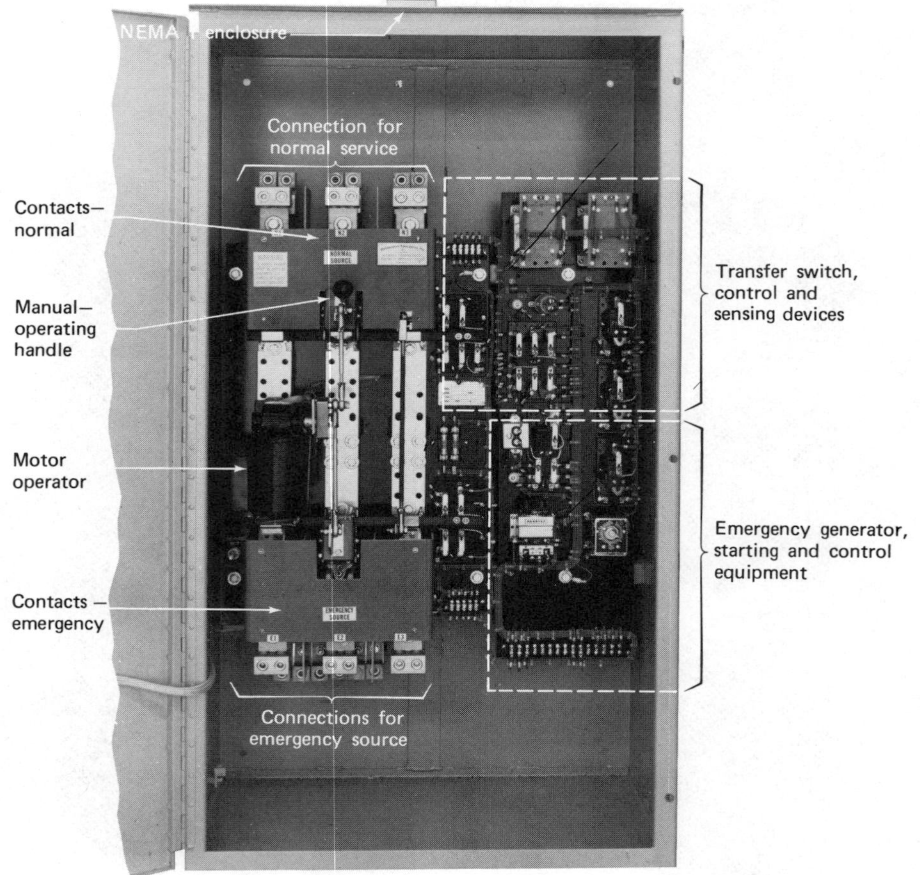

Fig. 16.17 Automatic transfer switch in a NEMA I, wall-mounted enclosure. This unit is rated 400 amps, 3-pole, 600 v, ac, which corresponds to approximately 150 kva at 208 v, 3-phase. The cabinet is 58 in. high, 32 in. wide, and 18 in. deep (147.3 × 81.3 × 43.2 cm) and the entire unit weighs 354 lb (160 kg). Courtesy of Russelectric, Inc.

Fig. 16.18 NEMA enclosure types.

Enclosure for S2500 line molded—case circuit breakers.

Fig. 16.18 (a) *Type 1 ventilated enclosure. It is apparent that this type enclosure is intended for a clean, dry environment. Figure 16.17 illustrates a nonventilated NEMA 1 enclosure. Note that no gasketing is used, and the enclosure serves only as physical protection for and from the enclosed equipment. Courtesy of General Electric, Circuit Protective Device Department.*

Air circuit breakers are available in two types; the molded case breaker and the "large air breaker." Molded case breakers consist of a complete mechanism encased in a molded phenolic case (see Table 16.8). In the large sizes (frames), the trips are both interchange-

Table 16.6 *Control Equipment Enclosures*

NEMA Designation: Type	*Description*	*Application*	*Figure Reference*
1	General purpose	Dry, indoor use.	16.18*a*
2	Drip-proof	Indoor, subject to dripping.	
3	Dust-tight, rain-tight, and sleet-resistant	Indoor/outdoor, where subject to wind-blown dust and water.	
3R	Rainproof and sleet-resistant	Outdoor, subject to falling rain, snow, and sleet.	16.18*b*
3S	Dust-tight, rain-tight, and sleet-proof	Outdoor, subject to windblown water, dust, and sleet. Most severe exterior duty.	
4	Watertight and dust-tight	Indoor/outdoor, subject to water from all directions. Not sleet-proof.	
4X	As Type 4, and corrosion-resistant	As Type 4, with corrosion resistance.	
5	Dust-tight	Superseded by Type 12.	
6	Submersible	Self-explanatory.	
7–9	Hazardous	Differing in application by class and group of hazardous use. See NEC.	16.18*c*
10	Bureau of Mines	Self-explanatory.	
11	Corrosion-resistant and oil-immersible	Special application for oil immersion in highly corrosive atmosphere.	
12	Industrial use, dust-tight and drip-tight	Indoor only, general use, industrial and other "dirty" environments.	16.14

Fig. 16.18 (b) *Type 3R outdoor enclosure. This is the type usually intended when "weatherproof" is specified. Courtesy of General Electric, Circuit Protective Device Department.*

GE's raintight NEMA 3B circuit breaker enclosure.

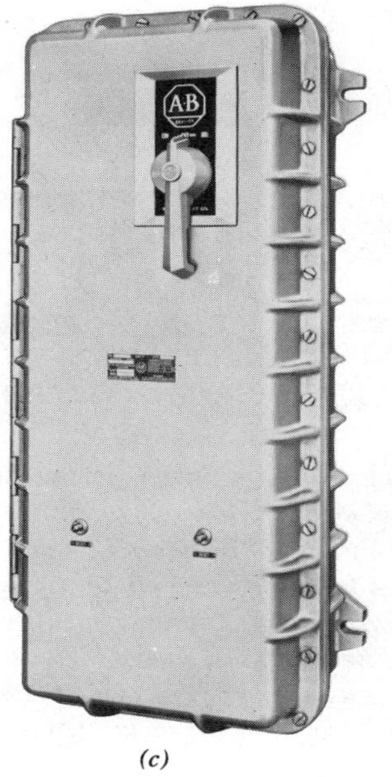

(c)

Fig. 16.18 (c) *Type 7 enclosure, suitable for Class 1, hazardous areas. Enclosures are built to contain any explosion, hence the extremely rugged construction. Courtesy of Allen Bradley Co.*

Table 16.7 *Fuse Ratings and Construction*

Current Rating	*Remarks*
0–10 15, 20, 25, 30	Plug fuse construction; max. 150 v to ground
0–10 15 20 25 30 35 40 45 50 60	Cartridge-type with ferrules; single and dual element; 250 and 600 v
70 80 90 100 110 125 150 175 200 225 250 300 350 400 450 500 600	Cartridge-type, knife blade contacts; 250 and 600 v
800 1000 1200 1600 2000 2500 3000 4000 5000 6000	Cartridge-type, bolted knife blade contacts; 600 v

able within a frame size and adjustable (see Fig. 16.20). The large air circuit breaker is a more complicated and widely adjustable mechanism and can be used in applications that preclude use of molded case breakers. All breakers can be equipped with remote trip and auxiliary contacts, and all good breakers have trip-indicating handles and are "trip-free" (i.e., will trip out harmlessly if closed in on a short-circuited line).

(*c*) ***Fuses and (versus) Circuit Breakers.*** Although both are circuit-protective devices, their characteristics differ markedly. Fuses are inherently instantaneous operating devices; time delay is provided by a second element, as in Figure 16.19*c*, *d*, and *e*. Due to the rapidity

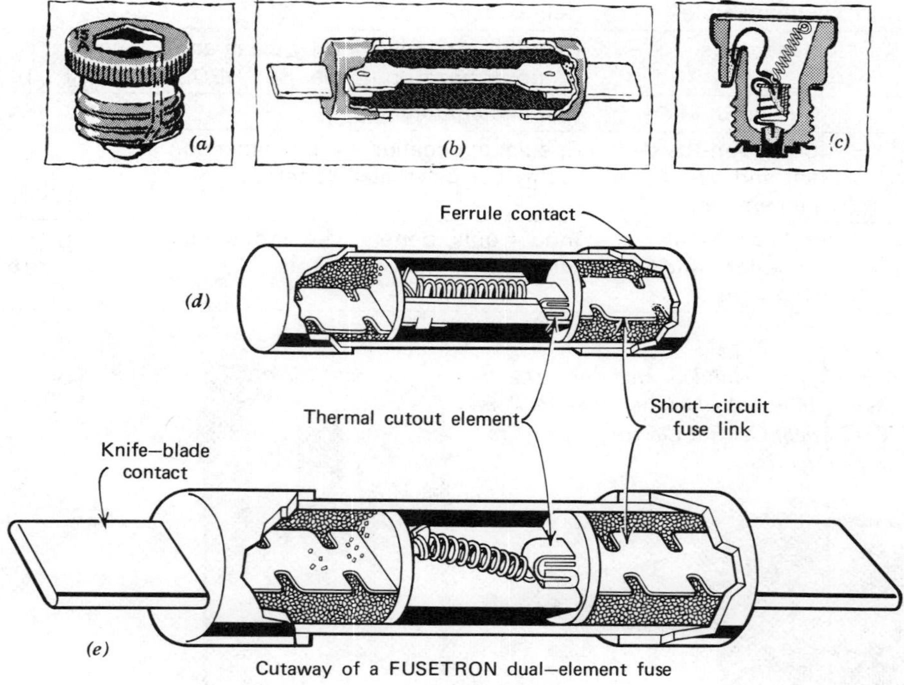

Fig. 16.19 *Standard types of fuses are:* (a) *common household plug fuse;* (b) *single-element, knife-blade fuse;* (c, d, *and* e,) *dual-element, time-delay fuses with Edison base, ferrule contacts, and knife-blade contacts, respectively. Courtesy of Bussman Division, McGraw Edison Co. Since fuses are inherently very fast-acting devices, time delay must be built into a fuse to prevent "blowing on short-time overloads, such as those caused by motor starting. A dual-element fuse, shown in* (c) *to* (e), *allows the heat generated by temporary overloads to be dissipated in the large center metal element, preventing fuse blowing. If the overload reaches dangerous proportions, the metal will melt, releasing the spring and opening the circuit. High capacity faults are cleared by the melting of the two end elements. The time to clear* (blow) *for fuses is inversely proportional to the amount of current.*

Table 16.8 ***Typical Molded Case Circuit Breaker Characteristics***

Frame Size (Amperes)[a]	*Trip Setting (Amperes)*	*Voltage (Volts)*	*Remarks*
50A	15 20 30 40 50	240	1–3 poles
100A	15 20 30 40 50 70 90 100	240 600	1–3 poles
225A	70 90 100 125 150 175 200 225	600	1–3 poles
400A/600A	125 150 175 200 225 250 300 350 400 500 600	600	1–3 poles
800A/1200A	250 300 350 400 500 600 800 1000 1200	600	1–3 poles
1600A	400 600 800 1000 1200 1600	600	2–3 poles 3 poles

[a]Available to 2500 amp, varies with manufacturer.

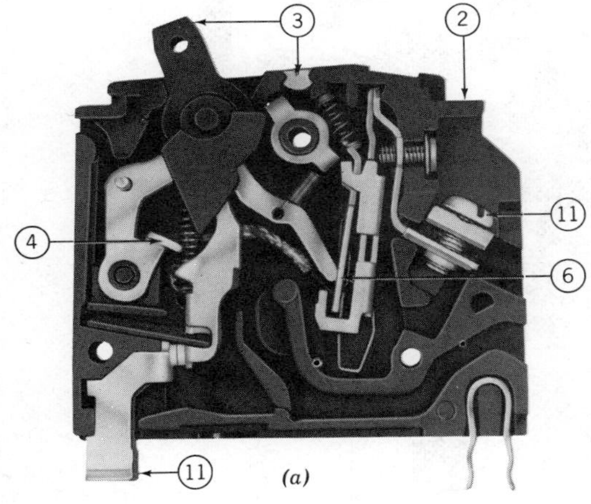

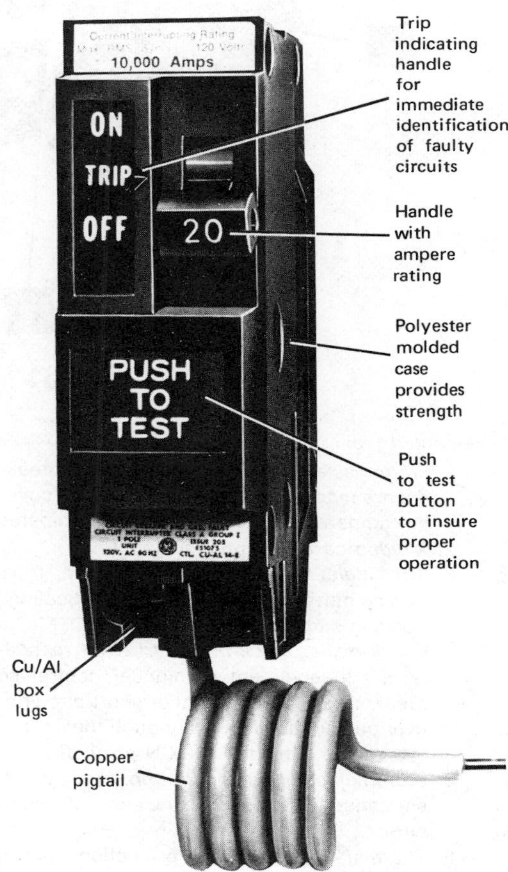

Fig. 16.20 *Molded-case circuit breakers.* (a) *Illustrates a conventional single-pole, 50A frame plug-in type c/b. Courtesy of the Square D Co. A unit that will provide ground-fault protection in addition to functioning as an ordinary circuit breaker is illustrated in* (b). *This unit is designed to fit into the same physical space as a conventional molded-case circuit breaker and therefore can be used as a replacement in an existing panelboard. The units are referred to as GFCI (ground fault circuit interrupters) or GFI. A 3-pole, 600-amp frame conventional c/b is shown in* (c). *Photos courtesy of Circuit Protective Devices dept of General Electric Co.*

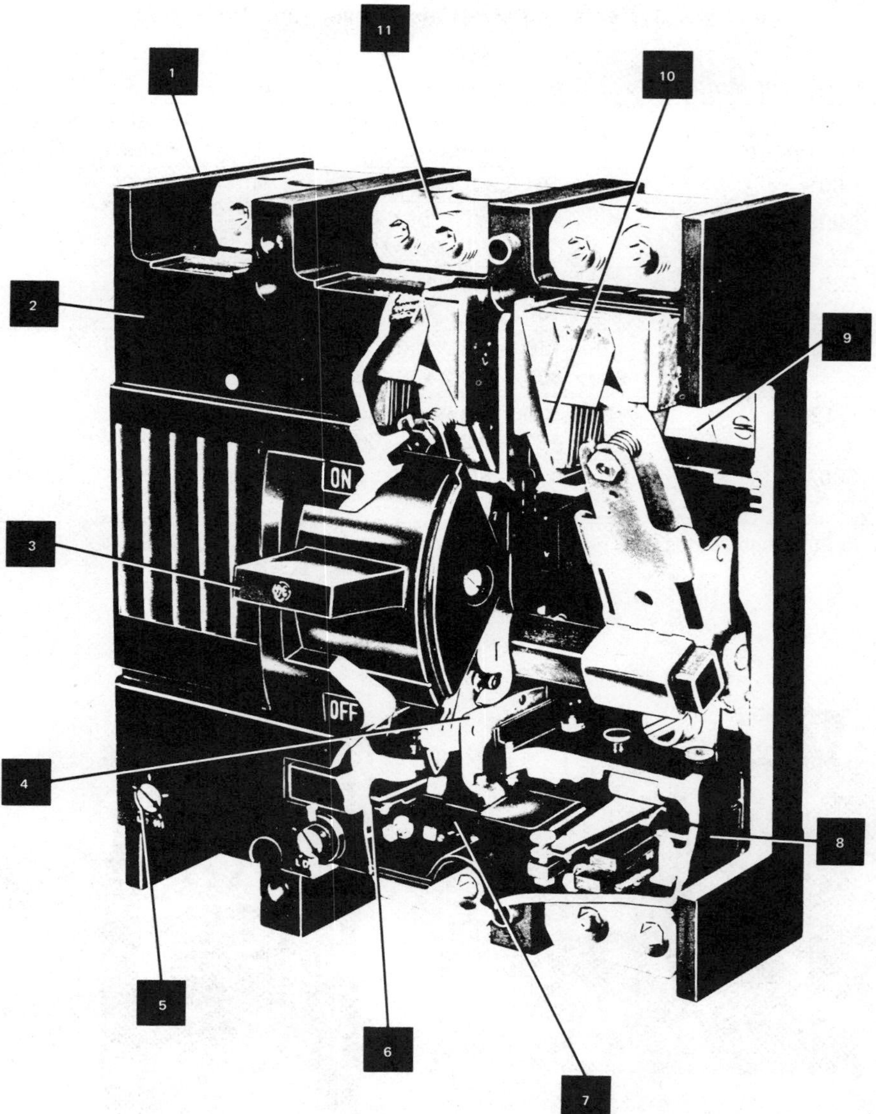

(Concluded from page 609)

(*c*)

1. *Frame sizes.* Molded-case circuit breakers are made in a few frame sizes that cover the ampere range from 0 to 2500 amperes
2. *Molded case.* Of phenolic material.
3. *Trip indication.* When mechanism trips, handle moves to middle, or "trip-indicating" position midway between ON and OFF.
4. *Quick-make, quick-break, trip-free mechanism.* Independent of manual control so breaker can trip open under short-circuit or overload conditions even though the operating handle is held in the ON position.
5. *Front-adjustable magnetic trip.* Provide instantaneous trip action in the event of a short circuit.
6. *Thermal trip.* Provides protection against sustained overloads. A bimetallic element reacts, "time-wise," in inverse proportion to the current, activating the trip mechanism and opening the circuit on overload.
7. *Common-trip bar.* Assures disconnection of all conductors when an overload or short circuit occurs on any one conductor in the circuit.
8. *Interchangeable trip units.* Trip units on frames above 100 amperes in rating can be interchanged.
9. *Silvered contacts.*
10. *Arc chute.* Provides fast arc interruption and guards against damage to the breaker.
11. *Line connections.* Designed to take appropriate sizes of copper or aluminum cable.

of operation, fuses are energy-limiting devices. This characteristic is of great importance in high-energy systems, where protection of the system's equipment from the enormous destructiveness of large short-circuit currents is obviously required. The inherent advantage of the breaker over the fuse lies in the fact that it can be reset after tripping by merely operating its handle; it is not self-destructive on operation, as is the fuse. Whereas fuses are single-pole devices, requiring a separate fuse for each ungrounded line, circuit breakers are available in multipole construction with a common trip so that an overload in any leg of a 3-phase circuit protected by a 3-pole breaker will cause the disconnection of the entire circuit. Like the fuse, the circuit breaker is rated in interrupting capacity.

Circuit breakers are complex mechanical devices that need periodic exercise, inspection, and maintenance. This is particularly true after opening on a large fault. The fuse, however, needs no maintenance and can be relied upon to operate, as designed, for many years after installation. Characteristics are tabulated below for ease of comparison.

Fuses—Switch and Fuse Combination

Advantages	*Disadvantages*
Simple and foolproof	Single pole only
Constant characteristics	Requires switch
Initial economy	Necessity for storage of replacements
Very high I.C. (interrupting capacity)	Self-destructive (one-time operation)
No maintenance	Nonadjustable
	Nonindicating
	No electric or remote control
	Not trip free

Circuit Breakers

Advantages	*Disadvantages*
Usable as switch	Low to medium I.C.
Multipole	Periodic maintenance
Small, convenient	High initial cost
No replacement	Complex construction
Resettable	Aging
Indicates trip	Not energy limiting
Trip free	
Remote control	
Adjustable	
Auxiliary contacts	

The above lists demonstrate that there is no all-inclusive answer to the oft-posed question "Which are preferable—fuses or breakers?" The answer depends on the specific application involved and is often based on highly technical factors, beyond this book's scope. Generally, breakers are used for all lighting and appliance panels and for main panels and switchboards where the I.C. requirement is not very high. Fuses are used in main switchboards and panels, where they are required for coordination with other fuses, or where the economics warrant their use.

16.18 Switchboards and Switchgear

Switchboards and switchgear are freestanding assemblies of switches, fuses, and/or circuit breakers, which normally provide switching and feeder protection to a number of circuits connected to a main source. A switchboard may be represented in a single-line diagram as in Figure 16.21. It serves in an electrical system to distribute, with adequate protection, large bulk power into smaller "packages." Thus, by hydraulic analogy, the main buswork of the switchboard is equivalent to a main header, the switches to on/off valves, the fuses to flow-limiting devices, and the feeders to subheaders connected to the main header. Modern switchboards (Figs. 16.22 to 16.24) are all dead-front; that is, they have all circuit breakers, switches, fuses, and live parts completely enclosed in a metal structure. The operator controls all devices by means of insulated handles in the front panel. When a switchboard has circuit breakers

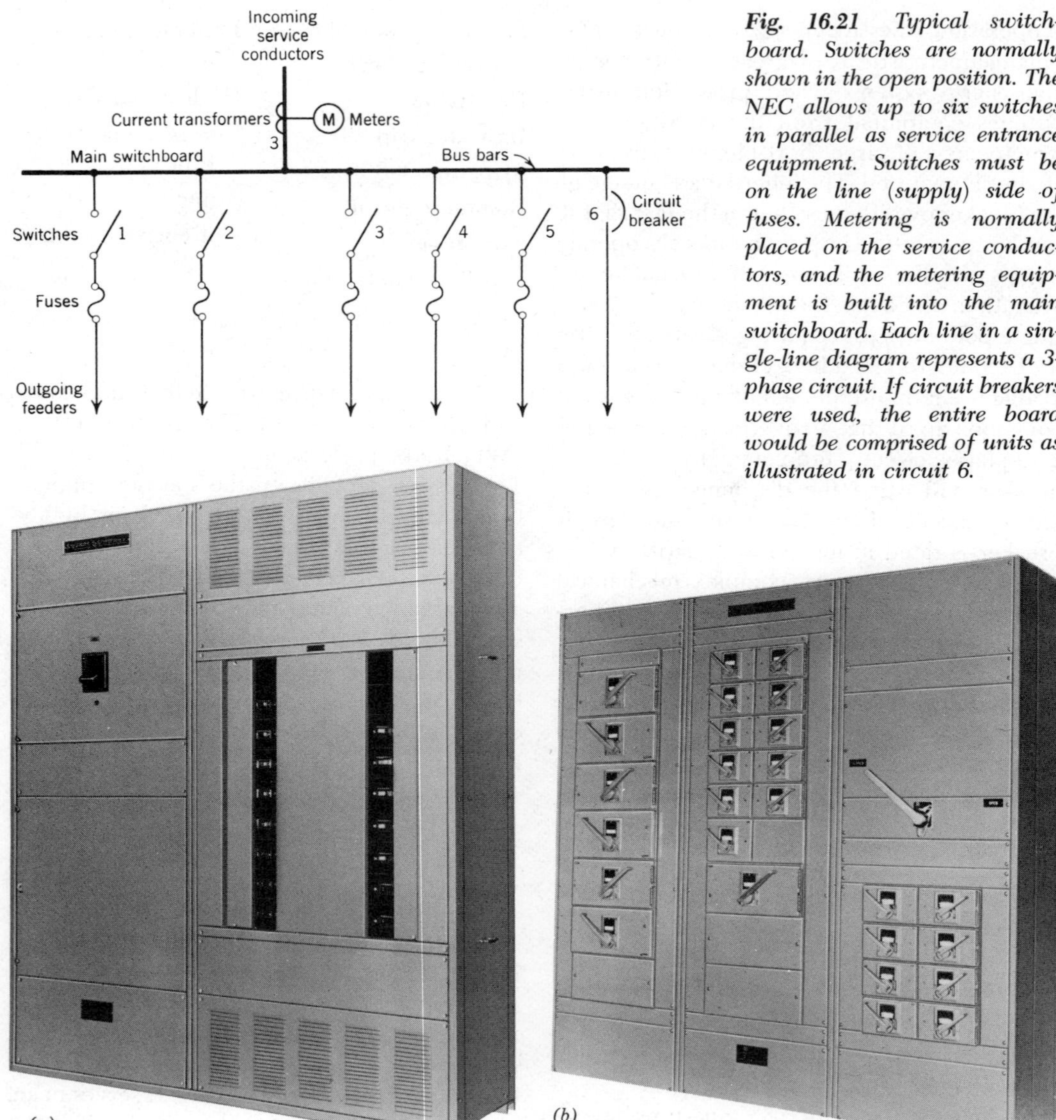

Fig. 16.21 Typical switchboard. Switches are normally shown in the open position. The NEC allows up to six switches in parallel as service entrance equipment. Switches must be on the line (supply) side of fuses. Metering is normally placed on the service conductors, and the metering equipment is built into the main switchboard. Each line in a single-line diagram represents a 3-phase circuit. If circuit breakers were used, the entire board would be comprised of units as illustrated in circuit 6.

***Fig. 16.22** Low-voltage switchboards are available in various designs.* (a) *Illustrates a NEMA 1, panel-type, freestanding, front-accessible unit with large air main circuit breaker and molded-case branch circuit breakers. The compartment below the main breaker can be used for metering, current transformers, or other accessories. The section across the top is for wiring between sections.* (b) *Illustrates a similar panel-type switchboard that is equipped with fusible switches. Here, too, the section above the large main switch can be utilized for metering, and so forth. Both these switchboards are constructed with a compact central bus in each section to which the section's circuit breakers and switches are connected. Since this is similar to panelboard construction, see Fig. 16.23, these switchboards are described as "panel-type." (Standard switchboard bus construction is similar to that shown in Fig. 16.39, with main horizontal buses feeding* all *switchboard sections and vertical buses tapped from them feeding the units in each section.) Units are from 14 to 20 in. deep and 90 in. high. This type of construction can also be used to house both circuit breakers and motor controllers, as in Figure 16.40. Courtesy of The Square D Co.*

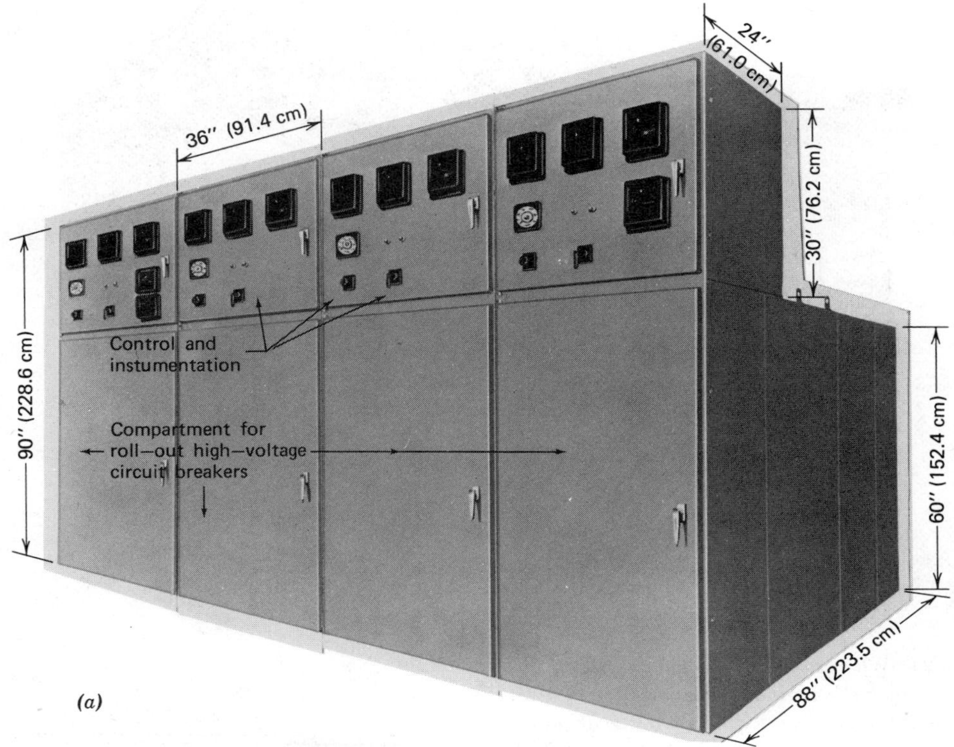

(a)

(b)

Fig. 16.23 Typical line-up of high-voltage (15 kv) metal-clad switchgear (a). *Length and width dimensions are typical, but vary with the item of equipment and with the manufacturer. A height of 90 in. (228.6 cm) is an industry standard. Each circuit breaker is mounted on a rolling "truck"* (b) *to facilitate removal, and plugs into stationary connectors in the switchgear. Enough space must be left in front of the switchgear to permit removal of the truck. Courtesy of The Square D Co.*

equipped with bayonet-type contacts and mounted in a movable drawer (like the drawers of a standard letter file), they are described as the "drawout" type. This draw-out arrangement facilitates emergency replacements, inspection, and repairs.

There is no clear distinction made between the terms "switchboard" and "switchgear," although often high-voltage equipment (above 600 v) is referred to as switchgear, as is equipment which comprises individual units rather than an assembly. When molded case circuit breakers are utilized in a switchboard it is

Fig. 16.24 When switchgear is to be installed outdoors, three methods may be employed: build a small house to enclose normal indoor gear, utilize weatherproof outdoor gear (illustrated), or utilize switchgear, which is built into its own metal house. Integral housings are equipped with heat and light and often prove the most economical choice. Courtesy of McGraw-Edison Co.

often known as a building-type switchboard. Space requirements for various type boards are shown in Figure 16.25. Main metal clad switchgear for commercial, industrial, and public buildings is almost invariably located in the basement, and housed in separate well-ventilated electrical switchgear rooms. The designer will realize that adequate lifting hooks, exits, hallways, and hatches should always be provided for the entrance and exit of the equipment of largest dimensions and weight to be moved. Therefore, the specifica-

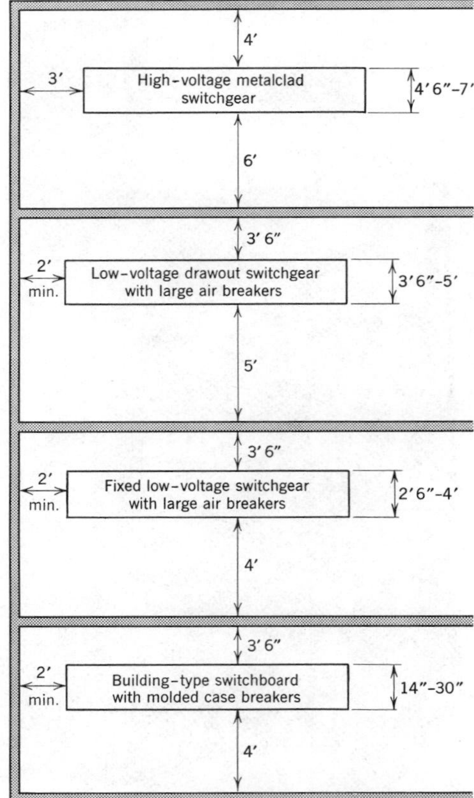

Fig. 16.25 Minimum switchgear space requirements. Clearance to ceiling should not be less than 3 ft. Each room should have two doors where switchgear is connected to high-capacity systems.

tions for switchgear should state the maximum number and overall maximum dimensions of sections to be bolted together as one portable section. Two, three, or four sections form the usual practical section. These sections may vary in length from about 8 to 12 ft. Smaller subdistribution switchboards require no special room enclosure, except to bar tampering or vandalism. In such instances a wire screen enclosure plus a large DANGER—HIGH VOLTAGE sign is usually adequate.

16.19 Unit Substations (Transformer Load Centers)

An assembly of primary switch-and-fuse or breaker, step-down transformer, meters, controls, buswork, and secondary switchgear is called a unit substation, or a load-center substation. It is available for indoor or outdoor use, to supply power from a primary voltage line to any large facility. The location in the building of the unit substation is governed by the type of transformer utilized, as explained above in the discussion on indoor transformer installations. For this reason, almost all indoor unit substations utilize dry-type (air-filled) transformers. Unit substations are utilized to effect the economies inherent in prefabricated construction with coordinated components. A basement location is most often selected, with ventilation requirements as detailed above. Access should be restricted to authorized persons only. Typical units are illustrated in Figures 16.26 and 16.27.

16.20 Panelboards

A panel, or panelboard, serves basically the same function as a switchboard, except on a smaller scale; that is, it accepts a relatively large block of power and distributes it in smaller blocks. Like the switchboard, it comprises main buses to which are connected circuit-protective devices (breakers or fuses), which feed smaller circuits. The panelboard

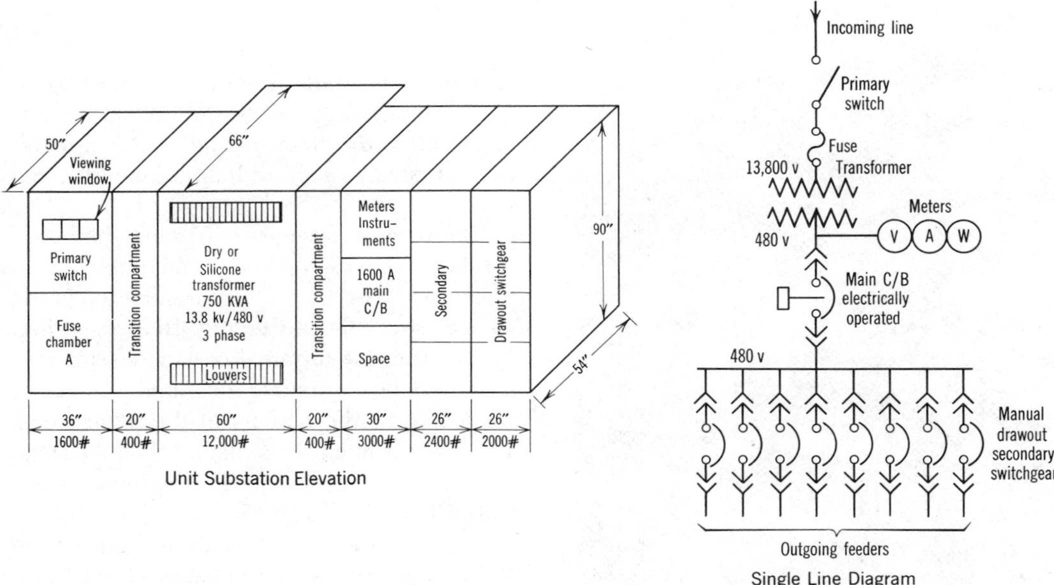

Fig. 16.26 *Unit substation. Approximate sizes and weights of a single-ended unit substation. Such a unit would supply a building with a maximum demand of 750 kva. The incoming 13,800 v cables enter cubicle* A *and connect to the switch and fuses. The load side of the fuses connects to the transformer, which in turn connects to the secondary switchgear. This main secondary switchgear in turn feeds various switchboards and panelboards distributed through the building.*

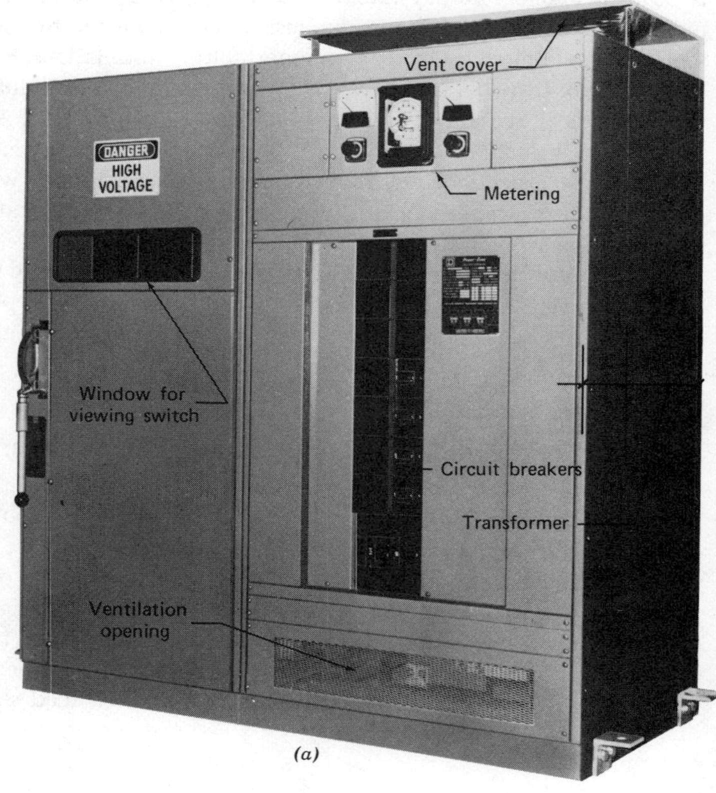

(a)

(b)

Fig. 16.27 *Shown front and back in* (a) *and* (b) *is a small unit substation, called a power center. It consists of a compact assembly of incoming high-voltage section, dry transformer, and secondary switchgear. Dry transformers and molded-case breakers are used for compactness. The pictured unit shows the high-voltage switch and fuses and the L.V. breakers. The entire unit is approximately 91 in. high, 90 in. long, and 48 in. deep. Note the window through which the position (open or closed) of the high-voltage switch can be verified. Note also the ventilating fan for cooling the dry-type transformer and the shallow enclosure for the molded-case circuit breakers. The transformer is behind the breakers. Courtesy of The Square D Co.*

level of the system is usually the final distribution point, feeding out to the branch circuits that contain the electrical utilization apparatus and devices, such as lighting, motors, and so on.

The panel components—that is, the buses, breakers, and so forth—are mounted on an insulating board that in turn is mounted inside an enclosing cabinet (see Fig. 16.28*a* and Fig. 16.29). The line terminal of each circuit-protective device (breaker or fused switch) is connected to the busbars of the panelboard. The load terminal of the device then feeds the outgoing branch circuit. This is shown schematically in Figure 16.28*c*. The busbars of the panelboard are energized by a feeder from some switchboard or load center. The cabinets are set in the walls and connected to the conduit system before the panelboards and trims are installed. The panelboards are then bolted into place in the cabinets, and the wires are pulled in through the conduits to make

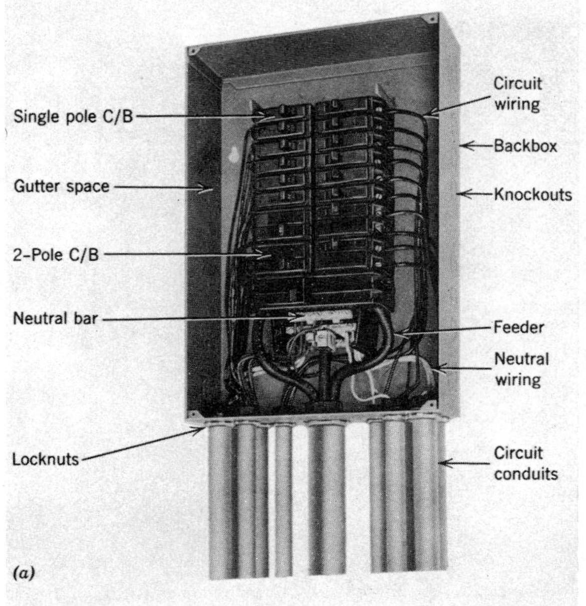

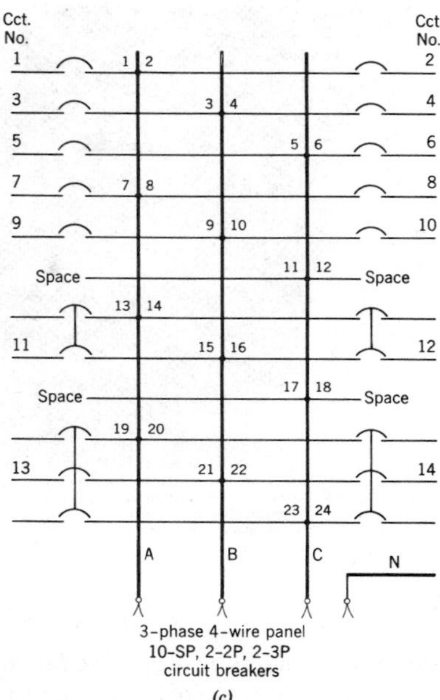

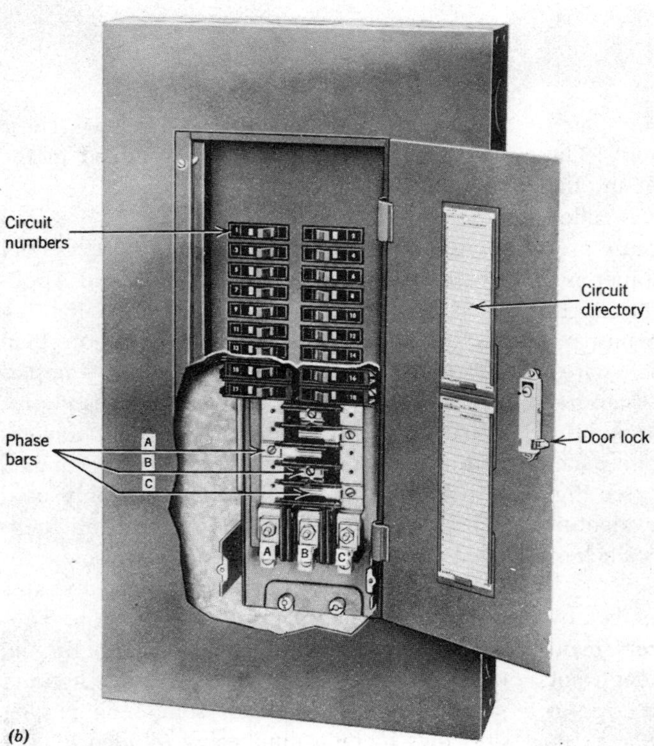

Fig. 16.28 Panelboards may be of the circuit breaker or fuse type. The illustrated panels contain 1, 2, and 3-pole branch circuits. Panels are provided with a minimum 4 in. gutter space, to allow routing of circuit wiring and feed-through conductors. Splices may also be made if space is sufficient. Lighting panels average 5 in. deep and 16 to 20 in. wide; power panels are 6 in. deep and 20 to 30 in. wide. Panels are mounted with the top circuit device no higher than 78 in. above finished floor (AFF), and the bottom device no lower than 18 in. AFF. The wired panel and backbox are illustrated in (a); *the panel front has been mounted in* (b). *A typical schematic diagram for a panel is shown in* (c). *Courtesy of The Square D Co.*

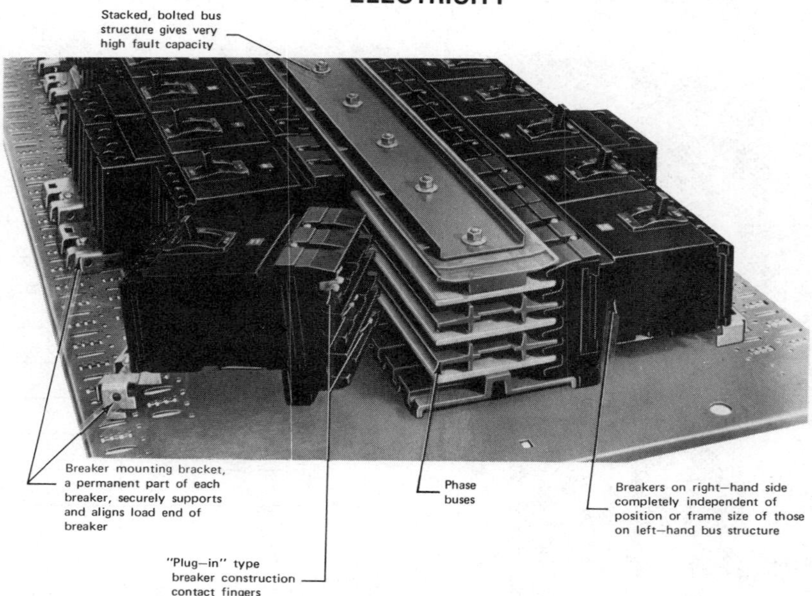

*Fig. 16.29 A design utilizing a busduct type of bus construction for a panelboard is shown here. This is the same type of bus configuration that is used in the switchboards illustrated in Figures 16.22*a *and* b, *and in the motor control center of Figure 16.40. The advantages of this type of bus construction are simplicity of panel assembly, high fault capacity, and complete independence of the devices on both sides. Thus no spaces (see Fig. 16.28*c*) need be left in order to make devices fit evenly. Courtesy of The Square D Co.*

connections to the terminals of the panel busbars and the outgoing branch circuits. The space between the edges of the panel and the sides, top, and bottom of the cabinet is called the wiring gutter. Wiring gutters must provide adequate space between the cabinet and the panelboard for making connections to branch circuits, the feeder, and the busbars, and for pulling through additional wires. When panels are stacked one above the other (on succeeding floors) and are fed from the same riser cable, a special construction panel known as a riser panel may be used (see Fig. 16.30). This type of installation is prevalent in multistory apartment houses where a stack of apartments are fed from a single riser.

Panelboards are described and specified by type, bus arrangement, branch breakers, main breaker, voltage, and mounting—though not necessarily in that order. A typical description might be: Lighting and Appliance Panel, 3-phase, 4-wire; 200A mains; main c/b, 225A frame, 150A trip. Branch breakers—all 100A frame; 8 ea. SP-20A, 4 ea. 2P-20A, 4 ea. 3P-20A; flush with hinged locked door. These terms and descriptions are explained in the following headings.

(*a*) ***Panel Type.*** The NEC recognizes only two types of panels—a "Lighting and Appliance Panel" and all other panels. A lighting and appliance panel is one having more than 10% of its overcurrent devices rated 30 amp or less, for which neutral connections are provided (single-phase loads). All other nomenclature—such as lighting panel, power panel, distribution panel, with the commonly used abbreviations of LP, PP, DP—is entirely unofficial. However, it is frequently convenient for the designer and subsequently for the operator to identify panels by use and location. Thus KP is frequently the kitchen panel, BP the basement panel, ELP the emergency lighting panel, and so on. Also, some engineering drawings include numbering to identify the floor location; thus LP3A means lighting panelboard "A" on the third floor. Such code numbers aid operators, electricians, and main-

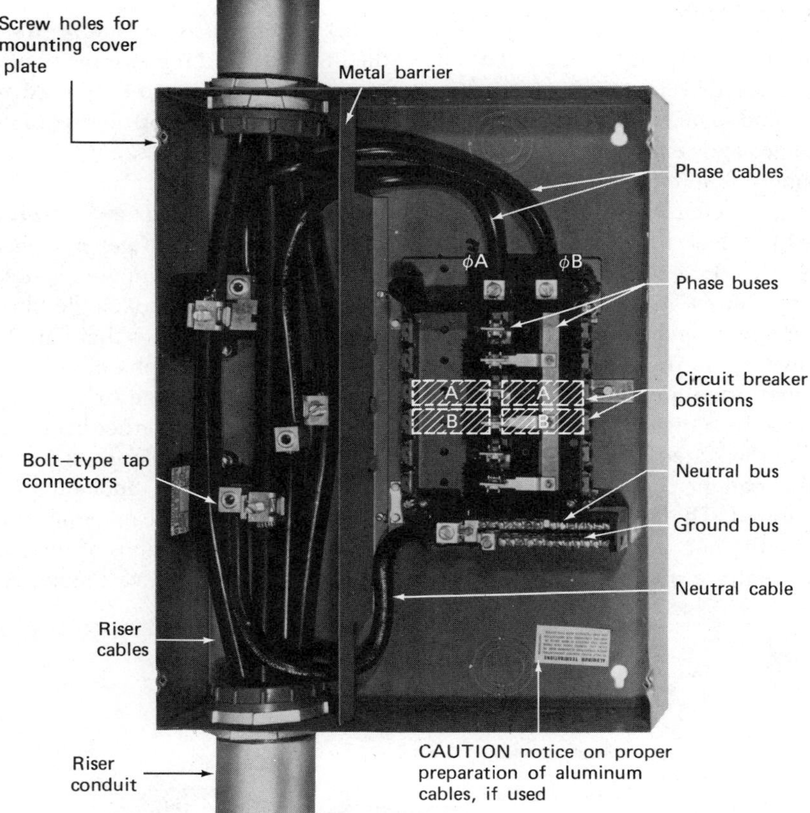

Fig. 16.30 *Twelve-pole 120/240 v riser panel, intended for use in multistory residences with identical apartments. The cable compartment is separated from the panel compartment by a metal divider. After the breakers and wiring are installed, the cover will be screwed on, which will give access only to the operating handles of the circuit breakers. Note the additional ground bus, intended for green-ground wire use. Courtesy of General Electric Company, Circuit Protective Device Department.*

tenance personnel in checking and maintaining circuits and equipment. These code numbers should be lettered on the outside of panelboard doors.

(*b*) ***Panel Electrical Description.*** This item gives the number of buses, their ampere rating, the panel voltage, and a complete description of the circuit-protective devices contained in the panel. If a contactor were built into the panel, its description would appear here.

(*c*) ***Panel Cabinet and Mounting.*** Panelboards are classified as flush-type or surface-type. Flush-type panelboards used in most buildings are those that have the trim and doors practically flush with the finished surface of the wall. Surface-type panelboards project into the room, the cabinets being bolted to wall surfaces or columns. The latter type is frequently used in industrial plants. Special features such as door locks, paint color, trim type, and the like, are also listed here (see Fig. 16.28*b*). As mentioned above, panels are the last "downstream" item of electrical power distribution equipment. Beyond the panel lies the power utilization equipment, which includes the motors and lighting fixtures and their control devices. These are discussed briefly below.

16.21 Electric Motors

Motors are very frequently supplied as adjuncts to specified driven equipment, such as fans, blowers, and so on, within the constraints of the voltage and enclosure. The actual choice of motor is left to the driven-equipment supplier, it being maintained that he is best qualified to select a motor that will optimally match the driven-equipment requirements and so supply a working whole for whose proper operation he is responsible. The supplier, however, is frequently guided primarily by the price motive. The specifier therefore should be sufficiently knowledgeable so that, within the stated constraints of applicability to the load, he can specify the particular motor desired. The following paragraphs are written with that purpose in mind and therefore concentrate on application data.

(*a*) ***Direct-Current Motors.*** As a result of the high cost and relative rarity of direct current, these motors are only used where continuous fine speed control is required, as in the case of elevator drives.

(*b*) ***Alternating-Current Motors.*** These motors fall into three general classifications: polyphase induction motors, polyphase synchronous motors, and single-phase motors. Within these categories there are further subdivisions. Thus polyphase induction motors are either squirrel cage or wound rotor; synchronous motors are either brush or brushless; and single-phase machines are manufactured in a host of designs, including split phase, repulsion-induction, capacitor, shaded pole, and various combinations of these designs. Of these many types, the vast majority are squir-

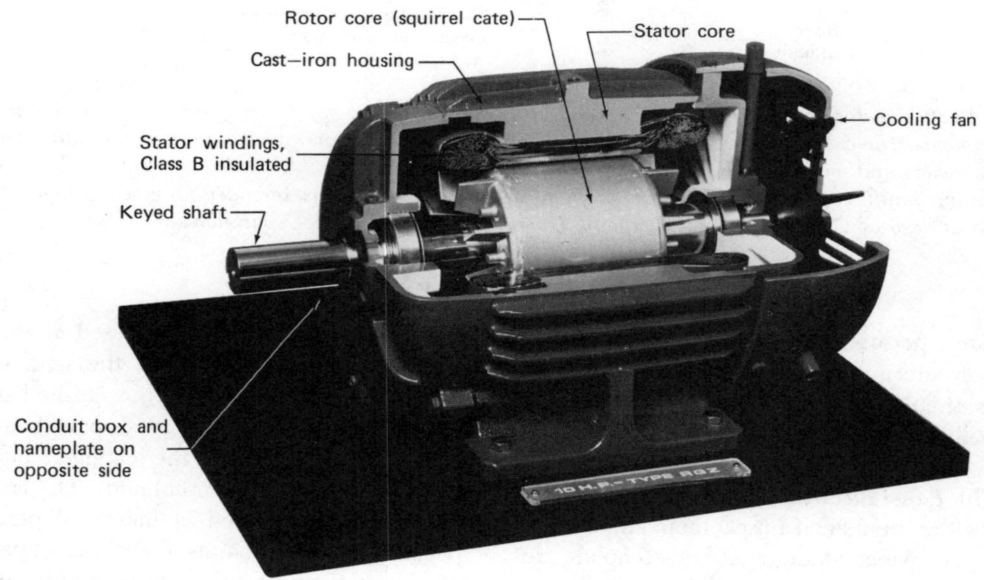

Fig. 16.31 *Totally enclosed, fan-cooled 10 hp polyphase induction motor. Note that the motor itself is totally enclosed; only the fan compartment is ventilated. The housing is ribbed to assist in heat dissipation and to maintain structural stiffness. Courtesy of Allis-Chalmers Corp., Motor Division.*

rel-cage induction machines; therefore, this type will be studied in some detail.

(*c*) ***Squirrel-Cage Induction Motors.*** This motor type owes its interesting name to an early design in which the rotor consisted of a group of bars welded together into a cylindrical cage-type shape. The design is basically unchanged today except for refinements. See Figure 16.31 for a cutaway view of a modern polyphase unit. Squirrel-cage motors are manufactured in four different NEMA designs to meet different application requirements. The characteristics of these four types are given in Table 16.9. A typical motor nameplate is shown in Figure 16.32.

Items on this nameplate that are not self-evident are:

1. *Type*—this is the manufacturer's designation and indicates primarily the enclosure. Thus in Figure 16.31 the type shown is RGZ, which is Allis-Chalmer's designation for a totally enclosed fan-cooled motor. Common enclosures are open drip proof, totally enclosed, fan cooled, and explosion proof. Special weather-protected designs are also available.
2. *Duty*—continuous or intermittent.
3. *Service factor*—indicates permissible overload, which generally runs 15%.
4. *kva code*—indicates by a letter the maximum starting current per horsepower. This is useful in selecting motor protective devices.
5. *Frame*—a NEMA standard number that indicates the motor's physical dimensions.
6. *Insulation class*—shows type of insulation used, which in turn indicates operating temperature. Insulation classes are B, F, and H corresponding to temperature rises of 80°, 105°, and 125° C, respectively.
7. *Motor voltage*—the standard motor voltages are 208, 230/460, and 575 v. Induction motors will generally operate satisfactorily ±10% voltage. Figure 16.33 indicates the effects of voltage change on a standard Design B, polyphase induction motor. Only 208-v motors should be used on 208-v systems, since actual line voltage may be as low as 200 v. Using a 230-v motor will result in sharply reduced torque and increased temperature rise, and poor overload capacity.

(*d*) ***Electric Motor Energy Considerations.*** Data compiled in 1976 by the Federal Energy Administration indicate that approximately one-half of all the electric energy generated in the United States is used to power electric motors! Thus an overall increase in motor

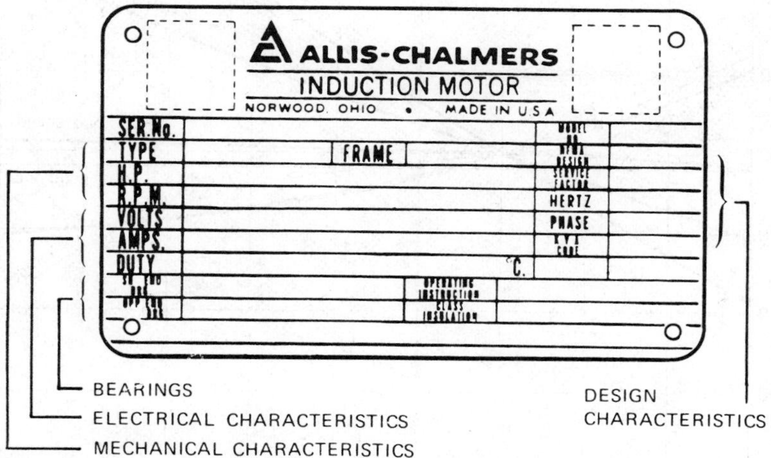

Fig. 16.32 *Typical motor nameplate. Courtesy of Allis-Chalmers Corp., Motor Division.*

Table 16.9 *Polyphase Motor Characteristics*

NEMA Design Letter	*Starting Current*	*Starting Torque*	*Break-down Torque*	*Description and Application*
A	High	Normal	High	Special design, not available except on special order.
B	Low	Normal	Normal	Standard design motor, widely used. High efficiency throughout load range. Power factor good at full load. Efficiency fair to good. Useful for constant speed, low torque loads, such as fans, blowers, centrifugal pumps, and machine tools.
C	Low	High	Normal	Lower efficiency and power factor than Design B. Also, slightly lower maximum torque in return for high starting torque. Useful on high starting torque loads, such as compressors, pumps, conveyors, crushers, etc. Constant speed.
D	Low	High	High	High slip, high torque design. Useful for heavy fluctuating loads, such as hoists, cranes, elevators, punch presses, shears, etc.
Wound rotor	Low	High	High	Used where some speed control is required, along with low starting current and high torque. Applicable to fans, pumps, compressors, etc.
Synchro-nous	Low to normal	Low to normal	Low to normal	Constant, low speed service. Large loads. High efficiency, high power factor, lower cost than comparable induction motors.

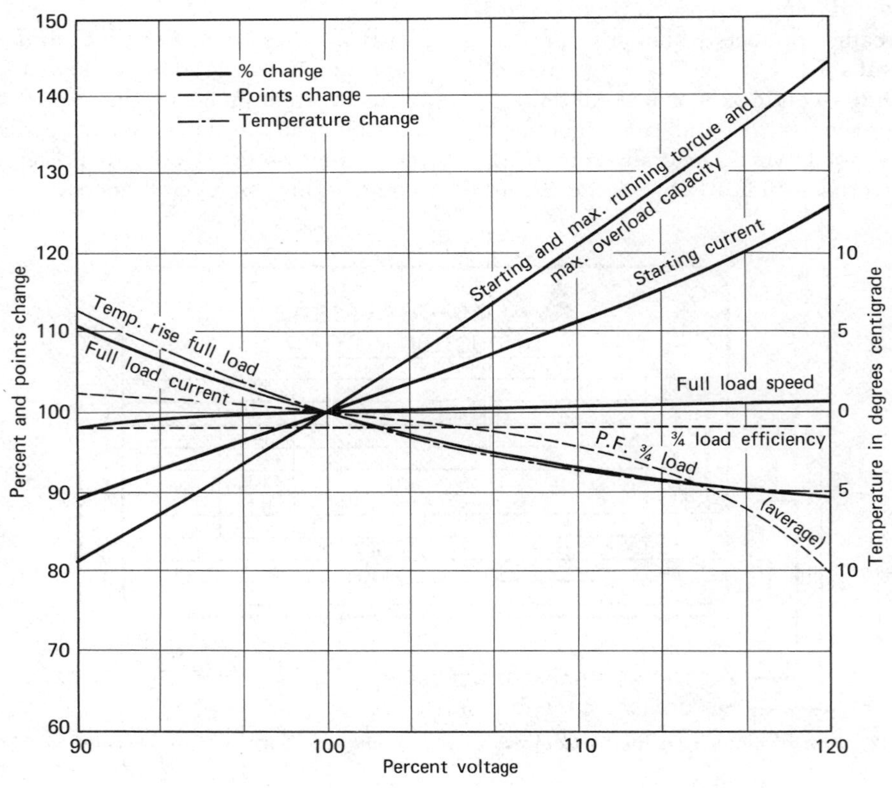

efficiency of only 1% would result in an energy savings of 2 to 3 billion kwh. As a result of these figures, legislation has been prepared that will require electric motors to meet certain energy standards, one of which is ASHRAE Standard 90-75. The energy losses associated with electric motors stem from two sources:

1. Losses in the motor itself, as indicated by the motor's efficiency.
2. Line losses caused by the motor's power factor.

Let us consider the latter item first.

The power factor (pf) of a device is an indication of the amount of reactive current that the device requires—the lower the power factor, the larger the reactive current. Reactive current does not draw power and does not therefore show up on the kilowatt-hour meter or the kilowatt-demand meter. It *does,* however, cause I^2R line losses throughout the distribution system, and it does tie up generating capacity with no financial return. Recognizing this, utility tariffs almost invariably contain a low power factor penalty clause. This additional revenue compensates the utility but does nothing to diminish the energy wasted in line losses. It is specifically this energy waste that the new regulations wish to reduce. ASHRAE Standard 90-75 specifies that "Utilization equipment, rated greater than 1000 w and lighting equipment greater than 15 w, with an inductive reactance load component, shall have a power factor of not less than 85% under rated load conditions."

To gain an appreciation of the magnitude of losses involved, consider a small industrial facility with a continuous load of 100 kw at 60% power factor. This low figure is readily attainable with low power factor fluorescent lighting, and single-phase motors whose power factor is inherently low (of the order of 40 to 50%).

Considering now the effect of an improved power factor, we would have

Fig. 16.33 (opposite) *Characteristics of Standard Design B Induction motors, as a function of applied voltage.*

	60% pf	*85% pf*
Load	100 kw	100 kw
kva $\frac{\text{kw}}{\text{pf}}$	166.67	117.65
Current	462A	326

Losses are proportional to I^2R. Assuming a 5% loss in the utility lines (5 kw) and another 5% loss in the building distribution system, the losses at the improved power factor are reduced to:

$$\text{loss at 85\% pf} = 10\text{ kw}\left(\frac{326}{462}\right)^2 = 5\text{ kw}$$

or the loss is cut in half. Assuming a 12-hr day, 5-day week, this amounts to 9000 kwh/year, a 4500 kwh savings to the utility and to the consumer (this is approximately an average residence's annual usage). Monetarily, the improvement is no gain to the utility which loses its surcharge. For the industrial facility, the outlay needed for capacitors to effect this improvement will generally pay for itself within a few years, depending on local energy rates, electrical installation costs, and other factors. For the economy, however, the energy savings is pronounced.

Some motor manufacturers have recently announced a line of high power factor, high-efficiency motors. Using these in lieu of standard motors and power factor correction with capacitors will markedly reduce the pay-back period. Also, since the power factor is very sensitive to load in motors, sizing motors close to the load requirement, in lieu of oversizing them as is common practice, will do much to improve the power factor and concomittantly will conserve energy. The newer high power factor design machines also have a better power factor versus load characteristic, which is particularly helpful in situations where motors handle variable loads; that is, where running at partial load is unavoidable.

Motor efficiency and its variation with load and motor size are shown in the curves of Figure 16.34. Here, unlike the power factor problem, any improvement in efficiency is reflected directly in the facility's electric energy bill, in addition to effecting a distinct savings in fuel. Unlike power factor, efficiency cannot be improved by the addition of an external device such as a capacitor. Thus for

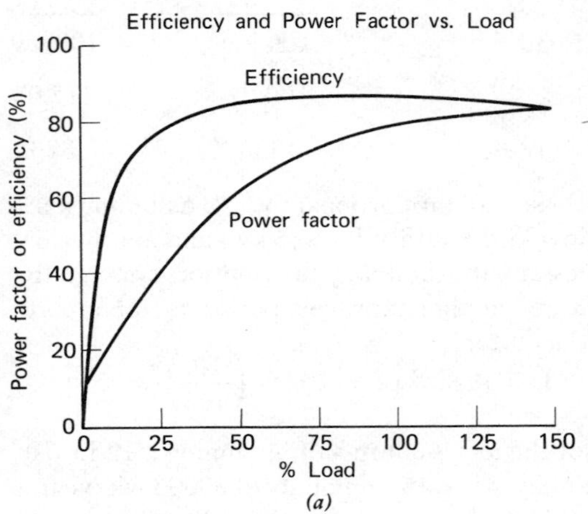

(a)

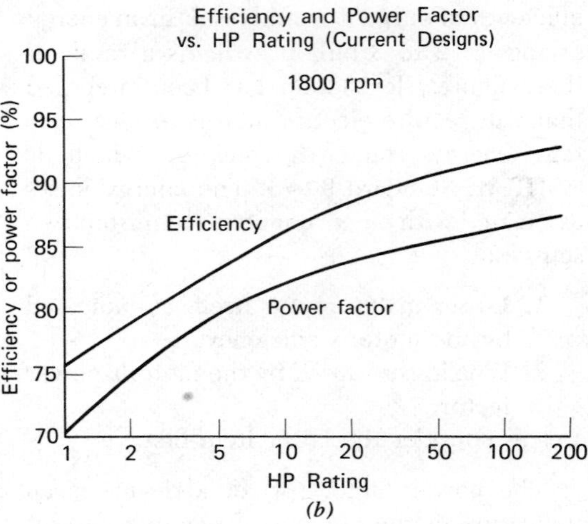

(b)

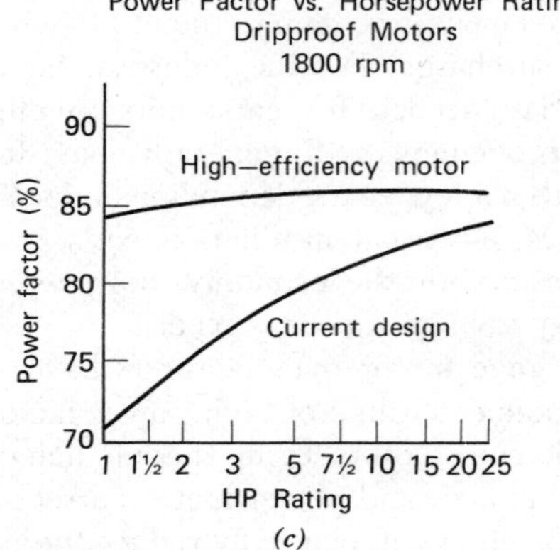

(c)

Fig. 16.34 Conventional motor characteristics. The curves shown represent averages and vary with individual manufacturers. Note particularly the extreme sensitivity of power factor to load in (a), *and the relation of both efficiency and pf to motor size in* (b). *The advantage of the high-efficiency, high-pf motor design, particularly in small motors, is clearly shown in* (c). *Courtesy of Cahners Publishing Co., Division of Reed Publishing Corp. Copyright © 1975.*

an existing facility a careful cost analysis would have to be made to determine whether replacing existing machines with newer, more expensive, high-efficiency machines is justifiable. For a new facility, the pay-back time for the cost differential of the high-efficiency machines is short. An added advantage of these machines is that the lower losses result in lower temperature operation and therefore extended life. A life-cycle cost analysis of the type given in Tables 15.6 and 16.4 should be performed before purchase of motors. The results will probably indicate a distinct advantage to the high-efficiency designs. A typical analysis of a 5 hp machine is given in Table 16.10, based on the published manufacturer's data, some of which is given in Figures 16.35 and 16.36.

16.22 Motor Control

Since d-c controllers are highly specialized and infrequently encountered, we will confine our remarks to a-c motor control. An a-c motor controller is basically a contactor (see Section 16.14) designed to handle the heavy inrush currents encountered in an a-c motor starting. Its function is twofold: to start and stop the motor and to protect the machine from overload. These two separate and distinct functions are accomplished by combining a set of contacts for on/off control with a set of thermal overload elements for overload

Fig. 16.35 (opposite) *The advantage of the high-efficiency design is most pronounced in the smaller sizes. Also, these motors all are less load-sensitive than conventional designs. Courtesy of Cahners Publishing Co., Division of Reed Publishing Corp. Copyright © 1975.*

Table 16.10 ***Life-Cycle Cost Comparison for 5 hp Motor***

Motor Type	*Conventional Motor*	*High-Efficiency Motor*
Efficiency at full load	83%	87%
Power output	3730 w	3730 w
Power input	4.494 kw	4.287 kw
kwh per year (12 hr/day, 5 day/wk, 52 wk/yr)	14,021	13,375
Annual kwh differential	*646 kwh*	
20-year kw differential	*12,920 kwh*	
Annual energy cost at $.03 per kilowatt-hour	$420.60	$401.26
20-year life-cycle energy cost (8% capital cost, 3% annual price escalation)	$6574	$6272
Cost differential	*$302*	
Initial motor cost	$150	$185
Motor life	10 years	20 years
Motor amortization cost	$69.47	$39.68
Cost differential	*$32.18*	
Total life-cycle cost	$6827	$6497
Total differential	*$330*	

Fig. 16.36 (below, right) *Cost payback of the price premium for high-efficiency motors can be taken directly from the curve, given energy cost and motor size. Thus a 3 hp motor will repay its price differential in less than half a year, with a $.03/kwh average energy rate. At the same energy rate (a very conservative estimate) a 25 hp machine will repay its premium cost in under two years. All figures are based on a particular manufacturer's product. Other manufacturers' products will yield somewhat different results. Courtesy of Cahners Publishing Co., Division of Reed Publishing Corp. Copyright © 1975.*

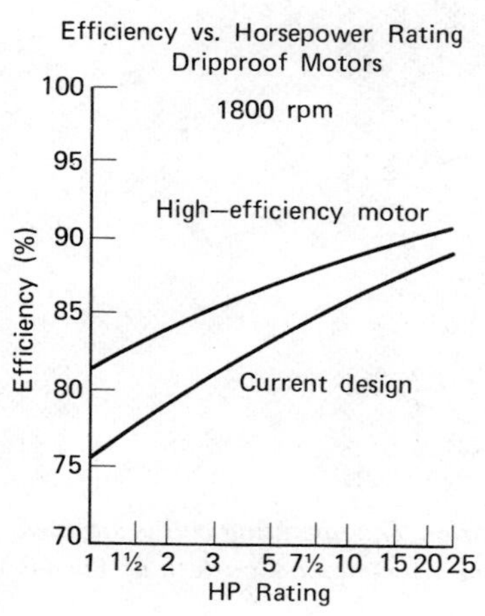

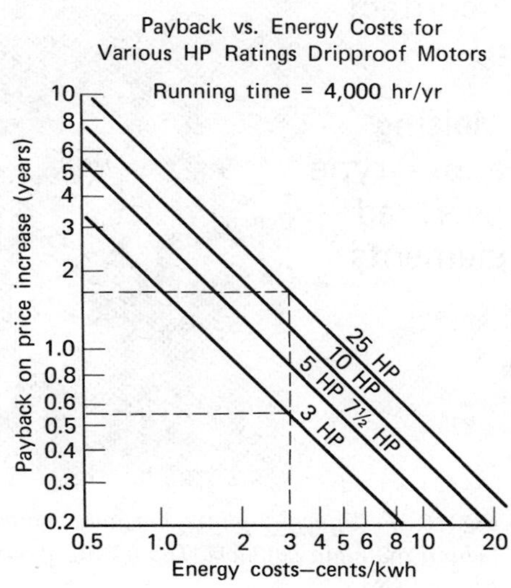

Table 16.11 ***Rating and Approximate Dimensions and Weights of a-c Full-Voltage Single-Speed Motor Controllers, 3-Phase Combination Circuit Breaker Type***

NEMA Size Designation	*Voltage*	*Maximum Horsepower*	*Width*	*Height*	*Depth*
				Inches	
0	208–230	3	10	24	7
	460	5			
1	208–230	$7\frac{1}{2}$	10	24	7
	460	10			
2	208–230	15	10	24	7
	460	25			
3	208–230	30	20	24	9
	460	50			
4	208–230	50	20	48	9
	460	100			
5	208–230	100	20	56	11
	460	200			

NOTE. All starters are housed in a NEMA 1 indoor ventilated enclosure.

Fig. 16.37 Typical 3-phase, magnetic, across-the-line (full voltage) starter. The operating coil is situated behind the main contacts. The wiring shown is for the holding (sealing-in) contact. Courtesy of Furnas Electric Co.

protection, in a single unit. When the contacts are operated by hand, the controller is called a manual starter; when the contacts are operated by a magnetic coil controlled by push buttons, thermostats, or other devices, the unit is known as a magnetic controller or simply and more commonly—a starter.

Motor overload protection normally comprises thermal relays that interrupt the magnetic coil circuit thus opening the starter contacts. Motors 1 hp or less are generally controlled by a manual switch which contains a thermal overload relay. It is advisable to utilize such a device for all fractional horsepower motors.

Starters are available in various sizes, voltages, and NEMA enclosures. Table 16.11 gives typical electrical and physical data for full-voltage starters, and Figure 16.37 shows construction details. Most starters are of the full-voltage across-the-line type; that is, the contacts place the motor directly onto the line and the motor starts up immediately. Where such a procedure is undesirable because of voltage dip and flicker caused by the large inrush current or because of utility company limitations, a reduced voltage starter, sometimes called a compensator, is used. These units apply reduced voltage to the motor and thus reduce starting inrush current. Every motor controller is required by the NEC to have a disconnecting means within sight of the controller. Where convenient, this disconnect switch may be combined with the starter into a single unit, known as a combination starter. A circuit breaker or fused switch is often used in such an arrangement, which then constitutes the branch circuit protection and disconnecting means (see Fig. 16.38).

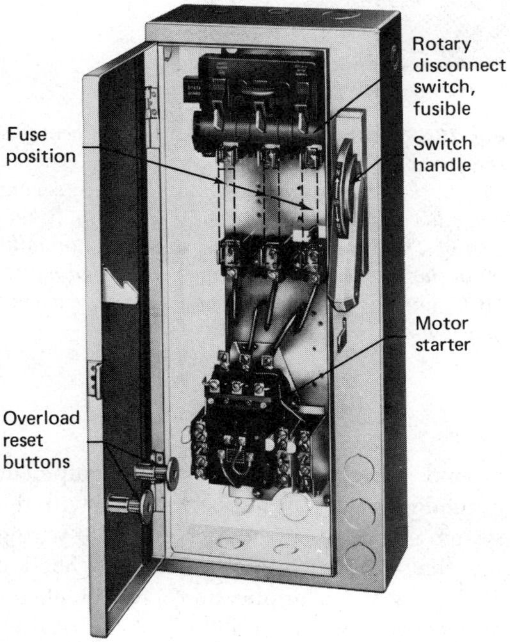

When starters are to be assembled for a group of motors, the motor starters, disconnect switches, motor controls, and indicating devices may be combined into a single large assembly for convenience and economy. Such an assembly is called a motor control center. Two typical motor control center construction types are shown in Figures 16.39 and 16.40. A typical, brief description of a motor controller would be similar to the following:

> Combination circuit-breaker type, across-the-line motor controller, NEMA size 2, 3 O.L. elements, 208 v, in a NEMA 1 enclosure. Starter shall contain integral on/off push buttons and integral control transformer for 24 v control circuit.

Fig. 16.38 Interior of a combination fused-switch type, across-the-line motor controller. Note that the unit is essentially a switch and a starter wired together and installed in a single cabinet. Courtesy of Allen-Bradley Co.

16.23 Wiring Devices

The general term "wiring devices" includes all devices that are normally installed in wall outlet boxes, including receptacles, switches, dimmers, and pilot lights. Attachment plugs, also called "caps," and wall plates are also included in any discussion of wiring devices. Devices are classified by grade in descending order of quality, as specification grade, intermediate grade, and standard or economy grade. In application, specification grade equipment is usually used in industrial and

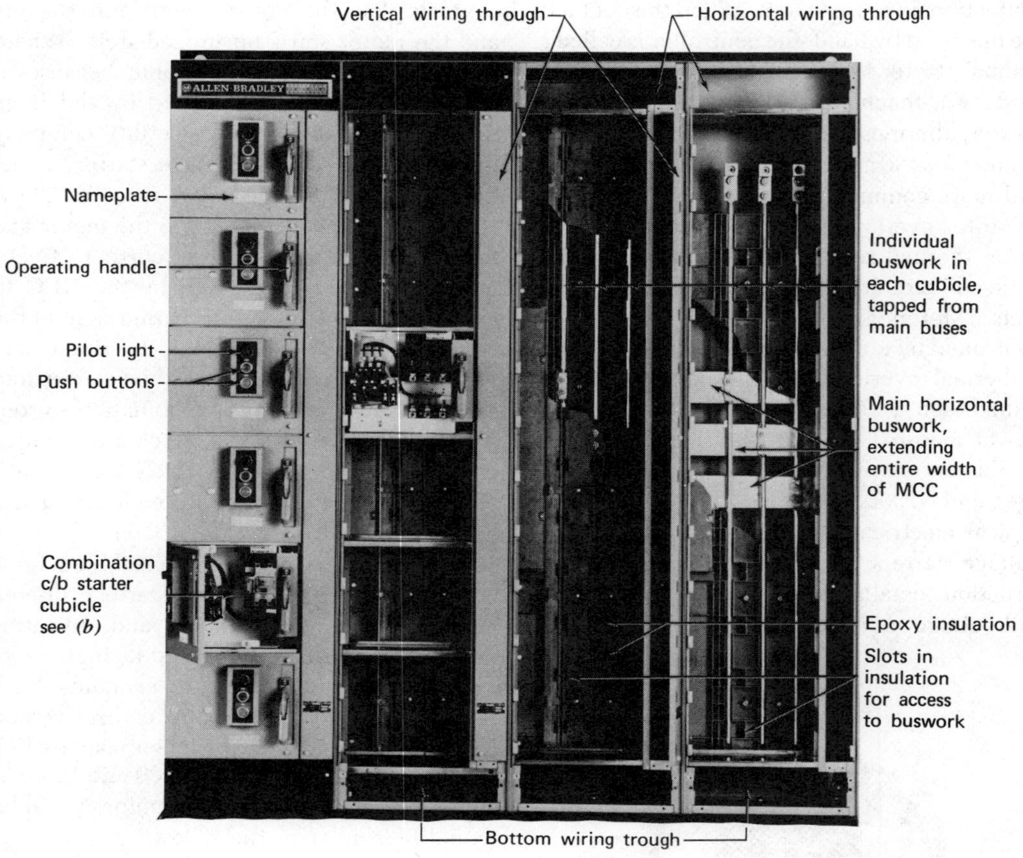

(a)

Fig. 16.39 A typical 600-v motor control center is shown in (a) *and* (b). (*Courtesy of Allen-Bradley Co.*) *Back-to-back construction* (c) *is space saving and adds only 5 in. to the basic 15 in. depth.* (*Courtesy of The Square D Co.*) *All units are normally 90 in. high and 20 in. wide per section, although MCCs with especially large components are deeper and wider.*

good commercial construction, intermediate grade in most educational and good residential buildings, and standard grade in low-cost construction of all types. The grade of wiring devices should, as with all electrical equipment, be consistent with the quality of construction in the entire facility.

16.24 Receptacles

In addition to the descriptive data specified above, receptacles are identified by the number of poles and wires, and whether or not the device is designed for connection of a separate grounding wire. The grounding pole *is not counted* in the number of poles (see Fig. 16.41) but is counted in "wires." The grounding pole is connected to the green ground wire where this is run, or to the conduit system where a ground wire is not run. The equipment grounding pole must not be confused with the system ground (neutral), nor must the wiring for the two be interchanged (see Section 17.3). In a typical application, a receptacle for an electric dryer with 4800 w, 208 v heating element and $\frac{1}{6}$ hp, 115 v motor would be NEMA 14-50R (see Fig. 16.41). The motor would connect across W and X, the heater across X and Y, and the appliance case to G.

Receptacles must be of the grounding type

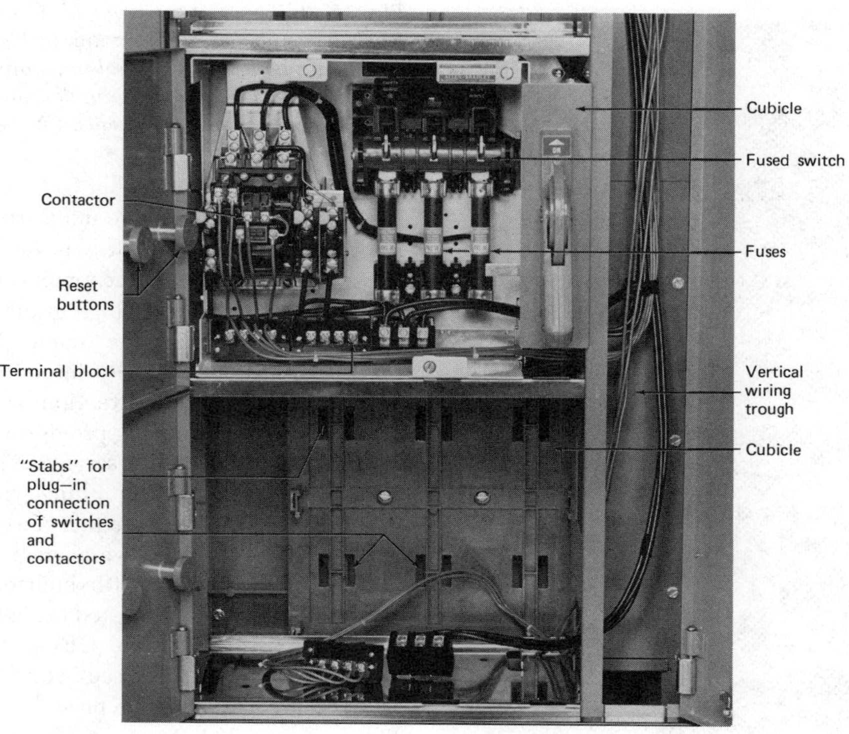
Cubicle
Fused switch
Contactor
Fuses
Reset buttons
Terminal block
Vertical wiring trough
Cubicle
"Stabs" for plug–in connection of switches and contactors

(b)

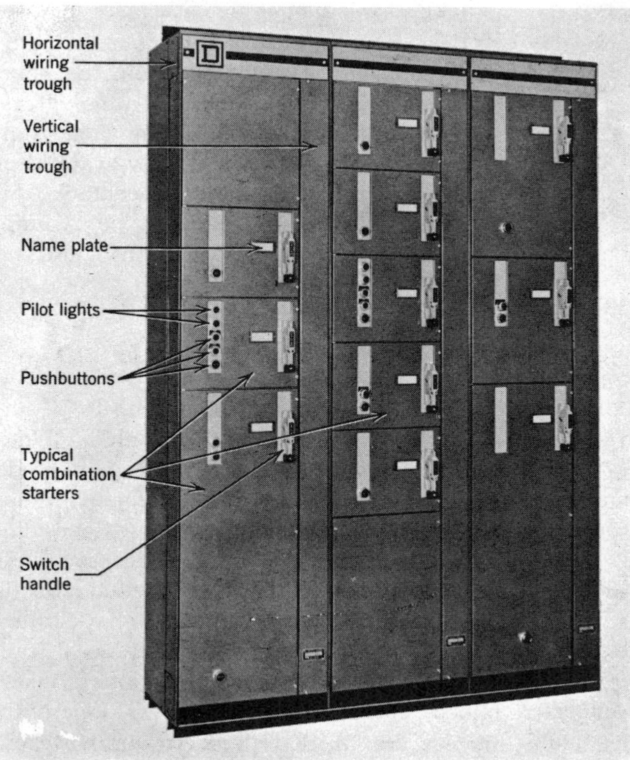
Horizontal wiring trough
Vertical wiring trough
Name plate
Pilot lights
Pushbuttons
Typical combination starters
Switch handle

(c)

Fig. 16.40 Shown here is an MCC with panelboard construction similar to that of Figure 16.29, containing both motor control apparatus and circuit breakers. Dimensions vary, depending on the number and size of components. Courtesy of The Square D Co.

where installed on standard 15 or 20 amp branch circuits. Receptacles connected to different voltages, frequencies, or current type (a-c or d-c) on the same premises must be polarized so that attachment plugs are not interchangeable. Figures 16.41 and 16.42 show some of the standard receptacle configurations and their ratings.

Receptacles are regularly available from 10 to 400 amp, 2 to 4 poles, and 125 to 600 v. In addition, special types are made such as locking types and miniature (interchangeable) types. Many specific usage units are available such as range receptacles, hazardous area types, and combination receptacle and switch assemblies. A line of very high-grade devices is manufactured under the name "hospital-grade" devices, specifically intended for this exacting use. Special attention is given in manufacture to grounding provisions, firmness of contact, and long-life service. These units are marked with a distinguishing symbol (usually a green dot) to denote this grade. In addition to these types, several manufacturers produce a receptacle with built-in ground-fault protection, as illustrated in Figure 16.43. For a discussion of these GFI (ground-fault interrupter) receptacles, see Section 17.4. All receptacles other than the normal 15/20 amp, 3-wire, parallel-slot type should be specified to be furnished with the required number of matching caps (plugs).

Receptacles are normally mounted vertically between 12 and 18 in. from the floor, except that in shops, labs, and other areas, where tables are used against the walls, 42 in. is the usual mounting height. A typical receptacle specification would be: Receptacle, duplex, 2-pole, 3-wire, grounding type, 20 amp, 250 v, specification grade, for indoor use.

16.25 Switch Devices

By NEC definition, switches up to 30 amperes that can be outlet-box mounted fall into this category. A typical specification grade toggle-type wall switch is shown in Figure 16.44*a*. Generally "a-c only" switches are preferable to the a-c/d-c type because of better construction. The usual a-c switch rating is 15, 20, or 30 amp at 120 or 120/277 v. Normal constructions are single-pole, 2-pole, 3-way, 4-way, momentary-contact, 2-circuit, maintained-contact SPDT and DPDT. Operating handles are toggle-type, key, push, rocker,

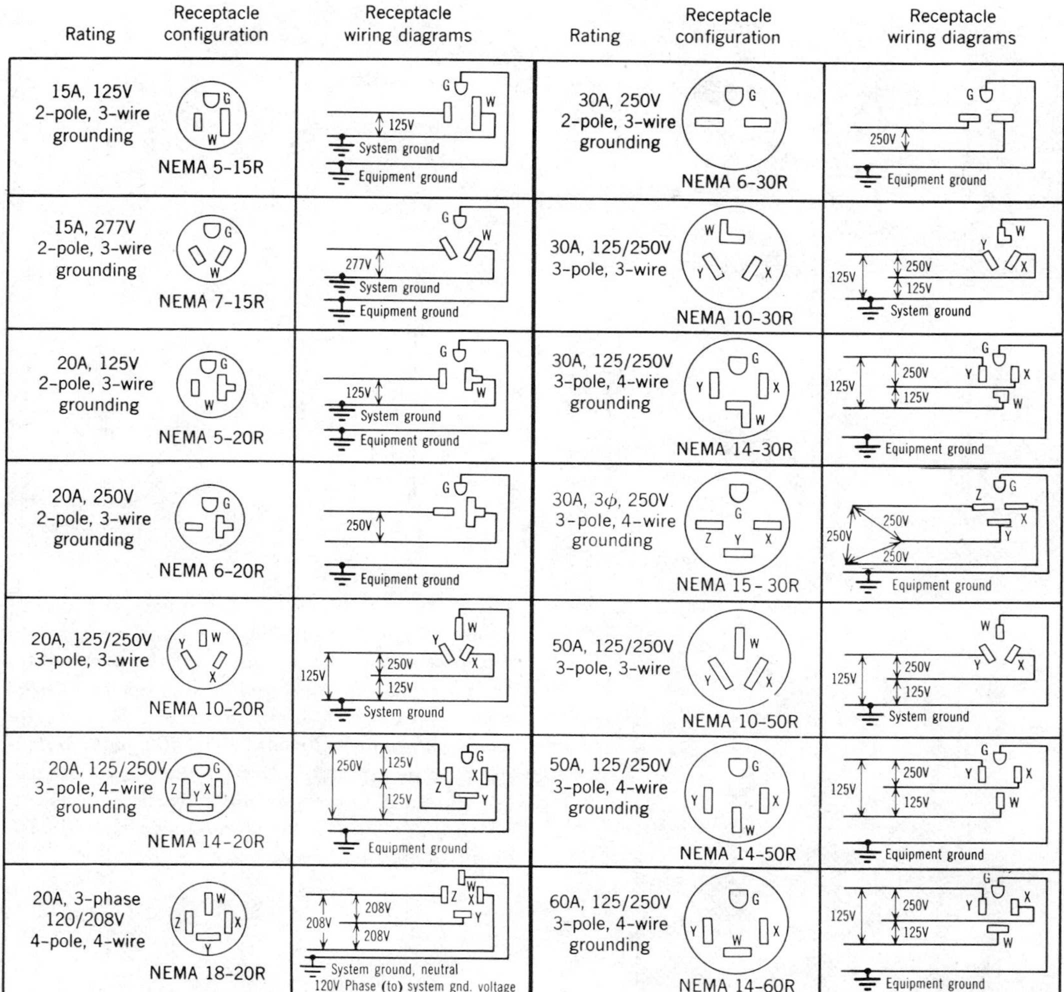

Note: "W" denotes white system ground or neutral
"G" denotes green equipment ground.

Fig. 16.41 *Receptacle configuration chart of selected common general purpose, nonlocking devices with related NEMA designations.*

rotary, and tap-plate types. The mercury and a-c quiet types are relatively noiseless; the toggle, tumbler, and a-c/d-c types are generally not. "Interchangeable" devices are miniature types, fully rated, which mount three to a strap in a normal outlet box. (Other devices available in this miniature construction are pilot lights, push buttons, and receptacles.) A typical switch specification would be: Switch, single-pole, a-c, quiet type, specification grade, 15 amp, 125 v, with press handle lighted when OFF, suitable for back or side wiring.

A switch incorporating a solid-state rectifier is readily available, which will give high/off/low control for *incandescent lamps,* within its wattage rating. These devices cost very little more than an ordinary switch and serve the purpose for which a more expensive dimmer is often used. Typical applications are

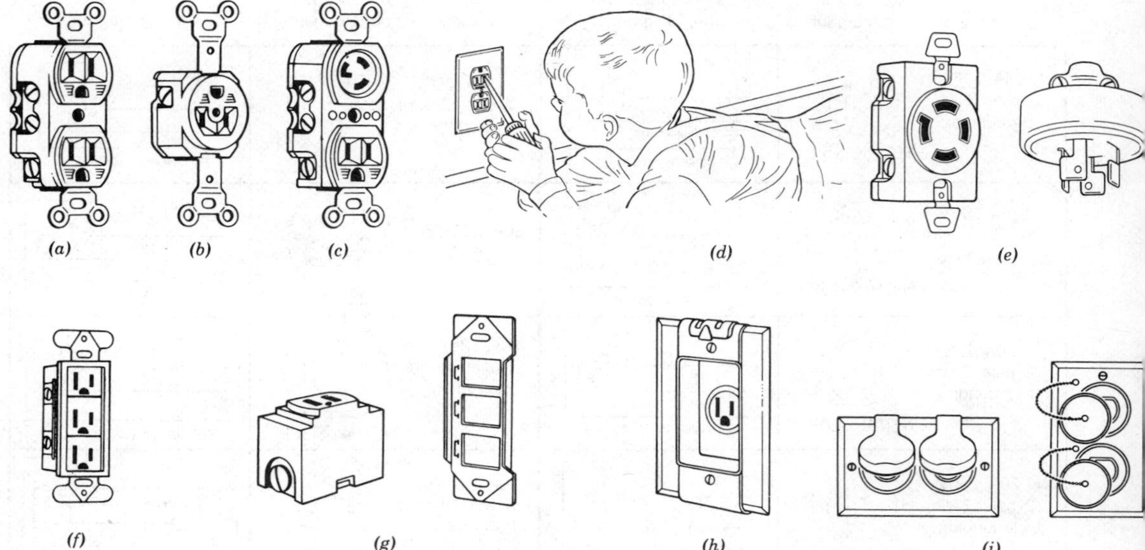

Fig. 16.42 Receptacle and cap types. (a) *A 2-pole, 3-wire 15 amp, 125 v, duplex grounding type;* (b) *the same as* (a) *except single;* (c) *the same as* (a) *except one normal and one locking device;* (d) *the same as* (a) *except safety type;* (e) *3-pole, 4-wire locking receptacle, 20 amp, 125/250 v, with matching cap;* (f) *triplex 15 amp grounding receptacle;* (g) *miniature (interchangeable) device with mounting strap;* (h) *clock outlet with hangar plate;* (i) *outdoor weatherproof receptacles.*

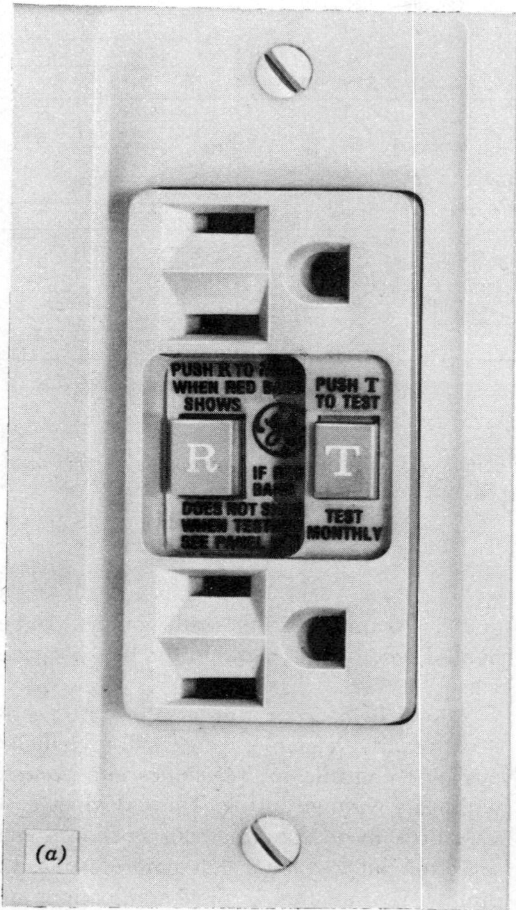

Fig. 16.43 Where it is desired to provide ground-fault protection at a single outlet, two choices are available—either to utilize a permanent, outlet-box installed unit as in (a), *or an add-on removable unit as in* (b). *The T and R buttons are for test and reset respectively. Courtesy of General Electric Company, Circuit Protective Device Department.*

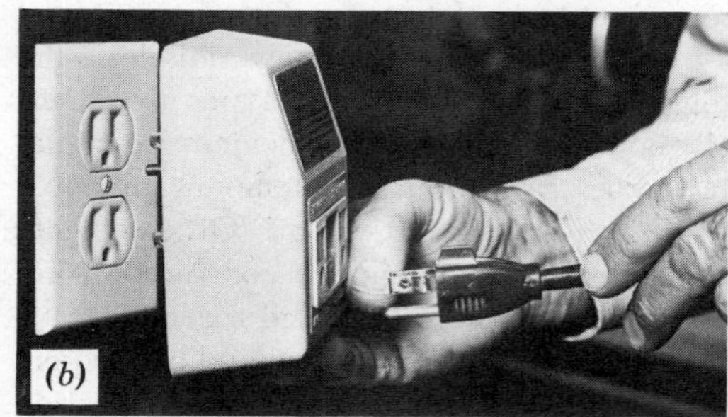

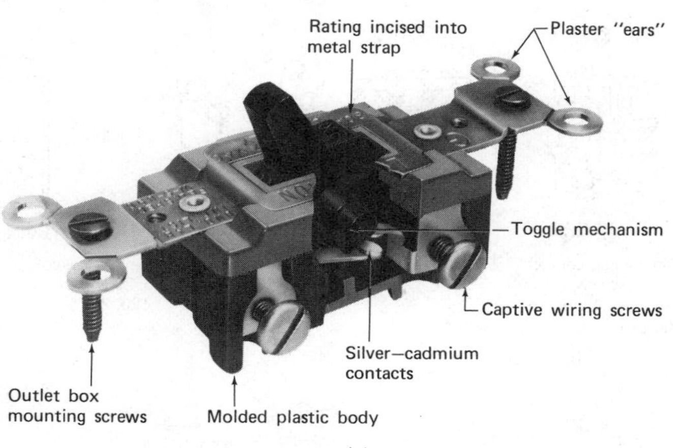

(a)

Fig. 16.44 *A cutaway of a specification grade, a-c, quiet toggle switch is shown in* (a). *This unit is rated 30 amps, 120/277 v, single pole, single throw. A Corbin lock tumbler unit intended for high-security areas is shown in* (b). *The rating is clearly visible on the front metal strap. A wall outlet box mounted time switch* (c) *is very useful for manual time control of load. The illustrated unit is rated 20 amp at 120 v. Photos are courtesy of:* (a) *Leviton Manufacturing Co.;* (b) *Arrow-Hart, Inc.;* (c) *M.H. Rhodes, Inc.*

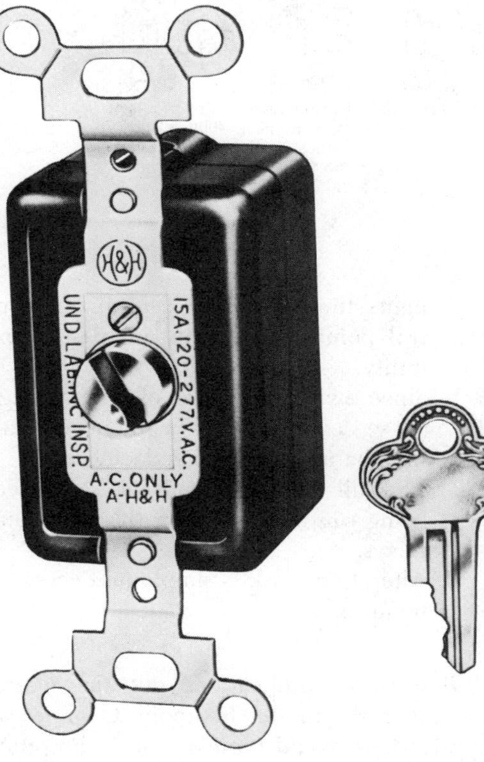

(b)

(c)

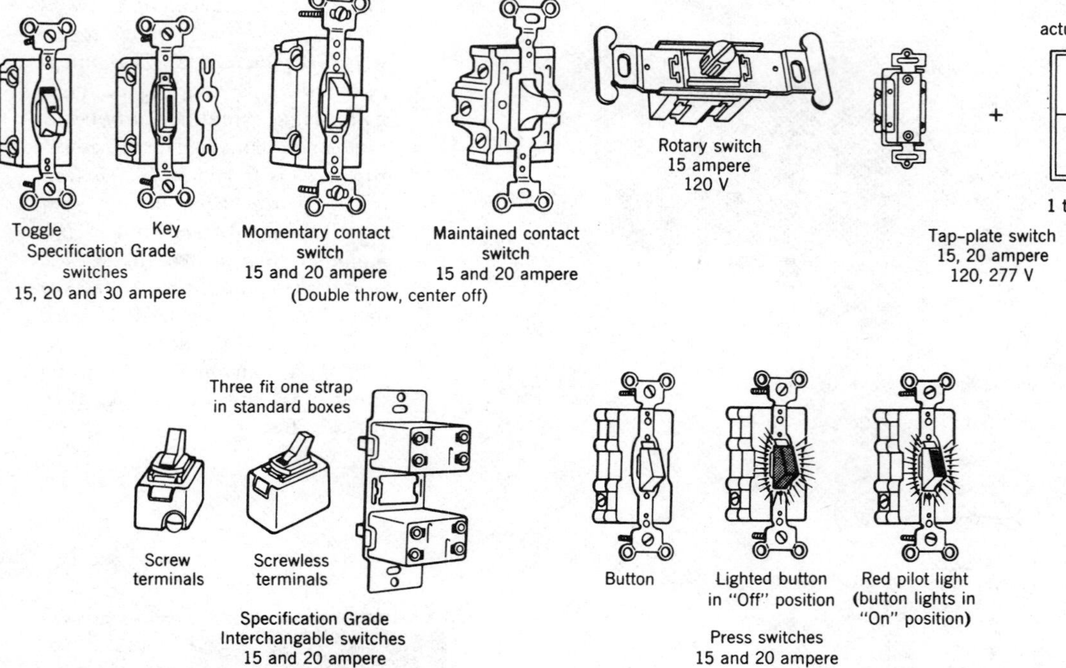

Fig. 16.45 *Typical branch circuit switch types.*

areas such as dining rooms, classrooms, and assembly rooms, where a lower illumination level is often acceptable and always desirable as an energy-conserving measure. In high-security areas, where the easily defeated normal key switch is inadequate, a tumbler lock controlled unit, similar to Figure 16.44*b*, can be used. Loads that can be timed-out, such as bathroom heaters and ventilating fans, can be controlled by the unit illustrated in Figure 16.44*c*. The common types mentioned above are illustrated in Figure 16.45.

16.26 Low-Voltage Switching

The switches illustrated above are all full-voltage types; that is, they are placed directly into the line and interrupt the load circuit. A system of switching that uses light-duty, low-voltage switches to control relays, which in turn control the load circuits, is shown in Figures 16.46 and 16.47. The advantages of this system are:

a. Permits local and remote control of lights, thus permitting centralizing control points and greatly increased flexibility.
b. Since switch legs are of low voltage (24 v), control wires without conduit may be used, greatly reducing costs.
c. The full voltage circuitry is shortened, being run only between the lights and relays.
d. Alteration work is simple and economical.

Residential applications put the master control in the master bedroom. Commercial applications would typically be in hospitals, schools, and industrial facilities where centralization of control is desirable, without the inconvenience of panel switching. The combination of local *and* centralized control is of great importance in energy conservation, since it permits lights to be controlled without the necessity of physically going to each switch point while at the same time retaining the local control so necessary for energy economy.

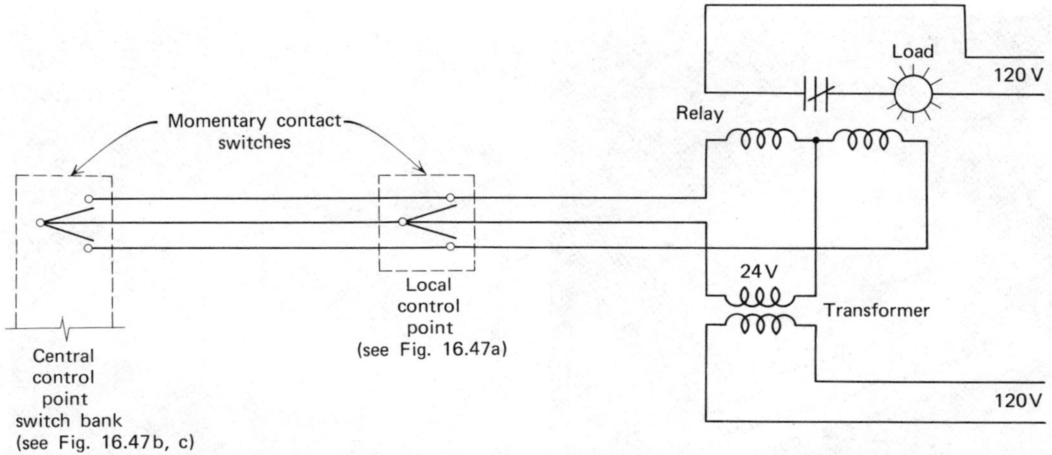

Fig. 16.46 *Low-voltage switching control. Multipoint control and central control are illustrated. The diagram shows the relays located at the load. Central relay cabinets are applicable to dense load areas.*

Fig. 16.47 *The basic low-voltage control device is a rocker-type illuminated momentary contact switch, with both contacts open* (a). *Switches can be ganged* (b), *or arranged with a rotary selector* (c).

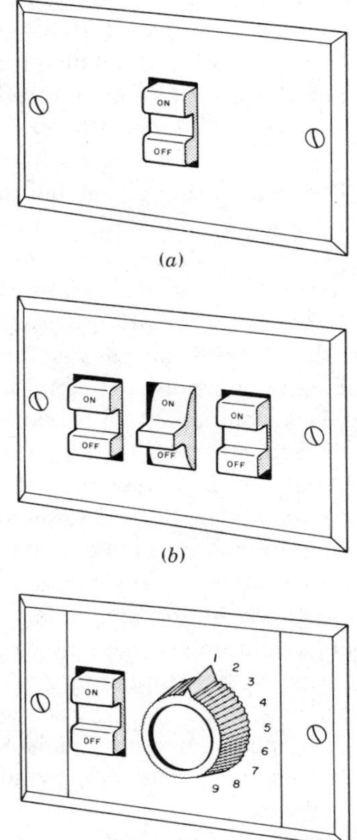

16.27 Outlet and Device Boxes

These boxes are generally of galvanized stamped sheet metal with multiple $\frac{1}{2}$ in., $\frac{3}{4}$ in., and several 1 in. conduit knockouts conveniently placed around the periphery and on the back. The most common sizes are the 4 in. square and 4 in. octagonal boxes used for fixtures, junctions, and devices, and the $4 \times 2\frac{1}{8}$ in. box used for single devices where no splicing is required. Box depths vary from $1\frac{1}{2}$ to 3 in. Generally, except for interchangeable devices, no more than one single device should be placed in a single-gang box. The number of wires permissible in various boxes is specified in the NEC. Nonmetallic boxes may be used with NM and NMC cable and with nonmetallic conduit installations. In wet locations and for outdoor work, cast-iron or cast-aluminum boxes are recommended.

Floor boxes for floor outlets are usually of cast-metal construction and are installed directly in the floor slab. An interesting design is shown in Figure 16.48 which alleviates the severe problem of outlet fouling during wet cleaning operations.

Fig. 16.48 A cleverly designed telescoping floor box locks in the "up" position during use and rests flush in the floor during periods of inactivity and when floors are cleaned. Courtesy of Maxicom Corp.

16.28 Lightning Protection Systems

The subject of lightning protection is a complex one, and the design of an adequate system is best left to specialists. Some of the initial considerations, NFPA requirements, and data on available materials will be discussed briefly below.

(*a*) The relevant standards are NFPA Bulletin No. 78, *Lightning Code* and Underwriters Laboratories Standard UL 96A. *Master Labeled Lightning Protection Systems.*

(*b*) The decision of whether or not to protect a structure depends on an evaluation of these factors:

1. Frequency and severity of thunder storms.
2. Value and nature of building and contents.
3. Hazard to building occupants.
4. Building exposure. Buildings in open and exposed areas are more susceptible to lightning than urban buildings.
5. Indirect effects. For example, loss of a water tower will seriously affect fire prevention and other services.

(*c*) If a decision is reached to protect a building, it should be done completely and properly, with UL label equipment (Label A and B) and UL approved installation (Label C). A partially protected building is in reality an improperly protected building, which may well be worse than one with no protection at all.

(*d*) The basic principle in lightning protection is to provide a metallic path to ground for the lightning stroke, since *there is no known method of protection that will prevent the occurrence of a lightning stroke.* This will prevent the stroke from passing through the nonconducting portions of a building. Such passage is accompanied by great heat and mechanical forces (see Fig. 16.49).

(*e*) Metal within a few feet of lightning conductors should be bonded to them to avoid arcing. Reinforcing bars in concrete construction should be electrically bonded and connected to the lightning conductors. If this is not done, a lightning stroke may cause severe damage due to the insulating gaps between the rods.

(*f*) A lightning arrester should be placed on all aerial service conductors, whether high or low voltage.

(*g*) Electrical and electronic equipment,

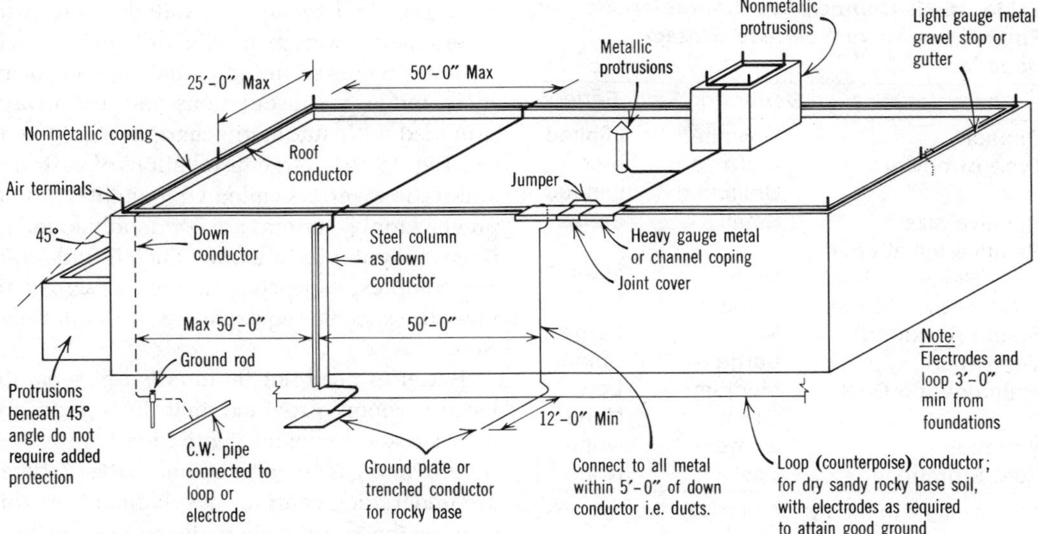

Fig. 16.49 *Typical lightning-protection diagram.*

Fig. 16.50 *Small ($2\frac{1}{2}$ in. sq × $1\frac{3}{8}$ in. deep) lightning arrester/surge protector useful in protecting electrical equipment from the destructiveness of electrical voltage surges. It allows the surge to bypass the protected equipment. Courtesy of Approved Lightning Protection.*

which is sensitive to voltage surges such as those accompanying lightning strokes, should be individually protected by surge arresters. A typical small unit of this type is illustrated in Figure 16.50. These units also protect equipment against voltage surges, transients, and spikes caused by phenomena other than lightning.

16.29 Emergency/Standby Power Equipment

The NEC makes a clear distinction between emergency power and standby power systems, covering the former in Article 700 and the latter in Article 750. The *equipment* involved in both is similar; the use is different. *Emergency systems* provide "power and illumination essential for *safety* to life and property, where such systems are legally required . . . ." Thus most states and cities require emergency power for exit lighting and egress lighting in places of assembly, plus power for equipment necessary for safety such as elevators, fire alarm systems, fire pumps, and so on, and extensive systems in health-care facilities.

Standby systems provide power "to selected loads (*other than those of Article 700, Emergency Systems*) in the event of failure of normal source . . . ." These systems are primarily intended to protect against property damage of financial loss and are not involved with safety considerations. For this reason the few authorities which legally require their installation do so only for water and sewage treatment plants. Private owners on the other hand do install standby systems when power interruption would cause serious damage to an

Table 16.12 *Comparative Characteristics of Engine-Generator Sets and Storage Batteries*

Characteristics	*Generator Set*	*Battery*
Capacity	Unlimited	Limited
Type of power	a-c or d-c	d-c[a]
Life	Unlimited	Limited
Relative size	Small	Large
Relative initial cost:		
(1) Below 50 kva	High	Medium
(2) Above 50 kva	Medium	High
Space Required	Medium	Large
Ventilation	Large	Small
Maintenance Cost	Medium	Low
Noise	Great	None
Vibration	Some	None
Fuel Storage	Yes[b]	None

[a] Alternating current can be provided, at high cost, by a static inverter.

[b] With natural gas engines, fuel storage is eliminated.

industrial process or a research project. System arrangements are discussed in Section 17.22. System equipment—which falls into two principal categories, that is, generators and battery installations—are discussed below and compared in Table 16.12.

(*a*) ***Engine-Generator Sets.*** An engine-generator set installation comprises basically three components: the fuel system including storage, if necessary; the set itself plus exhaust facilities; and the space housing the unit (see Fig. 16.51 and Table 16.13). The principal advantages of the engine-generator set are unlimited kva capacity, duration of power limited only by size of fuel tank, and, if properly maintained, indefinite life. Disadvantages are noise, vibration, nuisance of exhaust piping, need for constant maintenance and regular testing, and difficulties with fuel storage. Gasoline can only be stored for a year at most, and subsequent disposal is difficult. Diesel fuel keeps somewhat longer, but disposal is also difficult. Use of gas for a diesel or gas engine obviates the fuel storage problem but poses the alternate problem of availability of gas service during emergencies.

(*b*) ***Battery Equipment.*** Storage batteries are often used to supply limited amounts of emergency power, primarily for lighting. Such units are mounted in individual cabinets or in racks for larger installations and are always provided with automatic charging equipment (see Fig. 16.52). Large installations of batteries and static inverters called UPS (uninterrupted power supply) systems are commonly found in large computer installations. Such installations are complex, expensive, and are designed to meet the specific requirements of the installation.

Batteries need not be installed in separate battery rooms provided that live parts are guarded and sufficient ventilation is provided to prevent gas accumulation. Battery types are undergoing constant development. At this writing the types principally in use are lead acid, nickel cadmium, lead-antimony and lead-calcium cells. The choice depends on the application and is best left to a specialist. Batteries have the distinct advantage that they can be installed either in a central system with distribution by feeder of the battery power throughout the facility, or they can be installed in small package units around the building. Central systems are 24 to 125 v, d-c and normally feed nonfluorescent emergency lighting only. Individual packs are used often to supply a-c power via built-in inverters. The great disadvantage of battery systems is limited duration of power. The NEC requires that batteries maintain loads for a $1\frac{1}{2}$ hour *minimum,* but larger battery capacity is normally installed. Details of emergency lighting equipment are discussed in Section 22.27.

16.30 System Inspection

Each electric wiring system is inspected at least twice by the local inspection authorities; once after raceways (roughing) have been installed and before the wiring and closing in of walls, and once after the entire job is complete. The purpose of these inspections is to determine whether design, material, and installation techniques are meeting the national and local code requirements. Excellence of installation is the responsibility of the con-

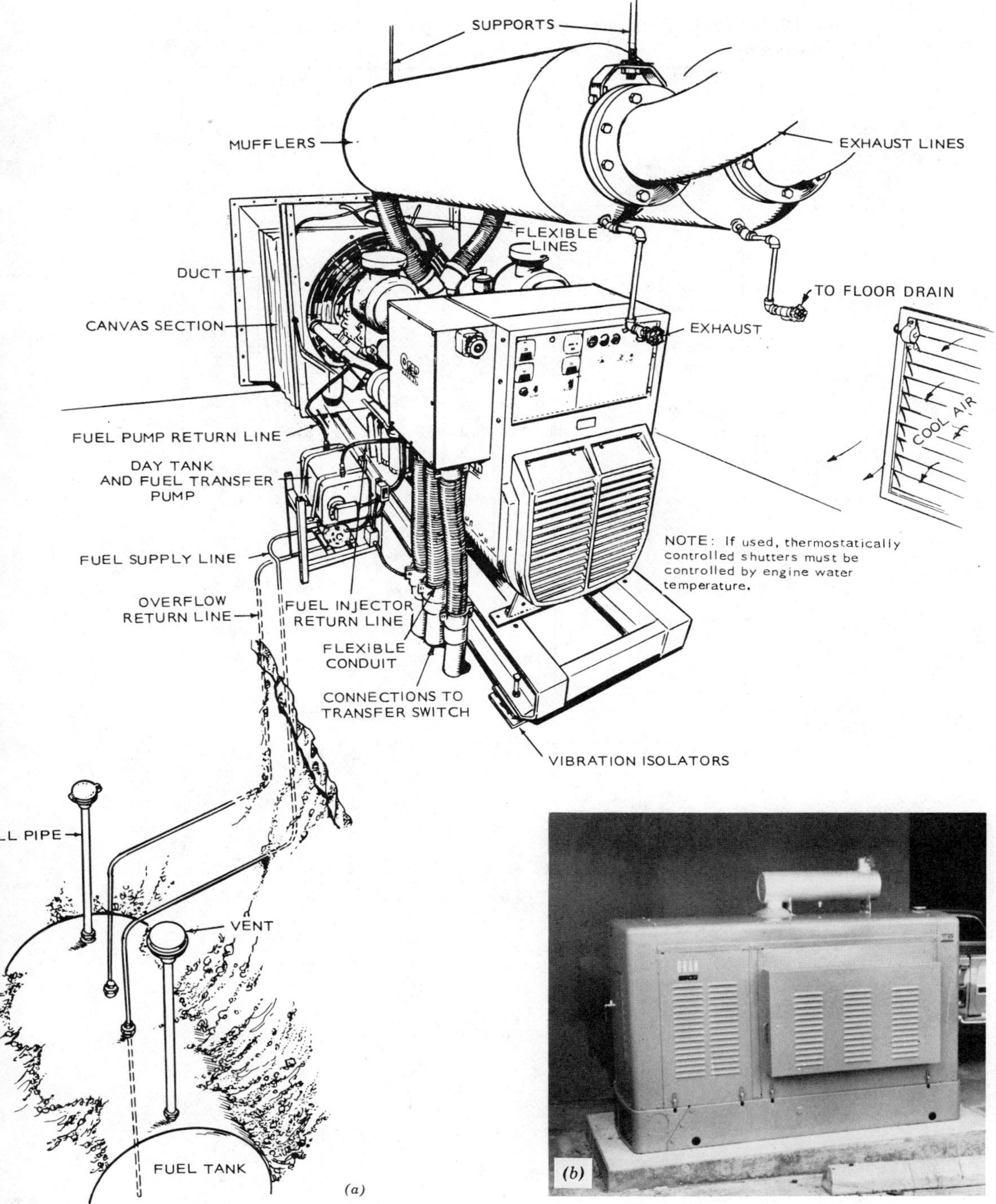

***Fig. 16.51** Typical engine-generator units and installation techniques. Illustrated in* (a) *is a radiator-cooled diesel installation. A housed 55 kw gasoline unit is shown in* (b)*, installed in an otherwise wasted space. Illustrations are Courtesy of ONAN Corp.*

Table 16.13 ***Basic Data, Representative Engine-Generator Sets***
Gas/Gasoline

Generator Ratings			Engine			Dimensions & Weight[b]			
Continuous Standby			*Fuel Consum. (gph)*	*Max. hp*	*Cool*	*Length*	*Width*	*Height*	*Net Weight*
kw	*kva @ 0.8 pf*	*60 Hz, volts a-c*[a]				*in./mm*	*in./mm*	*in./mm*	*lb/kg*
10	12.50	All packages are available in single- and three-phase from 120 through 6600 v on specification.	2.32	22.5	AIR	47/1194	27/686	27/686	950/431
15	18.75		3.21	32.0	AIR	47/1194	31/787	35/889	1150/522
23	29.00		3.90	49.5	RAD.	64/1626	24/610	36/914	1150/522
55	68.60		6.35	105.0	RAD.	82/2083	27/686	51/1295	2400/1089
85	106.00		9.85	172.0	RAD.	86/2184	37/940	58/1473	3600/1633
140	175.00		2130 CF/HR 1000 BTU	260.0	RAD.	114/2896	29/737	73/1854	5200/2359
170	212.50		21.00	290.0	RAD.	123/3124	81/2057	40/1016	7600/3447

Diesel

kw	kva @ 0.8 pf	60 Hz, volts a-c	Fuel Consum. (gph)	Max. hp	Cool	Length in./mm	Width in./mm	Height in./mm	Net Weight lb/kg
15	18.80		1.61	28.0	AIR	52/1321	26/660	40/1016	1150/522
35	43.70		3.42	61.0	RAD.	66/1676	25/635	45/1143	1800/817
60	75.00		5.31	105.0	RAD.	81/2057	24/610	51/1295	2800/1270
110	137.00		8.08	180.0	RAD.	101/2565	27/686	64/1626	4100/1860
250	312.50		18.37	420.0	RAD.	117/2972	45/1143	72/1829	7800/3538
500	625.00		40.00	750.0	RAD.	152/3861	69/1753	85/2159	10240/4645
725	908.00		62.00	1155.0	RAD.	178/4521	69/1753	90/2286	16780/7611
1000	1250.00		83.00	1540.0	RAD.	193/4902	69/1753	90/2286	20438/9271

Source. Data reproduced with permission from Katolight Corp.

[a] Electrical voltage ratings are: single-phase, 115/230 v; three-phase, 120/208 v; three-phase, 277/480 v.

[b] Dimensions are for unhoused units.

Fig. 16.52 *A large capacity installation of lead-calcium cells, rack mounted, is shown in* (a). *Cells can be added as desired to give the required amperage and voltage. A package unit consisting of an automatic battery charger and a nickel-cadmium battery pack is shown in* (b). *This unit will supply up to 50 amps at 24 v, d-c. The enclosure is 44 in. high, 25 in. wide, and 22 in. deep and contains knockouts for incoming a-c and outgoing d-c lines. A similar package unit except equipped with a d-c to a-c inverter is shown in* (c). *This unit will come on to the line automatically and will supply up to 2500 w,* a-c *at either 120 v or 277 v, thus making it suitable to feed fluorescent lighting. This unit, which utilizes lead-acid cells, also contains a battery charger. Photos are courtesy of:* (a) *Gould Inc., Industrial Battery Division;* (b) *LaMarche Mfg. Co.;* (c) *Standby Systems, Inc.*

tractor. The designer, however, must be completely familiar with installation work and the equipment's physical characteristics in order to properly design an electrical system. Such a design will not present the contractor with unwarranted difficulties. The designer must also be very wary of equipment substitutions by the contractor who, having submitted a bid on the basis of plans and specifications, should be required to supply the specified equipment.

17

Electric Wiring Design

17.1 General Considerations

In wiring design, as in other design, there are numerous possible solutions to each problem—some good, some fair, and some poor. Experience guides the designer to a solution that best suits the job, since it is his or her responsibility to establish the most economical design within the framework of the design criteria. Some of these criteria are discussed below.

(*a*) ***Flexibility.*** Every wiring system should incorporate sufficient flexibility of design in branch circuitry, feeders, and panels to accommodate all probable patterns, arrangements, and locations of electric loads. The degree of flexibility to be incorporated depends in large measure on the type of facility. Thus laboratories, research facilities, and small educational buildings require a great deal more flexibility than residential, office, and fixed-purpose industrial installations. As part of the design for flexibility, provision for expansion must be provided as experience has demonstrated that most facilities will grow, both physically and in electrical demand. It must, however, be emphasized that overdesign is as bad as underdesign, being wasteful of money and resources both initially and in operation.

(*b*) ***Reliability.*** The reliability of the electrical power within a facility is determined by two factors: the utility's service and the building's electrical system. The service record of the utility should be studied along with the economic impact of a power outage to determine whether, and to what extent, standby power equipment is justified (see Section 16.29). Emergency equipment, being concerned with the safety of a building's occupants, is determined by local, state, and national building codes and is largely independent of the electrical service arrangements. (In some cases two *independent* services can be used in lieu of emergency equipment backup for normal service.) Beyond the service point, the reliability of power is entirely dependent on the wiring system. Here, too, economic studies must be made to determine the quality of equipment and the amount of redundancy (duplicate equipment) to be installed. The subject of reliability is a complex one, and we can only state a few general principles here.

1. The reliability of an electric system is only as good as that of its weakest element. Therefore, it may be necessary to provide redundancy at selected weak points in the system.
2. The electrical service and the building's distribution system act together. An extremely reliable (and expensive) service is of little use if the power cannot reach the desired points.
3. Critical loads within the facility should be pinpointed to determine how best to reliably serve them, that is, by establishing reliable power paths to them or by furnishing individual standby power packages for them. The latter course is often chosen for critical health-care and computer installations.

4. The system must be so designed that equipment failures which will occur can be readily detected and corrected, in order that the cost and inconvenience of such unavoidable outages be minimized.

(*c*) ***Safety.*** Although rigid adherence to the requirement of the NEC and other applicable NFPA codes will assure an initially safe electrical installation, the designer must be constantly alert to such factors as electric hazards, caused by misuse or abuse of equipment or by equipment failure. Also, a thorough acquaintance with the size of equipment used will eliminate the oft-encountered physical hazard caused by obstruction of access spaces, passage, closets, and walls with electric equipment. Finally, lightning protection can be subsumed under the heading of safety; this topic is discussed in Section 16.28.

(*d*) ***Economic Factors.*** This item can readily be divided into two frequently interrelated items; first cost and operating cost. All other factors being equal, the first cost depends in large measure on whether the constructor is interested in minimum first cost or minimum owning cost. We have demonstrated that these two costs frequently stand in inverse relationship to one another (exceptions are mentioned below). Low first-cost equipment generally results in higher energy cost, higher maintenance cost, and shorter life. The decision, however, is not purely an economic one, inasmuch as the electrical energy cost factor in the operating-cost equation is directly related to energy consumption, with one exception. That exception is the utility's demand charge, which is discussed at length in Section 14.12. Means for minimizing this cost are covered in detail in Section 14.13, which is devoted to demand control equipment. Here there occurs a coincidence of reduction to both first cost and operating costs. Load leveling equipment permits the electrical distribution system to be sized without consideration of coincident load peaks, thus resulting in smaller equipment, operating more efficiently—near its full-load capacity. All other reductions in electric energy cost flow directly from the corresponding reduction in energy consumption.

(*e*) ***Energy Considerations.*** This factor is a complex one involving considerations of energy codes and budgets, energy conservation techniques, and energy control.

1. Energy codes and budgets. Buildings constructed with governmental participation will increasingly be subject to energy budget limitations expressed in Btu/square foot/year. Although the lion's share of this budget is taken by heating/cooling and lighting systems, the electrical distribution system will also be subject to conformity to stated codes. Among these are ASHRAE Standard 90-75, which is reproduced in part in Section 17.5(a).
2. Energy control. This item consists largely of providing facilities for metering current flow, voltages, power and energy consumption. These may take the form of built-in metering and/or meter points at selected locations throughout the electrical system, to permit accurate load analysis. Meters, both instantaneous reading and recording types, can provide data on equipment loading, load patterns, load coincidence, power factor, load voltage, power demand, and energy consumption.
3. Energy conservation techniques. These can effectively and economically be built into the design rather than being applied to any existing design. Thus, a study of the intended building use pattern may lead to a design that utilizes separate systems for daytime use and for night/weekend use. Such a design permits shutdown of whole systems rather than operate at very low load with concomitant high losses and low power factor. Also, providing relatively small items such as coffee-vending machines and eliminating coffee-pot outlets, connecting water coolers to time clocks, and the like, are included in this category and at this stage. All of the energy

considerations are discussed in detail in Section 17.5.

(*f*) ***Space Allocations.*** The general impression that electrical equipment is small and easily concealed is accurate only for wire and conduit. Panels, motor control centers, busduct, distribution centers, switchboards, transformers, and so on can be large, bulky, noisy, and highly sensitive to tampering and vandalism. Thus, space allocations must be concerned with maintenance ease, ventilation, expandability, centrality (to limit length of runs), limitation of access, and noise, in addition to the basic item of space adequacy.

(*g*) ***Special Considerations.*** These depend on the specialized nature of certain facilities and may include items such as security, central and/or remote controls, interconnection with other facilities, and the like.

17.2 Load Estimating

When initiating the wiring design of a building it is important to be able to estimate the total building load in order to plan such spaces as transformer rooms, chases, and closets. This information is also required by the local power company well in advance of the start of construction. An exact load total can be made after completing the design but, since this is often several months later, a good preliminary estimate is required. Such an estimate can be made from the figures given in Table 17.1. These figures are average. Where it appears that the building will have heavier or lighter loads because of lighting levels or other factors, the figures should be modified accordingly. Having established the figures in the individual categories, they should be added together without application of demand or diversity factors in order to obtain the maximum demand load for which the building service equipment must be sized, in the absence of electric load control (load leveling) equipment. At this point an analysis must be made to determine the feasibility of incorporating such equipment into the facility (see Section 14.13). Input to this study includes the utility's complete rate schedule, including all penalty clauses, a detailed analysis of the building's equipment load patterns, and any external constraints such as maximum loads imposed by power and energy budgets.

Equipment load patterns must be carefully analyzed because they determine a load's "sheddability." Thus, for kitchen equipment, load interruption may be undesirable but shifting of cooking time by a half hour is entirely feasible. On the other hand, for HVAC equipment, building thermal inertia and "stretching" maximum and minimum temperatures and humidities permit considerable load control without adverse effects. Also, as explained in Section 14.13, the degree and duration of load shedding is a function of the type of control equipment utilized. It is well to repeat here what is stated there—load control affects maximum demand, with only minor effect on total energy consumption.

The external constraints referred to are the energy budgets imposed by codes, legislation, and funding bodies. After the above load control analysis is completed, or simultaneously with it, a building energy consumption analysis must be performed. This may be done manually, although the numerous computer programs available are considerably more accurate. The results of this analysis will indicate whether the annual electrical energy budget is being exceeded. If so, loads will have to be modified by reconsideration of projected systems and system criteria, by incorporating energy-conservation devices and techniques into the electrical system and by drawing up energy-use guidelines that will be applied when the building is occupied. Since this last item depends for its success on the day-to-day voluntary actions of the building's occupants, it should not be considered a major conservation source. Conservation measures fall into three categories, and each one will be covered separately in the relevant discussions. They are:

a. Electric wiring system, Section 17.5.
b. Electric lighting, Sections 18.19, 19.7, 19.26, 20.5, 20.7, 21.26, and 21.28.

Table 17.1 ***Electric Load Estimating***[a]

I	II	III	IV	V	VI
	Volt-Amperes per Square Foot[b]				Ten Year Percent Load Growth
		Misc.	Air Conditioning[f]		
Type of Occupancy	*Lighting*[c,g]	*Power*[d]	*Electric*	*Nonelectric*	
Auditorium					
General	1.0–2.5	0	12–20	5–8	20–40
Stage	20–40	0.25			
Art gallery	4–6	0.5	5–7	2.0–3.2	20–40
Bank	2.5–4.5	2.0	5–7	2.0–3.2	30–50
Cafeteria	3–5	0.5	6–10	2.5–4.5	20–40
Church	1.5–3.0	0.5	5–7	2.0–3.2	10–30
Computer area	3–6	1.5[e]	12–20	5–8	50–200
Department store					
Basement	4–6	1.5			
Main floor	2.5–4.5	1.0	5–7	2.0–3.2	50–100
Upper floor	2–4	0.5			
Dwelling 0–3000 sq ft	3.0	0.5	—	—	
(not hotel) 3000–120,000	2.0	0.25	—	—	50–100
above 120,000	1.0	0.15	—	—	
Garage (commercial)	0.5	0.15	—	—	10–30
Hospital	2–3	1.0	5–7	2.0–3.2	40–80
Hotel					
Lobby	6–8	0.5	5–8	2.0–3.5	30–60
Rooms (no cooking)	1.0–2.5	0.5	3–5	1.5–2.5	
Industrial loft building	1.5–2.5	1.0	—	—	50–100
Laboratories	3–5	5–20	6–10	2.5–4.5	100–300
Library	2.5–4.5	0.5	5–7	2.2–3.2	30–40
Medical center	2.5–4.5	1.5	4–7	1.5–3.2	50–80
Motel	1.0–2.5	0.5	—	—	30–60
Office building	2.5–4.0	2.0	4–7	1.5–3.2	40–80
Restaurant	1.5–2.5	0.25	6–10	2.5–4.5	20–40
School	2–4	1.5	3.5–5.0	1.5–2.2	50–80
Shops					
Barber & beauty	3–5	1.0	5–9	2–4	
Dress	2–5	0.5			
Drug	3.0	0.5	4–7	1.5–3.2	40–80
Five and ten	3.0	0.5			
Hat, shoe, specialty	3.0	0.5			
Warehouse (storage)	0.25–1.0	0.25	—	—	10–30
In the above except single dwellings:					
Halls, closets, corridors,	0.5	—	—	—	
storage spaces	0.25	—	—	—	

[a] Figures assume energy-conservation techniques applied.

[b] These figures do not include allowance for future loads.

[c] The figures given in Article 220 of the NEC are minimum figures for calculation of electric feeder sizes.

[d] These figures are based on experience and must be applied judiciously.

[e] This figure does not include the power used by the computer.

[f] Includes loads of air-handling equipment and pumps.

[g] See Figure 20.3 for luminous efficacy of various systems of illumination.

c. Vertical transportation, Sections 23.16, 24.1, 24.2, 24.3, and 25.12.

The electrical loads in any facility can be categorized as follows:

a. Lighting.
b. Miscellaneous power, which includes convenience outlets and small motors.
c. Heating, ventilating, and air conditioning (HVAC).
d. Plumbing or sanitary equipment.
e. Vertical transportation equipment.
f. Kitchen equipment.
g. Special equipment.

Category (*a*) is self-explanatory and is covered by column II of Table 17.1.

Category (*b*), column III includes, in addition to receptacles and small motors, such items as small business machines, plug-in heaters, water fountains, and so on.

Category (*c*), column IV includes all loads imposed by the HVAC equipment. Included therefore are fuel pumps, boiler motors, condensate pumps, all heaters, blower motors, exhaust fans, and so forth. Also included in column IV, for air conditioning loads, is refrigeration compressors. This item is omitted in column V since the air conditioning utilizes absorption machines that do not use electricity for primary power. When air conditioning is not anticipated, the HVAC load is still appreciable because of heating and ventilating (H and V) requirements. A rough estimate for this load would be $\frac{2}{3}$ or 65% of the loads in column V.

Category (*d*) includes all loads associated with the water and sanitary system, including house water pumps, air compressors and vacuum pumps, sump pumps and ejectors, well pumps and fire pumps, water heaters and pneumatic tubes, plus such special items as display fountain pumps. Since these loads vary widely with local conditions and with facility design as much as type of facility, an estimate cannot be made on a volt-amperes per square foot basis, by type of building. For this reason, no figure is included in Table 17.1. If actual data cannot be used, it is helpful to remember that plumbing loads are relatively small, rarely exceeding 20% of the HVAC load, though, for the most part, they are unrelated to it.

Category (*e*), vertical transportation, is also obviously unrelated to square footage and therefore cannot be tabulated in Table 17.1. These loads, including elevators, moving stairs, and dumbwaiters, can be estimated from the data given in the vertical transportation chapters of this book.

Category (*f*), kitchen equipment, is also not included in Table 17.1, though obviously present in all restaurants, in most hospitals, and in some office and religious use buildings.

Table 17.2 ***Nominal Service Size in Amperes***

Nominal service sizes are 100A, 150A, 200A, 400A, 600A.

Facility	*Area in Square Feet* 1000	2000	5000	10000	*Remarks*
Single-phase, 120/240 v, 3-wire					
Residence	100A	100A	150A	—	Minimum 100A
Store[a]	100A	150A	—	—	
School	100A	100A	150A	—	
Church[a]	100A	150A	—	—	
3-phase, 120/208 v, 4-wire					
Apartment House	—	—	100A	150A	
Hospital[a]	—	—	200A	400A	
Office[a]	—	100A	400A	600A	
Store[a]	—	100A	400A	600A	
School	—	100A	100A	200A	

[a] Fully air conditioned using electric-driven compressors. Based on figures in Table 17.1.

The reason for this omission is that the primary power for the major load, the cooking equipment, may be either gas or electric. Other large energy-use equipment such as dishwashers can be electric, gas, or steam fed. Furthermore, no correlation can be made between facility type, area, and load, even if electrically powered, since population and schedule are also major factors. When kitchens are planned, a kitchen consultant or other experienced planner can usually supply an estimate of the electric power requirement, which can then be added to the other figures.

Category (g), special equipment, is so variegated that no figures can be listed. Under this title is subsumed such items as laboratory equipment, shop loads, display area loads, floodlighting, canopy heaters, display window lighting, and so on. This load data must be gathered for individual items of equipment and added to the foregoing totals.

Table 17.2 gives a tabulation of service entrance sizes in amperes, based on single- and 3-phase service for typical occupancies. These figures are for quick estimate purposes and should be adjusted after the design is completed.

17.3 System Voltage

There are several systems of voltage commonly available in the United States and Canada. The simplest of these is:

(*a*) 120 v, single-phase, 2-wire (see Fig. 17.1*a*). This is used for the smallest of facilities such as small residences, out-buildings, and isolated small loads up to 6 kva. Load is calculated by multiplying current and voltage. For 60-amp service, which this type service is normally limited to, no more than 50 amp are usually drawn. Thus

$$VI = 120 \times 50 = 6000 \text{ va} = 6 \text{ kva}$$

The nominal system voltage is 120 v, although it is also referred to as 110 v and 115 v (see Fig. 17.1*a*).

For somewhat heavier loads, the system normally used is:

(*b*) 120/240 v, single-phase, 3-wire (see Fig. 17.1*b*). The Code requires that all residences with five or more 2-wire circuits have a minimum of 100 amp, 3-wire service. Service disconnect for 100-amp service would be a 100 amp, 2-pole and solid neutral switch, fused at no more than 80% of rating, or 80 amp. This is usually written 100A, 2P & SN, 80AF. This service is used principally for residences, small stores, and other occupancies where the load does not exceed 80 amp or 19.2 kva. Load is calculated thus:

$$\text{kva} = \frac{V \times I}{1000} = \frac{240 \times 80}{1000} = 19.2$$

See Figure 17.1*b*-1. Although it may appear otherwise, the neutral carries no more than full-load current. Note that each "hot leg" of the 3-wire system carries line current. Thus, total load can also be calculated:

$$\text{Load kva} = \text{twice load on each line}$$

Assuming a balanced 80-amp load

$$\begin{aligned}\text{Total kva} &= 2 \times 80 \text{ amp} \times 120 \text{ v} \\ &\doteq 2 \times 9{,}600 \text{ va} = 19{,}200 \text{ va} \\ &= 19.2 \text{ kva}\end{aligned}$$

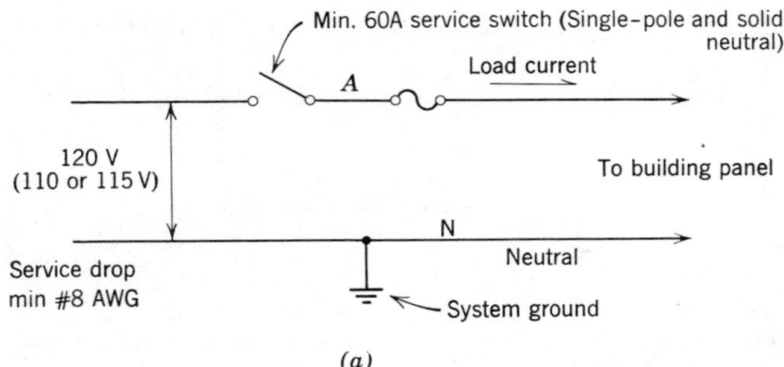

Fig. 17.1 (a) *120 v, single-phase, 2-wire service. This is also the arrangement of the usual branch circuit.*

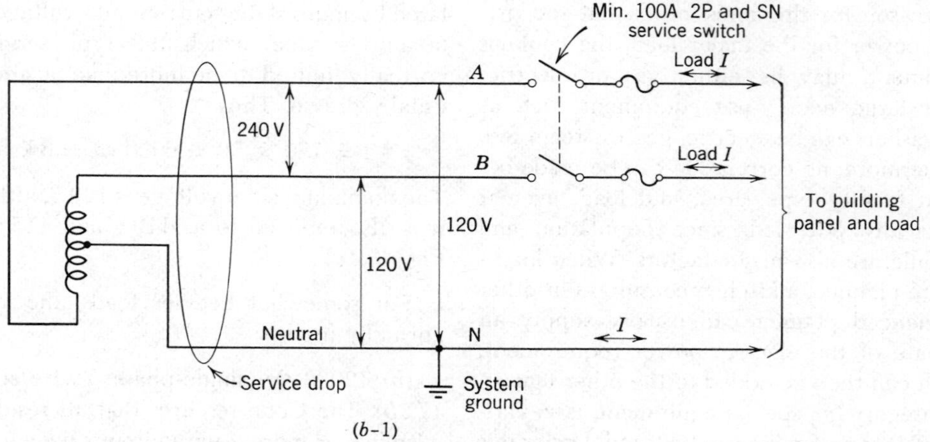

Fig. 17.1 (b-1) *120/240 v, single-phase, 3-wire service. The single phase transformer is center-tapped to establish a neutral.* The neutral connection is always grounded.

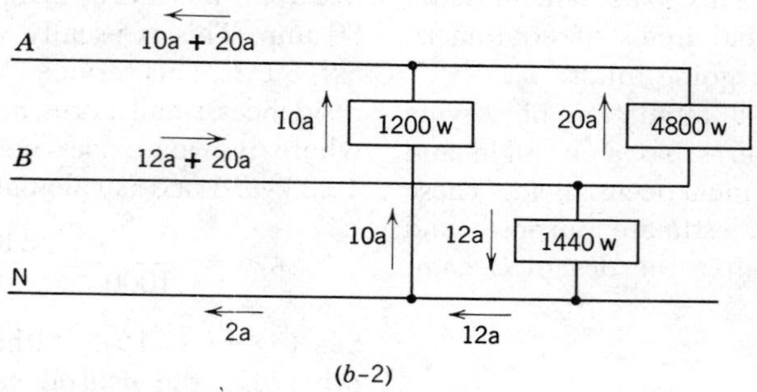

Fig. 17.1 (b-2) *Note that the neutral carries the difference in current between the* A *and* B *legs and therefore a maximum equal to the current in one of the legs (when the other is zero).*

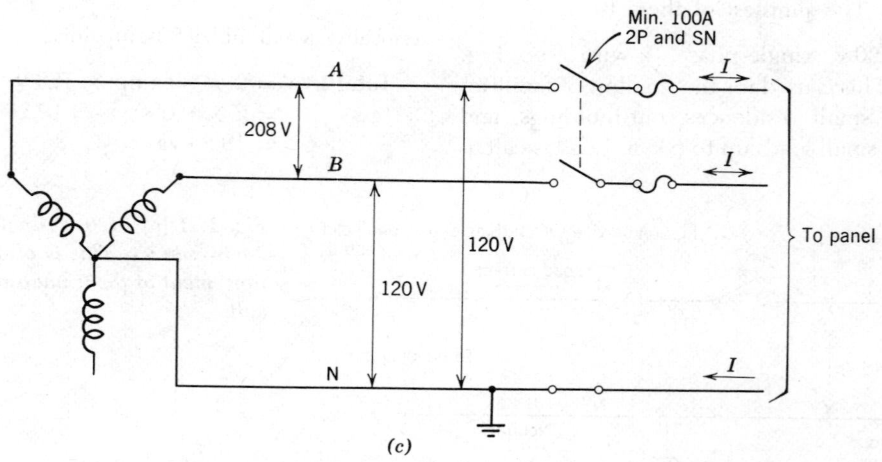

Fig. 17.1 (c) *120/208 v, single-phase, 3-wire service. This arrangement comprises two-thirds of the full 120/208 v, 3φ 4-wire connection shown in Figure 17.1*d-1, *below.*

If load is unbalanced, with say 30 amperes in one line and 50 amperes in the second, total load is

$$120 \times 30 + 120 \times 50 = 3600 + 6000$$
$$= 9600 \text{ va} = 9.6 \text{ kva}$$

Depending on the rating of the service transformer, system voltages can be 120/240, 115/230, or 110/220, although 120/240 is the accepted industry standard. Loads that are 120 v cause a current only in one line. Loads that are 240 v, such as a clothes dryer, cause current in both lines. For example, to find the line currents caused by the 3 loads shown on Figure 17.1*b*-2.

120 v, 1200 w iron on line *A*
120 v, 1440 w hair dryer on line *B*
240 v, 4800 w dryer on lines *A* and *B*

we calculate

$$I_A = \frac{1200}{120} = 10 \text{ amperes}$$

$$I_B = \frac{1440}{120} = 12 \text{ amperes}$$

$$I_{AB} = \frac{4800}{240} = 20 \text{ amperes}$$

Note that the neutral only carries the unbalance of 2 amperes.

Total current in $A = 20 + 10 = 30$ amperes
Total current in $B = 20 + 12 = 32$ amperes
Total load $= 120(30) + 120(32)$
$= 3600 + 3840 = 7440$ w
Loads are $1200 + 1440 + 4800 = 7440$ w

The 120/240 v, single-phase system comes from a center-tapped single transformer. When two hot legs and a neutral of a 3-phase system are taken, a similar system is obtained with a rating of:

(*c*) 120/208 v, single-phase, 3-wire (see Fig. 17.1*c*). This system is really part of a 3-phase system and is most often found *within* a building with 3-phase distribution, and is used to serve a small load that does not require 3-phase, 4-wire. Calculation of loads and line currents is considerably more complex than in (*b*) above because of the 120° phase displacement between phases A and B. Here, as before, the neutral will carry no more than line current whether the system is balanced or not.

When feeding large facilities, a 3-phase, 4-wire system is normally employed. This is usually:

(*d*) 120/208 v 3-phase, 4-wire (see Fig. 17.1*d*). This system is the most widely used 3-phase arrangement and is applicable to all facilities except the very largest ones. In those, lengths of feeders and sizes of loads become so great that a higher voltage is required. In this system, 120 v loads such as lighting, small machines, receptacles, and so on, are fed at 120 v by connection between each phase leg (see Fig. 17.1*d*-2) and neutral. Motors larger than $\frac{1}{2}$ hp and all 3-phase loads are fed at 208 v by connection between the 3-phase legs. Single-phase, 208 v loads such as heaters are accommodated as in (*c*) above, by connection between 2 phase legs. Such loads are often referred to as "2 pole" loads, alluding to the 2-pole current breakers used to feed them. Although this system is nominally 120/208 v, it is also used as 115/200. However, this is inadvisable because of nominal motor voltage, which will be discussed below.

Where buildings are large, either horizontally or vertically, lighting is principally fluorescent, and 120 v load does not exceed one-third of the total load, a more economical system of voltages is available. This system is:

(*e*) 277/480 v, 3-phase, 4-wire (see Fig. 17.1*e*). It utilizes 277 v fluorescent lighting, 480 v machinery, and small (3 to 25 kva) dry-type closet-installed transformers to step down from 480 to 120 v for supplying receptacles and other 120 v loads. This system is ideally suited to multistory office buildings and large single-level or multilevel industrial buildings. Savings are generated by the smaller feeder and conduit sizes and smaller switchgear, which more than offset the additional cost of step-down transformers for the 120 v load.

As an example of the savings possible with this system, let us consider the wiring required for a 15 kw heater.

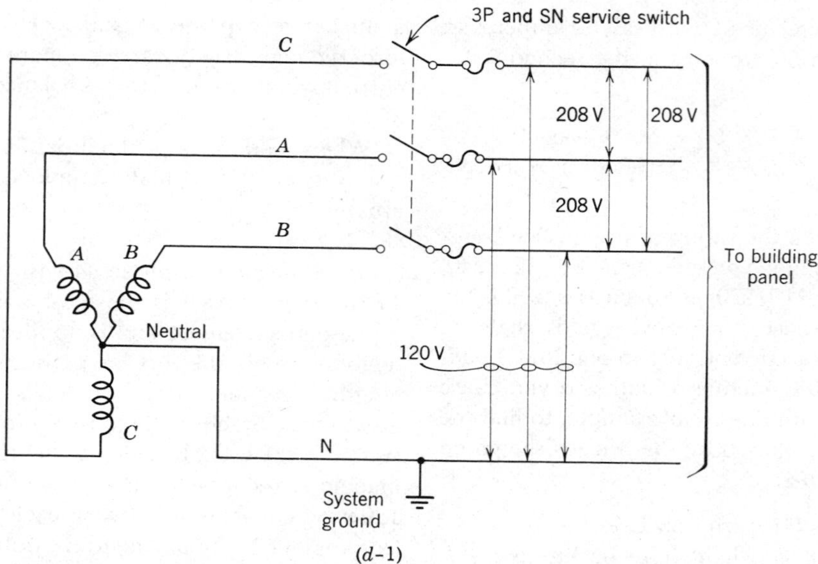

Fig. 17.1 (d-1) *120/208 v, 3-phase, 4-wire system. The neutral connection is connected to the system ground and is* not *broken by the service switch.*

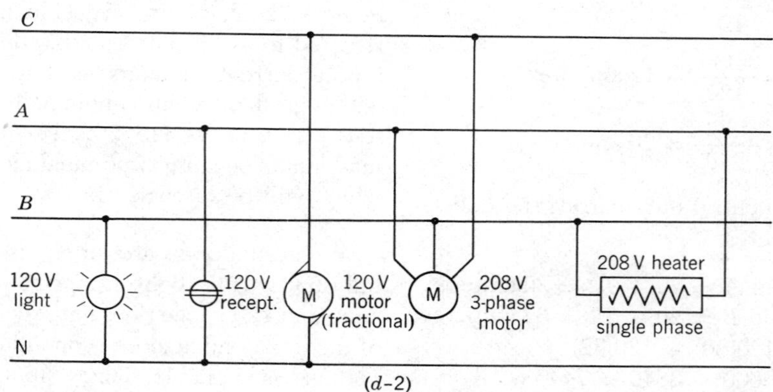

Fig. 17.1 (d-2) *The flexibility of the 120/208 v system is illustrated and accounts for its wide application. Loads are shown schematically. In practice, the loads are fed via protective devices in the panel. These are omitted here for clarity.* A, B, C, *and* N *represent the panel buses.*

At 3 ϕ, 208 v:

$$I = \frac{15{,}000 \text{ w}}{\sqrt{3} \times 208 \text{ v}} = 42 \text{ amperes}$$

requiring No. 8 RHW wire (45A capacity) at 208 v.

At 480 v:

$$I = \frac{15{,}000 \text{ w}}{\sqrt{3} \times 480 \text{ v}} = 18 \text{ amperes}$$

requiring only No. 12 RHW wire (20A capacity). This system is also frequently referred to as 265/460 v and 255/440 v.

(*f*) 2400/4160 v, 3-phase, 4-wire systems are only used in very large commercial buildings or in industrial buildings with machinery

Voltage above 150 v to ground is generally avoided in residential branch circuits, but may be used in commercial and industrial facilities, within the guidelines established by the NEC.

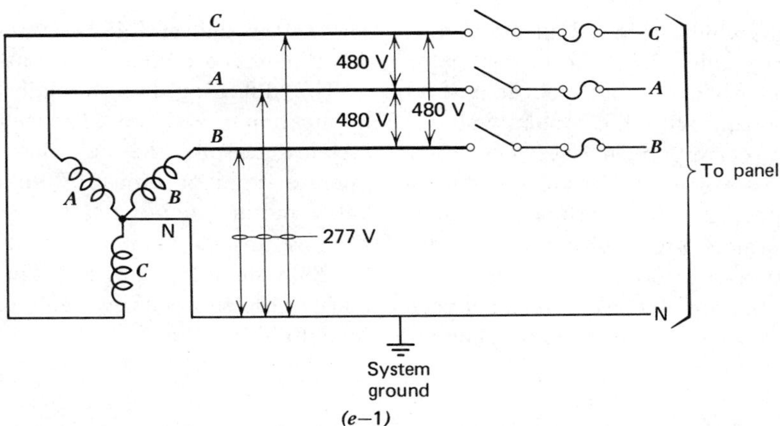

Fig. 17.1 (e-1) *277/480 v, 3-phase, 4-wire service system. The system is identical to the 120/208 v system shown in* (d-1) *except for voltages.*

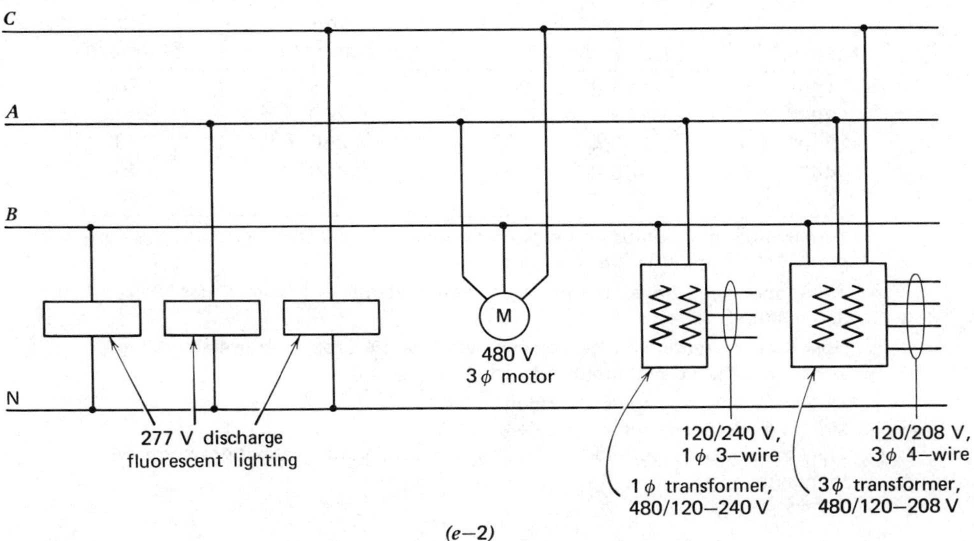

Fig. 17.1 (e-2) *Normal load arrangement for this system is illustrated. The lighting can be fluorescent or HID (mercury, metal-halide, sodium). Transformers, either single or 3-phase, supply 120 v for receptacles and 208 v for loads requiring that voltage.*

requiring these voltages. The cost of running this voltage feeder within a building is high because of NEC requirements and the inherent high cost of 5-kv equipment. A detailed cost and engineering analysis by a competent engineer is required for each case. Voltages above this level are widely used in large industrial plants and are beyond this text's consideration.

Reference was made above to the varied voltages assigned to the same voltage system. Thus, at the lowest level, we have 110, 115, and 120 v, at the next level 200, 208, 220, 230, and 240 v, at the next 255, 265, and 277 v, and finally 440, 465, and 480 v. These voltage differences arise because of the difference between transformer voltage standards, which establish the *system* voltage, and motor voltage standards, which govern *utilization* voltage level (see Table 17.3). The newer motor voltage standards are established at a level that is consonant with the system voltage. Note the close correspondence between motor standard voltage and system voltage with a normal 4% feeder voltage drop. Also, since most transformers are equipped with taps which allow voltage adjustment, very close correlation can normally be obtained between supply and device voltage ratings. Thus we see that on 240 and 480 v systems, 230 and 460 v motors are completely suitable.

The difficulty that normally arises is in application of 230 and 240 v motors to 208 v systems. Despite the fact that motors will operate at plus or minus 10% voltage, 230 and 240 v motors *should not* be used on 208 v systems. Instead, motors specifically wound for 200 v should be specified. For a brief summary of the effects of over and under voltage, refer to Table 17.4.

Table 17.3 ***System and Utilization Voltages***[a]

Standard Voltages[b]			
System Voltage (Transformers)		Utilization Voltage (Motors)	
Nominal	With 4% Drop[c]	New Standard	Old Standard
120	115.2	115	110
208[d]	199.7	200	208
240[d]	230.4	230	220
480	460.8	460	440
600	576.0	575	550

[a]To eliminate any confusion between system and utilization voltages, the current NEMA standards are tabulated above.

[b]When specifying transformers, use system voltages; for motors, use utilization voltage.

[c]Note that utilization voltage corresponds to a 4% drop from system voltage, well within the normal motor tolerance.

[d]Motors for 208 v systems are rated 200 v.
Motors for 240 v systems are rated 230 v.
They cannot be used interchangeably without seriously affecting motor performance.

Table 17.4 ***Effects of Overvoltage and Undervoltage***

	Condition		
Load	10% Undervoltage	10% Overvoltage	Remarks
Lighting			
Incandescent	Output reduced 30%	Life reduced 67%	See Fig. 19.21
Fluorescent	Output reduced, poor start	Ballast overheating	
Mercury	Low output, poor start	Ballast overheating	
Motors	20% lower torque, hotter operation, reduced life, overloading	High starting current, excessive starting torque, higher noise	See Fig. 16.33
Heaters	20% reduction in output	Overheating, shorter life	
Small tools	Stalling, low power	Reduced life, burned windings & bearings	

17.4 Grounding and Ground-Fault Protection

The vast majority of secondary wiring systems are grounded. The reasons for this arrangement are several and varied. Among them are

a. To prevent sustained contact between the low-voltage secondary system and the high-voltage primary system in the event of an insulation failure. Such contact could cause a breakdown of the secondary system insulation and severely endanger the system's users.
b. To prevent single grounds from going unnoticed until a second ground occurs, which will extensively disable the secondary system.
c. To permit locating ground faults with ease.
d. To protect against voltage surges.
e. To establish a neutral at zero potential for safety and for reference.

Points b to d are quite technical and a full explanation is beyond the purpose of this book. Point e requires that the neutral is

1. Never interrupted by switches or other devices.
2. Connected to ground only at one point—the service entrance.
3. Color coded white or natural gray for easy recognition.

A typical service-grounding diagram is given in Figure 17.2. Universal acceptance and use of grounded secondary 120 v systems *introduces* another shock hazard while eliminating the dangers described above. This is shown in Figure 17.3*a*. An accidental fault within an appliance can connect the metal case of the appliance to the line. This may readily occur with such common devices as an electric drill,

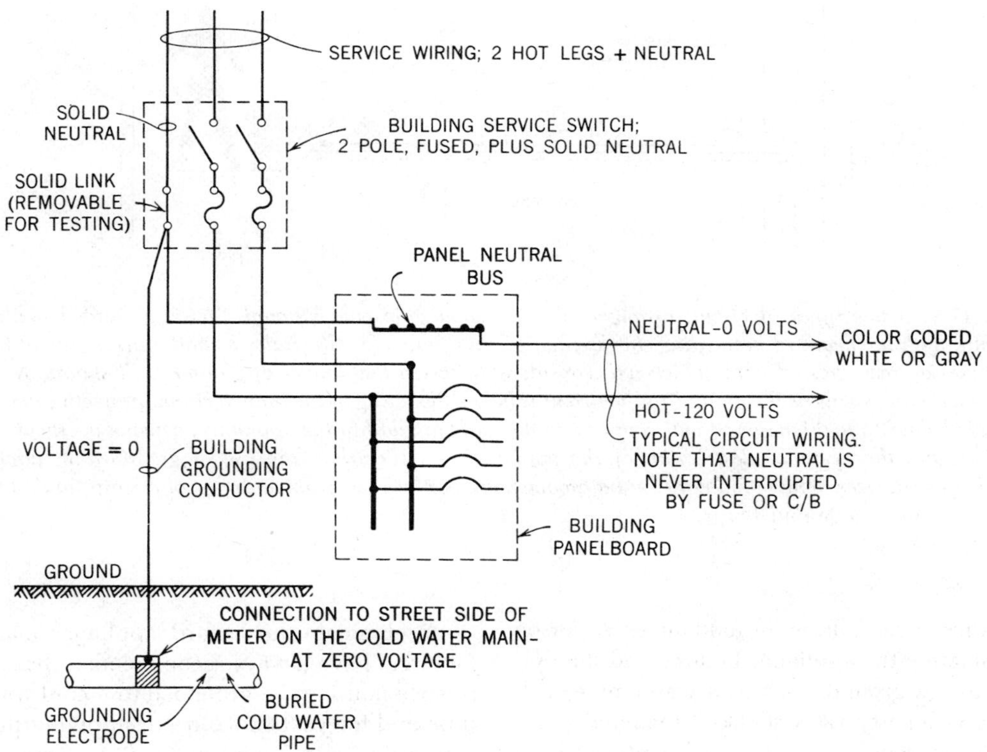

Fig. 17.2 *Typical service grounding arrangement. Note that the grounded neutral is continuous throughout. The ground bus, if present, is separate and distinct from the neutral bus, and both are grounded at the service entrance point.*

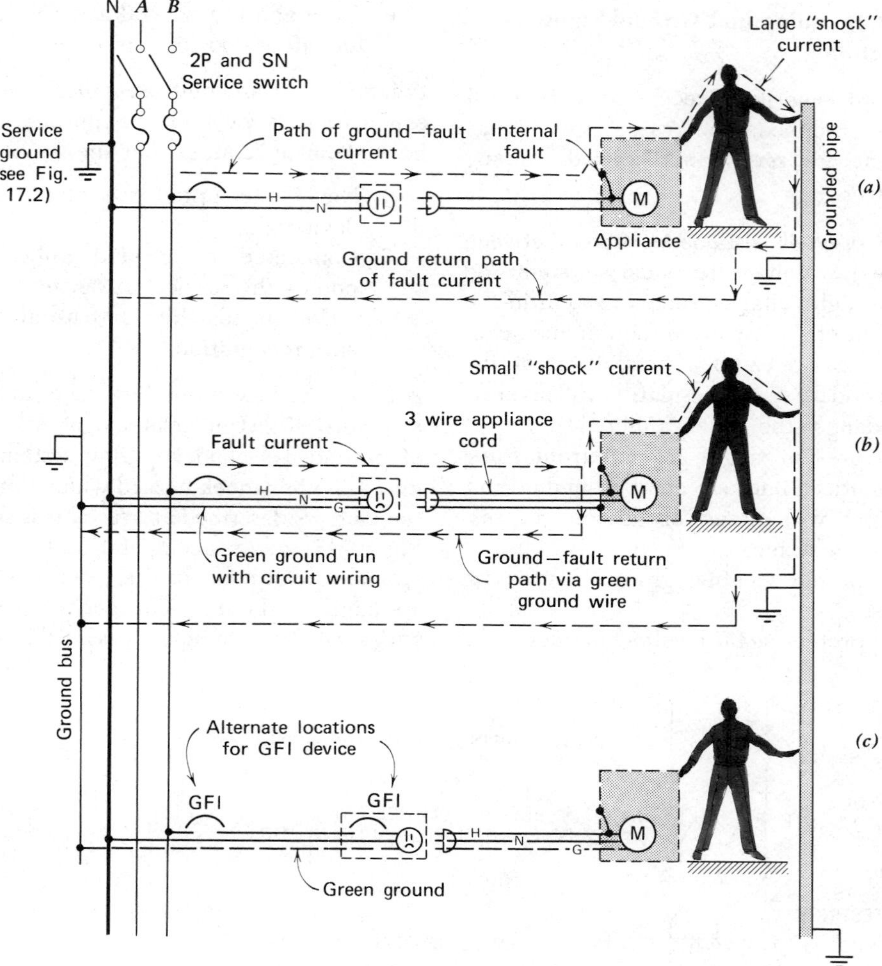

Fig. 17.3 Three types of circuit arrangements. (a) *Shows the conventional 2-wire, grounded neutral circuit, with no means of preventing shocks from ground faults.* (b) *Illustrates a similar arrangement but includes a green ground wire, which will considerably lessen the danger of ground-fault shocks. Note, however, that as long as the ground-fault current is below the rating of the branch circuit protective device (i.e., is below 15 to 20 amps), it will continue to flow and arc within the appliance, causing overheating, arcing, and the eventual destruction of the appliance.* (c) *Illustrates the use of ground-fault circuit interrupter devices, now mandatory for bathrooms, outdoor locations, swimming-pool circuits, and other areas sensitive to ground faults.*

clothes washer, dryer, or food mixer. A person contacting the appliance housing and simultaneously a ground, such as a water pipe, will receive a nasty 120 v shock. If the hands were wet, the shock could be fatal. Until such an incident occurred, however, the internal fault would remain an unnoticed but constant source of danger.

To eliminate this hazard, appliance manufacturers have always recommended that appliance housings be grounded to a cold water pipe and for years have been supplying their appliances with 3-wire plugs: two wires connected to the appliance and the third wire to the housing. To accommodate such plugs and to provide a ground path, the National Electric Code requires all receptacles to be of the grounding type and all wiring systems to provide a ground path, separate and distinct from the neutral conductor. The result of such wiring is shown in Figure 17.3*b* where the ground current passes harmlessly through the internal fault, along the ground wire path and back to the panel. A person contacting the appliance housing establishes a parallel ground path. However, since this path is usually of much higher resistance than the ground-conductor path, only a very small current will flow. Wet hands materially reduce contact resistance, and shock current can increase to a dangerous level. If the ground current is high, the branch circuit breaker or fuse will open, disconnecting the circuit.

When wiring systems are installed in metallic conduit, the conduit itself or the conduit plus a separate conductor within the conduit may be used as the grounding path. This latter method with the additional green ground wire is very much preferable, as explained in Figure 17.3. When nonmetallic or flexible metallic wiring is used ("Romex" or BX), a separate grounding conductor run with the regular circuit conductors *must* be used. All insulated ground conductors must have their covering colored green for identification as a grounding conductor. Many industrial installations have in recent years installed complete "green-ground" systems in an attempt to eliminate shock hazard and to reduce insulation failures. This has not been entirely successful for the reason alluded to above; that is, in order to clear the ground fault, current must be high enough to trip the branch circuit protective device. Otherwise, the ground fault continues to "leak," unnoticed by the system's protective devices. Unfortunately ground faults are by nature low current, leak-type faults, because they result from weak spots in insulation, dirt accumulation, and so on. Therefore, although the shock hazard is in large measure eliminated by the green-ground path, the fault continues to leak and arc until it becomes large enough to cause a major breakdown, generally accompanied by fire.

To eliminate this dangerous situation, which occurs anytime there is a leak of current to ground in an electric circuit, the ground fault circuit interrupter (GFCI or GFI) was developed (see Fig. 17.3*c*). This device compares, with extreme precision, the current flowing in the hot and neutral legs of a circuit; if there is a difference, it indicates a ground fault and the device trips out. The rapidity of this action—approximately 1/2 of a second—eliminates the possibility of dangerous shock hazard, which exists even in a properly grounded circuit as in Figure 17.3*b* and, all the more so, for the circuit arrangement of Figure 17.3*a*. The separate ground wire shown in Figure 17.3*c*, and required by the NEC, serves to minimize the shock current taken by a person before the GFCI operates. The device can be applied at the panel to replace a normal circuit breaker (see Fig. 16.20*b*, page 609) or at an individual outlet to replace a normal receptacle device (see Fig. 16.43, page 632). The GFCI finds a ready application in the old 2-wire circuits illustrated in Figure 17.3*a*. These aging circuit components are prone to ground faults and can best be protected with a GFI. It is also advisable to use GFI devices on all appliance circuits (see the NEC for locations where GFI use is mandatory, such as outdoors and in bathrooms). Application to lighting circuits is not essential, since fixtures are generally out of reach and are switch controlled. In mixed lighting and receptacle circuits the GFCI is best applied at the outlet that is to be protected.

17.5 Energy Conservation Considerations

Before proceeding with a detailed description of design procedure we will survey in this section many of the energy conservation ideas and techniques applicable to electrical distribution systems. This is done for ease of reference and cross-reference, since the individual items appear throughout the lengthy design procedure. An appropriate beginning would be to reproduce here (with permission) the relevant section from ASHRAE Standard 90–75, *Energy Conservation in New Building Design.*

(*a*) ASHRAE Standard 90-75

8.0 ELECTRICAL DISTRIBUTION SYSTEMS

8.1 Scope

This section deals with electrical distribution systems. Such systems *shall* be designed for efficient distribution of electrical energy from the service entrance to the points of use.

8.2 Power Factor

Utilization equipment, rated greater than 1,000 W and lighting equipment greater than 15 W, with an inductive reactance load component, *shall* have a power factor of not less than 85 percent under rated load conditions. Utilization equipment with a power factor of less than 85 percent *shall* be corrected to at least 90 percent under rated load conditions. Power factor corrective devices, installed to comply with this criterion, *shall* be switched with the utilization equipment, except where this results in an unsafe condition or interferes with the intended operation of the equipment. [See Section 16.21(d), page 612.]

8.3 Service Voltage

Where a choice of service voltages is available, a computation *shall* be made to determine which service voltage would produce the least energy loss, and that voltage *shall* be selected. [See Section 17.3.]

8.4 Voltage Drop

In any building, the maximum total voltage drop *shall not* exceed 3 percent in branch circuits or feeders, for a total of 5 percent to the farthest outlet based on steady state design load conditions.

8.5 Lighting Switching

Switching *shall* be provided for each lighting circuit, or for portions of each circuit, so that the partial lighting required for custodial or for effective complementary use with natural lighting *may* be operated selectively.

8.6 Electric Energy Determination

In any multi-tenant residential building, provisions *shall* be made to separately determine the energy consumed by each tenant. Where local codes and regulatory agencies permit, each tenant *shall* be made financially responsible for the energy he uses.

EXCEPTION. Hotels, college dormitories, and other transient facilities are excluded from the provisions of this paragraph.

(*b*) After establishing an energy budget based on projected loads and normal operation, set an energy reduction figure of 10 to 20% and accomplish it with the techniques given below. Annual energy consumption estimates are best made with the aid of the many computer programs available for that purpose.

(*c*) Recognize the energy-use characteristics of all materials and systems specified (see Table 15.6, page 554, and Table 16.4, page 593). In general, select high-efficiency equipment (motors, transformers, etc.). If not available as such, use materials and equipment with the lowest temperature rise, since these have lowest losses. Economic justification can be established with life-cycle cost analysis. To avoid making a detailed cost analysis on every item, utilize one of the many available shortcut calculations for pay-back time. This will generally indicate the material of choice. When comparisons are close, a detailed analysis may be necessary.

(*d*) Provide electric load control equipment (demand control), either as part of an overall building control system or separately. See Section 14.13, page 524.

(*e*) Avoid the use of electric heating elements if alternatives are available. Electric heat is an inefficient use of national resources because of the low efficiency of fuel-to-electricity conversion. See Section 1.5, page 13.

(*f*) Provide metering points (for fixed or plug-in meters) throughout the system to permit accurate analysis of power and energy use.

(*g*) Size equipment as close as possible to the load. This normally results in maximum efficiency and high power factor. Where load varies considerably, for example, between day and night or weekend, consider splitting the loads so that part of the equipment can be switched off when load is low, and the remaining load can be fed from a "night and weekend" feeder.

The design must be sufficiently flexible to permit shifting of loads between feeders if measurements on the operational facility indicate that this is possible. The purpose would be to operate equipment near rated load and deenergize lightly loaded sections.

(*h*) Include a provision for power factor correction in the system. Then, if metering [see (*f*) above] indicates the necessity, add capacitors as required. High power factor reduces line losses, permits maximum utilization of equipment capacity, and avoids utility penalty charges.

(*i*) Select the highest voltage in each class. This means 480 v in the 600 v class, 4 kv in the 5 kv class, and 13 kv in the 15 kv class. The result (depending on building size) will be low line losses, small panelboards at the branch circuit level, and generally lower electrical contract cost.

(*j*) Use the most efficient type of control. This means solid-state control for motors, remote switch control for blocks of lighting, electronic control systems for elevators, and so on.

(*k*) Arrange automatic time controls for 24-hr loads such as vent fans, water coolers, vending machines, and calculators.

(*l*) Seal all electric riser shafts to avoid heat loss by stack action.

(*m*) Generally select the coolest possible locations for electric equipment. Low ambient (below 40° C) permits use of smaller equipment for the same load with concomitant lower cost and losses. Thus, if below-grade space is available, it is well suited for this purpose.

(*n*) Provision for future expansion should be made by means of additional equipment in lieu of oversizing equipment initially. Here again the higher cost of two pieces of equipment can normally be justified by a detailed owning-and-operating cost analysis using realistic cost escalation and capital cost figures.

(*o*) Energy-conservation techniques in lighting and lighting control are found in Sections 18.19, 19.7, 19.26, 20.5, 20.7, 21.26, and 21.28.

(*p*) Energy conservation techniques in vertical transportation are found in Sections 23.16, 24.1, 24.2, 24.3, and 25.12.

17.6 Terminology and Definitions

Since much of the succeeding discussions will utilize terms common in the electrical field, a short list of these terms and definitions is being included here. This list is taken from the National Electric Code, 1975 edition.

Accessible: (As applied to wiring methods) Capable of being removed or exposed without damaging the building structure or finish, or not permanently closed in by the structure or finish of the building. (See *Concealed* and *Exposed.*)

Accessible: (As applied to equipment) Admitting close approach because not guarded by locked doors, elevation, or other effective means. (See *Readily Accessible.*)

Ampacity: Current-carrying capacity of electric conductors expressed in amperes.

Appliance: Utilization equipment, generally other than industrial, normally built in standardized sizes or types, which is installed or connected as a unit to perform one or more functions such as clothes washing, air conditioning, food mixing, deep frying, etc.

Appliance, Fixed: An appliance which is fastened or otherwise secured at a specific location.

Appliance, Portable: An appliance which is actually moved or can easily be moved from one place to another in normal use.

Appliance, Stationary: An appliance which is not easily moved from one place to another in normal use.

Bonding: The permanent joining of metallic parts to form an electrically conductive path which will assure electrical continuity and the capacity to conduct safely any current likely to be imposed.

Branch Circuit: The circuit conductors between the final overcurrent device protecting the circuit and the outlet(s).

Branch Circuit, Appliance: A branch circuit supplying energy to one or more outlets to which appliances are to be connected; such circuits to have no permanently connected lighting fixtures not a part of an appliance.

Branch Circuit, General Purpose: A branch circuit that supplies a number of outlets for lighting and appliances.

Branch Circuit, Individual: A branch circuit that supplies only one utilization equipment.

Branch Circuit, Multiwire: A branch circuit consisting of two or more ungrounded conductors having a potential difference between them, and an identified grounded conductor having equal potential difference between it and each ungrounded conductor of the circuit and which is connected to the neutral conductor of the system.

Cabinet: An enclosure designed either for surface or flush mounting, and provided with a frame, mat, or trim in which swinging doors are hung.

Concealed: Rendered inaccessible by the structure or finish of the building. Wires in concealed raceways are considered concealed, even though they may become accessible by withdrawing them. [See *Accessible* (As applied to wiring methods).]

Demand Factor: The ratio of the maximum demand of a system, or part of a system, to the total connected load of a system or the part of the system under consideration.

Disconnecting Means: A device, or group of devices, or other means by which the conductors of a circuit can be disconnected from their source of supply.

Enclosed: Surrounded by a case, housing, fence, or walls which will prevent persons from accidentally contacting energized parts.

Exposed: (As applied to wiring methods) On or attached to the surface or behind panels designed to allow access. [See *Accessible* (As applied to wiring methods).]

Feeder: All circuit conductors between the service equipment, or the generator switchboard of an isolated plant, and the final branch circuit overcurrent device.

Ground: A conducting connection, whether intentional or accidental, between an electrical circuit or equipment and the earth, or to some conducting body that serves in place of the earth.

Grounding Conductor: A conductor used to connect equipment or the grounded circuit of a wiring system to a grounding electrode or electrodes.

Identified: Identified, as used in this Code in reference to a conductor or its terminal, means that such conductor or terminal is to be recognized as grounded. See Article 200 in Code.

Isolated: Not readily accessible to persons unless special means for access are used.

Lighting Outlet: An outlet intended for the direct connection of a lampholder, a lighting fixture, or a pendant cord terminating in a lampholder.

Listed: Equipment or materials included in a list published by a nationally recognized testing laboratory, inspection agency, or other organization concerned with product evaluation that maintains periodic inspection of production of listed equipment or materials, and whose listing states either that the equipment or material meets nationally recognized standards or has been tested and found suitable for use in a specified manner.

Location:

Damp Location: Partially protected locations under canopies, marquees, roofed open porches, and like locations, and interior locations subject to moderate degrees of moisture, such as some basements, some barns, and some cold-storage warehouses.

Dry Location: One not normally subject to dampness or wetness.

Wet Location: Installations underground or in concrete slabs or masonry in direct contact with the earth, and locations subject to saturation with water or other liquids, such as vehicle washing areas, and locations exposed to weather and unprotected.

Outlet: A point on the wiring system at which current is taken to supply utilization equipment.

Panelboard: A single panel or group of panel units designed for assembly in the form of a single panel; including buses, automatic overcurrent devices, and with or without switches for the control of light, heat, or power circuits; designed to be placed in a cabinet or cutout box placed in or against a wall or partition and accessible only from the front. (See *Switchboard.*)

Raceway: Any channel for holding wires, cables, or busbars that is designed expressly for, and used solely for, this purpose.

Rainproof: So constructed, protected, or treated as to prevent rain from interfering with successful operation of the apparatus.

Rain-tight: So constructed or protected that exposure to a beating rain will not result in the entrance of water.

Readily Accessible: Capable of being reached quickly for operation, renewal, or inspections, without requiring those to whom ready access is requisite to climb over or remove obstacles or to resort to portable ladders, chairs, etc. (See *Accessible.*)

Receptacle: A receptacle is a contact device installed at the outlet for the connection of a single attachment plug.

Receptacle Outlet: An outlet where one or more receptacles are installed.

Service: The conductors and equipment for delivering energy from the electricity supply system to the wiring system of the premises served.

Service Conductors: The supply conductors that extend from the street main or from transformers to the service equipment of the premises supplied.

Service Drop: The overhead service conductors from the last pole or other aerial support to and including the splices, if any, connecting to the service-entrance conductors at the building or other structure.

Service Equipment: The necessary equipment, usually consisting of a circuit breaker or switch and fuses, and their accessories, located near the point of entrance of supply conductors to a building or other structure, or an otherwise defined area, and intended to constitute the main control and means of cutoff of the supply.

Service Lateral: The underground service conductors between the street main, including any risers at a pole or other structure or from transformers, and the first point of connection to the service-entrance conductors in a terminal box, or meter or other enclosure with adequate space, inside or outside the building wall. Where there is no terminal box, meter, or other enclosure with adequate space, the point of connection shall be considered to be the point of entrance of the service conductors into the building.

Switchboard: A large single panel, frame, or assembly of panels on which are mounted, on the face or back or both, switches, overcurrent and other protective devices, buses, and usually instruments. Switchboards are generally accessible from the rear as well as from the front and are not intended to be installed in cabinets. (See *Panelboard.*)

Thermal Cutout: An overcurrent protective device that contains a heater element in addition to and affecting a renewable fusible member which opens the circuit. It is not designed to interrupt short-circuit currents.

Thermally Protected: (As applied to motors) The words "thermally protected" appearing on the nameplate of a motor or motor-compressor indicate that the motor is provided with a thermal protector.

Utilization Equipment: Equipment which utilizes electric energy for mechanical, chemical, heating, lighting, or similar purposes.

Voltage, Nominal: A nominal value assigned to a circuit or system for the purpose of conveniently designating its voltage class (as 120/240, 480Y/277, 600, etc.).

Watertight: So constructed that moisture will not enter the enclosure.

Weatherproof: So constructed or protected that exposure to the weather will not interfere with successful operation.

17.7 Design Procedure

The steps involved in the electrical wiring design of any facility are outlined below. These may in some instances be performed in different order, or two or more steps may be combined, but the procedure normally used is that listed below.

(*a*) Make an electrical load estimate based on areas involved, building data, and any other pertinent data. See Section 17.2.

(*b*) In cooperation with the local electric utility, decide upon the point of service entrance, type of service run, service voltage, metering location, and building utilization voltage. See Sections 16.1 to 16.11 and Section 17.5(i).

(*c*) Determine with the client the usage of all areas, and type and rating of all client-furnished equipment including their specific electric ratings and service connection requirements.

(*d*) Determine from other consultants such as HVAC, plumbing, elevators, kitchen, etc., the exact electrical rating of all the equipment in their designs. This determination will often result after conferences during which the electrical consultant makes valuable recommendations to these other specialists about the comparative characteristics and costs of equipment. See Sections 17.5(a), (c), (e), and (g).

(*e*) Determine the location and estimate the size of all required electric equipment spaces including switchboard rooms, emergency equipment spaces, electric closets, and so forth. Panelboards are normally located in closets but may be located in corridor walls or elsewhere. This work is necessary at this point to enable the architect to reserve these spaces for the electrical equipment. Once the design is accomplished in detail, the estimated space requirements can be checked and necessary adjustments made.

(*f*) Design the lighting for the facility. This

step, as will be discussed in detail in Chapters 18 to 21, is complex and involves a continued interaction between the architect and the lighting designer.

(*g*) Depending on the type of facility, it may be necessary to separate the lighting layout from the receptacle and signal device layout for the sake of clarity of the plans. Once the decision has been made as to how this is to be handled, the lighting fixture layout can be made.

(*h*) On the same plan, or on a separate plan, as decided, locate all electrical apparatus including receptacles, switches, motors, and other power consuming apparatus. Underfloor duct and ceiling track systems would be shown at this stage. If extensive, a separate plan is made.

(*i*) On the plans, locate signal apparatus such as phone outlets, speakers, microphones, TV outlets, fire and smoke detectors, and so on. Since many of these systems are covered by special drawings or by specification alone, this step may be limited to showing phone outlets only. Since the material which follows deals with wiring design only, all further discussion of signal equipment will be reserved for Chapter 22.

(*j*) Circuit all lighting, devices, and power equipment to the appropriate panels, and prepare the panel schedules. Included in this step is the separate circuitry for emergency equipment.

(*k*) Compute panel loads.

(*l*) Prepare the riser diagram. This includes design of distribution panels, switchboards, and service equipment.

(*m*) Compute feeder sizes and all protective equipment ratings.

(*n*) Check the preceding work.

(*o*) Coordinate the electrical work with the other trades and with the architectural plans. This is not really a separate step, but a continuing process starting at (*i*) above and covering all subsequent stages of the work.

The material for (*a*), (*b*), (*c*), (*d*), (*f*), (*g*), and (*i*) is covered elsewhere in this text. The remaining steps, that is, (*e*), (*h*), (*j*), (*k*), (*l*), (*m*), and (*o*) will be discussed in order below.

17.8 Electric Spaces

The spaces required for electric equipment in a facility vary radically, depending on the design and nature of the building. The spaces required around major pieces of electrical switchgear and transformers are discussed above (see Fig. 16.25, page 614). The NEC in Article 110 further specifies the minimum working spaces required in front of electrical equipment and it should be consulted for details. In general, a minimum of 42 in. of clear space should be maintained in front of panels, switches, and other electrical apparatus.

(*a*) ***Residences.*** In private residences, other than the very largest, the service equipment and the building panelboard are incorporated into a single unit. The main disconnect(s) is usually installed as the main switch/breaker of the panel. A number of typical residential service-panel arrangements are shown in Figure 17.4. The panel is normally placed in the garage, utility room, or basement, since the building is rarely so large that length of branch circuits exceeds 100 ft. The panel should be placed as close to the load center as practicable, without sacrificing valuable space or making the panel inaccessible. Frequently a smaller panel can be subfed from the main panel to feed the kitchen and laundry loads. In apartments, panels are normally placed in the kitchen or the corridor immediately adjoining the kitchen. This location is chosen so that the panel circuit breaker may act as the required disconnecting means for fixed appliances larger than 300 volt-amperes or $\frac{1}{8}$ hp, as required by NEC.

(*b*) ***Commercial Spaces.*** The location of the required panelboards depends on their type and quantity, and availability of space. In the research building of which Figure 17.5 is a part plan, lighting panels are recessed into the corridor wall, since the building is only two stories high and the panels can be vertically stacked and fed by a single conduit. If this building were six or more stories high, an electric closet of the type shown in Figure 17.7 would be advisable to accommodate the

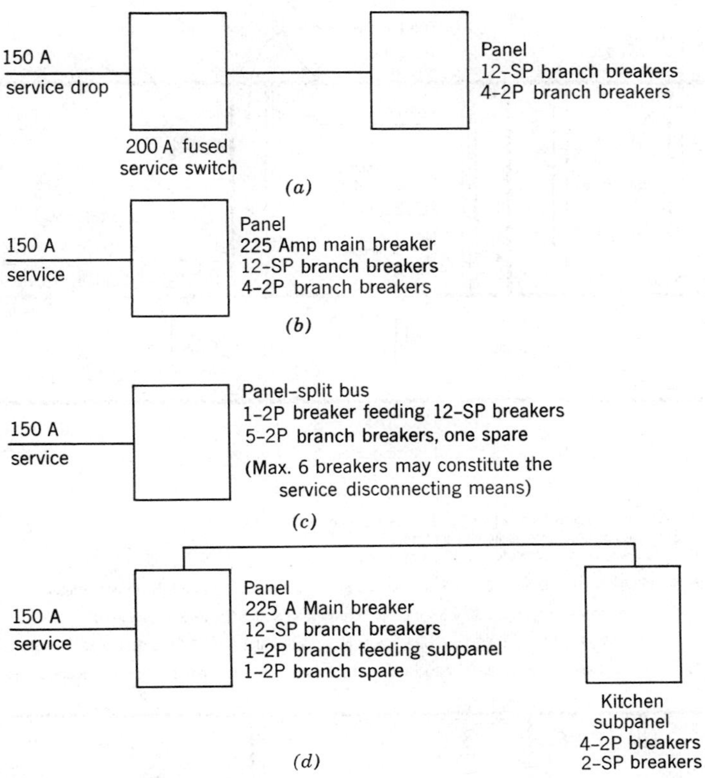

Fig. 17.4 *Typical residential service arrangements* (*all with 150 amp secondary service*).

panel and riser conduits. Of course, when panels are installed in finished areas such as corridors, flush mounting is required.

To limit the voltage drop on a branch circuit in accordance with the Code requirements, panelboards should be located so that no circuit exceeds 100 ft in length. If circuits longer than this are unavoidable, No. 10 AWG wire should be used for runs of 100 to 150 ft and No. 8 AWG for longer circuits. These wire sizes apply to 15 or 20 amp branch circuits, which are normally wired with No. 12 AWG wire.

The laboratory between the two offices of Figure 17.5 is intended to function as a self-contained unit and is therefore equipped with its own panel. Multioutlet assemblies, all wiring within the room, and the panel itself are surface-mounted to allow ready access to all components for the frequent rewiring encountered in laboratories. A main circuit breaker should be provided in such a panel to act as a main disconnect, whether required by Code or not. Where panels are convenient to the loads controlled, the panel circuit breakers may be used for switching.

Panels supplying large blocks of load simultaneously switched, such as auditorium house lights, lobby lights, large single-use office areas, store lighting, and the like, can be constructed with built-in contactors to switch the entire panel, with control at any desired remote location. These remote-control (RC) switches are discussed in Section 16.15. If only part of the panel's circuits are so arranged, a split bus panel is provided, partially contactor controlled (see Fig. 17.6). When lighting and appliance panels require more than 42 poles at a single location, a double panel is utilized comprising two panels in a single enclosing cabinet with a steel barrier between the two units.

Small offices, stores, and other small buildings have lighting panels mounted in a con-

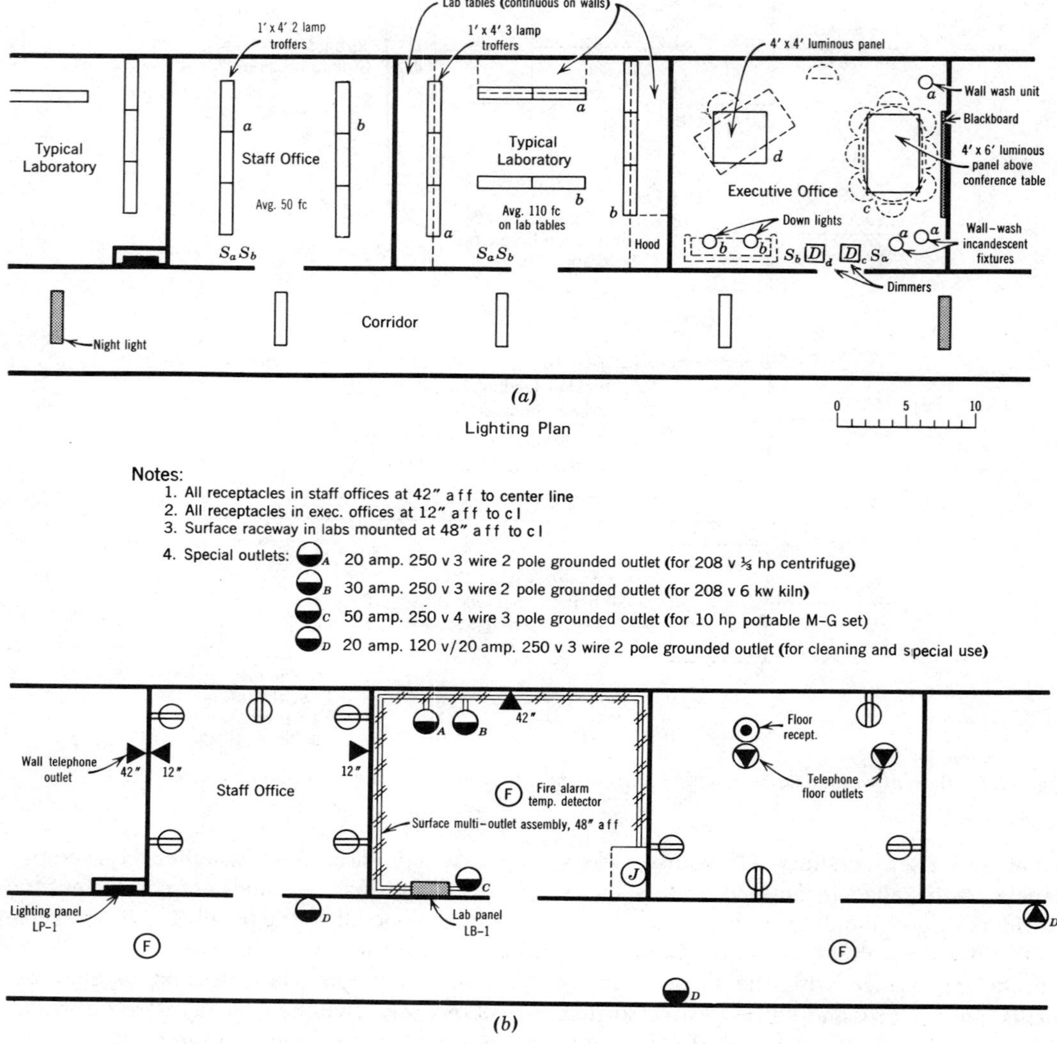

Fig. 17.5 Typical floor plans for lighting and power for a section of an office-laboratory building. Separate lighting and power plans are drawn for the sake of clarity. For circuited plan see Figure 17.17, page 688.

venient finished area and utilize the breakers for load switching. In large buildings, strategically located electric closets are provided to house all electrical supply equipment. Power panels and distribution panels are located as required by the loads fed through them.

In general, branch circuit panels, distribution panels, and switchboards are best located near the electrical load center. This minimizes feeder length and reduces voltage drop making it the most economical arrangement.

Every completely enclosed switchgear room, emergency generator room, or transformer vault should be equipped with an emergency light source. In generator rooms these should be battery operated to give illumination for generator repairs in the event of generator failure during a power outage.

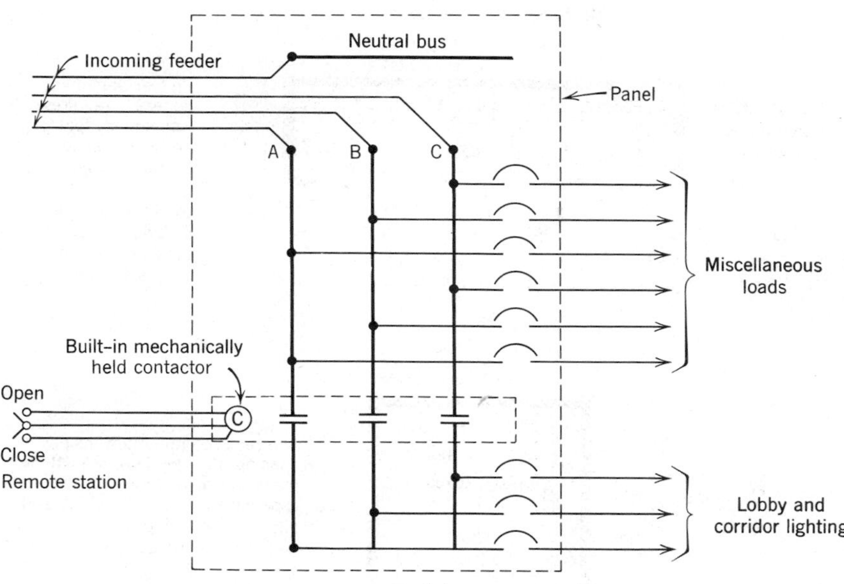

(a)

Fig. 17.6 (a) *Split bus panel. Lower section of panelboard is contactor that is operated in order to accomplish simultaneous control of the lobby and corridor lighting. Breakers in this section are kept normally closed.*

Fig. 17.6 (b) *Newly installed RC switches (mechanically held contactors) are shown to the left of a bank of existing panels. The contactors were installed to accomplish photocell and timer control of a department store's lighting. Courtesy of Automatic Switch Co.*

17.9 Electric Closets

In the design of a building electric system, particularly in multistory construction, it is often advantageous and convenient to group the electrical equipment in a small room called an electric closet. The shape of this space can be varied to fit the architectural and electrical demands, but it should provide the following:

(*a*) One or more locking doors.

(*b*) Vertically stacking, above and below other electric closets and located so as not to block conduits entering or leaving horizontally. Thus locations on outside walls and adjoining shafts, columns and stairs are poor.

(*c*) Space free of other utilities such as piping or duct, passing through the closet, either horizontally or vertically.

(*d*) Sufficient wall space to mount all requisite and future panels, switches, transformers, telephone cabinets, and signal equipment.

(*e*) Floor slots or sleeves of sufficient size for all present and future conduit or bus risers.

(*f*) Sufficient floor space so that an electrician can work comfortably and safely on initial installation and repair.

(*g*) Adequate illumination and ventilation (see Fig. 17.7).

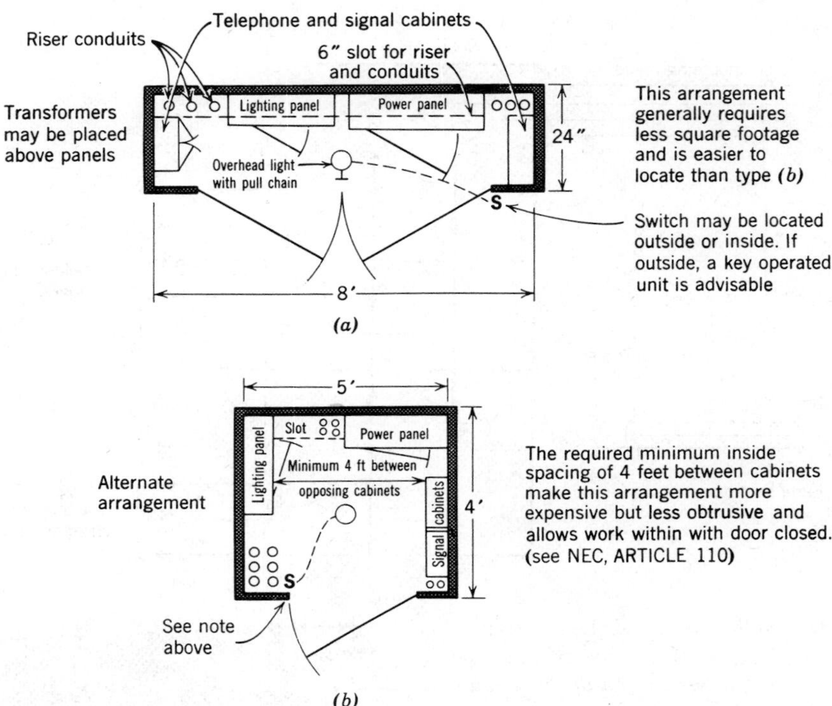

Fig. 17.7 Typical electric closets with some usual equipment. If warranted by amount of equipment, separate closets may be used for signal and telephone conduits and cabinets.

17.10 Equipment Layout

Wiring devices as defined above, but principally comprising receptacles and switches, are located as required by the known equipment to be served and by the anticipated area use. All 15 amp and 20 amp convenience receptacles must be of the grounding type.

Switches for control of lighting or receptacles are normally placed on the strike side of the door. Other devices such as plug-in strip on walls and special purpose receptacles are shown and identified. Signal outlet locations are often noted but generally remain uncircuited on floor plans, a riser being utilized to show interconnections. These include fire-alarm equipment, telephone and intercom equipment, radio and TV outlets, thermostats, and so on. These devices may be identified by a special symbol or note where a standard symbol is not available.

As mentioned above, lighting fixture outlets are normally placed on the same drawing as wiring devices unless the large number of the latter precludes showing the lighting, without undue cluttering of the drawings. In such event, the lighting is shown on one drawing and receptacles on another, with signals shown on the one least occupied. A ceiling raceway system (Section 15.33, page 576), as an underfloor duct system, would probably necessitate such separation. Motors, heaters, and other fixed and permanently wired equipment are shown and identified on the receptacle drawings (also called power drawings, in contradistinction to lighting drawings). Similar equipment—portable or fixed—is not shown if furnished with a cord and plug. However, the receptacle intended for supplying the particular device is shown and identified. A typical device layout is shown in Figure 17.5. A fairly complete symbol list is given in Figure 17.8.

SYMBOLS – PART I – RACEWAYS

NOTE 1	CONDUIT AND WIRING CONCEALED IN CEILING OR WALLS TICS INDICATE NO. OF CONDUCTORS EXCLUDING GROUNDS; 2 #12, ¾" CONDUIT UON.
	CONDUIT AND WIRING CONCEALED IN OR UNDER FLOOR
	CONDUIT AND WIRING EXPOSED
	CONDUIT AND WIRING TURNED UP
	CONDUIT AND WIRING TURNED DOWN
F–6	FEEDER F–6, SEE SCHEDULE, DWG NO.
	CONDUIT WITH ADJUSTABLE TOP AND FLUSH PLUG SET LEVEL WITH FINISHED FLOOR
—BX—	BX WIRING
—NM—	NON–METALLIC CABLE (ROMEX) WIRING
2, 4 2PLA 3# 12, ¾" C	HOME RUN TO PANEL 2PLA – NUMERALS INDICATE CIRCUITS, 3 #12 AWG, ¾" RIGID STEEL CONDUIT
TC2A	HOME RUN TO TELEPHONE CABINET TC2A
	FINAL CONNECTION TO EQUIPMENT IN FLEXIBLE CONDUIT
—EC—	EMPTY CONDUIT, SUBSCRIPT INDICATES INTENDED USE T – TELEPHINE, IC – INTERCOM, FA – FIRE ALARM ETC
A	SURFACE METAL RACEWAY, SEE NOTE ____ , DWG ____ SIZE AND RECEPTACLES AS SHOWN
	MULTI–OUTLET ASSEMBLY, SEE NOTE ____ (SEE PART III FOR ALTERNATE SYMBOL)

NOTE 1. IF THE COMPLETE WIRING SYSTEM IN A BUILDING IS OF A TYPE OTHER THAN CONDUIT AND WIRE, THIS SYMBOL MAY STILL BE USED WITH AN APPROPRIATE NOTE ON THE DRAWINGS OR SPECIFICATIONS, AND ELIMINATION OF THE WORD "CONDUIT" ABOVE.

Fig. 17.8 *Symbol lists for architectural-electrical plan; see parts (**II**) to (**IX**), following.*

SYMBOLS – PART II – OUTLETS

CEILING WALL NOTE 3

OUTLET AND INCANDESCENT FIXTURE; LETTER IN CIRCLE INDICATES TYPE – SEE SCHEDULE. SUPERSCRIPT NO. INDICATES CIRCUIT. SUBSCRIPT LETTER INDICATES SWITCH CONTROL.

OUTLET AND HID LAMP FIXUTURE; INCLUDES MERCURY, METAL HALIDE, SODIUM. LETTER INDICATES TYPE – SEE SCHEDULE. INTEGRAL BALLAST UON.

OUTLET AND FLUORESCENT FIXTURE; LETTER INDICATES TYPE, SEE SCHEDULE.

OUTLET AND EXIT SIGN FIXTURE, UPPER CASE LETTER INDICATES TYPE, ARROWS INDICATE REQUIRED SIGN ARROWS.

TYPE P, BARE LAMP FLUORESCENT STRIP

CONTINUOUS ROW FLUORESCENT FIXTURES, TYPE R, WITH SINGLE OUTLET AND WIRING IN FIXTURE CHANNEL. SEE NOTE 1

OUTLET BOX, BLANK COVER – NOTE 2

JUNCTION BOX, BLANK COVER.

OUTLET WITH FLEXIBLE CONDUIT CONNECTION TO DEVICE XY, NOTE 2

NOTE 1. A CONTINUOUS ROW OF FLUORESCENT FIXTURES WITH WIRING RUN IN THE FIXTURE WIRING CHANNEL IS CONSIDERED A SINGLE OUTLET.

2. IDENTIFYING LETTER 'E' CAN BE OMITTED WHEN THERE IS NO POSSIBILITY OF CONFUSION WITH COLUMNS, ETC.

3. SHOW MOUNTING HEIGHT OF ALL ITEMS.

SYMBOLS – PART III – WIRING DEVICES

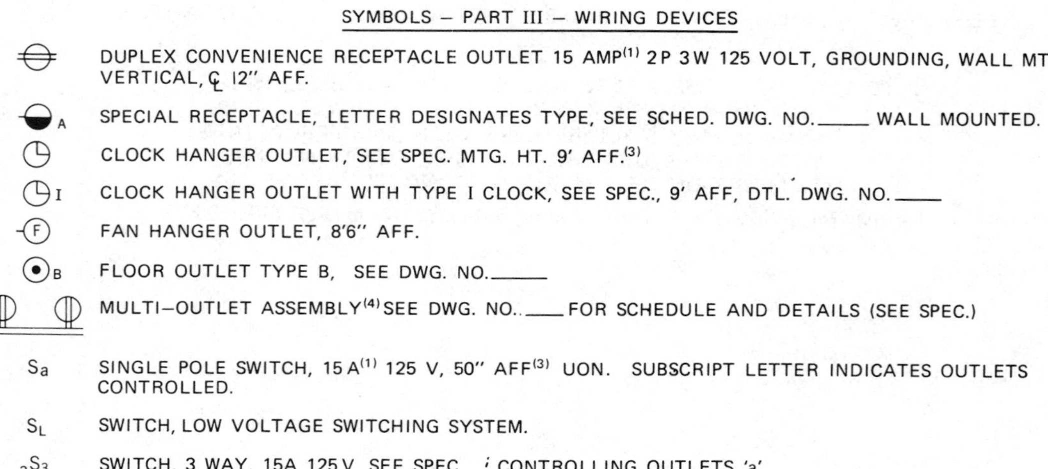

DUPLEX CONVENIENCE RECEPTACLE OUTLET 15 AMP[1] 2P 3W 125 VOLT, GROUNDING, WALL MTD.[2], VERTICAL, ℄ 12" AFF.

SPECIAL RECEPTACLE, LETTER DESIGNATES TYPE, SEE SCHED. DWG. NO.____ WALL MOUNTED.

CLOCK HANGER OUTLET, SEE SPEC. MTG. HT. 9' AFF.[3]

CLOCK HANGER OUTLET WITH TYPE I CLOCK, SEE SPEC., 9' AFF, DTL. DWG. NO.____

FAN HANGER OUTLET, 8'6" AFF.

FLOOR OUTLET TYPE B, SEE DWG. NO.____

MULTI–OUTLET ASSEMBLY[4] SEE DWG. NO..____FOR SCHEDULE AND DETAILS (SEE SPEC.)

S_a SINGLE POLE SWITCH, 15 A[1] 125 V, 50" AFF[3] UON. SUBSCRIPT LETTER INDICATES OUTLETS CONTROLLED.

S_L SWITCH, LOW VOLTAGE SWITCHING SYSTEM.

${}_aS_3$ SWITCH, 3 WAY, 15A 125 V, SEE SPEC. ; CONTROLLING OUTLETS 'a'

S_{DP} SWITCH, DOUBLE POLE, 15 A 125 V.

S_4 SWITCH, 4 WAY, 15 A 125 V.

S_K SWITCH, KEY OPERATED, 15A 125V.

S_D	DOOR SWITCH, SEE SPEC. FOR RATING AND TYPE.
S ⌀	SWITCH/RECEPTACLE COMBINATION IN 2 GANG BOX.
S_P	SWITCH, SP 15A 125V. WITH PILOT LIGHT.
S_{SA}	SWITCH, SPECIAL PURPOSE, TYPE A, SEE SPEC.; SEE DWG. NO._____
———	ABBREVIATIONS RELEVANT TO SWITCHES: SP – SINGLE POLE DP – DOUBLE POLE SPDT – SINGLE POLE DOUBLE THROW DPDT – DOUBLE POLE DOUBLE THROW RC – REMOTE CONTROL
S_{WP}	SWITCH, WEATHER PROOF ENCLOSURE, SEE SPEC.
S_{MC}	SWITCH, MOMENTARY CONTACT.
(M) A	SWITCH, REMOTE CONTROL, TYPE A, MECHANICALLY HELD. SEE SPEC.
(E) A	SWITCH, REMOTE CONTROL, TYPE A, ELECTRICALLY HELD. SEE SPEC.
[R]	OUTLET–BOX–MOUNTED RELAY
[D]	OUTLET–BOX–MOUNTED DIMMER.
[S/D]	OUTLET–BOX–MOUNTED SWITCH AND DIMMER.
⊖ GFCI	RECEPTACLE, RATED AS SHOWN, EQUIPPED WITH GROUND FAULT CIRCUIT INTERRUPTER.

NOTE 1. SPECIFY 20 AMP IF DESIRED.

2. ALL RECEPTACLES ARE WALL MOUNTED UON
3. ALL MOUNTING HEIGHTS ARE TO OUTLET ℄; SPECIFY MH. OF EACH OUTLET.
4. ALSO SHOWN IN SYMBOLS, PT. I, FOR COMPLETENESS, USING ALTERNATE SYMBOL.
5. REFER TO SPECIFICATIONS FOR DATA ON SWITCHES.

A,a	Amperes	N	Neutral
AFF	Above finished floor	NC	Normally closed
C	Conduit	NO	Normally open
C/B	Circuit breaker	NL	Night light
CCT	Circuit	NIC	Not in contract
DF	Drinking fountain	OH	Overhead
DN	Down	OL	Overload relay
EWC	Electric water cooler	OC	On center
EM	Emergency	PB	Push-button, Pull-box
EL	Elevation		
EC	Empty conduit	PC	Pull chain
F	Fuse	PL	Pilot light
FA	Fire alarm	RHC	Reheat coil
F-3	Fan No. 3	SW	Switch
FC	Fan coil unit	TC	Telephone cabinet
GND	Ground	T	Thermostat, Transformer
GFCI	Ground fault cct-interrupter	TEL	Telephone
GFI	Ground fault interrupter	TV	Television
		TYP	Typical
HOA	Hand-off-automatic selector switch	UON	Unless otherwise noted.
		UF	Unfused
HP	Horsepower	UG	Underground
L	Line	UH	Unit heater
LTG	Lighting	VP	Vaporproof
MCC	Motor control center	WP	Weatherproof
MH	Mounting height, Manhole	XP	Explosion proof
		XFMR	Transformer

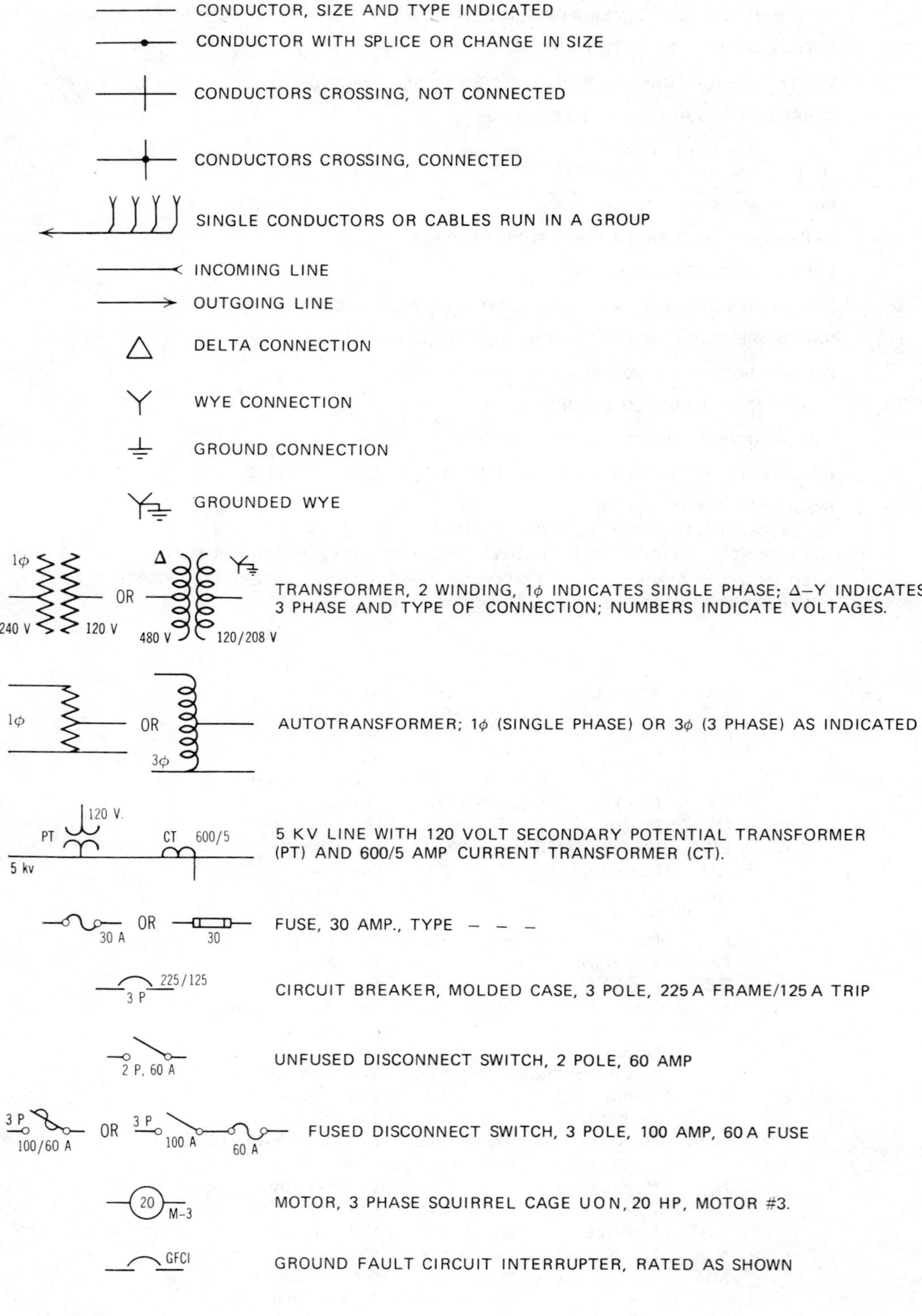
SYMBOLS – PART V – ONE LINE DIAGRAMS
CONDUCTOR, SIZE AND TYPE INDICATED
CONDUCTOR WITH SPLICE OR CHANGE IN SIZE
CONDUCTORS CROSSING, NOT CONNECTED
CONDUCTORS CROSSING, CONNECTED
SINGLE CONDUCTORS OR CABLES RUN IN A GROUP
INCOMING LINE
OUTGOING LINE
DELTA CONNECTION
WYE CONNECTION
GROUND CONNECTION
GROUNDED WYE
1φ
240 V
120 V
OR
Δ
480 V
120/208 V
TRANSFORMER, 2 WINDING, 1φ INDICATES SINGLE PHASE; Δ–Y INDICATES 3 PHASE AND TYPE OF CONNECTION; NUMBERS INDICATE VOLTAGES.
1φ
OR
3φ
AUTOTRANSFORMER; 1φ (SINGLE PHASE) OR 3φ (3 PHASE) AS INDICATED
120 V.
PT
CT
600/5
5 kv
5 KV LINE WITH 120 VOLT SECONDARY POTENTIAL TRANSFORMER (PT) AND 600/5 AMP CURRENT TRANSFORMER (CT).
30 A
OR
30
FUSE, 30 AMP., TYPE – – –
225/125
3 P
CIRCUIT BREAKER, MOLDED CASE, 3 POLE, 225 A FRAME/125 A TRIP
2 P, 60 A
UNFUSED DISCONNECT SWITCH, 2 POLE, 60 AMP
3 P
100/60 A
OR
3 P
100 A
60 A
FUSED DISCONNECT SWITCH, 3 POLE, 100 AMP, 60 A FUSE
20
M-3
MOTOR, 3 PHASE SQUIRREL CAGE UON, 20 HP, MOTOR #3.
GFCI
GROUND FAULT CIRCUIT INTERRUPTER, RATED AS SHOWN

SYMBOLS – PART VI – EQUIPMENT

	INDIVIDUALLY MTD. DISC. SW.; 30A 3P, SN (SOLID NEUTRAL) 250V., UNFUSED, TYPE ND, NEMA I ENCLOSURE, UON
3 P SN 60/30 3R	SW., 3 POLE AND SN., 60A/30 AF. NEMA 3R ENCL., ND (NORMAL DUTY) 250 V.
CB 100/60	ENCLOSED CIRCUIT BREAKER, 3 POLE, NEMA TYPE I ENCLOSURE UON, 100 A FRAME, 60 A TRIP. SEE SPEC.
E 2 P, 30 A	CONTACTOR, ENCLOSED, ELECTRICALLY OPERATED ELECTRICALLY HELD, 2 POLE, 30 AMP.
M OR RC 3 P, 60 A	CONTACTOR, ENCLOSED, ELECTRICALLY OPERATED, MECHANICALLY HELD, 3 POLE 60 AMP.
T S 3 P, 100 A	AUTOMATIC TRANSFER SWITCH, ENCLOSED, 3 POLE, 100 AMP.
	ITEM OF ELECTRICAL EQUIPMENT, AS INDICATED
XFMR A	TRANSFORMER; TYPE A, SEE SCHEDULE
LP-1	ELECTRIC PANELBOARD LP–1, RECESSED, SEE SCHEDULE ON DWG. ____________.
P-2	ELECTRIC PANEL (BOARD) P–2, SURFACE MTD.
TEL. 18 × 24 × 6	CABINET, MOUNTING, SIZE AND PURPOSE AS SHOWN.
PB 12 × 24 × 5″	PULLBOX, SIZE AND IDENT. SHOWN.

SYMBOLS – PART VII SIGNALLING DEVICES

F 1, A	BELL OR GONG, INSCRIBED LETTER INDICATES SYSTEM (SEE BELOW AND SUBSCRIPT LETTER OR NUMBER INDICATES TYPE e.g. A – 8″ VIBRATING BELL, 12 V. DC; B – 12″ WEATHER PROOF SINGLE STROKE GONG, 120 V. D C, etc.
W A	BUZZER, TYPE A.
F 1	FIRE DETECTOR, TYPE 1. } ETC.
I 1	INTRUSION DETECTOR, TYPE 1. } ETC.
SD 2	SMOKE DETECTOR, TYPE 2. } ETC.
F	MANUAL STATION – WATCHMEN TOUR, FIRE ALARM, ETC. LETTER INDICATES SYSTEM, SEE BELOW.
F 2	ANNUNCIATOR, LETTER AND NUMBER INDICATE SYSTEM AND TYPE .
SD A	CABINET OR CONTROL PANEL , SMOKE DETECTION, USE IDENTIFYING TYPE LETTER IF MORE THAN ONE TYPE IS USED ON THE PROJECT.
BATT	AUXILIARY DEVICE.
• •• PB	PUSH BUTTONS
A S 2	HORN OR LOUDSPEAKER, TYPE A, TYPE 2
B	TELEPHONE OUTLET, TYPE B
IC 2	INTERCOM OUTLET, TYPE 2
A	CLOCK SYSTEM OUTLET, TYPE A
TV	TV ANTENNA OUTLET

SYSTEM TYPES

F, FA	FIRE ALARM
S, SD	SMOKE DETECTION
I, IA	INTRUSION ALARM
W, WF	WATER, WATERFLOW
S, SP	SPRINKLER
W, WT	WATCHMAN'S TOUR
T, TEL	TELEPHONE
TV	TELEVISION
IC	INTERCOM
NC	NURSE CALL

AUXILIARY DEVICES

BATT	BATTERY
CT	CONTROL TRANSFORMER
DH	DOOR HOLDER
DO	DOOR OPENER
S, SP	SPEAKER, LOUDSPEAKER
TC	TELEPHONE CABINET
BT	BELL TRANSFORMER
DB	DOOR BELL
CH	CHIME

SYMBOLS – PART VIII
MOTORS AND MOTOR CONTROL

⊠ I	MOTOR CONTROLLER, 3 POLE ACROSS–THE–LINE (ATL) UON, NEMA SIZE I, SEE SCHED., DWG. __ __ __
⊠ II	COMBINATION TYPE MOTOR CONTROLLER; ATL STARTER PLUS FUSED DISCONNECT SWITCH, NEMA SIZE II, SEE SCHEDULE DWG. __ __ __
⊠ I CB	COMBINATION TYPE MOTOR CONTROLLER; ATL STARTER PLUS CIRCUIT BREAKER, NEMA SIZE I, SEE SCHEDULE DWG. __ __ __
—(5)— M – 1	MOTOR #1, 5 HP., 3ϕ SQUIRREL CAGE UON.
(T) OR [T]	DEVICE 'T', SEE LIST OF ABBREVIATIONS, SYMBOLS Part IX
S_T	MANUAL MOTOR CONTROLLER WITH THERMAL ELEMENT.
[• •]	PUSH BUTTON STATION – MOMENTARY CONTACT
[•][•]	PUSH BUTTON STATION – MAINTAINED CONTACT
△, ⧅, ⦶	SYMBOL INDICATING LOCATION OF AN EQUIPMENT ITEM

ABBREVIATIONS

ATL	ACROSS THE LINE STARTER – MAGNETIC
CATL	COMBINATION ACROSS–THE–LINE–MAGNETIC STARTER
FS	FUSED SWITCH

CB	CIRCUIT BREAKER
RV	REDUCED VOLTAGE
FV	FULL VOLTAGE
SR	STARTER RACK
MCC	MOTOR CONTROL CENTER
S	START BUTTON – MOMENTARY CONTACT
ST	STOP BUTTON – MOMENTARY CONTACT
S/S PB	START–STOP PUSH BUTTON
PL	PILOT LIGHT; COLOR INDICATED BY LETTER; A – AMBER, G – GREEN, B – BLUE, R – RED, Y – YELLOW
MER	MECHANICAL EQUIP. ROOM
NO	NORMALLY OPEN
NC	NORMALLY CLOSED
LO	LOCKOUT
R	RELAY
UV	UNDERVOLTAGE
OC	OVERCURRENT
REV	REVERSING

SYMBOLS – PART IX

CONTROL DIAGRAMS & WIRING DIAGRAMS

MOMENTARY CONTACT PUSH BUTTON – NO – (START)

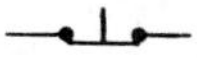

MOMENTARY CONTACT PUSH BUTTON – NC – (STOP)

MAINTAINED CONTACT START–STOP PUSH BUTTON ONE NC AND ONE NO CONTACT.

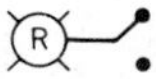

R PILOT LIGHT, R–RED, G–GREEN, Y–YELLOW (SWITCH INDICATES PUSH–TO–TEST).

OL THERMAL OL ELEMENT WITH NC OL CONTACT

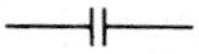

NORMALLY OPEN CONTACT – NO

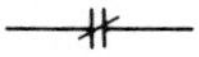

NORMALLY CLOSED CONTACT – NC

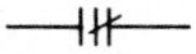

DOUBLE ACTION CONTACT; ONE NO AND ONE NC.

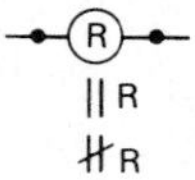

R OPERATING COIL FOR RELAY OR OTHER MAGNETIC CONTROL DEVICE. WITH ONE NO AND ONE NC CONTACT. LETTERS NORMALLY USED ARE C, R FOR CONTROL COIL AND RELAY.

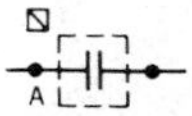

A PILOT CONTROL DEVICE TYPE A, ⧅ INDICATES REMOTE LOCATION.

POWER WIRING

CONTROL WIRING

WIRES CROSSING

WIRES CONNECTED

LIST OF ABBREVIATIONS

T	THERMOSTAT	MOM	MOMENTARY CONTACT
H	HUMIDISTAT	EP	ELECTRO–PNEUMATIC
SD	SMOKE DETECTOR	PE	PNEUMATIC–ELECTRIC
A,AQ	AQUASTAT	BG	BREAK–GLASS
R	RELAY	F, FL	FLOAT SWITCH
M	MOTOR	PS	PRESSURE SWITCH
MD	MOTORIZED DAMPER	HOA	HAND–OFF–AUTOMATIC SWITCH
PB	PUSH–BUTTON	LS, HS	LOW SPEED, HIGH SPEED
OL	OVERLOAD		

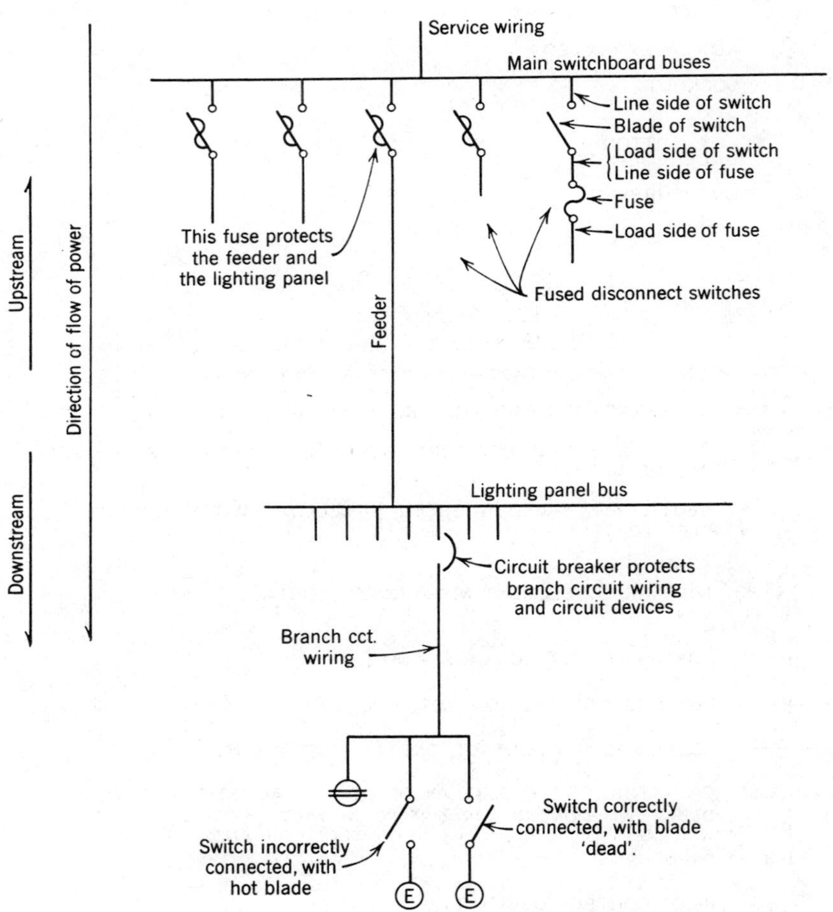

Fig. 17.9 Location of overcurrent protective equipment. Protective equipment should always be located at the point where the conductor receives its source of supply so that when it operates the current supply is cut off.

17.11 Application of Overcurrent Equipment

Before beginning an explanation of circuiting, it is necessary to explain the principles underlying overcurrent protection. As outlined in Chapter 16, the function of an overcurrent device is to open (interrupt) a circuit when the current rating of the equipment being protected is exceeded. These overcurrent devices are placed in circuits to protect wiring, transformers, lights, and all other equipment that can be damaged by excessive current. The following general rules govern the application of overcurrent protection:

(*a*) Overcurrent devices must be placed on the line or supply side of the equipment being protected (see Fig. 17.9).

(*b*) Overcurrent devices must be placed in all *ungrounded* conductors of the protected circuit (see Fig. 16.28*c*, page 617).

(*c*) All equipment should be protected in accordance with its current-carrying capacity (see Fig. 17.10).

(*d*) Conductor sizes shall not be reduced in a circuit or tap, unless the smallest size wire is protected by the circuit overcurrent devices (see Fig. 17.11).

(*e*) Overcurrent devices shall be located so as to be readily accessible, protected from physical damage, and away from easily ignited material.

17.12 Branch Circuit Design

A branch circuit by NEC definition refers only to the circuit conductors, although for our purposes and in trade parlance it includes the protective device and the outlets served. Such circuits may be multioutlet general-purpose type (Fig. 17.12*a*), multioutlet appliance type (Fig. 17.12*b*), or single-outlet type intended for a specific piece of equipment (Fig. 17.12*c*). The multioutlet types are limited to 50 amp in capacity, while the single-outlet type is governed in size only by the requirements of the item being served, and may be 200 or 300 amp in size.

In its simplest form, a branch circuit comprises only two wires. However, multiwire branch circuits carrying 2- or 3-phase wires plus a neutral are also widely used. Generally, each branch circuit should be sized for the load connected to it, plus the load expansion that is expected. These general rules of good practice should be followed:

(*a*) In all but the smallest installations, connect lighting, convenience receptacles, and appliances on separate groups of circuits, although this is not an NEC requirement.

(*b*) General purpose branch circuits should be 20 amp and wired with No. 12 AWG wire. Switch legs may be No. 14 AWG if the lighting loads permit.

(*c*) Limit the circuit load on 15 amp and 20 amp circuits to the values shown in Table 17.5.

(*d*) Use only grounding type receptacles on 15A and 20A branch circuits. This is an NEC requirement.

Since lighting and specific devices are circuited according to their nameplate rating, the only circuitry item left to the judgment of the designer is the number of convenience

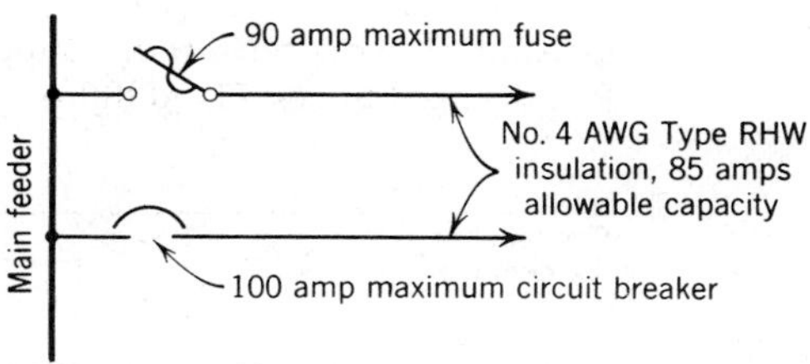

Fig. 17.10 *The overcurrent protection must correspond to the rating of the protected equipment. Where ratings do not correspond exactly, the next larger standard size may be used.*

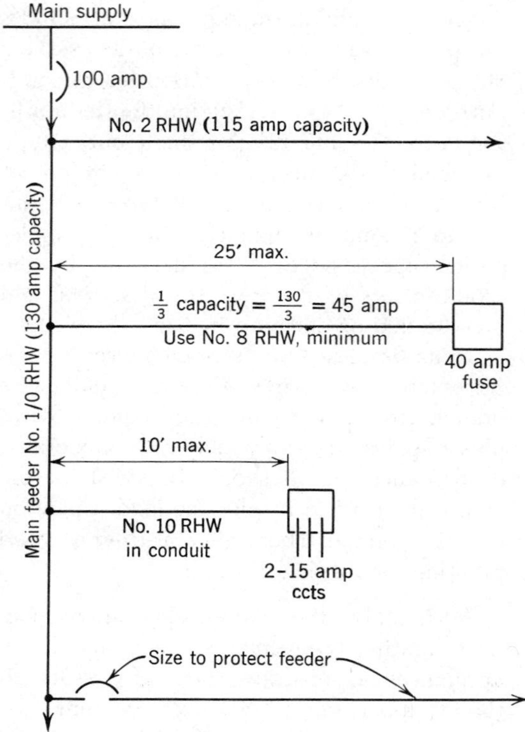

Taps may be made if smaller conductor is protected by main feeder protection (100 A C/B protects the No. 2).

Taps up to 25′ long are permitted it tap conductor has min. 1/3 ampacity of main, and terminates in a single C/B or set of fuses, which limits the current on the tap to the wire capacity. (40 A fuse limits current on 45 A capacity wire).

Taps up to 10′ long are permitted provided they have sufficient capacity for the circuits they feed; eg, if the wiring feeds 2–15 amp ccts, it must have 30 amp capacity and should be No. 10 AWG, RHW or equal.

Taps of any length may be made if the conductor is protected at the tap point by an appropriately sized overcurrent device.

Fig. 17.11 Permissible tap arrangements; see NEC Article 240-21.

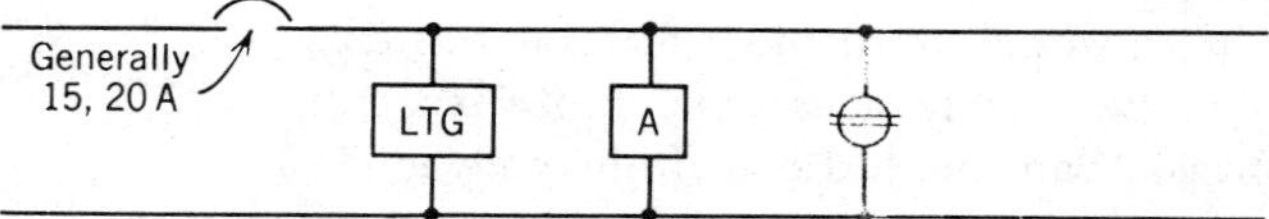

(a) General purpose branch circuit. Supplies outlets for lighting and appliances, including convenience receptacles.

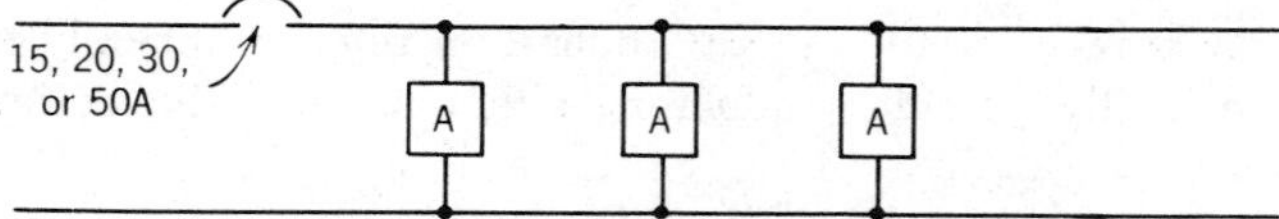

(b) Appliance branch circuit. Supplies outlets intended for feeding appliances. Fixed lighting not supplied.

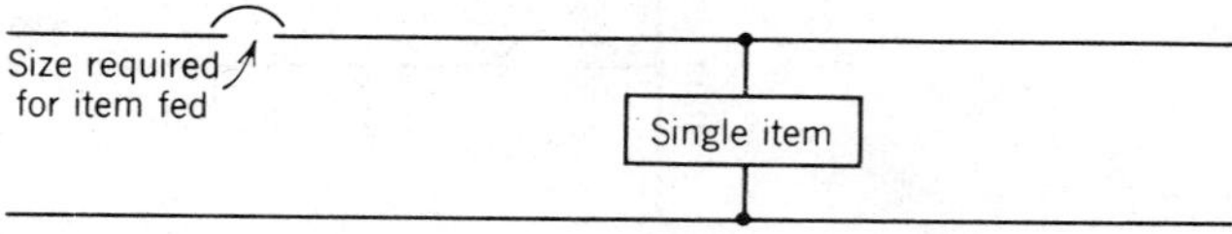

(c) Individual branch circuit, designed to supply a single, specific item.

Fig. 17.12 Branch circuit types.

Table 17.5 ***Recommended Branch Circuit Loads***

Size Circuit	Maximum Continuous Load (80% of Rating) Amperes	Volt-Amperes at 120 v	Volt-Amperes at 277 v	Anticipated 5 to 10 Year Load Growth (Table 17.1) — (A) 0–30%[a] Expansion: Initial Load	Volt-Amperes at 120 v (Approximate)	Volt-Amperes at 277 v (Approximate)	(B) 40% or More[b] Expansion: Initial Load	Volt-Amperes at 120 v (Approximate)	Volt-Amperes at 277 v (Approximate)
15A	12	1440	—	9.6	1150	—	8	960	—
20A	16	1920	4440	12.8	1520	3600	11	1300	3000

[a]Loading gives approximately 25% spare capacity,

15 amp cct at 120 v: $\frac{1440 - 1150}{1150} = \frac{290}{1150} = 25\%$

20 amp cct at 120 v: $\frac{1920 - 1520}{1520} = \frac{400}{1520} = 26\%$

[b]Loading gives approximately 50% spare capacity,

15 amp cct at 120 v: $\frac{1440 - 960}{960} = \frac{480}{960} = 50\%$

20 amp cct at 120 v: $\frac{1920 - 1300}{1300} = \frac{620}{1300} = 47\%$

at 277 v: $\frac{4440 - 3000}{3000} = \frac{1440}{3000} = 48\%$

Table 17.6 ***Branch Circuit Requirements***

	Branch Circuit Size				
	15 amp	*20 amp*	*30 amp*	*40 amp*	*50 amp*
Minimum size conductors	No. 14	12	10	8	6
Minimum size taps	No. 14	14	14	12	12
Overcurrent device rating	15 amp	20	30	40	50
Lampholders permitted	Any type	Any type	Heavy duty	Heavy duty	Heavy duty
Receptacle rating permitted (see note 7)	15 amp	15 or 20	30	40 or 50	50 amp
Maximum load (see note 6)	15	20	30	40	50

NOTES:

1. Wiring shall be types RHW, RHH, T, THW, TW, THWN, THHN, XHHW in raceway or cable.
2. On 15-amp circuit, maximum single appliance shall draw 12 amp. On 20-amp, circuit maximum single appliance shall draw 16 amp. If combined with lighting or portable appliances, any fixed appliance shall not draw more than 7.5 amp on a 15-amp circuit, and 10 amp on a 20-amp circuit.
3. On a 30-amp circuit, maximum single appliance draw shall be 24 amp.
4. Heavy-duty lampholders are units rated not less than 750 w.
5. 30, 40, and 50-amp circuits shall not be used for fixed lighting in residences.
6. When loads are connected long periods, actual load shall not exceed 80% of the branch circuit rating. Conversely, continuous type loads shall be figured at 125% of actual load in all load calculations.
7. A single receptacle on an individual branch circuit shall have a rating not less than the circuit, for example, 15 amps on a 15-amp circuit, etc. 15-amp receptacles on 20-amp circuit shall not supply a load greater than 12-amp for appliances. 20-amp receptacles on 20-amp circuit shall be limited to a 16-amp load.

receptacles per circuit (cct). The NEC specifies that plug outlets (convenience receptacles) be counted, in totaling loads, at 1.5 amp each unless included in the load for general lighting. Thus, following the guidelines stated above of 9 and 12 amp loading on 15 and 20 amp circuits, respectively, we would have by this method:

$$15 \text{ amp circuit } \frac{9}{1.5} = 6 \text{ outlets per cct}$$

$$20 \text{ amp circuit } \frac{12}{1.5} = 8 \text{ outlets per cct}$$

These figures must be used judiciously. If the devices to be energized are small, these quantities may be used. Such would be the case in a drafting room where only erasing machines (50 va) or a desk lamp (100 va) would be plugged in. However, for laboratory tables, office machines, or assembly benches, no more than two or three receptacles should be used or a 20 amp circuit. Of course, diversity of use is an all-important factor and the more closely it can be estimated, the better the design result.

A further note of caution is in order. Receptacles should be arranged, if at all possible, so that the loss of a single circuit does not deprive an entire area of power. That is, for the sake of reliability, circuitry should be alternated to give each space parts of different circuits.

The Code further specifies certain requirements for conductors, devices, and loads permissible on general purpose branch circuits. These are summarized in Table 17.6.

17.13 Alternative Wiring Techniques

The usual wiring system is a radial or "tree" type system in which conductors of progressively smaller size emanate radially from distribution points. These distribution points are the switchboards and panelboards throughout the system, which provide overcurrent protection for each of these radials (refer to Fig. 17.9). Note that at each step, power is tapped off in a smaller "package" and in accordance with the principles of overcurrent protection an appropriately sized protective device is placed at the point of tap. An alternate technique utilizes this tapping principle somewhat differently (refer to Fig. 17.11). The arrangement shown at the bottom of the figure—a tap protected at the

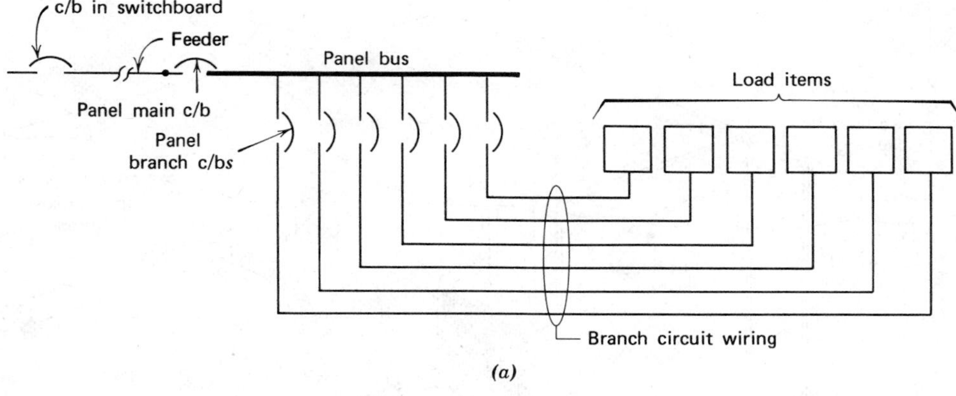

(a)

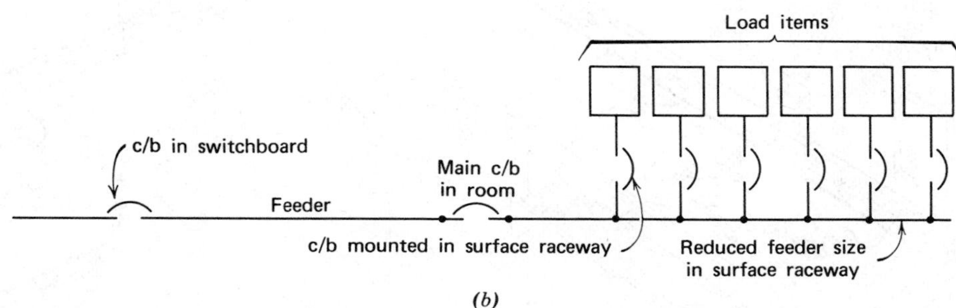

(b)

Fig. 17.13 (a) *Conventional radial wiring system. Branch circuits radiate from the panel to the loads and are protected at the panel by the panel circuit breakers. The panel itself is protected by a main c/b.* (b) *In this arrangement the heavy switchboard feeder is tapped by the room c/b that protects the room feeder. This feeder is in turn tapped at each load, thus eliminating a concentrated panelboard and almost all branch circuit wiring.*

source—is precisely what occurs at a panelboard, with the panel bus substituting for the feeder. This arrangement is shown in Figure 17.13*a*. If we *eliminate* the panel and tap the feeder directly as in Figure 17.13*b*, we have accomplished exactly the same end. The purpose of this second arrangement is to eliminate the branch circuit wiring, by placing the protective device at the load. Such an arrangement is shown in Figure 17.14. The advantage of this system, in addition to the advantages inherent in a surface raceway wiring system, is that by eliminating branch circuit wiring, installation costs are reduced, voltage drop and energy losses in branch circuit conductors are negligible, and loads are individually protected. Obviously, this system finds its best application in areas where surface raceways are desirable for flexibility and loads would benefit from this type of individual control and protection. As an exercise, the reader might examine the possibility of utilizing this system for the layout of Figure 17.5, justifying any conclusions from technical and economic points of view.

17.14 Branch Circuit Design Guidelines—Residential

(*a*) The NEC (1975 edition) requires for residences sufficient circuitry to supply a load of 3 w/sq ft in the building, excluding unfinished spaces such as porches, garages, and

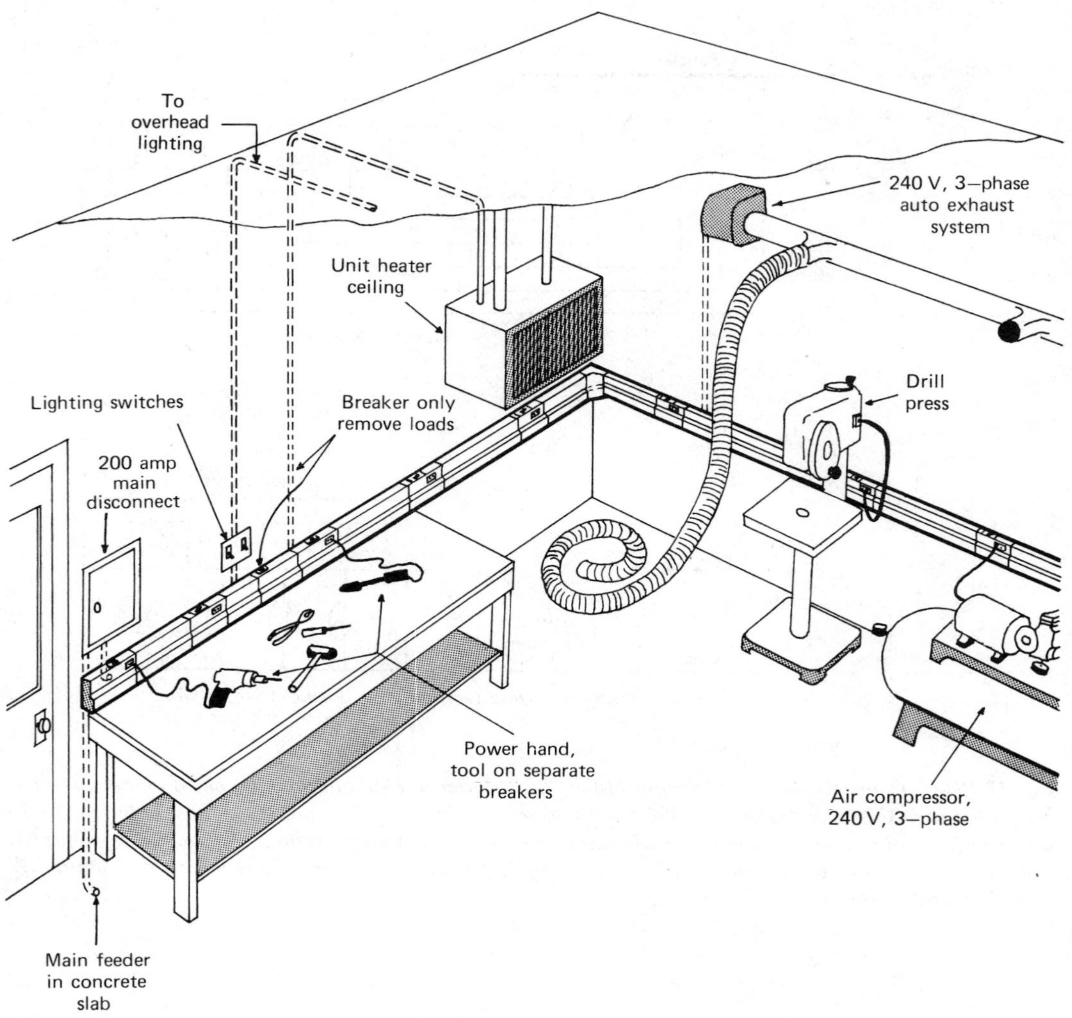

(a)

Fig. 17.14 (a) *Typical room layout using the feeder tapping wiring technique. The room feeder is tapped from the main feeder and protected by the 200 amp main disconnect that is either a circuit breaker or a fused switch. Beyond this point, the room feeder is run in surface raceway and is tapped at each device with a circuit breaker, which feeds an individual outlet. Overhead loads—such as the heater (shown) and lighting (not shown)—are fed from raceway circuit breakers.*

basements. This requirement of 3 w/sq ft works out to 800 sq ft per 20-amp circuit (2400 w) or 600 sq ft per 15-amp circuit (1800 w). Good practice, however, dictates a load of no more than 1600 w for a 20-amp circuit or 1200 w for a 15-amp circuit. This gives:

15-amp circuit, 1200 w max. 400 sq ft max.

20-amp circuit, 1600 w max. 530 sq ft max.

(*b*) The NEC requires a minimum of two 20-amp appliance branch circuits (see Fig. 17.12*b*) to feed all the small appliance outlets in the kitchen, pantry, dining room, and family room, and *only* these outlets. Furthermore, all kitchen outlets must be fed from at least two of these circuits (which may also feed

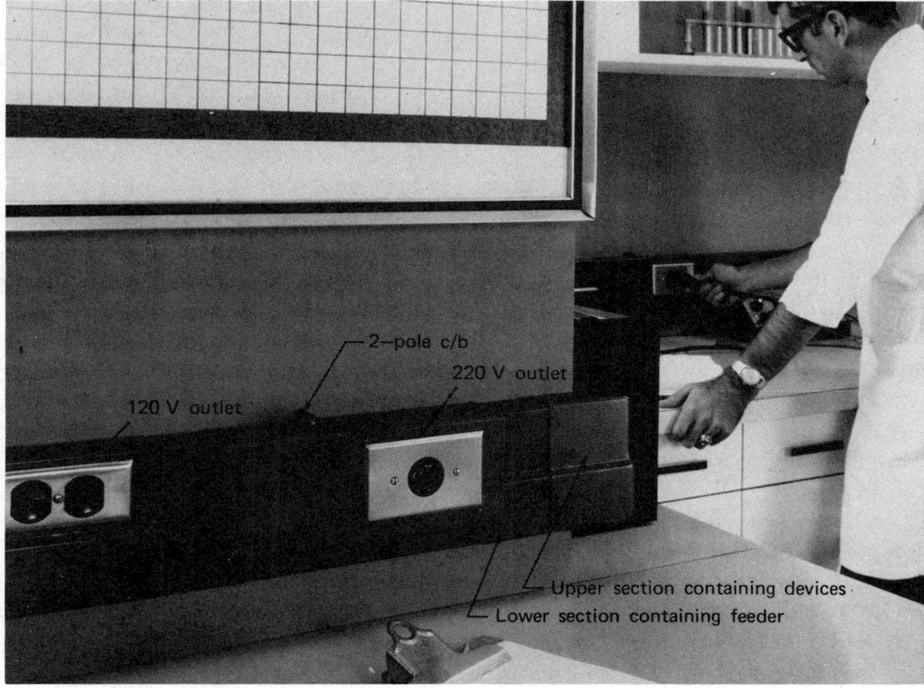

(b)

Fig. 17.14 (b) *The equipment as actually installed in a laboratory facility is neat, flexible, and highly practical. Courtesy of Versa-Tek, Inc.*

other appliance outlets). This NEC requirement needs clarification. The NEC requires that at least two circuits be reserved for *appliance outlets,* but it does not specify what these appliance outlets are except to say, by inference, that all kitchen outlets are appliance outlets. In effect, then, according to the NEC all the receptacles are potential appliance outlets, and at least two circuits should be supplied to serve them. Good practice dictates that certain receptacles in each room be designated as appliance outlets even though they do not differ from the other outlets in appearance. These outlets are:

1. All kitchen receptacles.
2. One dining room receptacle.
3. One receptacle in the family (or living) room.

These receptacles should be circuited with preferably two, but no more than four, such outlets on a 20-amp circuit, and the circuits should be arranged so that the kitchen has part of at least two circuits feeding its outlets.

(*c*) Additional circuits similar to appliance circuits should be furnished to supply one outlet in each bedroom of a house that is not centrally air-conditioned. Such outlets are intended for window air conditioners. (Good architectural and HVAC design will provide window arrangement, attic ventilation, insulation, sun-screening, and the like to obviate the necessity for these noisy energy users.) Also, on circuits of this kind place basement workbench outlets. These additional circuits must not be mixed with the appliance branch circuits discussed above, as they are not strictly appliance circuits by NEC definition.

(*d*) The NEC requires that at least one 20-amp circuit supply the laundry outlets. This requirement satisfies good practice. If an electric clothes dryer is anticipated (and it should be unless it is definitely known that a gas dryer will be used), an individual branch circuit should be supplied to serve this load, via a heavy-duty receptacle. (Obviously, facilities for hanging clothes must be provided for those who prefer not to waste energy.)

(*e*) Lay out convenience receptacles so that no point on a wall is more than 6 ft from an outlet. Use 20 amp, grounding-type receptacles only. Do not combine receptacles and switches into a single outlet except where convenience of use dictates high mounting of receptacles, as above counter spaces.

(*f*) Circuit the lighting and receptacles so that each room has parts of at least two circuits. This includes basements and garages.

(*g*) Avoid placing all the lighting in a building on a single circuit.

(*h*) Supply at least one receptacle in the bathroom and one outside the house. Both must be GFCI types. This is an NEC requirement. An additional convenience is switch control of the outside receptacle from *inside* the house. Also, a timer-controlled outlet for a plug-in bathroom heater is a welcome convenience, but obviously means additional expense.

(*i*) In rooms without overhead lights, provide switch control for one-half of a strategically located receptacle that is intended to supply a lamp. See Figure 17.15 for the wiring arrangement in such a case.

(*j*) Provide switch control for closet lights. Pull chains are a nuisance (but are considerably cheaper).

(*k*) In bedrooms supply two duplex outlets at each side of the bed location to accommodate electric blanket, clocks, radios, lamps, and other such appliances.

(*l*) Since receptacles are counted as part of general lighting and no additional load is included for them, no limit is placed on the number of receptacle outlets that may be wired to a circuit. For good practice they should be limited to 6 on a 15-amp circuit and 8 on a 20-amp circuit.

(*m*) Kitchens should have a duplex appliance outlet every 36 in. of counter space, but no less than two in addition to the normal wall outlets.

(*n*) A disconnecting means, readily accessible, must be provided for electric ranges, cook tops, and ovens. Plugs accessible by removal of a drawer or by reaching across a unit meet the Code requirements but are not really "readily" accessible. It is better practice to utilize a small kitchen panel recessed into a corner wall to control the large kitchen appliances and to provide completely safe, accessible disconnecting means. Such an arrange-

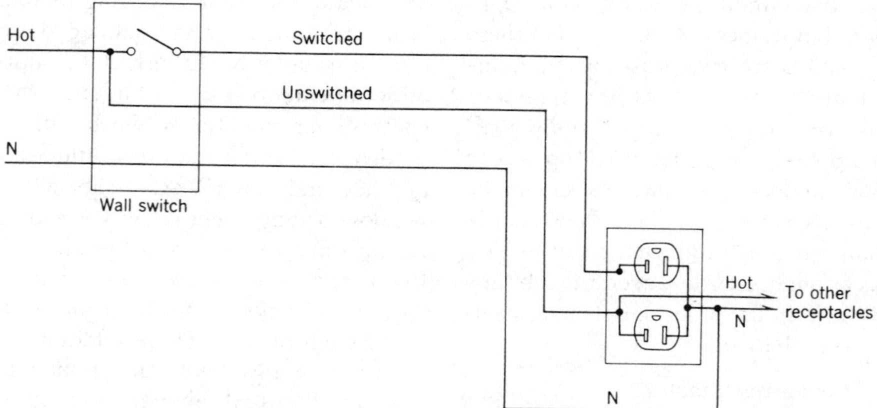

Fig. 17.15 *Split wiring of a duplex receptacle. Upper half is switch controlled; lower half is "hot" all the time. This allows wall switch control of a lamp or other device while maintaining part of receptacle line for independent use. Notice that the receptacle is mounted with the grounding pole* at the top. *This is the safest way to install receptacles, since a metallic item such as a paper clip falling on to an inserted tap will contact the ground only.*

ment can also be cheaper if the length of run between the main panel and the kitchen is appreciable.

(*o*) Perimeter lighting, inside switch controlled, can do much to lessen vandalism and discourage prowlers, in addition to illuminating the house at night.

(*p*) A tabulation of residential electrical equipment, including recommended circuits and receptacles is shown in Tables 17.7 and 17.8. A complete residential wiring plan for a small house is shown in Figure 17.16. Although residential plans are normally left uncircuited, a completely circuited design is shown here for didactic purposes.

17.15 Branch Circuit Design Guidelines—Nonresidential

(***a***) ***Schools.*** Since schools comprise an assembly of varied use spaces, including instruction, lab, shop, assembly, office, gymnasium, plus special areas such as swimming pools, photographic labs, and so on, it is not possible to generalize on branch circuit design considerations except for the following:

(1) To accommodate the opaque and film projectors frequently used in the classroom, 20-amp outlets wired two to a circuit are placed at the front and back of each such room. A similar receptacle, wired 6 or 8 to a circuit is placed on each remaining wall.

(2) Light switching should provide:

a. High-low levels for energy conservation and to permit low-level lighting for film viewing. With fluorescent lighting this can be accomplished by alternate ballast wiring and switching, thus avoiding the high cost of dimming equipment.
b. Separate switching of the lights on the window side of the room, which is often lighted sufficiently by daylight.

(3) Provide appropriate outlets for all special equipment in labs, shops, cooking rooms, and the like.

(4) Use heavy-duty devices and key-operated switches for public area lighting (corridors, etc.), plastic instead of glass in fixtures, and vandal-proof equipment wherever possible. All panels *must* be locked and should be in locked closets.

(5) The NEC requires sufficient branch circuitry to provide a minimum of 3 w/sq ft for general lighting in schools. Refer to the NEC Article No. 220. Unlike residential occupancy this figure does *not* include receptacles. Receptacles are calculated separately at 180 w each for ordinary convenience outlets.

(6) Keep lighting and receptacles completely separate when circuiting.

(***b***) ***Office Space***

(1) In small office spaces (less than 400 sq ft) provide either one outlet for every 40 sq ft, or one outlet for every 10 linear ft of wall space, whichever is greater. In larger office spaces, provide one outlet every 100 to 125 sq ft beyond the initial 400 sq ft (10 outlets). These should comprise wall outlets spaced as above plus floor outlets sufficient to make up the required total. In view of the increasingly heavy loads of office machines, these receptacles should be circuited at no more than 6 to a 20-amp branch circuit, and less if the equipment to be fed so dictates. Figure 17.17 shows one possible circuiting arrangement for the room layouts shown in Figure 17.5. Although other arrangements are possible, the net result is the same.

(2) Corridors should have a 20-amp, 120-v outlet every 50 ft, to supply cleaning and waxing machines.

(3) As with all nonresidential buildings, convenience receptacles are figured at 180 w each.

(4) Only specification grade equipment should be used.

(***c***) ***Industrial Spaces.*** These areas are so specialized that no meaningful guidelines can be given.

(***d***) ***Stores.*** In stores, good practice requires at least one convenience outlet receptacle for every 300 sq ft in addition to outlets required for loads such as lamps, show windows, and demonstration appliances.

Table 17.7 ***Load, Circuit and Receptacle Chart for Residential Electrical Equipment***

Appliance	*NEC Type*[a]	*Typical Connected Volt-Amperes*	*Volts*	*Wires*[b]	*Circuit Breaker or Fuse*	*Outlets on Circuit*	*NEMA Device*[c] *and Configuration See Fig. 16.41*	*Remarks*
					Kitchen			
Range	(F)	12000	115/230	3 #6	60A	1	14–60R	Use of more than one outlet is not recommended.
Oven (built in)	(F)	4500	115/230	3 #10	30A	1	14–30R	May be direct connected.
Range top	(F)	6000	115/230	3 #10	30A	1	14–30R	May be direct connected.
Range top	(F)	3300	115/230	3 #12	20A	1	14–30R	May be direct connected.
Dishwasher	(F)	1200	115	2 #12	20A	1	5–20R	May be direct connected.
Waste disposer	(F)	300	115	2 #12	20A	1	5–20R	May be direct connected.
Broiler	(P)	1500	115	2 #12	20A	1 or more	5–20R	See note e.
Fryer	(P)	1300	115	2 #12	20A	1 or more	5–20R	See note e.
Coffeemaker	(P)	1000	115	2 #12	20A	1 or more	5–20R	See note e.
Refrigerator	(S)	300	115	2 #12	20A	1 or more	5–20R	Separate circuit serving only refrigerator and freezer is recommended.
Freezer	(S)	350	115	2 #12	20A	1 or more	5–20R	Separate circuit serving only refrigerator and freezer is recommended.
					Laundry			
Washing machine	(S)	1200	115	2 #12	20A	1 or more	5–20R	Grounding-type receptacle on circuit is required.
Dryer	(S)	5000	115/230	3 #10	30A	1	14–30R	Appliance may be directly connected—must be grounded.
Hand iron; Ironer	(P)	1650	115	2 #12	20A	1 or more	5–20R	
Water heater	(F)	4500	115/230	2 #10	30A	1	—	May be direct connected.

					Living Areas			
Workshop	(P)	1500	115	2 #12	20A	1 or more	5–20R	Separate circuits recommended.
Portable heater	(P)	1300	115	2 #12	20A	1	5–20R	Should not be connected to circuit serving heavy-duty loads.
Television	(S)	300	115	2 #12	20A	1 or more	5–20R	Should not be connected to circuit serving appliances.
Portable lighting	(P)	1200	115	2 #12	20A	1 or more	5–20R	
					Fixed Utilities			
Fixed lighting	(F)	1200	115	2 #12	20A	1 or more	—	
Air conditioner $\frac{3}{4}$ hp	(F)	1200	115	2 #12	20A or 30A	1	5–20R	Separate circuit recommended.
Air conditioner $1\frac{1}{2}$ hp	(F)	2400	115/230	3 #12	20A or 30A	1	—	Connect through disconnect switch.
Central air conditioner	(F)	5000	115/230	3 #10	40A	1	—	Connect through disconnect switch.
Sump pump	(F)	300	115	2 #12	20A	1 or more	5–20R	Use of 1 pole thermal disconnect recommended.
Heating plant, i.e., forced-air furnace	(F)	600	115	2 #12	20A	1	—	Direct connected. Some local codes require separate circuit.
Fixed bathroom heater	(F)	1500	115	2 #12	20A	1	—	Direct connected.
Attic fan	(F)	300	115	2 #12	20A	1 or more	5–20R	Connect via thermal element 1 pole switch.

[a]Appliance types: (F) Fixed; (S) Stationary; (P) Portable.

[b]Number of wires does not include equipment grounding wires. Ground wire is #12 AWG for 20A circuit and #10 AWG for 30A and 50A circuits.

[c]Equipment ground is provided in each receptacle.

[d]For a discussion of disconnect requirements, see NEC Article 422.

[e]Heavy duty appliances regularly used at one location should have a separate circuit. Only one such unit should be attached to a single circuit at the same time.

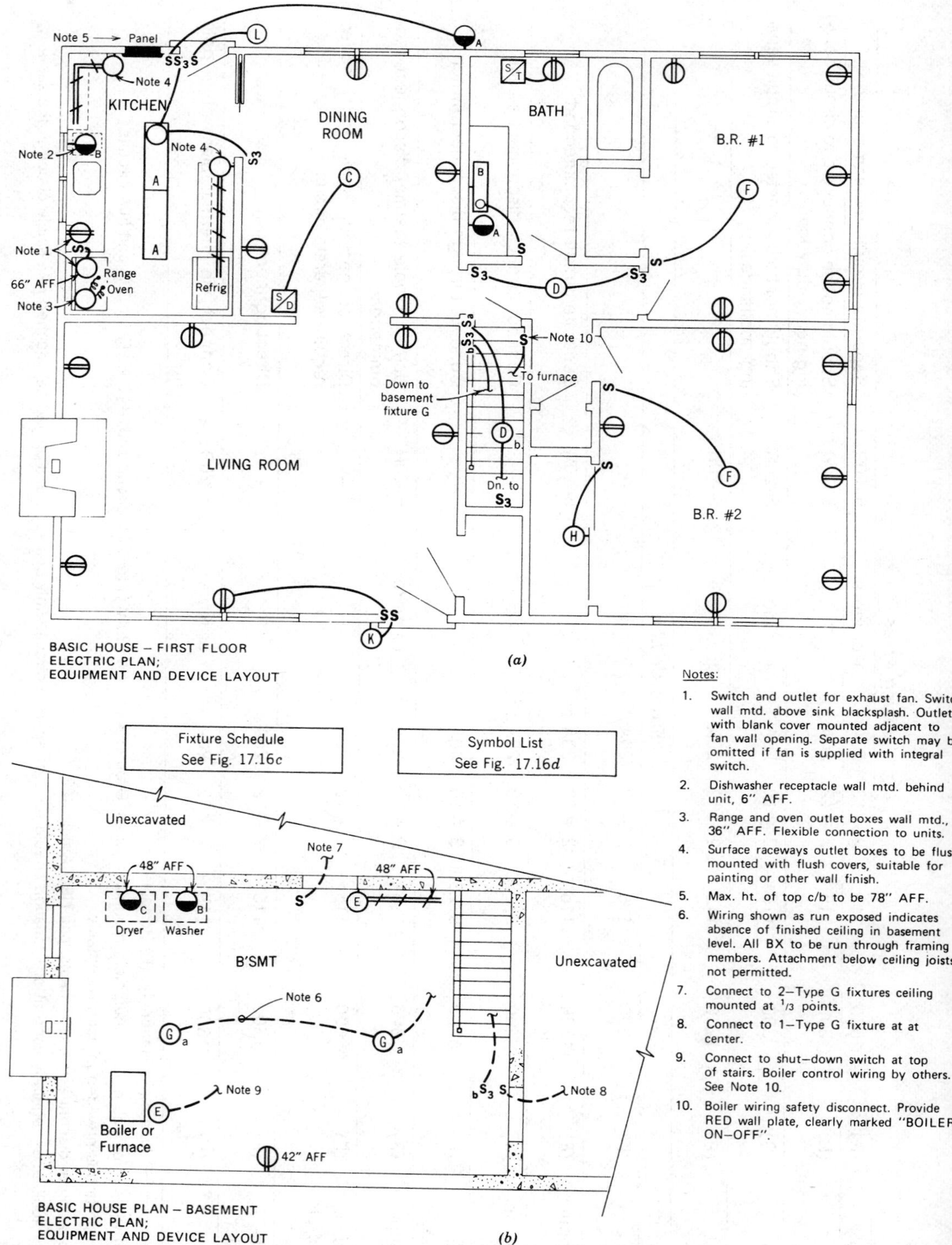

Fig. 17.16 *Electrical plan of a small house. Devices and lighting are shown on main floor and basement plans* (a) *and* (b). *A detailed fixture schedule* (c) *and symbol list* (d) *is also prepared. If circuitry is added, the completed plans for both levels will appear as in* (e) *and* (f). *A detailed panel schedule is given in* (g).

LIGHTING FIXTURE SCHEDULE			
TYPE	DESCRIPTION	MANUFACTURER	REMARKS
A	48" L X 12" W X 4" DEEP NOMINAL, 2 LAMP/FLUORES– CENT, WRAP–AROUND ACRYLIC LENS, F 40 WW/LAMPS. SURFACE MTD.	BRITE–LITE CO. CAT. #2/40/KFF OR EQUAL	4" DEPTH MAXIMUM
B	24" L, 1 LAMP 20W FLUOR. FIXTURE, WRAP–AROUND WHITE DIFFUSER, WITH SINGLE–SWITCHED RECEPTACLE. MOUNT ABOVE MEDICINE CABINET.	BRITE–LITE CO. CAT. #1/20/BFF OR EQUAL	MAX. MTG. HT. 78" TO ℄.
C	ADJUSTABLE HEIGHT PENDANT INCANDESCENT, 3–75W MAX., BUILT–IN 3–POSITION SWITCH.	HOMELAMP CO. CAT. #3/75/DRP OR EQUAL	———
D	10" D. DRUM–TYPE FIXTURE, WHITE GLASS DIFFUSER, CENTER LOCK–UP, 2–60W INCAND. MAX., SURF. MTD.	BRITELITE CO. CAT. #2/60/HF OR EQUAL	6" MAX. DEPTH.
F	12" D. DRUM FIXTURE, CONCEALED HINGE ON OPAL GLASS DIFFUSER FOR RELAMPING WITHOUT GLASS REMOVAL, 2–75W INCAND. MAX. SURFACE MTD.	DENMARK LIGHTING SPECIAL UNIT #374821	NO SUBSTITUTION WILL BE ACCEPTED.
G	PORCELAIN LAMPHOLDER, PULL CHAIN WITH WIRE GUARD, 100W. INCAND. SURF. MTD.	———	———
H	SAME AS TYPE G, EXCEPT W/O GUARD.	———	———
K	DECORATIVE OUTDOOR LANTERN, MAX. 150W INCAND., WALL MTD. 84" AFF TO ℄.	TO BE CHOSEN BY OWNER	———
L	UTILITY OUTDOOR LIGHT, ANODIZED ALUMINUM BODY AND CYLINDRICAL OPAL GLASS DIFFUSER. 1–100W INCAND. MAX. 84" AFF TO ℄.	UTIL–LITE CO. CAT. #1/100/BP OR EQUAL	IF VANDALISM IS OF CONCERN, SUBST. PLASTIC DIFFUSER.

(c)

SYMBOLS AND ABBREVIATIONS

BX CABLES RUN CONCEALED; TICS INDICATE NUMBER OF CONDUCTORS EXCLUDING GROUND WIRES. 2 #12 + BARE GROUND, UON

SAME AS ABOVE EXCEPT RUN EXPOSED.

WIRING RUN TURNING DOWN; WIRING TURNING UP

1 3 HOME RUN TO PANEL; ARROWS AND NUMERALS IDENTIFY CIRCUITS; TICS INDICATE WIRING – AS NOTED ABOVE.

OUTLET BOX AND FINAL CONNECTION TO EQUIPMENT WITH FLEXIBLE CONDUIT (OR BX).

OUTLET WITH SECTION OF SURFACE RACEWAY, 2 WIRE, SINGLE CIRCUIT, AND SEPARATE GREEN GND. 15A, 2P, 3 WIRE, RECEPTACLES ON 12" CENTERS.

D a — CLG. OUTLET WITH INCANDESCENT LTG. FIXTURE D, SWITCH CONTROL – 'a'.

H — WALL OUTLET W/INCAND. FIXT. 'H', M.HT. SHOWN.

A — CLG. OUTLET W/FLUOR. FIXT. 'A'.

B — WALL OUTLET W/FLUOR. FIXT. 'B', M.HT. SHOWN.

J — JUNCTION BOX

DUPLEX CONVENIENCE RECEPTACLE, 15A, 2P, 3W, 125V. GROUNDING, WALL MTD., VERTICAL, ℄ 12" AFF NEMA 5–15R.

A — DUPLEX CONVENIENCE RECEPTACLES, 15A, 2P, 3W, 125V, GROUNDING. W/INTEGRAL GFCI AND GASKETED W.P. SELF–CLOSING COVER.

B — SINGLE RECEPTACLE, 20A, 2P, 3W, GND'G., NEMA 5–20R.

C — SINGLE RECEPTACLE, 30A. 125/250V. 3 POLE–4 WIRE GROUNDING NEMA 14–30R; (NOTE 1)

S_a — SINGLE POLE SWITCH, 15A, 125V, ℄ 50" AFF, UON, CONTROLLING OUTLET(S) 'a'.

$_aS_3$ — SWITCH, 3 WAY, 15A, 125V, ℄ 50" AFF, UON, CONTROLLING OUTLETS 'a'.

S/T — MANUAL TIMER SWITCH, 1 SET 15 AMP N.O. CONTACTS.

S/D — OUTLET BOX MTD. SWITCH AND DIMMER, INCAND. LOAD ONLY, 600 WATTS MAXIMUM. ℄ 50" AFF.

FLUSH MTD. PANELBOARD;

AFF ABOVE FINISHED FLOOR
MHT MOUNTING HEIGHT
T THERMOSTAT
UON UNLESS OTHERWISE NOTED
GFCI GROUND FAULT CIRCUIT INTERRUPTER
WP WEATHER–PROOF
NO NORMALLY OPEN

NOTE 1. CONTRACTOR TO SUPPLY MATCHING CAP.

(d)

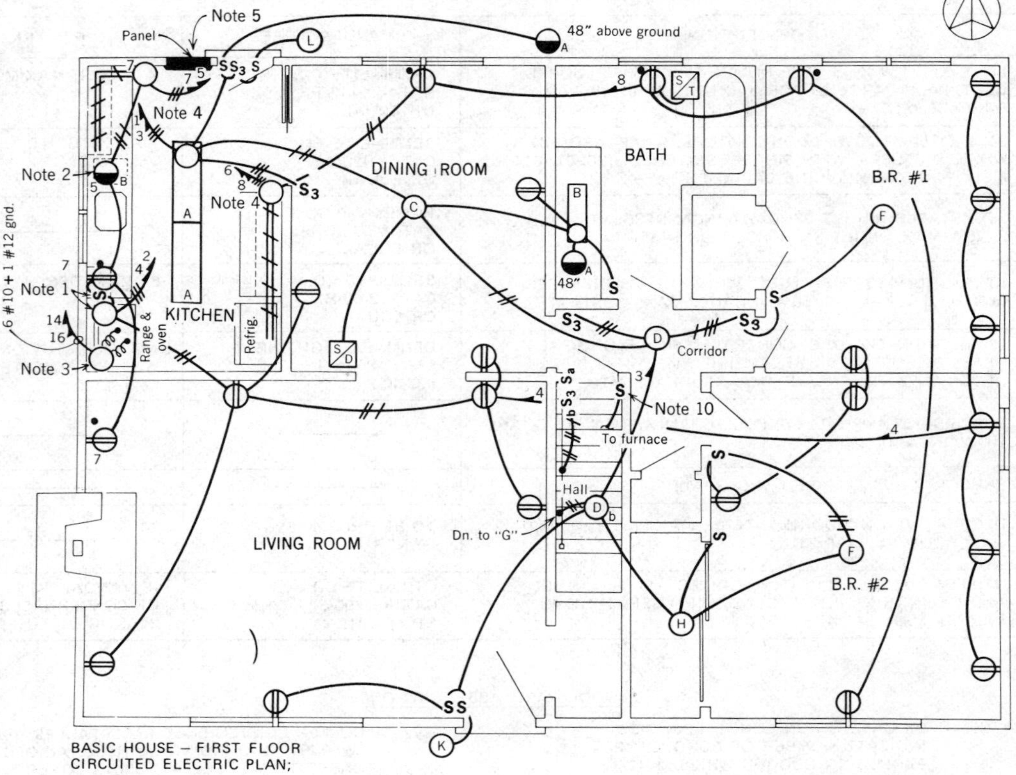

BASIC HOUSE – FIRST FLOOR
CIRCUITED ELECTRIC PLAN;

(e)

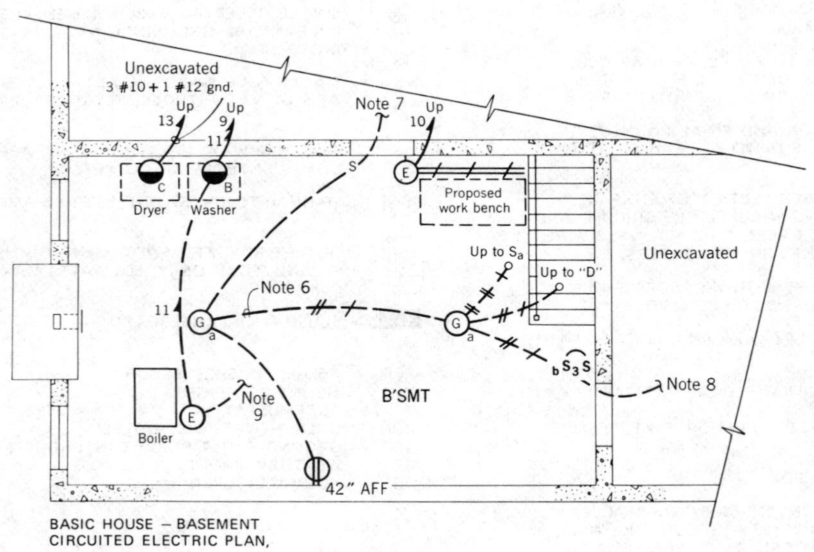

BASIC HOUSE – BASEMENT
CIRCUITED ELECTRIC PLAN,

(f)

PANEL SCHEDULE FOR BASIC HOUSE								
CIRC. NO.	DESCRIPTION	LOAD VA.	CIRCUIT BREAKERS			DESCRIPTION	LOAD VA.	CIRC. NO.
1	LTG – {KIT. DR., BR. #1 / OUTSIDE, BATH + ⏀	820 3R	20	1 | 2	20	OUTLETS – LR. & DR. + EXH. FAN	30 6R	2
3	LTG – {OUTSIDE / LR., HALL, BR. #2, BSMT. + ⏀	935 4R		3 | 4		OUTLETS – BR. 1 & 2	6R	4
5	DISHWASHER	1500		5 | 6		APPLIANCE OUTLETS – KIT., DR.	—	6
7	APPLIANCE OUTLETS – KITCHEN, LR.	—		7 | 8		OUTLETS – BATH, BR's.	—	8
9	LAUNDRY OUTLET – BSMT.	—		9 | 10		OUTLETS – BSMT.	—	10
11	HOT WATER BOILER	1300		11 | 12		SPARE		12
13	ELECTRIC CLOTHES DRYER	5000	30 A 2 P	13 | 14 / 15 | 16	30 A 2 P	RANGE	6000	14
15	SPACE FOR 2 – 1 P	—		17 | 18	30 A	OVEN	4800	16
	OR 1 – 2 P	—		19 | 20	2 P			

PANEL DATA
MAINS, GND. BUS: 150 A MNS., 60 A GND. BUS
MAIN C/B ~~OR SW/F~~ 150/100
BRANCH C/B INT. CAP. 5000 AMP.
MOUNTING – ~~SURF~~/RECESS
REMARKS: FRONT SUITABLE FOR PAINTING

VOLTAGE 120/240 1 PH.

(g)

Table 17.8 *Average Volt-Ampere Ratings for Residential Appliances*

Item	*Volt-Amperes*	*Item*	*Volt-Amperes*
Air conditioners (room)		Heating pad	60
½ ton	880	Heater	up to 1650
¾ ton	1200	Ice cream freezer	115
1 ton	1540	Iron	1320
Blanket	175	Knife sharpener	50
Blender	275	Odorizer	11
Bottle warmer	440	Power tools	up to 1000
Casserole	510	Projector	up to 1000
Clock	2	Radio	30
Corn popper	440	Recorder	95
Dehumidifier	185	Record player	50
Door chime	15	Roaster	1320
Egg cooker	440	Sandwich grill	960
Fans		Sewing machine	75
Floor circulator	120	Serving tray	600
Attic	345	Shaver	11
Kitchen exhaust	75	Sun lamp	275
Portable	50	Tea kettle	550
Floor polisher	475	Toaster	1130
Food warmer	310	Trivet	50
Frying pan	1085	Vacuum cleaners	
Food mixer	130	Bag type	340
Hair dryer	415	Canister type	725
Heating equipment		Tank type	555
Warm air furnace fan	320	Hand type	310
Oil burner	230	Vaporizer	385
Humidifier	185	Waffle baker	960
Heat lamp	250		

All these items utilize a 5–20R receptacle.

17.16 Load Tabulation

While circuiting the loads, a panel schedule is drawn up which lists the circuit numbers, load description and wattage (actually volt-amperes), and the current rating and number of poles of the circuit-protective device feeding each circuit. Spare circuits are included to the extent that the designer considers them necessary and consonant with economy, but normally no less than 20% of the number of active circuits. Finally, spaces are left for future circuit breakers, in approximately the same quantity as the number of spare circuits, but always to round off the total number of circuits.

Panels (other than the type shown in Fig. 16.29, page 618) are normally manufactured with an even number of poles. Thus if a panel had, with spares, 21 poles, the designer would probably require 3 spaces, to give a 24-circuit box. A typical panel schedule is shown in Figure 17.18, which includes the laboratory of Figure 17.17.

In calculating panel loads, the following rules apply.

(*a*) Each specific appliance, device, lighting

Fig. 17.17 (a) *Here alternate methods of circuiting are shown in* (a). *Room 205 shows the actual junction box location, with flexible connections to the box at each fixture. Room 207 shows circuit numbers and switch designations only; the placement of junction boxes is understood, and conduit runs are omitted for clarity, and because they most often are not representative of actual installation. Room 209 shows an outlet box at each fixture, with schematic conduit connections. All these systems are in common use.*

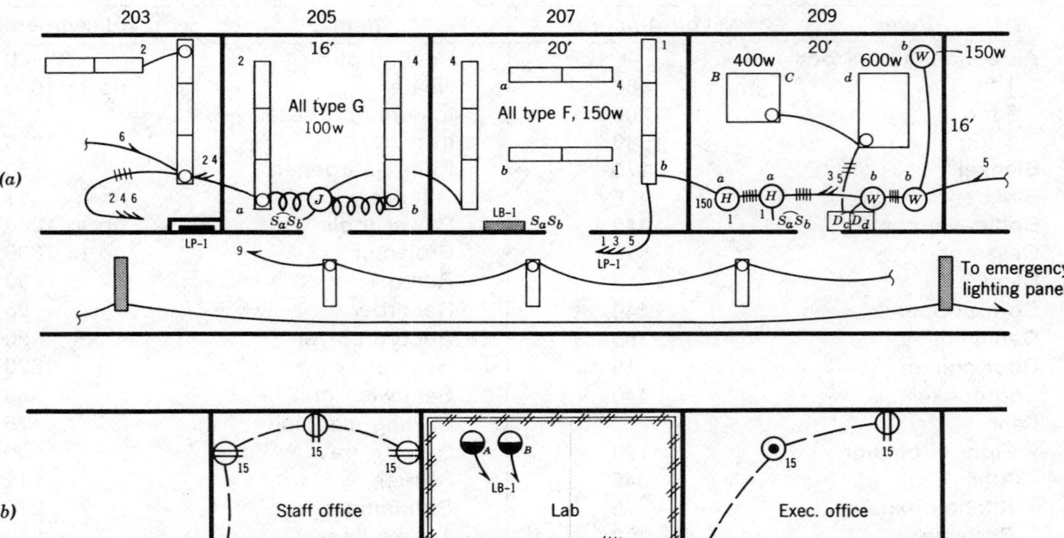

Fig. 17.17 (b) *Typical circuiting of several rooms in an office-lab building. Lighting and power (receptacles) are shown on separate plans* (a) *and* (b) *to avoid crowding (see Fig. 17.8 for symbols and Fig. 17.5 for notes). Lighting in offices is recessed; lighting in labs is surface mounted for flexibility. Note the double circuiting of the Type D receptacles.*

fixture, or other load is taken at its nameplate rating, except certain kitchen and laundry appliances for which the NEC allows a demand factor. (See NEC Article 220.)

(*b*) Each convenience outlet, in other than residential spaces, is counted as 1.5 amp (180 w).

(*c*) Loads for special areas and devices such as show window lighting, heavy-duty lampholders, and multioutlet assemblies are taken at the figures given in NEC Article 220.

(*d*) Spare circuits are figured at approximately the same load as the average active circuits (1200 to 1500 w).

ELECT. PANEL— *LP-1* *120/208 V 3φ 4W*

No.	SERVES	LOAD IN WATTS φA	φB	φC	BRANCH CIRCUIT Poles	Frame	Trip
1	*Lighting*	*1050*			*1*	*50*	*20*
2	*Lighting*	*1050*			↓	↓	↓
3	*Lighting*		*1450*				
4	*Lighting*		*1050*				
5	*Lighting*			*1100*			
6	*Lighting*			*1200*			
7	*Lighting*	*800*					
8	*Lighting*	*1100*					
9	*Lighting – Corridor*		*700*				
10	*Lighting*		*1050*				
11	*Lighting*			*1000*			
12	*Lighting*			*1200*			
13	*Receptacles 5 @ 1.5 amp*	*900*					
14	*Receptacle – Corridor (single pole)*	*900*					
15	*Receptacle*		*900*				
16	*Receptacle*		*900*				
17	*Spare*			*1200*			
18	*Spare*			*1200*			
19	*Spare*	*1200*			↓		
20	*Receptacle – Corridor 2-pole portion*		*1000*	*1000*	*2*		
21	*Spare*	*1200*			*2*	↓	↓
22			*1200*				
23-26	*Spaces only*				*1*	*50*	
	Phase totals	*8250*	*8250*	*7900*			
	Panel total	*24,350*					
	Max. φ current		*70 amp*				
	25% spare capacity	~	*20.*	*(Future loads)*			
	Total I		*90 amps*				

Main breaker *225A 3 pole (see text)*
Trip *100 A*
Feeder size *4 #2 RHW in 2" C.*

***Fig. 17.18** Schedule for lighting panel LP-1.*

(*e*) Spaces are not added into the load.

In calculating total panel loads, as shown in Figure 17.18, no demand factor may be applied except as specifically stated in the NEC. This is true despite the knowledge that most often the usage will be such that average load will be lower than the maximum demand. However, since feeders are calculated for *maximum* loads to be carried, the 100% demand is used. If it is known that certain loads will not or cannot be used simultaneously, the load total should reflect only the larger of the two. Thus, heating and cooling loads are *generally* not concurrent. Nor is building night floodlighting concurrent with the business machine load, but it may be with general interior lighting. Note in Figure 17.18 that 2-pole loads (208 single-phase) appear in two columns. Similarly, 3-phase loads would appear in three columns. Also note that the phase loads are *not equal*. It is the responsibility of the designer (or contractor) to circuit the loads so that the phases are as closely balanced in load as possible. If this is not done, one phase will carry considerably more current than the others. Since the panel feeder must be sized for the maximum phase current, this may lead to an oversized feeder and therefore a waste of money.

Having tabulated and balanced the loads and totaled them by phase, the maximum current is calculated. A portion of the spare capacity available in the branch circuits is added to the above total, as the basis for the calculation of the feeder load. This spare capacity, shown in Table 17.5, is something between 25 and 50%. The exact amount to be added initially in feeder sizing is developed below.

17.17 Spare Capacity

Load calculations for dwelling occupancies are detailed in NEC Article 220 and examples given in Chapter 9 of the NEC. Since these calculations are specialized but routine, and are covered there in detail, they will not be repeated herein. Again it is emphasized that possession of a current copy of the National Electrical Code (NFPA Bulletin 70) is a sine qua non of proper electrical design. Having arrived at the panel load totals as detailed above, the next step is to size the conductors feeding the panel. To do this, an examination of the spare capacity of the panel and of feeders is necessary, in order that the system design be consistent, giving equal capacity for future growth in all its components. Considering the panel circuitry first, let us examine the effect of load expansion, including spares and spaces (see Table 17.9).

(*a*) For buildings expecting a load expansion of 25% (column A, Table 17.5)

(1)	Initial load L =	1.0 L	
	Additional load due to 20% spare circuits =	0.2 L	
	Initial capacity, including spares		1.2 L
(2)	Spare capacity of active circuits = 25% of 1.0 L =	0.25 L	
	Spare capacity of spare circuits = 25% of 0.2 L =	0.05 L	
	Total spare capacity of all circuits, active and spare		0.3 L
(3)	Total panel capacity, without utilizing spaces		1.5 L
(4)	Capacity of 20% spaces 0.2 × 1.25 L		0.25 L
(5)	Maximum panel capacity, including full load on all circuits, active spare and spaces		1.75 L

(*b*) For buildings expecting a load expansion of 50% (column B, Table 17.5)

Table 17.9 ***Panel Initial and Expanded Loads***

	Panels in Facilities Expecting Limited Expansion; Circuits Initially Loaded to Give 25% Expandability (See Col. A, Table 17.5)	*Panels in Facilities Expecting Extensive Expansion; Circuits Initially Loaded to Give 50% Expandability (See Col. B, Table 17.5)*
Initial load	1.0	1.0
Initial plus spares	1.2	1.2
Load after all circuits including spares are loaded to maximum allowable	1.5	1.8
Load after utilizing 20% spaces also	1.75	2.1

(1) Initial load L =	1.0 L	
Additional load due to 20% spare circuits =	0.2 L	
Initial capacity, including spares		1.2 L
(2) Spare capacity of active circuits = 50% of 1.0 L =	0.5 L	
Spare capacity of spare circuits = 50% of 0.2 L =	0.1 L	
Total spare capacity of all circuits, active and spare		0.6 L
(3) Total panel capacity, without utilizing spaces		1.8 L
(4) Capacity of 20% spaces 0.2 × 1.50 L		0.3 L
(5) Maximum panel capacity, including full load on all circuits, active spare and spaces		2.1 L

These results can be summarized as follows: For panels in buildings expecting limited expansion, for which branch circuits are loaded to 80% of capacity (25% expansion), the ultimate panel load without new conduit work, merely by filling out circuits, is 1.5 L or 50% beyond the initial load. By adding breakers in the spaces this load can be expanded to 75% beyond the initial load. The corresponding figures for panels that are lightly loaded (66% capacity), in anticipation of considerable expansion, are 80 and 110%, respectively. These results are summarized in Table 17.9.

17.18 Feeder Capacity

To achieve economy, the panel feeder must accommodate the initial load plus some portion of the future load. Spare capacity in feeders is provided by one or more of the following procedures:

(*a*) Provide feeder (and conduit) capacity initially, to handle the entire eventual load. This method is most expensive—requiring initial outlay for no return—and is rarely used.

(*b*) Provide feeder for initial plus spare with properly sized conduit. Conduit is sized for

1. Type RHW with outer covering.
2. Type THW or RHW without covering.

This method, as we shall see, is generally most economical.

(*c*) Provide feeder for initial plus spare, with conduit oversized by one size. As above, size conduit for type A wire. Some additional cost is entailed here. This is only used where large load expansion is anticipated.

Table 17.10 ***Maximum Number of Conductors in Trade Sizes of Conduit or Tubing***

Type Letters	Conductor Size AWG, MCM	Conduit Trade Size (Inches)									
		$\frac{1}{2}$	$\frac{3}{4}$	1	$1\frac{1}{4}$	$1\frac{1}{2}$	2	$2\frac{1}{2}$	3	$3\frac{1}{2}$	4
TW, T, XHHW	12	7	12	19	35						
(14 to 8)	10	5	9	15	26	36					
	8	2	4	7	12	17	28				
RHW and RHH	12	4	8	13	24						
(without outer	10	4	6	11	19	26					
covering), THW	8	1	3	5	10	13	22				
TW, T, THW,	6	1	2	4	7	10	16				
RHW, and RHH	4	1	1	3	5	7	12	17			
(without outer	2	1	1	2	4	5	9	13			
covering)	1		1	1	3	4	6	9			
	0		1	1	2	3	5	8	12		
	00		1	1	1	3	5	7	10		
	000		1	1	1	2	4	6	9	12	
	0000			1	1	1	3	5	7	10	
	250			1	1	1	2	4	6	8	10
	300			1	1	1	2	3	5	7	9
	350				1	1	1	3	4	6	8
	400				1	1	1	2	4	5	7
	500				1	1	1	1	3	4	6
	600					1	1	1	3	4	5
	700					1	1	1	2	3	4
	750					1	1	1	2	3	4
	12	10	18	29							
THWN	10	6	11	18	32						
	8	3	5	9	16	22					
THHN	6	1	4	6	11	15	26				
	4	1	2	4	7	9	16				
	2	1	1	3	5	7	11	16	25		
	1		1	1	3	5	8	12	18		
XHHW (4 to 500	0		1	1	3	4	7	10	15		
MCM)	00		1	1	2	3	6	8	13		
	000		1	1	1	3	5	7	11	14	
	0000		1	1	1	2	4	6	9	12	
	250			1	1	1	3	4	7	10	12
	300			1	1	1	3	4	6	8	11
	350			1	1	1	2	3	5	7	9
	400				1	1	1	3	5	6	8
	500				1	1	1	2	4	5	7
	600				1	1	1	1	3	4	5
	700					1	1	1	3	4	5
	750					1	1	1	2	3	4

(*d*) Provide for initial load plus spare, with an empty conduit for future. This method is expensive because of high conduit cost, and it is infrequently advisable.

In (*b*) and (*c*) the future capacity beyond that initially supplied is handled by the use of larger gauge wire in the existing conduit, as allowed by the NEC. To examine exactly what these alternatives provide in spare capacity we have tabulated in Table 17.11 the maximum ampacity of various size conduits and the future capacity obtainable. Table 17.10 is taken directly from the NEC.

Note from Table 17.11 that simply by rewiring we can obtain additional capacity of from 16 to 138%, and if the conduit had been oversized one size, this rewiring capacity would be from 37 to 261%.

Returning now to the question of how large

Table 17.10 *Maximum Number of Conductors in Trade Sizes of Conduit or Tubing (Continued)*

Type Letters	Conductor Size AWG, MCM	Conduit Trade Size (Inches)									
		$\frac{1}{2}$	$\frac{3}{4}$	1	$1\frac{1}{4}$	$1\frac{1}{2}$	2	$2\frac{1}{2}$	3	$3\frac{1}{2}$	4
XHHW	6	1	3	5	9	13	21				
	600				1	1	1	1	3	4	5
	700					1	1	1	3	4	5
	750					1	1	1	2	3	4
	12	3	5	9	15	21					
RHW	10	2	4	7	13	18	29				
	8	1	2	4	7	9	16	22			
RHH (with outer covering)	6	1	1	2	5	6	11	15			
	4	1	1	1	3	5	8	12			
	2		1	1	3	4	6	9	14		
	1		1	1	1	3	5	7	11	14	
	0		1	1	1	2	4	6	9	12	
	00			1	1	1	3	5	8	11	14
	000			1	1	1	3	4	7	9	12
	0000			1	1	1	2	4	6	8	10
	250				1	1	1	3	5	6	8
	300				1	1	1	3	4	5	7
	350				1	1	1	2	4	5	6
	400				1	1	1	1	3	4	6
	500				1	1	1	1	3	4	5
	600					1	1	1	2	3	4
	700					1	1	1	1	3	3
	750						1	1	1	3	3

Source. Extracted from the *National Electrical Code, 1978.*

to make the feeder for a given panel load, we must balance future panel load, initial cost of feeder, and future capacity of existing conduit. Prime consideration must be given to avoid the installation of empty conduits since this is extremely expensive. However, rewiring is relatively inexpensive.

Referring to Table 17.11, note that normal design uses Type 2 cable and that design with Type 1 is, in effect, a first step in oversizing conduit. The second step is a deliberate oversizing of conduit that results in much increased conduit ampacity, as reflected by the figures in Table 17.11. Using these figures in actual practice requires that the designer juggle cable and conduit cost against anticipated load growth to arrive at the most economical long-term solution. Applying these numbers to concepts previously developed we have:

(*a*) Buildings expecting 25% expansion have a panel capacity of 1.75 times the load. The feeder can be matched to this by

1. Using Type 2 feeder (normal design). Calculate the feeder on the basis of panel load plus 20% spares, and oversize the feeder by 20%. This gives a feeder capacity of
   $$1.20 \times 1.2 = 1.44 \times \text{initial load}$$
   Table 17.11 indicates that rewiring will add another 20% on the average. This gives
   $$1.2 \times 1.44 = 1.73 \times \text{initial load}$$
   which corresponds closely to the 1.75 desired.
2. Using a Type 1 feeder, and oversizing it initially by only 10%, gives approximately the same result, that is, 1.10 × 1.2 plus 30% rewiring expansion or
   $$1.32 \times 1.3 = 1.72$$
   which is the desired figure.

Table 17.11 ***Maximum Wire and Ampacity of a Conduit, and Ampacity Gain on Rewiring***

Conduit	Type 1[a] Cable RHW with Covering		Type 2[b] Cable THW & RHW Without Covering		Rewiring with Types THHN or XHHW; Expansion over Type 1 & Type 2 Cable					
					Same Conduit Size				Having Initially Oversized the Conduit One Size[l]	
Size (Inches)	Max.[c,e] Wire	Max.[d] Amperes	Max.[c] Wire	Max.[d] Amperes	Max.[c] Wire	Max.[d] Amperes	Type 1	Type 2	Type 1	Type 2
$1\frac{1}{2}$	6	65	1	130	1/0	155	138%	19%	261%	81%
2	1/0	150	3/0	200	4/0	235	57%	18%	100%	50%
$2\frac{1}{2}$	4/0	230	250	255	300	300	30%	18%	76%	59%
3	350	310	400	335	500	405	31%	21%	58%	46%
					2–4/0	376[f,g]				
$3\frac{1}{2}$	500	380	600	420	700	490	29%	17%	52%	37%
			2–250	408[f,h]	2–300	480[f,i]				
4	600	420	750	475	750	500	37%	16%	—	—
	2–250	408[f,h]	2–350	496[f,j]	2–400	576[f,k]				

[a]Type 1 cable is 75° C with outer covering, that is, RHW with covering.

[b]Type 2 cable is 75° C cable without covering, that is, THW & RHW uncovered.

[c]See Table 17.10 for conduit capacities, sizes AWG and MCM.

[d]See Table 15.2, page ■■■, for ampacities.

[e]Each conduit is assumed to be carrying a 3-phase load, with full neutral.

[f]Parallel feeders; ampacity is double that of a single feeder, reduced by derating factors given in Table 15.2*b*.

[g]$2 \times 235 \times 0.8 = 376$ amperes.

[h]$2 \times 255 \times 0.8 = 408$ amperes.

[i]$2 \times 300 \times 0.8 = 480$ amperes.

[j]$2 \times 310 \times 0.8 = 496$ amperes.

[k]$2 \times 380 \times 0.8 = 576$ amperes.

[l]For wire size and ampacity see columns 6 & 7, but one line down. For example, a 2″ conduit holds #4/0 THHN or XHHW, with 235A capacity. Expansion is therefore: $\frac{235A - 65A}{65A} = 261\%$.

(*b*) For a building with 50% expansion, utilizing full space capacity gives us a feeder size of 2.1 times the initial load (see Table 17.9). This is accomplished with

1. Type 2 feeder. Oversize the feeder by 15% and oversize the conduit by one size. This latter step gives an expansion of approximately 50%. Therefore,
   $$\text{feeder capacity} = 1.15 \times 1.2 \times 1.5 = 2.07$$
   which is the desired figure. Alternatively
2. Using a Type 1 feeder, simply oversize the conduit by one size. This gives
   $$\text{feeder capacity} = 1.0 \times 1.2 \times 1.75 = 2.1$$
   having assumed a 75% expansion due to the conduit oversizing (see Table 17.11, last column; since the figures in this column vary considerably, the actual number should be used in lieu of the 75% assumed above).

If, as in the case of laboratories, more than 100% expansion is anticipated (see Table 17.1), conduit should be oversized and initial wiring oversized by 50%. Feeders thus arranged will handle the new panels required to meet the load expansion.

Initial feeder capacity (in oversize conduit) $= 1.5 \times 1.2 = 1.8 \times$ load

Rewired capacity $= 1.7 \times 1.8 = 3.0 \times$ load (assuming 70% increase in capacity with rewiring per Table 17.11)

Two factors should be carefully noted here. First, note that the smaller conduits offer the largest expandability although, in dollars per amperes, they are more expensive. Second, in order to take advantage of spaces in a panel, conduit stubs should be taken from the panel and extended into hung ceilings or another procedure used to make the panel circuitry accessible in the future.

17.19 Panel Load Calculation

EXAMPLE 17.1. Refer to Figure 17.18. The panel is for a laboratory/office area. Since large expansion is anticipated, circuitry follows column B of Table 17.5. Ultimate panel load would be 26 cct at 1900 w = 50 kva = 138 amp. Thus the initial feeder is sized for 90 amp but rewiring will allow as much as 235 amp in a 2 in. conduit (see Table 17.11). A 225-amp frame c/b, is chosen initially, since eventually the trip will be raised to 150 amp.

EXAMPLE 17.2. Assume a single floor of an office building 100 ft × 200 ft. Assume also that 15% of the area is corridor and storage. Calculate the required number of panels, circuits, and feeder size. Assume a good grade speculative construction venture, utilizing Code minima, consistent with good practice.

SOLUTION

$$\text{Office space} = 85\% \text{ of } 20{,}000 \text{ sq ft} = 17{,}000 \text{ sq ft}$$

$$\text{Corridor and storage} = 15\% \text{ of } 20{,}000 \text{ sq ft} = 3000 \text{ sq ft}$$

With respect to minimum loads, the Code specifies that they be increased by 25% if loads are continuous (3 or more hours). This requirement allows for breakers to heat up in panels while carrying continuous load, and is waived for c/b—which are ambient compensated, that is, are rated to carry 100% load. Since we have established 80% of the breaker rating as maximum load (see Table 17.5), *we have already accounted for this factor in circuitry,* but must keep it in mind in feeder calculation.

Office load

17,000 sq ft @ 5 w/sq ft = 85 kw

Storage

3000 sq ft @ 0.5 w/sq ft = 1.5

TOTAL LOAD = 86.5 kw

Minimum feeder capacity $1.25 \times 86.5 = 108$ kw. The 25% additional capacity is for continuous load, as stated above.

Since this building is of good grade office construction, we anticipate 40 to 80% expansion, and utilize an initial branch circuit loading of 1300 w per cct.

$$\text{Number of cct} = \frac{86.5}{1300} = 68$$

Because of the size of the building, three panels are required to keep branch circuits below 100 ft in length. In practice, *though not required by Code,* additional circuits are provided for receptacles and spares.

Receptacles:

First 400 sq ft	10 receptacles
Next 16,600 @ 110 sq ft per	151 receptacles
	161 receptacles

Circuited at 6 per 20A cct

$$\text{No. of ccts} = \frac{161}{6} = 27 \text{ ccts}$$

Lighting plus receptacles 68 + 27 =	95 ccts
Spares = 20%	20
Total	115 ccts

Each panel would then have 115/3 or 38 circuits plus 4 spaces, for a maximum of 42 poles.

Thus, with initially three panel locations we proceed to circuit the lighting and receptacles according to the actual tenant requirements. If the actual number of circuits is less than 68, then 68 must be provided; if greater as in our calculation above, the actual number required must be used. In either case, 20% spare should be used. In calculating the panel load for feeder sizing, the actual load as determined by adding the lighting and other loads on the panel is compared to the load by square foot calculation. The larger figure is used in determining the required panel feeder size. Thus actual load would be compared to 108 kw and the larger used.

Feeder would be sized thus:

Assuming even distribution of load, and actual load greater than minimum

Panel load: 38 cct @ 1300 w	= 49.5 kw
25% future	12.5
Feeder load	62 kw
	= 172 amp

Use 4 #2/0 THW, $2\frac{1}{2}$ in. C.

The feeder current is calculated in terms of the panel 3-phase kva thus:

$$I = \frac{\text{kva}}{0.360} \text{ amps}$$

Thus, above feeder current is

$$I = \frac{62 \text{ kw}}{0.360} = 172 \text{ amps}$$

As an aid in computing currents, Table 17.12 lists various current, voltage, and wattage relationships.

Table 17.12 *Current and Wattage Relationships* [a]

Load (Watts)	120 v Single-Phase	120/240 v 3-Wire	120/208 v Single-Phase, 3-Wire	120/208 v 3-Phase	277/480 v 3-Phase	277 v Single-Phase
100	0.83	0.41	—	—	—	0.362
200	1.6	0.8	—	—	—	0.72
500	4.2	2.1	—	—	—	1.8
1,000	8.3	4.2	4.8	2.77	1.2	3.6
2,000	16.6	8.3	9.6	5.5	2.4	7.2
5,000	41.7	20.8	24.0	13.9	6.0	18.0
10,000	83.2	41.6	48.0	27.7	12.0	36.0
20,000	—	—	96.0	55.6	24.0	72.0
50,000	—	—	240.0	139.0	60.0	180.0
100,000	—	—	480.0	277.0	120.0	362.0
	$I = \frac{W}{120}$	$I = \frac{W}{240}$	$I = \frac{W}{208}$	$I = \frac{W}{360}$	$I = \frac{W}{830}$	$I = \frac{W}{277}$

[a] Assuming 100% power factor.

17.20 Riser Diagrams

When all devices are circuited and panels are located and scheduled, we are ready to prepare a riser diagram. A typical diagram, shown in Figure 17.19, represents a block version of a single-line diagram except that, as the name implies, vertical relationships are shown. All panels, feeders, switches, switchboards, and major components are shown up to, but not including, branch circuiting. This diagram is an electrical version of a vertical section taken through the building.

EXAMPLE 17.3. Feeder F10 of Figure 17.19 serves lighting panels 1A, 2A, and 3A. Calculate the required feeder size, considering loads, future expansion, and voltage drop.

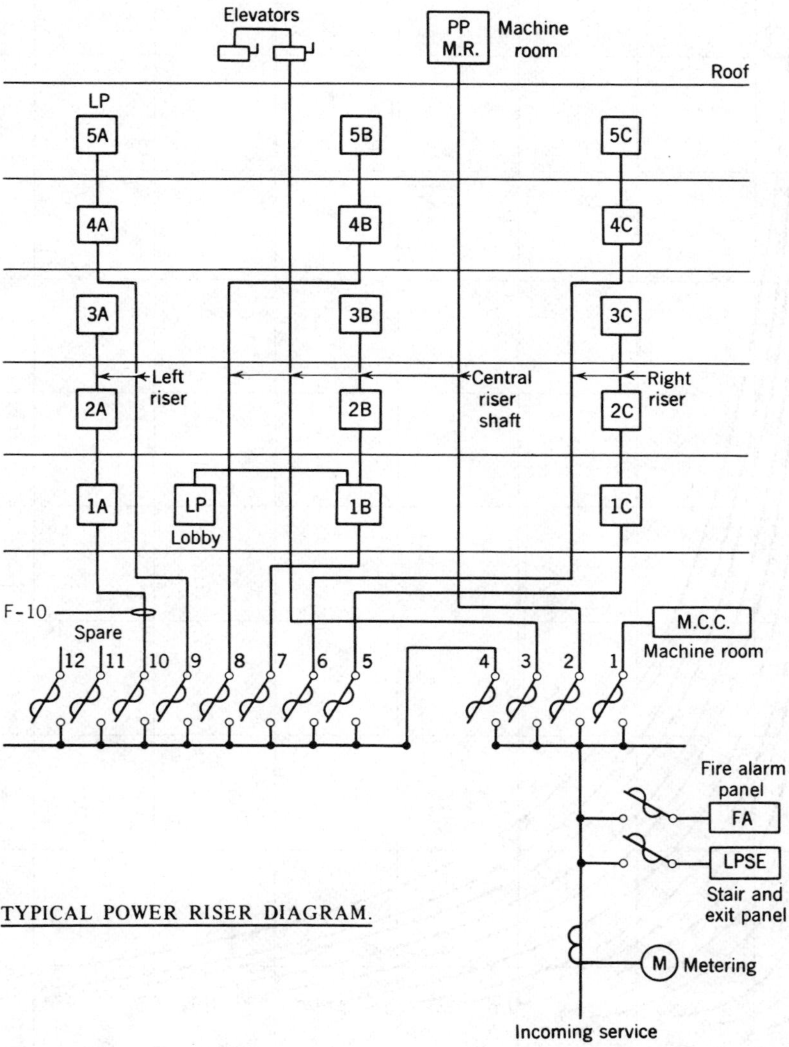

Fig. 17.19 Typical power riser diagram. Ordinarily the main switchboard would be shown as a large rectangle with the feeder emanating from it, and a switchboard schedule would detail the contents. Here, because of the unusual bus arrangement, we have shown the main switchboard as it would appear on a single-line diagram.

SOLUTION. The loads on these panels have been computed in accordance with the above considerations and are:

Connected Load	LP-1A—125 amp
	LP-2A—150 amp
	LP-3A—140 amp
	415 amp

These figures include connected load, spares, and a 25% future factor. *The NEC requires that demand be taken per panel at 100% only for load calculated by square feet, which is lower than actual circuited loads.* If actual panel circuit loads are utilized, any reasonable demand may be used, provided that at no time do the loads drop below the

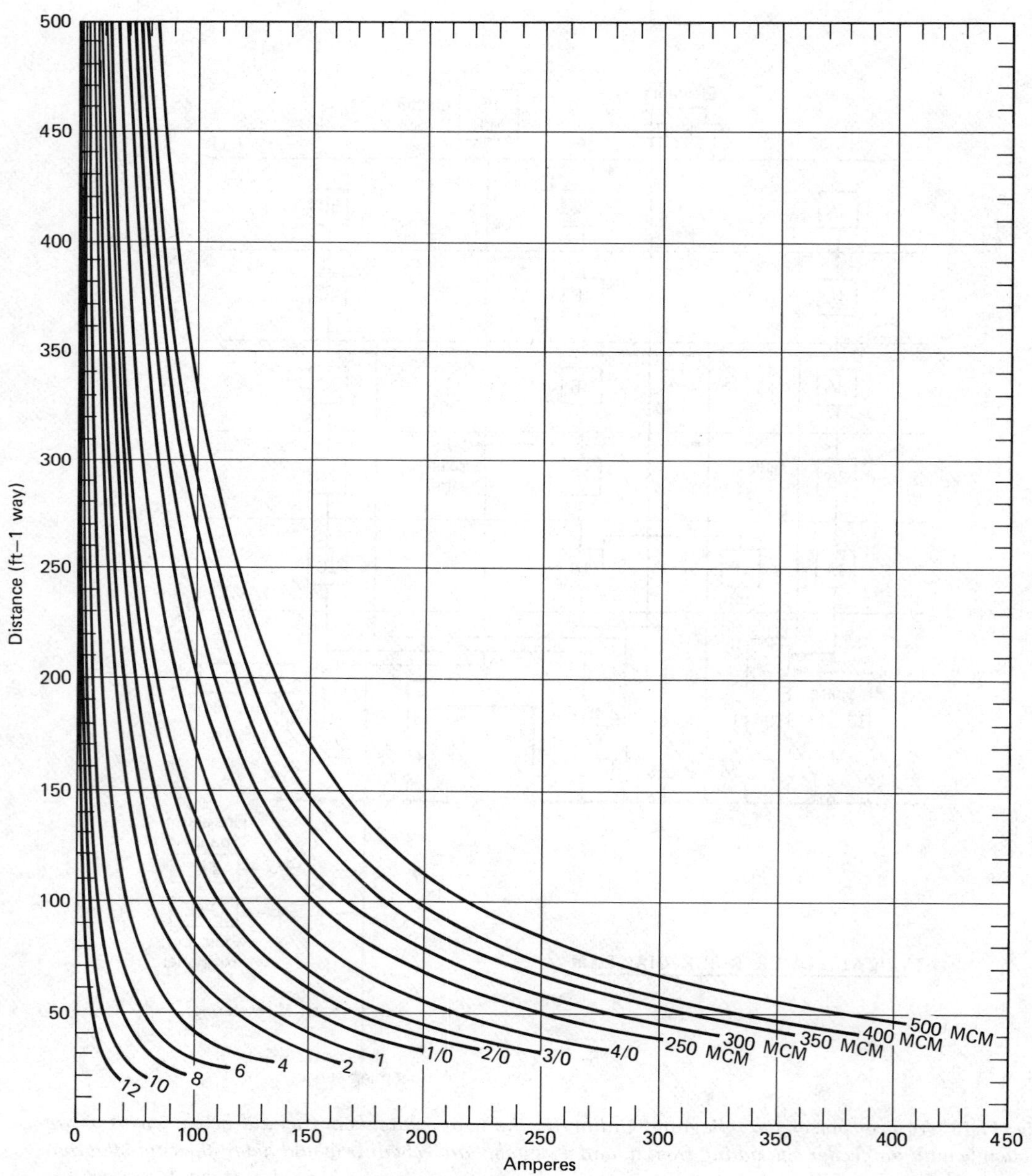

Fig. 17.20 Curves for determining voltage drop in copper cables.

minima specified in the NEC. Furthermore, we may apply diversity factors between panel loads in a judicious manner.

In office work, typical diversity factors are

Lighting Panels Fed from a Single Feeder	*Diversity Factors*
1,2	1.00
3,4	1.09
5,6, or 7	1.18
8,9, or 10	1.33

Thus the load on feeder F-10, using 100% demand per panel and 1.09 diversity between panels, would be 415 × 1.0/1.09 = 380 amps.

Methods for handling future expansion were discussed above. In this case the feeder, before voltage drop considerations, would be (from Tables 15.2 and 17.10) 4-500 MCM THW in $3\frac{1}{2}$ in. C. Note from Table 17.11, however, that $3\frac{1}{2}$ in. C gives the lowest percent increase in rewiring. In this case, the wise choice would be to increase the feeder conduit to 4 in., giving an eventual rewired capacity of 576 amps, a 52% increase.

The final consideration in sizing a feeder is voltage drop. The NEC specifies that the drop from source to final panel should be no more than 1% for lighting and combination lighting and power feeders and 3% for power and heating load feeders. These figures should be adhered to as closely as possible. In instances of long runs where these restrictions will cause excessive cost, lighting feeders up to the branch panel may be run with 2% drop and power feeders up to 4% voltage drop. Total voltage drop to the last outlet in the circuit shall not exceed 5%, to comply with ASHRAE Standard 90-75. See Section 17.5(a). Many tables and curves are published by manufacturers from which voltage drop can be obtained. Such a set of curves is shown in Figure 17.20. Applying these curves to our last example:

Allowable voltage drop
= 1% of 208 v = 2.08 v
Distance—assume 80 ft run

From the curves, 380 amps on 500 MCM cable will give a 1% drop in 40 ft. Therefore, the allowable 2% drop will be reached in 80 ft. Since our run is 80 ft in length, 500 MCM is adequate.

In summary, then, feeders are sized in accordance with load (actual or sq ft, whichever is larger) and voltage drop. Conduit may be oversized for large future load expansion.

17.21 Service Equipment and Switchboard Design

The main switchboard shown in Figure 17.19 constitutes a combination of service equipment and feeder switchboard. The service equipment portion of the board comprises the metering and the 4 main switches feeding risers, motor control center (MCC), roof machine room, and elevators. The feeder board comprises switches 5 through 12. Such an arrangement is permissible inasmuch as the NEC allows up to 6 fused switches or circuit breakers to serve as the service disconnect means. This arrangement was chosen in order to separate to the largest extent possible the motor loads (elevators, air-conditioning equipment, basement power, etc.) from the lighting. Such a procedure minimizes lighting fluctuations resulting from motor starting and yields simpler maintenance. Also, the size of the main switch is reduced. This switchboard would be of the metal clad dead-front type with switches or circuit breakers, as desired. For protection of Feeder F10 designed above, an 800-amp circuit breaker or 600A switch, with 400-amp trip or fuse, respectively, would be chosen, assuming the initial installation to be a set of No. 500 MCM cables.

Other considerations and general rules affecting service equipment are listed below.

(*a*) A building may be supplied at one point by either a single set or parallel sets of service conductors.

(*b*) Service drops may generally be not less than No. 8 AWG and service entrance conductors or underground service conductors not less than No. 6 AWG.

(*c*) All equipment used for service including cable, switches, meters, and so on, shall be approved for that purpose.

(*d*) It is recommended that a minimum of 100-amp, 3-wire, 120/240 v service be provided for all individual residences.

(*e*) No service switch smaller than 60 amp or circuit breaker frame smaller than 50 amp shall be used.

(*f*) In multiple occupancy buildings tenants must have access to their own disconnect means.

(*g*) All building equipment shall be connected on the load side of the service equipment except that service fuses, metering, fire alarm, and signal equipment and equipment serving emergency systems may be connected ahead of the main disconnect (see Fig. 17.19).

In computing a size for the service equipment *bus* a total is taken of the various feeder loads. Although application of a Diversity Factor to this total is permissible, good practice dictates the use of a unity Diversity Factor in order to provide a measure of spare capacity in the service equipment.

The connected loads on the machinery feeders are normally taken at 70 to 85% Demand Factor in comparison to the 90 to 100% Demand Factor normally used on lighting feeders.

17.22 Emergency Systems

Some of the considerations relevant to power reliability were discussed in Section 17.1(b), and a brief review of the equipment available to supply emergency power was presented in Section 16.29. Emergency lighting equipment is covered in Section 22.27. In this section, we will discuss possible arrangements of emergency power supply. The choice of arrangement and the size and type of equipment depends in large measure on the requirements of local codes, which determine the loads to be fed from the emergency system. The reader should note that, although we are using the term "emergency," the concepts involved are equally applicable to "standby" systems (see Section 16.29, page 637), remembering that the NEC requires emergency and standby systems to be kept entirely separate.

In general, when emergency power is discussed, it is assumed to be replacing "normal" power; that is, the assumption underlying governmental codes and ordinances is that power must be supplied to selected loads within the building because of a utility power outage. Cognizance *is not* taken of situations where normal power has not failed and the outage is localized because of an equipment failure. That aspect of design—reliability—is left to the designer. Some of the arrangements that will be discussed below differentiate between the nature of outages, that is, a utility or general outage versus an equipment or local outage. An exception to the above generalization occurs with health-care facilities, where the NEC in Article 517 specifies an internal electrical design that will in large measure cover both types of outages, down to the distribution level. The interested reader should refer to the referenced NEC article for further reading.

The emergency system includes all devices, wiring, raceways, and other electrical equipment, including the emergency source that is intended to supply electric power to the selected loads. These loads normally include egress lighting (stair, corridor, and exit and lobby lights), signal equipment such as public address and fire alarm that must remain functional during an emergency, and one or more elevators as required by Code. The recognized arrangements are discussed below.

(*a*) Where emergency loads are light, a storage battery arranged to be connected automatically on power outage is used. Where all emergency loads can be supplied with direct current, the arrangement of Figure 17.21*a* is used. (Note that a-c lighting can accept d-c emergency power if equipped with a local inverter, as discussed in Section 22.27.) If a-c is required, the arrangement of Figure 17.21*b* is utilized. When the emergency equipment is entirely separate from the normal equipment and is *normally deenergized,* the system of Figure 17.21*c* is used. This arrangement is used in small facilities requiring egress lighting only, where it is found that supplying a completely separate emergency

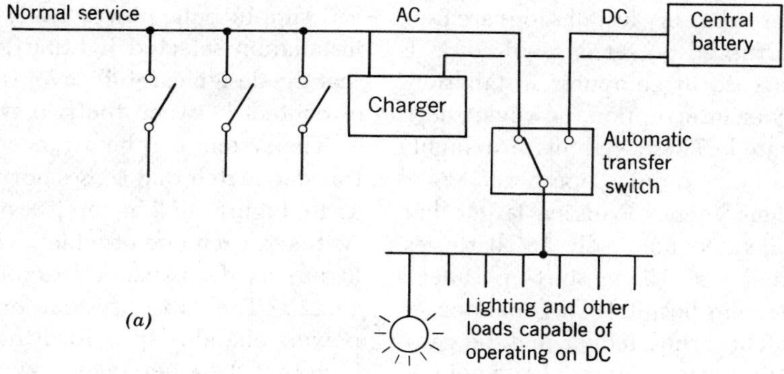

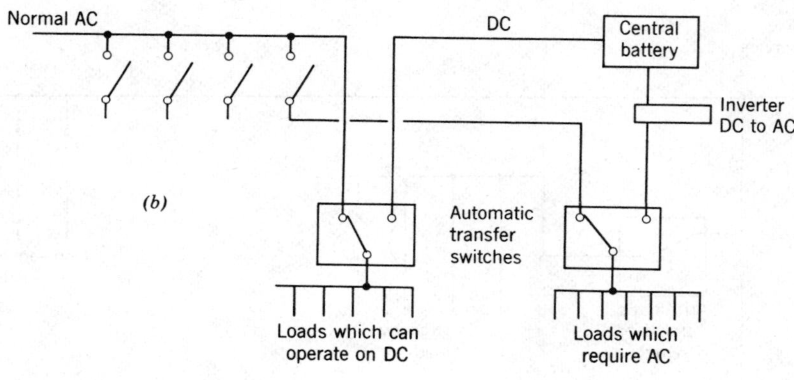

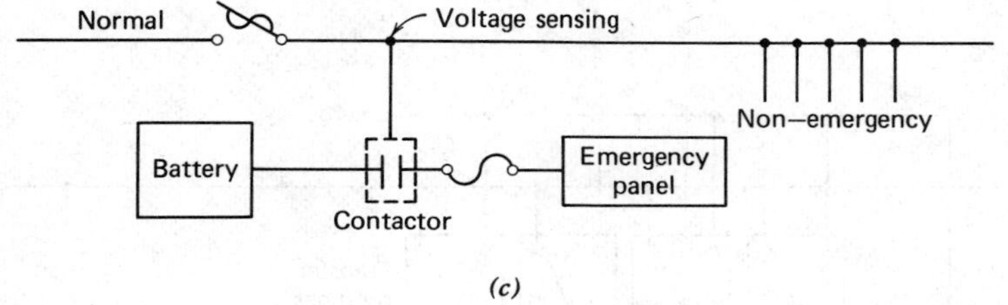

(c)

Fig. 17.21 *Use of a central battery for emergency power supply. Where all loads can be energized with d-c the arrangement of* (a) *is satisfactory. When a-c as well as d-c must be distributed, a central inverter is added as in* (b), *and the a-c and d-c emergency loads fed separately. In* (c) *the emergency loads are normally de-energized and are activated through the contactor when it senses power loss.*

system is the preferred economic or engineering choice. Large battery installations are used where uninterrupted power is required, as is generally the case in computer installations where no power interruption, however short, can be tolerated. These systems are highly technical and beyond our scope here.

(*b*) Where emergency loads are larger than can be supplied economically by batteries, and where the 8 to 15 sec start-up time is tolerable (10 sec in hospitals), a generator set is employed. The prime mover may be gasoline, diesel, steam, or gas. (It should be pointed out that a combination of sources can be used in a single building. For instance, a generator can supply bulk power loads and a battery installation selected lighting loads, provided that the design carefully avoids any possibility of contact between the two systems.)

The system can be arranged with a single transfer switch that senses normal power loss, as in Figure 17.22*a*, or it can use multiple switches, each one of which will sense power loss *at its downstream location*, as in Figure 17.22*b*. The latter system provides greater power reliability, provided that the design is such that the emergency power uses an independent power path to the transfer switches.

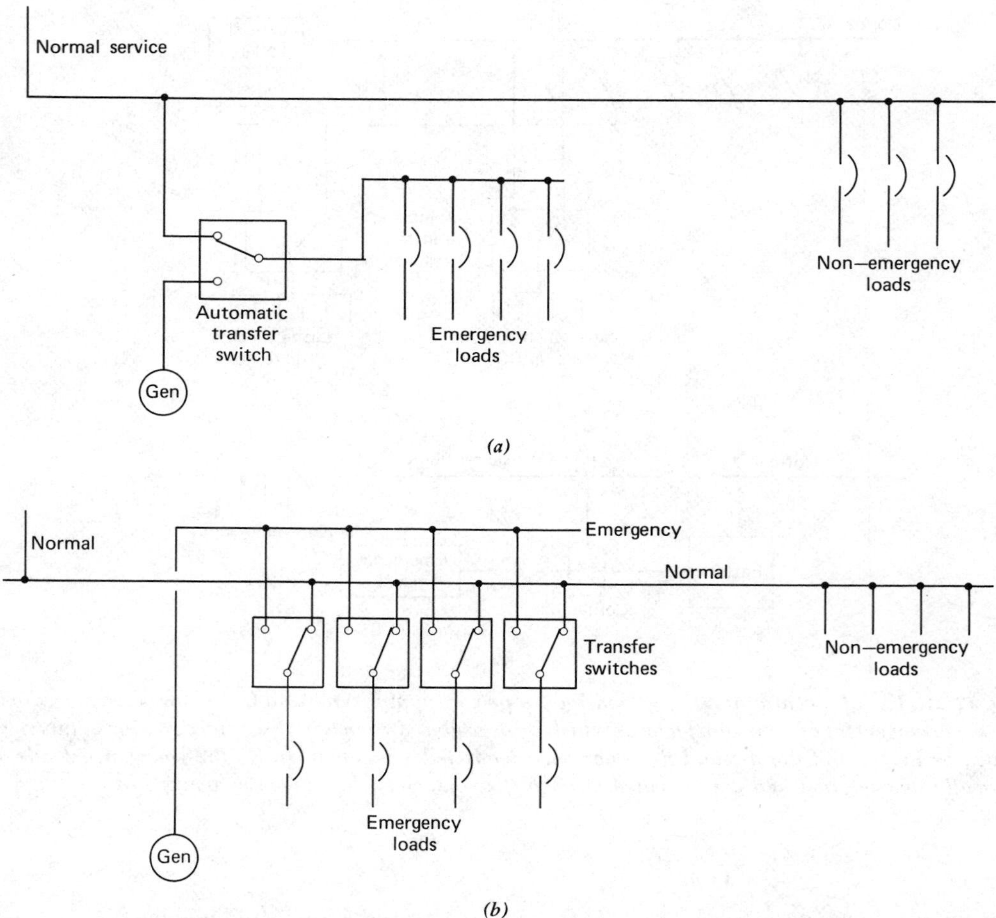

Fig. 17.22 *Alternate arrangements of emergency/normal power feed. In* (a) *a single transfer switch serves the normal power and transfers to the generator on power failure. In* (b) *the transfer switches are smaller, thus reducing the chance of a single equipment failure faulting out the entire emergency power system.*

Otherwise, a faulted piece of equipment that will interrupt normal power downstream will also prevent emergency power from reaching that point.

(*c*) Many codes permit the use of two separate electric services in lieu of a normal service plus an emergency source, provided that the two sources are independent, that is, come from different utility transformers or feeders, enter the building at different points and preferably from different directions, and use separate service drops or laterals. The point is, of course, that the type of reliability desired can only be obtained by minimizing the possibility of a single event interrupting both services. The usual arrangement is for one service to be "normal" and the other "standby," as in Figure 17.23*a*. A much less frequent case utilizes both feeders as "normal," each carrying part of the normal load and each acting as a standby for the other. This is shown in Figure 17.23*b*.

(*d*) The least reliable arrangement is one in which the emergency loads are connected ahead of the main disconnects and are so arranged that a downstream fault within the building will not affect these items. This situation is illustrated in the riser diagram of Figure 17.19 where the stair and exit panel, which supplies egress lighting, and the fire alarm panel are connected ahead of the building main disconnect and protected with their own fuses. Such an arrangement obviously can do nothing in the event of a power outage and, although once very popular, it is now falling into disuse as a result of more stringent codes. A portion of a typical stair and exit riser is shown in Figure 17.24. Circuits connected to this panel are not switched, being of the constant burning type, unless switches are accessible to authorized persons only. Emergency system wiring must be kept entirely independent of all other wiring and equipment and should not occupy the same enclosure or conduit as normal system wiring, except in dual fed units such as transfer switches.

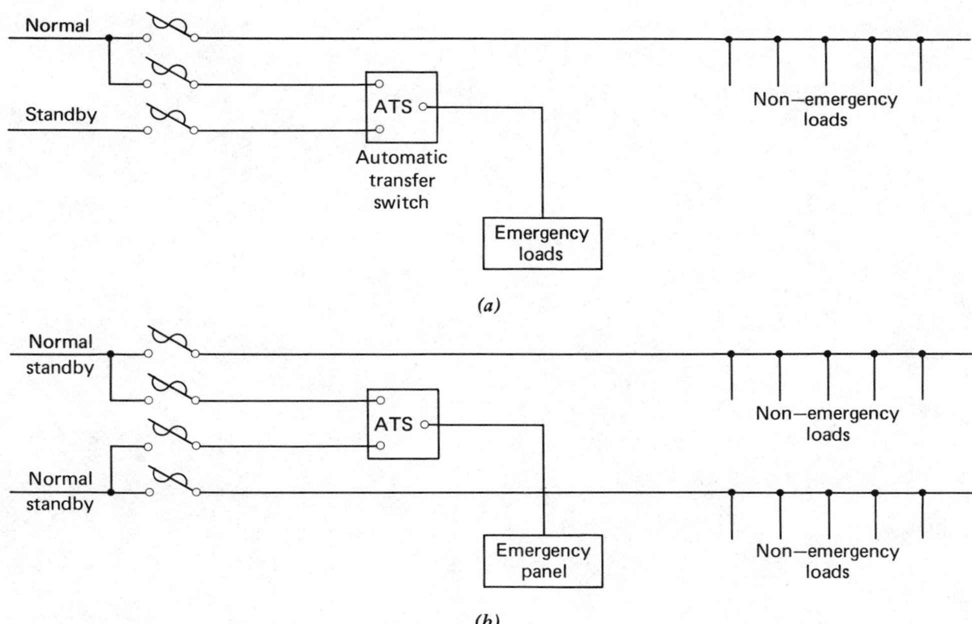

Fig. 17.23 *Emergency power is supplied by dual service. The arrangement shown in* (a), *where one service acts only as a standby, is more common than that of* (b), *where both supply normal loads and each acts as a standby for the other.*

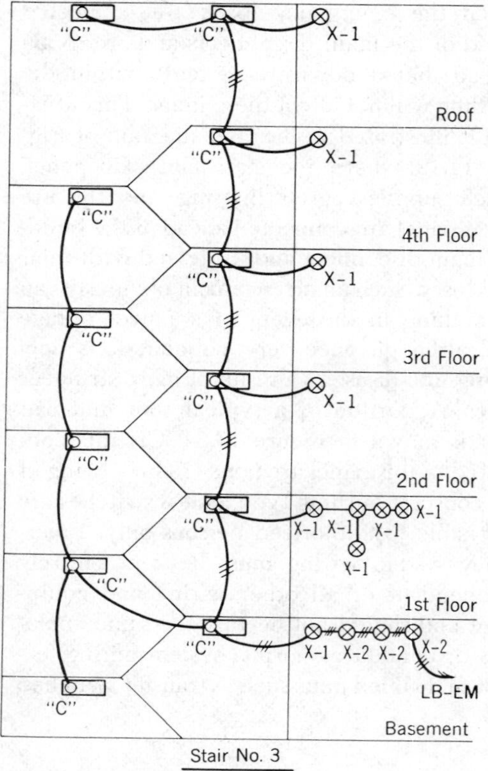

Where individual battery units are installed in a space to provide emergency lighting, they should be permanently wired and not plug connected. Also, the panel device feeding these outlets should be capable of being locked or so arranged as to be accessible to authorized personnel only.

***Fig. 17.24** Portion of a typical stair-and-exit riser. Panel LB-EM represents Lighting panel, Basement, EMergency.*

PART SIX

illumination

18

Lighting Fundamentals

Architecture is the masterly, correct and magnificent play of masses brought together in light. Our eyes are made to see forms in light; light and shade reveal these forms; cubes, cones, spheres, cylinders or pyramids are the great primary forms which light reveals to advantage. *Le Corbusier*

18.1 Introductory Remarks

For many years an artificial dichotomy existed in the field of lighting design, dividing it into two disciplines: architectural lighting and utilitarian design.

The former trend found expression in architectural building design that took little cognizance of vision needs, but that displayed an inordinate penchant for incandescent wall washers and architectural lighting elements, while regarding the added-on utilitarian lighting with partially justified asperity. The latter trend saw all spaces in terms of room or cavity ratios and designed lighting with footcandles and dollars as the ruling considerations. That both these trends have in large measure been eliminated is due in large measure to the work of the Illuminating Engineering Society (IES), and its members and new found energy consciousness that followed the 1973 Arab oil embargo. The latter spurred research into satisfying real vision needs within a framework of minimal energy use, and convinced architects that in addition to seeing the building, it must be possible to see *within* the building. The architectural designer must then take cognizance of these factors:

1. The manifold ramifications of daylighting, some of which are discussed in detail in Chapter 19.
2. The intimate interrelation between the energy aspects of artificial and natural lighting, heating, and cooling.
3. The effect of lighting needs on interior space arrangement, for example, the desirability of grouping similar lighting requirement tasks.
4. The characteristics, means of generation, effects, and utilization techniques of artificial lighting.

As a result of the need to consider these and other interrelated factors, many of which are mutually incompatible, the architect is faced with many trade-off type decisions. The purpose of the lighting chapters in this book is then twofold: to provide the background that will help the architect make these decisions correctly and to make him or her proficient in the use of lighting as a design material.

PHYSICS OF LIGHT

18.2 Light as Radiant Energy

The Illuminating Engineering Society, hereafter referred to as the IES, defines light as “visually evaluated radiant energy” or, more simply, a form of energy which permits us to see. If light is considered as a wave, similar to a radio wave or an alternating current wave, it has a frequency and a wave length. Figure 18.1 shows the position of light in the wave spectrum with relation to other wave phenomena of various frequencies.

From the chart we see that even the longest wavelength light (red) is a much higher frequency than radio and radar, and that visible light comprises only a very small part of the

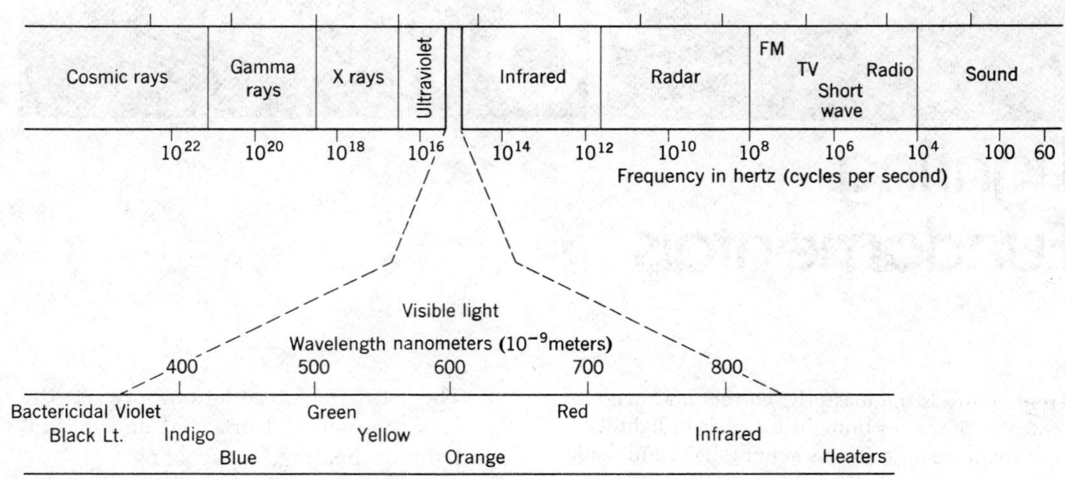

Fig. 18.1 Electromagnetic spectrum. See Figure 19.45, page 847, for spectrum of natural light.

wave energy spectrum. Yet it is this energy that makes possible our sight and with which we are here concerned. Color is determined by wavelength. Starting at the longest wavelengths with red, we proceed through the spectrum of orange, yellow, green, blue, indigo, and violet to arrive at the shortest visible wavelengths (highest frequency).

When a light source produces energy over the entire visible spectrum in approximately equal quantities, the combination of colored light produces white as is the case with the sun, whereas a source producing energy over only a small section of the spectrum produces its characteristic colored light. Examples are the blue-green clear mercury lamp and the yellow sodium lamp.

18.3 Fundamental Laws of Light

Design of lighting installations is possible because light is predictable, that is, it follows certain laws and exhibits certain fixed characteristics. Although some of these are so well known as to appear self-evident, a review is in order.

The *luminous transmittance* of a material such as a fixture lens or diffuser is a measure of its capability to transmit incident light. By definition, this quantity known variously as *transmittance, transmission factor,* or *coefficient of transmission* is the ratio of the total emitted light to the total incident light. In the case of incident light containing several components passing through a material that displays selective absorption, this factor becomes an average of the individual transmittances for the various components and must be used cautiously. A piece of frosted glass and a piece of red glass may both have a 70% transmission factor but obviously affect the incident light differently. In general then, transmission factors should be used only when referring to materials displaying nonselective absorption. Clear glass, for instance, displays a transmittance between 80 and 90%, frosted glass between 70 and 85%, and solid opal glass between 15 and 40%.

Similarly, the ratio of reflected to incident light is variously called *reflectance, reflectance factor,* and *reflectance coefficient.* Thus if half the amount of light incident on a surface is bounced back, the reflectance is 50% or 0.50. The remainder is absorbed, transmitted or both. The amount of absorption and reflection depends on the type of material and the angle of light incidence (see Fig. 18.3), since light impinging upon a surface at small (grazing) angles tends to be reflected rather than absorbed or transmitted. An example of almost perfect reflection from an opaque surface

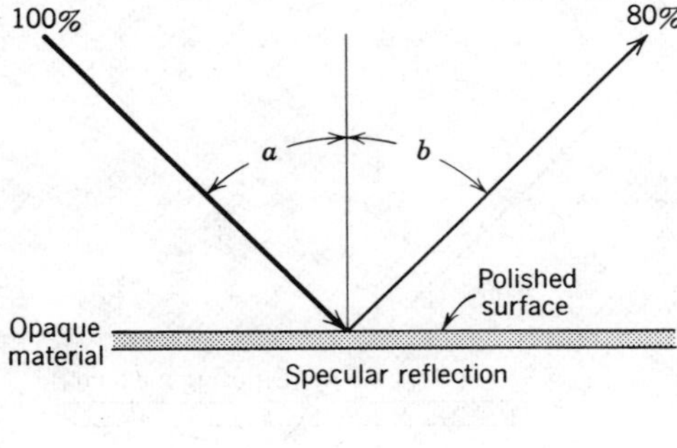

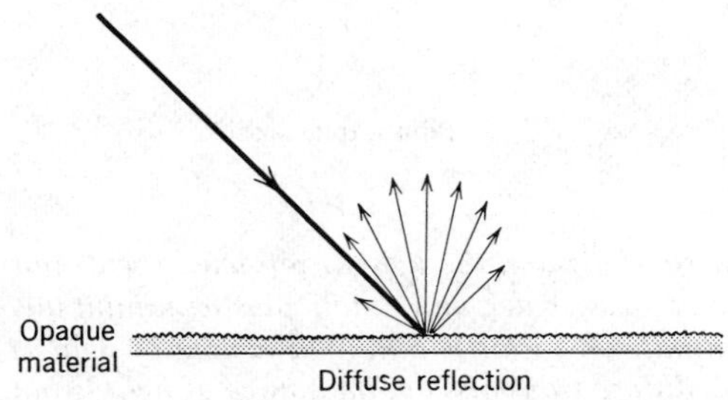

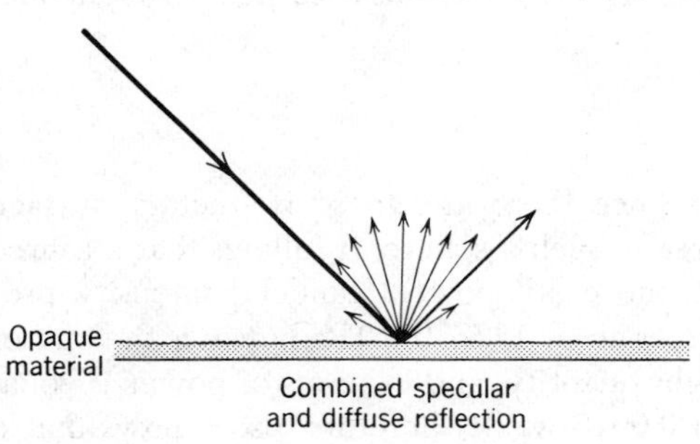

Fig. 18.2 Reflection characteristics. (a) *In specular reflection, angle of incidence equals angle of reflection* (a–b). *Since 80% of light is reflected, reflectance is 80%; 20% of light is absorbed.* (b) *In diffuse reflection, incident light is spread in all directions by multiple reflections on the unpolished surface. Such surfaces appear equally bright from all viewing angles.* (c) *Most materials exhibit a combination of specular and diffuse reflection. Such a surface will mirror the source while producing a bright background.*

would be that from a well-silvered mirror while almost complete absorption takes place on an object covered with lamp black or matte finish black paint. The effect of the material finish on reflection is shown in Figure 18.2*a, b, c.*

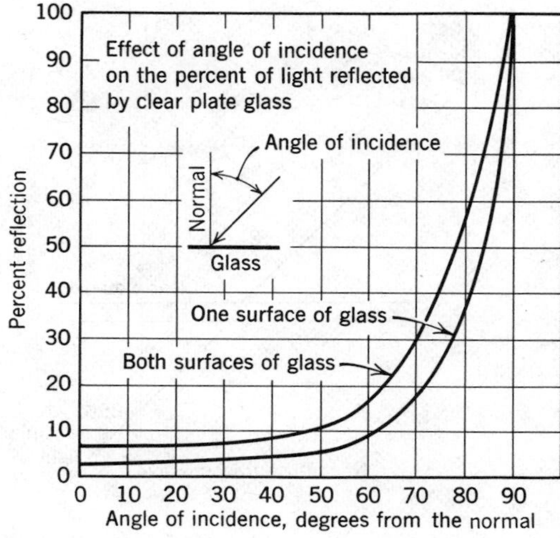

Fig. 18.3 Relation between angle of incidence and percent reflectance.

If the reflection takes place on a smooth surface such as polished glass or stone it is called specular reflection, as in Figure 18.2*a*. If the surface is rough, multiple reflections take place on the many small projections on the surface, and the light is diffused as in Figure 18.2*b*. Since the reflection factor is a measure of total light reflected, it does not depend on whether the reflection is specular or diffuse, or a combination of both, as shown in Figure 18.2*c*. Diffuse transmission takes place through any translucent source such as frosted glass, white glass, milky plexiglas, tissue paper and so on. This diffusing principle is widely employed in lighting fixtures to spread the light generated by the bulb or tube within the fixture. Diffuse and nondiffuse transmission are illustrated in Figure 18.4*a* and *b*.

18.4 Terminology and Definitions

Before beginning any discussion of lighting studies, techniques, and effects it is important to have a thorough understanding of the physical concepts and terminology involved, and their interrelations. We will use and explain both the conventional American Standard (AS) and the SI systems of units. The latter,

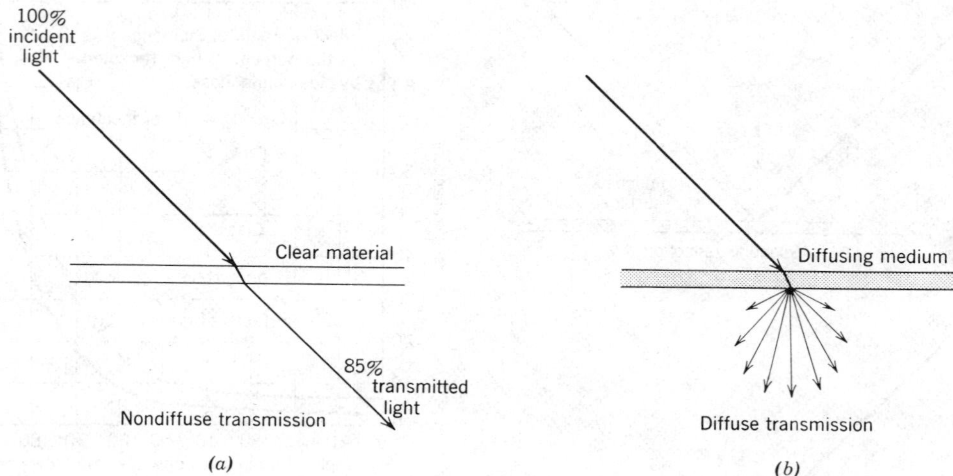

Fig. 18.4 *Transmission characteristics.* (a) *In nondiffuse transmission, the light is refracted (bent) but emerges in the same beam as it enters. Clear materials such as glass, water, and certain plastics exhibit this type of transmission. In the instance illustrated the transmission factor is 85%. The source of light is clearly visible through the transmitting medium.* (b) *With diffuse transmission, the source of light is not visible and, in the case of multiple sources, the diffusing surface will exhibit generally uniform brightness if the spacing between the light sources does not exceed $1\frac{1}{2}$ times their distance from the material.*

also called the metric system, is gradually coming into use in the United States and will, as with other SI units, eventually replace the American Standard system. To familiarize the reader with the SI system, we will frequently use duplicate units, with the SI unit enclosed in square brackets [ ] following the conventional unit.

The candlepower [candela], abbreviated cp [cd], is the unit of luminous intensity. It is analogous to pressure in a hydraulic system and voltage in an electric system and represents the force that generates the light that we see. An ordinary wax candle has a luminous intensity horizontally of approximately one candlepower [candela], hence the name. The candela and candlepower have the same magnitude. A candle radiates light equally in all directions. If we imagine such a source surrounded by a transparent sphere of one foot [meter] radius (see Fig. 18.5*a*), then *by definition* the amount of luminous energy (flux) emanating from one square foot [meter] of surface on the sphere is one *lumen* [lumen], abbreviated lm.

Since there are 4π sq ft [meters] surface area in such a sphere, it follows that a source of one candlepower [candela] intensity produces 4π or 12.57 lm. The lumen is the unit of light quantity, and in terms of power is equal to 0.0015 w. (It therefore also follows that a 1-cp [cd] source produces 12.57×0.0015 w, that is, 0.0189 w or approximately 1/50 w of luminous energy.) The lumen, as luminous flux, or quantity of light is analogous to flow in hydraulic systems and current in electric systems.

One lumen of luminous energy incident on one square foot of area produces an *illumination* of one *footcandle,* (fc). Restated, illumination is the density of luminous energy, expressed in terms of lumens per unit area. If we were to consider a lightbulb to be analogous to a sprinkler head, then the amount of water released would be the lumens and the amount of water per square foot of floor area would be the footcandles. When the area is expressed in square feet the resulting illumination is footcandles; when the area is in square meters, the illumination is expressed in *lux* (*lx*). Thus the

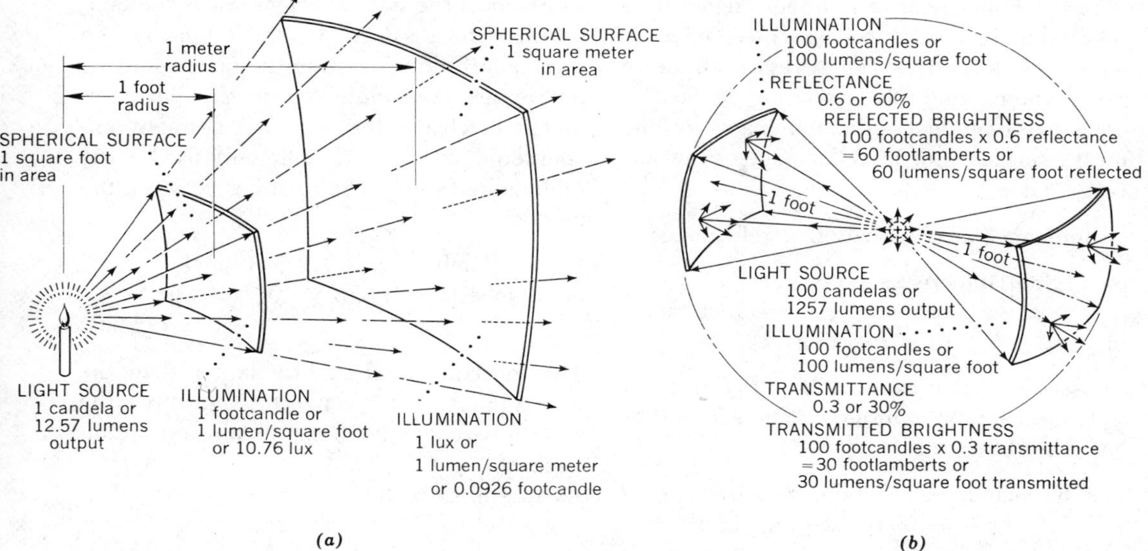

Fig. 18.5 (a) *The relationship between a unit source of one candela and the illumination produced, expressed in conventional and SI units (footcandles and lux, respectively).* (b) *Luminance (brightness) is either reflected or transmitted. In the former case the reflected brightness is calculated as the product of the incident lumens and the reflection factor; in the latter, as the product of the incident lumens and the transmission factor.*

SI unit, lux, is smaller than the corresponding unit, footcandles, by the ratio of square feet to square meters, that is

$$10.764 \text{ lux} = \text{one footcandle.}$$

or

multiply footcandle by 10.764 to obtain lux.

Restating the above mathematically,

$$\text{footcandles} = \frac{\text{lumens}}{\text{square feet of area}}$$

$$\text{fc} = \frac{\text{lm}}{\text{sq ft}} \tag{18.1}$$

and

$$\text{lux} = \frac{\text{lumens}}{\text{square meters of area}}$$

$$\text{lx} = \frac{\text{lm}}{\text{sq m}} \tag{18.2}$$

As an *approximation* (8% error)

$$10 \text{ lx} \simeq 1 \text{ fc} \tag{18.3}$$

EXAMPLE 18.1. A 40-w, 430-ma (milliampere), 48-in. fluorescent tube produces 3200 lm. What is the illumination on the floor of a 10-sq ft room assuming 40% overall efficiency and uniform illumination?

SOLUTION

$$\text{useful lumens} = 0.4 \times 3200 = 1280$$

$$\text{fc} = \frac{1280}{10 \times 10} = 12.8$$

$$\text{lx} = 12.8 \times 10.76 = 137.7$$

Calculating lux directly:

$$\text{lx} = \frac{1280}{10 \times 10} \times (3.28 \text{ ft/m})^2 = 137.7$$

By approximation:

$$\text{lx} \simeq 10 \cdot \text{fc} \simeq 128$$

Footcandle illumination at a point can also be computed from intensity as shown in Section 18.8. A *footlambert,* the conventional unit of luminance or brightness, is defined as the luminance of a surface reflecting, transmitting, or emitting 1 lumen of illumination per square foot of area, *in the direction being viewed.* This latter qualification is important since many surfaces (fabrics, for instance) exhibit different luminances at different angles. This unit has no readily conceivable mechanical or electrical analogy. The brightness of a *nonluminous* diffusely reflecting surface is equal to the product of the illumination falling on the surface and the reflectance of the surface, that is,

$$\text{luminance} = \text{illumination} \times \text{reflectance}$$

or

$$\text{footlamberts} = \text{footcandles} \times \text{reflectance factor}$$

or

$$\text{fL} = \text{fc} \times \text{RF} \tag{18.4}$$

Thus a source that causes 100 lm to fall on a 1 sq ft surface with a diffuse reflectance of 60% has a brightness of

$$\begin{aligned}\text{luminance} &= 100\ \text{lm} \times 60\%\ \text{RF}\\ &= 60\ \text{fL (see Fig. 18.5}b\text{)}\end{aligned}$$

Reflecting surfaces that derive their brightness from incident illumination are known as secondary sources. Primary sources are those that generate light and transmit directly to the eye. The largest and best known primary and secondary brightness sources are the sun and the moon, respectively. The same unit of luminance, the footlambert (fL) is used for a luminous surface that either emits (self-luminous) or transmits light. As stated above, a surface emitting or transmitting an average of 1 lm/sq ft has a luminance of 1 fL. Therefore the luminance of a transilluminated source is equal to the product of the illumination and the transmission factor, that is

$$\text{luminance} = \text{illumination} \times \text{transmission factor}$$

or

$$\text{fL} = \text{fc} \times \text{TF} \tag{18.5}$$

Thus a source of 100 lm behind a translucent diffusing material one square foot in area, with a transmission factor of 30%, exhibits a brightness of

$$\begin{aligned}\text{luminance} &= 100\ \text{lm/sq ft} \times 0.3\\ &= 30\ \text{fL (see Fig. 18.5}b\text{)}\end{aligned}$$

For an emitting surface:

luminance = lumens emitted per square foot. Thus a surface emitting uniformly 100 lm/sq ft has a brightness of 100 fL.

In SI units, the unit of brightness is the lambert (L) which is defined as the luminance of a surface reflecting, transmitting, or emitting one lumen *per square centimeter.* Since this unit is much greater than that normally encountered, the SI unit commonly used is the millilambert (mL). The relations between the units are

$$\begin{aligned}\text{lamberts} &= 1.076 \times 10^{-3}\ \text{fL}\\ \text{millilamberts} &= (1.076 \times 10^{-3})(1000)\\ &= 1.076\ \text{fL}\end{aligned}$$

The footlambert is slightly larger than the millilambert. To convert, multiply the footlambert by 1.076 to obtain the millilambert.

As an approximation,

$$1\ \text{mL} \simeq 1\ \text{fL}$$

EXAMPLE 18.2. A fixture with a 2 × 4 ft plastic diffuser having a transmittance of 0.8 and illuminated by four 3200 lm lamps, would have (assuming 100% use of the light flux), a luminance of:

$$\begin{aligned}\text{luminance} &= \frac{\text{transmitted lumens}}{\text{area}}\\ &= \frac{\text{generated lumens} \times \text{transmission factor}}{\text{area}}\\ \text{luminance} &= \frac{4 \times 3200\ \text{lm} \times 0.8}{2 \times 4\ \text{sq ft}}\\ &= 1320\ \text{fL}\end{aligned}$$

Each fluorescent tube has a luminance of 3200 lm divided by the tube surface area. The T-12, 4-ft tube has an area of π × diameter × length:

$$\pi \times 1.5 \times 48 = 226.5\ \text{sq in} = 1.58\ \text{sq ft}$$

This gives a luminance for a 48-in., T-12 tube of 3200/1.58 or approximately 2000 fL (see Table 18.3). If it is desired to express these results in SI units conversion factors would be used rather than direct calculation, since the dimensions 2 × 4 ft and 1.5 in. are clumsy in metric units. Using conversion factors we have:

luminance of the fixture

$$1320\ \text{fL} \times 1.076 = 1420\ \text{mL}$$

Table 18.1 *Lighting–Conversion Factors*

	Multiply	*By*	*To Obtain*
Illumination	Footcandle	10.764	Lux
	Lux[a]	0.0929	Footcandle
Luminance	Footlambert	1.076	Millilambert
(Brightness)	Millilambert[a]	0.929	Footlambert
	Footlambert	0.00221	Candela/in.^{2}
	Candela/in.^{2}	452.0	Footlambert
Intensity	Candlepower	1.0	Candela[a]

[a]SI Units.

luminance of the fluorescent tube

$$\frac{3200}{1.58} \times 1.076 = 2180 \text{ mL}$$

Reviewing, we have established that a source of one candlepower [candela] produces 4π lumens of luminous flux which produce illumination at the rate of one footcandle [lux] per lumen per sq ft [meter]. The luminous flux also produces luminance at the rate of one footlambert [millilambert] per lumen transmitted or reflected, per square foot [cm^2]. Conversion factors are given in Table 18.1, which for convenience also contains less used factors for luminance expressed in candela per unit area.

Fig. 18.6 *Typical light flux meters.*

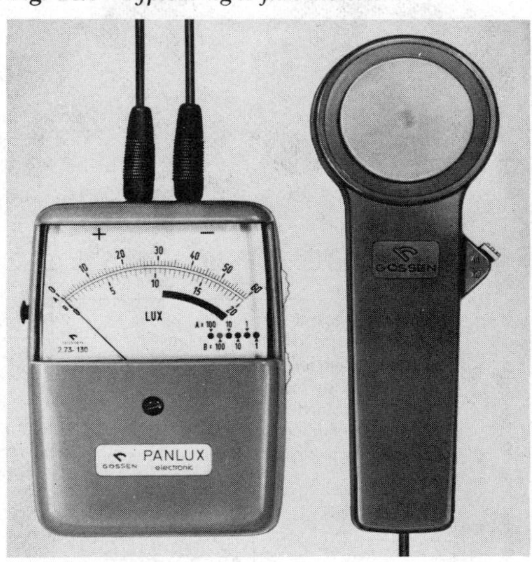

Fig. 18.6 (a) *Meter calibrated in lux and equipped with remote cable-connected cell unit, multiple ranges, and electronic color and cosine correction. Ranges cover 0 to 120,000 lux with maximum ±7.5% error. Unit is small and lightweight (380 grams). Courtesy of Gossen.*

18.5 Footcandle Measurements

Measurements of illumination levels are most commonly made with one or another of the available portable footcandle meters, some of which are illustrated in Figure 18.6. These cells comprise a light-sensitive material connected to a microammeter, and calibrated in footcandles. The smaller portable units are convenient to use but accuracy is generally not better than ±5% and fairly frequent recalibration is advisable. If greater accuracy is desired, the larger units should be employed.

The human eye is not equally sensitive to the various wavelengths, that is, colors. Maxi-

Fig. 18.6 (b) *Multirange, color and cosine-corrected footcandle meter. Range is 0 to 1200 fc (equivalent to 0 to 13,000 lux). Accuracy is approximately the same as unit* (a). *Courtesy of Weston Instruments, Division of Sangamo Weston Inc.*

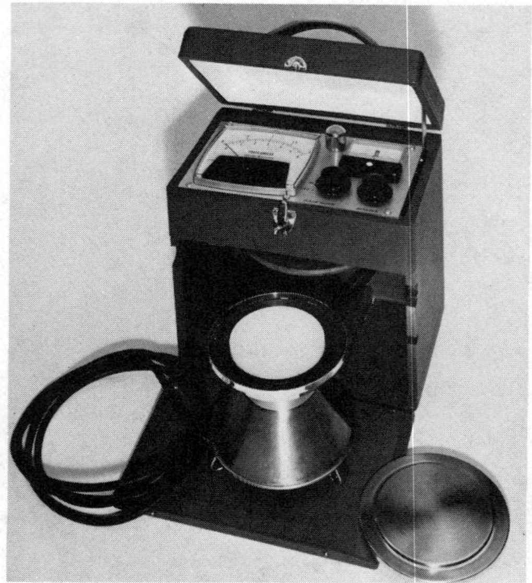

Fig. 18.6 (c) *Low-range, highly sensitive illumination meter with range of 0 to 2, 0 to 6, and 0 to 20 fc. This type of meter is useful for checking outdoor lighting levels in areas such as streets, walkways, parking lots, and highways. Courtesy of General Electric Corp.*

Fig. 18.7 *Curves showing response of uncorrected (glass and quartz filters) and corrected light flux meters, superimposed on the response curve of a normal light-adapted human eye. (The night/dark response curve of the eye is similar in shape but peaks at 500 nanometers, that is, blue-green). The corrected curve shown corresponds to the meter of Figure 18.6b. The meter in Figure 18.6a shows a very similar corrected response curves, accomplished primarily electronically. Courtesy of Weston Instruments, Division of Sangamo-Weston Inc.*

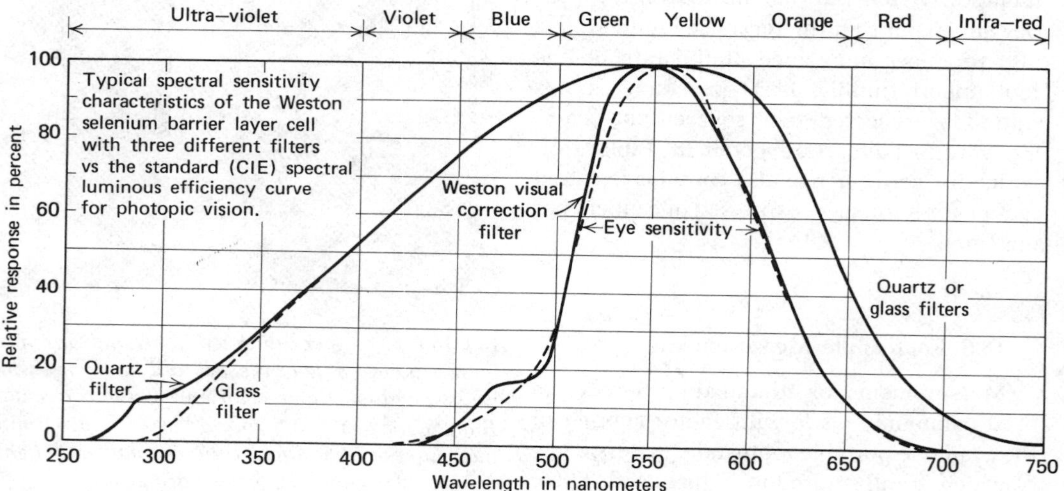

mum sensitivity is in the yellow-green area (wavelength of 555 nanometers) while sensitivity at the red and blue ends of the spectrum is quite low. This effect is so pronounced that 10 units of blue energy are required to produce the same visual effect as 1 unit of yellow-green. Therefore, if a meter is to be useful in terms of human eye response, its inherent response, which is quite different from that of the human eye must be corrected to correspond to the eye. For this reason most modern meters are "color corrected." The effects described above are shown graphically in Figure 18.7.

The cells (meters) must also be corrected for light incident at oblique angles that does not reach the cell due to reflection from the surface glass and shielding of the light-sensitive cell by the meter housing. This correction is known as cosine correction. A good meter must therefore be (and it will plainly so indicate) color and cosine corrected.

When taking actual readings, meters should be placed on a stable surface and readings taken after the needle stabilizes. For determining average room illumination a number of readings should be taken and an average computed. Where no definite height is specified, readings are taken at 30 in. above the floor. The meter must always be held with the cell parallel to the plane of the test. Thus to measure wall illumination, the meter must be held with the cell parallel to the wall. If nighttime illumination readings are desired and the test is being conducted during daylight hours, readings should be taken with and without the artificial illumination and the results subtracted. Detailed instructions for conducting field surveys are contained in the IES publication "How to Make a Lighting Survey." Briefly a survey of an existing indoor lighting installation should establish:

1. Type, rating, and age of sources.
2. Type, design, and model of luminaire.
3. Maintenance schedule.

It should also measure:

1. Mounting height.
2. Spacing and pattern of luminaires.
3. Reflectances of walls, floor, ceiling and major items of furniture and equipment.
4. Footcandle levels throughout the area plus levels at walls and columns, all at the working plane elevation.

18.6 Luminance (Brightness) Measurements

In many respects, luminance measurements are more important than illumination measurements since the eye sees brightness and not illumination. That the usual measurement is of illumination is due in large measure to the difficulty of measuring luminance easily. Luminance meters are of three types, all of which are illustrated in Figure 18.8. The first is a comparator type, which requires that the operator make a brightness equivalence judgment of target or background (see Fig. 18.8*a*). This meter requires frequent recalibration and because it is judgmental, it is advisable to take a number of readings. The second type (Fig. 18.8*b*) is a direct reading type. Basically it is an illumination meter equipped with a hooded cell arranged to block oblique light and calibrated in units of luminance. The third type is an accurate laboratory instrument, obviously unsuitable for field work.

An approximation of the brightness of a reflecting or luminous source can be obtained using a footcandle meter of the type shown in Figure 18.6. For diffuse reflecting surfaces the cell of the meter is placed against the surface and then slowly retracted 2 to 4 in. until a constant reading is obtained. The brightness is then 1.25 times the reading in footcandles, the 1.25 factor compensating for wide angle losses For a luminous source, the cell of the meter is placed directly (see Fig. 18.8*d*) against the surface; the reading in footcandles is the brightness in footlamberts (because footlamberts = lumens per area = footcandles).

18.7 Reflectance Measurements

It is often desirable to know the reflectance of a given surface since brightnesses can then be readily computed (fL = fc × RF). Two methods of measuring reflectance are shown in Figure 18.9: the known sample method and the light ratio method. If a sample of known reflectance factor (RF) is available, this method should be used since it yields more accurate results than the ratio method. The sample should be no smaller than 8 in. square.

It is well for a budding lighting designer to determine the reflectance, brightnesses, and illumination levels of spaces and surfaces familiar to him or her, such as his office desk, adjoining wall, and the like—even to the extent of marking these figures on the respective surfaces in order to develop an appreciation of and a memory for these parameters. This will enable the designer to visualize the result of a lighting design and should be of considerable assistance.

18.8 Inverse Square Law

We have already learned in Section 18.4 that, by definition, a source of 1 cd produces an illumination of 1 fc on the inside surface of a surrounding sphere of 1 ft radius (r). Also, since the surface area of this sphere is 4π sq ft, a 1-cd source produces 4π lumens of luminous flux. Now assume a sphere of 2 ft radius surrounding this same source (see Fig. 18.10*a*). Since the same amount of flux is spread over a larger area, the illumination on the larger

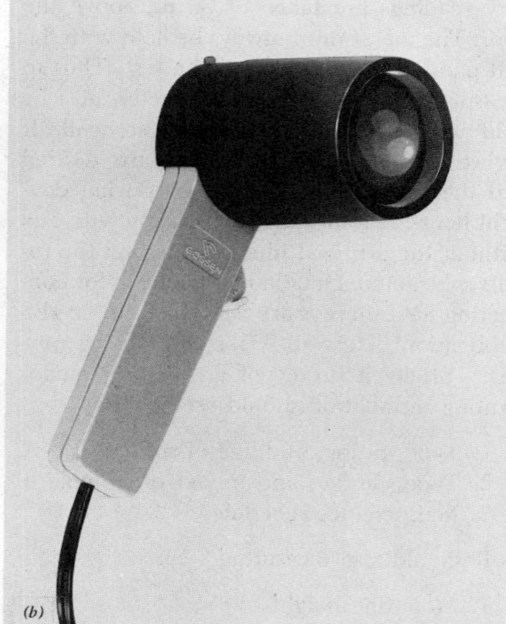

Fig. 18.8 *Brightness measurements of luminous sources can be made with* (a) *a direct-reading electronic brightness meter and* (b) *a hooded cell designed to exclude oblique angle light and connected to an illumination meter. The illustrated unit is an attachment for the lux meter illustrated in Figure 18.6*a. (c) *A direct-reading electronic brightness meter.* (d) *A simple footcandle meter. The latter, when held in contact with the luminous source, will read brightness in footlamberts directly on the footcandle scale. In* (a) *the author is checking brightness of a louver, which utilizes parabolic wedges to form the individual cells, giving low brightness above 45°. This is readily apparent in comparison to the adjacent fixture, which utilizes a prismatic lens diffuser.* (b) *Courtesy of Gossen and* (c) *Courtesy of General Electric.*

Reflected/Incident Light Method

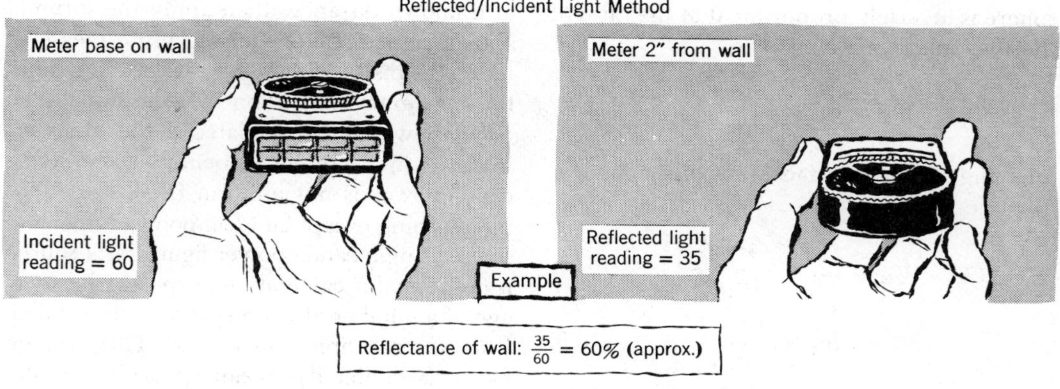

Known Sample Comparison Method

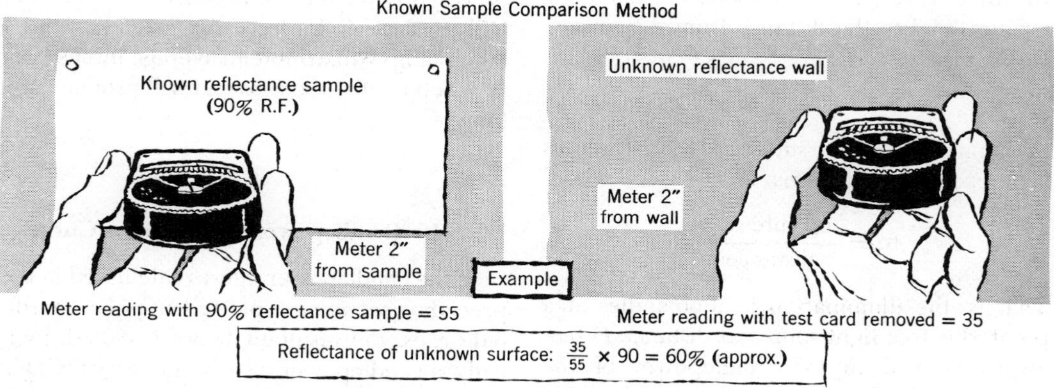

Fig. 18.9 Two methods of measuring reflectance factor.

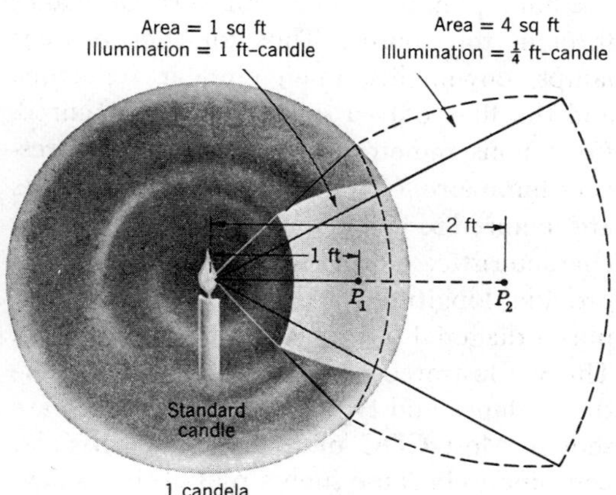

Fig. 18.10 (a) *Relations between candlepower, lumens, and footcandles defined with reference to a standard light source of 1 mean spherical candlepower (1 candela) located at the center of a sphere.*

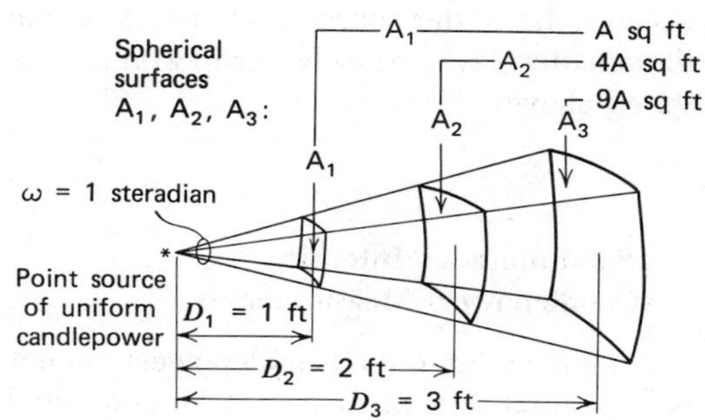

Fig. 18.10 (b) *Demonstration of inverse square law properties using a solid angle of unit size. Note that the surfaces are necessarily spherical, since points on a planar surface are not equidistant from the source.*

sphere is inversely proportional to the ratio of the two sphere areas, that is,

$$fc_2 = fc_1 \times \frac{\text{area}_{\ 1}}{\text{area}_{\ 2}}$$

knowing that the surface area of a sphere is $4\pi r^2$ we have

$$fc_2 = fc_1 \times \frac{4\pi r_1{}^2}{4\pi_1 r_2{}^2}$$

$$= fc_1 \times \frac{r_1{}^2}{r_2{}^2} \qquad (18.6)$$

In other words, the illumination is inversely proportional to the distance from the source. In the case stated

$$fc_2 = 1 \times \tfrac{1}{4} = \tfrac{1}{4}\,\text{fc}$$

of illumination on the surface of the 2-ft radius sphere. In general terms

$$\text{fc} = \frac{\text{cp intensity}}{\text{distance}^2} \qquad (18.7)$$

That is, the illumination in footcandles at a point, due to a light source at distance D feet away is equal to the candlepower of the source divided by the square of the distance. (This holds true for surfaces normal to a source. For other situations see Section 20.38, page 924.)

This relationship can also readily be derived by using any solid angle and the area it intercepts, as in Figure 18.10*b*. A glance at this figure shows clearly that the area intercepted is proportional to the square of the distance from the source and therefore the illumination is inversely proportional, as stated above.

18.9 Luminous Intensity (Candlepower) Measurements

Luminous intensity (candlepower) cannot be measured directly but must be computed from its illumination effects. The simplest way of doing this is to use the inverse square relationship developed in the preceding section. We measure the footcandle illumination produced on a plane at right angles to the source, at a known distance; then apply the formula above, cp = fc $\times D^2$. For accurate measurement, the distance should be at least five times the maximum dimension of the source. The candlepower thus calculated is the luminous intensity in the direction being viewed. Since candlepower is not uniform in all directions for anything except an ideal point source, and since a single candlepower figure for a source is desirable for calculation purposes, the average of a number of candlepower figures taken from several directions is used. This average figure is called the mean spherical candlepower (mscp), and represents a source that will produce 4π lm for every candela. Thus a 10-cp lamp will exhibit an average intensity of $\pm$ 10 cp in all directions and will produce 40π lumens.

18.10 Candlepower Distribution Curves

If the candlepower figures calculated in the preceding section are plotted on polar coordinate axes, the resultant figure is called, logically, a candlepower distribution curve (CDC) for the particular source involved. The procedure for making this curve is illustrated in Figure 18.11. If the sources' distribution is symmetrical as shown in Figure 18.11*a*, then only a single set of values are required, and the resultant plot is valid in all vertical planes through the source. Thus for incandescent lamps, downlights, open circular reflectors, and the like, only a single CDC is required. For a nonsymmetric source such as a fluorescent luminaire, CDC curves in several planes are required to define the fixture's distribution characteristic. Normally manufacturers will provide longitudinal and crosswise curves, plus a diagonal (45°) plane curve on request. This is illustrated in Figure 18.12, where the three planes and typical resultant curves are shown. Most CDC plots are made on polar coordinates because such a plot clearly shows directions and magnitudes. Nevertheless, polar plots tend to crowd near the nadir and accurate magnitude readings at the cutoff angle are difficult to make. For this reason it is occasionally desirable to obtain a plot on rec-

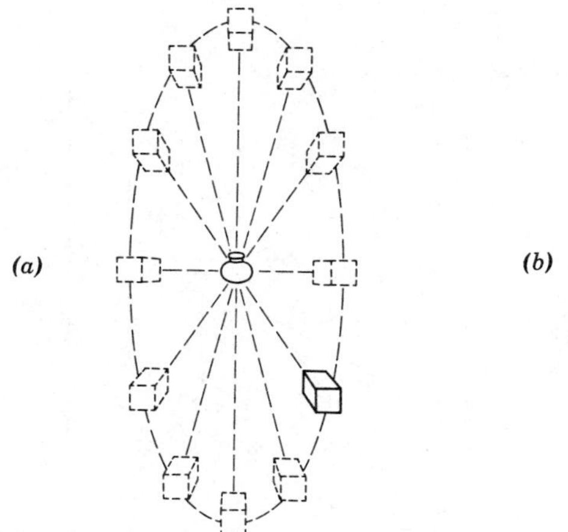

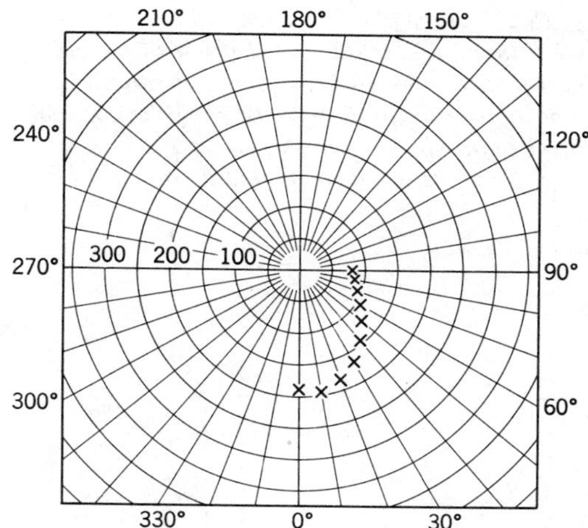

Fig. 18.11 (a) *Photocell is rotated around luminaire at a specific distance to measure footcandles at specified angles. The device that performs the testing function is called a ganio-photometer.*

Fig. 18.11 (b) *Candlepower values, computed from footcandle values at each angle, are plotted on polar coordinate graph.*

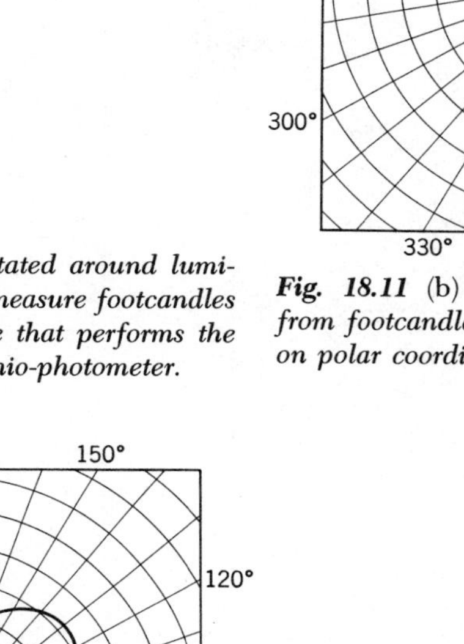

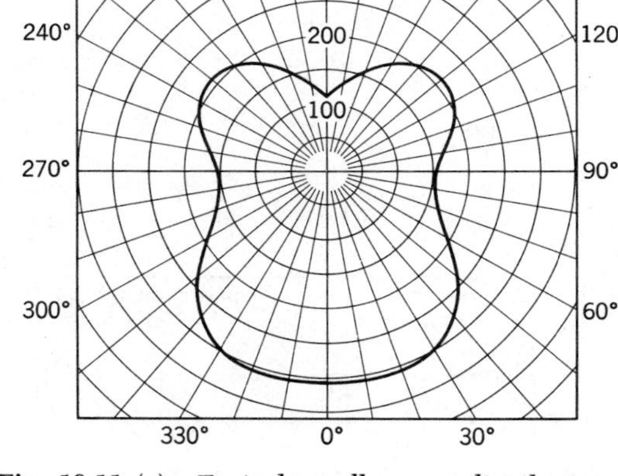

Fig. 18.11 (c) *Typical candlepower distribution curve for a general diffuse-type luminaire. Since the unit is symmetrical about the vertical axis, only one curve need be shown.*
Steps in producing a candlepower distribution curve. Courtesy of GTE Sylvania Inc.

tangular coordinate. One such plot is shown in Figure 18.13. The usefulness of candlepower distribution curves will become clear in our subsequent discussions on fixture diffusers (Section 20.19), point-by-point calculations (Section 20.38), and direct and reflected glare (Section 18.20 to 18.22).

18.11 Lumen Calculations from Candlepower Distribution Curves

It is important to be able to determine the total lumen output of a luminous device. Light source manufacturers must know the total output of their lamps and fixture manufacturers the efficiency of their products. The latter is calculated as

$$\text{efficiency} = \frac{\text{fixture lumen output}}{\text{lamp lumens}}$$

Two methods are in common use. The first utilizes an integrating sphere into which the device is placed. The second involves use of the candlepower distribution curves described in the previous section. Refer to Figure 18.14. We know that 1 sq ft of area on an enclosing 1 ft radius sphere intercepts 1 lm given a 1-cp source. Note from Figure 18.14*a* that this 1 sq ft of area may have any shape, for example, a

Fig. 18.12 *Three planes are used in testing a fluorescent luminaire, within each of which a candlepower distribution curve* (a) *is measured and plotted at a given uniform distance from the common central point* (p) *of plane intersection. The three resulting curves are then plotted as shown in* (b).

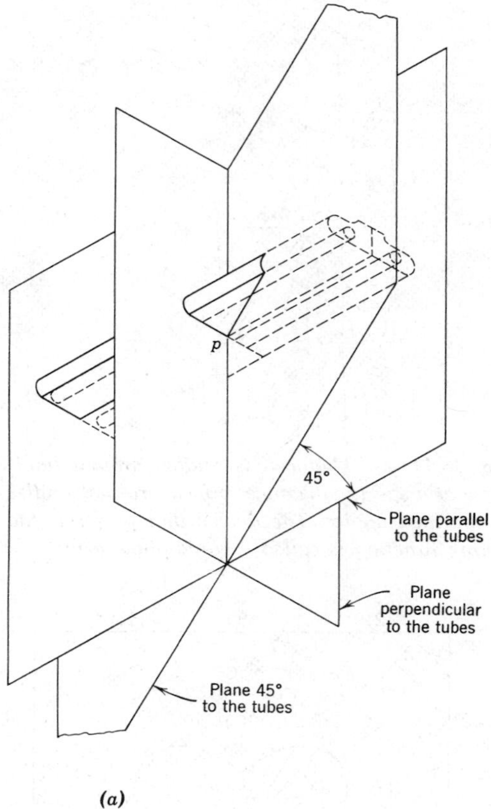

(a)

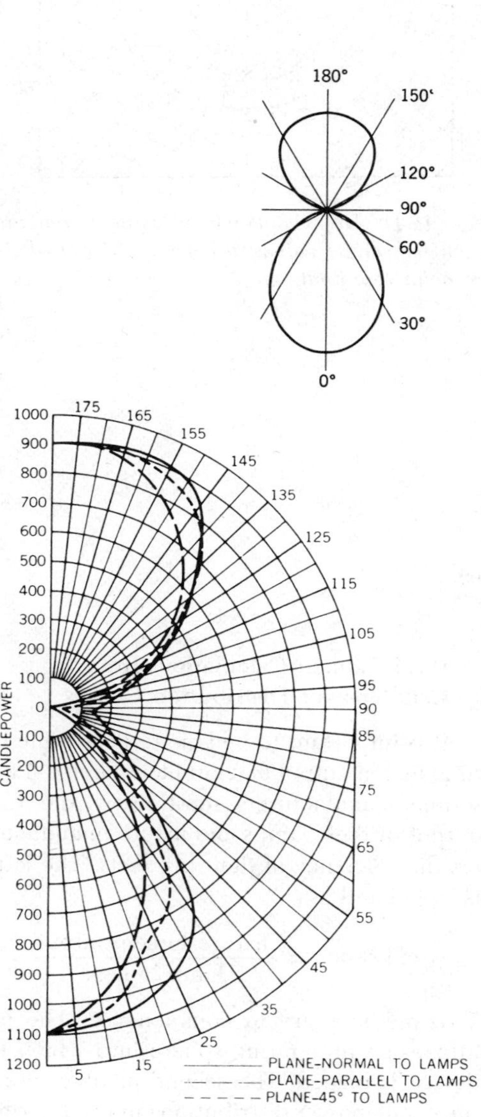

(b)

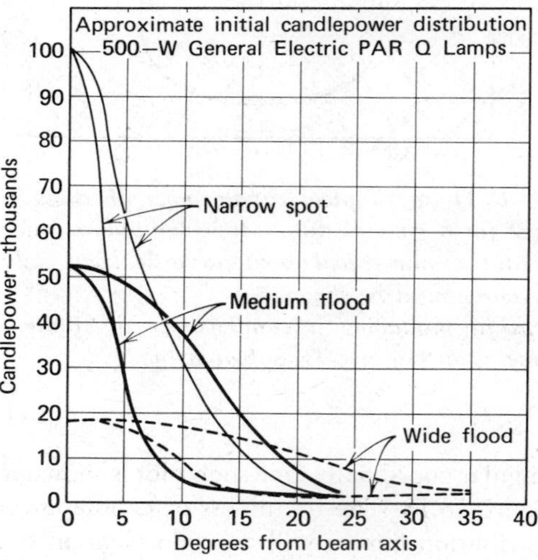

Fig. 18.13 *Candlepower distribution curves plotted in rectangular coordinates. Note that candlepower values near the cutoff angles are easily read, which is not the case in polar plots.*

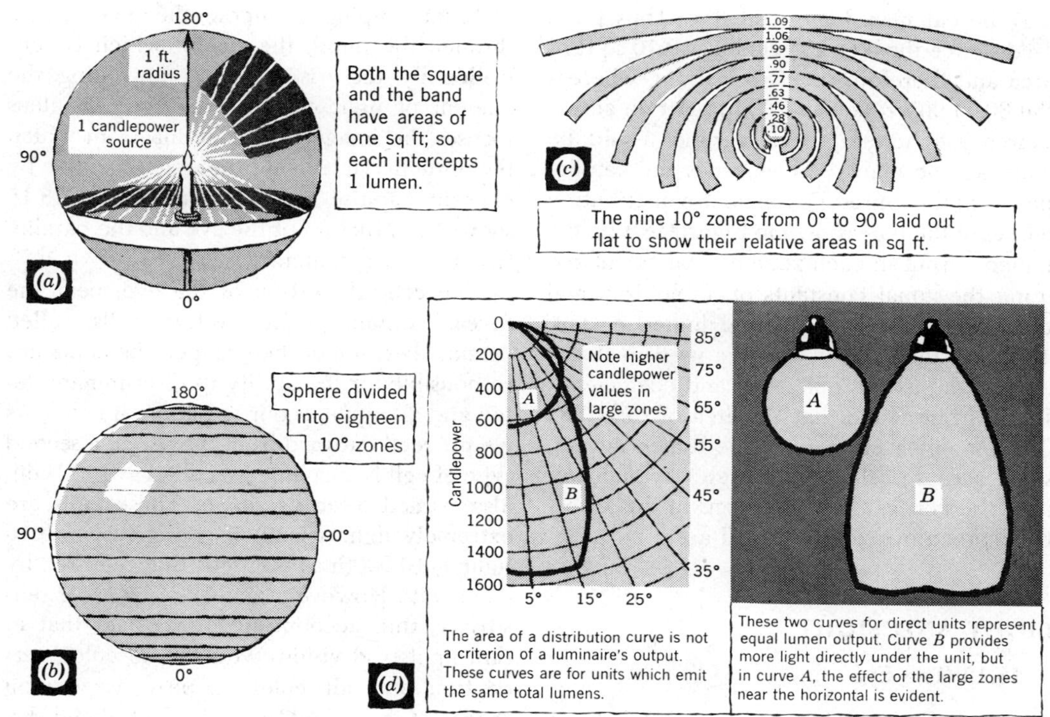

Fig. 18.14 Standard method of dividing a 1-ft sphere to obtain zonal bands and factors.

Table 18.2 *Zonal Factors*

Zone, Degrees	*Zonal Constant*	*Zone, Degrees*	*Zonal Constant*	*Percent of Lumens in each Band*
0–10	0.10	170–180	0.10	0.8
10–20	0.28	160–170	0.28	2.2
20–30	0.46	150–160	0.46	3.7
30–40	0.63	140–150	0.63	5.0
40–50	0.77	130–140	0.77	6.1
50–60	0.90	120–130	0.90	7.2
60–70	0.99	110–120	0.99	7.9
70–80	1.06	100–110	1.06	8.4
80–90	1.09	90–100	1.09	8.7
Total	6.28	Total	6.28	50.0
		Grand total	$12.56 = 4\pi$	100% in both halves

square on the surface or a band around the sphere. If the sphere is divided into 10° bands (see Fig. 18.14*b*), the actual number of lumens intercepted by each band is equal to the candlepower of the source times the area in square feet of that 10° zone. The candlepower is taken from the corresponding candlepower distribution curve at the midzone angle (45° for the 40–50° zone, etc). Note that zone areas near the 90° positions have a larger area than those near the 0° and 180° positions, which become very small. The value of these areas may be calculated by geometry. Thus for a 1-cd source the 0 to 10° zone has 0.10 sq ft of area and therefore receives 0.10 lm, whereas the 80 to 90° zone has 1.09 lm. For an actual source the lumens in each zone would be equal to the above values times the candlepower of the source. Given fixture distribution curves of the type shown in Figure 18.12*b*, the lumen output in each zone can be calculated using the zonal constants of Table 18.2 and then summed to obtain total lumen output. This plus the zone figures are used in determining the fixture efficiency and coefficients. It should be noted that the area of the CDC curve is not a measure of the lumen output, since a source that gives high candlepower near the vertical has its lumens in this direction spread over only a small area.

LIGHT AND SIGHT

18.12 The Eye

Since all discussion of light and lighting techniques is irrelevant to our purposes unless ultimately related to vision, we turn to a cursory examination of the human eye before proceeding further with discussions of lighting.

Light impinging upon the eye enters through the pupil, the size of which is controlled by the iris, thereby controlling the amount of light entering the eye. The lens focuses the image on the retina from which the optic nerve conveys the visual message by electric impulse to the brain. Figure 18.15 shows the structure of the eye and the parallel structure of a camera.

The central portion of the eye, near the fovea, contains light-sensitive cells called "cones" because of their shape. The cones are responsible for the ability to discriminate detail and also give us our sensation of color. As we proceed outward from the fovea, a second type of cell is encountered called a "rod" cell, also named after its shape. These cells are extremely light sensitive, giving response to light 1/10,000th as bright as that required by cone cells. However, rod cells lack color sensitivity, thus accounting for the fact that in dim light (rod vision) we have no color perception and all colors appear as varying shades of gray. Rod cells also lack detail discrimination, making "night vision" quite coarse. Finally, rod cells are slower acting than cone cells and therefore have a low degree of flicker fusion, or, stated conversely, they are highly motion sensitive. Since these

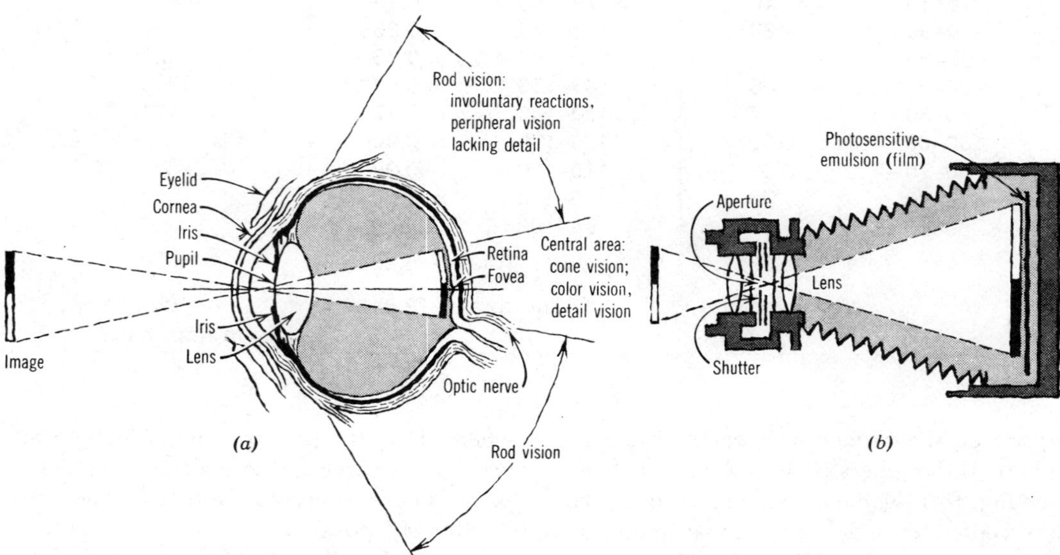

Fig. 18.15 *The human eye and the camera operate on the same optic principles.*

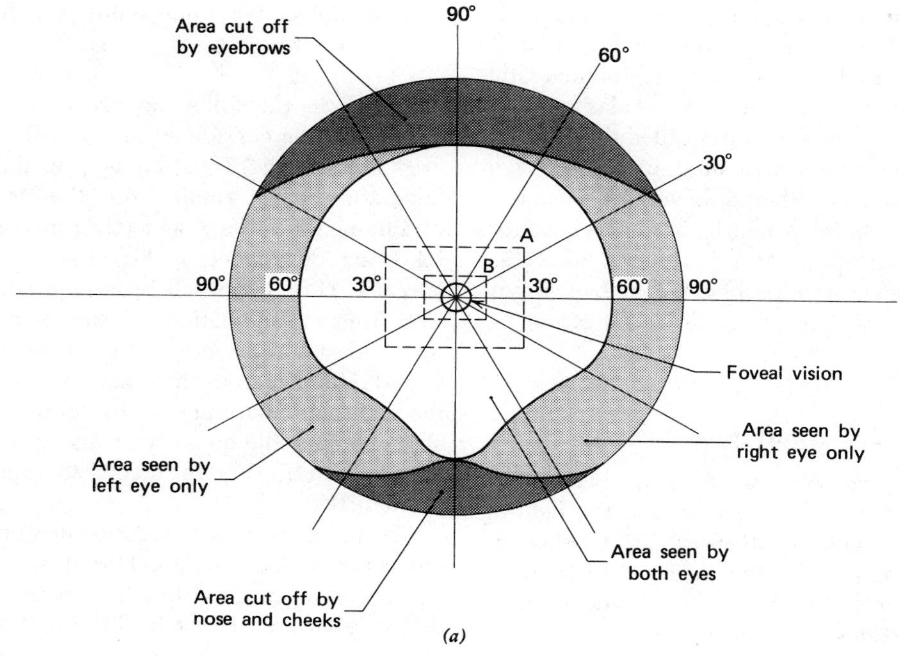

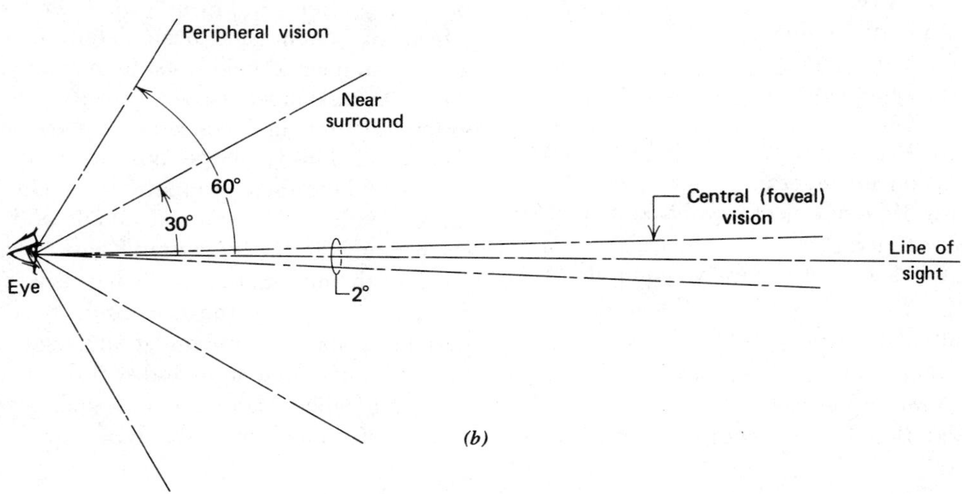

Fig. 18.16 (a) *The field of view of a normal pair of human eyes, and the angles subtended* (b). *The rectangles A and B superimposed on the field of vision* (a) *represent a magazine* (*A*) *and a small book* (*B*).

cells occur at the outer portions of the retina, their motion sensitivity results in our being best able to detect movement when looking out of the "corner of the eye." Looking at a fluorescent tube directly and then obliquely will demonstrate this effect.

Figure 18.16 shows a map of the binocular visual field, with a sketch illustrating the angles involved. Of particular interest is the extreme narrowness of the cone of central (foveal) vision, in which acute perception of detail takes place. This area is so small that the eye must refocus on successive letters in the words you are now reading if you wish to examine each individually. Surrounding this central area is a cone of binocular vision of 60° half-angle, sometimes called the near surround, in which area most of coarser sight information is gathered. Beyond this cone we have peripheral, primarily horizontal, monocular vision. It is this peripheral area that largely gives us our subjective, ambience-type reactions, which will be discussed below.

18.13 Factors in Visual Acuity

The three components of any seeing task are obviously the object or task, the lighting conditions, and the observer. Listed below are the variables affecting each of these three components. Based on the results of many investigations they can be categorized as of primary or secondary importance.

I. The Task
  Primary Factors
  (a) Size
  (b) Luminance (brightness)
  (c) Contrast
  (d) Exposure time—needed or given
  Secondary Factors
  (e) Type of object—required mental activity
  (f) Degree of accuracy required
  (g) Task; moving or stationary
  (h) Peripheral patterns

II. The Lighting Condition
  Primary Factors
  (a) Illumination level
  (b) Disability glare
  (c) Discomfort glare
  Secondary Factors
  (d) Luminance ratios
  (e) Brightness patterns
  (f) Chromaticity

III. The Observer
  Primary Factors
  (a) Condition of the eyes
  (b) Adaptation level
  (c) Fatigue level
  Secondary Factors
  (d) Subjective impressions; psychological reactions

Although in the following discussions these factors will be considered individually, many are interrelated. Thus luminance I(a) and adaptation III(b) result from the presence of illumination II(a); subjective impressions III(d) are dependent on brightness patterns II(e) and chromaticity II(f); fatigue III(c) results from a combination of many of the factors, and so on. In the literature it is common to find reference to the quantity and quality of the lighting environment. In terms of the above factors, the quantity of light has reference to item II(a) and the quality to items II(b) through II(f).

The basic visual tasks are the perception of low contrast, fine detail, and brightness gradient. Assuming a good lighting environment, that is, low glare, acceptable brightness ratios, and white light plus a normal pair of unfatigued eyes, visual acuity is primarily dependent on items I(a) to I(d), the interrelated effects of which have been determined by a large number of field tests. Remember that the seeing task under discussion involves foveal vision focusing and concentrating on small area detail. This is a vastly different task than reading, where the eye rapidly scans familiar images without focusing on details and the brain immediately understands even when much of the information is missing, as in poor reading copy. Here the task could be compared to reading an unfamiliar language written with unfamiliar signs and symbols, necessitating detailed examination of each symbol or part of symbol, individually.

18.14 Size of Visual Object

Visual acuity is generally proportional to the physical size of the object being viewed given fixed brightness, contrast, and exposure time. Since the actual parameter is not physical size but subtended visual angle, visual ability can be increased by bringing the object nearer the eye (see Fig. 18.17). It is assumed that we are dealing with a pair of *young* eyes,

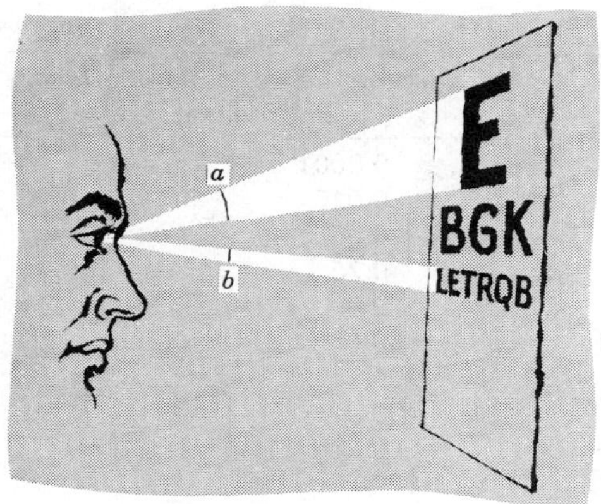

Fig. 18.17 Relationship between object size and visibility is demonstrated by comparison of subtended angles a *and* b.

since at ages above 40 the accommodation ability of the eye becomes limited and bringing the object closer blurs the focus.

18.15 Luminance (Brightness)

The terms *brightness* and *luminance* are almost entirely interchangeable, with the latter being the newer term. The difference is slight. Brightness is the subjective perceived light coming from an object. Luminance is, by definition, "photometric brightness" and is the luminous flux (light) being emitted, transmitted, or reflected from a surface, as discussed at length in Section 18.6. For the purposes of this text, the two terms are identical, and, although luminance is the preferred term, brightness is the more common and will also be used.

The sensation of vision, as explained above, is caused by light entering the eye. This light may be thought of as a group of convergent rays, each ray coming from a different point in space and therefore carrying different visual information. The composite of these rays comprises the entire visual picture that the eye sees and the brain comprehends. The individual rays differ from each other in intensity and chromaticity depending on the part of the viewed object from which they were reflected. The intensity of these cones of light determines and describes the perceived brightness of the object being viewed (see Fig. 18.18).

If the surface reflectance of the object being viewed is uniform and the illumination is also uniform, then the reflected rays of light will be equal in intensity and we will see an object of uniform brightness. If, however, as is generally the case, either the object or the illumination is nonuniform we will see an object of varied luminance.

The human eye detects brightness over an astonishing range of more than 100 million to 1, the lower levels being accomplished after an accommodation period, called adaptation time. This period varies from 2 min for cone vision to up to 40 min for rod vision for dark

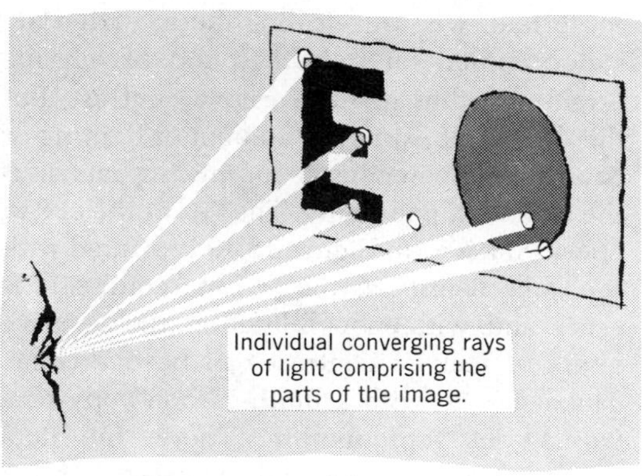

(a)

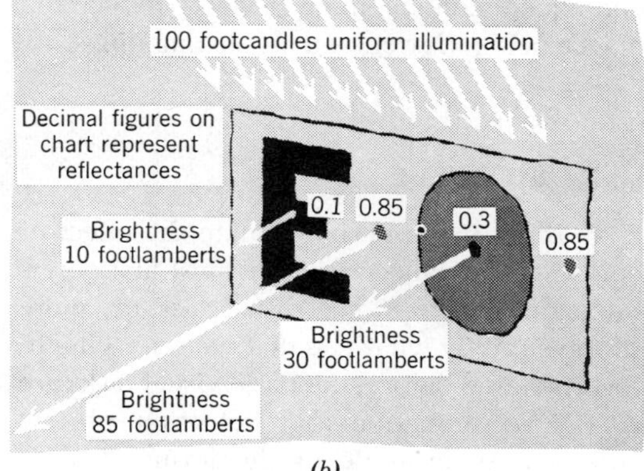

(b)

Fig. 18.18 (a) *Composition of the visual image.* (b) *brightness of a nonuniform surface.*

Table 18.3 ***Typical Luminance Values***

Object	Luminance in Footlamberts (Brightness)		
Black glove on cloudy night	0.0001		
Snow in moonlight	0.015		
Asphalt road—street lighting	0.05		
This sheet of paper lit by a candle	0.75		
Floor brightness in a poorly lighted office	2.0	↑ Full color	
Wall brightness in a well-lighted office	50.0		
Luminous ceiling	200		
Asphalt paving—overcast day	400		
North sky	1,000		
Moon, candle flame	1,500		
Asphalt paving—sunny day	2,000		
Fluorescent tube	2,000		
Kerosene flame	2,500		↑ Human eyes / Blink or squint
Hazy sky or fog	4,500		
Snow in sunlight	10,000		
40-w I.F. lamp	15,000		
500-w inside frost incandescent lamp	95,000		
Sun	450,000,000	↓	↓

adaptation, but is much faster for both types for bright adaptation (going from dark to light). The effects of adaptation on apparent brightness are discussed in the following section. Table 18.3 lists some measured brightnesses of every day visual tasks (regardless of eye adaptation level).

Returning then to the primary consideration of visual acuity as affected by luminance, we can state that, in general, visual performance increases with object brightness. However, a great deal depends on the background against which an object is viewed and the consequent contrast in brightness between the object being viewed and its surroundings.

18.16 Contrast

In order to properly evaluate the effect of contrast (luminance ratio) we must first determine the nature of the visual task or, more simply, exactly what it is that we are trying to see. As stated before, the basic visual tasks are detail discrimination and detection of low contrast. An example of the former task is reading of fine print, while of the latter would be examination of surface textures.

High contrast is helpful in delineating outline, size, and silhouette detail as shown in Figure 18.19. Note that the black-on-white letters can be read with ease even in moonlight, which is, at best, 0.01 fc whereas the latter letters of "performance" require an illumination of up to 100 fc (depending on visual acuity) and that because we expect the letter "e." Were it an unknown sign, illumination of the magnitude of direct sunlight, that is, 3000+ fc, would be required.

It is important to emphasize here what should be obvious but is often overlooked, or forgotten: we are dealing here with data superimposed on a background of varying brightness, thus giving variable contrast. But the data itself exists. No amount of lighting or contrast will compensate for missing data, as is the case with poor copy. The "e" at the end of "performance" in Figure 18.19 is printed with the same density as the "c" in "contrast." It exists, and, with enough lighting, the negative effect of lack of contrast can be overcome. This is not so with a ninth-carbon copy or a washed-out reproduction. There, the data simply does not exist, and increased lighting will simply make this fact plainer.

High background brightness makes the

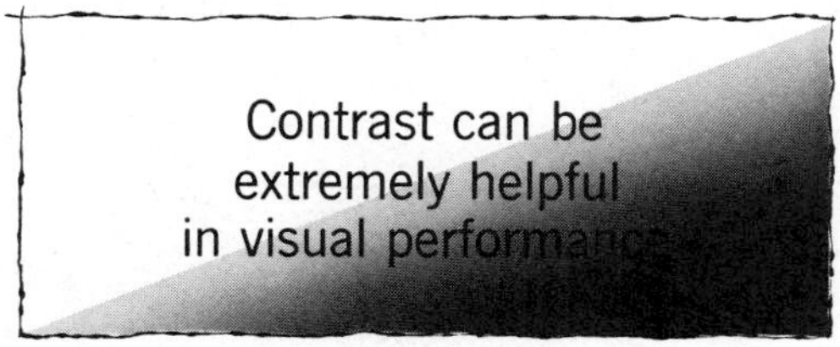

Fig. 18.19 High contrast is helpful when the seeing task involves detection of silhouette detail.

object viewed look darker and therefore assists in outline detail discrimination, as shown in Figure 18.20. Figure 18.21 shows a chart of actual visual performance in a high contrast (black on white) situation, for unknown material. The data used to prepare this curve demonstrates that smaller contrasts can be seen with higher background luminance, that is, the eye is more sensitive to brightness differ-

Fig. 18.20 *The positive effect of high-background brightness on silhouette delineation is demonstrated by the trees. In order to see surface detail in the presence of high-background brightness (high adaptation level), an equally high or higher object brightness is required. This is supplied here by the light-colored, high-reflectance building.*

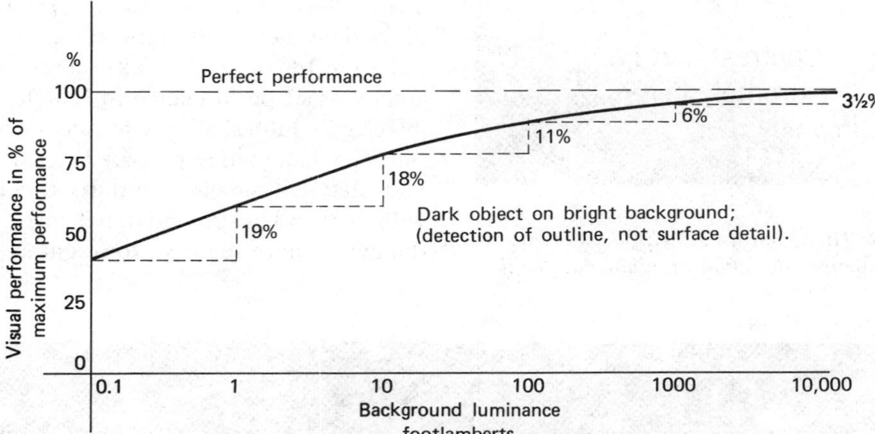

Fig. 18.21 Visual performance as a function of background luminance. Note diminishing return as brightness is increased.

ences at higher brightness levels. The important fact is that to maintain comfort, luminance ratios must be small as luminance levels increase. Thus a 10:1 ratio in object to background luminances is much more annoying visually if the numbers are 700 and 70 fL than if the luminances are 70 and 7. Examples might be a two-lamp, 2 ft × 4 ft fluorescent fixture with plastic diffuser (700–1000 fL), a sheet of paper in a brightly lighted office (50–100 fL), and a dark floor in the same office (5–15 fL). However, when the task depends on surface contrast discrimination rather than silhouette, as would be the case in examining the bark of the tree in Figure 18.20, or seeing a person's facial features when he or she stands facing us with a window behind him or her, high background illumination is a definite hindrance since the eyes automatically adjust to the average brightness of a scene. It is well known that when using an automatic (electric-eye) camera to photograph a dark object on a light background such as a person in a snow scene, it is necessary to manually increase the camera aperture in order to obtain additional light to photograph the detail of the darker object. (In doing this we overexpose the rest of the scene.) Since we cannot easily control the aperture of our eyes, we must compensate for the detrimental effect of high background brightness in another way, by increasing the surface brightness of the visual task, for example. Limited visual compensation can be made by squinting; this reduces the field of vision and the scene brightness. *Ideally, the brightness of the task should be the same as that of the background,* but ratios of 3:1 are acceptable in most circumstances. Thus, in Figure 18.20 we see that the building that is light colored (high reflectance) and well illuminated has a brightness approaching that of the background, and therefore surface visibility is good. The detrimental effect of high background brightness is thus overcome by an increase in the surface brightness of the object being viewed.

Another way of understanding the above is to consider the adaptation characteristic of the human eye (refer to Figure 18.22). As we stated, the eye adapts to the brightness level of the overall scene and sees objects within its view with respect to that adaptation level. Thus at an adaptation level of 1 fL a measured brightness ratio of 10:1 only appears to be approximately 4:1, that is, the apparent ratio is smaller than the actual. At an adaptation level of 1000 fL (daylight conditions) the apparent and actual *ratios* correspond, that is, smaller ratios are recognizable. The second important conclusion that can be drawn is that at high adaptation levels apparent brightness is lower than actual and vice versa. Thus a

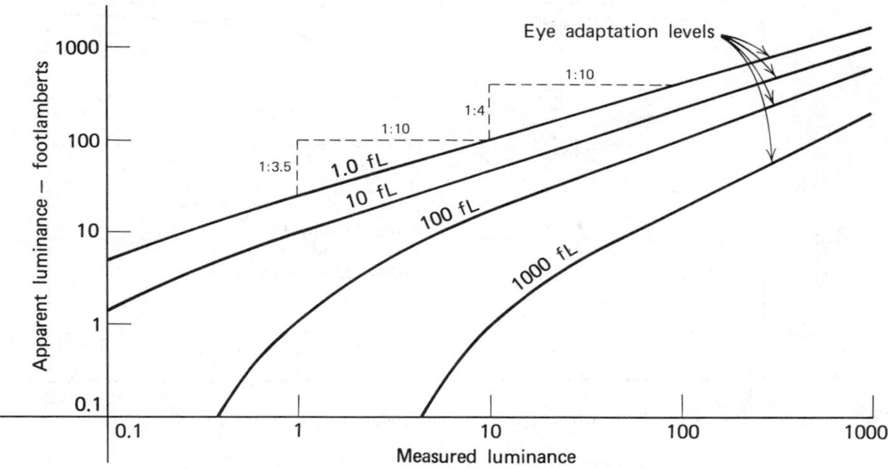

Fig. 18.22 Relation between apparent and actual measured brightness, for different eye adaptation levels. (Adapted from H. Cotton, Principles of Illumination, *Wiley, New York, 1960.)*

shadowed object near a window will look *darker* than it actually is, and, therefore, contrary to first expectation, must be lighted more brightly than a similar object further inside the room for equal visibility. This corresponds to the above conclusion with respect to viewing building surface detail in Figure 18.20, that the high background brightness (and therefore high eye adaptation level) requires high surface brightness for good surface detail visibility. That this is primarily important in daylight situations is also apparent from the curves since at a 100-fL adaptation level, which is approximately the indoor condition, apparent and actual brightness *levels* coincide (within visual ability to recognize differences).

18.17 Exposure Time

Registering a meaningful visual image is not an instantaneous process, but one that requires finite amounts of time. Just as a photograph can be taken in dim light by using a longer exposure, so can the human eye distinguish and discriminate fine detail in poor light given enough time (and neglecting eyestrain). Figure 18.23 is a plot of seeing time versus illumination for a given visual task. Of course, the time needed depends on the type of task, and different curves can be plotted for different tasks, but the principle of shorter time at higher illumination remains the same. This is particularly true when the object being viewed is not static but in motion. This curve shows that by increasing the luminance (brightness) by a factor of 6 we can halve the seeing time, whereas a further increase of sixfold in luminance only reduces the time approximately 20%. Thus, as in the case of improved contrast with increasing background brightness in Figure 18.21, we have a case of diminishing returns. Refer also to Table 18.5, which gives British IES recommendations. The table indicates that when eyestrain is not a consideration, that is, with tasks of very short duration, illumination can be lowered to the minimum, consistent with desired performance.

With this seeing parameter we must state the same reservations as with the forgoing: when dealing with material that does not require detail discrimination, improved performance does not necessarily result from improved illumination. It has been amply demonstrated[1] that speed of reading and comprehension are substantially independent of illumination levels above a minimum, but are very much dependent on the contrast quality of the material.

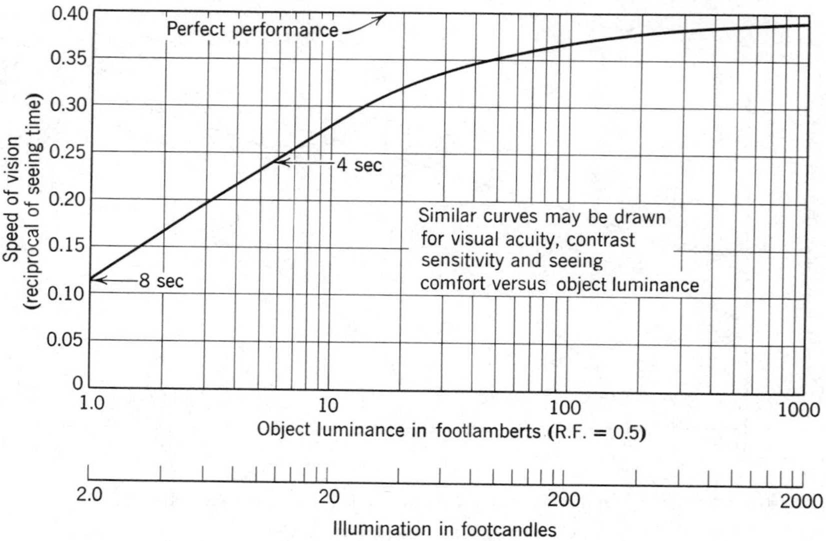

Fig. 18.23 Typical curve of visual speed versus luminance.

18.18 Other Factors In Visual Performance

The four factors considered in the preceding sections were listed in Section 18.13 as the four primary task-oriented factors. Before proceeding with the lighting factors that occupy the remainder of this chapter, a few remarks bearing on the secondary task factors and the observer-oriented considerations are in order. The preceding discussion on visual performance factors implied that optimum performance is synonymous with optimal seeing conditions, that is, the lighting conditions that yield optimal performance of certain tasks under laboratory-controlled conditions will, when extrapolated to actual field conditions, yield an ideal visual environment. Unfortunately, this may *not* be the case since a direct correlation between optimal performance and minimum fatigue has not been established, nor is it possible to extrapolate the laboratory tests to the field condition without considerable inaccuracy. An observer performing a lab test has a different level of concentration and performance than a person at an 8-hr-a-day task. The latter will compensate for unsatisfactory seeing conditions by:

1. Moving the work to a better viewing angle.
2. Moving his or her head and eyes to a more comfortable position.
3. Reducing the distance between eyes and task, to the extent that the eyes accommodate.
4. Complaining about a poor contrast task so that something is done about it (such as fixing the Xerox machine).
5. Taking more time to perform the seeing task involved. (This item, if it affects production, will frequently spur management to improve other factors.)

Furthermore, maximum performance may not be synonymous with maximum comfort or minimum fatigue. Indeed the reverse may be true. This field, which is under active research at this writing, has tentatively established that what is normally referred to as "eyestrain" is a condition of the eye muscles resulting from extensive and intensive eye use. [See Section 18.13, items I(e,f).] Thus, excellent performance under excellent lighting conditions can still produce fatigue because of the demanding nature of the task. Conditions will generally not produce fatigue if the observer is not "straining" to perform properly. In addition, as we discuss below, discomfort glare or even excessive lighting can cause fatigue without affecting performance. Thus the lighting designer who previously was concerned only with providing adequate uniform lighting levels should properly be concerned with all the factors involved. Many are indeed beyond his or her control. Some, such as task contrast, previously thought to be outside the lighting designer's province, should be examined by the designer in the framework of an overall lighting plus task plus observer problem, and recommendations made. Acceptance of these recommendations is a management decision, based on cost effectiveness. This means that management must weigh the costs involved against the expenses, including the cost of errors caused by poor seeing conditions, and reach its conclusions accordingly.

18.19 Quantity of Light; Illumination Levels

Returning to the list of factors in Section 18.13, and having discussed the task-oriented and observer-oriented items [except for item III(d), which is covered in Section 18.29], we turn now to item II, *the lighting condition.* This is frequently, if somewhat inaccurately, divided into two groups—quantity and quality of lighting—with item II(a) representing quantity and items II(b) to II(f) quality. That such a division is not accurate will become clear in our discussion of glare in Sections 18.20 to 18.22.

The factors in visual acuity discussed above have been widely known for many years and numerous attempts have been made to relate them to actual seeing tasks in order to establish brightness requirements and thereby illumination levels, so that practical design use can be made of the laboratory experimental data. The best known and most successful of these attempts from an acceptance viewpoint has been the continuing work of Dr. H. R. Blackwell at Ohio State University. The lighting level recommendations of the IES of North America are largely based on this work. Blackwell used the detection of a circular target on a field of uniform brightness as his basic experimental visual task, the variable parameters being the size, brightness, and exposure time of the target. The result of these tests was the determination of contrast requirements for various accuracies of sightings. Blackwell's problem was then to relate these laboratory findings to actual practical seeing tasks. By application of certain adjustment factors (previously called field factors), and use of a device known as a Field Task Simulator, he was able to extrapolate his laboratory results to actual task requirements. These derived figures are the basis of the current American IES recommendations. These are given in IES Handbook, 5th edition, Section 7. The figures in Table 18.4 are similar to these recommendations, as modified by the author. The British IES recommendations are generally lower than their American counterparts for two reasons:

1. The British standards are based on the work of Weston. The field factors resulting from his work are lower than that based on Blackwell's work in the ratio of 8:7.2. This small difference results in very significant illumination differences because the contrast/illumination curve is almost horizontal at the points of interest. Thus a very small difference in multiplying factor yields a large difference in required task brightness. (The present terminology for these curves is Visual Performance Reference Curves.)
2. The American IES recommendations are based on maximum performance. The British IES uses a graded system starting with a base illumination level and increasing it for poor task contrast, absence of daylight, and requirement for *maximum* rather than simply *good* performance (see Table 18.5). Contrariwise, illumination maybe reduced if seeing task is of unusually short duration.

The requirement for energy conservation has had a profound and salutary effect on illumination recommendations in several ways.

(*a*) It spurred the study of veiling reflections (see Section 18.22), which has led to increased use of Equivalent Sphere Illumination (ESI) in specifying illumination levels.

Table 18.4 *Illumination Levels*

	Recommended Minimum Footcandles
Industrial	
Airplane Manufacturing	
Parts manufacturing	
Drilling, riveting, and screw fastening	70
Final assembly	100
Airplane Hangars	
Repair service only	100
Assembly	
Rough easy seeing	30
Medium	100
Fine	500
Bakeries	
Mixing room	50
Oven room	30
Book Binding	
Cutting, punching, and stitching	70
Embossing and inspecting	200
Chemical Works	
Hand furnaces, boiling tanks, stationary driers, stationary and gravity crystallizers.	30
Clay Products and Cements	
Molding, pressing, cleaning, and trimming	30
Color and glazing—rough work	100
Cloth Products	
Cloth inspection	2000
Cutting	300
Sewing	500
Electrical Equipment Manufacturing	
Insulating: coil winding	100
Testing	100
Exterior Areas	
Entrances	
Active (pedestrian and/or conveyance)	5
Inactive (normally locked, infrequently used)	1
Building surrounds	1
Active shipping area	5
Storage areas—active	20
Storage areas—inactive	1
Loading and unloading platforms	20
Sheet Metal Works	
Miscellaneous machines, ordinary bench work	50
Presses, shears, stamps, spinning, me-	
Garages—Automobile and Truck	
Service garages	
Repairs	100
Active traffic areas	20
Parking garages	
Entrance	50
Traffic lanes	10
Storage	5
Inspection	
Ordinary	50
Difficult	100
Highly difficult	200
Most difficult	1000
Laundries	
Washing	30
Flatwork ironing, weighing, listing, and marking	50
Machine and press finishing, sorting	70
Leather Manufacturing	
Cleaning, tanning, and stretching, vats	30
Finishing and scarfing	100
Locker Rooms	20
Machine Shops	
Rough bench and machine work	50
Medium bench and machine work	100
Fine bench and machine work, fine automatic machines	500
Materials Handling	
Wrapping, packing, labeling	50
Picking stock, classifying	30
Loading, trucking	20
Inside truck bodies and freight cars	10
Paint Shops	
Dipping, simple spraying firing	50
Fine hand painting and finishing	100
Polishing and Burnishing	100
Printing Industries	
Printing plants	
Color inspection and appraisal	200
Composition	100
Presses	70
Proof reading	150
Receiving and Shipping (see Materials Handling)	
Exhibitions	30
Social activities	5
Banks (see also Offices)	
Lobby	50

Table 18.4 ***Illumination Levels (Continued)***

	Recommended Minimum Footcandles
dium bench work	50
Punches	50
Stairways, Corridors, and Other Service Areas	20
Storage Rooms or Warehouses	
Inactive	5
Active	
Rough bulky	10
Fine	50
Testing	
General	50
Extra fine instruments, scales, etc.	200
Toilets and Washrooms	30
Upholstering—Automobile, Coach, Furniture	100
Watch and Jewelry Manufacturing	500
Warehouse (see Storage)	
Welding	
General illumination	50
Precision manual arc welding	1000
Woodworking	
Rough sawing and bench work	30
Sizing, planing, rough sanding, medium quality machine and bench work, glueing, veneering, cooperage	50
Fine bench and machine work, fine sanding, and finishing	100
Stores, Offices, and Institutions	
Art Galleries	
General	30
On paintings (supplementary)	30
Dark paintings with fine detail may require two or three times as much illumination.	
On statuary	100
In some cases, much more illumination is necessary to reveal the beauty of statuary.	
Auditoriums	
Assembly only	15
General	10
Corridors, elevators, and stairs	20
Entrance foyer	30
Linen room	
Sewing	100
General	20
Writing areas in lobby	70[a]
Teller's stations, posting, keypunch	150[a]
Barber and Beauty Shops	100
Churches & Synagogues	
Altar, arc	100
Pews	15
Pulpit (supplementary)	50
Club Reading Rooms	30
Courtrooms	
Seating area	30
Court activity area	70[a]
Hospitals	
Autopsy	
General	100
Supplementary	1000
Corridors	20
Emergency Rooms	
General	100
Local	2000
Examination and Treatment Rooms	
General	50
Examining table	100
Laboratories	
General	50
Closework	100
Patients' Rooms	
General	20
Supplementary for reading	30
Supplementary for examination	100
Recovery Rooms	30
Surgery	
General	200
Supplementary on table	2500
Toilets	20
Waiting Rooms	20
Hotels and Motels	
Bars and cocktail lounges (see Restaurants)	
Bathrooms	
General	10
Mirror	30
Bedrooms	
Reading (books, magazines, newspapers)	30
Subdued environment	15
Quick service type	
Bright surroundings	100
Normal surroundings	50
NOTE: Footcandle levels in dining areas are highly variable. Variations	

Table 18.4 *Illumination Levels (Continued)*

	Recommended Minimum Footcandles
Lobby	
General lighting	10
Reading and working areas	30
Power Plant	
Boiler room	10
Equipment room	20
Storerooms	10
Libraries	
Reading rooms and carrells	70[a]
Stacks	30
Book repair and binding	70
Check-in and check-out, catalogs,	70[a]
Card files	100[a]
Offices	
General	
Cartography, designing, detailed drafting	200[a]
Accounting, auditing, tabulating, bookkeeping, business machine operation	150[a]
Regular Office Work	
Good copy	70[a]
Regular office work—reading, transcribing, active filling, mail sorting, etc., fair-quality copy	100[a]
Corridors, elevators, escalators, stairways	20
(Or, not less than $\frac{1}{5}$ the level in adjacent areas.)	
Post Offices	
Lobby, on tables	30
Sorting, mailing, etc.	100
Storage	20
Corridors and stairways	20
Restaurants	
Dining Areas	
Cashier	50
Intimate type	
Light environment	10
Subdued environment	3
Leisure type	
Light environment	30
Theaters	
Auditoriums	
During intermission	5
During performance or presentation	0.1
Foyer	5
Entrance lobby	20
***Residential* (*see Table 000 p. 000*)**	
Outdoor Floodlighting	
Building	

	Recommended Minimum Footcandles
depend on such factors as time of day, desired atmosphere, individuality, and attractiveness.	
Food Displays—twice the general levels but not under	50
Kitchen—commercial, hospital, hotel	
Inspection, checking, and pricing	70
Other areas	30
Schools	
Tasks	
Reading printed material	30[a]
Reading pencil writing	70[a]
Reading spirit duplicated material	
Good copy	30[a]
Poor copy	100[a]
Classrooms	
Chalkboards (supplementary illumination)	150
Drafting rooms	100[a]
Laboratories	100
Lecture rooms	
General	70[a]
Special exhibits and demonstrations	150
Lipreading classes	150
Shops	100
Sewing rooms	150
Sightsaving classes	150[a]
Study halls	70[a]
Corridors and stairs	20
Stores	
Store interiors	
Circulation areas	30
Merchandising areas	
Service stores	100
Self-service stores	200
Showcases and wall cases	
Service stores	200
Self-service stores	500
Feature displays	
Service stores	500
Self-service stores	1000
Stockrooms	30

Building Exteriors, and Monuments (Floodlighted)

	Surroundings Dark	Surroundings Bright
Light surfaces	5	15
Medium—dark surfaces	15	30
Dark surfaces	20	50

Flags	50
Parking Lots	

[a]Equivalent sphere illumination (ESI footcandles).

Table 18.4 *Illumination Levels (Continued)*

	Recommended Minimum Footcandles	
Construction	10	
Excavation	2	
Bulletins and Poster Panels		
	Surroundings	
	Dark	**Bright**
Light surfaces	20	50
Dark surfaces	50	100

	Recommended Minimum Footcandles
Self-parking	1
Attendant parking	2
Shopping centers (customer attraction device)	5

Table 18.5 *Flowchart*

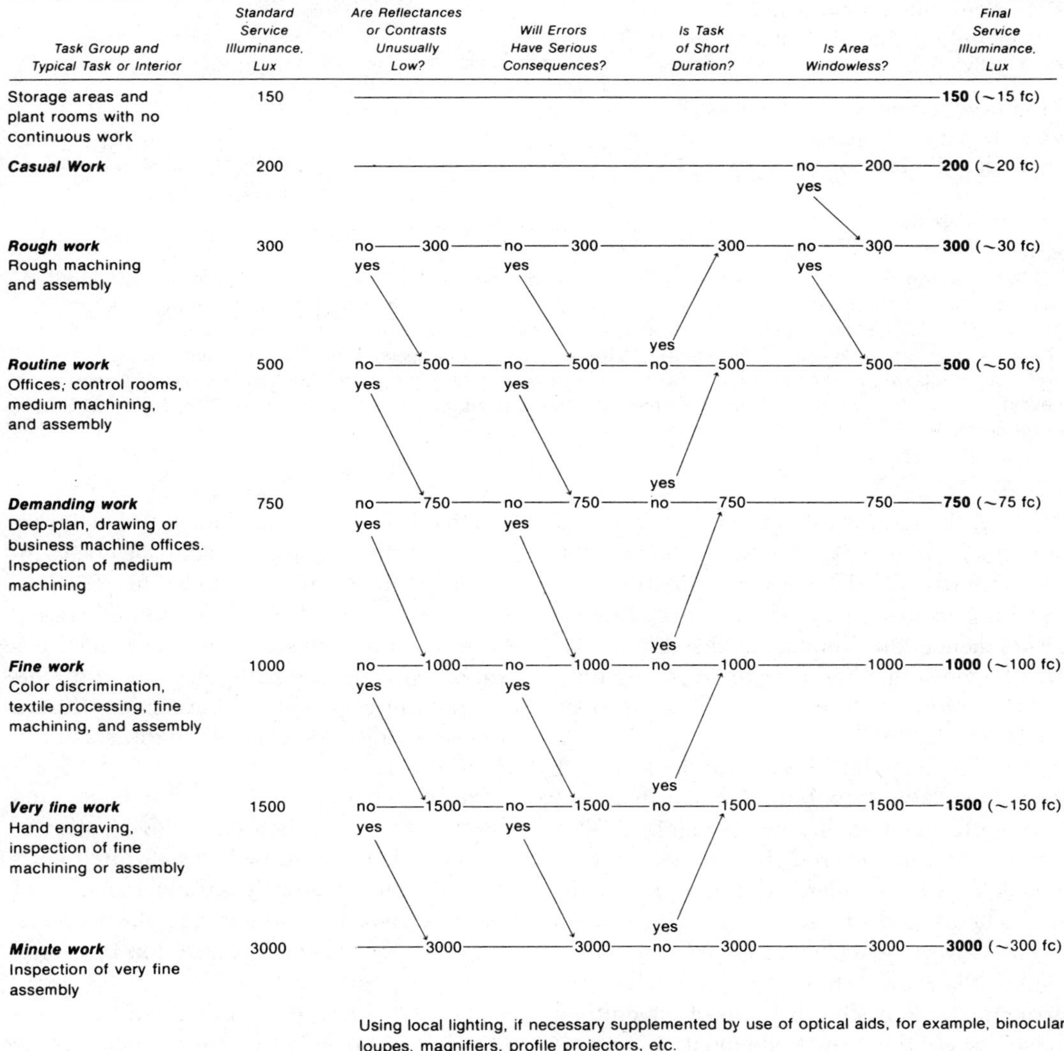

Source. British IES Code for Interior Lighting, 1977.

NOTES.

The standard service illuminance recommended for an application should be increased if:

1. Unusually serious consequences, in terms of cost or danger, could result from mistakes in perception.
2. Unusually low reflectances or contrasts are present in the particular task.
3. Tasks for which the recommended standard service illuminance is less than 500 lx are carried out in windowless interiors.

The recommended standard service illuminance may be decreased if, in the judgment of the designer, the duration of the task is unusually short. This flowchart gives the steps by which the recommended standard service illuminance should be modified when one or more of these conditions apply. The resulting final service illuminance derived from the flowchart should then be used as the design value.

Table 18.6a *Federal Energy Administration and General Services Administration Recommended Maximum Lighting Levels*

Task or Area	*Footcandle Levels*	*How Measured*
Hallways or corridors	10 ± 5	Measured average, minimum 1 fc
Work and circulation areas surrounding work stations	30 ± 5	Measured average
Normal office work such as reading and writing (on task only), store shelves, and general display areas	50 ± 10	Measured at work station
Prolonged office work which is somewhat difficult visually (on task only)	75 ± 15	Measured at work station
Prolonged office work which is visually difficult and critical in nature (on task only)	100 ± 20	Measured at work station as maximum
Industrial tasks	ANSI–A11.1-1973	

NOTE.

Table 18.6*a* shows levels for office work that are recommended when work stations are occupied; otherwise, consideration should be given to turning lights off or to switching to 30 fc if other workers remain nearby. For tasks requiring levels higher than 50 fc, switching to lower levels is desirable if the work changes to less critical tasks. Illumination at the task should be reasonably free of veiling reflections and body shadows. Refer to Table 18.6*b* for guidance in determining visual difficulty of office tasks. Levels for industrial work are from the American National Standards Institute All. 1-1973, June 1973, Practice for Industrial Lighting.

(*b*) Although levels were always specified for *tasks*, common practice was to provide the maximum required level for the entire area resulting in extensive energy waste. Present recommendations emphasize the differentiation between task and surroundings, resulting in nonuniform lighting layouts (see Sections 21.15 and 21.16).

(*c*) The inapplicability of a single set of recommendations to both familiar material (reading) and unfamiliar was recognized. This has resulted in lowered illumination recommendations for standard reading tasks in offices, schools, and so on.

(*d*) Governmental agencies have commissioned their own studies[1,2,3] and trial construction buildings on the basis of which they have issued their own illumination recommendations. These are binding in many types of government-owned and sponsored projects (see Tables 18.6, 18.7, 18.8). A distinct advantage in these recommendations is the recognition that task duration as well as task nature is important, that is, a time factor has been introduced into the recommendations. Furthermore, the General Services Administration (GSA) recommends following the British system of avoiding dogmatic obedience to footcandle-level recommendations, since the eye does not readily detect small brightness differences (see Table 18.9).

(*e*) In lieu of assuming a fixed task and observer to which lighting is added, the energy-conscious design includes all three in its purview. This necessarily introduces factors of cost effectiveness and energy effectiveness. The latter finds detailed expression in energy budgeting (see Section 20.5).

(*f*) Daylight is to be utilized to the maximum extent possible, (recognizing the inherent connection with heat loss and heat gain through windows) (see Section 19.2). This means that illumination levels *include* daylight contributions.

Table 18.6*b* *Relative Visual Task Difficulty for Common Office Tasks*

Task Description	*Visual Difficulty Rating*
Large black object on white background	1
Book or magazine, printed matter, 8 point type and larger	2
Typed original	2
Ink writing (script)	3
Newspaper text	4
Shorthand notes, ink	4
Handwriting (script) in No. 2 pencil	5
Shorthand notes, No. 3 pencil	6
Washed-out copy from copying machine	7
Bookkeeping	8
Drafting	8
Telephone directory	12
Typed carbon, fifth copy	15

Source. Ross and Baruzzini, *Energy Conservation Applied to Office Lighting,* Federal Energy Administration, April 1975.

NOTE.

Table 18.6*b* may be used as a guide in evaluating the degrees of visual difficulty for office work. It is based on the concept that visual difficulty for this kind of work is not only a function of the intrinsic characteristics of the task and the lighting system, but also of the length of time in which the task must be performed. To use this table, multiply the difficulty rating, as shown in the table, for each task performed at a given work place by a single worker by the number of decimal hours per day it is performed; for example: 3 hours, 15 minutes = 3.25 decimal hours. Add the products for each task. If the sum is greater than 40, provide 75 fc on the work station. If the sum is greater than 60, provide 100 fc on the work station. Multiply the difficulty factors by 1.5 if the operator is over 50 years of age, or if he has uncorrectable eyesight problems.

(g) ASHRAE Standard 90-75, Section 9, has been legally adopted by numerous states and will probably become legally binding throughout the United States. Among its provisions are a procedure for developing a power budget for a project (see Section 20.5). In calculating the lighting loads, ASHRAE Standard 90-75 specifies that nonuniform layouts be used as follows:

1. Task lighting to accord with IES recommendations.
2. General area lighting to be one-third of task lighting.
3. Noncritical (circulation) lighting to be one-third of general area lighting.

That is, a room with task lighting of 100 fc would have area lighting of 30 to 35 fc and circulation lighting of 10 to 12 fc. Albeit this standard is binding only for budget calculation within which budget the designer is free to do as he or she pleases, it is difficult to see how a designer could avoid nonuniform design even if he so desired. In effect, then, ASHRAE 90-75 (reprinted from IES Standard EM S-2) makes nonuniform layout the standard required technique.

As a result of this multiplicity of sources, agencies, and authorities the lighting designer cannot simply refer to the American IES recommendations as *the* authoritative list. Furthermore, there are contradictions between different agency recommendations and con-

Table 18.7 ***General Services Administration Recommended Electric Lighting Levels for Materials Handling and Storage Applications (Levels shown are average maintained horizontal footcandles, measured 4 ft above the floor.)***

Task	*Nominal Footcandle Level*[a]	*Equivalent Footcandle Range*[c]
Bulk storage (active, medium)	10	8–12
Bulk storage (inactive)	5	4–6
Bulk storage, aisles	15	12–18
Bin storage (active)	15	12–18
Bin packing	30	24–36
Weighing and marking	30	24–36
Inspection	50/75	40–60/60–90
Parcel post packing	30	24–36
Bin replenishment holding	5	4–6
Bin assembly	15	12–18
Shipping and receiving (bulk)	10	8–12
Shipping and receiving, inspect and check	50	40–60
Flammable storage	10	8–12
Truck wells	3	2.4–3.6
Truck docks[b]	15	12–18
Rail docks[b]	5	4–6
Rail dock entrances[b]	15	12–18
MHE shop	30	24–36
Battery charging	15	12–18
Shop storage	5	4–6
Exterior packing	10	8–12
Exterior truck maneuvering	3	2.4–3.6
Exterior storage	0.2	0.16–0.24
Exterior security lighting	0.2	0.16–0.24

Source. Ross and Baruzzini, *Lighting Systems Study,* General Services Administration, Public Buildings Service, March 1974.

[a]These are levels to be achieved for a task when personnel are present. When personnel are absent, lighting should be switched off or be reduced to a level suitable only for safety requirements.

[b]Supplement the truck and rail dock general lighting with bracket-mounted lighting that can be directed into the interior of the truck or rail car. These bracket lights should have local switching and should be turned off when not in use.

[c]Specific tasks within the areas described may require substantially different levels of lighting. See, for example, Table 18.6 for office work.

siderable vagueness in some. This problem is under active consideration by the lighting design profession at this writing. The only firm recommendation that can be made is that the designer consult with the authorities having jurisdiction before establishing the criteria by which the job will be designed.

18.20 Considerations of Lighting Quality

Quality of lighting is a term used to describe all the factors in a lighting installation not directly concerned with quantity of illumination. Certainly it is obvious that if a given room is alternatively lighted with a bare bulb

Table 18.8 ***Energy Conservation Guidelines***

Illumination[a]
Carefully select the tasks and duration of tasks. Illuminate those tasks only to the following standards: 1. Corridors, lobbies, and means of egress. Minimum of 10 raw fc with the average not to exceed 18 raw fc. 2. Storage areas: $\frac{1}{2}$ raw fc average with luminaires located strategically where needed in the area. 3. Waiting rooms and lounge areas: 12 to 18 raw fc. 4. Office areas: selective lighting with maximum background lighting of 10 to 20 raw fc. 5. Desk tops: 50 ESI fc. 6. Conference tables: 30 ESI fc. 7. Secretarial desks: 50 ESI fc with auxiliary localized (lamp) task lighting directed at paper holder (for typing) as needed. 8. Over open drawers of filing cabinets: 30 raw fc. 9. Over vertical surfaces as bookshelves: 30 raw fc. 10. Courtrooms and auditoriums: over 30 raw fc. 11. Kitchens: selective, but average 50 raw fc. 12. Cafeterias: 30 raw fc. 13. Testing labs: as required. 14. Drafting: full time, 100 ESI fc; part time, 70 raw fc. 15. Accounting offices: at work surfaces, 100 ESI fc. 16. General clerical: 50 ESI fc. 17. Toilet rooms: 20 raw fc.

Source. Dubin-Mindell-Bloome Associates, *Energy Conservation Design Guidelines for Office Buildings,* GSA/PBS, January 1974.

[a] The above illumination levels are possible with a maximum of 2.3 w per net square foot.

Table 18.9 ***Schedule of Illumination Levels—GSA***

Each level represents a significant subjective change. (Changes in levels of substantially lesser magnitude are of little consequence to the eye.)

Footcandles	*Lux*
0.2	2
0.5	5
1.0	10
2.0	20
3.0	30
5.0	50
7.5	75
10.0	100
15.0	150
20.0	200
30.0	300
50.0	500
75.0	750
100.0	1000
150.0	1500
200.0	2000
300.0	3000

Source. Adapted from Note B of the "General Schedule" of the *British IES Code for Interior Lighting,* 1977.

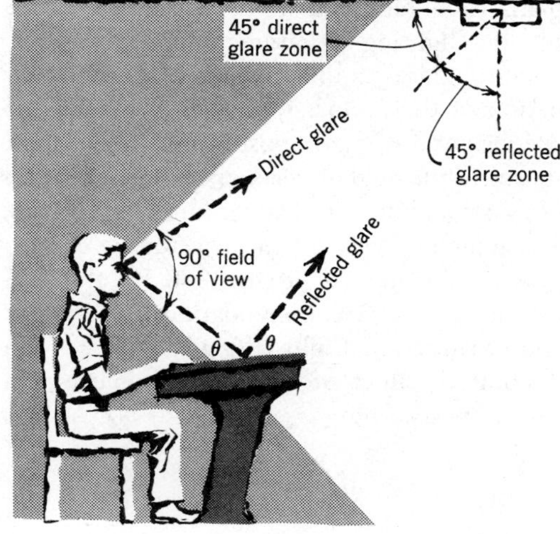

Fig. 18.24 Glare zones with observer in a head-up position. The direct and reflected glare light paths are delineated on the diagram although reflected glare is usually studied with the eyes down at a reading angle. Placement of fixtures, room size, ceiling height, paint finishes, windows, and so on also affect brightness ratios, and therefore glare.

and with a luminous ceiling, both giving the same average quantitative illumination (in terms of lumen output), there is a vast difference in the two lighting systems. This difference is in the quality of the lighting, a term which describes the luminance ratios, diffusion, uniformity and chromaticity of the lighting. Since uncomfortable brightness ratios, where background luminance exceeds object luminance, are commonly referred to as glare, the quality of the lighting system is also a description of the visual comfort and visual adequacy of the system.

When the discomfort glare is caused by light sources in the field of vision it is known as *direct glare.* When the glare is caused by reflection of a light source in a viewed surface it is known as reflected glare or "veiling reflection" (see Fig. 18.24).

18.21 Direct Glare

Glare by definition produces discomfort and interference with vision. This is a general qualitative statement. To determine whether a specific lighting situation is producing discomfort glare requires a quantitative examination of the factors involved. These factors are brightness, size, position of each light source in the field of vision, plus consideration of surrounding or background brightness. These factors were given quantitative expression in a study issued in 1966 by the IES Committee on Recommendations for Quality and Quantity of Light (RQQ-Report 2). The discomfort effect of a single luminaire was determined to be

$$M = \frac{LQ}{PF^{0.44}}$$

M = individual source glare factor
L = apparent source luminance
Q = function of solid angle subtended by the eye when viewing the source
P = position index factor, relating discomfort to position of the glare source in the field of view
F = room luminance factor, that is, the related eye adaptation level

The formula is a mathematical statement of logical conclusions. Obviously glare is proportional to the source luminance. The size of the source defined with respect to the viewer by the angle subtended by the source at the eye, is the second parameter to which glare is directly proportional (see Fig. 18.25). It is necessary to consider both these parameters together to understand why a small, very bright source is not a serious problem whereas a large, low brightness source, such as a luminous ceiling, may be. Indeed, a small bright source adds sparkle to the field of vision, and many observers find it pleasant in an otherwise monotonous lighting environment.

The remaining two factors are less self-evident. Glare decreases rapidly as the brightness source is moved away from the direct line of vision and thus the glare produced by a source depends on its position in the field of view. We indicated in Section 18.15 that one way to decrease the detrimental effect of high background brightness was to increase the task brightness. This technique is also effective in reverse with a source of glare, since the amount of discomfort glare produced by a source is inversely proportional to the background or field brightness (also known as eye adaptation level). Thus a ceiling fixture with a luminance of 1200 fL at 65° might easily constitute a source of discomfort glare in a space with an eye adaptation level of 50 fL. The same fixture would not be objectionable in a daylight condition, where the eye adaptation level might be 500 fL (see Fig. 18.22). A more striking example is that of an automobile's headlights, which at night are so severe a source of glare as to be described as "disabling glare," whereas in daylight, with its concomitant high eye adaptation level, these lights are barely noticeable.

The sum of the individual glare source contributions yields a quantitative value for glare, called the "discomfort glare rating." Since this factor is not readily usable, it is converted to a criterion called "Visual Comfort Probability," or VCP, which is defined as the percent of normal vision observers who will be *comfortable* in that specific visual environment. Since calculating VCP for each space

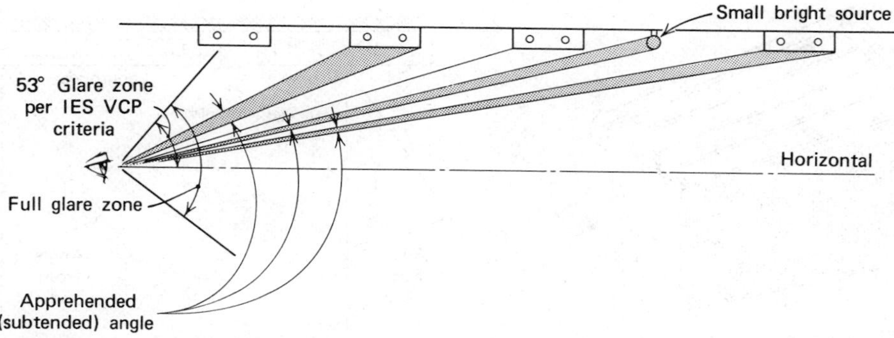

Fig. 18.25 Glare determination. The glare contribution of each source depends on its size (subtended or apprehended angle), luminance, and location in the field of view. Note that the apprehended angle of a small source is such that even with high brightness it is not objectionable. Such sources are normally called "sparkle." Glare will be much more objectionable with dark background than with a light one; therefore, light-colored paints on ceilings and upper walls are recommended.

being lighted is an extremely time-consuming job even with computer assistance, the IES has established a set of standard conditions for which the VCP for different sources can be calculated. These conditions are:

- Initial illumination level of 100 fc (high for most uses; see Tables 18.4 to 18.7).
- Room surface or effective cavity reflectances: ceiling 80%, walls 50%, floor 20%.
- Mounting heights above the floor: 8.5, 10, 13, and 16 ft.
- A range of room dimensions (floor area) to include square, long-narrow, and short-wide rooms.
- A standard layout involving luminaires uniformly distributed throughout the space.
- An observation point 4 ft in front of the center of the rear wall and 4 ft above the floor.
- A horizontal line of sight, directly forward.
- A field of view limited to 53° above and directly forward from the observer.

With these conditions, direct glare will not be a problem if all three of the following conditions are satisfied:

1. The VCP is 70 or more.
2. The ratio of maximum-to-average luminaire luminance does not exceed 5:1 (preferably 3:1) at 45, 55, 65, 75, and 85° from the nadir crosswise and lengthwise.
3. *Maximum* luminaire luminances crosswise and lengthwise do not exceed:

Angle above Nadir, Degrees	Maximum Luminance, Footlamberts
45	2250
55	1605
65	1125
75	750
85	495

A typical set of manufacturer's brightness and VCP data is shown in Figure 18.26 for an actual ceiling-mounted fluorescent fixture with 4-40WT12 lamps. Note that all VCP values are considerably above the 70 minimum criteria. If full VCP data of this type is not available, an earlier criteria commonly called the "scissors curve criteria" can be used. This criteria, which is intended for application to a *single fixture*, by extrapolation can be applied to an entire space; that is a room with fixtures that meet the scissors curve criteria will generally not fall below a VCP of 70. A brightness data check sheet using scissors curve criteria and typical fixture data is shown in Figure 18.27. If we replace the scissors shown in the figure with the eye, this

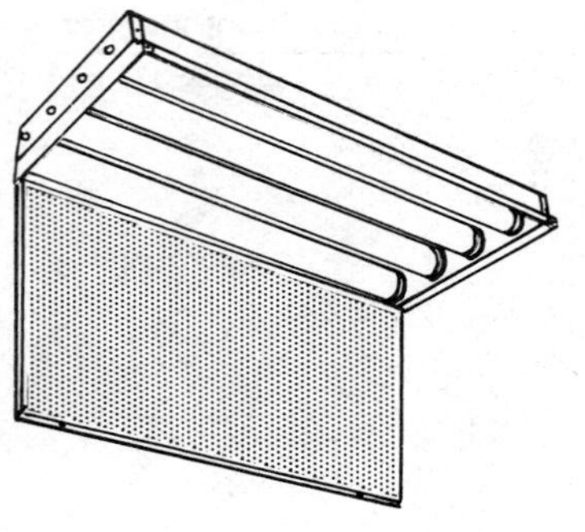

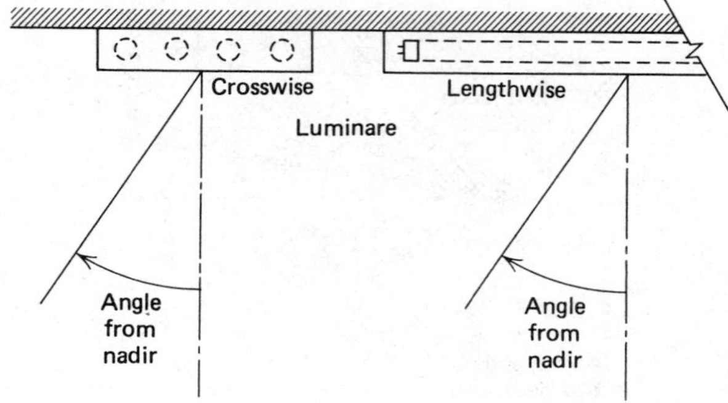

AVERAGE LUMINANCE DATA

In Footlamberts
Holophane No. 8224G4L1.

Vertical Angles	*Across Axes*	*45° Plane*	*Along Axes*
60°	587	571	512
65°	310	302	317
70°	149	168	165
75°	121	120	112
80°	140	120	100
85°	162	136	122

IES VISUAL COMFORT PROBABILITY DATA

Holophane No. 8224G4L1
Reflectance:
Wall 50%,
Ceiling Cavity 80%,
Floor Cavity 20%.
Work Plane.
Illumination: 100 fc.

Room Size		*Luminaires Lengthwise*				*Luminaires Crosswise*			
W	*L*	Ceiling Height (in feet)							
(in ft.)		8.5	10.0	13.0	16.0	8.5	10.0	13.0	16.0
20 x	20	81	77	76	78	79	75	72	74
	30	81	78	76	74	79	76	72	70
	40	82	79	77	74	79	77	74	71
	60	81	80	78	76	79	77	75	72
30 x	20	84	80	77	76	82	78	74	74
	30	83	80	77	73	82	79	74	70
	40	83	81	78	73	81	79	75	70
	60	82	80	78	74	80	78	76	72
	80	82	80	79	76	80	78	76	73
40 x	20	86	83	80	77	85	81	77	75
	30	85	83	79	74	84	81	76	71
	40	84	82	79	74	83	81	77	72
	60	83	82	79	75	82	80	77	73
	80	83	81	79	76	81	80	77	73
	100	82	81	80	76	80	79	77	74
60 x	30	86	84	81	76	85	83	79	74
	40	85	83	81	76	84	82	79	74
	60	84	82	80	76	82	81	79	74
	80	83	82	80	77	81	80	78	75
	100	82	81	80	77	81	80	78	75
100 x	40	87	86	83	79	86	85	82	78
	60	85	84	82	79	84	83	81	78
	80	84	83	82	79	83	82	80	77
	100	83	82	81	79	82	81	79	77

Fig. 18.26 A typical set of manufacturers' published VCP and brightness data. Data at 45, 50, and 55° are not presented. (Courtesy of Holophane, a Johns-Manville Division.)

criteria tells us the limiting comfort brightnesses of the eye, that is,

Angle above Nadir, Degrees	*Average Brightness Well Tolerated (⅓ Max.), Footlamberts*
45	750
55	400
65	375
75	250
85	165

If we compare the VCP limiting brightnesses and cavity reflectance to the previous scissors criteria limiting brightnesses and recommended reflectances, we see that the newer criteria is directly derived from its predecessor.

Manufacturer's Company Name ABC Company Luminaire Catalog No. 12345RS240
Laboratory Report A67413
Lamp Description F40WW No. of Lamps 2 Rated Lumens Each 3200

Maximum to Average Luminaire Brightness Ratios (Typical Fixture Data)

(Normal) Crosswise

Angle	3 Times Value[1]	Maximum	Average	Ratio Max/Avg
45°	2250	1320	710	1.9
55°	1605	985	385	2.6
65°	1125	620	250	2.5
75°	750	430	215	2.0
85°	495	250	185	1.4

Lengthwise (Parallel)

Angle	3 Times Value[1]	Maximum	Average	Max/Avg
45°	2250	1320	535	2.5
55°	1605	920	335	2.7
65°	1125	530	220	2.4
75°	750	330	195	1.7
85°	495	265	145	1.8

[1]Three times the value of the sloped limiting line, in footlamberts.

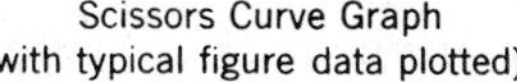
Scissors Curve Graph
(with typical figure data plotted)

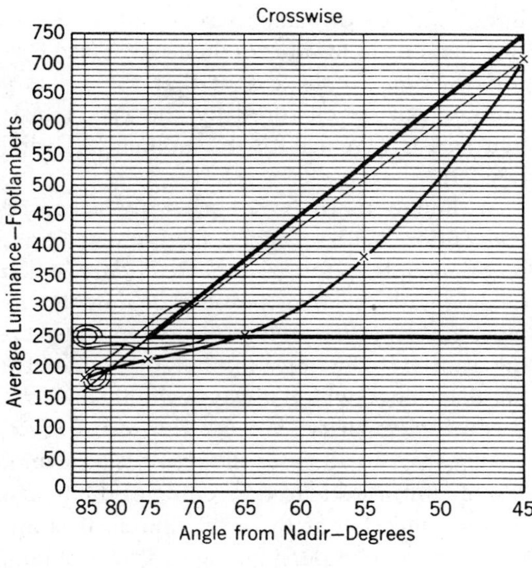

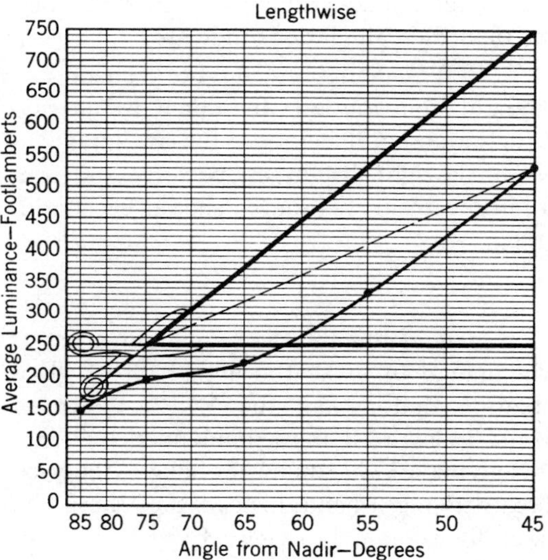

***Fig. 18.27** Luminaire brightness check sheet.*

Despite the accuracy and excellence of the above VCP criteria, it is inherently limited by its own standard conditions that are not easily applied to other situations, as follows:

1. In small spaces, VCP has little significance.

2. Tabulated VCP figures are given for the worst case in the room, assuming a uniform layout. Since VCP varies dramatically with observer position and aspect with respect to fixture alignment, the VCP values given are generally lower than the actual ones. Thus, in a room lighted with continuous rows of fixtures and an observer facing along the lengthwise fixture axis, the VCP is much higher than for an observer facing the crosswise axis. The only way to determine the actual VCP is to select a few observer positions and run (via computer) a few calculations.

3. VCP is only approximately proportional to illumination. Thus in rooms with other than 100 fc illumination, an educated guess is required, depending on layout.

4. VCP calculations are based on *overhead* lighting. The increasing trend toward perimeter lighting because of its salutory effect on veiling reflections makes VCP judgment more difficult. However since perimeter lighting is in the direct line of sight, the VCP will worsen (decrease).

5. VCP calculations are based on uniform layouts. Recent criteria effectively require nonuniform arrangements. To minimize veiling reflections, the closest fixtures in such arrangements are placed *outside* the frontal line of sight, thus decreasing direct glare and increasing VCP.

6. Most office tasks are performed in a head-down position. Since tabulated VCP is calculated for a glare zone of 53° *above* the horizontal, actual VCP will be markedly better than the calculations indicate.

7. VCP calculations do not reflect any daylight contribution. Inasmuch as energy-oriented design encourages the advantageous use of daylight, brightness patterns and contributions from this source should be considered.

In view of the above reservations, most of which tend to make the actual direct glare situation better than the VCP calculation would indicate, it is recommended that layouts giving a VCP somewhat below 70 not be discarded out of hand. Instead, they should be carefully examined in view of the above and, if possible, several observer positions calculated. By so doing the designer will develop a "feel" for glare sources that will enable him or her to make the intelligent judgments necessary. If the job warrants the expenditure of the additional computer time necessary, there are programs available that will compute and plot VCP values in the space being studied and compute minimum or average VCP. With these as a guide the designer can rearrange and substitute equipment to obtain the desired condition. As explained in detail in Section 18.22, a trade-off is frequently necessary between direct and reflected glare since the two are in large measure mutually incompatible.

18.22 Reflected Glare

(*a*) ***Nature of the Problem.*** The problem of reflected glare is much more complex than that of direct glare because it involves both the source *and* the task and is inherent in the act of seeing (refer to Figure 18.28). Vision is produced by light being reflected from the object seen. The object mirrors the source(s) of the light into the eye of the observer, that is, if the object being viewed were replaced by a mirror we would see the source(s) of light in the room (see Fig. 18.28*a*). In an interior space there are multiple sources of light. The primary sources are usually one or more lighting fixtures near the observer. Secondary sources are other, more remote fixtures in the room, *and all the room surfaces* that obtain their light from the primary sources and by reflection become light sources themselves (see Figure 18.28*b*). To the extent that the *primary* sources can be mirrored by the vision task, glare exists. (We are for the moment ignoring daylight.)

Although there is no generally accepted convention with respect to nomenclature, many people refer to *reflected glare* when dealing with specular (polished or mirror) surfaces and to *veiling reflections* when considering source reflections in dull or semimatte finish surfaces, which always exhibit some degree of specularity. We use the terms interchangeably. In all cases the result is a distinct loss of contrast due to the veiling of the image by the reflection of the light source. It is imperative to an understanding of this problem, to appreciate the importance of the nature of the object being viewed, that is, the task. If the object were prefectly absorbent, that is, if it had a reflection coefficient of 0%, it would appear completely black as no light would be reflected into the eye (see Fig. 18.28*c*). Conversely, if the object were perfectly specular, as a clean mirror, and no light source were within the geometry of reflection, *we would not see it* (see Figure 18.20*d*). Thus if we took a mirror out on a cloudy night and shined a light on it from over our shoulder, it would be practically invisible since no light would be reflected into the eye.

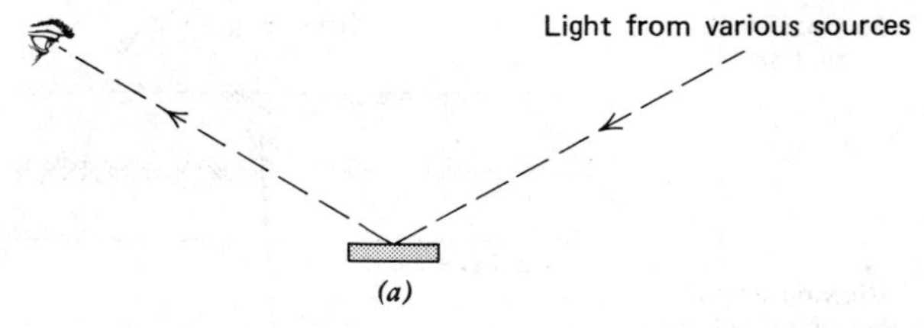

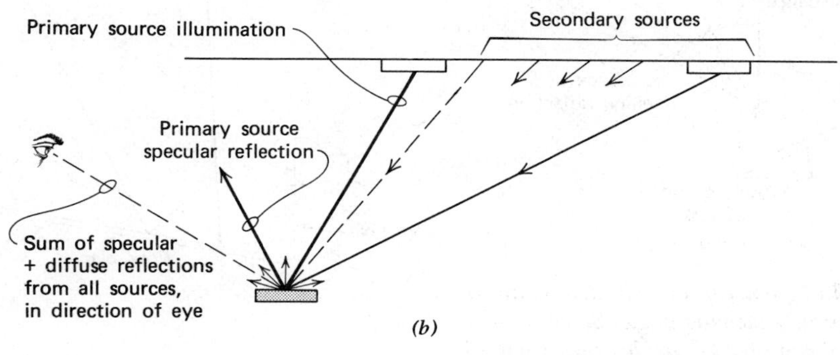

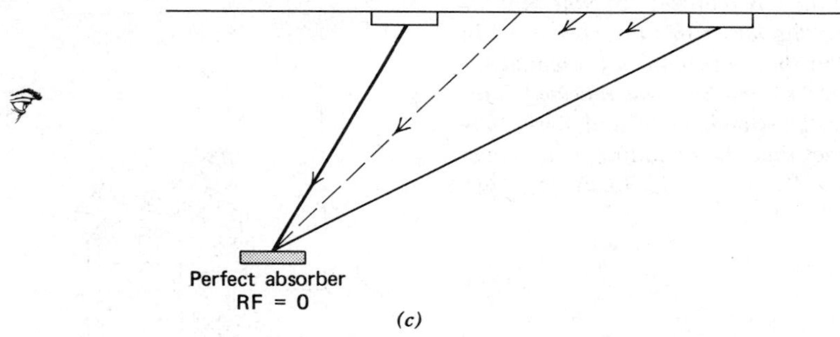

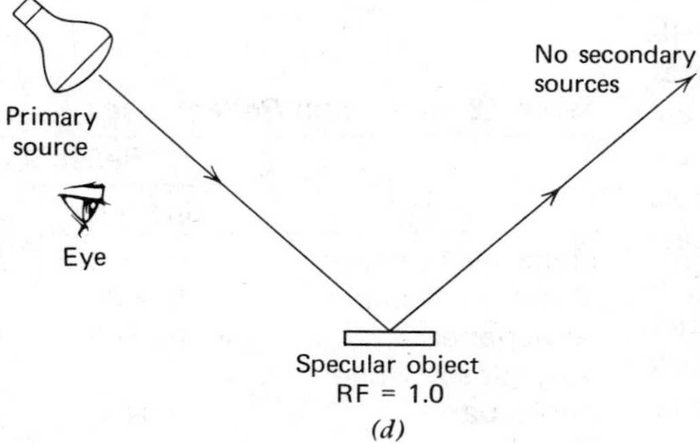

Fig. 18.28 (a) *The nature of the seeing process requires that light be reflected from the source(s) by the task into the eye.* (b) *The light entering the eye is the sum of all the reflected light, specular and diffuse, from all sources in the direction of the eye. If the task is specular all the sources will be seen reflected in the task.* (c) *A perfectly absorptive object is jet black.* (d) *A perfectly reflective one is invisible since geometrically it cannot reflect light into the eye.*

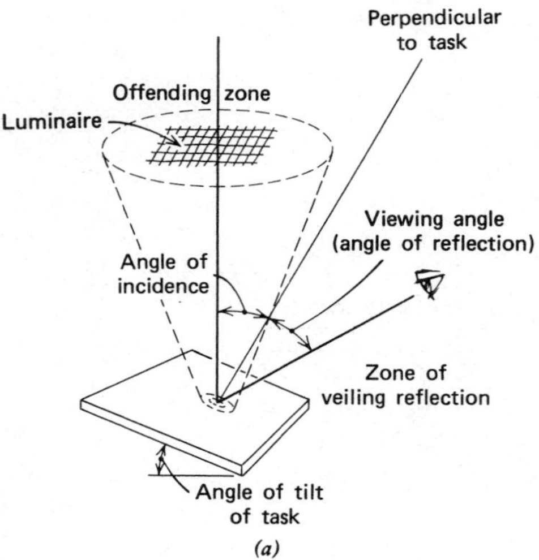

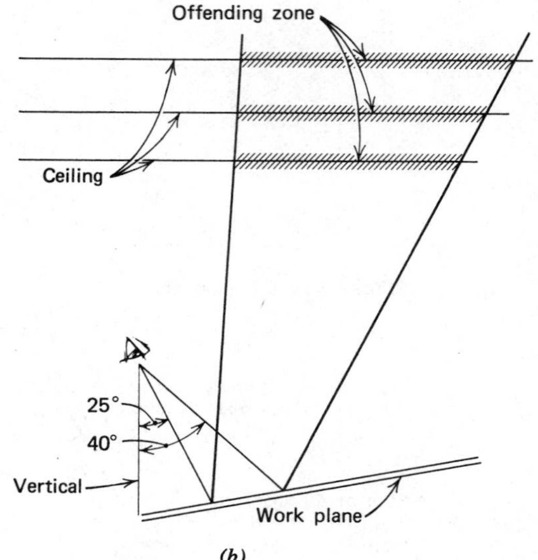

Fig. 18.29 The geometry of reflected glare is shown in (a) *for any viewing angle. Since normal desk-top, head-down viewing angles vary from 25 to 40 from the vertical, the offending zone is the area on the ceiling corresponding to specular reflection between these two angles.* (b) *Note that the higher the ceiling the larger this area becomes. In an office situation the draftsman in* (c) *would see ceiling fixtures in the offending zone reflected in his instruments, parallel straight edge, and work. Note the important fact that the offending zone moves* back *as the table tilts up (See Fig. 18.35).* (a) *Courtesy of the IES.*

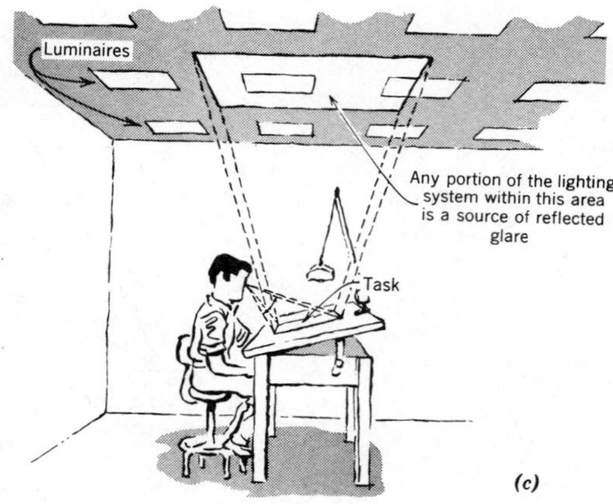

The reader might try this experiment: in an inside space with overhead luminaire, try to examine the surface of a clean mirror. You will find that the best angle to hold it is *almost* at the angle at which the light source is seen. This is because the mirror is *almost* completely specular, and it is the slight diffuse reflection near the viewing angle which permits us to see the surface. If the ceiling is relatively dark, holding the mirror at other angles results in a dark image (i.e., of the dark ceiling) and no detail of the mirror surface itself. Thus we understand that reflected glare is due to task surface specularity whereas object definition (i.e., surface detail) is due to task surface diffuseness. As stated above, task contrasts that enable us to see outline, form, silhouette, and so on (as in reading) are dimin-

Table 18.10 *Typical Reflectances*

	Reflectance	
Material	*Specular*	*Diffuse*
Matte black paper	0.0005	0.04
Matte white paper	0.0030	0.77
Newspaper	0.0065	0.68
Very glossy white photo paper	0.048	0.83
Metallic paper—copper	0.11	0.28
Dull black ink	0.006	0.045
Super gloss black ink	0.039	0.016

Source. Courtesy of the IES.

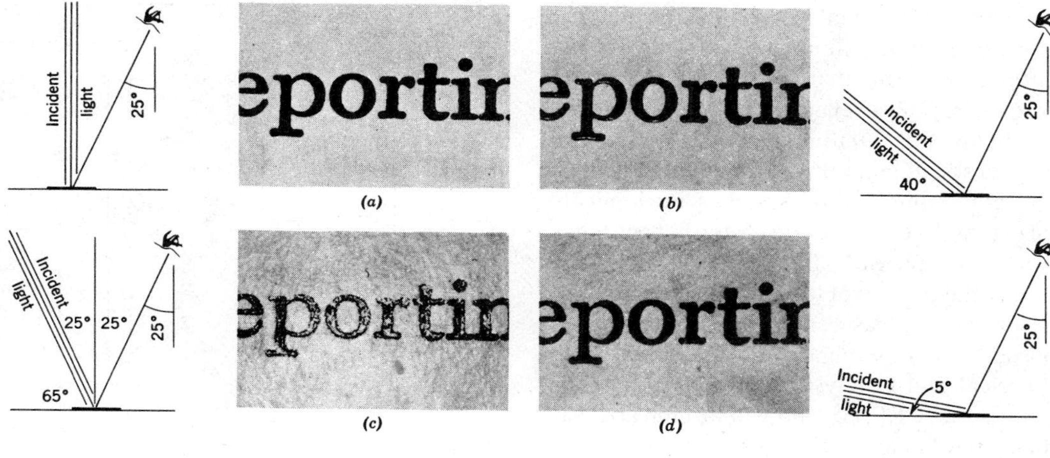

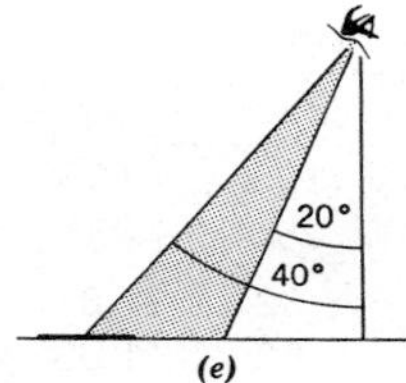

Fig. 18.30 Veiling reflection increases the difficulty of the seeing task. In (a), (b), and (d), it is diffuse reflection that permits us to see, since the specular reflection angle does not correspond to the viewing angle except in (c). *Veiling is pronounced at the 5° glancing angle even on a matte-finish surface. The normal viewing zone is from 20 to 40°, as shown in* (e). *Courtesy of the IES.*

ished to the extent that brightness (sources) are reflected in the task. These brightnesses within the geometry of reflected vision are shown in Figure 18.29 and the effects are shown in Figure 18.30. Table 18.10 lists a few sample reflectance figures to demonstrate that most materials exhibit both a specular and a diffuse reflectance. In studying Figure 18.29 it is important to note that a majority of visual work is done in the zone of 20 to 40° from the vertical, below the eye, bracketing the 25° angle shown.

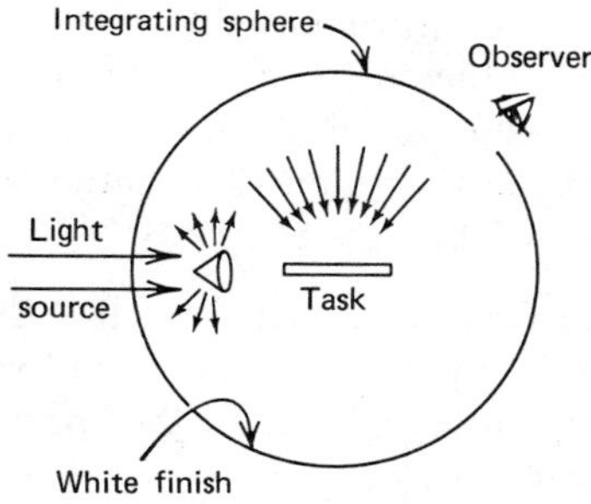

Fig. 18.31 Sphere illumination is produced by illuminating an object by reflection from the inside walls of an integrating sphere. The light source and observer are normally external.

(b) *Equivalent Spherical Illumination (ESI), Contrast Rendition Factor (CRF), and Lighting Effectiveness Factor (LEF).* As with direct glare, a scientific approach to the solution of reflected glare problems required a means for accurately defining the loss of contrast due to glare. This requires a reproducible, measurable light quality, that is, a reference lighting system in which seeing ability, which is defined by degree of contrast for a given task, can be measured and to which other lighting systems' visability for the same task can be compared (refer to Fig. 18.31). In order to achieve a lighting system almost free of reflected glare, it is necessary to construct an enclosed volume whose surfaces are uniformly reflective and whose primary source is obscured to the maximum extent possible. As illustrated, the integrating sphere is such a device. Light is introduced from the outside, split by a deflector and evenly distributed throughout the sphere by the multiple reflections from the white painted walls. The result is an evenly illuminated volume. When a task is introduced, the illumination falling on it is *entirely* uniform, that is, there are no high brightness points reflected in it. It is therefore termed *spherically* illuminated. (Note the parallel to sky illumination.) The extent to which any other illumination system can duplicate this glare-free environment is that system's equivalent spherical illumination (ESI), and is simply the portion of its total illumination that is spherical. Measurement of ESI is accomplished by comparing contrast rendition in the spherical and test systems. A contrast rendition factor (CRF) of 1.00 would indicate that the system under test gives the same contrast rendition as the integration sphere and that all its illumination is spherical illumination. With a lower CRF, the ESI drops sharply. The lighting effectiveness factor (LEF) is the ESI footcandles divided by the test illumination in "raw" footcandles, and is therefore a measure of the lighting effectiveness of the total system. A well-known study[4] of school lighting gave the illustrated results for four viewing positions in a classroom lighted with ceiling-mounted continuous rows of 2 by 4 ft, four-lamp, 40-w fluorescent fixtures with lens type wraparound diffusers, on 10 ft centers (see Fig. 18.32). Carefully note that:

1. The CRF, and therefore, the ESI, depends entirely on position and viewing angle, other factors in the space being equal.
2. In an ostensibly very well lighted (215 fc) position (M1), the useful illumination is only 28 fc!
3. The CRF can exceed 1.00, that is, the integrating sphere does not produce perfectly glare-free illumination but only nearly so.

Thus it is possible for ESI footcandles to be greater than raw footcandles. (Note also that referring to ESI footcandles as glare-free footcandles as is common in the literature is not entirely accurate.)

The results could have been anticipated, at least qualitatively, by examination of the observer positions vis-à-vis the layout. Positions M1 and M3 have bright sources in the offending zone; M1 more so than M3, as is borne out by the CRF figures. M2 is an excellent position in that it receives light contributions from the two sides; its footcandle value being lower than the others due to wide row spacing. M4 is ideally placed with no glare sources in the offending zone but with a row of fixtures positioned behind it, so that it is geometrically impossible to act as a glare source. The CRF/ESI analysis gives quantitative expression to our qualitative prejudgment and as such is an extremely valuable design tool.

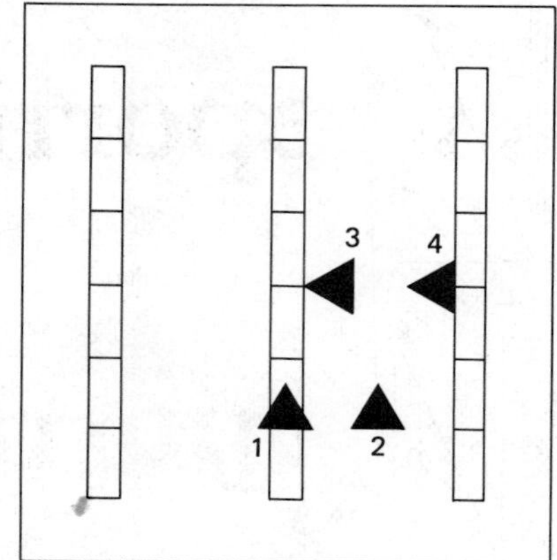

		Position			
		M1	*M2*	*M3*	*M4*
TI	2L	108	92	125	118
	4L	215	185	250	235
CRF	2L	.75	1.00	.82	1.01
	4L	.76	1.00	.83	1.03
ESI	2L	17.8	91.9	31.5	27.8
	4L	28.4	185.3	58.1	308.3
LEF	2L	.165	1.0	.25	1.08
	4L	.132	1.00	.23	1.31

TI—Task Illumination
2L—2 lamps (inside pair)
4L—4 lamps

Fig. 18.32 *A test classroom illuminated by three widely spaced rows of four-lamp fixtures. Observer positions are shown by arrows. (Reproduced from Foster K. Sampson,* Contrast Rendition in School Lighting, *Educational Facilities Laboratories, 1970.*

As with VCP, the CRF/ESI calculation system has the following limitations:

1. CRF, ESI, and LEF are critically dependent on observer position and viewing direction and viewing angle. Although position is generally fixed by chair location, observers can and do change viewing angle and head aspect to correct for glare situations.
2. The nature of task (i.e., its specularity) is assumed to be fixed and unique. In many situations the task nature varies from hour to hour and thus also the CRF. When tasks *are* constant, a poor CRF will frequently lead to an improvement in the task, thus changing the CRF and ESI. (This, as the item above, is inherent in the nature of the problem, that is, the definition of glare condition is dependent on the system components.)
3. Determination of task and lighting system characteristic data requires complex, expensive, and bulky equipment. These data are necessary to predetermine the system ESI performance.
4. CRF/ESI calculations require a computer with associated expenses.
5. The CRF/ESI relationship is nonlinear. Thus a change in illumination level does *not* produce a proportional change in CRF or ESI as can readily be seen from the tabulation in Figure 18.32.
6. No ESI meter is available as of this writing, although they are under active development and should be available soon.

18.23 Control of Reflected Glare

Since the causes of veiling reflections are well understood it would seem that a solution to the problem should by now have been adduced. Unfortunately, this is not the case. Although there is no known lighting method or material that will completely eliminate veiling reflections, there are a number of techniques that will minimize contrast loss due to veiling reflections while maintaining adequate illumination. These are:

Physical arrangement of sources, task, and observer so that reflected glare is minimal. See Section (a) below.

Adjusting brightnesses (eye adaptation level) so that objectionable brightness is minimized. See Section (b) below.

Design of the light source so that it causes minimal reflected glare. See Section (c) below.

Changing the task quality. See Section (d) below.

These techniques are discussed individually as follows:

(a) Physical Arrangement of System Elements. Since reflected glare is caused, as the name states, by reflection from a specular surface, the simplest and *most effective* technique is to arrange the geometry of the system so as to avoid the possibility of reflection. That is, we must remove the source from the offending zone, as in Figure 18.33. Unfortunately this is only totally effective when a single luminaire is involved and when its placement with respect to the observer is completely adjustable—a rare combination. As should be clear from Figure 18.32 and the related discussion, in a larger space utilizing multiple sources, particularly in continuous

Fig. 18.33 *Lighting fixture at* (a) *will produce more glare than one at* (b) *because of the geometry of the light rays. Desk finish and luminaire brightness can be chosen to minimize loss of contrast.*

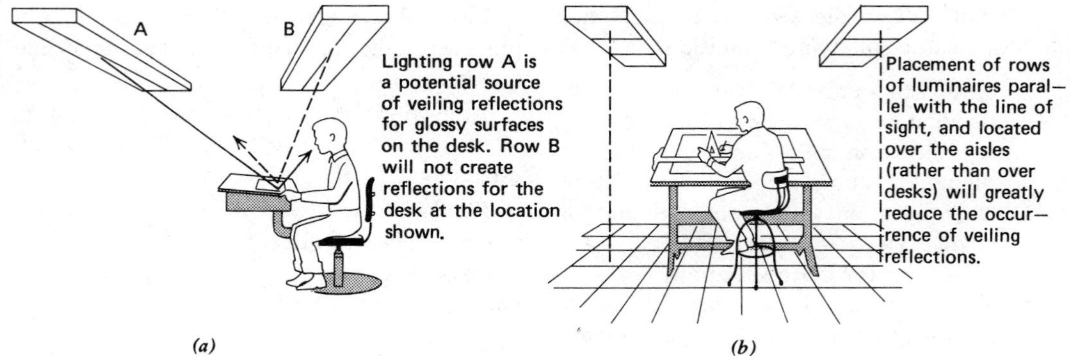

Fig. 18.34 *Continuous rows of lighting run across the offending ceiling zone* (a), *creating a severe glare problem. Crosswise rows* (b) *are much more desirable and effective, particularly where exacting tasks such as drafting are involved. Courtesy of General Electric Co.*

rows, placing the work between rows with the line of sight parallel to the long axis of the units is a very effective technique (see Fig. 18.34). It is important to remember, however, that the offending zone is also dependent on the tilt of the desk, assuming the work is to be kept flat on it. Thus for a horizontal 3 × 5 ft standard desk, the offending zone is forward of the desk, as in Figure 18.35*a;* with an elevated drafting table the ceiling glare source zone may well be behind the source, as in Figure 18.35*b*. This being so, it is often possible to reduce glare simply by tilting the work and/or the work surface to such an angle that glare is eliminated.

All of the above geometric solutions presuppose a known detailed fixed-furniture layout, a situation that obtains in many but certainly not all cases. In the absence of such data two alternatives are possible: to do a uniform layout and adjust the furniture to it, or vice versa. In practice, a combination of both is the

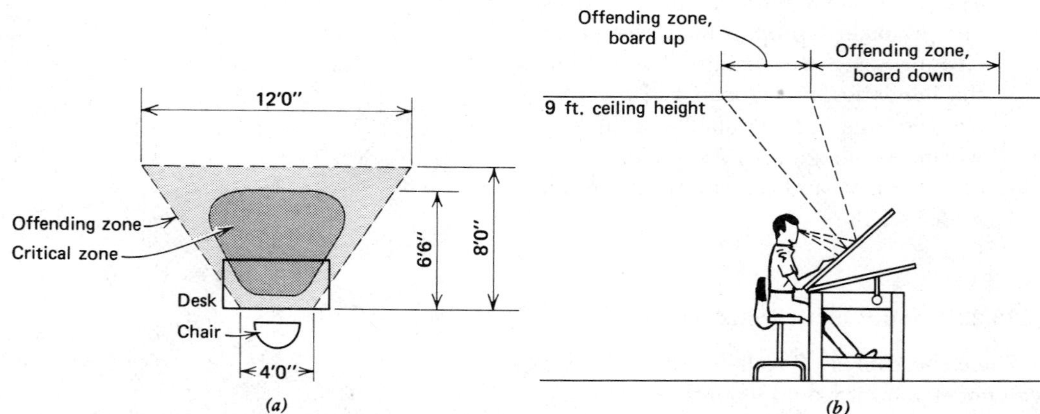

Fig. 18.35 (a) *If luminaires are kept out of the trapezoidal offending zone, an LEF of 1.0 or better should be obtained. If part of a luminaire projects into this area, LEF will drop to between 0.8 and 0.9. If the bulk of one or more luminaires projects into this zone and, in particular into the critical zone, LEF will drop to between 0.4 and 0.7. These figures are for specular tasks. The dimensions shown are for a flat desk 3 ft x 5 ft and a 9-ft ceiling height. (From Ross and Barruzini,* Lighting Systems Study, *General Services Administration, Public Buildings Service, 1974.*
(b) *The dependence of glare zone on table tilt is illustrated. The offending zone becomes small as the table is raised so that with a table near vertical position glare is all but eliminated.*

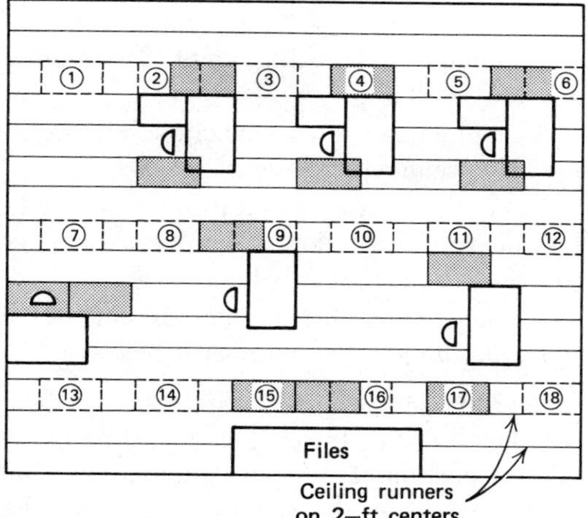

Fig. 18.36 The original uniform fixture layout utilized three rows of six 2 x 4, 4-lamp fixtures, giving a total load of 3600 w, a load density of 3.3 w/sq ft, and a uniform illumination level of approximately 90 (raw) fc. The original layout is shown dotted and numbered. The rearranged layout uses 13 fixtures (shown shaded) for a total 2600 w, a load density of 2.4 w/sq ft, and better than 100 ESI fc on each work surface. In addition, five fixtures are saved.

most practical approach. Indeed, with the energy-oriented criteria discussed above requiring a nonuniform layout (to give task lighting rather than maximum levels throughout), after-the-fact rearrangement of fixtures is mandatory. Since low watts per square foot budgets have made ducted lighting fixture heat removal systems (air troffers) much less necessary, fixtures are easily shifted. This mobility is further enhanced by the extreme flexibility of lighting fixtures fed from ceiling plug-in raceways as discussed in Sections 15.33 and 15.34. Thus ceiling fixtures of the lay-in type in hung ceilings can readily be shifted to yield the nonuniform layout required by task illumination criteria and by the necessity to reduce veiling reflections. The latter is all the more important in view of the reduced lighting levels being specified *and* the strong trend toward specifying ESI footcandle illumination rather than raw footcandles. Figure 18.36 shows such a rearrangement, which results in saving five fixtures, a load reduction of 1 kw and an *improvement* in lighting. The load density of 2.4 w/sq ft is very close to the recommended values (see Section 20.5).

(b) Control of Area Brightness and Eye Adaptation Level. As discussed in Section 18.15, loss of contrast can be compensated for (and glare eliminated) by increased overall illumination. We are simply making the task brighter to override the detrimental veiling reflection. Figure 18.37 is a curve that quanti-

Fig. 18.37 A loss in contrast can be compensated for by increasing the background brightness. Here a 14% loss in contrast (0.7 to 0.6) due to glare may be eliminated by a 300% increase in brightness. (Contrast in this curve is quantitatively defined as the ratio of minimum detectable change in brightness of the task to background brightness).

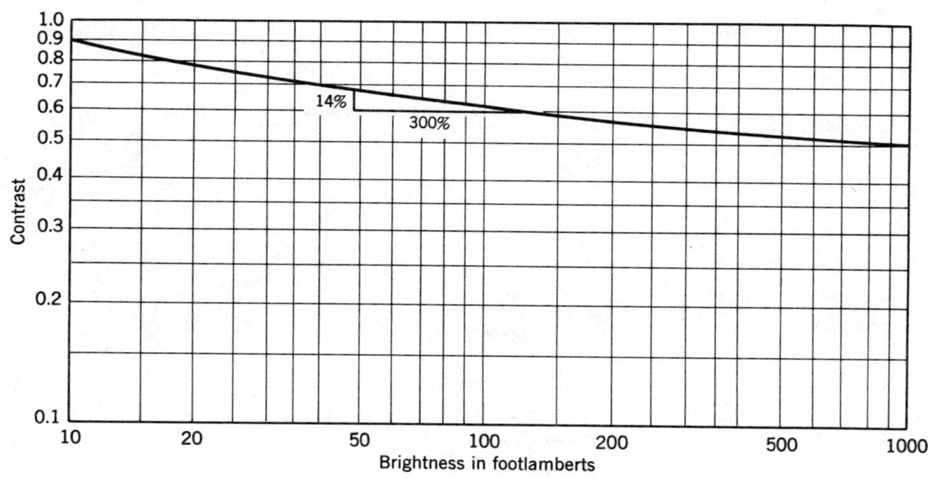

tatively relates the two parameters. In the particular instance shown in the curve, a 300% brightness increase is required to compensate for the loss of contrast experienced. This can, in many instances, be most practically accomplished not by increasing overall room illumination with the associated extremely high energy consumption but by adding a supplementary source so arranged as to be free of reflected glare. By making this supplementary source's position adjustable (as in Figure 18.29*c*) we accomplish three things.

1. Overcome veiling reflection.
2. Provide the high level of illumination needed for exacting tasks, with minimum energy expenditure.
3. Grant the observer complete control with resultant optimum lamp placement plus psychological satisfaction that will generally prevent worker complaints. (It is advisable to instruct users in the proper placement of supplementary sources, since there is a natural tendency to place the light immediately in front of the task and to thus aggravate rather than improve a glare situation. Proper positioning is generally above and to the left of the task.)

(*c*) ***Control of Source Characteristics.*** The reflected brightness that causes loss of contrast is proportional to the luminaire brightness. It is apparent then that glare may be reduced by reducing luminaire brightness at the reflection angle. This can be accomplished in four ways:

Dimming or switching lamps. See number 1 below.

Using luminaires with lower overall brightness. See number 2 below.

Using the luminaire as a primary source to illuminate a large, low-brightness secondary source. See number 3 below.

Reduce the luminaire brightness *only at the offending angles.* See number 4 below.

1. Although it may seem to be a trivial solution, reducing the output of a fixture in the critical portion of the ceiling glare zone can actually *increase* the ESI footcandles on the task. As noted in Section (b), number 3 above, adding light to a glare situation is only beneficial when the brighter source is at a nonglare angle; it then "overpowers" the glare source. Increasing the light of the glare source itself simply aggravates the problem, sometimes to the extent of adding "negative" footcandles. In these instances reducing overall illumination (raw footcandles) will *increase* ESI footcandles. (In one test situation reducing TI from 250 to 125 fc *increased* ESI footcandles from 104 to 108 fc, and increased LEF dramatically.)

2. In lieu of using a few, small, high-brightness sources, utilize larger-area low-brightness sources (see Fig. 18.38). This has the effect of reducing the source brightness in the ceiling glare zone while increasing the illumination contribution from outside the glare zone, resulting in better CRF at the same or lower LEF. The disadvantage of this technique is the increased fixture cost.

Fig. 18.38 *Largest concentration of light downward in the glare zone (left) produces largest amount of reflected glare. As number of light sources is increased (center) in the glare zone and brightness is decreased, reflected glare is decreased. Least glare is from all-luminous ceiling, which also has lowest luminance (right). Courtesy of (IES Standard on School Lighting).*

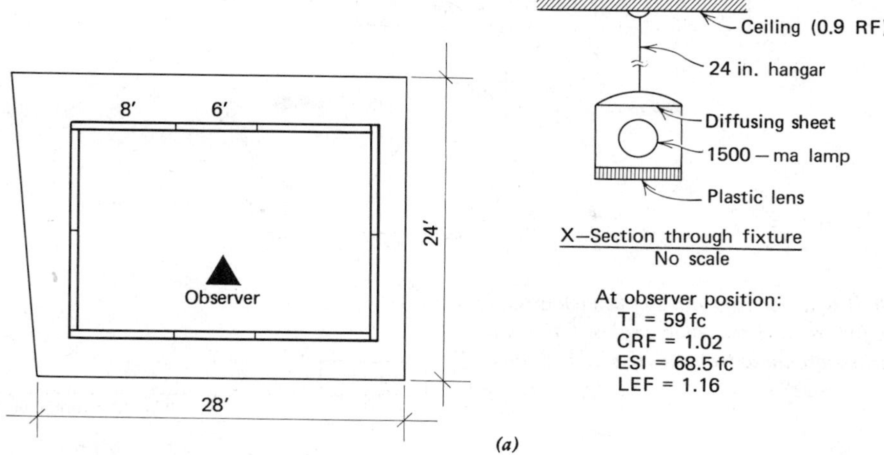

(*a*)

Fig. 18.39 (a) *With a high-reflectance, matte-finish ceiling this installation yields* more *ESI than raw footcandles (CRF > 1.0). Load is approximately 3 w/sq ft. (Reproduced from Foster K. Sampson,* Contrast Rendition in School Lighting, *Educational Facilities Laboratories), 1970.*

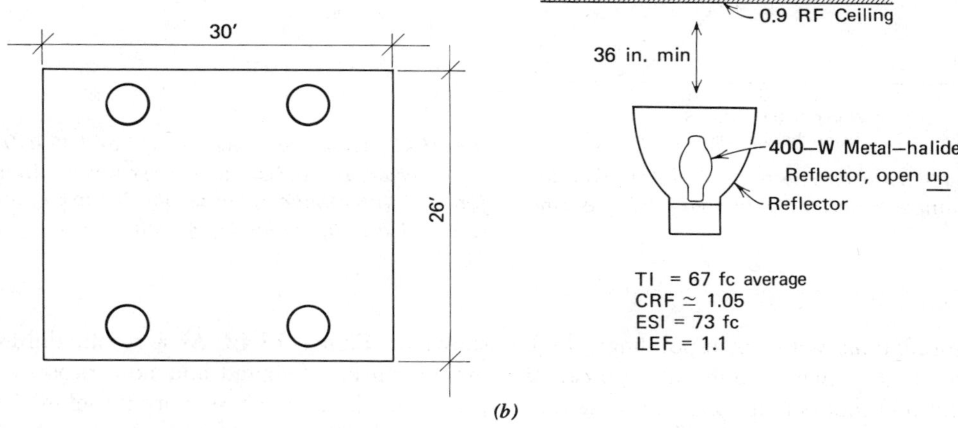

(*b*)

Fig. 18.39 (b) *Utilizing industrial reflectors to produce indirect lighting yields approximately the same results as in* (a) *above with only 2.2 w/sq ft. Due to compactness and high intensity of the source, a minimum of 3 ft between fixture and ceiling is desirable.*

3. To overcome the economic disadvantage of multiple low-brightness sources, the ceiling can be used as a secondary source illuminated from high brightness indirect or semiindirect fixtures (see Section 20.11). These high brightness sources, which can be fluorescent or HID (mercury, metal-halide, sodium), have the additional advantage of high efficacy. The space's ceiling height must be sufficient to permit suspending the unit at least 18 in. down, to avoid "hot spots" on the ceiling. The minimum suspension length depends on the luminaire characteristic and is normally provided by the manufacturer. To assure high efficiency the ceiling should be painted with a high reflectivity matte white paint, and kept clean. A semiindirect installation using 1500 ma very-high-output lamps and the results are shown in Figure 18.39*a*. Another utilizing 400 w indirect metal halide lighting unit is illustrated in Figure 18.39*b*. Of extreme importance is the CRF in excess of 1.00 in

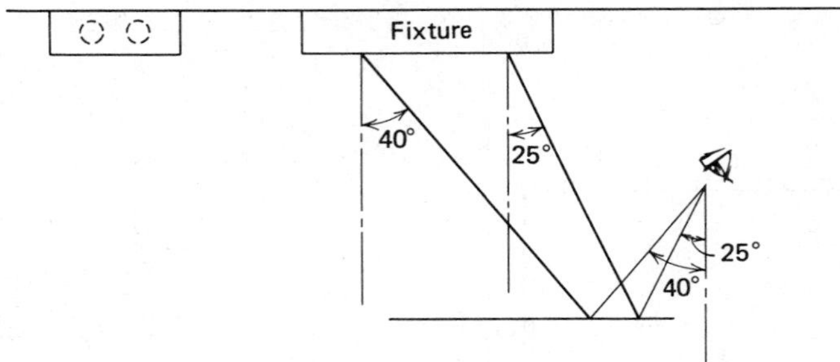

Fig. 18.40 *Light entering the eye at vision angles leaves the fixture at those same angles. For a tilted desk the emission angle from the fixture is reduced.*

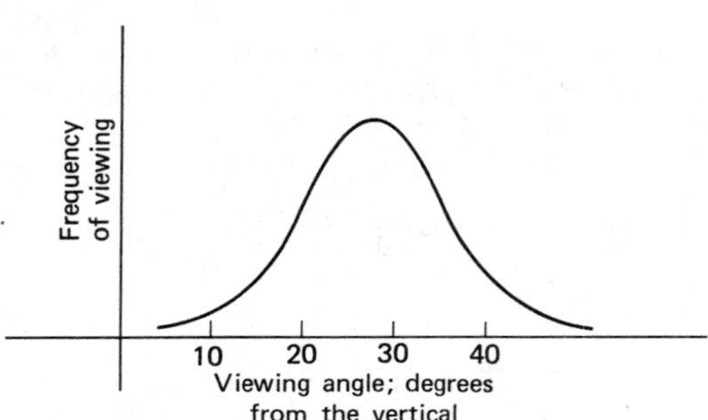

Fig. 18.41 *Extensive tests have shown that the viewing angle most commonly used is 25° from the vertical.*

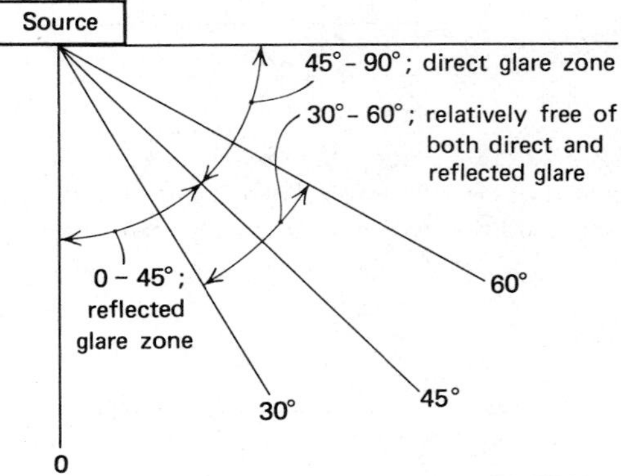

Fig. 18.42 *Glare zones are 0 to 45° and 45 to 90° for reflected and direct glare, respectively. Therefore, a diffuser that emphasizes the 30° to 60° zone will be least objectionable on both counts.*

both installations with correspondingly high ESI and LEF. These results are typical of well-designed indirect lighting installations.

4. To control luminaire brightness by straightforward geometric construction, it is simple to demonstrate (see Fig. 18.40) that the light causing veiling reflections is emitted by the source at the same angle as the sight angle. Since most vision takes place at about 25° (see Fig. 18.41) the light emitted by the fixture at 25° angle is the offending light. Therefore, any fixture whose light control element emits little or no light below 40° (which is the maximum vision angle) *cannot* produce veiling reflection regardless of its position in the field of view. This conclusion has been borne out by numerous studies[5] that compared many different types of diffusers as to the direct and reflected glare produced. The results are shown in Figure 18.42. As a result, diffuser manufacturers designed and now produce a prismatic diffuser whose output below 30° and above 60° are diminished, in order to minimize reflected and direct glare, respectively. Due to the characteristic shape of the distribution curve, elements that are so designed are known industrywide as "batwing" diffusers or lenses. If observers can be positioned so that their sight lines are perpendicular to the longitudinal axis of the ceiling fixtures, lenses with linear (side-to-side) batwing characteristics will perform well. If the observing position varies in aspect with respect to the fixture, a radial batwing curve (in all directions) is required. Figure 18.43 shows typical batwing distribution curves and Figure 18.44 shows photographs of the prismatic action involved.

Fig. 18.43 (a) *Comparison of typical distribution curves for common fluorescent fixture diffuser elements. The white plastic diffuser characteristic is almost circular, giving essentially the same distribution as the bare lamps, since it contains no light control elements. The curve of a typical prismatic diffuser shows high-angle cutoff (low direct glare) but high output in the 0 to 45° reflected glare area. Batwing distribution concentrates output in the 30 to 60° area while reducing the 0 to 30° zone to minimize reflections. All curves are across the axis of a 2 ft x 4 ft recessed fluorescent fixture with four 40-w lamps. Courtesy of Daybrite Lighting Div., Emerson Electric Co.*

Fig. 18.43 (b) *Linear batwing distribution with extremely sharp cutoff in the upper and lower ranges. Curve taken across lamp axis for Holophane Percepta (registered Trademark) single lamp unit. Courtesy of Holophane Division, Johns-Mansville Corp.*

Fig. 18.43 (c) *Complete photometric data for a radial bating distribution lens. Note that the perpendicular, parallel, and diagonal curves are almost identical. Zonal flux is maximum in the 30 to 60° range and drops off at both extremes, as desired. Courtesy of J. W. Carroll and Sons.*

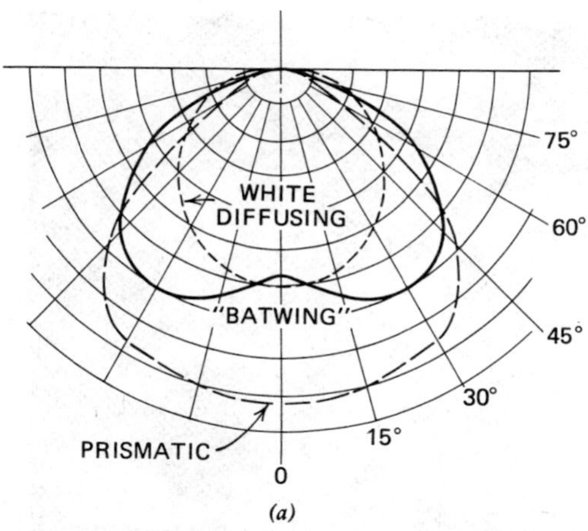

(a)

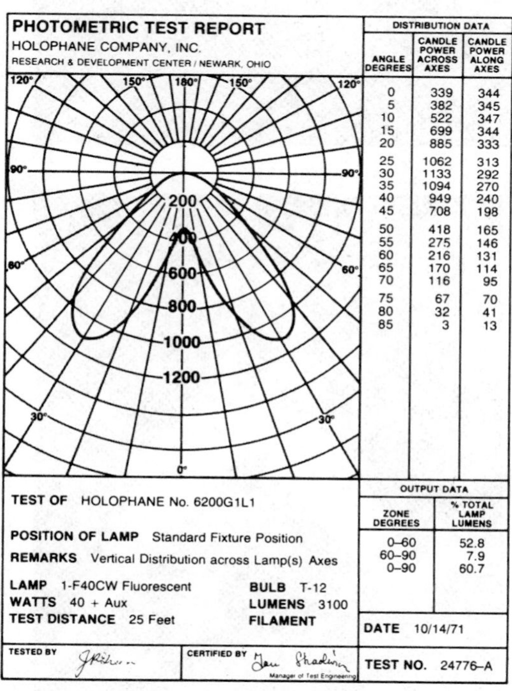

PHOTOMETRIC TEST REPORT
HOLOPHANE COMPANY, INC.
RESEARCH & DEVELOPMENT CENTER / NEWARK, OHIO

DISTRIBUTION DATA		
ANGLE DEGREES	CANDLE POWER ACROSS AXES	CANDLE POWER ALONG AXES
0	339	344
5	382	345
10	522	347
15	699	344
20	885	333
25	1062	313
30	1133	292
35	1094	270
40	949	240
45	708	198
50	418	165
55	275	146
60	216	131
65	170	114
70	116	95
75	67	70
80	32	41
85	3	13

TEST OF HOLOPHANE No. 6200G1L1

POSITION OF LAMP Standard Fixture Position

REMARKS Vertical Distribution across Lamp(s) Axes

LAMP 1-F40CW Fluorescent **BULB** T-12
WATTS 40 + Aux **LUMENS** 3100
TEST DISTANCE 25 Feet **FILAMENT**

OUTPUT DATA	
ZONE DEGREES	% TOTAL LAMP LUMENS
0–60	52.8
60–90	7.9
0–90	60.7

DATE 10/14/71

TESTED BY [signature] **CERTIFIED BY** [signature] Manager of Test Engineering

TEST NO. 24776–A

(b)

PHOTOMETRIC DATA:

DISTRIBUTION CURVES

2′ x 4′ two lamp luminaire
Radialens with opal acrylic overlay
——— PERPENDICULAR
— — — PARALLEL
-- ------ 45°

250
500
750
1000

CANDLEPOWER AT 20 FT.				ZONAL FLUX
ANGLE	PERP.	PAR.	45°	
90	0	0	0	–
85	48	38	51	51
75	282	229	288	287
65	553	462	561	526
55	808	687	813	700
45	1013	892	1018	762
35	1152	1045	1130	698
25	1110	1065	1090	505
15	929	929	927	263
5	883	881	883	83
0	880	880	880	–

LUMINAIRE EFFICIENCY – 62.5%

Brightness values based upon 3100 lumen lamps

(c)

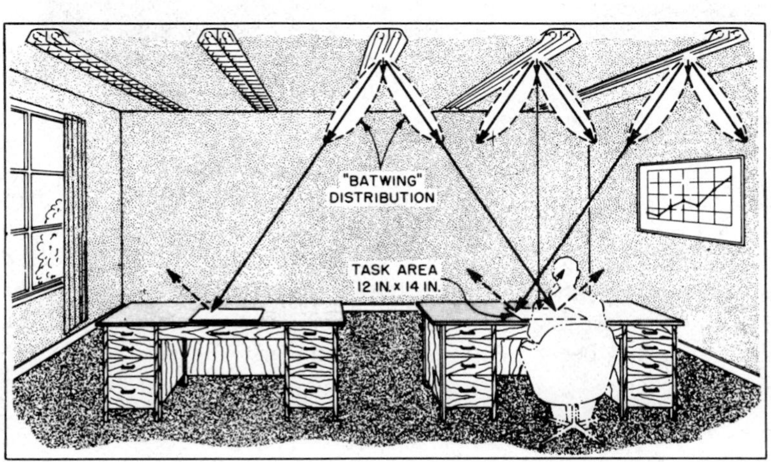

(d)

Fig. 18.43 (d) *The action of the batwing is illustrated here. Note that the principal light output cannot produce veiling reflections. See also Figure 20.31, page 896. Courtesy of Illuminating Engineering Research Institute of the IES.*

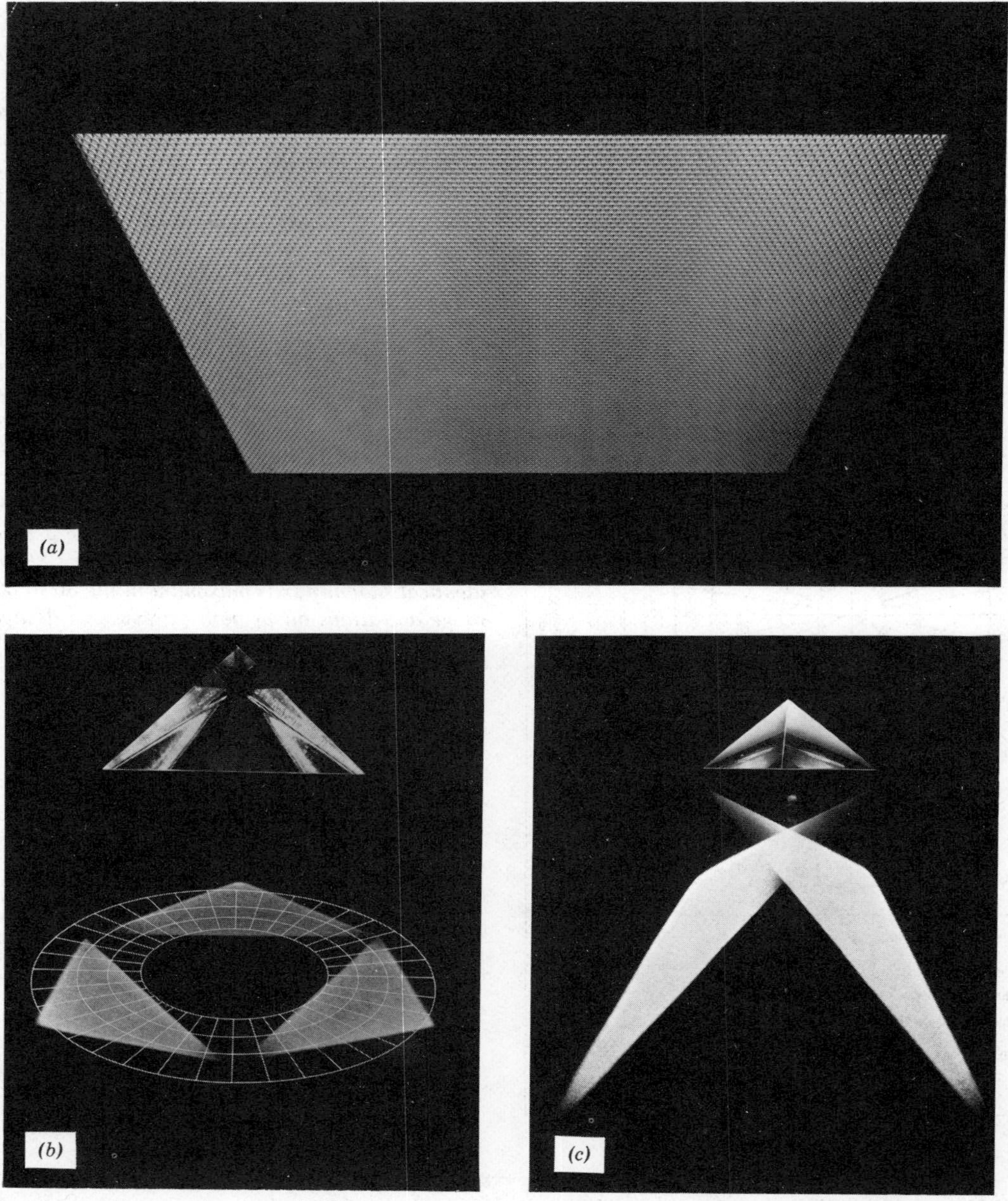

Fig. 18.44 Photographs of a radial batwing lens showing (a) *the low brightness of a lighted lens,* (b) *radial light distribution pattern through a single prism, and* (c) *vertical light distribution. All photos are of the RADIALENS (TM), Courtesy of J. W. Carroll and Sons.*

(d) Changing the Task Quality. At this point it is abundantly clear that reducing the task specularity is at least as effective a means of reducing veiling reflections as changing the lighting system characteristics, if not more so. Traditionally, however, this option has been outside the purview of the lighting designer. With energy limitations having legal force, no area of consideration should be considered untouchable, least of all the nature of the work. It is therefore recommended that task contrast and specularity be actively considered and recommendations made in a framework of energy effectiveness and cost effectiveness. Thus to produce adequate visibility it will often be cheaper to upgrade the task (in the visibility sense) than to upgrade the lighting system. Similarly, it will almost always be energy-economical to improve the task quality. A few suggestions are:

1. Use felt pens and nonspecular paper for office tasks.
2. Use paper in lieu of pencil cloth, nylon, or photo reproductions in drafting work. (In the latter instance, the money saved by using photo reproductions may well be offset by reduced production and increased fixture and energy cost.)
3. Provide means for tilting the work and changing the observers' viewing angle or position.
4. Institute a testing program to determine task characteristics and determine possible substitutions.

18.24 Other Glare Control Considerations

The batwing distribution is a useful tool in reducing glare problems but it is primarily applicable to fluorescent sources and to specific indoor situations. It is always desirable to be able to predict, at least qualitatively, the amount of glare that will be caused by a given fixture position and by a given fixture distribution. The former is a matter of geometry, and Figure 18.45 should be of assistance in this regard.

The ability to predict both glare problems and ESI efficiency is obviously desirable, and much effort has been expended in this direction. IES Report QQ3 analyzed inherent glare at various angles and determined, for example, that 1 fL at 65° (from the vertical) produced as much glare as 2 fL at 55°, and so on. Using this principle of geometric analysis the Illuminating Engineering Research Institute (IERI) produced an analysis table that can be applied to any light distribution curve to determine incipient direct and reflective glare problems (see Fig. 18.46). The system is limited, as are all such analyses, by their assumptions. Thus small rooms, higher or lower illumination levels, and suspended rather than surface fixtures would affect the results considerably. Still, such approaches are useful if applied with full knowledge and appreciation of the assumptions and limitations.

The analysis table introduces a concept called effective lumens per watt. This is simply an empirical way of judging the overall effectiveness of a fixture's output *including direct glare*, just as LEF is a measure of effective wattage considering only the deleterious effect of reflected glare.

Polarizing panels, when introduced, held great promise because they produce a marked decrease of veiling reflection at an angle of 60°, but much less at other angles. Since most viewing is in the 20 to 40° range, using these panels does not result in any appreciable reduction in reflected glare.

From personal experience in a drafting room equipped with luminaires utilizing high-efficiency multilayer polarizers, it can be stated that visual discomfort from reflected glare, as personally experienced, and as reported by a large staff, is not noticeably reduced.

18.25 Brightness Ratios

As explained above, visual performance increases with contrast, that is, with difference in brightness between the object being viewed and its immediate surrounding brightness. Conversely, however, the difference between the average luminance of the visual field (task) and the remainder of the field of vision, should be low. Restated, contrast is desirable in the object of view but undesirable in the field of view. The room for which the basic scissors curve data was developed used reflectances of 50, 30, and 80% for walls, floor, and ceilings, respectively, and 35% for furniture, in order to establish a fairly high eye adaptation level so that direct glare which results from excessive brightness ratios, is minimized. This is the fourth factor referred to above. The IES rec-

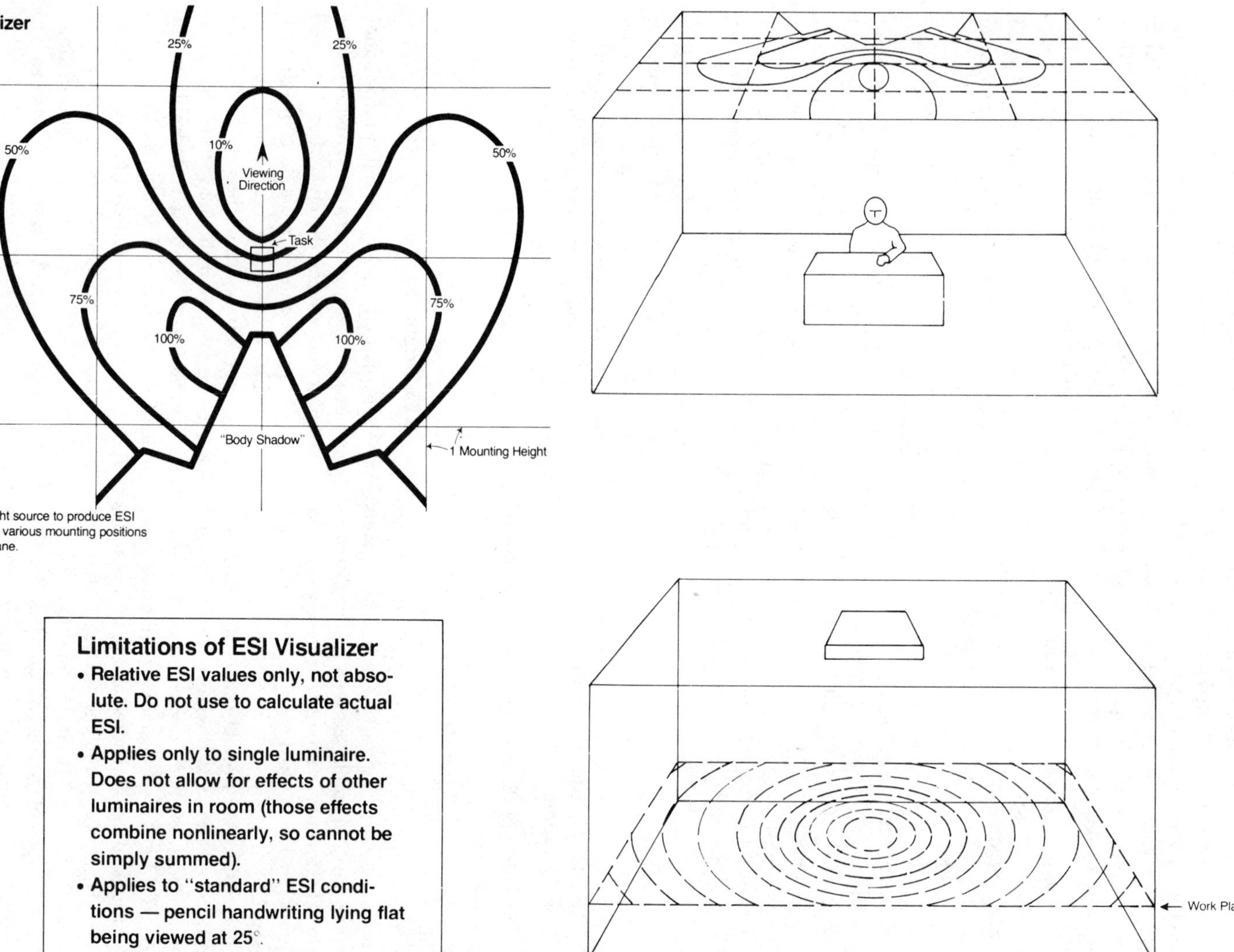

Limitations of ESI Visualizer

- Relative ESI values only, not absolute. Do not use to calculate actual ESI.
- Applies only to single luminaire. Does not allow for effects of other luminaires in room (those effects combine nonlinearly, so cannot be simply summed).
- Applies to "standard" ESI conditions — pencil handwriting lying flat being viewed at 25°.

Fig. 18.45 *These diagrams should assist the designer in visualizing reflected glare situations, since they show the work plane projected onto the ceiling. Courtesy of Widelight Corp.*

Zone	Sphere	*Lumens*	*Equiv. Sphere Factor*	*Equiv. Sphere Lumens*	*Glare Factor*	*Equiv. Glare Lumens*
0–10°	Vert. Polarized Horiz. Polarized Unpolarized		0.37 0.37 0.37		0	
10–20°	Vert. Polarized Horiz. Polarized Unpolarized		0.35 0.35 0.35		0	
20–30°	Vert. Polarized Horiz. Polarized Unpolarized		0.46 0.41 0.43		0	
30–40°	Vert. Polarized Horiz. Polarized Unpolarized		0.68 0.62 0.65		0	
40–50°	Vert. Polarized Horiz. Polarized Unpolarized		1.02 1.10 1.06		0	
50–60°	Vert. Polarized Horiz. Polarized Unpolarized		1.28 1.66 1.47		1	
60–70°	Vert. Polarized Horiz. Polarized Unpolarized		1.46 2.44 1.95		2.5	
70–80°	Vert. Polarized Horiz. Polarized Unpolarized		1.70 2.90 2.30		6	
80–90°	Vert. Polarized Horiz. Polarized Unpolarized		0.80 0.92 0.86		5	
90–180°			0.62			

Total ESL ______ Total EGL ______

Factors calculated on the basis of a 60 × 60 × 10 ft room; room surface components and body shadow included: room surfaces, ceiling—80%, walls—60%, and floors—30%.

NOTES

1. From the photometric curve of the luminaire, enter zonal lumens in "Lumens" column. If vertically and horizontally polarized data are available, insert on appropriate lines; if polarized curves were not run, enter total lumens on "Unpolarized" line.
2. Multiply "Lumens" by corresponding "Equivalent Sphere Factor," and enter the result under "Equivalent Sphere Lumens": multiply "Lumens" by "Glare Factor" and enter under "Equivalent Glare Lumens." Add "Equivalent Sphere Lumens" and "Equivalent Glare Lumens" columns. (Per IES RQQ Report #3 one fL at 65° causes as much glare as 2 fL at 55°. The factors in the "Glare Factor" column relate *lumens* in the various 10° bands to lumens in the 55° zone, and hence direct glare. Equivalent lumens at 55° are termed "Equivalent Glare Lumens").
3. Multiply luminous area of luminaire (square feet) by 400; this is the *allowable* Equivalent Glare Lumens.
4. If total Equivalent Glare Lumens (EGL) is less than the allowable value, divide Equivalent Sphere Lumens (ESL) by total luminaire watts, including ballast. The result is *Effective Lumens per Watt.*
5. If EGL is greater than the allowable value, multiply the difference by 1.47 and subtract from ESL before dividing by watts to get Effective Lumens per Watt.

Fig. 18.46 *Fixture analysis table enabling the user to analyize a distribution curve to determine ESI lumens, direct glare lumens, and effective fixture wattage. Refer to the text for additional notes on application of this table. (Reproduced from Ross and Baruzzini,* Lighting Systems Study, *GSA, 1974.*

Fig. 18.47 *Recommended reflectances for surfaces and furnishings in the classroom. (Note control media used at windows to reduce exterior brightness so that they are in balance with interior brightness.) Courtesy of the IES.*

Table 18.11 *Recommended Luminance Ratios*

To achieve a comfortable brightness balance, it is desirable and practical to limit brightness ratios between areas of appreciable size from normal viewpoints as follows:

1 to $\frac{1}{3}$	Between task and adjacent surroundings
1 to $\frac{1}{10}$	Between task and more remote darker surfaces
1 to 10	Between task and more remote lighter surfaces
20 to 1	Between luminaires (or fenestration) and surfaces adjacent to them
40 to 1	Anywhere within the normal field of view

Source. Courtesy of the IES.

These ratios are recommended as maximums; reductions are generally beneficial. See also the discussion on energy reduction and its relation to brightness ratios, Section 20.7, point G.

ommendations for luminance ratios to achieve a comfortable environment are tabulated in Table 18.11. To achieve these luminance ratios it is obviously necessary to carefully control the reflectances of the major surfaces in a room. The recommended reflectances for a typical schoolroom and office are shown graphically in Figures 18.47 and 18.48. The marked difference between a background with proper reflectance and one with excessive brightness ratios caused by the low surrounding reflectances is shown in Figure 18.49.

18.26 Patterns of Luminance

Returning to the list of characteristics in Section 18.13, we note among the secondary factors in illumination the existence of *"patterns of luminance."* This is a way of describ-

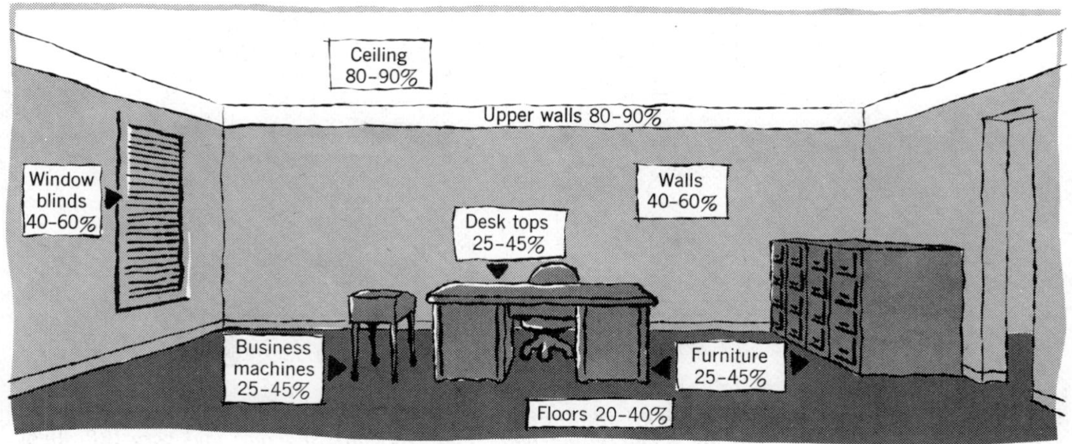

Fig. 18.48 *Recommended reflectences for room and furniture surfaces in offices. (Adopted from the IES.)*

Fig. 18.49 *The reflected glare from luminaires disappears when a piece of light, diffuse linoleum is placed over the dark, polished desk top. Light-colored desk tops with a 25 to 40% reflectance result in task-to-background ratios within the 3 : 1 recommended range. Prior to placing the linoleum, a reflection similar to the one seen above also existed on the left of the desk due to another luminaire. (Courtesy of the IES).*

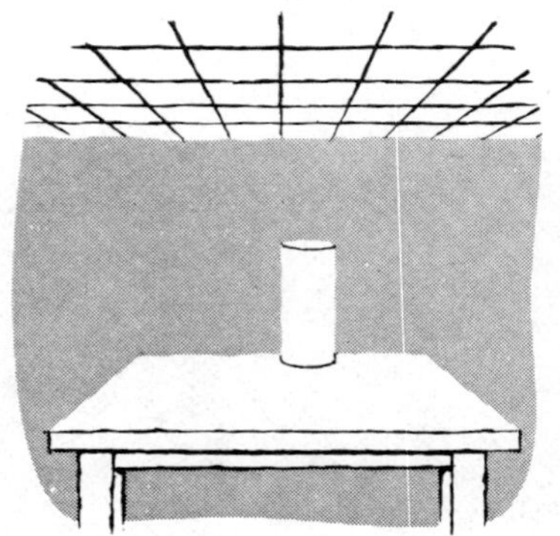

Fig. 18.50 *Diffuse illumination. The luminous ceiling installation provides shadowless, almost perfectly diffuse lighting. By comparison, the single-ceiling bulb produces sharp shadows and very little light diffusion.*

ing the patterns of light and shadow in a space as they result from the method of illumination in that space. Thus a single source produces sharp shadows while a luminous ceiling or a completely indirect illumination system produces almost completely diffuse light (see Fig. 18.50). Diffusion is the degree to which light is shadowless and is therefore a function of the number of directions from which light impinges on a particular point and the relative intensities.

Perfect diffusion, rarely obtainable, would have equal intensities of light impinging from all directions, therefore yielding no shadows. Not being readily mathematically expressible, diffusion is generally judged by the depth and sharpness of shadows. A room with well-diffused illumination resulting from multiple sources and high room surface reflectances, yields soft multiple shadows that do not obscure the visual task. The only naturally occurring example of perfectly diffuse lighting is a daytime fog, which we know to be extremely disturbing to the eye, demonstrating that some directivity is desirable.

There is a widely held but erroneous belief that diffuse lighting is better than directional lighting for all installations. Although this is frequently true for offices, schoolrooms, machine shops, and drafting rooms where shadows would be highly disturbing and could be dangerous (as in the case of a machine shop), it is decidedly not the case where texture must be examined, surface imperfections detected by grazing angle reflections, or in any installation where the flat monotony of diffuse lighting is undesirable. For this reason, some directional lighting is often introduced as an adjunct to diffuse general lighting to lend interest by producing shadows and high brightness variations.

Indeed, as seen in Figure 18.51, directional light is what creates shape and is precisely the characteristic best used to influence architectural space and form.

Section 20.10 and the sections following, which deal with systems of lighting, illustrate a few of the light/dark patterns produced by different lighting arrangements. The combinations of uplighting and downlighting, perimeter lighting and ceiling are legion; each produces its own shadows and modeling, and each has a quality of its own. It is very much in the interest of the lighting designer to be familiar with these effects so that he or she can mentally visualize them as the design progresses. Indeed it would be well for a designer to prepare a reference sketchbook of such shadow diagrams. It is these patterns of light and darkness that give the ambience and the subject reactions of sociability/isolation, clarity/fuzziness, spaciousness/crampedness, simplicity/clutter, formality/informality, boredom/excitement, definition/shapelessness, and so on. (Color has a great deal to do with subjective reactions and is discussed separately

Fig. 18.51 *Totally diffuse lighting* (a) *destroys texture, whereas a combination of diffuse and directional lighting* (b) *produces the required modeling shadows. Courtesy of Holophane Company, a Johns-Manville Division.*

below.) The subject of psychological reactions to lighting environment is extensive and complex and can only be touched upon here to the extent of mentioning a few of the salient lighting techniques and their usual subjective responses.

In addition to modeling and texture accent, spots of high brightness, which can be called "glare" or "sparkle" depending on one's point of view, create interest and visual excitement. Lighting installations generally yield a sense of vividness or activity proportional to the level of illumination. However, this is not the case with very diffuse lighted areas, which even at high footcandle levels are tedious. This is particularly noticeable in large, low, luminous ceiling installations, which are completely lacking in visual interest. Small exposed incandescent lamps, a brightly lighted, rough-textured wall, and pendant fixtures with pierced reflectors are some of the techniques used to create this visual interest.

Visual attention can be drawn by high brightness. This well-known fact is used constantly in displaying merchandise. Note the following usual reactions.

A 3:1 brightness ratio will be noticed but will usually not affect behavior or draw attention.

A 10:1 brightness ratio will attract attention and, if interesting, will hold it.

A 50:1 brightness ratio or larger will highlight the object thus illuminated, practically to the exclusion of all else in the field of view.

Since brightnesses draw the eye's attention, all the individual brightness sources in the field of view produce an overall impression. If there is some form or order or pattern to these brightnesses (as a pattern of lighting fixtures) then the overall impression is not disturbing—it can be thought of as visually harmonious. If, on the other hand, they are in disarray they produce a discordancy precisely as sound produces discordancy in the ear. This visual "noise" is frequently referred to as visual clutter and can be very disturbing. The designer is well advised to keep this important fact in mind when arranging light sources that are the primary sources of luminance in an enclosed space. A few aspects of fixture patterning are shown in Section 20.16.

18.27 Color Temperature

A light source is often designated with a "color temperature" such as 3400 K for quartz iodine lamps, 4500 K for cool white fluores-

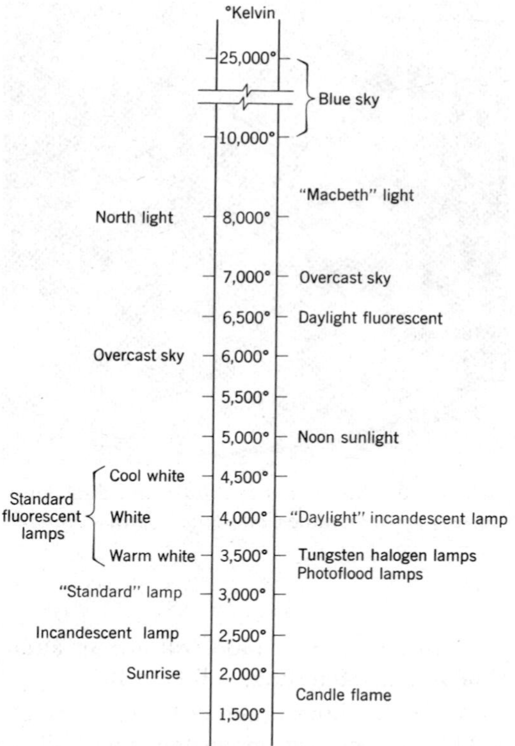

Fig. 18.52 *Color temperature scale.*

cent tubes, and so on. This nomenclature is derived from the fact that when a light-absorbing body (called a black body) is heated, it will first glow deep red, then cherry red, then orange until it finally becomes blue-white hot. The color of the light radiated—red from a red-hot body, white from a white-hot body—is thus related to its temperature. Therefore, by developing a black-body color temperature scale, we can compare the color of a light source to this scale and assign to it an approximate "color temperature," that is, the temperature to which a black body must be heated to radiate a light approximating the color of the source in question. Temperature is measured in degrees Kelvin, which is a scale that has its zero point at minus 456° F. Figure 18.52 shows the assigned color temperature of some common light sources.

It must be understood, of course, that color temperature of a light source is an indication of the color of the light produced and not of the actual source temperature. This is so since in many sources the light is produced by fluorescence and not incandescence. Certain colors, such as green, are not radiated by an incandescent black body as its temperature is raised, and therefore green lamps are not assigned a color temperature. However, since most common light sources are designed to deliver "white light," an approximate color temperature may be given. When a composite light such as white falls on a surface other than black or white, selective absorption occurs. The component colors are absorbed in different proportions so that the light reflected or transmitted is composed of a new combination of the same colors as had impinged on the surface. Thus a white light reflected from a red wall acquires a red tint since the component colors of the white light other than red were absorbed in greater proportion than the red. When reflected, the red light took prominence thus giving the reflected light a red tint. This is illustrated in Figure 18.53*a*. Similarly, a white light when passed through a piece of red glass emerges as a reddish light since the other components were absorbed in much greater proportion than the red. This well-known phenomenon is illustrated in Figure 18.53*b*.

It is this phenomenon that allows us to see color at all; the individual object pigmentation absorbs all other colors of light and reflects or transmits to the eye only its own hue.

18.28 Color

The color of the illuminant (light) and correspondingly the coloration of the objects within a space constitute an important facet of the lighting quality. The two factors, however, must not be considered separately since by definition the color of an object is its ability to modify the color of light incident upon it. It does this, as stated above, by a process of selective absorption, absorbing most of the light and reflecting or transmitting a spectrally modified light rich in a single hue, as shown in Figure 18.53*a*. The color reflected or transmitted is apprehended by the eye as the

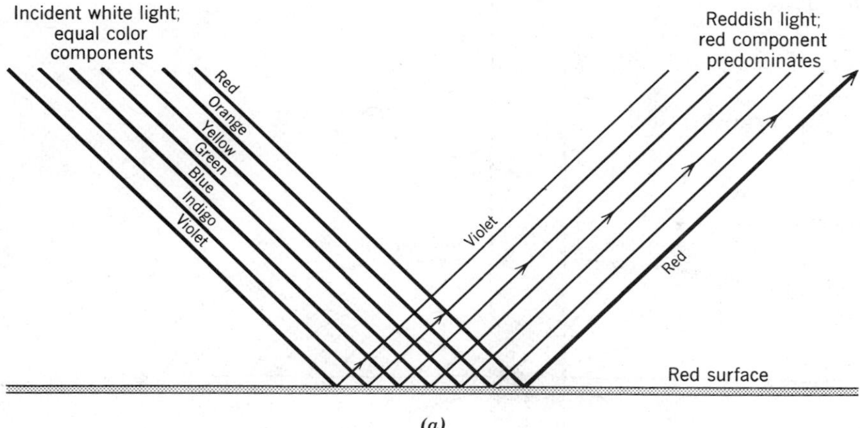

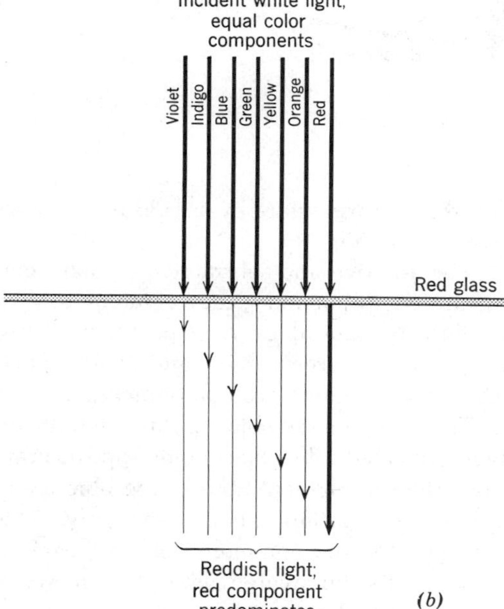

Fig. 18.53 (a) *Selective absorption of reflected light.*

Fig. 18.53 (b) *Selective absorption of transmitted light.*

color of the object. An object is technically said to be colorless (not transparent) when it does not exhibit selective absorption, reflecting and absorbing the various components of the incident light nonselectively. Thus, white, black, and all shades of gray are colorless, neutral, achromatic or, more precisely, lack hue.

Hue is defined as that attribute by which we recognize and therefore describe colors as red, yellow, green, blue, and so on. Just as it is possible to form a series from white to black with the intermediate grays, so it is also possible to do the same with a hue.

The difference between the resultant colors of the same hue so arranged, is called *brilliance or value.* White is the most brilliant of the neutral colors and black the least, pink is a more brilliant red hue than ruby, and golden yellow a more brilliant (lighter) yellow hue than raw umber.

Colors of the same hue and brilliance may still differ from each other in saturation, which is an indication of the vividness of hue or the difference of the color from gray. Thus pure gray has no hue; as we add color we change the saturation without changing the brilliance. The three characteristics then that define a particular coloration are *hue, brilliance,* and *saturation.* Using these terms we may define "bay" as a color red-yellow in hue of low brilliance and low saturation; while carmine is a color red in hue, of low brilliance and very high saturation.

Various systems of color classification have been devised, including the ISCC-NBC color system, the Munsell Color System, the Ostwald Color System, C.I.E., and the Chromaticity Diagram. In the Munsell color system (see Fig. 18.54) brilliance is referred to as *"value"* and saturation as *"chroma";* thus a color is defined by hue, value, and chroma.

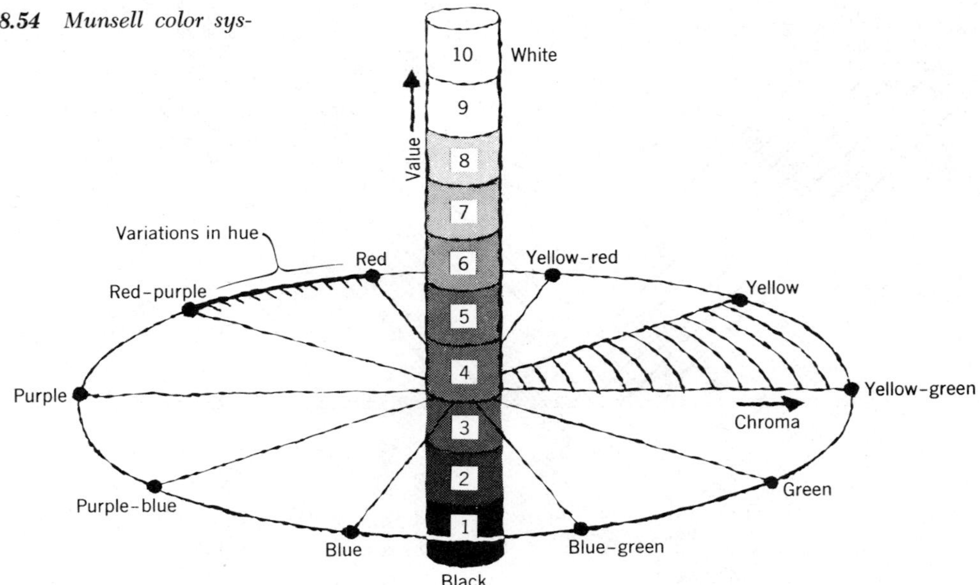

Fig. 18.54 Munsell color system.

The brilliance (value) of a pigment or coloration is related to its reflectance to white light. The higher the brilliance or value, the higher the reflectance factor as might be predicted when one considers that white and black are the poles of brilliance. Chroma or saturation may be thought of as either the difference from gray or the purity of the color. Spectral colors have 100% purity and therefore maximum chroma.

When white is added to a pigment, it produces a tint; adding black produces a shade. When pigments are mixed to produce a particular color, we create this color by a subtractive process. That is, each pigment absorbs certain proportions of white light; when mixed, the absorptions combine to subtract (absorb) various colors of the white spectra, and leave only those colors that finally constitute the hue, value, and chroma of the pigment. This subtractive effect is also utilized when producing colors by filtering white light. The filter selectively absorbs component colors, transmitting only the component desired. Thus a blue filter transmits only blue, etc. (see Fig. 18.53*b*).

Conversely, when light of the three primary colors of red, green, and blue are combined they form white by an additive process (see Fig. 18.55).

The additive and subtractive primary colors are complimentary; they combine to give a white or neutral gray, respectively. Thus, red and blue-green, blue and yellow, and green and magenta are complimentary.

Therefore if a red object is illuminated with blue-green light the object color appears gray, since the red pigment absorbs the blue-green and reflects nothing; hence the gray. This accounts for the common "lost red car" in parking lots illuminated with clear mercury lamps with their characteristic blue-green color. Similarly, a blue filter on a yellow light would transmit nothing.

18.29 Reactions to Color

Light of a particular hue (other than white) is rarely used for general illumination except to create a special atmosphere. When a space is lighted with colored light, the eye adapts by a phenomena known as "color constancy" so that it can to a degree recognize colors of objects despite the spectral quality of the illuminant. However, the eyes become more sen-

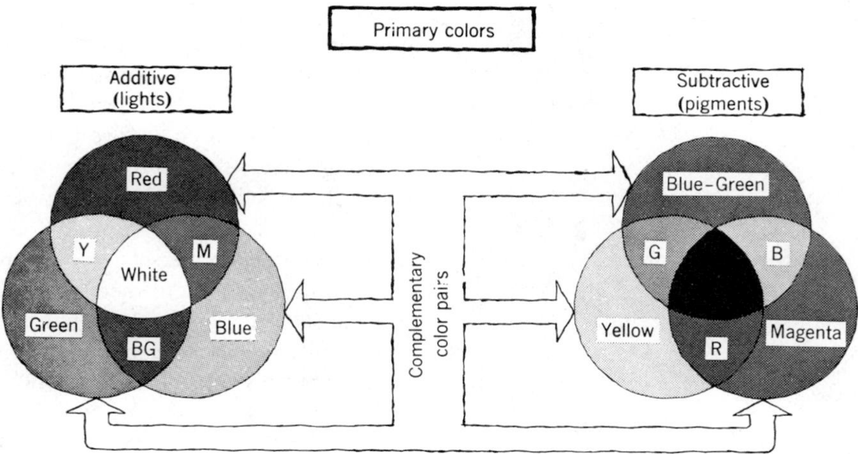

Fig. 18.55 *Primary and complementary colors. Complementary color pairs are shown by arrows. Pigments form color by an absorptive (subtractive) process; colored lights form colors by a reflective (additive) process.*

sitive to the missing colors that would make up white light. This phenomenon could be used to make meat look redder on a butcher counter by using blue-rich, red-poor, cool white lighting in the remainder of the store. A similar phenomenon occurs when the eye is exposed to a monochromatic scene where the chromaticity is due to coloration of the objects, rather than the illumination. The eye in such a situation becomes sensitized to the complimentary color; thus if after looking at a green surface one shifts the gaze to a white surface, one sees the complimentary red color. Returning to our meat market, the use of green paint on the walls also enhances the redness of the meat. This effect in reverse also accounts for the extensive use of green for paints, linens, and gowns, and so on, in operating rooms. The eyes of the surgeons and nurses when diverted from the redness of the surgical area will be more comfortable seeing green on a green background than on a white one. A similar effect is apparent object color differences when background color is changed. Thus a green object looks somewhat blue-green on a yellow background because the eye is supplying the complimentary color to yellow—that is, blue. Similarly, the same green object looks slightly yellow-green when on a blue background, the eye supplying the yellow.

Apparent brightness of a color is a function of its hue, in that light colors appear lighter than dark colors even when measured brightness is the same. Thus spaces may be defined by color within an area of equal illumination. Also, all colors tend to appear less saturated, that is, they appear "washed out" when illumination is high. Thus pigments of high saturation (chroma) must be used in well-lit spaces if they are to be effective although extensive use of saturated colors is generally best avoided.

Furthermore, there is an observed relationship between the color of the light in a space and the range of acceptable levels of illumination. The curves of Figure 18.56 indicate that cool illuminant color is desirable at high levels and warm at low levels, corresponding to blue sky and the light of a fire, respectively.

Other well-known psychological effects of colors are the coolness of blues and greens and the warmth of reds and yellows. Similarly, red and yellow are "advancing" colors because objects lit with them tend to "advance" toward the observer, giving the appearance of becoming larger. The opposite effect is noted with blue and green, accounting for their

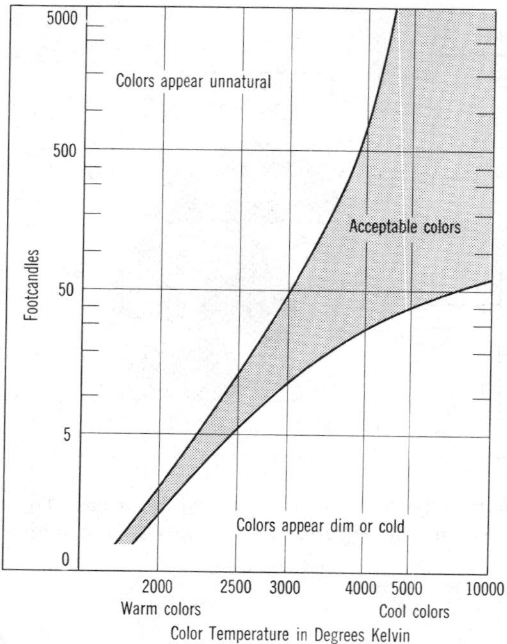

Fig. 18.56 *Note that color preference seems to be warm tones (flame) at low intensity and cool (blue sky) at high intensity.*

being known as receding colors. Thus cool colors might well be used in a fur salon and warm colors in a display of summer wear. A practical, energy-saving application of these color phenomena would be to use warm colors to compensate somewhat for lowered thermostats in the winter, and cool colors for the opposite effect in summer. How to accomplish this without the expense of repainting twice a year is left to the ingenuity of the architect and interior designer. In an atmosphere designed to be calm and restful, greens should generally predominate either in illuminant color, object color, or both, except in eating areas, which should be lighted with reds and yellows since cool colors are generally unappetizing. Yellows and browns emphasize motion sickness, whereas blues and greens tend to the reverse. Warm and saturated colors produce activity; conversely cool, unsaturated colors are conducive to meditation. Cool colors also seem to shorten time passage and are well applied in areas of dull repetitive work.

A further discussion of color control, source colors, and color matching will be found in Section 19.23 dealing with spectral energy distribution of sources.

References and Bibliography

1. *Energy Conservation Applied to Office Lighting,* April 1975, Ross and Baruzzini, Federal Energy Administration.

2. *Lighting Systems Study,* March 1974, Ross and Baruzzini, General Services Administration, Public Buildings Service.

3. *Energy Conservation Design Guidelines for Office Buildings,* January 1974, Dubin-Mindell-Bloome Associates, General Services Administration, Public Buildings Service.

4. *Contrast Rendition in School Lighting,* 1970, Foster K. Sampson, a report from Educational Facilities Laboratories.

5. "Comparison of the Energy Effectiveness of Office Lighting Systems," September 1976, Noel Florence, *Lighting Design & Application.*

6. Special Report #5, Architectural Engineering, "Lighting: Tools That Suit Architectural Objectives," March 1968, *Architectural Record.*

19

Light Sources and Their Characteristics

19.1 General Remarks

Long before the dawn of recorded history, the double blessing of fire, heat and light, was discovered. Even today in our sophisticated space age, fire is still used almost universally as the source of heat and, in a large proportion of the world's dwellings, as the source of light. Electrical lighting had its real beginning in about 1870 with the development of commercially usable arc lamps and was given greater impetus nine years later by Edison's first practical incandescent lamp. Today's electric light sources fall into three generic classifications: the incandescent lamp, including the tungsten-halogen types; the gaseous discharge lamp, which includes the well-known fluorescent, neon, and mercury lamps plus the more recent metal-halide and sodium lamps; and the electroluminescent sources. The efficiency of a light source is termed its *efficacy* and is measured in lumens per watt. Table 19.1 lists efficacies of modern light sources, including ballast losses where applicable.

Table 19.1 ***Efficacy of Various Light Sources***

Source	*Efficacy (Lumens per Watt)*
Candle	0.1
Oil lamp	0.3
Original Edison lamp	1.4
1910 Edison lamp	4.5
Modern incandescent lamp	14–20
Tungsten halogen lamp	16–20
Fluorescent lamp[a]	50–80
Mercury lamp[a]	30–60
Metal-halide lamp[a]	60–80
High-pressure sodium[a]	90–100
Low-pressure sodium	120–140

[a] Including ballast losses.

It is misleading to use the efficacy of the light sources alone, as is often done in the literature, because the ballasts are inseparable from the lamp. With development of low-loss electronic ballasts—a field now very active—the efficacy of the lamp/ballast combination will obviously rise. Most experts forsee a 150 lumens per watt luminous efficiency within a decade. Note that Table 19.1 gives a range of efficacies for each lamp type. In general, efficacy increases with wattage; therefore it is energy-economical to use a small number of higher-wattage lamps than the reverse. (It is also usually more economical with respect to fixtures.) Since electric lighting in American nonresidential buildings consumes 25 to 60% of the electric energy utilized, any attempt to reduce this must necessarily include integration of the cheapest (insofar as energy is concerned), most abundant and, in many ways, most desirable form of lighting available—daylight.

DAYLIGHTING

19.2 Daylighting as a Lighting Design Factor

The provision of daylight in structures in the United States has in the recent past largely been considered an amenity rather than a necessity. As such its provision has been the province of architecture rather than lighting design. The reasons for this are clear. Daylight is indeed an amenity. Windows provide visual contact with the outside and the resultant daylight provides a bright, pleasant, airy ambience. When daylight enters through windows (side lighting, as opposed to toplighting), its horizontal directivity provides good modeling shadows, minimal veiling reflections, and excellent vertical surface illumination. Furthermore, the continual variation of daylight which is one of its prominent characteristics, provides a constantly changing pattern of space illumination; one that is unattainable with artificial light. Since these changes are gradual, the eyes adapt easily (see Section 18.16) and the effect is one of visual interest. Undoubtedly, as a result of these effects, numerous studies have conclusively demonstrated a marked preference for daylight over any other form of lighting.

On the other hand, no ill effects have conclusively been demonstrated to have been caused by lack of daylight, that is, by working in an artificially lighted space. Since an artificial lighting system must be installed in any event, to furnish interior illumination during periods when daylight is insufficient, the practice arose in the United States of ignoring daylight and even of shutting it out deliberately. Careful design of an electric-lighting system can provide a good visual atmosphere as these chapters explain. Further, unlike daylight, control of such systems is relatively simple. Perhaps most important, an interior electric-lighting system has minimal impact on the building architecture, least of all on the all-important building facade. Finally, the energy to power electric-lighting systems *was* cheap.

The option of ignoring daylight in our high-energy-cost and energy-resource-poor society is no longer available. That being the case, the American designer must learn to cope with the special problems that daylight use presents in order to reap its benefits. Since daylight is variable it creates special problems of glare control, direct sunlight control, and heat-gain limitation. In large measure the science (and art) of daylighting is not so much how to provide daylight, as how to do so without the attendant undesirable effects. Put otherwise, the American designer must adapt and adopt the British technique of PSALI (Permanent Supplementary Artificial Lighting in Interiors), which is almost universally applied in Europe. This technique, which is really a design approach, views artificial lighting as supplementary to daylighting and not vice versa.

The PSALI technique recognizes that nonresidential buildings are principally used during daylight hours and that sufficient daylight is generally available during these hours to provide *much* of the structure's lighting needs. Understood with this statement are the well-founded assumptions that:

a. The same visual performance can be achieved with less daylight than artificial light, when compared on a footcandle basis.
b. Current American footcandle recommendations can be reduced appreciably without noticeable depreciation of visual performance of most tasks (see Section 18.19).
c. Daylight and artificial light can be readily and successfully combined, that is, that artificial light can *supplement* daylight when the latter is insufficient.

Reference was made above to nonresidential occupancies since residential spaces are also in use during periods of complete absence of daylight, that is, at night. That being so, the artificial lighting system must be designed to furnish all of the required lighting, and therefore PSALI is not applicable. The same is true, of course, for other structures that are expected to be in use at night.

Since interior daylighting is obviously directly dependent on exterior lighting levels, an understanding of the latter is a clear prerequisite to designing for the former.

19.3 Characteristics of Outdoor Illumination

(*a*) ***Factors.*** The most prominent characteristic of daylight is its variability. Obviously the source of all daylight is the sun. The level of exterior illumination, at a particular place and time, depends on

a. Altitude and azimuth (latitude, date, time of day).
b. Weather conditions (cloud cover, smog).
c. Effects of local terrain (natural and man-made obstructions and reflections).

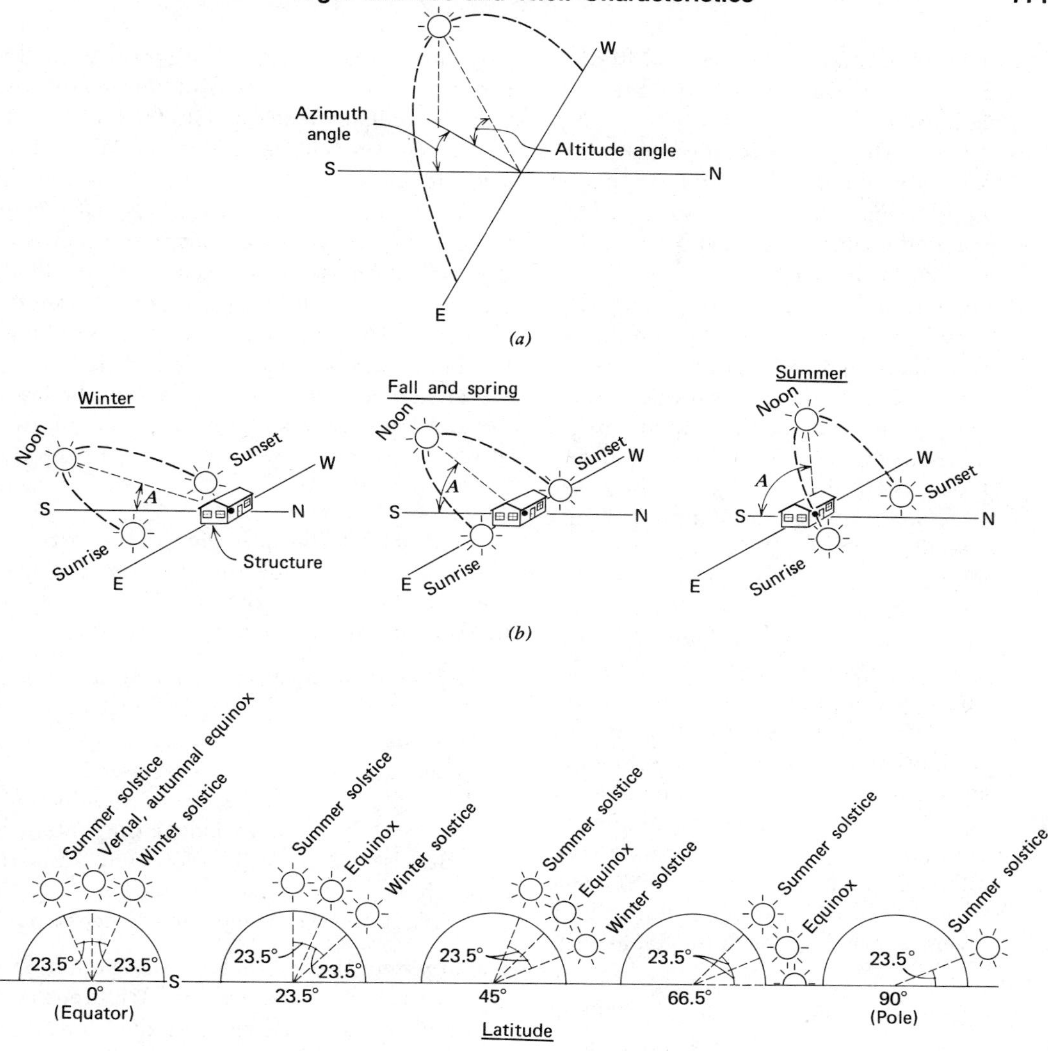

Fig. 19.1 (a) *The position of the sun is expressed in terms of vertical angle above the horizon (altitude) and horizontal angle, measured from the south (azimuth).* (b) *Approximate position of the sun in each of the seasons, at a midnorthern latitude (approximately 45°). Note that altitude angle is maximum in summer, minimum in winter, and in-between in spring and fall. Note too the length of daylight hours: maximum in summer, minimum in winter, and in-between in spring and fall.* (c) *Maximum sun altitude at various latitudes, for both solstices and equinoxes. Maximum summer sun altitude is 90° minus latitude plus $23\frac{1}{2}$°. Minimum winter sun altitude is 90° minus latitude minus $23\frac{1}{2}$°. Thus for all latitudes the yearly difference between maximum and minimum altitudes is twice $23\frac{1}{2}$°, or 47°, as shown.*

The position of the sun in the sky is expressed in terms of its altitude above the horizon and its azimuth angle. The latter is defined as its horizontal position angle, measured from the *south*. Both are normally expressed in degrees (see Fig. 19.1*a*). It is assumed that the reader is familiar with the basic astronomical phenomena governing the motion of the earth, which produce seasonal and latitudinal variations in the position of the sun. For our purpose, we will simply state the important facts, which are:

1. For all latitudes, the sun's altitude is highest in summer, lowest in winter, and in-between in spring and fall (see Fig. 19.1*b*). This fact, coupled with the shorter winter day, leads to the apparent rapid motion of the sun across the horizon in the winter and the apparent slow motion in the summer.
2. As a location approaches the equator (low latitude, either north or south), the sun's daily maximum altitude increases. The seasonal altitude variation, however, is the same for all latitudes (except at those extreme north and south latitudes where the sun is above or below the horizon for extended time periods). See Figure 19.1*c*. This factor not only affects exterior illumination levels but also has a pronounced effect on the design and efficacy of sun-shading devices.
3. The sun's azimuth angle is entirely dictated by the time of day, since the sun by definition transverses the sky between sunrise and sunset. The principal significance of the azimuth angle is encountered when discussing building orientation, exposures and shading angles, although obviously the azimuth angle strongly affects exterior illumination levels as can be seen from Table 19.2.

The factor of cloud cover, unlike that of solar position, is predictable only statistically. On the basis of extensive U.S. Weather Bureau observations it is possible to estimate that at a particular location, sky illumination will exceed a given figure for a specified number of hours per month. The figures have been derived from observations of cloud cover at numerous weather stations throughout the United States. At locations other than those for which these data are available, interpolation or extrapolation is necessary, with all the inaccuracies inherent in such procedures. Outside the United States, the designer must rely on locally available data, which are frequently meager or entirely unavailable. Since cloud cover, which can vary from scattered clouds in a blue sky to a solid overcast, is so very important, and the data establishing it generally so unsatisfactory, designers normally avoid accurate calculation of this factor (as will be explained below), preferring use of prevailing weather conditions to establish the sky brightness. The third factor—that of local terrain and construction conditions that either reduce illumination by shadowing or increase it by reflection—is so particular and individual that it can only be considered on a case-by-case basis. This does not imply, however, that this factor can be ignored. Most of the calculation methods consider the effects of shadowing and obstruction and, to a lesser extent, the effects of reflections (see Sections 19.4 to 19.6).

(*b*) ***Sky Conditions.*** In view of the extreme variability of exterior illumination and the difficulty involved in determining it, several fundamental questions face the designer:

1. Should an accurate calculation be attempted, based on a specific location, or should prevailing weather and its attendant sky conditions be used?
2. What value(s) of exterior illumination should be used in calculating interior levels in view of the daily and seasonal changes?
3. What degree of accuracy is necessary?

It is apparent that all three of these questions are related, as are the answers. Different design philosophies answer these questions differently, as do different designers. Nevertheless, all agree that when using hand methods (as opposed to computers), it is not practical to calculate hourly, daily, or even monthly variations. It is sufficient to establish four basic sky conditions, which can then be used with the specific design approach desired. These are:

1. Completely overcast sky.
2. Clear sky, without sun.
3. Clear sky, with sun.
4. Partly cloudy sky.

1. Completely Overcast Sky. This condition, which obtains for much of the year in northerly climates such as England, Scandanavia, and the Pacific northwest, is also called

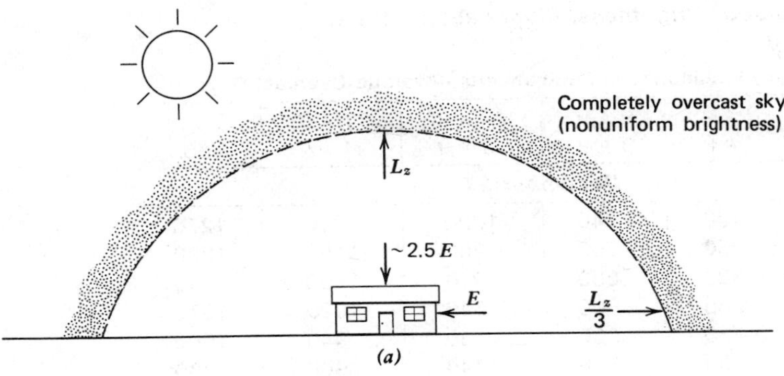

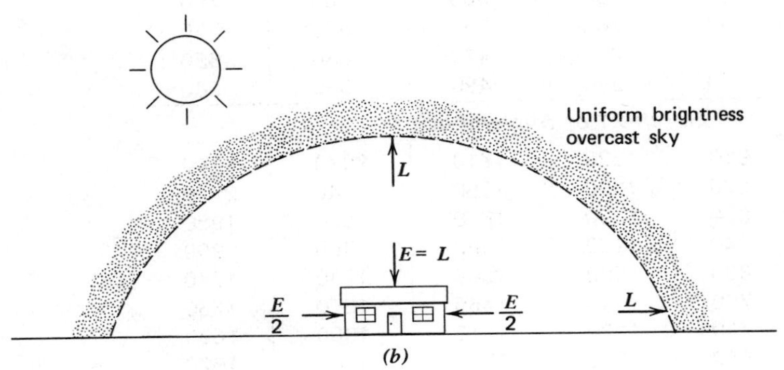

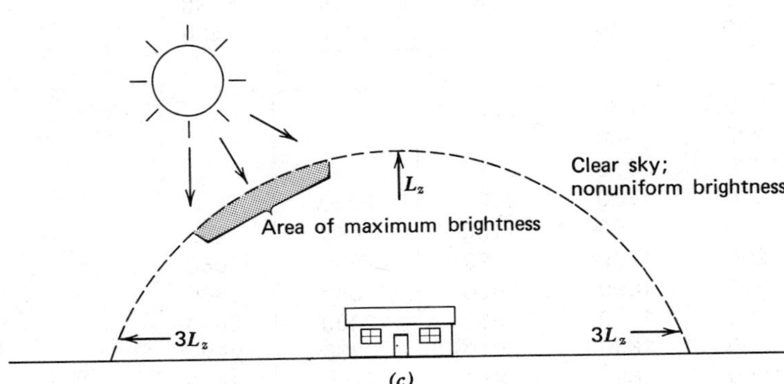

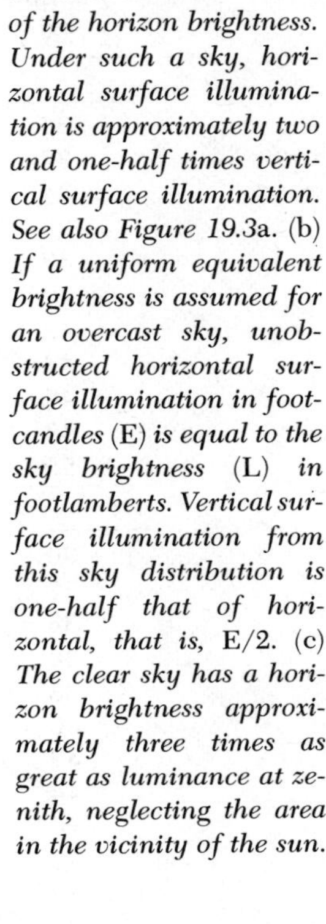
Fig. 19.2 (a) *The completely overcast sky has a zenith-brightness* L_Z, *which is three times that of the horizon brightness. Under such a sky, horizontal surface illumination is approximately two and one-half times vertical surface illumination. See also Figure 19.3a.* (b) *If a uniform equivalent brightness is assumed for an overcast sky, unobstructed horizontal surface illumination in footcandles* (E) *is equal to the sky brightness* (L) *in footlamberts. Vertical surface illumination from this sky distribution is one-half that of horizontal, that is,* E/2. (c) *The clear sky has a horizon brightness approximately three times as great as luminance at zenith, neglecting the area in the vicinity of the sun.*

the CIE sky, since it was adopted by the Commission Internationale de l'Éclairage (CIE) as the standard design sky for daylighting calculations.[16] This sky has a nonuniform brightness distribution, increasing from horizon to zenith in approximately the ratio of 1 : 3. Sky luminance at an altitude angle above the horizon is defined as

$$L_\theta = L_z \frac{1 + 2\sin\theta}{3} \tag{19.1}$$

where L_θ = luminance at θ^0 above horizon
L_z = luminance at zenith.

Thus at the horizon, where $\theta = 0^0$,

$$L_\theta = \frac{L_z}{3}$$

as stated above. The illumination (footcandles) produced by this distribution on horizontal and vertical surfaces stands in the approximate ratio of 2.5:1 (see Fig. 19.2*a*).

Table 19.2 ***Outdoor Brightness, Illumination Levels, and Solar Angles***

(*a*) Equivalent Sky Luminance in Footlamberts—Average Overcast Day

Latitude (°N.)	8 A.M. 4 P.M.	9 A.M. 3 P.M.	10 A.M. 2 P.M.	11 A.M. 1 P.M.	Noon
			December 21		
30	420	740	1020	1210	1270
32	350	700	960	1150	1200
34	320	650	910	1100	1140
36	260	600	840	1020	1070
38	230	550	790	940	1000
40	190	500	740	900	930
42	150	450	660	820	860
44	100	380	600	760	790
46	60	340	550	680	730
48	40	290	470	630	650
50	0	240	420	560	580
			March 21 or September 21		
30	910	1320	1710	2010	2140
32	880	1290	1650	1940	2070
34	860	1250	1600	1870	1980
36	840	1220	1560	1800	1900
38	800	1200	1500	1740	1840
40	790	1140	1460	1670	1760
42	760	1120	1410	1600	1690
44	740	1080	1340	1540	1620
46	710	1030	1229	1470	1550
48	690	990	1240	1410	1480
50	650	940	1180	1330	1400
			June 21		
30	1270	1730	2250		
32	1280	1730	2240		
34	1290	1730	2220		
36	1290	1730	2200	2960	
38	1290	1720	2160	2840	
40	1290	1700	2120	2650	3060
42	1300	1690	2080	2540	2860
44	1290	1670	2050	2430	2660
46	1290	1640	2010	2330	2520
48	1290	1620	1960	2250	2400
50	1260	1590	1900	2160	2280

Source. From *Recommended Practice of Daylighting*, IES, 1962.

To perform accurate calculations with this nonuniform distribution one requires some sort of sky protractor with which to measure sky angles and from which luminances and illumination can be calculated. Since this is an involved and time-consuming procedure, three types of approximation are in common use.

(*a*) *Uniform equivalent sky brightness, overcast sky.* This approximation takes cognizance of the sky luminance distribution and uses it to establish a *uniform equivalent sky brightness,* which is independent of altitude angle. The advantage of this simplification is the elimination of angle-luminance calculations. Remembering that an infinite plane of

Table 19.2 *Outdoor Brightness, Illumination Levels, and Solar Angles (Continued)*

(*b*) Overcast Sky, Outdoor Horizontal Illumination, Major Cities, United States

Location	Footcandles Illumination
Boston	700
Chicago	750
Detroit	750
Los Angeles	800
Minneapolis	700
New Orleans	900
New York City	750
Portland, Oregon	700
Salt Lake City	750
San Francisco	800

NOTE. Footcandle values given represent levels available for a minimum of 85% of working day.

uniform brightness, L fL, produces a surface illumination of L fc (see Fig. 20.25, page 889) we can appreciate the value of this simplification [see Table 19.2(*a*)]. The sky luminances there given in footlamberts translate directly to an equal number of footcandles on a horizontal plane, and one-half of that figure on the vertical plane (see Fig. 19.2*b*).

(*b*) *Simplified nonuniform distribution.* A second approximation maintains the brightness variation with altitude given in equation 19.1, and gives exterior surface illumination values directly, maintaining the 2.5:1 horizontal to vertical illumination ratio shown in Figure 19.2*a*. Solar altitude is determined by latitude, data, and time. Many designers consider this approximation adequate for most design purposes (see Fig. 19.3*a*).

(*c*) *Single-value exterior illumination—overcast sky.* This is the roughest of the approximations and should only be used for initial calculations. Single daylong values of 1250 to 1700 fc horizontal illumination and 500 to 700 fc vertical illumination are the values most frequently utilized when a specific location is not known. A few typical values for major American cities are listed in Table 19.2(*b*).

2. *Clear Sky.* This daylight condition was referred to above in two forms, that is, with and without sun. Obviously a clear sky cannot exist without sun. This form, however, is necessary for vertical *northerly* building exposures, which obtain all their daylight from the sky only (since they are not exposed to direct sun radiation) and for eastern and western exposures, which obtain daylight from sky alone and from sun and sky, depending on the time of day. Southern exposures obtain their daylight from a combination of sun and sky. Clear-sky, no-sun, illumination values are also required for toplighting installations not exposed to direct sunlight. The brightness of a clear sky is obviously highest in the area of the sun. If for the reason stated above, we separate the two factors, that is, the clear sky and direct sunlight, and neglect scattered clouds, we obtain an approximate brightness distribution for clear sky *alone*, as shown in Figure 19.2*c*. This figure shows that its brightness distribution is exactly opposite that of the completely overcast sky in that horizon brightness is three times that of the zenith. There is no simple relation between horizontal and vertical surface illumination, since the illumination on a vertical surface depends on solar azimuth as well as altitude. Accurate clear-sky brightness distribution charts are available for many areas from the Weather Bureau. However, here too, as with the overcast sky condition, use of these charts is time-consuming and laborious, and therefore simplifications are in common use. These are:

Table 19.2 ***Outdoor Brightness, Illumination Levels, and Solar Angles*** **(Continued)**

(*c*) Equivalent Sky Luminance in Footlamberts—Clear Days (No Sun)[a]

Latitude (°N.)	December 21					March and September 21					June 21				
	8 A.M.	10 A.M.	Noon	2 P.M.	4 P.M.	8 A.M.	10 A.M.	Noon	2 P.M.	4 P.M.	8 A.M.	10 A.M.	Noon	2 P.M.	4 P.M.
North															
30	450	600	600	600	450	700	1000	1050	1000	700	1550	1400	1000	1400	1550
34	350	550	550	550	350	800	800	900	800	800	1350	1400	950	1400	1350
38	300	550	550	550	300	750	800	900	800	750	1350	1300	950	1300	1350
42	250	500	500	500	250	700	750	800	750	700	1300	1300	950	1300	1300
46	150	450	500	450	150	700	750	750	750	700	1300	1250	950	1250	1300
South															
30	1100	1950	2250	1950	1100	1700	2300	2800	2300	1700	1200	1600	2400	1600	1200
34	1100	1900	2200	1900	1100	1700	2650	2900	2650	1700	1350	1650	2300	1650	1350
38	900	2300	2200	2300	900	1700	2700	2950	2700	1700	1350	1650	2300	1650	1350
42	600	2100	2150	2100	600	1700	2700	2450	2700	1700	1350	2000	2500	2000	1350
46	400	1900	2100	1900	400	1700	2700	2900	2710	1700	1350	2100	2700	2100	1350
East															
30	1550	1500	1000	700	400	2000	2500	1500	900	700	2800	2650	1400	1000	700
34	1350	1400	950	700	400	2400	2600	1600	950	650	2800	2700	1450	1000	700
38	1200	1300	900	650	350	2500	2600	1500	900	600	2800	2700	1400	1050	700
42	750	1200	850	600	250	2400	2400	1450	800	600	2900	2600	1400	1000	700
46	500	1100	800	500	150	2300	2100	1400	700	600	2850	2600	1400	1000	700
West															
30	400	700	1000	1500	1550	700	900	1500	2500	2000	700	1000	1440	2650	2800
34	400	700	950	1400	1350	650	900	1600	2600	2400	700	1000	1400	2700	2800
38	350	650	900	1300	1200	600	900	1500	2600	2500	700	1050	1400	2700	2800
42	250	600	850	1200	750	600	800	1450	2400	2400	700	1000	1400	2600	2900
46	150	500	800	1100	500	600	700	1400	2100	2300	700	1000	1400	2600	2850

Source. From *Recommended Practice of Daylighting,* IES, 1962.

[a]Average values, direct sunlight excluded.

(*d*) Average Solar Illumination as a Function of Altitude

Latitude (°N.)	Plane	Illumination (footcandles)								
		December 21			March, September 21			June 21		
		8 A.M. 4 P.M.	10 A.M. 2 P.M.	Noon	8 A.M. 4 P.M.	10 A.M. 2 P.M.	Noon	8 A.M. 4 P.M.	10 A.M. 2 P.M.	Noon
30	Perp.[a]	4200	7000	7700	6400	8300	8600	7700	8600	8900
	Horiz.	700	3400	4400	2600	5900	7000	4400	7200	8500
34	Perp.[a]	3100	6500	7100	6300	8100	8400	7600	8600	8900
	Horiz.	400	2700	3700	2400	5600	6700	4700	7100	8400
38	Perp.[a]	2500	6000	6900	6100	8000	8300	7600	8500	8900
	Horiz.	100	2000	3000	2100	5400	6200	4400	7000	8300
42	Perp.[a]	2000	5500	6400	6000	7800	8200	7600	8400	8800
	Horiz.	100	1600	2700	2000	4800	5800	4700	6800	7900
46	Perp.[a]	500	4500	5800	5800	7600	8100	7600	8100	8800
	Horiz.	—	1000	1800	1800	4400	5500	4400	6700	7400

Source. From *Recommended Practice of Daylighting*, IES, 1962.

[a]Perpendicular to sun's rays.

(*e*) Daylong Average Illumination; Approximation

Latitude	Illumination—Footcandles				
		Vertical			
	Horizontal	N	S	E	W
28–33° N					
Sky only	—	1300	1600	900	900
Sun and sky	9000	—	4500	4000	3500
34–38° N					
Sky only	—	1100	1400	800	800
Sun and sky	8000	—	4000	3500	3000
39–43° N					
Sky only	—	900	1200	700	700
Sun and sky	7000	—	3500	3000	2500

Table 19.2 *Outdoor Brightness, Illumination Levels, and Solar Angles* (*Continued*)

(*f*) Solar Altitude and Azimuth for Different Latitudes

Latitude (°N)		Date	A.M.: 6 / P.M.: 6	7 / 5	8 / 4	9 / 3	10 / 2	11 / 1	Noon
			Solar Time						
30	Altitude	June 21	12	24	37	50	63	75	83
		Mar.–Sept. 21	—	13	26	38	49	57	60
		Dec. 21	—	—	12	21	29	35	37
	Azimuth	June 21	111	104	99	92	84	67	0
		Mar.–Sept. 21	90	83	74	64	49	28	0
		Dec. 21	—	60	54	44	32	17	0
34	Altitude	June 21	13	25	37	50	62	74	79
		Mar.–Sept. 21	—	12	25	36	46	53	56
		Dec. 21	—	—	9	18	26	31	33
	Azimuth	June 21	110	103	95	90	78	58	0
		Mar.–Sept. 21	90	82	72	61	46	26	0
		Dec. 21	—	—	54	43	30	16	0
38	Altitude	June 21	14	26	37	49	61	71	75
		Mar.–Sept. 21	—	12	23	34	43	50	52
		Dec. 21	—	—	7	16	23	27	28
	Azimuth	June 21	109	101	90	83	70	46	0
		Mar.–Sept. 21	90	81	71	58	43	24	0
		Dec. 21	—	—	54	43	30	16	0
42	Altitude	June 21	16	26	38	49	60	68	71
		Mar.–Sept. 21	—	11	22	32	40	46	48
		Dec. 21	—	—	4	13	19	23	25
	Azimuth	June 21	108	99	89	78	63	39	0
		Mar.–Sept. 21	90	80	69	56	41	22	0
		Dec. 21	—	—	53	42	29	15	0
46	Altitude	June 21	17	27	37	48	57	65	67
		Mar.–Sept. 21	—	10	20	30	37	42	44
		Dec. 21	—	—	2	10	15	20	21
	Azimuth	June 21	107	97	88	74	58	34	0
		Mar.–Sept. 21	90	79	67	54	39	21	0
		Dec. 21	—	—	52	41	28	14	0
48	Altitude	June 21	17	27	37	47	56	63	65
		Mar.–Sept. 21	—	10	20	29	36	40	42
		Dec. 21	—	—	1	8	14	17	19
	Azimuth	June 21	106	95	85	72	55	31	0
		Mar.–Sept. 21	90	79	67	53	38	20	0
		Dec. 21	—	—	52	41	28	14	0

(*a*) *Uniform equivalent sky luminance—clear sky.* As with the overcast sky, this approximation recognizes the nonuniformity of clear-sky brightness in arriving at an *equivalent* uniform brightness value. This luminance figure is dependent on latitude, data, time, and exposure but is an approximation in that it assigns a single luminance value to the entire section of the sky seen by that exposure (N, S, E, W). Table 19.2(*c*) gives these average values, *excluding* direct sunlight.

(*b*) *Average illumination curves—clear sky.* The data presented in Figure 19.3*b* to *e* are adapted from U.S. Weather Bureau observations and are approximate in that single values are given for an entire season; and that interpolation is required between curves. Note that in curves (*b*) to (*d*) the term *horizontal angle* rather than *azimuth angle* is used to avoid confusion with the above defined azimuth angle of the sun, which determines its position with respect to South. Intended in these curves is the horizontal angular offset between the sun's position and a line perpendicular to the vertical plane. Thus, a zero offset angle indicates that the sun is positioned directly in front of the vertical surface. Depending on exposure, these curves will be used with or without direct sun contribution, which will be discussed below.

(*c*) *Single value approximations—clear sky.* As in 1(*c*) above, single-number average values for horizontal and vertical surface illumination for clear sky can be used. These figures, given in Table 19.2(*e*), are rough and therefore appropriate for first approximations and comparisons only.

3. *Partly Cloudy Sky.* There is no known way of mathematically expressing the sky luminance distribution of this type sky because of its infinite variability. There are, however, Weather Bureau statistical data on percent cloud cover for number of days per year, for given locations. This can be related to tables relating percent cloud cover to surface illumination. Most frequently, cloud cover data are used to select the type of sky brightness distribution to be used, that is, completely overcast or clear sky.

4. *Solar Illumination.* As explained above, illumination from a clear sky and that from direct sunlight were separated for conven-

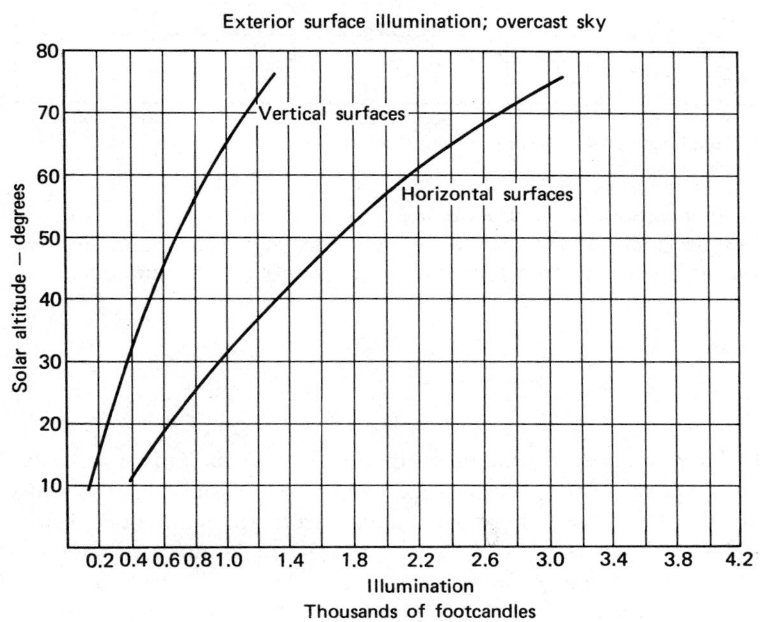

Fig. 19.3 (a) *Graphs based on nonuniform brightness overcast sky, giving unobstructed exterior surface illumination directly. The ratio between horizontal and vertical surface illumination is 2.5:1 as shown in Figure 19.2*a. *These data are based on U.S. Weather Bureau observations.*

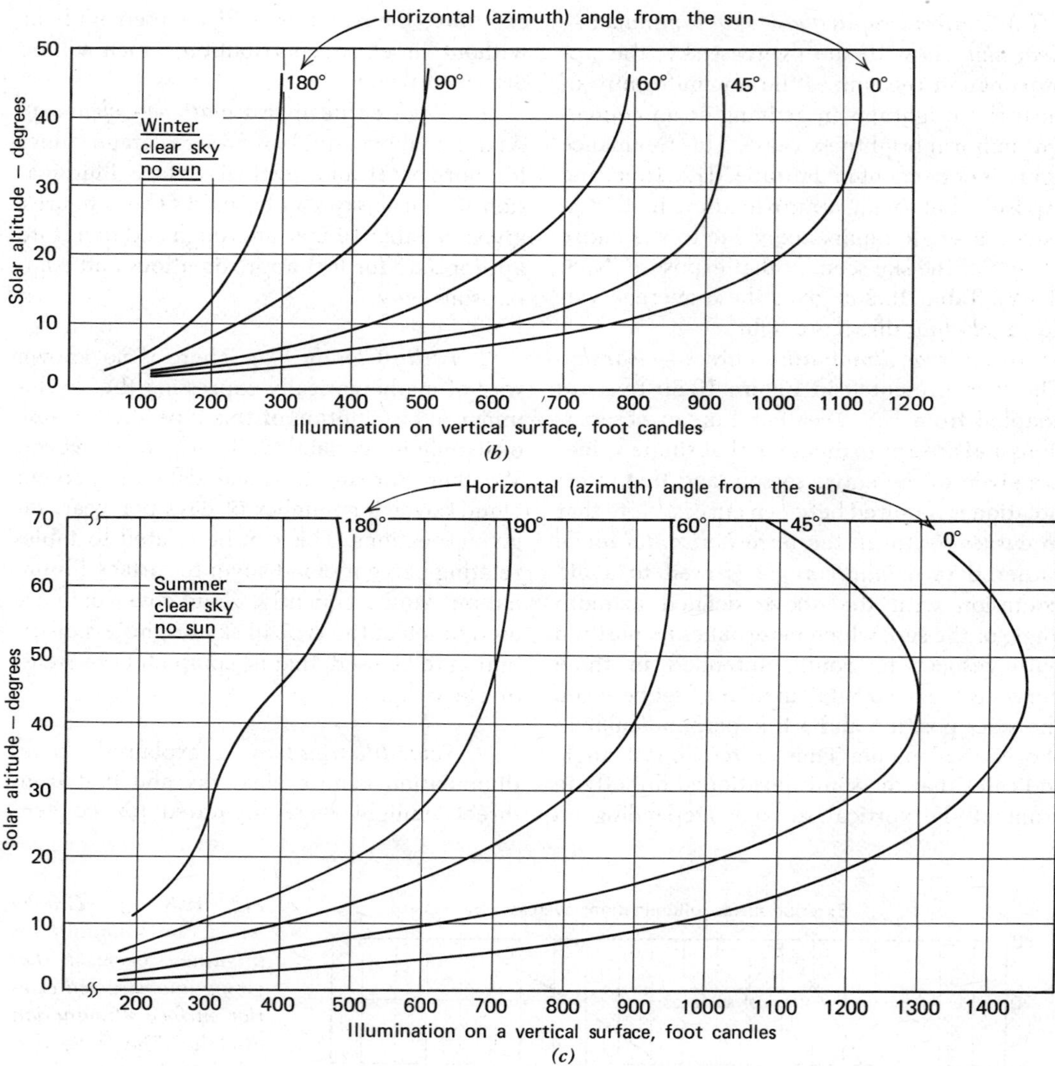

Fig. 19.3 (b)–(d) *Vertical surface illumination from a clear sky, with no sun contribution, for seasons of the year. See sketch on Figure 19.3*f *for graphical representation of horizontal angle between the surface and the sun's rays.* (e) *Horizontal surface illumination from a clear sky, with no sun contribution, for various seasons of the year.*

ience of calculations. Table 19.2(*d*) gives *average* solar illumination as a function of solar altitude for two planes—perpendicular to the sun's rays and on the horizontal—for different latitudes, seasons, and times of day. Since most practical application involves vertical surfaces, rather than those perpendicular to the sun's radiation, Figure 19.3*f* gives sun-only illumination values for such surfaces at various horizontal angles from the sun's rays. Table 19.2(*e*) gives rough single-number, daylong illumination values for both clear sky and clear sky plus sun, for the four vertical-surface exposures, and for horizontal surfaces. As cau-

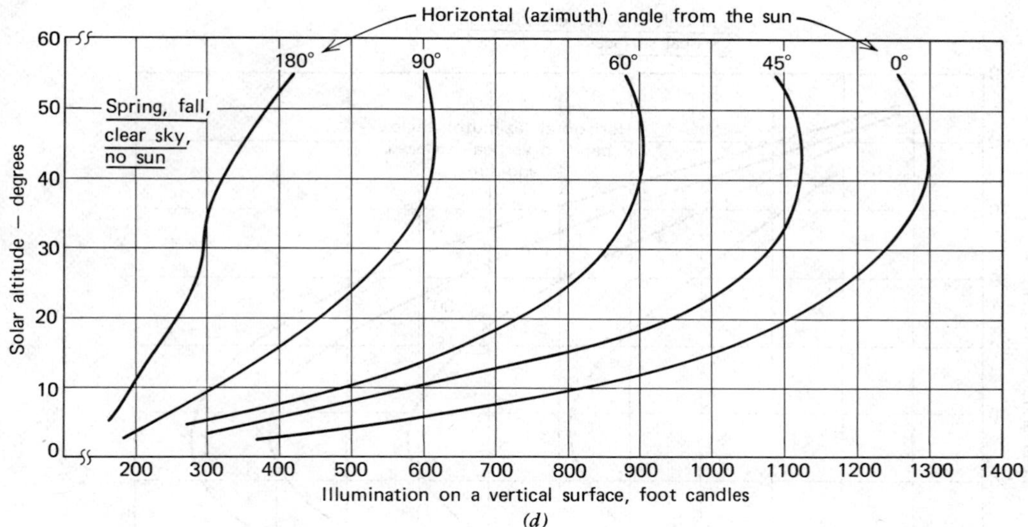

(d)

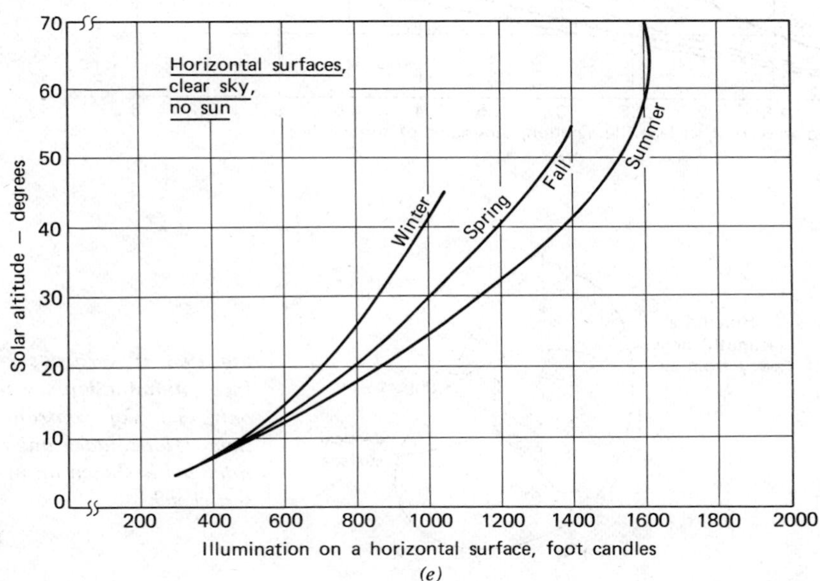

(e)

tioned above, these figures should be used for first approximation and comparative schemes only.

Table 19.2(*f*) gives values for solar altitude and azimuth for various latitudes. The solar azimuth figures (horizontal offset from South) are used to calculate the horizontal angle between a vertical surface and the sun, as required for use in Figure 19.3*b* to *f*.

(c) ***Selection of Outdoor Illumination Levels.*** At this point we must return to the very basic and important questions posed in Section 19.3(b) above; what value(s) of sky illumination should be used, and what degree of accuracy is possible and/or required? The answer to these questions depends on the calculation method adopted, and will be detailed in the description of three principal hand

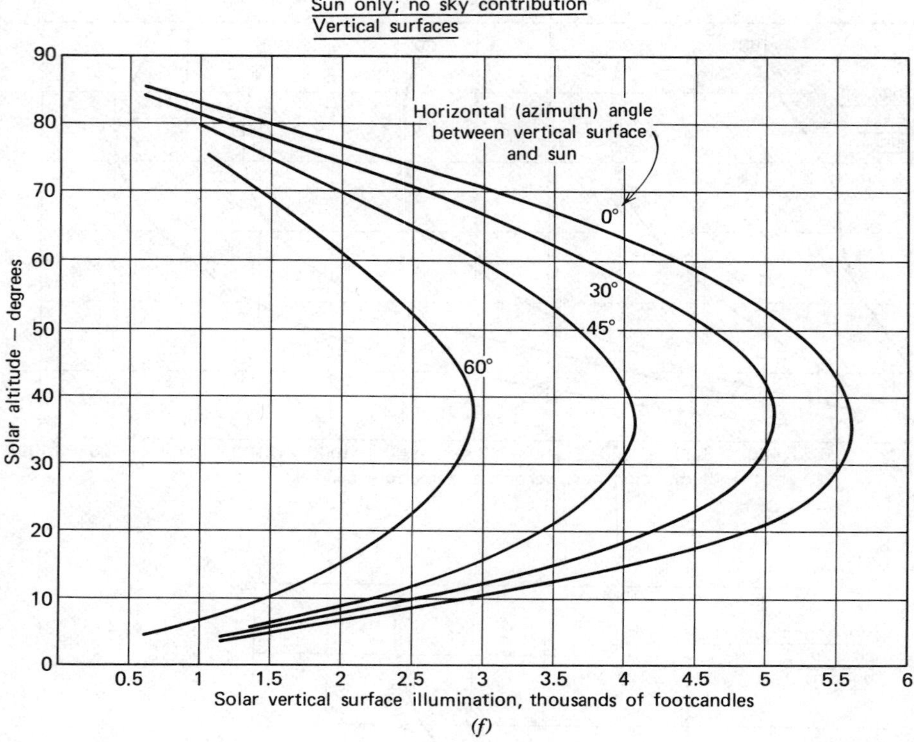

(f)

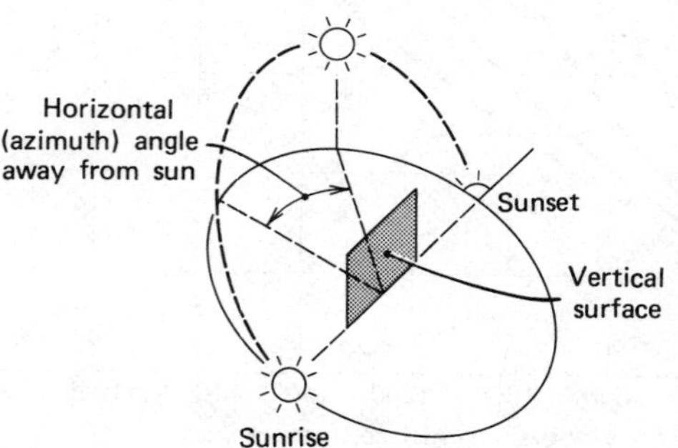

Fig. 19.3 (f) *Vertical surface illumination, sun only, no sky contribution. Horizontal angle intended is shown on insert sketch.*

methods in use today (contrasted with computer techniques). A general remark is in order here, however, with respect to accuracy and precision. These two terms are not synonymous in our usage. The former refers to correctness, the latter to degree of minuteness. Thus an answer to a calculation can be highly precise yet inaccurate, and quite accurate yet imprecise. In other words, no amount of precision can alter a degree of inaccuracy in data. Therefore, if sky brightness is used with an accuracy of ±50%, *that* will be the best degree of accuracy of the resultant daylight calculation, and little is gained by using coefficients to three decimal places. Computer programs are now becoming available that will detail sky brightness for a particular location hour-by-hour, and statistically throughout the year. These can then be used with interactive programming to yield plots, charts, or tables of hour-by-hour daylight levels, given all the variables involved. Here, a high degree of precision in all data will yield a high degree of accuracy. The methods described below are, as stated, hand calculation techniques with their inherent limitations.

19.4 Daylight Design Using the Daylight Factor Method

(*a*) ***General.*** This method was developed in England and is based on the solid overcast sky that prevails for much of the year in that country. It is also usable with the equivalent brightness, clear-sky figures as in Table 19.2(*c*) and (*e*). The approach of this method is to determine the *minimum* daylight that will be available within a space during most of the space's use. This light can then be augmented by artificial illumination; hence the PSALI (Permanent Supplementary Artificial Lighting for Interiors) approach. Since minimum figures will be achieved by using the solidly overcast sky, and since that sky condition obtains for so large a portion of the year in Britain, the method is eminently applicable there, and to other such climates. The error introduced when using an *equivalent* clear-sky luminance is not large, since obstructions are always at low angles, where clear-sky brightness is highest (see Fig. 19.4). This will become clearer in the discussion below.

(*b*) ***Daylight Factor and Its Components.*** The daylight factor expresses the interior illumination as a fraction of that available outdoors, generally in a horizontal plane. This fraction, usually written as a percentage, can be expressed as

$$\text{DF} = \frac{\text{indoor illumination at a given point}}{\text{outdoor illumination}} \times 100\% \qquad (19.2)$$

The factor is a constant because the sky *distribution* is fixed for the standard overcast sky, or the uniform bright sky, and therefore interior daylight will correspond exactly to fluctuations in sky brightness. This assumes minimal effect from obstructions and ground reflections, as is the case with a dull overcast sky. Eye adaptation levels will tend to smooth out these variations, making the *apparent* room

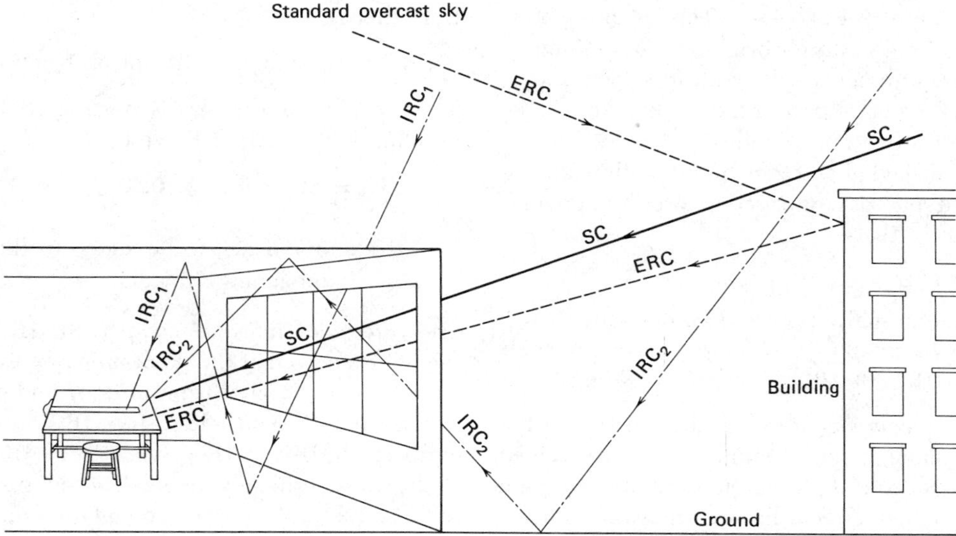

Fig. 19.4 *Total daylight factor (DF) is composed of SC (sky component), ERC (externally reflected component), and IRC (internally reflected component). The latter in turn is subdivided into reflected skylight and reflected ground light. Note that surfaces deep in the room are lighted in large part with rereflected light.*

brightness due to daylight fairly uniform. Thus the Daylight Factor method permits study of interior daylight distribution for varying fenestration, architectural arrangement, and building orientation. The overcast sky in the DF method, as stated, represents minimum exterior illumination, since the sky is frequently brighter. Hence actual interior daylight levels will frequently exceed the design minima. Since the overcast sky is *not* a glare source, the large window areas utilized in overcast sky design locales are not a glare source even when exterior levels are considerably above the design minima.

It is apparent that the DF varies within an interior space. This variation, however, is constant for a given architectural configuration. Therefore, knowing the DF variation for a given space and the exterior illumination as derived from sky brightness data, the actual interior daylight levels throughout the space can easily be calculated. Going one step further, minimum DF levels (as well as minimum artificial illumination levels) can be *specified* for different occupancies. These assume a minimum average daylight figure in a given geographic location, and the architecture is then designed to meet these requirements [see subsection 19.4(c) below]. Thus a combination of minimum exterior brightness and minimum DF requirements will result in sufficient daylight for a large part of the working day under almost all exterior conditions.

The daylight at any point within an enclosed space is comprised of three components (see Fig. 19.4):

1. Sky component (SC).
2. Externally reflected component (ERC).
3. Internally reflected components ($IRC_1 + IRC_2$).

The *sky component* (SC) is that portion of the total daylight at a point, which is received directly from the area of sky visible through the window. Since the brightness of an overcast sky varies, this factor must be computed. (For a uniformly bright sky this component is proportional only to the area of sky viewed, and is called the *sky factor*.) Since the sky component is *received* light, it takes into account light reduction due to window obstructions (mullions, etc.) and losses in transmission; that is,

$$\text{SC} = \text{incident skylight} - \text{window losses}$$

Incident skylight can be calculated with the use of instruments[3] and tables[4] prepared for the purpose. Losses in the window vary from 10% for small mullioned windows with clear glass to 50% for heavy window construction with dirty or frosted glass.

Note from Figure 19.4 the importance of obstructing buildings. Since the obstructions are at low angles, they obstruct the dullest part of an overcast sky and the brightest part of a clear sky. This has the effect of roughly equalizing the two conditions when considering a building in a built-up area and further justifying the use of overcast sky figures for all calculations with this method.

The externally reflected component (ERC) is light reflected from exterior obstructions onto the point under consideration. This *does not include ground-reflected light.* ERC is only of significance in heavily built-up areas and can be estimated as the portion of sky component for that area of obstructed sky, reduced by the reflectance factor (RF) of the obstruction; that is,

$$\text{ERC} = \text{sky factor} \times \text{RF (of obstruction)}$$

Thus if 25% of the sky is obstructed by a building with a 20% RF, we have

$$\text{ERC} = \text{SC} \times 0.25 \times 0.20$$

or

$$\text{ERC} = 5\% \text{ of SC, to be added to the remaining } 75\% \text{ SC.}$$

The *internally reflected component* (IRC) is the light received at the point under consideration that is reflected from interior and exterior surfaces. Nevertheless, since IRC_2 is generally small, $IRC \cong IRC_1$ (see Fig. 19.4). IRC is therefore critically dependent on surface reflectances and on the amount of window glazing, and becomes a large portion of DF deep within the interior space. See Table 19.3 and Figure 19.5. IRC is normally calculated using published interreflectance tables, as direct calculation is extremely complex. The DF

Table 19.3 ***Effect of Wall Reflectance Factor on Proportion of Internally Reflected Component (IRC) in Daylight Factor***

Distance from Window, Feet	*30% Wall Reflectance*		*60% Wall Reflectance*	
	Total DF	$\frac{IRC}{DF}$ %	*Total DF*	$\frac{IRC}{DF}$ %
0	30	1	31	3.5
5	16	1.9	17	6.5
10	5.5	5.5	6.3	16.9
15	2.1	14.3	2.9	37.9
20	1.3	23	2.1	52.4

ROOM DATA:
Room 24 × 28
Window on 28-ft wall—one side only
Window area = 20% of floor area
70% ceiling RF
20% floor RF

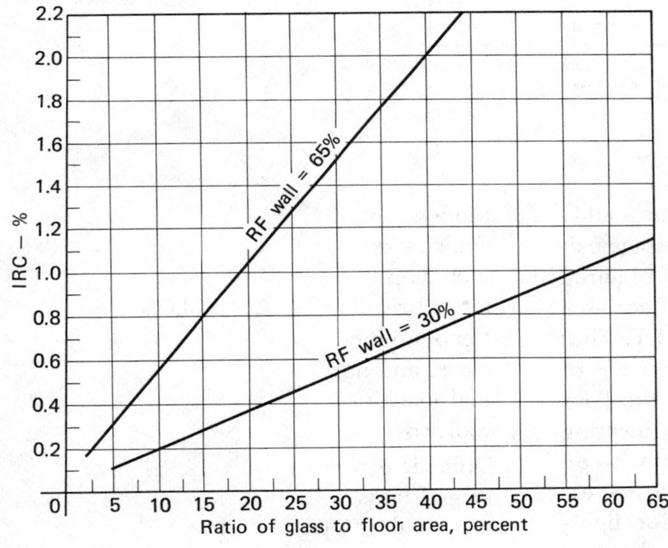

Fig. 19.5 *Variation of internally reflected component of DF as a function of wall reflectance. As might be expected the effect is more pronounced as the fenestration area increases.*

then is the sum of the three components:

$$DF = SC + ERC + IRC$$

all calculated individually for each location being considered.

(*c*) ***Characteristics of Daylight Factor.*** A typical DF curve, as produced by the longhand, daylight-protractor-aided techniques required, is shown in Figure 19.6. The three components—SC, IRC, and ERC—are calculated separately and added to produce the total daylight factor. Any change in parameters, such as window dimensions or design, ceiling height, surface reflectance, ground reflection, and obstructions, alters these curves and requires recalculation and replotting. As must be obvious from the above, exact calculation of even a few variants in a space is a tedious, time-consuming procedure when done longhand. For this reason two alternative approaches are available:

1. Use of a computer program.
2. Use of simplifications, such as standard curves, tabular data, or the CIE method described in Section 19.5.

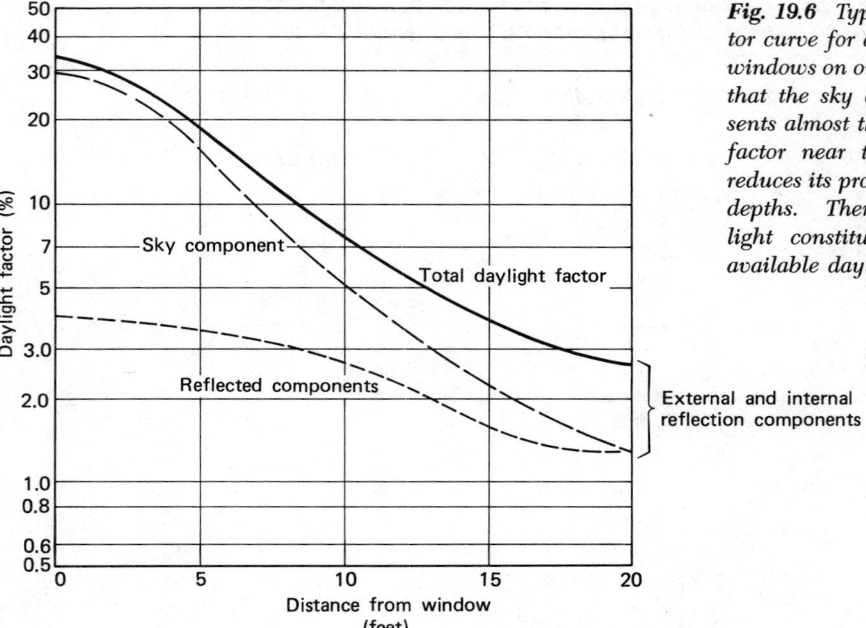

Fig. 19.6 Typical daylight factor curve for a room with large windows on one side only. Note that the sky component represents almost the entire daylight factor near the window, but reduces its proportion at greater depths. There, interreflected light constitutes 50% of the available daylight.

When curves and tabular data are not readily available, or when quick rough first-stage calculations are desired, calculate the required net clear glazing area in percent of floor area as 10 times the minimum required DF. Thus for a room with a required minimum DF of 1.4, the required glazing is 14% of the floor area. This rule-of-thumb is within engineering accuracy for rooms with single fenestration on the long side, length to depth ratio not exceeding 10:6 in proportion, and for light-colored walls. Other formulas and techniques for calculating DF are found in the literature.[6-9]

Despite the fact that DF is a ratio and not an absolute value such as footcandles, it is applied as an absolute, that is, recommendations are made of minimum daylight factor for specific seeing tasks. Recommended minimum levels of daylight factor vary, but the following are average English recommendations:

Ordinary seeing tasks, such as reading, filing, and easy office work	1.5 to 2.5%
Moderately difficult tasks, such as prolonged reading, stenographic work, and normal machine tool work	2.5 to 4.0%
Difficult prolonged tasks, such as drafting, proofreading poor copy, fine machine work, and prolonged fine inspection	4.0 to 8.0%

Since seeing tasks require minima of actual light, these DF recommendations are obviously based on an assumed *minimum* exterior illumination level. This level was, in Britain, historically 5000 lux, or approximately 500 fc. Thus the above recommendations would translate:

Ordinary seeing tasks	7.5 to 12.5 fc

Moderately difficult tasks	12.5 to 20 fc
Difficult tasks	20 to 32 fc

It is apparent that during periods when exterior illumination was indeed as low as 5000 lux, supplementary artificial illumination would no doubt be applied.

19.5 Daylight Design—CIE Method

(*a*) ***General.*** This method was the result of a search for a simple, rapid, straightforward, and fairly accurate daylight calculation method that would yield reliable results without the time-consuming constructions and calculations necessitated by other methods. After a study of considerable length and intensity, the CIE adapted and adopted a system developed in Australia by Dr. A. Dresler.[5] The current CIE recommendations were published in 1970.[16]

This system is based on the daylight factor described above. Dr. Dresler developed a set of more than 100 curves covering rooms of varying proportions and fenestration. One such typical curve is shown in Figure 19.7. The curves relate minimum DF (at a point 2 ft from the back of the room, in depth) to the maximum permissible room depth, for given reflectances and a standard window design, thus establishing the room's proportions. The curves were calculated using the techniques developed in England, as described in the preceding section. The published recommendations[16] make a clear strong case for the minimalist approach in selecting sky illumination and in design. It states simply that the number of design variables is so large and daylight itself so variable, that a simple routine method can only be based on minimal conditions. Therefore, the diagrams will give the *lowest* level of daylight that can reliably be expected for a given percentage of normal working hours in side-lighted rooms, and the *average* level in toplighted spaces.

(*b*) ***Characteristics of the Method.*** Advantages of the system are:

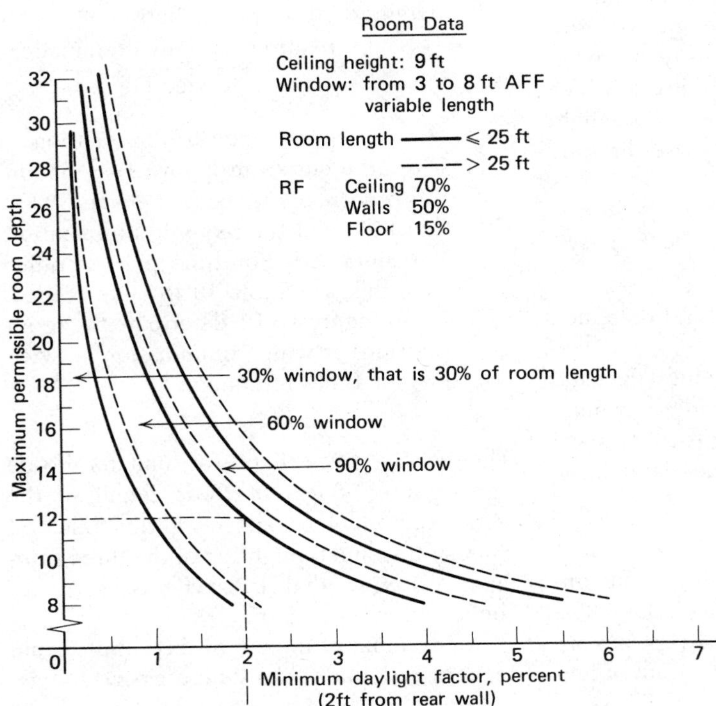

Fig. 19.7 *Maximum room depth to maintain minimum DF is proportional to window size. Thus for a room less than 25 ft in length, depth cannot exceed 12 ft if 2% DF is to be maintained. From* Daylight Design Diagrams, *published by Service Division, Commonwealth Department of Labor and National Service, Melbourne, Australia.*

1. Consideration of obstructions, reflections, and interior reflections.
2. Use of sky brightness that varies with latitude rather than a standard 5000 lux.
3. Applicability to a very wide range of side and top fenestration designs.
4. Establishment of required room proportions is architecturally more useful than solving for specific dimensions.

Limitations of the system are:

1. Inapplicable to clear-sky and sun conditions.
2. Calculation of minima only.
3. Inapplicable to other than rectangular-shaped rooms.
4. Unusable with sun-shading or high-reflectance ground.
5. Results give points of minimum, twice minimum, and four times minimum daylight factor only. Other points must be interpolated or extrapolated.
6. Window proportions and position in a wall are fixed.

Overall, the system accomplishes what it intended. The limitations listed are inherent in any rapid, simplified daylight calculation technique. They are listed here not as a criticism, but to advise the reader of these necessary limitations. The standard itself, which should be in the library of every daylight designer, also clearly states these limiting characteristics.

(*c*) ***Calculation Procedure.*** The CIE system is usable in two modes:

1. Given complete architectural data, find resultant daylight.
2. Given incomplete architectural data and required daylighting find maximum room depth and/or other room proportions that will satisfy the daylighting requirement.

Obviously the former mode is simpler since it leads directly to an answer. It will be demonstrated in an illustrative example below. The latter mode, since it has many answers, is more complex. That is, various combinations of fenestration and room dimensions will yield adequate daylight. For this reason, the designer should fix room length and percent fenestration of the window wall leaving depth as a variable, or fix room length and depth with fenestration as the variable. Ceiling height is usually fixed by other considerations. The procedure in mode 1 is as follows (the reader can work out mode 2 procedure):

1. Express room depth in terms of window height. From Figure 19.10 determine *design* daylight factor, that is, the daylight factor in the room.
2. The design daylight factor is larger than the actual service daylight factor because it is based on clear, clean glass and unobstructed horizon. Since this is rarely the case, the actual (service) daylight factor must be calculated using the correction factors of Figure 19.11 and Table 19.4; that is,

   $$\text{DF}_{\text{service}} = \text{DF}_{\text{design}} \times \text{correction factors}$$

3. Using the service daylight factor in eq. 19.2,

   $$\text{Required exterior illumination} = \frac{\text{required interior illumination}}{\text{service DF}}$$

   obtain required exterior illumination.
4. From the curves in Figure 19.8, obtain the percentage of hours between 0900 and 1700 that the required illumination is maintained. For time periods other than this, see Table 19.4(*c*).
5. From Figure 19.12, determine the room positions at which illumination is twice or four times minimum.

The curves given here for unilateral side lighting are typical of those found in the standard,[16] which covers, in addition, bilateral side lighting and toplighting of the three principal varieties, that is, skylights, sawtooth roofs, and monitor roof.

An example of the use of this simple rapid technique should make its use clear.

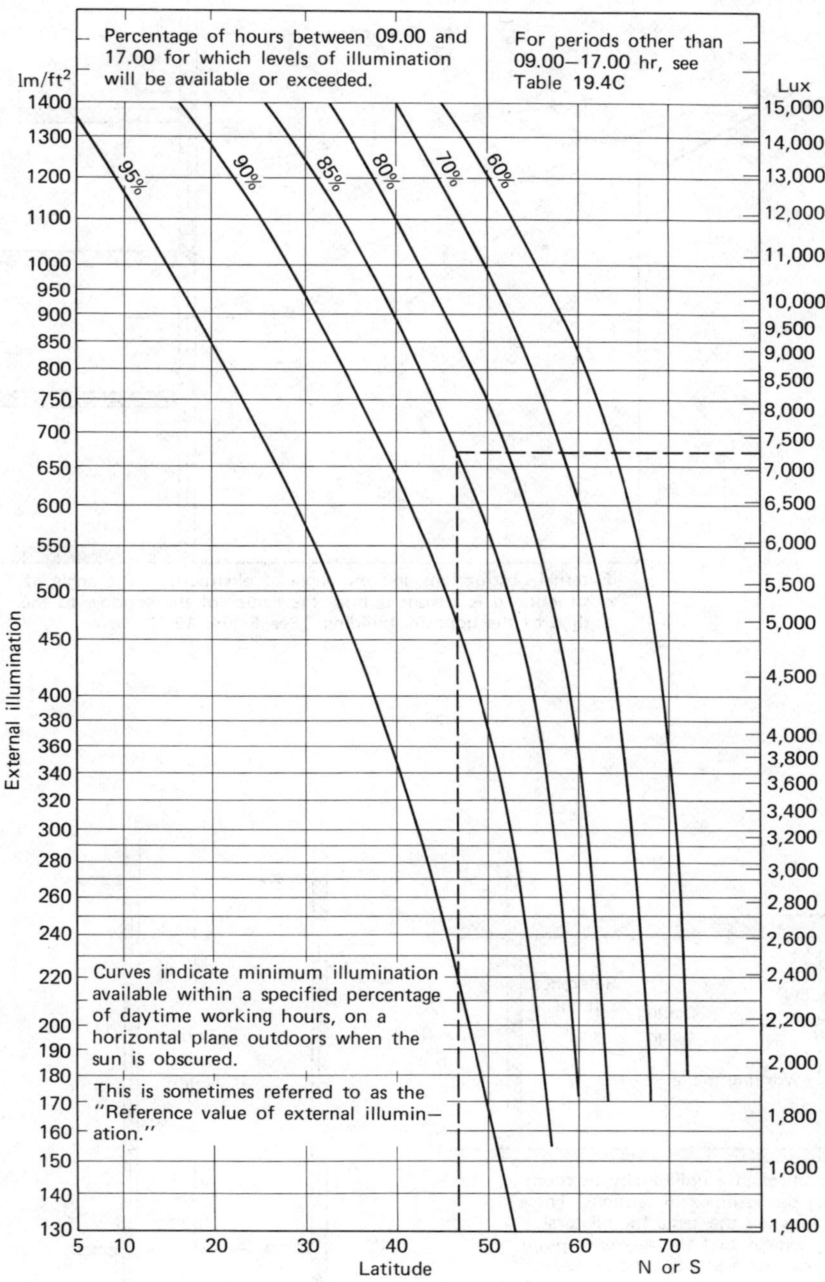

Fig. 19.8 Chart giving minimum maintained external illumination as a function of latitude, for a given percentage of the normal working day. From Daylight, International Recommendations for the Calculation of Natural Daylight, *Publication CIE No. 16 (E-3.2), 1970; Commission Internationale de l'Éclairage.*

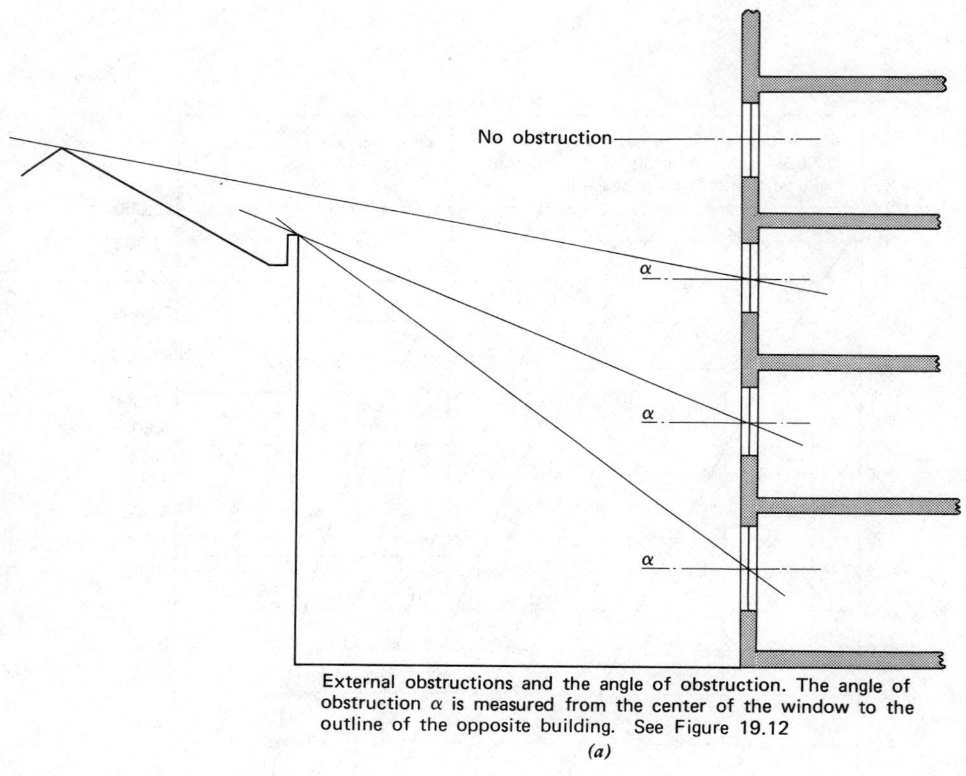

External obstructions and the angle of obstruction. The angle of obstruction α is measured from the center of the window to the outline of the opposite building. See Figure 19.12

(a)

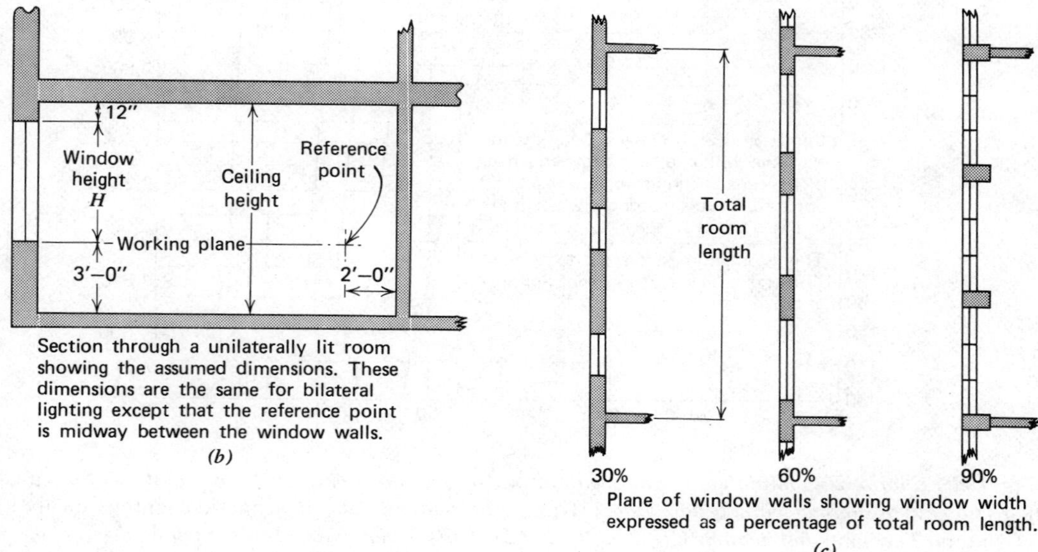

Section through a unilaterally lit room showing the assumed dimensions. These dimensions are the same for bilateral lighting except that the reference point is midway between the window walls.

(b)

Plane of window walls showing window width expressed as a percentage of total room length.

(c)

Fig. 19.9 *Sketches indicating the parameters of the CIE calculation systems.* (a) *Indicates how obstruction angle α is calculated.* (b) *Shows a vertical section through a room with dimensional data relevant to this system.* (c) *Indicates how size (length) of windows is calculated with respect to overall room length. From* Daylight, International Recommendations for the Calculation of Natural Daylight, *Publication CIE No. 16 (E-3.2), 1970; Commission Internationale de l'Éclairage.*

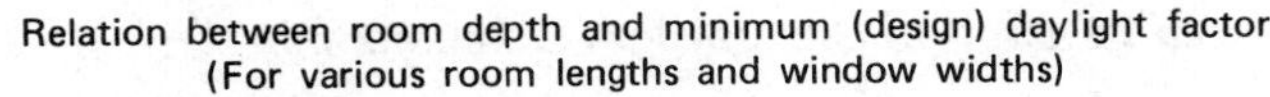

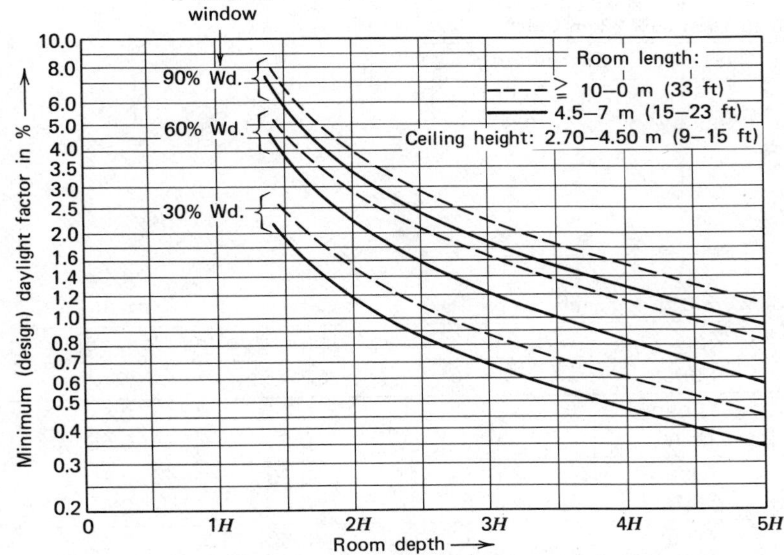

Fig. 19.10 Basic design diagram that relates minimum daylight factor to room width. Inasmuch as room width is expressed in terms of window height, the curves effectively relate minimum daylight factor (2 ft from back wall) to room proportion. From Daylight, International Recommendations for the Calculation of International Daylight, *Publication CIE No. 16 (E-3.2); Commission Internationale de l'Éclairage.*

EXAMPLE 19.1. A classroom in a single-story, Seattle elementary school is 25 ft long, 18 ft deep, and has a 10-ft ceiling. It receives daylight unilaterally from windows totalling 16 ft in length net. A building across the road obstructs the horizon to an angle of 15° above the window centerline. Window glazing is wired glass having a transmittance of 80%. The school is situated in a built-up residential area with a satisfactory cleaning schedule. Determine for what portion of the year tasks requiring a minimum illumination of 10 fc (108 lux) can be carried out by daylight, and at what distances from the window, levels of 20 and 40 fc are maintained, for this same portion of the year.

SOLUTION

General. The latitude of Seattle is 47.6° N, and its climate is appropriate for use of an overcast sky technique.

1. Window height H is:

$$\text{10-ft ceiling} - \text{3-ft sill height} - \text{1-ft soffit} = 6 \text{ ft}$$

Room depth in terms of window height:

$$\frac{\text{18-ft depth}}{\text{6-ft window height}} = 3H$$

From Figure 19.10, for room length of 25 ft, ceiling height of 10 ft, and window coverage of 90% (16 ft/18 ft), design daylight factor at $3H$ is 2.25.

2. Using Table 19.4 and Figure 19.11, we find these correction factors:

Glass transmission—0.95
Glass cleanliness—0.90
Obstruction—0.78

Therefore

$$\text{DF}_{\text{service}} = 2.25 \times 0.95 \times 0.9 \times 0.78 = 1.50$$

3. Required exterior illumination = 108 lux/1.50 × 100 = 7200 lux.
4. From Figure 19.8, 7200-lux exterior illumination at a latitude of 47.6° N will be maintained for 85% of the time between 0900 and 1700. Nevertheless, since the school day only extends to 1500, this figure of 85% can be increased to 95%, as shown in Table 19.4(*c*).

Table 19.4 ***Correction Factors to be Used in CIE Daylight Calculations***

(*a*) Correction Factor to be Used if the Glazing Has a Diffuse Transmittance of Less Than 85% When Clean

Diffuse Transmittance of Glass	*Correction Factor*
80%	0.95
70%	0.8
60%	0.7
50%	0.6
40%	0.45
30%	0.35

(*b*) Correction Factors to Allow for Dirt Accumulation on Glass

Locality	*Class of Industry*	*Angle of Slope (Measured to the Horizontal)* 90–75°	60–45°	30–0°
Country or outer-suburban area	Clean	0.9	0.85	0.8
	Dirty	0.7	0.6	0.55
Built-up residential area	Clean	0.8	0.75	0.7
	Dirty	0.6	0.5	0.4
Built-up industrial area	Clean	0.7	0.6	0.55
	Dirty	0.5	0.35	0.25

(*c*) Percentages to Use When Figure 19.8 Curves Are Applied to Periods Other Than 09.00–1700

Curve in Figure 19.8	*95%*	*90%*	*85%*	*80%*	*70%*	*60%*
Alternative period	Percentage of alternative period					
07.00–15.00	95	90	85	80	70	60
08.00–16.00	100	100	95	85	70	60
07.00–17.00	95	85	75	65	55	45
06.00–18.00	75	70	65	60	50	40

Source. From *Daylight, International Recommendations for the Calculation of Natural Daylight,* Publication CIE No. 16 (E-3.2), 1970; Commission Internationale de l'Éclairage.

5. From Figure 19.12, we find that for this room (depth = $3H$), twice the minimum level (20 fc) of illumination will occur at $1.75H$ or 9 ft from the window, and four times the minimum level (40 fc) will occur at $1.1H$ or 6.6 ft from the window. (The reader should plot these figures on a curve shaped as in Figure 19.6 to get a complete picture of daylight in this space.)

The CIE standard[16] also includes an extensive bibliography of daylight calculation methods that can be applied in instances where the above method is inapplicable.

19.6 Daylight Design—American IES Method

(*a*) ***General Remarks.*** The CIE method is in large measure parallel to the IES method for calculating artificial lighting, in that it treats the window or toplight as a large area lighting source—essentially a transilluminated lighting fixture. It then applies coefficients to the light output of this "daylight fixture," similar in usage to the familiar coefficient of utilization and light loss factor of a lighting fixture (see Sections 20.28 to 20.30), and arrives at an interior daylight level. The lighting levels calculated for side lighting are minimum levels, at predetermined points in the

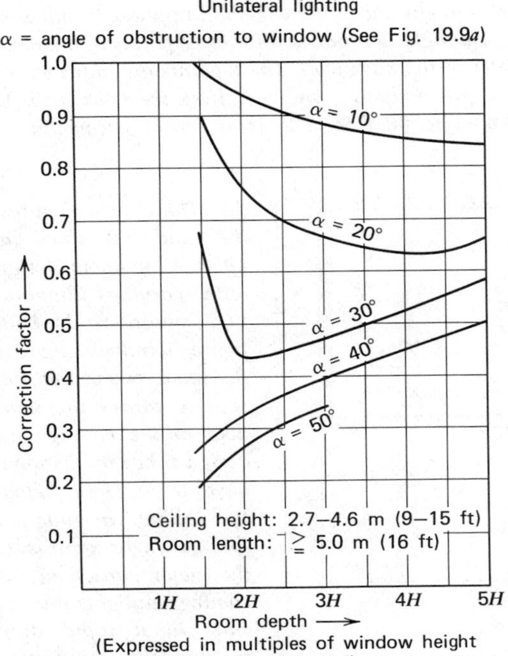

Fig. 19.11 Correlation factors to account for the influence of external obstructions on minimum daylight factor. See Figure 19.9(a) *for definition of obstruction angles.*

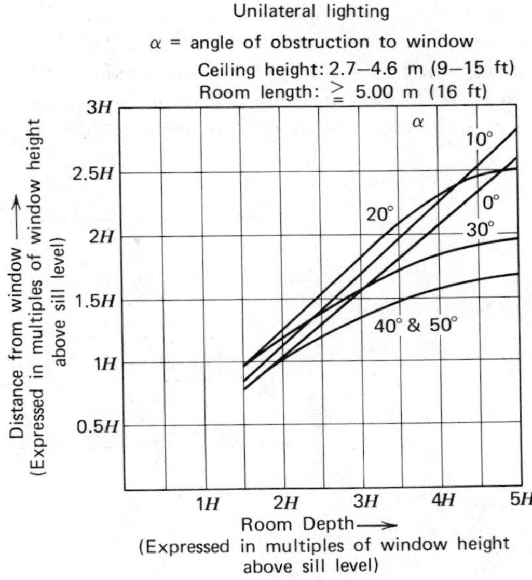

Fig. 19.12 (a) *Distance from window at which daylight factor is twice minimum daylight factor.*

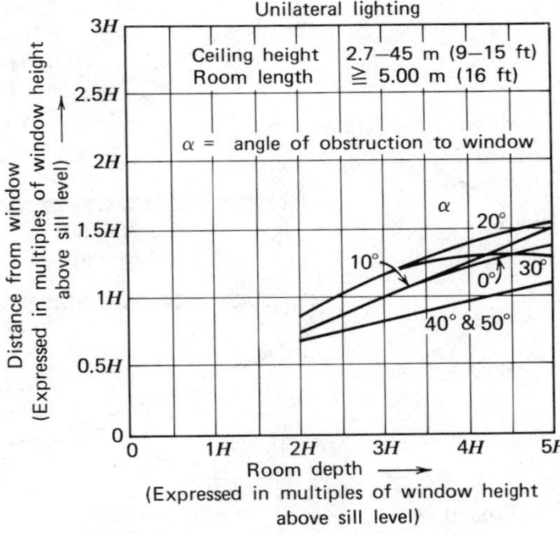

Fig. 19.12 (b) *Distance from window at which daylight factor is four times minimum daylight factor.*

room. For toplighting it is an average overall level, assuming proper spacing of toplights. The method is completely described in the relevant current IES publication.[2]

(b) Characteristics of the Method. Before entering into a detailed description of the technique, and demonstrating it with an illustrative example, a few general remarks are in order. The IES method is probably the most flexible technique available. It recognizes the importance of reflected light (see Fig. 19.13), can accommodate shading devices that will almost always be used on direct sun exposure, can readily account for various types of glazing—blinds, drapes, screens, and so on—is applicable to unilateral and bilateral side lighting plus the usual varieties of toplighting, and is well documented with tabular and graphic data. Nevertheless, it does have, to our mind, several distinct limitations or shortcomings of which the reader should be aware. These are:

1. The method is only usable in one mode, that is, given location and full architectural data, daylighting can be calculated. It cannot be used to determine room proportions, given the other data, as can the CIE method.

Fig. 19.13 (a) *Standard room section for which IES daylight calculation method applies. Window is assumed to extend from a 36-in.-high sill to the ceiling. Daylight is caluclated for three points in the room along a line from the center of the window to the center of the back wall. These points are indicated as Max., Mid., Min. and are 5 ft from the window, at depth midpoint, and 5 ft from the back wall.* D *represents room depth from window to back wall. Room height from 8 to 14 ft can be accomodated.*

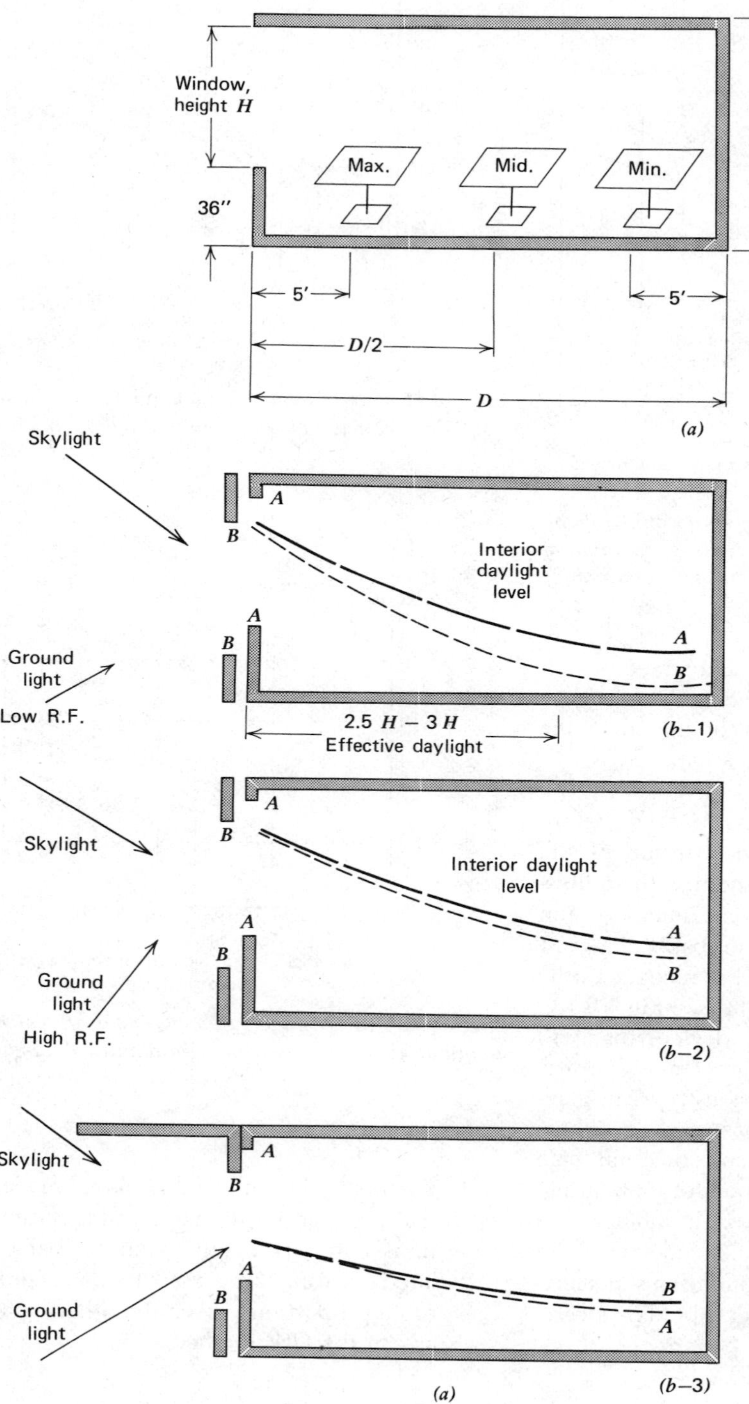

(b) *These two figures, (b-1) and (b-2), show the effect of window design with varying illumination conditions. In both figures window height is the same, except that design* A *carries the window to the ceiling and design* B *has the window head 18 to 24 in. below the ceiling. In diagram (b-1) skylight represents the major portion of incoming daylight; this results in a rapid drop in daylight penetration and a large difference between conditions* A *and* B. *In diagram (b-2) reflected ground light, resulting from a high-reflectance surface outside the window, comprises a large portion (30–50%) of the incoming light. Since ground light is rereflected from the ceiling into the depth of the room, the result is higher daylight levels within the room, particularly at the back, and only a small difference between the interior daylight levels for window designs* A *and* B. *Effective daylight can be expected to a depth of* $2\frac{1}{2}$ *to 3 times window height. Diagram (b-3) represents a design where direct skylight is blocked by an overhang, resulting in interior daylight being principally rereflected ground light. Here a lower sill height* (B) *results in better penetration.*

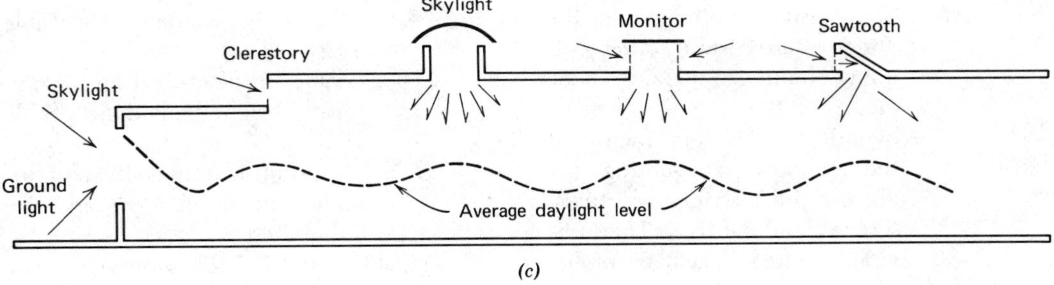

Fig. 19.13 (c) *This figure demonstrates the effect of various toplighting devices. The choice of design depends on prevailing sky conditions. Refer to Reference 2 for recommendations and design detail.*

2. Since it considers the daylight source essentially as a fixture, which necessarily has constant brightness over its entire surface, *a single figure* of equivalent sky brightness must be used. Variable sky luminance, which corresponds to the actual sky condition, seen through the window, cannot readily be used. This introduces a degree of approximation, making the four-place precision of the coefficients in Table 19.6 questionable. Different coefficients, however, are available for *clear* sky than for *uniform* sky, which compensates to an extent for variable sky luminance.

3. A corollary to the above limitation is that obstructions such as buildings and trees are difficult to take into account. This leads to their being ignored, and a further reduction in accuracy.

4. The standard unilaterally side-lighted room on which the calculations are based (Fig. 19.13*a*) calculates minimum lighting for a position 5 ft from the back wall. To our mind the CIE reference point of 2 ft from the back wall is a more useful figure. Similarly this standard room assumes windows extending to the ceiling. This is infrequently the case, and is important because the upper portion of a window contributes to deep daylight penetration much more than an equivalent lower area (see Fig. 19.13*b*). Thus minimum levels calculated with this window configuration are overoptimistic.

(c) ***Calculation Procedure.*** The calculation procedure for side lighting will be discussed, since it is more frequently encountered than toplighting. The reader is referred to Reference 16 for a discussion and presentation of the latter. It should present no difficulties since it is similar to, and simpler than, the side-lighting calculation procedure.

The daylight illumination on the workplane, for the reference point(s) shown in Figure 19.13*a*, can be expressed as the sum of two components: that resulting from sky illumination, which we can call E_S and that resulting from light reflected into the room from the ground, which we call E_G. Thus

$$E_D = E_S + E_G \qquad (19.3)$$

where

E_D = total daylight on the workplane, fc
E_S = workplane daylight resulting from skylight, fc
E_G = workplane daylight resulting from groundlight, fc

These can be calculated as follows:

$$E_S = [E_W \times AF \times TF \times LLF] \times CU \times K \qquad (19.4)$$

where

E_{SG} = daylight illumination on the workplane, for the point selected (max., mid., min.) resulting from skylight, in fc (see Section 19.3)
E_W = daylight illumination incident on the window, from sky, in fc

AF = net transilluminated area of the window after subtracting columns, mullions, glazing bars, etc. in sq ft

TF = overall transmission factor of the window, accounting for glazing plus shades, curtains, screens, and the like. (The individual factors of each are multiplied to obtain the overall factor.) Expressed as a dimensionless decimal. See Table 19.5(*a*)

LLF = light loss factor, expressing the degree of cleanliness of the glazing. Dimensionless decimal. See Table 19.5(*b*)

CU = coefficient of utilization selected for the point being calculated (max., mid., min.), the sky condition (clear, overcast), the room characteristics, and the type of sun control device in use, if any. See Table 19.6

K = second utilization coefficient, selected with the same criteria as CU above. See Table 19.6

The quantity enclosed in square brackets represents the daylight entering the room through the window. Factor CU modifies this entering light in accordance with room *length* and depth. Factor K further accommodates the entering light in accordance with room *height* and depth. Thus it requires two factors to introduce the effect of all room dimensions and proportions.

The second factor in total daylight, E_G, is

$$E_G = \frac{E_{IG} \times RF}{2} \times AF \times TF \times LLF \times CU \times K \quad (19.5)$$

where

E_G = daylight illumination on the workplane, for the point selected (max., mid., min.), resulting from reflected ground light, in fc

E_{IG} = illumination incident on the ground outside the window, in fc. See Section 19.3

RF = reflectance factor of the ground outside the window. See Table 19.7

CU = coefficient of utilization for uniformly bright ground. See Table 19.6

K = second utilization coefficient, for uniformly bright ground. See Table 19.6

AF, TF and LLF are same as above.

Equation 19.5 results from the observed fact that the illumination on a vertical surface (window) caused by adjacent ground reflection is approximately one-half of the ground luminance. Ground luminance is simply the sky brightness it sees, multiplied by its reflection factor. This will become clear in the illustrative examples below.

EXAMPLE 19.2. For the sake of comparison, we will use the building of Example 19.1. Refer to page 971 for the data. Find E_{max}, E_{mid}, and E_{min} for a spring day and a winter day, at 10 A.M. (2 P.M.). Assume the sky to be overcast so that a direct comparison with Example 19.1 can be made.

SOLUTION (see eq. 19.4).

A. *Sky Contribution* E_D

1. E_W: From Table 19.2(*f*) we find solar altitude to be

   March 21—36°
   Dec. 21—14°

   From Figure 19.3*a*, vertical surface illumination E_W from overcast sky is

   March 21—500 fc
   Dec. 21—200 fc

   (Note that approximately the same figures can be derived from Table 19.2(*a*) by applying a factor of 50% for conversion to a vertical surface.)

2. *AF:*

   window area = 25 ft long × (10 − 3) ft high = 175 sq ft
   net area = 0.8 × (175) = 140 sq ft

3. *TF*= 0.8 (given)
4. *LLF* = 0.9 (given)

Table 19.5 ***Transmission Data and Light Loss Factors for Windows—IES Method***

(*a*) Transmittance Data of Glass and Plastic Materials

Material	*Transmittance (In Percent)*
Polished plate window glass	80–90
Sheet drawn window glass	85–91
Heat-absorbing plate glass	70–80
Heat-absorbing drawn sheet glass	70–85
Neutral low-transmission glass	10–60
Configurated glass	70–90
Corrugated glass	80–85
Glass block	60–80
Clear plastic sheet	80–92
Neutral low-transmission plastic	9–42
Colorless patterned plastic	80–90
White translucent plastic	10–80
Glass fiber reinforced plastics	5–80

Source. From *Recommended Practice of Daylighting,* IES, 1962.

(*b*) Average Window Maintenance Factors Expressed as Percentage of Clean Glass Transmission

	Window Position				
	Office[a]	*Factory*[b]			
	Vertical	*Vertical*	*30° from Vertical*	*60° from Vertical*	*Horizontal*
Avg. value over 6-month period	83%	71%	65%	58%	54%
Value end of 3-month period	82%	69%	62%	54%	50%
Value end of 6-month period	73%	55%	45%	39%	34%

Source. From *Recommended Practice of Daylighting,* IES, 1962.

[a]Typical clean location. [b]Typical dirty location.

5. *CU:* From Table 19.6(*a*)

$$CU_{max} = 0.021 \text{ (by interpolation)}$$
$$CU_{mid} = 0.0117 \text{ (by interpolation)}$$
$$CU_{min} = 0.0067 \text{ (by interpolation)}$$

6. *K:* From Table 19.6(*a*)

$$K_{max} = 0.121$$
$$K_{mid} = 0.107$$
$$K_{min} = 0.0951$$

We can now calculate the three levels of illumination for March 21:

$$E_s\text{max} = 500 \times 174 \times 0.8 \times 0.9 \times 0.021 \times 0.121 = 160 \text{ fc}$$

$$E_s\text{mid} = 500 \times 175 \times 0.8 \times 0.9 \times 0.0117 \times 0.107 = 79 \text{ fc}$$

$$E_s\text{min} = 500 \times 175 \times 0.8 \times 0.9 \times 0.0067 \times 0.0951 = 40 \text{ fc}$$

The corresponding figures for Dec. 21 are 64 fc, 32 fc, and 16 fc.

B. Ground Light Contribution (eq. 19.5)

1. E_{IG}: From Figure 19.3*a*, horizontal surface illumination from overcast sky is

March 21—(36° solar alt.)—1200 fc
Dec. 21—(14° solar alt.)—500 fc

Table 19.6 ***Coefficients of Utilization and K Factors***

DATA: 75% Ceiling Reflectance[a]; 30% Floor Reflectance, No Window Controls

(*a*) Overcast Sky

CU							
Length		*20 ft*		*30 ft*		*40 ft*	
Wall Reflectance		70%	30%	70%	30%	70%	30%
Depth (ft)							
	20	.0248	.0226	.0172	.0156	.0129	.0121
Max	30	.0245	.0223	.0169	.0155	.0123	.0118
	40	.0242	.0221	.0164	.0154	.0120	.0117
	20	.0143	.0105	.0091	.0078	.0073	.0064
Mid	30	.0052	.0045	.0049	.0036	.0031	.0030
	40	.0035	.0024	.0027	.0021	.0020	.0017
	20	.0078	.0048	.0057	.0039	.0045	.0033
Min	30	.0029	.0017	.0026	.0015	.0018	.0013
	40	.0017	.0008	.0014	.0008	.0011	.0007

K									
Ceiling Ht.		*8 ft*		*10 ft*		*12 ft*		*14 ft*	
Wall Reflectance		70%	30%	70%	30%	70%	30%	70%	30%
Width (ft)									
	20	.125	.129	.121	.123	.111	.111	.0991	.0973
Max	30	.122	.131	.122	.121	.111	.111	.0945	.0973
	40	.145	.133	.131	.126	.111	.111	.0973	.0982
	20	.0908	.0982	.107	.115	.111	.111	.105	.122
Mid	30	.156	.102	.0939	.113	.111	.111	.121	.134
	40	.106	.0948	.123	.107	.111	.111	.135	.127
	20	.0908	.102	.0951	.114	.111	.111	.118	.134
Min	30	.0924	.119	.101	.114	.111	.111	.125	.126
	40	.111	.0926	.125	.109	.111	.111	.133	.130

(*b*) Clear Sky

CU							
Length		*20 ft*		*30 ft*		*40 ft*	
Wall Reflectance		70%	30%	70%	30%	70%	30%
Depth (ft)							
	20	.0185	.0156	.0129	.0111	.0099	.0088
Max	30	.0183	.0156	.0123	.0108	.0088	.0083
	40	.0180	.0151	.0118	.0107	.0086	.0082

K									
Ceiling Ht.		*8 ft*		*10 ft*		*12 ft*		*14 ft*	
Wall Reflectance		70%	30%	70%	30%	70%	30%	70%	30%
Width (ft)									
	20	.145	.155	.129	.132	.111	.111	.101	.0982
Max	30	.141	.149	.125	.130	.111	.111	.0954	.101
	40	.157	.157	.135	.134	.111	.111	.0964	.0991

Mid	20	.0138	.0094	.0090	.0071	.0075	.0060	Mid	20	.110	.128	.116	.126	.111	.111	.103	.108
	30	.0074	.0049	.0056	.0039	.0041	.0033		30	.106	.125	.110	.129	.111	.111	.112	.120
	40	.0047	.0029	.0036	.0025	.0026	.0021		40	.117	.118	.122	.118	.111	.111	.123	.122
Min	20	.0095	.0054	.0071	.0044	.0060	.0039	Min	20	.105	.129	.112	.130	.111	.111	.111	.116
	30	.0049	.0025	.0042	.0021	.0029	.0019		30	.0994	.144	.107	.126	.111	.111	.107	.124
	40	.0028	.0013	.0024	.0012	.0019	.0011		40	.119	.116	.130	.118	.111	.111	.120	.118

(c) Uniform Brightness Ground

CU								K									
Length		*20 ft*		*30 ft*		*40 ft*		*Ceiling Ht.*		*8 ft*		*10 ft*		*12 ft*		*14 ft*	
Wall Reflectance		*70%*	*30%*	*70%*	*30%*	*70%*	*30%*	*Wall Reflectance*		*70%*	*30%*	*70%*	*30%*	*70%*	*30%*	*70%*	*30%*
Depth (ft)								Width (ft)									
Max	20	.0132	.0101	.0092	.0079	.0073	.0064	Max	20	.124	.206	.140	.135	.111	.111	.0909	.0859
	30	.0127	.0101	.0088	.0079	.0069	.0063		30	.182	.188	.140	.143	.111	.111	.0918	.0878
	40	.0123	.0101	.0084	.0077	.0065	.0062		40	.124	.182	.140	.142	.111	.111	.0936	.0879
Mid	20	.0115	.0081	.0085	.0064	.0066	.0054	Mid	20	.123	.145	.122	.129	.111	.111	.100	.0945
	30	.0075	.0051	.0056	.0043	.0045	.0037		30	.0966	.104	.107	.112	.111	.111	.110	.105
	40	.0050	.0033	.0040	.0030	.0038	.0023		40	.0790	.0786	.0999	.106	.111	.111	.118	.118
Min	20	.0095	.0064	.0074	.0049	.0060	.0040	Min	20	.0994	.108	.110	.114	.111	.111	.107	.104
	30	.0046	.0023	.0037	.0021	.0030	.0019		30	.0816	.0822	.0984	.105	.111	.111	.121	.116
	40	.0026	.0016	.0023	.0011	.0020	.0010		40	.0700	.0656	.0946	.0986	.111	.111	.125	.132

[a]To convert CU figures for a ceiling reflectance of 80%, divide by 0.9. *K* figures do not change.

[b]Information in this table was supplied by Professor B. Evans of V.P.I. & S.U., and is reproduced with his permission.

Table 19.7 ***Reflectances of Building Materials and Outside Surfaces***

Material	Reflectance (In Percent)	Material	Reflectance (In Percent)
Bluestone, sandstone	18	Asphalt (free from dirt)	7
Brick		Earth (moist cultivated)	7
light buff	48		
dark buff	40		
dark red glazed	30	Granolite pavement	17
Cement	27	Grass (dark green)	6
Concrete	55	Gravel	13
Granite	40	Macadam	18
Marble (white)	45	Slate (dark clay)	8
Paint (white)		Snow	
new	75	new	74
old	55	old	64
		Vegetation (mean)	25

Source. From *Recommended Practice of Daylighting,* IES, 1962.

Note correspondence with figures in Table 19.2(*a*)

2. *RF:* Assume vegetation immediately outside the window. From Table 19.7, $RF = 0.25$
3. *TF, LLF,* and *K* are the same as in *A* above.
4. *CU:* From Table 19.6(*c*)

$$CU_{max} = 0.0112 \text{ (by interpolation)}$$
$$CU_{mid} = 0.0100 \text{ (by interpolation)}$$
$$CU_{min} = 0.0085 \text{ (by interpolation)}$$

For March 21

$$E_{G\max} = \frac{1200 \times 0.25 \times 175 \times 0.8 \times 0.9 \times 0.0112 \times 0.121}{2} = 26 \text{ fc}$$

$$E_{G\text{mid}} = \frac{1200 \times 0.25 \times 175 \times 0.8 \times 0.9 \times 0.0100 \times 0.107}{2} = 20 \text{ fc}$$

$$E_{G\min} = \frac{1200 \times 0.25 \times 175 \times 0.9 \times 0.9 \times 0.0085 \times 0.0951}{2} = 15 \text{ fc}$$

The corresponding figures for Dec. 21 are 11 fc, 8 fc, and 6 fc.

C. Total Daylight

March 21

$$E_{D\max} = 160 + 26 = 186 \text{ fc}$$
$$E_{D\text{mid}} = 79 + 20 = 99 \text{ fc}$$
$$E_{D\min} = 40 + 15 = 55 \text{ fc}$$

Dec 21

$$E_{D\max} = 64 + 11 = 75 \text{ fc}$$
$$E_{D\text{mid}} = 32 + 8 = 40 \text{ fc}$$
$$E_{D\min} = 16 + 6 = 22 \text{ fc}$$

The reader is advised to compare the above method and results with the CIE method in Example 19.1. It should be immediately apparent that the IES method is more time-consuming but gives much more information as a result.

It is our recommendation that each daylighting problem should be worked through assuming three different sky conditions:

1. Completely overcast sky. These results will determine the maximum amount of artificial lighting (PSALI) to be designed.
2. Equivalent sky luminance. These results will assist in determining room propor-

tions, window size and type, average HVAC loads, and yearly energy consumption.
3. Clear sky with sun. These results will indicate need for shading and shielding devices, and will give maximum HVAC loads.

Spaces lighted with multiple sources such as bilateral side lighting, windows and toplighted, or daylight plus artificial light are calculated by superposition. In the case of bilateral lighting the description of the three calculation points as maximum, midpoint, and minimum obviously loses significance.

The reader is referred to Reference 2 for a full discussion of the many aspects of daylighting, which we could not cover here such as toplighting, calculation with overhangs, daylight color, sun controls, and similar topics.

19.7 Factors in Interior Daylighting

(*a*) ***Horizontal and Vertical Surfaces.*** Since the sky component of daylight enters side fenestration at an angle, it can be resolved into horizontal and vertical components, as shown in Figure 19.14. The vertical component that illuminates horizontal surfaces is proportional to the sine of the angle of incidence, and the horizontal component that illuminates vertical surface is proportional to the cosine of this angle. Therefore, for horizontal tasks, windows should be *as high as possible* and, for vertical tasks, as low as possible. Since most tasks are horizontal, tall narrow windows will give better, deeper, penetration than short wide windows of the same area.

(*b*) ***Window Details.*** The effect of window construction on total fenestration area reduction is often neglected. Even windows with narrow mullions and light metal frames have 8 to 10% obscuration; heavy window supports and small glass lights can result in 12 to 15% obscuration with proportional daylight reduction. Further obscuration readily results from dirt accumulation, wired glass, and mechanical system items such as pipes and ducts inside the room, adjacent to windows.

(*c*) ***Surface Reflectances.*** Interior reflectances are very important in daylight design. In addition to determining the magnitude of the internally reflected light component (IRC) within the room, they determine in large measure the eye adaptation level. A high adaptation level is desirable to avoid a sensation of glare when the window and its immediate surround are in the field of vision. Furthermore, the internally reflected light component contributes largely to the diffuseness of the room light. With low IRC the sky component

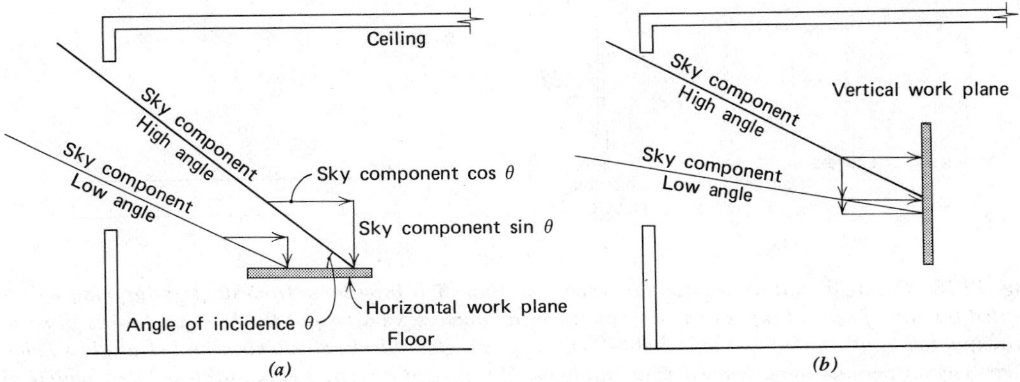

Fig. 19.14 *Effect of daylight incidence angle on illumination components. High-angle skylight is more effective for illumination on the horizontal plane* (a) *whereas low-angle skylight is more effective on vertical surfaces* (b).

of daylight is the essential illuminant, and diffuseness and room penetration are reduced. Floors receive the sky component directly and should have at least 20% reflectance. Ceilings receive the ground component and should have at least 70% reflectance. Walls receive rereflected light. Since they are the surfaces seen at normal vision angles, they are responsible for eye adaptation levels and should have at least 50% reflectance.

Exterior surface reflection can provide the deep daylight penetration that is required for effective daylighting. Thus a concrete or light-painted ground surface (RF of 50 to 70%) will furnish one-quarter to two-thirds of the light incident on a window, depending on shading and orientation. When combined with a high-reflectance ceiling, optimal interior distribution is achieved.

(*d*) ***Glare and Heat Control.*** These are among the most difficult problems to overcome in daylighting. The following recommendations will be helpful:

1. As stated above, provide high-reflectance surfaces, particularly toward the back of the room where daylight factor is low.
2. Building orientation is the factor that determines which areas are exposed to natural glare and heat extremes. Bear these facts in mind:
   a. Southern exposure receives maximum overall daylight but without the extremes associated with low sun angles, except at high latitudes.
   b. Northern exposure receives minimum daylight and no direct sun. As a result levels are low but relatively constant throughout the day. (An associated advantage is color constancy of approximately 8000 K for clear sky and 5500 K for overcast sky. This compares with approximately 4500 K for direct sunlight).
   c. East and west exposures receive extremes of light and heat because of low sun angles in early morning and

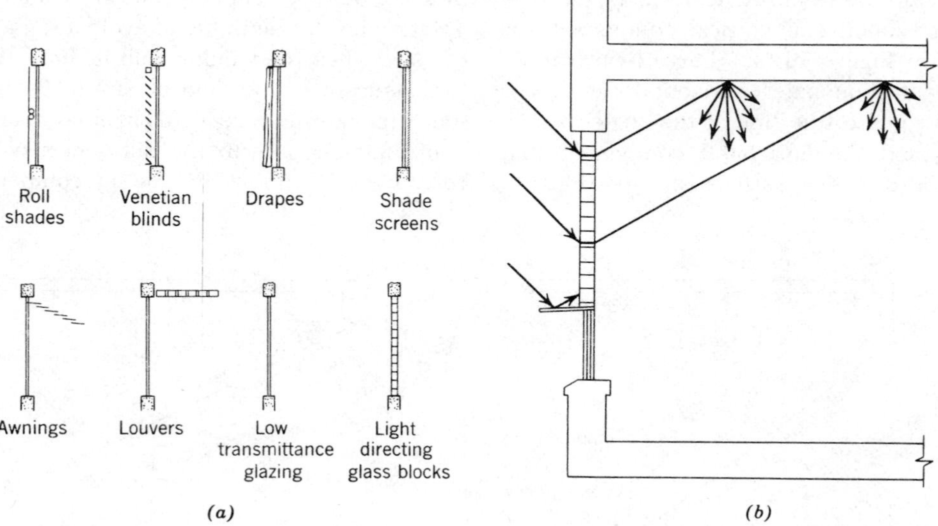

Fig. 19.15 *Daylight control devices.* (a) *Uncomfortably high brightness from daylighting may be prevented by any of several types of fenestration control devices. Courtesy of the IES.* (b) *Arrows illustrate principal light paths into room with light-directing prism glass blocks above visor and clear glass below. Clear section permits unobstructed view outdoors. When light-directing glass or glass block panels are used, they should begin approximately 6 ft above floor level. They are particularly effective in low-ceiling rooms, since the lower band of blocks near the 6-ft sill delivers most of the light to the deeper portions of the room. Courtesy of the IES.*

late afternoon. Heat buildup on west exposures in winter can be almost as severe as summer because of the low sun angle. Furthermore, the low sun angle in winter makes the sun's relative motion appear extremely rapid, necessitating either a fixed sun control device, or one that is easily and rapidly adjustable.

3. Provide fixed sunshades on sun exposures at low latitudes; and operable sun control devices on sun exposures at all latitudes. A few of the common types are shown in Figure 19.15*a*. Vertical devices are effective for low sun angles; horizontal at high sun angles. The latter are also useful to reflect incoming skylight onto the ceiling.
4. Translucent, limited-brightness, glass or plastic fenestration, including light-directing glass block just below the ceiling line and above the vision panels (clear windows), provides maximum penetration and minimum glare (see Fig. 19.15*b*).
5. Tinted windows and heat-reflective films are not usually desirable, except in retrofit installations, because they affect the quality of daylight. (The day/night appearance of the structure is also affected. During daylight hours vision out is possible and vision in is blocked. The reverse is true at night.)
6. Orient furniture so that daylight comes from the left side or the rear of the line of sight. Never face a window except one with northern exposure and no exterior glare sources in the line of sight.
7. Sunlight reflection from adjacent structures can be a source of intense glare and heat. Orientation, vegetation, and fixed shading are possible solutions to this problem.

Some of the principles involved in the use of daylight in construction are demonstrated in Figure 19.16.

19.8 Daylight and Energy Considerations

It must be assumed for the purposes of this book that the architect/lighting designer is intimately familiar with the direct relationships among daylight, sunlight, heat loss and gain, building siting, window aspect, sunshading devices, and so on. A discussion of these would take us far afield into architecture, and this extensive study is well covered in other sources. The following are specialized

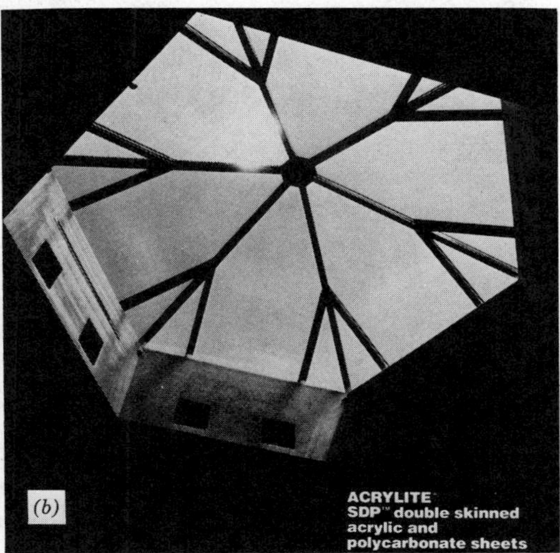

Fig. 19.16 (a) *When oriented East or West, expanses of glass must be provided with sun control devices. This unshielded, west-facing, 15-ft, glass-door facade creates intolerable disabling glare, the more so for the dimly lit interior corridor and associated low eye-adaptation level. Photo by Stein.* (b) *It is desirable from an architectural viewpoint to have the ceiling skylight area also incorporate artificial light sources, to be used when daylighting is insufficient. These alternate sources can be automatically modulated by the available daylight. Photo courtesy of Cy/Ro Industries.*

Fig. 19.16 (c) *This shopping mall uses suspended, acrylic-enclosed mercury luminaires that blend with the large skylights. They illuminate the skylight well with light of similar color to daylight, giving the ceiling area a daylight appearance at all times. Note too the stepped ceiling construction, which serves to reduce brightness gradually. (Maximum lighting power in this installation is 2 w/sq ft). Photo Courtesy of* LD&A Magazine, *from article by J. Wilson, May 1974.*

(d) *Daylight factor here has a large ground component due to cutoff of the sky component by deep overhang, which also acts as a reflector for the ground light. Low-reflectance surfaces are suitable for a corridor that has no demanding visual task. Photo courtesy of Libby-Owens-Ford.*

considerations that will be of value for their own sake and in making other related decisions.

1. Wall and ceiling designs that admit daylight without excessive attendant heat gain and direct glare are most desirable. These rely principally on utilizing reflected daylight (see Fig. 19.16g).

(f) *(opposite) Unusual application of skylighting provides daylighting to an inner court of the International Monetary Fund Headquarters, Washington D.C. Translucent acrylic panels give sun control while limiting heating and glare. Photo courtesy of Roper IBG.*

Fig. 19.16 (e) *Full-height clear glazing with negligible direct sun shading is only applicable on North elevations or climates with continual overcast sky, as in this Pacific Northwest installation. Note that work positions are not placed facing the glass, to avoid excessive brightness ratios in the field of view. Exterior planting reduces any incipient glare. Photo Courtesy of Libby-Owens-Ford.*

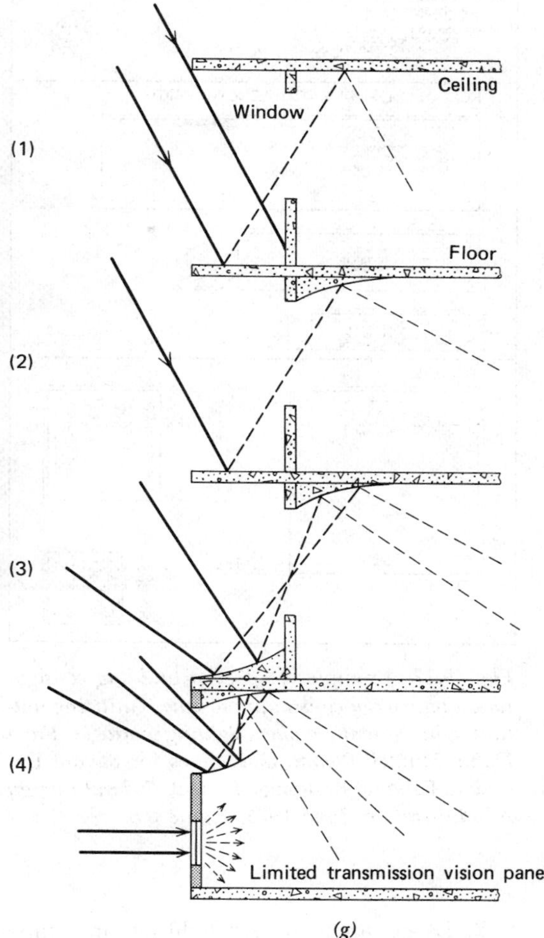

(g)

(g) *Various window treatments that permit reflected daylight to enter the room but block direct sunlight.*

(1) *Simple overhang acts as shield and reflector. Penetration is poor.*
(2) *Addition of high-reflectance, curved section at ceiling increases room penetration.*
(3) *Using a curved collector and reflector increased daylight factor deep into the room.*
(4) *Reflector can be placed inside the building wall, using building wall for sun shield. Daylight penetration can be very deep, since sky factor angle is large.*

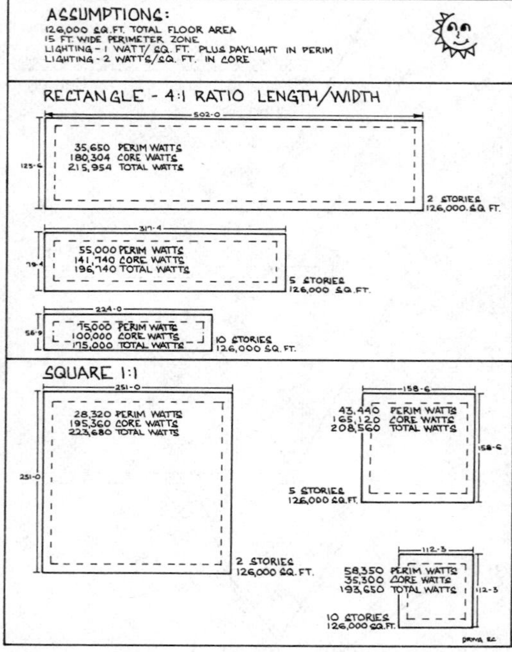

Fig. 19.17 *Example of effect of building configuration on energy consumption, where utilizing natural light. Note the modest lighting wattages. From Dubin-Mindell-Bloom,* Guidelines for Saving Energy in Existing Buildings, *ECM-2, Federal Energy Administration, June 1975, Figure 9.5.*

2. Large areas of sidelighting and toplighting can seriously affect heat transfer effect for the space. To minimize these effects while maintaining the advantage of incident light, double and triple glazing can be used, and operable thermal barriers can be added during nondaylight hours (see Fig. 3.12, page 69).
3. Since daylight penetration from side fenestration is limited and artificial lighting must be provided for areas with an insufficient daylight factor, it is apparent that building configuration affects the proportion of daylight use and hence the overall energy consumption. This is shown graphically in Figure 19.17.
4. The interrelationships between heat gains and losses due to fenestration, lighting energy economies due to daylight, and heating and cooling energy changes due to reduced lighting are extremely complex, and the use of one of the available energy consumption computer programs is strongly recommended. The results of one such study performed for a simulated 20-story, 312,000-sq ft building in St. Louis, Missouri, are shown in Table 19.8. The three cases shown represent:

Case 1. No allowance made for daylight; that is, electric lighting operated continuously during working hours. Blinds were operated any time solar heat gain exceeded 70 Btu/sq ft.

Case 2. Perimeter lighting was reduced in response to daylight as follows.

Natural Lighting Contribution	*Amount of Scheduled Lighting Turned Off*
Under 20 fc	None
Over 20 fc	1/3
Over 30 fc	2/3
Over 50 fc	All

Case 3. Windows were equipped with modulating shading devices that limited the daylight contribution to 50 fc. Perimeter lighting responded as in case 2. Shading device was given a maximum shading coefficient of 0.6.

The figures in Table 19.8 show a very significant (25%) reduction in lighting energy between cases 1 and 2; despite the increase in the space-heating energy requirement due to the loss of lighting heat, the building demonstrates a total annual energy reduction of 490,000 kwh or 6.6% of the *total* building energy input. No further overall energy savings is accomplished with the automated system of case 3. In terms of energy resources, the savings are much greater than shown. Lighting energy is electrical, with an overall fuel-to-usable energy efficiency of 20 to 30%, whereas heating energy is direct fuel combustion with an overall efficiency of 40 to 70%. Thus a 25% savings in lighting energy is

Table 19.8 ***Energy Trade-Offs by Using Daylighting in Office Buildings (St. Louis, Missouri, 3 w/sq ft Lighting)***

	Case No. 1	*Case No. 2*	*Case No. 3*
	No Daylight Allowance Artificial Light Only	*Partial Utilization of Daylight*	*Optimum Use of Daylight*
Space Heating:			
Therms	98,191	103,600	104,865
kwh (3413 Btu/kwhr)	2,877,000	3,035,500	3,072,500
Building Lighting, kwhr	2,392,077	1,805,801	1,788,851
	−25% →		
Chiller, kwh	699,094	644,508	630,830
Fans and Auxiliary, kwh	1,451,752	1,448,576	1,447,720
	−6.6% →		
TOTALS:			
kwh/yr	7.42×10^6	6.93×10^6	6.94×10^6
kwh/(yr)(sq ft)	23.74	22.17	22.21

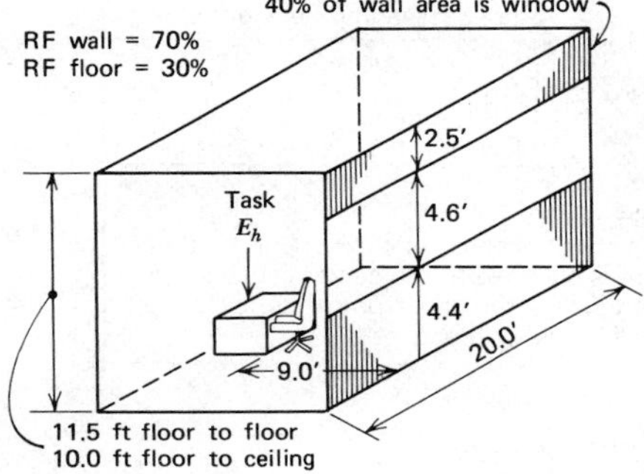

Source. From Ross and Baruzzini, *Energy Conservation Applied to Office Lighting,* Federal Energy Administration, April 1975, Section III.

NOTE. *Ku* (average, for clear sky, overcast sky, and ground reflectance) = 0.00121.

equivalent to approximately 50% in terms of energy resources.

It must be emphasized that, although the figures presented are relevant to a specific model, the model is sufficiently accurate to allow valid generalized conclusions to be drawn. Other studies, both theoretical[12] and actual,[13] support the figures presented, within the framework of their own situation.

A number of manufacturers have made available results of extensive studies that show the effect of daylight utilization on building energy usage for various parts of the United States. All of these studies draw the same conclusion that careful, intelligent design can result in good daylight penetration, avoidance of glare and undesirable heat gain, and very significant energy reductions that rapidly repay the additional construction costs involved.

INCANDESCENT LAMPS

19.9 The Incandescent Filament Lamp

(*a*) ***Construction.*** This lamp consists simply of a tungsten filament inside a gas-filled, sealed glass envelope (see Fig. 19.18). Current passing through the high-resistance filament heats it to incandescence, producing light. Gradual evaporation of the filament causes the familiar blackening of the bulbs and eventual filament rupture and lamp failure. Incandes-

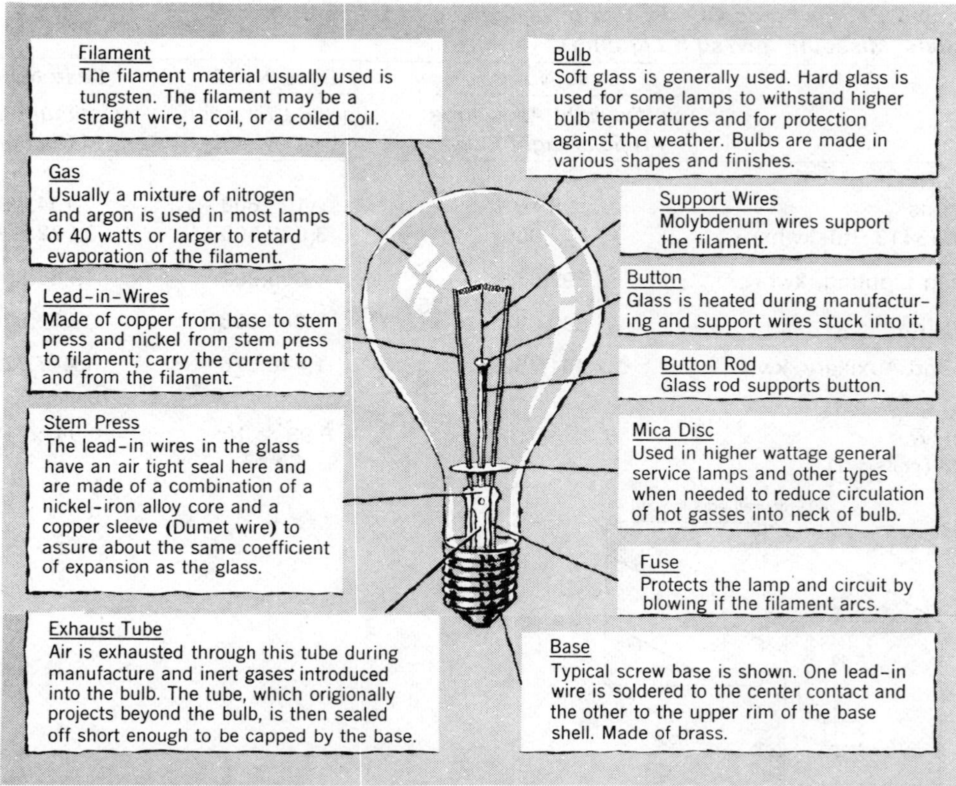

Fig. 19.18 Typical general service incandescent lamp construction.

cent lamps are available in many bulb and base types and special designs for particular application (see Figs. 19.19 and 19.20). To diffuse the light, most bulbs are either etched on the inside (inside-frosted) or are coated inside with white silica. The silica coating provides almost complete light diffusion at a cost of approximately 2 to 3% of the light output, whereas inside-frosted bulbs provide only partial diffusion but do not reduce light output. Inside-frosted bulbs are normally supplied for general service use unless other types are specified. Colored light is also readily available from either coated bulbs or bulbs of colored glass.

The lamp base is the means by which connection is made to the socket and thereby to the source of electric current. Most lamps are made with screw bases of various sizes, the most common being the medium screw base. General service lamps, of 300 w and larger, use the mogul screw base. Where exact positioning of the filament is important, as it is when lamps are placed in precise reflectors or in lens systems, a screw base cannot be used. Lamps designed for such use are furnished with one of the special bases illustrated in Figure 19.19.

(*b*) ***Operating Characteristics.*** These are critically dependent on the voltage at the lamps; therefore the life, output, and efficiency of a lamp can be markedly altered by even a small change in operating voltage, as illustrated by Figure 19.21.

For example, burning a 120-v lamp at 125 v (104.2%) means approximately:

16% more light (lumens).

7% more power consumption (watts).

8% higher efficacy (lumens per watt).

42% less life (hours).

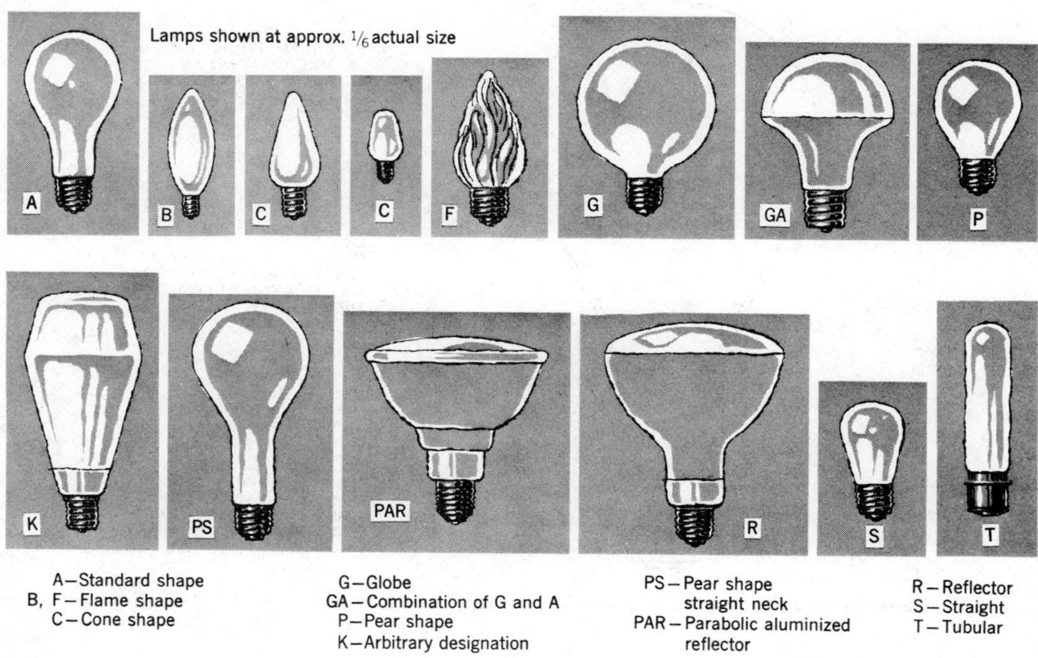

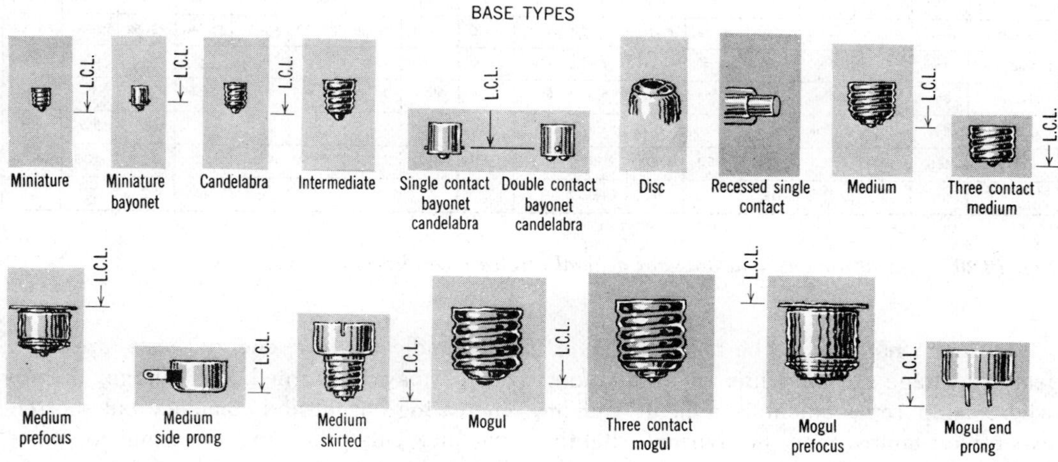

Fig. 19.19 *Incandescent lamp bulb and base types with nomenclature. The bulb nomenclature indicates type and size; the letter being an abbreviation of the shape and the number equal to the maximum diameter in eighths of an inch. Thus a PS-52 is a Pear-Shaped bulb, 52/8 (6½) in. in diameter and an R-40 is a Reflector lamp 40/8 (5) in. in diameter.*

Burning a 120-v lamp at 115 v (95.8%) means approximately:

15% less light (lumens).

7% less power consumption (watts).

8% lower efficacy (lumens per watt).

72% more life (hours).

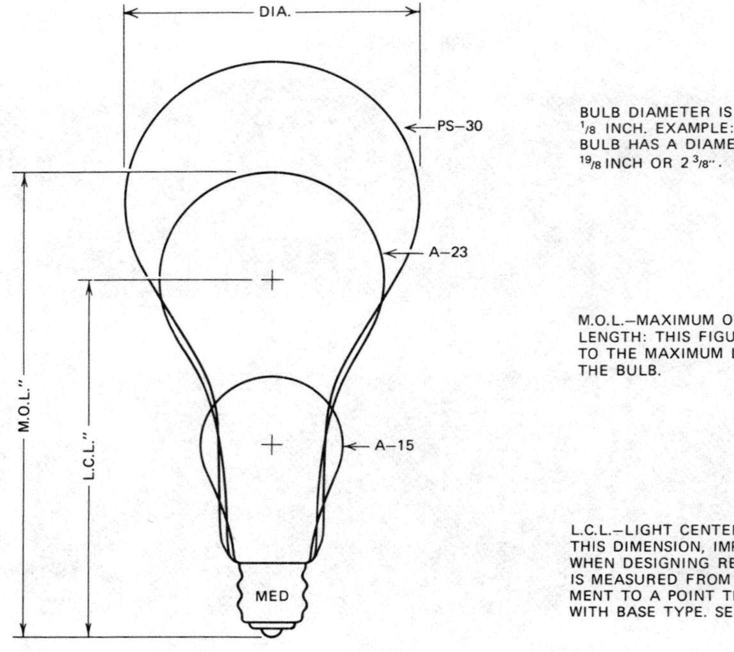

	A – STANDARD SHAPE									PS – PEAR SHAPE							
WATTS	15	25	40	60	75	100	100	150	150	150	200	300	300	500	750	1000	1500
BULB	A–15	A–19_1	A–19_2	A–19_3	A–19_3	A–19_3	A–21_1	A–21_2	A–23	PS–25	PS–30	PS–30	PS–35	PS–40	PS–52	PS–52	PS–52
DIAMETER"	1 7/8	2 3/8	2 3/8	2 3/8	2 3/8	2 3/8	2 5/8	2 5/8	2 7/8	3 1/8	3 3/4	3 3/4	4 3/8	5	6 1/2	6 1/2	6 1/2
M.O.L."	3 1/2	3 7/8	4 1/4	4 7/16	4 7/16	4 7/16	5 1/4	5 1/2	6 3/16	6 15/16	8 1/16	8 1/16	9 3/8	9 3/4	13	13	13
L.C.L."	2 3/8	2 1/2	2 15/16	3 1/8	3 1/8	3 1/8	3 7/8	4	4 5/8	5 3/16	6	6	7	7	9 1/2	9 1/2	9 1/2
BASE	MED	MED	MED	MED	MED	MED	MED	MED	MED	MED	MED	MED	MOG	MOG	MOG	MOG	MOG
STANDARD FINISH	IF	IF	IF W	IF	IF	IF	IF	IF	CL IF	CL IF	IF	IF	IF	IF	CL IF	CL IF	CL IF

CL – CLEAR IF – INSIDE FROSTED

Fig. 19.20 Typical dimensional data for general service incandescent lamps.

Particular note should be taken of the effect of voltage on lamp life. In installations where lamp replacement is difficult and/or expensive, lamps may be burned slightly under voltage and life prolonged, thereby decreasing the frequency of replacement. However, since efficiency is decreased by this procedure and since energy cost is normally a major cost in any lighting installation over the life of the installation, a detailed cost analysis should be made by the consulting engineer involved. Conversely, where lamps are replaced before burnout on a group replacement system and initial installation cost per footcandle and/or energy costs are high, lamps may be burned overvoltage, thereby increasing output and efficiency but shortening life. This procedure is normal in sports-lighting installations because of the high cost of tower-mounted floodlights, making it mandatory to extract the maximum light from each unit. In stadium installations that have yearly burning schedules averaging less than 200 hours, 10% overvoltage operation doubles the light output but still allows a once-a-year, off-season relamping and is therefore a highly economical procedure.

In general, however, it is advisable to oper-

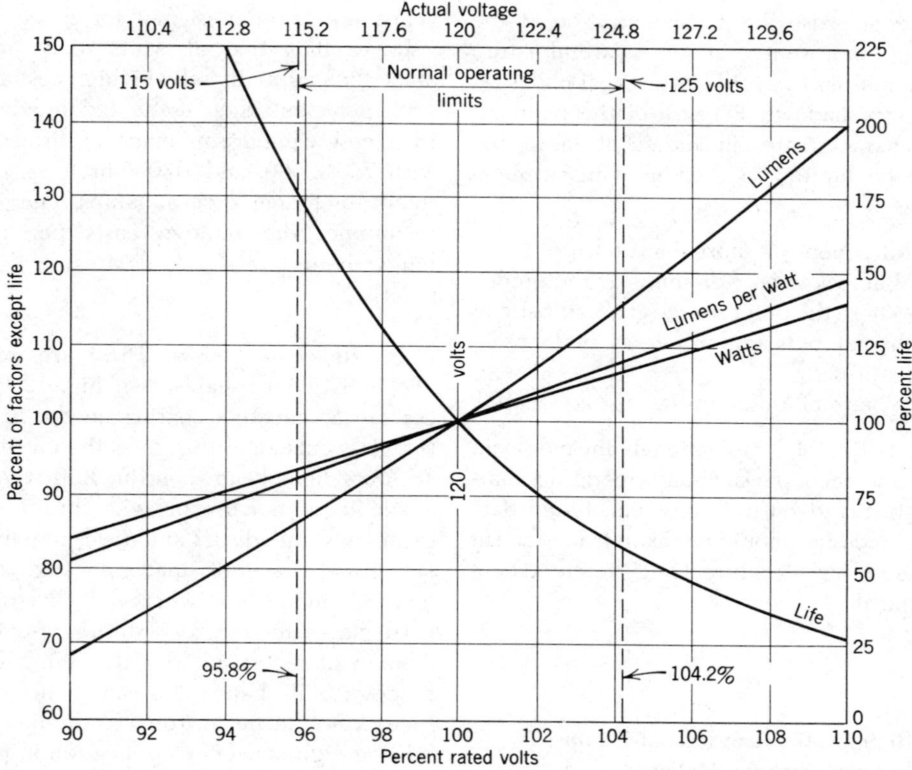

Fig. 19.21 Operating characteristics of a standard 120-v, general-service incandescent lamp.

ate incandescent lamps at rated voltage, accepting balanced efficiency, output, and life.

(c) ***Other Characteristics:***

1. *Lumen maintenance.* Light output decreases slowly with lamp life as the bulb blackens. Position during burning and bulb temperature affect this characteristic.
2. *Color.* White, with large, yellow-red component and therefore highly flattering to the skin. Since color depends on temperature, high-wattage lamps are bluer, low-wattage lamps are yellower. Dimmed lamps give yellow-red light.
3. *Surroundings.* Generally impervious to external heat, cold, or humidity. Starting completely unaffected.
4. *Lamp efficiency.* Since incandescent lamps produce light as a by-product of heat, they are inherently inefficient. Efficiency increases with wattage, varying from 8% for a 25-w lamp to 13% for a 1000-w unit. This increased efficiency can also be noted from the lumen output figures in Table 19.9. Thus a 100-w lamp produces the same 1750 lm as two 60-w lamps, representing a 20% increase in efficacy.

(*d*) ***Summary.*** The principal advantages of incandescent lamps are low cost, instant start and restart, simple inexpensive dimming, simple compact installation requiring no accessories, cheap fixtures, focusable as a point source, high power factor, life independent of number of starts, and good color.

The principal disadvantages are low efficacy (see below), short lamp life, and critical voltage sensitivity. Low efficacy results in a large number of fixtures, high maintenance costs, and large heat gain. Short lamp life results in high replacement labor cost. Voltage sensitivity requires careful and expensive circuit design. Also, light concentration at the filament (point source) requires careful fixture design in order to avoid glare and, if undesirable, sharp shadows. Because of the poor energy characteristic, incandescent lamp use should be limited to the following applications.

1. Infrequent or short duration use.
2. Where low-cost dimming is required.
3. Where the point source characteristic of the lamp is important, as in focusing fixtures.
4. Where minimum initial cost is essential.

A brief list of conventional incandescent lamps and their physical and operating characteristics is given in Table 19.9. Lamp data for use in design should be taken from current manufacturers' literature. Data presented here are typical.

19.10 Special Incandescent Lamps, Including Tungsten-Halogen

In the field of incandescent lamps other than the tungsten-halogen lamp, which will be discussed separately, numerous special types are available. Some of the more important types are covered briefly in the following pages.

Rough service and *vibration* lamps are built to withstand rough handling and continuous vibration, respectively, both of which conditions are extremely hard on general service lamp filaments. Neither type is intended for general use, and both types have lower luminous efficacy than a general service lamp (see Table 19.9).

Extended service lamps are designed for 2500-hr life and are useful, as mentioned, in locations where maintenance is irregular and/or relamping is difficult. The lamp is really designed for slightly higher voltage than that at which it is applied, and therefore efficacy is reduced. See Table 19.9 and Figure 19.21.

So-called "long-life" lamps, which are guaranteed to burn for two, three, or five years, are lamps designed for much higher voltages than that at which they operate. Since they normally sell at a high cost and are very inefficient, their use is seldom advisable. In a cost comparison made of three lamps with 750, 2500, and 10,000-hr lives, respectively, including cost of lamps, energy and relamping, the relative costs per million lumen hours were 1.0, 0.94, and 1.17, respectively.

(a) Reflector Lamps. These are made in "R" and "PAR" shapes (see Fig. 19.19) and contain a reflective coating on the inside of the glass envelope; this gives the entire lamp accurate light beam control. Both types are available in narrow or wide beam design, commonly called *spot* and *flood*, respectively. R lamps are generally made in soft glass envelopes for indoor use, whereas PAR lamps are hard glass, suitable for exterior application. Also available is a lamp with an elliptical reflector bulb shape. This causes the beam to focus a few inches in front of the lamp, permitting high-efficiency application in pinhole downlights or deep baffle units where use of ordinary R lamps causes trapping and loss of most of the lamps' output. This elliptical reflector action *in a fixture* is illustrated in Figure 20.21. Typical reflector lamp dimensional and photometric data are given in Figure 19.22. Illumination patterns resulting from typical PAR spot and flood lamps are shown in Figure 19.23. When using R and PAR lamps the fixture acts principally as a lampholder, since beam control is built into the lamp.

(b) Interference (Dichroic) Filters. Such filters, which had been previously used only in specialized applications such as projection lamps to remove heat from the light beam, are now available in PAR lamps. The basic filter is a thin film that operates on the interference principle rather than absorption. Thus the surface remains relatively cool.

In one design that is utilized to limit the heat in the light beam, the film is applied to the inside back of the lamp. It acts by transmitting infrared heat out the lamp back while reflecting light out the lamp front (see Fig.

Table 19.9 ***Typical Incandescent Lamp Data (Listing a Few of Many Sizes and Types of 115-, 120-, and 125-v Lamps)***

Watts and Life		*Approx. Color Temp. (K)*	*Lumens*		*Physical Data*		
Lamp Watts[a]	*Average Rated Life (Hours)*		*Initial Lumens*	*Lumens per Watt*	*Shape of Bulb*[b]	*Base*	*Description*
6 Cl	1500	2370	58	9.7	S-6	Cand.	Indicator
6	1500	2370	40	6.7	S-14	Med.	—
7½ Cl	1400	—	45	6.0	S-11	Med.	—
10 Cl	1000	—	115	11.5	S-14	Med.	Sign
15	2500	—	126	8.4	A-15	Med.	—
15	1000	—	120	8.0	A-17	Med.	—
15 Cer	400	—	138	9.2	S-11	Med.	Refrigerator
15	1000	—	115	7.7	T-7	Cand.	Appliance
25	1000	—	228	9.1	A-17	Med.	Rough service
25	1000	2550	357	14.3	A-19	Med.	—
25	1000	—	238	9.5	A-19	Med.	Vibration service
25 Cl	1000	—	242	9.7	T-6½	Inter.	Showcase
40	1000	2770	452	11.3	A-15	Med.	Appliance
40	1500	—	460	11.5	A-19	Med.	—
40 SW	1500	—	445	11.1	A-19	Med.	—
40	1500	—	—	—	G-25	Med.	Decorative
50 W	1000	2790	830	16.6	A-21	Med.	—
60	1000	2800	890	14.8	A-19	Med.	—
60	2500	—	760	12.7	A-19	Med.	Long life
60	1000	—	575	9.6	A-19	Med.	Rough surface
75	750	—	1210	16.1	A-19	Med.	—
75 SW	750	—	1180	15.7	A-19	Med.	—
100	750	2870	1740	17.4	A-19	Med.	—
100	750	—	1690	16.9	A-21	Med.	—
100	2500	—	1460	14.6	A-19	Med.	Long life
100	1000	—	1220	12.2	A-21	Med.	Rough surface
150	750	2900	2810	18.7	A-21	Med.	—
150	750	—	2580	17.2	PS-25	Med.	—
200	750	2930	3940	19.7	A-23	Med.	—
200 SBIF	1000	—	3320	16.6	PS-30	Med.	—
300	750	2940	6000	20.0	PS-30	Med.	—
300	1000	—	6000	20.0	PS-30	Mogul	—

[a]Figures in this column designate the input watts, and the letters identify the treatment of the glass bulb; thus: 60 means 60 w. All inside frosted unless otherwise noted. Other letters have these meanings: W, white; SBIF, silver bowl, inside frosted; Cl, clear; SW, soft white; Cer, ceramic.

[b]*Bulb Designations.* Bulb designations consist of a letter to indicate its shape and a figure to indicate the approximate maximum diameter in eighths of an inch (see Fig. 19.19).

19.24). Typical applications are in window displays, over food counters, and in any location where a "cool beam" is desirable. Of course, provision must be made for removal of the heat from the fixture if the lamp is housed.

In a second design, multiple-layer filters are applied to the front of the lamp. Each film acts to transmit one color and reflect its complement (two color, hence dichroic). These dichroic filter lamps produce a purer, more saturated color at high efficacy than is possible with selective absorption filters (see Fig. 19.25).

(c) ***Low-Voltage Lamps.*** These lamps, in PAR shape and for 6-v operation, are available in extremely narrow beam spread (5–10°) for special precision control floodlighting. The low voltage makes their application to exterior work simpler.

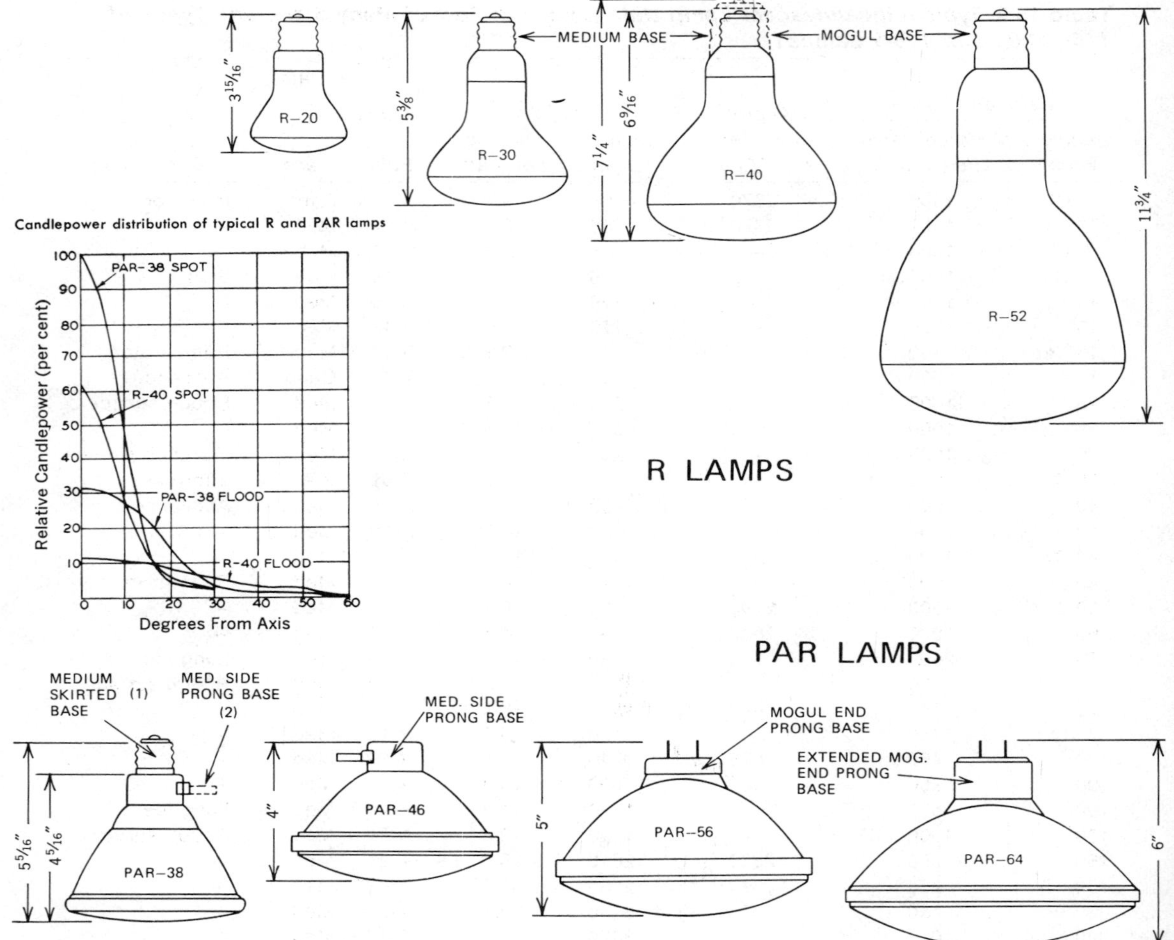

Approximate Initial Output of Typical PAR and R Lamps

Wattage	Bulb	SPOT LAMPS				FLOOD LAMPS			
		Central Candle-power	Beam Spread (1)	Beam Lumens	Total Lumens	Central Candle-power	Beam Spread (1)	Beam* Lumens	Total Lumens
75	R-30	1,850 (2)	50°	430	860	415 (2)	78°	610	860
150	R-40	7,400 (2)	20°	910	1,950	1,100 (2)	110°	1,530	1,950
300	R-40	13,500 (3)	35°	1,660	3,700	2,200 (2)	115°	3,250	3,700
500	R-40	22,000 (3)	60°	4,240	6,500	4,700 (2)	120°	5,930	6,500
500	R-52	—	—	—	—	—	—	—	7,850
750	R-52	—	—	—	—	—	—	—	13,000
150	PAR-38	9,100 (3)	30°	960	1,730	3,500 (2)	60°	1,220	1,730
200	PAR-46	36,000 (3, 4)	17° x 23°	1,200	2,250	10,500 (2, 5)	20° x 40°	1,300	2,250
300	PAR-56	70,000 (3, 4)	15° x 20°	1,800	3,840	24,000 (2, 5)	20° x 35°	2,000	3,840
300	PAR-56	—	--	—	—	10,000 (2, 6)	30° x 60°	2,100	3,840
500	PAR-64	104,000 (3, 4)	13° x 20°	3,000	6,500	35,000 (2, 5)	20° x 35°	3,400	6,500
500	PAR-64	—	—	—	—	11,000 (2, 6)	35° x 65°	3,500	6,500

(1) To 10% (approx.) of maximum candlepower. (2) Average in 10° cone. (3) Average in 5° cone. (4) Narrow spot.
(5) Medium flood. (6) Wide flood. *Estimated. All lamps have 2000 hour life.

Fig. 19.22 *Typical dimensional and photometric data for reflector spot and flood lamps. For accurate current data, consult large lamp and specialty catalogs. Courtesy of General Electric Company, Lamp Division.*

The beam of the PAR lamp is cone—like in shape. Each type of PAR lamp has a distinct illumination pattern which varies in size and light intensity – depending on the angle at which the lamp is aimed and on its distance from the area illuminated.

When centered directly on the surface to be lighted (at right angles or zero degrees) the small PAR 38 sizes give a round lighting pattern. The concentric rings show the amount of light measured in footcandles at various distances from the beam center. The round lighting pattern changes to oval or elliptical when the lamp is aimed at an angle.

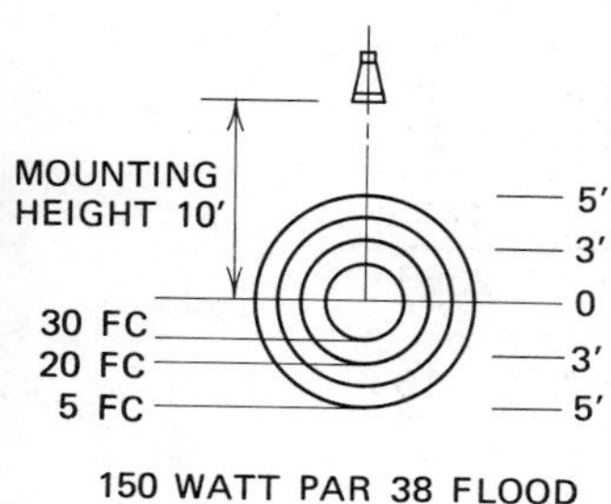

150 WATT PAR 38 FLOOD

70°
45°
0°

300 WATT PAR 56 NARROW SPOT

300 WATT PAR 56 MEDIUM FLOOD

The lighting pattern of the larger PAR 46, 56, and 64 lamps is oval or elliptical, whether centered directly on the surface or aimed at an angle. As shown in the two diagrams, aiming PAR lamps at progressively greater angles – proportionately increases both the length and width of the area illuminated. In general, for spotlights the length of the lighting pattern becomes proportionately greater – and in the case of floodlights, the width.

Fig. 19.23 *Typical illumination patterns for PAR spot and flood lamps. Courtesy of GTE Sylvania Inc.*

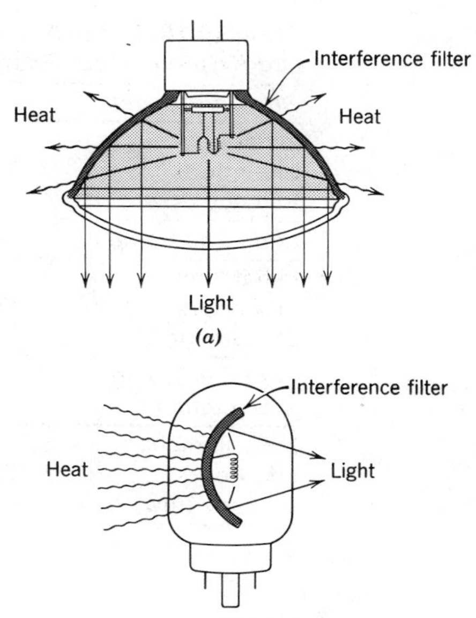

Fig. 19.24 *The reflectors in the cool-beam PAR lamp* (a) *and the projection lamp* (b) *reflect most of the visible light while transmitting the infrared energy, which constitutes two-thirds of the total heat energy. In an incandescent lamp, about 90% of the input energy is converted to heat. Of course, provision must be made for heat removal in the fixture.*

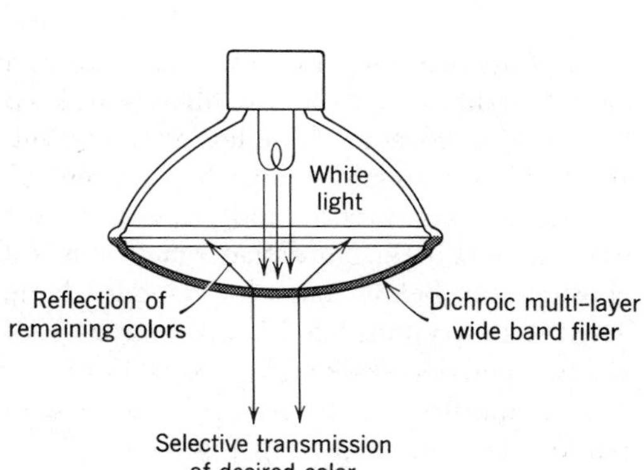

Fig. 19.25 *Action of dichroic filter is one of selective interference rather than absorption. Each film layer transmits one color while reflecting its complement. Desired color is obtainable by action of several films.*

Table 19.10 *Comparison of Characteristics of Standard, Long-Life and Krypton-Filled, Energy-Saving Lamps*

Standard Lamps	*40 w*	*60 w*	*75 w*	*100 w*	*150 w*
Initial lumens	452	890	1180	1740	2810
Life hours	1200	1000	750	750	750
Efficacy (lpw)	11.3	14.8	15.7	17.4	18.7
Long-Life Lamps	*40 w*	*60 w*	*75 w*	*100 w*	*150 w*
Initial lumens	420	760	970	1460	2380
Life hours	2500 →				
Efficacy (lpw)	10.5	12.7	12.9	14.6	15.9
Energy-Saving Lamps	*36 w*	*54 w*	*69 w*	*93 w*	*143 w*
Initial lumens	420	760	970	1460	2380
Life hours	2500 →				
Efficacy (lpw)	11.7	14.1	14.1	15.7	16.6

(*d*) ***Krypton Gas.*** This gas in lamps in place of the usual nitrogen-argon mixture conducts heat more slowly from the filament and results in approximately 10% higher efficacy, longer life, and a smaller envelope. The cost premium for krypton lamps is approximately 50%. Applications are in long-life lamps to increase efficacy (see below) and in exterior spots and floods to increase life and output.

(*e*) ***Energy-Saving Lamps.*** These are basically long-life lamps that are filled with krypton to raise efficacy. These lamps can be substituted for standard lamps for appreciable savings in energy costs and relamping costs (see Table 19.10). Note that efficacy is still considerably below that of a standard lamp. The use of krypton-filled lamps increases initial cost and decreases energy costs. For a true cost comparison, life cycle costs must be compared. These are shown in Figure 19.47.

(*f*) ***Tungsten-Halogen (Quartz) Lamp.*** The IES defines this lamp as a "gas-filled tungsten incandescent lamp containing a certain proportion of halogens." The halogens are iodine, chlorine, bromine, and fluorine. Thus the quartz-iodine-tungsten filament lamp is a member of this class. The lamp is basically an incandescent lamp, producing light and heat from the incandescence of its coiled filament. Unlike the normal inert gas-filled incandescent lamp, the lamp envelope, which is quartz to withstand high temperature, is filled with an iodine vapor that prevents the evaporation of the tungsten filament. This evaporation is what normally occurs in the incandescent lamp, resulting in blackening of the bulb, light output deterioration, and eventual burnout. The mechanism of the regenerative halogen cycle is shown in Figure 19.26, along with a graphic comparison of light loss between a normal incandescent and a tungsten-halogen lamp. Although the lamp has approximately the same efficacy as an equivalent normal incandescent, it has the advantages of longer life, low lumen depreciation (98% output at 90% life), and a smaller envelope for a given wattage. Some typical lamp data are given in Table 19.11. Since quartz lamps are not normally used for general lighting, comparative life cycle cost analyses are not appropriate.

FLUORESCENT LAMPS

19.11 The Fluorescent Lamp—Construction

The second major category of light sources is that of electric discharge lamps, of which the fluorescent lamp is the best known and most widely used type. It has become so popular since its major introduction in 1937 that it

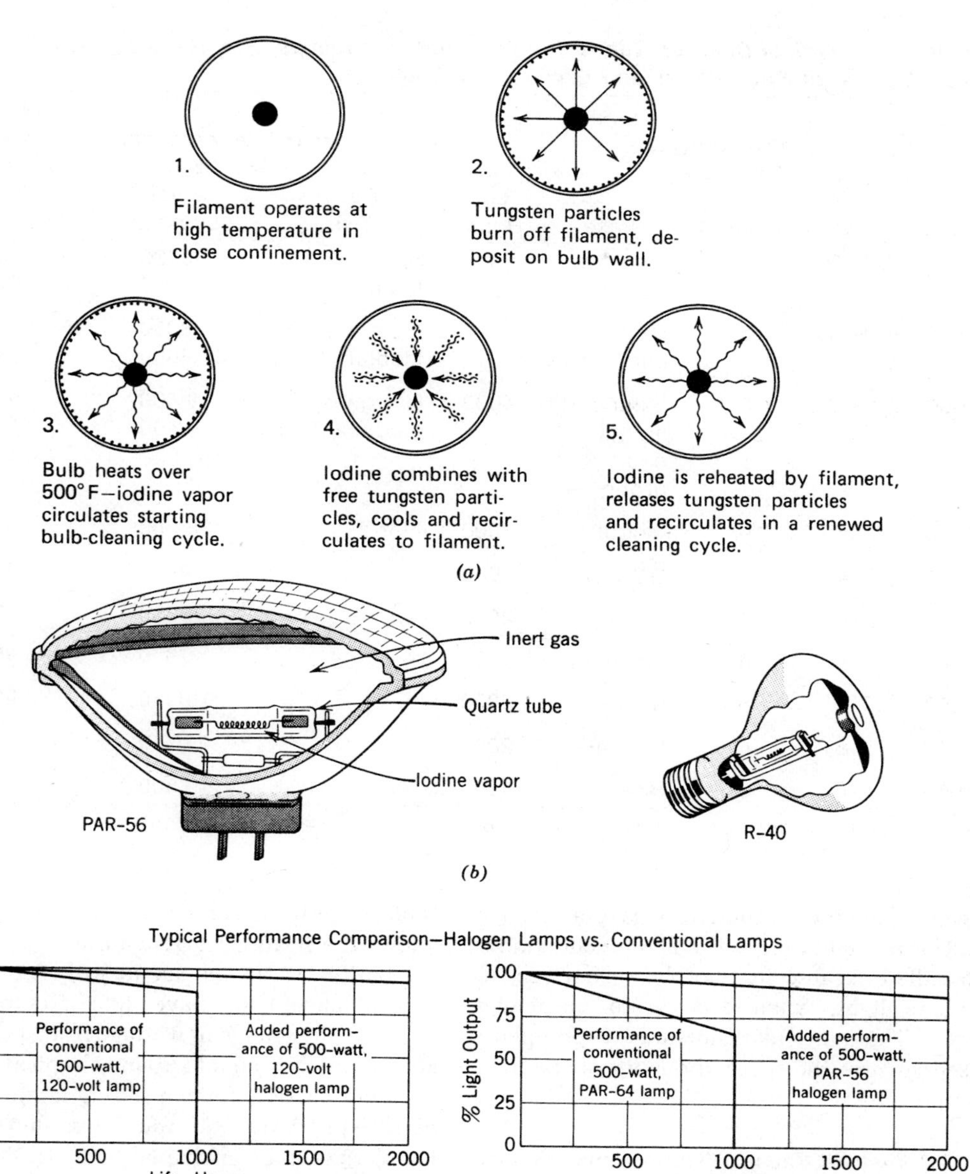

***Fig. 19.26** The self-regenerative halogen cycle* (a) *results in low light depreciation* (c) *whether the lamp is bare or in an enclosure* (b).

has almost completely supplanted the incandescent lamp in all fields except specialty lighting and residential use. The typical fluorescent lamp comprises a cylindrical glass tube sealed at both ends and containing a mixture of an inert gas, generally argon, and *low-pressure* mercury vapor. Built into each end is a cathode that supplies the electrons to start and maintain the mercury arc, or gaseous discharge. The short-wave ultraviolet light, which is produced by the mercury arc, is absorbed by the phosphors with which the

Table 19.11 ***Typical Data for Quartz Tungsten-Halogen Lamps, Par, Reflector, and Tubular 120-v Lamps for Spot, Flood, and General Lighting***[a]

Watts	Bulb	Maximum Overall Length (Inches)	Base	Rated Life (Hours)	Beam Type	Approximate Initial Total Lumens	Mean Lumens Through Life (Percentage)
250	PAR-38	$5\frac{5}{16}$	Medium skirted	6000	Spot flood	3220 3220	94 94
500	PAR-56	5	Mogul end prong	4000	Narrow spot Medium flood	8000 8000	94 94
1000	PAR-64	6	Extended mogul end prong	4000	Narrow spot Medium flood	19,400 19,400	94 94
1000	R-60	$10\frac{1}{8}$	Mogul	3000	Flood	18,300	95
250	T-4	3	DC bay	2000	—	4850	95
300	T-4	$3\frac{1}{8}$	RSC	2000	—	5650	95
400	T-4	$3\frac{5}{8}$	Mini-can	2000	—	7970	95
500	T-4	6	Med.-PF	2000	—	10,450	95
750	T-6	6	Med.-PF	2000	—	15,750	95
1000	T-6	$5\frac{5}{8}$	RSC	4000	—	19,800	95

[a]For a complete listing, see manufacturer's catalog.

inside of the tube is coated and is reradiated in the visible light range. The fluorescent lamp is so called because its phosphors fluoresce, or radiate light, when exposed to ultraviolet light. The particular mixture of phosphors used governs the spectral quality of the light output.

(*a*) ***Preheat Lamps.*** The original fluorescent lamp was of a preheat design. Construction of a typical hot cathode lamp (preheat and rapid start) is shown in Figure 19.27*a;* the basic preheat circuit is shown in Figure 19.28*a*. The circuit utilizes a separate starter, which is a small cylindrical device that plugs into a preheat fixture. When the lamp circuit is closed, the starter energizes the cathodes; after a 2- to 5-sec delay, it initiates a high-voltage arc across the lamp, causing it to start. Most starters are automatic, although in desk lamps the preheating is accomplished by depressing the start button for a few seconds and then releasing it. This closes the circuit and allows the heating current to flow; releasing the button causes the arc to strike.

All preheat lamps have bipin bases (see Fig. 19.27*b*). They range in wattage from 4 to 90 w and in length from 6 to 90 in. A typical ordering abbreviation for a preheat lamp would be F15T12WW. This translates: fluorescent lamp, 15 w, tubular-shaped bulb, 12/8-in. diameter (number represents diameter in one-eighths of an inch), warm white color (see Table 19.12). In large measure preheat lamps have been supplanted by rapid-start and instant-start types.

(*b*) ***Rapid-Start Lamps.*** These are similar in construction to the preheat lamps; the basic difference is in the circuitry (see Fig. 19.28*b*). This circuit eliminates the delay inherent in preheat circuits by keeping the lamp cathodes constantly energized (preheated). When the lamp circuit is energized, the arc is struck

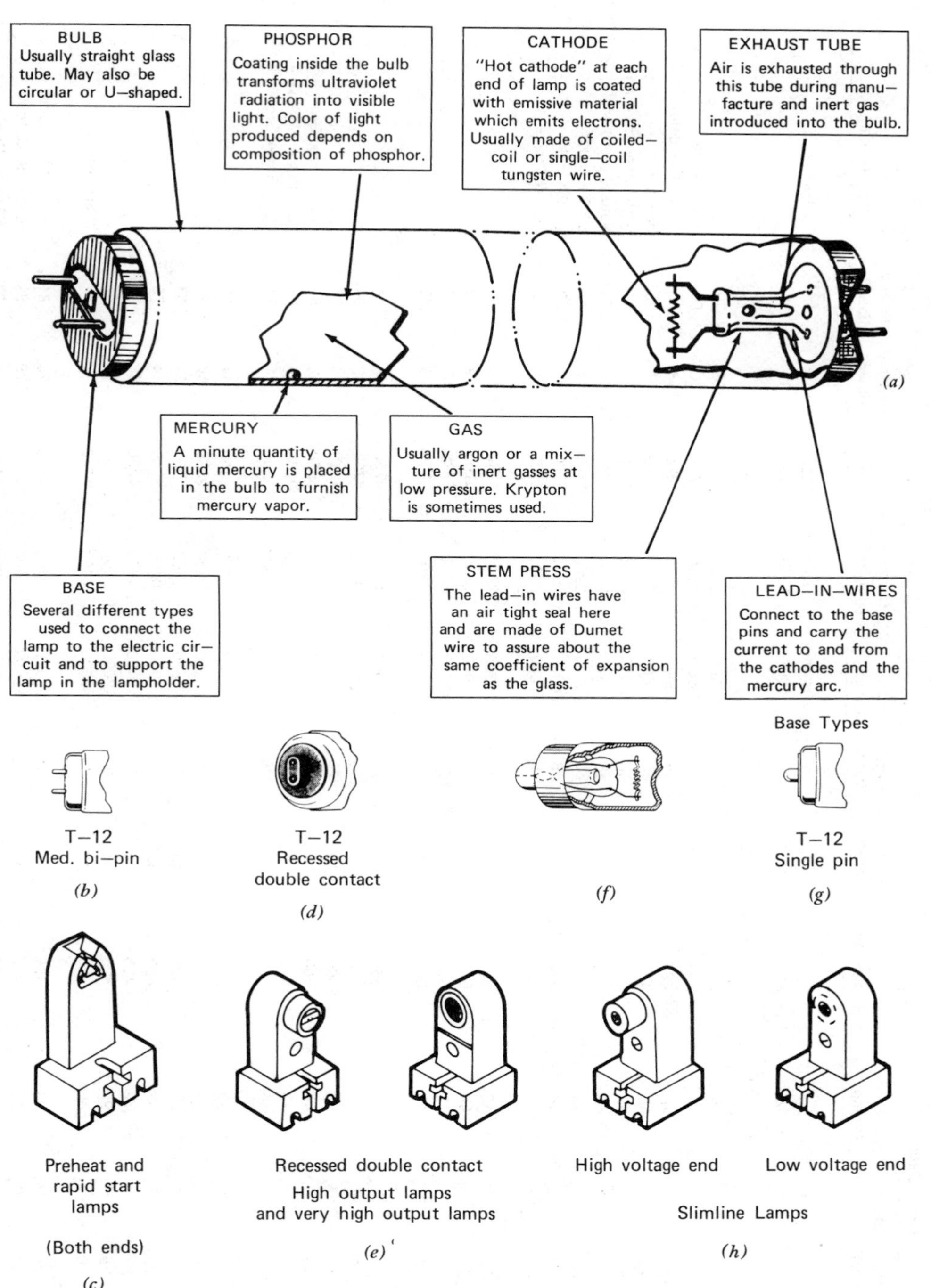

Fig. 19.27 *Details of typical fluorescent lamps and associated lampholders.* (a) *Construction of preheat/rapid-start bipin base lamp. This type lamp has type* (b) *base and is held in type* (c) *lampholder. High output HO and VHO rapid-start lamps use recessed dc base* (d) *and lampholders* (e). *Instant start lamps are similar in construction to* (a) *except with cathode construction* (f), *have a single pin base* (g), *and use single pin lampholders* (h), *which are different for each end.* (a) *Courtesy of GTE Sylvania Inc.*

Table 19.12 *Typical Fluorescent Lamp Data*

Lamp Abbreviation	Lamp Data: Lamp (Watts)	Lamp Data: Diameter (Inches)	Lamp Data: Length (Inches)	Lamp Current (milliamperes)	Ballast Watts[b,c]	Total[k] Watts	Lamp Life (Hours)[d]	Initial Output Lumens[e]	Lumens at 40% Life	Initial: Actual Efficacy (lpw)[f]	Initial: Lamp Efficacy (lpw)[g]	Remarks
Preheat lamps[a]												
F15 T8 CW	15	8/8	18	425	8	23	7500	870	750	38	58	Cool white
F20 T12 CW	20	12/8	24	425	10	30	9000	1300	1155	43	65	
Rapid start—preheat lamps[h]												
F30 T12 CW	30	12/8	36	425	7.5	37.5	18000	2300	1955	61	77	
F40 T12 CW	40	12/8	48	425	6	46	20000	3150	2770	68	79	
F40 T12 WW	40	12/8	48	425	6	46	20000	3200	2770	70	80	Warm white
F40 T12 CWX	40	12/8	48	425	6	46	20000	2200	1825	48	55	Cool white deluxe
F40 T12 D	40	12/8	48	425	6	46	20000	2600	2290	57	65	Daylight
F40 T12 CW/S	40	12/8	48	425	6	46	15000	3250	2960	71	81	High lumen maintenance
F40 T12 H	40	12/8	48	425	6	46	20000	2100	1745	46	53	Natural (soft) white
F40 T12/C50	40	12/8	48	425	6	46	20000	2200	1890	48	55	5000 K color
F40 T12/C75	40	12/8	48	425	6	46	20000	2000	1720	44	50	7500 K color
F40 T12/U	40	12/8	—	425	6	46	12000	2900	2525	55	63	"U" shape[i]
Rapid start—high output												
F48 T12 CW/HO	60	12/8	48	800	12.5	72.5	12000	4300	3740	55	72	
F60 T12 CW/HO	75	12/8	60	800	15	90	12000	5400	4700	60	72	
F72 T12 CW/HO	87	12/8	72	800	13	100	12000	6650	5785	67	76	
F96 T12 CW/HO	112	12/8	96	800	14	126	12000	9200	8005	73	82	
Rapid start—very high output												
F48 T12 CW/VHO	116	12/8	48	1500	5	121	9000	6250	4750	52	54	
F72 T12 CW/VHO	168	12/8	72	1500	5	173	9000	9900	7720	57	59	
F96 T12 CW/VHO	215	12/8	96	1500	10	225	9000	14500	11310	64	67	
F48 PG17 CW	116	12/8	48	1500	5	121	12000	7450	5950	62	64	G.E. Power Groove®
F72 PG17 CW	168	12/8	72	1500	5	173	12000	11500	9200	66	68	G.E. Power Groove®
F96 PG17 CW	215	12/8	96	1500	10	225	12000	16000	13000	71	74	G.E. Power Groove®
Instant start (Slimline) lamps												
F42 T6 CW	25	6/8	42	200	13	38	7500	1750	1490	44	70	
F64 T6 CW	38	6/8	64	200	13	51	7500	2800	2350	56	70	
F24 T12 CW	20	12/8	24	425	14	34	7500	1150	990	34	58	
F48 T12 CW	39	12/8	48	425	12	51	9000	3000	2760	54	75	
F72 T12 CW	57	12/8	72	425	13	73	12000	4600	4280	51	65	
F96 T12 CW	75	12/8	96	425	13	88	12000	6300	5800	68	84	
F36 T12	30	12/8	36	425	13	43	7500	2000	1740	47	67	Warm white
Energy-conserving lamps												
F40 CW/RS[j]	35	12/8	48	425	6	41	20000	2850	2510	70	81	Rapid Start, Cool white
F96 CW/HO[j]	95	12/8	96	800	14	109	12000	8500	7395	78	89	RS, High Output
F96 CW[j]	60	12/8	96	425	13	73	12000	5600	5150	76	93	Slimline, Cool white
F96 PG CW[j]	185	—	—	—	13	198	12000	14000	—	71	76	G.E. Power Groove®

[a] Data given for a preheat circuit.
[b] Figures are for a two-lamp circuit.
[e] After 100-hr burning.
[f] Includes ballast loss.
[i] U-shaped lamps available with $3\frac{5}{8}$ or 6-in. leg spacing; all other characteristics equal.
[j] Designations for these lamps vary between manufacturers; thus G.E.—Wattmiser—

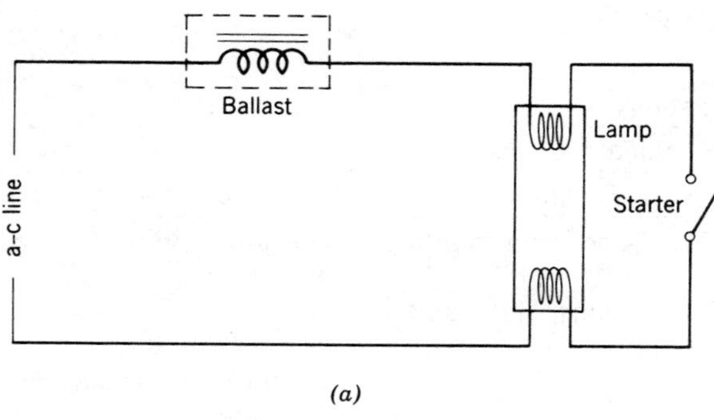

Fig. 19.28 *Basic circuits for preheat, rapid-start and instant-start (Slimline) are illustrated in* (a), (b), *and* (c). *For the sake of clarity only single-lamp circuits are shown, and power-factor correcting capacitors, autotransformers, and compensators are omitted.*

(a) *Basic preheat circuit. Starter may be any of several types, manual or automatic. The circuit does not show compensators or other detailed elements for the sake of clarity. Most preheat lamps are T-12 and operate at 425 ma.*

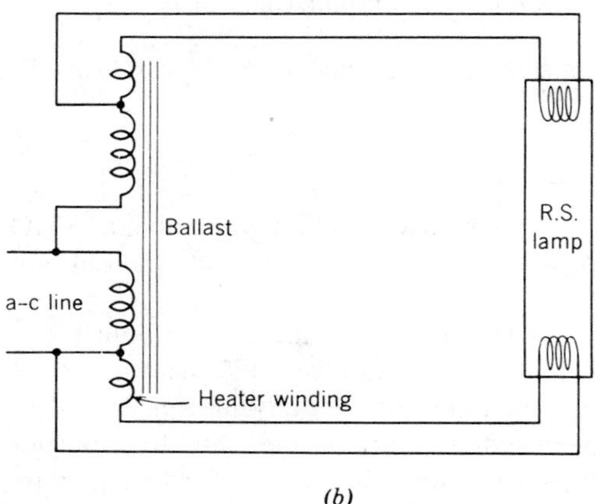

(b) *Basic rapid-start circuit. Note the special end windings used to supply voltage to heat the cathode continuously. To assure proper starting all standard RS lamps* must *be mounted within ½ in. of a grounded metal strip extending the full length of the lamp (1 in. for HO and VHO lamps). Normal output lamps operate at 425 ma, HO at 800 ma, and VHO at 1500 ma.*

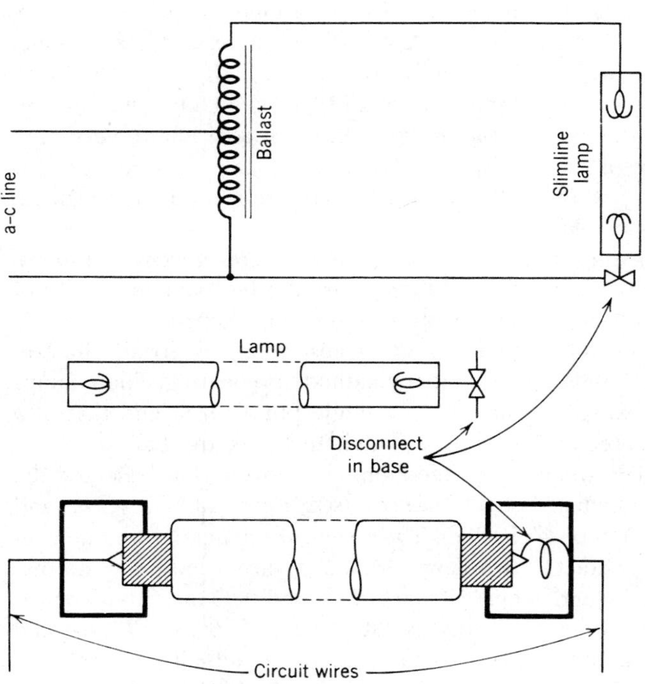

(c) *Basic instant-start circuit. Voltage from ballast transformer is high enough to strike an arc directly. Note that unlike preheat and rapid-start lamps, these are single pin, since cathodes are not preheated. T-6 and T-8 lamps normally operate at 200 ma, T-12 lamps at 425 ma.*

(Lower portion of figure) *Because of the high voltage involved, the lampholder at one end is a disconnecting device that opens the circuit when the lamp is removed.*

Table 19.13 ***Fluorescent Lamp Interchangeability***

Lamp Type	Ballast/Circuit Type		
	Preheat	Rapid-start	Instant-start
Preheat	OK	Not good, poor starting	Not good, poor starting short life[a]
Instant-start (Slimline)	Won't start, not good[b]	Won't start, Not good[b]	OK
Rapid-start	OK	OK	Not good, poor starting short life[a]
Preheat/ rapid-start	OK	OK	Not good, poor starting short life[a]

[a]Normally no possibility of interchange because of incompatible base and lampholder: preheat has bipin base, instant-start lampholder is single pin.

[b]Normally no possibility of interchange. Instant-start lamp is single pin base; preheat/rapid-start lampholders are for bipin bases.

immediately. No external starter is required. Because of this similarity of operation, rapid-start lamps will operate satisfactorily in a preheat circuit. The reverse is not true, because the preheat requires more current to heat the cathode than the rapid-start ballast provides (see Table 19.13 for interchangeability of lamps in the various circuits). By far the most popular lamp is the 40-w T-12 lamp. A standard ordering abbreviation for a lamp would be F40T12WW/RS which indicates fluorescent, 40 w, T-12 bulb, warm white color, rapid start.

Most rapid-start lamps operate at 425 ma. If this current is increased, the output of the lamp also increases. Two special types of higher output rapid-start lamps are available. One operates at 800 ma and is called simply high output (HO). The second, which operates at 1500 ma (1.5 amp), is called by different manufacturers—very high output (VHO), super-high output or, simply, 1500-ma, rapid-start lamp. There is also a 1500-ma special lamp that uses what looks like a dented or grooved glass tube. This lamp, called Power Groove by General Electric, has somewhat higher output than the standard VHO tube. All high-output lamps use double contact bases and special ballasts (see Figs. 19.27*d* and 19.27*e*). This lamp is used in applications where high output is required from a limited size source such as outdoor sign lighting, street lighting, and merchandise displays. Because of the serious heat problems involved, VHO lamps are frequently operated without enclosing fixtures.

The HO and VHO lamps are slightly less efficient than the standard 425-ma, rapid-start lamp and have considerably shorter life. Typical ordering abbreviations for high-output lamps are similar to the standard rapid-start lamps except that the number indicates length, not wattage. For instance, F72T12/CW/HO is fluorescent, 72-in. long, T-12 bulb, cool white, high output (800 ma). Similarly, F72T12/CW/VHO is fluorescent, 72-in. long, T-12 bulb, cool white, very high output (1500 ma). Typical characteristics for all rapid-start lamps are given in Table 19.12.

(*c*) *Instant-Start Fluorescent Lamps.* Slimline lamps are the best-known variety of instant-start fluorescent lamps. They use a high-voltage transformer to strike the arc without any cathode preheating. These lamps have only a single pin at each end that also acts as a switch to break the ballast circuit when the lamp is removed, thus lessening the shock hazard (see Figs. 19.27*f*, *g*, *h*, and 19.28*c*). The lamps are generally operated in two-lamp circuits at various currents; normal currents are 200 and 425 ma, and normal lengths are 24, 36, 42, 48, 60, 64, 72, 84, and 96 in. These lamps are actually hot cathode instant-start lamps, which differentiates from the high-voltage cold cathode type. Slimline

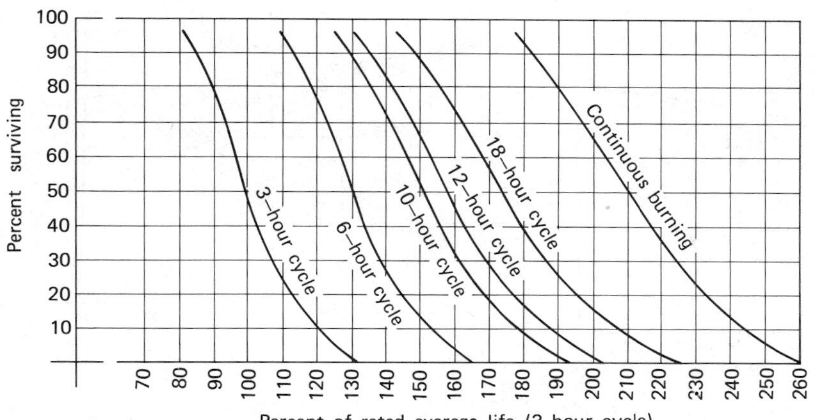

Fig. 19.29 Mortality curves of fluorescent lamps. Data Courtesy of GTE Sylvania, Inc.

lamps and ballasts are more expensive than rapid-start and are somewhat less efficient. However, they are manufactured in certain sizes and currents not made in rapid-start (e.g., 96 in. 430 ma), and they have the additional advantage of being able to start in much lower ambient temperatures (below 50° F) than rapid-start circuits. This starting characteristic makes the instant-start circuit particularly applicable to outdoor use. A typical ordering description for such a lamp would be: F42T6CW Slimline, which means: fluorescent, 42-in. length, tubular, 6/8-in. diameter, cool white, instant start. The T-6 narrow tube indicates a low-current, 200-ma lamp, in lieu of T-12 for the 425-ma lamp. Note also that in instant-start lamps the number following F indicates length, not wattage. This is true of all lamps that operate at other than 425 ma, which is the normal current. Typical characteristics appear in Table 19.12.

(*d*) ***Cold Cathode Tubes.*** The true cold cathode tube uses a large, thimble-shaped cathode and a high-voltage transformer that literally tears the electrons out of the large cathode to strike the arc. These lamps have a very long life which, in contradistinction to hot cathode lamps, is virtually unaffected by the number of starts. Cold cathode lamps have a lower overall efficiency than the hot cathode types and are normally used where long continuous runs are required, as in architectural-type lighting rather than in lighting fixtures. Cold cathode lamps are readily dimmed and also operate well at varying ambient temperatures.

19.12 The Fluorescent Lamp—Characteristics and Operation

(*a*) ***Lamp life.*** The lamp life of a standard fluorescent tube is greatly dependent on the burning hours per start. One manufacturer now produces lamps whose life is independent of burning hours per start, but at the cost of overall life. The figures listed in Table 19.12 and in the lamp catalogs for lamp life are based on a burning cycle of 3 hr per start and represents the average life of a group of lamps; that is, half the lamps of any group will have burned out at this time. Typical lamp mortality curves are shown in Figure 19.29, and the effect of burning hours per start is shown in Figure 19.30.

The significance of this item is connected with energy costs and utilization. From an energy source viewpoint, if an area is not utilized for periods of 15 min or more, fluorescent lamps should be shut off. This takes into account the resource energy required to replace a tube as a result of shortening its life. From a cost viewpoint, the break-even point depends on these factors:

1. Lamp life reduction as a function of burning hours per cycle.
2. Cost of energy.

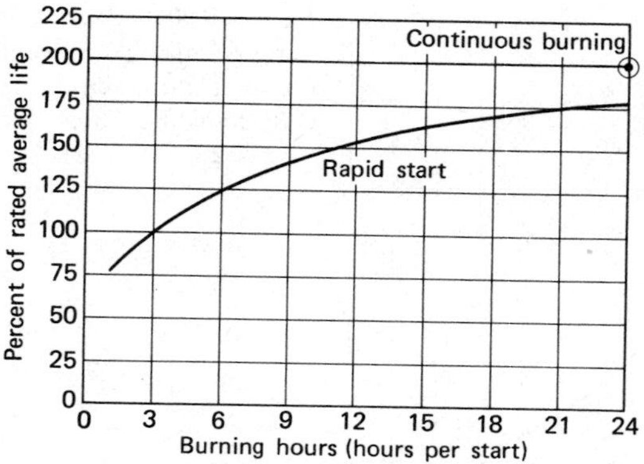

Fig. 19.30 Effect of burning hours on fluorescent lamp life. Note that at three burning hours per start, the average lamp life is 100% of the nominal catalog figure. Data courtesy of GTE Sylvania, Inc.

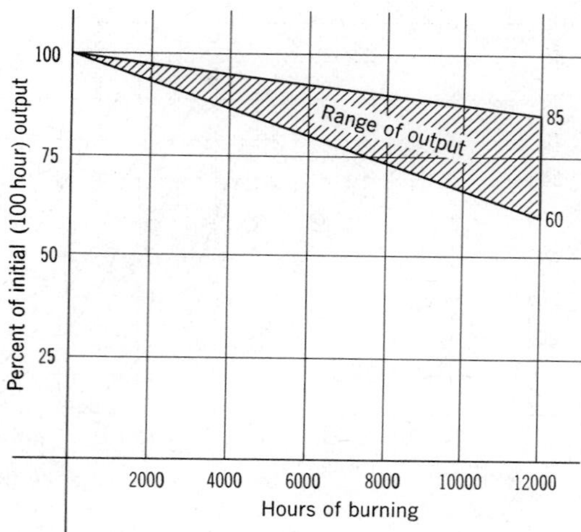

Fig. 19.31 Fluorescent lamp output depreciates with life. The majority of lamps fall in the upper part of the band. Lumen depreciation is unaffected by burning hours per start.

3. Cost of lamp and lamp replacement.
4. Amount of time lamp remains off when shut off.
5. Cost of switching equipment (if any).
6. Life of the building.

With this number of variables it is not possible to give general solutions, and an individual analysis is required. However, several analyses by the author have shown that: assuming ordinary office conditions, using life data as given in Figure 19.30, 20-year fixture life, \$0.035/kwh energy cost escalating 3% annually, \$1.25 lamp cost, 15-min relamping time, and \$5 switching cost per lamp (one switch per 2-2 lamp fixtures), lamps should be switched off anytime they are not in use for 10 to 15 min or more. (The spread is caused primarily by variation in local labor rates.) It is thus clearly an economic fallacy to leave lamps burning in order to achieve longer lamp life.

(*b*) ***Lumen Output.*** Lumen output of a fluorescent tube decreases rapidly during the first 100 hr of burning and thereafter much more slowly. For this reason the tabulated initial lumen figures represent output after 100 hr of burning. Data are also generally published on the lumen output at 40% of average rated life. This figure is approximately 85 to 90% of the 100-hr initial value (see Fig. 19.31 and Table 19.12). A lamp that is specifically designed for low-lumen depreciation is available; it has a 91% output at 40% of life (see Table 19.12, item F40-12/CW/S). This lamp has shorter overall life than a standard lamp (15,000 hr as compared to 20,000 hr), and is therefore best applied in installations using group relamping. In these, lamps are replaced at approximately 70% life; the result is that the shorter life to burnout is insignificant and the higher output throughout life is clearly advantageous. Too, these lamps are well applied where illumination is borderline and lumen depreciation would adversely affect performance of the tasks in the area.

(*c*) ***Efficacy.*** The efficacy of a fluorescent lamp depends on operating current and the phosphors utilized. Figure 19.32 shows the energy distribution of a typical fluorescent lamp alone, not including ballast losses. Normally warm white lamps are most efficient, followed closely by cool white, white, daylight, and colored lamps. Specialty colors such as "natural" white, or lamps designed to produce specific Kelvin temperatures, are low in output, with *lamp* efficacies in the 40 to 50 lpw region. The range of efficacy for stand-

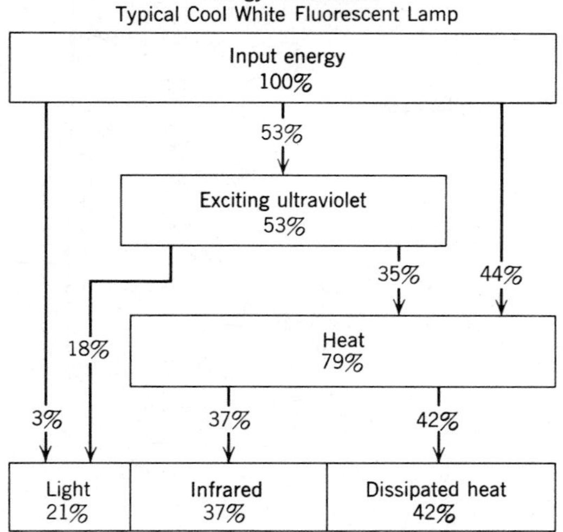

Fig. 19.32 *Fluorescent lamps with efficacies of up to 85 lpw are among the most efficient light sources available, yet still convert less than one-quarter of their energy to useful light. Ballast losses are not included in the figures shown.*

ard lamps is 40 to 85 lpw, *including* ballast losses in the wattage figure (see Table 19.12). This is important, since discharge lamps are inoperative without ballasts, and neglecting ballast losses yields an artificially high and therefore misleading efficacy.

Generally, standard 425-ma lamps are most efficient, followed by HO 800-ma lamps, then VHO 1500-ma lamps. Specialty lamps such as reflector and low-wattage units are discussed in following paragraphs. Ballast losses, which constitute 5 to 12% of lamp wattage, depend on ballast type, circuit, manufacturer, type of fixture, ambient temperature, and number of lamps connected. Figures given in Table 19.12 are average.

To make a proper comparison between lamp types, it is not sufficient to compare cost to produce a given quantity of lumens (see Fig. 19.47), since high-output lamp installations use a smaller number of fixtures. A meaningful cost comparison requires a full life cycle or annual owning cost analysis.

(*d*) *Temperature.* The temperature of the tube, which is also an important factor in light output, is affected by the ambient temperature. Maximum efficiency occurs with the tube operating at a bulb temperature of 100 to 120° F, with output reduction above and below these values. Special all-weather and jacketed lamps are available that will maintain fairly constant lumen output over a wide ambient temperature range. For outdoor use where starting below 50° F is a requirement, rapid-start lamps require special low-temperature ballasts. Slimline lamps with normal ballasts will start readily down to 20° F and, by using the next higher voltage ballast, starting down to −20° F is possible.

(*e*) *Voltage.* Voltage either above or below rating adversely affects life, unlike the effect of low voltage on the incandescent lamp. The results of operation at other than rated voltage are shown graphically in Figure 19.33. Normal operating voltage range for ballasts is 110 to 125 v on 120-v circuits, 200 to 215 v on 208-v circuits, and 250 to 290 v on 277-v circuits.

(*f*) *Dimming and Low-Output.* Dimming and low-output operation are accomplished by the use of special one- and two-lamp ballasts, with appropriate controls in the case of dimming. Smooth dimming control down to 1% output is possible with solid-state electronic dimming, with the lamp starting at any level. When specifying dimming equipment, care should be exercised to select high-quality SCR controls, so as to avoid causing radio frequency interference (RFI). The cost of dimming equipment is very high and is only justified when smooth changes and unlimited choice of light level are mandatory. Where it is desired simply to be able to reduce the lighting level, as in classrooms, lecture halls, or multipurpose areas, two- and three-level ballasts are available for one or two 40-w, 48-in. RS lamps. Performance is given in Table 19.14, along with possible energy and cost economies. Flashing greatly reduces fluorescent lamp life unless a special flashing ballast is used that keeps the cathode heated while the lamp arc is quenched. A flasher control device is also required.

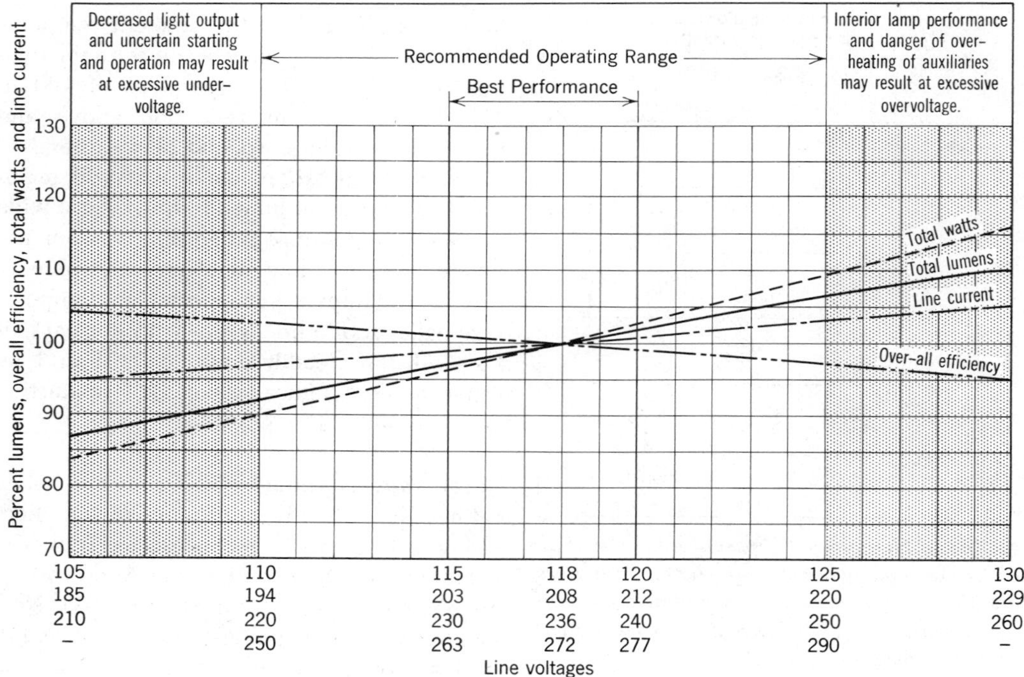

Fig. 19.33 Recommended operating ranges of circuit voltages for most satisfactory operation. The curves indicate the percentage changes in output lumens, efficiency, total watts, and current for line-voltage changes from the rated value. The nominal circuit voltages are 120, 208, and 277 v.

Table 19.14 *Effect of Multilevel Ballast Use with 2- to 40-w, 48-in. T12 Rapid-Start Lamps*

			2500-Hr Annual Operation	
Level	*Watts Input*	*Light Output (Percent)*	*Energy Reduction*	*Cost Savings*
1	98.7	100	—	—
2	55.7	55	107 kwh	@ 3¢/kwh—$3.21 @ 4¢/kwh— 4.28 @ 5¢/kwh— 5.35
3	37.0	37.0	154.25 kwh	@ 3¢/kwh—$4.63 @ 4¢/kwh— 6.17 @ 5¢/kwh— 7.71

Source. Ballast data courtesy of Advance Transformer Co.

(g) *Other Characteristics.* Fluorescent light color is discussed in Sections 19.23 and 19.25. Fluorescent lamps are large and therefore necessitate a relatively expensive fixture both to hold the lamps and control the light. Since the tubes emit light throughout their considerable length, accurate beam control is not possible, making fluorescent units best applicable to area lighting. The advantages of fluorescent lamps are long life, low cost, high output and efficacy, availability in an extremely wide range of sizes, colors, and brightnesses, and relative insensitivity to voltage fluctuation (important in brownout areas). Disadvantages are large size, which creates storage, handling, and relamping problems, and the fixture situation previously referred to.

19.13 Special Fluorescent Lamps

(*a*) ***"U"-Shaped Lamps.*** U-shaped lamps were developed to answer the need for a high-efficacy fluorescent source that could be utilized in a square fixture, since the normal fluorescent lamp shape is frequently not architecturally suitable (see Fig. 19.34). The U lamp is basically a standard 40-w, 48-in. fluorescent tube bent into a U shape and available with $3\frac{5}{8}$ or 6 in. leg spacing; the former can be accommodated three to a 2-ft sq fixture and the latter two to a 2-ft sq fixture. The lamps operate on standard ballasts and have slightly lower output than the corresponding straight tube. Insofar as energy is concerned, their use is much more desirable than using 2-ft lamps, as can readily be seen from the following data.

Two foot square fluorescent fixture with

Four 2-ft CW lamps	110 w	5200 lm	9000-hr life
Two U-shaped, CW lamps	100 w	5800 lm	12,000-hr life

Fig. 19.34 Each 5 ft square module uses a 2 ft square fixture with two 40-w U-shaped fluorescent lamps, of the type shown in the insert. Photo courtesy of GTE Sylvania, Inc.

[The panel fluorescent lamp that was produced in the late 1960s, for the same purpose (to provide a nonrectilinear source) is no longer readily available.]

(b) Reflector and Aperture Lamps. These lamps contain an internal reflector that performs in the same fashion as the more common reflector in the incandescent R and PAR lamp. The reflector lamp is completely phosphor-coated, while the aperture lamp has a clear "window" resulting in very high luminance of this slot (see Fig. 19.35). Both types have lower efficacy than a normal tube and are generally applied where an enclosing fixture is uneconomical or impractical, as in handrails or for sign illumination.

Tests using 235° reflector lamps in normal fluorescent fixtures intended for standard tubes indicate that the fixture coefficient of utilization increases up to 50%, depending on the fixture design. This is because the light normally trapped between the tubes and the fixture is saved, since almost no light is radiated above 62.5° from the vertical (cutoff of the internal reflector). Thus, using reflector lamps for general illumination can result in considerable savings in energy costs. If the use of these lamps is contemplated for normal lighting, the fixture and diffuser must be carefully selected to provide a sufficient degree of brightness control, since the lamp aperature brightness is approximately 5500 fL approximately $2\frac{1}{4}$ times the brightness of a normal 425-ma lamp (2400 fL). A life cycle cost analysis of a single fixture using standard lamps and reflector lamps is given in Table 19.15; it is based on the assumption that task illumination is approximately the same with three normal lamps as with two reflector lamps. This conclusion is reached as follows.

1. The illumination levels maintained are the same for all lamp types. Therefore, the number of fixtures is in direct proportion to the maintained lumens.
2. Standard lamp (STD) maintained lumens:

   $$3150 \text{ lm} \times \text{CU} \times 0.7 \text{ LLF} = 2205 \text{ CU}$$

3. Reflector lamp (RFL) maintained lumens:

   $$2600 \text{ lm} \times 1.4 \text{ CU} \times 0.9 \text{ LLF} = 3276 \text{ CU}$$

4. Ratio of maintained lumens:

   $$\frac{\text{lumens of RFL lamp}}{\text{lumens of STD lamp}} = \frac{3276}{2205} = 1.5$$

   (see Section 20.20 for definition of terms)

Therefore, a fixture with two reflector tubes will give the same output as a fixture with three standard lamps. Note that the results of this analysis will change markedly if a single three-lamp fixture is used with standard lamps and/or a higher-priced fixture is used with the reflector lamps, as may be necessary to control lamp brightness.

(c) Energy-Conserving Lamps. These

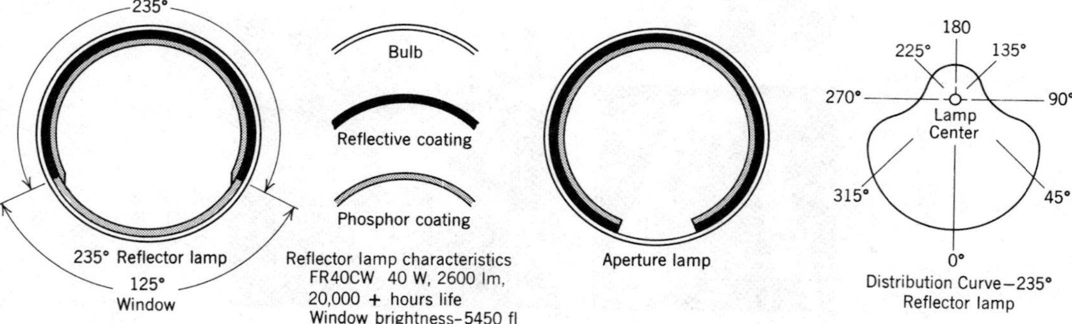

Fig. 19.35 Characteristics of reflector and aperture fluorescent lamps.

Table 19.15 ***Life-Cycle Cost Analysis of Installations with Standard and Reflector Lamps***

Initial Cost	*Three Standard Lamps*	*Two RFL Lamps*
Initial Cost		
Fixture ($20 per fixture)	$30.00[a]	$20.00
Lamps	5.00	10.50
Installation ($30 per fixture)	45.00	30.00
Totals	$80.00	$60.50
Annual Operating Cost (3000-hr operation)		
kwh, 1 year	450	300
Energy cost (@ 3.5¢/kwh)	$15.75	$10.50
Lamp Replacement—lamp cost	.72	1.50
Lamp Replacement—labor cost[b]	2.25	1.50
Fixture cleaning	22.50	7.50[c]
Annual operating cost	$46.22	$21.00
20-Yr Life-Cycle Cost		
(Initial cost and present value of annual costs, assuming 8% interest rate and 3% annual power cost escalation.)	$485.00	$325.00

[a]Assuming 2 lamp fixtures; 3 lamps require 1.5 fixtures.

[b]Assume individual lamp replacement, 15 min per lamp.

[c]Since reflector lamps require only clean diffusers but not clean fixtures, cleaning cost is one-half that of standard lamp unit. Assume 1 man-hour per year for total cleaning job at $15 per man-hour.

lamps are produced by all three major manufacturers and have trademarked names.

Watt-Miser by General Electric.
Econo-Watt by Westinghouse.
Super-Saver by Sylvania.

These lamps are intended as lower-wattage replacements for standard lamps. Comparative wattages follow.

Standard Lamps	*Energy-Conserving Lamp*
48 in., 40 w	34 w
96 in., 112 w HO	95 w
96 in., 215 w VHO	185 w
96 in., 75 w Slimline	60 w

It is important to know that although the efficacy of these energy-saving lamps is higher than their standard counterparts, their output is lower and life is shorter (except for the 40-w unit). This means that application is most appropriate where direct substitution for standard lamps is possible and the corresponding reduction in lighting levels is acceptable. This is frequently the case in stores, corridors, walkways, and many offices. Life cycle analyses based on this assumption are given in Tables 19.16 and 19.17; the first analysis uses a single four-lamp, 48-in. fixture of the type found in lay-in hung ceilings in offices; the second uses a two-lamp 96-in. Slimline unit. This type of analysis should be performed instead of relying on oversimplified charts and graphs, which show savings as a function of energy cost for the life of the lamps or per year. Most of these charts ignore relamping costs and interest rates, which results in a grossly distorted cost picture. Low-energy lamps are clearly marked by the manufacturer, as can be seen in Figure 19.36.

(*d*) ***Zero-Output Lamps.*** These lamps were developed to replace those removed from fixtures in energy-saving programs. As discussed

Table 19.16 ***Comparative Life-Cycle Cost Analysis for Relamping a Single 4-Lamp 4-Ft Fluorescent Fixture with Standard or Energy-Conserving Lamps***

Initial Cost	*Standard Lamps*	*Energy-Saving Lamps*
4 Lamps, CW	$6.68	$10.00
Annual Costs (3000 hr)		
Kilowatt-hours per year[a]	480	420
Energy cost @ $0.035/kwh	$16.80	$14.70
Replacement costs[b]	—	—
20-Year Costs		
(8% interest rate and 3%		
annual energy cost increase)	$218.70	$195.50

[a] Losses in ballasts are not included, since they are the same for both types of lamps.

[b] Replacement costs are the same for both lamps since life is the same. This is not true for other energy-saving lamps.

Table 19.17 ***Comparative Life-Cycle Cost Analysis for Relamping a Single 2-Lamp, 8-Ft Slimline Fluorescent Fixture with Standard Slimline or Low-Energy Slimline Lamps***

Initial Costs	*75-w Standard*	*60-w Low Energy*
2 Lamps—WW	$9.00	$9.30
Annual Costs (3000 hr)		
Kilowatt-hours per year		
(incl. ballast loss)	528	438
Energy cost @ $0.035/kwh	$18.48	$15.33
Lamp Replacement costs		
Lamp	2.25	2.33
Labor	2.50	2.50
	$23.23	$20.16
20-Year Costs @ 8% interest		
rate and 3% annual escalation		
of energy costs.	$302.16	$263.71
20-Year Energy Savings		1800 kwh

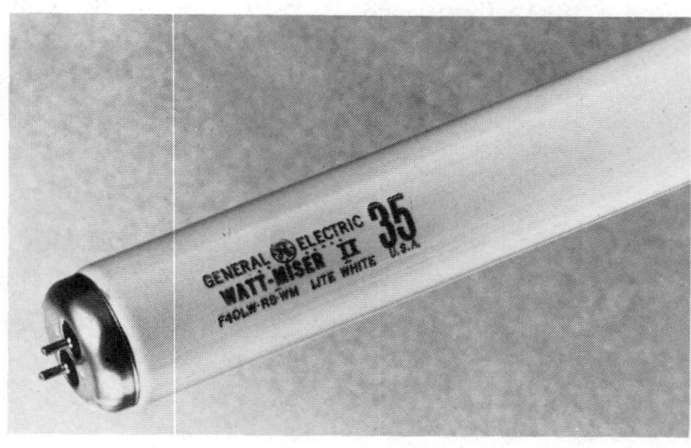

Fig. 19.36 *Energy-saving lamps are clearly so marked, as shown. Illustrated is General Electric's 35-w, T12 rapid start unit; intended to replace the standard 40-w lamp. The illustrated lamp has substantially the same output as a standard lamp, thus resulting in no foot candle loss when replacement is made. Courtesy of General Electric, Lamp Division.*

in Section 19.14, removal of a single lamp from a two-lamp fixture *may* have a deleterious effect on the ballast. To avoid this, dummy tubes are available that maintain balanced load on the ballasts.

(*e*) ***Compact Fluorescent Lamps.*** Compact fluorescent lamps that can directly replace incandescents have long been in development, and at least two types are presently in the testing stage. The desirability of such a lamp is obvious; acceptance will depend largely on initial cost and electric power rates.

19.14 Fluorescent Lamp Ballasts

The fluorescent lamp ballast is a complex assembly of electrical components producing considerable heat and noise, and it is responsible for the proper starting and continued operation of the fluorescent lamps in a fixture. Some of the important considerations in selection and application of ballasts are discussed below.

(*a*) ***Ballast Temperature.*** Normal ballasts are designed to operate in the fixture at a maximum temperature of 90° C in a 25° C ambient. Ballast temperature is important, since ballast life is directly affected by it. At normal temperature a ballast life of 12 to 15 years can be expected. Ballast life is halved for every 10° C above 90° C operation and conversely is doubled for every 10° C reduction in operating temperature. Ballast temperature rises 0.9° C/1° C of ambient temperature above 25° C. Where ambient conditions above 25° C (77° F) are anticipated, special cool operation ballasts, suitable for use in ambients up to 50° C (122° F) should be specified. These "cool" ballasts are more expensive but, when it is remembered that ballast replacement takes up to 1 hr, without considering inconvenience, work stoppage, and possible fixture damage due to dripping of the ballast compound, the additional cost is more than offset.

(*b*) ***Ballast Labels.*** There are several organizations involved with ballast standards and testing. They are:

CBM—Certified Ballast Manufacturers Association.

ANSI—American National Standards Institute. Originates standards on a national level.

ETL—Electrical Testing Laboratories, Inc. A private, independent organization and a recognized authority in measurements and testing of lamps and lighting equipment.

UL—Underwriters' Laboratories, Inc. An independent, nonprofit organization testing for public safety.

CSA—Canadian Standards Association.

Ballasts should be UL labeled and CBM/ETL certified. The UL (Underwriters Laboratories) label assures intrinsic safety. CBM (Certified Ballast Manufacturers) establishes high-quality design criteria, and ETL (Electrical Testing Laboratories) tests the ballasts to determine that the design standards have been met.

(*c*) ***Ballast Protection.*** The National Electrical Code (NEC) requires that all ballasts for *indoor* fixtures be protected by an integral thermal-sensing device that will disconnect the ballast in the event of overheating. Overheating is caused by excessive voltage, excessive ambient temperature, or failure of a ballast component. These devices are either thermostatic (self-resetting) or fuse-type (self-destructive). Since two of the three conditions that cause overheating are usually correctable, our recommendation is to specify the self-resetting type of protector. Thermally protected ballasts are designated "Type P" by the UL.

(*d*) ***Ballast Heat.*** Ballast heat is transferred to the fixture body by direct metal-to-metal contact (which must be unimpeded) and is then dissipated by radiation and convection from the fixture. Obviously, therefore, the location and method of fixture installation affect the heat transfer from the fixture and consequently the ballast temperature. Pendant fixtures (more than 6 in. below the ceiling) and fixtures recessed into ventilated suspended ceilings do not generally present a temperature problem. Fixtures mounted on insulating surfaces such as low-density acoustic tile, or into unventilated or heated ceiling spaces, or when boxed by a fire-rated enclosure and recessed into a fire-rated ceiling, *do* present serious heat dissipation problems. For such installations, as for high ambient temperature installations, cool operation ballasts must be specified. For surface-mounted fixtures, an air space between the fixture and the ceiling material will markedly reduce ballast temperature. The space should be a minimum of 1.5 in. Since each installation situation represents almost a unique case because of the variables of fixture, ballast, ceiling material, and ambient temperature, the designer should require a temperature test of the specific fixture in the installation situation involved before acceptance of the unit. This can be specified along with shop drawings, photometric data, and a sample unit.

(*e*) ***Normal Ballasts.*** Normal ballasts are designed to start fluorescent lamps in an ambient temperature range of 50 to 105° F. If conditions outside these limits are expected, special ballasts must be specified. For outdoor installations, ballasts suitable for starting at +20, 0, and −20° F are available.

(*f*) ***Ballast Power Factor (pf).*** This factor is determined by unit design and is either high (above 0.9) or low (0.5–0.6). High pf ballasts are more expensive than low pf units, but the additional cost is readily repaid by lower line losses, smaller circuit conductors in long runs, and larger number of fixtures per circuit.

(*g*) ***Radio Noise.*** This is frequently referred to as radio frequency interference (RFI). It is *not* produced by the ballast, but by the arc discharge in the fluorescent tube. (Occasionally a defective ballast does cause RFI.) To minimize RFI, ballasts are available with integral RF noise suppressors. In extreme cases additional suppression can be obtained by installation of RF noise attenuators in the fixture.

(*h*) ***Stroboscopic Effect.*** This is no longer a problem with single-lamp ballasts due to the use of long persistence phosphors in all lamps.

(*i*) ***Energy Considerations.*** Ballasts have a heat loss of their own due to the current passing through them. Recently, low-loss ballasts have been placed on the market as part of the general energy-saving design direction. Ballasts *are* affected by lamp failure or removal as follows.

1. In *rapid-start* circuits, operation of a two-lamp ballast with one lamp removed or burned out will not damage the ballast. Ballast power loss will remain approximately the same.
2. In two-lamp *preheat circuits,* one-lamp operation will cause ballast overheating and shortened life. In such circuits a dummy lamp is well applied.
3. In two-lamp *instant-start* circuits, deactivation of one lamp will normally deactivate the entire ballast. In general, two-lamp ballasts operate at their lowest temperature with two active lamps, other conditions being equal.

(*j*) ***Noise.*** Ballasts produce a hum or buzz when operating that is transmitted to the fixture because of the integral contact required for heat transfer. This contact generally amplifies the noise. Total noise depends on sound rating of the ballast, fixture design, and acoustical characteristics of the room.

Ballasts are sound rated by a letter that indicates not actual sound developed, but performance in a space. The ballast selected should be suitable for the lowest sound level likely to be encountered in the subject space. Generally, "A"-rated ballasts are selected for offices and other normally quiet areas where ballast noise would be objectionable. Since no manufacturer's standardization of ratings exists, recommendations should be obtained from ballast and fixture manufacturers or from an acoustic consultant. Where ballast noise or

Table 19.18 ***Acoustic Criteria for Selection of Fluorescent Lamp Ballasts***

For an Installation in:	*Average Ambient Noise Level of Interior*	*First Choice* Use of ballasts with this rating will be satisfactory within the noise levels shown.	*Second Choice* Use of ballasts with this rating will be satisfactory if care is exercised in selecting the acoustic environment.
TV or radio station, church, synagogue	20–24 decibels	A	B
Offices, residence, night school	26–30 decibels	B	C
Library, reception or reading rooms, school study hall	31–36 decibels	C	D
Noisy office, doctor or dentist office, classroom	37–42 decibels	D	E
Stores and factories	43–49 decibels	F	

heat buildup may present a problem, ballasts can be remotely mounted if provision is made for heat dissipation and noise control (see Table 19.18). Typical ballast data are given in Table 19.19.

(*k*) ***Electronic Ballasts.*** Electronic ballasts comprising a solid-state control package are presently being developed. They are small, light, almost lossless, and permit full-range dimming. Assuming that present problems of RFI and low pf can be overcome, these ballasts would prove a desirable replacement for today's heavy, high-loss unit.

19.15 Fluorescent Lamp High-Frequency Operation

Operation of fluorescent lamps, particularly rapid-start tubes, at frequencies above 60 Hz has many beneficial effects. It increases lamp efficiency (see Fig. 19.37), decreases ballast cost, size, weight, and heat losses (by as much as 90%), and lowers maintenance. Typical *relative* data for a standard 40-w, rapid-start lamp and accessories are as follows.

	60 Hz	*3000 Hz*
Lamp efficacy	1.0	1.08
Lamp life	1.0	1.0
Ballast loss	1.0	0.07
Ballast sound	A	—
Ballast weight, lb	4	2

The principal drawback to wide adoption of this lighting supply system is the high initial cost of the static conversion system. With lower lighting levels the payback period from energy savings is lengthened, and a detailed life cycle cost analysis is required, as it is for all other system design choices.

19.16 Neon Lamps

Neon-vapor lamps consist of exhausted glass tubes filled with neon gas that is ionized

Table 19.19 ***Fluorescent Lamp Ballasts***

Lamp Data		Circuit Voltage	Min. Degree Temp. °F	Sound Rating	Dimensions	
Description	Watts				Length	Cross Sect. Reference
RAPID START—430 ma						
(2) F30T12/RS **Two-Lamp**	30	120	0	B	$9\frac{1}{2}$	A
		120	50	A	$9\frac{1}{2}$	D
(2) F40T12/RS **Two-Lamp**	40	120	0	B	$9\frac{1}{2}$	A
		120	50	A	$9\frac{1}{2}$	A
		277	50	A	$9\frac{1}{2}$	A
HIGH OUTPUT—800 ma						
(1) F72T12/HO or	87	120	−20	C	$11\frac{3}{4}$	C
(1) F96T12/HO **Single Lamp**	112	120	−20	D	$11\frac{3}{4}$	C
(2) F24T12/HO **Two-Lamp**	32	120	−20	C	$11\frac{3}{4}$	C
(2) F36T12/HO **Two-Lamp**	45	120	50	C	$11\frac{3}{4}$	B
(2) F48T12/HO **Two-Lamp**	60					
(2) F72T12/HO **Two-Lamp**	87	120	−20	C	$11\frac{3}{4}$	C
		120	−20	B	$11\frac{3}{4}$	C
		277	−20	B	$11\frac{3}{4}$	C
(2) F96T12/HO **Two-Lamp**	112	120	−20	B	$11\frac{3}{4}$	C
		277	−20	B	$11\frac{3}{4}$	C
		277	50	B	$11\frac{3}{4}$	C
POWERGROOVE SHO/VHO—1500 ma						
(2) F72PG17 **Two-Lamp** or	168	120	−20	D	$16\frac{11}{16}$	C
(2) F72T12 **Two-Lamp**		277	−20	D	$16\frac{11}{16}$	C
(2) F96PG17 **Two-Lamp** or	218	120	−20	D	$16\frac{11}{16}$	C
(2) F96T12 **Two-Lamp**		277	−20	D	$16\frac{11}{16}$	C
SLIMLINE AND INSTANT START—200 ma						
(1) F24T12 **Single Lamp**	10	120	0	B	$11\frac{3}{4}$	B
(1) F36T12 **Single Lamp**	15					
(1) F42T12 **Single Lamp**	17					
(1) F48T12 **Single Lamp**	20					
(1) F64T6 **Single Lamp**	37	118	0	D	$14\frac{5}{16}$	C
(1) F72T8 **Single Lamp**	37.5					

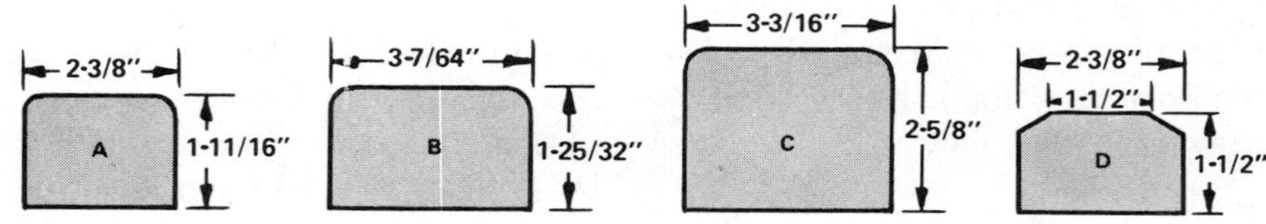

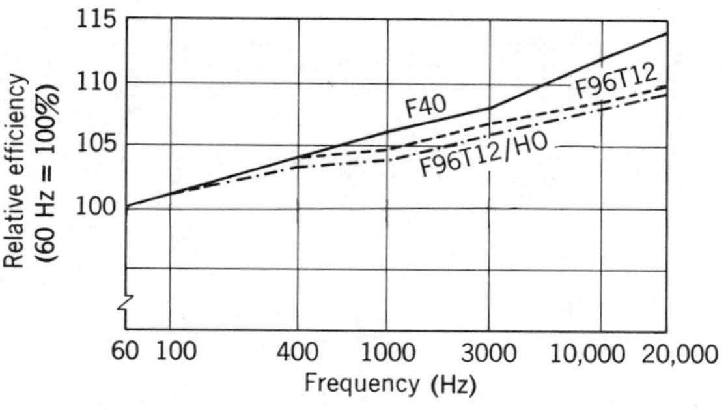

Fig. 19.37 The efficiency of most fluorescent lamps increases when frequency increases. On this curve, the efficiency at 60 cps is taken as the 100% value.

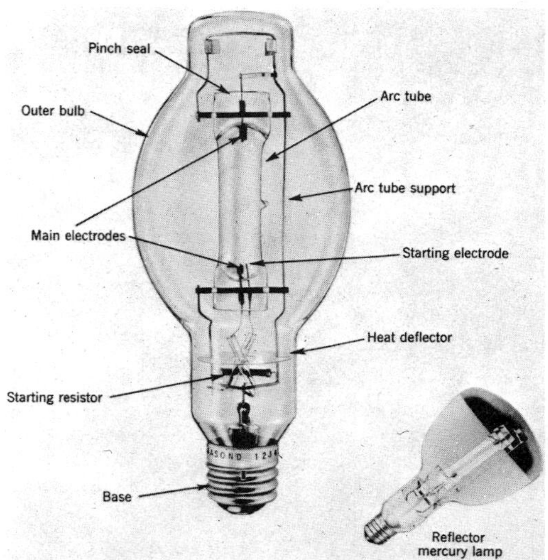

Fig. 19.38 Typical construction of a clear mercury lamp.

and conducts an electric current through the tube. A high voltage is required because of the large voltage drop at the cathode; consequently, a transformer is a necessary part of the equipment. A step up from 115 to 6000 or 10,000 v may be required. Neon light has a pink to dark red color, depending on the gas pressure. The tubes are commonly used in street, window, and indoor signs. Different colors may be obtained by using helium gas instead of neon, by using mixtures of the two, or by using colored glass tubing.

HIGH-INTENSITY DISCHARGE (HID) LAMPS

19.17 General Characteristics of HID Lamps

Subsumed under this heading are mercury, metal-halide, and sodium-vapor lamps. These lamps have inherently high efficacy and, with appropriate color correction, can be utilized in any application, indoor or outdoor, that does not have critical color criteria.

19.18 Mercury Lamps

These lamps operate by passing an arc through a *high-pressure* mercury vapor contained in an arc tube made of quartz or glass (see Fig. 19.38). This action produces light in both the ultraviolet region (as in the low-pressure fluorescent lamp tube) and in the visible region, principally in the blue-green band. This color is characteristic of the clear mercury lamp. Details of spectral distribution are given in Section 19.24.

(*a*) ***Lamp Designations.*** Designations for mercury lamps are more complex than for incandescent or fluorescent lamps and have been a source of confusion for years, because manufacturers invented their own systems in an attempt to simplify specifications. The American National Standards Institute (ANSI) adopted a simplified code some time ago that is now used by all manufacturers and is shown in Table 19.20. This code has five parts and is best illustrated by example. Lamp designation H 38 MP 100 DX indicates:

H—Mercury lamp
38—Ballast number
MP—indicates lamp physical characteristics
100—lamp wattage
DX—Identifies phosphor, glass coating, or coloring. Optional with each manufacturer. Lack of a letter indicates a clear lamp.

Table 19.20 ***Typical Data for Mercury Vapor Lamps***[a]

Lamp Watts	Bulb	Base	ANSI Ordering Abbreviation	Description (See below)	Light Center Length (In.)	Max. Overall Length (In.)	Rated Aver Life (Hr)	Approximate Lumens Initial	Approximate Lumens Mean[c]	Average[b] Ballast Loss (w)	Efficacy (LPW) Lamp & Ballast	Efficacy (LPW) Lamp
40	B-17	Med.	H46DL-40-50/DX	G	$3\frac{1}{8}$	$5\frac{1}{8}$	16000+	1100	800	8	23	28
50	B-17	Med.	H46DL-40-50/DX	G	$3\frac{1}{8}$	$5\frac{1}{8}$	16000+	1550	1150	10	26	31
75	B-21	Med.	H43AZ-75	G, S	$3\frac{3}{4}$	$6\frac{1}{2}$	16000+	2800	2350	12	32	37
			H43AY-75/DX	G, S	$3\frac{3}{4}$	$6\frac{1}{2}$	16000+	2800	2200		32	37
			H43AY-75/N	G, S	$3\frac{3}{4}$	$6\frac{1}{2}$	16000+	2050	1600		24	27
			H43AY-75/R	G, S	$3\frac{3}{4}$	$6\frac{1}{2}$	16000+	2800	2200		32	37
100	A-23	Med.	H38LL-100	G	$3\frac{1}{2}$	$5\frac{7}{16}$	24000+	4100	3450	15	36	41
			H38MP-100/DX	G	$3\frac{1}{2}$	$5\frac{7}{16}$	24000+	4300	3200		37	43
			H38MP-100/N	G	$3\frac{1}{2}$	$5\frac{7}{16}$	24000+	3600	2650		31	36
	BT-25	Mog.	H38HT-100	G, S, B	5	$7\frac{1}{2}$	24000+	4100	3450	15	36	41
			H38JA-100/R	G, S	5	$7\frac{1}{2}$	24000+	4400	3300		38	44
	R-40	Med.	H38BP-100/DX	RF, FF, VW		$7\frac{1}{2}$	24000+	2850	2280	15	25	29
			H38BP-100/N	RF, FF, VW		$7\frac{1}{2}$	24000+	2450	1950		21	25
			H38BP-100/R	RF, FF, VW		$7\frac{1}{2}$	24000+	2850	2280		25	29
175	BT-28	Mog.	H39KB-175	G, S, B	5	$8\frac{5}{16}$	24000+	7700	6600	20	40	44
			H39KC-175/DX	G, S	5	$8\frac{5}{16}$	24000+	8500	6800		44	49
			H39KC-175/N	G, S	5	$8\frac{5}{16}$	24000+	7000	5600		36	40
			H39KC-175/R	G, S	5	$8\frac{5}{16}$	24000+	8500	6800		44	49
	R-40	Med.	H39BM-175	RF, FF, W		$7\frac{1}{2}$	24000+	6100	5150	20	31	35
			H39BP-175/DX	RF, FF, VW		$7\frac{1}{2}$	24000+	5750	4600		30	33
250	BT-28	Mog.	H37KB-250	G, S, B	5	$8\frac{5}{16}$	24000+	12100	9850	25	44	48
			H37KC-250/R	G, S	5	$8\frac{5}{16}$	24000+	13000	9750		47	52
300	BT-37	Mog.	H33CD-300	G, S	7	$11\frac{1}{2}$	16000+	14000		40	41	47
	(Econ-o-watt)		H33GL-300/DX	G, S	7	$11\frac{1}{2}$	16000+	15700			46	52

400	BT-37	Mog.	H33CD-400	G, S, B	7	$11\frac{1}{2}$	24000+	21000	18300	40	48	53
			H33GL-400/DX	G, S	7	$11\frac{1}{2}$	24000+	23000	18400		52	58
	R-57	Mog.	H33FY-400	G, B, RF, FF, W		$12\frac{3}{4}$	24000+	18500	16400	40	42	46
			H33DN-400/DX	G		$12\frac{3}{4}$	24000+	23000	18400		52	58
700	BT-46	Mog.	H35NA-700	G, S	$9\frac{1}{2}$	$14\frac{1}{2}$	24000+	41000	35700	70	53	59
1000	BT-56	Mog.	H34GV-1000	G, B	$9\frac{1}{2}$	$15\frac{3}{8}$	16000+	55000	44000	100	50	55
			H34GW-1000/DX	G	$9\frac{1}{2}$	$15\frac{3}{8}$	16000+	56000	36400		51	56
			H36GV-1000	G, S, B	$9\frac{1}{2}$	$15\frac{3}{8}$	24000+	57500	47100	100	52	58
			H36GW-1000/DX	G, S	$9\frac{1}{2}$	$15\frac{3}{8}$	24000+	63000	44700	↓	57	63

Source. Westinghouse Corporation.

[a]For accurate, current data consult the manufacturers' catalogs.

[b]Losses vary widely with type of ballast; figures given are average.

[c]At two-thirds life.

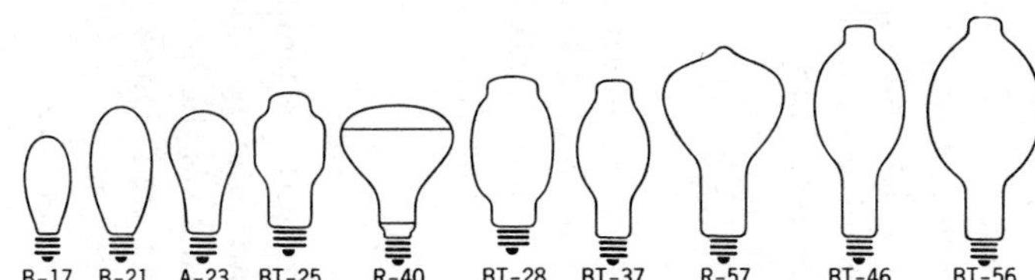

• Explanation of color suffix in ordering abbreviation:

/DX	Deluxe White	/R	Beauty Lite
/N	Style-Tone		No suffix—Clear (non-phosphor coated)

Descriptive Symbols

B	Black Light	RF	Reflector Flood
FF	Frosted Face	S	Street Lighting
G	General Lighting	VW	Very Wide Beam
W	Wide Beam		

(*b*) ***Lamp Life.*** Lamp life is extremely long, averaging 24,000+ hours based on 10 burning hours per start. Mercury lamps are not suitable for applications that are subject to constant switching; therefore, a relatively long period of burning per start was selected. A typical lamp mortality curve is shown in Figure 19.39. Life is affected by ambient temperature, line voltage, and ballast design. Mercury lamps are not as sensitive to short burning cycles as fluorescents but, because of accelerated lumen depreciation near the end of life, they are normally replaced before burnout.

(*c*) ***Lumen Maintenance.*** This depends on the type of lamp and its burning position. Manufacturers publish data on *each* of their lamp types. In general, clear lamps have the best lumen maintenance, followed by color-improved and phosphor-coated units. A curve showing the average data is given in Figure 19.40.

(*d*) ***Color Correction and Efficacy.*** These are added because the blue-green light distorts almost all colors. The outer bulb is coated with phosphors that are excited by the UV light and reradiate generally in the red band, which is entirely absent in the basic lamp color. Depending on the arc tube design and the phosphors used, the color of the emitted light can be corrected to make it acceptable for general indoor use. Lamps are available in clear, white, color-corrected, and white-deluxe, in ascending order of color improvement. The deluxe lamp also uses a stain on the envelope to filter out some of the blue-green, which obviously reduces lamp output. Efficacy of selected typical units, with and without ballast loss, is given in Table 19.20. Note that in general efficacy is lower than that of fluorescent lamps.

(*e*) ***Ballasts.*** Ballasts are required, as with all arc discharge lamps, to start the lamp and thereafter to control the arc. The basic ballast is simply a reactor that controls the arc after the discharge has been initiated. Three to six minutes are required for the lamp to reach full output, since heat must be generated by elec-

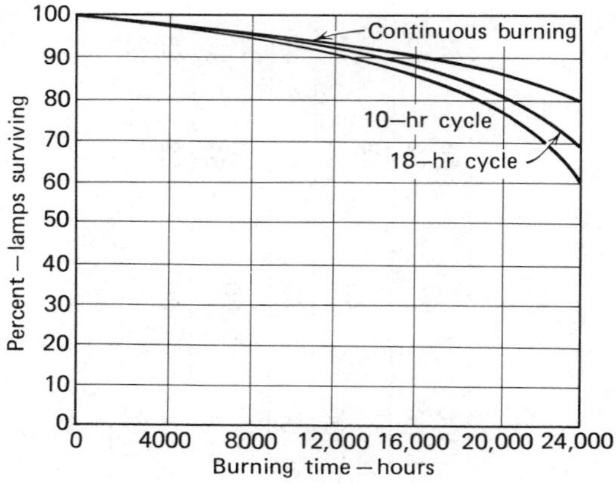

Fig. 19.39 *Typical life expectancy or survival curves for 175-, 400-, and 1000-w mercury lamps at various burning cycles.*

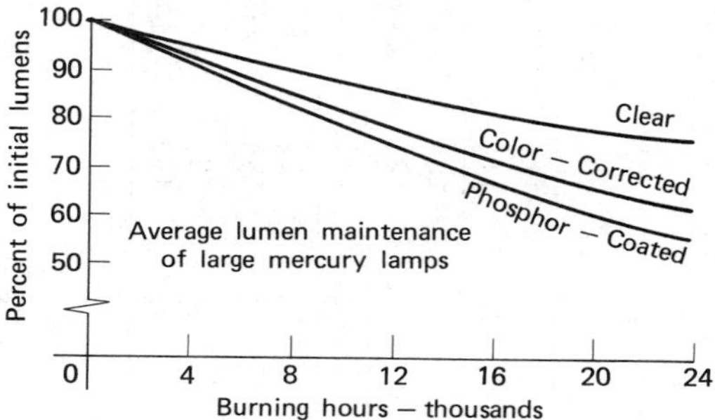

Fig. 19.40 *Average lumen maintenance for mercury lamps 175-w and larger.*

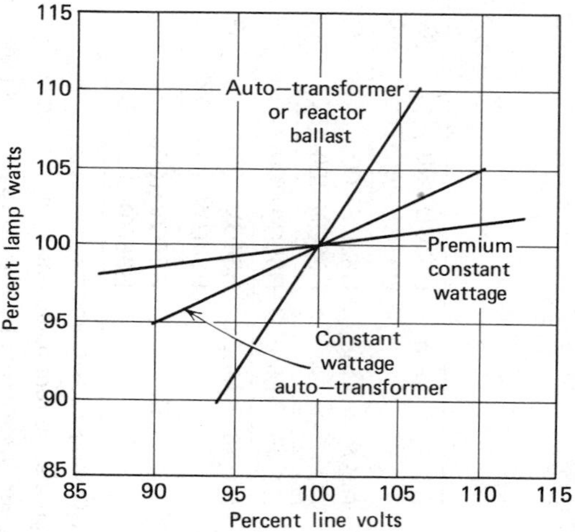

Fig. 19.41 *Effect of line voltage variation on lamp watts with various ballast types.*

tron flow to vaporize the mercury in the arc tube before the arc will strike. Once extinguished, the lamp must cool and the pressure must be reduced before restrike is possible. This restart delay amounts to 3 to 8 min, depending on the ballast type, and is an important consideration in design, since a momentary outage will extinguish all lamps, leaving an interior area in the dark. Special fixtures now available utilize small quartz lamps to supply light during such outages. Alternatively, some incandescent lighting can be utilized that will maintain minimum illumination. Principal types of ballasts are:

1. Reactor ballast is low power factor, provides no voltage regulation, and should only be used where line voltage fluctuation does not exceed ±5%.
2. Autotransformer ballast is a reactor unit with a transformer to match line voltage to lamp voltage. It also is low pf and nonregulating.
3. High pf autotransformer ballast. Same as type 2, except with additional capacitor to improve pf.
4. Constant wattage autotransformer ballast, also called a lead circuit ballast, is a regulating high pf unit that maintains lamp voltage and hence wattage and lumen output fairly constant. Lamp wattage will vary 5% with a 10% voltage change.
5. Premium constant wattage or stabilized ballast provides lamp isolation, voltage regulation, high pf, low extinction voltage and low inrush current.

Ballast prices increase from type 1 to type 5 in the ratio of 1.0 to approximately 2.5. When line voltage varies, lamp voltage is affected also, depending on the type of ballast (see Fig. 19.41). Lamp wattage and lumen output are directly proportional to lamp voltage. Lamp operation at overwattage is inadvisable, since lamp temperature increases while life and lumen maintenance decrease. Mercury ballasts are normally quite noisy. Where this may be a problem, remote mounting should be considered. Typical ballast data are given in Table 19.21.

(*f*) ***Dimming.*** Dimming of mercury lamps is possible and entirely practicable with the use of a dimming ballast and solid-state dimming control. These are available for 400-, 700-, and 1000-w units and, unlike the case of fluorescents, dimming is a desirable and economical control means. Mercury lamps have so large an output that shutting off a unit creates an imbalance in the lighting coverage—a problem readily solved by dimming. Cost analyses indicate the economic feasibility of such control, including sensing equipment that will automatically maintain illumination levels at preset levels. The payback period depends on power rates and local labor rates. One analysis performed by the writer for a large installation, with 50–1000-w lamps, a \$.035/kwh energy rate, and 8% capital cost, indicated a $4\frac{1}{2}$ year payback period, assuming 4000-hr annual operation (see Tables 19.16 and 19.17 for calculation techniques).

A little used but very effective and economical output reduction technique is simply to change the circuit capacitance by an amount, depending on lamp size and ballast type. By doing this, the lamp wattage and output can be reduced by approximately 50% with no deleterious effect on lamp or ballast. This technique is by far the cheapest method of accomplishing an overall, even reduction in output.

(*g*) ***Application.*** Mercury-vapor lamps are applicable to indoor and outdoor use with proper attention to color and fixture brightness. Indoor application is generally limited to mounting 10 AFF or higher to avoid glare problems and to permit adequate area coverage. Use in industrial spaces and stores is common as is discussed in detail in Chapter 21.

19.19 Special Mercury Lamps

In an attempt to satisfy the desire for a small lamp to take the place of incandescents in interior fixtures, manufacturers have made available mercury lamps in 40-, 50-, 75-, 100-, and 175-w sizes, in deluxe white and other color-corrected designs (see Table 19.20). For

Table 19.21 ***Typical Mercury Vapor Lamp Ballast Data. Constant Wattage Ballasts—Indoor Service***[a,b]

Lamp type and Wattage	Nominal Line Voltage	Line Voltage Range	Watts Loss	Line Current	Min. Start. Temp (°F)	Ship. Wgt. (lb)	Dimension in Inches: Over. Lgth. A	Case Width B	Case Hgt. C
1–75 w H-39	120 × 240	108–132 × 216–264	35	1.83 0.94	0	17	$13\frac{3}{8}$	$6\frac{1}{8}$	$3\frac{3}{8}$
	240 × 277	216–264 × 250–300		0.94 0.79					
1–250 w H-37	120 × 240	108–132 × 216–264	35	2.56 1.27	0	20	$13\frac{3}{8}$	$6\frac{1}{8}$	$3\frac{3}{8}$
	240 × 277	216–264 × 250–300		1.27 1.08					
1–400 w H-33	120	108–132	55	4.0	0	22	$15\frac{3}{4}$	$6\frac{1}{8}$	$3\frac{3}{8}$
	240 × 277	216–264 × 250–300		2.0 1.7					
	208	190–230		2.3					
	480	432–528		1.0					
1–1000 w H-36	120 × 240	108–132 × 216–264	90	9.30 4.65	−20	37	$16\frac{1}{16}$	$6\frac{1}{8}$	$5\frac{15}{16}$
	277	250–300		4.05					

Source. Sola Electric Co.

[a]All constant-wattage ballasts have minimum power factor of 90%.

[b]For accurate, current data refer to the manufacturers' catalogs.

the smaller sizes, screw-in ballasts are available, so that replacing an incandescent is simply a matter of screwing in a ballast and a small mercury lamp. In the 175-w size the ballast must be separately mounted. Available also are self-ballasted lamps (require no separate ballast), which can be used where ballast mounting is inconvenient, expensive, or undesirable for other reasons. In both cases, there is no doubt that the extremely long life, good color, and reliability of these incandescent substitutes make their use attractive in locations where relamping is difficult and expensive (see Fig. 19.42). But it also is true that their efficacy is only slightly better than that of the incandescent. Indeed many of the *self-ballasted* units have efficacies *lower* than that of the incandescent, making their use, from an energy standpoint, questionable. Furthermore, the high cost of self-ballasted units and the loss of the ballast when the lamp burns out makes this lamp questionable economically, except in very special cases. Similarly the long delay (5 to 10 min) in starting and restarting severely restricts applicability.

In all instances where self-ballasted or small mercury units are contemplated, consideration should be given to fluorescents, whose low price, long life, very high efficacy, and good color make them attractive economically and from the energy-resource aspect. Table 19.22 gives an analysis of four alternatives for lighting an *existing*, difficult-access location, as in Figure 19.42:

1. Continue using 150-w incandescent lamps.
2. Use a screw-in ballast and a 75-w deluxe white mercury unit.
3. Install directly a self-ballasted 160-w mercury lamp.

Fig. 19.42 Application of self-ballasted mercury lamp in a difficult access location. Photo courtesy of Westinghouse Electric Corp., Lamp Division.

4. *Replace* the existing fixture with a simple, 48-in. open-reflector strip fluorescent for one F40 T12 CW lamp.

From the table it is clear that alternative 4 is the economic choice and also, by far, the energy-resource choice. If installation of a fluorescent is impossible, the next choice is clearly the externally ballasted 75-w mercury unit.

19.20 Metal-Halide Lamps

The metal-halide lamp is basically a mercury lamp that has been altered by the addition to the arc tube of halides of metals such as thallium, indium, or sodium. The addition of these salts causes light to be radiated at frequencies other than the basic mercury colors and increases efficacy, but reduces the life and reduces lumen maintenance to 60% at two-thirds life. The color produced is much warmer than that of the mercury light. Clear lamps are recommended for exterior use and phosphor-coated units for all indoor application including food displays.

Table 19.22 ***Cost Analysis of Four Alternative Methods of Relighting a Stairwell***

Lamp Data and Fixture	*150-w Incandescent*	*75-w Mercury Separate Ballast*	*160-w Self-Ballasted Mercury*	*New Fluorescent Fixture with 40-w T12-CW Lamp*
Initial Lamp Cost	$1.10	$18.00	$35.00	$1.67
Initial Ballast/Fixture				
Cost Incl. Labor	—	24.00	—	60.00
Total Initial Cost	$1.10	$42.00	$35.00	$61.67
Lamp Wattage Including				
Ballast	150W	90W	160W	50W
Lamp Initial Lumens	2250	2800	2700	3150
Lamp Life-Hours	750	24000	20000	20000
Annual Data—4000 hr				
Cost of Lamp Replacement				
Lamp	$5.90	$6.00	$4.50	$0.35
Labor	110.00	3.50	5.00	5.00
Total—Lamps	$116.90	$9.50	$9.50	$5.35
Cost of Energy:				
Kilowatt-hours	600	360	640	200
@ 0.035/kwh	$21.00	$12.60	$22.40	$7.00
Total Annual Cost	$137.90	$22.10	$31.90	$12.35
10-Year Life Cycle Cost	$1036.00	$214.00	$275.00	$153.00
20-Year Life Cycle Cost	$1741.00	$321.00	$438.00	$217.00
10-Year Energy Use—kwh	6000	3600	6400	2000
Relative Energy Use	3.0	1.8	3.2	1.0

A brief comparison of these two lamps shows:

	Mercury	*Metal-Halide*
Life	16,000 to 24,000 hr	7500 to 15,000 hr
Color	Poor to fair	Good to excellent
Lamp efficacy	50 to 60 lpw	80 to 100 lpw

Since the color of the metal-halide lamp depends on the amount of ionized halide salt in the arc, lamp performance is extremely sensitive to voltage, temperature, and burning position. Mortality and lumen maintenance curves are similar to those for mercury lamps except for lower values as noted above. Strike time is shorter than that of the mercury lamp, being 2 to 3 min, but restrike time is up to 10 min, making it necessary to supply an instant start source in indoor areas lighted with these lamps. Certain metal-halides are usable with mercury ballasts (see Table 19.23) although generally they require special ballasts with higher voltage and better regulation than mercury lamps. Common trade names for these lamps are Metal-Arc (Sylvania), Multi-Vapor (G.E.), and Metal Halide (Westinghouse). Typical data are given in Table 19.23.

19.21 High-Pressure Sodium (HPS) Lamps

The most recently developed HID source is the high-pressure sodium lamp, marketed by its developer, General Electric, under the trade name *Lucalox.* (Other trade names for this product are Westinghouse's Ceramalux and Sylvania's Lumalux.) Details of the lamp are shown in Figure 19.43, and life and lumen maintenance data are given in Figure 19.44. Construction is quite different from that of mercury and metal-halide lamps and, although it operates as an arc discharge unit, its excellent characteristics stem from the spectral absorption phenomenon of the contained sodium under high pressure. The resultant light is a yellow-tinted color, similar to that of warm white fluorescent lamps. Typical characteristics are:

Lamp efficacy	85 to 140 lpw
Efficacy including ballast losses	55 to 125 lpw
Life	16,000 to 24,000 hr
Lumen maintenance	80 to 90%
Warm-up time	3 to 4 min
Restrike time	$\frac{1}{2}$ to $1\frac{1}{2}$ min

Unlike the metal-halide lamp, the HPS unit is not voltage sensitive and is color constant. Its lumen maintenance is outstanding as is its efficacy, both of which are the highest available for any general-use light source. Details of light color are discussed in Section 19.23, and comparative characteristics with other HID sources and fluorescent lamps are given in Section 19.25. As with all discharge lamps, a ballast is required to supply the high voltage to strike the arc, and to control the arc once struck. HPS ballasts are quite different than those for mercury or metal-halide lamps because of the high voltages necessary. However, in order to make the changeover from existing mercury installations to HPS attractive financially, a line of special HPS units is available that can be substituted directly for existing mercury lamps. This gives considerable improvement in illumination and reduction in energy costs, since on the average HPS lamps will supply double the efficacy of mercury lamps. These direct mercury replacements are marketed under the trade names of E-Z Lux and Unalux by G.E. and Sylvania, respectively. HPS lamps are available in clear and coated designs. The former is effectively a point source and, because of its extreme brightness, must be enclosed in a fixture. The latter is intended to substitute photometrically for coated mercury lamps and to constitute a lesser glare source, since lamp surface brightness is correspondingly reduced. Table 19.24 gives characteristics of typical HPS lamps that are presently available. Since HPS sources are the highest-efficacy general lighting sources in use, their energy-saving implications are obvious. Therefore, this lamp should be considered primarily, or as a substitute, for all existing or new HID applications. Its yellow color, which becomes less noticeable as the eye color-adapts, can be made even more acceptable by using white sources in conjunction. Application recommendations are found in Chapter 21.

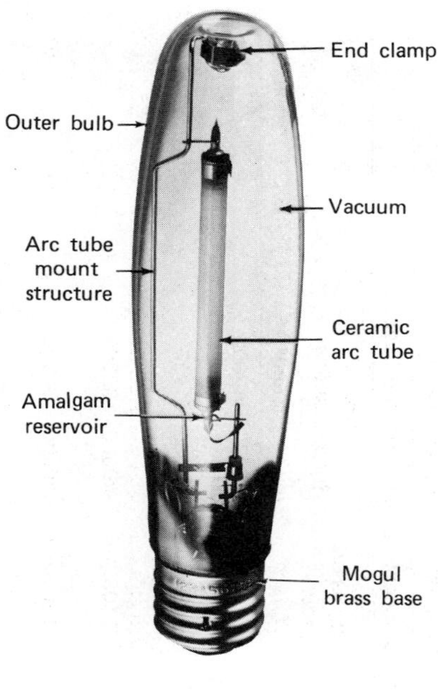

(a)

Fig. 19.43 (a) *The main features of the HPS lamp are the alumina ceramic arc tube, amalgam reservoir, and rigid arc tube mount structure. The illustrated unit is rated 310 w and yields 37,000 initial lumens and 24,000 hr life.*

End cap
Ceramic arc tube
Electric discharge through sodium vapor
Coated tungsten electrodes
Exhaust tube with amalgam reservoir
Mogul base
Ballast
Line
Voltage to lamp

(b)

(b) *Electrically, the HPS lamp is a simple device with the base contacts and internal wiring serving only to provide a current path between the ballast and arc tube electrodes. Illustrations courtesy of General Electric, Lamp Division.*

19.22 Low-Pressure Sodium Lamps

This lamp, also referred to as SOX, produces light of sodium's characteristic monochromatic deep yellow color, making it inapplicable for general lighting. Because of its very high efficacy of over 150 lumens per watt *including* ballast loss (compare HPS, Table 19.24), it can be applied wherever color is not an important criteria. Thus SOX is widely used for street, road, and area lighting, as well as for emergency or after-hours indoor lighting. Another desirable aspect of SOX lamps is their 100% lumen maintenance. This, coupled with the discharge lamp's typically long life (18,000+ hrs), makes SOX lamps the most economical source available today in terms of cost per million lumens produced.

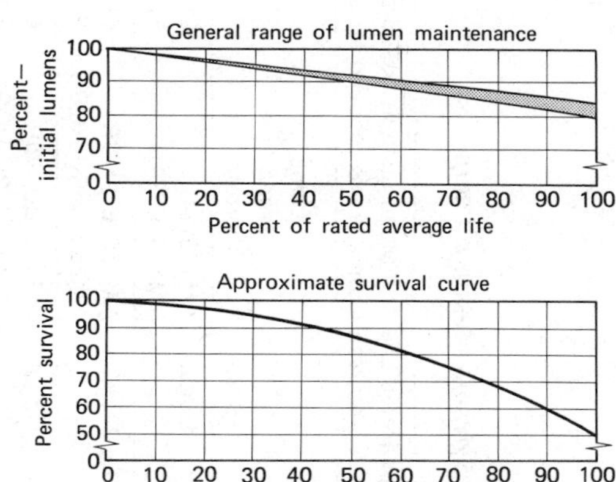

Fig. 19.44 *Characteristics of HPS lamps. Its lumen maintenance of 90% at 50% life is by far the best of all the HID sources.*

Table 19.23 *Sylvania Metalarc Lamps*

Special Ballasts, except as noted, are required; will not operate on most Mercury ballasts. For operation on some 400-w Mercury ballasts, see Metalarc Swingline. Above 50° F all Lamps will operate on 480-v Mercury Reactor; 175-w and 250-w will operate down to −20° F. For lumen output performance at angles other than vertical or horizontal, see Metalarc Lamp Engineering Bulletin 0-344.

Metalarc Lamps Rated Life and Mean Lumens based on minimum of 10 hr per start operation for 175-, 250-, 400-, and 1000-w lamps and 4 hr start on 1500-w lamps. Rated Life and Mean Lumens are reduced for shorter burning cycles.

Watts	*Bulb*	*Base*	*Lamp Ordering Abbreviation*	*Description*	*Avg. Rated Hours Life*	*Approx. Vert. Lumens*		*L.C.L.*	*M.O.L.*	*Lamp Efficacy*
						Initial	*Mean*[b]			
175[a]	BT-28	Mogul	M175/BU	Clear Metalarc, base up, vertical only[c]	7500	14000	10800	5	$8\frac{5}{16}$	80
			M175/C/BD	Phosphor coated, base down, vertical only[c]	7500	14000	10200	5	$8\frac{5}{16}$	
250[a]	BT-28	Mogul	M250/BU-HOR	Clear Metalarc, base up to horizontal	7500	20500	17000	5	$8\frac{5}{16}$	82
			M250/C/BU-HOR	Phos. coated Metalarc, base up to horizontal	7500	20500	16000	5	$8\frac{5}{16}$	
400[d]	BT-37	Mogul	M400/BU-HOR	Metalarc, base up to horizontal	15000	34000	25600	7	$11\frac{1}{2}$	85
			M400/C/BU-HOR	Phos. coated Metalarc, base up to horizontal	15000	34000	24600	7	$11\frac{1}{2}$	
1000[d]	BT-56	Mogul	M1000/BU-HOR	Metalarc, base up to horizontal	10000	100000	82500	$9\frac{1}{2}$	$15\frac{3}{8}$	100
			M1000/C/BD	Phosphor coated, base down	10000	100000	79000	$9\frac{1}{2}$	$15\frac{3}{8}$	
1500[a]	BT-56	Mogul	M1500/BU-HOR	Metalarc, base up to horizontal	1500	155000	142500	$9\frac{1}{2}$	$15\frac{3}{8}$	103

Super-Metalarc Lamps Lamps must be operated within ±15° of horizontal. Require special socket to accept position-oriented base.

Watts	*Bulb*	*Base*	*Lamp Ordering Abbreviation*	*Description*	*Avg. Rated Hours Life*	*Approx. Horiz. Lumens*		*L.C.L.*	*M.O.L.*	*Lamp Efficacy*
						Initial	*Mean*[b]			
175[a]	BT-28	Position oriented	MS175/HOR	Clear Metalarc, horizontal only	7500	15000	12000	5	$8\frac{5}{16}$	
		Mogul base	MS175/C/HOR	Phos. coated, horizontal only	7500	15000	11300	5	$8\frac{5}{16}$	86
400[a]	BT-37	Position oriented	MS400/HOR	Clear Metalarc, horizontal only	15000	40000	32000	7	$11\frac{1}{2}$	
		Mogul base	MS400/C/HOR	Phos. coated, horizontal only	15000	40000	31000	7	$11\frac{1}{2}$	100

Metalarc Swingline Designed for operation on 400-w Constant Wattage (CW) or Constant Wattage Autotransformer (CWA) mercury ballasts.

Watts	Bulb	Base	Lamp Ordering Abbreviation	Description	Avg. Rated Hours Life	Approx. Vert. Lumens		L.C.L.	M.O.L.	Lamp Efficacy
						Initial	Mean[e]			
400[d]	BT-37	Mogul	MM400/BU-HOR	Metalarc Swingline, base up to horizontal	12000	34000	24320	7	$11\frac{1}{2}$	
			MM400/C/BU-HOR	Phosphor coated, Metalarc Swingline, base up to horizontal	12000	34000	23370	7	$11\frac{1}{2}$	85

Source. Courtesy of GTE Sylvania, Inc.

[a]Operation is restricted to suitably enclosed fixtures only.

[b]Taken at 40% of rated life. Mean lumens are established on ballasts with crest factors of 1.8; higher crest factors reduce values.

[c]Lamps must be operated within $\pm 15°$ of vertical.

[d]Horizontal operation of Metalarc Lamps or operation within 60° of horizontal is restricted to suitably enclosed fixtures only.

[e]Taken at 40% rated life.

Table 19.24 ***Sylvania High Pressure Sodium Lamps***

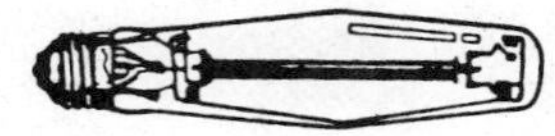

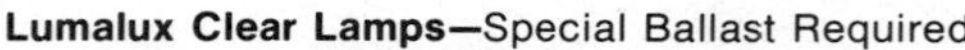

Lumalux Clear Lamps—Special Ballast Required

Watts	Bulb	Base	Avg. Rated Hours Life	Approx. Lumens		L.C.L.	M.O.L.	Lamp Efficacy (LPW)	Lamp and Ballast Efficacy
				Initial	Mean				
70	BT-25	Mogul screw	20000[a]	5800	5220	5	$7\frac{1}{2}$	83	58
100	BT-25	Mogul screw	20000[a]	9500	8850	5	$7\frac{1}{2}$	95	66
150	BT-25	Mogul screw	24000[a]	16,000	14,400	5	$7\frac{1}{2}$	106	79
150	BT-28	Mogul screw	24000[a]	16,000	14,400	5	$8\frac{5}{16}$	106	79
250	E18	Mogul screw	24000[a]	27,500	24,750	$5\frac{3}{4}$	$9\frac{3}{4}$	110	90
400	E18	Mogul screw	24000[a]	50,000	45,000	$5\frac{3}{4}$	$9\frac{3}{4}$	125	104
1000	E25 (*)	Mogul screw	24000[a]	140,000	126,000	$8\frac{3}{4}$	$15\frac{1}{16}$	140	124

Lumalux Coated Lamps Use in open-bottomed fixtures or where glare is a problem; Special Ballast Required

Watts	Bulb	Base	Avg. Rated Hours Life	Approx. Lumens		L.C.L.	M.O.L.	Lamp Efficacy (LPW)	Lamp and Ballast Efficacy
				Initial	Mean				
70	BT-25	Mogul screw	20000[a]	5400	4860	5	$7\frac{1}{2}$	77	54
100	BT-25	Mogul screw	20000[a]	8800	7920	5	$7\frac{1}{2}$	88	62
150	BT-25	Mogul screw	24000[a]	15,000	13,500	5	$7\frac{1}{2}$	100	74
400	BT-37	Mogul screw	24000[a]	47,500	42,750	7	$11\frac{1}{2}$	119	99

[a]Based on operation on proper auxiliary equipment for 10 hr or more per start.

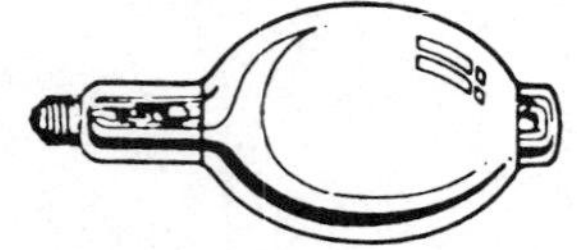

- Operate on existing Mercury lag type auto transformer ballasts or 240-277 volt reactors.
- Will operate on most mercury series circuits.

Unalux Clear Lamps

Watts	Bulb	Base	Avg. Rated Hours Life	Approx. Lumens		L.C.L.	M.O.L.	Lamp Efficacy (LPW)	Lamp and Ballast Efficacy
				Initial	Mean				
150	BT-28	Mogul screw	12000[a]	13,000	11,700	5	$8\frac{5}{16}$	87	70
360	BT-37	Mogul screw	16000[a]	38,000	34,200	7	$11\frac{1}{2}$	106	93

Unalux Coated Lamps Use in open-bottomed fixtures or where glare is a problem.

Watts	Bulb	Base	Avg. Rated Hours Life	Approx. Lumens		L.C.L.	M.O.L.	Lamp Efficacy (LPW)	Lamp and Ballast Efficacy
				Initial	Mean				
150	BT-28	Mogul screw	12000[a]	12,000	10,800	5	$8\frac{5}{16}$	80	65
360	BT-37	Mogul screw	16000[a]	36,000	32,400	7	$11\frac{1}{2}$	100	88

Source. Courtesy of GTE Sylvania, Inc. [a]Based on operation on proper auxiliary equipment for 10 hr or more per start.

COLOR OF LIGHT SOURCES

19.23 Spectral Distribution of Light Sources

In Sections 18.27 to 18.29 we discussed color as an absolute characteristic of both pigments and light; we defined it, classified it, and reviewed some of the salient optical and psychological characteristics. At this point we are ready to discuss the interrelation between illuminant and pigment since it has been truly observed that "color is how you light it." It behooves the lighting designer to choose the light source with the particular color characteristic that will accomplish his or her purpose. To do this, it is necessary that the designer be familiar with the color composition or spectral distribution of the various sources, and their apparent effect (with respect to the eye) on various pigments. If we were to separate each illuminant under consideration into its component colors and plot the relative amounts of energy on a wavelength axis, the results would be somewhat as shown in Figure 19.45. The results are not surprising; we know that north light is very blue, noon daylight yellow-white, light from an incandescent lamp yellow-orange, and that light from fluorescent tubes, mercury lamps, and metal-halides varies with the type of lamp. Light from HPS lamps is fixed in spectral composition by the strong sodium lines in the spectrum.

Fig. 19.45 *Spectral energy distribution curves for typical illuminants, plus their correlated color temperature and color rendering indices. Since only a radiating blackbody has a true color temperature, a source with mixed color illuminants is assigned a color temperature that* correlates *most closely with blackbody radiator, that is, it expresses the illuminant's dominant color. This is its* correlated color temperature.

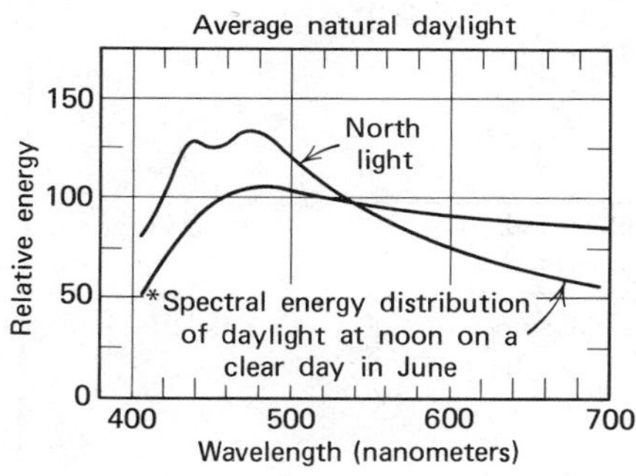

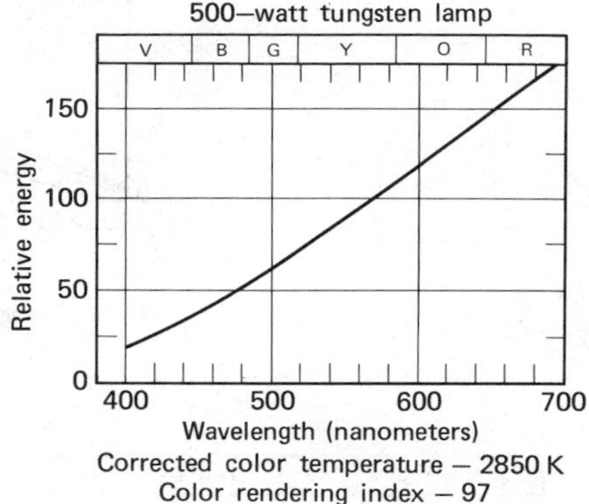

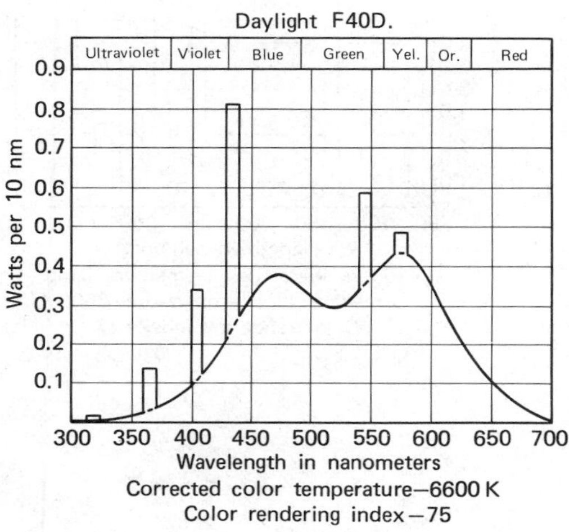

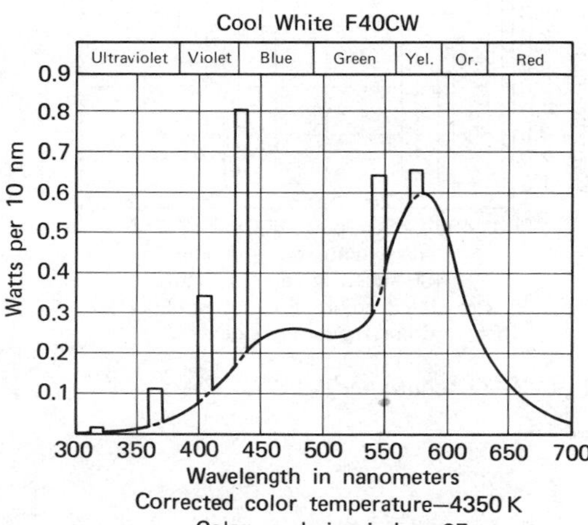

Warm White Deluxe F40WWX.

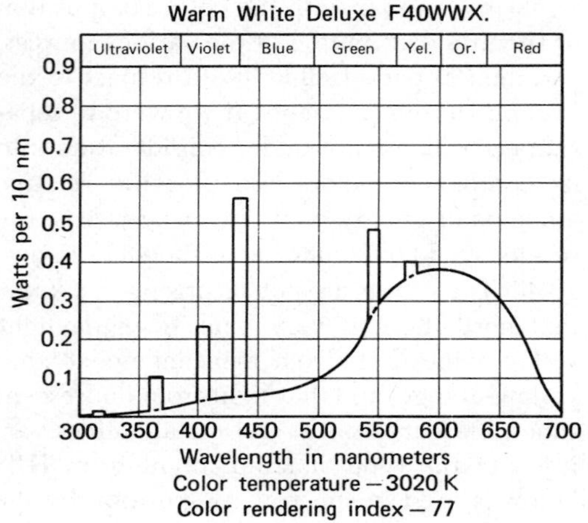

Wavelength in nanometers
Color temperature – 3020 K
Color rendering index – 77

Warm White F40WW

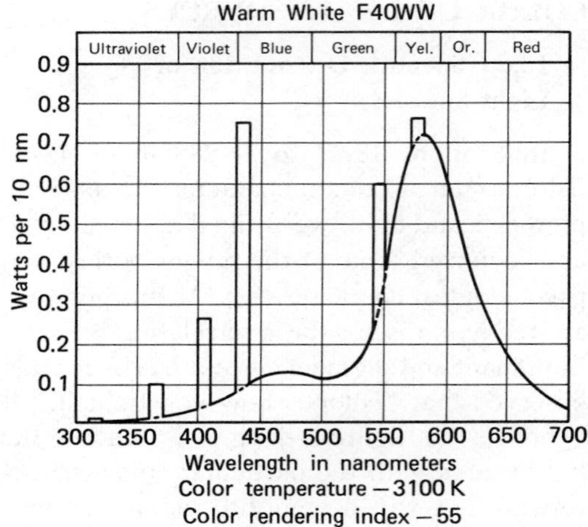

Wavelength in nanometers
Color temperature – 3100 K
Color rendering index – 55

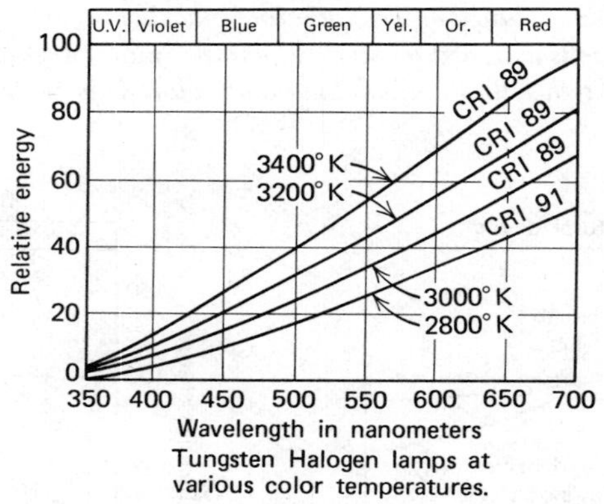

Wavelength in nanometers
Tungsten Halogen lamps at various color temperatures.

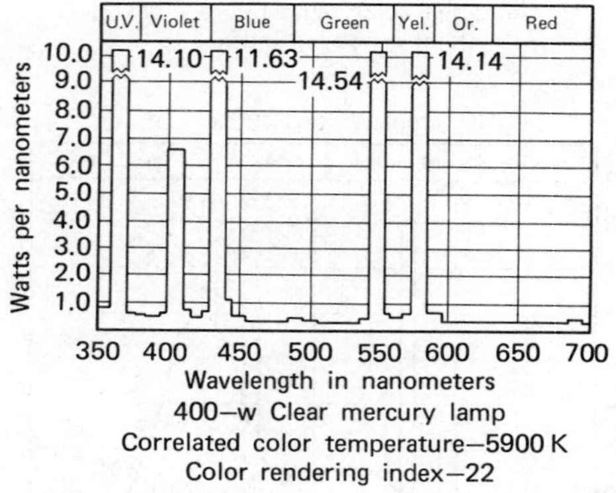

Wavelength in nanometers
400–w Clear mercury lamp
Correlated color temperature–5900 K
Color rendering index–22

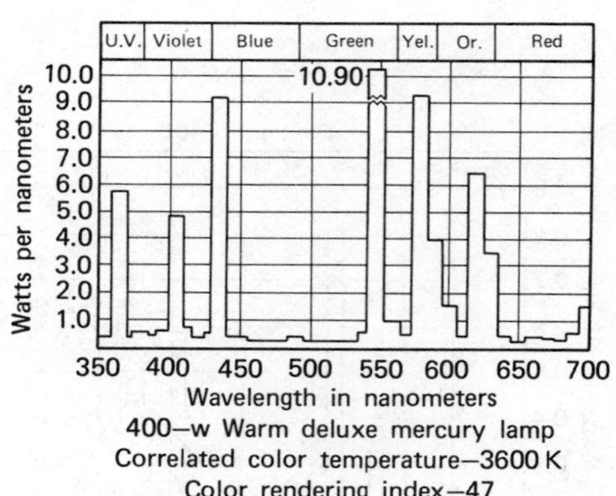

Wavelength in nanometers
400–w Warm deluxe mercury lamp
Correlated color temperature–3600 K
Color rendering index–47

Fig. 19–45 (continued)

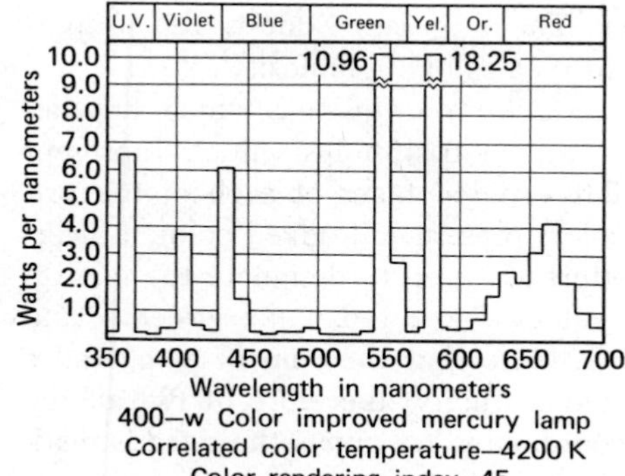

400—w Color improved mercury lamp
Correlated color temperature—4200 K
Color rendering index—45

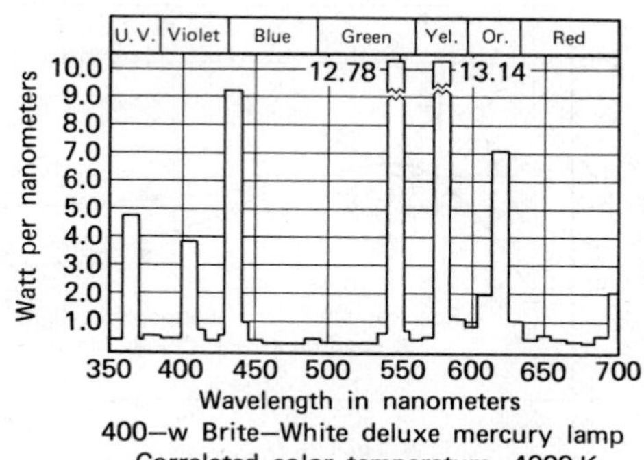

400—w Brite—White deluxe mercury lamp
Correlated color temperature—4000 K
Color rendering index—43

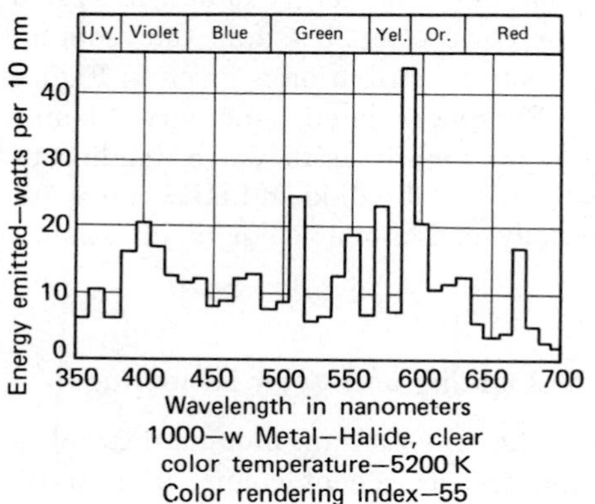

1000—w Metal—Halide, clear
color temperature—5200 K
Color rendering index—55

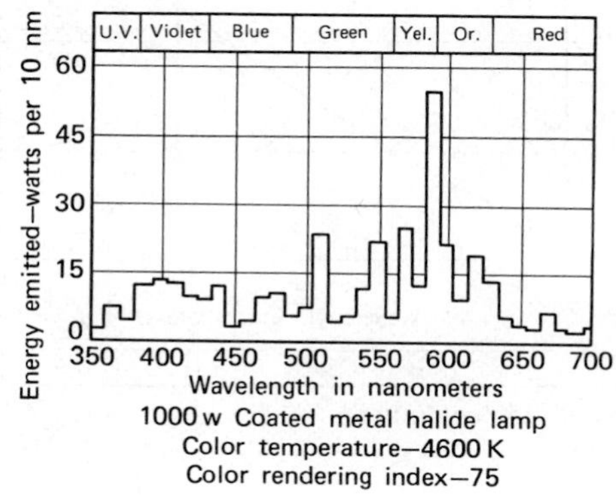

1000 w Coated metal halide lamp
Color temperature—4600 K
Color rendering index—75

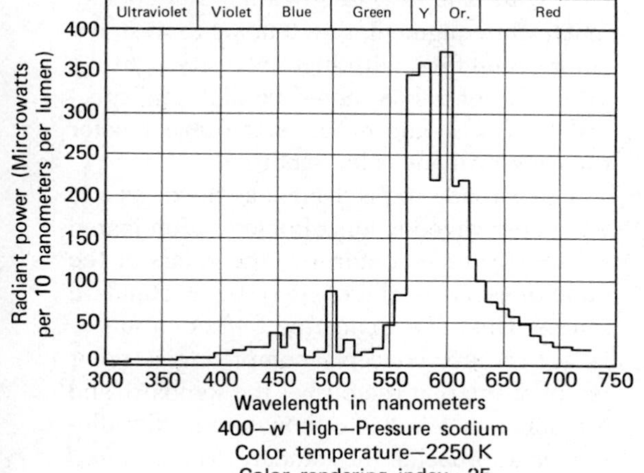

400—w High—Pressure sodium
Color temperature—2250 K
Color rendering index—25

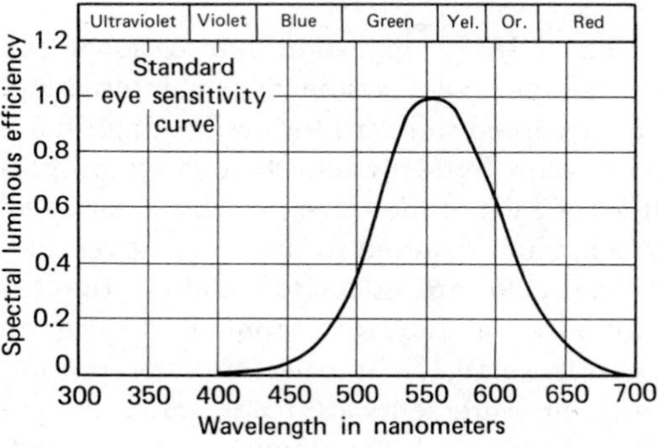

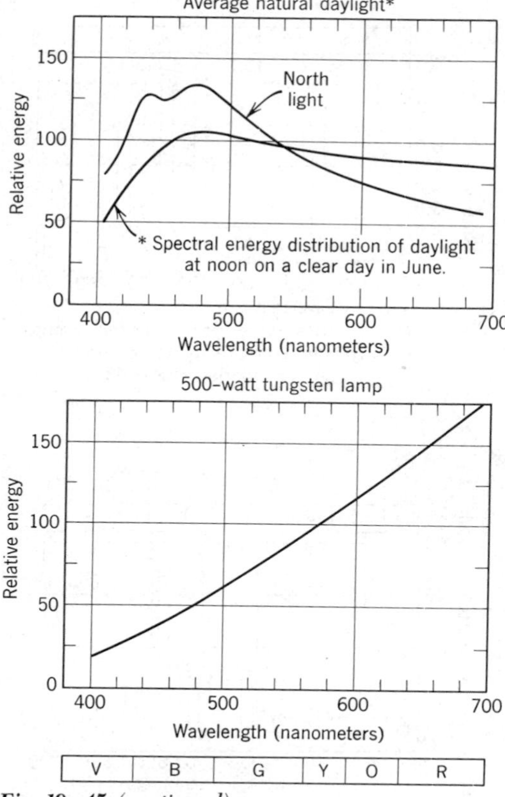

Fig. 19–45 (*continued*)

19.24 Chromaticity

The CIE (Commission Internationale de l'Éclairage) color system is the internationally accepted standard for designating illuminant color. In this system the relative proportions of each of the three primary colors (red, green, blue) required to produce a given illuminant color are calculated, and the result is plotted on a standard chromaticity diagram (see Figure 19.46). To calculate these proportions, measurements are made across the entire spectrum of the illuminant under test. These figures are then weighted in terms of the three primary colors. The resulting figures represent the proportions of red, green, and blue required to produce the spectrum color at that wavelength. These values are called the *tristimulus values* for that color and are designated by capital letters: X (red), Y (green), and Z (blue). The Y (green) value is also proportional to that color's luminosity.

Chromaticity coordinates, which are derived from the tristimulus values, represent the relative percentages of each of the primary colors present in a given color. Lowercase letters are used to designate the coordinate values: x = red, y = green, and z = blue. Note that the sum of x, y, and z always equals unity; therefore, if two of the three coordinates are given, the third is readily computed. Normally only the x and y values are published for a lamp's color coordinates. The $x = y = z = 0.333$ point is known as the point of equal energy. Table 19.25 gives chromaticity coordinates for the sources whose spectral distribution is given in Figure 19.45. Phosphor-coated mercury lamps change color slightly as they age, tending to cooler light. Metal-halide and HPS lamps are substantially constant in color of illuminant.

19.25 Lighting and Color Rendering

Since the intensity of illumination influences the appearance of colors, it must be considered in choosing object colors. As intensity is increased, reflection increases, particularly with pale tints (high value) that contain much white pigment, thus tending to wash out color. Therefore, with high-intensity lighting, saturation of colors should be high and spectral quality should enhance the object color for true, brilliant color rendition.

Color rendering is defined as the degree to which perceived colors of objects illuminated by a test source conform to the colors of the same objects as illuminated by a standard source. The *Color Rendering Index* of a source is a two-part concept, comprising a color temperature that establishes the standard, and a number that indicates how closely the illuminant approaches the standard. *The standard is always daylight at that color temperature.* Therefore, the color rendering index of a light is really a measure of how closely it approximates daylight of the same color temperature. Two lights cannot be compared unless their

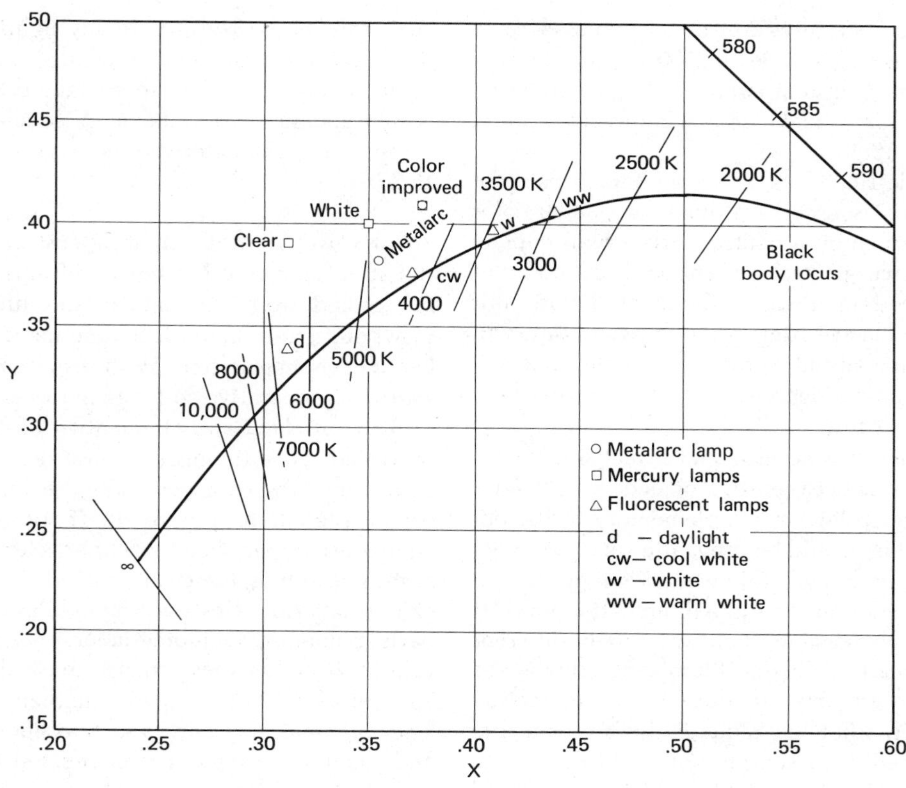

Fig. 19.46 Chromaticity diagram of Metalarc (metal-halide), mercury and fluorescent lamps. Courtesy of GTE Sylvania, Inc.

Table 19.25 *C.I.E. Color Coordinates (Chromaticity) for Typical Light Sources, plus Correlated Color Temperature and Color Rendering Index (CRI)*

Lamp	*Chromaticity* *X*	*Y*	° *K*	*CRI*
500-w Incandescent	0.448	0.407	2850	97
Tungsten Halogen, 3000°	0.437	0.404	3000	89
Fluorescent				
Cool White	0.367	0.375	4350	67
Warm White	0.431	0.406	3100	55
Daylight	0.313	0.332	6600	75
Deluxe WW	0.434	0.400	3020	77
Mercury				
Clear	0.32	0.379	5900	22
Color Improved	0.383	0.419	4200	45
WW Deluxe	0.400	0.380	3600	47
White Deluxe	0.382	0.385	4000	43
Metal Halide				
Clear	0.341	0.377	5200	55
Coated	0.358	0.380	4600	75
High-Pressure Sodium	0.515	0.435	2250	25

color temperatures are equal or quite close. A color rendering index of 100 is equal to daylight of that temperature. Color rendering indices for typical common lamps are given in Figure 19.45.

Table 19.26 lists the color characteristics of the major sources. An illuminant's own color appearance on a neutral surface will depend on its own spectral content but if the observer is placed in a space illuminated with this source, the eye after a short exposure time will become adapted to the source color and will detect only a degree of whiteness rather than an actual tint.

Where it is necessary to detect small color differences between two objects, a light poor in object color, or complementary to the object color, should be used, at a relatively high illumination level, followed by a light high in object color, at the same illumination level. If this is not possible, then two widely different but broad spectrum illuminants should be used, preferably at the same illumination level. Another technique is the use of a special, fixed color source. For a full discussion, see IES Handbook. It should be remembered in all considerations of color comparison, matching and rendering, that object color depends on the spectral energy distribution of the light source (illuminant), and therefore any change in the spectral content will change the object appearance. Two sources of the same color temperature and therefore apparent whiteness can have quite different spectral content and will therefore render object colors differently. A case in point would be a 3000 K warm-white fluorescent tube and an incandescent lamp (500-w photoflood) of approximately the same color temperature. Color temperature is an expression of dominant color, not spectral distribution.

19.26 Choice of a Light Source

The previous sections familiarized the reader with the color characteristics of the various sources so that these characteristics can be intelligently considered and weighed in the choice of a source. This choice, which is one of the early decisions in any lighting design, involves a number of factors of which color is only one. The others have been discussed and tabulated above. A summary of comparative characteristics is given in Table 19.27.

(*a*) ***Cost.*** The total lighting cost of an installation should, as has been said repeatedly, be studied from the annual and life-cycle viewpoints, and should include impact on the heating/cooling system. With respect to *light sources*, comparative costs can most readily be studied on the basis of quantity of lumens produced, since illumination also depends on fixture and type of space, and these vary with each installation. Figure 19.47 presents a comparative bar chart of life-cycle source costs, per million lumens produced, for a specific installation. Costs are divided into three parts: relamping including labor, energy, and capital. If any of these change markedly, the balance may shift to favor another source. Thus high-efficacy sources such as fluorescent and sodium are *relatively* unaffected by energy cost, and so on.

(*b*) ***Noise.*** Ballasts produce a hum that in certain cases can be objectionable. This sound problem can be mitigated by the use of special ballasts, soundproof cases, or by remotely mounting the ballasts. All of these procedures, of course, involve additional expense (see Chapter 27).

(*c*) ***Radio Interference.*** In addition to audible noise, a discharge type fixture also produces radio noise, or radio interference. This interference is of two kinds, direct radiation from the lamp and wiring, and conducted interference induced in the power-line wiring. Although this interference is usually small to negligible and can be mitigated by suppressors, as discussed above, it is sufficient to rule out the use of fluorescent lighting in some commercial, industrial, and military electronic laboratory installations. Cold cathode fluorescent lighting produces considerably less radio noise than conventional fluorescent tubes and is therefore applicable where the radio noise

Table 19.26 ***A Guide for Lamp Selection Based on General Color Rendering Properties***

Type of Lamp	Efficacy (lpw)	Lamp Appearance Effect on Neutral Surfaces	Effect on "Atmosphere"	Colors Strengthened	Colors Grayed	Effect on Complexions	Remarks
			Fluorescent Lamps				
Cool[a] white CW	High	White	Neutral to moderately cool	Orange, yellow, blue	Red	Pale pink	Blends with natural daylight—good color acceptance
Deluxe[a] cool white CWX	Medium	White	Neutral to moderately cool	All nearly equal	None appreciably	Most natural	Best overall color rendition; simulates natural daylight
Warm[b] white WW	High	Yellowish white	Warm	Orange, yellow	Red, green, blue	Sallow	Blends with incandescent light—poor color acceptance
Deluxe[b] warm white WWX	Medium	Yellowish white	Warm	Red, orange, yellow, green	Blue	Ruddy	Good color rendition; simulates incandescent light
Daylight	Medium-high	Bluish white	Very cool	Green, blue	Red, orange	Grayed	Usually replaceable with CW
White	High	Pale yellowish white	Moderately warm	Orange, yellow	Red, green, blue	Pale	Usually replaceable with CW or WW
Soft white/natural	Medium	Purplish white	Warm pinkish	Red, orange	Green, blue	Ruddy pink	Tinted source usually replaceable with CWX or WWX
			Incandescent Lamps, Tungsten Halogen				
Incandescent filament	Low	Yellowish white	Warm	Red, orange, yellow	Blue	Ruddiest	Good color rendering
			High-Intensity Discharge Lamps				
Clear mercury	Medium	Greenish blue-white	Very cool, greenish	Yellow, blue, green	Red, orange	Greenish	Very poor color rendering
White mercury	Medium	Greenish white	Moderately cool, greenish	Yellow, green, blue	Red, orange	Very pale	Moderate color rendering
Deluxe white[a] mercury	Medium	Purplish white	Warm, purplish	Red, blue, yellow	Green	Ruddy	Color acceptance similar to CW fluorescent
Metal halide[a]	High	Greenish white	Moderately cool, greenish	Yellow, green, blue	Red	Grayed	Color acceptance similar to CW fluorescent
High-pressure sodium[b]	High	Yellowish	Warm, yellowish	Yellow green, orange	Red, blue	Yellowish	Color acceptance approaches that of WW fluorescent

Source. Courtesy of General Electric Co., Lamp Division.

[a]Greater preference at higher levels.

[b]Greater preference at lower levels.

restrictions are stringent. Direct radiation can be reduced by distance separation or by metallic shielding; conducted interference can be mitigated by special filters that are built into the fixtures. The problem of radio noise does not exist with incandescent fixtures.

(d) Pattern. Small lamps mounted in unobtrusive fixtures do not dominate the architecture. This cannot be said of tubular sources, which must be carefully applied to blend with the architecture.

Table 19.27 *Comparative Characteristics of Light Sources*

Characteristic	*Incandescent*		*Fluorescent*		*High-Intensity Discharge (HID)*				
					Mercury			*Metal Halide*	*High-Pressure Sodium*
	Filament	*Tungsten Halogen*	*Standard*	*High-Output*	*Clear*	*Phos.*	*Self-Ball.*		
Wattages	10–1500	100–1500	15–100	60–200	40–1500	—	150–750	175–1500	70–1000
Efficacy—lamp	15–25	15–25	50–80	55–85	30–65	—	—	80–105	85–140 lpw
Lamp and ballast	—	—	40–70	45–75	25–55	—	15–25	65–85	55–125 lpw
Life-hours	750–1000	2000–6000	12000–20000		16000–24000 hr			—	—
Lumen maintenance @ 50% life	80–90%	95%	85–90%		70–80%		—	75–85%	80–90%
Color acceptability	Excellent		VG	VG	Poor	Fair to good		VG	Fair to good
Start time	Instant	Instant	Instant	Instant	3–6 minutes		—	2–3 min	3–4 min
Restrike time	—	—	Instant	Instant	3–8 minutes		—	5–10 min	$\frac{1}{2}$–$1\frac{1}{2}$ min
Light control	VG	VG	Poor	Poor	Good	Fair	Good	Good	Good
Glare control	VG	VG	VG	Fair to Good	Fair	Good	Good	Fair	Fair
Fixture cost	Low	Low	Low to moderate		Moderate to high			High	Highest
Cost to produce 10^6 lm	See Figure 19.46.								

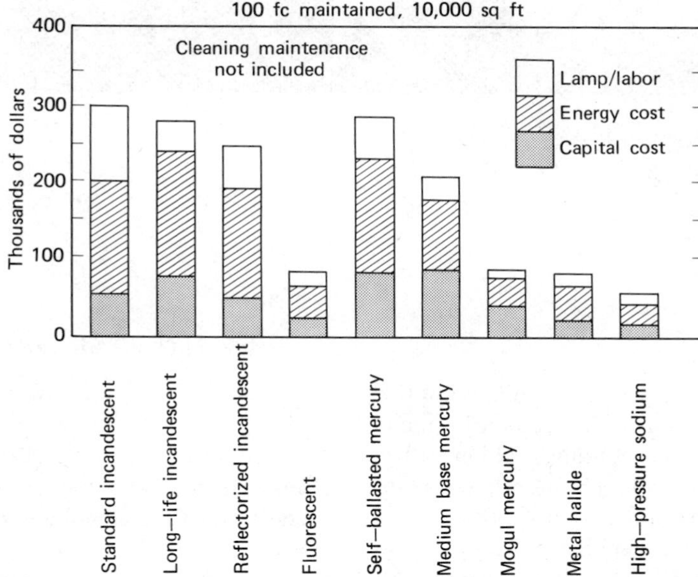

Fig. 19.47 *Ten-year, life cycle costs of producing 1 million usable lumens. (Although there are many types of fluorescent lamps, all types generally follow the costs shown). From Dubin-Mindell-Bloom,* Guidelines for Saving Energy in Existing Buildings, *ECM-2, Federal Energy Administration, June 1975.*

References and Bibliography

1. "Sky Brightness and Daylight Illumination Measurement," H. H. Kimbal and I. F. Hand, *Trans. of IES,* Vol *16,* No. 7.

2. *Recommended Practice of Daylighting,* 1979 IES.

3. Building Research Station (BRS) Daylight Factor Protractors, H. M. Stationery Office, London.

4. *Simplified Daylight Tables,* 1958, Hopkinson, Longmore, and Graham, National Building Studies, Report #26, H. M. Stationery Office, London.

5. *Daylight Design Diagrams,* published by Service Division, Commonwealth Department of Labor and National Service, Melbourne, Australia.

6. *Lighting of Buildings,* 1972, Hopkinson and Kay, Faber, London.

7. *Daylighting,* 1966, Hopkinson, Petherbridge, and Longmore, Heinemann, London.

8. *Daylighting, British Standard Code of Practice,* British Standard Institution, Pt. 1, 1964.

9. *Principles of Natural Lighting,* 1968, J. A. Lynes, Elsevier, New York.

10. *IES Handbook,* 5th ed., 1972, Section 7.

11. *Energy Conservation Applied to Office Lighting,* Ross & Baruzzini, Inc., Federal Energy Administration, April 1975, Section III.

12. "Procedure for calculating the potential savings in lighting energy from the use of skylights," J. B. Merdock, *Journal of the IES,* July 1977.

13. Peery, R., "Daylighting and energy conservation," *Lighting Design & Application,* October 1974.

14. Dubin-Mindell-Bloom, *Energy Conservation Design Guidelines for Office Buildings,* January 1974, General Services Administration, Public Buildings Service.

15. Dubin-Mindell-Bloom, *Guidelines for Saving Energy in Existing Buildings,* June 1975, ECM-2, Federal Energy Administration.

16. *Daylight, International Recommendations for the Calculation of Natural Daylight,* Publication CIE No. 16 (E-3.2), 1970; Commission Internationale de l'Éclairage.

20

Lighting Design

20.1 General

Lighting design is a combination of applied art and applied science. There can be many solutions to the same lighting problem, all of which will satisfy the minimum requirements, yet some will be dull and pedestrian while others will display ingenuity and resourcefulness. The competent lighting designer approaches each problem afresh, bringing to it a knowledge of current technology and years of background and experience, yet rarely being satisfied with a carbon copy of a previous design. And it is these years of background with their successful and not-so-successful designs coupled with a constant striving for improvements that are the characteristics differentiating the lighting consultant, designer, or engineer from the person who attempts to force each new job into the unwilling mold of a previous design.

Because of the large number of interrelated factors in lighting, no single design is the correct one, and for this very reason it is not entirely desirable to solve a lighting problem with a step-by-step technique. However, since this technique is the only avenue of approach open to the uninitiate who lacks the experience necessary to view an entire solution, we have adopted this approach.

20.2 Goals of a Lighting Design

Simply stated, the goal of lighting is to create an efficient and pleasing interior. These two requirements, that is, the utilitarian and esthetic, are not antithetical as is demonstrated by every good lighting design. Light can and should be used as an adjunct architectural material. We elaborate on these goals below.

a. Lighting levels should be adequate for efficient seeing of the particular task involved. Variations within acceptable brightness ratios in a given field of view are desirable to avoid monotony and to create perspective effects.

b. Lighting equipment should be unobtrusive, but not necessarily invisible. Fixtures can be chosen and arranged in various ways to complement the architecture or to create dominant or minor architectural features or patterns. Fixtures may also be decorative and thus enhance the interior design.

c. Lighting must have the proper quality as discussed previously. Accent lighting, directional lighting, and other highlighting techniques increase the utilitarian as well as architectural quality of a space.

d. The entire lighting design must be accomplished efficiently in terms of capital and energy resources; the former determined principally by life-cycle costs and the latter by operating energy costs and resource-energy usage. Both the capital and energy limitations are, to a large extent, outside the control of the designer, who works within constraints in these areas. Obviously, these constraints are maxima.

With these goals before us we can write a lighting design procedure, keeping in mind that the order of steps shown is not necessarily the same in each lighting problem and that, since all of the factors are closely interrelated, it is often necessary to apprehend several of the stages simultaneously before arriving at a decision.

20.3 Lighting Design Procedure

(*a*) ***Project Constraints.*** *Refer to Fig. 20.1.* This flowchart, which represents the design procedure and its interactions, should be referred to throughout the necessarily lengthy discussion that follows, in order to maintain perspective. It is important that the reader be aware of job constraints and of the interactions between the lighting designer and the remainder of the design group. We deliberately emphasize this to demonstrate the interdisciplinary nature of lighting design in general and its particular connection with HVAC and daylighting (fenestration). This approach, which is most often referred to as the systems design approach, will be followed throughout the discussion.

In (d) of the preceding section we referred to constraints. These can alternatively be classified as the owner-architect-user team and the jurisdictional authorities. In some detail these are:

1. *Owner-designer-user group. The owner* establishes the cost framework, both initial and operating. As part of both of these may be a rent structure, which in turn determines and is determined by the space usage. If the owner is also the occupant the cost factors change somewhat but remain in force. The architect determines the amount and quality of daylighting and the architectural nature of the space to be lighted. Much of these data are detailed in the building program. Obviously the architect and lighting designer (who may be one person) should interact in this aspect of building design.
2. The jurisdictional authorities *may* include:

   FEA—Federal Energy Administration.

   ERDA—Energy Research and Development Administration.

   GSA—General Services Administration.

   NCSBCS—National Conferences of States for Building Codes and Standards.

   ASHRAE—American Society of Heating, Refrigeration and Air Conditioning Engineers.

   IES—Illuminating Engineering Society.

   NBS—National Bureau of Standards.

   CIE—International Commission on Illumination.

Most of these are jurisdictional by reference, that is, the actual authorities will specify that the lighting system meet the requirements of ASHRAE 90-75, IES EMS-1, and so on. If public money is involved, FEA/GSA standards will probably be involved. The principal area of involvement is that of energy budgets and lighting levels, both of which affect every aspect of lighting design including source type, fixture selection, lighting system, fixture placement, and even maintenance schedules. For this reason the first step in the lighting design procedure is to establish the *project lighting cost framework and the project energy budget.*

(*b*) ***Task Analysis.*** As shown in Figure 20.1 this step essentially determines the needs of the task. Factors to be considered in addition to the nature of the task are its repetitiveness, variability, who is performing it (i.e., condition of the occupant's eyes), task duration, cost of errors, and special requirements. Several of these factors have been discussed in the preceding sections dealing with quality of light. The reader will be referred to the appropriate sections in the following analysis.

(*c*) ***Design Stage.*** This is the active consideration stage during which detailed suggestions will be raised, considered, modified, accepted, or rejected. This is also the most interactive stage as is clearly seen from Figure 20.1. At its completion, a detailed, workable design is in hand. The critical interactions here are with the architect in daylighting and with the HVAC group in power loads. The former may result in relocating a space within the building; the latter in making a change in a lighting system or HVAC system. In brief, this stage consists of the following steps:

1. Select the lighting system. Select type of

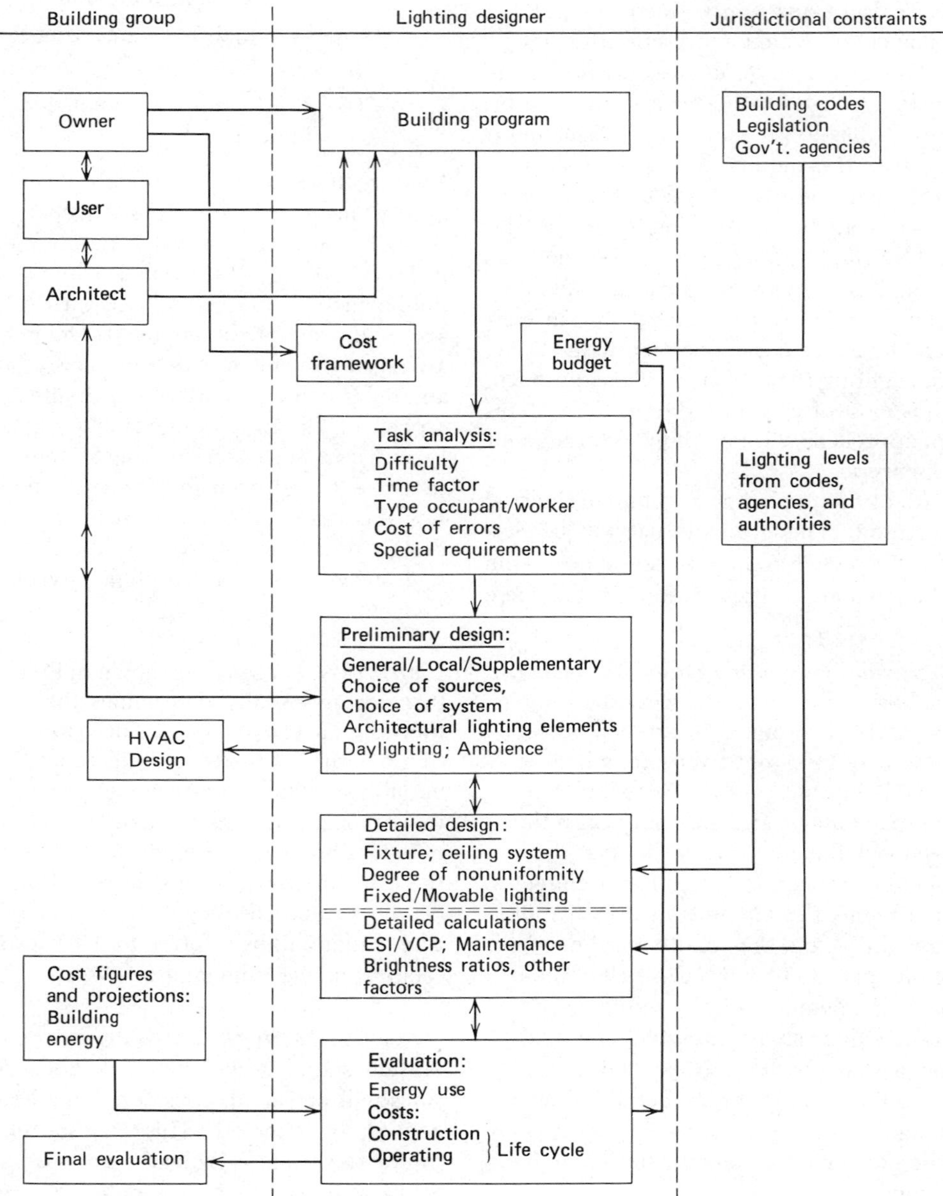

Fig. 20.1 Lighting design procedure chart.

light source, distribution characteristic of fixture(s) or area source, consider effects of daylighting, economics, and electric loads.

2. Calculate the lighting requirements. Use the applicable calculation method and establish the fixture pattern, considering the architectural effects.
3. Design the supplemental lighting.
4. Review the resultant design. Check the design for quality, quantity, aesthetic effect, and originality.

(*d*) ***Evaluation Stage.*** With the design on paper, it can now be analyzed for conformance to the principal constraints of cost and energy. If the design stage has been carefully accomplished, with due attention to these factors, the result of the final evaluation should be gratifying. The results of this stage are fed to the architectural group for use in the final overall project evaluation. In the following sections we consider in detail each of the steps in the design procedure.

20.4 Cost Factors

This is a particularly difficult item for a novice lighting designer because it requires experience in the field and an acquaintance with commercially available equipment. Also, the inevitable trade-offs between first cost and operating cost cannot intelligently be made unless the cost structure is clearly understood. The following guidelines should be of considerable assistance both in avoiding unpleasant surprises when a job is estimated and in preparing cost analyses:

1. Decide at the outset what cost criteria will be applied, that is, the relative importance of first cost, operating costs, annual owning costs, and life-cycle costs.
2. Trade-off decisions are required between first cost and operating costs. For example, incandescent lamps and fixtures are low in first cost and high in operating cost, and so on (see Table 19.22). Dimming and switching equipment falls into this area of decisions.
3. Manufacturer's catalog items are *always* cheaper than specials, and can be priced more readily.
4. To assist in comparing annual owning costs of two systems or methods, Table 20.1 is included. Conversion of these data to life-cycle cost comparisons is straightforward. With the continued increase in energy costs it is suggested that even in an *annual* owning cost comparison, two different energy costs be used and the impact of a sharp increase studied.
5. The impact of lighting energy on the operating cost of the entire building must be studied, and the apportionment of costs determined. As explained in Section 19.8, the only practical means of accomplishing this is by computer program. Programs can be readily adjusted, to reflect the impact of the lighting system on building costs, and in particular on HVAC first cost and operating costs. This technique is most useful in comparing systems and should be used, since it is incorrect to artificially separate the lighting system from the HVAC system with which it intimately interacts. Even without resort to detailed calculation it can be stated unequivocally that the lower the lighting system's energy level, the lower the building's overall operating cost. The argument that heat from a lighting system is fully utilized to heat the building and is therefore not wasted is a specious one, which has been refuted on many counts:
   a. HVAC system first cost is higher.
   b. HVAC year-round cost is higher.
   c. Lighting energy cost is higher.
   d. Life-cycle costs are higher.
   e. Energy resource use is higher.

This subject is discussed in detail in Section 21.18.

20.5 Energy Budgets

The requirement to establish a project's energy budget in accordance with a specified procedure, has now been incorporated into the building codes of most states. As such it is effectively the "law of the land." This procedure and its parameters, however, are at present in a state of development and change. For this reason the text is not reproduced here, and the reader is referred to the *latest issue* of *IES Recommended Lighting Power Budget Determination Procedure; EMS-1.* (ASHRAE 90-75 is a verbatim reprint of EMS-1, and its Chapter 9 paragraph headings can be made to

Table 20.1 ***Standard Form for Calculation of Lighting System Costs on an Annual Owning Cost Basis***

Lighting Cost Comparison

	Lighting Method #1	*Lighting Method #2*
Installation Data		
Type of installation (office, industrial, etc.)		
Luminaires per row		
Number of rows		
Total luminaires		
Lamps per luminaire		
Lamp type		
Lumens per lamp		
Watts per luminaire (including accessories)		
Hours per start		
Burning hours per year		
Group relamping interval or rated life		
Light loss factor		
Coefficient of utilization		
Footcandles maintained		
Capital Expenses		
Net cost per luminaire		
Installation labor and wiring cost per luminaire		
Cost per luminaire (luminaire plus labor and wiring)		
Total cost of luminaires		
Assumed years of luminaire life		
Total cost per year of life		
Interest on investment (per year)		
Taxes (per year)		
Insurance (per year)		
Total capital expense per year		
Operating and Maintenance Expenses		
Energy expense		
Total watts		
Average cost per kwh		
Total energy cost per year[a]		
Lamp renewal expense		
Net cost per lamp		
Labor cost each individual relamp		
Labor cost each group relamp		
Percent lamps that fail before group relamp		
Renewal cost per lamp socket per year[b]		
Total number of lamps		
Total lamp renewal expense per year		
Cleaning expense		
Number of washings per year		
Man-hours each (est.)		
Man-hours for washing		
Number of dustings per year		
Man-hours per dusting each		
Man-hours for dustings		
Total man-hours		
Expense per man-hour		
Total cleaning expense per year		
Repair expenses		
Repairs (based on experience, repairman's time, etc.)		
Estimated total repair expense per year		
Total operating and maintenance expense per year		
Recapitulation		
Total capital expense per year		
Total operating and maintenance expense per year		
Total lighting expense per year		

Table 20-1 (Continued)

Source. Courtesy of the IES.

[a] Total energy cost per year $= \dfrac{\text{Total watts} \times \text{burning hours per year} \times \text{cost per kwh}}{1000}$

[b] The following formulas give the annual cost per socket for lamps and replacement, and can be used for determining the most economical replacement method.

$$\text{Individual replacement} = \frac{B}{R}(c + i) \text{ dollars/socket/year.}$$

$$\begin{array}{c}\text{Group replacement}\\ \text{(early burnouts replaced)}\end{array} = \frac{B}{A}(c + g + cKL + Ki) \text{ dollars/socket/year.}$$

$$\begin{array}{c}\text{Group replacement}\\ \text{(no replacement of early burnouts)}\end{array} = \frac{B}{A}(c + g) \text{ dollars/socket/year.}$$

where B = burning hours per year
R = rated average lamp life, hours
A = burning time between replacements, hours
c = net cost of lamps, dollars
i = cost per lamp for replacing lamps individually, dollars
g = cost per lamp for replacing lamps in a group, dollars
K = proportion of lamps failing before group replacement (from mortality curve)
L = the portion of the cost of early burnouts that is charged against group replacement.

No general rule can be given for the use of group replacements; each installation should be considered separately. In general, group replacement should be given consideration when individual replacement cost i is greater than half the lamp cost c and when group replacement cost g is small compared to i.

coincide by placing the digit '9' before those appearing in the IES text. The IES publishes material corresponding to Attachments A and B of the ASHRAE standard plus a special section on power budget guidelines for residential kitchens and bathrooms. These are not an integral part of the standard itself. The IES also publishes a detailed example of the use of the IES/ASHRAE procedure.)

The purpose of this budget determination procedure is not to dictate design procedure. Indeed this standard explicitly so states. Instead it presents development of an overall maximum power budget *within* which the designer is free to do as he or she wishes. Obviously prodigality in one area is necessarily at the expense of another area, since maximum power is inflexible, and the entire budget is built on reasonable design techniques. Still, there is enough leeway in the budget and enough exceptions (see its Section 3.4) so that the designer is not overly restricted. It is suggested that the reader obtain and scan through the referenced IES or ASHRAE material at this point but reserve detailed study of it and its application guidelines, since they contain references to terminology and techniques with which he or she will not be familiar until completion of the chapter. However, this material is presented here in order to maintain the proper order of events in the design procedure. Several important comments are in order:

1. Because of the proliferation of standards and guidelines, the lighting designer may find a contradiction between authorities. Thus the above budget calculation uses lighting levels for task lighting, general lighting, and noncritical lighting that may contradict FEA/ESA requirements (Tables 18.6 to 18.8). In such event, written clarification from the appropriate authorities is advisable.

2. This same type of conflict may arise with recommended lighting load figures stated in watts per square foot. The overall figure arrived at by the IES/ASHRAE procedure will probably be higher than the 2.3 w/sq ft maximum used by the FEA/GSA. Remember that the FEA/GSA proposals are primarily intended for office-type spaces.

3. The IES/ASHRAE technique does not consider daylight contributions at all. This is an area that can be used to great advantage by

the lighting designer to achieve flexibility and "breathing space" in other areas.

4. The footcandle levels are to be taken from the IES recommendations. Many of these specify ESI footcandles, for example, in drafting rooms, 200 ESI footcandles are specified. Wattage calculations are based on a coefficient of utilization (CU) of 0.55. (See Section 20.28 for a definition of this term.) As we have seen, a CRF of greater than 1.0 is possible, which would substantially reduce actual wattages by giving an effective CU considerably above 0.55. The budget there represents a challenge to the designer to produce efficient, low-wattage solutions.

5. Exceptions to the budget criteria include residential, theatre, and exhibition spaces. Since these are specifically areas that utilize low-efficiency sources, no difficulties should be encountered in their lighting design.

6. The budget criteria considers tasks as they are stated. Subsequent interaction with the user in terms of task changes can reduce lighting requirements. Thus the hard-pencil copy can be changed to soft-pencil copy, reducing the lighting requirement, and so forth.

7. Overall, the designer should have no difficulty staying within the energy budget if the techniques already referred to are used: nonuniform lighting, high-efficiency sources, high reflectances, efficient fixtures, and supplemental task lighting.

20.6 Task Analysis

Refer to Figure 20.1. This is the stage at which the quantity and quality of lighting required for the tasks are decided. The factors affecting this choice as shown in the figure are difficulty, time factor, occupant, cost of errors, and special requirements.

(*a*) ***Difficulty.*** The components of visual difficulty were discussed at length in Sections 18.13 to 18.19, and the results in terms of lighting levels appear in Tables 18.4 to 18.7. Essentially, the designer will examine the type of task involved, on the basis of which he will select brightness/footcandle levels, the applicable authority, that is, IES, FEA/GSA, and so on. In the absence of specific instructions to the contrary the designer will use IES recommendations. If there are several tasks to be performed at the same point, the most difficult, subject to time considerations, will be selected; that is, if the more difficult task occurs infrequently it may be reasonable to provide supplementary portable lighting or even to suggest moving to another brighter location. If the more difficult task is frequent but not dominant, fixed supplementary lighting is a reasonable solution. If it is the major task, lighting should be based on it and provision made for intensity reduction for less demanding work. Notice that this solution may be identical to the previous case depending on systems, ceiling arrangement, local lighting, and so on.

Variation in task difficulty is particularly common in spaces in public buildings. Thus a school gym can be used for athletics, band concerts (despite the acoustics), and town meetings—three totally disparate lighting requirements. In these and similar instances it is common practice to treat the space as essentially three different spaces and to design lighting for each with a careful eye to maximum common equipment usage. Similar problems are encountered in basements, multipurpose rooms, and conference/meeting/lecture/exhibition rooms. Fortunately most such spaces do not have severe seeing tasks. The task variation referred to here is the variation that occurs in one very specific location and is not to be confused with task variation in an area, however restricted. Thus a small private office of say 8×8 ft has a desk, file cabinet, and circulation space; three tasks of differing but constant difficulty in one small space. The corresponding lighting for these is also fixed and varies with the task severity. As explained in Chapter 18, visual task studies that have been performed by the researchers in this area have resulted in tabulations of required task luminances for hundreds of usual tasks. If these are categorized and acceptable luminances tabulated, the results are approximately these (assuming good quality, that is, high VCP and CRF):

Category of Visual Task	*Required Luminance (fL)*
Casual	3–6
Ordinary	6–30
Moderately difficult	30–60
Difficult	60–120
Severe	Above 120

Since the relation between required luminance and incident illumination is known, fL = fc × reflectance, a determination of the required illumination for the listed tasks can readily be accomplished. The crucial dependence of illumination on task reflectance can be seen by a glance at Table 20.2, which shows quantitatively the illumination requirements in the above categories for tasks of radically different reflectance and points out why a single illumination scheme is often inadequate for an area containing widely differing visual tasks. Note that a 10% RF makes all tasks difficult and that casual seeing would be only outline recognition as discussed at length in Chapter 18. The values listed in Table 18.4 et al represent the required illumination on the surface in question whether horizontal, vertical, or in between. Since the flux method of calculating illumination normally yields the 30-in., horizontal-plane illumination level, it is necessary to be cognizant of the ratio of horizontal to vertical illumination for various lighting systems. This ratio is approximately 3:1 for narrow-distribution direct and semidirect lighting; 2.5:1 for wide-distribution luminaires of the same type; and 1.5:1 for indirect, general diffuse lighting.

Since the illumination values listed assume adherence to both recommended brightness ratios and reflectance (see Figs. 18.47, page 760, and 18.48, page 761), it is necessary to select, in conjunction with the interior designer, finishes and reflectances for surfaces within the area. Where this choice is not left to the lighting designer, as is often the case, he must at least be fully aware of the surface finishes in order to calculate luminances and to make adjustments in illumination levels, if necessary. If for instance in a private office a dark wall finish of 10% reflectance is chosen by others, it will be necessary for the lighting designer to compensate for this by additional wall lighting to maintain the recommended maximum 10:1 brightness ratio (see Table 18.11, page 760, and the discussion of IES point 6 in Section 20.7). This assumes, of course, that a working office quality is desired. The atmosphere created by vertical surface luminances will be discussed below. Table 20.3 lists reflectances of some common interior paint finishes.

(*b*) ***Time Factor.*** As discussed in Section 18.18, the length of time for which the task must be accomplished is important in difficult work. Thus in Table 20.2, beginning at the fourth level, that is, 50 fL nominal, prolonged intensive application to the seeing task would require illumination to be raised one level or

Table 20.2 ***Illumination Requirements for Similar Tasks of Differing Reflectances***

Category of Visual Task (see text)	*Average Required Luminance, fL*	*Required Illumination Footcandles for Reflectances of*	
		50%	*10%*
Circulation, casual, infrequent	10 (5–15)	20	100
Rough work	20 (15–25)	40	200
Routine work	35 (25–35)	70	350
Moderate difficulty	50 (35–60)	100	500
Difficult	70 (60–80)	140	700
Very difficult	100 (80–120)	200	1000
Most difficult	150	300+	1500+

Table 20.3 *Approximate Reflection Factors*

Medium Value Colors	
White	80–85
Light gray	45–70
Dark gray	20–25
Ivory white	70–80
Ivory	60–70
Pearl gray	70–75
Buff	40–70
Tan	30–50
Brown	20–40
Green	25–50
Olive	20–30
Azure blue	50–60
Sky blue	35–40
Pink	50–70
Cardinal red	20–25
Red	20–40

quality to be markedly improved by using a large daylight component or high ESI component system. If, on the other hand, this worker readily learns the work and it becomes routine, illumination can be reduced one level because, effectively, the degree of difficulty has been reduced. This is essentially what occurs with printed material and is the reasoning behind the FEA/GSA 50-fc recommendation for office work.

(*c*) ***Occupant.*** Since ordinarily the age and other specific characteristics of the worker are not known, a standard distribution is assumed and the recommendations as tabulated take account of this. On the other hand if there is a high percentage of older workers, as is the case in certain industries, lighting should be raised one level. This compensates for inability of the eyes to accommodate and for the tendency to tire easily.

(*d*) ***Cost of Errors.*** This involves an economic trade-off between cost of errors, or restated, cost of improving seeing accuracy as against the cost of the additional lighting (see Fig. 18.23, page 730). Performance can be brought as close as desired to perfection but the cost of so doing increases much more rapidly than the proportional increase in performance. (Standard lighting should provide 90%± seeing efficiency). This is partially due to the diminishing returns shown by the curve and partially because the data upon which this curve is based assume high-quality, low-glare lighting even in the upper range. To accomplish this practically is difficult, and very costly. Thus this step is basically an economic calculation, the criteria for which must come from the owner or user. Tasks in which this problem is encountered include all types of inspection, proofreading, textile matching, very fine machining, and jewelry manufacturing.

(*e*) ***Special Requirements.*** These include any nonstandard task lighting requirements. Some of these are specific illuminant color, directionality for shadowing and reflections as required for inspection, polarization, and controlled variations, as required in a space with varied tasks or varying daylight factor. In addition to these the physical dimensions of the task often create special requirements of their own. We tend to assume a small object in the horizontal plane as the task, since that is the normal office task. However, there are exceptions such as a drafting board, a machine shop, an inspection bench, or a cutting table. Consequently these special requirements arise:

1. *Large tasks.* In large tasks, the angle of seeing changes from 20 to 70° from the vertical, resulting in radically changed glare angles and reflection from the task.
2. *Three-dimensional tasks.* These tasks shadow themselves, particularly when containing undercuts and reveals. An architect's model shop presents such tasks. When it is necessary to see into an opening, an intense narrow beam is required.
3. *Tools.* Tools cast shadows below and in front when lighted from above and behind. A fabric cutter must see ahead of and below the cutting machine.
4. *Nonhorizontal tasks.* These must be calculated for the plane in which they stand. As stated earlier, the ratio between horizontal and vertical illumination varies between 1.5:1 to 3:1, de-

pending on the system. Task lighting requirements are stated in the plane of the task. This can have a pronounced effect on the lighting system selected and its arrangement.

5. *Task observed from various positions.* There are instances where a fixed task is observed from several angles, as a drawing in a conference room. Task illumination must be adequate for all viewing angles.

PRELIMINARY DESIGN

20.7 Energy Considerations

Energy considerations must pervade every aspect of the design process in order that the completed design satisfy the energy budget criteria. Some background material is in order here, to place the lighting energy subject in proper perspective. Best estimates indicate that at this writing lighting consumes approximately 25% of the electric power generated in the United States. In terms of *resources* this amounts to approximately 4 million barrels (bbl) of oil per day. The usage by occupancy is approximately:

Residential—20%
Industrial—20%
Stores—20%
Schools and offices—15%
Outdoor and other—25%

In commercial buildings lighting consumes about 20 to 30% of the building's electric energy; more in residences and less in industrial facilities. By judicious design a reduction of 40 to 50% in lighting energy is attainable. Translated into resources and put in terms of buildings under the designer's control, that is, excepting outdoor use and much of residential use, this reduction can readily amount to more than 1 million bbl of oil per day. Few will disagree that such a goal is well worth the effort. In this connection the reader is advised to be sceptical of articles purporting to demonstrate that a reduction in lighting energy will result in either negligible overall energy economy or even an increase in oil consumption. A cursory investigation of these arguments will quickly reveal their speciousness and the author's lack of disinterest. In point of fact every watt per square foot reduction in lighting energy results in 1.25 w/sq ft savings in air-conditioned buildings (see Section 19.7). It has been demonstrated by actual designs that offices and schools can be *well* lighted with 2.5 w/sq ft in lieu of the 4 to 5 w/sq ft in common use. The question to be answered then is: What design guidelines can be followed to effect this energy-conscious design?

In 1972–1973, the IES prepared and documented 12 recommendations for optimizing the use of energy for lighting. Since then, they have become so well known as to essentially constitute a standard; therefore, they are presented here in a group, in addition to appearing at appropriate points throughout the description of the design process. Comments on implementation follow each recommendation:

1. *Design lighting for expected activity.* This is the task lighting approach to which we have already devoted considerable space. Phrased otherwise, this point states that it is wasteful, of energy, to light any surface at a higher level than it requires. Since most spaces contain varied seeing tasks, nonuniform lighting is recommended. In order to accomplish this for areas where exact furniture layout is not available, it may be necessary to furnish readily movable fixtures as discussed in Sections 15.33, 15.34, and 18.23(a). A trade-off is involved here between the additional first cost of movable fixtures and the lowered operating cost. Providing overall high-level illumination with provision for switching to reduce levels is not advisable because of increased first cost and the psychological impetus to operate at maximum levels. Also, use of movable fixtures makes heat removal via ducted air-troffers impractical [see (d) below]. A compromise solution here is fixed fixtures for general low-level lighting and movable ones for supplementary task lighting. Other factors and techniques to be borne in mind are:

a. Grouping of tasks with similar lighting

requirements will generally increase efficiency.

b. Place most severe seeing tasks at best daylight locations.

c. Fixed-position tasks involve nonuniform lighting and vice versa.

d. Heat removal fixtures (air-troffers) increase efficiency of the units 10 to 20% but make the fixtures immobile. Trade-off decision involved.

e. Advantages of nonuniform lighting increase as the space between work stations increases.

2. *Design with more effective luminaires and fenestration.* By effective is meant providing useful light with high ESI component and minimum direct glare. In selecting fixture diffusers a trade-off is often encountered between these two usually mutually incompatible factors. In cases where much of the viewer's time is spent in a head-up position, as in schools, or where the viewer can compensate for veiling reflections as discussed in Section 18.18, the decision should lean toward high VCP. Where work position and viewers are fixed, most of the viewer's time is spent head-down, viewer's room position reduces direct glare, or illumination levels are well below the 100 fc level for which VCP is calculated, the decision should lean toward low-reflected glare. Fenestration too must be considered as a light source in all respects, with particular attention to the deleterious effects of excessive window brightness and the beneficial effects, that is, the excellent quality of daylighting.

3. *Use efficient light sources.* This point is self explanatory. Trade-offs involved here are:

a. Between first cost and life-cycle costs.

b. Between desired illuminant color and efficiency. Some of the highest-efficacy sources have less than ideal color but, since the eye adapts quickly, off-white color, as from HPS lamps, should be considered for most indoor uses.

c. Between light control and efficiency. Fluorescent sources, which are highly efficient, do not lend themselves to good beam control and are principally useful for area coverage.

d. Between architectural requirements and efficiency. Fluorescent sources are efficient and have good color but require a large fixture, which tends to dominate the space. A possible compromise is the U-shaped lamp in a modular ceiling (see Fig. 19.34, page 827).

Figure 20.2 will assist the designer in determining roughly what the various sources represent in terms of watts per square foot load. Thus, with a target of 2 to 2.5 w/sq ft for office-building lighting, the use of incandescent downlights is obviously severely restricted. Daylight must be considered as a regular light source subject to weather variations and time of building use. Obviously a three-shift industrial plant cannot use daylight on all shifts, but it can for at least one shift, and design should reflect this fact. Spill light and borrowed light are often neglected sources. Glass in upper wall sections can provide sufficient corridor lighting from borrowed office lighting. Sources with high lumen maintenance such as tungsten halogen and high-pressure sodium should be given preference.

Another aspect of light source efficiency is the ability to reduce levels easily. Use multilevel ballasts and switching to give flexible lighting control. Unused light is wasted energy.

4. *Use more efficient luminaires.* This point is connected to the one above in that only by using an efficient source in an efficient enclosure can best results be obtained. By efficient enclosure we mean a fixture with high coefficient of utilization CU (see Section 20.27) and *high* light loss factor (LLF) (see Section 20.30), since useful lumens are defined as

$$\text{Useful lumens} = \text{generated lumens} \times \text{CU} \times \text{LLF}$$

Care must be exercised here since simple lampholders have high CU and LLF but of course create direct glare problems. Therefore, here also a trade-off is involved between high CU and LLF and high VCP.

Other factors being equal, select luminaires with a high-maintenance category. Enclosed

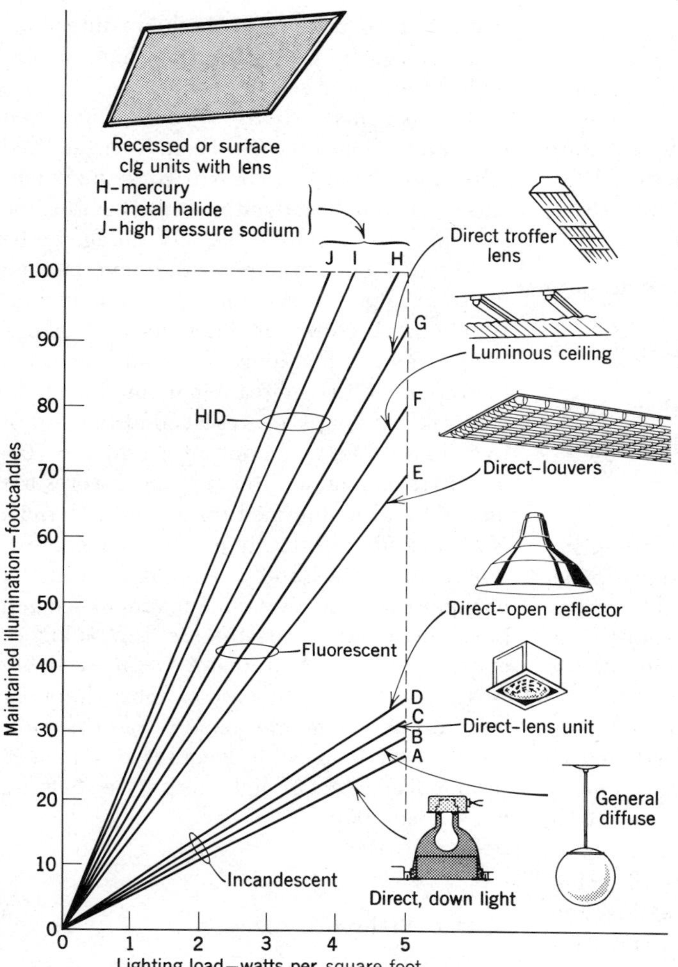

Fig. 20.2 Estimating chart for lighting load and related illumination levels for different lighting sources. The chart was calculated for a fairly large room—approximately classroom size. Although all the figures are necessarily approximate because of the variables involved, the chart gives figures close enough for a first approximation. Notice the increase in output as the sources change from incandescent to fluorescent to HID.

nongasketed fixtures collect dirt and yield a lower overall LLF. Also, the quality of the luminaire finish will have considerable bearing on its performance after 8 to 10 years. Permanent finishes such as Alzac have better long-range properties than painted finishes. With the latter, the quality of the fixture will make itself known in the long run.

5. *Use thermal-controlled luminaires.* In other words, do not waste the heat generated by lighting and, conversely, avoid placing additional load on the air-conditioning system with lighting heat. With lower lighting levels this point has considerably less significance than it did when 6 to 10 w/sq ft lighting loads were common. Disposition of this point depends, to a large extent, on the type of HVAC system, the lighting heat load, and the type of fixtures employed. Detailed analysis of this point involves HVAC considerations and the overall impact of lighting energy on the building.

6. *Use lighter finishes on ceilings, walls, floors, and furnishings.* This point is self-explanatory and is examined in a number of sections. A brief summary of recommendation ranges would be:

Ceilings—80 to 92%

Walls—40 to 60%

Furniture, office machines, and equipment—25 to 45%

Floors—20 to 40%

In addition to higher footcandle levels in the room, high reflectances minimize uncomfortable high-brightness ratios, as between fixture and upper wall or task and background. With respect to brightness ratios (see Table 18.11, page 760):

a. Between task and near surround—aim for 3:1, accept 5:1.
b. Between task and immediate area—aim for 10:1, accept 20:1.
c. Between luminaires and their background—aim for 20:1, accept 40:1.
d. Anywhere in the normal field of view—aim for 40:1, accept 80:1.

Note that the targets themselves are maximums and acceptance of values above maximum should only be for very good reasons.

7. *Use efficient incandescent lamps.* This point is self-explanatory. See Sections 19.8 to 19.10.

8. *Turn off lights when not needed.* Self-explanatory. The trade-off involved here is between cost of control equipment and energy savings (see Section 19.12). Here too, as in item 3 above, switching includes reduction of levels by use of multilevel ballasts, dimming, and partial switching. Coordinate this item with item 10 below so that daylight and artificial light complement each other.

9. *Control window brightness.* Excessive brightness causes severe and even disabling glare. A corollary of excessive brightness is excessive heat gain. Both are manageable with common control devices, manual and automatic.

10. *Utilize daylighting as practicable.* Self-explanatory. See Sections 19.2 to 19.7.

11. *Keep lighting equipment clean and in good working order.* This item is not as self-evident as it might seem. Lighting equipment must be selected with life-cycle costs in mind, a large portion of which are maintenance and relamping costs. Fixtures in relatively inaccessible locations such as high ceilings must be designed for low maintenance, and maintenance should be on a fixed schedule. Trade-off here is between higher cost of low-maintenance units and high maintenance costs. A corollary is the use of high-lumen-maintenance sources.

12. *Post instructions covering operation and maintenance.* By operation, we mean fixed scheduling based on known work shift. This can be accomplished automatically with time switches or manually if the scheduling can be enforced. Time switches should have manual override to permit accommodation to unusual situations. Lighting maintenance is defined here as fixed scheduling of maintenance and relamping. A 20% increase in maintained light is possible if lamps are replaced at the end of their useful life, when output is down to 70% of initial maintained lumens, and fixtures are cleaned and maintained on a fixed schedule. No trade-off is generally involved, since periodic maintenance and relamping is normally cheaper than one-at-at-time maintenance and burn-out replacement, *and* yields 20% higher average lumens. A trade-off *is* involved between the additional cost of more frequent maintenance and the energy savings produced. Maintenance procedures can have a pronounced effect on light output (see Section 20.30, page 000).

20.8 Preliminary Design

Again, referring to Figure 20.1, the preliminary design phase is the time during which ideas crystallize, but in terms of areas, patterns, as well as light and shadow; not yet in terms of hardware. At this stage the quality of the system is decided upon, that is, the brightness ratios, diffuseness, chromaticity, and proportion of vertical to horizontal lighting are determined. The latter establishes in large measure the room "mood" or lighting ambience. In preceding sections these items were discussed in some detail. In the sections that follow on lighting systems (direct, indirect, etc.) the quality of each will be considered and applications suggested. In the overall view, however, the ultimate quality of the lighting system, its visual pleasantness, centers of visual attention, highlights and shadows, as well as texture and forms, will be a deft and perhaps artistic combination of the above considerations, and will establish, as the term implies, the quality of the lighting design. A few observations, not covered elsewhere, are mentioned below.

Planes other than the "working plane" must always be considered. The ratio of vertical to horizontal illumination of the chosen lighting system will determine wall brightnesses, while the floor finish will have a pronounced effect on the ceiling illumination for direct lighting systems.

The chromaticity of the room lighting depends primarily on the source but secondarily on the luminaire and surface finishes. A "white" source can be tinted slightly by the use of a colored reflector in the luminaire. Of course, the effect on luminaire output of such a change must be considered. In the case of semi-indirect and indirect lighting this same effect can be accomplished by the use of colored ceiling and upper wall surfaces, which serve as secondary reflectors and become the actual luminous source for the room. Recommendations in Tables 19.26, page 853, and 19.27, page 854, cover choice of source.

20.9 Illumination Methods

There are three methods of illumination: general, local and supplementary, and combined general and local.

(*a*) ***General Lighting.*** This is a system designed to give uniform and generally, though not necessarily, diffuse lighting throughout the area under consideration. The method of accomplishing this result varies from the use of luminous ceiling to properly spaced and chosen downlights, but the resultant lighting on the *horizontal working plane* must be the same, that is, reasonably uniform. It may be, but is not necessarily, task lighting.

(*b*) ***Local and Supplementary Lighting.*** These are two terms that are used interchangeably but have slightly different meanings. By definition, *local lighting* provides a small, high-level area of lighting without contributing to the general lighting. *Supplementary lighting* also provides a restricted area of high intensity, but supplements the general lighting. In actual practice it is difficult to differentiate between the two. A desk lamp, a high-intensity downlight on a merchandising display, and a track light illuminating wall displays, all seem to answer both definitions, and in practice are referred to as local, supplementary, or local-supplementary lights. Typical of this genre are the units illustrated in Figure 20.3.

(*c*) ***Combined General and Local Lighting.*** This illumination method is used in areas where the general visual task is low, but local high-intensity lighting, in such instances called supplementary lighting, is required. An excellent example is the department store where circulation requires low-level lighting while merchandising areas and showcases require up to 500 fc. The indicated solution for this problem is a general lighting system that will provide 30 fc of uniform, diffuse lighting of the proper color and supplementary local lighting in restricted areas. The quality of the local lighting supplied depends on the particular item being displayed. A nonuniform lighting layout (see Fig. 18.36, page 751) that is arranged to localize lighting but does not have a uniform overall pattern is difficult to classify. We would think of it as a combination of general lighting in some areas and local in others, that is, a combined system.

These three *methods* of illumination can be accomplished in many ways by the use of luminaires and luminous sources of different types, since the illumination method is a function of *both* fixture placement and arrangement as well as the inherent fixture lighting distribution. The term used to describe the effect of the combination of a particular fixture type applied in a particular way is the *lighting system.* Thus a reflector-type fixture when aimed down gives *direct* light. The same fixture beamed up at the ceiling gives *indirect* light. The following section describes the systems that constitute the vast majority of lighting installations.

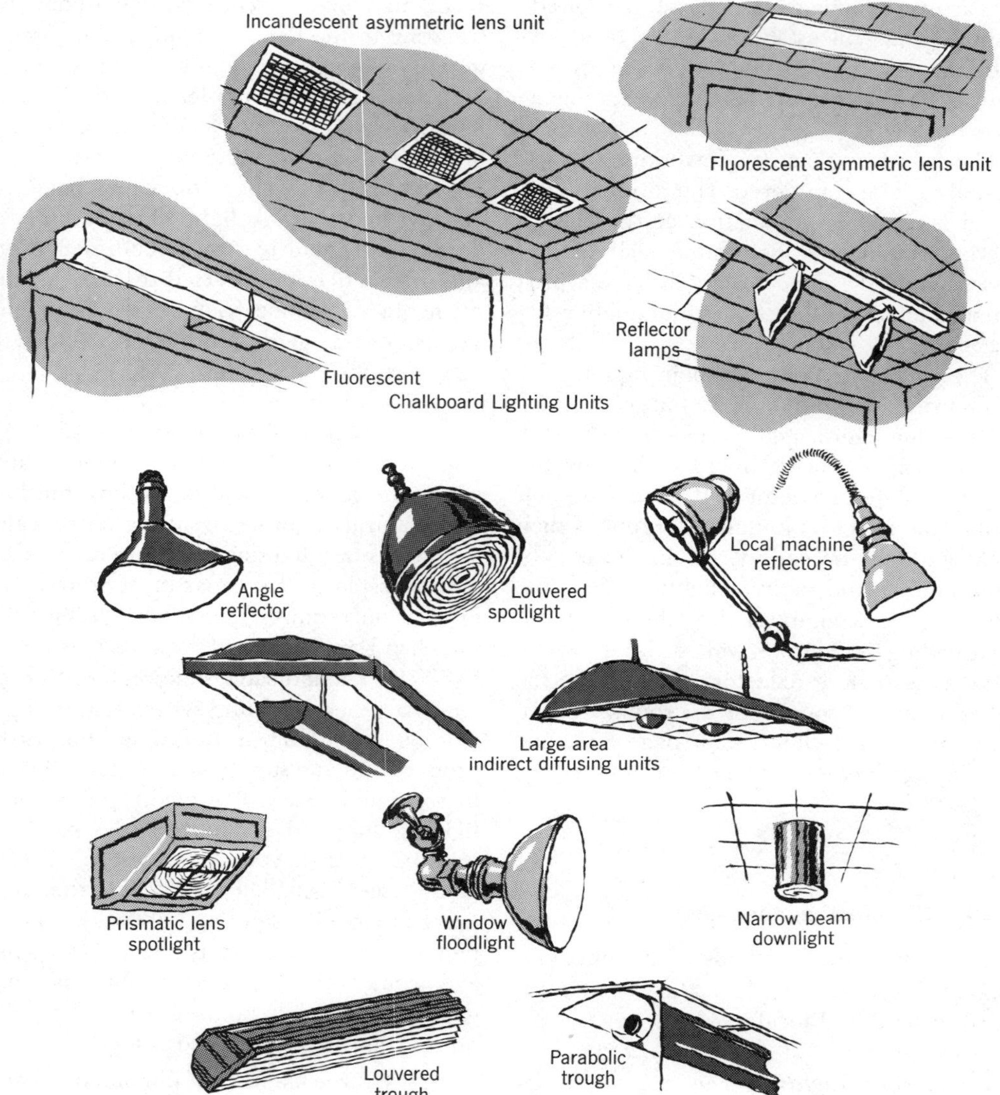

Fig. 20.3 Typical supplementary lighting units for incandescent and fluorescent sources.

20.10 Types of Lighting Systems

No one lighting system can be said to be the single choice in a given instance; on the contrary, the designer will normally have a choice of at least two systems that will, if utilized properly, yield illumination of adequate quantity and good quality. However, other factors, such as harmonization with the architecture and economics, will usually tip the balance in favor of one or the other. The five generic types of lighting systems are: indirect, semi-indirect, diffuse or direct-indirect, semi-direct, and direct.

20.11 Indirect Lighting

See Figure 20.4*a*. Ninety to one-hundred percent of the light output of the luminaires is directed to the ceiling and upper walls of the room. The system is called indirect because

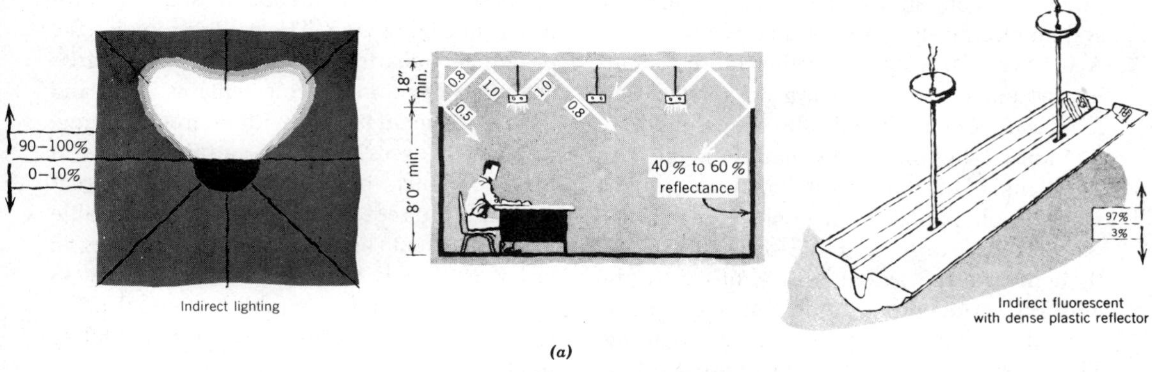

(a)

Fig. 20.4a *Indirect lighting. The white surfaces of the drawings represent the areas that are illuminated by the indirect fixtures, and in turn illuminate the room. Indirect luminaires give uniformly bright ceiling.*

(b)

Fig. 20.4b *Use of architectural coves gives brightness gradient on ceiling and, if properly designed, uniform illumination in the room.*

practically all of the light reaches the horizontal working plane indirectly, that is, via reflection from the ceiling and upper walls. *Therefore, the ceiling and upper walls in effect become the light source* and, if these surfaces have a high-reflectance finish, the room illumination is quite diffuse (shadowless). Since the source must be suspended at least 18 in. (depending on the unit's brightness) in order to avoid excessive ceiling brightness, this system requires a minimum ceiling height of 9 ft, 6 in.

In addition to diffuseness, the resultant illumination is generally uniform, and direct and reflected glare are low. A CRF in excess of 1.0 is common, with associated high ESI footcandles. See Figure 18.39*b*, page 753, and Section 18.23(c)3, page 753.

In order to avoid an unacceptable (greater than 20:1) brightness ratio between the luminaire and its surrounding field, the luminaire is made translucent, at least on the bottom surfaces and sometimes on the sides. This type is known as "luminous indirect" in contradistinction to metal reflectors, which are totally indirect. Architectural coves are classed as indirect lighting (see Fig. 20.4*b*).

In all cases, but most particularly in the instance of high-level illumination, the luminaire spacing and suspension length and the cove or valance dimensions must be carefully chosen to avoid excessive ceiling brightness. We consider 75 raw fc to be the maximum horizontal-plane illumination attainable without exceeding ceiling brightness limits of 400 fL. With a CRF in excess of 1.0, this is sufficient for all but difficult tasks. The lack of shadow, low source brightness, and highly diffuse quality created by the indirect lighting system give a very quiet, cool ambience to this type of lighted space, suitable for private offices, lounges, and plush waiting areas. Areas having specular visual tasks use this system to advantage.

When properly designed, particularly when the source of light is architectural coves, the ceiling has a floating, almost infinitely deep or skylike quality, which is pleasant and can be used to give an impression of height in a large room of low ceiling. This system is not to be confused with the self-luminous transilluminated ceiling, which is a direct lighting system of entirely different quality and effect. A further characteristic of the indirect lighting system is loss of texture on vertical surfaces as is common to all fully diffuse lighting.

Indirect lighting is by nature inefficient, since much of the useful light reaches the working plane only after double reflection—within the fixture and off the ceiling. Although, to some extent, this is offset by the high ESI, applications to difficult seeing tasks normally require supplementary lighting. Thus an indirectly lighted drafting room having tables equipped with supplementary lamps would take advantage of both systems—the local high-intensity light at about 200 fc for the restricted area being worked on, and overall table lighting of 40 to 50 fc ESI. The latter would also solve any reflected glare problems arising from the many viewing angles required by large tasks such as drawing.

20.12 Semi-Indirect Lighting

Sixty to ninety percent of the light is directed upward to the ceiling and upper walls. This distribution is similar to that of indirect, except that it is somewhat more efficient and allows higher levels of illumination without undesirable brightness contrast between fixture and surroundings along with lower ceiling brightness. A typical fixture employs a translucent diffusing element through which the downward component shines, and is illustrated in Figure 20.5. The ceiling remains the principal radiating source and the diffuse character of room-lighting remains. VCP and ESI both remain high as with indirect lighting. A CRF in excess of 1.0 is attainable, as in Figure 18.39*a*, page 753. In both indirect and semi-indirect systems, it is often desirable to add accent lighting or downlighting in order to break the monotony inherent in these systems, and to establish a visual point of interest, or to create required modeling shadows.

The quality of the semi-indirect system is somewhat different than indirect when using fixtures (semi-indirect cannot use architectural coves) because attention is not drawn to the fixtures, since they exhibit less contrast with the background ceiling brightness. Some of the feeling of ceiling cavity depth can be achieved by using well-shielded luminaires with luminous sides.

In both indirect and semi-indirect lighting systems the light undergoes a number of ceiling and wall reflections before reaching the horizontal working plane. Greater illumination can be achieved if these surfaces are colored than if they are grays of the same luminous reflectance. With an 80% reflectance paint, the difference is approximately 2% per reflection. Also, the use of colored paints serves to tint the room illumination slightly because of selective absorption as explained previously.

Fig. 20.5 Semi-indirect lighting.

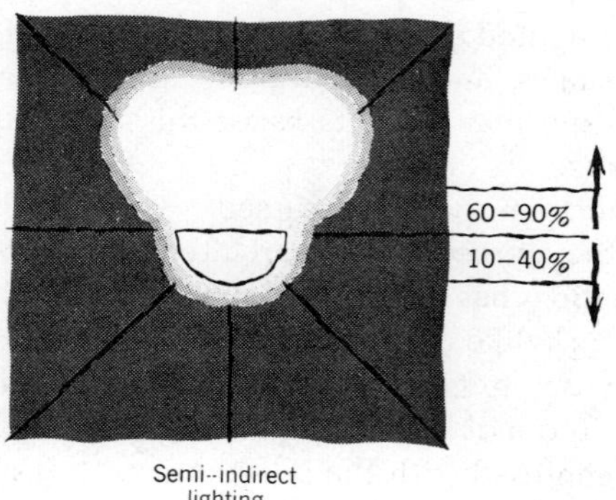

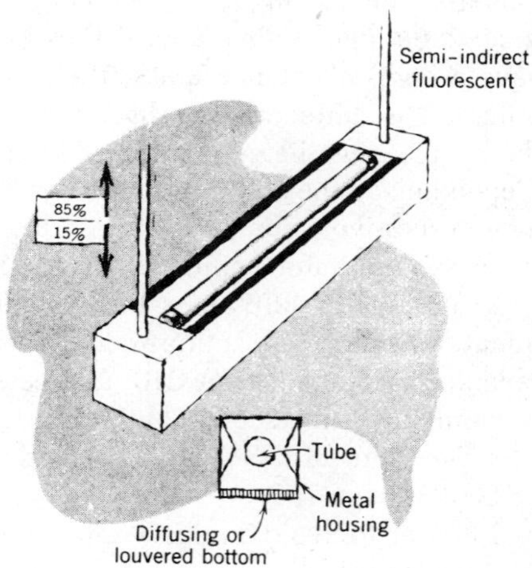

20.13 General Diffuse and Direct-Indirect Lighting

This type provides approximately equal distribution of light upward and downward, resulting in a bright ceiling and upper wall background for the luminaire. For this reason brightness ratios in the upper-vision zone are usually not a problem, although direct and reflected glare may be troublesome at high-illumination levels (100 fc and above). Since the ceiling is a major though secondary source of room illumination, diffuseness will be good, with resultant satisfactory vertical-plane illumination. Light falling on a horizontal surface will derive principally (65 to 75%) from the luminaire and secondarily from the ceiling (25 to 35%), the exact ratio being dependent on the ceiling reflectance and the fixture characteristic.

The difference between general diffuse (e.g., an opal diffusing globe) and direct-indirect (e.g., an open-top luminous side and bottom luminaire) lies in the fixture characteristic; diffuse fixtures give light in all directions, whereas direct-indirect have little horizontal component (see Figs. 20.6 and 20.7). Stems should be of sufficient length to avoid excessive ceiling brightness; generally not less than 12 in.

Since the impression of illumination depends to a large extent on *wall brightness* because this is the surface we see most often, a space with general diffuse illumination will *appear* lighter than one with direct-indirect because of the darker walls in the latter (see Fig. 20.7*b*). This effect is most pronounced with highly directional downlights, as discussed below. If this effect is not desired, fixtures must be placed near the walls or other wall illumination provided (see Fig. 20.11).

Quality of the lighting depends in large measure on the layout and on the tasks involved. By avoiding excessively bright units and giving attention to positioning of sources and viewing angles, VCP and CRF can both be kept high. Fixture brightnesses are interest points and the space will not appear dull and monotonous. Efficiency of these two systems is good. Both are well applied in spaces requir-

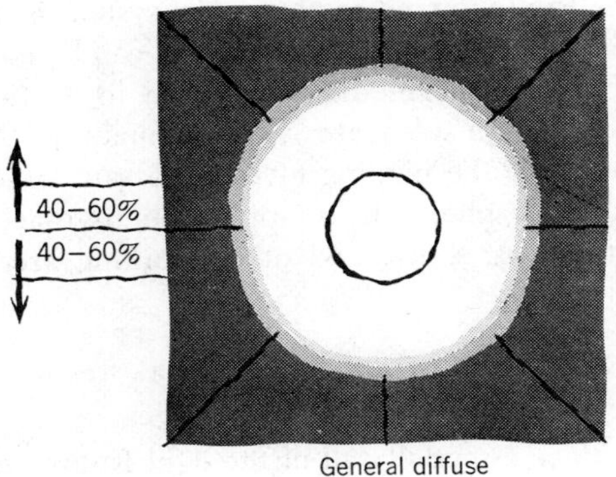

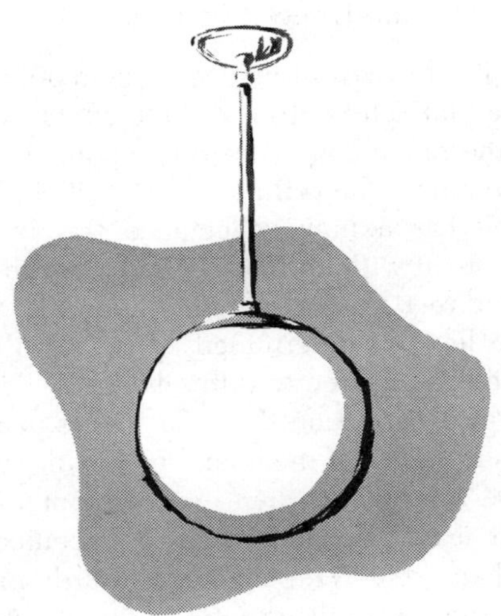

Fig. 20.6 *General diffuse lighting. Note that all room surfaces are illuminated and become secondary sources; the primary source of illuminant is the direct radiation from the fixture. The floor contribution is low due to its normally low reflectance.*

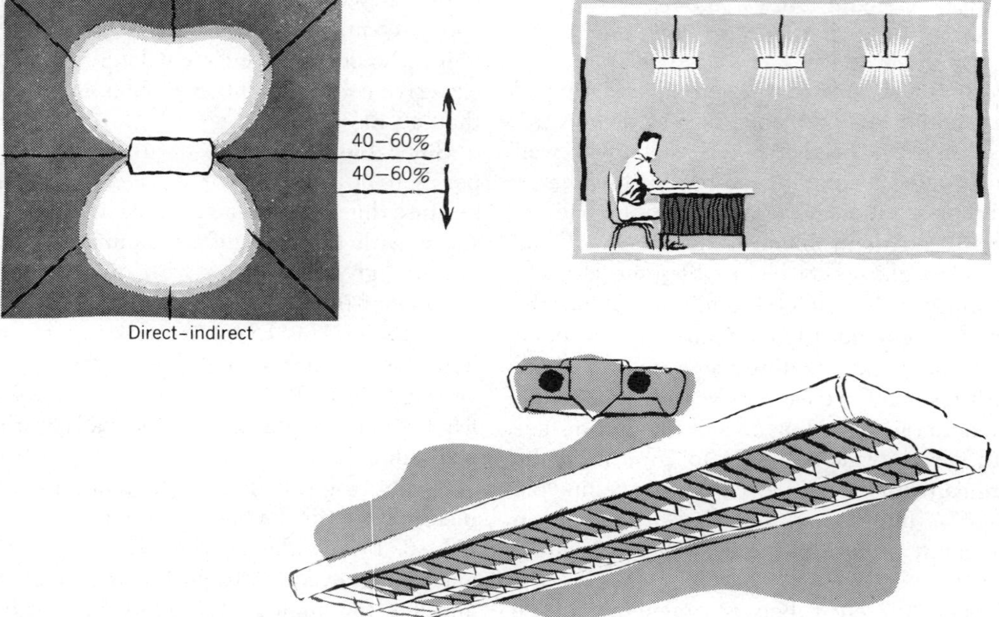

Fig. 20.7 Direct-indirect lighting. Upper and lower room surfaces are luminous but center of walls is not because of the lack of horizontal light from fixtures. Principal light on working plane comes from luminaire directly.

ing overall uniform lighting at moderate levels such as classrooms, standard office work spaces, and merchandising areas.

20.14 Semi-Direct Lighting

With this type of lighting system 60 to 90% of the luminaire output is directed downward and the remaining upward component serves to illuminate the ceiling (see Fig. 20.8). If the ceiling has a high reflectance this upward component will normally be sufficient to minimize direct glare, depending upon eye adaptation level as determined by overall illumination. The degree of diffuseness will depend in large measure on the reflectances of room furnishings and of the floor. Shadowing should not be a problem when upward components are at least 25% and ceiling reflectance not less than 70%. With smaller upward components the system is essentially direct lighting (see the following section). The system is inherently efficient. Reflected glare can be controlled by the methods discussed in Section 18.23. With adequate wall illumination, the quality of the lighting gives a pleasant working atmosphere. It is applicable to offices, classrooms, shops, and other working areas.

20.15 Direct Lighting

Since essentially all of the light is directed downward, ceiling illumination is entirely due to light reflected from the floor and room furnishings. This system then, more than any other, requires a light, high-reflectance, diffuse floor unless a dark ceiling is desired from an architectural or decorative viewpoint. Occasionally the ceilings are deliberately painted a dark color and pendant direct fixtures used in order to lower the apparent ceiling of a poorly proportioned room or to hide unsightly piping, ductwork, and so on.

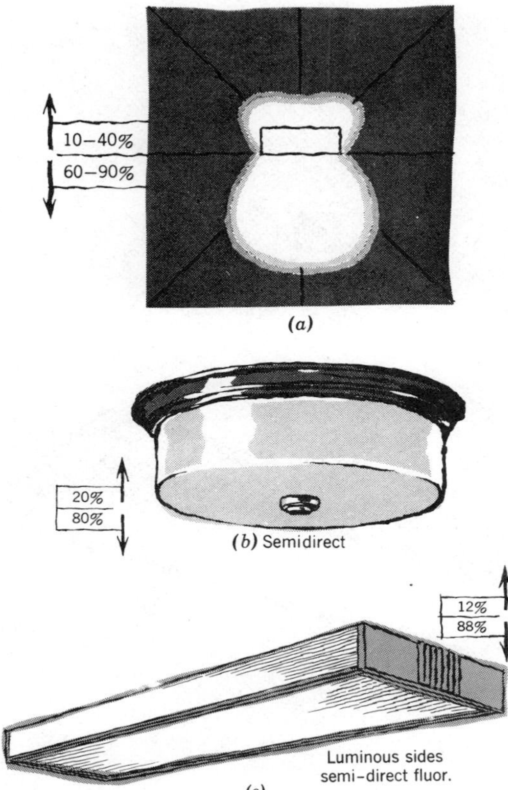

Fig. 20.8 *Semi-direct lighting provides its own ceiling brightness* (a), *with surface-mounted fixtures* (b), *or pendant/surface units* (c). *Other characteristics are similar to direct lighting.*

The effect of direct lighting depends greatly on whether the luminaires are spread or concentrating (Figs. 20.9 and 20.10). In the former case considerable diffusion results from reflections on floor, furniture, and walls of light emitted at high angles from the nadir. The result is a working atmosphere with slightly darkened walls and ceiling. This type of lighting, which is most widely represented by the recessed fluorescent troffer in a suspended ceiling, is standard for general office lighting. The fixtures themselves form a ceiling surface of light and dark areas, and the quality of the entire system is pleasant. Difficulties associated with low VCP and veiling reflections can be controlled by proper use of reflectances, use of low-brightness units (see Section 20.27), and judicious arrangement of viewing positions (see Fig. 18.32, page 748). When direct lighting units are used in an overall pattern, this latter option is radically diminished, and the need for particularly low-brightness units and high ceiling reflectivity, or specialty diffusers like the batwing design, is increased. For this reason (and from the energy viewpoint), lighting of *difficult* seeing tasks with uniform layouts is discouraged in favor of a nonuniform layout, or general lighting with spread-direct sources, supplemented by concentrating direct sources.

Direct lighting gives little vertical surface illumination, requiring the addition of perimeter lighting in business atmospheres (see Fig. 20.11). Concentrating downlights used alone are appropriate in restaurants and other areas where the privacy type of atmosphere generated by limited-area horizontal illumination and minimal vertical-surface illumination is desired. When these fixtures are designed with black cones or baffles or other devices that are nonreflecting at the viewing angle, the fixture appears dark. It is our opinion that installations providing high-horizontal-surface illumination, with no apparent source of brightness, such as those using black-cone downlights, are disturbing to the eye and to our normal bright-sun-and-sky orientation and should therefore be used cautiously and only in limited areas. This same comment, but to a lesser extent, is applicable to very low-brightness diffusers such as the parabolic wedge type (see Fig. 18.8*a*). There, however, the unit has the redeeming characteristic of low reflected glare, which is not the case with downlights.

In summary then, spread direct lighting is suitable and appropriate for general lighting while concentrating direct lighting, which reduces vertical illumination, is appropriate for highlights, local and supplementary lighting, and specialized or casual viewing.

20.16 Size and Pattern of Luminances

Because of its luminance each luminaire or other luminous source is a point of visual attention. To the extent that luminaires are

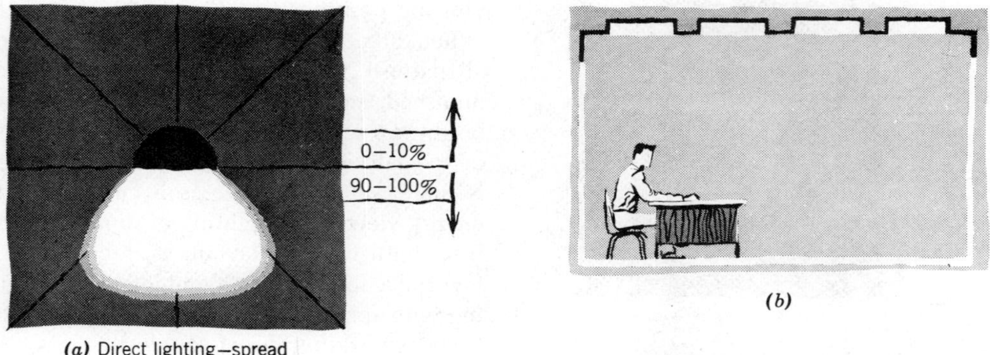

(a) Direct lighting—spread

(b)

Fig. 20.9 *Spread-type direct lighting* (a), *illuminates all room surfaces except the ceiling* (b), *which is only illuminated by reflection from the floor. Some diffuseness is evident. The most common type of unit in this category is the direct fluorescent either surface-mounted* (c) *or troffer type in a hung ceiling.*

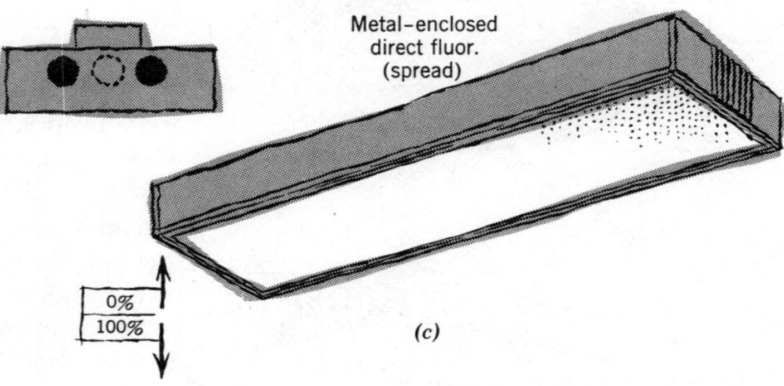

(c)

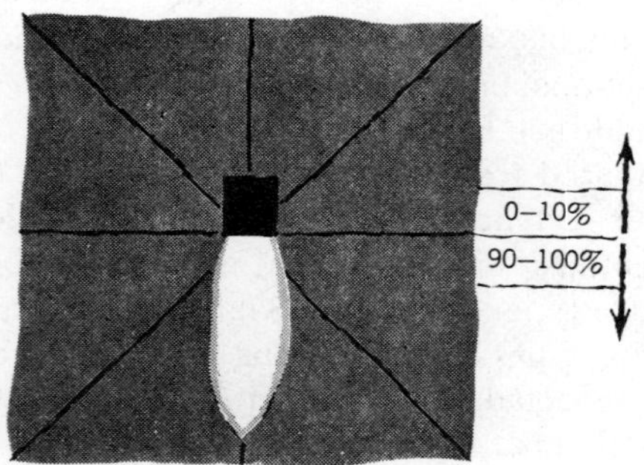

(a) Direct lighting—concentrating

(b)

Fig. 20.10 *With concentrating direct distribution* (a), *the floor is the only luminous surface* (b), *other than the ceiling fixture. Diffuseness is absent. Walls are dark. Incandescent downlights* (c) *are of this type unless equipped with spread-type lenses.*

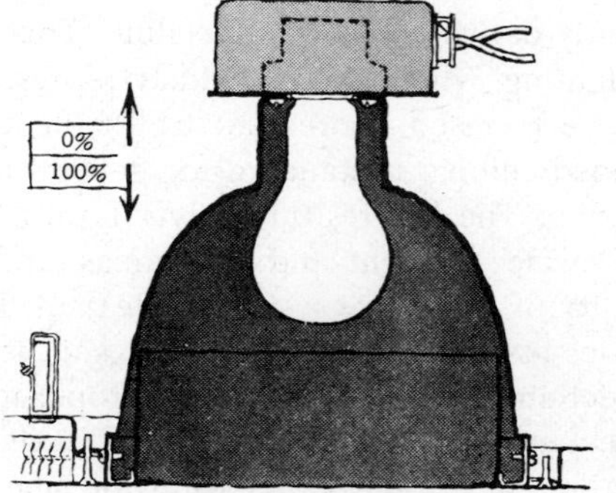

(c) Direct concentrating

Fig. 20.11 *The use of large-dimension, lighting fixtures in a low-ceiling room is possible if the apparent size of the unit is reduced. Here at a 7 ft, 6 in. mounting height, 4 ft x 4 ft units are acceptable because the lattice on the face of each unit gives the impression of reduced fixture size. Note also that the apparent illumination in* (b) *is greater than in* (a) *(though both are exactly equal on the table surface) due to the wall wash in the background. The eye perceives vertical surface illumination more readily than horizontal, and retains the impression for the entire space.*

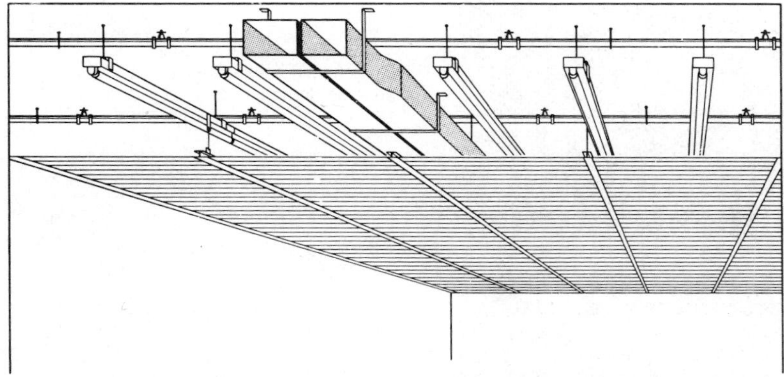

Fig. 20.12 *Luminous ceiling installation. Note that in properly designed installation, piping and ductwork do not affect the light distribution.*

numerous, large, very bright, or arranged in striking patterns, attention will be drawn to them and away from other surfaces. In the absence of these characteristics it may be desirable to add accent color or accent lighting to avoid monotony. Rigid rules cannot be set down covering these criteria, but examples can demonstrate the principles.

Luminaire size should correlate with room size and ceiling height. Fluorescent fixtures larger than 2 × 4 ft should not be used in ceilings below 10 ft unless their size is minimized by some sort of surface pattern (see Fig. 20.11). Transilluminated ceilings are *all* fixture and therefore require a minimum of 12 ft mounting height. When installed below this level, particularly in large rooms, the effect is oppressive, as if the sky were lowered on us. To offset this effect, the use of colored, shaped, or dark panels is of some help. In place of a luminous ceiling (Fig. 20.12) a large-area, coffer-type fixture can be utilized, which gives the impression of great depth (see Fig. 20.13).

In order to achieve the uniformity of illu-

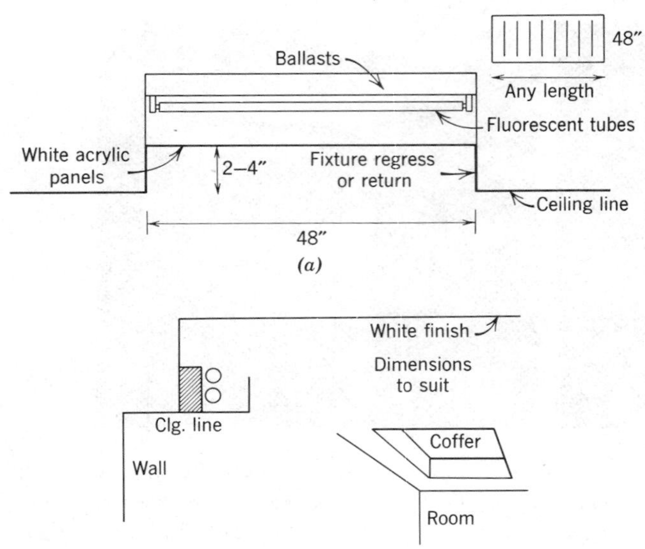

Fig. 20.13 *Types of coffer.* (a) *Coffer fixture, generally available in 4-ft width and any length.* (b) *Architectural coffer, designed as required. Both types give an illusion of great depth and soft, glare-free illumination.* (a) *is direct diffuse and* (b) *is indirect lighting. Coffers are also ideal daylight sources when designed in conjunction with skylights as in Figure 19.10.*

(a) Longitudinal lines increase apparent length, direct traffic flow, decrease direct glare

(b) Horizontal lines create a plane, increase apparent width but also increase direct glare

(c) Diagonal lines minimize shadows and break rectangular patterns. They are architecturally dominant

Fig. 20.14

(*d*) Rectangular pattern is architecturally dominant and therefore must be used carefully. It is a poor choice in stores where attention downward is desired

(*e*) Cornices, valances and coves are luminous ceiling borders. In large rooms suspended coves achieve uniform ceiling brightness and when designed with a downward component or combined with local lighting, as illustrated, give a pleasant, intimate atmosphere

(*f*) Coffers create a decorative architectural effect and can be designed to resemble skylights or can be built into actual skylights

(*g*) Luminous ceiling system utilizing louvers or translucent material suspended beneath rows of fluorescent lamps provides high illumination, low brightness and high diffusion. The system is architecturally oppressive, and monotonous, requiring some accent of either color or lighting

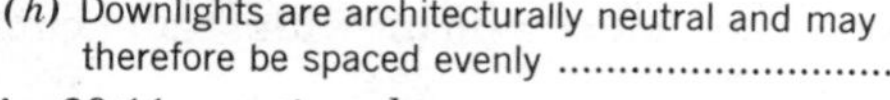

(*h*) Downlights are architecturally neutral and may therefore be spaced evenly .... (*j*) or unevenly

Fig. 20.14 *continued on next page*

(k) The large can-shaped surface-mounted downlights dominate the area's appearance despite the high ceiling. The pimpled look could be eliminated by recessing the units. Or, taking advantage of the prominence of these units, interest could be added by altering the regular pattern. [See *(l)*.]

(l) The lines of lights in the center converge optically to produce a directional flow of traffic toward the escalator. The remaining floor lighting is provided by visually pleasing circular patterns around columns (Penn Station, N.Y.C.)

(m) That patterns of lighting are plainly visible even during the daytime is apparent from this photo. The attractiveness of uniformity of fixture pattern can readily be seen. Photo by Stein.

mination necessary for general lighting, regular spacing is desirable. However, various effects may be obtained within the regularity, to accomplish an architectural purpose, as shown in Figure 20.14. The pattern of lights must never be at cross-purpose with any dominant architectural pattern; rather it should either reinforce an architectural form to be neutral. If a strong architectural element is absent, a dominant lighting pattern may be desirable. Conversely, a strong architectural element can either be reinforced (Fig. 20.14*p*) or utilized to carry a neutral lighting pattern (Fig. 20.14*q*).

Generally, continuous row installations are more attractive than individual units, and eliminate the dominating checkerboard effect of the latter. Coves and cornices give the ceiling a floating or light effect. Geometric patterns can be used to add interest or break monotony of large areas, such as department stores. Generally, incandescent downlights are not dominant, and regularity of placement is not essential. Nonuniform layouts with large sources create a distinct pattern problem inasmuch as they are too large to be neutral, and the nonuniformity can create visual confusion (see Fig. 20.15). The only cure for this problem is to minimize the source brightness by using low-brightness diffusers (see Section 20.27). The incorporation of daylight into the luminous sources of a space should take cognizance of size and pattern as well. Large windows are not consonant with small ceiling

(n) Since fixtures are readily visible even when unlit during daylight hours, their outline can be accentuated and the resultant pattern utilized as an architectural motif. Courtesy of Welton Becket & Assoc.

(o) Lighting can be utilized as a medium to connect the inside and outside of a building. The simple expedient of continuing the lighting pattern beyond the window or wall glass provides visibility from inside out as well as outside in. Care must be exercised to avoid fixture placement which will reflect in the glass. Photo by Marc Neuhof.

(p) Treatment of dominant architectural patterns. In each case the lighting designer was faced with essentially the same problem, viz., a low-level seeing task in a large space with a dominant architectural ceiling. Different solutions were arrived at, each of which is consonant with the architectural ceiling. Both accomplish this by following the dominant line; *(q)*, next page, follows it with a neutral pattern. Photo by M.B. Warren.

(q) See *(p)* on previous page. Photo by I. Reens, and TAC.

(r) The lighting in this extremely strong architectural pattern was hormonized deftly by recessing fixtures deeply into the lattice motif. Metal halide HID units and tungsten-halogen units were used. Courtesy of GTE/Sylvania Inc.

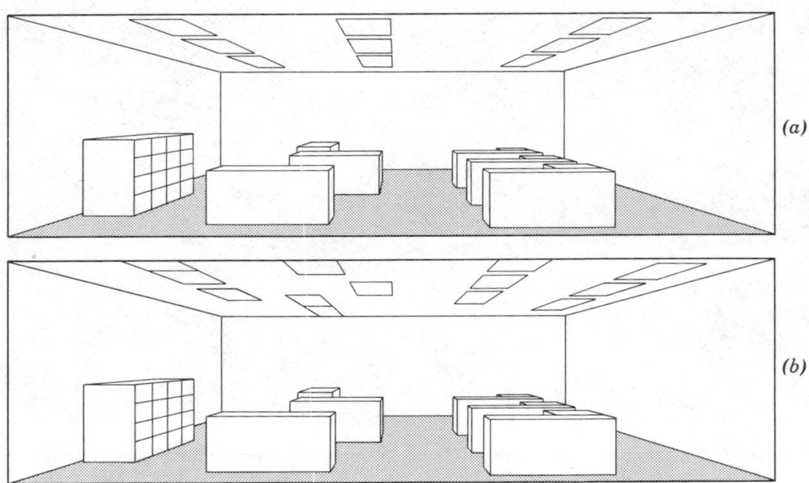

Fig. 20.15 (a) *The uniform layout of Figure 18.36 is neutral in that it does not dominate the space or draw the eye.* (b) *The nonuniform layout can be dominant in the pejorative sense if the eye is drawn by the lack of pattern or symmetry.*

sources, whereas skylights are readily integrated with other ceiling units. A frequently neglected consideration is the appearance of a source when de-energized. With proper daylight and energy-conserving design, many sources will be "dark" during the normal-use hours of the space. Obviously, low-brightness sources will change least in appearance, which is a factor in their favor.

DETAILED DESIGN PROCEDURES

20.17 Design Decisions

Refer again to Figure 20.1. At this point in the design process the lighting hardware is chosen on the basis of the considerations adduced in the preliminary design stage and the appropriate calculations performed. Some spaces will require overall, uniform illumination. These spaces are calculated by the lumen method, which yields average illumination. Other spaces will utilize local lighting alone (or local lighting in addition to general), requiring point-by-point illumination calculations or some other method for restricted-area calculation. Part of these calculations are those of VCP and ESI, either exact or estimated. Additional considerations at this design stage are type of ceiling system, for example, modular, movable fixture, and integrated service, and ancillary considerations of ballast noise, fixture heat distribution, brightness ratios, and maintenance. Also decided here is whether to utilize work station mounted or built-in lighting, both of which are principally applicable to open-plan spaces. The sections immediately following are a discussion of the photometric and other characteristics of lighting fixtures, without which proper equipment selection is obviously impossible.

20.18 Direct-Lighting Luminaire Characteristics

It is important to understand the action of luminaire reflectors. The basic shapes and beam patterns are illustrated in Figures 20.16 to 20.19, while shielding methods are shown in Figure 20.20. Although the illustrations use point sources (incandescent or HID lamps), the principle illustrated is applicable to reflectors for linear (fluorescent) sources, when considered in section. See also Figure 21.1, page 940. An interesting development in the area of elliptic reflectors was produced recently by one of the major lamp manufacturers. Note from Figure 20.17*a* and *b* that the so-called pin-hole downlight requires an elliptic reflector to focus the light through this hole at point "*x*" in order to maintain even minimal fixture efficiency. Elliptic reflectors are large and frequently space above the ceiling is too restricted for their use. A lamp with an integral elliptic reflector, which can therefore be utilized in a standard baffled reflector without severe losses, is illustrated in Figure 20.21.

20.19 Lighting Fixture Distribution Characteristics

When selecting a ceiling fixture for general coverage, the designer is faced with a choice of quite literally hundreds of units, most of which purport to be ideal for the purpose intended. Efficiency and economics will be discussed below; glare considerations were covered previously; distribution will be our concern in this section. A review of Sections 18.10, 18.11, and 18.23(c) by the reader would also be helpful.

The two distribution curves shown in Figure 20.22 are actual test results of two 2-lamp, 1 ft wide by 4 ft long, semi-direct fluorescent fixtures, with prismatic enclosures. The flat bottom of curve (*a*) indicates even illumination over a wide area and therefore a high spacing to mounting height ratio (1.5), whereas the rounded bottom of curve (*b*) indicates uneven illumination and closer required spacing for uniformity (1.2 ratio of spacing to mounting height above working plane for a maximum of 20% illumination variation). The straight sides of curve (*a*) show a fairly sharp cutoff, and the small amount of light above 45° means high efficiency, sufficient wall lighting, adequate diffuseness, and very little direct glare prob-

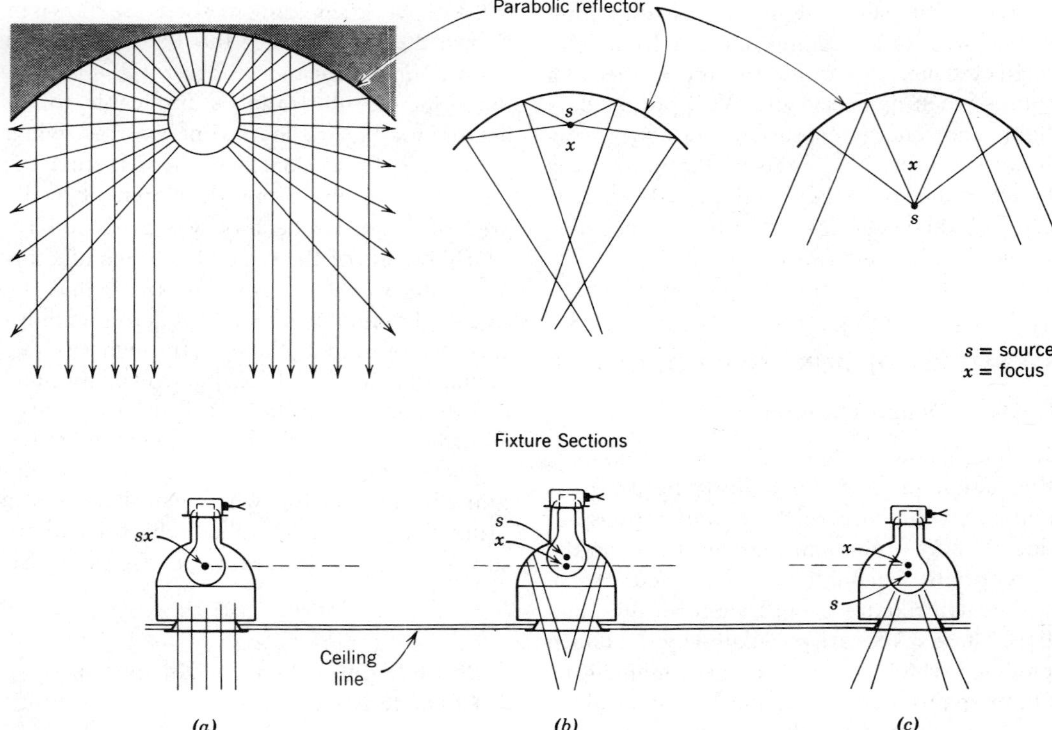

Fig. 20.16 Parabolic reflector action. With source at focal point (a) *rays are parallel. Above* (b) *they converge and below* (c) *they diverge. This focusing action is illustrated by fixtures correspondingly designated.*

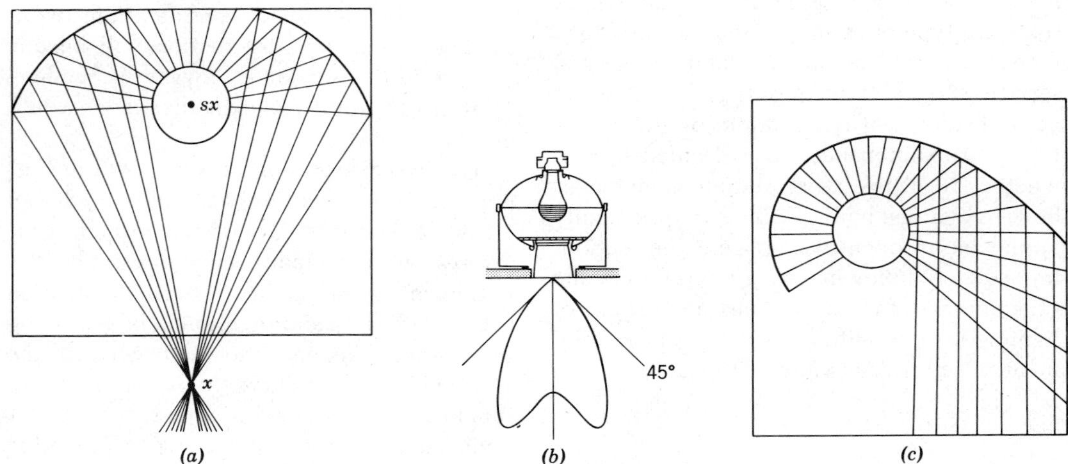

Fig. 20.17 Action of elliptical and extended reflector sections. In the ray diagram (a), *with the light source at one focal point, the light converges at the other focal point. This effect is useful in fixture design as in* (b). *By projecting light* up *only, through use of a silvered bowl lamp, the output light can be redirected through a constricted aperture at the other focal point, with little loss. Beyond the aperture, the light diverges. The cone in the bottom of the fixture provides cutoff at high angles of viewing (see Fig. 20.20). The extended section reflector in* (c) *allows the source to be concealed (shielded) and yet projecting the light downward.*

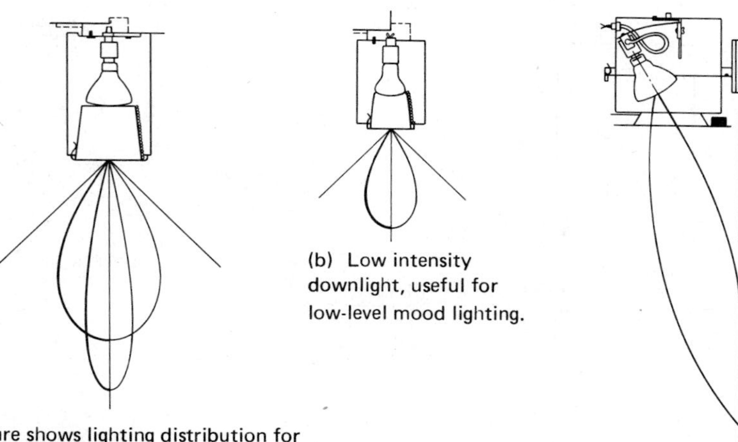

Fig. 20.18 Fixtures utilizing lamps with integral reflectors serve generally as lamp holders, and to provide beam shielding.

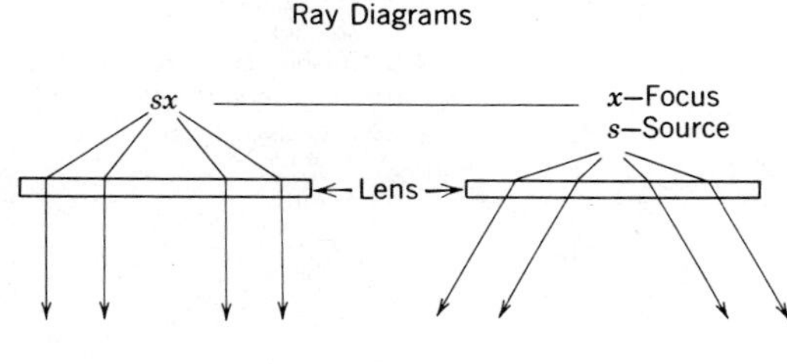

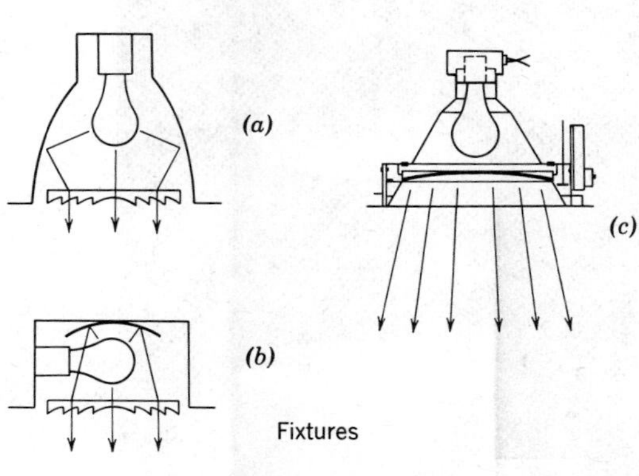

Fig. 20.19 *Action of a Fresnel lens. The lens performs the same function as a reflector, controlling the beam as a function of source placement. By utilizing a lens fixture, the curved reflector can be largely eliminated yielding a smaller fixture, while maintaining accurate beam control. Common design* (c) *uses regressed lens to provide shielding, although lens brightness is not normally objectionable.*

lem, but a distinct possibility of veiling reflections. Conversely, curve (*b*) shows a large amount of horizontal illumination (above 45°), with resultant direct glare, diffuseness, and relative inefficiency, since horizontal light is attenuated by multiple reflections before reaching the horizontal working plane. Here, however, low output below 45° minimizes reflected glare. The uplight component of fixture (*a*) is directed outward to cover the ceiling and will not cause hot spots; the corresponding light from fixture (*b*) is concentrated above the fixture and will give uneven illumination of the ceiling. Thus we see that a rapid inspection of a fixture curve performed by an informed person can yield a large amount of data on the fixture's performance.

PAR Lamp

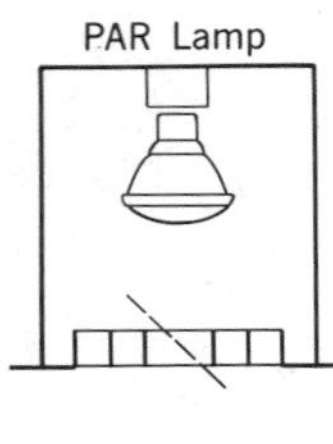

R Lamp

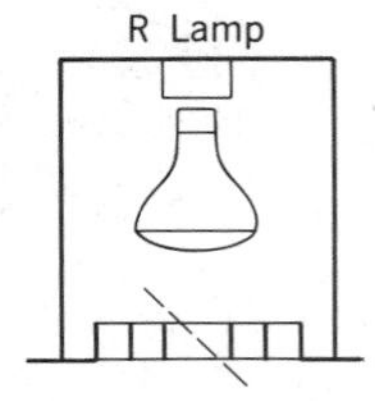

General Service Lamp

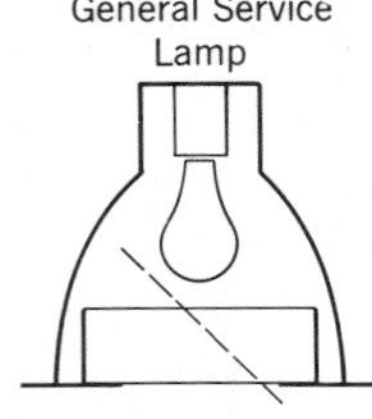

45° Shielding with vertical baffles

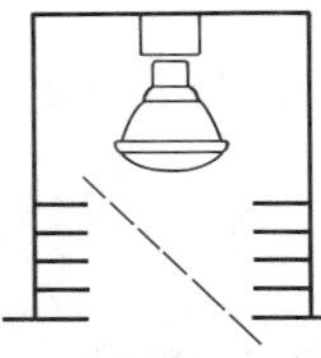

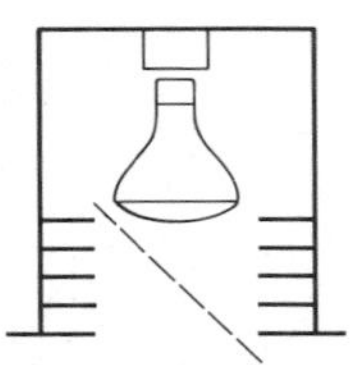

T Lamp

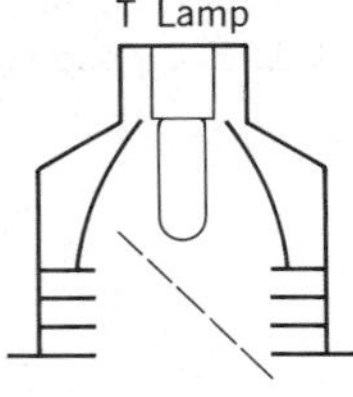

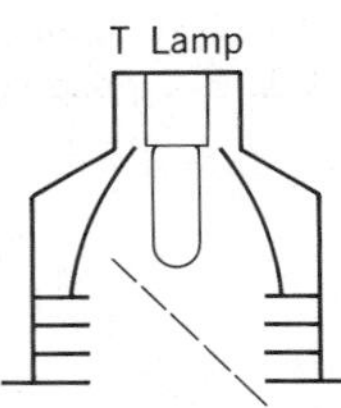

45° Shielding with horizontal baffles

Baffled downlights control unwanted high angle light by cutoff as illustrated. Black baffles aid by absorbing and appearing dark.

Other colors give a ring of light at the baffle edge.

R-Lamp

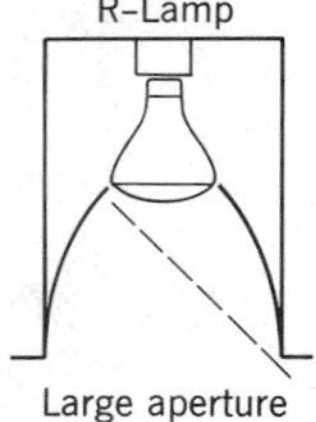

Large aperture

General Service Lamp

Small aperture

Cones control brightness by cutoff and by redirection of light due to shape. They are either parabolic or elliptical. A light specular finish appears dull; a black specular finish appears unlighted.

Black finishes require high quality maintenance since dust shows as a bright reflection.

Fig. 20.20 *Downlight shielding methods.*

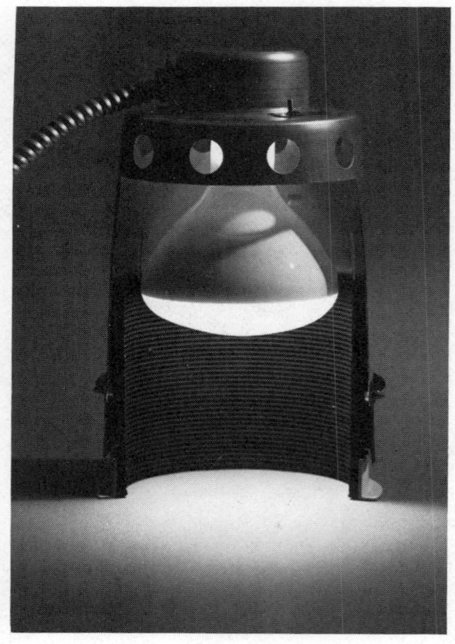

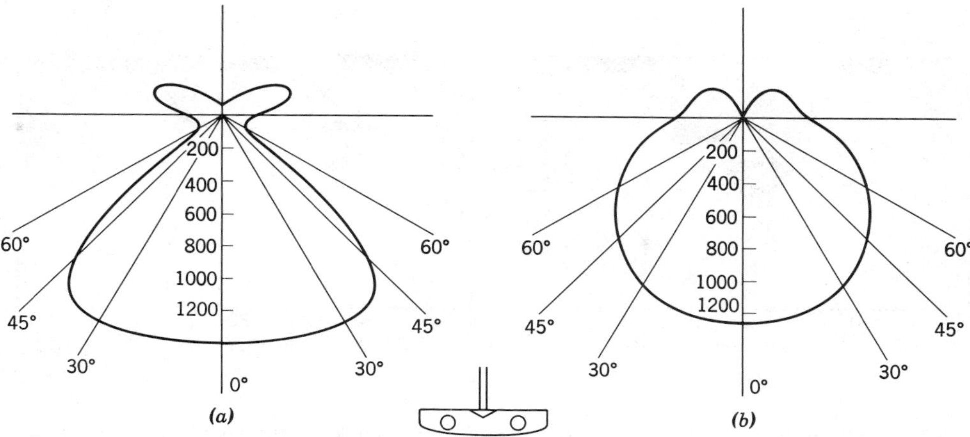

Fig. 20.22 *Semi-direct fluorescent fixture distribution (2-lamp, 40-w, prismatic enclosure). Note the sharp cut-off and wide horizontally even distribution of* (a) *in contrast to the diffuse, broad, and horizontally uneven distribution of* (b).

20.20 Uniformity of Illumination

In any space intended to be lighted uniformly with multiple discrete ceiling-mounted light sources (rather than a luminous ceiling), it is necessary to establish a fixture spacing that will give acceptable uniformity of illumination. A ratio of maximum to minimum illumination on the working plane of 1.2 is readily acceptable and 1.3 is tolerable. For general background or circulation lighting, up to 1.5 is acceptable. The data given by manufacturers (see figures immediately above the distribution curves for each fixture in Table 20.7) are generally based on this 1.2 figure, which should not be exceeded in a quality design. See Figure 20.23. We mentioned above that the fixture of Figure 20.22*a* had a high spacing to mounting-height ratio because of its flat-bottomed curve. This ratio, when not given by the manufacturer, may be approximated from the figures in Table 20.4. The distribution types shown in Table 20.4 are generic and therefore may not be readily applicable in some cases. The curves of Figure 20.24 are an approximation for direct distribution incandescent fixtures.

A much more accurate method of calculating maximum to minimum illumination ratios is available (see Reference 1) for critically accurate application. The concept of Spacing to Mounting-Height ratio in this more accurate method is replaced by the term Luminaire Spacing Criterion, abbreviated *SC*. For fixture distributions other than those shown, interpolation will be necessary.

It is well known that illumination levels near walls drop off at least 30% even in a well-designed installation. To counteract this effect, particularly when placement of furniture is such that visual tasks will occur near walls, the designer should arrange to provide additional illumination in these areas. This may readily be accomplished by additional fixtures, higher-

Fig. 20.21 (opposite) *Baffled downlights are low in brightness at normal viewing angles because the black baffles trap and absorb the lamp's output. The result is low efficiency, as in the fixture on the right. The lamp on the left contains an elliptical reflector that focuses the light below the fixture, resulting in even lower fixture brightness with high efficiency. This principal is also useful in pinhole downlights as illustrated in Figure 20.17. Courtesy of General Electric, Lamp Division.*

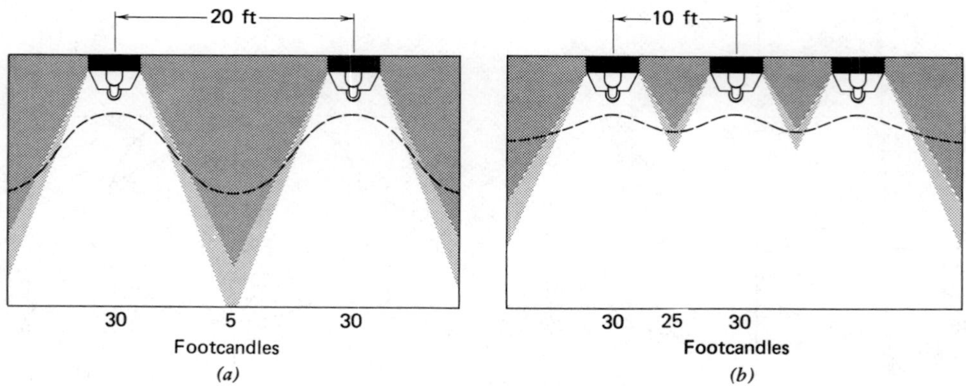

Fig. 20.23 *Uniformity of illumination. The ratio of maximum to minimum illumination should not exceed 1.20 in areas requiring uniform illumination.* (a) *Roller coaster lighting results when spacings are too great.* (b) *Closer spacings result in uniform lighting by overlapping the distributions from individual luminaires.*

Table 20.4 ***Spacing and Mounting-Height Relationship of Luminaires for Illumination Uniformity***[a]

	Light Distribution							
	Indirect		*Semi-indirect*	*General Diffusing*	*Direct-Indirect*	*Spread Direct*	*Semiconcentrating Direct*	*Concentrating Direct*
Ceiling Height	*Distance[b] from Walls*	*Length of Suspension*	*Maximum[c] Spacing of Luminaires*	*Mounting[d] Height of Luminaires*	*Distance[b] from Walls*	*Maximum[c] Spacing of Luminaires*	*Maximum[c] Spacing of Luminaires*	*Maximum[c] Spacing of Luminaires*
8	2	*	9	8	2	$7\frac{1}{2}$	$5\frac{1}{2}$	$2\frac{1}{2}$
9	2		$10\frac{1}{2}$	9	2	9	6	3
10	$2\frac{1}{2}$		$12\frac{1}{2}$	10	$2\frac{1}{2}$	$10\frac{1}{2}$	7	4
11	$2\frac{1}{2}$		$13\frac{1}{2}$	11	$2\frac{1}{2}$	12	8	$4\frac{1}{2}$
12	3		15	12	3	$13\frac{1}{2}$	9	5
13	4		17	13	4	15	10	$5\frac{1}{2}$
14	5		19	14	5	$16\frac{1}{2}$	11	6
15	5		20	15	5	18	12	$6\frac{1}{2}$
16	6		22	16	6	20	13	7
18	6		24	18	6	22	$15\frac{1}{2}$	8
20 or more	7		28	20 or more	7	25	$17\frac{1}{2}$	9

[a] All dimensions in feet.

[b] These spacings apply where desks and benches are next to wall, otherwise one-third the spacing between units is satisfactory.

[c] The actual spacing of luminaires is usually less than the maximum spacing to suit bay or room dimensions.

[d] For mounting height of general diffusing and direct-indirect fixtures, see Table 20.5.

*See Table 20.5.

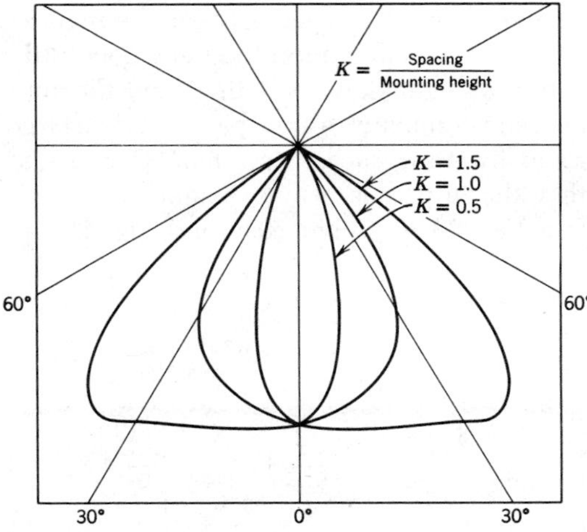

Fig. 20.24 Determination of spacing to mounting height ratio for direct incandescent sources with various distribution curves. To determine S/MH of a specific unit, compare its curve to those given and interpolate to obtain K. *(From the* IES Journal, *January 1963, after Odle & Smith).*

Fig. 20.25 Variation of illumination intensity, vertically, from different sources.

output units, perimeter lighting, or some type of wall-washing arrangement. Particular stress should be placed on this type of local lighting where wall reflectances are low, such as at walls covered with book shelves, equipment racks, low-reflectance paint, or dark wood paneling. Fixture end should be no more than 1 ft and fixture sides no more than 2 ft from walls.

The foregoing discussion of illumination uniformity concerned itself with uniformity on a horizontal plane. Occasionally, it is necessary to know the degree of uniformity vertically (on horizontal planes at different elevations). Four different lighting situations are normally encountered. They are: point sources such as incandescent downlights, line sources such as continuous-row fluorescent fixtures, infinite sources such as luminous ceilings—whether transilluminated or indirect—and parabolic reflector beams such as from PAR lamps. The vertical uniformity of each type is shown graphically in Figure 20.25.

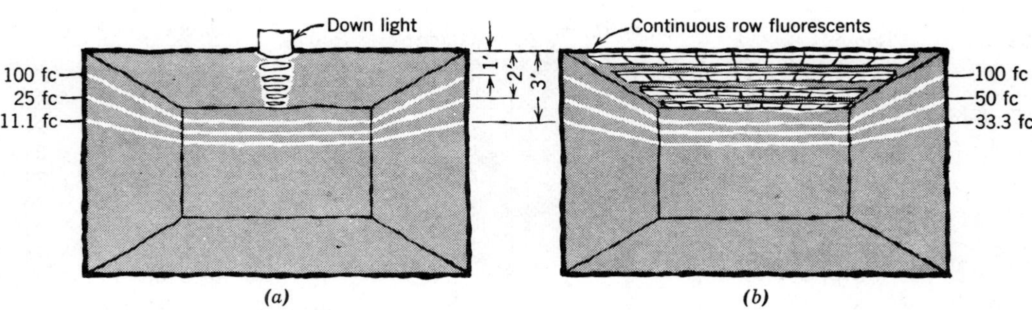

(a), (b). Illumination varies inversely with square of distance for a point source and inversely as distance for a line source.

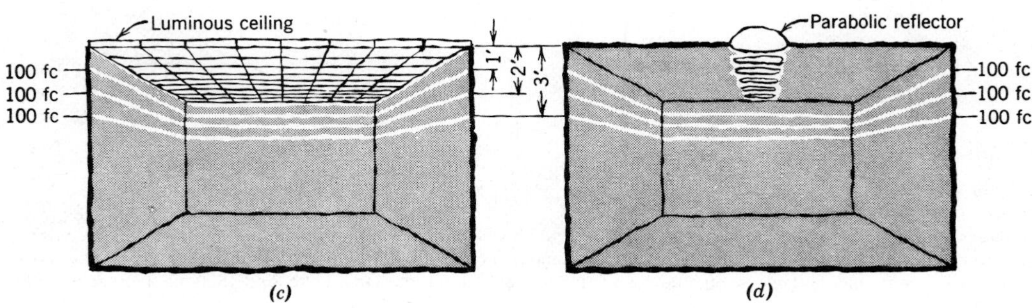

(c), (d) Illumination remains constant at all distances from either an infinite (or nearly) source or a parabolic reflector.

20.21 Luminaire Mounting Height

The mounting height of luminaires is normally established before spacing, and uniformity requirements govern spacing as explained above. In arriving at a mounting height for fixtures with an upward component, a balance must be struck between the requirement of low-ceiling brightness and good utilization of light (low mounting) and the reticence to dominate an area, particularly a large room by using such a low mounting height that the apparent ceiling height is affected (see Fig. 20.26, Table 20.5, and Fig. 20.24).

Fig. 20.26 *Mounting height of fixtures may be lower in a small room than in a large room because of the illusion of lowness created in a large room.*

Table 20.5 ***Practical Hanger Lengths for Suspended Luminaires in Offices and Schools***[a] (See Figs. 20.26 and 20.27.)

Ceiling Height (Feet)	*Room Width (Feet)*	*Hanger Length (Inches)*		*Ceiling Height (Feet)*	*Room Width (Feet)*	*Hanger Length (Inches)*	
		[b]*Offices and Classrooms*	*Drafting Rooms*			[b]*Offices and Classrooms*	*Drafting Rooms*
	7	*	*		13	21 or 24	24
7	14	*	*	13	26	21 or 24	24
	28 and up	*	*		52 and up	21 or 24	24
	8	*	6		14	30	30
8	16	*	6	14	28	24	24
	32 and up	*	6		56 and up	24	24
	9	6	12		15	36	36
9	18	6	12	15	30	30	30
	36 and up	*	6		60 and up	24	24
	10	18	21		16	42	42
10	20	12	18	16	32	36	36
	40 and up	6	12		64 and up	30	30
	11	21	21		18	42	42
11	22	18	21	18	36	36	36
	44 and up	12	18		72 and up	30	30
	12	21	21 or 24		20	54	54
12	24	21	21 or 24	20	40	42	42
	48 and up	21	21 or 24		80 and up	36	36

Source. From *Westinghouse Lighting Handbook,* 1963.

[a] Indirect and semi-indirect luminaires provide low utilization when suspended less than 18 in. If mounted too close to the ceiling, luminaires having a substantial upward component may cause excessively high ceiling brightness. Two-lamp fluorescent units normally should be suspended at least 12 in.; four-lamp units, and 1500-ma fluorescent, 18 in.

[b] Also for libraries, reception rooms, etc.

* Luminaires should be surface-mounted.

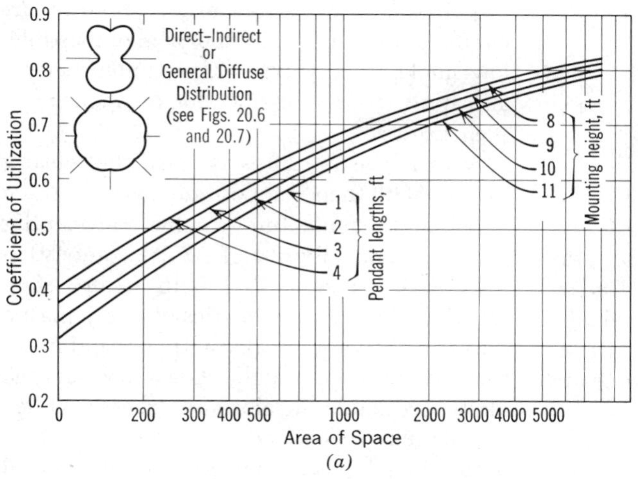

(a)

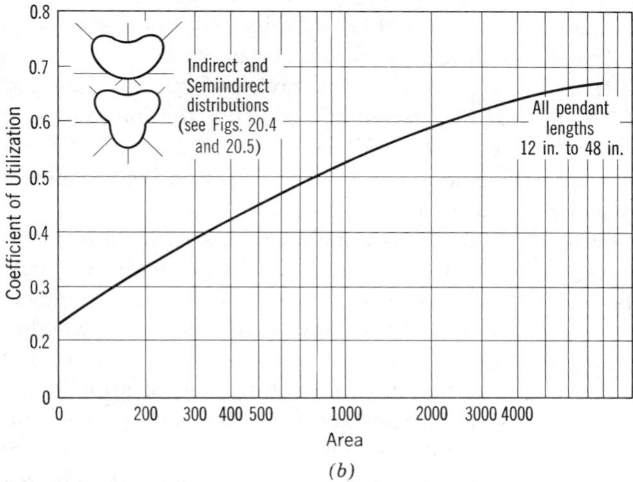

(b)

***Fig. 20.27** Lighting system efficiency (CU) as a function of pendant length. With a substantial downward component as in direct-indirect or general diffuse* (a), *system efficiency rises slowly as the fixture descends (pendant length increases). Where the ceiling is the light source as in indirect and semi-indirect systems* (b), *the pendant length does not change the room illumination (see also Fig. 20.25c).*

The effect of pendant length on coefficient of utilization is given in Figure 20.27 for semi-indirect and direct-indirect distributions.

20.22 Lighting Fixtures

Before proceeding further with design, we now present an excursus on the principal item of lighting hardware, the fixture itself. This and the following three sections will cover fixture construction, installation, and appraisal. The architect need simply stop to consider that lighting fixtures constitute 25 to 30% of the electrical budget or 4 to 5% of the overall building budget, to appreciate their importance. Since the difference between a quality unit and an inferior one is often not readily visible to the casual observer, particular care must be taken in the specification of lighting fixtures and in examination of shop drawings *and* samples. All fixtures if applied properly will give a sufficient quantity of light, but only a good unit will combine quantity with good quality, ease of installation, facility of maintenance, and indefinite life. In addition, regardless of the manufacturing details of a lighting fixture, installation procedure must be proper to insure mechanical rigidity and safety, electrical safety, freedom from excessive temperatures, and requisite accessibility of component parts and of the fixture outlet box. The material below is a combination of NEC minimum requirements plus factors that the author has found important beyond these minima.

20.23 Lighting Fixture Construction

a. All fixtures shall be wired and constructed to comply with local codes, NEC and Underwriters Laboratories Standard for Lighting Fixtures, and shall bear the Underwriters Label, where label service is available. Reflector Luminaire Manufacturers (RLM) standards shall be adhered to for all porcelain enameled fixtures.

b. Fixtures shall generally be constructed of 20 gauge (0.0359 in.) thick steel minimum. If other metals are used they shall be of the required thickness to have the same mechanical strength as 20-gauge steel. Cast portions of fixtures shall be no less than 1/16 in. thick.

c. All metals shall be coated. Unless specified otherwise, the coating shall be a baked-enamel white paint of minimum 85% reflectance, on top of a rust-inhibiting priming process. Unpainted surfaces shall be finished with a clear lacquer except for anodized or "Alzac" surfaces. All hardware shall be cadmium-plated or otherwise rust-proofed.

d. Lampholders in incandescent fixtures shall be porcelain, with nickel-plated screw shell.

e. An incandescent fixture shall not cause a temperature exceeding 90° C on any outside surface.

f. No point on the outside surface of a fluorescent fixture shall exceed 90° C.

g. Each fixture shall be identified by a label carrying the manufacturer's name and address and the fixture catalogue number.

h. Glass diffuser panels in fluorescent fixtures shall be mounted in a metal frame. Plastic diffusers shall be suitably hinged. "Lay-in" plastic diffusers should not be used.

i. Plastic diffusers should be of the slow-burning or self-extinguishing type with low smoke density rating and low heat distortion temperatures. This latter shall be low enough so that the plastic diffuser will distort sufficiently to drop out of the fixture before reaching ignition temperature.

j. It is *imperative* that plastics used in air-handling fixtures be of the noncombustible, low-smoke-density type. These requirements also apply to other nonmetallic components of such fixtures.

k. All plastic diffusers shall be clearly marked with their composition material, trade name, and manufacturer's name and identification number. Results of ASTM combustion tests shall be submitted with fixture shop drawings. The characteristics of many plastic diffusers change radically with age and exposure to ultraviolet light. Glass and acrylic plastic are stable in color and strength. Other plastics may yellow and even turn brown, thus diminishing light transmission radically as well as changing the fixture appearance. Some plastics that are initially very tough and "vandalproof" embrittle with age and exposure to weather or the ultraviolet light of a mercury or fluorescent source. Thus the long-range as well as initial characteristics of all diffuser elements must be investigated before specification and approval.

l. Ballasts shall be mounted in fixtures with captive screws on the fixture body, to allow ballast replacement without fixture removal.

m. All fixtures mounted outdoors, whether under canopies or directly exposed to the weather, shall be constructed of appropriate weather-resistant materials and finishes, including gasketing to prevent entrance of water into wiring, and shall be marked by the manufacturer, "Suitable for Outdoor Use."

20.24 Lighting Fixture Installation

a. Fixtures may not be used as a conductor raceway unless specifically so designed and labeled by the manufacturer.

b. Fixtures mounted on combustible surfaces shall not subject these surfaces to a temperature exceeding 90° C, in a 40° C ambient space. To this end insulating-material spacers or simply an air space may be used.

c. Although most codes allow fluorescent fixtures less than 40 lb in weight to be mounted directly on the horizontal metal members of hung-ceiling systems, experience has shown that vibration, member deflection, routine maintenance operation on equipment in hung ceilings, and poor workmanship can cause such fixtures to fall, endangering life. It is therefore strongly recommended that all fixtures, surface, pendant, or recessed, whether mounted individually or in rows, must be supported from the black channel iron supporting the ceiling system (purlins) or directly from the building structure, but in no case by the ceiling system itself. This is particularly important in the case of an exposed "Z" spline ceiling system.

d. Fixtures installed in wet plaster ceilings shall utilize plaster frames installed for that purpose.

e. Fixtures installed in bathrooms shall *not* have an integral receptacle and when installed on walls, shall have nonmetallic bodies. These are safety precautions.

f. The voltage to ground on branch circuits supplying lampholders and lighting fixtures shall not exceed 150 v except for:

1. Mogul-base lampholders mounted not less than 8 ft above the floor, and accessible only to authorized personnel.
2. Fluorescent lamp fixtures, permanently installed with appropriate ballasts, in which case the voltage may be up to 300 v. Such fixtures may not have integral switching but must use wall switches. This rule allows the use of 277-v fluorescent and mercury lighting. Use of lighting circuits above 150 v is not permitted in residences.

20.25 Standards for Recessed Fluorescent Luminaires

The prevalence of fluorescent troffers, that is, fluorescent fixtures recessed into a suspended-ceiling system, has necessitated industry standardization to reduce fixture-ceiling incompatibility. The five basic types are shown in Figure 20.28. This information is extracted from NEMA Standard LE-1. In order to assure coordination between ceiling type and fixture, the designer should insist that the NEMA ceiling designation appear in the fixture catalogue code and on the fixture.

20.26 Lighting-Fixture Appraisal

As mentioned above, the intense competition in the lighting field necessitates close scrutiny of the characteristics of fixtures. In order to compare the relative merits of similar lighting fixtures as manufactured by different companies, complete test data plus a sample in a regular shipping carton from a normal manufacturing run are necessary.

The following list should be used as a basic guide, with additional items added according to job requirements:

A. *Photometric Performance*
   1. Distribution (ETL tests).
   2. Coefficient of utilization.
   3. Average brightness and brightness distribution.
   4. Spacing-to-mounting-height ratio (calculated by IES method).

B. *Construction*
   1. Engineering.
   2. Workmanship (metal and paint; edge and end metal finish).
   3. Damage susceptibility.
   4. Rigidity in all directions, including twisting.
   5. Rattle prevention.
   6. Quality of materials and finishes.

C. *Installation*
   1. Instruction sheets.
   2. Variety and quality of mounting accessories.
   3. Mounting ease.
   4. Results of ballast temperature tests.
   5. Provision for alignment and leveling.
   6. Wiring ease.
   7. Diffuser attachment (lay-in unacceptable).

D. *Maintenance Ease*
   1. Relamping ease.
   2. Hinging diffuser from either side.
   3. Resistance to dirt collection.
   4. Ease of cleaning.
   5. Individually replaceable lampholders.

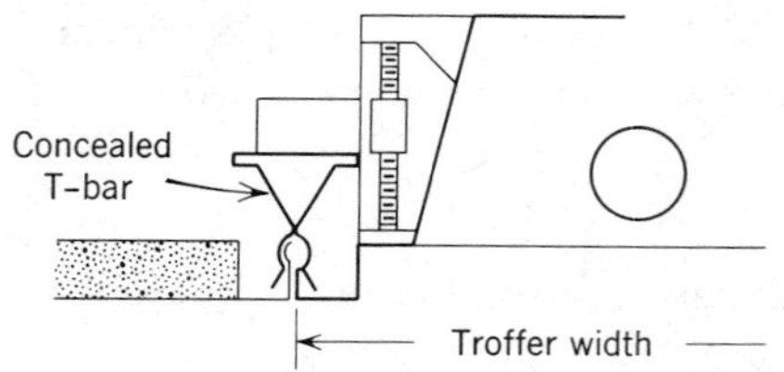

Type M luminaire is one having vertical turned-up edges which are parallel to the lamp direction and intended to "snap-in" or otherwise align the luminaire with a concealed T-bar suspension system, the center opening of the Tees being located on modular or other symmetrical dimensional lines.

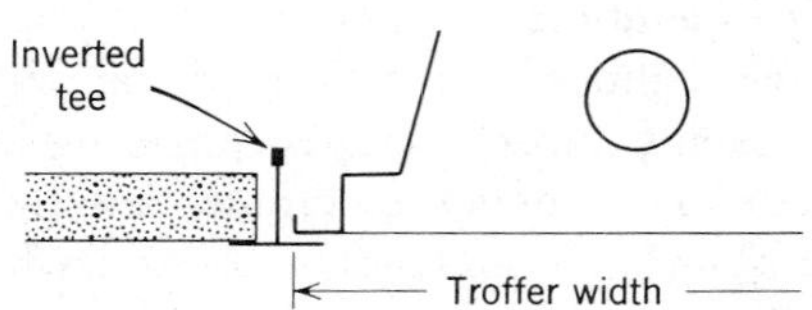

Type G luminaire is one having edges which are designed to rest on or "lay-in" the exposed inverted T of a suspension system (customarily described as a grid ceiling system) with the webs of the Tees being located on modular or other symmetrical dimensional lines.

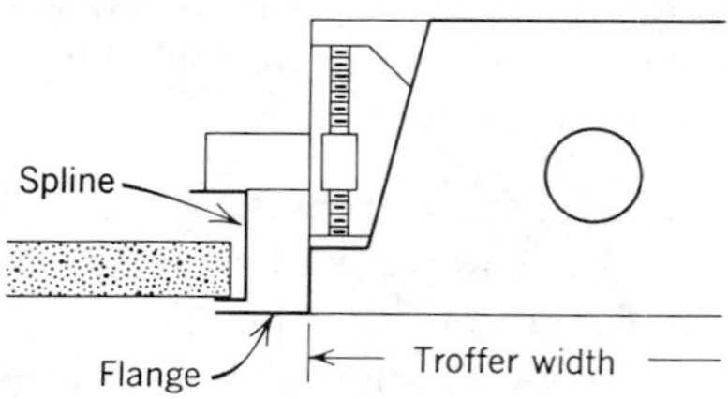

Type F luminaire is one having horizontal flanges which are parallel to the lamp direction and designed to conceal the edges of the ceiling opening above which the luminaire is supported by concealed mechanical suspension.

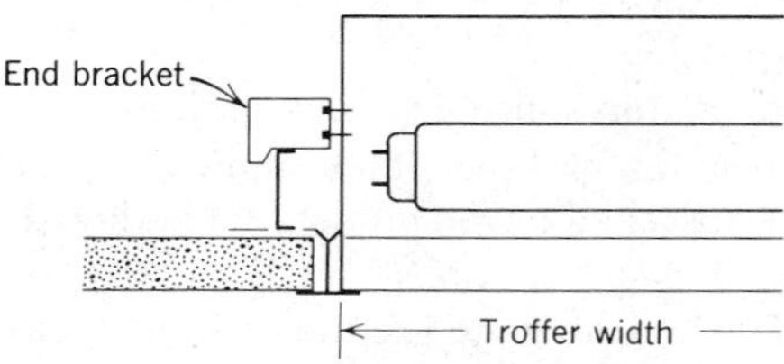

Type H luminaire is one having end brackets, hooks or other attachments and designed to be supported at the ends by "hooking-on" to some member of the ceiling suspension system.

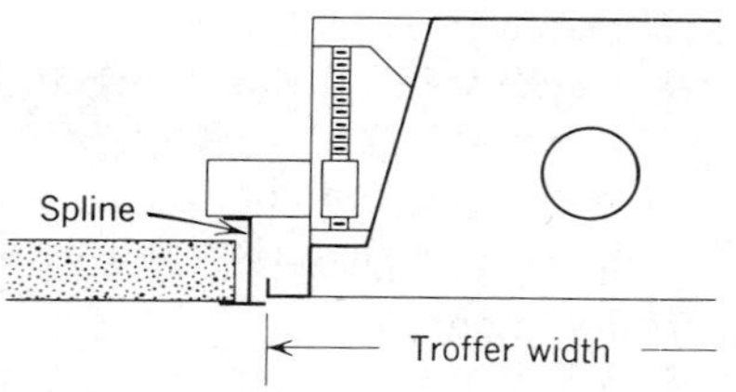

Type S luminaire is one which is designed for mechanical suspension from exposed splines and dependent on splines parallel to the lamp direction for concealment of the edges of the luminaire.

Type HS luminaire is a Type H luminaire having edges parallel to the lamp direction and dependent on splines of the ceiling suspension system for concealment of the edges of the luminaire.

Type HF luminaire is a Type H luminaire having edges parallel to the lamp direction and designed to conceal the edges of the ceiling opening in which the luminaire is recessed.

Fig. 20.28 *Standard fluorescent troffer details.*

6. Nonturning ballast screws.
7. Thermal contact of ballast with channel.
8. Ballast replacement.

E. *Appearance*
   1. Light-tightness.
   2. Styling.

F. *Price*
   1. Relative annual cost (unit cost per footcandle more pertinent than unit cost).
   2. Service availability and parts replacement (company reputation is important).

20.27 Luminaire Diffusing Elements

The diffusing elements usually considered include white plastic, striped and prismatic glass, prismatic plastic, high-reflectance aluminum (alzac), baffles, and miniature eggcrate louvers. In addition to these types there are various sizes and shapes of metal and plastic louvers and baffles, white glass, ribbed glass, etc. Each of these diffusers must be considered on its merits, and a decision arrived at based on photometric characteristics, cost, ease of maintenance, appearance, and fire safety. A rapid review of the photometric characteristics of this most important element of a lighting fixture yields the following:

(*a*) ***Plain White Diffusers.*** These have a circular distribution giving equal light in all directions. Diffusion is good, VCP is poor, and ESI is generally poor. Similar to Figure 20.22*b*.

(*b*) ***Prismatic Lens.*** There are dozens of designs that vary one from another considerably. Figure 20.22*a* can be taken as typical of this genre. They produce good diffusion, high VCP, and fair ESI, depending on viewing angles and positions.

(*c*) ***Louvers and Baffles.*** These are made of metal or plastic, in various shapes and sizes. They generally have circular to egg-shaped distribution, providing good diffusion, good VCP, and poor ESI.

(*d*) ***Parabolic Louvers.*** These are available in small and large designs, the principal characteristic of which is extremely low brightness yielding very high VCP. One such unit with miniature cells is shown in Figure 18.8*a*, page 716. A section through the unit is shown in Figure 20.29. Another design is shown in Figure 20.30 along with its distribution curves. Notice the low value of high-angle *and* low-angle brightness of the crosswise curve. This unit therefore gives very high VCP, good ESI, and good efficiency.

(*e*) ***Batwing Diffusers.*** These diffusers are made in a number of designs. They are generally prismatic lenses modified to give this special characteristic. Refer to Section 18.23(c) and Figures 18.42 to 18.44, pages 754-6. A comparison of batwing distribution with that of louvers and white plastic is shown in Figure 20.31. Table 20.6 presents a study drawn from Reference 3 comparing similar fixtures with different diffusers, the basis of comparison being *effective lumens per watt*. See Figure 18.46, note 5, page 759, for definition of the term. It is basically a figure of merit derived by positive weighting for ESI and negative weighting for glare. Note that since ESI depends on viewing position, the results shown should be used as a method rather than fixed results on which to base design.

20.28 Luminaire Efficiency

A luminaire, variously called a fixture, lighting unit, or reflector, comprises a device for physically supporting the light source and usually for directing or controlling the light output of this source. Because of internal reflections, some of the generated lumen output of the lamp is lost within the fixture. The ratio of output lumens to lamp (input) lumens, expressed as a percentage, represents the luminous efficiency of the fixture. Although this information is normally available from the manufacturer it is readily calculated from the fixture distribution curve by application of the zonal factors to the fixture curve. However, this characteristic has little meaning by itself, since the actual overall efficiency of a luminaire depends on the space in which it is used.

To illustrate, let us consider the case of a large high-ceiling room in which the ceiling is dark and is covered with dirty piping and ductwork. If we were to use a high-efficiency (say 80%) indirect lighting unit in such a room, most of the light directed upward would be lost (absorbed), and the actual lighting on the working plane 30 in. above the floor would be very low. If, however, this room were illuminated with low-efficiency direct lighting units (e.g., 50%) utilizing the same wattage, the

Fig. 20.29 Section through an egg-crate type of parabolic wedge louver. Most such units are aluminized plastic, which gives exceptionally low brightness and high VCP.

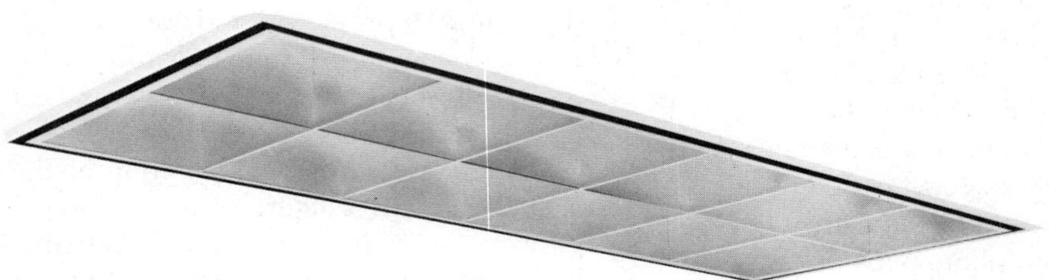

Fig. 20.30 Fixture with large aluminum parabolic baffles. As can be seen from the inset photometric data, this type unit gives low brightness, high VCP, and good ESI because of reduced output below 30° crosswise. Courtesy of Columbia Lighting Inc.

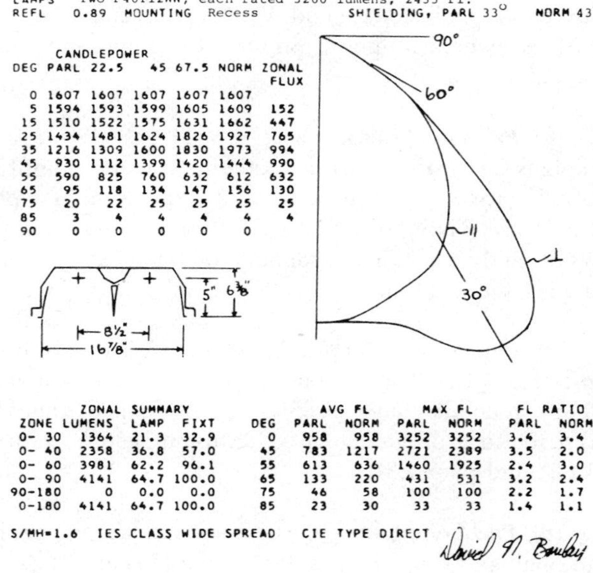

Columbia Test # 4137-A

PHOTOMETRIC REPORT PREPARED FOR COLUMBIA LIGHTING, INC.
CATALOG NUMBER 4542G-43-042-20" REPORT NO. 18899
REPORT OF CANDLEPOWER DISTRIBUTION IN 5 PLANES. DATE 3-28-75
LUMINAIRE Metal troffer, synthetic enameled. Large cell semi-specular aluminum parabolic louver.

LAMPS Two F40T12WW, each rated 3200 lumens, 2435 fl.
REFL 0.89 MOUNTING Recess SHIELDING, PARL 33° NORM 43°

CANDLEPOWER

DEG	PARL	22.5	45	67.5	NORM	ZONAL FLUX
0	1607	1607	1607	1607	1607	
5	1594	1593	1599	1605	1609	152
15	1510	1522	1575	1631	1662	447
25	1434	1481	1624	1826	1927	765
35	1216	1309	1600	1830	1973	994
45	930	1112	1399	1420	1444	990
55	590	825	760	632	612	632
65	95	118	134	147	156	130
75	20	22	25	25	25	25
85	3	4	4	4	4	4
90	0	0	0	0	0	

ZONAL SUMMARY					AVG FL		MAX FL		FL RATIO	
ZONE	LUMENS	LAMP	FIXT	DEG	PARL	NORM	PARL	NORM	PARL	NORM
0- 30	1364	21.3	32.9	0	958	958	3252	3252	3.4	3.4
0- 40	2358	36.8	57.0	45	783	1217	2721	2389	3.5	2.0
0- 60	3981	62.2	96.1	55	613	636	1460	1925	2.4	3.0
0- 90	4141	64.7	100.0	65	133	220	431	531	3.2	2.4
90-180	0	0.0	0.0	75	46	58	100	100	2.2	1.7
0-180	4141	64.7	100.0	85	23	30	33	33	1.4	1.1

S/MH=1.6 IES CLASS WIDE SPREAD CIE TYPE DIRECT

David N. Bowlay

Fig. 20,31 The light distribution of the luminaire itself can also be effective in reducing veiling reflections. Luminaire (a) *has a strong downward component of light that is likely to produce most serious losses of visibility. Luminaire* (b) *has a diffuse distribution that is considerably better than the first. Luminaire* (c) *has a controlled, wide-angle distribution that emits little light directly downward, and minimizes veiling reflections by linear, wide-angle distribution. All three luminaires have the same total light output. See also Figures 18.32 and 18.44, page 748 and page 756. Courtesy of the IES, from* Lighting Design and Application, *May 1973, p. 33.*

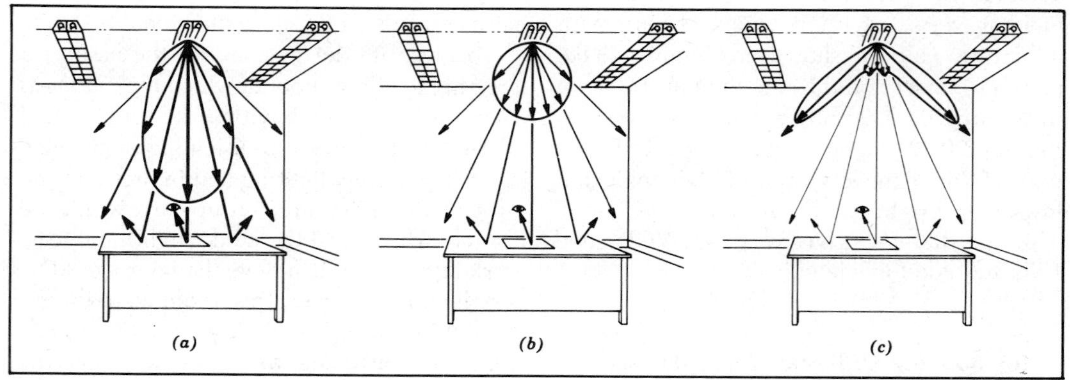

Table 20.6 ***A Lighting Efficiency Comparison for Common Fluorescent Fixtures***

Fixture Description	*Flux Distribution*	*Effective[a] Lumens per Watt*
1. Recessed, 2 × 4 ft, four-lamp, static, metal troffer; synthetic enameled, 0.91 reflectance, with clear prismatic acrylic lens, without diffusing overlay.	Radial batwing	15.1
2. Recessed, 2 × 4 ft, four-lamp static, metal troffer; synthetic enameled, 0.88 reflectance, with clear acrylic prismatic lens, with 0.40-in. diffusing overlay.	Linear batwing	15.3
3. Recessed, 1 × 4 ft, two-lamp, air-handling, metal troffer with semispecular, anodized aluminum parabolic reflector and louvers (crosswise only).	Linear batwing, low brightness	27.5
4. Recessed, 2 × 4 ft, four-lamp, air-handling, metal troffer, with baked enamel reflector, 0.91 reflectance, with semispecular anodized aluminum parabolic louvers (crosswise and lengthwise).	Linear batwing	28.3
5. Recessed, 2 × 4 ft, four-lamp, static, metal troffer; baked enamel, 0.87 reflectance, with clear acrylic prismatic lens, without overlay.	General prismatic	30.3
6. Same as 5 above except 2 × 4 ft two-lamp.	General prismatic	30.5
7. Same as 5 with clear, acrylic, low-brightness prismatic lens without overlay.	Low brightness	32.5
8. Recessed, 1 × 4 ft, two-lamp, static, metal troffer enameled; 0.88 reflectance, with specular Alzac parabolic louvers.	Low brightness	37.1

Source. From Ross & Baruzzini, Inc., *Lighting Systems Stads,* GSAIPBS, March 1974.

[a]This analysis based on data in Figure 18.46, page 759, for standard pencil tasks, at 70 fc TI, in 60 ft × 60 ft × 10 ft room. See this figure for definition of "effective lumens per watt."

illumination on the working plane would be considerably higher than in the first case.

Similarly, if we consider a small room with dark walls and ceiling, lighted alternatively by diffuse lighting and by direct lighting units of the same wattage and unit efficiency, the horizontal-plane illumination will be higher in the case of the direct units because of the large loss of the horizontal and upward components of the diffuse lighting on the walls and ceiling. It should be obvious then that the fixture efficiency *alone* is not a meaningful factor but that *the overall luminous efficiency of a particular unit in a particular space* is the figure that merits our attention. This figure, since it describes the utilization of the fixture output in a specific space, is known as the *coefficient of utilization* (CU). It is defined as the ratio between the lumens reaching the working horizontal plane to the generated lumens. Since each luminaire will have a different coefficient for every different space in which it is used, a system of standardization has been evolved utilizing room cavities (explained below) of certain proportions and various surface reflectances. The fixture coefficients are then computed and tabulated as shown in Table 20.7. It should be emphasized that the figures given in this table are for the generic fixture type only; in an actual job, actual fixture data should be used. The CU then, is a factor that combines fixture efficiency and distribution with room proportions, mounting height, and surface reflectance.

A few remarks about Table 20.7 are in

898

Table 20.7a *Coefficients of Utilization for Typical Luminaires with Suggested Maximum Spacing Ratios and Maintenance Category*

To obtain a Coefficient of Utilization:

1. Determine cavity ratios for the room, ceiling, and floor from Table 20.8.
2. Determine the effective ceiling and floor cavity reflectances from Table 20.9. Use initial ceiling, floor, and wall reflectances.
3. Obtain Coefficient of Utilization (CU) for 20% effective floor cavity reflectance from appropriate table below for luminaire type to be used. Interpolate, when necessary, to obtain CU for exact room cavity ratio for nearest effective ceiling cavity reflectances above and below reflectance obtained in Step 2; interpolate between these CUs to obtain CU for Step 2 ceiling cavity reflectance.
4. If effective floor cavity reflectance differs significantly from 20%, obtain multiplier from Table 20.10 and apply this to the CU obtained in Step 3.

NOTE. In some cases, luminaire data in this table are based on an actual typical luminaire; in other cases, the data represent a composite of generic luminaire types. Therefore, whenever possible, specific luminaire data should be used in preference to this table of typical luminaires.

The polar intensity sketch (candlepower distribution curve) and the corresponding spacing-to-mounting height guide are representative of many luminaires of each type shown. A specific luminaire may differ in perpendicular plane (crosswise) and parallel plane (lengthwise) intensity distributions and in S/MH guide from the values shown. However, the various coefficients depend only on the average intensity at each polar angle from nadir.

Typical Luminaire	Typical Distribution and Percent Lamp Lumens		ρ_{cc}[a] →	80			70			50			30			10			0	WDRC[e] ↓
	Maint. Cat.	Maximum S/MH Guide[d]	ρ_w[b] →	50	30	10	50	30	10	50	30	10	50	30	10	50	30	10	0	
			RCR[c] ↓	Coefficients of Utilization for 20% Effective Floor Cavity Reflectance (ρ_{fc} = 20)																
1	V	1.5	0	.87	.87	.87	.81	.81	.81	.69	.69	.69	.59	.59	.59	.49	.49	.49	.44	
	45%		1	.71	.67	.63	.66	.62	.59	.56	.53	.50	.47	.45	.43	.39	.37	.35	.31	.35
			2	.61	.54	.49	.56	.50	.46	.47	.43	.39	.39	.36	.33	.32	.29	.27	.23	.27
			3	.52	.45	.39	.48	.42	.37	.41	.36	.31	.34	.30	.26	.27	.24	.22	.18	.22
			4	.46	.38	.33	.42	.36	.30	.36	.30	.26	.30	.26	.22	.24	.21	.18	.15	.19
	45%		5	.40	.33	.27	.37	.30	.25	.32	.26	.22	.26	.22	.19	.21	.18	.15	.12	.16
			6	.36	.28	.23	.33	.26	.21	.28	.23	.19	.23	.19	.16	.19	.15	.13	.10	.14
			7	.32	.25	.20	.29	.23	.18	.25	.20	.16	.21	.16	.13	.17	.13	.11	.09	.13
Pendant diffusing sphere with incandescent lamp			8	.29	.22	.17	.27	.20	.16	.23	.17	.14	.19	.15	.12	.15	.12	.09	.07	.12
			9	.26	.19	.15	.24	.18	.14	.20	.15	.12	.17	.13	.10	.14	.11	.08	.06	.11
			10	.23	.17	.13	.22	.16	.12	.19	.14	.10	.16	.12	.09	.13	.09	.07	.05	.10
3	IV	1.3	0	.99	.99	.99	.97	.97	.97	.92	.92	.92	.88	.88	.88	.85	.85	.85	.83	
	10%		1	.88	.85	.82	.86	.83	.81	.83	.80	.78	.79	.78	.76	.77	.75	.73	.72	.29
			2	.78	.73	.68	.76	.72	.67	.73	.69	.66	.71	.67	.64	.68	.65	.63	.61	.28
			3	.69	.62	.57	.67	.61	.57	.65	.60	.56	.63	.58	.55	.61	.57	.54	.52	.26
			4	.61	.54	.49	.60	.53	.48	.58	.52	.48	.56	.51	.47	.54	.50	.46	.45	.24
	85%		5	.54	.47	.41	.53	.46	.41	.51	.45	.41	.50	.44	.40	.48	.43	.40	.38	.23
			6	.48	.41	.35	.47	.40	.35	.46	.39	.35	.44	.39	.34	.43	.38	.34	.32	.21
			7	.43	.35	.30	.42	.35	.30	.41	.34	.30	.39	.34	.30	.38	.33	.29	.28	.20
Porcelain-enameled ventilated standard dome with incandescent lamp			8	.38	.31	.26	.38	.31	.26	.37	.30	.26	.36	.30	.26	.35	.30	.26	.24	.19
			9	.35	.28	.23	.34	.27	.23	.33	.27	.23	.32	.27	.23	.31	.26	.22	.21	.17
			10	.31	.25	.20	.31	.24	.20	.30	.24	.20	.29	.24	.20	.29	.23	.20	.18	.16

8	IV 0.7	0	.53	.53	.53	.52	.52	.52	.49	.49	.49	.47	.47	.47	.45	.45	.45	.44	
	0%	1	.51	.50	.49	.50	.49	.48	.48	.47	.47	.46	.46	.45	.45	.44	.44	.43	.04
		2	.48	.47	.46	.48	.46	.45	.46	.45	.44	.45	.44	.44	.44	.43	.43	.42	.03
		3	.47	.45	.44	.46	.45	.43	.45	.44	.43	.44	.43	.42	.43	.42	.41	.41	.03
		4	.45	.43	.42	.44	.43	.42	.43	.42	.41	.43	.41	.41	.42	.41	.40	.40	.03
	45%	5	.43	.41	.40	.43	.41	.40	.42	.40	.39	.41	.40	.39	.41	.40	.39	.38	.03
		6	.42	.40	.39	.41	.40	.38	.41	.39	.38	.40	.39	.38	.40	.39	.38	.37	.03
		7	.40	.38	.37	.40	.38	.37	.39	.38	.37	.39	.38	.37	.38	.37	.36	.36	.03
		8	.39	.37	.36	.38	.37	.36	.38	.37	.35	.38	.36	.35	.37	.36	.35	.35	.03
Reflector downlight with baffles		9	.37	.36	.34	.37	.35	.34	.37	.35	.34	.36	.35	.34	.36	.35	.34	.33	.03
and inside frosted lamp		10	.36	.34	.33	.36	.34	.33	.36	.34	.33	.35	.34	.33	.35	.34	.33	.32	.03
16	III 1.5	0	.92	.92	.92	.90	.90	.90	.86	.86	.86	.82	.82	.82	.79	.79	.79	.77	
	10%	1	.85	.82	.80	.83	.81	.79	.79	.78	.76	.76	.75	.74	.74	.72	.71	.70	.19
		2	.77	.73	.70	.75	.72	.69	.73	.70	.67	.70	.68	.66	.68	.66	.64	.63	.19
		3	.70	.65	.61	.68	.64	.60	.66	.62	.59	.64	.61	.58	.62	.59	.57	.56	.18
		4	.63	.58	.53	.62	.57	.53	.60	.56	.52	.58	.55	.52	.57	.54	.51	.49	.18
		5	.57	.51	.47	.56	.51	.47	.55	.50	.46	.53	.49	.46	.52	.48	.45	.44	.17
	65%	6	.51	.45	.41	.51	.45	.41	.49	.44	.40	.48	.43	.40	.47	.43	.40	.38	.16
		7	.46	.40	.35	.45	.39	.35	.44	.39	.35	.43	.38	.35	.42	.38	.34	.33	.16
		8	.41	.35	.31	.41	.35	.31	.40	.34	.31	.39	.34	.30	.38	.33	.30	.29	.15
Wide-distribution ventilated re-		9	.37	.31	.27	.37	.31	.27	.36	.30	.27	.35	.30	.27	.34	.30	.26	.25	.15
flector with clear HID lamp		10	.33	.27	.24	.33	.27	.23	.32	.27	.23	.31	.27	.23	.31	.26	.23	.22	.14
21	II 1.5/1.3	0	.94	.94	.94	.90	.90	.90	.82	.82	.82	.75	.75	.75	.69	.69	.69	.66	
	15%	1	.85	.82	.80	.82	.79	.77	.75	.73	.72	.69	.68	.66	.64	.63	.62	.59	.18
		2	.76	.72	.68	.74	.70	.66	.68	.65	.62	.63	.61	.58	.58	.56	.55	.52	.17
		3	.69	.63	.59	.66	.61	.57	.62	.58	.54	.57	.54	.51	.53	.51	.48	.46	.17
		4	.62	.56	.51	.60	.54	.50	.56	.51	.47	.52	.48	.45	.48	.45	.43	.41	.16
		5	.55	.49	.44	.53	.48	.43	.50	.45	.41	.47	.43	.39	.44	.40	.38	.36	.15
		6	.50	.43	.39	.48	.42	.38	.45	.40	.36	.42	.38	.35	.40	.36	.33	.31	.15
	65%	7	.45	.38	.34	.43	.37	.33	.41	.36	.32	.38	.34	.30	.36	.32	.29	.27	.14
		8	.40	.34	.29	.39	.33	.29	.37	.31	.28	.34	.30	.26	.32	.28	.25	.24	.13
Diffuse aluminum reflector with		9	.36	.30	.25	.35	.29	.25	.33	.28	.24	.31	.26	.23	.29	.25	.22	.20	.13
35°CW shielding		10	.33	.26	.22	.32	.26	.22	.30	.25	.21	.28	.23	.20	.26	.22	.19	.18	.12
23	II 1.5/1.1	0	.83	.83	.83	.79	.79	.79	.71	.71	.71	.65	.65	.65	.59	.59	.59	.56	
	25%	1	.75	.72	.70	.72	.69	.68	.65	.64	.62	.60	.59	.58	.55	.54	.53	.50	.15
		2	.67	.63	.60	.65	.61	.58	.59	.57	.54	.55	.53	.51	.50	.49	.47	.45	.14
		3	.61	.56	.52	.58	.54	.51	.54	.50	.48	.50	.47	.45	.46	.44	.42	.40	.14
		4	.55	.49	.45	.53	.48	.44	.49	.45	.42	.45	.42	.40	.42	.39	.37	.36	.13
		5	.49	.44	.40	.47	.42	.39	.44	.40	.37	.41	.38	.35	.38	.35	.33	.31	.12
	55%	6	.45	.39	.35	.43	.38	.34	.40	.36	.33	.37	.34	.31	.35	.32	.30	.28	.12
		7	.40	.35	.31	.39	.34	.30	.36	.32	.29	.34	.30	.27	.32	.29	.26	.25	.11
		8	.36	.31	.27	.35	.30	.26	.33	.28	.25	.31	.27	.24	.29	.25	.23	.21	.11
Diffuse aluminum reflector with		9	.33	.27	.23	.32	.26	.23	.29	.25	.22	.28	.24	.21	.26	.22	.20	.19	.10
35°CW x 35°LW shielding		10	.30	.24	.21	.29	.24	.20	.27	.22	.19	.25	.21	.19	.23	.20	.18	.16	.10

Table 20.7a ***Coefficients of Utilization for Typical Luminaires with Suggested Maximum Spacing Ratios and Maintenance Category (Continued)***

Typical Luminaire	Maint. Cat.	Maximum S/MH Guide[d]	ρ_{ce}[a] →	80			70			50			30			10			0	
			ρ_w[b] →	5	30	10	50	30	10	50	30	10	50	30	10	50	30	10	0	WDRC[e] ↓
			RCR[c] ↓	Coefficients of Utilization for 20% Effective Floor Cavity Reflectance (ρ_{fc} = 20)																
28	VI	1.5 (75% up, 5% down)	0	.77	.77	.77	.67	.67	.67	.49	.49	.49	.33	.33	.33	.18	.18	.18	.11	
			1	.67	.64	.62	.59	.57	.54	.44	.42	.41	.30	.29	.28	.17	.16	.16	.10	.04
			2	.59	.54	.50	.51	.48	.45	.38	.36	.34	.26	.25	.23	.15	.14	.13	.09	.04
			3	.51	.46	.42	.45	.41	.37	.34	.31	.28	.23	.21	.20	.13	.12	.12	.07	.04
			4	.45	.40	.35	.40	.35	.31	.30	.27	.24	.20	.18	.17	.12	.11	.10	.06	.03
			5	.40	.34	.30	.35	.30	.27	.26	.23	.20	.18	.16	.14	.10	.09	.08	.05	.03
			6	.36	.30	.26	.32	.27	.23	.24	.20	.18	.16	.14	.12	.09	.08	.07	.05	.03
			7	.32	.26	.22	.28	.23	.20	.21	.18	.15	.15	.12	.11	.08	.07	.06	.04	.03
			8	.29	.23	.19	.25	.21	.17	.19	.16	.13	.13	.11	.09	.08	.06	.06	.03	.03
Luminous bottom suspended			9	.26	.20	.17	.23	.18	.15	.17	.14	.12	.12	.10	.08	.07	.06	.05	.03	.02
unit with extra-high-output lamp	Max. S/CH$_{wp}$ = 1.5		10	.24	.18	.15	.21	.16	.13	.16	.12	.10	.11	.09	.07	.06	.05	.04	.03	.02
30	V	1.5/1.2	0	.80	.80	.80	.77	.77	.77	.71	.71	.71	.66	.66	.66	.60	.60	.60	.58	
			1	.71	.69	.66	.69	.66	.64	.64	.62	.60	.59	.58	.56	.55	.54	.53	.50	.20
			2	.64	.59	.56	.61	.58	.54	.57	.54	.51	.53	.51	.49	.49	.48	.46	.44	.18
			3	.57	.52	.48	.55	.50	.47	.51	.48	.45	.48	.45	.42	.45	.42	.40	.38	.17
			4	.51	.46	.41	.49	.44	.40	.46	.42	.39	.43	.40	.37	.41	.38	.35	.34	.16
			5	.46	.40	.36	.44	.39	.35	.41	.87	.34	.39	.35	.32	.37	.33	.31	.29	.15
			6	.41	.35	.31	.40	.35	.31	.38	.33	.30	.35	.31	.28	.33	.30	.27	.26	.14
			7	.37	.31	.27	.36	.31	.27	.34	.29	.26	.32	.28	.25	.30	.27	.24	.23	.13
Two-lamp prismatic wrap-			8	.33	.28	.24	.32	.27	.23	.30	.26	.22	.29	.25	.22	.27	.24	.21	.19	.12
around—multiply by 0.95 for			9	.30	.24	.20	.29	.24	.20	.27	.23	.19	.26	.22	.19	.24	.21	.18	.17	.12
four lamps			10	.27	.22	.18	.26	.21	.18	.25	.20	.17	.23	.19	.16	.22	.18	.16	.15	.11
35	IV	1.0 (0% up, 45% down)	0	.59	.59	.59	.58	.58	.58	.55	.55	.55	.53	.53	.53	.51	.51	.51	.50	
			1	.54	.52	.50	.52	.51	.49	.50	.49	.48	.48	.47	.46	.47	.46	.45	.44	.15
			2	.48	.45	.43	.47	.44	.42	.45	.43	.41	.44	.42	.40	.42	.41	.39	.39	.14
			3	.43	.40	.37	.42	.39	.37	.41	.38	.36	.40	.37	.36	.39	.37	.35	.34	.13
			4	.39	.35	.32	.38	.35	.32	.37	.34	.32	.36	.33	.31	.35	.33	.31	.30	.13
			5	.35	.31	.28	.35	.31	.28	.34	.30	.28	.33	.30	.28	.32	.29	.27	.26	.12
Four-lamp, 2-ft-wide troffer with			6	.32	.28	.25	.32	.28	.25	.31	.27	.25	.30	.27	.25	.29	.26	.24	.23	.11
45° plastic louver—multiply by			7	.29	.25	.22	.29	.25	.22	.28	.25	.22	.27	.24	.22	.27	.24	.22	.21	.10
1.05 for two lamps and 0.95 for			8	.26	.22	.20	.26	.22	.20	.25	.22	.20	.25	.22	.19	.24	.21	.19	.18	.10
six lamps			9	.24	.20	.17	.24	.20	.17	.23	.20	.17	.23	.19	.17	.22	.19	.17	.16	.09
	Max. S/MH$_{wp}$ = 0.9		10	.22	.18	.16	.22	.18	.16	.21	.18	.16	.21	.18	.15	.20	.17	.15	.15	.09

41	V 0% 1.4/1.2 65%	0	.73	.73	.73	.72	.72	.72	.68	.68	.68	.66	.66	.66	.63	.63	.63	.62	
		1	.66	.64	.62	.65	.63	.61	.62	.60	.59	.60	.58	.57	.57	.56	.55	.54	.20
		2	.59	.55	.52	.58	.54	.52	.56	.53	.50	.54	.51	.49	.52	.50	.48	.47	.19
		3	.53	.48	.45	.52	.48	.44	.50	.46	.44	.48	.45	.43	.47	.44	.42	.41	.18
		4	.47	.42	.39	.46	.42	.38	.45	.41	.38	.43	.40	.37	.42	.39	.37	.36	.17
		5	.42	.37	.33	.41	.37	.33	.40	.36	.33	.39	.35	.32	.38	.35	.32	.31	.16
		6	.38	.33	.29	.37	.32	.29	.36	.32	.29	.35	.31	.28	.34	.31	.28	.27	.15
		7	.34	.29	.25	.33	.29	.25	.33	.28	.25	.32	.28	.25	.31	.27	.25	.23	.14
Fluorescent unit with flat pris-		8	.30	.25	.22	.30	.25	.22	.29	.25	.22	.28	.24	.21	.28	.24	.21	.20	.13
matic lens, 4-lamp, 2 ft wide—		9	.27	.22	.19	.27	.22	.19	.26	.22	.19	.25	.21	.19	.25	.21	.18	.17	.12
multiply by 1.10 for 2-lamp	Max. $S/MH_{wp} = 1.2$	10	.25	.20	.17	.24	.20	.16	.24	.19	.16	.23	.19	.16	.23	.19	.16	.15	.12
45		1	.42	.40	.39	.36	.35	.33	.25	.24	.23	Coves are not recommended for lighting areas having low reflectances.							
		2	.37	.34	.32	.32	.29	.27	.22	.20	.19								
		3	.32	.29	.26	.28	.25	.23	.19	.17	.16								
		4	.29	.25	.22	.25	.22	.19	.17	.15	.13								
		5	.25	.21	.18	.22	.19	.16	.15	.13	.11								
		6	.23	.19	.16	.20	.16	.14	.14	.12	.10								
		7	.20	.17	.14	.17	.14	.12	.12	.10	.09								
Single row fluorescent lamp cove without		8	.18	.15	.12	.16	.13	.10	.11	.09	.08								
reflector, mult. by 0.93 for two rows and		9	.17	.13	.10	.15	.11	.09	.10	.08	.07								
by 0.85 for three rows.		10	.15	.12	.09	.13	.10	.08	.09	.07	.06								
48	ρ_{cc} from below ~45%	1							.51	.49	.48				.47	.46	.45		
		2							.46	.44	.42				.43	.42	.40		
		3							.42	.39	.37				.39	.38	.36		
		4							.38	.35	.33				.36	.34	.32		
		5							.35	.32	.29				.33	.31	.29		
Louvered ceiling.		6							.32	.29	.26				.30	.28	.26		
1. Ceiling efficiency ~50%; 45° shielding		7							.29	.26	.23				.28	.25	.23		
opaque louvers of 80% reflectance. Cavity		8							.27	.23	.21				.26	.23	.21		
with minimum obstructions and painted		9							.24	.21	.19				.24	.21	.19		
with 80% reflectance paint—use $\rho_c = 50$.		10							.22	.19	.17				.22	.19	.17		

Source. Selected and reprinted with permission from *ICS Handbook,* 5th ed.

[a] ρ_{cc} = percent effective ceiling cavity reflectance.

[b] ρ_w = percent wall reflectance.

[c] RCR = Room Cavity Ratio.

[d] = Maximum *S/MH* guide—ratio of maximum luminaire spacing to mounting or ceiling height above work plane.

[e] WDRC = Wall Direct Radiation Coefficient.

Table 20.7b *Wall Luminance Coefficients and Ceiling Cavity Luminance Coefficients for Typical Luminaires*

To obtain a luminance coefficient follow the procedure detailed at the head of Table 20.7*a* to find a coefficient of utilization.

Although it is recommended that luminance coefficients and wall direct radiation coefficients be published to three decimal places, only two are shown here for these typical luminaires. Three decimal place data should be obtained from manufacturers of actual luminaires used.

Wall luminance coefficients for 20% effective floor cavity reflectance ($\rho_{fc} = 20$)

Typical Luminaire	$\rho_{cc} \rightarrow$	80	70	50	30	10
	$\rho_w \rightarrow$	50 30 10	50 30 10	50 30 10	50 30 10	50 30 10
	RCR ↓					
1 Pendant diffusing sphere with incandescent lamp	1	.32 .18 .06	.30 .17 .05	.27 .15 .05	.24 .14 .04	.21 .12 .04
	2	.27 .15 .05	.25 .14 .04	.23 .13 .04	.20 .11 .04	.18 .10 .03
	3	.24 .13 .04	.22 .12 .04	.20 .11 .03	.17 .09 .03	.15 .08 .03
	4	.21 .11 .03	.20 .10 .03	.17 .09 .03	.15 .08 .02	.13 .07 .02
	5	.19 .10 .03	.18 .09 .03	.16 .08 .02	.14 .07 .02	.12 .06 .02
	6	.18 .09 .03	.16 .08 .02	.14 .07 .02	.13 .07 .02	.11 .06 .02
	7	.16 .08 .02	.15 .08 .02	.13 .07 .02	.12 .06 .02	.10 .05 .02
	8	.15 .07 .02	.14 .07 .02	.12 .06 .02	.11 .05 .02	.09 .05 .01
	9	.14 .07 .02	.13 .06 .02	.12 .06 .02	.10 .05 .01	.09 .04 .01
	10	.13 .06 .02	.12 .06 .02	.11 .05 .01	.09 .05 .01	.08 .04 .01
3 Porcelain-enameled ventilated standard dome with incandescent lamp	1	.23 .13 .04	.23 .13 .04	.21 .12 .04	.20 .12 .04	.20 .11 .04
	2	.22 .12 .04	.22 .12 .04	.21 .11 .04	.20 .11 .03	.19 .11 .03
	3	.21 .11 .03	.21 .11 .03	.20 .11 .03	.19 .10 .03	.18 .10 .03
	4	.20 .10 .03	.19 .10 .03	.19 .10 .03	.18 .10 .03	.17 .09 .03
	5	.19 .09 .03	.18 .09 .03	.18 .09 .03	.17 .09 .03	.16 .09 .03
	6	.18 .09 .03	.17 .09 .03	.17 .09 .02	.16 .08 .02	.16 .08 .02
	7	.17 .08 .02	.16 .08 .02	.16 .08 .02	.15 .08 .02	.15 .08 .02
	8	.16 .08 .02	.15 .08 .02	.15 .07 .02	.14 .07 .02	.14 .07 .02
	9	.15 .07 .02	.15 .07 .02	.14 .07 .02	.14 .07 .02	.13 .07 .02
	10	.14 .07 .02	.14 .07 .02	.13 .07 .02	.13 .06 .02	.13 .06 .02
8 Reflector downlight with baffles and inside frosted lamp.	1	.06 .03 .01	.06 .03 .01	.05 .03 .01	.05 .03 .01	.04 .02 .01
	2	.05 .03 .01	.05 .03 .01	.05 .03 .01	.04 .02 .01	.04 .02 .01
	3	.05 .03 .01	.05 .03 .01	.04 .02 .01	.04 .02 .01	.04 .02 .01
	4	.05 .02 .01	.05 .02 .01	.04 .02 .01	.04 .02 .01	.04 .02 .01
	5	.05 .02 .01	.04 .02 .01	.04 .02 .01	.04 .02 .01	.04 .02 .01
	6	.04 .02 .01	.04 .02 .01	.04 .02 .01	.04 .02 .01	.04 .02 .01
	7	.04 .02 .01	.04 .02 .01	.04 .02 .01	.04 .02 .01	.03 .02 .01
	8	.04 .02 .01	.04 .02 .01	.04 .02 .01	.04 .02 .01	.03 .02 .01
	9	.04 .02 .01	.04 .02 .01	.04 .02 .01	.03 .02 .00	.03 .02 .00
	10	.04 .02 .00	.04 .02 .00	.04 .02 .00	.03 .02 .00	.03 .02 .00

Ceiling cavity luminance coefficients for 20% effective floor cavity reflectance ($\rho_{fc} = 20$)

Typical Luminaire	$\rho_{cc} \rightarrow$	80	70	50	30	10
	$\rho_w \rightarrow$	50 30 10	50 30 10	50 30 10	50 30 10	50 30 10
	RCR ↓					
1 Pendant diffusing sphere with incandescent lamp	0	.42 .42 .42	.36 .36 .36	.25 .25 .25	.14 .14 .14	.05 .05 .05
	1	.42 .40 .37	.36 .34 .32	.25 .23 .22	.14 .14 .13	.05 .04 .04
	2	.42 .38 .35	.36 .33 .30	.24 .23 .21	.14 .13 .12	.05 .04 .04
	3	.41 .37 .33	.35 .32 .29	.24 .22 .20	.14 .13 .12	.04 .04 .04
	4	.41 .36 .32	.35 .31 .28	.24 .22 .20	.14 .13 .12	.04 .04 .04
	5	.40 .35 .31	.34 .30 .27	.24 .21 .19	.14 .12 .11	.04 .04 .04
	6	.39 .34 .31	.34 .30 .27	.23 .21 .19	.14 .12 .11	.04 .04 .04
	7	.39 .34 .31	.33 .29 .27	.23 .21 .19	.13 .12 .11	.04 .04 .04
	8	.38 .34 .30	.33 .29 .26	.23 .20 .19	.13 .12 .11	.04 .04 .04
	9	.38 .33 .30	.33 .29 .26	.23 .20 .19	.13 .12 .11	.04 .04 .04
	10	.37 .33 .30	.32 .29 .26	.22 .20 .19	.13 .12 .11	.04 .04 .04
3 Porcelain-enameled ventilated standard dome with incandescent lamp	0	.15 .15 .15	.13 .13 .13	.09 .09 .09	.05 .05 .05	.02 .02 .02
	1	.15 .13 .11	.13 .11 .10	.09 .08 .07	.05 .04 .04	.02 .01 .01
	2	.14 .11 .08	.12 .09 .07	.08 .07 .05	.05 .04 .03	.02 .01 .01
	3	.13 .10 .06	.12 .08 .06	.08 .06 .04	.05 .03 .02	.01 .01 .01
	4	.13 .08 .05	.11 .07 .04	.08 .05 .03	.04 .03 .02	.01 .01 .01
	5	.12 .08 .04	.11 .07 .04	.07 .05 .02	.04 .03 .01	.01 .01 .00
	6	.12 .07 .03	.10 .06 .03	.07 .04 .02	.04 .02 .01	.01 .01 .00
	7	.11 .06 .03	.10 .05 .02	.07 .04 .02	.04 .02 .01	.01 .01 .00
	8	.11 .06 .02	.09 .05 .02	.06 .04 .01	.04 .02 .01	.01 .01 .00
	9	.10 .05 .02	.09 .05 .02	.06 .03 .01	.04 .02 .01	.01 .01 .00
	10	.10 .05 .02	.09 .04 .02	.06 .03 .01	.03 .02 .01	.01 .01 .00
8 Reflector downlight with baffles and inside frosted lamp.	0	.08 .08 .08	.07 .07 .07	.04 .04 .04	.03 .03 .03	.01 .01 .01
	1	.08 .07 .07	.06 .06 .06	.04 .04 .04	.03 .02 .02	.01 .01 .01
	2	.07 .06 .05	.06 .05 .05	.04 .04 .03	.02 .02 .02	.01 .01 .01
	3	.06 .05 .04	.05 .04 .04	.04 .03 .03	.02 .02 .02	.01 .01 .01
	4	.05 .04 .04	.05 .04 .03	.03 .03 .02	.02 .02 .01	.01 .01 .00
	5	.05 .04 .03	.04 .03 .03	.03 .02 .02	.02 .01 .01	.01 .00 .00
	6	.05 .03 .03	.04 .03 .02	.03 .02 .02	.02 .01 .01	.01 .00 .00
	7	.04 .03 .02	.04 .03 .02	.03 .02 .01	.01 .01 .01	.00 .00 .00
	8	.04 .03 .02	.03 .02 .02	.02 .02 .01	.01 .01 .01	.00 .00 .00
	9	.04 .02 .02	.03 .02 .01	.02 .02 .01	.01 .01 .01	.00 .00 .00
	10	.04 .02 .01	.03 .02 .01	.02 .01 .01	.01 .01 .01	.00 .00 .00

Luminaire	RCR						RCR					
							0	.15 .15 .15	.13 .13 .13	.08 .08 .08	.05 .05 .05	.02 .02 .02
	1	.18 .10 .03	.17 .10 .03	.16 .09 .03	.15 .09 .03	.14 .08 .03	1	.14 .13 .11	.12 .11 .10	.08 .08 .07	.05 .04 .04	.02 .01 .01
	2	.17 .09 .03	.16 .09 .03	.15 .09 .03	.15 .08 .03	.14 .08 .02	2	.13 .11 .09	.11 .09 .08	.08 .06 .05	.04 .03 .03	.01 .01 .01
	3	.16 .09 .03	.16 .08 .03	.15 .08 .02	.14 .08 .02	.14 .08 .02	3	.12 .09 .07	.11 .08 .06	.07 .06 .04	.04 .03 .03	.01 .01 .01
	4	.16 .08 .02	.15 .08 .02	.14 .08 .02	.14 .07 .02	.13 .07 .02	4	.12 .08 .06	.10 .07 .05	.07 .05 .03	.04 .04 .02	.01 .01 .01
	5	.15 .08 .02	.15 .07 .02	.14 .07 .02	.13 .07 .02	.13 .07 .02	5	.11 .07 .05	.10 .06 .04	.07 .04 .03	.04 .03 .02	.01 .01 .01
	6	.14 .07 .02	.14 .07 .02	.14 .07 .02	.13 .07 .02	.13 .07 .02	6	.11 .07 .04	.09 .06 .03	.06 .04 .02	.04 .02 .01	.01 .01 .00
	7	.14 .07 .02	.14 .07 .02	.13 .07 .02	.13 .07 .02	.12 .06 .02	7	.10 .06 .03	.09 .05 .03	.06 .04 .02	.04 .02 .01	.01 .01 .00
	8	.13 .07 .02	.13 .06 .02	.13 .06 .02	.12 .06 .02	.12 .06 .02	8	.10 .06 .03	.09 .05 .03	.06 .04 .02	.03 .02 .01	.01 .01 .00
Wide-distribution ventilated reflector with clear	9	.13 .06 .02	.13 .06 .02	.12 .06 .02	.12 .06 .02	.11 .06 .02	9	.10 .05 .03	.08 .05 .02	.06 .03 .02	.03 .02 .01	.01 .01 .00
HID lamp	10	.12 .06 .02	.12 .06 .02	.12 .06 .02	.11 .06 .02	.11 .06 .02	10	.09 .05 .02	.08 .04 .02	.06 .03 .01	.03 .02 .01	.01 .01 .00
21							0	.28 .28 .28	.24 .24 .24	.16 .16 .16	.09 .09 .09	.03 .03 .03
	1	.20 .11 .04	.19 .11 .03	.17 .10 .03	.15 .09 .03	.13 .08 .02	1	.27 .26 .24	.23 .22 .21	.16 .15 .15	.09 .09 .08	.03 .03 .03
	2	.19 .10 .03	.18 .10 .03	.16 .09 .03	.15 .08 .03	.13 .07 .02	2	.27 .24 .22	.23 .21 .19	.16 .14 .13	.09 .08 .08	.03 .03 .03
	3	.18 .10 .03	.17 .09 .03	.15 .08 .03	.14 .08 .02	.13 .07 .02	3	.26 .22 .20	.22 .19 .17	.15 .13 .12	.09 .08 .07	.03 .03 .02
	4	.17 .09 .03	.16 .09 .03	.15 .08 .02	.13 .07 .02	.12 .07 .02	4	.25 .21 .18	.22 .18 .16	.15 .13 .11	.09 .08 .07	.03 .02 .02
	5	.16 .08 .02	.16 .08 .02	.14 .07 .02	.13 .07 .02	.12 .06 .02	5	.25 .21 .17	.21 .18 .15	.15 .12 .11	.08 .07 .06	.03 .02 .02
	6	.15 .08 .02	.15 .07 .02	.14 .07 .02	.12 .06 .02	.11 .06 .02	6	.24 .20 .17	.21 .17 .15	.14 .12 .10	.08 .07 .06	.03 .02 .02
	7	.15 .07 .02	.14 .07 .02	.13 .07 .02	.12 .06 .02	.11 .06 .02	7	.24 .19 .16	.20 .17 .14	.14 .12 .10	.08 .07 .06	.03 .02 .02
	8	.14 .07 .02	.13 .07 .02	.12 .06 .02	.11 .06 .02	.11 .05 .02	8	.23 .19 .16	.20 .16 .14	.14 .11 .10	.08 .07 .06	.03 .02 .02
Diffuse aluminum reflector with 35° crosswise	9	.13 .06 .02	.13 .06 .02	.12 .06 .02	.11 .06 .02	.10 .05 .01	9	.23 .19 .15	.20 .16 .13	.14 .11 .10	.08 .07 .06	.03 .02 .02
shielding	10	.13 .06 .02	.12 .06 .02	.11 .06 .02	.11 .05 .01	.10 .05 .01	10	.23 .18 .15	.20 .16 .13	.14 .11 .09	.08 .07 .06	.03 .02 .02
23							0	.26 .26 .26	.22 .22 .22	.15 .15 .15	.09 .09 .09	.03 .03 .03
	1	.17 .10 .03	.16 .09 .03	.14 .08 .03	.13 .07 .02	.11 .06 .02	1	.26 .24 .23	.22 .21 .20	.15 .14 .14	.09 .08 .08	.03 .03 .03
	2	.16 .09 .03	.15 .09 .03	.14 .08 .02	.12 .07 .02	.11 .06 .02	2	.25 .23 .21	.21 .20 .18	.15 .14 .13	.08 .08 .07	.03 .03 .02
	3	.15 .08 .02	.15 .08 .02	.13 .07 .02	.12 .06 .02	.10 .06 .02	3	.24 .21 .19	.21 .19 .17	.14 .13 .12	.08 .08 .07	.03 .02 .02
	4	.15 .08 .02	.14 .07 .02	.13 .07 .02	.11 .06 .02	.10 .05 .02	4	.24 .21 .18	.20 .18 .16	.14 .12 .11	.08 .07 .07	.03 .02 .02
	5	.14 .07 .02	.13 .07 .02	.12 .06 .02	.11 .06 .02	.10 .05 .02	5	.23 .20 .17	.20 .17 .15	.14 .12 .11	.08 .07 .06	.03 .02 .02
	6	.13 .07 .02	.13 .06 .02	.11 .06 .02	.10 .05 .02	.09 .05 .01	6	.23 .19 .17	.20 .17 .14	.14 .12 .10	.08 .07 .06	.03 .02 .02
	7	.13 .06 .02	.12 .06 .02	.11 .06 .02	.10 .05 .01	.09 .05 .01	7	.23 .19 .16	.19 .16 .14	.13 .11 .10	.08 .07 .06	.03 .02 .02
	8	.12 .06 .02	.12 .06 .02	.11 .05 .02	.10 .05 .01	.09 .04 .01	8	.22 .18 .16	.19 .16 .14	.13 .11 .10	.08 .07 .06	.02 .02 .02
Diffuse aluminum reflector with 35° crosswise and 35°	9	.12 .06 .02	.11 .05 .02	.10 .05 .01	.09 .05 .01	.08 .04 .01	9	.22 .18 .15	.19 .16 .13	.13 .11 .10	.08 .06 .06	.02 .02 .02
lengthwise shielding	10	.11 .05 .01	.11 .05 .01	.10 .05 .01	.09 .04 .01	.08 .04 .01	10	.22 .18 .15	.19 .15 .13	.13 .11 .09	.08 .06 .06	.02 .02 .02
28							0	.65 .65 .65	.55 .55 .55	.38 .38 .38	.22 .22 .22	.07 .07 .07
	1	.20 .12 .04	.18 .10 .03	.13 .08 .02	.09 .05 .02	.05 .03 .01	1	.65 .63 .62	.55 .54 .53	.38 .37 .37	.22 .22 .21	.07 .07 .07
	2	.19 .10 .03	.17 .09 .03	.12 .07 .02	.08 .05 .01	.04 .03 .01	2	.64 .61 .59	.55 .53 .51	.38 .37 .36	.22 .21 .21	.07 .07 .07
	3	.17 .09 .03	.15 .08 .02	.11 .06 .02	.08 .04 .01	.04 .02 .01	3	.64 .60 .58	.55 .52 .50	.37 .36 .35	.22 .21 .21	.07 .07 .07
	4	.16 .08 .02	.14 .07 .02	.11 .06 .02	.07 .04 .01	.04 .02 .01	4	.63 .59 .57	.54 .51 .49	.37 .36 .35	.22 .21 .21	.07 .07 .07
	5	.15 .08 .02	.13 .07 .02	.10 .05 .02	.07 .04 .01	.04 .02 .01	5	.63 .59 .56	.54 .51 .49	.37 .36 .34	.21 .21 .20	.07 .07 .07
	6	.14 .07 .02	.12 .06 .02	.09 .05 .01	.06 .03 .01	.03 .02 .01	6	.06 .58 .55	.54 .50 .48	.37 .35 .34	.21 .21 .20	.07 .07 .07
	7	.13 .07 .02	.12 .06 .02	.09 .04 .01	.06 .03 .01	.03 .02 .01	7	.62 .58 .55	.53 .50 .48	.37 .35 .34	.21 .21 .20	.07 .07 .07
	8	.12 .06 .02	.11 .05 .02	.08 .04 .01	.06 .03 .01	.03 .02 .00	8	.61 .57 .55	.53 .50 .48	.37 .35 .34	.21 .21 .20	.07 .07 .07
Luminous bottom suspended unit with extra-high-output	9	.12 .06 .02	.10 .05 .01	.08 .04 .01	.05 .03 .01	.03 .02 .00	9	.61 .57 .55	.53 .50 .48	.36 .35 .34	.21 .21 .20	.07 .07 .07
lamp	10	.11 .05 .01	.10 .05 .01	.07 .04 .01	.05 .02 .01	.03 .01 .00	10	.61 .57 .54	.52 .49 .47	.36 .35 .34	.21 .21 .20	.07 .07 .07

Table 20.7b Wall Luminance Coefficients and Ceiling Cavity Luminance Coefficients for Typical Luminaires (Continued)

Typical Luminaire	$\rho_{cc} \rightarrow$	80	70	50	30	10	$\rho_{cc} \rightarrow$	80	70	50	30	10
	$\rho_w \rightarrow$	50 30 10	50 30 10	50 30 10	50 30 10	50 30 10	$\rho_w \rightarrow$	50 30 10	50 30 10	50 30 10	50 30 10	50 30 10
	RCR ↓	Wall luminance coefficients for 20% effective floor cavity reflectance ($\rho_{fc} = 20$)					RCR ↓	Ceiling cavity luminance coefficients for 20% effective floor cavity reflectance ($\rho_{fc} = 20$)				
30							0	.22 .22 .22	.18 .18 .18	.12 .12 .12	.07 .07 .07	.02 .02 .02
	1	.19 .11 .03	.18 .10 .03	.17 .10 .03	.15 .09 .03	.14 .08 .03	1	.21 .20 .18	.18 .17 .16	.12 .12 .11	.07 .07 .06	.02 .02 .02
	2	.18 .10 .03	.17 .09 .03	.15 .09 .03	.14 .08 .02	.13 .07 .02	2	.21 .18 .16	.18 .16 .14	.12 .11 .10	.07 .06 .06	.02 .02 .02
	3	.16 .09 .03	.16 .08 .03	.14 .08 .02	.13 .07 .02	.12 .07 .02	3	.20 .17 .14	.17 .15 .13	.12 .10 .09	.07 .06 .05	.02 .02 .02
	4	.15 .08 .02	.15 .08 .02	.14 .07 .02	.12 .07 .02	.12 .06 .02	4	.19 .16 .13	.17 .14 .12	.11 .10 .08	.07 .06 .05	.02 .02 .02
	5	.14 .07 .02	.14 .07 .02	.13 .07 .02	.12 .06 .02	.11 .06 .02	5	.19 .15 .13	.16 .13 .11	.11 .09 .08	.06 .05 .05	.02 .02 .02
	6	.14 .07 .02	.13 .07 .02	.12 .06 .02	.11 .06 .02	.10 .05 .02	6	.18 .15 .12	.16 .13 .10	.11 .09 .07	.06 .05 .04	.02 .02 .01
	7	.13 .06 .02	.12 .06 .02	.12 .06 .02	.11 .05 .02	.10 .05 .02	7	.18 .14 .11	.16 .12 .10	.11 .09 .07	.06 .05 .04	.02 .02 .01
	8	.12 .06 .02	.12 .06 .02	.11 .05 .02	.10 .05 .01	.10 .05 .01	8	.18 .14 .11	.15 .12 .10	.11 .08 .07	.06 .05 .04	.02 .02 .01
Two-lamp prismatic wraparound—multiply by 0.95 for 4 lamps	9	.12 .06 .02	.11 .05 .02	.11 .05 .01	.10 .05 .01	.09 .05 .01	9	.17 .13 .11	.15 .12 .09	.10 .08 .07	.06 .05 .04	.02 .02 .01
	10	.11 .05 .01	.11 .05 .01	.10 .05 .01	.09 .05 .01	.09 .04 .01	10	.17 .13 .11	.15 .11 .09	.10 .08 .07	.06 .05 .04	.02 .02 .01
35							0	.09 .09 .09	.08 .08 .08	.05 .05 .05	.03 .03 .03	.01 .01 .01
	1	.13 .07 .02	.12 .07 .02	.12 .07 .02	.11 .06 .02	.10 .06 .02	1	.09 .08 .07	.08 .07 .06	.05 .05 .04	.03 .03 .02	.01 .01 .01
	2	.12 .07 .02	.12 .06 .02	.11 .06 .02	.11 .06 .02	.10 .06 .02	2	.08 .07 .05	.07 .06 .05	.05 .04 .03	.03 .02 .02	.01 .01 .01
	3	.11 .06 .02	.11 .06 .02	.11 .06 .02	.10 .06 .02	.10 .05 .02	3	.08 .06 .04	.07 .05 .03	.05 .03 .02	.03 .02 .01	.01 .01 .00
	4	.11 .06 .02	.10 .05 .02	.10 .05 .02	.10 .05 .02	.09 .05 .02	4	.07 .05 .03	.06 .04 .03	.04 .03 .02	.03 .02 .01	.01 .01 .00
	5	.10 .05 .01	.10 .05 .01	.10 .05 .01	.09 .05 .01	.09 .05 .01	5	.07 .04 .03	.06 .04 .02	.04 .03 .02	.02 .02 .01	.01 .01 .00
	6	.10 .05 .01	.09 .05 .01	.09 .05 .01	.09 .04 .01	.08 .04 .01	6	.07 .04 .02	.06 .04 .02	.04 .02 .01	.02 .01 .01	.01 .00 .00
	7	.09 .04 .01	.09 .04 .01	.09 .04 .01	.08 .04 .01	.08 .04 .01	7	.06 .04 .02	.06 .03 .02	.04 .02 .01	.02 .01 .01	.01 .00 .00
	8	.09 .04 .01	.08 .04 .01	.08 .04 .01	.08 .04 .01	.08 .04 .01	8	.06 .03 .02	.05 .03 .01	.04 .02 .01	.02 .01 .01	.01 .00 .00
Four-lamp, 2-ft-wide troffer with 45° plastic louver—multiply by 1.05 for 2 lamps and 0.95 for 6 lamps.	9	.08 .04 .01	.08 .04 .01	.08 .04 .01	.08 .04 .01	.07 .04 .01	9	.06 .03 .01	.05 .03 .01	.04 .02 .01	.02 .01 .00	.01 .00 .00
	10	.08 .04 .01	.08 .04 .01	.07 .04 .01	.07 .04 .01	.07 .04 .01	10	.06 .03 .01	.05 .03 .01	.03 .02 .01	.02 .01 .00	.01 .00 .00
41							0	.11 .11 .11	.10 .10 .10	.06 .06 .06	.04 .04 .04	.01 .01 .01
	1	.16 .09 .03	.16 .09 .03	.15 .09 .03	.14 .08 .03	.14 .08 .03	1	.11 .10 .09	.09 .08 .07	.06 .06 .05	.04 .03 .03	.01 .01 .01
	2	.15 .08 .03	.15 .08 .03	.14 .08 .02	.14 .08 .02	.13 .07 .02	2	.10 .08 .06	.09 .07 .06	.06 .05 .04	.03 .03 .02	.01 .01 .01
	3	.15 .08 .02	.14 .08 .02	.14 .07 .02	.13 .07 .02	.13 .07 .02	3	.10 .07 .05	.08 .06 .04	.06 .04 .03	.03 .02 .02	.01 .01 .01
	4	.14 .07 .02	.13 .07 .02	.13 .07 .02	.12 .07 .02	.12 .06 .02	4	.09 .06 .04	.08 .05 .03	.05 .04 .02	.03 .02 .01	.01 .01 .00
	5	.13 .07 .02	.13 .07 .02	.12 .06 .02	.12 .06 .02	.11 .06 .02	5	.09 .06 .03	.08 .05 .03	.05 .03 .02	.03 .02 .01	.01 .01 .00
	6	.12 .06 .02	.12 .06 .02	.12 .06 .02	.11 .06 .02	.11 .06 .02	6	.09 .05 .03	.07 .04 .02	.05 .03 .02	.03 .02 .01	.01 .01 .00
	7	.12 .06 .02	.12 .06 .02	.11 .06 .02	.11 .06 .02	.10 .05 .02	7	.08 .05 .02	.07 .04 .02	.05 .03 .01	.03 .02 .01	.01 .01 .00
	8	.11 .05 .02	.11 .05 .02	.11 .05 .02	.10 .05 .02	.10 .05 .01	8	.08 .04 .02	.07 .04 .02	.05 .03 .01	.03 .02 .01	.01 .01 .00
Fluorescent unit with flat prismatic lens; four-lamp, 2 ft wide—multiply by 1.10 for 2 lamps.	9	.11 .05 .01	.10 .05 .01	.10 .05 .01	.10 .05 .01	.10 .05 .01	9	.08 .04 .02	.07 .03 .01	.05 .02 .01	.03 .01 .01	.01 .00 .00
	10	.10 .05 .01	.10 .05 .01	.10 .05 .01	.09 .05 .01	.09 .05 .01	10	.07 .04 .01	.06 .03 .01	.04 .02 .01	.03 .01 .01	.01 .00 .00

Source. Selected and reprinted with permission from *IES Handbook*, 5th ed.

order. The spacing to mounting-height (S/MH) guide should be used with care, with attention given to the direction when crosswise and lengthwise data are given. The S/MH figure shown is conservative. Field tests have shown no adverse reaction to light variations of 1.3:1 particularly in general illumination. A S/MH ratio in excess of 1.5 will generally not give satisfactory results except with special diffusers specifically designed for wide spacing. See also the previous discussion on lighting uniformity in Section 20.20. The Wall Direct Radiation Coefficient (WDRC) in the last column is an indication of wall illumination, that is, the WDRC figure can be used in similar fashion to the CU, to obtain wall illumination. Light loss factor should be used in any such calculation. In all cases actual fixture data are preferable to the generic data of this table. The table is principally useful for preliminary calculations.

20.29 Interior Illumination Calculations

Having selected a luminaire on the basis of all the foregoing criteria, it remains only to calculate the number of such units required in each space for *general* illumination, and to properly arrange them. Although a number of calculation methods are available, the lumen method is simplest and most applicable to our needs for area lighting calculations. Intensity calculation from point, line, or area sources is covered in Sections 20.37 to 20.41.

20.30 Light Loss Factor (LLF) in Lumen Calculations

The lumen method of calculation is a procedure for determining the *average* maintained footcandle illumination intensity on the working plane in a room. The method presupposes that luminaires will be spaced so that uniformity of illumination is provided in order that an *average* calculation have validity. The method is based on the definition of one footcandle as one lumen incident on one square foot of area. Or

$$\text{footcandles (fc)} = \frac{\text{lumens (lm)}}{\text{area in sq ft}}$$

As explained above, the ratio between the lumens reaching the working plane and the lumens generated is the coefficient of utilization, CU. Or

$$\text{lumens on the working plane} = \text{lamp lumens} \times \text{CU}$$

Therefore

$$\text{fc} = \frac{\text{lamp lumens} \times \text{CU}}{\text{area}}$$

The footcandle figure so calculated is *initial* average illumination. This initial level is reduced by the effect of temperature and voltage variations, dirt accumulation on luminaires and room surfaces, lamp output depreciation, and maintenance conditions. All of these effects are cumulatively referred to as the Light Loss Factor, LLF:

$$\text{Maintained fc} = \text{initial fc} \times \text{LLF}$$

(This factor was previously known as the maintenance factor, MF.) In order to arrive at the LLF to be used, consideration must be given to the following component factors:

(*a*) ***Luminaire Ambient Temperature.*** Light output is reduced when a fixture operates at other than its design temperature. With normal indoor installation use 1.0, that is, no depreciation. For other conditions, refer to technical data on the source involved.

(*b*) ***Voltage.*** As (a) above. Details of source sensitivity to voltage are given in Chapter 19.

(*c*) ***Ballast Factor.*** Depreciation of output due to use of a ballast other than that used in the test photometry of the fixture and source.

(*d*) ***Luminaire Surface Depreciation.*** This results from deterioration in the fixture's reflecting surfaces (not dirt), and is proportional to age. Exact figures depend on type of surface involved.

All of the above factors are nonrecoverable, that is, they cannot be improved by normal maintenance. When no exact data are available and no special conditions apply, use 0.9 for the product $a \times b \times c \times d$. The factors below are *recoverable,* that is, can be improved to initial conditions by maintenance.

(*e*) ***Room Surface Dirt.*** This factor is self-explanatory. Use the following approximations for direct lighting units (for exact data see Reference 4):

Clean room, frequent maintenance: 0.95.

Medium conditions and frequency of cleaning: 0.9.

Dirty room, infrequent maintenance: 0.85.

(*f*) ***Lamp Lumen Depreciation.*** This factor depends on the type of lamp and the replacement schedule. Use the following when exact data are unavailable:

	Group Replacement	*Replacement on Burnout*
Incandescent	0.94	0.88
Tungsten Halogen	0.98	0.94
Fluorescent	0.90	0.85
Mercury	0.82	0.74
Metal-Halide	0.87	0.80
High-pressure Sodium	0.94	0.88

(*g*) ***Burnouts.*** Self-explanatory. Factor depends on maintenance schedules and method of replacement. Use the following:

Group replacement procedures: 1.0.

Individual replacement on burnout: 0.95.

(*h*) ***Luminaire Dirt Depreciation*** (***LDD***). This factor depends on luminaire design, atmosphere conditions in the space, and maintenance schedule. The maintenance category is obtained from manufacturer's data or from Table 20.7. The type of atmosphere is determined by considering the space involved. LDD is then selected from the curves in Figure 20.32. Note that the difference between a 12-month cleaning cycle and one of 36 months is as much as 30%. If fixtures containing long-life lamps are cleaned only when burned-out lamps are replaced, that is, in 36 months or more, LDD frequently drops below 0.7.

Total LLF is the product of all the depreciation factors above, that is,

$$\text{LLF} = a \times b \times c \times d \times e \times f \times g \times h$$

For example, a fluorescent troffer in a regularly maintained group-lamp-replacement, air-conditioned office might typically have an LLF of

$$\text{LLF} = 1 \times 1 \times 1 \times 0.92 \times 0.95 \times 1.0 \times 0.9 \times 0.95 = 0.75$$

The same fixture in the same office, but with fixture cleaned only when replacing burned-out lamps, would typically have an LLF of

$$\text{LLF} = 1 \times 1 \times 1 \times 0.92 \times 0.90 \times 0.95 \times 0.85 \times 0.85 = 0.57$$

Thus, if in the first case the maintained illumination is A fc, in the second case it is 0.57/0.75 or 0.76A fc, that is, a reduction of 24% as a result of poor maintenance.

Our final expression for footcandle illumination as calculated by the lumen method is

$$\text{fc} = \frac{\text{lamp lumens} \times \text{CU} \times \text{LLF}}{\text{area}} \tag{20.1}$$

Lamp lumens is the total within the space, and equal to

$$\text{No. of fixtures} \times \text{lamps/fixture} \times \text{lumens/lamp}$$

Our formula then becomes

$$\text{fc} = \frac{\text{luminaires} \times \text{lamps per luminaire} \times \text{lumens/lamp} \times \text{CU} \times \text{LLF}}{\text{area}} \tag{20.2}$$

or, conversely,

$$\text{No. of luminaires} = \frac{\text{fc} \times \text{area}}{\text{lamps per luminaire} \times \text{lumens per lamp} \times \text{CU} \times \text{LLF}} \tag{20.3}$$

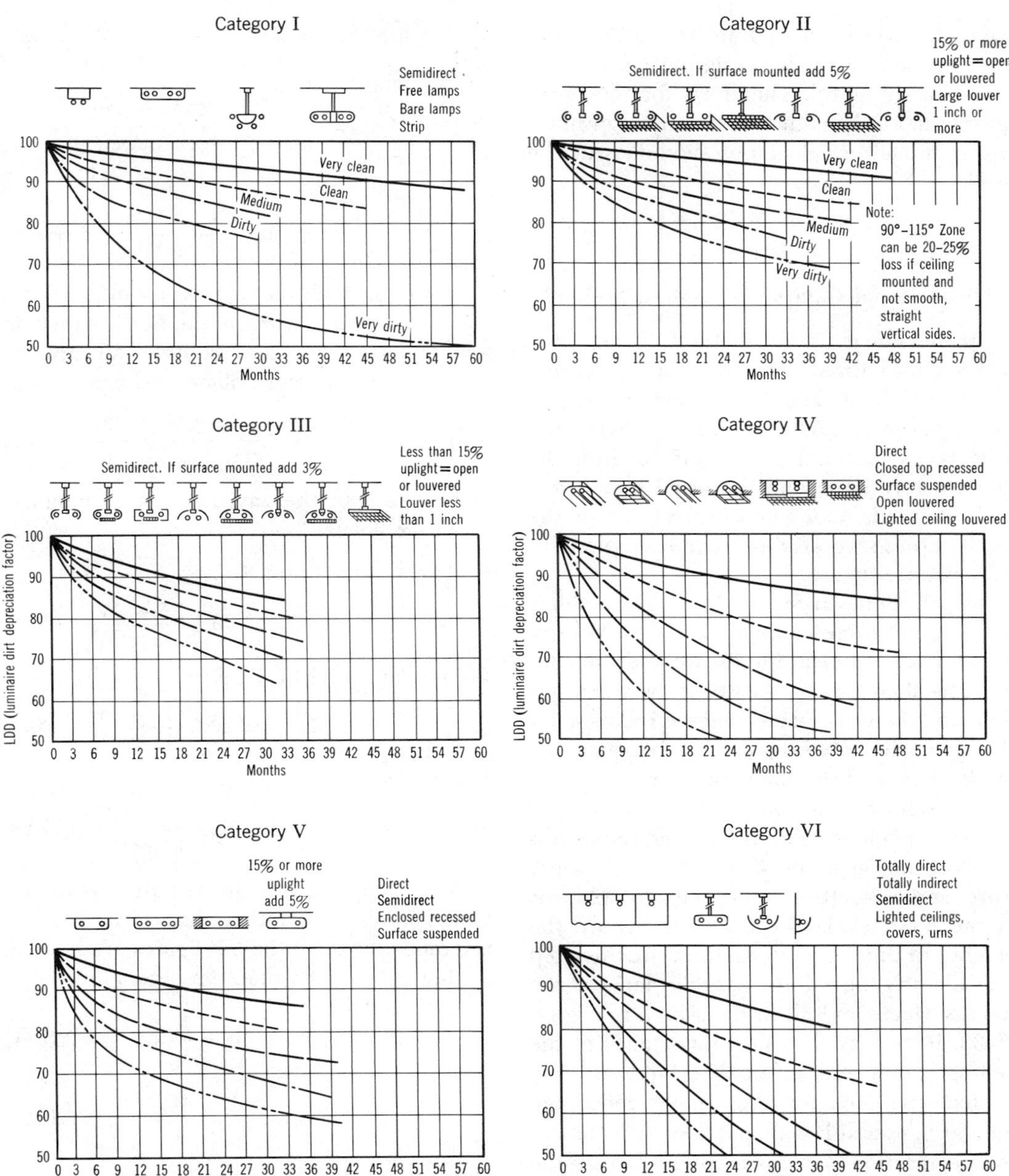

Fig. 20.32 Luminaire dirt depreciation factor (LDD) for six luminaire categories and five degrees of dirtyness. (Courtesy of the IES).

If we wish to determine the area per luminaire that is required to give a specified illumination, we have

$$\text{Area/luminaire} = \frac{\text{lamps} \times \text{lumen per lamp} \times \text{CU} \times \text{LLF}}{\text{fc}} \qquad (20.4)$$

This is usually a much more useful procedure in large spaces than calculating for the entire room. For instance, it is much more convenient to know that to maintain say 60 fc with a given luminaire, 100 sq ft per unit are required than it is to know that for an 18,000 sq ft floor, 180 fixtures are necessary. The former figure allows us to establish a pattern, say 10×10; the latter figure is too large to be immediately useful. Therefore, for rooms requiring more than a nominal number of luminaires, the latter calculation should always be used.

20.31 Zonal Cavity Calculation Method

In the Zonal Cavity method, the room is divided into three cavities, that is, ceiling, room, and floor, and the "effective" reflectance of each cavity calculated. With the reflectance figures, the CU is selected from the given fixture data, and the usual lumen formula applied. A detailed explanation of the method plus several illustrative examples will demonstrate its use.

Step 1. Refer to Figure 20.33. The room is divided into three cavities; the ceiling cavity is the space between the fixture and the ceiling, the floor cavity the space between the floor and the work plane, and the room cavity the space in between, that is, between the work plane and the luminaire center line. In offices, schools, and many other occupancies the work plane is 30 in. In drafting rooms it is 36 to 38 in., in shops 42 to 48 in., in carpet stores and sail-cutting rooms the work plane is at the floor level. The 3 "h" terms are the heights of the various cavities. In this step also, identify the *maintained* reflectance of the room surfaces and fill in the sketch in Figure 20.33. If the initial surface reflectance of the ceiling is 90% and a 10% deterioration is expected, use 80% for the ceiling reflection. Similarly, establish wall and floor reflectances by specification or by assumption. Utilize the nearest reflectance given in Table 20.9. Interpolation between values is *not* required.

Step 2. See Table 20.8 and Figure 20.33. This step involves determining the cavity ratios of the room, either by calculation or from Table 20.8. The basic expression for a cavity ratio (CR) is

$$CR = 2.5 \times \frac{\text{area of cavity wall}}{\text{area of work plane}} \qquad (20.5)$$

In a rectangular space the area of the cavity wall is $h \times (2L + 2W)$ or $2h(L + W)$; therefore

$$CR = \frac{2.5 \times 2h \times (L + W)}{\text{area of work plane}}$$

or

$$CR = 5h \times \frac{L + W}{L \times W} \qquad (20.6)$$

For other than rectangular rooms, the area can be calculated as required. For instance, in a circular room, the cavity wall area = $h \times 2\pi r$ and the work plane area is πr^2. Thus

$$CR = \frac{2.5 \times h \times 2\pi r}{\pi r^2} = \frac{5h}{r} \qquad (20.7)$$

For each of the cavities in a rectangular room we have:

Ceiling Cavity Ratio, CCR

$$= 5h_{cc} \frac{L + W}{L \times W} \qquad (20.8)$$

Room Cavity Ratio, RCR

$$= 5h_{rc} \frac{L + W}{L \times W} \qquad (20.9)$$

Floor Cavity Ratio, FCR

$$= 5h_{fc} \frac{L + W}{L \times W} \qquad (20.10)$$

Most people prefer, for rectangular rooms, to use Table 20.8 rather than formulas. For reference, since all the CR figures are related, having determined one, the others are

$$CCR = RCR \frac{h_{cc}}{h_{rc}} \qquad (20.11)$$

$$FCR = RCR \frac{h_{fc}}{h_{rc}} \qquad (20.12)$$

and

$$CCR = FCR \frac{h_{cc}}{h_{fc}} \qquad (20.13)$$

Step 3. See Table 20.9 and Figures 20.33 and 20.34. This step involves obtaining the effective ceiling reflectance (ρ_{cc}) from Table 20.9. Note that the wall reflectance remains as selected in Step 1. If the fixtures are surface mounted or recessed, then CCR = 0 and ρ_{cc} = selected ceiling reflectance.

Step 4. See Table 20.9 and Figures 20.33

and 20.34. This step involves obtaining the effective floor reflectance ρ_{fc} as above in Step 3, for ρ_{cc}. If the floor is the working plane, FCR = 0 and ρ_{fc} = selected floor reflectance.

Step 5. Select CU from manufacturer's data. Note that some interpolation may be necessary, since all CU data are given for an effective floor cavity reflectance of 20%. Factors for this correction are given in Table 20.10.

Step 6. Calculate footcandles and number of fixtures or area per luminaire in the usual fashion.

Illustrative examples and shortcut methods are demonstrated below. CU coefficients are listed in Table 20.7.

20.32 Zonal Cavity Calculations; Illustrative Examples

EXAMPLE 20.1. Refer to Figure 20.35, which will be filled in as the solution progresses.

GIVEN. *Classroom:* 20 × 25 × 12 ft, elementary school.
Initial Reflectances: ceiling 80, wall 50, floor 20. Provide adequate lighting using fluorescent lamp fixtures. Assume yearly maintenance, lamp replacement at 70% life, proper voltage and ballasts, medium clean atmosphere.

SOLUTION. From Table 18.4, page 734, the required illumination level for pencil writing is 70 fc ESI. We select from Table 20.7 fixture no. 23 for two lamps, which has semi-direct distribution, low brightness for high VCP (students spend a large proportion of their time in heads-up position, requiring high VCP), batwing-type crosswise distribution for high CRF and low veiling reflection, and good CU, which means high efficiency and low energy use. Furthermore, maintenance category II is excellent as is the wide lateral spacing possible with the 1.5 S/MH listed. Arrangement of fixtures will be in rows, front-to-back. *Based on experience, we would here design for 80 raw fc to obtain the requisite 70-fc, ESI average.*

Calculations

Step 1. From Table 20.6 assume 24-in. stem length (in lieu of 21 in., for ease of calculation). Also assume standard working plane as 30 in. AFF. We will fill in the blanks in the calculation sheet, Figure 20.35, as we proceed. Therefore:

$$h_{cc} = 2.0 \qquad h_{rc} = 7.5 \qquad h_{fc} = 2.5$$
$$\rho_C = 80\% \qquad \rho_w = 50\% \qquad \rho_F = 20\%$$

Step 2. From formulas

$$\text{RCR} = 5h_{rc}\left(\frac{L + W}{L \times W}\right)$$
$$= 5(7.5)\left(\frac{45}{500}\right) = 3.375$$

Similarly

$$\text{CCR} = 5(2)\left(\frac{45}{500}\right) = 0.9$$
$$\text{FCR} = 5(2.5)\left(\frac{45}{500}\right) = 1.125$$

Step 3. From Table 20.9.

$$\rho_{cc} = 68\%$$

Step 4. By interpolation between 10 and 30% floor reflectances and 1.1 and 1.2 cavity ratio, we obtain

$$\rho_{fc} = 19\% \text{ and } \rho_w = 50\% \text{ as given.}$$

Step 5. By double interpolation

	CU		
ρ_{cc}	*70*	*68*	*50*
RCR = 3	0.58		0.54
3.375	*0.549*	*0.545*	*0.509*
RCR = 4	0.53		0.49
Therefore, CU = 0.545			

Step 6. Light loss factors as shown on Figure 20.35:

$$\text{LLF} = 0.9 \times 0.9 \times 0.9 \times 0.9 = 0.656$$

Step 7. Footcandle calculations:

Number of luminaires

$$= \frac{80 \times 20 \times 25}{2 \times 3200 \times 0.545 \times 0.656}$$
$$= 17.48 \text{ fixtures}$$

or

GENERAL INFORMATION

Project identification: ______________________________
(Give name of area and/or building and room number)

Average maintained illumination for design: ____footcandles

Luminaire data:

Manufacturer: ______________________________

Catalog number: ______________________________

Lamp Data:

Type and color: ______________________________

Number per luminaire: ______________________________

Total lumens per luminaire: ______________________________

SELECTION OF COEFFICIENT OF UTILIZATION

$\rho =$ __ %

$\rho =$ __ %

$\rho =$ __ %

Work plane

$\rho =$ __ % $\rho =$ __ %

$h_{CC} =$ __

$h_{RC} =$ __

$h_{FC} =$ __

L = __

W = __

Step 1: Fill in sketch at right.

Step 2: Determine Cavity Ratios from Fig. 20.34, or by formulas.

Room Cavity Ratio, RCR = ______________________________

Ceiling Cavity Ratio, CCR = ______________________________

Floor Cavity Ratio, FCR = ______________________________

Step 3: Obtain effective ceiling cavity reflectance (ρ_{CC}) from Table 20.9 ρ_{CC} = __________

Step 4: Obtain effective floor cavity reflectance (ρ_{FC}) from Table 20.9 ρ_{FC} = __________

Step 5: Obtain coefficient of utilization (CU) from manufacturer's data. CU = __________

SELECTION OF LIGHT LOSS FACTORS

Unrecoverable		Recoverable	
Luminaire ambient temperature	______	Room surface dirt depreciation	______
Voltage to luminaire	______	Lamp lumen depreciation	______
Ballast factor	______	Lamp burnouts factor	______
Luminaire surface depreciation	______	Luminaire dirt depreciation LDD	______

Total light loss factor, LLF (product of individual factors above): ____

CALCULATIONS

(Average Maintained Illumination Level)

$$\text{Number of Luminaires} = \frac{(\text{Footcandles}) \times (\text{Area in square feet})}{(\text{Lumens per luminaire}) \times (\text{CU}) \times (\text{LLF})}$$

$$= \frac{\qquad\qquad}{\qquad\qquad} =$$

$$\text{Footcandles} = \frac{(\text{Number of luminaires}) \times (\text{Lumens per luminaire}) \times (\text{CU}) \times (\text{LLF})}{(\text{Area in square feet})}$$

$$= \frac{\qquad\qquad}{\qquad\qquad} =$$

Calculated by: ______________________________ Date: ______________

Fig. 20.33 Zonal Cavity Method calculation Form (this and opposite page). |Courtesy of the IES.

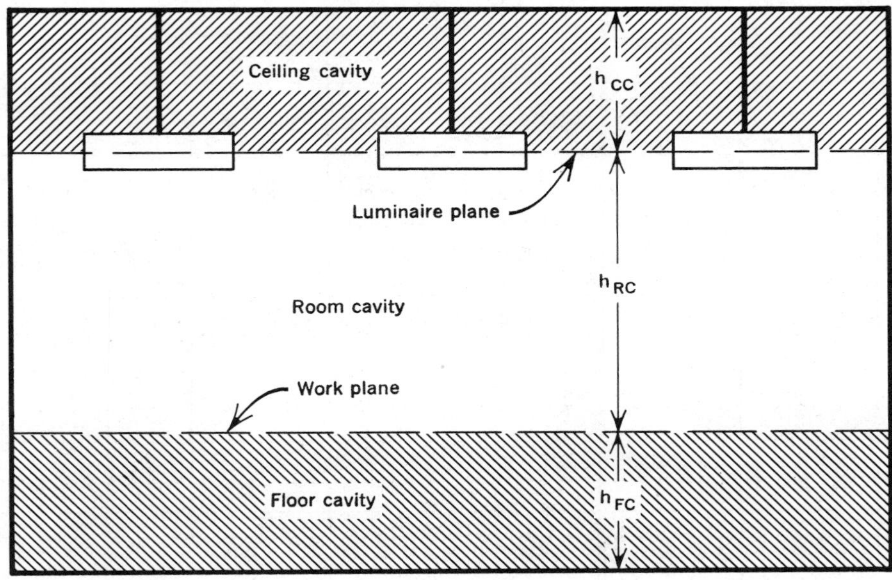

(a)

Notes:

1. Obtain Effective Ceiling Cavity Reflectance (ρ_{CC}) for combination of ceiling and wall reflectance to be employed from Table 20.9. Note that for surface-mounted or recessed luminaires, CCR = 0 and the ceiling reflectance may be used as the effective cavity reflectance. Since LLF accounts for room surface depreciation, use initial reflectances.
2. Obtain Effective Floor Cavity Reflectance (ρ_{FC}) for combination of floor and wall reflectances to be employed from Table 20.9.
3. Obtain Coefficient of Utilization for 20 percent effective floor cavity reflectance condition from table for luminaire, interpolating between tabulated values as required to match room size and ceiling and wall reflectance combinations. If Effective Floor Cavity Reflectance (ρ_{FC}) obtained in Step 3 differs significantly from 20 percent, obtain the multiplier from Table 20.10 (or by interpolating between the tabulated multipliers and the 1.00 multiplier for 20 percent effective floor cavity reflectance). Multiply the Coefficient of Utilization by this multiplier.
4. If *initial illumination* is desired, CU selection should be obtained by above procedure and the light loss factor omitted from the formula.
5. Fig. 20.34, which is a graphic plot of the data in Table 20.9 may be used in its stead.

$$\text{Area/luminaire} = \frac{2 \times 3200 \times 0.545 \times 0.656}{80}$$

$$= 28.6 \text{ sq ft/fixture}$$

Therefore, we can run three rows of fixtures front to back and obtain excellent wall lighting and blackboard illumination. An 8-ft spacing between rows gives an S/MH of less than 1.0, which is obviously excellent. The row adjacent to the window would be individually switched. Depending on daylight conditions, it might well be possible to eliminate one row of fixtures entirely and utilize two rows of six fixtures at 10-ft centerline spacing, 6 to 7 ft from the window and 3 to 4 ft from the inside wall. Actual ESI can be calculated by one of the computer programs available.

EXAMPLE 20.2. Large clerical business office 60 × 100 × 8 ft; initial reflectances 80, 30, 30. Provide 50 fc general lighting with fluorescent, four-lamp lay-in troffers. Space is air-conditioned. Lamps are replaced on a burn-out basis and fixture then cleaned.

SOLUTION. From Table 20.7 we select fixture no. 41 because of high S/MH ratio and good CU and VCP.

Calculations. (The reader can fill in a copy of Figure 20.33.) Assume working plane at 36

Table 20.8 *Cavity Ratios*

Room Dimensions		Cavity Depth																			
Width	Length	1.0	1.5	2.0	2.5	3.0	3.5	4.0	5.0	6.0	7.0	8	9	10	11	12	14	16	20	25	30
8	8	1.2	1.9	2.5	3.1	3.7	4.4	5.0	6.2	7.5	8.8	10.0	11.2	12.5	—	—	—	—	—	—	—
	10	1.1	1.7	2.2	2.8	3.4	3.9	4.5	5.6	6.7	7.9	9.0	10.1	11.3	12.4	—	—	—	—	—	—
	14	1.0	1.5	2.0	2.5	3.0	3.4	3.9	4.9	5.9	6.9	7.8	8.8	9.7	10.7	11.7	—	—	—	—	—
	20	0.9	1.3	1.7	2.2	2.6	3.1	3.5	4.4	5.2	6.1	7.0	7.9	8.8	9.6	10.5	12.2	—	—	—	—
	30	0.8	1.2	1.6	2.0	2.4	2.8	3.2	4.0	4.7	5.5	6.3	7.1	7.9	8.7	9.5	11.0	—	—	—	—
	40	0.7	1.1	1.5	1.9	2.3	2.6	3.0	3.7	4.5	5.3	5.9	6.5	7.4	8.1	8.8	10.3	11.8	—	—	—
10	10	1.0	1.5	2.0	2.5	3.0	3.5	4.0	5.0	6.0	7.0	8.0	9.0	10.0	11.0	12.0	—	—	—	—	—
	14	0.9	1.3	1.7	2.1	2.6	3.0	3.4	4.3	5.1	6.0	6.9	7.8	8.6	9.5	10.4	12.0	—	—	—	—
	20	0.7	1.1	1.5	1.9	2.3	2.6	3.0	3.7	4.5	5.3	6.0	6.8	7.5	8.3	9.0	10.5	12.0	—	—	—
	30	0.7	1.0	1.3	1.7	2.0	2.3	2.7	3.3	4.0	4.7	5.3	6.0	6.6	7.3	8.0	9.4	10.6	—	—	—
	40	0.6	0.9	1.2	1.6	1.9	2.2	2.5	3.1	3.7	4.4	5.0	5.6	6.2	6.9	7.5	8.7	10.0	12.5	—	—
	60	0.6	0.9	1.2	1.5	1.7	2.0	2.3	2.9	3.5	4.1	4.7	5.3	5.9	6.5	7.1	8.2	9.4	11.7	—	—
12	12	0.8	1.2	1.7	2.1	2.5	2.9	3.3	4.2	5.0	5.8	6.7	7.5	8.4	9.2	10.0	11.7	—	—	—	—
	16	0.7	1.1	1.5	1.8	2.2	2.5	2.9	3.6	4.4	5.1	5.8	6.5	7.2	8.0	8.7	10.2	11.6	—	—	—
	24	0.6	0.9	1.2	1.6	1.9	2.2	2.5	3.1	3.7	4.4	5.0	5.6	6.2	6.9	7.5	8.7	10.0	12.5	—	—
	36	0.6	0.8	1.1	1.4	1.7	1.9	2.2	2.8	3.3	3.9	4.4	5.0	5.5	6.0	6.6	7.8	8.8	11.0	—	—
	50	0.5	0.8	1.0	1.3	1.5	1.8	2.1	2.6	3.1	3.6	4.1	4.6	5.1	5.6	6.2	7.2	8.2	10.2	—	—
	70	0.5	0.7	1.0	1.2	1.5	1.7	2.0	2.4	2.9	3.4	3.9	4.4	4.9	5.4	5.8	6.8	7.8	9.7	12.2	—
14	14	0.7	1.1	1.4	1.8	2.1	2.5	2.9	3.6	4.3	5.0	5.7	6.4	7.1	7.8	8.5	10.0	11.4	—	—	—
	20	0.6	0.9	1.2	1.5	1.8	2.1	2.4	3.0	3.6	4.2	4.9	5.5	6.1	6.7	7.3	8.6	9.8	12.3	—	—
	30	0.5	0.8	1.0	1.3	1.6	1.8	2.1	2.6	3.1	3.7	4.2	4.7	5.2	5.8	6.3	7.3	8.4	10.5	—	—
	42	0.5	0.7	1.0	1.2	1.4	1.7	1.9	2.4	2.9	3.3	3.8	4.3	4.7	5.2	5.7	6.7	7.6	9.5	11.9	—
	60	0.4	0.7	0.9	1.1	1.3	1.5	1.8	2.2	2.6	3.1	3.5	3.9	4.4	4.8	5.2	6.1	7.0	8.8	10.9	—
	90	0.4	0.6	0.8	1.0	1.2	1.4	1.6	2.0	2.5	2.9	3.3	3.7	4.1	4.5	5.0	5.8	6.6	8.3	10.3	12.4
17	17	0.6	0.9	1.2	1.5	1.8	2.1	2.3	2.9	3.5	4.1	4.7	5.3	5.9	6.5	7.0	8.2	9.4	11.7	—	—
	25	0.5	0.7	1.0	1.2	1.5	1.7	2.0	2.5	3.0	3.5	4.0	4.5	5.0	5.5	6.0	7.0	8.0	10.0	12.5	—
	35	0.4	0.7	0.9	1.1	1.3	1.5	1.7	2.2	2.6	3.1	3.5	3.9	4.4	4.8	5.2	6.1	7.0	8.7	10.9	—
	50	0.4	0.6	0.8	1.0	1.2	1.4	1.6	2.0	2.4	2.8	3.1	3.5	3.9	4.3	4.5	5.4	6.2	7.7	9.7	11.6
	80	0.4	0.5	0.7	0.9	1.1	1.2	1.4	1.8	2.1	2.5	2.9	3.3	3.6	4.0	4.3	5.1	5.8	7.2	9.0	10.9
	120	0.3	0.5	0.7	0.8	1.0	1.2	1.3	1.7	2.0	2.3	2.7	3.0	3.4	3.7	4.0	4.7	5.4	6.7	8.4	10.1
20	20	0.5	0.7	1.0	1.2	1.5	1.7	2.0	2.5	3.0	3.5	4.0	4.5	5.0	5.5	6.0	7.0	8.0	10.0	12.5	—
	30	0.4	0.6	0.8	1.0	1.2	1.5	1.7	2.1	2.5	2.9	3.3	3.7	4.1	4.5	4.9	5.8	6.6	8.2	10.3	12.4
	45	0.4	0.5	0.7	0.9	1.1	1.3	1.4	1.8	2.2	2.5	2.9	3.3	3.6	4.0	4.3	5.1	5.8	7.2	9.1	10.9
	60	0.3	0.5	0.7	0.8	1.0	1.2	1.3	1.7	2.0	2.3	2.7	3.0	3.4	3.7	4.0	4.7	5.4	6.7	8.4	10.1
	90	0.3	0.5	0.6	0.8	0.9	1.1	1.2	1.5	1.8	2.1	2.4	2.7	3.0	3.3	3.6	4.2	4.8	6.0	7.5	9.0
	150	0.3	0.4	0.6	0.7	0.8	1.0	1.1	1.4	1.7	2.0	2.3	2.6	2.9	3.2	3.4	4.0	4.6	5.7	7.2	8.6
24	24	0.4	0.6	0.8	1.0	1.2	1.5	1.7	2.1	2.5	2.9	3.3	3.7	4.1	4.5	5.0	5.8	6.7	8.2	10.3	12.4
	32	0.4	0.5	0.7	0.9	1.1	1.3	1.5	1.8	2.2	2.6	2.9	3.3	3.6	4.0	4.3	5.1	5.8	7.2	9.0	11.0
	50	0.3	0.5	0.6	0.8	0.9	1.1	1.2	1.5	1.8	2.2	2.5	2.8	3.1	3.4	3.7	4.4	5.0	6.2	7.8	9.4
	70	0.3	0.4	0.6	0.7	0.8	1.0	1.1	1.4	1.7	2.0	2.2	2.5	2.8	3.0	3.3	3.8	4.4	5.5	6.9	8.2
	100	0.3	0.4	0.5	0.6	0.8	0.9	1.0	1.3	1.6	1.8	2.1	2.4	2.6	2.9	3.1	3.7	4.2	5.2	6.5	7.9
	160	0.2	0.4	0.5	0.6	0.7	0.8	1.0	1.2	1.4	1.7	1.9	2.1	2.4	2.6	2.8	3.3	3.8	4.7	5.9	7.1

30	30	0.3	0.5	0.7	0.8	1.0	1.2	1.3	1.7	2.0	2.3	2.7	3.0	3.3	3.7	4.0	4.7	5.4	6.7	8.4	10.0
	45	0.3	0.4	0.6	0.7	0.8	1.0	1.1	1.4	1.7	1.9	2.2	2.5	2.7	3.0	3.3	3.8	4.4	5.5	6.9	8.2
	60	0.3	0.4	0.5	0.6	0.7	0.9	1.0	1.2	1.5	1.7	2.0	2.2	2.5	2.7	3.0	3.5	4.0	5.0	6.2	7.4
	90	0.2	0.3	0.4	0.6	0.7	0.8	0.9	1.1	1.3	1.6	1.8	2.0	2.2	2.5	2.7	3.1	3.6	4.5	5.6	6.7
	150	0.2	0.3	0.4	0.5	0.6	0.7	0.8	1.0	1.2	1.4	1.6	1.8	2.0	2.2	2.4	2.8	3.2	4.0	5.0	5.9
	200	0.2	0.3	0.4	0.5	0.6	0.7	0.8	1.0	1.1	1.3	1.5	1.7	1.9	2.0	2.2	2.6	3.0	3.7	4.7	5.6
36	36	0.3	0.4	0.6	0.7	0.8	1.0	1.1	1.4	1.7	1.9	2.2	2.5	2.8	3.0	3.3	3.9	4.4	5.5	6.9	8.3
	50	0.2	0.4	0.5	0.6	0.7	0.8	1.0	1.2	1.4	1.7	1.9	2.1	2.5	2.6	2.9	3.3	3.8	4.8	5.9	7.2
	75	0.2	0.3	0.4	0.5	0.6	0.7	0.8	1.0	1.2	1.4	1.6	1.8	2.0	2.3	2.5	2.9	3.3	4.1	5.1	6.1
	100	0.2	0.3	0.4	0.5	0.6	0.7	0.8	0.9	1.1	1.3	1.5	1.7	1.9	2.1	2.3	2.6	3.0	3.8	4.7	5.7
	150	0.2	0.3	0.3	0.4	0.5	0.6	0.7	0.9	1.0	1.2	1.4	1.6	1.7	1.9	2.1	2.4	2.8	3.5	4.3	5.2
	200	0.2	0.2	0.3	0.4	0.5	0.6	0.7	0.8	1.0	1.1	1.3	1.5	1.6	1.8	2.0	2.3	2.6	3.3	4.1	4.9
42	42	0.2	0.4	0.5	0.6	0.7	0.8	1.0	1.2	1.4	1.6	1.9	2.1	2.4	2.6	2.8	3.3	3.8	4.7	5.9	7.1
	60	0.2	0.3	0.4	0.5	0.6	0.7	0.8	1.0	1.2	1.4	1.6	1.8	2.0	2.2	2.4	2.8	3.2	4.0	5.0	6.0
	90	0.2	0.3	0.3	0.4	0.5	0.6	0.7	0.9	1.0	1.2	1.4	1.6	1.7	1.9	2.1	2.4	2.8	3.5	4.4	5.2
	140	0.2	0.2	0.3	0.4	0.5	0.5	0.6	0.8	0.9	1.1	1.2	1.4	1.5	1.7	1.9	2.2	2.5	3.1	3.9	4.6
	200	0.1	0.2	0.3	0.4	0.4	0.5	0.6	0.7	0.9	1.0	1.1	1.3	1.4	1.6	1.7	2.0	2.3	2.9	3.6	4.3
	300	0.1	0.2	0.3	0.3	0.4	0.5	0.5	0.7	0.8	0.9	1.1	1.3	1.4	1.5	1.7	1.9	2.2	2.8	3.5	4.2
50	50	0.2	0.3	0.4	0.5	0.6	0.7	0.8	1.0	1.2	1.4	1.6	1.8	2.0	2.2	2.4	2.8	3.2	4.0	5.0	6.0
	70	0.2	0.3	0.3	0.4	0.5	0.6	0.7	0.9	1.0	1.2	1.4	1.5	1.7	1.9	2.0	2.4	2.7	3.4	4.3	5.1
	100	0.1	0.2	0.3	0.4	0.4	0.5	0.6	0.7	0.9	1.0	1.2	1.3	1.5	1.6	1.8	2.1	2.4	3.0	3.7	4.5
	150	0.1	0.2	0.3	0.3	0.4	0.5	0.5	0.7	0.8	0.9	1.1	1.2	1.3	1.5	1.6	1.9	2.1	2.7	3.3	4.0
	300	0.1	0.2	0.2	0.3	0.3	0.4	0.5	0.6	0.7	0.8	0.9	1.0	1.1	1.3	1.4	1.6	1.9	2.3	2.9	3.5
60	60	0.2	0.2	0.3	0.4	0.5	0.6	0.7	0.8	1.0	1.2	1.3	1.5	1.7	1.8	2.0	2.3	2.7	3.3	4.2	5.0
	100	0.1	0.2	0.3	0.3	0.4	0.5	0.5	0.7	0.8	0.9	1.1	1.2	1.3	1.5	1.6	1.9	2.1	2.7	3.3	4.0
	150	0.1	0.2	0.2	0.3	0.3	0.4	0.5	0.6	0.7	0.8	0.9	1.0	1.2	1.3	1.4	1.6	1.9	2.3	2.9	3.5
	300	0.1	0.1	0.2	0.2	0.3	0.3	0.4	0.5	0.6	0.7	0.8	0.9	1.0	1.1	1.2	1.4	1.6	2.0	2.5	3.0
75	75	0.1	0.2	0.3	0.3	0.4	0.5	0.5	0.7	0.8	0.9	1.1	1.2	1.3	1.5	1.6	1.9	2.1	2.7	3.3	4.0
	120	0.1	0.2	0.2	0.3	0.3	0.4	0.4	0.5	0.6	0.8	0.9	1.0	1.1	1.2	1.3	1.5	1.7	2.2	2.7	3.3
	200	0.1	0.1	0.2	0.2	0.3	0.3	0.4	0.5	0.5	0.6	0.7	0.8	0.9	1.0	1.1	1.3	1.5	1.8	2.3	2.7
	300	0.1	0.1	0.2	0.2	0.2	0.3	0.3	0.4	0.5	0.6	0.7	0.7	0.8	0.9	1.0	1.2	1.3	1.7	2.1	2.5
100	100	0.1	0.1	0.2	0.2	0.3	0.3	0.4	0.5	0.6	0.7	0.8	0.9	1.0	1.1	1.2	1.4	1.6	2.0	2.5	3.0
	200	0.1	0.1	0.1	0.2	0.2	0.3	0.3	0.4	0.4	0.5	0.6	0.7	0.7	0.8	0.9	1.0	1.2	1.5	1.9	2.2
	300	0.1	0.1	0.1	0.2	0.2	0.2	0.3	0.3	0.4	0.5	0.5	0.6	0.7	0.7	0.8	0.9	1.1	1.3	1.7	2.0
150	150	0.1	0.1	0.1	0.2	0.2	0.2	0.3	0.3	0.4	0.5	0.5	0.6	0.7	0.7	0.8	0.9	1.1	1.3	1.7	2.0
	300	—	0.1	0.1	0.1	0.1	0.2	0.2	0.2	0.3	0.3	0.4	0.5	0.5	0.6	0.6	0.7	0.8	1.0	1.2	1.5
200	200	—	0.1	0.1	0.1	0.1	0.2	0.2	0.2	0.3	0.3	0.4	0.5	0.5	0.6	0.6	0.7	0.8	1.0	1.2	1.5
	300	—	0.1	0.1	0.1	0.1	0.1	0.2	0.2	0.2	0.3	0.3	0.4	0.4	0.5	0.5	0.6	0.7	0.8	1.0	1.2
300	300	—	—	0.1	0.1	0.1	0.1	0.1	0.2	0.2	0.2	0.3	0.3	0.3	0.4	0.4	0.5	0.5	0.6	0.7	0.8
500	500	—	—	—	—	0.1	0.1	0.1	0.1	0.1	0.1	0.2	0.2	0.2	0.2	0.2	0.3	0.3	0.4	0.5	0.6

Source. Courtesy of the IES of North America.

Table 20.9 ***Percent Effective Ceiling or Floor Cavity Reflectance for Various Reflectance Combinations***

Percent Ceiling or Floor Reflectance	90				80				70			50			30				10		
Percent Wall Reflectance	*90*	*70*	*50*	*30*	*80*	*70*	*50*	*30*	*70*	*50*	*30*	*70*	*50*	*30*	*65*	*50*	*30*	*10*	*50*	*30*	*10*
Ceiling or Floor Cavity Ratio																					
0	90	90	90	90	80	80	80	80	70	70	70	50	50	50	30	30	30	30	10	10	10
0.1	90	89	88	87	79	79	78	78	69	69	68	59	49	48	30	30	29	29	10	10	10
0.2	89	88	86	85	79	78	77	76	68	67	66	49	48	47	30	29	29	28	10	10	9
0.3	89	87	85	83	78	77	75	74	68	66	64	49	47	46	30	29	28	27	10	10	9
0.4	88	86	83	81	78	76	74	72	67	65	63	48	46	45	30	29	27	26	11	10	9
0.5	88	85	81	78	77	75	73	70	66	64	61	48	46	44	29	28	27	25	11	10	9
0.6	88	84	80	76	77	75	71	68	65	62	59	47	45	43	29	28	26	25	11	10	9
0.7	88	83	78	74	76	74	70	66	65	61	58	47	44	42	29	28	26	24	11	10	8
0.8	87	82	77	73	75	73	69	65	64	60	56	47	43	41	29	27	25	23	11	10	8
0.9	87	81	76	71	75	72	68	63	63	59	55	46	43	40	29	27	25	22	11	9	8
1.0	86	80	74	69	74	71	66	61	63	58	53	46	42	39	29	27	24	22	11	9	8
1.1	86	79	73	67	74	71	65	60	62	57	52	46	41	38	29	26	24	21	11	9	8
1.2	86	78	72	65	73	70	64	58	61	56	50	45	41	37	29	26	23	20	12	9	7
1.3	85	78	70	64	73	69	63	57	61	55	49	45	40	36	29	26	23	20	12	9	7
1.4	85	77	69	62	72	68	62	55	60	54	48	45	40	35	28	26	22	19	12	9	7
1.5	85	76	68	61	72	68	61	54	59	53	47	44	39	34	28	25	22	18	12	9	7
1.6	85	75	66	59	71	67	60	53	59	52	45	44	39	33	28	25	21	18	12	9	7
1.7	84	74	65	58	71	66	59	52	58	51	44	44	38	32	28	25	21	17	12	9	7
1.8	84	73	64	56	70	65	58	50	57	50	43	43	37	32	28	25	21	17	12	9	6
1.9	84	73	63	55	70	65	57	49	57	49	42	43	37	31	28	25	20	16	12	9	6
2.0	83	72	62	53	69	64	56	48	56	48	41	43	37	30	28	24	20	16	12	9	6

Ceiling or Floor Cavity Ratio																					
2.1	83	71	61	52	69	63	55	47	56	47	40	43	36	29	28	24	20	16	13	9	6
2.2	83	70	60	51	68	63	54	45	55	46	39	42	36	29	28	24	19	15	13	9	6
2.3	83	69	59	50	68	62	53	44	54	46	38	42	35	28	28	24	19	15	13	9	6
2.4	82	68	58	48	67	61	52	43	54	45	37	42	35	27	28	24	19	14	13	9	6
2.5	82	68	57	47	67	61	51	42	53	44	36	41	34	27	27	23	18	14	13	9	6
2.6	82	67	56	46	66	60	50	41	53	43	35	41	34	26	27	23	18	13	13	9	5
2.7	82	66	55	45	66	60	49	40	52	43	34	41	33	26	27	23	18	13	13	9	5
2.8	81	66	54	44	66	59	48	39	52	42	33	41	33	25	27	23	18	13	13	9	5
2.9	81	65	53	43	65	58	48	38	51	41	33	40	33	25	27	23	17	12	13	9	5
3.0	81	64	52	42	65	58	47	38	51	40	32	40	32	24	27	22	17	12	13	8	5
3.1	80	64	51	41	64	57	46	37	50	40	31	40	32	24	27	22	17	12	13	8	5
3.2	80	63	50	40	64	57	45	36	50	39	30	40	31	23	27	22	16	11	13	8	5
3.3	80	62	49	39	64	56	44	35	49	39	30	39	31	23	27	22	16	11	13	8	5
3.4	80	62	48	38	63	56	44	34	49	38	29	39	31	22	27	22	16	11	13	8	5
3.5	79	61	48	37	63	55	43	33	48	38	29	39	30	22	26	22	16	11	13	8	5
3.6	79	60	47	36	62	54	42	33	48	37	28	39	30	21	26	21	15	10	13	8	5
3.7	79	60	46	35	62	54	42	32	48	37	27	38	30	21	26	21	15	10	13	8	4
3.8	79	59	45	35	62	53	41	31	47	36	27	38	29	21	26	21	15	10	13	8	4
3.9	78	59	45	34	61	53	40	30	47	36	26	38	29	20	26	21	15	10	13	8	4
4.0	78	58	44	33	61	52	40	30	46	35	26	38	29	20	26	21	15	9	13	8	4
4.1	78	57	43	32	60	52	39	29	46	35	25	37	28	20	26	21	14	9	13	8	4
4.2	78	57	43	32	60	51	39	29	46	34	25	37	28	19	26	20	14	9	13	8	4
4.3	78	56	42	31	60	51	38	28	45	34	25	37	28	19	26	20	14	9	13	8	4
4.4	77	56	41	30	59	51	38	28	45	34	24	37	27	19	26	20	14	8	13	8	4
4.5	77	55	41	30	59	50	37	27	45	33	24	37	27	19	25	20	14	8	14	8	4
4.6	77	55	40	29	59	50	37	26	44	33	24	36	27	18	25	20	14	8	14	8	4
4.7	77	54	40	29	58	49	36	26	44	33	23	36	26	18	25	20	13	8	14	8	4
4.8	76	54	39	28	58	49	36	25	44	32	23	36	26	18	25	19	13	8	14	8	4
4.9	76	53	38	28	58	49	35	25	44	32	23	36	26	18	25	19	13	7	14	8	4
5.0	76	53	38	27	57	48	35	25	43	32	22	36	26	17	25	19	13	7	14	8	4

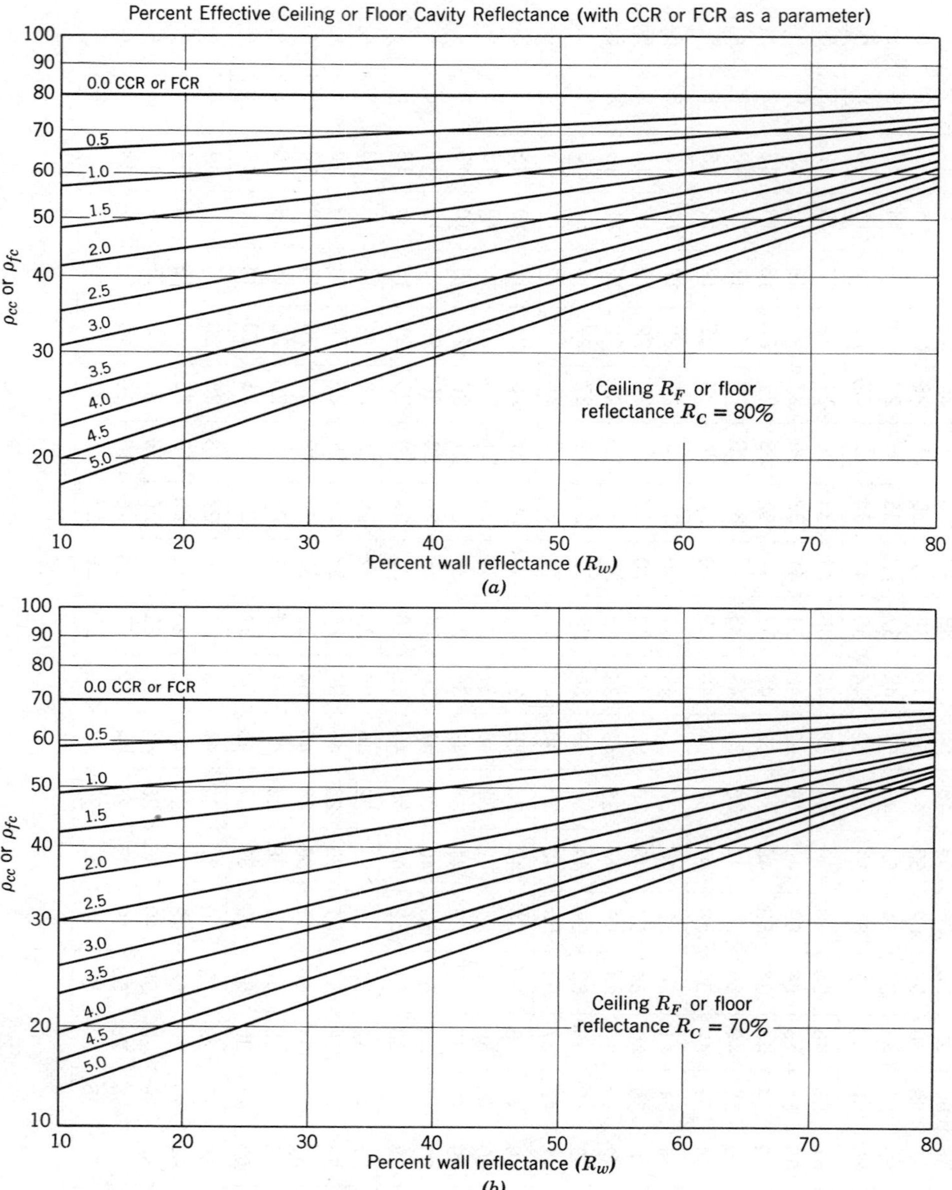

Fig. 20.34 *Graphic plot of Table 20.9.*

in. AFF; although usually at 30 in., the large number of business machines in use raises the elevation here.

Step 1. $h_{cc} = 0$; $h_{rc} = 5$; and $h_{fc} = 3$.

Step 2. CCR = 0; RCR = 0.666; and FCR = 0.4.

Step 3. From Table 20.9, $\rho_{cc} = 80\%$ (recessed fixture; use ceiling reflectance) and $\rho_W = 30\%$.

Step 4. From Table 20.9, $\rho_{fc} = 27\%$ and $\rho_W = 30\%$.

Step 5. From Table 20.7, an interpolation is again necessary:

	CU
ρ_{cc}	*80%*
RCR = 0	0.73
0.666	*0.70*
RCR = 1.0	0.64
CU for 20% floor cavity reflectance = 0.70	

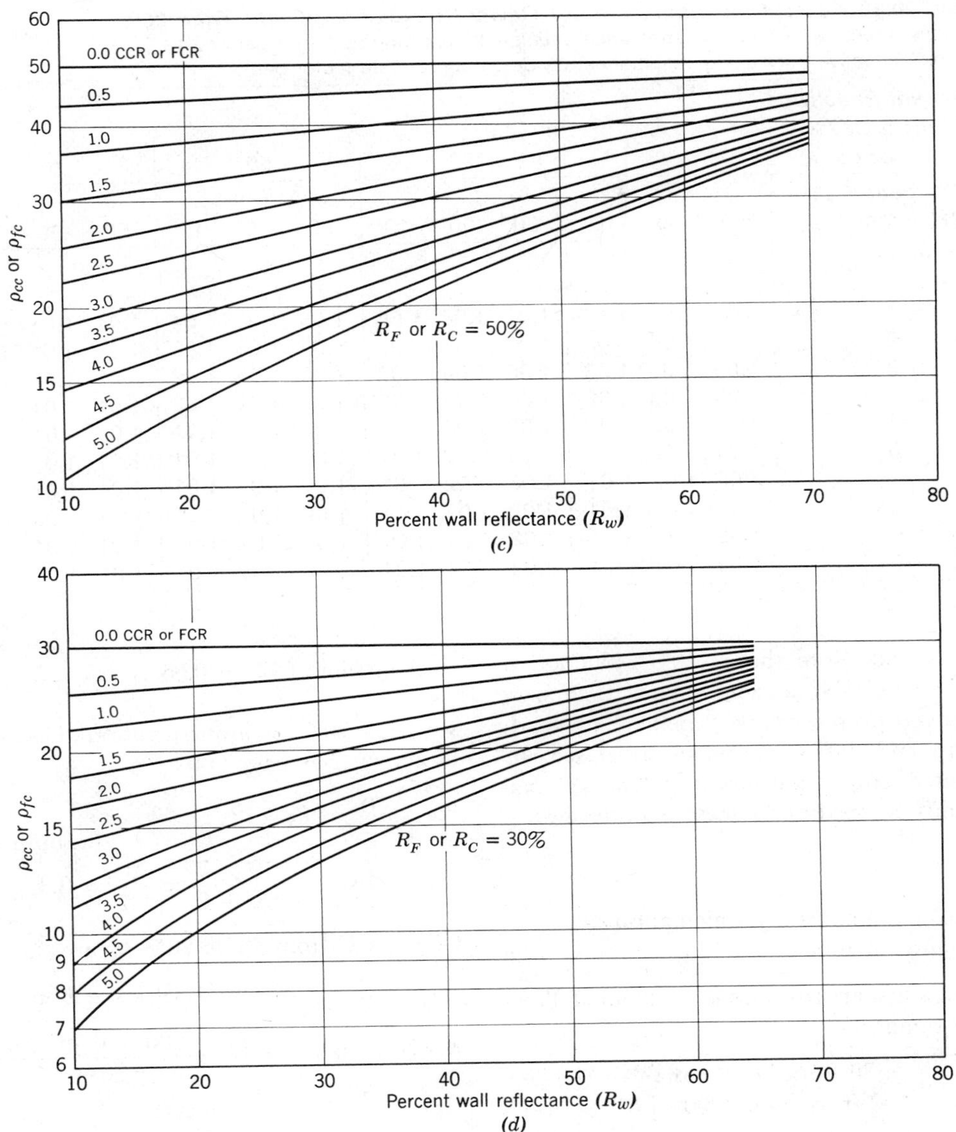

(c)

(d)

Since our ρ_{fc} is 20%, we obtain a correction factor of 1.09 by extrapolation from Table 20.10.

Therefore, final CU is

$$\text{CU} = 0.70\ (1.09) = 0.763$$

Step 6. Light loss factors, per Figure 20.33:

1.0	0.92
1.0	0.85
1.0	0.95
0.9	0.78

$$\text{LLF} = 0.9 \times 0.92 \times 0.85 \times 0.95 \times 0.78 = 0.51$$

Step 7:

$$\text{Area/luminaire} = \frac{4 \times 3200 \times 0.763 \times 0.51}{50} = 100 \text{ sq ft}$$

that is, fixtures can be mounted in a 12 × 8 or 10 × 10 grid and still meet the maximum spacing requirement of 1.4 side to side and 1.2

Table 20.10 *Factors for Effective Floor Cavity Reflectances Other Than 20%*
For 30% effective floor cavity reflectance, *multiply* by appropriate factor below
For 10% effective floor cavity reflectance, *divide* by appropriate factor below

Percent Effective Ceiling Cavity Reflectance, ρ_{cc}	*80*			*70*			*50*			*10*		
Percent Wall Reflectance, ρ_w	50	30	10	50	30	10	50	30	10	50	30	10
Room Cavity Ratio												
1	1.08	1.08	1.07	1.07	1.06	1.06	1.05	1.04	1.04	1.01	1.01	1.01
2	1.07	1.06	1.05	1.06	1.05	1.04	1.04	1.03	1.03	1.01	1.01	1.01
3	1.05	1.04	1.03	1.05	1.04	1.03	1.03	1.03	1.02	1.01	1.01	1.01
4	1.05	1.03	1.02	1.04	1.03	1.02	1.03	1.02	1.02	1.01	1.01	1.00
5	1.04	1.03	1.02	1.03	1.02	1.02	1.02	1.02	1.01	1.01	1.01	1.00
6	1.03	1.02	1.01	1.03	1.02	1.01	1.02	1.02	1.01	1.01	1.01	1.00
7	1.03	1.02	1.01	1.03	1.02	1.01	1.02	1.01	1.01	1.01	1.01	1.00
8	1.03	1.02	1.01	1.02	1.02	1.01	1.02	1.01	1.01	1.01	1.01	1.00
9	1.02	1.01	1.01	1.02	1.01	1.01	1.02	1.01	1.01	1.01	1.01	1.00
10	1.02	1.01	1.01	1.02	1.01	1.01	1.02	1.01	1.01	1.01	1.01	1.00

end to end. Note that a 20% reduction in fixtures could be accomplished if LLF were improved by the simple expedient of yearly fixture and diffuser cleaning. Enclosed ungasketed fixtures are dirt traps and are particularly in need of frequent maintenance.

20.33 Zonal Cavity Calculation by Approximation

For a first approximation, make the following assumptions:

a. $\rho_{cc} = 80$ for a large room $\simeq$ 100 ft sq ($\sim$ 10,000 sq ft)

   $\rho_{cc} = 70$ for a medium room $\simeq$ 30 ft sq ($\sim$ 1000 sq ft)

   $\rho_{cc} = 60$ for a small room $\simeq$ 12 ft sq ($\sim$ 150 sq ft)

b. $\rho_{fc} = 20$

c. Assume all rooms are square. To do this for a rectangle, take 1/3 the difference in dimensions and add to the smaller dimension. For square rooms

$$\text{RCR} = \frac{10h_{rc}}{W}$$

d. Assume LLF = 0.65.

EXAMPLE 20.3 (by approximation). Classroom as above. (See page 909).

$\rho_{cc} = 70 \qquad \rho_w = 50 \qquad \rho_{fc} = 20$ (assumptions)

$$\text{RCR} = \frac{10 \times 7}{20 + 2} = \frac{70}{22} = 3.3$$

Obtain CU from Table 20.7, fixture 23.

CU = 0.56 by visual inspection

$$\text{Area/luminaire} = \frac{2 \times 3200 \times 0.56 \times 0.65}{80} = 29.1 \text{ sq ft}$$

Thus, the result is within 2% of the accurate calculation. Let us also check Example 20.4.

EXAMPLE 20.4 (by approximation).

$\rho_{cc} = 80 \qquad \rho_w = 30 \qquad \rho_{fc} = 20$

$$\text{RCR} = \frac{10 \times 5}{60 + 40/3} = \frac{50}{73} = \frac{2}{3} = 0.66$$

CU = 0.70 by inspection
LLF = 0.65

$$\text{Area/luminaire} = \frac{4 \times 3200 \times 0.7 \times 0.65}{50} = 116 \text{ sq ft}$$

GENERAL INFORMATION

Project identification: BRIGHT CHILD ELEMENTARY SCHOOL
(Give name of area and/or building and room number)

Average maintained illumination for design: (1) footcandles

Luminaire data:

Manufacturer: ANY GOOD

Catalog number: 000

Lamp Data:

Type and color: F40T12 CW

Number per luminaire: 2

Total lumens per luminaire: 6400

SELECTION OF COEFFICIENT OF UTILIZATION

Step 1: Fill in sketch at right.

$\rho = 80\%$ (ceiling); $\rho = 80\%$; $\rho = 50\%$ (walls); Work plane; $\rho = 50\%$; $\rho = 20\%$ (floor)

$h_{CC} = 2.0$; $h_{RC} = 7.5$; $h_{FC} = 2.5$; L = 25; W = 20

Step 2: Determine Cavity Ratios from Fig. 20.34, or by formulas.

Room Cavity Ratio, RCR = 3.375

Ceiling Cavity Ratio, CCR = 0.9

Floor Cavity Ratio, FCR = 1.125

Step 3: Obtain effective ceiling cavity reflectance (ρ_{CC}) from Table 20.9 ρ_{CC} = 68%

Step 4: Obtain effective floor cavity reflectance (ρ_{FC}) from Table 20.9 ρ_{FC} = 19%

Step 5: Obtain coefficient of utilization (CU) from manufacturer's data. CU = 0.545

SELECTION OF LIGHT LOSS FACTORS

Unrecoverable		Recoverable	
Luminaire ambient temperature	1.0	Room surface dirt depreciation	0.9
Voltage to luminaire	1.0	Lamp lumen depreciation	0.9
Ballast factor	1.0	Lamp burnouts factor	1.0
Luminaire surface depreciation	0.9	Luminaire dirt depreciation LDD	0.9

Total light loss factor, LLF (product of individual factors above): 0.656

CALCULATIONS

(Average Maintained Illumination Level)

$$\text{Number of Luminaires} = \frac{\text{(Footcandles)} \times \text{(Area in square feet)}}{\text{(Lumens per luminaire)} \times \text{(CU)} \times \text{(LLF)}}$$

$$= \frac{30 \times 20 \times 25}{6400 \times 0.545 \times 0.656} = 17.48$$

$$\text{Footcandles} = \frac{\text{(Number of luminaires)} \times \text{(Lumens per luminaire)} \times \text{(CU)} \times \text{(LLF)}}{\text{(Area in square feet)}}$$

= See text =

Calculated by: B. STEIN Date: MAR. '78

(1) 70 ESI, 80 raw

Fig. 20.35 Zonal Cavity Calculation Form for Example 20.1.

Noting the category of maintenance V, we would arbitrarily decrease this by 10% making it 105 sq ft or within 5% of the accurate calculation. We thus see that these simple approximations give answers sufficiently accurate for most uses and are therefore recommended.

In conclusion, then, with respect to zonal cavity calculations, we can make the following statements:

1. For everyday calculations of rectangular rooms, with assumed reflectances, use the assumptions listed as a first approximation in (a) to (d) above.
2. For rooms where a high degree of accu-

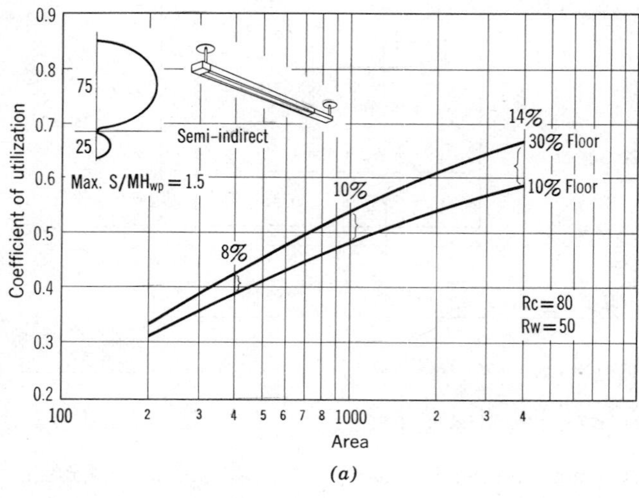

(a)

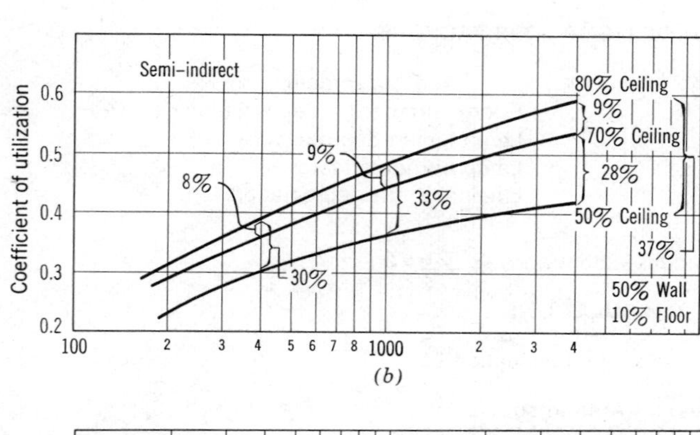

(b)

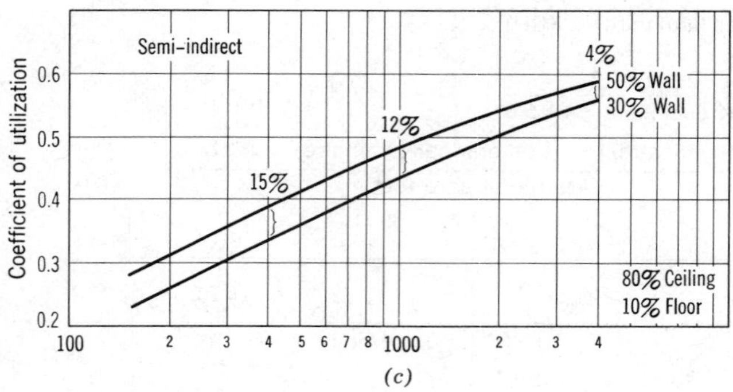

(c)

Fig. 20.36 *Effect of surface reflectances—semi-indirect distribution. As expected, since the ceiling becomes the light source, its reflectance has the most pronounced effect* (b). *With this particular unit having a 25% downward component, floor finish also has an appreciable effect* (a). *Wall reflectance has appreciable effect* (c), *in small rooms where wall surface is large.*

racy is desired and actual reflectances are known, use the long method, with visual interpolation.

3. For unusually shaped rooms or rooms with special conditions such as coffered ceilings, mixed-material walls, and partial height partitions, use the long method. This should not be necessary for more than 5% of calculations.

20.34 Effect of Cavity Reflectances

It is obvious to every reader at this point in our study that the reflectances of the various room cavities have a marked effect on the CU, because of light reflections within the room. In order to demonstrate this graphically, we have plotted in Figures 20.36, 20.37, and 20.38 the effect of varying cavity reflectances on the three principal types of fixture distribution: indirect, direct-indirect, and direct-spread. Note that, as expected, the ceiling cavity reflectance has the most pronounced effect with indirect fixtures, and the floor reflectance with direct units. Since lighting costs amount to 3 to 5% of *total construction* cost for many types of buildings such as offices, a 20% differential in lighting units can amount to as much as 1% of the total cost of the facility. This amount would not only pay for the increased cost of higher-reflectance finishes and materials but also would reduce both initial and operating cost. These data clearly indicate the necessity for the lighting designer to have considerable influence on the choice of room materials and finishes, a situation which unfortunately does not usually occur.

20.35 Application of Computer Calculation

As pointed out in previous sections the use of a computer in lighting calculation is a necessity rather than a convenience if accurate results are to be produced. Thus in the classroom of Example 20.4 experience led us to the fixture and layout choice and the ratio of raw to ESI footcandles assumed (80/70) for calculation. Once the calculation is done, only a computer analysis will give us the VCP and ESI values for selected viewing locations and directions. In addition, the computer gives the designer a degree of flexibility he would not otherwise have in that:

a. It performs the calculations accurately and rapidly.
b. It frees the designer for other, less routine work.
c. It gives the designer the ability to change parameters repeatedly, without making the amount of calculation excessive as would be the case in hand calculation.

It is this last characteristic that gives the designer greatest flexibility. The ability to run a series of calculations for a pendant fixture with pendant lengths varying from 12 to 36 in. in 3-in. intervals, or to change paint colors and reflectance on various surfaces and to note the effect, gives the designer a freedom urgently needed. In addition the computer can consider related items such as first costs, energy use, operating costs, and impact on HVAC, items whose complexity because of interrelations put them well beyond the pencil-and-hand-calculator's ability.

20.36 Modular Design

An increasingly large number of buildings are being designed on a modular system resulting in a need for flexible lighting to fit the module utilized. In such buildings, having established the general lighting scheme and the fixture involved, it is convenient to draw a family of curves for the fixture chosen, thereby facilitating the utilization of the modular unit in various spaces. "Area" may readily be replaced with multiples of modular areas, as shown in Figure 20.39.

20.37 Calculating Illumination at a Point

The lumen method calculation technique discussed earlier is appropriate for spaces in

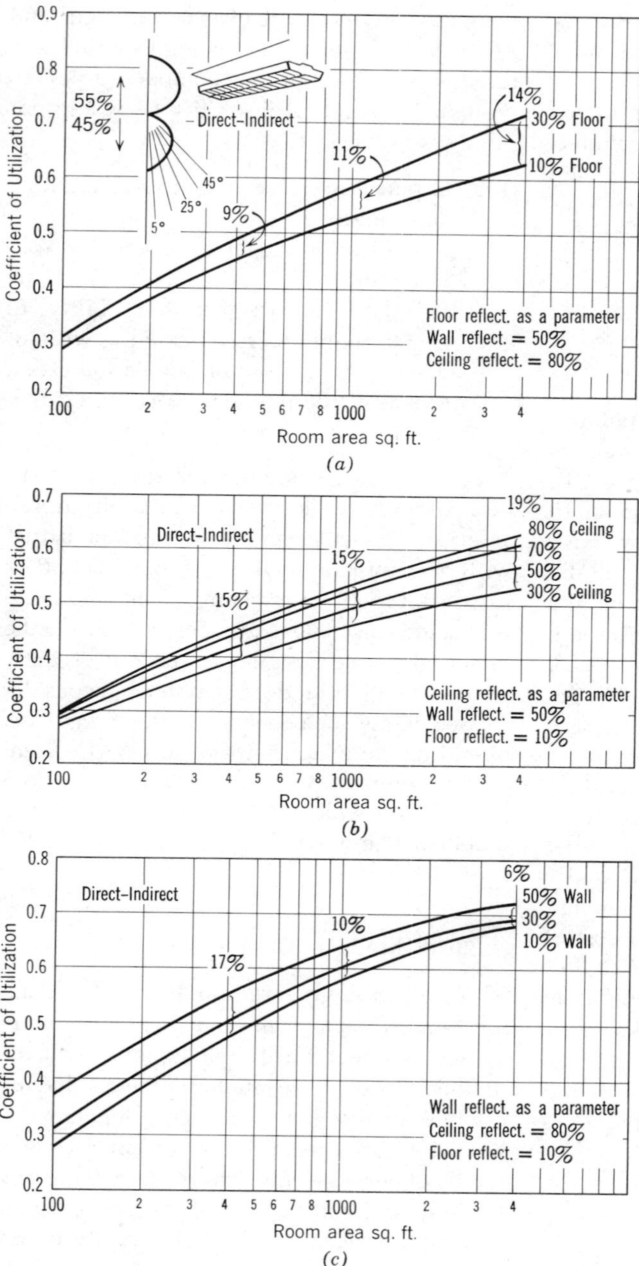

Fig. 20.37 *Effects of surface reflectances for direct-indirect distribution. With this distribution the effect of ceiling and floor are most pronounced, with a wall effect only in small rooms.*

which illumination is essentially uniform throughout. Where illumination is intentionally nonuniform, that is, where we purposefully light a restricted area from one or more discrete sources, other methods are required to enable us to calculate the illumination at a point within the restricted area. It is well to note at the outset that all the methods to be discussed below do just that—calculate illumination at *a point.* The answer to the constant query, "How much light will I have on my desk from a luminaire at this location?", is arrived at by taking several points on the desk and calculating illumination at each one.

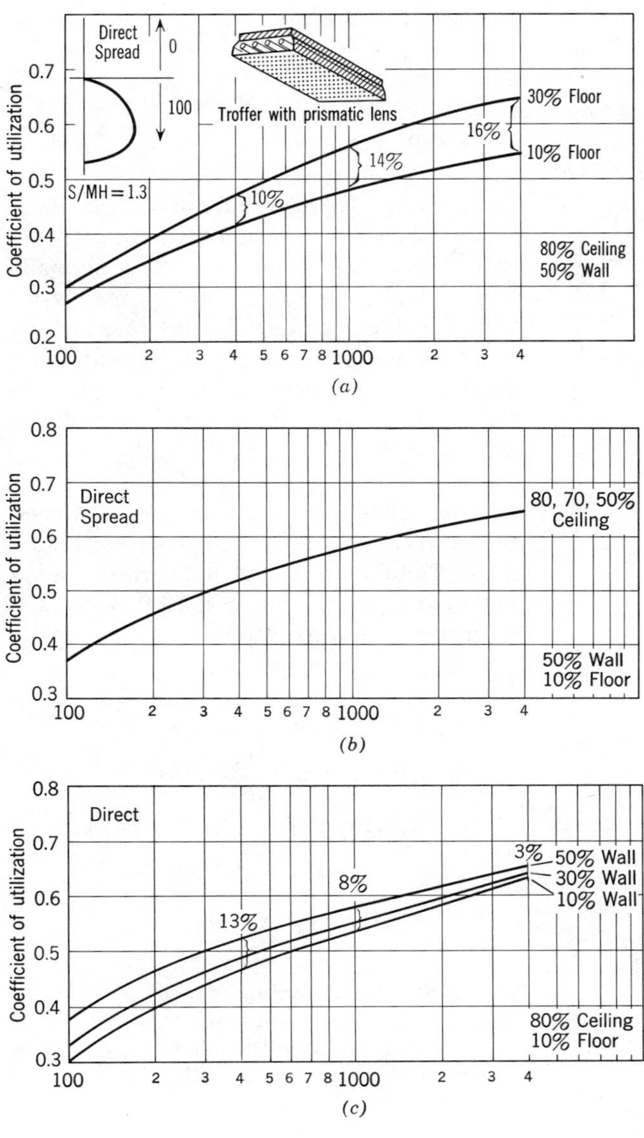

Fig. 20.38 Effect of surface reflectances—direct (spread) distribution. Floor finish is most important (a), *with wall only important in small rooms* (c). *There being no ceiling component, and with a 10% floor reflectance* (b), *ceiling finish has no effect on room illumination.*

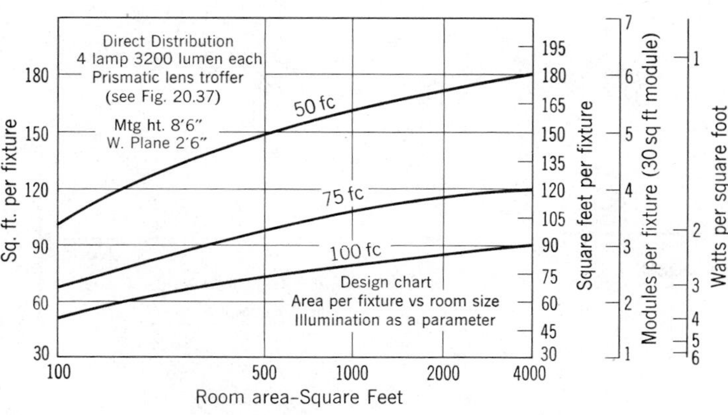

Fig. 20.39 Luminaire design chart. For frequently used fixtures, this type chart gives a rapid design figure for various size rooms. As seen from the ordinates, the figures can be translated into number of modules and watts per square foot.

Shortcuts can be made for symmetry and so on. The reader should also understand that most of these techniques are laborious and are generally performed by the consulting engineer or lighting specialist rather than the architect. They are presented here as source material for the technically oriented reader.

20.38 Point Sources

The basis of point source calculations is the inverse square law developed in Section 18.8:

$$fc = \frac{cp}{D^2} \tag{18.7}$$

where *fc*, *cp*, and *D* are footcandle illumination, candlepower intensity, and distance, respectively. Refer to Figure 20.40. The horizontal illumination at a point *P* as shown in Figure 20.40*a* is

$$\text{Horiz. } fc_P = \frac{cp}{D^2} \cos\theta \tag{20.14}$$

and the vertical illumination at that same point is

$$\text{Vert } fc_P = \frac{cp}{D^2} \sin\theta \tag{20.15}$$

however, since

$$\cos\theta = \frac{H}{D} \quad \text{and} \quad \sin\theta = \frac{R}{D}$$

we have then at point *P*

$$\text{Horizontal illumination} = \frac{cp}{H^2}\cos^3\theta \tag{20.16}$$

and

$$\text{Vertical illumination} = \frac{cp}{R^2}\sin^3\theta \tag{20.17}$$

Since the candlepower intensity in the direction of point *P* is normally taken from a candlepower distribution curve, θ is normally known, and these expressions are readily usable. Very few commercial sources are actually point sources. However, *when the maximum dimension of the source is less than five times the distance to point P*, the equations will give satisfactory results.

EXAMPLE 20.5. Referring to Figure 20.40 and the candlepower distribution curve of Figure 20.41, find the horizontal and vertical illumination at a point *P*, which is 10 ft below and 12 ft away from the source.

SOLUTION

$$H = 10 \text{ ft} \qquad R = 12 \text{ ft}$$

$$\theta = \tan^{-1}\frac{12}{10} = 50°$$

$$\sin\theta = 0.766 \qquad \cos\theta = 0.643$$

$$cp \text{ at } 50° = 6600$$

Fig. 20.40 Relationships for point calculations when source can be considered a point source, that is, when inverse square law applies. Source major dimension must not exceed 0.2D.

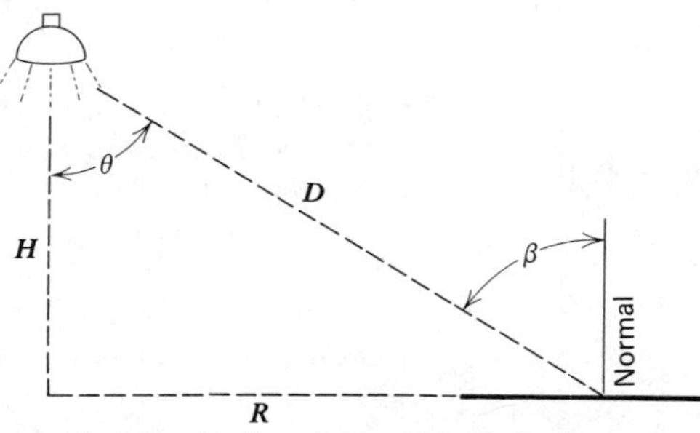

Footcandles (on the horizontal plane)

$$= \frac{\text{Candlepower} \times \cos\theta}{D^2} = \frac{CP}{H^2}\cos^3\theta$$

(*a*)

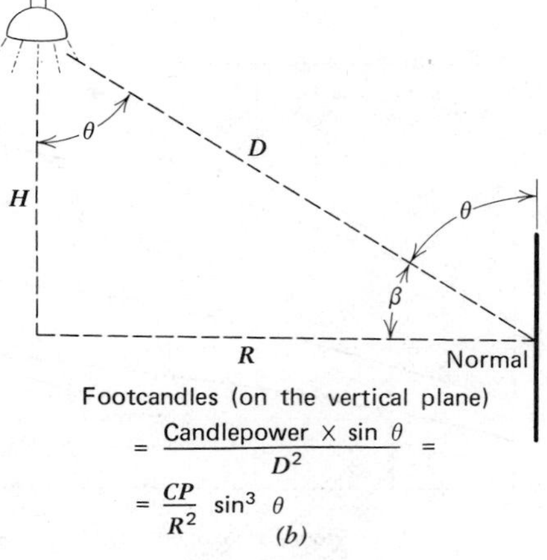

Footcandles (on the vertical plane)

$$= \frac{\text{Candlepower} \times \sin\theta}{D^2} = \frac{CP}{R^2}\sin^3\theta$$

(*b*)

$$\text{Horizontal illumination} = \frac{6600}{10^2}(0.643)^3$$

$$= 17.4 \text{ fc}$$

$$\text{Vertical illumination} = \frac{6600}{12^2}(0.766)^3$$

$$= 20.8 \text{ fc}$$

To assist in performing these calculations, tables have been prepared that give the illumination at a point P from a point source in terms of H and R, and to a 100-cp base. One such table for *horizontal* illumination is given in Table 20.11. (The table may be used for vertical illumination by reversing the headings). Entering the table at horizontal distance 12 ft and vertical distance 10 ft, we find 50° and 0.263 fc per 100 cp. Therefore,

$$\text{Horizontal illumination} = \frac{6600}{100}(0.263)$$

$$= 17.4$$

which checks. Using the table in reverse, that is, entering the table at 10 ft/12 ft, we read 0.315 or

$$\text{Vertical illumination} = 6.6(0.315) = 2.08$$

which also checks. As a further convenience Table 20.12 is included, which gives actual illumination figures for two of the most common types of reflector lamps. These data can be calculated from curves similar to those given in Figure 19.22, page 814, and are based on the output data of that figure.

20.39 Line Sources

When the maximum source dimension exceeds 20% of the distance to the point illuminated, it can no longer be considered a point source. A source that is long and narrow such as a fluorescent tube can be considered as a line source. For a linear source where length is much greater than width the illumination varies *inversely with distance* as opposed to inversely with the square of the distance for a point source, that is,

$$fc = \text{constant} \times \frac{cp}{D} \tag{20.18}$$

Thus if illumination at point P_1 is known, that at P_2 can be calculated as

$$fc_2 = fc_1 \times \frac{D_1}{D_2} \tag{20.19}$$

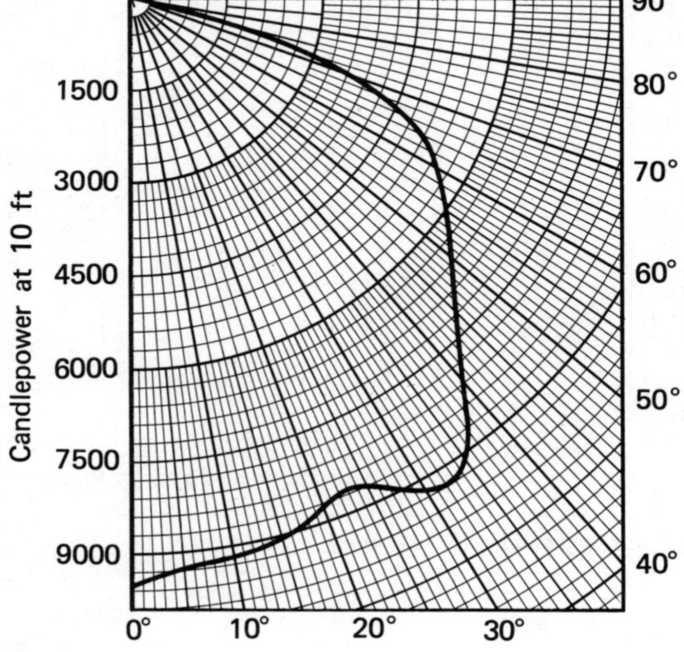

Fig. 20.41 *Typical candlepower distribution plot for a point source.*

Table 20.11 *Point-By-Point Footcandle Calculation Table*

Top figures—Angle between direction of light and vertical axis.

Bottom figures—Footcandles on the horizontal plane for each 100 cp of the source in that direction.

Height of Light Source Above Surface—Feet	Horizontal Distance from Axis of Light Source—Feet																										
	0	1	2	3	4	5	6	7	8	9	10	11	12	13	14	15	16	18	20	22	24	26	28	30	35	40	50
2	0° 25.000	27° 17.850	45° 8.850	56° 4.275	63° 2.245	68° 1.298	71° 0.802	74° 0.528	76° 0.355	78° 0.255	79° 0.190	80° 0.142	81° 0.113	81° 0.090	82° 0.070	82° 0.058	83° 0.048	84° 0.038	84° 0.025	85° 0.020	85° 0.015	86° 0.013	86° 0.008	86° 0.007	87° 0.004	87° 0.000	87° 0.000
3	0° 11.110	18° 9.500	34° 6.400	45° 3.933	53° 2.400	59° 1.522	63° 1.000	67° 0.680	69° 0.477	72° 0.356	73° 0.264	75° 0.205	76° 0.161	77° 0.126	78° 0.100	79° 0.084	80° 0.070	81° 0.050	81° 0.036	82° 0.027	83° 0.021	83° 0.016	84° 0.012	84° 0.011	85° 0.007	86° 0.004	87° 0.002
4	0°0′ 6.250	14° 5.707	27° 4.472	37° 3.200	45° 2.210	51° 1.524	56° 1.066	60° 0.764	63° 0.559	66° 0.419	68° 0.320	70° 0.249	72° 0.198	73° 0.159	74° 0.130	75° 0.107	76° 0.090	78° 0.064	79° 0.047	80° 0.037	81° 0.028	81° 0.022	82° 0.018	82° 0.015	84° 0.009	84° 0.006	86° 0.003
5	0°0′ 4.000	11° 3.771	22° 3.202	31° 2.522	39° 1.904	45° 1.414	50° 1.050	54° 0.785	58° 0.595	61° 0.458	63° 0.358	66° 0.283	67° 0.228	69° 0.185	70° 0.152	72° 0.126	73° 0.106	74° 0.077	76° 0.057	77° 0.044	78° 0.034	79° 0.027	80° 0.022	81° 0.017	82° 0.010	83° 0.008	84° 0.004
6	0°0′ 2.778	9° 2.673	18° 2.372	27° 1.987	34° 1.600	40° 1.260	45° 0.982	49° 0.766	53° 0.600	56° 0.474	59° 0.378	61° 0.305	63° 0.249	66° 0.205	67° 0.170	68° 0.142	69° 0.120	71° 0.088	73° 0.066	75° 0.051	76° 0.050	77° 0.032	78° 0.026	79° 0.021	80° 0.013	81° 0.009	83° 0.005
7	0°0′ 2.041	8° 1.980	16° 1.814	23° 1.585	30° 1.336	36° 1.100	41° 0.893	45° 0.722	49° 0.583	52° 0.473	55° 0.385	58° 0.316	60° 0.261	62° 0.218	63° 0.183	65° 0.154	66° 0.131	69° 0.097	71° 0.074	72° 0.057	74° 0.045	75° 0.036	76° 0.029	77° 0.024	79° 0.016	80° 0.010	82° 0.006
8	0°0′ 1.563	7° 1.527	14° 1.427	21° 1.283	27° 1.118	32° 0.953	37° 0.800	41° 0.666	45° 0.552	48° 0.458	51° 0.381	54° 0.318	56° 0.267	58° 0.225	60° 0.191	62° 0.163	63° 0.140	66° 0.105	68° 0.080	70° 0.063	72° 0.050	73° 0.040	74° 0.032	75° 0.026	77° 0.018	79° 0.012	81° 0.007
9	0°0′ 1.235	6° 1.212	13° 1.148	18° 1.054	24° 0.943	29° 0.825	34° 0.711	38° 0.607	42° 0.515	45° 0.437	48° 0.370	51° 0.314	53° 0.267	55° 0.228	57° 0.196	59° 0.168	61° 0.146	63° 0.110	66° 0.085	68° 0.067	69° 0.053	71° 0.043	72° 0.035	73° 0.029	76° 0.019	77° -0.013	80° 0.008
10	0°0′ 1.000	5°43′ 0.985	11° 0.943	17° 0.879	22° 0.801	27° 0.716	31° 0.631	35° 0.550	39° 0.476	42° 0.411	45° 0.354	48° 0.305	50° 0.263	52° 0.227	54° 0.196	56° 0.171	58° 0.149	61° 0.115	63° 0.089	66° 0.071	67° 0.057	69° 0.046	70° 0.038	72° 0.032	74° 0.021	76° 0.014	79° 0.008
11	0°0′ 0.826	5°12′ 0.816	10° 0.787	15° 0.742	20° 0.686	24° 0.623	29° 0.559	32° 0.496	36° 0.437	39° 0.383	42° 0.335	45° 0.292	48° 0.255	50° 0.223	52° 0.195	54° 0.171	56° 0.150	59° 0.117	61° 0.092	63° 0.074	65° 0.060	67° 0.049	69° 0.040	70° 0.034	73° 0.023	75° 0.015	78° 0.009
12	0°0′ 0.694	4°46′ 0.687	9° 0.668	14° 0.634	18° 0.593	23° 0.546	27° 0.497	30° 0.448	34° 0.400	37° 0.356	40° 0.315	43° 0.278	45° 0.246	47° 0.217	49° 0.191	51° 0.169	53° 0.150	56° 0.119	59° 0.094	61° 0.076	63° 0.063	65° 0.051	67° 0.043	68° 0.036	71° 0.024	73° 0.017	77° 0.009
13	0°0′ 0.592	4°24′ 0.587	9° 0.571	13° 0.547	17° 0.517	21° 0.481	25° 0.447	28° 0.404	32° 0.366	35° 0.329	38° 0.295	40° 0.263	43° 0.235	45° 0.209	47° 0.187	49° 0.166	51° 0.148	54° 0.119	57° 0.096	59° 0.078	62° 0.064	63° 0.053	65° 0.044	67° 0.037	70° 0.025	72° 0.017	76° 0.010
14	0°0′ 0.510	4°5′ 0.506	8° 0.495	12° 0.477	16° 0.454	20° 0.426	23° 0.396	27° 0.365	30° 0.334	33° 0.304	36° 0.275	38° 0.248	41° 0.223	43° 0.201	45° 0.180	47° 0.162	49° 0.146	52° 0.118	55° 0.096	58° 0.079	60° 0.065	62° 0.054	63° 0.046	65° 0.039	'68° 0.026	71° 0.018	75° 0.011
15	0°0′ 0.444	3°49′ 0.442	8° 0.433	11° 0.419	15° 0.401	18° 0.380	22° 0.356	25° 0.331	28° 0.305	31° 0.280	34° 0.256	36° 0.233	39° 0.212	41° 0.192	43° 0.174	45° 0.157	47° 0.142	50° 0.117	53° 0.096	56° 0.079	58° 0.066	60° 0.055	62° 0.047	63° 0.040	67° 0.027	69° 0.019	73° 0.011
16	0°0′ 0.391	3°35′ 0.388	7° 0.382	11° 0.371	14° 0.357	17° 0.339	21° 0.321	24° 0.300	27° 0.280	29° 0.259	32° 0.238	35° 0.219	37° 0.200	39° 0.183	41° 0.167	43° 0.152	45° 0.138	48° 0.115	51° 0.095	54° 0.080	56° 0.067	58° 0.056	60° 0.048	62° 0.041	66° 0.028	68° 0.020	72° 0.012
17	0°0′ 0.346	3°22′ 0.344	7° 0.339	10° 0.331	13° 0.319	16° 0.306	19° 0.290	22° 0.274	25° 0.256	28° 0.239	30° 0.222	33° 0.205	35° 0.189	37° 0.174	39° 0.159	41° 0.146	43° 0.134	47° 0.112	50° 0.094	52° 0.079	55° 0.069	57° 0.057	59° 0.048	60° 0.042	64° 0.029	67° 0.021	71° 0.012
18	0°0′ 0.309	3°11′ 0.307	6° 0.303	9° 0.297	13° 0.287	16° 0.276	18° 0.264	21° 0.250	24° 0.236	27° 0.221	29° 0.206	31° 0.192	34° 0.178	36° 0.165	38° 0.152	40° 0.140	42° 0.129	45° 0.109	48° 0.092	51° 0.079	53° 0.067	55° 0.057	57° 0.049	59° 0.042	63° 0.030	66° 0.021	70° 0.012

Height of Light Source Above Surface—Feet																											
19	0°0′ 0.277	3°1′ 0.276	6° 0.273	9° 0.267	12° 0.260	15° 0.251	18° 0.240	20° 0.229	23° 0.217	25° 0.205	28° 0.192	30° 0.180	32° 0.167	34° 0.156	36° 0.145	38° 0.134	40° 0.124	43° 0.106	46° 0.090	49° 0.077	52° 0.066	54° 0.057	56° 0.049	58° 0.042	62° 0.030	65° 0.022	69° 0.013
20	0°0′ 0.250	2°51′ 0.249	5°43′ 0.246	9° 0.242	11° 0.236	14° 0.228	17° 0.219	19° 0.210	22° 0.200	24° 0.190	27° 0.179	29° 0.168	31° 0.158	33° 0.147	35° 0.137	37° 0.128	39° 0.119	42° 0.103	45° 0.088	48° 0.076	50° 0.066	52° 0.057	54° 0.049	56° 0.043	60° 0.030	63° 0.022	68° 0.013
21	0°0′ 0.227	2°44′ 0.226	5°26′ 0.224	8° 0.220	11° 0.215	13° 0.210	16° 0.201	18° 0.194	21° 0.185	23° 0.176	25° 0.167	28° 0.158	30° 0.144	32° 0.139	34° 0.131	36° 0.122	37° 0.114	41° 0.099	44° 0.086	46° 0.075	49° 0.065	51° 0.056	53° 0.049	55° 0.043	59° 0.031	62° 0.023	67° 0.014
22	0°0′ 0.207	2°36′ 0.206	5°10′ 0.205	8° 0.201	10° 0.196	13° 0.192	15° 0.185	18° 0.179	20° 0.171	22° 0.164	25° 0.155	27° 0.148	29° 0.140	31° 0.132	33° 0.124	34° 0.114	36° 0.109	39° 0.096	42° 0.084	45° 0.073	47° 0.064	50° 0.056	52° 0.049	54° 0.043	58° 0.031	61° 0.023	66° 0.014
23	0°0′ 0.189	2°29′ 0.189	4°58′ 0.187	7° 0.184	10° 0.181	12° 0.176	15° 0.171	17° 0.165	19° 0.159	21° 0.153	24° 0.146	26° 0.139	28° 0.132	29° 0.125	31° 0.118	33° 0.111	35° 0.105	38° 0.092	41° 0.081	44° 0.071	46° 0.063	49° 0.055	51° 0.049	53° 0.043	57° 0.031	60° 0.023	65° 0.014
24	0°0′ 0.174	2°23′ 0.173	4°45′ 0.172	7° 0.170	10° 0.166	12° 0.163	14° 0.158	16° 0.154	18° 0.148	21° 0.143	23° 0.137	25° 0.130	27° 0.124	28° 0.118	30° 0.112	32° 0.106	34° 0.100	37° 0.089	40° 0.079	43° 0.070	45° 0.061	47° 0.054	49° 0.048	51° 0.042	56° 0.031	59° 0.024	64° 0.014
25	0°0′ 0.160	2°17′ 0.160	4°34′ 0.158	7° 0.157	9° 0.154	11° 0.151	14° 0.147	16° 0.143	18° 0.138	20° 0.133	22° 0.128	24° 0.123	26° 0.117	27° 0.112	29° 0.106	31° 0.101	33° 0.096	36° 0.086	39° 0.076	41° 0.068	44° 0.060	46° 0.053	48° 0.047	50° 0.042	55° 0.031	58° 0.024	63° 0.015
27	0°0′ 0.137	2°7′ 0.137	4°14′ 0.136	6° 0.135	8° 0.133	10° 0.130	12° 0.128	15° 0.124	17° 0.121	18° 0.117	20° 0.113	22° 0.109	24° 0.105	26° 0.100	27° 0.096	29° 0.092	31° 0.087	34° 0.079	37° 0.071	39° 0.064	42° 0.057	44° 0.051	46° 0.046	48° 0.041	52° 0.031	56° 0.024	62° 0.015
30	0°0′ 0.111	1°54′ 0.111	3°50′ 0.111	5°43′ 0.109	8° 0.108	9° 0.107	11° 0.105	13° 0.103	15° 0.100	17° 0.098	18° 0.095	20° 0.092	22° 0.089	23° 0.086	25° 0.083	27° 0.080	28° 0.077	31° 0.070	34° 0.064	36° 0.058	39° 0.053	41° 0.048	43° 0.043	45° 0.039	49° 0.031	53° 0.024	59° 0.015
33	0°0′ 0.092	1°44′ 0.092	3°28′ 0.091	5°12′ 0.091	7° 0.090	9° 0.089	10° 0.087	12° 0.086	14° 0.084	15° 0.082	17° 0.080	18° 0.078	20° 0.076	22° 0.074	23° 0.072	24° 0.069	26° 0.067	29° 0.062	31° 0.058	34° 0.053	36° 0.049	38° 0.045	40° 0.041	42° 0.037	47° 0.030	50° 0.024	57° 0.015
36	0°0′ 0.077	1°36′ 0.077	3°11′ 0.077	4°46′ 0.076	6° 0.076	8° 0.075	9° 0.074	11° 0.073	13° 0.072	14° 0.070	16° 0.069	17° 0.067	18° 0.066	20° 0.064	21° 0.062	23° 0.061	24° 0.059	27° 0.055	29° 0.052	31° 0.048	34° 0.044	36° 0.041	38° 0.038	40° 0.035	44° 0.029	48° 0.023	54° 0.015
40	0°0′ 0.063	1°26′ 0.062	2°52′ 0.062	4°17′ 0.062	5°43′ 0.062	7° 0.061	9° 0.060	10° 0.060	11° 0.059	13° 0.058	14° 0.057	15° 0.056	17° 0.055	18° 0.054	19° 0.053	21° 0.051	22° 0.050	24° 0.047	27° 0.045	29° 0.042	31° 0.039	33° 0.037	35° 0.034	37° 0.032	41° 0.027	45° 0.022	51° 0.015
45	0°0′ 0.049	1°16′ 0.049	2°33′ 0.049	3°49′ 0.049	5°5′ 0.049	6° 0.049	8° 0.048	9° 0.048	10° 0.047	11° 0.047	13° 0.046	14° 0.045	15° 0.045	16° 0.044	17° 0.043	18° 0.042	20° 0.041	22° 0.040	24° 0.038	26° 0.036	28° 0.034	30° 0.032	32° 0.030	34° 0.028	38° 0.025	42° 0.021	48° 0.013
50	0°0′ 0.040	1°9′ 0.040	2°17′ 0.040	3°26′ 0.040	4°34′ 0.040	5°43′ 0.039	7° 0.039	8° 0.039	9° 0.039	10° 0.038	11° 0.038	12° 0.037	14° 0.037	15° 0.036	16° 0.036	16° 0.035	18° 0.035	20° 0.033	22° 0.032	24° 0.031	26° 0.029	27° 0.028	29° 0.027	31° 0.025	35° 0.022	39° 0.019	45° 0.013
55	0°0′ 0.033	1°2′ 0.033	2°5′ 0.033	3°7′ 0.033	4°10′ 0.033	5°9′ 0.033	6° 0.032	7° 0.032	8° 0.032	9° 0.032	10° 0.032	11° 0.031	12° 0.031	13° 0.031	14° 0.030	15° 0.030	16° 0.029	18° 0.028	20° 0.027	22° 0.026	24° 0.025	25° 0.024	27° 0.023	29° 0.022	33° 0.020	36° 0.018	42° 0.013
60	0°0′ 0.028	0°57′ 0.028	1°55′ 0.028	2°52′ 0.028	3°50′ 0.028	4°16′ 0.027	5°43′ 0.027	7° 0.027	8° 0.027	9° 0.027	9° 0.027	10° 0.026	11° 0.026	12° 0.026	13° 0.026	14° 0.025	15° 0.025	17° 0.024	18° 0.024	20° 0.023	22° 0.022	23° 0.021	25° 0.021	27° 0.020	30° 0.018	34° 0.016	40° 0.013
70	0°0′ 0.020	0°9′ 0.020	1°38′ 0.020	2°34′ 0.020	3°16′ 0.020	4°5′ 0.020	4°54′ 0.020	5°43′ 0.020	7° 0.020	7° 0.020	8° 0.020	9° 0.020	10° 0.020	11° 0.019	11° 0.019	12° 0.019	13° 0.019	14° 0.019	16° 0.018	17° 0.018	19° 0.017	20° 0.017	22° 0.016	23° 0.016	27° 0.015	30° 0.013	36° 0.011

NOTE. Footcandles on the vertical surface—at a point that lies in a vertical plane that also includes the light source—may be determined by using the multiplying factor four when the table headings are reversed, that is, the height of the light source is read on the horizontal distance scale, etc.

Table 20.12 ***Illumination from 150-Watt, PAR-38 Lamps***

(a) Approximate footcandles from a single lamp at various mounting heights and lateral distances.

Height in Feet		Horizontal Distance from Lamp in Feet: 0		1		2		3		4		5		6	
		Spot	*Flood*	*Spot*	*Flood*	*Spot*	*Flood*	*Spot*	*Flood*	*Spot*	*Flood*	*Spot*	*Flood*	*Spot*	*Flood*
4	A	690	220	113	175	18	36	10	13	4	5	—	—	—	—
	B	690	220	627	200	490	156	355	113	248	79	262	83	259	82
	C	—	—	159	51	249	79	263	83	248	79	331	116	348	111
6	A	306	128	140	114	25	80	8	20	5	8	4	7	—	—
	B	306	128	291	121	251	109	218	91	181	75	138	57	110	46
	C	—	—	49	20	89	37	110	46	125	48	115	48	110	46
8	A	172	70	118	66	28	55	8	53	4	11	3	7	2	3
	B	172	70	168	68	157	64	143	58	122	50	105	43	88	36
	C	—	—	21	9	40	16	52	21	62	25	64	26	65	26
10	A	110	36	90	33	29	30	10	24	4	13	3	5	2	4
	B	110	36	108	35	103	34	97	32	88	29	78	26	70	23
	C	—	—	11	4	21	7	30	10	35	12	40	13	41	13
12	A	76	25	65	24	34	23	11	20	6	17	3	6	2	4
	B	76	25	75	25	73	24	69	23	65	21	60	20	54	18
	C	—	—	6	2	12	4	18	6	22	7	25	8	27	9
14	A	56	18	50	18	33	17	12	14	6	13	3	9	2	5
	B	56	18	55	18	54	17	52	17	50	16	47	15	42	14
	C	—	—	4	1	8	3	11	4	14	5	16	5	18	6
16	A	43	14	40	14	30	13	11	12	7	11	4	9	2	6
	B	43	14	43	14	42	14	41	13	39	13	38	12	36	12
	C	—	—	3	1	5	2	8	3	10	3	12	4	13	4

A—Footcandles on horizontal plane with lamp aimed straight down.
B—Footcandles on horizontal plane with lamp aimed at point of measurement.
C—Footcandles on vertical plane with lamp aimed at point of measurement.

Distances and planes of measurement

Illumination from 300-Watt, R-40 Lamps

(b) Approximate footcandles from a single lamp at various mounting heights and lateral distances.

Height in Feet		Horizontal Distance from Lamp in Feet: 0		1		2		3		4		5		6	
		Spot	*Flood*	*Spot*	*Flood*	*Spot*	*Flood*	*Spot*	*Flood*	*Spot*	*Flood*	*Spot*	*Flood*	*Spot*	*Flood*
4	A	900	185	140	148	44	62	26	29	17	19	11	13	—	—
	B	900	185	820	169	637	131	463	95	324	67	342	70	337	69
	C	—	—	207	42	324	66	342	70	324	67	432	89	531	109
6	A	400	83	175	75	36	50	20	28	12	16	10	11	7	8
	B	400	83	380	79	340	70	384	59	236	49	180	37	144	30
	C	—	—	64	13	116	24	144	30	150	31	150	31	144	30
8	A	220	47	140	43	34	37	16	24	10	16	9	11	6	7
	B	220	47	215	43	200	43	183	39	156	33	132	28	113	24
	C	—	—	27	6	50	11	66	14	79	17	82	16	84	18
10	A	144	30	110	29	49	25	14	20	10	14	7	10	5	7
	B	144	30	141	29	135	28	127	26	115	24	102	21	92	19
	C	—	—	14	3	27	6	39	8	46	10	52	11	53	11
12	A	100	21	83	20	45	19	15	17	9	13	6	10	4	7
	B	100	21	99	21	95	20	91	19	85	18	78	16	71	15
	C	—	—	8	2	16	3	23	5	29	6	33	7	36	7
14	A	73	15	64	15	39	15	18	13	8	11	5	9	4	7
	B	73	15	72	15	71	14	68	14	65	13	61	13	56	11
	C	—	—	5	1	10	2	15	3	18	4	22	4	24	5
16	A	56	11	52	11	36	10	17	9	8	8	5	7	4	5
	B	56	11	56	11	55	10	53	10	51	10	49	10	46	9
	C	—	—	3	—	7	1	11	2	13	3	15	3	17	3

Source. Reprinted, with permission from IES Publication ED-4.

For 75-w spot, use 10% and for 75-w flood, use 15% of above values.

For 100-w lamps, use 30% of above values.

For 150-w lamps, use 40% of above values.

For 200-w lamps, use 67% of above values.

For 500-w lamps, increase above values by 50%.

Table 20.13 K_H and K_V Values

(*a*) Horizontal $F_C = K_H \times$ Lamp Lumens per foot Broad Distribution—White-Enameled Channel

Vertical Distance from Centerline of Unit	K_H Values			
		Horizontal Distance from Centerline of Unit		
	0 ft	*1 ft*	*2 ft*	*3 ft*
1 ft	0.087	0.045	0.017	0.007
2 ft	0.040	0.033	0.020	0.011
3 ft	0.025	0.022	0.017	0.011
4 ft	0.018	0.015	0.014	0.009

(*b*) Vertical $F_C = K_V \times$ Lamp Lumens per foot Broad Distribution—White-Enameled Channel

Vertical Distance from Centerline of Unit	K_V Values				
	Horizontal Distance from Centerline of Unit				
	3 in.	*6 in.*	*9 in.*	*12 in.*	*18 in.*
1 ft	0.014	0.030	0.037	0.043	0.041
2 ft	—	0.007	0.017	0.014	0.019
3 ft	—	—	—	0.005	0.008
4 ft	—	—	—	—	—

(*c*) Horizontal $F_C = K_H \times$ Lamp Lumens per foot Broad Distribution—White-Enameled Reflector

Vertical Distance from Centerline of Unit	K_H Values			
		Horizontal Distance from Centerline of Unit		
	0 ft	*1 ft*	*2 ft*	*3 ft*
1 ft	0.438	0.127	0.008	0.001
2 ft	0.233	0.150	0.061	0.017
3 ft	0.145	0.120	0.077	0.041
4 ft	0.106	0.095	0.072	0.048

(*d*) Vertical $F_C = K_V \times$ Lamp Lumens per foot Broad Distribution—White Painted Cornice, No Reflector

Vertical Distance from Centerline of Unit	K_V Values				
	Horizontal Distance from Centerline of Unit				
	3 in.	*6 in.*	*9 in.*	*12 in.*	*18 in.*
9 in.	0.185	0.159	0.175	0.165	0.129
1 ft, 9 in.	0.011	0.028	0.044	0.057	0.068
2 ft, 9 in.	0.004	0.010	0.017	0.023	0.032
3 ft, 9 in.	0.002	0.005	0.008	0.012	0.018

(*e*) Horizontal $F_C = K_H \times$ Lamp Lumens per foot
Narrow Distribution—Polished Aluminum Reflector aimed at bottom shelf in each column.

Vertical Distance from Centerline of Unit		K_H Values — Horizontal Distance from Centerline of Unit		
	0 ft	*1 ft*	*2 ft*	*3 ft*
1 ft	0.753	0.079	—	—
2 ft	0.330	0.165	0.035	0.006
3 ft	0.212	0.161	0.066	0.022
4 ft	0.153	0.131	0.086	0.038

Source. Courtesy of General Electric.

(*f*) Vertical $F_C = K_V \times$ Lamp Lumens per foot
Narrow Distribution—Polished Aluminum Reflector aimed at bottom of each vertical plane.

Vertical Distance from Centerline of Unit	K_V Values — Horizontal Distance from Centerline of Unit				
	3 in.	*6 in.*	*9 in.*	*12 in.*	*18 in.*
9 in.	0.121	0.125	0.135	0.096	0.080
1 ft, 9 in.	0.028	0.056	0.077	0.086	0.090
2 ft, 9 in.	0.010	0.028	0.036	0.044	0.059
3 ft, 9 in.	0.006	0.013	0.020	0.031	0.037

When the point being considered is directly below the source and the source is very long, eq. 20.18 becomes

$$fc = \frac{fL \times W}{2D} \tag{20.20}$$

where

fc = illumination at distance D in footcandles
fL = source brightness in footlamberts
D = distance in feet.

When the point is offset from the source, the equation becomes complex (see Reference 4, pages 9–57). When the source is very long with respect to width, the effect of longitudinal offset (in the same direction as the length of the source) diminishes and the only offset that need be considered is a lateral one—away from the axis of the source. Table 20.13 provides illumination values for this situation. Obviously the longer the row the more accurate the results. Lumens per foot data for fluorescent lamps are available from Table 19.12, page 820, by simply dividing lumen output by length. By applying the linear relationship of eq. 20.19, the K values shown can be extrapolated.

EXAMPLE 20.6. Find the horizontal illumination on a line 8 ft below and 2 ft offset from a continuous row of two 800 ma HO lamps in a narrow-distribution reflector.

SOLUTION. From Table 20.13, K_H at 4 ft vertical distance and 2 ft offset is 0.086. At 8 ft and 2 ft:

$$K_H = 0.086\left(\frac{\sqrt{20}}{\sqrt{68}}\right) = 0.047$$

From Table 19.12, output of an HO lamp is approximately 1100 lm/ft. Therefore, horizontal illumination on the line indicated is

$$fc = 0.047 \times 2 \times 1100 = 103 \text{ fc}$$

20.40 Linear Sources

When the source width is too large with respect to length for it to be considered a line, several other calculation techniques can be used. Among these are the British IES Aspect-Factor method (see Reference 4, pages 9–52), the Configuration-Factor method, and the Angular Coordinate System. The latter two are more widely applicable and are therefore presented here. Basic application is to continuous rows of fluorescent fixtures. To obtain illumination at a point, we combine a direct illumination component obtained from tables of typical fixture types (Angular Coordinate method) with a reflected component obtained by a calculation similar to the zonal cavity method explained in preceding sections (Configuration Factor method). The following description is extracted with permission from the IES Handbook, Fifth Edition:

Angular Coordinate—DIC (Direct Illumination Component) Method

The Angular Coordinate Method is most applicable to continuous rows of fluorescent luminaires. Two angles are involved in this calculation, a longitudinal angle α and a lateral angle β. See Fig. 20.42. Angle α is the angle between a vertical line (perpendicular to the ceiling) passing through the seeing task (Point P), and a line from the seeing task to the end of the rows of the luminaires. If the seeing task is not in the vertical plane of a row of luminaires, a parallel reference plane is created for the specification of angle α.

Angle α is easily determined graphically from a chart showing angles α and β for various combinations of V, the vertical distance from the seeing task to the plane of the luminaire, and H, the horizontal distance parallel to the luminaires from the seeing task to the end of the row of luminaires. See Fig. 20.42. Usually, all rows of luminaires have the same α coordinates, one coordinate for each end of the row.

Angle β is the angle between the vertical plane of the row of luminaires and a tilted plane containing both the seeing task and the luminaire or row of luminaires. This angle is determined from the same chart as angle α, again using V, the mounting height of the luminaires above the seeing task, and H, which in determining angle β, is the horizontal distance from the seeing task to the row of luminaires, measured perpendicular to the luminaires. Angle β is different for each row of

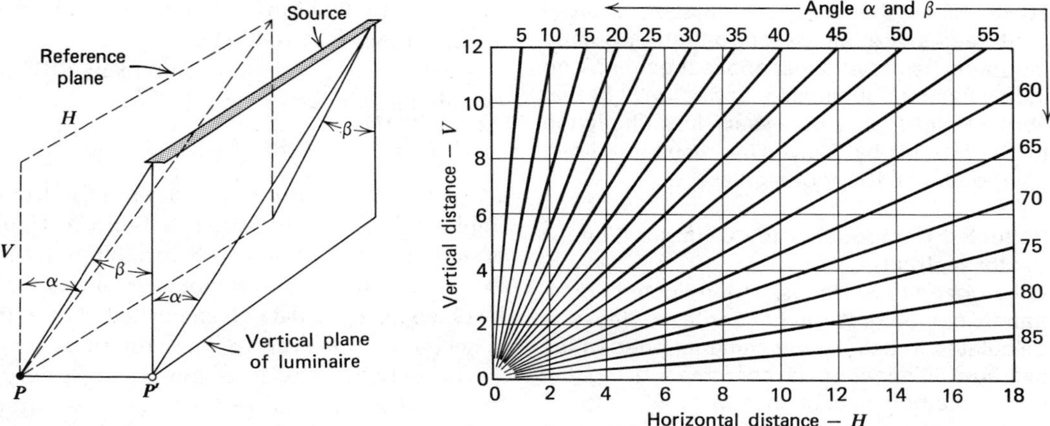

Fig. 20.42 Relationships of angles and distances used in the Angular Coordinate method. Courtesy of the IES.

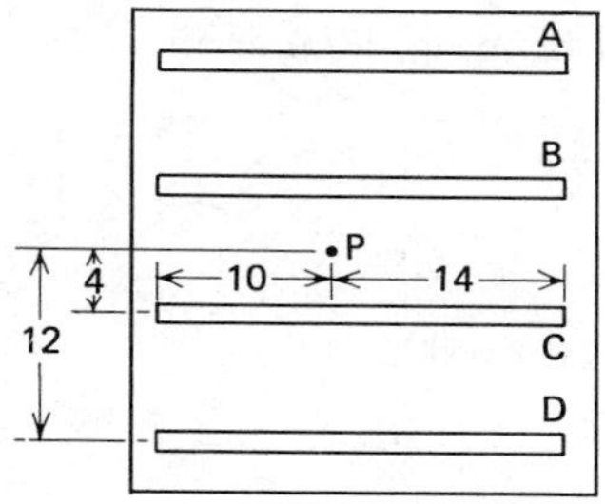

Row	α_1	α_2	β	Direct Illumination Component From Left End	Direct Illumination Component From Right End	Total
A	50	60*	55	3.9	4.4	8.3
B	50	60*	25	30.8	32.5	63.3
C	50	60*	25	30.8	32.5	63.3
D	50	60*	55	3.9	4.4	8.3
					Total	143.2

*Actually, α_2 is 59° but is rounded off to 60°.

Fig. 20.43 Layout and data obtained using the Angular Coordinate method. Courtesy of the IES.

luminaires. Each row has only one β coordinate.

The direct illumination component for each luminaire or row of luminaires is determined by referring to a table of direct illumination components for the specific luminaire. Tables such as shown in Table 20.15 list direct illumination components for specific luminaires for determining the illumination on the horizontal and on vertical planes either parallel to, or perpendicular to the luminaires. The direct illumination components are based on the assumption that the luminaire is mounted six feet above the seeing task. If this mounting height above the task is other than six feet, the direct illumination component shown in Table 20.15 must be multiplied by 6/V where V is the mounting height above the task. Thus, the total direct illumination component would be the product of 6/V and the sum of the individual direct illumination components for each row.

As an example of the method, assume four rows of six four-foot luminaires for which data are shown in Table 20.15(b). They are surface mounted on 8-foot centers in a room 28 by 30 feet. Assume the ceiling reflectance is 80 per cent and that of the walls is 50 per cent. Floor cavity reflectance is 20 per cent. The mounting height of the luminaires is $8\frac{1}{2}$ feet above the work-plane. The initial illumination on the horizontal work-plane at Point *P* in Fig. 20.43 is desired.

It is first necessary to determine angle α for both ends of the rows of luminaires and angle β. For angle α, H is 10 feet for α_1; and 14 feet for α_2. The vertical distance, V, is $8\frac{1}{2}$ feet. For angle β, H is 12 feet for rows A and D and is four feet for rows B and C. The vertical distance, V, still is $8\frac{1}{2}$ feet. Fig. 20.42 is used to determine the angles and values shown in Fig.

20.43. Since the direct illumination component tables are all based on an assumed mounting height of 6 feet above the point, and in this case the luminaires are actually $8\frac{1}{2}$ feet above Point *P*, it is necessary to multiply the total of 143.2 by $6/8\frac{1}{2}$. The resultant direct component is 101 footcandles.

Reflected Components (by Configuration Factor method)

For Horizontal Surfaces. The reflected illumination component on horizontal surfaces is calculated in exactly the same manner as the average illumination is computed using the lumen method except that the RRC, the reflected radiation coefficient, is substituted for the CU, coefficient of utilization.

$$\text{Reflected Illumination (Horizontal)} = \frac{\text{Lamp Lumens per Luminaire} \times \text{RRC}}{\text{Area per Luminaire (on Work-Plane)}} \quad (20.21)$$

where

$$\text{RRC} = \text{LC}_{\text{w}} + \text{RPM}\ (\text{LC}_{\text{cc}} - \text{LC}_{\text{w}}) \quad (20.22)$$

LC_{w} = wall luminance coefficient (see Table 20.7)
LC_{cc} = ceiling cavity luminance coefficient (see Table 20.7)
RPM = room position multiplier (see Table 20.14)

The wall luminance coefficient, LC_{w}, and ceiling cavity coefficient, LC_{cc}, are selected from Table 20.7 in exactly the same fashion as the CU is chosen. The room position multiplier (RPM) is selected from Table 20.14 after determining the position of the point being calculated and the room's RCR. The position coordinates are obtained from the chart in Table 20.14 and, together with RCR, determine the room position multiplier RPM. An example should help clarify the calculation procedure.

EXAMPLE 20.7. Determine the reflected component of illumination at point *P* in Figure 20.43 (for which the direct illumination component was there calculated as 101 fc).

SOLUTION. The reflected component can be calculated from eq. 20.21, which necessitates establishing the factors in that equation.

(a) Lamp lumens per luminaire = 2 lamps at 3200 ln each, or 6400 ln.

(b) Reflected radiation component RRC is calculated from eq. 20.22, that is,

$$\text{RRC} = \text{LC}_{\text{w}} + \text{RPM}\ (\text{LC}_{\text{cc}} - \text{LC}_{\text{w}}).$$

LC_{cc} and LC_{w} are selected directly from Table 20.7. Since the fixture in Table 20.15(*b*) is similar to that shown as fixture 30 in Table 20.7, the two required luminance factors will be taken from the data given for that fixture. It is necessary, however, to know the room cavity ratio RCR as well as the ceiling cavity, floor cavity, and wall reflectances. The latter are given above as

$$\rho_{\text{cc}} = 80\%$$
$$\rho_{\text{w}} = 50\%$$
$$\rho_{\text{fc}} = 20\%$$

RCR is calculated from eq. 20.9 as

$$\text{RCR} = 5h_{\text{rc}} \frac{\text{L} + \text{W}}{\text{L} \times \text{W}} = 5 \times 8.5 \times \frac{28 + 30}{28 \times 30} = 2.93, \text{ say } 3$$

from Table 20.7, fixture 30,

$$\text{LC}_{\text{w}} = 0.16$$

and

$$\text{LC}_{\text{cc}} = 0.20$$

Room position factor RPM is read from the tabular data in Table 20.14 after having established the grid coordinates of the desired point *P*. By a simple calculation the equivalent location of point *P* on the 10 × 10 grid diagram of Table 20.14 is found to be F4. With that information and an RCR of 3, RPM is found to be 0.75. Using this figure we can now calculate RRC.

$$\begin{aligned} \text{RRC} &= \text{LC}_{\text{w}} + \text{RPM}\ (\text{LC}_{\text{cc}} - \text{LC}_{\text{w}}) \\ &= 0.16 + 0.75\ (0.20 - 0.16) \\ &= 0.16 + .03 \\ &= 0.19 \end{aligned}$$

(c) Area per luminaire on work plane:

$$\text{Area per luminaire} = \frac{\text{room area}}{\text{number of luminaires}} = \frac{28 \times 30}{24} = 35$$

Table 20.14 ***Room Position Multipliers***

(For All Room Cavity Ratios and for All Points Designated by a Number and a Letter as Illustrated)

Points Along Width of Room	RCR	A	B	C	D	E	F
		Points Along Length of Room					
0	1	.24	.42	.47	.48	.48	.48
	2	.24	.36	.42	.44	.46	.46
	3	.23	.32	.37	.40	.42	.42
	4	.22	.28	.32	.35	.37	.37
	5	.21	.25	.28	.31	.33	.33
	6	.20	.23	.26	.28	.29	.30
	7	.18	.21	.23	.25	.26	.27
	8	.17	.18	.21	.22	.22	.23
	9	.15	.17	.18	.19	.20	.20
	10	.14	.16	.16	.17	.18	.18
1	1	.42	.74	.81	.83	.84	.84
	2	.36	.51	.60	.63	.66	.68
	3	.32	.40	.48	.51	.53	.57
	4	.28	.33	.40	.42	.44	.48
	5	.25	.29	.33	.36	.38	.42
	6	.23	.26	.29	.31	.33	.36
	7	.21	.23	.26	.28	.29	.30
	8	.18	.20	.23	.25	.26	.26
	9	.17	.18	.20	.21	.22	.23
	10	.16	.17	.18	.19	.19	.20
2	1	.47	.81	.90	.92	.93	.93
	2	.42	.60	.68	.72	.78	.83
	3	.37	.48	.58	.61	.64	.67
	4	.32	.40	.48	.50	.52	.57
	5	.28	.33	.40	.42	.44	.48
	6	.26	.29	.35	.37	.38	.40
	7	.23	.26	.30	.32	.33	.34
	8	.21	.23	.26	.27	.28	.29
	9	.18	.20	.23	.24	.25	.25
	10	.16	.18	.19	.21	.22	.22
3	1	.48	.83	.92	.94	.95	.95
	2	.44	.63	.72	.77	.82	.85
	3	.40	.51	.61	.65	.69	.71
	4	.35	.42	.50	.54	.58	.61
	5	.31	.36	.42	.46	.49	.52
	6	.28	.31	.37	.39	.41	.43
	7	.25	.28	.32	.34	.35	.36
	8	.22	.25	.27	.29	.30	.30
	9	.19	.21	.24	.25	.26	.26
	10	.17	.19	.21	.22	.23	.23
4	1	.48	.84	.93	.95	.96	.97
	2	.46	.66	.78	.82	.85	.86
	3	.42	.53	.64	.69	.73	.75
	4	.37	.44	.52	.58	.62	.64
	5	.33	.38	.44	.49	.52	.54
	6	.29	.33	.38	.41	.43	.45
	7	.26	.29	.33	.35	.37	.37
	8	.22	.26	.28	.30	.31	.32
	9	.20	.22	.25	.26	.26	.27
	10	.18	.19	.22	.23	.23	.24
5	1	.48	.84	.93	.95	.97	.97
	2	.46	.68	.83	.85	.86	.87
	3	.42	.57	.67	.71	.75	.77
	4	.37	.48	.57	.61	.64	.66
	5	.33	.42	.48	.52	.54	.56
	6	.30	.36	.40	.43	.45	.47
	7	.27	.30	.34	.36	.37	.38
	8	.23	.26	.29	.30	.31	.32
	9	.20	.23	.25	.26	.27	.27
	10	.18	.20	.22	.23	.24	.25

Source. Courtesy of the IES of North America.

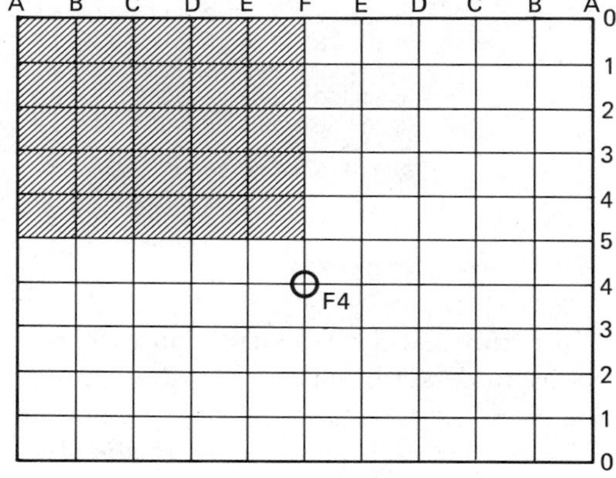

We now have all the factors required for eq. 20.21.

Reflected horizontal illumination

$$= \frac{\text{Lumens per luminaire} \times \text{RRC}}{\text{Area per luminaire}}$$

$$= \frac{6400 \times 0.19}{35}$$

$$= 34.74, \text{ say } 35$$

Thus total horizontal illumination at point P equals the sum of the direct and reflected components, that is,

$$\text{Illumination at point } P = 101 + 35$$
$$= 136 \text{ fc}$$

This, of course, is initial illumination, which must be reduced by light loss factor LLF to determine maintained illumination.

Table 20.15 ***Typical Angular Coordinate Method Data***

	β	5	15	25	35	45	55	65	75	
	α	*FC at a Point on Workplane*								
(a)	0-10	24.2	21.6	17.0	11.6	5.3	1.6	0.3	—	Category V
	0-20	47.6	41.9	33.2	22.7	10.5	3.2	0.6	—	
	0-30	67.4	59.8	47.8	32.6	15.2	4.8	.9	—	
	0-40	82.7	73.6	59.1	40.6	19.1	6.3	1.2	—	
	0-50	91.9	82.0	66.0	45.3	21.6	7.6	1.6	0.1	
	0-60	96.5	86.4	69.8	48.3	23.5	8.4	2.0	0.2	
	0-70	98.1	87.9	71.2	49.6	24.6	9.0	2.3	0.3	
	0-80	98.5	88.3	71.6	50.0	25.0	9.3	2.5	0.4	
	0-90	98.6	88.4	71.7	50.1	25.1	9.4	2.6	0.4	Four T-12 Lamps—430 MA Prismatic Lens 2 ft Wide—For T-10 Lamps—CU × 1.02
	α	*FC at a Point on Workplane*								
(b)	0-10	11.5	10.6	8.1	4.9	2.0	0.8	0.2	—	Category V
	0-20	22.7	20.4	15.7	9.6	4.0	1.6	0.4	—	
	0-30	31.9	28.7	22.5	13.7	5.9	2.4	0.6	—	
	0-40	39.1	35.1	27.5	17.1	7.5	3.2	0.8	—	
	0-50	43.5	39.1	30.8	19.4	8.7	3.9	1.1	—	
	0-60	45.6	41.1	32.5	20.8	9.6	4.4	1.4	—	
	0-70	46.3	41.8	33.2	21.5	10.1	4.8	1.7	—	
	0-80	46.5	42.0	33.4	21.7	10.3	5.0	1.9	—	
	0-90	46.5	42.0	33.4	21.7	10.3	5.0	1.9	—	Two T-12 Lamps—430 MA 1 ft Wide Prismatic Wrap-Around
	α	*FC at a Point on Workplane*								
(c)	0-10	10.6	9.5	7.6	5.5	3.3	1.3	0.3	—	Category III
	0-20	20.6	18.5	14.9	10.9	6.6	2.6	0.7	—	
	0-30	29.4	26.5	21.6	16.0	9.8	4.0	1.1	—	
	0-40	36.5	33.1	27.4	20.6	12.9	5.4	1.5	—	
	0-50	41.8	38.1	31.9	24.3	15.7	6.7	2.0	0.1	
	0-60	45.2	41.3	34.8	26.8	17.6	7.9	2.6	0.2	
	0-70	46.9	43.0	36.4	28.3	18.9	8.9	3.2	0.3	
	0-80	47.4	43.6	36.9	28.8	19.4	9.3	3.5	0.4	
	0-90	47.5	43.7	37.0	28.8	19.4	9.3	3.5	0.4	Two T-12 Lamps—Any Loading For T-10 Lamps—CU × 1.02
	α	*FC at a Point on Workplane*								
(d)	0-10	6.9	6.5	5.4	4.0	2.5	1.3	0.5	0.1	Category III
	0-20	13.5	12.6	10.6	7.9	5.0	2.5	0.9	0.2	
	0-30	19.2	17.9	15.3	11.5	7.4	3.8	1.4	0.3	
	0-40	23.6	22.2	19.2	14.7	9.7	5.2	2.0	0.4	
	0-50	26.9	25.4	22.2	17.3	11.8	6.5	2.6	0.6	
	0-60	28.9	27.4	24.1	19.1	13.4	7.8	3.2	0.8	
	0-70	29.9	28.4	25.1	20.1	14.3	8.7	3.9	1.0	
	0-80	30.2	28.7	25.4	20.8	14.6	9.0	4.2	1.3	
	0-90	30.2	28.7	25.4	20.8	14.6	9.0	4.2	1.3	Two Lamp Strip—Any Loading

Source. Courtesy of Westinghouse.

20.41 Area Sources

To perform accurate calculation with area sources, the exact brightness distribution over its surface must be known and calculations performed for each point. Since this is obviously impractical except by computer, most hand methods assume a lambertian distribution, i.e., that the surface of the source has uniform luminance. The direct illumination from such a *rectangular* source can be calculated by use of the curves in Figure 20.44, and the reflected component by the technique explained in the preceding section.

These curves are valid for illumination at a point distant d from one corner of the source. Therefore, in order to calculate illumination at a point beyond the source corner, it is necessary to calculate for an enlarged source and subtract the illumination components not desired. Refer to Figure 20.45. In order to compute the illumination at point *P* from source D, we must calculate the illumination from ABCD and then subtract the contribu-

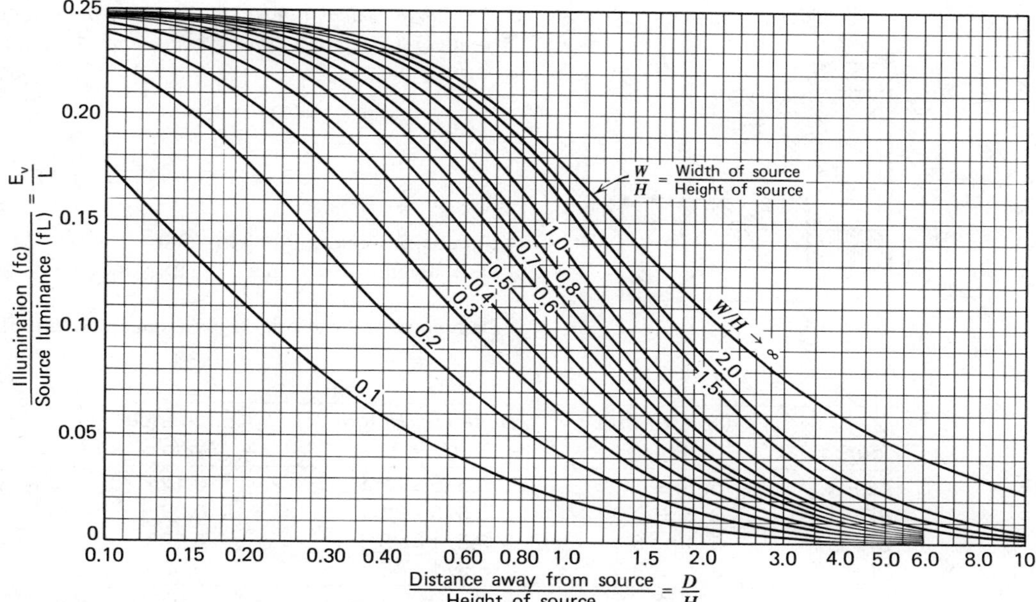

Fig. 20.44 Illumination from a rectangular uniform source parallel to illuminated plane. Courtesy of the IES.

tions from A, B, and C. Thus illumination from D is

$$I_D = \text{ABCD} - \text{AB} - \text{AC} + \text{A}$$

An example should clarify this.

EXAMPLE 20.8. Refer to Figure 20.46 (which is a section of Figure 18.36). Calculate the illumination at point P at the center of the desk, resulting from a single four-lamp troffer placed as shown. This type of problem is a practical one for nonuniform layouts. Assume surface brightness of the fixture to be 1280 fL.

SOLUTION. Using the technique shown above, with data from Figure 20.44:

fc from ABCD =
AEPG − BEPH − CFPG + DFPH

The calculation on Figure 20.46 gives the direct illumination at the center of the desk as 41 fc. Coupled with general and reflected illumination of 15 to 30 fc the desk illumination is adequate for all but the most difficult tasks.

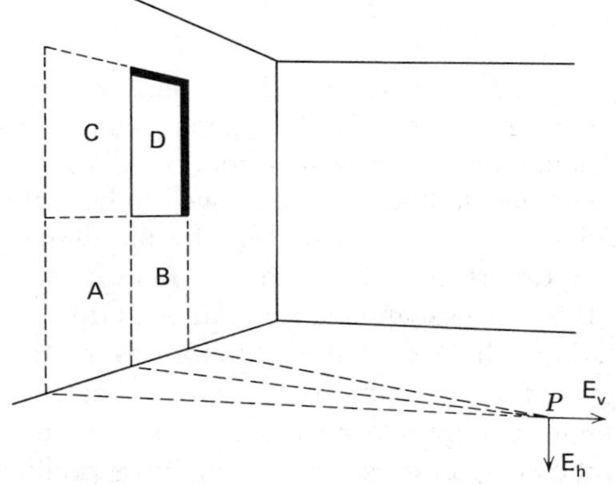

Fig. 20.45 Example of illumination calculation from source D. Courtesy of the IES.

EVALUATION STAGE

20.42 Design Evaluation

Having completed the design stage the final step is evaluation of the design from the three aspects—lighting, costs, and energy. The lighting aspects include quantity, quality, detailed ESI calculations, brightness ratios, mood, ambience, texture, color, variation, psychological impressions, orientations, and daylight use—in short, a review of all of the

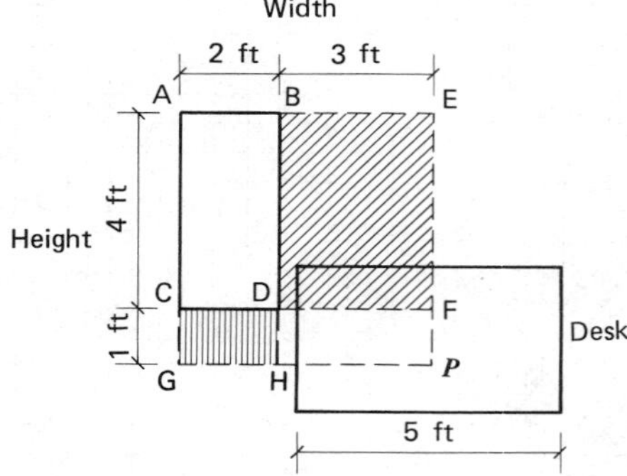

Fig. 20.46 Calculation of illumination at a point resulting from an area source above and to the side of the point.

Luminaire mounting height—8 ft
Desk height—2 ft 6 in.

Luminaire uniform brightness assumed at 1280 fL

Source	*Aspect Ratio* $w/H(L)$	*Distance/Height* (D/H)	*Illuminance Ratio*
AEPG	5/5 = 1	5.5/5 = 1.1	0.125
BEPH	3/5 = 0.6	5.5/5 = 1.1	−0.09
CFPG	5/1 = 5	5.5/1 = 5.5	−0.028
DFPH	3/1 = 3	5.5/1 = 5.5	0.025
			0.032

Illumination at point P = 0.032 (1280 fL) = 41 fc

lighting factors discussed in detail above. Since a good deal of experience is required to visualize the actual results from the drawings, the novice designer would do well to have the review done, at least in part, by one having such experience. The other two aspects of evaluation, cost and energy, can be performed readily with the aid of contractor's estimating figures for cost and a straightforward calculation for energy. The results are then compared to the budget figures developed at the preliminary design stage. As we have repeatedly stressed, the useful cost figures are life cycle, annual operating and first cost, for economic comparisons, operating budgets and construction budgets, respectively. In the following chapter we will present recommendations for specific occupancies, accompanied by actual cost studies and energy analyses. Detailed cost studies including the impact of lighting on air-conditioning, the proportional cost of the wiring system, and the proper apportionment of costs involves the entire building and, as we have repeatedly stated, can only be accurately performed by computer. Studies of this type are generally made by consulting engineers rather than architects, and then only after initial, operating, and total costs have been set in proper perspective for the particular job, by the architect and client. This is necessary because often, as in the case of speculative construction, the client's overriding consideration is first cost, thereby rendering a complete cost analysis unnecessary. Any attempt to completely separate costs for lighting, HVAC, structure, and so on is arbitrary because of the intimate interaction between these factors. Lighting designers are well advised to keep themselves and the construction team aware of this, if they are to fully fulfill their responsibility.

References and Bibliography

1. "Recommended Practice for Classification of Interior Luminaires by Distribution: Luminaire Spacing Criteria," IES, August 1977, *Lighting Design & Application.*

2. "Revision of the S/MH Concept," August 1977, R. E. Levin, *Lighting Design & Application.*

3. *Lighting Systems Study,* March 1974, Ross & Baruzzini, Inc., General Services Administration, Public Buildings Service.

4. *IES Handbook,* 5th ed.

21

Lighting Application

21.1 Introduction

In the preceding three chapters, we examined lighting fundamentals, sources, and design procedures. In this final lighting chapter we shall consider application of lighting principles to specific situations. The facilities to be covered in some detail will include residential, educational, commercial, institutional, and industrial occupancies. Each will be examined from the viewpoint of its special requirements and suggested approach. The latter will include lighting materials, sources, as well as comparative economics and energy considerations. The chapter will conclude with consideration of special types of indoor lighting plus a short section on exterior lighting. Part of the data presented are adapted from IES publications.

RESIDENTIAL OCCUPANCIES

21.2 Residential Lighting

Residential lighting offers to the lighting designer great opportunity for originality and ingenuity, since a residence combines more functions and needs than any other building. Furthermore, it generally requires that all work be done at minimal cost, and that the end result please a range of tastes. The designer approaches the problem with a list of requirements and two basic tools: the fixture and the architectural lighting element. The former was covered in Sections 20.18 through 20.27 and is augmented by the fixture data in Figure 21.1. The latter is discussed below.

(*a*) ***Energy Considerations.*** Only kitchens, bathrooms, and laundry areas in private residences are subject to the lighting-power budget limitations of ASHRAE 90-75. See ASHRAE 90-75 paragraph 93-4.1. Nevertheless, in view of the large portion of national lighting energy used in residential spaces—estimated at 20%—conservation measures are in order:

1. Provide means for reducing light levels in all areas. A kitchen during food preparation does not have the same lighting requirements as a kitchen being entered for a "refrigerator raid." Low-level lighting provision should be made in *all* rooms, including bathrooms. To accomplish this use high-low switches, simple dimmers, multilevel ballasts, and multilevel switching. An ancillary benefit is that ambience can be changed thereby in multiuse rooms such as dining rooms, family rooms, and finished basements.
2. Provide local task lighting for difficult tasks such as the location at which family accounts are handled.
3. Provide switching for accent lighting.
4. Provide time switches for exterior lights.
5. In large residences consider low-voltage control for its ease of remote control. Considerable energy savings can be effected in this way.
6. Use daylight in areas normally occupied during daylight hours such as kitchens and living rooms. Consider skylights with built-in artificial lighting for these areas.

(*b*) ***Sources.*** When using fluorescent, choose proper color for space (see Table 19.26, page 853). Despite their lower efficacy, use of daylight fluorescents as the artificial source in lighted skylights is very effective.

Wide Profile

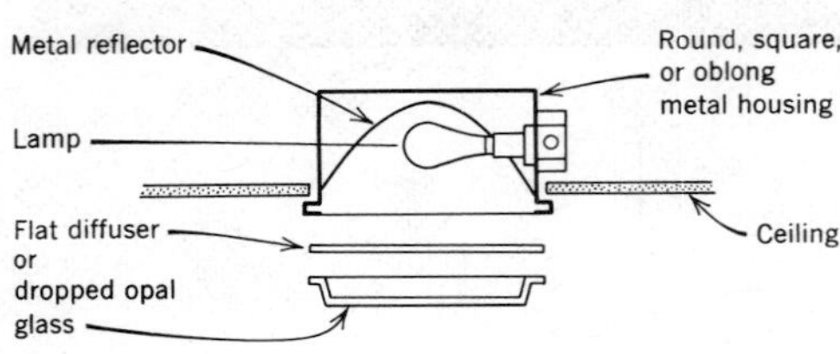

For general illumination (almost always used in multiple). Basement, reacreation rooms, kitchens, laundries, halls (service). Used singly or in small groups for small areas such as walk-in closets, garage, entry doors, overhangs in porches. Because of high luminance of diffuser, seldom used in living or social areas.

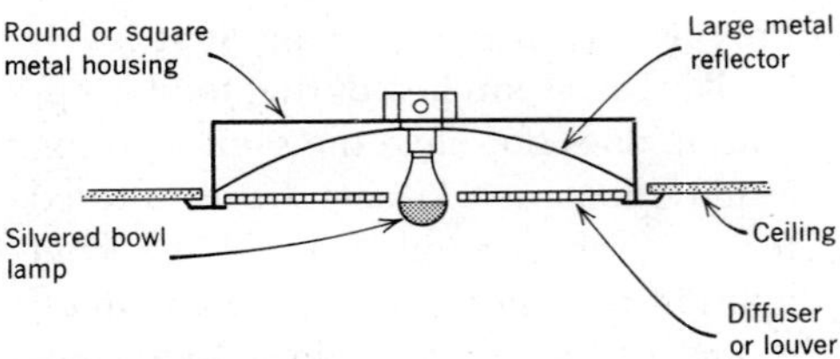

Used in multiple for general lighting in kitchens, baths, laundries, recreation rooms and family rooms. Used singly or in groups over game tables. Lower luminance allows this type of luminaire to be used in living areas if styling permits.

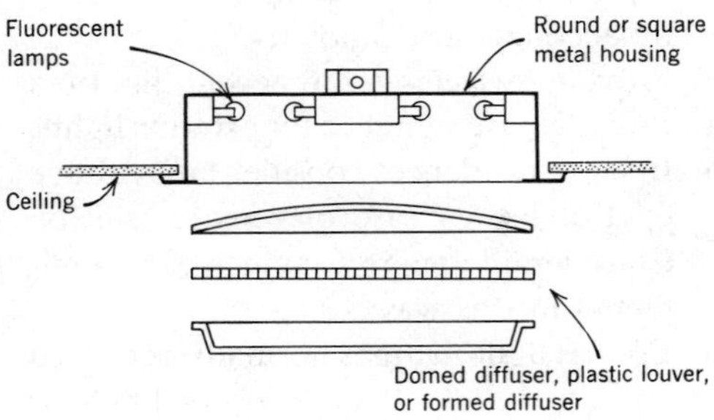

Very wide distribution, excellent as general lighting for kitchen, laundry, recreation room and bath. Because of large size and high lumen output, fewer units required. Often used signly or in pairs for entry halls and foyers and for skylight effect in interior halls and stairways.

Medium Profile

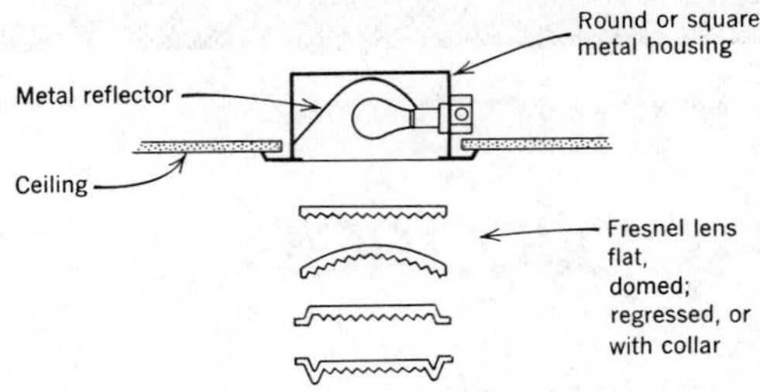

Used for specific task lighting where task area is large, such as kitchen sink, kitchen island counter or range, laundry tubs and ironing, game table, workbench, hobby area. Used for general lighting in restricted areas such as halls, entries and baths. Multiple groupings are satisfactory for the general lighting of kitchens and recreation rooms. If weatherproof, appropriate for outdoor uses, including overhangs, porches, entries.

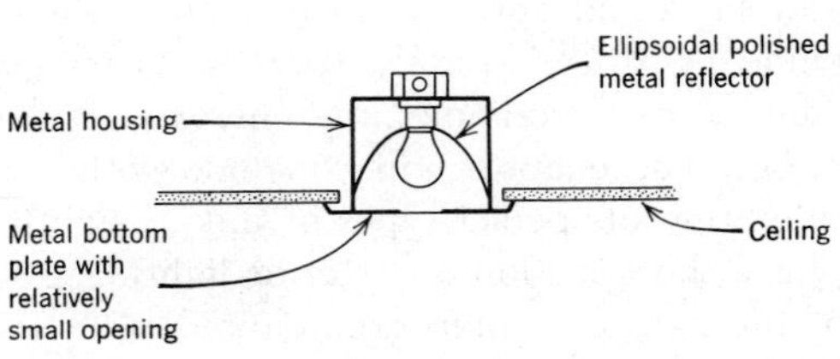

Uses basically the same as the Fresnel unit listed above, except that the lower luminance makes this type of equipment more usable in living and dining areas.

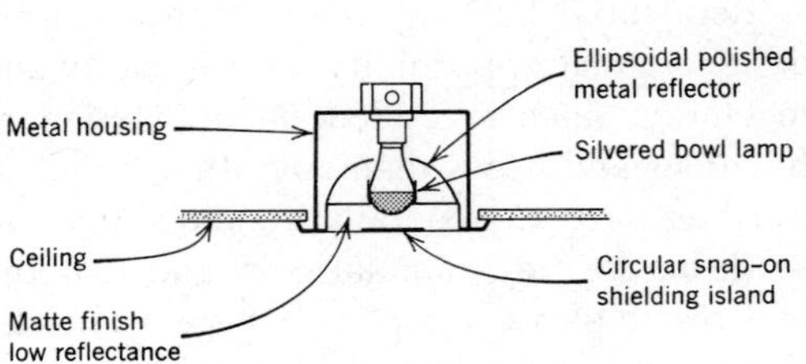

Same uses as open ellipsoidal unit listed above but has a far better control of lamp luminance.

Fig. 21.1 *Design features of recessed luminaires having wide, medium, narrow, and asymmetric profiles. See also Figures 20.17 through 20.21, pages 884 to 886. Courtesy of the IES.*

Narrow Profile

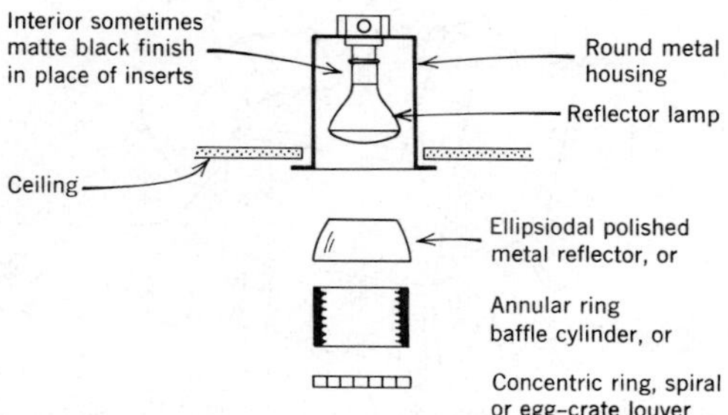

Accent lighting over plants, cocktail tables, etc. Wall lighting–mounted close to textured surfaces such as brick, stone, rough wood and fabrics. Task lighting–food preparation areas (may cause specular reflections), in multiple on quite close spacing for general lighting. Most effective when used near perimeter of room so that some light spills onto wall. Dramatic effects for family rooms, recreation rooms, and formal living areas. Supplementary stair lighting–shadow patterns define treads and risers. Dining tables–provide functional light on dining table to supplement decorative effect from hanging luminaire.

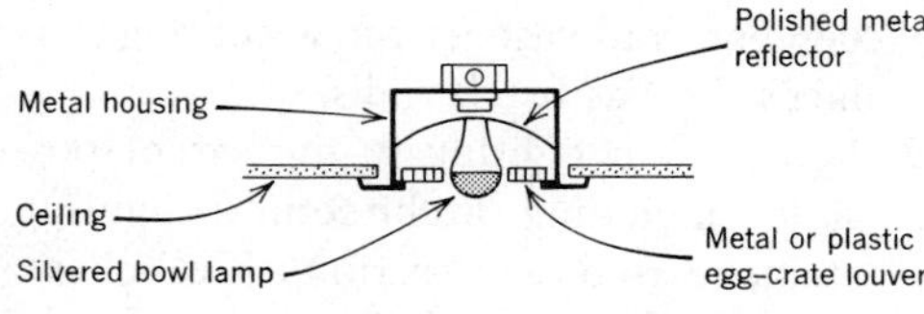

Same uses as type immediately above. Low luminance very desirable, but larger size sometimes prohibits use in highly styled interiors.

Special Asymmetric Profile

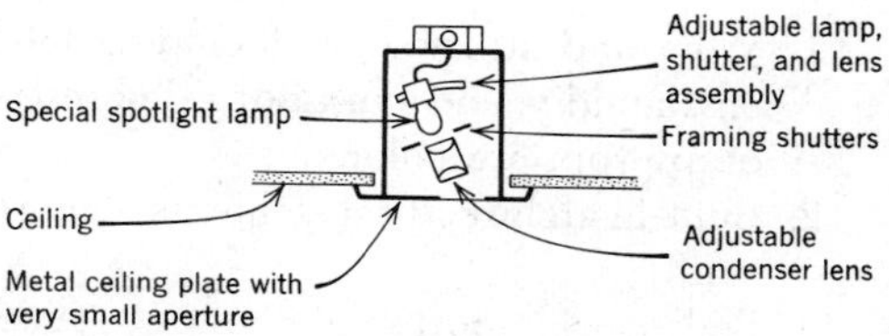

Adjustable beam can be framed precisely to outline paintings, pictures, maps, and niches. When aimed directly down, shutters can frame dining tables, cocktail or coffee tables, or other horizontal elements. Special high-light-output lamps providing controlled beams often used in these units. Should have easy access for the relamping required. Many come equipped with top-access openings for relamping from above.

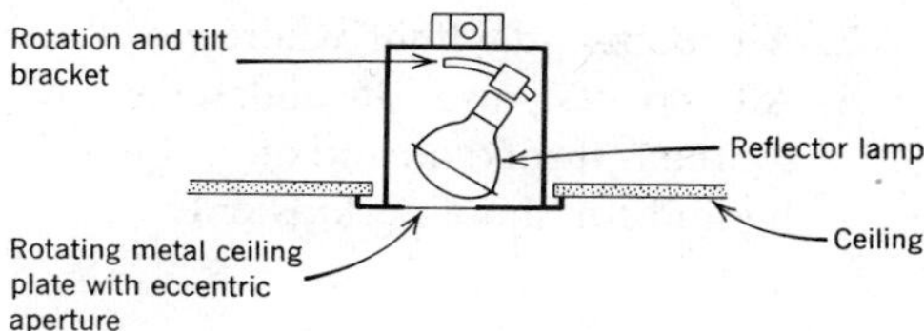

Useful for gallery or picture lighting and to light sculpture. If scalloping effect is acceptable can be used for wall lighting and to accent fireplace surfaces. Large size of bottom aperture sometimes make this unacceptable for highly styled interiors. May also be used for lighting piano music and sewing machines.

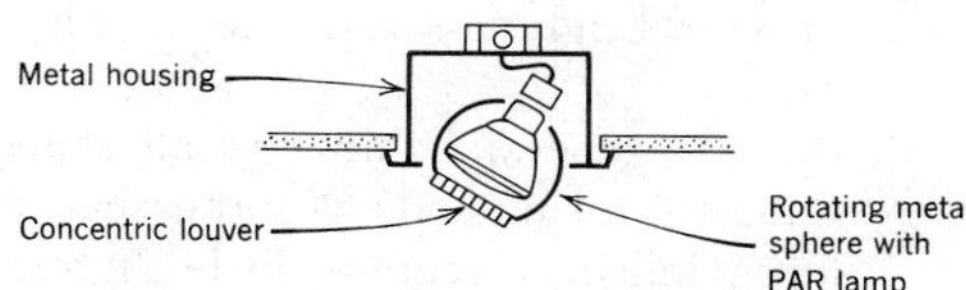

Same uses as recessed adjustable luminaire immediately above.

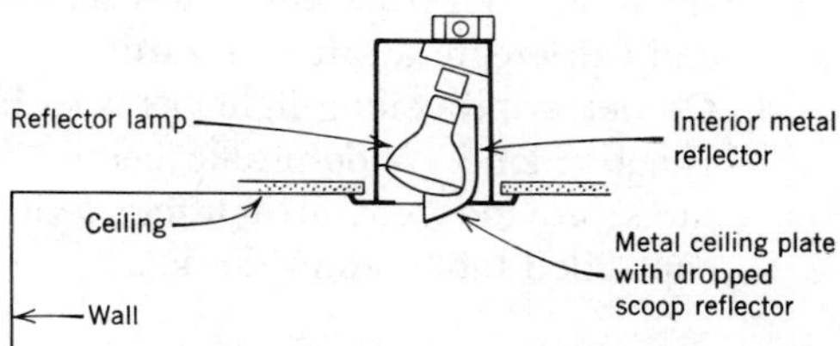

For uniform illumination of plane wall surfaces. Extremely effective for lighting murals and for minimizing wall imperfections. *Not* generally to be used for lighting textured wall surfaces because it directs no grazing light at wall. Spacing of these units is critical–follow manufacturer's recommendations closely.

Listed below are the appropriate sources to be used in different areas of a residence:

1. Work and utility areas including kitchens, laundry, and workshop—fluorescent of appropriate color.
2. Built-in architectural elements—fluorescent.
3. Bedrooms, portable lamps, accent lights—incandescent, tungsten-halogen.
4. Circulation areas, stairwells, closets—incandescent.
5. Exterior: for short periods—incandescents; for long periods—HID.
6. Bathrooms: general incandescent or warm white fluorescent; mirror lighting—incandescent.
7. All rooms—daylight where possible.
8. All spaces—use incandescent when source is turned on and off frequently or lighted for short periods only.

Zone 2 The immediate surroundings (area adjacent to the visual task)	
Desirable ratio	1/3 to equal to task*
Minimum acceptable ratio	1/5 to equal to task*

Zone 3 The general surroundings (not immediataly adjacent to task)	
Desirable ratio	1/5 to 5 times task*
Minimum acceptable ratio	1/10 to 10 times task*

*Typical task luminance range is 12 to 35 footlamberts (seldom exceeds 60 footlamberts)

Fig. 21.2 *Seeing zones and recommended luminance ratios for residential visual tasks. Courtesy of the IES.*

(*c*) ***Recommendations.*** These are design recommendations, which are applicable to residential occupancies of all types.

1. Use general/task-lighting concept with recommended levels as in Tables 21.1 and 21.2.
2. Provide brightness ratios as in Figure 21.2.
3. Provide general lighting in all spaces, sufficient for movement and casual seeing. Hallways require little lighting; stairs require more. Light stairs from directly above or ahead to create a shadow directly below the tread front. Lighting from front eliminates shadows and can create a safety hazard.
4. Do *not* avoid ceiling lights as is so frequently done. Wide-profile ceiling fixtures provide general lighting; switch-controlled table lamps do not.

(*d*) ***Fixtures and Luminous Elements.*** See Figure 21.1. Tungsten-halogen sources are applicable to many of the fixtures illustrated.

1. Utilize diffuse distribution for general lighting, narrow-distribution downlight for area and furniture accents, and narrow-distribution, ceiling-recessed incandescent wallwashers for accenting surfaces such as brick walls.
2. Use built-in lighting to the extent possible, including architectural lighting elements (see next section). We believe that this demonstrates integrity of concept. For this reason, we recommend that the flexibility of track lighting be utilized for accent and task lighting but not for general lighting or as the lighting system throughout the residence. Track lights and applications are shown in Figures 21.3 and 21.4.
3. Private residences are the exception to the rule of selecting off-the-shelf items in preference to specials. The lighting should compliment the architecture and furnishings, and frequently this can best be accomplished by original designs.

21.3 Architectural Lighting Elements

Reference to architectural lighting elements is usually made when dealing with

Table 21.1 ***Current Footcandle Recommendations for General Lighting***

Activity or Area	Typical American Recommendation: Footcandles[a] Minimum at Any Time	Other Authorities: Average Footcandles
Conversation and relaxation	10[a]	5–10
Passage areas	10[a]	5–10
Areas, other than kitchen	30	10–20
Kitchen	50	30

[a]General Lighting in these areas need not be uniform.

Table 21.2 ***Current Footcandle Recommendations for Specific Visual Tasks***[a]

Seeing Task	Typical American Recommendation: Minimum Footcandles	Other Authorities: Average Footcandles
Dining	15	10–15
Grooming, Makeup	50	50
Handcraft		
Ordinary seeing tasks	70	20–40
Difficult seeing tasks	100	50–70
Critical seeing tasks	200	>125
Ironing	50	20–30
Kitchen Duties		
Food preparation and cleaning involving difficult seeing tasks	150	70–90
Serving and other noncritical tasks	50	20–30
Laundry Tasks		
Preparation, sorting, inspection	50	20–30
Washer and dryer areas	30	10–15
Reading and Writing		
Handwriting, reproductions, and poor copies	70	70
Books, magazines, and newspapers	30	30
Reading Piano Scores		
Advanced	70	50
Simple	30	30
Sewing, Hand or Machine		
Dark fabrics	200	>125
Medium fabrics	100	70–90
Light fabrics	50	30–50
Occasional—high contrast	30	10–20
Study	70	50–70
Table Games	30	30

[a]These levels are based on young eyes with 20-20 vision. Older eyes, even when properly corrected by glasses, have reduced visual acuity, a longer period of adaptation, and decreased resistance to glare. To state it simply, older persons need more light, and special precautions against glare.

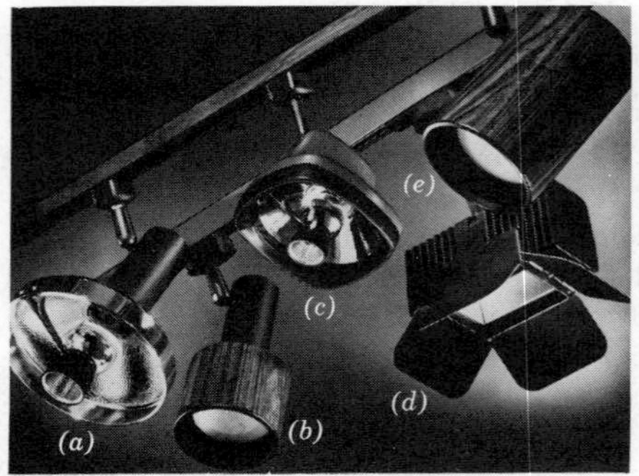

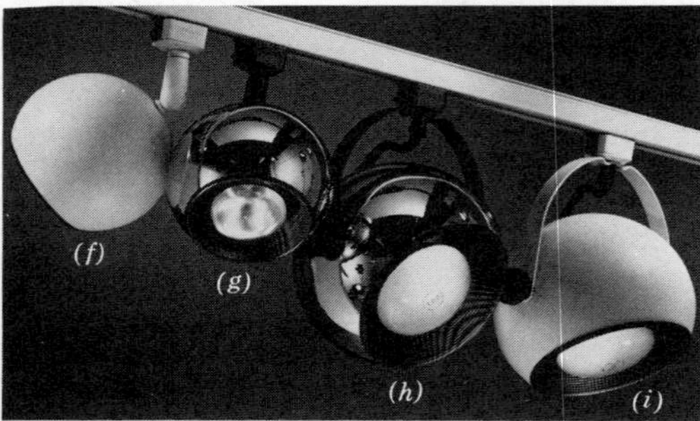

Fig. 21.3 *Track lights are available in a wide variety of practical and decorative designs. Types* (a) *and* (c) *use standard A-shape incandescent lamps;* (b) *and* (e) *are wood-grain finished and hold an 'R'-shape lamp;* (d) *is an adjustable wall wash unit using a 500-w tungsten-halogen lamp.* (f) *through* (i) *are decorative ball units holding R-lamps of different sizes. Courtesy of Lightcraft/Scovill.*

Fig. 21.4 (a) *Since the primary objective of this lighting is display, track units offer the desired flexibility. Spill light is sufficient for walking through the corridor. When display is not required, one unit should be separately controlled to provide needed passage lighting.*

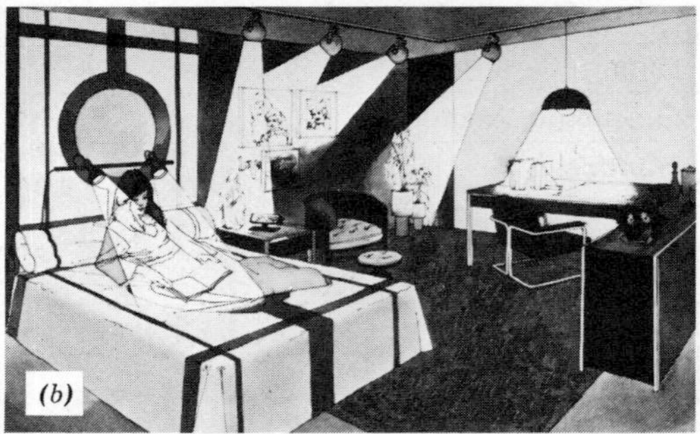

Fig. 21.4 (b) *Track light provides needed task lighting for reading in bed or working at the desk. The remaining units should be separately controlled, since display lighting in a bedroom is hardly needed constantly. Illustrations Courtesy of Lightcraft/Scovill.*

coves, cornices, valances, coffers, skylights, or other luminous surfaces not normally comprising a lighting fixture. Although such units are inherently relatively inefficient, their use is often indicated by architectural consideration, since they generally create an attractive indirect-lighting source. Empirical design data are given in Figure 21.5.

Using fluorescent tubes, it is possible to avoid dark spots between lamps by placing lamps at a slight angle rather than end-to-end, thus enabling ends to overlap. Reflectors, when used, should be aimed 15 to 25° above the horizontal and field-adjusted for best ceiling coverage. When using double strips they should be stacked vertically as shown in Figure 21.5*m*. Coefficients for double-lamp installation rarely exceed 0.75 times the single-lamp coefficient. Interiors of cove surfaces should be painted with a high-reflectance white paint with diffuse (flat) rather than specular (gloss) finish.

Figures 21.5*k* and *n* give dimensions of typical wall-washing cornices and valance. Reflectors, though not required, increase the efficiency of the installation. As in the case of coves, finish of the valance interior should be flat white of approximately 30 to 40% RF. Incandescent fixtures may also be effectively employed in architectural lighting.

Although not always considered as an architectural element, sections of luminous ceil-

(a) Lighted Cornices

Cornices direct all their light downward to give dramatic interest to wall coverings, draperies, murals, etc. May also be used over windows where space above window does not permit valance lighting. Good for low-ceilinged rooms.

(b) Lighted Valances

Valances are always used at windows, usually with draperies. They provide up-light which reflects off ceiling for general room lighting and down-light for drapery accent. When closer to ceiling than 10 inches use closed top to eliminate annoying ceiling brightness.

(c) Lighted Coves

Coves direct all light to the ceiling. Should be used only with white or near-white ceilings. Cove lighting is soft and uniform but lacks punch or emphasis. Best used to supplement other lighting. Suitable for high-ceilinged rooms and for places where ceiling heights abruptly change.

(d) Lighted High Wall Brackets

High wall brackets provide both up and down light for general room lighting. Used on interior walls to balance window valance both architecturally and in lighting distribution. Mounting height determined by window or door height.

Fig. 21.5 *Residential architectural lighting elements. Courtesy of the IES, except as otherwise noted.*

(e) Lighted Low Wall Brackets

Low brackets are used for special wall emphasis or for lighting specific tasks such as sink, range, reading in bed, etc. Mounting height is determined by eye height of users, from both seated and standing positions. Length should relate to nearby furniture groupings and room scale.

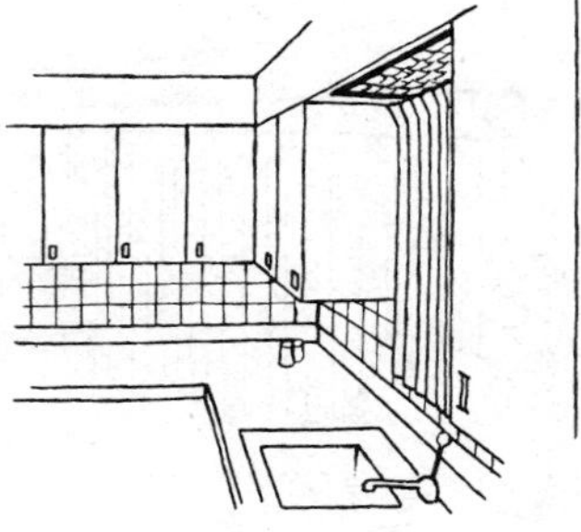

(f) Lighted Soffits

Soffits over work rease are designed to provide higher level of light directly below. Usually they are easily installed in furred-down area over sink in kitchen. Also are excellent for niches over sofas, pianos, built-in desks, etc.

Bath or dressing room soffits are designed to light user's face. They are almost always used with large mirrors and counter-top lavatories. Length usually tied to size of mirror. Add luxury touch with attractively decorated bottom diffuser.

(g) Lighted Canopies

The canopy overhang is most applicable to bath or dressing room. It provides excellent general room illumination as well as light to the user's face.

Fig. 21-5 (continued)

(*h*) Luminous Ceilings

Totally luminous ceilings provide skylight effect very suitable for interior rooms or utility spaces, such as kitchens, baths, laundries. With attractive diffuser patterns, more decorative supports, and color accents they become acceptable for many other living spaces such as family rooms, dens, etc. Dimming controls desirable.

(*j*) Luminous Wall Panels

Luminous wall panels create pleasant vistas; are comfortable background for seeing tasks; add luxury touch in dining areas, family rooms and as room dividers. Wide variety of decorative materials available for diffusing covers.

(*k*) Typical Valance

This "typical" dimensional drawing applies only to commonly encountered window valance situations. Obviously, other window treatments could necessitate modifications in these critical dimensions; i.e., vertical blinds, double track situations, curved bay windows, etc.

The same "job-tailored" variations can occur in the design of any type of structural lighting device. Therefore no other dimensional drawings have been included here.

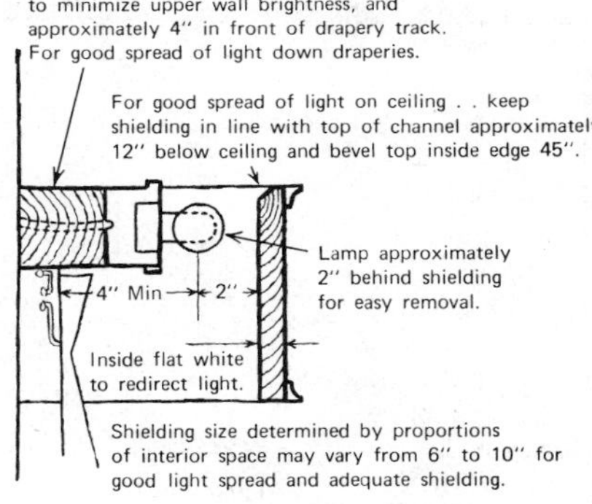

ing are frequently employed in kitchens to some advantage. The difficulty lies in lack of sufficient ceiling height. Lamp spacing should not exceed 1.5 times the cavity depth, for uniformity. An illusion of ceiling height may be obtained by blurring the ceiling-wall joint with light via a valance illuminating a light color joint.

EDUCATIONAL FACILITIES

21.4 Institutional and Educational Buildings

The lighting requirements for the various spaces in educational facilities are many and varied and, to a considerable extent, coincide with requirements for commercial (office) and

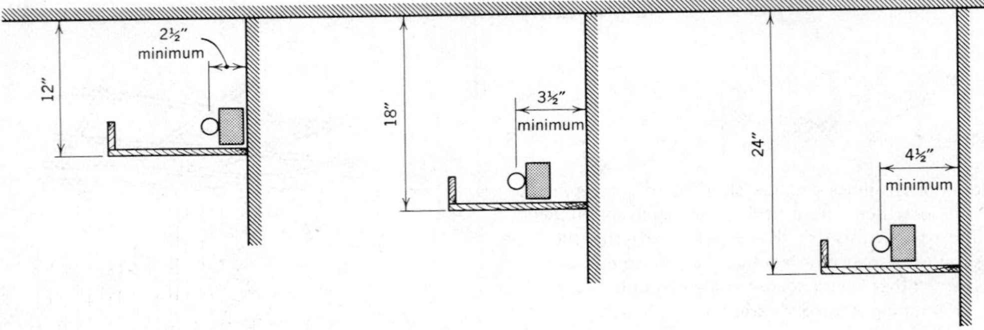

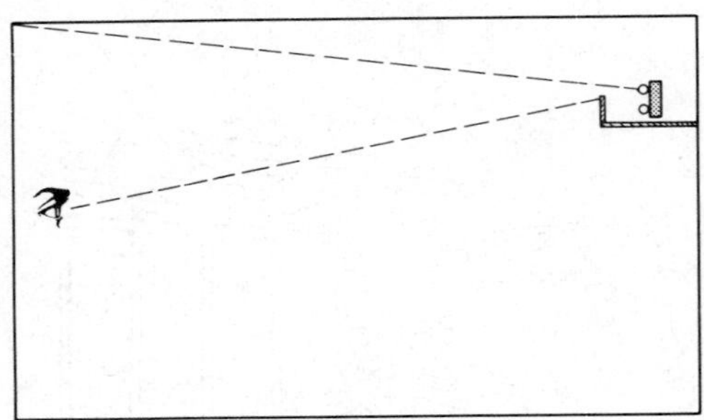

(l) and *(m)* Cove Installations

Proper cove proportions: Height of front lip of cove should shield cove from the eye yet expose entire ceiling to the lamp. Orientation of fluorescent strip as shown is preferable to upright arrangement. ***Westinghouse Lighting Handbook.***

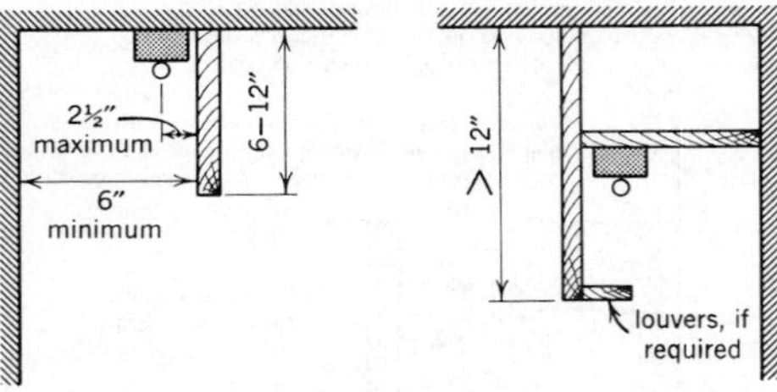

(n) Typical Cornices

Wall washing equipment mounted in valances and cornices provide improved brightness ratios and may be used for lighting desks against walls, or vertical illumination of walls and objects mounted thereon. ***Westinghouse Lighting Handbook.***

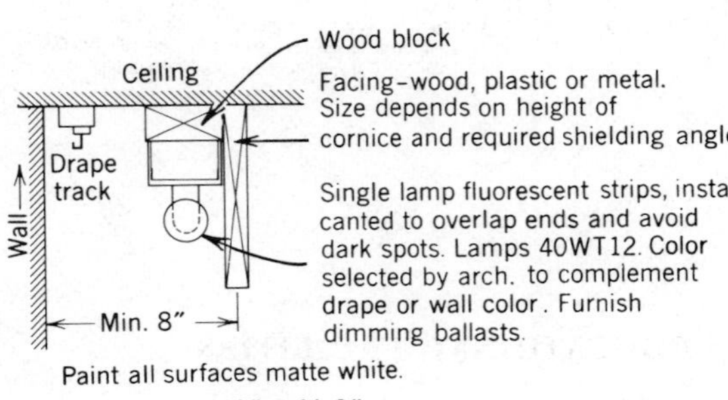

Fig. 21-5 (continued)

institutional buildings. To that extent the remarks herein are applicable there also. Generally school buildings are constructed from capital budget funds and maintained from operating funds. The former is often based on sale of bonds, and makes a fixed amount available. The latter is financed through taxes and is *always* tight. Therefore, all equipment in public buildings must be extremely hardy, long-lived, punishment-proof, as maintenance-free as possible, and low in energy consumption. Maintenance in such buildings is generally poor and on a repair rather than preventive basis. With this in mind as overall criteria, the following remarks apply to lighting equipment.

(*a*) Use source with highest possible efficacy. Remember that daylight has the highest efficacy, followed by HPS, fluorescent, and other HID sources.

(*b*) Where specific color lamps are called for, such as deluxe white, the requirement should be permanently stencilled in large letters on the lighting fixture.

(*c*) Long-life sources should always be given preference because of lower maintenance. Thus corridor and stair lighting should be fluorescent. This is also important in locations where relamping is difficult as in high-ceiling rooms such as gyms and assembly rooms. In such spaces, relamping should be possible by stick, and extended-life lamps are recommended, with preference to HID sources.

(*d*) In calculating levels, low figures for LLF (light loss factors) should be used to allow for aging of paints and dirt accumulation. Cleaning of lighting fixtures in schools is virtually unknown. A figure of 0.5 to 0.6 is reasonable.

(*e*) Most schools are *not* air conditioned. The masking air noise being absent, careful control must be exercised on noise and vibration from ballasts, diffusers, etc. Ballast noise increases with current rating, that is, 430-ma rapid start, 800-ma high output, and 1500-ma very high output lamps. The latter two must therefore be used with caution, particularly in locations that amplify sounds, or where low NC obtains.

(*f*) Lighting equipment must be designed for an absolute minimum of maintenance. This means captive screws, rust-preventive plated parts, captive-hinged diffusers requiring only one man to maintain, ballast replacement without demounting fixtures (plug-in ballasts are available), nonyellowing plastics, and high-quality finish and assembly.

Considerations will be adduced as applicable to the different usage spaces normally found in schools. Levels of illumination are tabulated in Table 18.4.

21.5 Art Rooms

The primary requirement here is for constant color daylight. Thus north windows and skylights are virtually indispensable. For artificial lighting, since color is so important, deluxe fluorescent tubes are recommended. General illumination should be augmented by user-adjustable supplementary lighting in the form of ceiling-mounted accent lights. If modeling is anticipated, spotlights for this purpose are required. For display of artwork, adjustable wall illumination is required. Ceiling track-mounted incandescent units are an excellent choice (see Fig. 21.6). These ceiling tracks can also be utilized to provide outlets for desk lamps (see Section 15.33).

21.6 Assembly Rooms, Auditoriums, and Multipurpose Spaces

The varied activities in these rooms make flexible lighting imperative. For performances, low-level dimmed incandescent lighting is required. Here incandescent is the recommended source because of the lower cost of dimming and short burning periods. For assembly, this can be augmented by architectural elements along walls and drapes, and in the ceiling. For study, additional ceiling fluorescents or HID units can be switched on. The

Fig. 21.6 Art exhibition room, illustrating good and bad techniques. Upper wall fenestration is excellent for deep daylight penetration. Track lighting is ideal for display of art. The mixture of incandescent downlights for general lighting is excessive and an eyesore. Also, the positioning of the track lights can create both direct and reflected glare problems unless the sources are selected properly and ceiling height is above 9 ft.

Fig. 21.7 Schools frequently utilize spaces for multiple functions. This space, normally used as a dining area, doubles as an assembly room. The architecture did not lend itself to conventional fixtures. High-intensity, indirect tungsten-halogen units, in concrete beam junctures, provide sufficient light for both uses.

combinations are legion; the different usages are the critical consideration (see Figs. 21.7 and 21.8). Also, the inaccessibility of high ceilings makes the use of very long-life sources such as HID imperative. With incandescent, 130-v or extended-life lamps are recommended. An additional consideration is step lighting. These should be of very low brightness and mounted to the side of or in risers. Baffles and louvers must be provided in these units to cast light down only, to illuminate the tread, and particularly its leading edge. Acoustic considerations are acute because of the low NC criteria. Thus the generally noisy ballasts of HID sources should be located with care.

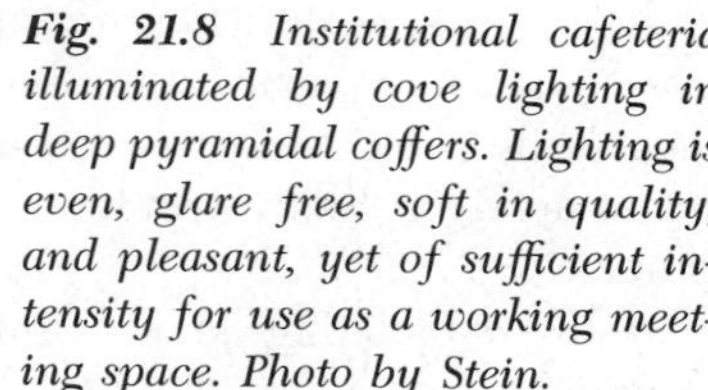

Fig. 21.8 Institutional cafeteria illuminated by cove lighting in deep pyramidal coffers. Lighting is even, glare free, soft in quality, and pleasant, yet of sufficient intensity for use as a working meeting space. Photo by Stein.

Fig. 21.9 This indoor tennis court uses 112 1000-watt metal-halide luminaires directed at the ceiling and reflecting shadow-free light of over 100 fc maintained on to the playing surface. Fixtures are mounted 20 ft above the floor. Sealed luminaires were chosen for both photometric performance and the low maintenance provided by their dustproof construction. Courtesy of Widelite Corp.

21.7 Gymnasium Lighting

Gyms present a situation similar to auditoriums in that they have widely varying usages. All fixtures should be sturdy and guarded. For gyms deluxe-white mercury, HPS, and metal-halide are excellent choices for color, life, control, and efficiency. More than one level should be available by switching [see Section 19.18(f)]. For dance and assembly use, other fixtures can be lamped with long-life incandescent or tungsten halogen, which provide good color for low-intensity lighting and also provide illumination during mercury startup or restart after an outage. Ratio of HID to incandescent can be between 4:1 and 5:1. All fixtures should be designed for relamping from the floor by means of stick or pole. Locker rooms should use guarded strip fluorescents. An interesting application of HID lighting in an indirect system is shown in Figure 21.9. Recognition of a possible problem with dirt accumulation and relamping is necessary with such an arrangement.

21.8 Classrooms

The essential room in the school is the classroom. Refer to Figure 18.47, page 760, for recommended surface reflectances. The modern classroom utilizes extensive audiovisual teaching aids and therefore requires multiple lighting levels. This is most economically accomplished by multiple switching and multilevel ballasts.

(*a*) ***Energy Considerations and Sources.*** Use fluorescent, standard 430-ma lamps, cool and warm white, for direct and direct-indirect fixtures; 800- and 1500-ma lamps, same colors, for semi-indirect and indirect lighting. Incandescent sources should not be used. Daylight, to the maximum extent, is desirable. Lights adjacent to daylight sources should be separately switched.

(*b*) ***Choice of Lighting System.*** With proper design, adequate lighting can be provided with 2.5 w/sq ft. Since ESI is so sensitive to viewing position and the recommended level is 70 fc ESI, the designer must be inherently familiar with the type of system, fixtures, and arrangements that will yield the requisite high ESI without exceeding energy limitations. To this end careful study of Reference 1 and a reading of References 2 to 5 will be of considerable assistance. Some general guidelines are:

1. Run rows of fixtures front to back and, if possible, between rows of students.
2. Use high VCP units.
3. Use semi-indirect lighting with HO or VHO fluorescent; direct-indirect, parabolic reflector, or batwing distribution fixtures with standard 430-ma lamps.
4. When using HO or VHO take special precautions against excessive ballast noise.

(*c*) ***Costs.*** Refer to Table 20.1, page 860. Convert annual costs to life-cycle costs using data given in Section 21.25. *Do not* compare costs on the basis of dollars per footcandle,

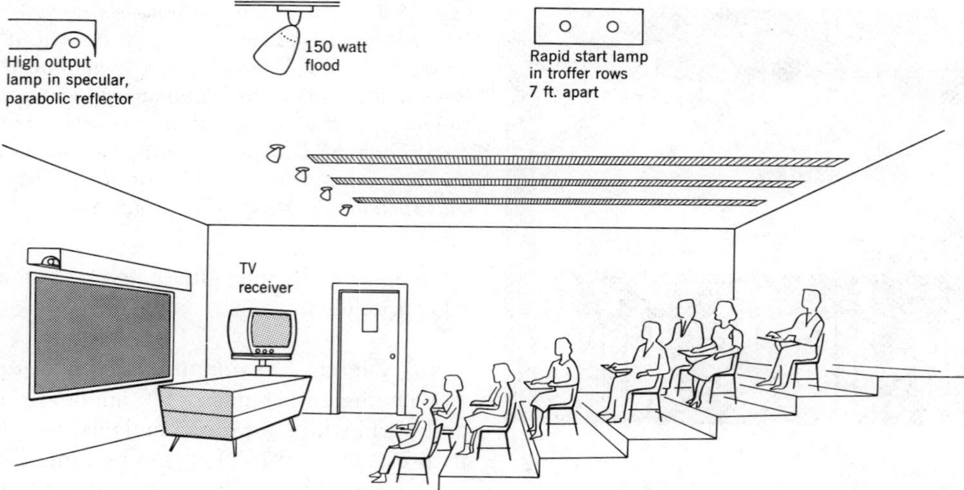

Fig. 21.10 *Lengthwise section through lecture room shows elevated seats toward rear, troffer general lighting in a suspended ceiling, surface-mounted adjustable R40 lamp units over demonstration table, and fluorescent chalkboard lighting.*

that is, by dividing maintained illumination by cost, since this leads to preference for higher footcandle levels and higher wattage levels. *Do* compare life-cycle costs of alternate adequate illumination systems. Note that the analysis of Table 20.1 does not include wiring costs and the effect on air conditioning. The latter, though rarely important in schools, is very important in office occupancies.

21.9 Lecture Hall Lighting

Lecture hall lighting is similar with respect to sources and other considerations to classrooms. Three-level fluorescent lighting which is readily accomplished via switching and multilevel ballasts, eliminates energy-wasting incandescents. Low levels are necessary during demonstrations, films, and the like. Auxiliary lighting for demonstration table and chalkboard complete the design. Step lights are not normally utilized. High-ceiling installations can utilize mercury or metal-halide for general lighting. Controls for lighting should be at the demonstration table (see Fig. 21.10).

21.10 Laboratory Lighting

Laboratories differ from classrooms in that tables are fixed, bench surfaces are frequently very dark, many of the items used exhibit specular reflection, vertical surface illumination is important, and visual tasks are not normally prolonged or severe. With low ceilings, use direct fixtures located parallel to and slightly behind the edge of tables to avoid reflected glare. These fixtures should have as wide a distribution as possible. If ceiling height is sufficient, indirect lighting is highly desirable for the same reason. Indirect lighting will also provide a high degree of diffuseness necessary for vertical surface illumination (see Fig. 21.11).

21.11 Library Lighting

Libraries comprise several different seeing tasks, each of which requires its own lighting solution.

(*a*) ***General Reading Room.*** Here two solutions are possible and both are in common use. In the first general lighting is supplied over the entire area, which is sufficient for reading tasks. For this purpose fluorescent or HID sources such as deluxe mercury or metal-halide (see Fig. 21.12) are normally ap-

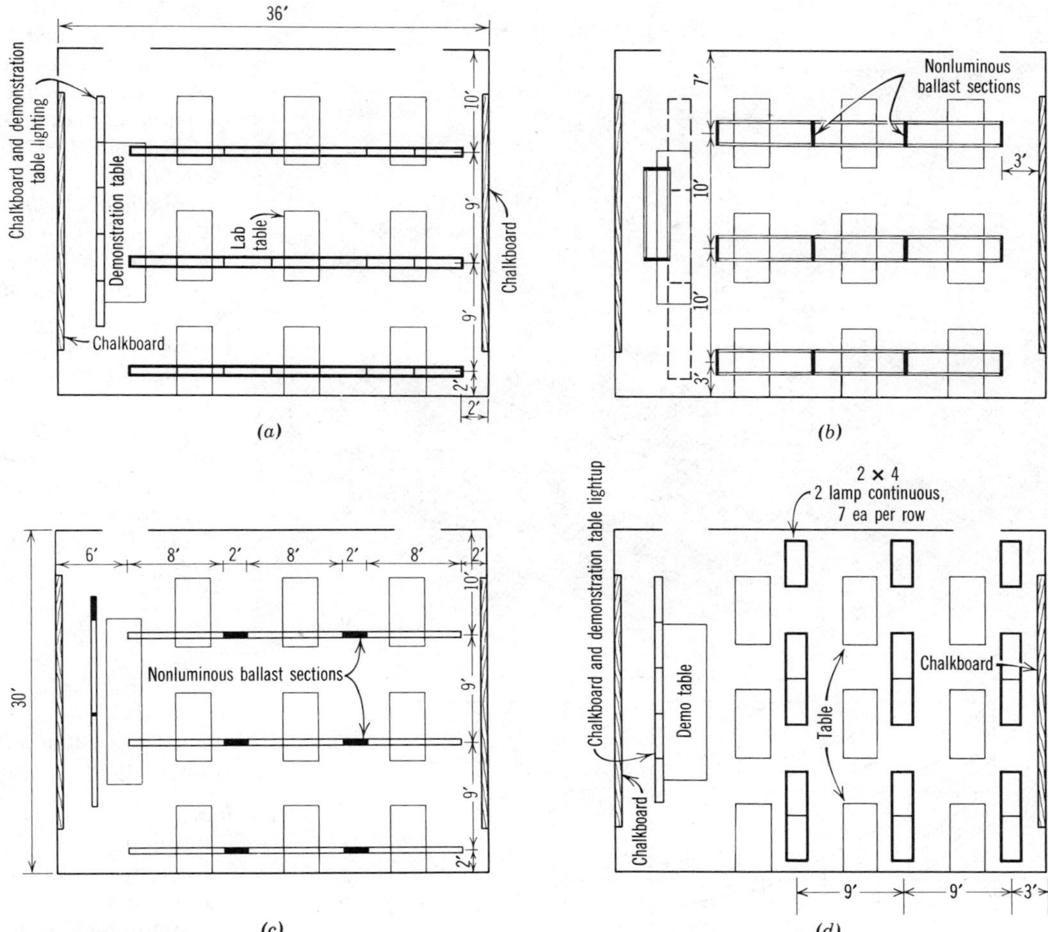

Fig. 21.11 *Laboratory lighting schemes. Running lights across tables or in aisles is generally preferable from a reflected-glare aspect. Scheme* (a) *utilizes pendant direct-indirect units, while* (b) *and* (c) *are variations of the single, semi-indirect HO and VHO unit design, illustrated in Table 20.7, Fixture No. 28. Relatively high noise level of ballasts is generally not objectionable in labs. Sketch* (d) *utilizes direct units in low ceiling, positioned for minimum glare.*

plicable, the latter two with ceiling heights at least 10 ft. The long life and good color and efficacy of these sources are suited to the long burning hours found in libraries. The second solution involves low-level general lighting supplied by fluorescent or HID sources supplemented by local reading lighting on the tables or carrels. Reading lights should be fluorescent, arranged to avoid veiling reflections (see Section 21.16).

Wherever HID sources are used, an additional instant restart source must be supplied to supply minimal lighting after an outage. The units illustrated in Figure 21.12 contain a small tungsten-halogen source for this purpose. Ballast noise can be a problem in low NC criteria spaces such as libraries. Special low-noise ballasts and enclosures are available and should be employed (see Fig. 21.13).

(*b*) *Stack Areas.* In stack areas required vertical surface illumination is best supplied

Fig. 21.12 Reading room in foreground is illuminated with 400-w deluxe white mercury at elevations of 10.5 and 14.5 ft. Metal-halide would be equally applicable. Stacks in background are lighted with fluorescent strips designed to give good vertical surface illumination required for scanning books titles on shelves. (Note shape of light scallop on wall and compare to distribution curves in Figure 21.14) Courtesy of Widelite.

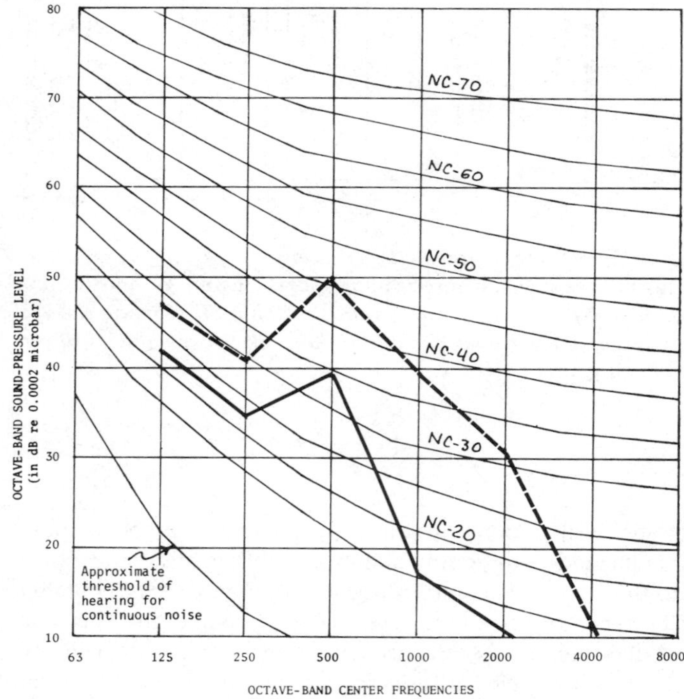

------- SilentGuard ballast in special mounting box

——— UltraQuiet ballast in special mounting box

Fig. 21.13 Acoustic Analysis of HID (Ballast) installation.

Total ballast noise for a complete installed lighting system is predicted by supplying ballast noise data, lighting-system parameters, all relevant acoustical characteristics such as noise reflection and absorption properties for room materials and configuration plus occupants. The plots here show standard Wide-Lite "SilentGuard" ballasts, mounted in the special box employed in the Columbia Falls library, and the improved performance provided by special UltraQuiet ballasts. Industry standards recommend an N-C noise level of 30 to 40 db for public libraries. See Table 27.11, page 1255. Courtesy of Widelite.

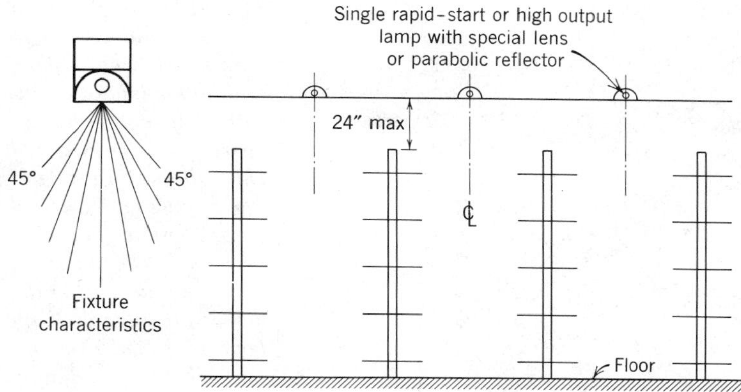

Fig. 21.14 Stack lighting is best accomplished by fixtures with lenses specifically designed for the purpose. Baffles and plastic diffusers generally do not give adequate vertical surface illumination.

by one of the special fluorescent units designed for this purpose. These are mounted between, and no higher than 24 in. above stacks for best results (see Fig. 21.14).

(*c*) ***Card Index Files.*** Vertical surface illumination is best supplied here by fluorescent or HID sources directly above and in front of the files.

(*d*) ***Work Desks and Checkout Areas.*** The lighting requirements for this area are the same as those for the reading room except that levels are somewhat higher. Supplementary lighting is recommended.

21.12 Specialty Room Lighting

Rooms with specific tasks such as sewing and typing rooms and shops must be carefully considered. Where moving parts are present, adjustable local lighting may be necessary to aid seeing by reflection from specular work. No general rules can be suggested other than those previously explained in detail. Particular attention must be given to often forgotten vertical illumination requirements, which are vital in areas such as storage, stacks, and shops. Kitchens and cafeterias are color sensitive, and sources must be carefully chosen. If dining areas serve other functions they must be considered in the total lighting.

21.13 Corridors and Stairways—All Buildings

Corridors intended only for circulation need only be lighted to 10 fc± unless a specific seeing task requires higher levels, for example, bulletin boards and lockers (see Fig. 21.15). Stairs, as previously noted, require clear delineation of the treads, generally by shadowing. Lighting can also be used to give direction by longitudinal arrangement. Wall-mounted or recessed wall lighting is particularly effective in corridors furnishing walk illumination plus lighting for posters, bulletins, and so on. Fluorescent luminaires mounted across corridors, particularly when corridors are long, are effective in reducing the tunnel impression. Incandescent sources are not recommended, because of low efficacy, high maintenance and frequency of relamping. Fluorescent and HID sources are also suggested for stairwells. Care must be exercised here, however, to avoid direct glare, which causes attention to shift from the stairs to the light and may thereby cause a hazard (see Fig. 21.16).

COMMERCIAL INTERIORS

21.14 Office Lighting

The following information applies primarily to offices in commercial buildings and sec-

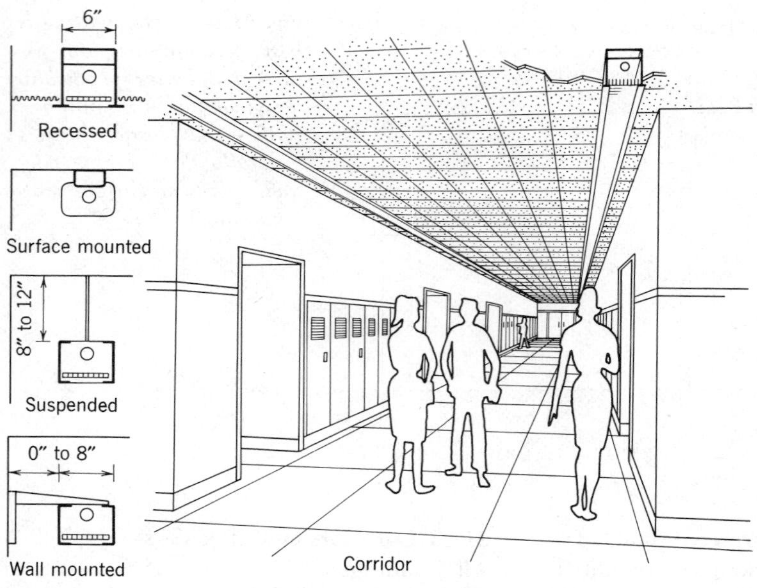

Fig. 21.15 Lighting of school corridors. High-reflectance walls, floor, and ceiling improve utilization of light and increase the feeling of cheerfulness. The lighting technique illustrated is appropriate for school corridors. The rows of luminaires at each sidewall illuminate bulletin boards, special displays, and the faces and interiors of lockers more effectively than units centered in the ceiling.

Fig. 21.16 Corridor illuminated by large cans producing strong pools of light on the floor. Walking the length of such a corridor is not a pleasant experience, nor is reading a room number easy because of the extreme brightness ratios.

ondarily to similar spaces in other occupancies, such as in educational and industrial buildings. In these latter cases, the general remarks applicable to those type facilities take precedence.

(*a*) ***Sources.*** Use fluorescent in low-ceiling areas and HID or pendant fluorescent in high-ceiling spaces. Color is not critical. Warm-white and cool-white fluorescent and deluxe mercury or metal-halide are usable. HPS can be used to advantage when mixed with deluxe mercury or metal-halide to reduce possible objections to the yellow color. Because of high output, HID sources may create direct glare problems, which can be minimized by use of a low-brightness, lens-type diffuser (see Fig. 21.26). Use low-wattage, energy-saving lamps as detailed in Section 19.13(c) for fluorescents. Incandescents may be used for storage areas, closets, and other short-burning period uses. Incandescent and tungsten-halogen track lighting is used to illuminate displays of all sorts.

(*b*) ***Illumination Levels.*** These are given in Tables 18.4 through 18.9, pages 732-9, and recommended finishes for room surfaces are given in Figure 18.48, page 761. Notice in Figure 18.48 that the upper-wall dado has a lighter finish than the remainder of the wall. This serves the double function of increasing ceiling cavity brightness, particularly with suspended fixtures, and increasing vertical illumination due to reflection from this surface. The basic design approach should be for non-uniform lighting. See Example 20.2 for general lighting design and Sections 20.37 through 20.41 for supplementary lighting design.

(*c*) ***Vertical Surface Illumination.*** This is required for visual tasks in offices, such as files, desk drawers, card files, and secretarial copy stands. Large area luminaires and a high degree of diffuseness are desirable. This is especially true in large offices where wall reflections are absent. Light-finish furniture surfaces, surface-mounted fixtures, and high-reflectance floors will also assist in this.

(*d*) ***Private Office.*** Here the lighting should be carefully designed for the specific areas and work involved; general and supplementary lighting with variable levels for desks, downlights in sitting areas, wallwash for accent and brightening dark walls, special lighting for display boards, paintings, and so on. If the ceiling is high, pendant fixtures can create a horizontal plane to correct poor room proportions.

(*e*) ***General Offices.*** These areas can use higher ceilings and mounting heights for fixtures. Direct-indirect distribution pendant units provide good diffusion, low direct and reflected glare, and good vertical light component but are dirt catchers and are unattractive in long pendant lines. Since desks and tables are frequently set near walls, additional peripheral lighting is frequently required.

(*f*) ***Office Lighting Equipment.*** Office lighting equipment should not be handled roughly. Fixtures may have touch latches, light hinges, and adjustable devices without fear of breakage or vandalism.

(*g*) ***Maintenance.*** In most offices maintenance is provided on a trouble call basis. Lamps are replaced on burnout, and the fixture is then cleaned. Because of the long life of fluorescents and HID sources, this generally means a 3- to 5-year cleaning cycle. Since most offices use lay-in troffers, an LLF of 0.6 to 0.65 is reasonable in air-conditioned spaces; lower in open-window offices.

(*h*) ***Noise.*** Noise from air conditioning plus adjacent street, traffic, and process noise, makes the use of higher-noise-level ballasts, as are found on HO and VHO lamps, frequently feasible. For private offices, A-rated ballasts for 430-ma lamps are the best choice. (For more detail, see Chapter 27.)

(*i*) ***Fenestration.*** When fenestration is absent, a lighted valance is recommended around the room. This will remove the wall-ceiling line and will partially compensate for the lack of windows. It will also brighten the walls and increase illumination on desks placed adjacent to the walls.

(*j*) ***Supplementary Lighting.*** Supplementary lighting can be mounted on ceiling–track-fed poles of the type described in Section 15.33. These are, in effect, track lights, mounted on vertical tracks.

(*k*) ***Switching.*** Provide switching to allow for:

1. Small groups of lights to remain lighted while the remainder are off.
2. Path lighting through large spaces to permit traverse without turning on all lights.
3. At least two levels via alternate lamp switching in areas where task severity changes or where daylight contribution varies. This is definitely preferable to entire fixture switching.
4. Central control to permit switching of large blocks of lighting.

With these general guidelines in mind, the following sections will discuss specific topics in office lighting.

21.15 Nonuniform Office Lighting Design Using Ceiling-Mounted Units

As stated in Section 21.14(b), modern office lighting design is predicated on nonuniform layout. This approach has three advantages:

a. Higher ESI.
b. Lower first cost.
c. Lower power and energy, yielding lower operating costs.

The general or *ambient* lighting can either be a uniform layout at a low level, the spill light from local lighting, or a combination of both. Figure 18.36, page 751, is an example of the former; Figure 21.17 of the latter.

Refer to Figure 21.17*a*, which is an office layout for a small technical office. In accordance with current recommendations, 50 fc is to be furnished on all work surfaces except the drafting table DT, and two small offices where close work on poor copy will be performed. Figure 21.17*b* shows a standard uniform layout, generally using three-lamp, 2 × 4 fluorescent troffers, with high-quality lens diffusers. VCP is good due to low lens brightness. Illumination on the drafting table is also good because of the fortuitous position of the fixture above it. Other work surfaces receive more than adequate raw footcandles but in many instances low ESI due to bad veiling reflections. In Figure 21.17*c*, fixtures have been arranged to supply task lighting. Ambient lighting comes from spill light and three fixtures placed in a relatively large, open circulation space. The results are good VCP, much higher ESI footcandles on the tasks, and somewhat lower energy use. As pointed out in Figure 20.15, a disadvantage of this system is the jumbled appearance of the ceiling. For this reason, nonuniform layouts are well suited to coffer-type ceilings, as in Figures 21.18 and 21.24, where the presence or absence of a fixture is not as prominent. This is all the more important where large glass wall areas make the ceilings readily visible from the outside.

This appearance problem can be mitigated considerably by using a uniform arrangement with adequate switching to provide the levels

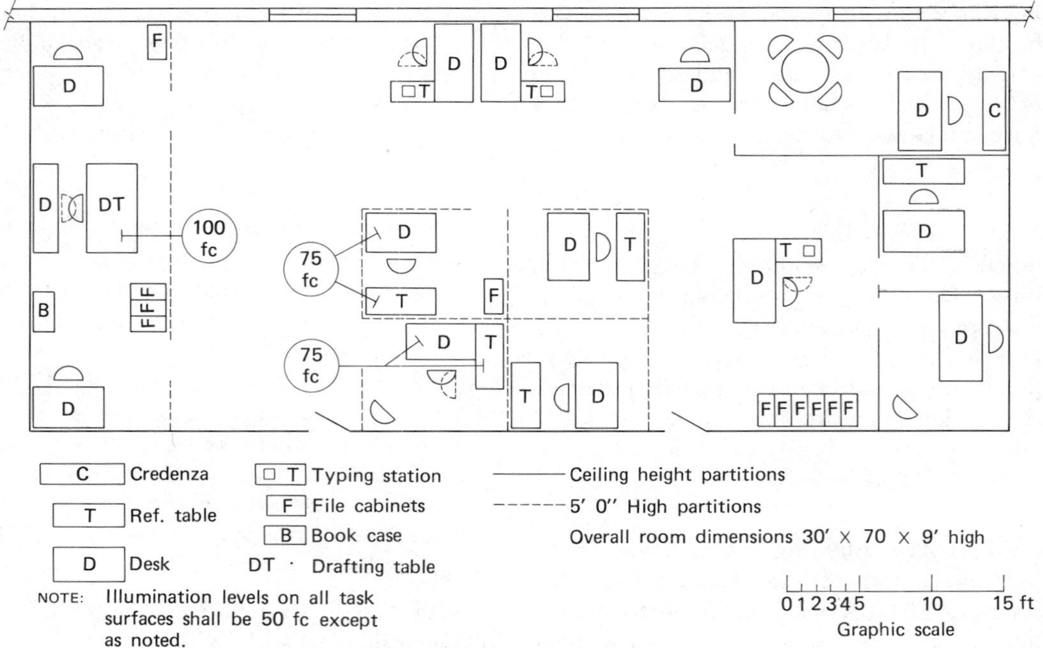

Fig. 21.17 (a) *Typical layout for a small technical office (from Reference 6).*

desired. This type of system, illustrated in Figure 21.20*a* to *d*, has the added advantage of usefulness when the furniture layout is not known in advance—a common occurrence. Its disadvantage is higher cost for fixtures, multi-level ballasts and switching, and lack of fixture-position control, so important in achieving low reflected glare. Figure 21.20*e* illustrates another technique in minimizing the appearance problem of nonuniform lay-

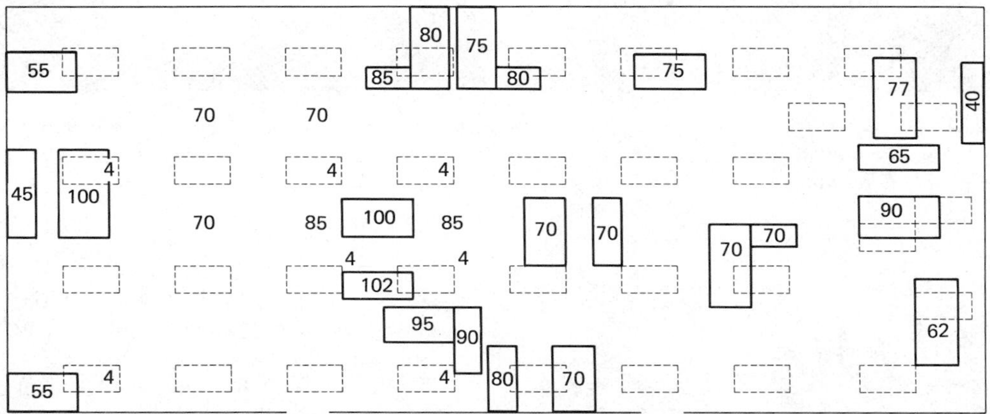

All 3–lamp F40 cw fixtures except where marked "4".
Footcandles are average over the desk.
35 fixtures; 5600 w total; 2.66 w/sq ft.
VCP > 80.
ESI – position dependent; generally poor to fair.
Average task illumination – all task; 75 fc raw, approx 45 in. ESI.

Fig. 21.17 (b) *Uniform lighting layout using 3-lamp 2 x 4 troffers except where "4" indicates a 4-lamp unit.*

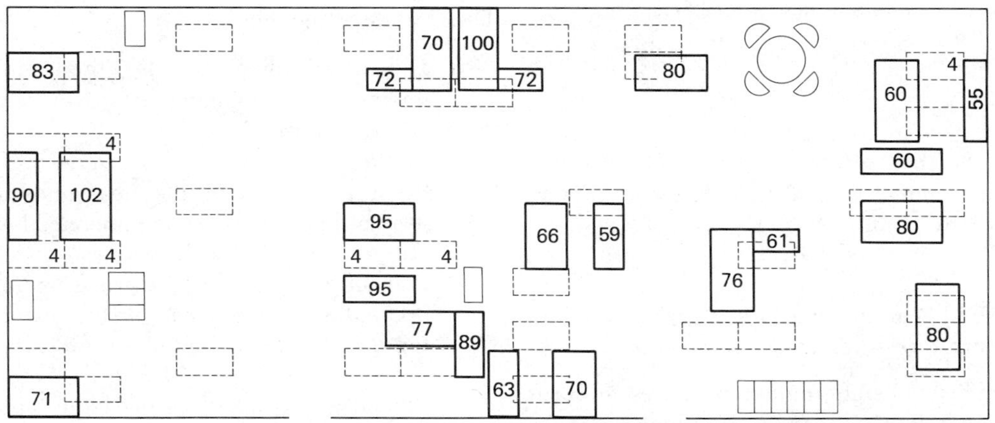

All 3–lamp F40 cw fixtures unless noted "4".
Footcandle figures are average over the desk, raw.
34 fixtures; 5400 w; 2.6 w/sq ft.
VCP > 80; ESI – good to excellent.
Average task illumination, all tasks – 76 fc raw; ~70 fc ESI.

Fig. 21.17 (c) *Nonuniform layout for the same space using 3-lamp and 4-lamp fixtures as indicated. All footcandles valued are* raw *footcandles, calculated by point-by-point methods.*

Fig. 21.18 A coffer ceiling room (a) *is illuminated with 1 x 4, 2-lamp units in alternate coffers. the appearance unlighted* (a) *and lighted* (b) *is symmetrical and pleasing. A nonuniform layout would be less objectionable here than in a flat hung ceiling. Photos by Stein.*

Fig. 21.19 A square modular ceiling utilizing square lighting units demonstrates architectural symmetry, blending, and incidentally that lighting is easily visible from the outside, even during the day. Photo by Stein.

out. Smaller sources are less objectionable when symmetry is absent. Refer to Figure 21.17*c* to confirm this.

21.16 Nonuniform Office Lighting Using Furniture-Integrated Luminaires

In lieu of ceiling-mounted fixtures, ambient and task lighting can be supplied by furniture-mounted units. This system has these advantages:

a. The problem of furniture layout and layout changes is eliminated.
b. Initial construction cost is reduced.
c. Energy requirements are lowered because of short distances between light source and task.
d. Each occupant has ON-OFF control of his or her task lighting and, in some designs, positioning control as well.
e. Maintenance is very much simplified, since fixtures are readily accessible from the floor.
f. Floor-to-floor height can frequently be reduced.
g. Tax advantages normally accrue due to higher depreciation rates on furniture than on the building.

Disadvantages include:

a. Difficulty in dissipating heat and minimizing ballast noise due to proximity of sources to user.
b. Veiling reflections can be severe.
c. Brightness ratios in the near and far surround may exceed recommended levels.
d. Difficulty in lighting a free-standing open desk, since most of the fixture types are undercounter or side-wall mounted. See Figure 21.22*e*.
e. Difficulty in evenly lighting large table or L-shaped desk areas because of the concentrating nature of the lighting units.

Figure 21.21 graphically shows the problem of local desk lighting. Figure 21.22 is a

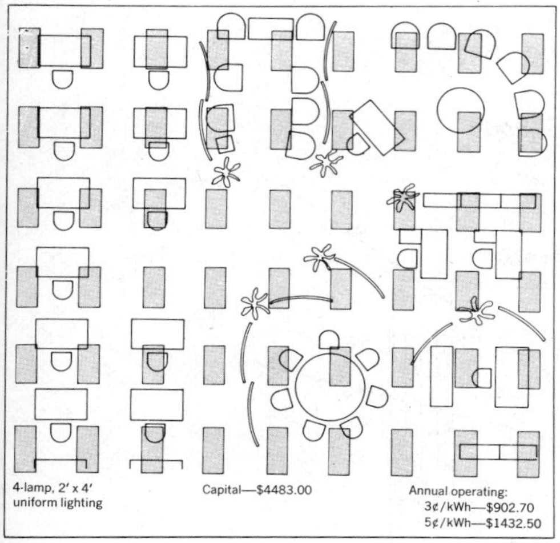

4-lamp, 2′ x 4′
uniform lighting
Capital—$4483.00
Annual operating:
3¢/kWh—$902.70
5¢/kWh—$1432.50

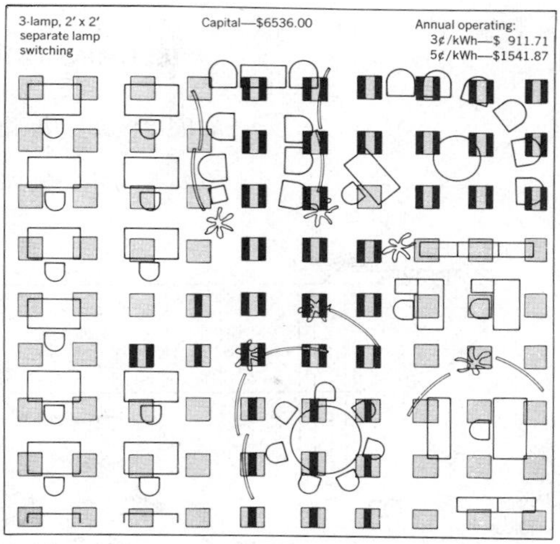

3-lamp, 2′ x 2′
separate lamp
switching
Capital—$6536.00
Annual operating:
3¢/kWh—$ 911.71
5¢/kWh—$1541.87

(d)

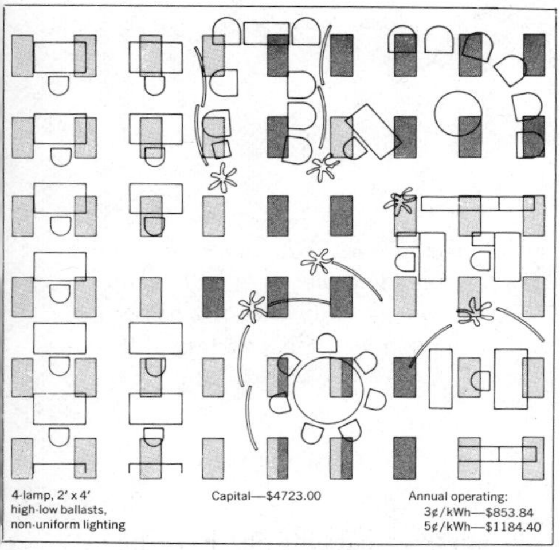

4-lamp, 2′ x 4′
high-low ballasts,
non-uniform lighting
Capital—$4723.00
Annual operating:
3¢/kWh—$853.84
5¢/kWh—$1184.40

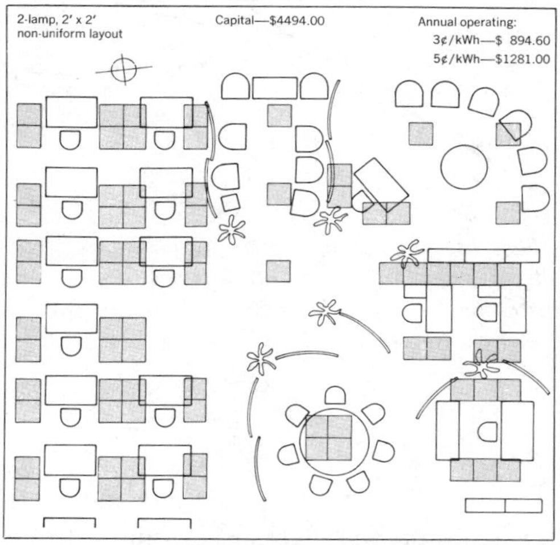

2-lamp, 2′ x 2′
non-uniform layout
Capital—$4494.00
Annual operating:
3¢/kWh—$ 894.60
5¢/kWh—$1281.00

(e)

3-lamp, 2′ x 2′
U-tube, uniform
Capital–$6376.00
Annual operating:
3¢/kWh—$1331.70
5¢/kWh—$2037.30

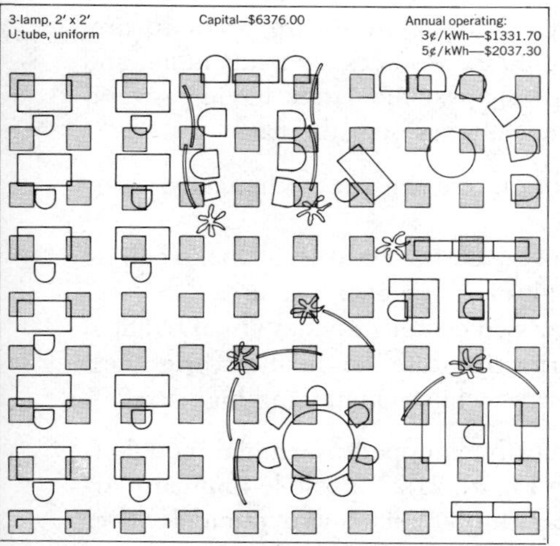

Fig. 21-20

(*a*) One hundred footcandles is supplied uniformly on the workplane by a very typical uniform layout of 2 x 4 ft troffers using four 40-w fluorescent lamps.
(*b*) A modification of Figure (*a*) using high/low ballasts that operate lamps at full or half output. Such switching can be accomplished locally in the luminaire or possibly from a remote, central location. Lighting is reduced for casual seeing, kept at 100 fc (or nearly so) for desk work. The shaded luminaires indicate low-level operation.
(*c*) A regular pattern of 2 x 2 ft troffers with three U-shaped lamps, as they might be used in a building module approach to supply 100 fc.
(*d*) A modification of Figure (*c*) that involves switching lamps individually to provide illumination appropriate for the areas. Darkened portions at the luminaire indicate either one or two lamps extinguished.
(*e*) A nonuniform array of 2 x 2 ft troffers in a flexible 2 x 2 ft ceiling system has the advantage of reducing veiling reflections, but the disadvantage of requiring the movement of lighting equipment whenever work positions are moved. The owning costs shown include the cost for a single move for all the luminaires during 10 years. One hundred footcandles is provided on the desks. *Assumptions:* Wiring cost: $100/kw. Luminaire installation: $20 to 30 each. Maintenance: 3-yr group relamping. Replacement labor: $0.25/lamp. Cleaning cost: $2.00 per fixture. Operation: 3000 hrs/yr

Fig. 21.20 *Nonuniform lighting designs. Reprinted with permission of the IES from* Lighting Design and Application, *October 1974.*

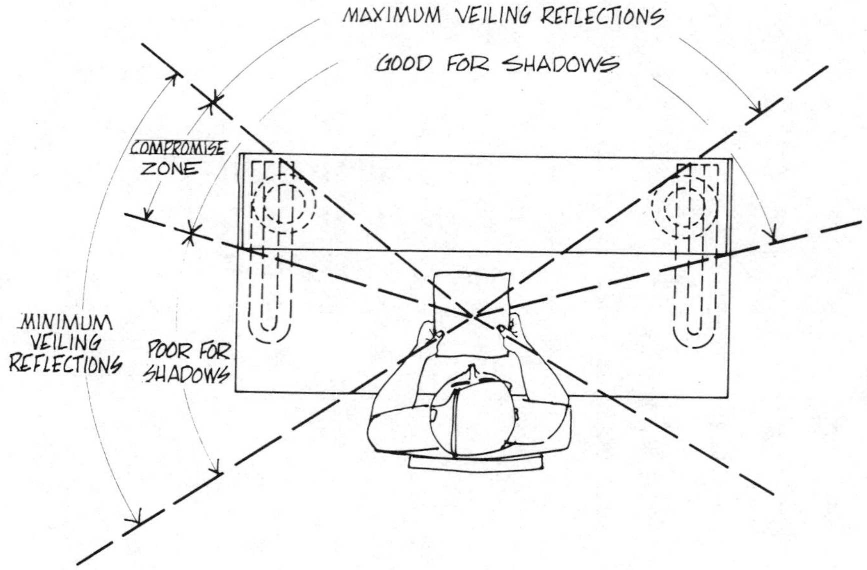

Fig. 21.21. Task lighting can create severe veiling reflections or hard shadows if the luminaire location is not carefully selected. Further constraints are location in elevation for shielding (needs to be low) and for good distribution (needs to be high), appearance considerations, and compatibility with the working physical movements. Reproduced from Lighting Design and Application, *September 1976, courtesy of IES.*

plan view of a prototype office space using this system, with views of this office mock-up, taken at the points indicated. The reader may judge the effectiveness of this particular design, which is one of many now appearing commercially. Table 21.3 is an economic analysis of the illustrated system versus an equivalent fluorescent-ceiling lighting system, including air-conditioning costs. To convert total annual operating and owning costs to life-cycle costs requires a knowledge of furniture-fixture life, which is beyond the range of our study. See Section 21.25 for the necessary equations and technique.

21.17 Integrated and Modular Ceilings

The cost, appearance, and design-flexibility advantages of an integrated ceiling design over field-assembled and coordinated systems has long been known. As a result ceiling systems with integrated lighting, acoustic control, and air-handling capabilities have been made commercially available in modular sizes, among which are 60 in. square, 48 in. square, and 30 × 60 in. Modules are made in flat and pyramidal shapes, the latter having several distinct advantages over the flat:

a. More interesting and esthetically pleasing.
b. More acoustic absorbency due to ceiling angles and large surface area.
c. Recessed center provides visual baffling, permitting use of higher brightness sources while maintaining high VCP.

A typically equipped pyramidal module is shown in Figure 21.23. Possible luminaire arrangements for both flat and pyramidal shapes are given in Figure 21.24, with examples of installations. In addition to the design flexibility available, electrified track can be integrated into the system runners, to supply both the lighting fixtures and power poles, as in Figure 15.44, page 580. This increases the

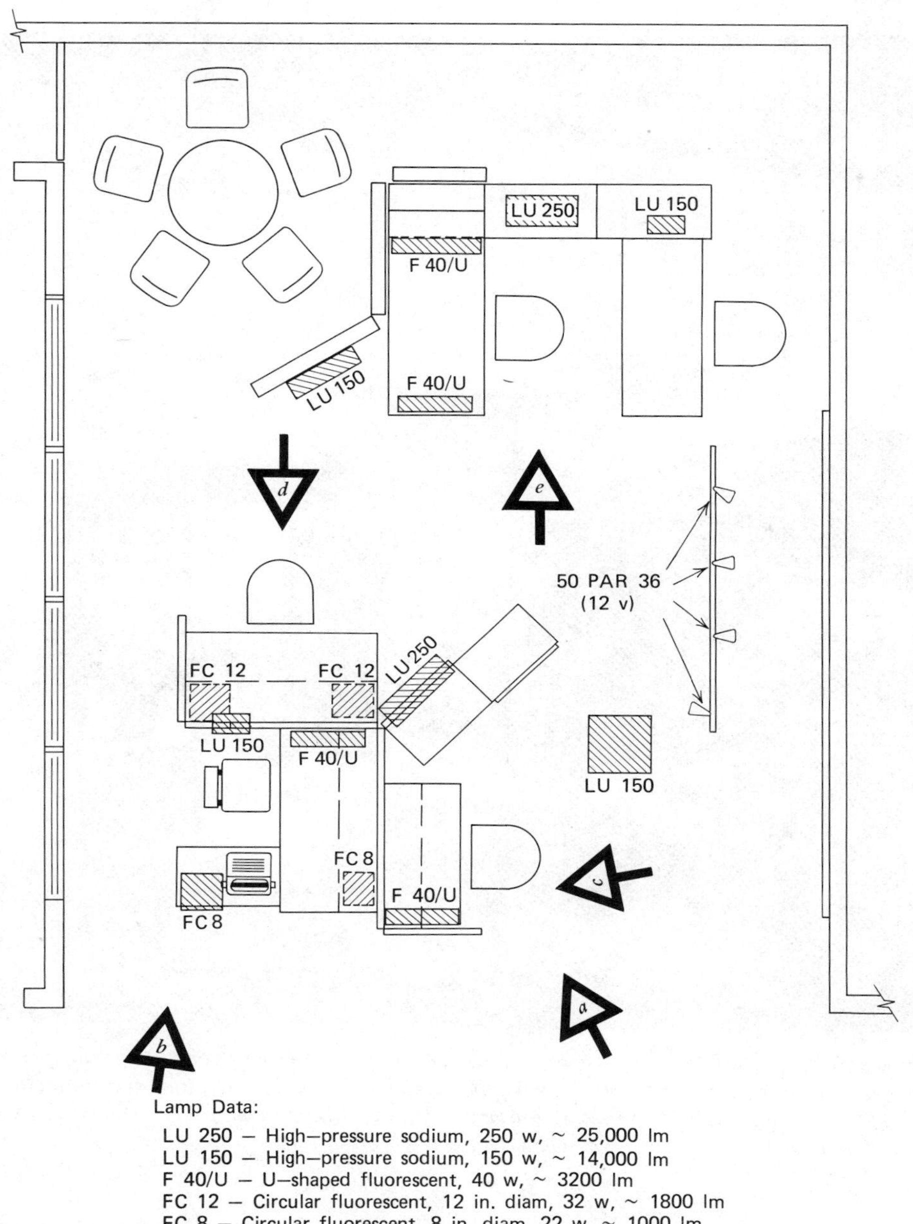

Fig. 21.22. *The plan of the prototype office space utilizing furniture-integrated lighting. The three work stations at the bottom are based on a Y shaped panel system 61 in. high; the two work stations at the top relate to lateral file units. These two stations are the most common forms of work stations used in offices today. The conference area at the top left is formed by 84-in.-high panels. The floor-standing pylon lighting unit is shown at lower light. The ceiling-mounted track system at right lights a tackboard on the right wall and a smaller one attached to the file cabinet at bottom center. Photographs* (a) *to* (e), *which follow, correspond to arrows lettered* (a) *to* (e) *above and show the results of the lighting. Reproduced by permission of IES and General Electric, who supplied photos.*

(a) *Normal view of office furniture work station arrangement as seen from observer entering the space. This is a prototype installation used for evaluating the approaches investigated for furniture-integrated lighting. The indirect illumination utilizes 150-w and 250-w lamps. The task lighting is supplied by compact (U-shaped and circular) fluorescent lamps in the warm-white color.*

(c) *Task lighting for the sales work station from the U-lamp unit at the left provides light with minimal veiling reflections. It is reinforced by the under-shelf two-lamp, 20-w, side mount unit located above the lateral file storage unit in the center of photo. There can be opportunities such as this for lighting such tasks that do not lie on the work station work surface; but where there is not, a sufficient amount of ambient illumination is then needed.*

(b) *Task lighting arrangement for the secretarial work station is a U-lamp fluorescent luminaire and a circular fluorescent lamp luminaire to light tasks at the desk with minimal-size equipment, and minimum veiling reflections. Another open bottom unit utilizing the 8-in circular fluorescent lamp is in a laterally tracking unit to locate to the right or left of the typewriter for copy illumination.*

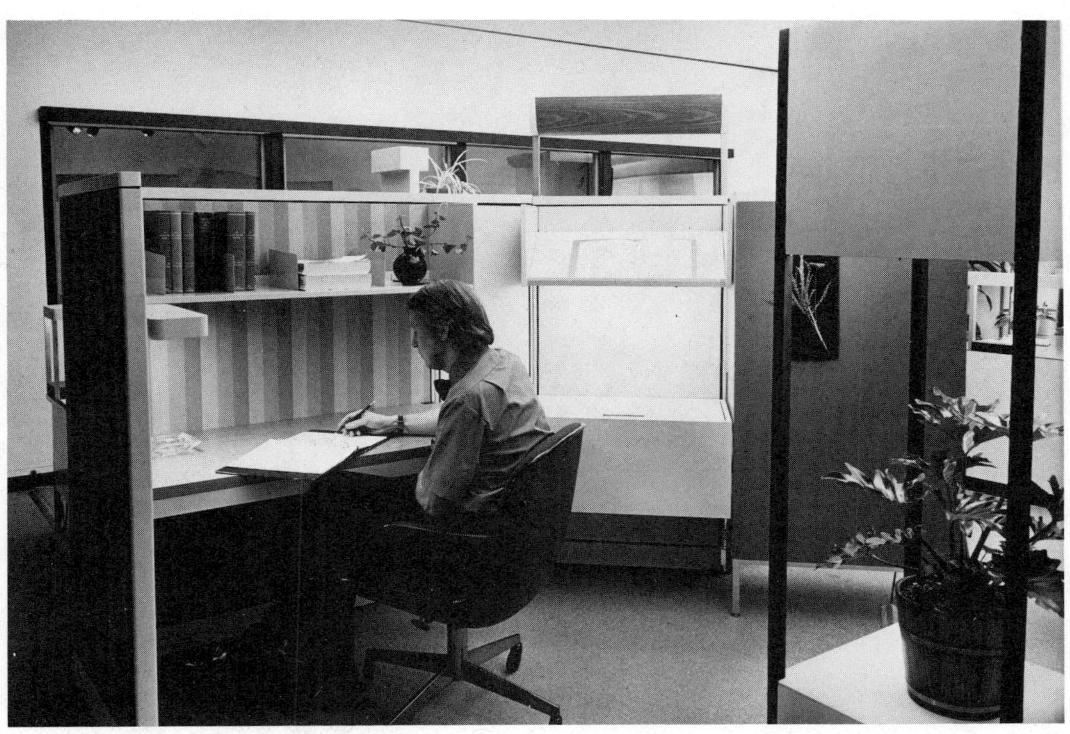

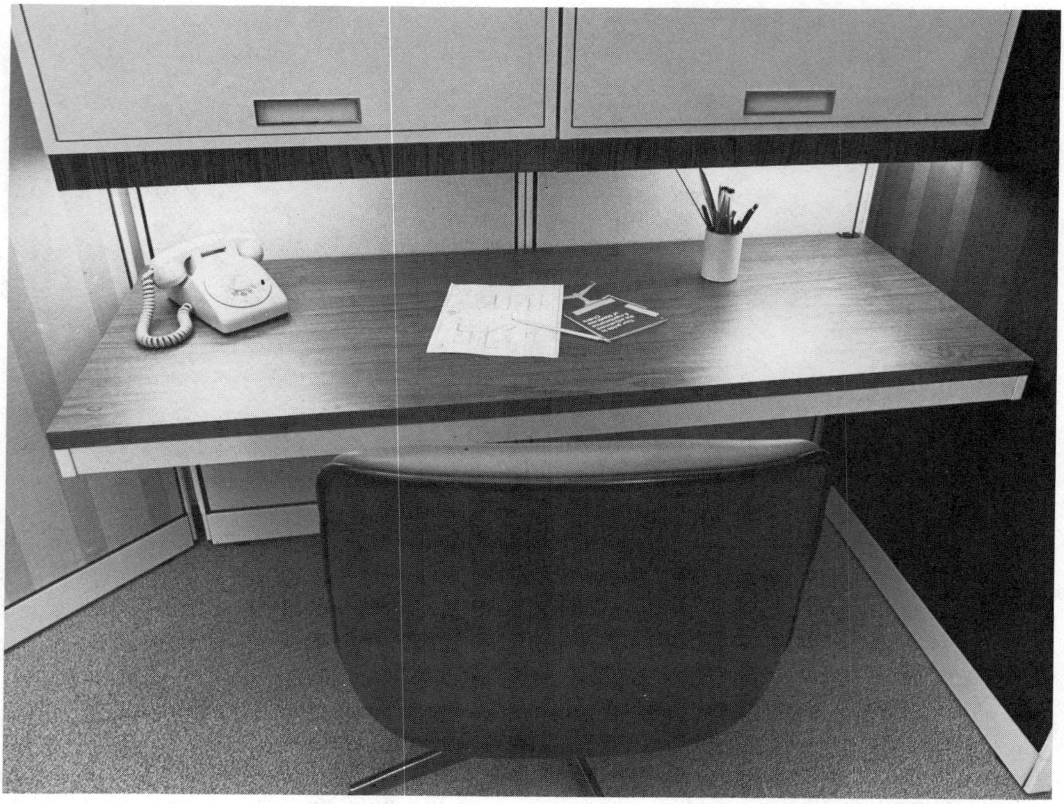

(d) *Work station with double side illumination for minimum shadowing and veiling reflections, supplied by two open bottom fixtures with 8-in. circular fluorescents.*

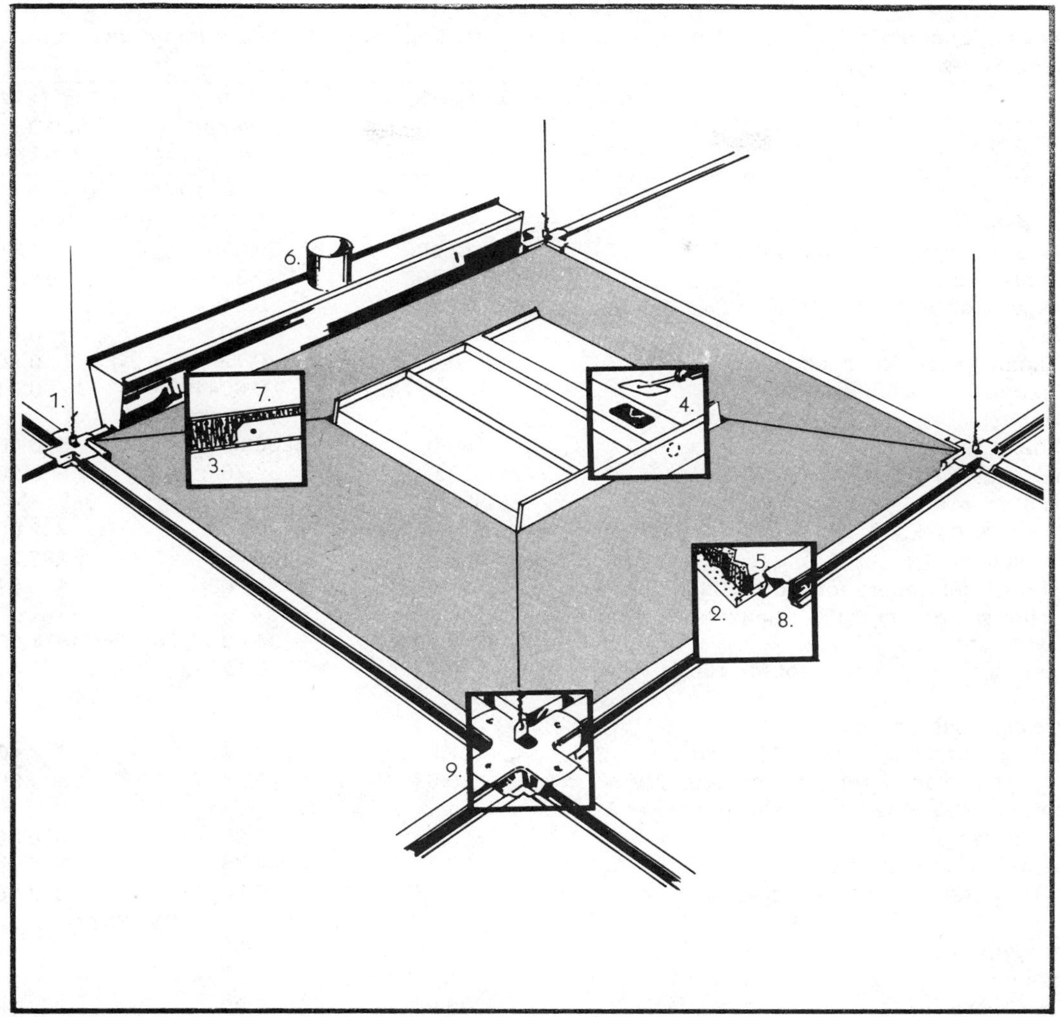

Fig. 21.23 *Plenum view of typical pyramidal module. Courtesy of Guth Lighting.*

1. *Typical hanger wire connection.*
2. *Perforated metal acoustical coffer.*
3. *Fiberglass sound-absorbing insulation material.*
4. *Wiring access plate for fast electrical connection of lighting fixtures.*
5. *Sound-attenuating panel prevents noise travel from room to room through plenum.*
6. *Air diffuser available in insulated or noninsulated versions.*
7. *Preassembled splay is ribbed for strength and rigidity.*
8. *Truss channel sides painted black to give splay a floating appearance from floor level.*
9. *Truss channel screws or rivets to intersection for fast accurate assembly of grid.*

(e) *A special "furniture-filler" unit is shown that uses a 24-in U-shaped 40-w fluorescent in the extended hood (background, supporting books and plant) and a special reflector to deliver light to the center of the desk. This unit could also be used at the open end of the desk and has facilities for telephone and file storage.*

Table 21.3 ***Economic Comparison of 2 × 4 ft Troffer Lighting with a Furniture-Integrated Lighting System, Shown in Figure 21.22***

	4-F40CW 2 × 4 ft Prismatic	*Lu250/S Indirect (50 fc)*	(100 fc, 100 sq ft per work station)	*2-fc 12T10/CW Task Lighting (50 fc)*
Basic data				
Initial lamp lumens per luminaire	12,600	30,000		3600
Luminaire watts	190	307		70
Coefficient of utilization (RCR $= 2\rho_{cc} = 0.8$)	0.60	0.55		—
Lamp lumen maintenance	0.88	0.90		0.83
Dirt factor (cleaning at relamping)	0.84	0.75		0.90
No. fixtures per square foot for 100 fc	0.01789	0.004489		0.01
Initial cost				
Luminaire (net)	$40.00	$200.00		$70.00
Lamps (net)[a]	$ 3.61	$ 37.76		$ 8.20
Installation at $16 per man-hour	$24.00	—		—
Wiring at $100/kw	$19.00	$ 30.70		$ 7.00
Total initial cost	$86.61	$268.46		$85.20
Initial cost per square foot for 100 fc	$ 1.549	$ 1.205		$ 0.852
Annual owning cost 0.15 × (initial—lamps) per luminaire	$12.99	$ 34.60		$11.55
Owning cost per square foot for 100 fc	$ 0.2324	$ 0.1553	$0.2708	$ 0.1155
Operating cost, annual				
Electricity at 4¢/kwh and 3000 hr/yr	$22.80	$ 36.84		$ 8.40
No. lamps replaced per year at 3000 hr/yr	0.6	0.2		0.667
Lamp and labor costs for replacements at $2.00 labor per lamp	$ 3.37	$ 7.95		$ 6.80
Total annual operating cost	$26.17	$ 44.79		$15.20
Annual operating cost per square foot for 100 fc	$ 0.467	$ 0.2011	$0.3531	$ 0.152
Total lighting cost				
Total annual owning and operating cost per luminaire	$36.99	$ 79.39		$26.75
Total annual cost per square foot for 100 fc	$ 0.6997	$ 0.3564	$0.6239	$ 0.2675
Air conditioning				
Tonnage required per luminaire to remove lighting heat	0.0538	0.0870		0.0198
Initial cost per luminaire ($1500 per ton)	$80.75	$130.50		$29.75
Initial cost per square foot for 100 fc	$ 1.445	$ 0.5858		$ 0.2975
Owning cost per square foot for 100 fc (0.1 × initial cost)	$ 0.144	$ 0.0586		$ 0.0298
Annual operating cost per square foot for 100 fc (1 kw per ton, 1000 hr/yr, 4¢/kwh, no maint.)	$ 0.0385	$ 0.0156		$ 0.0079
Total annual A.C. owning and operating costs	$ 0.1825	$ 0.0742	$0.1119	$ 0.0377
Total costs				
Total lighting and A.C. costs per square foot for 100 fc	$ 0.8822	$ 0.4306	$0.7358	$ 0.3052

Source. Reprinted with permission of the IES, from *Lighting Design & Application,* September 1976.

[a] Typical net lamp prices for large quantities.

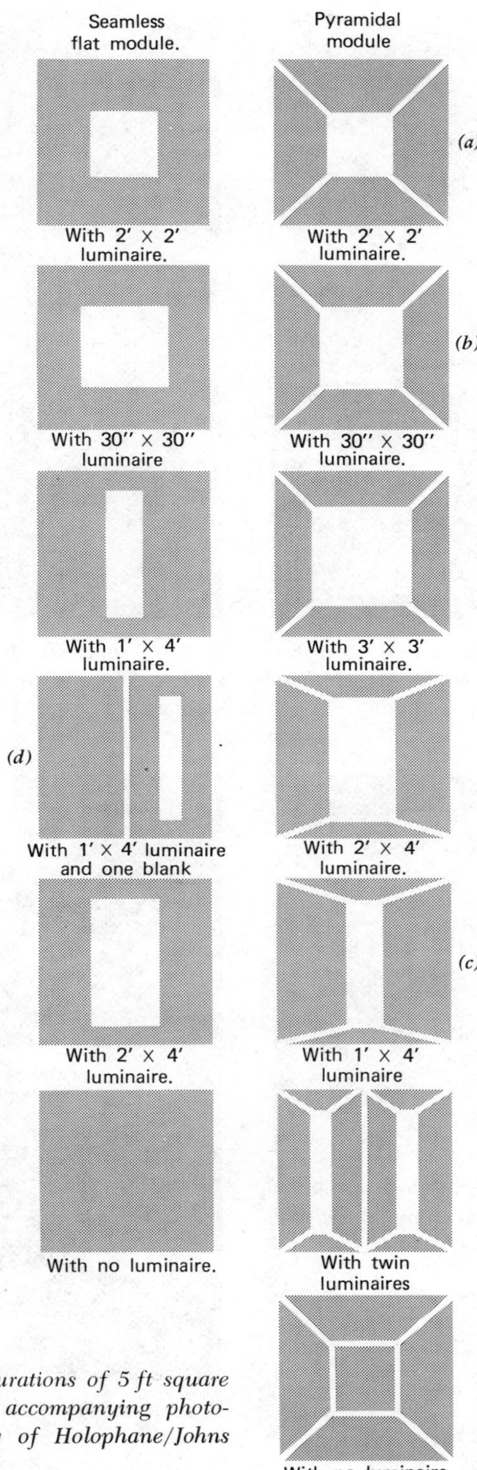

Fig. 21.24 *Various configurations of 5 ft square modules. Letters refer to accompanying photographs. (Sketches courtesy of Holophane/Johns Manville).*

Fig. 21.24 (a) *2 ft square fixture with 'U'-tubes. Note low brightness of the miniature parabolic egg-crate diffuser.*

(b) *30-in. square luminaires alternating with blank section. Note the air intake along the module perimeter.*

(c) *Note the shielding provided by the recess in the pyramidal shape. The effect is most pronounced with a low ceiling. Compare to* (b) *above.*

(d) *Splitting the module in half, rather than centering the fixture as in* (c), *creates the dominant checkerboard pattern shown. Photos* (a) *to* (d) *Courtesy of Armstrong Cork Company.*

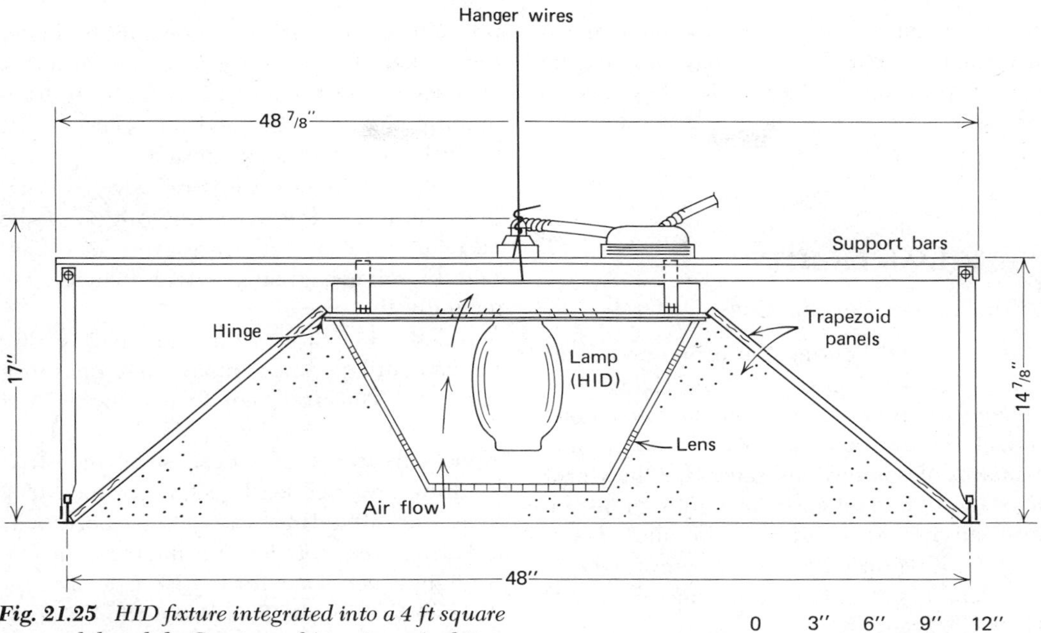

Fig. 21.25 HID fixture integrated into a 4 ft square pyramidal module. Courtesy of Armstrong Architectural Ceiling Systems.

Fig. 21.26 Office installation with HID source in a pyramidal modular fixture. The lamp used here is a 150-w, high-pressure sodium unit. Because of its high output and the relatively low ceiling, shadows are sharper than is desirable. Use of 100-w HPS lamps with an efficacy of more than 80 fc/w would considerably lessen this problem, as would an additional 2 ft of ceiling height. Photo courtesy of Westinghouse.

mobility of the fixtures, so necessary for adequate task lighting. The partial recessing accomplished with pyramidal modules is particularly useful when applying high-brightness HID sources to commercial interiors (see Figs. 21.25 and 21.26).

21.18 Lighting and Air Conditioning

The reduction of lighting power density levels to below 3 w/sq ft in all but special areas has considerably reduced the impact of lighting-generated heat on a building's HVAC system. In non air-conditioned buildings, the lighting heat contribution is directly applicable to building heating. Fixture efficiency is directly affected by its temperature. Fluorescent units operate at an optimum temperature of 77° F. Temperatures above and below this decrease output and fixture efficiency. Thus heat removal from units is desirable even at low lighting-energy levels. The most effective method of fixture heat removal is by duct connection to the unit itself. This method, however, is relatively expensive, and immobilizes the fixture. Alternatively the plenum can be exhausted with air passing through the fixtures, picking up excess heat. These details are essentially part of the HVAC design and are covered in Part Two.

INDUSTRIAL LIGHTING

21.19 Industrial Lighting—General

In industrial lighting the prime and overriding consideration of all work, lighting included, is its profitability; that is, its economic impact on the company. Given acceptable standards of comfort and safety for the working staff, additional costs for lighting must be self-justifying economically. In one case a good lighting installation was improved at considerable cost. Production jumped 15%, of which 3% was sufficient to amortize the cost of the lighting alteration. In another case an outlay for new inspection lighting reduced product failures and proved economically sound. In a third, improved lighting reduced accidents, improved employee morale, and consequently improved production. The cases studied are far too numerous to mention; general principles will be adduced instead.

21.20 Industrial Lighting—Levels and Sources

Levels of illumination are detailed in the IES, *American National Standard Practice for Industrial Lighting* (A 11.1), latest issue. Similar values are found in Table 18.4, page 732. Other applicable recommendations are found in Tables 18.5 and 18.7, pages 735 and 738. Where levels above 50 to 75 fc are required, ambient illumination must be supplemented by local illumination. For details on supplementary lighting the IES, *Recommended Practice for Supplementary Lighting*, may be consulted, which also contains data on inspection lighting. Both of these subjects are too specialized to be treated here. It is worthwhile to note that with increased lighting levels, and control of glare problems, industrial workers are approaching the lighting condition under which their forebears worked—out-of-doors. That enviable situation, with 1000+ fc horizontally and vertically and low glare, is not attainable with artificial lighting at its present technology without grossly excessive cost and energy use—nor is it necessary. A lesson to be learned, however, is the need to utilize daylight to a much greater extent than is now generally the case.

Industrial facilities lend themselves readily to daylighting, since many are one-story structures. Thus roof monitors, skylights, and clerestories are readily applicable and extremely desirable. However, since industrial facilities are frequently sited in industrial areas with attendant heavy atmospheric soot and dirt, a frequent cleaning and maintenance program is necessary if the LLF is to be kept at reasonable levels. This observation is obviously also applicable to indoor light facilities (to the extent that the indoor activity warrants).

Sources for industrial application should be high-efficacy, low-maintenance types, that is, HID and fluorescent. Where color is not critical, HPS is the recommended source. Adaptation to its warm yellow color is rapid and, as stated above, if it is mixed with metal-halides or mercury sources, no problem should be encountered. HID sources are easier to maintain, store, clean, and relamp than fluorescent and have equal or better efficacy, but have the disadvantage of delayed restrike time. Because of their concentrating nature, HID sources are more applicable to high-bay (>25 ft) and medium-bay (15–25 ft) installations, while fluorescent are suited for low-bay (<15 ft), although specially designed low-bay HID fixtures are available, which minimize the inherent glare and distribution problems involved (see Fig. 21.28).

21.21 Industrial Illumination Brightness Ratios

For reasons explicated at length in preceding sections, brightness ratios in industrial situations must be controlled. Recommendations are given in Table 21.4. In many situations it is

Table 21.4a *Industrial Lighting*

Recommended Maximum Luminance Ratios

	Environmental Classification		
	A	*B*	*C*
1. Between tasks and *adjacent darker* surroundings	3 to 1	3 to 1	5 to 1
2. Between tasks and *adjacent lighter* surroundings	1 to 3	1 to 3	1 to 5
3. Between tasks and more *remote darker* surfaces	10 to 1	20 to 1	*
4. Between tasks and more *remote lighter* surfaces	1 to 10	1 to 20	*
5. Between luminaires (or windows, skylights, etc.) and *surfaces adjacent* to them	20 to 1	*	*
6. Anywhere within normal field of view	40 to 1	*	*

*Luminance ratio control not possible.

A—Interior areas where reflectances of entire space can be controlled in line with recommendations for optimum seeing conditions.

B—Areas where reflectances of immediate work area can be controlled, but control of remote surround is limited.

C—Areas (indoor and outdoor) where it is completely impractical to control reflectances and difficult to alter environmental conditions.

Table 21.4b *Industrial Lighting*

Recommended Minimum Reflectances of Surfaces, Applicable to Classifications A and B of Table 21.4a

	Reflectance
Ceiling (all area above fixture line)	80 to 90%
Walls	40 to 60%
Desk and bench tops, machines and equipment	25 to 45%
Floors	Not less than 20%

Source. Reprinted from *American National Standard Practice for Industrial Lighting A11.1,* 1972.

impossible to control the surrounding brightness, as is shown in the tables. Note from Table 21.4 that ceilings that so frequently are covered with piping, ducts, and other equipment must be *light.* In other words, the above equipment must be painted with high-reflectance finishes, maintenance and cleaning must be good, and fixtures should have an upward component of light to avoid more than a 20:1 ratio of task to ceiling luminance.

Use of bright saturated colors for general surface painting should be avoided, however, since they draw attention and frequently have special significance. In addition to color-coded piping (banding is preferable), red frequently means fire equipment; green, first aid; orange, danger; etc. White is also to be avoided, being excessively bright and susceptible to dirt. Light, unsaturated colors are preferable.

21.22 Industrial Lighting Glare

The problem of direct glare can be acute in low-bay installations, and that of reflected glare in high-bay designs, when both use point

Lighting Application

Average Luminance and Output Comparison

Holophane No. 575 Vs. 575-ELB
400W., H33-1GL/C Improved Color Mercury Lamps
21000 Lumens

Average Luminance in Footlamberts

Vertical Angles	575 Luminance	575-ELB Luminance	% Reduction
65	3220	1570	51%
70	1330	405	69%
75	780	190	76%
80	670	235	65%
85	535	285	47%

Output Comparison

Zone Degrees	575 % Output	575-ELB % Output	% Reduction
0-45	43.6	40.6	7%
0-60	61.2	53.7	12%

sources. One method of reducing direct glare is the use of low-brightness prismatic lens units that utilize a black aluminum reflector behind the prismatic lens. The pronounced reduction in high angle brightness of such luminaires (shown with 400-w mercury lamp) is shown in Figure 21.27. This reduction in brightness is accomplished with only a 10%

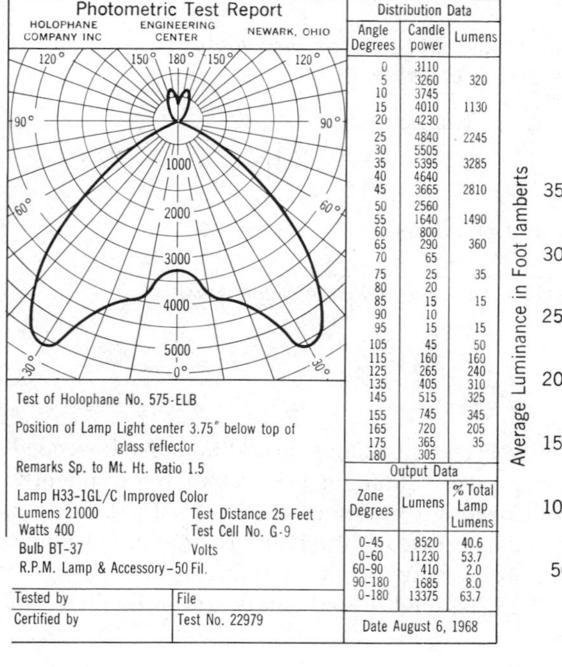

Photometric Test Report

HOLOPHANE COMPANY INC — ENGINEERING CENTER — NEWARK, OHIO

Test of Holophane No. 575-ELB

Position of Lamp Light center 3.75" below top of glass reflector

Remarks Sp. to Mt. Ht. Ratio 1.5

Lamp H33-1GL/C Improved Color
Lumens 21000 — Test Distance 25 Feet
Watts 400 — Test Cell No. G-9
Bulb BT-37 — Volts
R.P.M. Lamp & Accessory–50 Fil.

Tested by	File
Certified by	Test No. 22979

Distribution Data

Angle Degrees	Candle power	Lumens
0	3110	
5	3260	320
10	3745	
15	4010	1130
20	4230	
25	4840	2245
30	5505	
35	5395	3285
40	4640	
45	3665	2810
50	2560	
55	1640	1490
60	800	
65	290	360
70	65	
75	25	35
80	20	
85	15	15
90	10	
95	15	15
105	45	50
115	160	160
125	265	240
135	405	310
145	515	325
155	745	345
165	720	205
175	365	35
180	305	

Output Data

Zone Degrees	Lumens	% Total Lamp Lumens
0-45	8520	40.6
0-60	11230	53.7
60-90	410	2.0
90-180	1685	8.0
0-180	13375	63.7

Date August 6, 1968

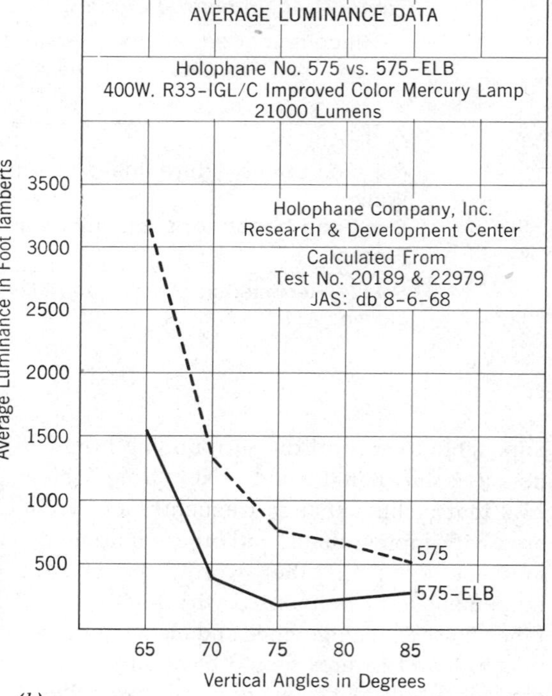

(b)

Fig. 21.27 (a) By utilizing a black reflector surface behind the prismatic glass lining in lieu of a polished aluminum one, fixture brightness can be dramatically reduced without great light loss. Courtesy of Holophane/Johns Manville. (b) Photometric comparison of the two lamps shown in (a).

Fig. 21.28 (a) *Low-bay installation of 400-w HPS fixtures in an industrial installation. Note particularly the low brightness of the unit; the fluorescent fixtures whose purpose is to provide minimal lighting during strike or restrike periods of the HPS; and the method of connecting the fixtures via flexible cords and plugs into structure-mounted receptacle boxes. See photometric data, next page.*

reduction in useful light. Another low-brightness unit with its photometric characteristics and an application using an HPS source are shown in Figure 21.28. Methods of minimizing veiling reflection from all sources have been discussed previously.

21.23 Industrial Lighting Equipment

The cost of maintenance increases with labor rates. For this reason, high-quality lighting equipment will yield lowest owning and life-cycle costs. For instance, the cost of replacing a ballast for an HID lighting unit frequently *exceeds* the cost of the ballast. It is thus obvious that it is more economical to utilize long-life, high-quality ballasts, particularly where luminaires are mounted high or are otherwise not readily accessible. This is graphically demonstrated in Figure 21.29, which relates annual savings for a 1000-fixture installation to the labor cost for ballast replacement, when using higher priced, long-life ballasts. Thus for a medium-bay installation, at a replacement *labor* cost of $30, *not including production loss,* an annual saving of $2600 is realized. Other suggestions for lowering costs, both initial and operating, include using ventilated luminaires that tend to be self-cleaning by convection (see Fig. 21.30) in addition to giving the needed upward light component, using bus-mounted fixtures for rapid installation and repair (see Fig. 21.31), using lowering mechanisms on high-bay units to avoid catwalk or platform relamping with concomitant *extremely* high cost, using fixtures arranged for "stick" relamping from the floor in medium- and low-bay work, and generally incorporating the most modern equipment into the plant. Typical industrial reflector fixture designs are given in Figure 21.32.

Proper maintenance is of paramount importance in industrial facilities because of the

Fig. 21.28 (b) *Photometric data on the fixture of* (a), *see insert. The fixture uses a refractor to give most of its output in the 20 to 50° range, thus limiting both types of glare. The fixture is equipped with filters and gaskets to minimize internal dirt accumulation and yield high maintained illumination. Courtesy of General Electric Co.*

(b)

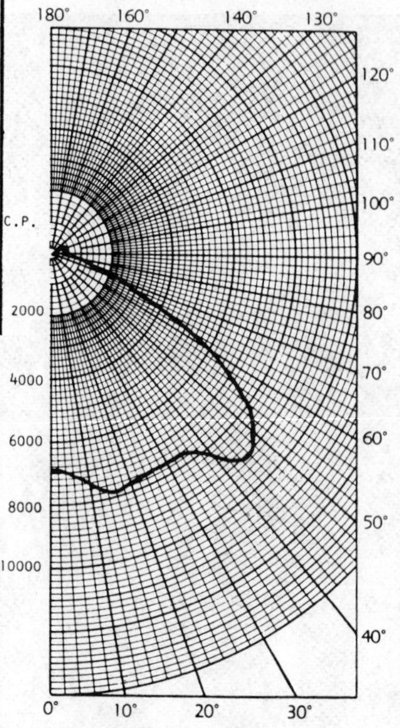

COEFFICIENTS OF UTILIZATION – ZONAL CAVITY METHOD*

	EFFECTIVE CEILING CAVITY REFLECTANCE – PERCENT															
	80			70			50			30			10			0
% WALL REFLECTANCE	50	30	10	50	30	10	50	30	10	50	30	10	50	30	10	0
ROOM CAVITY RATIO (RCR) 0																
1	.739	.705	.682	.718	.692	.666	.680	.657	.639	.647	.629	.613	.617	.601	.590	.570
2	.641	.597	.557	.624	.584	.547	.593	.560	.529	.564	.538	.514	.539	.520	.496	.479
3	.559	.506	.461	.547	.496	.457	.520	.478	.442	.497	.461	.432	.475	.445	.417	.403
4	.483	.429	.391	.470	.421	.384	.452	.409	.374	.433	.396	.368	.415	.384	.359	.343
5	.423	.369	.328	.414	.362	.325	.398	.352	.319	.382	.341	.311	.366	.331	.303	.288
6	.350	.304	.273	.344	.301	.268	.329	.292	.265	.319	.285	.257	.307	.276	.251	.242
7	.305	.262	.228	.294	.254	.224	.288	.249	.225	.277	.242	.215	.267	.234	.215	.201
8	.274	.228	.195	.263	.224	.191	.258	.222	.190	.248	.212	.184	.237	.206	.181	.171
9	.215	.184	.160	.207	.181	.160	.203	.175	.156	.197	.173	.154	.191	.170	.152	.144
10	.192	.160	.139	.185	159	.138	.180	.154	.135	.176	.152	.133	.169	.148	.131	.123

BASIS OF CALCULATIONS:
SPACING TO MOUNTING HEIGHT RATIO
EFFECTIVE FLOOR CAVITY REFLECTANCE 20 % (ρ_{fc} = 0.20)

* FOR "INTERFLECTANCE" OR "LUMEN" METHODS SUBSTITUTE ROOM INDEX (RI) RATIO IN PLACE OF ROOM CAVITY RATIO (RCR) WHERE RI = $\frac{5}{RCR}$

MAXIMUM RECOMMENDED SPACING: 1.9 TIMES HEIGHT ABOVE WORKING PLANE

PHOTOMETRIC DATA

AVERAGE	I BRIGHTNESS	
VERT ANGLE	FOOT LAMBERTS	CANDELAS/ SQUARE IN
85°	2282	
80°	2644	
75°	3463	
70°	4274	
65°	5542	

DISTRIBUTION DATA

VERTICAL ANGLE	AVERAGE CANDELAS	ZONE	ZONAL LUMENS
180°			
175°		170°-180°	
165°		160°-170°	
155°		150°-160°	
145°		140°-150°	
135°	52	130°-140°	40
125°	238	120°-130°	213
115°	542	110°-120°	538
105°	369	100°-110°	390
95°	399	90°-100°	435
85°	732	80°-90°	799
75°	1785	70°-80°	1889
65°	3987	60°-70°	3959
55°	7676	50°-60°	6885
45°	9163	40°-50°	7092
35°	7687	30°-40°	4827
25°	7687	20°-30°	3559
15°	7854	10°-20°	2223
5°	7081	0°-10°	673
0°	6902		

SUMMARY DATA

ZONE	LUMENS	% BARE LAMP	% LUMINAIRE
90°-180°	1616	3.0	4.8
0°-90°	31906	68.0	95.2
0°-60°	25259	53.7	75.4
0°-30°	6455	13.7	19.3
0°-180°	33522	71.0	100.0
TOTAL EFFICIENCY:		71.0	

CANDELA AND LUMEN DATA ARE SHOWN FOR 47000 LUMENS AT THE SOURCE. EXPECTED PERFORMANCE VALUES FOR ANY OF THESE ITEMS MAY BE DETERMINED BY MULTIPLYING GIVEN VALUES BY AN APPROPRIATE FACTOR SUCH AS: LAMP LUMEN RATING × BALLAST FACTOR × MAINTENANCE FACTOR 47000 LUMENS

C.P.
180° 160° 140° 130° 120° 110° 100° 90° 80° 70° 60° 50° 40°
2000 4000 6000 8000 10000
0° 10° 20° 30°

TEST DISTANCE FT. 25
LAMP LUMENS 47,000

IF DATA IS DESIRED FOR OTHER LAMP LUMENS MULTIPLY CANDELA AND LUMEN VALUES GIVEN BY RATIO: DIFFERENT LAMP RATING DIVIDED BY 47,000 LUMENS.

LAMP CANNOT BE VIEWED DIRECTLY

LUMINAIRE DESCRIPTION

GENERAL ELECTRIC LOWMOUNT IND. LUMINAIRE
FACETED SPEC. REFLECTOR
STD. SOCKET POSITION

LAMP – 400 WATT LUCALOX
G.E. No. LU-400/BD

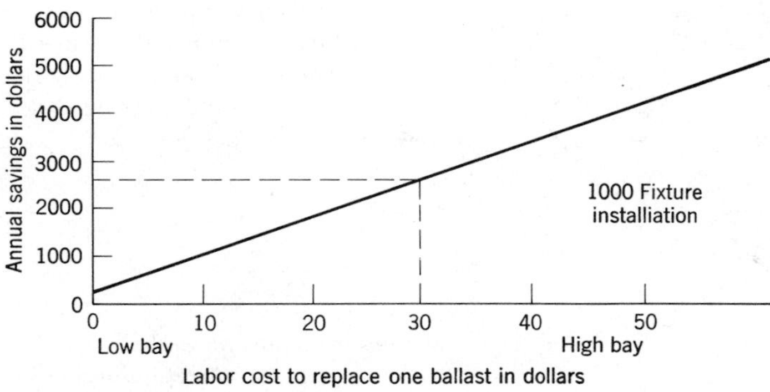

Fig. 21.29 A graphic representation of the savings possible with the use of long-life ballasts in HID fixtures.

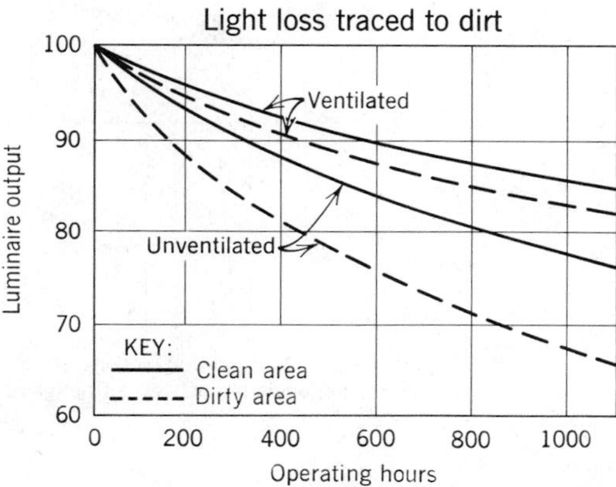

Fig. 21.30 The advantage of ventilated fixtures is demonstrated by this graph.

prevalence of dirt, vibration, and rough service. Under maintenance is subsumed cleaning, relamping, inspection, and preventive maintenance. Relamping on a burnout basis is extremely uneconomical because of disruption of production and lowered production due to lumen depreciation before burnout. Relamping is normally done on a planned group basis. Similarly, if the specific facility has a high dirt accumulation rate, cleaning must also be done on a planned group basis, rather than only at relamping time.

Ballast noise, including the high levels of HID ballasts, is not usually a factor in industrial facilities because of high ambient noise. In relatively quiet installations and/or where fluorescent fixtures are mounted a short distance above the work bench as in inspection and fine assembly, this is not true and noise ratings and conditions must be examined carefully.

21.24 Vertical-Surface Illumination

In industrial facilities more than any other, the illumination of vertical surfaces is crucial. This is a result of the nature of the work; machines, storage, gages, etc., all require high-level vertical-surface illumination. Examining Figure 21.33 and the derivation given there, we note that maximum vertical illumination (illumination resulting from the horizontal-lighting component) is obtained when the angle between the fixture's vertical axis and the work is approximately 35°. Hence, we should select a fixture whose candlepower distribution curve demonstrates such a characteristic most closely. The fixture of Figure 21.34 shows this characteristic and would be a good choice. Of course, the derivation is for a single location and fixture. For good vertical and angular illumination over a large area, select a fixture as above, and arrange fixtures with considerable overlap.

ECONOMIC ANALYSIS

21.25 Life-Cycle Costing

The economic comparison of Table 20.1, page 860, as that of Table 21.3, page 968, are accurate within their own assumptions, but have distinct shortcomings that can only be overcome by life-cycle costing. These short-

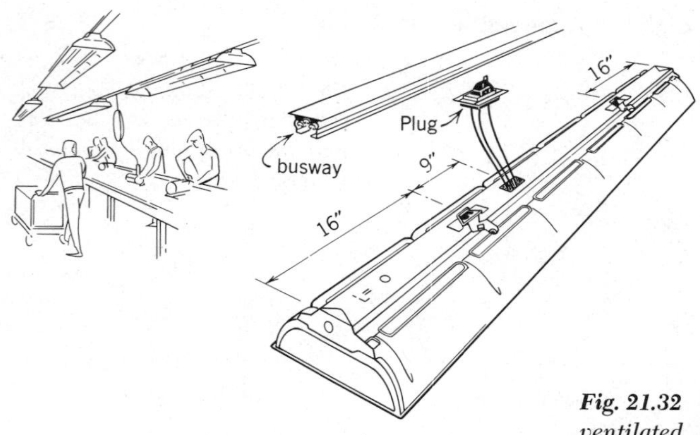

Fig. 21.31 *Industrial lighting design using plug-in lighting bus to feed and support fixtures. This allows fixtures to be installed, moved, and maintained with great ease, lowering initial and operating costs. Courtesy of Daybrite.*

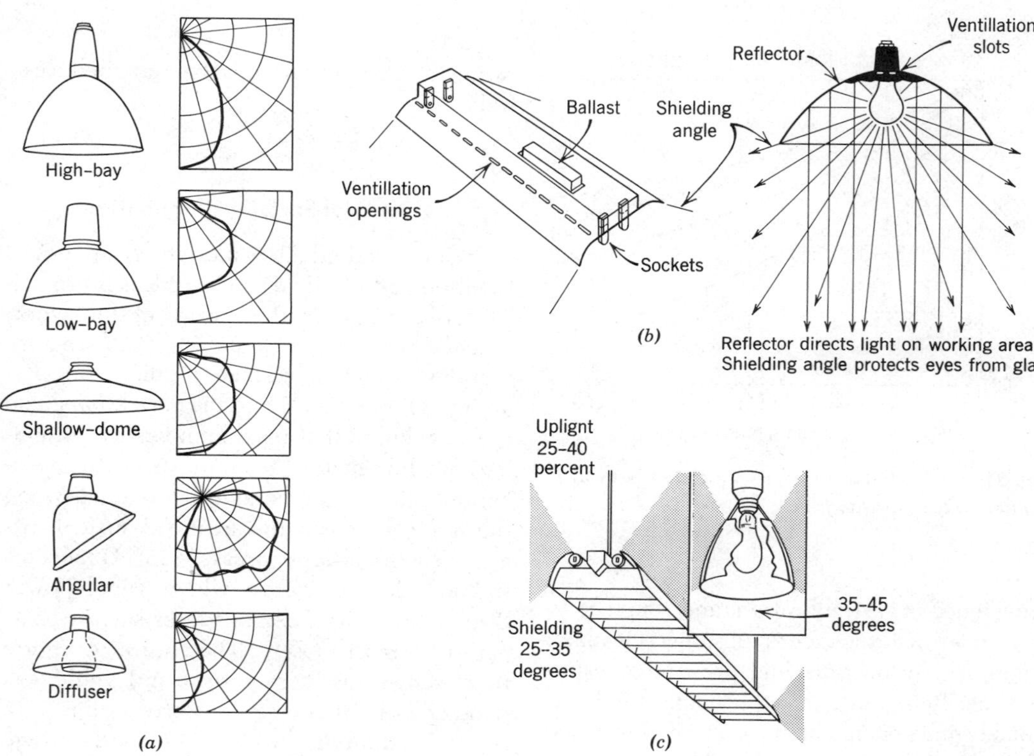

Fig. 21.32 *Fixtures in* (a) *are typical designs, in* (b) *ventilated variety, and* (c) *direct-indirect distribution units.*

Fig. 21.33 Vertical surface illumination. *The illumination on the vertical surface is the result of the* horizontal *component of the lighting. This is*

$$fc = \frac{\text{cp}}{\text{D}^2} \sin\theta = \frac{\text{cp} \times \text{R}}{\text{D}^3} = \frac{\text{cp} \times \cos^2\theta \sin\theta}{\text{H}^2}$$

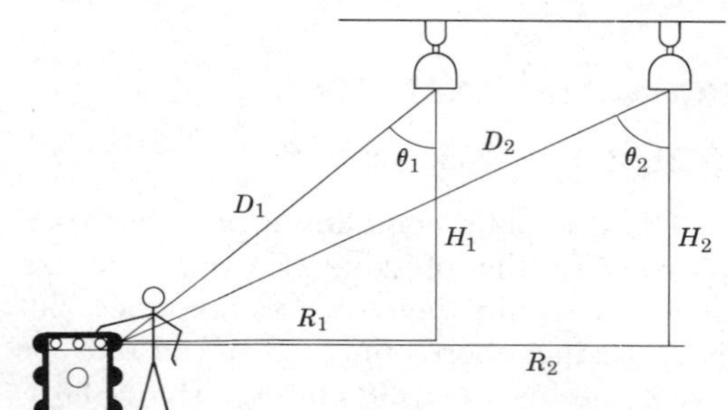

To maximize the horizontal component, we set the derivative of FC with respect to θ, to zero. Thus

$$\frac{dfc}{d\theta} = \frac{\text{cp}}{\text{H}^2}(-2\cos\theta\sin^2\theta + \cos^3\theta) = 0$$

or

$$2\sin^2\theta = \cos^2\theta; \qquad \tan^2\theta = \tfrac{1}{2};$$

$$\tan\theta = 0.707 = \frac{\text{R}}{\text{H}}; \qquad \theta \cong 35°$$

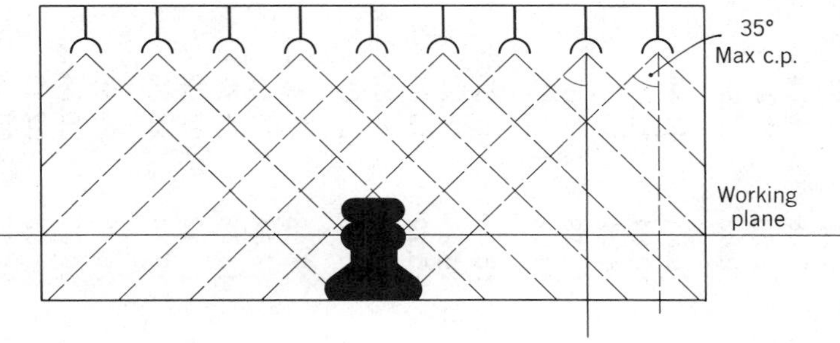

Fig. 21.34 Layout for good vertical surface illumination.

comings result from the assumption in both tables that all costs will remain stable through the operating life of the equipment. This is obviously inaccurate for most items and is grossly inaccurate for labor and energy costs, as recent history and the forseeable future will attest. Since these two factors weigh heavily in all economic comparisons, and generally exist in different proportions in the systems being considered, escalation in their costs will obviously change the economic balance when costs are projected over the equipment life, that is, life-cycle costs are calculated. The most accurate approach is therefore to project such escalation as can reasonably be expected. This was the approach taken in our analysis of Table 15.6, page 554, Table 16.4, page 593, Table 19.15, page 829, and Table 19.16, page 830.

To recapitulate the method (see Section 15.19), the life-cycle cost of a system is the sum of the first cost plus the present value of all future operating costs, over the life of the system:

$$C_{LC} = C_I + PV_{oc} \tag{21.1}$$

Where

C_{LC} = life-cycle cost
C_I = initial cost

and

PV_{oc} = present value of future operating costs over the life of the equipment.

The initial cost C_I is readily calculated by adding the costs of equipment, installation, wiring system, and related HVAC systems costs. This last factor may simply be the cost of tonnage or a more complex figure if the HVAC system is designed to utilize lighting heat. This cost must be relevant to each lighting system being considered, since the HVAC system may be different. The present value of future costs is found in most engineering economics texts, for *fixed* annual expenditures, but is not readily available for escalating costs. A brief excursus into this area is necessary in order to demonstrate the development of this very important data, presented in Table 21.5. The present value of future payments of 1.00, made at the *end* of a series of *n* periods, is

$$a_{\overline{n}|} = \frac{(1+i)^n - 1}{i} \tag{21.2}$$

where

$a_{\overline{n}|}$ = present value
i = interest rate expressed as a decimal

and

n = number of periods

If, however, the future payments escalate, the expression becomes

$$a_{\overline{n}|} = \frac{1 - \left(\frac{K}{1+i}\right)^n}{\left(\frac{1+i}{K}\right) - 1} \tag{21.3}$$

where K is the escalation rate per period, and the remaining items are as in equation 21.2. Thus for a 5% annual escalation, $K = 1.05$, and so on. The values of $a_{\overline{n}|}$ are tabulated in Table 21.5 for interest rates of 6 to 9% and escalation rates of 3, 5, 8, and 10%. Other values can be calculated from the equations. To demonstrate the effect of these escalative

Table 21.5a *Present Value of n Future Payments Beginning at 1.00 and Escalating at K per Period*

Periods	Interest Rate—6%				
	Escalation Rate K				
n	*1*	*1.03*	*1.05*	*1.08*	*1.10*
1	0.94	0.97	0.99	1.02	1.04
2	1.83	1.92	1.97	2.06	2.11
3	2.67	2.83	2.94	3.11	3.23
4	3.47	3.72	3.91	4.19	4.39
5	4.21	4.59	4.86	5.29	5.60
6	4.92	5.43	5.80	6.41	6.84
7	5.58	6.25	6.74	7.55	8.14
8	6.21	7.05	7.67	8.71	9.49
9	6.80	7.82	8.59	9.89	10.88
10	7.36	8.57	9.50	11.10	12.33
11	7.89	9.30	10.40	12.33	13.83
12	8.38	10.01	11.29	13.58	15.39
13	8.85	10.69	12.17	14.85	17.01
14	9.29	11.36	13.05	16.15	18.69
15	9.71	12.01	13.92	17.48	20.43
16	10.11	12.65	14.78	18.82	22.24
17	10.47	13.26	15.63	20.20	24.12
18	10.83	13.86	16.47	21.60	26.07
19	11.16	14.44	17.31	23.03	28.09
20	11.47	15.00	18.13	24.48	30.19
21	11.76	15.55	18.95	25.96	32.36
22	12.04	16.08	19.76	27.47	34.62
23	12.30	16.59	20.57	29.01	36.97
24	12.55	17.10	21.36	30.57	39.40
25	12.78	17.58	22.15	32.17	41.92
26	13.00	18.06	22.93	33.79	44.55
27	13.21	18.52	23.71	35.45	47.26
28	13.41	18.97	24.48	37.14	50.08
29	13.59	19.40	25.24	38.86	53.01
30	13.76	19.82	25.99	40.61	56.05

Table 21.5b *Present Value of n Future Payments Beginning at 1.00 and Escalating at K per Period*

Periods	Interest Rate—7%				
	Escalation Rate K				
n	*1*	*1.03*	*1.05*	*1.08*	*1.10*
1	0.93	0.96	0.98	1.01	1.03
2	1.81	1.89	1.94	2.03	2.08
3	2.62	2.78	2.89	3.06	3.17
4	3.39	3.64	3.82	4.09	4.29
5	4.10	4.47	4.73	5.14	5.44
6	4.77	5.26	5.62	6.20	6.62
7	5.39	6.03	6.50	7.27	7.83
8	5.97	6.77	7.36	8.34	9.08
9	6.52	7.47	8.20	9.43	10.36
10	7.03	8.16	9.03	10.53	11.68
11	7.50	8.82	9.84	11.64	13.03
12	7.94	9.45	10.64	12.75	14.43
13	8.36	10.06	11.42	13.88	15.86
14	8.75	10.64	12.19	15.02	17.33
15	9.11	11.21	12.94	16.17	18.85
16	9.45	11.75	13.68	17.33	20.40
17	9.76	12.28	14.41	18.50	22.00
18	10.06	12.78	15.12	19.69	23.65
19	10.34	13.26	15.82	20.88	25.34
20	10.59	13.73	16.50	22.08	27.08
21	10.84	14.18	17.18	23.30	28.87
22	11.06	14.61	17.84	24.53	30.70
23	11.27	15.03	18.48	25.77	32.59
24	11.47	15.43	19.12	27.02	34.53
25	11.65	15.82	19.74	28.28	36.53
26	11.83	16.19	20.36	29.55	38.58
27	11.99	16.55	20.96	30.84	40.69
28	12.14	16.89	21.55	32.13	42.86
29	12.28	17.22	22.12	33.44	45.09
30	12.41	17.54	22.69	34.77	47.38

Table 21.5c ***Present Value of n Future Payments Beginning at 1.00 and Escalating at K per Period***

Periods	Interest Rate—8%				
	Escalation Rate K				
n	*1*	*1.03*	*1.05*	*1.08*	*1.10*
1	0.93	0.95	0.97	1.00	1.02
2	1.78	1.86	1.92	2.00	2.06
3	2.58	2.73	2.84	3.00	3.11
4	3.31	3.56	3.73	4.00	4.19
5	3.99	4.39	4.60	5.00	5.28
6	4.62	5.10	5.44	6.00	6.40
7	5.21	5.82	6.26	7.00	7.54
8	5.75	6.50	7.06	8.00	8.70
9	6.25	7.15	7.84	9.00	9.88
10	6.71	7.78	8.59	10.00	11.08
11	7.14	8.37	9.33	11.00	12.30
12	7.54	8.94	10.04	12.00	13.55
13	7.90	9.48	10.73	13.00	14.82
14	8.24	9.99	11.41	14.00	16.11
15	8.56	10.48	12.06	15.00	17.43
16	8.85	10.95	12.70	16.00	18.77
17	9.12	11.40	13.32	17.00	20.13
18	9.37	11.82	13.92	18.00	21.53
19	9.60	12.23	14.51	19.00	22.94
20	9.82	12.62	15.08	20.00	24.39
21	10.02	12.99	15.63	21.00	25.86
22	10.20	13.34	16.17	22.00	27.35
23	10.37	13.68	16.69	23.00	28.88
24	10.53	14.00	17.20	24.00	30.43
25	10.67	14.30	17.69	25.00	32.01
26	10.81	14.59	18.17	26.00	33.62
27	10.94	14.87	18.64	27.00	35.27
28	11.05	15.14	19.10	28.00	36.94
29	11.16	15.39	19.54	29.00	38.64
30	11.26	15.63	19.97	30.00	40.37

Table 21.5d ***Present Value of n Future Payments Beginning at 1.00 and Escalating at K per Period***

Periods	Interest Rate—9%				
	Escalation Rate K				
n	*1*	*1.03*	*1.05*	*1.08*	*1.10*
1	0.92	0.94	0.96	0.99	1.01
2	1.76	1.84	1.89	1.97	2.03
3	2.53	2.68	2.79	2.95	3.06
4	3.24	3.48	3.65	3.91	4.09
5	3.89	4.23	4.48	4.86	5.14
6	4.49	4.94	5.27	5.81	6.20
7	5.03	5.62	6.04	6.75	7.26
8	5.53	6.25	6.79	7.68	8.34
9	6.00	6.85	7.50	8.60	9.42
10	6.42	7.42	8.19	9.51	10.52
11	6.81	7.96	8.85	10.41	11.62
12	7.16	8.46	9.49	11.31	12.74
13	7.49	8.94	10.10	12.20	13.87
14	7.79	9.40	10.70	13.07	15.00
15	8.06	9.82	11.27	13.94	16.15
16	8.31	10.23	11.82	14.81	17.31
17	8.54	10.61	12.35	15.66	18.47
18	8.76	10.97	12.86	16.51	19.65
19	8.95	11.31	13.35	17.35	20.84
20	9.13	11.63	13.82	18.18	22.04
21	9.29	11.94	14.28	19.00	23.25
22	9.44	12.23	14.72	19.82	24.48
23	9.58	12.50	15.14	20.63	25.71
24	9.71	12.76	15.55	21.43	26.96
25	9.82	13.00	15.94	22.23	28.21
26	9.92	13.23	16.32	23.01	29.48
27	10.03	13.44	16.68	23.79	30.76
28	10.12	13.65	17.04	24.57	32.05
29	10.20	13.84	17.37	25.33	33.35
30	10.27	14.03	17.70	26.09	34.67

Table 21.6 ***Life-Cycle Economic Comparison of Lighting Systems Described in Table 21.3 and Figure 21.22***

Item	Ceiling System	Integrated System Lighting		
		Total	*Ambient*	*Task*
Basic Data see Table 21.3				
Initial Cost per square foot				
Lighting	1.549	2.057	1.205	0.852
Air conditioning	1.445	0.883	0.5858	0.2975
Total	2.994	2.940		
Operating Costs				
Energy:				
Lamp	0.4071		0.1654	0.0840
Air conditioning	0.0385		0.0156	0.0079
Total annual	0.4456	0.2729	0.1810	0.0919
Present value life cycle	4.69	2.87		
Labor:				
Annual	0.0214	0.0168	0.0035	0.0133
Present value life cycle	0.25	1.96		
Lamp Replacement:				
Annual	0.0387	0.0814	0.0267	0.0547
Present value life cycle	0.272	0.572		
Total present value of costs	5.212	3.638		
Total life-cycle cost per square foot	8.206	6.58		
Relative	1.25	1.00		

considerations, Table 21.6 is a reworking of Table 21.3, with these assumptions:

a. Equipment life—10 years.
b. Capital cost (interest rate)—7%.
c. Annual labor escalation—10%.
d. Annual energy cost escalation—8%.

The short span of 10 years is taken, since it is assumed that within that period new developments in lighting sources will markedly alter the energy factors involved. The results of Table 21.6 show a result more heavily in favor of the furniture-mounted lighting system. This figure might swing in favor of the ceiling-mounted system if the study were taken over a longer period and the probable shorter life of the integrated equipment taken into account. Table 21.7 is a typical life-cycle economic analysis of three alternate solutions to an industrial-lighting problem. Notice how heavily the energy costs weigh when considered over the life of the installation, particularly in view of the assumed price escalation.

OTHER DESIGN TOPICS

21.26 Automatic Energy Control

Daylighted buildings in clear-sky climates are subject to large changes in lighting due to sun and cloud variations. Photoelectric control of lights is not entirely satisfactory, since it generally operates in an on-off mode. This is particularly undesirable for HID lighting because of:

a. Restrike time delay.
b. Lamp-life reduction due to repeated restarts.
c. Inability to maintain a desired illumination level.

Table 21.7 *Life-Cycle Economic Cost Comparison for a Light Industry Space, Difficult Seeing Tasks*[a]

Basic Data: Illumination level—100 fc maintained.
Capital costs 8%.
Labor cost escalation 5%.
Energy cost escalation 10%, non-air-conditioned.
System A—250-w deluxe mercury.
System B—4- to 40-w fluorescent, rapid start.
System C—170-w metal-halide.
3000 hr/yr usage; Initial energy cost 0.03/kwh.

	System		
Item	*A*	*B*	*C*
Initial Costs			
Fixtures	8100	5400	6840
Wiring	1500	900	900
Outlets	1125	1125	900
Fixture installation	1350	1350	1080
Lamps	698	234	1087
Total	12773	9009	10807
Operating Costs			
Energy annual	2309	1490	1426
Present value, 20-year life cycle	56316	36341	34780
Lamps—annual	87	39	311
Present value, 20-year life cycle	854	383	3054
Maintenance, labor—annual	363	810	316
Present value, 20-year life cycle	5474	12214	4765
Present Value of Life Costs— Operating	62644	48938	42599
Life-Cycle cost	75417	57947	53406

[a] All costs in dollars.

More recently photoelectric control has been applied to HID and fluorescent dimmers in a modulating mode, to maintain preset illumination levels. The dimmer controls can be preset to maintain a certain light level, arranged to dim or extinguish selected lamps during "off" hours and can be regulated manually. The resulting energy-cost savings are usually sufficient to amortize the system first cost in a short time (see Fig. 21.35). These controls are available for all HID sources, that is, mercury, metal-halide, and HPS.

One frequently ignored factor in lighting design, which is corrected by level control of this type, is the initial overlighting of a space. As the reader is aware, lighting-level calculations include a light loss factor for lumen depreciation, dirt accumulation, and so on. This means that a space is initially *overlighted* to the extent of this factor. Example 20.2 can be utilized as a case in point; the LLF there is 0.51. Therefore, to maintain 50 fc, the initial level is 98 fc! It will take approximately 3 years for the level to drop to 50 fc. During this period, automatic-level control could hold the level to 50 fc uniformly. The energy savings in this particular installation can be calculated as follows:

Number of fixtures—60
Wattage per fixture—200
Total wattage—$60 \times 200 = 12$ kw

Assuming smooth light depreciation over 3 years and 3000 burning hours per year, the total amount of energy saved is

$$6\text{ kw} \times 3\text{ yr} \times \frac{3000\text{ hr}}{\text{yr}} = 54000\text{ kwh}$$

Fig. 21.35 Simple control units for automatic light-level control allow operation mode to be selected—automatic, preset, dimmed, or continuously adjustable dimmed—and meter indicates percent of "full output" wattage. Courtesy of Widelite.

At an average cost of \$0.05/kwh including escalation, this amounts to

Total cost = 0.05 (54000) = \$2700

This amounts to \$45 per fixture, which is more than sufficient to pay for all the dimming equipment. Similar results are obtainable by using multilevel ballasts and switching. The disadvantage there is that control is not smooth, that is, levels "jump" by 25 to 30% and savings are much reduced.

SPECIAL LIGHTING APPLICATION TOPICS

21.27 Emergency Lighting

Emergency lighting is required when the normal source is interrupted for any of three reasons:

1. Interruption of current flow through operation of a circuit disconnect. For example, inadvertent de-energizing of a circuit at a panel or switchboard.
2. Failure of the building's electrical system.
3. General power failure.

Systems that cover the above three situations are normally defined as follows:

Type 1—all three situations.
Type 2—situation 2.
Type 3—situation 3.

Therefore, sensors for Type 1 systems are highly localized, those for Type 2 less so, and those for Type 3 are only at the service entrance point.

(*a*) ***Codes.*** Since this item involves safety, it is covered in various codes, all of which have jurisdiction and unfortunately not all of which agree in their requirements. The principal authorities are:

1. Life Safety Code.
2. National Electrical Code.
3. OSHA regulations.

(*b*) ***Occupancies Requiring Emergency Lighting.*** The Life Safety Code and the NEC require that emergency lighting be provided automatically for means of egress, in specific occupancies. These are usually taken to mean:

1. Places of assembly for 300 or more persons; Type 1; 2, or 3. With more than 1000 persons; Type 1.
2. Structure into which daylight does not penetrate, with occupancy of 100 persons or more: Types 1, 2, or 3.
3. All educational and institutional occupancies: Types 1 or 2.
4. Multiple residences with 25 or more dwelling units: Types 1, 2, or 3.
5. Merchandising spaces with areas greater than 3000 sq ft, with levels above or below the street, and those on one level that are so large that daylight will not penetrate sufficiently: Types 1 or 2.
6. Office building with occupancy in excess of 1000 persons: Types 1, 2, or 3.

Where the requirements of codes differ, clarification must be obtained from the authorities.

Table 21.8 ***Condensed General Criteria for Preliminary Consideration for Emergency Lighting Applications***

Specific Need	*Maximum Tolerance Duration of Power Failure*	*Recommended Minimum Auxiliary Supply Time*	*Type of Auxiliary Power System*		*System Justification*
			Emergency	*Standby*	
Evacuation of personnel	Up to 10 seconds, preferably not more than 3 seconds	2 hours	X		Prevention of panic, injury, loss of life Compliance with building codes and local, state, and Federal laws Lower insurance rates Prevention of property damage Lessening of losses due to legal suits
Perimeter and security	10 seconds	10 to 12 hours during all dark hours	X	X	Lower losses from theft and property damage Lower insurance rates Prevention of injury
Warning	From 10 seconds up to 2 or 3 minutes	To return to prime power source	X		Prevention or reduction of property loss Compliance with building codes and local, state, and Federal laws Prevention of injury and loss of life
Restoration of normal power system	1 second to indefinite depending on available light	Until repairs completed and power restored	X	X	Risk of extended power and light outage due to a longer repair time
General lighting	Indefinite; depends on analysis and evaluation	Indefinite: depends on analysis and evaluation		X	Prevention of loss of sales Reduction of production losses Lower risk of theft Lower insurance rates
Hospitals and medical areas	0.1 second to uninterruptible. Life Safety Code (NFPA No. 101) allows 10 seconds for engine to start and power to be available	To return of prime mover	X	X	Uninterruptible service to patients by surgeons, medical doctors, nurses, and aids Compliance with all codes, standards, and laws Prevention of injury or loss of life Lessening of losses due to legal suits
Orderly shutdown time	0.1 seconds to 1 hour	10 minutes to several hours	X		Prevention of injury or loss of life Prevention of property loss by a more orderly and rapid shutdown of critical systems Lower risk of theft Lower insurance rates

Source. Extracted from "Recommended Practice for Emergency and Standby Power Systems," IEEE Standard 446-1974, as reprinted from *Lighting Design & Application,* May 1975.

Table 21.9 ***Typical Emergency and Standby Lighting Recommendations***

Standby[a]	*Immediate, Short-Term*[b]	*Immediate, Long-Term*[c]
Security Lighting	Evacuation Lighting	Hazardous Areas
Outdoor perimeters	Exit signs	Laboratories
Closed circuit TV	Exit lights	Warning lights
Night lights	Stairwells	Storage areas
Guard stations	Open areas	Process areas
Entrance gates	Tunnels	
	Halls	Warning Lights
Production Lighting		Beacons
Machine areas	Miscellaneous	Hazardous areas
Raw materials storage	Standby generator areas	Traffic signals
Packaging	Hazardous machines	
Inspection		Health-Care Facilities
Warehousing		Operating rooms
Offices		Delivery rooms
		Intensive care areas
Commercial Lighting		Emergency treatment areas
Displays		
Product shelves		Miscellaneous
Sales counters		Switchgear rooms
Offices		Elevators
		Boiler rooms
Miscellaneous		
Switchgear rooms		
Landscape lighting		
Boiler rooms		
Computer rooms		

Source. Extracted from "Recommended Practice for Emergency and Standby Power Systems," IEEE Standard 446-1974, as reprinted from *Lighting Design & Application,* May 1975.

[a] An example of a standby lighting system is an engine-driven generator.

[b] An example of an immediate short-term lighting system is the common unit battery equipment.

[c] An example of an immediate long-term lighting system is a central battery bank rated to handle the required lighting load only until a standby engine-driven generator is placed on line.

(*c*) ***Type, Amount and Duration of Emergency Lighting.*** The codes and most authorities accept 1.0 fc of illumination as sufficient to avoid panic and permit orderly egress. This level is usually understood to be the *average* illumination but none of the codes makes uniform coverage mandatory. This point will be discussed below under equipment arrangement. Duration of emergency lighting varies with the codes, from 1/2 hour to $1\frac{1}{2}$ hours for egress, and up to indefinite periods (at higher levels) for facilities that cannot be evacuated. Tables 21.8 and 21.9 give general criteria and typical recommendations for emergency lighting for egress as well as for other purposes.

(*d*) ***Types of Emergency Lighting Systems.*** Section 16.29 and Figure 16.52, page 641, cover emergency power systems including generators and central battery systems. The former supplies selected portions of the normal lighting system through special emergency-lighting panels. The latter can supply a d-c distribution system or, if equipped with an inverter, can supply a-c as well. The availability of efficient inverters has practically eliminated central d-c systems. These have the added disadvantage that the incandescent fixtures they supply are not part of the normal system, and may obtrude upon the architecture, even when recessed and attractively finished. This is also true of package units with

Fig. 21.36 (a) *Completely self-contained, maintenance-free package. Each fixture contains a light, battery, and inverter, good for 1.5 hr of light.* (b) *Remote heads supplied from a d-c emergency power system. Test switches indicate unit's condition.* (c) *Slim, attractive, wood-grained case contains a maintenance-free lead calcium battery, inverter, and controls. Units such as this can be recessed and heads* (b) *mounted remotely. Photos courtesy of* (a) *Dual-Lite and* (b, c) *Tork, Inc.*

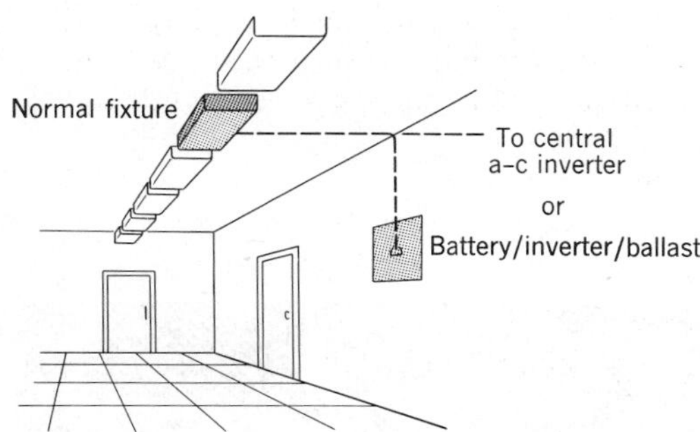

Fig. 21.37 *The use of an integral or remote (shown) concealed battery/inverter allows instantaneous emergency lighting of areas blacked out by local or general power failure. The use of fluorescent sources provides seven times the illumination possible with incandescent, for the same battery size.*

spotlight-type heads (see Fig. 21.36). Such units are best applied in individual rooms and isolated locations. The central battery with inverter, like the central generator, supplies lighting units that are usually (although not necessarily) part of the normal system. This arrangement has the advantages of economy, neatness, ability to use a-c sources (fluorescent), and reliability (see Fig. 21.37). Reliability can be further enhanced by using completely integral battery-charger-inverter packages mounted entirely within a fluorescent fixture (see Fig. 21.38). These units are usually designed to provide the $1\frac{1}{2}$ hours of illumination required by the code and to be completely maintenance-free for 7 to 10 years, after which the battery is simply replaced. Since high temperatures seriously affect battery life, it is advisable to mount these integral packages in a location other than that occupied by the fixture ballast (see Fig. 21.39).

(*e*) ***Arrangement of Fixtures.*** When the illumination level in an interior space drops sharply from a level of 30 to 150 fc to 1.0 fc, the eyes require up to 5 minutes to fully ac-

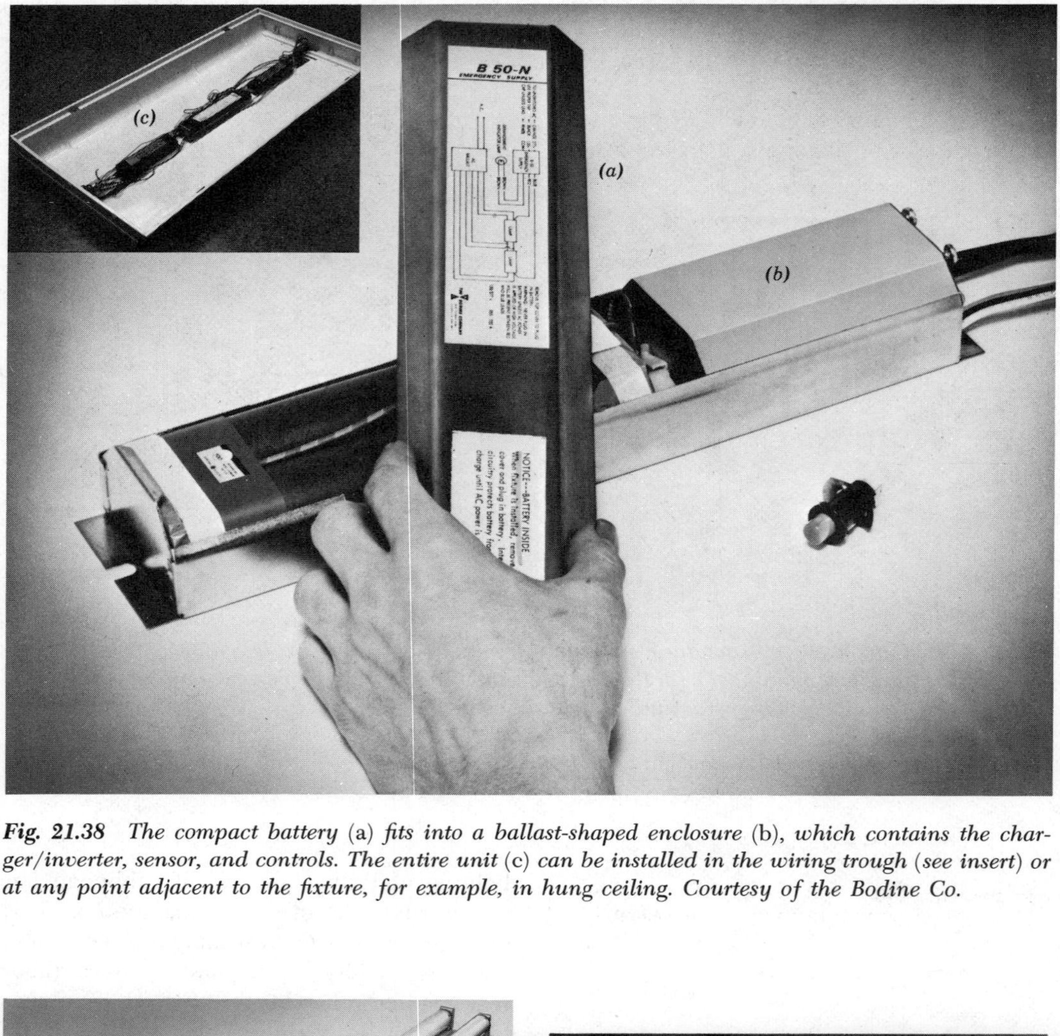

Fig. 21.38 *The compact battery* (a) *fits into a ballast-shaped enclosure* (b), *which contains the charger/inverter, sensor, and controls. The entire unit* (c) *can be installed in the wiring trough* (*see insert*) *or at any point adjacent to the fixture, for example, in hung ceiling. Courtesy of the Bodine Co.*

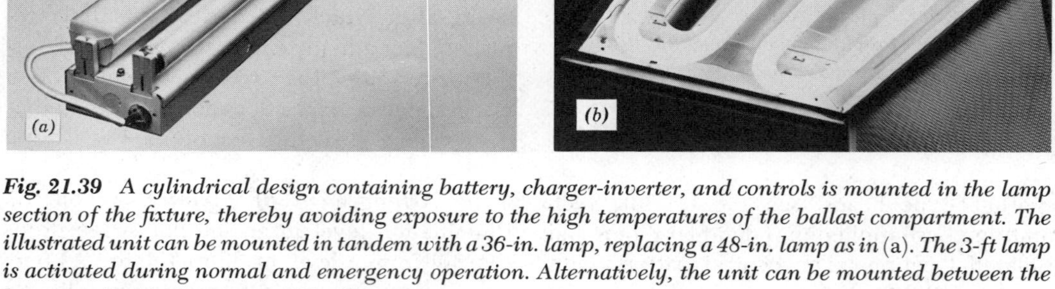

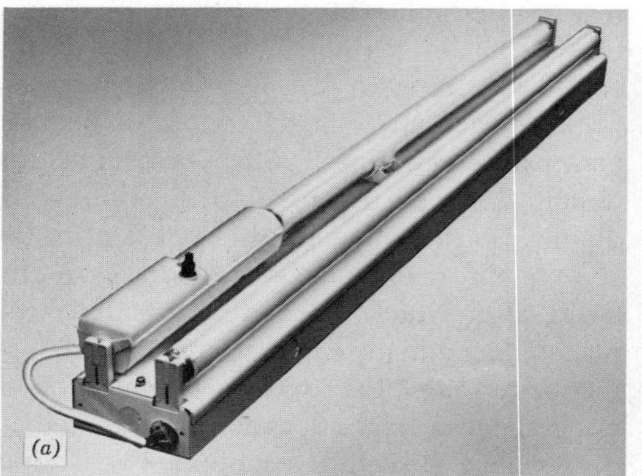

Fig. 21.39 *A cylindrical design containing battery, charger-inverter, and controls is mounted in the lamp section of the fixture, thereby avoiding exposure to the high temperatures of the ballast compartment. The illustrated unit can be mounted in tandem with a 36-in. lamp, replacing a 48-in. lamp as in* (a). *The 3-ft lamp is activated during normal and emergency operation. Alternatively, the unit can be mounted between the lamps as in* (b). *Courtesy of Dual-Lite.*

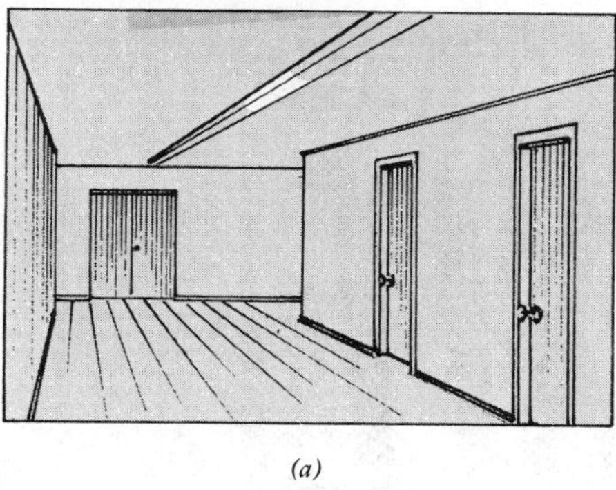

(a)

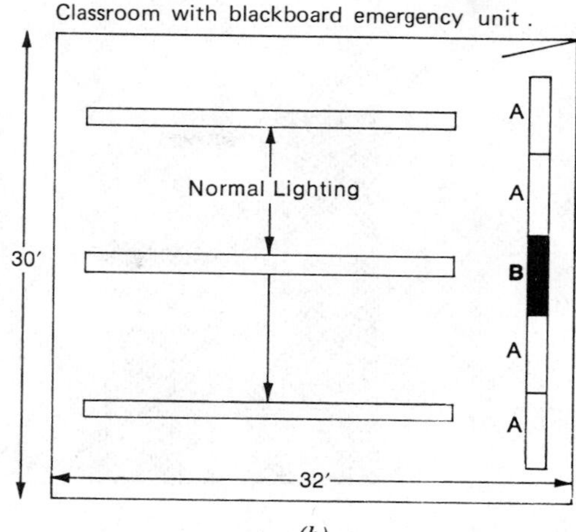

(b)

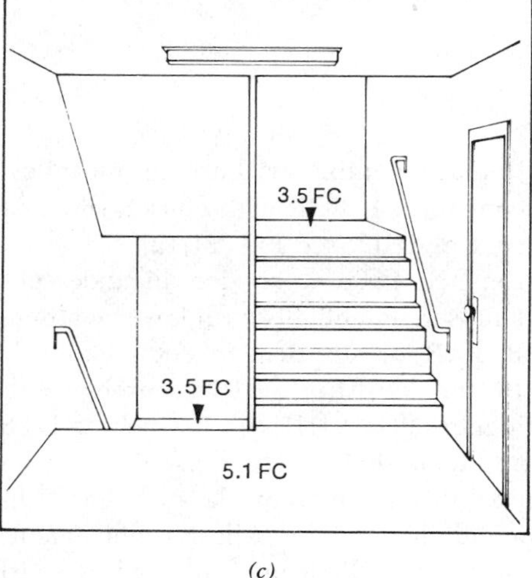

(c)

Fig. 21.40 *Typical emergency-lighting installations using dual-function fluorescent units with self-contained power packs. Courtesy of Holophane/J. Manville.*

commodate. During this long period the space's occupants are essentially sightless—a condition that lends itself readily to panic. For this reason, we recommend that the emergency lighting units be designed to give 1.0 fc uniform, not average, illumination. The difference is that the latter can be accomplished with bright, spotlight-type heads which, *unless carefully arranged,* can create disabling glare, distorting shadows, and impede eye accommodation. In such an event they do little to fulfill their primary function. A requirement for uniform illumination will normally result in a design using selected normal fixtures or lamps within such fixtures. This will result in altering only the level of lighting during a power failure—not the quality. This allows the eye to adapt rapidly, eliminates disabling glare, and assists in creating the atmosphere necessary for orderly, safe egress (see Figure 21.40).

Similarly, all exits must obviously be well illuminated so that traffic will flow smoothly. The well-known exit sign can also be equipped with an integral battery and control equipment to illuminate itself (normally with incandescent lamps) and the area immediately below it (see Figure 21.41).

21.28 Building Retrofit

Existing building lighting systems can be modified to increase efficiency, decrease glare, and decrease energy consumption by applying the procedures, techniques, and knowledge gained in the preceding four chapters.

(a) *Sources*

1. Replace standard fluorescent tubes with low-energy units.

Fig. 21.41 Exit light with built-in battery, charger, and controls. Insert shows typical application. Note that the bottom of the unit illuminates the area immediately in front of the exit. Photo of exit light courtesy of Dual-lite. Insert photo by Stein.

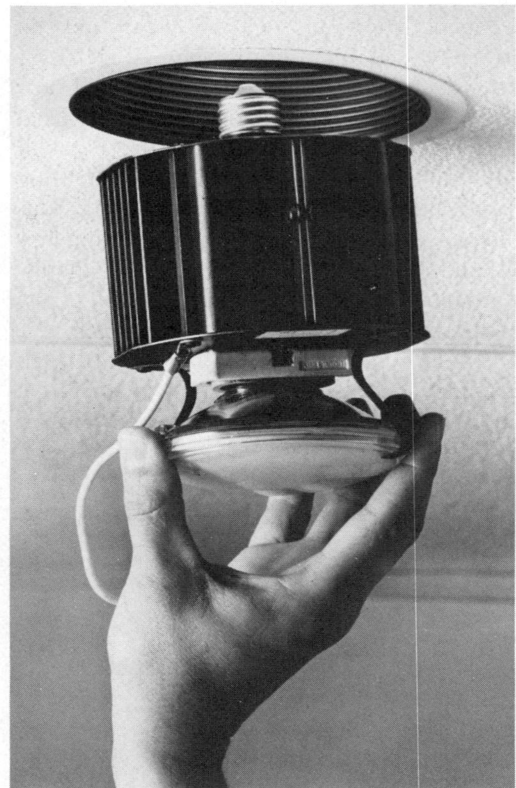

Fig. 21.42 Direct conversion of incandescent downlight to a mercury unit. Illustrated unit is 75-w PAR 36, which will replace a 150-w R or PAR incandescent lamp and generally pay back the investment within 3 years. Photo courtesy of Crouse Hinds.

2. Replace continuous burning incandescent sources with high-efficacy fluorescent or HID (see Fig. 21.42).
3. Replace general service incandescent lamps in downlights with lower wattage R, PAR, or elliptical reflector lamps.
4. Replace existing HID sources with higher-efficacy HID units such as HPS and metal-halide. Use those that can be operated on mercury-lamp ballasts, to avoid the cost of ballast replacement.
5. Increase daylight use—add reflectors to increase room penetration.

(*b*) ***Fixtures***

1. Replace diffusers with more efficient ones—reduce glare, increase CRF and ESI footcandles; this will permit reducing lighting levels.
2. Install multilevel ballasts.
3. Add two-level control to HID units.
4. Institute a program of maintenance that will permit decreasing energy use by at least 20% while maintaining output.
5. Modify fixture locations to give minimum direct and reflected glare.
6. Add task lighting.
7. Reduce overall lighting to appropriate ambient level by fixture removal.

(c) *Other*

1. After making the above adjustments check that levels do not exceed recommendations. Reduce levels by removing two of four lamps in a fixture. Also remove the associated ballast. With two-lamp fixtures, remove one tube and replace with a dummy tube if the ballast will be adversely affected by one-lamp operation.
2. Rearrange tasks so that the most difficult ones benefit most from daylight.
3. Install low voltage or sonic switching to give requisite local light control.
4. Repaint to give requisite reflectances.
5. Install time switches to automatically coordinate lighting with tasks and time, for example, reduce lighting during lunch hours.
6. Operate closet lights with door switches.
7. Install pilot lights outside storage rooms to indicate lighting "ON."

21.29 Floodlighting

Floodlighting, both interior and exterior, is extensively used for such diverse locations as are listed in Table 21.10, in addition to the more common sports lighting, which is not listed. At the designer's disposal are a variety of sources with respect to output, color, life, efficiency, and wattage (see Chapter 19).

Table 21.10 ***Lighting Application Guide***

Application	Minimum Footcandles Maintained[a]	Watts Per Square Foot Generally Required: Tungsten Halogen	Mercury Units	Metal-Halide	High-Pressure Sodium
Automobile Parking					
Attendant parking	2	0.38	0.17	0.11	0.075
Industrial lots	1	0.13–0.15	0.06–0.07	0.037–0.044	0.026–0.03
Self-parking lots	1	0.13–0.15	0.06–0.07	0.037–0.044	0.026–0.03
Shopping Centers					
Neighborhood	1	0.13–0.19	0.06–0.09	0.037–0.055	0.026–0.038
Average commercial	2	0.26–0.3	0.12–0.135	0.075–0.087	0.052–0.06
Heavy traffic	5	0.65	0.29	0.19	0.13
Automobile Sales Lots					
Front row (Front 20 ft)	50	10.	4.5	2.9	2.0
Remainder	10	1.5–1.8	0.68–0.81	0.44–0.52	0.3–0.36
Building					
Construction	10	1.5–1.8	0.68–0.81	0.44–0.52	0.3–0.36
Excavation	2	0.26–0.3	0.12–0.14	0.075–0.09	0.052–0.06
Buildings up to 50 ft High	*Adj. Area* Light / Dark				
Light surfaces	15 / 5	3.3 / 1.2	1.5 / 0.54	0.96 / 0.35	0.66 / 0.24
Medium light	20 / 10	4.3 / 2.2	1.94 / 1.0	1.25 / 0.64	0.86 / 0.44
Dark surfaces	50 / 20	10.0 / 4.3	4.5 / 1.94	2.9 / 1.25	2.0 / 0.86
Billboards and Signs	*Adj. Area* Light / Dark				
Good contrast	50 / 20	10.0 / 4.3	4.5 / 1.94	2.9 / 1.25	2.0 / 0.86
Poor contrast	100 / 50	20.0 / 10.0	9.0 / 4.5	5.8 / 2.9	4.0 / 2.0
Protective Lighting					
Gates and vital area	5	1.2	0.54	0.35	0.24
Building surrounds	1	0.15–0.19	0.07–0.09	0.044–0.055	0.03–0.04
Roadways					
Along buildings	1	0.24	0.11	0.07	0.05
Open areas	0.5	0.08–0.1	0.036–0.045	0.023–0.029	0.02
Storage yards (active)	20	3.6–4.3	1.6–1.94	1.04–1.25	0.72–0.86
Storage yards (inactive)	1	0.15–0.19	0.07–0.09	0.044–0.055	0.03–0.04
Shopping Centers					
Parking areas (attraction)	5	0.65	0.29	0.19	0.13
Buildings (attraction)			(See Buildings)		
Used Car Lots			(See Automobile Parking)		

[a] All footcandle levels for ground area applications are *horizontal* values.

Fig. 21.43 (a) *The Statue of Liberty was relighted for the American Bicentennial. It was found that 58- 1000 watt metal halide units give 40 to 50 fc of white light on the statue. Eleven 400-w HPS units were selected to compliment the color of the granite base, which they light to 10 fc. Courtesy of Crouse-Hinds Company.*

Fig. 21.43 (b) *Floodlighted section of wall surrounding the Old City of Jerusalem, Israel, adjacent to the Jaffa gate. Light sources are 400-w, high-pressure sodium units, giving an average illumination level of 50 lux. Courtesy of City of Jerusalem and J. Stroumsa, Chief Engineer.*

Fig. 21.43 (c) *Church of All Nations, Mount of Olives, Jerusalem, Israel. Floodlight sources are 250- and 400-watt mercury and metal-halide units, giving an average illumination of 70 lux. Sources were selected to compliment the colors in the mosaic at the top of the facade. Courtesy of City of Jerusalem, Israel and J. Stroumsa, Chief Engineer.*

Although a detailed floodlighting design involves complex calculations beyond the scope of this work, it is often sufficient for the designer to utilize a watts per square foot table such as Table 21.10 in order to determine the approximate floodlighting requirements.

Thus, if one is concerned with lighting a self-service parking lot at a neighborhood shopping center, and metal-halide is selected, Table 21.10 tells us that approximately 0.055 w/sq ft will suffice. If the lot is 200×500 ft or 100,000 sq ft, then $0.055 \times 100{,}000 = 5500$ w are required.

Arrangement and choice of equipment remains then, before the problem can be considered solved. Considerable assistance on this score can be obtained from either the lighting engineer involved or from representatives of the equipment manufacturers.

Although most floodlight installations use a single type, the installation of Figure 21.43*a* used a combination of metal-halide and HPS to obtain the desired effect.

21.30 Street Lighting

Although detailed street-lighting calculations and design considerations are beyond our scope (see appropriate IES standards) a few remarks are in order. New installations now use HID sources almost exclusively. The low

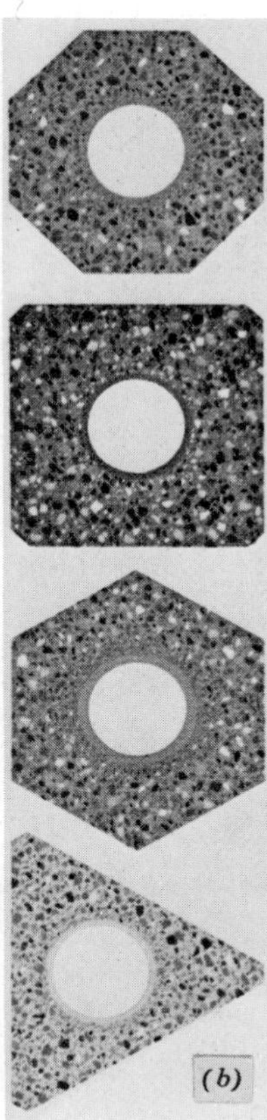

Fig. 21.44 (a) *Concrete pole with aluminum arm and mercury luminaire.* (b) *Other concrete pole sections. Courtesy of American Concrete Div./ Union Metal.*

Fig. 21.45 Street-lighting pole with base embedded in the ground. The appearance is artificial, since poles are not trees and do not grow from the ground. (*Photo by Stein*).

Fig. 21.46 The "lollypop" fixture, even if aesthetically pleasing, gives poor illumination downward (note large collar). Narrow pole is weakened by large handhole. (Photo by Stein).

efficacy and short life of incandescent sources and the bulkiness of fluorescents make them obsolete. Furthermore, high street-lighting levels reduce vandalism and crime, improve night merchandising, and add to an area's attractiveness. Some typical designs of street lighting and other outside luminaires are shown in Figures 21.44 to 21.47.

References and Bibliography

1. *Contrast Rendition in School Lighting*, 1970, F. K. Sampson, Educational Facilities Laboratories, New York.
2. "Creative Classroom Lighting," March 1976, A. J. Rudd, *Lighting Design & Application* (LDA), page 12.
3. "Comfortable, Economic, Effective Classroom Lighting," February 1974, J. J. Niedhart, *LDA*, page 16.
4. "Saving Energy in the Classroom," February 1974, J. F. Finn, *LDA*, page 29.
5. "Better Lighting with Less Energy," February 1974, N. Falk, *LDA*, page 4.
6. *Lighting System Study*, March 1974, Ross & Baruzzini, Inc., General Services Administration, Public Buildings Service.

Fig. 21.47 Built-in walk and stair lights should be unobtrusive as in (a), *and arranged to illuminate front of treads as in* (b). *(Photos by Stein.)*

PART SEVEN

signal equipment

22

Signal Systems

22.1 Introduction

No area of equipment design and application to buildings has seen such great and rapid changes as the field of sound and signal equipment. Under this rather vague title is subsumed all signal, communication, and control equipment, the function of which is to assist in effecting proper building operation. Included are surveillance equipment such as fire and interior alarm; audio and visual communication equipment such as telephone, intercom, and television, both public and closed circuit; time equipment such as clock and program. These systems are no longer limited in application. Clock and program equipment, which once were the exclusive interest of schools and some industrial facilities, are now incorporated into building mechanical equipment control systems. Closed circuit TV, which was once limited to classroom and college use, is commonplace in mercantile areas as part of surveillance systems. The hundreds of signals generated throughout a large facility are logged, channeled, and applied by means of specially programmed computers. All the signal systems that once were separate and distinct are now frequently combined and serve multiple purposes.

Obviously a detailed study of the design and application of such equipment is beyond the scope of this book or, for that matter, of any single book. We shall attempt, however, to discuss the basic operation of the various systems, some of the equipment available, application to different types of facilities, and the impact of these systems on the spaces within a structure. The types of facilities that will be considered are single and multiple residences, schools, stores, office buildings, and industrial facilities. Hospitals and laboratories are combinations of the above area types plus facilities too highly specialized to be discussed herein.

The systems to be covered will include surveillance, communication, and time-based signal arrangements. Antenna systems will be included as a special case of a communication (reception) system.

PRIVATE RESIDENTIAL SYSTEMS

22.2 General

Modern private residences utilize a variety of signal apparatus that greatly enhance their functional value. Figure 22.1 shows a residence that has been provided with what would be considered adequate but by no means excessive sound and signal equipment for a house of this size. In general, all signal systems require a source of signal, equipment to process the signal including transmitting it, and finally a means of indicating the signal, either audibly, visually, or on permanent record "hard copy." The complex system still falls into this threefold category except that the individual items of equipment and their functions proliferate and complicate. Thus, the difference between a small residential intercom system and a huge hospital communication system is really of degree and not of kind. Of course, circuits vary as do functions, but the basic line of signal initiation, transmission-and-process, and reception remains unchanged. To illustrate this we have listed in Table 22.1 the systems and equipment found in the residence of Figure 22.1, by this threefold classification. Note that the fire alarm,

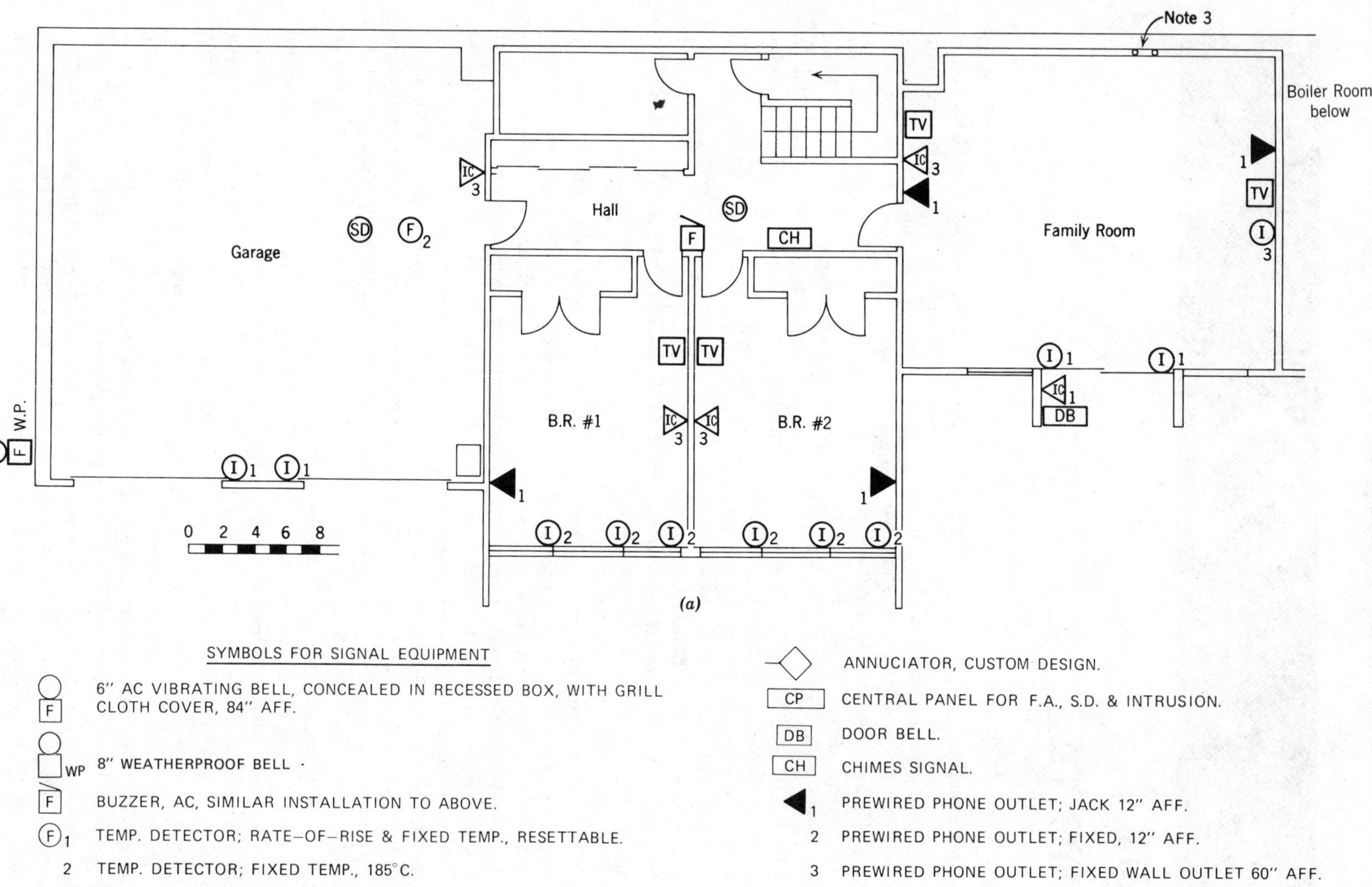
Note 3
Boiler Room below
Garage
Hall
Family Room
B.R. #1
B.R. #2
W.P.
SD
CH
TV
DB
0 2 4 6 8
SYMBOLS FOR SIGNAL EQUIPMENT
6" AC VIBRATING BELL, CONCEALED IN RECESSED BOX, WITH GRILL CLOTH COVER, 84" AFF.
8" WEATHERPROOF BELL
BUZZER, AC, SIMILAR INSTALLATION TO ABOVE.
TEMP. DETECTOR; RATE-OF-RISE & FIXED TEMP., RESETTABLE.
2 TEMP. DETECTOR; FIXED TEMP., 185°C.
ANNUCIATOR, CUSTOM DESIGN.
CP CENTRAL PANEL FOR F.A., S.D. & INTRUSION.
DB DOOR BELL.
CH CHIMES SIGNAL.
PREWIRED PHONE OUTLET; JACK 12" AFF.
2 PREWIRED PHONE OUTLET; FIXED, 12" AFF.
3 PREWIRED PHONE OUTLET; FIXED WALL OUTLET 60" AFF.

(a)

(SD)	SMOKE DETECTOR WITH RESETTABLE FIXED TEMP. DETECTOR.
(I)$_1$	INTRUSION DETECTOR, MAGNETIC DOOR·SWITCH.
2	INTRUSION DETECTOR; MAGNETIC WINDOW SWITCH.
3	INTRUSION DETECTOR; ELECTRONIC, MOTION DETECTOR.
IC$_1$	INTERCOM OUTLET, OUTDOOR, W.P. 60″ AFF.
2	INTERCOM OUTLET, MASTER STATION 60″ AFF.
3	INTERCOM OUTLET, REMOTE STATION 60″ AFF.
TV	PREWIRED TV ANTENNA OUTLET, 12″ AFF.

(c)

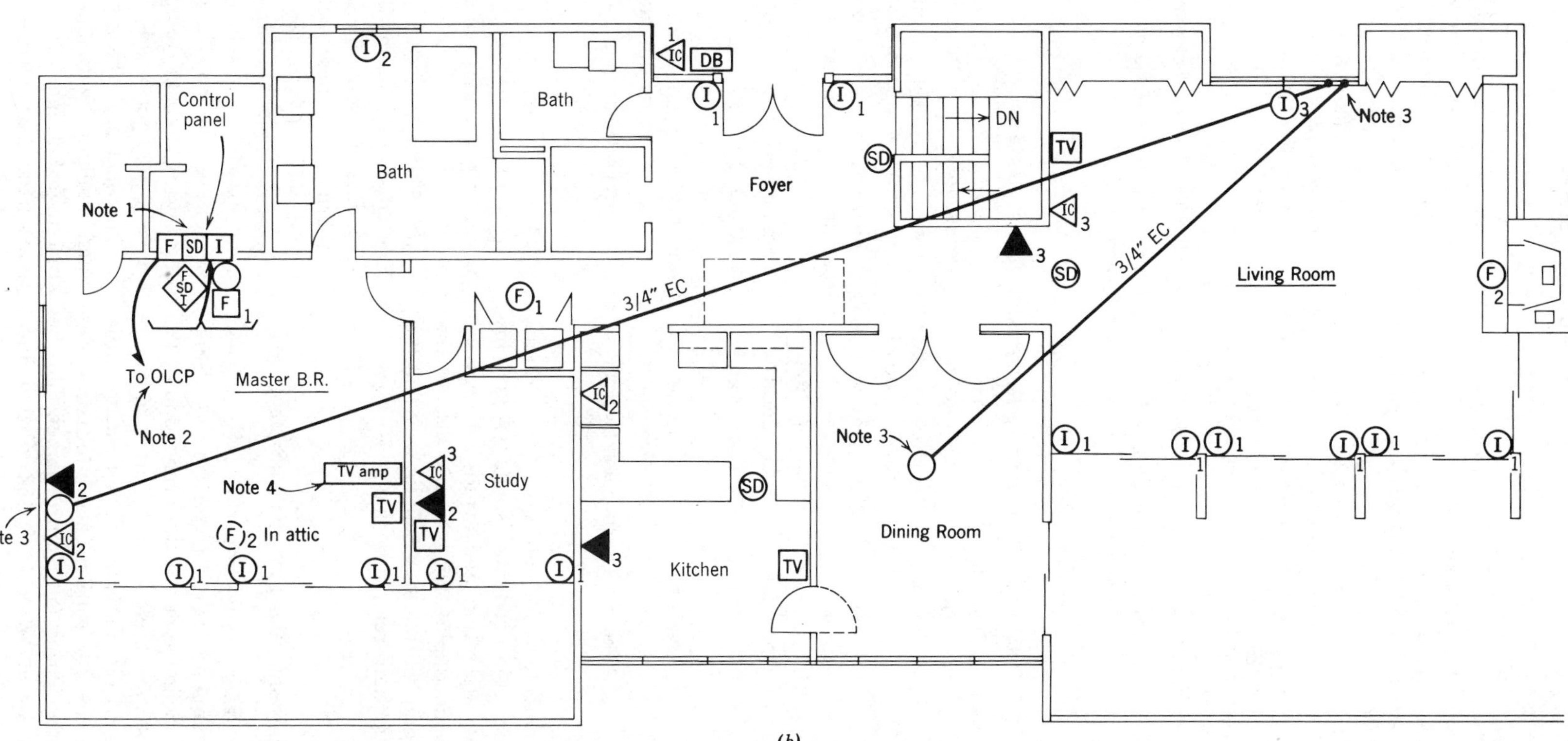

(b)

Fig. 22.1 (a) *Electrical plan, lower level, signal devices.* (b) *Electrical plan, upper level, signal devices.* (c) *Symbols for signal equipment.* (d) *Notes for signal systems: (on next page).*

Fig. 22-1 (continued)

1. *The fire detection, smoke detection, and intrusion alarm devices all operate from a single control panel. The alarm bell is common. The annunciator indicates the device operated and its location.*
2. *The connection between the signal control panel and OLCP (outside lighting control panel) activates all outside lights when a signal device trips. Selected lights inside the house can also be connected to go on.*
3. *Two $^3/_4$ in. empty plastic conduits, extending from 2 to 10 4 in. boxes in living room wall down to family room and terminating in 4 in. flush boxes. Boxes to be 18 in. AFF and fitted with blank covers. Extend a $^3/_4$ in. plastic EC from one 4 in. box in living room to 12 in. speaker backbox recessed in dining room ceiling. Locate in the field. From the second 4 in. box in living room extend a $^3/_4$ in. empty plastic conduit to an empty 4 in. box in the master bedroom, 18 in. AFF. Finish with blank cover.*
4. *Provide television antenna amplifier, recessed in wall box, with hinged ventilated cover, 18 in. AFF.*

   Connections to antenna and to all television outlets by television antenna subcontractor. Provide 120 v outlet at the amplifier, with switch to disconnect.

Table 22.1 ***Elements of Residential Signal Systems***

System Type	Signal Generator	Signal Processor[a]	Signal Transducer
Fire alarm	Temperature and smoke detectors	Control cabinet(s)	Bells, annunciator buzzer
Intrusion alarm	Door and window switches, motion detector	Control cabinet	Bells, buzzer, annunciator
Door bell	Push button	Transformer	Buzzer, chime
TV antenna	TV station and house antenna	Amplifier	TV set
Intercom	Microphone, speaker—mike	Amplifier	Speakers in various stations

[a]The proper wiring and switching is included under this title in all cases.

smoke detection, and intrusion alarm systems have been combined into a single system. This simplifies operation and avoids unnecessary equipment duplication. As we discuss the more complex systems it will be seen that the basic functions remain unchanged.

22.3 Residential Fire Alarm Systems

(*a*) ***General.*** If properly designed and functioning normally, the system should provide sufficient time for the evacuation of the residents and then for appropriate countermeasures to be initiated. The elements of the system are the various alarm-initiating devices, the wiring and control panel, and the audible alarm devices. Systems utilizing automatic sensing equipment are referred to as automatic fire alarm systems, even if manual signal units are also used.

(*b*) ***Alarm Devices.*** The most common residential alarm initiating device is the temperature detector, sometimes called (erroneously) a thermostat. These detectors are of two types; the fixed temperature unit and the rate-of-rise unit. The former operates a set of contacts when a preset (nonadjustable) temperature is reached. The latter operates when the rate of ambient temperature change exceeds a predetermined amount, indicative of an incipient fire. The rate-of-rise unit is normally combined with a fixed temperature unit in a single housing. The fixed temperature unit is available in a one-time nonrenewable design that utilizes a low melting point alloy plug and an automatic resetting unit of design similar to a thermostat. For most applications, the resettable unit is preferred. Two typical units are illustrated in Figure 22.2.

Smoke detectors, once a rarity in residences, despite it being well known that most

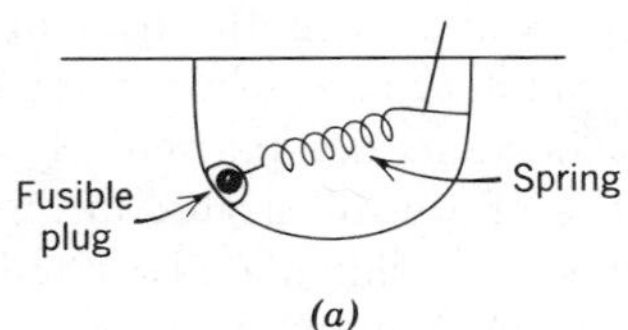

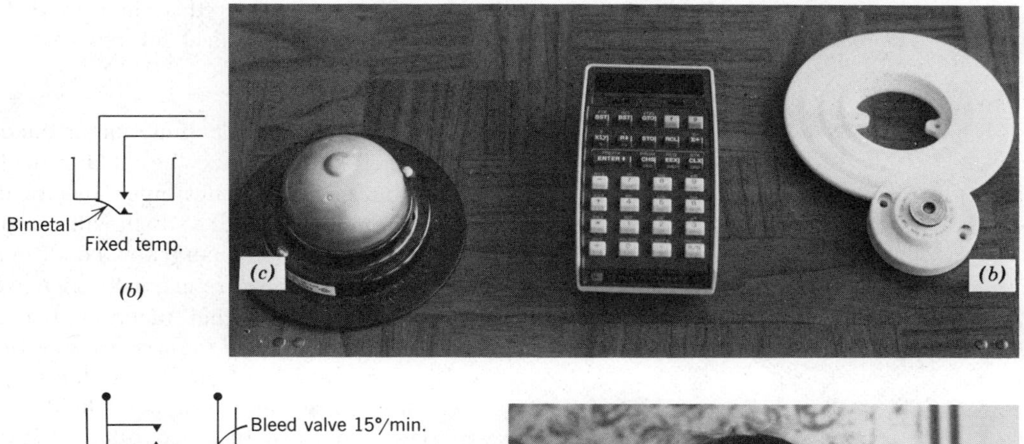

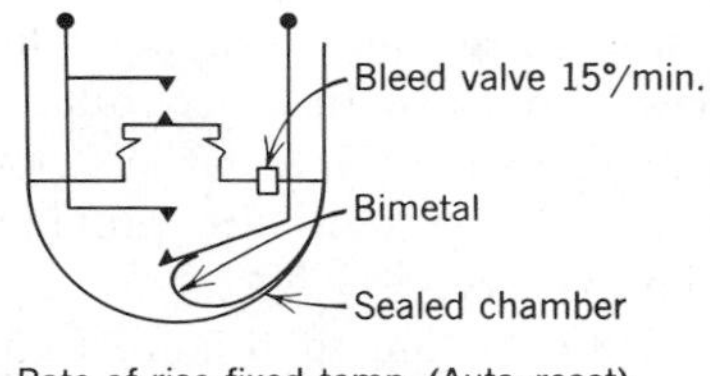

Bleed valve 15°/min.
Bimetal
Sealed chamber
Rate of rise fixed temp. (Auto-reset)
(c)

Fig. 22.2 *Fire alarm temperature detectors.* (a) *Fusible plug melts out at predetermined temperature. Unit is indicating and nonrenewable.* (b) *Bimetallic unit action is similar to thermostat and is self-restoring.* (c) *Rate-of-rise unit comprises an air chamber with restricted bleed valve. Rapid temperature rise causes expansion of bellows before loss of air by bleeding and consequent alarm. Unit is combined with bimetallic fixed temperature unit, similar to* (b). (d) *This unit is combined with a photocell smoke detector and operates on house current.*

(a)

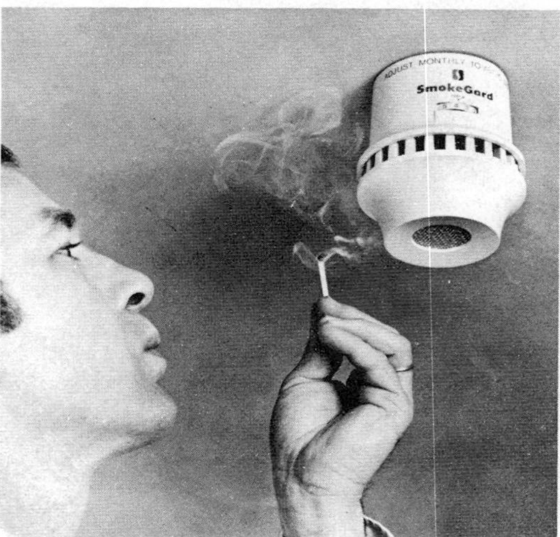

(b)

Fig. 22.3 Types of self-contained residential smoke detectors. (a) *Contains a photocell detector, control equipment, and an audible alarm. It operates from house current. Photo courtesy of Edwards.* (b) *Contains an ionization detector, controls, audible device, and batteries, making it independent of wiring systems. Photo courtesy of Statitrol.*

fire fatalities are caused by smoke poisoning, have become quite common, because

1. The Federal Housing Authority (FHA) requires at least one in all residences it finances.
2. Local building codes are increasingly mandating them.
3. Price has dropped to reasonable levels and availability is widespread.

The various types of smoke detectors available and their application are covered in Section 22.12. Most fire authorities agree that home fires smolder before bursting into flame. Photoelectric units are particularly effective for this type of fire. Illustrated in Figure 22.3 are two types of residential alarm units—both completely self-contained, since most residences are *not* prewired as in Figure 22.1. Also, such units are ideal for retrofitting an existing residence.

(*c*) ***Control Unit.*** The function of the central panel is to energize the audible devices (bells, buzzers, and gongs) upon receipt of a signal from the detectors, which will continue to sound until the emergency condition is cleared or until they are manually silenced at the control panel. Other functions that the panel may be designed to serve are shut-off of oil and gas lines, shut-off of attic fan to prevent fire spread, and turn-on of lights, both inside and outside. In order to assure system operation even in the event of a power interruption, most systems are provided with a standby battery.

As with the unit of Figure 21.1, a single control panel can serve a multiplicity of residential systems. Typical units of this type are illustrated in Figure 22.4. The panel of Figure 22.4*b* is designed to display the nature and location of the alarm device which has "tripped." A riser diagram for the system of Figure 22.1 is shown in Figure 22.5. The alarm devices themselves are not shown in this diagram, because they appear on the plans and duplication serves no useful purpose.

(*d*) ***Audible Alarm Devices.*** The audible devices comprise conventional bells, gongs, buzzers, and horns, some of which are illustrated in Figure 22.6. Most commonly used is the a-c vibrating bell, since gongs are normally only used in coded nonresidential systems. A weatherproof external bell, to alert neighbors and passersby, is also desirable.

(*e*) ***System Design.*** Detectors are rated by temperature and coverage; for example, 135° F and 200 sq ft and are normally located

(a)

Fig. 22.4a Typical residential control panel for fire and intrusion alarm systems. The illustrated unit is 15 in. wide, 14 in. high, and mounts semi-flush. Courtesy of NUTONE/SCOVILL.

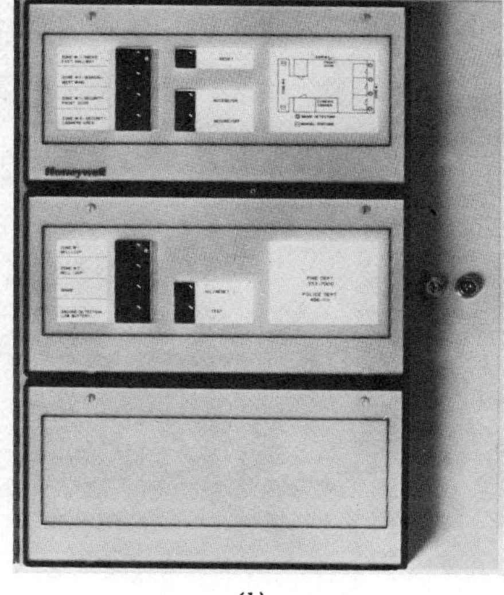

(b)

Fig. 22.4b A modern control panel and annunciator unit shows the zone and actual device which operated. On a plan of the house, a light indicates the location of the tripped device. Such an annunciator can show all the devices and distinguish between types by use of different color light. Courtesy of Honeywell.

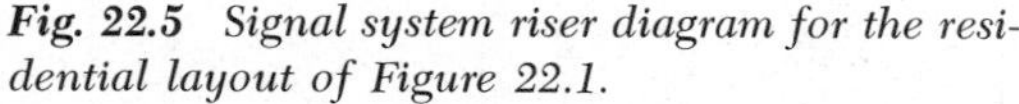

Fig. 22.5 Signal system riser diagram for the residential layout of Figure 22.1.

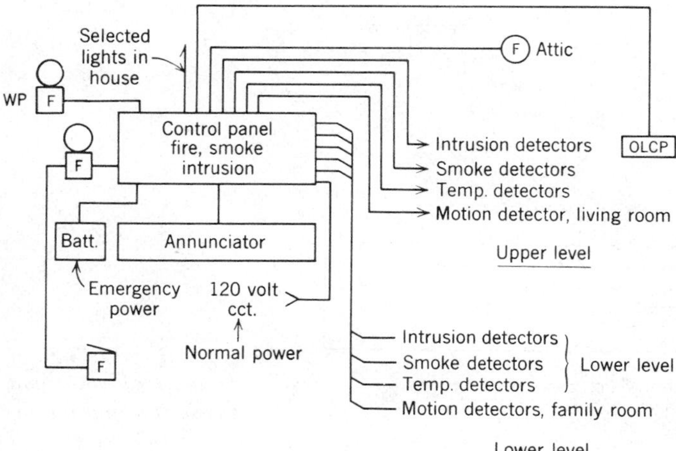

on the ceiling in all rooms and stairwells, including the attic and the basement. Because of high ambient temperatures, the units installed in the kitchen, attic, and basement near the heating unit are normally rated at 190 to 200° F, whereas the units in the other rooms are usually set at 135° F. Smoke detectors are normally placed adjacent to the house heating unit in the bedroom area, in the kitchen, and occasionally in the garage and in unoccupied areas such as the attic.

(*f*) ***Circuit Design.*** A system which is normally deenergized and carries no current ex-

Fig. 22.6 Audible signals, both indoor (a) *and outdoor type* (b), *alert occupant and/or neighbors to actual or incipient fire. Courtesy of NUTONE/SCOVILL.*

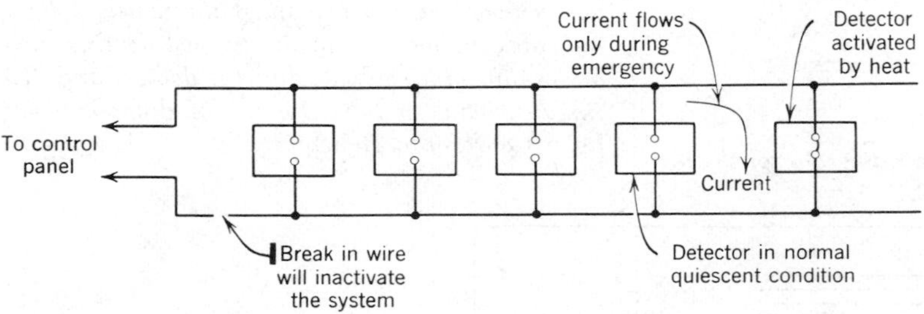

(a) Wiring of an open-circuit fire-alarm system

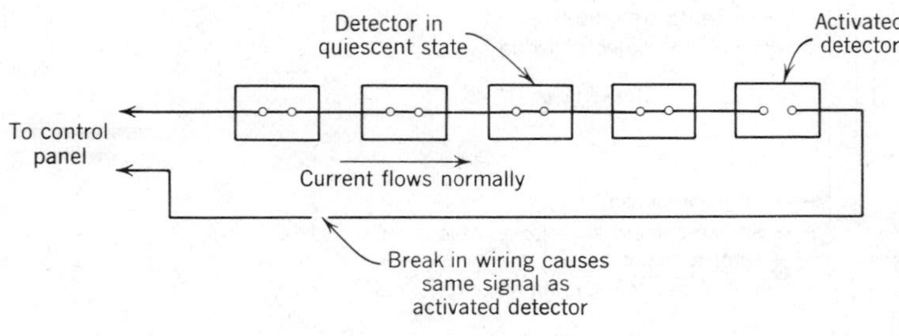

(b) Wiring of a closed-circuit fire-alarm system

Fig. 22.7 Wiring of open- and closed-circuit systems, showing the inherent advantage of closed-circuit wiring.

cept when functioning is called an open circuit system (see Fig. 22.7*a*). Such a system is the simplest and most economical type but has the disadvantage of not indicating a broken wire or other malfunction that will render the system inoperative.

Figure 22.7*b* shows a closed circuit system. This arrangement will set off the alarm bells in the event of trouble in the equipment, but since this type of "false alarm" is to an extent undesirable, a further refinement in the form of a trouble bell and (or) light can be added (at approximately 50% cost increase) that will then indicate to the occupants an equipment failure without ringing the fire alarm bells. This feature is known as supervision, and such a system is known as a supervised system. Supervised systems can utilize open or closed circuit devices, depending upon circuit arrangement. Furthermore, by special wiring and circuit design, the system can be arranged so that a single break or ground in the wiring to the devices will not prevent the operation of the system.

If the system becomes extensive as in Figure 22.1, a desirable feature to be incorporated in the circuitry arrangement is the grouping of the devices into zones, as in the panel in Figure 22.4. Each zone is covered by a separate panel section or module, permitting extremely rapid identification of an alarm signal.

In summary, then, the basic residential system is an open circuit, temperature, and smoke detector type of evacuation alarm. Additional cost is incurred by adding desirable features such as closed circuit operation and supervision.

22.4 Residential Intrusion Alarm Systems

An increasing number of private residences are utilizing intrusion (burglar) alarm systems, often in conjunction with fire alarm equipment.

Basically, an intrusion alarm system is similar to a fire detection system except that instead of thermal detection, devices such as metallic tape and micro- and magnetic switches are used to detect door and window motion, glass breakage, and so on. More sophisticated devices such as motion detectors are also used occasionally. A manual switch at the end of a long cord is also often provided so that the resident may at will set off the alarm in the event an intruder is heard. The system may employ the same audible signals as the fire system or its own components. Although done infrequently, intrusion alarm systems can be continuously supervised by connection with central stations of companies whose business such supervision is, and who will either respond directly to an alarm call or notify local police authorities of any illegal entry.

22.5 Residential Television Antenna Systems

The increasingly large number of multiset American homes have made the central television antenna system a desirable feature of the modern residence. Systems with more than two outlets generally require a booster amplifier (except in strong signal areas), and are known as amplified systems.

The function of the system is to supply a television signal at each wall outlet, so that a receiver may be operated at any location and so that two or more receivers may operate simultaneously.

The functioning of the system is simply to amplify the signal received by the antenna and by means of special cable to distribute these amplified signals in a concealed cable to the various wall outlets. The type and location of antenna, gain (amplification) of the amplifier, and type of cable are variables that, being dependent on the specific installation, are best left to a competent and reliable local television company or design engineer.

22.6 Residential Intercom and Sound Systems

The public demand for step-saving conveniences has resulted in the wide acceptance of the home "intercom" (see Fig. 22.8). Although available with various features, the basic system comprises one or more masters and several remote stations, one of which monitors the front door allowing it to be answered from various points within the home. In general, master stations allow selective calling, whereas remote stations operating through the masters are nonselective. The systems are particularly useful when left in the open (monitor) position for remote "baby sitting." The applicability of such systems to residences with outbuildings should be immediately apparent. Since wiring is low voltage

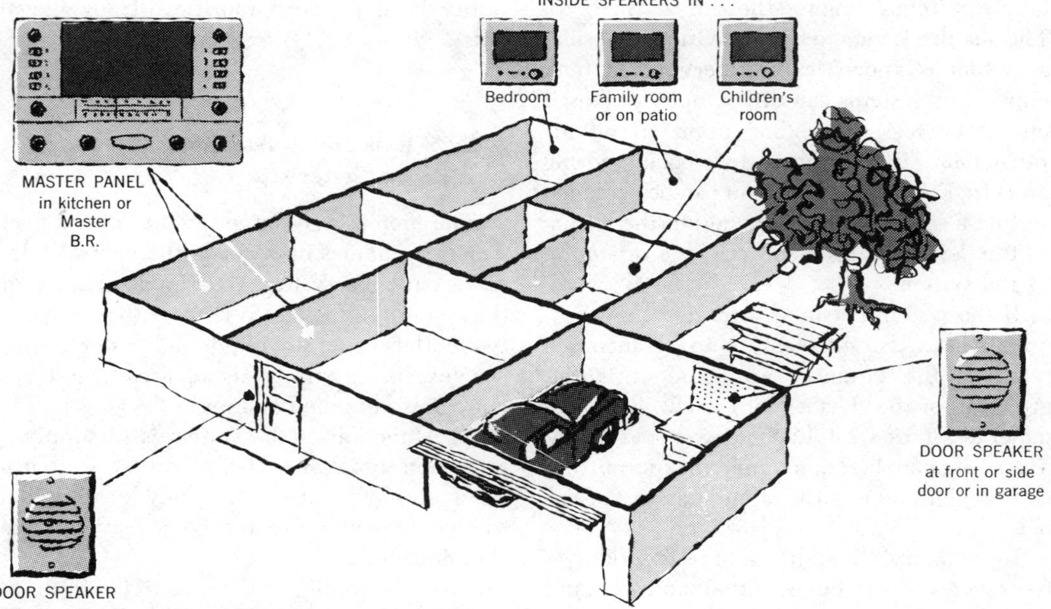

Fig. 22.8 Typical residential intercom equipment.

and low power, multiconductor color-coded intercom cable is generally used, run concealed within walls, attics, and basements.

Systems are also available that impose the signals onto the house power wiring. This has the advantage of eliminating separate wiring and making remote stations portable—they are connected simply by plugging into a power outlet. Many manufacturers have incorporated a tuner (AM, FM, or both) into the home intercom system so that music can be "piped" to each of the stations within the home. Since these stations generally utilize inexpensive 5 to 7 in. speakers, the results can only be construed as high-fidelity by an unknowledgeable home owner or an overzealous manufacturer.

A music system can be readily installed in the home as a separate entity. The design, and arrangement of such a system is a highly specialized field and the advice of experts should be sought if the results are to be gratifying.

22.7 Residential Telephone Systems

Telephone system work differs from all other types inasmuch as we are normally *only* making provisions for work to be done by others, namely the local telephone company. In residential work the telephone company normally follows the route of the electric service, entering the building overhead or underground as desired. In both cases a separate service entrance means must be provided; if aerial, a sleeve through the wall; if underground, a separate entrance conduit. Unless a residence has many entering lines, no source of power is required for the telephone equipment.

Wiring of telephone instruments when installed *after* completion of the residence consists of a single surface mounted $\frac{1}{8}$ in. diameter, 4-conductor cable that, even if skillfully installed, is unsightly at best and completely objectionable at worst. Most phone companies are happy to prewire a house at any stage prior to closing in of walls. Prewiring consists of running the cables on the wall framing and into empty device boxes. This results in a completely concealed installation that is desirable from the owner's viewpoint and, depending upon the number of outlets, cheaper for the phone company to install. Instruments can be wall or desk type, the latter also being available for jacking into outlets around the house. Most phone companies will also provide a number of other services including intercom, by special arrangement.

FIRE ALARM SYSTEMS

22.8 General

Fire alarm systems in other than small single residences are complex assemblies of equipment that should be understood at least to the extent of application. Our discussion will digress at this point to a survey of nonresidential fire alarm equipment and will then be followed by data on the application of such systems to multiple dwellings.

22.9 Fire Alarm Definitions and Terms

The following list will serve as a reference to terminology used herein which the reader will encounter in fire alarm work.

Alarm Initiating Device, Automatic. Automatic alarm initiating devices, such as fire detectors, smoke detectors, and water flow switches, are devices that automatically transmit an alarm signal when a condition indicative of a fire to which they respond occurs.

Alarm Initiating Device, Manual. A fire alarm station that will transmit an alarm signal when manually operated.

Automatic System. A system in which an alarm initiating device operates automatically to transmit or sound an alarm signal.

Auxiliary Fire Alarm System. A system that is maintained and supervised by a responsible person or corporation and having alarm initiating devices that, when operated, cause an alarm to be transmitted over a municipal fire alarm system to the fire station or to the fire alarm headquarters for retransmission to the fire station.

Bell, Single-Stroke. A device whose gong is struck only once each time operating energy is applied to the bell. This item is usually called a single-stroke gong.

Box, Fire Alarm. See *Station, Fire Alarm.*

Breakglass. Refers to a false-alarm deterrent available in fire alarm stations; a glass rod is placed across the pull-lever and breaks easily when the lever is pulled.

Coded Alarm Signal. An alarm signal that represents a 1, 2, 3, or 4-digit number indicative of the location of the fire alarm station operated.

Coded System. One in which not less than 3 rounds of coded alarm signals are transmitted, after which the fire alarm system may be manually or automatically silenced.

Common Code. See *Dual Coded System, Master Coded System, Noncoded System, Selective Coded System, Zone Coded System.*

Continuous Ringing. Refers to a continuous alarm. In coded systems so arranged, it refers to the signal that sounds after the completion of the normal number of rounds (usually 4) of identifying coded alarm signal.

Control Unit (Fire Alarm Panel). Comprises the controls, relays, switches, and associated circuits necessary to (1) furnish power to a fire alarm system, (2) receive signals from alarm initiating devices and transmit them to indicating devices and accessory equipment, and (3) electrically supervise the system circuitry.

Double Supervised System. A system in which the source of power for the trouble signal is supervised, in addition to the circuitry.

Dual-Coded System. A system in which a unique coded alarm is sounded for each separate fire box or fire zone to notify owner's personnel of the fire location, while noncoded or common coded alarm signals are sounded on separate signals to notify other occupants to evacuate the building.

Local Fire Alarm System. An electrically operated system producing signals at one or more places at the premises served, primarily for the notification of the occupants.

Local Noninterfering Coded Station. A fire alarm station that, once actuated, will transmit not less than 4 rounds of coded alarm signals and cannot be interfered with by any subsequent actuation of that station until it has transmitted its complete signal.

Manual System. One in which the alarm

initiating device is operated manually to transmit or sound an alarm signal.

Master Coded System. One in which a common coded alarm signal is transmitted for not less than 3 rounds, after which the fire alarm system may be manually or automatically silenced. The same code is sounded regardless of the location of the alarm initiating device. This system can be arranged to ring continuously.

Noncoded System. One in which a continuous audible alarm is transmitted for a predetermined length of time, after which it may be manually or automatically silenced.

Positive Noninterfering and Succession Coded Station. A fire alarm station that once actuated, will transmit not less than 4 rounds of coded alarm signals without interference from any other station on the circuit. One or more of these stations, if subsequently operated, will transmit not less than 4 rounds of their coded signals without interference with each other or with the first station actuated.

Presignal System. One in which the operation of an automatic detector or the first operation of a manual fire alarm station actuates only a selected group of alarm indicating devices for the purpose of notifying key personnel. A general alarm may be sounded on these same indicating devices and on an additional group of devices from any manual station, to warn all occupants.

Proprietary Fire Alarm System. A proprietary fire alarm system is a system with supervision by competent and experienced observers and operators in a central supervising station at the property to be protected.

Recorder (Punched Tape). A device for tape recording the actuation of a fire alarm initiating device. The station or zone code is punched on the tape. May be used with a TIME STAMP.

Remote-Station Fire Alarm System. A system of electrically supervised devices employing a direct-circuit connection between alarm initiating devices or a control unit in protected premises and signal-indicating equipment in a remote station, such as a fire or police headquarters.

Selective Coded System. A system in which each manual fire alarm station and each *group* of automatic detectors has its own individual code, which sounds on all alarm indicating devices in the system when the manual station or automatic detector is actuated.

Station, Fire Alarm. A manually operated alarm initiating device; may be equipped to generate a continuous signal (noncoded station) or a series of coded pulses (coded station).

Supervised System. A system in which a break or ground in the wiring which prevents the transmission of an alarm signal will actuate a trouble signal.

Trouble Signal. A signal indicating trouble of any nature, such as a circuit break or ground, occurring in the device or wiring associated with a fire alarm system.

Zone Coded System. A system in which the building has been divided into zones. Alarm initiating devices in each zone activate a zone code that indicates only the location of the affected zone.

22.10 Noncoded Manual Stations

In contrast to the automatic detectors that constitute the most common alarm initiation devices in residential systems, the manual station is operated by hand. The detector monitors automatically and is therefore best applied in buildings with sleeping residents and in unoccupied spaces in other buildings. Manual stations serve to spread the alarm, which has already been detected by other means, either human or automatic (see Fig. 22.9). If it is desired to know which manual noncoded station has been operated, an annunciation panel can be added to the system. This is equivalent to using each station as a noncoded indicating zone (see Section 22.16 and Fig. 22.10). Because of wiring costs, such an annunciated system becomes expensive. Beyond 10 stations, coding should be considered.

22.11 Coded Manual Stations

When the system design is such that immediate aural identification of the operated sta-

Fig. 22.9a A small manual fire alarm station with break glass rod and single set of contacts. Applicable to noncoded evacuation-type systems.

Fig. 22.9b If desired, the station can be enclosed in a cabinet that is readily opened and well marked.

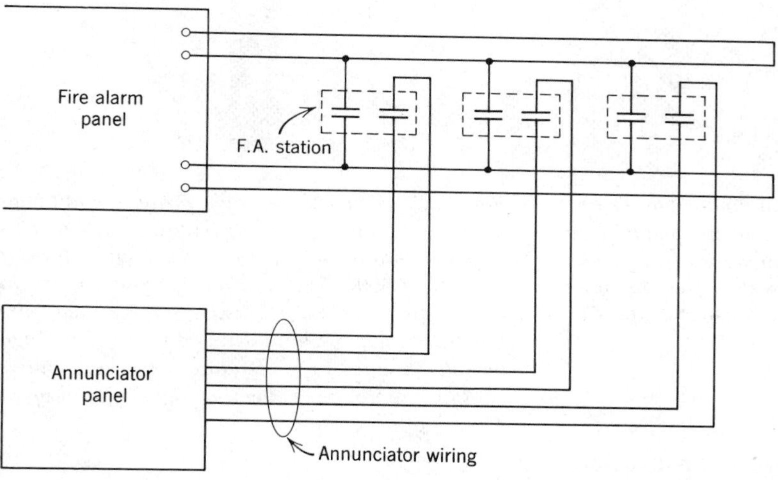

Fig. 22.10 Wiring of noncoded fire alarm manual stations. An additional set of contacts in each station provides annunciation for that station. Note that the pair of wires required for each station rapidly increases the cost of such a system.

tion is necessary, a coded station is used. The code sent out by the station is received at the control panel, processed, and then transmitted audibly on the system gongs. Not less than 3 rounds of code, and normally 4 rounds, are transmitted.

The code usually comprises 3 or 4 digits, for example, 2-3-2 with a pause between the ringing groups and a longer pause between the rounds. The first number may identify the building floor, the second digit the wing, and the third digit the individual station. Establishment of codes is left to the user. Wiring of coded stations is similar to that shown in Figure 22.10. Stations are places in the normal path of egress from a building so that an alarm may be turned in by a person as he or she exits. It is *imperative* therefore that stations be well marked and easily found. Architects who place fire alarm stations in nooks and corners and in camouflaged cabinets because they spoil the decor of the lobby are defeating the purpose of the system. Similarly, placement of bells *inside* hung ceilings because they are unattractive is not only foolish but dangerous, and should *never* be done, regardless of the circumstances. Loss of property and even of life may result from such ill-conceived aesthetic considerations.

22.12 Sprinkler Alarms

Various circuit arrangements are possible by use of water flow switches that are placed in sprinkler pipelines and operate when a sprinkler head goes off (see Fig. 22.11). In electrical terms, a water flow switch is a set of

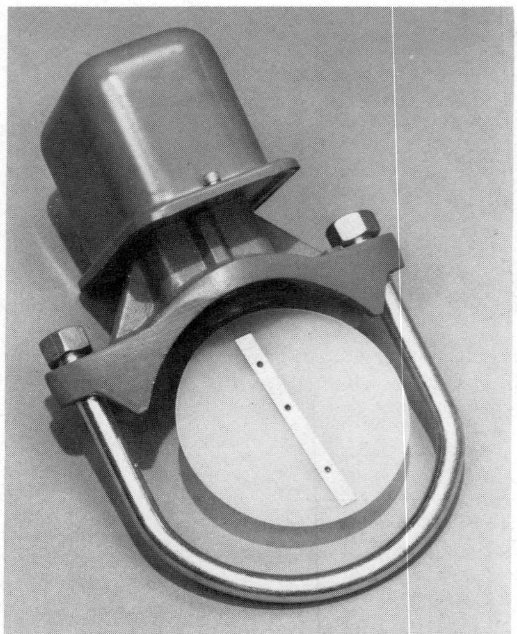

Fig. 22.11 *Typical waterflow indicator. The unit bolts onto a sprinkler pipe with the paddle inside the pipe. Any water motion deflects the paddle, causing a signal to be transmitted from the microswitch mounted in the box on top of the pipe. Courtesy of Notifier Company.*

Fig. 22.12 *Commercial grade surface-mount temperature detector of the self-resetting variety. This unit will alarm at a preset fixed temperature of 135° F or 200° F, and when a rapid change in temperature occurs, as when an incipient fire bursts into flame. The illustrated unit is 6 in. high overall, mounts on a standard 4-in. box, and is equipped with an indicating light. Courtesy of Pyrotronics.*

contacts, similar to a temperature detector. It can be used to trip a coded transmitter, setting off a sprinkler code, to show up on sprinkler annunciator board called a sprinkler alarm panel, or to act as a zone in a noncoded system. Wiring of water flow switches is the same as for stations in Figure 22.10.

22.13 Fire Detection

There are four basic types of fire detectors—temperature, photoelectric, ionization particulate, and flame type. A brief description of the characteristics and application of each follows.

(*a*) ***Temperature Detectors.*** These devices, as their name states, respond to elevated temperatures that accompany *actively* burning fires [see Section 22.3(b)]. As such, they act much like the fusible link in a sprinkler head. Effective applicability therefore is restricted to locations where the subsequent alarm will result in adequate countermeasures being taken in time to prevent injury or loss (see Fig. 22.12).

(*b*) ***Photoelectric Detectors.*** These detectors react to obscuration of a light beam by smoke. There are several designs available that vary in sensitivity and applicability (see Fig. 22.13). These detectors are best applied where fires, either active or incipient, produce considerable quantities of smoke and gas. This is the case with many types of smoldering fires, such as those caused by the burning of plastics and fabrics. These same materials can also burn in an open free mode, in which

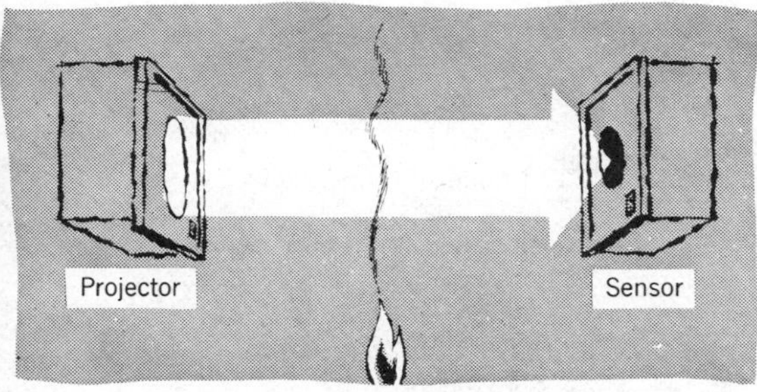

Fig. 22.13 Principle of smoke detection. Smoke detection is useful in spaces where occupants may be asleep, since smoke poisoning (asphyxiation) may occur before temperature-sensitive detectors operate.

instance these detectors are less effective because of the dispersal of smoke by heat. Because the smoldering stage often precedes the active burning stage, photoelectric units can be classified as early warning devices (in contradistinction to temperature detectors).

(c) ***Ionization Detectors.*** These devices, of which many designs are now commercially available, all purport to react to "products of combustion." In reality, they operate by detecting ionized *particles* in the air, hence the generic name—ionization particulate detectors. They do *not* detect smoke. Fires that produce heavy smoke but few particles—such as many types of smoldering fires, alcohol fires, and plastic fires—are not detected by these devices until they are about to enter the active flaming stage. On the other hand, even very small fires that do produce particle matter are rapidly detected. For this reason, many knowledgeable fire experts use a combination of obscuration-type and ionization-type detectors in applications that may produce either high-smoke, low-particle or high-particle, low-smoke conditions, depending on ignition circumstances. Ionization detectors are generally classed as early-warning, smoke-detection devices (see Fig. 22.14).

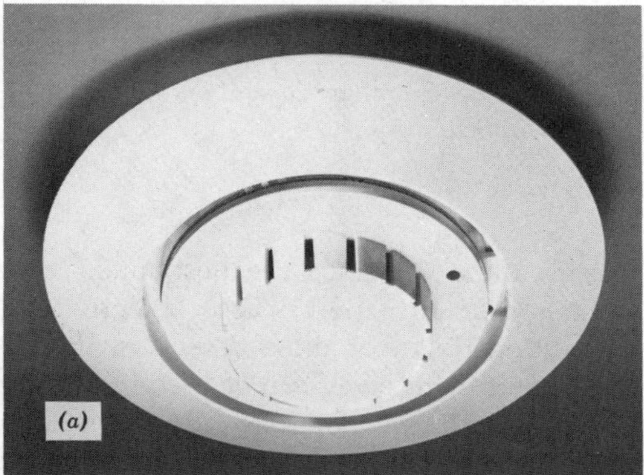

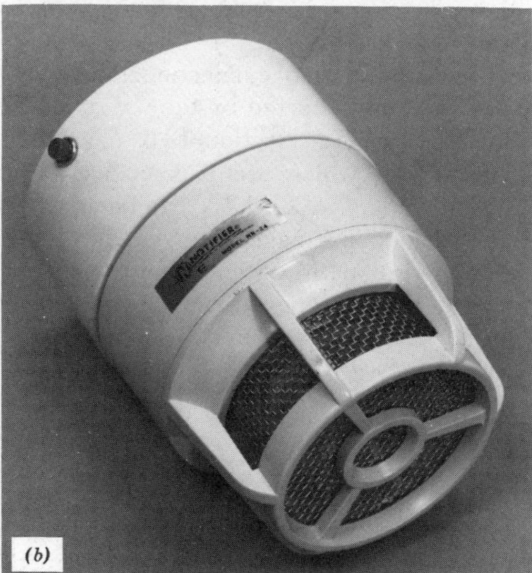

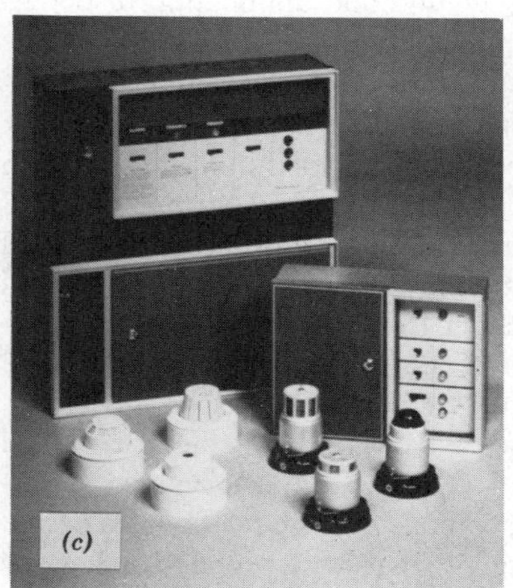

Fig. 22.14 Ionization-type fire detectors are available in many shapes and sizes, but basically operate on the same principle, that is, detection of the ionized particles usually generated by an incipient fire. Unit (a) *is approximately 8 in. in diameter;* (b) *approximately 5 in.* (c) *Illustrates a variety of ionization-detector designs, and some of the related control equipment. Coverage varies with unit* and *conditions of application. Photos courtesy of Honeywell* (a), *Notifier* (b), *and Pyrotronics* (c).

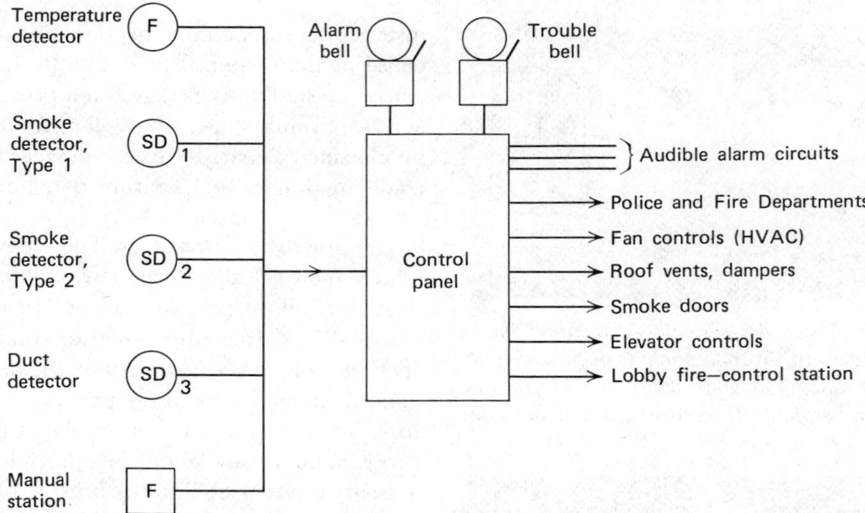

Fig. 22.15 Typical control functions for a fire alarm panel. The exact functions depend on the nature of the building, local fire codes, type of HVAC system, and the budget.

(*d*) ***Flame Detectors.*** These are of two types—infrared (IR) and ultraviolet (UV). The former reacts to IR radiations and the latter to ultraviolet radiated from active flames. Both are extremely sensitive, but can hardly be termed "early-warning." Applications are generally industrial, normally arranged to immediately activate protective measures.

Actual selection and application of detection units require intimate knowledge of the characteristics of the detectors and the environment into which each will be placed. For instance, because the operation of ionization detectors depends on "capturing" ionized particles, anything that will disturb expected air convection patterns or block the units will seriously affect their operation. Thus, placement of a detector in the path of air discharge from a diffuser, or on an air discharge ceiling, or adjacent to baffle in a beamed or baffled ceiling will depreciate performance. Similarly, a hot ceiling caused by deliberate design or trapping of heat will produce a hot air blanket adjacent to the ceiling, preventing the ionized particles from reaching the detector. Also, excessive or insufficient air motion will have a negative effect on performance unless anticipated and compensated. For the above reasons these decisions are best made with assistance of an expert whose objectivity is unquestioned. All of these detectors, though complex of themselves, can be used to sound a general alarm, a coded alarm (by tripping a code transmitter in the panel), and perform other functions such as shut down the building fan system (see Fig. 22.15).

In addition to smoke detection, smoke control is partially effected by the use of smoke control doors. (See Section 13.6, page 498). These doors, which are to corridors what smoke dampers are to ducts, can be arranged to be held open magnetically and released when the fire alarm system is actuated. Door releases can be arranged in zones to repond only to smoke or fire detection in a specified area.

22.14 Noncoded Systems

Noncoded systems are *continuous ringing* evacuation types using manual and automatic alarm initiation. If desired, the devices can be zoned and, if sufficiently large, annunciation can be provided. Audible devices are continuous ringing vibrating bells and horns (see Figs. 22.16*a* and 22.17).

22.15 Master Coded Systems

This system, also called Common Coded and Fixed Coded, generates 4 rounds of code when any signal device operates. It utilizes a single code transmitter at the panel. Normally, the system stops after 4 rounds of code, although it can readily be arranged to sound continuously thereafter. When the code is set to ring the bells at an even 108 strokes per minute it is known as "march time" because of the rhythmic cadence. This beat aids in the rapid panic-free evacuation of a building and therefore is frequently used in schools (see Fig. 22.16*b*).

22.16 Zone Coded Systems

In a system where it is desired to identify a zone, one can do so by utilizing a noncoded system with zone lights, an annunciator, or a coded system. In the first two cases, it is necessary to *go to the panel or annunciator* to determine the location of the operated device entailing possible critical delay. All coded systems obviate this necessity by sounding the code on all the gongs in the building, thus immediately identifying the station and permitting the building staff to quickly investigate the cause of the alarm and take appropriate measures.

Therefore, if a *coded* system is desired but by zone rather than device (and this is less expensive by far), *noncoded* manual stations are used, along with automatic detectors, grouped by circuit into zones. These trip zone transmitters *in the panel*, which in turn ring the zone's code on the single stroke gongs or chimes (see Fig. 22.16*c*). As with all coded systems, 4 rounds of coded signal are sounded after which the system is silenced. In all coded systems it is advisable to include a device that records in plain English all alarms, including time of receipt and the code sounded.

22.17 Dual Coded Systems

This arrangement is a combination of noncoded and zone coded systems. When an alarm device operates it initiates two separate functions—an identifying coded alarm and a continuous ringing evacuation alarm. The alarms are sounded simultaneously; the coded alarm in the building's maintenance office and the evacuation alarm on separate audible devices throughout the building. A requisite to the application of this system is a continuously manned office in which the coded identifying signal can be received and acted upon (see Fig. 22.16*d*).

22.18 Selective Coded Systems

This is a fully coded system in which all manual devices are coded and all automatic devices are arranged to trip code transmitters at the panel. Each manual station can be immediately identified by its distinctive code. Automatic devices may be grouped in any fashion desired, and annunciated if desired. The combinations and circuitry are entirely in the hands of the designer. In large systems, which fully selective coded systems usually are, sprinkler transmitters and smoke detectors operate as integral subsystems of the main fire alarm panel (see Fig. 22.16*e* and *f*).

22.19 Presignaling

When it is desired to alert only key personnel a system called presignaling is used. Small bells or chimes are activated only in their vicinity. Since these systems are always selectively coded, the personnel alerted can immediately investigate and, if necessary, manually turn in a general alarm by key operation of a station (see Fig. 22.16*g*). Because of the delay involved, this type of system is used only in buildings where evacuation is difficult and sufficient staff is available to immediately investigate the cause of an alarm.

MULTIPLE-DWELLING SYSTEMS

22.20 General

The types and design of the various sound, signal, and alarm systems employed in multi-

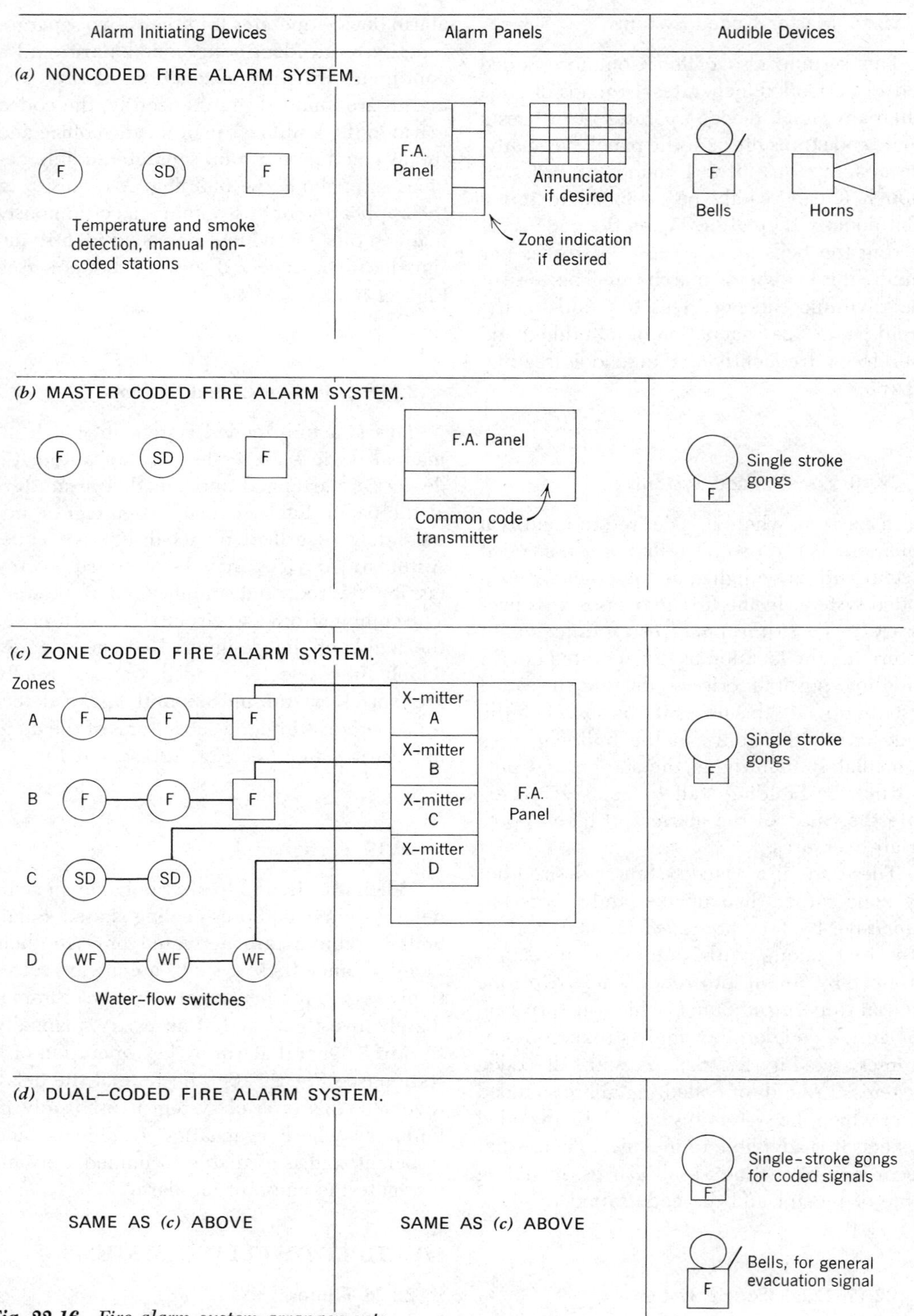

Fig. 22.16 *Fire alarm system arrangements.*

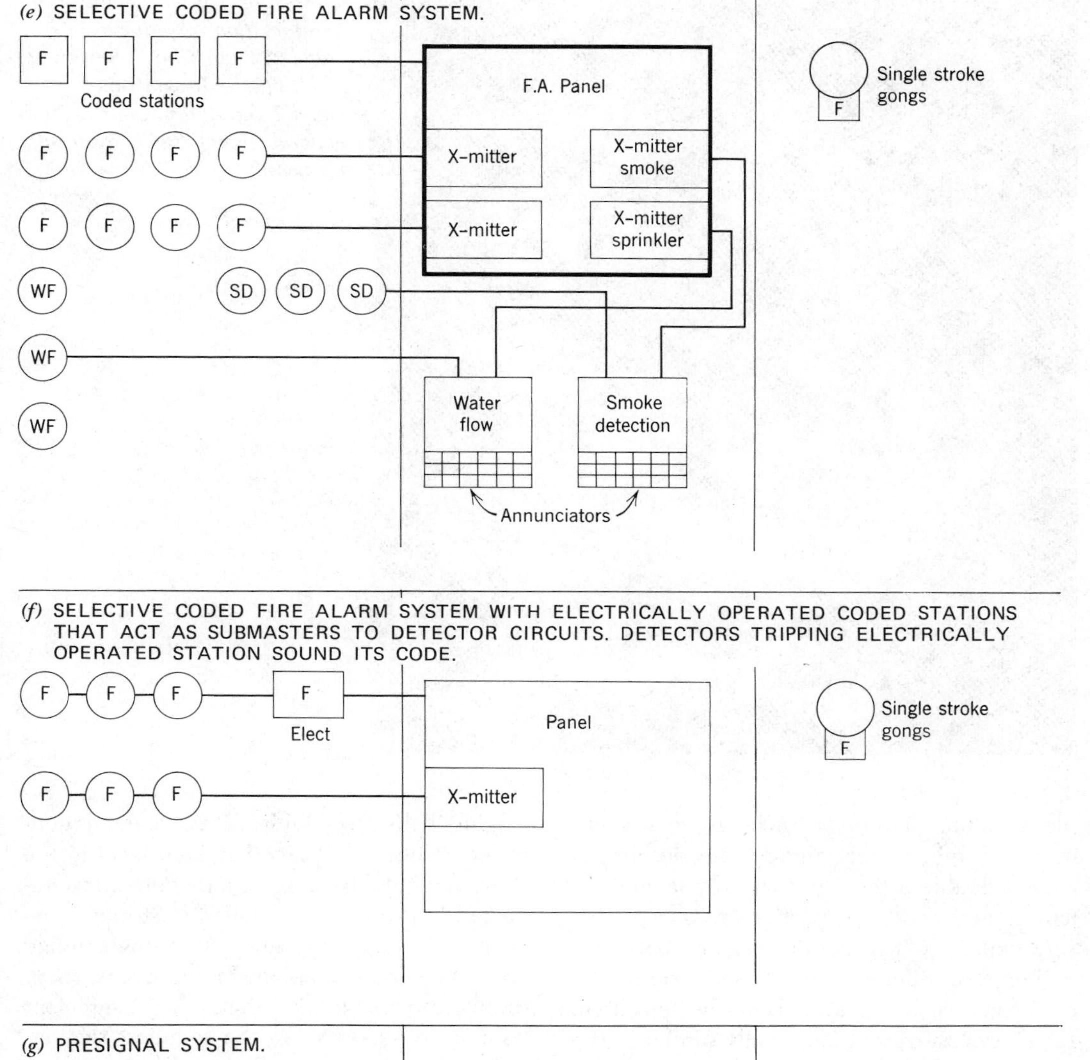

(g) PRESIGNAL SYSTEM.

SAME AS (e) ABOVE

SAME AS (e) ABOVE

F

Single-stroke gongs or chimes for coded presignal

CH

Bells for evacuation alarm

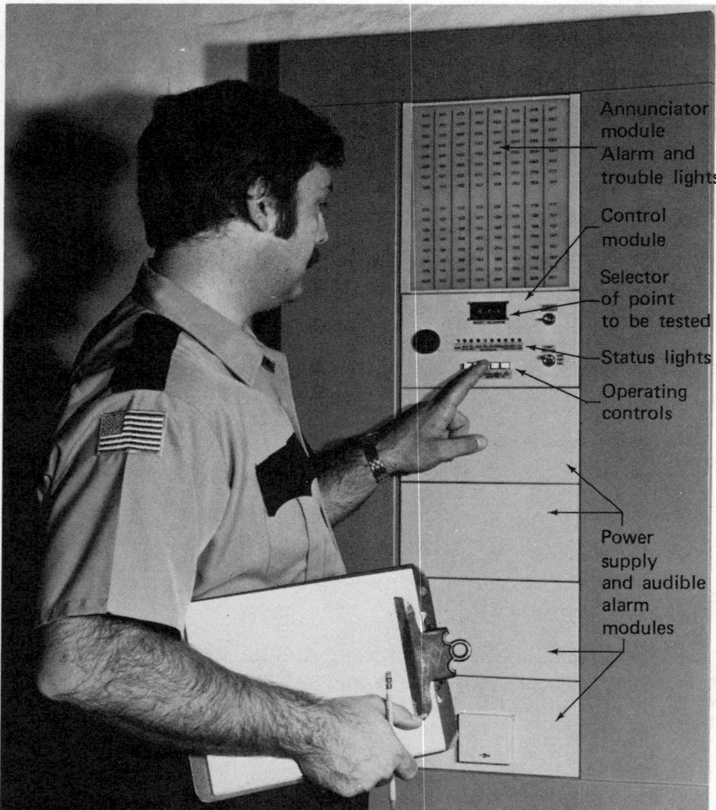

Fig. 22.17 Control panel for a noncoded general evacuation alarm-type system with full annunciation. Applicable (see Table 22.2) to schools, multiple dwellings, industrial and commercial buildings. Courtesy of Federal Signal Corp., Autocall Division.

ple-dwelling buildings such as apartment houses, hotels, motels, and dormitories are, to a considerable extent, governed by local ordinances and codes. In particular, NFPA 101, *Life Safety Code,* details requirements for buildings of different sizes, types, and number of occupants, and should always be consulted. In general, they are functionally similar to the systems discussed for individual residences.

Since multiple dwellings range in size from small 4-family homes and 10-student dormitories to huge hotel and apartment complexes, obviously the fire alarm system applicable to the former is not suited to the latter and vice versa.

22.21 Multiple-Dwelling Fire Alarm Systems

Dormitories normally utilize supervised, zoned, noncoded systems with continuously ringing bells (see Table 22.2). Manual noncoded stations are placed at each level in fire stairs and in lobbies. Automatic detectors connected in groups are installed in storage, mechanical equipment, and other unoccupied areas. Bells must be placed in corridors, study rooms, and alcoves and should be of sufficient volume to waken even the soundest sleeper. Since dormitories are designed with soundproofing in mind, to allow for ideal study and sleeping conditions, bells and horns of high sound intensity must be selected. The panel is generally placed in the mechanical equipment room. If annunciation is used to assist fire fighters in locating the cause of trouble, the annunciator should be placed in the *entrance lobby.* Since most dormitories are built under the aegis of some public agency, the applicable fire codes should be consulted in all cases. If none is extant, the fire-fighting arrangements should be investigated and the person in charge consulted.

Table 22.2 *Fire Alarm System Recommendations*

	Schools[b]	*Hospitals & Nursing Homes*[b]	*Colleges*[b] *(For Dorms, see Multiple Dwellings)*	*Industrial*[d]	*Commercial*	*Libraries & Record Storage*[b]	*Multiple Dwellings*[b] *(Dorms, Hotels, Motels, Apts.)*
Single story building—Under 20,000 sq ft	Noncoded, annunciated general alarm system[a]	Coded, annunciated general or presignal alarm system[c]	Noncoded, annunciated general alarm system	Noncoded, annunciated general alarm system	Noncoded, annunciated general alarm system[a]	Noncoded, annunciated general alarm system	Noncoded, annunciated general alarm system
Single story building—Over 20,000 sq ft	Dual coded, annunciated general alarm system[a]	Dual coded, annunciated general or presignal alarm system[c]	Coded, annunciated general alarm system	Coded, annunciated general alarm system	Coded, annunciated presignal alarm system[c]	Coded, annunciated general alarm system	Coded, annunciated presignal alarm system[c]
Multiple story building	Dual coded, annunciated general alarm system[a]	Dual coded, annunciated general or presignal alarm system[c]	Coded, annunciated general alarm system	Coded, annunciated general alarm system	Coded, annunciated presignal alarm system[c]	Coded, annunciated general alarm system	Coded, annunciated presignal alarm system[c]

[a] A common coded, annunciated general alarm system may be desirable in these occupancies.

[b] Complete smoke detection of area or entire building is recommended for these applications.

[c] The use of presignal system presupposes that trained personnel are awake and on duty at one location 24 hr/day. If this is not the case, use noncoded, annunciated general alarm in multiple dwellings and coded, annunciated general alarms in nursing homes and hospitals.

[d] Special attention should be given to the selection of signals due to the high ambient noise usually present in these areas.

NOTE. Based on recommendations of Autocall, Federal Signal, with emendations.

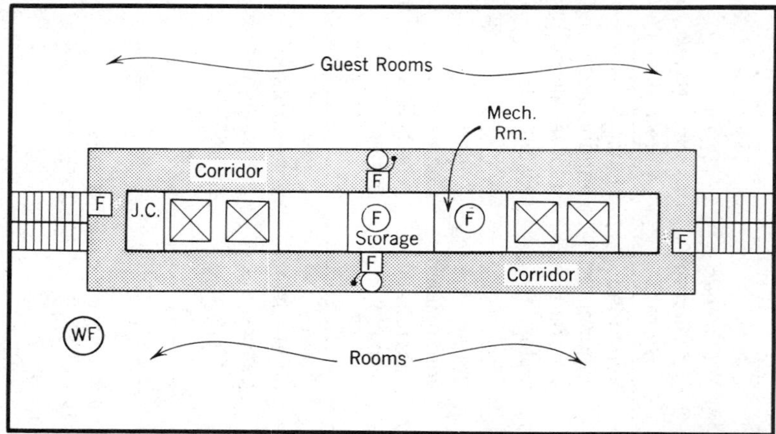

Fig. 22.18 Typical hotel floor plan. Bells are placed so that an alarm will be audible in all parts of the building. Stations are normally installed at all points of egress, such as stairwells and main floor exits. A typical location for sprinkler alarm transmitter would be in a janitor's closet (J.C.) or electric closet. Automatic temperature detectors are properly located in storage and mechanical spaces.

Apartment houses should have a line-voltage-powered smoke detector in each apartment. Medium-size buildings require a manual evacuation-type system. High-rise units require automatic detection, annunciation, central panel with voice communication, and—depending on local codes—elevator control.

A hotel of any magnitude would utilize a supervised, presignal or dual coded system with automatic stations in storerooms, kitchen, boiler room, and other unsupervised areas. Sprinkler transmitters and annunciation are common, as are smoke detectors in the ducts, activating a fan shutdown system. This latter device has as its obvious purpose the prevention of smoke spreading through the building. Presignal bells would be installed in the office of the building engineer as well as that of the hotel managers. The fire alarm panel is most often placed in the mechanical equipment area. An auxiliary circuit can be added to the system that trips a city fire alarm box or, in large cities, connects to a fire supervision company, which in turn notifies the city fire department. All such connections are rigidly controlled by city ordinances. Figure 22.18 shows a typical hotel floor plan.

22.22 Multiple-Dwelling Lobby Intercom and Security Systems

Apartment houses, residences, and hotels combine the functions of the intrusion alarm and doorbell systems in the familiar lobby-to-apartment communication system. The most basic system is a series of push buttons in the lobby and an intercom speaker or telephone with which to communicate with residents. At the other end, the tenant has a speaker microphone plus a lobby door-opener button (see Fig. 22.19*a*). This system can also be arranged to utilize the tenants' regular telephones, as in Figure 22.19*b*. When the number of tenants is large, an alphabetical roster is added to the apartment-button panel to avoid the nuisance of scanning all the apartment names when the sought party's apartment number is not known. When the number is larger yet, a simple push button per apartment arrangement becomes cumbersome, and is usually replaced by an alphabetical tenant register plus a dial or button phone. Closed-circuit television can be added to the lobby-tenant system, enabling the occupant not only to converse with, but actually to see, the caller. Such a system will increase the electrical contract cost for an average apartment house from 7 to 10%.

In addition to the security provisions provided by the apartment-to-lobby audio and video connections, additional security and alarm devices have been used such as emergency call buttons within the apartment at desired locations. These will light alarm lights, ring bells, and perform any other alarm functions required, to cover the situation of an intruder who manages to bypass the lobby security check. In geriatric designs, these buttons serve to *unlock* the apartment door to allow aid to enter, being summoned by lights and alarms. In luxury apartments, apartment doors can be monitored from a central security desk and any unscheduled door movement subjected to immediate investigation. These systems are custom designed to the needs and requirements of the owner.

22.23 Multiple-Dwelling Television Antenna Systems

All modern multiple residences supply each room with one or more TV/FM jack outlets. The master antenna systems feeding these are similar to the residential type discussed except for size and electronic design of the components. The antenna should be placed only after a survey has been made to determine signal strength patterns on the roof. Space below the roof and a single 15-amp circuit should be allowed for the amplifier equipment. Coaxial cables are run through floor sleeves and tapped at various points to provide good signals in each apartment.

22.24 Multiple-Dwelling Telephone Systems

As for the small residence, the telephone service normally follows the same entrance

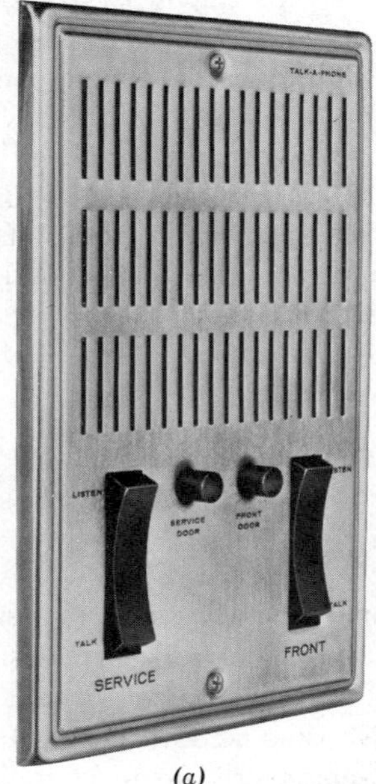

(a)

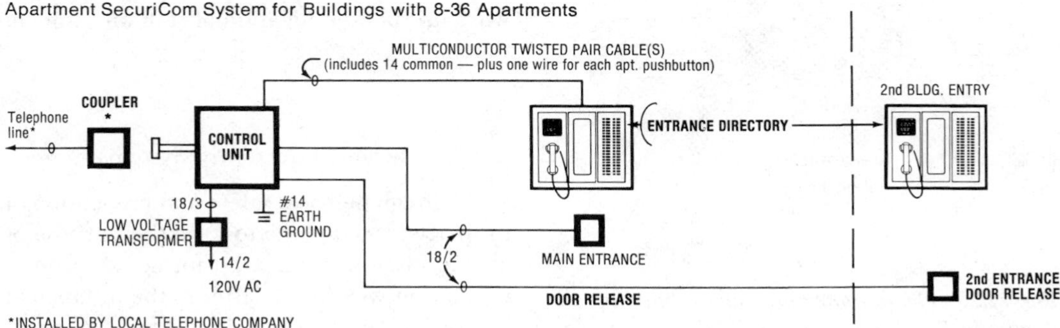

(b)

Fig. 22.19 *Apartment house communication-security equipment.* (a) *Typical apartment unit contains intercom controls and door-opening buttons for main and auxiliary entrances. Courtesy of Talk-a-Phone Co. The lobby and service entrance units contain a tenant registry and intercom instrument as in* (b), *which is a block diagram of a system utilizing the normal telephone for intercom. Courtesy of NUTONE / SCOVILL.*

path and method of entrance as the electric power service. For the sake of economy in underground construction, the two services often share the same trench albeit in different raceways, and utilize twin manholes where such are required. Typical entrance arrangements for any large building, residential or other, are shown in Figure 22.20.

The service entrance space requirements vary with the size of the building and telephone capacity. For a small apartment house of the garden or three-storey type, a clear wall space of 2 to 4 ft is sufficient. A terminal room is only required in very large buildings.

Apartment buildings and dormitories differ from commercial structures in that the floor plans of all floors are similar, so that the arrangement of risers is relatively simple. It is common practice to utilize cable only, in risers that extend through vertically aligned closets in apartments. To accommodate these cables, a sleeve through the floor between closets is necessary. If a riser is located in a shaft other than a closet, conduit is normally utilized to allow for easy installation, protection, and repair. If the location is accessible, as in an alcove, only a sleeve is provided. When the riser is located outside the apartment, each dwelling unit is connected to the riser by a conduit with a junction box at either end. These conditions are illustrated in Figure 22.21.

Beyond the apartment service point, the individual rooms can normally be prewired entirely without conduit, or with only a few short sleeves.

Prewiring has demonstrated itself to be the most economical arrangement for the owner and the phone company, in addition to eliminating unsightly wires, and should be employed wherever possible.

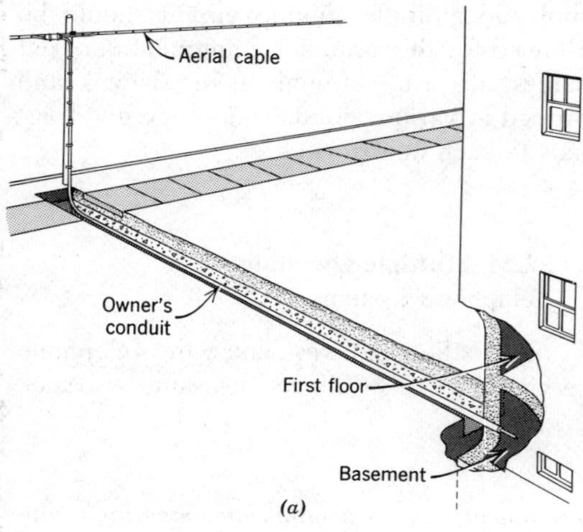

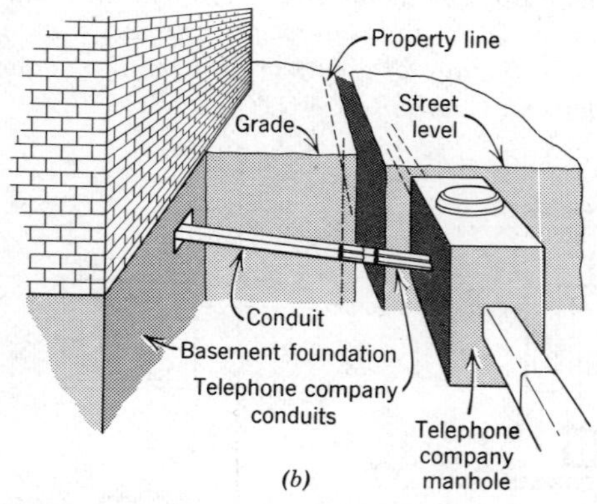

Fig. 22.20 *Phone cable may enter a building underground, originating on overhead lines* (a) *or in manholes* (b).

SCHOOL SYSTEMS

22.25 General

The proper operation of a modern school requires that flexible and efficient signal and communications equipment be available to the administrative and teaching staff. Such equipment, engineered to meet the needs of the individual institutions, will do much toward optimum utilization of staff and student time.

22.26 School Fire Alarm Systems

Although personal safety and prevention of property loss combine to form the purpose of all fire alarm systems, the former consideration far outweighs the latter in the instance of school buildings, particularly of the elementary grades. For this reason a general school-

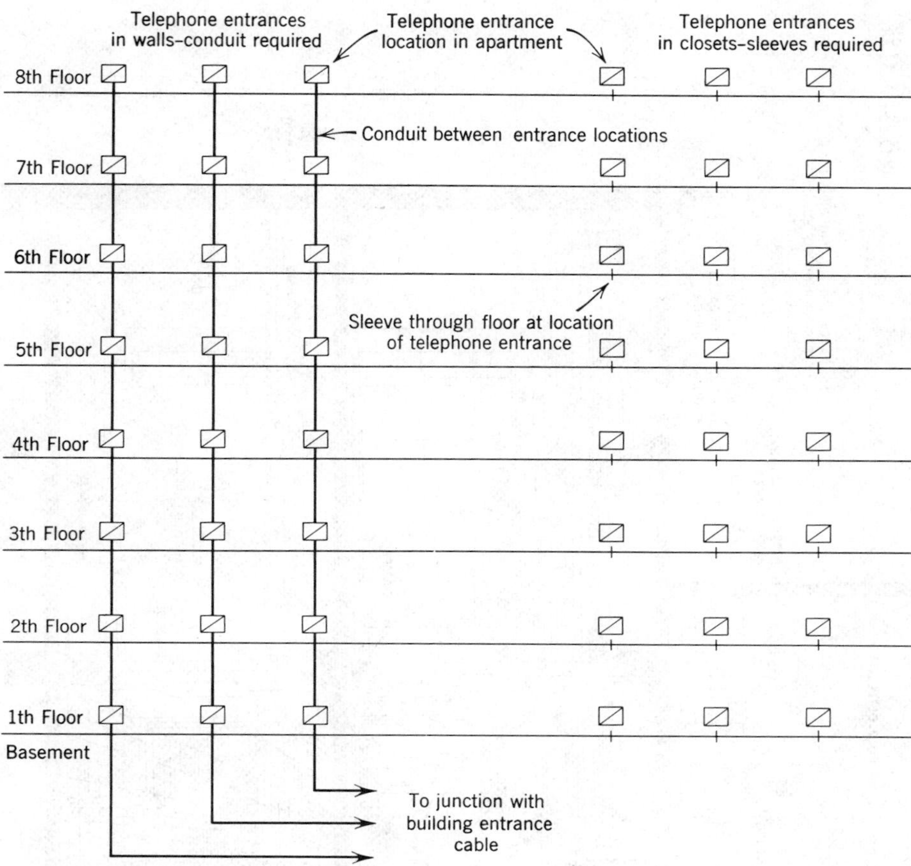

Fig. 22.21 *Typical telephone riser diagram. Note the need for conduit between apartments when installation is made inaccessible, as in a wall.*

wide alarm causing an immediate evacuation of the premises is the primary requirement of the system (see Table 22.2). Consideration also must be given to maintaining the uniqueness of the sound of the fire alarm gongs to allow no possibility of confusion with the program gongs, where the latter are used.

The system employed almost universally calls for a closed-circuit, supervised arrangement, noncoded or master coded in the case of smaller schools, and dual coded in the instance of large or multibuilding institutions. For a multiple-building school, the circuitry is generally arranged to sound an evacuation alarm in the affected building only. The signal also is transmitted to administrative areas in other buildings.

Since regular fire alarm drills are mandatory in all schools, the system circuitry must be arranged to allow for this type of testing. As with other systems, manual stations are placed on each floor at exit points, such as stairways, with automatic stations in the boiler room, kitchen, some laboratories, shop classrooms, and selected storage areas (see Fig. 22.22). It is also advisable to connect any sprinkler flow switches to the alarm system to effect building evacuation while utilizing, in larger schools, a central sprinkler annunciator panel, which will indicate the particular water flow switch involved. This accomplishes the same purpose as coding of fire alarm stations.

Most states have adopted statutes requiring the application of automatic detection in ad-

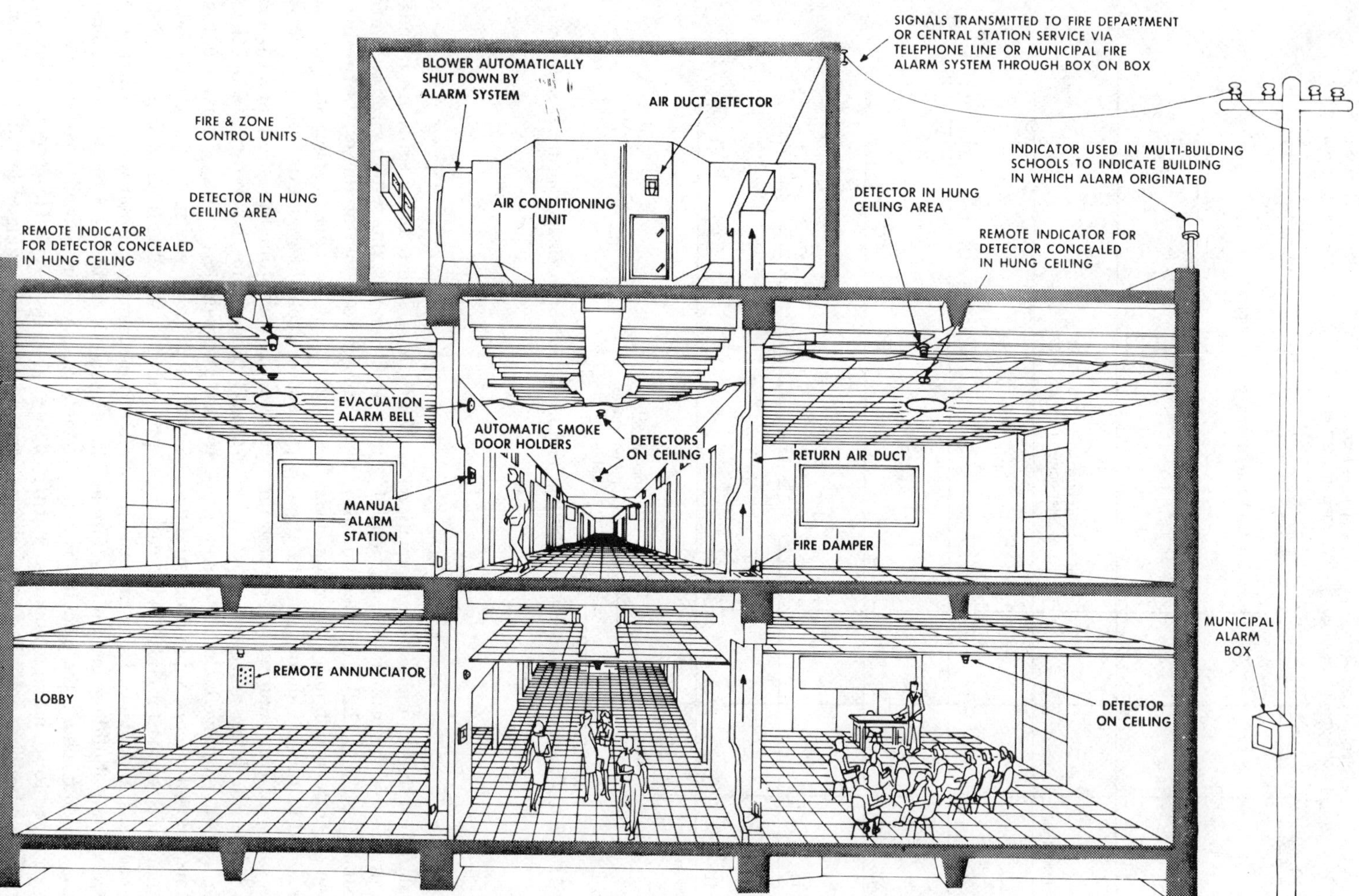

Typical system affording good protection of students in a school building.

Fig. **22.22** *Courtesy of Pyrotronics.*

dition to the previously universal requirement for manual systems. In an unoccupied building this will reduce property loss, and in an occupied building it may give the short alarm margin time required to avoid serious injury.

22.27 School Security Systems

Although intrusion alarms and security systems have not been a normal school requirement, this situation has unfortunately changed. Sensing devices on doors and windows can be arranged to trip both local alarm devices and, via auxiliary circuits, to notify police headquarters. Often, vandals can be frightened off by having the alarm system actuate a protective lighting system that will illuminate the building exterior and any interior areas desired, such as record rooms. Malicious vandalism and arson are other compelling reasons for the installation of sensitive automatic-device coverage in the fire alarm system.

22.28 School Clock and Program Systems

The clock and program system commonly used in educational institutions is in reality a combination of two separate and distinct systems, that is, a clock system and a program system. They are normally combined because a program system requires timing facilities that the clock system offers. However, for the sake of clarity in our discussion, we will consider the functions separately.

The function of the clock system is simply to indicate the correct time on the various room and corridor clocks in the system. This can be accomplished in four ways—three employ a master clock and one does not.

The synchronous, wired, master clock system utilizes a master clock that transmits over separate wiring (3 wires) power and correcting signals to the secondary clocks that are driven by synchronous motors. The correcting signal is generally sent hourly and every twelfth hour. The entire system, as with all systems employing masters, depends on the proper function of the master clock for accuracy.

The electronic system utilizes the power wiring for the transmission of correcting signals from the master clock to the secondary clocks. The individual clocks are powered locally and receive a high frequency correcting signal over the power wires. The outstanding advantage of this system is the elimination of a physical wiring connection between master and secondary clocks, thus making it particularly useful in renovation work or in multiple-building schools.

The minute impulse system also utilizes master and secondary units, the latter being driven by a d-c signal from the master, transmitted at minute intervals, rather than by individual synchronous motors. A correcting signal is also normally transmitted hourly. This system uses a 2-wire connection between units and master. A typical unit of this type is shown in Figure 22.23.

The *dual motor* system does not employ a master clock. Instead it utilizes clocks with dual motors; one synchronous for normal drive and one high-speed motor for correction. These clocks depend on their synchronous motors and the constancy of the 60 cycle power supply for accuracy. The high-speed correcting motors merely speed up the clock hands to correct for loss of time because of power interruption. This resetting is centrally controlled, either manually or automatically. The system is separately wired, thus allowing for ease of intentional corrections, such as is required for daylight-saving time.

Program systems in schools serve primarily to delineate audibly the various time periods into which the school day and week are divided. A single-circuit unit is utilized in the instance of an institution which operates entirely on one schedule, such as an elementary school on a morning period-lunch-afternoon period regimen. However, for schools employing different schedules for its various parts, program instruments are available offering up to twelve different program schedules.

The most basic type of program instrument is simply a timing device that drives one or

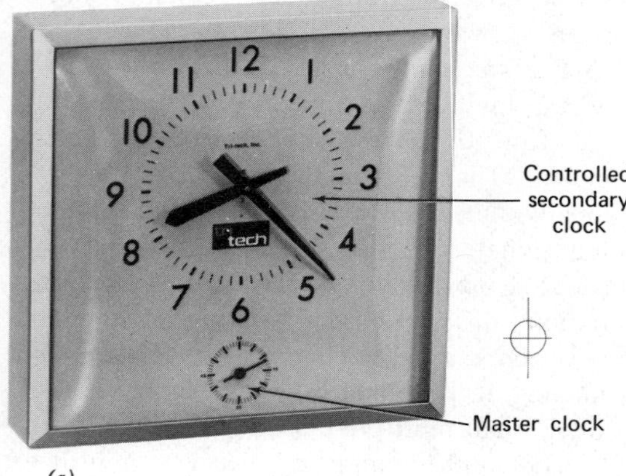

Fig. 22.23 Typical modern master clock and control for a minute impulse system (a). *The master clock receives second and minute impulses and transmits minute impulses to secondary clocks.* (b) *Shows the clock's depth dimensions and its connections and operating control. Courtesy of Tri-Tech, Inc.*

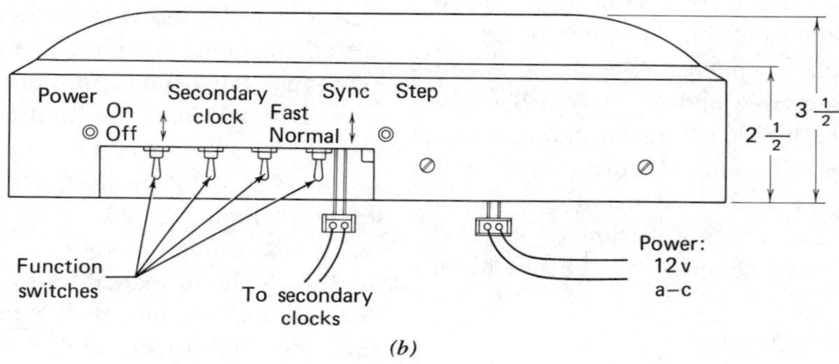

(b)

more drums, providing individual minute-interval, 24-hour, and weekly control. Plugs or pins inserted into the various drums activate contacts and relays, which in turn control the audible devices. Provision is usually made as well for manual control of individual circuits. This type of device, which is illustrated in Figure 14.11, page 525, is programmed by placing the pins into the appropriate holes in the drum, according to the desired schedule. A more sophisticated unit is shown in Figure 22.24. This device, which incorporates a master clock control as well, is programmed by punching paper memory tapes, which in turn control the 12 program circuits. Manual control of these circuits is provided by the 12 push buttons below the clock. A completely solid-state unit is shown in Figure 22.25. This 6-circuit program control is entirely electronic and does not use wheels, drums, pins, or escapements. Programming is accomplished by operating the front controls. This unit also can be equipped to serve as a master clock to control remote clock circuits. All modern units are equipped to operate for 12 to 24 hours without power supply, and without the necessity for reprogramming.

The audible devices in a program system may be bells, gongs, buzzers, horns, or a tone reproduced on a classroom loudspeaker. This latter system has the following advantages:

1. Clear audibility in each classroom, with adjustable volume to cover quiet and noisy areas.
2. No possibility of confusion between program tone and other signals such as fire alarm gongs.
3. Multiple use of the speaker unit, for classroom sound as well as program tone.
4. Complete flexibility of programming that is not possible with hall gongs. This is particularly desirable in schools with special programs for groups of students.

A logical extension of the clock and program

Fig. 22.24 Modern 12-circuit master time and program controller. Programming is accomplished with punched paper tape. Courtesy of Standard Electric Time Corp.

Fig. 22.26 A modern school intercom station that permits in-house and outside communications in addition to a number of other functions as described in the text. Switchboards are eliminated, and all control and switching equipment is solid state. Courtesy of Rauland-Borg.

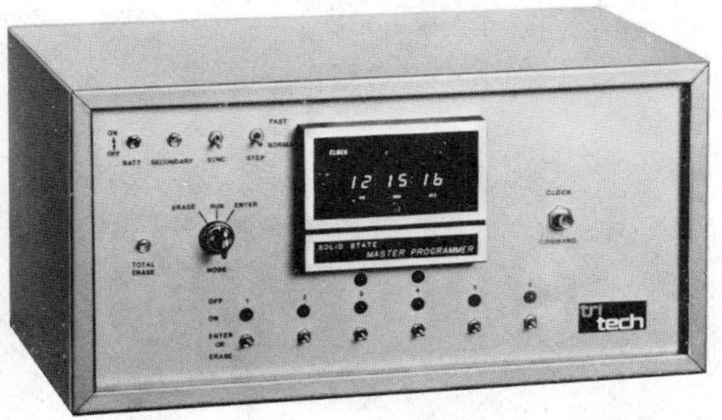

Fig. 22.25 Completely solid-state master clock/program controller. Programming of its six-circuit capacity is accomplished by operating the front controls. Courtesy of Tri-Tech, Inc.

equipment is the control of building utilities according to preset schedules. This is particularly effective and important in an energy-conservation and load-control program.

22.29 School Intercom Systems

Various types of systems are available depending on the needs of the building involved. In small schools, a simple wired intercom system connecting the various building offices is usually sufficient. This is supplemented by outside telephones in the administrative offices and a functional paging system that is normally part of the school's sound system arrangement. With larger buildings and correspondingly larger numbers of extensions and multiple-function demands, more sophisticated equipment is required. The unit illustrated in Figure 22.26 is typical of modern school intercom equipment, which is in actuality a private telephone system of considerable flexibility. Such a system is generally interfaced with the school sound system, and provides these functions:

1. Intercom between staff members and offices.
2. Direct communication with classrooms, including selective and all-call capability.
3. Zone, group call, and conference call functions.
4. Interconnection with the outside phone system.

These systems use direct push-button "dialing," eliminating the necessity for switchboards and operators. All stations are coded with three-digit alpha-numeric codes, and all switching is solid state, minimizing maintenance problems. Such systems are adequate for all but the large institutions. For these, the computer-logic circuitry of electronic PBX

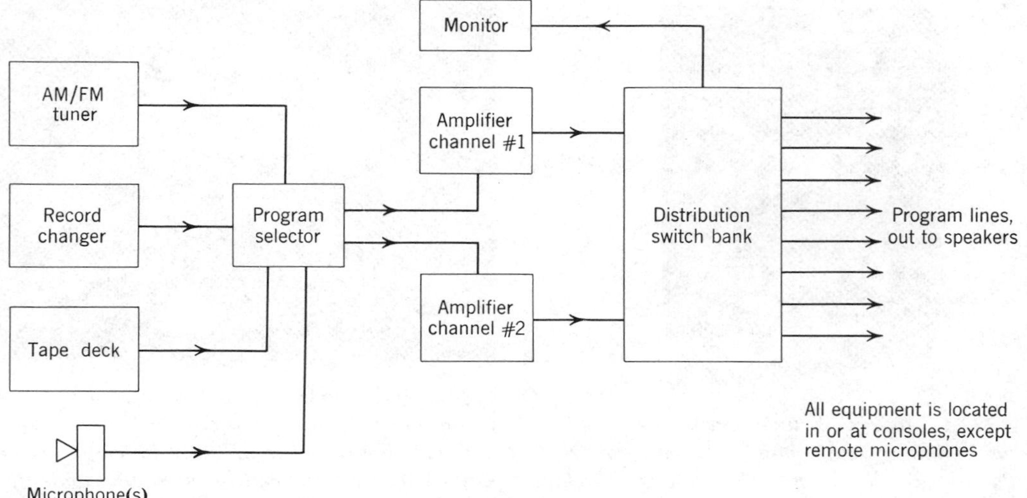

Fig. 22.27 (a) *Block diagram of a two-channel sound system.* (b) *Typical school-type sound control center. Optional facilities that can be added are tape deck, private telephone communications, and equipment for rebroadcast of signals between areas. Intercom and master clock and program can also be incorporated. Courtesy of Rauland-Borg.*

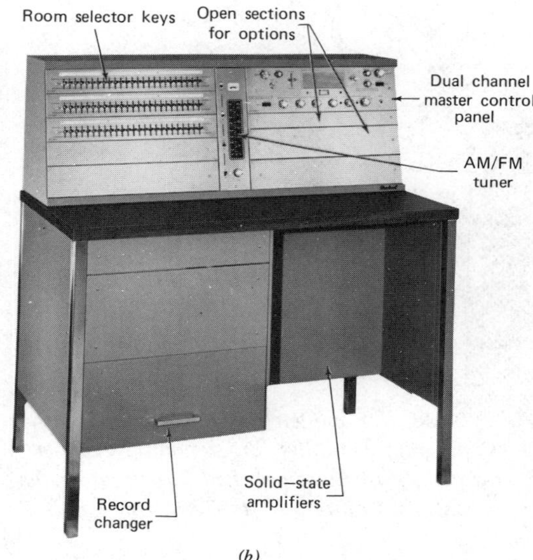

systems as described in Section 22.36 is necessary.

22.30 School Sound Systems

The integrated sound-paging-radio system designed for school use offers several modes of operation and considerable flexibility. Its function is to provide a means for distributing recorded (records, tapes), broadcast (AM/FM), or live sound to preselected areas of the school. Thus, a simple system might provide a record player and single microphone input, and a single channel to all of the speakers in the school, whereas a complex system can be arranged to operate with three simultaneous input signals distributed to six different areas of the school. As might be expected, increasing flexibility appreciably increases the cost of the system (see Fig. 22.27).

The system consists of a desk size control-console containing most of the input units, amplifiers, and switching devices, connecting to the remote loudspeakers, with related volume controls where required. The input units may comprise one or more AM/FM tuners, multispeed turntables or record changers, tape deck, and microphones. One microphone is normally located at the console with others in the principal's office, auditorium, school office, or other selected location. If desired, mike outlets can be spotted around the school and a spare mike and stand supplied to be plugged in at any of these points.

Loudspeakers, located in classrooms, gymnasium, auditorium, cafeteria, and outdoors receive the amplified signal through the switching mechanisms located in the console. It is the function of these switches to deliver the program material to the various loudspeaker circuits, which are also called program lines. Thus, using a system with multiple amplifiers, music can be piped to the cafeteria, an important radio address to senior classrooms and teachers' lounge, and instructions to an outdoor gym class or team practice. An all-call feature also allows announcements to reach all speakers in the system simultaneously. The intercom system discussed above can be incorporated into the sound system to allow conversation between classrooms and the console or other points, although it is often kept entirely separate.

The console is usually built in a desk arrangement, and it is advisable to provide sufficient space for it and for the person who operates it. Often an alcove of 30 to 50 sq ft is reserved for it and a library of recordings.

Loudspeakers may be in flush or surface baffles, at the discretion of the designer. Gymnasium, cafeteria, and auditorium units are normally flush-mounted in the ceiling. For large areas such as these it is well to provide a volume control, enclosed in a recessed wall box with a locking cover. A common variation of the above-described system uses separate subsystems for the cafeteria, auditorium, ball field, or other areas utilizing sound equipment frequently. These smaller systems have their own input, amplification, and control devices but utilize speakers in common with the central console. Normally the console has an override feature that allows it to override local systems.

For a discussion of high-quality sound systems required for recital halls in music schools and the like, refer to Sections 26.29 through 26.31, beginning on page 1195.

22.31 School Television Antenna Systems

Their principal function at present is to bring into the classroom or auditorium applicable portions of broadcast educational or cultural material. Classroom outlets obviate the necessity for congregating large groups of students in assembly halls, and increase the system cost only slightly in proportion to their usefulness. The system design is similar to that discussed above for dwellings, except that the equipment is designed for the UHF educational channels in addition to the usual VHF channels. These remarks, of course, apply only to antenna systems which receive broadcast material and not to internal, closed-circuit TV systems.

22.32 School Electronic Teaching Equipment

Asked to describe an ideal learning situation, one would probably include at least the following criteria:

Best possible teachers.

Each student learning at his or her optimum rate.

Comfortable learning environment.

Teacher accessible to students at all times, including retrieval of past classwork or access to classes held during enforced absences such as illness.

Elimination of limitations imposed by class and room sizes, such as poor audibility or visibility.

Elimination of equipment or other failures and consequent time losses.

Access to library information without time-consuming procedures.

The foregoing list generally reflects the student's viewpoint, although it is obviously concomitantly beneficial to the instructor. We might then add to this list, from the teacher's viewpoint, these criteria:

Facilities to dry-run procedures.

Access to equipment when it is required, without scheduling problems.

Improved ability to schedule specific material for a specific time.

Facilities for giving individual attention on a much increased scale.

Elimination of such time-consuming procedures as set-up, roll call and, even, blackboard erasure.

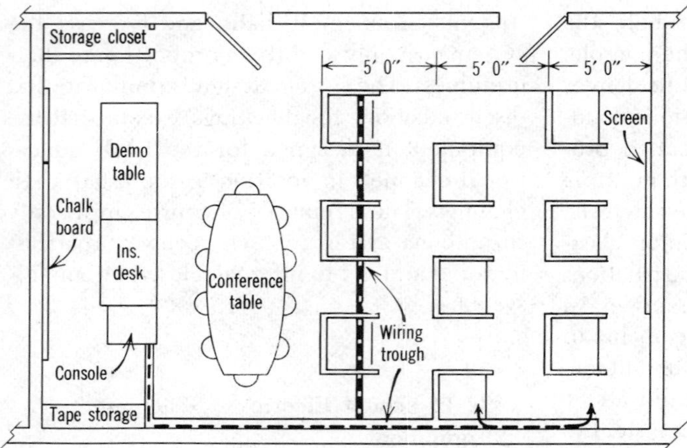

Typical classroom layout for utilizing computer terminals

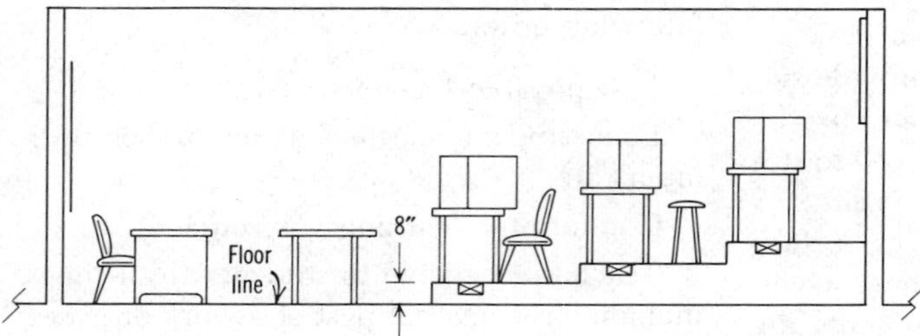

Fig. 22.28 Classrooms using computer terminals must be architecturally preplanned for this function. Note the wiring troughs, tape closet, and elevated rows which allow visibility.

Access to demonstration material and teaching aids that can be utilized without leaving the classroom.

Facilities to see oneself or other teachers actually teaching a class, in order to learn how to teach better.

Ability to conduct many simultaneous teaching procedures involving students singly and/or in groups, thus utilizing to a maximum the teacher's potential.

The above are practical arrangements; the basic hardware is available. The hardware consists of a library of recordings, a closed-circuit TV terminal, and an interactive computer terminal programmed with learning material. The first two devices are useful with a student in a passive mode. The computer terminal permits the student to actively participate in the lesson on a one-to-one basis, with the terminal acting as tutor (see Fig. 22.28). An auxiliary function of the computer terminal is that of information retrieval for more advanced students working on projects.

The entire subject of electronic teaching aids is complex and rapidly developing. The well-designed school will have adequate space and electrical and HVAC provisions for the extensive expansion into this field, which will undoubtedly occur.

OFFICE BUILDING SYSTEMS

22.33 General

Under this category we include systems found in all office, professional, and sales-type buildings. Such buildings house tenants with varying schedules and requirements and, unless large, do not have a full-time custodian. These factors must then be considered in the design of the signal systems for such buildings.

Although in many of the medium and large size buildings control, alarm, and security functions are combined in multiuse apparatus and consoles which we will discuss, the basic systems are essentially separate despite shared-use equipment. We will discuss them individually to demonstrate function and equipment, and combined to show economies and modern practice.

22.34 Office Building Fire Alarm Systems

(*a*) ***Buildings with Maximum Height of Seven Stories, Maximum Population of 500 Persons.*** Since these buildings are unoccupied for considerable periods of time, fire alarm systems in office buildings normally utilize automatic detection equipment connected to a double supervised selectively coded system. Temperature detectors of the combination thermal and rate-of-rise type, plus smoke detectors are located in critical areas unless sprinklers are required. Smoke detection in return-air ducts is often arranged to shut down all building supply and exhaust fans in order to avoid both feeding the fire and spreading smoke. A fan restart panel at an accessible location near the lobby should then be provided for selective restart of the *exhaust* fans only. Since this arrangement of fan shutdown and restart is not universal, it is well to consult local fire authorities and to follow their recommendations. Coded stations are required at egress points, plus a lobby station. Additional circuits can be arranged to release smoke control doors.

Waterflow switches on the sprinkler system and smoke detectors in their system should run to individual annunciators, which in turn connect to the fire alarm control panel. This panel then may be auxiliarized for outside supervision if the building is unattended for periods of time. Presignal systems may be used in fully attended buildings to avoid unnecessary building evacuation and the risk of confusion and even panic, since frequently building tenants and certainly transient occupants are unfamiliar with the alarms (see Table 22.2). This is particularly important in public-type structures where most of the occupants are transient. To avoid the possibility of the presignal going uninvestigated, a timing arrangement is possible that will turn in a general alarm if the alarm device causing the signal is not reset within a predetermined time.

(*b*) ***High-Rise Office Buildings.*** Sad experience (see Chapter 13, page 483) has demonstrated that high-rise buildings, once thought to be "fireproof," are emphatically not so. Indeed, due to their size, they have particularly severe fire protection problems, one of which is the need to relay reliable communications during fire emergencies. As a result, a new type of fire-alarm-and-communications system has appeared which is specifically intended for high-rise building use. In addition to the usual fire alarm system functions, this type of system can provide from the lobby fire command post:

1. Two-way communication with at least one station per floor, all mechanical equipment rooms, elevator machine rooms, and air-handling (fan) rooms.
2. Control of alarm signals.
3. Visual display of alarms, by floor location.
4. Selective and group control of all audible devices in the building.
5. Communication with the Fire Department.
6. On-off indication of all air-handling units in the building, plus override manual control of all fans.

The exact equipment supplied depends on the building and the local fire code. The outstanding characteristics of this system are the communications system, the visual display of alarm locations, and the remote control of air-handling equipment. A unit of this type is shown in Figure 22.29.

22.35 Office Building Security Systems

Although automatic surveillance systems *are* applicable to office and mercantile occupancies, they are more frequently found in industrial facilities and are discussed under that heading. Office buildings normally utilize some type of manual watchmen's tour system

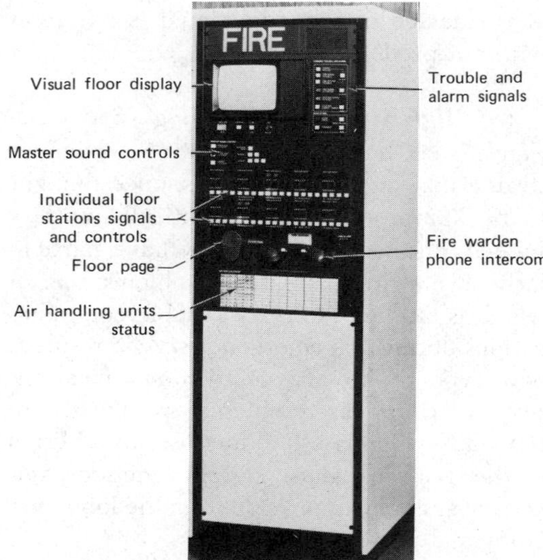

Fig. 22.29 Typical lobby fire command station for a high-rise building. Essential features are alarm pinpointing and 2-way communications with upper floors. Courtesy of Gamewell/Alarmtronics.

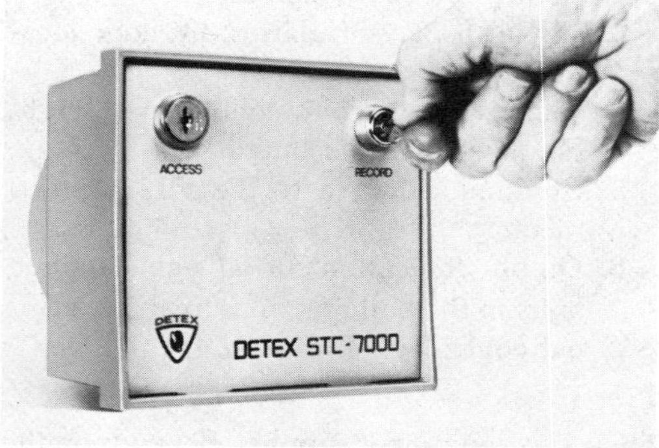

Fig. 22.30 Watchmen's tour station. Here the clock and recording tape are inside. The station is operated by simply inserting a key, as shown. Courtesy of Detex Corp.

so that surveillance of unoccupied areas be conducted on some regular basis.

The simplest type that is nonelectric comprises a number of small cabinets, each containing a key, placed at intervals around the interior and exterior of the building. The watchman uses these keys to operate a special clock that he carries about, thus recording the exact time at which he "clocked-in" at any specific location. Alternatively, the clock is wall-mounted and the guard carries only a key (see Fig. 22.30).

Electrical systems are available that permit constant supervision and are particularly effective where more than one person is on duty. Such systems show on a panel the location and progress of the watchman, by means of lights which glow when the device at each location is operated. Since part of the effectiveness of these systems lies in the timing of the tour, a system can be arranged to sound an alarm if a particular station is not operated within a specific time period. Telephone jacks spaced at points along the guard's route allow him to communicate with the supervising office or other point, without interrupting the scheduled tour. For protection of areas housing extremely valuable items or documents, an intrusion alarm system utilizing the radar principle may be employed. This equipment detects any motion within the protected area. Such spaces are obviously excluded from the watchman's tour.

22.36 Office Building Communications Systems

This planning item is composed of two parts—*intra*office communication or intercom, and *outside* communication via telephone. The former includes equipment; the latter, space only, since the telephone company provides all of the equipment. It is possible to combine the two functions into a single system, and indeed many offices choose that arrangement; that is, all communication equipment is provided by the local telephone company. Such an arrangement has these advantages:

1. Minimal initial cost since only space is provided.
2. Zero maintenance cost—the phone company maintains its own equipment.
3. Minimum capital costs to expand, change, or modernize the system.

The alternative approach is to purchase or lease a private telephone system for interoffice communication, and interface with the telephone company for outside communications.

This is the route taken by an increasing number of offices, both initially and by retrofitting since studies have shown that the majority of calls made in an average office are internal. These can be handled better and cheaper by a private, internal system. Obviously each case is different and must be studied individually. Still, large numbers of actual case studies show that an average office with more than 5 trunk lines and more than 25 extension phones can repay the initial cost of a private phone system within five years (often less), after which net savings of 10 to 40% on the telephone bill can be expected. Our discussion below will therefore cover both systems, that is, private telephone and space planning for telephone company equipment.

22.37 Office Building Private Telephone and Intercom Systems

The usual differentiation between intercom and private telephone is interface with phone company equipment, which is found on the latter and absent on the former. Modern office intercom systems are similar to those described in Section 22.28, for schools. Master and remote stations are frequently called administrative and staff, respectively, and their functions are the same. Instruments may look like telephones (see Fig. 22.26) or more traditional, executive-style, push-button stations. A typical complete system will provide:

1. Direct-dialing, two-way telephone communications between all master stations.
2. Direct communications between all masters and remotes.
3. Staff station call origination to the centrally located LED readout display panel(s) from which it can be routed by a master station to any other point.

In addition, options are available for conference call capability, all-call mode, and even automatic time signaling. Control equipment is based on integrated electronic circuitry and is therefore extremely compact, occupying no more space than a file cabinet. Power requirements are readily provided by local branch circuitry. Heat dissipation is minimal, and no special equipment room is required. The disadvantage of a separate intercom is the necessity for a separate instrument on the desk. This inconvenience is removed with a private telephone system.

Private electronic branch exchange systems (PABX) are, like the modern intercom, based on advanced solid-state technology. Thus, the entire switching system for a system that will handle up to 500 lines and trunks, 4 operator consoles, and 140+ simultaneous conversations, plus full intercom facilities, occupies a cabinet 2 ft × 2.3 ft × 6 ft high. The controls are completely automated, and attendants do little more than route incoming calls and provide information. Typical of this type of equipment are the two stations illustrated in Figure 22.31 and Figure 22.32. Operational features of this system include:

1. Direct internal and external dialing.
2. Consultation hold, that is, ability to hold an outside call while making an inside call.
3. Conference call capability, including internal and external units.
4. Call transfer and camp-on feature.
5. Automatic call back.
6. Call forwarding.
7. Distinctive ringing for different functions.
8. Paging, executive priority, dictation access, personnel location, plus other options as desired.

Manufacturers and suppliers of PABX equipment will furnish technical consultation and complete written proposals so that prospective users can compare operational and financial aspects among suppliers and with telephone company proposals. As stated, operational and long-term financial considerations will usually favor the private system.

22.38 Office Building Telephone Planning

Planning for the telephone equipment in an office building is of prime importance because of the large amounts and critical locations of required space, and is therefore necessarily done simultaneously with other space planning. Exact phone requirements for office space are generally unknown at design time and, even if they were known, planning must account for changes in space usage as well as increased phone and phone-related services. For this reason, all planning is based on square foot areas. Planning is essentially for spaces only, from incoming service to final instrument, since all cabling and equipment are furnished and installed by the telephone company. Space is required for:

1. Service entrance including terminal space, cabinet, or room.

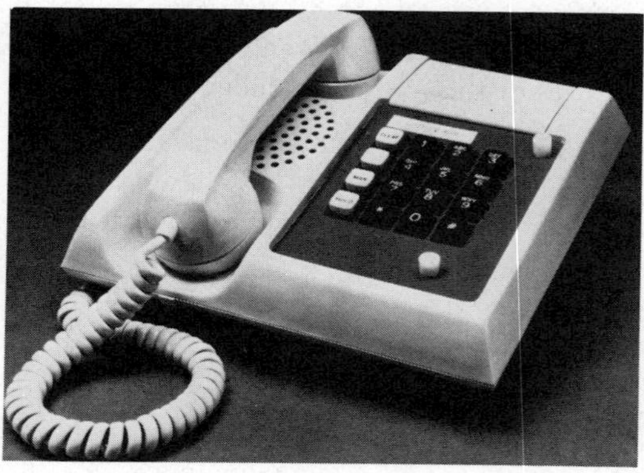

2. Riser spaces, shafts, conduits, and cabinets.
3. Apparatus closets for equipment.
4. Satellite locations for interconnections.
5. Equipment rooms for specific-use equipment.
6. Distribution system including conduits, boxes and underfloor duct, and raceway systems.

When solid-state switching is used, as with electronic PBX (Private Branch Exchange) equipment and Centrex 11, space requirements for items 3, 4, and 5 can often be reduced. It is in the interest of the client to be aware of this, in order that precious rental space not be wasted. The information that follows is for *standard switching equipment* (not electronic), except where specifically stated otherwise. A cutaway drawing illustrating most of the above items is shown in Figure 22.33 and in the single-line diagram in Figure 22.34.

(*a*) ***Service Entrance.*** Adjacent to the service conduit a spare sleeve placed in the foundation wall and sealed will provide for future service expansion. Inside the building the telephone cable is terminated in a wall box,

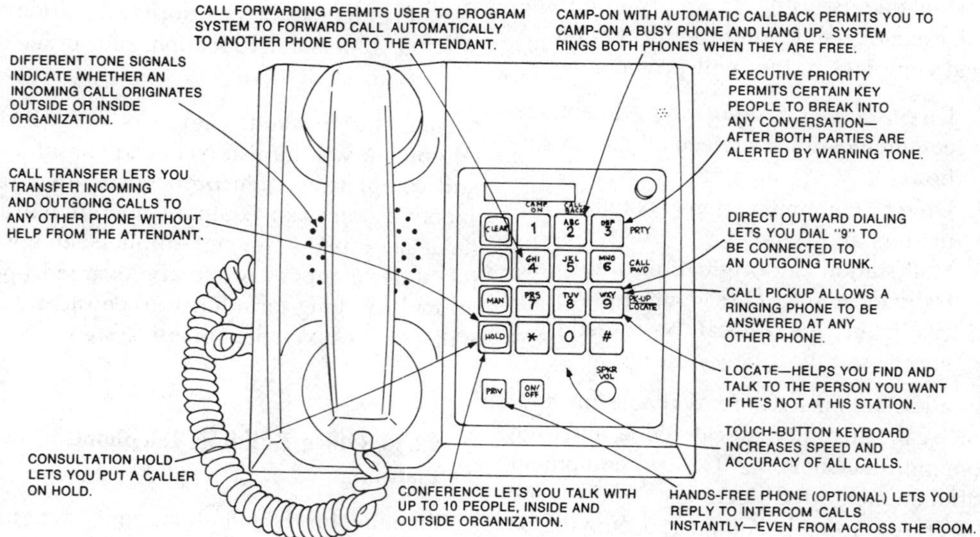

Fig. 22.31 *Typical modern PABX system station. Courtesy of Executone, Inc.*

cabinet, or terminal room depending on the cable size. For buildings up to 70,000 sq ft in rentable area, wall-mounted terminal cabinets are normally sufficient. In any case, the area should be dry and requires light, ventilation, one or more 120 v outlets on a separate circuit, and a good ground connection. In special cases more power is required.

Surface-type wall installations comprise simply a sheet of $\frac{3}{4}$ in. marine plywood on to which the phone company mounts its cabinet. Flush installations in finished areas require a cabinet with $\frac{3}{4}$ in. plywood back.

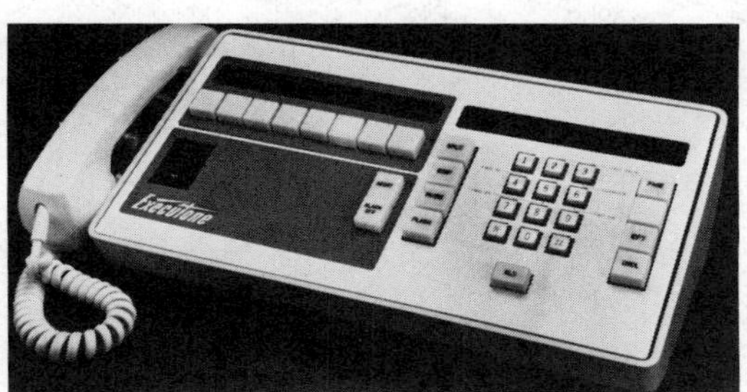

(*b*) ***Riser Shafts.*** These can accept the cables extending beyond the terminal room and carry them vertically through the building. Connection between terminal space and risers is preferable in conduit. A typical riser shaft is illustrated in cutaway in Figure 22.35.

The riser shafts provide means for the cables to extend vertically and to terminate at each floor. Ideally, the risers comprise a series of vertically aligned closets connected by $3\frac{1}{2}$ in. sleeves set in the floors and extending 1 in. above the floor. If electrical closets are adequately sized, telephone risers and cabinets may be installed therein. Where multiple risers are used, shafts should be interconnected by one or more $1\frac{1}{2}$ in. or 2 in. conduits to allow for interconnection of systems.

(*c*) ***Apparatus Closets.*** Within the apparatus closet, cables from the riser system are interconnected by means of connecting blocks to relay and power equipment, as well as to

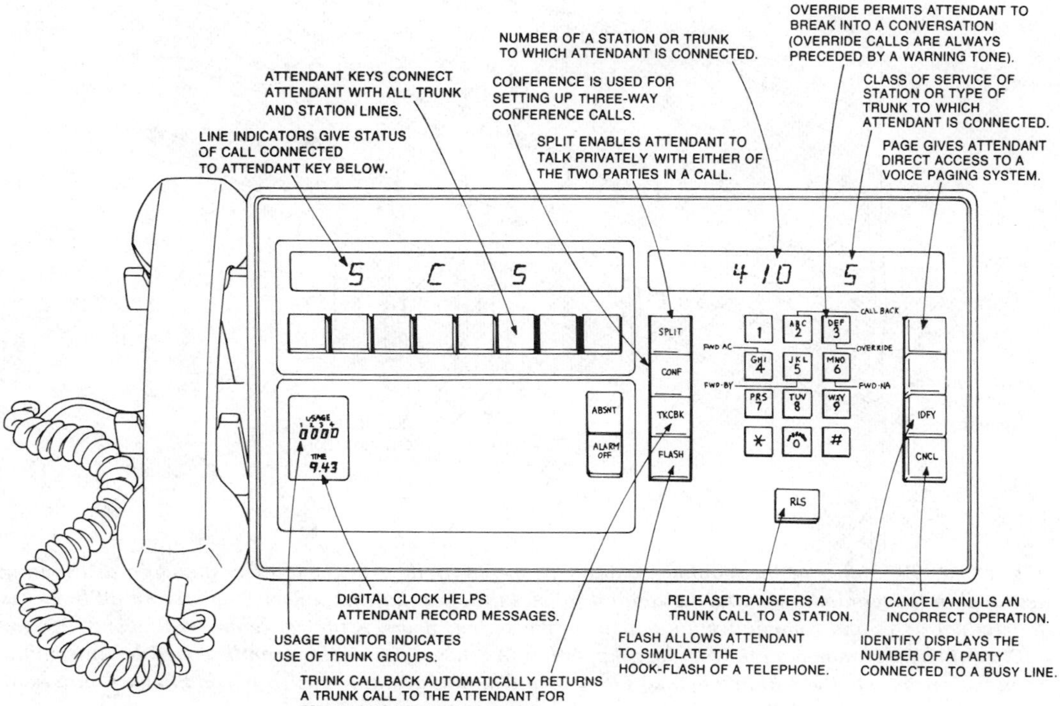

Fig. 22.32 *Attendant console for electronic PABX system. Courtesy of Executone, Inc. and* Specifying Engineer, *March 1977.*

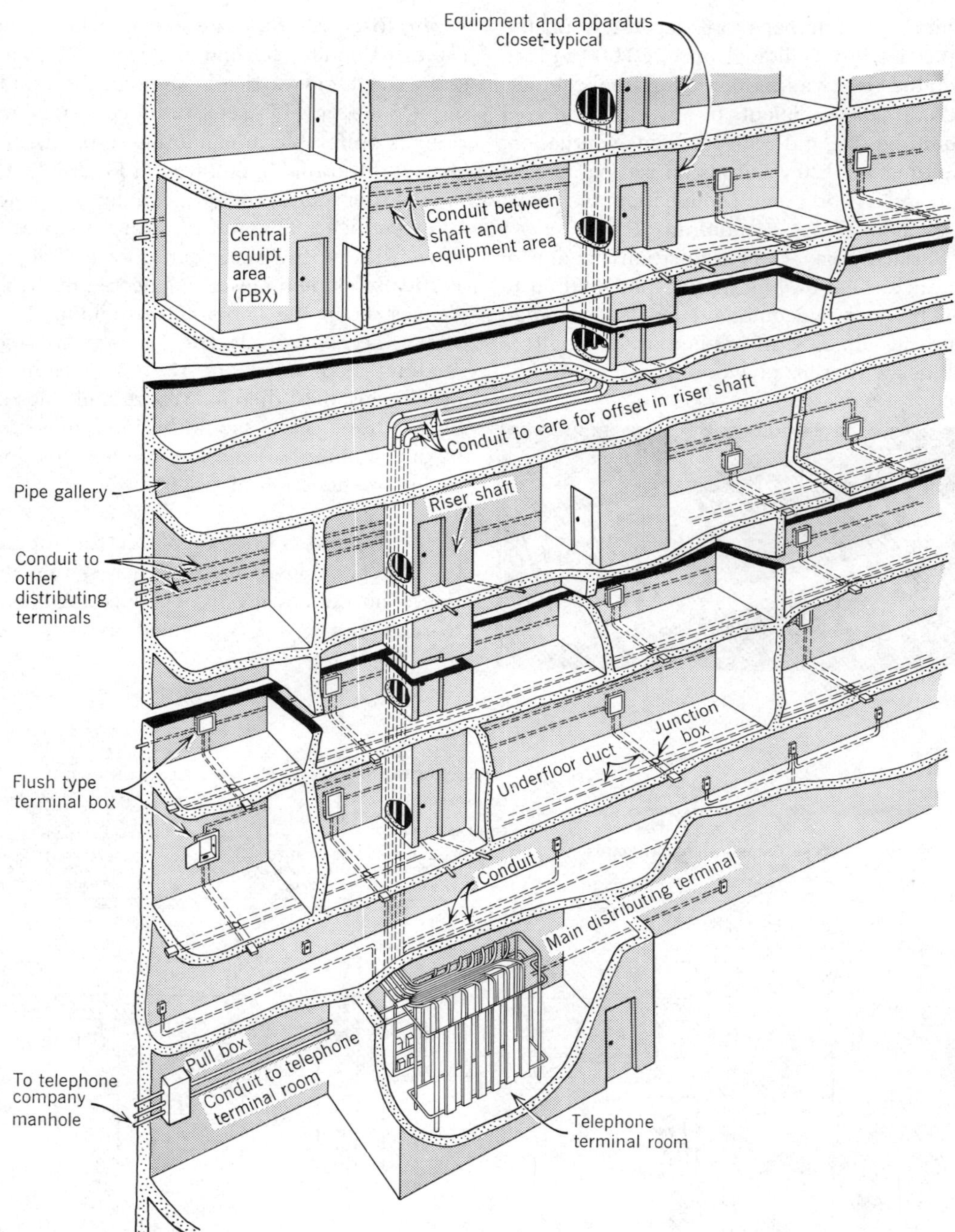

Fig. 22.33 *Beginning at the manhole, lower left, we can follow the conduit to the main distributing terminal in the terminal room. As the conduit follows the riser shaft from floor to floor, conduit branches to terminal boxes and communication outlets on the various floors. A typical central equipment area for PBX equipment is shown on the top floor, upper left. As shown here, cables can be concealed in conduit from the entrance to the building to their ultimate outlet, thus avoiding unsightly conditions. Courtesy of American Telephone and Telegraph Co.*

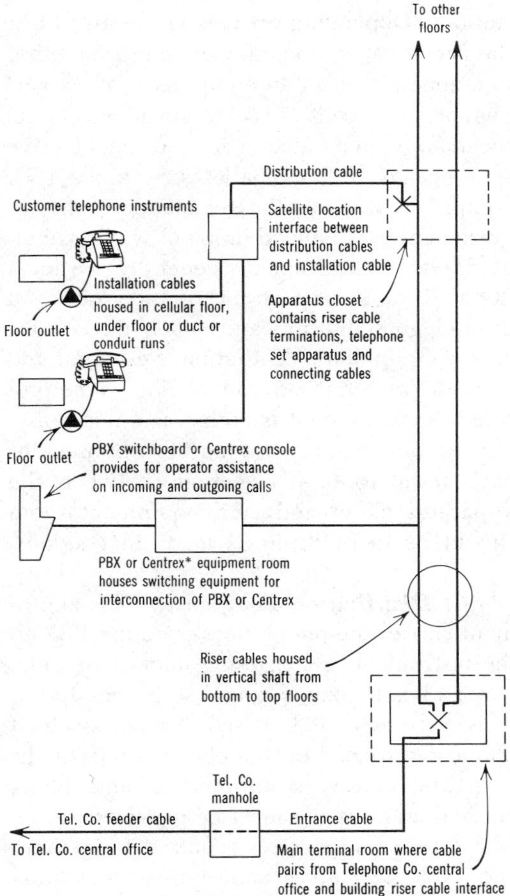

Fig. 22.34 Diagramatic view of major components of a commercial building's telephone system. Compare to Figure 22.33. Courtesy of American Telephone and Telegraph Co.

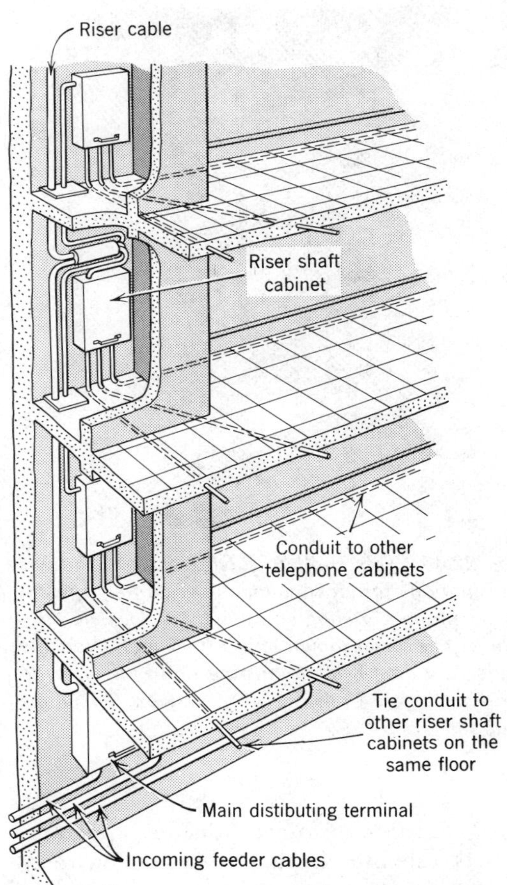

Fig. 22.35 Typical telephone riser arrangement. Closets provide terminal and splicing space and may double as equipment rooms if conditions permit. Courtesy of American Telephone and Telegraph Co.

the cable pairs that radiate from the closet to station locations throughout the floor. These closets may be called *zone closets* or *riser closets* if they function with an underfloor raceway system and/or a cable riser system. A single apparatus closet may serve a floor area of as much as 36,000 sq ft. Walls of the closet should be lined with plywood at least $\frac{3}{4}$ in. thick to support the weight of relay panels and circuit units, power equipment, terminals, connecting blocks, and other hardware.

The following table gives an idea of the space required for typical, centralized telephone system apparatus closets.

Square Feet of Rentable Floor Area Served	*Linear Feet of Required Wall Space*
Up to 9000	11
9001 to 18,000	22
18,001 to 27,000	33
27,001 to 36,000	44

Each apparatus closet must be provided with a switched ceiling light, and a separate 20-amp, 120 v circuit with 2 duplex receptacles. A source of emergency power is desirable to avoid curtailment of telephone service during outages.

Fig. 22.36 This shallow satellite closet contains terminations for cross-connecting relay equipment to telephones. Relay equipment is located elsewhere. The closet shown serves 7000 sq. ft. Note the underfloor duct headers turning up in the closet to receive cables. Courtesy of American Telephone and Telegraph Co.

(*d*) ***Satellite Closets.*** Unlike apparatus closets, satellite closets do not contain relay and power equipment. Their primary use is to provide cable terminating facilities, that is, connecting blocks, for key telephone system services, station (telephone) and central office, PBX, or centrex lines (see Fig. 22.36). The following table gives examples of the dimensions of satellite closets:

Square Feet of Rentable Floor Area Served	Linear Feet of Required Wall Space
Up to 9000	6
9001 to 18,000	12
18,001 to 27,000	18
27,001 to 36,000	24

(*e*) ***Equipment Rooms.*** These are sometimes required when tenants utilize a PBX system. This is the familiar private switchboard that accepts numerous incoming lines and manually redistributes them to interior stations. Depending on the size of the PBX, the necessary equipment can be placed either in a zone closet, in an equipment room of its own or, as a result of recent advancements in technology, in an alcove or other unobtrusive spot because of its smaller size. Since PBX equipment is generally free standing, the requirement for plywood lining of walls is obviated. Other requirements generally are for a 208 v 20 to 30 amp circuit, a 120 v outlet, a ground, good lighting, and sufficient space for all the equipment. Ventilation is essential and acoustic material on the ceiling is helpful since the equipment is both warm and noisy. Exact space requirement must be obtained for each installation. Connections between the apparatus closet and PBX equipment room should be in multiple $3\frac{1}{2}$ in. rigid conduits.

(*f*) ***Distribution Equipment.*** This equipment carries the phone lines from the PBX to the instrument. When using underfloor raceways, a header capacity of 1 sq in. per 100 sq ft is recommended, based on an assumed phone density of 1 button phone per 100 sq ft. If greater density is anticipated, and this is unusual, more space must be provided. Table 22.3 gives conduit sizes required for phones and Table 22.4 gives some typical telephone cabinet and plywood backboard sizes.

22.39 Office Building Supervisory Control Center

As the modern office building's mechanical and electrical systems increased in complexity, the need arose for a central point of supervision, control, and data collection from which an operator could survey an entire building's functioning. From such a point the water, air conditioning, heating, ventilating, electrical, and other systems could be controlled manually and automatically with much greater accuracy than if no such central control point were available. Data on temperatures, pressures, flow, current, voltage, and all the many parameters of mechanical systems could be made instantly available, so that operational decisions could be made more accurately. Also

Table 22.3 ***Telephone Design Data***

Type of Outlet	*Minimum Conduit Home Run Size (in Inches)*	*Maximum Number of Outlets on Home Run*
Pay phone	$\frac{3}{4}$	3
	1	6
Standard phone	$\frac{3}{4}$	4
	1	7
Unknown phone (general rule)	$\frac{3}{4}$	1 or 2
	1	3 or 4
	$1\frac{1}{4}$	5 to 7
Six-button phone	$\frac{3}{4}$	1
	1	2
	$1\frac{1}{4}$	3
Call director phones	$1\frac{1}{4}$	1

Table 22.4 ***Telephone Cabinet Sizes***

Number of Telephone Pairs	*Size of Cabinet* *W*	*H*	*D*
Min.–26	12	30	3
27–76	12	36	4
77–152	18	42	4
153–303	18	66	6
304–606	36	72	6
or	54	54	6
607–1212	54	72	6

Plywood board sizes (W × H)

all systems could be monitored here, and all alarms instantly acted upon, automatically or manually.

Such control centers, generally called *supervisory control and data centers,* are now installed as a matter of course in office buildings. Most are equipped with computers that process the huge amount of data received to arrive at operational decisions intended to optimize system performance. As such they result in a considerable savings in operating costs, in addition to the workforce savings generated by their remote monitoring and control functions. These units normally perform these functions:

a. Remote monitoring, recording, and logging.
b. Remote start-stop.
c. Remote controls, resets, changes.
d. Alarm functions.

The computer is programmed to respond to the data input of (a) under normal circumstances and (d) under unusual ones with the responses of (b) and (c), automatically. Operator intervention is needed to change programming, override automatic responses, check alarms, and periodically run through the system.

Typical energy-saving functions of a computer center would be:

a. Programmed lighting—reduce levels, turn off unused area lights.
b. Electric load control—as described in Section 14.13.
c. Optimized HVAC operation—based on preprogrammed procedures and continuous response control.

A typical performance specification for a control system might read as follows:

Furnish and install a Central Control System capable of performing the following basic functions:

1. Start and stop fans, pumps, and systems according to program, with provision for manual override.
2. Verify running conditions.
3. Provide alarm for fire, smoke, high or low temperature, high or low pressure, humidity or dew point, flow, and dirty filters from multiple points.
4. Read temperatures and pressures and reset control points as necessary.
5. Move dampers as necessary to control temperatures, and upon fan failure.
6. Isolate and identify malfunctions by building, system, or zone.
7. Measure flow and temperature and compute Btu of chilled water and high temperature water.
8. Measure all flows and temperatures necessary to determine boiler efficiencies.

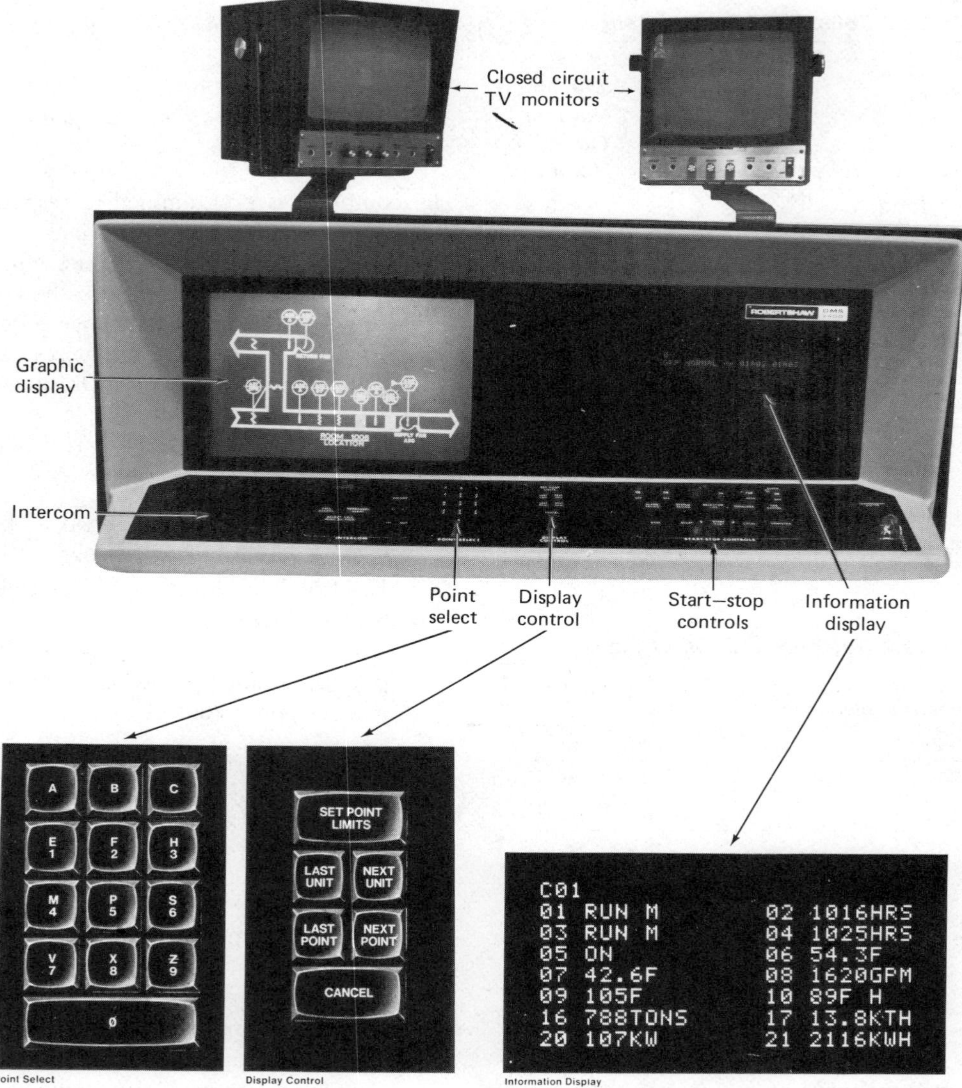

Fig. 22.37 Typical supervisory data center control console with functions displayed. Control equipment in adjacent cabinets are not shown. Courtesy of Robertshaw Controls Company.

9. Accumulate flow and Btu readings and log on demand.
10. Log kilowatt hours and the time and value of maximum loads by feeder and panel.
11. Cut off low priority loads and control timing of the starting of major loads, *in order to hold demand to a minimum.*
12. Provide control of chillers including starting, monitoring, selection of equipment for optimum plant efficiency, and shutting down.
13. Provide visual display by slide projection of any selected system, and automatically display a system in which an alarm exists, and hold the display until manually released.

This last item is illustrated on the display screen of the control console in Figure 22.37.

Note that in addition to the graphic display there is a simultaneous information display giving the data requested. Economic studies readily demonstrate the financial justification for the considerable first cost involved in acquiring a control unit both for new construction and for retrofitting an existing building.

Architecturally the unit requires good lighting and ventilation, extensive raceway space, but little area. Systems are tailored to the building so that no guidelines can be stated for space requirements.

INDUSTRIAL BUILDING SYSTEMS

22.40 General

All industrial facilities ranging from the taxpayer loft, housing a small hand assembly plant, to the immense steel manufacturing plant require a variety of signal and alarm equipment. Although, as in the case of commercial structures, a detailed analysis of the equipment is out of place here, it behooves the building designer to be sufficiently familiar with this area to know generally the function, operation, and availability of this type of equipment.

22.41 Industrial Building Fire Alarm Systems

Industrial building systems are normally selective coded and fully supervised (refer to Sections 22.8 through 22.18 and Table 22.2). Presignaling is utilized in structures where for any reason an evacuation alarm is undesirable. In addition to manual stations at points of egress, these devices may be used:

1. Temperature and smoke detectors in storage areas and laboratories.
2. Smoke and flame detectors in record rooms and continuous process laboratories.
3. Waterflow switches on all sprinklers.

Annunciators, control panel, and alarm register are best placed in the guard-room as will be discussed below. If none is available, an auxiliarized circuit should be added to allow remote supervision.

Because of the high ambient noise level in many plants, horns are substituted in such areas for bells and gongs which might be inaudible (see Figs. 22.38 and 22.39).

22.42 Industrial Building Security Systems

Among the most important signal functions in this type of facility is the protective one. Although the control point may be a common one, these systems are varied and perform separate functions.

(*a*) ***Door and Exit Controls.*** Outside doors and doors to restricted areas are supervised by electrified security hardware that triggers an alarm when a door is opened without authorization. The alarm mechanism may be concealed, or openly displayed as in Figure 22.40, to act as a deterrent. Annunciation can be provided by a separate panel as shown, or tied into a central alarm panel as in Figure 22.43.

(*b*) ***Personnel Entry Control.*** Several levels of security are available and can be applied in accordance with security needs. Beyond a simple lock, the first level is an electric card reader (Fig. 22.41*a*), which grants entry to the card holder. The next level requires the encoding of a 3-digit number (Fig. 22.41*b*) simultaneously with presentation of a card, thus barring entry to unauthorized card holders (found or stolen cards). The third level involves an attendant. In Figure 22.41*c* the attendant compares the card data and the person's appearance with stored data displayed on an adjacent screen, providing a triple check and negating the effect of forged cards.

(*c*) ***Watchmen's Tour Equipment.*** When tour equipment is used, it is frequently of the combination-alarm type, as illustrated in Figure 22.42. This type of station allows guards to call in, alerts them to waiting calls, permits a general alarm to be turned in by key operation, and is available as a manual fire alarm

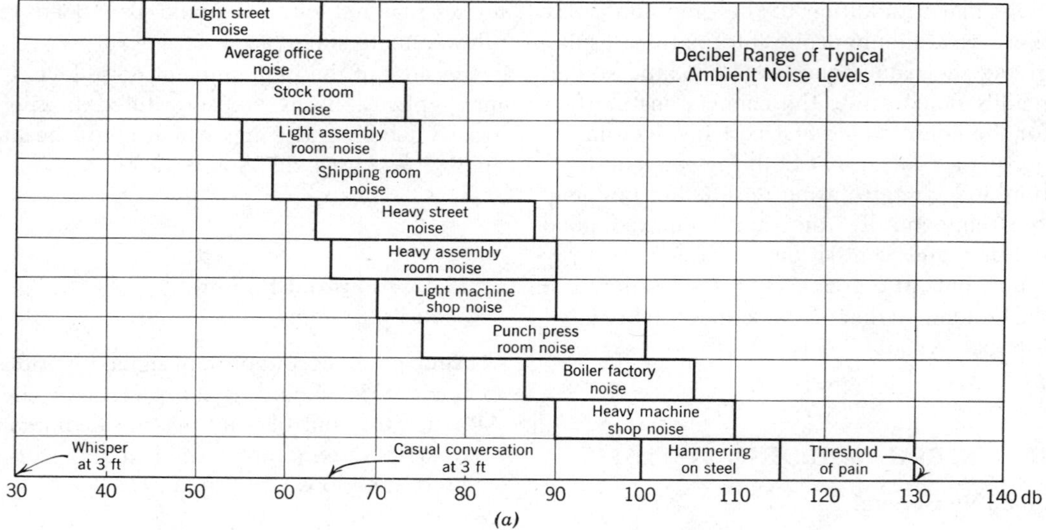

(a)

Fig. 22.38 (a) *Decibel range of typical industrial noise levels.*

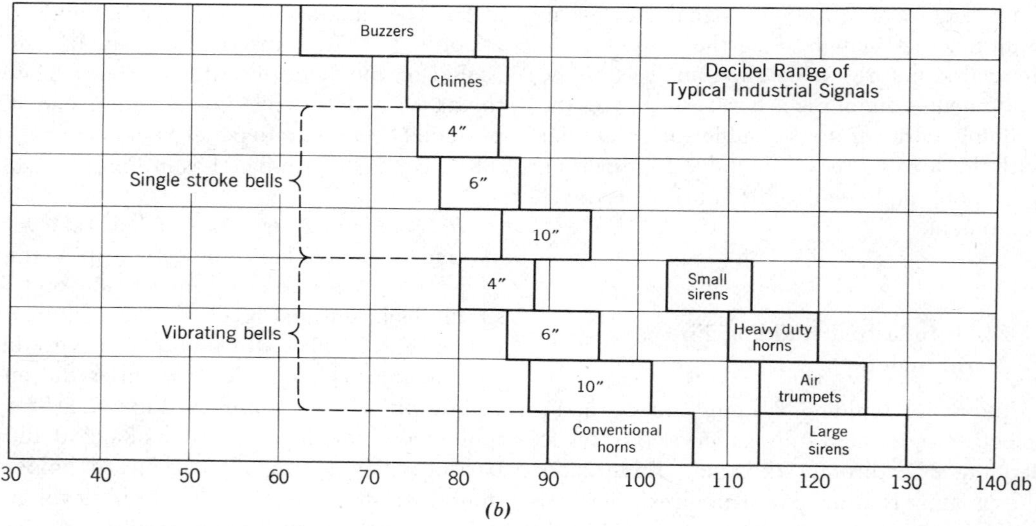

(b)

Fig. 22.38 (b) *Decibel rating of typical industrial signals may be compared with ambient noise levels* (a) *to facilitate selection.*

station. The station may be coded or noncoded as desired. The tour is normally timed so that any delay automatically sends an alarm and brings help.

(d) Interior Security Systems. In addition to the usual photoelectric arrangements, magnetic and contact devices, and closed-circuit television viewing, motion detectors and proximity alarms are now becoming common. The latter system utilizes a generator that emits a signal and a receiver that receives or detects the signal. When *any motion* occurs in the protective area the alarm is initiated. The

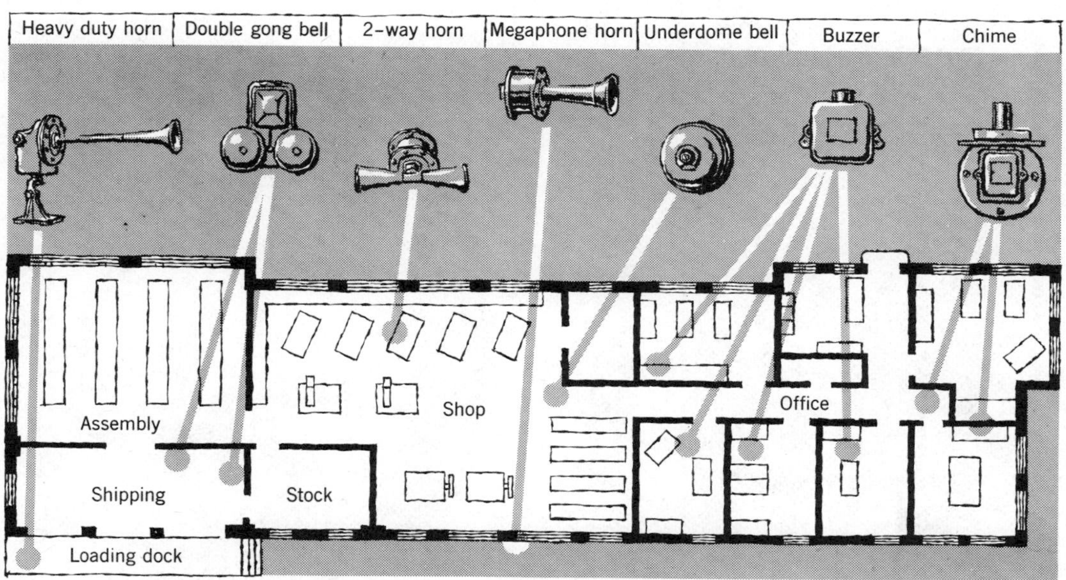

Fig. 22.39 *Suggested locations of typical signals are shown in this plan of a small industrial facility.*

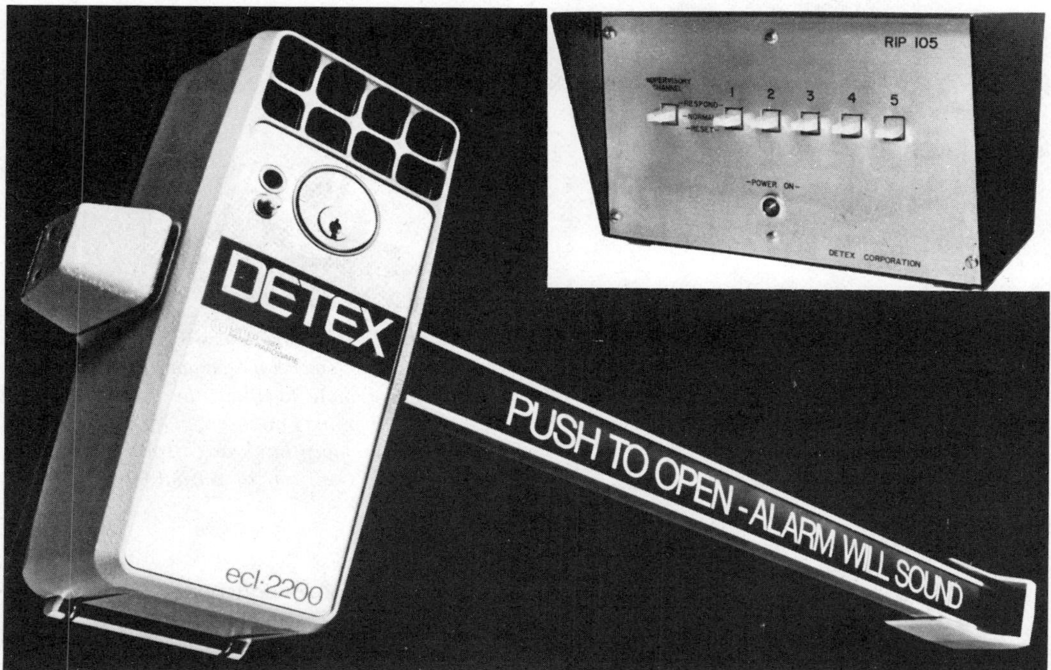

Fig. 22.40 *Typical security door hardware. Audible alarm is sounded locally and annunciated at the control panel (inset). Clearly written warning on the hardware acts as a deterrent. Courtesy of Detex Corp.*

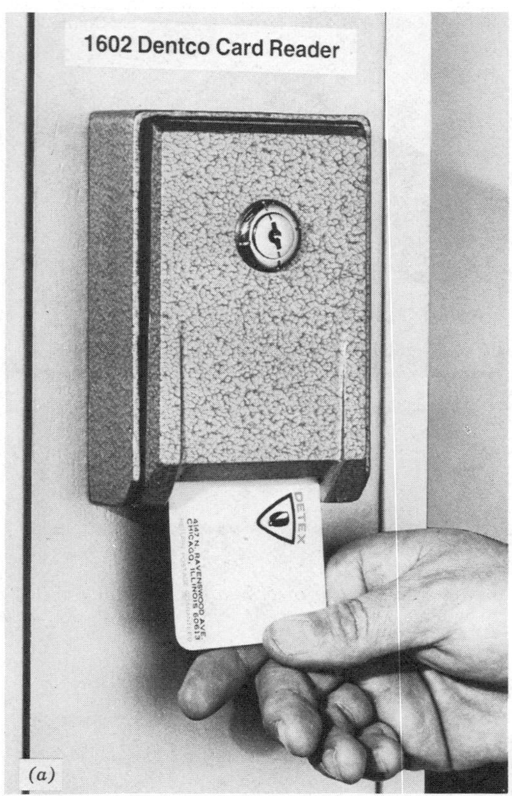

Fig. 22.41 (a) *Access is gained by holder of identification card.*

Fig. 22.41 (b) *In addition to card presentation, a corresponding code must be entered. Courtesy of Detex Corp.*

Fig. 22.41 (c) *The person desiring entry inserts an ID card and is viewed by a camera at the entrance (inset). At a remote location, the subject and the ID card are displayed on a screen (right), while ID data from a memory bank are displayed for comparison (left). Courtesy of Visual Methods, Inc.*

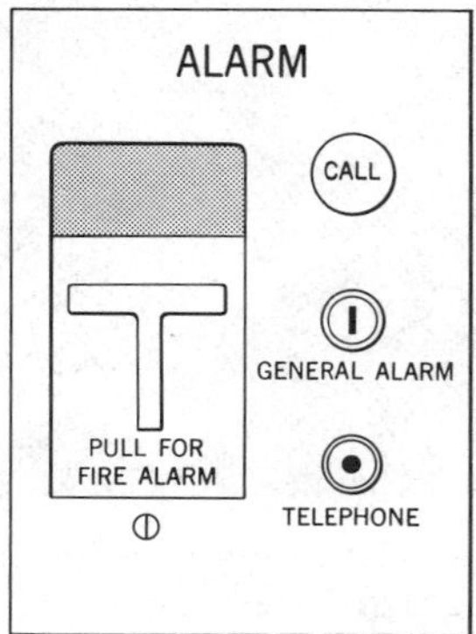

Fig. 22.42 A typical combination alarm station combines fire and general alarm with an intercom station.

signal may be of several types. An ultrasonic signal fills the area protected (indoors) with a standing wave pattern, which when disturbed is detected. This type is extremely sensitive and is used in museums and other areas of extreme value. A disadvantage is that, unless carefully adjusted, the extreme sensitivity will cause false alarms in response to curtain flutter, rodents, and the like.

(a)

Fig. 22.43a *Central monitoring and control unit for all types of hazard signal systems. The unit prints and displays station status on demand. An alarm or other change of status will cause it to print location, flash emergency light, transmit alarm signal, and sound an audible device if desired. Courtesy of Gamewell/Alarmtronics.*

Another variation is a tone radar system utilizing microwave frequencies. Motion within its antenna range causes a change in reflected signal and trips an alarm.

A third variation, intended for outdoor application, establishes a perimeter fence comprising an electromagnetic field between antennas. When the field is disturbed by an intruder, the system unbalances and an alarm is initiated.

Fig. 22.43b *Sketch of composite alarm system. All alarms can be monitored by a single person at this type of console. These would include fire, intrusion, and sprinkler alarms, and special units that detect liquid levels, flame failure, gas leakage and so on.*

(*e*) ***Proximity Alarm.*** This system utilizes the electrical capacitive effect of a human to trip an alarm. It is particularly useful in protecting a single item, for example, a file or safe within a room. Since the protective field extends only a few inches, normal use of the space can be maintained while protecting a single unit within the area.

All of the above, plus others not discussed, can be connected to a central monitoring console of either the automatic variety, as in Figure 22.43*a*, or annunicated at a control console, as in Figure 22.43*b*. The latter is employed where an attended guard post is required for such other functions as personnel access control and closed-circuit TV monitoring. The automated alarm center transmits its signal to an observed location such as the mechanical room or the plant offices.

22.43 Industrial Building Control Center

The control center for an industrial facility contains its program equipment, utility controls, and data logging equipment plus circuits that may be utilized to monitor and/or control process work within the facility. Program audible devices must be carefully selected as explained above for fire alarm devices.

22.44 Industrial Building Paging Systems

All of the time-saving and efficiency potential that result from the proper use of the various signal and communication equipment available can be lost if there is no recipient for the information transmitted. Furthermore, sometimes a decision must be reached quickly to avoid costly delay, reruns, and so on. These factors combine to make rapid and accurate paging an extremely important function in a manufacturing operation, particularly of the larger variety.

Paging systems fall into two general categories and several subcategories: they are either visual, audible, or both, and are either common or selective. The simplest visual and audiovisual types comprise flashing lights, which may be combined with buzzers or bells, either or both of which are generally coded. Such systems are nonselective in that they impinge upon the senses of all the building occupants—an obvious disadvantage.

More sophisticated systems utilize a small pocket device that is carried by each person likely to be paged; maintenance personnel, plant engineers, executives, and so forth. By means of either direct radio transmission or of electric fields induced by induction loops installed throughout the building, an individual pocket device can be alerted by a buzz. In some systems, the alerted person then listens to the message directly. On others, it is necessary for this person, once having been paged, to go to a phone and call in to a central paging desk to receive the message. Others utilize small hand-held, two-way radio transmitters with paging, to enable conversation between the page originator and the recipient.

In any of these systems, it is necessary to have a paging operator and a coding device at which the paging calls originate. Often, in a small factory, paging is handled by the regular phone operator.

References and Bibliography

Life Safety Code (latest code issue). Published by the National Fire Protection Association.

Design for Security, 1968, R. J. Healy, John, Wiley & Sons, New York.

PART EIGHT

transportation

23

Vertical Transportation: Passenger Elevators

GENERAL INFORMATION

23.1 Introduction

Among the many important functional decisions that must be reached by the designer of a multistory building, probably none is more important than the proper selection of the vertical transportation equipment, specifically the passenger elevators, service and freight elevators, and escalators (moving electric stairways). Not only do these items represent a major building expense, being in the case of a 25-story office building as much as 10% of the construction cost, but also the quality of elevator service is an important factor in a tenant's choice of space in competing buildings.

Although the final decision as to the type of equipment rests with the architect, the factors affecting it are so numerous that it behooves the building designer to consult with an elevator expert. Such consultation service is readily available from consultants in the field and to an extent from the major elevator and escalator manufacturers. It is the function of this chapter to familiarize the architect and engineer with the nature and application of vertical transportation equipment so as to enable them to make preliminary design decisions before consulting manufacturer's representatives or consultants.

23.2 Passenger Elevators

Passenger elevators are normally subdivided by application into four general classifications: general purpose or commercial, residential, institutional, and store. Our discussion will be principally concerned with the general purpose type, though much of the material also is applicable to the other types. Furthermore, our initial discussion of passenger elevators is restricted to traction types. Hydraulic units are covered in Section 24.8.

Ideal performance of an elevator installation provides minimum waiting time for a car at any floor level, comfortable acceleration, rapid transportation, smooth and rapid retardation, automatic leveling at landings, and rapid loading and unloading at all stops. Furthermore, the system must provide quick and quiet power operation of doors, good visual floor indication both in the cars and at landings, easily operated car and landing call buttons (or other devices), smooth quiet and safe operation of all mechanical equipment for all conditions of loading, comfortable lighting and generally pleasant car atmosphere.

In addition to the above passenger-oriented service considerations, the elevators have architectural aspects as well. The cars and shaftway doors must be treated in a manner consonant with the architectural unity of the building. More important though, the shaftways are major space elements whose integration into the building is a prime factor in composition.

23.3 Codes and Standards

Perhaps more than any other item of construction, elevators are governed by strict installation codes. The "bible" of the industry is the American National Standards Institute (ANSI) code A 17.1, *Safety Code for Elevators, Dumbwaiters, Escalators and Moving Walks*, an up-to-date copy of which should be an integral part of every architect and engineer's working library. This code has legal force in most parts of the United States, and its requirements—particularly in regard to pit depth and height of overheads (runby)—are strictly enforced. As with other items, some states and municipalities have their own elevator codes (Massachusetts, Wisconsin, Pennsylvania, New York City, Seattle, Boston, among others) that are generally based on, and more stringent than, the ANSI code.

In addition to the elevator code, other construction and installation codes have an influence on elevator work. Thus NFPA No. 101, *Life Safety Code,* states certain fire safety requirements, NFPA No. 70 (The National Electric Code) governs some of the electrical aspects of elevator construction, and state and local laws add a multitude of requirements and restrictions bearing on fire safety, emergency power, security regulations, and special accommodations for handicapped persons. Provisions for the handicapped are covered by a recent ANSI code, a special industry code, and in most locations by local law. As with most large industries, the elevator industry is self-regulating and standardized. The National Elevator Industry, Inc. (NEII) publishes standard elevator layouts for traction and hydraulic installations, and as mentioned above its own elevator standard for the handicapped. Local elevator consultants and companies are normally knowledgeable as to all the codes and standards in force, but this does not relieve the architect-engineer of legal responsibility for the installation. Therefore, we strongly recommend that in the preliminary planning stage all pertinent regulations concerning vertical transportation be assembled—this should be a high-priority item.

ELEVATOR EQUIPMENT

23.4 Arrangement of Principal Parts

The car, cables, elevator machine, control equipment, counterweights, shaft of hoistway, rails, penthouse, and pit make up the principal parts in any traction elevator installation. An idea of the functioning and orientation of these units of equipment can be obtained from an inspection of Figure 23.1. Specific installations will vary somewhat; in sophisticated group supervisory systems, logic modules may be used in lieu of the more conventional relay-operated control and selector panels, or solid-state equipment may be used in lieu of a motor-generator set. On the whole, however, the basic functional components remain the same for all installations.

The cars, with their equipment for safety, convenience, comfort, and finish, are the only items with which the average passenger is familiar. Indeed, some of the building's prestige depends on proper design of the car. Essentially, the car is a cage of some fire-resistant material supported on a structural frame, to the top member of which the cables are fastened. By means of guide shoes on the side members the car is guided in its vertical travel in the shaft. The car is provided with safety doors, operating-control equipment, floor-level indicators, illumination, emergency exits, and ventilation. It is designed for long life, quiet operation, and low maintenance.

The cables raise and lower the car. Four to eight cables, depending on car speed and capacity, are placed in parallel, the weight of the car being equally distributed among them. The cables that are fastened to the top of the car pass over a motor-driven cylindrical sheave at the traction machine (grooved for the cables) and then downward to the counterweight. Useful cable life depends on car speed, frequency of operation, and type of roping.

The elevator machine turns the sheave and lifts or lowers the car. It consists of a heavy structural frame on which are mounted the sheave and driving motor, the gears (if any), the brakes, the magnetic safety brake, and certain other auxiliaries. The governor that limits the car to safe speeds is mounted on or near the elevator machine. In most installations the elevator-driving motor receives its energy from a separate motor-generator set (m-g set), which is in operation during the period that the particular elevator is available for handling traffic. This m-g set is properly considered a part of the elevator machine,

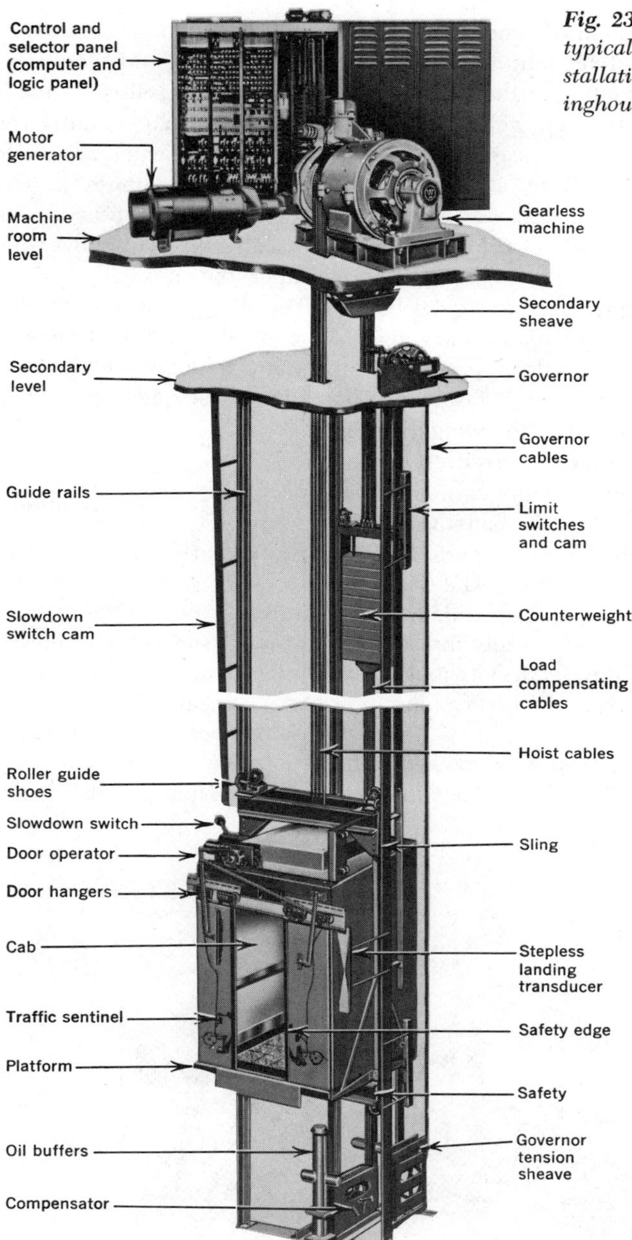

Fig. 23.1 *Components of a typical gearless elevator installation. Courtesy of Westinghouse Elevator Co.*

although it may be located some distance from it. In some recent installations, solid-state power control equipment replaces the traditional m-g set. Installations of this type will increase because of the inherent advantages of such an installation as discussed in Section 23.16.

The control equipment, in a general sense, is the combination of push buttons, contacts, electronic equipment, relays, solid-state switching, cams, and devices that are operated manually or automatically to initiate the door operation, starting, acceleration, retardation, leveling, and stopping of the car. These auxiliaries are interrelated in such a way that the major apparatus functions to produce the maximum of safety, comfort, and convenience. Electrical limit switches automatically stop the car from overrunning at the top and bottom of the hoistway. The well-known floor indicators, floor pilot lights, car panels, lobby control panel, call buttons at floor levels, floor-leveling devices, and up and down indicating lamps are all parts of the coordinated control equipment.

The counterweights are rectangular blocks of cast-iron or cut steel plates stacked in a frame that is supported at the opposite ends of the cables to which the car is fastened. The counterweight is related to the weight of the car and its load so that the required energy input to the elevator machine (which moves the car) is relatively low. In fact the energy required is large only during the periods of car acceleration and retardation. The counterweight is guided in its travel up and down the shaft by two guide rails typically installed on the back wall of the shaft. Obviously the counterweight travels in the reverse direction to the car.

The shaft is the vertical passageway for the car and counterweights. On its side walls are the car guide rails, door frames, and certain mechanical and electrical auxiliaries of the control apparatus. At the bottom of the shaft are the car and counterweight buffers. At the top is the structural platform on which the elevator machine rests. The elevator machine room which houses the elevator machine is usually directly above the hoistway. It contains the m-g set or solid-state control that supplies energy to the elevator machine, the control board, and other control equipment. All machinery and control equipment are designed for quiet operation.

23.5 Gearless Traction Machines

A gearless traction machine, Figure 23.2, consists of a d-c motor, the shaft of which is directly connected to the brake wheel and driving sheave. The elevator hoist ropes are placed around this sheave. The absence of gears means that the motor must run at the same speed as the driving sheave. Since it is not practical to build d-c motors for operation at very low speeds, this type of machine is

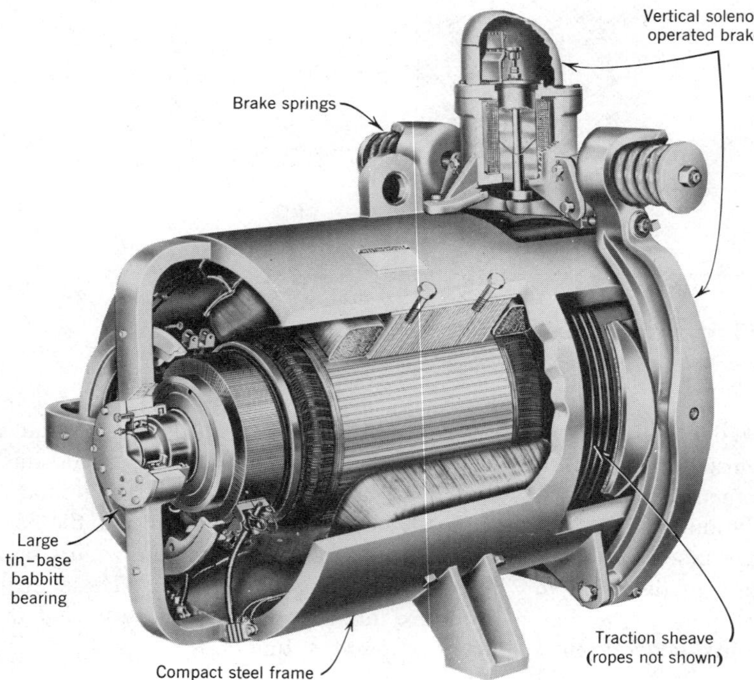

Fig. 23.2 Cutaway view of a typical high-speed gearless traction machine. Courtesy of Westinghouse Elevator Co.

utilized for medium- and high-speed elevators, that is, speeds above 350 ft per min (fpm). The motors are built in ranges from 20 to 375 hp. Gearless machines are generally utilized for passenger service, with minimum speed of 400 fpm and normal capacities of 2000 to 4000 lbs, although specials of up to 10,000 lbs as at the World Trade Center, have been built. Below this speed geared machines are used. In the range of 400 to 700 fpm, a 2:1 roping (see Section 23.7) is generally used. This reduces motor size and increases sheave speed, thus reducing the cost. Above 600 fpm, motor speed is high enough for 1:1 roping to be applied economically.

The gearless traction machine is generally considered superior to the geared machine. Since there are fewer moving parts, the gearless machine is more efficient, is quieter in operation, requires less maintenance, and has longer life. The decision as to whether these advantages are worth the additional cost involved is made only after a careful analysis. Generally, a gearless machine is chosen where rise is more than 150 ft and smooth, high-speed operation is desired.

23.6 Geared Traction Machines

This type of traction machine (Fig. 23.3) employs a worm and gear interposed between the driving motor and the hoisting sheave. The driving motor can therefore be a smaller high-speed unit, running at speeds of 600 to 1800 rpm, depending on the elevator speed and gear ratio. The motor itself is either a-c or d-c, in contradistinction to the gearless unit which is always d-c. The a-c unit is usually used on low-speed applications of 25 to 150 fpm, utilizing single- or two-speed a-c motors with rheostatic control. The a-c traction elevator has only limited application since, unlike the d-c unit multivoltage machine, speed cannot be varied smoothly—this results in rough operation. On installations of

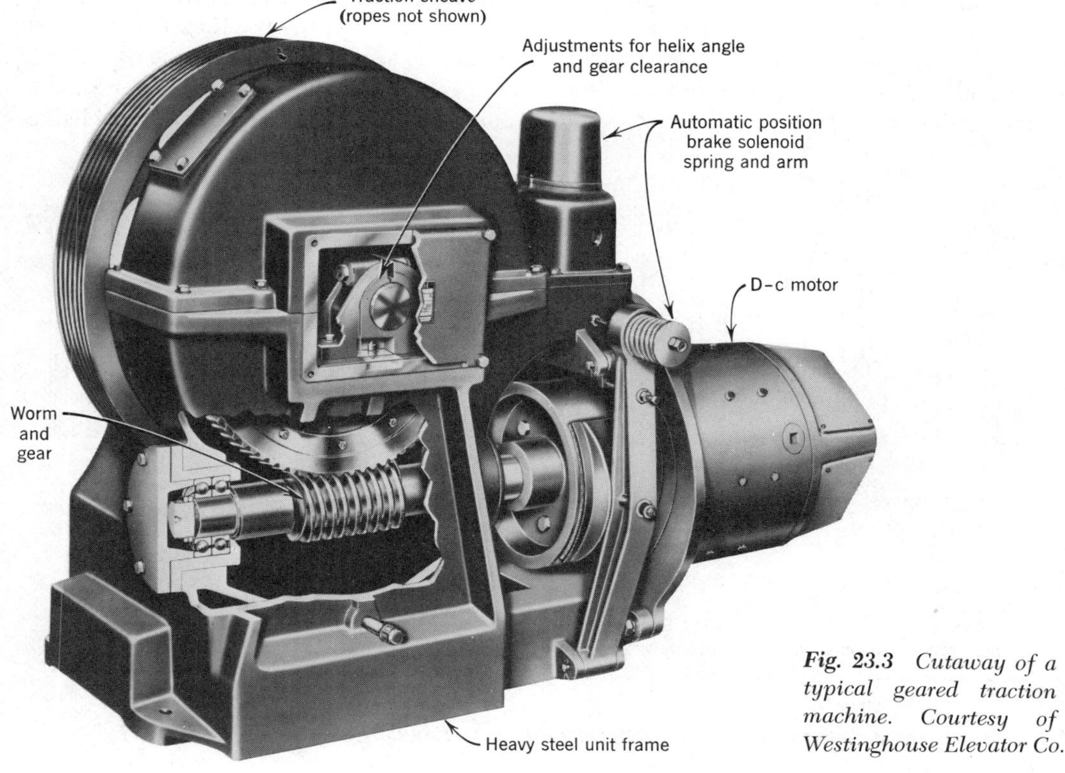

Fig. 23.3 *Cutaway of a typical geared traction machine. Courtesy of Westinghouse Elevator Co.*

Table 23.1 ***Comparative Table of Geared and Gearless Elevators***

	Rise (Feet)	*Speed (fpm)*	*Control*	*Life*	*Maintenance*	*Initial Cost*	*Smoothness*
Geared	50–150 50–175	50–200 50–350[a]	Rheostatic, variable voltage	30–40 years for gear and worm	Medium	Low medium	Low medium
Gearless	Above 100	400 and up	Variable voltage	Indefinite	Low	High	High

[a]The current trend is to raise the upper speed limit to 400 fpm.

150 to 350 fpm, unit multivoltage control (umv) is preferable, utilizing a d-c motor.

The geared traction machine is used for some passenger and most freight elevators, with motor horsepower ratings ranging from 3 to 100 hp (see Table 23.1).

23.7 Arrangement of Elevator Machines, Sheaves, and Ropes

The simplest method of arranging vertical travel of a car is to pass a rope over a pulley and to counterbalance the weight of the car by a counterweight. Then, by rotating the pulley, the car would move up or down and require very little energy to move it. This is essentially the scheme that is used on a majority of high-speed passenger elevators, as illustrated in Figure 23.4*a*.

The pulley referred to above is made in the form of a cylindrical sheave containing grooves for the several ropes that support the weight of the car.

When the supporting ropes merely pass over the sheave (in the grooves) and connect directly to the counterweights, the lifting power is exerted by the sheave through the traction of the ropes in the grooves. This system is referred to as the single-wrap traction elevator machine. The function of sheave *S* is merely that of a guide pulley; usually it is called the deflector sheave. Each of the four or more supporting ropes lies in a groove cut parallel to all other grooves on the sheave.

In Figure 23.4*b* the ropes from the car are first wrapped over the traction sheave *T*, then around the secondary or idler sheave *S*, once more around sheave *T*, and back over *S* to the counterweights. This arrangement is charac-

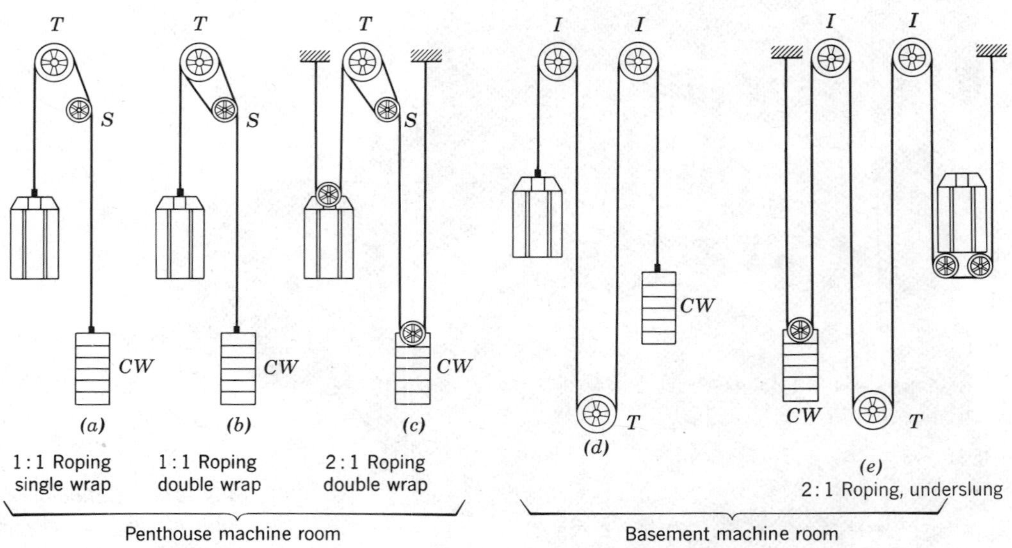

Fig. 23.4 *Elevator roping and sheave arrangements (see Section 23.7).*

teristic of the one-to-one, double-wrap traction machine. It provides greater traction than the single-wrap machine and is used in many automatic high-speed installations.

A 1:1 roping arrangement (Fig. 23.4*a*, *b*, and *d*) gives no mechanical advantage; that is, the drive must supply sufficient power to move the unbalanced load. The 2:1 roping (Fig. 23.4*c*) has a mechanical advantage of 2, which results in a high-speed, low-power and, therefore, low-cost traction machine. This arrangement is used for a wide variety of installations varying from medium-speed (500 to 700 fpm) gearless passenger elevators to low-speed, heavy-duty freight units.

In types *a*, *b*, and *c*, the elevator machines are located at the top of the hoistway. When the elevator machines are placed in the basement, a very different arrangement of cables and sheaves must be utilized to secure the same results. Figure 23.4*d* and *e* show such an arrangement. Much more rope is required when elevator machines are located in the basement, and consequently the problems of rope maintenance are increased. These systems, however, obviate the necessity for a tall penthouse; and where this is desirable for architectural or other reasons, a basement machine is used (see Fig. 23.34). This arrangement uses geared traction equipment, with speeds up to 400 fpm. The roping of Figure 23.4*e* is unusual and finds application where lifting the car from the top is not desirable. This system may be used in residential or commercial application with limited rise (100 ft) and speed (200 fpm). All the illustrated ropings are applicable to the full range of car capacities up to 4000 lbs.

23.8 Hoisting Ropes and Counterweights

The ropes that are connected to the crosshead (top beam of the elevator) and which carry the weight of the car and its live load are made of groups of traction steel wires especially designed for this application.

Although multiple ropes are used primarily to increase traction area, they also increase the elevator safety factor since each rope is normally capable of supporting the entire load. The minimum factor of safety varies from 7.6 to 12 for passenger elevators and 6.6 to 11 for freight elevators.

In order to compensate for rope weight which in high-rise elevators becomes an important factor, cables are attached to the bottom of the car and the counterweight, thus equalizing loads regardless of the cab position. These cables can be seen in Figure 23.1.

Counterweights comprise a steel frame in which weights are placed, totaling car dead weight plus 40% of live load. Use of the counterweight provides lowered power costs and adequate traction for empty or loaded cars.

23.9 Safety Devices

The main brake of an elevator is mounted directly on the shaft of the elevator machine (see Fig. 23.2). When d-c machines are used, the elevator is first slowed down by dynamic braking action of the motor and the brake then operates to clamp the brake drum, thus holding the car still at the floor.

A safety is designed to stop an elevator car automatically before the car's speed becomes excessive. The action of one such device is controlled by a centrifugal governor, which is independent of the other elevator machinery. At normal speeds the safety system has no effect on the operation of the elevator. On overspeed the governor will cut off the power to the d-c motor and set the brake. This usually stops the car, but, should the speed still increase, the governor actuates the two safety rail clamps, which are mounted at the bottom of the car, one on each side. These devices clamp the guide rails by wedging action, bringing the car to a smooth stop (see Fig. 23.5*a*).

Oil or spring buffers are usually placed in the elevator pit. Their purpose is not to stop a falling car but to bring it to a partially cushioned stop if it should overtravel the lower terminal (see Fig. 23.5*b* and *c*).

Electrical final-limit switches are located a few feet below and above the safe travel limits of the elevator car. If the car overtravels

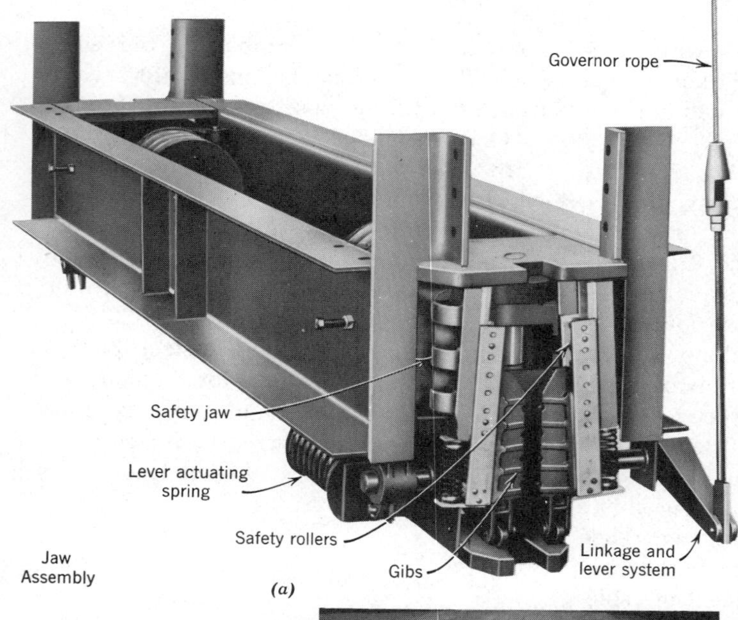

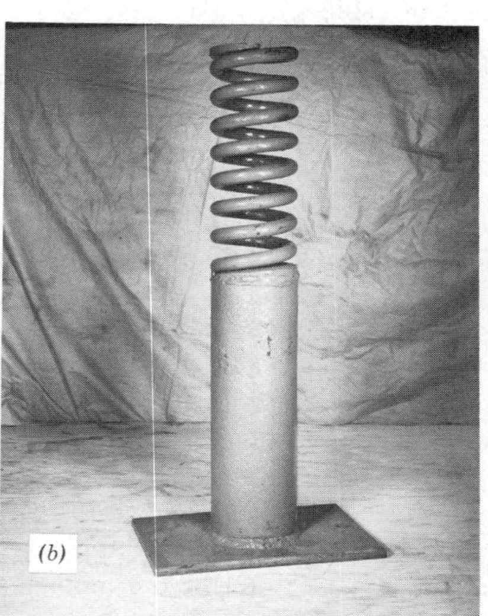

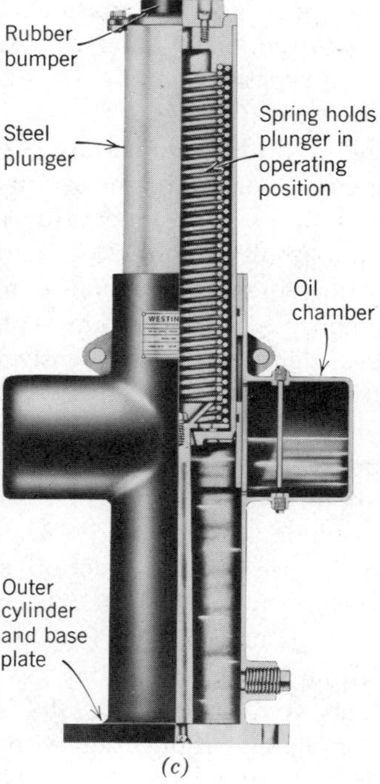

Fig. 23.5 Elevator safety devices. The governor trips, clamping the governor rope (a) *and releasing the safety jaws that exert a constant retarding force on the car rails, thus bringing the car to a gradual and safe stop. Buffers at the bottom of the car shaft are either of the spring type* (b) *or the oil type* (c). *Courtesy of Westinghouse Elevator Co.*

(down or up), these switches deenergize the traction motor and set the main brake. Safety arrangements under emergency condition of fire or power failure are discussed in Section 24.4.

23.10 Elevator Doors

The choice of car and hoistway door affects the speed and quality of elevator service considerably. Doors for the most modern passenger elevators are power operated and synchronized with the leveling controls so that the doors are fully opened by the time a cab comes to a complete stop at the landing. The closing time, however, varies with the type of door and size of opening. The kinetic energy of an automatic door, for safety reasons, is limited by the ANSI Code to 7 ft-lb. To provide fastest closing within this energy limitation a center-opening door is used. Also, in order to reduce passenger transfer time and avoid discomfort, a 3 ft 6 in. (106.7 cm) clear opening is used in most commercial installations, which permits simultaneous loading and

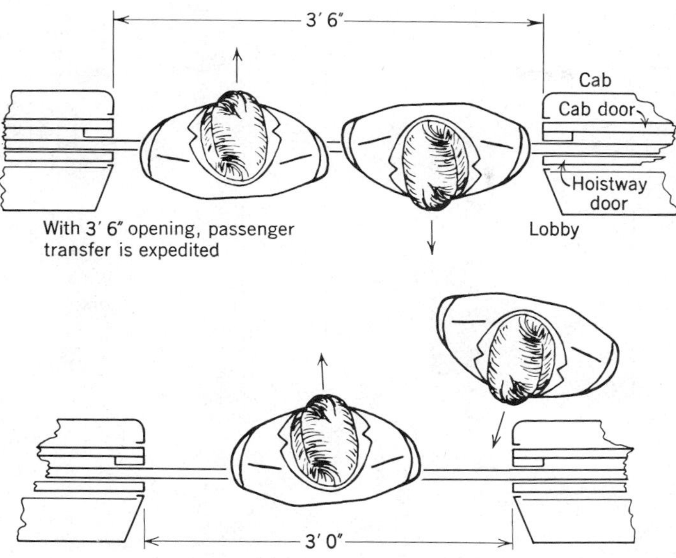

Fig. 23.6 *Transfer of passengers with different width door openings. With openings smaller than 3 ft. 6 in. simultaneous loading and unloading is difficult and transfer time is lengthened.*

unloading without passenger contact (see Fig. 23.6). When a narrower opening is used, loading will be delayed until unloading is complete; therefore, speed and quality of service will be markedly reduced. Such small doors are applied only in residential or small, light-traffic buildings. The available types are shown in Figure 23.7.

A two-speed door design is used where space conditions dictate or where a wide opening is required. The two-speed nomenclature results from the fact that the two halves of the door must travel at different speeds in order to complete their travel simultaneously (see Fig. 23.7*c*).

Detection of passengers on the car threshold can be accomplished electrically by beams across the doorway. Interruption of these beams will prevent a door from closing or will cause a door to reopen which has begun to close. All automatic elevators, whether equipped with detection beams or not, are required by ANSI to have a safety edge device on the car doors that will cause the car and hoistway doors, which operate in synchronism, to reopen when the safety edge meets any obstruction (see Fig. 23.8).

23.11 Cabs and Signals

Possibly the only area in which the architect has a free hand in selection of equipment is in the decor of the cabs and the styling of hallway and cab signals. A normal elevator specification is a functional one that describes the intended operation of the equipment, and normally includes an amount to cover optional decor of the cabs. The type and functioning of signal equipment is also specified, but finish and styling are optional. Cab interiors may be finished in wood paneling, plastic (Micarta or Formica), stainless steel, or almost any material desired. Floors may be tile, wood, or carpeting as selected. Illumination may be from ceiling fixtures, coves, or completely illuminated luminous ceiling, of standard or special design (see Figs. 23.9 and 23.10). For each bank of elevators, it is wise to furnish at least one set of wall mats, to protect wall finishes when cars are being used to move tenant furniture. This is especially important where no separate service car has been provided.

The purpose of the hallway lantern is to signal the approach of a car. This is normally

Fig. 23.7 Typical hoistway doors and applications. (a) *Single slide door, 24 to 36 in. wide, for small commercial building or residential use.* (b) *Standard commercial door, 42 in. center opening, for office building use.* (c) *Two-speed 42 in., general commercial use door.* (d) *48 to 60 in. center opening, for hospital or service car.* (e) *Two-speed center opening, 60 in. department store door, freight and passenger, nonautomatic service.*

accomplished by a two-color lantern to which may be added an audible signal to draw the attention of waiting persons. If desired, the hallway lantern can be equipped with indication as well, which will visually show the exact position of the car in the hoistway, and the direction of travel.

The hallway car-call device, normally a push button, serves to signal the car and, when furnished with illumination, also can indicate the direction of car motion if the lantern device is not so equipped.

Although many designs are commercially standard, architects may at their discretion design the lanterns, indicators, and push buttons to complement the building architecture. Several of the common types are illustrated in Figure 23.11.

ELEVATOR CONTROL

23.12 Elevator Control Systems

The equipment, arrangement, and interconnections that determine the movement and performance of a single car is designated by us as elevator control. Subsumed under this heading is the equipment that controls travel, door operation, leveling, call buttons, and floor signals (lanterns). This control is separate and distinct from the control that governs the response of a single car to calls and the response and interaction of a group of cars—the latter is designated as operating or supervisory systems. To an extent, and particularly with a single car, the two designations overlap.

Elevator car acceleration and deceleration are accomplished by controlling the speed of

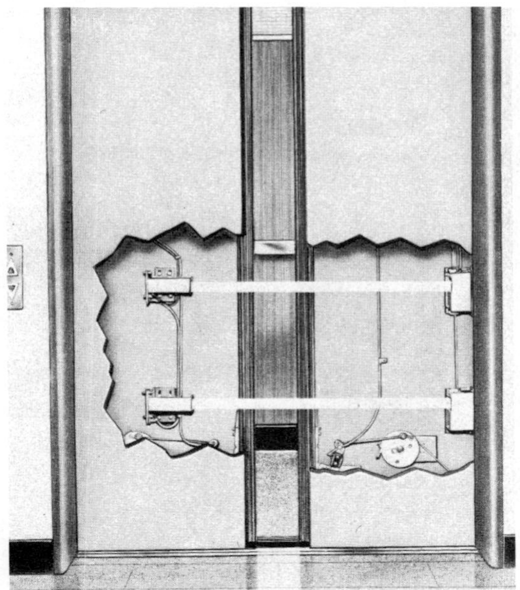

Fig. 23.8 Electric eye door protection is used to avoid the annoyance of being bumped by the door's safety edge.

the motor that drives the elevator traction machine. This speed control is performed in a number of ways, which will be described below. The extent to which this motor control can be performed smoothly, rapidly, and accurately is the principal figure of merit of the control system. The other functions of the control system mentioned above are, in general, independent of the motor control system and are normally furnished to match the quality of the drive system. Thus, a highly sophisticated leveling system would not usually be found with a rheostatic motor control system, and so forth.

23.13 Rheostatic Elevator Control

In installations where economic considerations, low rise and low speeds (25 to 150 fpm), or the type of traffic do not justify the use of the more costly variable voltage control (umv), the traction machine may be driven by an a-c motor whose speed is controlled by rheostat (variable resistance; see Fig. 23.12*a*).

(a)

(b)

***Fig. 23.9** Typical cab interior.* (a) *Residential with single slide door. Courtesy of Burlington Elevators, Division of Dover Elevator Co.* (b) *Commercial, with standard 42 in. center opening door. Courtesy of Westinghouse Elevator Co.*

Fig. 23.10 Interior finishes for elevator cars are as much the province of the architect as other lobby fixtures. Note the coordination of wall finish, car interior, and even lantern direction indicators. Parravano Office Building, Los Angeles. Courtesy of Montgomery Elevator Co.

Such elevators are usually operated by an operator's hand wheel or lever in the car rather than being automatic. The ride is not smooth. Use of a two-speed motor can improve this deficiency somewhat, although it remains much inferior to variable voltage installations. Automatic leveling is available on some a-c, two-speed, rheostatically controlled elevator motors, operating at car speeds up to 100 fpm.

23.14 Variable Voltage Elevator Control

This system of elevator control, also known as umv, requires that a variable voltage be impressed on a d-c traction machine motor, thus varying the speed of the drive. A d-c traction machine is employed because of the extremely fine speed control obtainable with it, yielding very smooth elevator operation. Such speed control is not possible with an a-c traction motor. The variable voltage required to drive the d-c elevator motor is obtained either from a motor-generator set (m-g set), comprising an a-c motor and a d-c generator (see Fig. 23.12*b*), or from a completely solid-state control system (see Fig. 23.12*c*). The latter is a recent development that has distinct advantages over the traditional m-g set, as will be discussed below. Each elevator requires a separate m-g set or solid-state control panel. All gearless elevator machines and most geared machines use the umv system of control, because it provides smooth and accurate speed regulation and efficient operation—in addition to lending itself readily to automatic supervisory systems. Automatic leveling is also furnished, which compensates for thermal expansion, elevator loading, and normal rope stretch.

23.15 Elevator Operation

When considering operation, whether the variable voltage is supplied by an m-g set or a solid-state installation is immaterial. Differences in other areas are discussed below. The following is a description of normal elevator operation.

Assuming the system to be energized and at rest, registration of a call from a station in the lobby or an upper floor corridor will activate the system. The particular elevator that will answer the call is selected by the supervisory

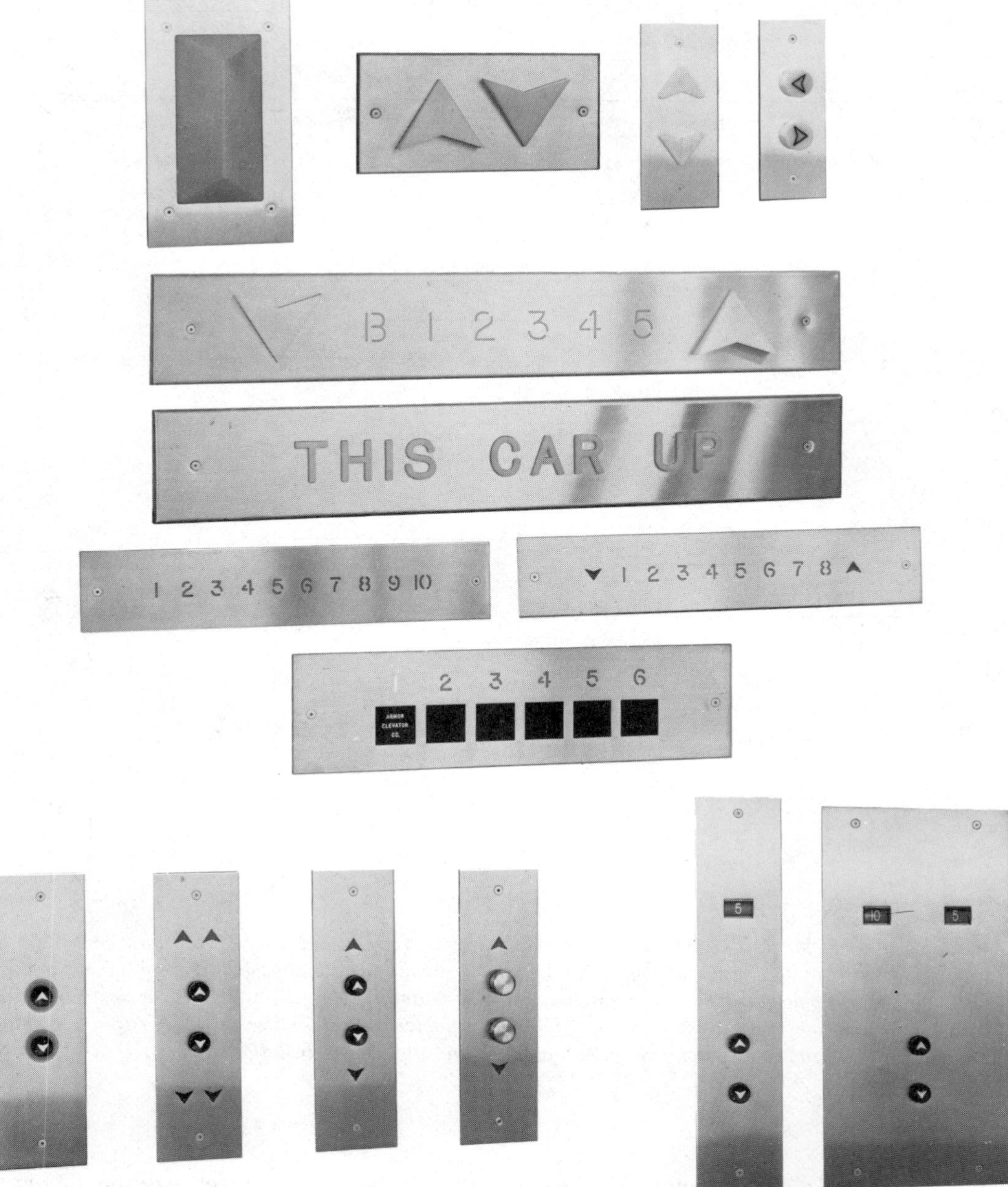

Fig. 23.11 Typical design of lanterns, push buttons, and indicators. Courtesy of Armor Elevator Co.

system. The m-g set related to that elevator car is started and comes up to a speed slowly through a reduced voltage starter. In the case of solid-state control, voltage is available immediately, and the m-g set starting time is eliminated. This is the single difference between m-g set and solid-state control, and it, too, disappears in the more usual case of an m-g set that is already running. When the d-c generator has developed full voltage, the elevator control is cut in and the car is ready for operation. Automatic signals initiate the automatic sequence for accelerating. Various relay devices in the control panel release the brake

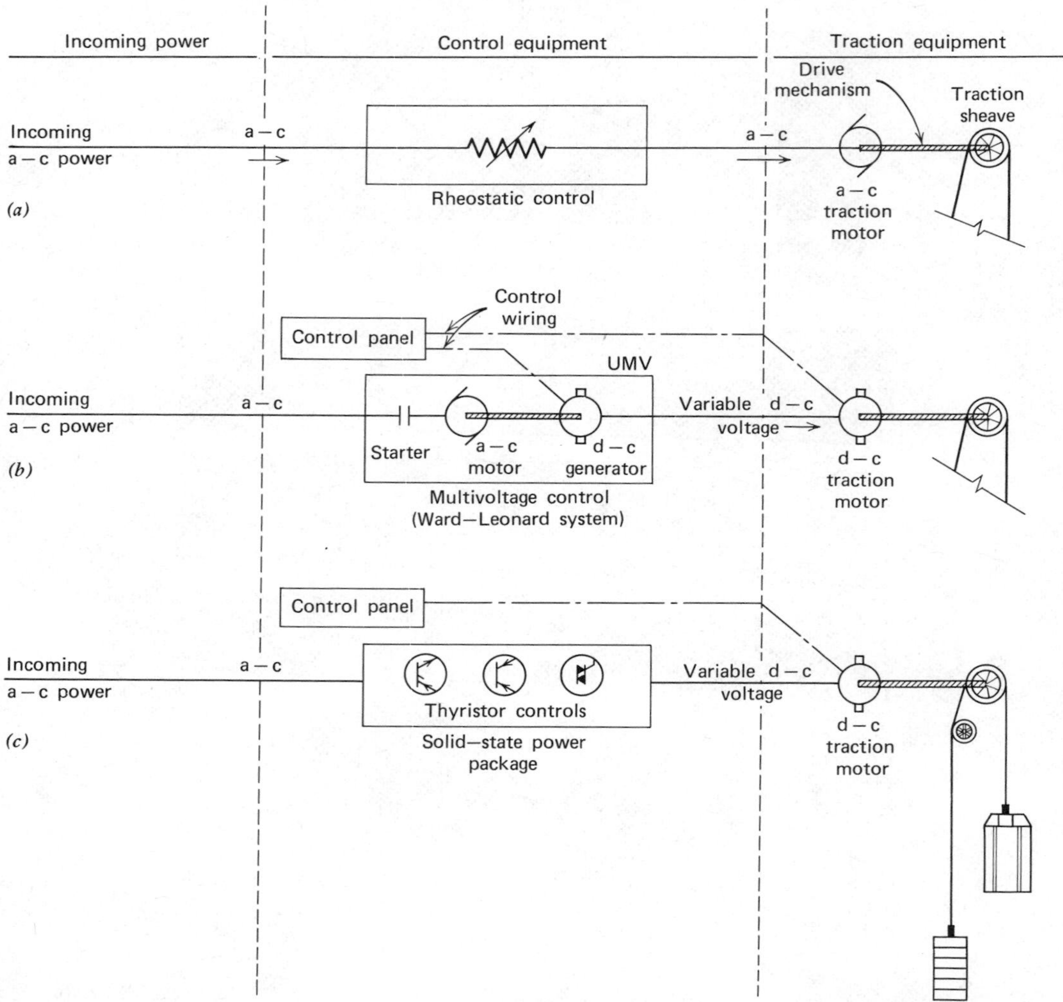

Fig. 23.12 *Arrangement of electric speed-control equipment for elevator drives. Rheostatic control* (a) *is the simplest variety and is used in slow speed, low-rise economy installations.* (b) *Illustrates the standard Ward-Leonard system, which produces finely controlled variable voltage d-c. This is the system in use today on the vast majority of installations. Solid-state thyristor control* (c) *also produces variable voltage d-c and is finding increasing application in medium- and high-speed installations.*

and energize the elevator motor, and the elevator accelerates to its rated speed. Reverse operations are initiated when decelerating and finally stopping (landing) the car. When the car stops, the brake holds the sheave and elevator stationary.

The motion of a single car is determined by the action of three principal items of equipment: the controller, the motor controls, and the system supervisory equipment. The action of this latter equipment will be discussed under the section covering operating systems.

The function of the controller is generally to duplicate the action of the elevators and to anticipate hall calls. This is accomplished in some controllers by the action of a sort of miniature elevator system that contains sets of contacts over which small sliding brushes move in synchronism with the corresponding car movements in the hoistway. The main carriage duplicates car movement; the advance carriage precedes the main, and records hall calls. This information, that is, the car position and the waiting calls, is then fed to the (supervisory or relay) panel, which initiates the proper procedures. In the case of group (system) control, the system panel operates to control the individual car control panel. In the final analysis however, the control panel controls the direction of car travel, starting, running, leveling and stopping, and operation of doors. In the more sophisticated systems solid-state switching devices are substituted for conventional electrical relays, but the action remains the same.

23.16 A Comparison—M-G Sets and Solid-State Control

Solid-state equipment will in time replace the traditional m-g set as the supply of variable voltage, just as solid-state switching is replacing relay switching because of its inherent advantages. The principal drawback at present is the higher initial cost of solid-state equipment. With increasing power costs this factor will lose importance.

(*a*) ***Energy Considerations.*** The principal advantage of the solid-state elevator control lies in the energy savings possible. A considerable portion of the energy consumed by an elevator system is wasted when it is not used to perform the fundamental traction work of the elevator system. Sources of energy waste are:

1. The m-g set idling while the elevator is at standstill—during loading, unloading, and waiting periods. Studies indicate this time to be approximately 50% of the total time the m-g set is operating.
2. The m-g set startup after being automatically shut down when a car is not in use.
3. Forced ventilation equipment in machine rooms, necessitated by the high heat losses of the m-g sets. (See Sections 24.1 and 24.2.)

Since the m-g set is eliminated with solid-state (thyristor) control, all m-g set losses are eliminated and ventilation requirements are sharply reduced. Depending on the particular elevator design, electrical energy costs are reduced by 10 to 25%. The actual cost impact can be seen in the cost study of (*e*) below.

(*b*) ***Physical Considerations.*** A thyristor control panel occupies approximately the same space as a standard m-g set control board. Elimination of the m-g set thus reduces the required machine room size with concomitant construction costs savings. Also, the absence of m-g sets reduces the machine room floor loading, yielding possible savings in construction. Comparative layouts for the same elevator installation are shown in Figure 23.13, graphically illustrating the space economy involved.

(*c*) ***Car Operation.*** No appreciable improvement in car operation is obtained by utilizing solid-state control except in installations without good group supervisory control. In those instances, m-g starting time does indeed affect system performance during light traffic periods when m-g set shutdown is common. With proper group control, a car already in service is utilized to answer a call, without the necessity for starting up an out-of-service car. Furthermore, this situation would normally occur at night or during another very light traffic period at which time an additional lobby waiting period is well tolerated.

(*d*) ***Other Factors.*** Elimination of the m-g set has other desirable effects:

1. Improvement of the facility power factor, since the m-g power factor is low due to starting and no-load operation. This may result in a reduction in the facility's electric energy bill.
2. Reduction in noise and vibration in the machine room that is frequently transmitted to other parts of the building.
3. Reduction in maintenance costs because the periodic maintenance required for

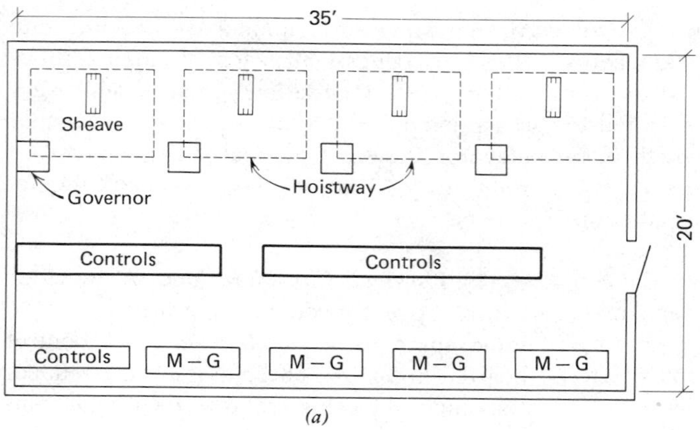

(a)

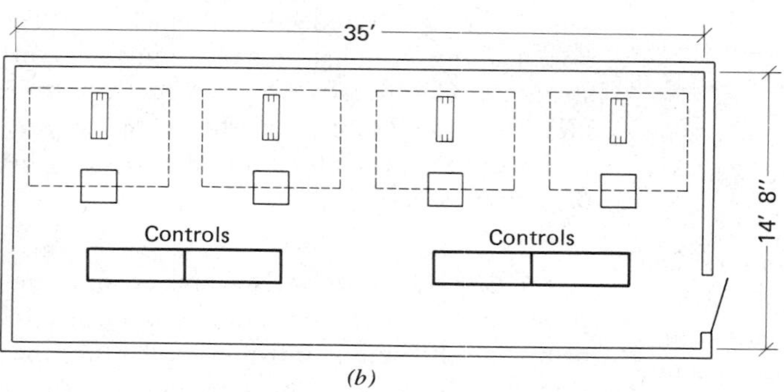

(b)

Fig. 23.13 Space savings possible with solid-state control (b) *over conventional controls* (a) *for a bank of four 3500 lb, 800 fpm, gearless traction elevators.*

continuously rotating equipment is eliminated. Solid-state equipment maintenance is minimal.

4. Possible reduction in the size of the emergency generator required to operate an elevator, with concomitant cost reduction. This reduction is possible since the very high current inrush taken on m-g starting is eliminated. The unit must, however, be large enough to handle the regenerative braking current produced by a full car traveling down.

(*e*) ***Operating Costs.*** A brief life-cycle cost analysis of a five-car elevator bank can serve as an illustrative example of the cost figures involved. Input data to the study are:

Five 3500 lb, 600 fpm, gearless cars, group supervisory control operation: 10 hrs/day, 22 days/month.

Overall system losses at 30%.

25-year equipment life.

Power cost \$.035 per kwh, increasing 10% per year.

Maintenance cost \$250 per car month for m-g set equipment, escalating 6% per year.

(*Note:* In major cities this cost will run nearer to \$400 per month.)

Cost of capital—8%.

Analysis: The two major operating costs that can be substantially reduced by the use of solid-state control are energy costs and maintenance costs. Overall savings possible are the sum of these two factors.

1. *Energy Cost*

From Section 24.2, page 1107, the average energy monthly cost of this *bank* is \$275. Assuming that a solid-state control system can affect a 20% savings in energy cost, yearly savings would be

$$\frac{\$275}{\text{month}} \times 20\% \times \frac{12 \text{ months}}{\text{year}} = \$660$$

The present value of the electric energy savings over 25 years, assuming 10% annual escalation, can be taken from Table 21.5*c*, page 000

$$\textit{Present value} = \$32.01 \times \$660 = \$21{,}125$$

2. *Maintenance Cost*
Assume that only a 15% economy in maintenance costs is possible. We then have

$$\text{Annual savings per car} = \frac{\$250}{\text{month}} \times 15\% \times 12 \text{ months} = \$450$$

$$\text{Annual savings per 5-car bank} = 5 \times \$450 = \$2250$$

Present value (*PV*) of life cycle maintenance cost *savings* (using Table 21.5*c*, page 981)

$$PV = \$19.78 \times \$2250 = \$44{,}505$$

The total present value of combined savings for energy and maintenance is

$$\$21{,}125 + \$44{,}505 = \$65{,}630$$

This can then be compared to the increased first cost of the solid-state equipment, in order to reach a decision.

SYSTEMS OF ELEVATOR OPERATION AND SUPERVISION

23.17 System Control Requirements

An operating system is that control, as stated above, which governs the automatic (operatorless) response of a single car or a group of cars to calls for service. A competent system therefore must take cognizance of all hall calls and car calls, car travel directions, and car positions—in relation to each other and in relation to the call requirements, plus the relationships between up and down traffic, and the trends of traffic. This latter is required in order that the system anticipate demand rather than react to it since only by anticipation can the system operate at maximum efficiency. The operating system considers all of the above data and dispatches cars accordingly. Obviously, since traffic and calls are never static, the controls that will satisfy all of the above demands in a large elevator system is an extremely sophisticated one. Without such control however, large elevator systems will render poor service to the dissatisfaction of owner and rider. On small systems the operating control is much simpler, as will be described below.

23.18 Single Automatic Push-Button Control

This system is the simplest of the passenger-operated automatic control schemes since it handles only one call at a time, providing an uninterrupted trip for each call. A single corridor button at each level can register a call only when the car is not in motion. To indicate availability of the car an "in use" light is placed over the hall call button. When out, calls can be placed. This control scheme is applicable only to a short-rise inactive elevator, that is, one making five or fewer trips per hour. Such elevators are found in small apartment houses, residences, small professional buildings, and industrial buildings using freight elevators.

23.19 Collective Control

This system is no longer used in new installations in the United States, although it is common in other countries. With only a slight increase in traffic beyond very light, single automatic push-button control becomes unsatisfactory, because no call storage provision is made and waiting periods can become extremely long, particularly with slow cars and moderately high rise. A control system therefore evolved that provides a single call button at each landing. The elevator stops at each floor that has registered a call to "collect" the waiting passenger. Hence the "collective" nomenclature. The car cannot differentiate between up and down calls since only a single button is provided. A car direction light is frequently placed adjacent to the call button. This will tell the prospective passengers whether the car which stops in response to their call is going in the right direction. As can readily be seen, this type of operating system is acceptable only for light residential or industrial service.

23.20 Selective Collective Operation

This type of collective operation is "selective" in that it is arranged to collect all waiting "up" calls on the trip up and all hall "down" calls on the trip down. The control system "stores" all calls until they are answered, and automatically reverses the direction of travel at the highest and lowest calls. When all calls have been cleared the car will remain at the floor of its last stop awaiting the next call, and its motor-generator set will stop after several minutes. Pressure on any "up" or "down" button at any landing, or on any floor button in the car, will start the m-g set and set the car into operation.

Collective control is standard in locations where service requirements are moderate such as in apartment houses, small offices, and hospitals. Since these locations often require more than one car, collective control is available for groups of one to three cars. This control scheme automatically assigns each landing call to the car best situated to answer it, prevents more than one car from answering a call, allows one car to be detached for freight duty while others serve passengers, and automatically parks all but one car at the ground floor, that car acting as a free car until service calls require the use of the parked cars.

Although collective control is standard for most residential and other light-to-moderate service requirement buildings, its inherent characteristics frequently result in long waiting periods for elevators. These characteristics are:

1. Highest call reversal.
2. Shutdown of the m-g set after calls have been answered.
3. Strong tendency toward bunching of cars.

This last characteristic is particularly annoying in groups of three cars. Frequently a passenger will arrive at a landing to find that all three cars have just passed going in the same direction. The result is that service is only slightly better than that which would be rendered by a single car, except that load (handling) capacity is greater. Operation of more than three cars with this system is not feasible.

Although collective control furnishes adequate and dependable elevator service where requirements are light to moderate, it is still basically a signal-controlled system that weighs all calls equally and takes no cognizance of traffic patterns. To overcome this shortcoming elevator engineers have developed numerous group supervisory control systems that can be grouped into three categories of increasing complexity and sophistication: multiple zoning, programmed traffic patterns, and computer-operated continuous demand analysis. Because of diverse development by different companies, these categories overlap somewhat but generally adhere to the system descriptions below.

23.21 Multiple Zoning Systems

This system has the basic selective, collective characteristic within predetermined zones in the building plus, in most versions, a simplified two-program traffic pattern recognition system. This two-program system, as with all preprogrammed traffic pattern arrangements, operates fundamentally as a time dispatching system. In the multiple zoning control system the building is divided into as many zones as there are elevators, with the lobby and floors near it acting as one zone. One car is assigned to each zone within which it operates, reversing travel at the upper and lower limits of its zone. The reasoning behind this arrangement is the desire to avoid bunching and excessive car travel to answer a hall call. When a car is taken out of its assigned zone by a car call, it loses its zone assignment after a predetermined time and another car is assigned to the vacated zone. The free car will then seek out another vacated zone and will acquire a new zone assignment. When the system recognizes a heavy lobby demand, cars in upper zones will be reassigned to lobby duty. In the lobby they will be released by time dispatch or when the elevator is fully loaded, whichever occurs sooner. This situation, which corresponds to an "up-peak" program of a traffic pattern system, will continue until lobby demand drops off, at which point the cars will be reassigned to zones. The reverse situation, "down-peak," is also recognized by the controls. In down-peak, the lobby zone is eliminated and its car is released to seek an unoccupied upper zone. This it will readily find because upper zone cars are leaving their zone to travel down to the lobby. As with up-peak, this arrangement lasts only as long as traffic demands, after which the system returns to normal zone operation. During up-peak and down-peak periods the maximum waiting periods for calls in directions opposite to the principal traffic flow will be automatically lengthened.

23.22 Programmed Traffic Pattern Control

The next step in automatic group supervisory control operates primarily with fixed dispatch times based on recognized programs but modified by continuous attention to system demand. Requirements can be best understood by referring to Figure 23.14. Inspection of this figure shows that there are two peak periods; one up and one down, corresponding to the incoming and outgoing crowds. During the day, traffic is balanced except for the noon lunch-hour peaks. Before and after working hours the traffic is light in both directions. This particular traffic pattern chart lends itself readily therefore to a six-program system: up-peak, balanced, heavy-down, heavy-up, down-peak, and off-hours, corresponding to the incoming rush, morning and afternoon off hours, lunch-out, lunch-return, homeward-rush, and nonworking hours, respectively.

Where a system is designed for four-program control, the noon peaks corresponding to *heavy-down* and *heavy-up* are omitted. A five-program system includes the basic four above, plus a *concentrated-down* program. This down program is provided since at quitting hour the elevator system is literally overwhelmed by a short time mass exodus. This program, also called *zone-down,* automatically zones the building in order to increase the elevator system handling capacity. Such zon-

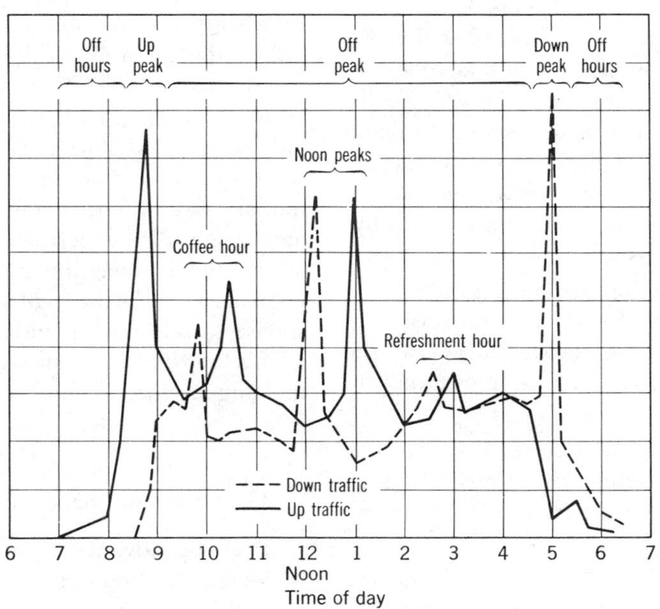

Fig. 23.14 Typical office building traffic chart-diversified tenancy.

Table 23.2 ***Elevator Supervisory Program Systems***

Description	*4 Program*	*5 Program*	*6 Program*	*7 Program*
Morning incoming	Up-peak	→		
Midmorning and afternoon two-way traffic	Balanced	→		
Out to lunch			Heavy-down	→
Return from lunch			Heavy-up	→
Outgoing	Down-peak	→		
Down rush		Zone-down		Zone-down
Off hours (nights, weekends, etc.)	Intermittent	→		

ing is also occasionally provided in the up direction, but less frequently, since up traffic is usually spaced over a greater time interval. A summary of the available programs is given in Table 23.2.

The action of the control system when placed in a particular mode (program) is essentially one of time control; that is, it regulates car dispatch from upper and lower terminals in accordance with predetermined schedules. Characteristics of the different programs vary between manufacturers and are adjustable in a system, but the following are basic.

Up-peak. Cars leave lobby when loaded or on schedule—whichever occurs first. A car leaving ahead of schedule advances the dispatch time of following cars so that time lags do not occur. Cars respond to up hall calls, reverse at highest call, and respond to down calls only if lobby load permits. If bypassed, down hall calls will be answered before waiting time becomes excessive, but will be bypassed by most cars.

Balanced. Elevators are spaced throughout the building so that calls can be served in shortest time, with at least one car at the street level at all times. Cars move only in response to calls.

Heavy-up and Heavy-down. Cars are operated on timed trips with timed dispatch from upper and lower terminals. Cars collect calls in both directions.

Down-peak and Zone-down. Cars are spaced and arranged to allow downward express when loaded. Certain cars may operate with no up-stop, and means are provided to avoid favoring upper floors. When demand is greatest, zones will automatically be established. Landing calls in the "up" direction are limited and timed.

Intermittent. The system parks cars at the lobby or spaced throughout the building, with closed doors and with timed m-g set shutdown. The car nearest a call is then activated on demand in the latter case. In the former, one car remains in service until demand requires the use of the others parked at the lobby.

In older systems, programs were selected by a starter at the lobby control panel who judged the condition of the system by observing the panel.

In a building with six or more cars, the judgment of the starter, however good, is insufficient. To obviate this human intervention, modern systems utilize a computer to analyze and weigh all the pertinent information and automatically select the appropriate program, regardless of the time of day, purely in response to service demands. Thus, the system avoids the rigidity of pure time dispatching by melding it with demand response.

23.23 Computerized System Control

The most advanced type of control system is one that continuously monitors demand and controls car motions in response to demand

HLL	HCS	TIME	AVL
00	00317	02367	007
01	00056	00496	009
02	00072	00533	007
03	00105	00954	009
04	00105	01097	010
05	00178	01486	008
06	00090	00850	009
07	00154	01945	013
08	00065	01047	016
09	00045	00822	018
10	00005	00119	024
11	00004	00086	022
07	00042		

This is a teletyped printout from a system in operation.

- HLL—Hall (floor) number
- HCS—No. of hall calls
- TIME—Total time at floor
- AVL—Average wait time

Fig. 23.15 *Printout produced by computer controlling an elevator bank. With data of this type operation of the control system can be adjusted and checked without laborious manual timing. Courtesy of Cubic Corp./U.S. Elevators.*

only; that is, it analyzes all the possibilities and answers each call in optimum fashion. Such a system is only possible with the aid of a computer, since the number of factors that must be considered, weighed, and computed is enormous. The factors that must be considered in determining which car will answer which call are:

1. Condition of each car including its load, registered calls, priority status, if any, and status of its m-g set, if any.
2. Distance of each car from a registered hall call.
3. Car unloading calls ahead of and behind each hall call.
4. Special conditions including delays, priority calls, and timing of registered lobby calls.

On the basis of all this data the computer calculates not only which car is in the best position to answer a call, but also travel time for the hall call *and* the calls already registered in that car. The fundamental criterion is optimum service, that is, minimum total trip time including waiting. The computer also analyzes the traffic trends every 1 to 3 sec and reevaluates the importance of all factors in order to anticipate traffic trends. Some computers are arranged to print out system performance in response to hall and lobby calls as a means of testing system operation (see Fig. 23.15). A typical group control panel is shown in Figure 23.16. These panels are installed in the elevator machine room and are interconnected with individual elevator control panels; These individual control panels are also normally solid state rather than relay units when this type of group control is used.

Fig. 23.16 *Integrated-circuit electronic control system of an advanced design, for an elevator bank in a 30-story building. This system considers 13 factors in .01 sec to decide how to answer a call, and resurveys the entire system every 2 to 3 sec. Courtesy of Reliance Electric/Haughton Elevators.*

23.24 Lobby Control Panel

A lobby control panel (also called a traffic director's panel) is furnished with each bank of elevators and is usually mounted in a readily

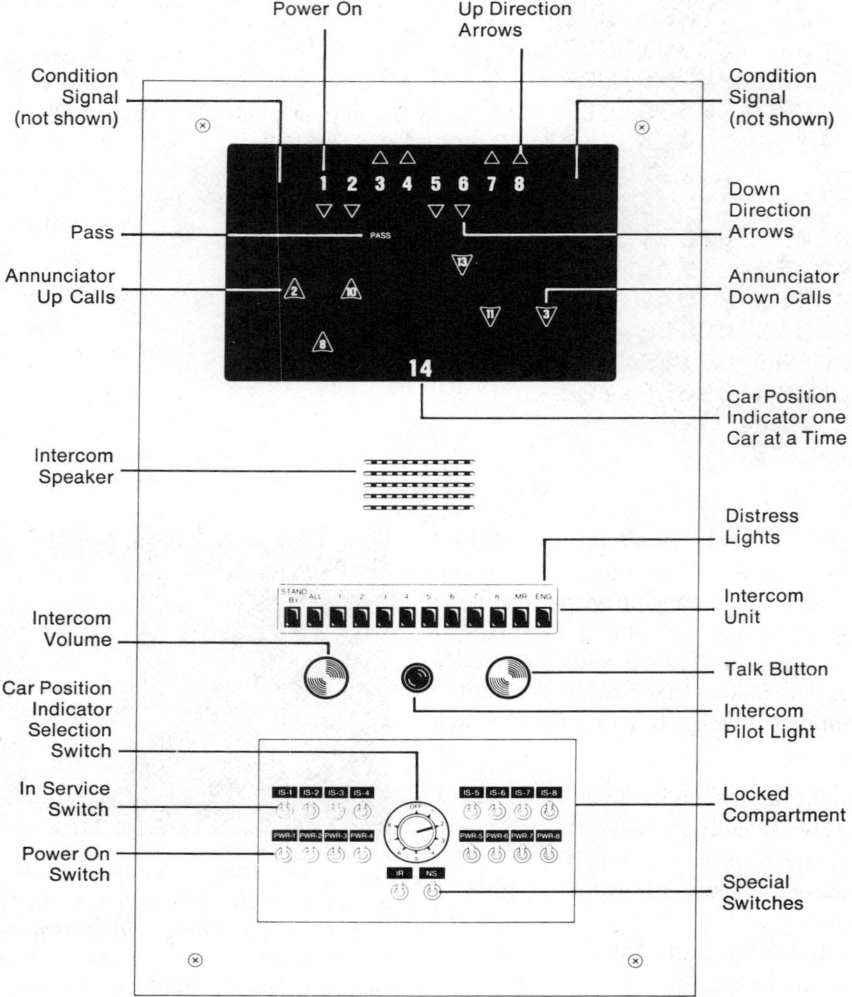

Fig. 23.17 Typical lobby panel for a modern group supervisory system. Westinghouse Selectomatic Mark IV control. Courtesy of Westinghouse Elevator Co.

accessible position in the building lobby. Obviously the type and function of the panel is dependent on the type of control system in use. One such panel, suited to a highly sophisticated demand controlled group supervisory system, is shown in Figure 23.17. Once the system is activated by operating the switches in the locked compartment, no manual intervention is necessary to keep the system in operation. Car location is *not* normally shown on the panel since this leads to the undesirable "tracking" of cars by people waiting in the lobby, with associated group movement, crowding, and impatience. It is much more desirable to encourage complete reliance on the automatic control system.

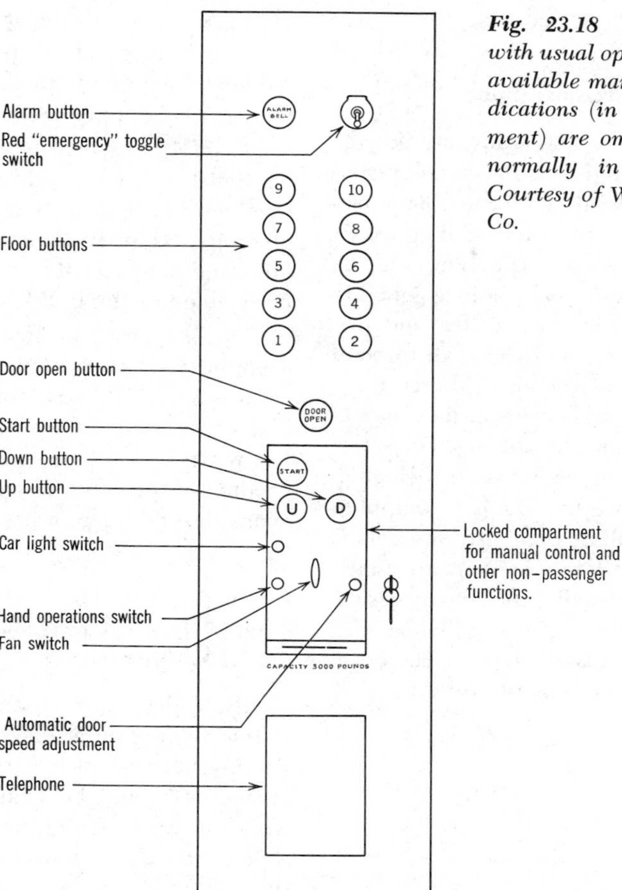

Fig. 23.18 Typical car station with usual options provided. Other available manual controls and indications (in the locked compartment) are omitted, since cars are normally in automatic control. Courtesy of Westinghouse Elevator Co.

The panel does provide for manual intervention to permit special types of operation. Thus a car can be

1. Arranged to travel without operating the usual audible and visual signals (*inconspicuous riser*).
2. Taken out of supervisory control and operated manually (*attendant service*).
3. Selected for night or weekend service while the other cars are shut down.
4. Assigned to a particular floor on a fixed or priority basis call (*convention feature or priority*).

Other operational features that can be provided are concerned with emergency service, including the "Firemen's Return" feature required by ANSI and the controls related to switching of power between cars in the event that operation on an emergency generator is necessary.

23.25 Car Panel

The car panel for a Westinghouse Selectomatic car is illustrated in Figure 23.18. When control is transferred from automatic to manual-attendant, a buzzer sounds every time a corridor call is registered, to alert the operator. The door closing is then manually initiated and the doors operate at a higher speed than when on automatic, thus reducing time delays. Provision for manual operation even on completely automatic equipment is occasionally made in outstanding buildings for prestige and in high-grade residential buildings for security and service.

ELEVATOR SELECTION

23.26 General Considerations

The selection of elevators for any but the simplest buildings requires the simultaneous consideration of several factors: adequate elevator service for the intended building usage, the economics of elevator selection, and the architectural integration of spaces assigned to elevators, including lobbies, shafts, and machine rooms. As must be obvious, all three of these factors are interdependent; therefore, in large complex buildings dozens and even hundreds of combinations are possible. The selection of a single optimum system for such cases is most practical with the aid of a computer, the use of which will be demonstrated below. For most buildings however, certain guidelines can yield entirely satisfactory results with hand computation. These guidelines will be developed and explained below. The criteria of elevator service quality are:

1. Interval and average waiting time.
2. Handling capacity.
3. Travel time.

23.27 Definitions

A clear definition of terms, including variant usages, is imperative for the proper study of a subject. To that end an abbreviated list of important definitions for our study follows.

Interval or Lobby Dispatch Time is the average time between departure of cars from the lobby.

Average Lobby Time or Average Waiting Time is the average time spent by a passenger between arriving in the lobby and leaving the lobby in a car.

Registration Time is the waiting time at an upper floor after registering a call.

Round Trip Time is the average time required for a car to make a round trip, starting from the lower terminal and returning to it. The time includes a statistically determined number of upper floor stops in one direction and an express return trip.

Travel Time or Average Trip Time is the average time spent by passengers from the moment they arrive in the lobby to leaving the car at an upper floor.

Handling Capacity is a figure of merit for an elevator system, indicating the maximum number of passengers that can be handled in a given period, usually 5 minutes; thus "5 minute handling capacity." It is expressed as a percentage of the building population.

Zone is a group of floors in a building that is considered as a group, with respect to elevator service. It may consist of a physical entity—a group of upper floors above and below which are blind shafts—or it may be a product of the elevator group control system, changing with system needs.

23.28 Interval or Lobby Dispatch Time, and Waiting Time

In an ideal installation, at least from the riding public's point of view, a car would be waiting at the lower terminal on the rider's arrival or would be available after a short wait. Since cars leave the lobby separated in time by the *interval* and passengers arrive at the lobby in random fashion, the average waiting time in the lobby should be half the interval. Field tests show, however, that it is actually higher than this, and the figure most used in the industry is 60%; that is,

$$\text{Average waiting time} = 0.6\, I$$

Excellent office building design provides a 15 to 18 sec average wait during up-peaks, with 22 sec considered good and 26 borderline.

Since modern control systems automatically zone the building with the result that some cars do not return to the lobby, the interval as a figure of merit can be somewhat misleading. As a basis of comparison, however, a 30 sec interval is excellent, 35 borderline, and 40 just acceptable in other than center-city office spaces. Table 23.3 lists acceptable intervals for various building types.

With interval in this range, riders will not be conscious of any irksome delay in elevator service. Consciousness of delay is considered a

Table 23.3 ***Suggested Elevator Intervals***

Facility	*Interval (in Seconds)*
Office Buildings	
Center city	25 to 30
Investment	30 to 40
Residential	
Prestige apartments	50 to 70
Middle-income apartments	60 to 80
Low-income apartments	80 to 120
Dormitories	60 to 80
Hotels—1st quality	40 to 60
Hotels—2nd quality	50 to 70

Table 23.4 ***Car Passenger Capacity***

Elevator Capacity (Pounds)	*Maximum Passenger Capacity*	*Normal Passenger[a] Load per Trip*
2000	12	10
2500	17	13
3000	20	16
3500	23	19
4000	28	22

[a]The number of passengers carried on a trip during peak conditions is approximately 80% of the car capacity.

major drawback in rental desirability and should be sedulously avoided for all conditions of traffic except up-peak and zone-down when a certain delay is expected and therefore tolerated, however grudgingly. Even in both these cases, a modern group supervisory system will recognize any *timed-out* call, that is, a call with registration time exceeding 50 sec, as a priority call. Priority calls are answered by the first available car, usually within 15 sec. If a considerable amount of interfloor traffic is expected during peak periods, as may be the case when a large company occupies several upper floors, elevator service must be increased by 20 to 40% over that calculated to maintain proper intervals.

23.29 Handling Capacity

The frequency, or interval, with which a car appears at the lobby is one of the two factors that determine the passenger capacity of an elevator system. The other is obviously the size or capacity of the elevator car. The system's *handling capacity* is completely determined by these two factors—car size and interval. A common difficulty in understanding the operation of elevator systems comes from the failure to appreciate that handling capacity is completely defined by car size and interval, and is independent of the number of cars. This can be best understood by visualizing the system as a single set of doors that opens periodically (interval) to remove a given number of passengers (car capacity) from the waiting group. Whether that set of doors represents a single car which returns quickly or many cars which take turns is immaterial. The only factors that fix handling capacity are passenger load (car capacity) and frequency of loading (interval). See Table 23.4.

Note that cognizance is taken of the fact that during peak traffic periods cars are not loaded to maximum capacity, but only to about 80%—a figure determined by actual count in many existing installations.

As a convenient measure of capacity, the handling capacity of a system for 5 min is taken as a standard. This is due to a 5-minute rush period being used as a measure of a system's ability to handle traffic. This may be expressed thus:

Handling capacity (HC)

$$= \frac{5 \text{ min} \times \dfrac{60 \text{ sec}}{\text{min}} \times \dfrac{\text{passengers}}{\text{car}}}{\text{Interval}}$$

or

$$HC = \frac{300p}{I} \qquad (23.1)$$

where p is individual car loading. When the interval is 30 seconds, the system handling capacity is $10p$, a convenient figure to remember.

In order to relate system capacity HC to the building size, to establish a criterion of merit for service, we establish a minimum percentage of building population that the system will handle in 5 minutes. Thus a good system for a diversified office building will

Table 23.5 ***Minimum Handling Capacities (HC)***

Facility	*Percent of Population to be Carried in 5 Minutes*
Office Buildings	
Center city	13 to 15
Investment	12 to 14
Single purpose	15 to 18
Residential	
Prestige	5 to 7
Other	6 to 8[a]
Dormitories	10 to 11
Hotels—1st quality	12 to 15
Hotels—2nd quality	10 to 12

[a]Due to more urgent traffic demands, particularly at the school and work exodus.

handle no less than 12% of the building population. Similar figures are shown in Table 23.5 for various types of facilities.

In planning a building the population must, of course, be estimated. This is particularly difficult in speculative-type, diversified-use buildings, where the occupancy can only be guessed. However, based on rental cost, area, and building type, a fair estimate can be made. Population estimates for office buildings are based on net area, that is, actual available area for tenancy. Table 23.6 gives suggested density figures while Table 23.7 gives average office building efficiency figures for use in calculating net area.

Table 23.6 ***Population of Typical Buildings for Estimating Elevator and Escalator Requirements***

	Net Area
Office Buildings	Square foot per person
Diversified:	
Large lower floors	110 to 120[a]
Upper floors	120 to 145
Average use	130
Single purpose	120
Hotels	Persons per sleeping room
Normal use	1.3
Conventions	1.9
Hospitals	Visitors per bed[b]
General private	1.5
General public (large wards)	3 to 4
Apartment Houses	Persons per bedroom
High rental	1.5
Moderate rental housing	2.0
Low-cost housing	2.5 to 3.0

[a]Density may vary for different floors. Clerical and stenographic area may have a population density as high as 50 sq ft per person.

[b]If visiting hours are restricted, visitor population will determine elevator requirements. If visiting is not restricted to only a certain few hours, staff requirements may determine elevator design. Where traffic is heavy, a combination of passenger cars and larger "hospital" cars should be used to provide optimum service.

Table 23.7 ***Office Building Efficiency***

Building Height	*Net Rentable Area as Percent of Gross Area*
0 to 10 floors	Approximately 85%
0 to 20 floors	Floors 1 to 10 approximately 80%
	11 to 20 approximately 85%
0 to 30 floors	Floors 1 to 10 approximately 75%
	11 to 20 approximately 75%
	21 to 30 approximately 85%
0 to 40 floors	Floors 1 to 10 approximately 75%
	11 to 20 approximately 80%
	21 to 30 approximately 85%
	31 to 40 approximately 90%

NOTE. Applicable to buildings with 15,000 to 20,000 gross sq ft per floor.

Source. Reprinted from G. R. Strakosch, *Vertical Transportation, Elevators and Escalators.* New York: Wiley, 1967.

23.30 Travel Time or Average Trip Time

The average trip time or time to destination is the sum of the lobby waiting time plus travel time to the median floor stop. Car round trip time is also used as a criterion but is not as relevant or meaningful as trip time. In a commercial atmosphere a trip of less than 1 minute is highly desirable; a 75-second trip acceptable, a 90-second trip annoying, and a 120-second trip (2 minutes!) the limit of toleration. Obviously in the more relaxed atmosphere of a residence where interval alone can account for a minute or more of trip time, these maxima are revised upwards. From Figure 23.19 we see that the 2000 lb and 2500 lb car used in residential buildings can have a 17-story rise, even with 60 second interval without excessive trip time. On the other hand the 3500 lb car, which is almost universal in office buildings (Fig. 23.20*d*), is limited to a maximum of 18 floors local run before exceeding the 90-second limit and about 8 to 10 floors to stay within the 75-second criterion. This, of course, is for the lower zone, which does not include any time for express run. Upper zones must be smaller, as will be demonstrated in examples below.

An important reservation on the above statements must be noted. The curves of Figures 23.19 and 23.20 are based on statistical calculations, empirical data, and field observations, as discussed in the next section. This being so, the average values that these figures give should be considered to be $\pm 10\%$, and borderline cases can be shifted either way. Designs that show high travel time on paper frequently work out well in the field, because lobby loading is often less than 80%, upper floor stops are less than the statistical figure due to a group of people going to the same floor, and staggered working hours relieve traffic peaks. Also sophisticated control systems do much to provide acceptable service in these borderline cases.

23.31 Round Trip Time

The figure for round trip time is composed of the sum of four factors, that is, time to accelerate and decelerate, time to open and close doors at all stops, time to load and unload, and running time (see Figs. 23.21 to 23.23). It is physically the time consumed by a car from door opening at the lower terminal to door opening at the same terminal at the end of a round trip. Since the actual number of stops made by a car is unknown, a statistical probability figure is used, based on the passenger capacity of the car and number of local floors above the lower terminal. In calculating this round trip time (*RT*) it is assumed that a car will depart the lower terminal when loaded. No intentional delay is included at either lower or upper terminal. The *RT* thus calculated is a median figure, with any single actual round trip taking more or less time. In detail, round trip time consists of the time expended in:

a. Loading at the lobby.
b. Door closing at the lobby.
c. Accelerating from the terminal.
d. Decelerating at each stop.
e. Passenger transfer at each stop.
f. Door operation at each stop.
g. Running time at rated speed between stops.
h. Return express run from the last stop.

These figures are obtained as follows:

1. *Field observations*—items a and e are based on a 3 ft, 6 in. door opening. This permits two people to pass each other without bumping. A smaller door opening increases passenger transfer time.
2. *Calculations*—items b and f, and items c, d, g, and h.

Door closing time is based on a 3 ft, 6 in. center-opening door with adjustable speed.

Acceleration and deceleration times are calculated with a maximum of 4 ft/sec/sec as anything beyond that results in physical discomfort to the passengers.

Running time at rated speed takes place after the car has accelerated and before it begins to decelerate. If we consider that it takes between 20 and 30 ft to accelerate to 700 fpm, depending on rate of acceleration, we see that in local runs the car never gets to

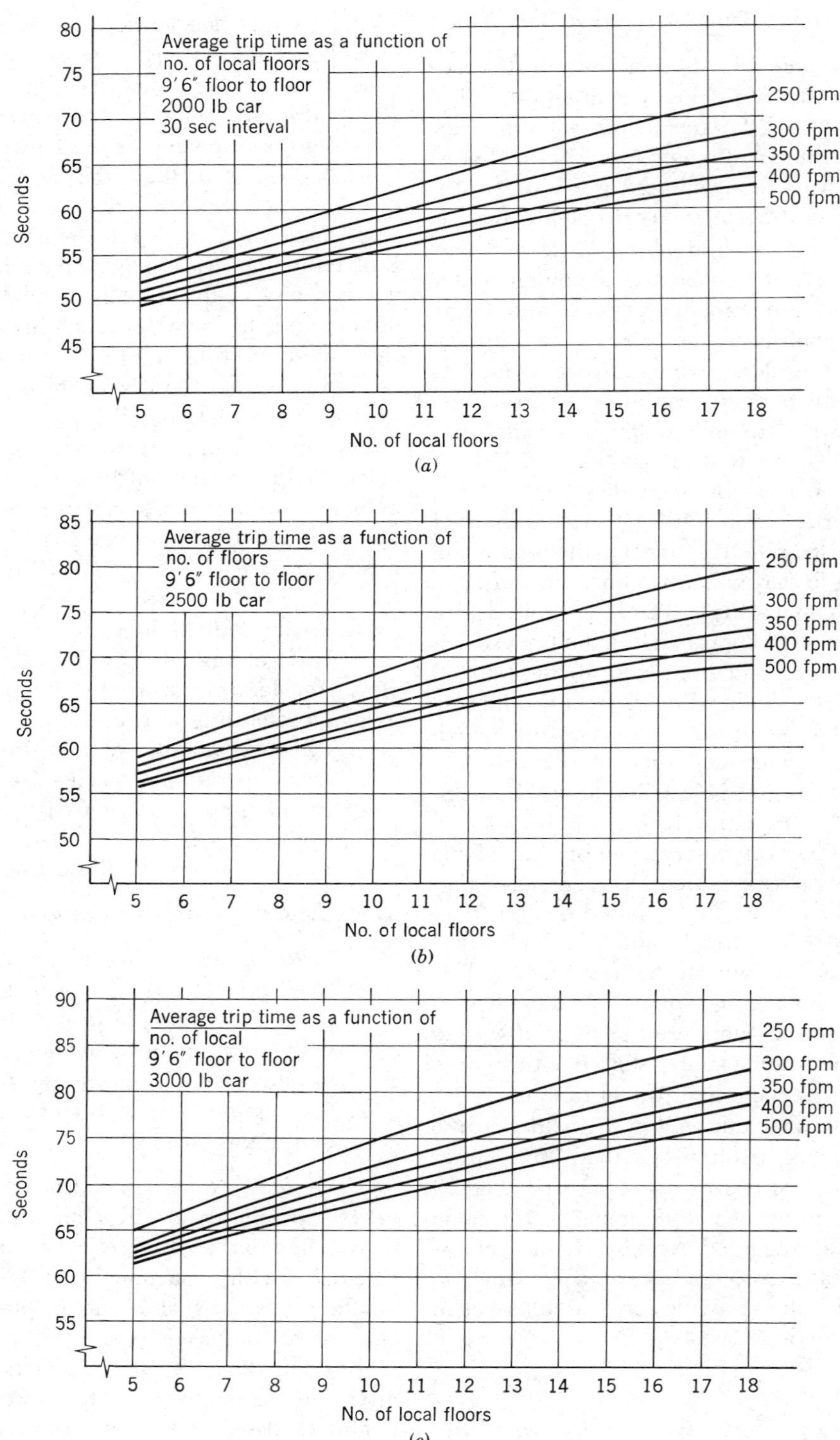

Fig. 23.19 *Plots of average trip time for various car speeds and capacities with 9 ft 6 in. floor height.*

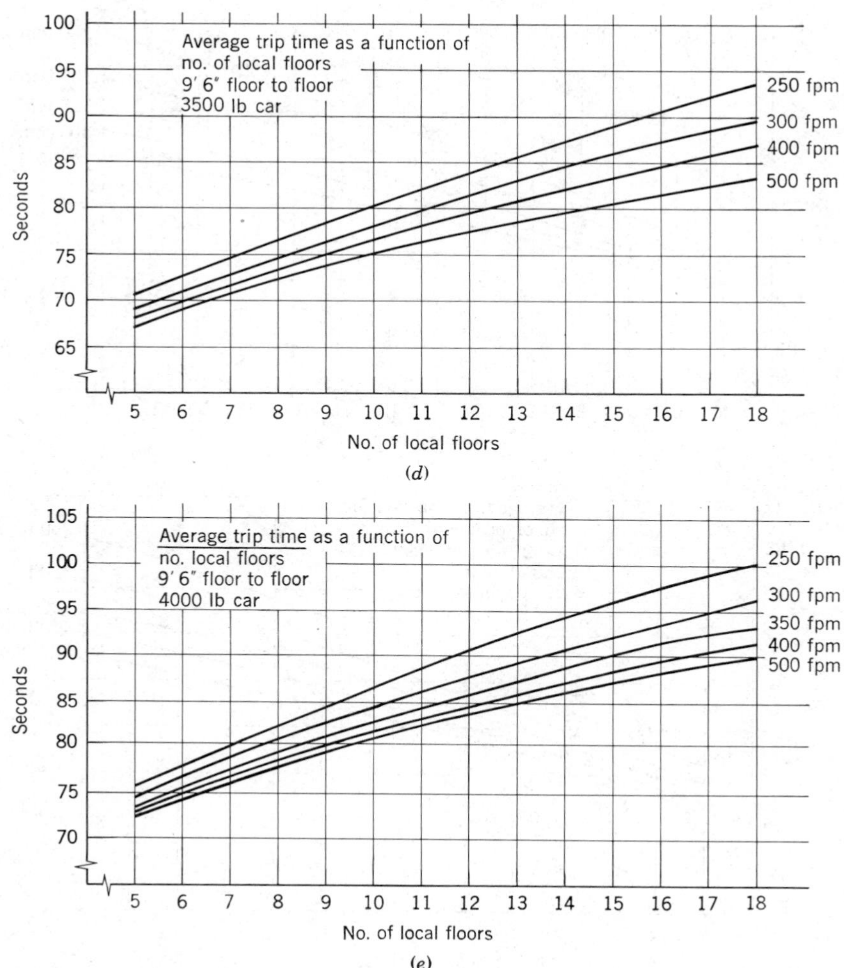

Fig. 23.19 *(continued)*

rated speed. It simply accelerates and decelerates. The higher speed equipment with its larger motor accelerates more quickly and gives some time advantage on the return express run, but has no great time advantage overall. This accounts for the bunching of curves above 500 fpm in Figures 23.21 and 23.22.

In calculating round trip time for cars in upper zones, it is necessary to know the time required to traverse the express floors. This may be obtained from Figure 23.23. The times given therein are for *one-way* express runs. Thus to calculate *RT* for an upper zone car, take the *RT* corresponding to the upper local floors and add *twice* the figure obtained for express run time from Figure 23.23.

23.32 System Relationships

The symbols that will be used in all elevator discussions are:

p	Individual car capacity, equal to 80% of maximum during peak hours.
h	5-minute capacity of a single car.

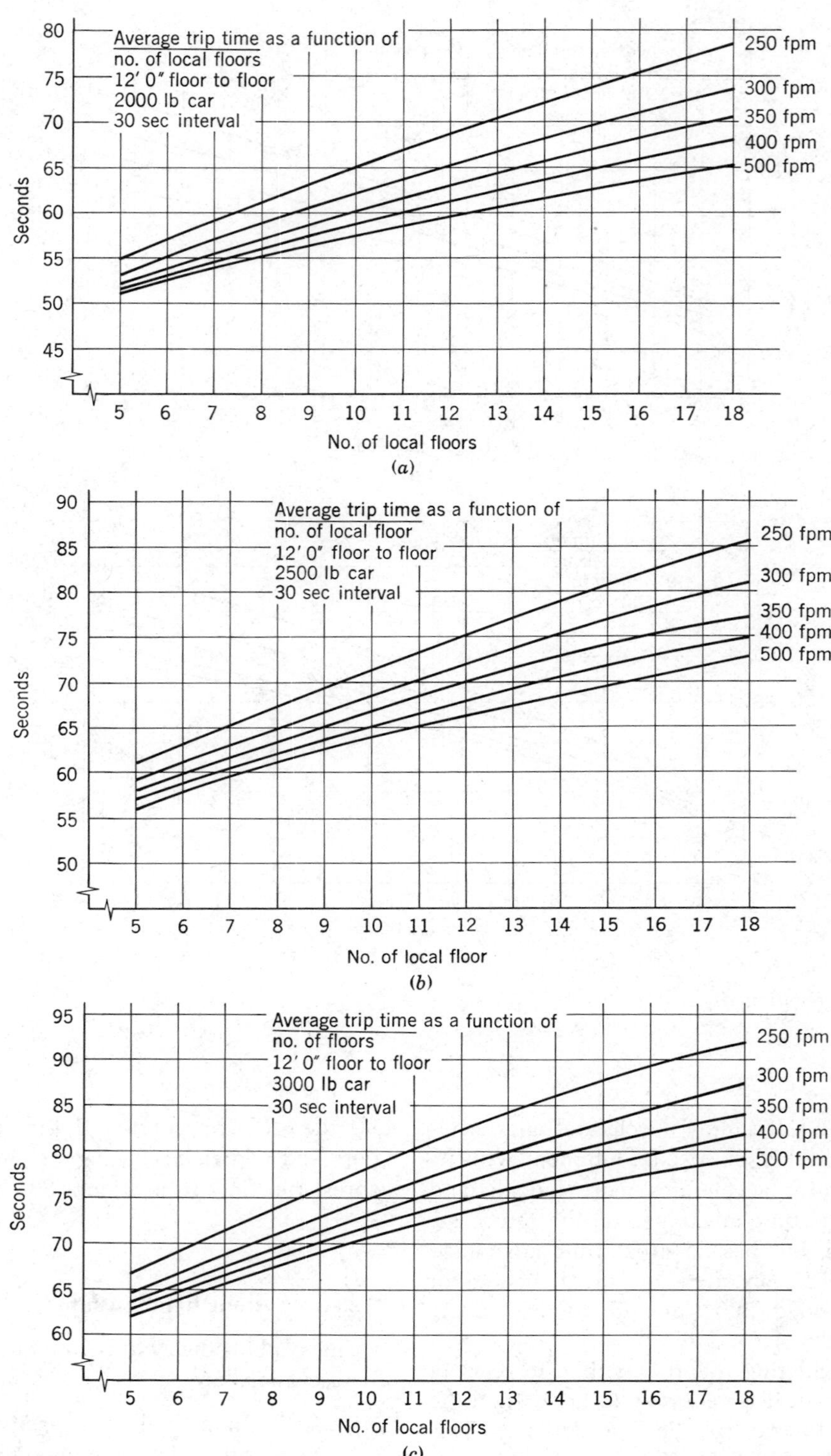

Fig. 23.20 *Plots of average trip time for various car speeds and capacities with 12 ft 0 in. floor height.*

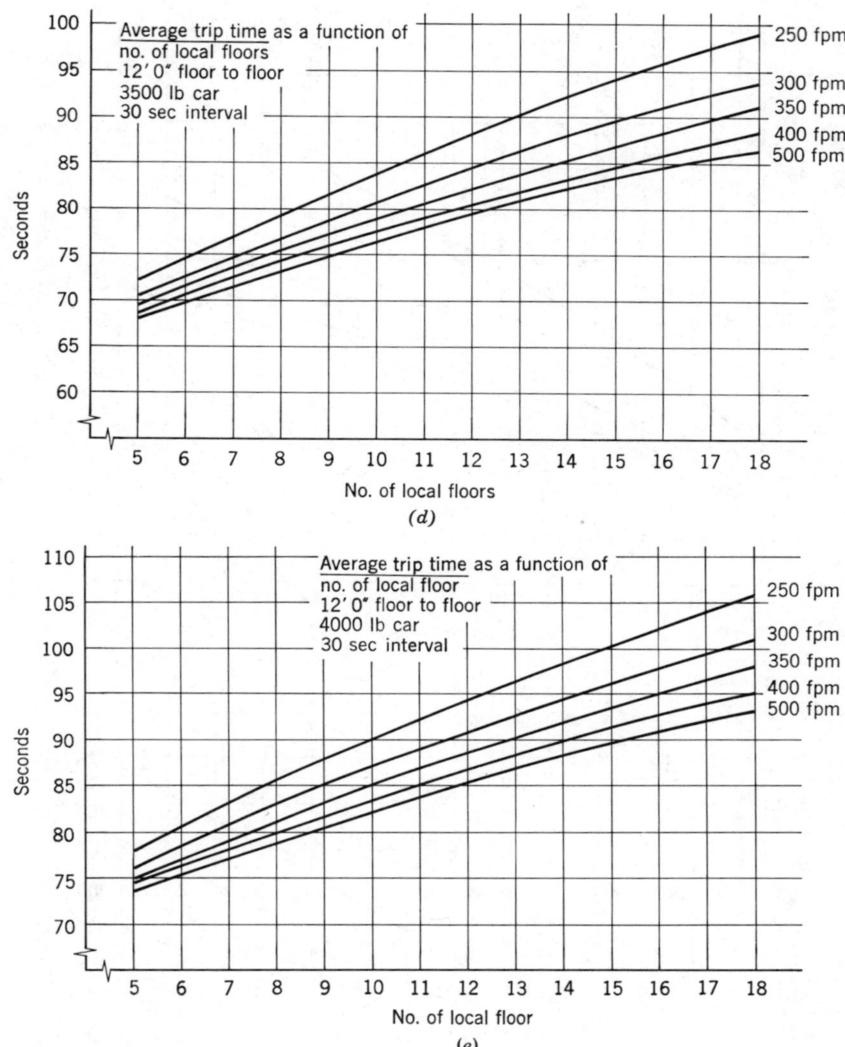

Fig. 23.20 (continued)

N	Number of cars in a system.
HC	System 5-minute handling capacity, expressed in number of persons.
RT	Round trip time, in seconds.
$AVTRP$	Average trip time, in seconds.
I	Interval, in seconds.
D	Population density, in square feet per person.
PHC	Percent of population to be moved in 5 minutes; expressed as a percentage figure.

Having considered the definition of interval, handling capacity, average and round trip time, it is well to demonstrate at this point the interrelationships between these quantities, and the other equations governing the remaining factors that define elevator systems. We have previously established that handling capacity is determined by car capacity p and interval I

$$HC = \frac{300p}{I} \tag{23.1}$$

In a system comprising a single car the

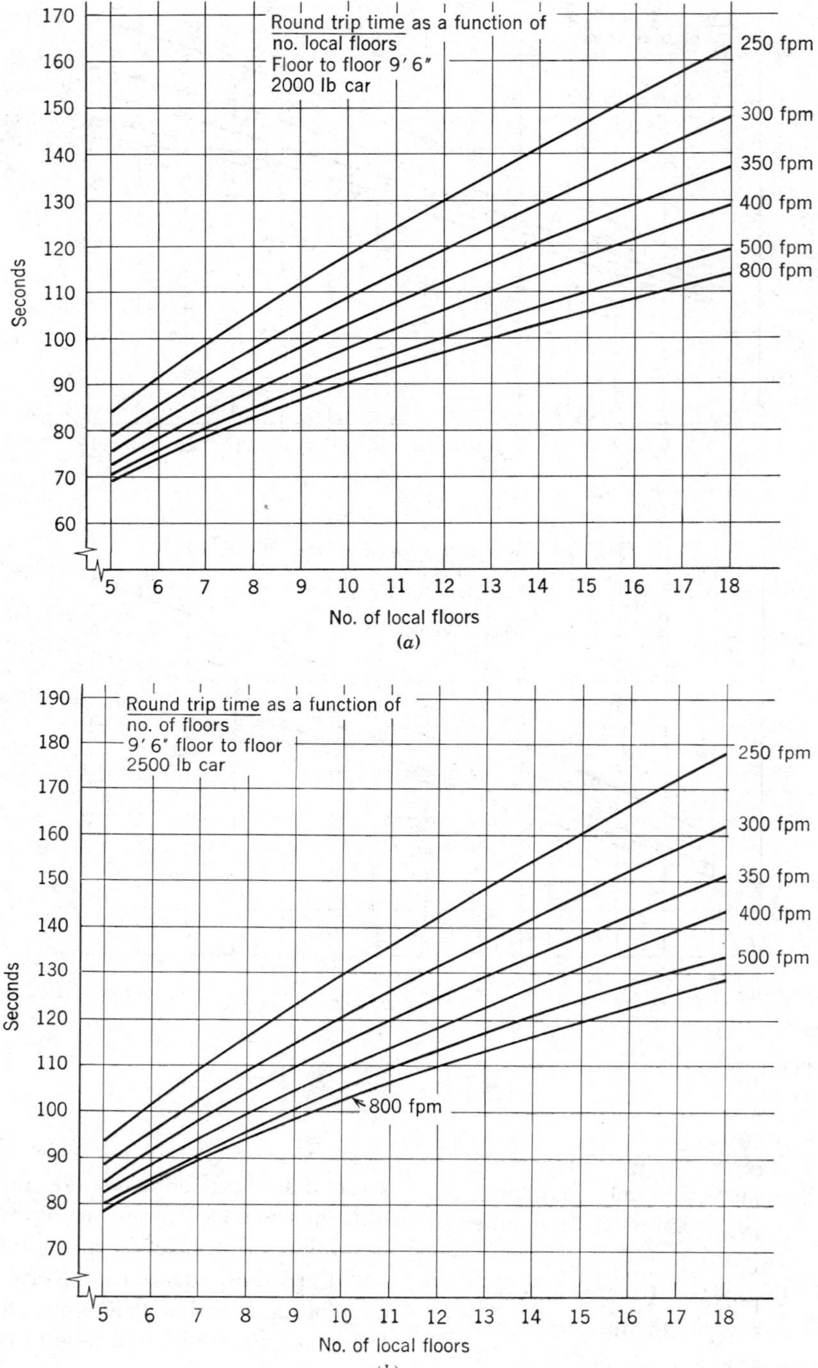

Fig. 23.21 *Plots of round-trip time for various car speeds and capacities with 9 ft 6 in. floor height.*

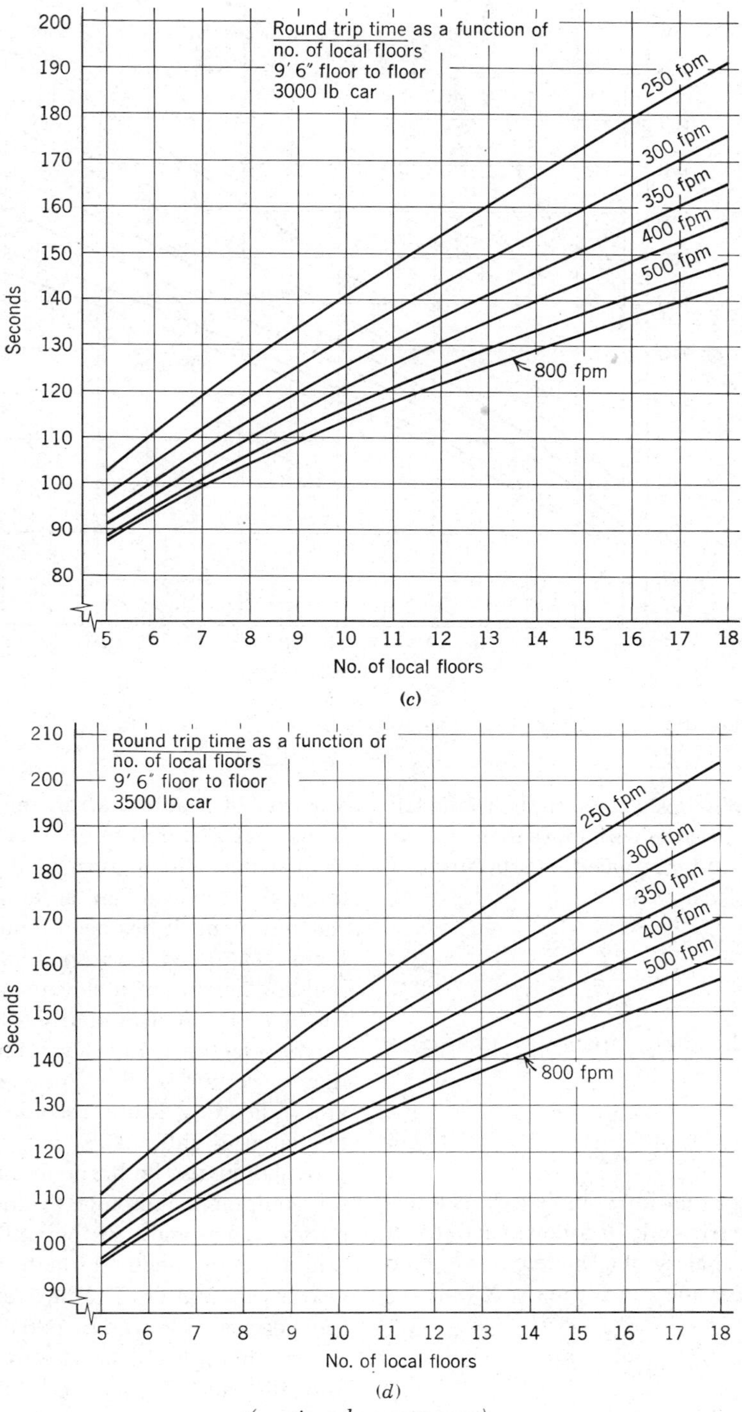

(*c*)

(*d*)

(continued on next page)

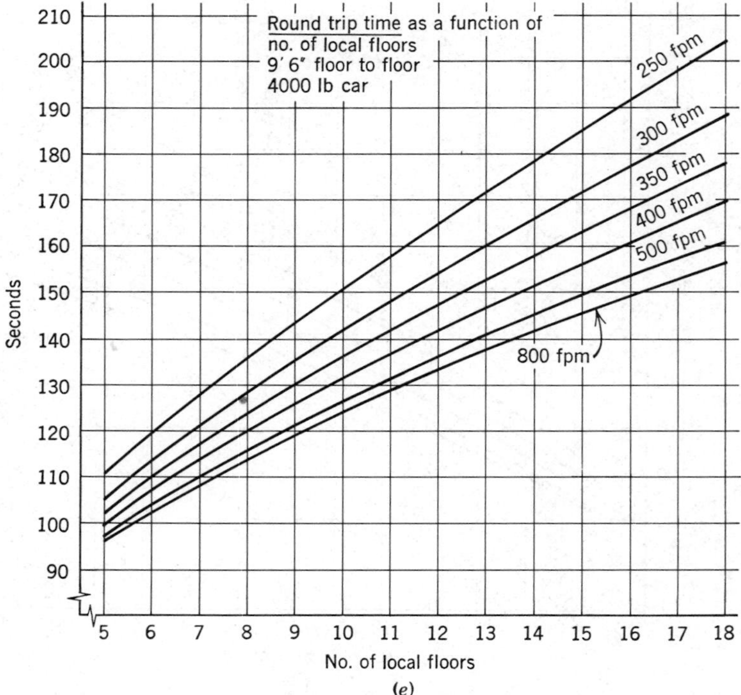

Fig. 23-21 (continued)

interval (I) is obviously equal to the round trip time (RT). In a system with more than one car, the interval will be reduced in proportion to the number of cars.

Thus

$$I = \frac{RT}{N} \tag{23.2}$$

The 5-min handling capacity (h) of a single car is then

$$h = \frac{300p}{RT} \tag{23.3}$$

if we remember that for a single car, its interval is its round trip time. It follows then that if the handling capacity of a single car is h, then the handling capacity of N cars is N times as much. Thus

$$HC = N \times h$$

or

$$N = \frac{HC}{h} \tag{23.4}$$

23.33 Car Speed

The selection of the car speed to be used is a matter of trial and error; the final selection being that required to give a round trip time (RT) which will in turn give an acceptable interval. However, in order to establish a starting point, it has been found that a minimum car speed corresponding to a given building height, or in elevator parlance—rise, can be established. Similarly, although car size can be selected at any value, it has been shown that for certain facility types, specific size cars are indicated. These recommendations are tabulated in Table 23.8.

It should also be borne in mind that elevator equipment falls into distinct speed categories. Thus most manufacturers will use geared equipment through 400 fpm and gearless thereafter. The next category is 500 fpm gearless, followed by 600 to 700 fpm, and so on. Therefore, it is wise to avoid moving into the next higher and more expensive

Fig. 23.22 *Plots of round-trip time for various car speeds and capacities with 12 ft 0 in. floor height.*

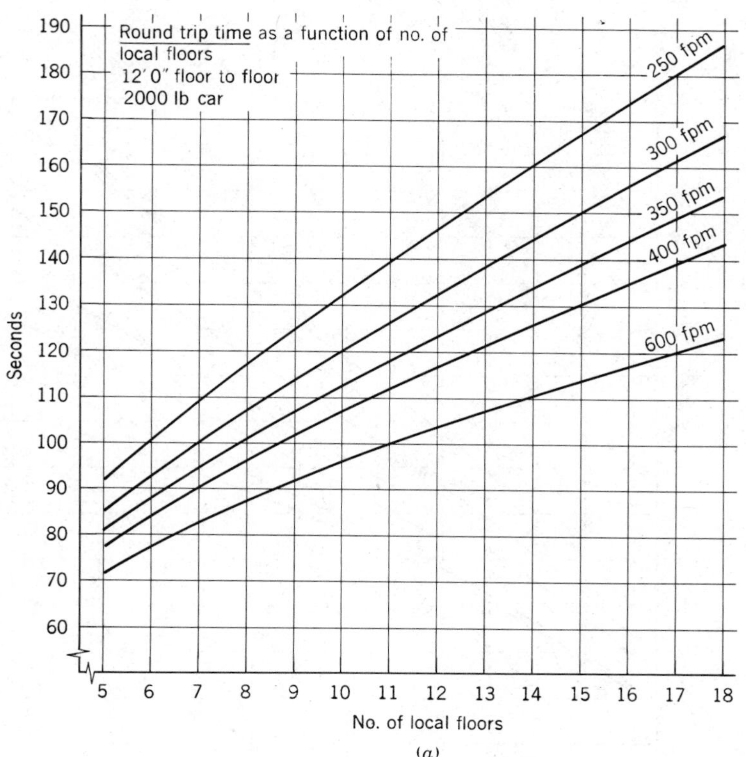
Round trip time as a function of no. of local floors
12′ 0″ floor to floor
2000 lb car
250 fpm
300 fpm
350 fpm
400 fpm
600 fpm
Seconds
190
180
170
160
150
140
130
120
110
100
90
80
70
60
5 6 7 8 9 10 11 12 13 14 15 16 17 18
No. of local floors
(*a*)

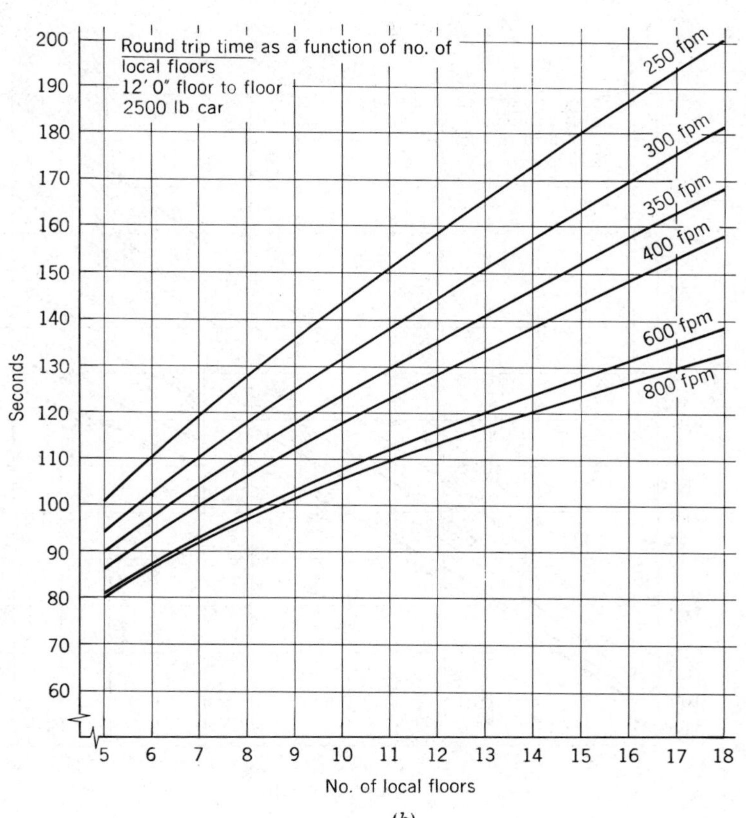
Round trip time as a function of no. of local floors
12′ 0″ floor to floor
2500 lb car
250 fpm
300 fpm
350 fpm
400 fpm
600 fpm
800 fpm
Seconds
200
190
180
170
160
150
140
130
120
110
100
90
80
70
60
5 6 7 8 9 10 11 12 13 14 15 16 17 18
No. of local floors
(*b*)

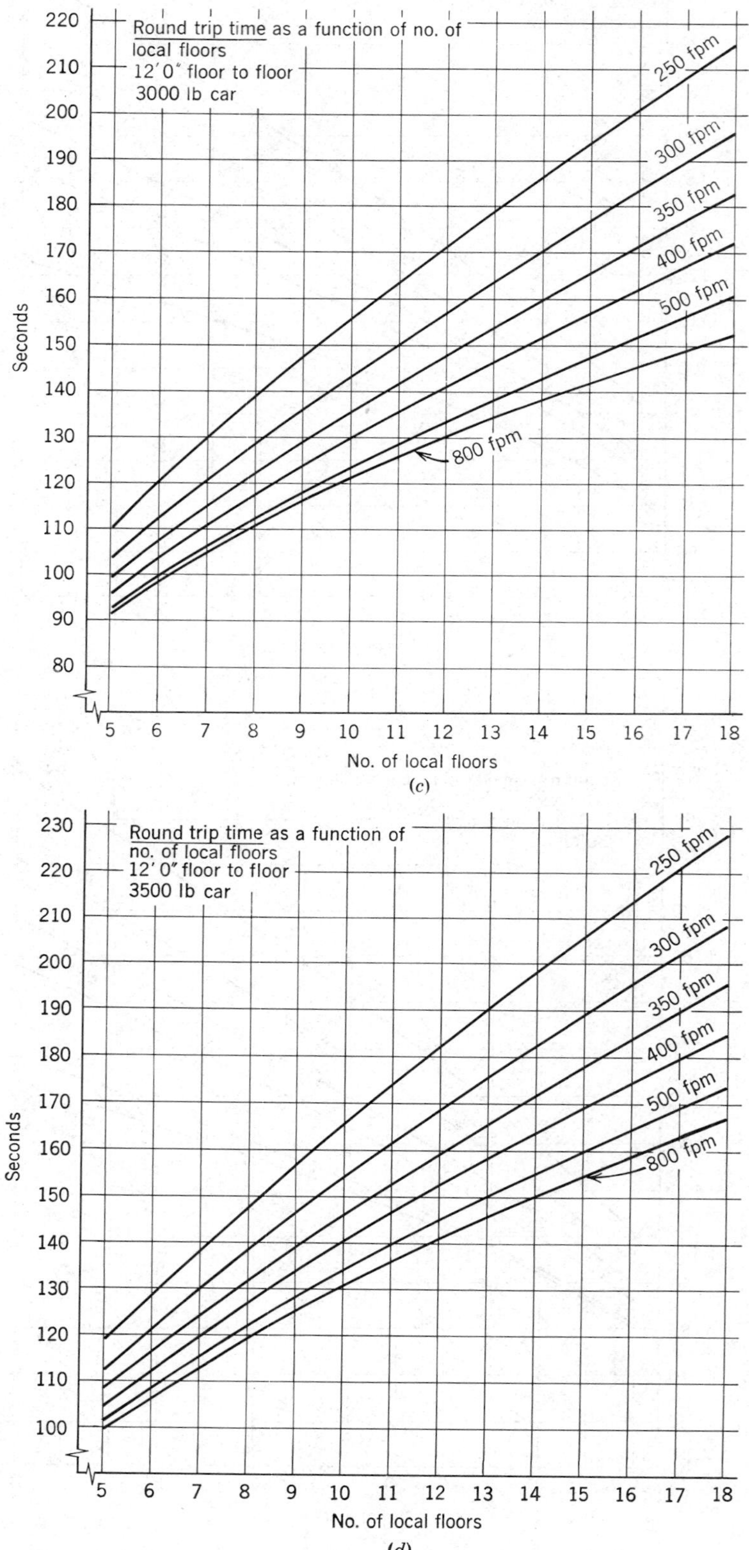

(*c*)

(*d*)

Fig. 23-22 (continued)

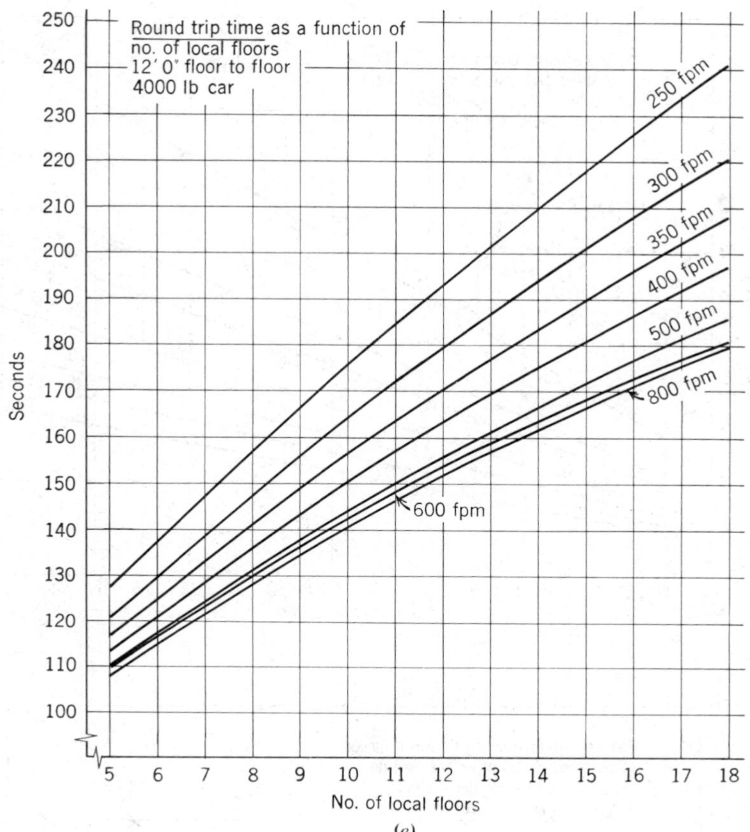

(e)

equipment category, if possible. This may mean exceeding recommended interval or dropping slightly low in handling capacity. It will be found, however, that this can be done without injury to the elevator system, provided a good group supervisory control system is employed.

23.34 Single Zone Systems

Having established the relationships that govern the design and performance of an elevator system, comprising a single zone, it would be helpful to follow through an illustrative example before proceeding to a discussion of multiple zone systems.

EXAMPLE 23.1. Office building, downtown, diversified use, 14 rentable floors above lobby, each 12,000 sq ft net. Floor to floor height—12 ft. Determine a workable elevator system.

SOLUTION. From Table 23.5, minimum HC is 13%. From Table 23.3, maximum interval is 30 sec. From Table 23.6, average population density is 130 sq ft per person. Due to rise only gearless units are considered.

Building Population

$$= \frac{14 \text{ floors at } 12{,}000 \text{ sq ft}}{130 \text{ sq ft per person}} = 1292 \text{ persons}$$

Minimum handling capacity $PHC = 13\%$

$$HC = 0.13 \times 1292 = 168 \text{ persons}$$
$$\text{Rise} = 14 \text{ floors at } 12 \text{ ft} = 168 \text{ ft}$$

From Table 23.8 we select a recommended car size of 3000 lb at 500 fpm.

From Figures 23.22*c* and 23.20*c*

$$RT = 143 \text{ sec}, \; AVTRP = 76 \text{ sec}$$

$$\text{Single car capacity } h = \frac{300p}{RT}$$

(see Table 23.4 for p)

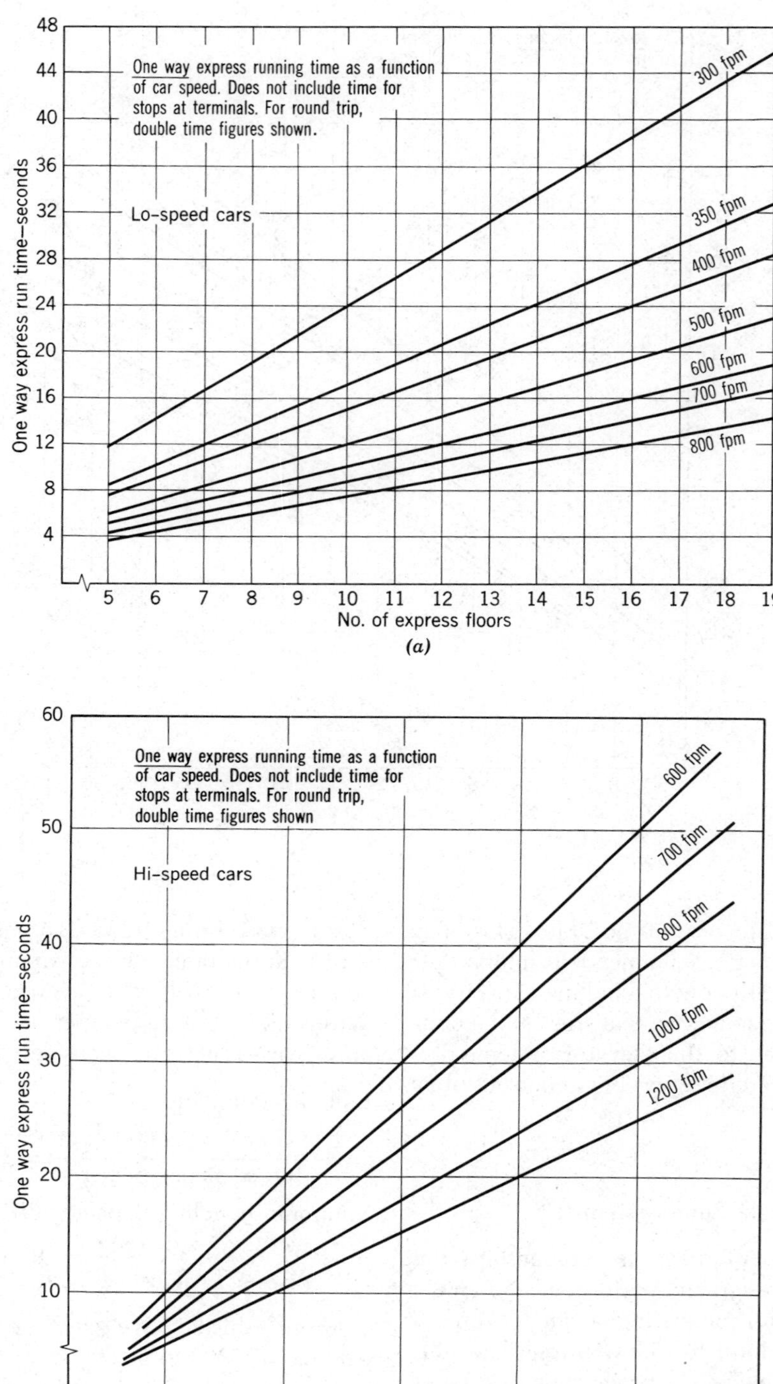

Fig. 23.23 One-way express running time, not including terminal time.

Table 23.8 ***Elevator Equipment Recommendations***

Usage		Car Capacity (Pounds)	Minimum Car Speed[a] (Feet per Minute)	Car Travel (Feet)
Office buildings			350 to 400	0 to 125
	Small building	2500	500 to 600	126 to 225
	Medium building	3000	700	226 to 275
	Large building	3500	800	276 to 375
			1000	Above 375
Hotels		2500		
		3000	As above	As above
Hospitals			150	0 to 60
			200	61 to 100
		3500	250 to 300	101 to 125
		4000	350 to 400	126 to 175
			500 to 600	176 to 250
			700	Above 250
Apartment houses[b]			100	0 to 75
		2000	200	76 to 125
		2500	250 to 300	126 to 200
			350 to 400	Above 200
Retail stores			200	0 to 100
		3500	250 to 300	101 to 150
		4000	350 to 400	151 to 200
		5000	500	Above 200

[a]Geared equipment is used up to 400 fpm, gearless at higher speeds.

[b]FHA minimum requirements call for full-collective variable voltage control; minimum of two cars; and approximately 120 bedrooms per car, for all buildings exceeding seven stories in height.

$$h = \frac{300(16)}{143} = 33.5 \text{ persons}$$

$$N = \frac{HC}{h} = \frac{168}{33.5} = 5.0 \text{ cars}$$

$$\text{Interval } I = \frac{RT}{N} = \frac{143}{5} = 28.3 \text{ sec}$$

This figure is excellent.

We should also try faster cars in order to reduce the number of cars, or slower cars to reduce car cost. We select 600 fpm and 400 fpm.

3000 lb

600 fpm

$$RT = 137;\ AVTRP = 74 \text{ sec}$$

$$h = \frac{300(16)}{137} = 35.0$$

$$N = \frac{168}{35} = 4.8 \text{ say } 5$$

$$I = \frac{137}{5} = 27.4 \text{ sec}$$

$$\text{Actual } PHC = \frac{5(13)}{4.8} = 13.5\%$$

This is no better than the previous solution, and is more expensive.

3000 lb

400 fpm

$$RT = 152;\ AVTRP = 77 \text{ sec}$$

$$h = \frac{4800}{152} = 31.6$$

$$N = \frac{168}{31.6} = 5.3 \text{ or } 5$$

$$I = \frac{152}{5} = 30.4$$

$$\text{Actual } PHC = \frac{5(13)}{5.3} = 12.26\%$$

which is somewhat low.

Thus in the 3000 lb car class we have an acceptable solution of

Five cars at 500 fpm

We try a larger car, say 3500 lb, to determine whether a smaller number of cars is possible.

3500 lb

500 fpm

$$RT = 155$$

$$h = \frac{300(19)}{155} = 36.8$$

$$N = \frac{168}{36.8} = 4.56$$

yielding no advantage since we still require five cars.

3500 lb

600 fpm

$$RT = 152$$

$$h = \frac{300(19)}{150} = 37.5$$

$$N = \frac{168}{37.5} = 4.48 \text{ say } 4$$

$$I = \frac{168}{4} = 42; \text{ excessive}$$

$PHC = \dfrac{4(13)}{4.42} = 11.7$, which is insufficient; and this solution is discarded.

3500 lb

700 fpm

$RT = 150$; no significant improvement.

We thus have no advantage in this larger car and may discount it as an economic solution. Turning to a smaller car, we try the 2500 lb size.

2500 lb

400 fpm

$$RT = 139 \text{ sec}$$

$$h = \frac{300(13)}{139} = 28.0$$

$$N = \frac{168}{28.0} = 6$$

$$I = \frac{139}{6} = 23.2 \text{ sec}$$

$$PHC = 13\%$$

This is a possible solution. Now trying a higher speed gearless unit:

2500 lb

700 fpm

$$RT = 122 \text{ sec}$$

$$h = \frac{3900}{122} = 32 \text{ persons}$$

$$N = \frac{168}{32} = 5.25 \text{ or } 5$$

$$I = \frac{122}{5} = 24.4 \text{ sec}$$

$$PHC = \frac{5}{5.25}(13) = 12.38\%$$

This is also a possible solution, since the *PHC* is close enough to 13% to be acceptable, and the interval is not too low.

We thus have three possible solutions:

Five 3000 lb cars at 500 fpm
Six 2500 lb cars at 400 fpm
Five 2500 lb cars at 700 fpm

The final selection should at this point be made on the basis of cost and core layout. When considering cost we should note that first cost is the governing factor only in a speculative venture or when the maximum building dimensions are not fixed. When dimensions are fixed, cost figures should reflect the impact of elevator space requirements on net rentable area in the building.

Comparative cost figures are given in Figure 23.24. These cost figures are for comparison only and *must not be used* to estimate job costs. Utilizing the cost data in Figure 23.24 for the solutions arrived at above, we obtain

a. Five 3000 lb cars at 500 fpm, gearless

Cost of each car	\$ 88,000
Additional floors, 14 × \$2870	40,180
	\$128,180
Total 5 cars	\$640,900

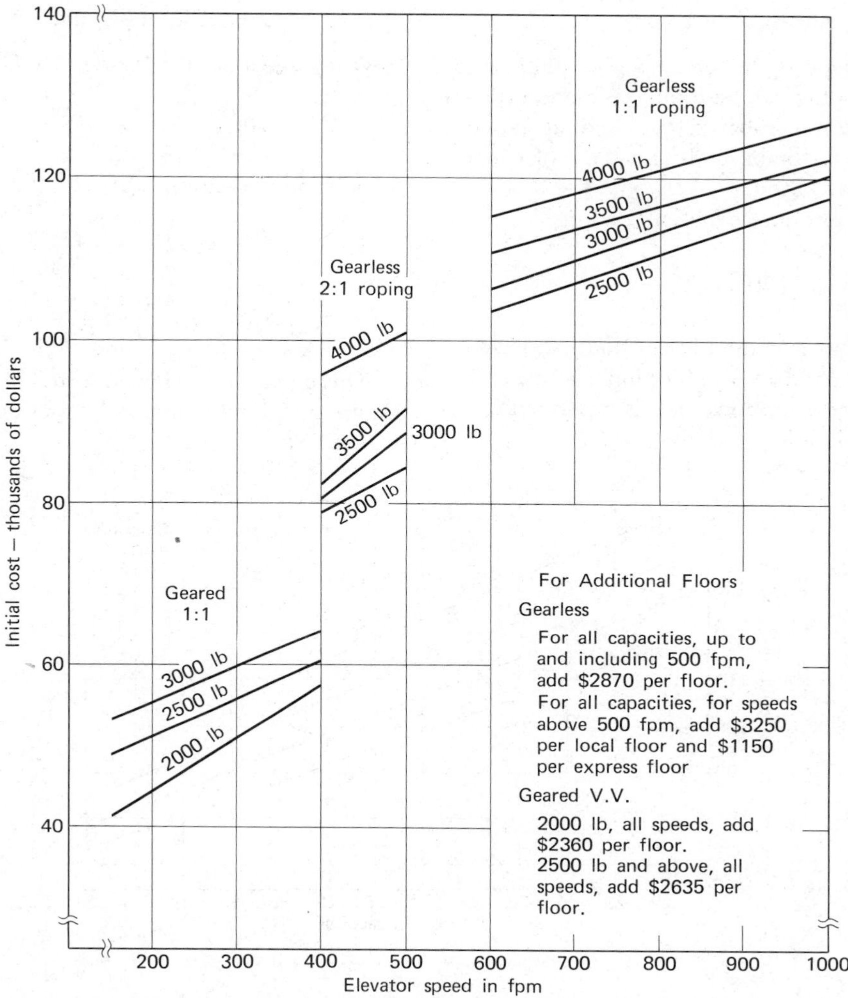

Fig. 23.24 *Comparative cost figures for variable voltage controlled elevator equipment. Cost data is for geared and gearless elevators in office buildings, general purpose buildings, hospitals, hotels, and industrial buildings with speeds of 200 to 800 fpm and with elevator passenger capacities from 2000 to 4000 lb.*

b. Six 2500 lb cars at 400 fpm, gearless

Cost of each car	$ 79,000
Additional floors, 14 × $2870	40,180
	$119,180
Total 6 cars	$715,080

c. Five 2500 lb cars at 700 fpm, gearless

Cost of each car	$107,000
Additional floors, 14 × $3250	45,500
	$152,500
Total 5 cars	$762,500

The choice is clearly in favor of the 3000 lb cars. The recommendations found in Table 23.8 are based on numerous analyses of this type and extensive field experience.

23.35 Multizone Systems

To properly understand the method by which zoning can be established and utilized, it must be remembered that handling capacity *HC* is fully established when size of car and interval are fixed.

If we examine equation 23.1

$$HC = \frac{300p}{I}$$

we see that it demonstrates that handling capacity is fixed by establishing *I* and *p*. Continuing by definition, *PHC* is a percentage of the population.

$$HC = PHC \times Population$$

If we assume a population density *D*, we have

$$HC = PHC \times \frac{Area}{D} \tag{23.5}$$

Substituting equation 23.1

$$\frac{300p}{I} = PHC \times \frac{Area}{Density}$$

$$Area = \frac{300 \times p \times D}{PHC \times I} \tag{23.6}$$

This equation has been plotted in curves of Figure 23.25, with area as a function of interval and *PHC*.

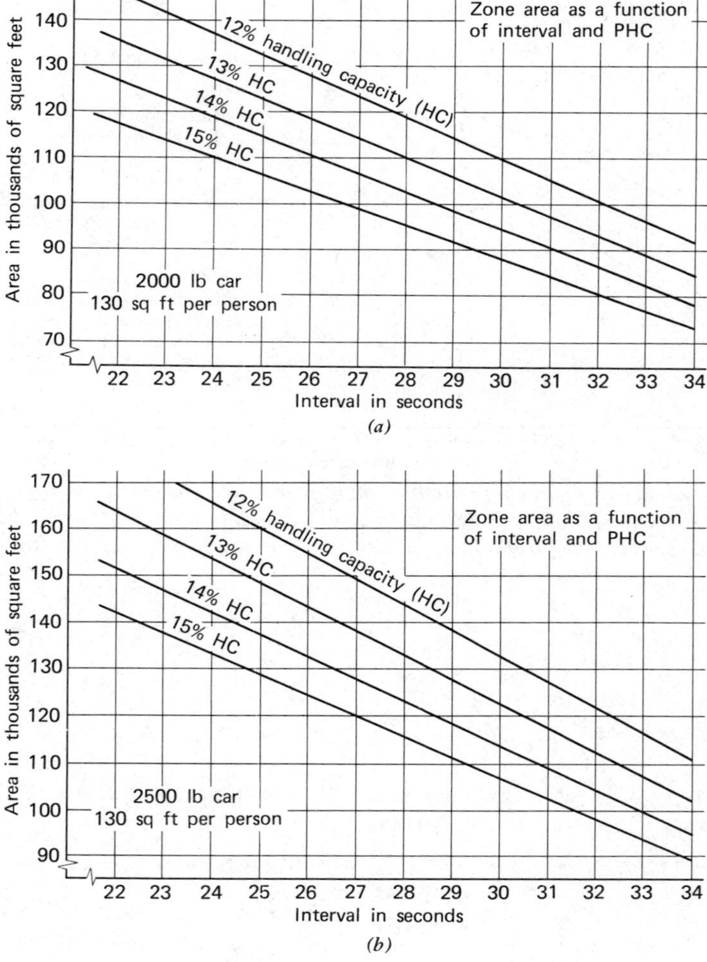

Fig. 23.25 *Zone area as a function of interval and percent handling capacity, for different capacity cars.*

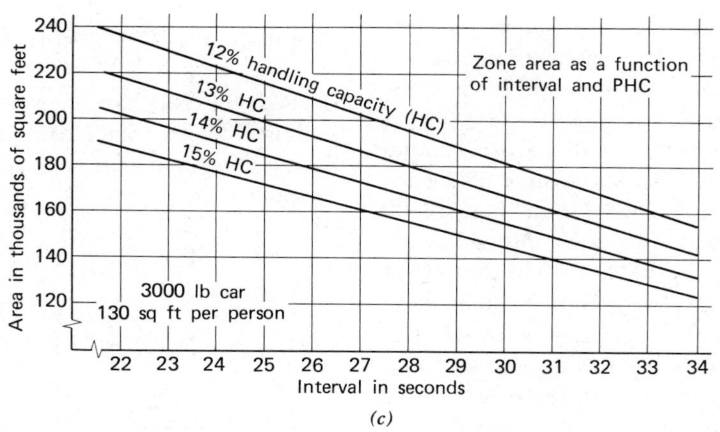
Zone area as a function of interval and PHC
12% handling capacity (HC)
13% HC
14% HC
15% HC
3000 lb car
130 sq ft per person
Area in thousands of square feet
240
220
200
180
160
140
120
Interval in seconds
22
23
24
25
26
27
28
29
30
31
32
33
34
(c)

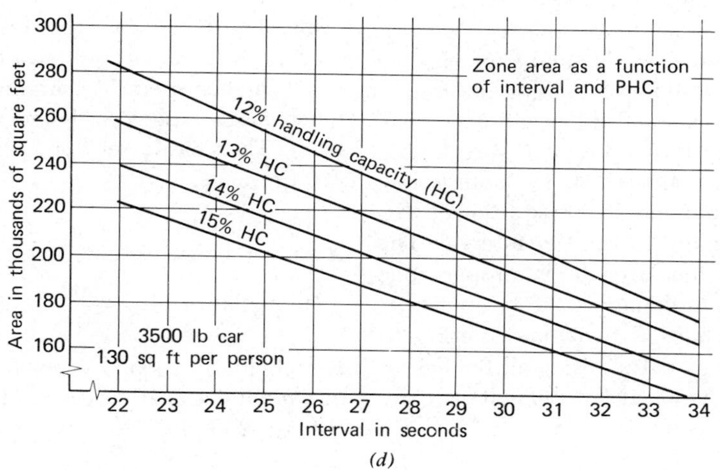
Zone area as a function of interval and PHC
12% handling capacity (HC)
13% HC
14% HC
15% HC
3500 lb car
130 sq ft per person
Area in thousands of square feet
300
280
260
240
220
200
180
160
Interval in seconds
22
23
24
25
26
27
28
29
30
31
32
33
34
(d)

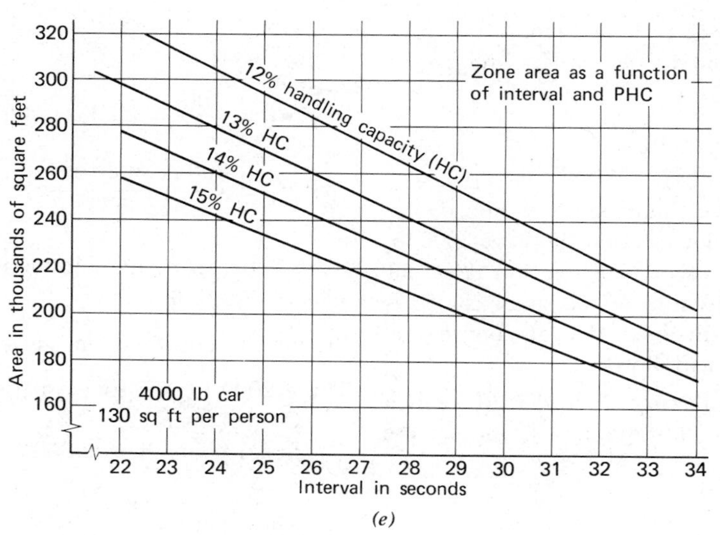
Zone area as a function of interval and PHC
12% handling capacity (HC)
13% HC
14% HC
15% HC
4000 lb car
130 sq ft per person
Area in thousands of square feet
320
300
280
260
240
220
200
180
160
Interval in seconds
22
23
24
25
26
27
28
29
30
31
32
33
34
(e)

A glance at these curves shows the area that can be handled by a specific combination of car size and interval. In the problem of the previous paragraph, an area of 168,000 sq ft with a minimum *PHC* of 13% and a maximum 30 sec *I* is readily located on curve 23.25*c* for a 3000 lb car. Curve 23.25*d* shows that a 3500 lb car will necessitate excessive interval, while curve 23.25*b* shows that a 2500 lb car cannot readily handle the load. Thus an immediate selection of a 3000 lb car can be made. The remaining calculations are routine.

Applying these curves to the results of the previous problem we can see immediately that the statement previously made—that zoning increases handling capacity—is justified. Since we have an odd number of cars, we would split the building into an upper 6-floor zone with two cars and a lower 8-floor zone with three cars. Inspection of Figure 23.25*c* for 3000 lb cars shows that for both zones, 72,000 sq ft upper and 96,000 sq ft lower, *HC* is very high, at the expense of interval. The 3000 lb car is too large for a proper zone solution to this building.

Is thère perhaps a two-zone solution to this building? An analysis using all the curves previously developed is not difficult.

EXAMPLE 23.2

Office building, average use

14 stories of 12 ft, above lobby

Population density $D = 130$ sq ft per person

Desired $PHC = 13\%$

Desired interval = 30 sec

SOLUTION. The single-zone solution developed in the previous section gives five 3000 lb, 500 fpm cars. A two-zone solution does not appear likely because of the small car sizes and high speed involved. We can confirm our first impression as follows.

TWO-ZONE SOLUTION. In order to keep a reasonable interval in the upper zone, we use an 8/6 split.
Thus

Lower Zone

Area = 8 floors × 12,000 sq ft = 96,000 sq ft

$$\text{Population} = \frac{96{,}000}{130}\text{ sq ft} = 738 \text{ persons}$$

Since 2000 lb cars have more than enough *HC* (see Fig. 23.25*a*), try a 2000 lb car at 400 fpm. This is the maximum speed with geared equipment.

From Figure 23.21*a*

$RT = 96$ sec

Using 3 cars

$$I = \frac{96 \text{ sec}}{3} = 32 \text{ sec, } OK$$

$$PHC = \frac{300\ (10)}{32(738)} \times 100 = 12.7\%,\ OK$$

(This figure can also be taken from Fig. 23.25*a*.) A larger car will give excessive interval; a higher speed is marginally better but much more expensive, being gearless equipment. Thus the lower zone is three 2000 lb, 400 fpm cars.

Upper Zone

Area = 6 floors × 12,000 sq ft = 72,000 sq ft

$$\text{Population} = \frac{72{,}000}{130} = 553 \text{ persons}$$

We obviously require a small, fast car. The *minimum* size gearless car available is 2500 lb. Using a 2500 lb, 700 fpm car we have:

$$RT = 2 \times 7 \text{ sec express} + 86 \text{ sec local} = 100 \text{ sec}$$

Using 3 cars

$$I = \frac{100}{3} = 33.3 \text{ sec, } OK$$

$$PHC = \frac{300\ (13 \text{ persons})}{33.3\ (553)} \times 100 = 21\%!$$

Thus, the upper zone solution is three 2500 lb cars at 700 fpm.

Pricing the entire system gives

Three 2000 lb, 400 fpm cars, geared

Cost of each car	$55,000
Local floors, 8 × $2360	18,880
	$73,880
Three-car bank	$221,640

Three 2500 lb, 700 fpm cars, gearless

Cost of each car	\$107,000
Local floors, 6 × \$3250	19,500
Express floors, 8 × \$1150	9,200
Cost per car	\$135,700
Three-car bank	\$407,100
Total system	\$628,740

Comparing this to the figure of \$640,900 for the single-zone system shows that the advantage is clearly for the *one-zone* system when rental value of the sixth car space is considered. The control system in this case would establish two zones during peak periods to increase handling capacity.

Summarizing the comparison between one- and two-zone systems, we find that the two-zone system has higher *HC* and interval, but lower trip time; it also has less shaftway door openings and lower maintenance cost. We are assuming no machine room in the middle of the building. If one were contemplated there, its cost impact on the building would be compared to the additional rental space saved on the floors above. Also, the architectural affect of an oversize floor in the building center must be considered.

23.36 Large Multizone Buildings

When faced with the problem of elevatoring a large office-type building, all of the above curves and tables can be applied to allow a solution by hand methods, although the use of a computer is helpful. The approach to such an example follows. Net floor area is defined as rentable space. Although this varies by zone, a first calculation can assume an average figure. Later detailed calculation should use refined area figures.

Parameters:

22 rentable floors above the lobby

Floor to floor height—12 ft

Average net floor area—24,000 sq ft

Population density—120 sq ft per person

Maximum interval—30 sec

Minimum *PHC*—13%

Our choice of cars would be 3000 lb and 3500 lb. Maximum interval has been established at 30 sec. Minimum interval should be not less than lobby loading time, which is approximately 1.2 sec per passenger, but in any case not less than 22 sec.

3000 lb car; see Figure 23.25*c*, modified for population density of 120 sq ft per person

Area for 13% *PHC* and 30 sec *I*

$$= 140{,}000 \text{ sq ft}$$

Minimum number of floors in a zone

$$= \frac{140{,}000}{24{,}000 \text{ per floor}} = 6 \text{ floors}$$

Area for 13% *PHC* and 22 sec *I*

$$= 200{,}000 \text{ sq ft}$$

Maximum floors in a zone

$$= \frac{200{,}000}{24{,}000} = 8 \text{ floors}$$

3500 lb car; see Figure 23.25*d*

Area for 13% *PHC* and 30 sec *I*

$$= 180{,}000 \text{ sq ft}$$

Minimum floors in a zone

$$= \frac{180{,}000}{24{,}000} = 7 \text{ floors}$$

Area for 13% *PHC* and 22 sec *I*

$$= 240{,}000 \text{ sq ft}$$

Maximum floors in zone

$$= \frac{240{,}000}{24{,}000} = 10 \text{ floors}$$

On the basis of these figures plus examination of the average trip time curves (Fig. 23.20) a figure of 8 or 9 floors suggests itself for the first zone. This gives these possible practical zonings:

	Scheme 1	*Scheme 2*	*Scheme 3*
Zone 1	8 floors	8 floors	9 floors
Zone 2	7 floors	8 floors	7 floors
Zone 3	7 floors	6 floors	6 floors

In all schemes a machine room level with no rental area would be placed above the first zone. This level is selected more from air conditioning requirements than for elevator needs. The computer output for this problem is given in Figure 23.26.

The cost figures in the program are more

```
ELEVATOR DESIGN PROGRAM -- 8080


COST DATA INPUT JAN 1967
COSTS ARE FOR PURPOSES OF DESIGN SELECTION NOT PROJECT ESTIMATING.


JOB TITLE--BOOK PROBLEM
JOB NUMBER--101-29
DATE--9/8/68
ENGINEER--STEIN


SCHEME NUMBER -- I


DESIGN CRITERIA --
---------------
MAXIMUM INTERVAL =  30.0 SEC.
MINIMUM INTERVAL =  20.0 SEC.
MINIMUM HANDLING CAPACITY = 13.0 PCT OF POPULATION IN 5 MIN.
BUILDING DENSITY =  120.0 SQ. FT. PER PERSON
TYPICAL FLOOR TO FLOOR DIMENSION =  12.0 FT.
GROUND TO 2ND FLOOR DISTANCE =  12.0 FT.
TOTAL NO. OF FLOORS(INCL. GRD. FL.) =  24.
TOTAL NO. OF ZONES IN THE BUILDING = 3.


SYMBOLS --
-------
N ELEV -- NO. OF ELEVATORS
SPEED -- SPEED OF ELEVATOR IN FT./MIN.
CAPACITY -- ELEVATOR CAPACITY IN LBS.
PHC -- PERCENTAGE HANDLING CAPACITY
INT -- INTERVAL IN SEC.
RT -- ROUND TRIP TIME IN SEC.
AVTRP -- AVERAGE TRIP TIME IN SEC.
COST -- COST OF ELEVATOR AND INSTALLATION IN $
NRSF/ELEV -- RENTABLE AREA PER ELEVATOR IN SQ. FT.


ZONE 1
------
NO. OF EXPRESS FLOORS= 0.
NO. OF LOCAL FLOORS= 8.
NET RENTABLE AREA PER FLOOR= 24000. SQ. FT.
POPULATION =      1600.
TOTAL RENTABLE AREA =  192000. SQ. FT.
```

N ELEV	SPEED	CAPACITY	PHC	INT	RT	AVTRP	COST	NRSF/ELEV
5	300	3500	13.0	27.6	138.	76.	207750.	38400.
5	300	4000	14.1	28.9	144.	81.	222750.	38400.
5	350	3500	13.6	26.3	131.	74.	216500.	38400.
5	350	4000	14.7	27.7	138.	79.	231500.	38400.
6	300	3000	14.5	21.1	127.	66.	231300.	32000.
6	300	3500	15.6	23.0	138.	73.	249300.	32000.
6	300	4000	16.9	24.1	144.	78.	267300.	32000.
6	350	3000	15.3	20.1	121.	64.	241800.	32000.
6	350	3500	16.3	21.9	131.	71.	259800.	32000.
6	350	4000	17.6	23.1	138.	76.	277800.	32000.

```
TOTAL SOLUTIONS EXPLORED =     380



THE LEAST EXPENSIVE SYSTEM WHICH WILL MEET DESIGN CRITERIA IS
5 -   300 FEET PER MIN 3500 LB ELEVATORS.
THE HANDLING CAPACITY IS 13.0 PCT OF POPULATION IN 5 MIN.
THE MAXIMUM INTERVAL IS  27.6 SECONDS.
TOTAL ESTIMATED COST = $ 207750.
OTHER SYSTEMS UP TO 50 PCT MORE EXPENSIVE ARE LISTED ABOVE.
```

Fig. 23.26 *Computer printout of solution to text problem in Section 23.36; zoning is input to the program.*

ZONE 2

NO. OF EXPRESS FLOORS= 9.
NO. OF LOCAL FLOORS= 7.
NET RENTABLE AREA PER FLOOR= 24000. SQ. FT.
POPULATION = 1400.
TOTAL RENTABLE AREA = 168000. SQ. FT.

N ELEV	SPEED	CAPACITY	PHC	INT	RT	AVTRP	COST	NRSF/ELEV
5	500	3000	13.3	26.4	132.	70.	394000.	33600.
5	500	3500	14.3	28.5	142.	78.	411500.	33600.
5	500	4000	15.6	29.9	149.	83.	429000.	33600.
5	600	3000	14.0	25.0	125.	68.	413000.	33600.
5	600	3500	15.0	27.2	136.	76.	430500.	33600.
5	600	4000	16.3	28.5	142.	81.	448000.	33600.
5	700	2500	13.3	22.2	111.	60.	414500.	33600.
5	700	3000	14.4	24.4	122.	68.	432000.	33600.
5	700	3500	15.4	26.6	133.	76.	449500.	33600.
5	700	4000	16.7	27.9	139.	81.	467000.	33600.
5	800	2500	13.4	21.9	109.	60.	433500.	33600.
5	800	3000	14.5	24.2	121.	68.	451000.	33600.
5	800	3500	15.5	26.4	132.	76.	468500.	33600.
5	800	4000	16.8	27.8	139.	81.	486000.	33600.
5	1000	2500	13.6	21.7	108.	61.	471500.	33600.
5	1000	3000	14.6	24.1	120.	70.	489000.	33600.
5	1000	3500	15.5	26.3	131.	78.	506500.	33600.
5	1000	4000	16.8	27.7	138.	83.	524000.	33600.
5	1200	2500	13.5	21.8	109.	62.	509500.	33600.
5	1200	3000	14.5	24.2	121.	71.	527000.	33600.
5	1200	3500	15.5	26.4	132.	79.	544500.	33600.
5	1200	4000	16.7	27.8	139.	84.	562000.	33600.
5	1400	2500	13.0	22.7	113.	66.	547500.	33600.
5	1400	3000	14.0	25.2	126.	75.	565000.	33600.
5	1400	3500	14.9	27.5	137.	83.	582500.	33600.
6	400	2500	13.4	22.0	132.	64.	429000.	28000.
6	400	3000	14.7	23.9	143.	71.	450000.	28000.
6	400	3500	16.0	25.6	153.	79.	471000.	28000.
6	400	4000	17.4	26.7	160.	84.	492000.	28000.
6	500	2500	14.6	20.2	121.	60.	451800.	28000.
6	500	3000	16.0	22.0	132.	68.	472800.	28000.
6	500	3500	17.2	23.8	142.	75.	493800.	28000.
6	500	4000	18.7	24.9	149.	80.	514800.	28000.
6	600	3000	16.9	20.9	125.	66.	495600.	28000.
6	600	3500	18.1	22.6	136.	73.	516600.	28000.
6	600	4000	19.6	23.8	142.	78.	537600.	28000.
6	700	3000	17.3	20.4	122.	66.	518400.	28000.
6	700	3500	18.5	22.1	133.	73.	539400.	28000.
6	700	4000	20.0	23.3	139.	78.	560400.	28000.
6	800	3000	17.4	20.2	121.	66.	541200.	28000.
6	800	3500	18.6	22.0	132.	74.	562200.	28000.
6	800	4000	20.1	23.1	139.	79.	583200.	28000.
6	1000	3000	17.5	20.1	120.	67.	586800.	28000.
7	400	3000	17.2	20.5	143.	69.	525000.	24000.
7	400	3500	18.6	21.9	153.	77.	549500.	24000.
7	400	4000	20.3	22.9	160.	82.	574000.	24000.
7	500	3500	20.1	20.4	142.	74.	576100.	24000.

TOTAL SOLUTIONS EXPLORED = 720

THE LEAST EXPENSIVE SYSTEM WHICH WILL MEET DESIGN CRITERIA IS
5 - 500 FEET PER MIN 3000 LB ELEVATORS.
THE HANDLING CAPACITY IS 13.3 PCT OF POPULATION IN 5 MIN.
THE MAXIMUM INTERVAL IS 26.4 SECONDS.
TOTAL ESTIMATED COST = $ 394000.
OTHER SYSTEMS UP TO 50 PCT MORE EXPENSIVE ARE LISTED ABOVE.

(continued on next page)

ZONE 3

NO. OF EXPRESS FLOORS=16.
NO. OF LOCAL FLOORS= 7.
NET RENTABLE AREA PER FLOOR= 24000. SQ. FT.
POPULATION = 1400.
TOTAL RENTABLE AREA = 168000. SQ. FT.

N ELEV	SPEED	CAPACITY	PHC	INT	RT	AVTRP	COST	NRSF/ELEV
5	700	3500	13.8	29.6	148.	81.	470500.	33600.
5	800	3000	13.1	26.9	134.	73.	472000.	33600.
5	800	3500	14.1	29.0	145.	81.	489500.	33600.
5	1000	3000	13.4	26.2	131.	73.	510000.	33600.
5	1000	3500	14.4	28.4	142.	81.	527500.	33600.
5	1000	4000	15.6	29.8	149.	86.	545000.	33600.
5	1200	3000	13.5	26.0	130.	74.	548000.	33600.
5	1200	3500	14.5	28.2	141.	82.	565500.	33600.
5	1200	4000	15.7	29.6	148.	87.	583000.	33600.
5	1400	3000	13.2	26.7	133.	77.	586000.	33600.
5	1400	3500	14.1	29.0	145.	86.	603500.	33600.
6	500	3000	13.8	25.5	153.	75.	498000.	28000.
6	500	3500	15.0	27.3	164.	82.	519000.	28000.
6	500	4000	16.4	28.4	170.	87.	540000.	28000.
6	600	2500	13.4	21.9	131.	64.	499800.	28000.
6	600	3000	14.8	23.8	143.	72.	520800.	28000.
6	600	3500	16.0	25.6	153.	79.	541800.	28000.
6	600	4000	17.4	26.7	160.	84.	562800.	28000.
6	700	2500	14.0	21.0	126.	63.	522600.	28000.
6	700	3000	15.4	22.9	137.	71.	543600.	28000.
6	700	3500	16.6	24.7	148.	78.	564600.	28000.
6	700	4000	18.0	25.8	155.	83.	585600.	28000.
6	800	2500	14.4	20.5	123.	63.	545400.	28000.
6	800	3000	15.7	22.4	134.	71.	566400.	28000.
6	800	3500	16.9	24.2	145.	78.	587400.	28000.
6	800	4000	18.4	25.3	152.	83.	608400.	28000.
6	1000	3000	16.1	21.8	131.	71.	612000.	28000.
6	1000	3500	17.3	23.7	142.	79.	633000.	28000.
6	1000	4000	18.7	24.8	149.	84.	654000.	28000.
6	1200	3000	16.3	21.6	130.	72.	657600.	28000.
6	1200	3500	17.4	23.5	141.	80.	678600.	28000.
6	1200	4000	18.9	24.7	148.	85.	699600.	28000.
6	1400	2500	14.6	20.1	121.	66.	682200.	28000.
6	1400	3000	15.8	22.2	133.	75.	703200.	28000.
7	400	2500	13.0	22.7	158.	70.	529900.	24000.
7	400	3000	14.5	24.2	169.	78.	554400.	24000.
7	400	3500	15.9	25.7	180.	85.	578900.	24000.
7	400	4000	17.4	26.7	187.	90.	603400.	24000.
7	500	2500	14.5	20.3	142.	66.	556500.	24000.
7	500	3000	16.1	21.9	153.	73.	581000.	24000.
7	500	3500	17.5	23.4	164.	80.	605500.	24000.
7	500	4000	19.1	24.3	170.	85.	630000.	24000.
7	600	3000	17.2	20.4	143.	70.	607600.	24000.
7	600	3500	18.7	21.9	153.	78.	632100.	24000.
7	600	4000	20.4	22.9	160.	82.	656600.	24000.
7	700	3500	19.3	21.1	148.	77.	658700.	24000.
7	700	4000	21.1	22.1	155.	81.	683200.	24000.
7	800	3500	19.7	20.7	145.	77.	685300.	24000.
8	400	3000	16.6	21.2	169.	76.	633600.	21000.
8	400	3500	18.2	22.5	180.	84.	661600.	21000.
8	400	4000	19.9	23.3	187.	88.	689600.	21000.
8	500	3500	20.0	20.5	164.	79.	692000.	21000.

TOTAL SOLUTIONS EXPLORED = 720

THE LEAST EXPENSIVE SYSTEM WHICH WILL MEET DESIGN CRITERIA IS
5 - 700 FEET PER MIN 3500 LB ELEVATORS.
THE HANDLING CAPACITY IS 13.8 PCT OF POPULATION IN 5 MIN.
THE MAXIMUM INTERVAL IS 29.6 SECONDS.
TOTAL ESTIMATED COST = $ 470500.
OTHER SYSTEMS UP TO 50 PCT MORE EXPENSIVE ARE LISTED ABOVE.

Fig. 23-26 (continued)

accurate than the ones obtainable from Figure 23.24, since they are arranged to recognize increments in both speed and height. They are based, however, on different data and are valuable as comparative costs only. The last column is net rentable square feet per elevator. The program is arranged to select the lowest cost system that will meet the stated requirements. Thus in Scheme 1, Figure 23.26*a*, Zone 2, the machine selection is five 500 fpm, 3500 lb cars with an average trip of 75 sec. The choice, however, would probably go to the five 600 fpm, 3000 lb cars with an *AVTRP* of 66 sec, for a cost differential of only $1500 or less than 1/2%. Interested students can select a specific zone and check the computer computations by the more conventional hand methods demonstrated above. In so doing, they will shortly come to appreciate the inestimable value of the computer (in this case an IBM 1130).

23.37 Other Elevator Selection Considerations

As in every design discipline certain generalizations, or rules-of-thumb, have grown up in elevator design. For commercial buildings rough costs can be estimated from Table 23.9.

A good rule-of-thumb for cost is 11% of total construction cost. Similarly, figures of 250 to 300 persons of population per elevator and 30,000 to 35,000 sq ft net, per elevator, are fairly accurate. Special considerations for various type buildings are given below.

(*a*) ***Office Buildings, Hotels, and Industrial Buildings.*** The expected population may be estimated from Table 23.6. An interval of 30 sec is desirable, although slightly longer is acceptable in all except large structures in congested areas. A 5-min capacity of 13% of building population is usual. The basic type of control is automatic electronic supervisory, although collective control is satisfactory for small buildings with up to three cars. Table 23.8 lists the characteristics of equipment usually used in these buildings.

Table 23.9 ***Office Buildings—Cost of Elevator and Electric Work***

Item	*Number of Stories* 20 Sty.	35 Sty.	60 Sty.
Elevator Work	10.9%	11.9%	12.2%
Electric Work	13.3%	12.6%	12.2%

Approximately one service car per 10 passenger cars should be provided, or alternatively one car for every 300,000 sq ft of net area. Service cars should be at least 3500 lb without dropped ceiling and, if also used for passenger service, equipped with wall pads. An oversized door, for example, 4 ft 0 in. or 4 ft 6 in. is particularly useful in handling furniture. Service elevators should have a shaftway door at every level plus easy access to the truck dock or other freight entrance as well as the lobby. These cars operate as freight cars normally but can serve as passenger cars in peak periods to alleviate congestion and delay. This fact is particularly useful in marginal service designs.

(*b*) ***Apartment Houses.*** Studies indicate that apartment house traffic depends not only on population but also upon location and type of tenant. Houses with many children experience a school-hour peak; houses in midtown with predominantly adult tenancy exhibit evening peaks due to the homecoming working group and outgoing amusement traffic. Normally a single elevator will suffice, although a second car functioning as a service and (or) passenger car is sometimes indicated, particularly in buildings taller than six stories. The cars may be banked or separated, as desired.

Self-service collective control is the general choice, with provision for attendant control in high-class buildings. With smaller cars and short rise, a swing-type manual corridor door is acceptable; in larger installations both the cab and corridor door should be the power-operated sliding type. Table 23.8 lists the preferred combinations of equipment for apartment applications.

Service elevators must be large enough to handle bulky furniture, and should therefore be at least 2500 lb and preferably 3000 lb, with a 48 in. door and high ceiling. Hoistways must be isolated from sleeping rooms by lobbies or other space. Similarly, machine rooms must be isolated since the starting and stopping of motors and other machine room noises are an effective barrier to sound sleep. Security arrangements are discussed in Section 24.5.

(*c*) ***Hospitals.*** As mentioned in Table 23.6 the governing factor in the determination of elevator requirements may be either normal hospital traffic or visitors traffic, depending on the visiting-hour schedule. Due to the large amount of vehicular traffic such as stretcher carts, wheel chairs, beds, linen carts, laundry trucks, and so on, the elevator cabs are much deeper than the normal passenger-type. This type of car when used for passenger service holds over 20 persons and therefore gives slow service. For this reason, it is occasionally advisable to utilize some normal passenger cars in addition to hospital-size cars, particularly in large hospitals.

The use of tray and bulk carts in food service imposes a considerable load on the elevators before, during, and after meals, and passenger service is seriously disrupted. To alleviate this congestion and delay, many architects and hospital administrators prefer the use of dumbwaiter cars or another of the many types of materials-handling systems that will handle a $15\frac{1}{2} \times 20$ in. food tray. These systems can also be used for transporting pharmaceuticals and other items and are discussed in Sections 24.19 and 24.20.

Elevators should be grouped centrally, although separated by type of use. Car control is normally collective, manually controlled, with optional self-service operation for the larger cars and self-service collective or fully automatic with manual option for the passenger size cars.

Population of the hospital may be estimated from Table 23.6. Experience has shown that a carrying capacity of 45 passengers in a 5-min period is adequate (estimating each vehicle as equivalent to 9 passengers).

Intervals should not exceed 1 min. Car and corridor doors should be power-operated, and cars must, of course, be self-leveling. Recommended combinations of hospital equipment are shown in Table 23.8.

(*d*) ***Retail Stores.*** Retail stores present a unique problem in vertical transportation inasmuch as the objective is partially to transport persons to a specific level and partially to expose these persons to a display merchandise. For this reason modern stores rely almost exclusively on escalators, with one or two small elevators intended for use by staff and handicapped persons. When for some reason it is desired to equip a store with elevators, use the recommendations shown in Table 23.8, calculated for a load of between 10 and 20% of store population. Control should be automatic, selective collective. Cars are arranged in a straight line to facilitate loading and waiting.

THE PHYSICAL PROPERTIES AND SPATIAL REQUIREMENTS OF ELEVATORS

23.38 Shafts and Lobbies

The elevator lobbies and shafts form one of the major space factors with which the architect is concerned. The elevator lobby on each floor is the focal point from which the corridors radiate for access to all rooms, stairways, service rooms, and so forth. Such lobbies obviously must be located one above the other. The ground-floor elevator lobby (also called the lower terminal) must be conveniently located with respect to main entrances. The modern equipment within or placed adjacent to this area should include public telephones, building directory, elevator starter service, elevator indicators, and control panels.

All lobbies should be adequate in area for the peak-load gathering of passengers to ensure rapid and comfortable service to all. The number of people contributing to the period of peak load (15- to 20-min peak) determines the required lobby area on the floor.

Approximately 4 sq ft of floor space per person should be provided at peak periods for waiting passengers at a given elevator or bank of elevators. The hallways leading to such lobbies should also provide about 4 sq ft per person approaching the lobby. This requires a check of human traffic through all approaches to elevator facilities.

The main lower terminal of elevator banks is generally on the street-floor level, although some buildings place this terminal on the basement or mezzanine level, particularly when the elevations of the street entrances vary around the building so that one side of the building is on the mezzanine level, while another main entrance is at a lower level. Such a situation is ideal for the use of escalators, which will economically and rapidly carry large numbers of persons between levels thus making practical and efficient a single main lower elevator terminal. The upper terminal is usually the top floor of the building. Typical dimensional data and lobby arrangements are shown in Figures 23.27 to 23.29.

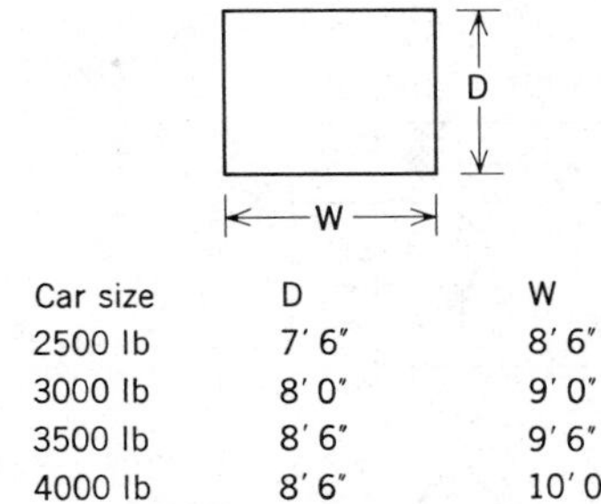

Car size	D	W
2500 lb	7′ 6″	8′ 6″
3000 lb	8′ 0″	9′ 0″
3500 lb	8′ 6″	9′ 6″
4000 lb	8′ 6″	10′ 0″

Fig. 23.27 Rough elevator dimensional data to be used in architectural single-line planning stage.

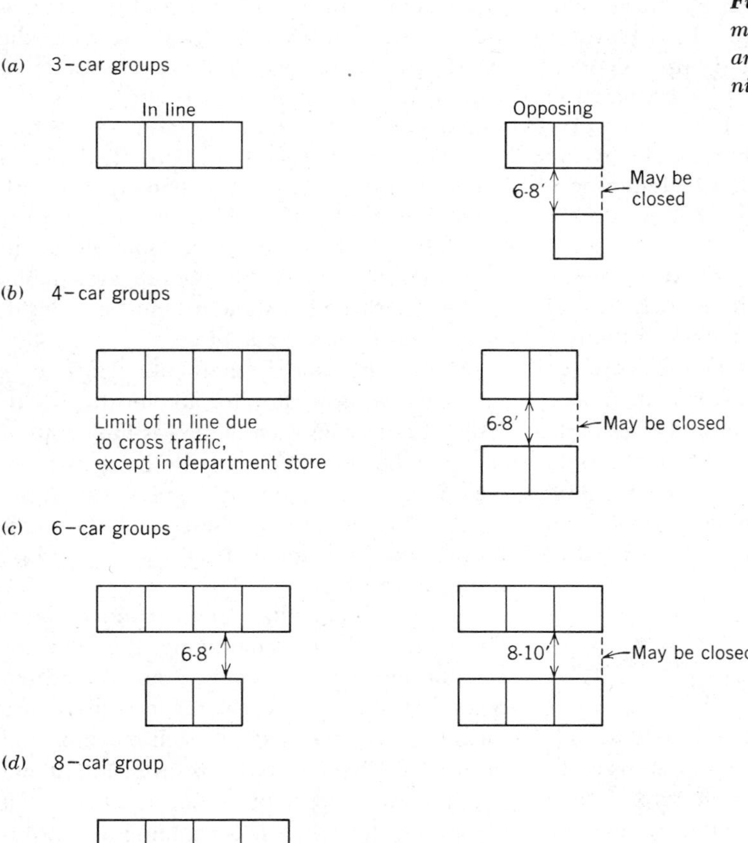

Fig. 23.28 Lobby groupings for single zones. Three, four, six, and eight-car groups.

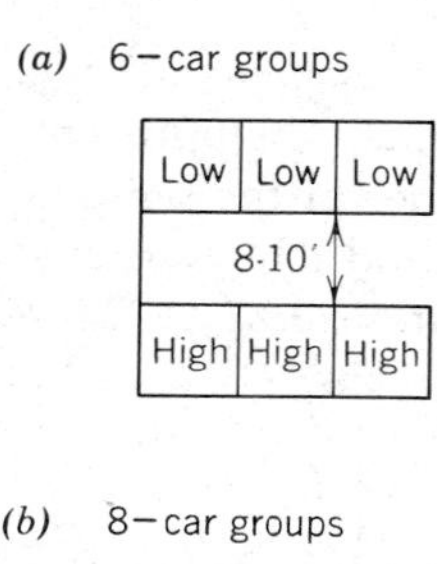

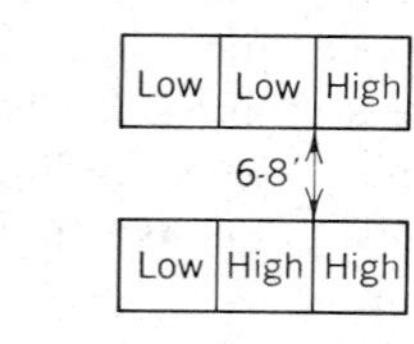

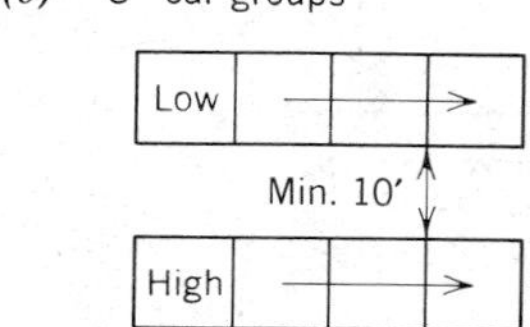

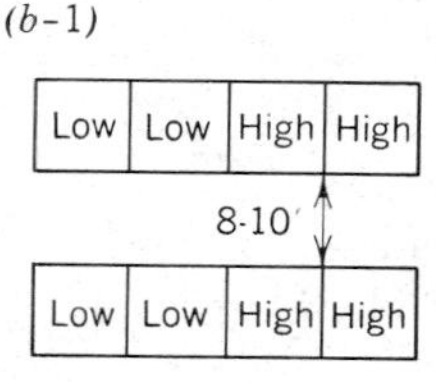

***Fig. 23.29** Lobby groupings for multiple zones. Arrangement* (a) *is preferable to* (a-1), *and* (b) *to* (b-1).

In the problem of Section 23.35, the solutions all point to three groups of five elevators each. One possible lobby and upper floor arrangement for these elevators is shown in Figure 23.30. It is suggested that as an exercise the reader prepare alternate lobby and upper floor core arrangements including service cars, utility shafts, and other utilities as desired.

No attempt has been made to present a completely integrated solution including all utilities. The purpose of the illustrations is simply to demonstrate workable core arrangements. If a second machine room level above Zone 2 is not architecturally desirable, the shafts for Zone 2 can be run through to the penthouse. Similarly, if a higher machine room is desired, it can be placed above Zone 2, with the shafts of Zones 1 and 2 terminating in it.

23.39 Dimensions and Weights

Manufacturers will supply standard layouts for elevators including dimensions, weights, and structural loads. One such layout for a preengineered 2500 lb geared unit is shown in Figure 23.31. The major manufacturers have agreed upon, and publish, a set of Standard Elevator Layouts via their trade organization, The National Elevator Industry, Inc. (NEII). One such standard, with applications, is reproduced in Figure 23.32 for 500 to 700 fpm gearless units in the full range of car capacities. These standards are available from the NEII. The NEII also publishes a recommendation for *Minimum Passenger Elevator Requirements for the Handicapped,* one section of which is reproduced in Figure 23.33. Inasmuch as many states and municipalities already have, or will have in the near future, requirements for accommodation of handicapped persons (particularly as regards public buildings), this NEII standard must be considered in all building planning.

As may be seen from Figure 23.31, it is necessary, in providing for an elevator installation, to consider such factors as the depth of the pit, the dimensions of the hoistway, the clearance from the top of the hoistway to the floor of the penthouse, the size of the penthouse, and the loads that must be carried by the supporting beams.

The penthouse floor and secondary level floor (respectively, containing the elevator traction machine and control panels, and the secondary sheave and selector tape drive) are located above the shaft of each elevator and need approximately two stories of additional height above the top of the support beam of a given elevator when it is standing at its top-floor location. The actual floor area required by the elevator traction machine, its motor-generator set, and control panels is roughly 2.0 times the area of the elevator shaft itself. The required area of the floor of the secondary level is no larger than the elevator shaft it

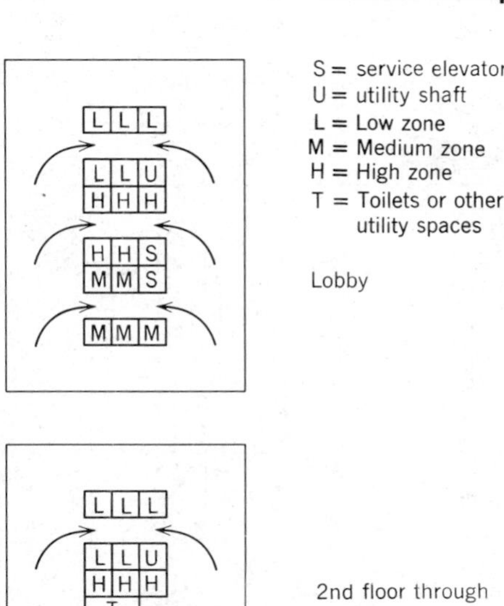

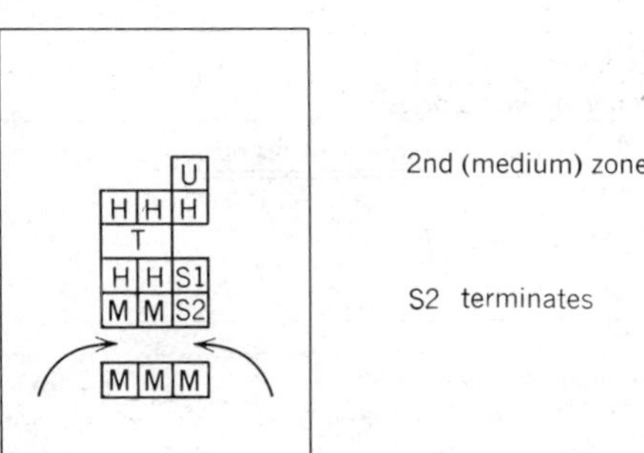

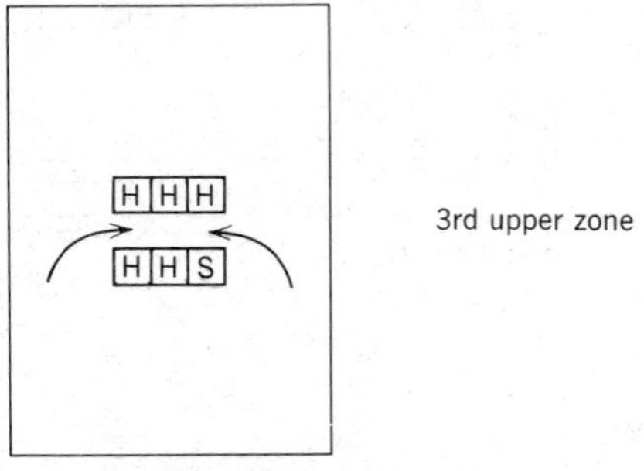

Suggested cores for 3-zone building. Assuming machine room above each zone.

Fig. 23.30 *High-rise core sections.*

serves. As seen in Figures 23.31 and 23.32, the machine room contains the bulk of the elevator machinery. Since some of this equipment will have to be moved for maintenance, it is advisable to furnish an overhead trolley beam that can be used during installation as well. The maximum beam load will be supplied by the elevator manufacturer.

When penthouse space is not available and where a hydraulic unit is not desired, a basement traction unit, also referred to as an underslung arrangement, can be used. These units are low speed (100 to 350 fpm) and are therefore applicable where rise is limited and traffic light to medium. Figure 23.34 shows a shaft section with car and dimensional data.

23.40 Structural Stresses

For the purpose of structural design it is necessary to know the footpounds of kinetic energy that must be supported by the foundations, by structural columns extending upward to the penthouse, and by the main beams which support the penthouse floor and subfloor. This kinetic energy is given for several typical elevators in Table 23.10. The weights given in columns *D*, *E*, *F*, *G*, and *H* include the actual dead weights of equipment when the elevator is not in motion, plus the added weight caused by the momentum of all moving parts and passengers when the elevator is at top speed and is suddenly caused to stop rapidly by the safety devices.

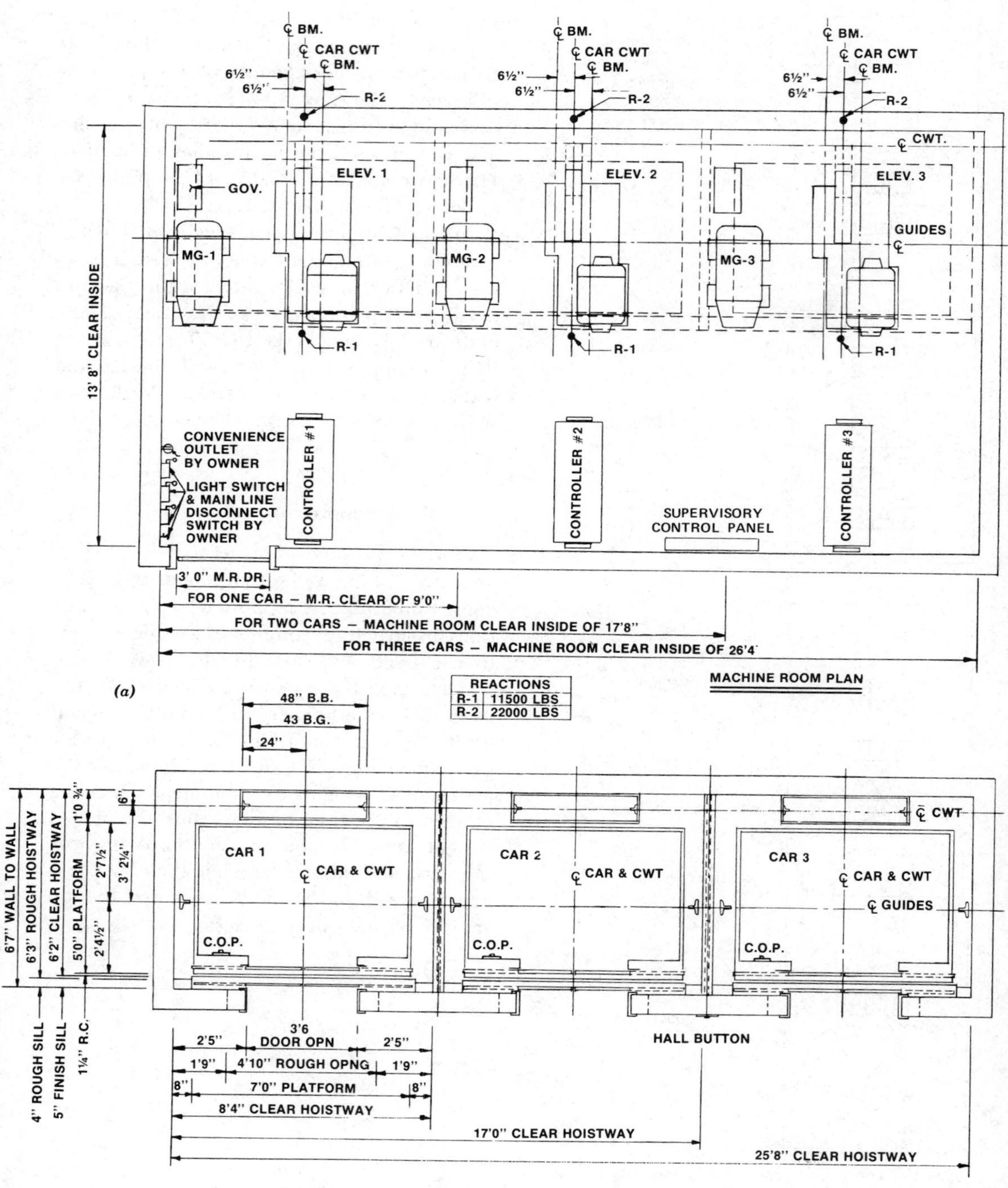

℄ BM.
℄ CAR CWT
℄ BM.
6½"
6½"
R-2
R-1
℄ CWT.
ELEV. 1
ELEV. 2
ELEV. 3
GOV.
GUIDES ℄
MG-1
MG-2
MG-3
13' 8" CLEAR INSIDE
CONVENIENCE OUTLET BY OWNER
LIGHT SWITCH & MAIN LINE DISCONNECT SWITCH BY OWNER
CONTROLLER #1
CONTROLLER #2
CONTROLLER #3
SUPERVISORY CONTROL PANEL
3' 0" M.R.DR.
FOR ONE CAR – M.R. CLEAR OF 9'0"
FOR TWO CARS – MACHINE ROOM CLEAR INSIDE OF 17'8"
FOR THREE CARS – MACHINE ROOM CLEAR INSIDE OF 26'4"
MACHINE ROOM PLAN
REACTIONS
R-1 11500 LBS
R-2 22000 LBS
48" B.B.
43 B.G.
24"
6'7" WALL TO WALL
6'3" ROUGH HOISTWAY
6'2" CLEAR HOISTWAY
5'0" PLATFORM
1'0 ¾"
2'7½"
2'4½"
6"
3' 2¼"
CAR 1
CAR 2
CAR 3
℄ CAR & CWT
℄ CWT
℄ GUIDES
C.O.P.
4" ROUGH SILL
5" FINISH SILL
1¼" R.C.
2'5"
3'6 DOOR OPN
2'5"
1'9"
4'10" ROUGH OPNG
1'9"
8"
7'0" PLATFORM
8"
8'4" CLEAR HOISTWAY
17'0" CLEAR HOISTWAY
25'8" CLEAR HOISTWAY
HALL BUTTON
PLAN VIEW

(a)

(b)

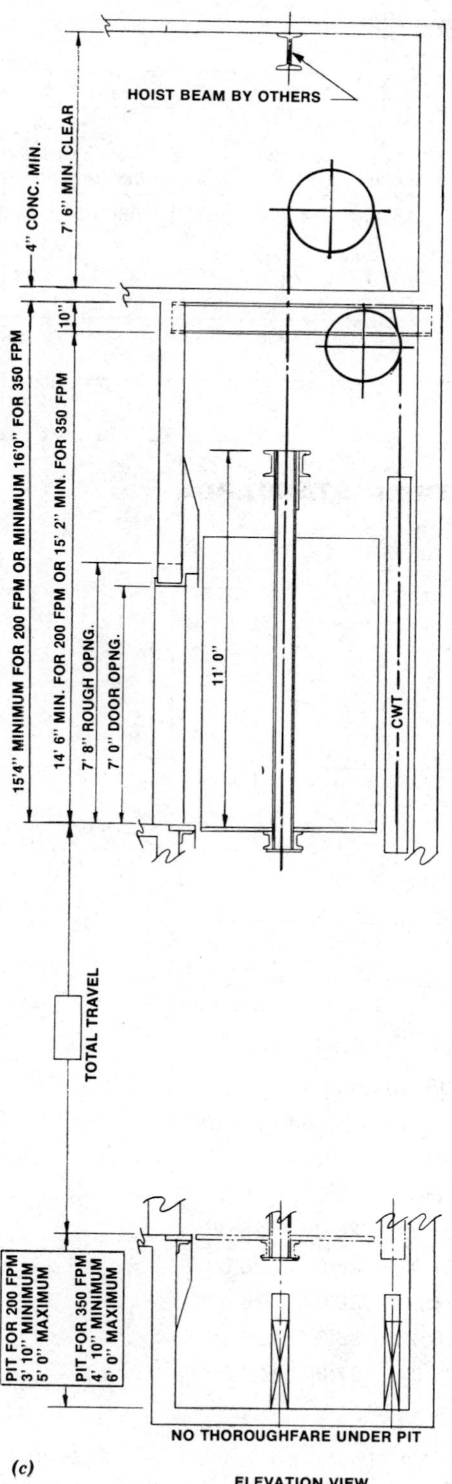

(c)

Fig. 23.31 *Manufacturer's typical layout drawing. Shown are machine room* (a), *hoistway plan* (b), *and hoistway section* (c), *for one, two, or three 2500 lb, 200 to 350 fpm geared premanufactured elevator cars. Courtesy of Montgomery Elevator Co.*

Table 23.10 ***Analysis of Kinetic Energy of Typical Elevators***

(Kinetic Energy at Rated Load and Speed, in Footpounds)

A	*B*	*C*	*D*	*E*	*F*	*G*	*H*	*I*
Traction-Machine Type, Hoist Motor and Control	*Rated Duty*	*Rise in Feet*	*Machine*	*Car*	*Counter-weight*	*Ropes*	*Live Load*	*Total of Columns D, E, F, G, H*
Gearless 1:1 umv, control	2500 lb at 800 fpm	420	4200	15,300	18,200	9800	6800	54,300
Gearless 2:1 umv, control	3000 lb at 500 fpm	200	2950	6675	7975	3925	3250	24,775
Geared 1:1 umv, control	500 lb at 250 fpm	130	6200	1450	1900	320	925	10,795
Geared 1:1 a-c motor, rheostatic control	2500 lb at 150 fpm	100	2600	500	600	140	310	4150

ELEVATOR ENGINEERING STANDARDS

1975 Edition

Passenger elevators,
rated speeds 500–700 fpm

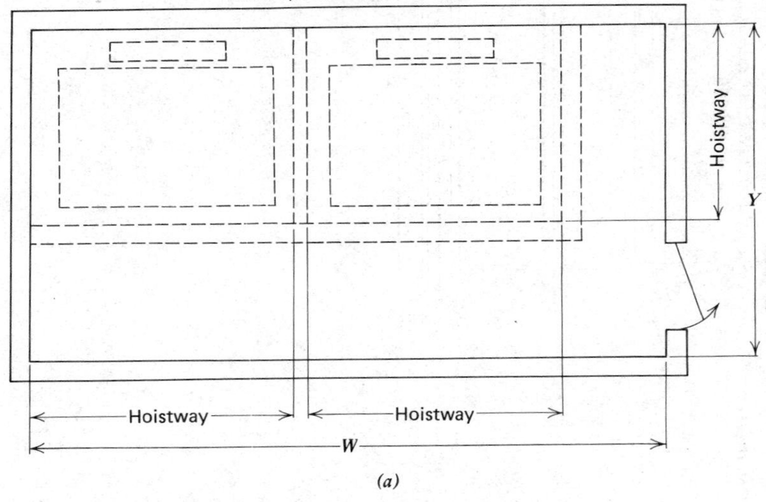

(*a*)

Min. Machine Room Dimensions

Rated Load	*W (Width)*	*Y (Depth) Rated Speeds 500*	*600*	*700*
2000	21′6″	18′0″	18′0″	18′0″
2500	21′6″	21′0″	26′0″	26′0″
3000	21′6″	21′0″	26′0″	26′0″
3500*	21′6″	21′0″	26′0″	26′0″
4000	24′6″	26′0″	26′0″	26′0″
5000	26′0″	26′0″	27′0″	27′0″

*For department stores, W = 24′0″.

Fig. 23.32 *Dimensional plans and application recommendations for two 500 to 700 fpm gearless cars at capacities of 2000 to 5000 lb. Dimensions for machine room* (a), *hoistway* (b), *and hoistway section* (c) *are minimum as recommended by the NEII. Reproduced with permission from Standard Elevator Layouts of Elevator Engineering Standards, 1975 edition, as published by National Elevator Industry, Inc.*

ELEVATOR ENGINEERING STANDARDS

1975 Edition

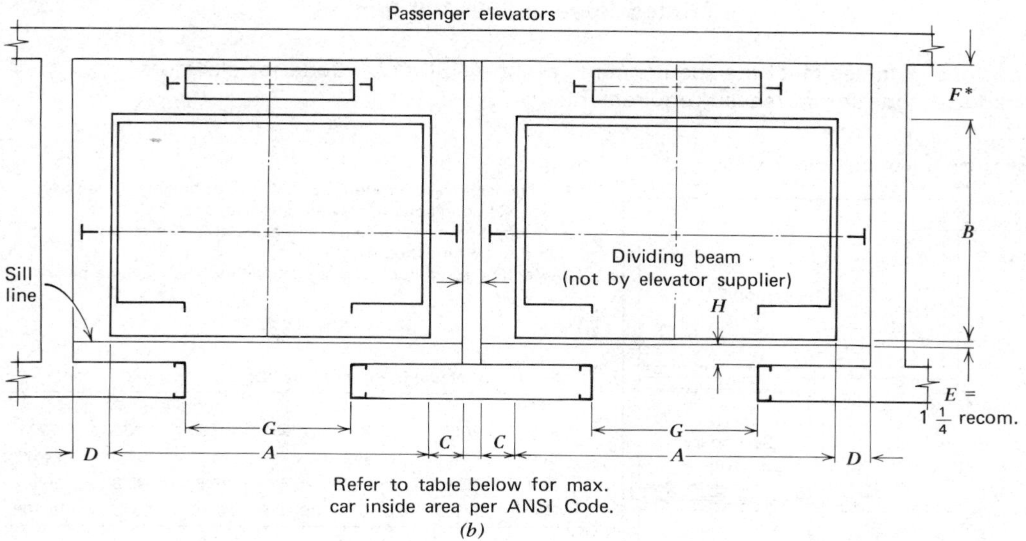

Refer to table below for max.
car inside area per ANSI Code.
(b)

Rated Speeds 500–700 fpm

Rated Load	Max. Area Car Inside	A	B	C	D	F	G	Entrances Type	H (Min.)
	** Application:* Office buildings, apartment houses, hotels, banks, etc.								
2000†	24.2	Dimensions available from National Elevator Industry, Inc., 600 Third Avenue, New York, N.Y. 10016						Single slide Center opening	5
2500†	29.1							Single slide Center opening	5
3000	33.7							Center opening	5
3500	38.0							Center opening	5
4000	42.2							Center opening	5
	Application: Department and other retail stores								
3500	38.0	See note above						Center opening	5
4000	42.2							Two-speed Center opening	6½
5000	50.0							Two-speed center opening	6½

Side platform clearances "C" and "D" do not include any allowance for rail backing. For max. allowable vertical spacing of rail supports, without backing, refer to ANSI-A17.1 Code, Rule 200.4*a*.

*Where pit floor is over occupied space, counterweight safeties are required and dimension "F" may be increased. Consult elevator supplier.

†Housing and Urban Development (H.U.D.) requirements:
2000 lb rated load—5 ft, 8 in. × 4 ft, 3 in. clear car inside, with 3 ft, 0 in. single slide entrance.
2500 lb rated load—6 ft, 8 in. × 4 ft, 3 in. clear car inside, with 3 ft, 6 in. single slide entrance.

(continued on next page)

ELEVATOR ENGINEERING STANDARDS

1975 Edition

Passenger Elevators, Rated Speeds 500–700 fpm

Dimensions in tables meet the requirements of the ANSI-A17.1 Code for elevators. Check local code for increased requirements.

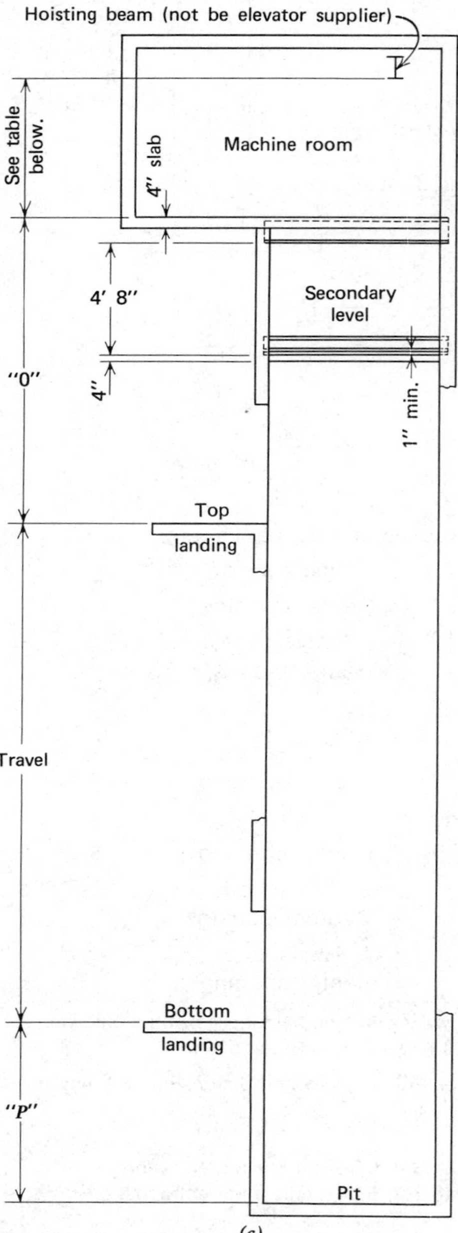

Notes:

* 1. Dimensions shown are for 8 ft, 4 in. cab heights. Refer any special conditions to elevator supplier.
2. Support for elevator machine beams at X-X in elevation are to be furnished by others.
3. Machine room floor slab is flush with top of machine beams.

**** Min. Top of Machine Room Floor "O"***

Speed				*Rated Load*		
fpm	*2000*	*2500*	*3000*	*3500*	*4000*	*5000*
500	24′4″	24′4″	24′4″	24′6″	24′6″	24′6″
600	25′8″	25′8″	25′8″	25′8″	25′8″	25′8″
700	26′10″	26′10″	26′10″	26′10″	26′10″	26′10″

Min. Pit Depth "P"

Speed					*Rated Load*	
fpm	*2000*	*2500*	*3000*	*3500*	*4000*	*5000*
500	9′4″	9′4″	9′4″	9′4″	9′4″	9′4″
600	9′4″	9′4″	9′5″	9′5″	9′6″	9′6″
700	10′3″	10′3″	10′4″	10′4″	10′5″	10′5″

Clear Height of Machine Room (*Min. Height Under Hoisting Beam*)

Rated			*Rated Speeds*
Load	*500*	*600*	*700*
2000	8′6″	8′6″	8′6″
2500	8′6″	8′6″	9′6″
3000	8′6″	9′6″	9′6″
3500	8′6″	10′6″	10′6″
4000	9′6″	10′6″	10′6″
5000	10′0″	10′6″	10′6″

Fig. 23-32 (continued)

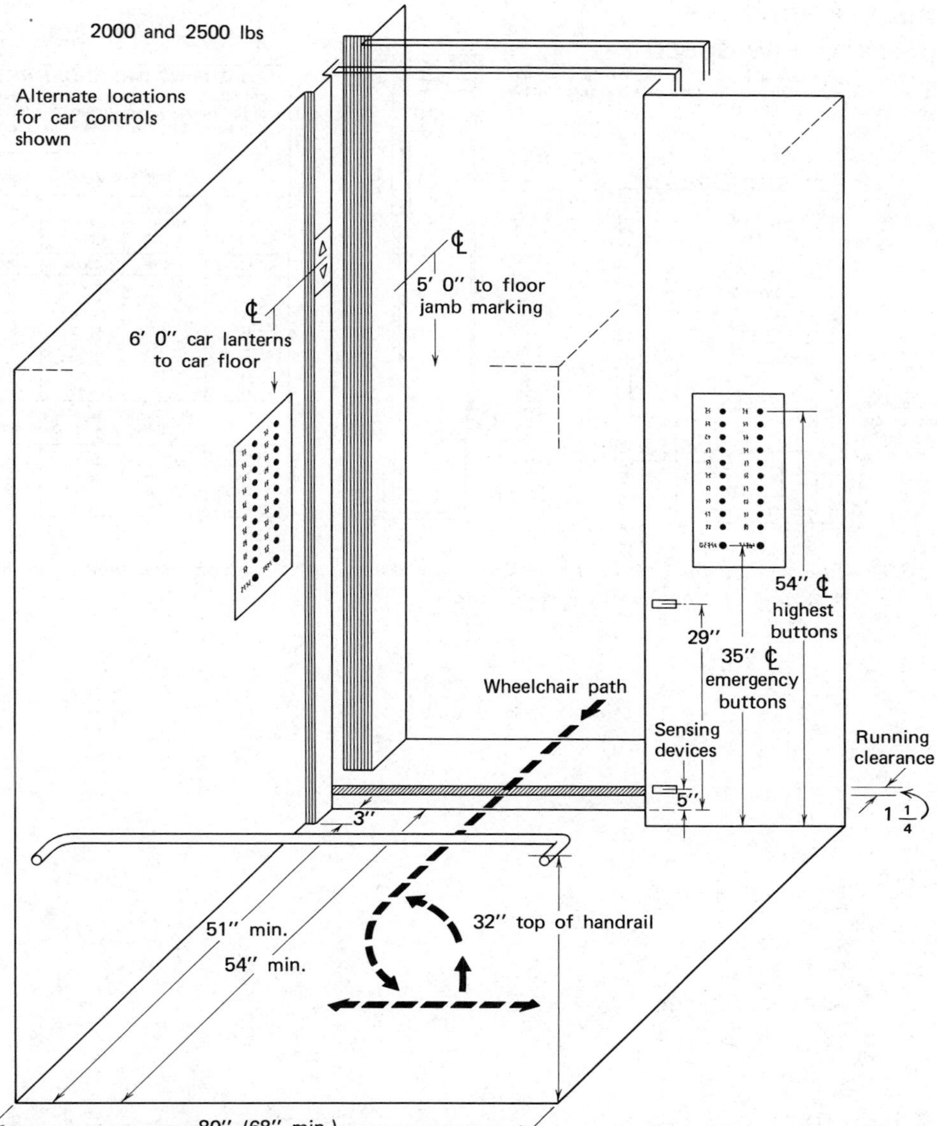

Fig. 23.33 *Composite sketch embodying recommendations of the National Elevator Industry, Inc., for 2000 and 2500 lb elevator car sizes, arrangements, and appointments to acommodate the handicapped. Reprinted with permission from NEII standards.*

Basement Traction — medium and low speed

Basement traction elevators are utilized for limited overhead conditions in new and existing buildings. This type of elevator facilitates future floor expansion.

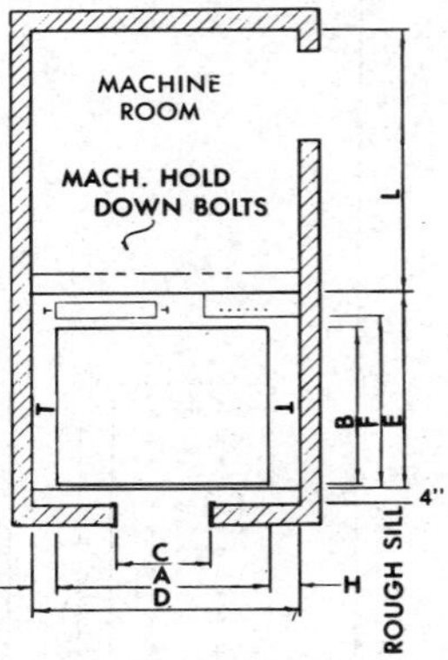

Machine room may be placed at side of elevator hoistway if dimension "D" is increased 4".

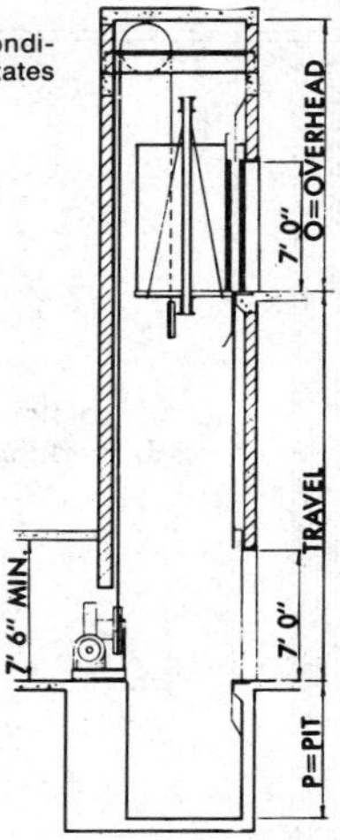

2:1 ROPING ARRANGEMENT permits a minimum overhead installation. Because of the sheave arrangement, it is necessary to have a greater pit depth than for a comparable 1:1 installation.

RECOMMENDED SIZES & CAPACITIES				
TYPE BUILDING	SMALL APARTMENT	SMALL OFFICE	AVERAGE OFFICE HOTEL	
CAPACITY	1200#	2000#	2500#	3000#
A	5' 0"	6' 4"	7' 0"	7' 0"
B	4' 0"	4' 5"	5' 0"	5' 6"
C	2' 6"	3' 0"	3' 6"	3' 6"
D	6' 10"	8' 2"	8' 10"	8' 10"
E	5' 4"	5' 9"	6' 6"	7' 0"
F	4' 5"	4' 10"	5' 5"	5' 11"
G	10"	10"	10"	10"
H	12"	12"	12"	12"
RECOMMENDED MACHINE ROOM OVERHEAD & PIT DIMENSIONS				
SPEED	100	200	250	300
L	13' 0"	15' 6"	15' 6"	15' 6"
O	12' 10"	13' 0"	13' 5"	13' 6"
P	5' 6"	6' 6"	6' 11"	7' 4"

Fig. 23.34 Typical data for basement traction or underslung arrangement, used where penthouse is unavailable. Courtesy of Montgomery Elevator Co.

24

Vertical Transportation: Special Topics

POWER AND ENERGY

24.1 Power Requirements

The power required by an elevator is that amount necessary to perform the necessary traction work and to overcome friction. Since power is equal to the rate at which work is done, that is

$$hp = (k)(\text{ft-lb})/\text{time}$$

the elevator motor size is directly proportional to the speed of the system, other factors remaining the same (k is a constant). In other words, it requires proportionately more power to lift a 3000 lb car at 700 fpm than at 200 fpm. This relationship is shown in Figure 24.1, which shows the size of the d-c elevator traction motor as a function of speed, for different capacity cars. This motor supplies the system friction in addition to the required traction power. (For power data on hydraulic elevators see Section 24.8.) Since friction is higher in geared machines than in gearless units, the traction motor must be larger. The m-g set motor size is approximately 20% larger than the value shown, to compensate for the set losses.

When solid-state control is used (and the m-g set eliminated) the values shown in Figure 24.1 are still valid inasmuch as the size of the traction motor is unaffected by the type of variable voltage supply. Some elevator manufacturers oversize the traction motor in order to provide rapid acceleration, resulting in the anomalous situation of the traction motor being larger than the m-g set motor. Figure 24.1, therefore, gives the *minimum* traction motor size. Since the current inrush is high when a motor is started, high-power demand occurs when starting the m-g set and when starting the traction motor under heavy load conditions. This high-current demand results in a high instantaneous power demand and a lowered power factor—both of which can result in penalty charges from the power company. If such occurs, it is possible to interconnect the controllers of the individual units in a bank to prevent simultaneous starting. The time lag introduced is very small, and the effect on system operation will be negligible.

As stated in Section 23.16, an elevator moves only about 50% of the time, the remainder being spent standing at various landings. As the number of cars in a bank increases, the probability of *all* the cars being in operation simultaneously decreases, resulting in a system demand factor. The factor is shown directly in Figure 24.1.

As an example of the use of the curves consider a bank of five 3500 lb, 600 fpm units. From Figure 24.1, each car requires 48 hp

$$\text{Group demand factor} = 0.67$$

Total instantaneous power required

$$= 5 \times 48 \times 0.67 = 160 \text{ hp}$$

Note that this is the *traction motor* power requirements. Assuming an overall efficiency (eff) of 80% for the m-g sets, the elevator system power requirement is

$$\text{System power} = \frac{160 \text{ hp}}{80\% \text{ eff}} = 200 \text{ hp}$$

which would have to be provided by the building power system.

24.2 Energy Requirements

The energy used by an elevator is essentially the system friction, including the heat generated by the brakes plus the electrical

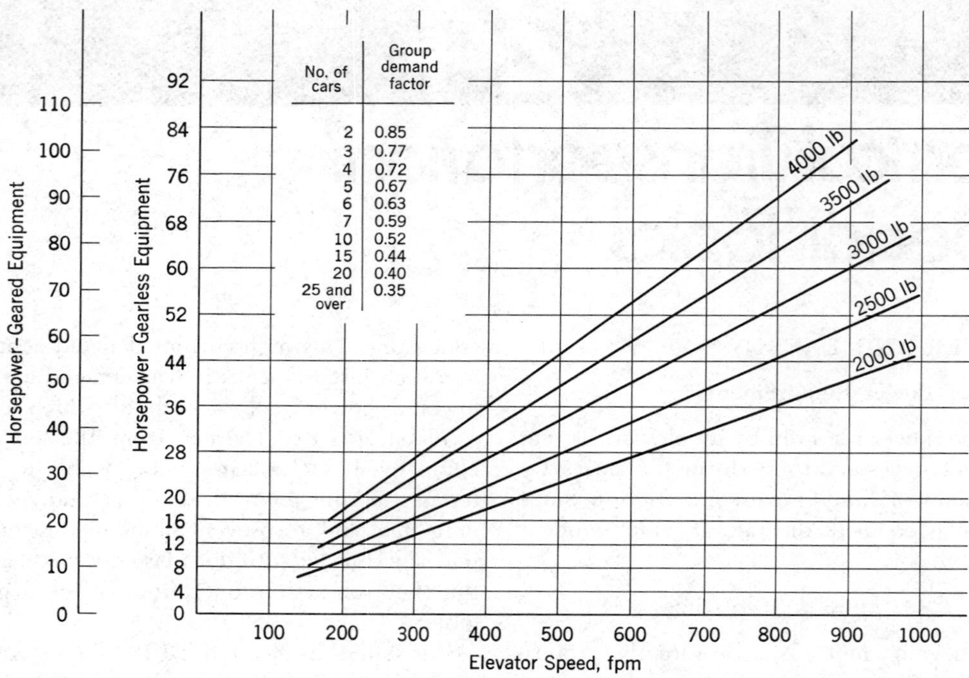

Fig. 24.1 Elevator traction motor power requirements per car. Size of M-G set motor is approximately 20% larger than the traction machine.

losses in the traction motor and m-g set, if any. The energy expended in raising the car and its passengers is simply stored as potential energy. It is *returned to the power system* when the car and passengers descend, via the system of regenerative braking in use in almost all elevator systems. Refer to Figure 24.2, which shows the approximate efficiencies of the components of a typical system. With this data, we are able to calculate a system's energy consumption.

EXAMPLE 24.1

Given a system of five 3500 lb, 600 fpm cars calculate:

a. Heat generated in the machine room during peak periods.
b. Approximate monthly energy cost using a combined demand/energy rate of $.035 per kwh.

SOLUTION

(a) During peak periods the m-g set operated continuously, idling about 50% of the time and loaded about 50% of the time. A combined figure of full load for 70% of the period would be sufficiently accurate for our purposes. The traction motor operates approximately 50% of the time and is at standstill the other half. Assume for both motors that while operating they draw 90% of full load. Therefore, for one car, from Figure 24.1

$$\text{Traction motor} = 48\text{ hp}$$

and

$$\text{m-g set motor} = \frac{48}{0.8} = 60\text{ hp}$$

Losses in m-g set (80% efficiency)

$$60\text{ hp} \times 90\%\text{ load} \times 70\%\text{ operation} \times 20\%\text{ loss} = 7.56\text{ hp}$$

$$\text{Btu} = 7.56 \times 2{,}545\,\frac{\text{Btu}}{\text{hp}}$$

$$= 19{,}243\text{ Btu/hr}$$

Losses in traction machine

$$48 \text{ hp} \times 90\% \text{ load} \times 50\% \text{ operation} \times 20\% \text{ loss} = 4.32 \text{ hp}$$

$$\text{Btu} = 4.32 \times 2{,}545 \frac{\text{Btu}}{\text{hp}} = 10{,}994 \text{ Btu/hr}$$

Total	30,237 Btuh
Say ±	30,000 Btu/hr/car

Since there are five elevators operating, the total heat generated is

$$5 \times 30{,}000 \text{ Btuh} = 150{,}000 \text{ Btuh!}$$

This is the rating of a large home furnace. No diversity is taken since all the machines are operating and the heating is additive; diversity is only applicable to calculate instantaneous load.

(b) To calculate the monthly energy cost it is required that an estimate be made of the total usage of the system. Assuming the system to be in an office building, a reasonable breakdown of operation during a 24 hr day would be

2 hrs peak use
2 hrs 70% of peak
6 hrs 50% of peak
14 hrs 10% of peak

This gives an average of 30% of peak load for the bank. Therefore, per car

$$\begin{aligned} \text{Energy} &= 30\% \times \text{total losses} \times 24 \text{ hr} \\ &= 0.3 \times (7.56 + 4.32) \text{ hp} \times 24 \text{ hr} \\ &= 85 \text{ hp-hr} = 64 \text{ kwh/day/car} \end{aligned}$$

Monthly cost would be

$$64 \frac{\text{kwh}}{\text{day}} \times 25 \text{ days} \times \$.035$$

$$\simeq \$55\text{/month/car}$$
$$\simeq \$275\text{/month for the bank}$$

See Section 23.16, page 1061, for an analysis of cost economies possible with solid-state variable voltage equipment.

In view of the large heat gain in an elevator machine room, it is well to equip the room with thermostatically controlled forced ventilation arranged to keep the room temperature from exceeding 90° F. Occasionally spill air from an air-conditioning system can be utilized for cooling purposes.

24.3 Energy Conservation

A reduction in energy consumption can be accomplished by implementing the following recommendations:

(*a*) For existing elevators

1. Increase interval during nonpeak hours.
2. Replace m-g sets with thyristor controls.
3. Recycle machine room waste heat.
4. Shutdown some units completely during off hours.

(*b*) For building in the planning stage

1. Design for maximum trip time.
2. Use the lowest speeds possible, within a type—that is, geared and gearless.
3. Use gearless equipment whenever possible (see Fig. 24.2).
4. Implement the recommendations of (*a*) above after construction.

Since elevator shafts have a powerful stack effect, measures should be taken to counteract this during the heating season.

24.4 Emergency Power

Recent major power failures have demon-

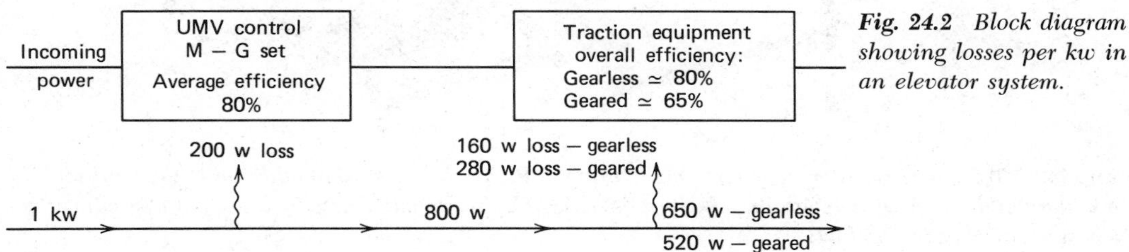

Fig. 24.2 Block diagram showing losses per kw in an elevator system.

strated forcefully the need for a standby or emergency power source of adequate size to operate a building's elevators. Few experiences are so harrowing as being trapped in the crowded narrow confines of a small box suspended in a long vertical shaft, with little or no light, and complete strangers for companions.

A common misconception relative to elevators is that on failure of power, the cars will automatically descend to the nearest landing where exit will be possible. In actuality, the brake is set immediately on power outage and the car remains stationary. Hydraulic cars can be lowered by operation of a manual valve; *small* traction cars can be cranked to a landing by hand; but large cars are fixed in position. This is particularly bad for cars in blind shafts, that is, express shafts with no shaftway doors. In such cases escape from the cars via hatchway is not practicable and, when emergency power is not available, breaking through the shaftway walls is the only recourse.

In addition to simple inconvenience, loss of elevator service in facilities such as hospitals, mental institutions, and jails constitutes a danger to life. For this reason most states require that emergency power be available in public buildings to operate one elevator at a time, and for lighting and communications. ANSI also requires that *every* elevator be equipped with emergency lighting and the ability to communicate with a central location. Most installations separate the emergency power functions, providing a diesel generator for traction power and separate battery packs for communications and lighting. The latter two items can be furnished by the elevator manufacturers with the cars, as an option. Typical self-contained emergency lighting units are shown in Figure 24.3.

The generator is normally sized to supply one elevator motor at a time, with manual or automatic switching arranged between unit controllers. Thus each car in turn can be brought to a landing and thereafter a single car retained in service. Obviously, if it is desired to operate more than one car, a larger generator can be installed. This might well be the case in a multiwing building with critical service requirements, such as a hospital.

The amount of power required, the size of the emergency generator, and the equipment size necessary to absorb regenerative power are all data that can be furnished by the consulting engineer and the elevator manufacturer.

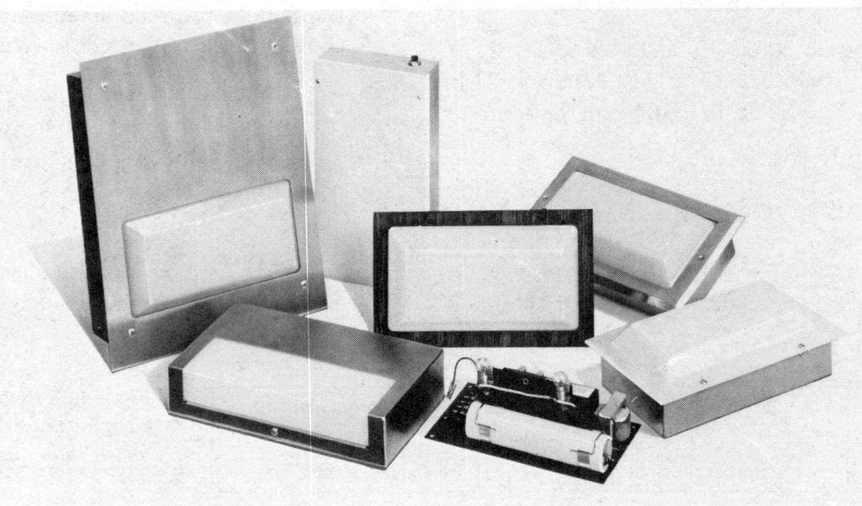

Fig. 24.3 *Typical self-contained emergency lighting units. Each contains, in addition to the light source, a battery and the requisite controls, which in the case of a fluorescent source includes a d-c/a-c inverter. Courtesy of Nylube Products.*

OTHER CONSIDERATIONS

24.5 Safety Considerations

(*a*) ***Fire Safety.*** As a result of sad experience, the ANSI elevator code (A 17.1) now requires what is commonly known as the Firemen's Return feature. This requires that a key switch be provided at the lower terminal which, when actuated, will return all cars to the terminal *without stopping* for any hall or car calls. After returning, the car parks with doors open until returned to normal service. The individual cars will then be operable manually only, from inside the car.

Many installations are equipped with automatic fire and smoke detection systems. Their action in general will

1. Cause all cars above the lower terminal to return to the terminal without stopping (action similar to the key action described above).
2. Sound an alarm in the selected location.
3. In certain instances, alert the fire department or a security organization.

In buildings equipped with automatic systems of this type, the lobby key switch has a third position (in addition to *on* and *off*) that overrides the action of the automatic detection system.

(*b*) ***Elevator Security.*** The problem of physical security in elevators is a serious one, inasmuch as the traveling elevator is an enclosed space which can be rendered inaccessible by the simple expedient of pressing the emergency stop button. To ameliorate this danger to an extent, many residential elevators are equipped with alarm buttons, which alert residents and security personnel, if any. Every elevator, by Code, must be equipped with communication equipment of some sort, but not necessarily voice equipment. Also, a hand-held phone is not practical in a security-emergency situation. More effective is an open two-way communication system with "no hands" operation in the car. When a closed-circuit TV monitor is added to this, utilizing a wide-angle camera in each car (see Fig. 24.4), the security problem will have been, in large measure, solved. Obviously, using a communication system presupposes continuous manning of the building security desk.

24.6 Elevator Specifications

Two basic types of specifications for elevator equipment, as for other types of equipment, are utilized. The performance specifications describe job conditions and invite interested contractors to submit detailed proposals including full engineering. The burden of comparing proposals then falls on the

Man Trying To Hide In Corner Under Camera

Fig. 24.4 *Wide-angle TV camera intended for elevator cab surveillance. The prominent, printed warning is an integral part of the system's effectiveness. Courtesy of Visual Methods, Incorporated.*

owner, who—if competent to properly perform such an evaluation—would most probably do better to utilize the equipment-type specification.

In recent years the use of performance specifications has increased because of the advent of preengineered, premanufactured systems. These are supplied by the major manufacturers and have these advantages

1. Approximately 10% lower cost than a custom-designed system.
2. A complete engineered and tested system is provided whose performance and cost is known exactly.
3. Rapid delivery.
4. Minimum supervision required by the owner and architect.

If architects decide to use a custom-designed system, they must prepare detailed drawings and specifications. The specifications must include:

Elevator rated load
Elevator speed
Maximum travel
Number of landings and openings
Platform and door size
Type of machine, for example, gearless
Type of control
Type of supervisory system
Type of car and door
Shaft doors
Signal equipment
Characteristics of power supply
Finishes

This last item can be left as a dollar allowance for architectural treatment of the car interior. Since the selection of, and technical specifications for, elevators are specialized and complex, the services of an elevator consultant are usually required.

In addition to the technical portions of the specifications, it is imperative that the following items be covered in detail:

(*a*) Owner's responsibility—The *construction* contractor normally provides:

1. The hoistway, including properly designed, lighted, drained, waterproofed, and ventilated machine room and pit.
2. Access doors, ladders, and required guards.
3. Guide rail bracket supports, support for machine and sheave beams, and grouting for hoistway sills.
4. Electric feeder terminating in switch in machine room.
5. Hoistway outlets for light, power, and telephone.
6. Temporary light and power during construction.
7. Concrete machine foundations.
8. Vents, holes, and other work to satisfy fire codes.
9. All cutting, patching, and chasing of walls, beams, masonry, and so on.
10. Coordination of all work.
11. Any special work, as negotiated.

(*b*) Elevator contractor's responsibility—A complete, working, tested, and approved system in accordance with specifications, plus any special work such as painting, special tests, scheduling of work, so as to provide temporary elevator service.

(*c*) Special job conditions such as work restrictions, scheduling, penalties or bonuses, test reports, and the like.

In alteration and modernization work the problems of coordination are more complex, and an elevator contractor experienced in this type of work must be selected. To this end, in all elevator contract work, bids should be solicited by consulting qualified bidder lists. Part of an elevator contract comprises maintenance of the installation for a specific period after completion. Contractors with poor maintenance facilities should be avoided.

SPECIAL ELEVATORS

24.7 Unique Traction Designs

Elevator engineers have in the past decade provided interesting and original solutions to unusual problems. A detailed analysis of these solutions is beyond our scope here; however, a rapid review is in order.

(*a*) ***Sky-Plaza System.*** For skyscraper buildings such as the World Trade Center and multiple-use buildings such as the John Hancock Tower—both of which are, in effect, stacked multiple buildings—the elevator solution involves transporting large groups of people from the street lobby to an upper lobby, called a sky-plaza. At this point the passengers transfer to another elevator to continue their upward journey (see Fig. 24.5). The advantages of this system are

1. Reduction in the space occupied by elevators since the shafts do not extend the entire height of the buildings.
2. Interrupting the otherwise lengthy vertical trip by the horizontal break at the sky lobby.

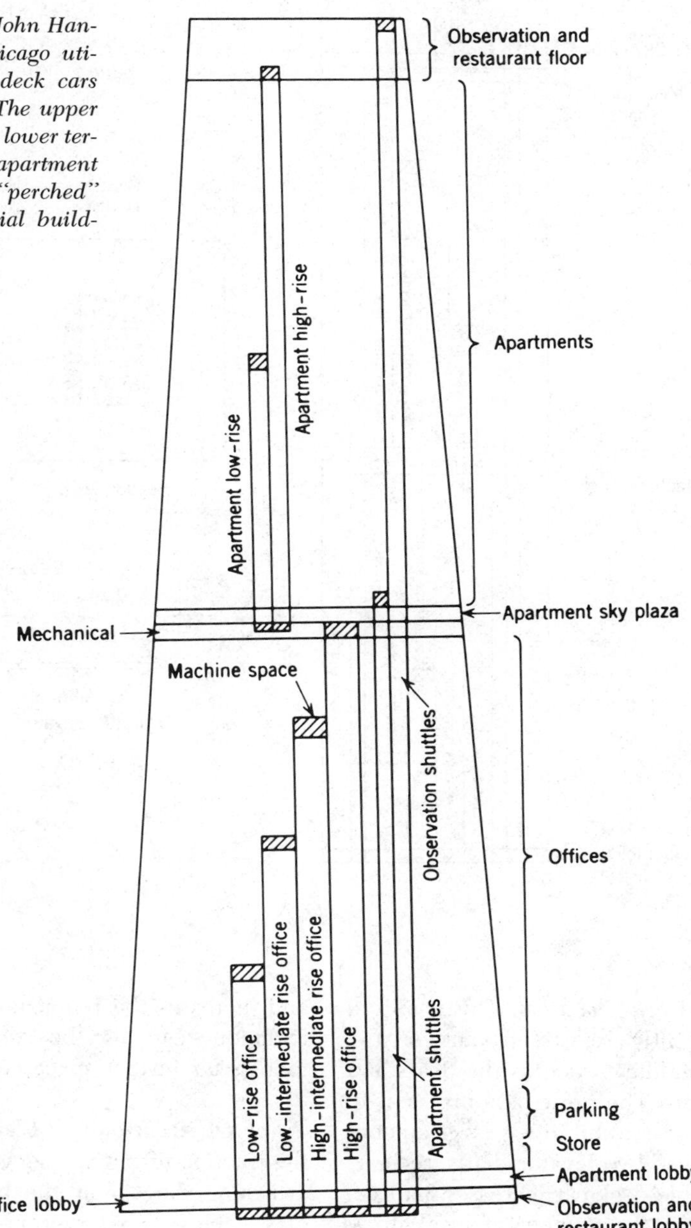

(*a*)

Fig. 24.5 (a) *The John Hancock Tower in Chicago utilizes both double-deck cars and a sky lobby. The upper lobby serves as the lower terminal for the apartment building which is "perched" atop the commercial building.*

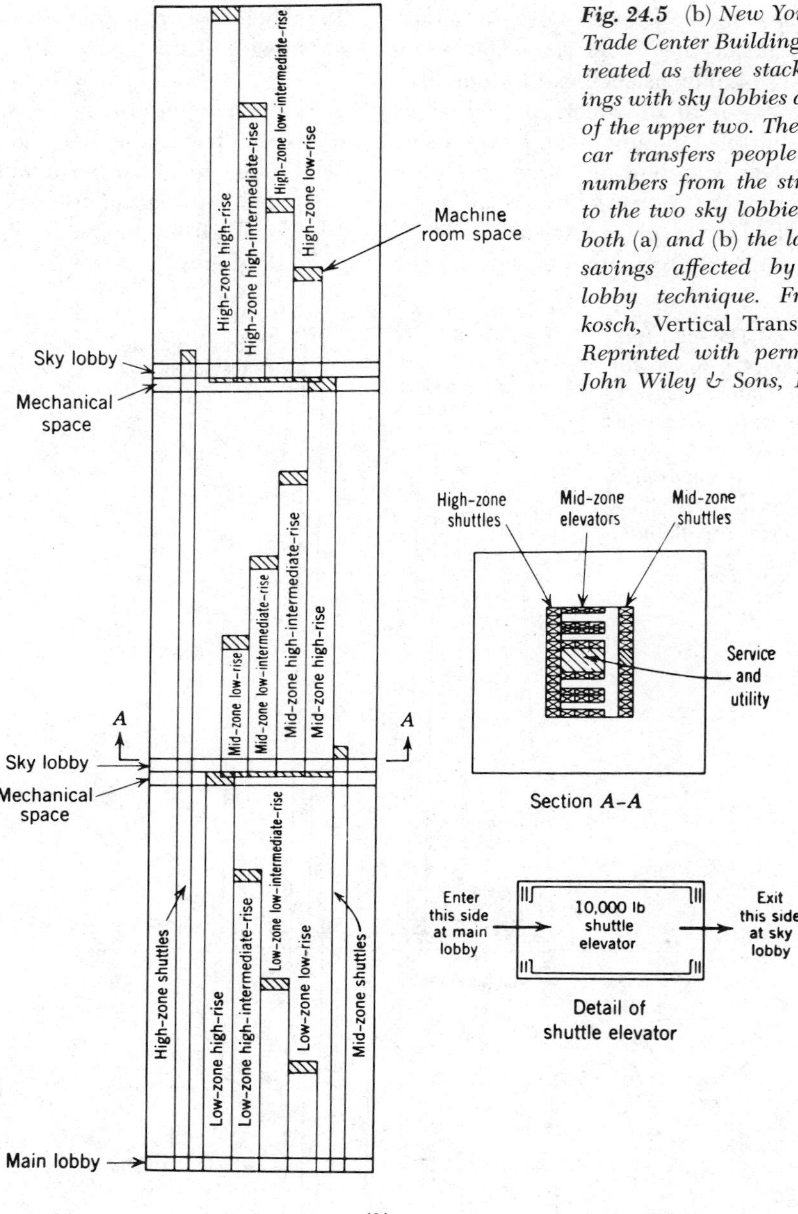

(b)

Fig. 24.5 (b) *New York's World Trade Center Buildings are each treated as three stacked buildings with sky lobbies at the base of the upper two. The 10,000 lb car transfers people in large numbers from the street lobby to the two sky lobbies. Note in both* (a) *and* (b) *the large space savings affected by the sky lobby technique. From Strakosch,* Vertical Transportation. *Reprinted with permission of John Wiley & Sons, 1967.*

(*b*) ***Double-Deck Elevators.*** This is an old technique, recently revised to answer the needs of tall buildings such as the Sears and Citicorp. Towers. The principal purpose is to limit the otherwise prohibitively large amount of space occupied by elevator shafts (see Fig. 24.6). The double-deck car increases shaft capacity, decreases the number of local stops, and increases the rental area available. This technique can also be combined with sky-lobbies for further space economy.

(*c*) ***Observation Car Elevators.*** By placing the traction lifting mechanism *behind* the car, attaching the car at the back, and using a glass-enclosed, observation-style car a spec-

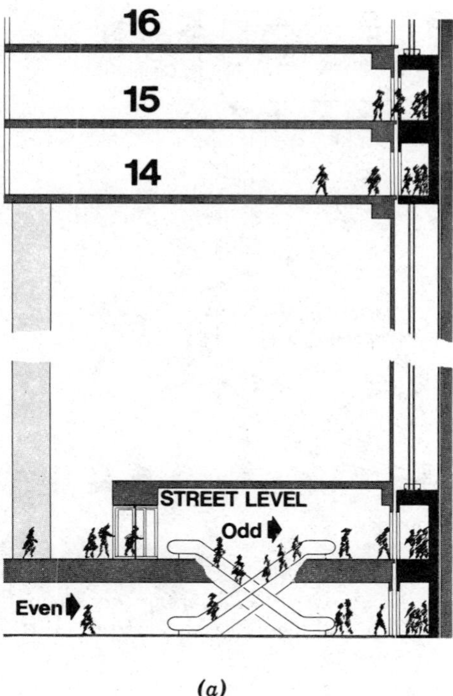

(a)

Fig. 24.6 (a) *The double-deck car serves to increase cab capacity and decrease shaft space. Coincidence of calls in the upper and lower cabs reduces the number of local stops made by the double car.*

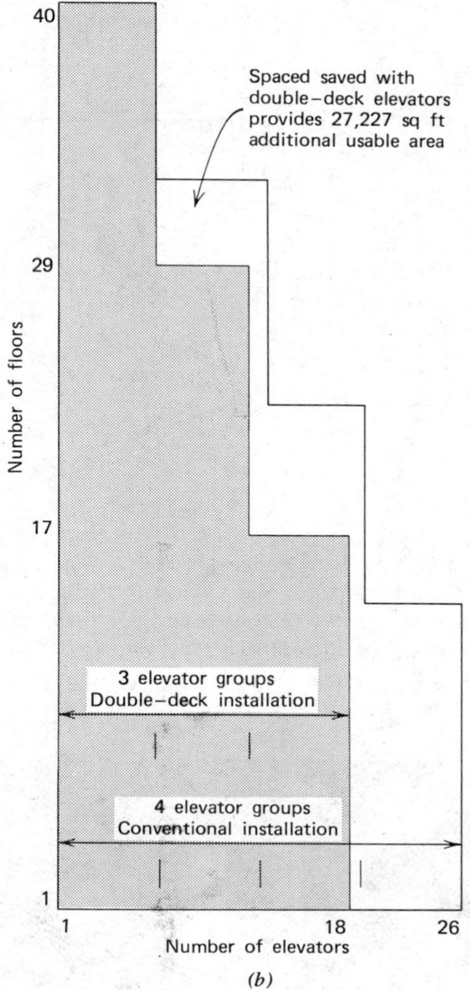

(b)

Fig. 24.6 (b) *Graphical representation of the space saved in a 40-story building by the use of double-deck elevators. Courtesy of Otis Elevator Co.*

tacular unit can be constructed that becomes an attraction in itself. The basic construction is shown in Figure 24.7, and two well-designed examples are seen in Figure 24.8. If the back screen is treated properly the car gives the impression of movement without any apparent motive force or machinery.

(d) Slant Elevators. Although elevators are normally conceived as traveling vertically, this is not necessarily so. Slant or inclined elevators have been constructed in numerous locations. The design varies depending on the angle of incline. One design system is shown in Figure 24.9 along with an extremely well-known application of an inclined elevator.

24.8 Hydraulic Elevators

All of the foregoing elevators are traction types, that is, they are raised and lowered as a result of the tractive force of cables attached to, or passing under, the car. In contradistinction to these, the hydraulic or plunger elevator is raised and lowered quite simply, by means of a movable rod (plunger) rigidly fixed to the bottom of the elevator car. The absence of cables, drums, m-g sets, elaborate controllers and safety devices, and penthouse equipment make this system inherently inexpensive and often the indicated choice for low-speed (up to 200 fpm), low-rise (up to 75 ft) applica-

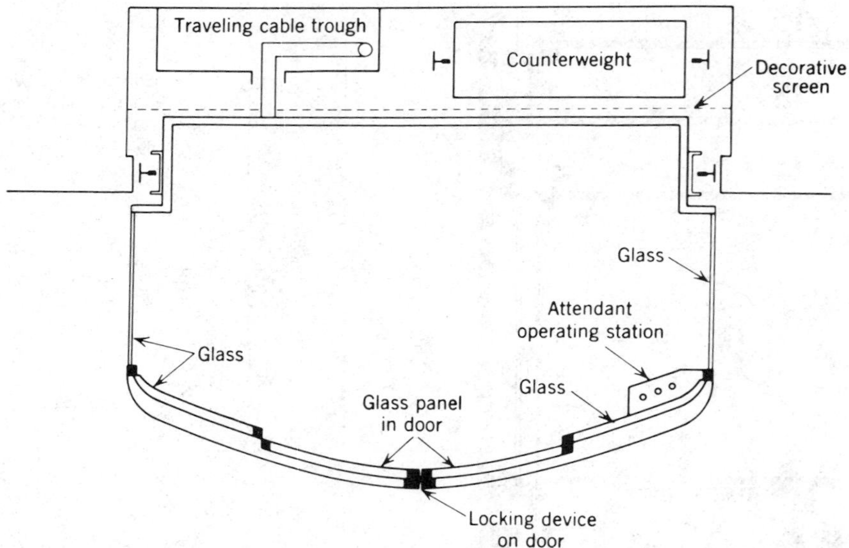

Fig. 24.7 *The basic design of a traction-type observation elevator is illustrated. Cars are small to permit all the occupants to enjoy the view through the cab's glass walls. An inside railing (not shown) is frequently used to prevent passengers from contacting the glass cab walls. From Strakosch,* Vertical Transportation. *Reprinted with permission of John Wiley & Sons, 1967.*

Fig. 24.8 (a) *One of the earliest observation-type elevator installations, at the Hyatt Regency Hotel in Chicago. Note the effectiveness of the screening, making the traction machinery barely visible. Photo courtesy of Bethlehem Steel Corporation, which supplied the elevator rope for this installation.* (b) *A twin installation with unusual car design, at the Hilton Hotel in Atlanta. Photo courtesy of Hauenstein and Burmeister, Inc., which supplied the cars for this installation.*

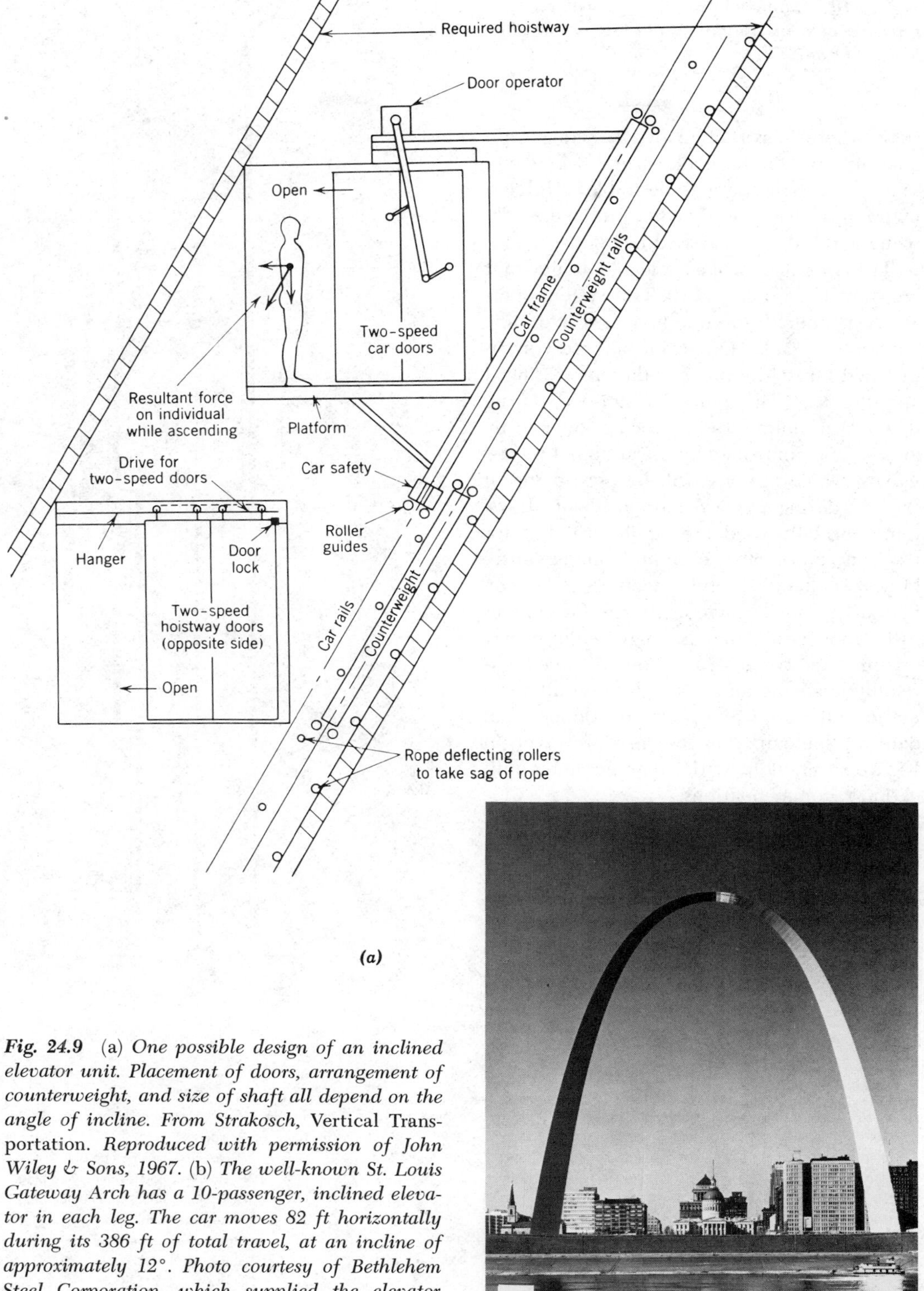

Fig. 24.9 (a) *One possible design of an inclined elevator unit. Placement of doors, arrangement of counterweight, and size of shaft all depend on the angle of incline. From Strakosch,* Vertical Transportation. *Reproduced with permission of John Wiley & Sons, 1967.* (b) *The well-known St. Louis Gateway Arch has a 10-passenger, inclined elevator in each leg. The car moves 82 ft horizontally during its 386 ft of total travel, at an incline of approximately 12°. Photo courtesy of Bethlehem Steel Corporation, which supplied the elevator rope for this installation.*

Fig. 24.10 *Phantom view of a typical plunger-type elevator installation. Courtesy of Westinghouse Elevator Co.*

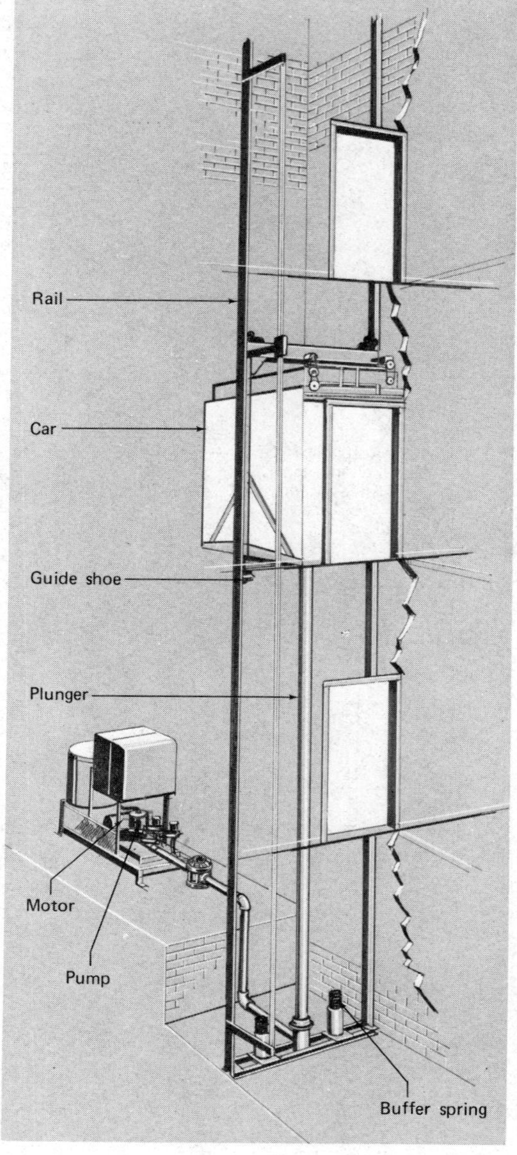

tions where construction of the plunger pit does not present difficulties and where absence of a penthouse is desirable. (Holeless hydraulics have recently come on the elevator scene and will be discussed below.)

The components of a typical hydraulic unit are shown in Figure 24.10. This system operates very much the same way as a hydraulic automobile jack. Oil from a reservoir is pumped under the plunger thereby raising it and the car. The pump is stopped during downward motion, the car being lowered by gravity and controlled by the action of bypass valves, which also control the positioning of the car during upward motion. Control systems normally used are similar to that for traction types, for example, single push-button, collective, and selective collective. Similarly, door arrangements are the same as in traction types, that is, single slide, center opening, and two-speed. Automatic leveling is readily available and is standard on all automatic units. Typical layout and dimensional data for standard plunger units are given in Figure 24.11, along with capacities and application recommendations.

Where drilling a plunger hole presents difficulties, a hydraulic installation using a tele-

Fig. 24.11 (below and opposite) *Typical dimensional, capacity, and layout data for conventional plunger-type hydraulic elevator. Extracted with permission from published data of Westinghouse Elevator Co.*

			Dimensions							
			Platform			*Hoistway*				
Capacity in Pounds	*Speed Feet Per Minute*	*Door System*	*A Width*	*B Depth*	*C Clear Hatch*	*D Clear Hatch*	*J Clear Opening*	*M Wall to Wall*	*No. of Passengers*	*Application*
1500	75	SS	5′4″	4′6″	6′8″	13′8″	2′8″	5′3″	10	Residential
2000	125	SS	6′4″	4′5″	7′8″	15′8″	3′0″	5′2″	13	Residential, small office bldg.
2500	125	CO	7′0″	5′0″	8′4″	17′0″	3′6″	5′9″	17	Residential, hotel, small office bldg.
2500	125	SS	7′0″	5′0″	8′4″	17′0″	3′6″	5′9″	17	Residential, hotel, small office bldg.
3000	100	CO	7′0″	5′6″	8′4″	17′0″	3′6″	6′3″	20	Store, small office bldg.
3500	100	CO	7′0″	6′2″	8′4″	17′0″	3′6″	6′11″	23	Store, small office bldg.
4000	100	CO	8′0″	6′2″	9′4″	19′0″	4′0″	6′11″	27	Store, small office bldg.

SS = Single slide
CO = Center opening

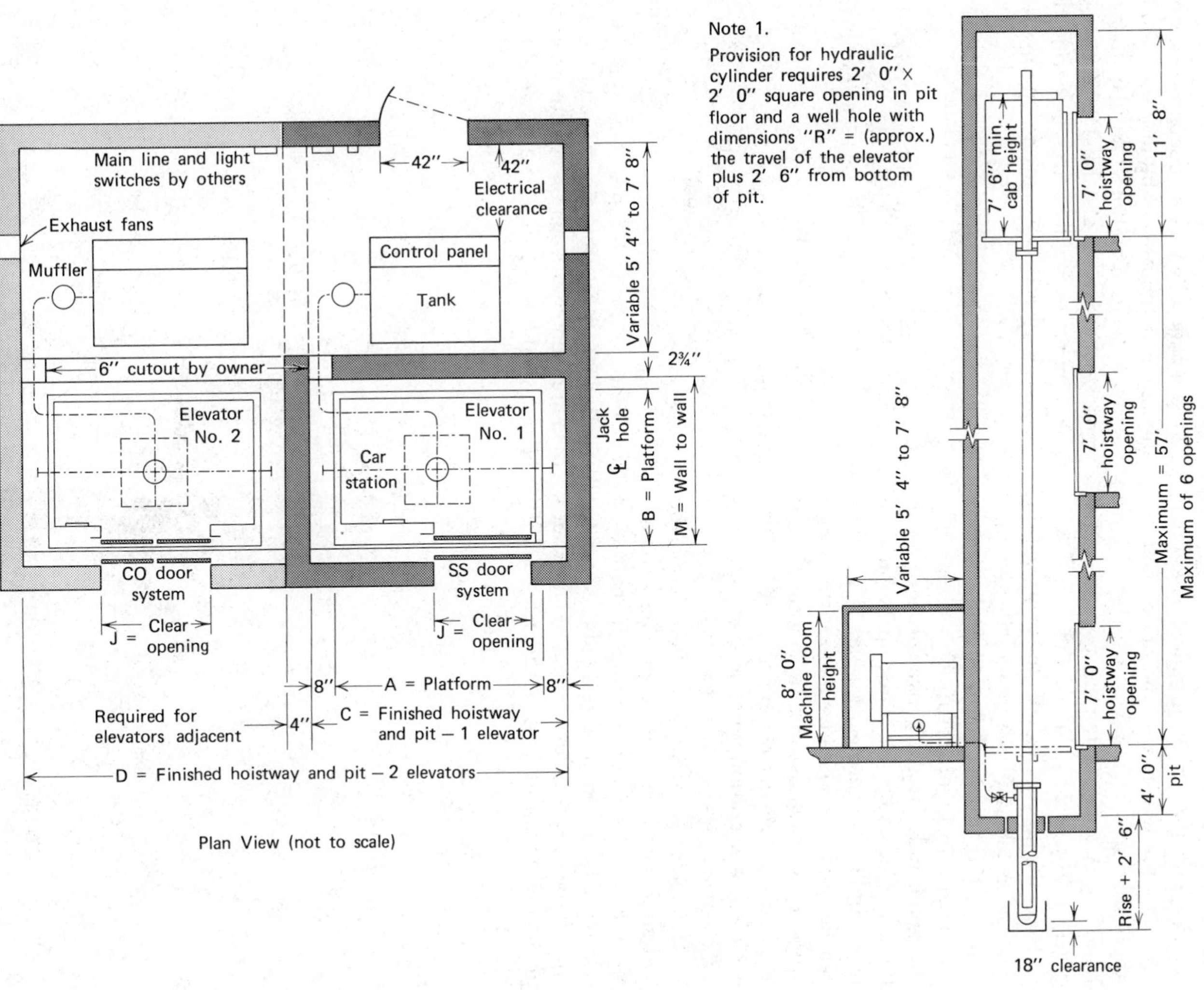
Main line and light switches by others
42"
42"
Electrical clearance
Exhaust fans
Muffler
Control panel
Tank
Variable 5' 4" to 7' 8"
2¾"
6" cutout by owner
Elevator No. 2
Elevator No. 1
Car station
℄ Jack hole
B = Platform
M = Wall to wall
CO door system
SS door system
Clear J = opening
Clear J = opening
8"
A = Platform
8"
Required for elevators adjacent
4"
C = Finished hoistway and pit – 1 elevator
D = Finished hoistway and pit – 2 elevators
Plan View (not to scale)
Note 1.
Provision for hydraulic cylinder requires 2' 0" × 2' 0" square opening in pit floor and a well hole with dimensions "R" = (approx.) the travel of the elevator plus 2' 6" from bottom of pit.
7' 6" min. cab height
7' 0" hoistway opening
11' 8"
7' 0" hoistway opening
Maximum = 57'
Maximum of 6 openings
Variable 5' 4" to 7' 8"
8' 0" Machine room height
7' 0" hoistway opening
4' 0" pit
Rise + 2' 6"
18" clearance

scoping plunger can be installed. These cars are very limited in rise and speed and are applicable only to small two to three story buildings. A cutaway for this type of unit is given in Figure 24.12.

From the point of view of the construction, the major inherent advantage of hydraulic units is the absence of overhead traction equipment. In Figure 24.10 we see that only the guide rails project above the car and, if these are camouflaged, the impression of a free-standing elevator car is given. This effect can be used to good advantage *inside* large, open spaces such as exist in shopping malls and stores; when combined with glass-enclosed, observation-type cabs, the effect is striking (see Figs. 24.13*a* and 25.4).

The major inherent *disadvantage* of the hydraulic elevator is its operating expense. Since it is not counterweighted it requires a relatively large motor to drive the oil pump, and *all* the energy is lost in heat. As an example of the operating cost consider a 3500 lb, 125 fpm hydraulic unit in a department store. Such a unit requires a 40-hp motor. Assuming the unit to be in operation 10 hr/day, 6 days a week, and assuming a normal 60% time-in-operation figure, we have (remembering that the motor operates only in the up-direction)

$$\text{Energy used/day} = \frac{40\text{ hp}}{0.82\text{ efficiency}} \times \frac{0.746\text{ kw}}{\text{hp}} \times 60\% \times 10\text{ hr} \times 1/2 = 110\text{ kwh/day}$$

At \$.035/kwh we have

$$\text{Monthly energy cost} = 110 \times \frac{6\text{ days}}{\text{week}} \times \frac{4.33\text{ week}}{\text{month}} \times \$.035 = \$100/\text{month}$$

Compare this to the previously calculated (Section 23.16) monthly energy cost of \$55 for a 3500 lb, 600 fpm traction car, for an appreciation of the value of a counterweight.

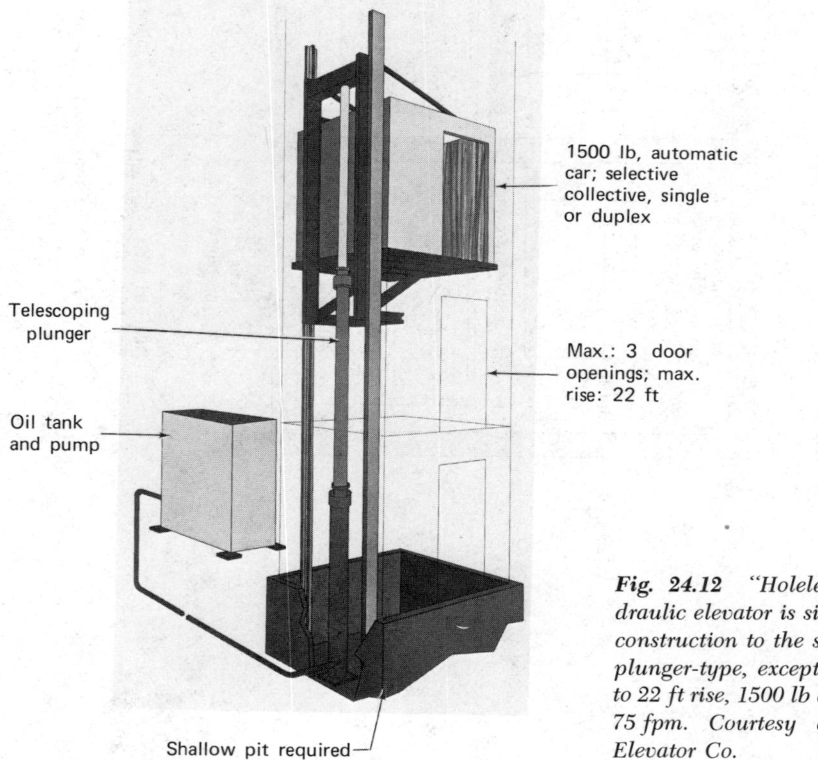

Fig. 24.12 *"Holeless" hydraulic elevator is similar in construction to the standard plunger-type, except limited to 22 ft rise, 1500 lb car, and 75 fpm. Courtesy of Otis Elevator Co.*

SCARBOROUGH TOWN CENTER Toronto, Ontario
Fig. 24.13 (a) *This hydraulic unit is more decorative than utilitarian, particularly, if we consider that it is bracketed by two sets of moving stairs. The design is especially interesting and attractive in that the elevator seems to rise out of a fountain.*

RAPID TRANSIT DISTRICT Los Angeles, California
Fig. 24.13 (b) *This 4 story high, exterior plunger installation fulfills both practical and viewing functions. Courtesy of Montgomery Elevator Co.*

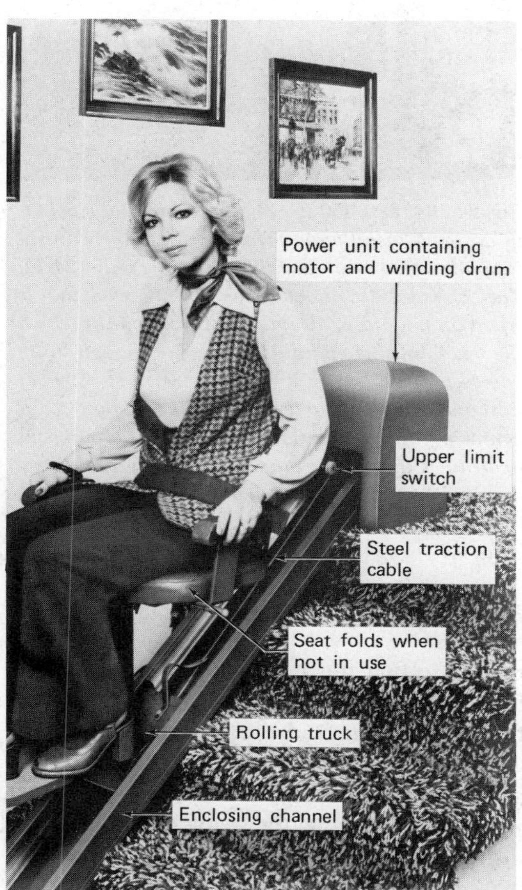

24.9 Residential Elevators and Chair Lifts

Although recognition of the special needs of the handicapped has only of late been made official through legislation, and only for public buildings, the elevator industry has been providing for the handicapped for years, on a private basis. Chair lifts shown in Figure 24.14, wheel chair lifts as in Figure 24.15, and private elevators as in Figure 24.16, are widely used to overcome the stair barrier in private homes. These items are also covered by the elevator code and must be equipped with the safety devices there specified. All units operate on household electric current and require minimal maintenance.

Fig. 24.14 *Typical layout for a single seat, folding chair lift. The seat is rigidly attached to a rolling truck mounted inside an enclosed steel track. The track is pulled by a steel cable operated from a winding drum in the power unit at the top of the stairs. Courtesy of the Inclinator Company of America.*

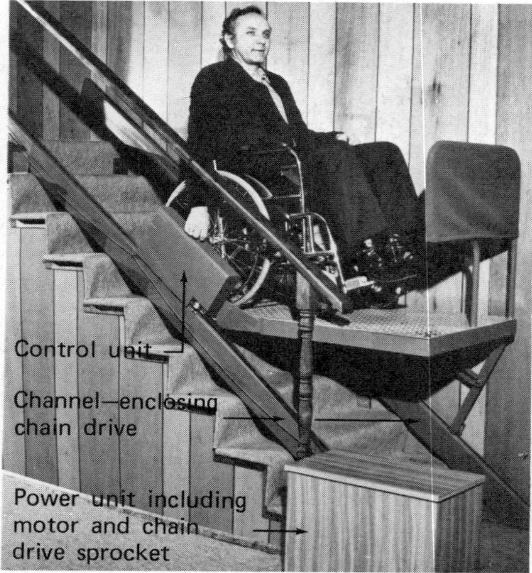

Fig. 24.15 A wheelchair lift installed relatively unobtrusively on a stair. The platform forms a bottom step, leaving the stair open for normal use. This unit is arranged to descend to the lower terminal in the event of a power failure. It is roller chain driven and operates at 22 fpm. Courtesy of Flinchbaugh/Murray Corp.

Fig. 24.16 (a) *A 450 lb, 30 fpm residential elevator in an open installation. The operating mechanism is similar to that of the chair lift in Figure 24.14. The cab is rigidly attached to a rolling truck that is lifted by a winding drum. The track, within which the truck rolls, is readily seen here, although in an enclosed installation it is concealed. The power unit and drum can be located at the top, bottom, or center of the installation. Limit switches prevent overrun. Control is manual or automatic, as selected. Some equipment details are shown in* (b) (opposite). *Courtesy of Inclinator Company of America.*

FREIGHT ELEVATORS

24.10 Freight Elevators—General

The preceding material, which dealt with passenger traffic, had as its prime consideration the most economical solution to the problem of vertically transporting a given number of persons in the most expeditious way. The problem with respect to freight elevators is similar; to transport a given tonnage of freight efficiently, economically, and quickly. The service car in a facility can be considered as a freight car but, if utilized for passenger duty at all, it must meet passenger service requirements. If passenger duty is not required, or if much freight is to be handled, a straight freight car is used.

Factors to be considered in freight elevator selection, in addition to tonnage movement per hour, are size of load, method of loading, travel, type of load, type of doors, and speed and capacity of cars. These factors are interrelated so that the actual process of selection involves making assumptions on the basis of recommendation and then by trial, deciding on a solution, very much as was done for passenger elevators.

Since it is beyond the scope of this book to discuss in the same detail the selection of material-handling elevators as was done for

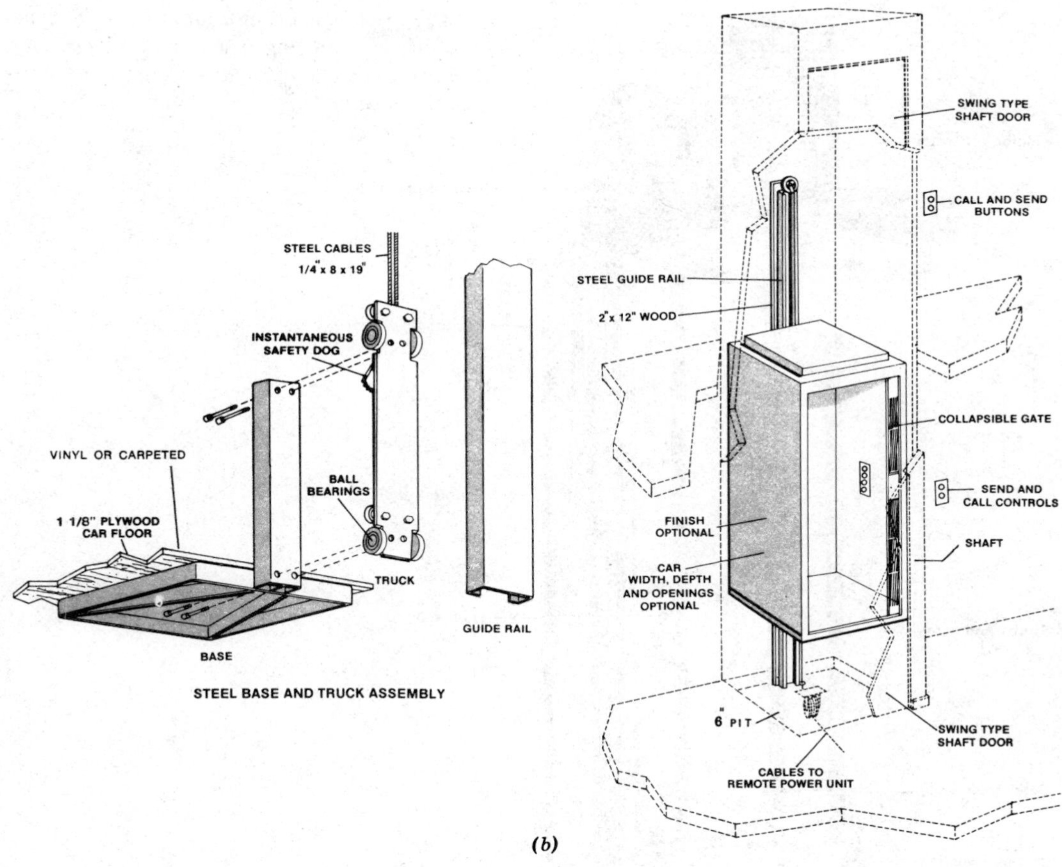

(b)

passenger elevators due to the large number of considerations involved, we shall restrict ourselves in the following paragraphs to descriptive material and recommendations. Also, since freight elevators form such an important link in industrial processes, a careful and detailed material-flow study should be made before the freight elevators are selected. Elevator manufacturer's representatives can be very helpful in this regard. Figure 24.17 shows a section through a typical freight car shaft.

24.11 Freight Car Capacity

The varying capacities shown for a specific platform size in Tables 24.1 and 24.2 are due to the varying square foot loads that are possible. Cognizance of this is taken by the ANSI Code for Elevators, which has established three load classifications for freight elevators.

Class A. General Freight Loading, by hand truck. Single items may not exceed 25% of the car-rated load. Rated load is based on 50 pounds per square foot (psf) of net inside platform area.

Class B. Motor Vehicle Loading. Car will carry automobiles or automobile trucks. Rating is based on a load of 30 psf of platform area.

Class C. Industrial Truck Loading. Maximum loading, 150% of rated capacity, based on a figure of 50 psf of net inside platform area.

24.12 Freight Elevator Description

Since speeds are generally between 50 and

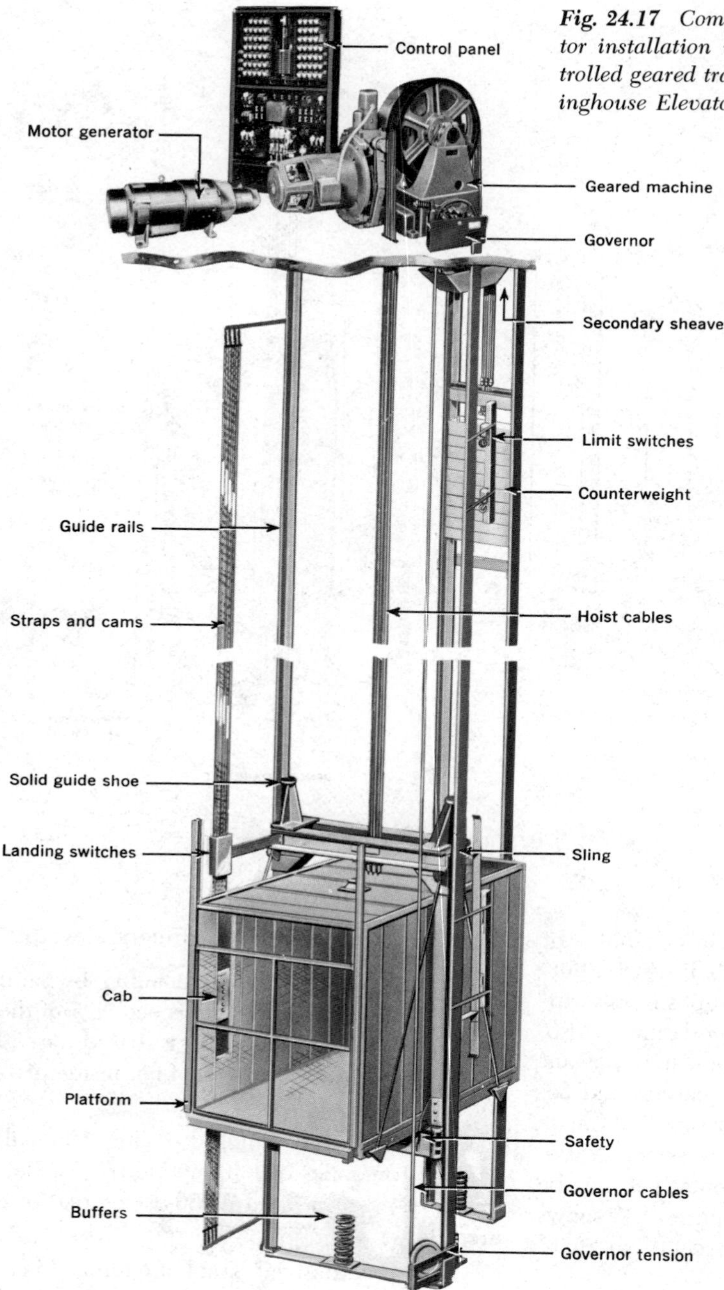

Fig. 24.17 Components of a typical freight elevator installation utilizing a variable voltage, controlled geared traction machine. Courtesy of Westinghouse Elevator Co.

200 fpm, a geared-type machine is used almost universally.

The preferred system of control is collective and variable voltage, d-c supply. However, if the car is used infrequently (less than five trips a day), economy is very important; and, if a rougher ride is tolerable, then a-c rheostatic control may be used.

If accurate automatic leveling is required, that is, $\pm \frac{1}{4}$ in., then umv control must be used. Rheostatic a-c control can provide at best only ± 1 in. automatic leveling. For

Table 24.1 ***Loading by Hand or by Hand Truck***

Capacity (Pounds)	*Platform (Width)*	*Size (Depth)*
2500	5′4″	7′0″
3000	6′4″	8′0″
3500	6′4″	8′0″
4000	6′4″	8′0″
5000	8′4″	10′0″
6000	8′4″	10′0″
8000	8′4″	12′0″
10,000[a]	8′4″	12′0″
12,000[a]	10′4″	14′0″

[a]Elevators of this size should always be considered for industrial truck loading.

Table 24.2 ***Loading by Industrial Trucks***

Capacity (Pounds)	*Platform (Width)*	*Size (Depth)*
10,000	8′4″	12′0″
12,000	10′4″	14′0″
16,000	10′4″	14′0″
18,000	10′4″	16′0″
20,000	12′4″	20′0″

low-rise jobs a hydraulic unit is most often employed. These, as the umv traction units, provide accurate control, smooth operation, and very accurate automatic leveling. Hydraulic units do not normally exceed 60 ft in height, but do operate at speeds up to 125 fpm. Accessories such as governors, safeties, brakes, and so on, are similar to those for passenger elevators previously described. These apply to both hydraulic and traction types.

Fig. 24.18 A typical counterweighted freight car with manual overhead gate. Note that the sling that lifts the car is arranged with double sheaves over which the hoisting ropes pass. This roping arrangement increases the mechanical advantage of the lifting ropes. Courtesy of Armor Elevator Co.

24.13 Freight Elevator Cabs, Gates, and Doors

Cabs for freight service are normally built of heavy gauge steel with a multilayer wooden floor, the entire unit being designed for hard service. Guarded ceiling light fixtures are required. Cab gates slide up vertically and are a minimum of 6 ft high. Hoistway doors are normally vertical lift, center-opening manual or power-operated. Both cab gate and hoistway doors are counterweighted and open fully to give complete floor and head clearance (see Figs. 24.18 and 24.19).

Fig. 24.19 A pair of large hydraulic freight elevators for automotive use. These units are automatic, self-leveling, and equipped with power-operated bi-parting shaftway doors and cab gate. As with many freight installations, both ends of the cab are open. Note that two control stations are provided in the cab—one at each gate. Photo courtesy of Harris-Preble Co., which supplied the doors in this installation. Hydraulic lifts are by Becker.

Freight Elevators — Traction

LIGHT AND MEDIUM DUTY FREIGHT ELEVATORS						
CAPACITY	2500#	3000#	4000#	6000#	8000#	10,000#
A	5'-4"	6'-4"	6'-4"	8'-4"	8'-4"	10'-4"
B	7'-0"	8'-0"	8'-0"	10'-0"	10'-0"	14'-0"
C	5'-0"	6'-0"	6'-0"	8'-0"	8'-0"	10'-0"
D	7'-4"	8'-4"	8'-4"	10'-4"	10'-10"	12'-10"
L	13'-0"	14'-0"	14'-0"	14'-0"	14'-0"	15'-0"

HEAVY DUTY POWER TRUCK FREIGHT ELEVATORS					
CAPACITY	10,000#	12,000#	16,000#	18,000#	20,000#
A	8'-4"	10'-4"	10'-4"	10'-4"	12'-4"
B	12'-0"	14'-0"	14'-0"	16'-0"	20'-4"
C	8'-0"	10'-0"	10'-0"	10'-0"	12'-0"
D	11'-4"	13'-6"	14'-0"	14'-2"	16'-6"
L	14'-0"	15'-0"	15'-0"	17'-0"	21'-0"

MINIMUM PIT & OVERHEAD DIMENSIONS FOR LIGHT & MEDIUM DUTY FREIGHT ELEVATORS				
CAR SPEED	50	75	100	200
O	16'-0"	16'-0"	16'-0"	16'-0"
P	5'-6"	5'-6"	5'-6"	6'-0"

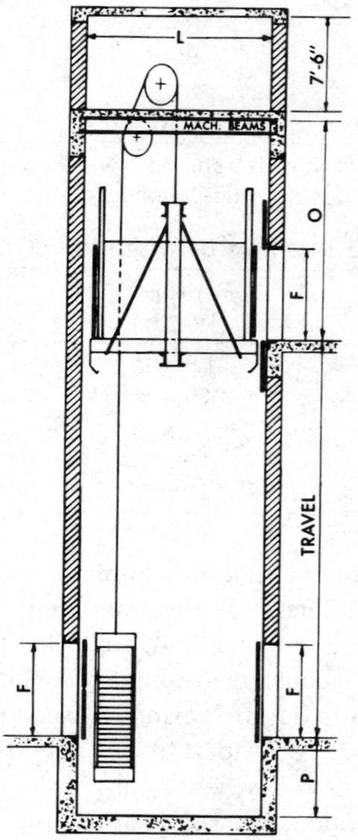

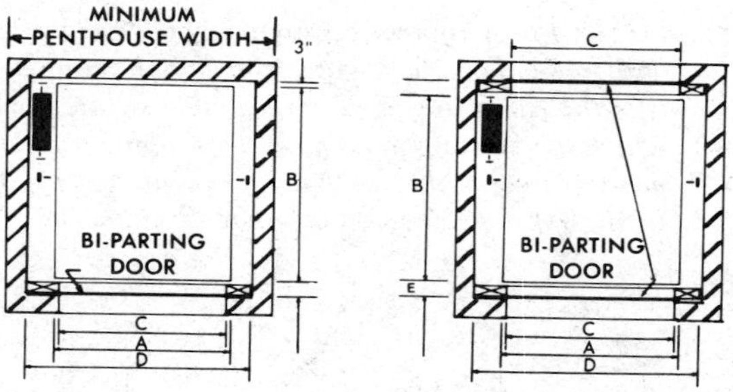

Fig. 24.20 Typical dimensional data for traction-type freight elevators. Courtesy of Montgomery Elevator Co.

24.14 General-Purpose Freight Elevators

These units, whether traction (Fig. 24.20) or hydraulic (Fig. 24.21) in load ranges up to 20,000 lb, are standard design items applicable to all types of commercial and industrial buildings. Heavier units are individually engineered. As with passenger elevators, structural reactions are supplied by the manufacturer to the architect, who is responsible for adequate structional supports. This item is of great importance in larger car installations, since traction unit rails must be supported every few feet and additional steel provided to accomplish this.

24.15 Freight Elevator Cost Data

The cost of a freight elevator installation, as with passenger elevator installation, is dependent on many factors, principally: capacity, type of control, use, and type of door operation.

Since exact pricing, as actual selection, is not covered in detail herein, being outside our scope, it is our recommendation that a reputable manufacturer or elevator consultant be consulted for such information. We can, however, make some general remarks on pricing, as follows:

Freight Elevators — Oil Hydraulic

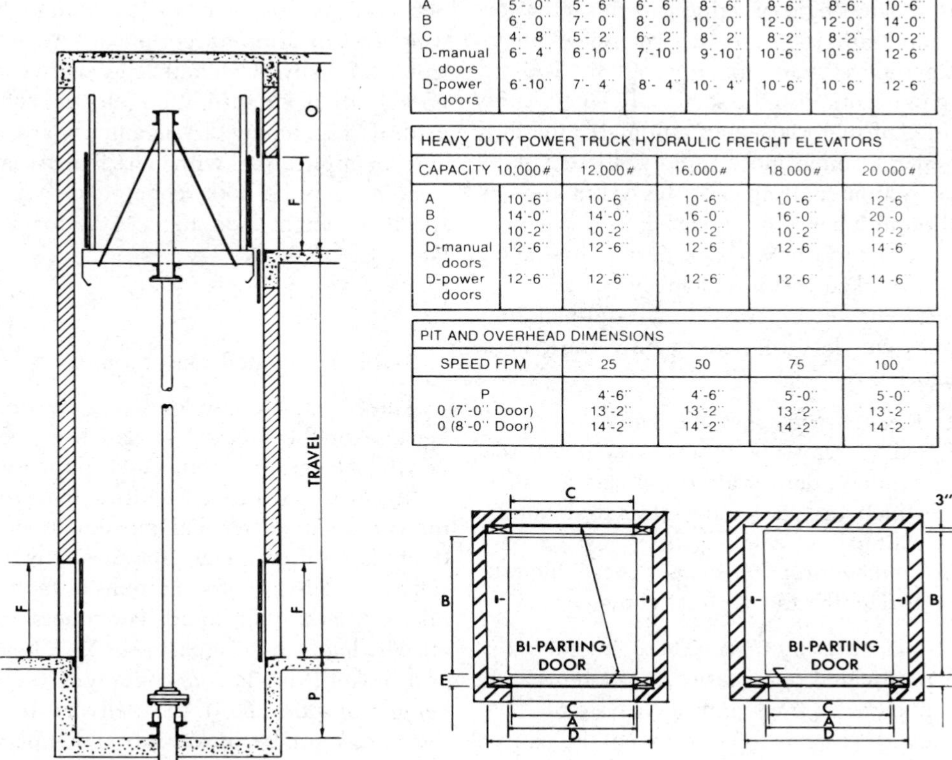

LIGHT AND MEDIUM DUTY HYDRAULIC FREIGHT ELEVATORS							
CAPACITY	2000#	3000#	4000#	5000#	6000#	7500#	10.000#
A	5'-0"	5'-6"	6'-6"	8'-6"	8'-6"	8'-6"	10'-6"
B	6'-0"	7'-0"	8'-0"	10'-0"	12'-0"	12'-0"	14'-0"
C	4'-8"	5'-2"	6'-2"	8'-2"	8'-2"	8'-2"	10'-2"
D-manual doors	6'-4"	6'-10"	7'-10"	9'-10"	10'-6"	10'-6"	12'-6"
D-power doors	6'-10"	7'-4"	8'-4"	10'-4"	10'-6"	10'-6"	12'-6"

HEAVY DUTY POWER TRUCK HYDRAULIC FREIGHT ELEVATORS					
CAPACITY	10.000#	12.000#	16.000#	18.000#	20 000#
A	10'-6"	10'-6"	10'-6"	10'-6"	12'-6
B	14'-0"	14'-0"	16'-0"	16'-0"	20-0
C	10'-2"	10'-2"	10'-2"	10'-2	12-2
D-manual doors	12'-6"	12'-6"	12'-6	12'-6	14-6
D-power doors	12'-6"	12'-6"	12'-6	12-6	14-6

PIT AND OVERHEAD DIMENSIONS				
SPEED FPM	25	50	75	100
P	4'-6"	4'-6"	5'-0"	5'-0"
0 (7'-0" Door)	13'-2"	13'-2"	13'-2"	13'-2
0 (8'-0" Door)	14'-2"	14'-2"	14'-2"	14'-2"

Fig. 24.21 *Typical dimensional data for hydraulic-type freight elevators. Courtesy of Montgomery Elevator Co.*

1. Variable voltage controlled equipment is 20 to 30% higher in cost than rheostatically controlled equipment.
2. Above a basic two-floor rise, the cost increases linearly with rise.
3. Electric door operation can increase a car installation cost between 10 and 25%. Thus, a set of manual doors might cost \$2500, the same doors motorized \$5000 to \$7000. A similar relation exists for the car gate.

As an example of cost, and for the purpose of *comparative* pricing only, an 8000 lb, 75 fpm, four-floor, manual door car with a-c rheostatic control and automatic leveling might cost \$30,000; the same car with variable voltage control, 150 fpm, and electrically operated doors would cost approximately \$55,000.

MATERIAL-HANDLING EQUIPMENT

24.16 Material Handling—General

The material-handling equipment that will be discussed briefly below is that which finds application in commercial and institutional buildings. Industrial materials handling is an entirely separate subject not germane to our purpose here. The need to transport material within a building has always existed and until approximately a decade ago was done largely manually, with mechanical assistance. Thus offices used messengers; hospitals used dumbwaiters, service elevators, conveyors, and chutes. The single exception to this situation was (and still is) the extensive use of pneumatic tube systems in large stores. Today's systems accomplish the same end—that is, the transfer of materials—but automatically and, in general, much more rapidly. First cost of these systems is frequently high, but the reduction in labor and increase in speed generally yield a short pay-back period combined with a marked rise in efficiency.

Modern, commercial material-handling systems can be grouped roughly into three categories:

1. *Elevator-type systems.* These are vertical lift car-type systems including the common dumbwaiter and ejection lifts, which are basically automated dumbwaiters.
2. *Conveyor-type systems.* These include horizontal and vertical conveyors.
3. *Pneumatic systems.* These include sophisticated pneumatic tube systems and pneumatic trash and linen systems.
4. *Other systems.* These do not fit easily into any of the above, and include automated messenger carts and automatic track-type container delivery systems.

24.17 Manual Load/Unload Dumbwaiters

The use of dumbwaiters in various types of structures often provides the most convenient and economical means of transporting relatively small articles between levels. In department stores such units transport merchandise from stock areas to selling or pickup counters; in hospitals dumbwaiters are often utilized for transporting food, drugs, linens, and other necessary small items. In multilevel restaurants, office dining rooms, and the like, dumbwaiters are almost always used for delivery of food from the kitchen and for return of soiled dishes.

Dumbwaiter cars are limited to a platform area of 9 sq ft and a maximum height of 4 ft. The car may be, and frequently is, compartmented by shelves. Normal speed ratings are 45 fpm to 150 fpm, with a capacity of up to 500 lb. Cars may be of the traction (counterweighted) or drum (direct pickup) type. Control is normally "Call and Send" between two floors, although multibutton selector switch or central dispatching arrangements are available for applications with more than two floors. Loading may be floor, counter, or any other specified height (see Fig. 24.22 for typical layouts).

24.18 Automated Dumbwaiters

These units are also known as ejection lifts because of the method of delivery (see Fig. 24.23). They find their best application in institutions and other facilities that require rapid scheduled vertical movement of relatively large items. Thus, the device is ideally suited for delivery of such items as food carts, linens, dishes, bulk liquids containers, and so on. The load can be a cart (Fig. 24.23*a* and *b*) or a basket (Fig. 24.23*c*) containing the items being transported. At the delivery terminal the item must be picked up and transferred horizontally to its final destination if remote from the delivery point. Payload capacity is available up to 1000 lb and car speeds up to 300 fpm. Round trip time for a 200-fpm unit with 5 loading stations is approximately 2 min, with 10 stations about $2\frac{1}{2}$ min. Major considerations for these units are their relatively high cost and the large shaft area required.

24.19 Horizontal Conveyors

Although horizontal conveyors find their best application in industrial facilities, they are also usable in commercial buildings such as mail order houses where there is a continuous flow of material required. Restrictions in application stem from inflexible right-of-way requirements, noise generation, and a degree of danger if left unprotected or exposed to unauthorized persons. Cost is relatively low, and capacity is virtually unlimited.

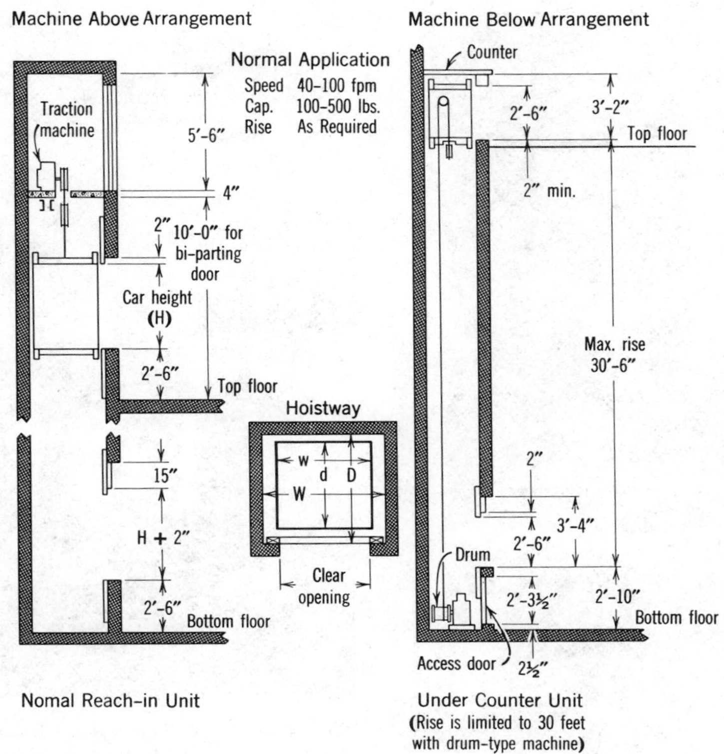

RECOMMENDED SIZES OF DUMBWAITERS

MAX. DUTY			CAR		HOISTWAY		
traction type machine 1:1 roping	traction type machine 2:1 roping	*drum type machine 2:1 roping	(w) width	(d) depth	(W) width	(D) depth without car gate	(D) depth with car gate
400 lb @ 100 fpm	500 lb @ 50 fpm	400 lb @ 45 fpm	2′-0″	2′-6″	3′-2″	2′-11″	3′-0¼″
↑	↑	↑	2′-0″	3′-0″	3′-2″	3′-5″	3′-6¼″
			2′-6″	2′-6″	3′-8″	2′-11″	3′-0¼″
			2′-6″	3′-0″	3′-8″	3′-5″	3′-6¼″
			3′-0″	2′-6″	4′-2″	2′-11″	3′-0¼″
↓	↓	↓	3′-0″	3′-0″	4′-2″	3′-5″	3′-6¼″
400 lb @ 100 fpm	500 lb @ 50 fpm	400 lb @ 45 fpm	3′-6″	2′-6″	4′-8″	2′-11″	3′-0¼″
Under-Counter Dumbwaiter		300 lb @ 50 fpm	2′-6″	1′-8½″	3′-5″	2′-1½″	——

Standard car heights—3′-0″, 3′-6″, 4′-0″

Fig. 24.22 *Typical layout data for manual-load dumbwaiters.*

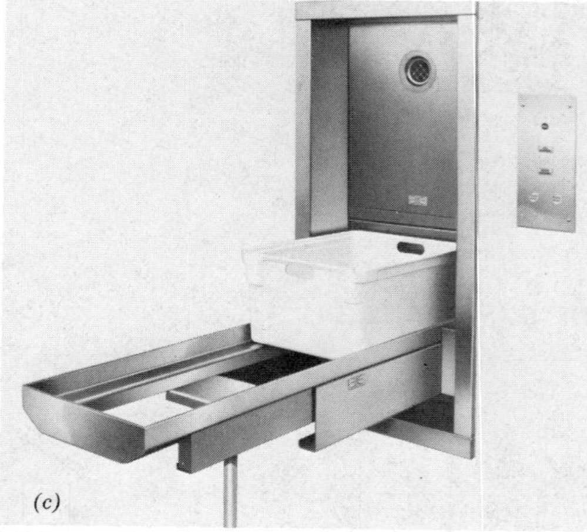

Fig. 24.23 (a) *Open ejection-lift unit showing cart ejection mechanism. Shaftway doors are vertical bi-parting.* (b) *The unit shown in* (a) *is being loaded with food carts and dispatched to the various floors of the hospitals. At the upper floors the carts are rolled away by attendants. Later the lifts are used to return soiled dishes and trays. Courtesy of Courion Industries, Inc.*(c) *Baskets can also be used in a similar system for handling smaller loads. Photo courtesy of AMSCO/American Sterilizer Co., Erie, Penn.*

24.20 Vertical Conveyors

The action of this system is similar to the automated dumbwaiter in that the system transfers vertically and automatically loads and unloads, but the similarity ends there. Vertical conveyors are constructed with a moving continuous-loop chain to which are attached carriages that pick up and deliver tote boxes. At sending and receiving stations the operator places the items to be moved (up to 40 lb) in the tote box, "addresses" the box in one of several ways depending on the system, and places it at a pickup point (see Fig. 24.24). The first empty carriage on the chain will pick up the box and deliver it to its address. Drawbacks of this system are the large shaft required, noise, and cumbersome arrangements when interfacing with horizontal conveyors. Cost is moderate.

24.21 Pneumatic Tubes

This well-tried system will undoubtedly continue in wide use until electronic data reproduction completely replaces the transfer of pieces of paper between two points. PT systems are available with $2\frac{1}{4}$ to 6 in. ranges of tube diameters (special shapes are also used)

Fig. 24.24 *Selective vertical conveyor. Basket is shown being placed into the sending terminal from which it will be picked up and delivered to its address. At the right is the receiving terminal. Photo courtesy of Standard Conveyor Company.*

and with single or multiple loops. Older systems were positive pressure; newer systems are generally vacuum. Compressors run 25 to 35 hp, require substantial machine room and air manifold space, and are noisy. Piping of the standard $2\frac{1}{4}$ in. lines is not difficult, although larger size systems present some space problems due to large minimum bending radii. Overall, pneumatic tube systems (also called pneumatic dispatch systems) are cheap, reliable, and very fast.

24.22 Pneumatic Trash and Linen Systems

Where applications require rapid movement of bagged or packaged trash and linen from numerous outlying stations to a central collecting point, this system is employed. Linen systems are found generally in hospitals; trash systems in various facilities, frequently in conjunction with compactors. The system is basically a network of large pipes, negatively pressurized, with numerous loading stations throughout the building. Pipes are 16, 18, or 20 in., operating at high-static pressure. A system normally can handle only one unit load at a time, but moves it so quickly (20 to 30 ft/sec) that system capacity is large and delays are not encountered. Material placed into a loading station is picked up as soon as the previous load clears. Compressors are large and very noisy, requiring considerable space allocation and acoustical isolation. In addition to the main vacuum system, a high-pressure air line is required to operate the doors, and sprinkler heads must be installed every few floors. Overall costs are low to moderate. For the specific task performed, a cheaper and more efficient transfer technique is difficult to find.

24.23 Automated Container Delivery Systems

This ingenious arrangement employs captive and secure containers locked onto a motorized carriage that, in turn, is locked onto the track system. Power for the motor in the carriage is picked off a third rail at 24 volts d-c. The entire assembly moves horizontally or vertically, with equal ease, at a constant 100 fpm. The container, which can be opened only with a proper key and only at a station, carries a maximum load of 20 lb, which is adequate for most uses. Addressing the unit is done simply by moving magnetic devices on the container. Right-of-way requirements are occasionally a problem but, because of the ease with which the unit moves and the small cross section of the container (about 13 in. × 14 in.), the problems are not serious. The equipment is quiet, unobtrusive, easy-to-use, and attractive. It is also expensive and proprietary as of this writing, being manufactured only by Mosler Airmatic and Electronic Systems under the trademarked name, Telelift (see Fig. 24.25).

24.24 Automated Self-Propelled Vehicles

These robot self-powered vehicles follow a predetermined route on a single level. One type stops at scheduled points for pick up and

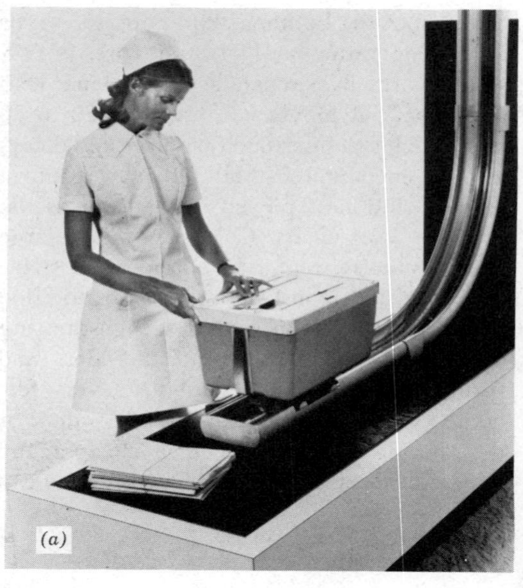

Fig. 24.25 (a) *Sending/receiving station for Mosler's "TELELIFT" system. User is shown adjusting the encoding magnets that address the car. The illustrated car has a capacity of 1700 cubic inches and 20 pounds. Car* (b) *is smaller, with a capacity of 850 cubic inches and 11 pounds. Mounted on the wall above the car in this library installation are a directory of stations and a network monitor panel that controls and monitors the entire system. Courtesy of Mosler Airmatic and Electronic Systems.*

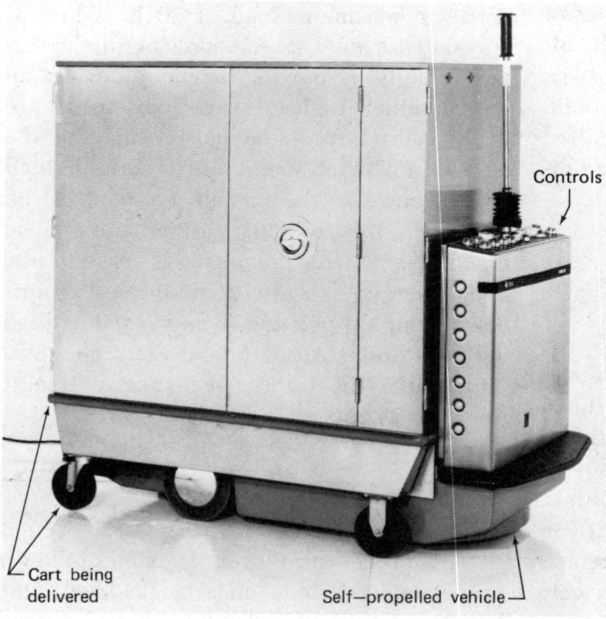

Fig. 24.26 *This automated, self-propelled vehicle follows a predetermined route, guided by wires buried in the floor. Illustrated unit is carrying a bulk cart, which it picked up from an automated dumbwaiter system (see Section 24.18 and Fig. 24.23). The cart will be delivered to a predetermined destination. These vehicles can also be operated manually. Photo courtesy of AMSCO/American Sterilizer Co., Erie, Penn.*

delivery of interoffice papers. In effect, it is an automated "office boy." It can be held at a stop as long as desired and, when released, will continue its scheduled trip. For offices with a large volume of incoming and outgoing mail and interdepartmental paper, this device works admirably. Another type (see Fig. 24.26) is an automated distribution system. Since it is restricted to a single level, it interfaces well with a selective vertical conveyor (Section 24.20), to give both horizontal and vertical capability.

24.25 Conclusion

The foregoing very brief summary simply describes the types of equipment available. For each facility being planned, the architect must study the material transfer problems, remembering that buildings not only handle and process but also generate material; an office building generates about a pound of waste per 100 sq ft per day—a prodigious amount in today's large office structures. This type of dry waste can be compacted, bailed, and sold, unlike garbage and wet waste. In addition to considerations of the type of material being handled, there are factors of speed, scheduling, location of stations, labor and material costs, space requirements, noise generation, and energy requirements. To consider and evaluate all of these factors in a large and complex facility is generally beyond the ability of the architect without expert advice from consultants who specialize in materials handling and from manufacturers' representatives.

25

Moving Stairways and Walks

MOVING ELECTRIC STAIRWAYS

25.1 General

The moving stairway is also referred to as an escalator or as an electric stairway. Throughout this section all three names will be used. The escalator was first operated at the Paris Exposition in 1900. In this country moving stairways now provide comfortable and rapid vertical transportation daily for many millions of persons.

The escalator not only delivers passengers comfortably, rapidly, and safely, but it also continuously receives and discharges its live load at a constant speed with practically no waiting periods at any landing. The annoying waiting for elevators is not present on traveling stairways. Also, time is not lost by acceleration, retardation, leveling, door operation, or operator's reactions, nor by pressing hall buttons, by passenger interferences in getting in or out of the cars, and so on. One seldom sees a waiting passenger or congestion of passengers at the lighted comb plate of an escalator.

Instead of formal lobbies and hallways leading to a bank of elevators on each floor, the electric stairway is always in motion and inviting passengers to "take a ride." The corridors, aisles, and other passageways in existing buildings usually provide space for floor openings adequate for the installation of escalators. In contrast, it would in most cases be almost impossible to install an adequate bank of elevators in an existing building to meet the growing needs for vertical transportation. Elevator hoistways must be vertical from bottom to top floors; an escalator installation can be "staggered" at various appropriate locations (see Fig. 25.1). Figure 25.2 shows a schematic view of a modular escalator. These are designed differently from standard units and are discussed separately in Section 25.6.

A standard stairway is assembled from three separate sections of structural truss—an upper section, a middle section, and a lower section similar to the upper one (see Fig. 25.6*a*). The middle straight section may be any desired length to provide rises for floor heights from 10 to 23 ft, for example. When the rise is over 20 ft an intermediate support is located between the two end supports of the stairway. Generally, the upper corners of the bottom and top ends of the truss, after assembly, carry the complete weight of the stairway mechanism and its live load.

25.2 Location

Because escalators are constantly moving and are generally part of a horizontal and vertical trip, they must be placed directly in the main line of circulation. This is in contrast to the elevator bank which, being a vertical transportation unit, can be set off as an element on its own, for people to approach and utilize. Escalators must therefore be placed in, and with a total and even dominating view of the area served. This means that potential riders must be able to immediately:

(*a*) Locate the escalators.

(*b*) Recognize the individual escalator's destination.

(*c*) Easily and comfortably move toward the escalator.

One of the most effective ways to disorient traffic movement is to inadequately or poorly mark escalator destinations. The resultant milling about, false starts, and general unhap-

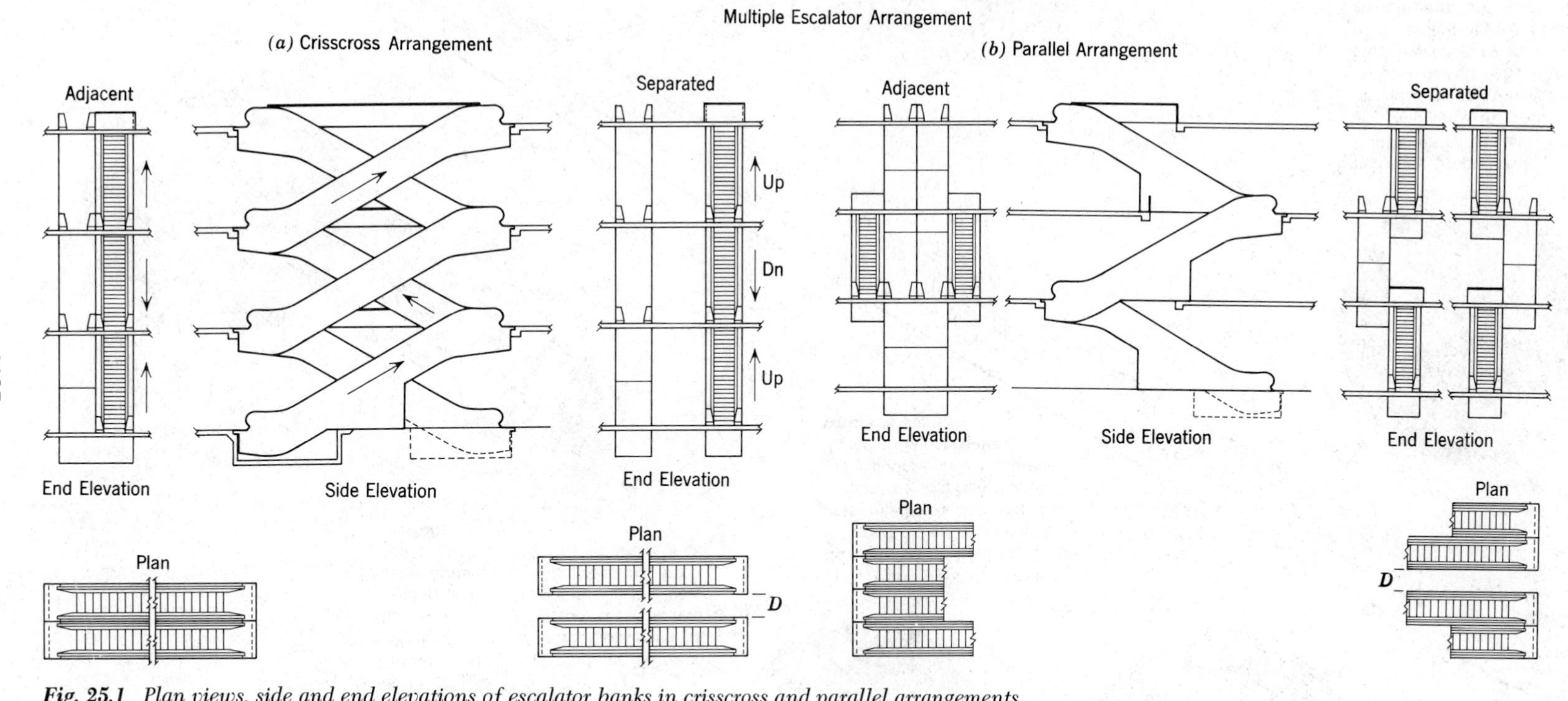

Fig. 25.1 *Plan views, side and end elevations of escalator banks in crisscross and parallel arrangements. Distance* D *between stairs may be selected as desired.*

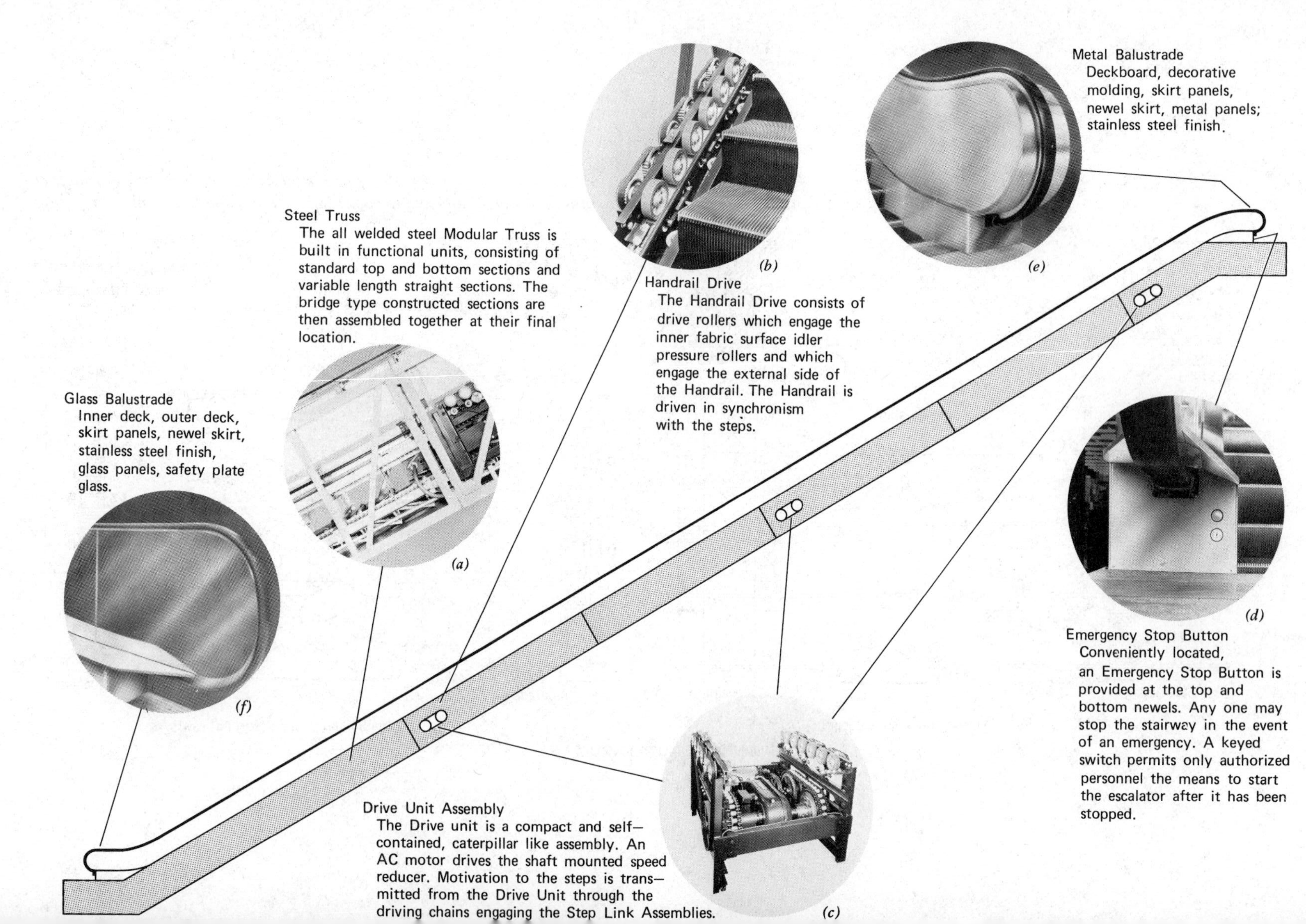

Steel Truss
The all welded steel Modular Truss is built in functional units, consisting of standard top and bottom sections and variable length straight sections. The bridge type constructed sections are then assembled together at their final location.
(a)
(b)
Handrail Drive
The Handrail Drive consists of drive rollers which engage the inner fabric surface idler pressure rollers and which engage the external side of the Handrail. The Handrail is driven in synchronism with the steps.
(e)
Metal Balustrade
Deckboard, decorative molding, skirt panels, newel skirt, metal panels; stainless steel finish.
Glass Balustrade
Inner deck, outer deck, skirt panels, newel skirt, stainless steel finish, glass panels, safety plate glass.
(f)
(d)
Emergency Stop Button
Conveniently located, an Emergency Stop Button is provided at the top and bottom newels. Any one may stop the stairway in the event of an emergency. A keyed switch permits only authorized personnel the means to start the escalator after it has been stopped.
Drive Unit Assembly
The Drive unit is a compact and self–contained, caterpillar like assembly. An AC motor drives the shaft mounted speed reducer. Motivation to the steps is trans–mitted from the Drive Unit through the driving chains engaging the Step Link Assemblies.
(c)

Fig. 25.2 Schematic of a modular design escalator. Note that length is varied by inserting additional center sections between the top and bottom sections, with additional drive units as required. Courtesy of Westinghouse Electric Corp., Electric Stairway Division.

piness can be observed in numerous otherwise well-designed buildings.

Sufficient lobby space must be provided at the base for queuing where anticipated, and most particularly at discharge points. A restricted area here, or again a poorly marked one, will cause passenger hesitation and traffic backup. Since the escalator discharges continuously, backup of traffic is usually dangerous and therefore intolerable. This is particularly crucial in theaters, stadiums, and wherever traffic backup could be disastrous. To avoid this, three design steps, in descending order of importance, are taken:

(*a*) Provide escalators with sufficient traffic-carrying capacity.

(*b*) Provide collecting space at intermediate landings so that pressure can be relieved.

(*c*) Provide a slight setback for the next escalator so that the necessary 180° turn can be readily negotiated (see Fig. 25.3).

At the bottom landing (or the top) an escalator should discharge into an open area with no turns or choice of direction necessary. Where such is unavoidable, large clear signs should make hesitation unnecessary. Landing space in front of an escalator terminal should be 6 to 8 ft *minimum* for a 90-fpm unit, and 10 to 12 ft for a 120-fpm unit.

25.3 Parallel and Crisscross Arrangements

Escalators may be installed so that the up and down stairs crisscross each other as in Figure 25.1*a* or may be arranged in parallel as in Figure 25.1*b*. Both arrangements may have the up and down stairs separated by any desired distance. Separating the stairways gives the advantage of easier mixing of riders entering at the various levels with riders making a continuous trip. Also, by separating the escalators in the crisscross arrangement or by stacking them in the parallel arrangement (see Fig. 25.4), passengers making a multifloor trip can be forced to traverse a specific area on each floor. This area can obviously be used to

LORD & TAYLOR Oakbrook Center, Chicago, Illinois

Fig. 25.3 Single, 32 in. crisscross escalator in an elegant department store. Note the setback of the descending escalator that is helpful in making a smooth turn to the next escalator. By separating the two escalators horizontally, an enforced walk-through can be created. Courtesy of Montgomery Elevator Co.

STIX BAER & FULLER Independence Mall, Kansas City, Missouri

Fig. 25.4 *The stacked parallel arrangement of escalators shown here requires riders to walk around when traveling more than one floor. To avoid the possible annoyance of excessively long trips, riders are provided with the alternative of glass-cab hydraulic elevators. The glass escalator balustrades are architecturally consonant with the glass cab and the glass-protective barriers around the center well of the store. Courtesy of Montgomery Elevator Co.*

Fig. 25.5 *The largest moving stairways in western Europe are installed in the Stockholm subway system. Illustrated is a parallel bank of three escalators, 70 m (230 ft) long with a 33 m (108 ft) rise. Unlike American standards of 30° incline, 90 or 120 fpm speed, and 32 or 48 in. width, these units are at 27.3° incline, 45 m/min (147.6 fpm) speed, and 1000 m (40 in.) width. Photo courtesy of Orenstein & Koppel, West Germany.*

advantage for display of impulse-buying merchandise, and this is indeed the major consideration in favor (from the store's point of view) of these arrangements. A negative reaction can be produced with these arrangements when

(*a*) Insufficient floor space is provided for the transit between escalators, causing crowding, pushing, and general annoyance.

(*b*) Insufficient elevator service is provided for passengers wishing to travel at least three floors. This forces people to make a multistory escalator trip, which in itself can be wearying, particularly when carrying parcels. If such a trip is further lengthened by an enforced walk-around at each floor, it becomes a source of severe irritation, often sufficient to keep customers away from the store.

The crisscross arrangement is generally favored because of lower cost, minimum floor space occupied, and lowest structural requirements. The parallel arrangement, being less efficient and more expensive, has as its only virtue a very impressive appearance which strongly draws people to it. For this reason it is frequently employed, particularly in multiple banks of three or four, in transportation terminals (see Fig. 25.5). In such installations, flexibility is maintained by operating all but one escalator in the direction of heaviest traffic. Reversibility of escalators provides this most desirable feature.

25.4 Size, Capacity, and Speed

All escalators in the United States are installed at an angle of 30 degrees from the horizontal. Thus the rise is equal to 57% of the projected floor area. The Safety Code limits escalator speed to 125 fpm along the axis of rise. In actual practice two speeds are available: 90 fpm and 120 fpm. Installations are frequently two-speed, with the higher speed utilized during rush hours and the lower speed during "off" hours. Where no rush is encountered, the lower speed is utilized since the 120 fpm speed presents some difficulty to the less agile passenger.

Table 25.1 ***Escalator Passenger Capacity***

Step Width	*Speed (fpm)*	*Passengers per Hour*	
		Maximum	*Actual*
32 in.	90	5000	3750
	120	6666	5025
48 in.	90	8000	6000
	120	10,665	8025

Moving stairways are generally available in widths of 32 in. and 48 in., both being measured at the hip level between the balustrades. (One manufacturer presently offers a unit in a 40 in. width that can carry two passengers per tread.) These two principal sizes correspond to tread widths of 24 in. and 40 in., respectively. All treads have a 16 in. depth and 8 in. rise. Table 25.1 lists the maximum and actual capacities of escalators. The 32 in. unit is rated for $1\frac{1}{4}$ persons per step. A more realistic figure assumes one person per step, or about 75% of the maximum figure. The 48 in. width assumes 2 persons per step. Here also a derating to about 75% of maximum is more realistic, taking into account empty steps, briefcase and package carrying persons, and so on.

25.5 Components

The major components of a standard (as opposed to a modular) escalator installation are shown in Figure 25.6

The truss (Fig. 25.6*a*) is a welded steel frame that supports the entire apparatus. The tracks are steel angles attached to the truss on which the step rollers are guided, thus controlling the motion of the steps. The sprocket assemblies, chains, and machine provide the motive power for the unit, much like the simple chain drive of a bicycle. An emergency brake located on the top sprocket will stop a loaded escalator safely in the event of a break in the chain. The controller, which consists of contactors, relays, and a circuit breaker, is normally located near the drive machine. An emergency stop button wired to the controller and placed near or on the escalator housing will stop the drive machine and apply the brake (see lower left escalator, Fig. 25.3).

Key-operated control switches at the top and bottom newels will start, stop, and reverse the stairway. The handrail is driven by two sheaves and is powered from the top sprocket assembly. It is synchronized with the tread motion to provide stability to riding passengers and a support for entering and leaving passengers. Handrails disappear at inaccessible points at newels. The balustrade assembly is designed for maximum safety of persons stepping on or off the escalators.

A particularly attractive design utilizing a transparent balustrade made of tempered glass is illustrated in Figure 25.4. They are frequently referred to as crystal balustrades. In these units, the handrail is pinch-driven within the truss. In addition to metal and glass as balustrade materials, back-illuminated fiberglass and wood are also used.

Fig. 25.6 (a) *This truss is a 3-section, built-up, welded steel unit that supports the moving stairway equipment.*

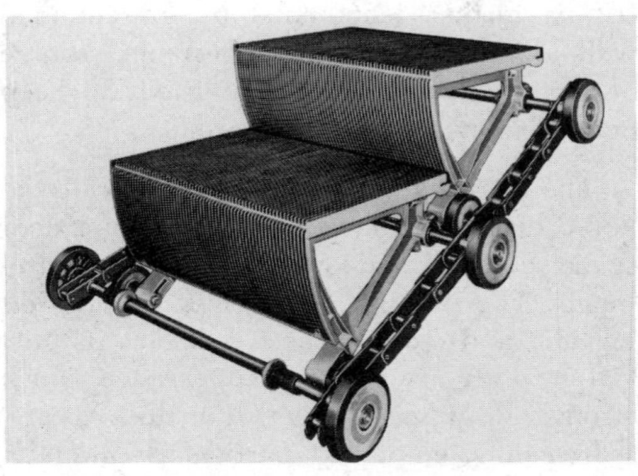

Fig. 25.6 (b) *Typical step assembly for electric stairway. Riser is 8 in. high. Comb pattern is designed to prevent materials from being caught between riser and tread. Courtesy of Westinghouse Electric Corp., Electric Stairway Division.*

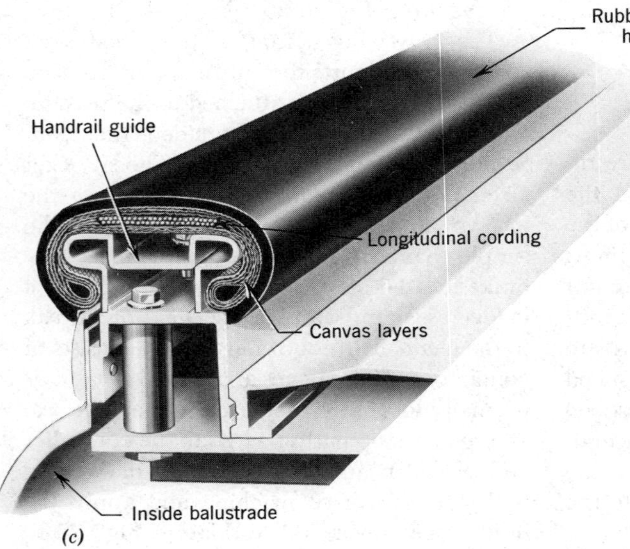

Fig. 25.6 (c) *The constant flexure of the handrail necessitates the layered construction illustrated here. Courtesy of Westinghouse Electric Corp., Electric Stairway Division.*

25.6 Standard Versus Modular Designs

In the conventional design escalator (see Fig. 25.7*a*) all the motive power is delivered at one point; that is, the drive motor drives the main chain, which drives the top sprocket which drives the step chain which pulls up the steps, causing the entire assembly to move. This arrangement is suitable for moderate rises of up to approximately 25 ft. Beyond that the design becomes increasingly inefficient. As the rise increases, the loads on all of the drive components including chains and sprockets increase sharply. Furthermore, to accommodate the heavier equipment necessitated, truss width increases as does wellway size and balustrade decks. For rises above 25 ft the drive motor is too large to fit inside the truss and requires a separate machine room below the truss, with attendant ventilation problems. All of these factors combine to limit conventional design units to a maximum rise of 60 feet (varies slightly between manufacturers).

To overcome these limitations, in the face of demands for higher rises, Westinghouse developed a radically new design that was introduced in 1973 under their trade name, Modular Escalators. This design has unlimited rise capability because it is constructed with additional drive motors along the length of the unit, in a modular design pattern (see Fig. 25.7*b* and Fig. 25.2). By spreading the drive load throughout the length of the unit the inherent limitations listed above, which are caused by a single drive location, are eliminated (refer to Fig. 25.7*c*). Note that this distributed drive principle simplifies the mechanism considerably and increases efficiency greatly. Chain loads are constant regardless of length, as is truss size. The modular drive machine units are all identical and a machine room is never necessary. Since tensions are held to low levels, maintenance is low. Furthermore, the helical gear drive in the modular units is 10 to 15% more efficient than the worm gear of the traditional design. In summary, then, the Westinghouse modular design offers distinct advantages in all sizes, plus rise capability beyond the previous 60-ft limit. Table 25.2 lists the necessary drives for various rises and stair widths.

Table 25.2 *Modular Escalator Motor Drives*

Escalator Size	*32 in.*	*48 in.*
Speed	90 to 120	90 to 120
Motor hp	10	10
One drive, nominal, max. rise	30 ft-0 in.	20 ft-0 in.
Two drives, nominal, max. rise	60 ft-0 in.	40 ft-0 in.
Three drives, nominal, max. rise	90 ft-0 in.	60 ft-0 in.

Source. Data Courtesy of Westinghouse, Escalator Division.

25.7 Safety Features

Protection to passengers during normal operation is ensured by a number of safety features associated with moving stairways:

(*a*) Handrails and steps travel at exactly the same speed (90 or 120 fpm) to ensure steadiness and balance on up or down travel and to aid naturally in stepping on or off the combplates.

(*b*) The steps are large, steady, and are designed to prevent slipping.

(*c*) Step design and their leveling with the combplates at each landing ensure against tripping as one enters or leaves the escalator. This is accomplished with two or three (depending on manufacturer) horizontal steps at each end of the escalator.

(*d*) The balustrade includes all enclosures as furnished by the escalator manufacturer, including the deckboards, inside panels, skirt guards, handrail guards, handrails, and combplates. Details of these parts are designed to prevent catching clothing or packages being carried by passengers. Close clearances provide safety features near the combplates and step treads.

(*e*) Automatic controls of a service brake will bring the stairway to a smooth stop if

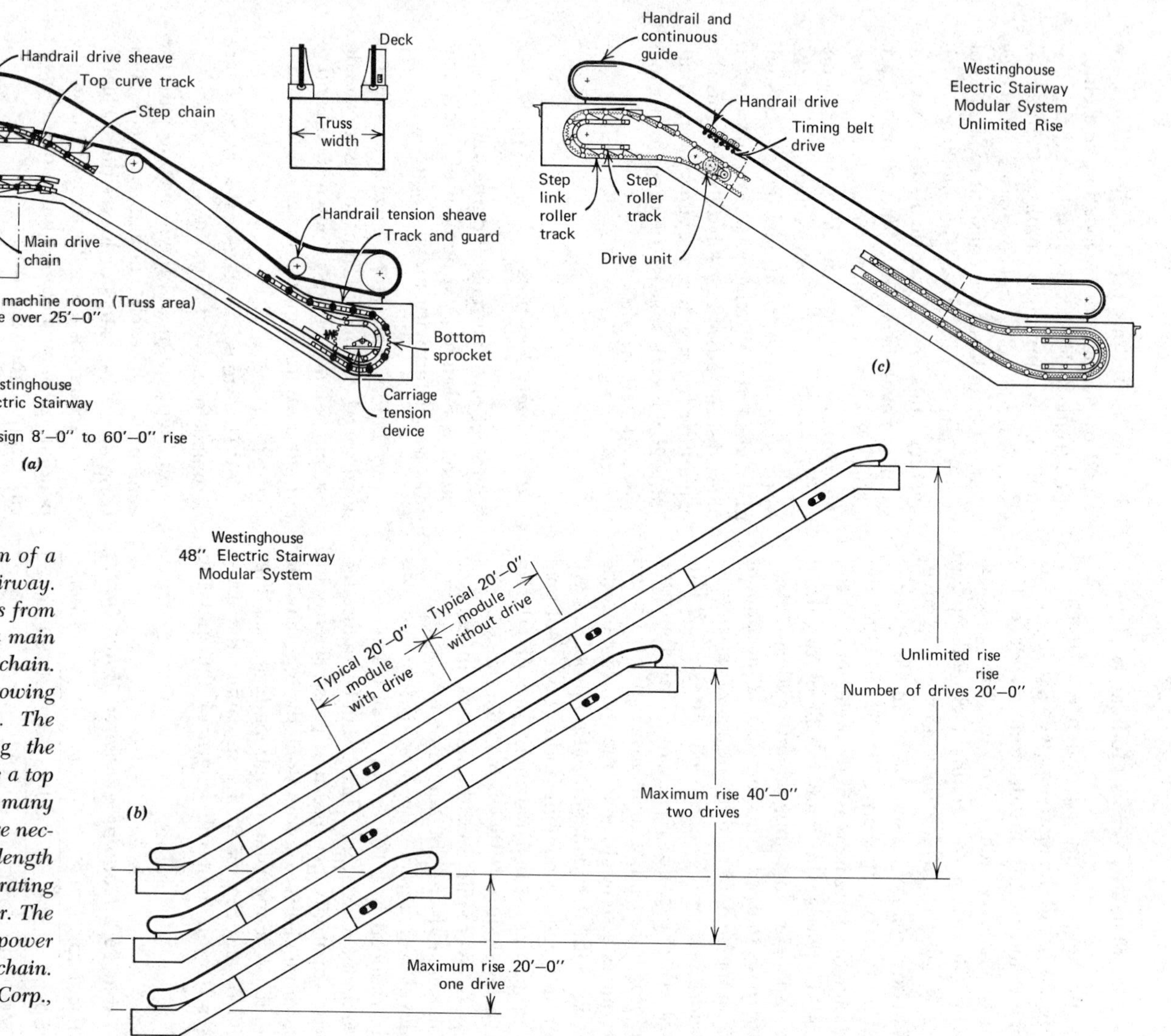

Fig. 25.7 (a) *Operating mechanism of a traditionally designed electric stairway. Note that all motive power proceeds from the drive at the top of the stair via main chain, top sprocket, and stair chain.* (b) *Schematic sectional drawing showing modularity of the newer design. The drive system is distributed along the length of the stair, which comprises a top section, a bottom section, and as many modular intermediate sections as are necessary to accomplish the requisite length and rise (see also Fig. 25.2).* (c) *Operating mechanism of the modular escalator. The distributed drive adds motive power throughout the length of the drive chain. Courtesy of Westinghouse Electric Corp., Electric Stairway Division.*

electric power or mechanical parts should fail. Passengers would then walk the steps as they would any stationary stairway.

(*f*) In case of overspeed or underspeed an automatic governor shuts down the escalator, prevents reversal of direction (up or down), and operates the service brake.

(*g*) An emergency stop switch is located near the combplate or in some unobtrusive location. Building employees and adult passengers may operate the switch to stop the escalator. The electric controls also are arranged to shut down the stairway if by some accident it is caused to reverse its direction.

(*h*) Adequate illumination must be provided by the building at all landings, at the combplates, and completely down all stairways. Some designs of escalators are provided with built-in lighting, as will be discussed in Section 25.9 below.

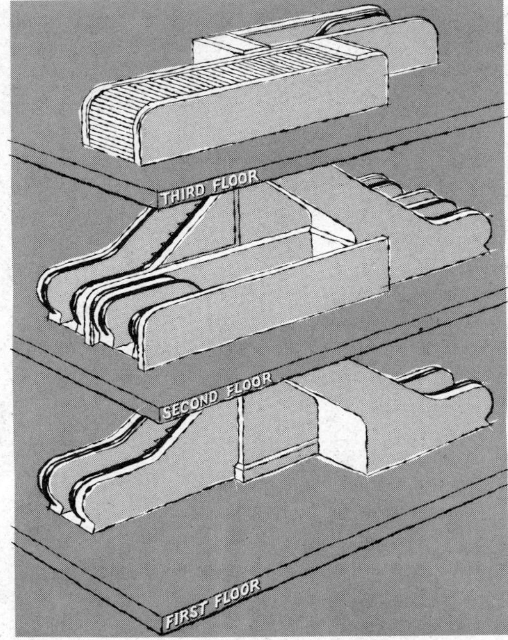

Fig. 25.8 *Rolling-shutter method of wellway fire protection.*

25.8 Fire Protection

Four methods of affording protection in case of fire near escalators are available: the rolling shutter, the smoke guard, the spray-nozzle curtain, and the sprinkler vent. One of these methods is required by Code when more than two floors are pierced. Figure 25.8 illustrates clearly how the wellway at a given floor level may be entirely closed off by the fire shutter, thus preventing draft and the spread of fire upward through escalator wells. The movement is actuated by temperature and smoke relays that automatically start the operation of the motor-driven shutters. The shutter in Figure 25.8 is shown at the third floor level, but other shutters may be installed at the tops of horizontal wellway openings at any floor.

Figure 25.9 illustrates the smoke-guard method of protection. It consists of fireproof baffles surrounding the wellway and extending downward about 20 in. below the ceiling level. Smoke and flames rising upward to the escalator floor opening meet a curtain of water automatically released from the usual type of sprinkler heads shown at the ceiling level. The baffle is a smoke and flame deflector. The vertical shields between adjacent sprinklers ensure that the spray from one will not cool the nearby thermal fuses and prevent the opening of adjacent sprinklers.

The spray-nozzle curtain of water (not shown) is quite similar to the above smoke-guard protection. Here closely spaced, high-velocity water nozzles form a compact water curtain to prevent smoke and flames from rising through the wellways. Automatic thermal or smoke relays open all nozzles simultaneously.

The sprinkler-vent fire control is shown in Figure 25.10. The fresh air intake housed on the roof contains a blower to drive air downward through escalator floor openings, while the exhaust fan on the roof creates a strong draft upward through an exhaust duct; this duct in turn draws air from the separate ducts just under the ceiling of each moving stairway floor opening. Three such separate wellway ducts are shown. Each duct has a number of smoke-pickup relays that automatically start the fresh air fans. The usual spray nozzles on the ceiling near the stairways aid in quenching the fire.

Fig. 25.9 (a) *Smoke-guard method of fire protection for a 32 in. moving stairway, crisscross type. Approximate dimensions are shown. The escalator floor opening (per floor) is approximately 4 ft 4 in. by 14 ft 6 in.*

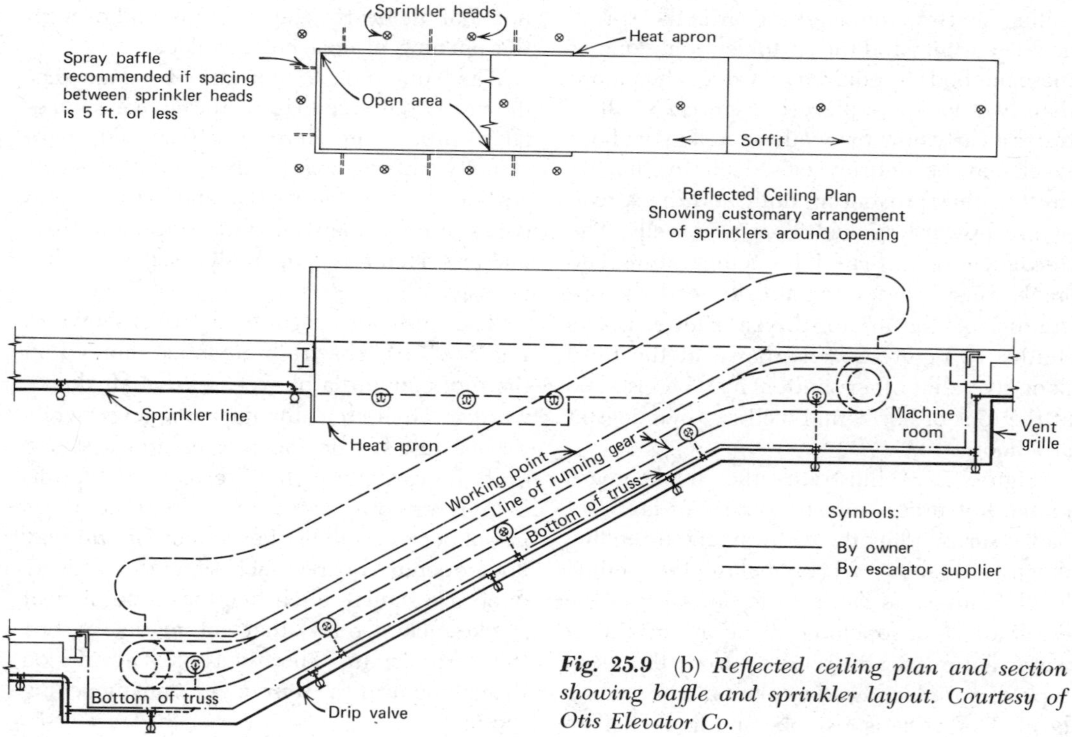

Sprinkler heads
Heat apron
Spray baffle recommended if spacing between sprinkler heads is 5 ft. or less
Open area
Soffit

Reflected Ceiling Plan
Showing customary arrangement of sprinklers around opening

Sprinkler line
Heat apron
Machine room
Vent grille
Working point
Line of running gear
Bottom of truss
Symbols:
By owner
By escalator supplier
Bottom of truss
Drip valve

Fig. 25.9 (b) *Reflected ceiling plan and section showing baffle and sprinkler layout. Courtesy of Otis Elevator Co.*

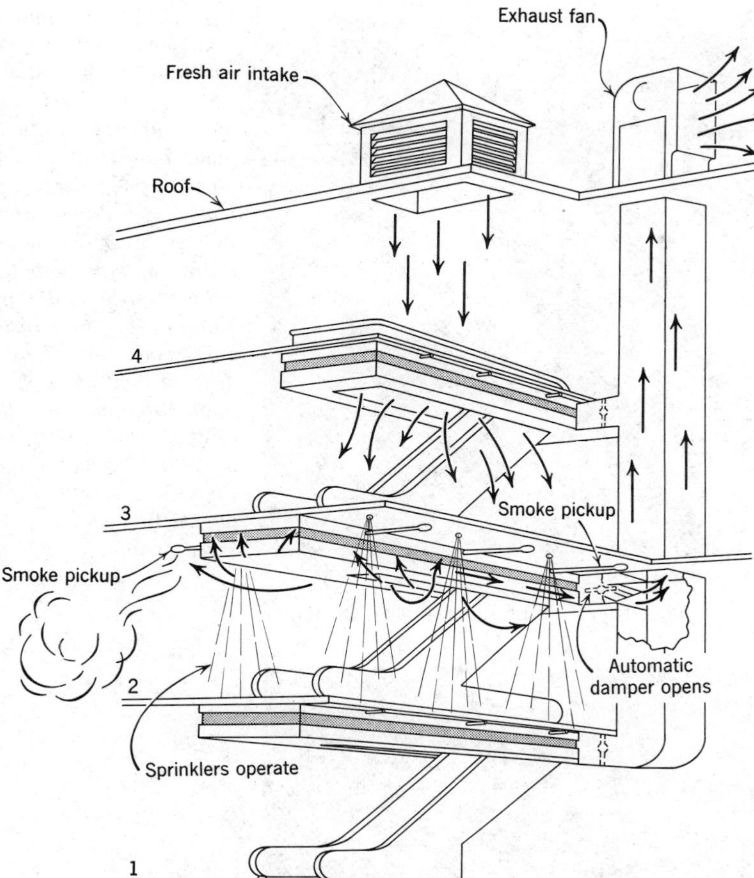

Fig. 25.10 Sprinkler-vent fire protection for escalator openings. An exception (with control) to the rule against perforations in floors.

25.9 Lighting

Adequate illumination of a moving stairway, particularly at the landings, is important from decorative as well as safety standpoints, since it is usually desirable to highlight the moving stair installation. In a stairwell-type installation, where general-area lighting does not provide sufficient illumination for the escalator, lighting consonant with the adjacent illumination is installed on the ceiling above the stairway, with special emphasis on lighting the combplate. Thus, in Figure 25.3 the general illumination is supplemented by downlights above the stairs. In Figure 25.5 banks of fluorescent lights are placed across the escalator bank at frequent intervals along the rise. Note the additional concentrated light at the combplate. Two different lighting treatments of similar type installations are shown in Figure 25.11.

25.10 Application

(*a*) Main floor locations should be chosen in the direct flow of traffic to assure maximum use.

(*b*) Vertical arrangements should be made to accomplish specific purposes, such as exposure of merchandise, maximum passenger capacity, maximum accessibility to various areas, and so on.

(*c*) The aspect of reversibility of an electric stairway should be considered in applications where major traffic flow is unidirectional. Light traffic in the reverse direction can be

FOX HILLS MALL Culver City, California

West Germany

Fig. 25.11 *Two lighting treatments of similar shopping mall installations.* (a) *In the United States; utilizes an illuminated balustrade to supplement the general high level of illumination.* (b) *In West Germany; utilizes a continuous fluorescent strip under the handrail of the crystal balustrade. Note the additional light at the base for illuminating the combplate.* (a) *Courtesy of Montgomery Elevator Co.* (b) *Courtesy of Orenstein & Koppel, West Germany.*

handled by a normal fixed stair, adjacent to the escalator (see Fig. 25.12). Similarly, a bank of two escalators can operate either both up, both down, or one up and one down to handle variable traffic conditions in such areas as office buildings and transportation terminals.

(*d*) Exterior escalators can provide an attractive, interesting, and economical solution to transporting people to selected entry points in a building without the necessity of extending the building to cover the entrance (see Fig. 25.13).

25.11 Elevators and Escalators

The use of elevators and escalators should be considered together as a single problem in

PLACE DES JARDIN Montreal, Quebec

Fig. 25.12 *A conventional arrangement provides a parallel bank of two electric stairways separated by a fixed stair. Reversibility of the escalators provides the desired flexibility, although normal operation is one UP and one DOWN. The fixed stair is provided for persons who cannot or will not use a moving stair. Courtesy of Montgomery Elevator Co.*

CANDLESTICK PARK San Francisco, California

Fig. 25.13 *An attractive exterior escalator installation avoids the necessity of interior stairs and escalators, while providing an item of architectural interest. Courtesy of Montgomery Elevator Co.*

vertical transportation, as applied to the particular facility being designed. In this connection, particularly in case of modernization, Figure 25.14 provides an interesting comparison.

In certain facilities there are times during which such a large demand for vertical transportation occurs that elevators are not a feasible solution. Primary among such facilities is the school building. During class change, virtually the entire building population moves, with as much as 80% moving between floors. Since class change time is at most 10 min, the only reasonable solution is the combined use of fixed and moving stairs. In other buildings such as multifloor stores, the escalator provides for short trips of one or two floors and the elevator generally transports passengers traveling three or more stories.

A comparison of travel time between esca-

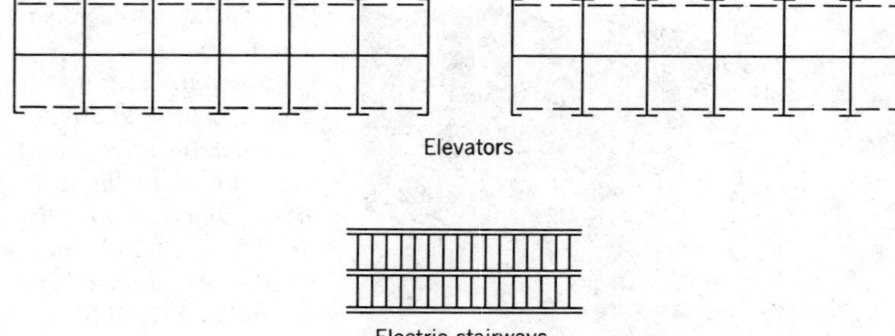

Fig. **25.14** *Comparative space requirements. This diagram shows the marked savings in space requirements for electric stairways as compared to elevators. Both means of vertical transportation provide equivalent capacity.*

lator and elevator is of interest. Using a normal speed of 90 fpm, a 12-ft floor requires 16 sec for travel plus about 5 to 6 sec to turn, for a total of approximately 22 sec. Thus a five-story trip would take approximately 110 sec. A similar elevator trip would take at most 60 sec. The additional escalator time is not noticed in the activity of boarding, turning, and riding. However, a trip of more than five stories would become tiresome.

25.12 Electric Power Requirements

Standard American electric stairways are driven by 3-phase, 60-hertz, a-c induction motors at standard voltages (208–230–460 v). Horsepowers of driving motors are shown in a tabulation in Table 25.3. A comparison with the figures in Table 25.2 shows that the power requirements for standard elevators in rises beyond 20 ft exceed those of the modular design. This is due to the higher friction losses in the gears and chain mechanisms of the standard drive.

It is recommended that no more than four escalators be served by a single electric feeder, and further that not all the escalators of an installation, whatever the number, be served from the same feeder.

Since obviously one cannot be trapped on an escalator, emergency power is rarely required. Ventilation for the machinery should be supplied for approximately 40% of the power to be dissipated as heat. Thus a 10-hp motor would require the dissipation of $0.40 \times 10 \times 2500$ or approximately 10,000 Btu/hr.

Table 25.3 ***Typical Escalator Motor Sizes***

Size (Inches)	*Speed (fpm)*	*Rise (Feet)*	*Motor (hp)*
32	90/120	14	5
	90/120	17	$7\frac{1}{2}$
48	90	17	$7\frac{1}{2}$
	90	21	10
	90/120	25	15

WIDTHS

	V	W	X	Y	Z
32	24″	4′-0¼″	3′-0¼″	32″	4′-2″
48	40″	5′-4¼″	4′-4¼″	48″	5′-6″

REACTIONS

RISE	32″ UNIT			48″ UNIT		
	A	B	C	A	B	C
6	9620	11730		11530	13610	
7	10090	12260		12130	14260	
8	10560	12780		12720	14910	
9	11030	13300		13320	15550	
10	11500	13820		13920	16200	
11	11980	14330		14520	16840	
12	12460	14850		15130	17470	
13	12940	15360		15730	18110	
14	13420	15870		16340	18750	
15	13900	16870		16940	19380	
16	14380	16890		17550	20000	
17	14870	17400		18160	20650	
18	15350	17910		18770	21280	
19	15840	18410		19380	21910	
20	16330	18920		20000	22540	
21	16810	19420		21610	23170	
22	17300	19930		8280	10830	25920
23	17790	20430		8520	11080	26660

Fig. **25.15** *(above and opposite) Typical escalator layout with dimensions and reactions, as furnished by the manufacturer. The illustrated unit is an Otis glass balustrade unit, Escal-aire Type, J series. Courtesy of Otis Elevator Co.*

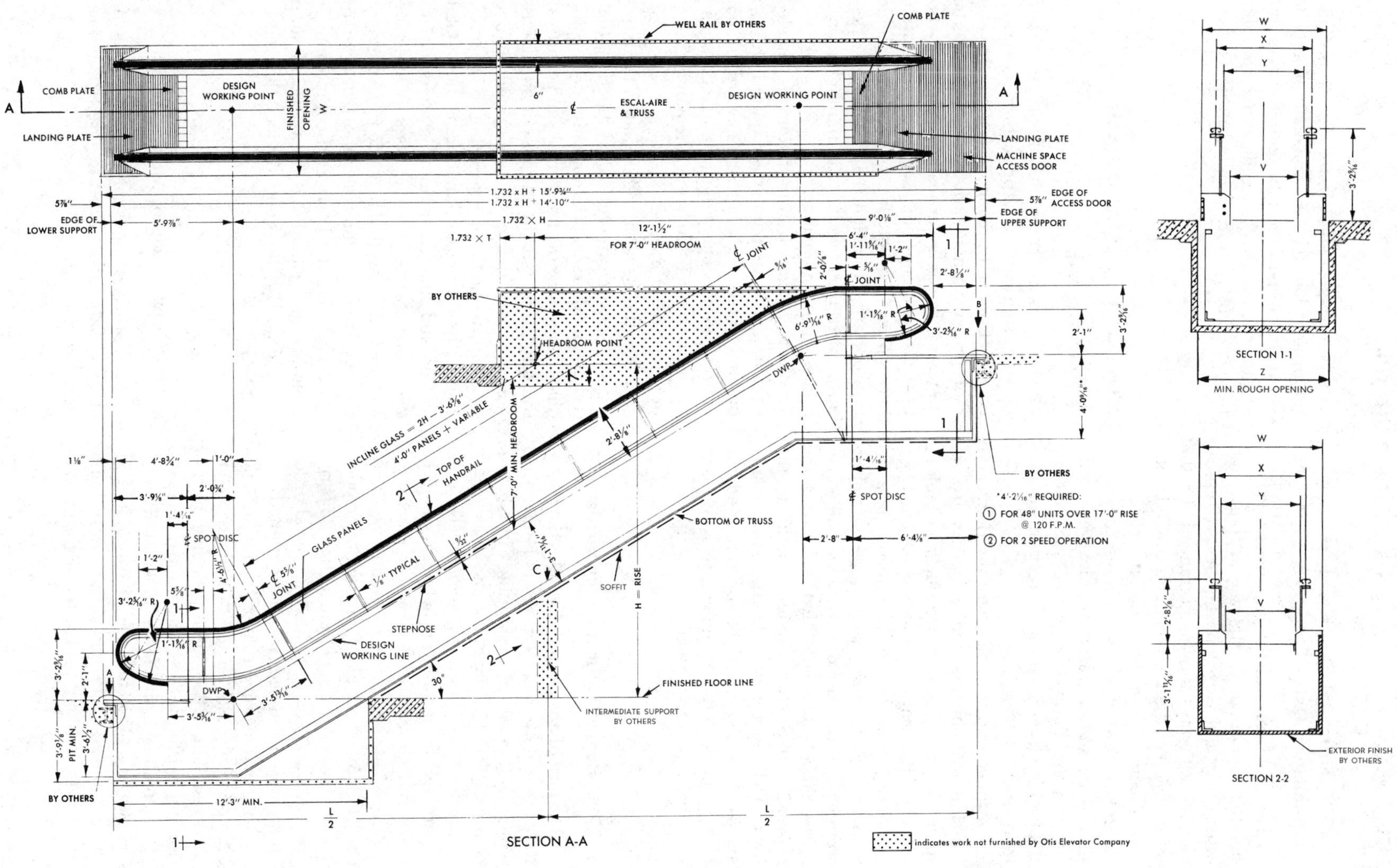
WELL RAIL BY OTHERS
COMB PLATE
DESIGN WORKING POINT
FINISHED OPENING W
ESCAL-AIRE & TRUSS
LANDING PLATE
MACHINE SPACE ACCESS DOOR
EDGE OF ACCESS DOOR
EDGE OF UPPER SUPPORT
EDGE OF LOWER SUPPORT
1.732 x H + 15'-9¾"
1.732 x H + 14'-10"
1.732 × H
1.732 × T
FOR 7'-0" HEADROOM
BY OTHERS
HEADROOM POINT
INCLINE GLASS = 2H − 3'-6⅝"
4'-0" PANELS + VARIABLE
TOP OF HANDRAIL
7'-0" MIN. HEADROOM
GLASS PANELS
⅛" TYPICAL
STEPNOSE
DESIGN WORKING LINE
BOTTOM OF TRUSS
SOFFIT
H = RISE
FINISHED FLOOR LINE
INTERMEDIATE SUPPORT BY OTHERS
℄ SPOT DISC
℄ JOINT
PIT MIN.
12'-3" MIN.
"4'-2⅙" REQUIRED:
① FOR 48" UNITS OVER 17'-0" RISE @ 120 F.P.M.
② FOR 2 SPEED OPERATION
SECTION 1-1
MIN. ROUGH OPENING
SECTION 2-2
EXTERIOR FINISH BY OTHERS
SECTION A-A
indicates work not furnished by Otis Elevator Company

25.13 Structural Design and Installation Data

The architect and the engineers must design the floor openings, stairway supports, and other structural work and finishes. A typical moving stairway drawing, arranged for the information of the architect and trained stairway erectors, is shown in Figure 25.15. A careful review of all the details shown on these plans and indicated specifications exhibits the coordination that is necessary among the architect, engineer, and erection superintendent.

Outlined on these plans are the details for which the architects and engineers are responsible, including structural, mechanical, and electrical features. It will be seen that two "working points" are identified, between which a very strong steel wire is tightly stretched. From these two points all other measurements are made, that is: locating the center line of the truss sections; to placing the lower and upper landing truss support beams, and so on. Such plans as this one are available from all escalator manufacturers for any standard type of stairway.

25.14 Budget Estimating for Escalators

The cost of an escalator includes the cost of the associated mechanical and electrical equipment, plus the shipping installation charges. The manufacturer provides expert engineering and a union field erector who supervises the installation which is done by unionized elevator and escalator mechanics.

The 32 in. and the 48 in. electric stairways are considered standard production models. These may be furnished from a 10 to 25 ft rise, operating at 90 or 120 fpm. On special orders other rises and speeds may be obtained.

A general average rule-of-thumb for quick estimating of costs for a 32 in. electric stairway, installed, is $40,000 for the first 10 ft of rise, and $1000 for each additional foot; the corresponding figure for a 48 in. stairway is $42,000, and $1250 for each additional foot. To these figures must be added the cost of builders' work, wellway protection, lighting, outside balustrades, and plaster. It is especially important that the assistance of experts in vertical transportation be employed in the engineering analysis, specifications, installation, and costs involved on all projects.

MOVING WALKS AND RAMPS

25.15 General

Moving walks and ramps are different from moving stairways in application, function, construction, and capacity. Escalators have as their primary function the movement of large numbers of people vertically, when such vertical distance does not exceed approximately five stories, as noted above. This specific transportation function the moving stair performs extremely well with minimum cost, space, and maintenance.

When vertical transportation of wheeled vehicles and large parcels is required, the use of an electric stairway is awkward, if not entirely impossible. For such functions and others to be discussed below, the moving ramp may best be utilized.

Unlike the elevator and escalator, the moving walk or ramp serves a dual function, that is, horizontal transportation only, or a combined function of horizontal and vertical transportation. For the purpose of our discussions, we will define a moving *walk* as one with an incline not exceeding 5 degrees, where the principal function is horizontal motion and inclined motion is incidental to the horizontal. A moving *ramp* is a device with an incline limited by Code to 15 degrees, where vertical motion is as important or more important than the horizontal component. It should be understood that the walk and ramp are physically the same device, differently applied.

25.16 Application of Moving Walks

The principal uses of moving walks or moving sidewalks, as they are sometimes called, are to:

Fig. 25.16 *Twin autowalks at Manchester Airport, England. These are the pallet type with glass balustrades and continuous built-in fluorescent lighting. These walks, which are each 1 m (40 in.) wide and 55 m (180 ft) long, travel in opposite directions and are both driven by a single-loop drive. They cannot, therefore, both be operated in the same direction. Courtesy of Orenstein & Koppel, West Germany.*

(*a*) Eliminate and/or accelerate burdensome walking.

(*b*) Eliminate congestion.

(*c*) Force movement.

(*d*) Easily transport large and bulky objects. Any person who has walked the seemingly endless distances in many of our major airports, carrying a heavy suitcase can appreciate the near absolute necessity for a moving walkway. It is for this reason that transportation terminals have become major users of this item (see Fig. 25.16). Other transportation terminals, such as rail and ship terminals, also can often find excellent applications for the moving walk, since much heavy and bulky luggage is moved in these areas.

Application of the apparent distance compression which moving walks provide, permits placement of parking areas remote from the pedestrians' destination. Thus, a store can extend its parking area with no annoyance to patrons who must make the long trip to their cars with bulky packages or shopping carts. These advantages are all the more appreciated by persons with a walking impediment.

A second application of walks, as in (*b*) above, is the routing of traffic to avoid congestion, milling about, and lost time and motion. This is particularly applicable in transportation terminals where persons are always traveling in opposite directions through the same and often restricted area, such as in "fingers" leading from airplanes to the main air terminal.

Movement of persons past a display window or some other point is used where the moving walk prevents congestion by preventing stopping. The application in (*d*)—movement of objects—demonstrates clearly that the moving walk is in function simply a large conveyor belt, regardless of its construction.

25.17 Application of Moving Ramps

The moving ramp that combines horizontal and vertical movement is principally applicable as follows:

(*a*) To move persons and wheeled vehicles vertically.

(*b*) To move persons without the agility required for an escalator.

(*c*) To vertically move large and bulky objects.

Ramps have found a fertile field of application in multilevel stores where escalators are not feasible for persons with shopping carts. Such stores may also utilize roof-top parking that is made accessible to persons with a cart via a moving ramp. Since luggage carriers do

Fig. 25.17 *Use of moving ramp between store levels. Courtesy of Sandvik Conveyor, Canada, Ltd.*

Fig. 25.18 *Moving ramps in weatherproof design provide direct access to upper floors. Illustrated is a 12° ramp leading to the second floor of a store in Copenhagen. Note the glass balustrade and continuous fluorescent lighting fixtures below the handrail. Courtesy of Orenstein & Koppel, West Germany.*

not easily adapt to usage on escalators, transportation terminals, which are almost always multilevel, also find extensive application for the moving ramps (see Figs. 25.17 and 25.18).

25.18 Size, Capacity, and Speed

The speed, physical dimensions, and therefore passenger capacity of walks and ramps are not standardized as is the case with escalators. Manufacturers utilize many different tread widths, combined with various speeds and ramp angles of incline. The combinations are designed to suit the situation. Most installations, however, are 26 in. (660 mm) 1 passenger or 40 in. (1000 mm) 2 passenger, these

Table 25.4 *Maximum Operating Speeds of Moving Ramps*

	Maximum Speed	
Angle of Incline	*Level Entrance (fpm)*	*Sloping Entrance (fpm)*
0° to 3°	180	180
3° to 5°	180	160
5° to 8°	180	140
8° to 12°	140	130
12° to 15°	140	120

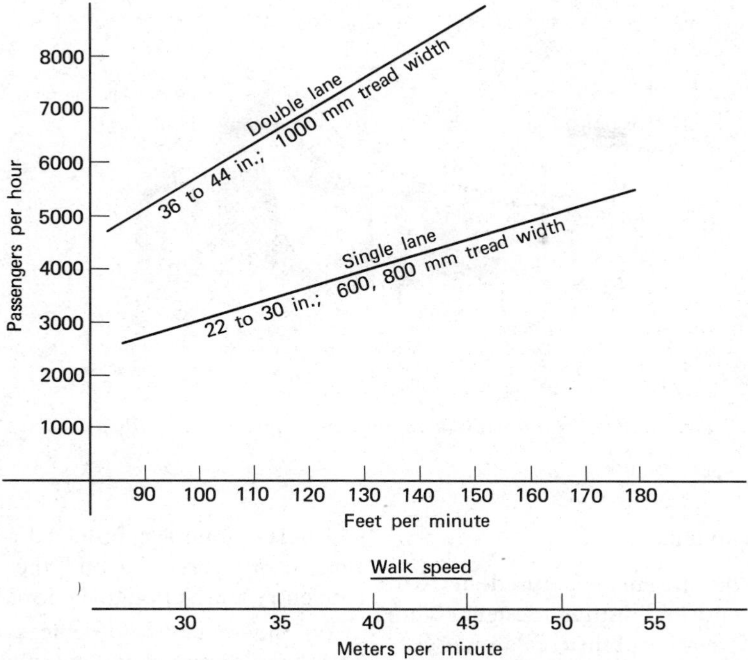

Fig. 25.19 *The capacity of moving walks varies with speed, angle of incline, and tread width. Because of the requirement for handrail support, tread widths greater than double lane are not utilized.*

figures being derived from stair pallet dimensions. The Code allows wider units on horizontal runs.

Furthermore, since the maximum ramp speed varies with angle of slope, and with design of entering point, passenger capacity ratings vary with each design. Higher speeds are allowed by Code where entrance is level but not with sloping entrance, for the obvious reason that the level entrance is easier to board. Table 25.4 and Figure 25.19 give maximum speeds and sample handling capacity for specific units. Figure 25.20 shows a typical section.

Since capacity varies with width, speed, and type of entrance, exact capacity figures must be obtained for each specific design. Maximum practical lengths at present are approximately 1000 ft, with longer units in design.

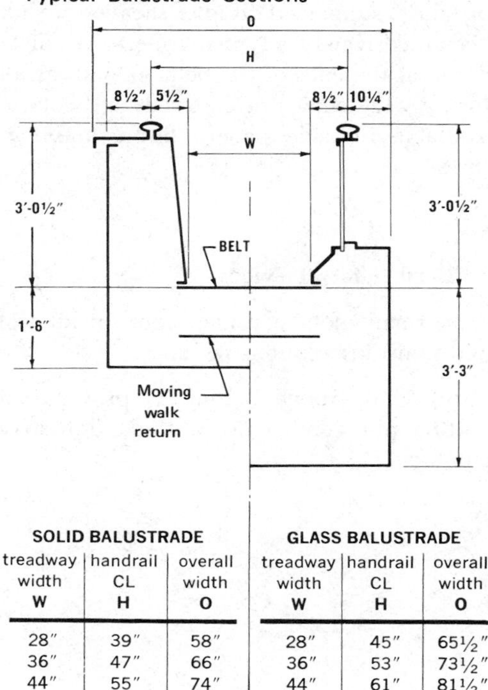

SOLID BALUSTRADE			GLASS BALUSTRADE		
treadway width W	handrail CL H	overall width O	treadway width W	handrail CL H	overall width O
28″	39″	58″	28″	45″	65½″
36″	47″	66″	36″	53″	73½″
44″	55″	74″	44″	61″	81½″

Fig. 25.20 *Typical moving walk balustrade sections. Courtesy of Sandvik Conveyor, Canada, Ltd.*

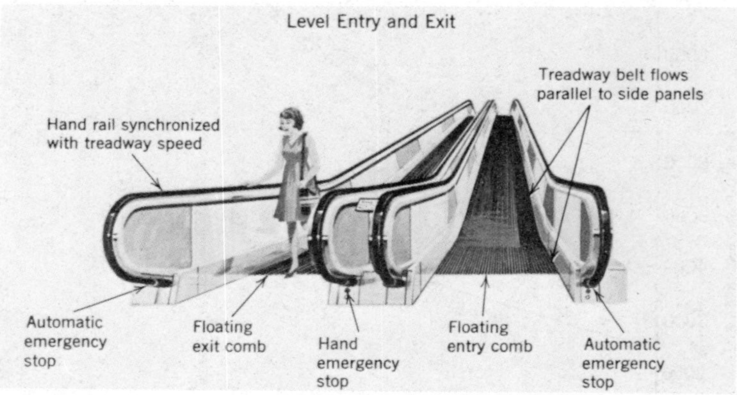

Fig. 25.21 *Moving walk or ramp, with safety features illustrated.*

25.19 Components

Moving walks are manufactured in two separate and entirely distinct designs. The first, which is derivative of the escalator, uses a flattened pallet in place of a step. In all other respects—the drive mechanism, safeties, brake, handrails, and so on—the unit is similar to an escalator. The second design is based on a conveyor belt and utilizes a continuous belt, variously constructed of rubber-covered steel or fabric, supported on idler sheaves or slider bed and driven by a typical roller-drive mechanism at the ends of the belt. As with escalators, a wide choice of materials and colors is available for side panels, drum, and balustrades.

25.20 Safety Devices

Normal safety considerations on all walk and ramp installations include:

(*a*) Thin groove belting to provide sure footing, plus comb action at the ends to avoid jamming.

(*b*) Handrails synchronized with tread motion, cantilevered beyond the treadway to ease entry and exit on to a normally level area.

(*c*) Emergency stop devices are provided, both hand operated and automatic; the latter is activated by disturbances on the tread or in the drive machinery, causing the walk to shut down.

As with escalators, handrails disappear at inaccessible points at newels, and special lighting is generally provided at combplates (see Fig. 25.21).

25.21 Conclusion

Since moving walks and ramps, unlike moving stairs, are nonstandard and are specifically tailored to each application, no general data can be given on power requirements, space, and structural design. Each job must be referred to the manufacturers and the details developed. Rough budget prices at this writing are $750 and $1000 per linear foot for the 26 in. and 40 in. units, respectively.

PART NINE

acoustics

26

Fundamentals of Architectural Acoustics

SOUND THEORY

26.1 General

Architectural acoustics may be defined as the technology of designing spaces, structures, and mechanical systems to meet hearing needs. With proper design, "wanted" sounds can be heard properly and "unwanted" sounds, or "noise," can be attenuated to the point where it does not cause annoyance. However, achieving good acoustics has become increasingly more difficult for a variety of reasons. To cut costs, the weight of construction materials used in many of today's buildings is reduced. Since light structures generally transmit more sound than heavy ones, this practice poses major acoustical problems. Forty percent or more of a building budget may be allocated for mechanical systems—most of which make noise. Outside noise sources such as cars, trucks, trains, and airplanes present problems in isolating interior spaces from exterior sound.

Building owners and tenants are aware that good acoustic environments in buildings are possible and the architect is expected to provide them. A clear understanding of the principles explained in this and the following chapter will assist the architect in accomplishing straightforward designs alone and, in more complex instances, cooperating knowledgeably with the project's acoustic consultant. The importance of proper acoustic design is all the more critical, since after-the-fact acoustic "repair" is often difficult and frequently impossible, without substantially altering the structure.

All acoustics situations have three common elements—source, transmission path, and receiver. The source can be made louder or quieter and the path can be made to transmit more or less sound. The listener's reception of sound also may be influenced. This chapter presents essential aspects of architectural acoustics to assist a designer in defining acoustic goals. Moreover, it describes the achievable goals as well as various methods for reaching them through design.

26.2 Sound: Definition and Generation

Sound can be defined in a number of different ways depending on the aspect we desire to study. Thus sound is a physical wave, or a mechanical vibration, or simply a series of pressure variations, in an elastic medium. For airborne sound, the medium is air. For structure-borne sound the medium is concrete, steel, wood, glass, and combinations of all of these. A much more limited definition is probably more appropriate to our study, that is, to define sound simply as an audible signal. This does not mean that subsonic or supersonic signals are not sound, nor does it mean that we are taking a stand on the existential question of whether unheard sound exists. It simply means that the science of architectural acoustics is concerned with the building occupant, and sounds which he or she cannot detect are generally not our concern. To further clear the air, it is always assumed that the hearer has a pair of healthy young ears with a detection range of 20 to 20,000 Hz. With these givens, it

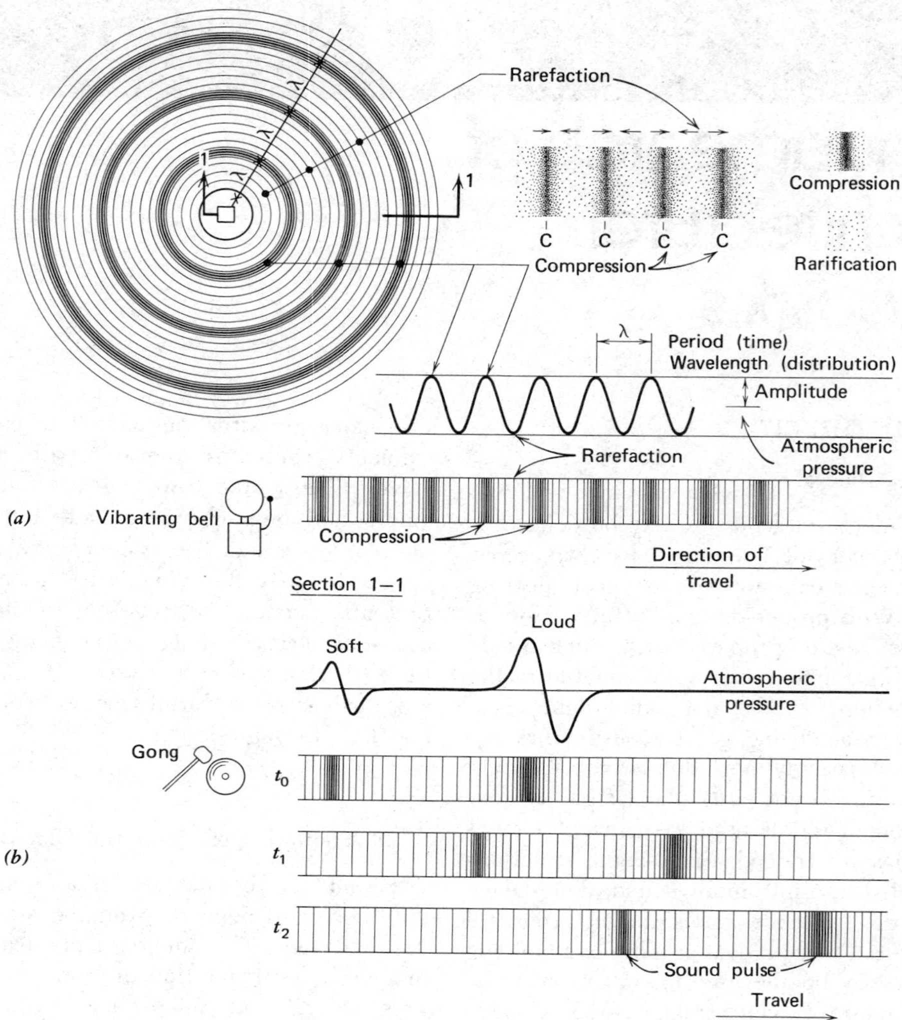

Fig. 26.1 *Sound pressure waves.* (a) *This is a continuous vibration that causes a series of compressions and rarefactions to travel outward longitudinally from the source. Amplitude information is carried by pressure, that is, greater amplitude means greater compression and greater rarefaction.* (*Compression and rarefaction are expressed diagramatically as line density, although they are actually molecular phenomena as shown in the upper drawing*). (b) *Two single impulses of different magnitude (amplitude) are shown traveling away from the source. Note how amplitude information is carried by difference in pressure.*

is probably best to view sound as a series of pressure variations. In air, these pressure variations take the form of periodic compressions and rarefactions (see Fig. 26.1). The bell radiates a pure tone in all directions equally, that is, it creates a circular wave front. As the bell vibrates it sets up vibrations in the air, of the same frequency, which can best be seen in the sectional view. Notice that the pressure changes containing the sound information travel in the same direction as the wave front—longitudinally. This is unlike a radio signal for instance in which the wave travels longitudinally but the information, that is, the wave height and shape, is transverse. Sound is therefore *longitudinal* mechanical wave motion.

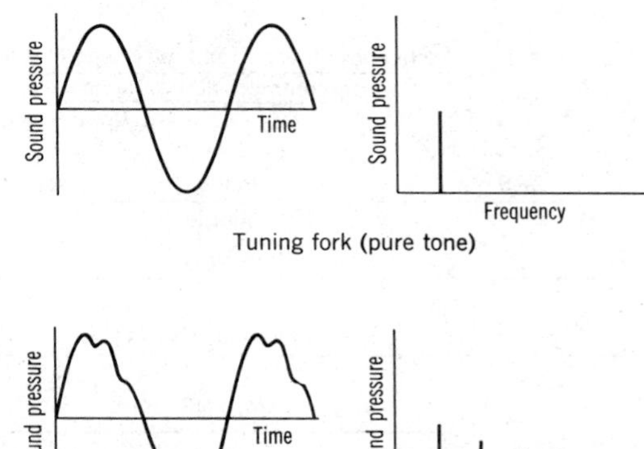

Tuning fork (pure tone)

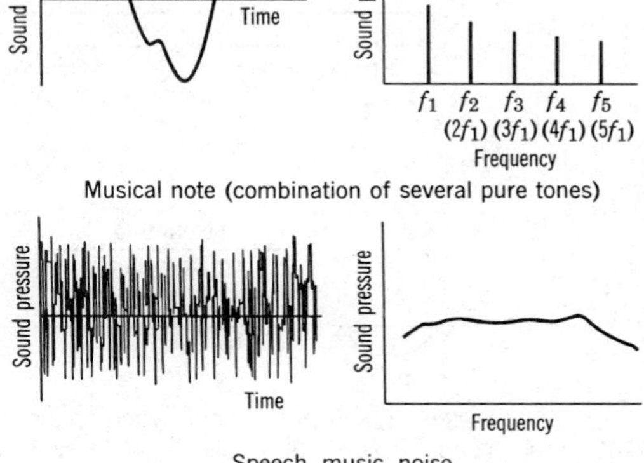

Musical note (combination of several pure tones)

Sound pressure
Time
Sound pressure
Frequency

Speech, music, noise

Fig. 26.2 *Schematic representations of a pure tone, a musical note, and more complex sounds (speech, music, and noise), showing the variation of sound pressure with time and frequency.*

26.3 Frequency

The number of times the cycle of compression and rarefaction of air occurs in a given unit of time is described as the frequency of a sound. For example, if there are 1000 cycles in one second, the frequency of the sound is 1000 cps [1000 hertz (Hz) in the standard nomenclature]. Thus in Figure 26.1, higher frequencies would be shown by compressions and rarefactions that are closer together and lower frequencies by those that are further apart. In sound, frequency is often referred to by a term borrowed from musical concepts—pitch. The higher the frequency the higher the pitch, and vice versa. As stated, the approximate frequency range of a healthy young person's hearing is 20 to 20,000 Hz. This upper limit decreases with age as a result of a process called presbycusis. Recognition of this phenomenon can be of importance in schools, since very high-pitched sounds that are inaudible to most adults (more pronounced in men than women) can be a source of extreme annoyance to students. [Dentists report that high-speed turbine drills and teeth-cleaning devices cause extreme auditory discomfort to many young patients. These devices produce sounds in the 15 to 20-Kilohertz (kHz) range.]

The human speaking voice has a range of approximately 100 to 600 Hz in *fundamentals*, but harmonics (overtones) reach to approximately 7500 Hz. Most speech information, as will be discussed below, is carried in the upper frequencies while most *energy* exists in the lower frequencies. The critical range of speech communication is 300 to 4000 Hz. Overtones outside these frequencies give the voice its characteristic sound and specific identity.

A sound composed of only one frequency is a pure tone. Except for the sound generated by a tuning fork, few sounds are truly pure. Musical sounds are composed of a fundamental frequency and integral multiples of the fundamental frequency (harmonics). Most common sounds are complex combinations of frequencies. Figure 26.2 shows examples of pure tones, musical notes, and common sounds; Figure 26.3 shows the frequency ranges of some common devices and phenomena. The frequencies shown in Figure 26.3 all stand in the ratio of 2:1 to each other, that is, 16:32:63:125:250, and so on. Borrowing again from musical terminology, they are one *octave* apart. These particular frequencies are also accepted internationally as the center frequencies of octave bands used for the purpose of sound specification. For technical reasons, a geometric mean is used. Thus 250 is the center frequency of an octave band ranging from $\frac{250}{\sqrt{2}}$ to $250\sqrt{2}$, and that particular octave is known as the 250-Hz octave. If finer division is required, 1/2 octave and 1/3 octave bands are used.

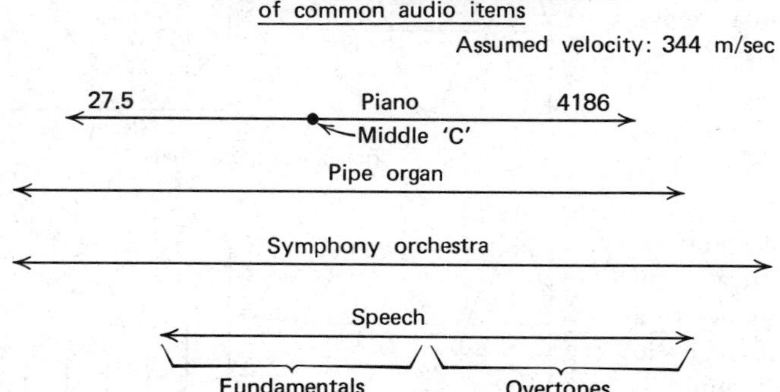

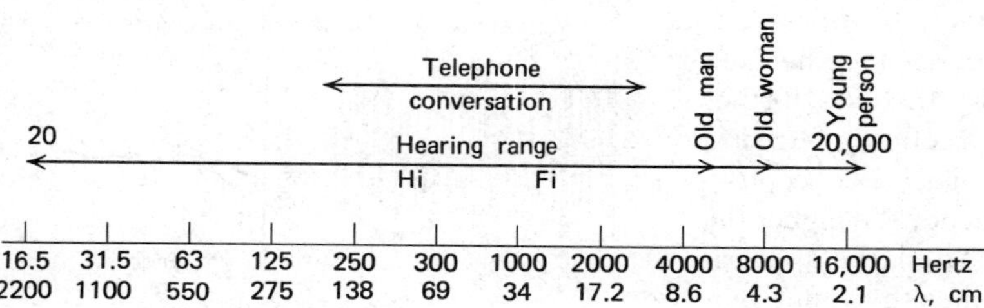

Fig. 26.3 Frequency ranges of common instruments.

26.4 Velocity of Propagation

Sound travels at different velocities depending upon the medium. In air, at sea level, sound velocity is 344 m/sec or 1130 fps. This corresponds to 770 miles per hour (mph) or 478 kilometers per hour (kmph)—slow indeed deed when compared to light at 186,000 miles per *second.* Since sound travels not only in air but also through parts of a structure, it is of interest to know the velocities in other media. See Table 26.1. For our purposes velocity changes due to temperature and altitude (atmospheric pressure) may be ignored and, for rough calculation, 1100 fps and 350 m/sec may be used as velocity in air since both are within 3% error.

26.5 Wavelength and Types of Propagation

The wavelength of a sound may be defined as the distance between similar points on successive waves or the distance the sound travels in one cycle. The relationship between wave-

Table 26.1 ***Sound Propagation Velocity in Various Media***

	Velocity	
Medium	*Meters per Second*	*Feet per Second*
Air	344	1130
Water	1410	4625
Wood	3300	10,825
Brick	3600	11,800
Concrete	3700	12,100
Steel	4900	16,000
Glass	5000	16,400
Aluminum	5800	19,000

NOTE. These figures are approximate, since the listed materials vary in density. Average frequency is used.

length frequency and velocity of sound is expressed as

$$\lambda = \frac{c}{f} \tag{26.1}$$

where

λ = wavelength, ft (m)
c = velocity of sound, fps (m/sec)
f = frequency of sound, Hz

Low-frequency sounds are characterized by long wavelengths and high-frequency sounds by short wavelengths. Sounds with wavelengths ranging from $\frac{1}{2}$ in. to 50 ft can be heard by humans. A simple nomograph is given in Figure 26.31, which permits rapid determination of wavelength, given frequency, and vice versa.

26.6 Sound Magnitude

The magnitude of a sound signal is a much more complex subject because of the different terms in use and because of the great range of numbers involved. When we speak of sound magnitude, we think of loudness, which is a subjective, ear-oriented reaction not linearly related to the physical quantity of sound. The level (quantity) of sound is described variously as sound power, sound pressure, sound pressure level (SPL), sound intensity, and sound intensity level (IL), all of which are different from each other, and from subjective loudness. To clearly understand these concepts, and it is imperative that they be so understood, a comprehension of how we hear and how sound is propagated in free space is necessary.

26.7 Sound Intensity; Free Field Propagation

A point sound source of constant power (watts) radiating in free space, that is, at a distance far from the effects of any reflecting surface, is represented in the drawing of Figure 26.4. The *sound intensity* at any distance from the source is expressed as

$$I = \frac{P}{A} \tag{26.2}$$

where

I is the sound intensity in w/cm^2 or w/m^2
P is acoustic power in watts
A is the area in cm^2 (or m^2)

(It is traditional in the architectural acoustics field to express area in square centimeters although the MKS (SI) system would require area to be stated in square meters. We will continue to use square centimeters. Conversion data are supplied in Table 27.13.) Since the sound radiates freely in all directions,

$$I = \frac{P}{4\pi r^2} \quad \text{w/cm}^2 \tag{26.3}$$

Where r is the radius of an imaginary enclosing sphere. (In English units this is

$$I = \frac{P}{930 \times 4\pi^2} \quad \text{w/ft}^2 \tag{26.4}$$

since there are 930 cm^2 in one sq ft.) The intensities at distances r_1 and r_2 from the source stand in the ratio of

$$\frac{I_1}{I_2} = \frac{{r_2}^2}{{r_1}^2} \tag{26.5}$$

which is the formula for the classic *inverse square law,* stating that *intensity is inversely proportional to distance from the source.* Note the exact correspondence of these relations to the derivations for illumination from a point source, found in Sections 18.4 and 18.8 and illustrated in Figures 18.5 and 18.10, pages 711 and 717. It may be helpful to the reader to review those sections at this point although analogies are not easily drawn between the quantities involved. Remember that the preceding derivation is based on a *point* source, that is, a source that is small relative to the wavelength produced. This type of source produces spherical waves. Line sources such as strings produce cylindrical waves. Large vibrating surfaces such as walls produce plane waves. The importance of these facts will become clear in the discussions on barriers and diffraction in Chapter 27. Figure 26.5 shows graphically how a sound pulse is attenuated in *strength* (but not in waveform) as it travels outward from the source by action of distance.

The threshold of hearing, that is, the mini-

Fig. 26.4 *The same total energy passes through* A_1 *and* A_2. *Since* A *is proportional to the square of* r, *the energy density or intensity is inversely proportional to* r.

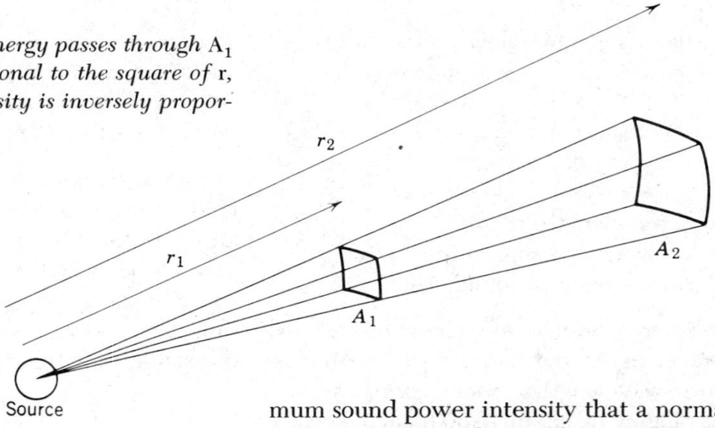

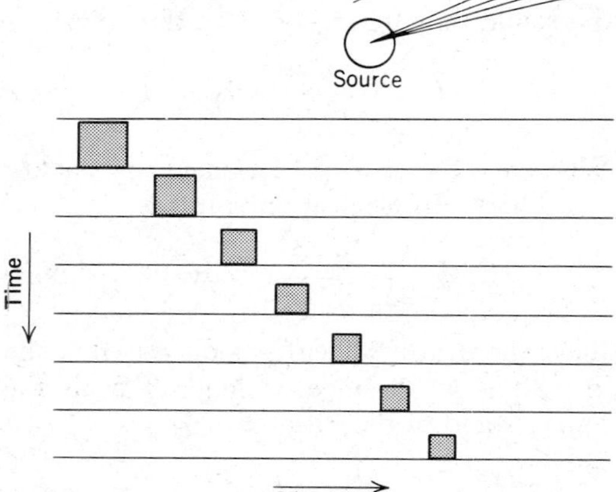

Fig. 26.5 *Attenuation of a sound signal as it travels away from the source. Note that the shape (information) remains constant when traveling in a nondispersive medium such as air. This is not the case with travel in solids where different frequencies travel at different velocities, and therefore cause a wave-shape change with time and distance.*

mum sound power intensity that a normal ear can detect, is 10^{-16} w/cm^2. (Actually, the ear responds to pressure, as will be explained below). The maximum sound intensity that the ear can accept without damage is approximately 10^{-3} w/cm. This gives a range of 10^{13}, or 10 million million to one (10,000,000,000,000:1). Table 26.2 gives the reader an idea of the physical significance of these numbers. Two problems arise immediately when dealing with quantities of this type; the numbers themselves are very small and the ratios are very large. Furthermore, the human ear responds logarithmically, not arithmetically to sound pressure (and intensity); that is, doubling the intensity of a sound does not double its loudness—the change is barely perceptible. To solve these problems it would be much more convenient if we were to construct a scale that:

Table 26.2 ***Comparison of Decimal, Exponential, and Logarithmic Statements of Various Acoustic Intensities***

Intensity (w/cm²)		*Intensity Level—Logarithmic Notation*	*Examples*
Decimal Notation	*Exponential Notation*		
0.001	10^{-3}	130 db	Painful
0.0001	10^{-4}	120 db	
0.00001	10^{-5}	110 db	75-piece orchestra
0.000001	10^{-6}	100 db	
0.0000001	10^{-7}	90 db	Shouting at 5 ft
0.000000001	10^{-9}	70 db	Speech at 3 ft
0.00000000001	10^{-11}	50 db	Average office
0.0000000000001	10^{-13}	30 db	Quiet unoccupied office
0.00000000000001	10^{-14}	20 db	Rural ambient
0.000000000000001	10^{-15}	10 db	
0.0000000000000001	10^{-16}	0 db	Threshold of hearing

a. Started at zero for the minimum sound (intensity or pressure) that we could hear.
b. Used whole numbers rather than negative powers of 10.
c. Had some fixed relationship between an arithmetic difference and a loudness change; say 10 units equals a doubling (or halving) of loudness. Thus, on such a scale, the difference between 20 and 30, and 60 and 70, would always be a doubling of loudness.

Such a scale is the decibel scale.

26.8 Intensity *Level* (IL), the Decibel (db)

The word "level" indicates a quantity relative to a base quantity. Intensity *level* is the ratio between a given intensity and a base intensity. If we express intensity level as

$$IL = 10 \log \frac{I}{I_0} \qquad (26.6)$$

where

IL = intensity level in decibels
I = intensity in watts per square centimeter
I_0 = base, that is, 10^{-16} w/cm^2 (threshold of hearing)

then we have established a scale that satisfies the three conditions set forth in the previous section. The quantity *IL*, intensity level, is dimensionless, since it indicates simply a ratio between two numbers. It is measured in decibels (db) for convenience in expressing the large numbers involved. Table 26.2 shows the great convenience of using the logarithmic decibel scale as compared to either decimal notation or exponential notation. Table 26.3 gives a short listing of subjective loudness changes expressed in db. Note that 10 db indicates a doubling of loudness, as specified; 20 db is loudness doubled twice, that is, four times as loud. The *difference* in db between any two intensity levels, expressed as a function of these respective intensities, is

$$IL_2 - IL_1 = 10 \log \frac{I_2}{I_0} - 10 \log \frac{I_1}{I_0}$$

$$= 10 \left(\log \frac{I_2}{I_0} - \log \frac{I_1}{I_0}\right)$$

$$\Delta IL = IL_2 - IL_1 = 10 \log \frac{I_2}{I_1} \qquad (26.7)$$

A few examples using db notation and logarithmic calculations should help the reader become comfortable with this excellent system.

Table 26.3 *Subjective Loudness Changes and Corresponding Intensity Level Changes*

Change in Level, Decibels	*Subjective Change in Loudness*
3	Barely perceptible
6[a]	Perceptible
7	Clearly perceptible
10	Twice or half as loud
20	Four times or one-quarter as loud

[a] Six decibels corresponds to the change encountered when distance to the source in a free field is doubled (halved).

EXAMPLE **26.1.** Two sound sources produce intensity levels of 50 and 60 db, respectively, at a point. When functioning simultaneously, what is the total sound intensity level? (We assume identical frequency content and random phase; that is, the phase relationship between the two sources changes in a random manner.)

SOLUTION. Note that we are dealing with intensity level, not intensity, since intensity itself has little significance for us. The technique involved in adding two sound intensity levels is to (a) convert both to actual intensity, (b) add arithmetically, and (c) reconvert to decibels. Thus

(a) $$IL = 10 \log \frac{I}{I_0}$$

so $$60 = 10 \log \frac{I_1}{10^{-16}}$$

$$6.0 = \log \frac{I_1}{10^{-16}}$$

$$10^6 = \frac{I_1}{10^{-16}}$$

$$I_1 = (10^{-16})\ 10^6 = 10^{-10}\ \text{w/cm}^2.$$

and $$50 = 10 \log \frac{I_2}{10^{-16}}$$

$$5.0 = \log \frac{I_2}{10^{-16}}$$

$$10^5 = \frac{I_2}{10^{-16}}$$

$$I_2 = 10^{-11}\ \text{w/cm}^2$$

(b) $$\begin{aligned} I_1 + I_2 &= 10^{-10} + 10^{-11} \\ &= (10 \times 10^{-11}) + 10^{-11} \\ &= 11 \times 10^{-11}\ \text{w/cm}^2 \end{aligned}$$

(c) $$\begin{aligned} IL \text{ combined} &= 10 \log \frac{11 \times 10^{-11}}{10^{-16}} \\ &= 10\ (\log 11 + \log 10^5) \\ &= 10\ (1.04 + 5) \\ &= 10\ (6.04) \\ &= 60.4 \text{ db} \end{aligned}$$

which is a fraction larger than the original 60 db of the stronger sound.

EXAMPLE 26.2. Assume two noise signals of 60 db each. What is the combined strength in decibels?

SOLUTION. One method would be to calculate levels as in Example 26.1. A shorter method is to find the difference between the two signals and to add it to either one. Using equation 26.7,

$$\begin{aligned} \Delta\ IL = IL_{\text{comb.}} - IL_1 &= 10 \log \frac{I_2}{I_1} \\ &= 10 \log \frac{I_{\text{comb.}}}{I_1} \\ &= 10 \log \frac{2I_1}{I_1} \\ &= 10 \log 2 \\ &= 10\ (0.3010) \\ &= 3 \text{ db} \end{aligned}$$

This answer (which is independent of any particular level) gives us the extremely important fact that doubling a signal intensity raises the intensity level by *3 db.* (In our case, the combined intensity level would obviously be 60 db + 3 db or 63 db.) Similarly, quadrupling a signal's intensity raises the received level by 6 db (see Table 26.3). Returning to the inverse square law expressed in equation 26.5, we are now in a position to determine the effect on sound intensity level of moving away from the sound source.

EXAMPLE 26.3. Given a sound source that produces sound intensity IL at a distance d_1 from the source (the reader can substitute any numbers desired, or follow the problem with symbols), what are the intensities at twice the distance? Three times? Four times?

SOLUTION. We know from equation 26.7 that

$$\Delta\ IL = 10 \log \frac{I_2}{I_1}$$

and we also know from equation 26.5 that

$$\frac{I_1}{I_2} = \frac{d_2^{\,2}}{d_1^{\,2}}$$

or

$$\frac{I_1}{I_2} = \frac{(2d_1)^2}{(d_1)^2} = 4$$

Substituting in equation 26.7, we have

$$\begin{aligned} \Delta\ IL &= 10 \log \frac{I_1}{I_2} \\ &= 10 \log \tfrac{1}{4} \\ &= 10\ (-0.6) \\ &= -6 \text{ db} \end{aligned}$$

which tells us that sound intensity level (not pressure) is reduced by 6 db. Similarly, when distance is tripled, intensity level is reduced by 9.5 db, and when distance is quadrupled, it is reduced by 12 db.

To summarize, then, intensity level increases (decreases) 3 db with every doubling (halving) of power and decreases (increases) 6 db with every doubling (halving) of distance. Figures 26.6, 26.7, and 26.8 illustrate these relationships. The ear responds to sound pressure, not intensity. As will be explained below, sound pressure *level* (SPL) is equal numerically to intensity *level* (at least for normal temperature and pressure, i.e., for our use in

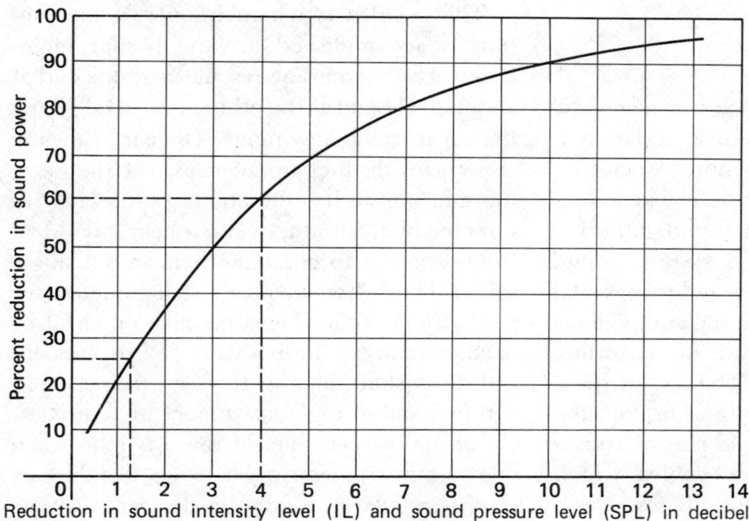

Fig. 26.6 Level decrease as a function of power (intensity) decrease.

EXAMPLE. A 25% reduction in sound power will result in an SPL reduction of $1\frac{1}{4}$ db. A 60% reduction in sound power will result in a 4-db decrease in sound intensity level and sound pressure level.

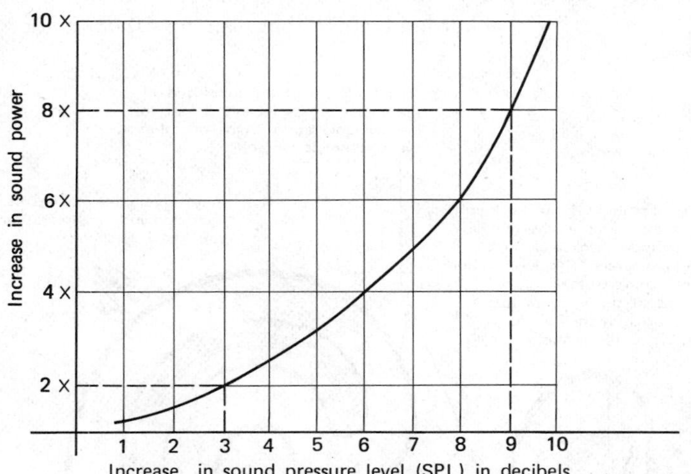

Fig. 26.7 Level increase as a function of power intensity increase.

EXAMPLE. A doubling of sound power produces a 3-db increase, and an eightfold increase in power produces a 9-db SPL increase (3 db × 3).

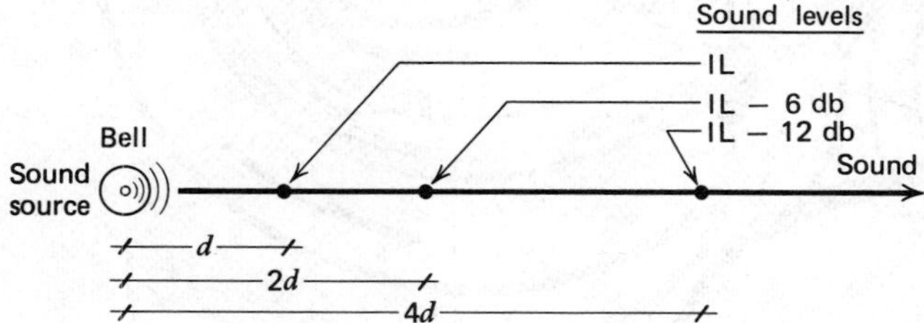

Fig. 26.8 Sound energy levels at varying distances from the source. Each doubling of distance reduces the level by 6 db. These relationships hold true only in a free field.

architectural acoustics), so that the foregoing examples and manipulations of intensity level are equally applicable to sound pressure level.

As will be explained in some detail in the following section, the combined effect of two sounds depends upon their frequency content. In the above examples we assumed signals either of identical frequency and random phase, or of very wide-frequency spectrum—so wide that phase phenomena are not significant. In architectural acoustics work, such an assumption is generally valid.

HEARING PHENOMENA

26.9 Human Hearing

Refer to Figure 26.9, which is a *schematic diagram* of the human ear, since explanation of the functioning would be more difficult on an anatomical drawing. The outer ear is funnel shaped and serves as a sound-gathering input terminal to the auditory system. Sound energy travels through the auditory canal (outer ear) and sets in motion the components of the middle ear, comprising the eardrum, hammer, anvil, and stirrup. The last of these acts as a piston to transmit vibrations into the fluid of the inner ear. This fluid motion causes movement of hair cells in the cochlea, which in turn stimulates nerves at their bases, which in turn transmit electrical impulses along the eighth cranial nerve, to the brain. These impulses we understand as sound.

The cochlea is where frequency recognition is accomplished, by the basilar membrane. This membrane resonates at one end at about 20 Hz and at the other at 20 kHz, giving the ear its frequency range. The ear hears and recognizes distinct frequencies, yet the hearing mechanism has the ability, apparently as directed by the brain, to either hear individual frequencies *or* to combine them into a single sound. Thus when we hear a string quartet we can, *at will*, either hear the entire quartet or each instrument individually. With concentration (vision helps in this), a "trained" ear can pick out a single instrument in an orchestra of 120 pieces, even if there is more than one such instrument in the group. Conductors do this regularly. Similarly, the ear can perform the selection known as the "cocktail party effect," that is, pick out one voice in background noise that can be 20 db louder

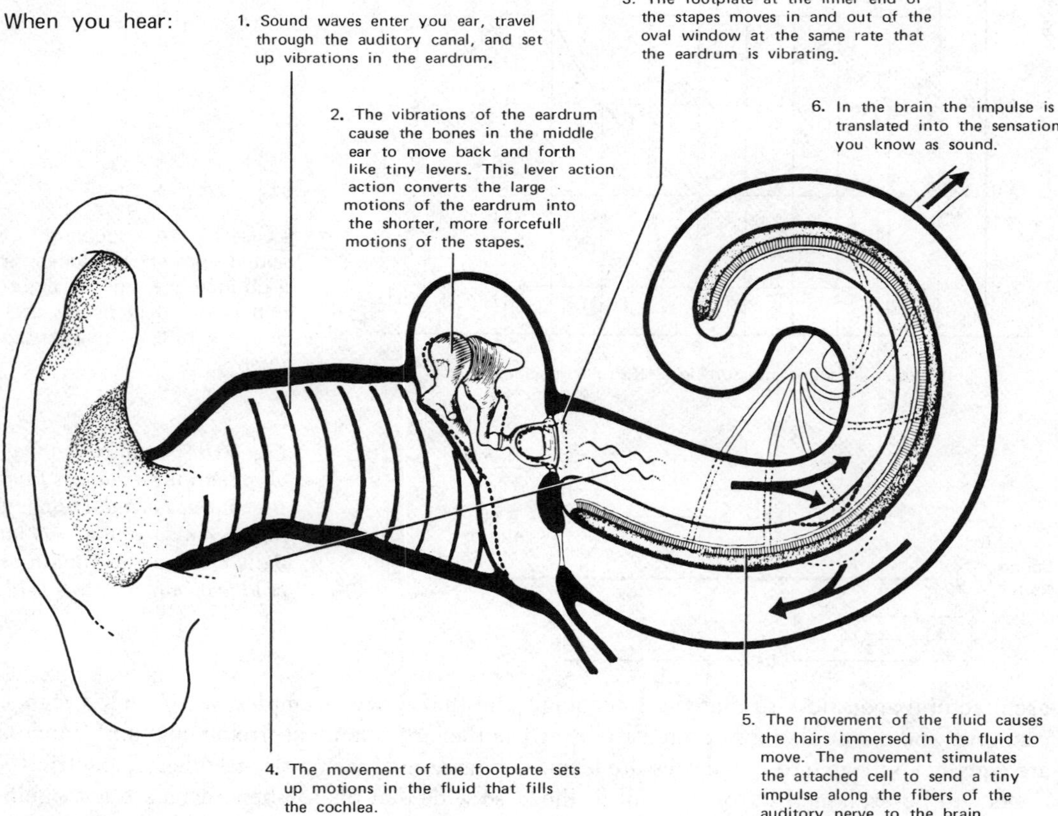

Fig. 26.9 *Schematic diagram showing functioning of the human ear. From* Quieting: A Practical Guide to Noise Control, *National Bureau of Standard Handbook 119, July 1976.*

than the wanted signal. In effect, the ear is attenuating the unwanted signals. Normally, however, the ear does precisely the opposite. It combines sounds that are clearly distinct from each other in frequency and phase. (Actually, the ear is not capable of detecting phase differences). The three tones in a chord struck on a piano are different in frequency and, if played as a very rapid triplet, out of phase. Yet the ear combines them, and hears a single sound, despite the fact that the maxima of the three tones do not occur simultaneously. In the preceding decibel calculations (Section 26.8), as well as those that will follow, we are assuming the latter action of the ear—that it ignores phase differences and combines frequencies. Nevertheless, this may not always be the case, particularly when the frequencies are very far apart. For this reason the type of information found in Table 26.4, that is, single-number representations of complex sounds, can be misleading and must be used with caution. This problem is discussed in Section 26.13, which explains measurements and weighting networks. For most purposes, information of the type shown in Figures 26.10 and 26.11 is more useful.

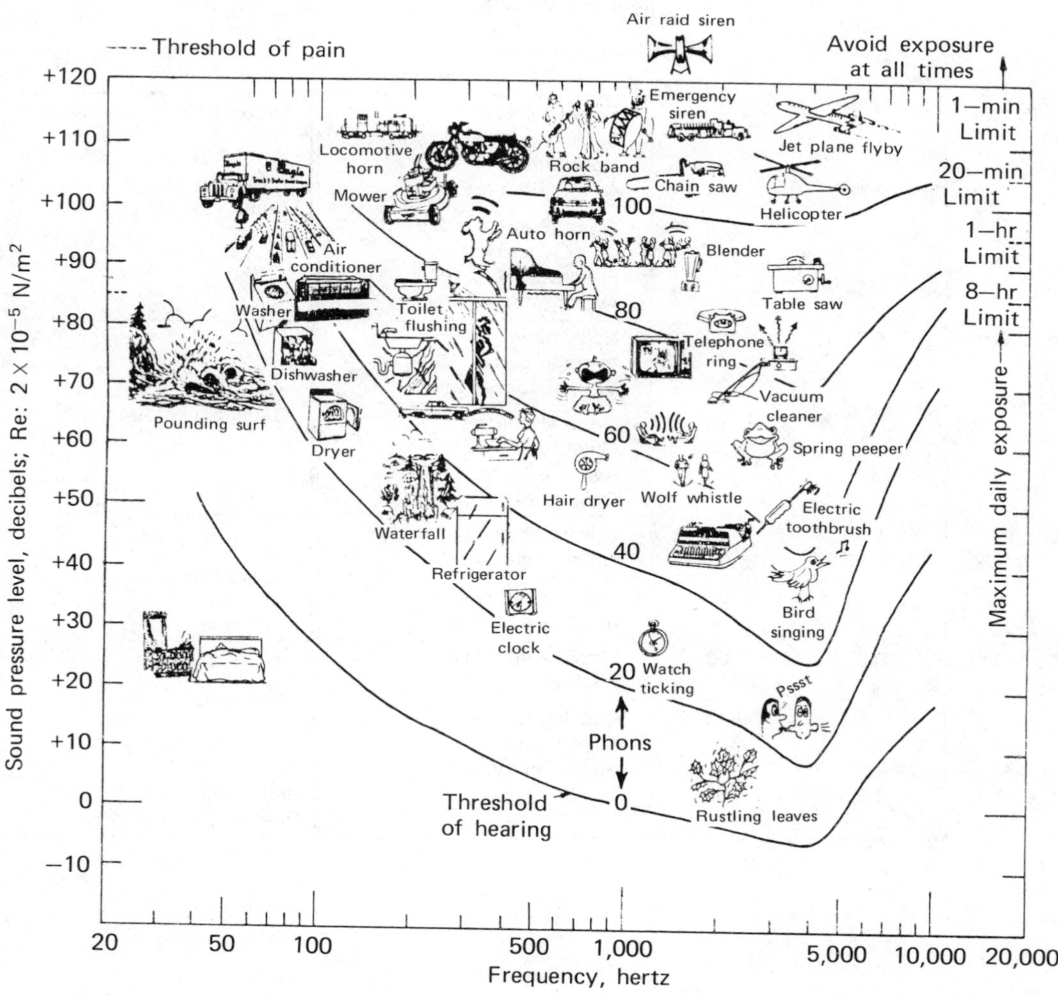

Fig. 26.10 Common sound sources plotted at their dominant frequencies and levels as typically heard by the observer. The equal loudness curves (see Section 26.11) show why certain sounds seem louder than others despite the pressure levels that would indicate the contrary. From Quieting: A Practical Guide to Noise Control, *National Bureau of Standard Handbook 119, July 1976.*

Table 26.4 *Common Noise Levels*

Sound Pressure		*Pressure Level, Decibels*	*Typical Sound*	*Subjective Impression*
Pascals	*Microbars (μbar)*			
		150		(Short exposure can cause hearing loss)
2×10^4	2000	140	Jet plane takeoff	
	2×10^3	130	Artillery fire Machine gun Riveting	Deafening (Threshold of pain)
2000 2×10^3	200	120	Siren at 100 ft Jet plane (passenger ramp) Thunder—Sonic boom	
		110	Woodworking shop Accelerating motorcycle Hard rock band	(Threshold of discomfort)
200	20	100	Subway (steel wheels) Loud street noise Power lawnmower Outboard motor	Very loud
		90	Truck unmuffled Train whistle Kitchen blender Pneumatic jackhammer	
20	2	80	Printing press Subway (rubber wheels) Noisy office Average factory	(Intolerable for phone use)
		70	Average street noise Quiet typewriter Freight train at 100 ft Average radio	Loud
2.0	0.2	60	Noisy home Average office Normal conversation	Usual background
		50	General office Quiet radio Average home Quiet street	
2×10^{-3} 0.002	0.02 2×10^{-2}	40	Private office Quiet home	Noticeably quiet
		30	Quiet conversation Broadcast studio	
2×10^{-4}	0.002	20	Empty auditorium Whisper	Very quiet
		10	Rustling leaves Soundproof room Human breathing	
2×10^{-5}	0.0002 2×10^{-4}	0 db		Intolerably quiet Threshold of audibility

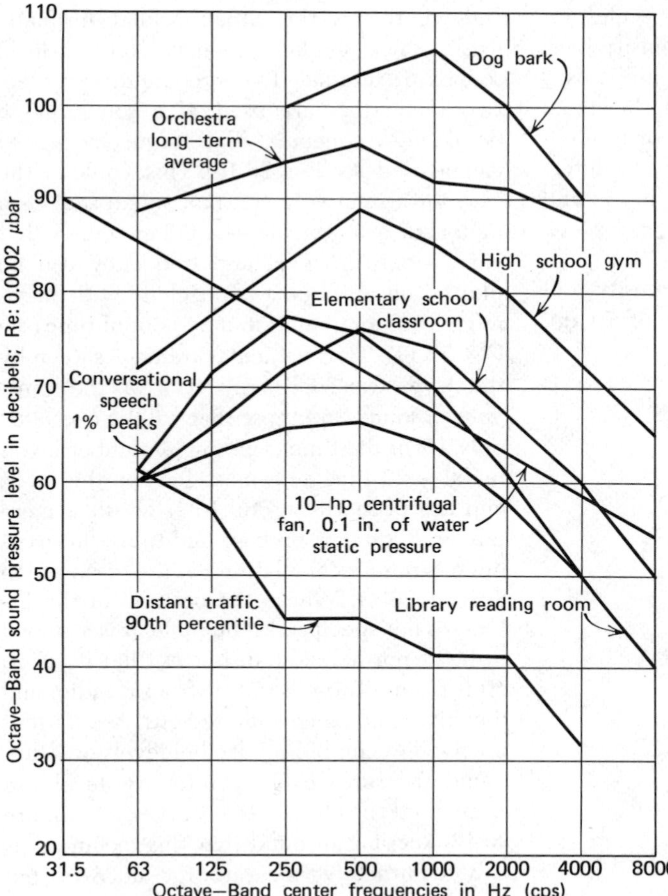

Fig. 26.11 *Sound pressure levels of eight typical sounds with different time characteristics.*

26.10 Sound Pressure and Sound Pressure Level (SPL)

At the threshold of hearing (approximately 0 db) the displacement of air molecules impinging on the eardrum, and the eardrum excursion, are approximately one Angstrom unit (10^{-8} cm), which is approximately the diameter of an *atom*. Were the ear an order of magnitude more sensitive, it would hear thermal noise, which is approximately at 1 micropascal (μPa). The human ear is thus close to the practical limit of sensitivity. At the other end of the pressure spectrum, the threshold of pain corresponds to 130 db, 10^8 μPa, and a molecular and eardrum motion of approximately $\frac{1}{4}$ mm. An astonishing range indeed! The pressure that corresponds to the threshold of hearing is taken to be 20 μPa or 2×10^{-4} microbars (μbar). As with intensity, this pressure is established as 0 db, for sound pressure *level*. Since the ear responds logarithmically to intensity and since pressure varies as the square root of intensity, we can write the expression

$$SPL = 10 \log \frac{p^2}{{p_0}^2}$$

or

$$SPL = 20 \log \frac{p}{p_0} \qquad (26.8)$$

where

SPL = sound pressure level in decibels

p = pressure in Pascals or bars

and

p_0 = reference base pressure in Pascals or bars

We have thus equalized the db scales for sound pressure level (SPL) and intensity level (IL) and the db values of the two can be used interchangeably. Obviously though, the actual intensity and the actual pressure corresponding to a particular decibel level are different—completely different, in magnitude and units.

EXAMPLE 26.4. Find the sound intensity and pressure corresponding to a level of 70 db.

SOLUTION

(a)

$$IL = 10 \log \frac{I}{I_0} \qquad (26.6)$$

where $I_0 = 10^{-16}\ \text{w/cm}^2$

Therefore

$$70 = 10 \log \frac{I}{I_0}$$

$$7 = \log \frac{I}{I_0}$$

$$10^7 = \frac{I}{I_0}$$

$$I = 10^7\ (10^{-16}) = 10^{-9}\ \text{w/cm}^2$$

(b)

$$SPL = 20 \log \frac{p}{p_0}$$

$$70 = 20 \log \frac{p}{p_0}$$

$$3.5 = \log \frac{p}{p_0}$$

$$10^{3.5} = \frac{p}{p_0}$$

since $p_0 = 2 \times 10^{-4}\ \mu\text{bar}$

$$p = (2 \times 10^{-4})\ (10^{3.5})$$

$$= 2 \times 10^{-0.5}$$

$$= \frac{2}{\sqrt{10}} = 0.63\ \mu\text{bar}$$

$$= 0.063\ \text{Pa}$$

Assigning a single-number decibel level to a sound presents two difficulties:

1. The sound pressure level varies with time.
2. The different components of the sound vary in pressure level.

Indeed, this is the situation that normally exists, since single-frequency, constant-level sounds are unusual. To overcome this problem two techniques are used. If a sound has a dominant frequency, that frequency's level can be used (see Fig. 26.10). This would be the case for a relatively constant sound such as a motor, or a fan or blower. Other sounds that vary widely in level and frequency can be plotted on an octave band chart with maximum level for a minimum percent of time (see Fig. 26.11). Thus vehicular traffic is shown by the level exceeded only 10% of the time. Speech sounds are those that will be exceeded only 1% of the time. The single-number levels in Table 26.4 have been assigned on this basis, and are primarily useful to establish a mental-aural comparison base and to use in maximum exposure calculations, as are discussed in Section 26.21. Where the position of the listener is not specified in the table, it is assumed to be at normal *close* distances, that is, 10 to 20 ft from a train, 3 to 5 ft from a radio, and the like. The somewhat tedious calculations involved in combining decibel levels of either sound pressure level or intensity level are eliminated by using the curves of Figure 26.12, keeping in mind that this technique is only completely accurate for uncorrelated sound sources.

26.11 Loudness Level—The Phon Scale

The human ear is not uniformly sensitive over its entire frequency range of 20 Hz to 20 kHz. At the upper limit the 120 to 130-db pain threshold occurs at all frequencies. However, at the lower limit, the 0-db threshold only occurs at 1000 Hz. The ear is in fact most sensitive at 3000 to 4000 Hz, at which frequencies the threshold is about −5 db. This nonlinear response exists throughout the ear's entire range. To determine the nature of this nonlinearity a great number of tests were conducted with pure tones at different frequencies, in which the listener was asked to equate loudness of signals. These tests resulted in a family of curves called *equal loudness level contours* or, alternatively, *Fletcher-*

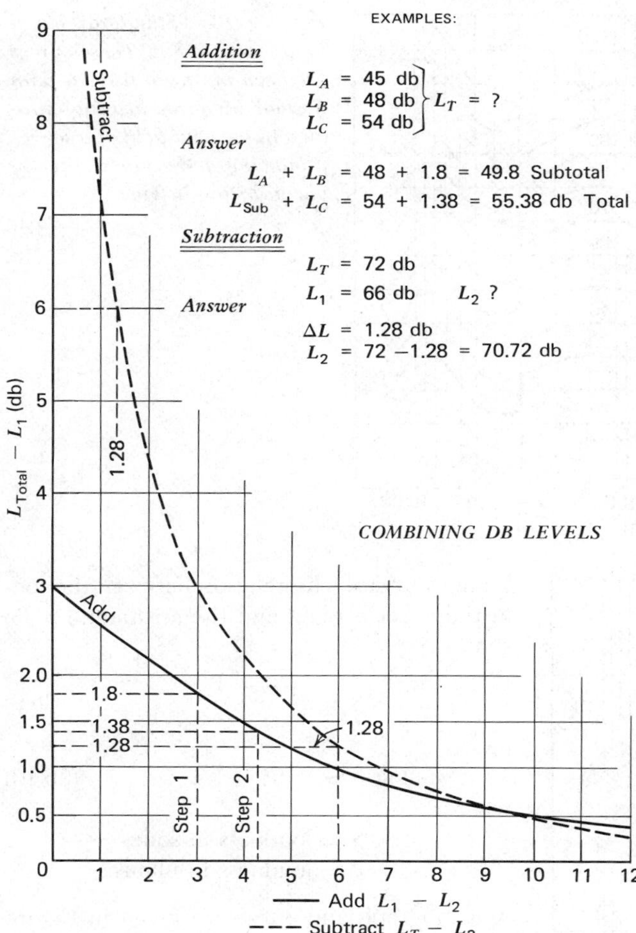

Fig. 26.12 Curves for combining decibel levels of uncorrelated sound sources.

Munson equal loudness contours (after two of the principal researchers). These curves (see Fig. 26.13) are internationally recognized and standardized, and are used as the reference for normal hearing response. They are also used to "weight" measuring devices as is explained in Section 26.13. Note that *by definition:*

1. All points on a single contour have the same subjective sensation of loudness.
2. The loudness level in *phons* is defined as the db level of that contour at 1000 Hz.

The curves demonstrate some interesting phenomena:

1. Sensitivity drops off sharply at low frequencies, particularly at low db levels. (For this reason most HI-FI amplifiers provide automatic bass boost at low-volume levels.)
2. Maximum sensitivity occurs between 3 and 4 kHz—precisely the frequencies that contain most information in human speech. (Note the dip between 3 to 4 kHz). See Figure 26.14.
3. In the normal listening range of 45 to 85 db, and the normal frequency range of 150 Hz to 6 kHz, the contour is substantially flat, that is, the ear's response is linear [with the exception of the dip noted in (2) above]. It is only at extremes of sound level and frequency that nonlinearity occurs.

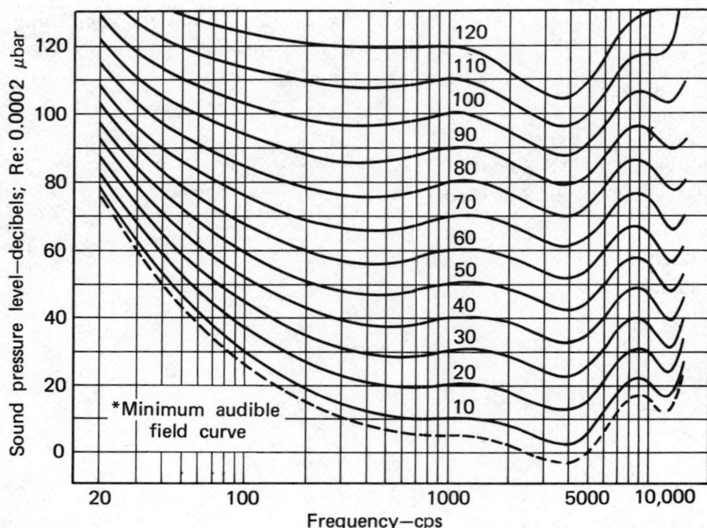

Fig. 26.13 Standard equal loudness contours. These curves are accurate for a listener with normal binaural hearing, situated in the near field of a source producing pure tones directly ahead of the listener.

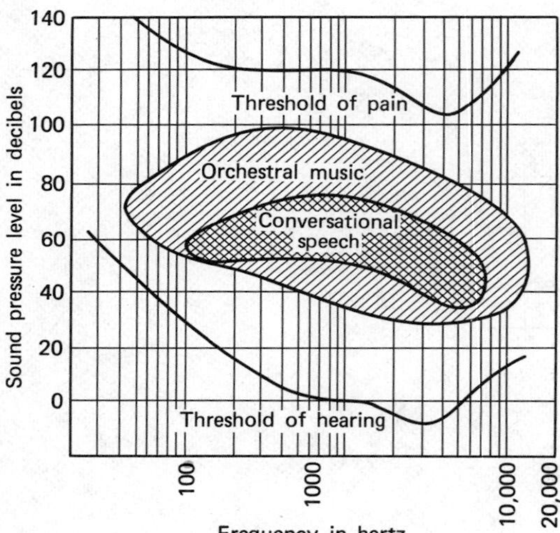

Fig. 26.14 The positions of speech and wide-range music in the human ear's aural field are illustrated. Speech is in the nominally linear response area of the ear as is most music (see Fig. 26.13). Beyond these frequencies the ear's action is effectively to attenuate the signal.

26.12 The Sone Scale

Since the phon scale is based on decibel levels at 1000 Hz, it too is a logarithmic scale in which a *doubling* of loudness corresponds to an *increase* of 10 phons. The *sone* scale was developed because most people much prefer an arithmetic scale, that is, one where doubling loudness would result in doubling the loudness index. The relation between the logarithmic scale phon and the arithmetic scale sone is

$$S = 2^{\left(\frac{P-40}{10}\right)} \tag{26.9}$$

or

$$\log S = 0.3P - 1.2 \tag{26.10}$$

where

S = loudness in sones
P = loudness in phons

A nomograph and curve are given in Figure 26.15 for conversion between the two quantities. Although both are in use, single-number decibel A ratings (see Section 26.13) are most frequently found in the literature.

26.13 Measurement of Sound Pressure Level (SPL)

The need for a means of measuring sound levels is obvious. One such instrument is the integrating sound-level meter illustrated in Figure 26.16*a*. In order to correlate meter readings with subjective loudness impressions, most such single-reading instruments are furnished with three weighting networks, the characteristics of which are given in Figure 26.16*b*. The A network corresponds to the 40-phon contour and discriminates against low frequencies (see Figure 26.13). The B network corresponds similarly to the 70-phon contour.

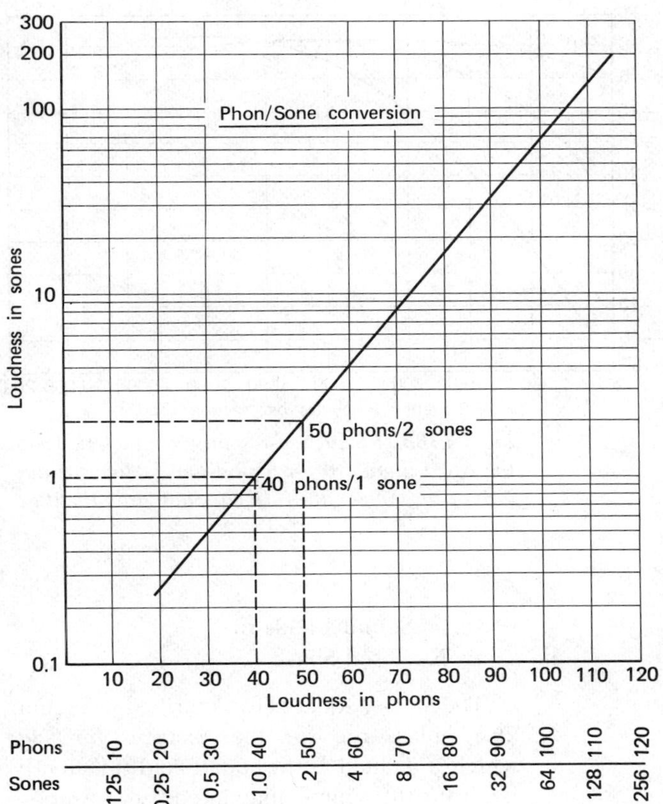

Fig. 26.15 Conversion chart between phons and sones. For convenience a nomograph with the same data is given below the X axis.

The C network gives substantially flat uniform response. In practice it was found that the B and C networks did not correspond well to subjective loudness reports. This was because the loudness contours on which they were based were taken for pure tones whereas most field measurements are of very complex sounds. The A network, however, did correspond well, and has therefore become the standard single-number measuring scale for loudness of sounds at any frequency and intensity. Measurements taken in this way should be labelled "dbA" to identify the scale. The sound meter is equipped with two different damping characteristics—fast and slow. The former will follow reasonably rapid fluctuations. The latter is more heavily damped and will give an average reading of sounds with rapid changes of more than 4 db. Extremely rapid noise pulses are impulse noises and present an entirely different problem.

It is known that the ear "averages" sounds over a minimum period, and that sound impulses shorter than this period sound quieter than they would sound as steady-state noise. This minimum sound length is taken to be between 50 and 200 milliseconds ($\frac{1}{2}$ to $\frac{1}{5}$ second) depending upon which researcher's work is used. A compromise figure of 70 ms ($\frac{1}{4}$ sec) is frequently found in the literature.

More accurate measurements of complex sounds than are possible with a standard meter are accomplished with sophisticated instruments that measure intensity in octave bands and plot the results as in Figure 26.11. Such measurements are necessary for accurate application of sound absorption and attenuation materials whose characteristics are also nonlinear over the frequency spectrum. Single-number dbA readings are known as *overall* levels, and are useful as preliminary data and for broad spectrum design.

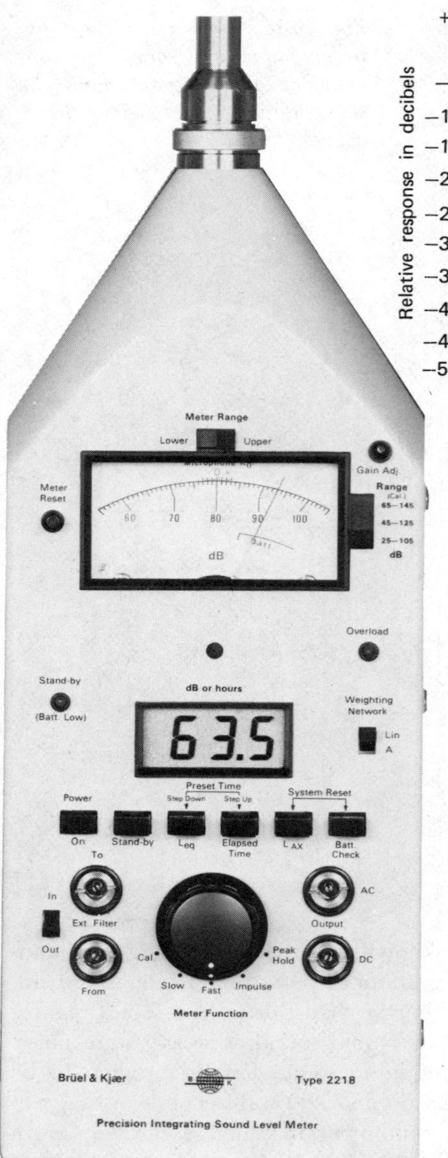

Fig. 26.16a *Precision-integrating sound level meter. This unit will measure either instantaneous sound level or* L_{eq} *(equivalent continuous sound level). The latter is important when sound at a location varies with time, and an equivalent level for design purposes is required. The unit takes level and duration into account and displays* L_{eq} *directly. It is arranged to measure dbA or linear (dbC), fast or slow time constant, adapts with filter networks to read octave or* ⅓ *octave bands, and reads out digitally. It also has peak-hold settings, which measure maximum sound during a time exposure. Courtesy of Brüel and Kjaer.*

Fig. 26.16b *Frequency-response characteristics in the ANSI Standard for Sound-Level Meters. Courtesy of the American National Standards Institute.*

26.14 Sound Fields in an Enclosed Space

The inverse square law described in Section 26.7 holds true for the acoustic *far-field,* which is defined as the sound field sufficiently far from the source that intensity is proportional to pressure squared. This field is quickly developed in open, free-field circumstances. Propagation in an enclosed space is quite different. When a sound propagated in an enclosed space reaches a wall or other large (with respect to wavelength) obstruction, part is reflected and part absorbed. The sound at any point in the room is then the combination of direct sound from the source plus reflected sound from walls and other obstructions. If the reflections are so large that the sound level becomes uniform throughout the room, the field within the room is termed a *diffuse* one (no shadows), and intensity measurements with respect to a specific source are meaningless. Of course, if it is our intention to measure sound pressure level at a specific point, such as a seat in an auditorium, then the type of field in the room is irrelevant. Most rooms do not have such a high level of reflection that a diffuse field is created. Instead, there is a near field near the source, a *free field* at a distance, and a *reverberant* field near the walls (see Fig. 26.17). They can be recognized as follows:

1. The near field is generally within one wavelength of the lowest frequency of sound produced by the source. Within this distance sound-pressure-level measurements vary widely and are not meaningful. (The maximum wavelength for the human male voice is about 11 ft.)

2. Near large obstructions such as walls, the *reverberant field* is dominant and approaches a diffuse condition. In auditoriums the reverberant (diffuse) field predominates and sound pressure level remains relatively constant beyond the free field area.

3. The *free field* exists between the near and reverberant fields, and there intensity varies as pressure squared and inversely with distance. In *this* field, sound pressure level drops 6 db with each doubling of distance from the source, and it is in this field that meaningful sound-pressure-level measurements can be made.

Many enclosed spaces are so small or so reverberant that no free field exists. In such instances it is possible to measure acoustic power output from which intensity levels can be calculated. Indeed, because accurate measurement of sound pressure level due to a single source is difficult, manufacturers very frequently supply sound power and/or sound-

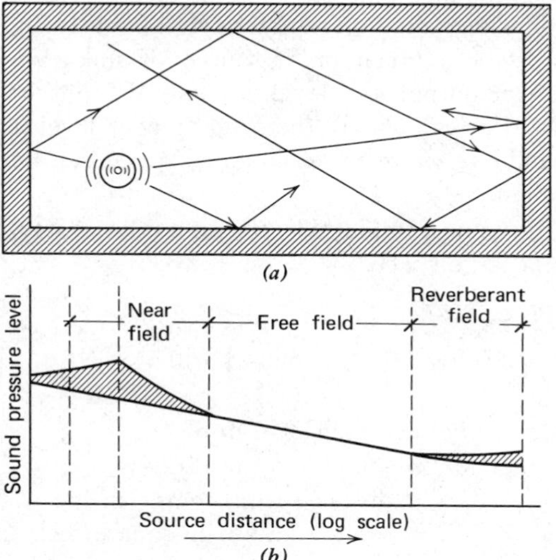

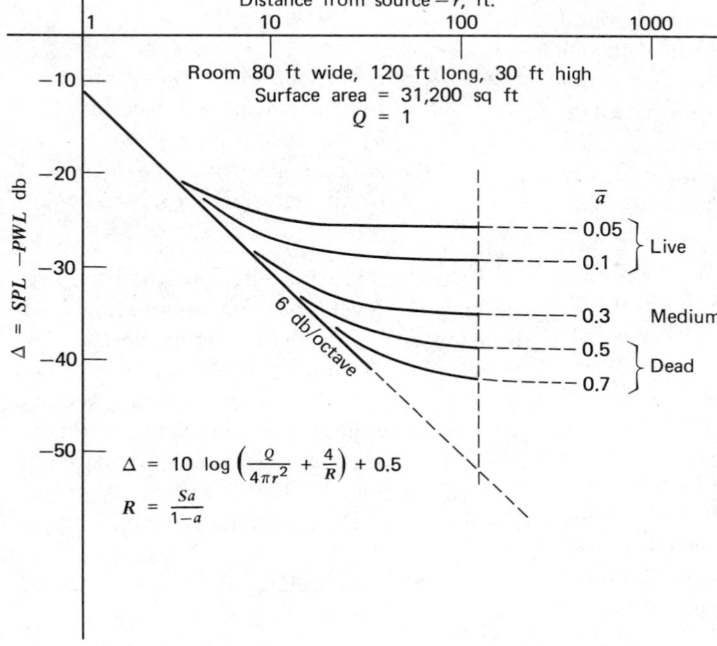

Fig. 26.17 *Type of sound field in an enclosed space depends in large measure on reflections (reverberations). In a typical room* (a) *there is near field adjacent to the sound, a free field beyond that, and a reverberant field adjacent to the walls.* (b) *In a large hall or auditorium the reverberant field dominates and level remains constant. Curves in* (c) *show that development of reverberant field (constant SPL, or flat curve) is dependent on the room's absorption characteristics. In a live room,* ($\bar{a}$ = *0.1*) *the reverberant field begins 20 ft from the source. In a dead room* ($\bar{a}$ = *0.5*) *it begins at 120 ft from the source, that is, at the back wall. Effectively then, there is no reverberant field in such a room. Slope of asymptote is 6 db per octave, that is, inverse square attenuation.*

power-level data for their equipment. These figures are derived by suspending the equipment in a reverberant chamber so as to establish a spherical wave in a diffuse field.

Then

$$W = \frac{13.8p^2V}{c^2T_R} \quad (26.11)$$

where

W = total sound power in watts
V = volume of the *reverberant* chamber
c = 344 m/sec, sound velocity in air
T_R = reverberative time in seconds (measured)
p = sound pressure in Pascal (Newtons/m^2)

Quantity p is calculated from measured sound pressure level:

since $$SPL = 20 \log \frac{p}{p_0} \quad (26.12)$$

and since p_0 is base pressure of 2×10^{-5} Pa,

$$SPL = (20 \log p + 94) \text{ db} \quad (26.13)$$

To avoid confusion, it will be helpful to the reader to think of sound pressure level (SPL) as the resultant "noise" or sound in an enclosed space, resulting from a source in that space, and affected by the characteristics of the space and the position of the listener. It is thus an end effect. The sound *power* level (PWL) is a measure of the amount of sound generated by a source, independent of its environment. In free space the two quantities are simply related by the inverse square law (see eq. 26.18 below), whereas in enclosed spaces the room characteristics come into play (see eqs. 26.15 and 26.17). The discussion below qualifies these relationships.

Roughly speaking, an analogy to lighting can be drawn: SPL corresponds to room illumination, that is, footcandles, and PWL corresponds to the lumen output of the source.

Reverberation time is defined as the time necessary for the sound to drop 60 db, that is, to effectively become inaudible, after power is shut off. (Reverberation time can also be calculated in terms of a room's dimensions and absorption coefficient, as is discussed in Section 26.24). Once W, the sound source's power level is determined, it in turn can be used to calculate the sound power level, *PWL*, in decibels:

$$PWL = 10 \log \frac{W}{W_0} \text{ db} \quad (26.14)$$

where

PWL = sound power level in decibels
W = sound power of source, in watts
W_0 = base power, 10^{-12} w.

In general, manufacturers will furnish W and PWL for equipment. It is only necessary to calculate these parameters when measurements are taken on a source of unknown power output and level. Given *PWL* (either listed or calculated) the sound power level it would generate in an enclosed space can be calculated as:

(a) For a sound source with no directivity, in a large space,

$$SPL = PWL + 10 \log \left(\frac{1}{12.5r^2} + \frac{4}{R} \right) + 0.5 \quad (26.15)$$

where SPL = sound level in db
PWL = sound level in db
r = distance from source in feet
R = room constant in square feet

the factor R can be calculated from

$$R = \frac{S\bar{a}}{1 - \bar{a}}$$

where R = room constant in square feet
S = total room surface area in square feet
$\bar{a}$ = average absorption coefficient of all materials, in sabins (see Section 27.3)

To understand the effect of room absorption on the development of reverberant fields in a large space refer to Figure 26.17*c*. The field is plotted from equation 26.15 for a room 80 ft × 120 ft × 30 ft high using average absorptions $\bar{a}$ ranging from 0.05 to 0.7, that is, from very live to very dead. ($\bar{a}$ is calculated by first calculating the room's entire absorption in sabins, and then dividing by total area:

$$\bar{a} = \frac{S_1\alpha_1 + S_2\alpha_2 + \cdots + S_n\alpha_n}{S_1 + S_2 + \cdots + S_n} \quad (26.16)$$

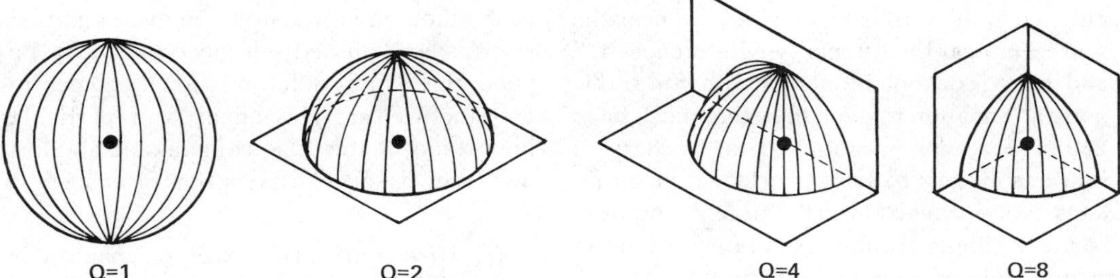

Fig. 26.18 Diagrams illustrating directivity factors for nondirective sources placed adjacent to large reflecting surfaces. Courtesy of Barry Blower Co.

see Section 26.24.) The importance of these curves will become clearer in the discussion on directional loudspeakers in Section 26.31 and room absorption in Section 27.3.

For a sound source with directional characteristics, the applicable equation is

$$SPL = PWL + 10 \log \left(\frac{Q}{12.5r^2} + \frac{4}{R} \right) + 0.5 \text{ db} \quad (26.17)$$

where Q is a directivity factor and all other factors are as in equation 26.15 above. The directional constant is either inherent in the sound source, and as such will be part of the given data, or can be obtained from Figure 26.18 for a nondirectional source made directional by adjacent reflecting surfaces. In a room containing more than one sound source, the sound pressure levels can be combined using the curves of Figure 26.12.

26.15 Sound Power and Pressure Levels in Free Space (Outdoors)

The equations of the preceding section are not applicable to outdoor propagation in which the large reflective component of the indoor condition is absent. Although the propagation of sound outdoors may not appear to be of immediate importance in architectural acoustics, outdoor noise sources such as traffic, cooling tower, and aircraft are frequently loud enough to disturb activities within or immediately adjacent to a building. Conversely, the noise made by building equipment such as cooling towers, heat pumps, and even window air conditioners may be sufficiently loud to disturb neighbors in a nearby building. For this reason it is desirable to have some basic understanding of outdoor sound propagation.

For preliminary evaluation of an outdoor noise problem, assuming a nondirectional source, one needs to know only the power level radiated by the source as a function of frequency and time; from this one can establish the intensity level of sound at the appropriate distance as follows:

$$SPL = PWL - 20 \log r + (Q\text{-}1) \text{ db} \quad (26.18)$$

where

SPL = sound pressure level
PWL = equipment power level
r = distance from the source, in feet
Q = directivity factor (see Figure 26.18)

This formula is fairly accurate for a *small* source. For large sources such as cooling towers and traffic, which do not exhibit inverse square properties, sound level estimates are best made on the basis of experience and empirical data beyond the scope of this book (see References 2 and 3). For small outdoor sources, the equipment power level can be estimated by measuring the sound power level at 5 ft and adding 15 db. Other factors, such as moisture in the air, the presence of trees, wind, and temperature gradients, will affect outdoor sound propagation to some extent but they can be ignored except when great distances (i.e., over 1000 ft) are involved. Barrier walls (such as solid fences) are an influence only when the wall is high, wide, and near the source, or near the listener (see References 4, 5 and 6). A convenient nomograph for traffic noise attenuation is given in Reference 2, page 239. The reader is cautioned to be chary of simple solutions to barrier attenuation problems. Noise travels through, around, and over barriers. Simple graphical solutions most often assume point sources and infinitely long barriers; situations that are not even approximated by actual field situations, and that yield misleadingly high attenuation figures, particularly at low frequencies.

26.16 Other Factors in Hearing

(*a*) ***Masking.*** When two separate sources of sound are perceived simultaneously, the perception of each is made more difficult by the presence of the other. This is known as masking, which is defined as the number of decibels a sound has to be raised above its threshold when perceived alone, to be perceived in the presence of another sound. Effectively then, masking is an upward shift of audibility threshold. Masking is greatest when two sounds are close in frequency or frequency content, since the ear has greater difficulty separating them. Also, a low frequency will more effectively mask a high frequency than the reverse, for the same decibel levels. With broad-frequency sounds the masking effect is difficult to predict, since it depends in part on how "hard" the listener is listening. Masking is an extremely important technique in noise control where background noise levels are deliberately manipulated to mask unwanted sounds. The background noises used for this purpose are of the broadband continuous variety, which are themselves noninformation bearing. They serve to obliterate lower-level, information-bearing sounds that would cause annoyance.

(*b*) ***Time.*** As stated above, impulse sounds are apprehended at lower levels than the same sound intensity over a longer period. Similarly, because of the time-constant of the ear's mechanical linkages, sounds closer than 10 msec apart cannot be distinguished from each other, and those up to 50 msec apart are poorly distinguished (see Section 26.25). Beyond this point differentiation becomes increasingly clear. This effect is of particular importance in the design of halls and auditoriums, with respect to reception of echoes.

(*c*) ***Directivity.*** The exact mechanism by which the binaural aspect of hearing detects direction is not clearly understood. The single ear is not phase sensitive, but it may be that binaurally, phase sensitivity exists, at least at low frequencies, and that this assists in detection of direction. At high frequencies, phase detection is clearly nonexistent and sense of direction may be due to diffraction effects around the head. These effects would also explain the accuracy of detection in a horizontal plane, which research indicates to be in the order of 5° change. It would not explain how the ears detect changes in the vertical plane immediately in front of the listener. This latter situation, although much less accurate than horizontal plane detection, definitely exists. In enclosed space, reverberation will blur most phase differences and "stereo" information will be almost completely dependent upon high frequencies in the near field.

(*d*) ***Concept of Reverberation.*** Although this is not strictly a hearing phenomenon it is one of the most pronounced hearing reactions in an enclosed space. Simply, it is the ear's reaction to echoes in an enclosed space, giving an impression of "liveness" or "deadness." We can obtain a good approximation of the subjective feeling of liveness of a room, for purposes of *speech*, from the relation

$$T = 0.3 \log \frac{V}{10} \tag{26.19}$$

where T = optimum reverberation time in seconds, for speech

V = room volume in cubic meters

For instance, a typical classroom might have a volume of 150 m^3 (5300 cu ft). Optimum reverberation time is

$$T = 0.3 \log 15 = 0.35 \text{ sec}$$

Reverberation times longer than this would sound live; shorter ones dead and flat. Indeed an increase of 20% in reverberation time would make the room excessively live and boomy and would negatively affect speech intelligibility. See Section 26.24 and 26.25 for a fuller discussion of this subject.

SOUND SOURCES

26.17 Speech

As can be seen from Figure 26.14 the ear's sensitivity is maximum in the speech frequency and normal energy range. Speech sounds vary in time length between 30 and 300 msec so that the ear perceives them individually and clearly. Speech is comprised of *phonemes*, which are individual and distinctive sounds that to an extent vary from language to language, that is, certain ones exist in one language and not in another. Since certain phonemes carry more information than others, it is these which good architectural acoustics must be particularly careful to preserve intact, to preserve intelligibility. In English, consonants carry much more information than vowels, as can readily be demonstrated by writing a sentence first without consonants and then without vowels:

Most speech energy is concentrated in the 100- to 600-Hz range
o ee ee i oeae i e 100–600 e ae
Mst spch nrgy s cncntrtd n th 100–600 Hrtz rng (see Fig. 26.19).

The male voice centers its energy around 500 Hz; the female about 900 Hz. It is, however, in the high frequencies that consonants have most of their energy. Phonemes such as s and sh have most of their energy above 2 kHz and both are particularly important in conveying intelligence.

Normal speech averages between 40 and 50 db sound pressure level at 3 to 4 ft, with a dynamic range of from about 30 db for a soft speech to about 65 db for loud speech at the same distance. Extremes of speech are 10 db for a whisper and 80 db for a shout, but in both these instances intelligibility is sharply reduced because of lack of consonant power. Indeed, in shouting, emphasis is necessarily on vowels so that it is generally accepted that 70 db SPL is about the upper limit of fully intelligible human speech. (Note that singers who frequently exceed 90 db do so at great loss of intelligibility). Another result of the high-frequency content of consonants and hence intelligibility is its directiveness. The higher the frequency the greater its directivity and the less its diffraction (ability to turn corners). Therefore, intelligibility of speech is greatest directly in front of the speaker and least behind him. The high-frequency tones are most easily absorbed and least reflected and diffracted.

26.18 Other Sounds

Music is much broader and complex than speech in frequency and dynamic range. It has

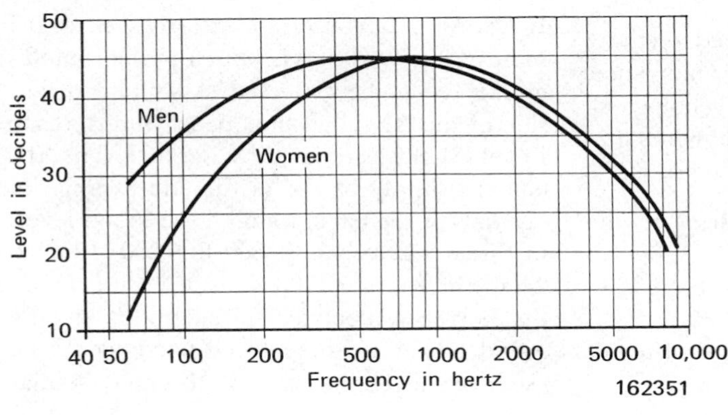

Fig. 26.19 Frequency distribution of typical male and female voices. From Architectural Acoustics, *Brüel and Kjaer, 1963.*

no direct parallel to intelligibility. "Reception" of music is a combination of physiological and psychological phenomena. As such, it is beyond most of the purposes of this study, but will be briefly examined in our discussion of room acoustics, auditoriums, and halls. *Noise* is variously defined as unwanted sound, sound with no intelligence content, and broadband sound depending on the listener and the situation. All three definitions are correct at various times and situations. For our purpose we must assume that any sound can be referred to as noise, as will be discussed in various sections below.

NOISE CRITERIA

26.19 Negative Effects of Noise

Although noise effects and their control are the subject of Chapter 27, noise criteria and their background are discussed here as part of our overall study of hearing and sound sources. There are two basic approaches to the negative effects of noise, a pyschological-practical one and a purely physiological one. The latter is concerned with the physical impact of noise on the body including hearing loss and other deleterious conditions (see Section 26.21). The former is concerned with noise levels that cause annoyance and disturbance to daily activities including work, relaxation, and rest, and is discussed in the following sections.

26.20 Noise and Annoyance

Although research has developed accurate data on loudness, the concept of annoyance, being primarily subjective and psychological, is much more elusive. Tests have shown that *in general,* annoyance as a result of noise is:

a. Proportional to the loudness of the noise.
b. Greater for high-frequency than low-frequency noise.
c. Greater for intermittent than continuous noise.
d. Greater for pure-tone than for broadband noise.
e. Greater for moving or unlocatable (reverberant) noise than for a fixed-location sound.
f. Much greater for an intelligence-bearing noise (neighbor's radio) than for a nonsense noise.

In order to establish criteria for acceptable background noise, certain of these effects must be neglected for the sake of simplicity. [They can and are considered in construction design and in establishing levels of masking noise (see Chapter 27)]. We thus ignore:

Factor (*c*), since design is based on continuous sounds.

Factor (*d*), since broadband noise is assumed.

Factor (*e*), since the noise source is assumed to be fixed in location.

Factor (*f*), since we consider noise level rather than content.

Thus the particularity and special characteristics of noises such as a barking dog (*c*), a whistle (*d*), a single passing vehicle (*e*), and intelligible sounds (*f*) are not considered. Consideration of the remaining factors (*a*) and (*b*) on interference with speech communication resulted in concepts called the Articulation Index (AI) and Speech Interference Level (SIL). This was determined by reading a carefully selected set of phonetically balanced nonsense syllables to a test audience in the presence of different levels and compositions of background noise. The ratio of correct answers to total syllables was the Articulation Index. An Articulation Index[11] in excess of 0.5 indicated a condition in which perfect intelligibility could be expected. A simplified version of the AI called the Speech Interference Level (SIL) was devised by Beranek. The SIL consists simply of the arithmetic average in decibels of the background-noise sound levels in three octave bands, 600 to 1200, 1200 to 2400, and 2400 to 4800 Hz, since it was found that correlation between intelligibility and the sound power in these three bands could be established. The results of this and similar

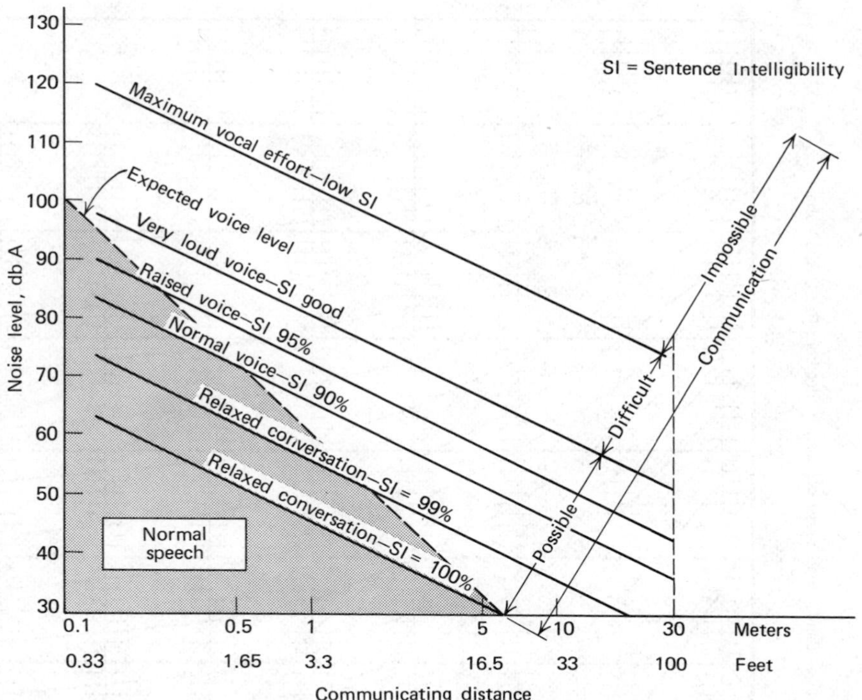

Fig. 26.20 Quality of speech communication as a function of voice and noise level and speaker-to-listener distance. Expected voice level *is the level that a speaker would normally use at that distance, in the presence of noise. Based on:* Information on Levels of Environmental Noise, *Report 550/9-74-004, USEPA, Washington, D.C., March 1974.*

Table 26.5 *Speech Interference Levels (SIL) for Male Voice*[a]

Distance Between Conversants		Voice Effort			
		Normal	Raised	Loud	Shout
Feet	Meters		SIL dbA		
1	0.3	65	71	77	83
2	0.6	59	65	71	77
4	1.2	53	59	65	71
6	1.8	49	55	61	67
12	3.7	43	49	55	61
24	7.3	37	43	49	55

[a]Subtract 5 db for female voice.

studies are shown in Figure 26.20 and Table 26.5.

On the basis of SIL data and loudness level (LL) information[7,8,9] (which is similar to equal loudness contour information of Fig. 26.13, except for broadband noise in lieu of pure tones), Beranek developed the well-known and widely accepted Noise Criteria (NC) curves shown in Figure 26.21. These curves take cognizance of the field-determined fact that most people prefer to speak at a level no greater than 22 phons above the background noise. By combining the SIL levels in decibels with this fact, the

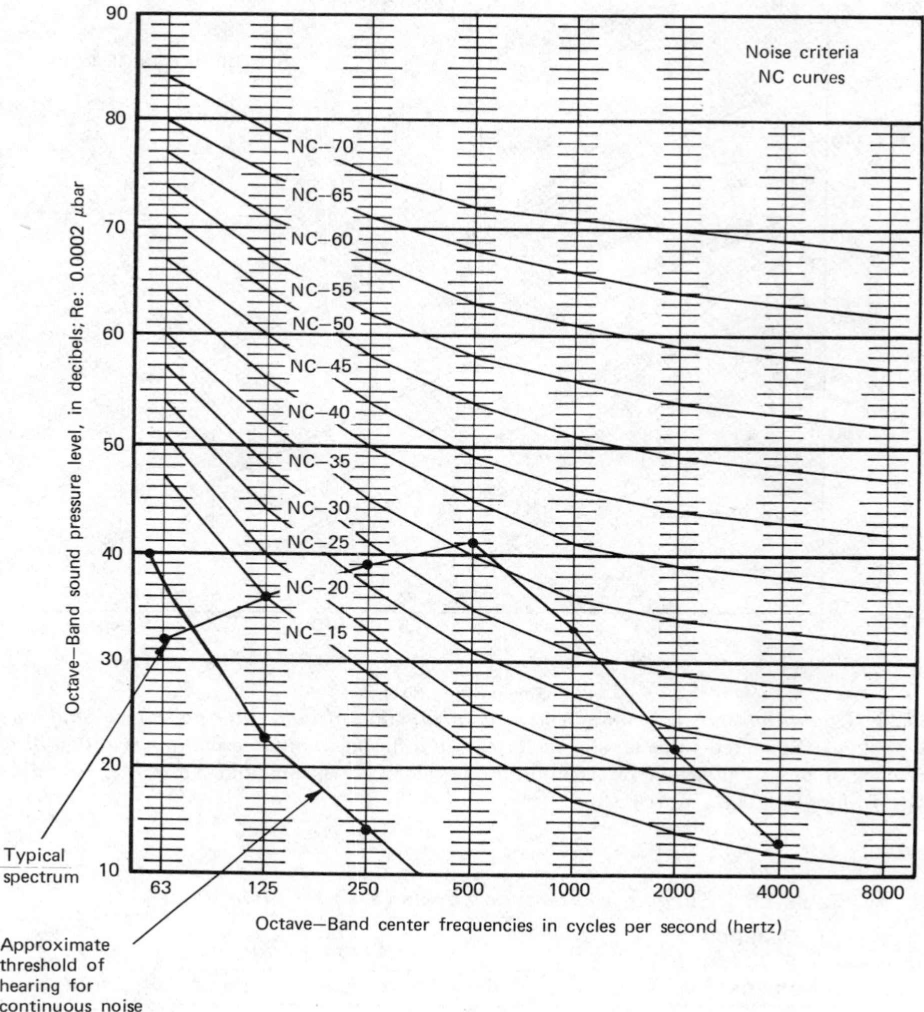

Fig. 26.21 Noise criteria curves. Typical spectrum plotted would be rated NC-36 as it exceeds NC-35 at 500 Hz by one db (see Table 27.11, page 1255, for specific recommendations for interior spaces.).

NC curves are derived, that is, they represent a loudness level 22 phons higher than the SIL in db. These contours represent then the maximum *continuous* background noise that will be considered acceptable in the environment specified and correspond fairly accurately to background noise level in commercial environments. A similar set of curves called Noise Rating (NR) curves has been proposed by the International Standards Organization (ISO), see Figure 26.22, and finds considerable application outside of the United States. These curves are less stringent than NC curves in the low frequencies and more stringent in the high frequencies.

In application, the octave-band spectrum of a noise over the range of 63 to 8000 Hz is measured and plotted on an NC curve sheet. The lowest NC curve that is not exceeded by any portion of the plot is the NC rating of the noise. Thus a specification of maximum noise levels of NC-30 means that no portion of

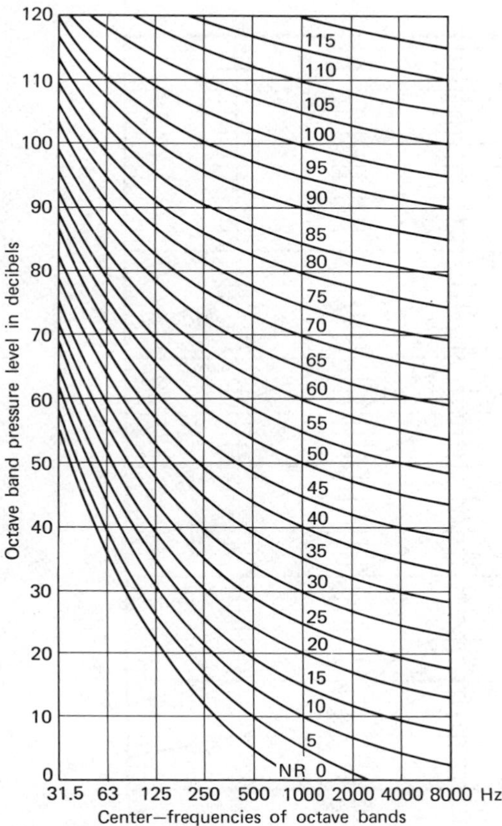

Fig. 26.22 *Noise rating (NR) curves.*

sound power level of any continuous background noise in the area may cross the NC-30 contour. Conversely, a piece of equipment rated NC-35 has an octave-band spectrum completely below NC-35. A fan rated NC-53 would indicate that at some point on its frequency-spectrum plot it exceeded NC-50 by 3 db (see Fig. 26.21). The NC number of a noise varies between 5 to 10 db below the measured dbA. NR numbers correspond closely to NC numbers. The virtue of the NC curve system is that it is a single-number specification for an entire frequency spectrum. Its disadvantage is that it was derived for, and is primarily accurate for, speech conversation in an office setting.

With the adoption of a new set of preferred frequencies for analysis in 1960 (centering on 500, 1000, and 2000 Hz), the SIL levels were replaced with Preferred Speech Interference Levels (PSIL), which are 3 db higher than the SIL levels of Table 26.5, and the NC curves of Figure 26.21 were replaced with Preferred Noise Criteria (PNC) curves as shown in Figure 26.23. It had previously been found that a background noise with a frequency spectrum conforming in shape to an NC curve was neither pleasant nor unnoticed (neutral), but instead sounded emphasized at both ends, that is, rumbled and hissed. PNC curves introduced by Beranek in 1971[12] overcame this objection. Table 27.11 gives both NC and PNC recommendations for different environments.

26.21 High Noise Levels; Hearing Protection; OSHA

It has long been recognized that continuous exposure to high noise levels causes a degree of temporary deafness in a majority of people and that long periods of such exposure, even on an intermittent 8-hr workday basis, appears to produce permanent hearing impairment. Many experts place the safe upper limit at 85 db. In addition, studies have indicated that continual exposure to noise levels as low as 75 to 85 dbA level can produce or contribute to numerous physical and psychological ailments including headache, digestive problems, tachycardia, high blood pressure, anxiety, and nervousness; in short, an almost complete catalog of human illnesses. Since continuous noises are most severe in industry, regulatory legislation in the United States has been most stringent and effective in this area. In 1969 the Walsh-Healy Public Contracts Act was passed, and thereafter its provision for maximum permissible exposure to noise levels was incorporated into the Occupational Safety and Health Act, known simply as OSHA. The relevant provisions of this act are reproduced in Figure 26.24. In order to avoid complex regulations, limitations of exposure are given in terms of single-number dbA values. The value is determined either by direct A-scale measurement, or by octave-band measurements. In the latter case, sound-pressure-level readings are taken with a linear C scale instrument and plotted on the curves given in Figure 26.24. These

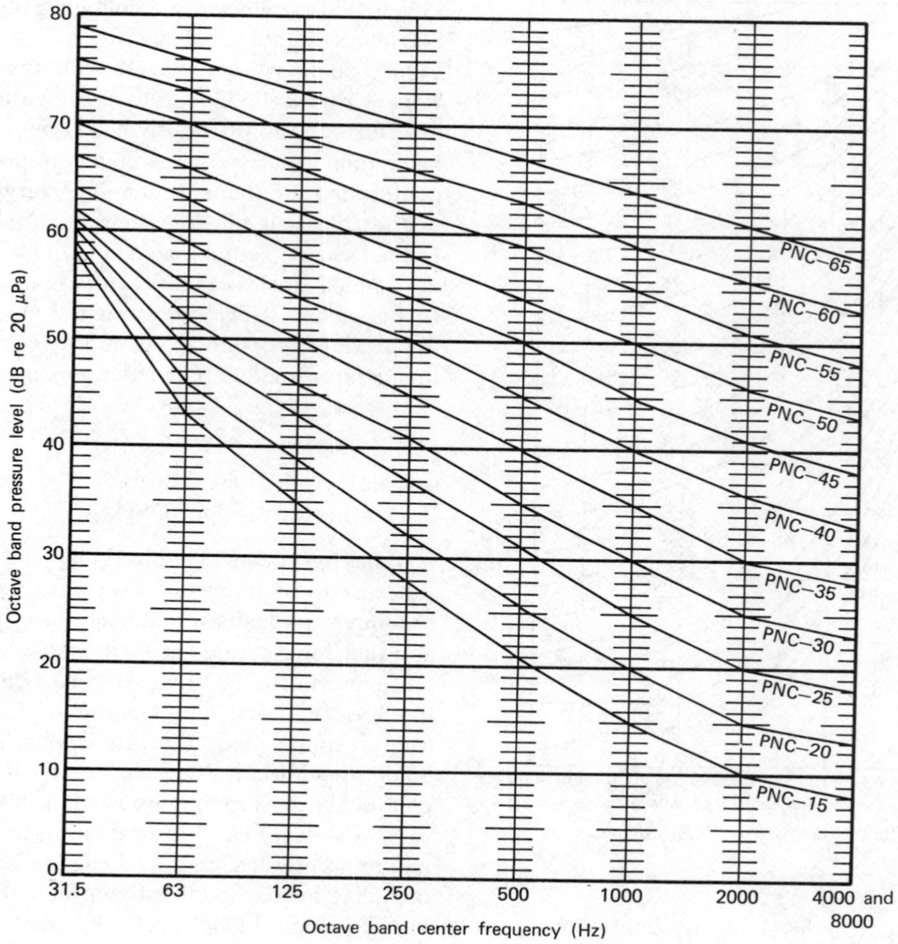

Fig. 26.23 *Prefered noise criteria (PNC) curves. From E. B. Magrab,* Environmental Noise Control, *Wiley, New York, 1975.*

curves represent A-scale response, that is, a meter with an A-weighted scale would read the values shown when exposed to a given sound pressure level at a given octave-band center frequency. The maximum penetration of the octave-band plot represents the dbA value of that noise, which is then applied as in Table G-16 of Figure 26.24. Typical industrial noise levels are given in Table 26.6. When these levels are exceeded, it is necessary for the management to take steps to reduce the exposure, either by reducing the noise or by providing hearing protectors of some type. Typical characteristics of a few types are given in Figure 26.25.

Note that OSHA does not deal extensively with impulse noises, except to state that they shall not exceed 140-db peak sound pressure level. The area of impulse noise is quite different from continuous noise, since apprehended levels depend on duration and specification is difficult. Much work has been done in this area by the military for obvious reasons, and the interested reader is referred to the literature[2,14] since this subject is substantially outside the area of architectural acoustics.

.95 Occupational noise exposure.) Protection against the effects of exposure shall be provided when the levels exceed those shown in Table when measured on the A scale of a ard sound level meter at slow re- e. When noise levels are determined ctave band analysis, the equivalent ighted sound level may be deter- d as follows:

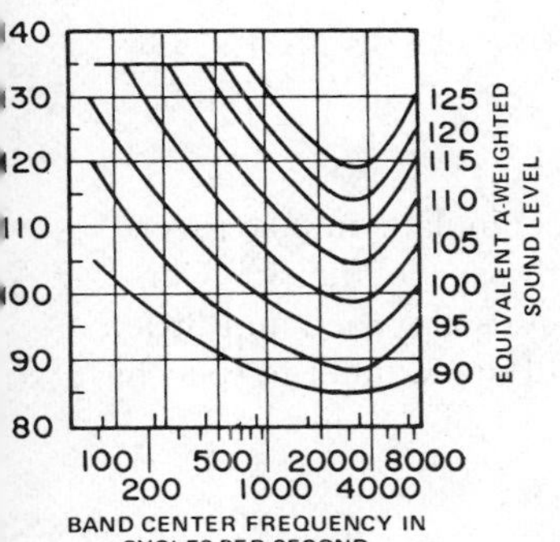

Equivalent sound level contours. Octave band sound pressure levels may be converted to the equivalent A-weighted sound level by plotting them on this graph and noting the A-weighted sound level corresponding to the point of highest penetration into the sound level contours. This equivalent A-weighted sound level, which may differ from the actual A-weighted sound level of the noise, is used to determine exposure limits from Table G-16. [1910.95 amended at 39 FR 19468, June 3, 1974]

(b)(1) When employees are subjected to sound exceeding those listed in Table G-16, feasible administrative or engineering controls shall be utilized. If such controls fail to reduce sound levels within the levels of Table G-16, personal protective equipment shall be provided and used to reduce sound levels within the levels of the table.

(2) If the variations in noise level involve maxima at intervals of 1 second or less, it is to be considered continuous.

(3) In all cases where the sound levels exceed the values shown herein, a continuing, effective hearing conservation program shall be administered.

Table G-16 Permissible Noise Exposures

Duration per day, hours	*Sound level dBA slow response*
8	90
6	92
4	95
3	97
2	100
1½	102
1	105
½	110
¼ or less	115

[1] When the daily noise exposure is composed of two or more periods of noise exposure of different levels, their combined effect should be considered, rather than the individual effect of each. If the sum of the following fractions: $C_1/T_1 + C_2/T_2 + \ldots C_n/T_n$ exceeds unity, then, the mixed exposure should be considered to exceed the limit value. Cn indicates the total time of exposure at a specified noise level, and Tn indicates the total time of exposure permitted at that level.

[1910.95 Table G-16 amended at 39 FR 19468, June 3, 1974]

Exposure to impulsive or impact noise should not exceed 140 dB peak sound pressure level.

Fig. 26.24 Provision of the Occupational Safety and Health Act (OSHA).

ROOM ACOUSTICS

26.22 Sound in Enclosures

When a continuous sound is generated in an enclosure, fields are set up as described in Section 26.14. When the sound is not a continuous tone or noise but a series of discrete sounds, following one upon the other and containing intelligence, as in speech or music, the room must be designed to maintain and enhance this intelligibility. That is what is meant by design of room acoustics.

The generated sound radiates out from the source until it strikes a room boundary or other large surface. Before reaching this surface the sound intensity is attenuated by distance (inverse square law) and by absorption in the air. This latter is only appreciable in large rooms, and at frequencies above 2000 Hz. (For detailed data on air absorption, see Reference 2. In the absence of exact data, assume a figure of 0.40 sabin/100 m^3 and see Table 27.1, page 1206). When the sound reaches the wall it is partially reflected and partially absorbed. A small portion is also transmitted into adjoining spaces. The energy transmitted is so small that it has little effect on the space within which the sound originates although, as will be discussed in Chapter 27, it may be very important in the surrounding spaces. The ratio between the energy absorbed and reflected will significantly affect what one hears within

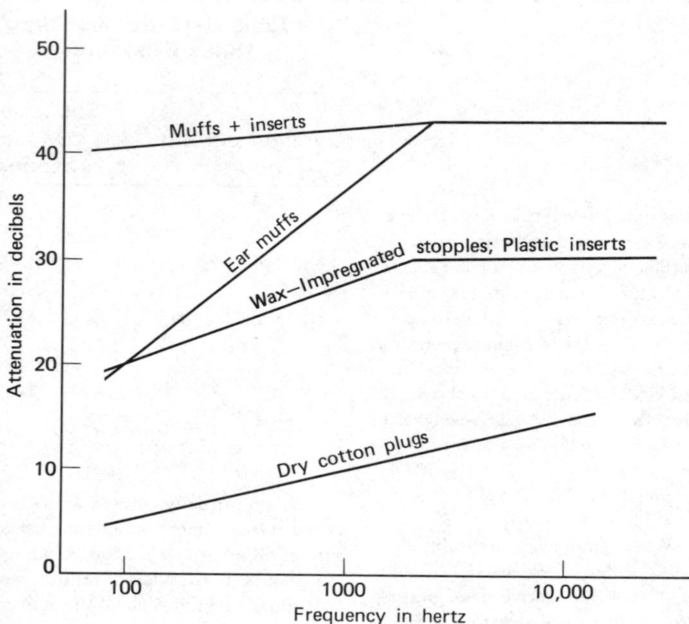

Fig. 26.25 Sound attenuation characteristics of various types of ear protectors. From Quieting: A Practical Guide to Noise Control, *National Bureau of Standard Handbook 119, July 1976, p. 13.*

Table 26.6 *Typical Industrial Noise Levels*[a]

Equipment	*dbA*
Printing press plant (medium size automatic)	86
Heavy diesel-propelled vehicle (about 25 ft away)	92
Heavy-duty grinder	93
Air compressor	94
Plastic chipper	96
Cutoff saw	97
Multiple spot welder	98
Turbine condenser	98
15 cu ft air compressor	100
Drive gear	103
Banging of steel plate	104
Magnetic drill press	106
Air chisel	106
Positive displacement blower	107
Air hammer	107
Vacuum pump	108
Jolt squeeze hammer	122

[a]These are approximate values for typical generic equipment types and should not be used as design values.

the space. Specifically, if little energy is absorbed and much is reflected, two effects will be noticeable. Intermittent sounds will be mixed together (which may make speech *less* intelligible or music *more* pleasant), and steady sounds will accumulate into a reverberant field, making the space "noisy." Conversely, if much energy is absorbed and little reflected, the room will sound quiet to speech and "dead" to music.

26.23 Sound Absorption

It is useful to express the above effects quantitatively. Most materials are neither perfect reflectors nor perfect absorbers. The coefficient of absorption (α) is defined as

$$\alpha = \frac{I_a}{I_i} \tag{26.20}$$

where I_i = intensity impinging on the material, watts/cm^2

I_a = intensity absorbed by the material, watts/cm^2

α = absorption coefficient (no units).

Thus α is a measure of absorption efficiency. If $\alpha = 1.0$ all the impinging energy is absorbed. Since open space has this characteristic, α has also been defined as the ratio between the absorption of a given material and that of an *open window* of the same area. Obviously then for an open window,

$$\alpha = 1.0$$

The total absorption (A) provided by a surface (S) is expressed in sabins, as

$$A = S\alpha \qquad (26.21)$$

where A = total absorption, sabins
S = Surface area, square feet
α = Coefficient of absorption

Since α is a ratio and thus unitless, and S is in square feet, $S\alpha$ should be in square feet. Actually, sound absorption units are called sabins in honor of W. C. Sabine, a pioneer in architectural acoustics.

Most rooms are constructed of several materials, each having different absorption coefficients α, and thus it becomes necessary to use

$$\Sigma S\alpha = S_1\alpha_1 + S_2\alpha_2 + \cdots + S_n\alpha_n$$
$$\text{or} \quad \Sigma \bar{A} = \bar{A}_1 + \bar{A}_2 + \cdots + \bar{A}_n \qquad (26.22)$$

where $\Sigma S\alpha$ = the total absorption in the room, sabins
S_1, S_2, etc. = the areas (ft^2) of each material
α_1, α_2, etc. = the coefficients of each material
$\bar{A}_1$, $\bar{A}_2$, etc. = total absorption of each different material

Room absorption is discussed in detail in Section 27.3 in connection with noise reduction techniques. Absorption coefficients for common materials, acoustic materials, and auditorium furnishings are tabulated in Table 27.1, page 1206. It is important to note that the coefficient varies with frequency and therefore must be calculated individually for the frequencies being studied.

26.24 Reverberation

Reverberation is the persistence of sound after the cause of sound has stopped—a result of repeated reflections. Reverberation time (T_R) describes the period required for the sound level to decrease 60 db after the sound source has stopped producing sound. For most rooms, the reverberation time at a specific frequency may be found by the formula:

$$T_R = K \times \frac{V}{\Sigma S\alpha} \text{ sec} \qquad (26.23)$$

where K = a constant, equal to 0.049 when V is in cubic feet and 0.16 when V is cubic meters
V = room volume, ft^3 (m^3)
$\Sigma S\alpha$ = total absorption, sabins, at the frequency

In most room acoustics studies, reverberation times are calculated at 125, 500, 1000, and 2000 Hz, although the midfrequency (500 to 1000 Hz) range is generally the reference used in specifying the reverberation time of a room. Equation 26.23 is only accurate for most usual size and shape rooms. Other equations exist that are more accurate for special situations (see References 15 to 18).

Reverberation can be considered as a mixture of previous and more recent sounds. The converse of reverberation or reverberance is articulation. An articulate environment keeps each sound event separate rather than running them together. Spaces for speech activities should be less reverberant—more articulate—than those designed for performance of romantic music. Figure 26.26 compares the reverberance/articulation requirements of performance activities and spaces.

Figure 26.27 shows average optimum reverberation time for various types of auditoriums as a function of size. The values are average because recommendations vary as much as 100% between respected sources. In addition to an optimum reverberation time in the central 500-Hz to 1-kHz range, the reverberations at other frequencies should exhibit a slight drop in the higher frequencies and a rise at the lower frequencies to compensate for the sharp drop in ear sensitivity. Thus T_R at 100 Hz should be, according to most authors, between 25 to 50% longer than T_R at the center frequencies. As a matter of interest, concert halls judged to be excellent by musical experts have a center frequency reverberation time between 1.6 and 1.8 sec.

26.25 Criteria for Speech Rooms

The overriding criteria for speech is intelligibility. Since speech consists of short dis-

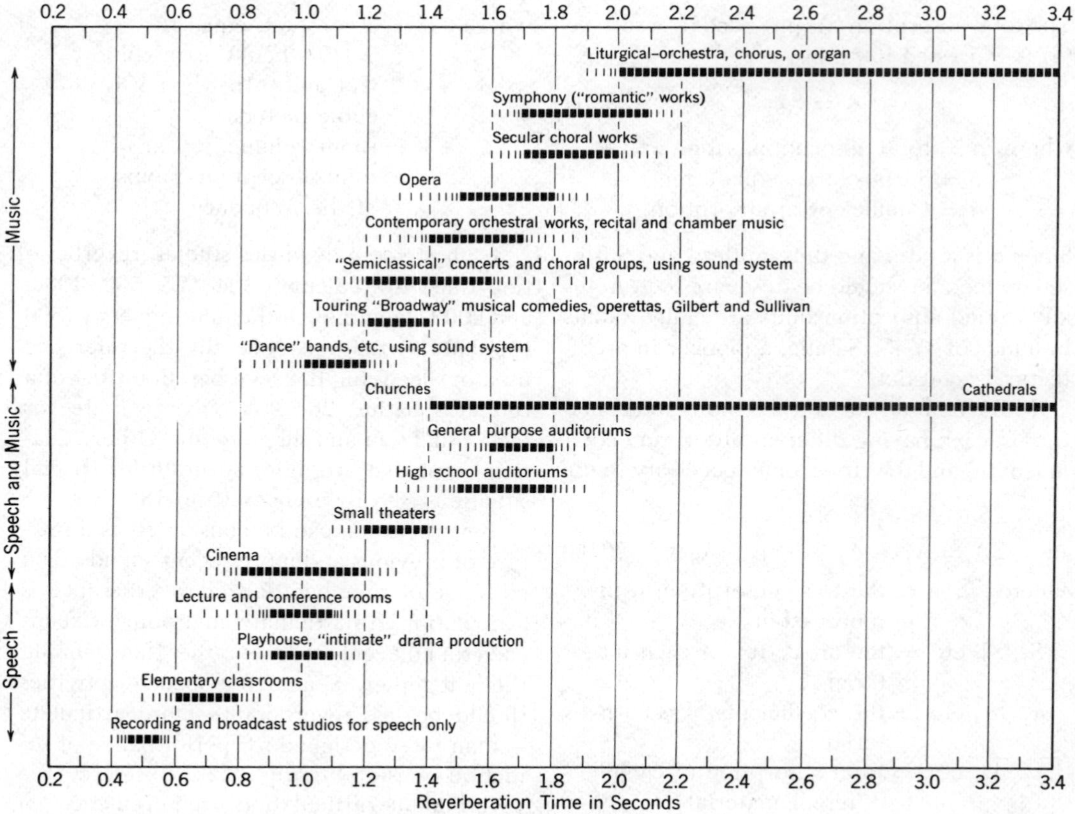

Fig. 26.26 *Optimum reverberation (500–1000 Hz) for auditoriums and similar facilities.*

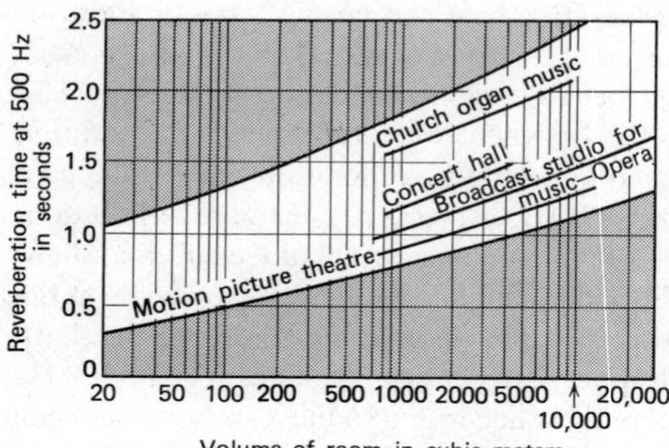

Fig. 26.27 *Optimum reverberation time for various sizes of music rooms. The curves shown are compromises between the recommendations of different authors. Data given for 500 Hz. From* Architectural Acoustics, *Brüel and Kjaer, 1963.*

connected sounds, among which are high-frequency, low-energy phonemes, the ideal room must assure the ear's reception of these phonemes as they are given. Since a slow decay rate is the equivalent of a masking noise, reverberation must be kept to a minimum. Maximum T_R is given in Figure 26.28, as determined by one well-known source. Too low a reverberation time (very high absorption, minimum reflection) is also undesirable because:

a. It limits the size of the room to that which can be covered by direct sound only.

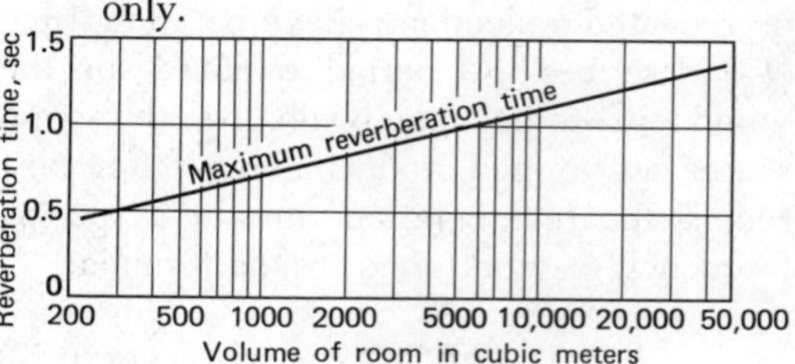

Fig. 26.28 *Maximum recommended reverberation time for speech in auditoriums and lecture halls. From P. V. Brüel,* Sound Insulation and Room Acoustics.

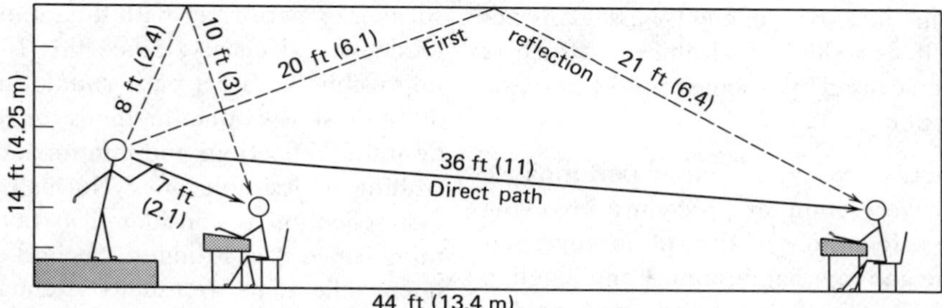

Fig. 26.29 Sound paths in a typical medium-size lecture room. Note that for both extremes of listener position, the maximum path-length difference between direct and first reflection is 11 ft. Thus signal is reinforced and intelligibility should be excellent if room absorption is provided to limit reverberation time to about ½ sec maximum (see Fig. 26.28). Numbers in parentheses are dimensions in meters.

b. It is disturbing to the speaker since absence of reflection prevents him or her from gauging proper voice level and tends to cause excessive effort (shouting).

Thus proper design of a room for speech is a compromise between the need for some reflection and the desire to minimize reflection to preserve intelligibility.

Another important factor must be considered. The reflections associated with reverberation can have either a salutory or a deleterious effect. The ear cannot distinguish between sounds that arrive within a maximum of 50 msec $\mp$ of each other. (Most authorities use the figure 35 msec as a conservative figure). Sounds arriving within this time *reinforce* the direct path signal and appear to come from the source. Sounds arriving after this time are apprehended as a fuzzy echo or elongation of the sound, reducing intelligibility and directiveness. Since 35 msec corresponds at 344 m/sec to 12 m or 40 ft, the room must be so arranged that the difference between the first reflection path and the direct path is no greater than 40 to 55 ft or 12 to 17 meters (see Fig. 26.29). For more detail concerning this and related factors on reflected paths, refer to Section 26.27. Optimum reverberation time *for speech,* according to one leading writer, was given in Section 26.16, in equation 26.18 as

$$T = 0.3 \log \frac{V}{10} \qquad (26.19)$$

where T = time in seconds
V = room volume in m^3

A good figure for lecture rooms is between 0.35 and 0.4 sec.

26.26 Criteria for Music Performance

Adequate design for a music space requires recognition of the following:

a. Large-volume spaces require direct-path sound reinforcement by reflection.
b. Relatively long reverberation time is needed to enhance the music—the exact amount depending on the type of music (see Fig. 26.26). This reverberance must, however, not include clear echoes.
c. Directivity declines if the reinforcing signal is excessively delayed. With large ensembles, directivity gives the sense of depth and instrument location necessary for proper appreciation. This is often referred to as clarity or definition in music. With a solo instrument this problem is diminished.
d. Brilliance of tone is primarily a function of high-frequency content. Since these frequencies are most readily absorbed, a good direct path must exist between sound source and listener. Since our eyes and ears are close together, a good sound path exists when a good vision path exists. At the other end of the spec-

trum, lack of sufficient bass expresses itself as a loss of "fullness," which is often caused by resonant absorption (see Section 27.3).

The actual design of music performance space is a very complex procedure involving extensive calculations of absorption, reverberation time and ray diagramming, and juggling of materials, dimensions, and wall angles. Simulation techniques and acoustic models are also employed. Most modern design also uses movable reflector panels and other active variables. After construction is completed extensive tests are conducted and field adjustments are made.

26.27 Ray Diagrams and Sound Paths

Ideally, every listener in a lecture hall, theater, or concert hall should hear the speaker or performer with the same degree of loudness and clarity. Since this is obviously impossible by direct-path sound, the essential design task is to plan methods for reinforcing desirable reflections and minimizing and controlling undesirable ones. Normally only the first reflection is considered in ray diagramming, since it is strongest. Second and subsequent reflections are usually attenuated to the point that they need not be considered except for the special situations of flutter, echoes, and standing waves discussed below.

(*a*) ***Reflections.*** These occur when sound hits a surface that is neither predominantly absorptive nor sound transparent at the specific frequency. Nonabsorptive surfaces that are large, as compared to the wavelength at a given frequency, reflect sounds. Thus, 3-in.-wide pickets in a fence reflect frequencies with wavelengths that are less than three in. (4500 Hz and higher). If the fence is 50%

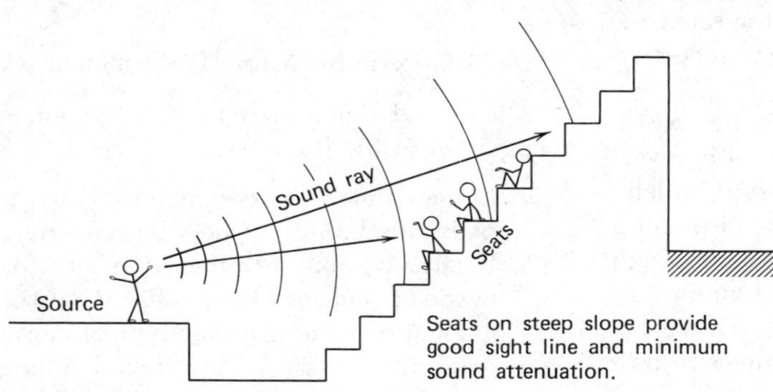

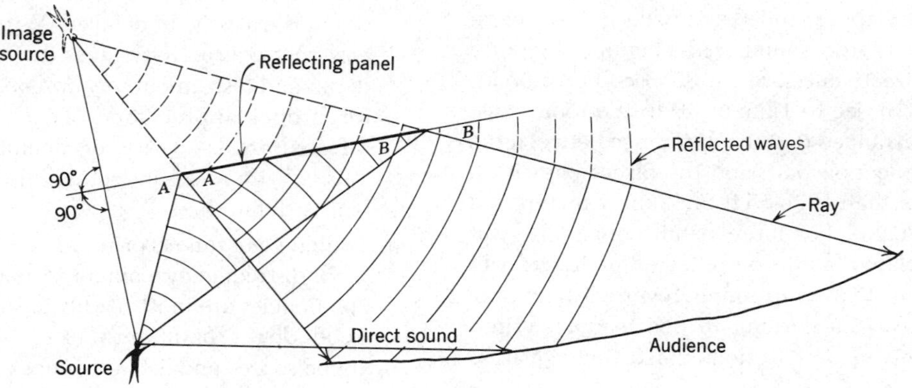

(*b*)

Fig. 26.30 *Use of an angled reflector panel* (b) *creates an image source that stands in approximately the same relation to the audience as the performer in the classic Greek theatre* (a).

open, it reflects about 50% of the high-frequency energy. However, it does not significantly affect the lower frequencies. These lower frequencies (longer wavelengths) simply diffract around the obstruction. A facing that transmits sound while serving as a visual barrier may be designed through careful application of these principles in selection of element sizes. Most designs for useful reflections involve large surfaces such as sections of walls or ceilings.

(*b*) ***Specular reflection.*** Specular reflection occurs when sound reflects off a hard polished surface. This characteristic can be used to good advantage to create an effective image source. In ancient Greek and Roman theatres, seats were arranged on a steep, conical surface around the performer. The virtue of the arrangement (see Fig. 26.30*a*) is that the sound power travels to each location, with minimal attenuation. This same effect can be accomplished by placing the sound source above the seats. This is not practical physically, but it can be accomplished effectively by the use of a reflecting panel (see Fig. 26.30*b*). Panel dimension must be at least one wavelength at the lowest frequency under consideration. Figure 26.31 is a conversion chart from frequency to wavelength in feet and meters.

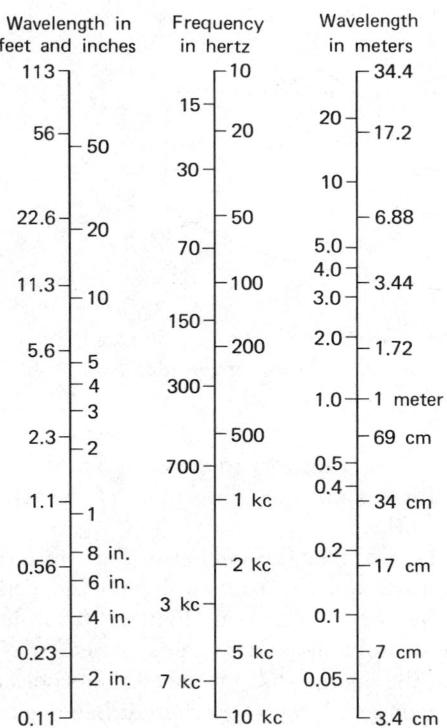

Fig. 26.31 *A nomograph for determining wavelength in feet or meters, given frequency in hertz, or vice versa. Speed of sound is taken as 344 m/sec and 1128 ft/sec. To use, hold a straightedge horizontally across the nomograph, and read the figures directly.*

(*c*) ***Ray Diagrams.*** Ray diagramming is a design procedure for analyzing reflected sound distribution throughout a hall, using the first reflection only. Figure 26.32 shows a ray diagram. The rays are drawn normal (perpendicular) to the spherically propagating sound waves. Specular reflection is assumed, that is, the angles between the reflecting panel and the incident and reflected rays are always equal. Thus, in addition to the direct sound, each listener is receiving reflected sound energy. It is as though there were additional sound sources, the *real* one and numerous image sources. Figures 26.32*a*, *b*, and *c* show the application of a ray diagram to a lecture hall. In Figure 26.32*a* the stage height and seating slope are arranged to provide good sight lines, and the ceiling height is established by reverberation requirements, esthetics, cost, etc. It can be seen that less than half of the ceiling is providing useful reflection. By dividing the ceiling into two panels (Fig. 26.32*b*), people in the rear of the room perceive the direct source plus two image sources, and the useful reflecting area is increased by 50%. In Figure 26.32*c*, the shape has been further refined to include a lighting slot and a loudspeaker grille.

Although they are a useful design tool, ray diagrams have certain restrictions. For example, the hall is three dimensional and the diagrams two dimensional, that is, sectional. To properly ray diagram, depth (width) must also be considered and this unduly complicates the diagrams. Also design must always be a compromise between ray diagrams for various "speaking positions" on the stage. Thus a

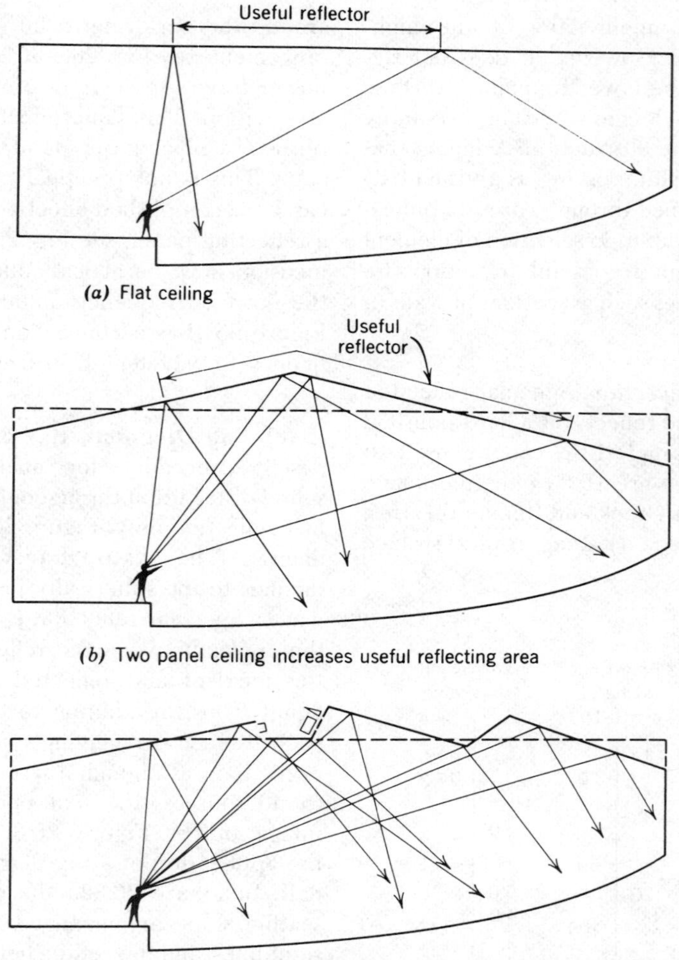

(a) Flat ceiling

(b) Two panel ceiling increases useful reflecting area

(c) Multifaceted ceiling incorporates lights and loudspeakers

Fig. 26.32 *Section through a typical lecture room showing use of ray diagrams.*

paraboloid may be a perfect shape for one source position but will be very poor for other positions.

(*d*) ***Echoes.*** A clear echo is caused when reflected sound at sufficient intensity reaches a listener approximately 70 msec or more after he hears the direct sound. This occurs whenever the reflected sound path is more than 70 ft longer than the direct path.

Echoes, even if not distinctly discernable, are undesirable in rooms. They are annoying and make speech less intelligible. The relative annoyance is dependent on the time delay, and loudness relative to the direct sound which, in turn, are dependent on the size, position, shape, and absorption of the reflecting surface.

Typical echo-producing surfaces in an auditorium are the back wall and the ceiling above the proscenium. Figure 26.33 shows these problems and suggests remedies. Note that the energy that produced the echoes can be redirected to places where it becomes useful reinforcement. If echo control by absorption alone were used on the ceiling and back wall, that energy would be wasted. The rear wall, since its area cannot be reduced too far, may have to be made more sound-absorptive to reduce the loudness of the reflected sound.

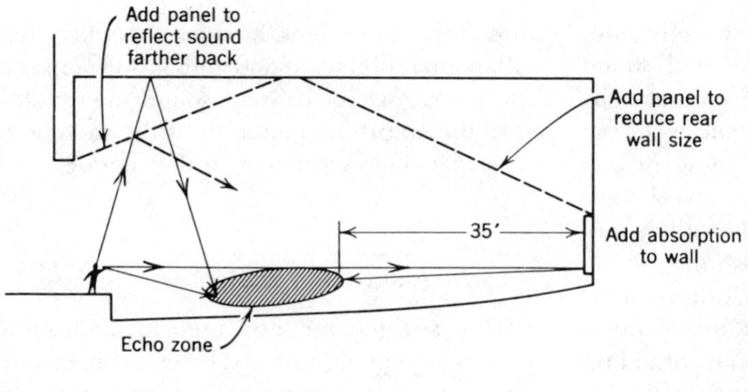

Fig. 26.33 Auditorium section showing the causes and remedies for two typical echoes.

Fig. 26.34 (below) Sound diffusion can be created with different shape reflectors, ranging from the horizontal flat (a), *inclined flat* (b), *or convex* (c). *Diffusion improves from* (a) *to* (c).

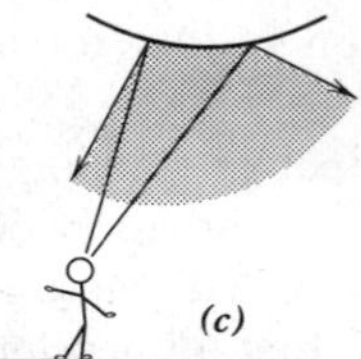

(*e*) ***Flutter.*** A flutter is perceived as a buzzing or clicking sound, and it is comprised of repeated echoes traversing back and forth between two nonabsorbing parallel (flat or concave) surfaces. Flutters often occur between shallow domes and hard, flat floors. The remedy for a flutter is either to change the shape of the reflectors, their parallel relationship, or add absorption. The solution chosen will depend on reverberation requirements, cost, or esthetics.

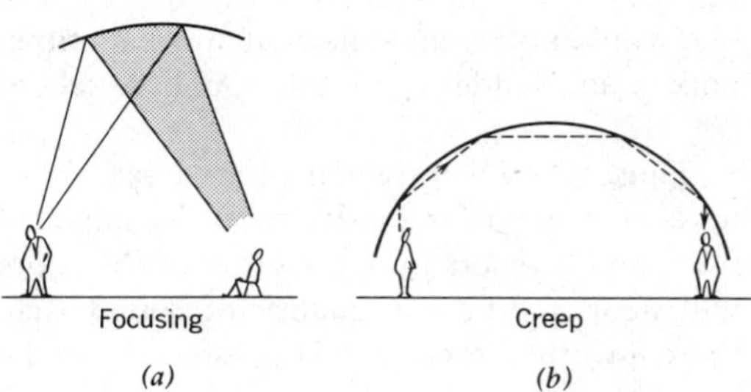

Fig. 26.35 Two undesirable phenomena in room acoustics.

(*f*) ***Focusing.*** Concave domes, vaults, or walls will focus reflected sound into certain areas of rooms. This has several disadvantages. For example, it will deprive some listeners of useful sound reflections and cause hot spots at other audience positions (see Fig. 26.35*a*).

(*g*) ***Diffusion.*** This is the converse of focusing and occurs primarily when sound is reflected from convex surfaces. A degree of diffusion is also provided by flat horizontal and inclined reflectors (see Fig. 26.34). In a diffuse sound field the sound level remains relatively constant throughout the space and as such is extremely desirable for musical performances.

(*h*) ***Creep.*** This describes the reflection of sound along a curved surface from a source near the surface. Although the sound can be heard at points along the surface, it is inaudible away from the surface. Creep is illustrated in Figure 26.35*b*.

(*i*) ***Standing Waves.*** Standing waves and flutters are very similar in principle and cause, but are heard quite differently. When an impulse (such as a hand clap) is the energy source, a flutter will occur between two parallel walls. When a steady pure tone is the source, a standing wave will occur, but only when the parallel walls are spaced apart at some integral multiple of a half wavelength.

When the parallel walls are exactly one-half wavelength apart, the tone will sound very loud near the walls and very quiet halfway between them. This is because at the center, the reflected waves traveling in one direction are exactly one-half wavelength away from those traveling in the other, and thus equal *but opposite in pressure,* which results in total cancellation. In other rooms standing waves are noted as points of quiet and maximum sound in the room. Standing waves are important only in rooms small with respect to the wavelengths generated (smallest room dimension <30 ft for music or <15 ft for speech).

Another effect of standing waves, or *resonance,* is the accentuation of the particular frequency, which will cause a standing wave in a room of that dimension. Thus, if one speaks (or plays a musical instrument) standing near a wall of a room, about 8 ft by 8 ft, one would notice an abnormal and sometimes unpleasant loudness in the sound at about 280 Hz.

Thus, when a musician plays a scale, one note may seem far louder than the adjacent ones, and listeners in one section of the room will hear a different quality of sound than those in other sections. This effect *must* be avoided for music performance but is merely an annoyance in rooms designed for speech use. This is one of the reasons that one finds music rehearsal rooms, broadcast studios, etc., with nonparallel walls and undulating ceilings. These irregularities direct sound energy toward the absorbing materials of the room and cause the standing waves to degenerate.

26.28 Auditorium Design

This is a general term used to describe a space where people sit and listen to speech or music. A large lecture hall, a multipurpose space, and a concert hall are auditoriums. Before beginning design of an auditorium, its potential use must be determined. If planned activities range from lectures to symphony orchestra concerts, the design approach for acoustics will differ, significantly, from a design approach for a space that would house only one of these activities. Therefore, the first step in the acoustical and architectural design must be determining the program. If the program for the auditorium includes activities that need different acoustical environments, it must be decided early whether the acoustics will be a compromise between the program extremes or adjustable for various activities. Acoustical environments can be altered by changing volume, moving reflecting surfaces, and adding or subtracting sound-absorbing treatment. Figure 26.36 illustrates several examples of acoustical adjustability.

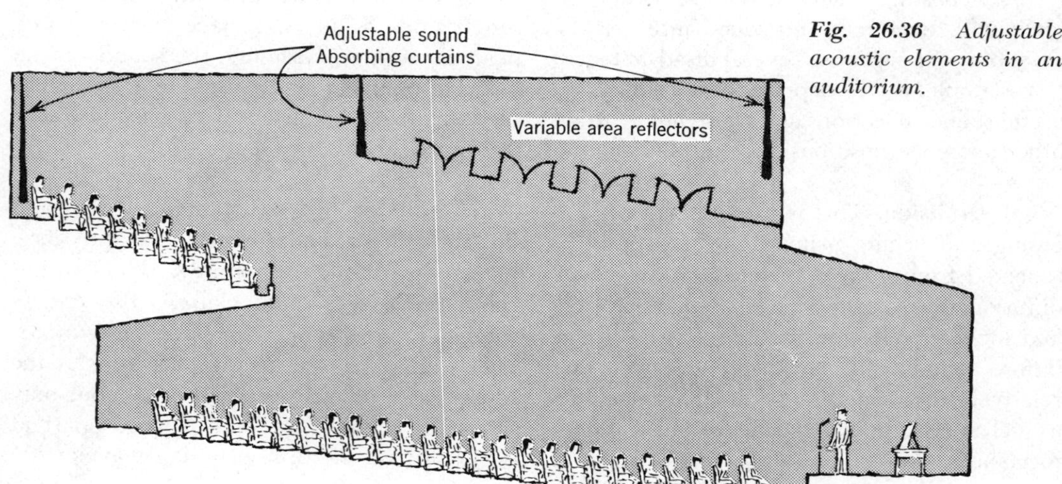

Fig. 26.36 *Adjustable acoustic elements in an auditorium.*

Factors that influence the acoustical design include audience size, range of performance activities, and sophistication of the potential audience. Obviously, an acoustically good 1200-seat theater is more difficult to design than an acoustically good 400-seat theater. In addition, the caliber of performance production and audience expectations are important design considerations. For example, a small school auditorium and a professional theater will have widely divergent demands from both audience and performers.

Acoustical design of an auditorium includes room acoustics, noise control, and sound system design. Noise control is covered in Chapter 27.

Room Acoustics. The audience size determines the basic floor area of an auditorium. Once this area has been fixed, the volume of the room is developed according to reverberation requirements of the space.

Figure 26.37 shows a typical auditorium in plan and section. The shape of wall and ceiling surfaces is developed to give proper distribution of sound and eliminate focusing or echoes. Essential characteristics of the design include:

1. Ceiling and side walls at the front of the auditorium distribute sound to the audience. These surfaces must be close enough to the performers to minimize time delays between natural sound and reflected sound.
2. Ceiling and side walls provide diffusion.

Acoustics must be considered in selection of materials used in an auditorium. Both sound-reflecting and sound-absorbing materials will be found in all auditoriums. Since the largest area of sound-absorbing material in any auditorium is the audience, the difference in acoustical characteristics that occur without an audience may be minimized by using fully upholstered seating.

Chairs with fully upholstered seats and backs, covered in an open-weave material, will have absorption characteristics closely approximating an audience. Using the auditorium in Figure 26.37 as an example, the reverberation characteristics of an auditorium with various materials may be examined. In the first example, the room use is assumed to be for music performance. The only sound absorption is that provided by the audience and seating. In the second set of calculations absorptive curtains were installed along the rear wall and a portion of the side wall. This configuration might be used for lectures in a room that is adjustable between speech and music configurations. A third configuration might use permanent sound-absorbing treatment installed on the ceiling and rear and side walls. Because of its low-reverberation time, this configuration would be appropriate only for movies and lectures, not for music activities.

These simple examples indicate the effect of changes in the amount of absorption on the characteristics of a room. Adjustable treatments permit the characteristics of the room to be modified to any point between the extremes to meet the program acoustic requirements of a multipurpose hall.

Existing spaces may require remedial treatment to eliminate unwanted phenomena such as focusing and echoes, as shown in Figure 26.38. In the first example, the surface of the dome was covered with sound-absorbing material to eliminate focusing; in the second, sound-absorbing treatment was applied to a curved rear wall to eliminate an echo. Such treatment also will affect the reverberation characteristics.

SOUND REINFORCEMENT SYSTEMS

26.29 Objectives and Criteria

The purpose of a sound reinforcement system is just what the name indicates—to reinforce the sound, which would otherwise be inadequate. Thus an ideal sound system will give the listener the same loudness, quality, directivity, and intelligibility as if the source of sound were immediately adjacent to him—a distance of 2 to 3 ft for speech and further for music depending on type and number of instruments. This situation must obtain for every position in the space except that a variation of ± 2 db for loudness is toler-

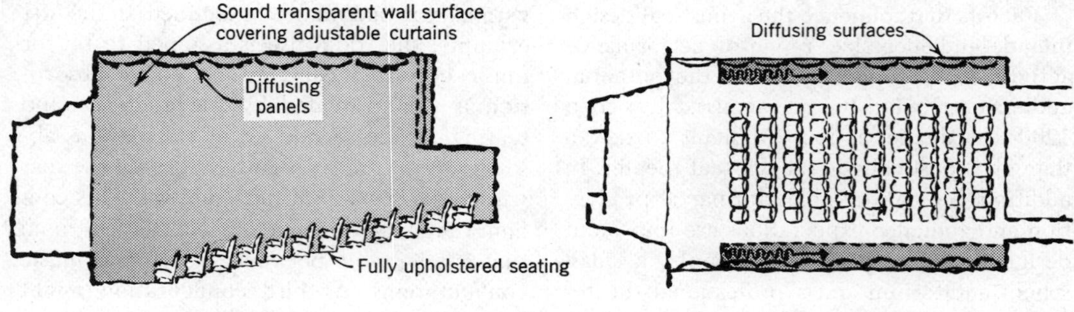

Fig. 26.37 Auditorium with surface treatments for control of reflections and reverberation.

Simplified calculations of midfrequency (500 Hz & 1000 Hz) average reverberation times.

$$\text{Reverberation Time (RT)} = \frac{0.049 \times \text{Volume (cu. ft.)}}{\text{Total Absorption (sabins)}}$$

Volume = 155,500 cu. ft.

More Reverberant Condition
(Curtains Retracted)

	Area	*a*	Absorption
Seating and Stage (with audience and performers)	3323	.92	3060
Wall Area—Concrete Block	8000	.2	1600
Lower Rear Wall—Permanent Sound Absorbing Treatment	450	.88	396
			5056

Total Absorption
More Reverberant
Condition

$$RT = \frac{0.049 \times 155,500}{5056}$$
= 1.48 sec.
= 1.5 seconds

Less Reverberant Condition
(Curtains Exposed)

	Area (sq. ft. =)	*a*	Absorption
Seating and Stage (with audience and performers)	3323	.92	3060
Wall Area—Concrete Block (Balance covered by curtains)	3600	.2	720
Curtains	4400	.45	1970
Lower Rear Wall—Permanent Sound Absorbing Treatment	450	.88	396
			6146

Total Absorption
Less Reverberant
Condition

$$RT = \frac{0.049 \times 155,500}{6146}$$
= 1.22 sec.
= 1.2 seconds

able. The other factors should remain constant. Of these factors, loudness and intelligibility have been discussed at length and should be well understood. By quality we mean that frequency response should be linear so that reproduced sound bears the same relation between its frequency components as the original sound. (Quality is then field-adjusted by "voicing" or "equalization," which will be discussed below).

Directivity is the characteristic whereby the sound appears to be coming from the originating source, that is, the loudspeakers should be directionally "invisible," and the listener must have the impression of actually hearing the source. It should be emphasized that sound systems cannot correct poor acoustic design completely although they can improve a bad situation.

Generally, sound systems will be required

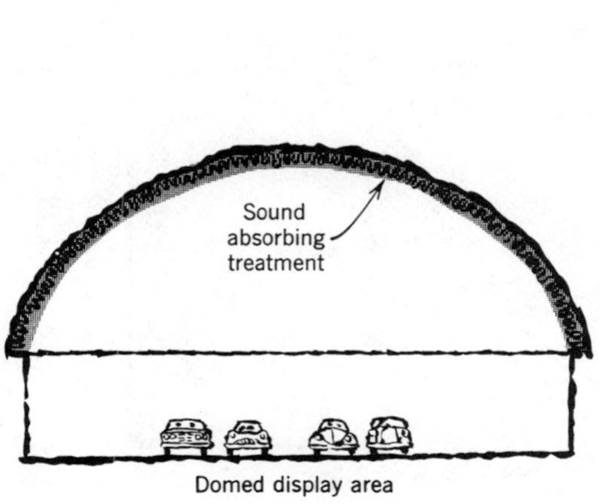

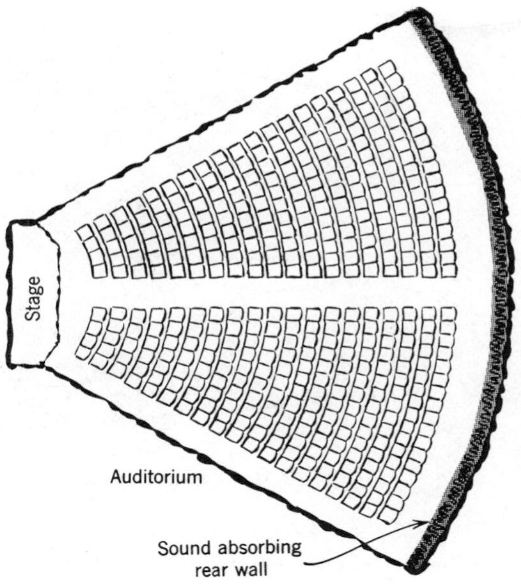

Fig. 26.38 Sound-absorbing treatment used to eliminate focusing from dome and curved auditorium wall.

in spaces larger than 50,000 cf (1500 m^3). In terms of population, this volume translates as 550 persons in lecture rooms (15 ft average ceiling height and 6 sq ft per person) and 325 persons in theatres (20 ft average ceiling height and 7.5 sq ft per person). In such a room (50,000 sq ft) a normal speaking voice can only maintain a volume of level of 55 to 60 db, depending on room design and voice strength. With background noise at PNC 35 (see Table 27.11, page 1255) the speaker will be heard; at higher noise levels intelligibility will suffer.

26.30 Components and Specifications

All sound systems consist of three basic elements: input devices, amplifier(s), and loudspeaker systems.

(*a*) *Input*

1. *Microphone.* If multiple mikes are required, mixing facilities must be provided.
2. Recorded material from phonograph, tape deck, cassette deck, and possibly commercial music source.
3. Radio source—AM/FM tuner.

(*b*) ***Amplifier and Controls.*** Amplifiers must be rated to deliver sufficient power to produce intensity levels of 80 db for speech, 95 db for light music, and 105 db for symphonic music. This assumes a *maximum* background noise level of 60 dbA. Thus 80-db speech intensity will be 20 db higher, or four times as loud as the noise level. If noise level is *known* to be below 60 db maximum, amplifier and loudspeaker power ratings can be reduced accordingly. The amplifier should carry technical specs for signal-to-noise ratio, linearity from x to y Hz $-$ 3 db, and low ($-$%) distortion. Exact figures here depend on application and are left to the acoustics specialist or sound engineer to supply. In addition to the usual volume, tone mixing, and input-output selector controls, the amplifier *must* contain special equalization controls for signal shaping. These are highly critical filter networks that are used to voice or equalize a system after installation. Equalization is a sine qua non of a good sound system, without which the system will howl, sound rough, give insufficient and poorly distributed gain and sound level, and generally sound poor. Essentially, voicing tailors the system to the acoustic properties of the space. A system not equipped for equalization is not a professional system and results will verify it. Furthermore, the specification must provide

Fig. 26.39 Loudspeaker system using delayed signal to underbalcony area.

for the services of a competent sound engineer to perform the equalization after installation and construction is complete.

Another control frequently required in theatre systems is a delay mechanism or circuit that can introduce a time delay into a signal being fed to a loudspeaker. Figure 26.39 shows a sound system that covers a majority of an auditorium from a central-loudspeaker cluster. The underbalcony seating areas are hidden from the central cluster and receive the reinforced sound from distributed loudspeakers in the underbalcony soffit. To provide directional realism, the signal to the underbalcony loudspeakers must be delayed to allow the weaker signal from the central speakers to arrive first. Delay is necessary since electrical signals travel at the speed of light whereas sound is much slower (one millionth the speed, approximately). With this arrangement sound will seem to come from the source, and the directivity so necessary to realism is maintained.

(*c*) ***Loudspeakers.*** These are the heart of any sound system and obviously must be of the same high quality as the remainder of the system. Indeed, system economies will show up much more quickly in loudspeaker performance than in any other component. Selection of speakers is a complex, technical task not within our scope. Nevertheless, a few general remarks are in order. The best systems use central-speaker arrays consisting of high-quality, sectional (multicell), directional, high-frequency horns and large cone woofers. These assemblies are very large, and the architect should be aware of the dimensions that must be accommodated. Units 6 ft wide, 8 ft high, and 3 ft deep, with a weight of 1000+ lbs are common in a large lecture hall or small theatre. This size is necessitated by the large wavelength of low-frequency sound (see Fig. 26.31). Smaller units using folded horns can be used, at a sacrifice in low-frequency response. If speech only is to be reproduced these will perform adequately. Distributed systems use small (4–12 in. diameter) low-level speakers, ceiling mounted, and firing directly down. To give adequate response these units must be mounted in at least a 2 cu ft enclosure. Smaller enclosures will usually seriously compromise performance.

26.31 Loudspeaker Considerations

Loudspeaker-system design and placement must be coordinated with the architectural design. The two principal types of loudspeaker systems are central and distributed. The loudspeakers in a central system are a carefully designed array of directional high-frequency units combined with less directional low-frequency units, placed above and slightly in front of the primary speaking position. In

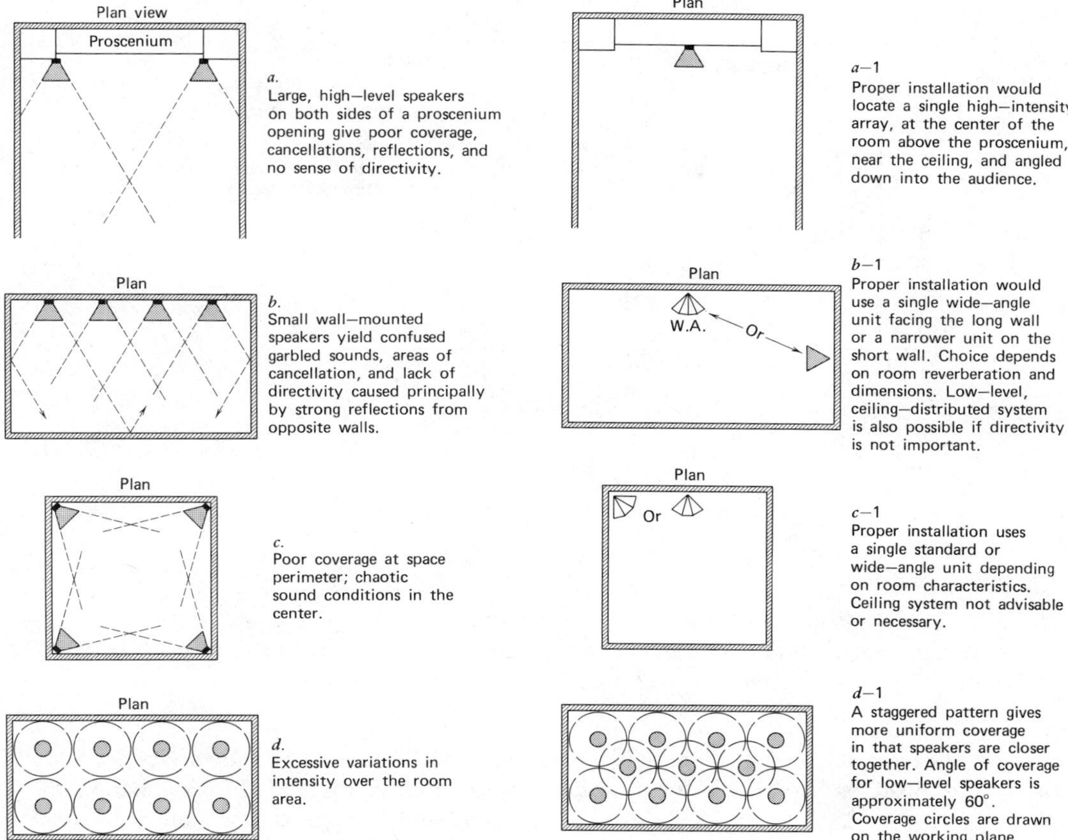

Fig. 26.40 *Poor layouts a to d and good layouts a-1 to d-1. Wide-angle speakers are available to fit most needs. Ceiling speakers give effective coverage on a 60° cone. Working plane is taken to be 4 ft AFF for seated audience and 6 ft AFF for standing listeners.*

most theaters, this location would be just above the proscenium on the centerline of the room. Located in this position, the system provides directional realism and is simple in its design.

The distributed loudspeaker system consists of a series of low-level loudspeakers located overhead throughout the space. Each loudspeaker covers a small area, in a manner similar to downlights. This type of system is used in low-ceilinged areas where a central-loudspeaker cluster could not provide proper coverage. It also can be used for public address functions if directional realism is not essential, in spaces such as exhibition areas, airline terminals, and offices. In addition, distributed loudspeaker systems provide flexibility for use in spaces where source and listener locations vary according to the use of the space, since loudspeakers can easily be switched to provide proper coverage. Combination loudspeaker systems that utilize both central and distributed units are used to solve special problems as discussed earlier and illustrated in Figure 26.39. In general, a listening position should receive sound from only one loudspeaker. Systems that cover seating areas with signals from several scattered loudspeakers usually increase the loudness of the sound but tend to produce garbled speech. This rule is the principal reason that the arrangements shown in Figure 26.40 will guarantee a bad job. The common practice of placing one loudspeaker on either side of a proscenium opening (Fig. 26.40*a*), or

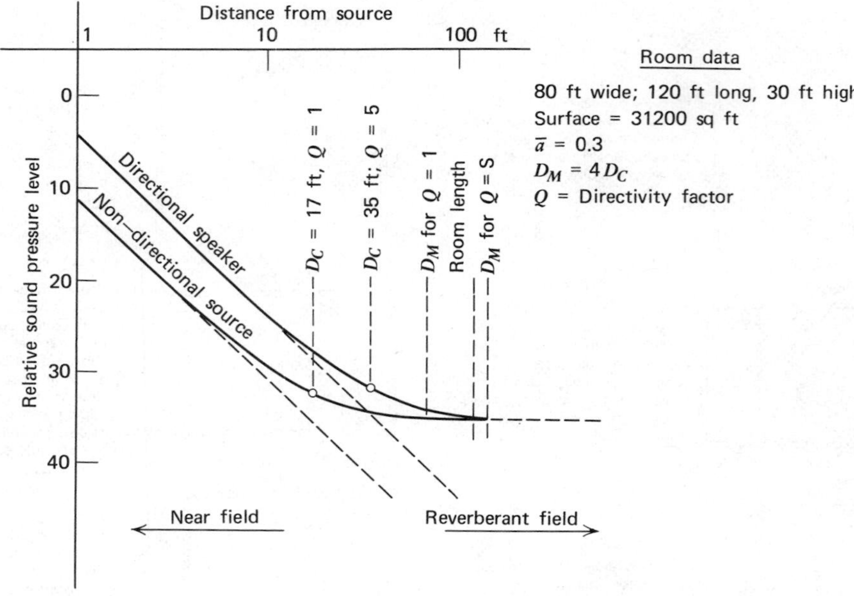

Fig. 26.41 *Plot of relative sound pressure level versus distance from sound source in a large, medium-absorption room. Lower curve is for a nondirectional source. Upper curve is for a directional loudspeaker.* D_c *is critical distance. Maximum distance Dm (throw) is four times* D_c.

rows of speakers on one or both sides of a room (Fig. 26.40*b*), is particularly to be deplored.

Essential parameters for location and design of the control position create problems for the architect. The sound system operator must be within the coverage pattern of the loudspeakers. For proper operation, he or she should be able to hear the sound as it is heard by the audience. Some current auditorium designs locate sound system controls within the audience seating pattern (see Fig. 27.49, page 1259). Others place a control room with a completely open wall or a large window at the rear of the auditorium. Monitor loudspeakers and earphones are inadequate substitutes for actual listening within the auditorium. In churches the simple control equipment can be located at the rear of the congregation area.

To understand selection and positioning of a high-level directional loudspeaker, it is necessary first to understand its action in a space. Refer back to Figure 26.41. We have transferred the curve $\bar{a} = 0.3$, that is, a medium (neither live nor dead) room, from Figure 26.17*c* to Figure 26.41 and have added a loudspeaker curve calculated from equation 26.16 for $Q = 5$, which is a common figure for multicell directional speakers. Note the result. First the level in the near field and free field areas has been raised 7 db. Second, the critical distance (*Dc*) has been doubled, that is, increased from 17 to 35 ft. *Critical distance* is defined as that point at which the distance between the asymptote and actual curve is 3 db. *At that point the direct and reverberant components are equal.* Beyond that point the reverberant field predominates. We have already stated that to maintain good directivity and clarity the direct component should predominate or at least constitute a large portion of the sound. It is widely accepted that *maximum* distance from loudspeaker to listener should not exceed four times the critical distance for good quality reproduction and comprehension (preferably less). Note then on Figure 26.41 that with *Dc* at 35 ft, the maximum distance becomes 140 ft. Since the room is only 120 ft long, a single high-level directional array is satisfactory. If the room were longer, the rear portions would require addi-

Table 26.7 ***Area of Coverage; 60° Cone Loudspeaker Firing Directly Down***

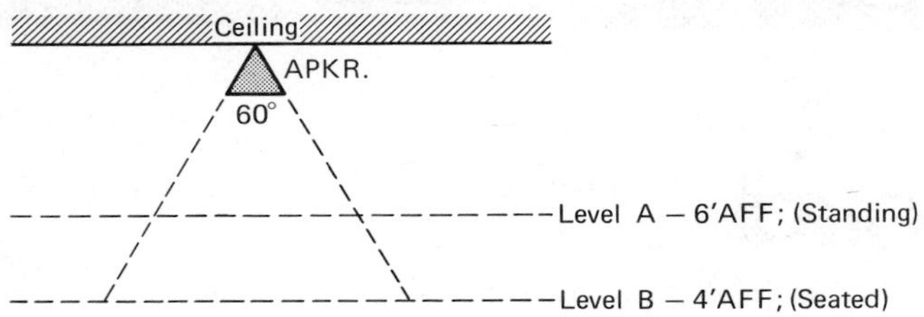

Ceiling Height, Feet	*At Level A*		*At Level B*	
	Diameter[a]	*Area*[b]	*Diameter*	*Area*
8	1.15	4.2	1.7	17
9	1.73	9.4	2.3	26
10	2.31	17	2.9	38
11	2.9	26	3.5	51
12	3.5	38	4.0	67
13	4.0	51	4.6	85
14	4.6	67	5.8	105
15	5.2	85	6.4	127
16	5.8	105	6.9	151

[a]Diameter of coverage circle (see Fig. 26.40).

[b]Area of coverage circle (see Fig. 26.40).

tional coverage. Alternatively room characteristics could be changed by adding absorption to reduce reverberance, or a more highly directive loudspeaker could be used. If it is not possible to use a central system, distributed speakers will perform adequately but without directivity (see Table 26.7). These decisions are the province of the acoustics expert, without whose expert advice complex or expensive installations should not be designed.

See References and Bibliography at the end of Chapter 27.

27

Building Noise Control

27.1 Introduction

Noise control in buildings is comprised of three components:

a. Reduction of noise generation at the source by proper selection and installation of equipment.
b. Reduction of noise transmission from point to point (along the transmission path) by proper selection of construction materials and appropriate construction techniques.
c. Reduction of noise at the receiver through acoustical treatment of the relevant spaces to meet NC criteria developed in Chapter 26.

Assurance of speech privacy is achieved by manipulation of all of the above plus the use of masking noise where necessary.

NOISE REDUCTION

27.2 Principles of Noise Reduction

Noise reduction is essentially the science of converting acoustical energy into another, less disturbing form of energy—heat. Since the amounts of energy involved are minute—130 db corresponds to 1/1000 of a watt or 0.003 Btu, the heat produced is completely negligible. This conversion is by absorption, by the room contents and wall coverings, and by the structure itself. The former controls noise levels *within* a space and the latter noise transmission between spaces. The reasons for this will become clear as our discussion proceeds, but it is important that it be understood at the outset. Noise control treatment within a room will affect the reverberant noise level within that room but will have minimal effect on the noise level in adjoining spaces. Refer to Figure 27.1 for a graphic presentation of this fundamental fact. The best that can be accomplished with acoustic room treatment is elimination of the reverberant field, that is, to make the intensity at the room boundaries what it would have been in free space, as in Figure 27.1*d*. (Even this is extremely difficult; the actual field at the wall would be above 72 db, except in a completely anechoic chamber). Adding further wall or other acoustic absorbent as in Figure 27.1*e* does nothing in the room itself and has minimal effect on the overall transmission loss, since the transmission loss of the acoustic material itself is very low, as can be seen in Figure 27.2*b*.

ABSORPTION

27.3 Mechanics of Absorption

We have already learned in Sections 26.23 and 26.24 the definition of sound absorption and its relevance to room reverberation characteristics. At this point it is appropriate to re-examine absorption as an acoustic phenomenon so that we may understand the application of absorptive material. Refer to Figure 27.2*a*. In an untreated room of normal construction, when the sound waves strike the walls or ceiling, a small portion is transmitted, a small portion is absorbed, and most of the sound is reflected. The exact proportions obviously depend on the nature of construction. When acoustical treatment is applied to the room surfaces as in Figure 27.2*c* some of the energy in the sound waves is dissipated before the sound reaches the wall. The transmitted

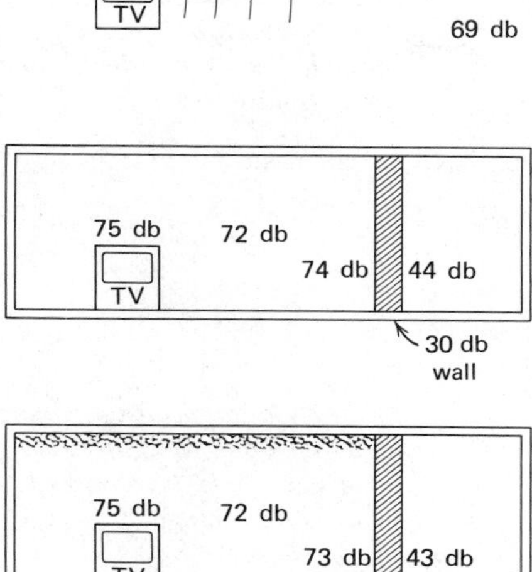

(*a*) TV set in free space produces 75-db sound level, which drops 6 db for each doubling of distance. Attenuation by inverse-square law (see Section 26.7).

(*b*) TV still produces 75 db. In the free field, sound drops to 72 db but builds up to 74 db at the wall due to reverberant field reinforcement (see Figure 26.17, page 1175). Wall attenuation is 30 db. Sound on other side of the wall is 74 − 30 = 44 db.

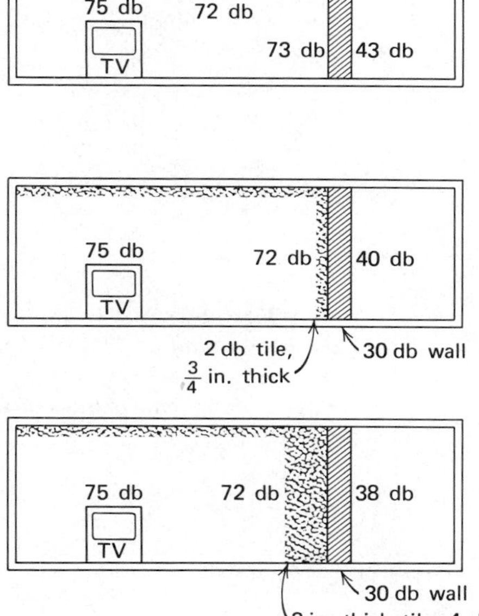

(*c*) Acoustic tile ceiling acts to reduce room reverberant field. Free field is extruded. Level at wall is 73 db. Level in second space is 73 − 30 = 43 db.

(*d*) Entire room is acoustically treated, effectively eliminating reverberant field. Room is 'dead'. Level on second side of wall is 72 db less acoustic tile loss, less wall loss, (that is, 72 − 2 − 30 = 40 db.

(*e*) Add another $2\frac{1}{4}$ in. of acoustic wall treatment. Room is dead. Level at wall 72 db. Level in second space = 72 − 4 − 30 = 38 db.

Fig. 27.1 *Action of acoustic absorbent material.*

portion is slightly reduced but the reflection is greatly reduced. The difference between the two situations is shown graphically in Figure 27.3. Refer back to Figure 26.17*c*, page 1175. Here the result of adding absorptive material to a room is shown in greater detail. The difference between a room with average absorption of 0.1 and the same room with $\bar{a}$ of 0.7 is 15 db, which is a reduction in *loudness* of approximately one and one-half times (see Table 26.3). We will now examine the acoustic materials themselves and the effect of varying type, quantity, thickness, and installation methods.

27.4 Absorptive Materials

There are three families of devices for sound absorption—fibrous materials, panel resonators, and volume resonators. All types absorb sound by changing sound energy into heat energy. Only fibrous materials and panel resonators are used commonly in buildings.

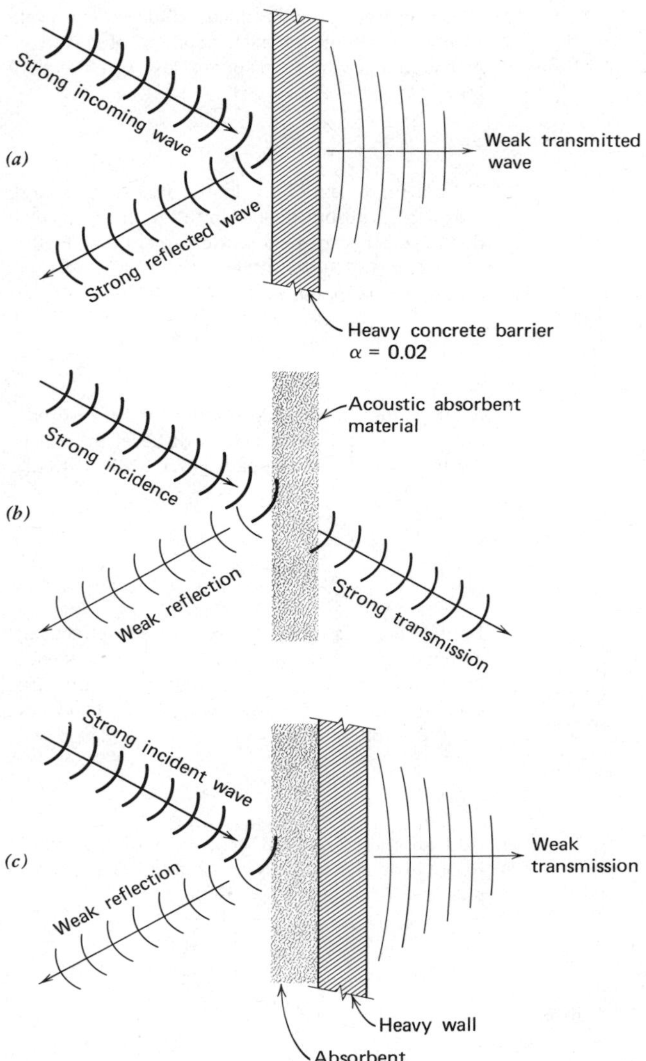

Fig. 27.2 (a) *Action of an incoming sound wave striking a heavy barrier. Much of the energy is reflected, some is absorbed, and a little is transmitted.*

(b) *Action of acoustic absorbent material alone. Very little energy is reflected, some is absorbed, and most is transmitted.*

(c) *When absorbent is applied to the heavy wall, it "traps" sound, preventing reflection, while wall mass acts to reduce transmission.*

Volume resonators, also known as Helmholtz resonators, are used principally as enclosures for absorbing a narrow band of frequencies. The following discussion refers to fibrous absorbers—the other two types are discussed separately in Section 27.8. The fibrous or porous materials absorb by the frictional drag produced by moving the air in small spaces within the material. The absorption provided by a specific material depends on its thickness, density, porosity, and resistance to air flow. For example, materials must be thick to absorb low-frequency sounds effectively. Since the action depends on absorbing energy by "pumping" air through the material, the air paths *must extend from one side to the other.* A fibrous material with sealed pores is useless as an acoustic absorbent. (Therefore, painting will generally ruin a porous absorber). A simple test is to blow smoke through the material. If the smoke passes through freely and the material is porous, fibrous, and thick it should be a good sound absorbent. Porosity, provided it is above 70%, does not much affect absorption. Below this figure sound absorbency decreases as porosity decreases. Table 27.1 gives

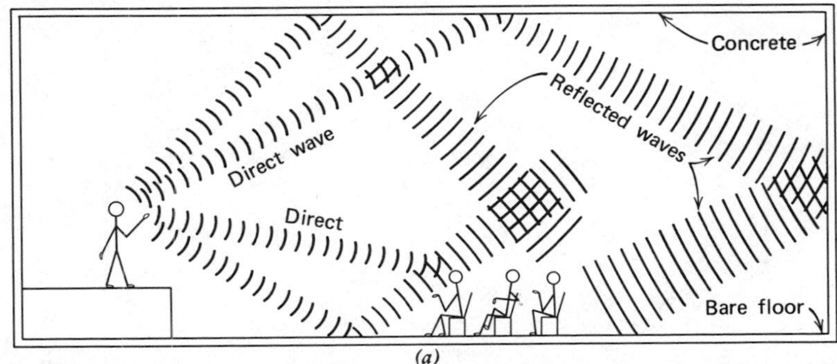

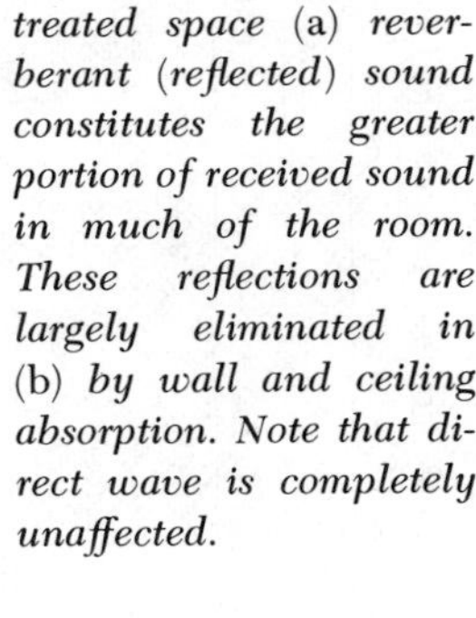
Fig. 27.3 In the untreated space (a) *reverberant (reflected) sound constitutes the greater portion of received sound in much of the room. These reflections are largely eliminated in* (b) *by wall and ceiling absorption. Note that direct wave is completely unaffected.*

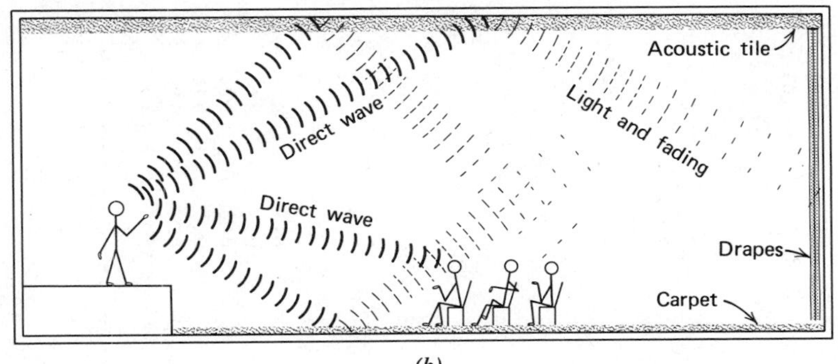

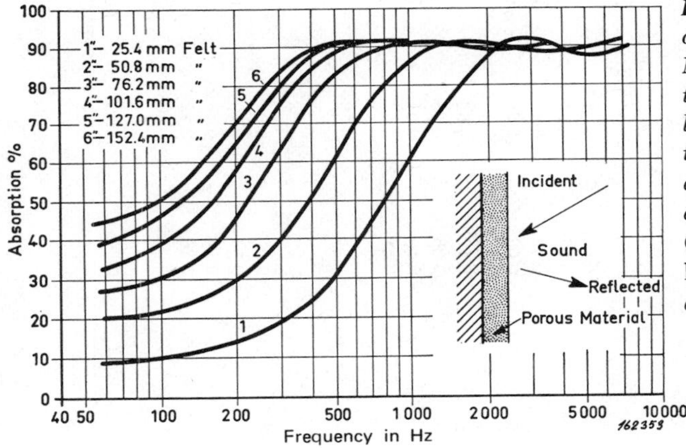

Fig. 27.4 Variation of absorption coefficient with thickness of felt absorbent. Note particularly that beyond 1 kHz, all thicknesses give the same α, whereas at low frequencies the absorption is proportional to thickness. Furthermore it requires a very heavy layer to give appreciable absorption at low frequency. Courtesy of Brüel and Kjaer Sound Insulation and Room Acoustics, *Chapman and Hall, London, 1951.*

absorption coefficients for fibrous absorbent materials and for building materials and furnishings. Several important conclusions can be drawn from examination of this table.

(*a*) For absorbent materials, absorption is normally higher at high frequencies than at low.

(*b*) Absorption is not always proportional to thickness, but depends on the type material being used and the method of installation (see Fig. 27.4). It is clear from this figure that beyond a nominal thickness, little is to be gained by adding thickness except at very low frequencies, or when installed discontinuously, as in (*c*) below.

(*c*) It is possible to obtain an α greater than 1.0 by using very thick blocks. See "Fiber Blocks" in the table. These are installed at a distance from each other and the edge absorption is very large, particularly at high frequencies.

Table 27.1 *Coefficients of Absorption—α*

Complete tables of coefficients of the various materials that normally constitute the interior finish of rooms may be found in the various books on architectural acoustics. The following short list will be useful in making simple calculations. Items are arranged in alphabetical order.

General Building Materials and Furnishings	Coefficients							
	125 Hz	250 Hz	500 Hz	1000 Hz	2000 Hz	4000 Hz	NRC[a]	Note
Brick, unglazed	0.03	0.03	0.03	0.04	0.05	0.07	0.005	
Brick, unglazed, painted	0.01	0.01	0.02	0.02	0.02	0.03	0.00	
Carpet, heavy, on concrete	0.02	0.06	0.14	0.37	0.60	0.65	0.29	
Same, on 40-oz hairfelt or foam rubber	0.08	0.24	0.57	0.69	0.71	0.73	0.55	
Same, with impermeable latex backing on 40-oz hairfelt or foam rubber	0.08	0.27	0.39	0.34	0.48	0.63	0.37	
Concrete block, coarse	0.36	0.44	0.31	0.29	0.39	0.25	0.35	
Concrete block, painted	0.10	0.05	0.06	0.07	0.09	0.08	0.05	
Fabrics								
Light velour, 10 oz/sq yd, hung straight, in contact with wall	0.03	0.04	0.11	0.17	0.24	0.35	0.15	
Medium velour, 14 oz/sq yd, draped to half area	0.07	0.31	0.49	0.75	0.70	0.60	0.55	
Heavy velour, 18 oz/sq yd, draped to half area	0.14	0.35	0.55	0.72	0.70	0.65	0.60	
Floors								
Concrete or terrazzo	0.01	0.01	0.015	0.02	0.02	0.02	0.00	
Linoleum, asphalt, rubber, or cork tile on concrete	0.02	0.03	0.03	0.03	0.03	0.02	0.05	
Wood	0.15	0.11	0.10	0.07	0.06	0.07	0.10	
Wood parquet in asphalt on concrete	0.04	0.04	0.07	0.06	0.06	0.07	0.05	
Glass								
Large panes of heavy plate glass	0.18	0.06	0.04	0.03	0.02	0.02	0.05	
Ordinary window glass	0.35	0.25	0.18	0.12	0.07	0.04	0.15	
Gypsum Board, $\frac{1}{2}$ in. nailed to 2 x 4's 16 in. o.c.	0.10	0.08	0.05	0.03	0.03	0.03	0.05	
Marble or Glazed Tile	0.01	0.01	0.01	0.01	0.02	0.02	0.00	
Openings								
Stage, depending on furnishings			0.25—0.75					
Deep balcony, upholstered seats			0.50—1.00					
Grills, ventilating			0.15—0.50					
Plaster, gypsum or lime, smooth finish on tile or brick	0.013	0.015	0.02	0.03	0.04	0.05	0.05	
Plaster, gypsum or lime, rough finish on lath	0.14	0.10	0.06	0.05	0.04	0.03	0.05	
Same, with smooth finish	0.14	0.10	0.06	0.04	0.04	0.03	0.05	
Plywood Paneling, $\frac{3}{8}$ in. thick	0.28	0.22	0.17	0.09	0.10	0.11	0.15	
Water Surface, as in a swimming pool	0.008	0.008	0.013	0.015	0.020	0.025	0.00	
Air, sabins per 1000 cu ft @ 50% RH				0.9	2.3	7.2	—	
Rough wood as tongue and groove cedar	0.24	0.19	0.14	0.08	0.13	0.10	0.14	

(Continued)

Acoustic Absorptive Materials	Coefficients							
	125 Hz	250 Hz	500 Hz	1000 Hz	2000 Hz	4000 Hz	NRC[a]	Note
Fiberglass Painted Ceiling Boards,[d] textured, $\frac{5}{8}$ in. thick	0.68	0.88	0.70	0.91	0.97	0.93	0.85	2
$\frac{3}{4}$ in. thick	0.66	0.85	0.72	0.94	0.99	0.98	0.90	
1 in. thick	0.69	0.91	0.79	0.99	0.99	0.99	0.90	
Random fissured, $\frac{5}{8}$ in.	0.64	0.82	0.68	0.86	0.83	0.57	0.80	
Perforated, $\frac{5}{8}$ in.	0.71	0.89	0.68	0.90	0.96	0.98	0.85	
Fiberglass Glass Cloth Ceiling Board[d]								
Nubby, $\frac{3}{4}$ in. thick	0.75	0.91	0.70	0.93	0.99	0.99	0.90	
1 in. thick	0.68	0.93	0.77	0.99	0.99	0.99	0.90	
Fiberglass prefinished ceiling tile[d]								
$\frac{3}{8}$ in. thick	0.70	0.83	0.62	0.78	0.91	0.92	0.80	
Celotex Mineral Fiber Tile[e]								
Natural fissured $\frac{3}{4}$ in. thick (Fig. 27.10*a*)	0.47	0.49	0.51	0.75	0.86	0.80	0.65	
Textured $\frac{3}{4}$ in. thick (Fig. 27.10*b*)	0.49	0.55	0.53	0.80	0.94	0.83	0.70	
Plaid design $\frac{3}{4}$ in. thick (Fig. 27.10*c*)	—	—	—	—	—	—	0.70	
LeBaron design, $\frac{3}{4}$ in. thick, (Fig. 27.10*d*)	—	—	—	—	—	—	0.70	
Striated design, $\frac{3}{4}$ in. thick, (Fig. 27.10*e*)	—	—	—	—	—	—	0.70	
Perforated lay-in panel $\frac{5}{8}$ in. thick (Fig. 27.10*f*)	0.27	0.26	0.52	0.75	0.68	0.53	0.55	
Gold Bond, National Gypsum[f] Mineral Fiber Tiles and Panels								
"Fire Shield" Solitude Panels, washable acrylic finish								
Perforated $\frac{5}{8}$ in. thick	0.25	0.29	0.60	0.83	0.71	0.53	0.60	
Fissured $\frac{5}{8}$ in. thick	0.28	0.32	0.65	0.73	0.73	0.75	0.60	
Textured $\frac{5}{8}$ in. thick	0.28	0.36	0.65	0.62	0.44	0.33	0.50	
Perforated Asbestos Panels, 1 in. thick								
Uniform	0.60	0.65	0.49	0.71	0.73	0.51	0.65	
Random	0.56	0.51	0.49	0.68	0.60	0.31	0.60	
"Acoustimetal" perforated metal panel, enameled, $1\frac{9}{16}$ in. thick								
Square pattern	0.59	0.85	0.88	0.99	0.97	0.79	0.90	
Diagonal pattern	0.63	0.84	0.86	0.99	0.99	0.91	0.90	
"Tectum" Sound Blocks								
3 in. thick x $15\frac{1}{2}$ in. square	0.32	0.60	1.43	2.36	2.32	2.41	1.68	3

Table 27.1 *Coefficients of Absorption—α(Continued)*

Absorption of Seats and Audience	*Values given are in sabins per square foot of seating area or per unit*[g] *125 Hz*	*250 Hz*	*500 Hz*	*1000 Hz*	*2000 Hz*	*4000 Hz*	*NRC*[a]	*Note*
Audience, seated in upholstered seats, per sq ft of floor area	0.60	0.74	0.88	0.96	0.93	0.85	—	
Unoccupied cloth-covered upholstered seats, per sq ft of floor area	0.49	0.66	0.80	0.88	0.82	0.70	—	
Unoccupied leather-covered upholstered seats, per sq ft of floor area	0.44	0.54	0.60	0.62	0.58	0.50	—	
Wooden Pews, occupied, per sq ft of floor area	0.57	0.61	0.75	0.86	0.91	0.86	—	
Chairs, metal or wood seats, each, unoccupied	0.15	0.19	0.22	0.39	0.38	0.30	—	
Students in tablet-arm chairs	0.30	0.42	0.50	0.85	0.85	0.84	—	

[a] Noise Reduction Coefficient is the arithmetic average of the α values at 250, 500, 1000, and 2000 Hz.
[b] Installed in hung ceiling with at least 16 in. to slab.
[c] Clipped or glued to wall; minimum 24 in. O.C.
[d] Courtesy of Owens-Corning Fiberglas.
[e] Courtesy of Celotex-Jim Walter Co.
[f] Courtesy of Gold Bond/National Gypsum.
[g] When the audience is randomly spaced, use an average of 5.0 sabins per person.

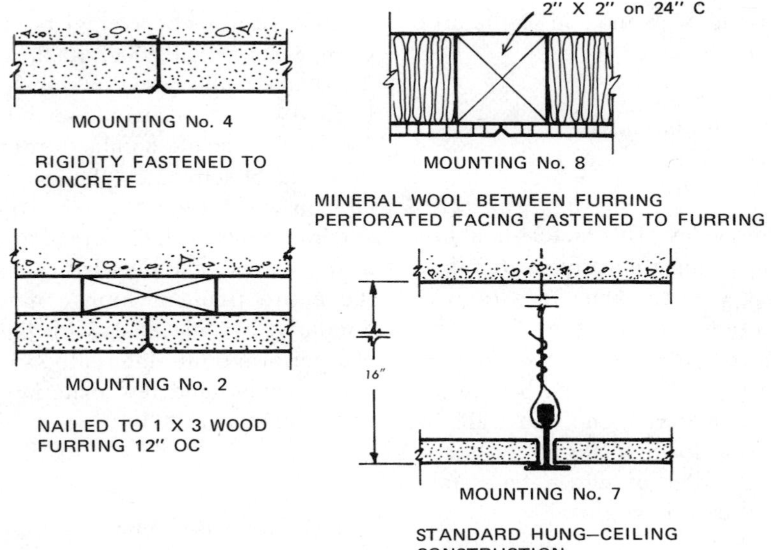

Fig. 27.5 Standard test-mounting methods for absorptive material, in accordance with which absorptive coefficients are given by manufacturers. Courtesy of Gold Bond/National Gypsum.

(*d*) Installation methods have a pronounced effect as discussed in the following section.

27.5 Absorptive Material Installation

Coefficient ratings for absorptive materials are always given with mountings corresponding to ASTM requirements. The most common standardized mounting methods are given in Figure 27.5. Installation of absorbent directly on a wall or ceiling is the least effective means, since exposure to sound energy is minimal (see Fig. 27.6). When an air gap is left between the porous layer and the rigid surface, the combination acts almost as well, in mid-frequencies, as a layer of absorbent equivalent in thickness to the air plus porous material. One problem with this technique is that at the $\lambda/2$ node of a standing wave there is a severe drop in absorption as can be seen from Figure 27.6*c*. Thus for instance at 1000 Hz, one-half wavelength is approximately 7 in. At that distance, α drops severely, but is a maximum at $\lambda/4$ or $3\frac{1}{2}$ in. For ceiling tile hung at 16 in.

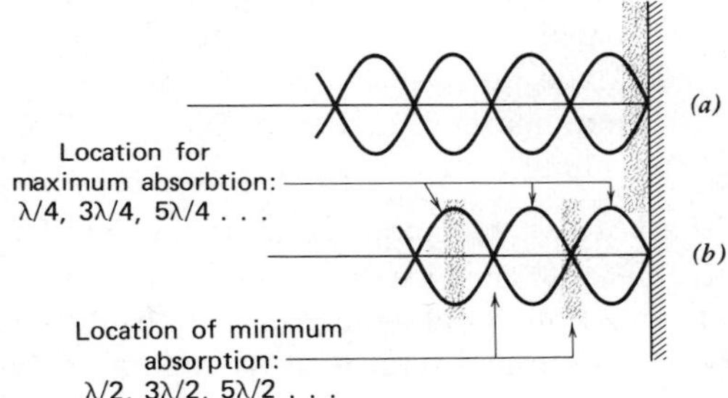

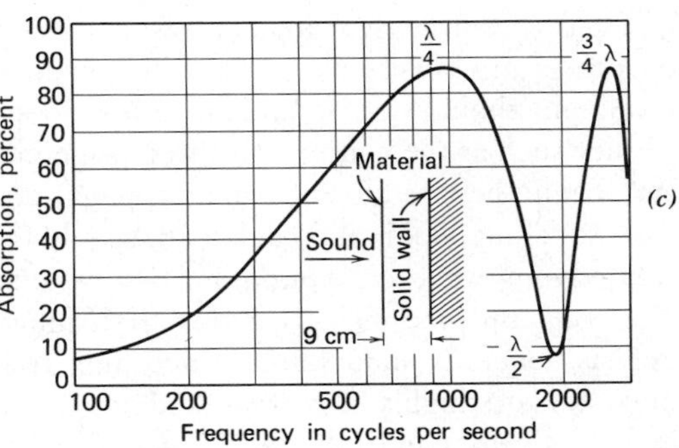

Fig. 27.6 Sound waves striking a surface of large mass will create standing waves at certain frequencies, depending on room dimensions (a). *If insulation is placed at the nodes of these waves* (b), *attenuation is drastically reduced* (c). *Illustration* (c) *from Brüel and Kjaer,* Sound Insulation and Room Acoustics, *Chapman and Hall, London, 1951.*

below the slab (Fig. 27.5, method 7), the drop in absorption occurs at

$$\lambda/2 = 16 \text{ in.}$$
$$\lambda = 32 \text{ in.} = 2.67 \text{ ft}$$
$$f = \frac{1128}{2.67} = 422 \text{ Hz}$$

which is midfrequency. This factor should be considered in applying absorptive material, to avoid a spacing corresponding to a drop in absorption at a sensitive frequency. To obtain good low-frequency absorption it is essential that a deep air space be provided behind the porous absorbent material, and that walls be treated in addition to the ceiling.

In increasing order of effectiveness, absorbent material can be applied:

a. Directly to room surface.
b. Hung below ceiling and supported away from the walls.
c. Hung from the ceiling as louvers or baffles.
d. Made up into shapes such as cubes, or tetrahedrons, and suspended from the ceiling.

The last two techniques are extremely effective because they expose a very large surface of porous material—much larger than could be accomplished with wall or ceiling covering. Of course, these suspended objects become architectural elements and must be handled accordingly. In contrast, surface coverings are relatively neutral. In general, treatment should not be limited to one room surface such as the ceiling. All three principal surfaces in the direction of sound propagation, that is, ceiling, floor, and back wall, should be treated *approximately equally* for best results. The common practice of treating the ceiling only is generally inadvisable, since high frequencies are highly directive and may not reach the ceiling until the third reflection. See specific recommendations in Section 27.36.

27.6 Noise Reduction Coefficient

The last column in Table 27.1 is labelled NRC—Noise Reduction Coefficient. This figure is the arithmetic average of the absorption coefficients at 250, 500, 1000, and 2000 Hz. The name is ill chosen inasmuch as it cannot be used as the name would seem to imply. It is useful as a single-number criterion for the *midband* effectiveness of a porous absorber. Obviously if careful design is required or high and low end frequencies are of interest NRC is nearly useless, and detailed calculations over the entire frequency range must be made. Similarly, two materials with the same NRC can perform quite differently, since the NRC is an average and few materials have a flat characteristic.

27.7 Noise Reduction by Absorption

As described earlier, the noise levels in a room are highest for a given source if the room's surfaces are primarily sound reflecting and lowest if there are large areas of sound-absorbing materials. Sound-absorbing ceilings, floors (carpets), and walls (permanent treatment or curtains) keep spaces such as restaurants relatively quiet. In the distant (reverberant) field of a noise source, the noise level in most rooms is

$$IL = PWL - 10 \log \Sigma S\alpha + 6 \text{ db}$$

or

$$IL = PWL - 10 \log \Sigma A + 6 \text{ db} \quad (27.1)$$

where

$\Sigma S\alpha = \Sigma A$
= total absorption in room, sabins
IL = Intensity level, decibels
PWL = Sound power level, decibels

Although increasing absorption decreases the noise level, the level never is reduced below the free field level for that distance from the source (see Fig. 27.1).

Noise reduction provided by additional absorption may be determined by

$$\text{Noise reduction} = IL_1 - IL_2$$
$$= 10 \log \Sigma A_2 - 2A_1$$
$$NR = 10 \log \frac{\Sigma A_2}{\Sigma A_1} \quad (27.2)$$

where

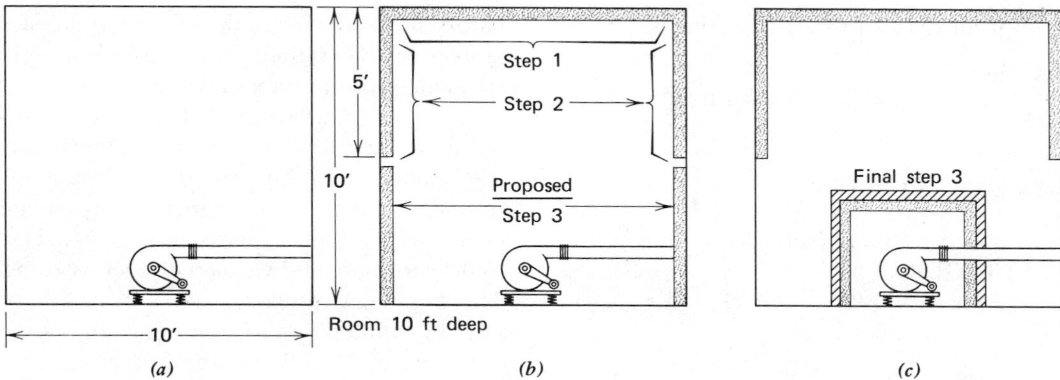

Fig. 27.7 Quieting the room (a) *by addition of absorptive material is only cost effective through step 2* (b). *Further quieting would be accomplished locally* (c), *which might obviate the necessity for wall treatment (step 2). Ceiling treatment should remain.*

ΣA_1 = absorption initial condition, sabins
ΣA_2 = absorption final condition, sabins

EXAMPLE 27.1. Referring to Figure 27.7, calculate the original noise level and the subsequent noise reduction by three steps of sound absorption treatment.

SOLUTION

(*a*) *Original Condition:* Painted Concrete Block Chamber—10 × 10 × 10 ft
Fan Power Level

At 500 Hz = 88 db
At 2000 Hz = 78 db

Absorption	Area	α	Total Absorption ($\Sigma S\alpha$)
500 Hz	600 ft²	0.06	36 sabins
2000 Hz	600 ft²	0.09	54 sabins

Sound Pressure Level Before Treatment

At 500 Hz:

$$\begin{aligned} IL &= \text{Sound Power} - 10 \log \Sigma S\alpha + 6 \text{ db} \\ &= 88 \text{ db} - 10 \log 36 + 6 \text{ db} \\ &= 88 \text{ db} - 15.6 \text{ db} + 6 \text{ db} \\ &= 78.4 \text{ db} \end{aligned}$$

At 2000 Hz:

$$\begin{aligned} IL &= 78 \text{ db} - 10 \log 54 + 6 \text{ db} \\ &= 78 \text{ db} - 17.3 \text{ db} + 6 \text{ db} \\ &= 66.7 \text{ db} \end{aligned}$$

(*b*) Ceiling Treatment Only

At 500 Hz; $\alpha = 0.82$

$$\begin{aligned} \text{Additional Absorption} &= 100\ (0.82 - 0.06) \\ &= 76 \text{ sabins} \end{aligned}$$

$$NR = 10 \log \frac{76 + 36}{36}$$

$$NR = 4.9 \text{ db}$$

At 2000 Hz:

$$\alpha = 0.94$$

$$\Delta A = 100\ (0.94 - 0.09) = 85$$

$$NR = 10 \log \frac{85 + 54}{54}$$

$$NR = 4.1 \text{ db}$$

(*c*) Ceiling and one-half wall treated

At 500 Hz:

$$\begin{aligned} \text{Added absorption} &= 300\ (0.82 - 0.06) \\ &= 228 \text{ sabins} \end{aligned}$$

$$NR = 10 \log \frac{228 + 36}{36}$$

$$= 8.7 \text{ db}$$

At 2000 Hz:

$$\begin{aligned} \text{Added absorption} &= 300\ (0.94 - 0.09) \\ &= 255 \text{ sabins} \end{aligned}$$

$$NR = 10 \log \frac{255 + 54}{54}$$

$$= 7.5 \text{ db}$$

(*d*) Ceiling and full wall treatment

At 500 Hz:

$$\Delta A = 500\ (0.82 - 0.06) = 380 \text{ sabins}$$

$$NR = 10 \log \frac{380 + 36}{36} = 10.6 \text{ db}$$

At 2000 Hz:

$$\Delta A = 500\ (0.94 - 0.09) = 425 \text{ sabins}$$

$$NR = 10 \log \frac{425 + 54}{54} = 9.5 \text{ db}$$

Summary

	IL	
	500 Hz	*2000 Hz*
Bare room	78.4	66.7
Ceiling treated	−4.9 db	−4.1 db
One-half wall treatment	−8.7 db	−7.5 db
Full wall treatment	−10.6 db	−9.5 db

We would conclude here that the third step is really not worthwhile, since only a negligible additional decibel of quieting is accomplished. This example is intended to indicate the law of diminishing returns in quieting by absorption. Starting with a live room, the initial application is effective. Beyond that, additional quieting by absorption is not economical, and the same outlay would be better applied in quieting the machine itself, probably with a machine enclosure as indicated in Figure 27.7*c*.

27.8 Panel and Cavity Resonators

Panel resonators are built with a membrane such as thin plywood or linoleum in front of a sealed airspace generally containing absorbent material. The panel is set in motion by the alternating pressure of the impinging sound wave. The sound energy is converted into heat through internal viscous damping. Panel resonators are used where efficient low-frequency absorption is required and middle-and high-frequency absorption is unwanted or provided by another treatment (see Fig. 27.8). Panel resonators are often used in recording studios.

Volume or cavity resonators (Helmholtz resonators, after their originator) are an air cavity within a massive enclosure, connected to the surroundings by a narrow neck opening. The impinging sound causes the air in the neck to vibrate, and the air mass behind causes the entire construction to resonate at a particular frequency. At that frequency absorption is very great—approaching unity, and dropping sharply above and below this frequency (see Fig. 27.8). By adjusting neck opening and cavity dimensions, the unit can be tuned to resonate at any desired frequency. This makes it extremely useful in instances where a major single frequency is present, as with 120-Hz transformer hum. Such an installation is shown in Figure 27.9 along with a typical concrete block resonator. Fibrous filler can be used in the block to increase high-frequency absorption.

27.9 Acoustically Transparent Surfaces

The soft porous material of which most absorbents are constructed is covered with perforated metal or wood panels, which act as physical protection and as stiffeners. These coverings are acoustically transparent except at higher frequencies. The frequency at which a noticeable reduction in absorption occurs can be estimated for circular holes as

$$f = \frac{40p}{d} \tag{27.3}$$

where f = frequency in hertz
p = percentage of open area
d = diameter of holes in inches

Thus for 1/4 in. holes and 60% open area, which is a typical commercial construction,

$$f = \frac{40(60)}{0.25} = 9500 \text{ Hz}$$

which is very high and generally not of major concern. It is always desirable for a given

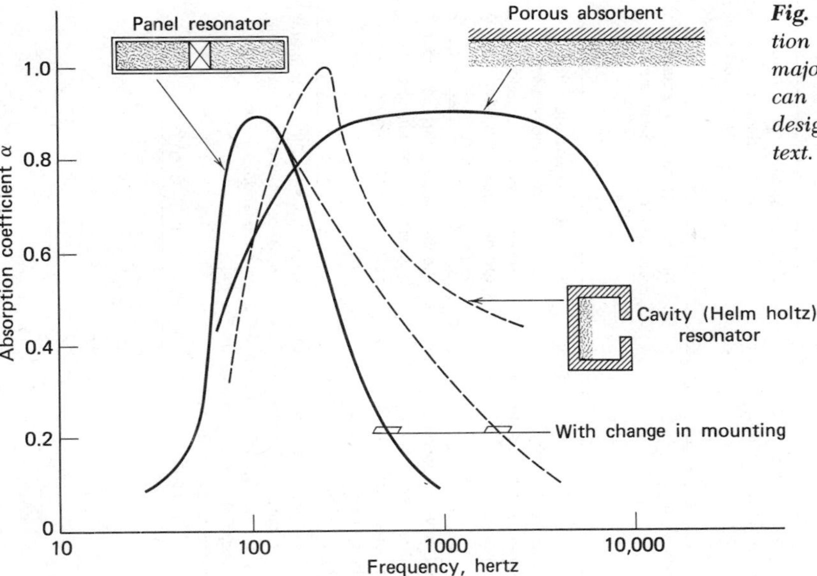

Fig. 27.8 Typical absorption curves for the three major types of absorbers. All can be changed by varying design, as discussed in the text.

Fig. 27.9 Typical application of cavity resonators to absorb transformer hum. Wall is constructed of masonry units constructed to act as resonators (see insert). (This installation is made with "SOUND-BLOX" as manufactured by the Proudfoot Co. through whose courtesy photo is supplied).

percentage of open area to have a larger quantity of small holes since, as is obvious from the formula, this raises the interference frequency. It is also desirable to stagger the holes as this improves absorption. An open-weave fabric is almost completely transparent to sound and can be used as a decorative cover on absorbent wall coverings (see Fig. 27.10).

27.10 Absorption Recommendations

To summarize the above discussion, absorption techniques are useful and effective:

a. To change room reverberation characteristics.
b. In spaces with distributed noise sources such as offices, schools, restaurants, and machine shops.
c. In spaces with a hard surface and little absorptive content.
d. Where listeners are in the reverberant field. (No amount of absorptive material can reduce intensity levels in the direct field.)

Concentrated noise sources are better handled by individual equipment enclosures than by room treatment, since enclosures reduce direct field noise which, as stated above, room surface treatment cannot do. Typical application recommendations are given in Table 27.2.

Table 27.2 *Typical Recommended Acoustical Treatments*

Acoustical Material	*Acoustical Tile*	*Metal-Faced Acoustical Units*	*Acoustical Plaster*	*Sprayed Mineral Asbestos*
Size and Form	Square tiles 12 in. x 12 in.; some up to 48 in. x 96 in. Roof deck 2 ft x 8 ft—thickness most common is $\frac{5}{8}$ in. & $\frac{3}{4}$ in.	12 in. x 24 in. up to 24 in. x 120 in. Thickness is controlled by acoustical pad backing. Either paper-wrapped mineral wool or cut or roll glass fiber. Units run 2 in. to 3 in. total thickness.	A plaster-like material of special fibrous or particulate aggregate $\frac{1}{2}$ in. to $1\frac{1}{2}$ in. thick.	Asbestos fibers sprayed with binder—$\frac{3}{4}$ in. to 3 in. thick.
Surface	Wide variety of perforated, textured, and sculptured surfaces with white, painted, vinyl, and glass cloth finishes.	Baked-enamel finish usually white, but available in color, perforated or slotted.	Fine-grain white texture, but may be spray painted.	Rougher grained than acoustical plaster, has deep texture and can be readily colored.
Method of Installation	Adhesive, nailing, or stapling to wood furring; lay-in grids; concealed spline-grid.	Attached to metal supports nailed to wood furring or hung in a proprietary metal suspension system.	Trowel or spray.	Sprayed on solid surfaces or immediately on metal lath.
Major Area of Application	All interior applications. Check specifications for application in high-humidity areas.	All interior applications. Check specifications for application in high-humidity areas.	Most interior applications. Especially useful for large curved surfaces. Not satisfactory in high humidity.	Architectural and industrial areas where combined fire-proofing and sound absorption are desired or rich texture is required.
Advantages	Provide widest range of finishes in high-absorption units. Mineral fiber tiles are rated incombustible. Lay-in units provide access to plenum above.	Incombustible. Easily maintained. Can be washed or painted; replacement units will match original job. Permits easy access to plenum.	Rapid, low-cost installation on large irregular or curved surface. Fine-grain texture. Incombustible—can be applied with (or as part of) fire protection.	Richly textured surface. Incombustible. Can be included as part of fire-proofing or thermal insulation.
Limitations	Allows transmission of sound over partition in hung ceiling unless selected for high-attenuation factor.	Allows transmission of sound over partition in hung ceiling unless backed by impervious layer.	Easily abused. Difficult to match finish in patching or repairing. Does not always perform as advertised. Performance limited to high-frequency absorption. Dusty in application.	Easily abused. Difficult to match patched areas to original. Dusty in application.

Acoustical Material	*Unit Sound Absorbers*	*Acoustical Form Board*	*Carpeting*	*Drapery*
Size and Form	From 12 in. x 12 in. x 2 in. units to 24 in. x 48 in. plastic coated glass fiber baffles.	Sizes and thickness vary to fit structural requirements (floor and roof slabs only) 1 to $2\frac{1}{2}$ in. for wall application.	Sized or seamed to fit any floor or wall area.	Acoustically transparent fabrics ranging from opaque velours to transparent/translucent glass fibers.
Surface	12 in. x 12 in. units in smooth to rough surface. Baffles are vinyl wrapped.	"Shredded Wheat" or smooth pattern.	Cut, looped, or combination to achieve any degree of "fuzzing."	Velveteen to boucle (rough).
Method of Installation	Units may be applied using special wall clips, pendents (ceiling only), or adhesive. Baffles are wire supported.	According to structural design (floor and roof slabs only). Adhesive or mechanical application on walls.	Tacking (floor only) or adhesive.	According to function and aesthetics.
Major Area of Application	12 in. x 12 in. units are useful in all interior areas. Baffles in industrial areas.	Floor and roof decks wherever applicable. Wall surfaces in most interior spaces.	Floors and walls.	Window or opaque wall areas and room dividers.
Advantages	Permit maximum flexibility. Hung units add sound absorption without requiring lowering lights or sprinkler heads.	Combines form board with thermal insulation and acoustical absorption.	Provides a relative degree of mid- and high-frequency absorption and luxurious appeal.	Provides a relative degree of overall sound absorption and luxurious appeal.
Limitations	Each application must be designed individually.	Not advisable for high-humidity areas.	Acoustical absorption increases with pile height, pad, and fiber density.	Acoustical absorption increases with fabric density percent, fold when drawn, and air space behind.

Source. Reprinted with permission of the authors. From Goodfriend and Sulewsky, "A Guide to Acoustic Materials," *Architectural and Engineering News,* February 1970.

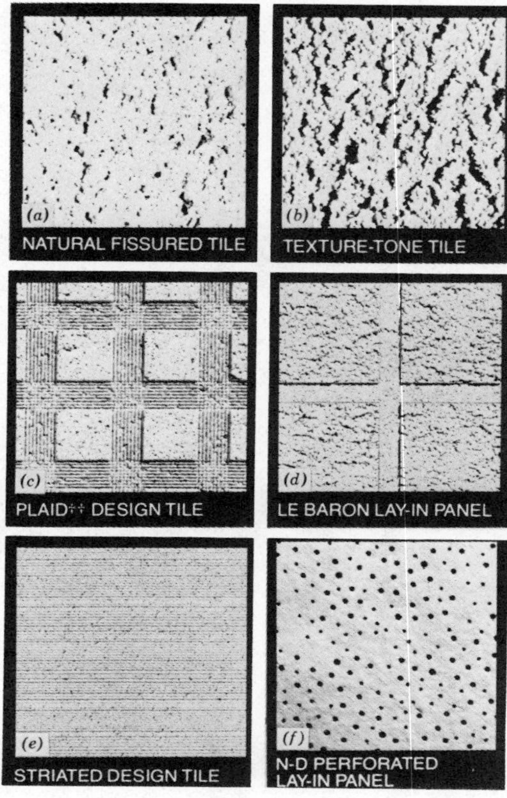

Fig. 27.10 (a to f) *Acoustic ceiling tiles and panels of varying designs, with varying types of perforations. Courtesy of Celotex Building Products Division, Jim Walter Corp.* (g) *Illustrates a fabric-covered mineral fiber, acoustic wall covering, coordinated with the acoustic ceiling tile pattern. Photo* (g), *courtesy of Armstrong Cork Co., illustrates their "Sound-Soak" Panels.*

SOUND ISOLATION

27.11 Airborne and Structure-Borne Sound

In contradistinction to the preceding sections, which were concerned with the phenomenon of in-room sound reduction by absorption, the following sections will discuss the characteristics of sound *transmission* between enclosed spaces. The distinction is often made between airborne and structure-borne sound, although in reality they differ only in sound origin. *Airborne sound* originates in a space with any sound-producing source and changes to *structure-borne* sound when the sound wave strikes the room boundaries. Structure-borne sound is generally understood as direct impact caused by a vibrating or impacting source directly contacting the structure. Hence a child crying in the adjoining apartment is air-borne sound; the same child bouncing a ball on the floor is creating structure-borne sound; in this case by impact. The building heating pumps that were installed without proper damping mounts create structure-borne sound by vibration. In reality all sound transmission is both airborne and structure-borne since, once having entered the structure, the sound travels along the structure and causes the structure to vibrate, generating airborne sound. Figure 27.11 should assist in understanding this action. In Figure 27.11*a* the sound is airborne, originating in the air at one side of the partition. The incidence of sound energy *causes the partition to vibrate,* generating sound on the other side. Sound does not "pass through"—it causes the structure to become a secondary source. The partition vibrates primarily in the direction of the sound, i.e., in the vertical plane. It also vibrates in other modes causing some sound energy to pass into the floor and ceiling, de-

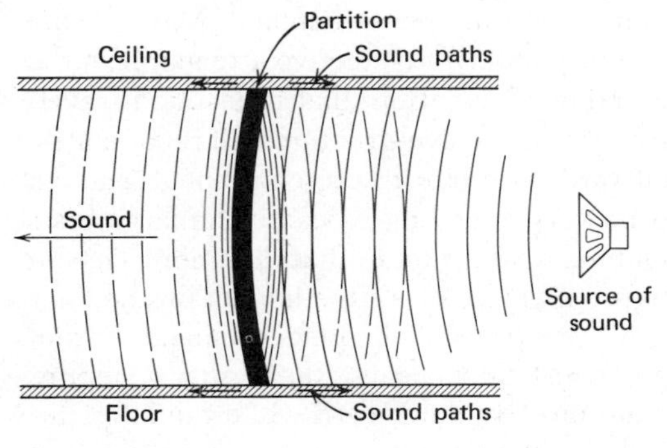

(a)

Fig. 27.11 (a) *Airborne sound is so called, since it originates in the air.* (b) *Structure-borne sound originates from mechanical contact between structure and vibrating or impacting sources. As such its energy level is usually much higher than airborne sound.* (c) *Illustrates the techniques available for controlling airborne sound.* A Guide to Airborne, Impact and Structure-Borne Noise Control, in Multi-family Dwellings, *U.S. Dept. of Housing and Urban Development, Washington, D.C., 1963.*

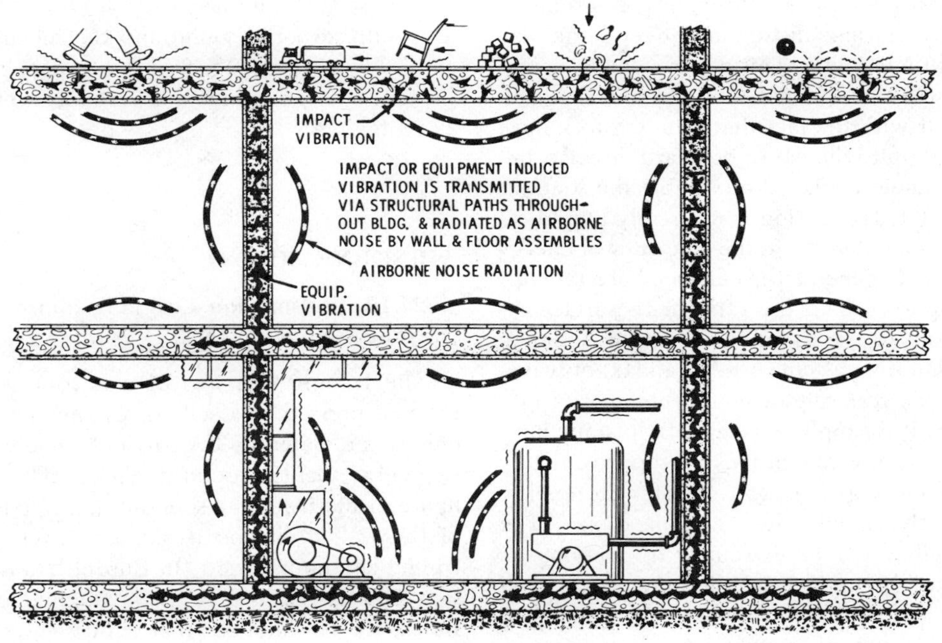

Transmission of Impact and Structure—Borne Noise.

(b)

A. MASSIVE WALL B. ISOLATED SURFACE C. DOUBLE WALL D. SOUND ABSORBER IN CAVITY E. SEAL HOLES AND CRACKS

(c)

pending on the mode of attachment. This energy becomes structure-borne sound.

In Figure 27.11*b* the process is similar, but reversed. Energy is introduced into the structure directly and efficiently by mechanical contact, that is, vibration and impact. Sound travels along the structure, as shown; and by causing the structure to vibrate, creates airborne sound. In a structure with rigid wall-to-floor connections, these sounds are clearly heard throughout the structure. Airborne sound is generally much less disturbing than structure-borne, since its initial power is very low and it attenuates rapidly at boundaries. Structure-borne sound is generally at a much higher energy level initially and attenuates slowly as it travels through the structure, thereby causing disturbance over large sections of a building. This disturbance is magnified by the "sounding board effect." We are all familiar with the fact that a tuning fork must be held up to the ear to be heard directly, but if its handle is placed on a table the sound is amplified. This action is not really amplification but an increase in the efficiency of energy transfer. In general, the efficiency of a radiator is proportional to the ratio of its surface dimensions to the sound wavelength. A tuning fork vibrating at concert A (440 Hz) of wavelength $2\frac{1}{2}$ *feet* cannot couple its energy into the air. It is simply too small. By placing it on a table whose dimensions are approximately one wavelength, it can transfer its energy efficiently; hence the "amplification." The same effect can be extremely troublesome in structure-borne sound. A vibrating pump itself makes little sound. However, it transfers a large amount of energy into the structure, which will appear as audible sound at each partition, floor, and wall that is rigidly coupled to the structure. Soft (damping) connections prevent energy transfer, thereby greatly attenuating the transmission of sound energy into connecting efficient radiating surfaces; hence the desirability of such flexible connections.

Airborne sound changes direction easily, with low frequencies most flexible in this regard. Thus sound can travel through long distances in corridors and ventilating ducts with attenuation of 6 db per doubling of distance. Structure-borne sound travels much more rapidly (see Table 26.1) and with attenuation as low as 1 db per kilometer. Mass of the structure is only an effective attenuator in the direction of radiation, that is, a sound traveling along a massive structure will radiate little outward from the structure (although enough to be very annoying) because the large mass minimizes vibration in that direction. Thus in Figure 27.11*b*, noise from impact on the floor above will probably be louder than noise from machines below because the former generates sound directly downward while the latter introduces energy into the entire network of parallel paths.

The sections immediately following will deal with airborne sound and the means for controlling it (see Fig. 27.11*c*). Impact noise (structure-borne sound) is covered in Sections 27.23 to 27.25.

AIRBORNE SOUND

27.12 Transmission Loss (TL) and Noise Reduction (NR)

The transmission loss of a barrier is the ratio, expressed in decibels, of the acoustic energy reradiated by the barrier to the acoustic energy incident upon it. This number is a figure of merit for the sound-isolating quality of the wall itself and is obtained from controlled laboratory tests. (In Europe *T*ransmission *L*oss is referred to as Sound Reduction Index, R). The number that is of greater importance to the building designer is the actual noise reduction between two spaces separated by a barrier of transmission loss TL, that is, the action of the barrier in situ. This isolation or noise reduction is defined as

$$NR = IL_1 - IL_2 \text{ db} \quad (27.4)$$

where NR = noise reduction between spaces, in db

IL_1 = sound intensity level in source room db

IL_2 = sound intensity level in receiving room db

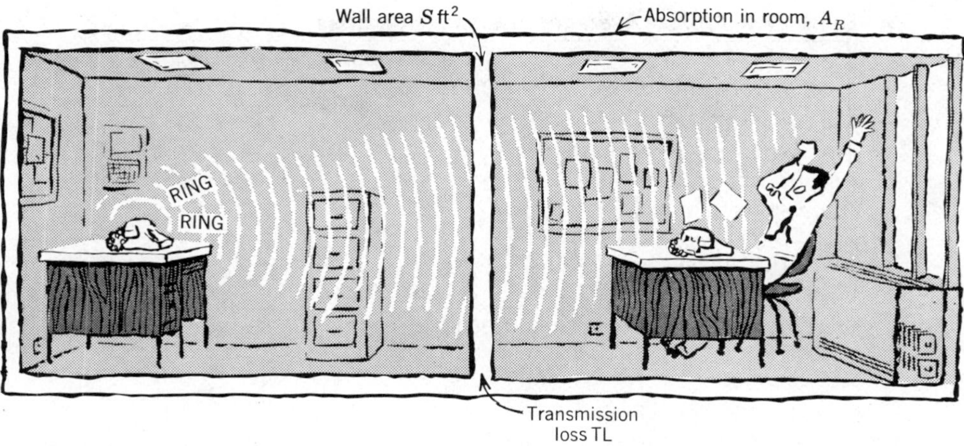

Fig. 27.12 *Illustration of the simple case of airborne sound transmission between adjacent rooms through a common barrier. With a sound source in one room the transmitted sound level is dependent not only on the transmission loss of the barrier, but also on the area of the barrier and the receiving-room absorption. The actual background "masking" noise levels determine whether or not the transmitted sound will be heard.*

and is related to the TL of the barrier by the expression

$$NR = TL - 10 \log \frac{S}{A_R} \qquad (27.5)$$

where NR = noise reduction in db
TL = barrier transmission loss in db
S = area of barrier wall in sq ft
A_R = total absorption of *receiving* room, in sabins

We see, therefore, that noise reduction and transmission loss are not equal but are related by the size of the dividing wall S and the absorption characteristic of the receiving room, A_R. A moment's thought will confirm the logic of this relation. When sound energy impinges on the wall, the wall in turn becomes the sound source, radiating into the receiving room. The sound level in the receiving room is related to its own reverberance (absorption characteristic A) as we have seen repeatedly. (Refer back to equation 26.15, for example.) Thus if the receiving room is a reverberant live space, A is low and the NR is considerably less than TL. Conversely if the receiving room is dead, A is large and noise reduction can be greater than transmission loss, depending on the ratio of barrier wall size to room area. Figure 27.12 illustrates the above relationships and demonstrates that the vital consideration is not transmission loss but noise reduction. In order to acquire facility with sound isolation techniques it is essential that the reader become familiar with the relationship of transmission loss to the barrier wall's physical characteristics, i.e., its mass, rigidity, material of which it is constructed, and method of construction and attachment. These considerations are the subject of the following sections.

Note: The reader is cautioned to be careful with the term noise reduction (NR), since a term Noise Reduction Coefficient (NRC) also exists, which is completely unrelated. The latter, as pointed out in Section 27.6, is very poorly named.

27.13 Barrier Mass

Sound transmission requires that the barrier be set into vibration by the incident sound energy. Although this was stated above, we

Table 27.3 ***Typical Average[a] Transmission Loss***

Barrier Construction	*Surface Weight (lb/sq ft)*	*Transmission Loss (db)*
$\frac{3}{16}$-in. Plate glass	2.5	20
$\frac{1}{4}$-in. Asbestos—cement sheet	3.0	21
2-in. Plaster on wire lath	15	34
3-in. Cinder block	25	39
4-in. Concrete	50	45
8-in. Brick, plastered	80	50
12-in. Brick, plastered	120	53

[a]Over the frequency range of 150 to 3000 Hz.

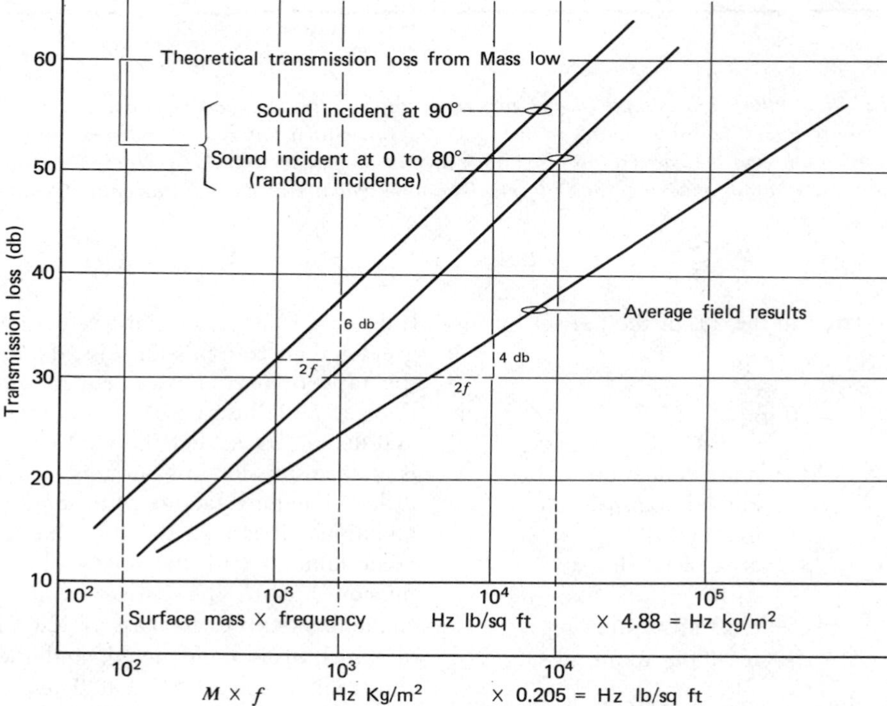

Fig. 27.13 *Graphic representation of Mass law action in attenuation of transmitted sound. Normal (90°) incidence results in maximum transmission loss. Field or random incidence (0 to 80°) is approximately 6 db lower, but maintains the 6 db per octave slope. Average field results are much lower due to flanking and stiffness effects.*

repeat it here to emphasize the fundamental importance of this simple statement. (We are of course referring to a barrier that is impervious to air, that is, a solid barrier. Otherwise the moving air molecules carrying the sound will simply pass through with minimal transmission loss.) The impinging energy acts as a force on the wall. Since

$$F = MA$$

the larger the mass the less it will vibrate. This is known in acoustics as the Mass law, which states that for a nonporous, homogeneous structure of low stiffness, the sound transmission loss is proportional to the logarithm of surface mass (the weight of the wall per square foot of area) and to the frequency of vibration (acceleration factor). Thus doubling the mass or frequency will, theoretically, cause an increase of 6 db in the transmission loss or, stated otherwise, the slope of the TL versus the frequency-mass (fM) curve is 6 db (see Fig. 27.12). Table 27.3 lists the average transmission loss of some common barriers, confirming the overall operation of the Mass law. Figure 27.13 is a graphic representation of Mass law operation. With sound incident at 90°, maximum energy is imparted to the barrier and the entire mass resists, resulting in maximum transmission loss. In practice, however, sound is incident from 0 to 80° (called field or random incidence) reducing the mass effect, but keeping the slope at 6 db/octave. Due to nonhomogeneity, porousness and stiffness, actual field results indicate transmission losses nearer 4 db per octave as shown by the lower curves in the figure.

27.14 Stiffness and Resonance

The stiffness of a barrier is a function of its material composition and the rigidity of its mounting. The former is dependent on its internal cohesiveness, that is, its modulus of elasticity, and the latter depends on its boundary restraints—whether the barrier is tightly or loosely held. A homogeneous material of high Young's modulus has great cohesiveness between its molecules. As soon as one is set in motion by the incident sound, the motion is passed to the next, and so on, making it an excellent sound conductor. Homogeneous materials with low modulus of elasticity have high internal damping—motion of molecules is not transmitted well and they are good sound insulators. Composite materials such as concrete and organic materials such as wood do not conform to these general rules.

The rigidity of mounting can be likened to a drumhead—the tighter it is stretched the better it resounds. Rigidity (stiffness) in a panel barrier resists damping and assists vibrations, making it a poor insulator. To recapitulate, stiffness, both internal and external, is an undesirable characteristic in sound isolation. A material such as lead, which has high mass, low modulus of elasticity, and resists rigid mounting is an excellent sound attenuator.

The effects of stiffness and mass both vary with frequency, unfortunately in opposite directions. Stiffness acts to reduce transmission loss as frequency increases while, as we have seen, mass acts to increase it; the combined effect is shown in Figure 27.14. At very low frequencies the mass and stiffness effects negate each other, giving the resonance dips shown. Beyond approximately 200 Hz most common wall construction enters the Mass law range and continues with it until the critical frequency. Deviations from a smooth 4 to 6 db per octave slope are due to the nonhomogeneous nature of most wall construction. At the critical frequency the phase of incident sound waves corresponds or "coincides" with the phase of vibration (shear wave) of the barrier in such a way as to pass a large portion of the incident energy. See insert on Figure 27.14. This shows in Figure 27.14 as the "coincidence dip." This effect is most pronounced in thin homogeneous partitions and light stiff ones. Critical frequency, f_c, as a function of panel thickness for common materials, is plotted in Figure 27.15. To avoid a coincidence dip in the audible range partitions can be either very heavy and stiff, which greatly decreases the critical frequency, or heavy and limp (resilient), which greatly increases the critical frequency. In practical terms, cost effectiveness is heavily in favor of the latter alternative, thus, for instance the TL of a wood partition can be improved by grooving it to increase flexibility. The dramatic improvement in TL, resonances, and coincidence dip by use of resilient mounting of a simple masonry partition is shown in Figure 27.16. Both walls have the same weight—21+ lb/ft^2. The solid wall A has better attenuation below 200 Hz in the stiffness-controlled range. Above that frequency the resilient-mounted partition is 10 db better. The sound transmission class (STC) (see Section 27.16) of wall A is 40; that of wall B is 51. Both show a coincidence dip at approximately 250 Hz (see Fig. 27.16) but that of wall B is shallow; that of A deep and wide. Furthermore, wall B is consistently better than Mass law attenuation; wall A consistently worse (due to stiffness).

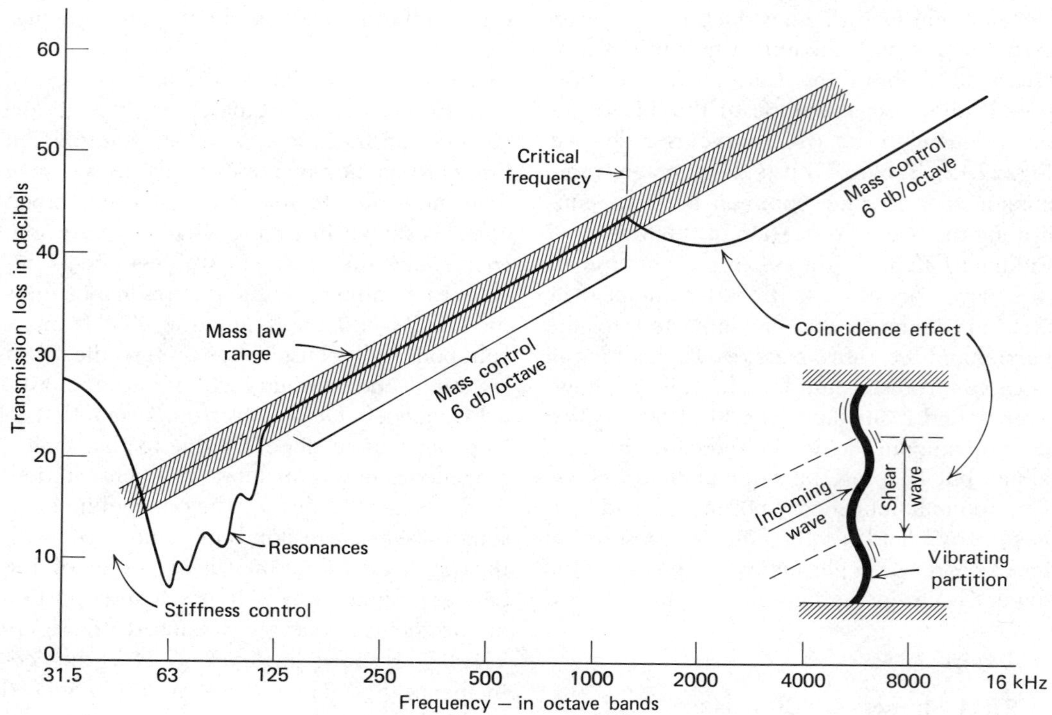

Fig. 27.14 Transmission loss as a function of frequency, assuming constant surface mass.

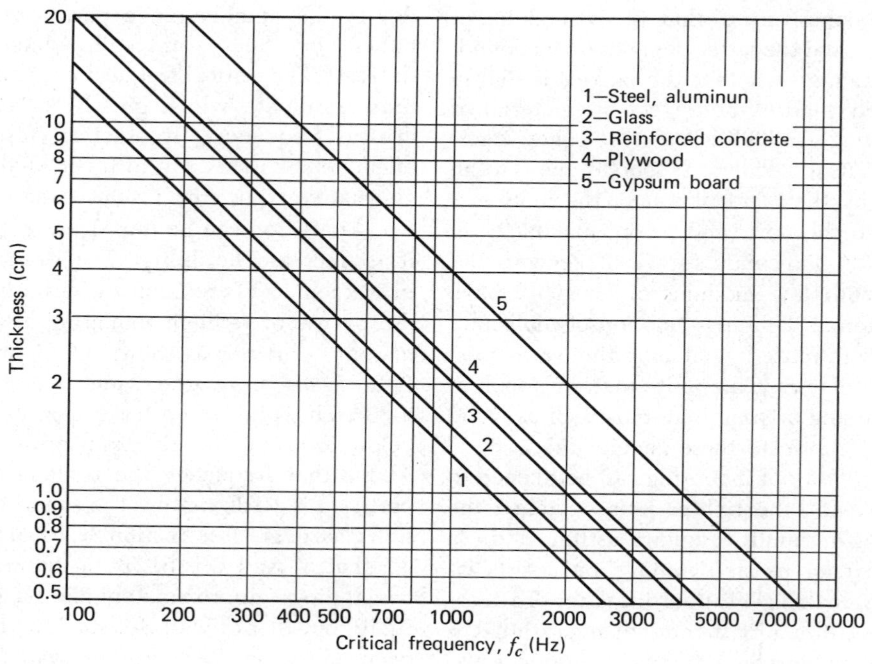

Fig. 27.15 Critical frequency as a function of thickness for several common materials. Reprinted with permission from E. B. Magrab, Environmental Noise Control, *Wiley, New York, 1975, p. 262.*

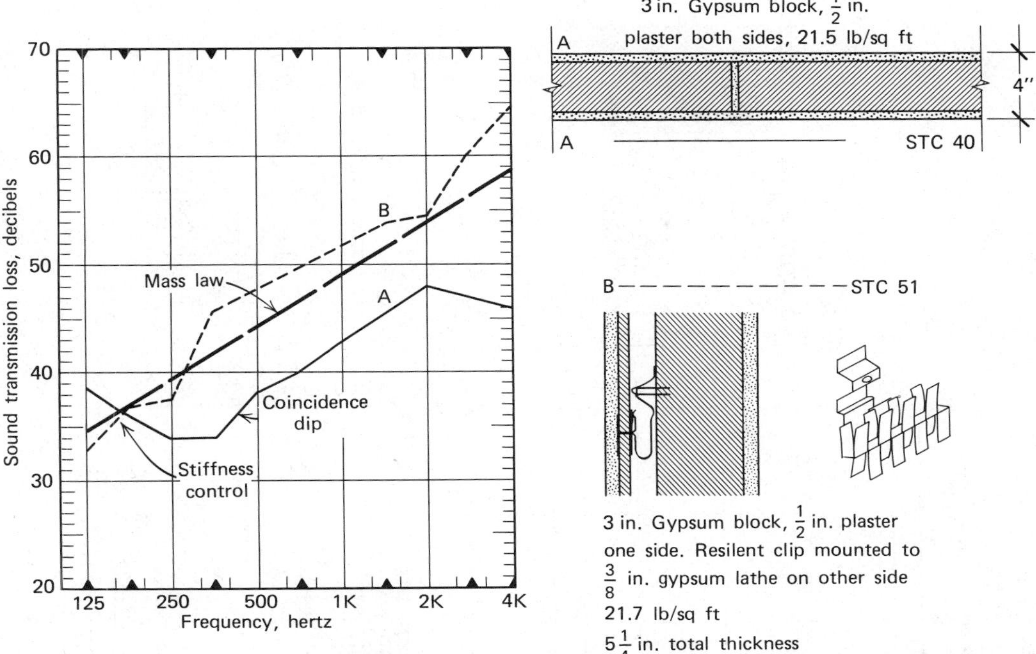

Fig. 27.16 Transmission loss characteristics of two equal weight partitions with similar boundary constraints. The solid partition A is worse than the Mass law due to stiffness. The resilient-mounted wall B performs better than the Mass law and much better than wall A, except at the lowest frequencies. Data extracted from A Guide to Airborne, Impact and Structure-Borne Noise Control in Multi-Family Dwellings. *U.S. Dept. of Housing and Urban Development, Washington, D.C., 1963, pp. 18, 19.*

27.15 Compound Barriers (Cavity Walls)

Since the maximum theoretical increase in transmission loss with mass increase is 6 db per doubling of mass, it is apparent that this method of transmission loss improvement rapidly reaches the limits of practicality. Indeed as we have seen, actual single homogeneous walls fall below the Mass law curve. This is because mass increase brings with it stiffness increase, which as we have seen acts to *reduce* transmission loss. If, however, a barrier is constructed of two separate layers without rigid interconnection, its performance is better than the calculated transmission loss based on mass alone. Note that even the nonrigid wire ties of wall B in Figure 27.17 lower its STC by five points. At low frequencies, where stiffness controls transmission loss (see Fig. 27.13), the cavity acts as a rigid connection between the layers, adding stiffness and increasing transmission loss. At higher frequencies, in the Mass law range, the air in the cavity cannot sustain the shear wave between the layers and therefore acts as a damping coupling to reduce stiffness. The net result is an improvement in performance throughout the frequency range. Transmission loss for the entire cavity wall increases with the size of the air space at the rate of approximately 5 db per doubling. Performance can be improved still further by filling the void with porous, sound absorbent material. This acts to further decrease the stiffness of the compound structure *and* to absorb sound energy that reflects back and forth between the two inside surfaces. The performance of cavity walls is reduced by any rigid interconnections between leaves. Thus a common stud wall with frequent rigid interconnections acts little better than a single homogeneous wall. However, a stud wall with staggered studs exhibits much improved per-

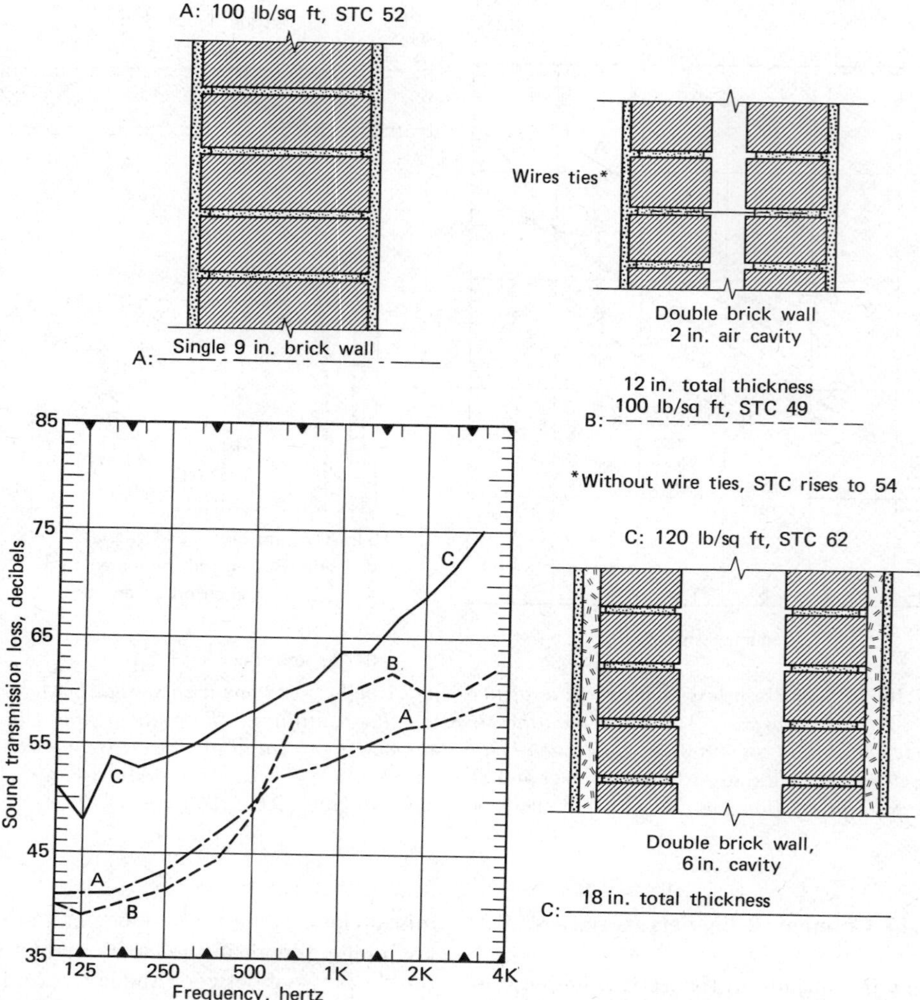

Fig. 27.17 Transmission loss curves showing the effect of air space on massive wall construction. All three walls are approximately the same mass. The 2 in. air space in wall B is not significant until until the higher frequencies, whereas the large, 6 in. air space of wall C is effective throughout the frequency spectrum. Data from A Guide to Airborne, Impact and Structure-Borne Noise Control in Multi-Family Dwellings, *U.S. Dept. of Housing and Urban Development, Washington, D.C., 1963, pp. W6, 22, 23.*

formance over a single-material wall or a common stud wall. The above effects are illustrated in Figures 27.17 and 27.18 (see also Appendix A).

27.16 Average Transmission Loss (TL); Sound Transmission Class (STC)

Various attempts at using a single-number average transmission loss to describe a barrier's characteristics have been tried but with little success. Indeed these averages can be misleading, since they ignore both deficiencies and proficiencies at particular frequencies. Their use, therefore, in all but rough work is to be discouraged.

To avoid the shortcomings of averages and yet to benefit from the indisputable convenience of single-number ratings, a system of

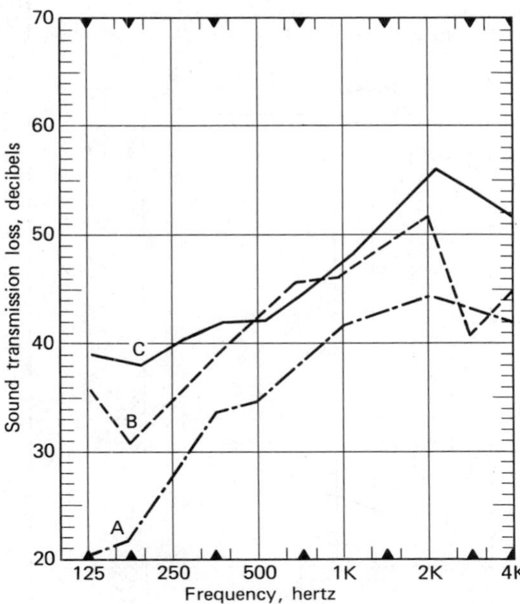

Fig. 27.18 Transmission loss curves illustrating the effect on lightweight walls of stiffness reduction and addition of absorptive material in the cavity. Curve A is a standard stud wall as found in frame construction. Curve B shows the advantage of staggered studs, over the entire frequency range. The dip at 3 kHz is a coincidence dip for a single leaf. Addition of absorption material C improves the attenuation characteristic at both ends, and is particularly useful for its low end improvement. Data from A Guide to Airborne, Impact and Structure-Borne Noise Control in Multi-Family Dwellings, *U.S. Dept. of Housing and Urban Development, Washington, D.C., 1963, pp. W20, 34, 37.*

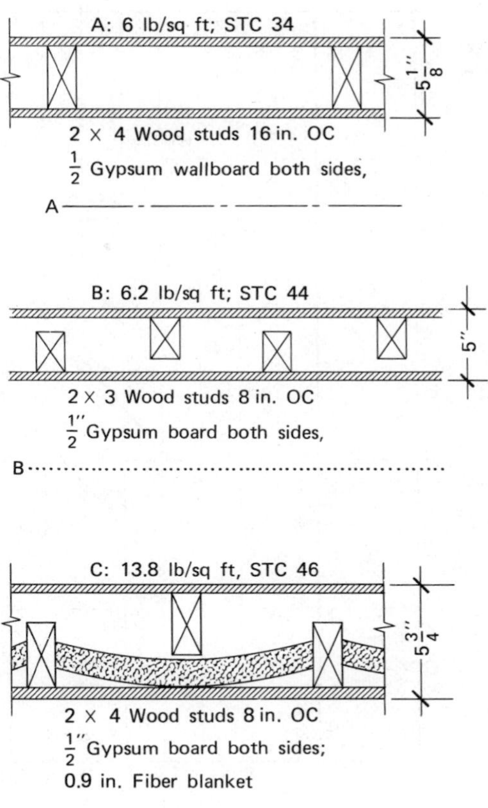

standard contours was developed, called Sound Transmission Class (STC) contours. Actual test results for a given construction, measured in a series of sixteen 1/3 octave bands, are compared to the standard STC contours according to a fixed procedure, and the STC number for that barrier is derived. The technique is illustrated in Figure 27.19. Figure 27.20 shows two TL curves, and the STC ratings of each. The nine-frequency average TL for each is approximately 40. However, there is a major dip in transmission loss of curve B. The STC of A is 42 and of B, 33. Thus, the STC approach considers a flaw in performance that is ignored by average TL. Nevertheless, since STC fails to give credit for performance *above* the established requirements, octave-band transmission-loss data, rather than STC ratings, should be used in all critical work such as music rooms or mechanical rooms where certain particular frequencies may be dominant. Figure 27.21 gives three standard STC contours that are of interest because they are used by the FHA to specify grades of construction. The criteria for their application is found in Section 27.36. An appreciation of the degree of sound isolation provided by walls with different STC ratings is given in Table 27.7. Since the subjective reaction on the quiet side depends on the background noise level there, the table gives this reaction for two NC curve levels. To assist

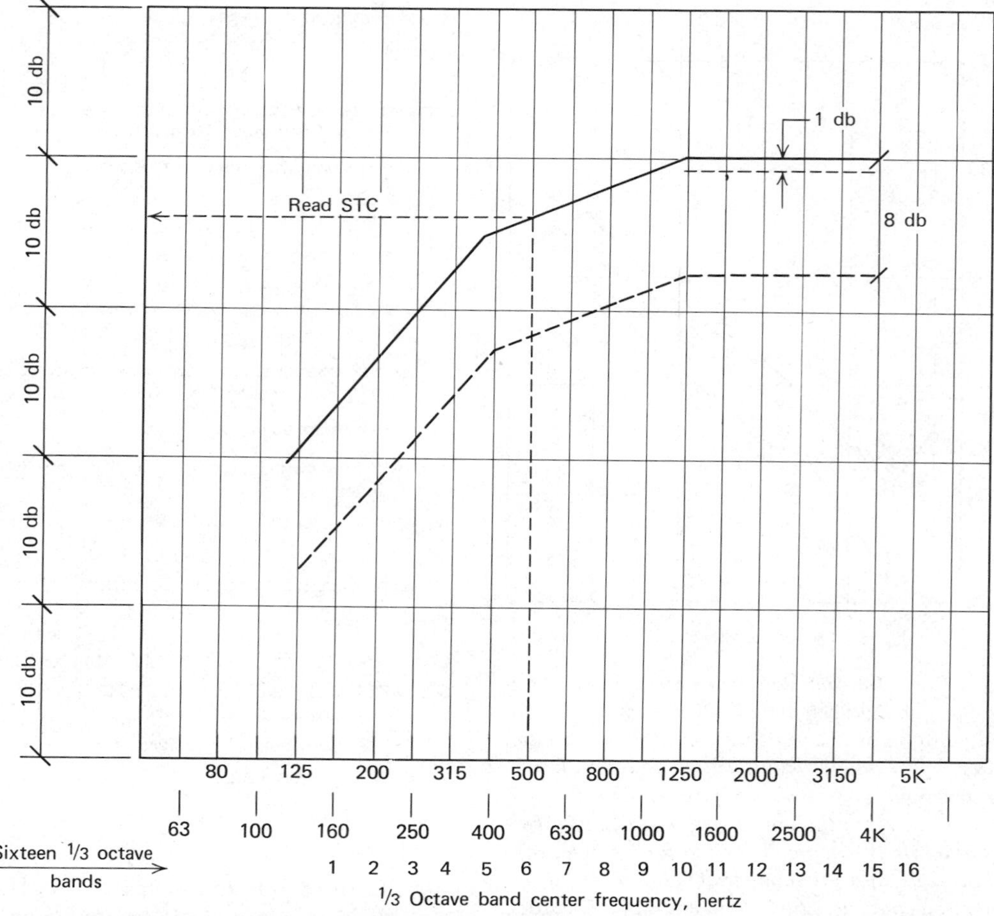

The STC is determined by comparison with a transparent overlay of Figure 27.19 on which the STC contour is drawn. The STC contour is shifted vertically, relative to the test curve, until some of the measured TL values for the test specimen fall below those of the STC contour (the solid line) and the following conditions are fulfilled:

1. The sum of the deficiencies (i.e., the deviations below the contour) shall not be greater than 32 db.
2. The maximum deficiency at a single test point shall not exceed 8 db [the broken (dashed) line beneath the STC contour].

When the contour is adjusted to the highest value (in integral db) that meets the above requirements, the sound transmission class for the specimen is the TL value corresponding to the intersection of the contour and the 500-Hz ordinate.

Fig. 27.19 *Overlay from which sound transmission class is determined graphically.*

the designer, extensive sound transmission testing has been performed on most types of standard wall and partition construction and the results published. Table 27.4 and Appendix A give descriptions of construction with typical details, transmission-loss data, STC ratings, and other pertinent data.

27.17 Composite Walls and Leaks

It is frequently necessary to determine the transmission loss of a composite wall, that is, a wall with a window, door, vent opening, and the like. It should be clearly appreciated that the two materials are "in parallel," to borrow

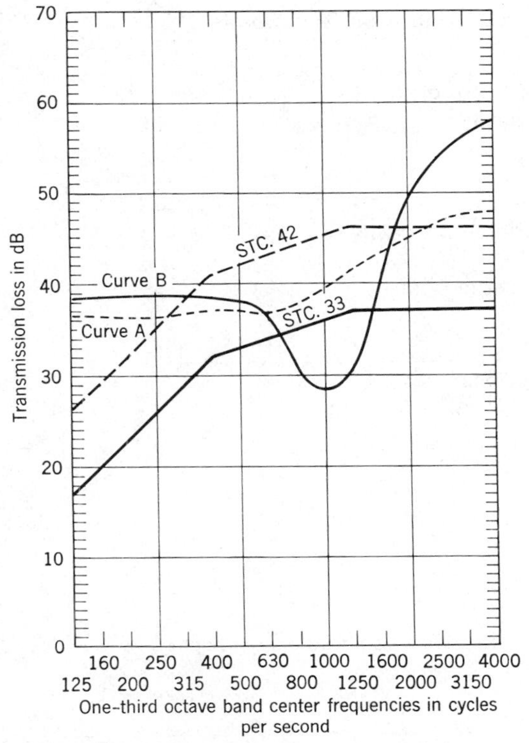

Fig. 27.20 Construction with the same average TL *yet different sound transmission class.*

an electrical concept, and the behavior is similar to that situation, i.e., the overall performance will be strongly affected by the poorer of the two, with some tempering of the degradation for small areas. Figure 27.22 enables us to calculate situations of this type.

EXAMPLE 27.1. Given a wall 9 ft × 18 ft with a transmission loss of 52 db at 1000 Hz, containing a 3 ft × 7 ft 6 in. hollow core door of 22-db transmission loss. Find the overall TL of the composite wall.

SOLUTION. Refer to Figure 27.22.

$$TL_1 - TL_2 = 30 \text{ db}$$

$$\frac{S_2}{S} = \frac{3 \times 7\frac{1}{2}}{9 \times 18} \times 100 = 13.9\%$$

From curves: $TL_1 - TL_c = 21.5$

$$TL_c = 52 - 21.5 = 30.5 \text{ db}$$

that is, a door with an area of only 14% of the entire wall reduces the TL of the structure from 52 db to 30.5 db—i.e., from excellent to *very poor.*

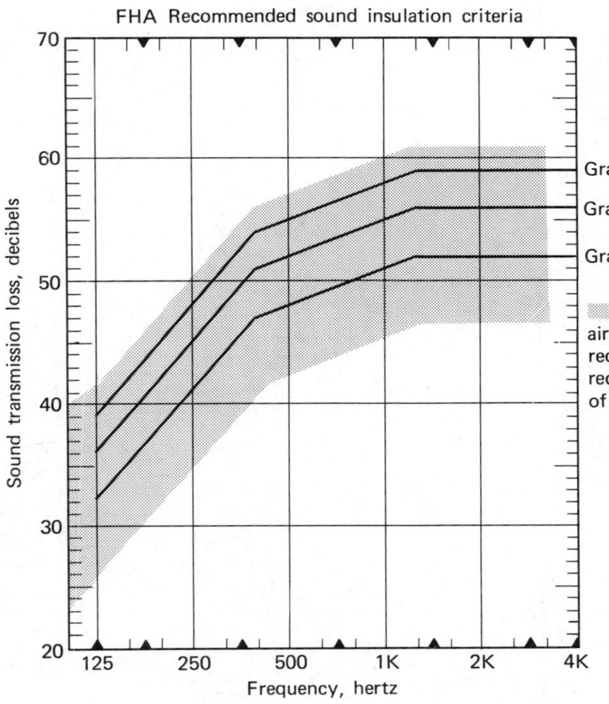

Fig. 27.21 FHA recommended sound insulation criteria. From A Guide to Airborne, Impact and Structure-Borne Noise Control in Multi-Family Dwellings, *U.S. Dept. of Housing and Urban Development, Washington, D.C., 1963, pp. 10–11.*

Table 27.4a *Improvements in STC Rating of Stud[b] Partitions*[a]

Description	*STC*
Basic Partition—single wood studs, 16 in. on centers, $\frac{1}{2}$-in. gypsum board each side, air cavity	35
Add to Basic Partition:	
Double gypsum board, one side	+2
Double gypsum board, both sides	+4
Single-thickness absorbent material in air cavity	+3
Double-thickness insulation	+6
Resilient channel supports for gypsum board	+5
Staggered studs	+9
Double studs	+13

[a]When using two improvements add additional +2.
When using three improvements add additional +3.

Example: Improvements to 35 STC basic partition:

Staggered wood studs	+9
Double gypsum board, one side	+2
Single insulation	+3
Adder	+3
Total	+17
Total STC	35 +17 = 52

For application to metal stud partitions, use adders as above except begin with STC = 40 for $3\frac{5}{8}$ in. basic partition.

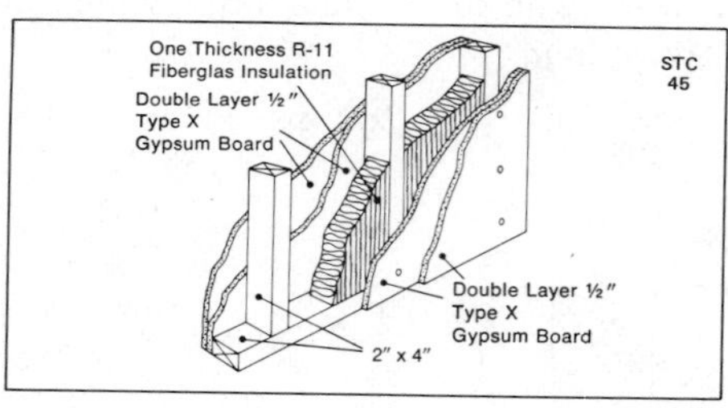

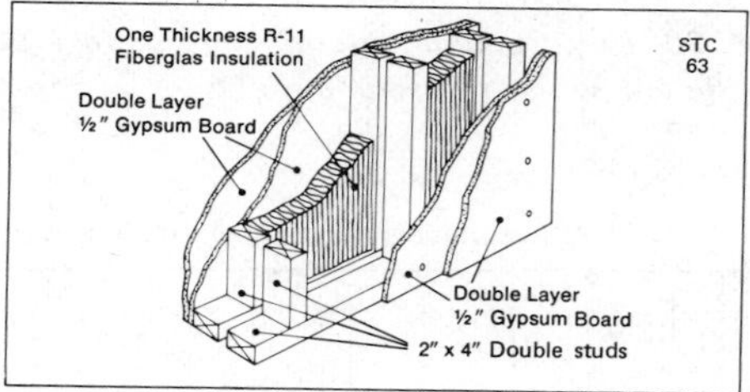

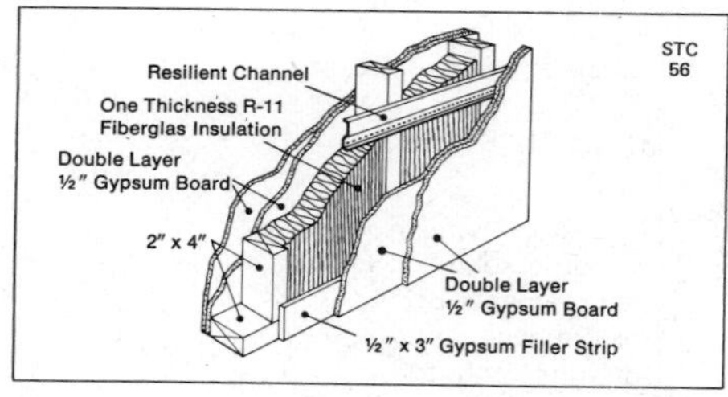

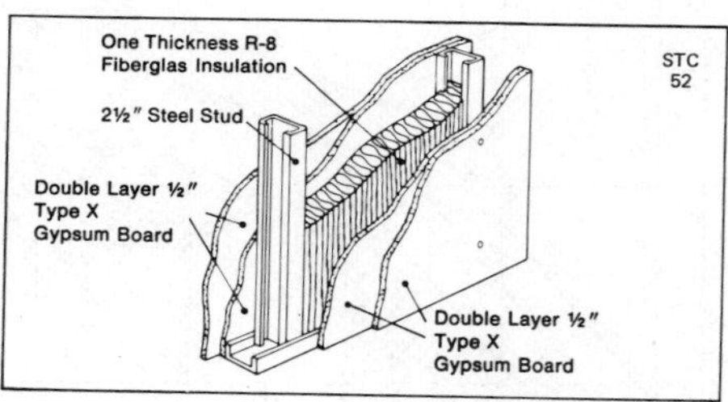

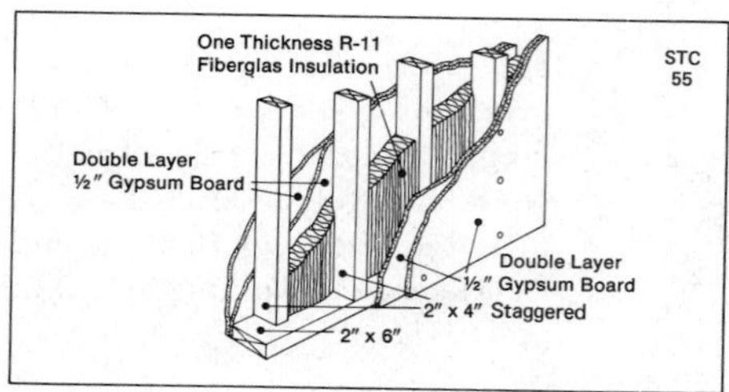

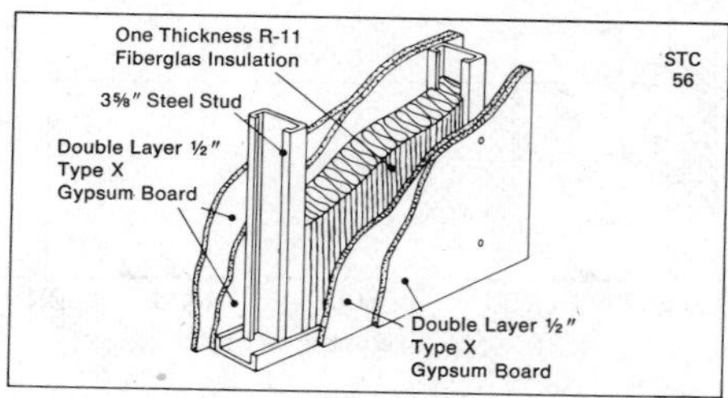

Table 27.4b *STC Ratings of Masonry Walls*

Description	*STC*
4-in. Lightweight[a] hollow block	36
4-in. Dense hollow block	38
6-in. Lightweight hollow block	41
6-in. Dense hollow block	43
8-in. Lightweight hollow block	46
8-in. Dense hollow block	48
12-in. Lightweight hollow block	51
12-in. Dense hollow block	53
4-in. Brick	41
6-in. Brick	45
8-in. Brick	49
12-in. Brick	54
6-in. Solid concrete	47
8-in. Solid concrete	50
10-in. Solid concrete	53
12-in. Solid concrete	56

NOTES. [a]All ratings of lightweight block assumes sealing with paint. Note that this reduces absorption.

MODIFICATIONS:

Add sand to cores of hollow blocks	+3
Add plaster to one side	+2
Add plaster to both sides	+4
Add furring strips, lath and plaster:	
1 side	+6
2 sides	+10
Add plaster via resilient mounting:	
1 side	+10
2 sides	+15

EXAMPLE 27.2. An exterior brick/frame wall of TL of 54 at 1000 Hz, measuring 8 ft × 16 ft, is pierced by two wood frame windows, each of area 3 ft × 4 ft, with single 1/8 in. glass, and a TL of 34 at 1000 Hz. Find the combined transmission loss.

SOLUTION.

$$TL_1 - TL_2 = 54 - 34 = 20 \text{ db}$$

$$\frac{S_2}{S} = \frac{2 \times 3 \times 4}{8 \times 16} \times 100 = 18.75\%$$

From Figure 27.22, $TL_1 - TL_c = 12.5$ db

$$TL_c = 41.5$$

Again the result is a reduction from an excellent wall to a poor one.

Since an opening in a wall is effectively a second material of $TL = 0$, the curves of Figure 27.22 can be replotted as in Figure 27.23. Note carefully that the curves very rapidly flatten out; thus any wall with 1% open area will

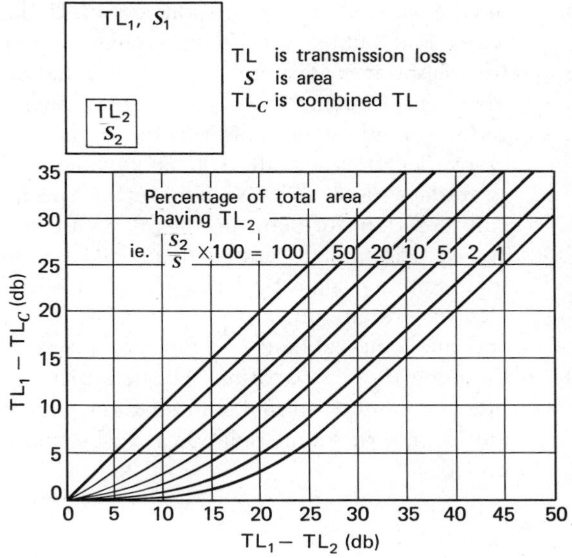

Fig. 27.22 Transmission loss of a two-element composite barrier as a function of the relative transmission loss of the components. From E. B. Magrab, Environmental Noise Control, *Wiley, New York, 1975, pp. 266, Fig. 7-45 and 268, Fig. 7-47.*

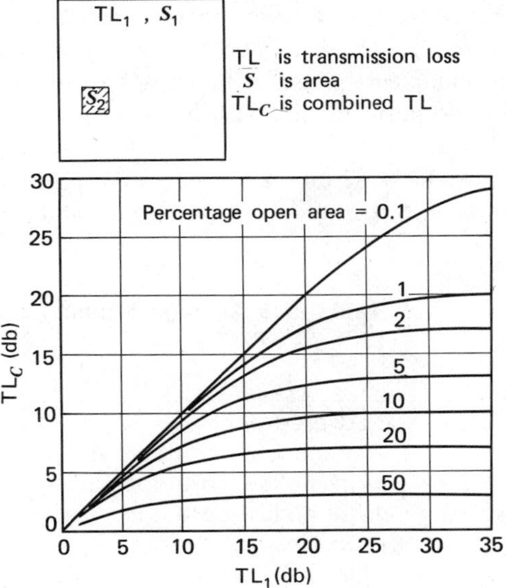

Fig. 27.23 The effect of a hole of a given percentage of the total area on a partition of TL, *db. From E. B. Magrab,* Environmental Noise Control, *Wiley, New York, 1975, pp. 266, Fig. 7-45 and 268, Fig. 7-47.*

have a *maximum* transmission loss of 20 db, which is all but useless as sound insulation. For this reason it is absolutely imperative that all openings be completely sealed, particularly those around doors and windows. A hairline crack degrades a wall 6 db; a keyhole will degrade a door 3 db, and so forth. Special considerations for doors and windows are discussed below. Care must also be taken with such common acoustic leaks as back-to-back electric outlets, pipes passing through walls, and medicine cabinets—in fact any break in the integrity of a partition. All such openings must be caulked to make an air-tight joint if any degree of sound isolation is to be maintained.

27.18 Doors and Windows

As can be appreciated from the preceding section, doors and windows in very large measure determine the overall transmission loss of a wall. Since in almost every instance the material of which doors and windows are constructed is of a lower acoustic rating than that of the wall in which they are mounted, particular care must be taken not to further degrade performance with leaks.

(*a*) ***Doors.*** Table 27.5 gives average TL figures for common wooden doors and for doors arranged in a sound lock, that is, two doors with a sufficient space between to allow door swing. The important conclusions to be drawn from this table are:

1. Louvered doors and doors undercut to permit air movement are useless as sound barriers.
2. The most important step in soundproofing doors is complete sealing around the opening. The door in the closed position should exert pressure on these gaskets, making the joints airtight.
3. Doors constructed of two leaves separated by a sound-absorbing material, that is, steel sheets with fiber batting inside, are more effective than a single solid material door of the same weight.

Commercial products are available with STC rating up to 65 for application in sensitive locations.

(*b*) ***Windows.*** Windows are critically important to block exterior noise, and all the more so, since exterior wall construction is generally of high STC, making the window the deciding factor in the composite transmission loss. Sound leaks through cracks in operable windows will normally establish the windows' rating regardless of type of glazing. Fortunately, the attention being given of late

Table 27.5 ***Average Sound Transmission Loss of Doors***

Construction	*Average TL (db)*	*STC*
Louvered door	10	15
Any door, 2-in. undercut	12	17
$1\frac{1}{2}$ in. hollow core door, no gasketing	15	22
$1\frac{1}{2}$ in. hollow core door, gaskets and drop closure	20	25
$1\frac{3}{4}$ in. solid wood door, no gasketing	25	30
$1\frac{3}{4}$ in. solid wood door, gaskets and drop closure	30	35
2 hollow core doors, gasketed all around with sound lock	38	45
2 solid core doors, gasketed all around, with sound lock	48	55
Special commercial construction, with lead lining and full sealing		45–65

Table 27.6 ***Sound Transmission Class (STC) of Window Construction***

Window Construction	*STC*
Operable wood sash, $\frac{1}{8}$ in. glass, unsealed	25
Operable wood sash, $\frac{1}{4}$ in. glass, unsealed	25
Operable wood sash, $\frac{1}{4}$ in. glass, gasketed	30
Operable wood sash, laminated glass, unsealed	28
Operable wood sash, double glazed, $\frac{1}{8}$ in. panes, $\frac{3}{8}$ in. air space, gasketed	35
Fixed sash, double $\frac{1}{8}$ in. panes, 3-in. air space, gasketed	44
Fixed sash, double $\frac{1}{4}$ in. panes, 4-in. air space, gasketed	48

to the sealing of windows for thermal purposes has had a salutory effect on their acoustic properties. As with doors, the importance of proper gasketing and sealing cannot be overstressed. Double glazing is only effective when the two panes are separated by a wide air gap (see Table 27.6). A narrow air gap acts as a stiff spring between the panes, reducing the transmission loss and causing a low-frequency resonance dip at 200 to 300 Hz. Figure 27.24 shows GSA recommendations for residential window placements, which are obviously also applicable to nonresidential building spaces with the same criteria.

27.19 Diffraction

The physical process by which sound passes around obstructions and through very small openings is called diffraction. Simply stated, diffraction is a process whereby any point on a wave front establishes a new wave front when passing an obstacle. Thus, although much of a sound wave is blocked by a small opening, the portion that does get through establishes a new wave front. The *amplitude* of the diffracted wave is affected by the relationship between the size of the opening and the wavelength. Hence for a small hole, short wavelengths (high frequencies) are attenuated less than long wavelengths (low frequencies) (see Fig. 27.25*a*). When sound meets a linear barrier such as a wall or half-high partition, it diffracts over it with little loss, but the interference patterns established create sound "shadows" close to the wall (see Fig. 27.25*b*). Generally, therefore, only an integral, nonporous barrier can effectively attenuate sound.

27.20 Flanking

Just as sound will pass through the acoustically weakest part of a composite wall, so it will also find parallel or flanking paths, that is, an acoustic short circuit. Proper design of window locations to avoid flanking paths has already been shown in Figure 17.24. The same situation obtains with respect to doors and any other openings between spaces. Thus in Figure 27.26 a high STC wall between the two spaces is in large measure defeated by flanking paths F5, F6, and F7. In office spaces the most common flanking path is via the plenum, as in Figure 27.26, path F1, and Figure 27.27*b* and *d*. Ductwork with registers or grills in various rooms acts as an excellent intercom system unless completely lined with sound absorptive material (see Section 27.28). Even then, low-frequency sound is only minimally attenuated, and special measures must be employed if good transmission loss is required. This subject is discussed further below, under mechanical equipment noise control.

27.21 Noise Reduction and Background Noise

Referring back to Section 27.12, we saw that

$$NR = TL - 10 \log \frac{S}{A_R} \qquad (27.5)$$

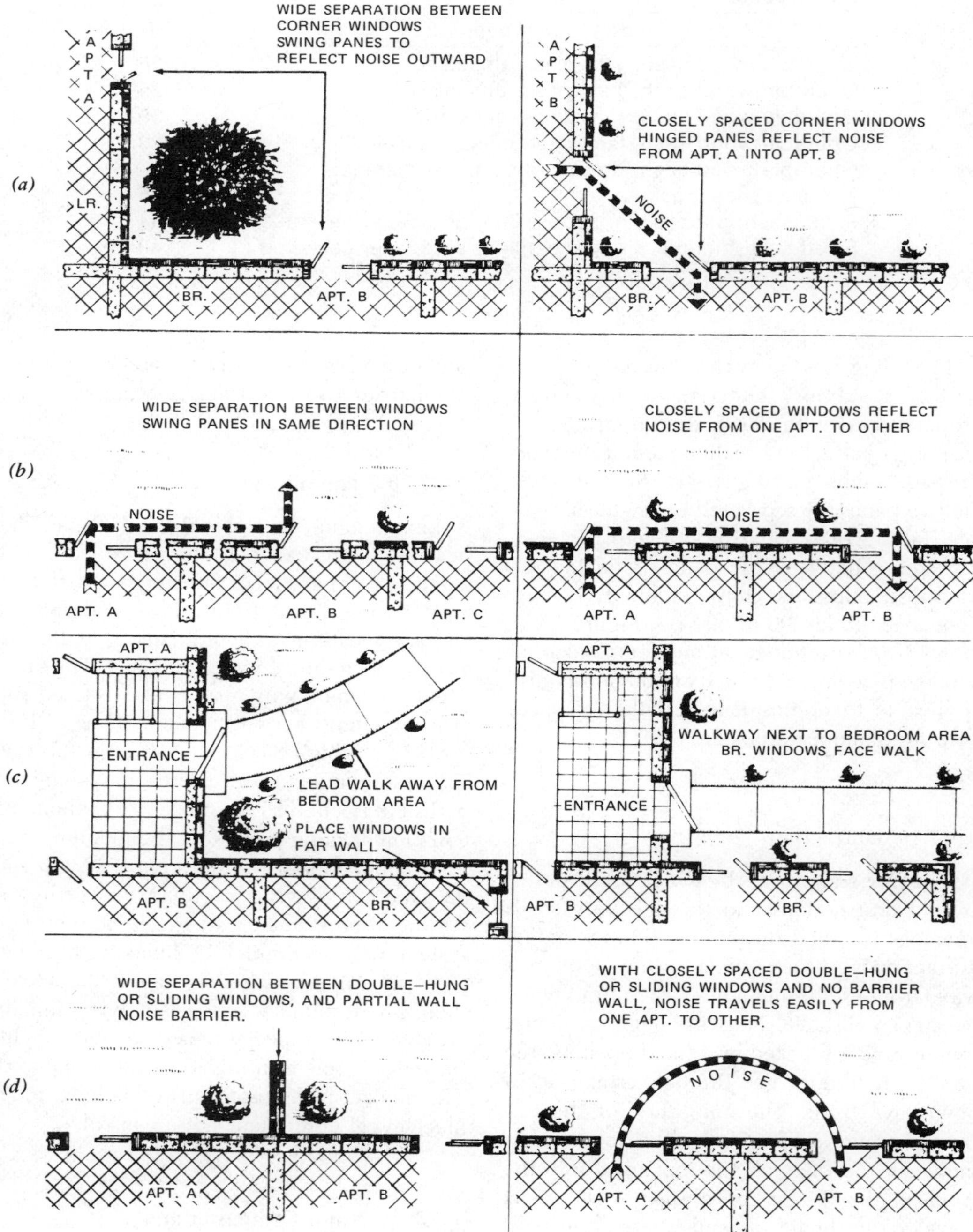

Fig. 27.24 *Recommended window placements. From* A Guide to Airborne, Impact and Structure-Borne Noise Control in Multi-Family Dwellings, *U.S. Dept. of Housing and Urban Development, Washington, D.C., 1963, pp. 8–53.*

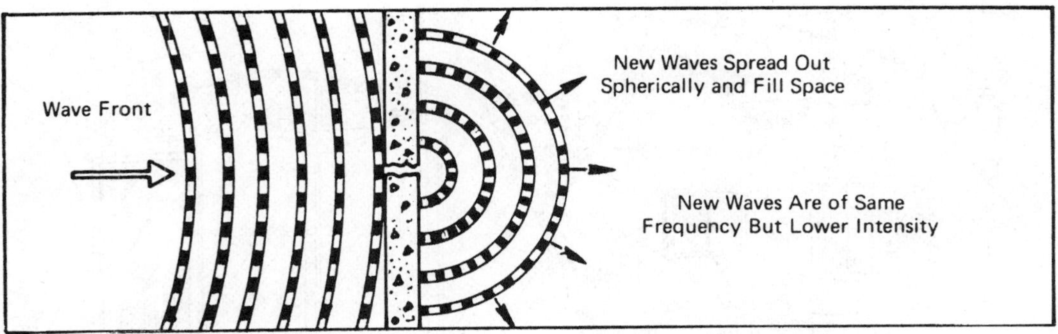

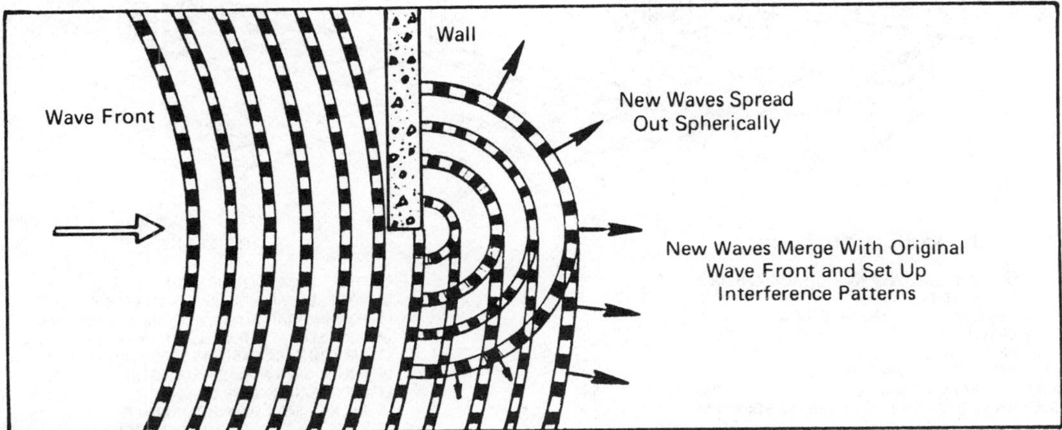

Fig. 27.25 (a) *Diffraction of a sound wave passing through a small hole.* (b) *Diffraction of a sound wave at the edge of a wall. From* Quieting: A Practical Guide to Noise Control, *National Bureau of Standard Handbook 119, July 1976.*

where *NR*, *TL*, *S*, and A_R are noise reduction, transmission loss, barrier area, and receiving-room absorption, respectively. In lieu of calculating, the following rule of thumb can be used:

a. For a live receiving room,
$NR = TL - 1$ db
b. For a medium receiving room,
$NR = TL + 4$ db
c. For a dead receiving room,
$NR = TL + 7$ db

Having established the noise reduction, we find the intensity level in the receiving room as simply

$$IL_2 = IL_1 - NR \qquad (27.4)$$

If the resulting IL_2 is below the background noise level, it will not be a source of annoyance; conversely, if it is greater than the background noise it will be heard and, depending on intensity and content, could be a source of disturbance. See Figure 27.28 for a graphic representation of this idea. With this in mind we can also tabulate the performance of a sound barrier, in terms of receiving-room background levels (see Table 27.7). Stated otherwise, the apparent isolation provided by a barrier may be greater than the actual noise reduction. Figure 27.29 shows two conditions of adjacent spaces. Although the source room level is uniform and partitions on both sides of the source room are identical, the background noise in the two receiving rooms is different. In A, the background is NC-35; in B, it is NC-25. The occupant of room A hears nothing

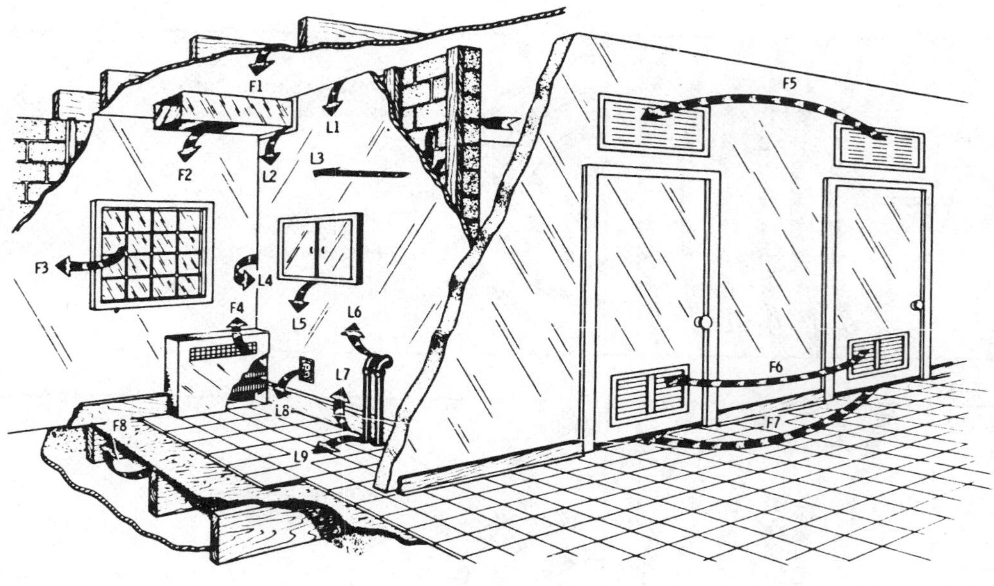

FLANKING NOISE PATHS

F1 OPEN PLENUMS OVER WALLS, FALSE CEILINGS
F2 UNBAFFLED DUCT RUNS
F3 OUTDOOR PATH, WINDOW TO WINDOW
F4 CONTINUOUS UNBAFFLED INDUCTOR UNITS
F5 HALL PATH, OPEN VENTS
F6 HALL PATH, LOUVERED DOORS
F7 HALL PATH, OPENINGS UNDER DOORS
F8 OPEN TROUGHS IN FLOOR-CEILING STRUCTURE

NOISE LEAKS

L1 POOR SEAL AT CEILING EDGES
L2 POOR SEAL AROUND DUCT PENETRATIONS
L3 POOR MORTAR JOINTS, POROUS MASONRY BLK
L4 POOR SEAL AT SIDEWALL, FILLER PANEL ETC.
L5 BACK TO BACK CABINETS, POOR WORKMANSHIP
L6 HOLES, GAPS AT WALL PENETRATIONS
L7 POOR SEAL AT FLOOR EDGES
L8 BACK TO BACK ELECTRICAL OUTLETS
L9 HOLES, GAPS AT FLOOR PENETRATIONS

OTHER POINTS TO CONSIDER, RE: LEAKS ARE (A) BATTEN STRIP A/O POST CONNECTIONS OF PREFABRICATED WALLS, (B) UNDER FLOOR PIPE OR SERVICE CHASES, (C) RECESSED, SPANNING LIGHT FIXTURES, (D) CEILING & FLOOR COVER PLATES OF MOVABLE WALLS, (E) UNSUPPORTED A/O UNBACKED WALL BOARD JOINTS (F) EDGES & BACKING OF BUILT-IN CABINETS & APPLIANCES, (G) PREFABRICATED, HOLLOW METAL, EXTERIOR CURTAIN WALLS.

Fig. 27.26 *Flanking transmission of airborne noise. From* A Guide to Airborne, Impact and Structure-Borne Noise Control in Multi-Family Dwellings, *U.S. Dept. of Housing and Urban Development, Washington, D.C., 1963.*

from the source room. The occupant of room B hears clearly. Occupant A probably will praise the partition while occupant B will complain. Although the level of reradiated sound is identical in the two receiving rooms, the intruding signal is masked by the background in A and it is audible in B. Thus, the apparent noise reduction is substantially higher in A than B. Section 27.22 investigates this phenomenon of speech privacy.

Sound isolation may be achieved very economically by careful planning. Storage and circulation can serve as buffers for noise-sensitive areas. Physical separation of noisy areas from quiet ones often eliminates the need for complicated and expensive multiple structures.

27.22 Speech Privacy

The subject of speech privacy received considerable additional study and emphasis with the advent of open-plan offices (office landscaping), although the same problem prevails in all spaces. Essentially the purpose of

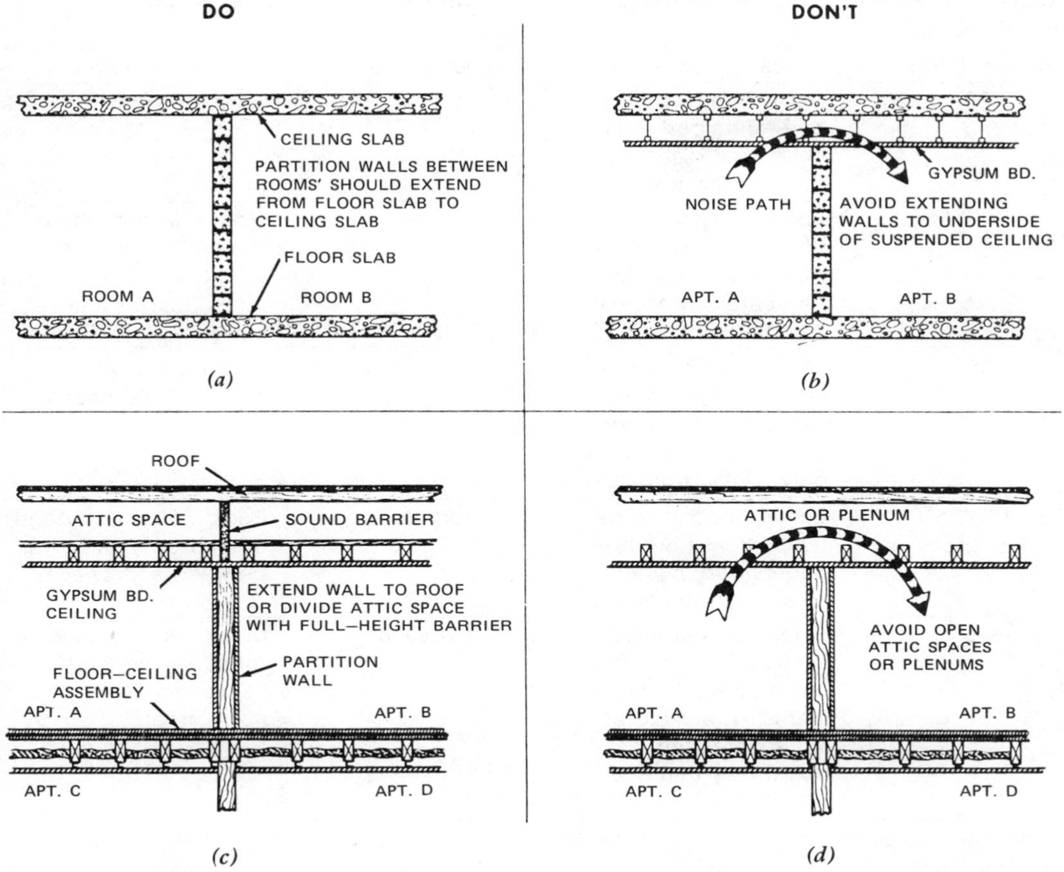

DO
DON'T
CEILING SLAB
PARTITION WALLS BETWEEN ROOMS' SHOULD EXTEND FROM FLOOR SLAB TO CEILING SLAB
FLOOR SLAB
ROOM A
ROOM B
(a)
GYPSUM BD.
NOISE PATH
AVOID EXTENDING WALLS TO UNDERSIDE OF SUSPENDED CEILING
APT. A
APT. B
(b)
ROOF
ATTIC SPACE
SOUND BARRIER
GYPSUM BD. CEILING
EXTEND WALL TO ROOF OR DIVIDE ATTIC SPACE WITH FULL-HEIGHT BARRIER
FLOOR–CEILING ASSEMBLY
PARTITION WALL
APT. A
APT. B
APT. C
APT. D
(c)
ATTIC OR PLENUM
AVOID OPEN ATTIC SPACES OR PLENUMS
APT. A
APT. B
APT. C
APT. D
(d)

Fig. 27.27 Construction techniques to avoid flanking paths. From A Guide to Airborne, Impact and Structure-Borne Noise Control in Multi-Family Dwellings, *U.S. Dept. of Housing and Urban Development, Washington, D.C., 1963.*

$$\text{Noise reduction, } NR = IL_1 - IL_2 = TL - 10 \text{ LOG } \frac{S}{A_2}$$

IL1
Source room level
Background noise level
IL2
Transmitted level

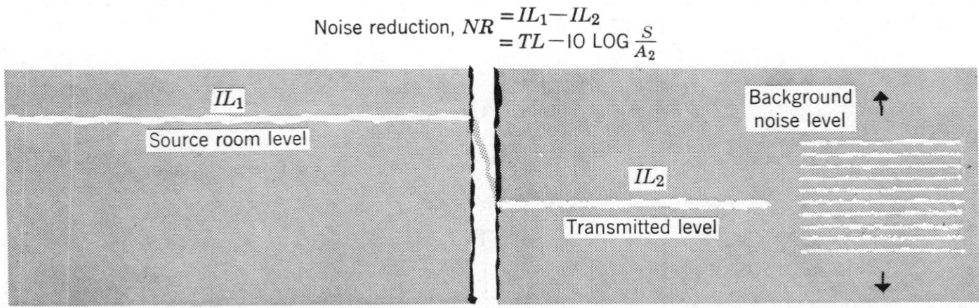

Fig. 27.28 The background noise level determines whether the transmitted sound will actually be heard.

Table 27.7 *Relations Between Barrier STC Rating and Hearing Condition on the Receiving Side*

Barrier STC[a]	*Hearing Condition*	*Description*	*Application*
25	Normal speech can be understood quite easily and distinctly through the wall.	Poor	Space divider
30	Loud speech can be understood fairly well. Normal speech can be heard but not easily understood.	Fair	Room divider where concentration not essential
35	Loud speech can be heard, but is not easily intelligible. Normal speech can be heard only faintly, if at all.	Good	Suitable for offices next to quiet spaces
42–45	Loud speech can be faintly heard but not understood. Normal speech is inaudible.	Very good	For dividing noisy and quiet areas; party wall between apartments
46–50	Very loud sounds, such as loud singing, brass musical instruments, or a radio at full volume can be heard only faintly or not at all.	Excellent	Music room, practice room, sound studio, bedrooms adjacent to noisy areas

[a]Assuming a background noise level of NC-25. With higher background, e.g., NC-35, STC can be degraded one step.

the investigations was to determine the factors affecting speech privacy and to quantify them with a degree of accuracy sufficient for design purposes. These studies indicate that there are six factors involved, which can be subsumed under two headings:

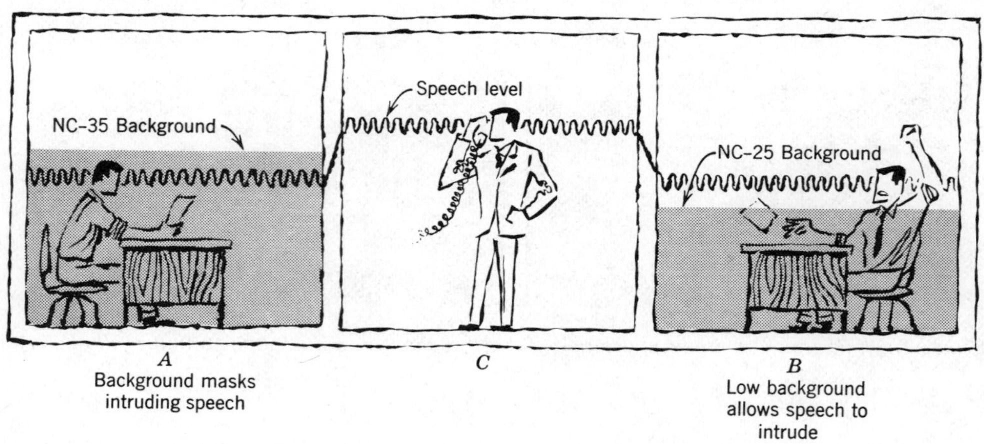

Fig. 27.29 *The occupant in room A with background noise NC-35 (≈ 45 dbA) is unaware of the noise that is so disturbing to occupant B whose NC-25 (≈ 36 dbA) is insufficient to mask C's loud speech.*

A. Speech Rating (Source Room, No. 1)
1. Speech effort—a measure of the loudness of speech.
2. Source room floor area—gives the approximate effect of room absorption.
3. Privacy allowance—What is the measure of privacy required?

B. Isolation Rating (Receiving Room, No. 2)
4. STC—rating of the barrier.
5. Noise reduction factor A_2/S (see Section 27.21); an indication of receiving-room absorption, that is, the difference between *NR* and *TL;* A_2 is area of receiving room; S the area of the barrier between rooms. Absorption is assumed to be average. For live rooms raise this factor two points; for a dead room lower two points.
6. Background noise level—in receiving room.

An analysis sheet for enclosed spaces is provided in Figure 27.30. See also References 24 and 25. The two examples of this analysis which follow should clarify its use.

EXAMPLE 27.3

(a) Source room: General clerical office $40 \times 60 \times 9$ ft
16-ft partition, STC 40

(b) Receiving room: Conference rooms, 16×24 ft
Background noise level 40 dbA (NC-30)

Privacy Analysis:

(a) 1. Speech effort: raised	66
2. Area A_1	0
3. Privacy—normal	9
Speech rating:	75
(b) 4. STC	40
5. A_2/S[a]	2
6. Noise level	40
Isolation rating:	82
$a - b =$	-7

Therefore the STC rating of the partition can be reduced to 33 without affecting speech privacy.

EXAMPLE 27.4

(a) Source Room: Drafting room 20×30 ft
Common Wall 12×8 ft high, STC: 26 ($\frac{1}{2}$ glass, with door)

(b) Receiving room: Supervisor's office $12 \times 14 \times 8$ ft
Background noise level 35 dbA

Privacy Analysis:

(a) Speech effort—conversational	60
Source room area A_1	2
Privacy—confidential	13
	75
(b) STC	26
A_2/S[b]	2
Noise level	35
	63
$a - b =$	12
or strong dissatisfaction	

The suggested corrections here are to increase STC to 36 by gasketing the door and increasing background noise level in the receiving room to 40 dbA (NC-30). This should result in a satisfactory condition. If still unsatisfactory, the glazing could be doubled.

A form similar to Figure 27.30 is available for open plan offices. We have not included it because these spaces require critically accurate design best left to an acoustic expert. In these open spaces where partial height partitions can provide a *maximum* of 15 db isolation, masking noise of fairly high level must be introduced. Too high a level will result in the noise itself being disturbing; too low a level results in speech interference, extreme dissatisfaction and impediments to work. Since in

[a] $\frac{16 \times 24}{16 \times 9} = 2.6$, corresponding to 2.0, see chart.

[b] $\frac{12 \times 14}{12 \times 8} = 1.8$, corresponds to 1.6; use 2.0 since system uses whole numbers only.

Anticipated response to privacy situation

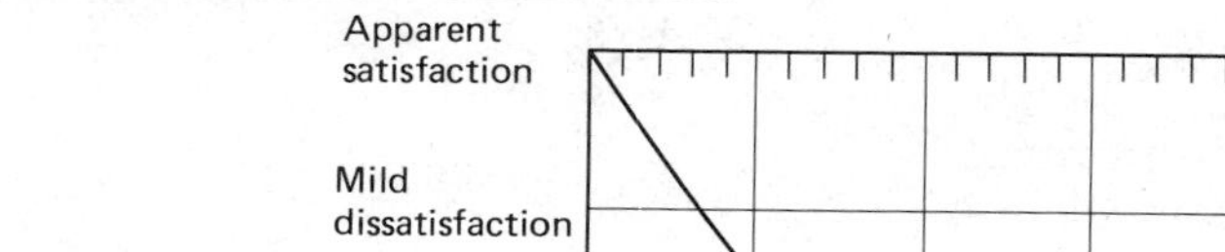
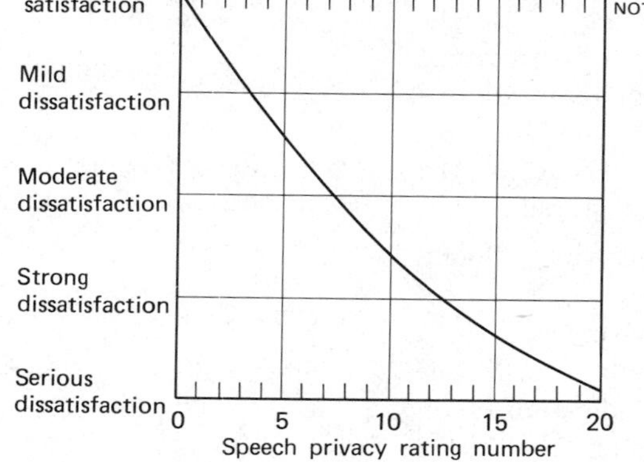

Speech rating

1. Speech effort — from source room
   Loud Raised Conversational
   72 66 60 ____

2. Source room floor area (A_1)— effect of source room absorption
   125 250 500 1000 (sq ft)
   10 6 3 0 ____

3. Privacy allowance — degree of privacy required
   Confidential Normal
   15 9 ____

Isolation rating

■ Speech rating total ____

4. Sound transmission class (STC)— common barrier ____

5. Noise reduction factor (A_2/S)— effect of receiving room absorption and barrier size

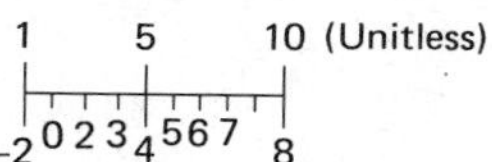

____

6. Adjacent room background noise level (dBA)— masking sound available ____

■ Isolation rating total ____

Speech privacy rating number ____

Find speech privacy rating number by subtracting isolation rating total from speech rating total. Then use graph at top of sheet to predict degree of satisfaction. ____

Fig. 27.30 Speech privacy analysis sheet. Based on "Speech Privacy in Buildings," by W. J. Cavanaugh, W. R. Farrell, P. W. Hirtle, and B. G. Wattern, J. Acoust. Soc. Amer., Vol. 38, No. 4, October 1965.

such areas, control of background noises such as office machines is extremely difficult, success depends in large measure on proper physical placement of noise sources and grouping of common acoustic need areas. The indiscriminate reliance on noise generators is to be discouraged (see Fig. 27.31). Generally speaking, open areas should use barriers at least 5 × 8 ft of very high absorbency; acoustic ceilings no higher than 9 ft with a minimum absorption coefficient of 0.7, backed by a plenum of at least 30-in. depth. The partial height barriers must be set so that the listener is shielded from line-of-sight noise, either direct or on first reflection. Obviously the taller the barrier, the better.

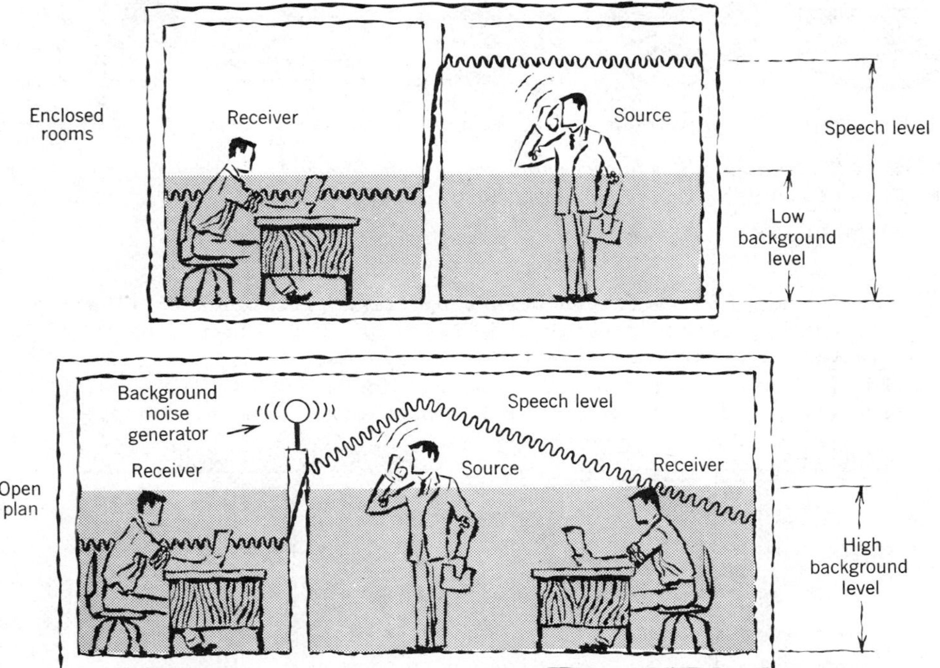

Fig. 27.31 *The open-plan background noise level must be high enough to mask adjacent speech but not so loud as to constitute a disturbance itself.*

STRUCTURE-BORNE SOUND

27.23 Structure-Borne Impact Noise

As stated in Section 27.11, structure-borne noise is at least as serious a problem as airborne noise:

(*a*) There is no air cushion between the source and the structure. Thus high-intensity energy is introduced into the structure, through which it travels with minimum attenuation and at great speed.

(*b*) Sound, once introduced into the structure, is attenuated well only by discontinuities in the structure. Since the structure must have integrity to carry the loads, discontinuities of the type that will stop noise are complex and expensive.

(*c*) The entire structure constitutes a network of parallel paths for sound. Therefore, partial solutions are useless, since sound will find flanking paths. The entire structure must be sound-proofed to yield good results.

(*d*) Unlike the case of airborne noise, additional mass does not usually alleviate floor-borne noise, particularly in long spans where the floor acts as a diaphragm.

(*e*) The increasing use of exposed structural ceiling eliminates the attenuation that can be introduced into a plenum above a hung ceiling. This is particularly bad, since most structure-borne noise is carried by floor structures (rather than walls), which radiate sound up and down.

The discussion that follows will be limited to impact noise. Refer to Section 27.27 for a brief treatment of vibration that is felt rather than heard, and is in effect a very low-frequency noise. Many of the recommendations and techniques that will minimize impact noise will also reduce vibration.

27.24 Control of Impact Noise

Impact noise problems can be controlled in

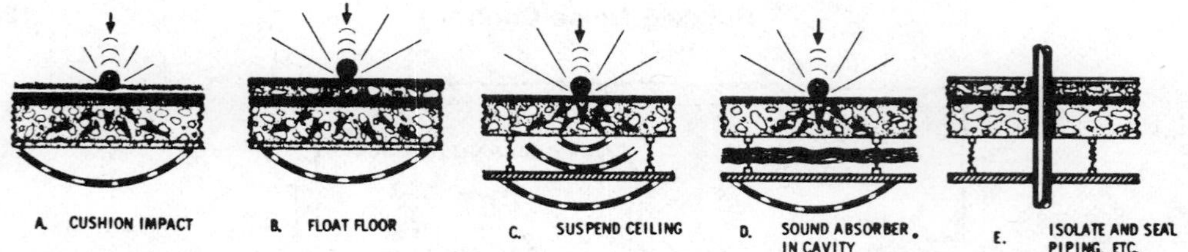

Fig. 27.32 Methods of controlling impact sound transmission through floors. From Quieting: A Practical Guide to Noise Control, *National Bureau of Standard Handbook 119, July 1976, p. 50.*

Caution must be exercised when supporting partitions on floating floors to prevent structural failures or short circuiting of the floating element, as illustrated.
For proper installation of floating floors and partition walls see other illustrations of such constructions in this chapter.

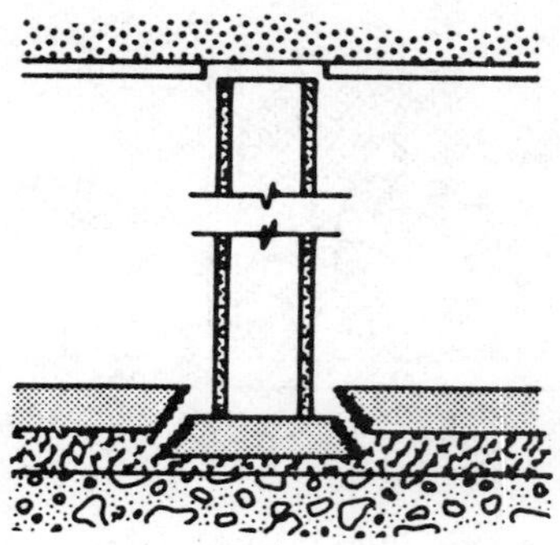

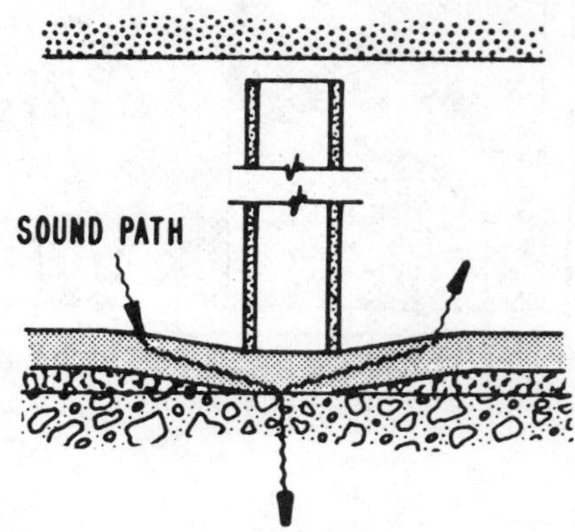

Fig. 27.33 Problems associated with floating floors. From A Guide to Airborne, Impact and Structure-Borne Noise Control in Multi-Family Dwellings, *U.S. Dept. of Housing and Urban Development, Washington, D.C., 1963, pp. 8–18.*

two ways—by preventing or minimizing the impact, and by attenuating it once it has occurred. Since we are concerned with structure, the latter problem will be discussed first. The former problem is covered in Section 27.26. Impact on floors is more serious than wall impact because the latter will be partially attenuated at the wall/floor joint whereas the former is introduced directly into the building framework (see Fig. 27.32).

(*a*) ***Cushion the Impact.*** This obvious solution will frequently eliminate all but severe problems. The resilient materials in common use are floor tiles of vinyl, rubber, and cork, or carpeting on pads, in ascending order of impact insulation. See Section 27.25 and Appendix B.

(*b*) ***Float the Floor.*** Since the key to elimination of structure-borne sound is *isolation,* separating the impacted floor from the structural floor by a resilient element is extremely effective. This element can be rubber or mineral wool pads, or blankets, or special spring metal sleepers. The effectiveness depends on the mass of the floating floor, compliance of the resilient support, and degree of isolation of the floating floor. This last element is extremely important, since flanking paths via end contacts with walls can short-circuit the floating element's sound impedance and defeat the system. With floating floors it is important that:

1. Mass of the floating floor be large enough to properly spread the loads. Otherwise, the pad will compress and deform sufficiently to transmit the impact.
2. Total construction must be airtight. Airtight is soundtight.
3. Particular care be exercised where par-

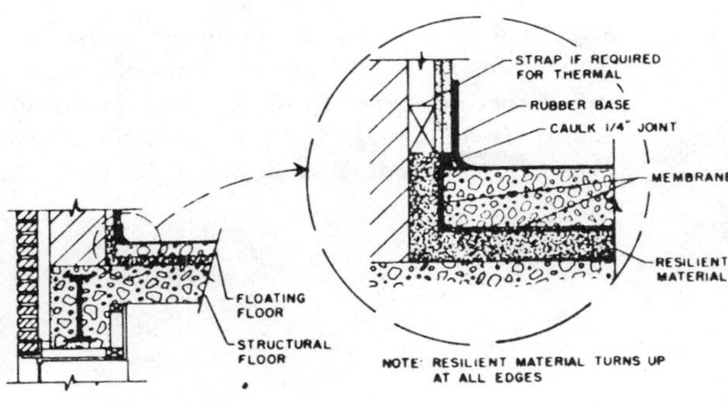

Fig. 27.34 *Constructional detailing of a floating floor. From* A Guide to Airborne, Impact and Structure-Borne Noise Control in Multi-Family Dwellings, *U.S. Dept. of Housing and Urban Development, Washington, D.C., 1963, pp. 8–18.*

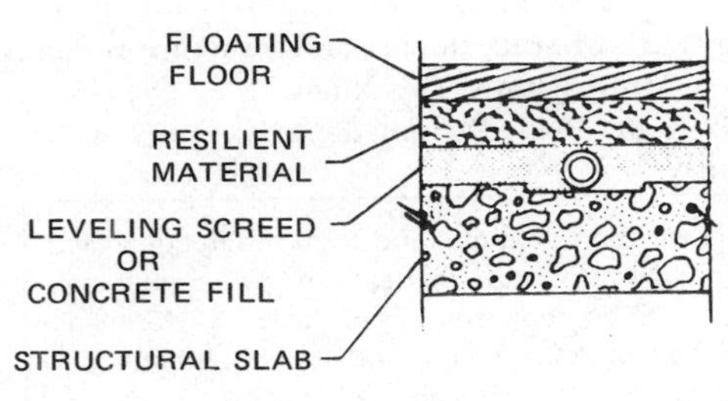

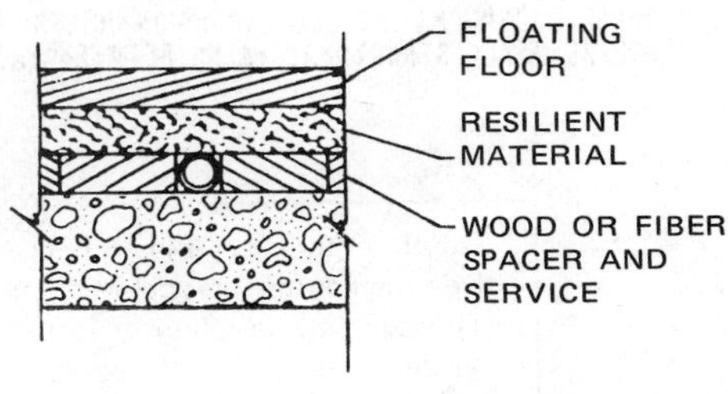

(a)

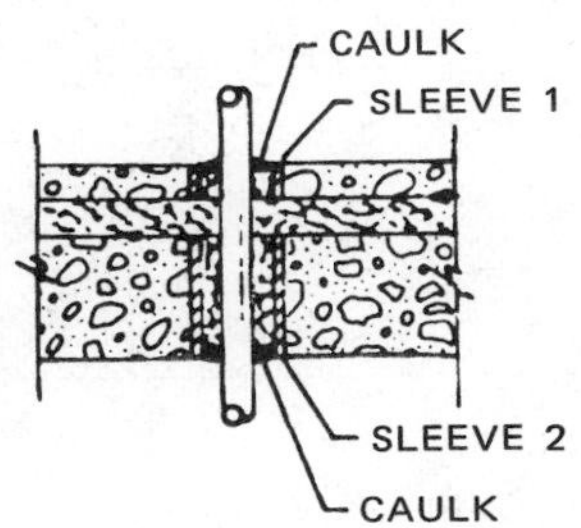

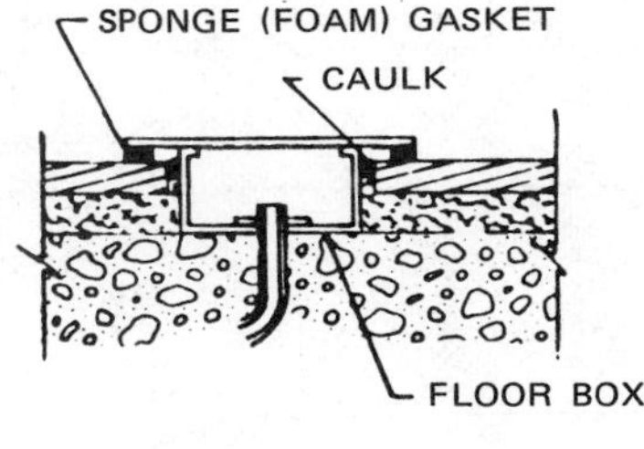

(b)

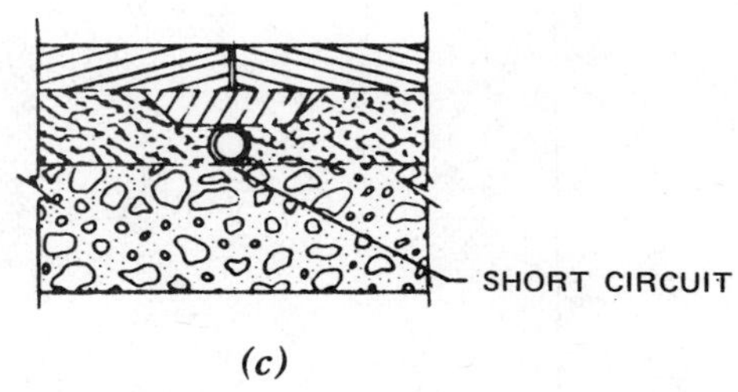

(c)

Fig. 27.35 (a) *Acceptable arrangement under floating floors.* (b) *Acceptable arrangement through floating floors.* (c) *Unacceptable arrangement under floating floor, since it provides a rigid sound path, thus "short circuiting" the floor. From* A Guide to Airborne, Impact and Structure-Borne Noise Control in Multi-Family Dwellings, *U.S. Dept. of Housing and Urban Development, Washington, D.C., 1963, pp. 8–16.*

titions rest on the floating floor (see Fig. 27.33).

4. Short circuits at walls (Fig. 27.34) or by penetrations be avoided (Fig. 27.35); see also Figure 27.11*b*.
5. Construction throughout must be consistent. Mixed-construction types invite flanking noise paths. (see Fig. 27.36).

(c) ***Suspend the Ceiling—and Use Ab-***

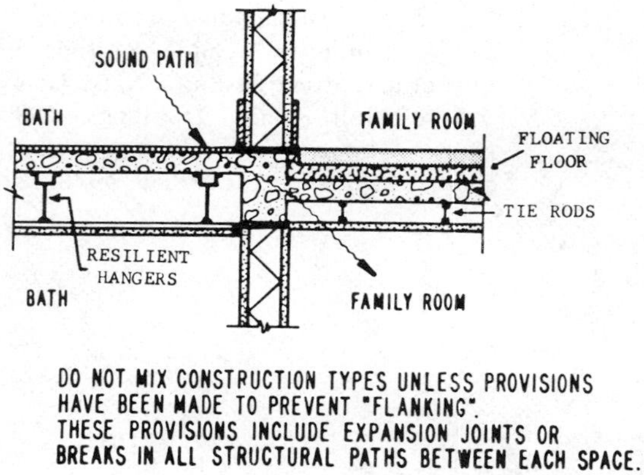

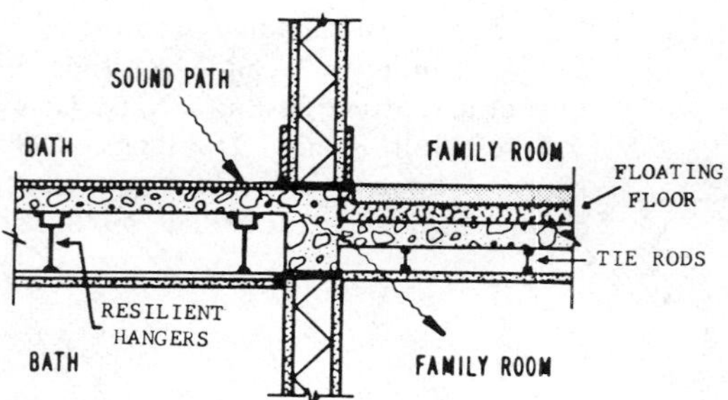

SOUND PATH
BATH
FAMILY ROOM
FLOATING FLOOR
TIE RODS
RESILIENT HANGERS
BATH
FAMILY ROOM
DO NOT MIX CONSTRUCTION TYPES UNLESS PROVISIONS HAVE BEEN MADE TO PREVENT "FLANKING".
THESE PROVISIONS INCLUDE EXPANSION JOINTS OR BREAKS IN ALL STRUCTURAL PATHS BETWEEN EACH SPACE.

Fig. 27.36 Flanking paths in mixed-construction types of floors. From A Guide to Airborne, Impact and Structure-Borne Noise Control in Multi-Family Dwellings, *U.S. Dept. of Housing and Urban Development, Washington, D.C., 1963, pp. 8–33.*

Fig. 27.37 Overlay from which impact isolation class can be determined graphically. From A Guide to Airborne, Impact and Structure-Borne Noise in Multi-Family Dwellings, *U.S. Dept. of Housing and Urban Development, Washington, D.C., 1963.*

Use of contour: Lay the contour over the isolation data plot and align the frequencies. Slide the contour vertically until it is at the lowest position that satisfies two conditions: (1) Maximum deviation of test curve over IIC contour (solid) is 8 db and (2) Sum of deviations at 16–½ octave frequencies is no more than 32 db. At that point, IIC rating is read on overlay corresponding to 60 db on the test curve abcissa (x–axis).
80
75
70
65
60
55
50
45
40
35
30
25
20
IIC
IIC
Contour
8 db
100 160 250 400 630 1K 1600 2500 4K 6300
1 2 3 4 5 6 7 8 9 10 11 12 13 14 15 16
16–Center frequencies.
125 200 315 500 800 1250 2K 3150 5K
1/3 Octave–band center frequency, hertz

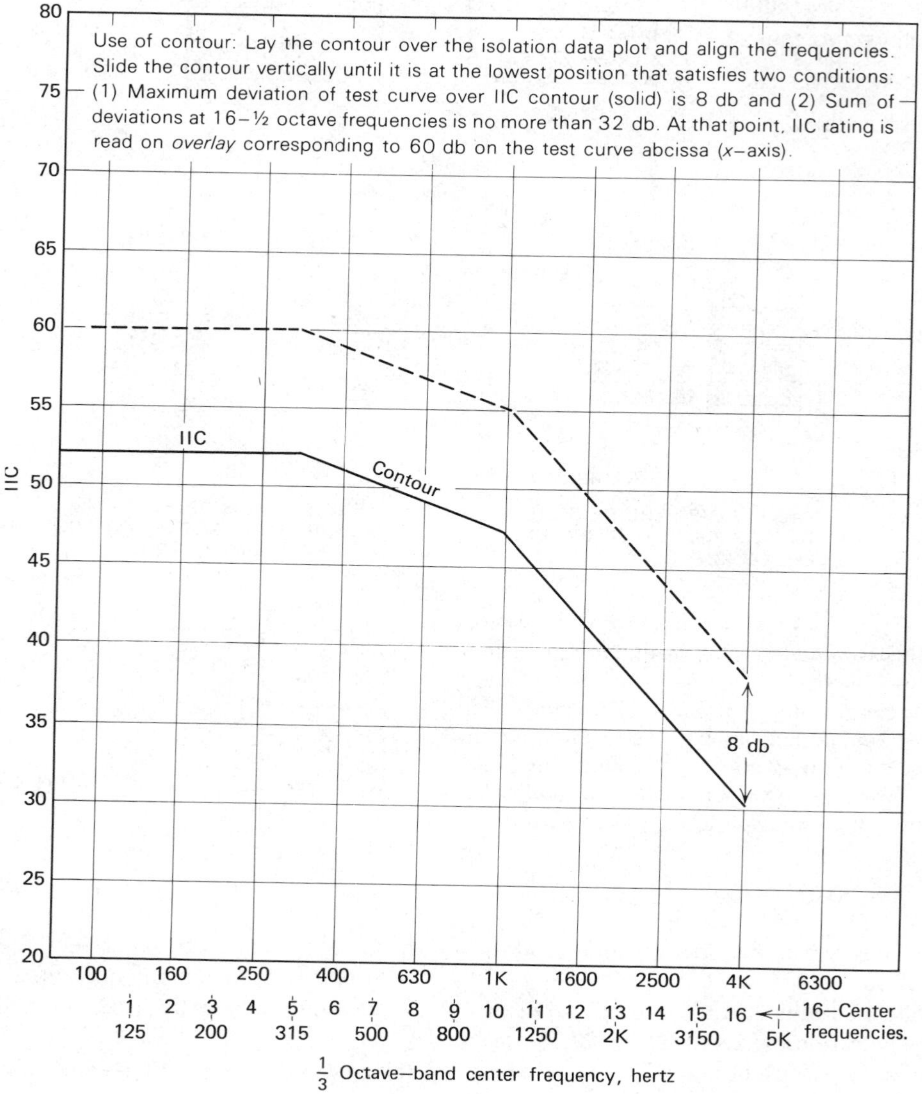

sorber in Cavity. As stated, the most disturbing noise is radiated down from the ceiling. A flexibly suspended ceiling with an acoustic absorbent layer suspended in it can be very effective if not flanked by paths leading into the walls and from there reradiating into the space below. It is imperative that the entire floor slab above be decoupled from the walls below by resilient separators.

(*d*) ***Isolate all Piping.*** All rigid structures such as piping must be isolated so as not to form a flanking path, and caulked with resilient sealing so as not to constitute an air-sound leak (see Fig. 27.36).

27.25 Impact Isolation Class, IIC

This is a single-number, impact isolation rating for floor construction, similar in intent and derivation to STC wall ratings. Tests are made with a standard tapping machine and the noise levels measured in 1/3 octave bands. These are plotted and compared to a standard contour, exactly as with sound transmission class. Figure 27.37 gives the standard contour. Figure 27.38 shows the IIC contours corresponding to the FHA grades of construction. The significance of these grades, as well as that of the STC grades, is given in Section 27.35. Details of typical floor constructions along with IIC ratings are given in Figure 27.39 and Appendix B. Resilient floor finishes on any of the floor constructions not provided with them will add to the IIC ratings approximately as follows:

1/16 in. vinyl tile	0
1/18 in. linoleum or rubber tile	4 ± 1
1/4 in. cork tile	10 ± 2
Low-pile carpet on fiber pad	12 ± 2
Low-pile carpet on foam rubber pad	18 ± 3
High-pile carpet on foam rubber pad	24 ± 3

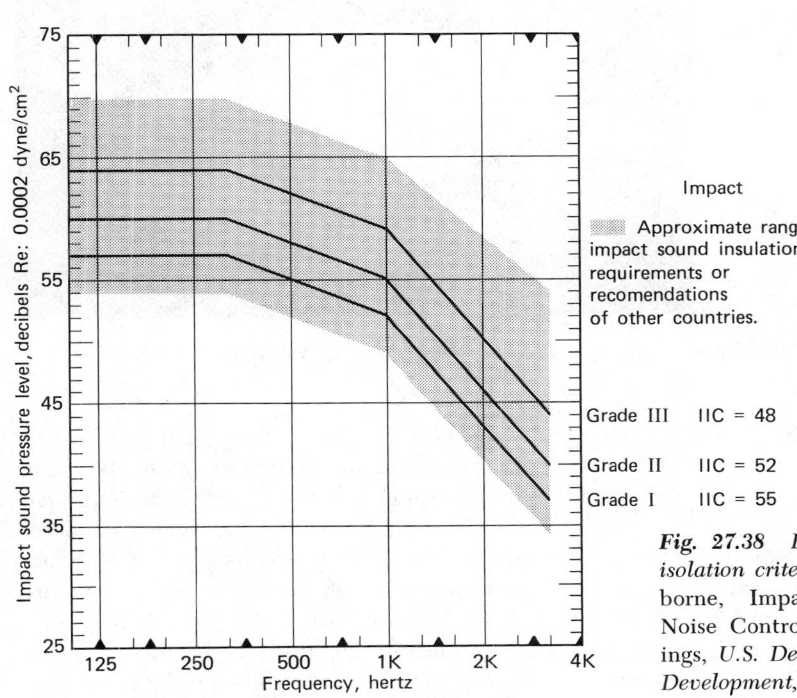

Fig. 27.38 *FHA recommended impact isolation criteria. From* A Guide to Airborne, Impact and Structure-Borne Noise Control in Multi-Family Dwellings, *U.S. Dept. of Housing and Urban Development, Washington, D.C., 1963, pp. 10–11.*

A. Wood Floors (all on 2″ x 10″ joists)
STC•53, IIC•73 Carpet and pad; particle board surface; plywood subfloor; single layer ½″ type x gypsum ceiling on resilient channel; one thickness R-11 Fiberglas insulation
STC•43, IIC•60 Carpet and pad; particle board surface; plywood subfloor; single layer ½″ type x gypsum ceiling attached directly to joists; one thickness R-11 Fiberglas insulation
STC•42, IIC•60 Carpet and pad; particle board surface; plywood subfloor; single layer ½″ type x gypsum ceiling attached directly to joists; no insulation

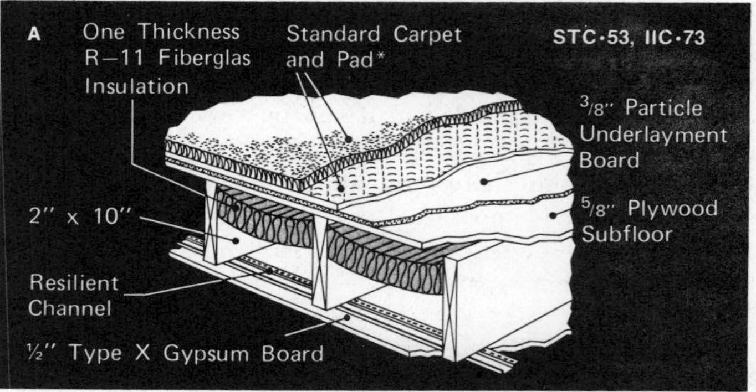

B. Lightweight (Cellular) Concrete Floors (1½″ thick over ⅝″ plywood subfloor on 2″ x 10″ joists)
STC•58, IIC•74 Carpet and pad; single layer ½″ type x gypsum ceiling attached to joists by resilient channel; one thickness R-11 Fiberglas insulation
STC•47, IIC•59 Carpet and pad; single layer ½″ type x gypsum ceiling attached directly to joists; no insulation

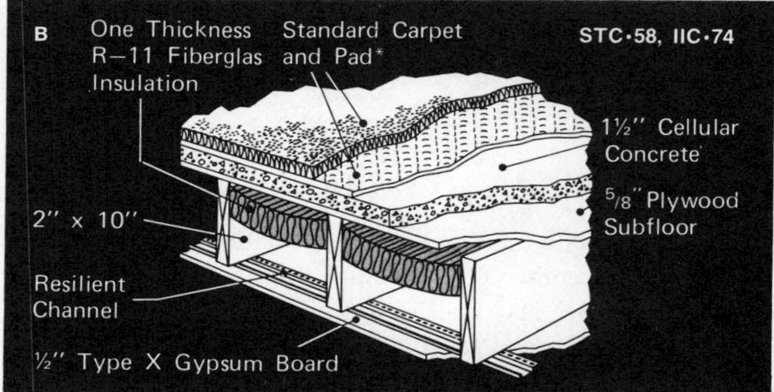

C. Steel Joist (7¼″ x 18 gage, @ 24″ o.c.) STC•56, IIC•71 Carpet and pad, ¾″ T & G plywood sub floor. ⅝″ type x gypsum board attached to ceiling joists by resilient channel; one thickness R-11 Fiberglas insulation
STC•43, IIC•57 Carpet and pad; ¾″ T & G plywood sub floor, ⅝″ type x gypsum board attached directly to the joists; no insulation

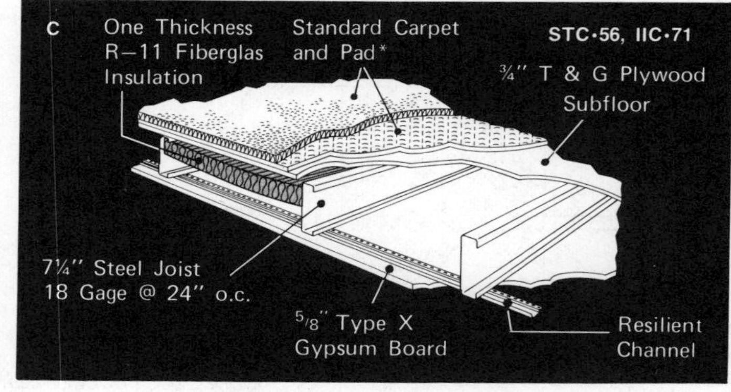

*The carpet and pad were selected to represent the type and quality commonly specified for new construction and was 20-oz. textured loop nylon carpet backed by a 40-oz. all-hair pad.

Fig. 27.39 *Typical floor/ceiling constructions. Courtesy of Owens-Corning Fiberglas.*

MECHANICAL SYSTEM NOISE CONTROL

27.26 Mechanical Noise Sources

Mechanical devices obviously make noise. And, generally, the more power they consume, the more noise they make. In many of today's buildings, 40% of the total cost is spent on mechanical systems. These systems are located throughout a building.

In most buildings, the primary sources of mechanical noise are the components of the air-conditioning and air-handling systems such as fans, compressors, cooling towers, condensers, ductwork, dampers, mixing boxes, induction units, and diffusers. The curve of Figure 27.40 depicts typical air-handling system noise and indicates the portions of the spectrum each group of components produces.

Pumps are another source of mechanical

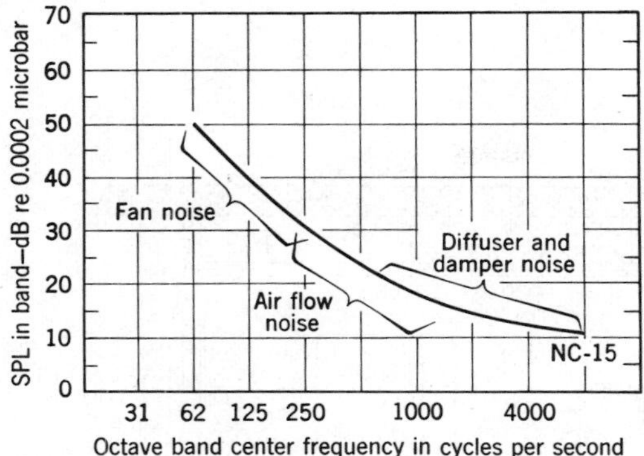

Fig. 27.40 HVAC system noise.

noise. Pump noise is frequently transmitted along pipes to remote points. Pumps for water-pressure boost systems may be located outside of mechanical rooms and be overlooked until problems occur in the completed building.

Elevators, escalators, and freight elevators also introduce mechanical noise into buildings. Escalators and freight elevators pose few problems, since they are localized in a specific area and have low-operation speeds. However, elevator-car operation is rapid and it affects large areas. In addition, the motors and switchgear are located on or above the prime upper floors of a building. Motor, shaftway, and switching noise must be properly controlled to prevent annoyance to building tenants located near the shaftways or mechanical penthouses. Vibration isolation of these major components is a specialized problem beyond the scope of this text.

27.27 Quieting of Machines

Machines cause noise by vibration. This noise is imparted directly to the surrounding air and by vibrational contact to the surrounding structure. Therefore, there are three ways to reduce this noise:

a. Reduce the vibration itself.
b. Reduce the airborne noise by decoupling the vibration from efficient radiating sources.
c. Decouple the vibrating source from the structure.

Refer to Figure 27.41: Items 1, 3, and 4 reduce vibration; items 4, 5, 6, and 7 reduce and decouple the vibration from the radiating

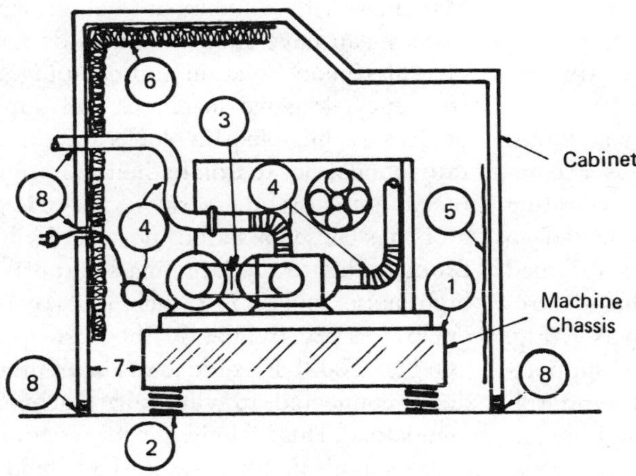

Fig. 27.41 Techniques to reduce the generation of airborne and structure-borne noise in machines and appliances. From Quieting: A Practical Guide to Noise Control, *National Bureau of Standard Handbook 119, July 1976, p. 25.*

1. Install motors, pumps, fans, etc. on most massive part of the machine.
2. Install such components on resilient mounts or vibration isolators.
3. Use belt drive or roller drive systems in place of gear trains.
4. Use flexible hoses and wiring instead of rigid piping and stiff wiring.
5. Apply vibration damping materials to surfaces undergoing most vibration.
6. Install acoustical lining to reduce noise buildup inside machine.
7. Minimize mechanical contact between the cabinet and the machine chassis.
8. Seal openings at the base and other parts of the cabinet to prevent noise leakage.

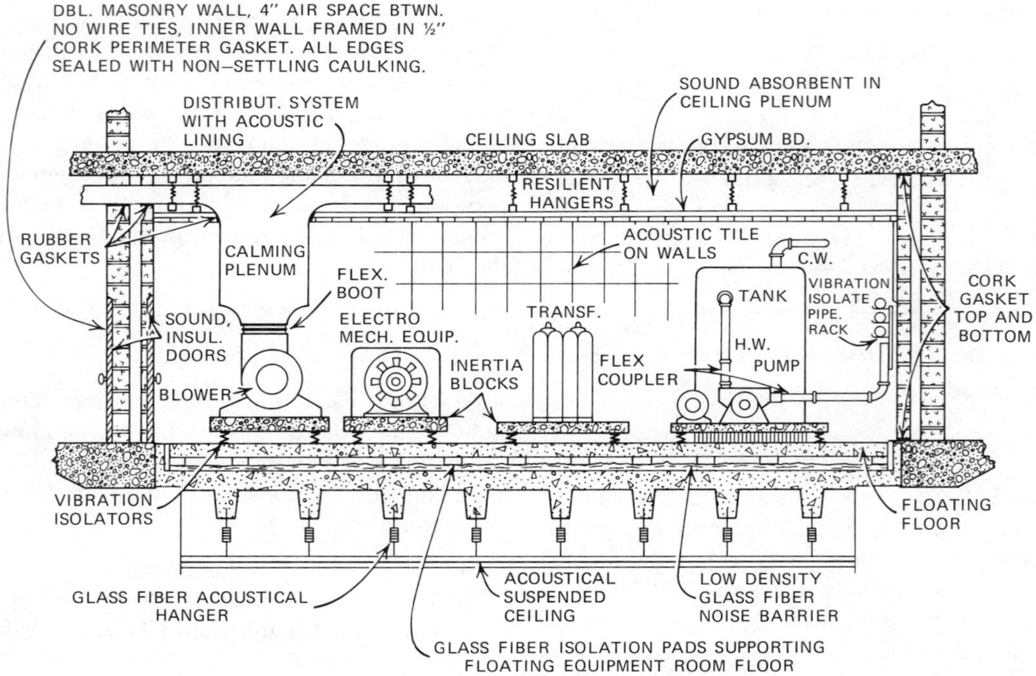

Fig. 27.42 *Sound-proofing a mechanical equipment room. From* A Guide to Airborne, Impact and Structure-Borne Noise Control in Multi-Family Dwellings, *U.S. Dept. of Housing and Urban Development, Washington, D.C., 1963.*

cabinet; and items 2 and 8 decouple the vibrating source from the structure. Once the noise becomes airborne or structure-borne, the isolation techniques studied above are employed.

Vibration reduction takes two forms, damping and isolation. Damping is accomplished by rigidly coupling the vibrating source to a large mass, frequently called an inertia block. Much of the energy is absorbed and dissipated as friction; the remainder results in lower-amplitude vibration (see Fig. 27.42). *Isolation* is accomplished by supporting the vibrating mass on resilient supports. These take many forms and are used in tandem. Thus machines are supported on fibrous, rubber, or spring steel vibration isolators, and the entire mass can be supported on a floating floor, which in turn rests on resilient vibration isolators as in Figure 27.42. Large machines are supported on special commerical "sandwiches" of asbestos, lead, cork, and other strong resilient materials. Piping is supported on cork pads and hung on resilient hangers. Machines with a dominant vibrational frequency can have special springs designed to give maximum isolation and damping at that frequency. Massive machines and impacting devices use huge inertia blocks and even separate foundations to isolate their vibration. The recent emphasis on emergency electric generators has caused a serious noise and vibration problem due to the large mass and extremely high noise levels. For such units complete enclosures are frequently the best approach (see Fig. 27.43). Flexible joints in all pipes and ducts connected to vibrating machines are mandatory. This includes flexible conduit connectors to all motors, transformers, and lighting fixtures using ballasts.

27.28 Duct-System Noise Reduction

Design of quiet duct-system operation entails more than specification of duct lining. Air

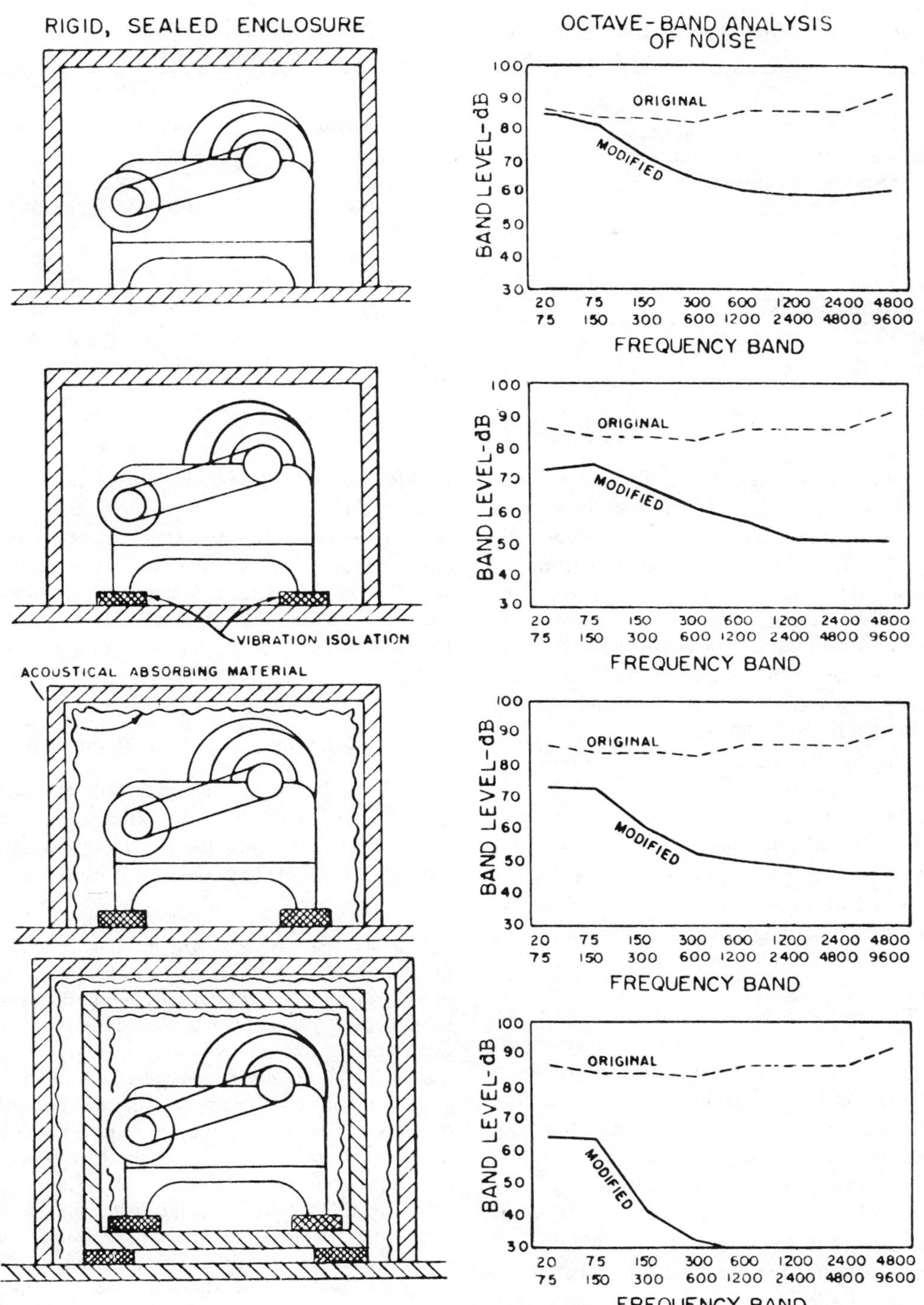

Fig. 27.43 The important aspect of this illustration is that each of the individual sound attenuation techniques is insufficient of itself, and only a combination of isolation and absorption will perform the requisite quieting. The last step—double isolation—is only required with very high noise level or very low background noise requirements. From A Guide to Airborne, Impact and Structure-Borne Noise Control in Multi-Family Dwellings, *U.S. Dept. of Housing and Urban Development, Washington, D.C., 1963, pp. 8–32.*

Table 27.8 ***Air Speeds in Ducts to Yield NC-15 or NC-25 Background Levels***
(with 1 to 2 in. inside duct lining, all duct sizes)

Location	Supply		Return	
	NC-15	*NC-25*	*NC-15*	*NC-25*
Slot speed at min. $\frac{1}{2}$ in. opening	250 fpm	350 fpm	300 fpm	420 fpm
10 ft of duct before opening	300	420	350	490
Next 20 ft	400	560	450	630
Next 20 ft	500	700	570	800
Next 20 ft	640	900	700	980
Next 20 ft	800	1120	900	1260
Next 20 ft	1000	1400	1100	1540
Next 20 ft	1300	1820	1450	2030
Next 20 ft	1600	2240	1800	2520

turbulence generates noise. Turbulence and noise levels increase as the velocity of air flow increases. Table 27.8 gives recommended design velocities for various distances from the terminal devices. Velocities should be as low as possible, since sound increases exponentially with velocity. Allowable velocities increase as the distance from the terminal device becomes greater. Return velocities may be slightly higher than supply.

Sound travels as easily against as with the air flow in ductwork. Therefore, both supply and return systems must be lined to control transmission of fan noise. Maximum noise reduction occurs at bends in the ducts. For maximum reduction in short runs, a pair of 90° bends often is designed into a system, since lining is not effective in runs shorter than 50 to 60 ft.

Other design approaches that are part of a quiet system include smooth transitions at changes of duct size and large radius bends with turning vanes. Attenuation drops rapidly with increasing size of duct, and therefore ducts should not be deliberately oversized. Cross talk between rooms and between ducts can be minimized by using lined ducts, separating adjacent ducts as much as possible, and gluing damping material on the outside and lining on the inside. Damping material is particularly effective in preventing the thin metal walls of the ducts from resonating. Mufflers and silencers are effective in reducing fan noise. The pressure drop they introduce must be compensated for in the fan. The *ASHRAE Guide and Data Book's* chapter on noise control in fans, plenums, housings, and ducts should be consulted for their recommendations. Figure 27.44 shows some of the ways by which cross talk, fan noise, and flanking noises can be reduced. Figure 27.45 shows some of the techniques employed for quieting duct noise.

27.29 Piping-System Noise Reduction

As with air flow, noise increases exponentially with velocity. Piping is not a major noise source normally, since the radiating diameter is small, although flow velocities much in ex-

Fig. 27.44 *Since ducts are extremely efficient sound transmission paths, considerable precaution must be taken to avoid cross-talk, ventilation, combustion, and equipment noise. Avoid running ducts as a common supply or return between rooms, unless such systems are properly baffled and lined with sound-absorbing material. The common practice, in wood frame structures, of using troughs between joists as a common return duct between dwellings results in very serious noise transmission problems. Caulk or seal around ducts at all points of penetrations through partitions. Use double-wall ducts, acoustical lining, flexible boots, and resilient hangers where required. Dwelling units should be serviced by separate supply and return ducts, which branch off a main duct system. From* A Guide to Airborne, Impact and Structure-Borne Noise Control in Multi-Family Dwellings, *U.S. Dept. of Housing and Urban Development, Washington, D.C., 1963, pp. 8–7, 8–8.*

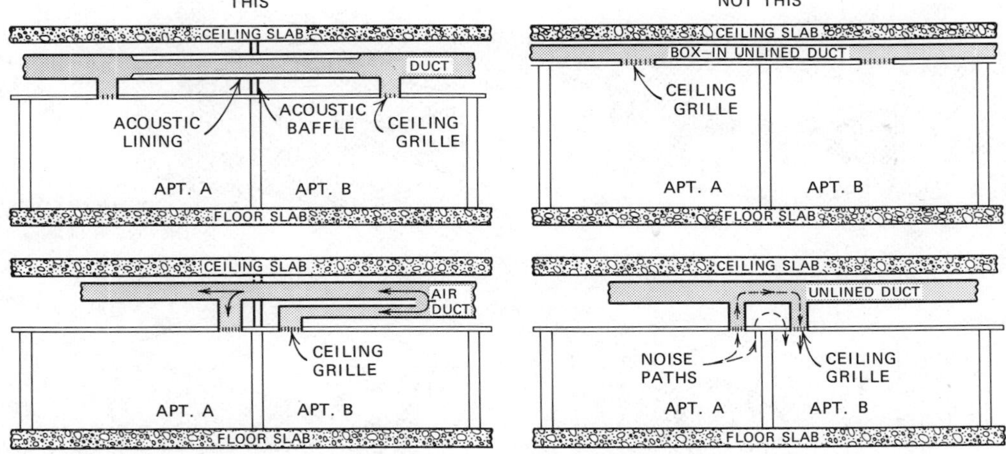

(a)

(a) *Method of controllong cross-talk and preventing flanking in ducts.*

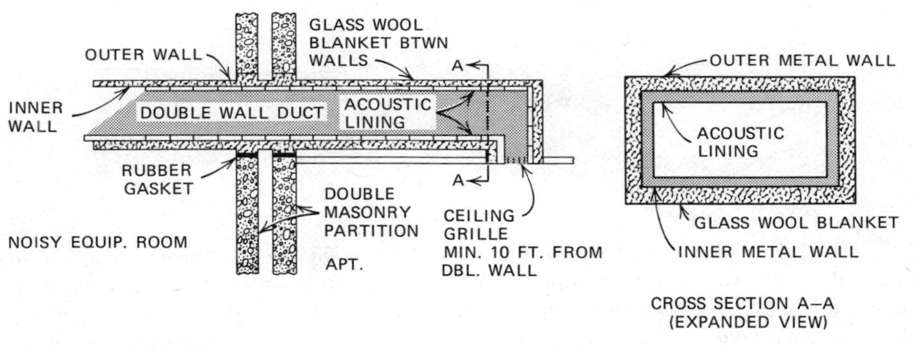

(b)

(b) *Proper installation of ductwork.*

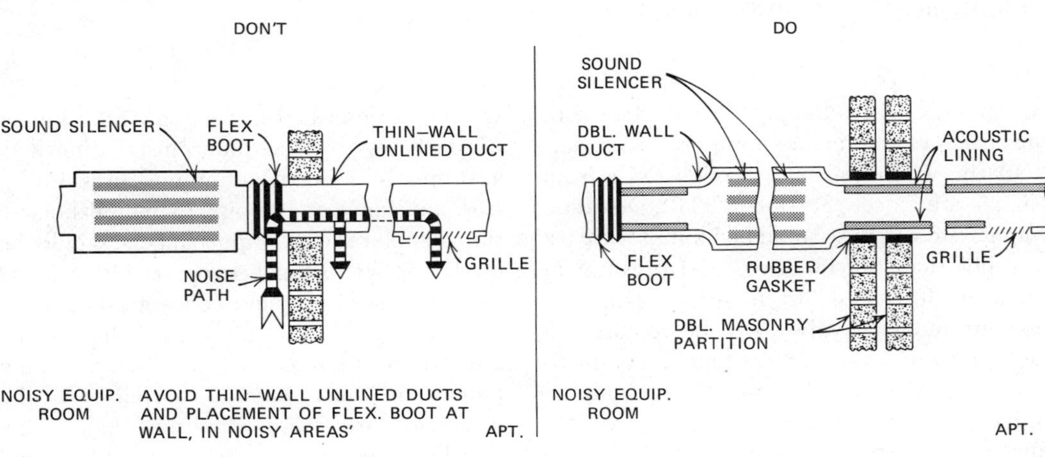

(c)

(c) *Installation of ductwork through equipment room wall.*

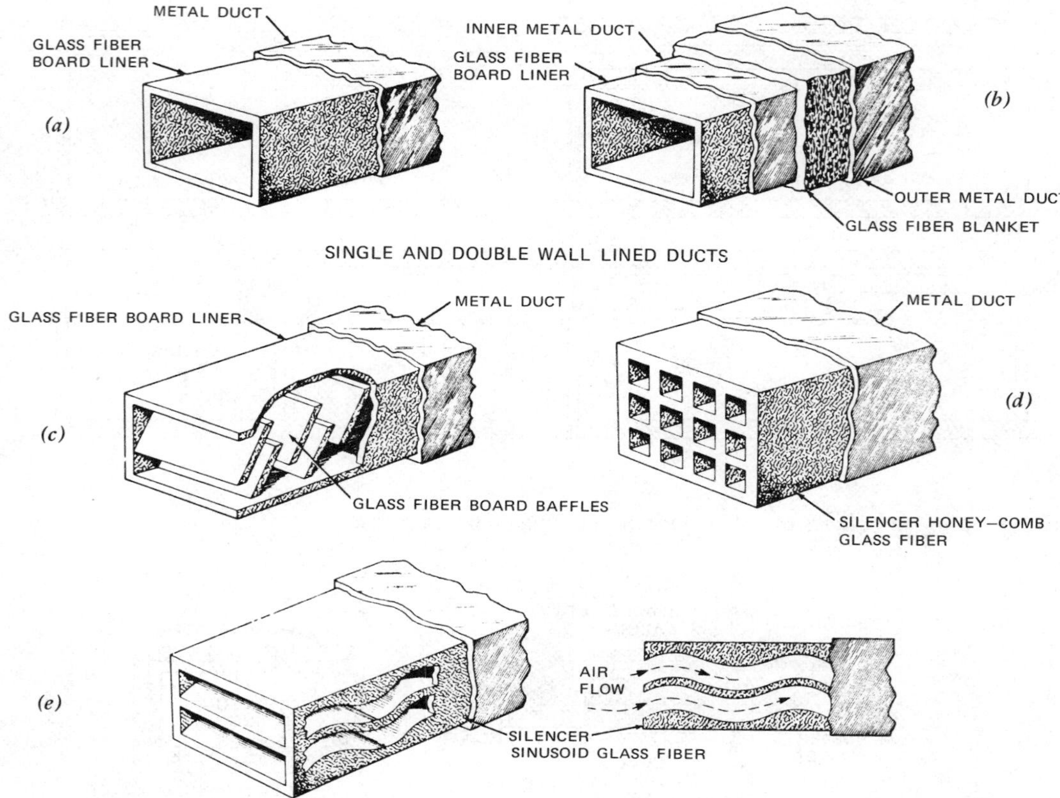

Fig. 27.45 *Unlined duct has negligible attenuation, and acts as an excellent speaking-tube. Inside lining* (a) *gives 2 to 3 db attenuation per foot in the 1 to 2 kHz range, dropping rapidly above and below those frequencies, and giving negligible low-frequency attenuation.* (b) *Double lining gives higher attenuation and reduces crosstalk between ducts. Duct silencers and baffles* (c–e) *give high broadband attenuation—maximum of 10 to 12 db/ft in the 1 to 2 kHz range, and lower above and below. They are useful to reduce fan noise in short runs but cause considerable pressure drop. From* A Guide to Airborne, Impact and Structure-Borne Noise Control in Multi-Family Dwellings, *U.S. Dept. of Housing and Urban Development, Washington, D.C., 1963, Figure 8–66.*

cess of 8 fps, where the pipe is in contact with the structure, can create noise problems, particularly when passing through NC-15 to NC-25 areas (see Section 27.34). Domestic water system mains should be limited to 50 psi in other than tall buildings and pressure in branches to 35 psi. In high-rise structures pressure-reducing valves will be required in high-pressure mains to meet these recommendations. Obviously piping must be designed to prevent water hammer, and noise sources must be located away from quiet areas.

Pumps, as with all rotating equipment, are sources of vibration and noise and should be treated as described in Section 27.27. Figure 27.46 shows a typical pump installation with appropriate noise reduction measures. For at least a distance of 100 pipe diameters beyond the pump, resilient pipe hangers should be used. With centrifugal pumps as with fan and blowers, machine sound concentrates in narrow bands and, if extremely disturbing, can be attenuated with resonant filters. Reciprocating pumps are more difficult to control as the pulsations are more vibration than noise. Flexible connections in the piping and U-joints in the piping will absorb much of this vibration.

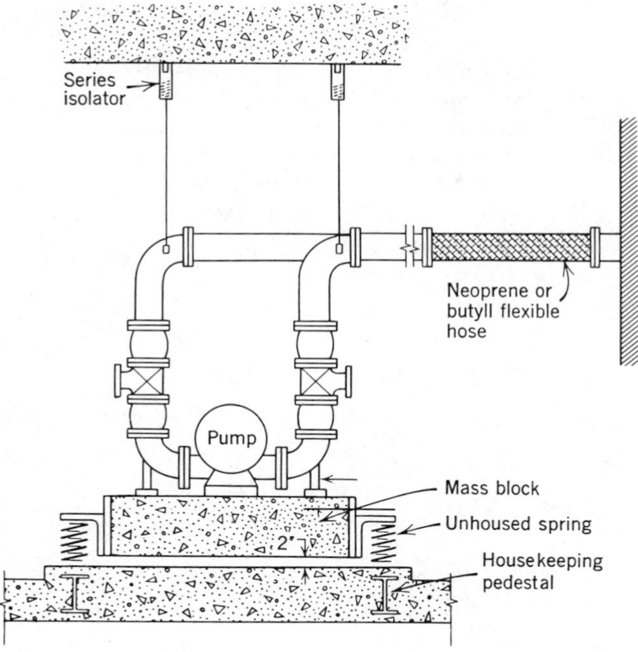

Fig. 27.46 Typical pump installation with appropriate isolation and damping measures.

27.30 Electrical Equipment Noise

Electrical equipment is generally overlooked as a noise source and this is unwise. Most electrical noise is 120-Hz hum (see Fig. 27.9). This can be very disturbing because it is so low a frequency and, as we have noted repeatedly, low-frequency noise is difficult to attenuate. Transformer noise levels are dictated by NEMA and ANSI standards. For a premium price lower noise units are obtainable. Table 27.9 lists maximum noise levels for dry-type units. Most manufacturers warranty noise levels below these. Oil- and silicone-filled units are normally quieter than these figures as are units designed for lower temperature rise. Transformer noise can be minimized by these steps:

a. Mount on vibration isolators.
b. If transformer is wall hung, use resilient hanging. If it is floor mounted, place on as massive a slab as possible.
c. Locate the unit so that reflections do not amplify the sound. Sound-absorbent material on the walls behind the units is not useful at 120 Hz. Only cavity resonators will absorb appreciable amounts of sound at that frequency.
d. Use only flexible conduit connections.
e. Avoid locating transformers adjacent to, or immediately outside of, quiet areas. A common error in this regard is placing a transformer pad immediately below the window of an NC 15-25 area.

Table 27.9 ***Maximum Sound Levels—Dry-Type Transformers***

KVa	*Decibels* (*NEMA STD*)
0–9	40
10–50	45
51–150	50
151–300	55
301–500	60

The second major source of 120-Hz hum is discharge light-fixture ballasts. This includes fluorescent plus all the HID sources. Table 27.10 lists the noise levels and the recommended application of fluorescent ballasts. HID ballasts were not a problem until re-

Table 27.10 ***Acoustic Criteria for Selection of Fluorescent Lamp Ballasts***

For an Installation in:	*Average Ambient Noise Level of Interior*	*First Choice Use of ballasts with this rating will be satisfactory within the noise levels shown.*	*Second Choice Use of ballasts with this rating will be satisfactory if care is exercised in selecting the acoustic environment.*
TV or radio station, church, synagogue	20–24 db	A	B
Offices, residence, night school	26–30 db	B	C
Library, reception or reading rooms, school study hall	31–36 db	C	D
Noisy office, doctor or dentist office, classroom	37–42 db	D	

cently when mercury, metal-halide, and sodium lamps moved from their former noisy industrial surroundings into quiet office spaces. These ballasts can be very noisy, and care must be excerised in their placement. With all ballasts, the method of mounting has a marked effect on the radiated noise. As pointed out earlier, when a small vibrating source is coupled rigidly to a larger body, noise is amplified because of increased source-to-air coupling. Since fluorescent ballasts are necessarily closely coupled to large metal fixtures for heat dissipation purposes, the sound radiation is much amplified. A large number of fluorescent fixtures mounted in a plenum can create a serious problem. Solution of the problem lies in use of absorptive material in plenums, flexible conduit connection to fixtures, and resilient fixture hanging. In severe cases ballasts can be remote-mounted. HID ballasts are inherently noisier than fluorescent but are generally less troublesome, being coupled to small radiating bodies and generally mounted higher.

27.31 Noise Problems Due to Equipment Location

Roof-mounted HVAC units have proven themselves very economical and very noisy. The vibration, short duct runs, and sound reflections are serious problems that can be solved with vibration isolators, sound mufflers, and careful location of equipment. Roof-mounted cooling towers are a particular problem when located adjacent to a taller building. That problem has led to a spate of lawsuits and noise-control legislation in many cities. For this reason particular attention should be paid in design to all exterior equipment.

In high-rise buildings, problems are engendered by the stringent noise requirements of the prime upper floor space conflicting with the near presence of elevator machine rooms, mechanical equipment rooms, and cooling towers. These problems are almost impossible to solve after construction and require the services of an acoustics expert during design.

Here especially the ounce of prevention is worth at least a pound of cure. Good planning with emphasis on proper location of noise sources will usually prevent serious, and very expensive, problems from arising.

27.32 Commercial Sound-Isolating Enclosures

In areas such as industrial plants with multiple high-noise sources, it is desirable to reduce noise at its source. To accomplish this many commercial products in the form of curtains, panels, and partial and full enclosures are available that will effectively isolate the noise source. Most of these operate on the mass principle and comprise sandwich construction of materials frequently including lead sheet. These enclosures are tailored to the specific noise and are not normally the responsibility of the building designer. It is, however, important to know that they will be used, as well as their characteristics, so that appropriate isolation can be designed into the structure for the sound that is radiated from the enclosures. One such product is illustrated in Figure 27.47.

BUILDING SITING; ROOM ARRANGEMENT

27.33 Building Siting

As important as interior structural design is building siting vis-à-vis exterior noise sources. Since this subject is somewhat beyond our scope the discussion will be brief. Buildings should be sited, with respect to noise sources:

a. To use natural terrain noise barriers—Figure 27.48*a*.
b. With respect to trees as noise barriers, rely only on thick wooded areas, Figure 27.48*b*.
c. To avoid naturally poor sites—Figure 27.48*c*.
d. To avoid sound reflection from other buildings—Figure 27.48*d*.

Factor (*d*) is also important in a multiwing building; in avoiding U-shapes or other configurations where a central court becomes an echo chamber.

(*a*) ***Room Arrangement.*** Where avoidance of an exterior noise source is impossible, quiet zones can be buffered from the noise by placing higher-noise areas on the noisy side of the building. Thus, in a school, classrooms and

Fig. 27.47 *Commercial multilayer acoustic noise barrier material composed of high-mass vinyl and absorbent blanket. It acts both as an absorbent and isolating barrier. When mounted on 1/4 plywood it forms a barrier of STC 33 rating. Courtesy of Ferro Corp.*

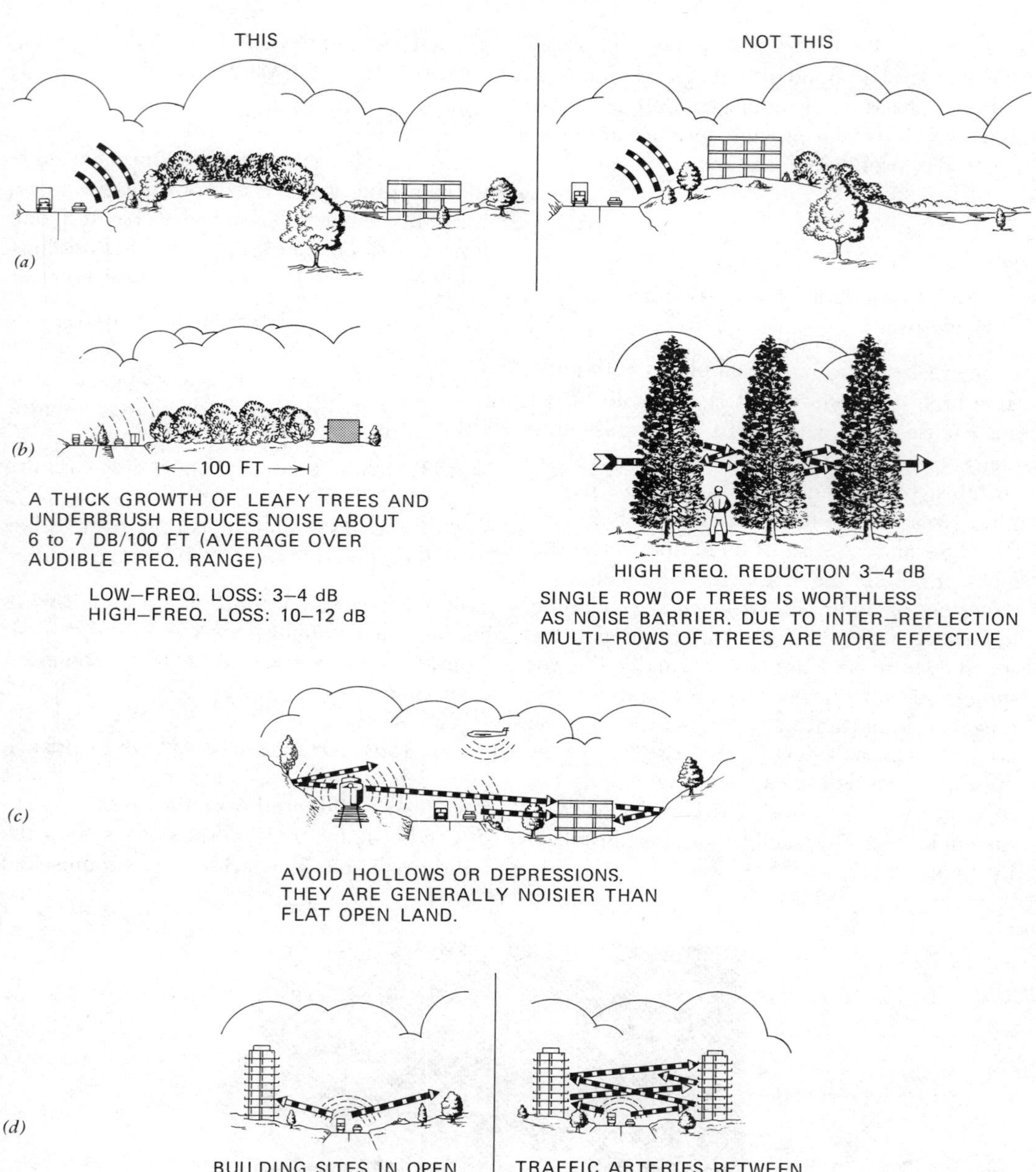

Fig. 27.48 (a) *Use of natural noise barriers.* (b) *Effectiveness of wooded areas as noise barriers. Noise reduction of trees.* (c) *An example of a poor building site.* (d) *Building sites near traffic arteries and other buildings. From* A Guide to Airborne, Impact and Structure-Borne Noise Control in Multi-Family Dwellings. *U.S. Dept. of Housing and Urban Development, Washington, D.C., 1963, p. 5–2.*

Table 27.11 ***Recommended Category Classification and Suggested Noise Criteria Range for Steady Background Noise***[a]

Type of space (and acoustical requirements)	*PNC Curve*	*NC Curve*	*Equivalent*[b] *dbA*
Concert halls, opera houses, and recital halls (for listening to faint musical sounds).	10–20	10–20	20–30
Broadcast and recording studios (distant microphone pickup used).	10–20	15–20	25–30
Large auditoriums, large drama theatres and houses of worship (for excellent listening conditions).	20 max.	20–25	30–35
Broadcast, television, and recording studios (close microphone pickup only).	25 max.	20–25	30–35
Small auditoriums, small theatres, small churches, music rehearsal rooms, large meeting and conference rooms (for good listening), or executive offices and conference rooms for 50 people (no amplification).	Not to exceed 35	25–30	35–40
Bedrooms, sleeping quarters, hospitals, residences, apartments, hotels, motels, and so forth (for sleeping, resting, relaxing).	25–40	25–35	35–45
Private or semiprivate offices, small conference rooms, classrooms, libraries, and so forth (for good listening conditions).	30–40	30–35	40–45
Living rooms and similar spaces in dwellings (for conversing or listening to radio and TV).	30–40	35–45	45–55
Large offices, reception areas, retail shops and stores, cafeterias, restaurants, and so forth (for moderately good listening conditions).	35–45	35–50	45–60
Lobbies, laboratory work spaces, drafting and engineering rooms, general secretarial areas (for fair listening conditions).	40–50	40–45	50–55
Light maintenance shops, office and computer equipment rooms, kitchens, and laundries (for moderately fair listening conditions).	45–55	45–60	55–70
Shops, garages, power-plant control rooms, and so forth (for just acceptable speech and telephone communication). Levels above PNC-60 are not recommended for any office or communication situation.	50–60	—	—
For work spaces where speech or telephone communication is not required, but where there must be no risk of hearing damage.	60–75	—	—

Source. Reprinted with permission from E. B. Magrab, *Environmental Noise Control,* Wiley, New York, 1975.

[a]After Reference 12.

[b]For information only.

offices can be buffered by a cafeteria and gym; in a residence, bedrooms by living rooms and corridors; in an office building, private offices by noisier clerical offices; and so on.

RECOMMENDATIONS AND CRITERIA

27.34 NC Criteria

Recommendations for background noise levels, structural isolation ratings, and impact isolation ratings are given in Table 27.11.

27.35 STC and IIC Criteria

(*a*) *Residential.* The most important and binding criteria for residential work in the United States are those issued by HUD/FHA. They are quoted here in full, as reprinted from Reference 19 (see Fig. 27.21 and Fig. 27.38):

F. Recommended Criteria

Descriptive definitions of three grades of acoustic environment are given in order to ascribe criteria suitable to the wide range of urban developments, geographic locations, economic conditions and other factors involved in the areas of concern of the FHA. Constructions which meet the criteria will provide good sound insulation and satisfy most of the occupants in the buildings which fit the conditions of each grade. Emphasis should be placed upon Grade II, as described below, for this category will be applicable to the largest percentage of multifamily dwelling construction and thus should be considered as the fundamental guide.

Grade I is applicable primarily in suburban and peripheral suburban residential areas, which might be considered as the "quiet" locations and as such the nighttime exterior noise levels might be about 35–40 dB(A) or lower, as measured using the "A" weighting network of a sound level meter which meets the current standards. The recommended permissible interior noise environment is characterized by noise criteria of NC20-25. In addition, the insulation criteria of this grade are applicable in certain special cases such as dwelling units above the eighth floor in high-rise buildings and the better class or "luxury" buildings, regardless of location.

Grade II is the most important category and is applicable primarily in residential urban and suburban areas considered to have the "average" noise environment. The nighttime exterior noise levels might be about 40–45 dB(A); and the permissible interior noise environment should not exceed NC25-30 characteristics.

Grade III criteria should be considered as minimal recommendations and are applicable in some urban areas which generally are considered as "noisy" locations. The nighttime exterior noise levels might be about 55 dB(A) or higher. It is recommended that the interior noise environment should not exceed the NC-35 characteristic.

In all cases, the partition structures should have STC and IIC ratings equal to or greater than the given criterion figures. For floor-ceiling assemblies, the criteria for both airborne and impact sound insulation must be met. A floor-ceiling structure which may provide adequate impact sound insulation but insufficient airborne sound insulation, or vice versa, will not assure freedom from occupant complaints.

The fundamental or key criteria of airborne and impact sound insulation of wall and floor assemblies *which separate dwelling units of equivalent function* are given in Table 1. *These criteria are based upon STC and IIC ratings derived from laboratory measurements,* since standard methods of test for field measurements have not as yet been formally adopted. Figures 27.21 and 27.38 illustrate the relationship of the fundamental FHA recommended criteria with the range of airborne and impact sound insulation requirements or recommendations of other countries.

Table 1 ***Key Criteria of Airborne and Impact Sound Insulation Between Dwelling Units***

	GRADE I	*GRADE II*	*GRADE III*
Wall Partitions	STC $\geq$ 55	STC $\geq$ 52	STC $\geq$ 48
Floor-Ceiling Assemblies	IIC $>$ 55	IIC $\geq$ 52	IIC $\geq$ 48

The following comprehensive tables show the criterion values related to partition function as applied in the separation of dwelling units. Indeed, these tables incude most of the typical separation combinations found in multifamily buildings, as well as some which are clearly undesirable for several reasons. The purpose of this detail is to illustrate the importance of the acoustical separation between sensitive and nonsensitive areas. Where the partition between dwelling units is common to several functional spaces, the partition must meet the highest criterion value.

Table 3 includes most of the floor-ceiling assembly combinations found in multifamily buildings as well as some which are clearly undesirable for several reasons. In addition, the importance of impact noise insulation is emphasized by giving separate criteria for reciprocal functional relationships.

Table 2 ***Criteria for Airborne Sound Insulation of Wall Partitions Between Dwelling Units***

Partition Function Between Dwellings			*Grade I*	*Grade II*	*Grade III*
Apt. A		*Apt. B*	*STC*	*STC*	*STC*
Bedroom	to	Bedroom	55	52	48
Living room	to	Bedroom[2,2]	57	54	50
Kitchen[3]	to	Bedroom[1,2]	58	55	52
Bathroom	to	Bedroom[1,2]	59	56	52
Corridor	to	Bedroom[2,4]	55	52	48
Living room	to	Living room	55	52	48
Kitchen[3]	to	Living room[1,2]	55	52	48
Bathroom	to	Living room[1]	57	54	50
Corridor	to	Living room[2,4,5]	55	52	48
Kitchen	to	Kitchen[6]	52	50	46
Bathroom	to	Kitchen[1]	55	52	48
Corridor	to	Kitchen[2,4,5]	55	52	48
Bathroom	to	Bathroom	52	50	46
Corridor	to	Bathroom[2,4]	50	48	46

NOTES,

1. The most desirable plan would have the dwelling unit partition separating spaces with equivalent functions, e.g., living room opposite living room, etc.; however, when this arrangement is not feasible, the partition must have greater sound insulating properties.
2. Whenever a partition wall might serve to separate several functional spaces, the highest criterion must prevail.
3. Or dining, or family, or recreation room.
4. It is assumed that there is no entrance door leading from corridor to living unit.
5. A common approach to corridor partition construction correctly assumes the entrance door as the acoustically weakest "link" and then incorrectly assumes that the basic partition wall need be no better acoustically than the door. However, the basic partition wall may separate the corridor from sensitive living areas such as the bedroom and bathroom without entrance doors, and must therefore have adequate insulating properties to assure acoustical privacy in these areas. In areas where entrance doors are used, the integrity of the corridor-living unit partition must be maintained by utilizing solid-core entrance doors, with proper gasketing. The most desirable arrangement has the entrance door leading from the corridor to a partially enclosed vestibule or foyer in the living unit.
6. Double-wall construction is recommended to provide, in addition to airborne sound insulation, isolation from impact noises generated by the placement of articles on pantry shelves and the slamming of cabinet doors. Party walls which utilize resilient spring elements to achieve good sound insulation may be used, providing wall cabinets are not mounted on them. It is not practical to use such walls for mounting of wall cabinets because the sound insulating performance of the walls can be easily short-circuited, unless specialized vibration isolation techniques are used.

Table 3 ***Criteria for Airborne and Impact Sound Insulation of Floor-Ceiling Assemblies Between Dwelling Units***

Partition Function Between Dwellings			*Grade I*		*Grade II*		*Grade III*	
Apt. A		*Apt. B*	*STC*	*IIC*	*STC*	*IIC*	*STC*	*IIC*
Bedroom	above	Bedroom	55	55	52	52	48	48
Living room	above	Bedroom[1,2]	57	60	54	57	50	53
Kitchen[3]	above	Bedroom[1,2,4]	58	65	55	62	52	58
Family room	above	Bedroom[1,2,5]	60	65	56	62	52	58
Corridor	above	Bedroom[1,2]	55	65	52	62	48	58
Bedroom	above	Living room[6]	57	55	54	52	50	48
Living room	above	Living room	55	55	52	52	48	48
Kitchen	above	Living room[1,2,4]	55	60	52	57	48	53
Family room	above	Living room[1,2,5]	58	62	54	60	52	56
Corridor	above	Living room[1,2]	55	60	52	57	48	53
Bedroom	above	Kitchen[1,4,6]	58	52	55	50	52	46
Living room	above	Kitchen[1,4,6]	55	55	52	52	48	48
Kitchen	above	Kitchen[4]	52	55	50	52	46	48
Bathroom	above	Kitchen[1,2,4]	55	55	52	52	48	48
Family room	above	Kitchen[1,2,4,5]	55	60	52	58	48	54
Corridor	above	Kitchen[1,2,4]	50	55	48	52	46	48
Bedroom	above	Family room[1,6]	60	50	56	48	52	46
Living room	above	Family room[1,6]	58	52	54	50	52	48
Kitchen	above	Family room[1,6]	55	55	52	52	48	50
Bathroom	above	Bathroom[4]	52	52	50	50	48	48
Corridor	above	Corridor	50	50	48	48	46	46

NOTES,

1. The most desirable plan would have the floor-ceiling assembly separating spaces with equivalent functions, e.g., living room above living room, etc.; however, when this arrangement is not feasible, the assembly must have greater acoustical insulating properties.
2. This arrangement requires greater impact sound insulation than the converse, where a sensitive area is above a less sensitive area.
3. Or dining, or family, or recreation room.
4. See text for proper vibration isolation of plumbing fixtures and appliances.
5. The airborne STC criteria in this table apply as well to vertical partitions between these two spaces.
6. This arrangement requires equivalent airborne sound insulation and perhaps less impact sound insulation than the converse.

(*b*) *Schools.* See Table 27.12.

(*c*) *Office Buildings.* See Table 27.12.

(*d*) *Motels, Hotels and Hospitals.* See Table 27.12.

27.36 Specific Occupancies

(*a*) ***Schools.*** School buildings house many kinds of spaces—classrooms, auditoriums, gymnasiums, cafeterias, shop areas, and music suites—which pose acoustics problems.

1. *Auditoriums.* Auditorium reverberation design is discussed in Section 26.28. Other aspects of auditorium design are noise control, sound systems, and multiple use of the space.

Noise Control. The two major noise control considerations for any auditorium are elimination of possible intrusion of extraneous sounds and design of the mechanical systems serving the space. Meeting the established background noise criterion is the design goal for both aspects.

If noise control problems are considered in

Fig. 27.49 *Auditorium sound system elements.*

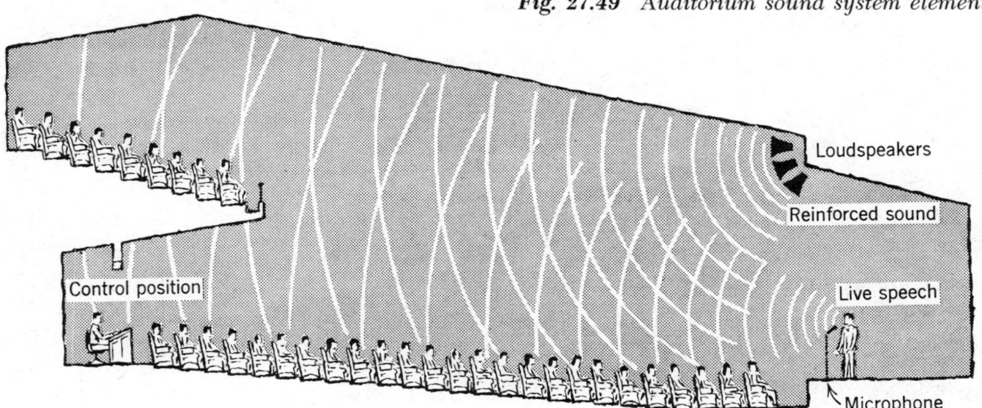

the early stages and basic building layouts that separate noisy and critically quiet areas are developed, the cost of achieving good acoustics will be minimized. Similarly noise control aspects should be considered during site selection.

Sound Systems. All auditoriums require a sound system for some of the activities accommodated. Figure 27.49 shows a typical sound system in an auditorium. Microphones, loudspeakers, and control locations are indicated. The most difficult aspect, architecturally, is integration of the loudspeaker system into the design. To provide proper sound reinforcement, loudspeakers must be located properly without obstructions. To accomplish this, the loudspeaker system should be incorporated in the earliest drawings.

Multiple Uses. In general, a school auditorium must be a multipurpose facility. It should be designed to meet speech requirements and also be suitable for the school's music activities.

Often a modified gymnasium (gymnatorium) or cafeteria (cafetorium) functions as an auditorium. Obviously, acoustics compromises occur in such facilities. The large areas of sound-absorbing treatment in either kind of space make them unsuitable as auditoriums, and most events require speech amplification.

2. *Classrooms.* Classrooms today range from the traditional enclosed spaces to open-planning designs that support innovative teaching methods.

Closed Classroom. Typical classrooms are approximately 30 ft square with 10-ft ceilings. Adequate speech communication is easily achieved in a room of this size. Classroom acoustic design usually involves:

a. Locating sound-absorbing treatment to reduce classroom noise levels.
b. Insuring adequate privacy between adjacent spaces.
c. Control of air-handling system noise.

Acoustic tile ceilings yield adequate sound absorption for most classrooms. An NRC of 0.7 is recommended. An alternate method involves a combination of sound-absorbing wall treatment and floor carpeting, which also produce adequate noise reduction. Carpeting, alone, is generally insufficient.

Partition systems must produce sufficient isolation to prevent disturbance from activities in other classrooms and corridors. Such partitions should run full height from floor to ceiling slab or roof construction. If return air-transfer ducts are needed, their noise reduction characteristics must be as good as the walls or doors that they penetrate (for NC data see Table 27.11). Unit ventilators commonly used for classrooms produce approximately this level of background noise.

Open Planning. Open-plan schools rely on partial height barriers and noise-reducing treatments, as well as properly controlled background noise for separation of activities. Experience with open-plan schools indicates that this design approach is feasible only when

Table 27.12 ***Recommended STC for Partitions; Specific Occupancies***

Type of Occupancy	Wall, Partition or Panel Between: Room being considered	AND Adjacent area	Sound Isolation Requirement (Background level in room being considered): Quiet	Normal
Normal school buildings without extraordinary or unusual activities or requirements	Classrooms	Adjacent classrooms	STC 42	STC 40
		Laboratories	STC 42	STC 40
		Corridor or public areas	STC 40	STC 38
		Kitchen and dining areas	STC 50	STC 47
		Shops	STC 50	STC 47
		Recreation areas	STC 45	STC 42
		Music rooms	STC 55	STC 50
		Mechanical equipment rooms	STC 50	STC 45
		Toilet areas	STC 45	STC 42
	Large music or drama areas	Adjacent music or drama rooms	STC 50	STC 45
		Corridor or public areas	STC 45	STC 42
		Practice rooms	STC 50	STC 45
		Shops	STC 50	STC 45
		Recreational areas	STC 50	STC 45
		Laboratories	STC 45	STC 42
		Toilet areas	STC 45	STC 42
		Mechanical equipment rooms	STC 50	STC 48
		Exterior of building	STC 45	STC 42
	Music practice rooms	Adjacent practice rooms	STC 55	STC 50
		Corridor and public areas	STC 45	STC 42
	Language laboratories	Same as for theaters, concert halls, auditorium, etc.		
	Counseling offices	Same as for executive offices		
Executive areas, doctors' suites; confidential privacy requirements	Office	Adjacent offices	STC 50	STC 45
		General office areas	STC 48	STC 45
		Corridor or lobby	STC 45	STC 42
		Washrooms and toilet areas	STC 50	STC 47

the various activities are relatively quiet. It would be impossible, for example, to achieve adequate separation of music practice and shop activities in open-plan design. Figure 27.31 shows how open planning relies on the interaction of noise reduction, partial height barriers, and background noise level. Closed- and open-plan isolation are compared also.

Open-plan school areas that are noisy (such as shops and music rooms) or require high degrees of privacy must be designed as traditional spaces using sound-isolating partitions. Successful open-plan school design utilizes its potential advantages while recognizing its drawbacks.

3. Music Suite. School music programs usually range from individual instruction to

Table 27.12 ***Recommended STC for Partitions; Specific Occupancies (Continued)***

Type of Occupancy	Wall, Partition or Panel Between: Room being considered	AND Adjacent area	Sound Isolation Requirement (Background level in room being considered): Quiet	Normal
Normal office areas; normal privacy requirements	Office	Adjacent offices	STC 40	STC 38
		General office areas	STC 40	STC 38
		Corridor or lobby	STC 40	STC 38
		Washrooms and toilet areas	STC 42	STC 40
		Exterior of building	STC 40	STC 38
		Kitchen and dining areas	STC 42	STC 40
Any normal occupancy, using conference rooms for group meetings or discussions	Conference rooms	Other conference rooms	STC 45	STC 42
		Adjacent offices	STC 45	STC 42
		General office areas	STC 45	STC 42
		Corridor or lobby	STC 42	STC 40
		Washrooms and toilet areas	STC 45	STC 42
		Exterior of building	STC 40	STC 38
		Kitchen and dining areas	STC 45	STC 42
		Manufacturing or other noisy interior areas	STC 45	STC 42
Normal business offices, drafting areas, banking floors, etc.	Large general office areas	Corridors or lobby	STC 38	STC 35
		Exterior of building	STC 38	STC 35
		Data processing area	STC 40	STC 38
		Manufacturing areas and mechanical equipment areas	STC 42	STC 40
		Kitchen and dining areas	STC 40	STC 38
Motels and urban hotels, hospitals and dormitories	Bedrooms	Adjacent bedrooms, separate occupancy	STC 52	STC 50
		Bathroom, separate occupancy	STC 50	STC 45
		Living rooms, separate occupancy	STC 45	STC 42
		Dining areas	STC 45	STC 42
		Corridor, lobby, or public spaces	STC 45	STC 42

Source. Courtesy of US Gypsum.

band and choral concerts. The teaching spaces required for such a program include practice rooms, ensemble rooms, and large rehearsal spaces. Both room acoustics design and sound isolation are critical in music suites. Privacy between adjacent spaces is particularly critical, since simultaneous use is necessary.

4. *Dining Areas.* The activity in cafeterias or lunchrooms generates noise. The kitchen and serving areas should be separated from the eating spaces. Ceilings and available wall areas in the cafeteria should be treated with sound-absorbing material such as acoustic tiles to reduce potential buildup of sounds produced by conversation, traffic, moving chairs, and dishes. Ideally, the area should be carpeted. Moreover, unless the ceiling is com-

pletely treated with a highly efficient sound-absorbing material, the environment will be unsatisfactory. Minimum NRC should be 0.8.

5. *Gymnasiums.* Activities in gymnasiums create so much noise that even extensive treatment will not quiet these spaces. A quiet gymnasium probably would be unsatisfactory since spectators are conditioned to consider the "noise" as an enjoyable aspect of athletic events. However, to provide a proper environment for normal sports activities, the ceiling area should be sound absorbing. In addition, if a sound amplification system is to be used, sound-absorbing wall treatment may be required to eliminate echoes that would reduce intelligibility of announcements. An NRC of 0.7 is suggested with sound absorbent material to be ceiling mounted.

If a gymnasium will also serve as an auditorium, loudspeaker system placement needs special consideration. For example, the loudspeakers should be located above the source location for speeches and plays. The need for flexibility in coverage also may be a design consideration. Flexibility may be achieved by switching loudspeakers on and off to fit the seating pattern.

6. *Swimming Pools.* The acoustic environment of swimming pools is often chaotic. Most sound-absorbing materials disintegrate in the high-humidity conditions prevalent in pool areas. Special combinations of materials, as well as some recently developed prefabricated sound-absorbing units, have moisture resistant properties. This ability to limit the amount of noise buildup creates a more pleasant acoustical environment for swimming pools.

7. *Shops.* Metal, woodworking, and scenery shops in schools contain many noise sources—saws, planers, drill presses, and manual tools. Each generates high airborne and structure-borne noise levels. Careful planning is essential for control of shop areas, as well as reduction of noise transmission into other spaces such as classrooms and auditoriums. Consolidating noisy areas and maximizing the distance between them and critically quiet spaces are essential. Ceiling and wall absorptive treatment with an NRC of at least 0.75 is recommended.

(*b*) ***Churches.*** The basic activities of most churches combine speech and music. Thus, the church environment must be hospitable, acoustically, to both. A nonreverberant chapel meets the requirement for speech communication but it is unsatisfactory for music. A cathedral usually will provide a magnificent acoustical environment for organ and choral works, while it is nearly impossible for intelligible speech communication. The architectural plan also must respond to religious requirements including the relative positioning of pulpits, lecterns, the altar, and choir.

Successful church acoustics can be achieved by designing the overall environment for music and providing special assist-

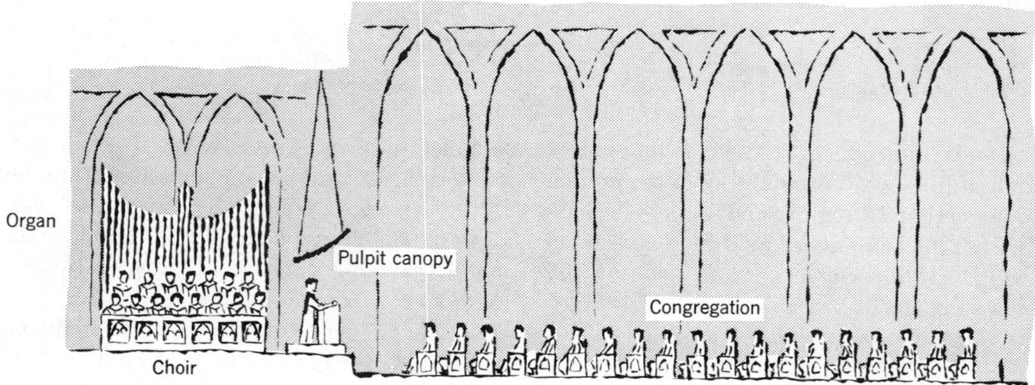

Fig. 27.50 *Church with pulpit canopy.*

ance for speakers. Figure 27.50 illustrates a church that is adequately reverberant for music and includes a sound-reflecting canopy over the pulpit to direct the minister's voice to the congregation. In some larger churches, a loudspeaker located above the canopy further reinforces speech from the pulpit. The choir and organ communicate with the entire volume of the church and, therefore, benefit from the reverberant environment. Certain traditional church designs divide the choir into two parts that are separated from the organ, thereby increasing difficulties for ensemble singing and coordination. Proper design should consolidate choir positions and also locate them as close to the organ as possible. Since concave geometry such as arches, vaults, and domes produces annoying focusing, it should be avoided in church design.

(*c*) ***Offices.*** Although office buildings contain public spaces, auditoriums, and restaurants, prime occupancy is in small office areas. Most acoustics problems in office buildings relate to privacy—either between spaces within a single firm or between adjacent firms. Speech privacy has been discussed at length in Section 27.22, including consideration of open-plan offices.

Before starting to design a series of offices, the architect must determine how much privacy is required in each space. Corporate officers and personnel departments, for example, may demand unusual degrees of privacy.

Satisfying speech privacy requirements does not necessarily create a good acoustics environment in an office building. As already pointed out, the operation of air-handling equipment, pumps, elevators, plumbing, and escalators can create excessive noise and vibration, particularly since executive offices or restaurants often are located on the top floor of multistory office buildings.

(*d*) ***Apartment Buildings.*** Large apartment buildings sometimes house thousands of residents. Privacy and freedom from annoyance are high on the list of tenant requirements. See HUD/FHA criteria in Section 27.35.

Satisfactory apartment acoustics design seldom requires special sound-isolating construction. With careful planning and supervision as well as avoidance of hard to construct details good housing can result.

Sound sources in apartments include conversation, phonographs, footsteps, telephones, doorbells, and incinerator chutes. Basic sound isolation between apartments can be provided by masonry or drywall partitions.

The performance of the partitions is compromised in many designs by careless planning of convenience outlets, medicine cabinets, and mechanical services. Moreover, direct-exhaust duct connections between apartments and back-to-back placement of medicine cabinets result in loss of privacy. Back-to-back convenience outlets must be avoided.

Installation of rugs or carpeting provides the best protection against footfall noise. Some leases now require that a tenant provide such impact-reducing floor covering over most of the floor area in his apartment. Good design also dictates that similar spaces in adjacent apartments be grouped—bedrooms next to bedrooms, for example. Absorptive material in bedrooms should be ceiling mounted. An NRC of 0.6 is recommended.

Apartment house site selection seldom includes consideration of acoustics. Nevertheless, truck routes, super highways, and airports can be annoying "neighbors." As mentioned above, cooling towers serving adjacent buildings must be considered during planning stages.

(*e*) ***Codes and FHA Criteria.*** Newer building codes and zoning ordinances usually include acoustic criteria. Zoning ordinances that previously were restricted to describing land use now include limitations on noise in the area. Preliminary research for a building project should include a detailed review of local codes and ordinances.

REFERENCE MATERIAL

27.37 Definitions

A-Scale. A filtering system that has characteristics that roughly match the response characteristics of the human ear at low sound levels (below 55 db sound pressure level, but frequently used to gauge levels to 85 db). A-scale measurements are often referred to as dbA.

Absorber, Sound. A device, panel, or material specifically designed to absorb sound energy. Such devices are usually constructed of porous materials composed of organic or mineral fibers.

Absorption Coefficient (a) (α). The absorption coefficient of a material or sound-absorbing device is the ratio of the sound absorbed to the sound incident on the material or device. The sound absorbed by a material or device is usually taken as the sound energy incident on the surface minus the sound energy reflected.

Acoustical Material. Any material considered in terms of its acoustical properties. *Commonly and especially*, a material designed to absorb sound.

Acoustical Treatment. The application or use of any sound absorbers, building materials, or structures and construction techniques for purposes of controlling noise and improving the acoustical environment.

Airborne Noise. Noise radiated initially into and transmitted through air.

Ambient Noise. The quiet-state noise level in a room or space, which is a composite of sounds from many external sources both near and far, over which one individually has no control.

Anechoic Room. An anechoic room provides a free-field acoustic testing environment like the out-of-doors. All of the sound emanating from a source is essentially absorbed at the walls of the anechoic room. Hence, there are no reflections and the spatial sound-radiation pattern of a source may be determined. An anechoic room may be described as "echoless" or acoustically "dead."

Attenuation, Sound. The reduction of the energy or intensity of sound.

Background Noise. The sound level present in a room or space at any given time above which speech, music, desired signal, or sound must be raised in order to be heard or made intelligible.

Baffle or Barrier, Sound. A shielding structure or partition used to increase the effective length of a sound transmission path between two locations. Such structures often are constructed or surfaced with sound-absorbing materials and are frequently used to seal open plenums above ceilings or below floors.

B-Scale. A filtering system that has characteristics that roughly match the response characteristics of the human ear at sound levels between 55 and 85 db. B-scale measurements are often referred to as dbB.

Bending Wave. A wave that travels in a structure by deforming the structure in bending. Bending waves may also be called flexural waves or transverse waves because their displacement is transverse to their direction of travel.

C-Scale. A filtering system that has characteristics that roughly match the response characteristics of the human ear at sound levels above 85 db. The filtering system in this case is flat with frequency. C-scale readings may be referred to as dbC.

Critical Frequency. The lowest frequency at which the wavelength of a bending wave, traveling in a structure, is the same as the wavelength in air at that frequency. Coupling between the air and the structure is very good at this point, and sound waves may move from the structure to the air and vice versa with ease.

Damping. Dissipation of *structure-borne noise* (usually traveling bending waves) by conversion to some other form of energy, usually heat. For the most part, this is accomplished by using a material with a high internal energy-absorbing capacity (i.e., high internal damping).

Decibel (db). See "Sound Pressure Level."

Diffuse Sound Field. A diffuse sound field is one in which the sound field at any given point is made up of sound waves of all angles of incidence.

Direct Field. The sound in a region in which all or most of the sound arrives directly from the source without reflection.

Flanking Transmission. The transmission of sound or noise from one room to another by indirect paths, rather than directly through an intervening partition.

Free Sound Field (Free Field). A free sound field is a field in a homogeneous, isotropic medium free from boundaries. In practice, it is a field in which the effects of the boundaries are negligible over the region of interest. In the free field, the sound pressure level decreases 6 db for a doubling of distance from a point source.

Frequency Spectrum. Usually a visual representation of a complex sound or noise that has been resolved into frequency components. The detailed nature of a complex sound may be studied by obtaining its frequency spectrum. Frequency spectra are commonly obtained in octave bands, $\frac{1}{3}$-octave bands, and various narrow bands.

Impact Insulation Class (IIC). A single-figure rating that provides an estimate of the impact sound-insulating performance of a floor-ceiling assembly.

Intensity Level (IL). A measure of the acoustic power passing through a unit area expressed on a decibel scale referenced to some standard (usually 10^{-12}) watt per square meter.

Longitudinal Wave. A longitudinal wave is a wave in which the direction of displacement at each point of the medium is normal to the wave front. Sound in air propagates as a longitudinal wave.

Loudness. Loudness is the subjective human definition of the intensity of a sound. Human reaction to sound is highly dependent on the sound pressure and frequency.

Loudness Level. A subjective method for rating loudness in which a 1000-Hz tone is varied in intensity until it is judged by listeners to be equally as loud as a given sound sample. The loudness level in phons is taken as the sound pressure level, in decibels, of the 1000-Hz tone.

Masking. The presence of a background noise increases the level to which a sound signal must be raised in order to be heard or distinguished. If the level of background noise is significantly higher than that of the sound signal, for instance a sound transmitted from another room, the transmitted sound signal cannot be heard. This effect is known as masking.

Mass Law. The law relating to the transmission loss of walls, which says that in a part of frequency range, the magnitude of the loss is controlled entirely by the mass per unit area of the panel. The law also says that the transmission loss increases 6 db for each doubling of frequency or each doubling of the panel mass per unit area.

Noise. Any undesired sounds, usually of different frequencies resulting in an objectionable or irritating sensation.

Noise Reduction (NR).

1. The reduction in sound pressure level caused by making some alteration to a sound source.
2. The difference in sound pressure level measured between two adjacent rooms caused by the transmission loss of the intervening wall.

Octave Band (OB). A range of frequency where the highest frequency of the band is double the lowest frequency of the band. The band is usually specified by the center frequency.

Phon. Loudness level, at a particular frequency, equal to the 1000-Hz db level of that equal-loudness contour.

Pink Noise. Wide frequency spectrum noise, whose amplitude drops 3 db per octave with increasing frequency. Particularly useful for masking.

Pitch. The pitch of a sound depends primarily on its frequency. In music, sounds of higher frequency are referred to as treble notes, while those of lower frequency are referred to as bass notes.

Pure Tone. A pure tone has a unique pitch and is characterized by a sinusoidal variation in sound pressure with time. The frequency spectrum of a pure tone shows a single line at a discrete frequency.

Radiation. The process of turning structure-borne noise into airborne (or some other fluid-borne) noise.

Random Noise. Random noise is a complex vibration made up of frequencies and amplitudes that vary with time in a random or statistical fashion.

Reverberation. Reverberation is the persistence or echoing of previously generated sound caused by reflection of acoustic waves from the surfaces of enclosed spaces.

Reverberation Room. Reverberation rooms are specially designed rooms in which the reverberation time is long (several seconds). When the reverberation room is properly designed, the sound field within the room is diffuse; and the sound pressure of an acoustic source is not a function of position within the room.

Sabin. The unit of acoustic absorption. One sabin is the absorption of one square foot of perfect sound-absorbing material.

Shear Wave. A shear wave is an elastic disturbance

within a medium in which the shape of the medium changes without a change in the volume of the medium. The speed of a shear wave is governed by the shear modulus and the shear mass.

Sone. A unit used in judging the loudness of sound. One sone is defined as the level, at 1000 Hz, that is 40 db above the subject's threshold of hearing. The loudness of a given sound is rated by the listener as some multiple of the sone.

Sound. Deformation waves that are traveling in the air or other elastic materials. It should be noted that sound can be defined as the disturbances themselves or the sensations they produce.

Sound Absorption

1. The process of dissipating or removing sound energy.
2. The property possessed by materials, objects, and structures such as rooms, of absorbing sound energy.
3. The measure of the magnitude of the absorptive property of a material, an object, or a structure such as a room.

Sound-Level Meter. An instrument for the direct measurement of sound pressure level. Sound-level meters often are made with various filtering networks that measure the sound directly on A, B, C, etc. scales. Sound-level meters may also incorporate octave-band filters for measuring sound directly in octave bands.

Sound Power Level (PWL). A measure of the total airborne acoustic power generated by a noise source, expressed on a decibel scale referenced to some standard (usually 10^{-12} w).

Sound Pressure. A fluctuating pressure superimposed on the static atmospheric pressure in the presence of sound. In analogy with alternating voltage, its magnitude can be expressed in several ways such as instantaneous sound pressure or peak sound pressure; but the unqualified term means root-mean-square (rms) sound pressure.

Sound Pressure Level (SPL). A measure of the air pressure change caused by a sound wave. Expressed on a decibel scale referenced to some standard (usually 0.0002 μbar).

Sound Transmission Loss (STL). The decrease or attenuation in sound energy (expressed in decibels) of airborne sound as it passes through a building construction. In general, the transmission loss increases with frequency, that is, the higher the frequency the greater the sound transmission loss.

Standing Wave. The interference pattern formed by two waves of the same wavelength traveling in opposite directions. The result is a wave pattern that does not move. A standing wave is a resonance within a medium as influenced by the boundaries of the medium (e.g., a standing wave in the air column enclosed within an organ pipe one end of which may be either open or closed). All standing waves are resonances. All resonances are not standing waves. For example, a rigid mass supported by a spring has a resonant frequency $\omega = \sqrt{\frac{k}{m}}$. This resonance does not represent a standing wave in the spring.

Structure-Borne Noise. A condition when the sound waves are being carried by a solid material. Sound waves in this state are inaudible to the human ear, since they cannot carry energy to it. Airborne noise can be created from the radiation of structure-borne noise into the air. Structure-borne noise may be propagated by shear waves, tension-compression waves, bending waves, or complicated combinations of waves.

Transmission Loss (TL). The reduction of airborne sound power that is caused by placing a wall or barrier between the reverberant sound field of a source and its receiver. Transmission loss is a property of the wall or barrier.

Wavelength (λ). The wavelength of a sound is the distance between a point of a given phase of one wave and a point of the same phase of an adjacent wave.

Wave Motion. A wave involves some physical quantity that is a function of both position and time and that (a) changes in magnitude more or less regularly with time at a given location, and (b) at a fixed time changes from place to place in a more or less regular manner.

White Noise. White noise is noise of wide frequency range in which the amplitude of the noise is essentially the same in all frequency bands.

27.38 Unit and Conversions—See Table 27.13.

27.39 Symbols—See Table 27.14.

Table 27.13 ***Acoustic Units and Conversions***

a: *Units*	*MKS*	*CGS*
Force	Kilogram-meter/sec^2 = Newton	gram-cm/sec^2 = dyne
Intensity	watts/meter^2	watts/cm^2
Pressure	Newtons/meter^2 = Pascal	dynes/cm^2 = microbars

b: *Conversion: Quantity*	*Multiply*	*By*	*To Obtain*
Force	Newtons	10^5	dynes
	dynes	10^{-5}	Newtons
Intensity	watts/cm^2	10^4	watts/m^2
	watts/m^2	10^{-4}	watts/cm^2
Pressure	Pascal	10	microbars
	microbar	10^{-1}	Pascal

NOTE. One atmosphere = 1 bar = 10^6 μ bar.

Table 27.14 ***Symbols and Abbreviations***

A =	Total absorption (sabins)
A_R =	Absorption in receiving room (sabins)
$\bar{A}$ =	Average absorption of a group of space, sabins
c =	Velocity of sound, feet per second
d =	Distance from source, centimeters or feet
db =	Decibel
f =	Frequency of sound, hertz (cps)
I =	Intensity, watts per square centimeter
I_a =	Absorbed energy (watts per square centimeter)
I_i =	Incident energy (watts per square centimeter)
I_0 =	Reference intensity, 10^{-16} watts per square centimeter
IIC =	Impact insulation class, no units
IL =	Intensity level, decibels
NC =	Noise criterion, no units
NR =	Noise reduction, decibels
p =	Pressure, Pascal or microbars
p_0 =	Reference base pressure 2×10^{-5} Pa
Pa =	Pascal, unit of pressure (SI)
PWL =	Sound pressure level, decibels
R =	Room constant, square feet
r,d =	Distance from source, meters or feet
S =	Surface area, square feet
SPL =	Sound pressure level, decibels
STC =	Sound transmission class, no units
T_R =	Reverberation time, seconds
TL =	Transmission loss, decibels
V =	Volume, cubic feet
W,P =	Sound power, watts
W_0 =	Reference base sound power 10^{-12} watts
a,α =	Absorption coefficient (no units)
$\bar{a},\bar{\alpha}$ =	Average absorption coefficient (no units)
λ =	Wavelength, feet or meters
Σ =	Sum of, or total (no units)
$\Sigma S_a = \Sigma A$ =	Total absorption, sabins
Δ =	Change in a quantity

NOTE. Where definitions are expressed in feet, centimeters or meters is also understood, with proper conversion factors.

References and Bibliography

1. *Quieting: A Practical Guide to Noise Control,* National Bureau of Standard Handbook 119, July 1976.

2. *Environmental Noise Control,* 1975, E. B. Magrab, John Wiley & Sons, Inc., New York.

3. *Sound, Man and Building,* 1976, L. H. Schaudinischky, Applied Science Publishers LTD, London.

4. *Acoustics,* 1954, L. L. Beranek, McGraw-Hill Book Company, New York.

5. *Noise Reduction,* 1960, L. L. Beranek, McGraw-Hill Book Company, New York.

6. *Handbook of Noise Control,* 1957, C. Harris, McGraw-Hill Book Company, New York.

7. *Handbook of Noise Rating,* April 1974, K. S. Pearsons and R. L. Bennet, NASA CR-2376, Washington, D.C., pp. 32–49.

8. *Fundamentals of Noise,* Report NTID 300.15 U.S. EPA, Washington, D.C., December 1971, pp. 55–56.

9. *Procedure for Computation of Loudness of Noise,* ANSI, USAS 53.4, 1968, New York.

10. *Music, Acoustics and Architecture,* 1962, L. L. Beranek, John Wiley & Sons, Inc., New York.

11. *Acoustic Measurements,* 1949, L. L. Beranek, John Wiley & Sons, Inc., New York.

12. L. L., Beranek, Blazer, and Figwer, "PNC Curves and Their Application to Rooms," *J. Acoust. Soc. Am.,* November 1971, pp. 1223–1228.

13. *Information on Levels of Environmental Noise,* Report 550/9-74-004, US EPA, Washington, D.C., March 1974.

14. *Proposed Damage-Risk Criterion for Impulse Noise, Report of Working Group 57,* July 1968, W. D. Ward et al, NAS-NRC Comm. on Hearing, Bioacoustics and Biomechanics (CHABA).

15. *Noise and Vibration Control,* 1971, L. L. Beranek, Ed., McGraw-Hill Book Company, New York.

16. R. W. Young, "Sabine Reverberation Equation and Sound Power Calculations," *J. Acoust. Soc. Am.,* Vol 31, #7 July 1959.

17. T. Embelton, "Absorption Coefficients of Surfaces Calculated from Decaying Sound Fields," *J. Acoust. Soc. Am.,* Vol 50, #3, 1971.

18. M. Gomperts, "Do the Classical Reverberation Formulas Still Have a Right for Existence?," *Acustica,* Vol 16, 1965/1966.

19. *A Guide to Airborne, Impact and Structure-Borne Noise Control in Multi-family Dwellings,* 1963, U.S. Dept. of Housing and Urban Development, Washington, D.C.

20. *Architectural Acoustics,* Brüel & Kjaer, 1963, Denmark.

21. *Measuring Sound,* Brüel & Kjaer.

22. *Application of B & K Measuring Systems to Acoustic Noise Measurements,* 2nd Ed., 1971, J. T. Broch, Denmark.

23. *Sound Insulation and Room Acoustics,* 1951, P. V. Bruel, Chapman and Hall, London.

24. W. J. Cavanaugh, W. R. Farrell, P. W. Hirtle, B. G. Watters, "Speech Privacy in Buildings," *J. Acoust. Soc. Amer.,* Vol 34, No. 4, April 1962.

25. R. W. Young, "Revision of the Speech Privacy Calculation," *J. Acoust. Soc. Amer.,* Vol. 38, No. 4, October 1965.

PART TEN

appendices

Sound Transmission Data for Walls

To use these tables:

a. Find desired construction Type Code in Index.
b. Find desired STC corresponding to the selected construction Type Code.
c. Refer to Table A.1 for details of wall construction.
d. Refer to Table A.2 for wall thickness weight, STC, fire rating, and the transmission losses at standard octave midpoints.

EXAMPLE. An interior masonry partition with an STC between 50 and 55 is desired. Of particular interest is the TL at 1000 Hz.

USE OF APPENDIX A

(a) From Index, we find Type Code "C."
(b) From Index, we note that constructions W25, W20, W6, W15b, and W22b have appropriate STC ratings.
(c) From Table A.1 we note that construction W15b is most suitable for the proposed use.
(d) From Table A.2 we find that construction W15b has a transmission loss of 55 db at 1000 Hz.

Index A *Sound Transmission Class, for Walls*

Type Code:

a. *wooden stud*
b. *metal stud*
c. *masonry*
d. *concrete*
e. *staggered stud*
f. *plaster*
g. *gypsum wallboard*
h. with resilient element
i. *absorbent blankets or fill*
j. *fiber board*
k. *lead*
l. *gypsum core board*
m. *double wall*

STC	*Type*	*Data Sheet No.*	*STC*	*Type*	*Data Sheet No.*
63	d,f	W4	46	a,f	W31a
62*	c,f,j,m	W23	46	a,f,k	W31c
56	c	W7	46	a,e,g,i	W37
56*	c,f	W8	45	c,d	W10b
56*	g,i,l,m	W87	45	c,f,h	W16
55*	b,g,i	W62b	45	a,f,i	W38
55*	b,g,i	W63	45	b,f,h	W50a
54*	c,f,m	W22b	45*	b,g,i	W62a
54	b,f,h	W52	45	g,i,l,m	W84
53*	d,f	W2	45	g,l,m	W85a
53	c,f,h	W15b	44	c	W12a
52*	c,f	W6	44	a,f	W30b
52*	c,f,h,i	W20	44	a,e,g	W34a
52*	c,f,m	W25	44	a,e,g	W35a
52*	b,g,j	W59	44	a,f,h	W40a
52	b,g,i	W61	43	c,d	W10a
51	b,f	W44a	43*	d,j,m	W21
51	b,g,h,i	W67	43	a,e,f	W35b
50	c,k	W12b	43	a,e,f	W36a
50	b,g,i	W60	43	b,f,k	W43b
50	g,i,l,m	W85b	43	b,f,h	W47
49*	c,f,m	W22a	42*	c,f	W5
48	c,d	W9	42	a,f	W14a
48	a,e,f,i	W36b	42	b,e,f	W71
48	b,f,k	W43c	41	b,f	W43a
48	b,g	W46	41	b,g	W55a
47	d	W1	40	c,f	W13
47	a,g,k	W28b	40	a,g	W32a
47	a,f,k	W31b	39	a,g	W28a
47	a,g,h	W39a	37	g,l	W82b
47*	b,f,h	W51a	36	a,g	W29
46	c,f	W11a	36	g,l	W80
46	c,f,h	W15a	34*	c,f	W14b
46	a,f	W30a			

*Field measurement.

Table A.1 ***Acoustic Characteristics of Walls***

Designation	Description	Section Sketch
	(for other data see Table A.2)	
Solid Concrete		
W1	3-in.-thick solid concrete wall poured in situ in test opening. All surface cavities were sealed with thin mortar mix.	
W2	6-in.-thick concrete wall with a $\frac{1}{2}$-in.-thick layer of plaster on both sides.	
W4	Wall of 4, 6, and 8- $\times$ 8 $\times$ 16 in. sand and gravel aggregate solid concrete blocks; on each side, $\frac{1}{4}$- to $\frac{1}{2}$-in.-thick layer of cement gypsum plaster and sand.	
W5	$4\frac{1}{2}$-in.-thick brick wall with a $\frac{1}{2}$-in.-thick layer of plaster on each side.	
W6	9-in.-thick brick wall with a $\frac{1}{2}$-in.-thick layer of plaster on each side.	
W7	12-in.-thick brick wall.	
W8	24-in.-thick stone wall with a $\frac{1}{2}$-in.-thick layer of plaster on both sides.	
Hollow Concrete Block		
W9	12-in. wall made of hollow 8 $\times$ 8 $\times$ 12 in. and 8 $\times$ 4 $\times$ 16 in. concrete blocks.	

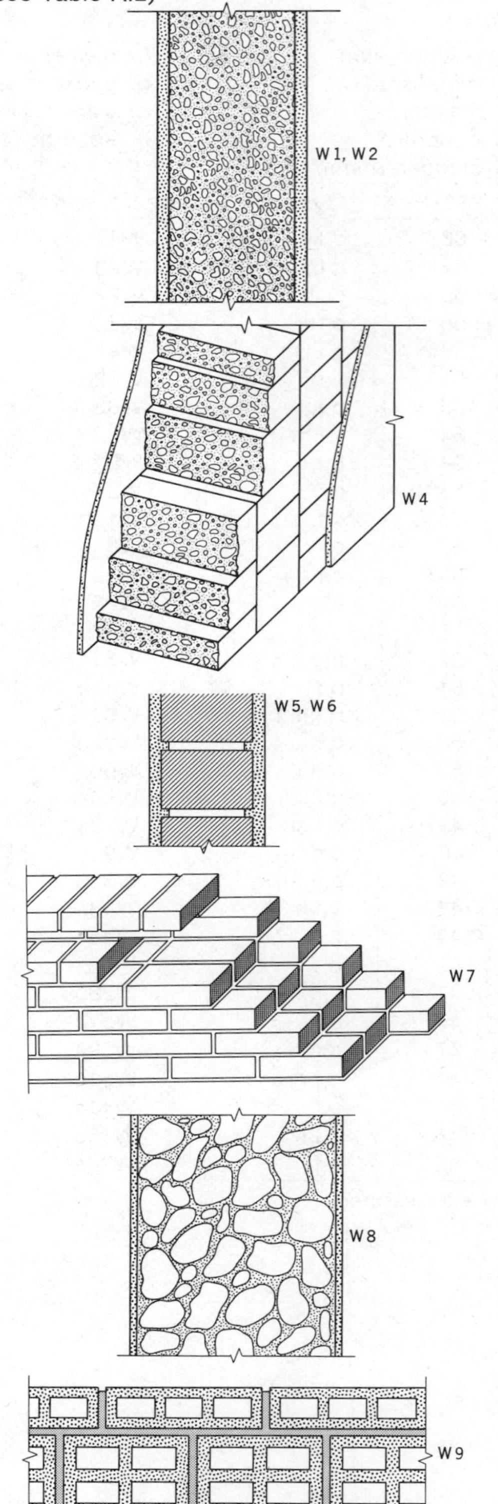

W10a 6-in. hollow concrete blocks constructed with vertical mortar joints staggered.

W10b Similar to (a) except wall was painted.

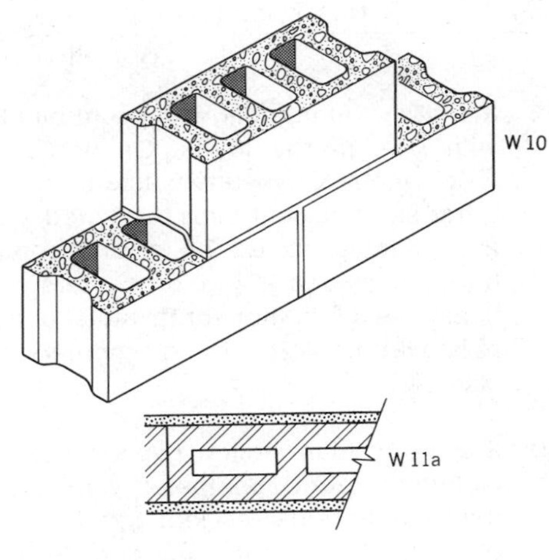

Cinder Block

W11a $4 \times 8 \times 16$ in. hollow cinder blocks; on each side $\frac{5}{8}$ in. of sanded gypsum plaster.

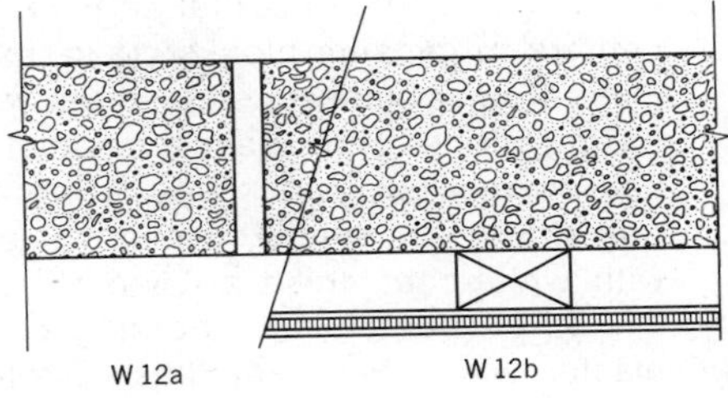

Cement Block

W12a $3\frac{5}{8} \times 7\frac{3}{4} \times 13\frac{1}{2}$ in. lightweight-aggregate cement blocks with $\frac{1}{2}$-in. mortar joints; three coats of masonry paint applied to each side of partition.

W12b Same as (a) except 1- by 2-in. furring strips were nailed vertically to partition on one side; $\frac{1}{16}$-in. layer of lead, 3.94 lb/ft^2, nailed to furring strips, $\frac{1}{4}$-in. plywood-covered lead with joints caulked.

W 12a W 12b

Hollow Gypsum Block

W13 3-in. hollow gypsum blocks cemented together with $\frac{3}{8}$-in. mortar joints; on each side, $\frac{1}{2}$-in. sanded gypsum plaster.

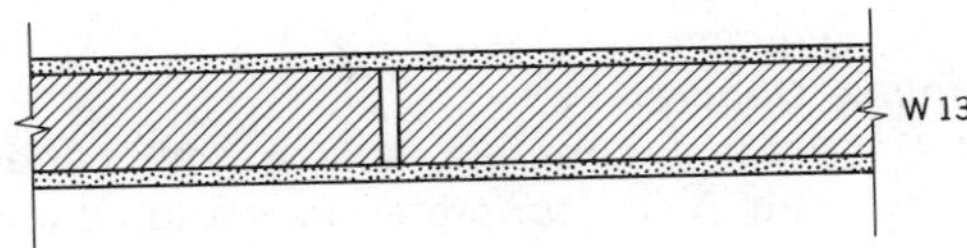

W14a 4-in. hollow gypsum blocks cemented together with $\frac{3}{8}$-in. mortar joints; on each side, $\frac{1}{2}$-in. sanded gypsum plaster.

W14b Similar to (a) except plaster coat was $\frac{5}{8}$-in. thick.

Hollow Gypsum Block, Resilient One Side, Plaster Both Sides

W15a $3 \times 12 \times 30$ in. hollow gypsum blocks with $\frac{1}{2}$-in. mortar joints. On one side $\frac{7}{16}$-in. sanded gypsum plaster; on the other side resilient clips, spaced 18 in. on centers vertically and 16 in. on centers horizontally, held to $\frac{3}{4}$-in. metal channels 16 in. on centers, to which expanded metal lath was wire-tied; $\frac{11}{16}$-in. sanded gypsum plaster. $\frac{1}{16}$-in. white-coat finish applied to both sides.

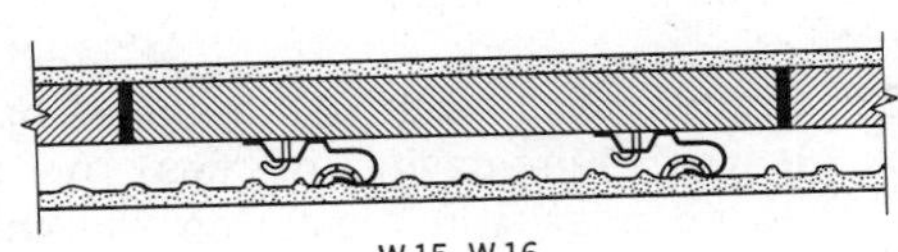

W 15, W 16

W15b Similar to (a), except $4 \times 12 \times 30$ in. gypsum blocks were used.

Table A.1 ***Acoustic Characteristics of Walls*** **(*Continued*)**

Designation	*Description*	*Section Sketch*
	(for other data see Table A.2)	
W16	3 × 12 × 30 in. hollow gypsum blocks with $\frac{1}{2}$-in. mortar joints. On one side $\frac{7}{16}$-in. sanded gypsum plaster; on the other side resilient clips, attached with 2-in. staples placed 24 in. on centers horizontally and $28\frac{1}{4}$ in. on centers vertically, held $\frac{3}{4}$-in. horizontal metal channels wire-tied $28\frac{1}{4}$ in. on centers to clips, $\frac{1}{2}$-in.	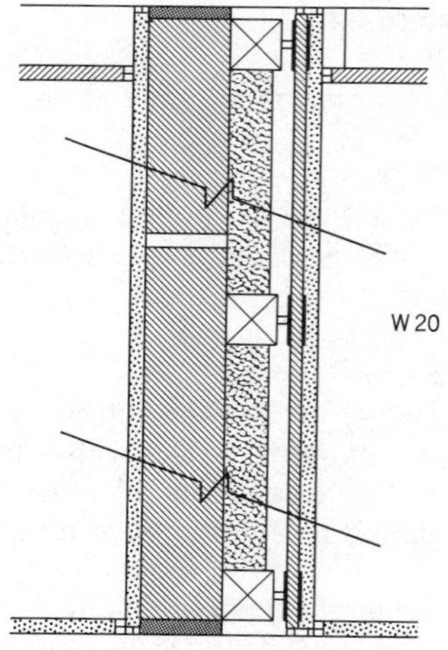
W20	3 × 12 × 30 in. hollow gypsum blocks isolated around perimeter with $\frac{1}{2}$-in.-thick continuous resilient gaskets. On one side, 2 × 2 in. wooden furring strips wire-tied horizontally 16 in. on centers to gypsum blocks. $1\frac{1}{2}$-in.-thick mineral fiber blankets stapled between furring strips. $\frac{3}{8}$-in. plain gypsum lath held by resilient clips nailed to the furring strips; $\frac{1}{2}$-in. sanded gypsum plaster with white-coat finish applied to lath. On other side, $\frac{5}{8}$-in. sanded gypsum plaster with white-coat finish applied directly to gypsum blocks. The $\frac{1}{4}$-in. clearance around perimeter closed with a nonsetting resilient caulking compound.	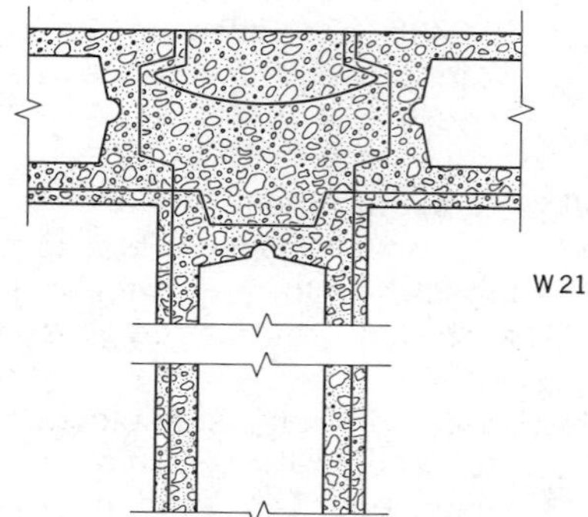
Hollow Concrete		
W21	Precast concrete hollow wall panels with in situ concrete posts and beams. The panels have $1\frac{1}{2}$-in.-thick concrete shells with a $6\frac{1}{4}$-in. airspace between them. A $\frac{1}{2}$-in.-thick layer of fiberboard is adhered to the exposed surfaces of the panel.	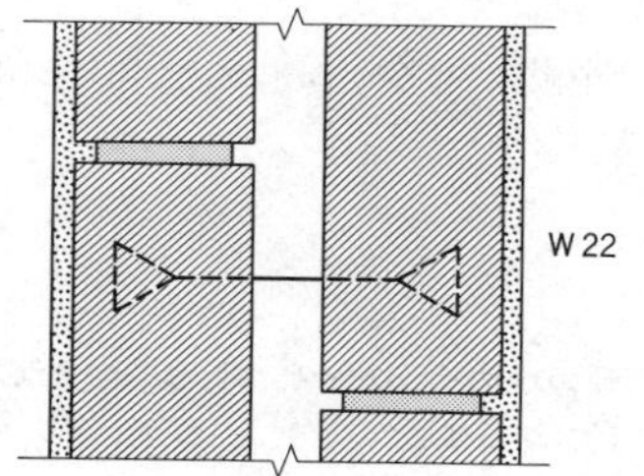
Double Walls		
W22a	Double wall with $4\frac{1}{2}$-in.-thick brick leaves separated by a 2-in. cavity (wire ties between leaves); $\frac{1}{2}$-in. plaster on exposed sides.	
W22b	Similar to (a), without wire ties between the leaves.	
W23	Double wall with $4\frac{1}{2}$-in.-thick brick leaves, 6-in. cavity (no ties); on exposed sides, $\frac{1}{2}$-in. plaster on 1-in.-thick wood-wool slabs mortared to the brick walls.	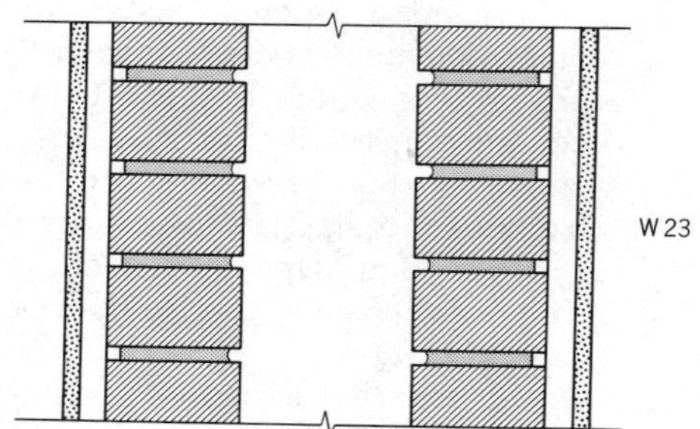

W25 Double wall with 4-in.-thick clinker block leaves, 2-in. cavity (no ties between leaves), $\frac{1}{2}$-in. plaster on exposed sides.

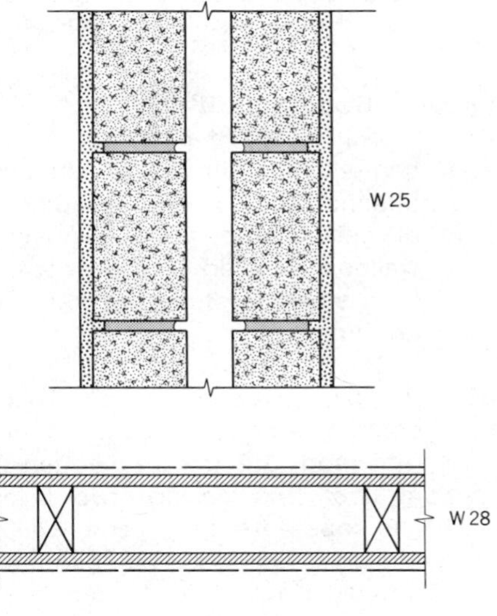

Wood Stud Walls

W28a 2×4 in. wooden studs 16 in. on centers, $\frac{1}{2}$-in. gypsum wallboard nailed to each side. All joints taped and finished.

W28b Similar to (a) except a layer of lead, 2.95 lb/ft^2, was laminated to each side of panel.

W29 2×4 in. wooden studs 16 in. on centers attached to 2×4 in. wooden floor and ceiling plates, $\frac{5}{8}$-in. tapered-edge gypsum wallboard nailed 7 in. on centers to both sides of studs. All joints taped and finished.

W30a 2×4 in. wooden studs 16 in. on centers attached to 2×4 in. wooden floor and ceiling plates, $\frac{3}{8}$-in. gypsum lath nailed to studs on both sides, $\frac{1}{2}$-in. sanded plaster with white-coat finish.

W30b Similar to (a) except the gypsum lath was perforated.

W31a 2×4 in. wooden studs 16 in. on centers, $\frac{3}{8}$-in. gypsum lath nailed to studs on both sides, $\frac{1}{2}$-in. sanded plaster with white-coat finish.

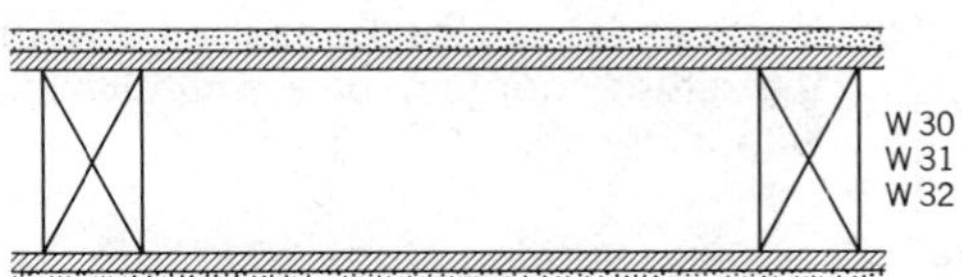

W31b Similar to (a) except a 0.065-in.-thick layer of lead weighing 3.85 lb/ft^2 was laminated to each side of panel.

W31c Similar to (a) except a 0.13-in.-thick layer of lead weighing 7.9 lb/ft^2 was laminated to one side of panel.

W32a 2×4 in. wooden studs 16 in. on centers; on each side two layers of $\frac{3}{8}$-in. gypsum wallboard cemented together; joints in exposed surfaces taped and finished.

Table A.1 ***Acoustic Characteristics of Walls (Continued)***

Designation	*Description*	*Section Sketch*

(for other data see Table A.2)

Staggered Wood Stud Walls

W34a 2 × 3 in. wooden studs 16 in. on centers, staggered 8 in. on centers, attached to 2 × 4 in. wooden plates at ceiling and floor; $\frac{1}{2}$-in. gypsum wallboard nailed 7 in. on centers on both sides to studs. All joints taped and finished.

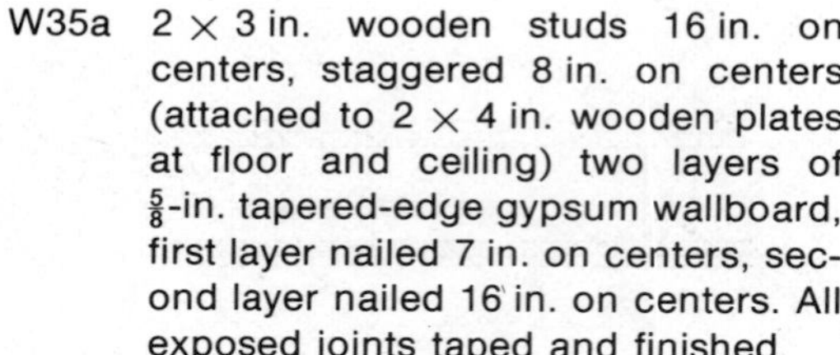

W35a 2 × 3 in. wooden studs 16 in. on centers, staggered 8 in. on centers (attached to 2 × 4 in. wooden plates at floor and ceiling) two layers of $\frac{5}{8}$-in. tapered-edge gypsum wallboard, first layer nailed 7 in. on centers, second layer nailed 16 in. on centers. All exposed joints taped and finished.

W35b Similar to (a) except the wall was constructed with $\frac{3}{8}$-in. perforated gypsum lath and $\frac{1}{2}$-in. sanded gypsum plaster with white-coat finish.

W36a 2 × 4 in. wooden studs 16 in. on centers, staggered 8 in. on centers, and offset $\frac{1}{2}$ in. On each side $\frac{3}{8}$-in. gypsum lath nailed to studs, $\frac{1}{2}$-in. gypsum vermiculite plaster, machine-applied, and a hand-applied white-coat finish.

W36b Same as (a) except the space between the studs contained vermiculite fill with a density of 6.3 lb/ft^3.

W37 2 × 4 in. wooden studs 16 in. on centers, staggered 8 in. on centers, attached to 2 × $4\frac{3}{4}$ in. wooden floor and ceiling plates; $\frac{1}{2}$-in. gypsum wallboard nailed on both sides to studs, 0.9-in. wood-fiber wool blanket stapled on the inside of one side of the wall. All joints taped and finished.

Slotted Wood Studs

W38 2 × 4 in. slotted wooden studs 16 in. on centers attached to 2 × 4 in. wooden floor and ceiling plates, $\frac{3}{8}$-in. gypsum lath nailed 7 in. on centers to studs, $\frac{1}{2}$-in. gypsum plaster with white-coat finish applied to both sides. 3-in.

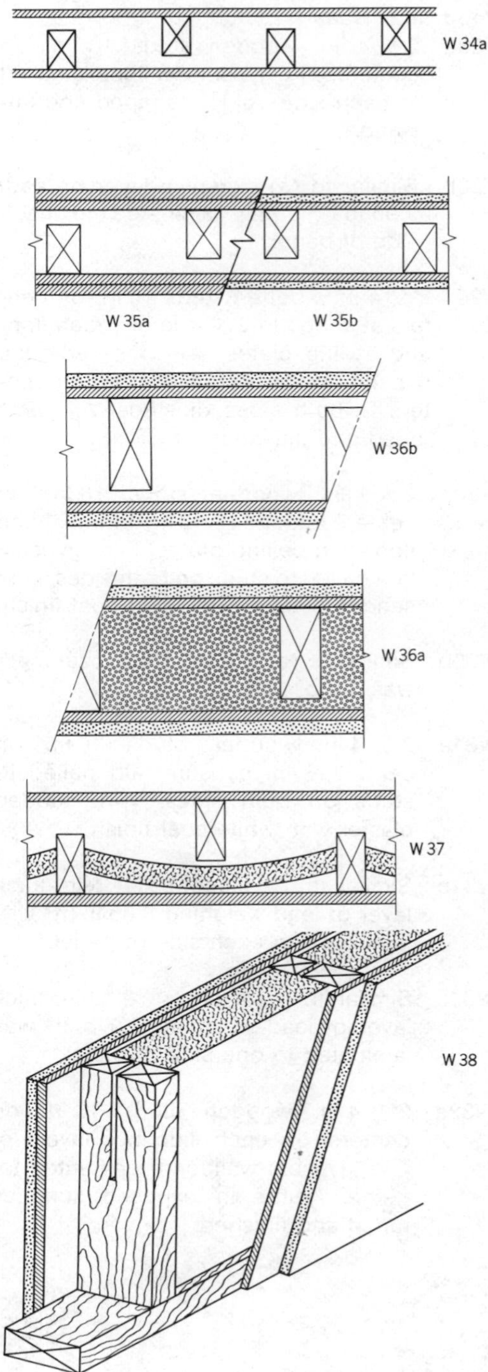

mineral fiber batts stapled between studs.

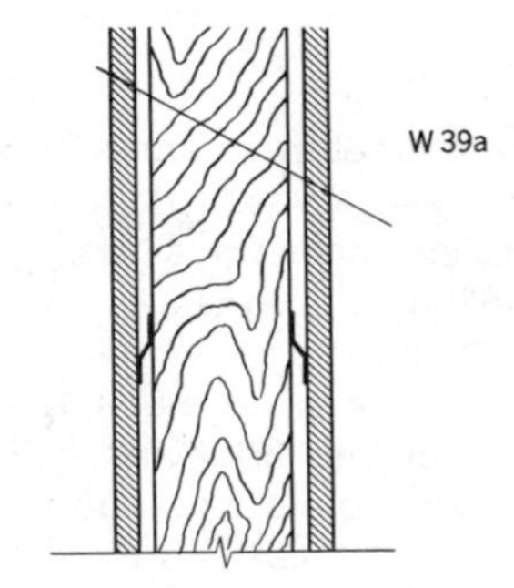

Wood Studs Resilient Mountings

W39a 2×4 in. wooden studs 16 in. on centers attached to 2×4 in. wooden floor and ceiling plates, resilient channels nailed horizontally to both sides of studs 24 in. on centers, $\frac{5}{8}$-in. gypsum wallboard screwed 12 in. on centers to channels. All joints taped and finished.

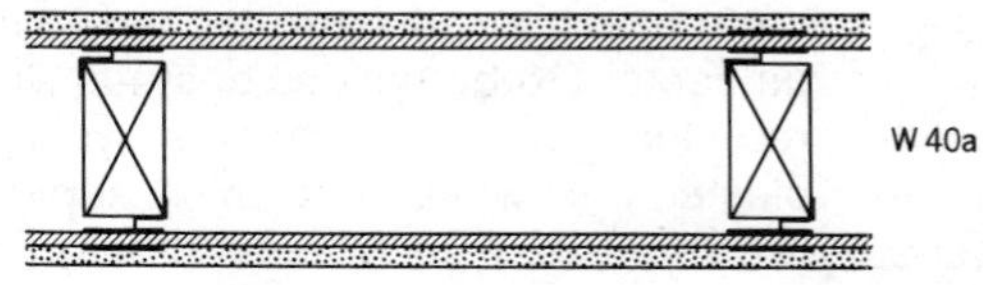

W40a 2×4 in. wooden studs 16 in. on centers; resilient clips, nailed to studs on both sides, held $\frac{3}{8}$-in. gypsum lath, $\frac{1}{2}$-in. sanded gypsum plaster with white-coat finish.

Steel Truss Stud Wall

W43a $1\frac{5}{8}$-in. steel truss studs; $\frac{3}{8}$-in. gypsum lath, $\frac{1}{2}$-in. plaster on both sides.

W43b Similar to (a) except a layer of lead, 2.95 lb/ft^2, was laminated to one side of partition.

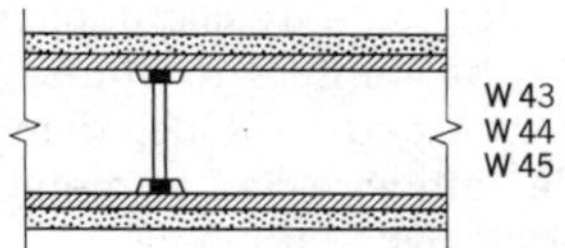

W43c Similar to (a) except a layer of lead, 2.95 lb/ft^2, was laminated to each side of partition.

W44a $3\frac{1}{4}$-in. steel truss studs 24 in. on centers attached to metal floor and ceiling tracks; on both sides $\frac{3}{8}$-in. perforated gypsum lath attached with wire clips wire-tied to studs, $\frac{1}{2}$-in. sanded gypsum plaster.

W46 $3\frac{1}{4}$-in. steel truss studs, 16 in. on centers, attached to top and bottom by stud shoes, starter clips, and stud tracks; $\frac{3}{8}$-in. gypsum wallboard (backer board) clipped to studs with galvanized wire clips; edges of wallboard held together by galvanized steel clips, $\frac{3}{8}$-in. gypsum wallboard laminated to the inner layer with joint cement.

Steel Truss Studs; Resilient Mountings

W47 $3\frac{1}{4}$-in. steel truss studs 16 in. on centers. $\frac{3}{8}$-in. gypsum lath attached on one side with resilient clips, and to the other side with galvanized wire clips; $\frac{1}{2}$-in. sanded gypsum plaster applied to both sides.

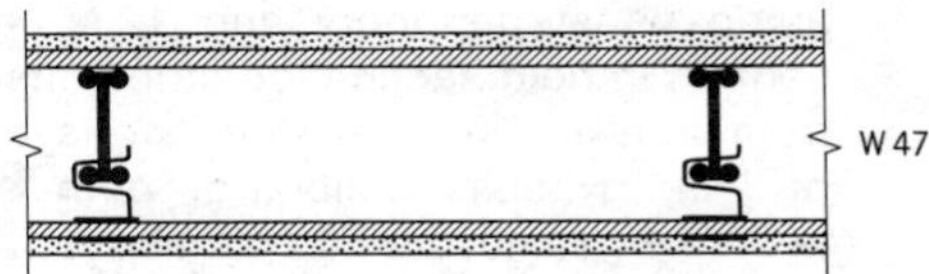

Table A.1 *Acoustic Characteristics of Walls (Continued)*

Designation	*Description*	*Section Sketch*

(for other data see Table A.2)

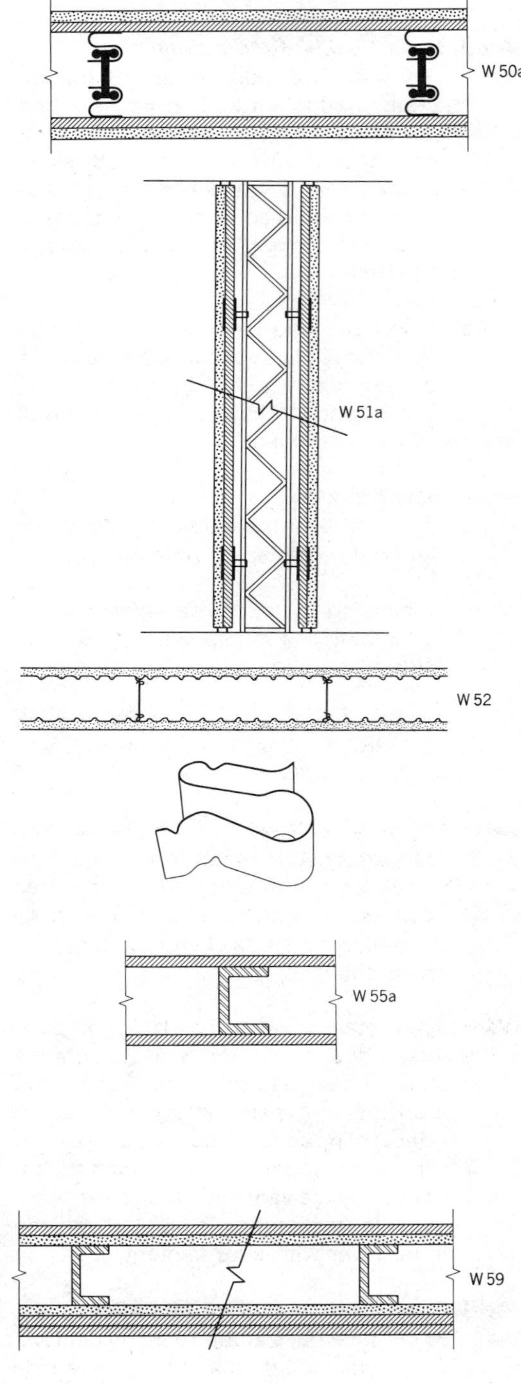

W50a $2\frac{1}{2}$-in. steel truss studs 16 in. on centers, $\frac{3}{8}$-in. gypsum lath attached with resilient clips to studs, $\frac{1}{2}$-in. plaster applied to both sides.

W51a $2\frac{1}{2}$-in. steel truss studs 16 in. on centers set into floor and ceiling tracks. Tracks isolated at floor and ceiling with $\frac{1}{4}$-in.-thick continuous resilient gaskets. Floor track attached 24 in. on centers to concrete slab. $\frac{3}{8}$-in. perforated gypsum lath attached 16 in. on centers to both sides of studs with resilient clips; $\frac{1}{2}$-in. sanded gypsum plaster with white-coat finish applied to lath. The $\frac{3}{16}$-in. clearance around perimeter closed with a nonsetting resilient caulking compound. One face of the wall primed with a pigmented sealer and the other face with shellac.

W52 $3\frac{1}{4}$-in. steel truss studs 16 in. on center; on each side resilient clips fastened 16 in. on centers to studs, $\frac{1}{4}$-in. metal rod wire-tied to clips, diamond mesh metal lath wire-tied to metal rods, $\frac{3}{4}$-in. sanded gypsum plaster.

Metal Channel Stud Wall

W55a $3\frac{5}{8}$-in. metal channel studs 24 in. on centers set into $3\frac{5}{8}$-in. metal floor and ceiling runners; $\frac{5}{8}$-in. gypsum wallboard screwed to studs on both sides. All joints taped and finished.

W59 $3\frac{5}{8}$-in. metal channel studs 24 in. on centers. Top, bottom, and side channels isolated from concrete floor and ceiling with a resilient caulking compound. $\frac{1}{2}$-in. mineral fiberboard screwed 24 in. on centers to each side. On one side, $\frac{1}{2}$-in. gypsum wallboard laminated and screwed 8 in. on centers along panel periphery and 12 in. on centers in field; lamination strips offset from screws. On other side, two layers of $\frac{1}{2}$-in. gypsum wallboard, both attached in same manner as above. All exposed joints taped and finished.

W60 2½-in. metal channel studs 24 in. on centers set in 2½-in. metal floor and ceiling runners; ½-in. vinyl-coated gypsum wallboard adhesively attached and screwed to studs on both sides. All joints sealed with caulking compound. Aluminum batten strips screwed 12 in. on centers to gypsum board at joints; top and bottom finished with aluminum ceiling and base trim. 2-in. mineral fiber blankets hung between studs.

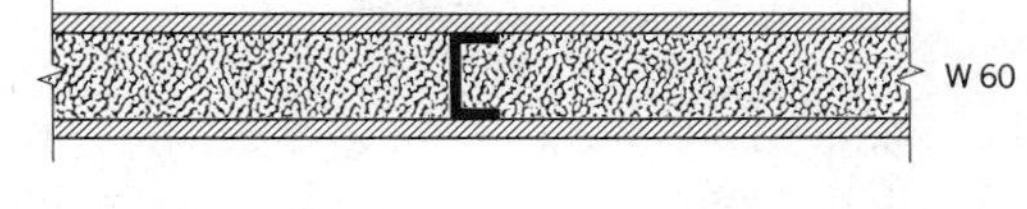

W61 2½-in. metal channel studs 24 in. on centers set into metal floor and ceiling runners, which were set on beads of nonsetting resilient caulking compound. Two layers of ½-in. gypsum wallboard attached to both sides of studs, both layers screwed 12 in. on centers with screws of each layer staggered 6 in. relative to each other. 3½-in.-thick glass fibered blankets, 2 lb/ft^3, stapled between studs. The ¼-in. clearance around the perimeter closed with a nonsetting resilient caulking compound.

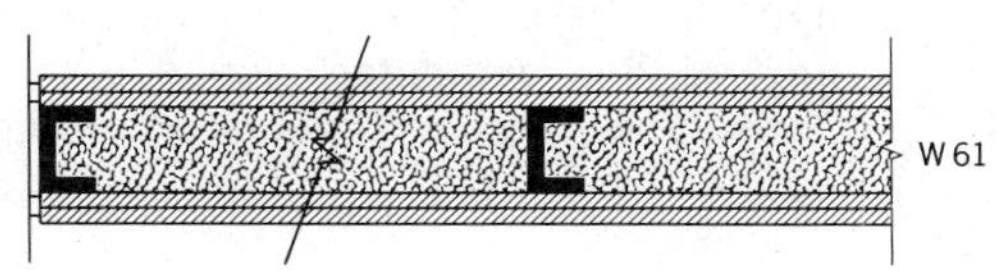

W62a 3⅝-in. metal channel studs 24 in. on centers set into floor and ceiling runners; stud at each adjoining wall and metal runners set on beads of nonsetting resilient caulking compound. On both sides, two layers of ⅝-in. gypsum wallboard; first layer screwed 12 in. on centers midway between joints and 8 in. on centers along joints, second layer glued to base layer with mastic spread so as to omit areas falling on the studs, screws 24 in. on centers at joints only. 1½-in. mineral wool blankets, 3 lb/ft^3, stapled between studs. The ¼-in. clearance around the perimeter closed with a nonsetting resilient caulking compound. All exposed joints taped and finished.

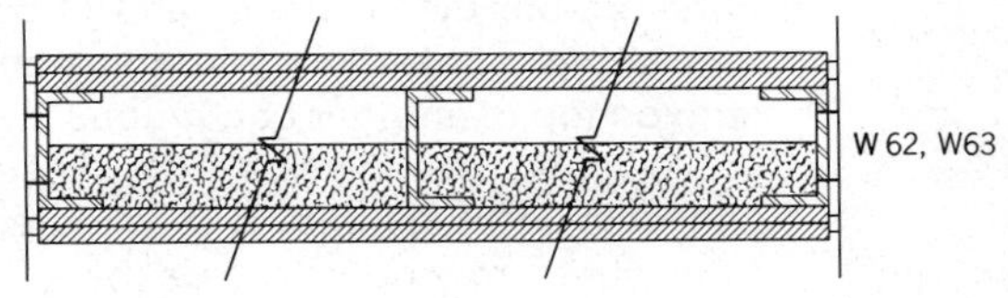

W62b Same as (a) except all possible flanking paths were eliminated. Therefore, the difference in sound transmission loss values may be attributed to the presence of flanking paths through the exterior window wall.

Designation	*Description*	*Section Sketch*
	(for other data see Table A.2)	
W63	$3\frac{5}{8}$-in. metal channel studs 24 in. on centers set into $3\frac{5}{8}$-in. metal runners, which were attached through continuous beads of nonsetting resilient caulking compound to floor and ceiling, respectively. Two layers of $\frac{5}{8}$-in. gypsum wallboard attached to both sides of studs; first layer screwed 8 in. on centers at joints and 12 in. on centers in field, second layer laminated and screwed 24 in. on centers to first layer with joints staggered 24 in. $1\frac{1}{2}$-in.-thick mineral fiber felt, 3 lb/ft^3, stapled between studs. All exposed joints taped and finished. The $\frac{1}{4}$-in. clearance around the perimeter closed with a nonsetting resilient caulking compound.	
W67	$3\frac{5}{8}$-in. metal channel studs 24 in. on centers set in $3\frac{5}{8}$-in. metal floor and ceiling runners; $\frac{5}{8}$-in. gypsum wallboard screwed to studs on both sides. On one side, resilient channels screwed horizontally 24 in. on centers to inner layer; $\frac{5}{8}$-in. gypsum wallboard screwed to channels. On the other side, $\frac{5}{8}$-in. gypsum wallboard laminated directly to inner layer. 3-in. mineral fiber blankets hung between studs. All exposed joints taped and finished.	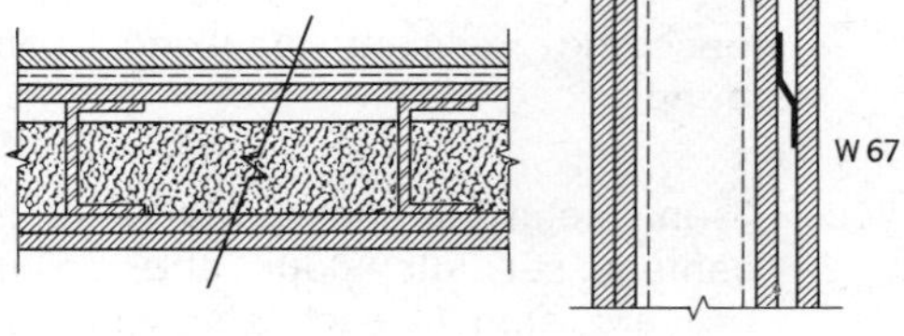
W71	Staggered $\frac{3}{4}$-in. cold-rolled steel channels, spaced 16 in. on centers, staggered $\frac{1}{2}$ in. and offset $\frac{1}{4}$ in. relative to opposite face. Channels held at top by punched-out metal runner and at bottom, set into holes in a $\frac{1}{4}$-in.-thick cork strip on top of another continuous layer of $\frac{1}{4}$-in. cork. On each side, $\frac{3}{8}$-in. gypsum lath held to studs with wire clips and from studs of opposite side by $\frac{3}{8}$-in.-thick, sponge-rubber dots; $\frac{1}{2}$-in. perlite gypsum plaster applied to lath.	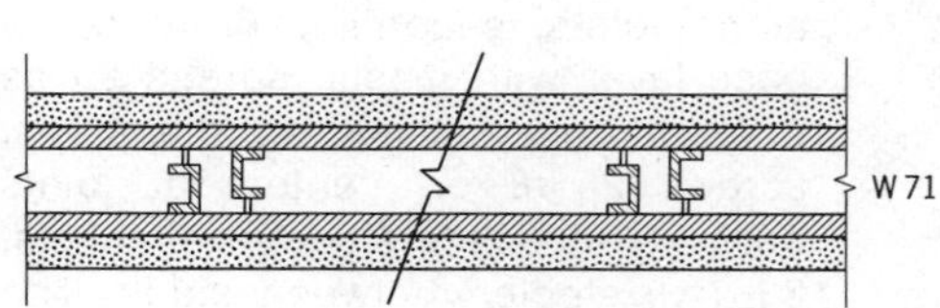

Gypsum Partitions

W80 24-in.-wide panels constructed of 1 × 24 in. gypsum core board offset $1\frac{1}{2}$ in. at edges to form tongue and groove edge; $\frac{5}{8}$-in. vinyl-faced gypsum wallboard laminated to both sides of core board. Panels inserted into two-piece metal floor and ceiling tracks. Gypsum to gypsum screws at $\frac{1}{4}$ and $\frac{1}{2}$ points along vertical edges of face boards.

W82b 24-in.-wide panels constructed of 1-in. gypsum core board strips, $7\frac{1}{2}$ in. and $4\frac{3}{8}$ in. wide, offset $1\frac{1}{2}$ in. at edges to form tongue and groove; $\frac{5}{8}$-in., vinyl-faced, gypsum wallboard laminated to both sides of core board strips. Panels inserted into two-piece metal floor and ceiling tracks. Gypsum to gypsum screws at $\frac{1}{4}$ points along vertical edges of face boards.

W84 1-in. tongue-and-groove gypsum core board attached to both sides of $2\frac{1}{2}$-in. metal channel runners at floor and ceiling, $\frac{1}{2}$-in. gypsum wallboard laminated to each core board. 2-in. mineral wool batts glued to inside surface of core board. All exposed joints taped and finished.

W85a Double wall with $1\frac{3}{8}$-in. airspace. Each leaf consisted of 24-in.-wide panels of $\frac{5}{8}$-in. gypsum core board strips, $7\frac{1}{2}$ in. and $4\frac{3}{8}$ in. wide, offset $1\frac{1}{2}$ in. at edges to form tongue and groove. $\frac{5}{8}$-in., vinyl-faced, gypsum wallboard laminated to both sides of core board strips. Panels screwed 12 in. on centers to $1\frac{1}{4}$ × 1 in. angle floor and ceiling runners.

W85b Similar to (a) except the space between leaves was $2\frac{1}{8}$ in. and contained 2-in. mineral fiber blankets stapled to one leaf. $\frac{1}{4}$-in. perimeter clearance closed with a nonsetting resilient caulking compound. Vertical face layer joints sealed with joint compound.

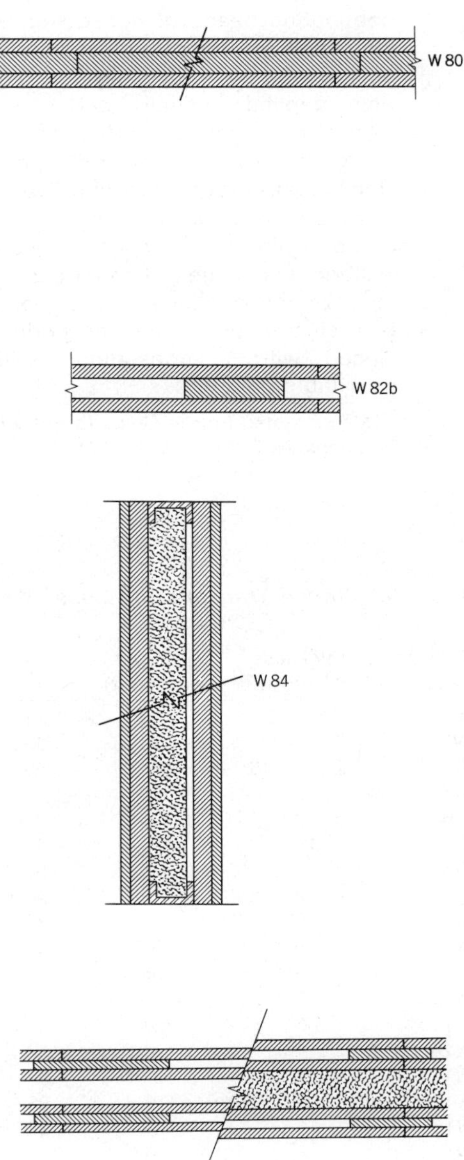

Table A.1 ***Acoustic Characteristics of Walls (Continued)***

Designation	*Description*	*Section Sketch*
	(for other data see Table A.2)	
W87	A pair of $\frac{3}{4}$ in. × 1 in. steel angle runners, with a 3-in. separation, set on continuous beads of nonsetting resilient caulking compound at floor and ceiling. 1 × 24 in. gypsum core board units applied vertically, and screwed 12 in. on centers to the angles with joints staggered 12 in. on opposite sides. $1\frac{1}{2}$-in.-thick mineral fiber blankets stapled to one inside surface of core boards. $\frac{1}{2}$-in.-thick by 48-in. gypsum wallboard laminated to core board with joints of opposite faces staggered. The $\frac{1}{4}$-in. clearances around the perimeter closed with a nonsetting resilient caulking compound.	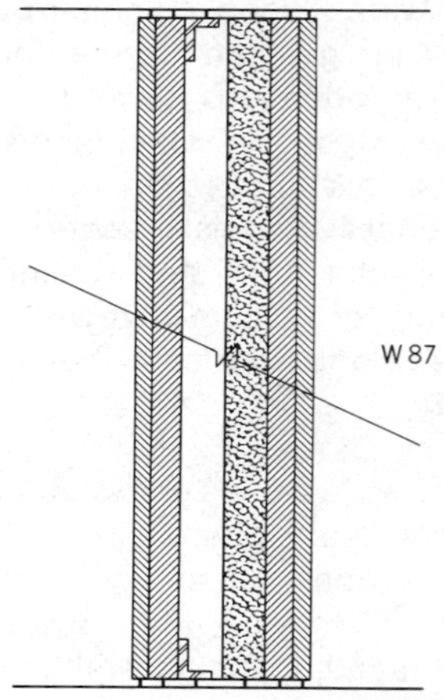

Source. Data extracted from *A Guide to Airborne, Impact and Structure-Borne Noise Control in Multi-Family Dwellings,* HUD/FHA/NBS, 1971.

Table A.2 ***Sound Transmission Loss—Walls***

Desig-nation	*Thickness (in.)*	*Weight (lb/sq ft)*	*Transmission Loss (db)* 125	250	500	1K	2K	4K	*STC*	*Fire Rating (hr)*
W1	3	39	35	40	44	52	58	64	47	$\frac{1}{2}$
W2	7	80	39	42	50	58	64	—	53	3
W4	approx. 16	184	50	54	59	65	71	68	63	4+
W5	$5\frac{1}{2}$	55	34	34	41	50	66	—	42	2.5
W6	10	100	41	43	49	55	57	—	52	4+
W7	12	121	45	45	53	58	60	61	56	4+
W8	25	280	50	53	52	58	61	—	56	4+
W9	12	79	46.5	44	46	52	54	56	48	4
W10a	6	34	32	33	40	47	51	48	43	1
W10b	6	34	37	36	42	49	55	58	45	1
W11a	$5\frac{1}{4}$	35.8	36	37	44	51	55	62	46	2
W12a	$3\frac{3}{4}$	26.1	40	40	40	48	55	56	44	1.5
W12b	5	31	41	46	46	56	63	67	50	1.5
W13	4	21.5	39	34	38	43	48	46	40	3
W14a	5	23.4	37	42	39	44	49	49	42	4
W14b	5	23.4	30	31	33	39	42	46	34	
W15a	5	27	38	37	44	51	56	59	46	3
W15b	6	31	45	44	50	55	56	59	53	4
W16	approx. $5\frac{1}{2}$	27	48	41	47	45	55	62	45	3
W20	7	22.9	32	46	49	53	58	66	52	3

Table A.2 ***Sound Transmission Loss—Walls (Continued)***

Designation	*Thickness (in.)*	*Weight (lb/sq ft)*	*Transmission Loss (db)*						*STC*	*Fire Rating (hr)*
			125	*250*	*500*	*1K*	*2K*	*4K*		
W21	$10\frac{1}{4}$	37	41	42	46	51	52	—	43	not available
W22a	12	100	37	41	48	60	60.5	—	49	4+
W22b	12	100	40	44	55	67.5	70	—	54	4+
W23	18	120	48	54	58	64	69	—	62	4+
W25	11	70	37	40	49	57	65	—	52	4+
W28a	5	6	21	28	35	42	45	41	39	0.5
W28b	approx. $5\frac{1}{8}$	12	27	37	43	52	56	—	47	0.5
W29	$5\frac{1}{4}$	7.2	30	31	37	44	39	52	36	1
W30a	$5\frac{3}{4}$	13.4 to 15.7	32	37	42	47	47	63	46	0.75
W30b	$5\frac{3}{4}$	14.2	42	32	42	49	50	62	44	1
W31a	$5\frac{3}{4}$	13.4 to 15.7	32	37	42	48	48	63	46	0.75
W31b	approx. $5\frac{7}{8}$		33	41	45	52	55	65	47	0.75
W31c	approx. $5\frac{7}{8}$	17–19	32.5	40	43	47	50	62	46	0.75
W32a	$5\frac{1}{2}$	8.2	27	31	39	45	52.5	48	40	1
W34a	5	6.2	36	36	40	47	52	45	44	0.5
W35a	$6\frac{1}{2}$	13.4	41	41	46	49	41	54	44	1.5
W35b	$5\frac{3}{4}$	15.6	48	46	48	48	48	59	43	1
W36a	$6\frac{1}{4}$	11.1	36	33	42	42	41	51	43	0.75
W36b	$6\frac{1}{4}$	12.8	37	37	49	50	52	66	48	1
W37	$5\frac{3}{4}$	13.8	39	40	42	47.5	55	51.5	46	0.5
W38	$5\frac{3}{8}$	14.2	39	45	48	50	44	54	45	1
W39a	$6\frac{1}{4}$	6.7	30	40	46	50	49	49	47	1
W40a	approx. $6\frac{1}{2}$	14.4	43	41	48	50	42	56	44	1
W43a	$3\frac{3}{8}$	12.3	27	37	43	46	39	47	41	0.75
W43b	$3\frac{1}{2}$	15.2	35	43	45	47	48	58	43	0.75
W43c	approx. $3\frac{1}{2}$	18.2	36	45	47	50	53	61	48	0.75
W44a	5	15.7	34	38	47	50	52	58	51	1
W46	$4\frac{3}{4}$	7.5	35	39	44	50	50	51	48	1
W47	$5\frac{1}{2}$	13	39	45	48	51	41	54	43	0.75
W50a	$5\frac{1}{4}$	13	35	46	48	51	48	43	45	0.75
W51a	5	13	33	40	45	53	47	57	47	1
W52	5	19	50	52	55	56	52	60	54	1
W55a	$4\frac{7}{8}$	6	29	36	40	46	40	46	41	1
W59	$6\frac{1}{8}$	8.2	31	42	54	61	62	65	52	1.5
W60	$3\frac{1}{2}$	5.4	34	40	47	50	53	54	50	1
W61	$4\frac{1}{2}$	8.3	38	44	48	53	51	58	52	1.25
W62a	$6\frac{1}{8}$	11.5	32	44	41	50	47	53	45	2
W62b	$6\frac{1}{8}$	11.5	36	47	51	56	57	62	55	2
W63	$6\frac{1}{8}$	11.5	36	47	51	57	57	62	55	2
W67	$6\frac{1}{2}$	11.3	41	46	49	51	50	60	51	2
W71	$2\frac{3}{4}$	8.6	27	31	38	41	47	56	42	NA
W80	$2\frac{1}{4}$	10.2	34	34	37	38	39	45	36	1
W82b	$2\frac{1}{4}$	8.3	31	24	38	41	38	45	37	1
W84	$5\frac{5}{8}$	14.3	45	49	56	44	51	59	45	2
W85a	$5\frac{1}{8}$	14.6	36	35	45	51	53	57	45	3
W85b	6	12.8	37	37	54	56	56	62	50	3
W87	6	12.5	34	46	52	60	63	70	56	2

B

Sound Transmission and Impact Insulation Data for Floor/Ceiling Construction

To use these tables:

a. Find desired construction Type Code in Index. Index B.1 lists the constructions by STC. Index B.2 lists the constructions by IIC.

b. Find desired STC/IIC ratings corresponding to selected Type Code.

c. Refer to Table B.1 for details of construction.

d. Refer to Table B.2 for thickness weight, STC, IIC, fire rating, and transmission loss, TL, at standard octave midpoints.

Index B.1 ***Sound Transmission Class, Floor-Ceiling Constructions***

Type Code:

a. *wooden joist*
b. *metal joist*
c. *concrete or masonry*
d. *plaster ceiling*
e. *gypsum board ceiling*
f. *with resilient elements*
g. *with absorbent blankets*
h. *with separate ceiling joists*

STC	*IIC*	*Type*	*Data Sheet No.*	*STC*	*IIC*	*Type*	*Data Sheet No.*
55*	57*	c,d,f	F14	47	37	b,c,d	F57a
55*	53*	c,d,f	F19	46	74	b,c,d	F60c
54*	64*	c,d,f	F17b	46*	42*	c,d	F23
54*	51*	c,d	F13a	46*	30*	c,d	F21a
52	80	a,e,g,h	F48	45	44	a,e,f	F44
51*	53*	c,d,f	F10	44*	48*	c,d	F28
51*	48*	c,d	F7a	44*	43*	a,e,g,h	F47
51*	47*	c	F4	44	25	c	F1a
50*	53*	c,e,f	F25	43*	43*	a,d	F32b
50*	51*	c,e	F27	42*	32*	c	F22
50*	49*	c,d,f	F29	40	32	a,e,g	F40a
50*	48*	c,d	F16	39	58	a,e,g	F42a
49*	48*	c,d	F9	39*	37*	a,e	F34
48*	47*	c,d	F12	37	33	a,e	F38a
48	33	b,c,d	F60a	37	32	a,e	F39a
48	32	b,c,d,f	F59a	35	39	a,e	F36a
47	69	b,e,f,g	F61	34*	32*	a,e	F30
47	62	b,c,e	F58	29*	32*	a,e	F35a
47*	42*	c,d	F24				

*Field measurement.

EXAMPLE. A standard wooden joist floor/ceiling construction is required with a minimum IIC of 55. Any STC above 35 is acceptable.

USE OF APPENDIX B

(a) Since IIC is the determining factor, refer to Index B.2. Basic Type Code is "a."
(b) From Index B.2, note that the only two suitable constructions are F42a and F48. Since the latter involves separated joists it is eliminated, leaving construction F42a.
(c) Refer to Table B.1, detail F42a, for construction details.
(d) From Table B.2 we obtain these data: thickness $12\frac{1}{2}$ in.; weight 10.5 lb/sq ft; TL at various frequencies as shown; STC—39; IIC—58; fire rating—1 hr.

Index B.2 ***Impact Insulation Class of Floor-Ceiling Constructions***

Type Code:

a. *wooden joist*
b. *metal joist*
c. *concrete or masonry*
d. *plaster ceiling*
e. *gypsum board ceiling*
f. *with resilient ceiling element*
g. *with resilient floor element*
h. *with carpeting*
i. *with absorbent blankets*
j. *with separate ceiling joists*

IIC	*STC*	*Type*	*Data Sheet No.*	*IIC*	*STC*	*Type*	*Data Sheet No.*
80	52	a,e,h,i,j	F48	44	45	a,e,f	F44
80	44†	c,h	F2-Ia	43*	44*	a,e,i,j	F47
74	46	b,c,d,h	F60c	43*	43*	a,d	F32b
69	47	b,e,f,h,i	F61	42*	47*	c,d	F24
64*	54*	c,d,g	F17b	42*	46*	c,d	F23
62	47	b,c,e,h	F58	42	44†	c,g	F3-II(d)
59	47†	b,c,d,h	F57b	41	44†	c	F3-I(a)
58	39	a,e,h,i	F42a	39	47	a,e,f	F39b
57*	55*	c,d,g	F14	39	35	a,e	F36a
56*	29†	a,e,h	F35b	37	47	b,c,d	F57a
53*	55*	c,d,g	F19	37*	39*	a,e	F34
53*	51*	c,d,g	F10	33	48	b,c,d	F60a
53*	50*	c,e,g	F25	33	37	a,e	F38a
51*	54*	c,d,g	F13a	32	48	b,c,d,f	F59a
51*	50*	c,e	F27	32*	42*	c	F22
49*	50*	c,d,g	F29	32	40	a,e,i	F40a
48*	51*	c,d	F7a	32	37	a,e	F39a
48*	50*	c,d	F16	32*	34*	a,e	F30
48*	49*	c,d,g	F9	32*	29*	a,e	F35a
48*	44*	c,d	F28	30*	46*	c,d	F21a
47*	51*	c,g	F4	29	44†	c	F1c
47*	48*	c,d	F12	25	44	c	F1a
45	44†	c,g	F3-II(a)				

*Field measurement.
†Estimated on the basis of similar structures.

Table B.1 ***Acoustic Characteristics of Floors***

*CODE** *Description* *Section Sketch*

(for acoustic data see Table B.2)

Reinforced Concrete Slab

F1a 4-in.-thick reinforced concrete slab, isolated from support structure. Concrete was reinforced with 6 × 6 in. number 6 AWG reinforcing mesh placed at the centerline horizontal plane of the slab. All surface cavities were sealed with a thin mortar mix.

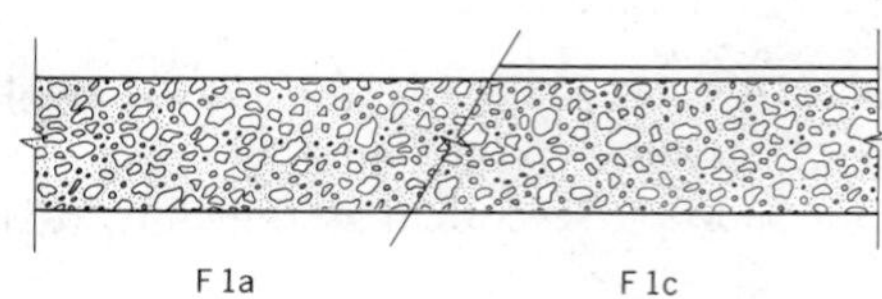

F1c Same as (a) except $\frac{1}{8}$-in.-thick vinyl tile was adhered to concrete.

Reinforced Concrete with Floor Coverings

See also F1-c above

F2-I(a) 4-in.-thick reinforced concrete slab with carpeting and pad. The carpeting was of $\frac{1}{4}$-in. wool loop pile with $\frac{1}{8}$-in. woven jute backing, 0.49 lb/ft^2; the foam rubber pad was $\frac{1}{4}$ in. thick and weighed 0.53 lb/ft^2.

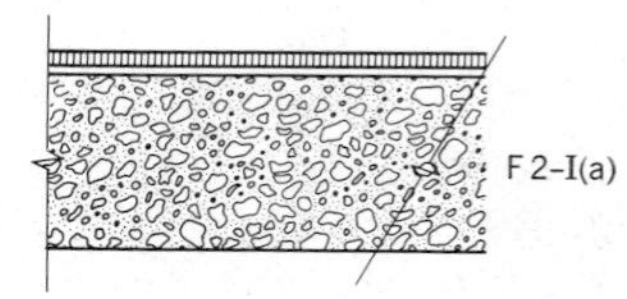

F3-I(a) 4-in. reinforced concrete slab with $\frac{1}{2} \times 9 \times 9$ in. oak blocks, 1.8 lb/ft^2, set in mastic.

F3-II(a) $\frac{1}{2}$-in. wood fiberboard, approximate density 21 lb/ft^3.

F3-II(d) $\frac{1}{8}$-in. cork.

F4 5-in.-thick reinforced concrete. On the floor side, linoleum on bitumen-felt underlayment. On the ceiling side, $\frac{1}{2}$-in.-thick papered fiberboard.

F7a $4\frac{3}{8}$-in.-thick reinforced concrete slab. On the floor side, $\frac{3}{4}$-in.-thick, sand-cement screed with $\frac{1}{8}$-in. linoleum floor covering. On the ceiling side, $\frac{3}{8}$-in. layer of plaster.

F9 $4\frac{3}{8}$-in.-thick reinforced concrete slab. On the floor side, $\frac{1}{2}$-in.-thick layer of bitumen with $\frac{1}{2}$-in.-thick soft wood fiberboard which was covered with a thin layer of bitumen with sand and a $\frac{3}{4}$-in.-thick sand-cement screed. On the ceiling side, $\frac{3}{8}$-in. layer of plaster.

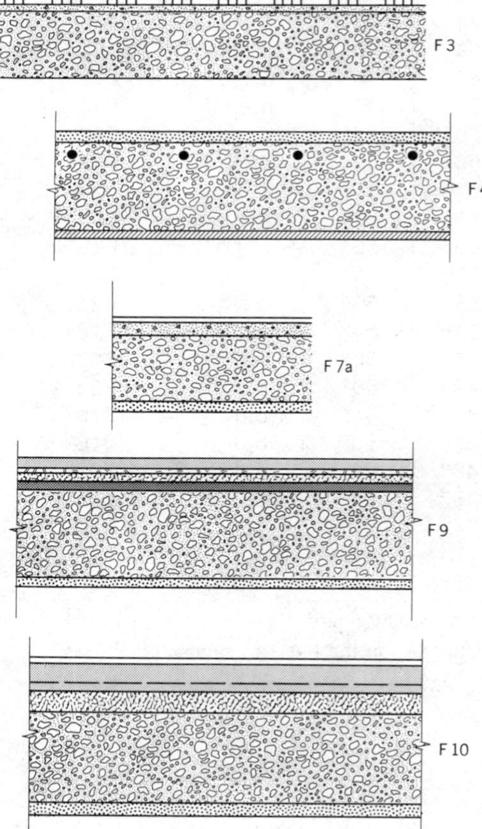

Reinforced Concrete Slab, Floating Floor

F10 5-in.-thick reinforced concrete. On the floor side, $1\frac{1}{2}$-in.-thick wire mesh reinforced sand-cement screed floating on

$\frac{1}{2}$-in.-thick bitumen-bonded, glass-wool quilt covered with building paper. On the screed, $\frac{1}{2}$-in.-thick pitch-mastic with a linoleum floor covering. On the ceiling side, $\frac{1}{2}$-in. layer of plaster.

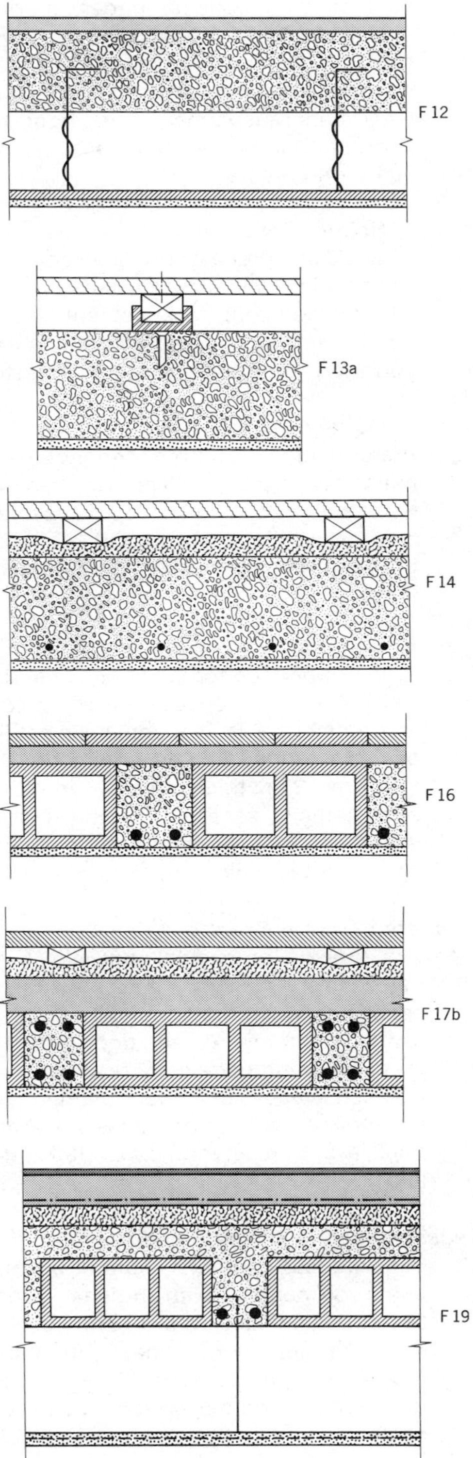

F12 $4\frac{3}{8}$-in.-thick reinforced concrete slab. On the floor side, $\frac{3}{4}$-in.-thick, sand-cement screed. On the ceiling side, brick wire mesh, suspended 4 in. with wire hangers, held $\frac{7}{8}$-in. gypsum plaster.

F13a $5\frac{1}{2}$-in.-thick reinforced concrete slab. On the floor side, $\frac{3}{4}$-in.-thick, tongue-and-groove wood flooring nailed to $1\frac{1}{2}$ in. $\times$ 2 in. wooden battens, which were held in asbestos-lined metal clips anchored to concrete slab. On the ceiling side, $\frac{1}{2}$-in. layer of plaster.

F14 6-in.-thick reinforced concrete slab. On the floor side, $\frac{3}{4}$-in.-thick, tongue-and-groove wood flooring nailed to $1\frac{1}{2} \times 2$ in. wooden battens, 16 in. on centers, floating on 1-in.-thick, glass-wool quilt. On the ceiling side, $\frac{1}{2}$-in. layer of plaster.

Concrete with Hollow Blocks

F16 5 $\times$ 10 in. hollow masonry blocks, 14 in. on centers, with spaces between blocks filled with 5-in.-thick reinforced concrete. On the floor side, $\frac{7}{8}$-in.-thick wood blocks adhered to $1\frac{1}{2}$-in.-thick, sand-cement screed. On the ceiling side, $\frac{3}{4}$-in. layer of plaster.

F17b 4 $\times$ $12\frac{1}{2}$ in. hollow masonry blocks, $15\frac{1}{2}$ in. on centers, with spaces between blocks filled with 4-in.-thick reinforced concrete. On the floor side, 2-in.-thick, sand-cement screed; linoleum on 1-in.-thick wood flooring nailed to 1 $\times$ 2 in. wooden battens, spaced $15\frac{1}{2}$ in. on centers, floating on glass-wool quilt, approximately 1 in. thick. On the ceiling side, $\frac{3}{4}$-in. layer of plaster.

F19 $5\frac{1}{2}$-in.-thick reinforced concrete with 4 $\times$ 12 in. hollow masonry blocks embedded 15 in. on centers. On the floor side, $1\frac{1}{2}$-in.-thick wire mesh reinforced sand-cement screed floating on 1-in.-

*CODE**	*Description*	*Section Sketch*

(for acoustic data see Table B.2)

thick, bitumen-bonded glass-wool quilt covered with building paper; thermoplastic tile floor covering. On the ceiling side, $\frac{1}{2}$-in. layer of plaster on ribbed expanded metal lath attached to $\frac{1}{4} \times 1\frac{1}{4}$ in. steel bars suspended 6 in. from the concrete slab by $\frac{1}{4}$-in. steel rods, spaced 48 in. on centers.

Precast Hollow Concrete Slab

F21a 6 × 20 in. precast pumice concrete slabs with cylindrical cavities. On the floor side, $\frac{3}{8}$-in.-thick cement mortar and $\frac{3}{4}$-in.-thick, sand-cement finish; on the ceiling side, $\frac{3}{8}$-in.-thick plaster.

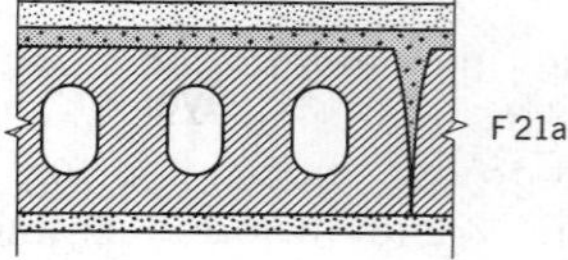

Concrete Channel Slab

F22 Prefabricated concrete channel slabs mortared together 20 in. on centers. Each slab had a 3-in.-deep trapezoidal channel with bases of 11 and $14\frac{3}{4}$ in. On the floor side, $\frac{3}{4}$-in.-thick, sand-cement finish.

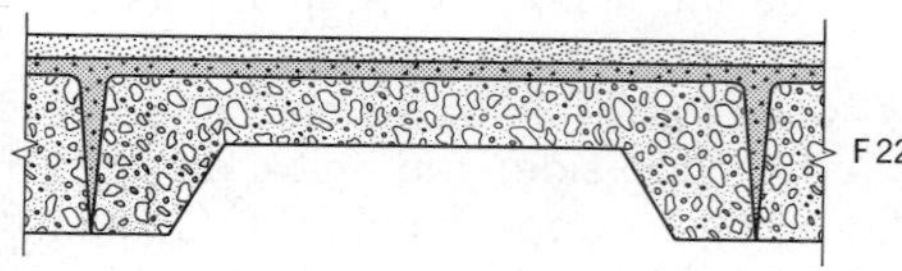

Ribbed Concrete

F23 $7\frac{1}{4}$-in. ribbed concrete floor. The ribs were $5\frac{1}{4} \times 3\frac{3}{4}$ in., spaced 21 in. on centers, with 1 × 2 in. wooden nailing strips cast into ends. On the floor side, the slab was 2 in. thick with a $\frac{3}{4}$-in.-thick, sand-cement screed. On the ceiling side, $\frac{5}{8}$-in.-thick wooden laths nailed to nailing strips, held $\frac{5}{8}$-in.-thick plaster.

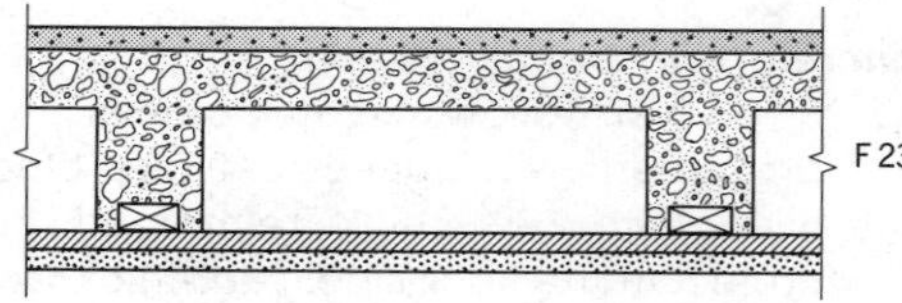

Concrete Channel Beam

F24 7-in. precast trapezoidal concrete channel beams, 14 in. on centers, with the spaces between the beams filled with a sand-cement mix. On the floor side, $1\frac{1}{2}$-in.-thick, sand-cement screed with 1-in.-thick, wood-block floor covering. On the ceiling side, approximately $\frac{3}{4}$-in.-thick layer of plaster on expanded metal lath.

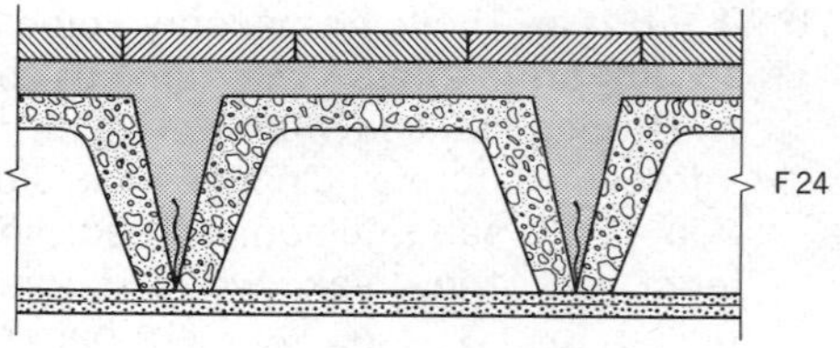

Precast Concrete Beam, Floating Floor

F25 5-in. precast concrete channel beams, $14\frac{1}{2}$ in. on centers, with the spaces between beams filled with a sand-cement mix. On the floor side, $\frac{7}{8}$-in.-thick, tongue-and-groove wood flooring nailed to 1 × 2 in. wooden battens, 20 in. on centers, on approximately 1-in.-thick,

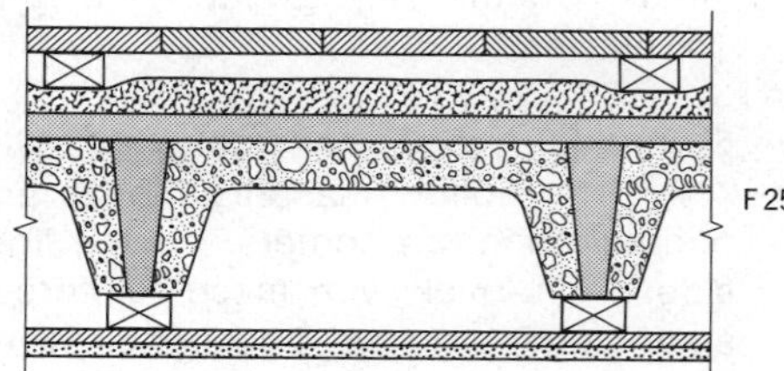

glass-wool quilt on $\frac{3}{4}$-in.-thick, sand-cement screed. On the ceiling side, $\frac{1}{8}$-in. layer of plaster on $\frac{3}{8}$-in. gypsum wallboard nailed to 1×2 in. wooden battens spaced $14\frac{1}{2}$ in. on centers.

Hollow Concrete Beam

F27 5-in. precast trapezoidal hollow concrete beams, $14\frac{1}{2}$ in. on centers, with bases of 14 in. and $12\frac{1}{2}$ in. The spaces between beams filled with a sand-cement mix. On the floor side, 1-in.-thick, sand-cement screed with $\frac{3}{16}$-in. cork tile floor covering. On the ceiling side, $\frac{3}{8}$-in.-thick gypsum wallboard attached to 1×2 in. wooden battens held by metal clips.

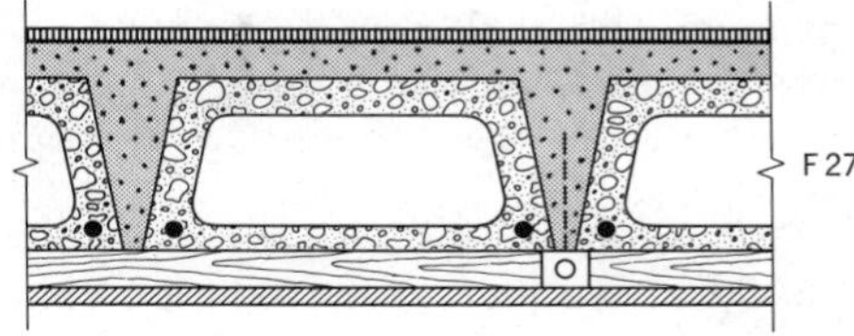

F28 7-in.-thick precast trapezoidal hollow concrete beams, $14\frac{1}{2}$ in. on centers, with bases of 14 in. and 12 in. The spaces between beams were filled with concrete. On the floor side, $\frac{7}{8}$-in.-thick, tongue-and-groove wood flooring nailed to 2×2 in. wooden battens, 18 in. on centers; linoleum floor covering. On the ceiling side, $\frac{3}{4}$-in. layer of plaster.

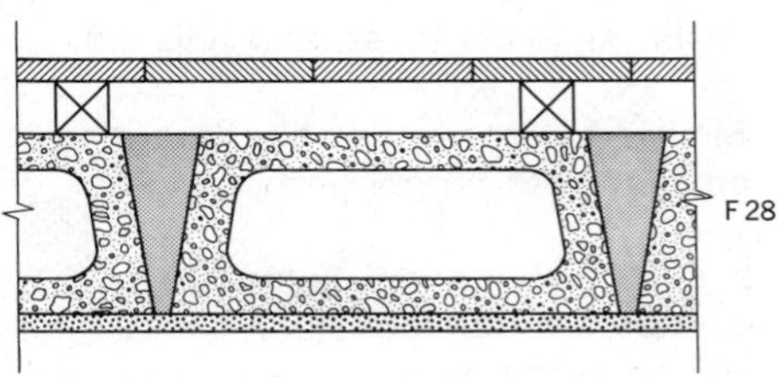

F29 5-in.-thick precast trapezoidal hollow concrete beams, $14\frac{1}{2}$ in. on centers, with bases of 14 in. and $12\frac{1}{2}$ in. The spaces between the beams were filled with a sand-cement mix. On the floor side, $\frac{7}{8}$-in.-thick, tongue-and-groove wood flooring nailed to $1\frac{1}{2} \times 2$ in. wooden battens, 20 in. on centers, floating on a glass-wool quilt, approximately 1 in. thick; linoleum floor covering. On the ceiling side, $\frac{5}{8}$-in. layer of plaster.

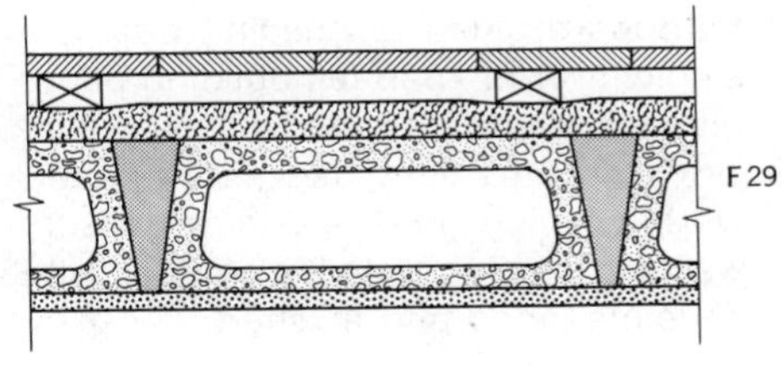

Wooden Joist

F30 2×8 in. wooden joists 16 in. on centers. On the floor side, $\frac{7}{8}$-in., tongue-and-groove flooring nailed to joists; on ceiling side, $\frac{3}{8}$-in. gypsum wallboard nailed to joists with the joints sealed.

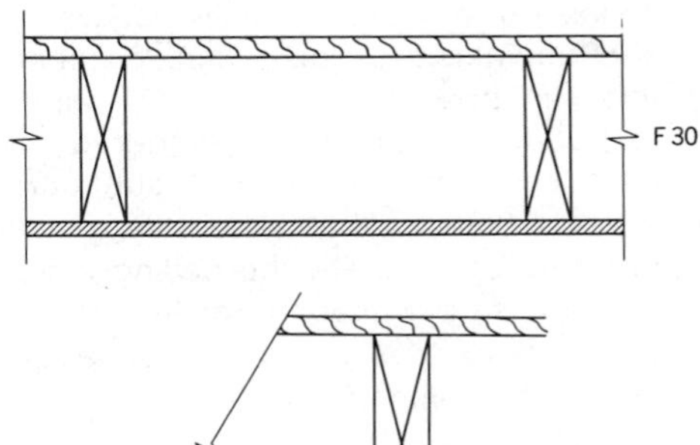

F32b 2×8 in. wooden joists 18 in. on centers. On the floor side, $\frac{7}{8}$-in., tongue-and-groove wood flooring nailed to joists. On the ceiling side, 1-in. battens nailed through glass-wool quilt, approximately 1 in. thick; $\frac{1}{2}$-in. layer plaster on $\frac{1}{4}$-in.-thick wood lath.

Table B.1 ***Acoustic Characteristics of Floors*** **(*Continued*)**

*CODE**	*Description*	*Section Sketch*
	(for acoustic data see Table B.2)	
F34	2 × 8 in. wooden joists 16 in. on centers. On the floor side, $\frac{1}{2}$-in.-thick C-D plywood nailed 8 in. on centers to joists, $\frac{25}{32}$-in.-thick hard wood flooring on plywood. On the ceiling side, $\frac{1}{2}$-in.-thick gypsum wallboard nailed 6 in. on centers to joists; all joints taped and finished; ceiling tile adhered to gypsum board.	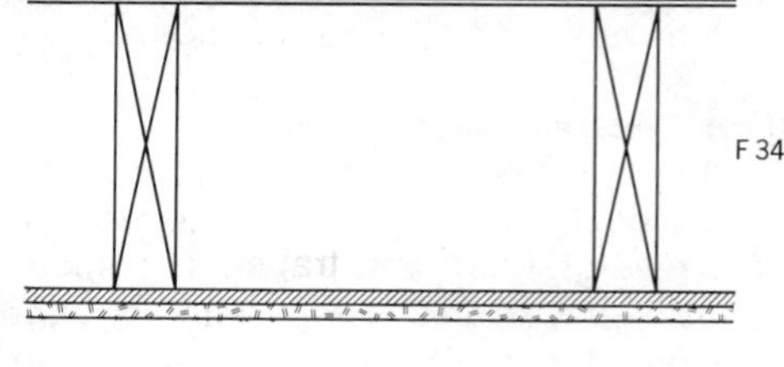 F 34
F35a	2 × 8 in. wooden joists 16 in. on centers. On the floor side, $1\frac{1}{2}$-in.-thick, tongue-and-groove wood fiberboard nailed to joists, vinyl tile floor covering. On the ceiling side, $\frac{1}{2}$-in.-thick gypsum wallboard nailed 6 in. on centers to joists. All joints taped and finished.	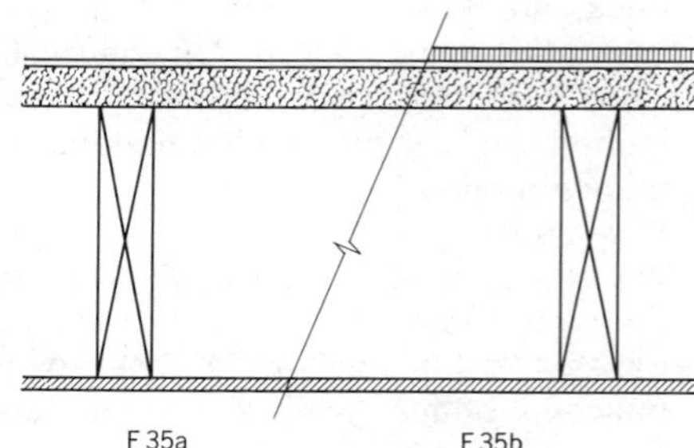 F 35a F 35b
F35b	Similar to (a) except fiberboard was covered with carpet and pad.	
F36a	2 × 10 in. wooden joists 16 in. on centers. On the floor side, $1\frac{11}{32} \times 23\frac{1}{4}$ in. compressed homogeneous paper pulp building board (approximate density 26.1 lb/ft³) nailed 8 in. on centers perpendicular to the joists, $\frac{1}{8}$-in.-thick hardboard glued to building board, a single layer of 15-lb felt building paper glued to hardboard, and $\frac{1}{8} \times 9 \times 9$ in. vinyl asbestos tile glued to felt paper. On the ceiling side, $\frac{1}{2}$-in.-thick gypsum wallboard nailed 12 in. on centers, with all joints taped and finished.	F 36a
F38a	2 × 10 in. wooden floor joists spaced 16 in. on centers. $\frac{5}{8}$-in. fir plywood subfloor nailed to joists 8 in. on centers; $\frac{1}{2}$-in. plywood underlayment nailed to subfloor with joints staggered to miss joints of the subfloor; $\frac{1}{8} \times 9 \times 9$ in. vinyl asbestos tile glued to underlayment. On the ceiling side, $\frac{1}{2}$-in. gypsum wallboard nailed 12 in. on centers with all joints and nailheads taped and finished.	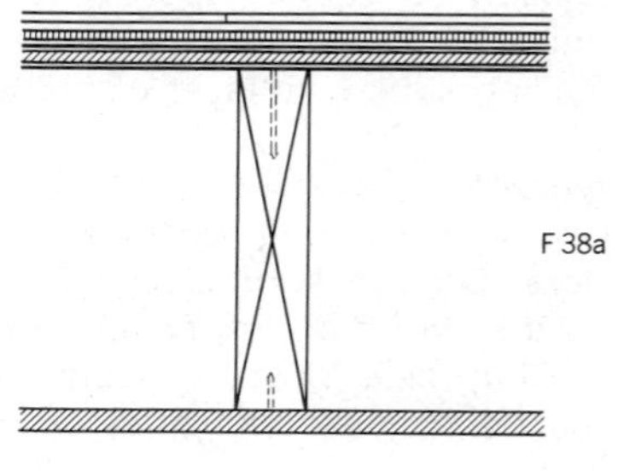 F 38a
F39a	2 × 10 in. wooden joists 16 in. on centers. On the floor side, $\frac{1}{2}$-in.-thick plywood subfloor nailed 6 in. on centers along edges and 10 in. on centers in field, building paper underlayment,	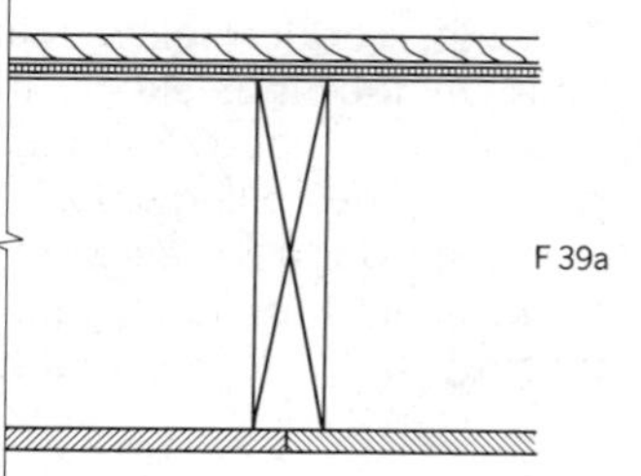 F 39a

$\frac{25}{32} \times 2\frac{1}{4}$ in. oak wood flooring nailed at each joist intersection and midway between joists. On the ceiling side, $\frac{5}{8}$-in.-thick gypsum wallboard nailed 6 in. on centers to joists; all joints taped and finished.

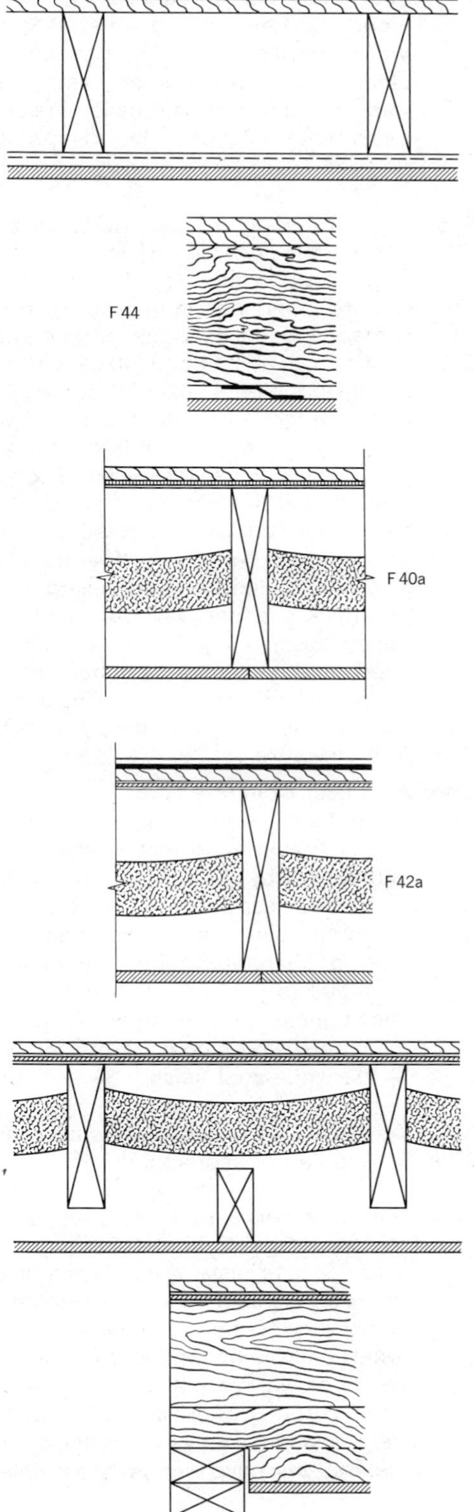

Wooden Joist, Resilient Ceiling

F44 2 × 8 in. wooden joists 16 in. on centers. On the floor side, $\frac{3}{4}$-in.-thick wood subfloor, a layer of building paper, and $\frac{3}{4}$-in.-thick tongue-and-groove fir finish flooring. On the ceiling side, resilient runners bridged across joists and nailed 12 in. on centers to the joists; $\frac{5}{8}$-in.-thick gypsum wallboard screwed to resilient runners, with all joints taped and finished.

Wooden Joist with Insulation

F40a 2 × 10 in. wooden joists 16 in. on centers with 3-in.-thick mineral fiber batts stapled between joists. On the floor side, $\frac{1}{2}$-in.-thick plywood subfloor nailed 6 in. on centers along edges and 10 in. on centers in field, building paper underlayment, $\frac{25}{32} \times 2\frac{1}{4}$ in. oak wood flooring nailed at each joist intersection and midway between joists. On the ceiling side, $\frac{5}{8}$-in.-thick gypsum wallboard nailed 6 in. on centers to joists; all joints taped and finished.

F42a 2 × 10 in. wooden joists 16 in. on centers with 3-in.-thick mineral fiber batts stapled between joists. On the floor side, $\frac{1}{2}$-in.-thick plywood subfloor nailed 6 in. on centers along edges and 10 in. on centers in field, building paper underlayment, $\frac{25}{32} \times 2\frac{1}{4}$ in. oak wood flooring nailed at each joist intersection and midway between joists; carpet, 44 oz/yd^2, with hair felt pad, 40 oz/yd^2, placed on flooring. On the ceiling side, $\frac{5}{8}$-in.-thick gypsum wallboard nailed 6 in. on centers to joists; all joints taped and finished.

F47 2 × 8 in. wooden joists 16 in. on centers with approximately 3-in.-thick fibered glass blankets stapled between joists. On the floor side, $\frac{1}{2}$-in.-thick plywood subfloor nailed 8 in. on centers to joists,

*CODE**	*Description*	*Section Sketch*

(for acoustic data see Table B.2)

$\frac{25}{32}$-in.-thick oak wood flooring on plywood. On the ceiling side, 2 × 4 in. wooden ceiling joists, 24 in. on centers, staggered between floor joists; $\frac{1}{2}$-in.-thick gypsum wallboard nailed to ceiling joists. Joints of gypsum board taped and finished.

F48 2 × 8 in. wooden joists 16 in. on centers. On the floor side, $1\frac{1}{8}$-in.-thick regular C-D rough plywood nailed 6 in. on centers along periphery and 16 in. on centers at other bearings, plywood covered with an all-hair pad (40 oz/yd²) and an all-wool pile (44 oz/yd²) carpet. The total weight of the carpet was 4.14 lb/yd² and the total thickness was $\frac{3}{8}$ in. On the ceiling side, 2 × 4 in. wooden joists 16 in. on centers staggered 8 in. on centers relative to the floor joists, 3-in.-thick fibered glass blankets stapled between ceiling joists, $\frac{5}{8}$-in.-thick gypsum wallboard nailed to ceiling joists. All joints taped and finished and the entire periphery of the panel caulked and sealed. The ceiling was supported independently from the floor structure.

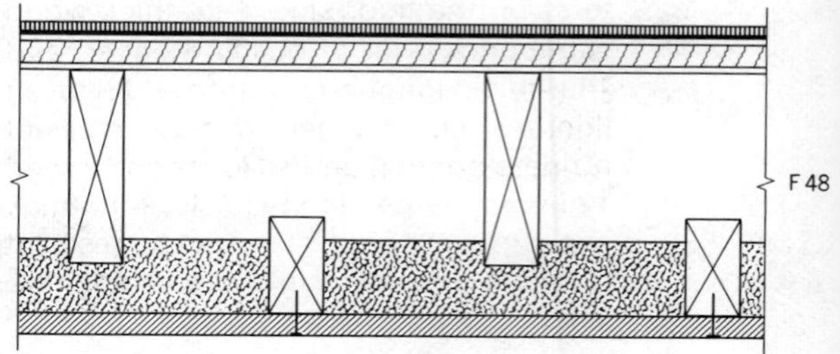

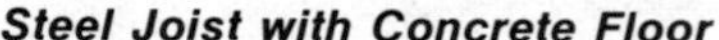

Steel Joist with Concrete Floor

F57a $2\frac{1}{2}$-in.-thick perlite concrete, 72 lb/ft³, on 28 gauge corrugated steel units supported by 14-in. steel bar joists; $\frac{1}{8}$-in.-thick asphalt tile cemented to concrete. On the ceiling side, $\frac{3}{4}$-in. furring channels, $13\frac{1}{2}$ in. on centers, wire-tied to joists, 3.4 lb/yd² diamond mesh metal lath wire-tied to furring channels, $\frac{9}{16}$-in. coat of plaster with $\frac{1}{16}$-in., white-coat finish.

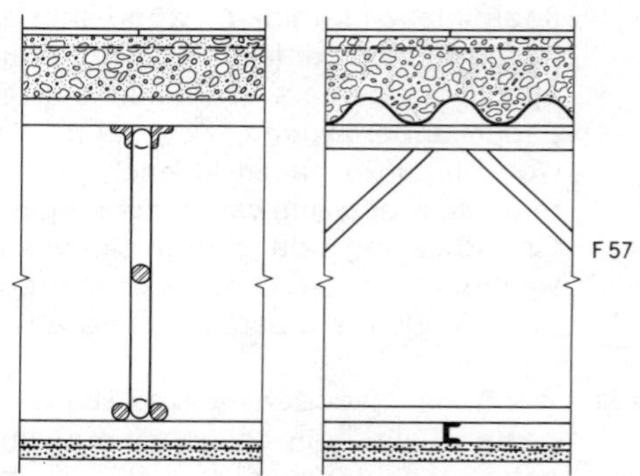

F57b Same as (a) except with carpet and pad in lieu of asphalt tile.

F58 18-in. steel joists 16 in. on centers. On the floor side, $\frac{5}{8}$-in.-thick C-D rough plywood nailed to joists, $1\frac{5}{8}$-in.-thick foamed concrete, 100 lb/ft³, slab constructed on the plywood; concrete covered with an all-hair pad (40 oz/yd²) and an all-wool pile (44 oz/yd²) carpet. The total weight of the carpet was 4.14 lb/yd² and the total thickness was $\frac{3}{8}$ in. On the ceiling side, $\frac{5}{8}$-in.-thick gypsum wallboard nailed

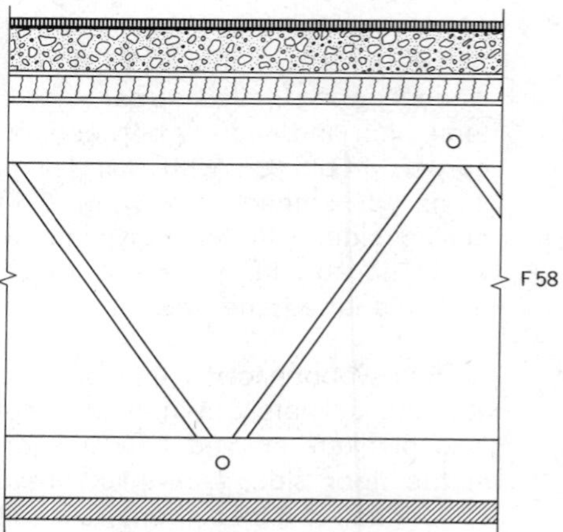

to joists. All joints taped and finished and the entire periphery of the panel caulked and sealed.

F59a 7-in. steel bar joists spaced 27 in. on centers. On the floor side, $\frac{3}{8}$-in. metal rib lath attached to top of joists, and 2-in.-thick poured concrete floor. On the ceiling side, resilient clips attached to joists held to $\frac{3}{4}$-in. metal furring channels 16 in. on centers; $\frac{3}{8}$ in. × 16 in. × 48 in. plain gypsum lath held with wire clips and sheet metal end joint clips; $\frac{7}{16}$-in. sanded gypsum plaster and $\frac{1}{16}$-in., white-coat finish.

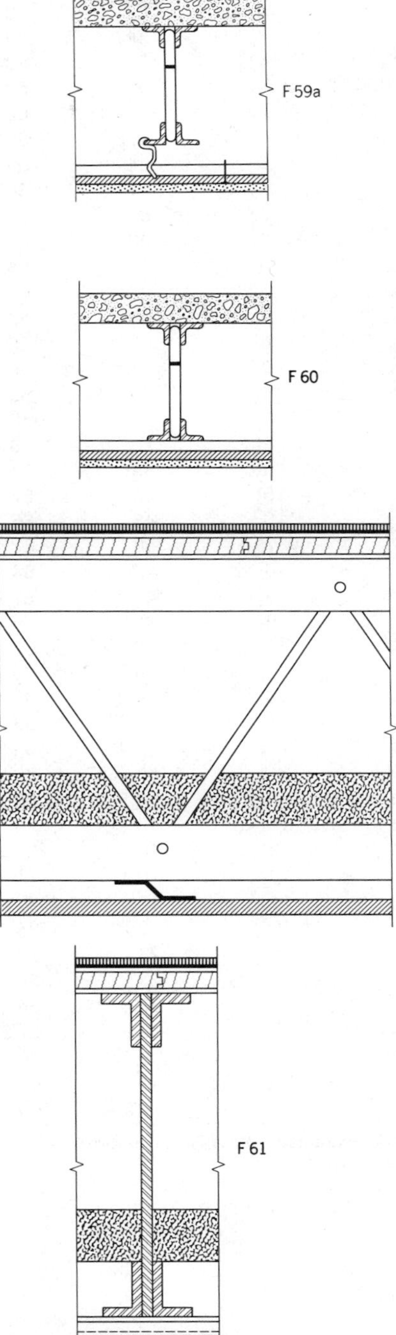

F60a 7-in. steel bar joists spaced 27 in. on centers. On the floor side, $\frac{3}{8}$-in. metal rib lath attached to top of joists, and 2-in.-thick poured concrete floor. On the ceiling side, $\frac{3}{4}$-in. metal furring channels wire-tied to joists 16 in. on centers; $\frac{3}{8}$ × 16 × 48 in. plain gypsum lath held with wire clips and sheet metal end joint clips; $\frac{7}{16}$-in. sanded gypsum plaster and $\frac{1}{16}$-in., white-coat finish.

F60c Structure (a) with nylon carpeting and foam rubber pad placed on the floor. The carpet pad had an uncompressed thickness of $\frac{1}{4}$ in., backed with a woven jute fiber cloth. The carpet had $\frac{1}{8}$-in. woven backing and $\frac{1}{4}$-in. looped pile spaced 7 loops per inch with a total thickness of $\frac{3}{8}$ in.

F61 18-in. steel joists 32-in. on centers with 3-in.-thick fibered glass blankets placed between joists. On the floor side, $1\frac{1}{8}$-in.-thick, tongue-and-groove plywood (grademarked 2-4-1) nailed to joists; plywood covered with an all-hair pad (40 oz/yd^2) and an all-wool pile (44 oz/yd^2) carpet. The total weight of the carpet was 4.14 lb/yd^2 and the total thickness was $\frac{3}{8}$ in. On the ceiling side, resilient furring channels, 24 in. on centers, screwed perpendicular to joists. $\frac{5}{8}$-in.-thick gypsum wallboard screwed 12 in. on centers to channels. All joints taped and finished and the entire periphery of the panel caulked and sealed.

*Material extracted from *A Guide to Airborne, Impact and Structure-Borne Noise Control in Multi-Family Dwellings,* HUD/FHA/NBS, 1971.

Table B.2 *Sound Transmission Loss—Floors*

*Designation**	*Thickness (in.)*	*Weight (lb/sq ft)*	*Transmission Loss (db)*						*STC*	*IIC*	*Fire Rating (hr)*
			125	*250*	*500*	*1K*	*2K*	*4K*			
F1a	4	53	59	67	75	78	77	75	44	25	1
F1c	$4\frac{1}{8}$	54	59	66	73	76	74	—	44	29	1
F2-I(a)	$4\frac{3}{8}$	54	38	36	32	23	—	—	44	80	1
F3-I(a)	$4\frac{1}{2}$	55	59	64	58	42	26	—	44	41	1
F3-II(a)	$4\frac{1}{2}$	54	63	67	69	62	47	32	44	45	1
F3-II(d)	$4\frac{1}{8}$	53	65	67	72	67	48	40	44	42	1
F4	6	64	59	64	64	63	56	—	51	47	2
F7a	$5\frac{1}{2}$	61	66	70	68	66	59	37	51	48	2
F9	$6\frac{5}{8}$	65	63	69	71	68	55	38	49	48	2
F10	$8\frac{1}{4}$	90	60	61	59	54	45	—	51	53	$2\frac{1}{2}$
F12	10	62	68	65	67	66	61	50	48	47	3
F13a	$9\frac{1}{4}$	75	61	67	62	55	43	—	54	51	3
F14	$9\frac{1}{2}$	83	59	58	55	48	42	—	55	57	3
F16	$8\frac{1}{8}$	65	56	60	60	62	54	—	50	48	2
F17b	$9\frac{1}{4}$	57	56	50	44	40	30	—	54	64	2
F19	$15\frac{1}{4}$	70	60	56	53	51	49	—	55	53	2
F21a	$7\frac{1}{2}$	43	70	78	77	77	75	—	46	30	2
F22	$6\frac{1}{4}$	28	76	69	80	79	75	—	42	32	$\frac{3}{4}$
F23	$9\frac{1}{2}$	45	71	73.5	72	71	66	54	46	42	$\frac{3}{4}$
F24	$10\frac{1}{4}$	65	67	66	67	66	62	—	47	42	$\frac{3}{4}$
F25	10	45	65	61	57	51	39	—	50	53	$\frac{3}{4}$
F27	$7\frac{5}{8}$	50	62	66	61	58	46	—	50	51	$\frac{3}{4}$
F28	$10\frac{5}{8}$	45	64	69	65	62	45	—	44	48	$\frac{1}{2}$
F29	$8\frac{3}{4}$	42	57	60	63	62	48	—	50	49	—
F30	$9\frac{1}{2}$	7	79	79	80	76	69	—	34	32	$\frac{1}{4}$
F32b	11	12	69	72	68	64	55	—	43	43	$\frac{3}{4}$
F34	$10\frac{1}{4}$	9.9	83	80	71	57	52	45	39	37	1
F35a	10	9.2	83	83	72	54	52	43	29	32	—
F35b	10	9.2	58	44	38	24	23	—	29	56	—
F36a	$12\frac{1}{4}$	8.4	83	82	76.5	63	51.5	—	35	39	$\frac{1}{2}$
F38a	$11\frac{3}{4}$	9	86.5	85.5	85	80	68.5	—	37	33	$\frac{1}{2}$
F39a	$11\frac{7}{8}$	9.5	80	80	81	71	66	58	37	32	1
F40a	$11\frac{7}{8}$	10	78	84	80	70	65	58	40	32	1
F42a	$12\frac{1}{2}$	10.5	61	47	43	25	16	14	39	58	1
F44	$10\frac{1}{2}$	10.1	71	74	70	63.5	61	—	45	44	$\frac{3}{4}$
F47	$11\frac{3}{4}$	13	77	71	68	61	53	42	44	43	$\frac{1}{2}$
F48	$12\frac{3}{8}$	10.7	34	28	20	17	18	—	52	80	$\frac{3}{4}$
F57a	$18\frac{9}{16}$	23.2	79	72	72	70	63.5	—	47	37	3
F57b	$18\frac{9}{16}$	23.2	59	43	32	18	—	—	47	52	3
F58	$21\frac{1}{2}$	20.4	56	41	33	23	17	—	47	62	1
F59a	12	40.2	67.5	69	72	75	76	—	48	32	$1\frac{1}{2}$
F60a	11	38.2	66	68.5	74	76	75	—	48	33	$1\frac{1}{2}$
F60c	$11\frac{5}{8}$	39	46	38	33.5	25	20	9	46	74	$1\frac{1}{2}$
F61	21	10.5	48	40	36	28	23	—	47	69	$\frac{3}{4}$

*Material extracted from *A Guide to Airborne, Impact and Structure-Borne Noise Control in Multi-Family Dwellings*, HUD/FHA/NBS, 1971.

Regional Weather Conditions

Table C.1 ***Average Yearly Degree Days for Cities in the United States and Canada***[a,b,c] ***(Base 65° F)***

State	Station		Avg. Winter Temp[d]	Degree Days Yearly Total
Ala.	Birmingham	A	54.2	2551
	Huntsville	A	51.3	3070
	Mobile	A	59.9	1560
	Montgomery	A	55.4	2291
Alaska	Anchorage	A	23.0	10864
	Fairbanks	A	6.7	14279
	Juneau	A	32.1	9075
	Nome	A	13.1	14171
Ariz.	Flagstaff	A	35.6	7152
	Phoenix	A	58.5	1765
	Tucson	A	58.1	1800
	Winslow	A	43.0	4782
	Yuma	A	64.2	974
Ark.	Fort Smith	A	50.3	3292
	Little Rock	A	50.5	3219
	Texarkana	A	54.2	2533
Calif.	Bakersfield	A	55.4	2122
	Bishop	A	46.0	4275
	Blue Canyon	A	42.2	5596
	Burbank	A	58.6	1646
	Eureka	C	49.9	4643
	Fresno	A	53.3	2611
	Long Beach	A	57.8	1803
	Los Angeles	A	57.4	2061
	Los Angeles	C	60.3	1349
	Mt. Shasta	C	41.2	5722
	Oakland	A	53.5	2870
	Red Bluff	A	53.8	2515
	Sacramento	A	53.9	2502
	Sacramento	C	54.4	2419
	Sandberg	C	46.8	4209
	San Diego	A	59.5	1458
	San Francisco	A	53.4	3015
	San Francisco	C	55.1	3001
	Santa Maria	A	54.3	2967
Colo.	Alamosa	A	29.7	8529
	Colorado Springs	A	37.3	6423
	Denver	A	37.6	6283
	Denver	C	40.8	5524
	Grand Junction	A	39.3	5641
	Pueblo	A	40.4	5462
Conn.	Bridgeport	A	39.9	5617
	Hartford	A	37.3	6235
	New Haven	A	39.0	5897
Del.	Wilmington	A	42.5	4930
D.C.	Washington	A	45.7	4224
Fla.	Apalachiocola	C	61.2	1308
	Daytona Beach	A	64.5	879
	Fort Myers	A	68.6	442
	Jacksonville	A	61.9	1239
	Key West	A	73.1	108
	Lakeland	C	66.7	661
	Miami	A	71.1	214
	Miami Beach	C	72.5	141
	Orlando	A	65.7	766
Fla. (Cont'd)	Pensacola	A	60.4	1463
	Tallahassee	A	60.1	1485
	Tampa	A	66.4	683
	West Palm Beach	A	68.4	253
Ga.	Athens	A	51.8	2929
	Atlanta	A	51.7	2961
	Augusta	A	54.5	2397
	Columbus	A	54.8	2383
	Macon	A	56.2	2136
	Rome	A	49.9	3326
	Savannah	A	57.8	1819
	Thomasville	C	60.0	1529
Hawaii	Lihue	A	72.7	0
	Honolulu	A	74.2	0
	Hilo	A	71.9	0
Idaho	Boise	A	39.7	5809
	Lewiston	A	41.0	5542
	Pocatello	A	34.8	7033
Ill.	Cairo	C	47.9	3821
	Chicago (O'Hare)	A	35.8	6639
	Chicago (Midway)	A	37.5	6155
	Chicago	C	38.9	5882
	Moline	A	36.4	6408
	Peoria	A	38.1	6025
	Rockford	A	34.8	6830
	Springfield	A	40.6	5429

Table C.1 *Average Yearly Degree Days for Cities in the United States and Canada*[a,b,c] *(Base 65° F) (Continued)*

State	Station	Avg. Winter Temp[d]	Degree Days Yearly Total
Ind.	Evansville . . . . A	45.0	4435
	Fort Wayne . . . . A	37.3	6205
	Indianapolis . . . . A	39.6	5699
	South Bend . . . . A	36.6	6439
Iowa	Burlington . . . . A	37.6	6114
	Des Moines . . . . A	35.5	6588
	Dubuque . . . . A	32.7	7376
	Sioux City . . . . A	34.0	6951
	Waterloo . . . . A	32.6	7320
Kans.	Concordia . . . . A	40.4	5479
	Dodge City . . . . A	42.5	4986
	Goodland . . . . A	37.8	6141
	Topeka . . . . A	41.7	5182
	Wichita . . . . A	44.2	4620
Ky.	Covington . . . . A	41.4	5265
	Lexington . . . . A	43.8	4683
	Louisville . . . . A	44.0	4660
La.	Alexandria . . . . A	57.5	1921
	Baton Rouge . . . . A	59.8	1560
	Lake Charles . . . . A	60.5	1459
	New Orleans . . . . A	61.0	1385
	New Orleans C	61.8	1254
	Shreveport . . . . A	56.2	2184
Me.	Caribou . . . . A	24.4	9767
	Portland . . . . A	33.0	7511
Md.	Baltimore . . . . A	43.7	4654
	Baltimore C	46.2	4111
	Frederich . . . . A	42.0	5087
Mass.	Boston . . . . A	40.0	5634
	Nantucket . . . . A	40.2	5891
	Pittsfield . . . . A	32.6	7578
	Worcester . . . . A	34.7	6969
Mich.	Alpena . . . . A	29.7	8506
	Detroit (City) . . . . A	37.2	6232
	Detroit (Wayne) . . . . . A	37.1	6293
	Detroit (Willow Run) . . . A	37.2	6258
	Escanaba C	29.6	8481
	Flint . . . . A	33.1	7377
	Grand Rapids . . . . . . A	34.9	6894
	Lansing . . . . A	34.8	6909
	Marquette C	30.2	8393
	Muskegon . . . . A	36.0	6696
	Sault Ste. Marie . . . . . A	27.7	9048
Minn.	Duluth . . . . A	23.4	10000
	Minneapolis . . . . A	28.3	8382
	Rochester . . . . A	28.8	8295
Miss.	Jackson . . . . A	55.7	2239
	Meridian . . . . A	55.4	2289
	Vicksburg C	56.9	2041
Mo.	Columbia . . . . A	42.3	5046
	Kansas City . . . . A	43.9	4711
	St. Joseph . . . . A	40.3	5484
	St. Louis . . . . A	43.1	4900
	St. Louis C	44.8	4484
	Springfield . . . . A	44.5	4900
Mont.	Billings . . . . A	34.5	7049
	Glasgow . . . . A	26.4	8996
	Great Falls . . . . A	32.8	7750
	Havre . . . . A	28.1	8700
	Havre C	29.8	8182
	Helena . . . . A	31.1	8129
	Kalispell . . . . A	31.4	8191
	Miles City . . . . A	31.2	7723
	Missoula . . . . A	31.5	8125
Neb.	Grand Island . . . . A	36.0	6530
	Lincoln C	38.8	5864
	Norfolk . . . . A	34.0	6979
	North Platte . . . . A	35.5	6684
	Omaha . . . . A	35.6	6612
	Scottsbluff . . . . A	35.9	6673
	Valentine . . . . A	32.6	7425
Nev.	Elko . . . . A	34.0	7433
	Ely . . . . A	33.1	7733
	Las Vegas . . . . A	53.5	2709
	Reno . . . . A	39.3	6332
	Winnemucca . . . . A	36.7	6761
N.H.	Concord . . . . A	33.0	7383
	Mt. Washington Obsv. . .	15.2	13817
N.J.	Atlantic City . . . . A	43.2	4812
	Newark . . . . A	42.8	4589
	Trenton C	42.4	4980
N.M.	Albuquerque . . . . A	45.0	4348
	Clayton . . . . A	42.0	5158
	Raton . . . . A	38.1	6228
	Roswell . . . . A	47.5	3793
	Silver City . . . . A	48.0	3705
N.Y.	Albany . . . . A	34.6	6875
	Albany C	37.2	6201
	Binghamton . . . . A	33.9	7286
	Binghamton C	36.6	6451
	Buffalo . . . . A	34.5	7062
	New York (Cent. Park) .	42.8	4871
	New York (La Guardia) . A	43.1	4811
	New York (Kennedy) . . A	41.4	5219
	Rochester . . . . A	35.4	6748
	Schenectady C	35.4	6650
	Syracuse . . . . A	35.2	6756
N.C.	Asheville C	46.7	4042
	Cape Hatteras	53.3	2612
	Charlotte . . . . A	50.4	3191
	Greensboro . . . . A	47.5	3805
	Raleigh . . . . A	49.4	3393
	Wilmington . . . . A	54.6	2347
	Winston-Salem . . . . . . A	48.4	3595
N.D.	Bismarck . . . . A	26.6	8851
	Devils Lake C	22.4	9901
	Fargo . . . . A	24.8	9226
	Williston . . . . A	25.2	9243
Ohio	Akron-Canton . . . . . . A	38.1	6037
	Cincinnati C	45.1	4410
	Cleveland . . . . A	37.2	6351

Table C.1 ***Average Yearly Degree Days for Cities in the United States and Canada***[a,b,c] ***(Base 65° F) (Continued)***

State	Station	Avg. Winter Temp[d]	Degree Days Yearly Total
	Columbus . . . . A	39.7	5660
	Columbus C	41.5	5211
	Dayton . . . . A	39.8	5622
	Mansfield . . . . A	36.9	6403
	Sandusky C	39.1	5796
	Toledo . . . . A	36.4	6494
	Youngstown . . . . A	36.8	6417
Okla.	Oklahoma City . . . . . . A	48.3	3725
	Tulsa . . . . A	47.7	3860
Ore.	Astoria . . . . A	45.6	5186
	Burns C	35.9	6957
	Eugene . . . . A	45.6	4726
	Meacham . . . . A	34.2	7874
	Medford . . . . A	43.2	5008
	Pendleton . . . . A	42.6	5127
	Portland . . . . A	45.6	4635
	Portland C	47.4	4109
	Roseburg . . . . A	46.3	4491
	Salem . . . . A	45.4	4754
Pa.	Allentown . . . . A	38.9	5810
	Erie . . . . A	36.8	6451
	Harrisburg . . . . A	41.2	5251
	Philadelphia . . . . A	41.8	5144
	Philadelphia C	44.5	4486
	Pittsburgh . . . . A	38.4	5987
	Pittsburgh C	42.2	5053
	Reading C	42.4	4945
	Scranton . . . . A	37.2	6254
	Williamsport . . . . A	38.5	5934
R.I.	Block Island . . . . A	40.1	5804
	Providence . . . . A	38.8	5954
S.C.	Charleston . . . . A	56.4	2033
	Charleston C	57.9	1794
	Columbia . . . . A	54.0	2484
	Florence . . . . A	54.5	2387
	Greenville-Spartenburg . A	51.6	2980
S.D.	Huron . . . . A	28.8	8223
	Rapid City . . . . A	33.4	7345
	Sioux Falls . . . . A	30.6	7839
Tenn.	Bristol . . . . A	46.2	4143
	Chattanooga . . . . A	50.3	3254
	Knoxville . . . . A	49.2	3494
	Memphis . . . . A	50.5	3232
	Memphis C	51.6	3015
	Nashville . . . . A	48.9	3578
	Oak Ridge C	47.7	3817
Tex.	Abilene . . . . A	53.9	2624
	Amarillo . . . . A	47.0	3985
	Austin . . . . A	59.1	1711
	Brownsville . . . . A	67.7	600
	Corpus Christi . . . . . . A	64.6	914
	Dallas . . . . A	55.3	2363
	El Paso . . . . A	52.9	2700
	Fort Worth . . . . A	55.1	2405
	Galveston . . . . A	62.2	1274
	Galveston C	62.0	1235
	Houston . . . . A	61.0	1396
	Houston C	62.0	1278
	Laredo . . . . A	66.0	797
	Lubbock . . . . A	48.8	3578
	Midland . . . . A	53.8	2591
	Port Arthur . . . . A	60.5	1447
	San Angelo . . . . A	56.0	2255
	San Antonio . . . . A	60.1	1546
	Victoria . . . . A	62.7	1173
	Waco . . . . A	57.2	2030
	Wichita Falls . . . . A	53.0	2832
Utah	Milford . . . . A	36.5	6497
	Salt Lake City . . . . . . A	38.4	6052
	Wendover . . . . A	39.1	5778
Vt.	Burlington . . . . A	29.4	8269
Va.	Cape Henry C	50.0	3279
	Lynchburg . . . . A	46.0	4166
	Norfolk . . . . A	49.2	3421
	Richmond . . . . A	47.3	3865
	Roanoke . . . . A	46.1	4150
Wash.	Olympia . . . . A	44.2	5236
	Seattle-Tacoma . . . . . A	44.2	5145
	Seattle C	46.9	4424
	Spokane . . . . A	36.5	6655
	Walla Walla C	43.8	4805
	Yakima . . . . A	39.1	5941
W. Va.	Charleston . . . . A	44.8	4476
	Elkins . . . . A	40.1	5675
	Huntington . . . . A	45.0	4446
	Parkersburg C	43.5	4754
Wisc.	Green Bay . . . . A	30.3	8029
	La Crosse . . . . A	31.5	7589
	Madison . . . . A	30.9	7863
	Milwaukee . . . . A	32.6	7635
Wyo.	Casper . . . . A	33.4	7410
	Cheyenne . . . . A	34.2	7381
	Lander . . . . A	31.4	7870
	Sheridan . . . . A	32.5	7680
Alta.	Banff C	–	10551
	Calgary . . . . A	–	9703
	Edmonton . . . . A	–	10268
	Lethbridge . . . . A	–	8644
B.C.	Kamloops . . . . A	–	6799
	Prince George* . . . . . A	–	9755
	Prince Rupert C	–	7029
	Vancouver* . . . . A	–	5515
	Victoria* . . . . A	–	5699
	Victoria C	–	5579
Man.	Brandon* . . . . A	–	11036
	Churchill . . . . A	–	16728
	The Pas C	–	12281
	Winnipeg . . . . A	–	10679

Table C.1 ***Average Yearly Degree Days for Cities in the United States and Canada***[a,b,c] ***(Base 65° F) (Continued)***

State or Prov.	Station	Avg. Winter Temp[d]	Degree Days Yearly Total	State or Prov.	Station	Avg. Winter Temp[d]	Degree Days Yearly Total
N.B.	Fredericton* . . . . A	—	8671		Kitchener C	—	7566
	Moncton C	—	8727		London . . . . A	—	7349
	St. John C	—	8219		North Bay C	—	9219
					Ottawa C	—	8735
Nfld.	Argentia . . . . A	—	8440		Toronto C	—	6827
	Corner Brook C	—	8978				
	Gander . . . . A	—	9254	P.E.I.	Charlottetown C	—	8164
	Goose* . . . . A	—	11887		Summerside C	—	8488
	St. John's* . . . . A	—	8991				
				Que.	Arvida C	—	10528
N.W.T.	Aklavik C	—	18017		Montreal* . . . . A	—	8203
	Fort Norman C	—	16109		Montreal C	—	7899
	Resolution Island C	—	16021		Quebec* . . . . A	—	9372
					Quebec C	—	8937
N.S.	Halifax C	—	7361				
	Sydney . . . . A	—	8049	Sasks.	Prince Albert . . . . A	—	11630
	Yarmouth . . . . A	—	7340		Regina . . . . A	—	10806
					Saskatoon C	—	10870
Ont.	Cochrane C	—	11412				
	Fort William . . . . A	—	10405	Y.T.	Dawson C	—	15067
	Kapuskasing C	—	11572		Mayo Landing C	—	14454

[a] Data for United States cities from a publication of the United States Weather Bureau, *Monthly Normals of Temperature, Precipitation and Heating Degree Days,* 1962, are for the period 1931 to 1960 inclusive. These data also include information from the 1963 Revisions to this publication, where available.

[b] Data for airport stations, A, and city stations, C, are both given where available.

[c] Data for Canadian cities were computed by the Climatology Division, Department of Transport from normal monthly mean temperatures.

[d] For period October to April, inclusive.

*The data for these normals were from the full ten-year period 1951–1960, adjusted to the standard normal period 1931–1960.

Table C.2a *Outside Design Conditions for United States*

State and City	Winter DB	Summer DB	Daily Range	Summer WB	Latitude, Deg.
ALABAMA					
Anniston	10	95	M	78	35
Birmingham	10	95	M	78	35
Gadsden	10	95	M	78	35
Mobile	20	90	L	80	30
Montgomery	20	95	M	78	30
Tuscaloosa	10	95	M	78	35
ALASKA					
Anchorage	−24	—	M	—	60
Barrow	−48	—	—	—	70
Bethel	−43	—	—	—	60
Cordova	−13	—	—	—	60
Fairbanks	−57	—	M	—	65
Juneau	−5	—	—	—	60
Ketchikan	4	—	—	—	55
Kodiak	4	—	—	—	55
Kotzebue	−46	—	—	—	65
Nome	−36	—	—	—	60
Seward	−4	—	—	—	60
Sitka	2	—	—	—	60
ARIZONA					
Bisbee	30	100	H	72	30
Flagstaff	−5	85	H	61	35
Globe	30	105	H	76	35
Nogales	30	105	H	72	30
Phoenix	35	105	H	76	35
Tucson	30	100	H	72	30
Winslow	−5	95	H	65	35
Yuma	40	110	H	78	35
ARKANSAS					
Bentonville	0	95	M	76	35
Fort Smith	5	95	M	76	35
Hot Springs	10	95	M	78	35
Little Rock	10	95	M	78	35
Pine Bluff	10	95	M	78	35
Texarkana	10	100	M	78	35
CALIFORNIA					
Bakersfield	30	105	H	70	35
El Centro	35	110	H	78	35
Eureka	30	90	M	65	40
Fresno	30	105	H	74	35
Long Beach	35	90	M	70	35
Los Angeles	40	90	L	70	35
Montague	15	95	M	70	40
Needles	25	115	H	—	35
Oakland	30	80	M	65	40
Pasadena	40	95	M	70	35
Red Bluff	15	100	H	70	40
Sacramento	30	100	H	72	40
San Bernardino	30	105	H	72	35
San Diego	45	80	L	68	35
San Francisco	35	80	M	65	40
San Jose	40	90	M	70	35
COLORADO					
Boulder	−15	95	M	64	40
Colorado Springs	−10	95	H	65	40
Denver	−10	95	H	64	40
Durango	−5	95	H	65	35
Fort Collins	−15	95	M	65	40
Grand Junction	−5	95	H	65	40
Leadville	−10	95	M	64	40
Pueblo	−15	95	H	65	40
CONNECTICUT					
Bridgeport	0	85	L	40	75
Hartford	0	90	M	40	75
New Haven	0	85	M	40	75
New London	5	85	L	40	75
Norwalk	0	85	L	40	75
Torrington	0	90	M	40	75
Waterbury	0	90	L	40	75
DELAWARE					
Dover	10	90	M	78	40
Milford	10	90	M	78	40
Wilmington	5	90	M	78	40
DIST. OF COLUMBIA					
Washington	10	90	M	78	40
FLORIDA					
Apalachicola	25	95	L	80	30
Fort Myers	40	95	M	78	25
Gainesville	30	95	M	78	30
Jacksonville	30	95	M	78	30
Key West	55	100	L	78	25
Miami	45	90	L	79	25
Orlando	35	90	M	78	30
Pensacola	25	95	L	78	30
Tallahassee	25	95	M	78	30
Tampa	35	95	M	78	30
GEORGIA					
Athens	10	95	M	76	35
Atlanta	10	95	M	76	35
Augusta	20	100	M	76	35
Brunswick	25	95	L	78	30
Columbus	20	100	M	76	35
Macon	20	95	M	78	35
Rome	10	95	M	76	35
Savannah	25	95	M	78	30
Way Cross	25	95	M	78	30
IDAHO					
Boise	−10	95	H	65	45
Idaho Falls	−15	90	H	65	45
Lewiston	−10	95	H	65	45
Pocatello	−15	90	H	65	45
Twin Falls	15	95	H	65	40
ILLINOIS					
Aurora	−10	95	M	75	40
Bloomington	−10	95	M	76	40
Cairo	0	100	M	78	35
Champaign	−10	95	M	77	40
Chicago	−10	95	M	75	40
Danville	−10	95	M	77	40
Decatur	−10	95	M	77	40
Elgin	−15	95	M	78	40
Joliet	−10	95	M	76	40
Moline	−10	95	M	76	40
Peoria	−15	95	M	76	40
Rockford	−15	95	M	78	40
Rock Island	−10	95	M	76	40
Springfield	−10	95	M	77	40
Urbana	−10	95	M	77	40
INDIANA					
Elkhart	−10	95	M	75	40
Evansville	−5	95	M	78	40
Fort Wayne	−5	95	M	75	40
Indianapolis	−10	95	M	76	40
Lafayette	−10	95	M	76	40
South Bend	−10	95	M	75	40
Terre Haute	−5	95	M	78	40
IOWA					
Burlington	−10	95	M	78	40
Cedar Rapids	−15	95	M	78	40
Charles City	−20	95	M	75	45
Clinton	−15	95	M	78	40
Council Bluffs	−15	100	M	78	40
Davenport	−10	95	M	78	40
Des Moines	−15	95	M	78	40
Dubuque	−15	95	M	78	40
Fort Dodge	−15	95	M	78	40
Keokuk	−15	95	M	78	40
Marshalltown	−15	95	M	78	40
Sioux City	−15	95	M	78	40
Waterloo	−15	95	M	78	40

Table C.2a *Outside Design Conditions for United States (Continued)*

State and City	Winter DB	Summer DB	Daily Range	Summer WB	Latitude, Deg.
KANSAS					
Atchison	−10	100	M	76	40
Concordia	−10	95	M	78	40
Dodge City	−10	95	H	78	40
Iola	−5	100	M	75	40
Leavenworth	−10	100	M	76	40
Salina	−10	100	M	78	40
Topeka	−10	100	M	78	40
Wichita	−5	100	M	75	40
KENTUCKY					
Bowling Green	0	95	M	78	35
Frankfort	0	95	M	78	40
Hopkinsville	0	95	M	78	35
Lexington	0	95	M	78	40
Louisville	0	95	M	78	40
Owensboro	0	95	M	78	40
Shelbyville	0	95	M	78	40
LOUISIANA					
Alexandria	20	95	L	78	30
Baton Rouge	20	95	M	80	30
New Orleans	25	95	L	80	30
Shreveport	15	95	M	78	35
MAINE					
Augusta	−15	85	L	73	45
Bangor	−20	85	L	73	45
Bar Harbor	−10	85	L	73	45
Belfast	−10	85	L	73	45
Eastport	−10	85	L	70	45
Lewiston	−10	85	L	73	45
Millinocket	−15	85	M	73	45
Orono	−20	85	M	70	45
Portland	−10	85	M	73	45
Presque Isle	−20	85	L	73	45
Rumford	−15	85	L	73	45
MARYLAND					
Annapolis	10	90	M	78	40
Baltimore	10	90	M	78	40
Cambridge	10	90	L	78	40
Cumberland	0	90	M	75	40
Frederick	5	90	M	78	40
Frostburg	−5	90	M	75	40
Salisbury	10	90	M	78	40
MASSACHUSETTS					
Amherst	−5	90	M	75	40
Boston	0	85	M	74	40
Fall River	0	85	L	75	40
Fitchburg	−5	90	M	75	45
Framingham	−5	85	L	75	40
Lawrence	−5	85	L	74	40
Lowell	−5	85	L	74	45
Nantucket	0	85	L	75	40
New Bedford	0	85	L	75	40
Pittsfield	−10	90	M	75	40
Plymouth	0	85	L	75	40
Springfield	−5	90	M	75	40
Worcester	−5	90	M	75	40
MICHIGAN					
Alpena	−10	90	M	75	45
Ann Arbor	−5	90	M	75	40
Big Rapids	−5	90	M	75	45
Cadillac	−10	90	M	75	45
Calumet	−20	80	M	73	45
Detroit	−5	90	M	75	40
Escanaba	−20	85	M	74	45
Flint	−10	90	M	75	45
Grand Haven	−5	90	M	75	45
Grand Rapids	−5	90	M	74	45
Houghton	−20	80	M	73	45
Kalamazoo	−5	90	M	75	40
Lansing	−10	90	M	75	45
Ludington	−5	90	M	75	45
Marquette	−15	80	M	73	45
Muskegon	−5	90	M	74	45
Port Huron	−10	90	M	75	45
Saginaw	−10	90	M	75	45
Sault Ste. Marie	−20	80	M	71	45
MINNESOTA					
Alexandria	−25	85	M	74	45
Duluth	−25	80	M	76	45
Minneapolis	−25	90	M	76	45
Moorhead	−30	95	M	75	45
St. Cloud	−25	90	M	76	45
St. Paul	−25	90	M	75	45
MISSISSIPPI					
Biloxi	25	90	L	80	30
Columbus	10	95	M	78	35
Corinth	5	95	M	78	35
Hattiesburg	20	95	M	80	30
Jackson	15	95	M	78	30
Meridian	15	95	M	79	30
Natchez	15	95	L	78	30
Vicksburg	15	95	L	78	30
MISSOURI					
Columbia	−10	100	M	78	40
Hannibal	−10	95	M	77	40
Kansas City	−10	100	M	76	40
Kirksville	−10	95	M	78	40
St. Joseph	−10	100	M	76	40
St. Louis	−5	95	M	78	40
Springfield	−5	100	M	77	40
MONTANA					
Anaconda	−30	85	H	59	45
Billings	−30	90	H	66	45
Butte	−30	85	H	59	45
Great Falls	−40	90	H	63	50
Havre	−40	95	M	70	50
Helena	−40	90	H	63	45
Kalispell	−50	90	H	63	50
Miles City	−35	95	H	69	45
Missoula	−30	90	H	63	45
NEBRASKA					
Grand Island	−15	100	H	75	40
Hastings	−15	100	M	75	40
Lincoln	−15	95	M	78	40
Norfolk	−15	95	M	78	40
North Platte	−15	100	H	73	40
Omaha	−15	100	M	78	40
Valentine	−20	95	M	78	45
York	−15	95	M	78	40
NEVADA					
Elko	−10	95	H	63	40
Las Vegas	10	110	H	71	35
Reno	5	95	H	65	40
Tonopah	5	90	M	63	40
Winnemucca	−10	95	H	65	40
NEW HAMPSHIRE					
Berlin	−15	85	H	73	45
Claremont	−15	85	M	73	45
Concord	−10	85	H	73	45
Franklin	−15	85	M	73	45
Hanover	−15	85	M	73	45
Keene	−10	85	L	73	45
Manchester	−10	85	M	74	45
Nashua	−10	85	L	74	45
Portsmouth	−5	85	L	74	45

Table C.2a *Outside Design Conditions for United States (Continued)*

State and City	Winter DB	Summer DB	Daily Range	Summer WB	Latitude, Deg.
NEW JERSEY					
Asbury Park	5	90	L	78	40
Atlantic City	10	90	L	78	40
Bayonne	0	90	L	75	40
Belvidere	0	90	M	75	40
Bloomfield	0	90	L	75	40
Bridgeton	5	90	L	78	40
Camden	5	90	L	78	40
East Orange	0	90	L	75	40
Elizabeth	0	90	L	75	40
Jersey City	0	90	L	75	40
Newark	0	90	M	76	40
New Brunswick	5	90	L	75	40
Paterson	0	90	L	75	40
Phillipsburg	0	90	M	75	40
Trenton	0	90	L	78	40
NEW MEXICO					
Albuquerque	10	95	M	65	35
El Morro	0	85	H	65	35
Raton	−5	95	H	65	35
Roswell	5	100	H	71	35
Santa Fe	5	90	M	65	35
Tucumcari	5	95	H	70	35
NEW YORK					
Albany	−10	90	M	74	45
Auburn	−10	90	M	74	45
Binghamton	−5	90	M	72	40
Buffalo	−5	85	M	73	45
Canton	−20	85	M	73	45
Cortland	−10	90	M	74	45
Elmira	−5	90	M	73	40
Glens Falls	−15	90	M	73	45
Ithaca	−5	90	M	73	40
Jamestown	−5	90	M	74	40
Lake Placid	−15	90	M	73	45
New York	5	90	M	76	40
Niagara Falls	−5	85	M	73	45
Ogdensburg	−20	85	M	73	45
Oneonta	−10	90	M	73	45
Oswego	−5	90	M	74	45
Port Jervis	0	90	L	75	40
Rochester	−5	90	M	74	45
Schenectady	−10	90	M	74	45
Syracuse	−10	90	M	74	45
Watertown	−15	85	M	73	45
NORTH CAROLINA					
Asheville	5	90	M	75	35
Charlotte	15	95	M	78	35
Greensboro	10	90	M	76	35
Hatteras	20	90	L	80	35
New Bern	20	95	L	78	35
Raleigh	15	95	M	78	35
Salisbury	10	90	M	78	35
Wilmington	20	90	M	81	35
Winston-Salem	10	90	M	76	35
NORTH DAKOTA					
Bismarck	−30	95	H	73	45
Devils Lake	−30	90	M	70	50
Dickinson	−30	95	H	70	45
Fargo	−30	95	H	75	45
Grand Forks	−30	90	M	72	50
Jamestown	−30	95	M	73	45
Minot	−35	90	M	71	50
Pembina	−35	90	M	73	50
Williston	−35	90	M	73	50
OHIO					
Akron	−5	90	M	75	40
Cincinnati	−5	95	M	78	40
Cleveland	−5	90	M	75	40
Columbus	−5	90	M	76	40
Dayton	−5	90	M	76	40
Lima	−5	90	M	75	40
Marion	−5	90	M	75	40
Sandusky	−5	90	M	75	40
Toledo	−5	90	M	75	40
Warren	−5	90	M	75	40
Youngstown	−5	90	M	75	40
OKLAHOMA					
Ardmore	5	100	M	78	35
Bartlesville	−5	100	M	77	35
Guthrie	0	100	M	77	35
Muskogee	0	95	M	79	35
Oklahoma City	0	100	M	77	35
Tulsa	0	100	M	77	35
Waynoka	−5	105	M	75	35
OREGON					
Arlington	5	95	M	68	45
Baker	−15	90	M	66	45
Eugene	15	90	H	68	45
Medford	20	95	H	68	40
Pendleton	−10	90	H	66	45
Portland	10	85	M	68	45
Roseburg	20	90	H	66	45
Salem	15	90	H	68	45
Wamic	0	90	H	66	45
PENNSYLVANIA					
Altoona	−5	90	M	75	40
Bethlehem	0	90	M	75	40
Coatesville	5	90	M	75	40
Erie	−5	85	M	74	40
Harrisburg	5	90	M	75	40
New Castle	−5	90	M	75	40
Oil City	−5	90	M	75	40
Philadelphia	5	90	M	78	40
Pittsburgh	−5	90	M	75	40
Reading	5	90	M	75	40
Scranton	0	90	M	75	40
Warren	−5	90	M	75	40
Williamsport	−5	90	M	74	40
York	5	90	M	75	40
RHODE ISLAND					
Block Island	5	85	L	75	40
Bristol	0	90	L	75	40
Kingston	0	85	L	75	40
Pawtucket	0	90	M	75	40
Providence	0	90	M	75	40
SOUTH CAROLINA					
Charleston	20	90	L	80	35
Columbia	20	95	M	78	35
Florence	20	95	M	79	35
Greenville	10	95	M	75	35
Spartanburg	10	95	M	78	35
SOUTH DAKOTA					
Aberdeen	−25	95	M	75	45
Huron	−20	100	H	75	45
Pierre	−20	95	M	73	45
Rapid City	−20	95	H	70	45
Sioux Falls	−20	95	H	75	45
Watertown	−25	95	M	73	45
TENNESSEE					
Chattanooga	10	95	M	76	35
Jackson	5	95	M	78	35
Johnson City	0	95	M	78	35
Knoxville	5	95	M	75	35
Memphis	5	95	M	78	35
Nashville	5	95	M	78	35

Table C.2a *Outside Design Conditions for United States (Continued)*

State and City	Winter DB	Summer DB	Daily Range	Summer WB	Latitude, Deg.
TEXAS					
Abilene	5	95	M	74	30
Amarillo	0	95	H	72	35
Austin	15	100	M	78	30
Brownsville	30	95	M	80	25
Corpus Christi	25	95	M	80	30
Dallas	10	100	M	78	35
Del Rio	20	100	H	78	30
El Paso	20	100	M	69	30
Fort Worth	10	100	M	78	35
Galveston	25	95	L	80	30
Houston	20	95	M	80	30
Palestine	10	100	M	78	30
Port Arthur	20	95	M	80	30
San Antonio	20	100	M	78	30
Waco	10	100	M	78	30
UTAH					
Logan	−10	95	H	65	40
Milford	−5	95	H	66	40
Ogden	−5	90	H	65	40
Salt Lake City	0	95	H	65	40
VERMONT					
Bennington	−10	90	M	73	45
Burlington	−15	90	M	73	45
Montpelier	−20	90	M	73	45
Newport	−20	85	M	73	45
Northfield	−20	90	M	73	45
Rutland	−15	90	M	73	45
VIRGINIA					
Cape Henry	15	90	L	78	35
Charlottesville	10	90	M	78	40
Danville	10	90	M	78	35
Lynchburg	10	90	M	76	35
Norfolk	15	90	L	78	35
Petersburg	10	90	M	78	35
Richmond	10	90	M	78	40
Roanoke	5	90	M	76	35
Wytheville	5	90	M	76	35
WASHINGTON					
Aberdeen	20	85	L	64	45
Bellingham	10	80	L	65	50
Everett	15	80	L	65	50
North Head	20	80	L	65	50
Olympia	15	80	M	64	45
Seattle	15	80	M	65	50
Spokane	−15	80	H	65	50
Tacoma	15	80	M	64	45
Tatoosh Island	20	80	L	65	50
Walla Walla	−10	90	H	65	45
Wenatchee	−10	90	M	65	50
Yakima	−5	90	H	67	45
WEST VIRGINIA					
Bluefield	0	95	M	75	35
Charleston	0	90	M	75	40
Elkins	−5	90	M	73	40
Fairmont	0	90	M	75	40
Huntington	0	90	M	76	40
Martinsburg	0	90	M	75	40
Parkersburg	0	90	M	75	40
Wheeling	−5	90	M	75	40
WISCONSIN					
Ashland	−25	80	M	71	45
Beloit	−15	95	M	78	45
Eau Claire	−20	90	M	75	45
Green Bay	−20	90	M	73	45
La Crosse	−20	95	M	75	45
Madison	−20	90	M	75	45
Milwaukee	−15	90	M	75	45
Oshkosh	−20	90	M	75	45
Sheboygan	−20	90	M	75	45
WYOMING					
Casper	−25	90	H	62	45
Cheyenne	−20	90	H	62	40
Lander	−30	90	H	65	45
Sheridan	−30	90	H	65	45
Yellowstone Park	−35	85	H	62	45

Table C.2b *Outside Design Conditions for Canada*

Province and City	Winter DB	Summer DB	Daily Range	Summer WB	Latitude, Deg.
ALBERTA					
Banff	−30	—	H	—	50
Camrose	−35	—	H	—	55
Calgary	−30	90	H	66	50
Cardston	−30	—	H	—	50
Edmonton	−35	90	H	68	55
Grande Prairie	−40	—	H	—	55
Hanna	−35	—	H	—	50
Jasper	−30	—	H	—	55
Lethbridge	−30	—	H	—	50
Lloydminster	−40	—	H	—	55
McMurray	−40	—	H	—	55
Medicine Hat	−35	90	H	65	50
Red Deer	−35	—	H	—	50
Taber	−35	—	H	—	50
Wetaskiwin	−35	—	H	—	55
BRITISH COLUMBIA					
Chilliwack	5	—	M	—	50
Courtenay	10	—	M	—	50
Dawson Creek	−40	—	H	—	55
Estevan Point	15	—	M	—	55
Fort Nelson	−40	—	H	—	60
Hope	0	—	M	—	50
Kamloops	−20	—	H	—	50
Kimberly	−25	—	H	—	50
Lytton	−5	—	H	—	50
Nanaimo	10	—	M	—	50
Nelson	−10	—	H	—	50
Penticton	−5	—	H	—	50
Port Alberni	10	—	M	—	50
Prince George	−30	—	H	—	55
Prince Rupert	10	—	L	—	55
Princeton	−15	—	H	—	50
Revelstoke	−25	—	H	—	50
Trail	−10	—	H	—	50
Vancouver	10	80	L	67	50
Vernon	−15	—	H	—	50
Victoria	15	—	L	—	50
Westview	10	—	M	—	50
LABRADOR					
Goose Bay	−25	—	L	—	55
MANITOBA					
Boissevain	−35	—	H	—	50
Brandon	−30	—	H	—	50
Churchill	−40	—	L	—	60
Dauphin	−35	—	M	—	50
Flin Flon	−40	—	M	—	55
Minnedosa	−35	—	H	—	50
Neepawa	−35	—	H	—	50
La Prairie	−30	—	M	—	50
Swan River	−35	—	M	—	55
The Pas	−40	—	M	—	55
Winnipeg	−30	90	M	71	50
NEW BRUNSWICK					
Bathurst	−10	—	L	—	45
Campbellton	−10	—	L	—	45
Chatham	−10	—	M	—	45
Edmunston	−15	—	M	—	45
Fredericton	−5	90	L	75	45
Moncton	−10	—	M	—	45
Saint John	−5	80	L	67	45
Woodstock	−15	—	M	—	45
NEWFOUNDLAND					
Corner Brook	0	—	L	—	50
Gander	−5	—	L	—	50
Grand Falls	−5	—	M	—	50
St. John's	0	—	L	—	50
NORTHWEST TERRITORIES					
Aklavik	−45	—	L	—	70
Fort Norman	−40	—	M	—	65
Frobisher	−50	—	L	—	—
Resolute	−40	—	L	—	—
Yellowknife	−50	—	L	—	60
NOVA SCOTIA					
Bridgewater	0	—	L	—	45
Dartmouth	0	—	L	—	45
*Halifax C	5	80	L	75	45
*Halifax A	0	80	L	75	45
Kentville	0	—	L	—	45
New Glasgow	0	—	L	—	45
Spring Hill	−5	—	M	—	45
Sydney	0	85	L	67	45
Truro	0	—	L	—	45
Yarmouth	5	—	L	—	45
ONTARIO					
Bancroft	−20	—	M	—	45
Barrie	−5	—	M	—	45
Belleville	−10	—	M	—	45
Brampton	−5	—	M	—	45
Brantford	−5	—	M	—	45
Brockville	−15	—	M	—	45
Chatham	0	—	M	—	45
Cobourg	−10	—	M	—	45
Collingwood	0	—	M	—	45
Cornwall	−15	—	M	—	45
Ear Falls	−35	—	M	—	50
Fort Frances	−30	—	M	—	50
Fort William	−25	85	M	70	50
Galt	−5	—	M	—	45
Geraldton	−35	—	M	—	50
Goderich	0	—	M	—	45
Guelph	−5	—	M	—	45
Hamilton	0	—	M	—	45
Haileybury	−25	—	M	—	50
Hanover	−5	—	M	—	45
Huntsville	−15	—	M	—	45
Kapuskasing	−30	—	M	—	50
Kenora	−35	—	M	—	50
Kingston	−10	—	M	—	45
Kirkland Lake	−25	—	M	—	50
Kitchener	−5	—	M	—	45
Lindsay	−15	—	M	—	45
London	0	—	M	—	45
Moonsonee	−35	—	M	—	50

Source. Reprinted by permission from *Load Calculations for Residential Winter and Summer Air Conditioning, Manual J* of National Warm Air Heating and Air Conditioning Association.

D

Metrication

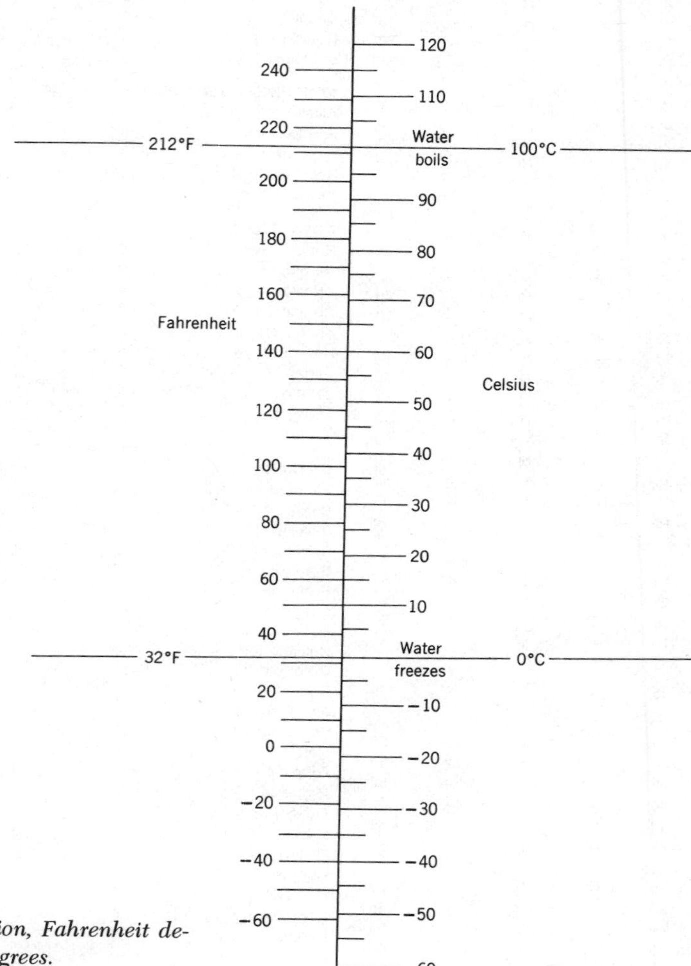

Fig. D.1 Conversion, Fahrenheit degrees to Celsius degrees.

D.2 Authors' Comments Regarding the Adoption of the Metric System

The schedule of change will be slower than anticipated in reference to the SI (Système Internationale), especially for buildings and their mechanical and electrical systems. Some of the professional engineering societies had hoped for 1979 as a date when all of their publications would be entirely in the metric system. Now the time span from 1982 to 1985 appears to be, tentatively, the period of transition. Manufacturers of equipment, understandably facing a tremendous expense in changing to metric sizes and to the metric "language," have been a strong influence in this reasonable delay.

The system is indeed a language. It is being

updated and there is not yet full agreement in regard to some of the more controversial engineering terms. For this reason, this book, the sixth edition of *Mechanical and Electrical Equipment for Buildings,* makes only casual reference throughout to metric equivalents. We have time, then, to study articles such as "Metrication is Coming: Prepare for the Transition," by Frank J. Versagi. It is with the permission of *Air Conditioning, Heating and Refrigeration News* that is reprinted here.

We look forward also to the publication by Industrial Press, Inc. of the book *Metric Units in Engineering—Going SI,* by Dr. Cornelius Wandmacher, Professor of Engineering at the University of Cincinnati.

D.3 METRICATION IS COMING: PREPARE FOR THE TRANSITION

by Frank J. Versagi

It is inevitable that the United States will officially adopt the metric system of measurements. This series of articles is being published to help *News* readers prepare for the transition.

The International System of Units, abbreviated SI, for *Systême Internationale,* is a modified version of the metric system, using modern fundamental discoveries to establish base units of measurement.

There are six, some sources list seven, fundamental physical quantities in SI:

length . . meter . . m
mass . . kilogram . . kg
time . . second . . s
electric current . . ampere . . A
thermodynamic temperature . . Kelvin . . K
luminous intensity . . candela . . cd
amount of substance . . mole . . mol

From these base units and a couple of supplementary units, are developed derived units, some with special names and several of specific interest to the construction industry and mechanical trades, to which we will get in a while.

Discussing things metric with persons unfamiliar with them, one realizes that an obstacle in the way of acceptance of metric measurement is that absence of a mental image and what one writer calls "recognition points." There are those, for example, who have no notion whether one meter is about as long as an inch or a mile.

It may be helpful, therefore, to pause long enough to attempt to establish a few mental images.

- A meter is equivalent to a yard (39 in. compared with 36 in.)

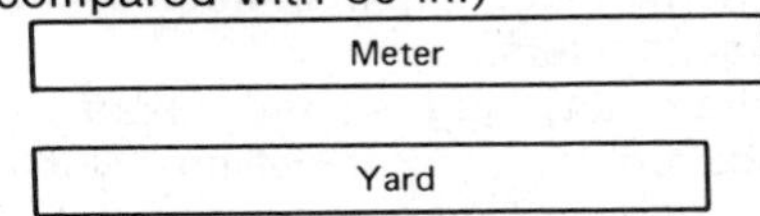

- A kilogram is a little over two pounds.

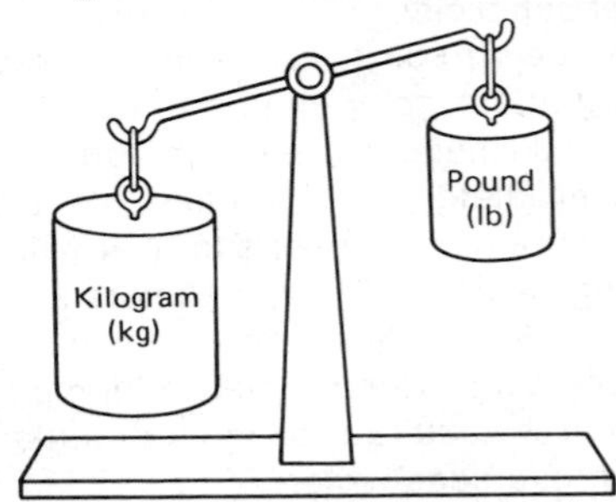

- A liter is just a bit larger than a quart (by the way, a kilogram is the mass of one liter of water).

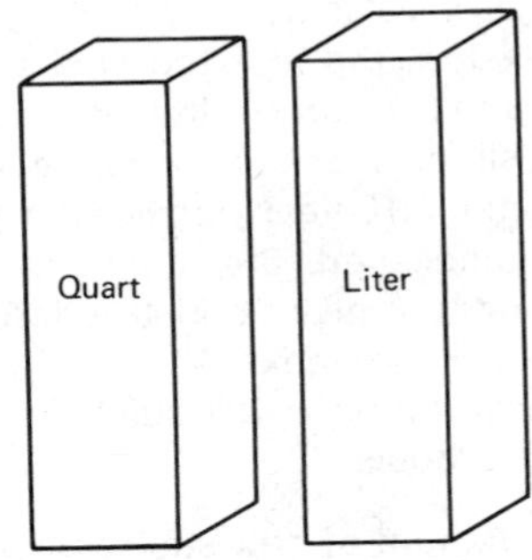

- A square meter is roughly equivalent to a square yard (11 sq ft compared with 9 sq ft).

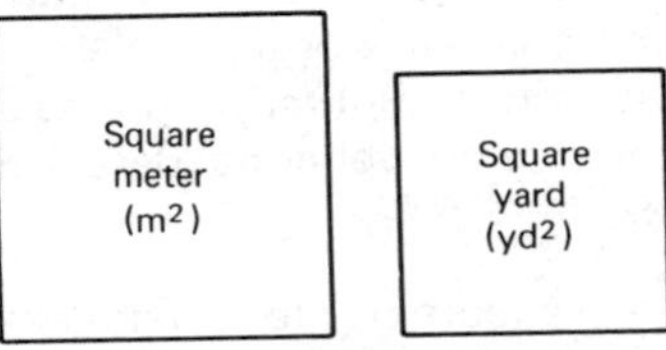

An Australian publication suggests recognition points called the "10-11-12-13 relationship."

10 meters = 11 yards

10 sq meters = 12 square yards

10 cubic meters = 13 cubic yards

Before we talk about conversion factors and dimensions specifically for our industry, let's touch on a couple of attitudinal and philosophical matters.

Is there anything sacred about, say, a specification for an 8-ft ceiling? Not really. There is no scientific or technological reason for that exact dimension. In most instances, such figures themselves were chosen for numerical simplicity.

That being so, it makes little sense when thinking about going metric to convert 8 ft to 2,438.4 millimeters or even to round it off to 2,440 millimeters. In practice, as the construction industry converts to SI, it is likely that a ceiling height of 2,400 millimeters will substitute for 8 ft. "There are very few instances in the building industry where a high degree of precision is required, or indeed achievable," says an overseas writer.

The experience of other nations has shown that the fear that there will be a period when errors are frequent and serious proved unfounded.

If U.S. law follows the example of other countries, it will also develop that (1) the faster the conversion, the less the cost and inconvenience and (2) many of the feared costs and problems simply do not materialize. Countries which planned 10-year conversion programs have accomplished the conversion in six years. The Australian construction industry has set a 5-yr timetable.

Design professions will quickly experience some added costs:

- Procurement of new scales, instruments, reference publications, and the like.
- Some temporary reduction in productivity caused by unfamiliarity, double-checking, etc.
- Conversion of necessary data where metric data are not available.
- Revision of computer programs and substitution of conventional data with metric data.

Further, according to Australia's Metric Conversion Board, the construction industry may see:

- Some manufacturing equipment will require modification and in some cases complete replacement to produce materials in metric sizes. Plant and equipment which have a weighing function will need to be recalibrated, in some cases to the original units for which they were designed. "Modification or recalibration may be combined with normal maintenance."
- Building regulations, codes of practice, design manuals, drawing scales, will have to be replaced or modified to give metric dimensions or properties.
- Data or records for estimating, costing, marketing, and other purposes will have to be rewritten in metric terms. Computer programs will have to be amended or rewritten in metric terms.
- Technical and marketing support literature will need to be stated in metric terms. "At the same time, advantage may be taken of this opportunity to update and review such literature."

SI, metrication, is inevitable.

An advantage claimed for the metric system is that all dimensions can be expressed as decimal multipliers or submultipliers of the basic unit. "Decimal" means proceeding by 10's.

Compare, for example, the English:

12 inches to one foot

3 feet to one yard

5,280 feet to one mile

1,760 yards to one mile

In metric:

10 millimeters to a centimeter

100 centimeters to a meter

1,000 meters to a kilometer

Multiplying or dividing by 10, 100, 1,000, 10,000 calls for merely moving the decimal point. Thus:

1.75 kilometers = 1,750 meters

300 millimeters = 0.300 meter

Again, in English:

16 ounces to one pound

2,000 pounds to one ton

And in SI:

1,000 grams to a kilogram

1,000 kilograms to a tonne

Even the words used are simpler, at least more uniform, when using metric. Using standard prefixes with the base unit makes all dimensions easily comparable. Again, compare the English and metric terms for length:

inch . . . foot . . . yard . . . mile

millimeter . . . centimeter . . . meter . . . kilometer

Those prefixes—milli, centi, and kilo—apply to all metric units: millileter and kilogram are examples. Table 1 shows the standard multiples and submultiples with their magnitudes and symbols. It is important to use the symbols correctly, capitalized or not, because there are different and specific meanings for "M", and "m", "G", and "g", sometimes used as prefixes, sometimes as the symbol for a base unit or a derived unit.

Special names for some derived units are already familiar: hertz for cycles-per-second and watt for joules-per-second are examples. For other derived units, unfamiliar special names will have to be learned: the newton as the unit of force (mass times length divided by time) is an example.

There are several general rules which apply to the use of metric units.

1. Unit symbols do not change in the plural. The designation for one meter is 1 m, for 570 meters is 570 m. For one millimeter is 1 mm, for 750 millimeters is 750 mm.
2. With one exception, all units and prefixes start with a small letter when written in full; the exception is "degree Celsius."
3. Where the prefix is combined with the unit name, the combination is written as one word: kilowatt, meganewton, milligram.
4. Prefix symbols should be written without spacing between the prefix symbol and the unit symbol: mm for millimeter, kW for kilowatt, MN for meganewton.

Here are the metric units with which the industries served by this newspaper will become involved (symbols in parentheses).

Linear measure . . . meter (m), millimeter (mm)

Area . . . square meter (m^2), hectare (ha)

Weight . . . gram (g), kilogram (kg)

Volume . . . cubic meter (m^3)

Capacity . . . liter (l)

Airflow . . . meter per second (m/s)

Volume flow . . . cubic meter per second (m^3/s), liter per second (l/s)

Temperature . . . degree Celsius (°C)

Force . . . newton (N), kilonewton (kN)

Pressure . . . kilopascal (kPa)

Energy, work . . . kilojoule (kJ), megajoule (MJ)

Table 1 ***These Prefixes May be Applied to All SI Units***

Multiples and submultiples	*Prefixes*	*Symbols*
1 000 000 000 000 = 10^{12}	tera	T
1 000 000 000 = 10^9	giga	G
1 000 000 = 10^6	mega	M*
1 000 = 10^3	kilo	k*
100 = 10^2	hecto	h
10 = 10	deka	da
0.1 = 10^{-1}	deci	d
0.01 = 10^{-2}	centi	c*
0.001 = 10^{-3}	milli	m*
0.000 001 = 10^{-6}	micro	μ*
0.000 000 001 = 10^{-9}	nano	n
0.000 000 000 001 = 10^{-12}	pico	p
0.000 000 000 000 001 = 10^{-15}	femto	f
0.000 000 000 000 000 001 = 10^{-18}	atto	a

*Most commonly used

Frequency . . . hertz (Hz)
Power . . . watt (W), kilowatt (kW)
Electric current . . . ampere (A)
Electric potential . . . volt (V), kilovolt (kV)
Electrical resistance . . . ohm (Ω)

A meter is slightly longer than a yard:

meters × 1.094 = yards
yards × 0.914 = meters

Using rounded conversion factors of 1.1 and 0.9, here are some mental images.

A 100-yard football field is 90 m long.

A 1,000 meter, or kilometer, run (1,000 m or 1 km) = 1,100 yards. (The reason that sprints in track competition are 110, 220, and 440 yards is that those English distances are the nearest to 100 meters, 200 meters, and 400 meters.)

A foot is about one-third of a meter.

feet × 0.305 = meters

An 8-ft ceiling, therefore, is roughly 2.4 m high. (Instead of speaking of "7-ft, 11$\frac{3}{4}$-in.," one might refer to "2,440 millimeters," for greater precision.)

A residential lot 75 feet wide and 200 feet deep would, in metric, be 22.5 m wide and 60 m deep.

The nearest equivalent to 1 inch is 25 millimeters (25 mm); the nearest equivalent to 1 foot is 300 mm.

inches × 25.4 = millimeters
inches × 2.54 = centimeters

An 8-in. diameter duct, is, in metric, a 203 mm diameter duct; a 6-in. duct = 152 mm.

A 36-in belt is expressed in metric as 91 cm or 914 mm.

On working drawings, preference is to use only meters and millimeters, not centimeters. Then, dimensions can be used without needing a unit symbol (m, cm, or mm), so long as all millimeter dimensions are expressed in whole numbers, like 3,600, and all meter dimensions are carried to three decimal places, like 3.600.

(Not all countries have agreed on (1) the use or not of a punctuation mark in large numbers—like 3600 or 3,600—and (2) whether the comma or period will be used to express decimal fraction—like 0.462 km or 0,462 km.)

A square yard contains 9 square feet; a square meter (m^2) contains about 11 ft^2.

$m^2 \times 10.77 = ft^2$
$m^2 \times 1.196 = yd^2$
$yd^2 \times 0.836 = m^2$
$ft^2 \times 0.093 = m^2$
$in.^2 \times 645.16 = mm^2$

A 100,000 ft^2 warehouse has an area of 9,300 m^2.

A 5,000 yd^2 field contains 4,180 m^2.

A 300 ft^2 window equals 28 m^2.

A television screen with a 200 $in.^2$ area has 129,000 mm^2, or 0.129 m^2.

An ounce, avoirdupois, contains about 28 grams and a pound contains about 450 grams.

Ounces × 28.35 = grams (g)
Pounds × 0.454 = kilograms (kg)
g × 0.035 = ounces
kg × 2.204 = pounds

An English ton, 2,000 pounds, is equivalent to 907 kilograms; the metric tonne contains 1,000 kg or 2,205 pounds. The seldom-used "long" ton contains 2,240 pounds, 1,016 kg.

Becoming accustomed to metric weights won't require as much of an adjustment as becoming familiar with some other metric units for two reasons: (1) most persons don't think much about fractional pounds when they read "12 oz" or "5 oz"; (2) many commodities, especially foods, already contain the weight in grams as well as in ounces.

The cubic meter is to the metric system what the cubic foot is to the English measurements of volume. A cubic meter (m^3) contains over 35 ft^3 and about one and one-third cubic yards.

$m^3 \times 1.307$ = cubic yards
$m^3 \times 35.31$ = cubic feet
$yd^3 \times 0.765 = m^3$
$ft^3 \times 0.283 = m^3$

A 10-by-12 room, 13 feet high, contains $10 \times 12 \times 13 = 1,560$ cubic feet. $1,560 \times 0.283 = 441$ m^3.

A concrete floor calling for 5,500 cubic yards of cement: $5,500 \times 0.765 = 4,208$ m^3.

For measuring capacity, as distinct from volume, the liter (l) is the customary unit. (The

Common Equivalents and Conversions

Approximate Common Equivalents		*Conversions Accurate to Parts per Million*	
1 inch	= 25 millimeters	inches × 25.4*	= millimeters
1 foot	= 0.3 meter	feet × 0.3048*	= meters
1 yard	= 0.9 meter	yards × 0.9144*	= meters
1 mile	= 1.6 kilometers	miles × 1.609 34	= kilometers
1 square inch	= 6.5 sq centimeters	square inches × 6.4516*	= sq centimeters
1 square foot	= 0.09 sq meter	square feet × 0.092 903 0	= sq meters
1 square yard	= 0.8 sq meter	square yards × 0.836 127	= sq meters
1 acre	= 0.4 hectare†	acres × 0.404 686	= hectares
1 cubic inch	= 16 cu centimeters	cubic inches × 16.3871	= cu centimeters
1 cubic foot	= 0.03 cubic meter	cubic feet × 0.028 316 8	= cu meters
1 cubic yard	= 0.8 cubic meter	cubic yards × 0.764 555	= cu meters
1 quart	= 1 liter†	quarts (liquid) × 0.946 353	= liters
1 gallon	= 0.004 cubic meter	gallons × 0.003 785 41	= cu meters
1 ounce (avdp)	= 28 grams	ounces (avdp) × 23.3495	= grams
1 pound (avdp)	= 0.45 kilogram	pounds (avdp) × 0.453 592	= kilograms
1 horsepower	= 0.75 kilowatt	horsepower × 0.745 700	= kilowatts
1 millimeter	= 0.04 inch	millimeters × 0.039 370 1	= inches
1 meter	= 3.3 feet	meters × 3.280 84	= feet
1 meter	= 1.1 yards	meters × 1.093 61	= yards
1 kilometer	= 0.6 mile	kilometers × 0.621 371	= miles
1 sq centimeter	= 0.16 sq inch	sq centimeters × 0.155 000	= sq inches
1 sq meter	= 11 sq feet	sq meters × 10.7639	= sq ft
1 sq meter	= 1.2 sq yards	sq meters × 1.195 99	= sq yards
1 hectare†	= 2.5 acres	hectares × 2.471 05	= acres
1 cu centimeter	= 0.06 cu inch	cu centimeters × 0.061 023 7	= cu inches
1 cu meter	= 35 cu feet	cu meters × 35.3147	= cu ft
1 cu meter	= 1.3 cu yards	cu meters × 1.307 95	= cu yards
1 liter†	= 1 quart	liters × 1.057	= quarts (liquid)
1 cu meter	= 250 gallons	cu meters × 264.172	= gallons
1 gram	= 0.035 ounces (avdp)	grams × 0.035 274 0	= ounces (avdp)
1 kilogram	= 2.2 pounds (avdp)	kilograms × 2.204 62	= pounds (avdp)
1 kilowatt	= 1.3 horsepower	kilowatts × 1.341 02	= horsepower

†common term not used in SI *exact

liter is slightly larger than a quart.) By international agreement the liter is considered to be one-thousandth of a cubic meter. The liter is subdivided into milliliters (ml) but is also spoken of as containing 1,000 cubic centimeters (cc). For practical purposes,

ml = cc

liters × 0.220 = gallons

1 × 1.760 = pints

ml × 0.035 = fluid ounces

Because a liter is one-thousandth of a cubic meter, and cubic meter-times-35.31 equals cubic feet, multiply liters by 0.03531 to get cubic feet. And milliliters × 0.610 = cubic inches.

When we work with units of volume and capacity to derive units for **volume rate of flow,** we encounter the phenomenon that the unit of time in the metric system is the second, followed by the hour and the day. **The minute is almost never used.**

Thus, flow rates are expressed in cubic meters per second (m^3/s), liters per second (l/s), or cubic meters per hour (m^3/h), instead of cubic meters per minute or liters per minute.

cubic feet per minute × 0.472 = liters per second

gallons per minute × 0.0758 = l/s

gallons per hour × 0.0013 = l/s

$m^3/s \times 35.31 = ft^3/s$

l/s × 13.20 = gal/min

l/s × 791.9 = gal/h

A design condition that calls for 600 cfm on cooling and 450 cfm on heating:

$600 \times 0.472 = 280$ l/s

$450 \times 0.472 = 212$ l/s

That comes from conversion. When the design is performed in metric, it is likely that the airflow would be specified at 300 l/s on cooling and 200 l/s on heating.

Simple velocity of air is expressed as meters per second (m/s) compared to the English feet per minute (fpm).

fpm $\times$ 0.00508 = m/s

So a terminal velocity of 75 fpm equals 0.4 m/s.

Although road speed is expressed in kilometers per hour (km/h), the correct SI unit is m/s.

miles per hour $\times$ 1.609 = km/h

km/h $\times$ 0.621 = miles per hour

So 70 mph = 113 km/h.
And 70 km/h = 43 mph.

The extent of American involvement with temperature conversions, till now, has been from Fahrenheit (°F) to Centigrade (°C). That stays the same in converting to metric, except that the temperature scale is called "degrees Celsius." The term "centigrade" means "by hundreds," just as the term "decimal" means "by tens," and refers to the arbitrary establishment by Celsius of a 100° differential between the freezing and boiling points of water. Water freezes at 0° C and 32° F; boils at 100° C, 212° F. Conversion is normally accomplished with a chart, but should calculations be required:

$$°C = \frac{(°F - 32)}{1.8}$$

Example:

$$\frac{(212° F - 32)}{1.8} = \frac{180}{1.8} = 100° C$$

$$°F = 1.8 \times °C + 32$$

Example:

$$100° C = 1.8 \times 100 + 32 = 180 + 32 = 212° F$$

An interesting sidelight is that the two scales cross at −40°.

$$°F = 1.8 \times °C + 32$$
$$= 1.8 \times (-40°) + 32$$
$$= -72 + 32 = -40° F$$

In science, though not often in engineering, one will encounter the "absolute" temperature scale, called "degrees Kelvin." Absolute zero, Kelvin, is −273.15° C or −459.67° F.

Water freezes at:

32° F
0° C
270° K

Water boils at:

212° F
100° C
373° K

When **temperature interval** is mentioned, one degree Celsius equals 1.8 degree Fahrenheit. When one says, for example, that a 14° C differential exists between two objects, the Fahrenheit temperature difference is 1.8×14, or 25°; NOT $1.8 \times 14 + 32$, or 57°.

When **heat flow rate** is discussed, there is still controversy because some want to retain traditional metric dimensions and not convert to SI standards. Purists would have all present dimensions, English and traditional metric, dropped in favor of watts (W) whether talking heating or cooling.

Gone would be not only British thermal units (Btu) and therms (100,000 Btu), but also kilocalories (kcal).

watts $\times$ 3.412 = Btuh

kW $\times$ 1.341 = horsepower

Btuh $\times$ 0.293 = watts

horsepower $\times$ 0.746 = kW

So, an 80,000-Btuh furnace would be rated: 80,000 $\times$ 0.293 = 23,440 W, or 23.44 kW.

And a 3-ton, 36,000-Btuh air conditioner: 36,000 $\times$ 0.293 = 10,548 W or 10.55 kW.

The relationships among work, energy, and heat are expressed in the following units:

megajoule (MJ) $\times$ 0.278 = kilowatthours (kWh)

kilojoules (kJ) $\times$ 0.948 = Btu

joules (J) $\times$ 0.7376 = foot-pounds.

There is argument, too, over **whether the newton (N) and/or kilopascal (kPa) will replace more familiar units** in expressions of force, inertia, and pressure. The pertinent conversions:

pounds per square inch (psi) $\times$ 0.068 = atmospheres

psi × 0.073 = kilograms per square centimeter (kg/cm^2)

psi × 703 = kg/m^2

psi × 6.895 = kPa

pounds per square foot × 47.88 = pascal (Pa)

kPa × 20.89 = pounds per square foot

Metric measurement can't be, needn't be, swallowed at one gulp. Most of us have need only for a few of the scores of units and learning those few isn't that difficult.

For reference, use the accompanying alphabetical list of most of the conversions which will be encountered in the refrigeration and hvac industries.

Ultimately, remember, we won't be converting to metric; we'll be designing in metric and we'll have forgotten what a Btu was.

Many companies, agencies, educational institutions, and associations are issuing metric-English conversion charts or booklets for specialized purposes, and many more will be doing so. Unfortunately, some early attempts in the United States contained errors, either in mathematics or in terminology. The best U.S. document we have so far encountered is the "ASTM Standard Metric Practice Guide," published by the American Society for Testing and Materials.

The International Organization for Standardization (ISO) has published "ISO Recom-

Typical Abbreviations

acre	no authorized abbreviation
atmospheres	atm
British thermal units	Btu
British thermal units per hour	Btuh
cubic feet	ft^3
cubic feet per minute	ft^3/min
cubic feet per second	ft^3/s
cubic inches	in^3
cubic meters	m^3
cubic millimeters	mm^3
cubic yards	yd^3
feet	ft
feet of water	ft H_2O
feet per second	ft/s
foot-pounds of force	lbf/ft
gallons	gal
gallons per hour	gal/h
gallons per minute	gal/min
grams	g
grams per square meter	g/m^2
hectares	ha
horsepower	hp
inches	in
inches of mercury	in Hg
inches of water	in H_2O
joules	J
kilocalories	kcal
kilograms	kg
kilograms per cubic meter	kg/m^3
kilograms per second	kg/s
kilograms per square meter	kg/m^2
kilojoules	kJ
kilojoules per cubic meter	kJ/m^3
kilojoules per kilogram	kJ/kg
kilometers	km
kilometers per hour	km/h
kilonewtons	kN
kilopascals	kPa
kilowatts	kW
kilowatt-hours	kWh
liters	l
liters per second	l/s
liters per minute	l/min
megajoules	MJ
meganewtons	MN
megapascals	MPa
meters	m
meters per second	m/s
miles	no abbreviation in metric
miles per hour	mile/h
millimeters	mm
millimeters of mercury	mm Hg
newtons	N
ounces	oz
ounces per square foot	oz/ft^2
pounds	lb
pounds of force	lbf
pounds of force per square foot	lbf/ft^2
pounds per cubic foot	lb/ft^3
pounds per second	lb/s
square feet	ft^2
square inches	in^2
square kilometers	km^2
square meters	m^2
square miles	$mile^2$
square millimeters	mm^2
watts	W
watts per square meter	W/m^2
yards	yd

mendation R786, Units and Symbols for Refrigeration."

The District Heating Association, of London, England, has published specialized metric information in its 1970/71 handbook.

Heating & Ventilating Publications Ltd., also of London, England, has published "Processed Calculations for Heating System Designs in SI Metric."

The Standards Association of Australia has published a metric handbook, "Metric Conversion in Building and Construction."

Scientific Notation

A majority of the public does not know the meaning of a term like 9.463×10^{-4}, so we have avoided using that format. Instead, the number is listed as 0.0009463.

That's the clue. The exponent (superior number) above the 10 tells one how many places to move the decimal point. If the exponent is preceded by a minus-sign, the decimal point is moved to the left. Examples:

$3.62 \times 10^3 = 3{,}620$

$3.62 \times 10^{-3} = 0.00362$

Useful Conversion Factors: Alphabetized

multiply	*by*	*to get*
acres	0.4047	hectares
acres	4,047	square meters
atmospheres	33.93	feet of water
atmospheres	29.92	inches of mercury
atmospheres	760.0	millimeters of mercury
atmospheres	1.058	tons per square foot
British thermal units	1,055	joules
British thermal units	0.2520	kilocalories
British thermal units	1.055	kilojoules
British thermal units per hour	0.2929	watts
British thermal units per pound	2.326	kilojoules per kilogram
cubic feet	0.02832	cubic meters
cubic feet	7.481	gallons
cubic feet	28.32	liters
cubic feet	29.92	quarts
cubic feet per minute	0.4719	liters per second
cubic feet per second	0.02832	cubic meters per second
cubic inches	16.39	cubic centimeters
cubic inches	16,387	cubic millimeters
cubic meters	35.32	cubic feet
cubic meters	1.308	cubic yards
cubic millimeters	0.00006102 or (6.102×10^{-5})	cubic inches
cubic yards	0.7646	cubic meters
feet	0.3048	meters
feet	304.8	millimeters
feet per second	0.3048	meters per second
foot-pounds of force	1.356	joules
foot-pounds of force per second	1.356	watts
gallons (liquid)	0.003785	cubic meters
gallons	3.785	liters
gallons per hour	0.001052	liters per second
gallons per minute	0.002228	cubic feet per second
gallons per minute	0.06308	liters per second
grams	0.03527	ounces (avoirdupois)
grams per square meter	0.003278	ounces per square foot
grams per square meter	0.02949	ounces per square yard

Useful Conversion Factors: Alphabetized (Continued)

multiply	*by*	*to get*
hectares	2.471	acres
horsepower	0.7460	kilowatts
horsepower	746	watts
inches	25.4	millimeters
inches of mercury	0.03342	atmospheres
inches of mercury	1.133	feet of water
inches of mercury	345.3	kilograms per square meter
inches of mercury (60°F)	3,377	newtons per square meter
inches of mercury	0.4912	pounds per square inch
inches of water	0.002458	atmospheres
inches of water	0.07355	inches of mercury
inches of water	25.40	kilograms per square meter
inches of water	0.03613	pounds per square inch
inches of water (60°F)	248.8	newtons per square meter
joules	0.7376	foot-pounds of force
kilocalories	3.968	British thermal units
kilocalories	4,190	joules
kilograms	2.205	pounds
kilograms per cubic meter	0.06243	pounds per cubic foot
kilograms per cubic meter	1.686	pounds per cubic yard
kilograms per second	2.205	pounds per second
kilograms per square meter	0.00009678	atmospheres
kilograms per square meter	0.003281	feet of water
kilograms per square meter	0.002896	inches of mercury
kilograms per square meter	0.2048	pounds per square foot
kilograms per square meter	0.001422	pounds per square inch
kilojoules	0.9478	British thermal units
kilojoules per cubic meter	0.02684	British thermal units per cubic foot
kilojoules per kilogram	0.4299	British thermal units per pound
kilometers	0.6214	miles
kilometers per hour	0.6214	miles per hour
kilonewtons	0.10036	tons of force
kilonewtons	224.8	pounds of force
kilopascals	20.89	pounds of force per square foot
kilowatts	1.341	horsepower
kilowatt-hours	3.6	megajoules
liters	0.03532	cubic feet
liters	61.02	cubic inches
liters	0.2642	gallons
liters	2.113	pints
liters	1.057	quarts
liters per minute	0.0005886	cubic feet per second
liters per second	2.119	cubic feet per minute
liters per second	951.0	gallons per hour
liters per second	15.85	gallons per minute
megajoules	0.2778	kilowatt-hours
meganewtons	100.36	tons of force
megapascals	145.04	pounds of force per square inch
megapascals	9.324	tons of force per square foot
megapascals	0.06475	tons of force per square inch
meters	3.281	feet
meters	1.094	yards

Useful Conversion Factors: Alphabetized (Continued)

multiply	*by*	*to get*
meters per second	2.237	miles per hour
miles	1.609	kilometers
miles per hour	1.609	kilometers per hour
miles per hour	0.4470	meters per second
milliliters	0.06102	cubic inches
milliliters	0.03520	fluid ounces
millimeters	0.0394	inches
millimeters of mercury	133.3	newtons per square meter
million gallons per day	0.005262	cubic meters per second
newtons	0.2248	pounds of force
ounces (avoirdupois)	28.35	grams
ounces (fluid)	28.41	milliliters
ounces per square foot	305.15	grams per square meter
ounces per square yard	33.91	grams per square meter
pounds	0.4535	kilograms
pounds of force	4.448	newtons
pounds of force per square foot	47.88	pascals
pounds of force per square inch	6.895	kilopascals
pounds per cubic foot	16.02	kilograms per cubic meter
pounds per cubic yard	0.5933	kilograms per cubic meter
pounds per second	0.4535	kilograms per second
pounds per square foot	4.882	kilograms per square meter
quarts	0.0009463	cubic meters
square feet	0.0929	square meters
square inches	645.2	square millimeters
square kilometers	0.3861	square miles
square meters	10.76	square feet
square meters	1.196	square yards
square miles	2.590	square kilometers
square millimeters	0.00155	square inches
square yards	0.8361	square meters
tons of force	9.964	kilonewtons
tons of force per square foot	107.25	kilopascals
tons of force per square inch	15.44	megapascals
torr (millimeters of mercury at 0°C)	133.3	newtons per square meter
watts	3.412	British thermal units per hour
watts	0.7376	foot-pounds of force per second
watts per square meter	0.3170	British thermal units per square foot
yards	0.9144	meters

Add your own conversion factors here.

multiply *by* *to get*

multiply *by* *to get*

multiply *by* *to get*

Reprinted from

AIR CONDITIONING HEATING & REFRIGERATION NEWS

P.O. Box 6000, Birmingham, Mich. 48012

E

Performance of Passively Solar-Heated Buildings

The following procedure was used in the 1978 Passive Residential Design Competition and Demonstration, sponsored by the U.S. Department of Housing and Urban Development. The version that appears here has been somewhat simplified.

E.1 Common-Basis, Solar-Heating Performance Estimate

The common-basis performance estimation technique, outlined below, follows a procedure devised at the Los Alamos Scientific

Table E.1 ***Passive Solar-Heating System Data***

A. Type of Solar Collection

(1) Direct Gain ______________________

(2) Indirect Gain ______________________

(3) Isolated Gain ______________________

B. Solar Collector

(1) *Aperture*	(2) *Net Area,* ft^2	(3) *No. Glazings*	(4) *Movable Insulation*	(5) *U-Value*	(6) *Shading*

(1) Windows, greenhouse, other; due south-facing vertical projected area.
(2) Actual glass area after mullions, etc., are subtracted.
(4) Yes or no; if yes, specify *U*-value when in place, Btuh/(ft^2)(°F) in Column (5).
(6) Percent shaded, June 21 at solar noon.

Table E.1 ***Passive Solar-Heating System Data* (*Continued*)**

C. Thermal Storage

(1) Storage Element	(2) Material	(3) Density lb/ft³	(4) Thickness, ft	(5) Area, ft²	(6) Percent Surface Exposed	(7) Exposed Mass (M), lb	(8) Specific Heat (CP) Btu/lb/°F	(9) MCP Btu/°F

(1) Floor, wall, ceiling, rockbed, etc.
(6) In direct sun, December 21 at solar noon, if applicable.
(7) Product of Columns (3), (4), (5), and (6).
(9) Product of Columns (7) and (8).

D. Collector/Storage Relationship

Divide Column (9) from category C, Thermal Storage, by Column (2) from category B, Solar Collector. The result should be approximately 30 Btu/(°F)(ft²) aperture.

Table E.2 ***Modified Building Loss Coefficient Method***

Surface Type	Area (ft²)	U-Value [Btuh/(°F)/(ft²)]	U × A (Btuh/°F)

Total (building skin conductance): ______ Btuh/°F

Laboratory (LASL). It has been developed from detailed hour-by-hour computer simulations for specific passive systems, and requires only readily available, monthly horizontal solar radiation and degree-day data. The method, strictly speaking, is only applicable for certain classes of thermal storage walls (Mass Trombe and Water Trombe) for which empirical curves have been determined. However, it can be used for all classes of passive solar-heating designs.

The key performance estimate obtained is the total annual heat energy required to maintain the home at 65° F. This number, when divided by the house floor area and divided by the heating degree-days, provides a common-basis figure of merit for comparing different buildings. Through use of solar gains, this number can usually be reduced to well below **5 Btu/(ft²)(DD)(yr).**

E.2 Modified Building Loss Coefficient

Determine the modified building loss coefficient. This can be calculated in either one of two ways.

Method A. If the design load has been determined, based on an assumed design temperature, then the modified building loss coefficient is calculated as follows:

Modified building loss coefficient Btu/DD

$$= \frac{24 \times \begin{pmatrix}\text{design heating}\\ \text{load}\end{pmatrix} \text{Btuh}}{\begin{matrix}\text{inside}\\ \text{temperature}\\ \text{assumed}\end{matrix}\ {}^\circ\text{F} - \begin{matrix}\text{outside}\\ \text{design}\\ \text{temperature}\end{matrix}\ {}^\circ\text{F}}$$

$$(\qquad)\ \text{Btu/DD} = \frac{24 \times (\qquad)\ \text{Btuh}}{(\qquad)\ {}^\circ\text{F} - (\qquad)\ {}^\circ\text{F}}$$

The design load used should be calculated without solar gains and without internal heat sources.

Method B. Alternatively, the modified building loss coefficient (see Table E.2) can be determined directly from the building outside envelope areas and U-values. Compute the area of each window, wall section, door, roof, etc. and list below. Then list the appropriate U-value [Btuh/(°F)(ft²)], multiply by the area, and tabulate the $U \times A$ products (Btuh/°F). The sum of the $U \times A$ products is the building skin conductance (Btuh/°F).

NOTE. IMPORTANT: In determining the building loss coefficient, the solar collection wall *must* be included assuming steady-state conduction through the wall and *no* solar gains. Solar gains are accounted for in the subsequent Solar/Load Ratio method in Table E.4, and must not be credited twice!

The infiltration load is now added, computed as the product of the building volume (ft³), the heat capacity of air [0.018 Btu/(ft³)(°F) at sea level], and the number of air changes per hour (ACH) used in the design calculations:

$$\text{Infiltration load} = \overset{\text{volume}}{[\qquad]} \times \overset{\text{specific heat}}{[\qquad]} \times \overset{\text{ACH}}{[\quad]} = \qquad \text{Btu/(°F)(hr)}$$

Finally:

$$24 \times \left[\begin{pmatrix}\text{building skin}\\ \text{conductance}\end{pmatrix} + \begin{pmatrix}\text{infiltration}\\ \text{load}\end{pmatrix}\right]$$

$$\begin{matrix}\text{Modified}\\ \text{building loss}\\ \text{coefficient}\end{matrix} = 24 \times \left[(\qquad) + (\qquad)\right] = \qquad \text{Btu/DD}$$

E.3 Building Thermal Load Profile

Table E.3 provides a *Building Thermal Load Profile* for each month of the year. This load profile should be based on a simple degree-day analysis, assuming a single building loss coefficient from the building-heated space to the outside, in the absence of solar gains.

To use Table E.3:

Step 1. Tabulate monthly heating degree-days in Column (1)

Table E.3 *Building Thermal Load Profile*

Month	(1) Monthly Degree-Days[a] °F	(2) Gross Monthly Load MBtu/mo[b]	(3) Internal Sources MBtu/mo	(4) Net Thermal Load MBtu/mo
Aug				
Sep				
Oct				
Nov				
Dec				
Jan				
Feb				
Mar				
Apr				
May				
Jun				
Jul				
TOTAL				

[a] List data only for months over 100 heating degree-days, base 65°F.
[b] MBtu = Btu $\times 10^6$.

Step 2. Multiply Column (1) by modified building loss coefficient (from above) to obtain gross monthly load (MBtu/month). List in Column (2).

Step 3. List monthly heating from internal sources other than heating systems such as lights, range, water heater energy retained in house, dryers, people, etc. (MBtu/month), in Column (3).

Step 4. Subtract Column (3) from Column (2) and list in Column (4). This is the net thermal load.

Step 5. Total up all columns.

Table E.4 provides an *Auxiliary Load Profile* based on the LASL Solar Load Ratio method. The total monthly solar energy transmitted through the solar collection surface is determined, and this is divided by the net monthly thermal load listed in Table E.3 to determine the monthly solar load ratio (SLR). The monthly solar heating fraction (SHF) is then determined from a selected curve, and the monthly auxiliary energy (backup heat required) is calculated.

E.4 Auxiliary Load Profile

To use Table E.4:

Step 1. List in Column (1) the net thermal load (MBtu/month) from Table E.3, Column (4).

Step 2. List in Column (2) the monthly solar radiation incident on a horizontal surface for the proposed site [Btu/(mo)(ft^2)].

Table E.4 ***Auxiliary Load Profile***

Month[a]	(1) Net Thermal Heating Load [from Table E.3, Column (4)]	(2) Solar Radiation on a Horizontal Surface	(3) Latitude Correction Factor	(4) Solar Radiation Absorbed	(5) Solar Load Ratio (SLR)	(6) Solar Heating Fraction (SHF)	(7) Auxiliary Energy
	MBtu/mo	Btu/(mo)(ft^2)		MBtu/mo			MBtu/mo
Aug							
Sep							
Oct							
Nov							
Dec							
Jan							
Feb							
Mar							
Apr							
May							
Jun							
Jul							
TOTALS							

[a]List data only for months over 100 heating degree-days.

NOTE. Do not use ASHRAE clear-day tables or cloud cover factors. Values listed must be based on National Weather Service measured data or other reliable measured data.

Step 3. Enter latitude of Site (L):

Step 4. Find the latitude correction factor from Fig. E.1, and list in Column (3).

Step 5. List in Column (4) the solar radiation absorbed. For vertical, south-facing double glazing (normal transmittance = 0.74), this can be determined from the following equation:

$$\text{Solar radiation absorbed Btu/mo} = \text{horizontal solar radiation [Column (2)] Btu/(mo)(ft}^2\text{)} \times \text{net solar collection area, ft}^2 \text{ (from Table E.1, Column B(2))} \times \text{solar absorptance} \times \text{latitude correction (from Table E.4, Column (3)}$$

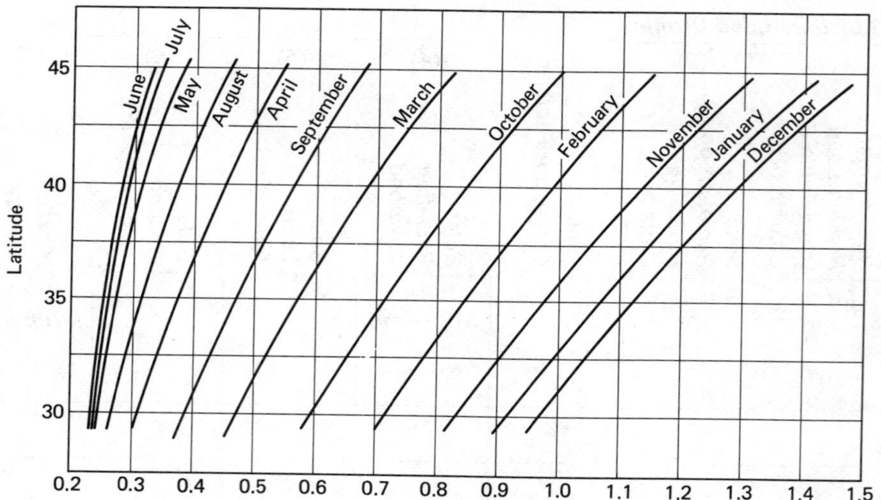

Fig. E.1 *Latitude correction.* [*Graphic solutions to value of* $0.226-(0.002512)(\text{L-D})+(0.0003075)(\text{L-D})^2$].

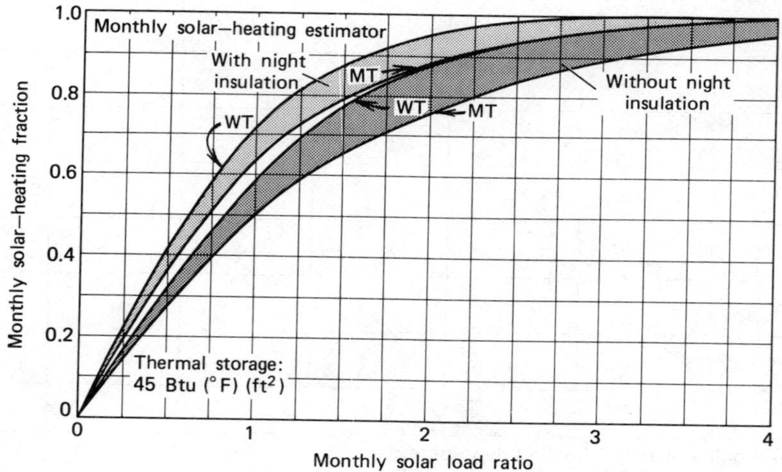

Fig. E.2 *Monthly solar load ratio (SLR):*

$$SLR = \frac{\text{Monthly solar energy absorbed}}{\text{Net monthly thermal load}}$$

including the static conduction through the solar wall, $A_w \times U_w \times DD$.
Note. *Use water trombe (WT) curves for direct gain and roof pond. Use mass trombe (MT) curves for sunspace.*

The *solar absorptance* is the absorptance of the surface inside the glazing. For a direct gain building, a value of 0.90 (light interior) to 0.95 (dark interior) is recommended.

Step 6. Divide Column (4) by Column (1) to obtain the solar/load ratio (SLR) and list this in Column (5).

Step 7. Using the solar load ratio from Step 6, obtain the value of monthly solar heating fraction (SHF) from Figure E.2. Enter these values in Column (6). Also *plot* these points on Figure E.2, and label each point with the month (useful for future reference).

For *direct gain* and *roof pond* buildings, it is recommended that the curves labeled "Water Trombe" on Figure E.2 be used.

For *sunspace* (or attached greenhouse) buildings, it is recommended that the curves labeled "Mass Trombe" on Figure E.2 be used.

For *Thermosyphon* buildings, determine the solar load ratio based on the type and location of the collector(s).

Step 8. List in Column (7) the monthly auxiliary energy (backup heating required) calculated from the formula:

$$\text{Monthly auxiliary energy, MBtu/mo} = (1 - \text{SHF}) \; \text{net monthly load [Column (1)], MBtu/mo}$$

To use Table E.5:

Add the total of Column (3) of Table E.3 (Internal Sources) to the total of Column (7) of Table E.4 (Auxiliary Energy), and divide this sum by the product of the heating degree-days [total Column (1), Table E.3, and building floor area].

Table E.5 ***Heating Energy***

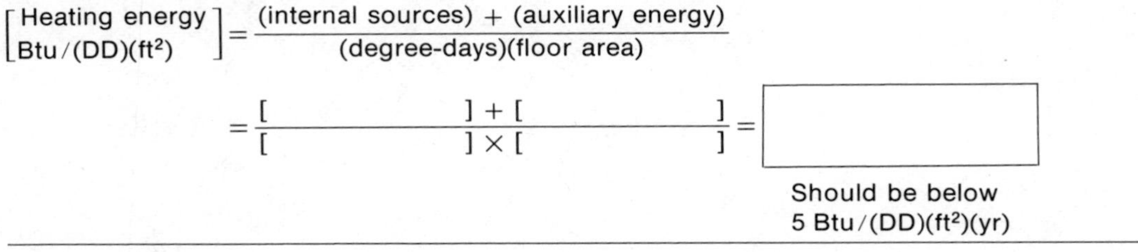

Index